# Encyclopedia of
# Mathematics Education

# Advisory Board

# Encyclopedia of
# Mathematics
# Education

*Editors*

**Louise S. Grinstein**
**Sally I. Lipsey**

**Routledge**
Taylor & Francis Group

NEW YORK AND LONDON

Published in 2001 by
Routledge
711 Third Avenue,
New York, NY 10017

Published in Great Britain by
Routledge
2 Park Square, Milton Park,
Abingdon, Oxfordshire OX14 4RN

First issued in paperback 2014

*Routledge is an imprint of the Taylor and Francis Group, an informa business*

**Library of Congress Cataloging-in-Publication Data**
Encyclopedia of mathematics education / editors-in-chief, Louise S. Grinstein, Sally I. Lipsey.
     p. cm.
   Includes bibliographical references and index.
   ISBN 978-0-8153-1647-3 (hbk)
   ISBN 978-0-415-76368-4 (pbk)
    1. Mathematics--Study and teaching--Encyclopedias. I. Grinstein, Louise S. II. Lipsey, Sally I.
(Sally Irene)

QA11 .E668 2001
510'.71--dc21

2001016599

# Contents

Alphabetical List of Entries                    vii

List of Contributors                            xiii

Preface                                         xxiii

Suggestions for the Reader                      xxv

Sample Entries by Category                      xxvii

Entries A–Z                                     1

Index                                           841

# Alphabetical List of Entries

## A

Abacus
Abel, Niels H.
Ability Grouping
Abstract Algebra
Actuarial Mathematics
Addition
Adult Education
Advanced Placement Program (AP)
Advising
African Customs as Applied to the Mathematics
    Classroom
Algebra, Introductory Motivation
Algebra Curriculum, K–12
Algebraic Numbers
Algorithms
American Mathematical Association of Two-Year
    Colleges (AMATYC)
American Mathematical Society (AMS)
American Mathematical Society (AMS), Historical
    Perspective
American Statistical Association (ASA)
Angles
Applications for the Classroom, Overview
Applications for the Secondary School Classroom
Archimedes
Aristotle

Arithmetic
Art
Assessment of Student Achievement, Issues
Assessment of Student Achievement, Overview
Association for Women in Mathematics (AWM)
Association of Teachers of Mathematics (ATM)
Asymptotes
Attribute Blocks
Awards for Students and Teachers
Axioms

## B

Babbage, Charles
Babylonian Mathematics
Bases, Complex
Bases, Negative
Basic Skills
Begle, Edward G.
Bernoulli Family
Binary (Dyadic) System
Binomial Distribution
Binomial Theorem
Biomathematics
Birkhoff, George D.
Board on Mathematical Sciences (BMS)
Bolyai, Janos

Books, Stories for Children
Boole, George
Boolean Algebra
Brooks, Edward
Brownell, William A.
Bruner, Jerome
Business Mathematics

## C

Calculators
Calculus, Overview
Calendars
Cambridge Conference on School Mathematics
    (1963)
Canada
Cantor, Georg
Career Information
Carroll, Lewis
Cauchy, Augustin-Louis
Chaos
Chemistry
China, People's Republic
Chinese Mathematics
Circles
Coalitions
Coding Theory
Cognitive Learning
Colburn, Warren
College Entrance Examination Board (CEEB)
    Commission Report (1959)
Combinatorics
Commission on Postwar Plans
Committee of Fifteen on Elementary Education
Committee of Ten
Committee on the Undergraduate Program in
    Mathematics (CUPM)
Competitions
Complex Numbers
Computer Assisted Instruction (CAI)
Computer Science
Computers
Congresses
Conic Sections (Conics)
Connected Mathematics Project (CMP)
Constructivism
Consumer Mathematics
Continued Fractions
Continuity
Cooperative Learning
Coordinate Geometry, Overview
Core-Plus Mathematics Project (CPMP)
Correlation

Creativity
Cryptography
Cuisenaire Rods
Curriculum, Overview
Curriculum Issues
Curriculum Trends, College Level
Curriculum Trends, Elementary Level
Curriculum Trends, Secondary Level
Curve Fitting

## D

D'Alembert, Jean
Data Analysis
Data Representation
Davies, Charles
Decimal (Hindu-Arabic) System
Dedekind, Richard
De Moivre, Abraham
De Morgan, Augustus
Descartes, René
Determinants
Developing Mathematical Processes (DMP)
Developmental (Remedial) Mathematics
Developmental Mathematics Program (DMP) and
    Apprenticeship Teaching
Dewey, John
Diagrams, Euler and Venn
Dienes, Zoltan P.
Differentiation
Diophantine Equations
Discovery Approach to Geometry
Discrete Mathematics
Divisibility Tests
Division

## E

$e$
Economics
Educational Studies in Mathematics (ESM)
Egyptian Mathematics
Einstein, Albert
Eisenhower Program
Engineering
England
Enrichment, Overview
Environmental Mathematics
Equalities
Equations, Algebraic
Equations, Exponential
Equations, Logarithmic
Eratosthenes

Errors, Arithmetic
Escher, Maurits C.
Estimation
Ethnomathematics
Euclid
Euler, Leonhard
Evaluation of Instruction, Issues
Evaluation of Instruction, Overview
Exponents, Algebraic
Exponents, Arithmetic
Extrapolation

# F

Factors
Family Math and *Matemática Para la Familia*
Fatou, Pierre
Federal Government, Role
Fehr, Howard F.
Fermat, Pierre de
Fibonacci Sequence
Field Activities
Fields
Finance
Formulas
Fourier, Jean Baptiste Joseph de
Fractals
Fractions
France
Functions
Functions of a Real Variable
Fundamental Theorems of Mathematics

# G

Gagné, Robert M.
Galilei, Galileo
Galois, Evariste
Game Theory
Games
Gauss, Carl F.
Geometry, Instruction
Germain, Sophie
Germany
Gifted
Gifted, Special Programs
Goals of Mathematics Instruction
Goals 2000
Gödel, Kurt F.
Goldbach's Conjecture
Golden Section
Graph Theory, Overview

Graphs
Greatest Common Divisor (GCD)
Greek Mathematics
Group Theory
Guidelines for Science and Mathematics in the Preparation Program of Teachers of Secondary School Science and Mathematics, Guidelines for Science and Mathematics in the Preparation Program of Elementary School Teachers

# H

Hadamard, Jacques S.
Health
Hexadecimal System
Hilbert, David
History of Mathematics, Overview
History of Mathematics Education in the United States, Overview
Homework
Humanistic Mathematics
Humor

# I

Index Numbers
Indian Mathematics
Individualized Instruction
Inequalities
Instructional Materials and Methods, Elementary
Instructional Methods, Current Research
Instructional Methods, New Directions
Integration
Integration of Elementary School Mathematics Instruction with Other Subjects
Integration of Secondary School Mathematics Instruction with Other Subjects
Interactive Mathematics Program (IMP)
Interest
International Comparisons, Overview
International Organizations
International Studies of Mathematics Education
Interpolation
Invariance
Inverse Functions
Irrational Numbers
Islamic Mathematics
Isomorphism
Israel

## J

James, William
Japan
Joint Commission Report (1940)
Joint Policy Board for Mathematics (JPBM)
Julia, Gaston

## K

Kepler, Johannes
Kindergarten
Klein, Felix
Kovalevskaia, Sofia

## L

*Ladies' Diary (Woman's Almanack)*
Lagrange, Joseph L.
Language and Mathematics in the Classroom
Language of Mathematics Textbooks
Laplace, Pierre S.
Learning Disabilities, Overview
Least Squares
Lebesgue, Henri
Legendre, Adrien M.
Leibniz, Gottfried W.
Length, Perimeter, Area and Volume
Lesson Plans
L'Hospital, Guillaume F.
Liberal Arts Mathematics
Lie, Sophus
Limits
Linear Algebra, Overview
Linear Programming
Liouville, Joseph
Lobachevsky, Nikolai I.
Logarithms
Logic
Logic Machines
Logo

## M

Maclaurin, Colin
Maclaurin Expansion
Magic Squares
Manipulatives, Computer Generated
Markov Processes
Mathematical Association of America (MAA)
Mathematical Association of America (MAA),
    Historical Perspective
Mathematical Expectation
Mathematical Literacy

Mathematical Sciences Education Board (MSEB)
*Mathematical Spectrum*
Mathematics, Definitions
Mathematics, Foundations
Mathematics, Nature
Mathematics Anxiety
Mathematics Education, Statistical Indicators
Mathematics in Context (MiC): A Connected
    Curriculum for Grades 5 through 8
Mayan Numeration
Mean, Arithmetic
Mean, Geometric
Mean Value Theorems
Measurement
Measures of Central Tendency
Media
Median
Middle School
Minorities, Career Problems
Minorities, Educational Problems
Mode
Modeling
Modern Mathematics
Money
Montessori, Maria
Moore, Eliakim H.
Mu Alpha Theta
Multicultural Mathematics, Issues
Multiplication
Music

## N

Napier, John
National Assessment of Educational Progress
    (NAEP)
National Center for Research in Mathematical
    Sciences Education (NCRMSE)
National Committee on Mathematical
    Requirements
National Council of Supervisors of Mathematics
    (NCSM)
National Council of Teachers of Mathematics
    (NCTM)
National Council of Teachers of Mathematics
    (NCTM), Historical Perspective
National Science Foundation (NSF)
National Science Foundation (NSF), Historical
    Perspective
Neutral Geometry
Newton, Isaac
Noether, Emmy
Non-Euclidean Geometry
Normal Distribution Curve

Number Sense and Numeration
Number Theory, Overview
Numerical Analysis, Methods
Nursing

# O

One-to-One Correspondence
Operations Research
Ordinary Differential Equations

# P

Paradox
Parametric Equations
Partial Differential Equations
Pascal, Blaise
Pascal's Triangle
Patterning
Peano, Giuseppe
Peirce, Charles S.
Percent
Pestalozzi, Johann H.
Philosophical Perspectives on Mathematics
Physics
Pi
Pike, Nicolas
Plato
Poetry
Poincaré, Jules H.
Poisson Distribution
Polar Coordinates
Pólya, George
Polygons
Population Growth
Power Series
Preschool
Prime Numbers
Probability, Overview
Probability Applications: Chances of Matches
Probability Density
Probability Distribution
Probability in Elementary School Mathematics
Probability Problems, Computer Simulations
Problem Solving, Overview
Problem-Solving Ability and Computer
    Programming Instruction
Programming Languages
Progressive Education Association (PEA) Report
Projections, Geometric
Projective Geometry
Projects for Students
Proof

Proof, Geometric
Proportional Reasoning
Psychology of Learning and Instruction, Overview
Pythagoras
Pythagorean Theorem

# Q

Quality Control
Questioning
Queueing Theory

# R

Ramanujan, Srinivasa
Random Number
Random Sample
Random Variable
Rates in Calculus
Rational Numbers
Readiness
Reading Mathematics Textbooks
Real Numbers
Recreations, Overview
Relativity
Remedial Instruction
Research in Mathematics Education
*Revolution in School Mathematics*
Riemann, George Friedrich Bernhard
Rounding
Russell, Bertrand
Russia

# S

School Mathematics Study Group (SMSG)
School Science and Mathematics Association
    (SSMA)
School Science and Mathematics Association
    (SSMA), Historical Perspective
Scientific Notation
Secondary School Mathematics
Series
Sets
Similarity
Skinner, Burrhus F.
Smith, David E.
Software
Space Geometry, Instruction
Space Mathematics
Spiral Learning
Sports
Standard Deviation

Standardization of Variables
Standards, Curriculum
State Government, Role
Statistical Inference
Statistical Significance
Statistics, Overview
Stern, Catherine
Stochastic Processes
Student Teacher Autonomy, Elementary School
Subtraction
Suppes, Patrick
Symmetry
Systems of Equations
Systems of Inequalities
Systems of Measurement

# T

Task Analysis
Teacher Development, Elementary School
Teacher Development, K–12: California
    Mathematics Project (CMP)
Teacher Development, Middle School
Teacher Development, Secondary School
Teacher Preparation, College
Teacher Preparation, Elementary School, Issues
Teacher Preparation, Middle School
Teacher Preparation, Secondary School
Teacher Preparation, Secondary School, Reform
    Project in California
Teacher Preparation in Statistics, Issues
Teaching Integrated Mathematics and Science
    Program (TIMS)
Techniques and Materials for Enrichment Learning
    and Teaching
Technology
Temperature Scales
Test Construction
Textbooks
Textbooks, Early American Samples
Third International Mathematics and Science Study
    (TIMSS)
Thorndike, Edward L.
Tiling
Time
Time Series Analysis
Topology
Transcendental and Algebraic Functions
Transcendental Numbers
Transformational Geometry
Transformations
Trigonometric Functions, Elementary

Trigonometric Identities
Trigonometry
Turing, Alan M.
Tutoring

# U

Underachievers
Underachievers, Special Programs
United States Department of Education
United States National Commission on
    Mathematics Instruction (USNCMI)
University of Chicago School Mathematics Project
    (UCSMP)
University of Illinois Committee on School
    Mathematics (UICSM)
University of Maryland Mathematics Project
    (UMMaP)

# V

Van Hiele Levels
Variables
Variance and Covariance
Vector Spaces
Vectors
Von Neumann, John

# W

Weierstrass, Karl
Weyl, Hermann
Whitehead, Alfred N.
Wiener, Norbert
Women and Mathematics, History
Women and Mathematics, International
    Comparisons of Career and Educational
    Problems
Women and Mathematics, Problems
Women and Mathematics Education (WME)
Writing Activities

# Y

Young, Jacob W.

# Z

Zeno
Zero

# List of Contributors

Albrecht H. Abele
*Paedagogische Hochschule*
*Heidelberg, Germany*

Francine F. Abeles
*Kean University*
*Union, New Jersey*

Geoffrey R. Akst
*Borough of Manhattan Community College*
*New York*

Henry L. Alder
*University of California*
*Davis*

Daniel S. Alexander
*Drake University*
*Des Moines, Iowa*

Gerald L. Alexanderson
*Santa Clara University*
*Santa Clara, California*

Lynne Alper
*San Francisco State University*
*San Francisco, California*

Alejandro Andreotti
*Rhode Island College*
*Providence*

George E. Andrews
*Pennsylvania State University*
*University Park*

Julia Anghileri
*Homerton College*
*Cambridge, England*

Alice F. Artzt
*Queens College*
*Flushing, New York*

Richard A. Austin
*University of South Florida*
*Tampa*

Ayoub B. Ayoub
*Pennsylvania State University, Ogontz Campus*
*Abington*

Steven J. Balasiano
*Sheepshead Bay High School*
*Brooklyn, New York*

Evelyne Barbin
*Université de Paris*
*Paris, France*

Sue Jackson Barnes
*De Queen High School*
*De Queen, Arkansas*

Michael T. Battista
*Kent State University*
*Kent, Ohio*

William J. Baumol
*Princeton University*
*Princeton, New Jersey*

Jay Becker
*University of Illinois*
*Chicago*

Joyce M. Becker
*Luther College*
*Decorah, Iowa*

George Berzsenyi
*Rose-Hulman Institute of Technology*
*Terre Haute, Indiana*

James K. Bidwell
*Central Michigan University*
*Mt. Pleasant*

Gary G. Bitter
*Arizona State University*
*Tempe*

Rolf K. Blank
*Council of Chief State School Officers*
*Washington, DC*

Derek I. Bloomfield
*Orange County Community College*
*Middletown, New York*

Martha J. Boles
*Bradford College*
*Bradford, Massachusetts*

Sylvia T. Bozeman
*Spelman College*
*Atlanta, Georgia*

John S. Bradley
*National Science Foundation*
*Arlington, Virginia*

Patricia A. Brosnan
*Ohio State University*
*Columbus*

Regina Baron Brunner
*Cedar Crest College*
*Allentown, Pennsylvania*

Pamela K. Buckley
*Eisenhower Math/Science Consortium*
*Charlestown, West Virginia*

James O. Bullock
*Cancer Research Center*
*Columbia, Missouri*

Robert J. Bumcrot
*Hofstra University*
*Hempstead, New York*

Grace M. Burton
*University of North Carolina*
*Wilmington*

Ralph W. Cain
*University of Texas*
*Austin*

Dennis Callas
*State University of New York*
*Delhi*

David William Carraher
*Technical Education Research Centers*
*Cambridge, Massachusetts*

Bettye Anne Case
*Florida State University*
*Tallahassee*

Kathryn Castle
*Oklahoma State University*
*Stillwater*

Rosalind Charlesworth
*Weber State University*
*Ogden, Utah*

Lily E. Christ
*John Jay College*
*New York, New York*

Mary Jo Cittadino
*University of California*
*Berkeley*

Robert G. Clason
*Central Michigan University*
*Mt. Pleasant*

Douglas H. Clements
*State University of New York*
*Buffalo*

Joan Countryman
*Lincoln School*
*Providence, Rhode Island*

Annalisa Crannell
*Franklin & Marshall College*
*Lancaster, Pennsylvania*

Lucille Croom
*Hunter College*
*New York, New York*

Frances R. Curcio
*Queens College*
*Flushing, New York*

Beatriz S. D'Ambrosio
*Indiana University—Purdue University*
*Indianapolis*

Ubiratan D'Ambrosio
*Universidade Estadual de Campinas*
*Campinas, Brazil*

Joseph W. Dauben
*Lehman College*
*Bronx, New York*

Donald M. Davis
*Lehigh University*
*Bethlehem, Pennsylvania*

David M. Davison
*Montana State University*
*Billings*

Richard M. Davitt
*University of Louisville*
*Louisville, Kentucky*

Joseph B. Dence
*University of Missouri*
*St. Louis*

Thomas P. Dence
*Ashland University*
*Ashland, Ohio*

Keith Devlin
*Saint Mary's College of California*
*Moraga*

Joan F. Donahue
*State Coalitions Project*
*Washington, DC*

Eileen F. Donoghue
*College of Staten Island*
*Staten Island, New York*

Willibald Dörfler
*Universität Klagenfurt*
*Klagenfurt, Austria*

Tommy Dreyfus
*Tel Aviv University*
*Holon, Israel*

David S. Dummit
*University of Vermont*
*Burlington*

David R. Duncan
*University of Northern Iowa*
*Cedar Falls*

Gail A. Earles
*Saint Cloud State University*
*Saint Cloud, Minnesota*

Betty M. Erickson
*Kearsage Regional Elementary School*
*Bradford, New Hampshire*

Paul Ernest
*University of Exeter*
*Exeter, Devon, England*

Mona Fabricant
*Queensborough Community College*
*Bayside, New York*

Daniel M. Fendel
*San Francisco State University*
*San Francisco*

Beverly J. Ferrucci
*Keene State College*
*Keene, New Hampshire*

William Fisher
*California State University*
*Chico*

James A. Fitzsimmons
*Ohio State University*
*Columbus*

Bernard A. Fleishman
*Rensselaer Polytechnic Institute*
*Troy, New York*

Peter Flusser
*Kansas Wesleyan University*
*Salina*

Robert J. Fraga
*Ripon College*
*Ripon, Wisconsin*

Richard L. Francis
*Southeast Missouri State University*
*Cape Girardeau*

Sherry Fraser
*San Francisco State University*
*San Francisco*

John E. Freund
*Arizona State University*
*Tempe*

Richard J. Friedlander
*University of Missouri*
*St. Louis*

Bernard A. Fusaro
*Salisbury State University*
*Salisbury, Maryland*

David Fuys
*Brooklyn College*
*Brooklyn, New York*

Stavroula K. Gailey
*Christopher Newport University*
*Newport News, Virginia*

Joseph Gani
*Australian National University*
*Canberra*

Robert A. Garden
*University of British Columbia*
*Vancouver, British Columbia, Canada*

Joan B. Garfield
*University of Minnesota*
*Minneapolis*

Lynn E. Garner
*Brigham Young University*
*Provo, Utah*

June L. Gastón
*Borough of Manhattan Community College*
*New York*

M. Katherine Gavin
*University of Connecticut*
*Storrs*

Dorothy Geddes (deceased)
*Brooklyn College*
*Brooklyn, New York*

William, J. Gilbert
*University of Waterloo*
*Waterloo, Ontario, Canada*

Alice J. Gill
*American Federation of Teachers*
*Washington, DC*

Lynda B. Ginsburg
*University of Pennsylvania*
*Philadelphia*

Terry Goodman
*Central Missouri State University*
*Warrensburg*

Ivor Grattan-Guinness
*Middlesex University*
*Enfield, Middlesex, England*

Mary W. Gray
*American University*
*Washington, DC*

Virginia Gray
*Columbia College*
*Sonora, California*

Judy Green
*Marymount University*
*Arlington, Virginia*

Louise S. Grinstein
*Kingsborough Community College*
*Brooklyn, New York*

Douglas A. Grouws
*University of Iowa*
*Iowa City*

Rod Haggarty
*Oxford Brookes University*
*Headington, Oxford, England*

Bettye C. Hall
*BCH Consultants*
*Houston, Texas*

Kathleen M. Harmeyer
*ExperTech Corporation*
*Timonium, Maryland*

Mary M. Hatfield
*Arizona State University*
*Tempe*

Anne Hawkins
*International Association for Statistical Education*
*London, England*

William A. Hawkins, Jr.
*Mathematical Association of America*
*Washington, DC*

Jerome D. Hayden
*McClean County Unit District 5*
*Normal, Illinois*

Christoper C. Healy
*El Monte School District*
*California*

Richard Herman
*University of Maryland*
*College Park*

Reuben Hersh
*University of New Mexico*
*Albuquerque*

Christian R. Hirsch
*Western Michigan University*
*Kalamazoo*

Peter Horák
*Slovak Technical University*
*Bratislava, Slovakia*

Donna D. Ignatavicius
*DI Associates*
*Hughesville, Maryland*

John Impagliazzo
*Hofstra University*
*Hempstead, New York*

Minoru Ito
*Teachers College, Columbia University*
*New York*

Linda Jackson
*Association of Teachers of Mathematics*
*Charvil, Reading, Berks, England*

William H. Jaco
*Oklahoma State University*
*Stillwater*

Judith E. Jacobs
*California State Polytechnic University*
*Pomona*

William D. Jamski
*Indiana University Southeast*
*New Albany*

George A. Jennings
*California State University*
*Dominguez Hills, Carson*

Zhonghong Jiang
*Florida International University*
*Miami*

Janet L. Johnson
*Wake County Public School System*
*North Carolina*

Graham A. Jones
*Illinois State University*
*Normal*

Vinetta C. Jones
*The College Board*
*New York, New York*

William Juraschek
*University of Colorado*
*Denver*

Jean Pierre Kahane
*Université de Paris*
*Paris, France*

Charity Vaughan Kahn
*High school teacher*
*San Francisco, California*

Robert Kalin
*Florida State University*
*Tallahassee*

Akihiro Kanamori
*Boston University*
*Boston, Massachusetts*

Bernice Kastner
*Towson State University*
*Towson, Maryland*

Herbert E. Kasube
*Bradley University*
*Peoria, Illinois*

Victor J. Katz
*University of the District of Columbia*
*Washington, DC*

Sandra Z. Keith
*Saint Cloud State University*
*Saint Cloud, Minnesota*

Patricia Ann Kenney
*University of Pittsburgh*
*Pittsburgh, Pennsylvania*

Alexander Kleiner
*Drake University*
*Des Moines, Iowa*

Israel Kleiner
*York University*
*North York, Ontario, Canada*

Dennis Kletzing
*Stetson University*
*Deland, Florida*

Peter Kloosterman
*Indiana University*
*Bloomington*

Eberhard Knobloch
*Technische Universität*
*Berlin, Germany*

Gerald Kulm
*Montana State University*
*Bozeman*

Carole B. Lacampagne
*U.S. Department of Education*
*Washington, DC*

Diana V. Lambdin
*Indiana University*
*Bloomington*

Glenda Lappan
*Michigan State University*
*East Lansing*

Carol Novillis Larson
*University of Arizona*
*Tucson*

Gilah C. Leder
*La Trobe University*
*Bundoora, Victoria, Australia*

Deborah B. Lee
*Troy State University*
*Phenix City, Alabama*

Lee Peng Yee
*National Institute of Education*
*Singapore*

Lisa Lefkowitz
*CNA Insurance Company*
*Chicago, Illinois*

Steven Leinwand
*Connecticut State Department of Education*
*Hartford*

Steven J. Leon
*University of Massachusetts*
*North Dartmouth*

Frank K. Lester, Jr.
*Indiana University*
*Bloomington*

Melvin D. Levine
*University of North Carolina*
*Chapel Hill*

Mary M. Lindquist
*Columbus College*
*Columbus, Georgia*

Sally I. Lipsey
*Brooklyn College*
*Brooklyn, New York*

Bonnie H. Litwiller
*University of Northern Iowa*
*Cedar Falls*

Richard Long
*Richard Long Associates*
*Washington, DC*

C. James Lovett
*Brooklyn College*
*Brooklyn, New York*

Sharon Lubkin
*University of Washington*
*Seattle*

Beatrice Lumpkin
*Malcolm X College*
*Chicago, Illinois*

David Lutzer
*College of William and Mary*
*Williamsburg, Virginia*

Daniel P. Maki
*Indiana University*
*Bloomington*

Elena A. Marchisotto
*California State University*
*Northridge*

George E. Martin
*State University of New York*
*Albany*

Edwin McClintock
*Florida International University*
*Miami*

Robert L. McGinty
*Northern Michigan University*
*Marquette*

William J. McKeough
*Hofstra University*
*Hempstead, New York*

Ruby Bostick Midkiff
*Arkansas State University*
*State University*

Sergei Mikhelson
*University of St. Petersburg*
*St. Petersburg, Russia*

Michael Mikusa
*Kent State University*
*Kent, Ohio*

Gayle M. Millsaps
*Ohio State University*
*Columbus*

Antonie F. Monna (deceased)
*Utrecht University*
*Netherlands*

Ann E. Moskol
*Rhode Island College*
*Providence*

Nitsa Movshovitz-Hadar
*Israel Institute of Technology*
*Haifa, Israel*

Marilyn S. Neil
*West Georgia College*
*Carrollton*

Ralph L. Nelson (deceased)
*Queens College*
*Flushing, New York*

Harald M. Ness
*University of Wisconsin Center*
*Fond du Lac*

Michael Neubrand
*Flensburg Universität*
*Flensburg, Germany*

Robert F. Nicely, Jr.
*Pennsylvania State University*
*University Park*

William J. O'Donnell
*Cherry Creek High School*
*Englewood, Colorado*

Enrique Ortiz
*University of Central Florida*
*Daytona Beach*

Anthony Orton
*University of Leeds*
*Leeds, England*

Douglas T. Owens
*Ohio State University*
*Columbus*

Vithal A. Patel
*Humboldt State University*
*Arcata, California*

A. Louise Perkins
*University of Southern Mississippi*
*Hattiesburg*

Teri Perl
*Teri Perl Associates*
*Palo Alto, California*

Benjamin M. Perles
*Author*
*Ft. Lauderdale, Florida*

Elizabeth Phillips
*Michigan State University*
*East Lansing*

Anthony V. Piccolino
*Montclair State University*
*Upper Montclair, New Jersey*

Eileen L. Poiani
*Saint Peter's College*
*Jersey City, New Jersey*

Henry O. Pollak
*Teachers College, Columbia University*
*New York*

Yvonne M. Pothier
*Mt. Saint Vincent University*
*Halifax, Nova Scotia, Canada*

Lee Etta Powell
*George Washington University*
*Washington, DC*

Donald L. Pratt
*Bloomsburg University*
*Bloomsburg, Pennsylvania*

James E. Pratt
*Oklahoma State University*
*Stillwater*

Michael H. Price
*University of Leicester*
*Leicester, England*

Libby Quattromani
*Western New Mexico University*
*Gallup*

Karin Reich
*Universität Hamburg*
*Hamburg, Germany*

Luetta Reimer
*Fresno Pacific College*
*Fresno, California*

Wilbert Reimer
*Fresno Pacific College*
*Fresno, California*

Joseph S. Renzulli
*University of Connecticut*
*Storrs*

Diane Resek
*San Francisco State University*
*San Francisco*

Catherine M. Ricardo
*Iona College*
*New Rochelle, New York*

Hadas Rin
*Sybase Inc.*
*Emeryville, California*

A. Wayne Roberts
*Macalester College*
*St. Paul, Minnesota*

David F. Robitaille
*University of British Columbia*
*Vancouver, British Columbia, Canada*

Thomas A. Romberg
*University of Wisconsin*
*Madison*

Mary Rouncefield
*Chester College*
*Chester, England*

Julie Sarama
*State University of New York*
*Buffalo, New York*

Mark E. Saul
*Bronxville Schools*
*Bronxville, New York*

Cathy G. Schloemer
*Indiana Area Senior High School*
*Indiana, Pennsylvania*

Joel Schneider
*Children's Television Workshop*
*New York, New York*

Benjamin L. Schwartz
*Consultant/author*
*Vienna, Virginia*

Diane Driscoll Schwartz
*Ithaca College*
*Ithaca, New York*

Elizabeth Senger
*Auburn University*
*Auburn, Alabama*

Chantal Shafroth
*North Carolina Central University*
*Chapel Hill*

Karen T. Sharp
*Mott Community College*
*Flint, Michigan*

Jean M. Shaw
*University of Mississippi*
*University*

Martha J. Siegel
*Towson State University*
*Towson, Maryland*

Murray H. Siegel
*Sam Houston State University*
*Huntsville, Texas*

Ronda Simmons
*Auburn School District*
*Auburn, Washington*

Leslie P. Steffe
*University of Georgia*
*Athens*

Marcy L. Stein
*University of Washington*
*Tacoma*

Bruce H. Stephan
*Webb Institute of Naval Architecture*
*Glen Cove, New York*

Lee V. Stiff
*North Carolina State University*
*Raleigh*

Wei Sun
*Towson State University*
*Towson, Maryland*

Frank J. Swetz
*Pennsylvania State University*
*Harrisburg*

Ronald J. Tallarida
*Temple University*
*Philadelphia, Pennsylvania*

Alan R. Taylor
*Coquitlam School District*
*British Columbia, Canada*

Lyn Taylor
*University of Colorado*
*Denver*

Lori A. Thombs
*University of South Carolina*
*Columbia*

Denisse R. Thompson
*University of South Florida*
*Tampa*

Maynard Thompson
*Indiana University*
*Bloomington*

Carol A. Thornton
*Illinois State University*
*Normal*

Ronald W. Towery
*Arkansas State University*
*State University*

Kenneth J. Travers
*University of Illinois, Urbana-Champaign*
*Champaign*

John R. Tucker
*Board on Mathematical Sciences*
*National Academy of Sciences*
*Washington, DC*

Zalman Usiskin
*University of Chicago*
*Chicago, Illinois*

Shlomo Vinner
*Hebrew University*
*Jerusalem, Israel*

Bruce R. Vogeli
*Teachers College, Columbia University*
*New York*

Erica Dakin Voolich
*Solomon Schechter Day School*
*Newton, Massachusetts*

Hiroko K. Warshauer
*Southwest Texas State University*
*San Marcos*

Max L. Warshauer
*Southwest Texas State University*
*San Marcos*

Ann E. Watkins
*California State University*
*Northridge*

Rhonda Neuborn Weissman
*Benjamin Cardozo High School*
*Bayside, New York*

Rosamund Welchman
*Brooklyn College*
*Brooklyn, New York*

Richard S. Westfall (deceased)
*Indiana University*
*Bloomington*

Alvin M. White
*Harvey Mudd College*
*Claremont, California*

Sharon Whitton
*Hofstra University*
*Hempstead, New York*

Connie Hicks Yarema
*Abilene Christian University*
*Abilene, Texas*

Rina Yarmish
*Kingsborough Community College*
*Brooklyn, New York*

Claudia Zaslavsky
*Author*
*New York, New York*

Daniel Zwillinger
*Zwillinger & Associates*
*Newton, Massachusetts*

# Preface

The purpose of this encyclopedia is to fulfill a need for an overall view of mathematics education, bringing together its most pertinent aspects at the elementary, secondary, and postsecondary levels. It is helpful to have a volume that brings together those aspects of mathematics education that are usually viewed separately: pure mathematics and a broad variety of applications, methods of teaching both, the history of both, and the history of the associated pedagogy. It is important to see modern attempts at reform in the light of reforms that were promulgated, and succeeded or failed, in earlier decades and centuries. Teachers and students, as well as other readers, will easily be able to find the connections to different levels and fields. Teachers can readily note connections of the curriculum, the history of mathematics, methodology, and the evolution of that methodology. The encyclopedia consists of over 450 signed articles, arranged alphabetically and written by experts in the field. The articles, varying in length from brief paragraphs to several pages, represent diverse perspectives and provide basic information. Selected references at the end of entries provide recommended sources for further research and in-depth study.

We hope that this volume will be useful not only to educators, researchers, and students, but also to scientists, government workers, policymakers, commercial and academic test developers, parents, and the public at large. Here is a source of information for answering unusual questions or questions that are simply outside a user's specialty. The encyclopedia should help to convey the beauty and power of mathematics and to counteract the tendency to view mathematics as "symbol-pushing." Although arranged alphabetically, entries may be grouped into the following major categories: (a) applications—realistic problems and materials to motivate and enhance learning, including such topics as the environment, health care, and statistics; (b) assessment—the evaluation of achievement, instructional programs, instructors, and textbooks; (c) content and instructional procedures—an overview of fundamental concepts of mathematics and methodology for teaching many areas such as algebra, arithmetic, calculus, and geometry; (d) issues—currently debated concerns such as the gifted student, multicultural instruction, underachievers, and women in mathematics; and (e) psychology of learning—theories and their classroom application including cognitive psychology, constructivism, disorders of mathematical learning, and psychology of instruction. Many entries fit into more than one category.

To the members of the advisory board and the 250 authors of the entries goes the credit for making this encyclopedia possible. It should be known that they were a uniformly cooperative group, understanding, patient, and helpful. It is impossible, of course, to fulfill every reader's needs or expectations, but we did our best to make judicious choices. We came together to produce this volume at the invitation of Marie Ellen Larcada, who had the original idea, an idea that we embraced because we thought that such a volume would be a useful contribution to the reference literature and fascinating to develop. We felt that our combined experiences in teaching, research, writing, and editing would enable us to meet the considerable challenges involved.

We appreciate the assistance of all the librarians at Kingsborough Community College in the task of crosschecking citations and references. Jeanne Galvin, Allan Mirwis, Coleridge Orr, Roberta Pike, and Angelo Trippichio were especially helpful. Our thanks and appreciation are also extended to the librarians at Brooklyn College, Hunter College, Columbia University, Teachers College, and New York University. They provided valuable help in answering bibliographic questions. Our thanks and appreciation are also tendered to Lucia De Natale and Helen and Masha Kovarsky for skilled assistance in disk format-

ting, and to Marinella Mosheni for resolving computer problems.

Special recognition must be given to Marie Ellen Larcada who inspired us to undertake this project and provided encouragement and invaluable guidance throughout the initial stages of its development. We thank Audrey Leung for her help and advice during a transition phase. Seema Shah was a very able editorial assistant. We are grateful to the anonymous reviewers for their insightful and wise suggestions.

Many thanks to Emily Autumn, production editor at Clarinda Publication Services, who has been invaluable in getting the manuscript to polished final form. We want to express our gratitude and appreciation to Andrew Bailis, production editor at RoutledgeFalmer for his excellent suggestions and conscientious supervision of the final stages of the project.

We want to thank our husbands, Jack and Bob, and our families, for their encouragement, optimism, and warm, generous helpfulness.

Louise S. Grinstein
Sally I. Lipsey

# Suggestions for the Reader

The following information and suggestions are presented to make it easy to use the encyclopedia. For help in locating appropriate entries in the main section, the reader can consult *Sample Entries by Category*, the *Alphabetical List of Entries*, and the *Index*.

## Alphabetical Entries in the Main Section (pp. 1–840)

The main section of the encyclopedia consists of entries arranged alphabetically without regard to particular subdivisions of mathematics education. Thus, for instance, under *A*, the reader will find *Assessment of Student Achievement, Overview;* followed by *Association for Women in Mathematics; Association of Teachers of Mathematics;* and *Asymptotes.* For the complete list of entries in the main section, see *Alphabetical List of Entries* (p. vii). The typical signed entry consists of an essay, cross-references, and recommended reading for further research. Cross-references direct the reader to additional entries, covering different but related topics or different perspectives on the same topic. Some entries also have lists of other resources, such as professional organizations or commerical contacts.

## Sample Entries by Category (p. xxvii)

Consult this section to gain an overall view of entries belonging to a particular subdivision of mathematics education. Thus, for example, to determine which entries focus on the role of the government in mathematics education, read the list under the category *Governmental Role*. The pertinent entries suggested there are *Federal Government, Role; Mathematics Education, Statistical Indicators;* and *State Government, Role.*

## Index (p. 841)

When interested in a specific topic that does not appear as an entry in the main section, consult the index for the entry heading and pages where the topic can be found. For example, "environmental mathematics" can be found directly under that heading in the main section. On the other hand, "absolute geometry" cannot be found as an entry heading, but the index refers the reader to the entry entitled "neutral geometry."

# Sample Entries by Category

## APPLICATIONS

*Sample entries:* Actuarial Mathematics; Applications for the Classroom, Overview; Economics; Environmental Mathematics; Health; Modeling; Numerical Analysis, Methods; Operations Research; Population Growth; Sports

## ASSESSMENT

*Sample entries:* Assessment of Student Achievement, Overview; Evaluation of Instruction, Overview; National Assessment of Educational Progress (NAEP)

## CONTENT and INSTRUCTIONAL PROCEDURES

*Sample entries:* Abstract Algebra; Addition; Algebraic Numbers; Algorithms; Arithmetic; Axioms; Calculus, Overview; Continuity; Data Representation; Decimal (Hindu-Arabic) System; Inequalities; Inverse Functions; Mathematics, Foundations; Mathematics, Nature; Number Sense and Numeration; Number Theory, Overview; Probability in Elementary School Mathematics; Statistics, Overview

## CURRICULUM

*Sample entries:* Addition; Curriculum Trends, Secondary Level; Discrete Mathematics; Integration of Elementary School Mathematics Instruction with Other Subjects; Standards, Curriculum

## ENRICHMENT

*Sample entries:* Art; Enrichment, Overview; Family Math and *Matemática Para la Familia;* Games; Projects for Students; Recreations, Overview

## GOVERNMENTAL ROLE

*Sample entries:* Federal Government, Role; Mathematics Education, Statistical Indicators; State Government, Role

## HISTORY OF MATHEMATICS

*Sample entries:* History of Mathematics, Overview; Abel; Archimedes; Bernoulli Family; etc.; African Customs as Applied to the Mathematics Classroom; Babylonian Mathematics; Islamic Mathematics; etc.

## HISTORY OF MATHEMATICS EDUCATION

*Sample entries:* History of Mathematics Education in the United States, Overview; Brooks; Dienes; Smith; etc.; Committee of Ten; School Mathematics Study Group (SMSG); etc.

## INSTRUCTIONAL METHODS and MATERIALS

*Sample entries:* Abacus; Ability Grouping; Adult Education; Attribute Blocks; Calculators; Computers; Homework; Instructional Methods, New Directions; Software; Technology; Textbooks; Van Hiele Levels

## INTERNATIONAL COMPARISONS

*Sample entries:* International Comparisons, Overview; International Studies of Mathematics Education; China, People's Republic; England; Japan; Russia; etc.

## ISSUES

*Sample entries:* Assessment of Student Achievement, Issues; Basic Skills; Developmental (Remedial) Mathematics; Ethnomathematics; Evaluation of Instruction, Issues; Gifted, Special Programs; Learning Disabilities, Overview; Mathematics Anxiety; Minorities, Career Problems; Minorities, Educational Problems; Multicultural Mathematics, Issues; Philosophical Perspectives on Mathematics; Underachievers; Women and Mathematics, Problems

## PSYCHOLOGY OF LEARNING and INSTRUCTION

*Sample entries:* Ability grouping; Advising; Cognitive Learning; Constructivism; Cooperative Learning; Goals of Mathematics Instruction; Humanistic Mathematics; Learning Disabilities, Overview; Psychology of Learning and Instruction, Overview

## RESEARCH and REPRESENTATIVE EXPERIMENTAL PROGRAMS

*Sample entries:* Research in Mathematics Education; University of Chicago School Mathematics Project (UCSMP)

## RESOURCES

*Sample entries:* American Mathematical Association of Two-Year Colleges (AMATYC); American Mathematical Society (AMS); American Statistical Association (ASA); Mathematical Association of America (MAA); National Council of Teachers of Mathematics (NCTM); United States Department of Education

## TEACHER PREPARATION and DEVELOPMENT

*Sample entries:* Teacher Development, Secondary School; Teacher Preparation, College; Teacher Preparation in Statistics, Issues

# Encyclopedia of
# Mathematics Education

# A

## ABACUS

Calculating device used for basic arithmetic operations, containing moving beads or counters. The Latin word *abacus* evolved from the Greek word *abax*, for a flat surface; this ancient Greek word evolved from a Semitic word meaning "dust." In ancient times, computations were completed by making easily erased impressions on sand or dust tables. Two specimens of the Roman hand abacus are on display in Paris and Rome. Each consists of two rows of grooves for counters or beads. Between the two rows of grooves are the Roman symbols for the integers 1, 10, 100, 1,000, 10,000, 100,000, and 1,000,000, and for the rational numbers $\frac{1}{2}$, $\frac{1}{4}$, and $\frac{1}{3}$.

These abaci were used for simple computations such as addition. Each upper row contained one five-group bead; each lower row contained four unit-group beads. To represent the number 60, a five-group bead in the upper row of the ten groove was moved downward to the bridge separating the upper and lower rows in the ten groove; then a unit-group bead in the lower row of the ten groove was moved upward to the bridge.

The Japanese abacus, called the *soroban,* is a rectangular frame with seventeen horizontal rods or wires. A long vertical strip separates one bead in the upper row from five beads in the lower row. The soroban does not have special rods for fractions. Both the Roman abacus and the Japanese soroban use quinary and decimal grouping of counters. Using a soroban to perform all four arithmetic operations is an art of efficient counter movement and of accurate mental mathematics related to place value rules, multiplication facts, and quick number combinations.

The Chinese *suan pan* has two five-group beads in the upper row and five unit-group beads in the lower row, so computations are easier to perform. The Russian *s'choty* abacus has ten beads on each wire and no five-group beads. The fifth and sixth bead are of a different color to help with ease of counting.

In classrooms, the abacus helps students concretely visualize abstract number concepts and processes such as place value, and model computations such as addition and subtraction. To add 142 to 60 depicted on the Roman abacus (see Figure 1), a student needs to move two unit-beads, four ten-beads, and one hundred-bead toward the bridge. First, the two unit-beads move toward the bridge above. The next attempt to move four ten-beads toward the bridge above is impossible, since there are only three ten-beads available. The six ten-beads already on the abacus summed to the four ten-beads required for the problem solution adds up to ten ten-beads. The student learns to move all the ten-beads away from the bridge and to move one hundred-bead

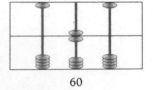 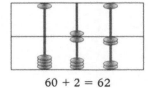 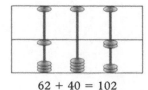 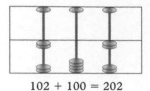

| 60 | 60 + 2 = 62 | 62 + 40 = 102 | 102 + 100 = 202 |

**Figure 1**

to the bridge. This is the process of carrying. Finally, the student moves a second hundred-bead to the bridge to compute the final sum of 202.

*See also* Chinese Mathematics; Japan; Techniques and Materials for Enrichment Learning and Teaching

### SELECTED REFERENCES

Davis, Harold T. "The History of Computation." In *Historical Topics for the Mathematics Classroom* 2nd ed. (pp. 91–93, 117–119), John K. Baumgart et al. eds. Reston, VA: National Council of Teachers of Mathematics, 1989.

Dirks, Michael K. "The Integer Abacus." *Arithmetic Teacher* 31(Mar. 1984):50–54.

Menninger, Karl. *Number Words and Number Symbols: A Cultural History of Numbers.* Paul Broneer, trans. Cambridge, MA: MIT Press, 1969.

Pullan, J. M. *The History of the Abacus.* New York: Praeger, 1970.

Reynolds, Barbara E. "The Algorists vs. the Abacists: An Ancient Controversy on the Use of Calculators." *College Mathematics Journal* 24(May 1993):218–223.

REGINA BARON BRUNNER

# ABEL, NIELS HENRIK
# (1802–1829)

Norwegian mathematician who made fundamental contributions in several areas. He is perhaps most famous for his studies of algebraic solutions to polynomial equations. On the one hand, he proved there exists no algebraic formula that can be used to solve all equations of a given degree if the degree is five or greater. On the other hand, he proved that if the group associated with a particular equation is commutative, then an algebraic formula does exist for that equation. It is because of this last result that commutative groups are generally called Abelian groups.

Abel also pioneered the theory of elliptic functions. A complex function $f(z)$ is elliptic if it is doubly periodic, that is, if there exist complex constants $u$ and $v$ such that for all $z$, $f(z) = f(z + u + v)$. Among his other accomplishments are the extension of the binomial theorem to complex exponents and one of the first rigorous examinations of the convergence of infinite series.

Because his birthplace, Norway, was far removed from the mathematical mainstream, Abel traveled to Paris and Berlin in 1825 in hopes of gaining the attention of the international mathematical community. His efforts were not entirely successful, and he returned to Norway in 1827 exhausted and in poor health. He died of tuberculosis in 1829.

*See also* Group Theory, Overview; History of Mathematics, Overview; Series

### SELECTED REFERENCES

Boyer, Carl B. *A History of Mathematics.* 2nd ed., rev. Revised by Uta C. Merzbach. New York: Wiley, 1991.

Cooke, Roger. "Abelian integrals." In *Companion Encyclopedia of the History and Philosophy of the Mathematical Sciences* (pp. 540–544). Ivor Grattan-Guinness, ed. London, England: Routledge, 1994.

Kline, Morris. *Mathematical Thought from Ancient to Modern Times.* New York: Oxford University Press, 1972.

Öre, Oystein. *Niels Henrik Abel.* Minneapolis, MN: University of Minnesota Press, 1957.

———. "Niels Henrik Abel." In *Dictionary of Scientific Biography,* Vol 1 (pp. 12–17). Charles C. Gillispie, ed. New York: Scribner's, 1973.

Stillwell, John. *Mathematics and Its History.* New York: Springer-Verlag, 1991.

DANIEL S. ALEXANDER

# ABILITY GROUPING

A method of dividing students into homogeneous groups. The many types of "grouping" or "tracking" methods have their roots in the 1920s in the work of forerunners such as Helen Parkhurst ("Dalton Plan"), Carleton Washburne ("Winnetka Plan"), Celestin Frenet (France), and Peter Petersen (Germany, "Jena Plan"). The varied development since then depended on the political and social affairs of the different countries. It is the aim of school to encourage each student to attain the highest possible level of his or her intellectual capability. To achieve this objective, students may be divided into ability groups, which are as homogeneous as possible. These employ three main forms, called streaming (or tracking), setting (or environment), and ability grouping within a class.

## STREAMING

Tracking the students in "streams" (such as "low," "middle," and "quick") according to their cognitive abilities, usually global criteria obtained from intelligence tests. The premise here is that: (a) differences in talent are inborn and thus stable and resistant to the environment; and (b) differences in inborn talent form a basis that is general and equally valid for differences in the results of learning in all subjects. Streams remain permanent for several years, often without any reevaluation.

## SETTING

Grouping the students according to their ability to learn in each subject. This subject-related ability grouping can be better than streaming with regard to optimal matching of course content to the particular level of development of the students from subject to subject. Research studies show that "the optimal learning environment, of course, for any student, would be when the level of material and pace of instruction are individually matched with ability . . . and slightly above the student's current level of performance . . ." (see Mills et al. 1994, 495). In general not all the teaching is done in subject-related ability groups. Some teaching is done in other groupings and other didactic forms such as "core teaching" or in the form of "projects." For example, all pupils of the same year who are weak mathematicians are collected in one group, and gifted mathematicians in another.

## ABILITY GROUPING WITHIN A CLASS

Forming groups for a limited time and for a particular purpose, such as to encourage weaker students, or to encourage especially gifted students in a defined area. This is based on the theory that in any particular learning situation intellectual ability depends on the following variables: (a) the subject-related level of development (e.g., knowledge and skills, linguistic competence, ability to apply rules of logical thinking); (b) the cognitive styles of processing information (e.g., ways of coding, storing and decoding

information, capacity for divergent thinking and creativity); and (c) motivation (e.g., level of attentiveness). For example, all pupils of a class who have difficulties in calculating with fractions are collected and form an ability group for one or two lessons.

Ability grouping makes the variables of the subject-related level of development more homogeneous in one learning group. There is then a narrower range of abilities in the learning group; it becomes more uniform. There is no upper ability group that is understretched and no lower ability group that is overstretched. According to the principle of optimal matching (i.e., the level of difficulty of the course content only continuously overstretches the level of ability of the students to a small extent), the teacher can direct his teaching at whole learning groups simultaneously. Homogeneous learning groups may be superior to nonhomogeneous learning groups, especially when the two following further conditions have been satisfied: (a) the structure of the subject develops in stages from general basic principles, as is particularly the case with mathematics and foreign languages; and (b) the subject matter is presented to the students over fairly long periods of time in the form of what is known as teacher-centered teaching. The latter condition implies that homogeneous learning groups, in a mathematics lesson structured as learning by doing, or discovery learning, have no advantage over heterogeneous learning groups.

Since the acquisition of social competence is now receiving more emphasis, other criteria for forming groups are becoming more important. It is no longer the homogeneous intellectual ability, which may actually hinder social maturity, but the social competence of the students that is becoming decisive in allocating groups. The ability to communicate, cooperate, argue, interact, and show responsibility are what is necessary for successful teamwork. Homogeneous ability groups are not superior to heterogeneous ability groups with regard to this objective.

*See also* Cooperative Learning; Gifted; Gifted, Special Programs; Underachievers; Underachievers, Special Programs

## SELECTED REFERENCES

Cobb, Paul. "Some Thoughts about Individual Learning, Group Development, and Social Interaction." In *Proceedings of the 15th PME Conference*, Vol. 1 (pp. 231–238). Fulvia Furinghetti, ed. Genoa, Italy: University of Genoa, 1991.

———, Erna Yackel, and Terry Wood. "Interaction and Learning in Mathematics Classroom Situations." *Educational Studies in Mathematics* 23(1)(1992):99–122.

Galton, Maurice, and John Williamson. *Group Work in the Primary Classroom*. London, England: Routledge, 1992.

Linchevski, Liora, and Bilha Kutscher. "Tell Me With Whom You're Learning, and I'll Tell You How Much You've Learned: Mixed-Ability Versus Same Ability Grouping in Mathematics." *Journal for Research in Mathematics Education* 29(5)(1998):533–554.

Loveless, Tom. *The Tracking Wars: State Reform Meets School Policy*. Washington, DC: Brookings Institution Press, 1999.

Maher, Carolyn A., and A. M. Martino. "The Construction of Mathematical Knowledge by Individual Children Working in Groups." In *Proceedings of the 15th PME Conference*, Vol 2 (pp. 365–372). Fulvia Furinghetti, ed. Genoa, Italy: University of Genoa, 1991.

Mills, Carol J., Karen E. Ablard, and William C. Gustin. "Academically Talented Students' Achievement in a Flexibly Paced Mathematics Program." *Journal for Research in Mathematics Education* 25(5)(1994):495–511.

Oakes, Jeannie. "Can Tracking Research Inform Practice?" *Educational Researcher* 21(4)(1992):12–21.

—— et al. *Multiplying Inequalities: The Effects of Race, Social Class, and Tracking on Opportunities to Learn Mathematics and Science*. Santa Monica, CA: Rand, 1990.

Slavin, Robert. *Cooperative Learning*. Englewood Cliffs, NJ: Prentice-Hall, 1989.

——. *A Practical Guide to Cooperative Learning*. Boston, MA: Allyn and Bacon, 1994.

Yates, Alfred, ed. *Grouping in Education*. New York: Wiley, 1966.

ALBRECHT H. ABELE

# ABSTRACT ALGEBRA

A branch of mathematics that deals with the construction and properties of algebraic structures and is occasionally called "modern" algebra. Mathematicians have created many different types of algebraic structures, depending upon which properties one wishes to model. Abstract algebra focuses primarily on three such structures, *groups, rings,* and *fields*. The concepts of group and field came about from nineteenth-century attempts to solve algebraic equations by means of radicals. Ring theory, on the other hand, arose from a need to better understand the many types of number systems that were being discovered and studied during the late nineteenth century.

## THE BEGINNINGS OF GROUP THEORY

Few centuries have had as profound an effect on the evolution of mathematical concepts as the nineteenth century. At that time, the foundations of arithmetic and algebra were formalized and axiomatized in much the same way that Euclid axiomatized the basic ideas of geometry. The nineteenth century also was a time when the fundamental concepts of group theory emerged. The concept of a group evolved out of early nineteenth-century attempts to solve algebraic equations by means of radicals; that is, to express the roots of an algebraic equation by means of addition, subtraction, multiplication, division, and extraction of roots applied to the coefficients of the equation. The two roots of the general quadratic equation $ax^2 + bx + c = 0$ are given by the formulas

$$x = \frac{-b \pm \sqrt{b^2 - 4ac}}{2a}$$

These formulas demonstrate that the roots may be expressed using the operations of addition, subtraction, multiplication, division, and extraction of roots when applied to the coefficients $a, b, c$ of the equation.

The search for formulas for the solution of algebraic equations had been an intriguing mathematical activity for centuries. The Greeks developed geometric algorithms for constructing the roots of certain linear and quadratic equations. Later, Renaissance mathematicians obtained formulas for solving the general algebraic equations of degree three and four, using clever algebraic manipulations. They transformed the original equation into another equation that was easily solved and whose roots could be transformed back to give the roots of the original equations.

During the next few centuries there were many attempts to imitate these methods for higher degree equations, especially for the general equation of degree five, the quintic equation, in hope of finding formulas for their roots. These efforts were largely unsuccessful, except in special cases. By the close of the eighteenth century there were still no formulas for the solution of the general quintic or higher degree equations—only a strong belief that such formulas existed and could be derived by a sufficient amount of ingenuity and algebraic manipulation.

This belief was shattered in 1826 when Niels Henrik Abel (1802–1829) showed that the general quintic is, in fact, not solvable by radicals. Then in 1832 Evariste Galois answered the entire question of solvability by creating a general theory for dealing with the solvability of any algebraic equation regardless of its degree. He first introduced the concept of a group, and then associated a group with every algebraic equation. By studying the relationship between the solvability of the equation and the properties of its group, Galois showed that the general algebraic equation of degree $n$ is not solvable by radicals whenever $n > 4$.

For Galois, the important issue was not the roots of the equation, but rather the possible ways in which the roots may be rearranged, or permuted, and still remain roots. He associated with any algebraic equation those permutations of the roots that leave the coefficients unchanged, showed that these permutations form a group, and that the properties of this group relate directly to the solvability of the equation by means of radicals. For example, the equation $x^4 - 1 = 0$ is first factored and written in the form

$$x^4 - 1 = (x^2 - 1)(x^2 + 1) = (x - 1)(x + 1)(x^2 + 1) = 0$$

Each of the three factors is irreducible over the rational numbers; that is, they cannot be further factored into a product of polynomials with rational number coefficients. The roots of $x^4 - 1 = 0$ are 1, $-1$, $i$, and $-i$. If the roots 1 and $i$ are interchanged, then the factor $x - 1$ becomes $x - i$, which is no longer a polynomial having rational number coefficients. Of the twenty-four possible permutations of the four roots, only the identity permutation and the permutation that interchanges $i$ and $-i$ leave the three irreducible factors unchanged. The Galois group of the equation consists of two permutations: the identity permutation and the permutation that interchanges $i$ and $-i$. By associating a group of permutations with an arbitrary algebraic equation, such as $x^4 - 1 = 0$, not just the general equation

$$ax^4 + bx^3 + cx^2 + dx + e = 0$$

and by using the properties of the group to study the solvability of the equation, Galois created a completely new and abstract approach to the question of solvability of algebraic equations. The central result of Galois theory states that an algebraic equation is solvable by radicals if and only if its Galois group has subgroups that satisfy a certain technical requirement. His work not only gave a complete solution of the solvability question, but also contained many of the basic concepts of group theory that would emerge later in the century: subgroup, cosets, normal subgroup, and quotient group.

## DEVELOPMENTS IN GROUP THEORY AFTER GALOIS

It would be nearly two decades before Galois's ideas were clarified and given a formal exposition. The first textbook to deal with the subject was the third edition of *Cours d'Algèbre Supérieure*, written in 1866 by Joseph-Alfred Serret (1819–1885). This was followed in 1870 by the popular text *Traité des Substitutions et des Équations Algébriques*, written by Camille Jordan (1838–1922). It is in Jordan's work that we find the term *simple group* used for the first time to describe a group that has no proper normal subgroups.

During the second half of the nineteenth century most of the basic concepts in group theory began to crystallize. The axioms defining a group were first formulated in 1854 by Arthur Cayley (1821–1895). Cayley is important in the development of group theory because of his abstract, axiomatic approach to the subject—an approach that was not popular at the time. Today we honor Cayley by referring to the multiplication table of a group as its Cayley table.

Research in group theory during the early years of the twentieth century turned to the question of determining the structure of groups—how are groups built up from subgroups, and what, if any, are the basic building blocks? These questions led to the development of extension theory. Extension theory showed that it is the simple groups that are the basic building blocks from which all finite groups are constructed. In much the same way that every positive integer can be written as a product of prime numbers, so every finite group may be constructed from finite simple groups.

The period from 1960 to 1980 saw an intense effort to find and classify all finite simple groups, much of it inspired and led by Daniel Gorenstein. In 1981 Gorenstein announced to the mathematical community what many believed to be impossible: "In February, 1981, the classification of the finite simple groups was completed, representing one of the most remarkable achievements in the history of mathematics. Involving the combined efforts of several hundred mathematicians from around the world over a period of thirty years, the full proof covered something between 5,000 and 10,000 journal pages, spread over 300 to 500 individual papers" (Gorenstein 1982, 1). Although the structure of every finite group is now known, at least theoretically, the work of understanding and interpreting the consequences of these results is just beginning.

## RINGS: NUMBERS AND MORE NUMBERS

By the middle of the nineteenth century, mathematicians began to axiomatize the foundations of arithmetic. What are integers? What are real numbers? Giuseppe Peano, Karl Weierstrass, Augustin-Louis Cauchy, Richard Dedekind, and others were developing axiomatic systems for constructing various number systems. Out of these efforts came the realization that the number systems themselves were remarkably similar from an algebraic point of view. Each had certain basic objects, the numbers, and two basic operations, addition and multiplication. The properties of the operations were essentially the same, regardless of the number system, although the properties of the numbers themselves changed from one system to the next. In an effort to unify these ideas, David Hilbert (1862–1943) prepared a report summarizing the current state of knowledge at that time. In this work the beginnings of ring theory are found.

The term *ring* was first used by Hilbert in 1897 to describe systems of real and complex numbers. It did not take on its modern, abstract meaning until 1914, when Abraham Fraenkel (1891–1965) gave a general definition that is essentially the same one used today. The concept itself stems from two distinct lines of investigation. The first goes back to early attempts by the British school of algebraists, notably George Boole (1815–1864), Augustus De-Morgan (1806–1871), and George Peacock (1791–1858), to formulate a general axiomatic approach to algebra, much the same way that Euclid formalized geometry. In his textbook, *A Treatise on Algebra,* published in 1830, Peacock suggested for the first time that the symbols of algebra need not refer to numbers, but may represent more general mathematical objects. His book represented an important step in the evolution of mathematical thinking because, up to this point, algebra was always discussed within the framework of numbers. The discovery of quaternions by William Rowan Hamilton (1805–1865) in 1843, and Arthur Cayley's introduction of matrix algebras in 1858 soon led to the general study of hypercomplex number systems and their eventual classification by Eli Cartan (1869–1951), Benjamin Peirce (1809–1880), and Peirce's son Charles Sanders Peirce (1839–1914). This branch in the evolution of noncommutative ring theory culminated with the work of Joseph Henry Maclagan Wedderburn (1882–1948), who developed the general structure theory for number systems over arbitrary fields.

## IDEALS: SAVING UNIQUE FACTORIZATION

The second line of investigation and the one that led to the evolution of commutative ring theory and the theory of ideals emerged during the period 1844–1897 from the work of Ernst Edward Kummer (1810–1893), Leopold Kronecker (1823–1891), and Richard Dedekind (1831–1916) on algebraic numbers. An algebraic number is any complex number that is the root of a polynomial whose coefficients are rational numbers. Numbers such as $\sqrt{2}$ and $i$ are algebraic numbers. The ring of Gaussian integers, which consists of all complex numbers of the form $n + mi$, where $n$ and $m$ are integers, and which are added and multiplied as complex numbers, is a ring of algebraic numbers. Numbers such as $\pi$ and $e$, however, are not algebraic; they are not the root of any algebraic equation with rational number coefficients. The original motivation for studying algebraic numbers goes back to attempts to prove Fermat's Last Theorem.

Kummer, Kronecker, Dedekind, and Hilbert were beginning to develop a general theory of commutative number systems in an attempt to extend the ideas of unique factorization from the integers to rings of algebraic numbers. If unique factorization remained valid in rings of algebraic numbers, they could prove Fermat's Last Theorem. Recall that, according to the fundamental theorem of arithmetic, every positive integer greater than 1 may be written as the product of uniquely determined prime numbers. Thus, $6 = 2 \times 3$ and $12 = 2 \times 2 \times 3$, where 2 and 3 are prime numbers. But the number 2 is no longer prime when viewed in the ring of Gaussian integers, since it may be factored as

$$2 = (1 - i)(1 + i)$$

Is it still true in the ring of Gaussian integers that every number is the product of primes? By extending the notion of prime number to the ring of Gaussian integers and other such rings, Kummer, Kronecker, Dedekind, and Hilbert hoped to develop a general theory of factorization similar to the fundamental theorem of arithmetic but applicable to more general number systems, thus saving unique factorization and proving Fermat's Last Theorem. But there are rings of algebraic numbers in which unique factorization is not valid. To save unique factorization required giving up numbers themselves and replacing them by a new object, called an ideal.

## DEDEKIND'S IDEALS

An *ideal* is a collection of elements $a, b, c, \ldots ,$ in a ring with the property that $a + b$, $a - b$, and $ra$ are in the collection whenever $a$ and $b$ are in the collection and $r$ is any element of the ring. For example, the set of even integers is an ideal of the ring of integers, since the sum and difference of two even integers is an even integer, and the product of any integer with an even integer is an even integer. Dedekind identified the concept of a prime ideal, and then showed that in number rings, every ideal may be expressed as a product of prime ideals that are essentially uniquely determined.

Dedekind's approach was revolutionary not only because it provided a framework in which to extend classical ideas about unique factorization but also because it represented a profound change in mathematical thinking. Infinite sets, in the form of ideals, were finally recognized as completed entities in and of themselves. Dedekind's theory of ideals was extended by twentieth-century mathematicians, notably Emmy Noether, Wolfgang Krull, and Oscar Zariski, and today forms the basis for the theory of commutative algebra, algebraic geometry, and algebraic number theory.

## THE GALOIS THEORY OF FIELDS

Field theory, on the other hand, remained closely associated with the solution of algebraic equations and Galois theory for nearly a century. In 1910 the German mathematician Ernst Steinitz (1871–1928) gave the more general, abstract definition of a field that is used today. Steinitz introduced many of the concepts that dominate current field theory, but it was Emil Artin (1898–1962) who reformulated Galois theory, and with it much of field theory.

Artin's approach was to view a field as a vector space over any of its subfields and then use vector space techniques to study the properties of the field extension. For example, a complex number is a number of the form $a + bi$. Such a number may be written as the sum $a(1) + b(i)$, where the numbers 1 and $i$ are regarded as vectors in the complex plane and $a$ and $b$ are coefficients. Vector addition and scalar multiplication correspond to addition of complex numbers and multiplication of a complex number by a real number, respectively. In this way the complex number field is a two-dimensional extension of the field of real numbers. Using these techniques for any field over any subfield, Artin successfully dissociated Galois theory from its classical origins in the solvability of algebraic equations and reformulated it as an abstract relationship between an extension of fields and its group of automorphisms. In the modern Galois theory of fields, the central focus is on the field extension $K/k$, its automorphism group $G(K/k)$, and the relationship between intermediate subfields of $K/k$ and subgroups of $G(K/k)$.

Galois theory not only provides the necessary mathematical tools for dealing with such classical problems as the solvability of algebraic equations and the impossibility of the classical Greek construction problems, but, when applied to finite fields, plays an important role in such contemporary areas of mathematics as combinatorics and coding theory.

*See also* Algebraic Numbers; Fermat; Fields; Group Theory, Overview; Modern Mathematics; Vector Spaces

## SELECTED REFERENCES

Boyer, Carl B. *A History of Mathematics.* 2nd ed., rev. Revised by Uta C. Merzbach. New York: Wiley, 1991.

Edwards, Harold M. "Dedekind's Invention of Ideals." In *Studies in the History of Mathematics* (pp. 8–20). Esther R. Phillips, ed. Washington, DC: Mathematical Association of America, 1987.

Gallian, Joseph. "The Search for Finite Simple Groups." *Mathematics Magazine* 49(4)(1976):163–179.

Gorenstein, Daniel. *Finite Simple Groups: An Introduction to their Classification.* New York: Plenum, 1982.

Kiernan, B. Melvin. "The Development of Galois Theory from Lagrange to Artin." *Archive for History of Exact Sciences* 8(1971–1972):40–154.

Kleiner, Israel. "The Evolution of Group Theory: A Brief Survey." *Mathematics Magazine* 59(4)(1986): 195–215.

———. "Thinking the Unthinkable: The Story of Complex Numbers (with a Moral)." *Mathematics Teacher* 81(7)(1988):583–592.

Reid, Constance. *Hilbert.* New York: Springer-Verlag, 1970.

Wussing, Hans. *The Genesis of the Abstract Group Concept.* Abe Shenitzer, trans. Cambridge, MA: MIT Press, 1984.

Zassenhaus, Hans. "Emil Artin, His Life and His Work." *Notre Dame Journal of Formal Logic* 5(1) (1964):1–9.

DENNIS KLETZING

# ACTUARIAL MATHEMATICS

Combines the mathematics of compound interest with probability concepts. Originally the word *actuary* meant a clerk or registrar. Now, however, it denotes a mathematically trained professional working mainly in the insurance industry. Actuaries translate uncertainties into financial terms. Using data on past experience, they predict the likelihood of future events and estimate the cost of paying for those events.

Actuaries work for life and health insurance companies, property and casualty insurance companies (e.g., auto and home insurance), pension plans, the Social Security Administration, consulting, accounting and investment firms, and state insurance departments. There are approximately fifteen thousand members of the Society of Actuaries (SOA; Life, Health and Pensions) and two thousand fellows of the Casualty Actuarial Society (CAS). Actuaries who determine the soundness of pension plans are licensed by the United States Government under the Employee Retirement Income Security Act of 1974 (ERISA) as Enrolled Actuaries. Most other countries have their own actuarial organizations and training requirements, although the SOA exams are offered in several other countries. For example, SOA sponsors a project to send members as teachers to Nanking University to train China's first actuaries. The International Actuarial Association organizes congresses every three or four years for the exchange of ideas between actuaries from different countries.

Within insurance companies, actuaries have two main roles: product development and financial reporting. To determine the price of new products, they evaluate new product features and decide appropriate levels of mortality, expenses, and expected interest rates to produce a product both attractive and useful to customers and profitable to the insurer. Financial actuaries work with accountants to report a firm's current financial health and project future financial events. Their major concern is the accurate calculation of *reserves,* money set aside to cover future expected payouts. Financial reports are used internally by management for planning purposes and also may be submitted to state insurance departments, the Securities Exchange Commission or credit rating agencies like Moody's and Standard & Poors. Recently, there has been a greater emphasis on how insurance companies invest their money. Actuaries now must confirm that future asset income will match expected future claim payments.

The basic tool used by actuaries to make these estimates and predictions is actuarial mathematics, the combination of the mathematics of compound interest with probability concepts. Compound interest or "the time value of money" depends on the principle that since money deposited in a bank or invested elsewhere earns income, a dollar today is worth more than a dollar a year from now. Cash payments to be received $n$ years into the future are *discounted:*

$$\text{present value (payment)} = \frac{\text{payment}}{(1 + i)^n}$$

and past receipts are *accumulated:*

accumulated value (payment) = payment $\cdot (1 + i)^n$

to produce the *present value* (where $i$ indicates the interest rate earned). These concepts can be used to calculate the amortization schedule for a home mortgage (the rate at which a loan is repaid and the interest and principal remaining at any date). A series of discounted or accumulated payments can be condensed into a simple formula through algebraic analysis. For example, the value of a savings account after ten years if $1,000 is deposited at the beginning of each year is equal to

$$1000 \cdot [(1 + i) + (1 + i)^2 + (1 + i)^3 + \cdots + (1 + i)^{10}]$$

$$= 1000 \cdot (1 + i) \cdot [\frac{(1 + i)^{10} - 1}{i}]$$

If an assumption is made about the likelihood of a risk occurring (for instance, death or sickness) during a specified period, it is possible to calculate the cost of making payments for that risk. For example, defining $z$ as the starting age of the annuitant, $\omega$ as the age beyond which there are assumed to be no survivors; the probability that someone aged $z$ will live $t$ more years as $_t\mathrm{p}_z = Pr[(x) = z \mid (x) = z + t]$; and the discount rate

$$v = \frac{1}{1 + i}$$

then the value of a life annuity payable at the end of each year during the recipient's lifetime is

$$_1\mathrm{p}_z \cdot v + _2\mathrm{p}_z \cdot v^2 + _3\mathrm{p}_z \cdot v^3 + \cdots + _\omega\mathrm{p}_z \cdot v^\omega$$

With available tables for mortality and simple probability one can price and calculate reserves for complicated insurance or annuity products. An important actuarial principle states that values calculated *retrospectively* (based on the past) or *prospectively* (based on the future) must be equal. Thus, life insurance reserves are equal to the accumulated value of previous premiums minus the accumulated value of benefits already paid out or the present value of future benefits minus the present value of future premiums. Another type of reserve, commonly used for health or disability insurance, is the estimate of claims that have occurred but have not yet been reported.

To become an actuary in the United States or Canada requires passing a series of exams given by SOA or CAS. Many actuaries study independently for these tests while they are employed. It is also possible to major in Actuarial Science at over fifty American and Canadian universities. The core exams cover differential and integral calculus, linear algebra, probability and statistics, operations research, numerical analysis, compound interest, life contingencies, demography, survival models, risk theory, mortality table construction, and graduation of data. Basic mathematical tools used every day are interpolation and extrapolation, and linear regression. Computer skills are very important to actuaries. A variety of computer software, some designed specifically for actuaries, and languages are used on personal computers, mainframes, and networks.

Crude estimates of life expectancy used for annuities or leases have existed since Roman times. Beginning in the seventeenth century, attempts were made to collect actual data from governmental and church records and create a theoretical framework with which to organize it. In 1693, Edmund Halley (the discoverer of Halley's comet) compiled a mortality table from birth and death statistics from the city of Breslau in Germany. The format for the table that he invented showing age, number of people living at that age, and the number of deaths during the year is still used. Thirty years later, Abraham De Moivre theorized that of eighty-six persons at birth, one will die each succeeding year. This implies an arithmetical progression where the probability of dying (the *mortality rate*) at age $x$ is equal to

$$\frac{1}{86 - x}$$

Later mortality laws by Gompertz (1825, $\mu_x = Bc^x$), Makeham (1860, $\mu_x = A + Bc^x$), and Weibull (1939, $\mu_x = kx^n$) assume exponential and geometrical increases in the mortality rate. Current research includes both statistical and theoretical topics, ranging from studies of AIDS-related death claims to risks of default in mortgages held by insurance companies to the application of Monte Carlo methods to the graduation of data. Emphasis has shifted from deterministic toward probabilistic results, producing a range of results derived from a probability distribution, rather than the application of *mean* analysis to produce a single answer.

*See also* Interest

SELECTED REFERENCES
*Associateship and Fellowship Catalog—Spring 1995.* Schaumburg, IL: Society of Actuaries, 1994.
Bowers, Newton L., Jr., et al. *Actuarial Mathematics.* Itasca, IL: Society of Actuaries, 1986.
Ingram, Thomas Allan, and Charlton Thomas Lewis.
"Insurance." In *Encyclopaedia Britannica.* Vol. 14 (pp. 656–673). 11th ed. New York: Encyclopaedia Britannica, 1910.
Kellison, Stephen G. *The Theory of Interest.* 2nd ed. Burr Ridge, IL: Irwin, 1991.
*1994 Society of Actuaries Yearbook.* Schaumburg, IL: The Society, 1994.
Society of Actuaries. *Actuaries Make a Difference.* Schaumburg, IL: The Society.
———. *The Actuary's Career Planner.* Schaumburg, IL: The Society.
———. Website ⟨http://www.soa.org⟩
*Society of Actuaries Speakers' Kit.* Schaumburg, IL: The Society, 1993.

LISA LEFKOWITZ

# ADDITION

One of the fundamental operations of arithmetic, building on the basic skills of counting, grouping objects, measuring and comparing quantities (Fuson 1992). On an elementary level, addition is a symbolic way of combining disjoint sets of items (classified as alike) to find the sum, that is, the total number of items combined. The number of items in a set is represented by an addend that may also represent a length or other measure. The sum is the total of the addends.

Addition of fractions and decimals also is commonly taught in the elementary grades. Addition of fractions involves two distinct types of problems, those with *like* denominators and those with *unlike* denominators (Stein et al. 1997). Fraction problems with *like* denominators are solved by keeping the denominator constant while adding the numerators (e.g., $\frac{3}{5} + \frac{1}{5} = \frac{4}{5}$). Fraction problems with *unlike* denominators require conversion to a common denominator. For example, $\frac{3}{4} + \frac{1}{3}$ is solved by changing each fraction to an equivalent fraction with a denominator of 12 (e.g., $\frac{3}{4} = \frac{9}{12}$ and $\frac{1}{3} = \frac{4}{12}$) and then adding (e.g., $\frac{9}{12} + \frac{4}{12} = \frac{13}{12}$).

Addition of decimals also can be divided into two distinct problem types. The first type includes problems in which each number in the problem has the same number of decimal places (e.g., 153.95 + 19.93). In problems such as these, lining up the decimals so that they are in a column is essential to accurately solving the problem. The second type of problem includes addends that have different numbers of decimal places (e.g., 1.8 + 4.32). To solve this type of problem, the addends must be rewritten as equivalent decimals (e.g., 1.8 is rewritten as the equivalent decimal, 1.80) before students can work the problem (Stein et al. 1997).

## VARIETY OF PEDAGOGICAL APPROACHES IN THE ELEMENTARY SCHOOL

Fairly sophisticated techniques for solving addition problems develop in children before formal schooling. Children assimilate mathematics instruction into their existing cognitive structures. Some students use methods taught in class. Others combine invented strategies with formal instruction while still others invent their own methods (Houlihan and Ginsburg 1981). Formal instructional approaches for teaching addition also vary. Some approaches involve explicit instruction in both memorization of basic addition facts (combinations of addends) and application of strategies for solving addition problems, for example, direct instruction approaches (Stein et al. 1997). Other more student-centered approaches encourage teachers to design instruction based on the existing knowledge of their students that is shown, for example, by the strategies they invent to solve word problems. An example of the latter approach is cognitively guided instruction (Carpenter et al. 1997; Carpenter et al. 1996).

### Addition Facts

Using two single nonzero digits at a time as addends, eighty-one different addition facts may be formed. Different types of instructional activities used to teach these facts include activities for understanding, building relationships, and mastery. Activities for understanding include concrete or pictorial explorations of how the answer (either the sum, $4 + 3 = \rule{1cm}{0.4pt}$, or a missing addend, $\rule{1cm}{0.4pt} + 3 = 7$) is derived. The relationship activities are designed to reduce the memory requirements by capitalizing on the relationships inherent in the number system. A common relationship activity involves number families. Students learn from the fact family consisting of 2, 4, and 6, the addition facts of $2 + 4 = 6$ and $4 + 2 = 6$. Note that two subtraction facts also can be formed from this fact family (e.g., $6 - 2 = 4$ and $6 - 4 = 2$), illustrating how addition is the inverse of subtraction. Another relationship activity involves teaching students a series of facts in order (e.g., $3 + 2 = 5, 3 + 3 = 6, 3 + 4 = 7$) and then practicing the facts randomly. Finally, students are often guided to recognize *doubles,* addition facts that have two like addends, such as $4 + 4 = 8$. Mastery activities include those activities that promote automaticity. Activities such as fact games, timings, and peer tutoring with flash cards are included in this category.

### Addition Problem-Solving Strategies

Addition is represented by two basic types of problems: combine or join ($4 + 2 = \rule{1cm}{0.4pt}$) and add-to ($4 + \rule{1cm}{0.4pt} = 6$). Students may solve combining problems by counting both addends ($1,2,3,4 + 5,6$) or by starting with one of the addends and counting forward ($4 + 5,6$). Add-to problems encourage students to begin with the first addend (4) and count forward until reaching the sum (6), keeping track of how many have been counted (2). Note that add-to problems also are frequently solved using subtraction strategies (e.g., to solve $4 + \rule{1cm}{0.4pt} = 6$, students subtract, $6 - 4 = \rule{1cm}{0.4pt}$). Both types of addition problems may require computational skills involving both single digit and multidigit problems or they may be represented as word problems.

*Simple Addition* Students may learn to solve simple addition problems using a variety of strategies (Geary 1994). These strategies include counting strategies, whereby students use manipulatives or pictures to solve problems. Students also may be introduced to addition through the use of the number line. Students taught using this approach locate the first addend on the number line and then count the number of line segments corresponding to the second addend to arrive at the sum. A third approach involves tally marks. Students draw tally marks corresponding to each addend and then count the total number of tallies drawn for the sum.

*Multidigit Addition* Multidigit addition is usually performed in columns, for example,

$$\begin{array}{r} 24 \\ + 15 \\ \hline 39 \end{array}$$

When the sum of numbers is greater than 10, it is necessary to regroup to the next column. This is called *exchanging* or *regrouping* (formerly called "carrying"). This skill is difficult for many students in that it requires both mental calculations and understanding of place value (Geary 1994). Students usually are introduced to multidigit problems without regrouping prior to being introduced to regrouping. Manipulatives are often used to introduce students to regrouping. For example, students are given colored cubes to represent units and when the sum of the units column exceeds 10, students trade the ten individual units of one color for a single unit of another color representing tens numbers (e.g., 40, 50, 60).

*Field Properties of Addition* (a) *Closure*—for any real numbers, the sum is a real number; (b) *Associa-*

*tive*—any method of grouping that preserves order may be used to obtain the sum of several addends, for example, $(3 + 4) + 5 = 3 + (4 + 5)$; (c) *Commutative*—the sum is the same regardless of the order in which the numbers are added, for example, $3 + 4 = 4 + 3$; (d) *Additive Identity*—when any whole number and zero are addends, the sum is the whole number, for example, $3 + 0 = 3$; and (e) *Additive Inverse*—for any real number, there is a real number (called the inverse), such that the sum of the number and its inverse is 0, for example, $-3 + 3 = 0$. Understanding these properties is helpful to students in several ways. For example, students use the Inverse Property when they use addition to check solutions to subtraction problems $(7 - 4 = 3; 3 + 4 = 7)$. Students also use the Inverse Property as they gain understanding of number families $(3, 4, 7)$ to help them remember math facts $(3 + 4 = 7; 4 + 3 = 7; 7 - 3 = 4; 7 - 4 = 3)$.

*Word problems* Word problems require a mixture of methods including combining (also referred to as joining), classification, and add-to. Simple combining problems include those problems in which two quantities are given and the total is requested. Some combining problems contain an action verb that indicates addition (e.g., Neil has six pencils. He *finds* three more. How many pencils does he have in all?) Combining problems also may require knowledge of classification, in that students must understand that two subordinate quantities are members of a superordinate class (e.g., There are eight *boys* in the class and nine *girls*. How many *children* are in the class?). Finally, add-to problems are those in which one quantity is given, the total is given and the amount of increase is requested (Marike had three cookies. After José gave her some cookies, she had seven. How many cookies did José give her? $3 + \_\_ = 7$.)

A variety of approaches is used to teach students to solve word problems. One approach is to have students act out the events in the problems. Another common approach is to use a number line to represent the quantities in the problem. Still other approaches involve teaching students to use manipulatives or draw pictures. Students often find word problems that are relevant to their daily lives more interesting. Teachers often have students work in groups or with partners to encourage dialogue that facilitates conceptual understanding of the problem (Fernandez et al. 1994).

Student difficulties with word problems may be due to errors in computation or a lack of understanding of the verbal language of the problem.

Teachers may need to assist students with specific vocabulary as well as help them solve problems mathematically. Moreover, it is important for teachers to serve as models for mathematical thinking and not just provide answers (Fernandez et al. 1994).

## RECOMMENDED MATERIALS FOR INSTRUCTION

Teachers of beginning addition often use a variety of materials to illustrate the concept of addition. These materials may include blocks or cubes (that often combine to demonstrate place value), unifix cubes, Cuisenaire rods, or any number of everyday objects used as counters (e.g., beads, beans, plastic chips). After being introduced to addition with manipulatives, students may be given pictorial representations including the number line. For practice with introductory as well as more advanced computation and problem solving, students may use technology available in the schools. Computerized drill and practice has been found to be related to a more positive attitude toward mathematics (Watkins 1989). Also, microcomputer programs can be beneficial in monitoring skill acquisition as well as by providing practice that leads to automaticity of math facts and teaching mathematics concepts (Woodward and Carnine 1993). Finally, students may practice their addition skills with a variety of mathematics games such as dominoes.

*See also* Basic Skills; Decimal (Hindu-Arabic) System; Fractions; Subtraction

## SELECTED REFERENCES

Carpenter, Thomas P., Elizabeth Fennema, and Megan L. Franke. "Cognitively Guided Instruction: A Knowledge Base for Reform in Primary Mathematics Instruction." *Elementary School Journal* 9(1)(1996):3–20.

Carpenter, Thomas P., Megan Franke, Victoria R. Jacobs, Elizabeth Fennema, and Susan B. Empson. "A Longitudinal Study of Invention and Understanding in Children's Multidigit Addition and Subtraction." *Journal for Research in Mathematics Education* 29(1997):3–20.

Carpenter, Thomas P., and James M. Moser. "The Acquisition of Addition and Subtraction Concepts in Grades One through Three." *Journal for Research in Mathematics Education* 15(Apr. 1984):179–202.

Fernandez, Maria L., Nelda Hadaway, and James W. Wilson. "Problem Solving: Managing It All." *Mathematics Teacher* 87(Mar. 1994):195–199.

Fuson, Karen C. "Research on Whole Number Addition and Subtraction." In *Handbook of Research on Mathematics Teaching and Learning* pp. 243–275. Douglas A. Grouws, ed. New York: Macmillan, 1992.

Geary, David C. *Children's Mathematical Development.* Washington, DC: American Psychological Association, 1994.

Houlihan, Dorothy M., and Herbert Ginsburg. "The Addition Methods of First and Second Grade Children." *Journal for Research in Mathematics Education* 12(Mar. 1981):95–106.

Stein, Marcy L., and Douglas Carnine. "The Design and Delivery of Effective Mathematics Instruction." In *Teaching in American Schools: Essays in Honor of Barak Rosenshine.* R. J. Stevens, ed. Columbus, OH: Prentice-Hall, in press.

Stein, Marcy L., Jerry Silbert, and Douglas Carnine. *Designing Effective Mathematics Instruction: A Direct Instruction Approach.* 3rd ed. Columbus, OH: Merrill/Prentice-Hall, 1997.

Watkins, Marley W. "The Computerized Drill-and-Practice and Academic Attitudes of Learning Disabled Students." *Journal of Special Education Technology* 9(Spring 1989):167–172.

Woodward, John, and Douglas Carnine. "Uses of Technology for Mathematics Assessment and Instruction: Reflections on a Decade of Innovation." *Journal of Special Education Technology* 12(Spring 1993):38–48.

MARCY L. STEIN
RONDA SIMMONS

# ADULT EDUCATION

Mathematical education designed to take into account the characteristics and needs of adult learners. Many adults leave school without gaining the mathematical skills needed to function well in everyday situations or to enter or advance in the job market. The National Adult Literacy Survey (Kirsch, Jungeblut, Jenkins, and Kolstad 1993) showed that almost half of U.S. adults perform at the lowest two levels of a five-level quantitative literacy scale of everyday functional tasks. (A sample task from the second level involved determining the difference in price between tickets for two shows.) Thus, a sizable segment of the U.S. population needs to improve basic mathematical and numeracy skills; millions of adults presently study mathematics in a variety of adult literacy, General Educational Development (GED) test preparation, and workplace-based programs (Gal and Schuh 1994).

While adult learners may need to acquire the mathematical skills typically learned by children in elementary and secondary school, the extensive experiences of adults in and out of school and their life situations must be taken into account in order to design effective instruction for them. When adult students return to learning, they often bring with them a history of frustrating, unsuccessful schooling experiences and accompanying negative beliefs about their own abilities to learn mathematics. In addition, adult learning time is constrained by responsibilities outside the classroom (e.g., raising children, holding jobs) as well as by the limited hours of instructional time offered in programs for adults. Most adults enter instructional programs with well-defined goals and will choose to continue participation only if they perceive instruction to be helping them attain those goals. Goals frequently mentioned include gaining job skills, becoming better able to support their children's education, and passing a gatekeeping test such as the GED or a test to enter a job training program.

Adults have usually developed some informal mathematical knowledge that they apply to everyday tasks in their lives. This knowledge may derive in part from school learning, but also may include ideas and strategies that were self-invented as situations arose for which available strategies were inadequate. While the ideas and strategies may be meaningful to the individuals who developed them, they often are limited to the contexts within which they were developed, may contain misconceptions, and are often not part of a mathematical system with meaningful connections and relationships between and among ideas (Ginsburg, Gal, and Schuh 1995).

A number of implications for adult mathematics instruction emerge from the characteristics and needs of adult learners. First, negative attitudes and beliefs should be identified, shared, analyzed, and frequently revisited so that they do not become a stumbling block to new learning. In addition, the curriculum may need to differ from the traditional linear model (e.g., four operations with fractions, then with decimals, followed by percents) to meet the goals of the learners and address the functional demands of adults in our society. Many, but not all, adults benefit from using manipulatives and hands-on activities, but these must be age-appropriate and connections between these activities and underlying mathematical ideas should be clarified.

Finally, mathematics instruction for adults should rely heavily on the use of real-world materials and problems. It is important that adult students see the relevance and applicability of what they are learning so that they see the value of continuing their study of mathematics. In addition, learning mathematics through and for functional tasks helps adults juxtapose their self-developed and intuitive ideas about mathematical procedures and concepts with new ideas so that earlier ideas can be examined and revised as necessary (Ginsburg and Gal in press). For example, adults periodically need to decide whether to buy a new car or continue upkeep on the old car. After gathering car and loan advertisements from newspapers, much meaningful mathematics can be explored as adult students evaluate the short- and long-term financial implications of the purchase, compare alternative financing arrangements, estimate the probable expenses of continuing to run the old car, and finally determine how to weigh all the collected data to inform a purchase decision. By grounding mathematics in everyday functional tasks, concerns about the ability of adult students to transfer knowledge from the domain of the mathematics class to the domain of the real world can be reduced.

*See also* Mathematical Literacy; Mathematics Anxiety

SELECTED REFERENCES

*Adult Numeracy Instruction: A New Approach.* Videotape No. VT94–02. Philadelphia, PA: National Center on Adult Literacy, University of Pennsylvania, 1994.

Coben, Diana, John O'Donoghue, and Gail E. FitzSimons, eds. *Perspectives on Adults Learning Mathematics: Research and Practice.* Dordrecht, Netherlands: Kluwer, 2000.

Gal, Iddo, and Alex Schuh. *Who Counts in Adult Literacy Programs? A National Survey of Numeracy Education.* Technical Report No. TR94–09. Philadelphia, PA: National Center on Adult Literacy, University of Pennsylvania, 1994.

Ginsburg, Lynda, and Iddo Gal. "Instructional Principles for Adult Numeracy Education." In *Numeracy Development: A Guide for Adult Educators.* Iddo Gal, ed. Cresskill, NJ: Hampton, in press.

Ginsburg, Lynda, Iddo Gal, and Alex Schuh. *What Does 100% Juice Mean? Exploring Adult Learners' Informal Knowledge of Percent.* Technical Report No. TR95–06. Philadelphia, PA: National Center on Adult Literacy, University of Pennsylvania, 1995.

Kirsch, Irwin S., Ann Jungeblut, Lynn Jenkins, and Andrew Kolstad. *Adult Literacy in America: A First Look at the Results of the National Adult Literacy Survey.* Washington, DC: National Center for Education Statistics, U.S. Department of Education, 1993.

LYNDA B. GINSBURG

# ADVANCED PLACEMENT PROGRAM (AP)

A cooperative endeavor between secondary schools and colleges and universities to enable students who are in secondary schools throughout the United States and elsewhere to pursue college-level courses and to receive college credit. Foreign countries participating in the program include Austria, Germany, Great Britain, and Norway. During the month of May, Advanced Placement final examinations are administered by the College Entrance Examination Board. The College Board contracts with the Educational Testing Service (ETS) in Princeton, New Jersey, an independent, nonprofit agency, for technical and operational educational services.

Many students in their junior or senior year of high school opt to take AP courses taught by highly qualified teachers who are experts in the subject area. ETS hires consultants to grade the AP examinations in June. The consultants are high school instructors as well as college instructors and professors. The College Board offers final examinations in Advanced Placement Calculus AB and BC, Computer Science A, and Computer Science AB, as well as many other subjects.

Each of the calculus AB and BC final examinations has two parts; the total time allotted is three hours and fifteen minutes. The first part of the examination has two sections. One section poses twenty-eight multiple choice questions with no use of a graphing calculator and a time limit of forty-five minutes. The next section consists of seventeen questions permitting use of a graphing calculator. The time limit for this part is sixty minutes.

The second part of the examination, devoted to free response questions, is also subdivided. There are two sections each consisting of three questions and having a time allotment of forty-five minutes. In one section, use of a graphing calculator is permitted. The free response part will require students to analyze and to write proofs and conclusions based upon given data.

Each AP examination score is based on a combination of the multiple choice section and the free response problems. Each free response question carries a weight of nine credits and the time limit is ninety minutes. The grading system is based on a scale from 1 to 5, where 3 is the passing grade. After the final examinations are graded, statisticians for ETS review the raw data scores and then decide upon the cut-offs for the grades of 1 through 5. Commencing in May 1998, there were changes in grading of the BC examination. The portion of the examination extended to BC-only topics was increased to 40%. In addition to a BC calculus grade, a Calculus AB subscore will be reported based on the portion of the BC examination directly related to the AB topics.

During the month of July, students throughout the country receive the AP examination results. Each individual college or university grants students credit for the AP final examinations. The AP program is a very rewarding educational experience for high school students who want to enter the first year of college with advanced standing.

*See also* Gifted; Gifted, Special Programs; Projects for Students

## SELECTED REFERENCES

College Entrance Examination Board and Educational Testing Service. *A Guide to the Advanced Placement Program.* New York: The Board and Service, 1995.

————. *Advanced Placement Course Description. Calculus.* New York: The Board and Service.

————. *Calculus AB.* New York: The Board and Service, 1998.

————. *Calculus BC.* New York: The Board and Service, 1999.

————. *The College Board.* New York: The Board and Service.

STEVEN J. BALASIANO

# ADVISING

Programs in colleges and universities that help students with curriculum and career decisions. Studies of exemplary undergraduate mathematics programs in the United States reveal a wide variety of successful advising systems, united by a common goal. They all aim to maintain personal contact between mathematics students and what might otherwise seem an impersonal department, and are central in departmental efforts to make the mathematics major an intellectual home for mathematics students (Association of American Colleges 1990).

Increased options available to mathematics students at every level, combined with changes in student demographics and backgrounds, make advising more challenging and simultaneously more important today. Entering freshmen need reliable help in choosing their first mathematics course. Students at all levels need advice about mathematics courses that prepare them for further study in mathematics and other disciplines. Two-year college students need advice about continued study at four-year schools and transfer students need advice about how courses taken elsewhere fit with local curricula and requirements. Mathematics majors need assistance in choosing among the many options available in today's mathematical sciences programs, and the advisor's perspectives can help students build coherence into their programs of study. Majors also need information to help them make postgraduation plans, whether for immediate employment or for continuing studies.

Counseling new students about their initial mathematics courses is sometimes the responsibility of the mathematics department, and sometimes of a central advising office. Whichever is the case, the mathematics department must play a central role in the design and interpretation of placement tests. Placement recommendations are usually based upon a combination of placement scores and high school background information, sometimes combined using complicated regression formulas (Cederberg 1994). But too often the success of the placement system is inadequately studied, given the dual goals of placing the student in a course that is both challenging and one in which the student is likely to succeed.

Students need reliable information if they are to make rational choices about their courses, majors, and postgraduation lives. Unfortunately, there is no guaranteed approach to providing such information. Recruiting brochures, open meetings, and informal after-class discussions are traditional methods. In recent years, many departments have begun to experiment with (electronic) publication of advising handbooks for mathematics students. These handbooks can answer the most frequently asked questions about the major so that individual advising sessions can be used more profitably. In addition, they can discuss courses and curricular options (e.g., Actuarial Science, Biostatistics, Computational Mathematics, and so forth) in greater detail than allowed by the condensed style of most college catalogs. Handbooks can discuss how to study mathematics and the value of group study. They can advise students about what employers will be looking for at the end of the students' academic programs and can present the local picture of careers for mathematics majors. Handbooks can give advice about further study options, for example, to help two-year college students prepare for study at four-year institutions, and to assist mathematics majors in choosing professional and graduate schools. They also can describe opportunities for undergraduate research and internships, and can introduce the department's faculty, providing information about each faculty member's personal and scholarly interests. Cooperating in the writing of the departmental handbook is a good project for local mathematics club members, if only because they know what students want to know. In addition, such cooperation builds ties between faculty and students. And while handbooks are student-oriented, they also can help newly appointed faculty members understand the goals and spirit of their departments (Lutzer 1994). Internet technology makes it possible for a department's advising handbook to be disseminated electronically, thereby decreasing publication costs and reaching a wider audience. Attached to the department's World Wide Web home page, a handbook can be a recruiting device for new students and can be used by high school counselors as they assist advisees in college searches.

Alumni surveys across the nation show that advising about postgraduation options needs improvement. Information on the national picture of careers for mathematics majors is available from professional societies such as the American Mathematical Association of Two Year Colleges (AMATYC), the Mathematical Association of America (MAA), the American Mathematical Society (AMS), the Society for Industrial and Applied Mathematics (SIAM), and the Association for Women in Mathematics (AWM). In addition, commercially available career guidebooks (e.g., Burnett 1993) and federally published data (e.g., Bureau of Labor Statistics 1994–1995) are usually available through the institution's Career Services Office (CSO). All of this data can help a department present information about what mathematics majors do nationally.

But it is the local, and not the national, picture that is most important to students. Students want to know what mathematics majors from their own department have done after graduation. The local CSO and the Alumni Association should be valuable sources of information for departments about what mathematics majors did after college. The CSO should be able to provide lists of corporations that interviewed or hired recent mathematics graduates, and also may have lists of mathematics alumni who are willing to serve as long-distance advisors for current students. Many departments make good use of alumni in advising. They ask their alumni to return to campus to conduct career advising sessions, and have found that current students relate well to recent graduates. In addition, mathematics alumni can sometimes assist today's students by arranging summer internships, or shorter term experiences in which the student closely follows the day-to-day work of a recent graduate.

Graduate or professional school is an attractive option for some mathematics majors. Advising

for prelaw and premedical students is usually not a departmental responsibility, but advising for mathematics-based graduate study is. A mathematics major with suitable supporting courses is strong preparation for applications-oriented masters programs in industrial mathematics, econometrics, finance, operations research, statistics, and so forth (Fink 1996; Herrmann and Lutzer 1995; SIAM 1995). That is sometimes overlooked by advisors who mistakenly identify graduate study with doctoral study and therefore advise students away from graduate school in the light of today's job market for new Ph.D.s. A recent SIAM study (SIAM 1995) confirms the viability of applications-oriented masters degrees for mathematics majors.

For those students whose interests and abilities orient them toward doctoral study in mathematics, the department is the primary information and advising source. Data on mathematical sciences graduate programs, and on the fellowships and assistantships that pay for most doctoral study in mathematics, are published annually by the AMS (1995). Information on national rankings of research doctoral programs is available from the National Research Council, and the AMS periodically publishes such listings (Connors et al. 1988, 532; Jackson 1995). Commercial graduate school guides also are available, and departments can assist students by collecting all of this information in a single place. Once again, alumni have a role to play in the advising process. Alumni who are now in graduate school may be able to answer undergraduates' questions, provided the department can furnish alumni telephone numbers or e-mail addresses. New, and visiting, faculty also are valuable sources of information. In addition, departments can help graduate-school-bound students feel less isolated by organizing GRE review sessions to bring such students together. In the end, however, graduate-school-oriented advising must match a given student with reasonable options for him/her. That requires personal knowledge of the student's abilities and interests, as well as data on the expectations of various graduate programs.

While some advising legitimately belongs elsewhere on campus, *academic* advising remains a departmental responsibility, central to the department's teaching mission. By enhancing students' relationship to their departments, advising can help to make the mathematics major an intellectual home for students and a lens through which students interpret their other college studies. Without the personal linkages that advising can build, there is a danger that the major will be little more than a transcript notation for its students.

*See also* Advanced Placement Program; Applications for the Classroom, Overview; Career Information

## SELECTED REFERENCES

American Mathematical Society. *Assistantships and Graduate Fellowships in the Mathematical Sciences.* Providence, RI: The Society, 1995.

Association of American Colleges. *The Challenge of Connecting Learning.* Washington, DC: The Association, 1990.

Bureau of Labor Statistics. *Occupational Outlook Handbook.* Washington, DC: U.S. Department of Labor, 1994–1995.

Burnett, Rebecca. *Careers for Number Crunchers and Other Quantitative Types.* Lincolnwood, IL: VGM Career Horizons, 1993.

Cederberg, Judith. "Administering a Placement Test, St. Olaf College." *Placement Newsletter of the Mathematical Association of America* (Fall 1994):2–5.

Connors, Edward A., et al. "1987 Annual AMS-MAA Survey." *Notices of the American Mathematical Society* 35(4)(1988):525–533.

Fink, James. "Can Mathematics Majors Become Engineers?" *Focus* 16(5)(1996):29.

Herrmann, Diane, and David Lutzer. "Advising Undergraduates about Graduate School in Mathematics: Methods that Work." *Focus* 15(3)(1995):15–16.

Jackson, Allyn. "New NRC Rankings of Graduate Programs Released." *Notices of the American Mathematical Society* 42(12)(1995):1535–1542.

Lutzer, David. "Career Advising for Mathematical Sciences Majors." *Focus* 14(5)(1994):3–4.

Schoenfeld, Alan. *A Sourcebook for College Teaching.* Washington, DC: Mathematical Association of America, 1990.

Society for Industrial and Applied Mathematics. *The SIAM Report on Mathematics in Industry.* Philadelphia, PA: The Society, 1995.

DAVID LUTZER

# AFRICAN CUSTOMS AS APPLIED TO THE MATHEMATICS CLASSROOM

Connections between African culture and elementary mathematics. All peoples have developed numeration systems to meet their needs, from the few simple words of the San (Bushmen) in southern Africa to the extensive systems of societies engaged in far-reaching trade. Many peoples of West Africa group numbers by 20, with 5 and 10 as subsidiary bases. The word for 20 might mean "a whole person," indicating that all twenty fingers and toes have been counted, as in the Mende language of Sierra Leone. Number words in eastern and southern Africa generally are based on groups of ten. Distinctive finger gestures often accompany or replace the number words in many cultures. Although the construction of words for large numbers usually involves the operations of multiplication and addition, subtraction also occurs, as in the Yoruba (Nigeria) word for 65: "five and ten from four twenties" (Zaslavsky 1999). As students analyze the construction of number words in other cultures, they improve their understanding of our base-10 system and may be inspired to invent their own words for numbers.

Cowrie shells and beads, grouped in various related denominations according to local customs, served as currency in many parts of Africa. Students might role-play market scenes, using strings of beads as money, as well as African number words and gestures. Ironically, West African numeration systems expanded in the nineteenth century when European traders dumped huge quantities of cowrie shells, causing inflation and the need for larger numbers (Zaslavsky 1999).

Generally lacking written systems of numerals until the introduction of the Hindu-Arabic numerals now used in most of the world, Africans south of the Sahara devised ways of recording numbers on tally sticks and strings. They amazed European traders by their skill in mental arithmetic, as exemplified by Thomas Fuller, who was brought to the American colonies as a slave in 1724 at the age of 14. Although he was not permitted to learn to read or write, he was known as the African Calculator and was celebrated by the abolitionist movement as an example of African intellectual ability (Fauvel and Gerdes 1990). Fuller's achievements in mental arithmetic might inspire students to develop that important skill. His contemporary, Benjamin Banneker, the free-born son and grandson of African slaves, achieved even greater fame. A self-taught mathematician and astronomer, he was the author of ten almanacs and served on the commission to survey the land for the new capital of the United States at the District of Columbia (Bedini 1972).

African architecture ranged from the temporary shelters of nomadic peoples to the elaborate stone buildings and walls of the centuries-old city-state known as Great *Zimbabwe,* or "great stone house." In some societies the round house is traditional—hemispheric, beehive-shaped, or cylindrical with a conical roof. Others have always constructed square or rectangular homes. Students might discuss the factors that lead to one style or another, and construct their own model homes. They may discover that the circle offers the largest floor space when a given amount of material is available for the walls (Zaslavsky 1991).

Most cultures incorporate characteristic designs into their basketry, cloth, carved objects, and buildings. Outstanding are Bakuba (Zaire) embroidered raffia cloth and wood carvings, with their intricate repeated patterns. Applications of transformational geometry to African art can range from stamping a motif on paper or cloth to the study of group theory as exemplified by African designs, and the creation of original patterns (Washburn and Crowe 1988).

The universal African game called *Mankala* in Arabic appears in various regional versions and has many names, such as *Owari, Bao, Avo,* and *Lela.* It is considered one of the world's best games of strategy. African children often play the game with pebbles in holes scooped out of the dirt, while Asante (Ghana) kings played on heirloom gold boards. The game is available commercially in the United States; it also can be played with an egg carton as the game board and beans as counters. For young children the game affords practice in counting and one-to-one correspondence, yet it can be so sophisticated as to foster national competitions for adults in some African countries. Once students have learned the game, they can vary it in several ways, thus developing critical thinking skills (Zaslavsky 1999).

Three-in-a-row games similar to Tic-Tac-Toe are played in many parts of Africa. In fact, workmen carved diagrams for such games into the roofing slabs of ancient Egyptian temples. One of the most complex versions, *Murabaraba,* originated in Lesotho (Zaslavsky 1982). An instructor in that country carried out a controlled study, and concluded that middle-grade students who knew the game performed better than nonplayers in certain standardized geometry tasks (Lepheana 1977).

Games of chance are popular among Africans. They might toss coins, or use the cowrie shells that were once the standard currency in many regions. While the outcomes of coin-tossing are predictable due to the symmetry of the coin, the asymmetric cowrie shell is not equally likely to land with the opening up or down. Students can compare outcomes, using cowries or macaroni shells—a good introduction to the study of probability theory (Zaslavsky 1999).

*See also* Ethnomathematics; Multicultural Mathematics, Issues

SELECTED REFERENCES

Bedini, Silvio A. *The Life of Benjamin Banneker.* Rancho Cordova, CA: Landmark, 1972.

Fauvel, John, and Paulus Gerdes. "African Slave and Calculating Prodigy: Bicentenary of the Death of Thomas Fuller." *Historia Mathematica* 17(1990):141–151.

Gerdes, Paulus. *Geometry from Africa: Mathematical and Educational Explorations.* Washington, DC: Mathematical Association of America, 1999.

Lepheana, Josiel N. "Spatial Relations through Morabaraba." Unpublished manuscript. Maseru, Lesotho: School of Education, National University of Lesotho, 1977.

Washburn, Dorothy, and Donald W. Crowe. *Symmetries in Culture.* Seattle, WA: University of Washington Press, 1988.

Zaslavsky, Claudia. *Africa Counts: Number and Pattern in African Culture.* 3rd ed. Chicago, IL: Hill, 1999.

———. "Multicultural Mathematics Education for the Middle Grades." *Arithmetic Teacher* 38(Feb. 1991):8–13.

———. *Tic Tac Toe and Other Three-in-a-Row Games from Ancient Egypt to the Modern Computer.* New York: Crowell, 1982.

CLAUDIA ZASLAVSKY

# ALGEBRA, INTRODUCTORY MOTIVATION

The use of letters for variables and constants, introduced slowly, with concrete problems to motivate new developments. One scenario is sketched here, which uses number games of steadily increasing difficulty to demonstrate the need for "unknowns" and for reliable ways of manipulating them. Certainly other scenarios are possible. In one instructional sequence of pre-algebra activities starting in the primary grades, it is suggested that "action situations or models are generally far easier" for younger pupils "to manage than static situations or models" (Nibbelink 1990). For example, a "missing number" idea is implemented via a mischievous Gerald Gerbil who eats numbers (exactly one per problem), thus transforming written arithmetic situations into story problems.

In the "number game" approach developed here the teacher might begin by posing the problem, "When 3 is added to a certain number, you get 17. What is the number?" Many students are sufficiently at ease with basic number facts that they can give the answer immediately. Some may not give the answer as readily; they may have to try a few guesses before hitting on the correct answer.

Mental calculations or guessing, then checking to see whether the answer is right, should be encouraged. It is pointless, in fact counterproductive, at this point to develop algebraic machinery. New or more difficult techniques will be respected and appreciated by the student to the extent that he or she recognizes the need for them. The teacher will create such a need by posing more difficult problems, as illustrated below.

After a few easier problems the class might be offered the following: "33 is added to 7 times a certain number, and the result is 117. What is the number?" Almost every student would have to write down the given fact in some form, to think about possible answers. The teacher may reasonably suggest (after the students have themselves had a chance to write down the information and play with the problem) that it might be helpful to write $33 + $ (7 times unknown number) $= 117$, or, more concisely, $33 + (7 \times$ unknown number$) = 117$. This apparently simple step of writing the clue in the form of a mathematical statement is important. It is now easy to think about, and experiment with, specific numbers that may solve the problem.

Again students should be encouraged to try different possibilities for "unknown number," to determine which of the suggested numbers, if any, makes the equation "balance." (A physical balance scale can be helpful here.) Such an exercise makes precise the meaning of a "solution;" also it helps the students realize that there is already available to them a legitimate method for attacking such problems—namely, trial and error.

After the students attack similarly several such problems, the teacher might suggest that an equation will look tidier and be easier to write down, if some symbol, say the letter $n$, is used to stand for "unknown number." This yields $33 + (7 \times n) = 117$.

At this stage of development, as at other stages, students should be asked to recall the problem represented by this so-called algebraic equation: Which value of $n$ (that is, which number substituted for $n$) will make the two sides of the equation have the same numerical value, 117? A few tries with different values of $n$ reveal that the so-called "solution" of the equation (and the original problem) is $n = 12$.

Here is a "real-life" problem leading to the same mathematical formulation: "I need 117 feet of fencing to protect my garden. From an earlier, smaller garden I have 33 feet of fencing. If additional pieces are available in 7-foot lengths, how many such lengths do I have to buy?"

Notice that algebraic manipulations have not yet been introduced. The emphasis has been on mean-

ing, the meaning of the new mathematical symbol (the letter) and the meaning of the equation. Much anxiety and confusion is generated by teachers passing too quickly through these preliminary stages.

Algebraic manipulations are certainly important. But their use can best be grasped and appreciated by the student if they are introduced when they serve a real need, that is, when problems are posed that are difficult to handle unless the algebraic expressions in them are simplified.

Suppose the students have been posing to each other the kinds of number problems discussed above. Someone (if necessary, the teacher) may well pose a problem such as the following: "A certain number is multiplied by 5, and the result is added to 11. From twice the new result we subtract three times the original number, and the final result is 64. What is the unknown number?" It is clear that no one can deal with this problem without writing it down. (For convenience here let us suppose that the students have already learned that $5n$ stands for "$5 \times n$" and that a number next to a parenthesis indicates multiplication.) Once the class is agreed that the problem may be formulated as: $2(5n + 11) - 3n = 64$, students can try guessing the unknown number and going through the calculation for each guess. It will soon be apparent, however, that for complicated equations trial and error is inefficient. This is the right moment to demonstrate the value of algebraic technique. The teacher, justifying briefly the legitimacy of the distributive law, reduces the equation above to a much simpler one. The students will see that without some knowledge of correct rules of manipulation, only very simple problems can be handled with any degree of efficiency.

If time and care are spent initially to motivate the need for nonnumerical symbols, and technique is introduced as needed, this early investment of effort will result in greater comfort on the part of the student with the new symbols, an intuitive feeling that the "rules" are based on common sense, and a consequent reduction in the time required later for repetitive drill.

*See also* Constructivism; Language and Mathematics in the Classroom; Psychology of Learning and Instruction, Overview; Variables

### SELECTED REFERENCES

Fleishman, Bernard A., and Stanley Kogelman. "Add Intuition, Subtract Anxiety." *NYSSBA Journal* (Dec. 1980):8–9.

Hiebert, James. "The Struggle to Link Written Symbols with Understandings: An Update." *Arithmetic Teacher* 36(Mar. 1989):38–44.

Nibbelink, William H. "Teaching Equations." *Arithmetic Teacher* 38(Nov. 1990):48–51.

BERNARD A. FLEISHMAN

# ALGEBRA CURRICULUM, K–12

Gradual development of classical and modern algebraic concepts beginning with exploration of patterns. What is algebra? When high school students are asked what algebra is, a common response is "finding $x$." Algebra, to these students, means solving equations to find an unknown quantity, more often than not expressed by the letter $x$. The interpretation held by the high school students is not so different from the definition of classical algebra: the science of solving equations. Diophantus, a Greek who taught at the University of Alexandria in Egypt, is generally held to be the Father of Algebra because of his work with equations in both first and second degree. Other mathematicians before Diophantus used algebraic principles in their problem solving. Many of them, such as Euclid, used a geometric approach to algebra. In an algebra that is regarded as the science of solving equations, not only must equations of different degrees be considered but also systems of equations to be solved simultaneously.

The idea that algebra is not only a method of problem solving but, in addition, is a mathematical discipline in which algebraic ideas and principles take precedence over problem solving, seems to have occurred first in nineteenth-century England with such notables as George Boole (1815–1864). The idea of laying down postulates for the manipulation of abstract symbols forms what we today refer to as abstract algebra. In an abstract algebra, mathematicians look at different parts of algebra, noticing common patterns of proofs that keep recurring. These patterns are then singled out as assumptions or axioms. All that is necessary for an axiomatic system to be an algebraic structure is for it to contain a set of elements and one or more operations defined on the set.

Whereas mathematicians and scientists in other branches of scientific learning have developed mathematics for their contemporary applications, many algebra concepts were developed before an application had been conceived. A thousand years ago, people saw no need for negative numbers. They either had a positive number of apples or no apples at all. A negative number of apples seemingly made no sense. The concept was too abstract for many of the greatest mathematicians of the time. Even Diophantus once described the solution of $4x + 6 = 2$ as being an absurd number (Pinter 1990). Another algebraic concept that was entirely useless in application for years was the complex number system. Jerome Cardan (1501–1576) was the first to introduce the complex numbers and even he had serious misgivings about it. In order to make their algebraic structure philosophically sound, Cardan and other mathematicians realized that they could not have an operation, in this case taking the square root, unless that operation could be defined over their entire set of numbers that included negatives. So $\sqrt{-1}$ existed for years in algebraic theory only. Now, in modern times, complex numbers are seen as a basic tool in physics.

Even more intriguing in their development have been the abstract algebras. Matrix algebra, first credited to Arthur Cayley, is a nineteenth-century mathematical discovery, now commonly used in business and scientific worlds where large amounts of data are to be recorded and manipulated. George Boole's algebra of logic has played a large role in modern computer design.

Have we reached the farthest limits to algebra as the science of solving equations? As a body of mathematical theory? The answer seems to be yes to the first question. The answer to the second question is much less definite. It has been estimated that over two hundred different algebraic systems have been developed and put into application for a particular aspect of scientific need within the last few years. Tensor algebra and the algebra of vectors are two such algebras. One hundred years ago, Boolean algebra and matrix algebra were considered abstract and theoretical. We may safely assume that theoretical algebra has not yet seen its limits, and that when a specific need or application arises, a new algebraic structure will be born to access it.

Until the 1980s in the United States, the formal study of algebra was usually begun on the ninth-grade level and covered solving linear equations, simple systems of equations, and simple quadratics. This was normally followed by a second year of algebra at the secondary level, which may or may not have been preceded by plane geometry. The algebra presented on the second level of high school usually included systems containing more than three equations where matrices and determinants are applied for solution, graphical representation of conic sections, and higher degree equations. It has been the norm in the mathematics curriculum of the United States to present the classical algebra to students at a secondary level and consider abstract algebras as college courses.

Educational research in the 1980s indicated stultifying effects of the mathematics curriculum for most seventh and eighth grades, where presentations of new concepts appeared almost nonexistent. Conceptual emphasis was placed on common fractions, decimals, and percentages in sixth grade, seventh grade, and again in eighth grade. The same research also showed a glut of new concepts presented in ninth-grade mathematics, normally the first year for algebra. Call for change in mathematics curriculum thrust many algebraic concepts down to seventh- and eighth-grade levels. The publication of the *Curriculum and Evaluation Standards for School Mathematics* (National Council of Teachers of Mathematics [NCTM] 1989) initiated a drive to introduce algebraic thought into the mathematics of the elementary schools as well. NCTM proposed that the K–4 mathematics curriculum should "include the study of patterns and relationships so that the student can explore the use of variables and open sentences to express relationships." Educators of fifth through eighth grades were asked to focus on informal explorations of algebraic concepts in order to build a foundation for the more formal study of algebra in later grades. Students at that educational level are expected to understand the concepts of variable, expression, and equation. They are expected to analyze tables, graphs, and charts, solve linear equations, investigate inequalities and nonlinear equations, and to apply algebraic methods to real-world situations. NCTM further suggested that a mathematics curriculum for grades 9–12 should now include the study of algebraic concepts and methods for all students, and that college-bound students be able to use matrices to solve linear systems and demonstrate technical facility with algebraic transformations. Some of the high school curriculums do not follow the traditional algebra 1, geometry, algebra 2 sequence but integrate all three courses over three years of instruction.

To bring the language of algebra out of an abstract setting into the real world of younger students, the methods by which it is taught have changed. Manipulatives, such as two-color counters, algebra tiles, equation balances, and graphing calculators are now found in mathematics classrooms, and several are proving beneficial when introducing algebraic concepts to lower elementary school students.

Mathematics has been called the handmaiden to the sciences. If this is so, the language she speaks is algebra. Much of the technological boom of the last part of the twentieth century has been made possible by mathematicians' and scientists' use of algebraic concepts. It is ironic that this technological thrust has brought about changes in the teaching of algebra. Access to the computer, the scientific calculator, and, finally, the graphing calculator has made it possible to free students from such manipulative drudgery as interpolation and the algorithm for extracting roots necessary for the solution of algebra problems two decades ago. Not only has technology freed algebra students from number crunching, it also has given access to creative new ways to teach and to learn algebra concepts. The first-year algebra student two decades ago would have had to graph enough linear equations to be convinced that lines with the same slope are parallel and represent a system of equations with no common solution. Now, in a matter of minutes, a student with access to a graphing calculator can be presented with enough equations of this nature to see and remember their graphic behavior. Once, only those students who not only understood the laws of logarithms but had learned how to interpolate logarithms could be asked to solve an exponential growth or decay problem. Now, with the recursive capabilities of the graphing calculator, which allows a process to be repeated with the push of a button, students can access answers to such problems without really having to know or understand what a logarithm is. Technology has freed mathematics educators from spending excessive amounts of time on algorithms per se and has empowered them with means of implementing algebraic concepts.

Technological tools in the realm of algebra have not been embraced by all educators. There are those who believe that the pencil and paper methods and algorithms used to access algebraic reasoning developed through the centuries from Diophantus to Descartes, are now so embedded within algebra as to be taken for concepts, themselves. For example, one of the topics chosen for deemphasis by NCTM in the algebra curriculum is factoring. Factoring had been a major topic in most elementary algebra textbooks up through the 1980s because it allowed access to solutions of nonlinear equations on a rudimentary basis. Now that there are technological means of solving nonlinear equations, some argue that factoring should still be taught as examples of algebraic patterning. The dilemma facing today's algebra teacher seems to be what algebraic methods to give up as archaic and what algebraic methods to keep for historical value. If students now have access to a tool that allows them to solve a quadratic equation by other means, should they spend an excessive amount of time in the classroom on the techniques of factoring? If a problem showing exponential decay can be solved on a graphing calculator with iterative or graphic techniques, is the student any the worse off for not understanding logarithms or being able to interpolate them?

In the book *Everybody Counts* (1989), the National Research Council says "Education reflecting only the mathematics of the distant past is no longer adequate for present needs." The Council goes on to refute the myth: "There is no algebra in my future," with the reality: "Over 75 percent of all jobs require proficiency in simple algebra and geometry." Students no longer have the option of not taking algebra. What now seems to be the question is, as more and more algebraic concepts, once considered for ninth graders only, are being pulled further and further down within the curriculum, and technology is accessing algebraic thought more quickly and easily, will algebra remain a course to itself, or become integrated and embedded within a larger mathematical framework? Will the abstract algebras move down from college level and find a place in the K–12 curriculum? "Algebra for All" has become a slogan of NCTM, which in the past few years has made this area a priority of study. Installed through the council, a task force on algebra in the nation's curriculum presented its comprehensive plan for action in the spring of 1994, and continues to move forward in offering opportunities for discussion and guidance for action (Lindquist 1994, 514–515).

*See also* Abstract Algebra; Curriculum, Overview; Curriculum Trends, Secondary Level; Manipulatives, Computer Generated; Technology

### SELECTED REFERENCES

Baumgart, John K., et al. *Historical Topics for the Mathematics Classroom.* 31st Yearbook. Washington, DC: National Council of Teachers of Mathematics, 1969.

Bell, Eric Temple. *Men of Mathematics.* New York: Simon and Schuster, 1986.

Lindquist, Mary. "President's Report: Linking Yesterday to Tomorrow." *Journal for Research in Mathematics Education* 25(5)(Nov. 1994):512–522.

Michalowicz, Karen Dee. "Episodes in the History of Algebra." *Mathematics Teaching in the Middle School* 1(Jan.–Mar. 1995):293–294.

National Council of Teachers of Mathematics. *Curriculum and Evaluation Standards for School Mathematics.* Reston, VA: The Council, 1989.

———. *Principles and Standards for School Mathematics.* Reston, VA: The Council, 2000.

National Research Council. *Everybody Counts: A Report to the Nation on the Future of Mathematics Education.* Washington, DC: National Academy Press, 1989.

Perl, Teri. "Manipulatives and the Computer: A Powerful Partnership for Learning." *Classroom Computer Learning* (Mar. 1990):20–27.

Pinter, Charles C. *A Book of Abstract Algebra.* New York: McGraw-Hill, 1990.

SUE JACKSON BARNES

# ALGEBRAIC NUMBERS

A proper subset of the set of complex numbers. A complex number $\alpha$ is said to be *algebraic* if $\alpha$ satisfies an *algebraic equation,* that is, if $\alpha$ is a zero of some polynomial equation with integer coefficients: $a_n\alpha^n + a_{n-1}\alpha^{n-1} + \cdots + a_1\alpha + a_0 = 0$ for some integers $a_0, a_1, \ldots, a_n$ with $a_n \neq 0$ $(n \geq 1)$. For example, $\sqrt{2}$ and $\sqrt{3}$ are both algebraic, since they are zeros of the polynomials $x^2 - 2$ and $x^2 - 3$, respectively. Not every complex number is algebraic and the complex numbers that are not the zeros of any polynomial are said to be *transcendental*. Loosely speaking (and as the name implies), the algebraic numbers can be defined "algebraically" in terms of the usual arithmetic operations of addition, subtraction, multiplication, and division (for example, $\sqrt{2}$ is defined as the positive real number that when multiplied by itself gives the integer 2). The transcendental numbers, in contrast, require "analytic" descriptions that involve limiting operations. The arithmetic operations of addition, subtraction, multiplication, and division take algebraic numbers to algebraic numbers (so the collection of all algebraic numbers has the algebraic structure of a *field* ): if $\alpha$ and $\beta$ are algebraic numbers, then also $\alpha + \beta$, $\alpha - \beta$, $\alpha\beta$, and $\alpha/\beta$ (provided $\beta \neq 0$) are algebraic. For example, $\sqrt{2} + \sqrt{3}$ is algebraic, in fact it is a zero of the polynomial $x^4 - 10x^2 + 1$. The set of polynomials with integer coefficients is countably infinite, that is, it is possible to give a $1:1$ correspondence between these polynomials and the integers 1, 2, 3, . . . . As a result, there also is a countably infinite number of algebraic numbers. Since there are uncountably many complex numbers, this implies that almost every complex number is not algebraic in the following sense: if $N$ complex numbers are chosen at random, then the probability that one of them is algebraic is very small and as $N$ tends to infinity this probability tends to zero. Loosely speaking, the "probability that a randomly chosen complex number is algebraic is zero" (a precise technical result is that the algebraic numbers have Lebesque measure zero in the set of complex numbers).

If $\alpha$ is algebraic, there is a polynomial of minimal degree having $\alpha$ as a zero. This polynomial is *irreducible,* that is, cannot be factored into two smaller polynomials with rational coefficients. Any other polynomial with rational coefficients having $\alpha$ as a zero is divisible by this polynomial. The degree of this polynomial, called the *degree of* $\alpha$, is an important invariant associated with $\alpha$ and is a measure of how "complicated" $\alpha$ is. For example, $\sqrt{3}$ is of degree 2 and $\sqrt{2} + \sqrt{3}$ is of degree 4.

The study of the behavior of the degree is a part of *field theory* in abstract algebra. There is a smallest field containing the rational numbers and the number $\alpha$, called the field *generated* over the rationals by $\alpha$. This is the smallest subset of complex numbers containing $\alpha$ that is closed under the arithmetic operations of addition, subtraction, multiplication, and nonzero division (i.e., the sum, difference, product, and quotient of two elements in the set is again in the set). For example, the field generated over the rationals by $\alpha = \sqrt{2}$ is the collection of numbers of the form $a + b\sqrt{2}$, where $a$ and $b$ are rational numbers. If $\alpha = \sqrt{2} + \sqrt{3}$, then the field generated by $\alpha$ is the collection of numbers of the form $a + b\sqrt{2} + c\sqrt{3} + d\sqrt{6}$ where $a$, $b$, $c$, and $d$ are rational numbers. The fact that these sets are closed under nonzero division amounts to "rationalization of denominators" familiar from elementary algebra: for example, $\frac{1}{3 + \sqrt{2}} = \frac{3}{7} - \frac{1}{7}\sqrt{2}$. The number of elements needed to describe the field generated by $\alpha$ over the rationals is called the *extension degree.* For example, if $\alpha = \sqrt{2}$, then two elements are needed (viz., the elements 1 and $\sqrt{2}$) whereas if $\alpha = \sqrt{2} + \sqrt{3}$ then four elements are needed (viz., the elements 1, $\sqrt{2}$, $\sqrt{3}$, $\sqrt{6}$. For both of these examples, the extension degree of $\alpha$ is the same as the degree of the minimal polynomial of $\alpha$. A fundamental result in abstract algebra is that this is always the case: the degree of $\alpha$ is the same as the extension degree of $\alpha$. In other words, if $\alpha$ is algebraic then it is possible to describe all the numbers obtained from $\alpha$ using the arithmetic operations in terms of a finite number of elements using only rational numbers as coefficients (and the number needed is precisely the degree of $\alpha$). Furthermore, if $\alpha$ is transcendental then it is not possible to describe the field generated by $\alpha$ using only a finite number of rational numbers as coefficients. Another consequence of this result is that it proves it is always possible to rationalize the denominator of any expression involving algebraic numbers (and generally not possible if transcendental numbers are involved).

Results on the degree of algebraic numbers also can be used to solve many classical problems in geometry. For example, it can be shown that any straightedge and compass construction leads to algebraic numbers whose degree is necessarily a power of 2. Substituting $\beta = \cos 20°$ into the triple angle formula for cosines, $\cos 3\beta = 4\cos^3 \beta - 3\cos \beta$, shows that $\beta = \cos 20°$ is a zero of the irreducible polynomial $8x^3 - 6x - 1$, so $\beta = \cos 20°$ is an algebraic number of degree 3. Since 3 is not a power of 2, this algebraic number cannot be constructed by straightedge and compass, and it follows that in general it is impossible to trisect an angle by straightedge and compass (although some angles, such as 180°, can certainly be trisected).

The other zeros of the irreducible polynomial satisfied by an algebraic number $\alpha$ are called the *algebraic conjugates* of $\alpha$. For example, the algebraic conjugates of $\sqrt{2} + \sqrt{3}$ are $\sqrt{2} - \sqrt{3}$, $-\sqrt{2} + \sqrt{3}$, and $-\sqrt{2} - \sqrt{3}$. The study of the field theory associated with $\alpha$ and its conjugates leads to the important and powerful *Galois Theory* in abstract algebra. Among other results, this theory proves that, in general it is impossible to express the zeros of a polynomial of degree 5 or greater in terms of *radicals,* or $n^{\text{th}}$ zeros. The quadratic formula states that the zeros of $ax^2 + bx + c$ are

$$\frac{-b \pm \sqrt{b^2 - 4ac}}{2a}$$

which gives these zeros in terms of the coefficients of the quadratic polynomial, the elementary arithmetic operations of addition, subtraction, multiplication, and division, and the extraction of a square root. There are similar formulas for the roots of a general cubic polynomial $ax^3 + bx^2 + cx + d$ and for a general quartic polynomial, called Cardano's formulas (Dummit and Foote 1991). These formulas are more complicated but, again, only require radicals (in these cases, extraction of square and cube roots) as well as the arithmetic operations. The theorem on the insolvability of the quartic in Galois Theory states that such formulas are *impossible* for polynomials of higher degree. The zeros of the polynomial $x^5 - x + 1$, for example, cannot be expressed in terms of radicals.

If $\alpha$ is an algebraic number that is the zero of a *monic* polynomial with integer coefficients (i.e., $a_n = 1$ above), then $\alpha$ is said to be an *algebraic integer,* generalizing the usual notion of the integers as a subset of the rational numbers. The sum, difference, and product (but not in general the quotient) of algebraic integers is again an algebraic integer, giving the collection of algebraic integers the structure of a *ring* in abstract algebra. The study of algebraic number fields and the arithmetic of algebraic integers, which originated in attempts to prove Fermat's Last Theorem (Weil 1984), is referred to as *algebraic number theory* and is an extremely active area of current mathematical research.

*See also* Fields

## SELECTED REFERENCES

Courant, Richard, and Herbert Robbins. *What Is Mathematics?* London, England: Oxford University Press, 1969.

Dummit, David S., and Richard M. Foote. *Abstract Algebra.* Englewood Cliffs, NJ: Prentice-Hall, 1991.

Hardy, Godfrey H., and Edward M. Wright. *An Introduction to the Theory of Numbers.* 4th ed. Oxford, England: Clarendon, 1965.

Van der Waerden, Bartel L. *History of Algebra.* New York: Springer-Verlag, 1985.

Weil, Andre. *Number Theory: An Approach through History from Hammurapi to Legendre.* Boston, MA: Birkhäuser, 1984.

DAVID S. DUMMIT

# ALGORITHMS

Step-by-step procedures for solving problems. An algorithm can produce yes or no answers to a decisional problem; numerical answers to a computational question; or a geometric solution, such as the shortest path for a traveling salesman to take in visiting a collection of cities.

There are two characteristics of an algorithm. First an algorithm should have a *finite description*. If one thinks of an algorithm as being described by steps, then the execution of the algorithm can be thought of as following a sequence of steps as in a computer program. Finiteness says that the number of steps in the description is finite. The second characteristic is that an algorithm should be *effective*. This means that each step should be mechanically performable.

In the study of algorithms, several properties are of interest. It is important to know if an algorithm is *terminating*. This means that the algorithm should produce a result after a finite number of steps. Note that even if an algorithm has a finite description, it is possible that the algorithm would not terminate if it proceeds in circles and never produces a result. Algorithms also are classified into two categories, *deterministic* and *nondeterministic*. In a deterministic algorithm, if we are in a given state, with a given input, then we will always move to a unique succeeding state. In short, the algorithm "determines" each subsequent step in a definite manner. An example of

a deterministic algorithm is Euclid's algorithm for finding the greatest common divisor of two integers. Each succeeding step is uniquely determined. In a nondeterministic algorithm, there can be choices as to what state the algorithm will move to at each step. One can think of nondeterminism as a game of dice, where the step that we follow each time depends on what we roll with the dice.

One of the most important measures of the usefulness of an algorithm is its *efficiency*. There are two measures of efficiency, the *time* and *space* complexity of the algorithm. The time complexity of the algorithm is measured by the number of steps the algorithm requires to produce an answer. The space complexity refers to the amount of storage space required by the algorithm during the course of its execution. If we are implementing an algorithm on a computer, then we could measure space complexity by the amount of memory required for the execution of the algorithm. If we try to implement an algorithm on a computer, but the algorithm requires a googol ($10^{100}$) steps, then even the fastest computer would not reach an answer for billions of years, longer than the estimated age of the earth. Similarly, if the algorithm requires more memory than the computer has, then the algorithm might be theoretically interesting but practically useless.

Algorithms have been studied since antiquity. Euclid developed a beautiful algorithm for finding the greatest common divisor of two integers. This algorithm is useful even today. For example, we might pick two numbers, each two hundred digits long. These numbers are so large that they could not be factored into primes even using high-speed computers for hundreds of years. Euclid's algorithm, however, would enable the same computers to find the greatest common divisor of these numbers in a matter of seconds.

Notational appearance of the algorithm is a consideration in handling algebraic formulations formally and simply. Flowcharts are often used to display steps and the relationships of the steps with great brevity. It is important to carefully design the algorithm used to solve a problem to manage the available time and space resources so that the algorithm will be useful.

Algorithms range from the basic algorithms for addition, subtraction, multiplication, and division, to finding the greatest common divisor of two whole numbers, to solving systems of equations, to algorithms for devising transport networks to manage the flow of oil or transportation of goods. A computer can be thought of as an algorithm machine, capable of taking data and performing operations to obtain a result through sequential, discrete, and well-defined steps.

The word "algorithm" and its variant "algorism" come from the ninth-century Arabian mathematician Mukhammad ibn Musa abu Djafar al-Khorezmi. Born in the area south of Lake Aral in what is now Uzbekistan, he lived in Baghdad, where the principal works of Greek mathematicians were being translated to Arabic. One of his lasting works is "Kitab hisab al-'adad allhindi," or the "liber algorithmi." His writings were instrumental in introducing present methods of numeration to the Western world. During the Middle Ages, algorithm referred to any use of the Hindu-Arabic numerals.

Algorithms arise in almost all fields in mathematics, algebra, differential equations, number theory, and graph theory. Computer scientists make a careful study of algorithms that can be implemented on a computer that operates sequentially (modeled classically by a Turing machine), as well as algorithms that are implemented by parallel machines. The study of the effectiveness and efficiency of algorithms to solve whole classes of problems contains numerous unsolved problems and is a fertile ground for ongoing research.

Algorithms to sort numbers, as well as to search for elements meeting certain criteria, depend on the representation of the data and cleverness of the designer of the algorithm. Equally difficult is the analysis of how well an algorithm performs in the best case, worst case, and average case. For example, a given algorithm to sort $n$ numbers may require only $n$ steps if the $n$ numbers are already nearly sorted (best case); up to $n^2$ steps if the numbers are in a particularly bad initial arrangement; and some number of steps between $n$ and $n^2$ on the average.

The challenge of inventing an algorithm involves carefully analyzing the problem and breaking the problem down into simple steps. Careful, precise logic and thinking are required. We can classify algorithms by how many steps the algorithm requires depending on the number of data ($n$) that are input. If the number of steps is a polynomial in $n$, such as $n^2$, then we say that the algorithm requires polynomial time. If the number of steps is exponential in $n$, such as $10^n$, then we say that the algorithm is exponential. In general, exponential algorithms take so long to execute for large amounts of data ($n$ large) that they are not practical to use. Even a polynomial algorithm such as one requiring $n^2$ steps can be very inefficient if $n$ is very large. Ideally, we try to devise algorithms using few steps, but this is not always possible for every problem. There are some problems that we can prove do not have simple (polynomial) solutions.

*See also* Addition; Algebra Curriculum, K–12; Arithmetic; Computer Science; Division; Greatest Common Divisor (GCD); Multiplication; Subtraction; Systems of Equations

## SELECTED REFERENCES

Artiaga, Lucio, and Lloyd D. Davis. *Algorithms and Their Computer Solutions.* Columbus, OH: Merrill, 1972.

Bauer, Friedrich L., and Hans Wossner. *Algorithmic Language and Program Development.* Berlin, Germany: Springer-Verlag, 1982.

Brassard, Gilles, and Paul Bratley. *Algorithmics: Theory and Practice.* Englewood Cliffs, NJ: Prentice-Hall, 1988.

Chabert, Jean-Luc, ed. *A History of Algorithms: From the Pebble to the Microchip.* New York: Springer-Verlag, 1999.

Cole, R. Wade. *Introduction to Computing.* New York: McGraw-Hill, 1969.

Machtey, Michael, and Paul Young. *An Introduction to the General Theory of Algorithms.* New York: Elsevier North-Holland, 1978.

Morrow, Lorna J., ed. *The Teaching and Learning of Algorithms in School Mathematics.* 1998 Yearbook. Reston, VA: National Council of Teachers of Mathematics, 1998.

HIROKO K. WARSHAUER

MAX L. WARSHAUER

# AMERICAN MATHEMATICAL ASSOCIATION OF TWO-YEAR COLLEGES (AMATYC)

Formed in 1974 due to the emerging issues in mathematics education and the needs of the mathematics faculty at the two-year college level, which fell between the interests of the major national mathematics organizations. Currently there are forty state or regional affiliate organizations; these affiliates are an integral part of the structure of AMATYC and provide important services to members (approximately 2,800).

AMATYC has published documents on the academic preparation of two-year college mathematics faculty, two-year college mathematics department guidelines, and standards for content and pedagogy of introductory college mathematics. *The AMATYC Review* is the official journal that is published each fall and spring, while *The AMATYC News* is the national newsletter published three times each year.

AMATYC sponsors a Student Mathematics League annual mathematics competition to give special recognition to excellence of two-year college mathematics students. The top qualified student receives the $3,000 Chuck Miller Memorial Scholarship Award to be used for further education at an accredited four-year institution. AMATYC also is a sponsor of the American Mathematics Competitions.

The AMATYC Convention is held each fall in a major North American city, at which two-year college mathematics education issues are discussed. In addition, AMATYC sponsors summer institutes for professional development of faculty.

AMATYC is governed by an executive board of five national officers and eight regional vice presidents and a delegate assembly composed of state and affiliate delegates. AMATYC has the following committees: The *Developmental Mathematics Committee* seeks to improve the quality of developmental mathematics programs in the two-year college; the *Education Committee* investigates the concerns of and promotes quality professional training of two-year college mathematics faculty and department chairs; the *Equal Opportunity in Mathematics Committee* seeks to enhance the position of women and minorities in mathematics; the *Grants Committee* serves as a resource for members and their institutions on matters regarding sources of grant funding and the preparation of grant proposals; the *Placement and Assessment Committee* provides information and serves as a resource; the *Student Mathematics League* encourages student excellence at the two-year college via an annual mathematics competition and other activities; the *Technical Mathematics Committee* supports mathematics courses in allied health and human services, business, computer/data processing, trade and industry, engineering, and emerging technologies; and the *Technology in Mathematics Education Committee* promotes the use of technology within the mathematics curricula and the interaction of mathematics and computer science curricula.

*See also* Awards for Students and Teachers; Competitions

## SELECTED REFERENCES

American Mathematical Association of Two-Year Colleges. *Crossroads in Mathematics: Standards for Introductory College Mathematics Before Calculus.* Memphis, TN: The Association, 1995.

———. *Guidelines for Mathematics Departments at Two-Year Colleges.* Memphis, TN: The Association, 1993.

———. *Guidelines for the Academic Preparation of Mathematics Faculty at Two-Year Colleges.* Memphis, TN: The Association, 1992.

KAREN T. SHARP

# AMERICAN MATHEMATICAL SOCIETY (AMS)

An organization whose central mission is the furtherance of scholarship and research in mathematics. The AMS fulfills this mission through programs and services that promote mathematical research and its uses, strengthen mathematical education, and foster awareness and appreciation of mathematics and its connections to other disciplines and everyday life.

AMS is one of the largest mathematics publishers in the world. Primarily a publisher of monographs, proceedings, and journals focusing on mathematical research, the Society has in recent years begun offering publications directed at a broader audience. Examples of book series include: *Issues in Mathematics Education,* which is published with the Conference Board of Mathematical Sciences and examines current ideas for improving the teaching and learning of mathematics; *Mathematical World,* which provides high quality mathematics exposition to mathematics students and teachers, and the interested lay reader; and *Graduate Studies in Mathematics,* which contains textbooks for upper-level undergraduate and graduate students in mathematics. In 1993, AMS launched *What's Happening in the Mathematical Sciences,* an annual publication aimed at mathematics students and interested lay readers that reviews current developments in mathematics. Many have found *What's Happening* to be an effective classroom tool.

Another major focus of AMS activity is meetings. Each year the AMS holds at least one major meeting. These meetings are the occasion for a wide range of lectures and presentations. In particular, sessions focusing on mathematics education have increased in number and in attendance. Examples of topics explored at recent meetings include the evaluation of mathematics teaching at the postsecondary level, calculus reform, and mathematics competitions. AMS also holds smaller meetings regionally and sponsors a wide range of research conferences each summer.

AMS policies and activities in education are overseen by its Committee on Education. This fifteen-member committee is made up of AMS members having interest and expertise in educational issues. In recent years, the committee has explored many issues, including evaluating teaching effectiveness, assessing student learning, recognition and rewards, undergraduate research experiences, and teacher preparation and enhancement. Through its office in Washington, DC, the AMS maintains links to federal agencies involved in education.

*See also* American Mathematical Association of Two-Year Colleges (AMATYC); American Mathematical Society (AMS), Historical Perspective; Mathematical Association of America (MAA); National Council of Teachers of Mathematics (NCTM)

## SELECTED REFERENCE

*Notices of the American Mathematical Society.* URL: http://www.ams.org.

WILLIAM H. JACO

# AMERICAN MATHEMATICAL SOCIETY (AMS), HISTORICAL PERSPECTIVE

Founded in 1894, the first national organization for mathematics in the United States. It grew out of the New York Mathematical Society founded in 1888 and is now the largest organization for research mathematicians in the United States. It created and maintains the primary research reporting journals in the United States: the *Bulletin,* founded in 1891, the *Transactions* (1900), and the *Proceedings* (1950). In its early years, before the Mathematical Association of America (MAA) was founded, the AMS was concerned with secondary school mathematics, college entrance requirements, and teacher training. AMS assisted in preparing, and sponsored, many reports on mathematics reform. It was active in World War II on military preparedness and mathematical skills; in the early work of the National Science Foundation (NSF) on secondary school mathematics; and in the founding of School Mathematics Study Groups in 1958. AMS now leaves precollege concerns to MAA.

*See also* American Mathematical Society (AMS); Mathematical Association of America (MAA); Mathematical Association of America (MAA), Historical Perspective

## SELECTED REFERENCES

Bidwell, James K., and Robert G. Clason, eds. *Readings in the History of Mathematics Education.* Washington, DC: National Council of Teachers of Mathematics, 1970.

Jones, Phillip S., ed. *A History of Mathematics Education in the United States and Canada.* 32nd Yearbook. Washington, DC: National Council of Teachers of Mathematics, 1970.

May, Kenneth O. *The Mathematical Association of America: Its First Fifty Years.* Washington, DC: The Association, 1972.

JAMES K. BIDWELL
ROBERT G. CLASON

# AMERICAN STATISTICAL ASSOCIATION (ASA)

Founded in 1839 in Boston and now based in Alexandria, Virginia, to promote statistics and its applications. Governed by a board of directors, it is organized both geographically, with an active local chapter system, and by professional subject matter, constituted as ASA sections. The local chapters of ASA share the Association's objectives and belong to the Council of Chapters, which publishes a system-wide quarterly newsletter, *LINK.* There currently are twenty subject matter sections, including Bayesian Statistical Science, Government Statistics, Physical and Engineering Sciences, Quality and Productivity, Statistical Survey Research Methods, and Statistical Education. Each section publishes proceedings of its meetings.

## MEETINGS

The ASA is the primary sponsor of the Annual Joint Statistical Meetings (held in August), which serve as a forum for the latest developments in statistical theory and applications. The Association's winter conferences, begun in 1987, are designed to be more application-oriented, and appeal to both ASA members and nonmembers. In addition, the local chapters hold regular meetings throughout the year.

## EDUCATION

ASA provides a forum for research in theoretical statistics as well as current educational practices. ASA has established the Center for Statistical Education, which serves as a national focal point for the coordination of resources, presentation of training materials, and development of K–12 curriculum materials. The Quantitative Literacy (QL) Program is directed toward improving junior and senior high school statistics instruction. The QL workshops for teachers in grades 7–12 are designed to promote professional development among secondary school teachers of mathematics and science, while preparing them for classroom instruction of statistical and probabilistic concepts. The Continuing Education (CE) Program provides videotaped short courses and tutorials. The ASA also provides a visiting lecturer program for colleges and universities.

*See also* Statistics, Overview; Teacher Preparation in Statistics, Issues

SELECTED REFERENCES

*The American Statistician* (1947–).

*Amstat News* (1974–).

*Chance* (1988–), with Springer-Verlag.

*Journal of Business & Economic Statistics* (1983–).

*Journal of Computational and Graphical Statistics* (1992–), with the Institute of Mathematical Statistics, and the Interface Foundation of North America.

*Journal of Educational Statistics* (1976–) with the American Educational Research Association.

*Journal of the American Statistical Association* (JASA)(1888–).

*Statistics Teacher Network* with the National Council of Teachers of Mathematics.

*Stats—The Magazine for Students of Statistics* (1989–).

*Technometrics* (1959–), with the American Society for Quality Control.

LILY E. CHRIST

# ANGLES

Defined mathematically in distinct but related ways. For example, an angle can be considered the figure formed by two rays extending from the same point. Angle also can be defined as the amount of turning necessary to bring one line or plane into coincidence with or parallel to another. The research on angles suggests a balanced approach to curriculum and teaching that includes, and more important, integrates, various conceptual frameworks for the angle concept.

## MATHEMATICAL BACKGROUND

An angle can be defined as the union of two rays, $a$ and $b$, with the same initial point, $P$. $P$ is called the vertex of the angle, the rays are called the arms. The rays can be made to coincide by a rotation about $P$, which determines the size of the angle between $a$ and $b$. That is, each arm defines a direction and the angle size is the measure of the difference of these directions. The orientation of this difference can be positive in the clockwise (as in surveying, or in "turtle math," described later) or counterclockwise (as in traditional Euclidean geometry).

Methods of measuring the size of angles are based on the division of a circle. The most common are measurement by degrees and by arc length. A degree is $\frac{1}{360}$ of the circumference of a circle. The length of an arc, $a$, between two radii is proportional to the angle between them and to the length of the radius. The following proportion holds. Circumference $(2\pi r)$:Arc::360°:Angle subtended at center. So, if the radius of a circle is known, the length of an arc on the circumference can be used to measure the corresponding angle at the center, $a/r$. The unit, called a *radian*, is thus the angle at the center of a circle subtended by an arc of length equal to the radius of the circle (Gellert et al. 1977).

On a unit circle, these proportions result in the measurement of the angle formed by two radii being equal to the measure of the length of the arc subtended by the angle. This notion can be generalized beyond the unit circle to a generalized arc, which can be defined as "a mapping of a directed line segment into the unit circle such that (1) it is single-valued, and (2) it is measure preserving for subsegments of length less than $\pi$." Then, the generalized angle $POQ$ can be defined as the "union of two rays $OP$ and $OQ$ with a common origin together with a directed generalized arc whose initial point lies on $OP$ and whose end point lies on $OQ$ and the signed measure of a directed generalized angle is the signed length of its directed generalized arc" (Allendoerfer 1965, 86–87). This definition is particularly useful in the study of trigonometric functions.

Angles are classified according to their measure. Categories by degrees are right (90°), acute ($< 90°$), obtuse ($> 90°$), straight (180°), reflex ($> 180°$), and full (360°). An interesting historical question remains: Why 360? The Babylonians created this measurement. They used a sexagesimal number system (base 60) rather than a decimal system (base 10). They knew that the perimeter of a hexagon is exactly equal to six times the radius of the circumscribed circle, in fact that was evidently the reason why they chose to divide the circle into 360. With a base of 60, 6 times 60 was a natural choice. Further, this coincided with their knowledge of astronomy of the last century B.C.: The year was divided into six equal parts, each having sixty plus some fractional number of days.

It is common to take degrees as a measure of rotation. One of the first sets of geometry books published in America, John Playfair's *Elements of Geometry* (1806), presents Euclid's Elements in a form that "renders them most useful" (Jones 1944, 4). Of the three original geometry works published in America, Playfair's was the only one to introduce the notion of an angle as formed by a rotation. He does this to present an alternate proof, presenting a definition of angle that, "if while one extremity of a line remains fixed at $A$, the line turns about that point from $AB$ to $AC$, it is said to describe the angle $BAC$ contained by the line $AB$ and $AC$" (Jones 1944, 7).

Angle concepts and angle measures play essential roles in analyzing and solving problems in a wide variety of geometric situations. For example, they aid in determining (a) heights that cannot be directly measured (of a redwood, for example), (b) a right angle from a hidden point, or (c) the altitude of the sun. To be able to solve such problems, students must learn about various aspects of the angle concept. To do so, students should work through the following.

Awareness of turns (rotations); for example, walking paths through school halls following directions such as "forward 10 steps, right turn"; or Logo turtle geometry activities.

Awareness of corners in surroundings and geometric figures; comparing angle size physically.

Estimation of turns and angles with nonstandard units.

Estimation of turns and angles with standard units.

Connection of turn and angle as the intersection of two rays with a common endpoint (middle school). This might involve, for example, discovering that a turn (e.g., 120°) and the angle produced by that turn (60°) are complementary.

Integration of regions, or the open convex part of a plane bounded by a pair of rays with a common endpoint, with rotations (beginning of secondary school).

Exploration of relationships between angles and other geometric figures, such as parallel lines, leading to mathematical abstraction of these relationships.

Formulation of a single, mathematically rigorous definition of angle.

## THEORETICAL BACKGROUND

Piaget's psychological definition contains aspects of both definitions of angle, although it emphasizes the second one. Piaget claimed that ". . . a general distinction is drawn between two major classes of shape, curvilinear or without angles, and rectilinear or with angles, though subdivisions within the two classes are hardly noticed. . . . There is no doubt that it is the analysis of the angle which marks the transition from topological relationships to the perception of Euclidean ones. It is not the straight line itself which the child contrasts with round shapes, but rather the conjunction of straight lines which go to form an angle" (Piaget and Inhelder 1967).

Thus, abstraction of shape is not a perceptual abstraction of a physical property, but is the result of a coordination of children's actions. Children "can only 'abstract' the idea of such a relation as equality on the basis of an action of equalization, the idea of a straight line from the action of following by hand or eye without changing direction, and the idea of an angle from two intersecting movements" (p. 43).

Furthermore, the child constructs his representation of angle not as two intersecting lines, but rather as the "outcome of a pair of movements (of eye and hand) which conjoin" (p. 31). In fact, "Euclidean shapes . . . are at least as much abstracted from particular actions as they are from the object to which the actions relate" (p. 31).

## RESEARCH

One does not have to look far for examples of children's difficulty with the angle concept. Many students believe that an angle must have one horizontal ray, a right angle is an angle that points to the right; the angle sum of a quadrilateral is the same as its area, and two right angles in different orientations are not equal in measure (Clements and Battista 1992).

This body of research indicates that students have many different ideas about what an angle is. These ideas include "a shape," a side of a figure, a tilted line, an orientation or heading, a corner, a turn, and a union of two lines (Clements and Battista 1990). Angles are not salient properties of figures to students (Clements et al. in press; Mitchelmore 1989). When copying figures, students do not always attend to the angles.

Students also hold many different schemes regarding not only the angle concept, but also the size of angles. They frequently relate the size of an angle to the lengths of the line segments that form its sides, the tilt of the top line segment, the area enclosed by the triangular region defined by the drawn sides, the length between the sides (from points, sometimes but not always, equidistant from the vertex), the proximity of the two sides, or the turn at the vertex (Clements and Battista 1989).

Intermediate grade students often possess one of two schemes for measuring angles. In the "45–90 schema," slanted lines are associated with 45° turns; horizontal and vertical lines with 90° turns. In the "protractor schema," inputs to turns are based on usage of a protractor in "standard" position (thus, to have a turtle at home position turn left 45°, students might use an input of 135°, which corresponds to a protractor's reading when its base is horizontal) (Kieran et al. 1986).

Moreover, such schemes may be resistant to change, especially through, for example, textbook definitions and examples. When they think, people do not use definitions of concepts, but rather concept images—a combination of all the mental pictures and properties that have been associated with the concept (Vinner and Hershkowitz 1980). Such images can even be adversely affected by inappropriate instruction. For example, the fact that, for many students, the concept image of an obtuse angle having a horizontal ray might result from the limited set of examples they see in texts and a "gravitational factor" (i.e., a figure is "stable" only if it has one horizontal side, with the other side ascending).

## PEDAGOGICAL APPROACHES

Students who know a correct verbal description of a concept but also have a specific visual image or concept image associated strongly with the concept may have difficulty applying the verbal description correctly. Instead, educators need to help them build a robust concept of angle.

One approach, researched by Mitchelmore (Mitchelmore in press), uses multiple concrete analogies. To develop the concept of an angle, the teacher must provide:

1. Practical experience in various situations, leading to an understanding of angular relationships in each individual situation.
2. Angle subconcepts develop when the common features of superficially similar situations are recognized. These types of situations (e.g., turns, slopes, meetings, bends, directions, corners, opening) form different angle contexts.
3. Superficial differences between contexts initially hinder children's recognition of such common features.
4. Less obvious similarities between contexts gradually become apparent, and angle subconcepts begin to emerge.
5. A full angle concept emerges when the same common features are recognized in all angle contexts.

Research on teaching activities based on these ideas revealed that most elementary-age students understood physical relations. Turns, or rotations, were a difficult concept to understand in concrete physical contexts. Other research supports the importance of integration of all types. Some children have only understood turns and angles in a meaningful way after months of work (Clements et al. in press). Initially, they gained experience with physical rotations, especially rotations of their own bodies. During the same time, they gained limited knowledge of assigning numbers to certain turns, initially by establishing benchmarks. A synthesis of these two domains (turn-as-body-motion and turn-as-number) constituted a critical juncture in learning about turns for many elementary students.

An implication of the Piagetian position and the emphasis on turns is that dynamic computer environments might be useful. Turns (and angles) are critical to the view of shapes as paths, and the intrinsic geometry of paths is closely related to real world experiences such as walking.

Computer games have been found to be marginally effective at promoting learning of angle estimation skills (Bright 1985). More extensive and promising are several research projects investigating the effects of Logo's turtle graphics experience on students' conceptualizations of angle, angle measure, and rotation. In one study, for example, responses of intermediate grade students in a control group were more likely to reflect little knowledge of angle or common language usage, whereas the responses of the Logo students indicated more generalized and mathematically oriented conceptualizations (including angle as rotation and as a union of two lines/segments/rays) (Clements and Battista 1989). A large group of studies has reported similar findings, although in some situations, benefits do not emerge until more than a year of Logo experience (Clements and Battista 1992). So, having these experiences over several years of elementary school is recommended.

Logo experiences may foster some misconceptions of angle measure, including viewing it as the angle of rotation along the path (e.g., the exterior angle in a polygon) or the degree of rotation from the vertical (Clements and Battista 1989; Clements et al., in press). In addition, such experiences do not replace previous misconceptualizations of angle measure (Davis 1984). For example, students' misconceptions about angle measure and difficulties coordinating the relationships between the turtle's rotation and the constructed angle have persisted for several years during their elementary schooling, especially if not properly guided by their teachers (Clements and Battista 1992). In general, however, Logo experience appears to facilitate understanding of angle measure. Logo children's conceptualizations of a "larger angle" are more likely to reflect mathematically correct and coherent ideas. If activities emphasize the difference between the angle of rotation and the angle formed as the turtle traced a path, misconceptions regarding the measure of rotation and the measure of the angle may be avoided. For example, students will understand that when the Logo turtle goes forward 50, right 120, forward 50, it draws an angle with a measurement of 60°.

To understand angles, students must understand the various aspects of the angle concept. They must overcome difficulties with orientation, discriminate angles as critical parts of geometric figures, and construct and represent the idea of turns, among others. Furthermore, they must construct a high level of integration between these aspects. This is a difficult task that is best begun in the elementary and middle school years, as children deal with corners of figures, comparing angle size, and turns. At the beginning of secondary school, regions should be integrated with rotations. The formulation of a single, mathematically rigorous definition of angle should follow (Mitchelmore 1989).

*See also* Geometry Instruction; Logo; Trigonometry

## SELECTED REFERENCES

Allendoerfer, Carl B. "Angles, Arcs, and Archimedes." *Mathematics Teacher* 58(Feb. 1965):82–88.

Bright, George. "What Research Says: Teaching Probability and Estimation of Length and Angle Measurements through Microcomputer Instructional Games." *School Science and Mathematics* 85(1985):513–522.

Clements, Douglas H., and Michael T. Battista. "The Effects of Logo on Children's Conceptualizations of Angle and Polygons." *Journal for Research in Mathematics Education* 21(1990):356–371.

———. "Geometry and Spatial Reasoning." In *Handbook of Research on Mathematics Teaching and Learning* (pp. 420–464). Douglas A. Grouws, ed. New York: Macmillan, 1992.

———. "Learning of Geometric Concepts in a Logo Environment." *Journal for Research in Mathematics Education* 20(1989):450–467.

Clements, Douglas H., Michael T. Battista, Julie Sarama, and Sudha Swaminathan. "Development of Turn and Turn Measurement Concepts in a Computer-based Instructional Unit." *Educational Studies in Mathematics* (in press).

Davis, Robert B. *Learning Mathematics: The Cognitive Science Approach to Mathematics Education.* Norwood, NJ: Ablex, 1984.

Gellert, Walter, H. Küstner, M. Hellwish, and H. Kästner, eds. *VNR Concise Encyclopedia of Mathematics.* New York: Van Nostrand Reinhold, 1977.

Jones, Phillip S. "Early American Geometry." *Mathematics Teacher* 37(1)(1944):3–11.

Kieran, Carolyn, Joel Hillel, and Stanley Erlwanger. "Perceptual and Analytical Schemas in Solving Structured Turtle—Geometry Tasks." In *Proceedings of the Second Logo and Mathematics Educators Conference* (pp. 154–161). Celia Hoyles, Richard Noss, and Roseland Sutherland, eds. London, England: University of London, 1986.

Mitchelmore, Michael C. "The Development of Children's Concepts of Angle." In *Proceedings of the Thirteenth Conference of the International Group for the Psychology of Mathematics Education* (pp. 304–311). Gerard Vergnaud, Janine Rogalski, and Michèle Artigue, eds. Paris, France: Paris University, 1989.

———. "The Development of Pre-angle Concepts." In *New Directions in Research on Geometry and Visual Thinking.* Annette R. Baturo, ed. Brisbane, Australia: Queensland University Press, in press.

Piaget, Jean, and Bärbel Inhelder. *The Child's Conception of Space.* New York: Norton, 1967.

Vinner, Shlomo, and Rina Hershkowitz. "Concept Images and Common Cognitive Paths in the Development of some Simple Geometrical Concepts." In *Proceedings of the Fourth International Conference for the Psychology of Mathematics Education* (pp. 177–184). Robert Karplus, ed. Berkeley, CA: Lawrence Hall of Science, University of California, 1980.

DOUGLAS H. CLEMENTS
MICHAEL T. BATTISTA
JULIE SARAMA

# APPLICATIONS FOR THE CLASSROOM, OVERVIEW

Components of the curriculum consisting of real-world problems solved by mathematics. Changes in mathematics education during the last half of the twentieth century have been caused by many factors, including changes in technology, new points of view in mathematics itself, and the growth of research in mathematics education. But no factor has had a greater influence than changes in the applications of mathematics.

What mathematics has significant applications? The historical position was that "applied mathematics" consists of classical analysis, such as calculus, differential equations both ordinary and partial, integral equations, and special functions. Given this position, the content of school mathematics also was defined: schools must aim for and teach all the prerequisites for calculus, and nothing else matters as much. Thus, the secondary schools emphasized plane and solid geometry, two years of algebra, trigonometry, and analytic geometry. A concentration on functions included both specific categories such as polynomial, rational, exponential, logarithmic, and trigonometric, and an attempt to get across the general notion of a function.

Classical analysis, in turn, was the basic mathematics of physics, chemistry, and astronomy, and of the various branches of engineering. No other applications were felt to be as important as these, and none had a comparable record of centuries of success. It is true that probability theory had begun to be important in physics and in various branches of engineering, such as communications. A path-breaking book was Thornton C. Fry's *Probability and Its Engineering Uses* (1928). But it is true that at least into the 1960s few departments advised students with great interests in either physics or mathematics to spend any time learning probability.

The fundamental change in our understanding of applications of mathematics, which pervades the second half of the twentieth century, is the demonstrated importance of mathematics to practically all fields of human endeavor, not just to the physical sciences and engineering. Many forces led in this direction. The social sciences and the biological sciences, for example, were rapidly being mathematized, operations research (usually called "OR") proved successful in World War II and rapidly spread across business and industry, and the unbelievable growth of both computer science and computer applications affected almost all disciplines. Currently, almost all fields of human endeavor tend to be examined in a systematic, structural, analytical, that is, mathematical way. Linguistics, art, and music, to name just a few, have not been mentioned explicitly, but have shared the tendency toward mathematization.

Along with this mathematization of almost everything came the realization that the areas of applicable mathematics were vastly broader than classical analysis. Probability theory, mentioned above, is basic to social sciences and to OR. Linear algebra was a graduate subject until the era under discussion, and its applicability was justified by systems of differential equations. Again, the social sciences and OR showed how universally important linear algebra is. Discrete mathematics is at the foundation of computer science and computer applications. Statistics, including its recently developed aspect of exploratory data analysis, turns out to be essential to almost everybody. Nonlinear mathematics is important not only to the physical sciences, but to an understanding of many aspects of human organization, history, and business. Even number theory, that cornerstone of allegedly "pure" mathematics, has become the foundation of modern cryptography. The term "Mathematical Sciences" became standard in the 1970s to reflect the much broader usefulness of mathematics, statistics, and computer science.

This has made the job of curriculum design enormously more difficult. How can all these topics with valid claims to time in the schools possibly be accommodated? In order to understand this issue, it is necessary to clarify the meaning and interrelation of such concepts as applied mathematics, mathematical modeling, and word problems.

## THE MEANING OF APPLIED MATHEMATICS

Applications in the classroom mirror applications in the real world to the extent possible. An application of mathematics in the real world can be outlined as follows:

1. Identify something we want to know, or do, or understand. The result of this step is a question in the real world.
2. Select relevant real-world "objects," and specify relations among them. The result is an identification of key concepts.
3. Decide what to keep and what to ignore among the properties of the objects and relations. The result is an idealized version of the original question.
4. Translate the idealized version into mathematical terms. The result is a mathematical version of the idealized question.
5. Identify the field(s) of mathematics involved. The result is the ability to bring to bear the instincts and knowledge of this field.
6. Use mathematical methods to get insights and answers. The results are techniques, interesting special cases, solutions, theorems, and algorithms.
7. Translate back to the original field. The result is a theory of the idealized question.
8. Confront the empirical data of reality with the theory of the idealized question. Do we believe what the mathematics is telling us? If we do, we write it up, present it to other people, and have done our job. If not, go back to the beginning, and start over.

It is possible for students to participate in this total process, either in following or recreating successful examples, or creating their own. Either way, they will be learning Mathematical Modeling (q.v.), which is the name for this total process. On other occasions, mathematics education begins at step 4, with the translation of the idealized question into mathematical terms, and ends with step 7, the translation back to the vocabulary of the idealized question. This is the traditional meaning of *Applied Mathematics*. Courses that teach the mathematics that arises in steps 5 and 6 when the real world under consideration is the physical sciences are called *Methods of Applied Mathematics*. We often have problems in the curriculum that use words from the outside world, but make no attempt to connect with that world. They begin with step 5 and end with step 6. These are *Word Problems*. One more category needs to be mentioned. There are word problems in which the context is not meant to be taken seriously, but serves only to lighten the mathematical task. These are *Whimsical* problems.

## THE RECENT HISTORY OF APPLICATIONS IN THE CURRICULUM

The history of applications in the classroom in modern times involved the teaching of arithmetic to all children, which culminated in grades 7 and 8 in the most important genuine applications of their time, namely shopkeeping arithmetic. Algebra was originally intended as a part of pure mathematics to be taught only to the privileged classes, with applications that were unreal and at best whimsical. Geometry was always highly practical in its collection of facts useful for agriculture and for many trades; the axiomatic development was again reserved for the elite. Thus algebra and geometry had a historic "gatekeeping" function by restricting access to college preparatory mathematics. The trigonometry course that was current in the middle of the twentieth century was essentially designed for future surveyors.

The swing of the pendulum away from the "New Math" of the 1960s included a greater emphasis on applications of mathematics. Shirley Hill (1975) chaired a committee of the Conference Board of the Mathematical Sciences called NACOME. Its report contains a long and valuable section on applications of mathematics. "The Interaction of Mathematics and Other School Subjects" (International Commission on Mathematical Instruction 1979) gives a good summary for its time of the state of applications in the classroom. The *Agenda for Action* (National Council of Teachers of Mathematics 1980) states as its first recommendation that problem solving must be the focus of school mathematics in the 1980s. *The*

*Mathematical Sciences Curriculum K–12: What is Still Fundamental and What is Not* (Conference Board of the Mathematical Sciences 1982) states that "the widespread availability of calculators and computers and the increasing reliance of our economy on information processing and transfer are significantly changing the ways in which mathematics is used in our society. To meet these changes we must alter the K–12 curriculum by increasing emphases on topics which are fundamental to these new modes of thought." In the meantime, Hugh Burkhardt's book *The Real World and Mathematics* (1981) was a milestone in advocating mathematical modeling for the schools by the use of everyday situations of interest to schoolchildren. The *Curriculum and Evaluation Standards* (National Council of Teachers of Mathematics 1989) took the next step beyond the problem solving of the *Agenda for Action* by stressing problem *finding*. All three of the K–4, the 5–8, and the 9–12 standards include statements on the formulation of problems from situations within and outside mathematics.

The Conference Board of the Mathematical Sciences report (1982) highlighted the uses of mathematics in society, which have become more and more pervasive in the thinking about mathematics education. Society gives us a large amount of time to teach mathematics because of its usefulness. It is used in practical everyday life, in employment, and as part of intelligent citizenship. Economic competitiveness as a motivation for changing school mathematics has in recent years replaced the earlier motivation of military and space competition.

## MATERIALS FOR THE TEACHING OF APPLICATIONS

The drive toward emphasizing applications in school mathematics is reflected in many activities in recent years. The first International Congress on the Teaching of Mathematical Modelling was held in 1983, and has continued regularly at two-year intervals. The published papers from these congresses may be found in the Horwood series, *Applications and Modelling in Learning and Teaching Mathematics* (Blum et al. 1989). Journals that specialize in the teaching of applications include various COMAP publications. Source books of genuine applications of mathematics have been published (Noble 1967; Bushaw et al. 1980). There have been many recent books for the teaching of mathematical modeling (see Giordano and Weir 1985).

The diversity of applicable mathematics and of areas of applications is reflected in many current textbook series at all levels. Series for grades K–8 have diversified beyond arithmetic into geometry, statistics, probability, and applications of technology; the problem-solving activities include a great variety of situations to which mathematics is applied. Unified comprehensive textbooks for high school also lend themselves to the inclusion of a variety of applications.

There have been a number of curriculum projects in recent years that have focused on applications and modeling. The pioneer was the USMES project (Unified Science and Mathematics in the Elementary School), which was active from the middle 1960s to the middle 1970s. This project produced thirty-two units, which combined mathematics, science, social studies, and English. Each unit began by gaining an understanding of a "challenge," developed the relevant mathematics, science, and social science, and ended by a proposal for action (USMES 1974). A similar project at the high school level was "The Man-Made World" (Engineering Concepts Curriculum Project 1969). It emphasized modeling, computers, and engineering concepts such as dynamical systems, feedback, and stability. *The Regional Math Network* was a teacher invigoration and curriculum project headed by Katherine K. Merseth. In 1987 it produced a series of junior high school materials entitled *Sports Shorts, Quincy Market,* and *Mathspace Mission.* Each book contained applications of mathematics with a particular theme, and took responsibility for the teaching of particular topics—decimals and percent, ratio and proportion, and estimation and geometry, respectively.

Several projects have emphasized modeling. For example, the Interactive Mathematics Project has produced a high school curriculum focused on a variety of applications; the ARISE project of COMAP is working on a curriculum for grades 9–11, which is based on mathematical modeling, and the Pacesetter project of the College Board has designed a high school capstone course for grade 12, which is based on modeling and is intended to serve a multiplicity of destinations such as calculus, two-year college, and the world of work.

What many of these projects have in common is the realization that all students need to learn how to use mathematics. The social goal of keeping students in high school together for more years may be more nearly achievable when we emphasize the common need to apply mathematics rather than the disparate mathematics requirements of various destinations.

Everyone gains when every student learns how to model situations in the real world mathematically.

## RESEARCH ON THE TEACHING OF APPLICATIONS

On the whole, there has not been much research in the teaching and learning of mathematics that deals with applications. What there is has dealt primarily with "problem solving." An excellent summary of what is known may be found in Alan Schoenfeld (1992). He points out how many different meanings mathematics education gives to "problem solving," "from remediation to critical thinking to developing creativity." The definition closest to applications is "to learn standard techniques in particular domains, most frequently in mathematical modeling." Much has been written about the many purposes of teaching applications and modeling. For example, Niss (1989) mentions furthering a creative, problem-solving attitude, developing critical potential, practicing applications and modeling, giving a balanced picture of mathematics, and aiding the acquisition and understanding of mathematics. But there appears to be no proof as yet that the teaching of applications and modeling brings mathematics education closer to these goals.

## FORCES THAT AFFECT THE TEACHING OF APPLICATIONS

A number of forces that help the teaching of applications of mathematics are the increasing mathematization of other fields, the fact that applications are now recognized as an integral part of mathematics, the ability of applications to motivate students, and the increasing mathematical needs of many jobs. The messy numbers that the real world is likely to produce no longer impede mathematical treatment. Both the calculator and the computer handle realistic numbers as easily as artificial ones. Some of the factors that impede the teaching of applications include the belief of some mathematics educators in the purity of the subject, and a possible ignorance and even fear of other disciplines. Teaching mathematics through modeling takes more time than dealing immediately with the mathematics, and the extra time may not be available. Some parents as well as educators believe that mathematics should be taught the same way they themselves learned it. They resist the influence of technology, of newly important subjects in the mathematical sciences, and the inclusion of applications because they are different from their own experience. Economic competitiveness is a key answer to such objections.

*See also* Applications for the Secondary School Classroom; Interactive Mathematics Program (IMP); Modeling

## SELECTED REFERENCES

Blum, Werner, et al. *Applications and Modelling in Learning and Teaching Mathematics.* Chichester, England: Horwood, 1989.

Burkhardt, Hugh. *The Real World and Mathematics.* Glasgow, Scotland: Blackie, 1981.

Bushaw, Donald, et al. *A Sourcebook of Applications of School Mathematics.* Reston, VA: National Council of Teachers of Mathematics, 1980.

COMAP. The series called HiMAP, HistoMAP, and UMAP. Lexington, MA: Consortium for Mathematics and its Applications, 1979 ff.

Conference Board of the Mathematical Sciences. *The Mathematical Sciences Curriculum K–12: What Is Still Fundamental and What Is Not.* A Report to the NSB Commission on Precollege Education in Mathematics, Science and Technology. Washington, DC: The Board, 1982.

Engineering Concepts Curriculum Project. *The Man-Made World.* New York: McGraw-Hill, 1969.

Fry, Thornton C. *Probability and its Engineering Uses.* New York: Van Nostrand, 1928.

Giordano, Frank R., and Maurice D. Weir. *A First Course in Mathematical Modelling.* Monterey, CA: Brooks/Cole, 1985.

Hill, Shirley, et al. *Overview and Analysis of School Mathematics Grades K–12.* Washington, DC: Conference Board of the Mathematical Sciences, 1975.

International Commission on Mathematical Instruction. *New Trends in Mathematics Teaching.* Vol. IV. Paris, France: UNESCO, 1979.

Merseth, Katherine K. *The Regional Math Network.* Cambridge, MA: Harvard Graduate School of Education, 1987. Available from Seymour, Palo Alto, CA.

National Council of Teachers of Mathematics. *An Agenda for Action: Recommendations for School Mathematics of the 1980s.* Reston, VA: The Council, 1980.

——. *Curriculum and Evaluation Standards for School Mathematics.* Reston, VA: The Council, 1989.

——. *Principles and Standards for School Mathematics.* Reston, VA: The Council, 2000.

Niss, Mogens. "Aims and Scope of Applications and Modelling in Mathematics Curricula." In *Applications and Modelling in Learning and Teaching Mathematics* (pp. 22–31). Werner Blum et al., eds. Chichester, England: Horwood, 1989.

Noble, Ben. *Applications of Undergraduate Mathematics in Engineering.* Washington, DC: Mathematical Association of America, 1967.

Schoenfeld, Alan H. "Learning to Think Mathematically: Problem Solving, Metacognition, and Sense Making in Mathematics." In *Handbook of Research on Mathematics*

*Teaching and Learning* (pp. 334–370). Douglas A. Grouws, ed. New York: Macmillan, 1992.

Unified Science and Mathematics for Elementary Schools (USMES). *Teacher Resource Books.* Newton, MA: Education Development Center, 1974 ff.

HENRY O. POLLAK

# APPLICATIONS FOR THE SECONDARY SCHOOL CLASSROOM

Word problems to enhance, motivate, and show relevance of topics in the secondary curriculum. Numerous federal and privately sponsored commissions have been actively promoting inclusion of applications (see Cohen 1993; National Council of Teachers of Mathematics 1989; National Research Council 1989, 1990); publishers have begun to develop new textbooks and supplementary materials focusing on applications. Sometimes the term *applications* is a term for traditional word problems; for example, the time it takes for two trains to collide, or the age of a relative.

Some of the newer applications in mathematics can be classified as job or game types. These types of applications show mathematics in broader terms, in ways that may help students see mathematics as more interesting and relevant to their lives than formerly. The first type of application demonstrates how the particular elements and style of the mathematics being used in the classroom are required on a job. For instance, the concept and formulas associated with volume can be related to pouring concrete at a construction site. Often, construction crews must set forms and pour concrete for bridge abutments, structures that support a bridge. Knowing the exact amount of concrete to be poured is important to assure the integrity of the structure and minimize cost. A low initial estimate requires a second pour. This is called a "cold joint," and is not permitted on most jobs. Cold joints cannot guarantee that the abutments will support the weight of the bridge. On the other hand, a high initial estimate leaves extra concrete, which is not only wasteful but difficult to discard.

An application that examines a practical situation may be simplified to match the level of student ability. Examples of this type of application illustrate how a system of equations is used for industrial-type linear programming problems or for understanding how a CAT scan functions. Since these applications involve hundreds, and sometimes thousands, of variables, any classroom presentation out of necessity would have to be "contrived." They also would have to be "contrived" because of the numerous prerequisites needed to understand both problems fully.

Consider the following "contrived" interpretation of the CAT scan (Malkevitch 1992). The machine takes many cross-sectional pictures of the body. For each of these cross-sectional slices, a detector on the opposite side of the body records the amount of the X-ray beam that passes through the body. As the scanner rotates around the body, the position of the X-ray beam changes many times, so for each cross-section there are many after-penetration X-ray measurements. At the same time, a photo is taken of the cross-section, and the various shades of grays are coded to represent various densities in the body. These densities might represent, for example, in the brain, the presence of water, bone, blood, air, gray matter, or tumors. With enough of these after-penetration X-ray measurements, a computer can determine an unknown internal region by computing its density value. Suppose regions $A$, $B$, $C$, and $D$ have different densities, and $W$, $X$, $Y$, and $Z$ measure the amount of $X$-ray passing through the brain. The resulting equations would be:

$$A + D + C = Y$$
$$B + A + D = X$$
$$C + B = Z$$
$$D + C = W$$

where regions $A$, $B$, $C$, and $D$ can be determined in terms of the known values $W$, $X$, $Y$, or $Z$.

Game-type applications may not be of any practical value in real life, but they are intriguing and charm some students into trying to solve them. A game type application with a practical value in real life is Conway's Game of Life (Dewdney 1988 and 1989; Von Neumann 1966). The game is based on the theory of cellular automata. This theory is used to model how organisms reproduce and die, how snowflakes or galaxies are formed, or how to encrypt messages. A simple example of how cellular automata can simulate real phenomena is to play Conway's Game of Life. Starting with a square matrix, containing only 1s and 0s, pretend the 1s represent living creatures and the 0s represent a region with no life. Since life forms utilize resources to grow and reproduce, we can imagine that too many 1s in a region represent "overcrowding" and too many 0s represent "not enough life forms to meet, mate and multiply." A simple rule to determine future genera-

tions might be (a) if the $3 \times 3$ neighborhood containing a cell sums to 3 or 4, in the next generation this cell will be alive; and (b) if the neighborhood sums to anything else, this cell gets a zero in the next generation.

Many educators believe that by engaging students in all types of applications, courses can become more attractive to them. Mathematics can be made to play to their interests, and to incorporate today's issues, such as the greenhouse effect or the AIDS epidemic; both issues can be modeled mathematically, the first using exponential functions (Fisher 1989) and the second using William Farr's Law of Ratios (Bregman and Langmuir 1990). Teachers are being encouraged by the many reports on the future of mathematics education to make these connections for their students, in order to show relevance. The hope is that this, along with other classroom innovations, will encourage more students to continue studying mathematics.

A federal initiative to incorporate applications in the secondary schools is Tech Prep. The program is targeted for the academically "middle 50%" of students, attempting to raise their sights and encourage them to undertake technology studies. Materials have been written in a number of areas, including mathematics. The mathematics materials were developed through a consortium of state educational agencies by the Center for Occupational Research and Development (CORD); its two-year program of thirty-six units meets standards established by the National Council of Teachers of Mathematics (NCTM). Each unit is divided into six activity sessions, including videos, hands-on exercises, and real-life applications. Learning is applied to problems in technology and careers (CORD 1994).

For teachers, inclusion of applications beyond those found in a textbook is dependent on their initiative. Teachers must find sources for such applications, determine what applications are appropriate, and how to incorporate them into course work. Fortunately there are many sources teachers can use to find applications. Two major sources are mathematics journals and materials produced by COMAP (Consortium for Mathematics and Its Applications). Membership in COMAP includes a quarterly newsletter of articles on job and game type applications, and discounts on high school modules in mathematics applications. These modules often include lesson plans, group and individual exercises, transparencies, and worksheets. Modules average eight pages and can be photocopied. COMAP also has a limited but excellent video application series in geometry and statistics; the videos come with a written summary and exercises.

Colleagues, newspapers, textbooks, and non-mathematical journals also are good sources of applications. A recurring story in newspapers is the use of computers to factor large numbers. Such stories describe in lay terms (suitable for classroom use) how factoring is used for coding messages to make military and banking secrets secure. Topics discussed in nonmathematical settings may be related to the study of a mathematical topic. For instance, the concept of ratio may be illustrated by showing how waist-hip ratios are used as an index of health (Folsom 1993). We may also show how linear and quadratic equations are being used to help farmers determine the amount of fertilizer needed to maximize corn yield (Klausner 1990).

Once teachers have found appropriate applications to present to the class, they must decide on the instructional format and develop lesson plans. While many teachers still prefer the lecture format, the reports on applications in the classroom encourage hands-on experiences for students (like the CORD project) and team projects (collaborative learning).

Lesson plans can take a lot of time, depending on the source of the application and how much additional information the teacher wants to include. A second and quicker option for teachers is simply to present the students with examples of applications of a particular topic, not to produce arguments through mathematical rigor or proof. A recently funded National Science Foundation Division of Undergraduate Education Grant, entitled *Snapshots of Applications in Mathematics*, has used this latter approach. Students are quickly introduced to an application, without all the details necessary to understand the problem thoroughly. These applications are not complete mathematical lesson plans. Preliminary results have shown improvement in student motivation, especially for the academically weaker students, and an improved appreciation for the relevance of mathematics (Callas 1994).

Teachers should use applications to help make the curriculum more relevant to students. How best to deliver the applications—quickly but with few details, or carefully but with lots of detail—is still open to debate. It is reasonable to conclude that both methods should be used.

*See also* Actuarial Mathematics; Applications for the Classroom, Overview; Art; Biomathematics; Business Mathematics; Chemistry; Consumer Mathematics; Cryptography; Economics; Environmental Mathematics; Finance; Health; Music; Nursing; Sports

## SELECTED REFERENCES

Bregman, Dennis, and Alexander Langmuir. "Farr's Law Applied to AIDS Projections." *Journal of the American Medical Association* (Mar. 16, 1990):1522–1525.

Callas, Dennis. *Snapshots of Applications in Mathematics.* Delhi, NY: State University, 1994.

Center for Occupational Research and Development (CORD). *A Closer Look at Applied Mathematics—Unit Objectives, Correlation to NCTM Standards, Equipment List.* Waco, TX: The Center, 1994.

Cohen, Don, ed. *Standards for Curriculum and Pedagogical Reform in Two-Year College and Lower Division Mathematics—Circulating Draft.* ED 362248. Columbus, OH: ERIC, 1993.

Dewdney, A. K. "Computer Recreations." *Scientific American* 259(Aug. 1988):104–107.

———. "Computer Recreations." *Scientific American* 261(Aug. 1989):102–105.

Fisher, Arthur. "Global Warming: Part Two—Inside the Greenhouse." *Popular Science* 235(Sept. 1989):63–70.

Folsom, Aaron R. "Waist-Hip Ratio Seen as an Index of Health." *New York Times* (Jan. 27, 1993):C16.

Klausner, Stuart. "The Status of Soil Testing for Nitrogen Recommendations." *Cornell Extension Farming Update* (Apr. 1990):10–11.

Malkevitch, Joseph. *Geometry: New Tools for New Technology.* Lexington, MA: COMAP, 1992.

National Council of Teachers of Mathematics (NCTM). *Curriculum and Evaluation Standards for School Mathematics.* Reston, VA: The Council, 1989.

———. *Principles and Standards for School Mathematics.* Reston, VA: The Council, 2000.

National Research Council (NRC). *Everybody Counts: A Report to the Nation on the Future of Mathematics Education.* Washington, DC: National Academy Press, 1989.

———. *Reshaping School Mathematics: A Philosophy and Framework for Curriculum.* Washington, DC: National Academy Press, 1990.

Von Neumann, John. *Theory of Self-Reproducing Automata.* Edited and completed by Arthur W. Burks. Urbana: University of Illinois Press, 1966.

DENNIS CALLAS

# ARCHIMEDES (287–212 B.C.)

The greatest mathematician of Greek antiquity and, together with Isaac Newton (1642–1727) and Carl Friedrich Gauss (1777–1855), one of the three greatest mathematicians of all time. He spent most of his life in the Greek city of Syracuse, but is known to have visited Egypt, which accounts for his extensive familiarity with the mathematical and scientific achievements of the Alexandrian school. Archimedes is credited with using his scientific prowess to help inflict serious losses on the Romans during their siege of Syracuse. It was during this siege that Archimedes was killed while trying to prevent a Roman soldier from disturbing a geometrical diagram that he had drawn in the sand. Although many of his original works have been destroyed, ten original treatises remain extant, of which three are devoted to plane geometry and two are devoted to solid geometry.

Archimedes' work can be classified into three areas: geometrical, arithmetical, and mechanical. He was the first to specify a method for computing pi by inscribing and circumscribing polygons about a circle. His analysis led to an approximation of pi between $3\frac{10}{71}$ and $3\frac{1}{7}$. Archimedes developed a spiral which he used to solve the problem of quadrature of a circle, that is, to determine a square whose area is equal to that of a circle. He also succeeded in solving quadrature problems for curvilinear plane figures and for solving quadrature and cubature problems for curved surfaces.

Archimedes often is described as an early pioneer in the development of the calculus, since he was able to devise methods equivalent to modern-day integration methods for determining the areas of parabolic segments and spirals, and computing the volume and surface area of a sphere. He discovered many of the laws governing the science of hydrostatics and invented the water-screw, which was used to irrigate fields and pump water out of mines. In his treatise *The Sand Reckoner,* Archimedes described a system of numeration that he devised to express numbers containing up to eighty thousand million million digits. To solve a problem that he posed asking for the number of grains of sand that it would take to fill the universe, Archimedes calculated that the number would be $10^{63}$. His system of representing large numbers laid the groundwork for the later development of logarithms in the seventeenth century.

*See also* Greek Mathematics

SELECTED REFERENCES

Archimedes. *The Works of Archimedes.* Thomas L. Heath, ed. New York: Dover, n.d.

Eves, Howard W. *Great Moments in Mathematics (Before 1650).* Washington, DC: Mathematical Association of America, 1980.

ANTHONY V. PICCOLINO

## ARISTOTLE (384–322 B.C.)

The great Greek philosopher of antiquity, whose greatest contribution was his codification and systematization of the laws of inferential thinking into a scientific body of knowledge. This achievement represents one of the greatest intellectual mileposts in the history of civilization. Although not a mathematician per se, Aristotle was instrumental in transmitting, through his writings, much of the known mathematical knowledge of his time and advancing what is known today as the scientific method. The elements of observation, hypothesis formation, prediction, and testing form the essential components of the scientific method. According to Aristotle, mathematics is one of the three theoretical sciences, the other two being theology and physics (natural philosophy). He utilized mathematics as the primary medium to illustrate the scientific method. With respect to the concept of infinity, Aristotle rejected the notion of an actual infinity, in favor of potential infinity. He distinguished the two types of infinity by contending that a straight line does not actually become infinite, but rather, whatever its length, it can always be extended. Therefore in Aristotle's view, a straight line can be made arbitrarily long, but will never actually be infinite in length. Although his writings indicate an awareness of several significant mathematical theories such as Eudoxus' theory of proportions and the Pythagorean view of numbers, Aristotle paid little attention to higher mathematics. Noticeably missing is any mention of the conic sections and the classical construction problems of duplicating the cube and trisecting an arbitrary angle.

*See also* Greek Mathematics

SELECTED REFERENCES

Burke, James. *The Day the Universe Changed.* Boston, MA: Little, Brown, 1985.

Calinger, Ronald, ed. *Classics of Mathematics.* Englewood Cliffs, NJ: Prentice-Hall, 1995.

Heath, Thomas L. *A Manual of Greek Mathematics.* New York: Dover, 1963.

ANTHONY V. PICCOLINO

## ARITHMETIC

A branch of mathematics that has three common definitions, two of which are current and one of which is historic: (a) From the pure mathematical point of view, arithmetic is one of the branches of mathematics (like probability, geometry, etc.) that deals with real numbers and computing with them. The four fundamental operations of arithmetic are addition, subtraction, multiplication, and division; (b) Throughout the history of education, the conventional definition of *arithmetic* has referred to the basic facts and algorithms; skills and rote procedures traditionally included in the elementary school curriculum; and (c) Common current usage of the term *arithmetic* refers to the elementary and middle school curriculum in the area of mathematics or school mathematics in the area of number. This latter definition has been further confirmed with the publication of the *Curriculum and Evaluation Standards for School Mathematics* (National Council of Teachers of Mathematics (NCTM) 1989). A change of emphasis from memorization and following directions to more cognitive tasks such as active use of intuition, number sense, and estimation has evolved (Leinhardt, Putnam, and Hattrup 1992). Historically, the image of arithmetic is one of memorization, drill, and algorithmic rules. Beginning with the September 1994 issue, the major journal in elementary mathematics education has changed its name from *The Arithmetic Teacher* to *Teaching Children Mathematics*. Barnett (1993) suggests a continuation of the use of the term *arithmetic* with, however, a change of focus toward its underlying concepts of number sense and patterns.

## THE HISTORY OF ARITHMETIC

Arithmetic, it is believed, developed in response to practical needs. In prehistoric times people needed to keep track of time and possessions. A person might add a pebble to a pile each time the sun came up or put a notch on a stick for each animal skin acquired. This became known as *one-to-one correspondence*. Gradually, terms such as *pair, couple, twin, trio, multitude,* and *heap,* still in use today, developed to identify groups of more than one (Baroody 1987). Human beings have a natural power of imaging, enabling them to accurately visualize up to amounts of four or five elements without actually counting them (Barrow 1992). Without the aid of *counting,* the capacity to further develop arithmetic concepts of quantity and measurement is critically limited. Eventually a need developed for more precise means of keeping track, and systems of counting were invented. The basic idea underlying arithmetic is counting. Some indigenous tribes, however, never developed counting and even today rely on visualization for their practical needs (Barrow 1992, 40).

While counting is the basis of the historical development of the fundamental arithmetic processes (addition, subtraction, multiplication, and division), *visualization,* the underlying basis of number sense, has renewed value in current thinking about school mathematics (NCTM 1989, 1991).

As the world became more complex, it became necessary to record amounts greater than the process of counting by ones could easily accommodate. As more efficient methods were needed, the process of grouping emerged. Probably due to the dependence on the ten fingers or digits, base ten evolved naturally as the basis of our Hindu-Arabic number system. Not all cultures used ten as a base. Around 3500 B.C., the first known numeral system developed. People discovered that by designating a symbol for a group of a certain culturally determined amount, they could easily record information. For example, a farmer might count off ten sheep and record the ten with a pebble. Upon reaching ten groups of ten, the farmer might use a larger stone to indicate one hundred sheep (Baroody 1987). Methods of calculation became more efficient as positional numeration systems were invented. In a symbol system with *place value,* the digit's position within the sequence of a numeral designates its worth. For example, in 3,333, the three on the right is three 1s, the next three stands for three 10s, the next for three 100s, and the next for three 1,000s. In this system, given the above four positions and the symbols 0 to 9, any whole number quantity between 0 and 9,999 can be represented. Around A.D. 500, Indian astronomers came up with the brilliant idea to use a small circle instead of an empty space to represent a place with no digit in it (Bunt, Jones, and Bedient 1976). This evolved into the zero we use to represent an empty place and was a milestone in transferring a number from a counting board, such as an abacus, to written symbols. Once zero was invented, arithmetic algorithms were developed that could be commonly understood and used.

The history of the development of arithmetic demonstrates how intuitive, informal numeration schemes developed that led to more formalized systems (Baroody 1987). Some mathematics educators view the development of children's knowledge of arithmetic as following a path similar to that of history. That is, children develop intuitive and informal arithmetic understandings and problem-solving methods which can then gradually lead them to an understanding of the more formal procedures (Baroody 1987; Kamii with DeClark 1985).

## THE HISTORY OF TEACHING ARITHMETIC

Mathematics has a rich history of instruction going back to ancient Egypt, where the first book on mathematics was written and its major principles taught to children (Kramer 1966). Specimens of children's arithmetic work have been found that date back to that early time period. The Greeks viewed both the study of number and calculation as essential elements in a liberal education. The Romans stressed the practical applications of calculations to everyday problems, such as building roads and carrying on business transactions (Kramer 1966).

In seventeenth-century America, there existed minimal mathematical studies. When mathematics was taught, it was in the form of simple arithmetic, often outside of school, since it was not considered a legitimate school subject. The arithmetic that was taught stressed the practical applications of the times, such as bookkeeping, navigation, and surveying (Rosskopf 1970). As early as the eighteenth century, Benjamin Franklin proposed changes to the core reading and writing curriculum by advocating the inclusion of arithmetic, science, and mechanical arts, thus establishing the "academy." Franklin's logic for the inclusion of mathematics into the curriculum was one based on practical utility and mental discipline. There was a strong influence of Hebart and Froebel's faculty psychology present, where the mind was considered a muscle to be developed through exercise. These and other singular attempts at innovation were not widespread and were short-lived, with mathematics viewed as mental discipline and learning as recitation (Rosskopf 1970). The "Committee of Ten" (1892–1894) recommended two major changes to the mathematics curriculum: (a) that topics of mathematics be integrated rather than taught separately, and (b) that algebra and geometry be integrated into elementary school arithmetic. Such recommendations are still considered innovative in more recent reform documents (National Research Council [NRC] 1989; NCTM 1989), and to this day are not considered common characteristics of the elementary school curriculum. John Dewey's influence on the mathematics curriculum consisted of the pedagogical theory of arithmetic content as measurement. He coauthored (with John McLellan) a text for teachers stressing problems in the form of measurement activities. This text, along with Herbert Spencer's theories of the spiraling nature of learning, were met with skepticism or avoidance by teachers in the schools (Rosskopf 1970).

The recursive appearance of drill and practice as a pedagogical theory is salient throughout history, and appears, again, in the Edward Thorndike period (1917–1935). Thinking of drill as an innovation may appear ludicrous even for teachers of that time period. Thorndike's "contribution" was more in the sense of a recent theory that added credence to this common rote pedagogy. His theories provided the field of pedagogy with another problem, however, the fragmentation of content into minute and discrete units that are taught and tested separately (Romberg and Carpenter 1986).

In the period from 1920 to 1945, the concept of junior high schools emerged and proliferation of these schools increased. Arithmetic was well established as the mathematics content of the K–6 schools. With little controversy on *what* to teach, attention was diverted to *how* and *when* to teach particular topics. Gestalt psychology had its effect on the curriculum through Leo Brueckner. In his book, *The Development of Ability in Arithmetic* (1939), Brueckner proposed a natural stage theory that affected the curriculum in the form of readiness tests, retention, and a postponement of formal arithmetic until the intermediate grades. The primary mathematics curriculum thus became, by default, an informal learning of the subject through projects or themes, also known as "incidental learning."

Another important happening during this same period was the introduction of meaning theory by William Brownell. In referring to the name for his theory, Brownell contended that it was "selected for the reason that, more than any other, this theory makes meaning, the fact that children shall see sense in what they learn, the central issue in the arithmetic instruction" (Brownell 1935, 520). Brownell considered his theory one of psychology and pedagogy, carefully delineating its relation to other theories of the time—drill theory and incidental learning theory. The translation of Brownell's theory into practice, however, varied widely in terms of content and procedures, perhaps, due to the paradoxical interpretation of the words "teaching for meaning" (Rosskopf 1970). Among the recommendations proposed by the 1945 National Council of Teachers of Mathematics Report were to "abandon the idea that arithmetic can be taught incidentally or informally," to "administer drill much more wisely," and to "conceive of arithmetic as having both a mathematical aim and a social aim."

After World War II, needs for and concerns about mathematics spread into new areas, such as statistics, practical problem solving, and consumer applications as well as an expansion of traditional skill areas in which students were perceived as deficient. Growing public awareness of the need for better schools sparked interest in the mathematics curriculum. This ended in a public realization that mathematics was a worthwhile discipline in its own right apart from its application to other areas of study. The public cry for specialization took the form of a proliferation of school programs labeled "enrichment" and "accelerated." Along with this growing need for better quality and broader mathematics programs was the concomitant shortage of qualified mathematics teachers at all levels of schooling. The period from the early 1950s to the late 1960s came to be known as the era of the "New Math," a top-down curriculum reform in mathematics. A move to promote understanding of theories behind arithmetic precipitated the inclusion of set theory and properties of arithmetic into the New Math curriculum even as low as first grade. The concern for building a mathematics curriculum with more meaning than the rote drill and memorization of earlier years had deep roots.

Nevertheless, the science explosion initiated with the launching of the space vehicle *Sputnik* by the U.S.S.R. in 1957 has been acknowledged as a direct cause of the New Math movement. This reform differed in several ways from earlier types; (a) it was public, drew public concern and discussion; (b) it was nationwide in character—had an economic and competitive base rousing national pride to meet the international challenge; (c) it was the first subsidized reform effort on a grand scale; (d) it was the first time in history that mathematicians in large numbers became involved in school curricular efforts; and (e) it was content-centered rather than child- or society-centered. Critics of the New Math varied from those who questioned its "correctness" to those who objected to it as an appropriate program for elementary school children. The most well-known critic of the time was Morris Kline, who criticized the content and pedagogy of the New Math reform efforts (Rosskopf 1970).

The influence of child psychologist Jean Piaget was not felt until the late 1950s, when several of his books were translated and published in this country. Though Piaget's stage theory has been questioned, there is no doubt concerning the impact that it had on the textbooks and teaching practices. He brought the curricular emphasis away from the subject area content back to the child. As in the earlier readiness theories, the curriculum was seen as driven by the development of the child. Though Piaget's works were available in English since the late 1950s, they became more popular in the late 1960s more as a reaction to the New Math with its strict content focus.

The "Back to Basics" movement in elementary school mathematics also was a reaction to, and dissatisfaction with, the New Math reform. Back to Basics was essentially a return to earlier drill and practice methods. It was a reaction to the weaknesses within New Math and failed to acknowledge its strength in attempting to make salient to students the meaning behind operations and procedures in arithmetic. The Back to Basics reform effort came from the public with some following in college education departments.

The NCTM document, *An Agenda for Action: Recommendations for School Mathematics of the 1980s,* suggested that problem solving be the focus of school mathematics and that basic skills in mathematics be defined to encompass more than computational facility (NCTM 1980, 1). "Problem Solving" became an ambiguous buzz word with multiple interpretations. Teachers' beliefs about mathematics and about how it could best be taught framed and reshaped efforts to change the curriculum (Romberg and Carpenter 1986).

Although the use of physical objects to represent abstract mathematical concepts had been a part of arithmetic since its inception, reliance on symbols and printed textbooks forced such concrete representations out of most classrooms until the 1970s and 1980s when such practices became more widespread. Perhaps the use of concrete objects or manipulatives was a reaction to the problem-solving movement with its emphasis on reasoning as abstract thought. *Cuisenaire Rods,* invented in Holland in the 1960s, and *Dienes' Blocks,* created in Canada not long after, came into widespread use during this latter period. Their use in schools increased with the popularity of workshops and in-service programs for teachers.

It appears that the goals and practices of arithmetic instruction have been following a roller coaster pattern throughout history. Pedagogical emphasis in arithmetic varied from lecture methods to student discovery methods. At certain times in history, abstract concepts and theories of arithmetic were stressed, while at other times, practical problems with concrete objects were valued. The arithmetic curriculum varied from child-centered to content-centered. The origins of the various reform movements differed. Sources of reform included: education departments, mathematics departments, the general public, and individual scholars.

## CURRENT BELIEFS ABOUT ARITHMETIC

NCTM and the National Research Council (NRC) have played a major role in redefining the goals of arithmetic as well as other areas of mathematics (NCTM 1980, 1989, 1991; NRC 1989). The introduction of the NCTM standards (1989, 1991) was an added stamp of approval to pockets of change nationwide in school mathematics. The current "perception of mathematics is shifting from that of a fixed body of arbitrary rules to a vigorous active science of patterns" (NRC 1989, 84). The move has been from a focus on strict accuracy through drill and practice to more cognitive and real-life–based curricula. Mathematics educators have recognized the need for new goals that better fit the needs of an information society. "Public attitudes about mathematics are shifting from indifference and hostility to recognition of the important role that mathematics plays in today's society" (NRC 1989, 82). The constant development of new technology has accelerated the pace of economic change. New demands are being made on the workforce for more than minimal competency in reading, writing and traditional "shopkeeper" arithmetic. NCTM has identified the following new goals for mathematics education: (a) mathematical literacy, (b) lifelong learning, (c) opportunity for all, and (d) an informed electorate. New employees in industry are expected to be able to "problem solve." That is, they will be required to (a) set up problems with the appropriate operations; (b) know a variety of techniques to approach and work on problems; (c) understand the underlying mathematical features of a problem; (d) have the ability to work with others on problems; (e) apply mathematical ideas to common and complex problems; (f) be prepared for open-ended problem situations; and (g) believe in the utility and value of mathematics (NCTM 1989, 4). The skills needed to meet the above competencies include reasoning abilities, communication skills, understanding of connections between mathematics and the real world, and problem-solving capability. These are different from basic arithmetic skills emphasized in the elementary schools of past centuries, but necessary for citizens in today's workplace.

## CURRENT BELIEFS ABOUT TEACHING ARITHMETIC

The pedagogical approaches in arithmetic today have their roots in the reform efforts of the past. The goals and activities of teaching and learning mathematics are being shaped by mathematics educators working with classroom teachers, mathematicians, school administrators, and parents. The "Standards" for curriculum, assessment, and teaching (NCTM 1989, 1991) and the research on mathematics teaching and learning (i.e., Grouws 1992; Jensen 1993) describe current trends. Arithmetic is integrated with other subject areas through problem solving, responding to literature, writing, dramatizing, investigating questions in science, and learning through thematic units. Elementary school mathematics curricula have broadened from a sole emphasis on arithmetic to include other topics so that children can be afforded opportunities for problem solving, communicating, reasoning, and making connections within and outside mathematics (NCTM 1989). Students communicate and reason through journal-writing, work on problems collaboratively with peers, reflect on their and others' work, make conjectures and defend their positions, collaborate on research, do independent and group investigations, and develop portfolios for assessment of their work.

Teachers model problem solving and move children gradually toward the generation of their own problems. Children work independently and in groups while the teacher serves as a facilitator and a guide. Instruction focuses on problem solving through interacting with materials and activities that promote logical thinking. Problems focus on areas that are important and valued by students (Charlesworth and Lind 1995).

## RECOMMENDED MATERIALS FOR INSTRUCTION

One way arithmetic concepts and skills are learned is by using hands-on materials for explorations. These materials include real objects, pictorial representations, two-dimensional cutouts, paper and pencil, calculators, and computers. Many ordinary materials can be collected for arithmetic instruction, including items that might otherwise be discarded. Items such as plastic lids from jars and bottles, thread spools, pine cones, sea shells, buttons, and seeds can be counted, sorted, graphed, and so forth. Egg cartons and frozen food containers can be used for sorting and organizing. String, ribbon, sticks, and so forth can be used for comparing lengths and for informal measurement.

A rich variety of commercial materials also is available. Picture books or "trade" books are becoming increasingly important as vehicles for suggesting problems to be solved and for integrating arithmetic and language arts. Basic materials include unit blocks, construction toys, *Unifix* cubes, *Lego* blocks, *Multilinks, Cuisenaire Rods,* pegboards and pegs, picture lotto games, beads and strings, attribute blocks, geoboards, balance scales, thermometers, flannel and magnet boards with accessory pieces, *Montessori* materials, manipulative clocks, *Dienes* or base-ten blocks, and fraction pieces. Calculators are available for use with complex problem-solving tasks and number theory investigations. Computer software provides opportunities for thinking beyond drill and practice (Charlesworth and Lind 1995).

Mathematics Learning Centers, Math Labs, or Math Menus (Burns 1992) build on the inquisitive nature of the child and create independent thinking. In such classroom settings, materials are available for exploration or specific tasks are designed for solution. For example, there might be a measurement center, a place value center, a graphing center, a trade book center, a mathematics writing center, or a center to integrate arithmetic into dramatic play such as a simulated grocery store or restaurant (Charlesworth and Lind 1995). A Math Lab might ask students to investigate the quality and economic value when comparing different name brand pencils, or discover winning strategies for the game Nim, and so on. A Math Menu is similar to a Math Center except it does not involve a physical space; materials are collected by the students from a storage area. Math Menus are similar to Math Labs, but allow groups of students to choose investigative tasks from a set of activities designed to achieve the same mathematical concept goals (Burns 1992).

## CONCLUSION

Arithmetic has been defined differently throughout time. The changes in arithmetic as a curricular discipline and in the way it is taught are numerous. With today's national and international organizations, however, current reform efforts have become more widespread. The use of technology and communication networks have brought together a community of people interested in the quality of mathematics education.

*See also* Addition; Decimal (Hindu-Arabic) System; Division; Multiplication; Number Sense and Numeration: One-to-one Correspondence; Subtraction

## SELECTED REFERENCES

Barnett, Carne S. "Arithmetic as Serious Mathematics." [Review of *Analysis of Arithmetic for Mathematics Teaching* by Gaea Leinhardt et al.] *Educational Researcher* 22(Oct. 1993):40–41.

Baroody, Arthur J. *Children's Mathematical Thinking.* New York: Teachers College Press, 1987.

Barrow, John. *Pi in the Sky: Counting, Thinking, and Being.* New York: Oxford University Press, 1992.

Brownell, William A. "Psychological Considerations in the Learning and Teaching of Arithmetic." In *The Teaching of Arithmetic* (pp. 1–31). 10th Yearbook of the National Council of Teachers of Mathematics. William D. Reeve, ed. New York: Bureau of Publications, Teachers College, Columbia University, 1935.

Brueckner, Leo J. *The Development of Ability in Arithmetic.* Philadelphia, PA: Winston, 1939.

Bunt, Lucas, Phillip Jones, and Jack Bedient. *The Historical Roots of Elementary Mathematics.* Englewood Cliffs, NJ: Prentice-Hall, 1976.

Burns, Marilyn. *About Teaching Mathematics: A K–8 Resource.* White Plains, NY: Burns, 1992.

Charlesworth, Rosalind, and Karen K. Lind. *Math and Science for Young Children.* 2nd ed. Albany, NY: Delmar, 1995.

Grouws, Douglas A., ed. *Handbook of Research on Mathematics Teaching and Learning.* New York: Macmillan, 1992.

Jensen, Robert J., ed. *Research Ideas for the Classroom: Early Childhood Mathematics.* Reston, VA: National Council of Teachers of Mathematics, 1993.

Kamii, Constance Kazuko with Georgia DeClark. *Young Children Reinvent Arithmetic.* New York: Teachers College Press, 1985.

Kramer, Klaas. *The Teaching of Elementary School Mathematics.* Boston, MA: Allyn and Bacon, 1966.

Leinhardt, Gaea, Ralph Putnam, and Rosemary A. Hattrup, eds. *Analysis of Arithmetic for Mathematics Teaching.* Hillsdale, NJ: Erlbaum, 1992.

National Council of Teachers of Mathematics. *An Agenda for Action: Recommendations for School Mathematics of the 1980s.* Reston, VA: The Council, 1980.

———. *Curriculum and Evaluation Standards for School Mathematics.* Reston, VA: The Council, 1989.

———. *Principles and Standards for School Mathematics.* Reston, VA: The Council, 2000.

———. *Professional Standards for Teaching Mathematics.* Reston, VA: The Council, 1991.

National Research Council. *Everybody Counts: A Report to the Nation on the Future of Mathematics Education.* Washington, DC: National Academy Press, 1989.

Reesink, Carole J., ed. *Teacher-Made Aids for Elementary School Mathematics: Readings from the Arithmetic*

*Teacher and Teaching Children Mathematics.* Vol. 3. Reston, VA: National Council of Teachers of Mathematics, 1998.

Reeve, William D., ed. *Multi-sensory Aids in the Teaching of Mathematics.* 18th Yearbook of the National Council of Teachers of Mathematics. New York: Bureau of Publications, Teachers College, Columbia University, 1945.

Romberg, Thomas, and Thomas Carpenter. "Research on Teaching and Learning Mathematics: Two Disciplines of Scientific Inquiry." In *Handbook of Research on Teaching* (pp. 850–873). 3rd ed. Merlin C. Wittrock, ed. New York: Macmillan, 1986.

Rosskopf, Myron F., ed. *The Teaching of Secondary School Mathematics.* 33rd Yearbook. Washington, DC: National Council of Teachers of Mathematics, 1970.

ROSALIND CHARLESWORTH
ELIZABETH SENGER

# ART

A problem-solving process that can support current pedagogical trends in mathematics. A mathematics teacher need not be an art expert to competently handle art projects in the mathematics classroom. Art and mathematics are not the opposites they are frequently thought to be. Art is a problem-solving process that involves rational as well as emotional components. Art teachers do not generally just hand out supplies and direct the students to do whatever they feel like doing. They set limits and the students create their artworks within those limits. The major difference between art and mathematics is the type of problem to be solved and the materials used to arrive at the solutions.

In mathematics there are usually very few right answers to a problem, whereas in art there can be infinitely many. That is not to say that all answers to an art problem are right answers, but, rather, that the opportunity for many successful outcomes exists. It is generally much easier to judge whether an answer is mathematically correct due to the objective nature of the discipline. A successful art project is one that has a unique and balanced blend of several elements such as composition, color, scale or size, craft, and use of materials, each of which will be discussed in detail later on. Because there is a subjective element to these components as well as to the notion of balance, and because human beings also are influenced by cultural norms and personal taste, not even artists or art critics will always agree as to the success of a particular artwork. In the art world the true test of success is whether the work endures over time.

Mathematics teachers should not be daunted by the subjective nature of the art discipline. The object of introducing art projects in the mathematics classroom is not to produce masterpieces, but, rather, to provide an opportunity for students to experience mathematical relationships in a different way and to enhance their ability to learn the subject. Some students, simply because of the setting, may take a little time to realize that there can be many correct solutions. "What does the teacher want?" will eventually be replaced by "What do I want?" as students gain confidence to experiment.

Because art is a personal activity, and one of the goals is to leave one's unique and personal stamp on a piece, there should be as many different answers as there are students in the room. For example, suppose the students in a class are building models of the Platonic solids (Pugh 1990). The "right answer" mathematically is that each student should produce a model of a tetrahedron, a cube, an octahedron, a dodecahedron, and an icosahedron, each with regular faces. Such models are useful for exploring mathematical properties of each of the polyhedra such as symmetries, numbers of faces, edges and vertices, and so forth. If, however, the models are displayed, they will all look alike. Now suppose the art problem is introduced: Add surface design to the faces. Without further instructions, students could draw or glue objects on the faces or develop patterns which might add a textural element, emphasize the type of polygon composing the face, or visually distort the model so that it is almost unrecognizable from a distance. If the mathematical problem is solved correctly, a comparison of all the students' models should yield sameness. If the art problem is solved correctly the same comparison should yield differences so that, with the exception of the original structures, each student's work is unique.

The previous example illustrates a natural place to use manipulatives in a mathematics class. The tactile and kinesthetic senses are valuable aids in helping students to grasp relationships, and the more opportunities there are for students to use their hands and literally "feel" an idea, the more successful their mathematics education is apt to be. Art, by its very nature, is a hands-on activity that requires processing visual and spatial information, a right brain function. Most mathematics education utilizes the analytic and sequential cognitive modes that are in the domain of the left brain (Springer and Deutsch 1981). Therefore, if a project requires students to create artworks based on the mathematical concepts that are being explored, both sides of the brain are exercised. More learning can take place because the two hemispheres of the brain are operating cooperatively rather than singly or in opposition to one another (Edwards 1979).

The introduction of art projects also fosters cooperative learning. When students sit around a table sharing art materials, the atmosphere is generally relaxed, and very naturally they begin to share ideas as well. They will look around to see what their classmates are doing, talk about the process, and an idea that one student is working with will inevitably spark other ideas for other students. If the relaxed atmosphere and desire to explore carries over into groups solving mathematical problems the benefits from the art experience are reaped yet another time.

Once the art/mathematics projects are completed, students should make oral presentations of their finished works to the class. A major part of these presentations should focus on the way the mathematical concepts are used in the work. These concepts may not be obvious in the finished work, so students should be encouraged to keep their sketches and to write about the methods they used to develop their artworks. This practice not only documents the art process but also documents the mathematical thinking that took place. If the teacher insists that students use the proper mathematical language, they will have more opportunities to practice by writing and doing these presentations. As a student's comfort level with the language increases, a major source of anxiety falls away, eliminating one of the stumbling blocks to learning the subject matter.

How can a mathematics teacher successfully grade an art project? To answer this question, it is important to remember that an artwork is a function of several variables. It is possible, and in fact desirable, for the teacher to assign the grade based solely upon those aspects of the work about which he or she is knowledgeable. Sometimes it is feasible to team up with an art teacher so that each may grade the projects in the areas of their respective expertise. Of course the students should be made aware of exactly how they will be graded, which aspects of the project figure into the grade, and which aspects may result in extra credit. In time, the mathematics teacher may develop more artistic expertise and adjust the grading criteria accordingly. A discussion of some of the art variables follows.

First, every artwork has structure or composition. This is the element that determines how the parts relate to one another and how they relate to the whole. All problems solved by visual artists have to do with creating relationships in space (Newman and Boles 1992). Since geometric problems all have to do

with describing spatial relationships, composition is one variable in the art arena that can be greatly enhanced by a knowledge of mathematics.

The way things relate to one another spatially has to do with ratio and proportion. Although mathematically specific, in art literature proportion is frequently presented as having to do with feelings. If the proportions are "good," the composition "feels right." This is hardly helpful for students of the arts. Suppose two people have different feelings; who is elected to be the one to decide what feels right? If it is always the teacher, does this not send a message to students that their feelings are either unimportant or must be denied? Surely, a more rational approach would be helpful.

The use of an armature is a valuable aid to help an artist place the objects in the composition whether the subject matter is representational, such as a still life or landscape or portrait, or is pure abstraction. In three-dimensional work, an armature is the wire structure upon which clay or other materials are layered to create a sculpture. In two-dimensional work it is the underlying geometry of the work. The geometry can become the subject matter of an abstract piece, such as in a quilt square, or can be the invisible scaffolding that acts as a guide for placement of objects in the composition.

Consider a drawing or painting. In subdividing the visual field, one should first think of the dimensions of the paper or canvas. If rectangular, the artist might want to repeat the length-to-width ratio in the interior spatial divisions so that the parts relate to the whole. For example, if the rectangle is twice as long as it is wide, dividing sides at midpoints and connecting them with line segments, vertical, horizontal, or diagonal, creates a simple but effective armature, for the 2:1 ratio is preserved.

Simple divisions of sides into halves, thirds, or fourths, with connecting segments, can also be effective. With this type of subdivision, however, each segment along a side of the rectangle is congruent to each of the others, so repetition is the result. Repetition is useful for certain kinds of artwork, particularly those involving patterning, but may be somewhat limiting. If the artist is interested in similar shapes instead, there are many ways to subdivide the sides into uneven parts, but the relationship to the whole may be problematic. The only way mathematically to divide a line segment unequally so that the parts relate proportionally to the whole is to find the golden cut. This is the point that divides the segment so that the ratio, $\phi$, of the whole segment to the longer part is the same as the ratio of the longer part to the shorter part. ɸ, the golden ratio, is found by solving the equation

$$\frac{x + 1}{x} = \frac{x}{1}$$

where 1 is the length given to the shorter segment, $x$ the length of the longer, and $x + 1$ the length of the original segment. The actual value of ɸ is

$$\frac{1 + \sqrt{5}}{2}$$

but 1.6 or $\frac{8}{5}$ are common approximations.

It is known that the classical Greek sculptors and architects used this ratio extensively in their work and that the Renaissance masters were schooled in mathematics. It is, in fact, difficult to imagine that any artist could simply begin to paint, for example, a large fresco without first creating an armature of simple geometric shapes to determine how the space will be subdivided. One of the most widely used art history texts (Janson 1986), although a wonderful resource with a vast array of information, only briefly mentions the golden ratio or the notion of an armature and with language that again relates to the emotions rather than to reason. *The Painter's Secret Geometry* (Bouleau 1963), unfortunately out of print, deals extensively with the concept of armatures, and will give the uninitiated a much clearer idea of the geometric structures of works from the history of art.

The golden ratio provides a wonderful focus for an exploration of the connections between art and mathematics. Armatures developed by golden ratio subdivisions work well to help achieve the goal of unity, as similarity is preserved. Far from providing a formula, it, rather, can be used as a tool in a myriad of ways to help the artist create an harmonious work of art; also, its mathematical properties are many and fascinating. In the series *The Golden Relationship: Art, Math & Nature, Book 1* (Newman and Boles 1992) examines natural patterns in light of their relationships to the golden ratio, and *Book 2* (Boles and Newman 1992) explores how these patterns can be played out on the plane. Both texts contain many different mathematical problems and art projects that can help strengthen students' understanding of the connections, yet neither presupposes any art background on the part of the reader. Consider the Fibonacci sequence, 1, 1, 2, 3, 5, 8, 13, . . . , for example. It has some fascinating mathematical properties, but the numbers translate readily to lengths that can be used to subdivide the sides of geometric figures to create "op art" patterns or to widths for paper strips to be used for a weaving.

Color is another variable of a work of art. Color theory, at its most elementary level, can be thought of in much the same way as a logical mathematical system. The primary colors—red, yellow, and blue—along with black and white, represent the undefined terms. All other colors can be mixed from these and, therefore, are analogous to the defined terms in a logical system. For example, green is the result of mixing equal parts of yellow and blue. The standard color harmonies (monochromatic, analogous, and complementary) can be thought of as the axioms and postulates. For example, any hue on the color wheel can be used to develop a monochromatic harmony where variations are introduced by tinting (adding white to the color), toning (adding gray), or shading (adding black). The more personal and sophisticated applications can be thought of as the theorems. Orange, blue-green, and blue-violet, together with their light and dark variations, can be used in harmony, because they lie at the vertices of an isosceles triangle on the color wheel, forming a triad. Of course the analogy breaks down at this point because of the nature of proof, and the fact that color theory is not an exact science.

It is important to remember that color harmony does not mean using every color available. Often, a most effective work results from using very few colors or from limiting a work to black and white and shades of gray. A teacher does not have to be an expert in color theory to bring art into the mathematics classroom.

If a group project is assigned, color must be considered ahead of time. Suppose the problem is for each student to create a square based on an harmonic armature that he or she designs. This will be a mini-artwork on its own, but the purpose will be to put all the squares together to make a paper quilt. Individually, many different color harmonies are possible. When put together, however, too many colors destroy the unity of the work. A simple solution in this case is to limit the choice to black and white and one color, where all students work with the same color and its light and dark variations. Students still have the opportunity to express their individuality, yet know that the finished piece will be a group effort where the separate pieces must all work together. Teachers may find that if they provide the materials for any group projects, it is easier to deal with a color choice that will be harmonious in the finished work.

Choice of materials is one more variable in the art process. Students should be encouraged to try new materials, as the only way to discover how to use them and which ones work best for which applications is actually to try them. They should not be penalized if their attempts with a new material are less than successful. Bonus points can be awarded for taking risks regardless of the outcome. Making mistakes is an important part of the art process. It is equally important in mathematics. For example, the discovery of non-Euclidean geometries was the result of experimenting with new assumptions concerning the parallel postulate.

Craft and neatness are related to materials. But neatness is not just an art skill; it is a life skill, and skills are honed with practice. In mathematics, learning to be neat can help a student with organization of both materials and thought processes, and, therefore, enhance learning. In art, simple ideas and materials can become elegant and sophisticated through exquisite craft. Neatness is a clearly observable factor in an artwork, and, therefore, can certainly figure into the grading process. Not all students, however, will be attracted to clean, sharp edges in their work. If the choice is to create a piece in which materials are used more loosely and with less precision, the intent should be clear in the finished work. It should not simply appear sloppy.

Scale or size is another consideration, but a little more difficult to objectify. Some ideas require execution in a large size to give them presence. The same ideas may appear insignificant interpreted in a size that is too small. Some ideas are more intimate and can be handled in a smaller format. For example, if a class is studying geodesics (Pugh 1990), a project might be to create a diorama of a neighborhood consisting of geodesic dome dwellings. This project would not work well in a very small size, partly because of the physical requirements and the enhancing details (addition of models of flora and fauna, topography, and so forth), and partly because of the scope of the idea. Attention would have to be paid to the plan of the overall layout, another subdivision problem where an armature would be helpful, as well as to the structure of the geodesic dwellings, which requires a different set of mathematical concepts. Since neighborhood planning also involves elements based on psychological and sociological choices, the scope of this interdisciplinary project is limited only by time and the physical constraints of the classroom.

Finally, time is a variable. Certainly students should receive credit for completing a project on time. It also is important, however, to remember that the art process is time-consuming. Here is an area in which students with little art experience but decent mathematics skills may gain respect for the length of time it takes for an idea, which may come quickly, to be translated into a finished physical expression. Involvement with the art process is most helpful in teaching students the importance of time management, as it is a process that cannot be rushed. Budgeting time to complete a project rather than trying to do it all at once is a good guideline whether the project is related to art, mathematics, or any other endeavor.

To discuss art projects in the light of a grading procedure is not to imply that the grade is the important outcome. Rather, it is to assure mathematics teachers that they can competently handle such projects regardless of their art backgrounds. The art "equation" is a complex, multivariable one, in which the unity of the artwork is a function of all the dependent and interdependent variables previously mentioned. As with mathematical subject matter, greater competence comes with practice. The desired outcome is for the class to have a meaningful interdisciplinary experience that enhances abilities in and respect for both subject areas.

Today, many students of the arts go to postsecondary schools, some hoping to avoid forever another experience in mathematics. Indeed, it is the case that few art schools include mathematics in their curricula. Many artists feel that the use of mathematical concepts necessarily means that somehow a formula is applied that will inhibit their creativity. They do not have the background to understand that mathematics is a tool much like a paint brush or a pen; a means to obtain a desired result that frees them to use their creativity in an optimal way. For example, if an artist wants to incorporate a regular pentagon into his or her composition, it is hardly a restriction of creativity to know that there are several compass and straightedge constructions that will accomplish the task easily. Nor does it restrict creativity to know the appropriate language so that one may quickly research methods to achieve the desired result if they are not in one's repertoire. To struggle trying to "eyeball" a regular pentagon might result in some creative thinking, but it will not necessarily be the type that enhances the artwork.

Some artists do use mathematics consciously, and a few others accidentally discover its value. In the summer of 1993, well-known artists, mathematicians, and educators eager to share or learn ways to make interdisciplinary connections congregated for a week at SUNY Albany for AM93, the second conference on Art and Mathematics. There were sculptors using variations on polygonal forms in works that ranged from very small pieces that could be held in one hand to extremely large ones that grace parks and commercial buildings around the world. There also were fiber artists and painters experimenting with tiling structures, and fabric designers using the computer and random number generators, to name a few. It was notable that several of them claimed not to have known that they were using mathematics in their work.

Artists who were in the "I hate mathematics" group as students are at a distinct disadvantage if later on they wish to learn more about the connections between art and mathematics. Because of limited backgrounds and lack of a fundamental knowledge of the mathematical language, they may not understand mathematics texts. They also may experience difficulties trying to communicate clearly and accurately about their own work if it relates to mathematics. If they attempt to generalize and explore related concepts in print, inaccuracies often occur or their use of the language is nonspecific, and, therefore, ambiguous. Because we tend to give greater credence to written concepts, misconceptions and errors proliferate.

Mathematics educators can do a great service for the artists-to-be as well as other students by introducing them to applications at an early level. Surely, if students understand the relevance of the connections, they will be more apt to discipline themselves to work diligently to understand the subject. We all play harder if we are excited or intrigued by the game. As success breeds success, these same students also might become encouraged enough to venture into the more abstract areas of mathematics with diminished anxiety.

*See also* Escher; Golden Section; Tiling

## SELECTED REFERENCES

Boles, Martha, and Rochelle Newman. *The Golden Relationship: Art, Math & Nature. Book 2: The Surface Plane*, Bradford, MA: Pythagorean, 1992.

Bouleau, Charles. *The Painter's Secret Geometry.* New York: Harcourt, Brace and World, 1963.

Edwards, Betty. *Drawing on the Right Side of the Brain.* Los Angeles, CA: Tarcher, 1979.

Janson, Horst Woldemar. *History of Art.* 3rd ed. Revised and edited by Anthony F. Janson. New York: Abrams, 1986.

Newman, Rochelle, and Martha Boles. *The Golden Relationship: Art, Math & Nature. Book 1: Universal Patterns.* 2nd ed, rev. Bradford, MA: Pythagorean, 1992.

Pugh, Anthony. *Polyhedra: A Visual Approach.* Palo Alto, CA: Seymour, 1990.

Springer, Sally P., and Georg Deutsch. *Left Brain, Right Brain.* San Francisco, CA: Freeman, 1981.

MARTHA J. BOLES

# ASSESSMENT OF STUDENT ACHIEVEMENT, ISSUES

Four prominent issues are (1) changing assumptions about mathematics learning and teaching, (2) clarifying the purposes of assessment, (3) adapting to technology and innovation, and (4) establishing criteria for judging the quality of assessments.

## CHANGING ASSUMPTIONS ABOUT MATHEMATICS LEARNING AND TEACHING

Educators are tending to move from viewing mathematics as a fixed collection of facts and skills to emphasizing processes of conjecturing, communicating, problem solving, and logical reasoning in mathematics learning. This tendency is related to a trend away from viewing learning as human information processing and toward seeing learning as model building. As a result, mathematics teaching is shifting from the demonstration, explanation, and practice of procedures taken out of context toward helping students construct their own knowledge through real-life mathematical investigations.

## CLARIFYING THE PURPOSES OF ASSESSMENT

Related to changing views of mathematics learning and teaching are changing views about the purposes of assessment. In general, "the aim of educational assessment is to produce information to assist in educational decision making, where the decision makers include administrators, policy makers, the public, parents, teachers, and students themselves" (Lesh and Lamon 1992, 4).

Proponents of changes in assessment use qualitative as well as quantitative data, and focus more on describing student progress than on categorizing individuals or predicting future success. There is a tendency away from short-answer or multiple choice tests to increased use of such alternatives as performance assessments, open-ended questions, group projects, portfolios, journal writing, oral reports, and observations. High stakes assessments (such as standardized or state-mandated tests, whose results have an impact on important decisions made about students, teachers, schools, or programs) also are undergoing some changes.

Traditionally, because tests were designed to compare students, efforts were made to standardize testing conditions by placing strict constraints on the time, resources, and tools that students could use. By contrast, the newest trend is elimination of as many artificial constraints as possible to create "authentic" assessments or "performance" assessments, although there is still debate over how to identify authentic tasks, how to characterize quality performance, and how to ensure valid uses of assessment findings.

## ADAPTING TO TECHNOLOGY AND INNOVATION

New technologies have resulted in some changes in the real-world problem-solving situations for which mathematics is useful and in the types of knowledge and abilities that are important today. As a result, technology is exerting an influence on assessment as well. For example, the availability of handheld calculators and notebook computers with graphing and symbol manipulation capabilities enables students to think differently, not just faster. Educators need first to determine the types of skills and understandings that prepare students for using technology optimally, and then identify ways to assess the new types of problem-solving made possible by technology.

## ESTABLISHING CRITERIA FOR JUDGING THE QUALITY OF ASSESSMENTS

Along with new assessments comes the need for explicit criteria by which to judge their appropriateness and effectiveness. Thus, the National Council of Teachers of Mathematics (NCTM) published *Assessment Standards for School Mathematics* (1995), a document that recommends six "standards" for judging assessments: mathematical topics, learning achievement, equity, openness, inferences, and coherence. Each of these standards raises important assessment issues.

### Standard I: Mathematics

Few would argue with the assertion that useful mathematics assessments must focus on important mathematics. Yet the trend toward broader concep-

tions of mathematics and mathematical abilities raises questions about the appropriateness of the mathematics reflected in most traditional tests, since that mathematics is generally distinct from the mathematics actually used in real-world problem solving. There is still much debate over how to define important mathematics and who should be responsible for doing so.

### Standard II: Learning

New views of assessment call for tasks that are embedded in the curriculum, the notion being that assessment should be an integral part of the learning process rather than an interruption of it. This raises the issue of who should be responsible for the development, implementation, and interpretation of student assessments. Traditionally, both standardized and classroom tests were designed—using a psychometric model—to be as objective as possible. By contrast, the alternative assessment movement affords teachers more responsibility and subjectivity in the assessment process. It assumes that teachers know their students best because teachers have multiple, diverse opportunities for examining student work performed under various conditions and presented in a variety of modes. When teachers have more responsibility for assessment, it can become almost seamless with instruction (Lesh and Lamon 1992).

### Standard III: Equity

Ideally, assessments should give every student optimal opportunities to demonstrate mathematical power. In practice, however, traditional standardized tests have sometimes been biased against students of particular backgrounds, socioeconomic classes, ethnic groups, or gender (Pullin 1993). Equity becomes even more of an issue when assessment results are used to label students or to deny them access to courses, programs, or jobs. More teacher responsibility means greater pressure on them to be evenhanded and unbiased in their judgments. Ironically, the trend toward more complex and realistic assessment tasks and elaborated written responses raises equity concerns since familiarity with contexts, reading comprehension, and writing ability may confound results for certain groups (Lane 1993). Similarly, efforts to establish national or state performance goals often are met with resistance because of fears that their use in high-stakes assessments will result in inequities. Thus, it is unclear whether recent trends will actually cause increased or decreased equity in mathematics assessment.

## Standard IV: Openness

Testing has traditionally been a secretive process in that test questions and answers were carefully guarded and criteria for judging performance were generally set behind the scenes by unidentified authorities. Today many believe that students are best served by open and dynamic assessments—assessments where expectations and scoring procedures are openly discussed and jointly negotiated. Traditionally, mathematics courses and tests have often been used as filters—to screen students for entry into programs, courses, and jobs; this helps to explain why test questions were kept secret. Many argue, however, that assessments today should be designed more to describe student proficiencies and deficiencies, to help in making instructional decisions, or to gauge the overall status of the educational system than to categorize individual students. For such purposes, it is argued, criteria certainly can be made more open.

## Standard V: Inferences

Changes in assessment approaches have tended to result in new ways of thinking about reliability and validity as they apply to mathematics assessment. For example, when assessment is embedded within instruction, it becomes unreasonable to expect a standard notion of reliability to apply (that a student's achievement on similar tasks at different points in time should be similar), since it is actually expected that students will learn throughout the assessment. Also, new forms of assessment prompt a reexamination of traditional notions of validity. Many argue that it is more appropriate to judge validity by examining the inferences made from an assessment than to view validity as an inherent characteristic of the assessment itself. It is difficult to determine how new types of assessments (e.g., student projects or portfolios) can be used for decision making without either collapsing them into a single score (thereby losing all their conceptual richness) or leaving them in their raw, unsimplified, difficult-to-interpret form.

## Standard VI: Coherence

The coherence standard emphasizes the importance of ensuring that assessments are appropriate for the purposes for which they are used. Assessment data can be used for monitoring student progress, making instructional decisions, evaluating student achievement, or program evaluation. The types of data that are appropriate for each purpose, however, may be very different. Policy makers and assessment experts often disagree on this issue: the former may have multiple agendas in mind and expect that all be accomplished by using a single assessment, while the latter warn against using assessments for purposes for which they were never intended.

## ISSUES FOR DEBATE

Recent trends in assessment raise many issues for debate. For example, the move away from objective, right-wrong test questions raises concerns about subjectivity and unclear expectations. Sharing assessment criteria more openly may counter some of these concerns, but such openness also may be difficult to accomplish and perhaps even subject to misuse. The trend toward relying more on teachers (and less on external assessments) for monitoring student progress may falter as a result of conscious or unconscious bias by teachers. Moreover, new assessment techniques (such as journals and portfolios) may be potentially beneficial yet practically unfeasible for overworked teachers in crowded schools.

*See also* Assessment of Student Achievement, Overview; National Assessment of Educational Progress (NAEP); Third International Mathematics and Science Study (TIMSS)

### SELECTED REFERENCES

Eisner, Elliott W. "Reshaping Assessment in Education: Some Criteria in Search of Practice." *Journal of Curriculum Studies* 25(1993):219–233.

Kulm, Gerald, ed. *Assessing Higher-Order Thinking in Mathematics.* Washington, DC: American Association for the Advancement of Science, 1990.

Lane, Suzanne. "The Conceptual Framework for the Development of a Mathematics Performance Assessment Instrument." *Educational Measurement: Issues and Practice* 12(2)(1993):16–23.

Lesh, Richard, and Susan J. Lamon, eds. *Assessment of Authentic Performance in School Mathematics.* Washington, DC: American Association for the Advancement of Science, 1992.

Lesh, Richard, Susan J. Lamon, Merlyn Behr, and Frank K. Lester, Jr. "Future Directions for Mathematics Assessment." In *Assessment of Authentic Performance in School Mathematics* (pp. 379–424). Richard Lesh and Susan J. Lamon, eds. Washington, DC: American Association for the Advancement of Science, 1992.

Mathematical Sciences Education Board. *Measuring Up: Prototypes for Mathematics Assessment.* Washington, DC: National Academy Press, 1993.

National Council of Teachers of Mathematics. *Assessment Standards for School Mathematics.* Reston, VA: The Council, 1995.

————. *Principles and Standards for School Mathematics.* Reston, VA: The Council, 2000.

National Research Council. *Measuring What Counts: A Conceptual Guide for Mathematics Assessment.* Washington, DC: National Academy Press, 1993.

Pullin, Diana C. "Legal and Ethical Issues in Mathematics Assessment." In *Measuring What Counts: A Conceptual Guide for Mathematics Assessment* (pp. 201–223). By National Research Council. Washington, DC: National Academy Press, 1993.

Romberg, Thomas. A., ed. *Mathematics Assessment and Evaluation: Imperatives for Mathematics Education.* Albany, NY: State University of New York Press, 1992.

Wiggins, Grant. "Assessment: Authenticity, Context, and Validity." *Phi Delta Kappan* 75(Nov. 1993): 200–214.

DIANA V. LAMBDIN

# ASSESSMENT OF STUDENT ACHIEVEMENT, OVERVIEW

Collecting, interpreting, and synthesizing information to assist in decision making (Airasian 1991). In its *Assessment Standards for School Mathematics* (1995), the National Council of Teachers of Mathematics (NCTM) defines assessment in mathematics as "the process of gathering evidence about a student's knowledge of, ability to use, and disposition toward mathematics and of making inferences based on that evidence for a variety of purposes" (p. 7). Assessing the achievement of students as individuals or as groups is not limited to standardized procedures as characterized by multiple-choice tests, but includes teacher-made assessments, informal observations, oral questioning, student projects, and portfolios. Information about student achievement is used to make a variety of decisions such as grading, placement, diagnosis of strengths and weaknesses, curriculum planning, and accountability. This entry provides a general overview of important topics associated with assessment of student achievement.

## BRIEF HISTORY OF ASSESSMENT OF STUDENT ACHIEVEMENT

The history of assessment of student achievement in school subjects, including mathematics, has its roots in the measurement of human behavior in general, and can be roughly divided into four overlapping periods: (1) from the beginning of historical record to the nineteenth century; (2) the nineteenth century; (3) the 1900s to the 1960s; and (4) the 1960s to the present (Romberg 1992). Although not much information exists about assessment during the first period, there is evidence that over three thousand years ago the Chinese used formal written examination systems to select civil servants. With the advent of the nineteenth century came notable contributions to assessment in education such as the use of written examinations as substitutes for oral tests in the Boston Schools (ca. 1845), George Fisher's use of benchmark standards as objective measures of student achievement in handwriting, spelling, mathematics, and other school subjects (ca. 1864), and spelling tests developed by Joseph M. Rice and used for program evaluation (ca. 1894). The third period brought with it major developments in mental ("aptitude") testing such as the Stanford-Binet Intelligence Quotient (IQ) test in the early 1900s and educational testing such as the development of a standardized test in arithmetic reasoning (ca. 1908), the introduction of "psychometrics" as a field of study (ca. 1920), and the introduction of the Scholastic Aptitude Test (SAT) in multiple choice format (ca. 1926). Since the 1960s, assessment of student achievement has been a firmly entrenched part of education in the United States. Especially in the 1980s and 1990s, "accountability" has become the key word in assessment, as measures of student achievement are used to document the need for reform in education.

The emphasis in national standards for student achievement in school subjects has focused attention on assessment. Specific to the assessment of student achievement in mathematics, the NCTM has articulated its position in the document *Assessment Standards for School Mathematics* (1995). Produced as a complement to its earlier Standards documents (NCTM 1989, 1991), the *Assessment Standards* proposed six standards as the criteria to be used for judging assessment practices: (1) assessment should reflect the mathematics that all students need to know and be able to do (The Mathematics Standard); (2) assessment should enhance mathematics learning and inform teaching (The Learning Standard); (3) assessment should promote equity by focusing attention on each student's learning (The Equity Standard); (4) assessment should be an open process in which information about the process is made available to those affected by it (The Openness Standard); (5) assessment should promote valid inferences about mathematics learning (The Inferences Standard); and (6) assessment should be a coherent process in which all phases fit together, the assessment matches the purpose for which it is being done, and the assessment is aligned with the curriculum and with instruction (The Coherence Standard). The *Assessment Standards* also suggests the applicability of these standards for different purposes, which include monitoring student progress, making instructional decisions, evaluating student achievement, and evaluating programs. Other NCTM publications (for example, Webb 1993) and materials produced under the auspices of the Mathematical Sciences Education Board (for example, National Research Council 1993) also address issues of assessment of student achievement in mathematics.

This very brief overview of educational assessment does not mention many important areas that have contributed to the current status of assessment of student achievement in mathematics as well as other school subjects. For a comprehensive treatment of educational assessment, see Linn (1989), and for a more complete treatment of the history of educational testing, see, for example, Dubois (1970) and Resnick (1982).

## PURPOSES OF ASSESSMENT

Assessment of student achievement has been used for a wide variety of purposes, including instructional guidance, grading, certification of competence in school subjects, placement into programs and courses, and accountability. These purposes are described briefly in the following section. More complete information on each purpose can be found in Linn (1989).

### Instructional Guidance

One of the most important purposes of assessment is to guide teachers' instructional decision making in order to improve teaching and learning for all students (Silver and Kenney 1995). The classroom teacher is the person closest to student performance, observing it on a daily basis and assessing it continually. It is therefore not surprising that teachers often rely on their own assessments rather than external measures such as standardized tests for the purposes of making instructional decisions. In addition to written tests and quizzes, teachers often use informal methods that include watching students as they work on tasks, observing students working collaboratively, asking appropriate questions, and listening to students present their solutions, explanations, and justifications. Each of these sources of information influences teachers' initial decisions

about the content of a particular unit (for example, inclusion of particular topics or selection of appropriate tasks) and the method of teaching (for example, a problem-based approach) as well as decisions concerned with the duration of the unit and the kinds of assessments used for summative evaluation. During the course of the unit, teachers use assessment techniques embedded within instruction that lead to modification of the original plans such as adjusting the pace of the lessons, restructuring group work, or selecting alternative or additional tasks.

### Grading

Grading is one way that teachers make judgments about their students' levels of achievement, and then report those levels in a consistent and fair manner. In general, grading is the process by which scores on tests and other descriptive information are converted into numerical quantities (often based on a scale in which 100 is a perfect score), letters as symbols (A through F), or performance scales (excellent to unsatisfactory) denoting the level of student achievement. Assigning grading is one of a teacher's most important professional responsibilities, yet it also is viewed as a difficult task which few teachers look forward to with pleasure (Airasian 1991). Grades can be assigned through a norm-referenced approach or a criterion-referenced approach. In the norm-referenced approach, a student's performance is compared to the performance of other students to determine a grade. Here, the distribution of grades is often expected to follow a pattern such as more Cs than any other grades and few As or Fs. In the criterion-referenced approach, a student's performance is compared to predetermined performance standards; this grading reflects a particular level of attainment. Because the comparison is between the individual student and the standard (and not between students), there may be no expected pattern of grades (depending on the group). In both grading approaches, it is assumed that the standards are meaningful and attainable by students, and teachers are consistent in their application of the approach.

### Certification of Competence
### in School Subjects

Assessment for the purpose of certifying student competence in school subjects saw an increase in the 1970s, as school districts and states began to implement mandatory examinations at benchmark points in a student's education, the most common being a high school exit examination. For the most part, the tests focused on minimal skills in core areas such as mathematics, reading, and writing, and were quite controversial. More recently, although externally mandated certification tests still exist in the 1990s, states such as California, Kentucky, and Vermont are experimenting with performance-based assessments that, in addition to traditional multiple choice questions, use open-ended tasks, essays, and portfolios to measure higher-order thinking skills.

## Placement into Programs or Courses

Perhaps assessment of student achievement for the purposes of placement is most commonly associated with special education students and those with a variety of handicapping conditions. Assessment results from a standardized test often form the basis for institutional decisions about placement of a student in a particular course. It also is common for classroom teachers to use results from their own assessments or from externally mandated tests to decide at what level in the instructional sequence a student should begin studies to avoid unnecessarily repeating what is already known. Placement decisions about individual students are often made by persons other than classroom teachers, such as school- or district-level administrators. While most placement decisions are made with the best interest of the student in mind, placements based on results from a single test and on homogeneous ability grouping, or "tracking," are sometimes counterproductive. In particular, academic tracking can reinforce students' negative self-perceptions by emphasizing their past failures; youngsters assigned to lower tracks may find themselves blocked from access to further educational opportunities.

## Accountability

As has been stated in a previous section of this entry, in the last two decades of the twentieth century, educational policy makers began to place increasing reliance on formal measures of student achievement for the purposes of accountability. The trend has been to use standardized or externally mandated tests to document the inadequacies in educational achievement and in the system of public education in the United States. An increase in testing was proposed not only to monitor the effects of educational reforms but also to encourage the implementation of reforms by creating a system

based on "high stakes testing." The current emphasis on assessment of student achievement for purposes of accountability raises salient issues such as the adequacy of currently available tests and the value of "teaching to the test" that will need to be addressed.

## TECHNIQUES FOR ASSESSING STUDENT ACHIEVEMENT

The following section presents some assessment techniques that can be used by classroom teachers. More complete information about suggested ways to assess student achievement, with a specific focus on assessment in mathematics for instructional guidance, can be found in Silver and Kenney (1995). It is understood that the selection of appropriate classroom assessment techniques will vary between teachers and will most likely be tempered by the many challenges teachers face daily in their classrooms.

## Observation

Observation is a basic classroom method of gathering information about students. Information gained from watching students while they are doing mathematics can provide insights into their strengths and weaknesses, and can be used by teachers to decide whether to move forward in a lesson, to allow more time for completing an activity, to provide an alternative explanation, or to modify the direction of instruction. Teachers can enhance the use of observation to assess student achievement through the use of observation instruments such as checklists. For example, an annotated class list consisting of a roster of student names with a blank space to the right of each name can be used to record information routinely about a student's understandings and attitudes. Another kind of recordkeeping instrument is the topical list, consisting of a set of predetermined categories (for example, selects appropriate strategy; reflects on work and answer; works collaboratively; uses mathematical language) as possible foci during the observation. Here, a teacher may choose one or more categories and then assess students on only those selected.

Teachers can use information from observations in a variety of ways, but teachers should also be aware that judgments based on observations are sometimes misleading. Since student observation typically represents just one facet of the teacher's many simultaneous activities, it is not unusual for the teacher to attend to more of what is wanted or expected rather than to interpret the information from observations in a completely objective manner. In some cases, however, students may in fact demonstrate more understanding while being observed than they show on a written test. Taking time to observe students at work in the classroom and then to interpret the observation data objectively has the potential to provide teachers with important information to be used to shape the direction of future classroom activities.

## Questioning

Whether questioning occurs informally as part of regular classroom activity or in a more structured individual or small group setting, teachers can gain insights into thinking and communication skills that may not be obvious from written work. Questioning can be a source of important information for student assessment related to areas such as cognitive performance, attitudes, self-assessment, and instructional decision making. Although teacher-generated questioning of students can enhance discourse, teacher questions may sometimes suppress discourse, particularly if the questions are posed directly and in ways that encourage students to respond with terse, factual statements rather than with complex, reflective responses. In light of recent developments in educational reform, particularly in mathematics education, as teachers adopt new ways to facilitate student learning, some direct questioning techniques will be less useful as assessment tools.

One situation in which teachers' use of direct questioning has the potential to retain its importance is in structured or semistructured student interviews. It is true that some situations such as large class sizes preclude the use of individual student interviews. In cases where formal interviews are a possibility, the key to using interviews to gather assessment information is a well-designed plan. Although such plans may vary depending upon the problem situation presented, they usually are composed of six steps: establishing rapport, presenting specific instructions, presenting the problem, probing for understanding of the problem, probing for a solution process, and coming to closure. During the course of the interview, sufficient time should be allowed for the student to formulate a response, and the student's thought processes should be of greater importance than the answer.

In addition to teacher-formulated questions, questions that students pose for themselves can be used as another way to gather information about achievement in mathematics. The most common use of student problem posing as a means of assessment occurs when the teacher invites students to identify their own questions or problems after they have worked through an assignment. The traditional "Any questions?" sets the stage for the next learning opportunity. Problem posing by students, however, can provide additional insights when it is employed on a different level. For example, teachers can ask students to create new problems based on a mathematical concept or a problem type currently being studied. Students who pose insightful, rich problems reveal their depth of understanding, whereas students who pose problems almost identical to examples provided in class or in the textbook may understand less. Additionally, the appearance of errors in mathematical language or the construction of implausible problem situations reveals gaps in student understanding for teachers to address during the course of instruction.

## Written Discourse

Not all important classroom discourse need be oral; there are various forms of written discourse that also can serve as sources of assessment information. For example, student journals or written responses to specific probes provide teachers with opportunities to consider students' ideas or their attitudes when making instructional planning decisions.

Journal writing can be instrumental in setting up a dialogue between student and teacher. The content and format of the journal depend upon its intended use in the classroom. For some purposes, journals might be used as a repository for writing in an expressive mode, in which students "think aloud on paper" and record their impressions of classroom activity or their learning. In other cases, students might be asked to use their journals for transactional writing, in which a journal serves as a place to record specific responses to a teacher's questions or to provide some designated kinds of information. In either case, the journal serves not only as a record of the students' thoughts but also as a medium for dialogue with the teacher. A dialogue between student and teacher can occur during the typical journal-writing activity sequence: student entries followed by teacher reading, reflecting, and commentary on the entries, followed by new student entries, and so on.

For a variety of reasons, but certainly because journals are perceived both to use valuable classroom instructional time and to require additional reading and commentary time from teachers, it is likely that relatively few teachers use journal writing as a major instructional activity in their classrooms. As an alternative to full-scale journal writing, teachers might regularly have students respond in writing to specific questions such as "Today in class I learned . . ." or "Of the classwork we've done lately, I'm most confused about . . ." Clearly, students' responses could serve as a valuable source of information to guide planning for the following lesson or week's work, and if students were asked to respond to such probes on a regular basis, this information might become an important component of assessment of student achievement.

## Projects, Investigations, and Extended Problems

Projects, investigations, and extended problems provide opportunities for teachers to assess their students' abilities to formulate problems, apply knowledge in novel ways, generate interesting solution approaches, and sustain intellectual activity for an extended period of time. By working on an extended investigation of an interesting mathematical problem, students participate in activities that are closely related to the activities that occur in the world of work. Such activities allow students to develop an understanding that the analysis of complex problems may take days or even weeks to explore, to learn to work independently or collaboratively on a large project, and to experience the process of producing a written and/or oral report of work over an extended period of time.

Results of projects, investigations, and extended problems may be reported orally rather than in writing. A combination of oral and written presentation allows for a more thorough evaluation of performances, and similar criteria could be used to evaluate oral and written presentations.

## Classroom Testing

Written tests prepared by teachers are commonly used in classrooms to assess each student's achievement. In addition to their role in providing summative information on achievement, classroom tests represent a major source of formative feedback that might be useful in guiding instructional decisions. Teachers interested in diversifying their classroom testing might include a project or an open-ended problem, as discussed above, as a "take-home" portion of a test. The inclusion of such an activity as a portion of a test would provide a teacher with information about aspects of student performance that could not be made available solely from classroom testing. Teachers, however, must weigh the benefits of an out-of-class assessment against the possibility that students, away from the teacher's supervision, may engage in unauthorized collaboration. Yet even within the time constraints characteristic of classroom testing, it is possible to include tasks that can provide information on reasoning, problem solving, and communication. This can be accomplished through the use of tasks that can be completed in five to fifteen minutes, in contrast to the thirty to forty-five seconds typically available for a response to a multiple choice question, and that bear a "family resemblance" to projects, and extended open-ended problems that require much longer to complete. Such tasks are being used in external testing programs in mathematics such as the College Board Advanced Placement (AP) Test in Calculus and in some state-level testing programs. Student responses to these open-ended tasks are typically scored holistically, using a scoring guide that provides detailed information about various levels of performance in solving a particular problem using categories such as mathematical knowledge, strategic knowledge, and communication.

Teachers who use open-ended tasks as part of formal assessment can derive a wealth of information from student responses. They can learn whether or not their students can recognize the main points of a problem, organize information, interpret results, use appropriate mathematical language, and express their reasoning processes. Reading lengthy responses to these tasks is far different, however, from checking multiple choice responses or scoring the more typical computational responses. Moreover, the development of detailed scoring rubrics that set forth requirements for varying levels of performance may be essential to ensure high degrees of interrater agreement when the tasks are used on external assessments, but classroom teachers are unlikely to have the time to create such detailed scoring guides. For classroom use, simple scoring rubrics can provide teachers with a mechanism for evaluating students' solutions and examining the evidence provided in the students' response to detect clues that might help guide instructional decision making. For example, a teacher might focus on strategy selection and how the strategy was implemented, and then use that information to plan additional instruction or to choose examples for the next unit.

## Portfolios

Much of the information obtained from classroom assessment can be especially useful to teachers if it is accumulated over time. Careful recordkeeping can help to ensure that longitudinal information is accumulated for examination. A technique that has been suggested as particularly appropriate for gathering assessment information over time is the portfolio. In its most general sense, a portfolio is a "container" of evidence of someone's knowledge, skills, and dispositions. Creation of a collection of work produced over time has long been an accepted form of assessment in the arts and humanities, and it has received some attention in mathematics and science in recent years. In general, attention to portfolios has focused on the use of this technique as an alternative or supplement to formal testing as a means for evaluating student achievement. Although portfolios could conceivably be assembled for a variety of purposes, the two purposes most often discussed are to serve as a display of "best work" or to demonstrate "growth over time." The purpose of the portfolio determines the criteria that will be used in selecting its contents. If the portfolio is meant to serve as a summative display of proficiency, then the samples representing a student's best work are most appropriate for inclusion. In contrast, if the purpose is for documentation of growth and progress over time, then it would be desirable for it to contain dated examples of student work, including drafts and final copies of projects, solution attempts (both successful and unsuccessful) for a particular problem, and perhaps some examples of the use of concepts or procedures early in a course as contrasted with their use later in the course.

Most discussions of classroom use of portfolios emphasize the benefits of involving students actively in the selection of items to include in their portfolios. In this way, they can engage in an important process of self-assessment. The portfolio can be further personalized, for example, when a student who is interested in art decides to include examples of pictorial representations of and solutions for mathematics problems. For young children, or for those who are compiling a portfolio for the first time, the teacher might maintain the portfolios and periodically review them with the students until they become more familiar with the process.

For the classroom teacher the portfolio provides a comprehensive view of a student's mathematical experiences over time that combines the advantages of other forms of instructionally embedded assessment. Virtually all of the classroom assessment information sources that have been discussed in this section—student journals, written responses to open-ended problems, summaries of group projects or independent investigations, homework papers, and tests—can be included in a portfolio. The coupling of portfolios with information from classroom discourse and activity (e.g., observations and interviews) constitutes a multifaceted approach to assessment of individual student achievement.

Implementation of portfolio assessment in classrooms and in large-scale assessment programs is an ongoing process. While most educators agree that portfolios are excellent sources of assessment information for instructional purposes, their usefulness as a source for summative information or for comparisons between students is still in debate. Unlike classroom tests or standardized assessment instruments, each student's portfolio is different from every other student's portfolio in important ways. Moreover, due to the array of materials that can be included in a portfolio, evaluation becomes a very complex, time-consuming task. In planning for the use of portfolios as an assessment method, teachers should be aware of the benefits and problems involved.

## CONCLUSION

In the 1990s, assessment of student achievement received national attention as educators, policy makers, and other stakeholders debated the issues involved in setting national standards and monitoring student progress toward those standards. Because of the continuing importance placed on accountability at all levels (national, state, local, and district), it is likely that issues concerned with assessment of student achievement will continue to be of major importance.

*See also* Advanced Placement Program (AP); Assessment of Student Achievement, Issues; National Assessment of Educational Progress (NAEP); Questioning; Test Construction

## SELECTED REFERENCES

Airasian, Peter W. *Classroom Assessment*. New York: McGraw-Hill, 1991.

Dubois, Philip H. *A History of Psychological Testing*. Boston, MA: Allyn and Bacon, 1970.

Linn, Robert L., ed. *Educational Measurement*. 3rd ed. New York: Macmillan, 1989.

National Council of Teachers of Mathematics. *Assessment Standards for School Mathematics*. Reston, VA: The Council, 1995.

———. *Curriculum and Evaluation Standards for School Mathematics.* Reston, VA: The Council, 1989.

———. *Principles and Standards for School Mathematics.* Reston, VA: The Council, 2000.

———. *Professional Standards for Teaching Mathematics.* Reston, VA: The Council, 1991.

National Research Council. *Measuring What Counts: A Conceptual Guide for Mathematics Assessment.* Washington, DC: National Academy Press, 1993.

Orpwood, G., and Robert A. Garden. *Assessing Mathematics and Science Literacy.* Vancouver, BC, Canada: Pacific Educational, 1998.

Resnick, Daniel P. "History of Educational Testing." In *Ability Testing: Uses, Consequences, and Controversies. Part II: Documentation Section* (pp. 173–194). Alexandra K. Wigdor and Wendell R. Garner, eds. Washington, DC: National Academy Press, 1982.

Romberg, Thomas A. "Evaluation: A Coat of Many Colors." In *Mathematics Assessment and Evaluation: Imperatives for Mathematics Educators* (pp. 10–36). Thomas A. Romberg, ed. Albany, NY: State University of New York Press, 1992.

Silver, Edward A., and Patricia Ann Kenney. "Sources of Assessment Information for Instructional Guidance in Mathematics." In *Reform in School Mathematics and Authentic Assessment* (pp. 38–86). Thomas A. Romberg, ed. Albany, NY: State University of New York Press, 1995.

Webb, Norman L., ed. *Assessment in the Mathematics Classroom.* 1993 Yearbook. Reston, VA: National Council of Teachers of Mathematics, 1993.

PATRICIA ANN KENNEY
CATHY G. SCHLOEMER

# ASSOCIATION FOR WOMEN IN MATHEMATICS (AWM)

An organization founded by six women in January 1971 following a meeting of the Mathematicians Action Group, a progressive political caucus within the mathematics community. The major goal of AWM is to encourage women to study and pursue careers in mathematics. As of 1995, AWM had about 4,000 individual members, men and women, students and teachers.

Although groups of women mathematicians had been meeting informally before 1971, the new organization, under the energetic leadership of Mary Gray, quickly established a national presence. Its first newsletter appeared in May 1971 and its first business meeting took place in August 1971 at the joint Summer Meetings of the American Mathematical Society and the Mathematical Association of America.

Through its bimonthly newsletter and the panels it presents at national and international mathematics meetings, AWM provides information to the mathematical community on women in mathematics. Its visibility in the 1990s has been enhanced by the *Alice T. Schafer Mathematics Prize*, awarded annually to an undergraduate woman for excellence in mathematics, the *Louise Hay Award*, honoring contributions to mathematics education, and the booklets it produces, such as *The Emmy Noether Lectures*, profiling women research mathematicians, and *Careers That Count* (Jackson 1991, 1994) describing careers in mathematics. In addition, AWM has published a membership directory and a directory of women mathematicians.

While AWM started within the community of research mathematicians, many of the women active in AWM also have been involved in designing and implementing educational programs to increase the participation of girls and women in mathematics. Articles describing successful education programs and strategies have appeared in the *Newsletter* and panels on education have been held at regional and national meetings.

Since the late 1980s, AWM has sponsored programs for high school women students (Sonia Kovalevsky High School Days) and women research mathematicians (workshops and travel grants) with support from government and private foundations.

In 1975 AWM hired a part-time secretary and opened a national office. Initially at Wellesley College, the office moved to the University of Maryland at College Park in 1993.

*See also* Women and Mathematics, History; Women and Mathematics, Problems

## SELECTED REFERENCES

*Association for Women in Mathematics (AWM) Newsletter* (1971–).

Blum, Lenore. "A Brief History of the Association for Women in Mathematics." *Notices of the American Mathematics Society* 38(1991):738–754. [Reprinted in *AWM Newsletter* 21(6)(1991):11–22; 22(1)(1992):12–25.]

*Directory of Women Mathematicians.* College Park, MD: Association for Women in Mathematics, 1995.

*The Emmy Noether Lectures.* College Park, MD: Association for Women in Mathematics, 1994.

Jackson, Allyn. *Careers that Count.* Wellesley, MA: Association for Women in Mathematics, 1991. [Reprinted, College Park, MD: Association for Women in Mathematics, 1994.]

*Membership Directory.* College Park, MD: Association for Women in Mathematics, 1994.

JUDY GREEN

# ASSOCIATION OF TEACHERS OF MATHEMATICS (ATM)

An organization, then called the Association for Teaching Aids in Mathematics (ATAM), started in 1953 in England. The aims of the original association were to promote the study of the teaching of mathematics and to improve the aids used in doing so, through weekend courses, exhibitions, demonstration lessons, and the publication of a bulletin. Its aims today have become more detailed, widening its basis (1) to encourage and enable increased understanding of the learning process, especially in relation to mathematics and its applications; (2) to encourage sharing of teaching and learning strategies and to promote the exploration of new ideas and possibilities; (3) to elucidate the previous two aims and their practical implications, for the benefit of those who have a working involvement with, or an interest in, mathematical education for all age groups. Members receive *MicroMath* and *Mathematics Teaching.* These journals provide the readers with information, such as useful teaching hints and reviews of new software and books, as well as an opportunity to express their points of view.

An annual conference is held each year in different locations in England. Members have also formed local branches, which run a wide variety of activities, including workshops, discussions, mathematical trails (in which students discover mathematics in their own environment), and working groups (where members meet to collate mathematical ideas and invited speakers pass on their expertise to local educators of mathematics). The first branches were initiated in 1957 to promote the aims of ATM within a geographical area, and the tradition has continued.

As new initiatives have been launched, ATM has sought to ensure that mathematics teaching has been strengthened and that the present government knows and understands the opinions of the ordinary teacher of mathematics, especially in view of the introduction of the national curriculum and standards of mathematics for school leavers and the new developments in teacher training. ATM has recently started to establish international links, and there is now a Polish ATM. It has also taken part in the International Congress for Mathematics Education, providing speakers for the seminars and workshops.

*See also* Congresses; England; Mathematical Association of America (MAA); National Council of Teachers of Mathematics (NCTM)

SELECTED REFERENCES

"A.T.A.M. Meetings." *Mathematics Teaching* 5(Nov. 1957):38.

Association of Teachers of Mathematics, 7 Shaftesbury Street, Derby, England DE23 8YB.

ATM. *ATM General Council Handbook.* Derby, England: The Association, 1993.

———. *Holes: An Activity Pack.* Derby, England: The Association, 1993.

———. *Mathematics Teaching.* Derby, England: The Association (four issues a year).

———. *MicroMath.* Derby, England: The Association (three issues a year).

———. *Teaching, Learning and Mathematics* Derby, England: The Association, 1995.

———. *Using and Applying Mathematics* Derby, England: The Association, 1993.

"Group Meetings." *Mathematics Teaching* 11(Nov. 1959):27.

LINDA JACKSON

# ASYMPTOTES

Auxiliary lines used when representing diverse rational and nonalgebraic functions graphically that, though not a part of the actual graph, reveal much concerning the pattern of the points. Functions of elementary mathematics are often of the polynomial type and thus have the appealing feature of being continuous everywhere. Illustrated by linear and quadratic forms, they are frequently the object of graphical treatment in fundamental algebra. As the field of discussion broadens, other functions appear, many of whose graphs have varying points of discontinuity. Notable among these are the hyperbola, $y = k/x$, and certain of the trigonometric functions such as $y = \tan x$ and $y = \sec x$.

Suppose that a point tracing a curve recedes infinitely and becomes ever nearer some line in the process. Such a line, called an *asymptote,* may be vertical, horizontal, or oblique. In this dynamic approach to curve definition, figures are thus generated by a moving point subject to some set of conditions. The dynamic point of view proves beneficial in providing the student a deeper understanding of conic sections and other notable curves of mathematics whenever infinite distances from the origin, especially in the case of asymptotes, are implied.

One of the earliest examples encountered of curves associated with asymptotes is the equilateral hyperbola $y = 1/x$ (a curve often arising in the study of inverse variation and applications such as those of the distance rate-time relationship or Boyle's Law) (see Figure 1).

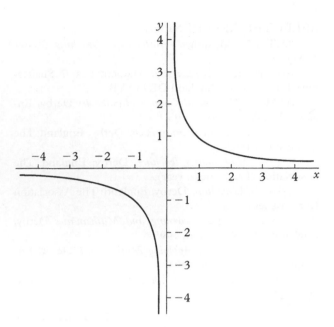

**Figure 1** *The equilateral hyperbola, $y = 1/x$, with vertical asymptote (the y-axis) and horizontal asymptote (the x-axis).*

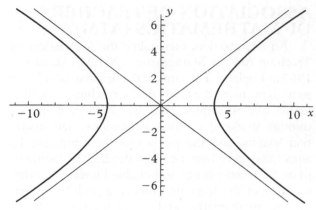

**Figure 2** *The hyperbola $\dfrac{x^2}{16} - \dfrac{y^2}{9} = 1$ with oblique asymptotes $y = \pm\dfrac{3}{4}x$.*

For positive $x$ values, it becomes immediately apparent that $y$ tends to zero as $x$ increases without bound. Moreover, $y$ increases without bound as $x$ tends to zero through positive values. This is made vivid by the first-quadrant portion of the graph. In particular, the $x$ and $y$ axes reflect this "infinite" characteristic in that they are approached by the hyperbola but never touched or crossed.

More rigorous accounts of asymptotes involve the limit concept, especially stressed in the calculus. Hence, the line $x = 0$ is a vertical asymptote in the preceding example because, as $x$ approaches 0 from the right, limit $1/x$ is infinity. Symbolically,

$$\lim_{x \to 0^+} \frac{1}{x} = \infty$$

Too, the line $y = 0$ is a horizontal asymptote because

$$\lim_{x \to \infty} y = 0$$

Oblique asymptotes often arise and are illustrated effectively by the hyperbola

$$\frac{x^2}{a^2} - \frac{y^2}{b^2} = 1$$

whose asymptotes are the lines

$$y = \pm\frac{b}{a}x$$

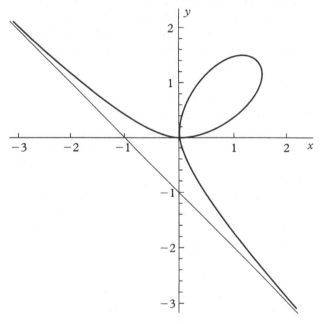

**Figure 3** *The Folium of Descartes, $x^3 + y^3 = 3xy$, and its asymptote, $x + y + 1 = 0$.*

(See Figure 2.) Another well-known illustration, and one of historical interest, is the folium of Descartes (see Figure 3). Transcendental functions provide a setting for asymptotes to appear. In addition to the familiar tangent and cotangent curves, the functions $y = e^x$ and $y = \ln x$ have this notable characteristic (as does the bell-shaped curve of probability, $y = \exp(-x^2)$).

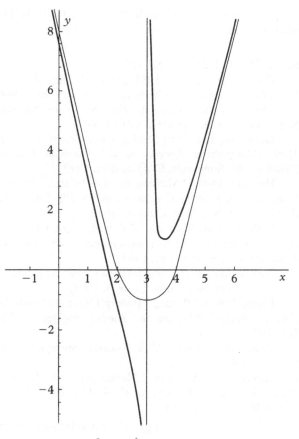

**Figure 4** $y = \dfrac{x^3 - 9x^2 + 26x - 23}{x - 3}$ *with linear*

*asymptote x = 3 and curved asymptote*

$y = x^2 - 6x + 8.$

Although the asymptote (which literally means "not falling together") is generally of the linear type, curved asymptotes are also possible (see Figure 4). In this case, the moving point that generates the desired graph approaches a curve other than a straight line. Such a variant reinforces the broad scope of asymptotes. They serve as auxiliary lines or curves that identify more thoroughly the pattern of points in the generating of various graphs. Asymptotes are indelibly associated with many of the most useful and appealing curves of mathematics and extend from antiquity and the works of Apollonius to the present day. Curve sketching, a significant classroom activity, is expedited by a consideration of asymptotes. Although asymptotes are not a part of the desired graph, they nevertheless are lines (or, in the general setting, familiar curves) that provide a better understanding of a graph's behavior.

*See also* Conic Sections; Limits

SELECTED REFERENCES

Barnett, Raymond A., and Michael R. Ziegler. *Analytic Trigonometry with Applications.* Boston, MA: PWS, 1995.

Boyer, Carl B. *A History of Mathematics.* 2nd ed. Revised by Uta C. Marzbach. New York: Wiley, 1989.

Hogben, Lancelot. *Mathematics for the Millions.* 4th ed. New York: Norton, 1968.

Larson, Roland E., and Robert P. Hostetler, *Calculus with Analytic Geometry.* Lexington, MA: Heath, 1986.

Penna, Michael A., and Richard R. Patterson. *Projective Geometry and Its Applications to Computer Graphics.* Englewood Cliffs, NJ: Prentice-Hall, 1986.

RICHARD L. FRANCIS

## ATTRIBUTE BLOCKS

Created by William P. Hull (1968) and used to develop logic, classification, and problem-solving skills. Sets consist of wooden blocks with three attributes: size (small, large), shape (circle, triangle, diamond, square), and color (red, blue, yellow, green). Thus, because there are two sizes, four shapes, and four colors, there are $2 \times 4 \times 4$, or thirty-two, blocks. Some sets also contain label cards and strings.

The concepts behind Hull's development of attribute blocks evolved during the 1950s and 1960s. Hull worked with the same students as they matured mathematically in nursery school and in grades 3 and 5. As a psychologist, he was interested in why "some of the students who scored highly in achievement tests were able to do so because they had developed an approach to learning in which they searched for 'recipes' to follow rather than facing up to the complexity of problems and staying with them until they developed a deeper understanding. When presented with unfamiliar problems they were often surpassed by students who had not been very successful in meeting 'production' expectations" (Hull, Mar. 1995). Thus, he developed a Block Sorting Test to identify students with proficient problem-solving skills.

Hull also worked with students aged 5, 6, and 7 in the British infant schools and interacted with Zoltan Dienes and also Caleb Gattegno, who discovered the work of Emile-Georges Cuisenaire in Belgium. In the Elementary Science Study (ESS), Hull refined the ideas behind attribute blocks by collaborating with physicists from the Massachusetts Institute of Technology (MIT) and David Hawkins, Director of ESS. The organization, ESS, was started by MIT scientists who had been working on curriculum for high school.

The umbrella group was Educational Services, Inc. (ESI), now Educational Development Center. ESS provided a stimulating environment in which scientists and teachers could take a fresh look at the complexities of children's thinking.

Sometimes included with a set of thirty-two attribute blocks are ten positive label cards that name each available attribute, for example, "red," "large," "circle," and ten negative label cards that negate each attribute, for example, "not blue," "not small," "not a triangle." The strings used to surround subsets of blocks may be elastic, twine, or shoelaces.

ESS developed the original attribute materials, commercially sold as Attribute Games and Problems. These included attribute blocks, label cards, colored elastic strings, color cubes, People Pieces (plastic tiles), Creature Cards (cardboard activity cards), and a teacher guide. Attribute Logic Blocks, available commercially, consists of sixty blocks in five shapes (circle, triangle, rectangle, square, hexagon), three colors (red, blue, yellow), two sizes (small, large), and two thicknesses (thin, thick).

Students explore various attributes of the blocks and communicate results verbally, in chart form, and by physically arranging the blocks in some systematic way. Some students form arrangements by color, some by shape, and others by size. An extension for student exploration is to determine how many blocks are in a set having six shapes, five colors, and three sizes. The Fundamental Counting Principle may be used to determine that this set has ninety attribute blocks. In groups, students can make such a set and use it for further exploration. Another recommended activity starts by choosing two blocks that differ by one attribute and continues by finding other blocks to create a sequence in which each block differs from the preceding block by one attribute. The ability to generate more complex sequences can develop from this simple beginning.

Attribute blocks are useful teaching tools for the development of logic, patterning and relationships, geometry, and reasoning concepts and skills. Students can use and communicate to each other similarities and differences in a set of blocks, solve circle logic problems, and learn the mathematical meaning of such words as *and, not, all, some,* and *none.* Additionally, students can describe in both verbal and written form relationships between objects in a pattern, ways to extend a given pattern, and explanations of their own patterns. Identifying and sorting activities help students to develop analytical, logical, and spatial reasoning skills, communication skills, and social interaction skills.

*See also* Cooperative Learning; Cuisenaire Rods; Dienes; Patterning

## SELECTED REFERENCES

Dienes, Zoltan P., and Edward W. Golding. *Learning Logic, Logical Games.* New York: Herder and Herder, 1966.

Elementary Science Study. *The ESS Reader.* Newton, MA: Education Development Center, 1971.

Goodnow, Judy, and Shirley Hoogeboom. *Overhead Math: Manipulative Lessons on the Overhead Projector (Grades 3–6).* Sunnyvale, CA: Creative, 1990.

Hawkins, David. "Messing about in Science." In *Open Education: The Informal Classroom* (pp. 58–70). Charles H. Rathbone, ed. New York: Citation, 1971.

Hull, William. "Concept Work with Young Children." *Bulletin of the I.S.G.M.L. (International Study Group for Mathematics Learning)* (Apr. 1963).

Hull, William P. Personal interview. (Feb. 7, 1995).

———. Personal Letter. (Mar. 3, 1995).

Pagni, David. "Playing 'Twenty Questions' with Attribute Blocks." *Mathematics Teacher* 86(Dec. 1993): 765–769.

Roper, Ann. *Junior High Jobcards.* Sunnyvale, CA: Creative, 1990.

Spikell, Mark A. *Teaching Mathematics with Manipulatives: A Resource of Activities for the K–12 Teacher.* Boston, MA: Allyn and Bacon, 1993.

REGINA BARON BRUNNER

# AWARDS FOR STUDENTS AND TEACHERS
## AWARDS FOR JUNIOR HIGH SCHOOL STUDENTS

There are two main types of awards based on competitions. Awards include certificates, conferring of titles, listing on honor rolls, invitations to participate in higher-level competition, and all-expense-paid trips.

### MATHCOUNTS Competition

This is a nationwide coaching and competition program to encourage and prepare students to pursue mathematics in high school and beyond by making it as exciting as a school sports event. MATHCOUNTS competitions involve a two-phase approach. The first phase begins in the fall with free distribution of MATHCOUNTS "coaching" materials to mathematics teachers nationwide. These materials focus on problem solving, analytical thinking, exploration, and teamwork. Throughout the fall, teachers use the specifically prepared "warm-ups" and "workouts" as supplemental classroom materials or as an extracurricular activity. The second phase gets underway in February, when participating schools have an opportunity to challenge other area schools in a head-to-head mathematics competition. Winning teams and individuals progress to state-level competitions in March, with top students from each state earning an all-expense-paid trip to the nation's capital in May to vie for national MATH-COUNTS titles, such as national individual champion or national team champion.

## American Mathematics Contest → 8 (AMC → 8)

Formerly known as the *American Junior High School Mathematics Examination (AJHSME)*, the AMC → 8 is a twenty-five question, forty-minute, multiple choice examination in junior high school (middle school) mathematics designed to promote the further development of problem-solving skills. Registration brochures are mailed in early September to all public and private schools with seventh and eighth grades, and the examination is given in November. The examination provides an opportunity to apply the concepts taught at the junior high levels to problems that not only range from easy to difficult but also cover a wide range of applications. Many problems are designed to challenge students and to provide problem-solving experiences beyond those provided in most junior high school mathematics classes. Calculators are allowed. High-scoring students are invited to participate in the AMC → 10 (formerly the American High School Mathematics Examination (see the following section)).

Participation in this examination is typically available to all interested students but is restricted in some schools to those most qualified. It provides an inexpensive way to recognize mathematical talent. Carefully written solutions provide an excellent opportunity for further discussion and investigation (see American Mathematics Competitions).

## AWARDS FOR HIGH SCHOOL STUDENTS
## AMC → 10 and AMC → 12 (Formerly AHSME)

The main competitions at the high school level are the American Mathematical Contests (formerly known as the American High School Mathematics Examination (AHSME)), sponsored by organizations including mathematical, actuarial, and statistical associations. The contests are multiple choice examinations in secondary school mathematics that contain twenty-five problems which can be under-

stood and solved with precalculus concepts. Calculators are allowed. Each examination is given in February. The main purpose of the AMC → 10 and AMC → 12 is to spur interest in mathematics and to develop talent through the excitement of solving challenging problems in a timed multiple choice format. The problems range from the very easy to the extremely difficult. Students who participate in the AMC → 10 or AMC → 12 should find that most of the problems are challenging but within their grasp. Each examination is intended for all students, not necessarily only the highest ranking.

Because the problems cover such a broad spectrum of knowledge and ability there is a wide range of scores. The National Honor Roll cutoff score, 100 out of 150 possible points, is typically attained or surpassed by fewer than 2% of all participants. For most students and schools, only relative scores are significant, so lists of top individual and team scores on regional and local levels are compiled. These regional lists and information on score distributions appear in the yearly summary sent to all participating schools. The most valuable comparison students can make is between their own current level of achievement and their levels in previous years. In particular, all students are encouraged to begin participating in the contests early in their secondary mathematics studies so that they can look back with pride each year on how they have learned to answer questions that they could not have answered previously.

A special purpose of the competition is to help identify those few students with truly exceptional mathematics talent. Students who are among the very best deserve some indication of how they stand relative to the other students in the country and around the world. The AMC → 10 and AMC → 12 provide one such indication, and are the first in a series of examinations (followed in the United States by the American Invitational Mathematics Examination (AIME) and the USA Mathematical Olympiad (USAMO)) that culminates in participation in the International Mathematical Olympiad (IMO), the most prestigious and difficult secondary mathematics examination in the world. In this way, the very best young mathematicians are recognized, encouraged, and developed by receiving prizes, given widespread publicity, brought to Washington, DC, for an awards ceremony, and so on.

## AIME

All students with a score of 100 out of a possible 150 on the AMC → 10 and AMC → 12 are invited to take the AIME. The AIME is intended to provide further challenge and recognition, beyond that provided by the AMC $\to$ 10 and AMC $\to$ 12, to the many high school students in North America who have exceptional mathematical ability. The students who receive the top scores (based on a weighted average) on AMC $\to$ 10 and AIME or AMC $\to$ 12 and AIME are invited to take the USAMO. The AIME is a fifteen-question, three-hour examination in which each answer is a three-digit integer. The questions on the AIME are much more difficult, and students are very unlikely to obtain the correct answer by guessing. All problems on the AIME can be solved by pre-calculus methods.

## USAMO

The USAMO provides a means of identifying and encouraging the talent of those who may become the next generation of leaders in the mathematical sciences. The USAMO is part of a worldwide system of national mathematics competitions, a movement in which both educators and research mathematicians are engaged to recognize and celebrate the imagination and resourcefulness of our youth. The top six scorers on the USAMO comprise the United States team that competes in the IMO. The USAMO is a five-question, three-and-one-half-hour essay/proof examination. Approximately 140 of the top scoring AMC/AIME students (based on a weighted average) are invited to take the USAMO. Participation is restricted to citizens of the United States and those who are permanent residents.

The top eight USAMO scorers are honored at ceremonies held in Washington, DC at the National Academy of Sciences and the U.S. Department of State. This gives our high school scholars a chance to meet some of the nation's best scientists and mathematicians, some of whom make presentations on advanced mathematical topics and others who talk to them individually. At the IMO, high school students from about eighty nations compete in a challenging two-day mathematical examination held in a different country each year. There, team members and leaders have a unique opportunity to interact with their counterparts from other countries in an exchange of mathematical ideas (see American Mathematics Competitions).

## OTHER STUDENT AWARDS

High school and junior college students are also eligible for election to the national mathematics honor society, Mu Alpha Theta, at high schools that have a chapter, and to participate in the annual Westinghouse Science Competition. It is relatively rare for a mathematician to win the top prize in this competition, but it did happen in 1996 when the $40,000 top prize went to a mathematics student.

## AWARDS FOR JUNIOR AND SENIOR HIGH SCHOOL TEACHERS
### Presidential Awards for Excellence in Mathematics Teaching

Kindergarten through twelfth-grade mathematics teachers who have taught mathematics for five years or more in public or private schools are eligible for this award. An award is granted to one mathematics teacher at both the elementary and secondary levels from each state, the District of Columbia, Puerto Rico, the Department of Defense Dependent Schools, and the U.S. territories. The Presidential awardees in each state and jurisdiction receive a presidential citation, an expense-paid trip to Washington, DC, a $7,500 National Science Foundation grant for their school, local and national recognition, and generous gifts from business and industry (see National Science Teachers Association).

### The Edyth May Sliffe Awards for Distinguished Junior High School Mathematics Teaching

These awards are given to teachers whose schools participate in the AMC $\to$ 8. Five awards are given in each of the ten regions associated with the AMC $\to$ 10 and AMC $\to$ 12. The ranking of a school in a given year for the purposes of this award is determined by the sum of the scores of the three top students at the school. Those schools that ranked in the top third of the schools within the region for each of the past three years are then ranked, by region, by summing their rankings for each of those years. The supervising teacher of each of the eligible five top-ranking schools on this list is declared an Edyth May Sliffe Award recipient for that region and receives a cash award of $100, an appropriate certificate, and a pin (see American Mathematics Competitions).

### The Edyth May Sliffe Awards for Distinguished High School Mathematics Teaching

These awards are given to teachers whose schools had the highest official team scores on the

AMC. The executive director of the AMC asks the three students on each of these teams for a recommendation on the teacher who, in the student's opinion, was most responsible for his or her success on the AMC. On the basis of the information received from the students, approximately twenty teachers are selected for the award, which consists of cash ranging from $750 for the first-place recipient to $350 for those ranking eleventh through twentieth. They also receive from the president of the Mathematical Association of America (MAA) a letter of congratulations, a certificate, a pin, and a one-year free membership in the MAA (see American Mathematics Competitions; Mathematical Association of America 1989).

## AWARDS FOR COMMUNITY COLLEGE TEACHERS

Community college teachers of mathematics are eligible for the Awards for Distinguished College or University Teaching of Mathematics. Eligibility requires an assignment of at least half time during the academic year to teaching a mathematical science in a public or private college or university in the United States or Canada, five years teaching experience in a mathematical science, and membership in the MAA. Each of the twenty-nine sections of the MAA can make such an award each year. The award consists of a certificate, recognition at a section ceremony, and, in some sections, a cash award. The section awardees then become eligible for the national Deborah and Franklin Tepper Haimo Awards for Distinguished College or University Teaching of Mathematics, of which at most three are conferred each year. Recipients of this award receive $1,000 and a certificate.

*See also* Advanced Placement Program (AP); Competitions

### SELECTED REFERENCES

American Mathematics Competitions. Information Booklet. Lincoln, NE: University of Nebraska. Tel.:(800) 527-3690. World Wide Web Page: http://www.unl.edu/amc.

MATHCOUNTS Foundation. *MATHCOUNTS School Handbook.* Alexandria, VA: The Foundation. Tel.: (703) 684-2828.

Mathematical Association of America. "The Edyth May Sliffe Award for Distinguished High School Mathematics Teaching." *Focus* (Jan.–Feb. 1989):12.

National Science Teachers Association. *Presidential Awards for Excellence in Science and Mathematics Teaching.* Arlington, VA: The Association.

HENRY L. ALDER

# AXIOMS

Statements accepted as true without proof. The axioms of any mathematical theory contain the undefined fundamental concepts from which all others are defined. From the axioms all other propositions are deduced. Synonyms for *axioms* include *postulates, assumptions,* and *primary propositions.* For centuries, authorities had distinguished between "axiom" and "postulate," regarding an axiom as a universal truth; for example, "The whole is greater than the part." Postulates, however, had a mathematical context. An example is Euclid's fifth postulate, now stated as follows: Given a line $l$ and a point $P$ not on $l$, there is exactly one line parallel to $l$ through $P$.

Euclid's *Elements* (ca. 300 B.C.), the first fully developed axiomatic system, distinguished between axioms and postulates by calling axioms "common notions" (Euclid 1956). From his assumptions and undefined terms, Euclid proved nearly 500 theorems, using deductive techniques set forth by Aristotle. The result became the basis for the geometry course offered in schools of many countries for centuries.

Originally, axioms were derived from the everyday world and were viewed as self-evident (empirical or intuitive) truths. In any axiomatic system of modern mathematics, no axiom need appeal to intuition; and neither may characteristics of any object (such as a geometric figure) nor any experiences with it (e.g., movement) ever be substituted for logic in any proof. Axioms are sentence functions in which the undefined terms are variables subject to interpretations; each interpretation satisfying the axioms automatically satisfies all theorems deduced from those axioms (Hempel 1956). Examples of such sentence functions are: If for elements $a$, $b$, $c$ of a set $E$, (1) a relation $R$ satisfies $aRa$; (2) $aRb$ implies $bRa$; and (3) if $aRb$ and $bRc$, then $aRc$, then the relation is an equivalence relation. Among the interpreting relations satisfying these axioms are congruence of segments, angles, or triangles, and equality within the real numbers, together with all of their theorems.

The only requirement for a system of axioms is consistency so that ensuing deductions shall not be contradictory. Independence and completeness are desirable (Wilder 1952). The history of Euclid's fifth postulate illustrates independence and consistency. For centuries, mathematicians questioned its independence, suspecting it could be proven from Euclid's other postulates. Their failure to find a proof led to acceptance of its independence and to the formulation of non-Euclidean geometries, where new assumptions contradictory to the fifth postulate were independent of and consistent with the remaining axioms. Early in the nineteenth century, Nikolai Lobachevsky did so by denying Euclid's parallel axiom; he essentially assumed that through point $P$ not on any line $l$, there were more than one parallel to $l$. Later in that century, Georg Friedrich Bernhard Riemann developed a non-Euclidean geometry from his denial of Euclid's parallel axiom; he in essence claimed that there was *no* line through $P$ parallel to $l$. Any set of independent and noncontradictory axioms guarantees only the validity of the resulting theorems.

Axioms are the basis of all mathematical systems, but until recently, only in geometry did axioms appear explicitly in school mathematics. The modern mathematics movement of the twentieth century led to reforms in the secondary school curriculum, such as the elimination of many of Euclid's weaknesses and the clarification and expansion of axiomatic language in both geometry and algebra (Commission on Mathematics 1959). Earlier, Harold Fawcett's experiment (1938) had indicated that thinking improved in geometry and other, even nonmathematical areas if students were led to develop their own axioms and undefined terms from their own experience. Since the early 1980s, many mathematics educators have subscribed to the pedagogical theory of Pierre van Hiele and Dina van Hiele-Geldof that students have difficulty with axioms, undefined terms, and deduc-tion unless they are first guided through preliminary thinking levels of visualization, analysis, and informal deduction (Shaughnessy and Burger 1985).

*See also* Euclid; Mathematics, Foundations; Non-Euclidean Geometry; Van Hiele Levels

## SELECTED REFERENCES

Commission on Mathematics. *Program for College Preparatory Mathematics.* New York: College Entrance Examination Board, 1959.

Euclid. *The Thirteen Books of Euclid's Elements.* 2nd ed. Thomas L. Heath, trans. New York: Dover, 1956.

Fawcett, Harold. P. *The Nature of Proof.* 13th Yearbook. National Council of Teachers of Mathematics. New York: Bureau of Publications, Teachers College, Columbia University, 1938.

Hempel, Carl G. "Geometry and Empirical Science." In *The World of Mathematics* (pp. 1635–1646). James R. Newman, ed. New York: Simon and Schuster, 1956.

Kalin, Robert, and Mary K. Corbitt. *Prentice Hall Geometry.* Englewood Cliffs, NJ: Prentice-Hall, 1993.

Shaughnessy, J. Michael, and William F. Burger. "Spadework Prior to Deduction in Geometry." *Mathematics Teacher* 78(Sept. 1985):419–428.

Wilder, Raymond L. *Introduction to the Foundations of Mathematics.* New York: Wiley, 1952.

ROBERT KALIN

# B

## BABBAGE, CHARLES
## (1792–1871)

The first person to appreciate fully the benefits of the mechanization of computation and best known for his *Difference* and *Analytical Engines*. In 1822, Babbage built an experimental model of the first *Difference Engine*, which used the method of finite differences to perform arithmetical calculations necessary for accurate mathematical tables. By 1834, he conceived of the *Analytical Engine,* a projected machine that would mechanize any mathematical operation—the machinery would be unchanged and only the instruction cards rewritten. Babbage continued to work on its design throughout his life. The modern computer differs from Babbage's design principally in that his computer was not based on the notion of a stored program. In 1842, Augusta Ada Lovelace provided a detailed description of the steps Babbage's machine took to produce solutions to particular mathematical problems.

In 1812, when Babbage was a student at Trinity College, Cambridge University, he and several other students founded the influential Analytical Society, whose purpose was to raise British mathematics to the level that existed on the European continent. From 1827 to 1839, as Isaac Newton did earlier, Babbage held the Lucasian chair at Cambridge. Babbage's interests were wide and to all of them he made important contributions: astronomy, cryptology, theory of life insurance, scientific manufacturing processes, and the design of tools (e.g., "On the Principles of Tools for Turning and Planing Metals" (1846)). His most successful book, *On the Economy of Manufactures and Machinery* (1832) foreshadowed operational research.

### SELECTED REFERENCES

Babbage, Charles, et al. *Charles Babbage on the Principles and Development of the Calculator: And Other Seminal Writings.* Philip Morrison and Emily Morrison, eds. New York: Dover, 1961.

Hyman, Anthony. *Charles Babbage: Pioneer of the Computer.* Princeton, NJ: Princeton University Press, 1982.

Swade, Doron D. "Redeeming Charles Babbage's Mechanical Computer." *Scientific American* 268(Feb. 1993): 86–91.

FRANCINE F. ABELES

## BABYLONIAN MATHEMATICS

Mathematics of ancient Babylonia (also called Mesopotamia), an area spanning the Tigris-Euphrates river valley from Baghdad to the Persian Gulf in what is now Iraq. The Babylonian era lasted from about 3000 B.C. to the invasion of Alexander the Great around 330 B.C. We know quite a lot about ancient Babylonia because its scribes kept detailed records by marking soft clay tablets with a stylus. Once dried, these tablets became rocklike, and thousands have survived. Many of the tablets concern mathematics, and through careful study of their contents, historians have pieced together a picture of Babylonian mathematics.

The Babylonians used a base 60 numeration system, complete with fractional parts, usually referred to as the *sexagesimal system*. Historians usually expressed a number sexagesimally by writing a sequence of numbers between 0 and 59, separating the various places by commas. A semicolon was then used as a "decimal" point to distinguish the integer part of a number from its fractional part. For example, the base 60 number 41, 38, 3; 18, 23 represents the base 10 number $41 \times 60^2 + 38 \times 60^1 + 3 \times 60^0 + \frac{18}{60} + \frac{23}{3600}$ or $149,883 \frac{1103}{3600}$. The Babylonian sexagesimal system became widespread throughout the Mediterranean rim as well as in Europe. Vestiges of this system survive in the units we use to measure time and angles. The Babylonians, however, did not seem to understand the concept of zero, although toward the end of the Babylonian era an empty space was sometimes used to indicate a zero. As was the case in most cultures up until the 1600s, negative numbers were not used, nor did the Babylonians use numerals, but rather a system of vertical and horizontal strokes. The Babylonians developed extensive multiplication, square root, and cube root tables as well as tables of reciprocals, which were used in computations involving fractions. Aside from the fact that they did not use zero, the Babylonians manipulated numbers much as we do, except that they relied on tables rather than a calculator for difficult computations.

Although they did not develop any systematic algebraic symbolism, the Babylonians knew how to solve simple equations such as a quadratic equation. Perhaps because there was no algebraic symbolism, algebraic techniques were not taught by the use of formulas. Instead, students would be given solutions to various specific problems and would, it was hoped, then be able to apply the techniques thereby learned to similar problems. Some have interpreted the Babylonian habit of teaching by example as evidence that they did not understand the general principles behind such skills as solving a quadratic. Although this may be the case, it is more likely that the general principles were indeed understood but that they were communicated by example, just as a present-day teacher might demonstrate a general principle through the use of a specific problem.

Tablets dating from approximately 1700 B.C. show that the Babylonians not only knew the Pythagorean theorem over 1,000 years before Pythagoras was born but also knew how to generate tables of Pythagorean triples, that is, natural numbers $a$, $b$, and $c$ such that $a^2 + b^2 = c^2$. The Babylonians also knew how to compute the area of basic geometric shapes such as rectangles or right triangles. However, their techniques for finding the areas of irregularly shaped objects generally yielded only approximate values, a fact they may not have understood. For example, one Babylonian formula for the area of a circle uses an implicit value of 3 for pi. It should be kept in mind, however, that the Babylonians do not appear to have abstracted the concept of pi per se, namely, the mathematical constant that is the ratio of the circumference of a circle to its diameter. Despite the fact that their formula for computing the area of a circle was not correct, they were aware that the volume of a cylinder was the product of its base and its height. And although the Babylonians initially used an incorrect formula to compute the volume of a truncated pyramid with a square base, they eventually discovered the correct one.

Many of the Babylonian clay tablets appear to be instructional texts containing word problems whose solutions require many of the skills just described. For example, a tablet from around 1600 B.C. asks the student to find the length and width of a rectangle if its area and semiperimeter are given. The solution of this problem involves the quadratic equation. It is interesting to observe that similar problems are found in modern-day algebra texts. Indeed, a young Babylonian might very well have studied many of the topics covered in a secondary school algebra course.

*See also* History of Mathematics, Overview

## SELECTED REFERENCES

Boyer, Carl B. *A History of Mathematics* (pp. 23–42). 2nd ed., rev. Revised by Uta C. Merzbach. New York: Wiley, 1991.

Friberg, Jöran. "Methods of Babylonian Mathematics." *Historia Mathematica* 8(1981):277–318.

Hoyrup, Jens. "Babylonian Mathematics." In *Companion Encyclopedia of the History and Philosophy of the Mathematical Sciences* (pp. 21–29). Ivor Grattan-Guinness, ed. London, England: Routledge, 1994.

Katz, Victor. *A History of Mathematics: An Introduction* (pp. 1–41). New York: Harper Collins, 1993.

Kline, Morris. *Mathematical Thought from Ancient to Modern Times* (pp. 3–13). New York: Oxford University Press, 1972.

Kramer, Edna. *The Nature and Growth of Modern Mathematics.* Princeton, NJ: Princeton University Press, 1981.

Neugebauer, Otto. *The Exact Sciences in Antiquity.* Princeton, NJ: Princeton University Press, 1951.

Van der Waerden, Bartel L. *Geometry and Algebra in Ancient Civilizations.* New York: Springer-Verlag, 1983.

———. *Science Awakening.* New York: Oxford University Press, 1961.

DANIEL S. ALEXANDER

# BASES, COMPLEX

Complex numbers in terms of which other complex numbers may be expressed. Complex numbers can be written as single numbers in positional notation using various complex bases, without separating the numbers into their real and imaginary parts. These representations yield some interesting fractal patterns in the complex plane.

A complex integer (or Gaussian integer) is a complex number $x + iy$, in which $x$ and $y$ are real integers. The complex integer $z$ is said to be represented in the complex integer base $b$, using a digit set $D$, if it can be written as $z = a_m b^m + a_{m-1} b^{m-1} + \cdots + a_2 b^2 + a_1 b + a_0$, where $a_m, a_{m-1} \ldots a_2, a_1, a_0$ are digits in the set $D$. This representation is denoted by the positional notation $(a_m a_{m-1} \cdots a_2 a_1 a_0)_b$. For example, all complex integers can be represented uniquely in the base $-1 + i$ using the binary digit set $\{0,1\}$. In this base, $-4 - i$ is represented by $(10111)_{-1+i}$ since $(-1 + i)^4 + (-1 + i)^2 + (-1 + i) + 1 = -4 = i$, and $2$ is represented by $(1100)_{-1+i}$ since $(-1 + i)^3 + (-1 + i)^2 = 2$. The complex integer $p_0 + iq_0$ can be converted to the base $-1 + i$ as follows. The right digit, $a_0$, is 0 if $p_0 + q_0$ is even, and 1 otherwise. Then calculate

$$p_1 + iq_1 = \frac{p_0 + iq_0}{-1 + i} - a_0$$

The next digit, $a_1$, is 0 if $p_1 + q_1$ is even, and 1 otherwise. Continue in this manner until $p_{m+1} + q_{m+1}$ is zero.

A good exercise for students in complex arithmetic is to have them determine which complex integers can be represented in the base $1 + i$ using the binary digit set $\{0, 1\}$ and expansions of length less than or equal to $m$. Then have them do the same exercise with the base $-1 + i$. This can lead to the construction, by hand or by computer, of various fractal sets in the complex plane (see Gilbert 1982). Let each square of a sheet of graph paper (or pixel on a computer screen) correspond to a complex integer.

If that complex integer can be represented in the base, color the square according to the length of the representation. One fractal obtained this way is called the *space-filling twin dragon curve.* The complex base $-1 + i$ can be used to analyze various dragon curves, which are obtained by repeatedly folding a strip of paper in half and then opening it up; see Davis and Knuth (1970) and Gardner (1990).

The first examples of complex numbers as bases were given in the 1960s; see Knuth (1973, Ch. 4.1) for the history of positional number systems. Only certain bases and digit sets can be used to represent all the complex integers. For each positive integer $n$, $b = -n + i$ can be used as a base for all the complex integers, using the natural number digit set $D = \{0,1,2, \ldots, n^2\}$. Hence the base $-3 + i$ yields a decimal representation of all the complex integers, using the digits $0, 1, \ldots, 9$, where, for example, $20 - 159i = (21073)_{-3+i}$. The digits could be complex; one interesting system is to use the base $2 + i$, with the symmetrical digit set $\{0, 1, -1, i, -i\}$.

If all the complex integers can be represented in a given base, then the noninteger complex numbers can also be represented, using expansions to the right of the radix point, in the form $(a_m \ldots a_1 a_0 . a_{-1} a_{-2} \ldots)_b = a_m b^m + \cdots + a_1 b + a_0 + a_{-1} b^{-1} + a_{-2} b^{-2} + \ldots$ where $a_m, \ldots, a_1, a_0, a_{-1}, a_{-2}, \ldots$ are digits in the set $D$. The usual arithmetic operations of addition, subtraction, and multiplication can be generalized, with some interesting twists, to complex bases; see Gilbert (1984). For example, in the base $-3 + i$ with decimal digits, $10 = (1540)_{-3+i}$, so that when the total in adding one column exceeds 10, 154 has to be carried to the next three columns. Sometimes these carry digits can accumulate into an infinite sequence of digits; however, the final answer is still finite.

*See also* Bases, Negative; Binary (Dyadic) System; Decimal (Hindu-Arabic) System; Fractals; Hexadecimal System

## SELECTED REFERENCES
Davis, Chandler, and Donald E. Knuth. "Number Representations and Dragon Curves—I and II." *Journal of Recreational Mathematics* 3(1970):66–81, 133–149.

Gardner, Martin. *Mathematical Magic Show.* Washington, DC: Mathematical Association of America, 1990. Reprinted from Gardner, Martin. "Mathematical Games." *Scientific American* 216(Mar. 1967):124–125; (Apr. 1967):118–120.

Gilbert, William J. "Arithmetic in Complex Bases." *Mathematics Magazine* 57(1984):77–81.

———. "Fractal Geometry Derived from Complex Bases." *Mathematical Intelligencer* 4(1982):78–86.

Knuth, Donald E. *The Art of Computer Programming*. Vol. 2. *Seminumerical Algorithms*. Reading, MA: Addison-Wesley, 1973.

WILLIAM J. GILBERT

# BASES, NEGATIVE

Amusing extension of the usual positional number systems. It is well known that all the positive integers can be represented in any positive base $n$ (if $n > 2$), using the digits $0, 1, \ldots, n - 1$; the decimal, binary, and hexadecimal systems being the most common. However, it is possible to represent all the integers in a negative base $-n$ (if $n > 2$), using the digits $0, 1, \ldots, n - 1$. Each positive and negative integer can be represented uniquely in each of these bases, without using a minus sign. For example, $-532$ can be represented as 1548 in the base $-10$, since $1 \times (-10)^3 + 5 \times (-10)^2 + 4 \times (-10) + 8 = -532$. The usual arithmetic operations can be extended to these bases, and the exploration of these bases would make a good project for a student. Numbers in the base $-10$ were first considered by Vittorio Grunwald in an obscure journal in 1885; see Knuth (1973, Ch. 4.1) for further details.

The general number $a_m a_{m-1} \ldots a_2 a_1 a_0$ in the base $-n$ represents $a_m \times (-n)^m + a_{m-1} \times (-n)^{m-1} + \cdots + a_2 \times (-n)^2 + a_1 \times (-n) + a_0$. It will represent a positive number if $m$ is even and a negative number if $m$ is odd. Examples in negative binary (base $-2$) and negative decimal (base $-10$) follow:

| Decimal | Negative Binary | Negative Decimal |
|---|---|---|
| −11 | 110101 | 29 |
| −10 | 1010 | 10 |
| −2 | 10 | 18 |
| −1 | 11 | 19 |

To convert a decimal to a negative base $-n$, repeatedly divide by the negative base, making sure that the remainder is a nonnegative number less than $n$. To convert $-237$ to negative decimal, for example, we perform the following divisions:

$$-237 = 24(-10) + 3$$

$$24 = (-2)(-10) + 4$$

$$-2 = 1(-10) + 8$$

$$1 = 0(-10) + 1$$

| Decimal | Negative Binary | Negative Decimal |
|---|---|---|
| 0 | 0 | 0 |
| 1 | 1 | 1 |
| 2 | 110 | 2 |
| 10 | 11110 | 190 |

Hence, the decimal $-237$ can be written as 1843 in negative decimal.

Arithmetic in these bases has some unusual features. In negative decimal, for example, 10 is represented by 190. Therefore, in negative decimal whenever 10 has to be carried, 19 has to be carried to the next two columns. Sometimes these carries accumulate into an infinite series of carry digits; this will always happen if a number is added to its negative. The following example shows the addition of the negative decimal numbers 25 and 58 to obtain the negative decimal 63; that is, $(-15) + (-42) = -57$ in decimal. The carry digits are written at the bottom.

```
            2   5
    +       5   8
    --------------------
    · · ·  0   0   0   6   3
                1   9
            1   9
    · · ·
```

In class, students should construct tables of positive and negative numbers in negative binary and negative decimal. They should then explore the arithmetic operations in these bases; see Nelson (1967) and Gilbert and Green (1979). They can then try to write fractional numbers in these bases.

*See also* Bases, Complex; Binary (Dyadic) System; Decimal (Hindu-Arabic) System; Hexadecimal System

## SELECTED REFERENCES

Gilbert, William J., and R. James Green. "Negative Based Number Systems." *Mathematics Magazine* 52(1979):240–244.

Knuth, Donald E. *The Art of Computer Programming*. Vol. 2. *Seminumerical Algorithms*. Reading, MA: Addison-Wesley, 1973.

Nelson, Allyn H. "Investigation to Discovery with a Negative Base." *Mathematics Teacher* 60(1967):723–726.

WILLIAM J. GILBERT

# BASIC SKILLS

Computational and other essential mathematical skills. Teaching basic mathematical skills may be one of the most misunderstood areas in public education. Basic mathematical skills, as well as concepts, are necessary elements in the instructional program, but difficulties have arisen from the lack of a common definition and an understanding of how students learn them, but not from the need to teach them.

## HISTORICAL BACKGROUND

The definition of "basic skills" in mathematics has changed as societal needs have changed. During the colonial period of American history, mathematics was limited to skills needed for clerks and bookkeepers. Addition, subtraction, multiplication, division, and limited use of fractions comprised the targeted skills. The expansion of business and commerce at the end of the nineteenth century required additional mathematical skills, such as percentage, ratio, and proportion. During the early 1900s, the social utility of mathematics still defined what the basic skills were and teaching concentrated on the actual mathematics needed in various occupations. During the 1950s and 1960s, a concern for national security created an increased emphasis on "curriculum development and research in mathematics" (Reys, Suydam, and Lindquist 1995, 4). Programs such as the "new math" emerged. However, the "Back to Basics" movement of the 1970s renewed the emphasis on computational skills.

The Back to Basics movement and state mandates for minimum performance testing during the 1980s required students to reach minimum levels of mastery of basic skills. In a number of states, students were tested in grades 3, 6, and 8, with teachers being held accountable for the success (or failure) of each class of students. When students in the eighth grade failed the Minimum Performance Test (MPT), they were given remediation and then retested to obtain a passing grade before being allowed to enter the ninth grade. As a result, teachers spent time teaching, reteaching, and reviewing computational skills using isolated drill activities that had little impact on the development of relationships among information, skills, and the real world (Eby 1993). Teachers often neglected higher-order thinking skills and problem solving in order to ensure successful student performance on the MPT.

In 1989, with the publication of standards by the National Council of Teachers of Mathematics (NCTM), numerous states began to revise their mathematics curriculum with a focus on mathematics as problem solving, mathematical communication, mathematical reasoning, and mathematical connections. Through the development of curriculum frameworks and a slowly evolving change in assessment, teachers in local districts were now encouraged to implement the NCTM standards in their classrooms. However, teachers recognized that students still needed to master computation skills to be successful "in school and in life" (Miller and Heward 1992, 98). Therefore, drill was still an integral part of mathematics classrooms, with "more than 70% of the time spent in independent practice, mostly with workbooks and paper-and-pencil tasks" (Kennedy and Tipps 1994, 15).

The phrase "basic skills" in this entry refers not only to computational skills but also to the additional essential skills described by the National Council of Supervisors of Mathematics (1989): problem solving, communicating mathematical ideas, mathematical reasoning, applying mathematics to everyday situations, alertness to reasonableness of results, estimation, algebraic thinking, measurement, geometry, statistics, and probability.

## DRILL AND PRACTICE

"Drill and practice" is a commonly used, often misunderstood phrase that usually refers to repetitive practice. In effective mathematics instruction, it refers to a different purpose for the activities.

Drill is a feature of the more traditional theories of learning based on Edward Thorndike's stimulus-response theory. However, Brownell, Piaget, Bruner, and Gagné discounted this theory and suggested more stress on understanding mathematical concepts (Kennedy and Tipps 1994). To develop understanding of basic skills, they said, meaningful tasks are needed. Practice with mathematics that grows out of clearly developed concepts is essential but need not be limited to workbook and text-based exercises. The use of computers, calculators, games, manipulatives, and problem solving can also develop accuracy. However, practice should only be utilized after understanding of the concept is developed.

Drill is utilized to enforce "accuracy and speed (with an emphasis on speed) . . . [including activities such as] flash cards, computerized games, races, timed tests" (Usnick 1991, 344). In other words, "drill is the repetitious activity used to make certain responses (e.g., finger placement) automatic" (344). The students simply become more proficient in using the techniques they employ to answer each fact, such as counting fingers, dots on the page, etc. Practice, however, involves developing meaningful understanding and accuracy of responses. Drill places more emphasis on speed, while practice places more emphasis on accuracy. The main difference between a drill activity and a practice activity may well be whether the activity is completed in an allowed time (five minutes) or students are simply given the assignment to complete. Drill should be limited to a few facts repeated at random on a page, while different, unrelated facts of an operation completed without a time limit would be more effective as practice (Usnick 1991). Furthermore, mathematics instruction that emphasizes timed tests and endless pencil-and-paper activities is often boring and frustrating to children. The result is a dislike of mathematics that may have long-lasting negative effects on the students (Kennedy and Tipps 1994).

## PEDAGOGICAL APPROACHES

Understanding of the mathematical concept is essential before progressing to practice activities. Instruction in any mathematical concept should progress through three distinct stages: (1) concrete, hands-on, or enactive; (2) representational, pictorial, graphic, or iconic; and finally (3) words or symbols and abstract information (Midkiff and Thomasson 1993; Kennedy and Tipps 1994). Thus, teachers should utilize manipulatives to provide concrete references for students as they model and explain the concept being taught. Students may process the information using all three modalities of learning. They can feel the manipulatives as they use them (kinesthetic), see what they are doing with the manipulatives and what the teacher models (visual), and hear the explanations of the material (auditory). This procedure will enable more students to grasp a concept before proceeding to higher levels of instruction. If abstract symbols are introduced before the concept has been developed with concrete experiences, these symbols can impede understanding instead of enhance learning. Although they often do, teachers should not usually introduce any new concept or skill simply by writing the problem (abstract symbols) on the board and demonstrating the solution (Carson and Bostick 1988; Midkiff and Thomasson 1993).

Furthermore, "informal exploratory activities, teacher-directed lessons, and mathematical investigations" (Kennedy and Tipps 1994, 92) should all be included in a well-balanced mathematics program when teaching the basic skills. While varying techniques accommodate many learning styles, abilities, and interests (Carson and Bostick 1988; Midkiff and Thomasson 1993; Kennedy and Tipps 1994), they also serve several distinct purposes. Free exploration often sets the stage for formal instruction to follow by developing essential background knowledge for the students. Structured experiences are provided through teacher-directed lessons for learning basic skills. Mathematical investigations or problem-solving approaches incorporate higher-order thinking with other basic skills in a single lesson. For example, instead of completing a worksheet on the values of various coins, students may list as many different ways as they can to make 25 cents. Resources available for solving problems include play money, cooperative learning groups, and calculators. Besides identifying the number of possible combinations, justifying responses is also encouraged.

Guided practice is an essential part of any basic skills lesson in mathematics. After free exploration and teacher modeling of the skill, the students should model or repeat the steps in the process. Integrating student modeling with direct instruction increases the amount of feedback received by the students. If the students lack understanding, the teacher simply models the procedure again, using a different explanation or different materials before having the students repeat the procedure. This is the best stage for remediation to occur instead of waiting until the homework assignment for the next day has been completed. Asking the students to verbalize the information presented in the lesson will allow the teacher to determine which students need additional assistance as well as provide closure for the lesson (Kennedy and Tipps 1994).

## MATERIALS FOR INSTRUCTION

A wide range of materials similar to the tools available to adults to complete mathematical tasks in their daily lives should be available for learning mathematics. Adults often use concrete objects to assist in analyzing problems, calculators to verify responses or complete large computations, and computers to manipulate large amounts of data. Students may be given personal, concrete experiences that fit their learning styles to provide a solid foundation for learning basic skills. The use of manipulatives to provide concrete experiences for students is necessary whatever the presented concept or skill. In fact, mathematical skills and concepts should always be introduced through concrete experiences, whether students are visual, auditory, or kinesthetic learners, kindergarten students, sixth graders, or high school students.

Good manipulatives are any concrete objects that model a mathematical concept or skill and that students can easily move and rearrange. These materials can be used to provide concrete ways for students to bring meaning to abstract mathematical ideas, learn new concepts, relate new concepts to previously learned ideas, devise problem-solving strategies, and master basic skills. Teachers should model the appropriate behavior, discuss and explain the concept or skill, and record the abstract version for the students. Additionally, students should be allowed to move gradually from using the concrete materials to the representational stage (pictures) before reaching the abstract level (working only with pencil and paper) of basic skills (Midkiff and Thomasson 1993).

Besides a wide variety of manipulatives, technology is also an excellent resource when teaching basic skills. Although calculators have been available for everyday use for over twenty years, their acceptance in learning mathematics has been slow. Many teachers and parents feel that if students are allowed to use calculators when learning mathematics, they will become dependent on the calculators and not learn basic arithmetic. However, research studies have documented that having the calculator as an integral part of instruction and learning improves performance of arithmetic and problem solving, as well as the attitudes of the students regarding mathematics (Hembree and Dessart 1992; Kennedy and Tipps 1994; Reys, Suydam, and Lindquist 1995; Holmes 1995).

Effective utilization of calculators begins in the primary grades with activities to practice counting and extends through secondary mathematics with graphing calculators. One necessary element of instruction includes teaching students when it is appropriate to use the calculator. For example, students could be divided into two groups; one group must use the calculator and one group cannot use the calculator but can use mental computations or pencil and paper. The teacher keeps score of the first group to answer each problem read to them. If the teacher structures the activity appropriately, the students soon see that not using the calculator is sometimes the best way to solve a problem. It is also often inappropriate to check class work or homework with a calculator.

Calculators should be only one of many tools used in teaching basic skills. They are not meant to replace manipulatives, but should be integrated into the mathematics curriculum to provide students with meaningful activities for studying numbers. Computers should also be available as aids to teaching and learning basic skills. Numerous types of computer-assisted instruction (CAI), ranging from tutorials to simulations, are available. Some of these programs introduce the material and allow students to practice a skill (tutorial); others provide repetitive exercises (drill and practice), re-create a real-world environment (simulation), or use motivational strategies with an element of competition for practice (instructional games). In selecting software, teachers should look for programs that provide motivational practice activities that keep the students actively involved for moderate amounts of time. Drill and practice activities that contain large amounts of repetition in a similar format may become boring and should be avoided. Teachers should also avoid programs that make a sound when students input the incorrect answer. Although the student needs to know when an incorrect response is given, the rest of the students in the class should not be alerted each time a class member makes an error.

Several software programs that pattern effective mathematics instruction (concrete, representational, and abstract stages) are available. One such program, *Hands-On Math* (Ventura Educational Systems 1993), provides graphics of a variety of manipulatives for integrating concrete and representational stages of instruction very effectively with abstract information. The program is available in three volumes and covers a wide range of activities for use in the elementary and middle school mathematics classroom.

## COMMITTING BASIC SKILLS TO MEMORY

Whether students are learning simple or more complicated basic skills, appropriate strategies for learning the material should be taught. As students continue to practice the strategies, they will commit many facts, procedures, and skills to memory. However, the amount of practice required will vary from a limited number of practice sessions to numerous activities required for remembering the information.

As teachers select activities for practice, attention should be given to the amount of time actually spent with basic skills. For example, when worksheets that require students to answer problems and then color different sections are used, students spend more time coloring than working the problems. A wide variety of games that students can play in pairs or small groups is available in various publications of the National Council of Teachers of Mathematics as well as other publishers. When students are engaged in playing games that involve practice of basic skills, they are actively involved in mathematics in a motivational way that can improve attitudes about mathematics as well as increase academic achievement in mathematics.

Principles that should be followed when teaching basic skills include: (1) review previous knowledge, (2) provide a few facts or problems at a time, (3) practice often for short periods, (4) use a variety of materials and activities that are developmentally appropriate for the students, (5) do not give speed drills each day, (6) recognize the success and progress of all students (Kennedy and Tipps 1994).

*See also* Bruner; Errors, Arithmetic; Estimation; Number Sense and Numeration

SELECTED REFERENCES

Carson, Joan, and Ruby Bostick. *Math Instruction Using Media and Modality Strengths.* Springfield, IL: Thomas, 1988.

Eby, Judy W. *Reflective Planning, Teaching and Evaluation for the Elementary School.* New York: Merrill, 1993.

Hembree, Ray, and Donald J. Dessart. "Research on Calculators in Mathematics Education." In *Calculators in School Mathematics.* 1992 Yearbook (pp. 23–32). James T. Fey, ed. Reston, VA: National Council of Teachers of Mathematics, 1992.

Holmes, Emma E. *New Directions in Elementary School Mathematics: Interactive Teaching and Learning.* Englewood Cliffs, NJ: Merrill, 1995.

Kennedy, Leonard M., and Steve Tipps. *Guiding Children's Learning of Mathematics.* 7th ed. Belmont, CA: Wadsworth, 1994.

Midkiff, Ruby, and Rebecca Thomasson. *A Practical Approach to Using Learning Styles in Math Instruction.* Springfield, IL: Thomas, 1993.

Miller, April D., and William L. Heward. "Do Your Students Really Know Their Math Facts?" *Intervention in School and Clinic* 28(Nov. 1992):98–104.

National Council of Supervisors of Mathematics. "Essential Mathematics for the 21st Century: The Position of the National Council of Supervisors of Mathematics." *Arithmetic Teacher* 37(Sept. 1989):44–46.

———. "Position Paper on Basic Skills." *Arithmetic Teacher* 25(Oct. 1977):19–22.

National Council of Teachers of Mathematics. *Curriculum and Evaluation Standards for School Mathematics.* Reston, VA: The Council, 1989.

———. *Principles and Standards for School Mathematics.* Reston, VA: The Council, 2000.

Reys, Robert E., Marilyn N. Suydam, and Mary Montgomery Lindquist. *Helping Children Learn Mathematics.* 4th ed. Boston, MA: Allyn and Bacon, 1995.

Usnick, Virginia E. "It's Not Drill AND Practice, It's Drill OR Practice." *School Science and Mathematics* 91(Dec. 1991):344–346.

Ventura Educational Systems. *Hands-On Math.* Grover Beach, CA: Ventura, 1993.

Willoughby, Stephen S. *Mathematics Education for a Changing World.* Alexandria, VA: Association for Supervision and Curriculum Development, 1990.

RUBY BOSTICK MIDKIFF
RONALD W. TOWERY

# BEGLE, EDWARD GRIFFITH (1914–1978)

Mathematician and mathematics educator at Yale (1942–1961) and Stanford (1961–1978) universities. He obtained his Ph.D. degree in topology from Princeton University. His interest in mathematics education began in 1954 with the publication of his innovative freshman calculus text. In 1958, he became director of the newly formed School Mathematics Study Group (SMSG). (He went into K–12 mathematics education because his child had difficulties with percentages.) Until he disbanded it in 1972, Begle was the driving force of that project. He directed the work of up to 100 writers, who were funded by National Science Foundation grants of $10 million. Begle molded SMSG into a national force for reform. He became internationally known and was characterized as "the foremost proponent of 'new math'" by the *New York Times* (1978). He initiated SMSG's five-year National Longitudinal Study of Mathematical Abilities in 1962. After 1972, Begle worked on research in mathematics instruction and materials at the school level (K–12).

*See also* School Mathematics Study Group (SMSG)

SELECTED REFERENCES

Begle, Edward G. *Introductory Calculus with Analytic Geometry.* New York: Holt, 1954.

———. "SMSG: The First Decade." *Mathematics Teacher* 61(Mar. 1968):239–245.

Jones, Phillip S., ed. *A History of Mathematics Education in the United States and Canada.* 32nd Yearbook. Washington, DC: National Council of Teachers of Mathematics, 1970.

*New York Times* (Mar. 3, 1978):B-2.

Zelinka, Martha. "Edward Griffith Begle." *American Mathematical Monthly* 85(Oct. 1978):629–631.

JAMES K. BIDWELL
ROBERT G. CLASON

# BERNOULLI FAMILY

Protestant refugees from Belgium who settled in Basel, Switzerland in 1622. Over the next six generations, more than a dozen members of the family achieved eminence in mathematics and physics. The most celebrated were the brothers Jakob (Jacques I, 1654–1705) and Johann (Jean I, 1667–1748). Johann's three sons, Nicolaus III, Daniel I, and Johann II, also made significant contributions to mathematics.

Jakob Bernoulli studied mathematics under Gottfried Leibniz and took the chair of mathematics at Basel University in 1687, where he taught until his death. By 1690, he was publishing papers on calculus and its applications, particularly to problems involving plane curves. Such was his fascination with curves that he had an equiangular spiral engraved on his tombstone, together with a Latin inscription, *Eadem mutato resurgo,* which translates as "I shall arise the same, though changed."

Jakob pioneered the use of polar coordinates and was the first mathematician to use the word *integral* in its calculus sense. He solved many of the popular problems of his era and, in common with other family members, published his results in the influential *Acta Eruditorum.* Jakob's major work on probability, *Ars Conjectandi,* was published posthumously in 1713. This work contains a proof (by induction) of the binomial theorem for positive integral powers; the Bernoulli numbers, which arise in the series expansions of trigonometric and hyperbolic functions; and Bernoulli's law of large numbers. The latter states that if an event occurs with probability $p$ and an infinity of trials are carried out, the proportion of successes is sure to be $p$.

Johann Bernoulli also studied under Leibniz and became a successful teacher. He published two books on the differential and integral calculus in 1691 and 1692 and succeeded his brother Jakob as professor at Basel in 1705. Johann is credited with inventing the calculus of variations from his work on the problem of the brachystochrone, the curve of quickest descent of a small body moving between two points in a gravitational field, more commonly known as a cycloid. Much of the material on differential and integral calculus and ordinary differential equations to be found in elementary texts today results from the collaborative work of Johann and Jakob Bernoulli and their fruitful correspondence with Leibniz.

Johann's three sons, Nicolaus III (1695–1726), Daniel I (1700–1782), and Johann II (1710–1790), all held chairs at St. Petersburg or Basel. The eldest, Nicolaus, died in a drowning accident only eight months after arriving at the St. Petersburg Academy. Together with his younger brother Daniel, he formulated the famous Petersburg paradox, which was widely debated throughout the eighteenth century. The paradox states that a person receiving a crown if a head is achieved on the first toss of a coin, two crowns if a head first appears on the second toss, four crowns if on the third, and so on (so that $2^{n-1}$ crowns are received if a head first appears on the $n$th toss) can expect to gain an infinite return. Intuition and experience suggest a more modest return.

Nicolaus was succeeded at St. Petersburg by Daniel, who subsequently returned to Basel in 1733. Daniel contributed to astronomy, mathematics, and physics. Together with Jean Le Rond D'Alembert, he pioneered the study of partial differential equations. He is best known for his work on hydrodynamics, *Hydrodynamica,* published in 1738. This treatise contained the Bernoulli principle of hydraulic pressure and laid the foundations for the development of the kinetic theory of gases. In the *Hydrodynamica,* Daniel explained the phenomenon of pressure by imagining a gas to consist of "very minute corpuscles," "practically infinite in number," "driven hither and thither with a very rapid motion." In recognition of his pioneering contributions, Daniel Bernoulli has been dubbed the father of mathematical physics.

The youngest son, Johann II, spent his later years as professor of mathematics at Basel. His main interest was in the mathematical theory of heat and light. The Bernoulli family holds a unique place in the history of mathematics: not only were so many of the family celebrated mathematicians, but they were also enthusiastic advocates who spread knowledge of calculus and its applications throughout Europe.

*See also* History of Mathematics, Overview

## SELECTED REFERENCES

Boyer, Carl B. *A History of Mathematics.* 2nd ed. Revised by Uta C. Merzbach. New York: Wiley, 1989.

Eves, Howard W. *An Introduction to the History of Mathematics.* 5th ed. Philadelphia, PA: Saunders College Publishing, 1983.

Newman, James R. *The World of Mathematics.* 4 vols. London, England: George Allen and Unwin, 1960.

Struik, Dirk J. *A Concise History of Mathematics.* New York: Dover, 1948.

ROD HAGGARTY

| Divide | Quotient | Remainder |
|--------|----------|-----------|
| 23/2 | 11 | 1 |
| 11/2 | 5 | 1 |
| 5/2 | 2 | 1 |
| 2/2 | 1 | 0 |
| 1/2 | 0 | 1 |

# BINARY (DYADIC) SYSTEM

A method for writing numbers in the base 2, using only 0 and 1 as digits. For example, 1101 represents $1 \times 2^3 + 1 \times 2^2 + 0 \times 2 + 1$, which is 13 in decimal notation. If different bases are being used, the base is often added as a subscript, so this binary number would be written as $1101_2$. The general binary number $r_m \cdots r_2 r_1 r_0$ represents the sum $r_m \times 2^m + \ldots + r_2 \times 2^2 + r_1 \times 2^1 + r_0 \times 2^0$. Every positive integer can be written in the binary system. For example, $1 = 1_2, 2 = 10_2, 3 = 11_2, 4 = 100_2, 5 = 101_2, 10 = 1010_2$, and $13 = 1101_2$.

Some ancient systems of measurements used a rudimentary form of the binary system; a remnant of these systems still in use today is the liquid measure 2 pints = 1 quart. In 1703, Gottfried Leibniz, the co-inventor of the calculus, was the first to make systematic use of the binary system. Leibniz said that binary arithmetic was not intended for practical calculations, but only for illustrating patterns in number systems. However, Leibniz did outline a design for a mechanical calculating machine using binary arithmetic, although it was never built. See Asimov (1977, Ch. 2), Knuth (1973, Ch. 4.1), and Resnikoff and Wells (1984, Ch. 1) for more about the history of binary and other positional number systems.

It was only with the advent of digital electronic computers in the 1950s that binary arithmetic became very practical. The binary system is used to represent numbers internally in a computer, as the digits 1 and 0 can correspond to current flowing or not flowing. Circuits can easily be built that will add and multiply numbers in binary. In computer science, the digits 0 and 1 are usually called bits. A byte often refers to eight bits; thus bytes are the binary numbers from 00000000 to 11111111, or from 0 to 255 in decimal. An alphanumeric character in a computer is usually encoded as one byte, so that 1 kilobyte will hold about a thousand characters, and 1 megabyte will hold about a million characters.

A positive integer is converted from decimal notation to binary notation by repeatedly dividing the number by 2, saving the remainders. For example, the following table converts 23 to binary. The binary representation is the sequence of remainders, written in the reverse order, namely $10111_2$.

Arithmetic can be performed in the binary system just as it is done in the decimal system. In fact, it is easier, since the addition and multiplication tables are much simpler (see Figure 1). The only disadvantage is that a number requires more digits when written in binary, as opposed to decimal. For example, consider the addition and multiplication of $110_2$ and $11_2$, that is, $6 + 3$ and $6 \times 3$. (See Figure 2.)

| + | 0 | 1 |
|---|---|---|
| 0 | 0 | 1 |
| 1 | 1 | 10 |

Addition Table

| × | 0 | 1 |
|---|---|---|
| 0 | 0 | 0 |
| 1 | 0 | 1 |

Multiplication Table

**Figure 1** *Tables showing addition and multiplication.*

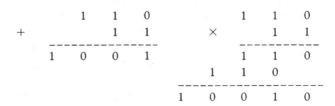

**Figure 2** *Illustration of binary addition and multiplication.*

Just as fractional numbers can be written in the decimal system using a decimal point, fractional numbers can be represented in binary using a binary (or radix) point, such as $.001_2 = 2^{-3} = \frac{1}{8}$. For example, converting from binary to decimal notation, $1.011_2 = 1 + 0 \times 2^{-1} + 1 \times 2^{-2} + 1 \times 2^{-3} = 1 + \frac{1}{4} + \frac{1}{8} = \frac{11}{8} = 1.375$.

A decimal number can be converted into binary by first converting the integer part, as shown above, and then converting the fractional part. The fractional part is converted by repeatedly multiplying by 2 and separating the integer and fractional parts of the result. The binary expansion will be the list of digits in the integer part, in the order generated. For example, the binary expansion of 0.69 to four binary places is found as follows:

| Multiply | Integer Part | Fractional Part |
| --- | --- | --- |
| $.69 \times 2$ | 1 | .38 |
| $.38 \times 2$ | 0 | .76 |
| $.76 \times 2$ | 1 | .52 |
| $.52 \times 2$ | 1 | .04 |

Hence 0.69 is $0.1011_2$, correct to four binary places. Its exact value will be an infinite binary expansion.

In class, students could construct tables of decimal and binary numbers, and they should compare, for example, the relationship between the representations of 0 to 15 with those from 16 to 31. They can construct punched cards for the first thirty-one numbers, with each hole corresponding to a digit 1. Students should also try to develop the rules for addition, subtraction, and multiplication in binary. Note that some calculators have a binary mode.

*See also* Bases, Complex; Bases, Negative

### SELECTED REFERENCES

Asimov, Isaac. *Asimov on Numbers.* Garden City, NY: Doubleday, 1977.

Knuth, Donald E. *The Art of Computer Programming.* Vol. 2. *Seminumerical Algorithms.* Reading, MA: Addison-Wesley, 1973.

Resnikoff, Howard L., and Raymond O. Wells. *Mathematics in Civilization.* New York: Dover, 1984.

WILLIAM J. GILBERT

## BINOMIAL DISTRIBUTION

The discrete probability distribution of a sum of $n$ independent Bernoulli random variables (r.v.s), where $n$ is a positive integer greater than 1. The Bernoulli r.v. $X$, named after Jakob Bernoulli (1654–1705) of the famous Swiss family of mathematicians, takes one of the two values 0 and 1, with the following probabilities

$$P\{X = 1\} = p, 0 < p < 1$$
$$P\{X = 0\} = 1 - p = q$$

(see Johnson and Kotz 1982). It may be thought of as representing success ($X = 1$) or failure ($X = 0$) in a single trial of a game, examination, or contest.

The binomial r.v.

$$Y_n = X_1 + \cdots + X_n$$

is the sum of a fixed number $n$ of independent Bernoulli random variables. It records the number of successes in $n$ independent trials $X_i, i = 1, \ldots, n$. The probability of $y$ successes in $n$ trials, where $y = 0, 1, \ldots, n$, is given by

$$P(Y_n = y) = \binom{n}{y} p^y q^{n-y}$$

These probabilities are, in effect, the terms in the binomial expansion

$$(p + q)^n = \sum_{y=0}^{n} \binom{n}{y} p^y q^{n-y}$$

whence the name of the distribution. The expectation and variance of $Y_n$ are, respectively,

$$E(Y_n) = np, \qquad \text{Var}(Y_n) = npq$$

Published tables of $P\{Y_n = y\}$ are available (Johnson and Kotz 1969) for various values of $n$ and $p$; for large $n$, this probability can be approximated using the standard normal distribution. For large $n$ and very small $p$, the Poisson distribution is a good approximation. A simple example may help to illustrate the use of this distribution. Suppose an archer hits the target with probability $p = 0.7$ in a single trial; he shoots $n = 5$ arrows, each trial being independent of the others. The probability that $y = 3$ arrows hit the target is then

$$P\{Y_5 = 3\} = \binom{5}{3}(0.7)^3 (0.3)^2$$
$$= 10(0.343)(0.09) = 0.3087$$

The expectation and variance of $Y_5$ are $5(0.7) = 3.5$ and $5(0.7)(0.3) = 1.05$, respectively.

*See also* Mathematical Expectation; Normal Distribution Curve; Random Variable; Variance and Covariance

### SELECTED REFERENCES

Johnson, Norman L., and Samuel Kotz, eds. *Discrete Distributions.* New York: Wiley, 1969.

———. *Encyclopedia of Statistical Sciences.* Vol. 1. New York: Wiley, 1982.

JOSEPH GANI

## BINOMIAL THEOREM

The theorem that gives the general form of the expansion of a binomial expression to a power. The rules for expansion of a binomial to several positive integral powers had been known as far back as A.D.

1300—for example, to Chinese mathematicians (Boyer 1991, 205).

$$(a + b)^1 = a + b$$
$$(a + b)^2 = a^2 + 2ab + b^2$$
$$(a + b)^3 = a^3 + 3a^2b + 3ab^2 + b^3$$
$$(a + b)^4 = a^4 + 4a^3b + 6a^2b^2 + 4ab^3 + b^4$$
$$(a + b)^5 = a^5 + 5a^4b + 10a^3b^2 + 10a^2b^3 + 5ab^4 + b^5$$

However, the general form of the expansion was first deduced by Isaac Newton (1642–1727) around 1665 (Boyer 1991, 393). The result he obtained is known as the Binomial Theorem, his first mathematical accomplishment (Rickey 1987). Newton did not give a proof of the theorem; this was done slightly later by Jakob Bernoulli (1654–1705).

When $n$ is a positive integer, the theorem reads (Dence and Dence 1994, 215)

$$(a + b)^n = \sum_{k=0}^{n} \binom{n}{k} a^{n-k} b^k$$

where the binomial coefficients $\binom{n}{k}$ are defined as

$$\binom{n}{k} = \frac{n!}{k!\,(n-k)!}, \; n! = n(n-1)(n-2) \ldots (2)(1)$$

and $0! = 1$. There are always $n + 1$ terms in the expansion of $(a + b)^n$, since $k$ runs from 0 to $n$. For example, if the theorem is used to work out the next expansion in the sequence shown at the beginning of this entry, the following seven-term expression is obtained:

$$(a + b)^6 = a^6 + 6a^5b + 15a^4b^2 + 20a^3b^3 + 15a^2b^4 + 6ab^5 + b^6$$

Newton also investigated (but did not prove) the expansion of a binomial when $n$ is a nonintegral rational number (positive or negative). In these cases, an infinite series is obtained (Dence and Dence 1994, 306). The expansion commonly given is that of $(1 + x)^n$, and in order for the infinite series to converge (have a finite sum), the variable $x$ must be restricted to the open interval $(-1, 1)$:

$$(1 + x)^n = 1 + \sum_{k=0}^{\infty} \binom{n}{k} x^k, \; |x| < 1$$

where the generalized binomial coefficient $\binom{n}{k}$ is defined as

$$\binom{n}{k} = \frac{n(n-1) \cdots (n - k + 1)}{k!}$$

The importance of the binomial theorem cannot be exaggerated; during the 300 years of its existence, it has been a key step in countless formal demonstrations. For example, Newton himself used it during his creation of the calculus to deduce the derivative of $f(x) = x^n$. Recently, it was employed to derive formulas for the sums of the powers of consecutive integers (Kelly 1984). The theorem is also useful computationally. For example, $\sqrt{3}$ can be estimated using only a few terms from the following infinite series:

$$\sqrt{3} = \left( \frac{676}{225} - \frac{1}{225} \right)^{\frac{1}{2}} = \frac{26}{15} \left( 1 - \frac{1}{676} \right)^{\frac{1}{2}}$$
$$= \frac{26}{15} \left\{ 1 - \frac{1}{1352} - \frac{1}{3655808} - \cdots - \right\}$$
$$\approx 1.732050809$$

The tabulated value is approximately 1.732050808.

The binomial coefficients in the binomial theorem are fascinating in their own right. They are given by the horizontal entries in Pascal's triangle (Krause 1991, 150), an oblique reading of which also yields the very interesting Fibonacci numbers, $\{F_n\}$ (Vernadore 1991).

A host of interrelationships connect the binomial coefficients (Leonard 1973), a reflection of the fact that the coefficients have a combinatorial interpretation (Elkin 1968). Thus, the binomial coefficients appear in the computation of probabilities of events in situations where there are two or more equally likely outcomes (Krause 1991, 499). More concrete uses include interesting applications to counting in geometry (Schielack 1991, 137) and to computing the number of ways of winning a playoff series in a sport (Litwiller and Duncan 1992).

*See also* Fibonacci Sequence; Pascal's Triangle

## SELECTED REFERENCES

Anderson, Ian. "Sums of Squares and Binomial Coefficients." *Mathematical Gazette* 65(1981):87–92.

Boyer, Carl B. *A History of Mathematics.* 2nd ed., rev. Revised by Uta C. Merzbach. New York: Wiley, 1991, pp. 205–206, 393–395.

Coolidge, Julian L. "The Story of the Binomial Theorem." *American Mathematical Monthly* 56(1949):147–157.

Dence, Joseph B., and Thomas P. Dence. *A First Course of Collegiate Mathematics.* Malabar, FL: Krieger, 1994, pp. 215–218, 306–308.

Elkin, Jack M. "Binomial Coefficient Formulas by General Reasoning." *Mathematics Teacher* 61(1968): 399–402.

Kelly, Clive. "An Algorithm for Sums of Integer Powers." *Mathematics Magazine* 57(1984):296–297.

Krause, Eugene F. *Mathematics for Elementary Teachers.* 2nd ed. Lexington, MA: Heath, 1991, pp. 150–151, 499–501.

Leonard, Courtney A. "Those Intriguing Binomial Coefficients Again." *Mathematics Teacher* 66(1973):665–666.

Litwiller, Bonnie H., and David R. Duncan. "Combinatorics Connections: Play-off Series and Pascal's Triangle." *Mathematics Teacher* 85(1992):532–535.

Rickey, V. Frederick. "Isaac Newton: Man, Myth, and Mathematics." *College Mathematics Journal* 18(1987): 362–389.

Schielack, Vincent P., Jr. "Combinatorics and Geometry." In *Discrete Mathematics Across the Curriculum, K–12* (pp. 137–142). 1991 Yearbook. Margaret J. Kenney, ed. Reston, VA: National Council of Teachers of Mathematics, 1991.

Vernadore, James. "Pascal's Triangle and Fibonacci Numbers." *Mathematics Teacher* 84(1991):314–316.

JOSEPH B. DENCE

# BIOMATHEMATICS

Also known as mathematical biology. Biomathematics is an ideal topic for the classroom, because it is a subject sufficiently broad to be accessible to everyone at some level. The mathematical modeling of a phenomenon, often from daily life, leads to insight into the phenomenon as well as to understanding of the mathematics itself. There are biological applications to be found in nearly all areas of mathematics, so students can practice applying their mathematics to real biological problems at all levels. Biomathematics employs algebra, geometry, probability, statistics, calculus, differential equations, recurrence relations, game theory, oscillations, dynamical systems, optimization, chaos, fractals, dimensional analysis (units of measurement), and even topology.

The first biomathematician was probably Galileo Galilei, who discovered/invented the biological laws of scaling, which he published in 1638 (Thompson 1961). He observed that the mass of an animal is proportional to the cube of its length, whereas the strength of its legs is proportional to the area of the legs' cross-section, which in turn is proportional to the square of the animal's body length. What this implies is that small animals have much thinner legs, proportionally, than do large animals (compare an insect, a dog, and an elephant) and that there is a limit to how large a land animal can get and still be able to support its own weight (Haldane 1985). (Could there ever be a land animal larger than a dinosaur?) Daniel Bernoulli, in 1760, may have been the first to apply mathematics to epidemiology, in this case to vaccination against smallpox. Another early biomathematician was Thomas Malthus, who in 1806 used very simple arguments to compare the growth rates of the human population and of its food supply, concluding that populations ultimately exceed their food supply. His argument hinged on the observation that human populations increase exponentially, as $P = P_0 e^{(b-d)t}$, where $P$ is the world population, $b$ is the birth rate and $d$ the death rate, yet their food supply only increases linearly, as $F = kt$, where $F$ is the amount of food produced per year and $k$ is the annual production improvement rate. These early examples are accessible to a high school audience.

A classic monograph on mathematical aspects of the form of organisms is D'Arcy Thompson's *On Growth and Form* (1917) (Thompson 1961). Thompson was fascinated by the form of natural objects and studied everything from the honeybee's comb to the spirals of seashells. Also in the early twentieth century, Alfred Lotka and Vito Volterra modeled the cycles of interacting populations, such as predators and prey, which formed the foundation of the huge field of population biology. Since then, mathematics has made important contributions to our understanding of epidemics, ecology, use of renewable resources such as forests, development, biorhythms such as the heartbeat and the circadian rhythm (Glass and Mackey 1988), and a host of other topics. Alan Hodgkin and Andrew Huxley, measuring the voltage response of squid neurons and fitting their data to a differential equation model, were able to predict the existence and shape of ion channels in cell membranes, for which they won the Nobel Prize. James Murray (1988) used more complicated scaling arguments than did Galileo, based on reaction-diffusion theory (nonlinear partial differential equations) to explain why small mammals are all one color, medium ones have patches, large ones have many spots and stripes, and the largest—elephants again—are all one color.

It is this usefulness which is at the core of biomathematics. Construction and analysis of a mathematical model of a biological phenomenon should lead not just to the understanding of the mathematics and of the modeling process, but to a greater understanding of the biology. The framework of mathematics, with its rigorous definitions and attention to detail, can ideally lead to a clearer, more logical understanding of real-world phenomena.

Excellent problems in biomathematics abound. At the simplest level, seventh-graders can get practice interpreting graphs and scaling maps by using a drawing of a whale as a template for drawing a life-size whale on a parking lot (with chalk). In the area of population biology, there is the classic example of bacterial growth (equivalent to the Malthus problem mentioned earlier), which can be used to introduce the concepts of powers of 2, exponential growth, and elementary differential equations. For example, if $N$ is the number of bacteria, and the growth rate is proportional to $N$, then $dN/dt = rN$, and $N = N_0\exp(rt)$. One can then introduce the concept of model limitations, as the bacterial population, as modeled, goes to infinity.

At the college level, the addition of bacterial migration as a diffusion term gives a simple and easily motivated example of a linear partial differential equation. Fibonacci's reproducing rabbits can lead to a discussion of recurrence relations, limits, continued fractions, geometry, and even Pascal's triangle (Huntley 1970). Fibonacci numbers also feature prominently in the study of phyllotaxis (Jean 1994), the arrangement of leaves on a plant that is often spiral. From simple exercises of counting the scales on a pinecone, one can, with very little mathematical background, explore phyllotaxis and be led through such exotic yet accessible mathematical topics as lattice geometry. Even in elementary school, a creative mathematics teacher can easily come up with biological applications of mathematical topics; for instance, if each cell divides in two, how many cells are there after three generations? If your bean seedling grows an inch a day, how long will it take to grow as tall as you are? How fast did you grow to get that tall, in comparison?

Biomathematics need not be confined to the mathematics classroom. There is certainly ample room for mathematical analysis in the biology classroom, and in schools with interdisciplinary programs, biomathematics is ideal for integrating learning across the curriculum. There are many biological examples that are ideal for physics classes as well: How much force/power does a grasshopper have to exert in order to jump one meter? What is the torque on your elbow if you are carrying a bag of groceries? How much force must your bicep exert to keep the bag elevated, and how does it depend on the angle of your bent elbow? Model the walking human leg as a pendulum; what is its natural period? What does this say about mice and giraffes? What is the difference between a mouse falling 10 feet and landing on its back and a human falling 10 feet and landing on its back?

*See also* Applications for the Classroom, Overview; Applications for the Secondary School Classroom

## SELECTED REFERENCES

Glass, Leon, and Michael Mackey. *From Clocks to Chaos: The Rhythms of Life.* Princeton, NJ: Princeton University Press, 1988.

Haldane, John B. S. "On Being the Right Size" and "The Biology of Inequality." In *On Being the Right Size* (pp. 1–8, 113–134). John M. Smith, ed. Oxford, England: Oxford University Press, 1985.

Huntley, Herbert E. *The Divine Proportion: A Study in Mathematical Beauty.* New York: Dover, 1970.

Jean, Roger. *Phyllotaxis: A Systemic Study in Plant Morphogenesis.* Cambridge, England: Cambridge University Press, 1994.

Malthus, Thomas. *An Essay on the Principle of Population; or, A View of Its Past and Present Effects on Human Happiness; with an Inquiry into Our Prospects Respecting the Future Removal or Mitigation of the Evils Which It Occasions.* 3rd ed. London, England: Bensley, 1806.

Murray, James D. "How the Leopard Gets Its Spots." *Scientific American* 258(3)(1988):80–87.

Thompson, D'Arcy Wentworth. *On Growth and Form.* Abridged version by John Tyler Bonner. Cambridge, England: Cambridge University Press, 1961.

SHARON LUBKIN

# BIRKHOFF, GEORGE DAVID (1884–1944)

One of the first Americans to achieve distinction in mathematics. A professor at Harvard University, Birkhoff specialized in divergent series, topology, and axiomatics. In 1933, he devised a system of postulates for geometry that used the real number system and the ruler and protractor. With the aid of Ralph Beatley, a mathematics educator at Harvard, Birkhoff wrote a high school text, *Basic Geometry,* published in 1941. This was one of the first texts to break from traditional Euclidean form and to utilize modern Riemannian concepts of geometry. The first and third of the five postulates used real numbers. The first was the "ruler postulate," that is: "The points on any straight line can be numbered so that number differences measure distances." The third, the "protractor postulate," states: "All half-lines having the same end-point can be numbered so that number differences measure angles" (Birkhoff and Beatley

1941, 40, 47). The School Mathematics Study Group used these postulates in its early geometry material.

## SELECTED REFERENCES

Bidwell, James K., and Robert G. Clason, eds. *Readings in the History of Mathematics Education.* Washington, DC: National Council of Teachers of Mathematics, 1970.

Birkhoff, George D., and Ralph Beatley. *Basic Geometry.* Chicago, IL: Scott, Foresman, 1941.

JAMES K. BIDWELL
ROBERT G. CLASON

# BOARD ON MATHEMATICAL SCIENCES (BMS)

Established by the National Research Council (NRC) in November 1984 to support and promote the quality of the mathematical sciences and their benefits to the nation. BMS carries out its mission by conducting and disseminating studies and technical assessments on mathematical science topics of national interest and representing the mathematical sciences to government, academic institutions, professional societies and communities, industry, and the public.

BMS consists of fifteen members, representing core and applied mathematics, statistics, operations research, and scientific computing. Members are nominated by the BMS and must be approved by the NRC. They serve a term of three years, with service limited to two consecutive terms. There are two standing committees, the Committee on Applied and Theoretical Statistics (CATS) and the U.S. National Commission on Mathematics Instruction. CATS addresses issues affecting research and education in the statistical sciences. Additional ad hoc committees, panels, and working groups are formed as needed to carry out individual projects, with members nominated by the BMS and also subject to NRC approval. The BMS, CATS, and ad hoc groups typically comprise over 100 mathematical scientists, scientists, engineers, and biomedical personnel, including many members of the National Academy of Sciences, the National Academy of Engineering, and the Institute of Medicine (IOM).

Recent BMS activities include: (1) producing the book *Calculating the Secrets of Life: Applications of the Mathematical Sciences in Molecular Biology,* which gives a state-of-the-art view of significant contributions, open problems, and likely future directions of mathematical sciences research and methods in molecular biology; (2) coproducing the report *Mathematical Challenges from Theoretical/Computational Chemistry,* which identifies areas of computational chemistry where the mathematical sciences have contributed and could contribute more; (3) coproducing the report *Mathematics, Physics and Emerging Biomedical Imaging;* (4) hosting in 1995 the tenth annual mathematical sciences department chairs colloquium, Managing While Science and Education Evolve, which focused on strategies for increasing the employment of young mathematicians, evolving graduate education, encouraging underrepresented groups, undergraduate programs and calculus reform, and faculty assessment issues; and (5) helping the mathematical sciences community through a period of major change via the project Actions for the Mathematical Sciences: Adapting to the Changed Environment, which examines what steps mathematics should take to build on existing connections with areas such as materials science (see National Research Council 1993), the biological sciences, and medical science, and identifies and commences beneficial actions the community should take in response to the changed research and education environment.

*See also* Federal Government, Role

## SELECTED REFERENCES

National Research Council. *Calculating the Secrets of Life: Applications of the Mathematical Sciences to Molecular Biology.* Washington, DC: National Academy Press, 1995.

———. *Educating Mathematical Scientists: Doctoral Study and the Postdoctoral Experience in the United States.* Washington, DC: National Academy Press, 1992.

———. *Mathematical Challenges from Theoretical/Computational Chemistry.* Washington, DC: National Academy Press, 1995. (Available on the World Wide Web at http://www.nas.edu.)

———. *Mathematical Research in Materials Science: Opportunities and Perspectives.* Washington, DC: National Academy Press, 1993.

———. *Mathematics, Physics and Emerging Biomedical Imaging.* Washington, DC: National Academy Press, 1995. (Available on the World Wide Web at http://www.nas.edu.)

JOHN R. TUCKER

# BOLYAI, JANOS (1802–1860)

A Hungarian mathematician known for his geometry of "absolute space," an alternative to Euclidean geometry with applications in the Einstein theory of relativity and other developments of modern physics. Euclid's parallel postulate states that in a plane there can be only one parallel to a given line through a point not on that line. After Euclid, mathematicians, doubting that it was necessary to *assume* this postulate, attempted for more than two millenia to *prove* this result as a *theorem* in plane geometry. No one thought to question it as being "intuitively true" until Carl F. Gauss, Janos Bolyai, and Nicholas Lobachevsky independently and simultaneously proposed an alternative plane geometry in which more than one parallel to a line can be drawn through any point not on the line. Bolyai had inherited his interest in the theory of parallels from his father, a friend of Gauss, and first published his theories in an appendix to one of his father's works in 1832. This classic essay of twenty-four pages was the only mathematics he published during his lifetime. Bolyai's geometry of absolute space was validated by Eugenio Beltrami and Felix Klein in the decade after his death and eventually became accepted and useful.

*See also* History of Mathematics, Overview; Non-Euclidean Geometry

### SELECTED REFERENCES

Bolyai, Janos. "The Science of Absolute Space." George B. Halstead, trans. In Robert Bonola, *Non-Euclidean Geometry.* New York: Dover, 1955.

Burton, David. *History of Mathematics, An Introduction.* 3rd ed. Dubuque, IA: Brown, 1995, pp. 532–535.

Eves, Howard W. *An Introduction to the History of Mathematics.* 6th ed. Philadelphia, PA: Saunders College Publishing, 1990, pp. 498–499.

Gillispie, Charles, ed. *Dictionary of Scientific Biography.* Vol. 2. New York: Scribner's, 1981–1990, pp. 259–265.

RICHARD M. DAVITT

# BOOKS, STORIES FOR CHILDREN

Children's trade books can be used to introduce, reinforce, or develop mathematical concepts and generally enrich mathematics learning. Especially in the early primary grades, mathematics and language skills develop together as the students listen, read, write, and talk. The children's books reflect the whole language philosophy of integrating literature throughout the curriculum by serving as a step between reading, using concrete manipulatives, and doing abstract paper-and-pencil activities.

## CATEGORIES OF CHILDREN'S BOOKS

The different children's trade books available as supplements to mathematics instruction may be separated into four broad categories: (1) counting books; (2) number books; (3) miscellaneous story books; and (4) concept or informational books. Counting books, in addition to developing and reinforcing counting and number concepts, can be used to introduce the four basic operations (addition, subtraction, multiplication, division) as well as the concepts of sets, subsets, and fractions. *Sea Squares* (Hulme 1991) is an example of a multipurpose type of counting book. A number book story emphasizes a specific number. This type of book helps the student to understand the meaning of that number. *My Six Book* (Moncure 1986) is an example. The miscellaneous story book may be a fairy tale, folk tale, or other story in which the author touches on a mathematics concept. For instance, *A Million Fish . . . More or Less* (McKissack 1992) can be used to reinforce the concept of large numbers. In contrast, concept or informational books are written purposely to investigate specific mathematical concepts. However, they are also written in a delightful manner to stimulate further exploration of the mathematical concepts. A concept book that can be used to introduce combinatorial analysis, permutations, combinations, and probabilities is *Socrates and the Three Little Pigs* (Anno 1986). Among the many excellent books in this category are the classic *Young Math Books* series and the *Let's Investigate* series (see Smoothey 1993).

In addition to the trade books just mentioned, a wealth of rhymes and poems either introduce a specific number or reinforce counting or an operation. Examples are *Hand Rhymes* (Brown 1985), which includes directions for finger play adaptation, and the poem *Over in the Meadow* (Wadsworth 1985).

## THINGS TO DO WITH THE STORIES

In order to make the literature available to all students, the teacher can create a mathematics book corner where students may go to read the books themselves or listen to audiotaped books. "Big Books" may also be used for group reading. After reading or listening, students can communicate their understanding by making mobiles, dioramas, murals, and posters; decorating the classroom door with work featuring the concepts that were reinforced in the book; creating a bulletin board or a book jacket; making quilt squares that use information from the book; making banners that depict book scenes that relate to mathematics; constructing bookmarks or making scrapbooks.

The students can also express their knowledge verbally, always concentrating on the mathematics concept, by writing newspaper advertisements for the books, newspaper articles, headlines, captions, and telegrams or by creating comic strips. The students may adapt the books into plays or write additional chapters for them. In order to reinforce selected vocabulary, the teacher may develop crossword puzzles and the students can create word problems based on the books. The students may pantomime, do finger plays, use sign language, or have puppet shows. They may write monologues or prepare mock television shows based on the books.

## ADDITIONAL EXAMPLES OF CHILDREN'S BOOKS

In *One, Two, Three* (Wildsmith 1995), a counting book, the author used geometric figures (circles, triangles, rectangles) to construct pictures that illustrate sets corresponding to given numbers. For example, to demonstrate the number 8 he formed a person by using seven rectangles (one for the main body, two for the arms, two for the feet, and two for a hat) and one circle (for the face). The page opposite the person shows the numeral 8 and the word name written in both uppercase and lowercase letters. In addition to introducing the numbers 1 through 10, the book can be used to reinforce shape concepts. *Anno's Magic Seeds* (Anno 1995) includes numerous concepts that can be used as the basis for an interdisciplinary unit to connect mathematics with social studies and science. It also has examples that reinforce operations with fractions, decimals, and percentages.

In the *Count Your Way Through . . .* series, the author counts, 1 through 10, in a given country's language. In addition, each number is associated with a certain historical, cultural, geographic, or economic aspect of that country. For example, when presenting the number 4, *Count Your Way Through Israel* (Haskins 1990) discusses the four questions that Jewish children ask during the Passover Seder, and 9 is associated with the nine main products that Israel exports. The book *Let's Investigate Circles* (Smoothey 1993) includes a section on the etymology of the word *circle* and descriptions of how circles can be used to form other shapes, such as squares, ellipses, cardioids, etc. It also discusses the discoveries of the Greeks and the Egyptians about measuring the area

of a circle.

The connections between children's literature and mathematics are enhanced by selecting books that encourage thinking on a number of different mathematical concepts. Talking about the books creates opportunities to talk about mathematics.

## SELECTED REFERENCES

Anno, Mitsumasa. *Anno's Magic Seeds.* New York: Philomel, 1995.

———. *Socrates and the Three Little Pigs.* New York: Philomel, 1986.

Brown, Marc, ed. *Hand Rhymes.* New York: Dutton, 1985.

Gailey, Stavroula K. "The Mathematics—Children's Literature Connection." *Arithmetic Teacher* 40(Jan. 1993):258–261.

Haskins, James. *Count Your Way Through Israel.* Minneapolis, MN: Carolrhoda, 1990.

Hulme, Joy. *Sea Squares.* New York: Hyperion, 1991.

McKissack, Patricia. *A Million Fish . . . More or Less.* New York: Knopf, 1992.

Moncure, Jane. *My Six Book.* Chicago, IL: Children's, 1986.

Smoothey, Marion. *Let's Investigate Circles.* North Bellmore, NY: Cavendish, 1993.

Thiessen, Diane, Margaret Matthias, and Jacquelin Smith. *The Wonderful World of Mathematics: A Critically Annotated List of Children's Books in Mathematics.* 2nd ed. Reston, VA: National Council of Teachers of Mathematics, 1998.

Wadsworth, Olive A. *Over in the Meadow: A Counting Out Rhyme.* New York: Viking Penguin, 1985.

Wildsmith, Brian. *One, Two, Three.* Brookfield, CT: Millbrook, 1995.

*Young Math Books.* New York: Crowell.

STAVROULA K. GAILEY

# BOOLE, GEORGE (1815–1864)

Developer of the system of mathematical logic named after him, Boolean algebra. Boole's life divides into two distinct parts. Until 1849, he worked largely in and around his home city of Lincoln, England, teaching and running his own school and learning mathematics for himself. Then, on the basis of his published mathematical researches, he took a post as founder professor of mathematics at Queen's College in Cork, Ireland.

Boole's research also divides into two parts, but they are closely linked. He was a leading figure in the strong English and Irish interest of the time: the (known) method of solving differential equations by symbolizing differentiation by the letter $D$ for $d/dx$

and developing an algebra in which $D$ was treated as an algebraic symbol (for example, $D^{-1}$ represented the inverse operation of integration). He also sought basic laws of operations with $D$ and took them over by analogy into his *Mathematical Analysis of Logic,* in which he developed a similar algebra to represent operations upon classes (such as the class of all men). He developed methods of deducing logical consequences from a given set of premises that was far more powerful than the syllogistic logic of his time. He also applied this logic to probability theory. The theory can partly be represented by the (misnamed) Venn diagrams, but Boole himself deliberately eschewed diagrams. The modern Boolean algebra is very important in computer science, but Boole was disinclined to study repetitive processes and took no interest in the calculating machines that Charles Babbage was developing.

*See also* Boolean Algebra

### SELECTED REFERENCE
MacHale, Desmond. *George Boole, His Life and Work.* Dublin, Ireland: Boole, 1985.

IVOR GRATTAN-GUINNESS

# BOOLEAN ALGEBRA

A system of mathematical logic developed by the British logician George Boole (1815–1864). It was first presented in 1854 in his "An investigation of the laws of thought on which are founded the mathematical theories of logic and probabilities."

A Boolean algebra is a set $B$ of elements $a$, $b$, $c$, . . . together with two binary operations $+$ and $\cdot$ (where $a \cdot b$ may be written $ab$) that for all elements $a, b, c, \ldots$ in $B$ satisfies the following properties:

1. The idempotent laws: $aa = a + a = a$
2. The commutative laws:
   $$ab = ba, \qquad a + b = b + a$$
3. The associative laws:
   $$a(bc) = (ab)c$$
   $$a + (b + c) = (a + b) + c$$
4. The absorption laws: $a(a + b) = a + (ab) = a$
5. The distributive laws:
   $$a(b + c) = ab + ac$$
   $$a + (bc) = (a + b)(a + c)$$
6. There exist *identity* elements 0 and 1 in $B$ that satisfy:
   $$0 \cdot a = 0, \quad 0 + a = a, \quad 1 \cdot a = a, \quad 1 + a = 1$$
7. For every element $a$ in $B$ there exists an element $a'$ (called "a prime") in $B$ which obeys the laws

$$a + a' = 1$$

and

$$aa' = 0$$

When presenting Boolean algebra as an abstract system, one first proposes a set of postulates and then derives theorems. The resulting mathematical system includes symbols that adhere to certain rules and that may be manipulated in certain ways. However, what the symbols *represent* may vary, and Boolean algebras may acquire various realizations, depending upon the contexts in which they are postulated or in which they arise. Thus, the set of elements and operations that comprise a Boolean algebra may embody different forms. Three common realizations are: (1) algebra of sets; (2) algebra of logic or of propositions (statements); and (3) algebra of switching circuits.

As an algebra whose elements are sets, the Boolean operation $+$ corresponds to the operation $\cup$ (union); the Boolean operation $\cdot$ (dot) corresponds to $\cap$ (intersection); and the Boolean operation $'$ (prime) corresponds to complementation. Thus, per the Boolean properties,

1. $a \cup a = a \cap a = a$
2. $a \cup b = b \cup a; \qquad a \cap b = b \cap a$
3. $a \cup (b \cup c) = (a \cup b) \cup c$
   $a \cap (b \cap c) = (a \cap b) \cap c$
4. $a \cap (a \cup b) = a \cup (a \cap b) = a$
5. $a \cup (b \cap c) = (a \cup b) \cap (a \cup c)$
   $a \cap (b \cup c) = (a \cap b) \cup (a \cap c)$
6. $0 \cap a = 0, \qquad 0 \cup a = a,$
   $1 \cap a = a, \qquad 1 \cup a = 1$
   where 1 denotes the universal set and 0 denotes the null or empty set
7. $a \cup a' = 1$
   $a \cap a' = 0$

As an algebra of logic, whose elements are propositions, the Boolean operation $\cdot$ (dot) corresponds to AND (conjunction); the Boolean operation $+$ corresponds to OR (disjunction); and the Boolean operation $'$ (prime) corresponds to NOT, or negation. Thus, if 1 represents a statement that is TRUE, and 0 represents a statement that is FALSE, and $=$ means "has the same truth value," then:

| | | |
|---|---|---|
| 1 AND 1 = 1 | 1 OR 1 = 1 | NOT 1 = 0 |
| 1 AND 0 = 0 | 1 OR 0 = 1 | NOT 0 = 1 |
| 0 AND 1 = 0 | 0 OR 1 = 1 | |
| 0 AND 0 = 0 | 0 OR 0 = 0 | |

Note that the Boolean properties may be interpreted as follows:

1.  $a$ AND $a$ = $a$ OR $a$ = $a$
2.  $a$ AND $b$ = $b$ AND $a$;     $a$ OR $b$ = $b$ OR $a$
3.  $a$ AND $(b$ AND $c)$ = $(a$ AND $b)$ AND $c$
    $a$ OR $(b$ OR $c)$ = $(a$ OR $b)$ OR $c$
4.  $a$ AND $(a$ OR $b)$ = $a$ OR $(a$ AND $b)$ = $a$
5.  $a$ AND $(b$ OR $c)$ = $(a$ AND $b)$ OR $(a$ AND $c)$
    $a$ OR $(b$ AND $c)$ = $(a$ OR $b)$ AND $(a$ OR $c)$
6.  0 AND $a$ = 0,     0 OR $a$ = $a$,
    1 AND $a$ = $a$,     1 OR $a$ = 1
7.  $a$ AND (NOT $a$) = FALSE (0)
    $a$ OR (NOT $a$) = TRUE (1)

As an algebra of switching circuits, whose elements are circuits, the Boolean operation + corresponds to circuits connected in parallel, the Boolean operation · (dot) corresponds to circuits connected in series, and the Boolean operation ′ (prime) corresponds to the change of state of a switch (closed to open or open to closed). 0 represents a circuit that is open (no current flows), and 1 represents a circuit that is closed (current does flow). The Boolean properties cited at the beginning of this entry apply in this context.

Boolean algebra may be presented on various levels. On an elementary level, sets would generally be presented first, since finite sets can be drawn or their elements listed, and simple examples abound. Boolean concepts in switching circuits are widely used in digital computer design; abstract Boolean concepts are applicable to the study of symbolic logic. Alternatively, Boolean algebra may be used to illustrate the structure of abstract mathematical systems, in which a set of postulates leads to derivation of all the relationships and operations that hold in the algebra, without reference to content or to what the symbols represent.

*See also* Abstract Algebra; Boole; Logic; Sets

SELECTED REFERENCES

Boole, George. *Collected Logical Works.* 2 vols. Chicago, IL: Open Court, 1916.

Frankiewicz, Ryszard. *Hausdorff Gaps and Limits.* New York: North-Holland, 1994.

Halmos, Paul R. *Lectures on Boolean Algebra.* New York: Van Nostrand, 1963.

Hohn, Franz Edward. *Applied Boolean Algebra.* New York: Macmillan, 1960.

Stone, M. H. "Applications of the Theory of Boolean Rings to General Topology." *Transactions of the American Mathematical Society* 41(1937):375–481.

Whitesitt, John Eldon. *Boolean Algebra and Its Applications.* Reading, MA: Addison-Wesley, 1962.

Young, Frederick H. *Digital Computers and Related Mathematics.* Boston, MA: Ginn, 1961.

RINA YARMISH

# BROOKS, EDWARD (1831–1912)

Influential mathematics educator and psychologist. His *Philosophy of Arithmetic* (1880) is recognized as the first arithmetic methods book for elementary teachers, and his *Mental Science and Methods of Mental Culture* (1883) is known as one of only five books on educational psychology published before the emergence of John Dewey. His arithmetic text series was on the market for over sixty years. Brooks was president of the Normal Department of the National Teachers' Association, superintendent of Philadelphia public schools, and one of the National Education Association's Committee of Fifteen. Brooks was an exponent of faculty psychology, a system that organizes mental functioning into an elaborate system of faculties and subfaculties, each of which must be exercised or "cultured." Brooks saw the study of mathematics as essential in the culture of various faculties, a point of view used to justify increased attention to mathematics in schools.

*See also* Committee of Fifteen on Elementary Education; History of Mathematics Education in the United States, Overview

SELECTED REFERENCES

Bidwell, James K., and Robert G. Clason, eds. *Readings in the History of Mathematics Education.* Washington, DC: National Council of Teachers of Mathematics, 1970.

Jones, Phillip S., ed. *A History of Mathematics Education in the United States and Canada.* 32nd Yearbook. Washington, DC: National Council of Teachers of Mathematics, 1970.

JAMES K. BIDWELL
ROBERT G. CLASON

# BROWNELL, WILLIAM ARTHUR (1895–1977)

Best known for his 1935 article, "Psychological Considerations in the Learning and the Teaching of Arithmetic." He was also lead author of a successful series of arithmetic texts. Brownell advocated a "meaning" theory of arithmetic learning that stressed understanding and that accepted the complexity of arithmetic learning. Brownell objected to the stimulus-response bond learning theory of the connectionist psychologists and to the incidental learning theory. He felt that connectionism was inadequate to explain higher-order arithmetic learning and that students would not gain a full understanding of arithmetic processes if they were exposed to arithmetic only incidentally, as it occurs in other settings.

*See also* History of Mathematics Education in the United States, Overview; Psychology of Learning and Instruction, Overview

## SELECTED REFERENCES

Bidwell, James K., and Robert G. Clason, eds. *Readings in the History of Mathematics Education.* Washington, DC: National Council of Teachers of Mathematics, 1970.

Brownell, William A. "Psychological Considerations in the Learning and the Teaching of Arithmetic." In *The Teaching of Arithmetic*, pp. 1–31. 10th Yearbook of the National Council of Teachers of Mathematics. William D. Reeve, ed. New York: Bureau of Publications, Teachers College, Columbia University, 1935.

Jones, Phillip S., ed. *A History of Mathematics Education in the United States and Canada.* 32nd Yearbook. Washington, DC: National Council of Teachers of Mathematics, 1970.

JAMES K. BIDWELL
ROBERT G. CLASON

# BRUNER, JEROME (1915– )

An educational psychologist and philosopher who has published extensively on theories of teaching and learning. He is best known for *The Process of Education* (1960) and *Toward a Theory of Instruction* (1966). In *The Process of Education,* he stated that "any subject can be taught effectively in some intellectually honest form to any child at any stage of development" (p. 33). Bruner (1971) eventually decided that this philosophical position, which regarded the "structure of academic disciplines" as the fundamental factor in understanding the educational process, was inadequate for the development of school curricula. Instead of focusing on the structure of disciplines in designing school curricula, Bruner later stated that it is better to deal with curricula concerns in the context in which teaching and learning problems occur. In *Toward a Theory of Instruction,* Bruner describes the nature of intellectual development and presents six characteristics of intellectual

growth. Earlier, he had also described three modes of representation of knowledge. These included the enactive mode (information is stored through habits of action), the iconic mode (images are used for storing information), and the symbolic mode (abstract systems such as language forms are utilized for information storage) (Bruner 1964).

Bruner observed mathematics classes and conducted studies on the teaching and learning of mathematics. As a result, he developed four "theorems" about how students learn mathematics, known as the *construction, notation, contrast and variation,* and *connectivity* theorems (Bruner 1963). These theorems help teachers formulate a variety of teaching strategies that can benefit students' ability to learn mathematics. For example, the construction theorem stresses the importance of students using concrete representations of mathematical concepts and operations from which they may formulate their own rules in mathematics. The notation theorem identifies the need to use mathematical notation that is appropriate for cognitive development. Thus, $y = 8x - 3$ may be appropriate for beginning algebra students, but $\triangle = 8 \,\square - 3$ is better for elementary-to-middle-grades students. Teaching strategies that compare and contrast critical attributes of a concept or skill, or strategies that provide a variety of examples of a concept or skill, adhere to the theorem of contrast and variation. Finally, the connectivity theorem states the importance of making mathematical connections among the areas of mathematics, such as algebra, geometry, and data analysis. These connections are critical to students' ability to reason and mature mathematically.

Bruner completed his undergraduate education at Duke University and obtained his Ph.D. degree in psychology in 1947 from Harvard University. He served as the director of the Center for Cognitive Studies at Harvard University and was a professor at Oxford University, the New School for Social Research in New York City, and the Institute for Advanced Study at Princeton. In 1963, he received the Distinguished Science Award of the American Psychological Association and became the Association's president in 1965.

*See also* History of Mathematics Education in the United States, Overview; Psychology of Learning and Instruction, Overview

## SELECTED REFERENCES

Bell, Frederick H. *Teaching and Learning Mathematics (in Secondary Schools).* Dubuque, IA: Brown, 1978.

Bruner, Jerome S. "The Course of Cognitive Growth." *American Psychologist* 19(1964):1–15.

———. "Observations on the Learning of Mathematics." *Science Education News* (Apr. 1963):1–5

———. *The Process of Education.* Cambridge, MA: Harvard University Press, 1960.

———. "The Process of Education Revisited." *Phi Delta Kappan* 53(Sept. 1971):18–21.

———. *Toward a Theory of Instruction.* Cambridge, MA: Belknap Press of Harvard University Press, 1966.

LEE V. STIFF

# BUSINESS MATHEMATICS

Primarily geared to business students and taught at most colleges, universities, and community colleges, with calculus or without (the latter course is generally called finite mathematics). At some colleges, business mathematics is a general education course in practical mathematics and admits more freedom in its design and implementation. But "business mathematics" usually refers to mathematics based on more fundamental computational skills, such as working with tables, computing percentages, and so forth.

Most comprehensive universities have had courses in college algebra for a long time. A course combining college algebra and finite mathematics began to emerge at many institutions as a required business course after 1950. In some cases, representatives from the mathematics and business departments developed the syllabus. Although this course also absorbs students in programs such as economics, nursing, and criminal justice, successfully directing it to both the business and life sciences audience is usually not a problem for mathematics departments, since most texts take a flexible approach.

Calculus-based mathematics for business students provides a different option and may be required of business majors in areas such as finance, accounting, or business computer information systems. This course is also increasingly required or recommended in fields such as engineering technology, premedicine, or pharmacology. Students and faculty alike often feel that such a course affords a broader overview of calculus than traditional calculus courses. As is the case with many introductory mathematics classes, the extent of students' algebra background can be a major problem. Tests such as the "Intermediate Algebra Skills" test by the Educational Testing Service may be administered prior to the first class to permit unprepared students to shift into preparatory algebra courses.

Features such as a packed syllabus and rapid pace in large sections contribute to the occasional reputation of business mathematics as a "weed-out" course. Students may question whether they will ever use the mathematics. It is true that in subsequent business courses, rote computer programs reduce the importance of some of the mathematics students learn in an introductory course; as a result, frequent consultation about the course content with other departments is advisable. However, departments requiring the course want students not only to be exposed to sound mathematical content but to learn how to absorb material conceptually, develop reasoning skills, and achieve a disciplined attitude to learning. These goals may be difficult to convey to students.

## COURSE CONTENT

An informal survey of twenty-five college catalogues and ten texts reveals a fairly standard curriculum: (1) an algebra preview with graphing; (2) linear and quadratic functions, with word problems focusing on cost and revenue; (3) exponential and logarithmic functions with problems on compound interest; (4) a brief overview of elementary linear algebra techniques, such as solving systems of equations by reducing matrices to row echelon form, determinants, and matrix inverses; (5) an introduction to linear programming; (6) some elementary probability, going as far as conditional probability and Markov chains; (7) some elementary statistics, including measures of central tendency, the normal distribution, the Central Limit Theorem, and possibly some more advanced topics; (8) sets, sequences, and arithmetic and geometric progressions; (9) an overview of other topics, such as game theory and difference equations.

Accrediting agencies and college administrators are seeking ways to demonstrate concretely that students are learning in ways that meet the institution's learning objectives, the needs of the departments served, and the needs of the students themselves. Current reforms in mathematics teaching have brought to the classroom the nonlecture format, group problem solving, learning through applications, writing assignments, and technology. Many faculty who are familiar with these techniques in other courses are finding success with these methods in this course. Calculators are becoming a staple of the course as is the computer, accompanied by a wide variety of software. The particularly applied nature of the course allows faculty to experiment with stimulating projects that can make the course more enjoyable, for example, classroom discussions and reports on up-to-date newspaper stories (see Snell 1995), video interviews of local businesspersons, and talks by business majors who have taken the course. But it is clear more incentives, such as classroom assessment techniques, are needed to help students learn to value and develop the methods of critical thinking, good study habits, and the work ethic the course demands of them.

*See also* Finance; Interest; Linear Algebra, Overview

## SELECTED REFERENCES

Barnett, Raymond A., and Michael R. Ziegler. *Finite Mathematics for Business, Economics, Life Sciences, and Social Sciences.* 6th ed. New York: Dellen, 1993.

Garfunkel, Solomon, Lynn Steen et al. 3rd ed. *For All Practical Purposes.* New York: Freeman, 1994.

Hoenig, Alan. *Applied Finite Mathematics.* 2nd ed. Boston, MA: Houghton Mifflin, 1995.

*Intermediate Algebra Skills.* Test provided by Multiple Assessment Programs and Services (MAPS). Princeton, NJ: Educational Testing Service, n.d.

Loacher, Mentkowski. "Creating a Culture in Which Assessment Improves Learning." In *Making a Difference* (pp. 5–24). Trudy Banta et al., eds. New York: Jossey-Bass, 1993.

Snell, J. Laurie. "Take a Chance on CHANCE." *UME Trends* 6(6)(1995):28–29.

SANDRA Z. KEITH
GAIL A. EARLES

# C

## CALCULATORS

Electronic devices that can perform mathematical computations and functions with speed and accuracy. Calculators can perform simple arithmetic tasks, such as adding, subtracting, multiplying, and dividing, as well as more complex tasks, such as the mathematical problems encountered by engineers and scientists requiring multiple steps using programs and several memories.

## HISTORY

The abacus is thought to be the earliest calculating device dating back to the early Babylonians around five thousand years ago. This device consists of a frame of movable beads on rods in rows of certain amounts and is still used by people in Asia. The first true mechanical calculator was created in 1642 by a French mathematician, Blaise Pascal. His machine required toothed gears with the digits 0 to 9 that interlocked to add and subtract. He found that building it was too expensive to be practical (Bitter 1984). In 1671 Gottfried Wilhelm Leibniz, a German physicist (atomist) and mathematician, invented a calculator called the *Reckoning Machine* that not only could add and subtract, but also could multiply, divide, and compute square roots. Multiplication was accomplished by repeated addition of the multi-

plicand. He was the first person to recognize the importance of the binary system (1 and 0) of notation (Bitter 1984).

As early as 1822, Charles Babbage began to envision more complicated calculating machines. In 1823, Babbage began a project to create a calculating machine with a twenty-decimal place capacity. His *Analytical Engine* was the forerunner of modern calculators and computers. The machine used punched cards to store partial answers as additional computations were performed. The punched card concept was originated by Joseph Jacquard, who developed an automatic loom with punched cards controlling the weaving sequence. The Babbage machine stored data in columns of wheels. By moving a lever forward or backward, problems could be computed in a sequential manner. Babbage persuaded the British government to invest in his project, but by 1842, since he was still unsuccessful in building the metal wheels to get the device to function properly, his financial support was withdrawn. The design and principle behind the *Analytical Engine* was sound and the mechanical system of wheels, cogs, levers, and cams on which his model was conceptualized was realized in later developments in electronics (Bitter 1992). It was not until the late 1880s that mechanical adding machines were available for purchase. An American inventor, William Burroughs, organized the American Arithmometer Company in 1886 and, after much trial and error, received a patent for an adding machine in 1892. Such calculators were handpowered by a crank handle. The first American patent on an adding machine using an electric motor for power was authorized in 1903.

## INTEGRATED CIRCUIT

In 1948 the transistor was invented and in 1958 the integrated circuit allowed the use of miniature electronic circuits on a single silicon chip. The transistor and the integrated circuit formed the basis for the electronic calculator revolution. The integrated circuit involved the fabrication and assembly of numerous transistors, capacitors, diodes, and/or resistors in a single integrated component called a chip. The integrated circuit could hold all the circuit elements needed to implement the algorithms, and numerous algorithms could be held in memory, to be retrieved when desired. These inventions created the "calculator on a chip" that was the basis for the electronic handheld or pocket-sized mini calculator (Bitter 1992).

During the 1960s, new products that decreased in size and increased in capabilities replaced earlier models. Transistorized models replaced the mechanical models. In 1972 calculators appeared with a liquid crystal display that provided a significant savings in power usage. By the mid-1970s, inexpensive handheld calculators using large-scale integrated circuits and costing less than $50 could outperform a $1,500 mechanical calculator. Estimates indicate around fifty million "pocket calculators" were owned in 1975. The changes in the four-function calculator (add, subtract, multiply, and divide) from 1974 to 1977 included the reduction of electronic parts from 82 to 2 and total parts from 119 to 17. During the late 1970s, many special features and numerous functions were incorporated including the long-life battery or solar power. The size of the calculator was reduced to small chips that could fit into watches, pens, and credit card-size calculators. By 1980, special features included music, a stopwatch, and numerous algorithms that could be placed into the memory of the calculator.

## PROGRAMMABLE CALCULATORS

Advanced calculators are "programmable" devices in which sequences of commands called programs have the ability to branch (to select and execute given parts of a program while excluding others, depending on some condition) and to loop (a group of instructions performed repeatedly until certain conditions are met) as in computer programs. Programmable scientific calculators accept programming by the operator in desired sequences of operations using separate memory systems for data and programs. Some scientific calculators have a liquid crystal display of four or five lines with an infrared printer interface. Such devices provide storage locations that are activated by the recall key and a location-designating key. Manufacturers offer preprogrammed cards for special calculations that customize the calculator and some units accept miniaturized magnetic cards to expand the nonvolatile storage capacity (storage that does not result in the loss of contents—program and data—when power is accidentally or purposely removed). The typical programmable electronic calculator uses a binary-coded decimal system and operates much like a computer. In a digital system, information is transmitted by switching conductors back and forth between two voltage states which are symbolized by 0 or 1. Each bit of information is stored as either 0 or 1.

## GRAPHING CALCULATORS

The 1990s ushered in the graphing calculators that educators use for interactive graphic analysis and discrete mathematics. A few of the special features of these calculators are menus with selection options, 32 K or more RAM (random-access memory) to store data, and split screen displays. The ability to compute in other number bases (hexadecimal, binary, octal), and to use built-in automatic constants, and Boolean logic operations make these calculators applicable for chemistry, physics, and calculus. The typical eight-line display allows up to ten functions traceable on a single graph, six polar equations, two recursively defined sequences, six parametric equations, and the simultaneous graphing of more than one function. Accessories allow information to be printed or stored on a disk.

Other graphing calculators graph, analyze, and store up to ninety-nine polar equations, solve thirty simultaneous equations, find the roots of a polynomial up to the thirtieth order, and can communicate with other calculators or computers. Students can solve problems graphically, numerically, and algebraically. The following equation-solving capabilities allow such calculators to be used in calculus classes: (a) plot and trace a function with its first and second derivative; (b) evaluate definite integrals and arc length; and (c) find maximums, minimums, points of inflection and zeros of functions.

Other versions display two graphs on a single screen for side-by-side comparison, or the formula and its graph can appear together. The animated graphing function illustrates the changes in any graph produced by a change in variables. The graphing capabilities of this type of calculator include rectangular coordinate graphs, polar coordinate graphs, parametric graphs, inequality graphs, integration graphs, statistical graphs, and function graphing.

The educational benefits of calculators have caused manufacturers to produce a wide variety of calculators emphasizing many mathematical concepts. Scientific calculators incorporate technology that allows students to enter equations in the order in which they appear in written form with operations displayed. This feature eliminates time typically spent on learning how to enter expressions and equations into the calculator. Additional available features include many levels of parentheses, constants for physics, fixed decimal capability, and a display that shows ten digits or more plus a two-digit exponent.

Although calculators using fractions already existed, Texas Instruments developed the Math Explorer and Math Explorer Plus for teaching fractions. These calculators have built-in fraction keys that calculate fractions, convert between mixed and improper fractions, and convert fractions to their decimal equivalents. Another feature of these calculators is the ability to perform integer division with the answer denoted as a quotient with whole number remainders rather than decimals.

## EDUCATIONAL ASPECTS

Hembree and Dessart's (1992) analysis of seventy-nine studies supports the use of calculators throughout the mathematics curriculum and the contention that the computational skills with traditional algorithms of low- and high-ability students were not harmed. Bitter and Hatfield ("Implementing . . ." 1992) reported increased performance scores for seventh and eighth graders on the computation, problem solving, and concepts subtests of achievement tests when using the calculator. They noted that effective ways to integrate calculators into the mathematics curriculum must be found. Mathematics educators recognize the importance of not relying on calculators when more appropriate ways to compute should be used. Students should be encouraged to select when it is appropriate to use the calculator, mental computation, estimation, or

paper-and-pencil computation. A balanced approach to using calculators as appropriate tools for problem-solving situations is needed.

Calculators are recommended as a valuable tool for learning mathematics by all students, beginning in kindergarten. "Calculators enable children to explore number ideas and patterns, to have valuable concept-development experiences, to focus on problem-solving processes, and to investigate realistic applications. The thoughtful use of calculators can ensure the quality of the curriculum as well as the quality of children's learning" (National Council of Teachers of Mathematics [NCTM] 1989, 7). Within the mathematics education community, there is a broad consensus that calculators have considerable power to improve levels of achievement in mathematics by providing access to more sophisticated mathematical problem solving and ideas. A calculator-based laboratory (CBL) allows students to collect real-world data, input it directly to the calculator, then generate graphs and analyze the results. With CBL probes, students measure temperature, light, motion, voltage, force, pH, and other information to be analyzed, to increase their understanding of science and mathematics (Brueningsen et al. 1995). Calculators facilitate the investigation of mathematical questions that involve computations that are too complex for paper-and-pencil solutions.

Even though many textbooks have not yet fully integrated calculator pedagogy, some recent textbooks center around the graphing calculator (Burzynski et al. 1995; Demana et al. 1994; Dennis and Neal 1995; Lund and Anderson 1995). Many support resources exist and are available from publishers and calculator manufacturers. Ancillary materials for calculators such as the *Math Explorer* (Bitter and Mikesell 1990) can be used in conjunction with classroom textbooks. Teachers' guides, newsletters, electronic mail networks, compatible overhead projectable models, loan systems for materials, and summer workshops are a few of the offerings for assistance *(Eightysomething!; It's About T.I.M.E.; T³ (Teachers Teaching with Technology); Using Calculators to Improve Your Child's Math Skills* [NCTM]*)*.

Appropriate calculators should be available to all students across all curriculum areas. Students can share calculators in collaborative groups if the logistics of the availability of an advanced tool, such as a graphing calculator, becomes an issue. A calculator study (Bitter and Hatfield, *Integration of . . .*, 1992; 1994) of over 600 middle grade students revealed the proliferation of typical four-function calculators. The data indicated that 97% of the students in the study reported calculators in their homes and a median value of "more than 3" calculators. Administrators should recognize that to facilitate instructional change in the use of calculators, teachers, students, and parents should be involved. Calculators may be viewed as a powerful tool to enhance children's thinking, rather than as a shortcut to arrive at answers that cannot be explained or interpreted. Implementation of calculators into a mathematics curriculum requires a solid commitment of resources, funding, and time. As we look at today's jobs, skills, and opportunities, most certainly calculators will play a role and students must be prepared to use them appropriately. Technological changes will continually modify calculators, increasing their speed and functionality, as well as incorporating new desirable features to allow new mathematical concepts and problems to be explored.

*See also* Abacus; Babbage; Technology

## SELECTED REFERENCES

Bitter, Gary G. *Computers in Today's World.* New York: Wiley, 1984.

Bitter, Gary G., ed. *Macmillan Encyclopedia of Computers.* New York: Macmillan, 1992.

Bitter, Gary G., and Mary M. Hatfield. "The Calculator Project: Assessing School-Wide Impact of Calculator Integration." In *Impact of Calculators on Mathematics Instruction* (pp. 49–65). George W. Bright, Hersholt C. Waxman, and Susan E. Williams, eds. Lanham, MD: University Press of America, 1994.

———. "Implementing Calculators in Middle School Mathematics: Impact on Teaching and Learning." In *Calculators in School Mathematics* (pp. 200–207). 1992 Yearbook. James T. Fey (ed.). Reston, VA: National Council of Teachers of Mathematics, 1992.

———. *Integration of the Math Explorer Calculator into the Mathematics Curriculum.* Tempe, AZ: Technology Based Learning and Research, 1992.

Bitter, Gary G., and Jerald Mikesell. *Using the Math Explorer Calculator.* Menlo Park, CA: Addison-Wesley, 1990.

Brueningsen, Chris, Bill Bower, Linda Antinone, and Elisa Brueningsen. *Real World Math with the CBL System.* Dallas, TX: Texas Instruments, 1995.

Burzynski, Denny, Wade Ellis, and Ed Lodi. *Precalculus with Trigonometry for Graphing Calculators.* Boston, MA: PWS, 1995.

Demana, Franklin, Bert K. Waits, Charles Vonder Embse, and Gregory D. Goley. *Graphing Calculator and Computer Graphing Laboratory Manual for Precalculus Series.* 2nd ed. Reading, MA: Addison-Wesley, 1994.

Dennis, Cynthia, and Linda M. Neal. *Trigonometry Activities for the T1-82 and T1-85 Graphing Calculators.* Boston, MA: PWS, 1995.

*Eightysomething!* Newsletter for Users of T1 Graphing Calculators. Dallas, TX: Texas Instruments.

Hembree, Ray. "Research Gives Calculators a Green Light." *Arithmetic Teacher* 34 (Sept. 1986):18–21.

———, and Donald J. Dessart. "Research on Calculators in Mathematics Education." In *Calculators in School Mathematics* (pp. 23–32). 1992 Yearbook. James T. Fey, ed. Reston, VA: National Council of Teachers of Mathematics, 1992.

*It's About T.I.M.E. (Technology in Math Education).* Dallas, TX: Texas Instruments.

Lund, Charles, and Edwin Anderson. *Graphing Calculator Activities: Exploring Topics in Precalculus.* Reading, MA: Addison-Wesley, 1995.

National Council of Teachers of Mathematics. *Curriculum and Evaluation Standards for School Mathematics.* Reston, VA: The Council, 1989.

———. *Principles and Standards for School Mathematics.* Reston, VA: The Council, 2000.

———. *Using Calculators to Improve Your Child's Math Skills.* Reston, VA: The Council, 1986.

$T^3$ (Teachers Teaching with Technology). Dallas, TX: Texas Instruments. Website: http://www.t3ww.org/

MARY M. HATFIELD

GARY G. BITTER

# CALCULUS, OVERVIEW

A branch of mathematics that provides computational (or calculating) rules for dealing with problems involving infinite processes. Though the subject clearly rests on ideas that go back at least to Archimedes, development of the subject as we know it today generally is traced to the work of Isaac Newton, Gottfried Leibniz, and others who worked in the second half of the seventeenth century.

Calculus draws upon all the algebra, geometry, and trigonometry commonly taught in U.S. high schools, making it a logical capstone of the secondary curriculum. Also, because calculus is the fundamental mathematical tool of natural science, of engineering, and increasingly of social science as well, it is the foundation of the college mathematics curriculum. It has found its place as both a culminating course for the high school mathematics program, and as an introductory course at the collegiate level. We shall try here to introduce the principal ideas, and to indicate their wide application. We also shall present ideas that have motivated lively discussion over the past decade of how the subject should be taught.

## NOTATION

Before stating the two basic problems of calculus, we need to discuss some convenient notation, the absence of which greatly handicapped such thinkers as Archimedes and other early investigators of infinite processes. The idea of locating a point in a plane by giving first its horizontal distance $x$ and then its vertical distance $y$ from a fixed reference point called the origin is commonly credited to René Descartes—hence the name *Cartesian plane*.

## FUNCTIONS

The next embellishment, seemingly simple, but needing hundreds of years to be developed, is to imagine a rule that assigns to a point $x$ on the horizontal axis a unique value $y$ on the vertical axis. The rule is called a *function* and is named with a single letter, say $f$ or $g$. We say that $y$ is a function of $x$, writing $y = f(x)$. Modern pedagogical reforms have advocated that students learn to think of this arrangement in three different ways, numeric, graphic, and symbolic.

### Numeric

It may be that the values of $x$ represent times at which a measurement was taken, and $y$ represents the measurement. A passenger on an automobile trip might, for instance, write down at each hour the reading of an odometer set to 0 when the trip started.

| $x$ (time in hours) | $y$ (distance in miles) |
|---|---|
| 0 | 0 |
| 1 | 50 |
| 2 | 88 |
| 3 | 126 |
| 4 | 176 |

### Graphic

The trip may be represented by plotting the points (0,0), (1,50), (2,88), (3,126) and (4,176).

### Symbolic

Sometimes one can find a formula relating $x$ to $y$ so that if the values of $x$ are substituted into the formula, the values of $y$ are produced. For example,

$$y = f(x) = 2x^3 - 12x^2 + 60x$$

has the property that $f(0) = 0, f(1) = 50, f(2) = 88, f(3) = 126, f(4) = 176$.

Information collected by a human observer, whether a passenger in a car or a scientist in a laboratory, is most likely to be available for analysis from a numeric list, and so some education reformers have argued that students need to learn how to work from this information. For analysis, one turns most naturally to a graphic display. If, for example, one wishes to know the distance that had been covered at time $x = 2.5$ or even $x = 2.1$, the most likely method would be to draw a reasonably smooth curve through the known data points, and then make estimates of the desired information. One might guess from a graph, for example, that when $x = 2.5, y = 110$; that when $x = 2.1, y = 92$. It should be noted that some monitoring devices, an electrocardiogram for instance, will provide original data in graphical form.

For analysis, the symbolic method certainly seems most attractive. In our example, it is easily seen that $f(2.5) = 106.25$ and $f(2.1) = 91.60$. The difficulty is, of course, that information very seldom comes to the analyst in symbolic form. The expression for $f$ is itself determined as part of the analysis, so that while $f(2.1) = 91.60$ carries with it an aura of great accuracy, one must ask how $f(x) = 2x^3 - 12x^2 + 60x$ was determined.

To simplify analysis, traditional calculus texts started a discussion of motion with, "Let $y = f(x) = 2x^3 - 12x^2 + 60x$ represent the distance $y$ that a car has traveled in time $x$." While the analysis that can follow is still of fundamental importance, it is now felt that clarity of understanding as well as experience in dealing with practical problems argues for presenting concepts with all three representations mentioned above.

## LIMITS

Suppose we wish to know at time $x = 2$ the speed in miles per hour of the car used in the illustration above. Since we only know distances and times at hour intervals, we might reason that the best we can do is approximate the speed (or rate) at $x = 2$ by finding the average rate over the hour from $x = 2$ to $x = 3$. Corresponding to the three representations of our information, we might proceed as follows.

*Numeric:* rate = distance/time = $(126 - 88)/(3 - 2) = 38/1$

*Graphic:* slope = change in $y$ values/change in $x$ values = $38/1$

*Symbolic:* rate = $[f(3) - f(2)]/(3 - 2) = (126 - 88)/(3 - 2) = 38/1$

None of these reliably give the reading of the speedometer at time $x = 2$. All of them give the average speed over the period of the third hour. We would clearly get a better estimate of the speed at $x = 2$ if we could use the average speed over, for instance, the tenth of an hour (the six-minute period) from $x = 2$ to $x = 2.1$. The distance traveled at $x = 2.1$ was in fact obtained for two of our representations above, affording us two ways to estimate the speed at $x = 2$.

*Graphic:* slope = change in $y$ values/

$$\text{change in } x \text{ values} = \frac{92 - 88}{0.1} = 40$$

*Symbolic:* rate $= \dfrac{f(2.1) - f(2)}{2.1 - 2} = \dfrac{91.60 - 88}{0.1} = 36$

These last calculations are significant for what they suggest. If using the six-minute interval after $x = 2$ is an improvement, would we not get still more improvement by using a smaller time interval? For a time interval $t$, the two methods lead to two very important representations.

*Graphic:*
slope = change in $y$ values/change in $x$ values

*Symbolic:* rate $= [f(2 + t) - f(2)]/[(2 + t) - 2]$

We observe that as $t$ gets smaller, the slope of the line (properly called a secant line, since it cuts the curve at two points) approaches the slope of a line tangent to the curve. Similarly, we define the rate analytically by looking for a limiting value, as $t$ gets smaller and smaller, to the expression: rate $= f(2 + t) - f(2)/t$.

We may estimate the slope (which is, remember, an approximation to the speed) as:

change in $y$ values/change in $x$ values = 38/1

Or, symbolically, $[f(2 + t) - f(2)]/t = 36 + 2t^2$. Now we can see that the closer $t$ gets to zero, the closer this expression gets to 36. We summarize this by writing:

$$\lim_{t \to 0} \frac{f(2 + t) - f(2)}{t} = \lim_{t \to 0}[36 + 2t^2] = 36$$

## THE DERIVATIVE

This problem leads us to one of the two fundamental concepts of calculus. For a given function $f$, the derivative $f'$ is defined by

$$f'(x) = \lim_{t \to 0} \frac{f(x + t) - f(x)}{t} \qquad (1)$$

This is the expression for an arbitrary value $x$, encountered above for $x = 2$. To carry out the computation of $f'(x)$ for our function $f(x) = 2x^3 - 12x^2 + 60x$, one calculates

$$\lim_{t \to 0} \frac{f(x + t) - f(x)}{t}$$

$$= \lim_{t \to 0} \frac{2(x + t)^3 - 12(x + t)^2 + 60(x + t) - [2x^3 - 12x^2 + 60x]}{t}$$

$$= 6x^2 - 24x + 60$$

If we set $x = 2$ in our expression for $f'(x)$, we get $f'(2) = 36$ which corresponds with the answer previously obtained. The value $f'(2) = 36$ is called the derivative of $f$ at $x = 2$.

In the discussion above $f'(2)$ not only represented the speed of the car at $x = 2$, but it also represented the slope of a line tangent to the graph of $y = f(x)$ at $x = 2$. This relationship, of utmost importance, is illustrated by the graph of

$$y = f(x) = 2x^3 - 12x^2 + 60x$$

and the graph of its derived function

$$y = f'(x) = 6x^2 - 24x + 60$$

Though we motivated the definition of the derivative by determining the velocity of a moving object from knowledge of its location at various times, we have noticed that the same concept gives us the answer to one of the two basic problems in calculus.

## First Basic Problem

For an arbitrary function $f$, how do we find the slope of a line tangent to the graph of $y = f(x)$ at a point $x_0$? We have seen that the required slope is given by $f'(x)$.

Definition (1) is a fundamental concept of calculus because, for given interpretations of the function $f$, the function $f'$ has very important meanings (Table 1).

Students of calculus have traditionally been required to learn, for certain familiar functions, how to find the derived function. As was already apparent in the computations needed above to find the derivative of the simple third degree polynomial $f(x) = 2x^3 - 12x^2 + 60x$, the calculation of derivatives requires good algebraic skills. Students also were expected to be able to graph functions and their derivatives. At one time, good graphing skills were taught in a course called analytic geometry before the study of calculus was undertaken.

Where the computer and the graphing calculator are readily available, there are teachers of calculus who argue for making less demands on student abilities to do algebra or graph functions. Whether or not one believes that a student should be required to develop facility with these skills, it is believed that a student should be able to explain in terms of slopes of tangent lines the relationship to the graphs of the de-

**Table 1** *Meaning of the derivative*

| Given meaning of $f(x)$ | Interpretation of $f(x)$ |
|---|---|
| *Distance* [$f(x)$ is the distance a moving object travels from some fixed starting point in time $x$.] | *Velocity* [$f'(x)$ gives the velocity at time $x$.] |
| *Amount* [$f(x)$ gives the measure of some changeable quantity (e.g., volume of a melting ice cube, length of a shadow) at time $x$.] | *Rate of change* [$f'(x)$ gives the measure in units/(time interval) of the rate at which the quantity is changing.] |
| *Graph* [The coordinates $(x,y)$ of points on a graph in the plane are related by $y = f(x)$.] | *Slope of a tangent line* [The line tangent to the graph at $(x,y)$ has slope $f'(x)$.] |

rivatives of the functions graphed. The reader should examine the three examples of Figure 1 in this way.

$$P(x) = x^3 \qquad T(x) = 2^x \qquad S(x) = \sin x$$

$$P'(x) = 3x^2 \qquad T'(x) = 2^x(.69) \qquad S'(x) = \cos x$$

Note that the first function $P(x) = x^3$ and its derivative $P'(x) = 3x^2$ illustrate a general rule. For any function $Q(x) = ax^n$ where $a$ is constant and $n$ is an integer, $Q'(x) = anx^{n-1}$. If you look at the function

$$f(x) = 2x^3 - 12x^2 + 60x$$

for which we actually computed the derivative, and think of using this rule term by term, you can quickly obtain

$$f'(x) = 6x^2 - 24x + 60$$

Of these functions, however, it is $T(x) = 2^x$ that we wish to single out for discussion. The derivative $T'(x)$ is computed from

$$T'(x) = \lim_{t \to 0} \frac{2^{x+t} - 2^x}{t} = 2^x \lim_{t \to 0} \frac{2^t - 1}{t}$$

The last expression can be estimated with a calculator. For instance, by setting $t = .001$, one gets $(2^{.001} - 1)/.001 = 0.69$ which is a close enough estimate for our purposes. Now consider $R(x) = 3^x$. Similar computations show

$$R'(x) = 3^x \lim_{t \to 0} \frac{3^t - 1}{t} \approx 3^x \frac{3^{.001} - 1}{.001} = 3^x(1.10)$$

Summarizing our results,

$$T(x) = 2^x \qquad T'(x) = 2^x(.69)$$

$$R(x) = 3^x \qquad R'(x) = 3^x(1.10)$$

It is natural, and it turns out to be extremely important, to ask if there is some number $e$, $2 < e < 3$, for which $E(x) = e^x$ will have $E'(x) = e^x(1)$. That is, we seek a function that is its own derivative. The value of $e$ that works can be found by experiment—another favorite learning device of the reform-minded teachers who, with calculators and computers at their disposal, can now teach calculus as an experimental science. Euler was the first mathematician to notice the importance of this number, hence its designation by $e = 2.71828 \ldots$.

## Three Important Applications of the Derivative

*Falling Bodies* If a ball is dropped from the top of the 555-foot Washington Monument, the distance it drops in $t$ seconds may be shown experimentally to vary proportionately to the square of the time $t$. Thus its height at time $t$ will be

$$y = h(t) = 555 - ct^2$$

The derivative of this function will give the velocity of the ball; and using the rule for derivatives of powers of a variable, we get

$$h'(t) = v(t) = -2ct$$

The rate at which velocity changes is commonly called acceleration. In the case under consideration

$$a(t) = v'(t) = -2c$$

In words, the acceleration due to gravity is constant. This constant is usually designated by $g$. Replacing $2c$ by $g$ makes $c = g/2$, and our first formula above is

$$y = h(t) = 555 - \left(\frac{1}{2}\right)gt^2$$

Suppose that instead of dropping the ball, we throw it with an initial velocity of $v_0$ feet per second. That would affect the velocity, making

$$v(t) = v_0 - 2ct$$

and it would add $v_0 t$ to the distance traveled in time $t$:

$$h(t) = 555 + v_0 t - \left(\frac{1}{2}\right)gt^2$$

The formulas we have developed would have perfect generality if instead of starting with 555, we started at a height of $y_0$. Since the acceleration $a(t)$ is the derivative of the velocity $v(t) = h'(t)$, it also is the

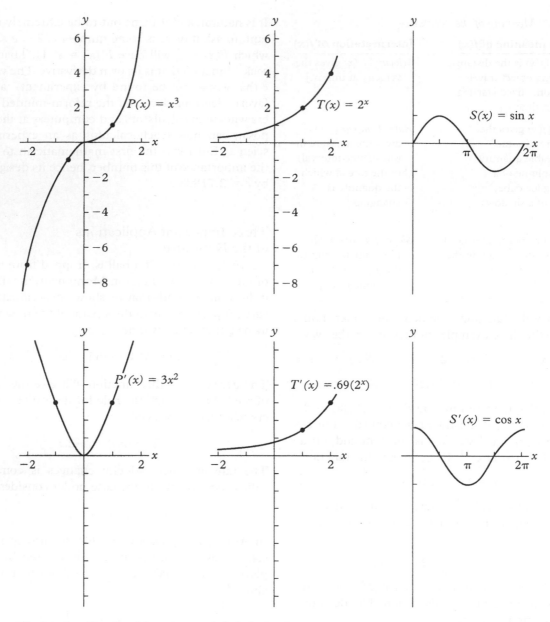

**Figure 1** *Graphs of functions and their derivatives.*

second derivative of $h(t)$, written $h''(t)$. Summarizing, we have three interrelated formulas.

$$y = h(t) = y_0 + v_0 t - \left(\frac{1}{2}\right) g t^2$$
$$v(t) = h'(t) = v_0 - gt$$
$$a(t) = h''(t) = -g$$

Note that this discussion could have proceeded from the bottom up; one begins with the hypothesis (experimentally verifiable) that the acceleration due to gravity is constant. What is differentiated to get the constant $-g$? That is, what is the antiderivative of $-g$? It is $-gt$, plus, perhaps, a constant. Since the constant will be the value of the velocity at $t = 0$, call it $v_0$, we get $v(t) = v_0 - g(t)$. In the same way, derive $h(t)$. There is much to be learned about differentiation and antidifferentiation by working back and forth between these formulae. There also is much to be learned about mathematical modeling and its dependence on the hypothesis of an orderly universe by pondering the fact that these formulae faithfully describe the behavior of a falling body.

*Growth and Decay* A large ice cube, with a lot of surface area, melts much more quickly than does a small ice cube. The mold that grows on food left too long in the refrigerator starts as a speck, grows slowly at first, and then grows at a much greater rate as the amount of mold increases. Populations, whether of bacteria or of people, grow at a rate that is proportional to the population present. These and many other examples obey the principle that *the rate at which something grows is proportional to the amount present.* This statement is easily translated to the language of calculus (see Table 1). If $A(t)$ represents the amount present at time $t$, then $A'(t)$ represents the rate of change. The statement translates to $A'(t) = kA(t)$. This is called a *differential equation.* Its solution is a function having a derivative that, except for a constant, is equal to the function itself, reminding one of $E(t) = e^t$. Modifying this slightly to take account of the constant $k$, the most general solution is $A(t) = A_0 e^{kt}$. The constant $A_0$ is the initial amount, being the value of $A(0)$, and $k$ will be positive according as the substance is increasing (like mold) or decreasing (like ice). This gives rise to the expression that something grows or decreases exponentially (Figure 2).

*Maxima and Minima* Many different cones of variable height $h$ can be inscribed in a sphere of radius 6. The volume of the cone is

$$v(h) = 4\pi h^2 - \left(\frac{\pi}{3}\right)h^3$$

a function having the graph pictured in Figure 3. One may ask what value of $h$ yields the maximum volume. The maximum we seek evidently occurs where the slope of the tangent line is 0. That occurs, however, where the derivative $v'(h) = 0$, i.e., where

$$v'(h) = 8\pi h - \pi h^2 = 0$$

Solutions of $h = 0$ and $h = 8$ give points where the slope is 0, and the volume is a minimum and a maximum, respectively. Thus, $v = 0$ when $h = 0$ (something that was obvious in the beginning); and when $h = 8$,

$$v = 4\pi(8)^2 - \left(\frac{\pi}{3}\right)(8)^3 \approx 268$$

## Second Basic Problem

For an arbitrary function $f$ defined for all values of $x$ between $a$ and $b$, what is the area under the graph of $y = f(x)$ for $a \leq x \leq b$?

This problem is approached by using sums of areas of approximating rectangles. Start by partitioning the interval $[a,b]$ into $n$ equal segments $a = x_0, x_1, x_2, \ldots, x_n = b$. Then note that $(x_1 - x_0)f(x_1)$ is the area of the leftmost shaded rectangle in Figure 4, and $(x_2 - x_1)f(x_2)$ is the area of the adjacent rectangle. If this process is continued for each of the $n$ rectangles, the sum of these areas gives an approximation of the desired area.

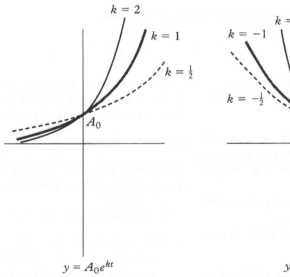

$$y = A_0 e^{kt}$$

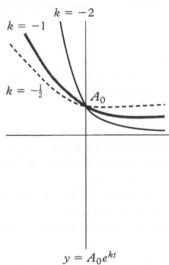

$$y = A_0 e^{kt}$$

**Figure 2**

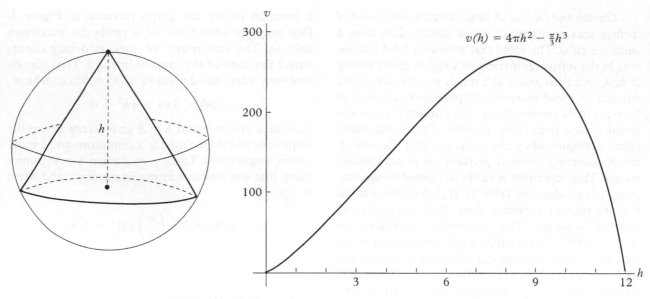

**Figure 3**   *The volume of a cone.*

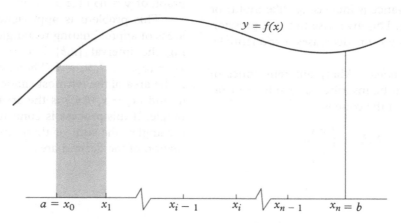

**Figure 4**   *Approximating the area under a graph.*

Since what we get is an approximation, one next asks how the approximation might be improved. At least three ways might be considered:

1.  The partition might use intervals of unequal length, making longer intervals where the graph is flat, shorter ones where it rises or drops quickly.
2.  We could choose a point $t_i$ in the interval $[x_{i-1}, x_i]$ different from the end point $x_{i-1}$.
3.  The number of segments in the partition might be increased.

For a broad class of functions, any of these methods, or combinations of these methods can be used to get

better and better approximations to the desired area. That area, being a sum of areas of rectangles between $a$ and $b$, with heights determined by the graph of $f$, is designated by $\int_a^b f$. The symbol is read, "the integral of $f$ from $a$ to $b$." It is common, when a formula is known for $f(x)$, to simply write that into the integral. For instance, if $f(x) = x^2$, we would write $\int_0^2 x^2$ to indicate the area under the graph of the parabola $y = x^2$ between 0 and 2 (Figure 5).

The definition just given goes to the heart of what an integral is and where it is likely to be useful (wher-

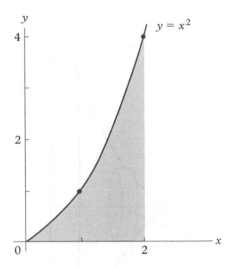

**Figure 5** *Area under the graph of the parabola.*

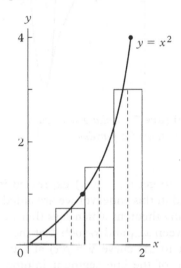

**Figure 6** *Approximating the area under the graph of the parabola.*

ever one needs to sum up a lot of pieces to approximate a whole). Until the introduction of electronic aids for calculating, however, the definition was not of much use in actually evaluating an integral. Clever ways, discussed at the end of this entry, had to be found to circumvent the laborious calculations.

With the aid of a handheld calculator, one can use the definition itself to approximate an integral to any level of accuracy that would be wanted for any practical purpose. Consider the problem introduced above of finding the area under $f(x) = x^2$ between 0 and 2. If we partition $[0,2]$ into four subintervals, and choose our $t_i$ to be the midpoint of each interval, we get the sum of the areas of the four rectangles in Figure 6. That sum is obtained from

$$(x_1 - x_0)f(t_1) + (x_2 - x_1)f(t_2) + (x_3 - x_2)f(t_3) + (x_4 - x_3)f(t_4)$$

$$= \frac{1}{2}f\left(\frac{1}{4}\right) + \frac{1}{2}f\left(\frac{3}{4}\right) + \frac{1}{2}f\left(\frac{5}{4}\right) + \frac{1}{2}f\left(\frac{7}{4}\right)$$

$$= \frac{1}{2}\cdot\frac{1}{16} + \frac{1}{2}\cdot\frac{9}{16} + \frac{1}{2}\cdot\frac{25}{16} + \frac{1}{2}\cdot\frac{49}{16} = \frac{21}{8} = 2.6250$$

Anyone who knows how to program a computer can easily write a program to carry out this same calculation for $n$ subintervals instead of 4. Indeed, most scientific calculators can be programmed to do the same thing. Even that, however, is made obsolete by calculators already programmed to evaluate sums that will approximate the integral to an accuracy that exceeds the display capabilities of the screen. One merely enters the function, the values of $a$ and $b$, and asks for an evaluation. For the problem above, calculators give 2.6667 (rounded accurate to four places beyond the decimal).

This raises a question for the teacher of a modern course in calculus. How many methods invented over the past three hundred years to evaluate integrals should still be taught? Many mathematics educators feel that much less time should be spent on ways to evaluate integrals, with more time spent on their applications.

Some of the methods remain essential to the subject of differential equations. Most important, however, the quest for easier ways to evaluate integrals leads to one of the most beautiful theorems in mathematics, a theorem that ties together the concepts of differentiation and integration, the fundamental result of calculus. Let $f$ be defined from $a$ to $b$. Then for any $x$ between $a$ and $b$, the area from $a$ to $x$ under the graph of $y = f(x)$ is represented by

$$F(x) = \int_a^x f.$$ What is the derivative of $F$? This, it turns out, is a key question to ask. It turns out that $F'(x) = f(x)$, a fundamental result. The function $F$ is called an antiderivative of $f$. Two antiderivatives of $f$ will differ by a constant. Let $G$ be any antiderivative of $f$. Then $G(x) = F(x) + c$. Since $F(a) = 0$, $G(a) = c$. Then $G(b) = F(b) + c = \int_a^b f + G(a)$ and

$$\int_a^b f = G(b) - G(a).$$ This is the fundamental theorem of calculus. It is the basis of what we referred to above as a clever way to avoid the calculations that seem inherent in the definition of $\int_a^x f$. Return, for example, to $\int_0^2 x^2$ which we calculated before. Since $G(x) = x^3/3$ is an antiderivative of $x^2$, the fundamen-

tal theorem tells us that $\int_0^2 x^2 = 2^3/3 - 0^3/3 =$ 2.6667. This method is dependent on knowing, for a given function $f$, how to find an antiderivative. Sometimes, as in the case of our example, this is easy. Sometimes it can be done using a variety of methods that have comprised large units in calculus courses taught for the last 300 years. And sometimes there is no antiderivative expressible in terms of familiar elementary functions.

## Applications of the Integral

*Volumes*  Many objects that we encounter in the world around us have cross-sections that are circles, and this observation gives rise to a way to find their volumes. Suppose the parabola pictured in Figure 5 is rotated about the $x$-axis to generate the solid pictured in Figure 7. The method we have in mind for finding the volume is suggested by the picture. Partition the interval from $x = 0$ to $x = 2$, and choose a point $t_i$ in each segment $[x_{i-1}, x_i]$. Then the volume of the cylinder pictured is $\pi[(t_i)^2]^2(x_1 - x_{i-1})$, and the sum of the volumes of these cylinders approximates the desired volume.

If we define $g(x) = \pi x^4$, then the sum takes the form $g(t_1)(x_1 - x_0) + g(t_2)(x_2 - x_1) + \cdots + g(t_n)$ $(x_n - x_{n-1})$. This is the sum that is designated by $\int_0^2 g$, and since an antiderivative of $g(x) = x^4$ is $x^5/5$, the fundamental theorem gives us the volume $\int_0^2 \pi x^4 = \pi\,(2^5/5) - \pi\,(0^5/5) = 32\,\pi/5$.

*Length of Arc*  Power lines suspended between supporting poles, cables from which a suspended bridge hangs, and the gentle bend of railroad tracks as viewed from the air all illustrate curves we see around us that can be described by an equation of the form $y = f(x)$, and in many cases, engineers will want to be able to find the length $L$ of such curves. Once again, we are led very quickly to partitioning the interval from $x = a$, where the curve starts, to $x = b$, where the curve ends, and then estimating the length we want from an approximating sum. Figure 8 suggests our idea. Focus on the interval $[x_{i-1}, x_i]$. This time the quantity we want, the length of the curve $y = f(x)$ from $x = x_{i-1}$ to $x = x_i$ is approximated by the length of the line segment that joins $(x_{i-1}, y_{i-1})$ to $(x_i, y_i)$. Its length is

$$L_i = \sqrt{(x_i - x_{i-1})^2 + (y_i - y_{i-1})^2}$$

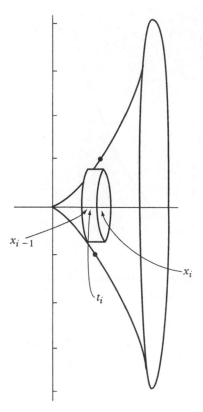

**Figure 7**  *Solid generated by rotation of a parabola.*

The trick is to get $(y_i - y_{i-1})$ expressed in terms of $x_i - x_{i-1}$, and in this endeavor we are aided by a plausible sounding theorem which says that we can find a point $t_i$ between $x_{i-1}$ and $x_i$ such that the slope of the line tangent to the curve $y = f(x)$ at $t_i$ will be equal to the slope of the line segment joining $(x_{i-1}, y_{i-1})$ to $(x_i, y_i)$. In symbols, $f'(t_i) = (y_i - y_{i-1})/(x_i - x_{i-1})$ or $y_i - y_{i-1} = f'(t_i)(x_i - x_{i-1})$. This enables us to write

$$L_i = \sqrt{(x_i - x_{i-1})^2 + [f'(t_i)]^2(x_i - x_{i-1})^2}$$
$$= \sqrt{1 + [f'(t_i)]^2}(x_i - x_{i-1})$$

If we define $g(x) = \sqrt{1 + [f'(x)]^2}$, then the approximating sum $L_1 + L_2 + \cdots + L_n$ takes the form

$$g(t_1)(x_1 - x_0) + g(t_2)(x_2 - x_1)$$
$$+ \cdots + g(t_n)(x_n - x_{n-1})$$

This is the sum that is designated by $\int_a^b g$, so

$$L = \int_a^b \sqrt{1 + [f'(x)]^2}$$

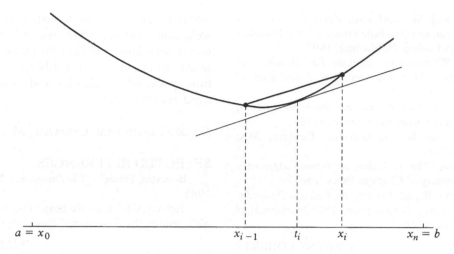

**Figure 8**   *Approximating the length of a curve.*

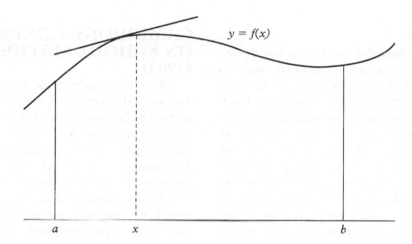

**Figure 9**   *Graph of a function.*

## SUMMARY

The two basic questions of calculus can both be asked about the graph of a function $f$ defined from $x = a$ to $x = b$ (Figure 9). They are: (1) what is the slope of the line tangent to the graph of $y = f(x)$ at $x$?, and (2) what is the area under the graph of $y = f(x)$ from $x = a$ to $x = b$? Attempts to answer these questions lead us to consider two infinite processes: the derivative $f'(x) =$

$$\lim_{t \to 0} \frac{f(x + t) - f(x)}{t}$$

and the integral $\displaystyle\int_a^b f$. Each of these concepts is useful in applications, and each can, with the aid of electronic calculation be determined with enough accuracy for any practical purpose. The Fundamental Theorem of Calculus is, nevertheless, that the two ideas are related by the fact that if $G'(x) = f(x)$, then

$$\int_a^b f = G(b) - G(a).$$

*See also* Differentiation; Integration; Rates in Calculus

## SELECTED REFERENCES

Apostol, Tom M. *Calculus.* 2 vols. Waltham, MA: Blaisdell, 1961.

Boyer, Carl B. *The History of the Calculus and Its Conceptual Development.* New York: Dover, 1959.

Ganter, Susan L. *Calculus Renewal: Issues for Undergraduate Mathematics Education in the Next Decade.* Dordrecht, Netherlands: Kluwer, 2000.

Hughes-Hallet, Deborah, Andrew M. Gleason, et al. *Calculus.* New York: Wiley, 1992.

Ostebee, Arnold M., and Paul Zorn. *Calculus from Graphical, Numerical, and Symbolic Points of View.* Philadelphia, PA: Saunders College Publishing, 1997.

Roberts, A. Wayne, ed. *Calculus: The Dynamics of Change.* Washington, DC: Mathematical Association of America, 1996.

Sawyer, Walter W. *What Is Calculus About?* Washington, DC: Mathematical Association of America, 1961.

Thomas, George B., Jr. *Calculus.* Reading, MA: Addison-Wesley, 1961.

Toeplitz, Otto. *The Calculus: A Genetic Approach.* Chicago, IL: University of Chicago Press, 1963.

Tucker, Thomas W., ed. *Priming the Calculus Pump: Innovations and Resources.* Washington, DC: Mathematical Association of America, 1990.

A. WAYNE ROBERTS

needed. Although this adjusted calendar serves us well, being off less than one half-minute per year, our descendants will have to make calendar adjustments to counteract this difference. If people move into space, other calendar and time issues will also need reconciliation.

*See also* Enrichment, Overview; Mayan Numeration

SELECTED REFERENCES

Boorstin, Daniel J. *The Discoverers.* New York: Vintage, 1985.

Taisbak, Christian Marinus. "Dionysius, Zero, and the Millennium: the Real Story." *Focus* 20(6)(2000):14–15.

WILLIAM D. JAMSKI

# CALENDARS

Systems for defining the beginning, length, and parts of a year for a planet, consistent with its movements through space. The need arose when developing civilizations wanted to coordinate fixed-date ceremonies on an earth completing one revolution about its sun every 365 days, 5 hours, 48 minutes, and 46 seconds, while spinning on its own axis and accompanied by a moon that completed twelve full cycles of phases in a lunar year of 354 days, 8 hours, and 48 minutes. Generally aware of this situation, our ancestors reconciled its incongruities quite well with their own calendars. Among many of special interest are: the old Chinese calendar based on six sixty-day cycles; the Moslem calendar, which still follows the lunar year; the Jewish calendar based on solar year adjustments to a lunar calendar; the Mayan 365-day calendar; and the early Roman calendar. Julius Caesar improved the Roman calendar in 46 B.C. by establishing a calendar (now called Julian) with a year of 365 days, except for every fourth one with 366, in effect making a year of 365 days and 6 hours. Despite what was a great short-term improvement, over time the seemingly minor, less than twelve-minute, annual discrepancy accumulated. By 1582 the planetary start of spring and the accepted calendar start were ten days apart, which Pope Gregory XIII rectified in Roman Catholic lands by just skipping forward the needed amount. This Gregorian calendar further established that centenary years not divisible by 400, such as 1700, would not be leap years, whereas those divisible by 400, such as 2000, would. In 1752, England and its colonies accepted this calendar, skipping eleven days as

# CAMBRIDGE CONFERENCE ON SCHOOL MATHEMATICS (1963)

With funding from the National Science Foundation (NSF), a group of twenty-five mathematicians created a discussion document on the K–12 mathematics curriculum that could be implemented "over the next few decades." Their report, *Goals for School Mathematics,* was a "new mathematics" effort to compress K–16 mathematics grade levels to K–12. The report further stated, "If this report, however, fulfills its purposes by providing general debate and bold experimentation, those guidelines [for curriculum] may ultimately emerge" (Cambridge Conference 1963, ii). The discussion for K–6 was organized under these headings: real number system, geometry, logic, set theory, functions, foundations, theory of real functions, and applications. For grades 7–12, the headings were: algebra, probability, geometry, geometry and topology of the complex plane, linear algebra, and analysis. This report caused widespread debate at the time, fueled by the review by Marshall Stone of the University of Chicago. In his review, Stone severely criticized the report for its lack of attention to the structure, and the practicability of implementation, of the suggested curriculum framework. He stressed the lack of realism in several areas, including teacher preparation and students' ability to learn. The report was too radical for general acceptance by the mathematics education community.

*See also* History of Mathematics Education in the United States, Overview; *Revolution in School Mathematics*

SELECTED REFERENCES

Cambridge Conference on School Mathematics. *Goals for School Mathematics*. Boston, MA: Houghton Mifflin, 1963.

Jones, Phillip, S., ed. *A History of Mathematics Education in the United States and Canada*. 32nd Yearbook. Washington, DC: National Council of Teachers of Mathematics, 1970.

Stone, Marshall H. "A Review of *Goals for School Mathematics*." *Mathematics Teacher* 58(Apr. 1965): 353–360.

JAMES K. BIDWELL
ROBERT G. CLASON

# CANADA

Mathematics education is influenced by complex issues such as political mandates, trends in the discipline, changes in the workplace, and the perceived needs of society. The impact of these issues on the educational system is tempered by the structure and governance of education in Canada, results of provincial and national assessments of mathematics, findings from research, and recommendations of professional groups such as provincial mathematics associations and the National Council of Teachers of Mathematics, based in the United States. To maintain a Canadian perspective on what mathematics is taught, how the curriculum is delivered, and what content and concepts are included in the curriculum, textbooks are either developed and produced in Canada or else revised from other jurisdictions.

Under the *Constitution Act* of 1867 (Gayfer 1991) education was placed under the sole jurisdiction of each provincial legislature. As a result, each province has centralized its decision making relative to four major components: the organization of the educational system, the development and prescription of curricula, the selection of learning resources, and the preparation and certification of teachers. In most of the ten provinces and in the two territories, the public education system begins either at the kindergarten level (age five) or grade 1 and finishes at the end of grade 12. Two exceptions are in the provinces of Quebec and Ontario, where public education ends after grades 11 and 13 respectively. In spite of centralized control of curriculum by each separate province and territory, more similarities than differences exist in the content and organization of mathematics courses across Canada.

## THE MATHEMATICS CURRICULUM

Most recent revisions to mathematics curricula across provinces have included greater focus on problem-solving skills and real-world applications. In the content area, changes have included increased emphasis on data analysis at all levels of the system, and on decimals in earlier grades at the elementary level. A greater interest has also been evident in the role of geometry, including spatial visualization and structured proof. An overriding theme calls for more authentic real-world applications (Jones 1991) and applications to technology (Burgess 1990).

A review of the mathematics content taught by each province at the secondary level, as described in *Secondary Education in Canada: A Student Transfer Guide* (Council of Ministers of Education 1991) identifies similarities across provinces. For example, at the grades 7–9 levels, several provinces organize content around the following themes: number and number operations, data analysis, geometry, measurement, algebra, and ratio and proportion. The rest include these topics as discrete units in courses offered at each level. In most cases problem solving and real-life applications are themes integrated across content areas. Greater diversity enters into the content and structure of courses beginning at grade 10. At this level, most provinces offer a variety of courses, focusing on consumer and/or computer applications and academic preparation. Each province and territory includes traditional precalculus topics among academic preparation courses offered in the upper secondary grades. At the elementary level, most jurisdictions include topics in number and operations; data analysis; geometric shapes, sizes, and transformations; measurement; and number sentences. There is an emphasis on the use of manipulative materials and the use of calculators is encouraged.

## TEACHING PRACTICES

Although direct instruction remains an important strategy, teachers have expanded the number and variety of teaching approaches and methods of classroom organization. For example, findings from the 1990 Provincial Assessment of Mathematics (Robitaille 1991) showed that in British Columbia, a majority of teachers of grades 4, 7, and 10 reported they were more likely to use cooperative learning groups than they did previously, and indicated a greater focus on problem solving now than in the past.

Liedtke (1990) felt that at the elementary level the shift in content emphasis and the adoption of new texts were of less importance than development of conceptual understanding and problem solving. The teaching strategy he felt particularly conducive to conceptual understanding was cooperative learning, due to the opportunities it provides students to learn in a group setting and to discuss ideas or strategies in their own words. Further cooperative learning techniques at the secondary level were reported by DeGroot (1990) and Carrodus (1993).

## ASSESSMENT PRACTICES

Teacher-made tests and quizzes are most commonly used in mathematics classrooms across Canada for student assessment purposes. The nature of these instruments is influenced to a great extent by exemplars in curriculum guides and sample questions in textbooks. In recent years, additional use has been made of student portfolios, observational techniques, and three-way conferencing at the elementary level, in which the student reviews and discusses his achievement with the parent and teacher. Four provinces administer final examinations in mathematics at the end of secondary school. Program level assessments in mathematics are administered every three or four years in five provinces at three different grade levels and one province administers commercially produced standardized tests in mathematics at grades 3, 6, and 9.

## CHANGES IN MATHEMATICS EDUCATION

As part of the curriculum analysis component for the Third International Mathematics and Science Study (Robitaille et al. 1993), provinces were surveyed on various aspects of the mathematics curriculum. Some of the questions asked dealt with major changes in the mathematics curriculum during the 1980s and anticipated changes. Based on responses to these questions it is clear that among the major changes in practice and focus during the 1980s were the increased use of manipulatives and an emphasis on problem solving. Content shifts have included the teaching of decimal fractions at earlier grades and a greater focus on data analysis, probability, and statistics. Provinces reported that it is expected that the focus on problem-solving skills will continue and the use of electronic aids will grow significantly. Further emphasis is expected in curriculum integration, mathematical modeling, and real-world applications. In addition, it is anticipated that assessment of student performance in activity oriented situations will grow in popularity.

*See also* International Comparisons, Overview; International Studies of Mathematics Education; Third International Mathematics and Science Study (TIMSS)

## SELECTED REFERENCES

Burgess, W. D. "New Mathematics Is an Essential Part of New Technology." *Ontario Mathematics Gazette* 28(3)(1990):6–7.

Carrodus, Jon. "Co-operative Learning: Transforming the Math Classroom." *Vector: Journal of the British Columbia Association of Mathematics Teachers* 35(1)(1993):48–51.

Council of Ministers of Education. *Secondary Education in Canada: A Student Transfer Guide.* 6th ed. Toronto, ONT, Canada: The Council, 1991.

DeGroot, Ian. "Co-operative Grouping and Assessment in a Mathematics Classroom." *Vector: Journal of the British Columbia Association of Mathematics Teachers* 32(1)(1990):26–28.

Gayfer, Margaret. *An Overview of Canadian Education.* 4th ed. Toronto, ONT, Canada: Canadian Education Association, 1991.

Hunter, Dave. "Where are we and Where do we go from Here?" *Ontario Mathematics Gazette* 28(1)(1989): 6–11.

Jones, M. Ann. "President's Message." *Ontario Mathematics Gazette* 30(2)(1991):3.

Liedtke, Werner. "Mathematics Curriculum Changes —Implications for Diagnosis and Remediation." *Vector: Journal of the British Columbia Association of Mathematics Teachers* 35(1)(1990):12–16.

Maloney, Jerry. "Specialization Years Committee Comments." *Ontario Mathematics Gazette* 31(3)(1993): 10–11.

Putz, Barry. "Elementary Mathematics in the 90's." *Saskatchewan Mathematics Teachers' Society Journal* 27(3)(1991):37–39.

Robitaille, David, ed. *The 1990 British Columbia Mathematics Assessment: Technical Report.* Victoria, British Columbia, Canada: British Columbia Ministry of Education and Ministry Responsible for Multiculturalism and Human Rights, 1991.

Robitaille, David, et al. *Curriculum Frameworks for Mathematics and Science.* Vancouver, British Columbia, Canada: Pacific Educational, 1993.

ALAN R. TAYLOR

# CANTOR, GEORG (1845–1918)

The originator of transfinite set theory. Cantor was born in St. Petersburg and raised in Germany. He studied at the universities of Göttingen and Berlin, where he earned his Ph.D. degree in 1866 with a dissertation on number theory, having studied with Ernst Eduard Kummer, Leopold Kronecker, and Karl Weierstrass. From Berlin he accepted a position at Halle University, where he spent the rest of his academic career. Following a series of innovative papers devoted to the question of whether for an arbitrarily given function, its representation by a trigonometric series was unique, Cantor was led from topological considerations of the domain of definition to a rigorous treatment of the real numbers (1872). This in turn prompted his first revolutionary discovery (in 1873) that the set of real numbers is nondenumerably infinite, which helped to stimulate his subsequent interest in developing a theory of transfinite ordinal numbers and later a theory of cardinal numbers.

In 1891, Cantor was elected the founding president of the Deutsche Mathematiker Vereinigung (Union of German Mathematicians). In 1895–1897 he published (in two parts) his definitive presentation of transfinite set theory, which included his famous Continuum Hypothesis. This asserts that the power or cardinal number, aleph one, of the set of real numbers $R$ is the second transfinite cardinal number immediately following the least transfinite power or smallest infinite cardinal number, aleph null, of the set of all natural numbers $N$. Cantor, unfortunately plagued by a life-long affliction of manic-depression, died in a mental hospital in Halle, Germany, in 1918.

Cantor's theory, despite its success in applications, especially in the theory of functions and in providing a general foundation for all of mathematics, also engendered paradoxes that led to various alternatives of set theory (which Gottlob Frege criticized sharply for its psychologism). Ernst Zermelo took an axiomatic approach to set theory, while Bertrand Russell and Alfred North Whitehead attempted to establish the consistency of all mathematics using symbolic logic in their *Principia Mathematica* (1910–1913). David Hilbert also had a similar goal in mind, hoping to establish formally the consistency of all mathematics. All such attempts were eventually shown to be impossible in a series of important papers published by Kurt Gödel just prior to World War II.

*See also* Gödel; Sets

SELECTED REFERENCES

Dauben, Joseph W. *Georg Cantor: His Mathematics and Philosophy of the Infinite.* Cambridge, MA: Harvard University Press, 1979. [Reprinted; Princeton, NJ: Princeton University Press, 1990.]

Faris, John Acheson. *Plato's Theory of Forms and Cantor's Theory of Sets.* Belfast, Ireland: Queens University, 1968.

Gödel, Kurt. *Collected Works: Vol. 1.* Solomon Feferman, et al., eds. New York: Oxford University Press, 1986.

———. *The Consistency of the Axiom of Choice and of the Generalized Continuum Hypothesis with the Axioms of Set Theory.* Princeton, NJ: Princeton University Press, 1940. [Reprinted with additional notes, 1951. Reprinted with more additional notes, 1966.]

———. "On Formally Undecidable Propositions of *Principia Mathematica* and Related Systems I" [in German]. *Monatshefte für Mathematik und Physik* 38 (1931):173–198. [In van Heijenoort 1967: 596–616; reprinted in Gödel 1986:145–195.]

Meschkowski, Herbert. *Probleme des Unendlichen: Werk und Leben Georg Cantors.* Braunschweig, Germany: Vieweg, 1967.

Purkert, Walter, and Hans J. Ilgauds. *Georg Cantor* [in German]. Basel, Switzerland: Birkhäuser, 1987.

Van Heijenoort, Jan, ed. and trans. *From Frege to Gödel. A Source Book in Mathematical Logic, 1879–1931.* Cambridge, MA: Harvard University Press, 1967.

JOSEPH W. DAUBEN

# CAREER INFORMATION

The first section of this entry describes the recent history and the current status of the job market in mathematics. The second section describes some of the most common mathematical professions and how to prepare for them. The third and last section deals with advice for students on how to prepare for a mathematical career: what classes to choose, how to learn more about mathematical professions, and how to develop a career portfolio.

## BRIEF HISTORY OF THE MATHEMATICS JOB MARKET

The end of the twentieth century has seen wild swings in the mathematical job market. The craze that began in 1957 with the launch of the Russian space vehicle, *Sputnik,* created many new jobs for physicists and mathematicians during the 1960s; this was followed by severe job shortages in the 1970s. During the 1980s, the market for mathematicians was relatively stable. Demographers of that decade took into account the increasing number of students who would need technically trained teachers, the growing use of technology in manufacturing and consulting, the large number of scientists and mathematicians preparing to retire, together with the decreasing number of American students actually pursuing technical careers. Considering all these factors, they concluded that the number of mathematically oriented jobs would soon far outweigh the available supply of technically trained people (Johnston and Packer 1987).

The 1990s proved that predictions of a shortage of mathematicians did not hold true—in fact, just the opposite happened. A global recession, widespread corporate restructuring, and cutbacks in the budgets of state universities combined to create record-setting unemployment rates among young mathematicians (Fulton 1995). There is always uncertainty about when and how the job market will improve. (It should be noted that the recession did not affect mathematics and science alone.)

Because of the increasingly technical nature of society, a lack of mathematical sophistication closes out job opportunities. Avoiding mathematics in school leaves an applicant with fewer career choices and fewer chances of promotion. Furthermore, mathematical careers are consistently rated among the most satisfactory by journals such as *Jobs Rated Almanac* (Krantz 1992): jobs which require a great deal of mathematical knowledge are likely to pay more, to be less hazardous to one's health, to have a more pleasant working atmosphere, and to provide greater personal satisfaction to the employee than nonmathematical jobs.

Students should take as many mathematics and English classes as they can, especially while in high school. College professors in the sciences note that the greatest shortcoming their students have is *not* a lack of scientific knowledge, but a lack of mathematical and communications skills. Students will find that many enjoyable fields of study—such as psychology, medicine, economics, chemistry, astronomy, and sociology—are closed to them unless they are able to master algebra and to have a reasonable facility with calculus and statistics.

## MATHEMATICAL JOBS

Mathematics careers can be successfully created in many fields, so it is difficult to provide a complete listing of jobs held by mathematicians. It is possible, however, to give a listing of the more popular and clearly defined mathematical professions. The professions described below have considerable overlap; nonetheless, they should help to provide some idea as to what mathematicians do.

There are many professional organizations that provide information on beginning and pursuing a career in their field. Many of these have information on fellowships and grants available to students; most have newsletters that are free to members of their organization (membership is usually inexpensive for students).

### Actuary

An actuary is a mathematician who does statistical and quantitative work for an insurance agency, for example, figuring out which factors make a person riskier than others to insure, or helping to set fees for individuals in different risk categories. An actuary must have at least a college degree in mathematics, with a good knowledge of statistics, probability, numerical analysis, and computer programming. Understanding medicine, economics, and business management helps, too. To become an actuary, the applicant takes a series of tests; each test successfully completed by the applicant promotes the applicant to a new professional level.

Summer internships are occasionally available from insurance firms or consulting firms; applications for these should be made in December, as such jobs are usually filled very quickly. For more information, contact the Society of Actuaries or the Casualty Actuarial Society.

### Consultant/Industrial Mathematician

Industrial mathematicians deal with problems that can be solved by quantitative methods: What is the most efficient way to do a certain job? How can we design this machine in the least expensive way? How can we simulate this design of the machine so we know what won't work *before* we build it?

Because it is difficult and expensive to hire an experienced, trained staff to deal with every temporary technical problem that arises, both government and industry frequently make use of technical consulting firms. These firms range in size from one to thousands of employees; their areas of expertise vary from firm to firm. In all of these, however, it is extremely important for the employee to be versatile and articulate: The job for which the employee was hired is seldom the same job that she or he will be expected to work on six months later. The ability to promote the firm's expertise and to communicate technical material to a nontechnical audience also is important.

Anyone wishing to work with a consulting firm or even to start a new firm should have, in addition to programming, mathematical, and writing skills, experience in consulting or industry. A visit to a consulting firm or a business is ideal for learning more about how local firms operate (Benkoski 1994). More information about working in industry can be obtained from the Society for Industrial and Applied Mathematics.

## Computer Scientist/Engineer

Computer scientists and engineers are not typically considered mathematicians, but both of these professions require a good working knowledge of mathematics. Although today it is possible to earn a degree in computer science, at first it was considered a subfield of mathematics. Computer scientists and engineers, because of their applied training, have less difficulty convincing potential employers that they will be productive employees.

A student can get a college degree in either of these fields; it also is possible to get a college degree in mathematics and then proceed to get an advanced degree in these fields.

## Cryptologist

Cryptology also is known as *coding theory;* it deals with creating codes that are very hard to break. Cryptologists used to work almost exclusively for the government—in particular, for the National Security Agency—but the widespread use of networked computers has impelled industries and financial institutions to make increased use of cryptology.

A person interested in pursuing a career in cryptology should take a good deal of mathematics—in particular, number theory and abstract algebra—and also courses in computer science. Because of the delay in obtaining security clearances, plans for summer internships with the government or consulting firms should be made early (possibly even the preceding summer).

## Finance

A career in finance can include working with stock exchanges, investment banks, securities dealers, trust companies, utilities and insurance companies, commodity exchanges, and brokerage firms. A broad mathematical background, a facility with computers, and an understanding of economics are important in the field of finance.

## Operations Research

Operations Research (OR) deals with making decisions about large, complex processes: It combines engineering, mathematics, and psychology. OR specialists work with companies to help them design their systems to be more efficient and also more useful to their customers. For example, it is easier to drive a friend's car than to program a friend's VCR, because cars are standardized from one to the next, and also the knobs and pedals and steering wheel of a car are easy to distinguish, whereas the buttons on a particular VCR are similar to each other. Knowing this, how would one design a remote control for a VCR? Like statisticians, operations researchers work in a variety of settings, including industry, academia, and government.

To prepare for a career in OR, a student should take probability and statistics, linear algebra, computer programming, and economics, and be prepared to pursue graduate level work after college. For more information, contact the Operations Research Society of America.

## Research Mathematician

Universities are the most common employers for research mathematicians, although it is not uncommon to find mathematical researchers working in technological industries. A research mathematician works at finding interesting mathematical problems and then at discovering solutions to those problems. Some of the more famous examples of questions that research mathematicians have considered are Fermat's Last Theorem, the Four Color Theorem, and the transition from stability to chaos. Researchers who work at universities or colleges will often be expected to publish their work, to teach classes, and to help with the governance of the institution. A Ph.D. in mathematics is the typical requirement for entering the field. For more information, contact the American Mathematical Society or the Association for Women in Mathematics.

## Retailing/Marketing

Working in marketing requires a great deal of general mathematical knowledge. Predicting the quantity of goods to be manufactured so that there will be neither a great surplus nor an undersupply; designing a timetable that minimizes storage and transportation costs; interpreting survey results in order to better understand the consumer market—all of these problems can be better analyzed within a mathematical framework.

To enter the field of retailing, a person should take a broad spectrum of courses in business, economics, and mathematics. As always, the ability to work with other people and to write well are important.

## Statistician

Statisticians work in a variety of settings, from academic to government to industry to marketing to media. Statistics is used to determine voting districts, landfill sites, the number of lawn mowers to make in December for the upcoming May, the effect of new drugs on AIDS, the environmental impact of new power plants, and so on. In fact, the abundant use of statistics in our society leads some professionals to suggest that students replace calculus with statistics as their first college mathematics course. Certainly any person contemplating a career in science should take at least one college course in statistics; enthusiasts can pursue a college degree or even a Ph.D. in the field. For more information about careers in statistics, contact the American Statistical Association.

## Teacher/Professor

Broad-sweeping educational reforms make this an exciting time to teach mathematics. There are many kinds of mathematics teachers, corresponding to the different kinds of institutions in which they work. Elementary and secondary school teachers who teach in public schools should have a college degree in mathematics or science—or possibly a degree in mathematics or science education—as well as a teaching certificate from the state in which they teach. Some schools require a master's degree. Those who wish to teach at the college level should plan to complete a Ph.D. in mathematics, although it is possible to find somewhat limited opportunities for teachers with a master's degree in mathematics or a doctorate in education.

There are many organizations that provide information about (and support for) mathematics teachers. These include: the American Mathematical Association of Two-Year Colleges, the Mathematical Association of America, and the National Council of Teachers of Mathematics.

## CAREER ADVICE

Preparing for a career starts long before the search for a particular job begins. This section focuses on choosing classes and designing a career portfolio.

## Choosing Classes

A student should choose classes that help to develop skills in communication and technical training with a wide, diverse background. Communication skills include the ability to write well, the ability to present information clearly and elegantly in front of an audience, and the ability to collaborate with co-workers. Whether the applicant aspires to be a teacher (who has to stand in front of a class), a business person (who makes presentations to clients, to co-workers, and to the boss), or a scientist (who presents her or his findings to the community at large meetings), the applicant will find that public speaking experience is a considerable asset in almost all professions. For similar reasons, it is important to get extensive experience writing and rewriting. Mathematicians in industry consistently emphasize that communications skills are essential to success in their profession (Davis 1994).

A technical background means knowing mathematics beyond the secondary school level (particularly calculus, probability and statistics, linear algebra, and differential equations), having familiarity with computers, and understanding how to make the connections between problems that one faces on the job and mathematical solutions.

There is a growing emphasis—both in industry and in academia—on interdisciplinary work, and so being able to communicate with people in other areas, or being able to cross the line between one field and another is a valuable skill.

Examples of interdisciplinary connections between mathematics and other fields abound. In music, mathematics is used to understand the acoustics of a music hall before it is built; computers help to fashion musical instruments; designing technological items such as CD players and stereos requires knowing a great deal about the mathematics of sound. In law, knowing mathematics is necessary for understanding statistical evidence (such as identifying a person by blood type or DNA); mathematics also is a great help to corporate lawyers, who must be familiar with accounting techniques. Medicine has seen recent technological discoveries such as ultrasound and artificial heart valves, both of which relied heavily on mathematical explorations; even daily medical practice uses mathematics in determining dosages of medicine and understanding the statistics in the most recent medical journals.

Keeping all this in mind, a student would be wise to choose a variety of classes that interest him or her and to avoid specializing too early. At the same time students should make an effort to include classes that emphasize writing, speaking, and mathematical skills, especially if these classes relate to the student's areas of interest.

## The Career Portfolio

One of the most difficult aspects of preparing for a career is actually remembering and organizing all one's past accomplishments, and then putting all this information together in a way that others can best appreciate it. A *portfolio* is an excellent way to begin this process. Artists and architects have long used portfolios to showcase their work. Teachers are beginning to use portfolios to document their teaching effectiveness, and job applicants can use portfolios to help them begin their own searches.

Into this portfolio, the potential applicant should place anything that makes the applicant proud or that explains what the applicant has been doing. Students should start their portfolios while still in school—for cxample, it might be kept by a student throughout high school or graduate school. Examples of what can go into a portfolio include descriptions of work or volunteer experiences; report cards, transcripts, and so forth; letters from others thanking the applicant for doing a good job; art, essays, published work; awards, honors, acceptance letters; self-evaluations; materials from conferences or summer workshops; and anything else that might go into a résumé or letter of recommendation.

The portfolio can be used to help put together a résumé; it also can be lent to letter writers (especially if it is well organized and visually appealing). It also can be taken along on job interviews, to serve as a reference and as a promotional tool. Teachers occasionally use portfolios as a means of evaluating students' performance in class; urging students to start a "Career Portfolio" is a natural extension of this idea that will serve students in good stead later on.

*See also* Actuarial Mathematics; Computer Science; Cryptography; Engineering; Finance; Operations Research; Statistics, Overview; Teacher Preparation, College; Teacher Preparation, Secondary School

## SELECTED REFERENCES

Benkoski, Stanley. "Preparing for a Job Outside Academia." *Notices of the American Mathematical Society* 41(Oct. 1994):917–919.

*Career Information in the Mathematical Sciences: A Resource Guide.* Washington, DC: Conference Board of the Mathematical Sciences, 1995.

Davis, Paul. *1994 SIAM Forum Final Report.* (Dec. 1994), available electronically at gopher://gopher.siam.org:70/00/reports/forum94.txt

Fulton, John. "1994 Annual AMS-IMS-MAA Survey." *Notices of the American Mathematical Society* 42(Aug. 1995):863–874.

Jackson, Allyn. *Careers that Count: Opportunities in the Mathematical Sciences.* College Park, MD: Association for Women in Mathematics, 1991.

*Job Choices in Science and Engineering.* Bethlehem, PA: National Association of Colleges and Employers (formerly College Placement Council, Inc.), 1995.

Johnston, William, and Arnold Packer. *Workforce 2000: Work and Workers for the Twenty-first Century.* Indianapolis, IN: Hudson, 1987.

Krantz, Les. *Jobs Rated Almanac.* New York: Pharos, 1992.

*Occupational Outlook Handbook.* Washington, DC: United States Department of Labor, Bureau of Labor Statistics, 1995, available electronically via http://stats.bls.gov:80/ocohome.htm

*Peterson's Job Opportunities for Engineering, Science, and Computer Graduates.* Princeton, NJ: Peterson's, 1993.

*Prentice-Hall Guide for Fellowships for Math and Science Students.* Des Moines, IA: Prentice-Hall, 1993.

*Seeking Employment in the Mathematical Sciences.* Providence, RI: Mathematical Sciences Employment Register, 1994.

Sterrett, Andrew, ed. *101 Careers in Mathematics.* Washington, DC: Mathematical Association of America, 1996.

ANNALISA CRANNELL

# CARROLL, LEWIS/CHARLES LUTWIDGE DODGSON (1832–1898)

Lewis Carroll was the pen name of Charles Dodgson, best known for his classic, *Alice in Wonderland.* Dodgson was actually a serious logician, and much of the fascination of his literary works, including *Alice,* comes from the delight he found in logical puzzles and games, both verbal and mathematical.

Dodgson graduated in 1854 from Christ Church College, Oxford University, where he was awarded first-class honors in mathematics (having placed first). Subsequently, he was appointed lecturer in mathematics, and several years later earned a Master of Arts degree. Dodgson also was ordained by the Church of England (his father was a clergyman), although he always remained at Oxford teaching mathematics. He was especially interested in determinants, geometry, and the mathematical analysis of tournaments and elections. In fact, his consideration of the seeding of tennis tournaments and how best to arrive at fair decisions by committees and juries led to his analysis of these subjects using matrix methods to represent multiple decisions. His results contributed greatly to understanding how preferential ballots could be used to reach a fair majority.

Dodgson's concern for education was reflected in a five-act comedy, *Euclid and His Modern Rivals* (1879), in which a mathematician (Minos) dreams that Euclid is debating with modern mathematicians, including lively discussion of the significance of the parallel postulate in Euclidean geometry. Among the textbooks he published was *An Elementary Treatise on Determinants* (1867), but in it he used unfamiliar notation and this prevented its widespread acceptance. Better known was an elementary work, *A Syllabus of Plane Algebraic Geometry* (1860).

Although Dodgson published his work in mathematics under his own name, he always published his work in logic using the well-known pseudonym Lewis Carroll. Dodgson was fascinated by logical, linguistic, semantic, and other conundrums, all of which he popularized most spectacularly in *Alice in Wonderland*. Subsequently, many of the problems in logic that Dodgson devised have stimulated serious research by mathematicians and logicians alike. He died in Surrey, England, January 14, 1898.

*See also* Determinants; Logic

## SELECTED REFERENCES

Abeles, Francine F. "Lewis Carroll's Method of Trees. Its Origins in Studies in Logic." *Modern Logic* 1(1990):25–35.

———. "Some Victorian Periodic Polyalphabetic Ciphers." *Cryptologia* 14(1990):128–134.

Collingwood, Stuart D. [Dodgson's nephew]. *The Life and Letters of Lewis Carroll.* London, England: Unwin, 1898.

Dodgson, Charles L. *The Annotated Alice.* Martin Gardner, ed. New York: Bramhall, 1960.

———. *The Annotated Snark.* Martin Gardner, ed. New York: Simon and Schuster, 1962.

———. *An Elementary Treatise on Determinants.* London, England: Macmillan, 1867.

———. *Euclid and His Modern Rivals.* London, England: Macmillan, 1879.

Eperson, D. B. "Lewis Carroll—Mathematician." *Mathematical Gazette* 17(1933):92–100.

Green, Roger L. *Lewis Carroll.* New York: Walck, 1962.

Kelly, Richard. *Lewis Carroll.* Boston, MA: Twayne, 1990.

JOSEPH W. DAUBEN

# CAUCHY, AUGUSTIN-LOUIS (1789–1857)

Contributed to all existing areas of mathematics as well as to astronomy, mechanics, optics, and elasticity, but his most influential works were in real and complex analysis. He was born in Paris, studied at the recently established *École Polytechnique,* and began his career as a military engineer. He was subsequently named professor of mathematics at both the *École Polytechnique* and the Sorbonne. He was a devout Catholic and an unyielding royalist. Cauchy was extremely productive—his collected works comprise twenty-seven thick volumes.

With Carl Friedrich Gauss, Niels Henrik Abel, and Bernhard Bolzano, Cauchy ushered in a spirit of scrutiny of the concepts and methods in various areas of mathematics, especially in analysis. He put the calculus on firm foundations, basing it on the concept of limit. This came about after two hundred years of the subject's vigorous, but unrigorous, development. And for over a quarter of a century Cauchy worked singlehandedly to establish the elements of complex function theory—one of the greatest creations of nineteenth-century mathematics.

*See also* Calculus, Overview; History of Mathematics, Overview

## SELECTED REFERENCES

Belhoste, Bruno. *Augustin-Louis Cauchy: A Biography.* Frank Ragland, trans. New York: Springer-Verlag, 1991.

Bell, Eric Temple. *Men of Mathematics.* New York: Simon and Schuster, 1937.

Grabiner, Judith V. *The Origins of Cauchy's Rigorous Calculus.* Cambridge, MA: MIT Press, 1981.

ISRAEL KLEINER

# CHAOS

Often called the theory of order within disorder. Its essential principles are illustrated with the following experiment (see Peitgen, Jürgens, and Saupe 1992, Chapter 11).

Enter into a programmable calculator the function $f_2(x) = 2x(1 - x)$. Next pick an arbitrary point $x_0$ from the interval $(0,1)$, and let $x_1 = f_2(x_0)$, $x_2 = f_2(x_1)$ and so forth. For example, if $x_0 = \frac{1}{3}$, then $x_1 = .4444\ldots, x_2 = .4938\ldots$, and $x_4 = .4999\ldots$. The result is an infinite sequence $S = \{x_0, x_1, x_2, \ldots\}$ contained in $(0,1)$, whose points are called the *iterates* of $x_0$. In our example, $S = \{\frac{1}{3}, .4444\ldots, .4938\ldots, .4999\ldots, \ldots\}$ are the iterates of $\frac{1}{3}$. Surprisingly, no matter which point $x_0$ from $(0,1)$ is chosen, $S$ converges to the point $p = \frac{1}{2}$. Since $f_2(\frac{1}{2}) = \frac{1}{2}$, $\frac{1}{2}$ is said to be a fixed point of $f_2$.

The particular function $f_2$ is an example from a family of functions defined on $(0,1)$ of the form $f_\lambda(x) = \lambda x(1 - x)$, which are called logistic functions. As $\lambda$ varies incrementally from 1 to 4, $f_\lambda$ changes from a well-behaved function to a chaotic one. To illustrate this transition to chaos, we will focus on the infinite sequence $S$ defined above for $f_2$, and defined analogously for any $f_\lambda$: let $x_0$ be any point in $(0,1)$ and let $S = \{x_0, x_1, x_2, \ldots\}$ where $x_1 = f_\lambda(x_0), x_2 = f_\lambda(x_1)$, and so forth.

When $1 < \lambda < 3$, $S$ behaves just as it does when $\lambda = 2$, that is, for any $x_0$ in $(0,1)$, $S$ converges to a fixed point of $f_\lambda$. When $\lambda$ becomes slightly greater than 3, $S$ no longer converges to a fixed point. Rather, it oscillates arbitrarily close to what are called periodic points of $f_\lambda$. In general, if $x_n$ is the first iterate of $x_0$ to equal $x_0$, then $x_n$ is a period $n$ point of $f_\lambda$. For example, if $\lambda = 3.1$ there are two periodic points and $S$ oscillates arbitrarily close to two period two points; when $\lambda$ becomes approximately 3.45, $S$ begins to oscillate between four period four points. Increase $\lambda$ to 3.55 and $S$ oscillates between eight period eight points. If we keep incrementally increasing $\lambda$ all periods eventually appear, as is the case when $\lambda = 3.83$.

Once $\lambda$ equals 4 the iteration of $f_\lambda$ becomes so complicated that (a) near any point $x_0$ in $(0,1)$ is a point $y_0$ whose iterates wander throughout $(0,1)$ in such a strange fashion that they come arbitrarily close to all points in $(0, 1)$ infinitely often; (b) iterates of nearby points generally separate from one another quite dramatically; and (c) arbitrarily close to any point in $(0,1)$ is a period $n$ point of $f_\lambda$. Functions with these three properties are called chaotic.

Once it became apparent that chaotic processes turn up in chemistry, biology, meteorology, physics, and geology, as well as in mathematics, chaos emerged as a field of study (Gleick 1987). It is generally agreed that the word "chaos" was first used in a precise mathematical fashion in "Period Three Implies Chaos" (Li and Yorke 1975). Nonetheless, the history of chaos serves to show that a mathematical object can exist long before anyone thinks to name it.

The first example of a chaotic set evidently occurred in 1871 in a paper by Ernst Schröder on the iteration of Newton's method over 100 years before Li and Yorke's paper (Alexander 1994, chap. 1). Examples of chaos also appear in studies of the stability of the solar system initiated by the great French mathematician Henri Poincaré in the 1880s (Ekeland 1988, chap. 2), as well as in Pierre Fatou and Gaston Julia's respective studies of the iteration of complex functions in the early 1900s (Alexander 1994).

The role of the computer in the development of the theory of chaos illustrates how technology can midwife the birth of a theoretical field of knowledge. In the early 1960s, Edward Lorenz's discovery of the so-called butterfly effect, namely that small perturbations in the weather in one locale can drastically alter the weather in another, is often cited as the beginning of the contemporary study of chaos and was made possible via the computer. Robert May's observation in the early 1970s that population changes can be chaotic was the result of an experiment with a pocket calculator very much like the one described at the beginning of this entry. Moreover, Benoit Mandelbrot's striking computer-generated images of fractal sets did much to popularize the theory of chaotic processes in the 1970s (Gleick 1987). During the same period, Steve Smale and others developed a mathematical theory of dynamical (i.e., time dependent) systems, including chaotic systems. This fusion of theoretical developments and experimental results in a number of fields gave rise to the theory of chaos, and it is this ongoing three-pronged unification of theory, experiment, and computational visualization that forms the backbone of the science of chaos.

*See also* Fractals

## SELECTED REFERENCES

Alexander, Daniel. *A History of Complex Analysis from Schroder to Fatou and Julia.* Wiesbaden, Germany: Vieweg, 1994.

Devaney, Robert. *Chaos, Fractals, and Dynamics: Computer Experiments in Mathematics.* Menlo Park, CA: Addison-Wesley, 1990.

Ekeland, Ivar. *Mathematics and the Unexpected.* Chicago, IL: University of Chicago Press, 1988.

Gleick, James. *Chaos: Making a New Science.* New York: Viking, 1987.

Hall, Nina. *Exploring Chaos: A Guide to the New Science of Disorder.* New York: Norton, 1993.

Li, Tien-Yien and James Yorke. "Period Three Implies Chaos." *American Mathematical Monthly* 82(1975): 985–992.

Mandelbrot, Benoit. *The Fractal Geometry of Nature.* New York: Freeman, 1983.

Peitgen, Heinz-Otto, Hartmut Jurgens, and Dietmar Saupe. *Chaos and Fractals: New Frontiers of Science.* New York: Springer-Verlag, 1992.

Peterson, Ivars. *Newton's Clock: Chaos in the Solar System.* New York: Freeman, 1993.

Stewart, Ian. *Does God Play Dice? The Mathematics of Chaos.* New York: Blackwell, 1989.

Turcotte, Donald. *Fractals and Chaos in Geology and Geophysics.* Cambridge, England: Cambridge University Press, 1992.

DANIEL S. ALEXANDER
ALEXANDER KLEINER

# CHEMISTRY

In chemistry, as in other branches of physical science, mathematics has become the core of its major theories (Kac and Ulam 1992, 145; Kline 1985, 210) and the primary language for the rigorous parts of the discipline, including stoichiometry, gas law problems, thermodynamics, equilibria, and pH problems. Mathematics is also, of course, excellent analytical training for coping with complex problems, whether these be scientific or not (Wilder 1973). A number of mathematical topics useful in elementary chemistry are functions, logarithms, and polynomials.

## FUNCTIONS

Most functions arising in mathematics or physical science are representable by formulas, but this need not always be so. There is no limit to the variety of representational forms that functions may assume. The following familiar formula is an elementary example from chemistry: $PV = nRT$.

*Example 1* For a purely hypothetical gas whose constituent particles possess no volume, exert no forces upon one another, and yet create a pressure at the walls of a confining vessel, the equation of state $PV = nRT$ holds. Here, $P$ is the gas pressure, $V$ is the volume of the container, $n$ is the number of moles of gas, $R$ is a universal constant, and $T$ is the Kelvin temperature. A hypothetical gas such as just described is called an *ideal gas;* an ideal gas is a close approximation to a real gas under nonextreme conditions (Ebbing 1990, 142).

If the point of view is the calculation of the pressure of an ideal gas, then write the above relation as $P = f(n, T, V) = nRT/V$. The domain $D$ is then the set of all triples $(n, T, V)$ in which each independent variable is a positive number.

## LOGARITHMS

*Example 2* The *acidity* of an aqueous solution is defined quantitatively by the logarithmic function $pH = -\log_{10}[H_3O^+]$, where $[H_3O^+]$ denotes the concentration in moles per liter of solution of the hydronium ion $H_3O^+$, a species present in any acidic aqueous solution (Ebbing 1990, 638). A related definition is that of pOH, given by $pOH = -\log_{10}[OH^-]$, where the hydroxide ion $OH^-$ is a species present in any aqueous solution of a *base.*

At room temperature, it is found experimentally that the product of the concentrations of $H_3O^+$ and $OH^-$ is nearly a constant for any aqueous solution whatsoever: $[H_3O^+][OH^-] \approx 1 \times 10^{-14}$. Hence, by using properties of logarithms, we obtain the operationally simple relationship $pH + pOH \approx 14$.

*Example 3* If generic substances $A$, $B$ are allowed to react at the Kelvin temperature $T$ in a closed vessel according to the chemical equation $aA + bB \rightleftarrows cC + dD$, then thermodynamics states that after sufficient time the four variable concentrations $[A]$, $[B]$, $[C]$, $[D]$ will attain constant (equilibrium) values, and the quantity $K$ given by

$$K = \left( \frac{[C]^c[D]^d}{[A]^a[B]^b} \right) \text{ at equilibrium}$$

will be fixed (so long as $T$ is constant). The number $K$ is called the *equilibrium constant.* On the other hand, if $a$ moles of $A$ and $b$ moles of $B$ are converted at temperature $T$ into $c$ moles of $C$ and $d$ moles of $D$, all four substances being in certain defined standard states, then there is an accompanying change in the thermodynamic *standard molar Gibbs free energy,* $\Delta \widetilde{G}°$. This quantity is related to $K$ by the remarkable formula, $\ln K = -\Delta \widetilde{G}°/RT$ (Ebbing 1990, 730), where $R$ is the universal ideal gas constant (see Example 1).

## POLYNOMIALS

The *Fundamental Theorem of Algebra* guarantees that every polynomial has a zero, which may itself be either real or complex (Churchill and Brown 1984, 118). Only real zeros are significant in chemistry. It may happen that one of these zeros in a specific application is to be rejected on purely physical grounds (e.g., a negative volume).

*Example 4* At some temperature T the equilibrium constant (see Example 3) for

$$N_{2(g)} + 3H_{2(g)} \overset{K}{\rightleftharpoons} 2NH_{3(g)}$$

is $K = 0.500$. Suppose the pressures of the three gases upon initial mixing are all 1 atm. These are not equilibrium pressures. Let $2x$ denote the decrease in pressure of $NH_3$ when equilibrium is reached. The numerical coefficients in the chemical equation above give us the corresponding changes in pressure of the other two ingredients, so that at equilibrium we have

$$N_{2(g)} \quad + \quad 3H_{2(g)} \overset{K}{\rightleftharpoons} 2NH_{3(g)}$$
equil. pressure $1 + x \quad 1 + 3x \quad 1 - 2x \qquad$ atm

Using gas pressure as a measure of concentration and using the definition of the equilibrium constant $K$ (see Example 3), we obtain

$$0.500 = \frac{(1 - 2x)^2}{(1 + x)(1 + 3x)^3}$$

which simplifies to the quartic equation $27x^4 + 54x^3 + 28x^2 + 18x - 1 = 0$.

We do not have a simple formula analogous to the quadratic formula for solving this quartic equation. Instead, we use the *Newton-Raphson method* (Kimberling 1985). Let $x_1$ be a close first guess to a real zero of $P(x)$. Then a closer estimate of the zero is given by

$$x_2 = x_1 - \frac{P(x_1)}{P'(x_1)}$$

where $P'(x_1)$ is the value of the derivative of $P(x)$ at $x = x_1$. Continuing, we use $x_2$ as a seed to generate a still closer estimate, $x_3$. The sequence is continued until consecutive estimates differ by less than some preassigned tolerance (Dence and Dence 1994, 281). If $x_1$ is not chosen close enough to a true zero, then the sequence just described may not converge. In this case one must experiment with other choices of $x_1$, but if any such sequence of estimates does converge, then it must converge to one of the zeros of $P(x)$.

Let $P(x) = 27x^4 + 54x^3 + 28x^2 + 18x - 1$, and therefore $P'(x) = 108x^3 + 162x^2 + 56x + 18$; let us set a tolerance of $\delta = 0.0001$. We have $P(0) = -1$ and $P(.1) \approx 1.13$, so we pick $x_1 = 0.05$. Then computation gives

$$x_2 = 0.05 - \frac{P(0.05)}{P'(0.05)} \approx 0.05109$$

$$x_3 = 0.05109 - \frac{P(0.05109)}{P'(0.05109)} \approx 0.05108$$

and $\qquad |x_3 - x_2| = 0.00001 < \delta$

Hence, $x = 0.05108$ is a zero, and the pressures of the three gases at equilibrium are: $NH_3 = 0.8978$ atm, $N_2 = 1.051$ atm, and $H_2 = 1.153$ atm. The Newton-Raphson method is easily implemented on a microcomputer or on a programmable calculator.

Other areas of mathematics that are pertinent to elementary chemistry are probability and statistics (Shulte 1981), and error analysis (Rasof 1968). These are especially relevant to laboratory work.

*See also* Equations, Algebraic; Functions of a Real Variable; Logarithms

## SELECTED REFERENCES

Bowen, John J. "Mathematics and the Teaching of Science." *Mathematics Teacher* 59(1966):536–542.

Churchill, Ruel V., and James W. Brown. *Complex Variables and Applications.* 4th ed. New York: McGraw-Hill, 1984.

Dence, Joseph B. "The Mathematics Needed in Freshman College Chemistry." *Science Education* 54(1970):287–290.

———, and Thomas P. Dence. *A First Course of Collegiate Mathematics.* Malabar, FL: Krieger, 1994.

Ebbing, Darrell D. *General Chemistry.* 3rd ed. Boston, MA: Houghton Mifflin, 1990.

Kac, Mark, and Stanislaw M. Ulam. *Mathematics and Logic.* New York: Dover, 1992.

Kimberling, Clark. "Roots: Newton's Method." *Mathematics Teacher* 78(1985):626–629.

Kline, Morris. *Mathematics and the Search for Knowledge.* New York: Oxford University Press, 1985.

Rasof, Bernard. "Error Analysis Without Calculus." *Mathematics Teacher* 61(1968):2–11.

Shulte, Albert P., ed. *Teaching Statistics and Probability.* Reston, VA: National Council of Teachers of Mathematics, 1981.

Wilder, Raymond L. "Mathematics and Its Relations to Other Disciplines." *Mathematics Teacher* 66(1973): 679–685.

JOSEPH B. DENCE

# CHINA, PEOPLE'S REPUBLIC

School education began in China in 1862. Prior to this, education was available only to the children of rich families through private tutoring, which was conducted in a way similar to classroom teaching. Even with the establishment of a school system, education still was available only to a small group of people. There was no age requirement at which children must begin school. From the 1850s to the end of the nineteenth century, some schools were established by foreign missionaries and included elementary schools, secondary schools, and colleges.

The school system of the Qing Dynasty that preceded the establishment of the Republic of China in 1912 was an imitation of the Japanese system. The Japanese school model included both elementary and secondary schools. The mathematics subjects taught in Japanese-type schools included arithmetic, algebra, geometry, and trigonometry. During the first ten years of the twentieth century, the textbooks used in schools were mostly translations of British, American, and Japanese textbooks. The first Chinese-compiled arithmetic textbook was published in 1904. Some schools also employed Western and Japanese teachers to teach mathematics (Wei 1987).

From 1912 until 1949, the Ministry of Education issued numerous revisions of the *Regulations of the School System*. In 1932 and 1933, the Ministry of Education published *Curriculum Standards for Elementary School Mathematics, Curriculum Standards for Junior High School Mathematics,* and *Curriculum Standards for Senior High School Mathematics*. They were the first complete national mathematics syllabi in the history of Chinese education. The concentration in elementary school mathematics was on whole number and rational number systems, the four basic operations, and simple applications. The use of the abacus was considered a very important topic in school mathematics. Mathematics teaching in the elementary and secondary schools was very "teacher centered." Lecturing was the main method employed.

During the first half of the twentieth century, mathematics education in China was strongly influenced by the development in the United States. The philosophy of John Dewey was especially influential and led to the consideration of "progressive" methodologies; however, educational reform ceased during World War II.

After 1949, following the Soviet example, China's educational authorities introduced national, unified teaching plans, syllabi, materials, and textbooks for every academic institution, from elementary school to higher education. All schools were obligated to follow the directions stated in these documents. Elementary education was compulsory, while secondary and higher education were available only to students who could successfully pass the entrance exams.

The educational system of the People's Republic of China is highly centralized—all policies concerning education are drafted, approved, and promulgated by the central government. From 1950 to 1980, there were many changes in the elementary and secondary mathematics curriculum. Most noticeable changes were the content of mathematics taught in schools. Modern mathematical concepts, such as set, probability, and statistics, and even calculus were introduced at different levels.

Current mathematics directives include two major documents that outline "mandatory directions" for all elementary and secondary school mathematics teachers—the *Teaching Plan,* now called the *Curriculum Plan,* and the *Teaching Syllabus*. These documents detail principles for the teaching of mathematics, for teacher preparation, and for the production of textbooks and reference books. The mathematics syllabus describes the topics taught in schools and includes separate subsyllabi for arithmetic, algebra, plane geometry, solid geometry, and analytic geometry. Calculus and probability and statistics are also included as nonrequired courses.

The overall objectives of mathematics education in the People's Republic of China are stated in the *Teaching Syllabus for Elementary School Mathematics* (State Education Commission 1989, 1): "To enable the students to grasp the basic knowledge of mathematics necessary for the participation in socialist construction, and for the study of modern science and technology, to acquire the ability of fast and accurate mathematical operations, logical thinking and spatial visualization, so that the students' ability to analyze and solve problems can be developed gradually. The students' enthusiasm in the study of mathematics for the realization of the nation's modernization program should be encouraged, and their scientific attitude and dialectical materialist viewpoint cultivated."

In August 1992, the State Education Commission issued a new *Curriculum Plan for the Nine-Year Compulsory Education System*. Based on this document, new teaching syllabi for elementary and secondary school mathematics were drafted. The new program emphasizes that the goals of mathematics instruction are to have children "understand the fundamental content of mathematics, to develop the ability to think mathematically and to reason mathematically, and to master the mathematical skills of computation and problem solving" (State Education Commission 1992, 5).

There are now two parallel school organizations in China. One is the Five-Four-Three System, that is, five years of elementary school, four years of middle school, and three years of high school. The other is the Six-Three-Three System, that is, six years of elementary school, three years of middle school, and three years of high school. Provincial School Boards decide which region within their borders can adopt these systems. National textbooks for both systems are provided primarily by the People's Education Press, which is the only publishing company in China that produces the national teaching and reference materials.

Since 1980, the emphasis in the teaching of mathematics has shifted gradually from teacher-centered instruction to classroom methodologies that promote communication, discussion, and inquiry involving students and the teacher. Classroom activities have changed from solving large numbers of traditional routine problems to using mathematical knowledge to solve real-world problems. Complicated traditional problems have been replaced with problems that are more challenging, and require more thinking. Practical application of mathematical knowledge and skills received special attention. Since China is a large country with a population of over 1.2 billion, educational development remains "unbalanced," with different regions participating in educational reforms to varying degrees. Mathematics education is, in particular, an area that appears to be receiving increased attention from the educational authorities, as well as mathematics educators and the Chinese public.

*See also* International Studies of Mathematics Education; Japan

## SELECTED REFERENCES

Ma, Liping. *Knowing and Teaching Elementary Mathematics: Teachers' Understanding of Fundamental Mathematics in China and the United States.* Mahwah, NJ: Erlbaum, 1999.

State Education Commission. *Curriculum Plan for the Nine-Year Compulsory Education System.* Beijing, China: People's Education Press, 1992.

———. *Teaching Syllabus for Elementary School Mathematics.* Beijing, China: People's Education Press, 1989.

———. *Teaching Syllabus for Secondary School Mathematics.* Beijing, China: People's Education Press, 1989.

Swetz, Frank J. *Mathematics Education in China, Its Growth and Development.* Boston, MA: MIT Press, 1974.

———, ed. *Socialist Mathematics Education.* Southampton, PA: Burgundy, 1978.

Wei, G. *The Secondary Mathematics Education in China.* Beijing, China: People's Education Press, 1987.

WEI SUN
BRUCE R. VOGELI

# CHINESE MATHEMATICS

Can be traced back to the Stone Age and may be divided into five periods: (1) from the Stone Age to the end of the West Han Dynasty (before A.D. 25); (2) from the East Han Dynasty to the mid-Yuan Dynasty (A.D. 25–1303); (3) from the late Yuan Dynasty to the mid-Qing Dynasty (1303–1840); (4) from the mid-Qing Dynasty to the end of the Qing Dynasty (1840–1911); and (5) from the end of the Qing Dynasty to the present (1911–present).

During the first period, the base-10 numeration system and the four basic operations were established. Mathematics was applied to real-life situations. One of the oldest publications involving mathematics is *Zhou Bi Suan Jing*. This book was written in the early West Han Dynasty (ca. 300–200 B.C.). In this book, the author was the first to indicate the "Gou Gu Theorem," now known as the Pythagorean Theorem for right triangles. During the second period (A.D. 25–1303), mathematics in China was systematized and reached its highest point.

Probably the most famous and successful mathematical work in ancient China was *Jiu Zhang Suan Shu (The Nine Chapters on Mathematical Art)*. This book was first written in the early East Han Dynasty. The author of the work is unknown and the original version of the book no longer exists. The work that people refer to is the annotation (A.D. 263) of the book by Liu Hui, a famous Chinese mathematician. It includes nine chapters and 246 problems organized in the form of questions and answers. This was the first work in the world to discuss positive and negative numbers in a formal way. It covers a variety of mathematical topics, including: Area of Geometric Figures and Land; Agricultural Product; Ratio; Square Root and Cube Root; Volume of Solids and Applications; Profit and Debt; Systems of Linear Equations; and the Gou Gu Theorem.

One of the most notable Chinese achievements was the calculation of the value of pi by Zhu Cong Zhi (429–500). Zhu studied *The Nine Chapters on Mathematical Art* and found that the value of pi used in the work ($157/50 = 3.14$) was far from accurate and needed to be calculated further. He found that pi could be approximated by $355/113$ and gave it as a positive number that is greater than 3.1415926 and less than 3.1415927. How he discovered this result has been a mystery since his original work was lost.

During the 1200s Chinese mathematics reached its peak, as reflected in the works of Qin Jiu Shao (1202 or 1209–1261) and Yang Hui. In 1247, Qin finished his book, *Shu Shu Jiu Zhang (Mathematical*

*Treatise in Nine Chapters*). It included eighteen volumes and had about 200,000 words. It, too, was organized in the form of questions and answers, but the depth of the content went much deeper than the earlier treatise. In this work, Qin discussed in detail the numerical solutions of higher-order equations. His work represented the most advanced worldwide achievement in mathematics at that time. He also improved the methods for solving systems of linear equations.

Yang Hui is well known in Chinese mathematical history for his discovery of the coefficients of binomial expansions. The "Yang Hui Triangle," known later in the West as "Pascal's triangle," was thoroughly discussed in *Si Yuan Yu Jian* (*Precious Mirror;* 1303) by another mathematician of the time, Zhu Shi Jie.

From the late Yuan Dynasty to mid Qing Dynasty (1303–1840), mathematics in China developed slowly. Emphasis shifted from the study of mathematical theory (pure mathematics) to the applications of mathematics (commercial mathematics). The study of the abacus replaced theoretical studies in mathematics. During this period, western mathematical works were brought into China. In the 1800s, some Chinese mathematicians began to study western mathematics. Li Shan Lan (1811–1882), in conjunction with Alexander Wylie and Joseph Edkins, translated many mathematical texts and treatises into Chinese. The content covered algebra, geometry, calculus, and probability. Although some research mathematicians found some interesting results, overall no significant discoveries were made by the Chinese mathematicians during this time.

Since the beginning of the twentieth century, many Chinese traveled to western countries to study mathematics. Upon their return, they offered many advanced courses in the universities. In 1921, the first mathematics department was founded in Beijing University. Some mathematicians began to investigate mathematical problems in different branches of the discipline. Representative mathematicians and their areas of concentration include: Cheng Jian Gong—mathematical analysis; Cheng Jian Gong, Xong Qing Lai—Theory of Functions; Zhou Wei Liang, Cheng Chuan Zhang—Differential Equations; Yang Wu Zhi, Hua Luo Geng, Min Si He—Number Theory; Hua Luo Geng, Wang Xiang Hao—Abstract Algebra; Yu Da Wei, Su Bu Qing, Cheng Xing Shen, Jiang Ze Han—Topology; Cheng Xing Shen, Su Bu Qing—Differential Geometry; Xu Bao Lu—Probability. These mathematicians published many research pa-

pers, and most of them continued their research after the establishment of the People's Republic of China.

*See also* History of Mathematics, Overview; Pascal's Triangle

## SELECTED REFERENCES

Boyer, Carl B. *A History of Mathematics.* 2nd ed. Revised by Uta C. Merzbach. New York: Wiley, 1989.

Compiling Group of Foreign and Chinese History of Mathematics. *History of Mathematics in China.* Jinan, China: Shandong Education Press, 1986.

Martzloff, Jean-Claude. *A History of Chinese Mathematics.* New York: Springer-Verlag, 1997.

Smith, David E. *History of Mathematics.* Boston, MA: Ginn, 1923.

Straffin, Philip D., Jr. "Liu Hui and the First Golden Age of Chinese Mathematics." *Mathematics Magazine* 71(3)(1998):163–181.

WEI SUN

# CIRCLES

Belong to the category of simple closed curves. These are curves that do not intersect themselves. A *circle* may be defined as the set of points in a plane that have the same distance from a fixed point in the plane. The fixed point is called the *center* of the circle and a line segment joining the center to any point of the circle is a *radius* of the circle. In the commonly used statement "a circle of radius *r*," the *r* represents the length of the radius. A line segment joining two points of a circle is called a *chord*. If a chord passes through the center it is called a *diameter*. The two ends of a diameter partition the circle into two semicircles. The theorem that any angle inscribed in a semicircle is a right angle (Figure 1(a)) is attributed to the Greek mathematician Thales (ca. 600 B.C.) (Boyer 1989). This theorem is a special case of a

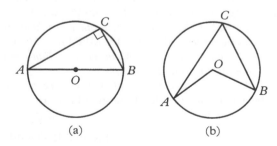

(a)                    (b)

**Figure 1**

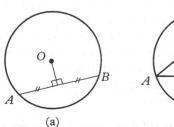 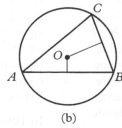 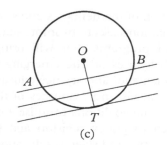

(a)    (b)    (c)

**Figure 2**

more general one that states that the measure of an inscribed angle in a circle is one-half the measure of the central angle with the same intercepted arc (Figure 1(b)). All theorems on circles taught in schools today can be found in the *Elements,* the book written by Euclid of Alexandria (ca. 300 B.C.) (Boyer 1989). A theorem that relates a chord of a circle to its center states that the line segment joining the center to the midpoint of a chord of a circle is perpendicular to the chord (Figure 2(a)). This theorem is used in constructing the circle that circumscribes a triangle. For the triangle *ABC*, the perpendicular bisectors of *AB* and *BC* intersect in the center *O* of the circle *ABC*, because $OA = OB = OC$ (Figure 2(b)). A *secant* is a straight line that contains a chord. If the secant moves parallel to itself, then in the limiting case the ends of the chord coincide to become one point and the secant becomes the tangent touching the circle at that point (Figure 2(c)). This motivates the following theorem: If a radius contains the point in which a tangent line touches a circle, then the radius is perpendicular to the tangent line. The circle may be described as "the perfect symmetric figure," because it is symmetric with respect to its center, with respect to each of its diameters and has infinitely many rotational symmetries around its center.

## CIRCUMFERENCE AND AREA OF A CIRCLE

The fact that all circles are similar, that is, have the same shape but not necessarily the same size, leads to the conclusion that the ratio between circumference and diameter is the same for all circles. This ratio is denoted by $\pi$. Hence the circumference of a circle of radius $r$ is $2\pi r$. It can be shown that the area of a circular region of radius $r$ is $\pi r^2$. Circles are associated with many three-dimensional figures such as circular cones, circular cylinders, spheres, toruses, and solids of revolution. They are widely used in the phys-

ical world, as seen in wheels, motors, and machines. They are vital to geometric construction and engineering design, and that is why a compass is found in almost every drawing set. The earliest mathematical records (ca. 2000 B.C.) indicate that the ancient Egyptians and the Babylonians dealt with circles. This is evident from the Egyptian Rhind papyrus and some Babylonian clay tablets (Boyer 1989).

In elementary schools, circles are introduced through pictures and models of discs, cylinders, cones, and toruses. Then, teachers get their students involved in activities such as drawing and matching games that deal with circles and other geometric figures. When students start to use the compass, they can intuitively establish some properties about circles. In secondary school, students are taught how to apply deductive reasoning to prove some theorems on circles. After introducing coordinate geometry, the equation $x^2 + y^2 = r^2$ of the circle of radius $r$ with the center at the origin is derived. Then, the students learn how to deal with problems involving circles analytically.

*See also* Conic Sections; Pi; Symmetry

### SELECTED REFERENCES

Billstein, Rick, Shlomo Libeskind, and Johnny W. Lott. *Mathematics For Elementary School Teachers.* 6th ed. Reading, MA: Addison-Wesley, 1997.

Boyer, Carl B. *A History of Mathematics.* 2nd ed. Revised by Uta C. Merzbach. New York: Wiley, 1989.

Euclid. *The Thirteen Books of Euclid's Elements.* Vol. 2. 2nd ed. Thomas Heath, trans. New York: Dover, 1956.

AYOUB B. AYOUB

## COALITIONS

A statewide alliance of leaders from the education, corporate, and public policy sectors working toward revitalization of mathematics and science education. Each coalition seeks to promote state education policies and programs that will bring about change that both focuses on the emerging national standards in mathematics and science education and is tailored to the needs and resources of the state. Working together, the coalitions promise a coherent national approach to policy revision and systemic change in mathematics and science education.

In 1989, the Mathematical Sciences Education Board (MSEB) of the National Research Council (NRC) launched the State Mathematics Coalition Project. Modeled on the MSEB, the coalitions were intended to serve as action-oriented, coordinating committees for mathematics education in the schools and colleges within each state. Although most coalitions were originally limited to mathematics, over the past six years the majority of them have expanded their focus to include science education. Between 1989 and 1991 coalitions were initiated in all U.S. states and the District of Columbia. Their boards include governors, legislators, parents, teachers, teacher educators, mathematics and science faculty, chief state school officers, mathematical scientists, other scientists, corporate executives, and leaders of state parent-teacher and school board associations.

A recent review of coalition activities revealed a variety of accomplishments, including the following. Coalitions have:

- selected leaders to serve on the boards of directors of state and regional initiatives in mathematics and science;
- served as the driving force behind the development of Statewide Systemic Initiative proposals to the National Science Foundation;
- developed a basic statewide public information plan regarding the reform of mathematics and science education;
- published a statewide directory of resources for improving mathematics and science education;
- secured an increase in the mathematics required for entry in teacher education programs;
- promoted legislation mandating Internet access for teachers through the state's regional education centers;
- secured a mandate for the use of calculators on statewide assessments;
- issued a policy statement on the preparation and professional development of teachers;
- produced a video course on the teaching and learning of mathematics aligned with the National Council of Teachers of Mathematics (NCTM) standards;
- sponsored conferences for school administrators on the substance of mathematics and science education reform; and
- conducted statewide conferences on equity issues.

In 1994 coalition leaders decided to wean themselves from the MSEB, and to establish the National Alliance of State Science and Mathematics Coalitions (NASSMC). NASSMC was established by the coalitions to serve as an independent, umbrella organization that provides a single voice for all state coalitions and that supports the coalitions in their efforts to reform mathematics and science education for all children.

Five basic tenets are the foundation of the goals and programs of NASSMC and its member coalitions:

- *all* children can learn more mathematics and science;
- both mathematical and scientific literacy and the opportunity to pursue careers in mathematics, science, and technology are essential objectives for *all* young people;
- standards-based systemic reform of mathematics and science education is the most promising strategy for achieving the learning goals implied by these literacy and career objectives;
- effective systemic reform, especially if it is to be sustained, requires the involvement of an active, discipline-centered, nongovernmental alliance of leaders from the mathematics and science education, business, and policy sectors (in some states, the NASSMC coalition will be that alliance; in other states, the NASSMC coalition will join in such an alliance); and
- the fundamental indicator of the success of a state coalition is whether the state's systemic revitalization process in mathematics and science education is effective in improving student learning.

In pursuing this agenda, NASSMC does not plan to undertake any project that some other organization can do (or is doing) as well. NASSMC will become part of the national infrastructure that supports educational reform by maintaining a national vision while building on the unique strengths of its member coalitions.

*See also* State Government, Role

JOAN F. DONAHUE

# CODING THEORY

A recent branch of mathematics used to design methods for reliable digital data transmission. These methods should detect, and possibly correct, errors that occur during transmission. The errors may occur when data is passed through an electrical circuit, such as a computer, or during transmission by modems on noisy telephone lines, or via satellite. Errors may occur when data is read from magnetic tape, compact discs, or bar codes. Errors also may occur when a human being copies a long identification number, such as a driver's license.

The need for coding theory resulted from the miniaturization of solid state devices that started in the 1960s. Coding theory is now a very active area of mathematical research that uses a variety of techniques from modern algebra including group theory, linear algebra, and polynomials over finite fields. (The term *coding theory* does not usually include cryptography, which is the science of sending secret messages that can only be read by an authorized user.)

There are two basic types of codes used: error-detecting and error-correcting codes. Error-detecting codes are simpler, and alert the machine, or user, to an error that has occurred. The data must then be reread or retransmitted. In some situations however, it is not possible to resend the data, and error-correcting codes are needed. For example, if data is stored on magnetic tape, and is found to be corrupt months later when the tape is read, the original data may no longer exist. Error-correcting codes also are used on compact discs so that the music or data can be read correctly, even if there is some dust or a scratch on the disc.

One example of coding in everyday use is the International Standard Book Number (ISBN). Each published book is given an ISBN consisting of ten digits, arranged in four groups, such as 0–123–45678–9. The first digit is a code for the language; 0 stands for English, 2 for French, etc. The second group of digits is a code for the publisher, and the third group is the publisher's number for the book. The last digit is one of 0, 1, 2, . . . , 9, X, and is a check digit that helps detect any errors when the ISBN is copied when ordering the book. This check digit is chosen so that for any valid ISBN, $a_1 \, a_2 \, a_3 \ldots a_9 \, a_{10}$, the sum $1a_1 + 2a_2 + 3a_3 + \cdots + 9a_9 + 10a_{10}$ is divisible by 11, or equivalently, congruent to 0 modulo 11. This means that the check digit is congruent to $1a_1 + 2a_2 + 3a_3 + \cdots + 9a_9$ modulo 11. The check digit X stands for the number 10. This code will detect any single error or a transposition error. Thus, if one digit is copied incorrectly, or if two digits are transposed, then the result will not be a valid ISBN, and a wrong book would not be ordered. See the two articles by Gallian (1991a, b) for descriptions of other checks used in identification numbers.

One simple way to detect errors in electrical circuits is to add a single parity check digit. Suppose that the basic unit inside a computer is a byte, consisting of eight binary digits. Then one additional digit can be attached to each byte to form nine binary digits called a code word. This extra digit is chosen so that there are an even number of ones in each code word, and so is called a parity check digit. For example, if 10110110 and 00110000 are two bytes, then the code words, including a parity check digit at the right end, would be 101101101 and 001100000. This coding will detect any error in a single digit, because the parity of the number of ones will change if a single 1 is changed to a 0, or if a single 0 is changed to a 1. The computer uses the first eight digits to determine which byte a word represents. There are circuits that check each word to determine if it contains an even number of ones. If the circuit detects an odd number of ones, it triggers an alarm called a parity check error, because there must have been a malfunction somewhere in the computer. This parity check cannot be used to correct the byte in case of an error, and it will not detect two simultaneous errors in one word. By adding further check digits, however, it is possible to correct some errors. See Salwach (1988) and Thompson (1983) for more information on error-correcting codes, and McEliece (1985) to see how these are applied to computer memory chips.

In a class on modular arithmetic, each student could find the ISBN from the back cover of any book, omit the check digit or any other digit, and hand it to another student who would determine the missing digit. Alternatively, two of the digits could be transposed and the other student would have to try to find the correct ISBN; in some cases, this is not always possible. Since most codes are linear, students can use matrix multiplication over the field of two elements to generate examples of error-detecting and error-correcting codes (Thompson 1983).

*See also* Binary (Dyadic) System

## SELECTED REFERENCES

Gallian, Joseph A. "Assigning Driver's License Numbers." *Mathematics Magazine* 64(1991a):13–22.

———. "The Mathematics of Identification Numbers." *College Mathematics Journal* 22(1991b):194–202.

McEliece, Robert J. "The Reliability of Computer Memories." *Scientific American* 252(Jan. 1985):88–95.

Salwach, Chester J. "Codes that Detect and Correct Errors." *College Mathematics Journal* 19(1988):402–416.

Thompson, Thomas M. *From Error-Correcting Codes through Sphere Packing to Simple Groups.* Washington, DC: Mathematical Association of America, 1983.

WILLIAM J. GILBERT

# COGNITIVE LEARNING

Encompasses all forms in which knowledge may be acquired and processed and all ways in which the individual orients him/herself. Cognitive processes may be understood as processes of mental orientation, carried out by the individual in perceiving, comprehending, sensing, thinking, imagining, knowing, meaning, believing, feeling, evaluating, expecting, and so on. The general characteristics of cognitive processes include four aspects: first, cognitive processes are similar to a searching movement. Motivations govern these processes and, conversely, cognitive processes influence motivation; second, cognitive processes are interpretations of reality, formed on a sensory, emotional, or rational basis; third, cognitive processes lead to objective decisions by the individual; and fourth, cognitive processes shape cognitive structures that, as organizational principles for other cognitive processes, may be stable and durable or unstable and transitory.

From different points of view the cognitive processes may be divided into classes in very different ways. The traditional category system, for example, recognizes cognition processes of *perceiving*, of *imagining,* and of *thinking.* Other approaches classify according to *target* orientation, *method* orientation, *means* orientation, and orientation to the *overall situation* (Newcomb 1959), or according to motivational orientation on the one hand and value orientation on the other hand. The relative, momentary totality of cognitive processes determines the "worldview" of the individual, his knowing, structuring, understanding of contexts, and determining of his own position in any overall situation he may experience. Restricting these overall situations to mathematics and learning mathematics has the corresponding effects on the network of relationships of cognitive processes.

## COGNITIVE ASPECTS IN THE LEARNING OF MATHEMATICS

The development of mathematical and intellectual ability is often described using Jean Piaget's phase theory of cognitive development as a base (sensory-motor and preoperative phase, phase of concrete logical thinking, phase of formal thinking) (Piaget 1950 and 1974). The intelligence model of Joy P. Guilford (Guilford 1971), with the three intelligence dimensions "product," "content," and "operation," is a more static analysis of mathematical (and nonmathematical) intellectual capacity. Of the total of $6 \times 4 \times 5$ factors, the most important here are the cognitive abilities, related to mathematics, which aid the acquisition of knowledge. Some examples are: the classifying of concrete or pictorial conditions (such as the formation of the concept of cardinal numbers); the forming of relations (such as $a < b$); and the execution of geometrical transformations (such as congruence mappings).

The general mathematical processes include: representing (e.g., enactivating, iconizing, symbolizing, verbalizing, formalizing); classifying (with abstracting, concretizing, generalizing, specializing); analyzing, synthesizing, combining, ordering (e.g., producing relationships, local ordering, global ordering, structuring, axiomatizing); defining, concluding, proving, heuristic thinking (e.g., experimenting, restructuring, seeking strategies); and mathematizing. They are often mentioned in connection with general educational objectives in curriculum theory.

Special educational objectives of mathematics teaching are divided in a taxonomy (e.g., according to Bloom 1956) of three fields: the cognitive, the affective, and the psychomotor. The taxonomy for the cognitive field contains six main classes: knowledge, comprehension, application, analysis, synthesis, and evaluation. For example, in teaching the Pythagorean Theorem, it is possible to formulate objectives relating to each of the above classes: To repeat the content of the theorem in one's own words (knowledge); to formulate the idea of a proof (comprehension); to apply the theorem within a word problem (application); to test an unknown proof for correctness (analysis); to find out a new proof (synthesis); and to compare several proofs (evaluation).

The cognitive theoretical background has been further investigated in numerous projects, and individual problem fields have also been specifically examined (e.g., Davis 1984; Dörfler 1988; Ginsburg 1983; Hasemann 1986; Minsky 1975; Nesher and Kilpatrick 1990; Piaget 1974; Skemp 1983; van Hiele 1981, 1986; Vergnaud 1982). New fields of research are being opened up by the inclusion of social components in the learning of mathematics and the use of computers as a "learning environment" or as an intelligent tutorial system.

*See also* Psychology of Learning and Instruction, Overview

## SELECTED REFERENCES

Bloom, Benjamin S., et al., eds. *Taxonomy of Educational Objectives.* New York: McKay, 1956.

Bruner, Jerome S., Rose R. Olver, and Patricia M. Greenfeld. *Studies in Cognitive Growth.* New York: Wiley, 1966.

Davis, Robert B. *Learning Mathematics. The Cognitive Science Approach to Mathematics Education.* Norwood, NJ: Ablex, 1984.

Dörfler, Willibald, ed. *Kognitive Aspekte mathematischer Begriffsentwicklung.* Vienna, Austria: Hölder-Pichler, 1988.

Gagné, Robert M. *The Conditions of Learning.* 2nd ed. New York: Holt, Rinehart and Winston, 1970.

Ginsburg, Herbert P., ed. *The Development of Mathematical Thinking.* New York: Academic, 1983.

Guilford, Joy P., "Intellectual Resources and Their Values as Seen by Scientists." In *The Third (1959) University of Utah Research Conference on the Identification of Creative Scientific Talent* (pp. 69–95). Calvin W. Taylor, ed. 1959.

———. *The Nature of Human Intelligence.* New York: McGraw-Hill, 1971.

Hasemann, Klaus. *Mathematische Lernprozesse. Analysen mit kognitionstheoretischen Modellen.* Braunschweig, Germany: Vieweg, 1986.

Minsky, Marvin L. "A Framework for Representing Knowledge." In *The Psychology of Computer Vision* (pp. 211–277). Patrick H. Winston, ed. New York: McGraw-Hill, 1975.

Nesher, Pearla and Jeremy Kilpatrick, eds. *Mathematics and Cognition.* Cambridge, England: Cambridge University Press, 1990.

Newcomb, Theodore M. "Individual Systems of Orientation." In *Psychology: A Study of Science* (pp. 384–422) Vol. 3. Sigmund Koch, ed. New York: McGraw-Hill, 1959.

———. Ralph H. Turner, and Philip E. Converse. *Social Psychology.* New York: Holt, Rinehart and Winston, 1965.

Piaget, Jean. *Psychologie der Intelligenz* (orig. *La Psychologie de l'Intelligence*) 6. Aufl. Olten: Walter, 1974.

———. *The Psychology of Intelligence* (orig. *La Psychologie de l'Intelligence*). New York: Harcourt Brace Jovanovich, 1950.

Skemp, Richard R. "The Functioning of Intelligence and the Understanding of Mathematics." In *Proceedings of the 4th International Congress on Mathematical Education* (pp. 533–539). Boston, MA: Birkhäuser, 1983.

Van Hiele, Pierre M. *Structure and Insight.* Orlando, FL: Academic, 1986.

———. *Struktuur.* Purmerend, Holland: Muusses, 1981.

Vergnaud, Gerard, "Cognitive and Developmental Psychology and Research in Mathematics Education: Some Theoretical and Methodological Issues." *For the Learning of Mathematics* 3(2)(1982):31–41.

Winston, Patrick H., ed. *The Psychology of Computer Vision.* New York: McGraw-Hill, 1975.

ALBRECHT H. ABELE

# COLBURN, WARREN (1793–1833)

Author of *First Lessons in Arithmetic on the Plan of Pestalozzi with Some Improvements* (1821), which in its many editions and revisions is credited with being the most popular arithmetic text ever published. Earlier American texts were not written with young children in mind, and were little concerned with how a child obtained fundamental number concepts. As schools became more concerned with teaching young children, Colburn, who was articulate and an able spokesperson, became the recognized authority in arithmetic teaching. Colburn also wrote *An Introduction to Algebra upon the Inductive Method of Instruction* (1825).

Following Johann Pestalozzi, Colburn saw object lessons, concrete materials, and applications to be fundamental in arithmetic learning. Today, he might be called a constructivist, since he believed that students must create their own mental structures and that what is a good mental structure for one person may not be for another. Colburn stated that "two persons never have exactly the same association of ideas," and ". . . when the scholar does not understand the question or proposition, he should be allowed to reason upon it in his own way, and agreeably to his own associations" (Bidwell and Clason 1970, 33). Colburn's ideas were recognized in American arithmetic teaching through the remainder of the nineteenth century, although they were sometimes challenged for obscuring the deductive structure of mathematics and modified to include deductive as well as inductive methods.

*See also* Constructivism; History of Mathematics Education in the United States, Overview; Pestalozzi

## SELECTED REFERENCES

Bidwell, James K., and Robert G. Clason, eds. *Readings in the History of Mathematics Education.* Washington, DC: National Council of Teachers of Mathematics, 1970.

Jones, Phillip S., ed. *A History of Mathematics Education in the United States and Canada.* 32nd Yearbook. Washington, DC: National Council of Teachers of Mathematics, 1970.

JAMES K. BIDWELL
ROBERT G. CLASON

# COLLEGE ENTRANCE EXAMINATION BOARD (CEEB) COMMISSION REPORT (1959)

The CEEB was founded in 1900 to design and administer standardized tests and provide the syllabi that defined the test content. Before 1940, these tests were taken by students primarily in the Eastern states. The Commission on Mathematics was appointed in 1955, chaired by Albert W. Tucker of Princeton University, to review existing curricula and make recommendations. The report offered a nine-point program for college-capable students: (a) strong preparation up to calculus, (b) deductive reasoning, (c) appreciation of structure, (d) unifying ideas (sets, relations, functions), (e) inequalities, (f) integration of plane, coordinate, and solid geometry, (g) trigonometry (grade 11), (h) elementary functions (grade 12) and (i) either probability and statistics or modern algebra (grade 12). The writers provided sample materials for the last two items. The Commission was the first body to recommend formal probability and statistics for precollege study. Pedagogy was not strongly emphasized. This report was an early and effective element in the curriculum reforms of the 1960s.

*See also* Advanced Placement Program; Cambridge Conference on School Mathematics (1963); Guidelines for Science and Mathematics; History of Mathematics Education in the United States, Overview; *Revolution in School Mathematics*

## SELECTED REFERENCES

Bidwell, James K. and Robert G. Clason, eds. *Readings in the History of Mathematics Education.* Washington, DC: National Council of Teachers of Mathematics, 1970.

College Entrance Examination Board. *Report of the Commission on Mathematics.* New York: The Board, 1959.

Jones, Phillip S., ed. *A History of Mathematics Education in the United States and Canada.* 32nd Yearbook. Washington, DC: National Council of Teachers of Mathematics, 1970.

JAMES K. BIDWELL
ROBERT G. CLASON

# COMBINATORICS

Probably one of the earliest types of mathematics used; sometimes referred to as combinatorial analysis. Combinatorics is sometimes described as "counting without really counting." The subject concentrates on techniques for determining how many elements there are in a set or how many functions exist over certain domains. These are counting problems but in more general form and far beyond what we normally think of as counting.

One of the oldest examples of combinatorial thinking is found in a Chinese book of about 2200 B.C. (Berge 1971, 4). Here a magic square is recorded, a three-by-three array of integers where each row, each column, and the two diagonals all sum to 15. In its earliest forms combinatorics is often indistinguishable from the theory of numbers. It involves finding properties of sums or arrays of squares, cubes, fourth powers and the so-called polygonal numbers (i.e., triangular numbers, squares, pentagonal numbers, etc.) The first book to treat combinatorics in a formal way was Gottfried Leibniz's *Dissertatio de Arte Combinatoria* (1666). Leibniz, who often sought to develop applications of mathematics in nonscientific fields, saw this book as having applications in history, metaphysics, and other areas removed from classical mathematics (Berge 1971, 6). The impact of this book, however, was not great. Problems and techniques that are now classified as combinatorial really derive from the work of Leonhard Euler in the eighteenth century. It was Euler, for example, who first investigated the function $p(n)$, the number of partitions of a positive integer into positive integer summands disregarding order. Thus $p(5) = 7$ because $5 = 4 + 1 = 3 + 2 = 3 + 1 + 1 = 2 + 2 + 1 = 2 + 1 + 1 + 1 = 1 + 1 + 1 + 1 + 1$. This function and many variants of it have proved to be of value in various branches of mathematics.

Techniques used by Euler included generating functions. These functions are formal power series such that the coefficients yield information about the exponents. For example, the partition function just described has a generating function that is an infinite product:

$$\frac{1}{1-x} \cdot \frac{1}{1-x^2} \cdot \frac{1}{1-x^3} \cdots$$
$$= 1 + p(1)x + p(2)x^2 + p(3)x^3 + \cdots +$$

Because it is easy to modify generating functions for various situations, such functions provide a powerful tool in the study of partition functions and elsewhere in combinatorics.

There are many counting functions that have been known for a long time. The binomial coefficients are found in the mathematical literature of China and India as far back as the twelfth century (Boyer 1989, 231). They have a combinatorial interpretation in that they count the number of possible combinations of $k$ objects that may be chosen from a set of $n$ objects, for example. The factorial function is a combinatorial function, since it counts the number of permutations of a fixed number of objects. Examples of other counting functions are Stirling numbers of the first and second kind, Bell numbers, and so forth.

A typical problem in elementary combinatorics is: determine the number of solutions in positive integers of the linear equation with unit coefficients, that is, $x_1 + x_2 + \cdots + x_k = n$. The elegant combinatorial solution is to observe that we could line up $n$ 1s in a row, with $n - 1$ spaces between them. Then the number of solutions is given by the number of ways one can place $k - 1$ "+" signs in the $n - 1$ spaces, that is, $_{(n-1)}C_{(k-1)}$ solutions (Niven 1965, 54–55). Generalized, this problem involves the distribution of objects into containers where the objects and containers may be distinguishable or indistinguishable (Hillman et al. 1987, 252–269).

In 1937 George Pólya proved what has since become known as the Pólya enumeration theorem. Motivated by a question about the benzene molecule, Pólya decided to investigate what would happen if one colored, say, the vertices of a polyhedral figure and then rotated that figure about various axes, always bringing the figure back into a position occupying the same space, but with vertices interchanged in some way. Would it be possible to count the number of distinct colorings? If one takes a cube, for example, and colors all the vertices around the top red and all the vertices around the bottom blue, then the rotation of 90° about the axis connecting the midpoint of the top face and the midpoint of the bottom face will leave the figure looking exactly as it did before the rotation. But a rotation about an axis connecting midpoints of opposite faces around the side would leave the colors in a different position. This question clearly has ramifications in chemistry since, instead of colorings of vertices, the vertices could be identified, for example, as hydrogen or oxygen or carbon atoms. This problem opens up a vast area of inquiry with many refinements of the theorem and many applications. The technique of solution involves a classic application of group theory (Tucker 1984, 333–361). It was later discovered that the problem had been solved by J. Howard Redfield in 1927, but the significance of his paper had not been recognized (Berge 1971, 6).

A subject closely related to combinatorics is graph theory. Many combinatorial problems are susceptible to a graph theoretic approach. The earliest significant problem of graph theory is probably the famous "seven bridges of Königsberg" problem solved in 1736 by Euler (Wilson and Watkins 1990, 123–124). The problem posed was whether one could take a walk in Königsberg leaving one's house, crossing each of the seven bridges of the city once and only once, and returning home. Euler showed that this was not possible given the configuration of the bridges. Another area in which graph theory and combinatorics are closely related is a class of optimization problems, the most famous of which is probably the traveling salesman problem: if we wish to visit, say, the capitals of Europe, can we find a route to follow in order to visit each of the capital cities once and only once and minimize the distance traveled? This turns out in general to be an extremely difficult problem and no complete solution of it is known at the present time.

In addition to applications of combinatorics as discussed above, basic combinations are essential in the study of probability, and configurations similar to magic squares come up in the design of experiments in statistics. The implications of combinatorics are therefore very broad. This has been even more dramatic perhaps in applications to the design of computers. Combinatorics is often referred to as "discrete mathematics," since the functions involved are over domains of discrete elements, integers, rather than over continuous domains. The connection with the design of digital computers then becomes clear.

Many problems in combinatorics remain unsolved. In the case of a number of problems it is necessary to be satisfied with approximations to the solution of a problem. For example, the best solution up to now may be certain bounds on the number of elements in a set. It might also be the case, for example, in getting an approximation to a solution of the traveling salesman problem. To find the best possible route involves too many calculations, even for the best computers. One can determine, however, something of the magnitude of the number of steps involved and thereby ascertain whether the problem is manageable or not.

A fundamental idea in combinatorics is the Dirichlet pigeonhole principle. It is an extremely simple idea but its applications are sometimes not so intuitively clear. The principle states that if $n + 1$ pigeons are to be placed in $n$ pigeonholes, then one of the holes must have two pigeons in it. For example, in any group of 367 people, two must have the same birthday (*note*: there are at most 366 days in a year). An extension of this principle is Ramsey's theory. The general statement of Ramsey's theorem is complicated but an example is the following problem: what is the least number of people you can have in a room so that there must be either three who are mutual friends or three who are mutual strangers? The answer in this case is six. For four friends and four strangers the answer is eighteen. Not known at the present time is the least number that would guarantee the presence of five mutual friends or five mutual strangers (Cohen 1978, 162–169). As with so many combinatorial problems, a question that sounds very simple can turn out to be very difficult to answer.

Because of the increased visibility of combinatorics as a subject, a number of new journals appeared in the 1960s and 1970s, specializing in papers on combinatorial problems. A number of combinatorial and discrete mathematics courses were introduced into the curriculum. A course in finite mathematics became a staple for students in the social sciences in the 1960s and has remained an important course in the curriculum. There was even a move to introduce a sequence in discrete mathematics that would to some extent replace the calculus for many students. In spite of optimistic expectations of this course, there is no evidence that it will replace calculus within any reasonable number of years. One of the problems involved is this: calculus is a cohesive sequence of courses that has a lot of internal structure to hold it together. Of course many applications of mathematics depend on the calculus. Discrete mathematics has proved to be more difficult to introduce into the curriculum in the sense that, although there are some unifying themes, such as generating functions and standard counting techniques, the subject is often perceived by students to be "a bag of tricks." The solution of a problem in combinatorics often requires ingenuity and insight rather than application of a standard algorithm or technique.

*See also* Discrete Mathematics; Graph Theory, Overview; Magic Squares; Pascal's Triangle

SELECTED REFERENCES

Berge, Claude. *Principles of Combinatorics.* New York: Academic, 1971.

Boyer, Carl B. *A History of Mathematics.* 2nd ed. Rev. by Uta C. Merzbach. New York: Wiley, 1989.

Cohen, Daniel I. A. *Basic Techniques of Combinatorial Theory.* New York: Wiley, 1978.

Hillman, Abraham P., Gerald L. Alexanderson, and Richard M. Grassl. *Discrete and Combinatorial Mathematics.* San Francisco, CA: Dellen (Macmillan), 1987.

Niven, Ivan. *Mathematics of Choice: How to Count without Counting.* Washington, DC: Mathematical Association of America, 1965.

Tucker, Alan. *Applied Combinatorics.* 2nd ed. New York: Wiley, 1984.

Wilson, Robin J., and John J. Watkins. *Graphs: An Introductory Approach.* New York: Wiley, 1990.

GERALD L. ALEXANDERSON

# COMMISSION ON POSTWAR PLANS

Established in 1944 by the National Council of Teachers of Mathematics; submitted four reports, between 1944 and 1947 on the postwar teaching of mathematics. The reports rejected the ideas that mathematics should be taught only as a tool subject and that mathematics should be taught incidentally and informally, ideas that had gained some prominence in the prewar years. For grades 1–6, the reports recommended more attention to meaning, evaluation, and the wise use of drill. For grades 7 and 8, the reports recommended a program essentially the same for all normal pupils, built around a few broad categories and aimed at both functional competence and building a foundation for subsequent mathematics courses. For high school, the report recommended a two-track system with algebra for some in grade 9 and general mathematics for the rest. In grades 10–12, it proposed sequential courses for those having the requisite ability, desire, or need for such work, with alternative courses centering on consumer and vocational mathematics. The reports moved away from the minimal mathematics advocated in the 1930s and anticipated to some extent the "New Math" reforms of the 1960s.

*See also* History of Mathematics Education in the United States, Overview

SELECTED REFERENCES

Bidwell, James K., and Robert G. Clason, eds. *Readings in the History of Mathematics Education.* Washington, DC: National Council of Teachers of Mathematics, 1970.

Jones, Phillip S., ed. *A History of Mathematics Education in the United States and Canada.* 32nd Yearbook. Washington, DC: National Council of Teachers of Mathematics, 1970.

Schorling, Raleigh, et al. "The Second Report of the Commission on Post-War Plans." *Mathematics Teacher* 38(May 1945):195–221.

JAMES K. BIDWELL
ROBERT G. CLASON

## COMMITTEE OF FIFTEEN ON ELEMENTARY EDUCATION

Appointed by the National Education Association in 1893. The Committee, which issued its report in 1895, included William T. Harris, U.S. Commissioner of Education, the president of the University of Illinois, the New Jersey State Superintendent, and the superintendents of schools of twelve large cities, including Edward Brooks of Philadelphia. The Committee proposed some deemphasis of arithmetic. They recommended that arithmetic be completed in grades 3 through 6 with a transition to algebra to follow in grades 7 and 8, breaking the tradition of a substantial amount of arithmetic in high school. They also recommended abandoning the practice of having separate daily lessons for mental and written arithmetic. The Committee further recommended a deemphasis on solving "conundrums," difficult problem-solving exercises that had gained popularity. The Committee warned against excess drill, especially on topics for which students would study simpler solutions in later work.

*See also* History of Mathematics Education in the United States, Overview

### SELECTED REFERENCES

Bidwell, James K., and Robert G. Clason, eds. *Readings in the History of Mathematics Education.* Washington, DC: National Council of Teachers of Mathematics, 1970.

Jones, Phillip S., ed. *A History of Mathematics Education in the United States and Canada.* 32nd Yearbook. Washington, DC: National Council of Teachers of Mathematics, 1970.

National Education Association. *Report of the Committee of Fifteen on Elementary Education.* New York: American Book, 1895.

JAMES K. BIDWELL
ROBERT G. CLASON

## COMMITTEE OF TEN

The Committee on Secondary School Studies, referred to as the Committee of Ten. Appointed by the National Education Association in 1892, the Committee issued a report in 1893 (reprinted in 1894) on the teaching of arithmetic—a subject then given considerable attention in secondary schools—on concrete geometry, on algebra, and on formal geometry. In arithmetic, the Committee recommended deletion of topics such as compound proportion, cube root, and obsolete units of measure, and suggested decreasing the amount of time spent on the subject. In concrete geometry, the Committee recommended both estimation and measurement of quantities and experimental verification of properties, but no formal proofs. In algebra, the Committee emphasized study of equations and an early introduction (in arithmetic) to literal expressions. In formal geometry, the Committee recommended that students give verbal as well as written proofs and that logical principles be taught with the geometry.

*See also* History of Mathematics Education in the United States, Overview

### SELECTED REFERENCES

Bidwell, James K., and Robert G. Clason, eds. *Readings in the History of Mathematics Education.* Washington, DC: National Council of Teachers of Mathematics, 1970.

Jones, Phillip S., ed. *A History of Mathematics Education in the United States and Canada.* 32nd Yearbook. Washington, DC: National Council of Teachers of Mathematics, 1970.

National Education Association. *Report of the Committee of Ten on Secondary School Studies.* New York: American Book, 1894.

JAMES K. BIDWELL
ROBERT G. CLASON

## COMMITTEE ON THE UNDERGRADUATE PROGRAM IN MATHEMATICS (CUPM)

A group of university mathematicians appointed by the Mathematical Association of America (MAA) in 1952. Under a grant from the National Science Foundation (NSF), in 1954, it produced *Universal Mathematics,* an early attempt to introduce to college freshmen the theory of functions based on the real number system. The text contained concepts that became "New Math" material, such as relations and functions; limits based on sequences; and Riemann sums and integrals. In 1961, CUPM began publishing recommendations and course guides for the preparation of mathematics teachers (K–12). These influenced mathematics education courses in most undergraduate programs. The original recommendations covered K–12 and college mathematics. In 1971 the program for prospective teachers (K–6) suggested four three-semester hour courses of specialized content. For junior high school teachers, CUPM recommended two courses in calculus and in algebra and single courses in geometry, probability and statistics, and computer science. In 1983 CUPM was succeeded by the Committee on the Mathematical Education of Teachers (COMET), which offered recommendations in 1988 and 1991. Recommendations for grades 5–8 were similar to those made in 1971, but required a minimum of five courses.

*See also* History of Mathematics Education in the United States, Overview

## SELECTED REFERENCES

Jones, Phillip S., ed. *A History of Mathematics Education in the United States and Canada.* 32nd Yearbook. Washington, DC: National Council of Teachers of Mathematics, 1970.

Mathematical Association of America. *A Call for Change: Recommendations for the Mathematical Preparation of Teachers of Mathematics.* Washington, DC: The Association, 1991.

JAMES K. BIDWELL
ROBERT G. CLASON

# COMPETITIONS

Even though there are no historical records of mathematical rivalries prior to the latter part of the Middle Ages, it is highly probable that much of the mathematics developed in Babylonia, Egypt, India, the Far East, and Greece was stimulated by rivalries among fellow mathematicians. Many of the problems posed by Diophantus could be used even today in most national mathematical olympiads. And it is very difficult to believe that the accuracy of the "trigonometric tables" developed by the Babylonians came into existence without supreme efforts to improve the results of one another. The rivalries for the solution of the general cubic and quartic equations are better known, and even more is known about the controversies surrounding the development of calculus and the creation of hyperbolic geometry (Eves 1976).

It is generally accepted that the first modern mathematical competition was held in Hungary in 1894. This competition (named first after Lórand Eötvös, and then after József Kürschák) became the prototype for all national, multinational, and international olympiads, as well as for a number of other competitions around the world. Even more important, the success of the Kürschák Mathematical Competition provides an ongoing impetus for the development and improvement of all other scholastic competitions as well. It kept its original format for over one hundred years, allowing four hours (and the use of any reference books the participants cared to bring along) for the solution of three extremely challenging problems, with the awarding of the top prize(s) only if the performance of the winner(s) is on the expected level. Interestingly, it took forty years before any other country was ready to follow Hungary's example (Freudenthal 1969).

## VARIATIONS ON THE THEME

In addition to the olympiadlike mathematical competitions, there are a number of different types of contests, which arose due to various circumstances. For instance, in the United States the American Mathematical Contests (AMC→10 and AMC→12), formerly known as the American High School Mathematics Examination (AHSME), consist of twenty-five problems of the multiple choice variety, since the aim of the AMC is to attract hundreds of thousands of students, and the grading would be impossible otherwise. Its time limit of seventy-five minutes and a built-in penalty for wrong answers also are appropriate for its intended audience. Similarly, the American Mathematics Contest (AMC→8), formerly the American Junior High School Mathematics Examination (AJHSME), is a forty-minute multiple choice test, consisting of twenty-five problems, with no penalty for wrong answers, to minimize the discouragement of younger participants.

Another development in Hungary (which also took place in 1894) probably should have attracted even more attention. That was the year that Hungary's famous high school mathematics journal, the *Középiskolai Matematikai Lapok* (*KöMaL*), was launched. In its ten issues per year, it features a multitude of problems at different levels, and its student readers are expected to submit solutions and accumulate points for them throughout their high school careers. These problems and their published solutions have become the training ground for all of Hungary's future scientists, and have been largely responsible for the incredible scientific accomplishments of that small country. Unfortunately, the evaluation of the solutions submitted is a huge task; hence only a few countries followed Hungary's example in this area. The U.S. Mathematical Talent Search (USAMTS), partially conducted through COMAP's *Consortium* by George Berzsenyi, is a relatively recent attempt to follow *KöMaL*'s example. The USAMTS encourages creative problem solving as a relaxed intellectual activity, rather than a speedy performance. In the USAMTS the emphasis is on well-written solutions rather than on the answers to the problems. For the problems and solutions of the USAMTS the reader is referred to the *Mathematics and Informatics Quarterly* (available from G. Berzsenyi), which features the international equivalent of the USAMTS.

In addition to the two extremes (competitions featuring many quickly solvable problems in a limited time frame or fewer and deeper problems with lots of time allotted for their solutions), there are several competitions that combine the virtues of each. The annual meets of the American Regions Mathematics League (ARML) are a prime example of such competitions. The components of the ARML meets also include relay-type competitions and team competitions. Moreover, at the ARML meets the students compete face to face, hundreds of them from all over the country at several selected sites linked via satellite communication. Australia's recent development of a competition conducted via a telecommunication network is even more spectacular, since one can follow the progress of the competition on huge screens set up in public areas throughout the country. In contrast, the equally excellent and in some ways similar Mandelbrot Competition allows the students to compete at their own schools. (This competition was started by three remarkable undergraduates in 1990.) The International Tournament of Towns is also relatively recent; its special feature is that the team scores are computed so that larger cities do not have an advantage over smaller ones. In yet other competitions, like MathCounts (for students in junior high schools), some of the problems need to be solved orally; the most famous competition of this type is the Leningrad Mathematical Olympiad, which was founded in 1934 (Fomin and Kirichenko 1995, Zimmerman and Kessler 1995, Taylor 1994, and Berzsenyi and Maurer 1997).

## PROBLEMS VERSUS EXERCISES

Ideally, competitions should feature problems rather than exercises. While it is true that even the simplest exercises are serious problems for the mathematically less capable students, ideally such students should not be subjected to competitions. In general, most problems featured in textbooks are exercises, since every student who masters the material in the book should be able to solve them. By contrast, the solution of a problem calls for at least one bright idea, an application of a procedure in an unexpected manner, a discovery of some relationship, or an unusual insight. As examples, we state below three

problems, which were used in the American Invitational Mathematics Examination (AIME) in 1983, 1984, and 1985, respectively. The reader should not be discouraged by their difficulty level; instead, we should all be encouraged and impressed by the fact that our students were able to cope with them.

*Problem 1* For $\{1, 2, 3, \ldots, n\}$ and each of its nonempty subsets a unique alternating sum is defined as follows: Arrange the numbers in the subset in decreasing order and then, beginning with the largest, alternately add and subtract successive numbers. For example, the alternating sum for $\{1, 2, 4, 6, 9\}$ is $9 - 6 + 4 - 2 + 1 = 6$ and for $\{5\}$ it is simply 5. Find the sum of all such alternating sums for $n = 7$.

To solve this problem for an arbitrary $n$, it is best to include the empty set and assign the alternating sum of 0 to it. Then one can set up a one-to-one correspondence between those subsets that contain $n$ and those that contain the same elements except for $n$, and show that the sum of the alternating sums of these pairs of subsets is always $n$. Thereby, since $\{1, 2, \ldots, 7\}$ has 128 subsets, and since these form 64 such pairs of subsets, the answer is 7 times 64 or 448. The most creative contestants were led to this approach by their insights; others had to experiment with smaller sets, recognize the pattern that governed the answers, conjecture the right conclusion, and then arrive at the above method of solution.

*Problem 2* What is the largest even integer that cannot be written as the sum of two odd composite numbers? (Recall that a positive integer is said to be composite if it is divisible by at least one positive integer other than 1 and itself.)

The answer to this problem is 38. To solve it, one has to recognize that all even positive integers greater than 38 can be expressed by adding odd multiples of 5 to the small composites 21, 33, 15, 27, and 9, whose last digits differ from one another. Alternately, one could partition the even integers into three residue classes modulo 6, use the composites 9, 25, and 35, and increase them by $6n + 9$, as $n$ ranges through the nonnegative integers. Both of these solutions require not only familiarity with the nature of the positive integers, but also a feel for patterns, which are basic to most mathematics.

*Problem 3* Three 12 cm by 12 cm squares are each cut into two pieces A and B as shown in Figure 1 by joining the midpoints of two adjacent sides. These six pieces are then attached to a regular hexagon, as shown in Figure 2, so as to fold into a polyhedron. What is the volume (in cubic centimeters) of this polyhedron?

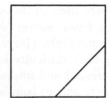

**Figure 1**

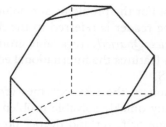

**Figure 3**

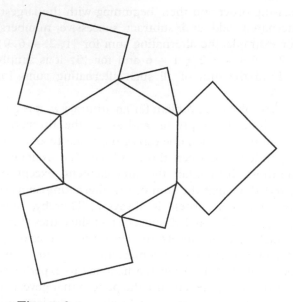

**Figure 2**

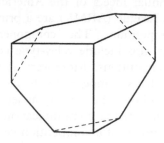

**Figure 4**

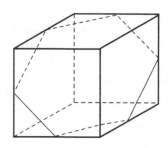

**Figure 5**

To solve this problem, the contestants had to have some spatial intuition, and recognize that the solid formed is one half of a cube; hence the answer was ½(12 × 12 × 12), or 864, as illustrated by Figures 3, 4, and 5. Geometric intuition is an extremely important ingredient to much of mathematics, physics, and engineering, hence its development through problems of this nature is to be encouraged.

The reader may question the lack of problems of a more applied nature. Their scarcity is due to several factors. In general, it is more difficult to state them without lengthy explanations, which would detract from the appeal of the problems. It also is more difficult to evaluate the solutions to them, since one must usually leave some room for different interpretations. Moreover, the areas of number theory, geometry, algebra, and combinatorics are proven grounds for excellent problems, which are easily understood by most of the contestants.

## DISCOVERY VERSUS DEVELOPMENT

The major purpose of mathematical competitions is the discovery of talents that may be undetected otherwise. Creative individuals do not always perform well enough in their regular classes to call attention to their abilities, but they often respond more positively to the challenges of competitions. Once their talent is discovered, however, it is time to develop that talent. To develop one's talents takes hard work, lots of attention, and encouragement. Further competing can allow for the monitoring of that progress, but cannot replace the development itself. Other competitions also can be used to reenforce one's self-confidence and to compare one's standing with respect to others, as well as for the selection of teams to represent one's school, region, or country. Nevertheless, these are all very secondary aims. Unfortunately, while the popularity of mathematical competitions is at its highest level, in most of the Western countries very little is done about the development of the talents discovered by these competitions. In this area we have a lot to learn from the countries of the former Communist Bloc, which developed a variety of special programs for the development of their most gifted students. They designated some of their high schools to specialize in the mathematical training of their most promising students, published scores of outstanding books and periodicals for their charges, and made the public aware of the importance of scientific inquisitiveness at an early age. Admittedly, the rationale of the communist bosses for promoting scientific development could be questioned. Nevertheless, the outcome greatly benefited millions of students hungry for knowledge.

## RECOMMENDATIONS

Every teacher should know about the variety of competitions available to their students, and guide them to take part in those most appropriate for them. Teachers also should be aware that tough competitions can have a bruising effect on not so tough egos, and should be prepared to minimize the harm. Some excellent students who are not "cut out" for competitions should be guided to other programs like the International Science and Engineering Fair (and its affiliated regional events), the Intel (formerly Westinghouse) Science Talent Search, and other events that are more suitable to them. Most importantly, all teachers should concentrate on programs that will best assist their students in the development of their talents discovered by competitions or otherwise.

*See also* Gifted; Gifted, Special Programs; Problem Solving, Overview; Projects for Students

## SELECTED REFERENCES

American Mathematics Competitions website: *http://www.unl.edu/amc*

Berzsenyi, George, and Stephen B. Maurer. *The Contest Problem Book, V*. Washington, DC: Mathematical Association of America, 1997.

Committee on the American Mathematics Competitions. *Mathematical Olympiads*. Washington, DC: Mathematical Association of America, 1976. . . . Booklets printed annually.

Eves, Howard W. *An Introduction to the History of Mathematics*. 4th ed. New York: Holt, Rinehart and Winston, 1976.

Fomin, Dmitry, and Alexey Kirichenko. *Leningrad Mathematical Olympiads 1987–1991*. Mansfield, OH: MathPro, 1995.

Freudenthal, Hans, ed. "ICMI Report on Mathematical Contests in Secondary Education (Olympiads) I." *Educational Studies in Mathematics* 2(1969):80–114.

Taylor, P. J. *Tournament of the Towns 1989–1993*. Belconnen, ACT, Australia: Australian Mathematics Trust, 1994.

Zimmerman, Lawrence, and Gilbert Kessler. *ARML-NYSML Contests 1989–1994*. Mansfield, OH: MathPro, 1995.

GEORGE BERZSENYI

# COMPLEX NUMBERS

Invented in the sixteenth century by Raphaello Bombelli (ca. 1526–1573) and others for the purpose of providing solutions to certain polynomial equations of the third degree (Ebbinghaus et al. 1990, 56; Kleiner 1988). The roots of lower-order equations also may be complex numbers; for example, the quadratic equation $x^2 + 1 = 0$ has no real roots. Although, at first, complex numbers were not accepted by all, their place in mathematics was secured around 1800 when Carl Friedrich Gauss (1777–1855) proved that a polynomial equation of any degree $a_0 x^n + a_1 x^{n-1} + \cdots + a_n = 0$, where the $a_i$'s may be real or complex, always has a root, which itself may be real or complex (Bühler 1981, 41; Churchill and Brown 1984, 118). This important result is known as the *Fundamental Theorem of Algebra* (Dence and Dence 1994, 282). The significance of complex numbers does not end with the Fundamental Theorem, for they are ubiquitous in modern mathematics and physical science. Complex numbers can be used to describe rotations of vectors in the plane, to represent solutions to certain differential equations, and in physics are invaluable in describing both electromagnetic and matter waves (Elmore and Heald 1985, 7).

One way to portray complex numbers was pioneered in 1835 by the Irish mathematician William Rowan Hamilton (1805–1865), who regarded them as ordered pairs (or "couples") of real numbers (Hankins 1980, 261). Ordered pairs $(a,b)$, $(a',b')$ are said to be *complex numbers* if, for all real numbers $a$, $b$, $a'$, $b'$, $k$, they obey these rules for addition and multiplication (Dubisch 1952, 83): (1) $(a,b) + (a',b') = (a + a', b + b')$; (2) $(a,b) \times (a',b') = (aa' - bb', ab' + a'b)$; and (3) $k(a,b) = (ka,kb)$. Alternatively, one can picture a complex number as a point in a plane with coordinates $(a,b)$. By convention, the ordered pair (0,1) is designated *i*. If the horizontal axis in the plane is the set of all points (*a*,0), and the vertical axis is the set of all points (0,*b*), then the point with coordinates *(a,b)* is the complex number (*a*,0) + *b* (0,1) = *a* + *bi*. The plane with a real horizontal axis and a vertical axis calibrated in multiples of *i* is called the Argand plane, after the Swiss mathematician Jean-Robert Argand (1768–1822). Two complex numbers *a* + *bi* and *a'* + *b'i* can be added and multiplied as algebraic binomials, with $i^2 = -1$. The results then agree with the statements given earlier for the ordered-pair formulation. Thus, (2 + 3*i*) + (−4 + *i*) = −2 + 4*i*, (2 + 3*i*)(−4 + *i*) = −11 − 10*i*, and 2(2 + 3*i*) = 4 + 6*i*.

The *modulus* of the complex number *a* + *bi*, denoted by |*a* + *bi*| is the nonnegative number $\sqrt{a^2 + b^2}$. For example, $|-4 + i| = \sqrt{17}$. The modulus possesses the following properties: (a) the product of two moduli equals the modulus of the product, and (b) the sum of the moduli is not less than the modulus of the sum. The modulus provides an alternative way to represent a complex number *a* + *bi*. Draw an arrow from the origin, (0,0), of the Argand plane to the point (complex number) with coordinates (*a*,*b*). Let *θ* be the angle this arrow makes with the positive real (horizontal) axis. Then the length of the arrow is *r* = |*a* + *bi*|, and *a* = *r* cos *θ*, *b* = *r* sin *θ*. The complex number *a* + *bi* can then be expressed in *polar form*, that is, in terms of *r* and *θ*, as *r*(cos *θ* + *i* sin *θ*). The expression cos *θ* + *i* sin *θ* can be shown to be $e^{i\theta}$, a result known as *Euler's relation*. Hence, the polar form for writing a complex number *a* + *bi* reduces to the exponential form $re^{i\theta}$. This form offers certain advantages, such as simplified multiplication:

$$re^{i\theta} \times r'e^{i\theta'} = rr'e^{i(\theta + \theta')}$$

*See also* Algebraic Numbers; Real Numbers; Vectors

### SELECTED REFERENCES

Bühler, Walter K. *Gauss: A Biographical Study.* New York: Springer-Verlag, 1981.

Churchill, Ruel V., and James W. Brown. *Complex Variables and Applications.* 4th ed. New York: McGraw-Hill, 1984.

Dence, Joseph B., and Thomas P. Dence. *A First Course of Collegiate Mathematics.* Malabar, FL: Krieger, 1994.

Dubisch, Roy. *The Nature of Number, an Approach to Basic Ideas of Modern Mathematics.* New York: Ronald, 1952.

Ebbinghaus, Heinz-Dieter, et al. *Numbers.* New York: Springer-Verlag, 1990.

Elmore, William C., and Mark A. Heald. *Physics of Waves.* New York: Dover, 1985.

Hankins, Thomas L. *Sir William Rowan Hamilton.* Baltimore, MD: Johns Hopkins Press, 1980.

Kleiner, Israel. "Thinking the Unthinkable: The Story of Complex Numbers (with a Moral)." *Mathematics Teacher* 81(1988):583–592.

JOSEPH B. DENCE

# COMPUTER-ASSISTED INSTRUCTION (CAI)

Most popular term describing the use of computers for instructional tasks. Educational software, courseware, CBI (computer-based instruction), and CMI (computer-managed instruction) also are used as terms referring to this type of computer use. Several distinct types of CAI exist: drill and practice, tutorial, simulation, computer-managed instruction, and problem solving. Educational programs often employ more than one of these techniques. A tutorial program, for example, typically includes drill and practice exercises after topics have been introduced. As CAI programs require only operational knowledge of microcomputers, no programming skills are necessary.

The first educational use of computers was in 1950 with a computer-driven flight simulator used to train pilots at MIT. In 1959, the IBM team of Rath, Anderson, and Brainerd reported a program to teach binary arithmetic to New York elementary school students via a typewriter inquiry station connected to an IBM 650 computer, thus becoming the first to use computers with schoolchildren. The use of computers to support school and college administrative functions has steadily increased over the years; however, it is the classroom application of computers that has seen the most dynamic and problematic growth (Bitter 1992).

Patrick Suppes is often referred to as the Grandfather of CAI, because of his original instructional computing efforts. His early contributions were the beginning of many CAI and CMI efforts in the late 1970s and early 1980s. The period from 1960 to 1980 was shaped by time-sharing activities offered by universities and major instructional computing projects, all of which were developed on and driven by large-scale mainframe or minicomputer systems. The 1980s were a time of such intense interest in and purchase of microcomputers for schools that it came to be known as the Microcomputer Revolution in education. In the 1990s microcomputer use was still widespread, but schools witnessed a strong interest in connectivity, electronic mail, and the World Wide Web.

## DRILL AND PRACTICE

Drill and practice software allows learners to perform a variety of exercises using the same facts, relationships, problems, and vocabulary until the material is committed to memory or until a particular skill has been refined. Such software possesses a format that encourages repeated practice by students, thus enabling mastery of the skill or establishment of the stimulus-response association required for memorization of certain facts. Drill and practice software incorporates sequential learning tasks that assist the student in mastering skills such as the mastery of the facts, relationships, problems, or vocabulary. Software is available for a great number of topics, from preschool learning to technical subjects.

## TUTORIAL

Tutorial software utilizes written explanations, descriptions, questions, problems, and graphic illustrations for concept development, much like a private tutor. Tutorials attempt to aid skill acquisition by careful presentation of ideas and feedback to the user's responses. Pretesting is often used to determine the placement of a student in a particular lesson of a tutorial courseware package. There are two common types of tutorial program designs. Tutorials that progress in a linear fashion present a series of screen displays to all users, regardless of individual differences among the students. Branching tutorials do not require all users to follow the same path, but direct users to certain lessons or parts of a lesson according to results of computerized pretests and posttests, or to the user's responses to questions embedded within the tutorial. Often computerized or traditionally written pretests are included with tutorial software to determine the most appropriate level of lessons for a particular student.

After the tutorial portion of a lesson has been presented, drill and practice exercises are often offered. Finally, a posttest for each objective or group of objectives determines mastery of concepts. Typically, scores are displayed at the end of lessons, as well as suggestions for further study and/or review. Tutorial courseware tends to require more complicated instructional design and programming techniques than drill and practice. The author of the tutorial must try to predict all possible correct responses and allow for errors. The program must respond with related additional instruction to incorrect answers, predict the most common incorrect answers, and offer specially tailored explanations and learning experiences based on the user's incorrect answers. In addition, tests generated by computers must be as valid and reliable as tests written by instructors, so that the user is not subjected to material already learned and so that learning can be evaluated fairly and accurately. Although they can supplement the curriculum, computer tutorials represent only one learning mode, and obviously will never match the intuition of a human teacher.

## COMPUTER-MANAGED INSTRUCTION

Computer-managed instruction (CMI) does not offer instruction of any type; rather, it manages the instruction in a classroom or school through computer-assisted testing and recordkeeping. By testing set objectives, the computer determines the student's mastery level to ascertain the educational needs of that student. Closely related to criterion-referenced testing, CMI programs are powerful tools which perform recordkeeping tasks and provide informational reports, thus allowing the teacher to choose the best individualized instructional plan (paired learning, whole group, textbook, etc.) for mastery of specific learning objectives.

Another type of management system includes not only objectives and mastery testing, but also provides drill and practice and/or tutorials related for each objective. In addition, teachers can request personalized worksheets or tests for specific students or groups of students according to demonstrated needs. Other useful options to a system may be student workbooks, teacher's guides, and individualized learning prescriptions. Such prescriptions not only list objectives yet to be mastered, but also list page numbers, locations, and titles of specific textual material, worksheets, workbooks, audio-visual aids, and other auxiliary materials matched to those objectives. This system helps teachers locate appropriate learning materials for students at the right time. Card readers are sometimes used to minimize data entry tasks. Results are stored and printed so that teachers can determine optimum grouping arrangements for instruction (paired learning, small or whole group). Generally, CMI is not considered the total learning package for students, but is integrated with other effective learning activities.

## PROBLEM SOLVING

Problem-solving software requires user strategy and input. A problem is presented for the user to solve. The software does not teach concepts; instead, the user applies concepts that have been mastered previously, learns from mistakes, and refines skills while gaining mastery of certain problem-solving techniques. Problem-solving software requires logical thinking on the part of students and assumes that some previous concept development has occurred. The continuing further development of logical thinking is an underlying principle of problem-solving software. Some educators refer to this software as educational games.

Bagels, a simple example of a problem-solving program, is available in slightly different formats from several publishers. In this case, the problem to be solved strengthens logical thinking and understanding of place value. The user can choose the difficulty level by indicating whether he/she wishes to guess a two-, three-, or four-digit number that the computer randomly selects. Then the user tries to determine the number in the least possible tries. After each guess, the computer gives certain clues regarding correct placement of digits in the guess. If none of the digits are correct, the computer responds with the word *bagels*. If one digit is correct, but is in the wrong value position, the computer prints *pico*. If one digit is correct and is in the correct value position, the computer responds *fermi*.

For example, the computer has chosen the secret number 287. The user guesses the number 345.

| | Guess | Clue |
|---|---|---|
| 1. | 345 | BAGELS (No digits are correct!) |
| 2. | 729 | PICO PICO (two correct digits, but in the wrong place value position) |
| 3. | 897 | PICO FERMI (one correct digit, one correct digit in the correct place value position) |
| 4. | 197 | FERMI (one correct digit in the correct place value position) |
| 5. | 687 | FERMI FERMI (two correct digits and correct place value positions) |
| 6. | 287 | THAT IS THE SECRET NUMBER! |

The clues must be interpreted and used to select the best guess possible during the subsequent turn. In the example above, a certain amount of logical thinking is apparent. The first clue was *bagels*, meaning that no digits were correct. Therefore the digits 3, 4, and 5 were not guessed again. In guess 4, the clue

logically indicates that 7 is the digit totally correct, and that neither the digits 1 nor 9 are present in the computer-chosen number.

Hurkle is another popular problem-solving program (available as shareware) that aids in learning to use grids and number lines of various types. A creature called a Hurkle hides behind the grid or number line. In the case of grids, the student enters where the coordinate pairs are guessed to be hiding. Unless the guess is exactly right, the computer responds by giving directional clues, such as North, South, East, West. The game continues until the Hurkle is revealed. The Hurkle program incorporates logical reasoning and coordinate geometry. Problem-solving programs are an integral part of many multimedia programs.

## SIMULATION

Simulations allow users to experience situations in which it would be difficult or impossible to make decisions and to be impacted by the outcome. Simulations offer opportunities for users to make and be affected by their own decisions. Guided by data provided by the simulations, the user selects certain options or risks, and then witnesses the result of the decision. Many simulations model some real or imaginary situation, but in most cases simulations emphasize specific skills. Since the user learns not only what the simulation is teaching but also how best to use the program, simulations represent more complex forms of computer instruction than drill and practice and tutorials. A simulation could be described as a high-level problem-solving program, because the user must evaluate situations to create strategies for successfully solving the problem. Simulations help the user learn through models that cannot occur in real situations due to unavailable materials, expense, time, or danger. Computer simulations offer the user opportunities to experience real-life events in safe environments. Decisions must be made to affect the outcome of the simulation, so the user tends to become an involved and interested learner.

## INTERACTIVE MULTIMEDIA

Multimedia includes video, text, sound, and animation to present learning. Simulation, drill and practice, tutorial, and problem solving are all part of an interactive multimedia program. The ability to engage the user's senses in learning situations makes interactive multimedia an exemplary CAI approach. Options include multilingual capability, voice response, hypertext notebook entries, video, and animation. These components can be an integral part of the program. Interactive multimedia CAI promotes user control and multiple pathways, incorporates development of higher-order thinking abilities, and facilitates development of decision-making and problem-solving abilities (Hatfield and Bitter 1994).

## TOOL SOFTWARE

Three popular software tools are word processing, databases, and spreadsheets. The word processing application provides capabilities to write memos, letters, and other professional and personal documents. The database is an electronic version of a file cabinet and can be used to store and retrieve test scores, prepare a calendar of events, and keep track of similar activities. A spreadsheet program resembles an electronic version of an accountant's ledger pad. This electronic worksheet can perform many computations, including calculations based on specified conditions. Spreadsheets can be used to prepare a budget for a school's activities, calculate weighted grade averages, and do many other mathematics applications. Often tool software includes function plotters, sketch pads, graphing packages, and various mathematical programs that include charting, graphing scroll, implicit plotting, trace mode, 2-D and 3-D graphics and plots, reflecting, zoom stretches, translating, rotating, dilating functions and/or relations, and symbolic algebra.

## CURRICULAR INTEGRATION OF CAI

Different types of CAI lend themselves to various classroom integration procedures. Drill and practice varieties should be used after some particular conceptual development has taken place. Tutorial software is most effective when students show varying levels of skill acquisition and conceptual understanding. CAI can provide for individual tutoring needs that may be difficult to satisfy through traditional instructional arrangements. Computer-managed instruction assesses specific behavioral objectives and suggests instructional materials and grouping arrangements. Simulations and problem-solving programs typically require that students have acquired certain prerequisite skills and knowledge. Such software allows students to practice skills and to develop strategies, and promotes logical thinking, since decisions determine outcomes.

## SUMMARY

Computer-assisted instruction describes the use of computers for instructional tasks. CAI programs require only operational knowledge of microcomputers. Some distinct types of CAI are drill and practice, tutorial, simulation, computer-managed instruction, and problem solving. Software can combine two or more of these approaches. The future of CAI will continue to emerge through connectivity and the World Wide Web. Video, animations, voice, text, sound, and interactivity will be an integral part of future computer-assisted instruction programs.

*See also* Computer Science; Computers; Software

### SELECTED REFERENCES

Bitter, Gary G., ed. *Macmillan Encyclopedia of Computers.* New York: Macmillan, 1992.

Bitter, Gary G., *Microcomputers in Education Today.* Watsonville, CA: McGraw-Hill, 1989.

————. *Understanding Teaching: Implementing the NCTM Professional Standards for Teaching Mathematics.* CD-ROM. Tempe, AZ: Technology Based Learning and Research, Arizona State University, 1996.

————, Ruth Camuse, and Vickie Durbin. *Using a Microcomputer in the Classroom.* Englewood Cliffs, NJ: Prentice-Hall, 1993.

Hatfield, Mary M., and Gary G. Bitter. "A Multimedia Approach to the Professional Development of Teachers: A Virtual Classroom." In *Professional Development for Teachers of Mathematics* (pp. 102–115). 1994 Yearbook. Douglas B. Aichele, ed. Reston, VA: National Council of Teachers of Mathematics, 1994.

National Council of Teachers of Mathematics. *Principles and Standards of School Mathematics.* Reston, VA: The Council, 2000.

GARY G. BITTER
MARY M. HATFIELD

# COMPUTER SCIENCE

Electronic computing machines have existed since the 1930s. Since then, many phenomena have emerged: the transistor in 1947, the first programming language (FORTRAN) in 1957, interactive computing in the 1960s, and the chip in the 1970s—all part of emerging computing technology. New courses in computing began to appear at some colleges and universities in the late 1950s and, by the 1960s, some colleges offered degree programs in the computing discipline.

The Association for Computing Machinery (ACM) formally defined the study of computing in

1968 when it published its first curriculum recommendation for four-year programs in computer science called *Curriculum '68* (ACM 1968). Curriculum '68 proposed a set of computing courses that included computer programming, computer systems, computer organization and architecture, algorithms, data structures, operating systems, programming languages, and numerical analysis. It also recommended study of discrete mathematics, calculus, linear algebra, and probability and statistics. In response to the rapidly changing field of computing, ACM published another report entitled *Curriculum '78* (ACM 1979). Curriculum '78 updated Curriculum '68 and advocated a stronger emphasis on programming. It became the model curriculum followed by many computing programs in the country.

Curriculum '78 focused on the mainstream of computing programs. The needs of the computer engineer, however, were not in its domain. So the Computer Society of the Institute of Electrical and Electronics Engineers (IEEE-CS) addressed the needs of the computer engineering professional. IEEE-CS published curriculum reports (IEEE-CS 1977, 1983); both these reports encouraged greater study and application in the engineering aspects of computing such as computer organization, architecture, and hardware. ACM published two recommendations (Koffman 1984, 1985) that encouraged the use of software design and implementation in the early stages of a computer science curriculum. They also supported the introduction and active use of data structures in the second course.

There were many other curriculum reports that emerged during the 1970s and 1980s. For example, among several other curriculum projects, ACM published recommendations for graduate programs in information systems (ACM 1972). The Data Processing Management Association (DPMA) also issued various curriculum reports. These reports focused on data processing and information systems in four-year programs (see DPMA 1986, Longenecker and Feinstein 1991) and in two-year programs (see DPMA 1992). Several computing societies joined their efforts and proposed the report *Information Systems IS'96*. The report addressed current needs and recommendations of information systems programs (see Longenecker et al. 1996).

By the early 1980s, it became evident that computer science needed some stability as an emerging discipline. In 1984 ACM and IEEE-CS created the Computing Sciences Accreditation Board (CSAB) that established the Computer Science Accreditation Commission (CSAC). The CSAC/CSAB set criteria for the accreditation of four-year computer science programs in the United States. The Accreditation Board for Engineering and Technology (ABET) was already serving the accreditation needs of computer engineering programs. By 1995 the CSAC/CSAB or the ABET accredited over 200 computing programs.

Computing curricula recommendations had become more reactive than proactive as they responded to the needs of a changing computing profession. They generally promoted an early study of a programming language in the curriculum. This did not necessarily fulfill the needs of the computer scientist. Among the various educational computing activities in the 1980s was a paper by Peter Denning that provided an evaluation of the issues surrounding the curriculum in computer science and represented a first step in a series of educational activities that were to follow (Denning 1985).

In the late 1980s a special task force chaired by Denning addressed the issue of curriculum reform. It published a report known as the *Denning Report,* in which computing is defined as ". . . the systematic study of algorithmic processes that describe and transform information: their theory, analysis, design, efficiency, implementation, and application. The fundamental question underlying all of computing is, 'What can be (efficiently) automated?'" (Denning et al. 1988). The report also established topic areas that define the discipline incorporating the paradigms of theory (a mathematical approach), abstraction (an experimental approach), and design (an engineering approach) within nine topic areas. These areas are: algorithms and data structures, programming languages, architecture, numerical and symbolic computing, operating systems, software methodology and engineering, database and information retrieval, artificial intelligence and robotics, and human-computer communication. Sensitivity to the social context of computing accompanies these nine areas so that computer manufacturers and operators are kept aware of the ethical, medical, and legal issues surrounding the use of computers in a global society.

ACM and IEEE-CS embraced the elements of the Denning Report. In 1991 both societies jointly published curriculum recommendations that model a variety of computing programs (Tucker et al. 1991). These recommendations did not prescribe courses for study as in previous curriculum reports. They decomposed the nine topic areas of computing into knowledge units arranged conveniently to serve the needs of an individual program or college. The nine topic areas reflect one or more of the paradigms of theory, abstraction, and design. They also formulated a broad range of topics for a program of study. Curricula '91 suggests the study of computing as a whole in the first stages of the discipline coupled with laboratory experiences. Such a pedagogical approach is much like that of biology, chemistry, and physics in the first year of study. This holistic manner of addressing the discipline of computing is known as the "breadth approach" to the study of computer science.

The professional societies also considered associate-degree programs. In a historic effort, the ACM published a four-volume set of five curriculum reports (ACM Two-Year College Computing Curricula Task Force 1993). These curriculum guidelines for associate-degree programs cover the areas of *Computer Support Services, Computing and Engineering Technology, Computing and Information Processing,* and *Computing Sciences.* The *Computing for Other Disciplines* report accompanies each of the previous reports in their respective volumes.

The discipline of computing with its breadth of study is becoming a part of secondary education also. Many governmental educational agencies and local school districts have sought guidance in establishing a well-balanced high school course in computing. In response to this need, the ACM published recommendations to foster the breadth of the discipline of computing—not just programming (ACM Task Force of the Pre-College Committee 1993). Following this program of study, high school students will not only learn how to program, but will gain knowledge in other areas of computing such as computer architecture, operating systems, and data communications.

Computer science is a changing discipline. New improvements, inventions, and achievements in this emerging technical field will have a dynamic effect on the growing number of educational programs. Curricula '91, the five reports addressing associate degree programs, and the high school model curriculum have had a positive influence to stabilize the volatility of curriculum changes in computing—at least for the near future. We can expect the discipline of computing to stay fluid. Today we face new and exciting challenges. Computer networking, multimedia, and communications on the information highway are just a few emerging areas. Instructors of computing will have to keep abreast of these emerging technologies such as using the Internet and learn new languages such as Ada, Java, and Perl, so their students will be better prepared to meet the changing world of computing.

*See also* Computers; Programming Languages

## SELECTED REFERENCES

ACM Curriculum Committee on Computer Science. "Curriculum '68—Recommendations for Academic Programs in Computer Science." *Communications of the ACM* 11(3)(1968):151–197.

———. "Curriculum '78—Recommendations for the Undergraduate Program in Computer Science." *Communications of the ACM* 22(3)(1979):147–166.

ACM Curriculum Committee on Information Systems. "Curriculum Recommendations for Graduate Professional Programs in Information Systems." *Communications of the ACM* 15(5)(1972):364–398.

ACM Task Force of the Pre-College Committee. *ACM Model High School Computer Science Curriculum.* New York: ACM, 1993.

ACM Two-Year College Computing Curricula Task Force. *Computing Curricula Guidelines for Associate-Degree Programs.* 4 vols. New York: ACM, 1993.

Data Processing Management Association. *CIS '86, The DPMA Model Curriculum for Undergraduate Computer Information Systems.* Park Ridge, IL: The Association, 1986.

———. *The DPMA Two Year Model Curriculum.* Park Ridge, IL: The Association, 1992.

Denning, Peter J. "The Science of Computing." *American Scientist* 73(Jan. 1985):16–19.

———, et al. *Computing as a Discipline—Report of the ACM Task Force on the Core of Computer Science.* New York: ACM, 1988. [Reprinted in part in *Communications of the ACM* 32(1)(1989):9–23, and in *Computer* 22(2)(1989): 63–70.]

IEEE Computer Society, Education Committee. *A Curriculum in Computer Science and Engineering.* Committee Report, IEEE Publ. EH0119–8. New York: The Society, 1977.

IEEE Computer Society Educational Activities Board/Model Program Committee. *The 1983 IEEE Computer Society Model Program in Computer Science and Engineering.* New York: The Society, 1983.

Koffman, Elliot, et al. "Recommended Curriculum for CS1: 1984." *Communications of the ACM* 27(10)(1984): 998–1001.

———. "Recommended Curriculum for CS2: 1984." *Communications of the ACM* 28(8)(1985):815–818.

Longenecker, Herbert E. Jr., and David L. Feinstein, eds. *IS '90: The DPMA Model Curriculum for Information Systems for Four Year Undergraduates.* Park Ridge, IL: Data Processing Management Association, 1991.

Longenecker, Herbert E. Jr., David L. Feinstein, John T. Gorgone, Gordon B. Davis, and J. Daniel Couger, co-chairs. *Draft Report: Information Systems—IS '96. Model Curriculum and Guidelines for Undergraduate Degree Programs in Information Systems.* Mobile, AL: University of South Alabama, 1996.

Tucker, Allen B., et al. *Computing Curricula 1991—Report of the ACM/IEEE-Computer Society Joint Curriculum*

*Task Force*. New York: ACM, 1991. [Reprinted in summary in *Communications of the ACM* 34(6)(1991):68–84.]

JOHN IMPAGLIAZZO

# COMPUTERS

To describe the role of computers in mathematics education is to attempt to put into words both exciting and frustrating observations about successes and failures thus far, and hopes and promises of the future. Computers have substantial potential for influencing the teaching and learning of mathematics at all levels. Yet, the instances of startling success are offset by instances of resistance, indifference, and even blocking behaviors as well as true difficulties. Likewise, computer technology provides a medium and tools for research into teaching and learning from which we can profit greatly. Although these tools are present and impressive, their use is impeded by incomplete knowledge about how to study learning and teaching of mathematics, and by a lack of real emphasis on and valuing of research.

The computer is an innovation of substantial magnitude, leaving few to doubt that computer technology's use in mathematics, science, and business is of immeasurable importance. Inventions as significant as computers are so rare as to occur only once in several decades or a century or two. "The emerging view is that computers are a cultural medium intimately connected to our identity and our function as human beings" (Kaput 1992, 547). Computers in the workplace are deeply embedded, not just a tool for use. They have become the main vehicle of communication as fax and telecommunication and serve as the basis of the Internet and the World Wide Web. Information sources like reference volumes, financial accounts, and design and art collections are processed through computers, and multimedia presentations, computer-based instruction, and learning laboratories are controlled by computers. Computers fundamentally alter the nature of work, not merely its efficiency, accessibility, or simplicity (Winograd and Flores 1986). Computers allow teachers and professors to pose problems that are more authentic than traditional class problems, and to expect a more well-researched or thoroughly validated response. They provide research mathematicians and mathematics educators with new problems and new ways to solve longstanding problems (e.g., the four-color problem). Yet it remains to be seen whether this invention of such substantial magnitude will have an equally substantial influence on the learning and teaching of mathematics, even in an era of educational reform that hails the technology's importance.

Just as mathematics education has been noted as central to the economic security and well-being of the individual (National Research Council [NRC] 1989) and of the nation, computers and their usage are central to any reform of mathematics education. How are changes in the technology likely to open up possibilities for improved learning and teaching of mathematics? To begin to answer this question, an examination of the changes that have occured and are occurring with computer technology is appropriate.

To use computer technology within the study of mathematics serves multiple goals: a better understanding of mathematics, tools for solving problems, tools for life-long learning, and the preparation for use of technology beyond mere computer and mathematical literacy. The types of uses of computers in mathematics are expanding, along with the growth of power of personal computers. Possibly the most important goal relates to the development of employability skills, for every occupation and profession requires increasingly greater facility with computer technology. A substantial quantity and quality of varied experiences with computers are necessary, particularly mathematically related experiences, to develop this requisite facility. Beyond employability skills, there are surely reasons to believe that computer technology will continue to grow in importance and power for many years. As Kaput says, "major changes are underway that are likely to produce substantial transformations in the 1990s, some based in the technology itself, some in the reform effort, and some in the interaction between the two" (Kaput 1992, 518).

## BRIEF HISTORY OF COMPUTER TECHNOLOGY'S GROWTH IN MATHEMATICS EDUCATION

From the entry of the microcomputer into the market (about 1978), the computer itself has been a limiting factor of the mathematics we could use it to study. For example, the first available "computer systems" were equipped with a central processing unit (CPU) for performing the operations and functions, a monitor (a TV-like device), 2–4 kilobytes (KB) of random access memory (RAM), and an unreliable audiotape recorder for external storage of programs. Within a few months, 16 K of RAM and a floppy drive made the computer a more functional machine. Programming, saving programs to 5¼" disks that had a capacity of 16 K, and using spreadsheets were evolving features mathematics educators could appreciate.

During the 1980s, the microcomputer characteristics included color graphics, 32–64 KB of RAM, and 5¼″ or the beginnings of the 3½″ disks, with capacities of upwards of 800 KB. Programs such as Green Globs (Dugdale 1982), the Geometric Supposer Series (Schwartz and Yerushalmy 1985), and function plotters proved to be viable alternatives to programming for mathematics. Earlier, writing computer programs was viewed as a good way of learning to think and reason mathematically, but the aforementioned programs get at some of these objectives more directly. Speed of operation was not so important an issue as were the quality of the graphics and the ease of use. On-screen menus began to offer options that engaged students in mathematics without having to use time learning how to get the program to operate. This stage of computer development began to attract mathematics teachers who did not want to study computers but did want to use computers as tools for teaching mathematics better. The software was consistent with course objectives, and in some cases, provided possibilities of investigating and discovering important relationships in mathematics. At this point the computer hardware had evolved so a teacher could see that there was potential for teaching mathematics better. To be able, in a single class period, to discuss graphs of $f(x) = \sin x$, $f(x) = \sin 2x$, $f(x) = \cos x$, as well as $f(x) = a \sin b(x + c) + d$ and $f(x) = a \cos b(x + c) + d$ was a power teachers valued. Each parameter could be studied individually and one graph could be superimposed on another for analysis, illustrating a better way of teaching mathematics. Rather than plotting one or two trigonometric functions of a particular type, being told the effects of a specific parameter and then asked to practice sketching several examples of that type, students could explore each parameter independently and in combinations to construct their own meaning.

## COMPUTERS AND THE CHANGING GOALS OF MATHEMATICS EDUCATION

The goals of mathematics education in a technological society are changing (Kaput 1992, 515; Cockroft 1982; Crust et al. 1994, 333; NRC 1989, 12) at all levels of education and research. These changes in the goals are necessary for several reasons: (a) the economic necessity that all citizens have a much more substantial mathematics education (e.g., the need for all students to have a mathematics education equivalent to that of the college bound student of previous years); (b) the changed role of mathematics in our technological society; and (c) the role that mathematics plays in computer technology, that is, the basis of its logic, applications, and research for its development. In addition, the role of the mathematically educated student has changed. Working in a computer technology environment, solving problems that may well have a combination of humanities, social sciences, and arts as well as scientific origin and data structure, and working with a team to make judgments, the goals for preparation for this workplace must be substantially reformed. Computer tools become, in this framework, a means of analysis and abstraction of data from sources, a current major mathematics ability we then need to develop. This is in contrast to that of years past, when the mathematical activities tended to involve the mind and its memory for the process of recalling and deducing results.

Computer technology and mathematics are affecting each other in important and inseparable ways. The following statement describes this quite well, and is highly suggestive of needed changes in the goals of mathematics and technology usage. "Although the public often views computers as a replacement for mathematics, each is in reality an important tool for the other. Indeed, just as computers afford new opportunities for mathematics, so also it is mathematics that makes computers incredibly effective. Mathematics provides abstract models for natural phenomena as well as algorithms for implementing these models in computer languages. Applications, computers, and mathematics form a tightly coupled system producing results never before possible and ideas never before imagined" (NRC 1989, 36–41).

The opportunity for the mathematics education community to influence the directions computer technology develops is important to consider. Computers that draw upon and support mathematical intuition, that utilize and develop the visualization aspects of mathematics, and that provide tools for building mathematical models, to mention a few possibilities, would surely be positive directions for computer technology and mathematics education. They would bind tightly applications, computers, and mathematics. An aggressive approach to creating the technology essential to these needs should be considered. A more reactive approach of realignment of curriculum and assessment to available technology and to remaining aware of possibilities presented by the rapid computer-technology introduced opportunities has been recommended (Crust et al. 1994, 333–337). This suggests making advances in curriculum and assessment that keep up with the most recent technology. (How? The author would recommend a proactive approach in which mathematics curriculum and assessment "drive" the development of software and hardware, rather than keeping up with it.) Envisioning the design of software and hardware that helps solve important learning and teaching problems in mathematics, as well as assessment problems, is a more appropriate position to take.

As Jurdak (1994, 199) indicates, the combination of computer technology and communication technology has transformed the workplace into an information-rich environment. It has transformed the social context in which we live and communicate into a much more highly interactive, social, and personally engaging surrounding. And it is changing our home into social-work-leisure mixes of lifelong learning and experiences. For example, computers are beginning to drive our home entertainment and media systems, our communication systems, and even our lighting and security systems.

The new directions are expressed well in a discussion of Information Age (IA) mathematical literacy by Jurdak: "The physical context requires the utilization of high-technology environment for learning, applying, or assessing authentic mathematical tasks. The social context makes it imperative that IA mathematical literacy be achieved by all, irrespective of social or ethnic divisions. IA mathematical literacy for all is assumed to be necessary for the economic survival and social harmony. The temporal context refers to the learning environment which provides the student with the power to learn for life" (Jurdak 1994, 202). The needed skills and thinking of the mathematically literate have expanded to include problem formulation, experimentation and data collection, drawing inferences, creating models, and verifying and making judgments. And, as Jurdak points out, these are essential mathematical literacy tools for all.

The views of mathematics educators and mathematicians are that computer technology must serve the objectives of their community (Kaput and Romberg 1989); that what society expects in a mathematics education for its students and citizens is undergoing considerable and rapid change (Lappan and Theile-Lubienski 1994), and that with increased power of computer technology comes a changing role of mathematics, with some parts taking on sub-

stantially more significance and others becoming obsolete or of much different import (NRC 1989, 45). The opportunity to use computers to teach mathematics more efficiently, with better student outcomes and understandings, and for more independent and lifelong learning is believed to be increasing with every improvement in computer technology. "As today's computer visions become tomorrow's verities, they will revolutionize the way mathematics is practiced and the way it is learned" (NRC 1989, 62).

Although the goals are changing and some changes in mathematics education are emerging, the rate of computer technology hardware and software development is much more rapid. The challenge of education is to iteratively evolve goals appropriate to needs and to narrow the gap between what is and what should be. It seems, however, that now is the right time and situation for finally realizing the substantial potential of computer technology throughout our mathematics education enterprise (Crust et al. 1994, 332).

## TEACHING MATHEMATICS BETTER AND TEACHING BETTER MATHEMATICS

Two important hopes for computer technology in mathematics education have been widely expressed: the ability to teach and have students learn mathematics better, and to teach better mathematics. Reformers suggest an emphasis on problem solving and reasoning as a means of teaching mathematics better. It is clear that computer technology potentially contributes to a learning environment in which students can explore, conjecture, test, validate, do mathematical modeling, and build mathematical understanding. This implies that introducing problems as a means of motivating and as a context for new mathematical ideas has an important place in mathematics. These descriptors characterize the reform documents' sense of mathematical reasoning. At the same time, several problem-solving researchers suggest that to pose problems for which the assumed essential prerequisite knowledge has not necessarily been "taught" is one good strategy in a problem-solving agenda. Computer technology provides a variety of tools for solving problems in creative, nontraditional ways, decreasing the prerequisite-dependency aspects of solving problems. For example, computers are understood to be facile as we create and analyze patterns, visualize relationships, or do inductive reasoning. They offer a thinking way around absence of "knowledge" or a way of constructing prerequisite knowledge. A simple example illustrates the point. Simultaneous equations have for decades provided the mathematical tools for solving several "types" of problems, so ninth grade was the year of their placement in the curriculum. With computer technology, these same problems can be solved much earlier and more easily, with more intuition and understanding. Basically the power of computer technology facilitates using graphic techniques to see, visually, the point of intersection of the graphs of equations simultaneously, and inductive approaches to find the solution through patterns, powerful alternatives to the traditional ninth-grade approach.

A concept, important for all of us to understand, is that with a computer, we can do much that was not within our grasp earlier, and we can achieve deeper understandings of those still important goals that are retained. Just how does the computer allow us to accomplish more? And, possibly more important, how are we now able to do a better job than we did before technology became widely available? The paragraphs that follow illustrate the possibilities.

*Mathematica* is a sophisticated and powerful tool for mathematics as well as a tool-maker (Wolfram, 1996). It offers manifold opportunities for teaching better mathematics. A sampling of the "better mathematics" that can be studied in *Mathematica* include: visualizing patterns in statistical data sets; transforming data sets and seeing the visual image of the transformation; visually analyzing trigonometric relationships, including trigonometric identities; and exploring multiple-linked representations as, for example, in probability problems, trigonometry, analytic geometry, and discrete mathematics. We now examine the computer hardware configuration essential to making use of technology to study better mathematics. *Mathematica* can only satisfactorily be used on mainframe, mini, and advanced microcomputer platforms (Centris, Quadra, 486 IBM or Compatible, Sun, and so on). The multiple-disk software requires installation on a hard drive. It is sluggish with processors of 33 MHz or less, but will operate at its own deliberate speed.

An approximating characteristic that represents the power of a computer is its speed in MHz. As is evident, power according to this measure is increasing rapidly. Several different approaches are used: the number of transistors is increased (e.g., from 3.3 million to 3.5 million with accompanying reduction in power consumption and less heat), or new chips are designed. Processing power is approximately following the "axiom" of Robert Moore (Glitman 1994, 67) that "processing performance (of micro-

processors) will double every 18 months." Critical specifications make a difference in reasonable performance of more sophisticated mathematics software like *Mathematica*. *Mathematica* and other advanced software need a fast processor, a top-of-the-line color monitor, a high-capacity hard drive, substantial internal memory, and an extra processor supplementing the central processing unit. In 1995, only the top 20% of the most powerful microcomputer systems met these specifications, and hence satisfactorily ran programs like *Mathematica*. Increasingly, hardware configurations require substantial RAM for reasonable operation. Particularly *Mathematica*, *Photoshop*, and graphics software require 32 MB or more RAM for reasonable operation, making memory a main cost in computer systems, with prices for memory remaining expensive and rather stable over time.

Far away from the more "obvious" uses of computers in mathematics, *Photoshop* draws heavily on mathematics and mathematical reasoning. The significant application of mathematics in *Photoshop* permits the artist, photographer, mathematics educator, or multimedia user to see the power of mathematics and the necessity for an increased, better mathematics education. After exploring this program, most would surely conclude that it is a unique mathematical environment exemplifying "better mathematics" because of its superb applications of mathematics and because it is a good example of a mathematical microworld.

As stated earlier, the computer configuration in use effectively determines the mathematics and applications possible. It also introduces mathematical problems and reasoning within daily uses of the technology and software. To exemplify these ideas we summarize some mathematical problem solving in *Photoshop*.

## Mathematical Optimization Problems

*Photoshop* used with a computer system (40–60 MHz, 16–32 MB RAM, 250–540 MB drive), poses a multitude of optimization problems. For example, assume time and disk space are at a premium for a presentation involving several high quality data plots (e.g., two- and three-dimensional sections of multidimensional data sets or of three-dimensional functions). For the presentation, six to ten of these graphs are to be created and stored on the disk drive before the presentation. It is clear that simply storing each graph produced for the presentation will consume 15–40 MB of disk space. How can the pixel density (number of dots per square inch), image size, and compression be considered while we seek to optimize disk space and time factors?

Other problems that require optimization of pixel density versus file size are faced daily in *Photoshop*. There is a given menu with a sliding scale to assist in the optimization, but the correspondence of increased file size with increased pixel density, and vice versa is usually an incomplete beginning to the solution of the problem. Thus, experimentation, decision making, and reasoning are required to finish the solution.

There are literally hundreds of optimization problems that are encountered within most hardware configurations while using *Photoshop*, each having to be solved relative to the situation at hand. Firsthand encounters of interesting mathematics with computer hardware and in software like *Photoshop*, require better mathematical preparation, quite different mathematics involving critical thinking that becomes situation-specific for larger numbers of people, many typically considered "outside" of those who need mathematics.

## Multimedia Computers

In the years 1993–1995, computer technology became powerful, integrated, and innovative, with multimedia computers becoming a major direction of development. The concept of multimedia embraces capabilities like the use of sound (music, vocal, narration), visual images (pictures, videos and animations), dynamics, and interactivity integrated into the same system. Thus, the computer system that serves as a multimedia machine typically connects to audio and video equipment or has these components built in. The video cassette recorder (VCR), video and audio disk players (and recorders), CD-ROM drives, TV, telephone, and so on become integrally tied to multimedia systems. Examples of products from using multimedia equipment are animated film, instructional video, or well-designed advertisements, such as typical TV advertisements or TV weather reports.

## EDUCATING FOR THE FUTURE: A TECHNOLOGICAL WORLD

As with computer technology, the social, economic, professional, and geopolitical world is rapidly changing. Our decisions today about what we do with technology will affect what our children will be able to do with their lives in fifteen or twenty years.

Children who are born in the 1990s and beyond come into a world that will likely change beyond our power to imagine. They will experience technologies and be employed in jobs that are not existent today. Those children who embrace current technologies and seek to master them will be the ones who will succeed in the vastly different world. On average children are able to adapt to changes and work with computers with greater ease than adults. The child who has frequent opportunities to use technology is better prepared for success.

At present, the use of computer technology is in the stage of transliteration of existing practices, quite immature by comparison to the potential the invention affords. This transliteration involves simply adapting current practices into the new form, without new practices emerging. Beyond transliteration, small glimpses of the possibilities are emerging, some of which are headed toward development. For example, in digital technology that is emerging, the integrity of the data is preserved through transformation and transmission. This contrasts markedly with current visual media such as videotape, telephone, and TV, all of which acquire electronic "noise" during normal transmission. Analog data used in current video is considerably more susceptible to degradation in image quality when manipulated, distributed, edited, or copied. The integration of TV, the communications medium, and computer technology into digital form typically produces a substantially improved product, of high quality, but with substantial cost and delay. For example, the storage capacity of the medium will far exceed what is currently conceived. A short video clip in digital form quickly exceeds our drive's capacities and also exceeds what should be considered commercially feasible costs.

The changing forms of computer technology make clear that those elements essential to the mathematics classroom will evolve. For example, duo-docking systems, notebook computers that plug into a platform that converts them into desktop computers, have capabilities and physical properties that create tools to take outside the classroom. They may evolve toward what is becoming known as "semicomputers," a computer-like technology that is highly portable. Such a "computer" could be taken home, to a research site, or to a location for data-gathering or other types of mathematical investigations. These semicomputers have capabilities and prices somewhere between a notebook computer and a palm-size computer. They hold promise, pricewise, for wide availability. At present, however, they tend to be more like a single-purpose tool, only able to do one type of application like word processing. In contrast, however, a duo-docking system is a full-purpose notebook computer that is lightweight and highly portable, with batteries and a built-in video screen. In the office it can be inserted into a docking system, a hardware component, that makes the portable computer into a desktop computer with the capabilities of most other desktops including a full-size monitor, keyboard, printer, mouse, and so on. At present, a duo-docking system with color monitor, 4 MB RAM and 250 MB hard drive and docking port is beyond the current price range of most school mathematics departments, although one or two may be found in several more technologically progressive mathematics departments.

Unique, evolving features can change this situation. Some duo-docking systems, for example, are available with modular structure and a second battery, and are exchangeable with an internal floppy or a "semihard drive" or other features like interchangeable parts (e.g., RAM cards). The plug-in semihard drive could be, for one hour, a drive containing *Mathematica*; for another *Geometer's Sketchpad* (Jackiw 1991); for another Data Desk, and so on. Thus, with one configuration, the duo-docking system could be used in a geometry class one hour, a probability and statistics class another, and an advanced mathematical applications, discrete mathematics, or calculus still another. And, since the capacity of the hard drive is a major cost factor, the total system's cost with these alternatives may be relatively low.

## Uses of Computers in Mathematics

Computers are used in K–16 mathematics education in a wide variety of ways. The simplest use, in the view of most who involve themselves with the study of technology, is the most inappropriate use: as a skill practice tool. The more sophisticated use is as a tool for inquiry and scientific investigations. Computers are having some effect on the mathematics curriculum, from no influence in many local schools to substantial use in technology magnet schools and some university mathematics departments. There are examples of elementary schools, middle schools, and senior high schools that have a computer for each two to five students and use these computers in each subject.

At one level, computers have had little effect on mathematics curriculum and teaching practices. Becker (1991, 6) speaks of the unfulfilled dreams and the minimal role computers have played in real schools. Yet he reminds us: "As we enter into the 1990s, it is important to understand how much the early limited reality still remains and to understand how much of the idea of transforming teaching and learning through computers remains plausible. We need to be aware of the 'old habits' and 'conventional beliefs' that are common among practicing educators and the 'institutional restraints' that impede even the best of intentions to improve schools through technology" (Becker 1991, 6). Yet from another perspective, there are several complete mathematics courses that utilize computers. These courses characteristically make substantial use of the computer for developing meaning of major mathematical ideas (e.g., the function concept) and for solving application problems. A notable example is *Computer-Intensive Algebra* (Fey et al. 1991).

Student use of computers in mathematics ranges from none at all to extensive; the greatest use is found in, for example, magnet technology schools, where computers and other technologies are available. Students use computers to write their papers and use software (such as *Sketchpad* or *Mathematica*) to solve problems, routine and nonroutine, and to experiment with simulations from which they are able to construct their own mathematical meaning. The technology is there, and software is becoming available that supports learning opportunities of varying quality.

Teachers also are at many different stages in the utilization of computers for mathematics teaching and learning. Their ability to use computers is quite varied. It is widely recognized that staff in-service and typical preservice approaches left much to be desired; and the evidence suggests that this condition has not changed appreciably in recent years. Yet there are teachers and even entire departments that have become such avid users of computers, and have brought their students to this stage as well, that one's hope remains alive that a revolution in computers to improve mathematics education is just a few years away.

## Research in Using Technology

A brief view of some research that examines the use of computers in teaching and learning mathematics will provide a flavor of the direction it is taking. Although not intended to be complete, it does provide a few of the typical directions such research is taking. One category of research involves computers as a means of examining student misconceptions in mathematics. In probability and in algebra, studies have pointed out the positive effects of certain computer software in exploring the depth and nature of student misconceptions in major ideas. In each case (Moschkovich 1989), student misconceptions were detected in ways not so obvious without technology, and more significantly, allowed students to detect their own misconceptions. In each case, multiple representations were available and served to provide students connections in their explorations. For example, using CHANCE (a multiple tools, multiple-linked representations software package) not only were students able to detect their own misconceptions, they were able to demonstrate an ability to construct sound concepts in sophisticated areas of probability, including independence, conditional probability, randomness, and sampling (Jiang and Potter 1994). Research by Biehler et al. (1988), found supporting evidence to the findings of Jiang and Potter, indicating that multiple representations and tools for representing data, particularly resulting from multiple-linked representations, tend to support the development of more complete conceptions of important ideas.

An important observation by Moschkovich (1989) is that while use of highly visual and interactive computer software helps to highlight misconceptions, it also can introduce other types of misconceptions. On the other hand, Bohren (1988, 218) describes research in which students and teachers were both misled to believe that the students were "sense making" of graphical representations and related tabular representations when further investigation showed little understanding of the relationship between the two variables, and in fact considerable misconception of the functional relationship. Furthermore, the point to be made is that teachers' beliefs are influenced by and do affect curricular change. Although research findings can contribute significantly to gains in understanding of uses of technology in the teaching and learning process, the entire research agenda to determine directions needs to be followed deliberately. It takes significant time to build the patterns of research that assure that there are not misconceptions in our interpretations much akin to the misconceptions students demonstrate in learning such important ideas of mathematics as major probability concepts or the function concept. Classroom interaction and teacher behaviors are quite different during periods of use of graphing or probability software, and this difference has an effect on the learning of students, as does students' direct work with technology.

*See also* Computer-Assisted Instruction (CAI); Computer Science; Programming Languages; Software; Technology

## SELECTED REFERENCES

Becker, Henry J. "Where Powerful Tools Meet Conventional Beliefs and Institutional Restraints." *Computing Teacher* 18(8)(1991):1–12.

Biehler, Rolf, Wolfram Rach, and Bernard Winkelmann. *Computers and Mathematics Teaching: The German Situation and Reviews of International Software*. Occasional paper #103. Bielefeld, FRG: University of Bielefeld, Institut für Didaktik der Mathematik, 1988.

Bohren, Janet L. "A Nine Month Study of Graph Construction Skills and Reasoning Strategies Used by Ninth Grade Students to Construct Graphs of Science Data by Hand and with Computer Graphing Software." Doctoral diss., Ohio State University, 1988.

Cockroft, Wilfred H., chairman. *Mathematics Counts: Report of the Committee of Inquiry into the Teaching of Mathematics in Schools*. London, England: Her Majesty's Stationery Office, 1982.

Crust, Rita, Richard Phillips, James T. Fey, Anthony Ralston, and Connie Widmer. "MiniConference on Calculators and Computers." In *Proceedings of the 7th International Congress on Mathematical Education* (pp. 331–337) Claude Gaulin, Bruce Hodgson, David Wheeler, and John C. Egsgard, eds. Sainte-Foy, Quebec, Canada: Les Presses de l'Université Laval, 1994.

Dugdale, Sharon. "Green Globs: A Microcomputer Application for Graphing of Equations." *Mathematics Teacher* 75(1982):208–214.

Fey, James T., ed. *Computing and Mathematics: The Impact on Secondary School Curricula*. Reston, VA: National Council of Teachers of Mathematics, 1984.

Fey, James T., M. Kathleen Heid, Richard Good, Charlene Sheets, Glendon Bhime, and Rose Mary Zbiek. *Computer-Intensive Algebra*. College Park, MD: University of Maryland Office of Technology Liaison, 1991.

Glitman, Russell. "Pentiums Speed Up to 100 MHz." *PC World* 12(May 1994):67.

Jackiw, Nicholas. *The Geometer's Sketchpad* [Software]. Berkeley, CA: Key Curriculum, 1991.

Jiang, Zhonghong and Walter D. Potter. "A Computer Microworld to Introduce Students to Probability." *Journal of Computers in Mathematics and Science Teaching* 13(2)(1994):197–222.

Jurdak, Murad. "Mathematics Education in the Global Village: The Wedge or the Filter." In *Selected Lectures from the 7th International Congress on Mathematical Education* (199–210). David F. Robitaille, David H. Wheeler, and Carolyn Kieran, eds., Sainte-Foy, Quebec, Canada: Les Presses de l'Université Laval, 1994.

Kaput, James. "Information Technologies and Affect in Mathematical Experiences." In *Affects and Mathematical Problem Solving* (89–103). Douglas B. McLeod and Verna M. Adams, eds. New York: Springer-Verlag, 1989.

———. "Technology and Mathematics Education." In *Handbook of Research on Mathematics Teaching and Learning* (pp. 515–556). Douglas A. Grouws, ed. New York: Macmillan, 1992.

———, and Thomas A. Romberg. *Exploiting New Technologies for Reform in Mathematics Education.* Madison, WI: National Center for Research in Mathematical Science Education, 1989.

Lappan, Glenda, and Sarah Theile-Lubienski. "Training Teachers or Educating Professionals? What Are the Issues and How Are They Being Resolved?" In *Selected Lectures from the 7th International Congress on Mathematics Education* (pp. 249–261). David F. Robitaille, David H. Wheeler, and Carolyn Kieran, eds. Sainte-Foy, Quebec, Canada: Les Presses de l'Université Laval, 1994.

Moschkovich, J. "Constructing a Problem Space through Appropriation: A Case Study of Tutoring during Computer Exploration." Paper presented at the 1989 meeting of the American Educational Research Association, San Francisco, March 1989.

National Research Council. *Everybody Counts: A Report to the Nation on the Future of Mathematics Education.* Washington, DC: National Academy Press, 1989.

Philipp, Randolph, William A. Martin, and Glen W. Richgels. "Curricular Implications of Graphical Representation of Functions." In *Integrating Research on the Graphical Representation of Functions* (pp. 239–278). Thomas Romberg, Elizabeth Fennema, and Thomas Carpenter, eds. Hillsdale, NJ: Erlbaum, 1992.

*PhotoShop* (3.1) [Software]. Mountainview, CA: Adobe Systems, 1994.

Schwartz, Judah L., and Michal Yerushalmy. *The Geometric Supposers.* (Software). Pleasantville, NY: Sunburst, 1985.

———. "Using the Microcomputer to Restore Invention to the Learning of Mathematics." In *Developments in Mathematics Education around the World* (pp. 623–636). Vol. 1. Izaak Wirszup and Robert Streit, eds. Reston, VA: National Council of Teachers of Mathematics, 1987.

Winograd, Terry, and Carlos F. Flores. *Understanding Computers and Cognition: A New Foundation for Design.* Norwood, NJ: Ablex, 1986.

Wolfram, Stephen. *Mathematica* [Software]. 3rd ed. Champaign, IL: Wolfram, 1996.

EDWIN McCLINTOCK

# CONGRESSES

In 1893 the citizens of Chicago organized a great fair, the World's Columbian Exposition, to celebrate the 400th anniversary of Columbus's discovery of America, albeit one year late. As part of this event, a series of scientific and scholarly congresses was scheduled, among them one in mathematics. One of the foremost mathematicians of the day, Felix Klein of Göttingen University, came from Germany to give the opening address and bring mathematical papers from some of his European colleagues. The fair and the accompanying congresses coincided with the founding of the University of Chicago; the university's first president, William Rainey Harper, was one of the organizers of the scholarly sessions (Parshall and Rowe 1994, 295–327).

This Chicago congress, though the first, was not truly an international meeting where representatives from many countries met to exchange information on the latest developments in mathematics. The first such congress was held in Zürich three years later and was the first of a long series of international congresses of mathematicians (ICM) that have been held roughly every four years since that time: in Zürich (1897); Paris (1900); Heidelberg (1904); Rome (1908); Cambridge, England (1912); Strasbourg (1920); Toronto (1924); Bologna and Florence (1928); Zürich (1932); Oslo (1936); Cambridge, Massachusetts (1950); Amsterdam (1954); Edinburgh (1958); Stockholm (1962); Moscow (1966); Nice (1970); Vancouver (1974); Helsinki (1978); Warsaw (1983); Berkeley (1986); Kyoto (1990); Zürich (1994); and Berlin (1998).

The history of the ICM has not been devoid of controversy, often being affected by world politics. But there have been some high points along the way. The Paris congress of 1900 was noteworthy for providing a setting for the famous lecture by David Hilbert in which he outlined the twenty-three problems "from the discussion of which an advancement of science may be expected" (Albers et al. 1986, 8–9; Browder 1976, 1–34).

The series of congresses was disrupted during World War I; when it was resumed in 1920 in Strasbourg, mathematicians from Germany and the other Central Powers were not invited. That policy continued through the congress in Toronto in 1924, but the Germans were again invited to the next congress held in Bologna, although some refused to attend. By the time the congresses were held in Zürich in 1932 and Oslo in 1936, politics was again making the organization of such meetings difficult; the 1940 congress was scheduled for the United States but had to be cancelled. The first postwar congress, held in Cambridge, Massachusetts, in 1950, demonstrated the extraordinary strength of American mathematics, not only from the rise in the number of internationally known native-born American mathematicians, but also through the migration of many outstanding European mathematicians who fled the Nazis in the 1930s and 1940s. The progress of subsequent congresses has been largely without crisis except for the Warsaw congress of 1982, which, due to political problems in Poland, had to be deferred to 1983. The first Fields Medals (often referred to rather inaccurately as the "Nobel Prizes" of mathematics) were awarded at the congress in Oslo; they were named for John Charles Fields who had organized the 1924 congress in Toronto and provided funding for the prizes (Albers et al. 1986, 46–53).

In 1969 a parallel set of congresses began with an International Congress on Mathematics Education (ICME) held that year in Lyons, followed by congresses in Exeter (1972), Karlsruhe (1976), Berkeley (1980), Adelaide (1984), Budapest (1988), Quebec City (1992), Seville (1996), and Tokyo (2000). Currently both sequences of congresses meet every four years, with ICMEs meeting in years divisible by four and ICMs in the other even numbered years. Both ICMs and ICMEs are organized under the auspices of the International Mathematical Union, a consortium of fifty-five national mathematical organizations. Extensive proceedings are published following each congress, giving full transcripts of most presentations.

*See also* International Organizations

### SELECTED REFERENCES

Albers, Donald J., Gerald L. Alexanderson, and Constance Reid. *International Mathematical Congresses: An Illustrated History 1893–1986.* New York: Springer-Verlag, 1986.

Browder, Felix E., ed. *Mathematical Developments Arising from Hilbert Problems.* Providence, RI: American Mathematical Society, 1976.

Parshall, Karen Hunger, and David E. Rowe. *The Emergence of the American Mathematical Research Community, 1876–1900: J. J. Sylvester, Felix Klein, and E. H. Moore.* Providence, RI: American Mathematical Society, 1994.

GERALD L. ALEXANDERSON

# CONIC SECTIONS (CONICS)

Curves obtained when a plane intersects a double-napped right circular cone. Assuming that the intersecting plane does not pass through the vertex of the cone, the resulting conic section will be a *circle, ellipse, parabola,* or *hyperbola,* depending on the inclination of the plane (Figure 1(a)). If the plane passes through the vertex, however, the section will be a degenerate conic, namely a point, or a pair of straight lines. Figure 1(b) depicts the case of two intersecting lines.

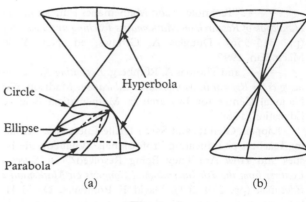

**Figure 1**

The Greek mathematician Menaechmus (ca. 350 B.C.) was the first to study the conic sections, and Apollonius (ca. 225 B.C.) was the one who wrote the celebrated *Conics,* in which he adopted the names ellipse, hyperbola, and parabola (Boyer 1989). In the *Conics,* Apollonius followed his predecessors in studying the conics as sections of a cone, but later he dispensed with the cone and studied the properties of the conics as plane curves. About eighteen centuries later, René Descartes invented analytic geometry and the topic of conic sections became an important part of it. The derivation of the Cartesian equation of each conic is based on a characteristic property of the conic.

The *parabola* is the set of all points in the plane that are equidistant from a fixed point (the *focus*) and a fixed line (the *directrix*). A line through the focus perpendicular to the directrix is the *axis* of the parabola, and the point where the axis cuts the parabola is called the *vertex.* In Figure 2(a), the focus is $F(p,0)$ and the directrix is $x = -p$. Since $PF = PD$, the equation of the parabola is $y^2 = 4px$.

The ratio $PF:PD = 1$ is the main characteristic of a parabola and is designated as its *eccentricity.* The parabola has the property that the tangent at a point $P$ makes equal angles with $PF$ and the line through $P$ parallel to the axis of the parabola (Figure 2(b)). This reflective property has many applications, for example, in headlight reflectors, searchlight mirrors, radars, solar energy devices, and telescopes. The parabola is crucial in designing suspended bridges and other constructions. In dynamics, the trajectory of a projectile is a parabola that is open downward.

The *ellipse* is the set of all points in the plane the sum of whose distances from two fixed points (the *foci*) is constant. The line through the foci is the *major axis* of the ellipse, the points where it cuts the ellipse

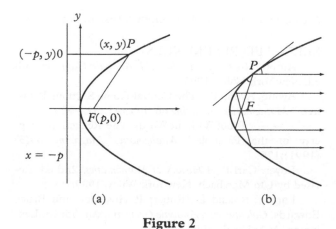

**Figure 2**

are the *vertices,* and the midpoint between the vertices is the *center* of the ellipse. In Figure 3(a), the foci are $F_1(-c,0)$ and $F_2(c,0)$. If $PF_1 + PF_2 = 2a$, $a > 0$, then the equation of the ellipse is $\frac{x^2}{a^2} + \frac{y^2}{b^2} = 1$, where $c^2 = a^2 - b^2$. This equation implies that the vertices of the ellipse are $V_1(-a,0)$ and $V_2(a,0)$ and the distance between the vertices $= 2a$.

Corresponding to the two foci $(+c,0)$ and $(-c,0)$ of the ellipse, there are two directrices $x = +a^2/c$ and $x = -a^2/c$ where the distance of any point $P$ of the ellipse from a focus to its distance from the corresponding directrix is a constant ratio $c/a < 1$. This ratio is called the *eccentricity* ($e$) of the ellipse. This property is used to derive the polar equation of the ellipse:

$$r = \frac{ed}{1 - e \cos \theta}$$

where $d$ is the distance $FD$ between the focus (at the pole) and the corresponding directrix (Figure 3(b)). This polar equation is essential for the proof of Kepler's first law, which states: Each planet in the solar system moves in an elliptical orbit with the sun at a focus.

The ellipse also has a reflective property, namely, the tangent line at a point $P$ of an ellipse makes equal angles with the lines through $P$ and the foci of the ellipse (Figure 3(c)). This property explains the phenomenon of whispering galleries. The cross-sections of the ceilings of these galleries are elliptic. Thus, when a person whispers at one focus, his soundwaves are reflected by the ceiling such that a person at the second focus hears the whispered sound (Anton 1995).

The *hyperbola* is the set of all points in the plane the difference of whose distances from two fixed points (the *foci*) is constant. The line through the foci is the *transverse axis* and the points where it cuts the hyperbola are the *vertices*. The midpoint between the vertices is the *center* of the hyperbola. In Figure 4(a), the foci are $F_1(-c,0)$ and $F_2(c,0)$. If $PF_1 - PF_2 = +2a$ or $-2a$, $a > 0$, then the equation of the hyperbola is

$$\frac{x^2}{a^2} - \frac{y^2}{b^2} = 1$$

where $c^2 = a^2 + b^2$. This equation implies that the vertices of the hyperbola are $V_1(-a,0)$ and $V_2(a,0)$ and the distance between the vertices is $2a$. Both branches of the hyperbola come arbitrarily close to the straight lines

$$y = \pm \frac{b}{a}x$$

which are called asymptotes of the hyperbola. The equations of the two asymptotes may be combined in the single equation

$$\frac{x^2}{a^2} - \frac{y^2}{b^2} = 0$$

Like the ellipse, the hyperbola has two directrices $x = \frac{+a^2}{c}$ and $x = \frac{-a^2}{c}$ corresponding to the foci $(c,0)$ and $(-c,0)$. However, the eccentricity of the hyperbola is $e = \frac{c}{a} > 1$.

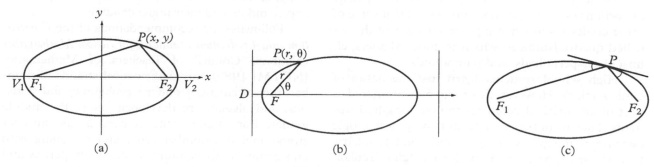

**Figure 3**

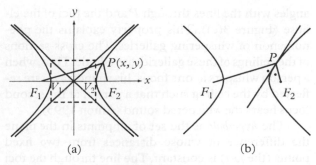

(a)                    (b)

**Figure 4**

The reflective property of the hyperbola also is similar to that of the ellipse, shown in Figure 4(b). This property is used in navigation to determine the location of a ship relative to two stations transmitting radio signals (Anton 1995).

The quadratic equation $ax^2 + bxy + cy^2 + dx + ey + f = 0$, where not all $a$, $b$, and $c$ are zeros, represents a conic section, whose axis need not be parallel to either of the coordinate axes. This can be shown by using rotation or translation of the coordinate axes to transform the quadratic equation to one of the forms: $Ax^2 + By^2 = 0$, $Ax^2 + By^2 = C$, or $y^2 = Ax$. However, the following rules, expressed in determinant form, may be used to identify the graph of the quadratic equation:

$$(1) \text{ If } \Delta = \begin{vmatrix} 2a & b & d \\ b & 2c & e \\ d & e & 2f \end{vmatrix} = 0$$

then the equation represents a degenerate conic section, that is, a pair of straight lines (Ayoub, 1993). (2) If $\Delta \neq 0$, then the discriminant $b^2 - 4ac$ of the quadratic equation may be used to classify the corresponding conic as follows: (a) a parabola if $b^2 - 4ac = 0$; (b) an ellipse if $b^2 - 4ac < 0$; (c) a hyperbola if $b^2 - 4ac > 0$ (Anton 1995).

While the conic sections were defined originally as sections of a cone, it should be noted that some of these conics result when a plane cuts any of the so-called quadric surfaces, which include cylinders, ellipsoids, hyperboloids, and paraboloids.

High school students learn just the basics of conic sections. However, this topic may be offered as part of an enriched geometry course for gifted students. They would appreciate and enjoy such a topic because it integrates analytic and solid geometry, loci, circles, spheres, and similar triangles (Teukolsky, 1987).

*See also* Circles; Space Geometry, Instruction

SELECTED REFERENCES

Anton, Howard. *Calculus with Analytic Geometry.* 5th ed. New York: Wiley, 1995.

Ayoub, Ayoub B. "The Central Conic Sections Revisited." *Mathematics Magazine* 66(5)(1993):322–325.

———. "Proof Without Words: The Reflection Property of the Parabola." *Mathematics Magazine* 64(3) (1991):175.

Boyer, Carl B. *A History of Mathematics.* 2nd ed. Revised by Uta Merzbach. New York: Wiley, 1989.

Larson, Roland E., Robert P. Hostetler, and Bruce Edwards. *Calculus with Analytic Geometry.* Alt. 5th ed. Lexington, MA: Heath, 1994.

Teukolsky, Roselyn. "Conic Sections: An Exciting Enrichment Topic." In *Learning and Teaching Geometry* (pp. 155–174). 1987 Yearbook. Mary Montgomery Lindquist, ed. Reston, VA: National Council of Teachers of Mathematics, 1987.

AYOUB B. AYOUB

# CONNECTED MATHEMATICS PROJECT (CMP)

National Science Foundation (NSF)-funded project to write a middle school curriculum for students and teachers (Lappan et al. 1995). The mathematical goal of CMP is: All students should be able to reason and communicate proficiently in mathematics. This includes knowledge and skill in using vocabulary, forms of representation, materials, tools, and techniques; and intellectual methods of the discipline of mathematics including the ability to define and solve problems. The goal includes a commitment to skill that is more than just proficiency with computation and manipulation with symbols, but also enables a student to deal with situations not previously encountered. The CMP curriculum is organized into units based on major mathematical ideas. Students explore problems included in each unit embodying these mathematical ideas; then practice, apply, and extend their understanding.

Following the recommendations of the *Curriculum and Evaluation Standards for School Mathematics* (National Council of Teachers of Mathematics [NCTM] 1989), the mathematical strands of number, measurement, geometry, probability and statistics, and algebra are developed across the middle grades. For example, the content in the number strand includes number sense and reasoning with and about numbers; number theory; properties and operations of number systems, with focus on integers and rational numbers; number estimation; ratio, proportion, and percentage; representation of numbers in concrete, graphic, and symbolic forms; scientific notation; and exponential notation. The four overarching goals in the NCTM *Standards*—Problem Solving, Communication, Reasoning, and Connections—serve as the major process goals for CMP. In addition, CMP has the following specific process goals: counting, visualizing, representing, comparing, measuring, estimating, modeling, reasoning, playing, and using tools.

The CMP curriculum supports the belief that curriculum and instruction are not distinct; the "what to teach" and "how to teach it" are inextricably linked (NCTM 1991). The CMP materials, for both student and teacher, are designed in ways that help students and teachers build a different pattern of interaction in the classroom. Traditionally, the general mode of instruction in mathematics classrooms has centered around the teacher as the source of knowledge. In the CMP curriculum students develop mathematical understanding by investigating mathematical problems under the guidance of the teacher. The CMP materials are designed to support a teacher and students in building a community of learners who are mutually supportive as they work together to make sense of the mathematics. This is done through the tasks that are provided, the justification that students are asked to provide regularly for their mathematical work, the opportunities to talk and write about ideas, and the help for the teacher in both alternative forms of assessment and using a problem-centered instructional model in the classroom.

Assessment in the CMP materials is intended as an extension of the learning process for students as well as an opportunity to check what students can do. For this reason the CMP assessment is multidimensional, giving students many ways to demonstrate how they are making sense of the mathematics in the units. The assessment tools include Check-Ups, Partner Quizzes, Projects, Unit Tests, Self-Assessment/Show What You Know and organizational tools that include Notebooks, Journals, Notebook Check, and Vocabulary Lists.

The CMP has collected comparative data on student achievement using both traditional and non-traditional methods. The Iowa Test of Basic Skills Survey Battery Form K, Levels 12 and 13, was used to understand how CMP students in the sixth to the eighth grades perform in traditional multiple-choice, timed, standardized tests. In addition, an hour-long individually administered test and three forty-five-minute performance tasks were developed based on

the *Assessment Standards for School Mathematics* (NCTM 1995). These last two assessment instruments focus on reasoning, mathematical communication, connections, and problem solving. The CMP students outperformed the non-CMP students on the *Standards*-based problem-solving test at all three grade levels. While the CMP students held their own on the Iowa Test of Basic Skills in the sixth and seventh grades, they scored significantly better than the non-CMP students by the eighth grade (Hoover 1997). In addition, a study of proportional reasoning by students in the seventh grade showed that CMP students scored significantly better than non-CMP students (Ben-Chaim et al. 1997).

## SELECTED REFERENCES

Ben-Chaim, David, James T. Fey, William Fitzgerald, C. Benedetto, and Jane Miller. *A Study of Proportional Reasoning among Seventh Grade Students.* Paper presented at the annual meeting of the American Educational Research Association, Chicago, 1997.

Connected Mathematics Project website: http://www.math.msu.edu/cmp/

Hoover, Mark N., Judith S. Zawojewski, and James Ridgeway. *Effects of the Connected Mathematics Project on Student Attainment.* Paper presented at the annual meeting of the American Educational Research Association, Chicago, 1997.

Lappan, Glenda, James T. Fey, William Fitzgerald, Susan Friel, and Elizabeth Phillips. *Connected Mathematics Project.* Menlo Park, CA: Addison-Wesley, 1995.

Lappan, Glenda, William Fitzgerald, Susan Friel, James T. Fey, and Elizabeth Phillips. *Connected Mathematics Project: Covering and Surrounding.* White Plains, NY: Seymour, 1998.

National Council of Teachers of Mathematics (NCTM). *Assessment Standards for School Mathematics.* Reston, VA: The Council, 1995.

———. *Curriculum and Evaluation Standards for School Mathematics.* Reston, VA: The Council, 1989.

———. *Principles and Standards for School Mathematics.* Reston, VA: The Council, 2000.

———. *Professional Standards for Teaching Mathematics.* Reston, VA: The Council, 1991.

GLENDA LAPPAN
ELIZABETH PHILLIPS

# CONSTRUCTIVISM

Theories of learning that claim that students learn by active meaning-making and by involvement in their own learning processes. These learner activities can be visible, such as in active problem solving, or not, such as when a learner is listening intently to understand an explanation. In each case, the learner is making sense of the mathematical situation presented, and drawing upon previously learned knowledge, skills, and understandings to do so. Since the 1980s, there has been a growing acceptance of constructivist theories of learning.

## PIAGET'S CONSTRUCTIVISM

It was the influence of the great Swiss psychologist Jean Piaget (1896–1980) that established constructivism as a leading theory of learning mathematics. His constructivism includes an epistemology, a structuralist view, and a research methodology. Piaget's epistemology has its roots in a biological metaphor. This says that just as an evolving organism must adapt to its environment in order to survive, so too the developing human intelligence undergoes a process of adaptation in order to fit in and remain viable in its environment. Personal theories or schemata are constructed to enable the person to make sense of his or her experiences. These theories are constellations of interconnected concepts, and are adapted by the twin processes of assimilation and accommodation in order to continue to fit with the person's experiences. Assimilation is the process by which new experiences are interpreted by means of existing schema. When existing schema cannot make sense of new experiences, a process of restructuring, termed accommodation, occurs. This also is triggered when there are conflicts between existing schema and attempts to resolve them. Piaget claims that the world as we experience it can only be known by us through our cognitive structures (an insight he takes from the philosopher Immanuel Kant).

Piaget's structuralism involves a belief that in organizing itself, the human intelligence necessarily constructs a characteristic set of logical and mathematical structures. He proposes that there is an invariant sequence of stages in constructing these structures through which an individual's cognition develops. Piaget's methodology centers on the use of the clinical interview, in which a subject is required to perform certain carefully designed tasks in front of, and with prompting from, an interviewer. For example, in a conservation of number task, the interviewer places a line of, for example, eight chips in front of a child. She then spreads them out to double the length and asks the child if the number of chips is now more, the same, or less. Several sessions are likely to be needed for the researcher to develop and test her model of the subject's understanding (see Piaget 1952, for an example of his interview transcripts). Piaget's clinical interview method is an important contribution to research methodology, and is one of the most widely used approaches by researchers.

## RADICAL CONSTRUCTIVISM

Ernst von Glasersfeld (1995) has developed a well-founded and elaborated constructivist epistemology that extends the foundational work of Piaget in a way many regard as significant. He bases this on the following two principles: (a) "knowledge is not passively received but actively built up by the cognizing subject," and (b) "the function of cognition is adaptive and serves the organization of the experiential world, not the discovery of ontological reality" (von Glasersfeld 1989). The adoption of the first principle alone results in a weak form of constructivism, which von Glasersfeld terms "Trivial Constructivism." Despite the disparaging title, it has important educational implications, for it states that knowledge is not transferred directly from the environment or other persons into the mind of the learner. Instead, any new knowledge has to be actively built from preexisting mental objects within the mind of the learner by extending and linking concepts, possibly in response to stimuli in the experiential world, to satisfy the needs and wants of the learner. An immediate consequence of this theory is that the transmission model of teaching and the passive reception model of learning are seen to be inadequate. Although lectures may work, the underlying mechanism is far more complex than that of simple information transfer. Individual learners construct unique and idiosyncratic personal knowledge representations, even when exposed to identical stimuli. For example, Herbert Ginsburg (1977) illustrates the variety of individual methods children invent in arithmetic.

Constructivism is widely cited as the underlying framework in research publications in mathematics education, and is claimed to underpin recent teaching reforms. It also generated the "alternative conceptions movement" in science education research (Duit 1995). However, although von Glasersfeld's first principle may help to reconceptualize the teaching of mathematics, it should be recognized that it does not strictly imply or disqualify any teaching approach. Rote learning, drill and practice, and passive listening to lectures do give rise to successful learning. The activity necessitated by this principle takes place cognitively, and so visible inactivity on the part of the learner is irrelevant. Constructivism does not equate to the "discovery method" or problem-solving teaching approaches, as Gerald Goldin (1990) argues, although it can be used to support them. In addition, it should be noted that the idiosyncratic nature of learners' mental constructions are not only the consequences of this principle. They also follow from other psychological approaches such as those of David Ausubel and George Kelly.

The second principle adds an epistemological dimension to constructivism, transforming it into "Radical Constructivism." This claims that all knowledge is constructed, and that it can tell us nothing certain about the world, or any other domain. This is not entailed by the first principle, which is consistent with the assumption that objective truth exists, but that the cognizing subject constructs idiosyncratic personal representations of it. Thus a radical constructivist teacher can never know what is behind a child's actions, but can only conjecture. Von Glasersfeld's second principle has been criticized for leading to the denial of the existence of the physical world. However, this is an incorrect conclusion. Von Glasersfeld (1995) has explicitly made the point that radical constructivism is ontologically neutral, and is consistent with the existence of the world. All that it denies is the possibility of any certain knowledge about it. The implication that there is no certain knowledge is very important, both philosophically and educationally. Current work in the philosophy of mathematics and philosophy of science questions the possibility of any absolute knowledge, and radical constructivism supports this view (Ernest 1991). For example "1 + 1 = 2" has been taken as an absolute truth. A new perspective is that its truth is relative to the context, with its underlying assumptions. In the alternate contexts of Boolean algebra and modulo 2 arithmetic, $1 + 1 = 1$ and $1 + 1 = 0$, respectively.

There is a growing body of constructivist research into children's learning of mathematics, and into the use of teaching approaches consistent with constructivism. Some very promising results have been obtained with learners of all ages, from young children's understanding of number (Steffe et al. 1983) to high school students' understanding of exponents (Confrey 1991).

## CRITICISMS OF CONSTRUCTIVISM

There have been criticisms from supporters of traditional expository teaching and basic skills-centered approaches who argue that all forms of constructivism are inefficient at raising standards of attainment in mathematics. Such criticisms often are based on an inadequate understanding of constructivism, and such critics have yet to produce evidence

from research for their claims. In addition, some constructivists reject both the labels "Radical" and "Trivial." For example, Goldin (1990) advocates a "moderate constructivist" view (p. 32), and criticizes radical constructivism for denying the existence of mathematical structures outside the mind of the learner. Some of the strongest criticisms of constructivism say that it overemphasizes the individual aspects of the learner and neglects social aspects of learning, including the important role of language. Radical constructivism views inner cognitive (and affective) experiences as necessarily preceding outer manifestations, and even other persons are understood to be the theoretical constructs of the knower. In contrast, Lev Vygotsky (1978) and other socially orientated psychologists emphasize the import of language and social experience, and regard much of learning as deriving from participation in shared social practices, before becoming part of the individual's conceptual structure. For example, the complex symbolism of mathematics must first be encountered in the external, social world before becoming part of the learner's thought processes. Contributions to both sides of the debate can be found in Ernest (1994), Steffe and Gale (1995), and Steffe et al. (1996). These also include versions of "Social Constructivism," which attempt to meet some of these criticisms by giving at least equal emphasis to the social dimension of learning.

## IMPLICATIONS FOR PRACTICE

All forms of constructivism suggest the need for and value of sensitivity and attentiveness to the learner's previous constructions. This means that the teacher needs to pay careful attention to students' mathematical methods and explanations. To develop students' incomplete or idiosyncratic conceptions (which may be the cause of errors), the teacher should use multiple representations of mathematical concepts to help extend conceptions, and also provide problems involving the use of discussion and collaboration between students, to help them confront the consequences of nonstandard conceptions and to self-correct them. Additional suggestions for practice are often associated with radical (and social) constructivism: (a) mathematical knowledge itself is understood to be fallible and corrigible, not just the learner's personal knowledge; (b) the methodology of research also becomes problematic and all inquiries must be conducted circumspectly and with self-criticism, as there is no truth to uncover, just accounts of the circumstances investigated of varying plausibility and viability; and (c) the focus of concern goes beyond the learner's cognitions and includes affects, beliefs, and personal views about the nature of mathematical knowledge and how it is validated. (The same holds for research on teachers.) For the classroom teacher, this should mean that all children's methods are valued and praised, but that the need for consistency and shared conventions is explained and consequently the class is led toward accepted knowledge. Also, the teacher should be confidently able to give students open-ended problem-solving tasks without knowing "the answers" in advance.

*See also* Psychology of Learning and Instruction, Overview

SELECTED REFERENCES

Confrey, Jere. "Learning to Listen: A Student's Understanding of Powers of Ten." In *Radical Constructivism in Mathematics Education* (pp. 111–138). Ernst von Glasersfeld, ed. Dordrecht, Netherlands: Kluwer, 1991.

Duit, Reinders. "The Constructivist View: A Fashionable and Fruitful Paradigm for Science Education Research and Practice." In *Constructivism in Education* (pp. 271–285). Leslie P. Steffe and Jerry Gale, eds. Hillsdale, NJ: Erlbaum, 1995.

Ernest, Paul, ed. *Constructing Mathematical Knowledge: Epistemology and Mathematics Education.* London, England: Falmer, 1994.

———. *Social Constructivism as a Philosophy of Mathematics.* Albany, NY: State University of New York Press, 1998.

———. *The Philosophy of Mathematics Education.* London, England: Falmer, 1991.

Ginsburg, Herbert. *Children's Arithmetic: The Learning Process.* New York: Van Nostrand, 1977.

Goldin, Gerald. "Epistemology, Constructivism, and Discovery Learning in Mathematics." In *Constructivist Views on the Teaching and Learning of Mathematics* (pp. 31–47). Robert B. Davis, Carolyn A. Maher, and Nel Noddings, eds. Reston, VA: National Council of Teachers of Mathematics, 1990.

Piaget, Jean. *Structuralism.* New York: Basic Books, 1970.

———. *The Child's Conception of Number.* New York: Norton, 1952.

Steffe, Leslie P., and Jerry Gale, eds. *Constructivism in Education.* Hillsdale, NJ: Erlbaum, 1995.

———, Ernst von Glasersfeld, John Richards, and Paul Cobb. *Children's Counting Types: Philosophy, Theory, and Application.* New York: Praeger, 1983.

———, Pearla Nesher, Paul Cobb, Gerald A. Goldin, and Brian Greer, eds. *Theories of Mathematical Learning.* Mahwah, NJ: Erlbaum, 1996.

Von Glasersfeld, Ernst. "Constructivism in Education." In *The International Encyclopedia of Education* (pp. 162–163). Suppl. Torsten Husén and T. Neville Postlethwaite, eds. Oxford, England: Pergamon 1989.

———. *Radical Constructivism: A Way of Knowing and Learning.* London, England: Falmer, 1995.

Vygotsky, Lev. *Mind in Society: The Development of Higher Psychological Processes.* Michael Cole et al., eds. Cambridge, MA: Harvard University Press, 1978.

PAUL ERNEST

# CONSUMER MATHEMATICS

Topics related to money and its uses in daily life. In school, primary-grade children explore these topics when they are learning to count money. At the secondary level, students apply the mathematics skills learned through middle school to the problems of daily life. Students should learn that use of numbers is inseparable from purchase, but other factors also enter into a decision to buy.

Topics for consumer mathematics should include how money comes into a consumer's hands. Earning and investing money, as well as saving money, are topics for applying the basic operations as well as exponentiation. Other items of concern for a consumer are housing and food. Purchasing clothing, traveling, and other consumer activities provide ample opportunity to strengthen mathematics skills.

Teaching good analysis skills for the selection of a home will help to save significant sums of money in the life of a consumer. For instance, students will be able to use mortgage tables to decide how to compare the amount of their down payment with the amount actually repaid on a house loan. The most important thing to teach students in consumer mathematics is that mathematics should enter into every decision they make about money. The cost may not be the deciding factor in the decision to purchase; however, it definitely must be a consideration.

## ESTIMATION

Being able to determine an approximate answer is an essential for incorporating number analysis in daily encounters with numbers. For example, department stores often have a sale where a certain discount, say 20%, will be deducted at the register, while the current price is marked on the item. Consumers need to realize that an exact price is not important in deciding if the sale is to their advantage. They can make an estimate with a small amount of mental calculation. Say the marked price is $34.99

and the touted discount is 20%. Rounding the price of the item to 35, doubling to 70, and mentally dividing by 10 is an easy way to approximate the savings. Then they may decide if $28 is a fair sale price. To many innumerates, 20% is a much bigger number than $7. The analysis of the final result is key to using mathematics to aid in consumer decisions. Students who have difficulty performing mental arithmetic should be coached in the use of a hand calculator while shopping. The marketing of sales to consumers takes advantage of the inability of many consumers to make that calculation at the point of investigation.

## CALCULATORS AND COMPUTERS

When a more precise answer is required, such as in income tax reporting, use of hand calculators is recommended. To that end, sufficient practice and instruction on the use of hand calculators should be included in a course in consumer mathematics. The important thing is that students determine the precision required of the answer for each situation. With the wide range of features and functions available to the public, consumers should purchase the calculator to suit their needs. The useful calculator functions for consumers are the basic four operations (that is, addition, subtraction, multiplication, and division), square root and exponentiation, fraction input, percent, and memory storage. These are generally available on relatively inexpensive calculators. Linking estimation skills with calculator use helps students to assess the value of the answer reported in the calculator display. A common misperception is that a calculator answer is the correct one. Students must be ready to challenge that answer if it disagrees with their estimated expected result.

Computers in the classroom have helped to make complex calculations easy. Use of a spreadsheet to compute, chart, and analyze provides an opportunity to study topics such as mortgage decisions. Helping students to construct a spreadsheet using the formulas for the monthly payment and the total repaid on a mortgage for various terms and interest rates helps students to realize how much a mathematical analysis of their spending can greatly impact their overall wealth.

## SIMULATIONS AND COMMUNITY SUPPORT

Generally, merchants in the vicinity of a school are pleased to help students develop consumer awareness. Field trips and price comparison discussions are facilitated by a project where students simulate a weekly food shopping trip. Determining unit prices, selecting the proper size of product, and balancing the cost of keeping a large package in inventory with the price saving of the larger size are all components of such a simulation. Local banks may support lessons on checkbook maintenance and savings instruments. Merchants are looking to these students as future customers; it is to the merchant's advantage that wise consumers are learning about their services.

Not only merchants but also real estate brokers can be helpful. A project on housing may be supported by newspaper ads, realtors, and other resources, and may include activities integrated to buying, renting, or furnishing a home. Students will learn the mathematics involved in buying carpet, wallpaper, curtains, arranging furniture, and so forth. Students are far more motivated to participate in a simulation project such as this when the tasks are defined in small, doable steps where they have made choices about how they will meet the objectives.

*See also* Calculators; Estimation; Finance; Interest; Money

### SELECTED REFERENCE

National Institute for Consumer Education (NICE). Housed at Eastern Michigan University. http://www.emich.edu/public/coe/nice/nice.html.

KATHLEEN M. HARMEYER

## CONTINUED FRACTIONS

Like any other *fractions* in possessing a numerator and denominator, but the denominator may be somewhat complex. To illustrate,

$$\frac{34}{15} = 2 + \frac{4}{15} = 2 + \frac{1}{\frac{15}{4}}$$

$$= 2 + \frac{1}{3 + \frac{3}{4}}$$

$$= 2 + \frac{1}{3 + \frac{1}{1 + \frac{1}{3}}} \qquad (\star)$$

the algebraic expression in $(\star)$ represents what is known as a finite, simple, continued fraction that is equal to 34/15. The "simple" refers to the succession of ones as numerators. Short-cut notation allows (*) to be written as [2; 3, 1, 3], where the integer before the semicolon is the whole number portion of the continued fraction, and the remaining integers are the whole number portions of the succession of denominators.

The partial convergents $[p_n/q_n]$ to 34/15 are given by the continued fractions

$$\frac{p_1}{q_1} = [2] = \frac{2}{1}$$

$$\frac{p_2}{q_2} = [2; 3] = \frac{7}{3}$$

$$\frac{p_3}{q_3} = [2; 3, 1] = \frac{9}{4}$$

$$\frac{p_4}{q_4} = [2; 3, 1, 3] = \frac{34}{15}$$

These partial convergents, as well as all others, follow two noteworthy principles (Eynden 1987):

1.  $\frac{p_1}{q_1} < \frac{p_3}{q_3} < \frac{p_5}{q_5} < \cdot \cdot \cdot < \frac{p_6}{q_6} < \frac{p_4}{q_4} < \frac{p_2}{q_2}$

2.  $p_i q_{i+1} - p_{i+1} q_i = (-1)^i$

The first rule says that the odd-numbered partial convergents form an increasing sequence, and the even-numbered convergents decrease, with each sequence converging to a common value. The second rule is quite useful in solving certain Diophantine equations, and serves as a springboard for solving many "word" problems in the classroom. For instance, to solve

$$34x - 15y = 1 \quad (\star\star)$$

in positive integers, the partial convergents to 34/15 are obtained, and since

$$\frac{p_3}{q_3} = \frac{9}{4}, \qquad \frac{p_4}{q_4} = \frac{34}{15}$$

it follows that $p_3 q_4 - p_4 q_3 = (-1)^3 = 9(15) - 34(4)$, so $x = 4$, $y = 9$ is a solution to ($\star\star$). Furthermore, the related equation

$$34x - 15y = 19$$

for example, must necessarily have the solution $x = 4(19)$, $y = 9(19)$.

Students today have an alternate method of solving ($\star\star$) by using their graphing calculators. With the $TI - 82$, for instance, the student can enter the function as $y_1 = (34x - 1)/15$, prepare a table of values with Tbl Min $= 1$ and $\Delta$Tbl $= 1$, and search for values where $x$ and $y_1$ are both integers, which will not take long as $x = 4$ and $y_1 = 9$.

Another area of fruitful study concerns infinite continued fractions. In this case, the sequence of denominators never ends, nor does the sequence of partial convergents. The constant $\pi$ has the representation

$$\pi = [3; 7, 15, 1, 292, 1, 1, 1, 2, \ldots]$$

with no apparent pattern at all. On the other hand, the infinite continued fraction

$$[1; 2, 2, 2, 2, \ldots]$$

is a representation for $\sqrt{2}$. A rational number that approximates $\sqrt{2}$ could be obtained quickly by selecting one of the partial convergents, say $p_5/q_5$, where

$$\frac{p_5}{q_5} = [1; 2, 2, 2, 2] = \frac{41}{29} \approx 1.4138$$

Infinite continued fractions sometimes occur quite ingeniously as solutions to equations. Consider the quadratic equation $x^2 - 4x - 1 = 0$. Isolating the quadratic term, dividing by $x$, and substituting gives

$$x = \frac{4x + 1}{x} = 4 + \frac{1}{x}$$

$$= 4 + \cfrac{1}{4 + \cfrac{1}{x}}$$

$$= 4 + \cfrac{1}{4 + \cfrac{1}{4 + \cdots}}$$

so one solution to the quadratic is $[4; 4, 4, 4, \ldots]$.

Many transcendental functions have continued fraction expansions. In particular, the tangent function can be represented (Olds 1963) by

$$\tan x = \cfrac{1}{1/x - \cfrac{1}{3/x - \cfrac{1}{5/x - \cfrac{1}{7/x - \cdots}}}}$$

From the late seventeenth century, when continued fractions first came under study as natural extensions of the Euclidean algorithm, and up to the present day, these fractions comprise an area of active research, especially pertaining to approximation theory. A typical lesson for the classroom is to find "good" rational approximations $p/q$ to some given irrational number $x$, where by "good" we mean $|x - p/q| < 1/q^2$. This degree of proximity is usually a standard measure of "closeness." The rational approximates $p/q$ are then determined as the partial convergents to the continued fraction expansion of $x$ (LeVeque 1990).

*See also* Diophantine Equations; *e*; Number Theory, Overview; Pi

## SELECTED REFERENCES

Eynden, Charles Vanden. *Elementary Number Theory.* New York: Random House, 1987.

LeVeque, William. *Elementary Theory of Numbers.* New York: Dover, 1990.

Olds, Carl D. *Continued Fractions.* New York: Random House, 1963.

THOMAS P. DENCE

# CONTINUITY

A property of a function whose graph has no breaks. With respect to a function $f(x)$, three different methods are commonly used to describe continuity at a point $x = a$. In an informal, intuitive approach, a value $x_0$ is selected that is close to some given $a$, perhaps equaling $a$, and another value $x_1$ that is close to $x_0$. If function $f$ is continuous at $x = a$, then the two function values $f(x_0)$ and $f(x_1)$ will be close. To illustrate, if $f(x) = x^2$, $a = 1.5$, $x_0 = 1.51$, and $x_1 = 1.49$, then because $f$ is continuous at 1.5 it must follow that the two function values $f(1.49)$ and $f(1.51)$ are close—and they are, since $1.49^2 = 2.2201$ and $1.5^2 = 2.25$. On the other hand, with $f$ given by

$$f(x) = 1 \quad \text{if } x \geq 0$$
$$= -1 \quad \text{if } x < 0$$

then $f$ is not continuous at $a = 0$ since $f(0.1)$ and $f(-0.1)$ differ by the relatively large amount of 2.

A more popular limit definition defines function $f$ to be continuous at $x = a$ if (1) $f$ has a definite (and hence finite) value $f(a)$ at $x = a$ and (2) as $x$ approaches $a$, $f(x)$ approaches $f(a)$ as its limit. Symbolically,

$$\lim_{x \to a} f(x) = f(a)$$

This definition proves especially helpful when trying to verify that a function $f$ is not continuous at $x = a$.

The third definition of continuity, developed by the German mathematician Karl Weierstrass (Dunham 1990), is the $\epsilon$–$\delta$ formulation that is fairly standard in most introductory calculus texts. In this approach, $f$ is continuous at $x = a$ if $f(a)$ is defined and if given any positive number $\epsilon > 0$, there exists a positive number $\delta > 0$ with the property that if $|x - a| < \delta$, then $f(x)$ is defined and $|f(x) - f(a)| < \epsilon$. The graphical interpretation of this (see Figure 1)

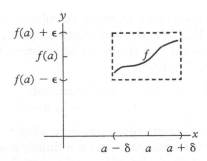

**Figure 1** $\epsilon$–$\delta$ *definition of continuity.*

shows that if $f$ is continuous at $x = a$ then the graph of $f$ lies entirely within the rectangle $a - \delta < x < a + \delta$ and $f(a) - \epsilon < y < f(a) + \epsilon$.

The last two definitions of continuity were the result of over a century of struggle by mathematicians to bring precision to the calculus of Isaac Newton and Gottfried Leibniz (Dunham 1990). Specifically, it was the limit concept that needed refinement and clarification. Credit is generally given to both the French mathematician Augustin-Louis Cauchy and to Weierstrass for pioneering a rigorous treatment of limits, continuous and differentiable functions, integrals, and series (Simmons 1992).

A function that is continuous at each point in its domain $D$ is said to be continuous on $D$. For example, $\sin x$ and $\cos x$ are continuous on the entire real line $(-\infty, \infty)$ as are all polynomials. Rational functions are continuous on $(-\infty, \infty)$ except at the points where the denominator function vanishes. For positive real $a$, the exponential functions $y = a^x$ are continuous on $(-\infty, \infty)$, while the logarithmic functions $y = \log_a x$ are continuous on $(0, \infty)$. Theorems regarding the arithmetic of continuous functions state that the sum, difference, and product of two continuous functions are again continuous functions, as is the composition of two continuous functions (Varberg and Purcell 1992). As in the case of rational functions, the quotient $f(x)/g(x)$ of two continuous functions is also continuous except when $g(x) = 0$.

Continuous functions are considered "nice" functions, not only because of the previously mentioned arithmetic properties but because continuity determines a measure of "smoothness" of a function. One should not think that all functions are "nice," for many abnormalities may occur. There are functions defined on the entire real line that are continuous everywhere, continuous everywhere except at a finite number of points, continuous only at a finite number (any finite number of your choosing) of points, and continuous absolutely nowhere! One only has to look on a graphing calculator at the function $y = \text{INT}(x)$, which is continuous at every real number except the entire set of integers. The "salt and pepper" function

$$f(x) = 1 \qquad \text{if } x \text{ is rational}$$
$$= -1 \qquad \text{if } x \text{ is irrational}$$

happens to be continuous nowhere. Finally, there are functions that are continuous at infinitely many points and, simultaneously, not continuous at infinitely many points. The classic example of a function (Kirkwood 1989) that is continuous at every irrational number and discontinuous at every rational number was a startling modern result that gave major impetus to Georg Cantor's classifications of infinite sets.

*See also* Calculus, Overview

SELECTED REFERENCES

Dunham, William. *Journey Through Genius: The Great Theorems of Mathematics.* New York: Wiley, 1990.

Kirkwood, James. *An Introduction to Analysis.* Boston, MA: PWS-Kent, 1989.

Simmons, George. *Calculus Gems.* New York: McGraw-Hill, 1992.

Varberg, Dale, and Edwin Purcell. *Calculus with Analytic Geometry.* 6th ed. Englewood Cliffs, NJ: Prentice-Hall, 1992.

THOMAS P. DENCE

# COOPERATIVE LEARNING

Teaching strategy that involves a small group of learners working together as a team to solve a problem, complete a task, or accomplish a common goal (Artzt and Newman 1997). It can be used in combination with other methods or as a total instructional system (Davidson, "Small-group . . .," 1990). Positive effects of cooperative learning under certain conditions have been documented (Slavin 1989). The goals of cooperative learning are improved learning of and attitudes toward mathematics, and the development of positive social attitudes and behaviors.

Cooperative learning requires more than a small group of students working together on a task. Careful attention must be given to structuring the group's work so that it qualifies as a cooperative learning experience. A number of components of cooperative learning are essential. *Task interdependence* requires the creation of a task that cannot be accomplished successfully without the individual contributions of each member of the group. That is, the individuals in the group must perceive themselves as dependent on one another's contributions. *Individual accountability* requires that each member of the group is responsible for learning the material and the burden for each person's learning is shared by other group members. Students must be taught *social skills* in order for them to function well in their groups. They must learn the *collaborative skills* that will enable them to work effectively with others, no matter what their abilities or personal characteristics. Finally they must engage in *group processing:* Teachers and their students must be given time to assess how well the students have worked together and to devise plans for improving the groups' effectiveness.

Cooperative learning is not a new idea. Scholars who studied group dynamics, Kurt Lewin (1935) and Morton Deutsch (1949), proposed theories of cooperative and competitive situations that have served as a foundation upon which research and discussions of cooperative learning have been based. Since that time, the cooperative learning literature documents many cooperative learning structures that teachers and researchers have studied and developed. Some of the most popular ones are *Using Student Team Learning* (Slavin 1980), *The Jigsaw Classroom* (Aronson 1978), *Learning Together and Alone* (Johnson and Johnson 1987), and *Numbered Heads Together* (described in Kagan 1989). In each of these structures the components that are described focus on the way the groups are formed, the way the tasks are designed, the reward structure used, and the way the group process is assessed. Applications of these strategies in mathematics instruction can be found in Neil Davidson (*Cooperative Learning . . .,* 1990).

Group formation is a critical component for successful group work. In general, it is recommended that groups consist of three to four students of heterogeneous ability. Recent recommendations suggest that, although the final decision on group arrangements should rest with the teacher, students' input regarding their evaluation of their own ability and their personal choices for people with whom they would like to work should be taken into account (Artzt 1994). To increase the feeling of cohesiveness among group members, team-building activities can be helpful. Students can be provided with opportunities to engage in tasks that will enable them to get to know one another better. For example, they can describe their past experiences in mathematics, their current attitudes toward mathematics, and their perceptions of their own mathematical ability. They can create names for their groups with group logos and they can photograph their group and post their pictures on the classroom bulletin boards.

Tasks for cooperative learning activities must be designed carefully so that all students have the opportunity and motivation to contribute their ideas. The tasks can be designed for different outcomes at different points in the lesson. For example, students may meet in groups for problem-solving sessions, for discovery sessions that contribute to the development of the lesson, for enrichment projects, for homework review, or for test review. The outcomes can range from agreeing on a set of solutions to discovering patterns and relationships and formulating and testing conjectures, to creating an enrichment project. The best-designed tasks are those that require students to come to the group having already thought about or done a part of the task. Other task designs that work well are those that define specific roles for each student in the group. For example, in mathematics, tasks can be designed where each student in the group examines a different case of a general property. When putting their work together they can notice a pattern in the results and arrive at a general statement. Specifically, each student may be given a different family of graphs to plot on his/her calculator. When they compare their results they may make a conjecture regarding the effects of different coefficients or constants. Also, mathematical projects can be designed whereby students must contribute a different part of the work. Specifically, a statistics project can be designed such that each person in the group must find one example of a specific kind of graph from magazines or newspapers. Each student must be prepared to explain the meaning of his or her graph to the other members of their group. Students must listen carefully, as one student may be selected randomly to explain the graphs to their group. The group grade is affected by the quality of the presentation.

Although most students are intrinsically motivated by the enjoyment they derive from working within a small group, not everyone, of course, thrives in this environment. Therefore, carefully designed reward structures can provide additional incentives for students to help and be helped by one another. For example, group scores that depend on the individual group members' performances work particularly well. Rewards for groups meeting certain preestablished criteria can take the form of publicity, stickers, handmade certificates, or points added to cooperation or participation grades.

For cooperative learning groups to be effective, teachers and students must engage in group processing, a form of assessment. That is, teachers must give students the opportunity to evaluate the quality of their own participation within the group and the quality of the participation between and among the group members. Students can record their assessment in written journals. They can be asked to describe the mathematical concepts they explained or received help with in their group that day. They also can record their assessment on evaluation forms that students can share with one another and discuss in their groups. The teacher must continually assess the behavior of the group members by observing the students at work, using evaluation forms to make notes about their participation, and reading and reacting to the journal entries of the students regarding their group work (see Dippong 1992 for sample evaluation forms).

*See also* Ability Grouping

## SELECTED REFERENCES

Aronson, Elliot. *The Jigsaw Classroom.* Newbury Park, CA: Sage, 1978.

Artzt, Alice. "Integrating Writing and Cooperative Learning in the Mathematics Class." *Mathematics Teacher* 87 (Feb. 1994):80–85.

———, and Claire M. Newman. *How to Use Cooperative Learning in the Mathematics Class.* 2nd ed. Reston, VA: National Council of Teachers of Mathematics, 1997.

Davidson, Neil, ed. *Cooperative Learning in Mathematics: A Handbook for Teachers.* Menlo Park, CA: Addison-Wesley, 1990.

Davidson, Neil. "Small-Group Cooperative Learning in Mathematics." In *Teaching and Learning Mathematics in the 1990s* (pp. 52–61). 1990 Yearbook. Thomas J. Cooney, ed. Reston, VA: National Council of Teachers of Mathematics, 1990.

Deutsch, Morton. "A Theory of Competition and Cooperation." *Human Relations* 2(1949):129–151.

Dippong, Jackie. "Two Large Questions: Teacher Challenges and Appropriate Student Tasks in Assessing and Evaluating Cooperative Learning." *Cooperative Learning* 13 (Fall 1992):6–8, 50–51.

Johnson, David W., and Roger Johnson. *Learning Together and Alone: Cooperative, Competitive, and Individualistic Learning.* 2nd ed. Englewood Cliffs, NJ: Prentice-Hall, 1987.

Kagan, Spencer. *Cooperative Learning Resources for Teachers.* San Juan Capistrano, CA: Resources for Teachers, 1989.

Lewin, Kurt. *Dynamic Theory of Personality.* New York: McGraw-Hill, 1935.

Slavin, Robert. *School and Classroom Organization.* Hillsdale, NJ: Erlbaum, 1989.

———. *Using Student Team Learning.* Rev. ed. Baltimore, MD: Center for Social Organization of Schools, Johns Hopkins University, 1980.

ALICE F. ARTZT

# COORDINATE GEOMETRY, OVERVIEW

Expresses the correspondence between algebra and geometry by the introduction of coordinates and graphing. In the late 1980s, the rise of the calculus reform movement and the widespread availability of computers and graphing calculators greatly encouraged the graphical approach to both discrete and continuous problems in mathematics (Douglas 1986; Steen 1988). While the study of analytic geometry as an end in itself has almost disappeared from the undergraduate curriculum, the use of coordinate geometry as a tool for both understanding and applying calculus appears to be on the increase.

## COORDINATE GEOMETRY IN THE PLANE

The Cartesian coordinate system for the Euclidean plane associates with each point $P$ in the plane an ordered pair $(x,y)$ of real numbers, as shown in Figure 1. This system, named for the French philosopher and mathematician René Descartes (1596–1650), is of great use in geometry because it allows statements of plane geometry to be translated into statements of algebra, and vice versa. Thus the ability to visualize is linked with the ability to analyze, resulting in improved understanding. Although many fundamentally important ideas of coordinate geometry are introduced by Descartes in an appendix on geometry to his famous *Discourse* of 1637, the system that bears his name was not fully presented by him. Many of the same ideas were discussed and more fully developed earlier in the correspondence of the noted French lawyer and mathematician Pierre de Fermat (1601–1665).

One of the most important examples of the correspondence between geometry and algebra is that between straight lines in the plane and linear equations. It can be shown that every line consists of all points whose coordinates satisfy an equation of the form $ax + by + c = 0$, where $a$ and $b$ are not both zero. Conversely, the *graph* of every such equation, that is, the set of points whose coordinates satisfy the

equation, consists of all points on a line. Thus, the geometrical study of lines in the plane corresponds to the algebraic study of linear equations in two unknowns. To illustrate the correspondence between algebra and geometry, consider a system of two linear equations in two unknowns. Assuming that neither equation is trivial, the system describes one of three possibilities: two distinct nonparallel lines intersecting at one point; two distinct parallel lines; or one line. This corresponds, respectively, to the system being *nonsingular,* with the coordinates of the point of intersection as its unique solution; *inconsistent,* with no solution; or *redundant,* with a one-parameter infinite family of solutions.

Comparable in importance to linear equations is the formula for the distance $|P_1P_2|$ between points $P_1$ and $P_2$ in the plane. If $P_1$ and $P_2$ have coordinates $(x_1,y_1)$, and $(x_2,y_2)$, respectively, then from the Pythagorean theorem, $|P_1P_2| = \sqrt{(x_1 - x_2)^2 + (y_1 - y_2)^2}$. From this it follows that every circle in the plane is the graph of an equation of the form $Ax^2 + Ay^2 + Bx + Cy + D = 0$, where $A \neq 0$ and $B^2 + C^2 - 4AD > 0$, and conversely that the graph of every such equation is a circle. It can be shown from the distance formula that the distance from a point $P_0$ with coordinates $(x_0, y_0)$ to a line $l$ with equation $ax + by + c = 0$ is given by $|ax_0 + by_0 + c|\sqrt{a^2 + b^2}$. From this it can be shown that every conic section has an equation of the form $Ax^2 + 2Bxy + Cy^2 + Dx + Ey + F = 0$, where $A$, $B$, $C$ are not all zero, and conversely that the graph of every such equation is a conic section. Thus the geometrical study of conics corresponds to the algebraic study of quadratic equations in two unknowns.

## CHANGE OF COORDINATES

Sometimes the equation for a curve in a given Cartesian coordinate system can be simplified by changing to another Cartesian coordinate system. The two most important types of coordinate system changes are *translation* and *rotation* of coordinates. Suppose an $xy$-coordinate system is translated to a new $x'y'$-coordinate system with the origin at the $xy$-coordinates $(h, k)$, as shown in Figure 2. Then a point with $xy$-coordinates $(x, y)$ will have $x'y'$-coordinates $(x', y')$, where $x' = x - h$ and $y' = y - k$. For example, suppose the graph of the equation $Ax^2 + Ay^2 + Bx + Cy + D = 0$ is a circle. If the coordinate system is translated to the center of the circle, this equation will simplify to $Ax'^2 + Ay'^2 + E = 0$, where the radius of the circle is $\sqrt{-E/A}$.

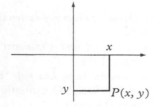

**Figure 1** *Cartesian coordinate system.*

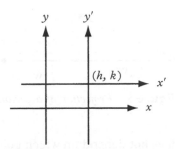

**Figure 2** *Translation of coordinate system.*

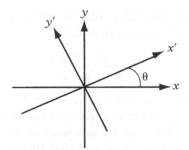

**Figure 3** *Rotation of coordinate system.*

Suppose an $xy$-coordinate system is rotated about the origin by angle $\theta$ (counterclockwise if $\theta > 0$, clockwise if $\theta < 0$) to a new $x'y'$-coordinate system, as shown in Figure 3. Then a point with $xy$-coordinates $(x, y)$ will have $x'y'$-coordinates $(x', y')$, where $x' = x\cos\theta + y\sin\theta$ and $y' = -x\sin\theta + y\cos\theta$. It can be shown, for example, that for any conic section with equation $Ax^2 + 2Bxy + Cy^2 + Dx + Ey + F = 0$, there is an angle through which the coordinates can be rotated so as to simplify the equation to $A'x'^2 + C'y'^2 + D'x' + E'y' + F' = 0$, and that this system can then be translated so as to further simplify the equation to $A''x''^2 + C''y''^2 + F'' = 0$, which is much easier to study.

## VECTOR GEOMETRY IN THE PLANE

A *vector* in coordinate geometry may be defined as a displacement. The length of the displacement is called the *magnitude* or *norm* of $\mathbf{v}$, denoted by $\|v\|$ or sometimes simply $|v|$. The displacement from a point $A$ to a point $B$ is often represented by a directed segment or arrow from $A$ to $B$: $\overrightarrow{AB}$ or sometimes $\vec{AB}$. Since a displacement is defined solely by its magnitude and direction, it follows that two arrows $\overrightarrow{AB}$ and $\overrightarrow{CD}$ that have the same length and direction represent the same vector $\mathbf{v}$, and we may write $\mathbf{v} = \overrightarrow{AB} = \overrightarrow{CD}$. A zero displacement, that is, a displacement with magnitude 0 and no direction, corresponds to the *zero vector*. This vector, denoted $\mathbf{0}$, may be represented by a virtual arrow $\overrightarrow{AA}$ for any point $A$.

Suppose $A$ and $B$ are points in the plane with Cartesian coordinates $(a_1, a_2)$ and $(b_1, b_2)$. Let $\mathbf{v} = \overrightarrow{AB}$. Recall that the *origin* of the coordinate system is the point $O$ with coordinates $(0,0)$. It it easy to show that $\mathbf{v} = \overrightarrow{OV}$, where the coordinates of $\mathbf{v}$ are $(v_1, v_2)$, if and only if $v_1 = b_1 - a_1$ and $v_2 = b_2 - a_2$. The ordered pair of numbers are called the *components* of $\mathbf{v}$ with respect to the coordinate system. Within a given coordinate system, each vector $\mathbf{v}$ may be identified with its components, $v_1, v_2$. This may be written as $\mathbf{v} = \langle v_1, v_2 \rangle$ or, if confusion with point coordinates is unlikely, simply as $(v_1, v_2)$.

The sum of two vector displacements $\mathbf{v}$ and $\mathbf{w}$, denoted $\mathbf{v} + \mathbf{w}$, is obtained by following one displacement by the other. In terms of representation by arrows, suppose $\mathbf{v} = \overrightarrow{AB}$. There is one and only one point $C$ such that $\mathbf{w} = \overrightarrow{BC}$. Then $\mathbf{v} + \mathbf{w} = \overrightarrow{AB} + \overrightarrow{BC} = \overrightarrow{AC}$. In terms of components, suppose $\mathbf{v} = \langle v_1, v_2 \rangle$ and $\mathbf{w} = \langle w_1, w_2 \rangle$. It can be shown that $\mathbf{v} + \mathbf{w} = \langle v_1, v_2 \rangle + \langle w_1, w_2 \rangle = \langle v_1 + w_1, v_2 + w_2 \rangle$. Given a vector $\mathbf{v}$, the solution $\mathbf{x}$ of the vector equation $\mathbf{v} + \mathbf{x} = \mathbf{0}$ is denoted by $-\mathbf{v}$. Clearly, $-\langle v_1, v_2 \rangle = \langle -v_1, -v_2 \rangle$.

In vector notation, ordinary numbers are called *scalars*. The *product* of a scalar $k$ and a vector $\mathbf{v}$, denoted $k\mathbf{v}$, is defined as the vector of norm $|k|\,\|v\|$ that points in the same direction as $\mathbf{v}$ if $k > 0$, and the opposite direction if $k < 0$. For all $k$ and all $\mathbf{v}$, the products $k\mathbf{0}$ and $0\mathbf{v}$ are defined to be the zero vector, $\mathbf{0}$. If $\mathbf{v} = \langle v_1, v_2 \rangle$, it can be shown that $k\mathbf{v} = k\langle v_1, v_2 \rangle = \langle kv_1, kv_2 \rangle$. Then, from the previous paragraph, $-\mathbf{v} = (-1)\mathbf{v}$.

Let $A$ and $B$ be distinct points. A point $X$ is on line $AB$ if and only if $\overrightarrow{AX} = t\overrightarrow{AB}$ for some scalar $t$. Point $A$ divides line $AB$ into two sides, called *rays* or *half-lines*. We have

$$t = \frac{|AX|}{|AB|}$$

if $X$ and $B$ are on the same side of $A$, and

$$t = -\frac{|AX|}{|AB|}$$

if they are on opposite sides. Let the coordinates of $A$, $B$, $X$ be $(a_1, a_2)$, $(b_1, b_2)$, $(x, y)$. Then $\overrightarrow{AB} = \overrightarrow{OB} - \overrightarrow{OA} = \langle b_1 - a_1, b_2 - a_2 \rangle$ and similarly $\overrightarrow{AX} = \langle x - a_1, y - a_2 \rangle$, from which we obtain *parametric equations* for line $AB$: $x = (1 - t)a_1 + tb_1$, $y = (1 - t)a_2 + tb_2$. Thus, for example, the value $t = \frac{1}{2}$

gives the coordinates of the midpoint of the segment with endpoints $A$ and $B$, namely,

$$\left(\frac{a_1 + b_1}{2}, \frac{a_2 + b_2}{2}\right)$$

Nonzero vectors $\mathbf{v}$ and $\mathbf{w}$ are said to be *parallel* if they point in the same direction or in opposite directions; otherwise, they are *independent*. Suppose that $\mathbf{v}$ and $\mathbf{w}$ are independent and that there are scalars $a$ and $b$ such that $a\mathbf{v} + b\mathbf{w} = \mathbf{0}$. Then $a\mathbf{v} = -b\mathbf{w}$. If $a$ is not zero, then neither is $b$. But then $a\mathbf{v}$ is parallel to $\mathbf{v}$ while $-b\mathbf{w}$ is parallel to $\mathbf{w}$; so these vectors cannot be equal. Thus, it must be that $a = b = 0$. From this follows the theorem of *uniqueness of representation*, which is of central importance to coordinate geometry: if $\mathbf{v}$ and $\mathbf{w}$ are independent and $a\mathbf{v} + b\mathbf{w} = c\mathbf{v} + d\mathbf{w}$, then $a = c$ and $b = d$.

A good illustration of the interplay between coordinate geometry and vectors is provided by Euclid's theorem that for any triangle $ABC$ the lines joining the vertices $A$, $B$, $C$ to the midpoints $A'$, $B'$, $C'$ of the opposite sides (the medians) meet at a point $G$ (the centroid), which is two-thirds of the way from each vertex to the opposite side (see Figure 4). Let $\overrightarrow{AB} = \mathbf{v}$ and $\overrightarrow{AC} = \mathbf{w}$. Then $\overrightarrow{A'A} = \frac{1}{2}(\mathbf{v} + \mathbf{w})$ and $\overrightarrow{B'B} = -\mathbf{v} + \frac{1}{2}\mathbf{w}$.

If these medians meet at $G$, then there are scalars $a$ and $b$ such that

$$\overrightarrow{AG} = \frac{a}{2}(\mathbf{v} + \mathbf{w}) \quad \text{and} \quad \overrightarrow{AG} = \mathbf{v} + b\left(-\mathbf{v} + \frac{\mathbf{w}}{2}\right)$$

Then, by the uniqueness of the representation theorem, $a/2 = 1 - b$ and $a/2 = b/2$, from which $a = b = \frac{2}{3}$. If the coordinates of $A$, $B$, $C$ are $(a_1, a_2)$, $(b_1, b_2)$, $(c_1, c_2)$, it follows that the coordinates of $G$ are $[(a_1 + b_1 + c_1)/3, (a_2 + b_2 + c_2)/3]$. Then, by symmetry, $C'C$ also passes through $G$.

The *angle between* two nonzero vectors is defined as the smallest angle between arrows having a common initial point that represents them. Clearly this

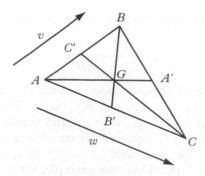

**Figure 4** *The centroid.*

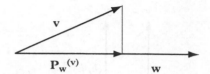

**Figure 5** *Projection of a vector.*

definition does not depend on which point is chosen as the initial point. If the angle between the nonzero vectors $\mathbf{v}$ and $\mathbf{w}$ is $\theta$, the *dot product*, also called the *scalar product* of $\mathbf{v}$ and $\mathbf{w}$, denoted $\mathbf{v} \cdot \mathbf{w}$, is defined as $\mathbf{v} \cdot \mathbf{w} = \|\mathbf{v}\| \|\mathbf{w}\| \cos \theta$. In terms of components, suppose $\mathbf{v} = \langle v_1, v_2 \rangle$ and $\mathbf{w} = \langle w_1, w_2 \rangle$. Using the trigonometric law of cosines, it can be shown that $\mathbf{v} \cdot \mathbf{w} = v_1 w_1 + v_2 w_2$.

Among the many uses of the dot product in coordinate geometry are the following. Two nonzero vectors are perpendicular if and only if their dot product is zero. The projection $\mathbf{P_w}(\mathbf{v})$ of a vector $\mathbf{v}$ onto a nonzero vector $\mathbf{w}$, defined as shown in Figure 5, is given by the formula

$$\mathbf{P_w}(\mathbf{v}) = \left(\frac{\mathbf{v} \cdot \mathbf{w}}{\mathbf{w} \cdot \mathbf{w}}\right)\mathbf{w}$$

From this we may obtain results on the altitudes of triangles and the formula for the distance from a point to a line that was given earlier.

## OTHER PLANE COORDINATE SYSTEMS

For certain objects of geometric interest, coordinate systems other than the Cartesian are useful. The system most used is the *polar coordinate system*, which is based on a half-line $i$, called the *initial ray*, and its endpoint $O$, called the *pole*. The polar coordinates of a point $P$ different from $O$ are $(r, \theta)$, where $r = |OP|$ and $\theta$ is a directed angle from $i$ to the half-line from $O$ through $P$, as shown in Figure 6. The value of $\theta$ is not uniquely determined by the point $P$. If $(r, \theta)$ are coordinates, so are $(r, \theta + 2n\pi)$ for all integers $n$. Negative values of $r$ are also used. If $(r, \theta)$ are coordinates of $P$, then so are $(-r, \theta + \pi)$ and $(-r, \theta + (2n + 1)\pi)$ for all $n$. The polar coordinates of the pole $O$ are $(0, \theta)$ for all $\theta$. The graph of an equation in polar coordinates $f(r, \theta) = 0$ is defined as the set of all points such that at least one pair of polar coordinates satisfies the equation. For example, the point $P$ with polar coordinates $(1, 2\pi)$ is on the spiral $r + \left(\frac{\theta}{\pi}\right) = 0$ because $(-1, \pi)$ are also polar coordinates for $P$.

If polar coordinates are introduced in a coordinate system with the origin as the pole and the positive side of the $x$-axis as the initial ray, then for

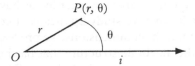

**Figure 6** *Polar coordinate system.*

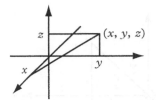

**Figure 7** *Cartesian coordinate system in 3-space.*

all polar coordinates $(r, \theta)$ of a point $P$, the Cartesian coordinates $(x, y)$ of P are given by $x = r \cos \theta$, $y = r \sin \theta$.

Another coordinate system used in location- and direction-finding technology is based on confocal ellipses and hyperbolas. Fix two distinct points $F_1$ and $F_2$ in the plane and assign as the coordinates of a point $P$ the pair $(u, v)$, where $u = |PF_1| + |PF_2|$, $v = |PF_1| - |PF_2|$. The curves $u = c$, where $c > |F_1F_2|$ are ellipses, and the curves $|v| = c$, where $c \geq 0$, are hyperbolas perpendicular to them.

## COORDINATE GEOMETRY AND LINEAR PROGRAMMING

Coordinate geometry deals with inequalities as well as equations. In the plane with Cartesian coordinates, the graph of every inequality of the form $ax + by \leq c$, where $a, b$ are not both zero, is a *half-plane,* consisting of all points on one side of the line $ax + by = c$.

The coordinate geometry of linear inequalities arises in problems of *linear programming,* in which the goal is to maximize or minimize a linear *objective function,* $z = ax + by$, subject to a set of linear *constraints,* $a_i x + b_i y \leq c_i, i = 1, 2, \ldots, n$. The graph of all solutions of the constraints, called the *feasible region* of the problem, is a convex polygonal region with at most $n$ sides and $n$ vertices, sometimes called *corner points.* It is not hard to show that a solution of the problem, if one exists, always occurs at a corner point. Moreover, it can be shown that a maximum corner point, if one exists, can always be found by starting at any corner point and moving along the edges of the feasible region so as to increase the objective function. When expressed in terms of the linear algebra of dual spaces, this result leads to the well-known *Simplex Method* for the solution of linear programming problems, which is frequently used in business and engineering.

## EUCLIDEAN THREE-DIMENSIONAL SPACE

Most of the ideas of coordinate geometry can be introduced in the Euclidean plane. The extension to three-dimensional Euclidean space, or *3-space,* is

then straightforward. We consider briefly some of the ideas introduced earlier.

The Cartesian coordinate system in 3-space assigns to each point an ordered triple of coordinates $(x, y, z)$ by means of three mutually perpendicular axes, as shown in Figure 7. Every plane in 3-space has an equation of the form $ax + by + cz = d$, where $a, b, c$ are not all zero, and the graph of every equation of this form is a plane. The distance between points $P_1(x_1, y_1, z_1)$ and $P_2(x_2, y_2, z_2)$ in space is given by

$$|P_1P_2| = \sqrt{(x_1 - x_2)^2 + (y_1 - y_2)^2 + (z_1 - z_2)^2}$$

From this, the formula for the distance from a point to a plane and descriptions of the quadric surfaces follow. In particular, every sphere has an equation of the form $Ax^2 + Ay^2 + Az^2 + Bx + Cy + Dz + E = 0$, where $A \neq 0$ and $B^2 + C^2 + D^2 > 4AE$; and the graph of every such equation is a sphere.

As in the two-dimensional case, translation and rotation of Cartesian coordinates in space can be used to simplify the equations of surfaces. Except for a few straightforward examples, this discussion is perhaps best left to a first course in linear algebra. The definitions of vector, vector equality, components, vector addition, multiplication by a scalar, and dot product are the same in space (of any dimension) as in the plane. In terms of components it can be shown that if $k$ is a scalar, $\mathbf{v} = \langle v_1, v_2, v_3 \rangle$, and $\mathbf{w} = \langle w_1, w_2, w_3 \rangle$, then

$$\mathbf{v} + \mathbf{w} = \langle v_1 + w_1, v_2 + w_2, v_3 + w_3 \rangle,$$
$$k\mathbf{v} = \langle kv_1, kv_2, kv_3 \rangle,$$
$$\mathbf{v} \cdot \mathbf{w} = v_1 w_1 + v_2 w_2 + v_3 w_3$$

Consider the plane through the point $A(a_1, a_2, a_3)$ perpendicular to the nonzero vector $\mathbf{n} = \langle n_1, n_2, n_3 \rangle$. A point $P(x, y, z)$ is on this plane if and only if $\overrightarrow{AP}$ is perpendicular to $\mathbf{n}$, that is, $(\overrightarrow{OP} - \overrightarrow{OA}) \cdot \mathbf{n} = 0$. It follows that if a plane has the equation $ax + by + cz = d$, then it is perpendicular to the vector $\langle a, b, c \rangle$, which is called a *normal* vector to the plane. The angle between two planes is then the angle between their normal vectors, which can be found with the dot product. If $A, B,$ and $C$ are non-

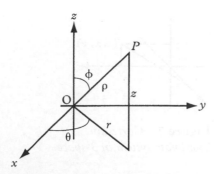

**Figure 8** *Cylindrical coordinate system.*

collinear points, then for every point $X$ in the plane containing $A$, $B$, $C$, we have $\overrightarrow{AX} = s\overrightarrow{AB} + t\overrightarrow{AC}$ for some scalars $s$, $t$. Using $\overrightarrow{OX} = \overrightarrow{OA} + \overrightarrow{AX}$, this yields a parametric description of the plane: $x = (1 - s - t)a_1 + sb_1 + tc_1$.

In the *cylindrical coordinate system* in 3-space, the Cartesian coordinate system in the $xy$-plane is replaced by the polar coordinate system. Then, a point with Cartesian coordinates $(x, y, z)$ has cylindrical coordinates $(r, \theta, z)$, where $x = r \cos \theta$, $y = r \sin \theta$. In this system, the surfaces with equation $r = k$, where $k$ is a constant, are cylinders centered on the $z$-axis. The *spherical coordinates* of a point $P$ different from the origin use the $\theta$ coordinate of the polar and cylindrical systems together with the distance $\rho = |OP|$ and the angle $\phi$ from the positive $z$-axis to the segment $OP$. From Figure 8 it is clear that $r = \rho \sin \phi$, $z = \rho \cos \phi$. Thus a point with Cartesian coordinates $(x, y, z)$ has spherical coordinates $(\rho, \theta, \phi)$, where $x = \rho \sin \phi \cos \theta$, $y = \rho \sin \phi \sin \theta$, $z = \rho \cos \phi$. In the spherical system, the graphs of the surfaces $\rho = k$, $k > 0$, are spheres centered at the origin, and for each such sphere, $\theta$ and $\phi$ are the longitude and colatitude.

In 3-space, the graph of each constraint $a_i x + b_i y + c_i z \leq d_i$ of a linear programming problem is a *half-space*, consisting of all points on one side of the plane $a_i x + b_i y + c_i z = d_i$. The feasible region is a (possibly unbounded) polyhedron, and optimum values of an objective function, if they exist, occur at corner points, where three or more boundary planes intersect. The Simplex Method is even more efficient in 3-space than in the plane.

## HIGHER-DIMENSIONAL COOORDINATE GEOMETRY

High-dimension geometry may be approached with a philosophical perspective different from that of geometry of low dimension. In low dimensions,

geometrical objects may be regarded as having an inherent reality that we study by means of coordinates and algebra. In high dimensions, we proceed by analogy with the algebraic structure that has been developed to *define* the geometrical objects. Thus a point in 3-space has an independent reality, which we may investigate by means of a coordinate system, whereas the points of $10^{100}$-space are simply the set of all $n$-tuples $(x_1, x_2, \ldots, x_n)$, where $n = 10^{100}$. A brief discussion of mathematical "meaning" and "reality" can sometimes serve to enliven a geometry class, but perhaps it should not be allowed to take much time away from the subject itself.

High-dimension coordinate geometry may be treated as a subfield of linear algebra. Let $V$ be a vector space, not necessarily finite-dimensional. The *affine geometry* of $V$ is the set $\mathbf{A}(V)$ of all *flats*, which are sets of the form $\mathbf{a} + U = \{\mathbf{a} + \mathbf{u}: \mathbf{u} \in U\}$, where $U$ is a subspace of $V$. The dimension of a flat $\mathbf{a} + U$ is the dimension of the subspace $U$. Flats of dimension 0, 1, 2, . . . are called *points, lines, planes,* . . . . A point $\mathbf{a} + \{0\} = \{\mathbf{a}\}$ may be identified with the vector $\mathbf{a}$. If $V$ is finite-dimensional, a flat of dimension $V - 1$ is called a *hyperplane*. For simplicity, let us assume that the vector space $V$ has finite dimension $n$. A *coordinate system* for the affine geometry $\mathbf{A}(V)$ is an ordered basis of $V$, that is, $B = (\mathbf{e}_1, \mathbf{e}_2, \ldots, \mathbf{e}_n)$. The coordinates of a point $\mathbf{a}$ relative to $B$ are the $n$-tuple $(a_1, a_2, \ldots, a_n)$, where $\mathbf{a} = a_1 \mathbf{e}_1 + a_2 \mathbf{e}_2 + \ldots + a_n \mathbf{e}_n$.

The *projective geometry* of $V$ is the set $\mathbf{P}(V)$ of all subspaces of $V$. The *projective dimension* of a member of $\mathbf{P}(V)$ is its subspace dimension minus 1. Thus the zero subspace has projective dimension $-1$. Members of projective dimension 0, 1, 2, . . . are called *points, lines, planes,* . . . . If $V$ is finite-dimensional, its subspaces of dimension one less are called *hyperplanes*.

Suppose $V$ has vector space dimension $n + 1$. A frame of reference for $\mathbf{P}(V)$ is an ordered $(n + 2)$-tuple of points, $(E_0 E_1 \ldots E_N | U)$ such that $\{E_0, E_1, \ldots, E_N\}$ span $V$ and the span of no proper subset of $\{E_0, E_1, \ldots, E_N\}$ contains $U$, which is called the *unit point* of the frame. Such a frame may be used to coordinatize the geometry as follows. If $A$ is any point, it can be shown that there are vectors $\mathbf{e}_i \in E_i$, $0 \leq i \leq n$, $\mathbf{u} \in U$ and scalars $a_i$, $0 \leq i \leq n$, such that $\mathbf{e}_0 + \mathbf{e}_1 + \cdots + \mathbf{e}_n = \mathbf{u}$ and $a_0 \mathbf{e}_0 + a_1 \mathbf{e}_1 + \cdots + a_n \mathbf{e}_n \in P$. Moreover, if $\mathbf{e}_i'$, $\mathbf{u}'$, $a_i'$ are another set of vectors and scalars satisfying these conditions, then there exists a scalar $t$ such that $a_i' = t a_i$, $0 \leq i \leq n$. The $(n + 1)$-tuples $(a_0, a_1, \ldots, a_n)$ are the *homogeneous coordinates* for $A$ relative to the frame of reference.

*See also* Conic Sections; Linear Programming; Polar Coordinates; Transformations; Trigonometry; Vectors

## SELECTED REFERENCES

Douglas, Ronald G., ed. *Toward a Lean and Lively Calculus.* Washington, DC: Mathematical Association of America, 1986.

Efimov, Nikolai V. *An Elementary Course in Analytic Geometry.* Oxford, England: Pergamon, 1966.

Glaubiger, Pearl, et al. *Modern Coordinate Geometry; A Wesleyan Experimental Curricular Study.* Boston, MA: Houghton Mifflin, 1969.

Kelly, Paul J. *Elements of Analytic Geometry.* Glenview, IL: Scott, Foresman, 1970.

Murdoch, David C. *Analytic Geometry with an Introduction to Vectors and Matrices.* New York: Wiley, 1960.

Steen, Lynn. *Calculus for a New Century.* Washington, DC: Mathematical Association of America, 1988.

Willmore, Floyd E., Donald R. Barr, and Donald Voils. *Analytic Geometry: A Vector Approach.* Boston, MA: Allyn and Bacon, 1971.

ROBERT J. BUMCROT

# CORE-PLUS MATHEMATICS PROJECT (CPMP)

A comprehensive curriculum development project funded by the National Science Foundation (NSF) to design, evaluate, and disseminate an innovative high school curriculum that interprets and implements the recommendations of the National Council of Teachers of Mathematics (1989, 1991, 1995). In contrast to the traditional practice of tracked programs that offer advanced mathematics for a few and minimal mathematics for the majority of students (National Research Council 1989), the CPMP curriculum is designed to make broadly useful mathematics accessible to all high school students. The curriculum consists of a single core sequence for both college-bound and employment-bound students during the first three years. This curriculum organization is intended to keep post-high school education and career options open for all students. A fourth-year course continues the preparation of students for college mathematics.

The CPMP curriculum builds upon the theme of *mathematics as sense-making.* Throughout, it extends the informal knowledge of data, shape, change, and chance that students bring to situations and problems (Hirsch, Coxford, Fey, and Schoen 1995). Investigations of real-life contexts lead to discovery of mathematics that makes sense to students and, in turn, enables them to make sense out of new situations and problems. The focus of the curriculum and instructional practices is on mathematical thinking and communication, both oral and written.

## UNIFIED MATHEMATICS

In each year of the CPMP curriculum, mathematics is developed along four strands: algebra and functions, geometry and trigonometry, statistics and probability, and discrete mathematics. Each course consists of seven units and a culminating capstone experience. For example, in Course 1, which is intended primarily for ninth-grade students, the first unit *(Patterns in Data)* and the seventh unit *(Simulation Models)* focus on ideas from the statistics and probability strand; the second unit *(Patterns of Change),* third unit *(Linear Models),* and sixth unit *(Exponential Models)* feature patterns and relationships from the algebra and functions strand; the fourth unit *(Graph Models)* introduces the discrete mathematics strand from the perspective of vertex-edge graphs; and the fifth unit *(Patterns in Space and Visualizations)* is centered around concepts and methods from the geometry and trigonometry strand. Across the strands and within the units, the emphasis is on mathematical modeling and modeling concepts of data collection, representation, interpretation, prediction, and simulation. The capstone at the end of each course is a thematic two-week project-oriented activity that enables students to pull together and apply the important modeling concepts and methods developed in the course. In the case of Course 1, the capstone evolves around mathematical problems and questions related to planning a benefits carnival. Graphics calculators help students to develop versatile ways of modeling realistic situations and remove barriers that have in the past prevented large numbers of students from continuing their study of mathematics.

The four strands of the curriculum are connected by fundamental ideas such as: symmetry, recursion, functions, data analysis and curve-fitting, and matrices. The strands are unified by mathematical habits of mind such as: visual thinking, recursive thinking, searching for and describing patterns, making and checking conjectures, reasoning with multiple representations, inventing mathematics, and providing convincing arguments. Overarching themes of data, representation, shape, and change serve to further unify the four strands.

## INTEGRATED INSTRUCTION AND ASSESSMENT

The CPMP curriculum materials were developed not only to reshape what mathematics all students have the opportunity to learn, but also to influence the manner in which learning occurs and is assessed. Each unit in the curriculum is developed around a series of four or five multiday lessons in which major ideas are developed through student investigations of applied problem situations. Lessons focus on several interrelated mathematical concepts and often span five or six days.

Each CPMP lesson is introduced as an entire class activity in which students are asked to think about a context such as that in Figure 1, which is used to launch the first lesson of the *Exponential Models* unit in Course 1.

Once launched, a lesson usually involves students working together collaboratively in small groups as they investigate focused problems and questions related to the launching situation. These small group investigations lead to (re)invention of mathematics that makes sense to students. Sharing, and agreeing as a class, on the mathematical ideas groups are developing is prompted by "Checkpoints" in the instructional materials. Each lesson is accompanied with a set of additional tasks to engage students in modeling with, organizing, reflecting on, and extending their mathematical understanding developed through the investigations. These "MORE" tasks are intended primarily for individual work outside of class.

Assessment is embedded in the CPMP curriculum materials and is an integral part of instruction. As students pursue the investigations that make up the curriculum, the teacher is able to assess student performance in terms of process, content, and disposition and use this information to help guide instruction. At the end of each investigation, the checkpoint and accompanying class discussion provide an opportunity for the teacher to assess levels of understanding that various groups of students have reached. Finally, the "On Your Own" problem situation as well as the tasks in the MORE sets provide further opportunities to assess the level of understanding of each individual student. Quizzes, in-class exams, take-home assessment activities, and extended projects are included in the CPMP teacher resource materials.

## EVALUATION

Each CPMP course goes through at least a three-year research, development, and evaluation cycle. After a year of initial development, the pilot version of a course is tested in nineteen Michigan high schools, revised as necessary, and then field tested the next year in thirty-six high schools in Alaska, California, Colorado, Georgia, Idaho, Iowa, Kentucky, Michigan, Ohio, South Carolina, and Texas. A broad cross-section of students from urban, suburban, and rural communities with ethnic and cultural diversity is represented. Achievement data from a nationally standardized test, *Ability to Do Quantitative Thinking*, indicates that CPMP students at the end of Course 1 and at the end of Course 2 grew in their ability to think and reason quantitatively more than the national norm group and more than a group of control students. Further information on the evaluation of the CPMP program may be found in Schoen and Ziebarth (1998).

*See also* Curriculum Trends, Secondary Level

## SELECTED REFERENCES

Core-Plus Mathematics Project website: http://www.wmich.edu/cpmp/

Coxford, Arthur F., James T. Fey, Christian R. Hirsch, Harold L. Schoen, Gail Burrill, Eric W. Hart, Ann E. Watkins, Mary Jo Messenger, and Beth Ritsema. *Contemporary Mathematics in Context, A Unified Approach.* Chicago, IL: Everyday Learning, 1996.

---

**Think About This Situation**

The graph below shows three possible patterns in the rate at which the school-closing rumor could spread.

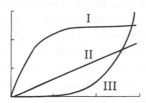

(a) How would you describe the rate of rumor spread in the case of each graph?

(b) Which pattern of spread is most likely if the students plant the story on the 5 o'clock television or radio news? Explain your reasoning.

(c) Which pattern of spread is most likely if the rumor spreads only by word of mouth around the community? Why?

---

**Figure 1** *Sample lesson launch from Exponential Models. [Courtesy of Everyday Learning Corporation]*

Hirsch, Christian R., Arthur F. Coxford, James T. Fey, and Harold L. Schoen. "Teaching Sensible Mathematics in Sense-making Ways with the CPMP." *Mathematics Teacher* 88(1995):694–700.

National Council of Teachers of Mathematics. *Assessment Standards for School Mathematics.* Reston, VA: The Council, 1995.

————. *Curriculum and Evaluation Standards for School Mathematics.* Reston, VA: The Council, 1989.

————. *Principles and Standards for School Mathematics.* Reston, VA: The Council, 2000.

————. *Professional Standards for Teaching Mathematics.* Reston, VA: The Council, 1991.

National Research Council. *Everybody Counts: A Report to the Nation on the Future of Mathematics Education.* Washington, DC: National Academy Press, 1989.

Schoen, Harold L., and Steven W. Ziebarth. "High School Mathematics Curriculum Reform: Rationale, Research, and Recent Developments." In *Annual Review of Research for School Leaders.* Peter S. Hlebowitsh and William G. Wraga, eds. New York: Macmillan, 1998.

CHRISTIAN R. HIRSCH

# CORRELATION

A relationship between two paired variables with the following property: if, as one increases, the other tends to increase (or to decrease) in a linear manner. The correlation coefficient, a number between −1 and 1, measures the strength of the linear relationship between the two variables. That is, it measures how closely the points on a scatterplot cluster about a line. The correlation is positive if the slope of the line is positive and negative if the slope of the line is negative. A perfect positive correlation of 1 or a perfect negative correlation of −1 means that all the points lie on a line. If there is neither an upward or downward trend on the scatterplot, the correlation is 0.

Three examples of scatterplots and their correlations, *r*, are given below:

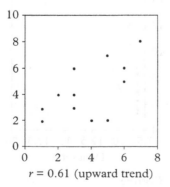

$r = 0.61$ (upward trend)

**Figure 1**

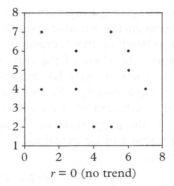

$r = 0$ (no trend)

**Figure 2**

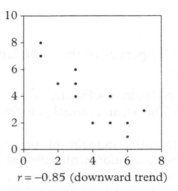

$r = -0.85$ (downward trend)

**Figure 3**

## PEARSON'S SAMPLE CORRELATION COEFFICIENT

Pearson's correlation coefficient, *r*, for a sample is given by the formula

$$r = \frac{1}{n-1} \sum \left( \frac{x - \bar{x}}{s_x} \right) \left( \frac{y - \bar{y}}{s_y} \right)$$

where *n* is the sample size, $\bar{x}$ and $s_x$ are the mean and standard deviation for one of the variables, and $\bar{y}$ and $s_y$ are the mean and standard deviation for the other.

To use this formula, convert each observed value to standard units (*z*-scores) by subtracting the sample mean from the value and dividing by the sample standard deviation. Multiply each standardized *x*-value by the corresponding standardized *y*-value. Finally, find the average (dividing by $n - 1$ if the data are from a sample) of these products. If *x* and *y* are both below or both above their respective sample means, a positive number will be contributed to the average product. If one is above and one is below, a negative number will be contributed. (See Freedman et al. 1991.) There are several equivalent versions of Pearson's formula (Weisberg 1980).

In addition to giving a measure of how linear the association between the variables is, the correlation coefficient is related to the regression line in two other ways. First, the slope of the regression line is $\left(\frac{s_y}{s_x}\right)r$. If the variables are standardized, the slope of the regression line is simply $r$. Second, the square of the correlation coefficient, $r^2$, called the *coefficient of determination*, is the proportion of the variability in the values of $y$ that can be explained by the regression line. That is,

$$r^2 = 1 - \frac{\Sigma(y - \hat{y})^2}{\Sigma(y - \bar{y})^2} = \frac{\Sigma(\hat{y} - \bar{y})^2}{\Sigma(y - \bar{y})^2}$$

where $\hat{y}$ is the value of $y$ predicted by the regression line.

## Additional Properties of the Correlation Coefficient

1.  The correlation coefficient is not resistant to outliers. That is, an unusual point can greatly affect $r$.
2.  Restricting the domain of one variable can make the correlation artificially low. A tragic example is the *Challenger* explosion data where scientists did not look at the (higher) air temperature at take-off of the flights that had no O-ring problems (Dalal, Fowlkes, and Hoadley 1989).
3.  Two data sets with the same correlation may look strikingly different. For example, the Ans-combe data sets below all have the same correlation coefficient, $r = 0.82$, and the same regression equation,

$$y = 3.0 + 0.5x$$

**Table 1**  *Anscombe Data Sets, where $r = 0.82$*

| Data set 1 | | Data set 2 | | Data set 3 | | Data set 4 | |
|---|---|---|---|---|---|---|---|
| $x$ | $y$ | $x$ | $y$ | $x$ | $y$ | $x$ | $y$ |
| 10 | 8.04 | 10 | 9.14 | 10 | 7.46 | 8 | 6.58 |
| 8 | 6.95 | 8 | 8.14 | 8 | 6.77 | 8 | 5.76 |
| 13 | 7.58 | 13 | 8.74 | 13 | 12.74 | 8 | 7.71 |
| 9 | 8.81 | 9 | 8.77 | 9 | 7.11 | 8 | 8.84 |
| 11 | 8.33 | 11 | 9.26 | 11 | 7.81 | 8 | 8.47 |
| 14 | 9.96 | 14 | 8.10 | 14 | 8.84 | 8 | 7.04 |
| 6 | 7.24 | 6 | 6.13 | 6 | 6.08 | 8 | 5.25 |
| 4 | 4.26 | 4 | 3.10 | 4 | 5.39 | 19 | 12.50 |
| 12 | 10.84 | 12 | 9.13 | 12 | 8.15 | 8 | 5.56 |
| 7 | 4.82 | 7 | 7.26 | 7 | 6.42 | 8 | 7.91 |
| 5 | 5.68 | 5 | 4.74 | 5 | 5.73 | 8 | 6.89 |

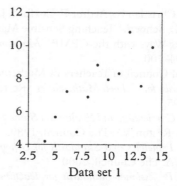

**Figure 4**

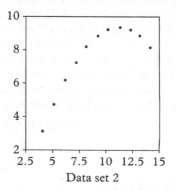

**Figure 5**

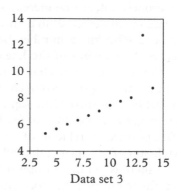

**Figure 6**

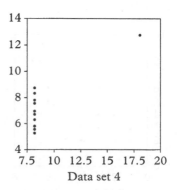

**Figure 7**

4. The correlation coefficient remains unchanged if each value of one variable is multiplied by the same positive number or has the same number added to it. For example, converting data from inches to centimeters does not change the value of $r$.

## Interpretation of the Correlation Coefficient

1. As the Anscombe data sets above illustrate, a high correlation does not necessarily mean that a linear model is appropriate.
2. A high correlation does not necessarily mean that one variable causes the other. A classic data set shows a high positive correlation between the number of people and the number of storks in a small German town (Box, Hunter, and Hunter 1978). A lurking variable is time—both the population of humans and of storks were growing over the years.
3. The correlation $\rho$ of all the points in a population and the correlation $r$ of the points in a random sample taken from that population generally will not be the same. Consequently, a nonzero correlation $r$ in a sample does not necessarily mean that there is a nonzero correlation in the population. Even if two variables are uncorrelated, there usually will not be zero correlation in a sample taken from that population. Many introductory statistics books give the procedure for testing the hypothesis that $\rho = 0$ and the procedure for constructing a confidence interval for $\rho$.
4. Variables with a strong nonlinear association may have zero correlation. For example, consider the points $(-3,9)$, $(-2,4)$, $(-1,1)$, $(0,0)$, $(1,1)$, $(2,4)$, and $(3,9)$.

## OTHER TYPES OF CORRELATION

*Rank correlation* formulas of Charles Edward Spearman and Maurice G. Kendall give the correlation for paired observations that are ranks, such as the ranking of ten gymnasts by Judge A paired with the ranking of the same ten gymnasts by Judge B (Noether 1976). If there are no ties in the ranks, Spearman's formula gives the same value as Pearson's.

The *partial correlation* between two variables is their correlation computed after removing the effect of other variables. For example, there is a positive correlation between the reading ability of children and the size of their feet. A lurking variable is the age of the child. Partial correlation will tell us if there is still a nonzero correlation between reading ability and foot size if we take age into account (Ferguson 1981).

Multiple linear regression is used to predict the value of one variable, $y$, from other variables: $\hat{y} = b_0 + b_1 x_1 + \cdots + b_n x_n$. For example, one might predict college grade point average by multiplying the high school grade point average by some factor, multiplying an achievement test score by another factor, and then adding the two products. The *multiple correlation* coefficient is the correlation between the actual values of $y$ and the predicted values $\hat{y}$. More useful is the square of this correlation, $R^2$, which, as in the two variable case, gives the proportion of the variation in the values of $y$ that can be explained by the multiple regression.

*Autocorrelation* occurs in time series data when part of the series is correlated with another part of the series (Cryer and Miller 1994). For example, autocorrelation occurs in the time series of the maximum daily temperature over the past ten years. The maximum temperature on a given date will be highly correlated with the maximum temperature on that same date in a previous year.

## HISTORY

Francis Galton is credited with first understanding the idea of correlation (1888), which he called *co-relation*. Galton was investigating the relationship between the heights of parents and their children. In 1896, Karl Pearson introduced his correlation coefficient, extending the idea to nonnormal populations. In 1915, Ronald A. Fisher published the probability distribution of the sample correlation coefficient (Stigler 1986).

*See also* Mean, Arithmetic; Numerical Analysis, Methods; Standard Deviation; Standardization of Variables

## SELECTED REFERENCES

Box, George E. P., William G. Hunter, and J. Stuart Hunter. *Statistics for Experimenters: An Introduction to Design, Data Analysis, and Model Building.* New York: Wiley, 1978.

Cryer, Jonathan D., and Robert B. Miller. *Statistics for Business: Data Analysis and Modeling.* 2nd ed. Belmont, CA: Duxbury, 1994.

Dalal, Siddhartha R., Edward B. Fowlkes, and Bruce Hoadley. "Risk Analysis of the Space Shuttle: Pre-Challenger Prediction of Failure." *American Statistician* 84(Dec. 1989):945–957.

Ferguson, George A. *Statistical Analysis in Psychology and Education*. 5th ed. New York: McGraw-Hill, 1981.

Freedman, David et al. *Statistics*. 2nd ed. New York: Norton, 1991.

Moore, David S. and George P. McCabe. *Introduction to the Practice of Statistics*. 2nd ed. New York: Freeman, 1993.

Noether, Gottfried E. *Introduction to Statistics: A Nonparametric Approach*. 2nd ed. Boston, MA: Houghton Mifflin, 1976.

Stigler, Stephen M. *The History of Statistics: The Measurement of Uncertainty before 1900*. Cambridge, MA: Harvard University Press, 1986.

Weisberg, Sanford. *Applied Linear Regression*. New York: Wiley, 1980.

ANN E. WATKINS

# CREATIVITY

"When speaking of creative potential we are looking at a process distinct from either academic ability or artistic talent which can be described as creative thinking ability. It is a composite of intellectual abilities, personality variables, and problem solving orientation which increases but does not guarantee a person's chance of behaving creatively. Both convergent and divergent thinking styles may be present but emphasis is clearly on the latter." (Carroll and Howieson 1991, 68).

## MATHEMATICAL CREATIVITY

Children who are mathematically creative have freedom of thought—they are not bound by traditional algorithms. They have the need and desire to look at things from a variety of perspectives and see what others may not. They enjoy taking the novelty of an idea and following it—wherever it leads. They value mathematics and receive intrinsic reward in pursuing a new idea. Mathematically creative students tend to have high self-concept and low anxiety regarding mathematics (Haylock, "Mathematical . . . ," 1987). They have an ability to form rich and elaborate visual and sensory images with little input from anyone (Carroll and Howieson 1991). Mathematical creativity may be seen when a talented student leaves the stereotyped means of solving a problem and finds a few different ways of solving it (Haylock, "Mathematical . . . ," 1987).

One reason that mathematical creativity is considered important by many mathematics educators is that it enables students to explore mathematics in a way that exposes the beauty of the discipline. With-

out creative thinking, students may be locked into a restricted mental state where the only way to solve a problem is to recall algorithms that someone else created. Teachers must provide students with a rich environment in terms of inquiry, analysis, decision making, intuition, flexibility, and creativity. It is important to foster more discovery learning and less rote learning by using different techniques such as questioning instead of lecture and having students invent their own problem-solving strategies as opposed to applying teacher-supplied algorithms. Students can be creative and resourceful in mathematics if appropriately taught (Davis 1983).

## THEORIES OF CREATIVITY

Research on creativity and its influence on personality and performance has been a topic of interest to psychologists and educators for more than three decades. This interest has led to attempts to develop theories of creativity. Balka (1974) and Krutetskii (1969) describe creativity as the ability to break through mental fixations or mind sets and use unique thinking processes. Torrance (1966) identifies creative individuals by a list of character traits including courageous in convictions, independent in judgment, and curious or willing to take risks. Getzels and Jackson (1962) and Torrance (1966) found that intelligence and creativity were somewhat different, while Kurtzman (1967) found that intelligence and creativity were directly related. Later, Hall (cited in McCabe 1991) showed complex interrelationships among intelligence, creativity, and achievement.

To assess the validity of these theories, measures of creativity have been devised. The most widely validated test of creative thinking is that developed by Torrance (1966). This test is designed to measure a range of creative indicators in both the verbal and the figural areas of performance. A framework for assessing creative ability has two main strands: creative process and creative product. The creative process in problem solving involves four stages: preparation, incubation, illumination, and verification (Haylock, "Mathematical . . . ," 1987). Assessment of creativity has conventionally been through a divergent production test, where the subject is given a problem with many appropriate responses. The divergent thinking revealed by such tests is contrasted with convergent thinking, in which the subject must seek (converge upon) the one and only one correct solution (Haylock, "A Framework . . . ," 1987).

Carroll and Howieson (1991), after testing for intelligence and creativity, conducted a study to compare the two. Forty-eight seventh-grade children were divided into four groups based on creative thinking scores and intelligence scores. On some measures of problem solving, imagery, and mathematics, highest scores were achieved by the high-intelligence/high-creativity group. For other school assessments, creativity did not add to performance and even appeared to depress results.

In another study, 210 female adolescents who were high achievers in English were more likely (than those who were not high achievers in English) to score high on tests of creative thinking and obtain high intelligence quotient (IQ) scores. Achievement in mathematics and art were not as highly correlated with creative thinking but were related to high IQ scores (McCabe 1991).

## PEDAGOGY FOR CREATIVITY

Goals in mathematics education should include divergent thinking, positive self-image, motivation, and creativity. Both high intelligence and high creativity maximize achievement in problem solving and imaginative situations. Creative problem-solving skills do not emerge suddenly; they must be encouraged and carefully shaped throughout schooling years. All classrooms should develop a climate where original thinking is rewarded (Carroll and Howieson 1991). Creative thinking can be developed in early grades through creative exploration activities including games (cards and tangrams), competitions, clubs, mathematical evenings (relays and treasure hunts), projects (patchwork quilting, art, and music) and camps (de Vries 1992). In problem solving, tasks that require generation of a variety of possible solutions rather than problems for which there is only one correct response appear to stimulate creative thinking (Torrance 1984). Problems should be immediately attractive, require data to be generated or gathered, appeal to students from elementary school to graduate school, involve fundamental mathematics concepts, have satisfying solutions, and suggest other problems. An exploratory process should include an inductive phase, a deductive phase, a creative phase, and use of technology where appropriate (Stevenson 1992).

An example of such a problem is the following: "There is a ship at sea. How far from shore is the ship? Find at least ten different ways to solve this problem. To solve the problem, you have only what you would normally find on a beach." This problem is rooted in experience, allows many methods of solution, and especially fosters creative solutions in-stead of cutting off students' initial intuitive reactions when a ready-made algorithm is provided.

Objectives regarding creativity, fostering independent thinking, using experiences, using mastery, reinforcing previously learned skills, emphasizing choices, following directions, encouraging divergent thinking, and developing concepts are central to quality education (Levy 1984). Research has shown that use of microcomputers, computer and calculator programming, and some training in logic and reasoning improves students' creative abilities (Bower 1985; Wagner 1984). The use of technology as a creative outlet is becoming more important as use of technology becomes more widespread. A study by Davis (1983) shows that students prefer creative computer programming work over drill and practice using a computer or calculator. Shumway (1984) also showed the importance of having students do computer programming in the classroom.

Teachers of all grades may encourage creative thinking by doing less telling and demonstrating and by posing situations to students and asking them to find different ways to solve the problems. This type of assignment will free students from thinking of that "one correct way" of solving problems and will actually encourage them to think, and invent. Teachers also may encourage creative thinking for brilliantly creative students by allowing choices in classroom assignments (e.g., allow students to solve the teacher-created problems or challenge them to create their own problems), providing enrichment activities for the students, asking the students to share the results of their creativity with the class, and asking them to generalize the problems on which they have been working and then to find generalized solutions.

*See also* Cooperative Learning; Gifted; Gifted, Special Programs; Problem Solving, Overview

## SELECTED REFERENCES

Balka, Donald S. "Creative Ability in Mathematics." *Arithmetic Teacher* 21(7)(1974):333–363.

Bernstein, Bob. *Math Thinking Motivators. A Good Apple Math Activity Book for Grades 2–7.* Carthage, IL: Good Apple, 1988.

Bower, B. "Computers and Kids: Learning to Think." *Science News* 127(5)(Feb. 1985):71.

Carroll, John, and Noel Howieson. "Recognizing Creative Thinking Talent in the Classroom." *Roeper Review* 14(2)(Dec. 1991):68–71.

Davis, Robert B. "Diagnosis and Evaluation in Mathematics Instruction: Making Contact with Students' Mental Representations." In *Essential Knowledge for Beginning Educators* (pp. 101–111). David C. Smith, ed. Washington,

DC: American Association of Colleges for Teacher Education, 1983.

De Vries, Marianne E. "Thinking beyond the Obvious Boundaries in Mathematics: An Exploration of Joyous Discovery." *Gifted Education International* 8(3)(1992): 163–179.

Getzels, Jacob W., and Philip W. Jackson. *Creativity and Intelligence: Explorations with Gifted Students.* New York: Wiley, 1962.

Haylock, Derek W. "A Framework for Assessing Mathematical Creativity in Schoolchildren." *Educational Studies in Mathematics* 18(1)(1987):59–74.

———. "Mathematical Creativity in Schoolchildren." *Journal of Creative Behavior* 21(1)(1987):48–59.

Krutetskii, Vadim A. "Mathematical Aptitudes." In *Soviet Studies in the Psychology of Learning and Teaching Mathematics* (pp. 333–363). Vol. 2. Jeremy Kilpatrick and Izaak Wirzup, eds. Chicago, IL: University of Chicago Press, 1969.

Kurtzman, Kenneth A. "A Study of School Attitudes, Peer Acceptance, and Personality of Creative Adolescents." *Exceptional Children* 34(3)(1967):157–162.

Levy, Barbara W. *The Math Master Levels 1–4. Ages Pre-school–12.* Carthage, IL.: Good Apple, 1984.

McCabe, Marita P. "Influence of Creativity and Intelligence on Academic Performance." *Journal of Creative Behavior* 25(2)(1991):116–122.

Shumway, Richard. "Young Children Programming and Mathematical Thinking." In *Computers in Mathematics Education* (1984 Yearbook; pp. 127–134). Viggo Hansen, ed. Reston, VA: National Council of Teachers of Mathematics, 1984.

Stevenson, Frederick. *Exploratory Problems in Mathematics.* Reston, VA: National Council of Teachers of Mathematics, 1992.

Taback, Stanley F. "The Wonder and Creativity in 'Looking Back' at Problem Solutions." *Mathematics Teacher* 81(6)(1988):429–434.

Torrance, Ellis P. "Some Products of Twenty Five Years of Creativity Research." *Educational Perspectives* 22(1984):3–8.

———. *Torrance Tests of Creative Thinking: Norms Technical Manual.* New Jersey: Personell, 1966.

Wagner, Paul. *Pre-College Philosophy: Will It Get Its Day in Court.* ERIC No. ED260975. Columbus, OH: Educational Resources Information Center, 1984.

PATRICIA A. BROSNAN
JAMES A. FITZSIMMONS

# CRYPTOGRAPHY

Concerned with secret codes, that is, codes intended to hide the content of the message from an unauthorized recipient. Codes, however, have many additional purposes. One, called *source coding*, is to compress a message before it is sent, and another is to protect a message against corruption by noise during its transmission. Both are the subject of algebraic coding theory. There are other codes that facilitate computer handling of information, such as zip codes for mail, bank-identification codes for processing checks, and International Standard Book Number (ISBN) codes for identifying books and their publishers. None of the latter is intended to hide information, although they share with algebraic coding theory the attempt to detect and correct errors. For an overview of this broad variety of codes see, for example, Malkevitch, Froelich, and Froelich (1991).

The known history of cryptography goes back to ancient times. Julius Caesar used a cipher, called a shift transformation, in which every letter of the alphabet is replaced by the letter that comes three letters later and x, y, and z are replaced by a, b, and c, respectively. This is an example of a substitution cipher, in which each letter of the alphabet is replaced by some other letter, that is, there is a one-to-one correspondence between letters in the original text and letters in the encoded message. Caesar's substitution rule was especially simple. A more general substitution cipher can be broken by making intelligent guesses using the known letter frequencies of the language of the message. Pure substitutions can be avoided by using several alphabets for a single message (a system called *polyalphabetic substitution*) or by using an $n \times n$ matrix to encode blocks of $n$ consecutive message symbols, called a polygraphic system (Sinkov 1976).

Cryptography has played a significant role in world history during the twentieth century. In 1917, the German foreign minister, Arthur Zimmermann, sent an encoded message to the German ambassador in Washington. In it, he instructed the ambassador to offer U.S. territory to Mexico in exchange for a Mexican alliance with Germany in the case of U.S. entry into World War I. The decryption by the British of this telegram helped to bring about just this entry. During World War II, the German ciphers were broken by the British, and the Japanese ciphers by the Americans. Thus Admiral Halsey obtained the itinerary of a forthcoming inspection tour by Fleet Admiral Yamamoto, and Yamamoto and his staff were shot down (Kahn 1967).

In all the cryptographic systems described so far, the sender and the recipient share the same secret key. This implies that they must both have the key before communication can take place, and that the problems of its distribution and security have been solved. In 1976, W. Diffie and M. Hellman proposed a totally new system, called public key cryptography. In this system, there are separate keys for encoding and for decoding. Person R who expects to receive messages publishes the encoding key that people should use to encode messages intended for R, but keeps secret a decoding key. The mathematical properties of the keys are such that no one can, in reasonable time, reconstruct the decoding key from knowing only the published encoding key. Such a system, called the RSA, actually was devised by A. Rivest, A. Shamir, and L. Adleman in 1977, and is in wide commercial use. It makes use of the fact that it is easy to construct a really large (say two hundred-place) number that is the product of two primes, but extremely difficult to find these primes if only the product is given (Wampler 1994).

Some school curricula use shift transformations to teach linear functions, and substitution ciphers to help with functions in general. Polygraphic systems make significant use of linear transformations and matrices. For example, Hill Codes, using matrix multiplication and inverses, are developed in Malkevitch and Froelich (1993).

*See also* Coding Theory; Prime Numbers

SELECTED REFERENCES
Kahn, David. *The Codebreakers.* New York: Macmillan, 1967.
Malkevitch, Joseph, and Gary Froelich. *Loads of Codes.* Histomap Module 22. Lexington, MA: COMAP, 1993.
———, Gary Froelich, and Daniel Froelich. *Codes Galore.* Histomap Module 18. Lexington, MA: COMAP, 1991.
Sinkov, Abraham. *Elementary Cryptanalysis.* Washington, DC: Mathematical Association of America, 1976.
Wampler, Joe F. "Elementary Cryptology." In *UMAP Models, Tools for Teaching 1993* (pp. 113–140). Paul J. Campbell, ed. Lexington, MA: COMAP, 1994.

HENRY O. POLLAK

# CUISENAIRE RODS

Invented by Emile-Georges Cuisenaire (1891–1976), a Belgian schoolmaster, as a teaching aid for self-discovery of basic mathematical concepts and principles. Cuisenaire had interests in both mathematics and music. He based the lengths of the rods on the lengths of pipe organs. Doubling the length of a pipe organ raises a note by an octave. He selected colors for the rods based on his own intuitive feelings for numbers. He chose hues of green and blue for 3, 6, and 9 and shades derived from reds and oranges for 2, 4, 8, and 10 (Hull 1995). Caleb Gattegno discovered Cuisenaire's work in Belgium and started the Cuisenaire Company in America to bring Cuisenaire's ideas to the United States.

Each rod is a rectangular prism of size 1 cm by 1 cm by n cm, where n is a whole number from one to ten inclusive. The lengths of the Cuisenaire rods range from 1 cm to 10 cm and each is color-coded by length as indicated: 1—white, 2—red, 3—green, 4—purple, 5—yellow, 6—dark green, 7—black, 8—brown, 9—blue, and 10—orange. To name the rods, the first letter of each color is used except for black, brown, and blue; then the last letter is used. The ordered list of colors in order of rod length is $w, r, g, p, y, d, k, n, e, and o$ (the last written in script to distinguish it from zero).

By moving Cuisenaire rods into various arrangements, students discover relationships among numbers. The simplest discovery is the arrangement of the rods in order from shortest to tallest. By using a white rod, students discover that one white rod added to a given rod equals the length of the adjacent next taller rod. Thus, $w + w = r, w + r = g, w + g = p$. By exploring rod relationships concretely, students build readiness for mathematical concepts, such as addition and "one more than."

At a higher level, students explore how many white rods are needed to build each rod and then express these relationships as fractions. Since $2 w = 1 r$, the length of a white rod is half the length of a red rod. Also, since $5 w = 1 y$, the length of a white rod is one-fifth the length of a yellow rod.

Cuisenaire rods are very versatile in teaching a variety of mathematical concepts and principles such as addition, subtraction, multiplication, and division of whole numbers, factorization, place value, ratio and proportion, perimeter, area, frequency distribution, signed numbers, and modular arithmetic. As an example, students can explore ratio and proportion by building one-color trains using rods. To build a one-color train equivalent to dark green, students may use two green rods, three red rods, or six white rods. Using these concrete materials, students may conclude that $6 w = 1 d, 3 r = 1 d,$ or $2 g = 1 d$. The same visualization could illustrate that $1 g = \frac{1}{2} d, 1 r = \frac{1}{3} d,$ or $1 w = \frac{1}{6} d$. Also, $2 r = \frac{2}{3} d, 4 w = \frac{4}{6} d = \frac{2}{3} d,$ and $3 w = \frac{3}{6} d = \frac{1}{2} d$ (see Figure 1).

| Dark Green | | | | | |
|---|---|---|---|---|---|
| Green | | | Green | | |
| Red | | Red | | Red | |
| W | W | W | W | W | W |

**Figure 1** *One-color trains for dark-green rod.*

*See also* Arithmetic; Fractions; Manipulatives, Computer-Generated

## SELECTED REFERENCES

Bradford, John. *Everything's Coming Up Fractions with Cuisenaire Rods.* New Rochelle, NY: Cuisenaire, 1981.

Davidson, Jessica. *Using the Cuisenaire Rods—A Photo/Text Guide for Teachers.* White Plains, NY: Cuisenaire, 1983.

Davidson, Patricia S. *Idea Book for Cuisenaire Rods at the Primary Level.* White Plains, NY: Cuisenaire, 1977.

———, and Robert E. Willcutt. *Symmystries with Cuisenaire Rods, Sets 1, 2, and 3.* New Rochelle, NY: Cuisenaire, 1990.

Hull, William P. Personal Interview. February 7, 1995.

Spikell, Mark A. *Teaching Mathematics with Manipulatives: A Resource of Activities for the K–12 Teacher.* Boston, MA: Allyn and Bacon, 1993.

REGINA BARON BRUNNER

# CURRICULUM, OVERVIEW

Goals and content of the mathematics curriculum in schools have evolved through many changes in the United States. As was the belief in Europe, in the 1700s, the belief in the United States that education was necessary to the welfare of society gave impetus to the content of the mathematics curriculum (Jones 1970). Benjamin Franklin established an academy in 1751 that included geometry and algebra in its curriculum. These secondary schools sprang up in competition with Latin schools of that time and were an attempt to satisfy the practical needs of seamen, merchants, artisans, and frontiersmen. From about 1820 until 1900, European educational philosophies and currricula were the strongest influences on mathematics education in the United States. The major issues during this period included defining the goals of mathematics education (Jones 1970). The content of instruction in schools most students attended consisted of only assorted arithmetic, since most students did not further their instruction beyond the common school.

## 1900–1950

During this period, issues debated about the curriculum included deciding on the goals of mathematical instruction and choosing the content of the mathematics curriculum to meet the background needs, interests, and abilities of students (Jones 1970). The need for efficiency in calculating and economic questions influenced the curriculum. Behaviorist psychology strongly influenced the mathematics curriculum, as did the changing economic status that emphasized technology, industry, and scientific development. In operant conditioning a small unit of information is presented to the learner, who makes a response simply by answering a question. The learner is informed if the response is correct. Mathematics content, which can be easily broken into small bits of information, therefore, became narrow and often irrelevant.

The introduction of a "meaning theory" of learning mathematics influenced the mathematics curriculum during this time; content to be taught was selected from situations familiar to students. Founded in 1920, the National Council of Teachers of Mathematics (NCTM) began publishing *The Mathematics Teacher* in 1921. The organization and its subsequent publications influenced the mathematics curriculum. Another influence was the entrance of the United States into World War II. An emphasis of the mathematics curriculum at the time was military preparation.

The National Education Association convened a Commission on the Reorganization of Secondary Education in 1918. The Commission identified seven goals for public education known as "Cardinal Principles." These principles, which lay the groundwork for a comprehensive high school, included two principles that influenced the mathematics curriculum: (a) the need for students to have a command of fundamental processes and (b) preparation for a vocation.

The junior high school concept grew out of suggestions by the Committee of Ten in the 1920s and 1930s that secondary education begin two years early to reduce elementary school from eight years to six; the junior high school became a separate housing facility. This again influenced the mathematics curriculum. Additionally, the progressive education movement supported the general goals of schools to respond to the needs and interests of individual students. This movement was influenced by the work of John Dewey. The mathematics curriculum was "adjusted" to fit the needs of students.

## 1950 TO PRESENT

A relaxation of college entrance requirements for mathematics, applications of mathematics and science in many fields based on knowledge gained in World War II; the launching by the Soviet Union of the space vehicle, *Sputnik*; and a greater concern for superior students have influenced the mathematics curriculum since the 1950s. Attempts were made to make the curriculum more strenuous by adding in-depth content and requiring more mathematics courses for graduation. The "new math" reform movement (1950–1975) changed and shaped the mathematics curriculum (Wirszup and Streit 1992). The priorities of the mathematics curriculum changed from teaching only skills, to teaching for understanding and emphasizing reasoning.

NCTM publications (1989; 1991; 1995) had a major influence on the mathematics curriculum. An updated version, *Principles and Standards for School Mathematics,* was published in 2000. These publications charted a course for changing the ways mathematics is taught, by describing changes needed in mathematics education and how the changes might be implemented. Many mathematics programs implemented recommended changes to varying degrees, and some of the recommendations met with controversy.

The NCTM presented five goals for the mathematics curriculum. All students should become mathematical problem solvers, and confident in their ability to do mathematics. They should learn to reason mathematically and to communicate mathematically as they learn to value mathematics (NCTM 1989). There is an emphasis on interactive learning, where students interact with other students as well as with the teacher. Attempts are being carried out to integrate mathematics into and with other areas of study. An emphasis is being placed on connecting mathematics with the real world. The goals of the general mathematics curriculum have changed as a result of the influences described above as well as the different needs of workers, changes in the ways mathematics is used, the availability of calculators and computers, and the emphasis being placed today on lifelong learning.

## MATHEMATICAL CONTENT

There has been an ongoing debate throughout history about what content should be included in the mathematics curriculum. The debate has been between those who advocate emphasizing drill, skills, and procedural knowledge, and those who believe that students should acquire understandings and conceptual knowledge. Is it better for students to learn a rule and then draw inferences, or for instruction to begin with examples and through the use of many examples, students arrive at a rule?

The *Curriculum and Evaluation Standards for School Mathematics* (NCTM 1989) has gained much

attention and seems to be serving as a model of reform in teaching practices and in changing the content of mathematics being explored at all levels of the curriculum. Estimation of discrete quantities and measurements are integral parts of the general curriculum and are used to help students solve problems. Content from kindergarten through high school includes exploration of number sense, geometry and spatial sense, patterns, and relationships. To help students develop higher-order thinking skills and become problem solvers, classification skills are fundamental. Classifying enables individuals to deal with all aspects of life in that we categorize and react to experiences in similar ways. Classification experiences in mathematics are necessary for students to become problem solvers in mathematics.

In the primary grades students collect and analyze data, formulate and solve problems, and explore probability and statistics. In the middle grades, these concepts are expanded and students begin to explore algebraic concepts. Making predictions, analyzing relationships, and identifying properties and relationships are integral parts of the middle school curriculum. The general curriculum for all high school students has been modified to include a core curriculum where all high school students study the same topics and those students who are college bound examine the same topics in greater breadth and depth.

## PEDAGOGICAL APPROACHES

Many changes are being recommended in the way mathematics is being taught. Interactive learning, in which students work in cooperative groups to produce both written and oral reports, is stressed. Less emphasis is placed on pencil-and-paper activities, and more emphasis is placed on hands-on activities, in which students make inquiries into relevant problems while they acquire a background of understandings and skills in mathematics. Less emphasis is placed on speed and more on competence in problem solving. Systematic programs, including a core curriculum, are planned for all students: students of both genders, minority students, those who have learning disabilities, and those who are gifted and talented in mathematics.

Methods that emphasize conceptual knowledge are being implemented. Students observe, discuss, write, and complete projects. Teachers use a less direct instructional approach and work to teach for higher-order thinking. Teachers serve as facilitators and guides, while students actively construct their own knowledge and understandings. "Authentic assessment" is being advocated. Authentic assessment is more than just testing. It uses a variety of strategies to evaluate the learning of students, both formally and informally. Authentic assessment is ongoing, covers a long time frame, and attempts to sample not only the mathematical skills and knowledge of students, but also their concepts and understandings in mathematics and whether they can make predictions, draw conclusions, and form generalizations. No longer are tests the only means of evaluation; students are being evaluated by many means, including all kinds of projects and quality of portfolios.

Integrated learning strands are planned and implemented. Integrated strands provide opportunities for students to explore topics in-depth and to make choices about what and how they learn. Mathematics concepts are included and extracted from other content areas. Strategies associated with integrated learning contrast with those of the time when mathematics was taught in isolation, only during a period set aside for mathematics. Depending on the philosophy adopted in schools, the grade levels of the school, and the extent to which the philosophy or beliefs of the teachers and the communities are accepted, some teachers plan activities where students integrate and apply their mathematical understandings in meaningful situations throughout the day.

## MATERIALS AND ACTIVITIES

Today, attempts are being made to provide rich learning environments for all students. In mathematics this means making calculators, computers, models, and manipulatives available to students in classrooms. Activities are generated that have a reasonable context in the everyday lives of students. The real experiences of students provide a basis for problems and activities. Tasks used in the teaching and learning of mathematics are carefully selected and require the use of many strategies for teacher and student. In many schools that have chosen to implement the 1989, 1991, and 1995 NCTM standards, students are allowed to use their own approaches and a variety of tools to solve problems.

Technology has had a major impact on the mathematics curriculum. NCTM suggested that all types of technology be made available for students in their mathematical explorations; for example, calculators can be used at all grade levels for concept learning, in problem solving, for skill development, and in testing. Computers also are becoming common in schools and are used as tutors for introducing new subject matter. Computer programs include those that allow drill and practice and those that simulate real-life events and allow students to play instructional games. Graphing programs, database programs, and spreadsheets allow computers to be used as tools for learning by which students retrieve information and make calculations using various computer programs.

*See also* Curriculum Trends, Secondary Level

## SELECTED REFERENCES

Armstrong, David D., Kenneth T. Henson, and Tom V. Savage. *Education: An Introduction.* New York: Macmillan, 1981.

Grouws, Douglas A., Thomas J. Cooney, and Douglas Jones, eds. *Effective Mathematics Teaching.* Reston, VA: National Council of Teachers of Mathematics, 1988.

Holmes, Emma E. *New Directions in Elementary School Mathematics: Interactive Teaching and Learning.* Englewood Cliffs, NJ: Merrill, 1995.

Jones, Phillip S., ed. *A History of Mathematics Education in the United States and Canada.* 32nd Yearbook. Washington, DC: National Council of Teachers of Mathematics, 1970.

National Council of Teachers of Mathematics. *Assessment Standards for School Mathematics.* Reston, VA: The Council, 1995.

———. *Curriculum and Evaluation Standards for School Mathematics.* Reston, VA: The Council, 1989.

———. *Principles and Standards for School Mathematics.* Reston, VA: The Council, 2000.

———. *Professional Standards for Teaching Mathematics.* Reston, VA: The Council, 1991.

Nelson, David, George G. Joseph, and Julian Williams. *Multicultural Mathematics: Teaching Mathematics from a Global Perspective.* New York: Oxford University Press, 1993.

Wertheimer, Richard D. "Issues of Implementation." *Mathematics Teacher* 88(Feb. 1995):86–88.

Wirszup, Izaak, and Robert Streit, eds. *Developments in School Mathematics Education Around the World.* Vol. 3. Reston, VA: National Council of Teachers of Mathematics, 1992.

MARILYN S. NEIL

## CURRICULUM ISSUES

Debates about what part of the existing body of mathematics should be presented to students. In the ninth century, Al-Khwarizmi (whose mathematical treatise gave us the word "algebra") selected "what is easiest and most useful in arithmetics, such as men constantly require in cases of inheritance" (Van der Waerden 1980, 20). In 1831, William Slocomb published *The American Calculator,* "containing all the rules necessary for transacting the common business of life" and "intended for the use of schools and young men who may be desirous of obtaining further knowledge of this science" (Slocomb 1831, title page). Slocomb explained in the preface how his text differed from those in current usage. With the latter, he argued, "much of the time of the pupil is wasted, without adding scarcely any thing to the stock of his useful knowledge."

Between 1955 and 1975, there were two forces driving a need for change: the space race with the Soviet Union and a sudden growth of technology. "New math" was one result; it included aspects of set theory that were beneficial in the dawning computer age and promoted understanding of concepts rather than rote learning of rules and algorithms. Interdisciplinary uses for mathematics were presented in areas other than and in addition to the physical sciences. The "new math" curriculum had a short life. Some educators believe that lack of public relations was a reason for its failure to catch on or last in the public schools. Parents did not accept the program (National Research Council [NRC] 1989, 79); and Americans at large became confused, not only about the changes, but the need for change. Another important factor was that unprepared elementary school teachers had to go back to school to learn and then teach unfamiliar concepts. In a backlash, some parents and teachers supported a "back-to-basics" movement. Others continued to support modernization of the curriculum, but recommended different selections.

Supporters of modernization believe that drills on calculation and traditional basic skills should be decreased. "Even in the absence of calculators, neither children nor adults make much use of the specific arithmetic techniques taught in school" (NRC 1989, 46–47). Instead, according to the NRC, the major objective of elementary school mathematics should be the development of number sense. In the secondary schools, the NRC recommended (1989, 49–50) that all students should study a common core of broadly useful mathematics and should study mathematics every year that they are in school. A primary goal is that all students learn the mathematics necessary to cope with the demands of their daily lives and to become productive citizens. The National Council of Teachers of Mathematics (NCTM) attempted to facilitate modernization by developing new recommendations for the curriculum, published in *Curriculum and Evaluation Standards for School Mathematics* (NCTM 1989). A major emphasis of the recommendations is the importance of teaching students to solve problems. According to Henry Pollak, as quoted in the introduction to the Standards,

the mathematics that new employees are expected to know includes the ability to state problems in mathematical terms, knowledge of concepts needed for solutions, ability to work with others, to see the applicability of mathematical ideas, preparation for open problem situations, and the belief in the utility and value of mathematics. As a result of positive and negative feedback from the educational community and the public, NCTM published a revision, *Principles and Standards for School Mathematics,* in 2000.

*See also* Cambridge Conference on School Mathematics (1963); College Entrance Examination Board (CEEB) Commission Report (1959); Curriculum Trends, College Level; Curriculum Trends, Secondary Level; Ethnomathematics; History of Mathematics Education in the United States, Overview; Multicultural Mathematics, Issues; *Revolution in School Mathematics;* School Mathematics Study Group (SMSG); Standards, Curriculum

## SELECTED REFERENCES

Bidwell, James K., and Robert G. Clason. *Readings in the History of Mathematics Education.* Washington, DC: National Council of Teachers of Mathematics, 1970.

National Commission on Excellence in Education. *A Nation at Risk: The Imperative for Educational Reform.* Washington, DC: Government Printing Office, 1983.

National Council of Teachers of Mathematics. *Curriculum and Evaluation Standards for School Mathematics.* Reston, VA: The Council, 1989.

———. *Principles and Standards for School Mathematics.* Reston, VA: The Council, 2000.

National Research Council. *Everybody Counts: A Report to the Nation on the Future of Mathematics Education.* Washington, DC: National Academy Press, 1989.

Slocomb, William. *The American Calculator.* Marietta, OH: William Davis and William Slocomb, 1831, reprinted extract. In *Readings in the History of Mathematics Education* (pp. 3–12). James K. Bidwell and Robert G. Clason, eds. Washington, DC: National Council of Teachers of Mathematics, 1970.

Van der Waerden, Bartel L. *A History of Algebra.* Berlin, Germany: Springer-Verlag, 1980.

SUE JACKSON BARNES

# CURRICULUM TRENDS, COLLEGE LEVEL

Changes in curricula in the areas of calculus and teacher education in particular. After the publication of *A Nation at Risk* (National Commission on Excellence in Education 1983), educational reform was advocated. Since that time, curricular reform in the discipline of mathematics has been promoted for levels K–14. In 1986, twenty-five mathematicians and mathematics educators met at Tulane University to discuss the calculus curriculum. In the final report of the conference, *Toward a Lean and Lively Calculus,* Ronald G. Douglas (1986) summarized some of the problems with the traditional calculus curriculum: (a) the curriculum had too many topics which hindered conceptual understanding of the subject; (b) the curriculum was not conducive to the use of technology; and (c) the curriculum was structured to expose mathematics majors to the rigors of the discipline and was not suitable for other students who were required to take calculus, such as those majoring in the fields of biology, chemistry, psychology, physics, sociology, and business. Thus, Douglas (1986) proposed rethinking and rewriting the calculus curriculum to meet the needs of today's students.

The resulting suggestions for a reformed curriculum include emphasis on problem solving in a real-world context, use of technology, collaborative work, and articulation by students of their reasoning processes (Johnson 1994). The approach has been designed to (a) allow students weak in algebraic skills to study the same level of calculus as the students who have stronger skills; and (b) improve attitudes about mathematics. In order to implement these goals, reform proposals emphasize conceptual understanding over symbolic manipulation. For example, the traditional curriculum stresses procedures such as how to take a derivative. In contrast, the new proposed curricula accentuate what the derivative means and use technology to teach the concept.

In addition, a long-term aim of the calculus reform movement is to enroll more students in mathematics or mathematics-related courses. In order to ensure better success in the calculus sequence, some schools have adopted programs modeled after the Emerging Scholars Program. This particular program is designed to assist minority students in their calculus studies through the use of collaborative work sessions. Through success in the calculus sequence, reformers hope more students will remain in mathematics-oriented fields. For further information on the Emerging Scholars Program, see Treisman (1992) and Bonsangue (1994).

By 1991 the National Science Foundation (NSF) ("NSF Awards" 1990; "NSF Calculus Awards" 1991) had funded over seventy calculus reformed curriculum projects. In comparing these curricula to a traditional calculus curriculum, C. E. Johnson (1994) indicated that the traditional calculus curriculum developed the skills needed to work a problem then considered application problems that

were typically in a separate section at the end of a chapter. In contrast, many reformed curricula first present a problem to catch the interest of the student and then develop the calculus concept needed to confront the problem. For example, in the Five Colleges' Calculus in Context curriculum, a problem about the spread of disease, such as measles, is introduced on the first class day. The rate of change for people succumbing to the disease and the rate of change for people recovering from the disease are considered. From the discussion about rates of change, the concept of the derivative is addressed.

Field testing of reformed curricula has been slow to occur. The project with the most field testing has been the Harvard's Calculus Consortium (CCH) curriculum. This particular curriculum is presently the most popular of all the reformed calculus curricula; however, it also has been cited as being the least reformed of all the reformed calculus materials (Tucker and Leitzel 1995). For an overall synopsis of calculus reform, see Tucker 1990, *UME Trends* (1995), and Tucker and Leitzel (1995).

About the same time that the calculus reform movement got underway, the National Council of Teachers of Mathematics (NCTM) published materials for curricular reform on other levels (NCTM 1989; 1991). NCTM published updated materials, *Principles and Standards for School Mathematics,* in 2000. NSF provided funding to mathematics departments desiring to educate teachers in accord with NCTM proposals. Projects, such as Professors Rethinking Options in Mathematics for Prospective Teachers (Project PROMPT), at Humboldt State University, focused on the use of manipulatives to teach mathematical concepts to preservice elementary education majors, and modeled teaching methods other than lecture.

Additionally, NSF provided funding for the professional development of inservice mathematics teachers. For example, funding was approved to provide professional development for inservice teachers in the states of Louisiana, Maryland, and Montana. Programs such as these educate teachers about curricular changes, the use of manipulatives, and the use of technology. Also, teachers are taught mathematical content such as algebra, statistics, or discrete mathematics.

Overall, there has been a trend by universities to change their teacher education programs to field-based teacher education that emphasizes professional development and technology. For instance, East Texas State University's teacher education program won the American Association of State Colleges and

Universities' 1994 Christa McAuliffe Showcase for Excellence Award. This particular program is a collaboration between the university and surrounding public school districts. Preservice teachers are required to spend an internship and residency at the public schools that may range in length from one public school semester to one academic year. The preservice teacher must be interviewed and accepted by a team of mentoring master teachers. If chosen, a preservice teacher is considered part of the faculty at the school and is exposed to many aspects of teaching in the public schools such as the first school day. The mentors of the preservice teachers are master teachers who must meet certain criteria and who also must participate in a professional development and technology program at the university. In return they are supplied materials and technology for their own classrooms. Since the trend in teacher education moved toward field-based programs, the professional development of master teachers may be seen as an important element in the preparation of preservice teachers.

In addition to curricular revision in the calculus sequence and in teacher education, other areas, such as introductory college mathematics and linear algebra courses, have begun curricular revisions, too. Introductory college mathematics courses include remedial, college algebra, and precalculus mathematics courses. Standards for the curricular revision for introductory college mathematics courses were released by the American Mathematical Association for Two-Year Colleges (AMATYC) in October 1994. A few revised curricula for these courses have been written and are in the field testing process. These new curricula emphasize problem solving, use of technology, and center upon patterns, relations, and functions. More emphasis is placed upon the development of algebraic thinking rather than manipulation skills. On the other hand, curricular revisions for linear algebra are limited, focusing mainly on the use of technology.

As technology became more and more important to society, student enrollment in mathematics (the language of technology) increased, but the failure rate in these courses brought about the need for reforming the undergraduate mathematics curriculum. Reform first began with curricular changes in calculus. Then reform branched into K–12 teacher education, both at the preservice and inservice levels. More recently, reformed curricular revisions at the college level has focused on courses other than calculus, such as introductory college mathematics, and linear algebra. Published research on undergraduate mathematics education remains sparse. See Dubinsky, Schoenfeld, and Kaput (1994) or Kaput and

Dubinsky (1994) for information about research on mathematics education at the undergraduate level.

*See also* Calculus, Overview; Teacher Preparation, College

## SELECTED REFERENCES

American Mathematical Association of Two-Year Colleges. *Standards for Introductory College Mathematics.* Memphis, TN: The Association, 1994.

Bonsangue, Martin V. "An Efficacy Study of the Calculus Workshop Model." In *Research in Collegiate Mathematics Education: I* (pp. 117–137). Ed Dubinsky, Alan H. Schoenfeld, and James J. Kaput, eds. Providence, RI: American Mathematical Society, 1994.

Douglas, Ronald G., ed. *Toward a Lean and Lively Calculus.* Washington, DC: Mathematical Association of America, 1986.

Dubinsky, Ed, Alan H. Schoenfeld, and James J. Kaput, eds. *Research in Collegiate Mathematics Education. I.* Washington, DC: American Mathematical Society, 1994.

Johnson, Charles E. "So What Precisely Is Going on in the Calculus Reformation?" *New York State Mathematics Teachers' Journal* 44(Sept. 1994):49–53.

Kaput, James J., and Ed Dubinsky, eds. *Research Issues in Undergraduate Mathematics Learning: Preliminary Analyses and Results.* Washington, DC: Mathematical Association of America, 1994.

National Commission on Excellence in Education. *A Nation at Risk: The Imperative for Educational Reform.* Washington, DC: Government Printing Office, 1983.

National Council of Teachers of Mathematics. *Curriculum and Evaluation Standards for School Mathematics.* Reston, VA: The Council, 1989.

———. *Principles and Standards for School Mathematics.* Reston, VA: The Council, 2000.

———. *Professional Standards for Teaching Mathematics.* Reston, VA: The Council, 1991.

"NSF Awards $2.4 Million for Calculus: Twenty Grants to Curriculum Projects." *UME Trends* 2(Oct. 1990):1, 5.

"NSF Calculus Awards 1991: $2.3 Million Plus for Calculus Curriculum Efforts." *UME Trends* 3(Oct. 1991):1.

Treisman, Uri. "Studying Students Studying Calculus: A Look at the Lives of Minority Mathematics Students in College." *College Mathematics Journal* 23(Nov. 1992):362–372.

Tucker, Alan C., and James R. C. Leitzel, eds. *Assessing Calculus Reform Efforts: A Report to the Community.* Washington, DC: Mathematical Association of America, 1995.

Tucker, Thomas W., ed. *Priming the Calculus Pump: Innovations and Resources.* Washington, DC: Mathematical Association of America, 1990.

*UME Trends.* 6(Jan. 1995). Special issue on the calculus reform movement.

CONNIE HICKS YAREMA

# CURRICULUM TRENDS, ELEMENTARY LEVEL

Evolving mathematical studies in elementary school grades, K–5. Typically includes the concept of number, the decimal number system, shapes and patterns, measurement, and arithmetic. Recommendations continue to be made urging greater attention to geometry and topics from probability and statistics. Flanders (1987) documented that the textbooks commonly used in elementary classes devote a rather large percentage of pages reviewing previously covered topics from earlier grades. It might be necessary for teachers to spend less time reviewing in order to include more attention to new topics.

## HISTORICAL PERSPECTIVE

The key single event impacting education in the late 1950s was the successful launch of the Soviet space vehicle, *Sputnik,* in October 1957. A significant reaction in the United States was an effort to reform mathematics and science education. The goal was to prepare a new generation of excellent scientists and engineers in order to "win" the race for space exploration. Several federal grants were made to projects for developing improved mathematics and science curricula. The programs emerging from these reform efforts became known as "new math." During the 1960s, many new texts incorporated the recommendations of the School Mathematics Study Group (SMSG) and provided discussion of the structure of mathematics and the discovery of patterns. Emphasis on computational skills was decreased. The term "new math" was never well defined in a mathematical sense. The adjectives logical, rigorous, and structured were frequently identified with the "new math." Formalism seemed to be a desired end in itself. The instructional focus was on teaching concepts and away from drilling basic computational algorithms. Attempts to have students correctly distinguish between a number and a numeral is a typical example of this formalism that did not have a lasting impact.

*Why Johnny Can't Add: The Failure of the New Math* (Kline 1973) presented a view that the "old" curriculum, pre-1960, was indeed in need of reform, but the "new math" movement made matters worse. Kline referred to formalism, rigor, and precise language as hindrances to student learning in mathematics. Kline also observed that mathematics had gotten away from its application to science and other areas, and had become a field of study itself. A growing number of critics questioned the value of "new math" programs, and noted that standardized test scores in mathematics were not impressive.

The "new math" was blamed for the decline of mathematics test scores. Reaction to this situation took the form of the "back-to-basics" movement. The problem, as popularly perceived, was that students were not spending enough time learning (memorizing) basic school mathematics. This "back-to-basics" movement was not any more well defined than was the "new math." Its emphasis was on a return to computational skills as taught prior to the "new math." During the later 1970s, textbook publishers changed their emphasis. An overall failure of students to perform computational skills at a level that the public felt to be acceptable attracted a great deal of attention in the press. Changes occurred in the books that were published in the late 1970s and early 1980s.

The popular perception within mathematics education was that the "back-to-basics" movement emphasized memorizing algorithms, and an almost exclusive arithmetic focus in the elementary grades. Both the National Council of Teachers of Mathematics (NCTM) and the National Council of Supervisors of Mathematics (NCSM) published position statements on basic skills in the October 1977 issue of the *Arithmetic Teacher.* While agreeing that there was a genuine concern over declining mathematical achievement measures, the NCTM paper pointed out the danger in going back to basics, if the basics meant strictly arithmetic computation. The paper further expressed concern that the "back-to-basics" movement might eliminate teaching for understanding. For students to have the ability to compute but not know what computations to perform when solving a problem would be of little benefit. That students may use a calculator for computation further emphasized the need for understanding: they must know when to push what button.

Following the NCTM lead, the NCSM position paper identified ten basic skill areas.

1. Problem solving
2. Applying mathematics to everyday situations
3. Alertness to the reasonableness of results
4. Estimation and approximation
5. Appropriate computational skills
6. Geometry
7. Measurement (both metric and customary systems)
8. Reading, interpreting, and constructing tables, charts, and graphs
9. Using mathematics to predict
10. Computer literacy

These topics later became the focus for *An Agenda for Action* (NCTM 1980). The publication of these topical lists was intended to broaden the scope of what was considered basic mathematics.

## REFORM EFFORTS

Many of the reform efforts in mathematics education have close ties with parallel reform efforts in science. Information from the Annenberg/CPB Math and Science Project, *The Guide to Math and Science Reform* (1995) lists seven reform efforts that had at least some focus on mathematics at the elementary school level prior to the publication of NCTM Standards in 1989. Many of the projects encompassed both the elementary and secondary mathematics curricula. One of these cited projects was *EQUALS: A Mathematics Equity Program*. The project was begun in 1977 in California, and continued into the 1990s. A main goal of this project was to attract and retain women and minority students in mathematics. The program also focused on the use of manipulative materials, cooperative learning, and assessment techniques as well as promoting student awareness of potential career opportunities. The *University of Chicago School Mathematics Project (UCSMP)* began in 1983 and also still continues. Its goal was to create a K–12 mathematics curriculum that upgraded the experiences of all students. The elementary level goal was to make a gradual transition from concrete manipulatives to abstractions while integrating use of mathematics with other school curricular topics. In 1987 both Colorado and Vermont initiated projects. The Exxon K–3 Mathematics Specialist Program, national in scope, also started in 1987. The Exxon program funded teacher projects designed to meet needs in their own schools.

The NCTM Yearbook for 1989, *New Directions for Elementary School Mathematics* (Trafton 1989), featured chapters by prominent mathematics educators who were supportive of NCTM recommendations on curriculum and evaluation standards, also published that year. Diana Kroll's chapter in the yearbook illustrates the change in the psychological perspective of mathematics education from a behaviorist position to a constructivist view.

### NCTM Standards

An influential publication in the area of reform in mathematics was the *Curriculum and Evaluation Standards for School Mathematics* (NCTM 1989). This document established a framework of what should be included in the school mathematics curriculum for various grade levels. A clear message, contained in the standards, is that all students need to be actively engaged in activities that promote problem solving and communicating about mathematics. At the elementary school level (grades K–4) the following recommendations were presented:

1. *Problem solving* should be viewed not as a topic to be studied, but rather as an approach to situations.
2. *Communication*. Students should be given the opportunity to talk about the mathematics that they are doing, including the strategies and materials they are using.
3. *Reasoning*. Students should be involved in making conjectures and evaluating their own ideas.
4. *Connections*. First, mathematics should be used in the study of other curricular topics. Second, previously learned (covered) mathematical topics should be connected to new mathematical studies under investigation.
5. *Estimation*. Students should learn when estimating is appropriate and that estimation is a valid mathematical process.
6. *Number Sense and Numeration*. Students should understand what a number means, and how to use the base-ten number system.
7. *Concepts of Whole Number Operations*. Concepts of addition, subtraction, multiplication, and division are essential for development of problem-solving skills.
8. *Whole Number Computation*. This is the part of mathematics teaching that has been traditionally done well. Unfortunately, students have not demonstrated that they can apply this computational ability to solving problems.
9. *Geometry and Spatial Sense*. Identification and classification of shapes should be combined with the development of spatial awareness.
10. *Measurement*. Both the concepts of measurement and the skills of measuring should be developed.
11. *Statistics and Probability*. This involves the collection, display, and interpretation of data. Very little is done with probability at the elementary level.
12. *Fractions and Decimals*. The concept of fraction should be developed.
13. *Patterns and Relationships*. Students should have activities dealing with recognizing and extending patterns.

An updated version of the NCTM standards, *Principles and Standards for School Mathematics*, became available in 2000.

## Reforms Following Publication of NCTM Standards

Shortly after the NCTM Standards were published, *Reshaping School Mathematics: A Philosophy and Framework for Curriculum* was published by the National Research Council (1990). This book was designed to provide support for the reformation of mathematics curriculum in general and for the NCTM Standards in particular. While many teachers welcomed the standards, many wondered how they could alter what they were currently doing in order to do a "better" job of teaching mathematics. To help fill the need, NCTM began developing its Addenda Series in 1990. These books were created to provide activities that would be consistent with the standards and involve students in hands-on activities.

The books for elementary teachers are organized in two ways. First, there is a booklet of activities for each grade, K–6. The activities also are organized by subject matter type. These subject lessons, identical to grade specific activities, are in *Patterns*, *Making Sense of Data*, *Number Sense and Operations*, and *Geometry and Spatial Sense*.

It is noteworthy that since the publication of the NCTM Standards, the number of funded curriculum projects has risen. Of the forty-eight curriculum projects with emphasis on elementary mathematics begun in 1990 or later, and cited in the Annenberg Guide, fifteen had an exclusive emphasis on elementary grades (Annenberg/CPB 1995). Not all of the projects that are currently active are cited in the Annenberg Guide. Materials from the Gateways III Conference of October 1994, funded by the National Science Foundation (NSF), included three elementary projects in addition to the UCSMP, which was mentioned earlier. The *TIMS Elementary Mathematics Curriculum Project*, at the University of Illinois at Chicago, is developing a K–5 curriculum. The *Cooperative Mathematics Project* (CMP), at Oakland, California, is developing K–6 curriculum materials, but not a full curriculum. Some of their units have been published by Addison-Wesley, Alternative Publishing Group under the name *Number Power*. There also is the *Investigations in Number, Data, and Space* project (Cambridge, Massachusetts). This is a complete K–5 mathematics curriculum that is directed toward the teacher; no student textbooks are being produced (Gateways III Conference, 1994).

NSF began funding a series of Statewide Systemic Initiatives (SSI) in 1991. There are currently twenty-four states and Puerto Rico with SSI programs designed to promote excellence in mathematics and science education (NSF 1994). While each of the states has its own particular set of needs and its own special focus, all of the funded projects require the collaboration of educators at all levels, parents, business and industry, policy-makers, and the community at large. Continuing the view that mathematics reform is for all students, Webb and Romberg edited *Reforming Mathematics Education in America's Cities* (1994), a study of collaboration as a process through *The Urban Mathematics Collaborative Project*. This continuing project began in 1984 with support from the Ford Foundation and has had an impact on eleven urban centers across the nation.

## Cognitively Guided Instruction (CGI)

While most of the areas covered so far were related to the reform of mathematics curriculum, CGI is a project dealing with instruction that has had an effect on curriculum. Begun at the University of Wisconsin in 1986 by Thomas Carpenter, Elizabeth Fennema, and Penelope Peterson (Knapp and Peterson 1995), the CGI model grew out of Carpenter's earlier work observing how five- and six-year-old students solved addition and subtraction problems. CGI has never been a set of procedures to be implemented, but rather a way of thinking about how children think about mathematics and how they learn mathematics. The initial teacher training involved forty first-grade teachers. Rather than telling students how to do a problem and model correct algorithms, the teacher's job is to pose problems and sometimes provide a structure so that students can develop their own understanding by working out the problems and discussing results with their peers. Often students use concrete manipulatives to model situations and talk about the mathematics they are using. These students are constructing their own mathematical concepts.

## Published Textbooks

The scope and sequences of many textbook publishers reveal that the expansion of topics as recommended by the NCTM Standards has been introduced in their new textbooks. In fact, the newest versions of elementary mathematics textbook series for elementary schools feature a close alignment with the recommendations of the NCTM Standards. It may be several years, however, before the newest textbooks and materials available today find their way into most of the nation's classrooms. Even then it will remain to be seen whether or not the teachers make use of the materials provided by publishers. Arithmetic continues to be the major component of the elementary mathematics curriculum, but teachers have never had so much resource material available as there is today to assist in covering other mathematics topics in addition to arithmetic. While no one can be sure what individual teachers cover in their own classes, the textbook, if used, provides a good indicator as to what will be covered. Certainly if topics are missing from a textbook they are very unlikely to be included in most elementary teachers' lesson plans. The inclusion of more word problems is an important positive feature of many new textbooks.

*See also* Arithmetic; Basic Skills; Curriculum, Overview; *Revolution in School Mathematics;* School Mathematics Study Group (SMSG); University of Chicago School Mathematics Project (UCSMP)

## SELECTED REFERENCES

Annenberg/CPB Math and Science Project. *The Guide to Math and Science Reform.* No. 3. Spring 1995.

Flanders, James R. "How Much of the Content in Mathematics Textbooks Is New?" *Arithmetic Teacher* 35(1)(1987):18–23.

Gateways III Conference. "Moving Forward." Washington, DC: National Science Foundation, October 6–9, 1994.

Kline, Morris. *Why Johnny Can't Add: The Failure of the New Math.* New York: St. Martin's, 1973.

Knapp, Nancy F., and Penelope Peterson. "Teachers' Interpretation of 'CGI' after Four Years: Meanings and Practices." *Journal for Research in Mathematics Education* 26(1)(1995):40–65.

Kroll, Diana L. "Connections between Psychological Learning Theories and the Elementary Mathematics Curriculum." In *New Directions for Elementary School Mathematics* (1989 Yearbook; pp. 199–211). Paul Trafton, ed. Reston, VA: National Council of Teachers of Mathematics, 1989.

National Council of Supervisors of Mathematics. "Position Paper on Basic Skills." *Arithmetic Teacher* 25(Oct. 1977):21–22.

National Council of Teachers of Mathematics. *An Agenda for Action: Recommendations for School Mathematics of the 1980s.* Reston, VA: The Council, 1980.

———. *Assessment Standards for School Mathematics.* Reston, VA: The Council, 1995.

———. *Curriculum and Evaluation Standards for School Mathematics.* Reston, VA: The Council, 1989.

———. "Position Paper on Basic Skills." *Arithmetic Teacher* 25(Oct. 1977):18.

———. *Principles and Standards for School Mathematics.* Reston, VA: The Council, 2000.

———. *Professional Standards for Teaching Mathematics.* Reston, VA: The Council, 1991.

National Research Council. *Reshaping School Mathematics: A Philosophy and Framework for Curriculum.* Washington, DC: National Academy Press, 1990.

National Science Foundation. *Foundation for the Future: The Systemic Cornerstone.* Washington, DC: The Foundation, 1994.

Trafton, Paul, ed. *New Directions for Elementary School Mathematics.* 1989 Yearbook. Reston, VA: National Council of Teachers of Mathematics, 1989.

Webb, Norman L., and Thomas Romberg, eds. *Reforming Mathematics Education in America's Cities: The Urban Mathematics Collaborative Project.* New York: Teachers College Press, 1994.

RICHARD A. AUSTIN

# CURRICULUM TRENDS, SECONDARY LEVEL

Evolving mathematical studies in secondary school. Major phases that have occurred in the United States during the latter part of the twentieth century are: the "new math" of the 1960s, the "back-to-basics" movement of the late 1970s and early 1980s, and the "standards reform" of the 1990s. During World War II, many recruits were found to lack the skills necessary for training in the technical areas that had developed such as radar, sonar, radio navigation, and aircraft piloting. National committees were appointed to begin discussions of curriculum reform, especially in science and mathematics. The space race that began with the 1957 launching of *Sputnik* by the Soviet Union generated the political will to allocate funds to develop a curriculum that would prepare future scientists and engineers. The National Science Foundation (NSF) was formed to fund curriculum projects and to train teachers to use the new materials.

An important characteristic of the curriculum projects of the 1960s was the leadership of research mathematicians. School textbook development was no longer left to mathematics teachers and those who prepared them in teachers' colleges. Concurrently, important developments took place in educational psychology. Behaviorism, which had been the dominant theory for mathematics learning, was challenged by "discovery" learning theory (Bruner 1966). These factors, added to the urgency to accelerate mathematical development, produced the beliefs that children could learn much more mathematical content, and learn it more quickly than previously thought. Most curriculum projects had the goal of having students complete at least a year or two of calculus, probability, and abstract algebra by the end of high school. Nearly all of the projects were aimed, at least implicitly, at college-bound students (Howson, Keitel, and Kilpatrick 1981).

More than fifty curriculum projects were funded by the NSF, most of them aimed at either the elementary grades or the secondary school. Early in the effort, the secondary school curriculum was the higher priority, since the quick production of scientists and engineers was important. There was a great deal of variation in the implementation of the new secondary school curriculum. Many schools used materials developed by the projects themselves, while the majority eventually used commercial textbooks adapted from these materials that may have lost some of the "spirit" of the curriculum. The content was difficult for many teachers. In many cases, only the new language and rules were emphasized rather than underlying concepts and structures. In spite of all of these problems, the general level of mathematics learning improved, helping the United States to meet the technical challenges of the late twentieth century (National Committee on Mathematical Education 1975).

During the 1970s, as larger percentages of students began applying to colleges and universities, test scores revealed poor mathematics achievement, even on relatively simple computational tests. Concurrently, a backlash occurred against the "new math," partly because many average and below average students found its approach difficult (Kline 1973). These factors, among others, produced a retrenchment in the mathematics curriculum, often referred to as "back-to-basics." Arithmetic was stressed in grades 1 through 8. Algebra focused on rules and procedures; geometry stressed the memorization of properties and standard proofs. Many states implemented "minimal competency" tests to assure that students learned basic arithmetic skills before graduating. Protests from the mathematics education community and recommendations for a curriculum based on problem solving (National Council of Teachers of Mathematics [NCTM] 1980) were largely ignored. The force of political conservatism, with its support of "back-to-basics" curricula, was too strong to overcome the warnings that mathematical reasoning and problem solving were losing ground.

In 1986, the Second International Mathematics Assessment was completed, comparing United States achievement with sixteen other countries. In grades 8 and 12 across all content topics, U.S. mathematics achievement was extremely poor, ranking near the bottom in comparison with other developed countries. Some of the curriculum taught in other countries was not covered in U.S. schools and the material that was taught was often at a rote level. Students had poor problem-solving and reasoning skills (McKnight et al., 1987). A report entitled *A Nation at Risk* (National Commission on Excellence in Education 1983) had suggested that current achievement levels placed the nation in jeopardy economically and militarily. The report of the international tests confirmed this fear. Other reports ushered in the reforms of the 1990s. At the center of these reforms was the idea that state and national standards needed to be set for curriculum, teaching, and assessment. The standards movement was led by the NCTM, which published the first standards document for any subject area, the *Curriculum and Evaluation Standards for School Mathematics* (1989), the model for future documents in other subject areas. At about the same time, the American Association for the Advancement of Science (AAAS) published *Science for All Americans* (1989), which outlined broad goals for literacy in science, mathematics, social sciences, and technology. A later document, *Benchmarks for Science Literacy,* provided specific outcomes expected of students, including objectives for mathematics (AAAS 1993).

The statement that best summarizes the philosophy and goals of the NCTM Standards is the following: "We are convinced that if students are exposed to the kinds of experiences outlined in the Standards, they will gain mathematical power. This term denotes an individual's abilities to explore, conjecture, and reason logically, as well as the ability to use a variety of mathematical methods to effectively solve nonroutine problems. This notion is based on the recognition that mathematics is more than a collection of concepts and skills to be mastered; it includes methods of investigating and reasoning, means of communication, and notions of contexts. In addition, for each individual, mathematical power involves the development of personal self-confidence" (NCTM 1989, 5). Unlike the previous two reform efforts, the current one had some important characteristics that sustained its life and its potential long-term impact: It was developed by a professional organization of teachers and supported by mathematicians, school administrators, parent organizations, and government agencies; it was based on findings from research on how children develop cognitively and learn mathematics; it took a long-term view of curriculum development rather than looking for a political "quick fix"; and it integrated curriculum development with learning, teaching, and assessment. An update of the NCTM Standards, *Principles and Standards for School Mathematics,* was introduced in 2000.

A number of curriculum projects (e.g., Connected Mathematics Project, Interactive Mathematics Program, and Teaching Integrated Mathematics and Science Program) were developed in the late 1990s that reflected the recommendations of the NCTM standards. Commercial textbooks began to respond by including applications of mathematics, attention to the use of technology, and a focus on meaning and understanding, since many states and school districts adopted curriculum guides that borrowed heavily from the directions suggested by the standards. Technology is an important influence on the current secondary curriculum. Most schools have relatively good access to calculators that not only have sophisticated graphing capabilities but also can do symbol manipulation, statistical calculations, and matrix operations. In addition, computers with spreadsheets, geometry tools, and communications networks are more readily available. These changes have challenged curriculum developers to rethink the relative emphasis on skill acquisition, concept learning, and problem-solving content. For example, most new curricula include an integration of topics, less attention to rote practice on outdated computational procedures, increased use of technology, and more focus on applications as the motivation and purpose for learning and understanding mathematics.

Teacher education and methods of assessing student learning have gradually become influenced by new perspectives on curriculum and testing. Many colleges of education have formed partnerships with schools to develop teacher preparation programs that begin earlier in the college years, integrate theory and classroom practice, and involve master teachers in program design and instruction. National, state, and local assessment programs include tests with open-ended items, the use of calculators, and greater attention to problem solving and other higher-order thinking abilities.

The capability of teachers is a critical issue. New curricula, which require significant shifts from the traditional school mathematics, can only succeed with similar changes in teaching. Teachers' roles include those of coaches, mentors, and guides to resources rather than only being purveyors of information. Lectures are sometimes less effective than group projects, investigations, and individualized practice. Much of the new curriculum demands new views of learning and teaching. Many students learn best in small, cooperative groups. Learning takes place by connecting new knowledge to previous experiences and information. Mastery takes place gradually through a process of integrating and applying concepts and skills. These changed ideas of teaching and learning have begun in the elementary and middle grades; further effort is needed at the secondary levels.

*See also* Algebra Curriculum, K–12; Begle; Cambridge Conference on School Mathematics (1963); Connected Mathematics Project (CMP); Curriculum, Overview; Geometry, Instruction; Interactive Mathematics Program (IMP); Teaching Integrated Mathematics and Science Program (TIMS); University of Chicago School Mathematics Project (UCSMP)

## SELECTED REFERENCES

American Association for the Advancement of Science. *Benchmarks for Science Literacy.* New York: Oxford University Press, 1993.

———. *Science for All Americans.* Washington, DC: The Association, 1989.

Bruner, Jerome S. *Toward a Theory of Instruction.* Cambridge, MA: Belknap Press of Harvard University Press, 1966.

Howson, A. Geoffrey, Christine Keitel, and Jeremy Kilpatrick. *Curriculum Developments in Mathematics.* New York: Cambridge University Press, 1981.

Kline, Morris. *Why Johnny Can't Add: The Failure of the New Math.* New York: St. Martin's, 1973.

McKnight, Curtis C., F. Joe Crosswhite, John A. Dossey, Edward Kifer, Jane O. Swafford, Kenneth J. Travers, and Thomas J. Cooney. *The Underachieving Curriculum: Assessing U.S. School Mathematics from an International Perspective.* Champaign, IL: Stipes, 1987.

National Commission on Excellence in Education. *A Nation at Risk: The Imperative for Educational Reform.* Washington, DC: Government Printing Office, 1983.

National Committee on Mathematical Education. *Overview and Analysis of School Mathematics Grades K–12.* Washington, DC: Conference Board of the Mathematical Sciences, 1975.

National Council of Teachers of Mathematics. *An Agenda for Action: Recommendations for School Mathematics of the 1980s.* Reston, VA: The Council, 1980.

———. *Curriculum and Evaluation Standards for School Mathematics.* Reston, VA: The Council, 1989.

———. *Principles and Standards for School Mathematics.* Reston, VA: The Council, 2000.

GERALD KULM

# CURVE FITTING

A method of approximating a curve that will contain a given set of points. Suppose we are given a set of points and do not have an analytic expression of a function, $f(x)$, whose graph contains these points. We may need the value of $f(x)$ at an arbitrary point other than the points of the given set. We can estimate the value of $f$ for an arbitrary $x$ by drawing a curve through the given points. There are various possibilities for the curve (the graph of $f$): it may be continuous or not, smooth (i.e., $f'(x)$ is continuous) or not. Assumptions can be imposed on the required $f$. For instance, we may want the $m^{th}$ derivative of $f$ to be continuous on $[a,b]$.

To generalize this problem, let us assume that we seek to fit a smooth curve (represented by a function, $f(x)$) to a set of points $(x_1, y_1), (x_2, y_2), \ldots, (x_N, y_N)$ and that we want the $m^{th}$ derivative of $f$ to be continuous on $[a,b]$. One way of constructing $f$ that ensures the continuity of all derivatives is to think of it as a polynomial $P(x)$ that passes through a given set of $(N + 1)$ points. A polynomial of degree $N$ can have $N - 1$ relative maxima and minima.

An alternate approach is to use a lower-degree polynomial in each subinterval. In the old days, one interpolated trigonometric, logarithmic, and exponential functions using linear interpolation in each subinterval. Thus, a function is replaced by a piecewise linear function whose values match the values of $f(x)$ at $x_0, x_i, \ldots, x_N$ and whose graph is a broken line. The line in each subinterval is not smooth at the interface. We can overcome this, however, by using a higher-degree polynomial in each subinterval and requiring it to be smooth at each interface. The piecewise polynomial is called a *spline*. This name is derived from a draftsman's tool (a flexible strip) that passes through each point smoothly.

The simplest spline function is a polygonal path consisting of line segments that passes through $(x_0, y_0), (x_1, y_1), \ldots, (x_N, y_N)$, and the resulting linear spline function is given by

$$S(x) = \begin{cases} y_0 + \dfrac{y_1 - y_0}{x_1 - x_0}(x - x_0) & x \in [x_0, x_1] \\[2mm] y_1 + \dfrac{y_2 - y_1}{x_2 - x_1}(x - x_1) & x \in [x_1, x_2] \\ \quad \vdots \\ y_N + \dfrac{y_N - y_{N-1}}{x_N - x_{N-1}}(x - x_{N-1}) & x \in [x_{N-1}, x_N] \end{cases}$$

The points $x_0, x_1, \ldots, x_N$ at which $S$ changes its character are known as *knots*. By the nature of its derivatives, $S$ is a continuous function on $[x_0, x_N]$; however, the derivative of $S$ is not continuous at the knots. A spline of degree one has "kinks" at each knot. To remove the kinks, the first derivative of $S$ must be continuous at each knot. Thus, we need $S$ to be a piecewise quadratic polynomial or a higher-degree polynomial. Quadratic splines have a smoother appearance but could have instantaneous changes in curvature at each knot. Therefore, cubic splines are most widely used in practice. More details can be found in Mathews (1992) and Patel (1994).

*See also* Data Analysis; Data Representation; Least Squares

## SELECTED REFERENCES

Mathews, John H. *Numerical Methods for Mathematics, Science, and Engineering.* 2nd ed. Englewood Cliffs, NJ: Prentice-Hall, 1992.

Patel, Vithal A. *Numerical Analysis.* Philadelphia, PA: Saunders College Publishing, 1994.

VITHAL A. PATEL

# D

## D'ALEMBERT, JEAN LE ROND (1717–1783)

One of the leading French mathematicians of the eighteenth century, best known for his contributions to kinetics and the theory of partial differential equations. D'Alembert was the son of an aristocratic sister of a cardinal and was abandoned as an infant near the church of St. Jean Le Rond in Paris. He was elected to the French Academy in 1741 and became its permanent secretary in 1754.

D'Alembert tackled many fundamental problems in mathematics; his *Traité de Dynamique* (1743) contained the principle of kinetics, now called "D'Alembert's principle," and in a treatise published in 1747 he determined the general equation of motion of a vibrating string. Between 1751 and 1772, he compiled most of the scientific entries in Denis Diderot's massive (twenty-eight volume) *Encyclopédie,* including an entry depicting his concept of a limit. In addition to his mathematical work, d'Alembert also was involved in politics (he was a friend of François Voltaire) and was one of the many intellectuals who helped pave the way for the French Revolution.

*See also* History of Mathematics, Overview

### SELECTED REFERENCES

Boyer, Carl B. *A History of Mathematics.* 2nd ed. Revised by Uta C. Merzbach. New York: Wiley, 1989.

Eves, Howard W. *An Introduction to the History of Mathematics.* 5th ed. Philadelphia, PA: Saunders College Publishing, 1983.

Newman, James R. *The World of Mathematics.* 4 vols. London, England: Allen and Unwin, 1960.

Struik, Dirk J. *A Concise History of Mathematics.* New York: Dover, 1948.

ROD HAGGARTY

## DATA ANALYSIS

A part of statistics, the science of gathering, displaying, and analyzing data (or information). Developments in statistics in recent years have revolutionized data analysis. Specifically, the invention of the stem-and-leaf plot and the box plot to help visualize distributions, and the use of computer simulations are significant developments. More recently, the advent of the graphing calculator has permitted high school students to examine the relationship between variables and to compare sets of data. Just a few years ago, these investigations would have required many tedious computations or the use of a large computer. These developments have permitted statistical problem solving to become a useful tool in the teaching of mathematical skills.

The use of median-based statistics in exploratory data analysis has greatly reduced the use of formulas and tedious computations. Median-based statistics is data analysis using the median instead of the mean to measure "average," and interquartile range (IQR) to measure variation. IQR is the distance between the lower quartile (25th percentile) and the upper quartile (75th percentile). The median fit line, which can be constructed by a seventh-grade student, has been used to fit bivariate data (i.e., data from two variables such as height and weight, or temperature and number of crimes). In the past, the "least squares best fit" line was used for this purpose and required numerous computations beyond the scope of a middle school student.

The median-based tools help develop conceptual understanding and facilitate the solution of important problems. The utilization of simulations in problem solving provides a means to develop the understanding of probability and can aid in the application of probability distributions. The box plot has proven to be a simple yet potent tool in comparing sets of data, analyzing distributions, and developing the underlying concepts in inferential statistics. Figure 1 is an illustration of the box plot. The line segment (or whisker) at the left represents the 1st quartile of the data. The two rectangles which form the box represent the 2nd and 3rd quartiles. The right-hand whisker represents the 4th quartile. The width of the box is the IQR. The vertical line segment in the box is located at the median. Finally, the graphing calculator has taken the drudgery out of data analysis by providing immediate computations and useful visual displays.

The following examples illustrate the use of data analysis in the teaching of mathematics.

*Example 1* Twelfth-grade students at Alan C. Pope High School (Marietta, GA) gathered the following data relating hours spent watching TV and working, to grade point average earned:

| Females | | | Males | | |
|---|---|---|---|---|---|
| GPA | TV | Work | GPA | TV | Work |
| 3.2 | 6 | 13 | 1.9 | 16 | 0 |
| 2.5 | 4 | 12 | 2.6 | 2 | 15 |
| 3.6 | 12 | 5 | 3.3 | 3 | 21 |
| • | • | • | • | • | • |
| • | • | • | • | • | • |

GPA = grade point average (on a four point scale) for the first seven semesters of high school

TV = the average number of hours of television watched per week

Work = the average number of hours worked for pay per week

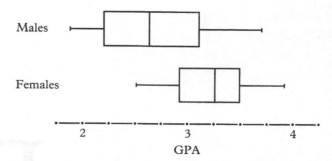

**Figure 1** *Box plot comparison: Male vs. female GPAs.*

Using graphing calculators, the students quickly determined the means (males = 2.67, females = 3.19), medians (males = 2.65, females = 3.25), standard deviations (males = .54, females = .415) and IQRs (males = .9, females = .6) for the GPA data. These statistics seemed to indicate that, on average, females had higher grades and were more consistent in their grades than the males. The use of box plots constructed by the class (some graphing calculators actually produce box plots on their screens) convinced the class that the females had better grades and had less variation in GPA than the males (see Figure 1).

The class was asked to determine why the female students had better grades than the males. It should be noted that the class originally gathered eight pieces of data on each person in the data set. Only the work and TV data are shown in the table above, as these were the data that the students decided intuitively were of significance in explaining the difference in GPAs. The males watched more hours of TV (male mean = 11.5, female mean = 6.61) and worked more hours (male mean = 11.7, female mean = 9.78). The class decided that the female students had more time to devote to their studies and thus received higher grades.

Example 1 illustrates the use of data analysis to reinforce problem-solving skills and simple arithmetic skills such as comparing numbers. Since the data were drawn from the students, they were interested in the analysis and were willing to state opinions that were to be justified by the mathematical analysis.

*Example 2* Lawyers for a convicted felon appeal their client's conviction by stating that the accused did not receive a trial by his peers. The client is left-handed, and, while 10% of the population is left-handed, all members of the jury were right-handed. The question that must be asked is how likely is it to have zero left-handed persons on a jury of twelve if

10% of the potential jurors are left-handed. The class conducted a simulation to assess the fairness of this trial. The students used tables of random digits (0 through 9). A student randomly selected twelve digits and counted the number of ones among the twelve digits.

The twelve digits represent twelve randomly selected jurors. Each time a one is obtained it represents a left-handed juror. The digits two through nine and zero represent right-handed jurors. Since the probability of randomly selecting a 1 from the digit table is 1 out of 10, this event can represent selecting a left-handed juror, the probability of which is also 10%.

The members of the class conducted one thousand simulations. The results of the simulations are as follows:

| # of left-handed jurors | 0 | 1 | 2 | 3 | 4 | 5 |
|---|---|---|---|---|---|---|
| frequency | 287 | 363 | 246 | 69 | 32 | 3 |

The simulation predicted that an unbiased selection process would result in a jury of twelve right-handed members 28.7% of the time. (Note that the binomial probability formula indicates that the probability of having zero left-handed jury members is 0.2824, which demonstrates that the results of a set of simulations can approximate the theoretical answer.) Since the chance of obtaining the jury with no left-handed members in a situation with no prejudice involved is 28.7%, it is reasonable to say that no prejudice was evidenced in the jury selection. Generally in a statistical test of this type, a probability of less than 5% would have to be found in order to conclude that prejudicial selection was present. The class then discussed how an appeals court judge would use this information to rule on the request for a new trial. Example 2 demonstrates the use of data analysis to reinforce the use of percentages. Because the example is relevant to students, they are motivated to determine whether the jury selection process was or was not fair.

*Example 3* A political pollster wishes to take a survey with a margin of error of plus or minus 3%. The pollster wants to report a 95% confidence interval on the proportion of the population that will answer "yes" to a question. That is, the interval, which will be 6% in width, will have a 95% likelihood of containing the actual percentage of the population who would answer in the affirmative. The pollster now must determine what size sample is needed.

The class has seen that the margin of error is a function of sample size. For a 95% confidence inter-

val the margin of error is approximately equal to the reciprocal of the square root of the sample size. Actually this is a reasonable approximation for what is called "the worst case scenario," which occurs when the proportion of people voting yes is one half. The class can now see a meaningful use for solving a square root equation. The students manipulate the equation so that the solution is equal to the square of the reciprocal of the margin of error $1/(0.03)^2$, which turns out to be 1,111 people.

The most effective use of data analysis in the teaching of mathematics is in the investigation of the equation of a line and the meaning of slope. Using actual data allows the teacher to provide a concrete basis for the development of mathematical concepts, such as slope, intercepts, and the equation of a line. The class is provided with the following data, which are the advertised prices and ages for a particular brand of automobile.

| age in years | 4 | 5 | 5 | 6 |
|---|---|---|---|---|
| price in dollars | 6500 | 7000 | 6000 | 5500 |
| age in years | 6 | 6 | 7 | 8 |
| price in dollars | 5000 | 4800 | 7250 | 3000 |
| age in years | 9 | 9 | 11 | 11 |
| price in dollars | 2200 | 1500 | 3000 | 1800 |
| age in years | 12 | 12 | 13 | 15 |
| price in dollars | 1000 | 900 | 1500 | 750 |
| age in years | 16 | 18 | | |
| price in dollars | 1000 | 750 | | |

Using age as the *X* variable and price as the *Y* variable, each ordered pair is plotted on graph paper. Each student draws a line that he or she feels best represents the eighteen points. This is termed an "eyeball fit line." The students find approximately how much value the car loses each year—this is the slope of the line. The students have a concrete example of the meaning of slope. There is even a name for the slope in this example. It is called *straight-line depreciation.*

The *Y*-intercept is found and the students realize this is the value of the car when new. The *X*-intercept is the age at which the car has no value. The class then discusses classic cars. In classes where the students have already been exposed to quadratic functions, a parabola can be sketched in to represent the appreciation as the car becomes a classic. A graphing calculator that provides a quadratic model can be used to find the equation of a quadratic function that will best fit the data. Points above the line represent cars whose value is greater than the value predicted by the car's age. This may be due to factors such as low mileage, extra options, mint condition, and so forth. When students investigate linear inequalities in two variables, this example can be used to verify that when the inequality involves "greater than," we shade above the line. If the line is extended below the $X$-axis, the $Y$ value is negative. This means that the car is so old that you have to pay someone to tow it away. This is an excellent concrete example of the meaning of a negatively valued function.

The use of data analysis equips the mathematics teacher with a valuable tool kit which allows students to tackle real-world problems while simultaneously developing critical thinking skills. Data analysis provides a rich background for the teaching of arithmetic, algebra, and calculus. Abstract concepts can be understood by students through the use of interesting, concrete problems.

*See also* Curve Fitting; Data Representation; Statistics, Overview

SELECTED REFERENCES

Burrill, Gail, et al. *Data Analysis and Statistics across the Curriculum.* Reston, VA: National Council of Teachers of Mathematics, 1992.

Gnanadesikan, Mrudulla, Richard Scheaffer, and James Swift. *The Art and Techniques of Simulation.* Palo Alto, CA: Seymour, 1987.

Landwehr, James, James Swift, and Ann E. Watkins. *Exploring Surveys and Information from Samples.* Palo Alto, CA: Seymour, 1987.

Landwehr, James, and Ann E. Watkins. *Exploring Data.* Palo Alto, CA: Seymour, 1986.

Moore, David S., and George P. McCabe. *Introduction to the Practice of Statistics.* 2nd ed. New York: Freeman, 1993.

MURRAY H. SIEGEL

# DATA REPRESENTATION

The recording and representation of quantitative data pictorially and symbolically, which has occurred since the dawn of civilization. The power of a visual display is that it provides a compact and concise way of representing data. This entry will focus on two aspects of representing quantitative data in tables, graphs, and plots. First, characteristics and features of some common displays are presented. Second, factors related to readers' ability to comprehend mathematical relationships expressed in visual displays are discussed.

## CHARACTERISTICS OF VISUAL DISPLAYS

Graphs and plots of quantitative data are visual representations of mathematical relationships and they make the comparison of data easier than other types of visual displays. For example, tables may contain estimated, exact, or rounded quantities, but trends and mathematical relationships may only be revealed when the data are displayed graphically.

The work of René Descartes, a seventeenth-century philosopher and mathematician, led to the development of the modern graph. The English political economist William Playfair (1759–1823) is credited with being one of the first to use graphic statistics to represent data. Traditional graphs used widely in newspapers, magazines, and reports employ the use of pictures, bars, lines, and circles. Some "new" statistical plotting techniques (i.e., stem-and-leaf, and box plots) created by John Tukey (1977) to facilitate data representation and analysis, have recently found their way into the mathematics curriculum (Curcio 1989; Landwehr and Watkins 1986; Russell and Corwin 1989). The purpose of the visual display, the nature of the data (e.g., discrete vs. continuous), and the nature of the audience to whom a message is to be conveyed must be considered when determining the type of visual display to select for recording and communicating information. Furthermore, the structure of tables, graphs, and plots must be understood.

A *table* is an abstract but compact display that contains organized lists of primary-source numerical data. In this type of display, understanding mathematical trends and relationships embedded in the data requires more than a simple surface reading of the quantities listed (MacDonald-Ross 1977).

The *picture graph* (also called picture chart, pictogram, pictograph, or pictorial graph) is used to display discrete (i.e., noncontinuous) or categorical data using uniform ideographs or symbols to depict quantities of objects or people. A set of labeled rectangular axes defines the field for the display. In simplest form the ideograph is in a one-to-one correspondence with the item it represents. In more complex displays, a many-to-one correspondence between the items represented and the ideograph requires the use of a legend. The ideographs must be uniform in size and shape to avoid misleading the reader (Huff 1954).

The *bar graph* (also called a bar chart) is used to display discrete or categorical data. Within rectangular axes that must be labeled, the heights of rectangular bars of uniform width are proportional to the

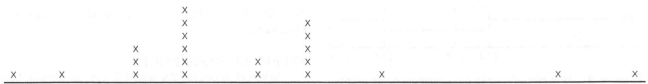

**Figure 1**  *Line plot of height (in centimeters) of 21 ninth-grade girls in one class.*

quantities they represent. The bar graph may be set up either horizontally or vertically. Discrete stratified data (i.e., data collected from particular groups) may be compared in double or multiple bar graphs.

The *histogram* contains grouped data arranged like a vertical bar graph. It satisfies three conditions: only one set of data may be represented, the class intervals must be equal, and both axes must contain a numerical scale (MacDonald-Ross 1977, 367). A frequency polygon may be constructed directly from a histogram.

The comparison of continuous data (e.g., change over a period of time or a linear functional relationship) is depicted in a *line graph* (also called a broken-line graph). The perpendicular axes that intersect at a common point, usually zero, are labeled and they define the field for the display. On each axis, the units of division must be consistent. After the points are graphed, they are connected by line segments. Two or more sets of continuous data may be graphed on the same set of axes to create a double or multiple line graph.

When data are to be compared to a whole or to different parts of a whole, a *circle graph* (also called an area graph, pie chart, pie diagram, or pie graph) is used. The field of display of a circle graph is defined by the circumference of a circle. Sectors of the circle created by line segments emanating from the center correspond proportionally to fractional parts of the unit being analyzed.

Having the appearance of a primitive bar graph, a *line plot* is created by representing numerical data in the form of $x$'s on a number line. This visual display is easy to construct. The shape and the spread of the data are revealed while having access to each individual datum. The median and the mode are easily identified from this display. The number of items plotted usually does not exceed twenty-five (see Figure 1).

A *stem-and-leaf plot* provides a display of data that is created by separating the digits in the data based on their place value. In a regular stem-and-leaf plot, two columns identified by place value are established to list each digit. The digit with the higher

place value is the stem, and the digit with the lower place value is the leaf. This type of plot usually contains more than twenty-five data entries. Similar to the line plot, it is easy to construct, and the shape and the spread of the data are revealed while having access to each individual datum. The median and the mode can be easily identified (see Figure 2). Back-to-back stem-and-leaf plots can be constructed to compare two sets of related data. For further descriptions and examples, see Tukey (1977), Curcio (1989), and Landwehr and Watkins (1986).

A *box plot* (also referred to as a box-and-whiskers plot) uses five summary points (i.e., the lower extreme, the lower quartile, the median, the upper quartile, and the upper extreme). The plot has one axis in the form of a number line, arranged either horizontally or vertically. A rectangle is used to represent the middle 50% of the data. The median is represented by a line segment that partitions the rectangle to show the spread of the upper and lower quartiles, proportional to the size of the rectangle. The extremes are represented by points, and they are connected to each end of the rectangle by line segments. Outliers are plotted as points unconnected to the rectangle and are identified in terms of the interquartile range. This plot is used when analyzing more than 100 pieces of data. Unlike the line plot and the stem-and-leaf plot, individual data are not identifiable (see Figure 3).

| 13 | 0 0 1 2 |
| 14 | 2 4 4 8 |
| 15 | 0 3 3 3 5 5 5 5 5 5 6 8 8 |
| 16 | 0 0 0 0 0 3 3 5 5 8 8 |
| 17 | 0 0 0 2 2 3 3 3 3 |

Key: 13 | 0 means 130 centimeters

**Figure 2**  *Stem-and-leaf plot of height (in centimeters) of 41 ninth-grade girls in two classes.*

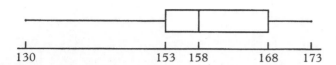

130               153  158      168    173

**Figure 3** *Box plot of height (in centimeters) of ninth-grade girls.*

A more comprehensive discussion of features and characteristics of tables and graphs can be found in Kosslyn (1989), MacDonald-Ross (1977), and Tufte (1983). A more in-depth explanation of plots can be found in Tukey (1977).

## COMPREHENSION OF QUANTITATIVE VISUAL DISPLAYS

Comprehending mathematical relationships expressed in visual displays has been studied from different perspectives: ability to construct various displays, level of difficulty of different displays, and cognitive factors related to comprehension. A review of the literature can be found in Curcio (1981).

Although some educators have recommended that students have experience constructing visual displays, the research is inconclusive as to whether ability to construct displays is a prerequisite for comprehension.

The more complex and technical a display, the more abstract it is for readers. When compared with a table, picture graph, or bar graph, the most difficult display to comprehend has been the line graph.

Comprehension has been defined as reading the data (i.e., literal), reading between the data (i.e., interpretation, interpolation), and reading beyond the data (i.e., prediction, extrapolation). In a study examining the effects of cognitive factors on graph comprehension, it was found that prior knowledge of both the topic of a graph and the graphical form may be predictors of comprehension above and beyond the general predictors of school success (i.e., reading and mathematics achievement). Knowledge of the mathematical content embedded in the graph contributes to understanding the mathematical relationships expressed in the graph (Curcio 1981, 1987).

Current recommendations for developing concepts related to data representation support a constructivist philosophy. Formulating questions, collecting and analyzing data, and inventing and designing visual displays (DiSessa et al. 1991; Whitin et al. 1990) can enable students in designing and directing their own learning.

*See also* Curve Fitting; Data Analysis; Statistics, Overview

### SELECTED REFERENCES

Arkin, Herbert, and Raymond R. Colton. *Graphs: How to Make and Use Them.* Rev. ed. New York: Harper, 1940.

Curcio, Frances R. "Comprehension of Mathematical Relationships Expressed in Graphs." *Journal for Research in Mathematics Education* 18(Nov. 1987):382–393.

———. *Developing Graph Comprehension.* Reston, VA: National Council of Teachers of Mathematics, 1989.

———. *The Effect of Prior Knowledge, Reading and Mathematics Achievement, and Sex on Comprehending Mathematical Relationships Expressed in Graphs.* Final Report. ERIC No. ED210185. Brooklyn, NY: St. Francis College, 1981.

DiSessa, Andrea, David Hammer, and Bruce Sherin. "Inventing Graphing: Meta-representational Expertise in Children." *Journal of Mathematical Behavior* 10(Aug. 1991):117–160.

Huff, Darrell. *How to Lie with Statistics.* New York: Norton, 1954.

Kosslyn, Stephen M. "Understanding Charts and Graphs." *Applied Cognitive Psychology* 3(1989):185–226.

Landwehr, James M., and Ann E. Watkins. *Exploring Data.* Palo Alto, CA: Seymour, 1986.

MacDonald-Ross, Michael. "How Numbers Are Shown." *AV Communication Review* 25(Winter 1977): 359–409.

Nuffield Foundation. *Pictorial Representation.* New York: Wiley, 1967.

Russell, Susan Jo, and Rebecca Corwin. *Statistics: The Shape of the Data.* Palo Alto, CA: Seymour, 1989.

Tufte, Edward R. *The Visual Display of Quantitative Information.* Cheshire, CT: Graphics Press, 1983.

Tukey, John W. *Exploratory Data Analysis.* Reading, MA: Addison-Wesley, 1977.

Wainer, Howard. "Understanding Graphs and Tables." *Educational Researcher* 21(Jan.–Feb. 1992):14–23.

Whitin, David, Heidi Mills, and Timothy O'Keefe. *Living and Learning Mathematics.* Portsmouth, NH: Heinemann, 1990.

FRANCES R. CURCIO

## DAVIES, CHARLES (1798–1876)

Introduced Adrien Marie Legendre's geometry and Louis Pierre Marie Bourdin's algebra to American students through translations that were part of his popular series of mathematics texts, a series that ranged from primary arithmetic through the calculus. He taught mathematics at the United States Military Academy, at Trinity College, and at Columbia College. Davies' *The Logic and Utility of Mathe-*

*matics with the Best Methods of Instruction Explained and Illustrated* (1850) is recognized as the first American book to include methods of teaching mathematics. While Davies recognized the value of "sensible objects" in early arithmetic learning—"manipulatives" in the current lexicon—and while he employed some precepts of faculty psychology by suggesting exercise of the abstraction, generalization, and judgment faculties, he found his primary framework for teaching mathematics in the structure of mathematics itself.

*See also* History of Mathematics Education in the United States, Overview

### SELECTED REFERENCES

Bidwell, James K., and Robert G. Clason, eds. *Readings in the History of Mathematics Education.* Washington, DC: National Council of Teachers of Mathematics, 1970.

Jones, Phillip S., ed. *A History of Mathematics Education in the United States and Canada.* 32nd Yearbook. Washington, DC: National Council of Teachers of Mathematics, 1970.

JAMES K. BIDWELL
ROBERT G. CLASON

# DECIMAL (HINDU-ARABIC) SYSTEM

Place-value system generally considered to have originated in India and to have been transmitted to the West by the Arabs. Symbols for the numbers 1 through 9 date back to the Brahmi system of writing in the third century B.C., in India. About the eighth century A.D., with the Moslem invasion of northern India, these digits were brought to Europe, first to Spain and later to the rest of Europe (Menninger 1969).

The ancient Egyptians used a base-10 system of numeration, with each higher power of 10 involving the creation of a new symbol. This was also true of the ancient Indian system. In the seventh century, it appears that the Indians began using the symbols 1 through 9 in the familiar place-value arrangement, perhaps due to interactions with Chinese traders and their counting boards. No mention is made of a symbol for zero, although it appears that a dot, and later the circle, was used for this purpose. Zero as we know it was being used by the Arabs by the ninth century A.D. (Menninger 1969, 399–403).

## MOTIVATION AND GOALS

The decimal place-value system is generally recognized as the most efficient form of notation for writing numbers. Important properties of the system include a base of 10, the sufficiency of the symbols 1, 2, 3, . . . , 9 for representing numbers, place value, and a symbol for zero. The use of base 10 is most likely a consequence of objects being placed in one-to-one correspondence with the fingers of both hands. While numbers in such ancient numeration systems as the Egyptian were typically written in hierarchical order, this was not really necessary because each power of 10 required the introduction of a new symbol. The use of only ten different symbols combined with the value of a digit being denoted by its position, and the use of zero as placeholder, made the Hindu-Arabic numeration system more efficient than any other which had been used.

## MATHEMATICAL EXPLANATION

The four properties of the decimal place-value system identified above mean that we can write numerals in expanded notation. Thus $3{,}016 = 3 \times 10^3 + 0 \times 10^2 + 1 \times 10 + 6$.

## VARIETY OF PEDAGOGICAL APPROACHES

Ross has pointed out: "To understand place value the student must coordinate and synthesize a variety of subordinate knowledge about our children's notational system for numbers and about numerical part-whole relationships. A student who understands place value knows not only that the numeral 52 can be used to represent 'how many' for a collection of fifty-two objects but also that the digit on the right represents two of them, the digit on the left represents fifty of them (five sets of ten), and that 52 is the sum of the quantities represented by the individual digits" (Ross 1989, 47). She developed a five-stage model of the interpretations children assign to two-digit numerals: (1) whole numeral, (2) positional property, (3) face value, (4) construction zone, and (5) understanding. She found that even with extensive experience with embodiments such as base-10 blocks, many students do not appear to gain an understanding of place value. In fact, if understanding is the goal, it does not matter which manipulative a teacher uses, if the students simply follow directions rather than construct their own knowledge of numbers and the relations among them (Ross 1989, 49–50).

Kamii (1985) and Kamii and Joseph (1988) showed how students invent their own strategies for working with place value. They claimed that traditional 10s and 1s materials are contrary to the child's natural system of order and hierarchical inclusion. As a preliminary place-value activity, learners typically experience grouping activities. Such activities include grouping sticks into bundles of ten, or sorting beans into egg cartons ten to a cup (using ten cups). More representational of the decimal place-value system are Dienes' multibase arithmetic blocks. Dienes (1960, 1966) argued that the concept of place-value is independent of base; he provided experiences in bases 2, 3, 4, 5, 6, and 10. While we can agree that a learner who understands the concept of base would be able to work with any such base, it is now common to work in the classroom only with base 10. Students investigating computer logic would also work with base 2. The dominant feature of base-10 blocks, that 10 units = 1 long, 10 longs = 1 flat, and 10 flats = 1 block, is that the model is verifiable visually. For example, 14 longs is readily seen to be equivalent to 1 flat and 4 longs.

At a more abstract level, exchange activities call for the learner to recognize that one blue chip has the value of ten yellow chips, for example. On a spike abacus, a disk on the 10s spindle has ten times the value of a disk on the 1s spindle. Thus, using base-10 blocks, 3,016 would be represented by 3 blocks 0 flats 1 long 6 units, a correspondence that does not necessarily imply an understanding of the position of a digit as a determiner of value. With exchange activities such as those involving a spike abacus, however, the same number would be represented by 3 disks on the 1000s spindle, 0 disks on the 100s spindle, 1 disk on the 10s spindle, and 6 disks on the 1s spindle.

Alternative approaches include such teacher-made exchange activities as using such fictitious characters as Oliver, Timothy, and Hercules to represent 1s, 10s, and 100s. Oliver, being young, can carry loads of no more than 9 objects; Timothy carries loads bundled in multiples of 10 up to a maximum load of 90 objects. Hercules carries loads bundled in multiples of 100 up to a maximum of 900 objects. A new character can be invented if a teacher wishes to represent numbers of 1,000 or more. Learners successfully completing such activities as those above should have a clear understanding of place-value and should readily transfer from exchange activities to symbolic representations.

## RECOMMENDED MATERIALS FOR INSTRUCTION

Grouping activities involve using such materials as tongue depressors in one-to-one correspondence with a set of objects to be counted. Groups of ten are bundled to facilitate counting. Base-10 blocks, mentioned above, are highly recommended as an introduction to place value. An economical substitute for 100s, 10s, and 1s, involves the use of beans: individually to represent 1s, ten glued to a tongue depressor (called a bean stick) representing 10, and ten bean sticks glued to a cardboard square (called a bean raft) representing 100.

Exchange activities include the spike abacus and chip trading, using a place-value mat (Davidson 1975). It is important, at this level, that a 6 in a 100s column be recognized as having a value ten times that of a 6 in the 10s column.

As a step leading to a symbols-only approach, learners may use numbered cards or may write numerals to denote the numbers represented by base-10 blocks or by a spike abacus. Again it must be emphasized that no matter what materials are used, the students need to be assisted to develop their own understanding of place value based on a suitable transition from a hands-on approach to a symbolic approach.

*See also* Abacus; Attribute Blocks; Egyptian Mathematics; Indian Mathematics; Islamic Mathematics

## SELECTED REFERENCES

Davidson, Patricia S. *Chip Trading Activities: Teacher's Guide.* Fort Collins, CO: Scott, 1975.

Dienes, Zoltan P. *Building Up Mathematics.* London, England: Hutchinson Educational, 1960.

———. *Mathematics in the Primary School.* Toronto, Ontario, Canada: Macmillan, 1966.

Heddens, James W., and William R. Speer. *Today's Mathematics. Part 1. Concepts and Classroom Methods.* Englewood Cliffs, NJ: Prentice-Hall, 1995.

Kamii, Constance. *Young Children Invent Arithmetic.* New York: Teachers College Press, 1985.

———, and Linda Joseph. "Teaching Place Value and Double-Column Addition." *Arithmetic Teacher* 35(Feb. 1988):48–52.

Menninger, Karl. *Number Words and Number Symbols: A Cultural History of Numbers.* Paul Broneer, trans. Cambridge, MA: MIT Press, 1969.

Ross, Sharon H. "Parts, Wholes, and Place Value: A Developmental View." *Arithmetic Teacher* 36(Feb. 1989): 47–51.

Smith, David E. *A Source Book in Mathematics.* Vol. 1. New York: Dover, 1959.

DAVID M. DAVISON

# DEDEKIND, RICHARD
## (1831–1916)

A pioneer in the use of the modern axiomatic method and set-theoretic modes of thinking—fundamental characteristics of twentieth-century mathematics. Dedekind was born in Brunswick, Germany, received his doctorate at Göttingen, under Carl Friedrich Gauss, at the age of twenty-one, and was professor at the Zürich Polytechnic for the next several years. He spent his last fifty years in Brunswick, teaching at the Technical University.

Dedekind was a prime exponent of the nineteenth century's new spirit of rigor and abstraction in mathematics. One of his major accomplishments was a rigorous definition of the real numbers, through "Dedekind cuts." This was needed, he recognized, in order to give his students a rigorous presentation of calculus. He is best known for his work in abstract algebra and (algebraic) number theory. He invented ideals, which enabled him to vastly generalize the theorem on unique factorization of integers into primes. He also defined axiomatically (in the context of algebraic numbers) the concepts of ring, field, and module.

*See also* Abstract Algebra

### SELECTED REFERENCES
Bell, Eric Temple. *Men of Mathematics.* New York: Simon and Schuster, 1937.

Biermann, Kurt-R. "Dedekind, (Julius Wilhelm) Richard." In *Dictionary of Scientific Biography* (pp. 1–5). Vol. 4. Charles C. Gillispie, ed. New York: Scribner's, 1981.

Dugac, Pierre. *Richard Dedekind et les Fondements des Mathématiques.* Paris, France: Vrin, 1976.

ISRAEL KLEINER

# DE MOIVRE, ABRAHAM
## (1667–1754)

Important contributor to the fields of probability and analytical trigonometry and known for De Moivre's Theorem. De Moivre, a French Huguenot who fled to England after the revocation of the Edict of Nantes in 1685, settled in London, where he earned his living as a private tutor of mathematics. He was elected to the Royal Society in 1697 and in 1712 was appointed to be one of the commissioners whose task was to report on the bitter dispute between Gottfried Leibniz and Isaac Newton as to which of them invented the calculus. De Moivre expounded the laws of chance in his treatise *Doctrine of Chances* (1718), later editions of which contained the first treatment of the probability integral:

$$\int_0^\infty e^{-x^2}\, dx = \frac{\sqrt{\pi}}{2}$$

Part of this work, concerned with actuarial mathematics, was published separately as *Annuities upon Lives.* Further contributions to probability and analytical trigonometry appeared in *Miscellanea Analytica* (1730), in which De Moivre dealt with imaginary numbers and circular functions. The formula $(\cos\theta + i\sin\theta)^n = \cos n\theta + i\sin n\theta$, where $i = \sqrt{-1}$ and $n$ is a positive integer, is used in this work. This result is still known as De Moivre's Theorem.

*See also* Complex Numbers; Probability, Overview; Trigonometry

### SELECTED REFERENCES
Boyer, Carl B. *A History of Mathematics.* 2nd ed. Revised by Uta C. Merzbach. New York: Wiley, 1989.

Eves, Howard W. *An Introduction to the History of Mathematics.* 5th ed. Philadelphia, PA: Saunders College Publishing, 1983.

Newman, James R. *The World of Mathematics.* 4 vols. London, England: Allen and Unwin, 1960.

Struik, Dirk J. *A Concise History of Mathematics.* New York: Dover, 1948.

ROD HAGGARTY

# DE MORGAN, AUGUSTUS
## (1806–1871)

Influenced the development of mathematics by being an outstanding teacher, a prolific writer of journal articles, and author of many fine textbooks on algebra, arithmetic, calculus, geometry, logic, and trigonometry. For most of his life De Morgan was professor of mathematics at what is now University College, University of London. *An Essay on Probabilities, and on Their Application to Life Contingencies and Insurance Offices* (1838) is the second important work in English on probability (after Abraham De Moivre's (1667–1754) *Doctrine of Chances.* He was the first president of the London Mathematical Society and served two terms as secretary of the Astronomical Society.

De Morgan's mathematical interests were primarily in algebra and in logic, and he stood at the brink of modern developments in both. Influenced by George Peacock's (1791–1858) work in the newly emerging symbolic algebra, De Morgan expounded the view that an algebraic system could be created from arbitrary symbols and the laws for combining those symbols. Although his contributions to logic were eclipsed by George Boole's (1815–1864) algebraic logic, De Morgan introduced new notation into the classical Aristotelian syllogism and the quantification of the predicate term, in which a numerically definite quantity ("all," "some") is applied to subject or predicate or to both. De Morgan's view of a relation was an extension of the subject-predicate concept, that is, $X$, $Y$ are subject and predicate and these names refer to the "mode of entrance" in the relation, not to the order in which they are used. His introduction of relations into logic foreshadowed modern developments attributed to Benjamin Peirce (1809–1880) by about twenty-five years.

*See also* Boole; Boolean Algebra; Logic

## SELECTED REFERENCES

Dubbey, John M. "De Morgan, Augustus." In *Dictionary of Scientific Biography* (pp. 35–37). Vol. IV. Charles C. Gillispie, ed. New York: Scribner's, 1971.

Heath, Peter, ed. *A. De Morgan. On the Syllogism and Other Logical Writings.* New Haven, CT: Yale University, 1966.

Macfarlane, Alexander. *Lectures on Ten British Mathematicians of the Nineteenth Century.* New York: Wiley, 1916, pp. 19–33.

Smith, David E., ed. *A Budget of Paradoxes by Augustus De Morgan.* 2nd ed. Two vols. Chicago, IL/London, England: Open Court, 1915.

FRANCINE F. ABELES

# DESCARTES, RENÉ (1596–1650)

French philosopher and mathematician who made significant contributions to analytic geometry and algebra. Descartes was a sickly child and therefore allowed to spend his mornings ruminating in bed, a habit he continued throughout his life. Indeed, it could be said that the abandonment of this habit killed him: in the winter of 1649–50, Descartes became the personal tutor of Queen Christina of Sweden. She insisted that Descartes begin her lessons at 5 A.M., and early in 1650 Descartes became ill and died soon afterward.

Although Descartes made significant mathematical contributions, he is perhaps best known for his philosophical works, in particular *Discourse on the Method of Rightly Conducting Reason,* published in 1637. Descartes was also a renowned scientist, publishing works on optics, meteorology, and physics.

Descartes was obsessed with the discovery of truth, and his interest in mathematics was prompted by his fervent hope that the axiomatic methods favored by mathematicians might also be applied to "all those things which fall under the cognizance of man" (Descartes 1931, 92). Not only did this belief influence the development of the scientific method, but it led Descartes to append to the *Method* a monograph entitled *The Geometry,* in which he introduced a system of analytic geometry (including the use of $x$ and $y$ as variables) that was in essence the first quadrant of the $xy$-plane.

Prior to the appearance of Descartes' *Geometry,* mathematicians generally used geometric constructions to solve a given problem, even those of an algebraic nature. Consequently, algebraic methods were not in general use. Descartes' analytic geometry changed all that. Even though François Viète introduced algebraic symbolism toward the end of the 1500s, Viète viewed variables such as $A$ as line segments (or areas, in the case of $AB$) and did not place his variables within a coordinate system. Descartes, however, viewed variables also as positions relative to an axis and thereby created a coordinate system. This innovation allowed him to express curves as algebraic equations, which gave mathematicians the means to solve geometric problems algebraically (and vice versa). It is in recognition of Descartes' accomplishments that the $xy$-plane is often referred to as the Cartesian plane.

In addition to his contributions to analytic geometry, Descartes made many other important algebraic discoveries in *The Geometry.* He introduced polynomial long division, stated a version of the fundamental theorem of algebra, and articulated the "rule of signs," namely, that the maximum number of positive zeros of a polynomial $p(x)$ is equal to the number of times the sign changes from $+$ to $-$ and $-$ to $+$, and the maximum number of negative zeros is equal to the number of times that two like signs appear in succession.

*See also* Coordinate Geometry, Overview

## SELECTED REFERENCES

Boyer, Carl B. "Analytic Geometry: The Discovery of Fermat and Descartes." *Mathematics Teacher* 37(1944): 99–105.

———. "The Invention of Analytic Geometry." *Scientific American* 180(Jan. 1949):40–45.

Descartes, René. *Discourse on Method, Optics, Geometry and Meteorology.* Indianapolis, IN: Bobbs-Merrill, 1965.
——. "*Discourse on the Method of Rightly Conducting Reason.*" In *The Philosophical Works of Descartes.* Elizabeth S. Haldane and G.R.T. Ross, trans. (pp. 79–130). Vol. 1. Cambridge, England: Cambridge University Press, 1931.
——. *The Geometry.* New York: Dover, 1954.
Scott, Joseph Frederick. *The Scientific Work of René Descartes.* London, England: Taylor and Francis, 1952.
Vrooman, Jack. *René Descartes: A Biography.* New York: Putnam's, 1970.

DANIEL S. ALEXANDER

# DETERMINANTS

Numbers associated with square matrices; often used to characterize the conditions under which matrices will have multiplicative inverses. Simply stated, a square matrix will have a multiplicative inverse if and only if its determinant is nonzero. Determinants are a pedagogical tool in the study of matrix theory.

## THE COFACTOR DEFINITION OF DETERMINANTS

Given a $2 \times 2$ matrix

$$A = \begin{pmatrix} a_{11} & a_{12} \\ a_{21} & a_{22} \end{pmatrix}$$

its determinant, denoted $\det(A)$, is defined by

$$\det(A) = a_{11}a_{22} - a_{21}a_{12}$$

Let $I$ denote the $2 \times 2$ identity matrix

$$I = \begin{pmatrix} 1 & 0 \\ 0 & 1 \end{pmatrix}$$

If $\det(A) \neq 0$ and

$$B = \frac{1}{\det(A)} \begin{pmatrix} a_{22} & -a_{12} \\ -a_{21} & a_{11} \end{pmatrix}$$

then $BA = AB = I$. Thus $B$ is the multiplicative inverse of $A$. If $\det(A) = 0$ then $A$ does not have a multiplicative inverse.

We can also determine whether or not a $3 \times 3$ matrix $A$ has a multiplicative inverse by computing a single number as a function of the entries of $A$. Indeed it can be shown that the matrix

$$A = \begin{pmatrix} a_{11} & a_{12} & a_{13} \\ a_{21} & a_{22} & a_{23} \\ a_{31} & a_{32} & a_{33} \end{pmatrix}$$

will have a multiplicative inverse if and only if

$$a_{11}a_{22}a_{33} - a_{11}a_{32}a_{23} - a_{12}a_{21}a_{33} \\ + a_{12}a_{31}a_{23} + a_{13}a_{21}a_{32} - a_{13}a_{31}a_{22} \neq 0$$

The determinant of $A$ is defined to be the expression on the left-hand side of this inequality.

To simplify the definition of the determinant of an $n \times n$ matrix $A$ we introduce some new notation. Let $A_{ij}$ be the $(n-1) \times (n-1)$ submatrix of $A$ obtained by deleting the $i$th row and $j$th column of $A$. The determinant of $A$ can be expressed in terms of the determinants of $n$ of these submatrices. To illustrate how this is done consider the $3 \times 3$ case.

$$\det(A) = a_{11}(a_{22}a_{33} - a_{32}a_{23}) - a_{12}(a_{21}a_{33} \\ - a_{31}a_{23}) + a_{13}(a_{21}a_{32} - a_{31}a_{22}) \\ = a_{11}\det(A_{11}) - a_{12}\det(A_{12}) + a_{13}\det(A_{13})$$

This method of representing the determinant of $A$ is called the *cofactor expansion* of the determinant of $A$ along the first row of $A$. Actually it is possible to represent $\det(A)$ as a cofactor expansion along any row or column of $A$. For example, if $A$ is a $4 \times 4$ matrix, then using the second column we can represent the determinant of $A$ by the cofactor expansion

$$\det(A) = -a_{12}\det(A_{12}) + a_{22}\det(A_{22}) \\ - a_{32}\det(A_{32}) + a_{42}\det(A_{42})$$

In general, if $A$ is an $n \times n$ matrix and we set $C_{ij} = (-1)^{i+j}\det(A_{ij})$ for each $i$ and $j$, then the cofactor expansion of $\det(A)$ along the $i$th row is given by

$$\det(A) = \sum_{j=1}^{n} a_{ij}C_{ij}$$

and the cofactor expansion along the $j$th column is given by

$$\det(A) = \sum_{i=1}^{n} a_{ij}C_{ij}$$

## PROPERTIES OF DETERMINANTS

Using cofactor expansions, it is not difficult to determine the effects of elementary row operations on the value of the determinant. These effects can be summarized as follows:

1. If the two rows of the matrix are interchanged, then the value of the determinant will be multiplied by $-1$.
2. If a row of a matrix is multiplied by a real number $c$, then the value of the determinant will be multiplied by $c$.

3. If a multiple of one row of a matrix is added to another row, the value of the determinant will remain unchanged.

As a consequence of these properties one can show that for any $n \times n$ matrices $A$ and $B$

$$\det(AB) = \det(A)\det(B)$$

If one considers the $n \times n$ matrix $A$ as a linear operator then one can gain valuable information about the operator by solving the *characteristic equation*

$$\det(A - \lambda I) = 0$$

for $\lambda$. The expression $\det(A - \lambda I)$ defines an $n$th degree polynomial in the variable $\lambda$. The roots of the polynomial are referred to as the *characteristic values* or *eigenvalues* of the operator $A$. Eigenvalues play an important role in many mathematical models. Often it is possible to represent physical phenomena such as the energy states of molecules or natural frequencies of a structure in terms of eigenvalues.

## THE HISTORY OF DETERMINANTS

Determinants first appeared implicitly in the 1693 letters from Gottfried Leibniz to Guillaume L'Hôspital. In the letters, Leibniz gave a formula characterizing the conditions under which a system

$$a_1 + b_1 x + c_1 y = 0$$
$$a_2 + b_2 x + c_2 y = 0$$
$$a_3 + b_3 x + c_3 y = 0$$

would be consistent. The formula was equivalent to requiring that the determinant of the $3 \times 3$ coefficient matrix be 0.

In 1750, Gabriel Cramer published his well-known theorem for expressing the solution to a linear system in terms of determinants. The rule was known to Colin Maclaurin probably as early as 1729 and it was used by Maclaurin in a treatise published in 1748.

One of the first important papers on determinants was by Alexandre T. Vandermonde in 1771. The following year, Pierre Laplace wrote a paper giving his rule for the expansion of a determinant. In 1775, Joseph L. Lagrange used determinants to derive formulas for the area of a triangle and the volume of a tetrahedron. The formula for the area of a triangle whose vertices are located at the coordinates $(0,0)$, $(x_1,y_1)$, $(x_2,y_2)$ is simply

$$\text{Area} = |x_1 y_2 - x_2 y_1|/2$$

The person most responsible for establishing the theory of determinants is Augustin-Louis Cauchy (1789–1857). Cauchy published a long memoir in 1812 on alternating symmetric functions, that is, functions such as $x_1 y_2 - x_2 y_1$. In this paper, he introduced the name "determinants." Cauchy gave general proofs for all the standard theorems and introduced the concepts of "characteristic polynomial and characteristic equation" of a matrix. Another notable paper on determinants, by Jacques P. M. Binet, also appeared in 1812. After the papers by Cauchy and Binet, determinants became a standard tool for all mathematicians.

In 1829, Carl G. J. Jacobi introduced functional determinants that were later named Jacobians by James J. Sylvester. Jacobi followed this up with a long landmark paper on the subject in 1841. The next major advance occurred in 1848, when Sylvester used arrays to generate determinants and introduced the term "matrix." In the following years a great number of papers appeared on the subject of determinants. During the period 1906–1930, Thomas Muir published a monumental five-volume history of determinants. It took Muir approximately twenty-five hundred pages just to summarize the work that others had done.

## THE DECLINING IMPORTANCE OF DETERMINANTS

In the latter half of the twentieth century, interest in determinants declined sharply. Determinants are no longer a major area of mathematical activity. The subject has been eclipsed by the growing importance of matrix theory. Determinants are now used primarily as a tool to help students to learn linear algebra.

One reason for the decline is that determinants have proved to be far less useful in modern applications than one would expect. Problems that traditionally were solved using determinants are now solved by alternative methods that are computationally superior. Today Cramer's rule is considered to be an exceedingly inefficient way to solve linear systems. Many linear algebra books now list it as an optional topic rather than a required topic, and many instructors choose not to discuss it.

The expression $p(\lambda) = \det(A - \lambda I)$ is called the *characteristic polynomial* of $A$. In a beginning linear algebra course students are taught to find the eigenvalues of $A$ by computing the roots of its characteristic polynomial. This is not a good technique, however, for computational applications. In general it is difficult to compute roots of polynomials accurately. Even the smallest rounding errors can result in relatively large changes in the computed values of the roots. For this reason, modern computational methods for eigenvalues avoid using the characteristic polynomial.

*See also* Linear Algebra, Overview

## SELECTED REFERENCES

Bell, Eric Temple. *The Development of Mathematics.* New York: McGraw-Hill, 1945.

Boyer, Carl B. *A History of Mathematics.* 2nd ed. Revised by Uta C. Merzbach. New York: Wiley, 1989.

Leon, Steven J. *Linear Algebra with Applications.* 4th ed. New York: Macmillan, 1994 [Chap. 2].

Muir, Thomas. *Contributions to the History of Determinants, 1900–1920.* London, England: Blackie, 1930.

————. *The Theory of Determinants in the Historical Order of Development.* 4 vols. London, England: Macmillan, 1906, 1911, 1920, 1923. [Reprinted, New York: Dover, 1960.]

STEVEN J. LEON

# DEVELOPING MATHEMATICAL PROCESSES (DMP)

Individualized elementary mathematics curriculum (K–6) developed by members of the mathematics education staff of the University of Wisconsin. Funding for its development was provided by the National Science Foundation and the U.S. Department of Education. Four features of the program are unique: approach to mathematics, flexible organization, assessment procedures, and recordkeeping.

## APPROACH TO MATHEMATICS

The mathematical content in DMP is approached through measurement. In this approach, children examine the objects in their world and focus on some attribute (length, numerousness, weight, capacity, area, time, etc.). They use various processes (describing, classifying, comparing, ordering, equalizing, joining, separating, grouping, and partitioning) to explore relationships between objects. Once they are familiar with each attribute, they symbolically represent (measure) it. Likewise, they represent the relationships between objects with mathematical sentences. In turn, they represent mathematical sentences with real objects to check their validity.

## FLEXIBLE ORGANIZATION

DMP is organized in the following way. Activities designed to promote attainment of closely related objectives are clustered to form ninety topics (twelve to fifteen topics per grade) in five mathematical strands (attributes, addition and subtraction, multiplication and division, geometry, and movement and direction). Since many objectives are hierarchically related, in each topic it is important both to review objectives that have been taught earlier and to prepare students for later objectives as well as instructing them to mastery. For planning instruction across topics the Resource Manual to DMP offers a number of suggestions, and within each topic considerable help is provided by describing suggested activities and alternates.

## ASSESSMENT PROCEDURES

Decisions regarding attainment of objectives are made in DMP at three levels. Mastery (M) means the teacher is convinced by the child's performance that, given a similar activity, the required behavior would be exhibited. Making Progress (P) means that the objective is not mastered but the child is making progress toward mastery. Needing Considerable Help (N) means that the child has not mastered the objective and will probably need individual attention and extra work. Also, there are three different times when the teacher is to assess student achievement on a topic: before beginning instruction, during the three or more weeks of instruction, and after it is completed.

## RECORDKEEPING

Records are kept so that the information is handy in making decisions relative to each child. Since children progress through DMP according to the rate of attaining objectives and not by year of school attendance, it becomes important to keep track of achievement of all objectives. In DMP there are three different forms. The Topic Checklist is designed for keeping records of students' performance during activities. The Group Record Card is similar, but it contains information on ten topics instead of one. It provides information about which students have completed which topics. The Individual Progress Sheet is designed for keeping substantial records on a single student.

In summary, DMP is a unique individualized mathematics program. It approaches mathematics from a measurement approach, it approaches instruction from a flexibly organized activity approach, and it approaches management of instruction by having coherent assessment procedures and methods of keeping records (Romberg 1977).

Since publication, the materials have been widely used and emulated. The activity approach has become commonplace in most elementary programs, and the emphasis on the processes involved in representing situations underlie, the K–4 Curriculum Standards published by the National Council of Teachers of Mathematics (1989). Evaluations of student performance in schools who used the program were very positive (Schall, Bauman, and Mohan, 1975; Schall, Mohan, and Hull 1974; Webb 1980). Like many experimental programs, however, it proved to be too innovative for most schools, and suffered from the "back-to-basics" emphasis of that era.

*See also* Individualized Instruction

## SELECTED REFERENCES

National Council of Teachers of Mathematics. *Curriculum and Evaluation Standards for School Mathematics.* Reston, VA: The Council, 1989.

————. *Principles and Standards for School Mathematics.* Reston, VA: The Council, 2000.

Romberg, Thomas A. "Developing Mathematical Processes: The Elementary Mathematics Program for Individually-Guided Education." In *Individually Guided Elementary Education: Concepts and Practices* (pp. 77–109). Herbert J. Klausmeier, Richard A. Rossmiller, and Mary Saily, eds., New York: Academic, 1977.

Schall, William E., Daniel Bauman, and Madan Mohan. "Developing Mathematical Processes (DMP): Field Test Evaluation, 1973–74." Working paper. Fredonia, NY: Teacher Education Research Center, 1975.

Schall, William E., Madan Mohan, and Ronald E. Hull. "Developing Mathematical Processes (DMP): Field Test Evaluation, 1972–73." Working paper. Fredonia, NY: Teacher Education Research Center, 1974.

Webb, Norman, L. "The Relation of Instructional Time and Classroom Process Variables to Achievement: An Evaluation of DMP." Paper presented at the Annual Meeting of the National Council of Teachers of Mathematics, April 1980.

THOMAS A. ROMBERG

# DEVELOPMENTAL (REMEDIAL) MATHEMATICS

Precollege mathematics such as arithmetic, algebra, or geometry taught in colleges. Remedial mathematics programs have become large-scale enterprises on many college campuses. In fact, a 1990 survey of college mathematics enrollments found that remedial courses accounted for 16% of these enrollments at four-year colleges and 58% of the enrollments at two-year colleges (Albers, Loftsgaarden, Rung, and Watkins 1992). Among the factors that contributed to the growth of these programs, most pronounced in the 1960s and 1970s, were the proliferation of open-admissions postsecondary institutions (notably two-year colleges) and the weakening of high school exit requirements.

On some campuses, remedial mathematics offerings are housed in the mathematics department; at others, in a department of developmental or basic skills. These programs typically make use of placement examinations to identify which students should enroll.

Developmental mathematics courses frequently are taught by part-time faculty. These courses often employ nontraditional pedagogical methods, such as computer-based instruction, collaborative learning, and mastery testing. Self-pacing is often employed, especially for students reviewing material or knowing a substantial part of course content. To further individualize learning, classes may be smaller than average, and supplemented by tutors in a learning center or mathematics laboratory.

In recent years, considerable pressure has been brought to bear on developmental mathematics programs to justify their legitimacy, to prove their effectiveness (often with mixed results), to reduce costs, and to get students through more quickly. Many programs have modernized the standard curriculum. Some have enhanced their coverage of geometry, usually following a descriptive rather than deductive approach. Other programs have adopted a graphics-calculator approach, stressing graphing at the expense of traditional algebraic algorithms.

*See also* Developmental Mathematics Program and Apprenticeship Teaching; Remedial Instruction

## SELECTED REFERENCES

Albers, Donald J., Don O. Loftsgaarden, Don C. Rung, and Ann E. Watkins. *Statistical Abstracts of Undergraduate Programs in the Mathematical Sciences and Computer Science in the United States, 1990–91 CBMS Survey.* Washington, DC: Mathematical Association of America, 1992.

American Mathematical Association of Two-Year Colleges. *Crossroads in Mathematics: Standards for Introductory College Mathematics Before Calculus.* Memphis, TN: The Association, 1995.

Davis, Ronald M., ed. *A Curriculum in Flux—Mathematics at Two-Year Colleges.* Washington, DC: Mathematical Association of America, 1989.

Maxwell, Martha, ed. *From Access to Success.* Clearwater, FL: H and H, 1994.

GEOFFREY R. AKST

# DEVELOPMENTAL MATHEMATICS PROGRAM (DMP) AND APPRENTICESHIP TEACHING

The program conducted at California State University, Northridge, to assist educationally disadvantaged students in acquiring the basic skills necessary to succeed in university-level mathematics. The DMP also provides support for graduate students in mathematics, preparing them for teaching positions at high schools or community colleges, or for teaching fellowships in Ph.D. programs. A coordinator from the Mathematics Department and a mathematics specialist from the university's Learning Resource Center oversee the program, supervising twenty graduate students, that is, teaching associates, who teach the remedial classes. A professor from the Mathematics Department serves as faculty advisor in charge of apprenticeship teaching.

Each remedial class in the DMP is conducted by a teaching associate with the assistance of two tutors. In order to qualify for teaching in the DMP, a graduate student must first tutor in the remedial classroom for at least one semester. Once hired as an associate, the graduate student then spends the first one or two semesters team-teaching remedial classes with a seasoned teaching associate. Graduate students must maintain at least B averages in their graduate courses, and generally serve as teaching associates, on the average, for about three years.

Weekly faculty meetings are the catalyst for apprenticeship teaching in the DMP. Graduate students learn to teach, not only by conducting classes within a highly structured program, but also by frequent collaboration with fellow instructors and supervisors. The faculty advisor meets every week with the coordinators and the teaching associates to discuss teaching strategies, to introduce new pedagogical research, and to monitor the progress of both teachers and students. The DMP is based on the philosophy that students learn in different ways. It seeks to provide a wide variety of options for remedial students to learn, giving them the primary responsibility for doing so. This focus on active student participation in the learning experience provides a wide variety of opportunities for apprenticeship teaching. Because the traditional lecture approach is not used, teaching associates must master an assortment of teaching strategies. As students themselves, the associates have an appreciation of what works in the classroom. As novices in the teaching profession, they are open to experimenting with different methodologies. The teaching techniques that the associates learn or develop are tested in the faculty meetings as well as in the classroom. These meetings not only provide an opportunity for the faculty advisor to instruct the associates, they serve as a forum for the associates to learn from each other.

The challenge in training graduate students as teachers is to provide a structured yet flexible environment in which they can work. Structure is provided by the use of a standardized text, fixed syllabus, and common final examination. Associates follow a teachers' manual that outlines the philosophy of the program as well as its various components. The faculty advisor and coordinators visit classrooms once or twice a semester to assess teaching. Flexibility is provided by the faculty meetings. Associates are encouraged to share teaching experiences, good and bad, to observe each other in the classroom, and to recommend new teaching strategies and improvements. In a very real sense, the DMP, which originated in 1978 and has changed considerably since that time, is the result not only of continued research in mathematics education, but also of the excellent suggestions of the teaching associates.

## PROGRAM COMPONENTS: STUDENT ACTIVITIES AND APPRENTICESHIP TEACHING

The DMP integrates several pedagogical components that engage the creativity of the teaching associates. The faculty meetings provide the environment in which they can share their ideas.

### Group Problem Solving and Collaborative Learning Projects

Group work is the primary method of instruction in all developmental courses. The associates design their own group projects to encourage students to see themselves as resources for learning. During faculty meetings, the associates share successful group projects with fellow instructors. Sometimes, a failed group project can engender a discussion of what pitfalls to avoid in designing collaborative learning experiences for students. Faculty meetings are often the setting for collaborative projects for the teaching associates themselves. These projects give them an understanding of the dynamics of group work, and, in some cases, act to convince the skeptic of the benefits of such activity. For example, in the remedial courses, ten-minute clips from videotapes showing real-life situations that involve mathematics are used to stimulate discussion of a topic. Associates preview these clips during faculty meetings, and then work in groups to design questions that will enable their students to focus on particular concepts. When the teaching associates break out of their groups to have a general discussion, they are able to see the benefits of their collaborative activity.

## Writing to Learn Mathematics

The writing component of the DMP has several goals: to enable students to learn mathematics by encouraging them to classify problems and then write procedures for solving them, or by having them write essays about topics they review in videotapes; to help students learn the language of mathematics; to assist students in overcoming their own anxiety about mathematics by having them confront it by writing their own "mathematics autobiography" or "best/worst experiences involving mathematics." Students are not the only ones who are assigned writing. The teaching associates write essays in order to accomplish certain tasks or to make suggestions for improving the DMP. For example, the associates assist in preparing final examinations. Preceding a discussion about choosing topics for a final examination, they are directed to write an essay in support of specific topics they believe are important. Then, in discussions of the examinations at a subsequent faculty meeting, the dialogue is focused and the group is much more likely to reach consensus.

## Testing as a Learning Experience

Students in developmental courses take weekly tests on a three-carboned form. Testing is conducted as a group project. After completing the twenty-minute test, the students tear off the top form to hand in to the instructor. Then, as a class, the students go over the quiz with each student grading her/his own paper. No partial credit is permitted. The students hand in their "graded" test to the instructor and leave the room with their own corrected copy of the test. This makes testing a learning experience for the students. It provides them with weekly assessments of their progress in the class, enabling them to see their errors immediately so that they can seek help. Since the teaching associates write their own weekly tests, they consult regularly with each other regarding appropriate questions. Testing, like group work, often becomes the focus of discussion at faculty meetings, and the associates learn how to assess student performance without testing everything.

## SUCCESS OF THE MODEL: STUDENT PERFORMANCE AND APPRENTICESHIP TEACHING

The DMP at California State University has been continually assessed since its inception in the late 1970s. Its effectiveness is evaluated primarily in two areas:

1. *Persistence:* About 93% of the students who enroll complete the remedial classes.
2. *Student success in subsequent mathematics courses:* Students who earn credit in remedial classes and then enroll in university-level mathematics classes the semester immediately following their developmental experience are consistently more successful than the general population.

The most outstanding characteristic of the program is the quality of instruction given by the teaching associates. They consistently earn high ratings on student evaluation forms (locally generated at the university to evaluate faculty teaching). Most graduate teaching associates have either chosen to pursue further graduate study or have obtained positions teaching in the community colleges or in high schools. Those who go on to Ph.D. programs are often awarded teaching fellowships. Many who are teaching report that they are successfully using the techniques they learned in the DMP. The DMP empowers the teaching associates by giving them a voice to discuss their experiences and a means to improve the structure of the program in which they teach. The weekly faculty meetings are the key to their success as teachers. These meetings help to create a feeling of community. They provide a vehicle for introducing new research in mathematics education and for stimulating continuous dialogue about teaching.

*See also* Developmental Mathematics; Test Construction; Writing Activities

## SELECTED REFERENCES

Birken, Marcia. "Using Writing to Assist Learning in College Mathematics Classes." In *Writing to Learn Mathematics and Science* (pp. 33–47). Paul Connolly and Teresa Vilardi, eds. New York: Teachers College Press, 1989.

Borrelli, Robert, and Stavros Busenberg. "Undergraduate Classroom Experiences in Applied Mathematics." *UMAP Journal* 3(1980):17–26.

Bruffee, Kenneth. "The Art of Collaborative Learning." *Change* (Mar./Apr. 1987):42–47.

Castro, Jim, Jerry Gold, Mark Schilling, Joel Zeitlin, and Elena Marchisotto. "Small Steps to a Student-Centered Classroom." *Primus* (Sept. 1991):253–274.

Connolly, Paul, and Teresa Vilardi, eds. *Writing to Learn Mathematics and Science.* New York: Teachers College Press, 1989.

England, David, and L. Diane Miller. "Writing to Learn Algebra." *School Science and Mathematics* 89(4) (1989):299–311.

Ganguli, Aparma. "Integrating Writing in Developmental Mathematics." *College Teaching* 37(4)(1988): 140–142.

Lesh, Kathryn. "Mathematical Problem Solving and Heuristics." *NLA News* 6(8)(1990):1–5.

Silver, Edward, ed. *Teaching and Learning Mathematical Problem Solving: Multiple Research Perspectives.* Hillsdale, NJ: Erlbaum, 1985.

Wolf, Dennis. "The Art of Questioning." *Academic Connections* (Winter 1987):1–15.

ELENA A. MARCHISOTTO

## DEWEY, JOHN (1859–1952)

Ranks among America's foremost educational thinkers. As a pragmatist he believed that learning is based on activity, social interaction, and real-world experience. Problem solving is fundamental for Dewey, and his five steps for problem solving or variations of them are enduring: (1) a felt need, (2) analysis of the problem, (3) discovery of alternative solutions, (4) experimentation with the alternatives, and (5) verification of a final solution. *Democracy and Education, The School and Society,* and *Experience in Education* are among Dewey's major works. John Dewey's view of mathematics is functional; he saw mathematics as a tool rather than a preexisting structure. He believed that in learning mathematics, students mentally construct numbers to solve real problems. Echoing William James, Dewey asserted that "*number* is not (psychologically) got *from* things, it is put *into* them." (McLellan and Dewey 1895, 71) His view of mathematics education can be found in *The Psychology of Number and Its Application to Methods of Teaching Arithmetic* with James A. McLellan, and in an arithmetic text series by McLellan and A. F. Ames.

*See also* Constructivism; Evaluation of Instruction, Overview; Psychology of Learning and Instruction, Overview

## SELECTED REFERENCES

Bidwell, James K., and Robert G. Clason, eds. *Readings in the History of Mathematics Education.* Washington, DC: National Council of Teachers of Mathematics, 1970.

Jones, Phillip S., ed. *History of Mathematics Education in the United States and Canada.* 32nd Yearbook. Washington, DC: National Council of Teachers of Mathematics, 1970.

McLellan, James A., and John Dewey. *The Psychology of Number and Its Application to Methods of Teaching Arithmetic.* New York: Appleton, 1895.

JAMES K. BIDWELL
ROBERT G. CLASON

## DIAGRAMS, EULER AND VENN

Geometrical representations of classes and syllogisms. Gottfried Wilhelm Leibniz (1646–1716) was the first person to use geometric diagrams to systematically represent syllogistic logic, the classical logic of Aristotle, which until the middle of the nineteenth century was thought to represent all forms of valid reasoning. In the eighteenth century, the use of circle diagrams to represent syllogisms was popularized by Leonhard Euler (1707–1783). Influenced by George Boole's (1815–1864) algebra of class logic, John Venn (1834–1923) constructed the most successful system of geometrical representations of classes and syllogisms in 1880.

Three classes are needed to represent a syllogism. Consider this example:

All *C* is *A*
No *A* is *B*
No *C* is *B*

In Figure 1 (an example of a Venn diagram), all the points in each of the three intersecting circles are members of the class represented by the given circle, while all points outside the circle are members of the complement (negation) of that class, e.g., *A'* denotes the complement of *A*.

The diagram shows the (unbounded) universe represented by the plane partitioned into eight compartments. The first premise of the syllogism, All *C* is

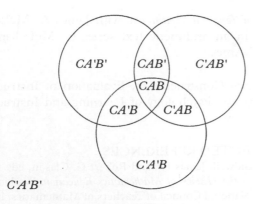

**Figure 1** *Example of a Venn diagram.*

*A*, means the class of things that are both *C* and *A'* is empty, that is, all compartments containing *C* and *A'* are empty. These are *CA'B'* and *CA'B*. The second premise, No *A* is *B*, means the class of things that are both *A* and *B* is empty. This class is represented by the compartments *CAB* and *C'AB*. Now we can see that all compartments in the diagram containing both *C* and *B* are empty, that is, compartments *CAB* and *CA'B*, leading to the conclusion of the syllogism that No *C* is *B*. Charles L. Dodgson (Lewis Carroll) improved Venn's indefinite treatment of the universe of discourse by enclosing each diagram in a square to bound the last compartment *C'A'B'*.

In principle, Venn's system can be extended to *n* classes, but for more than four classes the diagrams become too complicated. (Four classes require ellipses.) Venn diagrams also can be used to solve problems in propositional logic. Beginning in the 1870s, class logic gradually gave way to propositional logic. In propositional logic, terms represent statements (connected by "or," "and," and "if-then") that can be designated true or false. In Figure 1, each circle now represents a proposition rather than a class, and a compartment is empty if it contains an impossible combination of truth values of propositions. Consider this example: "If *A* is true, then *B* is true." This proposition is true in all cases except when *A* is true and *B* is false. So all compartments containing *AB'* are empty. These are *C'AB'* and *CAB'*.

*See also* Boole; Logic

### SELECTED REFERENCES

Baron, Margaret E. "A Note on the Historical Development of Logic Diagrams: Leibniz, Euler and Venn." *Mathematical Gazette* 53(May 1969):113–125.

Bartley, William W. III, ed. *Lewis Carroll's Symbolic Logic.* New York: Potter, 1977 [pp. 244–246].

Gardner, Martin. *Logic Machines and Diagrams.* 2nd ed. Chicago, IL: University of Chicago Press, 1982.

FRANCINE F. ABELES

# DIENES, ZOLTAN PAUL (1916– )

Mathematician and educator, born in Hungary, educated in England, and taught in England (1940–1959), Australia (1961–1965), and Canada (1966–1978). He believed that learners are brought to full awareness of the essential invariance of a mathematical structure by the presentation of multiple embodiments (perceptual variability) and to an understanding of the complete mathematical variability of the structure through activities with structured materials. Thus, he believed in concept construction before discussion.

Learning is accomplished in six stages: free play, games, search for commonalities, representation, symbolization, and formalization. He is perhaps best known for his development of multibase material (Dienes blocks) as a vehicle for understanding whole numbers and their operations. In his Canadian program, children worked for years with multibase material before turning to pencil-and-paper algorithms. Today only base-10 material is integrated with algorithm development. Dienes wrote in several languages; his most significant book is *Building Up Mathematics* (1971).

*See also* Cuisenaire Rods; History of Mathematics Education in the United States, Overview; Montessori

### SELECTED REFERENCES

Aichele, Douglas B., and Robert E. Reys. *Readings in Secondary School Mathematics.* 2nd ed. Boston, MA: Prindle, Weber, and Schmidt, 1977.

Dienes, Zoltan P. *Building Up Mathematics.* 4th ed. London, England: Hutchinson Educational, 1971.

Jones, Phillip S., ed. *A History of Mathematics Education in the United States and Canada.* 32nd Yearbook. Washington, DC: National Council of Teachers of Mathematics, 1970.

JAMES K. BIDWELL
ROBERT G. CLASON

# DIFFERENTIATION

If $y = f(x)$ is some function, the fraction

$$\frac{f(x) - f(a)}{x - a}$$

is the ratio of the change in the output variable to the change in the input variable. This ratio is the slope of the secant line joining $(a, f(a))$ to $(x, f(x))$ on the graph of the function. The derivative of this function, $f(x)$, if it exists at $x = a$, is the limit of this ratio as $x$ approaches $a$, that is

$$\lim_{x \to a} \frac{f(x) - f(a)}{x - a}$$

Symbols for this process are

$$f'(a), \frac{dy}{dx}\bigg|_a, \frac{df}{dx}\bigg|_a, y'(a), Df(a)$$

This derivative is the slope of the tangent line to the graph of $y = f(x)$ at the point $(a, f(a))$. The derivative may also be viewed as the instantaneous rate of change of $f(x)$ at $x = a$. If $f'(a)$ is positive, then $y$ increases as $x$ increases, while if $f'(a)$ is negative, then $y$ decreases as $x$ increases. If $f'(a) = 0$, then $f(x)$ has a horizontal tangent line and locally behaves as if it were the constant $f(a)$, neither increasing nor decreasing for small changes in $x$. If $f(x)$ is displacement of a particle and $x$ is time, then $f'(a)$ represents the velocity of the particle at $x = a$. If the moving object were a car, then $f'(a)$ is the speedometer reading at time $a$. In economics, $f'(x)$ is the marginal (cost, revenue, profit) function.

The property of $f(x)$ having a derivative at $x = a$ is distinctly different from that of $f(x)$ being continuous at $x = a$. Functions that have derivatives at $x = a$ are continuous at $x = a$, while functions continuous at $x = a$ may not have derivatives at $x = a$. For example, although $y = |x|$ is continuous at $x = 0$, it is not differentiable there. If a function has no derivative at $x = a$, its graph either has a vertical tangent or a break or a sharp turning point (cusp) at $(a, f(a))$ and no tangent line there. If $f'(a) = 0$, then the graph of $f(x)$ has a horizontal tangent at $(a, f(a))$, which may correspond to the largest or smallest value for $f(x)$.

The derivative of $f(x)$ is itself a function, $f'(x)$, which may have a derivative. Symbols for applying the differentiation process twice are $f''(x)$, $y''$ (if $y = f(x)$), $f^{(2)}(x)$, $d^2f/dx^2$. If $f(x)$ is displacement and $x$ is time, then $f'(x)$ is velocity and $f''(x)$ is acceleration. $f''(x)$ also measures the bending of the graph $y = f(x)$. Repeating the process $n$ times produces the $n$th derivative, whose symbols are $f^{(n)}(x)$, $d^ny/dx^n$.

If $f(x)$ and $g(x)$ are functions and $c$ is a constant, then the algebraic operations interact with the derivative as

$$(cf)'(a) = cf'(a), (f \pm g)'(a) = f'(a) \pm g'(a),$$
$$(fg)'(a) = f'(a)g(a) + f(a)g'(a)$$

and

$$(f/g)'(a) = \frac{g(a)f'(a) - f(a)g'(a)}{g(a)^2}$$

The derivative of $f(x) = x^n$ is $f'(x) = nx^{n-1}$. These theorems provide mechanical means for finding derivatives of a variety of functions efficiently. For instance, the derivative of a polynomial $p(x) = a_0 + a_1x + a_2x^2 + \cdots + a_nx^n$ is $p'(x) = a_1 + 2a_2x + \cdots + na_nx^{n-1}$, and the derivative of $(u(x))^n$ with respect to $x$ is $n(u(x))^{n-1}u'(x)$.

If $f(x)$ has a derivative at $x = a$ and $f'(x)$ is continuous at $a$, then the tangent line $y = f(a) + f'(a)(x - a)$ approximates the function, and $f(x) = f(a) + f'(c)(x - a)$ where $c$ is between $x$ and $a$. For example, if $f(x)$ is the displacement of a moving vehicle and $x$ is time, this says that the average velocity of the moving vehicle from $a$ to $x$ must actually be the speedometer reading at some previous time $c$. The essence of differentiability is that the function $f(x)$ may be approximated near $a$ by terms linear in the deviation from $a$. For $y = f(x)$, the differential of the dependent variable, $dy = f'(a)dx$, is just the change in $y$ that results along the tangent line to $y = f(x)$ at $x = a$ for a change in $x$ of $dx$.

In addition to the algebraic functions (ratios, sums, products, and differences of polynomials), the transcendental functions such as the trigonometric, hyperbolic, and exponential functions also have derivatives computed from the definition above and their special properties. The derivatives of the basic sine and cosine functions are $d\sin(x)/dx = \cos(x)$ and $d\cos(x)/dx = \sin(x)$. The quotient rule coupled with these two derivatives gives the derivatives of the remaining trigonometric functions. The most important transcendental function that has a derivative is $y = e^{ax}$, where $(d/dx)e^{ax} = ae^{ax}$. That is, the rate of growth of $y$ is proportional to the value of $y$. The fact that the rate of growth of a substance is proportional to the amount of the substance present is called the *mass action law* and is fundamental to financial, biological, and physical science calculations. For example, if $y$ is the amount of money on deposit at a bank and $x$ is time, and interest $dy$ is paid on this deposit at an annual rate of $a$ over a time period $dx$, then $dy = ay\,dx$ and $dy/dx = ay$.

If a function $w = f(x_1, x_2, \ldots, x_n)$ is a mapping from $n$-space into 1-space, the partial derivative of $w$ with respect to $x_i$ is found by treating all variables other than $x_i$ as constants and differentiating $w$ as if it were a function of the one variable $x_i$ alone. Symbols for this are $\partial w/\partial x_i$ or $w_{x_i}$ (sometimes $\partial f/\partial x_i$ or $f_{x_i}$ may also be used). The function $w$

may also have higher order partial derivatives $\partial^m w/\partial x_1^{a_1} \partial x_2^{a_2} \cdots \partial x_n^{a_n}$ where $a_1 + a_2 + \cdots + a_n = m$. This symbol represents the repeated differentiation ($m$ times) of the function $w$. This differentiation is performed $a_i$ times with respect to each of the variables $x_i$. As long as the resulting higher-order derivatives are continuous, the order of this repeated differentiation ($m$ times) does not matter.

A function $w = f(x,y)$ is differentiable at $(a,b)$ if it has partial derivatives $\partial w/\partial x$ and $\partial w/\partial y$, which are continuous. In this case the function has a tangent plane, $w = f(a,b) + f_x(a,b)(x - a) + f_y(a, b)(y - b)$, to the surface, $w = f(x, y)$, at $((a,b), f(a,b))$, which approximates this function. For $w = f(x,y)$, the differential of the dependent variable $w$ at $(a,b)$ is a function of four variables, $a$, $b$, $dx$, and $dy$ and is $dw = f_x(a,b)dx + f_y(a,b)dy$. Here $dx$ and $dy$ are respective arbitrary changes in the independent variables $x$ and $y$ away from $x = a$ and $y = b$.

The differentiation process is extended to cases where the input variable $x$ is a point in $n$-space and the output variable is $m$-space, for arbitrary positive integers $n$ and $m$, through the idea of linear approximation. The derivative of the mapping $(x_1, x_2, \ldots, x_n) \rightarrow (f_1(x_1, x_2, \ldots, x_n), \ldots, f_m(x_1, x_2, \ldots, x_n))$ is the $m$ by $n$ matrix $A$ ($m$ rows and $n$ columns) so that $\Delta f \approx A \Delta X$ where $\Delta x$ is the column vector $[h_1, h_2, \ldots, h_n]^T$ and $\Delta f$ is the column vector $[\Delta f_1, \Delta f_2, \ldots, \Delta f_m]^T$ with $\Delta f_i = f_i(x_1 + h_1, x_2 + h_2, \ldots, x_n + h_n) - f_i(x_1, x_2, \ldots, x_n)$. The entry in the $i$th row of $A$ and the $j$th column of $A$ is $\partial f_i/\partial x_j$. With this viewpoint, the derivative of $w = f(x,y)$ is the 1 by 2 matrix $[\partial f/\partial x, \partial f/\partial y]$. This matrix is a function of $(x,y)$. The differential of $w = f(x,y)$ mentioned above is then $[f_x(a,b), f_y(a,b)]$ multiplied by the column vector $[dx/dy]$.

*See also* Calculus, Overview

## SELECTED REFERENCES

Anton, Howard. *Calculus with Analytic Geometry*. 5th ed. New York: Wiley, 1995.

Hughes-Hallett, Deborah, Andrew M. Gleason, et al. *Calculus*. New York: Wiley, 1992.

Marsden, Jerrold E., and Anthony J. Tromba. *Vector Calculus*. 3rd ed. New York: Freeman, 1988.

Repka, Joe. *Calculus with Analytic Geometry*. Dubuque, IA: Brown, 1994.

Stewart, James. *Calculus*. 3rd ed. Monterey, CA: Brooks/Cole, 1995.

Thomas, George B., and Ross L. Finney. *Calculus and Analytic Geometry*. 8th ed. Boston, MA: Addison-Wesley, 1992.

BRUCE H. STEPHAN

# DIOPHANTINE EQUATIONS

Named in honor of Diophantus, equations whose only solutions of interest are rational numbers, usually integers. Diophantus, one of the early scientists who studied extensions of the Pythagorean Theorem, was a Greek mathematician who lived in Alexandria about A.D. 250 (LeVeque 1990).

Over the years, many people have studied a variety of Diophantine equations. A few of the more interesting and widely studied equations include the following:

(1) $x^2 + y^2 = z^2$,     $x, y, z > 0$
(2) $x^2 - dy^2 = 1$,     $x, y, d > 0$
(3) $ax + by = c$,     $a, b, c, x, y = $ integers

Any triplet of positive integers $(x, y, z)$ that satisfies (1) is known as a Pythagorean triple and geometrically represents the three sides of a right triangle. Examples include $(3, 4, 5)$, $(6, 8, 10)$ and $(5, 12, 13)$. There are, in fact, infinitely many examples of "really different" Pythagorean triples (i.e., one is not a multiple of another), and they are all generated by the equations (Eynden 1987)

$$x = 2st$$
$$y = t^2 - s^2$$
$$z = t^2 + s^2$$

where $s$, $t$ are relatively prime integers, one of $s$ and $t$ is even and the other is odd, and $t > s$. Setting $t = 7$ and $s = 4$, for example, produces the Pythagorean triple $(56, 33, 65)$.

Equation (2) is a special case of Pell's equation, named (a mistake made by Euler) after the seventeenth-century English mathematician John Pell (LeVeque 1990). The equation proves important in the study of the general quadratic equation in two variables $(ax^2 + bxy + cy^2 + dx + ey + f = 0)$. It will have infinitely many solutions in positive integers provided $d$ is not a perfect square. In this case, if $(x_0, y_0)$ is one particular solution, found by any method, then another solution $(x,y)$ can be found from the relationship (Niven and Zuckerman 1966)

$$x + y\sqrt{d} = (x_0 + y_0\sqrt{d})^k$$

for any positive integer $k$. With $d = 3$, the equation $x^2 - 3y^2 = 1$ has $(2, 1)$ for a solution. Thus another solution can be found by solving (with $k = 2$)

$$x + y\sqrt{3} = (2 + \sqrt{3})^2$$

which yields $x = 7$ and $y = 4$.

Linear Diophantine equations occur many times in the solution of algebra word problems. If, for instance, Billy spends $3.17 on a batch of pencils costing $0.11 each and some erasers costing $0.15 each, then this statement translates to the equation

$$11p + 15e = 317$$

This equation has infinitely many integral solutions (e.g., $p = 22$, $e = 5$), because the greatest common divisor of 11 and 15 happens to divide into 317. Unfortunately, not all the solutions are necessarily positive (e.g., $p = -1268$, $e = 951$), and this is typically determined by the sign of the coefficients in the equation (Long 1987). In the classroom, one need only graph the function $11x + 15y = 317$, which has slope $-11/15$ and $y$-intercept $317/15$ to see that the graph intersects quadrants I, II, and IV, and hence all integral solutions $(x,y)$ will consist of either two positive integers, or one positive and one negative.

*See also* Number Theory, Overview

### SELECTED REFERENCES

Devlin, Keith. "Fermat's Last Theorem." *Math Horizons* (Spring 1994):4–5.

Eynden, Charles Vanden. *Elementary Number Theory.* New York: Random House, 1987.

LeVeque, William J. *Elementary Theory of Numbers.* New York: Dover, 1990.

Long, Calvin T. *Elementary Introduction to Number Theory.* 3rd ed. Englewood Cliffs, NJ: Prentice-Hall, 1987.

Niven, Ivan, and Herbert Zuckerman. *An Introduction to the Theory of Numbers.* 2nd ed. New York: Wiley, 1966.

THOMAS P. DENCE

# DISCOVERY APPROACH TO GEOMETRY

A technique that minimizes the role of the teacher and textbook, allowing students to learn through inductive reasoning. It is a technique that encourages student involvement in the learning process, and provides practice in mathematical communication and independent thinking. Socratic questioning may be considered a precursor to (guided) discovery learning. Socrates (ca. 400 B.C.) used directed dialogue to elicit answers from students that would promote insights into the subject under discussion. Thus Socrates not only taught the subject but encouraged his students to think for themselves. As described by Socrates in the *Great Dialogues of Plato:* ". . . the object was that of a midwife to bring other men's thoughts to birth." In the late nineteenth century, John Dewey (1859–1952) introduced "reflective thinking," which challenged stu-

dents to consider actively any belief or supposed form of knowledge. This method was a forerunner of learning by discovery.

In 1938, William Reeve, Halbert Christofferson, and Harold Fawcett (Fawcett 1938) wrote of the importance of learning geometry through the discovery method. They found the technique not only taught geometric facts, but gave the students a practical approach useful in daily living. In the 1950s, advocacy of the discovery approach continued: *"Students should be encouraged to use inductive procedures to help them make discoveries and lead to conjectures concerning the problem at hand.* [authors' italics] . . . Mathematics is a deductive system, but induction certainly plays a great role in the discovery and creative aspects of the subject" (Henderson and Pingry 1953, 260). Educational reform leaders such as Max Beberman (1958) and Robert Davis (1964) recommended an atmosphere of discovery in the mathematics classroom. The theory was that with respect and intellectual challenge, students would not only discover facts but also would be learning to think creatively in ways the adult world demanded. In the 1980s, geometry textbooks (Chakerian, Crabill, and Stein 1986; Serra 1986) based on the discovery approach were published and geometry education moved in that direction. Many cognitive theorists believe that the nature of geometric knowledge does not require sequential learning, unlike other subjects where order is of greater significance. For this reason, geometry is well suited to the discovery approach. An experimental discovery learning technique that uses neither textbook nor traditional lecture was introduced in 1987 (Healy 1993). The teacher functions as a facilitator, advisor, and questioner for the students. Given three basic theorems at the start of the course, the students make geometric conjectures, debate their findings, and build their own text. This method is based on the premise that high school students not only can learn geometry through their own investigations, but will sharpen their critical thinking in the process.

Geometry is currently the subject at the secondary level that uses the most discovery learning. The number of teachers who use it is limited; reasons include the need to fulfill a mandated curriculum or the belief that some students benefit from more structure.

*See also* Constructivism; Cooperative Learning; Geometry, Instruction; History of Mathematics Education in the United States, Overview; Instructional Methods, New Directions; Moore

## SELECTED REFERENCES

Beberman, Max. *An Emerging Program of Secondary School Mathematics.* Cambridge, MA: Harvard University, 1958.

Chakerian, G.D., Calvin D. Crabill, and Sherman K. Stein. *Geometry: A Guided Inquiry.* Pleasantville, NY: Sunburst, 1986.

Davis, Robert. *Discovery in Mathematics.* Reading, MA: Addison-Wesley, 1964.

Dewey, John. *How We Think.* Boston, MA: Heath, 1910.

Fawcett, Harold P. *The Nature of Proof.* 13th Yearbook. National Council of Teachers of Mathematics. New York: Bureau of Publications, Teachers College, Columbia University, 1938.

Healy, Christopher. *Build-A-Book Geometry.* Berkeley, CA: Key Curriculum, 1993.

Healy, Jane M. *Endangered Minds: Why Our Children Don't Think—& What We Can Do About It.* New York: Simon and Schuster, 1989.

Henderson, Kenneth B., and Robert E. Pingry. "Problem-Solving in Mathematics." In *The Learning of Mathematics: Its Theory and Practice.* 21st Yearbook. Howard F. Fehr (ed.). Washington, DC: National Council of Teachers of Mathematics, 1953.

Plato. *Great Dialogues of Plato.* New York: New American Library, 1956.

Serra, Michael. *Discovering Geometry.* Berkeley, CA: Key Curriculum, 1986.

CHRISTOPHER C. HEALY

# DISCRETE MATHEMATICS

No precise definition exists, but most mathematicians agree that it is the mathematics of finite and countably infinite systems, including such topics as combinatorics, difference equations, graph theory, and linear programming. Mathematicians at the college and university level concur that discrete mathematics is an essential part of the mathematics curriculum. The curriculum standards of the National Council of Teachers of Mathematics (NCTM) support this view and recommend an expanded role for discrete mathematics at both the elementary and secondary levels (NCTM 1989). The power of discrete mathematics to contribute to the cognitive development of students in elementary, secondary, and collegiate settings has been recognized since the 1950s. An early, innovative college textbook for freshmen and sophomores was the *Introduction to Finite Mathematics* (Kemeny, Snell, and Thompson 1956). This book was developed at Dartmouth College at about the same time that John Kemeny, Thomas Kurtz, and their colleagues were introducing the programming language, BASIC, in a computer course required of all students at the college. In the more than forty years since the publication of that book, many more have been written. For other examples, see Goldstein, Schneider, and Siegel (1995) and Lial (1993).

Among the purposes of the "finite mathematics" course was an early introduction to concepts of modern mathematics and to applications in the social and biological sciences. The course included compound statements (logic), sets and subsets, partitions and counting, probability theory, vectors and matrices, linear programming and the theory of games, and applications to behavioral science. Many of these topics still appear in the typical syllabus of a discrete mathematics course.

Discrete mathematics is linked with the growth of computing and the sophistication of computers and communication networks. In fact, although courses in number theory, combinatorics, operations research, probability, and even graph theory were offered at the upper division at many colleges and universities in the 1950s, it was in the computer science major that a discrete mathematics course first was offered at the freshman-sophomore level as a separate and distinct field of study from mathematics. The driving force is the importance of the computer and computing to not only solving discrete mathematics problems, but to generating new problems linked by their very structure to the nature of a computing machine. This area of mathematics continues to be rich in its own right with interesting and challenging problems and applications to many disciplines.

Although even simply stated problems in discrete mathematics can be very difficult to solve, often it is possible for students, with few specific algebraic and geometric skills, to become engaged in solving interesting and elementary problems in this area quickly. Thus, cooperative learning and group work (pedagogical methods now recommended by many educators) can be applied to many aspects of discrete mathematics, such as graph theory and combinatorics, and can be used with students at every grade level (see Kenney 1991). Textbook and ancillary materials for grades K through 12 are now available for introducing students to these topics, both on a theoretical level and with applications that relate to their interests (Core-Plus Mathematics Project 1998; Usiskin et al. 1993), for example.

The cognitive techniques for handling many problems in this area are linked to algorithmics and hence are ideal for helping to develop good habits of thought and mathematical maturity. Recursion, induction, and construction (and eventual proof of correctness) of algorithms form the basis of many courses or extracurricular materials. This reenforces the relationship of mathematics to computing and teaches problem-solving techniques applicable across the curriculum.

In addition, difference equations and graphs can be used to model many phenomena and provide ways of introducing applications in biology, business, social behavior, sports, finance, as well as physical and chemical processes. Dynamical systems and fractals can be introduced as discrete mathematical topics and can provide fascination for students of all ages. Thus, discrete mathematics is useful, provides stimulating problems, and develops cognitive skills.

To put the present situation into perspective, we offer some history of recent curricular reform. In 1982, *The Future of College Mathematics,* a conference held at Williams College, focused on the curriculum of the first two years of college. As an outgrowth of that meeting, the Alfred P. Sloan Foundation provided funds for the development of elementary discrete mathematics courses in tandem with revision of the calculus sequence at six schools. The Mathematical Association of America (MAA) Committee on Discrete Mathematics in the First Two Years, also funded by Sloan, with representation from mathematics, engineering, and computer science, held many public forums on the role of discrete mathematics in the curriculum. The committee published its recommendations for such courses in 1985 (Siegel 1985). Numerous experiments began, with various ways of integrating discrete and continuous mathematics in the first two years, and they were in some ways the direct precursors of the calculus reform movement. These are summarized in Ralston (1989). While the 1985 report was the point of departure for many authors and schools, there has been continual evolution since then. There were scarcely any textbooks available for this level prior to 1983, but by 1994 there were more than forty books available for this market. Coupled with the changes in calculus that have added discrete methods to the continuous ones, we have seen a remarkable change in the way we approach these elementary courses and the way in which a discrete approach has been integrated into the undergraduate curriculum.

The emphasis on discrete mathematics at the collegiate level can be categorized roughly as being of several types. In addition to the influence on calculus itself, there are courses at the freshman-sophomore level designed primarily for mathematics, computer science, and electrical engineering majors. There the emphasis is on algorithmics, or on algebraic systems, or on difference equations and models, or on logic and proof using number theory, combinatorics, and recursion as possible vehicles. See representative texts such as Ross and Wright (1992), Epp (1995), and Johnsonbaugh (1993).

For students with less professional interest in the mathematical sciences, we find an emphasis on finite mathematics—matrices, linear models, discrete probability, graphs, and elementary difference equations—with applications to other disciplines.

At the junior-senior level, traditional courses from the 1950s and 1960s have been updated and serve mathematics, computer science, and electrical engineering majors with one-semester courses in several different concentrations, such as applied algebra, combinatorics, operations research, graph theory, logic and set theory, probability, discrete dynamical systems, and coding theory. Possible texts are Tucker (1995) and Grimaldi (1989).

For teacher preparation, collaborative efforts in the schools, colleges of education, and collegiate mathematics and science departments (National Science Foundation (NSF)-funded systemic initiatives and/or collaboratives) have given rise to new content courses that include more discrete models and their underlying mathematics. Courses for preservice elementary education majors have been designed to include elementary number theory, graph theory, probability, discrete models, logic, and sets so that teachers can meet currently accepted standards and feel comfortable with modern school materials.

Significantly, since the number of computer science majors has decreased through the 1990s, the number and variety of discrete mathematics textbooks have not continued to increase as they did in the 1980s. On the other hand, discrete mathematics has pervaded the curriculum in many ways, and quite distinct from its fundamental role as a computer science adjunct. For example, the general survey course in introductory college mathematics, *For All Practical Purposes* (Garfunkel 1994) is a popular way to offer a nontrivial "general education" course. Topics such as street networks, the traveling salesman problem, planning and scheduling, linear programming, statistics, coding and information transfer, geometries, symmetry and tilings, and social choice and decision making are handled in lively and applied settings, enhanced by a series of videotapes highlighting the applications. Discrete mathematics courses in secondary school are not uncommon (Dossey et al. 1992).

As an alternate to having mathematics and science majors take calculus as the first serious mathematics course, some colleges are requiring students (even those who have already had calculus) to begin with a course in discrete dynamical systems (Sandefur 1993). In addition, as the next step in the calculus-reform movement, begun in the 1980s, precalculus is now being modernized, and such "reform" courses focus on discrete models and difference equations, retaining the traditional emphasis on the concept of function (Gordon et al. 1995). As computers and mathematical modeling become even more prevalent and complex, discrete mathematics will help us route the snowplows, supply refugee camps, settle disputes, manage an airplane fleet, and finance and build long-term projects with added efficiency and optimal (or near optimal) results.

*See also* Combinatorics; Curriculum Trends, College Level; Graph Theory, Overview; Linear Programming; Tiling

## SELECTED REFERENCES

Core-Plus Mathematics Project. *A Balanced Approach to Mathematics Education: Contemporary Mathematics in Context.* Chicago, IL: Everyday Learning, 1998.

Dossey, John A., Albert D. Otto, Lawrence E. Spence, and Charles Vanden Eynden. *Exploring Discrete Mathematics.* Reading, MA: Addison-Wesley, 1992.

Epp, Susanna S. *Discrete Mathematics with Applications.* 2nd ed. Boston MA: PWS, 1995.

Garfunkel, Solomon, proj. dir. *For All Practical Purposes.* 3rd ed. New York: Freeman, 1994.

Goldstein, Larry, David Schneider, and Martha Siegel. *Finite Mathematics and Its Applications.* 5th ed. Englewood Cliffs, NJ: Prentice-Hall, 1995.

Gordon, Sheldon, et al. *Functioning in the Real World: A PreCalculus Experience.* Prel. ed. Reading, MA: Addison-Wesley, 1995.

Grimaldi, Ralph P. *Discrete and Combinatorial Mathematics: An Applied Introduction.* 2nd ed. Reading, MA: Addison-Wesley, 1989.

Johnsonbaugh, Richard. *Discrete Mathematics.* 3rd ed. New York: Macmillan, 1993.

Kemeny, John, J. Laurie Snell, and Gerald Thompson. *Introduction to Finite Mathematics.* 1st ed. Englewood Cliffs, NJ: Prentice-Hall, 1956. [3rd ed. 1974.]

Kenney, Margaret J., ed. *Discrete Mathematics across the Curriculum K–12.* 1991 Yearbook. Reston, VA: National Council of Teachers of Mathematics, 1991.

Lial, Margaret, Charles Miller, and Raymond Greenwell. *Finite Mathematics.* 5th ed. New York: HarperCollins, 1993.

National Council of Teachers of Mathematics. *Curriculum and Evaluation Standards for School Mathematics.* Reston, VA: The Council, 1989.

———. *Principles and Standards for School Mathematics.* Reston, VA: The Council, 2000.

Ralston, Anthony, ed. *Discrete Mathematics in the First Two Years.* Washington, DC: Mathematical Association of America, 1989.

Rosenstein, Joseph G., Deborah S. Franzblau, and Fred S. Roberts, eds. *Discrete Mathematics in the Schools.* Reston, VA: American Mathematical Society and National Council of Teachers of Mathematics, 1997.

Ross, Kenneth A., and Charles R. Wright. *Discrete Mathematics.* 3rd ed. Englewood Cliffs, NJ: Prentice-Hall, 1992.

Sandefur, James T. *Discrete Dynamical Modeling.* New York: Oxford University Press, 1993.

Siegel, Martha, ed. *Final Report of the MAA Committee on Discrete Mathematics in the First Two Years.* Washington, DC: Mathematical Association of America, 1985. Reprinted with *Afterthoughts, 1989* in Ralston (1989) and with *1989 Preface* in Steen (1989).

Steen, Lynn Arthur, ed. *Reshaping College Mathematics,* Washington, DC: Mathematical Association of America, 1989.

Tucker, Alan. *Applied Combinatorics.* 3rd ed. New York: Wiley, 1995.

Usiskin, Zalman, et al. *University of Chicago School Mathematics Project.* Glenview, IL: Scott, Foresman, 1993.

## ADDITIONAL RESOURCES

Committee on Calculus and the First Two Years (CRAFTY) of the Mathematical Association of America.

SIAM Special Interest Group in Discrete Mathematics, Society for Industrial and Applied Mathematics, 3600 University City Science Center, Philadelphia, PA 19104–2688, siam@siam.org.

DIMACS at Rutgers University under the direction of Fred Roberts, Rutgers University, New Brunswick, NJ 08903.

MARTHA J. SIEGEL

# DIVISIBILITY TESTS

Rules for determining whether or not a number is divisible by some given number. At the school level, the commonly used rules are for divisibility by 2 through 11. The rules for 7 and 11 are used less often as they are not readily justified at the elementary level. The divisibility tests are usually presented in treatments on number theory and are a testimony to the quest for patterns in number (Dantzig 1967).

## MOTIVATION AND GOALS

The primary motivation for including rules for divisibility in a mathematics curriculum is as a problem-solving tool. Learners can be encouraged to develop the rules from appropriately guided examples.

Two justifications for teaching divisibility rules are (a) for the expression of numbers as products of prime factors and (b) the reduction of fractions to simplest terms.

## MATHEMATICAL EXPLANATION

These rules are:

1. A number is divisible by 2 if the last digit is divisible by 2. Thus 3168 is divisible by 2 because 8 is divisible by 2. In alternate terminology, "2 divides 3168" or 2 is a factor of 3168.
2. A number is divisible by 3 if the sum of the digits is divisible by 3. Thus 3168 is divisible by 3 because $(3 + 1 + 6 + 8)$ is divisible by 3.
3. A number is divisible by 4 if the last two digits are divisible by 4. Thus, 3168 is divisible by 4 because 68 is divisible by 4.
4. A number is divisible by 5 if the last digit is 0 or 5. Thus 3175 is divisible by 5.
5. A number is divisible by 6 if it is divisible by 2 and by 3. Thus 3168 is divisible by 6 because it is divisible by both 2 and 3.
6. A number is divisible by 7 if the number represented without its units digit, minus twice the units digit of the original number, is divisible by 7. Thus 2968 is divisible by 7 because 7 divides $(296 - 2 \times 8)$ or 280.
7. A number is divisible by 8 if the last three digits are divisible by 8. Thus 8 divides 3168 because 8 divides 168.
8. A number is divisible by 9 if the sum of the digits is divisible by 9. Thus 9 divides 3168 because it divides $(3 + 1 + 6 + 8)$.
9. A number is divisible by 10 if the last digit is 0.
10. A number is divisible by 11 if the difference between the sums of the alternating digits is divisible by 11. Thus 11 divides 3168 because it divides $[(3 + 6) - (1 + 8)]$.

The rules for divisibility by 3 and 9 can be justified either numerically or visually by observing that dividing a power of 10 by 3 (or 9) leaves a remainder of 1; thus a number can be divided by 3 (or 9) if it divides the sum of the remainders. That is, consider $n = a \times 10^3 + b \times 10^2 + c \times 10 + d = a(999 + 1) + b(99 + 1) + c(9 + 1) + d = (a \times 999 + b \times 99 + c \times 9) + (a + b + c + d)$. Since 9 divides $(a \times 999 + b \times 99 + c \times 9)$, then 9 divides $n$ if 9 divides $(a + b + c + d)$, the sum of the digits.

An application of divisibility by 9 in mental arithmetic is "casting out 9s." The sum of the digits of the addends equals the sum of the digits of the computed sum. It may be used as a check on the accuracy of calculations performed mentally or by hand. Ready accessibility of calculators has made this application less important.

## VARIETY OF PEDAGOGICAL APPROACHES

Most curriculum presentations focus on the rules for divisibility as a mathematical curiosity. It certainly is a good problem-solving application to test whether 36 divides 1,728 or to find the missing digit, A, such that 36 divides 1,7A8. It is rare, however, for school textbooks to include the application of the tests in checking whether or not a given number is prime, or can be divided by a particular single digit integer. This has direct application to expressing a number as a product of its prime factors, an important pre-algebra skill. Equally important is the application to the simplification of fractions. For example, to simplify the fraction 30/72, we note that 6 is a factor of both the numerator and denominator, because both 30 and 72 are even, and for both, the sum of the digits is a multiple of 3.

## RECOMMENDED MATERIALS FOR INSTRUCTION

In addition to the mental exploration involved in studying number patterns, instruction in divisibility rules is enhanced if visual or tactile representations (illustrating numbers on grid paper or manipulating blocks) provide learners with concrete models to show why a given divisibility test works.

*See also* Factors; Fractions; Number Theory, Overview

SELECTED REFERENCES

Bezuszka, Stanley J. "A Test for Divisibility by Primes." *Arithmetic Teacher* 33(Oct. 1985):36–38.

Dantzig, Tobias. *Number, the Language of Science.* New York: Free Press, 1967.

Heddens, James W., and William R. Speer. *Today's Mathematics. Part 1: Concepts and Classroom Methods.* Englewood Cliffs, NJ: Prentice-Hall, 1995.

DAVID M. DAVISON

# DIVISION

First encountered in school as a procedure using *sharing* or *grouping* (repeated subtraction). Most elementary school texts will suggest experience with sharing and grouping before representing division in symbolic form. These two aspects of division arise from different types of concrete examples. The problem, "12 marbles are shared among 3 children. How many marbles does each child get?" suggests a sharing procedure while "Marbles are sold in packs of 3. How many packs can be filled with 12 marbles?" suggests a grouping or repeated subtraction approach.

(a) 12 marbles shared among 3 children

(b) 12 marbles boxed in 3s

Symbolic representation of both these problems as 12 ÷ 3 may present children with the first ambiguity in division where a single symbol "÷" is used to signify two different procedures that may be used to find a solution. (This symbol for division first appeared in print in 1659 in a book on algebra by the Swiss Johann Rahn (1622–1676) and became known more widely when the work was translated into English. Modeling the problem 12 ÷ 3 with materials can lead to both sharing and repeated subtraction procedures. In each procedure, the number 3 is formally referred to as the *divisor*, the number 12 as the *dividend* and the number 4 as the *quotient*. At a later stage in school mathematics the symbol "÷" is used less frequently as the expression 12 ÷ 3 is replaced by the ratio *a/b*, which has the same value as *a* divided by *b*.

As well as sharing and repeated subtraction, the problem "12 divided by 3" is effectively tackled by noting that "the sum for 3 fours is 12" and that division by 3 is the inverse operation to multiplication by 3. Once it is recognized that each division problem has a multiplicative counterpart, the symbolic problem 12 ÷ 3 takes on yet another meaning, "How many 3s add up to 12?" and the process of sharing or repeated subtraction may become implicit rather than explicit in solution procedures. Familiarity with the multiplication facts will enable children to solve some division problems by referring to the associated multiplication fact. An alternative symbolic presentation of division problems reverses the written position of the divisor and dividend (for example, $3\overline{)12}$) to enable extension to a written algorithm. This has implications for the language of division and is known to lead to confusion over whether "3 divided into

12" or "12 divided into 3" is the alternative to "12 divided by 3," and may suggest a more abstract approach to the calculation. Although this representation may simplify division of larger numbers, some problems will present substantial difficulties to the child who relies on a written algorithm or on a sharing procedure, for example, "68 divided by 17." But when it is related to the fact that "17 and 17 make 34, which is half of 68," its solution becomes clear.

## REMAINDERS

When children have substantial experience of division problems with whole number solutions, division with remainder is introduced. It should be noted that with ready availability of calculators in the classroom the idea of a remainder or items "left over" must now be reconciled with the result of a calculation using the division key, for example, 10 ÷ 4 = 2 remainder 2, or 10 ÷ 4 = 2.5. In real-life problems, the sharing procedure may be continued so that even "leftovers" are divided up. For children used to sharing at home, the idea of a remainder may be puzzling. Remainders also may be expressed as fractions; and the problem "10 ÷ 4" related to the problem "10 pizzas shared among 4 people" will result in the outcome of "2 and $\frac{1}{2}$ pizzas each." Certainly, the word *remainder* will not be familiar to young children, and ways to record a remainder will need to be discussed.

Some high school children assume that the decimal point is a "grown up" way of expressing remainders; thus, 13 ÷ 4 = 3 remainder 1 erroneously becomes 13 ÷ 4 = 3.1. Here it will be necessary to discuss what is meant by 3.1. Another error among children is to find a closely related result and use the remainder to indicate how far away the exact result is. For example, children sometimes use 23 ÷ 6 = 4 remainder 1 to indicate that 24 ÷ 6 is one unit away from the exact result. In both these cases, it will be helpful to use counters for sharing, or to draw illustrations to show how the answer is found.

When division is explored on a calculator, some problems will result in surprises. The outcome of a problem like 10 ÷ 3 on a calculator will be 3.3333333, and the result of 15/7 will be 2.1428571. This is due to the fact that some fractions have a nonterminating representation as decimals. In the problems above, one-third is represented as .3333333 and one-seventh as .1428571. Where such results are found in division problems, decisions will need to be made about a "sensible" solution and decimals may be "rounded" to the nearest decimal with fewer places or "truncated" by simply stopping after a given number of places. The answer to 20/3 may be 6.7 if it is rounded, or 6.6, if it is truncated. Since each of these answers is an approximation, either may be adequate, although 6.7 is nearer the exact answer.

## WORKING WITH LARGE AND SMALL NUMBERS

Although the idea of "sharing" may provide an initial base for understanding division, it also may develop in children the misconception that division will always lead to a smaller answer than the number at the start. It appears to be a common misconception that "multiplication makes bigger" and "division makes smaller," although multiplication and division by numbers smaller than 1 will not conform to such conclusions. In order to interpret the problem $6 \div \frac{1}{2}$ and find the solution 12, it is helpful to interpret the problem in terms of multiplication as "How many halves in 6?" rather than "6 shared by $\frac{1}{2}$. On the other hand, it may not be effective to use this interpretation for "How many 6s in 6,000?" Better in this example to visualize 6,000 partitioned into 6 sets each of 1,000 and read the question as 6,000 shared among 6.

Sometimes a distinction is made between *short division* where the divisor is a single digit—for example, $246 \div 6$—and the answer may be found mentally, and *long division,* where large numbers are involved—for example, $5632 \div 24$—and the calculation needs to be done with pencil and paper. The standard algorithm for dividing large numbers starts by representing the problem as $24 \overline{)5632}$ and can be very complex, involving estimating, multiplying, and subtracting partial sums with difficulties in recording answers appropriately at each stage. There has been much controversy about the relevance of practicing such problems when calculators are normally used in the workplace and children's invented methods enable them to reach a solution. Written methods can be developed, based on students' own understanding. One method is a *standard* pencil-and-paper method in which 23 is divided into 495 by considering first the 100s, then the 10s, and then the units. At each stage a subtotal is calculated, subtracted from the total, and the remainder is added to the remaining digits. Another method uses estimates that are known facts or easy calculations and the total (495) is reduced in stages by subtracting multiples of 23 until no further multiples can be subtracted. Where a number does not belong to a relevant multiplication table, for example, $23 \div 6$, approximations are helpful and decisions must be made by the child about the most appropriate solution procedure. Children may start by tackling an associated problem ($24 \div 6$) rather than using a direct procedure for the solution.

## FRACTIONS AND DECIMALS

Division using fractions and decimals presents considerable problems, particularly when children's concepts are built on sharing and grouping. Meanings need to be attached to such expressions as $6 \div \frac{1}{2}, \frac{1}{2} \div 6, 0.2 \div 6$, and $6 \div 0.2$ before expressions like $\frac{1}{2} \div \frac{3}{4}$ and $0.2 \div 0.3$ can be tackled effectively. Rules like "turn the second fraction upside down and then multiply" do not make sense to children who may lose confidence in their understanding and have difficulty memorizing such "tricks." This type of calculation is easier to consider if children are first introduced to *multiplicative inverses.* When two fractions multiplied together give the result 1, they are said to be multiplicative inverses, for example, $\frac{2}{3} \times \frac{3}{2} = 1, \frac{4}{5} \times \frac{5}{4} = 1$ so $\frac{2}{3}$ and $\frac{3}{2}$ are multiplicative inverses, $\frac{4}{5}$ and $\frac{5}{4}$ are multiplicative inverses. Now to calculate $\frac{2}{3} \div \frac{4}{5}$ first write it as

$$\frac{\frac{2}{3}}{\frac{4}{5}}$$

and then multiply the top and the bottom by $\frac{5}{4}$. The calculation then becomes

$$\frac{\frac{2}{3} \times \frac{5}{4}}{\frac{4}{5} \times \frac{5}{4}}$$

Now notice that the bottom has the value 1. So the calculation becomes $\frac{2}{3} \times \frac{5}{4}$.

## DIVIDING BY ZERO

A difficulty in division arises when 0 is involved. Even adults may be confused over problems like $0 \div 2$ and $2 \div 0$. The first is the easier to explain: if there is nothing to be divided then the result must always be nothing, that is, 0 divided by any nonzero number will always result in 0. To understand the second problem, $2 \div 0$, it is helpful to investigate what happens when 2 is divided by ever-decreasing numbers:

$$2 \div 2 = 1$$
$$2 \div 1 = 2$$
$$2 \div 0.5 = 4$$
$$2 \div 0.1 = 20$$
$$2 \div 0.01 = 200$$
$$2 \div 0.001 = 2000$$

As the divisor gets smaller, the quotient gets larger. The result of $2 \div 0$ is not finite. Mathematicians say that the result of dividing any nonzero number by 0 is undefined and division of 0 by 0 is indeterminate.

*See also* Fractions; Multiplication; Zero

## SELECTED REFERENCES

Anghileri, Julia. *Children's Mathematical Thinking in the Primary Years.* London, England: Cassell, 1995.

Boyer, Carl. *A History of Mathematics.* New York: Wiley, 1968.

Durkin, Kevin, and Beatrice Shire, eds. *Language in Mathematical Education: Research and Practice.* Philadelphia, PA: Open University Press, 1991.

Eves, Howard W. *An Introduction to the History of Mathematics.* Rev. ed. New York: Holt, Rinehart and Winston, 1964.

Post, Thomas R., ed. *Teaching Mathematics in Grades K–8: Research Based Methods.* 2nd ed. Boston, MA: Allyn and Bacon, 1992.

JULIA ANGHILERI

# E

## e

A number that serves as a base for the natural logarithm (ln $x$) or "Napierian" logarithm (although Napier did not invent this type of logarithm). Leonhard Euler in a letter (1731) to Christian Goldbach used the notation $e$. In 1744, Euler proved that $e$ is irrational, and in 1873, Charles Hermite proved it transcendental.

It is shown in calculus that

$$e = 1 + \frac{1}{1!} + \frac{1}{2!} + \frac{1}{3!} + \cdots$$

and

$$e = \lim_{n \to \infty} \left(1 + \frac{1}{n}\right)^n$$

Numerically, $e = 2.7182818284\ldots$ . Euler gave an expression for $e$ using continued fractions:

$$e = 2 + \cfrac{1}{1 + \cfrac{1}{2 + \cfrac{1}{1 + \cfrac{1}{1 + \cfrac{1}{4 + \cfrac{1}{1 + \ldots}}}}}}$$

$$= [2, 1, 2, 1, 1, 4, 1, \ldots, 1, 2n, 1, \ldots].$$

The number $e$ is closely connected with exponential and logarithmic functions. We use the exponential function to describe population growth and say that the population grows exponentially. For example, a colony of bacteria grows at a rate $f'(t)$ proportional to the number $f(t)$ present at time $t$, that is, $f'(t) = kf(t)$. Then $f(t) = f(0)e^{kt}$ for $t > 0$ and $k > 0$. If $k < 0$, then $f(t)$ represents negative growth, as, for example, in radioactive decay. Also, the amount $A$ after $t$ years accrued on a principal $P$, invested at an annual rate of interest $r$ compounded continuously, is $A = Pe^{rt}$. The problem was originally considered by Jacob Bernoulli.

Again, it is shown in calculus that if $f(x) = e^x$ then the derivative $f'(x) = f(x)$. Conversely, if $f'(x) = f(x)$ for all $x$ and $f(0) = 1$, then $f(x) = e^x$. The inverse function of $f(x) = e^x$ is the logarithmic function, that is, $f^{-1}(x) = \log_e x$ or ln $x$ for $x > 0$. Logarithms were used as an aid to computation but this use is now obsolete due to the invention of computers. Euler showed that $e^{ix} = \cos x + i \sin x$, providing a link between $e$ and the trigonometric functions. In particular, $e^{i\pi} = -1$.

*See also* Continued Fractions; Interest; Irrational Numbers; Logarithms; Napier; Transcendental and Algebraic Functions; Transcendental Numbers

SELECTED REFERENCES

Apostol, Tom M. *Calculus*. Vol. 1. 2nd ed. Waltham, MA: Blaisdell, 1967.

Hardy, Godfrey H. *A Course of Pure Mathematics*. Cambridge, England: Cambridge University Press, 1958.

Maor, Eli. *e: The Story of a Number*. Princeton, NJ: Princeton University Press, 1998.

Olds, Carl D. *Continued Fractions*. New York: Random House, 1963.

Struik, Dirk J. *A Concise History of Mathematics*. 4th ed. New York: Dover, 1987.

LEE PENG YEE

# ECONOMICS

A field that has made substantial use of mathematics for several centuries. Distinguished mathematicians, notably Daniel Bernoulli (1738) and Antoine A. Cournot (1838), were among the earliest and most substantial contributors. During most of its history, mathematical economics has relied heavily upon the differential calculus in its theoretical research. The most obvious and fruitful application is to analysis of the behavior of decision-making individuals or bodies whose goal is maximization or minimization of some variable. For example, in economic theory it is customary to assume that the only goal of the business firm is maximization of profits. Economists are well aware that the motives of those who make business decisions are more complex, but they believe the oversimple premise (that profit is the exclusive object of the enterprise) is sufficiently close to the truth to be capable of providing substantial insights into business behavior. This premise does, indeed, yield results that are often illuminating and sometimes surprising. An elementary example relates to business responses to changes in the magnitude of "fixed costs," that is, costs that are unaffected by a change in the quantity of output produced by the firm. Thus, suppose that a firm rents a piece of land that the landlord supplies on an all-or-none basis, initially renting it to the company for $1 million per year. This part of the firm's cost is fixed, because it remains the same $1 million whether the company produces one hundred or ten thousand units of output per year on the land.

Suppose now that the annual rent is suddenly doubled. How much will it pay the firm to change the price of its product in response? The precise answer *seems* hard to guess, but most observers think it will certainly pay the company to raise the price to some extent, to shift part of the added rent burden on to the customer. The differential calculus shows, however, that this guess is wrong and that it is best for the profit-maximizing firm to leave its price absolutely unchanged when this rent increase occurs. How does calculus show this? The volume of sales that maximizes profit is the one at which the first derivative of profit with respect to sales volume equals zero—the most basic result on maximization in the differential calculus. But the added rent simply amounts to the subtraction of a constant term from profit, and we know that the subtraction of a constant from any function does not change its first derivative. Consequently, the added rent does not change the derivative of the firm's profits—meaning that to maximize profits it should arrange to sell the same output as before. If, instead, the firm were to raise its product price, that would simply lead to a reduction in consumer purchases. That last observation explains our mathematical result—that it does not pay a profit-maximizing firm to raise its product price when its fixed costs increase. The price rise will just drive away customers and so must fail to preserve profit.

Mathematical analysis of the sort just described has been used to study how firms react to tax or to interest rate changes, how consumers react to changes in the price of goods they purchase, and so on. It has been used to guide the work of regulatory agencies such as the Federal Communications Commission (FCC) and the Interstate Commerce Commission (ICC). It also has had a wide variety of other applications. Mathematical economics uses many mathematical tools as well as differential calculus. Its tools range from elementary algebra to sophisticated analytic approaches such as set theory and topology (see, for example, Von Neumann and Morgenstern [1947] and Arrow [1963]).

Some decades ago a new mathematical analysis, game theory, was introduced, with economic behavior intended as its primary application. The theory of games was the joint product of mathematician John von Neumann and economist Oskar Morgenstern. The theory can be thought of as an investigation of the possible outcomes when two rival firms consider the strategies they can use in dealing with one another, just as one can analyze how two players in a fiercely competitive game can decide to respond to each other's moves. Since game theory was first introduced, its use by economists has expanded greatly and it is now a required subject for students who specialize in study of the behavior of firms and industries.

*See also* Finance; Game Theory; Interest

## SELECTED REFERENCES

Arrow, Kenneth J. *Social Choice and Individual Values.* New Haven, CT: Yale University Press, 1963.

Bernoulli, Daniel. "Exposition of a New Theory on the Measurement of Risk." *Econometrica* 22(Jan. 1954): 23–36. [Louise Sommer, trans., from the Latin original in *Papers of the Imperial Academy of Sciences in St. Petersburg* 5(1738):175–192.]

Cournot, Antoine A. *Researches into the Mathematical Principles of the Theory of Wealth.* Nathaniel T. Bacon, trans. New York: Macmillan, 1897. [French original. Paris, France: Hachette, 1838.]

Theochares, Reginos D. *The Development of Mathematical Economics.* New York: Macmillan, 1993.

Varian, Hal R. *Intermediate Microeconomics: A Modern Approach.* New York: Norton, 1990.

Von Neumann, John, and Oskar Morgenstern. *Theory of Games and Economic Behavior.* 2nd ed. Princeton, NJ: Princeton University Press, 1947.

WILLIAM J. BAUMOL

# EDUCATIONAL STUDIES IN MATHEMATICS (ESM)

International journal founded in 1970 by Hans Freudenthal in response to the emergence of mathematics education as an independent field of scientific research as indicated by dedicated institutionalizations at various universities. At about the same time, other journals were started as well. The general policy of ESM has remained essentially unchanged over the years: to present "new ideas and developments which are considered to be of major importance to those working in the field of mathematical education. It seeks to reflect both the variety of research concerns within this field and the range of methods used to study them. It deals with didactical, methodological and pedagogical subjects rather than with specific programmes for teaching mathematics. All papers are strictly refereed and the emphasis is on high-level articles which are of more than local or national interest" (ESM, inside front cover of every journal issue).

Leading editorial principles are: highest standards of articles selected to be published; breadth of themes, topics, and methods; openness to new developments and research paradigms; internationality of contributors and readers. The editors are supported by the members of an International Editorial Board. Since it began, the journal has grown considerably and its development reflects that of mathematics education with respect to content and scientific relevance. Special issues are dedicated to topics of great

importance or broad interest, but also try to promote areas of research that are felt to be in need of enhancement. Among topics recently featured are assessment, radical constructivism, mathematics and gender, theories of mathematics education, and the legacy of Hans Freudenthal. The growing number of contributions from many different countries indicates the dynamic growth of the science of mathematics education.

## SELECTED REFERENCE

*Educational Studies in Mathematics,* ISSN 0013-1954. 2 volumes of 4 issues per year. Dordrecht, Netherlands: Kluwer.

WILLIBALD DÖRFLER

# EGYPTIAN MATHEMATICS

This entry discusses ancient Egyptian mathematics of the Pharaonic period, 3100 B.C. to 332 B.C. Historical sources include surviving mathematical papyri, inscriptions on monuments, and bookkeepers' accounts, supplemented by reports of Greeks who studied in Egypt including Thales, Pythagoras, Plato, Democritus, Solon, and Eudoxus. By 3100 B.C., Egyptians had developed an additive, grouping system of numerals. Tally marks stood for 1 to 9, and a different symbol was used for each power of 10 from 10 to 1,000,000. Since these numerals did not have place value, there was no need for a placeholder. A zero symbol was used by 2600 B.C. (Arnold 1991, 17) as a reference for directed numbers. Leveling lines for construction purposes were labeled as "1 cubit above zero," "3 cubits below zero," and so forth. The same symbol was used for a zero remainder in a balance sheet that showed food items received and distributed (Scharff 1922). Educators who wish to infuse African contributions throughout the curriculum use many examples from ancient Egypt (Diop 1991, 231–307). Egyptian numerals have been used as manipulative aids to introduce African contributions and to teach mathematical concepts, such as transition from counting to addition; exchange of a 10 symbol for ten units or a 100 symbol for ten 10 symbols; subtraction as the inverse of addition; and division as the inverse of multiplication.

Besides unit fractions (such as $\frac{1}{3}$), Egyptians also used $\frac{2}{3}$, and rarely, $\frac{3}{4}$. Other fractional quantities were skillfully changed to a sum of unit fractions. The fraction symbol, the Egyptian letter r, which looks like a mouth, is written over the denominator, and is

*211*

an operator symbol giving instructions to "invert." Addition and subtraction were sometimes shown by a pair of feet walking the numeral toward or away from other numerals (Chace 1986, 99). Other symbols used were the word *aha* for the unknown; and symbols for "add," "subtract," and "answer." Egyptian multiplication is included in many textbooks to deepen understanding of the distributive property of multiplication. The modern multiplication of $17(35) = (7 + 10)35 = (7)(35) + (10)(35)$ can be compared with the Egyptian method of $17(35) = (1 + 16)35 = (1)(35) + (16)(35)$. Egyptians used a process of doubling multiplier and multiplicand. In effect, they were expressing the multiplier as a sum of powers of 2, then adding the required partial products.

First- and second-degree equations were usually solved by "false position," although examples exist of modern-type solutions (Gillings 1982, 157). A convenient solution value for an equation is tested; it usually turns out to be false. The resulting error leads to a proportional correction and the correct solution. An example by the scribe Ahmose asked, "What is the quantity, given that the sum of the quantity and $\frac{1}{4}$ the quantity is 15?" He assumed 4. But $4 + (\frac{1}{4})(4) = 5$ not 15. The correction factor is 3. Multiply the "false" value of 4 by 3 to get 12, the correct solution (Chace 1986, 37). These examples help in teaching the concept of proportion. The solution value was always checked, another feature of educational value. Problems using arithmetic and geometric series were elegantly solved in the RMP (*Rhind Mathematical Papyrus* written by Ahmose) and provide insight into the derivation of formulas for sums of series (Lumpkin 1995).

Development of Egyptian geometry was probably stimulated by the need to redraw farm boundaries after the annual flood, and to solve problems that arose in the construction of pyramids and other massive structures. Ancient Egyptians developed correct formulas for areas of triangles, rectangles, trapezoids, and a good approximation, $A = [(\frac{8}{9})D]^2$, for a circle of diameter $D$. Examples from RMP can be used to introduce these formulas. An Egyptian formula for surface area of a curved surface is believed either to apply to a hemisphere, or to half a cylinder (Gillings 1982, 194–201). Another advanced formula gave the volume of a truncated square pyramid with bases $a$, $b$, and height, $h$: $V = (h/3)(a^2 + ab + b^2)$. Rectangular coordinates were used as early as 2700 B.C. in an architect's plan at Saqqara. The height of the curve is given at horizontal coordinates spaced 1 cubit apart. Chicago

Public Schools have used this example to introduce graphing in a coordinate plane.

*See also* History of Mathematics, Overview

## SELECTED REFERENCES
Arnold, Dieter. *Building in Egypt.* New York: Oxford University Press, 1991.

Chace, Arnold Buffum. *The Rhind Mathematical Papyrus.* Reston, VA: National Council of Teachers of Mathematics, 1986.

Chicago Public Schools. *Algebra I Instructional Framework.* Chicago, IL: The Schools, 1991.

Diop, Cheikh Anta. *Civilization or Barbarism: An Authentic Anthropology.* Yaa Lengi Meema Ngemi, trans. Harold J. Salemson and Marjolyn de Jager, eds. New York: Hill, 1991.

Gillings, Richard J. *Mathematics in the Time of the Pharaohs.* New York: Dover, 1982.

Lumpkin, Beatrice. *African and African American Contributions to Mathematics.* Portland, OR: Portland Public Schools, 1995.

Scharff, Alexander. "Ein Rechnungsbuch des Königlichen Hofes aus der 13. Dynastie (Papyrus Boulaq Nr. 18)." *Zeitschrift für ägyptische Sprache und Altertumskunde* 57(1922):58–59.

BEATRICE LUMPKIN

# EINSTEIN, ALBERT (1879–1955)

Most outstanding physicist, perhaps even scientist, of the twentieth century. In his youth, theoretical physics was just being established. With the development of relativity theory, which Einstein originated, and quantum theory, to which he contributed, physics became the premier science. Einstein was born in Ulm, Germany, and grew up in Munich (1883–1894). In 1896, he began his studies in Zürich (Switzerland) at the Federal Technical College (Eidgenössische Technische Hochschule). He completed work for a diploma as a teacher of mathematics and physics on the secondary school level in 1900.

In the years immediately following, Einstein failed to obtain a position as an assistant. While studying for the Ph.D. degree at the University of Zürich, he had to withdraw his first dissertation in 1902. In that year, however, he became technical expert at the patent office in Bern, Switzerland (1902–1909). Einstein's first publications concerned statistical physics. His second dissertation, "A New Determination of Molecular Dimensions," was accepted and became a major contribution to the field of statistical physics despite its short length of only seventeen pages. In 1905, he published three papers, each of which would have promoted him to being ranked among the very best physicists. His most well-known paper introduced his theory of special relativity (Einstein 1905). Not until 1915 did Einstein himself call it "special relativity," after he finished his paper on general relativity (Einstein 1916), which was based on a new mathematical concept, tensor calculus.

By 1914 Einstein was professor at the Academy of Berlin. In 1919, his fame spread internationally. At the occasion of a solar eclipse, an English expedition was able to confirm the bending of light, which Einstein had predicted. In 1921, Einstein was awarded the Nobel Prize, not because of his relativity theory, but "for his services to theoretical physics and especially for his discovery of the photoelectric effect" (Pais 1982, 502–512).

Due to the threatening political climate in Germany, Einstein emigrated to the United States in 1932 and settled at the Institute for Advanced Study in Princeton. During his life in America, Einstein continued to contribute to physics, mainly to general relativistic field equations. He also was engaged in philosophy, politics, and Judaism. Though a convinced pacifist, in 1939 Einstein signed a letter to President Franklin D. Roosevelt in which he drew attention to the military implications of atomic energy. After World War II, Einstein warned against the development of the hydrogen bomb, atomic rearmament, and an uncontrolled arms race. He supported the use of science for peace and social good until his death.

*See also* Relativity

## SELECTED REFERENCES

Aichelburg, Peter C., and Roman U. Sexl. *Albert Einstein: His Influence on Physics, Philosophy and Politics*. Braunschweig, Wiesbaden, Germany: Vieweg, 1979.

Bucky, Peter A., with Allen G. Weakland. *The Private Albert Einstein*. Kansas City, MO: Andrews and McMeel, 1992.

Clark, Ronald William. *Einstein: The Life and Times*. New York: World, 1971.

Dukas, Helen, and Banesh Hoffmann. *Albert Einstein: The Human Side: New Glimpses from His Archives*. Princeton, NJ: Princeton University, 1979.

Einstein, Albert. *Collected Papers*. 5 vols. Princeton, NJ: Princeton University, 1987–1995.

———. "Electrodynamics of Moving Bodies." *Annalen der Physik* 17(1905):891–921.

———. *Essays in Humanism*. New York: Philosophical Library, 1983.

———. "General Relativity." *Annalen der Physik* 49(1916):769–822.

———. *Maric, Mileva: The Love Letters*. Jürgen Renn and Robert Schulmann, eds. Shawn Smith, trans. Princeton, NJ: Princeton University, 1992.

———. *The Meaning of Relativity*. Princeton, NJ: Princeton University, 1955; London, England: Methuen, 1956; Franklin Center, PA: Franklin Library, 1981.

———. *Out of My Later Years*. New York: Bonanza, 1989.

———. *Sidelights on Relativity*. New York: Dover, 1983.

Pais, Abraham. *Subtle Is the Lord . . . The Science and Life of Albert Einstein*. New York: Oxford University Press, 1982.

Sayen, Jamie. *Einstein in America: The Scientist's Conscience in the Age of Hitler and Hiroshima*. New York: Crown, 1985.

Schilpp, Paul Arthur, ed. *Albert Einstein, Philosopher-Scientist*. Evanston, IL: Library of Living Philosophers, 1949.

KARIN REICH

# EISENHOWER PROGRAM

Consortia established to assist states in developing and implementing standards-based curriculum frameworks, designing assessments to align with the new curriculum frameworks, creating partnerships, and disseminating successful classroom practices. The last decade has seen several large-scale national efforts to encourage comprehensive reform in mathematics and science education. Much of the emphasis has been on establishing standards for what students are expected to know and be able to do. In October 1992, following a national competition, the U.S. Department of Education awarded three-year grants to establish ten Eisenhower Regional Consortia for Mathematics and Science Education.

The consortia serve the same regions as the ten regional educational laboratories funded by the U.S. Office of Educational Research and Improvement. In October 1995, a second round of awards was made to continue the consortia for another five years. Eight of the original consortia received continuation grants and two are new grantees not connected with the regional educational laboratories. The ten regions served are: Appalachia, the Far West, the Pacific Islands, the Far Northwest, the Mid-Continent, North Central, the Mid-Atlantic, the Southwest, the Southeast, and the Northeast and Islands (including Puerto Rico, and the U.S. Virgin Islands).

The mission of the Eisenhower Regional Consortia is to (a) coordinate mathematics and science resources within each region, (b) disseminate exemplary mathematics and science education instructional materials, and (c) provide technical assistance for the implementation of teaching methods and assessment tools for use by elementary and secondary school students, teachers, and administrators. Resources provided by the consortia include databases, printed materials, training opportunities, and technical assistance to states and local schools.

As part of the Eisenhower National Program for Mathematics and Science Education, the ten consortia work together as a network. The Eisenhower National Clearinghouse (ENC) in Columbus, Ohio, a partner in the national network, works with the consortia to strengthen and improve mathematics and science through technology. The ENC's comprehensive electronic catalog and other electronic resources serve the ten consortia, and the consortia assist the ENC in identifying resources for the electronic catalog, and disseminating information. Each of the consortia also has established an ENC Technology Demonstration Site. Some of the consortia have chosen to locate the demonstration site in their facilities, while others have selected higher education institutions, department of energy laboratories, and museums.

Each consortium has its own individual evaluation plan. In addition, the ten consortia are collaborating in a national effort to assess their collective impact. Because the consortia have only been in operation since 1992, it will be several more years before the long-range impact on teachers and students can be determined.

*See also* Federal Government, Role; United States Department of Education

SELECTED REFERENCE
"ENC's Partners: Eisenhower Regional Consortia and ENC Demonstration Sites." *ENC Focus* 7(2)(2000): 8–9. URL: enc.org

PAMELA K. BUCKLEY

# ENGINEERING

A profession concerned with the application of scientific knowledge to the solution of practical problems. Typical problems include the design and construction of structures such as buildings and bridges, machinery, air and land vehicles, electrical circuits and devices, as well as computer hardware and software. Biomedical engineering is a newer discipline in which engineering methods are applied to solving problems of human beings and animals. The language of the engineer and, indeed, all of science, is mathematics.

Engineers use mathematics routinely and thus students of engineering study mathematics as a core course and apply it in the study of engineering principles. In these applications the physical principles and other information that underlie the engineering problem must be translated into mathematical forms, that is, mathematical models of the physical situation are created. These models may take the form of equations or other mathematical expressions that are treated by methods learned in formal mathematics courses. For example, equations are solved; complicated mathematical forms are simplified. Different coordinate systems may be used in order to aid in solving equations or in gaining new information or insights. Ultimately the mathematical results must be interpreted in physical or biological terms.

High school courses in algebra, geometry, and trigonometry are the usual prerequisites for the more advanced mathematics courses that are taken by college engineering students. The key mathematics courses at the collegiate level are the differential and integral calculus. These courses, often of three-semester duration, almost always include vector calculus. Most engineering students will also study linear algebra (sometimes called matrix algebra), probability, and differential equations in formal courses given by the mathematics department, and will have exposure to additional specialized mathematical topics in the context of certain engineering courses. For example, electrical engineering courses and those in control engineering (e.g., as in missile guidance systems) make much use of complex analysis, Fourier series, and Laplace transforms, whereas certain mechanical engineering courses that involve vibrations in membranes require Bessel functions. It is interesting to note that in the early part of the twentieth century, few engineers studied (or used) calculus in their work. Today, it is difficult to imagine how they got along without it.

Some insights on the use of mathematics in engineering may be gained by discussing certain applications. A standard and historically important application of calculus is concerned with straight-line motion. This application may be illustrated in an analysis of vertical motion. Relative to the surface of the earth, the height of a particle may be denoted by $y$; its instantaneous velocity $v$ is then the time derivative, $dy/dt$, whereas the acceleration is the derivative of velocity, $dv/dt$. If the acceleration is constant, as it is near the earth's surface, one may write this as $dv/dt = -g$, where $g$ has the magnitude 32 ft/sec$^2$. This is an example in which a physical law is expressed in mathematical form, in this case, a differential equation. The solution, obtained by integration, is given by $v = -gt + v_0$, where $v_0$ is the initial velocity. A further integration yields the expression for the coordinate $y$ at any time $t$: $y = -gt^2/2 + v_0t + y_0$, where $y_0$ is the initial vertical position ($y$-coordinate). The above expression for $y$ is an equation of motion, that is, it allows a determination of the vertical position at any time from initial conditions of speed and position. Like all models it is based on assumptions—in this case, that air resistance is negligible. The reasoning is typical of what goes into mathematical model-building in engineering situations.

The simple harmonic oscillator represents another model system that is fundamental in the engineering of mechanical systems. Newton's second law of motion, which relates force $F$, mass $m$, and acceleration $a$, by $F = ma$, is explicit in this model in which a piece of physical matter (e.g., a spring) is stretched in one direction and released. It was determined by experiment that a stretched spring experiences a restoring force whose magnitude is proportional to the change in length. This phenomenon is represented by the equation, $F = -kx$, where $F$ is the restoring force, $x$ is the change in length, and $k$ is a constant of proportionality that relates to the spring's stiffness. The algebraic sign is negative in this relation, since the direction of the force is opposite to the direction of stretch. Acceleration $a$ is the second derivative, $d^2x/dt^2$, so that Newton's law gives $-kx = md^2x/dt^2$, a second-order linear differential equation that leads to the oscillatory solution $x = A \cos wt + B \sin wt$, where $A$ and $B$ are constants that depend on initial conditions and $w = (k/m)^{1/2}$ is related to the frequency of oscillation.

Space travel represents one of the great engineering accomplishments of the twentieth century. In order for a spacecraft (or any other body) to escape the earth's gravitational force it is necessary that it attain an initial velocity whose magnitude is determined from the solution of a rather basic differential equation. This "escape velocity," obtained from the solution of the differential equation, is very nearly seven miles per second and, thus, space engineers were required to develop boosters that could attain this velocity. This is an outstanding example of how mathematics is used by engineers.

Radioactive substances such as radium and uranium are well known, the former for its use in the treatment of some malignancies, and the latter for its use in nuclear reactors and explosives. Engineers, utilizing the principles of physics and the equations describing these principles, have harnessed the energy from these natural elements. For example, it has been calculated that the nuclear fission of just 1 gram of uranium 235 would release energy equivalent to the explosion of twenty tons of TNT. Fundamental to an understanding of radioactive elements is the rate of their disintegration, that is, the loss of mass from the nucleus of the atom. A quantitative understanding of radioactive disintegration is well approximated by the mathematical model, $dN/dt = -kN$, where $N$ is the number of atoms of the original element and $k$ is the disintegration rate constant. The quantity $dN$ is an integer, but it is small compared with $N$, so that $N$ is approximately continuous. For an initial number $N_0$ the equation above leads to the following exponential relation for the number $N$ at any later time $t$: $N = N_0 e^{-kt}$. Analysis of this decreasing exponential shows that the half-life, the time to decay to one-half the original amount, is given by $(\ln 2)/k$.

Electrical and electronic devices have numerous uses and applications in communication systems, control systems, computers, laser technology, and so forth. Electrical engineers use mathematics extensively in their work and learn the most basic facts of electrical circuitry by using the language and methods of mathematics. For example, the basic relation among voltage, current, and resistance in a resistor is given by the well-known Ohm's law, $V = IR$, where $I$ is the current and $R$ the resistance. Other passive circuit elements have different relations among these; an inductor with inductance $L$ relates these with the derivative, $V = L \, di/dt$, and a capacitor connects them with an integral,

$$V = \frac{1}{C} \int i \, dt$$

Complex numbers, introduced in elementary algebra courses, and functions of a complex variable, studied at the college level, provide concepts and techniques that are fundamental to the activities of electrical engineers. Work with electric and magnetic fields has its basis in the classic equations of James Clerk Maxwell (1831–1879) and these use concepts of vector analysis. Computers, which are capable of doing enormous calculations rapidly, as well as storing and retrieving data of all kinds, do all of this in a binary (base 2) system consisting of just 0s and 1s, which correspond to electrical circuits being either opened or closed. A number in the binary system, such as 1001, has the following interpretation: $1 \times 2^3 + 0 \times 2^2 + 0 \times 2^1 + 1 \times 2^0$, which is equal to 9. Today, some students learn about different number bases early in their studies of arithmetic, since the concepts are fundamental to an understanding of how we count and also have a major application in the field of computers that now have a significant impact on our lives.

*See also e*; Modeling

## SELECTED REFERENCES

Tallarida, Ronald J. *Pocket Book of Integrals and Mathematical Formulas.* 2nd ed. Boca Raton, FL: CRC, 1992.

Wylie, Clarence R., and Louis C. Barrett. *Advanced Engineering Mathematics.* 6th ed. New York: McGraw-Hill, 1995.

RONALD J. TALLARIDA

# ENGLAND

Mathematics in English education has a long and complex history stretching back to the Middle Ages. Individuals, institutions, and the state have each, at various times, helped to shape this history. Various developments have been surveyed in the literature, which itself goes back to the nineteenth century (Ball 1889; De Morgan 1847). From the sixteenth century, Cambridge came to exert a profound influence on English mathematics education at both university and school levels. Cambridge produced many leading textbook writers, teachers, and administrators. Furthermore, the Cambridge system of competitive examinations by means of *written* papers in mathematics, which developed in the eighteenth century, became a paradigm for examinations at all levels in the nineteenth century (Price 1994).

Robert Recorde was educated at Oxford and Cambridge Universities in the sixteenth century and was the first writer to produce a series of mathematics textbooks in *English* (as opposed to Latin or Greek). The series covered arithmetic, algebra, geometry, and astronomy, and was designed for individual study at a time when institutional provision was limited. Recorde also has been referred to as "the first mathematics educator" (Howson 1982) on the strength of the explicit and implicit concern for pedagogy in his writings.

Until the nineteenth century, charitable secondary schools were required to devote themselves largely to classical studies, and mathematics was generally neglected. The teaching of mathematics, beyond arithmetic, was largely in the hands of private tutors, "philomaths" (lovers of learning) or "mathematical practitioners," whose range of interests in applicable mathematics was already considerable by 1700. Exceptionally, from the late seventeenth century, special mathematical *schools* were established for aspiring naval apprentices. Furthermore, mathematics education for military purposes was well served by the establishment of the Royal Military Academy, Woolwich, in 1741, with which the name of Charles Hutton, a leading mathematical practitioner and textbook writer, became closely associated (Howson 1982).

Through textbooks and private teaching alongside the basic skills of writing, accounts, and bookkeeping, utilitarian arithmetic became the most widely disseminated of all the branches of mathematics. Building on the early work of De Morgan (1847), a valuable survey of English arithmetic textbooks, 1535–1935, has been provided by Yeldham (1936). From the 1830s, when the state first turned its attention to the provision of elementary education for working-class boys and girls from the age of 6, and associated teacher training, arithmetic became firmly established as one of the "3 Rs" in what became effectively a national curriculum with national testing for elementary schools in the nineteenth century (Howson 1982).

The first half of the nineteenth century saw the demise of the tradition of the many-sided mathematical practitioner. The cause of practical mathematics was furthered through the nineteenth-century development of technical education in evening classes, schools, and colleges, supported by central and local government grants and a state system of examinations in "science and art" subjects. At university level, the monopoly of Oxford and Cambridge ended in the 1830s through the development of London University. A national *system* of universities, however, is a twentieth-century development.

By 1870, a firm place for mathematics in the traditionally classics-dominated secondary schools for boys was secured through two major developments: the rise of written examinations of a general secondary education and the competition of newly established private schools with more "modern" curricula emphasizing mathematics from the outset. All sectors of English education—elementary, secondary, technical, teacher, and university—came under the pervasive influence of examinations. Nineteenth-century education for girls normally stopped at arithmetic, and inequality of access to mathematics for girls is a persistent theme in the twentieth century (Howson 1982; Price 1994).

Recorde, Hutton, and De Morgan were three major figures in the development of English mathematics education (Howson 1982). From the 1860s, organizations and movements also are identifiable as part of an emerging professional consciousness in mathematics education. The unassailable position of pure Euclid as a textbook for secondary school and university examinations led to a movement for reform and the establishment, in 1871, of an Association for the Improvement of Geometrical Teaching. This Association became the Mathematical Association (MA) in 1897 and was by far the most important English professional organization for mathematics education—principally at the secondary school level—until the 1950s (Price 1994).

The development of central and local government involvement in English education is closely associated with the major Education Acts of 1870, 1902, 1944, and 1988. The 1870 act brought in a national system of elementary schooling up to thirteen years of age, administered by local school boards. These boards were replaced by local education authorities, following the 1902 act, and their responsibilities were extended to the provision of new grant-aided selective secondary schools, alongside the well-established private sector. For the majority who failed the selection tests, however, education still meant elementary schooling up to thirteen—raised to fourteen from 1918—until the 1944 act.

The 1944 act formalized a national system of primary followed by differentiated secondary schooling of three types—grammar (alongside the private sector), technical, and "modern"—for all pupils. The school leaving age was raised to fifteen in 1947 (and sixteen in 1972), and the tripartite school system was largely undermined by the comprehensive school movement of the 1960s. The 1988 act introduced a compulsory national curriculum for all boys and girls aged 5–16 in state schools. Mathematics as a core subject now involves national testing at ages 7, 11, and 14, alongside existing local examinations at 16. The 1988 act reduced the powers of local education authorities and encouraged schools to work on a direct-grant basis, independently of their local authority. The principle of partnership, since 1902, between central and local authorities has been fundamentally eroded.

Periods of major instability in English mathematics education are detectable in the years around 1870, 1900, and 1960 (Price 1985). The first reform movement was largely unsuccessful, and the dominance of Euclid in England was not ended until the 1900s, as part of the worldwide "Perry movement"

to make mathematics teaching more practical and relevant for the nonspecialist secondary pupil. The introduction of practical geometry, graphical methods in algebra, simple four-figure tables, and the beginnings of trigonometry in a general education were all products of this movement (Price 1994). The worldwide modern mathematics movement from the 1950s brought further broadening of the secondary curriculum to include some descriptive statistics and probability. The movement also produced collaborative projects in textbook production, such as the School Mathematics Project, which continued for decades to exert a major influence on mathematics for the 11–19 age range (Cooper 1985).

The general postwar movement toward *mathematics for all* gained strength from the international visual aids movement; and the establishment of the Association for Teaching Aids in Mathematics (ATAM) in 1952 was one result. The ATAM became the Association of Teachers of Mathematics (ATM), in 1962, and played a leading part both in the modern mathematics movement and in the promotion of pupil-centered learning (Price 1994). In addition to the MA and the ATM, from around 1960 a range of differentiated interests in mathematics and mathematics education surfaced at an organizational level:

1959   Mathematics Section of the Association of Teachers in Colleges and Departments of Education (now part of the Association of Mathematics Education Teachers [AMET])

1961   University Departments of Education Mathematics Study Group (now part of AMET)

1962   Joint Mathematical Council of the United Kingdom

1963   Institute of Mathematics and Its Applications

1971   British Society for the History of Mathematics

1974   National Association of Mathematics Advisers

1978   British Society for the Psychology of Learning Mathematics (now British Society for Research into Learning Mathematics)

1981   Girls and Mathematics Association (now Gender and Mathematics Association)

The state of organizational complexity in English mathematics education is proving problematic in relation to professional influence on a centrally controlled curriculum in the 1990s.

The modern mathematics movement was followed by a major reaction in the 1970s, which was fueled by public concerns about both standards of "numeracy" and the suitability of the reforms for the majority of pupils in comprehensive schools. A major government inquiry led to the publication of the influential Cockcroft Report (1982), which advocated differentiated curricula, on the basis of pupils' levels of attainment, and a broader range of teaching styles to promote active learning for all pupils. These principles were subsequently embodied in both the new sixteen-plus examinations for the General Certificate of Secondary Education, from 1985, and the National Curriculum arrangements from 1989 (Cooper 1994).

The development of the National Curriculum and associated national testing may be regarded as the largest and most controversial of all the curriculum development projects in England. The structure of this National Curriculum has been fittingly described as "unique" in a comparative study by Howson (1991, 217). The concern for national standardization and monitoring contrasts sharply with the emphasis on professional judgment, autonomy, and initiative that has characterized mathematics curriculum development in England for much of the twentieth century.

*See also* Association of Teachers of Mathematics (ATM); International Comparisons, Overview

## SELECTED REFERENCES

Ball, Walter W. Rouse. *A History of the Study of Mathematics at Cambridge.* Cambridge, England: Cambridge University, 1889.

Cockcroft, Wilfred H., Chairman. *Mathematics Counts.* London, England: Her Majesty's Stationery Office, 1982.

Cooper, Barry. "Secondary Mathematics Education in England: Recent Changes and Their Historical Context." In *Teaching Mathematics* (pp. 5–26). Michelle Selinger, ed. London, England: Routledge, 1994.

———. *Renegotiating Secondary School Mathematics: A Study of Curriculum Change and Stability.* Lewes, East Sussex, England: Falmer, 1985.

De Morgan, Augustus. *Arithmetical Books from the Invention of Printing to the Present Time.* London, England: Taylor and Walton, 1847.

Howson, A. Geoffrey. *A History of Mathematics Education in England.* Cambridge, England: Cambridge University, 1982.

———. *National Curricula in Mathematics.* Leicester, England: Mathematical Association, 1991.

Price, Michael H. "Historical Perspectives on English School Mathematics, 1850–1950." *Zentralblatt für Didaktik der Mathematik* 17(Jan. 1985):1–6.

———. *Mathematics for the Multitude? A History of the Mathematical Association.* Leicester, England: Mathematical Association, 1994.

Yeldham, Florence A. *The Teaching of Arithmetic through Four Hundred Years (1535–1935).* London, England: Harrap, 1936.

MICHAEL H. PRICE

# ENRICHMENT, OVERVIEW

Curriculum enhancement originally provided for the academically talented; now available for all students. Typical enrichment activities involve selected mathematical and historical topics, challenging problems, group experiments, applications, mathematical recreations, reports, competitions, and use of manipulatives, audiovisual materials, calculators, and computers. Early in the twentieth century, there were recommendations that problems posed should have a puzzle element; and that there should be more emphasis on graphs to illustrate ideas needed in science and business, and in the affairs of everyday life. Measurement and how it related to industry was also to receive increased attention. Teachers were encouraged to stress applications of geometry, trigonometry, and statistics, and to welcome multiple ways of finding solutions. Laboratory methods were suggested, also.

In the elementary schools in the 1920s there were recommendations to supplement the problems found in textbooks by making up problems using children's experiences outside of school; adult activities that interest children, such as the cost of clothing or the upkeep of an automobile; newspaper articles; advertisements; sports scores; or holiday experiences. Children also could make up their own problems using these same topics. There were supplementary projects in which students kept graphs of arithmetic scores from day to day; determined ways to speed up service at the cafeteria; calculated the cost of chicken raising; and worked out a budget. Playacting was recommended to help children learn mathematics. This included having stores in the classroom where students could act as shopkeepers and customers. Items such as groceries could be purchased and the amount due as well as the change would have to be calculated. Alternatively, the students could set up a bank to achieve similar results.

Concrete materials, of a variety of types, were to be used to help students understand large measures such as acre, ton, mile, or one million, as well as small, positive numbers less than one. Everyday objects such as checks, deposit slips, and coins were to be used in classroom activities. Also, because of the increased cost of housing after World War I, a larger portion of household income needed to be spent for shelter. Extra units were to be included, which took into account taxation, mortgages, upkeep, and owning versus renting.

There was a desire to have condensed and readable accounts of the history of measurement so these concepts and techniques could become an addition to the curriculum when teaching metric measure. These units included navigation, mapmaking, solar measures, time, industrial uses, and medical uses. Supplementary units involving these topics might have students in the class act as realtors and customers. Contracts could be drawn up with payments figured, diagrams of the land and buildings made, and records looked up in the local courthouse. Also, connections to architecture could be made, both past and present, including symmetry and geometric shapes. These projects could be tailored to the pupils' interest. For example, a city person would not be given a problem involving farms.

The enrichment materials for senior high school students during this period tended to be of a more mathematical nature. Extra units for algebra included topics such as solutions of higher degree equations, indeterminate equations, differentiation and integration, logarithms, and interpolation using engineering manuals. Additional topics for geometry included geometric interpretation of square roots of numbers, golden section, geometric construction of reciprocals, and mechanical drawing. Instruments used included surveying transits and slide rules.

In the 1930s, activities outside of the classroom were in vogue. Mathematics clubs were formed so that after school students could explore mathematical recreations. The Sunday section of newspapers contained mathematical puzzles that could be solved by club members. Also, historical topics provided sources of problems.

In the elementary schools teachers were encouraged to use field trips, exhibits, and projects to show children how mathematics was being used. The elementary school curriculum was to be enriched by having students study arithmetic through projects that looked at such topics as food processing, the story of measurement, or the beginning of the earth. Many projects had a social aspect to them, such as using maps and charts to compare urban versus rural populations, or a project in which a student or a group of students would research the local taxes and the amount taken in and how these monies were spent, or one in which students would have to use resources and research the concept of large numbers by looking at populations of schools, towns, cities,

and the nation. There needed to be more concrete introductions and more practical applications. Constructions, drawings, measuring, paper folding, and cutting should become commonplace, or seeing that a three-legged stool is stable; or folding a triangle to see the sum of the measures of the angles as 180°. The interest of the students was to be aroused by discussing how the shape of shelters had changed over the ages (from tepees to lean-tos to houses) and conjecturing as to why these were used; or by using symmetry in art, analytic geometry in building design, or geometric principles in astronomy.

Across the curriculum students were assigned special topics of interest, and then the student or a group of students would research the topic and report back to the class. One such topic was drawn from then current loan procedures. Because of the high unemployment rate, a loan or second mortgage had to be paid for in advance. Thus the person would get only a portion of the entire amount, but the full amount would be due at the end of the loan period.

With the impetus of the earlier emphasis on applications of mathematics, the early 1940s saw many supplementary materials that were designed to show a variety of uses of mathematics. A sourcebook (Olds et al. 1942) of applications was arranged in four main sections, arithmetic, algebra, geometry, and trigonometry, with topics listed alphabetically in each section. There were over six hundred supplementary problems. Pupils were encouraged to make "source books" by collecting items from newspapers, magazines, and books that related to mathematics. Local community personnel such as engineers, bankers, and businessmen could be brought in to the schools to tell students how mathematics was used in everyday life. Teachers were encouraged to take their classes on field trips to local industries and businesses to witness mathematics being used. Clubs and school newspapers devoted to mathematical topics were encouraged. In the early 1940s, an attempt was made to reorganize mathematics instruction around centers of interest to help students from diverse cultural backgrounds appreciate mathematics. These interest centers contained supplementary projects and activities.

After World War II, the improvements in the curriculum were rooted in the then current learning theory and the implied practices from this theory. This brought about the laboratory approach to mathematics, which included discovery learning, in which facts and generalizations were to be revealed through the manipulation of objects. Mathematics laboratories were to be used to supplement the curriculum and they were to be furnished with multisensory equipment. With the end of World War II, prosperity came to the United States. People worked to make up for the deficiencies of many previous years. This led to more money becoming available for multisensory aids. Also, multisensory aids made the classroom more enjoyable and helped to increase comprehension.

The list of aids was very extensive and included posters, graph paper of all types, geometric models, stereoscopes and views, slide rules, parallel rulers, vernier calipers, surveying transits, sextants, sundials, colored chalk, abacus, Napier's Bones, hypsometers, film strips, puzzle books, clay, calculating machines, drawing instruments, nomographs, opaque projectors, slides, lantern-slide projectors, motion pictures, measuring cups, rulers, tiles, work tables, cases to house the aids, and even separate rooms for laboratories (Reeve 1945). Other additions to the curriculum were centered around radio and television. This gave teachers a chance to see as well as hear others teaching various topics and units in mathematics. In 1952 the Federal Commerce Commission reserved 242 VHF channels for noncommercial use, and educational television was born. One of the first courses was on teaching the slide rule.

With the new emphasis on the structures of mathematics, the 1960s saw many enrichment materials that concentrated on the better students and upon helping them understand the underlying principles of mathematics. This included supplementary units on sets, Venn diagrams, properties of numeration systems, different bases for numeration, logic, topology, graph theory, and open sentences for elementary school students. The supplements for the junior high schools included topics such as challenging puzzles, number theory, repeating decimals, mathematical systems, logical connectives, and informal geometry. The high school topics included primes, modular arithmetic, Fibonacci Numbers, unsolved problems from arithmetic, linear programming, induction, functions, and non-Euclidean geometries. Also, historical topics for all levels were used as additions to the curriculum. This included ancient numeration systems, the Golden Mean, the four color problem, Pascal's Triangle, Greek numbers (perfect, pentagonal, etc.), trisection problems, history of mathematical terms, Horner's Method, Descartes' Rule of Signs, and infinitesimals. Of the seven yearbooks published by the National Council of Teachers of Mathematics (NCTM) during the 1960s, five were on enrichment.

The "new math" was not the only thing that led to innovative changes during the decade. The trend toward using concrete experiences continued in the early grades as did the use of film and visual aids for the later grades. In the high schools, programmed materials were used for instruction. These materials ranged from whole courses to supplementary units that individuals studied at their own pace. Outside of the school day, academic games and contests became popular. The overhead projector became a common fixture in the classroom, leading to the development of supplemental overhead transparencies to illustrate mathematical concepts. Late in the decade, textbook publishers included supplementary materials with each series. These included manipulatives, audiotapes, duplicating masters, extra problems for "experts," and a separate teacher's edition with suggestions for incorporating these supplementary materials into daily lessons. Related to the many curriculum projects from the 1960s, the 1970s witnessed an attempt to evaluate and find the projects or pieces of a project that seemed to work the best. This led to refining previously used supplemental materials.

It also was increasingly recognized that textbooks were the dominant force in determining the direction of the curriculum; the selection of textbooks became crucial. Many cities, school districts, and states devised criteria for evaluating textbooks and this included rating the texts on appropriate supplementary materials. Because textbooks per se did not always cover all the topics that teachers deemed important, selecting appropriate supplementary materials received increased attention. One of the resulting consequences was the continuing explosion of supplementary materials being offered by educational supply companies.

The role of calculators and computers was being debated. Initially calculators, because of their cost and teacher resistance to their use, were mainly used as an add-on to the curriculum. They were used for checking answers and playful activities such as calculating the answer to a problem and then turning the calculator upside down to read the answer as a word. Computers were very expensive and their use was limited by the dearth of appropriate materials. Supplementary software was being developed, however, for computer-assisted instruction. This included reinforcement activities, flow charting, and solving numerical problems that had previously been too time-consuming. A counter force was also developing to the "back-to-basics" movement, which indicated that appropriate use of inexpensive calculators could eliminate the need for some drill and practice. This development gave rise to a movement to provide extra materials for approximation and application. The applications were centered around everyday situations designed to arouse student interest. Problems involving wildlife, the environment, and symmetry were used.

Several forces set the tone for enrichment materials of the 1980s. The *Agenda for Action* (NCTM 1980) called for an emphasis on problem solving. Reactions to the poor scores of U.S. students compared to their counterparts in international studies, notably in Japan, sparked cries for reform, as did reactions to the report *A Nation at Risk* (National Commission on Excellence in Education 1983). There ensued renewed efforts at providing supplementary materials for problem solving and applications. One of the main differences from previous attempts at emphasizing problem solving and applications was that this time it was aimed at all students instead of a select few. There were recommendations that standard textbook problems be augmented with a wider variety of problems, including nonstandard problems.

The increasing use of calculators, graphing calculators, and computers was hastened because the costs dropped dramatically. More and more teachers also accepted them as an integral part of the curriculum rather than as an add-on. Because of this increasing use of calculators and computers and the concomitant uses of problems in statistics, probability, data processing, social sciences, and the other sciences, many of the supplementary materials for applications involved integration of mathematics with other subject areas.

During the 1990s, cooperative learning, writing tasks across the curriculum, and new assessment techniques combined to produce new types of supplementary materials and projects. The use of cooperative learning became more widespread, because it was a pedagogically sound way for students to learn. Students could discuss and grapple with approaches to a solution of a problem that one person alone might not reach, but their combined efforts could enhance the learning for all of them.

This situation has spawned an effort to find supplementary materials that are appropriate for group work. The problems are generally more complex and require using exploration, references, and group discussions. An integral part of the process is writing up the outcomes of these deliberations, which also has an impact on the types of problems that are developed. The increased use of technology also has af-

fected the topics that are being considered for enrichment of the curriculum. We are seeing an increasing use of supplementary topics chosen from discrete mathematics, combinatorics, graph theory, and networks.

What will be the driving forces that guide the enrichment for the next century?—certainly some of the same ones that have driven change during this century, namely, demands from business and industry, psychological considerations, motivation, application, integration, equity, diversity, technology, innovation from mathematics projects, and new physical devices.

*See also* Applications for the Classroom, Overview; Art; Family Math and *Matemática Para la Familia*; Games; Recreations, Overview; Techniques and Materials for Enrichment Learning and Teaching; University of Chicago School Mathematics Project (UCSMP)

## SELECTED REFERENCES

Berger, Emil J., ed. *Instructional Aids in Mathematics.* 34th Yearbook. Washington, DC: National Council of Teachers of Mathematics, 1973.

Hilgard, Ernest R. *Theories of Learning.* 2nd ed. New York: Appleton-Century Crofts, 1956.

Hilton Peter J., and Gail S. Young, eds. *New Directions in Applied Mathematics.* New York: Springer-Verlag, 1982.

National Commission on Excellence in Education. *A Nation At Risk: The Imperative for Educational Reform.* Washington, DC: Government Printing Office, 1983.

National Council of Teachers of Mathematics. *An Agenda for Action: Recommendations for School Mathematics of the 1980s.* Reston, VA: The Council, 1980.

Olds, Edwin G. et al., comps. *A Source Book of Mathematical Applications.* 17th Yearbook. National Council of Teachers of Mathematics. New York: Teachers College, Columbia University, 1942.

Reeve, William D. *Multi-Sensory Aids in the Teaching of Mathematics.* 18th Yearbook. National Council of Teachers of Mathematics. New York: Teachers College, Columbia University, 1945.

Seymour, Dale, comp. *Encyclopedia of Math Topics and References.* Palo Alto, CA: Seymour, 1996.

ROBERT L. McGINTY

# ENVIRONMENTAL MATHEMATICS

New subject emerging as a response to the environmental devastation in our time: overpopulation, pollution, and vanishing wilderness. Environmental mathematics can be defined analogically by saying it has the same relation to the environment as engineering mathematics has to engineering. It must be distinguished from subjects such as mathematical ecology or ecological modeling that tend to be at the research or graduate level. On the other hand, environmental mathematics addresses general problems of the environment dealt with at a level suitable for students in the range from high school seniors to college sophomores. Moreover, many mathematicians who are setting the agenda have the view that the subject calls for a certain amount of commitment. This sets a tone that is somewhat different from the conventional "white coat" or clinical approach to science.

A pioneer, Richard H. Schwartz, began teaching an elementary environmentally oriented course at Staten Island College, New York in 1975. In 1983 the author of this entry introduced a freshman course, "Environmental Mathematics," at Salisbury State University, Maryland. Marcia Sward, Executive Director of the Mathematical Association of America (MAA), and the author of this entry wrote an article for the Association newsletter that was issued on the twentieth anniversary of Earth Day (Fusaro and Sward 1990). Lynn Steen suggested that an electronic conference be held on mathematics and the environment. This was done in June–July 1991. He included the results as a chapter in an MAA publication that called for curriculum changes (Fusaro 1992). The term *environmental mathematics* made its first appearance in that publication.

The *de facto* theme of the MAA national meeting in 1992 was Environmental Mathematics. It was the focus of an extensive report in a national academic journal, *Chronicle of Higher Education* (Wheeler 1992). Environmental mathematics was chosen as the theme for Math Awareness Week 1992. Although interest has increased over the last several years, the level is still modest. Environmental awareness has swept through the country and the school system. The interest and activity are very strong in the lower grades. Many teachers weave environmental topics into their field trips and class projects. These activities drop off as one moves to graduate school. This pattern is reflected in the amount and variety of software packages with environmental themes. Activity is easy to check at the college level. Catalogs and curricula are replete with introductory courses with such titles as Environmental Chemistry, Environmental History, and even Environmental Psychology. Counterparts in mathematics are rare.

An important goal is to raise the *environmental awareness* of mathematicians. As mathematicians become more aware of the seriousness of environmental problems, changes such as the following may be expected. Teachers will begin using examples from the environment in their classrooms; applied mathematicians will be more receptive to problems of the environment; and authors will write on mathematical aspects of the environment. A popular and crucially important source is population growth and arable land. Some other sources of illustrations or accessible problems abound in such topics as recycling, energy use, and biodiversity. Kurt Kreith (1993) created a simple but powerful metaphor to suggest the radical change in outlook that is required. Just as Copernicus' civilization had a perspective clouded by the *geo*centric hypothesis, modern civilization has a view distorted by its *homo*centric bias. Mathematicians, especially, must make a special effort to reconnect with the natural world via applications and modeling.

How is environmental mathematics envisioned? It is associated with a pedagogy that emphasizes teamwork, hands-on projects, and written reports in a framework of common English. The emphasis is on exploratory learning and visualization. It is an interdisciplinary subject whose applications come more from such fields as environmental education, environmental engineering, and environmental science. The subject should be presented in a modeling context—the mathematics driven by the application. Charts, diagrams (Odum and Odum 1978), and other graphical illustrations are used extensively. Computational mathematics includes the use of graphing calculators, simulation packages (Odum and Odum 1989), and spreadsheets (Kreith 1993).

As an example, consider a small population of subterranean termites that tunnel up to a nicely laid-out dinner, a wooden house $W$. Let $P$ represent the energy in the termite colony and $H$ the energy in the house that is available to the termites. A qualitative analysis suggests $P$ will increase as $W$ decreases. When $W$ is depleted, the termites will find another food source. If they stay in place, they will eventually die. This qualitative analysis can be transformed to a quantitative model by drawing a simple diagram in which $P$ and $W$ are represented by energy storage tanks. An application of the commonsense "flow equation," net energy flow = energy inflow − energy outflow, to $P$ and to $W$ yields flow equations. An ordinary calculator can be used to solve these equations in a step-by-step manner. They also can be solved with a graphing calculator, a spreadsheet, or by writing a simple BASIC program. The values can then be plotted on $P$ versus time and $W$ versus time coordinate systems. It is noteworthy that this process allows a student with very little algebraic background to solve a problem using equations that are actually differential equations in disguise (Fusaro 1995).

*See also* Applications for the Classroom, Overview; Biomathematics; Modeling; Population Growth

### SELECTED REFERENCES

Fusaro, Bernard A. "Environmental Mathematics." In *Heeding the Call for Change: Suggestions for Curricular Action* (pp. 82–92). Lynn A. Steen, ed. Washington, DC: Mathematical Association of America, 1992.

———. *Environmental Mathematics.* Faculty Workshop. Thomas Nelson Community College, Hampton, VA, April 8, 1995.

———, and Marcia Sward. "Solving Environmental Problems: Where Are the Mathematicians?" *Focus* 10(2)(1990):9.

Kreith, Kurt. *Building a Mathematical Base for Environmental Studies Curricula.* Davis, CA: Department of Education, University of California, 1993.

Odum, Howard T., and Elizabeth Odum. *Computer Minimodels and Simulation Exercises.* Gainesville, FL: Center for Wetlands, University of Florida, 1989.

———. *Energy Basis for Man and Nature.* New York: McGraw-Hill, 1978.

Schaufele, Christopher, and Nancy Zumoff. *Earth Algebra: College Algebra with Applications to Environmental Issues.* New York: HarperCollins, 1993.

Schwartz, Richard H. *Mathematics and Global Survival.* 2nd ed. Boston, MA: Ginn, 1990.

Wheeler, David L. "Mathematicians Develop New Tools to Tackle Environmental Problems." *Chronicle of Higher Education* 38(Jan. 22, 1992):A7, A10–11.

BERNARD A. FUSARO

# EQUALITIES

Robert Recorde was the first to publish the modern equality symbol in his 1557 algebra book, *The Whetsone of Witte* (Eves 1983). Recorde used two extended equal parallel line segments because he felt that nothing could be more equal. In his 1693 *De Algebra Tractatus: Historicus and Practicus,* John Wallis shortened the equal sign to the form we use today (Baumgart 1969).

The fact that two expressions are equal is indicated by an equality sign. The simplest equalities are arithmetic such as $4 + 5 = 9$. Algebraic equalities true for all values of a variable are called identities. These include equalities such as: $x + 3x = 4x$; $(x - 1)^2 = x^2 - 2x + 1$; or $\sin^2 x + \cos^2 x = 1$. Sometimes the symbol "$\equiv$" is used to denote an al-gebraic identity. Conditional equalities are true for one or more values of $x$. For example, $3x + 1 = 7$ is true for $x = 2$, but false for all other values. The equation $|x + 2| = 3$ is true for $x = 1$ or $x = -5$. Some equalities, such as $x^2 + 1 = 0$, are true for no real values. Besides algebraic equalities, other mathematical entities can be expressed in terms of equalities. For example, two sets $A$ and $B$ are said to be equal if they consist of the same elements. Two matrices $A$ and $B$ are equal if they have the same numbers of rows and columns, and all corresponding elements are equal. For complex numbers, $a + bi = c + di$ if and only if $a = c$ and $b = d$.

The reflexive, symmetric, and transitive properties hold for equalities as follows:

- Reflexive property: $a = a$. For example, $2x + 1 = 2x + 1$.
- Symmetric property: If $a = b$, then $b = a$. For example, if $2x + 1 = 5$, then $5 = 2x + 1$.
- Transitive property: If $a = b$ and $b = c$, then $a = c$. For example, if $5 = 2 + 3$, and $2 + 3 = 6 - 1$, then $5 = 6 - 1$.

If two algebraic expressions are equal, it is possible to add, subtract, multiply by, or divide by equal quantities on both sides of the equation without destroying the equals relationship. It also is helpful to simplify by combining like terms and eliminating parentheses. Commonly introduced for solving linear equations, these steps result in equivalent equations. For example, if

$$2(x - 3) = 10$$
$$2x - 6 = 10$$

then

$$2x - 6 + 6 = 10 + 6$$
$$2x = 16$$
$$x = 8$$

The National Council of Teachers of Mathematics (NCTM) recommends that students in grades 5–8 "develop confidence in solving linear equations using concrete, informal and formal methods" (NCTM 1989, 102). Teeter-totters and balance beams are useful physical devices to help justify the "balance operations" (Bernard and Cohen 1988). *The Algebra Lab* (Picciotto 1990), a complete manipulative program for teaching algebra concepts from middle school through high school, helps students learn informal methods to solve linear (and later quadratic) equations.

In grades 9–12, the NCTM recommends that students be able to represent a physical situation in terms of an appropriate equality (NCTM 1989, 151). For example, to determine how speed ($s$) is related to a car's stopping distance ($d$), students can collect data on reaction distance, braking distance, and stopping distance for various speeds. From this data, they should be able to deduce that stopping distance ($d$) equals the reaction distance ($s$) added to the braking distance ($s^2/20$): that is, the equation $d = s + s^2/20$. The Project for Computer-Intensive Curricula in Elementary Algebra (1991) is a good source for other examples. The NCTM also recommends that high school students use a graphing calculator or a computer as a visual tool to help solve polynomial equations. If an equation has the form $f(x) = k$, students can solve this equation by rewriting the problem as $y = f(x) - k = 0$, graphing the function, and finding the values of $x$ where the graph crosses the $x$-axis.

*See also* Equations, Algebraic; Equations, Exponential; Equations, Logarithmic; Inequalities

SELECTED REFERENCES

Baumgart, John K. "The History of Algebra." In *Historical Topics for the Mathematics Classroom* (pp. 233–260). 31st Yearbook. John K. Baumgart et al., eds. Washington, DC: National Council of Teachers of Mathematics, 1969.

Bernard, John E., and Martin P. Cohen. "An Integration of Equation-solving Methods into a Developmental Learning Sequence." In *The Ideas of Algebra, K–12* (pp. 97–111). 50th Yearbook. Arthur F. Coxford, ed. Washington, DC: National Council of Teachers of Mathematics, 1988.

Eves, Howard W. *An Introduction to the History of Mathematics.* 5th ed. Philadelphia, PA: Saunders College Publishing, 1983.

National Council of Teachers of Mathematics. *Curriculum and Evaluation Standards for School Mathematics.* Reston, VA: The Council, 1989.

———. *Principles and Standards for School Mathematics.* Reston, VA: The Council, 2000.

Picciotto, Henri. *The Algebra Lab.* Sunnyvale, CA: Creative, 1990.

Project for Computer-Intensive Curricula in Elementary Algebra. *Computer-Intensive Algebra.* College Park, MD: University of Maryland, 1991.

ANN E. MOSKOL

# EQUATIONS, ALGEBRAIC

Equations involving polynomial expressions. Originally the science of equations, algebra began in Egypt about 1850 B.C. Early algebra was characterized by the gradual invention of symbolism and the solution of equations by various methods. Problems were found on cuneiform clay tablets dating back to the reign of King Hammurabi (ca. 1700 B.C.). The algebraic use of letters for numbers, however, began with Diophantus about A.D. 275. He employed a symbol for equality, but it never came into common use. About A.D. 825, a famous Arabian mathematician, Mohammed ibn Musa al-Khowarizmi, wrote a book on mathematics called *Al-jabr w'al muqabalah*, which used the Hindu numerals in the solution of equations, from which the term *al-jabr* became our word *algebra* (Baumgart 1969).

As early as the twelfth century, the Hindus wrote one member of an equation over the other to indicate that they were equal. In 1557 Robert Recorde, an Englishman, published a book called *The Whetstone of Witte,* in which he used the sign "$=$" for equality (Hawkes et al. 1951). Solutions of equations by various methods showed only minor improvements until the "general solutions of cubic and quartic equations" (ca. 1545) by François Viète (Baumgart 1969).

Equations show an equals relationship between sets of numbers and/or phrases. Equations are solved using the postulates (axioms) of equality and successive operations to produce a series of *equivalent* equations. Each equivalent equation in the series is simpler than the previous one. The series continues until only the unknown is in one set and its numerical value in the other. For example:

| | |
|---|---|
| $3(x + 2) - 4 = 10 + 2(x + 1)$ | given |
| $3x + 6 - 4 = 10 + 2x + 2$ | clear parentheses |
| $3x + 2 = 2x + 12$ | combine similar terms |
| $x + 2 = 12$ | subtract $2x$ from both sides |
| $x = 10$ | subtract 2 from both sides |

To help beginning students visualize the solutions of equations many teachers today use concrete materials to represent the algebraic expressions. Concrete materials generally are used for two purposes. First, they give the teacher and the student a common ground for conversation, and second, they furnish something on which the student can act. The goal is that students reflect on their action in relation to the concepts the teacher is working to establish (Thompson 1994).

Teachers in first-year algebra classes often use two- or three-dimensional tiles to help students visualize the structure of equations and polynomials. The

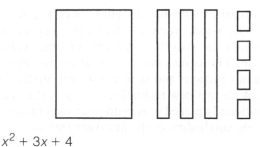

$$x^2 + 3x + 4$$

**Figure 1** *Quadratic polynomial with algebra tiles.*

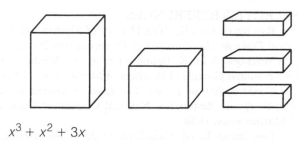

$$x^3 + x^2 + 3x$$

**Figure 2** *Cubic polynomial with algebra tiles.*

two-dimensional tiles are based on an area concept and the three-dimensional pieces on volume. (See Figures 1 and 2.)

Another set of concrete objects that teachers find helpful is small plastic cups, called portion cups, and chips that are painted a different color on each side. The chips represent integers, one color being +1 and the other being −1, and the cups represent the unknowns. When one counter of the positive color and another counter of the negative color occur, this represents +1 and −1. Since +1 + (−1) = 0, the pair of different color counters is called a "zero pair." When the cup is open and up, the variable is positive; if it is upside-down, the variable is negative. A positive and a negative cup also form a zero pair. A pencil or a pipe cleaner is used to divide the two members of the equation.

The cups and chips are manipulated in the same manner as in the steps of the algebraic solution. Students should write the symbolic notation for each step. This process allows first-year algebra students to use the senses of seeing and touching to relate to very abstract concepts. Manipulatives should appeal to as many of the senses as possible.

Some advantages of using the cups are:

- any number of chips can be placed in the cups— thus $x$ can have different values in different equations;

- the concept of the value of $-x$ can be illustrated by placing chips in the cup and inverting it, that is, $x = 5$ and $-x = -5$;
- the distributive property for polynomials is reinforced so that $3(x + 2) = 3x + 6$, rather than $3x + 2$;
- equations can be checked by building the original equation and filling the cups with the number of chips obtained in the solution;
- the value of expressions can be found by filling the cups with the given number of chips.

Using partitioning with manipulatives helps students visualize solutions involving fractions. To solve the equation $2x = 7$, represent the equation by setting two cups on the left and seven paper chips on the right. Partition each member to solve for *one x.* That is, take each chip and place it in one of the cups, dividing the group evenly. The last chip may be cut in half and each half placed in a cup, leading to the solution $3\frac{1}{2}$. Students have little or no trouble visualizing this concept.

From characters on cuneiform tablets to portion cups and chips is a long trip through history, but the manipulatives help unlock the mystery of algebraic equations for many students. Using concrete materials is not enough to guarantee success. "We must look at the total instructional environment to understand effective use of concrete materials . . . concrete materials do not automatically carry mathematical meaning for students" (Thompson 1994). Students must be taught to record their findings and to make generalizations from their use of concrete materials.

Tiles can be used to model first degree and second degree equations in one variable. The degree of an equation in one variable is the value of the highest exponent in any one term. The tiles are especially helpful to illustrate the method of completing the square used by the Greeks to solve quadratic equations, as shown in Figure 3 (Hall and Fabricant 1993).

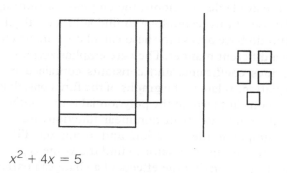

$$x^2 + 4x = 5$$

**Figure 3** *Quadratic equation with algebra tiles.*

This helps students understand why it is called *completing the square,* and why you must add the square of one half of the coefficient of the first degree term to both sides of the equation.

Many of the equations that occur in practical situations, such as scientific and banking formulas, may involve common fractions, decimal fractions, and/or algebraic fractions. An efficient method of working with equations containing fractions is to clear the equation of the fractions. Multiplying each term by the least common denominator of all the fractions contained in the equation will clear the equation of fractions so that it can be solved using the axioms, postulates, and properties of algebra. Example:

Given

$$\frac{6x}{2x-1} - \frac{x}{x-3} = 2$$

multiply the whole equation by the common denominator $(2x-1)(x-3)$ and simplify the result. If an equation contains decimal fractions as coefficients, the process is the same. Multiplying by the appropriate power of 10 will clear the denominator of the smallest decimal fraction and produce whole number coefficients.

Equations with degree greater than 2 are generally called *polynomial equations.* If the polynomial can be factored easily, the equation is solved by factoring and setting each factor equal to zero. For example, to solve $p(x) = x^3 + 3x^2 - 4x - 12 = 0$, rewrite the polynomial as a product of the factors; thus $(x+3)(x+2)(x-2) = 0$. Set each factor equal to 0. The roots of $p(x) = 0$ are $-3$, $-2$, and 2. For polynomial equations with real coefficients, complex roots appear in pairs and can be found when one of the factors is irreducibly quadratic. Real-valued solutions to polynomial equations also can be found by graphing the corresponding polynomial function and finding its intersection with the $x$-axis, that is, $\{(x,y): y = p(x) \text{ and } y = 0\}$.

As calculators with graphing capabilities become more available in schools, the emphasis is shifting to the use of this technology to find solutions. Teachers who do have access to these calculators are teaching in a different manner. They are emphasizing how the various coefficients and constants contained in the equations relate to the graphs of the functions, that is, the constant term of the polynomial is the $y$-intercept of its graph. Roots of nonlinear functions are found by graphing the equation and using the TRACE function of the calculator to find the $x$-intercept. Students are studying the effects of a change in parameters in relation to linear, quadratic, cubic, and other polynomial functions. The effect of folding the graph over the $x$-axis when absolute value appears in the equation is visually apparent on the calculator screen. The conics are visualized as the union of a pair of functions and their graphs are studied in detail. With this new technology many of the concepts traditionally placed in second-year algebra are entering the curriculum of the first-year classes.

*See also* Calculators; Equalities; Equations, Exponential; Equations, Logarithmic

SELECTED REFERENCES

Baumgart, John K. "The History of Algebra." In *Historical Topics for the Mathematics Classroom* (pp. 233–260). 31st Yearbook. John K. Baumgart et al., ed. Washington, DC: National Council of Teachers of Mathematics, 1969.

Coxford, Arthur F., ed. *The Ideas of Algebra K–12.* 50th Yearbook. Reston, VA: National Council of Teachers of Mathematics, 1988.

Fey, James T., ed. *Calculators in Mathematics Education.* 54th Yearbook. Reston, VA: National Council of Teachers of Mathematics, 1992.

Hall, Bettye C. *Developing Concepts Using Algebra Tiles.* Englewood Cliffs, NJ: Prentice-Hall, 1994.

———, and Mona Fabricant. *Algebra 2 with Trigonometry.* Englewood Cliffs, NJ: Prentice-Hall, 1993.

Hawkes, Herbert E., William A. Lubby, and Frank C. Touton. *First-Year Algebra, Elementary Course.* Boston, MA: Ginn, 1951.

Thompson, Patrick W. "Concrete Materials and Teaching for Mathematical Understanding." *Arithmetic Teacher* 41(May 1994):556–558.

Wah, Anita, and Henri Picciotto. *Algebra, Themes, Concepts, and Tools.* Mountain View, CA: Creative, 1994.

BETTYE C. HALL

# EQUATIONS, EXPONENTIAL

Equations involving expressions that have variables for exponents. René Descartes, the great French mathematician, is credited with introducing the use of Hindu-Arabic numerals as exponents on a given base in about 1637. Descartes was not the first to use exponents, only the first to use numbers as we know them today. The Swiss mathematician Jobst Burgi used Roman numerals as exponents. In 1798 Thomas Malthus, a British economist, published his theory of exponential growth in populations. The theory applies to human populations, as well as populations of cells, bacteria, and so forth.

Exponential equations are particularly useful in the solution of problems in physics, biology, and other sciences. To determine the number of days $t$ needed for a certain bacterial culture to grow to a given number $N$ of bacteria, a biologist may use the formula: $N = 5000(3)^t$ (Angel 1988). In physics, at a constant temperature, the atmospheric pressure $p$, in pascals, is given by the formula: $p = 101.3e^{-0.001h}$, where $h$ is the altitude in meters (Hall and Fabricant 1993). An equation with a variable for an exponent is called an *exponential equation*. Some further examples are $3^x = 27$, $2^x = \frac{1}{8}$, and $y = 3^{2x}$.

If $x$ and $a$ are real numbers with $a > 0$ and $a \neq 1$, then an equation in the form $y = a^x$ defines an *exponential function*, where $a$ is called the base and $x$ is called the exponent. To help students distinguish between the functions defined by linear and exponential equations, such as $y = 2x$ and $y = 2^x$, have the students cut strips from grid paper representing various integral values of the two functions, such as for $x = 1, 2, 3, 4, \ldots$, and then paste the strips on another grid to form the graphs. Remind students that there are values between those represented by the strips.

All exponential functions are one-to-one, that is, $a^m = a^n$ if and only if $m = n$. This fact is used in the solution of exponential equations. Examples:

$$3^x = 243 \qquad 5^x = \frac{1}{125}$$
$$3^x = 3^5 \qquad 5^x = 5^{-3}$$
$$x = 5 \qquad x = -3$$

The solution of exponential equations that are not expressed in terms of the same base may be found by using logarithms. Since $\log_a x = b$ is equivalent to $a^b = x$ and $\log_a M = \log_a N$ if and only if $M = N$, take the common (or base 10) logarithm of each side of the equation, and then solve for $x$. Example:

$$4^x = 19$$
$$\log 4^x = \log 19$$
$$x \log 4 = \log 19$$
$$x = \log 19/\log 4 = 2.12$$

You can use a graphing calculator to solve an exponential equation. To solve $3^x = 7$, for instance, graph the functions $y = 3^x$ and $y = 7$. Use the TRACE function and ZOOM features to read the value of $x$ where the lines intersect.

In 1947 William Lubby discovered that the approximate age of organic materials could be determined by measuring the amount of the radioactive isotope carbon-14 left in an organism and then using a so-called *exponential decay formula:* $A = A_0 \cdot 2^{-1/kt}$, where $A$ is the present amount of the isotope, $A_0$ is the original amount of isotope, $k$ is the half-life of the

isotope measured in the same units as $t$, and $t$ is the time to reduce the original amount of the isotope to the present amount (Hall and Fabricant 1993). If nuclear energy is used to produce electricity, the length of time for decay of nuclear wastes becomes very important. The exponential decay formula is used by nuclear engineers to determine how long nuclear wastes must be stored before the amount of radioactivity is negligible. A typical nuclear power plant produces about 10 pounds of Krypton 85 per year. Since the half-life of Krypton 85 is eleven years, it must be contained for about seventy-three years. The mathematical concepts of growth and decay figure prominently in many fields of science which influence our daily lives.

*See also* Engineering; Equations, Logarithmic; Exponents, Algebraic; Interest; Logarithms; Population Growth

SELECTED REFERENCES
Angel, Allen R. *Intermediate Algebra for College Students.* Englewood Cliffs, NJ: Prentice-Hall, 1988.

Fleming, Walter, and Dale Varberg. *College Algebra: A Problem-Solving Approach.* Englewood Cliffs, NJ: Prentice-Hall, 1988.

Hall, Bettye C., and Mona Fabricant. *Algebra 2 with Trigonometry.* Englewood Cliffs, NJ: Prentice-Hall, 1993.

National Council of Teachers of Mathematics. *Curriculum and Evaluation Standards for School Mathematics.* Reston, VA: The Council, 1989.

———. *Principles and Standards for School Mathematics.* Reston, VA: The Council, 2000.

BETTYE C. HALL

# EQUATIONS, LOGARITHMIC

Equations containing logarithms, that is, exponents. John Napier (1550–1617), a Scottish baron, amused himself by studying science and mathematics. In 1614 he published his discussion of logarithms, simplifying the methods of complicated calculations, under the title *Mirifici logarithmorum cannonis descriptio* ("A Description of the Wonderful Law of Logarithms") (see Baumgart 1969). Today, *logarithmic equations*, that is, equations containing logarithms, are found in real-world situations. For example, the pH of a solution is the concentration of hydrogen ions, $H^+$, in gram atoms per liter, that is, $pH = \log_{10} 1/H^+$. The concentration of hydrogen ions determines the acidity of a solution.

For all positive real numbers $x$ and $b$, where $b$ is not equal to 1, the inverse of $y = b^x$ is defined as

$\log_b y = x$. Notice that as $x$ increases arithmetically, $y$ increases geometrically. Every positive real number has a unique logarithm to the base $b$. Logarithms are used to compare the intensities of earthquakes. The Richter scale gives the magnitude of an earthquake as the logarithm of the intensity $I$ of its shockwaves: $R = \log_{10} I$, where $I$ is the number of times more intense the quake is, compared to a minimum level. For example, if the 1906 San Francisco earthquake was 8.25 on the Richter scale and the 1971 Los Angeles earthquake was 6.7, how much more intense was the 1906 quake? Using $R = \log_{10} I$, then $8.25 = \log_{10} I$ (SF) and $6.7 = \log_{10} I$ (LA); the antilog $I(\text{SF}) = 178{,}000{,}000$ and $I(\text{LA}) = 5{,}000{,}000$. Thus, SF is 35.6 times more intense.

Since by definition a logarithm is an exponent, operations with logarithms follow the laws of exponents. This fact can be used to solve logarithmic equations. For example, if

$$\log_4 y + \log_4(y + 6) = 2$$
$$\log_4 y(y + 6) = 2$$
$$y(y + 6) = 4^2$$

Then solve the quadratic equation for the positive root. Logarithms also are used to solve exponential equations when the bases in the equation are different. For example, if

$$2^{x-3} = 3^4$$
$$\log 2^{x-3} = \log 3^4$$
$$(x - 3) \log 2 = 4 \log 3$$
$$x - 3 = 4\left(\frac{\log 3}{\log 2}\right) = 6.3398\ldots$$
$$x = 9.3398\ldots$$

The graphing calculator also can be used to find the solution. Graph the two functions $y = 2^{x-3}$ and $y = 3^4$ and find the point of intersection.

*See also* Equations, Exponential; Exponents, Algebraic; Logarithms

### SELECTED REFERENCES

Angel, Allen R. *Intermediate Algebra for College Students*. Englewood Cliffs, NJ: Prentice-Hall, 1988.

Baumgart, John K. "The History of Algebra." In *Historical Topics for the Mathematics Classroom* (pp. 233–260). 31st Yearbook. John K. Baumgart et al., eds. Washington, DC: National Council of Teachers of Mathematics, 1969.

Fleming, Walter, and Dale Varberg. *College Algebra: A Problem-Solving Approach*. Englewood Cliffs, NJ: Prentice-Hall, 1988.

Stewart, Ian. "Change." In *On the Shoulders of Giants, New Approaches to Numeracy* (pp. 183–216). Lynn A.

Steen, ed. Washington, DC: National Academy Press, 1990.

BETTYE C. HALL

# ERATOSTHENES
## (ca. 276–ca. 194 B.C.)

Known primarily for developing an algorithm, now called the "Sieve of Eratosthenes," which is used to determine all primes up to a given number $N$. The method is based upon identifying primes $< \sqrt{N}$ and eliminating the multiples of each of these primes. Eratosthenes was a Greek scholar who excelled in several different areas, including mathematics, geography, and literature. After studying in Athens, he spent most of his life in Alexandria, at that time under Greek rule. Eratosthenes also is credited with using parallel lines and arcs of great circles to measure the circumference of the earth. His results differed from what we know to be the actual circumference by less than 160 miles! He also is known for developing a mechanical device that was used to solve the famous construction problem of doubling the volume of a given cube.

*See also* History of Mathematics, Overview; Prime Numbers

### SELECTED REFERENCES

Baumgart, John K. et al., eds. *Historical Topics for the Mathematics Classroom*. 2nd ed. Reston, VA: National Council of Teachers of Mathematics, 1989.

Heath, Thomas L. *A Manual of Greek Mathematics*. New York: Dover, 1963.

ANTHONY V. PICCOLINO

# ERRORS, ARITHMETIC

An important part of the evaluation of a mathematics program is the analysis of errors in arithmetic. A considerable amount of information can be gleaned by carefully studying children's erroneous procedures and strategies (Ashlock 1994). Roberts (1968) identified four major error categories in arithmetic computation: (1) *incorrect operation* (e.g., addition rather than subtraction); (2) *obvious computational error*—the student uses the correct operation, but makes a mistake in recalling basic facts (e.g., $8 + 7 = 13$), (3) *defective algorithm*—the student tries to use the correct operation but makes errors in carrying out the necessary steps (e.g., incorrect division procedure); (4) *random response*—the response the student gives has no connection to the problem.

Research indicated that the errors could be generalized. Carelessness (though not a separate category) and not knowing basic facts occurred at about the same frequency among all ability levels, but use of the wrong operation and random guesses occurred more often with low-ability students.

Researchers have recognized for years that studying students' errors is a good source of data about the learning process. In most schools, errors that students make are not analyzed but simply used to determine grades. Some teachers do analyze errors; their goal is diagnosis and remediation. However, using errors as a springboard for inquiry with students may provide a stimulus to learning. When using errors as a springboard for inquiry, the following goals of school mathematics may be advanced: (a) enabling students to better understand the nature of mathematics as a discipline; (b) facilitating the learning of significant mathematical content; (c) making the students more proficient in doing mathematics; and (d) making the students more confident in their ability to learn and use mathematics (Borasi 1994, 189–190).

## COMMON ERRORS

In a study of systematic errors, Cox selected 700 students in grades 2 through 6. The researcher defined a systematic error as "a repeatedly occurring incorrect response that is evident in a specific algorithmic computation," whereas random errors "give no evidence of a recurring incorrect process of thinking or recording" (Cox 1975, 203–204).

Systematic errors in addition of a two-digit and a one-digit number with no renaming were made by 73% of the students in the study, with the students adding each digit separately without regard to the places of the digits:

$$\begin{array}{r} 31 \\ +5 \\ \hline 9 \end{array}$$

In this example, the student added $3 + 1 + 5 = 9$.

When adding two-digit numbers with renaming, 67% of the children in the study made the error of adding and recording 1s and then 10s, without regard to place value in the sum:

$$\begin{array}{r} 38 \\ +96 \\ \hline 1214 \end{array}$$

When subtracting a two-digit number from a two-digit number with renaming, 83% of the children apparently considered each position as a separate subtraction problem, subtracting the smaller number from the larger number without regard to which number represented the subtrahend and which represented the minuend:

$$\begin{array}{r} 94 \\ -68 \\ \hline 34 \end{array}$$

Seventy-five percent of the children renamed a subtraction problem (such as $395 - 72$) whether or not it was necessary:

$$\begin{array}{r} 81 \\ 395 \\ -72 \\ \hline 3113 \end{array}$$

In this example, the student renamed the 5 to 15 and subtracted to get 13. The student then subtracted 7 from 8, getting 1, and then brought down the 3, for a final answer of 3113.

In multiplication of a two-digit or a three-digit multiplicand by a one-digit multiplier with no renaming, 90% of the children did not multiply the number in the tens place or the hundreds place; instead, they simply placed those digits in the product:

$$\begin{array}{rr} 32 & 423 \\ \times 4 & \times 2 \\ \hline 38 & 426 \end{array}$$

Sometimes the children add in the number carried before multiplying the tens digit. In Cox's study, 43% of the children displayed this particular error:

$$\begin{array}{r} 2 \\ 35 \\ \times 4 \\ \hline 200 \end{array}$$

In division, one of the most common difficulties experienced by children is determining when a zero is required in the tens place of the quotient. Seventy-two percent of the children failed to insert zero in the quotient:

$$\begin{array}{r} 7\ \ 8\ r2 \\ 5\overline{)3542} \\ \underline{35} \\ 42 \\ \underline{40} \\ 2 \end{array}$$

Cox also reported that students realized that patterns and structures were necessary for solving the problems, but they could not figure out the correct ones. The misconceptions were not caused by lack of knowledge of number facts, but by failure to understand the meaning of number, operation, place value, or renaming.

## SUGGESTED INSTRUCTIONAL STRATEGIES

In the teaching of computational procedures, the instruction should be combined with diagnosis. As soon as it is apparent a student is making a common error, that is the time to remediate, before the error becomes part of the student's conceptual understanding. One of the primary reasons children have difficulty learning computation is they do not understand the underlying mathematical concepts (Ashlock 1994).

The first step in helping children who are experiencing systematic arithmetic errors is to determine if the error is a component skill error or a strategy error. If it is a component skill error, this indicates the child does not understand the underlying concept behind the computation. A strategy error is simply confusion as to the correct steps necessary to solve a computation problem (Silbert, Carnine, and Stein, 1990).

One good strategy, whether component skill or strategy error, is to have the child estimate before beginning the computation. If this is done first, the child should have an idea if the result of the computation is reasonable. In order to be good estimators, students need to have good number sense. Number sense involves understanding numbers and their relationships, recognizing the relative magnitude of numbers, and understanding the relative effect of operating on numbers (National Council of Teachers of Mathematics (NCTM) 1989). Another good strategy is to have the child explain what he or she is thinking so that the correct remediation can be applied. Conversation also leads to increased understanding, since learning is an active process rather than a passive one (Greeno 1991).

Continuous, purposeful use of manipulatives is imperative in helping children gain conceptual understanding of computation. Children need many concrete experiences using base-10 blocks, counters, unifix cubes, Cuisenaire rods, and so on, setting up different problems so they can solve the problems physically. This creates a mental reference so they can eventually move from concrete to abstract solving of arithmetic problems. Concrete experiences lay the foundation for developing mathematical concepts. However, it is not wise to try to move a student directly from concrete experiences to abstract solving of problems. Students must be able to link their concrete experiences to abstract thought through learning bridges, which are connections between the concrete and the abstract. These bridges connect instructional models and formal mathematical ideas and symbols. These learning bridges work both ways. They need to be crossed and recrossed for each new concept the student learns (Reys, Suydam, and Lindquist 1992).

Carelessness, or inattention to detail, is not the same as failure to understand a concept. Carelessness is simply a mistake in computation. One strategy to help students eliminate careless errors is to have each child keep track of his or her mistakes by using a checklist such as the one shown here:

| Homework | Page | | |
|---|---|---|---|
| Did not regroup in subtraction | | | |
| Forgot to carry in addition | | | |
| Skipped problem | | | |
| Did not record zero in quotient | | | |
| Did not record zero in product | | | |

The use of the checklist makes the child aware of his or her mistakes, which helps eliminate them. It is also very useful during parent conferences so parents can see exactly where their child needs help (McAcy 1993).

In conclusion, arithmetic errors should not be ignored or passed over as just wrong answers. They need to be analyzed carefully for use in remediation, discovery, and inquiry in the mathematics classroom by teachers and students.

*See also* Addition; Basic Skills; Division; Multiplication; Subtraction

## SELECTED REFERENCES

Ashlock, Robert B. *Error Patterns in Computation*. 6th ed. New York: Macmillan, 1994.

Borasi, Raffaella. "Capitalizing on Errors as 'Springboards For Inquiry': A Teaching Experiment." *Journal for Research in Mathematics Education* 25(Mar. 1994): 166–208.

Cox, L. S. "Systematic Errors in the Four Vertical Algorithms in Normal and Handicapped Populations." *Journal for Research in Mathematics Education* 6(Nov. 1975): 202–220.

Greeno, James G. "Number Sense as Situated Knowing in a Conceptual Domain." *Journal for Research in Mathematics Education* 22(May 1991):170–218.

McAcy, Karen B. "Careless Mistakes." *Mathematics Teacher* 86(Apr. 1993):298–299.

National Council of Teachers of Mathematics. *Curriculum and Evaluation Standards for School Mathematics*. Reston, VA: The Council, 1989.

———. *Principles and Standards for School Mathematics*. Reston, VA: The Council, 2000.

Reys, Robert E., Marilyn N. Suydam, and Mary Montgomery Lindquist. *Helping Children Learn Mathematics*. 3rd ed. Boston, MA: Allyn and Bacon, 1992, pp. 44–59.

Roberts, Gerhard H. "The Failure Strategies of Third Grade Arithmetic Pupils." *Arithmetic Teacher* 15(May 1968):442–446.

Silbert, Jerry, Douglas Carnine, and Marcy Stein. "Direct Instruction." In *Direct Instruction Mathematics*. 2nd ed. Columbus, OH: Merrill, 1990.

DEBORAH B. LEE

## ESCHER, MAURITS CORNELIUS (1898–1972)

A graphic artist with a distinctive mathematical style born in the Netherlands. He became internationally known for his surrealistic woodcuts and lithographs based on tessellations spaced in a regular but striking manner. These tessellations were made from squares, equilateral triangles, and other polygons, which were systematically altered, shaded, and transformed by various symmetries into imaginative pattern pieces. These elements were then combined artistically into visual paradoxes and illusions that intrigue the mind and the eye. Escher's approach was strongly influenced by Moorish art, which he observed while moving through and studying art in southern Europe. Returning to the Netherlands, his distinctive mathematical-artistic style increasingly gained recognition and acceptance. During his successful career, he produced over 270 works, such as "Day and Night" (1938), "Drawing Hands" (1948), and "Circle Limit IV" (1960), which dramatically illustrate his tessellation effect and express artistic looks at the infinite.

*See also* Art; Tiling

### SELECTED REFERENCES
Escher, Maurits C. *The Graphic Work of M.C. Escher*. New York: Hawthorn, 1970.

Ranucci, Ernest R. "Master of Tessellations: M. C. Escher, 1898–1972." *Mathematics Teacher* 67(4)(1974): 299–306.

WILLIAM D. JAMSKI

## ESTIMATION

The process of obtaining, often mentally, a quick, commonsense approximation to the solution of a problem, rather than solving it exactly. One of the most important reasons for using estimation is to prevent computation errors. It enables the problem solver to determine if a solution is reasonable. It also helps children understand that not everything done in mathematics is exact. In many real-life situations, estimation is usually a more appropriate process than actual measurement or calculation. Research surveys suggest that people use estimation in over 80% of all mathematics applications in everyday life instead of exact computation (Harte and Glover 1993; Reys 1992). Learning to estimate helps children develop good number sense, but it should not be taught as an isolated page or unit in a textbook. Rather, it is a skill that relates to every facet of elementary school mathematics (from addition to measurement to geometry) and to higher-level mathematics as well. Children must be provided with experiences that help them realize that estimation is an essential and practical skill. Reys (1992) states that characteristics of good estimators include (a) producing an approximation to the answer quickly, (b) using mental computation to make estimates rather than writing out the details, (c) producing answers that are adequate but not exact in order to make a decision, (d) feeling comfortable with an estimate, and (e) having a variety of estimation strategies to choose from, depending on the problem.

Reys, Suydam, and Lindquist (1992) state that "it is critical in all stages of estimation that children think about the problem, the operation, and the numbers involved, and not rely on a fixed set of rules to produce an estimate" (p. 241). They also recommend instructional guidelines to use when teaching computational estimation:

*Instruction*—Computational estimation has to be taught because it does not always come naturally. The teacher should provide learning situations that require the children to estimate using various strategies.

*Practice*—The children should be provided with different types of estimation situations after they have been given specific instruction. This can be accomplished in short practice sessions, which helps children maintain and improve their basic facts, mental computation skills, and estimation skills.

*Testing*—Testing motivates children to develop computational estimation skills. The tests should be short and may focus on specific estimation strategies or a combination of strategies.

Different strategies can be used to estimate solutions to problems. Several of these strategies are: (a) ballpark guesses; (b) different types of rounding; (c) front-end estimation; (d) compatible numbers strategies; and (e) more sophisticated techniques used in statistics and economics (Usiskin 1986; Reys 1992).

Front-end estimation is a good strategy to use with addition. In front-end estimation, the problem is solved by adding the leftmost digits first and then adjusting by focusing on the other digits. Consider $98 + 37 + 63$. Adding the tens digits, $9 + 3 + 6 = 18$ tens. The unit digit of 98 (i.e., 8) is approximately 1 ten; the unit digits 7 and 3 sum to 1 ten. The total sum is about 20 tens or 200, a good estimate of the exact sum, 198.

Compatible numbers strategy is a more sophisticated estimation strategy. The student should round so the numbers are compatible and easy to compute. This strategy is good with division, such as $7158/22 = 7000/20$ or $7500/25$. Special numbers are numbers that are near values and can be estimated easily. For example, $24/47$ of 800 may be approximated by $1/2$ of $800 = 400$; 23.8% of 1200 by $1/4$ of $1200 = 300$.

Rounding is a good strategy to use with multiplication. It involves computing with the rounded figures and adjusting the final product. To multiply $47 \times 38$, for instance, round to $50 \times 40 = 2000$ and adjust to less than 2000, or $2000-$. For $33 \times 51$, multiply $30 \times 50 = 1500$ and adjust to more than 1500, or $1500+$. In many textbooks, the only estimation strategy taught is rounding. All of the different procedures need to be taught, because rounding is not an appropriate strategy in many situations (Reys 1992). There are times when just a ballpark guess or the use of compatible numbers is a reasonable way to solve a problem. Estimation is one of the most important skills that can and should be taught because it facilitates accurate computations and promotes number sense, mental computation, and development of various mathematical concepts (Reys and Reys 1990).

*See also* Basic Skills; Rounding

SELECTED REFERENCES

Harte, Sandra W., and Matthew J. Glover. "Estimation Is Mathematical Thinking." *Arithmetic Teacher* 40(Oct. 1993):75–77.

Reys, Barbara. "Estimation." In *Teaching Mathematics in Grades K–8: Research Based Methods*. 2nd ed. (pp. 279–300) Thomas R. Post, ed. Needham Heights, MA: Allyn and Bacon, 1992.

Reys, Barbara J., and Robert E. Reys. "Implementing the *Standards*. Estimation: Directions from the *Standards*." *Arithmetic Teacher* 37(Mar. 1990):22–25.

Reys, Robert E., Marilyn N. Suydam, and Mary Montgomery Lindquist. *Helping Children Learn Mathematics*. 3rd ed. Boston, MA: Allyn and Bacon, 1992.

Usiskin, Zalman. "Reasons for Estimating." In *Estimation and Mental Computation* (pp. 1–15). 48th Yearbook. Harold L. Schoen, ed. Reston, VA: National Council of Teachers of Mathematics, 1986.

DEBORAH B. LEE

# ETHNOMATHEMATICS

Research and education based on knowledge derived from quantitative and qualitative practices (such as counting, weighing and measuring, sorting and classifying) and accumulated through generations in distinct cultural environments.

Ethnomathematics (or Multicultural Mathematics) takes into account all the forces that shape a mode of thought, in the sense that it examines the generation, organization (both intellectual and social), and diffusion of knowledge. The research interrelates results from the cognitive sciences, epistemology, history, sociology, and education and assumes that mathematics is an intellectual construct by humans in response to survival needs (D'Ambrosio 1994). The history of mathematics contains many examples of ethnomathematics. Much of the finger counting and arithmetic of street and market vendors practiced by people of European ancestry have their origins in the Middle Ages. And today the mathematical practices of the peoples in the Americas are performed in a very distinctive way that results from the mutual exposition since colonial times of different cultural forms. It is common to see indigenous peoples in the Americas using Hindu-Arabic numerals but performing the operations from bottom to top, which they explain by saying that this is the way trees grow.

Mathematical achievement in everyday life is easily recognized. This is evident in the many professions that require mathematical abilities. But school tests, since these are attached to subjects and skills taught in the classroom, do not necessarily reveal mathematical abilities. However, there are alternative assessment techniques that take into account cultural components of mathematical knowledge. One such new assessment technique, having students write mathematics, is suitable for multicultural mathematics. Writing invites reflection on the nature of manipulation of symbols, codes, and rules, and the cultural context emerges (Drake and Amspaugh 1994).

Practices and perceptions of learners are the substratum upon which new knowledge is built. Thus new knowledge has to be based on the individual and cultural history of the learner. There are educational benefits in (a) recognizing the diversity of extant cultures, present in specific communities all over the world; and (b) an historical perspective that recognizes the contribution of past cultures to modern thought.

For example, pre-Columbian cultures had different styles of measurement and computation, and these practices are still prevalent in some native communities. Most Amazonian tribes have counting systems that use only "one, two, three, four, many." With these numbers alone they can satisfy all their needs (Closs 1986). We see ways of dealing with pottery, tapestry, and everyday knowledge with strong mathematical characteristics in several cultures (Ascher 1991). The same is true of African cultures (Gerdes 1995). The people from these cultures have no problem at all in assimilating the current European number system and can deal with counting, measurement, and money when trading with individuals from European-based cultures. Land measurement, as practiced by peasants in Latin America, comes from ancient geometry. Carpenters, bricklayers, and carpet layers all over the world use a very specific geometry in their work. They have to cut pieces produced in the usual geometrical forms, such as squares, rectangles, and regular polygons, and adjust them to the surface to be covered, practicing optimization techniques.

To respond adequately to multicultural demands, the traditional organization of the curriculum, subdivided into objectives, content, and methods, may take on a more dynamic approach. The current order of topics, seen in most curricula, is based entirely on tradition. In ethnomathematical pedagogy, there are no natural prerequisites. The history of mathematics tells us that the motivation for al-Khowarizmi's introduction of algebra was the societal need to make operational some of the precepts of the Koran. "The modern teacher is trained to do much more than follow a programme of work contained in a series of textbooks or commercial learning schemes. The ability to devise appropriate learning materials and situations is also essential. So also is the ability to guide pupils to investigative work and personal or group project work" (Nelson, Joseph, and Williams 1993, 43).

There are several strategies for modifying the curriculum to introduce a multicultural approach.

Of course, the support of the school administration, parents, and students is needed to introduce modifications. An important aspect of ethnomathematics is the possibility it offers to overcome the fear of mathematics, since it deals with familiar situations, present in daily life (see Zaslavsky 1994). If in some cases it is necessary to follow a traditional programmatic curriculum, there is always the possibility of introducing ethnomathematics as curriculum enrichment. For example, when dealing with measurement, it is rather easy to bring ethnomathematics into the curriculum. Several examples are given in Harris (1991); and Nelson, Joseph, and Williams (1993). In a relatively prosperous and culturally homogeneous community, dealing with interesting issues such as building a golf course or analyzing the daily routine of families or the cost of home energy leads to interesting ethnomathematics (Skovsmose 1994).

A dynamic curriculum can be achieved with a three-stage methodology: (1) *sensibilization,* (2) *instrumentation,* (3) *supporting subjects.* This approach was implemented in a graduate program for leadership in science and mathematics education sponsored by the Organization of American States at the University of Campinas, Brazil, from 1975 to 1980. The following is a brief description of the three stages.

1. *Sensibilization* (or *socialization*) is achieved by raising issues in the classroom related to a main theme of social concern. This can be achieved by examining newspapers, magazines, and television and by conducting interviews. Of course, these practices have a local character directly related to the social and natural environment of the school. The methodology at this stage is essentially ethnographic. The use of mathematical codes associated with writing depends on the level of the students and on the subject under examination. For example, in a small town it is difficult to quantify environmental conditions. Students could research the city archives to learn of new housing developments over the last fifty years in order to evaluate the elimination of woodlands to provide open space for dwellings. Many issues relating to daily concerns are not found in textbooks or in the traditional curricula but result from modern societal behavior and are reported in the media (Frankenstein 1990). Sociocultural issues may thus take priority in choosing classroom topics over predesigned content. How much mathematics is needed for this

stage? What students know at the moment is enough to undertake this initial stage, which is essentially motivation to increase their mathematical knowledge.

2. *Instrumentation* supplies the mathematical tools needed to translate problems into mathematical language. Of course, some students already possess these intellectual instruments, which are put into action according to need. The teacher becomes an important resource for students, providing the stimulus to develop practical capability in tackling problems and in using the various cognitive, social, and behavioral skills they are learning.

3. *Supporting subjects* provide the established mathematical instruments needed to deal with the issues now formulated in mathematical language in stage 2. Clearly, the listing of mathematical content in the program should not be defined *a priori*, but should rather be a result of the process. This is one of the strongest characteristics of the new curricular organization implicit in multicultural mathematics. Good textbooks try to produce updated motivations. But these are obviously rooted in the natural and cultural environment of their authors. Thus another typical issue to be dealt with in multicultural mathematics is the necessity, in order to achieve motivation, of treating each theme as a self-contained unit, such as was proposed in the 1960s in the modular approach. Of course, instead of a propaedeutic approach, background material should be frequently retrieved during the process. Technology—encyclopedias, videos, and software—makes learning possible as the motivation occurs.

*See also* Multicultural Mathematics, Issues

## SELECTED REFERENCES

Ascher, Marcia. *Ethnomathematics. A Multicultural View of Mathematical Ideas.* Pacific Grove, CA: Brooks-Cole, 1991.

———, and Ubiratan D'Ambrosio, eds. "Ethnomathematics in Mathematics Education." *For the Learning of Mathematics.* Special Issue. 14(2)(June 1994).

Closs, Michael, ed. *Native American Mathematics.* Austin, TX: University of Texas Press, 1986.

D'Ambrosio, Ubiratan. "Environmental Influences." In *Studies in Mathematics Education.* Vol 4. (pp. 29–46). Robert Morris, ed. Paris, France: UNESCO, 1985.

———. "Ethno-mathematics, the Nature of Mathematics and Mathematics Education." In *Mathematics, Education and Philosophy: An International Perspective* (pp.

230–242). Paul Ernest, ed. London, England: Falmer, 1994.

Drake, Bob M., and Linda B. Amspaugh. "What Writing Reveals in Mathematics." *Focus on Learning Problems in Mathematics* 16(3)(Summer 1994):43–50.

Frankenstein, Marilyn. *Relearning Mathematics.* New York: Columbia University Press, 1990.

Gerdes, Paulus. *Ethnomathematics and Education in Africa.* Stockholm, Sweden: Institute of International Education, Stockholm University, 1995.

Harris, Mary. *Schools, Mathematics and Work.* Bristol, England: Falmer, 1991.

National Science Foundation. *Science, Mathematics, Engineering, and Technology Education for the 21st Century.* Summer Symposium on Educating for Citizenship in the 21st Century, July 1992. Final Report. Washington, DC: The Foundation, 1993.

Nelson, David, George G. Joseph and Julian Williams. *Multicultural Mathematics. Teaching Mathematics from a Global Perspective.* Oxford, England: Oxford University Press, 1993.

*Newsletter of the International Study Group on Ethnomathematics (ISGEm).* Patrick J. Scott, ed. Albuquerque, NM: University of New Mexico, 1985–.

Powell, Arthur B., and Marilyn Frankenstein, eds. *Ethnomathematics: Challenging Eurocentrism in Mathematics Education.* Albany, NY: State University of New York Press, 1997.

Saxe, Geoffrey. *Culture and Cognitive Development. Studies in Mathematical Understanding.* Hillsdale, NJ: Erlbaum, 1991.

Skovsmose, Ole. *Towards a Philosophy of Critical Mathematics Education.* Dordrecht, Netherlands: Kluwer, 1994.

Vithal, Renuka, and Ole Skovsmose. "The End of Innocence: A Critique of 'Ethnomathematics'." *Educational Studies in Mathematics* 34(2)(1997):131–157.

Zaslavsky, Claudia. *Africa Counts: Number and Pattern in African Culture.* New York: Hill, 1979; Revised 1999.

———. *Fear of Math. How to Get Over It and Get On with Your Life.* New Brunswick, NJ: Rutgers University Press, 1994.

UBIRATAN D'AMBROSIO

# EUCLID (ca. 300 B.C.)

Best known for compiling most of the mathematical knowledge of his time and organizing it into the monumental work known as the *Elements.* Although little is known about Euclid's personal life, the historian Proclus places Euclid in the reign of Ptolemy I Soter of Egypt (304–285 B.C.). In the *Elements,* Euclid masterfully organized propositions from plane and solid geometry, number theory, and geometrical algebra (using geometric figures and the concept of area to describe algebraic truths) into a deductive system that became the forerunner of modern axiomatic systems.

The 465 propositions contained within the *Elements* are organized into thirteen books and are preceded by Euclid's inclusion of definitions, axioms, and postulates from which the propositions are derived. Euclid distinguished between an axiom and a postulate, a distinction not made in contemporary mathematics. Axioms referred to truths (considered self-evident) such as "If equals are added to equals, the sums are equal," whereas postulates were self-evident truths pertaining specifically to geometric entities. In addition to the familiar propositions of plane and solid geometry, the *Elements* contains propositions dealing with constructions, geometrical algebra, Eudoxus' theory of proportion, number theory, and the regular polyhedra. Conspicuously missing from the *Elements* are conic sections and spherical geometry.

Euclid's Fifth Postulate (Parallel Postulate), in its modern form, specifies that through a given point not on a given line, one and only one line can be drawn parallel to the given line. This postulate served as the basis of much controversy among mathematicians for over two thousand years. The recognition, in the nineteenth century, that the Fifth Postulate is indeed independent of Euclid's other postulates served as the basis for the development of non-Euclidean geometries by mathematicians such as Carl Gauss (1777–1855), Janos Bolyai (1802–1860), Nikolai Lobachevsky (1793–1856), and Georg Riemann (1826–1866). Since the first twenty-eight propositions in Euclid's *Elements* are independent of the Parallel Postulate, that collection of propositions is often referred to as neutral or absolute geometry.

In the *Elements*, Euclid also utilized the method of *reductio ad absurdum* (indirect proof) to prove that the number of primes is infinite and that $\sqrt{2}$ is irrational.

*See also* Greek Mathematics; History of Mathematics, Overview; Non-Euclidean Geometry

## SELECTED REFERENCES

Calinger, Ronald, ed. *Classics of Mathematics.* Englewood Cliffs, NJ: Prentice-Hall, 1995.

Euclid. *The Thirteen Books of Euclid's Elements.* 2nd ed. 3 vols. Thomas L. Heath, trans. New York: Dover, 1956.

Swetz, Frank, ed. *From Five Fingers to Infinity: A Journey Through the History of Mathematics.* Chicago, IL: Open Court, 1994.

ANTHONY V. PICCOLINO

# EULER, LEONHARD (1707–1783)

A most productive mathematician and one of the greatest. Born in Switzerland, he studied mathematics with Johann (Jean) Bernoulli and wrote his first mathematical paper at eighteen. Scientific activity in the eighteenth century centered around academies rather than universities, and Euler spent all his professional life at the academies of St. Petersburg and Berlin. He became totally blind in 1766 but continued his research until the day he died. His collected works will fill about eighty volumes (not all have been published to date). Euler made seminal contributions to all then-existing areas of mathematics as well as to mechanics and astronomy. As were Carl Gauss and Srinivasa Ramanujan, Euler was superb at calculation, an important experimental tool of the mathematician. He often arrived at beautiful results inductively and heuristically. To Euler (as to most mathematicians of his time), algorithms were at least as important as abstract proofs, and special problems were at least as weighty as general theories.

In the seventeenth century, calculus was geometric, dealing with *curves*. Euler made it algebraic, focusing on *functions*. His influential textbooks, which can still be read with great profit, systematized the subject. In the *Introductio in Analysin Infinitorum*, "Euler accomplished for analysis what Euclid (in the *Elements*) and al-Khowarizmi (in *Al-jabr*) had done for synthetic geometry and elementary algebra, respectively" (Boyer 1951). For example, Euler gave the earliest treatment of logarithms as exponents and of trigonometric functions as numerical ratios. He introduced and systematized notation and terminology that is standard today. He was the first to use the symbols $e$ and $i$ and the first to use $\pi$ and $f(x)$ systematically. He also discovered (and proved) the Euler-Cotes formula $e^{i\theta} = \cos\theta + i\sin\theta$.

Euler's contributions to number theory alone would earn him a place in the mathematics hall of fame. But his major work was in what later came to be independent fields of analysis: calculus of variations, differential equations, special functions, infinite series, and complex functions. He also touched on aspects of differential geometry and topology. "The study of Euler's works will remain the best school for the different fields of mathematics and nothing else can replace it," asserted Gauss, who was not given to excessive praise.

*See also* $e$; Exponents, Algebraic; Graph Theory; History of Mathematics, Overview

SELECTED REFERENCES

Bell, Eric Temple. *Men of Mathematics.* New York: Simon and Schuster, 1937.

Boyer, Carl. "The Foremost Textbook of Modern Times." *American Mathematical Monthly* 58(1951): 223 226.

Schattschneider, Doris, ed. *Mathematics Magazine* 56(5)(1983). A tribute to Euler on the 200th anniversary of his death.

Youschkevitch, Adolf Pavlovich. "Euler, Leonhard." In *Dictionary of Scientific Biography* (pp. 467–484). Vol. 4. Charles C. Gillispie, ed. New York: Scribner's, 1981.

ISRAEL KLEINER

# EVALUATION OF INSTRUCTION, ISSUES

Includes the major questions (1) what criteria should be employed? (2) who should do the evaluation? and (3) for what purposes should the results be used? The answers in each of these cases are of utmost importance to teachers, administrators, parents, and students. Decisions based on the results of such evaluations have effects on student achievement, employment and retention of personnel, organization of instructional teams, allocation of funds, and other matters.

For purposes of this entry, *instruction* will be defined as the activities of a teacher that are related to the attainment of curricular goals. It will include the planning and tactical activities of the teacher and emphasize classroom performance. It will not include textbooks or other commercial materials over which the teacher may have no control, but may include materials developed by the teacher for use by him or her in the classroom.

With respect to evaluation criteria, there are two major approaches. One may be thought of as a *process* approach, in which the focus is upon the performance of the teacher. The other is more of a *product* approach, in which the focus is upon the performance of the students in regard to the attainment of stated objectives. The former takes a position that the effectiveness of the teacher in terms of questioning, exposition, and demonstration, as well as motivating and keeping students on task, should be the basis for evaluation of instruction. The latter approach argues that if the students do not attain stated objectives, what the teacher does is almost irrelevant, so that achievement, including that on standardized tests, should be the primary criterion for evaluation.

Sets of performance criteria for "good" teaching have been established by teacher education researchers over the past decades (e.g., Borich 1977, Good and Brophy 1987, Koehler and Grouws 1992). They include the teacher's subject-matter knowledge, the quality of questioning strategies and activities, the nature of teacher-student interaction, the perceptions of students regarding the quality of their teachers' performances, affective factors, and "general atmosphere" of classrooms. At the elementary school level the utilization of manipulative materials could be an important criterion of instructional quality, while a middle school teacher might be evaluated more on the use of written and oral explanations of mathematical relations and concepts by students. A high school teacher could be evaluated on the use of higher-order thinking skills in both questioning and assessment activities.

One difficulty with some of these models is that they do not always allow for personality differences among teachers or different learning styles of students. If one includes in the definition of "good teaching" the ability and willingness to adapt instruction to each student's individual needs, then differences among teachers and students are covered, but the model then would require knowledge of individual student characteristics. Furthermore, it is highly unlikely that teachers have the time or resources to determine and respond to the needs of all the students they encounter on a daily basis. Thus the use of teacher-performance criteria in the evaluation of instruction is necessarily limited.

Using student achievement as the primary, or sole, criterion for the quality of instruction seems tempting. It focuses on the attainment of stated objectives, and, if that is the mission of the instruction, then satisfactory accomplishment of the mission would seem to be the ideal way to evaluate the instruction. There are, however, serious threats to the validity of such a model. One such threat is related to the validity of the procedures used in assessing the achievement of the objectives. The emphasis on alternative assessment procedures, such as advocated in the *Curriculum and Evaluation Standards for School Mathematics* (National Council of Teachers of Mathematics 1989), has raised serious questions about the validity of traditional pencil-and-paper testing as the primary method of assessing achievement. Therefore, some consensus on what constitutes achievement would be necessary before adopting it as a criterion of instructional effectiveness. Another threat is the failure to include student variables in making instructional evaluation decisions. Despite some attempts to minimize the effects on achievement of individual differences in ability, it is true that not all students are equally capable, for many different reasons, of reaching equal levels of achievement. Further, not all students come to a teacher with anywhere near equal prior performance data. Thus, to be fair to a teacher, one must factor in such variables if his or her instructional effectiveness is to be judged on the basis of student achievement.

An ideal model of evaluation of mathematics instruction should include both teacher and student performance dimensions. Teachers should be expected to exhibit knowledge and understanding of mathematical content, relate mathematical topics to each other and to other fields, provide opportunities to learn for all students, and be able to utilize resources effectively (National Council of Teachers of Mathematics 1989). It is also true that if teachers provide good instruction, then students should be able to perform well on tests and to provide other evidence of mathematics learning, such as portfolios of their work and successfully completed projects.

If a model of evaluation of instruction is formulated, then questions arise as to who should carry out the evaluations. Existing models include peer evaluations, evaluations by administrators, evaluations by specialists who are assigned evaluation duties as their role, and evaluations by students. Peer evaluations can be quite valid, but they are sometimes tainted by factors ranging from differences in instructional philosophy to simple jealousy. Evaluations by administrators are a common practice, but administrators are often ignorant of the nature of a specific subject area, in particular, mathematics, and, if using some type of generic evaluation form, may make judgments that are invalid. Evaluations by specialists trained for that purpose often suffer from the same problems as those performed by administrators, unless the specialist is also a subject-matter specialist. Evaluations by students should be the most valid, for they see the teacher on a regular day-to-day basis, but students often judge a teacher not by the quality of the instruction, but by how much they "enjoy" the class. Quite often, that "enjoyment" is based upon factors other than how much students feel they have learned. It would appear, then, that using a combination of sources would be the most likely way to develop a valid method of evaluating instruction.

If a model has been developed, and the evaluators have been selected, what is to be done with the results of the evaluations? Certainly, an important, if not the most important, use of evaluation results should be the improvement of instructional practice. That is, any performance by human beings is likely to be subject to improvement, and the results of an evaluation can point out the areas for improvement. For example, if a classroom of students, after planned instruction, performs poorly on a test of the application of some principles of probability to real-world situations, the teacher should consider where the specific difficulties arise and, preferably in conjunction with the students, make changes in instruction that would seem to improve future student performance. If all parties concerned know that the purpose is not to penalize, but to make a teacher a better instructor, then honest opinions are more likely to result, and defensiveness and suspicion are less likely to occur. Also, teachers are more likely to be willing to listen to the criticisms, both positive and negative, that may be included in an evaluation and take steps to act upon those criticisms in order to maximize the effectiveness of their instruction.

It is also possible to use results to reward the "good" and punish the "bad." Although poor instruction certainly should be improved or weeded out as soon as possible, it is imperative that both merits and demerits be based on valid evaluations and reasonable criteria. A sad state of affairs exists when teachers are evaluated solely on the performance of their students on externally mandated standardized tests. This approach, depending only on student performance, overlooks some of the factors just discussed, many of which are beyond the control of the teacher. Certainly, good instruction should be rewarded and poor instruction should be eliminated, but one must be careful in selecting the criteria on which such decisions are made.

Evaluation of instruction is a very complex issue. There are different schools of thought as to the most important criteria, best methods, and best uses of results. However, it is a task that needs to be done. Validity of results and the improvement of instruction must be the primary concerns in this activity.

*See also* Evaluation of Instruction, Overview

## SELECTED REFERENCES

Borich, Gary D. *The Appraisal of Teaching: Concepts and Process.* Reading, MA: Addison-Wesley, 1977.

Good, Thomas L., and Jere E. Brophy. *Looking in Classrooms.* 4th ed. New York: Harper and Row, 1987.

Koehler, Mary, and Douglas A. Grouws. "Mathematics Teaching Practices and Their Effects." In *Handbook of Research on Mathematics Teaching and Learning.* Douglas A. Grouws, ed. New York: Macmillan, 1992.

National Council of Teachers of Mathematics. *Curriculum and Evaluation Standards for School Mathematics.* Reston, VA: The Council, 1989.

———. *Principles and Standards for School Mathematics.* Reston, VA: The Council, 2000.

RALPH W. CAIN

# EVALUATION OF INSTRUCTION, OVERVIEW

Traditionally, an administrative activity for the purpose of making decisions about teacher retention or tenure or about teachers' advancement on a career ladder. Ideally, the main purpose of evaluation should be to improve teaching, leading to greater student achievement. Three main procedures are used to obtain data for the evaluation of instruction: (1) observation of the teacher, sometimes supplemented with a review of lesson plans or tests; (2) information gathered from students; and (3) teacher self-reflection and analysis. The objects for evaluation can be the act of teaching itself or the materials and products used to support instruction, such as lesson plans, tests, student worksheets, or laboratory activities. At the college level, evaluation also includes a review of the instructor's scholarly products, such as reports or articles on research in mathematics teaching or learning, textbooks, or descriptions of specific teaching methods or strategies. In secondary schools and colleges, the most typical approach to evaluation is through observation of actual teaching. Most often, current practice is that the evaluation itself is done by an administrator once or twice a year, and usually more often for a teacher in the first year or two of teaching. No specific steps are taken to accommodate modes of teaching in different subject areas, and the criteria used in the observation are often not well defined, existing mainly in the mind of the person doing the evaluation.

During the 1980s, in response to the calls for improved accountability in education, many states and school districts implemented more specific and expanded teacher evaluations. A few states began testing teachers on content and pedagogical knowledge. This practice was unpopular and not very successful in identifying weak teachers who were already in the classroom. Many states, however, do require such tests of those who are applying for initial certification. Typically, secondary mathematics teachers are required to pass a content test up to and including calculus; elementary teachers generally do not take a specific test on mathematics content.

Another innovation was the development of more focused and structured teacher-observation instruments. Some research on teaching had indicated that certain teacher behaviors were associated with student achievement. Among these behaviors were (a) providing students with the lesson objectives; (b) using a motivational activity; (c) presenting clear directions; (d) making smooth transitions between activities; (e) asking and encouraging questions; (f) providing students with appropriate examples and practice, and (g) providing closure and summary of the main points of the lesson (Grouws, Cooney, and Jones 1988). In addition to these teacher behaviors, certain student interactions or responses were often monitored, including whether students are "on-task" (paying attention or working), the number of students called on or volunteering to answer questions, and student ability to answer questions. The advantage of these observation instruments is that they provide objective criteria and identify areas in which teachers can improve. The distinct disadvantage is that they are often very prescriptive, giving the impression that there is only one correct way to be an effective teacher. They presume a very traditional teacher-centered classroom and lesson in which clear, simple teaching activities are structured to achieve a specific objective. Lessons such as this are often best adapted to the learning and development of skills and procedures. Teachers who attempt to use small-group instruction, inquiry-based approaches, or long-term interdisciplinary projects may find themselves penalized for not fitting into the structure of the observation form.

Two clear deficiencies in traditional evaluation of instruction are that it focuses only on the teacher rather than on the overall program and teaching environment, and that it is usually not tied to staff development with appropriate follow-up. Teachers are usually constrained by the setting and resources available. An algebra teacher who has access to a graphics calculator for every student and whose students attend class regularly may have very different evaluations from those that would be obtained in less desirable circumstances. Most school districts provide staff development workshops and classes for teachers, but it is unusual for these to be tied to areas identified by teacher evaluations to be in need of development. Even more rarely would follow-up teacher evaluations be performed to assess the effectiveness of the staff workshops. These approaches are not common because the teacher's day is organized so that teachers are in almost constant contact with students. Little time is provided for instructional evaluation, professional development, and improved teaching in the structure of the school day or budget in most school districts.

*Professional Standards for Teaching Mathematics* (National Council of Teachers of Mathematics (NCTM) 1991) includes a section on standards for evaluation of the teaching of mathematics. It sets out the role of evaluation in achieving the goal of improving teaching and enhancing professional growth. The first standard emphasizes that the evaluation process should be a cyclical one in which data on a teacher is collected and analyzed, professional development is done based on the analysis, and teaching is improved as a result of the focused professional development.

The second and third standards recommend that teachers should be involved as participants in evaluation, have opportunities to analyze their own teaching, discuss their teaching with colleagues, and confer with supervisors about their teaching. Evaluation should be seen not only as an external requirement but also as an ongoing opportunity for teachers to engage in self-improvement. When performance is evaluated for administrative purposes such as retention, tenure, or promotion, teachers should be involved throughout the process and have the opportunity to provide their own information and interpretations.

The actual data collected for evaluation may be from multiple sources and of different types to reflect various strengths and areas for improvement. Examples of sources for data should include the teacher's goals and expectations for student learning, the teacher's plans for achieving these goals, the teacher's portfolio consisting of a sample of lesson plans, student activities and materials and means of assessing student understanding of mathematics, analyses of multiple episodes of classroom teaching, and evidence of student understanding of and disposition to do mathematics (NCTM 1991). Since teaching is a complex activity, it is only reasonable that evaluation should reflect the many facets of instruction, including planning, teaching, and student outcomes. The process of data collection and analysis should be continual and integrated with an overall program of professional development.

Further recommendations focus on mathematical content, student attitudes, assessment, and the learning environment. The standards provide a kind of detailed structure for evaluating mathematics instruction. Traditionally, evaluation has been performed by administrators who are often not trained in mathematics and who rely primarily on their own past classroom experiences as learners of mathematics to judge what might be good teaching. As mathematics content and methods change as a result of

reforms, teachers who implement these approaches should be recognized and rewarded through the evaluation process. This may also help to develop teachers' abilities to meet the challenges of new curricula. The following are examples of expected observable teacher behaviors: (a) engages students in mathematical discourse that extends their understanding of mathematical concepts, procedures, and understandings; (b) models and emphasizes mathematical communication using written, oral, and visual forms; (c) demonstrates the value of mathematics as a way of thinking and its applications in other disciplines and in society; (d) matches assessment methods with the developmental level, the mathematical maturity, and the cultural background of the student; and (e) respects students and their ideas and encourages curiosity and spontaneity.

These standards "emphasize the importance of significant mathematics when evaluating the teaching of mathematics" (NCTM 1991, 119). A central belief is that evaluation of mathematics instruction must be specific to mathematics rather than general teaching strategies or generic classroom environments. This type of evaluation requires an intimate knowledge of mathematics content, instruction, and goals. These guidelines also provide an outline for developing observation instruments that are specific enough to provide feedback that is useful for teacher improvement yet general enough to allow for individual styles of effective mathematics teaching.

In order to accomplish the goal of improving instruction, data for evaluation must come from multiple sources. As suggested in the *Professional Standards* (NCTM 1991), some districts and teachers are beginning to consider the use of teaching portfolios, which would contain selected information of many types. Much like portfolios of student work, which are beginning to be used as more authentic evidence of performance, these teaching portfolios are structured to present comprehensive evidence of teaching abilities. Among the characteristics of a portfolio would be (a) clear organization of contents, including materials that are agreed upon by the teacher and administrators; (b) statement of personal teaching goals and philosophy; (c) data from multiple classroom episodes, using a variety of observational and recording instruments; (d) data from students, including work samples and statements about the teacher or class; (e) evidence of professional activity, such as work on committees and staff development; and (f) reflective statement about accomplishments and future goals for development. The review of portfolios is an opportunity for broad involvement, adding to the credibility of the teaching profession. A committee of master teachers and administrators could have the responsibility for reviewing and providing useful feedback to teachers. This process could help put in place an evaluation cycle that would lead to recommended professional development activities.

A comprehensive teaching portfolio should include information from students. Although used extensively at the university level, student evaluations of teaching are seldom obtained in high school. The obvious concern is whether students have sufficient maturity to do an objective evaluation. Even at the university level, instructor personality, expected course grade, and other extraneous factors are sometimes associated with evaluation scores. However, appropriate safeguards can be used to protect against these problems. For example, teachers can select some of the questions to be placed on the student evaluation, so that strengths can be reflected. Teachers need not be required to include student evaluations from all classes every year. In the spirit of the portfolio, student evaluations can be selected by the teacher according to agreed criteria. Another approach to obtaining student input is through their reactions to specific course topics, types of learning activities, and reflection upon their own learning progress. This type of information can provide indirect but very important information about a teacher's performance.

Finally, as a component of data from students, indicators of mathematical performance are important. In recent years, many states have implemented statewide assessments of student achievement, often focusing on mathematics. In many communities, newspapers have published scores from these state assessments or other standardized tests and have compared individual schools and grade levels. Administrators have sometimes used these scores as data to evaluate individual teachers. As a part of instructional evaluation, this approach has been questioned, especially in terms of individual teachers. More appropriate student achievement data can include work samples and class data that indicate how well students perform on specific mathematical objectives such as the ones outlined in the *Professional Standards* (NCTM 1991). Standardized tests designed for large populations are not useful tools for assessing class and individual performances. As teachers become more skilled at using student performance assessments, they can apply these multiple approaches and obtain information about student achievement that is appropriate for their own teaching portfolios (Kulm 1994).

Teachers' own continuous self-reflection is perhaps the most important and effective component of instructional evaluation. Unless a teacher reflects on methods and strategies, it is unlikely that any real improvement in instruction will result. Teaching is a highly complex and individualistic activity. Evaluation of teaching should take these characteristics into account in order to support and encourage teachers who reflect on what they do and take the inevitable risks associated with making changes to enhance their teaching.

*See also* Evaluation of Instruction, Issues

## SELECTED REFERENCES

Grouws, Douglas A., Thomas J. Cooney, and Douglas Jones, eds. *Effective Mathematics Teaching.* Reston, VA: National Council of Teachers of Mathematics, 1988.

Kulm, Gerald O. *Mathematics Assessment: What Works in the Classroom.* San Francisco, CA: Jossey-Bass, 1994.

National Council of Teachers of Mathematics. *Curriculum and Evaluation Standards for School Mathematics.* Reston, VA: The Council, 1989.

———. *Principles and Standards for School Mathematics.* Reston, VA: The Council, 2000.

———. *Professional Standards for Teaching Mathematics.* Reston, VA: The Council, 1991.

GERALD KULM

# EXPONENTS, ALGEBRAIC

The repeated product $xxx \ldots x$ of $a$ factors, denoted by $x^a$ (also called the $a^{th}$ power of $x$), where $x$ is the *base* and $a$ is the *exponent*. This definition can be broadened to include zero, negative, and rational exponents. Through the use of logarithms and series, $x^a$ can be understood for any real or complex number $a$. In the most general case, the exponent is an algebraic expression with one or more variables.

## HISTORICAL OVERVIEW
### Evolution of the Exponential Notation $x^a$ ($a$ rational)

François Viète (1540–1603) is generally acknowledged to be the first mathematician to employ letters for numbers as a general procedure (Bell 1945, 120). Descartes, in the appendix to his *Discours de la Méthode* (1637), introduced the modern $x$, $xx$ (he preferred $xx$ to $x^2$), $x^3$, $x^4$, etc. (Bell 1945, 129). Newton (1676) employed negative exponents to mean the inverse of a power (as in $x^{-5} = 1/x^5$) and fractional exponents to mean roots (as in $x^{\frac{1}{3}} = \sqrt[3]{x}$) (Dessart 1969, 331).

## Logarithms, Real Exponents, and Complex Exponents

The origin of the meaning of $x^a$, where $a$ is any real number, arose from the invention of logarithms by John Napier (1550–1617). Seeking to facilitate the calculation of products of sines, Napier devised a system in which for each number between 1 and $10^7$, he assigned a second number, the logarithm (Eves 1990, 308). This logarithm of Napier is closely related to "ln," the natural logarithm, or logarithm to the base $e$. Napier wrote extensive tables of logarithms, and since $A = \ln B$ is equivalent to $B = e^A$, these could also be read as exponentiation tables. However, few mathematicians understood the link between the two concepts. It took the genius of Leonhard Euler to accomplish this feat. In his *Introductio in Analysin Infinitorum* (1748), he realized the importance of the base of the natural logarithm, gave the name $e$ to that number, and found a series converging to it:

$$e = 1 + \frac{1}{1} + \frac{1}{1 \cdot 2} + \frac{1}{1 \cdot 2 \cdot 3} + \cdots$$

More importantly, he expressed $x^a$ as the series

$$x^a = 1 + ka + \frac{k^2 a^2}{1 \cdot 2} + \frac{k^3 a^3}{1 \cdot 2 \cdot 3} + \cdots$$

where $k = \ln(x)$, giving a solid foundation to the meaning of an irrational exponent. Finally, using this series and De Moivre's Formula, $(\cos \theta + i \sin \theta)^n = \cos n\theta + i \sin n\theta$, he established the following interpretation of imaginary exponents: $e^{i\theta} = \cos \theta + i \sin \theta$ (Coolidge 1950).

## PROPERTIES AND DEFINITIONS

Exponential expressions satisfy the following properties: (1) $x^a x^b = x^{a+b}$ (Product Rule); (2) $x^a/x^b = x^{a-b}$ (Quotient Rule); and (3) $(x^a)^b = x^{ab}$ (Power Rule).

## Rational Exponents

| Exponent | Meaning of $x^a$ |
|---|---|
| Positive integer | $xxx \ldots x$ {$a$ factors} |
| Zero | $x^0 = 1, \quad x \neq 0$ |
| Negative integer | $x^{-a} = 1/x^a, \quad x \neq 0$ |

| Fraction | $x^{a/b} = \sqrt[b]{x^a} \geq 0$    If $a$ is odd and $b$ even, $x \geq 0$; if both exist, $\sqrt[b]{x^a} = (\sqrt[b]{x})^a$ |

## Real Exponents

Real exponents are first defined when the base is the number $e$, using the series

$$e^a = \sum_{n=1}^{\infty} \frac{a^n}{n!}$$

When $a$ is a rational number, it can be proven that this definition agrees with the one given earlier and that it satisfies the properties of exponents (Rudin 1976, 178). For a general positive base $x$, $x^a$ is defined as follows: Since $\ln(x)$ and $e^x$ are inverse functions, $x = e^{\ln(x)}$. Hence, $x^a = (e^{\ln(x)})^a = e^{a \ln(x)}$, so

$$x^a = \sum_{n=1}^{\infty} \frac{(a \ln(x))^n}{n!} = \sum_{n=1}^{\infty} \frac{a^n \ln^n x}{n!}$$

which is Euler's formula. The base $x$ must be positive since otherwise $\ln(x)$ will not be real.

## Complex Exponents

Given any complex number $z = a + ib$, Euler's formula, $e^{i\theta} = \cos \theta + i \sin \theta$, is used to write $e^z = x^{a+ib} = e^a e^{ib} = e^a(\cos b + i \sin b)$. And, as above, for positive $x$: $x^z = e^{z \ln(x)} = e^{(a+ib)\ln(x)} = e^{a \ln(x)} e^{ib \ln(x)} = x^a[\cos(b \ln(x)) + i \sin(b \ln(x)]$.

## Algebraic Exponents

If $P(a_1, a_2, \ldots, a_n)$ is an algebraic expression where $a_i$ is real or complex, then Euler's formula is used to write

$$x^{P(a_1, a_2, \ldots, a_n)} = \sum_{n=1}^{\infty} \frac{[P(a_1, a_2, \ldots, a_n)]^n \ln^n(x)}{n!}$$

An equation with algebraic exponents can be solved for one or more of the variables through the use of logarithms of the appropriate base. For example, in the equation $3^{2x-1} = 81$, taking the $\log_3$ (logarithm base 3) of both sides gives $2x - 1 = \log_3 81 = 4$, or $x = \frac{5}{2}$.

## PEDAGOGICAL APPROACHES

An educator could introduce the concept of algebraic exponents by using continuity arguments, models of exponential growth and decay, or an inverse function approach through the logarithm, among others.

## Continuity Arguments

In this setting one introduces the possibility of irrational exponents by an interpolation process. This can be done both graphically and numerically. As an example of the graphic approach, one could start by plotting the pairs $(n, 2^n)$, with $n$ being an integer between $-4$ and $4$. Then more points could be added to this graph (by including square roots, fourth roots, etc.). By continuity, one obtains the graph of $2^x$ for a general $x$. In the numerical approach, one tries to answer the question, What could $2^\pi$ possibly mean? by considering the sequence of numbers $2^3$, $2^{31/10}$, $2^{314/100}$, $2^{3141/1000}$, etc.

## Models of Exponential Growth and Decay

Using these models, one first establishes the meaning of the exponential function when the independent variable (the exponent) is an integer. Then one lets the exponent become a continuous variable. The meaning of the resulting expression is given by the physical or real-life context.

As an example of exponential growth, if a colony of bacteria originally numbered at 100,000 grows by 8% every 20 seconds, the number of bacteria as a function of time $t$ (in seconds) is given by $B(t) = 100,000(1.08)^{t/20}$. We can start by asking for the number of bacteria after 20 seconds, 40 seconds, or other multiples of 20 seconds. This will involve calculating expressions with integer exponents. One can then ask for the number of bacteria for units of time that are not multiples of 20, thus introducing rational exponents. Moreover, it is natural in this context to calculate $B(t)$ for irrational $t$ and to consider algebraic exponents. For example, the question, What is the number of bacteria at time $t + 45$ seconds? leads to $B(t + 45) = 100,000(1.08)^{(t+45)/20}$. This is also an ideal setting for introducing equations with algebraic exponents. A traditional way of doing this is to ask, When will the number of bacteria double? which leads to the equation $1.08^t = 2$ (Project for Computer-Intensive Curricula in Elementary Algebra 1991, 5–2).

## Inverse of ln

This approach follows the historical development more closely. One starts by introducing the $\ln(x)$ function as

$$\ln x = \int_1^x \frac{1}{t}\, dt$$

Then $e^x$ is introduced as the inverse of $\ln(x)$, and the exponent properties are proven. This method is adequate for an advanced audience at the calculus level (Larson, Hostetler, and Edwards 1994, 344).

*See also* Equations, Algebraic; Equations, Exponential; Exponents, Arithmetic; Logarithms

### SELECTED REFERENCES

Bell, Eric T. *The Development of Mathematics.* 2nd ed. New York: McGraw-Hill, 1945.

Coolidge, Julian L. "The number *e.*" *American Mathematical Monthly* 57(1950):591–602.

Dessart, Donald J. "Exponential Notation." In *Historical Topics for the Mathematics Classroom* (pp. 327–331). 31st Yearbook. John K. Baumgart et al., eds. Washington, DC: National Council of Teachers of Mathematics, 1969.

Eves, Howard W. *An Introduction to the History of Mathematics.* 6th ed. Philadelphia, PA: Saunders College Publishing, 1990.

Larson, Roland E., Robert P. Hostetler, and Bruce H. Edwards. *Calculus.* Lexington, MA: Heath, 1994.

Phillips, Elizabeth, et al. *Patterns and Functions. Agenda Series, Grades 5–8.* Reston, VA: National Council of Teachers of Mathematics, 1991.

Project for Computer-Intensive Curricula in Elementary Algebra. *Computer-Intensive Algebra.* College Park, MD: University of Maryland, 1991.

Rudin, Walter. *Principles of Mathematical Analysis.* 3rd ed. New York: McGraw-Hill, 1976.

ALEJANDRO ANDREOTTI

ANN E. MOSKOL

# EXPONENTS, ARITHMETIC

Rational numbers, written as superscripts, to indicate multiplication of identical quantities or extraction of roots. A positive integer exponent (or power) indicates multiplication of identical quantities. For example, in $3^4$ the exponent 4 indicates that four threes are multiplied together, namely, $3^4 = 3 \times 3 \times 3 \times 3 = 81$. The numeral 3 is called the base. If the exponent is a rational number, but not equivalent to a positive integer, then the meaning of the exponent is as follows: $x^0 = 1$, $x^{-p} = 1/x^p$, $x^{1/p} = \sqrt[p]{x}$, for $p$ a positive integer and $x \neq 0$ when it appears in the denominator. For example, $5^0 = 1$, $3^{-2} = 1/(3^2) = 1/9$, and $8^{1/3} = \sqrt[3]{8} = 2$. If $n = p/q$, where $p$ and $q$ are both integers, and $q \neq 0$, then $x^n = x^{p/q} = (x^p)^{1/q} = (x^{1/q})^p$. For example, $8^{2/3} = 4$.

The idea of powers of numbers is an old idea. There is even a hint of it in the Egyptian Rhind papyrus which dates from 1650 B.C. However, our modern superscript notation first appeared in René Descartes' *Discours de la Méthode* (1637). For some unexplained reason, even though Descartes would use $x^3$ and $x^4$, he never used $x^2$, instead he used $xx$. The Pythagoreans (ca. 450 B.C.) classified numbers by shape. The square numbers were the ones that could be represented by arranging dots or squares in a square. Cubic numbers could be represented by building cubes. The Greeks had thought of $x^2$ and $x^3$ as areas and volumes of squares and cubes, whereas Descartes considered them as lengths of lines. Our reading $x^2$ as "x-squared" and $x^3$ as "x-cubed" is a legacy from the Pythagoreans. In 1676, Isaac Newton introduced the idea of fractional and negative exponents. The rules for multiplying and dividing quantities involving exponents can be found in Diophantus' *Arithmetica* (mid-third century). Nicole Oresme (fourteenth century) gave meaning to powers and roots of ratios.

Before formally teaching elementary school children about exponents, the groundwork for the concepts of square numbers and exponential growth can be laid. For example, books such as *Bunches and Bunches of Bunnies* (Mathews 1978) and *The King's Chessboard* (Birch 1988) can be read, modeled, and discussed. Using manipulatives such as Base Ten Blocks or Cuisenaire's Metric Blocks to develop place value concepts or building squares of different sizes with square tiles are effective activities. Another activity that could precede the formal introduction of exponents is to have students continue sequences and explain how to find the next term. For example, students are frequently asked to continue sequences based on repeated addition of a quantity such as 0, 2, 4, 6, . . . or 0, 5, 10, 15, . . . . Students could also be asked to continue sequences that are based on repeated multiplication, such as 1, 2, 4, 8, ——, ——, . . . or 1, 5, 25, ——, ——, . . . .

To introduce exponents, explain that mathematicians invented the shortcut for writing a product of identical quantities. For example, $2 \times 2 \times 2 \times 2 = 2^4$. You can ask what they think $3^5$ or $2^{50}$ or $(\frac{3}{4})^2$ means. The students can compute values of various exponential expressions. With materials such as Cuisenaire's Metric Blocks, you can visually model the growth of each power, or you can use the Base Ten Blocks to model the powers of 10. Exploring the patterns with the powers of 10, showing students the names for the various place values, and telling them about "googol" can be interesting to them. Also, finding ways to model 1 million and showing students the book *One Million* (Hertzberg

1970) as well as looking at the video (and/or book) *Powers of Ten* (Morrison and Morrison 1982) can be helpful in developing a concept of relative size. Letting students use a calculator for large numbers is fine, but initially they should not be told what the "$y^x$" key means, so that they get the feel of how many times one actually multiplies for each power.

The easiest way to introduce the meaning of the zero exponent and negative exponents is by means of patterns. Have the students look at the following sequence of equations, ask what has to be the exponent in the last equation in order to maintain the pattern, and then point out that to go from one row to the row below, you divide by 2; ask what $2 \div 2$ is.

$$2^3 = 8$$
$$2^2 = 4$$
$$2^1 = 2$$
$$2^? = \text{———}$$

Clearly, the exponent must be 0 and the result 1. You can do the same with powers of another number. After a sufficient number of examples, the students will accept the definition that the zero power of a number must be 1. Then the same sequence of equations can be put on the board and continued to develop the idea of negative exponents. Again, you get each answer by dividing the previous answer by 2, and if the answer is written in fractional form, the students will quickly see the pattern.

$$2^3 = 8$$
$$2^2 = 4$$
$$2^1 = 2$$
$$2^0 = 1$$
$$2^? = \text{———}$$
$$2^? = \text{———}$$

Later, when the students have learned the rules for dividing powers with the same base, you can use this to demonstrate the zero power:

$$\frac{2^3}{2^3} = \frac{2 \times 2 \times 2}{2 \times 2 \times 2} = \frac{8}{8} = 1$$
$$\frac{2^3}{2^3} = 2^{3-3} = 2^0$$

So $2^0 = 1$.

Middle school and high school students learn how to use the "$y^x$" key on the calculator. The following question can be discussed: if $2^3 = 8$ and $2^4 = 16$, is there a power of 2 between 8 and 16? A little bit of exploration on the calculator will lead to answers such as $2^{3.5}$, and these will then justify connecting the points on a graph of the exponent vs. the value of the power ($y = 2^x$). A similar discussion of powers of 10 can lead to an explanation of the Richter scale (Voolich 1995) and why an earthquake with a Richter reading of 6.8 is 100 times more intense than one with a reading of 4.8. Various theorems can be proved that lead to rules for operations with exponential quantities. These include $(x^y)(x^z) = x^{y+z}$, $(x^y)^z = x^{yz}$, $(xy)^z = x^z y^z$, and $(x/y)^z = x^z/y^z$.

*See also* Patterning

SELECTED REFERENCES

Birch, David. *The King's Chessboard.* New York: Dial Books for Young Readers, 1988.

Burton, David. *The History of Mathematics: An Introduction.* 2nd ed. Dubuque, IA: Brown, 1991.

Eames, Charles, and Ray Eames. *Powers of Ten.* Films, Vol. 1. Santa Monica, CA: Pyramid Film and Video, 1989.

Hertzberg, Hendrik. *One Million.* New York: Simon and Schuster, 1970.

Katz, Victor J. *A History of Mathematics: An Introduction.* New York: HarperCollins, 1993.

Mathews, Louise. *Bunches and Bunches of Bunnies.* New York: Scholastic, 1978.

Morrison, Philip, and Phyllis Morrison. *Powers of Ten.* New York: Scientific American, 1982.

Voolich, Erica Dakin. "Seize that Teachable Moment." *Mathematics Teaching in the Middle School* 1 (Jan.–Mar. 1995):268, 270.

ERICA DAKIN VOOLICH

# EXTRAPOLATION

The making of estimates or predictions beyond the range of data or information already available. Consider the following data (graphed in Figure 1), which show the percentage of a population using filtered water and the death rate due to typhoid recorded for various years.

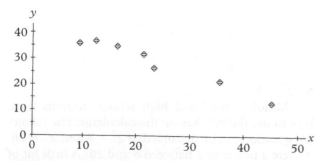

**Figure 1** *Percentage using filtered water vs typhoid death rate.*

| Percentage using filtered water ($x$) | Typhoid death rate per 100,000 ($y$) |
|---|---|
| 9 | 36 |
| 12 | 37 |
| 16 | 35 |
| 21 | 32 |
| 23 | 27 |
| 35 | 22 |
| 45 | 14 |

There is a very strong relationship between the two variables, as measured by the high correlation coefficient ($r = -0.98$) and the regression line (line of best fit) is given as

$$y = 43.93 - 0.649x$$

Although this equation can sensibly be used to find estimates for values within the indicated range of data, caution should be exercised in applying it to values outside the known range. A prediction for the death rate when $x = 90$ yields an impossible result using that equation (when $x = 90$, $y = -14.48$).

In this instance, it is likely that any decrease in the death rate will eventually level off, and that if more data were available, the resulting graph might actually form a nonlinear curve. Just as it is unwise to extrapolate too far beyond the range of data given in the sample, it is also inadvisable to extrapolate to different countries or to different time periods, where the environment and prevailing background conditions may be different.

Similarly, if predictions are to be made for time series data, any extrapolation beyond the time interval used for estimating the equation of the regression line should be interpreted cautiously. Even if it can be assumed that the relationship does remain linear, and that background conditions have not changed, the addition of a few extra points at either end may well change the estimate of the regression equation.

*See also* Correlation; Least Squares

SELECTED REFERENCES

Hudson, Brian, and Mary Rouncefield. *Handling Data Core Y9. Century Maths Series.* Cheltenham, England: Thornes, 1992.

Moore, David S., and George P. McCabe. *Introduction to the Practice of Statistics.* New York: Freeman, 1993.

MARY ROUNCEFIELD

# F

## FACTORS

Divisors or submultiples of given numbers. The fundamental theorem of arithmetic, attributed to Euclid, states that every number can be expressed uniquely as a product of prime factors. Another aspect of number theory, which has attracted the attention of mathematicians through the ages, is finding out whether a number is a factor of a given number. The idea of factor precedes the study of number theory. In elementary school, learners studying intuitive concepts of division are able to determine that 12 can be divided into 3 lots of 4. Thus, 4 divides 12 three times. It follows that 4 is a factor of 12. In the middle grades, learners study whether a number is divisible by given numbers and investigate the factor combinations of numbers. Being able to express a number as a product of prime factors is also important as a pre-algebra skill.

To give a formal definition, $a$ is a factor of $b$, where $a$, $b$ are integers, if there exists some integer $k$ such that $b = ka$. The study of factors involves finding values of $a$ and $k$ for given values of $b$. A variety of pedagogical approaches may be used. Individual unit cubes may be counted out to a given number—say, 24. To test if 4 is a factor, the cubes can be arranged in lines of 4. Since no cubes are left over, we can say that 4 divides 24, or that 4 is a factor of 24. Likewise, if the cubes are arranged in lines of 5, then there are 4 cubes left over. Therefore, 5 is not a factor of 24. This method can be used to show that the factors of 24 are 1, 2, 3, 4, 6, 8, 12, and 24. A link to the process of division is illustrated by having learners divide a given number of objects into groups by the sharing method. If no objects are left over, then the group size is a factor of the given number (Burns 1991).

An alternative approach would be to use Cuisenaire rods or similar materials (Chambers 1964). Thus, 24 would be represented by two orange (10) rods and a purple (4) rod. To test if 4 is a factor, children make a train of purple rods to check if it will match orange + orange + purple. Six purple rods do match; therefore, 4 is a factor of 24. Likewise, to test if 5 is a factor, a train of yellow rods is made. This train does not match orange + orange + purple; therefore, 5 is not a factor of 24.

Learners should be led to observe that to express a number as a product of factors, it is necessary to test no further than the square root of the given number. Note that $24 = 1 \times 24 = 2 \times 12 = 3 \times 8 = 4 \times 6$. Since the square root of $24 = 4.9$, it is evident that factors greater than the square root of 24, such as 6, have already been included in the factor pairs.

The most efficient method of expressing a number as a product of prime factors is to use a factor tree. For example, from $24 = 4 \times 6$ the factors 4 and 6 are obtained. Factoring further gives the factor tree:

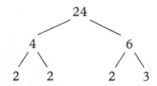

and $24 = 2 \times 2 \times 2 \times 3$. Note that if we begin by dividing by the smallest prime factor, we have $24 = 2 \times 12$, $12 = 2 \times 6$, $6 = 2 \times 3$, and $24 = 2 \times 2 \times 2 \times 3$, as before. This method is often used to find the greatest common factor of two or more numbers.

Interesting problem-solving applications follow from a search for a relationship between a number and the number of its divisors. For example, "What kinds of numbers have three and only three divisors?" and "Place all the numbers up to 100 in categories based on the number of divisors" (Brown 1984).

The concept of factoring extends to algebraic expressions. Applying the distributive property, $ab + ac = a(b + c)$, gives the following factor types, commonly taught in elementary algebra courses: $6x^2 + 8x = 2x(3x + 4)$; $x^2 - 6x + 8 = (x - 4)(x - 2)$; $x^2 - 16 = (x + 4)(x - 4)$.

Materials for instruction include cubes. These can be formed into rectangular arrays and are highly recommended as an introductory means of having learners express numbers as a product of two factors. Conceptually, this focuses the learner's attention on the inverse of the array model of multiplication. The use of Cuisenaire rods, as described earlier in this entry, provides more of a linear model of number. This method is a useful way to check if one number divides another and has a direct application to the set-of-divisors method of finding the greatest common divisor. The symbolic approach to factoring, expressing a number both as the product of a pair of factors and as a product of its prime factors, follows directly from the use of manipulatives.

*See also* Divisibility Tests; Number Theory, Overview; Prime Numbers

SELECTED REFERENCES
Brown, G. W. "Searching for Patterns of Divisors." *Arithmetic Teacher* 32(Dec. 1984):32–34.
Burns, Marilyn. "Introducing Division through Problem-solving Experiences." *Arithmetic Teacher* 38(Apr. 1991):14–18.
Chambers, C. E. *The Cuisenaire-Gattegno Method of Teaching Mathematics*. Vol. 1. Reading, England: Educational Explorers, 1964.
Dearing, Shirley A., and Boyd Holtan. "Factors and Primes with a T Square." *Arithmetic Teacher* 34(Apr. 1987):34–35.
Hawkins, Wayne. "Rectangles." *Australian Mathematics Teacher* 47(Dec. 1991):10–13.
Heddens, James W., and William R. Speer. *Today's Mathematics*. Part 1. *Concepts and Classroom Methods*. Englewood Cliffs, NJ: Prentice-Hall, 1995.

DAVID M. DAVISON

# FAMILY MATH AND *MATEMÁTICA PARA LA FAMILIA*

Programs whose primary focus is on providing ways for families of children in grades K through 8 to enjoy learning mathematics together. *MATEMÁTICA PARA LA FAMILIA* is the Spanish-language version of the program. Adaptations that reflect the diversity of the various Spanish-speaking cultures have enriched the FAMILY MATH program. FAMILY MATH and *MATEMÁTICA PARA LA FAMILIA* offer parents and other caregivers ideas and materials for doing mathematics with their children other than checking homework, using flashcards, or telling their children how to do a problem.

The programs exist as books (the original English version has been translated into Spanish, Swedish, and Chinese), as a series of classes for families (adults and children) in their communities, as workshops to prepare class leaders and staff developers, and as a network. Both programs are an outgrowth of the philosophy, goals, and methods of the EQUALS teacher education program at the Lawrence Hall of Science at the University of California at Berkeley.

Since 1977, EQUALS has worked with teachers who wanted to improve mathematics education to provide activities and methods to help all students succeed in mathematics, especially those who have been underrepresented in mathematics and science. It was these teachers who asked for help in responding to parents' requests for ideas and materials to use at home to help their children in mathematics. Many teachers reported that parents expressed frustration at not knowing enough about their children's mathematics program to help them or at not understanding the mathematics their children were studying. As a result, the FAMILY MATH program was created, and a grant from the Fund for the Improvement of Postsecondary Education (U.S. Department of Education) in 1981 allowed the dissemination of the program. Additional funding has been provided by the National Science Foundation, the Carnegie Corporation of New York, and the Women's Education Equity Act Program. Interest and enthusiasm for the programs have spread rapidly, and the programs are located in other areas of California, twenty-nine other states, Washington, DC, and in at least eight other countries. These sites offer locally the services available at the Lawrence Hall of Science.

As a part of EQUALS, FAMILY MATH and *MATEMÁTICA PARA LA FAMILIA* believe that all children can learn and enjoy mathematics. In addition, parents as well as other family members are their children's first and most influential teachers. Just as we do not expect children to develop all of their reading skills in school, we cannot expect them to develop a full understanding of mathematics without interactions outside class and formal homework settings. Family members are key to providing the time for these interactions outside school.

Awareness, confidence, and encouragement are three strands woven throughout the EQUALS program, and they underlie the goals of the FAMILY MATH and *MATEMÁTICA PARA LA FAMILIA* programs. In addition to providing families with activities to help their children with mathematics at home, the programs seek to inform families about the importance of mathematics in future schooling and work; build awareness among parents that mathematics consists of more than arithmetic and rote computations; build confidence by developing family problem-solving skills and ability to communicate about mathematics; encourage positive attitudes toward mathematics among parents and children; validate parents' roles in their children's mathematics education; and create an opportunity for all family members to enjoy doing mathematics together.

Both programs advocate the *doing* of mathematics. Instead of memorizing facts or listening to lectures, families participate in doing mathematics, either by using the book, *FAMILY MATH* or *MATEMÁTICA PARA LA FAMILIA* (Stenmark et al. 1986, 1987), at home and/or by attending a series of classes with other families in their respective communities. Asking questions rather than giving answers is presented as a very important way to help individuals find their own methods of working through any problem. Adults are urged to support their children in being more concerned with the

*processes* of doing mathematics than with only getting a correct answer. For example, parents are encouraged to ask questions that guide their children without necessarily telling them what to do—questions which cannot be answered with "yes" or "no": "What do you know now?" "How might you begin?" "Can you make a drawing to explain what you're thinking?" "Do you see any patterns?" "What did you try that didn't work?" "Can you explain that in a different way?" "Can you convince me your solution makes sense?" "Can you make a prediction?" These are just a few of the types of "good" questions that open up a problem and support different ways of thinking about it.

Opportunities to develop *problem-solving skills* and to build a *conceptual* understanding of mathematics with *hands-on* materials are the program's main characteristics. Problem-solving skills include building a repertoire of strategies, such as looking for patterns, drawing a picture, working backward, making the problem less complex (e.g., substituting smaller numbers), working with a partner, or eliminating possibilities. The rationale is this: having a supply of strategies allows you a choice of ways to tackle a problem, helping relieve the frustration of not knowing how or where to begin. Having more strategies at your command develops greater confidence about being able to solve problems, so there is a greater willingness to tackle new problems, leading you to becoming a better problem solver. Many research and applied mathematicians use manipulative materials and models as tools since much of mathematics can best be explained and understood that way. This is emphasized to families so that they will not perceive such aids to making abstract concepts more concrete as the domain only of the very young or intellectually immature. Families use inexpensive materials of all kinds—beans, blocks, bottle caps, toothpicks, coins—as aids to figuring out problems.

*Mathematics is more than arithmetic.* Today's elementary curriculum reflects this, and FAMILY MATH and *MATEMÁTICA PARA LA FAMILIA* give families overviews of the mathematics topics at their children's grade levels and explanations of how these topics relate to each other. The programs involve adults in exploring and learning about mathematics topics that may be new to them, while helping the whole family develop a broader picture of what mathematics is. The curriculum starts with traditional number-based activities familiar to adults and gradually introduces measurement, exploring geometry, studying the relationships between numbers and shapes, looking for patterns, organizing information, learning about probability and statistics, and the role of calculators and technology in the curriculum and workplace. Most of the activities are of interest across age groups, and often adults find themselves engaged, challenged, and learning with their children.

Whether they are done at home or in a class, FAMILY MATH and *MATEMÁTICA PARA LA FAMILIA* activities are meant to be fun. Suggestions for creating a nonthreatening and supportive environment are included in the book and are modeled by class leaders. Feeling comfortable while doing mathematics encourages the experimentation and risk taking necessary to develop problem-solving skills and the willingness to persist when encountering challenging problems. The goal of such an environment is to build self-confidence and enjoyment for both adults and children and to help families develop mathematical understanding. The programs have been professionally evaluated, in addition to being the subject of both masters' theses and doctoral dissertations. The book was written with families in mind, with over 100 activities, several lesson plans, and a rationale for a problem-solving approach to the mathematics curriculum and for family involvement in mathematics education. A special effort has been made to present the mathematical activities and directions clearly so families can use the book easily at home without attending a class.

Classes occur with many variations depending on the needs of the class leader(s) and the community. In general, they meet once a week for one and one-half to two hours in grade-level groupings (K–1, 2–3, 3–4, 4–6, and 7–8) and over a period of four to six weeks. They have been conducted in private homes, schools, libraries, churches, and community centers. Class leader workshops prepare educators, parents, and others to lead classes for families. Anyone who is enthusiastic, kind, and committed to helping others enjoy mathematics can attend a workshop of at least two days that includes materials and methods for leading FAMILY MATH or *MATEMÁTICA PARA LA FAMILIA* classes. A good mathematics background is helpful but not essential for a class leader. People with less mathematics experience usually team with someone with more experience. However, nonprofessional educators are able to offer excellent classes in any setting so that people of different ages, levels of experience, and knowledge are engaged in meaningful mathematics challenges. The programs have proven a positive ve-

hicle for promoting collegiality among parents, classroom teachers, and administrators in the interest of meeting the needs of students. Staff development workshops prepare experienced class leaders to organize and offer class leader workshops. Experienced class leaders can participate in a staff development workshop of at least two days that includes materials, methods, and strategies for preparing others to lead classes for families. Workshop presenters need a good background in mathematics. The *Network* is maintained to provide support to class leaders, staff developers, parents, teachers, and others interested in FAMILY MATH and *MATEMÁTICA PARA LA FAMILIA*. The network functions as a two-way conduit between the two programs at the Lawrence Hall of Science and class leaders and staff developers, as well as the general community.

*See also* Enrichment, Overview

## SELECTED REFERENCES

David, Jane L., and Patrick M. Shields. FAMILY MATH in Community Agencies. A report for the FAMILY MATH Project of the EQUALS Program. Berkeley, CA: Lawrence Hall of Science, University of California, October 1988.

Devaney, Kathleen. Interviews with Nine Teachers. Report for the FAMILY MATH Project of the EQUALS Program. Berkeley, CA: Lawrence Hall of Science, University of California, October 1988.

Kreinberg, Nancy, and Virginia Thompson. FAMILY MATH: Report of Activities, September 1983–September 1986. Report to The Fund for the Improvement of Postsecondary Education, U.S. Department of Education. Berkeley, CA: Lawrence Hall of Science, University of California, October 1986.

Ramage, Katherine, and Patrick M. Shields. *MATEMÁTICA PARA LA FAMILIA,* San Diego County. Evaluation report prepared for EQUALS. Berkeley, CA: Lawrence Hall of Science, University of California, 1994.

Shields, Patrick M., and Jane L. David. The Implementation of FAMILY MATH in Five Community Agencies. Report for the EQUALS Program. Berkeley, CA: Lawrence Hall of Science, University of California, 1988.

Stenmark, Jean Kerr, Virginia Thompson, and Ruth Cossey. *FAMILY MATH.* Berkeley, CA: Lawrence Hall of Science, University of California, 1986.

———. *MATEMÁTICA PARA LA FAMILIA.* Berkeley, CA: Lawrence Hall of Science, University of California, 1987.

———. *Matte hemma Matte i skolan.* Berkeley, CA: Lawrence Hall of Science, University of California, 1988.

Thompson, Virginia, and Mary Jo Cittadino. "Joining Home and School Through FAMILY MATH." *Educational Horizons* 69(4) (Summer 1991): 195–199.

———, and Nancy Kreinberg. "FAMILY MATH: A Report of an Intervention Program That Involves Parents in Their Children's Mathematics Education." Paper presented at American Association for the Advancement of Science annual meeting, Philadelphia, 1986.

"We All Count in FAMILY MATH." Film and video. Berkeley, CA: EQUALS, Lawrence Hall of Science, University of California.

Weisbaum, Kathryn Sloane. Families in FAMILY MATH. Research Project. NSF Proposal No. MDR-8751375. Final Report, December 1990.

MARY JO CITTADINO

SELECTED REFERENCES
Alexander, Daniel. "Civilized Mathematics." *Focus* 14(3)(Jun. 1994):10–11.
———. *A History of Complex Analysis from Schröder to Fatou and Julia.* Wiesbaden, Germany: Vieweg, 1994.
Devaney, Robert. *Chaos, Fractals, and Dynamics: Computer Experiments in Mathematics.* Menlo Park, CA: Addison-Wesley, 1990.
Nathan, Henry. "Pierre Fatou." In *Dictionary of Scientific Biography* (pp. 547–548). Vol. 4. Charles C. Gillispie, ed. New York: Scribner's, 1980.

DANIEL S. ALEXANDER

# FATOU, PIERRE (1878–1929)

Prominent French astronomer and mathematician who made significant contributions in several areas, including real and complex analysis. Fatou is best known for providing the foundation for the contemporary theory of the iteration of complex functions, often referred to as *complex dynamics.*

The iterates of a point $z$ in the complex plane under a function $f$ are the points $z$, $f(z)$, $f(f(z))$, $f(f(f(z)))$, etc. A simple example of iteration occurs with Newton's method for solving an equation of the form $f(z) = 0$. To use this method, let $N$ be the Newton's method function for $f$. Then pick a point $z$ near a suspected solution of $f(z) = 0$, evaluate $N(z)$, plug the result back into $N$, and continue repeating the process in hopes that the iterates of $z$ under $N$ converge to the solution.

Fatou apparently intended to submit his work in complex dynamics to the French Academy of Sciences, hoping to win the academy's 1918 Grand Prize of Mathematics, which the academy had previously announced would be given to the best paper it received concerning the iteration of complex functions. However, when Fatou published his initial findings in late 1917, another French mathematician, Gaston Julia (1893–1978), announced that he also had submitted many of the same results to the academy and demanded that it immediately judge who deserved priority.

Although both men achieved their results independently, the academy ruled in Julia's favor. Fatou never entered the contest, and the prize was awarded to Julia. However, Fatou eventually published his results, and today the work of each is held in equal regard.

*See also* Chaos; Julia

# FEDERAL GOVERNMENT, ROLE

U.S. government policy with respect to mathematics education over the last two decades of the twentieth century could be divided into three phases: wind-down, build-up, and systemic. These phases are very distinct and were significantly affected by publications of the National Council of Teachers of Mathematics (NCTM) in 1980 and 1989.

## THE WIND-DOWN PHASE

The U.S. government had provided funds for the development of curriculum and training through the National Science Foundation (NSF) sporadically through most of the 1960s and 1970s. By the end of the 1970s and the early 1980s, the government had decided to wind down many of its elementary and secondary education programs. This was done for several reasons. First, there was growing concern over the budget, and second, the NSF's leadership was less interested in developing programs in education and more interested in supporting "working scientists." But in the early 1980s, the need for more-qualified mathematics teachers and for people with a background in mathematics began to grow; accordingly, the government began shifting its policy.

## THE BUILD-UP PHASE

In the 1980s, the federal government allotted money to support teacher development, curriculum development, and teacher recruitment. These specifically targeted programs, supported by both parties, were passed and signed by President Reagan in a political environment of reducing categorical programs and cutting spending. The reason the programs were adopted under these circumstances is that the federal government believed that it was in the nation's economic interest to produce more engineers, scientists, and technicians. During the 1980s, government spending rose from $20 million to $500 million for mathematics education. This money was divided between the highly specific NSF grants and the formula-funded Dwight David Eisenhower Professional Development Act, which was enacted in 1988 for the purpose of providing high-quality professional development to classroom teachers. The formula for the Eisenhower act was based on the total student population and the number of disadvantaged students in a state.

## THE SYSTEMIC PHASE

In the early 1990s, changes started to take place. By that time, the NCTM had issued its standards for curriculum and evaluation, the NSF was working to coordinate the efforts of many education-related providers, formal as well as informal, and the nation was debating national education goals and the processes for supporting those goals. The impact on mathematics education was significant. Funding was no longer an annual fight, and the sense that progress was being made was present at all levels of government and the profession. By the mid-1990s, however, systemic reform, the national goals, and the standards were undergoing significant review. People questioned the control that the national goals gave to the federal government, and many educators in other disciplines who were working on standards raised controversial issues. One of the results of this controversy has been that spending on mathematics education has come under increasing scrutiny, and the future of funding is in doubt.

*See also* Eisenhower Program; Goals 2000; Mathematics Education, Statistical Indicators; National Assessment of Educational Progress (NAEP); National Science Foundation (NSF); National Science Foundation (NSF), Historical Perspective; State Government, Role; United States Department of Education; United States National Commission on Mathematics Instruction (USNCMI)

## SELECTED REFERENCES

National Council of Teachers of Mathematics. *Agenda for Action: Recommendations for School Mathematics of the 1980s.* Washington, DC: The Council, 1980.

——. *Curriculum and Evaluation Standards for School Mathematics.* Reston, VA: The Council, 1989.

——. *Principles and Standards for School Mathematics.* Reston, VA: The Council, 2000.

RICHARD LONG

# FEHR, HOWARD FRANKLIN (1901–1982)

Mathematics educator born in Pennsylvania, who received B.A. and M.A. degrees from Lehigh University and a Ph.D. degree from Columbia University. He taught at the Montclair State Normal School in New Jersey (1934–1948) and subsequently at Teachers College, Columbia University, retiring in 1967. Throughout his career, he advocated unifying courses in secondary mathematics about central concepts, and that the mathematics taught in elementary and secondary schools should be relevant to societal needs. He advocated a restructuring of the mathematics curriculum so that students would be prepared more adequately for the increase of knowledge in all branches of mathematics. Algebra, geometry, trigonometry, and analysis should, he believed, be taught not as separate subjects, but as parts of a more general mathematics program. The concepts of functions, variables, sets, and symbolic logic should be introduced in the primary grades to provide a basis for such topics as Diophantine equations and linear programming to be presented in the secondary school.

Fehr published numerous articles and textbooks expressing his ideas about content and curriculum as well as quality teacher preparation. Active in many national and international education societies, he was president (1956–1958) of the National Council of Teachers of Mathematics (NCTM). In the 1960s, he served as a mathematics education consultant for the Organization for Economic Development and Cooperation and the United Nations Educational Scientific and Cultural Organization (UNESCO). In 1965, Fehr established and subsequently directed the Secondary School Mathematics Curriculum Improvement Study (SSMCIS) at Teachers College, producing texts designed to unify the teaching of mathematics (Fehr 1971–1974).

## SELECTED REFERENCES

Anzovin, Steven. "Howard Franklin Fehr." In *The Annual Obituary 1982* (pp. 213–215). Janet Podell, ed. New York: St. Martin's, 1983.

Fehr, Howard F. "Mathematics Education: Some New Thinking." *Teachers College Record* 62(1961):456–464.

———. *Secondary Mathematics, a Functional Approach for Teachers.* Boston, MA: Heath, 1951.

———. *Unified Mathematics, Course I–IV.* Reading, MA: Addison-Wesley, 1971–1974.

———, and Lucas N. H. Bunt. *New Thinking in School Mathematics.* Paris, France: Organization for Economic Cooperation and Development, 1961.

Leef, Audrey J. V. "An Historical Study of the Influence of the Mathematics Department of Montclair State College on the Teaching of Mathematics, 1927–1972, in the Context of the Changes in Mathematics Education During This Period." Unpublished Ph.D. dissertation. Rutgers University, 1976.

LOUISE S. GRINSTEIN

# FERMAT, PIERRE DE (1601–1665)

Fermat was born in Beaumont de Lomagne, France, studied law in Toulouse, and spent his adult professional life in government service. "The prince of amateurs" (Bell 1937), Fermat did mathematical research in his spare time. He was also an accomplished classical scholar and knew Latin, Greek, Italian, and Spanish. Fermat published little, so his influence was less than it deserved to be. Many of his ideas were made public through correspondence with fellow mathematicians, and, in the case of number theory, through his notes in the margins of Diophantus' *Arithmetica.*

The seventeenth century saw the advent of "modern" mathematics. Although the initial inspiration for mathematical research came from recently translated works of Archimedes, Apollonius, Diophantus, and Pappus, mathematicians soon began to make fundamental changes. Old fields were completely transformed and new ones created, notably analytic geometry, calculus, number theory, probability, and projective geometry. Fermat made outstanding contributions to all but the last.

Number theory was Fermat's mathematical passion. In fact, he founded the subject in its modern form. His investigations of divisibility, representation of integers as sums of squares and of polygonal numbers, Pell's equation, and so on, inspired Euler and Legendre. No less a legacy was his claim to having proved Fermat's Last Theorem, namely, that for $n > 2$ the equation $x^n + y^n = z^n$ has no nonzero integer solutions. Subsequent attempts to prove the theorem motivated the creation of concepts and results far more important than that theorem.

Fermat's contributions to calculus—his methods of maxima and minima, tangents, and quadrature—

were conceptually the most advanced prior to the work of Newton and Leibniz. Also groundbreaking were his (and Descartes') invention of analytic geometry and his founding (in correspondence with Pascal) of probability theory. Fermat's mathematical output was remarkable for its originality and diversity. He has been called the greatest (pure) mathematician of the seventeenth century (*pace* Newton).

*See also* History of Mathematics, Overview; Number Theory, Overview

SELECTED REFERENCES

Bell, Eric Temple. *Men of Mathematics.* New York: Simon and Schuster, 1937.

Mahoney, Michael S. "Fermat, Pierre de." In *Dictionary of Scientific Biography* (pp. 566–576). Vol. 4. Charles C. Gillispie, ed. New York: Scribner's, 1981.

———. *The Mathematical Career of Pierre de Fermat, 1601–1665.* 2nd ed. Princeton, NJ: Princeton University, 1994.

ISRAEL KLEINER

# FIBONACCI SEQUENCE

Considered by many to be the premier mathematician to emerge from Europe during the thirteenth century. Leonardo Fibonacci (ca. 1175–1250) was born in Pisa, Italy. As a young man, Leonardo (known also as Leonardo of Pisa) lived for a while in Algeria, and later traveled to Greece, Sicily, Egypt, and Syria. In 1202, Fibonacci published the *Liber Abaci,* a book that "was instrumental in displacing the clumsy Roman numeration system and introducing methods of computation similar to those used today" (Hoggatt 1969, 1).

One of the most famous problems from the *Liber Abaci* is summarized as follows: Beginning the year with one pair of mature rabbits, how many pairs of rabbits will you have on December 31st if your original rabbits produce a new pair of rabbits on the first day of each month and each new pair of rabbits matures for two months and then produces a pair on the first of each subsequent month?

Because of its appearance in the solution of the rabbit problem, the sequence 1, 1, 2, 3, 5, 8, 13, . . . is known as the Fibonacci sequence. Each term of the sequence, beginning with the third, can be written as the sum of the two previous terms. The sequence can be defined recursively in the following manner: Let $F_1 = F_2 = 1$. Then $F_n = F_{n-1} + F_{n-2}$, where $F_n$ is the $n$th term.

For $n = 1, 2, 3, \ldots$, the Fibonacci numbers can also be generated from their Binet form

$$\frac{a^n - b^n}{\sqrt{5}}$$

In this form, $a$ and $b$ are the positive and negative roots, respectively, of the quadratic equation, $x^2 - x - 1 = 0$. The positive root of this equation,

$$a = \frac{1 + \sqrt{5}}{2}$$

is also known as the Golden Section (Brousseau 1973).

Elementary and middle school students may study patterns and develop identities involving Fibonacci numbers. Students might even discover previously unknown relationships. Advanced students may use mathematical induction to prove Fibonacci identities as well as identities involving both the Fibonacci sequence and other number-theoretic sequences. Fibonacci numbers also occur while studying the arrangement of leaves on a plant stem (also known as *phyllotaxis*), the ancestry of a male bee, electrical networks, light reflections, as well as other natural phenomena (Bicknell and Hoggatt 1973).

*See also* Enrichment, Overview; Golden Section; Recreations, Overview

SELECTED REFERENCES

Bicknell, Marjorie, and Verner E. Hoggatt, Jr., eds. *A Primer For the Fibonacci Numbers*. San Jose, CA: Fibonacci Association, 1973.

Brousseau, Brother Alfred. "Fibonacci Sequences." In *Topics for Mathematics Clubs* (pp. 1–9). LeRoy C. Dalton and Henry D. Snyder, eds. Reston, VA: National Council of Teachers of Mathematics, 1973.

Eves, Howard W. *An Introduction to the History of Mathematics*. 6th ed. Philadelphia, PA: Saunders College Publishing, 1990.

Ganis, Sam E. "Fibonacci Numbers." In *Historical Topics for the Mathematics Classroom* (pp. 77–78). 2nd ed. John K. Baumgart et al., eds. Reston, VA: National Council of Teachers of Mathematics, 1989.

Hoggatt, Verner E., Jr. *Fibonacci and Lucas Numbers*. Boston, MA: Houghton Mifflin, 1969.

WILLIAM J. O'DONNELL

# FIELD ACTIVITIES

In the United States, "hands-on science and mathematics" programs for exploring mathematics at the elementary level. Many of these programs are sponsored by the U.S. government, and many are widely available (cf. Pasterz 1992; Goin, et al. 1989; Lind 1991; and Markle 1988). Several projects organized at the national level are directly tied to government-sponsored research; for example, at Stennis Space Center in Mississippi, the "Digital Display Technology: A Comprehensive Tool for Education" project funded by the National Aeronautics and Space Administration (NASA) is being performed by schoolchildren and is part of the Mission to Planet Earth science project.

The goal of these programs is to show how to motivate students to see the world as a big laboratory for both educational progress and personal satisfaction. The focus is on how to teach students to *explore* the physical world around them. An example project is included in this entry to illustrate how an experiment-based mathematics project is defined, executed, and then followed up by a field expedition. Field trips should clarify the connections among the question asked, the experimental data collected or techniques learned, and the mathematical methods used to determine an answer to the posed question.

Hands-on mathematics is designed to provide students with first-hand experience using their knowledge and application abilities. They can design and perform experiment-based calculations. And from their own experimental observations, they may learn to derive logical conclusions. They may also augment their knowledge and application base and increase their comprehension and synthesis. Toward this end, it is nice to instill in children the concept of an experiment or a calculation as a means of discovery. Children ask questions about everyday life; how adults respond to these questions either reinforces children's natural tendency to figure out how things work or quenches their inquisitive nature.

For example, when the author was a child of six, she asked her mother how old she would have to grow to be half as old as her mother; my mother assured me that the answer could be found using the numbers that I knew. Equipped with the tool of counting, I was encouraged to determine for myself how to utilize it. I began by making a list of my age in one column and my mother's age in an adjacent column. I then checked each row until I located the correct age relationship. I saw that when the child reached the mother's age, the mother would have lived just twice as long. My mother then pointed out to me that I had posed a mathematical question and established a procedure to answer the question. Then she told me about another way to answer the question. She wrote the equation, Mom's age = Child's age + Mom's birth age, which is true on the child's birthday, at least. Since both the mother and child are aging at the same rate, she formalized my question by writing it as an equation, (Child's age) × 2 = Mom's age. By substituting the age equation (the relationship above) into the first equation (the "model") we directly solved for the answer: Child's age = Mom's birth age. With this simple example, my mother instilled in me a love of, and respect for, mathematics. It provided a way to pose and then answer questions, and I could do it by myself.

Posing questions is, perhaps, the *most important skill* a mathematician or scientist can acquire: science is the art of asking relevant questions that are simple enough to answer. This skill carries over to many other disciplines as well. Posing answerable questions is very difficult, so hands-on experiment-based mathematics projects should be simple to execute, logically direct, and relevant. If they also provide insight, that is even better. For widespread use, they should also be inexpensive. The subject of investigation should be relevant to most students' daily lives, not just future scientists. Everyday tools of different occupations are good technologies to adapt. For example, suppose a plumber has been asked to fix a stopped-up sink that is full of water, but his treatment works only in a specified volume of water that is about half the volume of the sink. The plumber needs to estimate the volume of the water that he wishes to leave and the amount to remove. The plumber then needs to select the appropriate-sized bucket to scoop the excess water into.

It is also quite important to understand what it means when a question "does not have an answer." It can either mean there were no cause and effect present, that there are multiple causes and effects offsetting each other, or that the method was not sensitive enough to distinguish between causes and effects. In the first case, a question is posed that presupposed a relationship that did not exist: an irrelevant question was asked. In the second case, a question is not basic enough to distinguish between the existing relationships: a question was posed that was not simple enough. In the third case, the method is not precise enough.

It is essential for students to understand the difficulties involved in deciding among these possibilities. If there are no cause and effect, you need to look in an entirely different area. If the question was posed badly, a simpler question must be developed. If the method is not precise, effort must be exerted to secure or invent a better method. There are many aspects of experiment-based mathematics that can fool

a researcher and lead to erroneous conclusions. This is why posing an answerable question is difficult, but so important. To prepare a conclusive analysis you must possess sufficient methodology to include relevant relationships. Some of the problems involved are illustrated in the following hands-on project.

## A TECHNOLOGY LIMITS PROJECT

A simple project illustrates the concept of engineering limitations. Collect several cardboard boxes and break them down into flat pieces, leaving at least one box as a three-dimensional object. This last box should be equilateral. Measure the length $L$ of any edge of the box and multiply this by the length of an adjoining edge. This will give the area ($A = L^2$) for this face of the box. This is an opportunity to point out why we call a square a square: it has area equal to its length squared. Demonstrate using a grid overlay that this is the area of all box faces.

The goal is to construct an equilateral box like the original. From the first stage, students will understand that they will need to construct six faces of equal area, with all edges the same length. If they also notice that the angles of the face are right angles, include this detail at this time. If they do not notice, do not mention it yet.

Once the definitions of equilateral box, and the concepts of length and area are assimilated into their level of understanding, have the students each take their own piece of cardboard and draw, freehand, six faces to construct their own box. The instructor should now use the best students' work, as chosen by the entire group, explaining that they are choosing to work with the best technician available. Cut out these six sides and tape the sides together to make a box. The constructed box will not fit together well because the six sides were not exactly equal. *The methodology (freehand drawing) is not exact enough for this geometric task.*

Repeat the construction, allowing students to use a straightedge and ruler. The box will still not fit together well, but it should be a better fit than the previous construction. Allow the students to figure out what is wrong now (the corners will not be right angles). Have the students suggest a new methodology to remedy this defect. They should conclude that they need to be able to make true corners. Repeat the construction a third time, allowing students to use a carpenter's square (a reference true corner). But before you assemble the box, ask the students if they can think of a way to check that the box has six equal sides before they waste time taping the box together. This can be done by overlaying the pieces to check alignment. Tracing and cutting errors should still be visible, but small. There are always limits to technological precision, and this is an important point for them to understand. Assemble this next box—it will not fit perfectly because of the small technological errors—but it should be a faithful reproduction of the original cube.

The students are now ready for a field visit to a carpenter's workshop. There they will be able to see the existing technology and discuss with an expert why these tools exist. It will also help them understand the technological limits faced by construction workers. Although this project was designed to illustrate the importance of application knowledge, it required critical thinking to identify the technical problems and overcome them. It also clarified the differences and similarities between the geometric concepts used, improving both geometric comprehension and synthesis. And it even teaches a useful skill.

*See also* Lesson Plans; Techniques and Materials for Enrichment Learning and Teaching

### SELECTED REFERENCES

Goin, Kenn, Eleanor Ripp, and Kathleen N. Solomon. *Bugs to Bunnies Hands-on Animal Science Activities for Young Children.* New York: Chatterbox, 1989.

Lind, Karen K. *Water, Stones, and Fossil Bones.* Washington DC: National Science Teachers Association, 1991.

Markle, S. *Hands-on Science.* Cleveland, OH: Instructor, 1988.

Pasterz, V. L. *Drill It: Science Workbook Level 2.* Bloomington, MN: Simon and Schuster, 1992.

Williams, Jack. *The Weather Book: An Easy-to-Understand Guide to the USA's Weather.* New York: Vintage, 1992.

A. LOUISE PERKINS

# FIELDS

Collections $K$ of elements $a, b, c, \ldots$, together with two operations called addition and multiplication that assign to each pair of elements $a, b$ in $K$ a sum $a + b$ and a product $ab$ in $K$ such that the following properties are satisfied:

1. Properties of addition: For all $a, b, c$ in $K$, $a + (b + c) = (a + b) + c$ (associativity) and $a + b = b + a$ (commutativity). There is an identity element 0 in $K$ such that $0 + a = a + 0$ for all $a$ in $K$. Given any element $a$ in $K$ there is an inverse element $-a$ in $K$ such that $a + (-a) = (-a) + a = 0$.

2. Properties of multiplication: For all $a, b, c$ in $K$, $a(bc) = (ab)c$ (associativity) and $ab = ba$ (commutativity). There is an identity element 1 in $K$ such that $a \times 1 = 1 \times a = a$ for all $a$ in $K$. Given any nonzero element $a$ in $K$, there is an inverse element $a^{-1}$ in $K$ such that $aa^{-1} = a^{-1}a = 1$.

3. For all elements $a, b, c$ in $K$, $a(b + c) = ab + ac$ (distributivity).

The set of rational numbers (fractions) with the operations of addition and multiplication is a field. The set of real numbers and the set of complex numbers are also fields under addition and multiplication.

Using modular arithmetic, we may construct a field containing a finite number of elements. Let $p$ be a prime number and let $K = \{0, 1, 2, \ldots, p - 1\}$. If $s$ and $t$ are any of the numbers in $K$, let $s + t$ (mod $p$) and $st$ (mod $p$) be the sum and product of $s$ and $t$ computed modulo $p$, that is, their remainder upon division by $p$. Then $K$ is a field containing $p$ elements under addition and multiplication modulo $p$. If $p = 2$ then $K = \{0, 1\}$ and $1 + 1 = 0$, since the remainder of $1 + 1$, or 2, upon division by 2 is 0. In this case, the arithmetic on $K$ is called binary arithmetic. If $p = 3$, then $K = \{0, 1, 2\}$ and the following tables summarize addition and multiplication in $K$:

| + | 0 | 1 | 2 | | × | 0 | 1 | 2 |
|---|---|---|---|---|---|---|---|---|
| 0 | 0 | 1 | 2 | | 0 | 0 | 0 | 0 |
| 1 | 1 | 2 | 0 | | 1 | 0 | 1 | 2 |
| 2 | 2 | 0 | 1 | | 2 | 0 | 2 | 1 |

Finite fields, such as those constructed using binary arithmetic, are especially important in applications of mathematics to coding theory and computer science, where sequences of zeros and ones form the basis for the digital transmission of information.

Field theory evolved out of eighteenth century attempts to solve algebraic equations. To solve the general linear equation $ax + b = 0$, where $a$ and $b$ are numbers in a field, $a \neq 0$, we first add $-b$ to both sides, getting $ax = -b$, then divide both sides by $a$ to obtain the solution $x = -b/a$. The general quadratic equation $ax^2 + bx + c = 0$ may be solved by a similar series of operations. In this case, however, the roots do not necessarily lie in the field since the solution requires the square root of $b^2 - 4ac$. If $b^2 - 4ac$ is negative, its square root is not in the field of real numbers and the field must be extended to a larger field, such as the field of complex numbers, in order that the roots exist. Working in the field of complex numbers, Carl Friedrich Gauss (1755–

1855) showed in 1799 that every algebraic equation having degree $n$ has $n$ roots in the field of complex numbers. It is not necessarily true, however, that the roots can be expressed using the four arithmetic operations of addition, subtraction, multiplication, and division, together with root extraction, as is the case for quadratic equations. When this is possible the equation is solvable by means of radicals. The general theory for dealing with the question of solvability by means of radicals was developed by Evariste Galois (1801–1833). By first associating a group with the equation, Galois showed that the equation is solvable by radicals precisely when its group has a series of subgroups that satisfy a certain technical requirement.

Field theory also provides the mathematical framework for solving four classical geometric construction problems. Elementary geometry dates back to Euclid of Alexandria (ca. 300 B.C.), whose book, the *Elements*, is the basis for standard geometry. Euclid gives elementary constructions for bisecting angles and many other similar constructions using only straightedge and compass. The Greeks proposed four special construction problems to be solved using only straightedge and compass: given a cube, to construct a cube having twice the volume; given an angle, to construct an angle whose measure is one-third that of the given angle; given a circle, to construct a square having the same area as the circle; given a circle and a positive integer $n$, to divide the circumference of the circle into $n$ equal parts.

The impossibility of these constructions is demonstrated by first transforming them into an algebraic framework. A basic unit of measure is selected and from this is built the set of rational points in the plane. The corresponding field is the field of rational numbers. At each stage in the construction process, either two lines intersect or a line and a circle intersect. The intersection point of two lines adds a new point but the coordinates of this point are still rational numbers, since solving two linear equations simultaneously that have rational coefficients gives a point with rational coordinates. However, when a linear and quadratic equation are solved simultaneously, the new point may involve radicals and hence enlarge the field of rational numbers by addition of that radical. In this way, each step in the construction process either leaves the field of numbers the same or enlarges it by a radical. The four classical construction problems are then shown to be impossible by showing that the required constructions result in points that do not have coordinates that lie in such an enlarged field.

*See also* Abstract Algebra

SELECTED REFERENCES

Fraleigh, John. *A First Course in Abstract Algebra.* Reading, MA: Addison-Wesley, 1989.

Gallian, Joseph. *Contemporary Abstract Algebra.* Lexington, MA: Heath, 1990.

Kletzing, Dennis. *Abstract Algebra.* San Diego, CA: Harcourt Brace Jovanovich, 1991.

DENNIS KLETZING

# FINANCE

Investing in a business corporation, or company, in one of two ways: by lending money to the company, usually by buying the bonds it sells, and becoming one of its creditors; or by buying the common stock the company sells and becoming one of its stockholders.

Buying the bond of a company, or otherwise lending money to it, gives the purchaser a legal claim to the interest on the bond or loan and a legal claim to the return of the bond's or loan's principal at the end of the loan period. Suppose a bond is bought from a company for its face value of $1,000 and a maturity of 15 years. This means that the company is obligated to pay back $1,000 at the end of fifteen years, at which time the bond is said to mature. Suppose also that the annual interest rate of the bond was 8%. This means that the company is obligated to pay an $80 interest payment each year for the next fifteen years. The $80 interest payment is 8% of the $1,000 face value of the bond. Most bonds pay interest two times a year, or semiannually. If this were the case, the bondholder would receive a $40 interest payment each six months for the next fifteen years.

Buying the common stock of a company makes the purchaser an owner of the company along with its other stockholders. Part of the return on common stock investment consists of cash dividends set by the company's board of directors. Dividends may be changed frequently and depend on how the directors feel about the prospects for future profit. The stockholders elect the board of directors and in this way may be able to influence the dividends received. However, as an owner of the company, stockholders have no legal guarantee of dividends. Also, unlike bondholders, stockholders have no legal claim to be paid back their original investment. The only way that a stockholder can get back the investment in the stock is to sell it to another investor. On rare occasions, a company will buy its stock back from its stockholders. In this case, however, the company will pay the stockholders the present market price for the stock and not the price the company received when it first sold its stocks to stockholders.

The annual return on common stock has two parts. The first is the *current dividend return*. These are the cash dividends paid for the year as set by the board of directors. The second is the *capital gains/loss return*. This is the change in the market price of the stock from the beginning to the end of the year. Together, the current dividend return and the capital gains/loss return add up to the *total return* for the year on the investment in the stock.

Suppose that on January 1 a share of stock in the PDQ Corporation was purchased at a price of $100. Over the year, PDQ paid four dividends of $2 each, for a total of $8 for the year. On December 31 the market price of PDQ stock had risen to $106. The total return would be $14, made up of $8 in current dividend return and $6 in capital gains return. Based on the January 1st investment of $100, the $14 total return amounts to an *annual rate of total return* of a *positive* 14%.

The annual rate of total return on a stock can be negative. This takes place in years when the stock's market price fell more than the cash dividends it paid. Suppose that on January 1 a share of stock in the IRK Corporation was purchased at a price of $50. IRK paid its first two quarterly dividends of $2 each, cut its third-quarter dividend to $1, and did not pay a fourth-quarter dividend, for a total annual current dividend return of $5. On December 31, the market price of IRK stock had fallen to $42. The total return for the year would be a negative $3, made up of $5 in current dividend return and $8 in capital loss return. Based on the January 1st investment of $50, the total return of a negative $3 amounts to an *annual rate of total return* of a *negative* 6%.

The value of a common stock can be expressed in mathematical form. Let $P_0$ be the present value to an investor of a share of common stock; let $D_0$ be the annual dividend the stock just paid; let $g$ be the projected growth rate of the dividend into the indefinite future; and let $k$ be the rate of return the investor requires on a common stock investment in this company. The highest price an investor would pay for the stock is given by the formula:

$$P_0 = D_0 \frac{1 + g}{k - g}$$

Suppose that the IRK Corporation was selling common stock and an investor was trying to determine how much to pay for a share. The company had just paid an annual dividend of $5 per share, stock analysts were projecting an indefinite growth rate of

6% per year, and the investor needs at least a 15% return on this investment. The highest price the investor would pay for a share of stock would be calculated as

$$\$5 \frac{1 + 0.06}{0.15 - 0.06} = \$58.89$$

Since stock market transactions are made in one-eighths of a dollar, the investor would offer $58.875 for the stock.

*See also* Interest

SELECTED REFERENCES

Brealey, Richard A., Stewart C. Myers, and Alan J. Marcus. *Fundamentals of Corporate Finance.* New York: McGraw-Hill, 1995, Chapter 3.

Ross, Stephen A., Randolph W. Westerfield, and Bradford D. Jordan. *Essentials of Corporate Finance.* Chicago, IL: Irwin, 1996, Chapter 5.

RALPH L. NELSON

# FORMULAS

Expressions for relationships between quantities. One type of formula is a compact, easy-to-use statement, in symbolic form, for finding the value of an unknown quantity when the values of the other quantities in the relationship are known. A formula is an alternative to a verbal statement, for example:

1.  The verbal statement "The area of a triangle is equal to one-half the product of its base and height" may be translated into the formula $A = \frac{1}{2}bh$, where $A$, $b$, and $h$ represent the measures of the area, base, and height, respectively.
2.  The amount of interest on a loan is given by the product of the principal, the yearly rate of interest, and the time in years for which the money is being lent. The corresponding formula is $I = prt$, where $I$ = yearly interest, $p$ = principal, $r$ = yearly rate, and $t$ = time in years.

In algebraic expressions, both variables and constants are used. Variables represent quantities that can assume different values, while constants are fixed in value. In the formula for the area of the triangle, $A$, $b$, and $h$ are variables, since their values change for different triangles, but $\frac{1}{2}$ is a constant—fixed in value for calculations of the area of any triangle.

Formulas are evaluated by substituting numbers for the letters. Suggestions for evaluating a formula are: (1) write the formula that expresses the given verbal statement; (2) replace each letter by its assigned numerical value; (3) do all arithmetic operations using the correct order of operations. The following examples illustrate the procedure:

1.  Find the area of a triangle if the length of the base is 16 in. and the length of the height is 6 in. After writing the formula $A = \frac{1}{2}bh$, replace each letter by its assigned value. Thus $A = \frac{1}{2}(16)(6)$, and $A = 48$ square inches.
2.  The total electrical resistance $R$ of two resistances $a$ and $b$ connected in parallel is their product divided by their sum. Find the total resistance $R$ if $a = 100$ ohms and $b$ is 400 ohms. (An ohm is a unit of electrical resistance.) Steps in the solution are:

$$\text{(i) } R = \frac{a \cdot b}{a + b}$$

$$\text{(ii) } R = \frac{100 \cdot 400}{100 + 400}$$

$$\text{(iii) } R = 80 \text{ ohms}$$

*See also* Equations, Algebraic; Equations, Exponential; Equations, Logarithmic; Variables

### SELECTED REFERENCE

Bloomfield, Derek I. *Introductory Algebra*. St. Paul, MN: West, 1994.

DEREK I. BLOOMFIELD

# FOURIER, JEAN BAPTISTE JOSEPH DE (1768–1830)

Author of *The Analytical Theory of Heat,* where the Fourier equation, series, and integrals were introduced. Perhaps no other mathematician's name appears in scientific literature as frequently: the Fourier equation, Fourier series, Fourier integrals, Fourier transforms, Fourier analysis, and Fourier formulas are terms familiar to many scientists and engineers. *The Analytical Theory of Heat,* his main work, initiated a series of impressive works on trigonometric series leading to the modern notions of function, integration, and sets. Moreover, Fourier worked on algebraic equations and on linear inequalities, anticipating modern developments in operations research.

Fourier was a product of the French Revolution and his life provides a very interesting cross-section of French history. His family was poor, and at age 10 he became an orphan. He was trained by Benedictine monks in a military college. After graduating, he wanted to serve in the artillery, the most scientific

arm, and was backed by Adrien-Marie Legendre, a famous mathematician. But it was a time of reaction. The minister of war answered Legendre that were he even a second Newton, Fourier could not enter the artillery since he was not a noble. Fourier had to choose the Church.

He was supposed to take his vows in 1789, at the exact time when the National Assembly decided to suspend religious vows. Fourier, already known as Abbé Fourier, gave up the vows and returned to his military school as a humanities teacher. He got involved in revolutionary activities in 1793, the year that began with the execution of Louis XVI and saw the formation of the European coalition against the French Republic, the mass levy of 300,000 men, and the founding of the revolutionary committees. He proved efficient in organizing food and military supplies in Orleans and moderating excesses in many places, but he was sent to prison several times, both before and after the fall of Robespierre.

In 1794, he became a student of the newly created *École Normale* in Paris, where the teachers were Joseph-Louis Lagrange, Gaspard Monge, and Pierre-Simon de Laplace. His intelligence recognized, he was elected to teach at the *École Polytechnique,* also founded in 1794. When Napoleon Bonaparte led the French expedition to Egypt in 1798, he established a duplicate of the French Institute in Cairo and installed Monge as president and Fourier as *secrétaire perpétual.* Fourier worked in many fields while in Egypt; he wrote papers on oases and monuments as well as on algebraic equations, conducted diplomatic negotiations, and prepared material for an extensive *Description de l'Égypte,* which he took back to France, that qualified him as an egyptologist. In 1801, Napoleon made him *Préfet de l'Isère,* a prominent political position. He was a good prefect and at the same time wrote his major scientific work, *The Analytical Theory of Heat.* His theory was not accepted immediately. Lagrange opposed his use of trigonometric series.

After the fall of Napoleon, who had made him "Baron" Fourier, he became director of the Bureau of Statistics in Paris and proved efficient as a statistician too. He was elected to the *Académie des Sciences* in spite of the opposition of King Louis XVIII and became *secrétaire perpétual* in 1822. He had competitors if not enemies, like Simeon-Denis Poisson and Augustin-Louis Cauchy. But he had friends among the younger people, Peter G. Lejeune-Dirichlet and Charles Sturm, his most famous successors. The importance of his work on trigonometric series was recognized in Germany more than in France and was fully established in the thesis of Georg F. B. Riemann on trigonometric series. The full recognition of his role as mathematician is quite recent in France.

SELECTED REFERENCES

Fourier, Joseph. *Théorie analytique de la chaleur.* Paris, France: Didot, 1822.

Grattan-Guinness, Ivor, with Jerome R. Ravetz. *Joseph Fourier, 1768–1830, A Survey of His Life and Work.* Cambridge, MA: MIT, 1972.

Herivel, John. *Joseph Fourier, The Man and the Physicist.* Oxford, England: Clarendon, 1975.

Kahane, Jean Pierre, and Pierre-Gilles Rieusset. *Fourier Series and Wavelets.* Reading, England: Gordon and Breach, 1995.

JEAN PIERRE KAHANE

# FRACTALS

Objects, either in pure mathematics or in the real world, that have the property of self-similarity, that is, small parts of them resemble the whole object after appropriate magnification. In the 1970s, IBM mathematician Benoit Mandelbrot launched the field of fractals by pointing out many fractal occurrences in nature and mathematics. Two familiar real-world examples are coastlines and trees. In the former, the wiggliness of a small section of the coast caused by small groups of rocks often resembles that of a broad expanse of the coast caused by huge inlets and bays. In the latter, the structure of tiny twigs often resembles that of the largest boughs.

Probably the most famous mathematical fractal is the Mandelbrot set (see Figure 1). For every point $(x,y)$ in the plane, a sequence of numbers can be generated according to a simple formula. (If the point is interpreted as the complex number $c = x + iy$, then the sequence is defined by $z_0 = 0$, $z_{n+1} = z_n^2 + c$.) The point $(x,y)$ is in the Mandelbrot set if no numbers in this sequence exceed 2. This yields a set of amazing intricacy, which

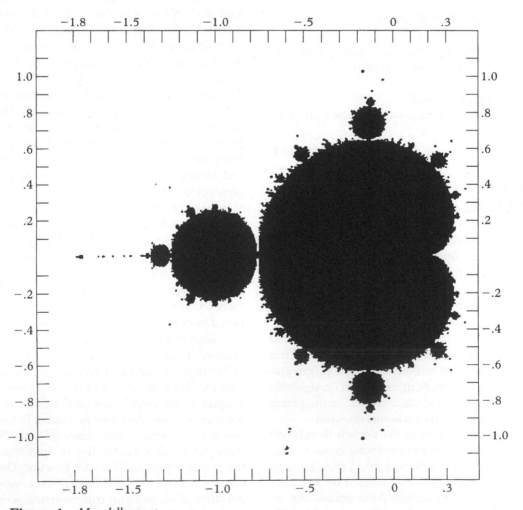

**Figure 1** *Mandelbrot set.*

is only hinted at by the computer-generated approximation in the illustration. Computer programs can easily magnify a small region of the picture; one finds warts appearing on warts at all levels of magnification. Moreover, it can be proved that the Mandelbrot set is a connected set. This means that what appear to be tiny islands disjoint from the main body are actually connected to the main body by filaments so fine that they cannot be seen completely by any computer depiction.

One important mathematical notion is fractal dimension, an idea developed by mathematicians such as Felix Hausdorff (1868–1942) in the early 1900s. A self-similar curve tends to fill up space more than a smooth curve, and fractal dimension gives a quantitative way of measuring this. The dimension of a smooth curve is 1, whereas that of a wiggly fractal curve may have a value such as 1.3. Important examples discovered before 1900 by Georg Cantor (Cantor set) and Karl Weierstrass (a curve nowhere differentiable) had much influence on the development of what is now called fractal geometry.

Fractals have been used since the mid-1980s in the work of many applied scientists and engineers. One example deals with drilling for oil. It has been found that the interface between oil and water is a fractal surface, and the relationship of the mathematical properties of this surface to the ease of separation of oil and water has been studied. Another example, developed by the mathematician Michael Barnsley, uses fractal generation of pictures to minimize the amount of data required for storage or transmittal of a picture. For example, an intricate fern can be completely described by just twenty-four numbers. Indeed, the fern can be partitioned into four sets, as in Figure 2, each of which is the image of the whole set under a translated linear transformation. Each such transformation is determined by six numbers, and iteration of these transformations will draw the fern. There was initial skepticism as to whether similar economy could be practically obtained in a fractal description of any picture, but Barnsley and his co-workers seem to have succeeded in making this a practical method of image compression. Finally, movie producers use computer-generated fractal landscapes to obtain scenery unavailable in the real world.

Fractals can be an excellent topic for exciting students about mathematics. Computer experimentation using software packages can be performed with minimal mathematical understanding, and the curious user can then be led into a study of the underlying mathematical notions. A thorough under-

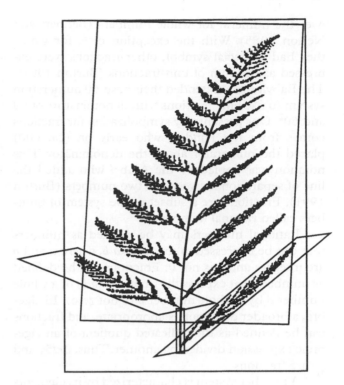

**Figure 2** *Fractal fern.*

standing of fractals requires graduate-level study of analysis and dynamical systems.

*See also* Chaos

SELECTED REFERENCES

Barnsley, Michael. *Fractals Everywhere*. San Diego, CA: Academic, 1988.

Mandelbrot, Benoit. *The Fractal Geometry of Nature*. New York: Freeman, 1982.

Peitgen, Heinz-Otto, Hartmut Jürgens, and Dietmar Saupe. *Fractals for the Classroom*. I and II. New York: Springer-Verlag, 1992.

DONALD M. DAVIS

# FRACTIONS

Term derived from the Latin word *fractum*, meaning "to break." From this literal meaning, an early concept of the fraction was of something "less than a whole." This primitive interpretation is frequently the one that is used when introducing fractions to children.

Some ancient numeration systems included fractions. The Egyptians used unit fractions, that is, fractions with a numerator of 1, which they symbolized by placing an elongated oval, meaning "a part," over the numeral for whole numbers (Bennett and Nelson 1985). With the exception of $\frac{2}{3}$, for which they had a special symbol, other fractions were expressed as the sum of unit fractions (Burton 1991). The Babylonians extended their base-60 numeration system to use only fractions with denominators of 60 and $60^2$. Our present-day symbolization for fractions comes from the Hindus, who early on (ca. 600) placed the numerator above the denominator. This notation was copied by the Arabs, who added the line of separation between the two numbers (Burton 1991). Fractions are a subset of the system of numbers called rational numbers.

Rational numbers may be defined as numbers that can be expressed in the form $a/b$, where $a$ and $b$ are integers and $b$ is not 0. Fractions can be defined as numbers that express the quotient ($a/b$) of a whole number $a$ by a whole number $b$ ($b$ not zero). In algebra, a broader definition is appropriate, and fractions can be defined as the indicated quotient of an algebraic expression divided by another. Thus, $2n/5y$ and $\frac{\pi}{2x^2}$ are fractions.

A number system is characterized by its elements (numbers), the basic operations that can be performed on those numbers, and the principles that hold true for the system. After the quantitative values of fractional numbers are understood, students use them in problem-solving activities. They learn that the basic operations of addition, subtraction, multiplication, and division can be performed with fractions as with whole numbers. They also learn that the principles that are true for whole numbers are also true for fractions. These include: commutativity, associativity, the closure property for addition and multiplication, and distributivity for multiplication over addition. In addition, the closure property holds true for division of fractions, except when the divisor has the value zero. Fractions also have the property of denseness, that is, between any two fractions, another fraction can be named. Thus, one cannot "count-on" by naming a fraction that is "next to" a given fraction. This changes substantively the process of computation from that with whole numbers, where counting-on enables one to arrive at a sum or product (Post 1992). In applications, fractions have various meanings: part of a whole, subset of a set, a ratio, an operator (multiplicative aspect), and a quotient. A mature knowledge of fractions involves an understanding of subconstructs and their interrelationships (Kieren, "The Rational Number Construct," 1980).

Introductory fraction instruction generally focuses on the part-whole interpretation. This interpretation depends directly on the ability to partition

a continuous quantity or a set of discrete objects into equal-sized parts (Post 1992). Length, area, and volume measures are part of the concept of continuous quantity. In this case, the unit or "the whole" is one object, such as a square piece of paper, a piece of ribbon, a pizza, or a candy bar. A unit can also be "part of a whole" (e.g., three-fourths of a large pizza to be shared among three friends) or it can be more than one whole (e.g., a pie and a half for the nine people at the dinner table).

When the unit consists of a set of objects, the unit is said to be discrete, and can be partitioned into different-sized groups of the (discrete) objects (Post 1992). Symbolically, two numbers are used to represent each fraction number: the numerator (first number) denotes the number of parts identified and the denominator (second number) indicates into how many equal parts the whole has been separated.

Research supports the use of the region model first for instruction (Hollis 1984; Payne 1984). However, some studies report better performance when a set model is used in instruction (Langford and Sarullo 1993). When using the region model, learning is enhanced when students engage in partitioning activities. That is, students should explore partitioning different regions or figures into different numbers of equal-sized parts. Figure 1 depicts children's

**Figure 1** *Children's partitioning attempts at attaining fractional parts of different regions.*

successful and unsuccessful attempts at partitioning geometric regions into fractional parts. Without this personal experience of partitioning figures, students are unable to use the model in problem-solving activities. (For additional children's attempts at partitioning geometric regions, see Pothier and Sawada 1984.) Students should also be involved in partitioning or separating activities with other kinds of physical units (e.g., sets) and in naming fractional parts. The language should be both oral and written ("two-thirds," "2 thirds," or "two out of three equal parts," not "two over three"). When these capabilities are well developed, the move can be made to symbolic representation of fractional parts.

The length measurement model is an effective one for comparing fractions. A set of "fraction strips" can easily be made and used to compare unit fractions and to find equivalent fractions. Cuisenaire rods and Fraction Bars are two commercial sets of materials that embody length measurement. Eventually, a number line can be used to name different points as fractions. Assessment results show that even in junior high school, students are not competent in their ability to show fractions on a number line (Kouba et al. 1988).

The challenge of the discrete quantity or part-of-a-set model is to name the result of partitioning. Research results show that whole-number thinking appears to dominate children's thinking and thus the results of a partition are named by the number of objects in the part rather than by the fractional parts constructed (Langford and Sarullo 1993; Pothier and Sawada 1990). For example, when sharing twelve candies among three friends, some students say that each friend gets "one-fourth" of the candies. An important idea to develop is that, for example, when "sharing for three," each one gets "one-third" of the set no matter how many things make up each person's share. Teachers should bear in mind that the leap from whole-number to fractional-number thinking is not trivial and develops over time with the aid of proper experiences (Langford and Sarullo 1993).

The concept of equivalent fractions distinguishes fractions from whole numbers in a significant way. The notion that a quantity can be expressed with different numbers (e.g., $\frac{1}{2}, \frac{2}{4}, \frac{3}{6}, \frac{4}{8}$, etc. all name the same amount) is not easy for students to grasp and mastery should not be expected until grades 5 or 6 (Driscoll 1984). To use equivalent fractions for comparison or addition of fractions, there are different ways to find common denominators: (a) represent each fraction concretely to "see" which de-

nominator is common; (b) list successive multiples of each denominator and then identify a common multiple; (c) list equivalent fractions until a common denominator is found; (d) list the prime factors of each denominator, then multiply together the minimum number of prime factors that are necessary so that the factors of each denominator are represented. For example, to find $\frac{5}{8} + \frac{3}{6}$, we first find a common denominator. By method (d), the prime factorization of 8 is $2 \times 2 \times 2$, and of 6 is $2 \times 3$. The lowest common denominator is $2 \times 2 \times 2 \times 3 = 24$. When equivalent fractions are understood, students are able to compare, rename, simplify, and add fractions. These ideas are prerequisites to problem solving involving fraction computation.

## COMPUTATION

In early problem-solving activities, students should use concrete and pictorial models to show how they arrived at a solution. The three types of models described earlier (i.e., region, measurement, and part-of-a-set models) can be used to show addition and subtraction of fractions. Computing with unlike fractions (e.g., $\frac{3}{4} + \frac{5}{6}$) should evoke an automatic move to write the equivalent fractions with the same denominators before proceeding with adding the number of parts.

Multiplication and division of fractions are surprisingly easy. In fact, learning simple rules (e.g., "multiply top numbers and then bottom numbers" or "change the sign and invert the second fraction") enables one to find answers to questions quickly. There is a danger that students may be satisfied with learning rules for computation. Studies show that students appear to rely on rules rather than on understanding of fractions as distinct quantities when computing (Post 1992). Thus, the orientation of instruction should be for understanding rather than for procedural knowledge and for students to mentally connect processes with their own knowledge from meaningful personal experiences (Besuk and Bieck 1993). In introductory fraction work involving multiplication and division, the fact that multiplication can yield a product that is less than either of the factors (when the factors are between 0 and 1) and division can give a quotient that is greater than the dividend (when the divisor is less than 1) must be discussed. A sequence for instruction could be (a) multiplying a whole number by a fraction (e.g., $\frac{3}{4} \times 8$) (b) multiplying a fraction by a fraction (e.g., $\frac{2}{5} \times \frac{1}{4}$); then (c) multiplying mixed numbers ($2\frac{1}{4} \times 3\frac{2}{3}$).

In learning an algorithm for division of fractions, the idea of a reciprocal or multiplicative inverse is helpful. For example, the reciprocal of $\frac{3}{5}$ is $\frac{5}{3}$ and $\frac{3}{5} \times \frac{5}{3} = 1$. Problem-solving experiences in fraction division could follow the sequence: (a) fraction divisor and whole-number dividend; (b) fraction divisor and dividend; then (c) mixed-number dividend. Calculators that display fractions can be used to advantage by students to develop concepts and computational competence.

Notwithstanding the numerous studies in fraction learning that have been reported in recent years, there are still many unanswered curricular and instructional questions. The consensus appears to be that learning is enhanced when the focus of instruction is on children building their own meaning for fractions and operations with them through problem-solving activities. This is achieved when varied models (region, part-of-a-set, length measurement) and different representations (real-world objects, representative concrete materials, pictures) are used, attention is given to children's oral and written language, and students are involved in constructing their own computation procedures. A best program sequence may only be judged from observing and listening to children as they connect new ideas to their existing knowledge about fractions.

*See also* Arithmetic; Cuisenaire Rods; Rational Numbers

SELECTED REFERENCES

Bennett, Albert B., Jr., and Leonard T. Nelson. *Mathematics, An Informal Approach.* 2nd ed. Boston, MA: Allyn and Bacon, 1985.

Besuk, Nadine S., and Marilyn Bieck. "Current Research on Rational Numbers and Common Fractions: Summary and Implications for Teaching." In *Research Ideas for the Classroom—Middle Grades Mathematics.* Douglas T. Owens, ed. (pp. 118–136). New York: Macmillan, 1993.

Burton, David M. *The History of Mathematics, An Introduction.* 2nd ed. Dubuque, IA: Brown, 1991.

Driscoll, Mark. "What Research Says." *Arithmetic Teacher* 31(Feb. 1984):34–35, 46.

Hollis, L. Y. "Mickey." "Teaching Rational Numbers—Primary Grades." *Arithmetic Teacher* 31(Feb. 1984): 36–39.

Kieren, Thomas E. "Knowing Rational Numbers: Ideas and Symbols." In *Selected Issues in Mathematics Education.* Mary M. Lindquist, ed. Berkeley, CA: McCutchen, 1980.

———. "The Rational Number Construct—Its Elements and Mechanisms." In *Recent Research on Number Learning.* Thomas E. Kieren, ed. Columbus, OH: ERIC/SMEAC, 1980.

Kouba, Vicky L., Catherine A. Brown, Thomas P. Carpenter, Mary M. Lindquist, Edward A. Silver, and Jane O. Swafford. "Results of the Fourth NAEP Assessment of Mathematics: Number, Operations, and Word Problems." *Arithmetic Teacher* 35(Apr. 1988):14–19.

Langford, Karen, and Angela Sarullo. "Introductory Common and Decimal Fraction Concepts." In *Research Ideas for the Classroom: Early Childhood Mathematics* (pp. 223–247). Robert J. Jensen, ed. New York: Macmillan, 1993.

Payne, Joseph N. "Curricular Issues: Teaching Rational Numbers." *Arithmetic Teacher* 31(Feb. 1984):14–17.

Post, Thomas R., ed. *Teaching Mathematics in Grades K–8 Research-Based Methods.* 2nd ed. Boston, MA: Allyn and Bacon, 1992.

Pothier, Yvonne, and Daiyo Sawada. "Some Geometrical Aspects of Early Fraction Experiences." *Recherches en Didactiques des Mathematiques* 5(2)(1984):215–226.

———. "Partitioning: An Approach to Fractions." *Arithmetic Teacher* 38(Dec. 1990):12–16.

YVONNE M. POTHIER

# FRANCE

In the 1970s, mathematics was considered as the preeminent discipline in French education. Curricular reform based on "modern mathematics" begun at that time was, however, abandoned in the 1980s. The reform had brought about a real examination of the existing mathematics program, but when the new programs were introduced, this did not recur. The ideology of "modern mathematics" still exists, and some teachers oppose the "spirit of the new programs," which has never been explained in all its epistemological dimensions, especially the nonformalist but constructivist interpretation of mathematics (Bkouche et al. 1991). Furthermore, the new programs were put in place at the same time as were new requirements, in particular, that of bringing 80% of a generation of students up to the baccalaureate level. Certain teachers also miss the time when students were chiefly selected for their skill in mathematics.

The political reality of education in France is such that successive ministers of education cannot resist proposing new programs. Therefore, since the end of the 1970s, there have been numerous program changes. The major current tendencies in French mathematics education are: more selectivity in the teaching of geometry and the learning of proof, new types of instruction in analysis, activity-based instruction, and the introduction of an historic perspective.

The teaching of geometry in the middle school (ages 11–15) is founded essentially on the idea of geometric transformation, which poses many learning difficulties. Moreover, this concept is conceived as part of a developmental approach to learning proof, beginning with "deductive blocks," meaning that the propositions are organized in subdivisions based on deduction from certain propositions accepted as true. Teachers tend to focus on the learning of proof, while neglecting the build-up of geometric understandings (Barbin 1993, 1994), because many consider proof as their chief responsibility. Actually, proof is one of the principal causes of failure in middle school. As a result, the trend is toward minimizing geometry instruction in favor of probability and statistics, strong components of the last programs in the middle school. This poses the risk of geometry instruction being decreased in the secondary school (ages 16–18) in the coming years, although many teachers continue to believe in the formative value of geometry (Audibert 1991).

The new method of teaching analysis in French secondary schools seeks to break with the formalism of $\epsilon$ and $\delta$, basing the studies of series and functions instead on diagrams. Many teachers are unsatisfied, however; they worry that their teaching is not rigorous enough. They feel the program should have been accompanied by an epistemological presentation on the concepts of analysis, conceived as means of understanding real phenomena. Upper-level secondary school teachers believe that the achievement level of students is falling.

The activity-based instruction (middle school) and the problem-solving activities (secondary school) proposed by the new programs have not always been understood by French teachers; many did not consider these activities worthwhile. These approaches demand the mastery of new teaching practices (Bach et al. 1992). Here again, a systematic training of teachers is necessary. Activity-based instruction, poorly understood and poorly managed, has some perverse effects: teachers suggest some introductory activities that are really either just pretenses or fragmented activities of such a nature that mathematical learning is no more significant than with traditional methods.

The introduction of an historic perspective into the teaching of mathematics is motivated by a revival of interest among French teachers in the history of their discipline. The approach is not to give just dates or relate anecdotes to the students; it is more concerned with presenting a cultural and interdisciplinary view of mathematics by proposing, for instance, that the students (chiefly in secondary school) read original texts (Fauvel 1990).

*See also* Constructivism

## SELECTED REFERENCES

Audibert, Gerard. "La Géométrie dans l'Enseignement." *Repères IREM* 4(July 1991):21–52.

Bach, Marie J., Dominique Gaud, Jeannette Gay, Jean P. Guichard, Madeleine Marot, and Claude Robin. "Enseigner par les Activités." *Repères IREM* 8(July 1992): 5–32.

Barbin, Evelyne. "The Epistemological Roots of a Constructivist Interpretation of Teaching Mathematics." In *Constructivist Interpretations of Teaching and Learning Mathematics* (pp. 49–68). J. A. Malone and P. C. S. Taylor, eds. Perth, Australia: Curtin University, 1993.

———. "The Meanings of Mathematical Proof: On Relations between History and Mathematical Education." In *Eves' Circles* (pp. 41–52). John M. Anthony, ed. Washington, DC: Mathematical Association of America, 1994.

Bkouche, Rudolf, Bernard Charlot, and Nicolas Rouche. *Faire des Mathématiques: le Plaisir du Sens.* Paris, France: Armand Colin, 1991.

Fauvel, John, ed. *History in the Mathematics Classroom, The IREM Papers.* Vol. 1. Leicester, England: Mathematical Association, 1990.

EVELYNE BARBIN
Muriel K. Sperling, translator

# FUNCTIONS

A concept characterized as "the keynote of Western culture" (Schaaf 1930), its history can be traced (Kleiner 1989) from rather vague notions in the precalculus and early calculus era through Euler's view of a function as an analytical expression composed of a (single) variable and numbers or constants, to its various modern generalizations. These include not only functions of several real variables but also mappings defined on any set (the Dirichlet-Bourbaki definition), probability distributions, and elements of abstract functional spaces. As one "climbs" this ladder of abstractions, the nature and then even the very existence of the variable become less and less important. From being a rule prescribing an action or process, a function becomes an object to be acted upon in its own right.

On a level accessible to high school students, this change in the role of function can be illustrated by means of function transformations. For example, using shift, stretch, and flip transformations, any quadratic function can be transformed into any other quadratic function. From this point of view, the (quadratic) functions involved are objects acted upon by transformations. Moreover, the transformations can be viewed in turn as functions whose domain is a set of functions. This provides a unifying view: unification of all quadratic functions on the one hand; a functional view at two different levels on the other hand.

More generally, the importance of the notion of function in mathematics stems from its unifying role. The function concept appears and reappears like a thread throughout school mathematics from grade 1 (e.g., addition as a function from $R \times R$ to $R$) to grade 12 (e.g., calculus) and beyond. In many school curricula it ties algebra, trigonometry, and geometry together. For example, trigonometry is usually first introduced as a means to describe and compute triangles; with the transition to trigonometric functions, trigonometry becomes integrated into a functional view of mathematics as a search for pattern and generalization that grows naturally out of algebra. Another manifestation of their unifying role is the appearance of functions in dynamic geometry problems in which attention is focused on how one geometric quantity varies as a function of another one; such a dynamic view of geometry arises naturally in many software-based geometry curricula.

Moreover, most applications of mathematics in the physical and social sciences use functions; phenomena in domains from physics (motion, waves, electric current, and so on) to economics (price, demand, rate of inflation, and so on) are usually modeled by differential or other functional equations. Day-to-day situations as well as states of quantum mechanical systems are commonly and easily described by functions. The use of functions to generate and describe chaotic systems is but the latest manifestation of this general pattern. Indeed, the emergence of chaotic behavior can easily be demonstrated by repeatedly applying a nonlinear function to a number $x_0$. For example, if one chooses $f(x) = cx(1-x)$ and $0 < x_0 < 1$, the sequence of numbers $x_0, x_1 = f(x_0), x_2 = f(x_1), \ldots,$ $x_k = f(x_{k-1}), \ldots$ converges as long as $c$ is smaller than 3; but as $c$ grows from 3 to 4, the sequence becomes periodic, with ever-increasing periods, and thus shows less and less regularity. Finally, at about $c = 3.57$, any regularity disappears and the sequence cannot be distinguished any more from a random sequence; it thus models the behavior typically observed in natural chaotic systems.

This state of affairs immediately raises a number of questions concerning the teaching of functions, including to what extent the function concept can and should be used as a unifying idea for school mathematics, to what extent such unification should be prepared from the early grades, to what extent applications of the function concept should be prepon-

derant, and to what extent an abstraction should be attempted. Traditionally, functions are introduced in many curricula in grade 9 or 10. Function is presented as a new idea that uses some topics introduced earlier (e.g., algebraic expressions, graphs) but which is quite independent and novel for the students. The teaching of functions serves as a first step to higher mathematics (precalculus, calculus) rather than as unifying what was taught in earlier grades. Often, a rather high degree of abstraction is required from the students in two respects: integration between different representations for functions from $R$ to $R$, and conception of function in terms of the set-theoretic (Dirichlet-Bourbaki) definition. In the remainder of this entry it will be shown that research has highlighted the difficulties that such an approach causes for many students. As a consequence, new curricula are beginning to emerge that do justice to the unifying role of the notion of function.

Since the 1970s, students' learning of the function concept has been a central topic of mathematics education research. Much of this research is reviewed in Harel and Dubinsky (1992) and in Romberg, Fennema, and Carpenter (1993). Some important issues addressed by the research are students' concept images of function (Vinner 1992), the development of a sense for functions (Eisenberg 1992), and students' ability to integrate between different representations of a function (Artigue 1992; Schwarz and Dreyfus in press). Functions have also served as prototypical examples for the investigation of the complexities of learning processes (Moschkovich, Schoenfeld, and Arcavi 1993). Finally, the role of computers in the learning of functions and the concurrent added importance of the graphical representation of functions have been addressed from various points of view (Dugdale 1993; Goldenberg, Lewis, and O'Keefe 1992; Kieran 1993; Schwartz and Yerushalmy 1992).

On the basis of this research as well as the added availability of computers in schools, fundamentally different approaches to teaching functions have been proposed and are being developed and implemented on a trial basis in several countries and in various settings. This development is consistent with a reconceptualization of the entire school mathematics curriculum in terms of strands extending from the early grades through college (Steen 1990). Change is the proposed theme of one such strand (Stewart 1990), and the function concept forms the central focus of this strand. The leading idea is thus the introduction of functional thinking early in the curriculum—some talk about calculus in elementary school (Kaput 1994). This can be done by means of appropriate applied situations and the use of representations, including graphical ones, that are accessible to young children, especially in computer-based learning environments. For example, graphs describing the change of temperature during a day, month, or year may be used to qualitatively introduce functional notions such as dependence, increase, decrease, slope, and local and global extrema in the early grades. The same notions may then be shown to appear in other situations, such as supermarket prices, traffic, water flow in a river, and so on. On the other hand, functions can also be used to describe arithmetic operations (e.g., multiplication, percentages), geometric relationships (how does one edge of a rectangle with fixed area depend on the other edge?), and statistical concepts (how does the average grade of a class change when the teacher raises one student's grade?).

On the basis of such early experience with functions, the function concept can be progressively refined, developed, generalized, and abstracted during the junior and senior high school years. Doing this consistently, curriculum developers may use functions as a unifying thread throughout the school mathematics curriculum.

*See also* Functions of a Real Variable

## SELECTED REFERENCES

Artigue, Michele. "Functions from an Algebraic and Graphic Point of View: Cognitive Difficulties and Teaching Practices." In *The Concept of Function: Aspects of Epistemology and Pedagogy* (pp. 109–132). Guershon Harel and Ed Dubinsky, eds. Washington, DC: Mathematical Association of America, 1992.

Dugdale, Sharon. "Functions and Graphs—Perspectives on Student Thinking." In *Integrating Research on the Graphical Representation of Functions* (pp. 101–130). Thomas A. Romberg, Elizabeth Fennema, and Thomas P. Carpenter, eds. Hillsdale, NJ: Erlbaum, 1993.

Eisenberg, Theodore. "On the Development of a Sense for Functions." In *The Concept of Function: Aspects of Epistemology and Pedagogy* (pp. 153–174). Guershon Harel and Ed Dubinsky, eds. Washington, DC: Mathematical Association of America, 1992.

Goldenberg, Paul, Philip Lewis, and James O'Keefe. "Dynamic Representations and the Development of a Process Understanding of Function." In *The Concept of Function: Aspects of Epistemology and Pedagogy* (pp. 235–260). Guershon Harel and Ed Dubinsky, eds. Washington, DC: Mathematical Association of America, 1992.

Harel, Guershon, and Ed Dubinsky, eds. *The Concept of Function: Aspects of Epistemology and Pedagogy*. Washington, DC: Mathematical Association of America, 1992.

Kaput, James J. "Democratizing Access to Calculus: New Routes to Old Roots." In *Mathematical Thinking and Problem Solving* (pp. 77–156). Alan H. Schoenfeld, ed. Hillsdale, NJ: Erlbaum, 1994.

Kieran, Carolyn. "Functions, Graphing and Technology: Integrating Research on Learning and Instruction." In *Integrating Research on the Graphical Representation of Functions* (pp. 189–238). Thomas A. Romberg, Elizabeth Fennema, and Thomas P. Carpenter, eds. Hillsdale, NJ: Erlbaum, 1993.

Kleiner, Israel. "Evolution of the Function Concept: A Brief Survey." *College Mathematics Journal* 20(1989): 282–300.

Moschkovich, Judith, Alan H. Schoenfeld, and Abraham Arcavi. "Aspects of Understanding: On Multiple Perspectives and Representations of Linear Relations and Connections among Them." In *Integrating Research on the Graphical Representation of Functions* (pp. 69–100). Thomas A. Romberg, Elizabeth Fennema, and Thomas P. Carpenter, eds. Hillsdale, NJ: Erlbaum, 1993.

Romberg, Thomas A., Elizabeth Fennema, and Thomas P. Carpenter, eds. *Integrating Research on the Graphical Representation of Functions*. Hillsdale, NJ: Erlbaum, 1993.

Schaaf, William L. "Mathematics and World History." *Mathematics Teacher* 23(1930):496–503.

Schwartz, Judah, and Michal Yerushalmy. "Getting Students to Function in and with Algebra." In *The Concept of Function: Aspects of Epistemology and Pedagogy* (pp. 261–289). Guershon Harel and Ed Dubinsky, eds. Washington, DC: Mathematical Association of America, 1992.

Schwarz, Baruch, and Tommy Dreyfus. "New Actions upon Old Objects: A New Ontological Perspective on Functions." *Educational Studies in Mathematics*, in press.

Steen, Lynn A., ed. *On the Shoulders of Giants—New Approaches to Numeracy*. Washington, DC: National Academy Press, 1990.

Stewart, Ian. "Change." In *On the Shoulders of Giants—New Approaches to Numeracy* (pp. 183–216). Lynn A. Steen, ed. Washington, DC: National Academy Press, 1990.

Vinner, Shlomo. "The Function Concept as a Prototype for Problems in Mathematics Learning." In *The Concept of Function: Aspects of Epistemology and Pedagogy* (pp. 195–214). Guershon Harel and Ed Dubinsky, eds. Washington, DC: Mathematical Association of America, 1992.

TOMMY DREYFUS

# FUNCTIONS OF A REAL VARIABLE

A real-valued function of a real variable is a correspondence between two sets of real numbers such that to each element of the first set there corresponds one and only one element of the second set. This concept may be traced to the Parisian scholar, Nicole

Oresme (1323?–1382), who eventually became the Bishop of Lisieux. He devised what mathematicians today would call the graph of the velocity of an object moving with constant acceleration. This idea yields a representation of a linear function. It does not seem likely, however, that the extension of this notion to other functions of a real variable occurred to Oresme (Boyer 1989). The work of the French mathematician and philosopher René Descartes (1596–1650) in analytic geometry suggests that he had an intuitive grasp of the notions of a variable and a function of a real variable, although a full-blown presentation of the theory was to wait another 150 years. It was the German mathematician Gottfried Leibniz (1646–1716) who introduced the term *function* into the mathematical vocabulary. His use of the term was primarily restricted to certain kinds of mathematical formulas (Apostol 1967). The basis for the theory of a function of a real variable is attributed to Joseph-Louis Lagrange (1736–1813), the leading French mathematician of the eighteenth century. Lagrange's book *Théorie des Fonctions Analytiques* (1797), regarded as a classic of rigorous mathematics (Boyer 1989), took the approach of expanding functions of a real variable in Taylor series. Many French and German mathematicians of the eighteenth and nineteenth centuries made contributions to the theory of functions, especially the German analyst Karl Weierstrass (1815–1897), who made rigorous the theory of functions begun by Lagrange.

## MATHEMATICAL EXPLANATION

Given two sets $X$ and $Y$, a function from $X$ to $Y$ is a correspondence that associates one and only one element of $Y$ to each element of $X$ (Apostol 1967). The set $X$ is called the *domain* of the function, and the set of elements of $Y$ associated by the function to the elements of $X$ is called the *range* of the function. (It need not be all of $Y$.) If $f$ denotes the function from $X$ to $Y$, it is common to write $f(x)$ for the element $y$ that is associated with $x$ by $f$. If $X = Y = R$, the real line (set of all real numbers), then $f$ is said to be a real-valued function of a real variable. The graph of such a function is the set $\{(x, f(x)) \mid x$ in the domain of $f\}$. It is commonly referred to as a *curve*. If $X = R$ and $Y = R^n$ for some natural number $n$, then $f$ is said to be *vector-valued*. The function $f$ from $R$ into $R^2$ that is defined by the rule $f(t) = (\cos t, \sin t)$ and whose range is a circle is an example of such a function. The variable in $X$ is called a *parameter,* and the graph of $f$, which is now a space curve in $n$-space, $R^n$, is said to be given *parametrically.*

## MOTIVATION AND GOALS

Students of elementary analysis, up to and including calculus, almost always see functions defined by formulas such as $f(x) = x^2$, $g(x, y) = e^y \sin x$. Functions, however, arise more naturally from sets of data or, when data points have been plotted, as graphs. For example, temperature is given as a function of time by readings made at one location over a period of a day or a week. Special functions of a real variable play a vital role in engineering, the physical sciences, and mathematics. Functions in elementary analysis are categorized as algebraic, trigonometric, or exponential functions. These categories offer a wealth of variety to illustrate the properties of functions, an examination of which occupies the bulk of students' time when they first encounter the concept of a function.

Before calculus is taught, a number of properties of functions can be studied. For example, determining the range of a function or whether a function is one-to-one are two such questions. More significant, perhaps, is the notion of periodicity and the way this quality distinguishes trigonometric functions, which are periodic, from algebraic functions, which are not. Another subject of study is the asymptotic behavior of functions. Examples of this can be found in all three of the categories listed earlier. To explain these properties for students who are not intrinsically motivated, it is wise to choose examples with which they can identify or which interest them. Real-world examples—electrocardiograms, the modeling of animal populations, associating the members of a class with their ages or scores on examinations—are frequently the most effective in engaging students' attention and in stimulating their understanding of functional properties.

Software packages for microcomputers and inexpensive calculators with a graphics capability have expanded the intellectual horizons of students of elementary analysis. Exploration of the effect of changes in the formulas that define functions, for example, is now possible at the push of a button or the click of a mouse. Parametrically given curves in 2- or 3-space can also be generated easily. Teachers who avail themselves of the technological aids that are now on the market will need to reconsider what their students are to study about functions and even how they are to pursue their study. In the study of calculus, new issues arise and old issues, like the question about the range of a function, can be reexamined. Among the new issues are a determination of where a function is continuous and where it is differentiable. Both these ideas involve limits: a function $f$ is continuous at a point $a$ in its domain if the limit of $f$ as $x$ approaches $a$ exists and if that limit is equal to $f(a)$. Similarly $f$ is differentiable at a point $a$ in its domain if

$$\lim_{x \to a} \frac{f(x) - f(a)}{x - a}$$

exists.

New properties, like concavity, can be introduced in terms of differentiability properties. Students will certainly be shown that a function that is differentiable at a point is necessarily continuous at that point. They will also see that a function that is continuous at a point need not be differentiable there. They probably need *not* see an example of a function that is continuous on an interval but that fails to have a derivative at any point of that interval. An example of such a function was found by the Czech priest Bernhard Bolzano (1781–1848), but his example went largely unnoticed. (Weierstrass is generally given credit for producing the first example of a continuous, nondifferentiable function.) Students may profit from hearing that the German analyst Georg Bernhard Riemann (1826–1866) produced an example of a function $f$ that is discontinuous at infinitely many points in any interval $[a,b]$. This function is defined as follows: let $d(x)$ denote the difference between $x$ and the nearest integer, with $d(x) = 0$ if $x$ lies halfway between two integers. Then $f$ is given by the formula

$$f(x) = \frac{d(x)}{1} + \frac{d(2x)}{4} + \frac{d(3x)}{9} + \ldots$$

This series converges for all values of $x$. Moreover, the function $F$, defined by the rule $F(x) = \int_a^x f(t)\,dt$, is continuous and nondifferentiable at the points where $f$ is discontinuous. It is recommended, however, that teachers adopt as a principal goal in early courses in analysis the development of their students' intuitive feel for what functions do and what they look like. Examples as pathological as Riemann's are best left for more advanced work.

## PEDAGOGICAL APPROACHES

Most functions arise from data sets, although until the calculus reform movement of the late 1980s, few calculus texts dealt in any significant fashion with this fact. Nonetheless, students' conceptual grasp of functions is greatly aided by work with examples in which a function is given through a data collection (Fraga 1993). Once data points are plotted, a function can be studied graphically. This, too, has the potential to enhance students' understanding of functional behavior. Any detailed study of how graphs are obtained from (finite) data sets, however, is best deferred to a course in numerical analysis. Underlying the remarks above is the assumption that students of elementary analysis courses learn from examples. Population data, to cite one instance, offer the opportunity to work with a real-world problem at the same time as they present a context to ask questions about concavity and extrema. (Given population $P(t)$ as a function of time $t$, when is the population greatest? least? Over what periods of time is the rate of growth of population increasing? decreasing?) In dealing with such situations, students should be encouraged to defend their reasoning in clear, well-constructed sentences. Studying functions that are defined by formulas has undergone a significant change since the appearance of technological aids. Use of technology requires a recasting of mathematical pedagogy, but this same technology offers opportunities as well as challenges. The transition from formula to graph, for example, can now be done by keying in the formula and domain specifications on a computer or a calculator, but this does not mean that the sort of questions that teachers have traditionally posed about that transition need be abandoned. These questions will have to be posed differently, of course. Instead of just asking *where* a function has asymptotes, or what these asymptotes are, teachers may now ask *why* a function has the asymptotic behavior exhibited on their students' monitors and to check how well students can explain their answers. In fact, one result of the technological evolution of the late 1980s and early 1990s is some shifting away from questions that demand only brief, numerical answers to questions whose answers involve persuasive written communication.

## RECOMMENDED MATERIALS FOR INSTRUCTION

In light of the direction that mathematics instruction in American high schools and colleges has taken in the 1990s, it now seems inescapable that the study of functions of a real variable will be done with the assistance of some kind of technology. The selection of what technology to use, not whether to use it at all, is the question teachers need to address. A variety of computer algebra systems that run on microcomputers and graphing calculators are available. All teachers of elementary analysis should require their students to familiarize themselves with and to use suitable technology in their study of functions.

*See also* Calculus, Overview; Functions

## SELECTED REFERENCES

Apostol, Tom M. *Calculus.* Vol. I. Waltham, MA: Blaisdell, 1967.

Bell, Eric Temple. *The Development of Mathematics.* New York: McGraw-Hill, 1940.

Boyer, Carl B. *A History of Mathematics.* 2nd ed. Revised by Uta C. Merzbach. New York: Wiley, 1989.

Fraga, Robert, ed. *Calculus Problems for a New Century.* Washington, DC: Mathematical Association of America, 1993.

Minty, George J., et al. "Functions." In *A Century of Calculus* (pp. 62–91) Pt. 2. Tom M. Apostol et al., eds. Washington, DC: Mathematical Association of America, 1992.

Stewart, James. *Calculus.* Monterey, CA: Brooks/Cole, 1987.

ROBERT J. FRAGA

# FUNDAMENTAL THEOREMS OF MATHEMATICS

Basic, essential, and/or significant theorems. The recognition of some theorems in mathematics as fundamental is reflected in their names, such as the Fundamental Theorem of Arithmetic, the Fundamental Theorem of Algebra, and the Fundamental Theorem of Calculus. Some are fundamental because they are basic, such as Euclid's theorem on the number of primes. Some are fundamental because they are essential components to the development of mathematics, such as the Pythagorean Theorem and the Binomial Theorem. Others are fundamental because of their major significance, such as the Euler Relation and Gödel's Theorem. And, of course, some fit into more than one category.

The Fundamental Theorem of Arithmetic deals with prime numbers. A prime number is an integer greater than 1 that is divisible only by itself and 1. This theorem establishes that every integer greater than 1 is either a prime or is a unique (excluding order) product of primes. Most authorities credit Euclid (ca. fourth century B.C.) for the proof of this theorem (proposition 14, book IX of Euclid's *Elements*). However, Carl Friedrich Gauss (1777–1855) is credited with stating it in its modern form and extending the concept to the field of complex numbers (Kline 1972). A theorem basic to the study of number theory establishes the existence of an infinite number of primes. Proof of this theorem is found in Euclid's *Elements* (Thomas 1956).

The Fundamental Theorem of Algebra deals with roots (solutions) of polynomial equations. It states that every polynomial equation of degree equal to or greater than 1 with coefficients from the field of complex numbers has at least one root (solution) that is a complex number. This conjecture was first made by Albert Girard (1595–1632). Several mathematicians, including René Descartes and Jean le Rond d'Alembert, attempted to prove it; but Gauss was the first to produce a rigorous deductive proof. It appeared in his dissertation for a doctor's degree from the University of Helmstedt in 1799 (awarded in absentia) (Bell 1956).

Calculus, developed essentially by Isaac Newton (1643–1727) and Gottfried Leibniz (1646–1716), independently and simultaneously, dealt with rates of change of variables (differential calculus) and calculation of areas (integral calculus). While Newton calculated areas by antidifferentiation and Leibniz used sums of rectangles to approximate areas, both recognized the inverse relationship between differentiation and integration. However, neither they nor any other seventeenth-century mathematician was able to express formally nor derive deductively this relationship. By defining the definite integral and using the mean value theorem as a critical element in his argument, Augustin Louis Cauchy (1789–1857) was able to establish deductively this relation, called the Fundamental Theorem of Calculus (Kline 1972). The mean value theorem assures that, in an interval for a continuous function, there is a point where the instantaneous rate of change (derivative) is equal to the average rate of change of the function in that interval.

The modern interpretation of the familiar Pythagorean Theorem is a relationship between numbers representing measures of the sides and hypotenuse of right triangles. This relationship was known well before the time of Pythagoras. The early Greek interpretation, however, is reflected in Euclid's statement that the sum of the squares constructed upon the sides equals the square constructed upon the hypotenuse; the proof is based on areas of these squares (Thomas 1956).

The Binomial Theorem, which is useful in many and diverse areas of mathematics and science, deals with the expansion of a binomial to a power. The binomial expansion for positive integral powers was known to the Arabs in the thirteenth century (Kline 1972). The array of binomial coefficients known as Pascal's triangle and used by Pascal to determine binomial coefficients was known to the Persian poet and mathematician Omar Khayyam (ca. 1050–1130) and the ancient Chinese mathematicians. James Gregory (1638–1675) and Newton, independently, formulated the Binomial Theorem, and Newton extended it to negative and fractional powers (a very significant extension). Gauss was the first to prove the theorem rigorously about a century later. The Euler relation, $e^{ix} = \cos x + i \sin x$, established by Leonhard Euler (1707–1783), is important because of its great unifying nature. It shows how seemingly diverse concepts—exponential functions, imaginary numbers, and trigonometry—are related; and it does so in a single short statement.

Kurt Gödel's theorem, proved in 1931, is significant because of the doubt it raised regarding the deductive process, which had become the linchpin of formal mathematics. Consistency of the set of axioms of a system, meaning they cannot lead deductively to contradictory statements, was considered essential if the conclusions were to be irrefutable. Gödel proved that it is impossible to establish consistency within an axiomatic system.

*See also* Binomial Theorem; Calculus, Overview; Gödel; Number Theory, Overview; Proof; Pythagorean Theorem

## SELECTED REFERENCES

Bell, Eric T. "The Prince of Mathematicians." In *The World of Mathematics* (pp. 295–339). James R. Newman, ed. New York: Simon and Schuster, 1956.

Kline, Morris. *Mathematical Thought from Ancient to Modern Times.* New York: Oxford University Press, 1972.

Thomas, Ivor. "Greek Mathematics." In *The World of Mathematics* (pp. 189–209). James R. Newman, ed. New York: Simon and Schuster, 1956.

HARALD M. NESS

# G

## GAGNÉ, ROBERT MILLS (1916– )

Educational psychologist at Florida State University since 1969. In the 1960s, he advocated a learning theory that knowledge must be organized hierarchically and that subordinate knowledge must be learned before higher principles. Thus, Skinnerian stimulus-response learning through reinforcement preceded later concept, principle, and problem-solving learning. Gagné's theory strongly influenced texts and research at that time. His theory led to instruction in which tasks are arranged according to the knowledge structure hierarchy. The learner works through a program of tasks arranged in a hierarchy which may be "linear" or "branching" (where each learner is directed to different parts of the program according to his or her responses). Gagné's theory of structure led to greater care in sequencing instruction. Presenting mathematics in this way was sometimes called *guided* learning, a way of learning in opposition to the less-structured approach of discovery learning, also popular at that time. Gagné's book *The Conditions of Learning* (1965) was common reading in mathematics education courses.

*See also* Computer-Assisted Instruction (CAI); History of Mathematics Education in the United States, Overview; Task Analysis

### SELECTED REFERENCES

Aichele, Douglas B., and Robert E. Reys. *Readings in Secondary School Mathematics.* 2nd ed. Boston, MA: Prindle, Weber, and Schmidt, 1977.

Gagné, Robert M. *The Conditions of Learning.* New York: Holt, Rinehart and Winston, 1965.

Jones, Phillip S., ed. *A History of Mathematics Education in the United States and Canada.* 32nd Yearbook. Washington, DC: National Council of Teachers of Mathematics, 1970.

JAMES K. BIDWELL
ROBERT G. CLASON

## GALILEI, GALILEO (1564–1642)

Often called the "Father of the Scientific Method." He rejected any speculations about the physical universe that could not be checked by observation and experience. To use a truly "scientific method," he taught, one must experiment, think about the results, and then try to come up with a law or principle, performing more experiments to test it. Galileo was a fine musician and excellent painter and impressed his early teachers with his literary ability and mechanical ingenuity.

Born near Pisa, Galileo taught at the Universities of Pisa and Padua. He created considerable controversy by dropping objects from high towers to disprove Aristotle's theories on falling bodies. He discovered the laws of the pendulum and was the first to show that the path of a projectile in a vacuum is a parabola. Galileo also helped to develop the telescope and, with it, discovered Jupiter's moons.

Galileo believed, like Copernicus, that the sun—not the earth—was the center of the universe. This opinion cost Galileo dearly: he was excommunicated from the church he loved and forced to recant. Under house arrest, he became blind and died unappreciated by most of the scientific world.

## SELECTED REFERENCES

Baumgart, John K., et al. *Historical Topics for the Mathematics Classroom.* 2nd ed. Reston, VA: National Council of Teachers of Mathematics, 1989.

Bell, Eric Temple. *Men of Mathematics.* New York: Simon and Schuster, 1965.

Burton, David M. *The History of Mathematics.* Boston, MA: Allyn and Bacon, 1985.

Eves, Howard W. *An Introduction to the History of Mathematics.* 6th ed. New York: Saunders College Publishing, 1990.

Kramer, Edna E. *The Nature and Growth of Modern Mathematics.* Princeton, NJ: Princeton University Press, 1981.

LUETTA REIMER
WILBERT REIMER

# GALOIS, EVARISTE (1811–1832)

Created seminal body of work in the theory of equations whose richness and insights were only appreciated in the latter part of the nineteenth century, long after his death. Born at Bourg-la-Reine, near Paris, France, Galois died twenty-one years later from wounds suffered in a duel at Paris. Few mathematicians have been more romanticized. Unfortunately, Galois' genius was largely unrecognized within the mathematical community, due to a combination of bad luck and his own contentious personality. Still, he established a definitive answer to the problem of precisely when polynomial equations have solutions that can be obtained by arithmetic formulas analogous to the familiar quadratic formula. Simultaneously, he introduced many concepts that today are fundamental in the study of modern algebra whereby basic number systems such as the rational numbers are generalized and analyzed. Some of these structures are finite in size and lend themselves to computational and computer applications.

*See also* Abstract Algebra

## SELECTED REFERENCES

Burton, David. *History of Mathematics: An Introduction.* 3rd ed. Dubuque IA: Brown, 1995.

Eves, Howard W. *An Introduction to the History of Mathematics.* 6th ed. Philadelphia: Saunders College Publishing, 1990.

Gillispie, Charles, ed. *Dictionary of Scientific Biography* (pp. 259–265). Vol. V. New York: Scribner's, 1981–1990.

RICHARD M. DAVITT

# GAME THEORY

Mathematically based strategies for determining optimal behavior in competitive situations, such as games. A *strategy* is used when one determines to act in a particular manner under specified circumstances. Game theory has also contributed to the knowledge of how to apply various other mathematical theories (particularly algebraic and topological methodologies) to the social sciences. The forging of game theory as a discipline is credited to John von Neumann (1903–1957). He introduced the concept in a paper discussing optimal behavior in situations involving conflict of interest (von Neumann 1928).

## MATRIX GAMES

Game theory is used in the analysis of conflict situations, such as games (say chess or cards), business (for example, labor and management at a bargaining table), war (opposing armies on a battlefield). The theory analyzes strategies available to opposing sides. The outcome of games of chess or of cards, for example, depends on more than mere chance. It requires certain strategies in anticipation of what the opponent will do.

Consider the following simple game played by two players, A and B. Each player simultaneously shows either 1 or 2 fingers. If the sum of the number of fingers is 2, player A gets 4 points from player B; if the sum is 4, player B gets 4 points from player A. If the sum is odd, no points are exchanged. Notice that the "payoffs" are such that a win for one player results in a corresponding loss for the other player. Such a situation is called a *zero-sum game*. This game can be summarized by the $2 \times 2$ *payoff matrix:*

|  |  | **Player B** | |
|---|---|---|---|
|  |  | 1 | 2 |
| **Player A** | 1 | 4 | 0 |
|  | 2 | 0 | −4 |

The payoff matrix summarizes the payoffs for each pair of strategies, for each player. The rows summarize the alternatives for Player A, and the columns summarize the alternatives for Player B. Each element of the matrix represents a payoff amount to Player A. Positive payoff implies transfer of that amount from Player B to Player A. A negative payoff implies transfer from Player A to Player B.

To develop a strategy (a determination to act in a particular manner), examine the payoff matrix. Player A's best strategy is to show 1 finger, since (looking at the first row) the worst possible outcome is breaking even, Player B's best strategy is to show 2 fingers, since (looking at the second column) the worst possible outcome for Player B is to break even. When players play their best strategies, the amount of the payoff is called the *value of the game*. A game is said to be *fair* if its value is 0.

Note that for teaching purposes, one generally begins, as above, with a *matrix game*, in which the entire mathematical content of the game is described by its *payoff matrix G*, with entries $g_{ijk}$ . . ., known as *payoffs*. The dimension of the matrix is given by the number of players.

## Number of Players

*1. Two-person Games*. Pedagogical presentation generally begins with a *two-person* game, involving exactly two players, as illustrated above. Pure strategies for players R and C are given by matrix rows and columns, respectively. A *positive payoff $g_{ij}$* implies a transfer of $g_{ij}$ units from Player C to Player R. A *negative payoff* implies a transfer from player R to C. The *solution* of a matrix game is a pair of optimum strategies for players R and C and a *value v* for the game.

*2. N-person Games*. Games with more than two (*n*) players are *n-person* games. They involve additional possibilities:

*Cooperative games:* Formation of *coalitions* of players is possible and permitted (some groups agree to cooperate to the advantage of coalition members and to the consequent disadvantage of nonmembers).

*Noncooperative games:* Formation of coalitions is either not possible or not permitted.

Other concepts may be expanded. An example: each player may have his/her own payoff matrix, each describing positive or negative payments from an independent "bank." A *bimatrix* or *multimatrix* (for 2 or *n* players, respectively) describes all strategies. Each matrix entry is an ordered *n*-tuple, each member of which represents the payoff to a particular player.

## Zero-sum and Nonzero-sum Games

In a *zero-sum* game,

- gains by the winner(s) imply loss to the loser(s) (the sum of the winnings is zero),
- the payoff matrix is visible to all players,
- players may not communicate.

In a *nonzero-sum* game, the sum of player winnings is nonzero for at least one play.

## Solution

Matrix games are solved using *linear programming* techniques, requiring knowledge of the simplex algorithm and the Duality Theorem of Linear Programming for sufficient treatment of the topic.

## GAMES OF CHANCE

Game theory also considers analyses in which part or all of the play is determined by chance. A particular game may have no chance moves (as in chess); chance may play a contributory role (as in bridge); or the outcome may be entirely chance-dependent (as in roulette). Additionally, there exists the possibility of incomplete knowledge by player(s) of the current position of the game. Games of chance may be zero-sum or nonzero-sum, cooperative or noncooperative.

## Solution

When chance is a factor, a foundation of knowledge in probability theory is necessary for meaningful instruction. If the pedagogical presentation is to include proofs, some knowledge of the calculus is generally necessary, as is familiarity with properties of convex sets and functions.

*See also* Games; Linear Programming

## SELECTED REFERENCES

Adelson-Velskii, Georgii M. *Algorithms for Games*. New York: Springer-Verlag, 1988.

Bacharach, Michael. *Economics and the Theory of Games*. Boulder, CO: Westview, 1977.

Casson, Mark. *The Economics of Business Culture: Game Theory, Transaction Costs, and Economic Performance*. New York: Oxford University Press, 1991.

Dresher, M., Albert W. Tucker, and Philip Wolfe, eds. *Contributions to the Theory of Games III*. Princeton, NJ: Princeton University Press, 1957.

Driessen, Theo. *Cooperative Games, Solutions, and Applications*. Boston, MA: Kluwer, 1988.

Koopmans, Tjalling C., ed. *Activity Analysis of Production and Allocation.* New York: Wiley, 1951.

Kuhn, Harold W., and Albert W. Tucker, eds. *Contributions to the Theory of Games I.* Princeton, NJ: Princeton University Press, 1950.

———. *Contributions to the Theory of Games II.* Princeton, NJ: Princeton University Press, 1953.

Luce, Robert Duncan, and Howard Raiffa. *Games and Decisions.* New York: Wiley, 1957.

Maital, Shlomo. *Economic Games People Play.* New York: Basic Books, 1984.

Tucker, Albert W., and Robert D. Luce, eds. *Contributions to the Theory of Games IV.* Princeton, NJ: Princeton University Press, 1959.

Von Neumann, John. "Zur Theorie der Gesellschaftsspiele." *Mathematische Annalen* 100(1928): 295–320.

———, and Oskar Morgenstern. *Theory of Games and Economic Behavior.* Princeton, NJ: Princeton University Press, 1944.

RINA YARMISH

# GAMES

The distinguishing feature of a mathematical game, in contrast to other topics from mathematics, is competition. Generally, there are two opposing players who take turns making moves according to some set of rules, and any gain for one player is a loss for the other. (Usually "player" refers to just one person, but in some games the "player" may be a team of more than one. In bridge, for example, each team of two partners is considered one "player.") Some games have a theoretical mathematical basis for the best way to play. Several such games, like tic-tac-toe, are trivially simple. Even beginning players quickly discover the best way to play, and the outcome is not in doubt. The best strategy may involve random choice. Stone-scissors-paper is a game in which on the count of three, each player extends a hand, with either a fist (stone) or two extended fingers (scissors) or an open palm (paper). Stone beats scissors; scissors beats paper, and paper beats stones. Each player should choose moves at random, with the three choices occurring equally often in the long run. The outcome of any one play then depends on chance, but the average result over a large number of games is that each player wins about half the time. In other games, like tic-tac-toe, there is not even the temporary uncertainty and these cases have little or no mathematical or educational interest.

Another group of games is the "hard" ones for which there is no mathematical theory. Chess, backgammon, and Go are examples. Although supposedly a general mathematical "theory of games" identifies the best move (including random choices, with known probability weights) in every possible situation, in most practical cases it turns out to be too complicated and computationally top-heavy. Mathematical game theory was invented before 1930 by John von Neumann; it is beyond the scope of most secondary school students.

The focus here is on the unfortunately rather small number of games for which there is some theory that specifies the best way to play but it is not trivial or obvious. Each of these cases gives rise to its own specific mathematical features. Educationally, many of these are within the reach of students, whose interest in the mathematics can be stimulated by the example of application in the game. A few of these are Number War, Nim, Morra, Bridge-It, and SIM.

Number War comes in many versions; for example, "Count to 50 by 7s." In the first move, a player announces any positive integer not greater than 7. The opponent then likewise chooses such an integer, adds it, and announces the new total. A player might start with 5; the opponent could then choose 3 and announce 8 (i.e., 5 + 3). Play continues with each player in turn choosing a positive integer not greater than 7 and adding it on. The winner is the first player to reach or pass 50. A sample game might go like this:

|  | First Player | Opponent | |
|---|---|---|---|
| Start | 5 | 8 | |
| | 15 | 16 | |
| | 21 | 24 | |
| | 29 | 33 | |
| | 37 | 42 | |
| | 49 | 50 | Opponent wins |

On first acquaintance with this game, even a mathematically naïve player will notice that a winning strategy is to be able to announce 42. Then the other player cannot reach 50 and cannot prevent the first player from reaching 50 on the next turn. In effect, the game is then reduced from counting to 50 to counting instead to 42. And in a few more plays, it will be discovered that not 42 but instead 34 should really be the goal, since announcing 34 assures that the player can reach 42 next time, and thus win in two more turns. But from 34, you can back up again to 26, then to 18, then 10, and finally 2. So if the starting player chooses 2, he or she can be sure to win. When both players know the theory, the outcome depends only on who goes first. In practice, with an opponent who has not yet figured out the set of winners, a player can generally wait until late to grab a winner, perhaps 34 or even 42.

This game illustrates some general properties, known as Sprague-Grundy rules in honor of the mathematicians who first developed them. Beasley (1989) has a good presentation. The numbers in the set {2, 10, 18, 26, 34, 42, 50} are called "winners." All others are "losers." There are three critical properties that determine the set of winners:

(1) The final goal is a winner.
(2) From any winner, any permitted choice must lead to a loser.
(3) From any loser, there is at least one permitted choice that leads to a winner.

Sprague-Grundy theory states that when the positions of a game can be separated into winners and losers, the player who first captures a winning position (e.g., with the first move) can always win the game. Finding the set of winners for any particular game is considered a significant achievement.

Other versions of Number War might be counting to 100 by 8s or to 75 by 3s. In those and similar cases, if the maximum increase permitted is (say) $n$, the winners will be the goal minus multiples of $n + 1$. It is a game in which elementary school students can get to practice all the operations of arithmetic.

Another famous game that falls in the class of Sprague-Grundy games is Nim. It is a well-known early example of a "solved" game. The theory dates from the early 1800s. The educational aspect of Nim is that its set of winners is described by use of binary notation. Nim is played using some large number of identical pieces, like pennies or pebbles. At the start of the game, the supply of pieces is divided up into a number of piles (usually three or four). Each pile may start with any number of pieces. A sample starting position with four piles is 33, 12, 18, 5. Players in turn select one of the piles and remove some or all of its pieces. (Pieces removed from a pile are simply put aside and have no further effect on play or scoring.) If in this case the third pile is selected, the player could remove any number of pieces up to and including 18. Say the player decided to remove 14 pieces, leaving the new position 33, 12, 4, 5 for the opponent. The opponent then does the same. The total number of pieces in play, initially 68 (i.e., 33 + 12 + 18 + 5), decreases with each move and must eventually become 0. The winner is the player who makes the last move. A complete game might go like this:

| Start 33 12 18 5 | Position after first player's move | Position after opponent's move |
|---|---|---|
| | 33 12 4 5 | 8 12 4 5 |
| | 4 12 4 5 | 4 12 4 (0) |
| | 4 3 4 | 1 3 4 |
| | 1 3 2 | 1 2 2 |
| | (0) 2 2 | 2 (0) |
| All 0's, player wins | | |

To determine if any position is a "winner," write the binary representation of the number of pieces in each pile. In the example this is

$$33 = 1\ 0\ 0\ 0\ 0\ 1$$
$$12 = 0\ 0\ 1\ 1\ 0\ 0$$
$$18 = 0\ 1\ 0\ 0\ 1\ 0$$
$$5 = 0\ 0\ 0\ 1\ 0\ 1$$

With the binary representations arrayed this way in columns, scan each column to see if the number of "1" bits in the column is even—yes (y) or no (n). If the array has y's in all the columns, then the position is a winner; otherwise, a loser. (The position here is a loser. The rightmost column has two "1" bits and so is a "y." But the second rightmost column has only one "1," so it is a "n." And some of the other columns are also "n," with an odd number of "1" bits, but even one such column is enough to make the position a loser.) To prove that this rule characterizes the set of winners, it is necessary to show Sprague-Grundy properties 1., 2., and 3. While that proof is not easy, with a teacher's help a high school student should be able to do it.

There are several other games in which the theory is known. SIM is played using the vertices of a hexagon and two differently colored pencils. Blue starts by drawing a line between any two vertices. Then Green draws another. The objective is to *avoid* completing a triangle all of one color on the hexagon vertices and forcing the opponent to do so. The game is so simple in appearance that it is surprising that it takes a computer to find the winning line to play. (Green, the second player, can always force a win.) Schwartz (1980) has a collection of articles on SIM showing how a mathematical theory can evolve. Most are accessible to school-level readers.

A rather similar game is Bridge-It. It was invented by David Gale. It is played on a pair of interlaced lattices, as shown in Figure 1. A player's move is to join two adjacent points of his or her own color, with the final objective being to connect his or her

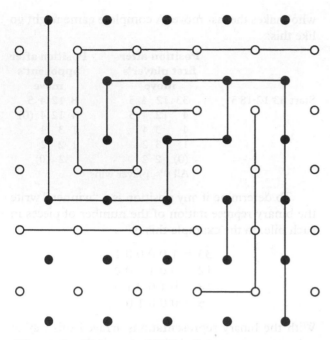

**Figure 1** *The game of Bridge-It.*

own two sides of the board. There cannot be a tie; if a player cannot get across the board, it must be because the other player has completely blocked the way, so that player must have a win. The first player can always win but the winning strategy is hard to find. However, it is easy to show by contradiction that there must be a win for the first player (Beasley 1989, 142). Temporarily assume that the second player had a winning set of moves. Then the first player could make his or her initial move at random and then switch to the winning scheme available for playing second. The extra line does the first player no harm, so this procedure must assure that he or she wins. And that contradicts the assumption. This argument illustrates that in games, as often elsewhere in mathematics, we can prove rigorously that something exists (like a winning strategy) without being able to produce it. This argument, called "strategy stealing," is credited to John Nash, the Nobel laureate.

For games in which random moves occur, like stone-scissors-paper, the Sprague-Grundy theory does not apply. In such games, the best move will not lead to a winning position, but rather to a new position that has some randomness which over a large number of trials will lead to the highest probability of winning for the player. Since the opponent is also trying to win, the best result may be 50/50 for a win, or some other mix. One game like that is an old Italian game called Morra. The rules are: On a signal

like a nod of the head, both players put up either 1 finger or 2. If the choices are the same, you win 2¢. If you choose 1 and your opponent chooses 2, you pay the opponent 3¢. And if you choose 2 and the opponent chooses 1, you pay 1¢. The game can be summarized in a payoff table.

*Payoff table for Morra*

| Your move | Opponent's move | |
|---|---|---|
| | 1 | 2 |
| 1 | +2 | −3 |
| 2 | −1 | +2 |

With Morra, it turns out that you should randomly choose 1 finger with probability 3/8, and 2 fingers with probability 5/8. With that strategy mix, you are sure to win, on the average, 1¢ in every eight times the game is played (more if your opponent blunders), even though some of those plays will be cases in which you lose. This strategy certainly is not intuitively obvious. It can be calculated using the methods of von Neumann.

Another game with the random strategy feature is called the three-coin game. Each player begins with three pennies or other coins, and initially chooses, unknown to the other, to put 0, 1, 2, or all 3 in his closed right hand. Then the first player tries to guess the total number of coins in both hands. He guesses a number between 0 and 6 inclusive. The second player, having heard the guess, makes his own best guess, which must be a different number. With both guesses announced, the players open their hands, and inspect to see if either of them has gotten the total number of coins right. The player who does so wins; if neither of them is a winner, they play again.

The winning line of play is for each player to select at random, with equal probability for each option, the initial number of coins put in the hand. Then the first player must guess "3" for the sum; any other choice gives the second player a useful clue. The second player may guess any other number at least as large as the number of coins held. (Obviously, to guess "0" or "6" would be ridiculous if he held 2 coins, but 2, 4, or 5 are all equally good strategies for that case.) When both players make these optimal choices, each wins half the time. The proof of that result in Schwartz (1959) uses no advanced mathematics, but it is tightly reasoned and will stretch most readers. However, as an example of the kind of result that has been found, the game still has educational value.

One variation on a game is to reverse the rule defining who is the winner. This is called the *misere* version. For instance, misere Nim makes the player who is forced to take the last piece (because the opponent has left only one) the loser. The mathematical impact of switching from a game to its misere version may range from trivial to profound. Some games have known solutions when their misere versions do not, and vice versa. In Number War, Counting to 50 by 7s, it is easy to see that to be a winner in the misere battle, you need to announce 49. Then the opponent cannot avoid going on to 50. So the only change is that, in effect, the game has become Counting to 49 by 7s, and the winning first move is to announce 1. In Nim, the change is much more complicated. Misere Nim involves the idea of a *singleton pile,* which is a pile with just one remaining piece. Until all piles but one are reduced to singletons, the winning strategy at misere Nim is the same as straight Nim, using the binary representations. Then there is a change. And for tic-tac-toe, the misere version is much more interesting and challenging than the straight version, although it still turns out that properly played, every game will still end in a draw with neither player forced to complete a line of three. Silverman (1973) has an extended and entertaining discussion of several misere cases.

*See also* Enrichment, Overview; Game Theory

## SELECTED REFERENCES

Beasley, John D. *The Mathematics of Games.* Oxford, England: Oxford University Press, 1989.

Berlekamp, Elwyn R., John Horton Conway, and Richard K. Guy. *Winning Ways for Your Mathematical Plays.* New York: Academic, 1982.

Falkener, Edward. *Games Ancient and Oriental and How to Play Them.* New York: Dover, 1961. Reprint of 1892 edition.

Holt, Michael. *Math Puzzles and Games.* New York: Dorset, 1992.

Sackson, Sid. *The Book of Classic Board Games.* Palo Alto, CA: Klutz, 1991.

Schwartz, Benjamin L. *Mathematical Solitaires and Games.* Farmingdale, NY: Baywood, 1980.

———. "Solution of a Set of Games." *American Mathematical Monthly* 66(8)(Oct. 1959):693–701.

Silverman, David L. *Your Move.* London, England: Kaye and Wood 1973.

Von Neumann, John, and Oskar Morgenstern. *The Theory of Games and Economic Behavior.* 3rd ed. Princeton, NJ: Princeton University Press, 1953.

BENJAMIN L. SCHWARTZ

# GAUSS, CARL FRIEDRICH (1777–1855)

Made groundbreaking contributions in all areas of mathematics to which he turned: algebra, analysis (both real and complex), geometry (differential and non-Euclidean), number theory, probability, and statistics. He was the Prince of Mathematicians to his contemporaries and is, by universal acknowledgment, one of the three greatest mathematicians of all time (the other two are Archimedes and Isaac Newton); see Bell (1937).

Born in Germany, Gauss was a precocious child and joked later in life that he could count before he could talk. In 1792 he entered the Collegium Carolinum, studying classical languages and, on his own, the works of Isaac Newton, Leonhard Euler, and Joseph-Louis Lagrange. In 1795, when he enrolled at the University of Göttingen, he was still undecided about which of his two intellectual loves—philology or mathematics—he would pursue as a career. He opted for mathematics the following year, when he showed that the regular polygon of seventeen sides is constructible with straightedge and compass. This was not just a personal triumph; it was the first discovery of a constructible regular polygon in over 2,000 years (the ancient Greeks knew how to construct regular polygons of three, four, five, and fifteen sides). Another early first was Gauss's proof in 1799 of the Fundamental Theorem of Algebra, which had eluded Jean d'Alembert, Euler, and Lagrange. This earned him a Ph.D. degree from the University of Helmstedt.

Number theory, the queen of mathematics according to Gauss, was his first and greatest mathematical love. The *Disquisitiones Arithmeticae,* arguably his greatest work, was completed in 1798 (when he was twenty-one!) but not published until 1801 (Gauss 1986). In the seventeenth and eighteenth centuries, number theory consisted of a collection of isolated, though brilliant, results. In the *Disquisitiones* Gauss systematized the subject, solved a number of its central and difficult problems, and pointed directions for future researchers.

The *Disquisitiones* begins with the definition of congruence (another first). This offers an excellent example of the power of a felicitous notation, here used for the familiar idea of divisibility. A major achievement is the proof of one of the central theorems of number theory: the *quadratic reciprocity law,* already conjectured by Euler and Adrien-Marie Legendre and rediscovered by Gauss at age fifteen. This describes the relationship between the solvability of

$x^2 \equiv p \pmod{q}$ and $x^2 \equiv q \pmod{p}$ for (odd) primes $p$ and $q$. Another fundamental accomplishment is the beautiful but intricate theory of *binary quadratic forms,* which investigates the representation of integers by forms $f(x, y) = ax^2 + bxy + cy^2$. (Pierre de Fermat's problem of the representation of integers as sums of two squares, $n = x^2 + y^2$, is a very special case.)

The final section of the *Disquisitiones,* a beautiful blend of algebra, geometry, and number theory, deals with *cyclotomy*—the division of a circle into $n$ equal parts. Algebraically, it asks for the solution of $x^n - 1 = 0$. Using number-theoretic ideas, Gauss shows that this so-called *cyclotomic equation* is *solvable by radicals* for every positive integer $n$; that is, its solution reduces to a sequence of rational operations and extractions of roots (cf. the quadratic formula). This was a most important result in the program, initiated by Lagrange in 1770 and brought to fruition by Evariste Galois about 1830, of determining which polynomial equations are solvable by radicals. An important "by-product" of Gauss's results on cyclotomy was the characterization of regular polygons constructible with straightedge and compass: a regular $n$-gon is constructible if and only if $n = 2^k p_1 p_2 \ldots p_s$, where the $p_i$ are *distinct* primes of the form $2^{2^t} + 1$ (Bell 1937; Bühler 1981; Dunnington 1955). Little wonder the *Disquisitiones* made Gauss an instant celebrity. Its wealth and profundity of ideas are still being mined.

Students of astronomy and physics salute Gauss as one of their own. In 1807, for economic reasons, he accepted the directorship of the Göttingen Observatory, a position he held for the rest of his life. He thenceforth made important contributions to both the theoretical and observational aspects of astronomy and to various branches of physics, including mechanics, optics, acoustics, and geomagnetism. But he always sought the mathematical connection, and in one instance in particular his efforts bore exceptional fruit. In 1820 Gauss was asked by the kingdom of Hanover (to which Göttingen belonged) to supervise a geodesic survey, which lasted several years. A major task was the precise measurement of large triangles on the earth's surface. The stimulus (presumably) provided to Gauss's fertile mind gave birth in 1827 to his famous paper on curved surfaces, in which he formulated the fundamental notion of (Gaussian) *curvature* and founded the study of the *intrinsic geometry* of curved surfaces. Bernhard Riemann built on these ideas in the 1850s to found the theory of $n$-dimensional manifolds, which later proved indispensable in Albert Einstein's general theory of relativity (Bühler 1981; Dunnington 1955).

Related to Gauss's astronomical work, particularly his calculation of orbits of asteroids and comets, were achievements in probability and statistics. In 1809, in the paper "Theory of Motion of Heavenly Bodies," he invented the *method of least squares* (independently found by Legendre) for obtaining the "best fit" to a series of experimental observations. In this connection, he devised what came to be known as *Gaussian elimination* for the solution of a system of linear equations. In the same work Gauss also showed that the distribution of errors when using the least-squares method is "normal." This is the source of the *Gaussian (normal) distribution,* represented graphically by a bell-shaped curve (Bühler 1981; Hall 1970).

Another groundbreaking idea appeared in Gauss's 1831 paper on *biquadratic reciprocity,* which investigates the relation between the solvability of $x^4 \equiv p \pmod{q}$ and $x^4 \equiv q \pmod{p}$. Just to *state* the law of biquadratic reciprocity Gauss found that he had to introduce what came to be known as the *Gaussian integers,* defined as $G = \{a + bi : a, b$ real integers$\}$. He carefully analyzed the arithmetical structure of $G$, showing that its elements can be written uniquely as products of "primes" (of $G$). Here was an important contribution to the founding of a new subject, *algebraic number theory.* This was also the paper in which Gauss defined the complex numbers as points in the plane, making them at long last respectable (Stewart 1977).

Ideas no less profound and far-reaching than those present in Gauss's published works were found in his mathematical diary (Gray 1984). This is a remarkable nineteen-page document of 146 brief (often cryptic) entries dealing with discoveries covering the years from 1796 to 1814. The diary became public only in 1898. The first entry, dated March 30, 1796, notes Gauss's discovery of the constructibility of the regular 17-gon: "The principles upon which the division of the circle depend, and geometrical divisibility of the same into seventeen parts, etc." (Gray 1984). The publication of almost any of the 146 entries would have made mere mortals famous. But some of the entries also anticipated major creations of nineteenth-century mathematics: complex analysis, elliptic function theory, and non-Euclidean geometry. Had Gauss published these in his lifetime, it has been speculated, the development of mathematics would have been advanced by half a century. (For speculation about why Gauss did *not* publish the discoveries listed in his diary, see Bühler 1981; Stewart 1977.)

The nineteenth century witnessed fundamental transformations in mathematics, among them a growing insistence on rigor. Gauss was a leading exponent of this emerging spirit, which began to permeate all areas of mathematics. For example, in his important 1812 work on the hypergeometric series, he was the first to insist on a rigorous treatment of the convergence of series (Bühler 1981; Dunnington 1955). "It is demanded of a proof that all doubt become impossible," he wrote to a friend. And he practiced what he preached. His proofs were elegant and polished, often to the point where all traces of his method of discovery were removed. "He is like the fox, who erases his tracks in the sand with his tail," deplored Niels Henrik Abel (Stewart 1977).

The finished product of Gauss's researches gives, in particular, no indication of his great skill in, and love of, calculation. Some of his deepest theorems in number theory were inspired by calculation. For example, he conjectured the *Prime Number Theorem,* namely that $\pi(x)$ *is asymptotic to* $\frac{x}{\log x}$ where $\pi(x)$ is the number of primes $\leq x$, by first putting together a table of all primes up to 3,000,000. It was likely a striking (and possibly unique) combination of remarkable insight, formidable computing ability, and great logical power that produced a mathematician whose ideas are still bearing rich fruit today.

*See also* History of Mathematics, Overview; Least Squares; Number Theory, Overview; Prime Numbers

### SELECTED REFERENCES

Bell, Eric Temple. *Men of Mathematics.* New York: Simon and Schuster, 1937.

Bühler, Walter Kaufmann. *Gauss: A Biographical Study.* New York: Springer-Verlag, 1981.

Dunnington, G. Waldo. *Carl Friedrich Gauss: Titan of Science.* New York: Hafner, 1955.

Gauss, Carl Friedrich. *Disquisitiones Arithmeticae.* Arthur A. Clark, trans. New York: Springer-Verlag, 1986.

Gray, Jeremy. "A Commentary on Gauss's Mathematical Diary, 1796–1814, with an English Translation." *Expositiones Mathematicae* 2(1984):97–130.

Hall, Tord. *Carl Friedrich Gauss: A Biography.* Cambridge, MA: MIT Press, 1970.

May, Kenneth O. "Gauss, Carl Friedrich." In *Dictionary of Scientific Biography* (pp. 298–315). Vol. 5. Charles C. Gillispie, ed. New York: Scribner's, 1981.

Stewart, Ian. "Gauss." *Scientific American* 237(July 1977):122–131.

ISRAEL KLEINER

# GEOMETRY, INSTRUCTION

Has the goal of teaching students about spatial objects, relationships, and transformations, along with the language and axiomatic mathematical systems that have been constructed to discuss and represent them. Students develop both their spatial and logical reasoning. Traditionally, secondary school geometry has been axiomatic in nature, and elementary school geometry has emphasized measurement and informal development of those basic concepts needed in high school (although in recent years, more informal approaches to high school geometry have become increasingly popular). Analytic geometry is usually studied at the beginning of the calculus sequence. Although mathematics majors in universities may study non-Euclidean geometries, the term "school geometry" usually refers to Euclidean geometry, despite the fact that there are numerous approaches to the study of the topic (e.g., synthetic, analytic, transformational, and vector). Students also encounter geometric ideas in discussions of other areas of mathematics, such as finding solutions to equations or systems of equations or using vectors for velocity and force problems.

## STUDENT PERFORMANCE

According to extensive evaluations of mathematics learning, students in the United States are failing to learn basic geometric concepts and geometric problem solving (Carpenter et al. 1980; Kouba et al. 1988). Only about half of all high school students enroll in a geometry course, and only about 30% of those enrolled in a course for which proof is a goal exhibit any understanding of the meaning of proof (Senk 1985; Suydam 1985). The primary explanation given for this poor performance is the traditional curriculum, in terms of both what and how topics are treated. Standard elementary and middle school curricula focus on a hodgepodge of low-level skills such as recognizing and naming geometric shapes, writing proper symbolism for simple geometric concepts, using measurement and construction tools, and applying formulas (Porter 1989). Furthermore, traditional school geometry curricula have not been designed to properly engender systematic progression to higher levels of geometric thought (as described in the next section on learning). In addition, teachers often do not teach even the impoverished geometry curriculum that is available to them.

## LEARNING

Van Hiele's theory describes how geometric thought develops through several levels of sophistication under the influence of a school curriculum

(Clements and Battista 1992). Such thought progresses from reasoning about shapes based on their appearance; to establishing the properties of shapes by experimenting, measuring, drawing, and model-making; to reasoning logically, forming abstract definitions, and classifying figures hierarchically; to reasoning formally within axiomatic systems and constructing proofs; and finally to reasoning formally about axiomatic systems rather than just within them, analyzing and comparing these systems. Student thinking follows the Piagetian progression from being nonreflective and unsystematic, to empirical, and finally to deductive. Both the van Hiele and Piagetian theories take physical and mental action as the source from which geometric knowledge arises. These theories suggest that students learn geometry best when they participate in curricula that accommodate these differing levels of thinking and when they are actively involved in solving geometric problems. In particular, students need plenty of visual, intuitive experiences with shapes before they begin to view those shapes logically in terms of properties.

## Spatial Reasoning

Spatial reasoning involves constructing and manipulating mental representations for spatial objects. According to Harris (1981), the U.S. Employment Service estimates that most technical-scientific occupations require persons having high ability to reason spatially. Furthermore, numerous mathematicians and mathematics educators have suggested that spatial reasoning plays a vital role in mathematical and scientific thought (Battista 1994; Wheatley 1990). It is conjectured that such reasoning is used to represent and manipulate information in learning and problem solving. Thus, it has often been suggested that spatial activities be included in the mathematics curriculum not only because spatial reasoning is important in its own right but because it is related to meaningful mathematical thinking (Wheatley 1990).

## Use of Computers

Because of their graphic capabilities, computers are uniquely suited for studying geometric ideas. Appropriately designed software permits students to actively manipulate and explore geometric objects in a dynamic environment unmatched by other media (Clements and Battista 1994). For example, certain computer environments are quite effective in helping to develop the ability to reflect on the properties of classes of geometric objects and to think in a more general and abstract manner about those objects. The path perspective taken in *Logo* software builds on the action-based intuitions students have constructed in moving about in the physical world and can facilitate their progress to higher levels of thinking about shapes (Clements and Battista 1992). Construction programs such as the *Geometer's Sketchpad* help students to make and test conjectures and can be used by teachers to promote a more inquiry-based approach to geometry.

*See also* Logo; Software; Van Hiele Levels

## SELECTED REFERENCES

Battista, Michael T. "On Greeno's Environmental/Model View of Conceptual Domains: A Spatial/Geometric Perspective." *Journal for Research in Mathematics Education* 25(Jan. 1994):86–94.

Carpenter, Thomas P., Mary K. Corbitt, Henry S. Kepner, Mary M. Lindquist, and Robert E. Reys. "National Assessment." In *Mathematics Education Research: Implications for the 80s* (pp. 22–38). Elizabeth Fennema, ed. Alexandria, VA: Association for Supervision and Curriculum Development, 1980.

Clements, Douglas H., and Michael T. Battista. "Computer Environments for Learning Geometry." *Journal of Educational Computing Research* 10(2)(1994): 173–197.

———. "Geometry and Spatial Reasoning." In *Handbook of Research on Mathematics Teaching* (pp. 420–464). Douglas A. Grouws, ed. New York: Macmillan, 1992.

Harris, Lauren J. "Sex-related Variations in Spatial Skill." In *Spatial Representation and Behavior Across the Life Span: Theory and Application* (pp. 83–125). Lynn S. Liben, Arthur H. Patterson, and Nora Newcombe, eds. New York: Academic 1981.

Kouba, Vicky L., Catherine A. Brown, Thomas P. Carpenter, Mary M. Lindquist, Edward A. Silver, and Jane O. Swafford. "Results of the Fourth NAEP Assessment of Mathematics: Measurement, Geometry, Data Interpretation, Attitudes, and Other Topics." *Arithmetic Teacher* 35(May 1988):10–16.

National Council of Teachers of Mathematics. *Curriculum and Evaluation Standards for School Mathematics.* Reston, VA: The Council, 1989.

———. *Principles and Standards for School Mathematics.* Reston, VA: The Council, 2000.

Porter, Andrew. "A Curriculum Out of Balance: The Case of Elementary School Mathematics." *Educational Researcher* 18(1989):9–15.

Senk, Sharon L. "How Well Do Students Write Geometry Proofs?" *Mathematics Teacher* 78(1985): 448–456.

Suydam, Marilyn N. "The Shape of Instruction in Geometry: Some Highlights from Research." *Mathematics Teacher* 78(1985):481–486.

Wheatley, Grayson W. "Spatial Sense and Mathematics Learning." *Arithmetic Teacher* 37(6)(1990):10–11.

MICHAEL T. BATTISTA
DOUGLAS H. CLEMENTS

# GERMAIN, SOPHIE (1776–1831)

Growing up in Paris during the French revolution, Sophie Germain occupied herself with learning mathematics from the books in her father's library. Not permitted to enroll in the *École Polytechnique* because she was female, Germain secured lecture notes and submitted papers under the name of LeBlanc, a student there. Joseph Louis Lagrange came to know of her work under that name, and it was as LeBlanc that she originally corresponded with Carl Friedrich Gauss. In 1809, at the behest of Napoleon, a *prix extraordinaire* was established for the mathematical analysis of the vibration of elastic surfaces. In the first competition the prize was not awarded; in the second, Germain received an honorable mention. Finally, in 1816 Germain won the prize, although there were still some deficiencies in her exposition. Work on the curvature of surfaces, including a concept of mean curvature, the other area of mathematics in which Germain published, arose out of her elasticity studies.

Inroads, substantial for the time, were made by Germain on the solution to Fermat's Last Theorem: the equation $x^n + y^n = z^n$ has no solution in integers if $n > 2$. She was able to prove that there is no solution if $n$ is a prime less than 100 and neither $x, y$, nor $z$ is divisible by $n$.

Germain never held a formal position and led a relatively reclusive life. For a period of time she corresponded productively with Gauss; at times, Adrien-Marie Legendre, Jean Baptiste Fourier, and others were helpful. She suffered, however, from gaps in her basic knowledge and from lack of opportunity for prolonged interchange with other mathematicians. Even those disposed to be helpful appeared not to take her efforts seriously. She is perhaps best remembered for her later philosophical work, which fitted with Auguste Comte's development of positivism. Germain died of cancer in 1831 as France again entered into a revolutionary period.

*See also* Women and Mathematics, History

## SELECTED REFERENCES

Bucciarelli, Louis L., and Nancy Dworsky. *Sophie Germain: An Essay in the History of the Theory of Elasticity.* Boston, MA: Reidel, 1980.

Gray, Mary W. "Sophie Germain." In *Women of Mathematics* (pp. 47–56). Louise S. Grinstein and Paul J. Campbell, eds. New York: Greenwood, 1987.

Libri, Guillaume. "Notice sur Mlle. Sophie Germain." In *Considérations sur l'État des Sciences et des Lettres*, Sophie Germain, and Jacques-Armand Lherbette, ed. Paris, France: Lachevardi é re, 1833.

Stupuy, H., ed. *Sophie Germain, Ouevres Philosophiques.* New ed. Paris, France: Firmin-Didot, 1896.

MARY W. GRAY

# GERMANY

Some of the current characteristics of the German school system today can be traced back through the history of Germany, both the history of the political organization of the country as well its social structures. Since Germany, from the Middle Ages on, was not strongly united in a central state but rather a conglomeration of independent states, cultural affairs and educational policy are still not a matter of the federal administration but of the sixteen states (*Laender*) into which Germany is divided today. Therefore, major differences appear among the educational systems of the several states in various respects, such as the traditions of the schools, the organization of the syllabi and the curricula, the organization of teacher education, and so on. Schools originally developed, on the one hand, from the clerical system, which provided an academic education, and, on the other hand, from the various needs of trade, crafts, industry, and so forth for a vocationally oriented education. This distinction is still in a certain sense reflected in the essentially three-division system that is generally a main characteristic of the German school system despite the differences among the states.

School starts with primary school (*Grundschule*), consisting of grades 1 to 4 (sometimes to grade 6) for all children aged six to ten (or twelve). Then the school system splits into three parts, as a result of choices made by the students and their parents: *Hauptschule* (grades 5–9 or 10), *Realschule* (grades 5 or 7–10), and *Gymnasium* (grades 5–12 or 13). Today, the number of students in these different schools is nearly equally distributed, which is a result of a development of the past thirty years when *Gymnasium* increased its share from 5%–10% to 30%. To a smaller extent, less then 10%, there are also comprehensive schools (*Gesamtschule*, grades 5–13) which are intended, based on political goals of equal opportunity, to avoid the separation of the students in different schools at such an early age.

*Hauptschule* (or "major school") was formerly attended by the majority of children. It provides a basic education that normally is completed by two or three years of vocational training. *Realschule* prepares children for the more advanced employment as technicians, administrators, medical assistants, and so on. *Gymnasium* offers, as the final examination, the *Abitur*, which is the (almost only) entrance qualification to university studies. Mathematics lessons are compulsory in all school types, except in the last year of *Gymnasium*. The number of lessons in mathematics varies from three to six a week.

## THE CONTENT OF SCHOOL MATHEMATICS

Although the curriculum at different types of schools and in different states may differ somewhat, the main topics of school mathematics are generally as follows:

| Grades | Content |
|---|---|
| 1–2 | Basic numeracy with numbers 1–100; basic geometric experiences. |
| 3–4 | Place value and numeration up to 1,000,000, including the written forms of the algorithms of the operations of whole numbers; basic ideas of measurement: length, weight, time, money; simple geometric properties of solids and shapes. |
| 5 | Because the students are not equally prepared by the different primary schools they attended before, there occurs at this point mainly a recapitulation, homogenization, and gradual extension of the contents of primary school. |
| 6 | Factorization of whole numbers, primes; meaning of the system of whole numbers and the concepts of representation and operations of common and decimal fractions; preparatory phase of geometry. |
| 7–8 | Negative numbers, both integers and rationals; proportionality and percentages; linear equations and representations of lines in the plane; geometric constructions, transformations, and congruence; area of triangles and circles. |
| 9–10 | Real numbers; square roots and quadratic equations; rational exponents and their properties (not in *Hauptschule*); equations and functions; geometric calculations, like area and volume, theorem of Pythagoras; concept of similarity; trigonometry and logarithms (in *Realschule* and *Gymnasium*). |
| 11–13 | Selected topics from analysis, linear algebra, and stochastics, among them differentiation and integration of functions, ideas of probability, lines, planes, and vectors in two- and three-dimensional geometry. To a greater extent than in lower grades, teachers, on their own responsibility, may choose various other topics, such as complex numbers, mathematical modeling, advanced combinatorics, etc. |

## MATHEMATICS PEDAGOGY AS A SCIENTIFIC DISCIPLINE

There are at present about 150 professors of mathematics education in German universities, nearly all of whom are charged with teacher preparation. These faculty members have developed the scientific field of "didactics of mathematics." Some of the professional organizations and activities in this field of mathematics education include: (a) a special scientific society, the *Gesellschaft fuer Didaktik der Mathematik (GDM)*, whose members also include the mathematics educators from Austria, Switzerland, and abroad; (b) a tradition since 1967 of annual meetings, the *Tagungen fuer Didaktik der Mathematik;* (c) a considerable number of scientific journals devoted to mathematics education, such as *Journal fuer Mathematik-Didaktik, Der Mathematikunterricht, Praxis der Mathematik, Mathematik in der Schule;* (d) a special journal for international documentation and reviewing, the *Zentralblatt fuer Didaktik der Mathematik;* (e) a research institute on mathematics education, the *Institut fuer Didaktik der Mathematik der Universitat Bielefeld (IDM);* and (f) the possibility to earn at most universities a doctoral degree (Ph.D.) and the *venia legendi (Habilitation)*, which is a special qualification to allow a person to teach didactics of mathematics at a university.

A relatively broad spectrum of research on issues of mathematics education is undertaken. This ranges from self-reflections about the discipline itself (Wittmann 1995) to a variety of concrete proposals for classroom use. Perhaps one can best characterize the "paradigm" of the present research in mathematics education in Germany as follows: one tries to understand mathematics learning and teaching as a human activity, in all its aspects. Research runs from observations of classroom interactions (Bauersfeld 1988) to epistemological considerations of the status of mathematical knowledge in school (Steiner 1987); from the investigation of the mathematical abilities of individual students (Lorenz 1994) to the content analysis of several subjects; from application (Blum 1991) to pure mathematics; and from curriculum studies and development (Keitel 1987) to the study of the impact of the computer. Diversity seems to be a characteristic in present mathematics education studies in Germany. See Schupp et al. (1992).

*See also* International Comparisons, Overview; International Studies of Mathematics Education; Third International Mathematics and Science Study (TIMMS)

## SELECTED REFERENCES

Bauersfeld, Heinrich. "Interaction, Construction and Knowledge—Alternative Perspectives for Mathematics Education." In *Effective Mathematics Teaching* (pp. 27–46). Thomas Cooney and Douglas A. Grouws, eds. Reston, VA: National Council of Teachers of Mathematics, 1988.

Blum, Werner. "Applications and Modelling in Mathematics Teaching—A Review of Arguments and Instructional Aspects." In *Teaching of Mathematical Modelling and Applications* (pp. 10–29). Mogen Niss et al., eds. Chichester, England: Horwood, 1991.

Keitel, Christine. "What Are the Goals of Mathematics for All?" *Journal of Curriculum Studies* 19(1987): 393–407.

Lorenz, J. H. "Mathematically Retarded and Gifted Students." In *Didactics of Mathematics as a Scientific Discipline* (pp. 291–302). Rolf Biehler et al., eds. Boston, MA: Kluwer, 1994.

Schupp, H., et al., eds. "Mathematics Education in Germany." *Zentralblatt fuer Didaktik der Mathematik* 24(1992). Special Issue.

Steiner, Hans G. "Philosophical and Epistemological Aspects of Mathematics and Their Interaction with Theory and Practice in Mathematics Education." *For the Learning of Mathematics* 7(1)(1987):7–13.

Wittmann, Erich C. "Mathematics Education as a 'Design Science'." *Educational Studies in Mathematics* 29(1995):355–374.

MICHAEL NEUBRAND

# GIFTED

"All students deserve the opportunity to achieve their full potential; talented and gifted students in mathematics deserve no less. It is a fundamental responsibility of all school districts to identify mathematically talented and gifted students and to design and implement programs that meet their needs" (House 1987, 100). Rarely do educators encounter students whose mathematical ability meets or exceeds their own in precollege education. Students of this ability often need intensive personalized attention to ensure that their unique talents are fostered. In most cases, regular classroom mathematics does not fulfill their insatiable appetite for more mathematics, particularly since most school mathematics curricula are so repetitive. This entry provides a starting place for work with students who are gifted learners of mathematics. First, identifying characteristics of these students and the psychological theories dealing with their extraordinary abilities will be explored. Then a variety of programs for such students will be presented, followed by a discussion of possible social effects. Finally, a list of some available instructional materials for the gifted will be provided.

## CHARACTERISTICS OF THE MATHEMATICALLY GIFTED

In identifying highly gifted students, there are usually two competing schools of thought: one school differentiates between innate and acquired ability; the other between creative ability (ability to create mathematics) and school ability (ability to master assigned mathematics). These differentiations are not necessarily mutually exclusive; overlap is sometimes apparent in highly gifted mathematics students. In order to look at these schools of thought, we must have an idea of what constitutes mathematical ability. Krutetskii (1976) defines the ability to learn mathematics as "individual psychological characteristics (primarily characteristics of mental activity) that answer the requirements of school mathematical activity and that influence, all other conditions being equal, success in the creative mastery of mathematics as a school subject—in particular, a relatively rapid, easy, and thorough mastery of knowledge, skills, and habits in mathematics" (74–75).

Another idea of what constitutes mathematical ability can be seen in teachers' most frequent statements about what characteristics they believe constitute high mathematical ability (note: a student need not display all of the characteristics in order to be considered of high ability): "Rapid mastery of mathematics; logical and independent thinking; resourcefulness; rapid and stable memory (memory here refers to the ability to store generalized relationships and structures. High ability students tend to have selective memory (they store only generalizations, not specific instances).); high level of generalization, abstraction, analysis, and synthesis: can generalize from very small sets of examples; reduced fatigue during lessons: this may be due to students' partiality toward certain activities at which they do well; flexible thought patterns/rapidly switch direction of thought (High ability students switch directions of thought and methods of solution many times in order to glean information about the problem at hand.); use of visualization; good spatial conceptualization; economy of reasoning ability and mental powers (High ability students often curtail their reasoning process (literally skipping explicit steps in logic as though they were looking at the solution as a whole instead of linearly-connected logical steps).); simple, elegant solutions (High ability students often strive for the 'best' (in some sense) solution to a problem.)" (Krutetskii 1976).

Highly talented mathematics students often benefit from what is termed "inspiration" when solving a problem. It is termed thus because the solver is unable to explain how he or she came upon the solution. This could be due to two of the abovementioned characteristics: the ability to generalize easily, which could cause unconscious generalizations, or curtailed reasoning processes, which would make it difficult for the solver to retrace his or her logical steps. Gifted students generalize problem situations to a certain kind of problem. For a simple example, students might generalize a given real-world problem to a general quadratic equation and then solve it using the quadratic formula. They then sort information from the problem into one of three types: information essential to the kind of problem situation; information essential in the current situation, but not essential to all problems of this kind; and superfluous information. Once the students sort the information, the solution is nearly immediate because they had already generalized. The difficulties (if there are any) for the high-ability students come during the generalization phase. These students encounter far fewer difficulties than average students because they generalize rapidly and broadly; generalize methods of solution; and notice small variations between similar problems (Krutetskii 1976).

Not all components of high mathematical ability form at the same time. The first stage is generalizing,

followed by curtailing of the reasoning process, generalization and relationship memory, and striving for simple, elegant solutions. Students look at the world through the lens of logic, interpreting all incoming data through this filter with three different types of minds. The first type of mind is the analytic type, which tends to be very abstract; the second is geometric, which tends to visualize problems pictorially; and the third is the harmonic type, which mixes parts of both the analytic and the geometric (Krutetskii 1976).

Davidson and Sternberg (1984) promote the view that exceptional mathematical insight is the key to high mathematical ability. Speed in doing mathematics is important, but it is subordinate to insight. One can be exceptionally insightful and not speedy and be highly able, but one cannot be considered highly able when speed is the only element present. Three characteristics of insight are believed to be used by highly able mathematics students: (1) selective encoding: this involves sifting out relevant information from a problem situation; (2) selective combination: here the student usefully synthesizes the relevant information; (3) selective comparison: the student compares the information synthesis to other similar problems. Davidson and Sternberg conducted studies on fourth-, fifth-, and sixth-grade gifted and normal students. The results indicated that all three characteristics of insight indeed seemed to be of importance, as they significantly aided in differentiating gifted from normal students.

Other observations of general characteristics of highly mathematically gifted students include mature and highly motivated; more interested in relationships and structure than irrelevant detail in problems; preference for mathematics and science to other school subjects; and having a higher probability of further education.

## PSYCHOLOGICAL THEORIES OF GIFTED DEVELOPMENT

Mathematical ability stages are very similar to Jean Piaget's stages of cognitive development. Mathematical reasoning goes through a critical development level at about the same time as Piaget said students would pass from a concrete operational to a formal operational stage. High-ability students reach this at age ten, while average students do so at age twelve (Kirschenbaum 1992). Students move from the concrete operational stage to the formal operational stage at different ages, and each student at different ages for different subject areas.

Stanley, Lupkowski, and Assouline (1990) caution against giving every gifted elementary student algebra and other high school topics on which to work because they may not have yet reached the Piagetian formal operational thinking needed for such study. Some elementary students, however, may be ready for abstract work, but specific testing is needed to be sure. Identification of gifted students should include testing of mathematical, verbal, mechanical, spatial, and nonverbal skills as well as IQ level. The Scholastic Aptitude Test (mathematical and verbal) and the Stanford-Binet IQ test are two of the most often-used tests. Of these, mathematical testing should carry the most weight, but other scores could signal latent aptitude.

Testing should be only one way of determining which students are highly gifted. Other methods have also been suggested and implemented. Brown (1984) states that "by demonstrating creative/productive behavior, we will do away with a traditional mode of identifying gifted children eligible for enrichment services which favored proficient lesson learners and test-takers at the expense of persons who may score somewhat lower on tests but who more than compensate for such scores by having high levels of task commitment and creativity" (126).

## PROGRAMMING FOR THE MATHEMATICALLY GIFTED

The National Council of Teachers of Mathematics (NCTM) recommends that vertical acceleration be used sparingly for mathematically gifted creative students, especially when interests and attitudes clearly indicate that they will succeed just as well with a carefully designed sequential curriculum. Records show that many accelerated students discontinue the study of or experience extreme difficulty with mathematics before graduation from high school. For most gifted students (with the exception of some highly gifted mathematics students), a strong, expanded program emphasizing mathematics enrichment is preferable (House 1987, 99–100). For the extremely gifted students, acceleration must be carefully designed and monitored on an individualized basis. Mentors would be essential in these extreme cases.

### Acceleration versus Enrichment

The most important difference in programming approaches for highly gifted mathematics students is that between acceleration and enrichment. Acceleration involves providing higher grade-level mathematics at a younger age, while enrichment stresses a deeper understanding of current material as well as a broad introduction to mathematics that may not be included in higher grade-level courses. Experts argue whether programming for highly gifted mathematics students should include enrichment in areas outside their expertise and outside school subjects. Many feel that enrichment should naturally take place in the student's area of interest. The essential question is whether gifted students should expand their horizons to other areas or concentrate on the fields in which they are most gifted or most interested. Enrichment is not, however, viewed as the only choice, but just the preferred one. Wolfle (1986) advocates a mixture of enrichment and acceleration activities in a gifted program.

Benbow, Perkins, and Stanley (1983) found that young highly gifted participants in advanced placement programs showed no gaps in mathematical knowledge and tested at or above the scores of nonparticipants on mathematics achievement tests. In fact, participants had a greater interest in mathematics and attended more selective colleges.

### General Aspects of Programs

Many different opinions exist about what should be included in gifted mathematics programs. The NCTM gives the following position statement on "Provisions for Mathematically Talented and Gifted Students" in House (1987): "These students need enriched and expanded curricula that emphasize higher-order thinking skills, nontraditional topics, and application of skills and concepts in a variety of contexts. All mathematically talented and gifted students should be enrolled in a program that provides a broad and enriched view of mathematics in a context of higher expectation" (100).

Experts recommend in-depth learning of mathematical topics, high expectation levels, and continuity in programs, explaining that gifted students learn not only at a different rate from average students but also in different ways, thus necessitating novel types of instruction. Highly talented students need to learn how to learn on their own, to assess themselves, and to become independent thinkers. Most young gifted mathematics students can reason very well but need to acquire knowledge. Many do not score as highly as they could on some tests, not because they cannot solve problems, but because they lack the background knowledge necessary for the situation.

Group work among gifted mathematics students, in a program separated from regular school, is an important part of any program (Wolfle 1986). These students can work together to become more creative and develop divergent thinking. Students can be asked to form their own groups for an activity or the teacher can create his or her own prepared groups. Open-ended problems that foster mathematical discovery and invention can then be worked on and discussed within each group and between groups. That is not to say, however, that individual work is not important. Solo work promotes self-sufficiency in a student's pursuit of knowledge and understanding.

Important aspects of a gifted mathematics program should steadily increase achievement while fostering deep, intellectual satisfaction. This can be accomplished by offering mathematics, at an appropriate mental level, on which the students enjoy working. In some cases, however, this may cause the student to outgrow his or her future high school courses, and some provision for further study must be made when that happens.

Technology is an important part of programming for highly gifted mathematics students. Grandgenett (1991) describes three roles of technology: as a tutor to the student (through computer-aided instruction (CAI)), as a tool for the student (as an aid for mathematical calculations and exploration), and as a tutee for the student (as the computer or calculator is taught or directed by the student). Technology should be used with gifted students for information processing and interactive learning, planning (for projects and presentations), communications (among fellow students and the mentor or teacher), and problem solving (both for calculations and for exploration). Students can program computers and/or calculators, use prepared computer-assisted instruction programs, communicate through an electronic mail system, and explore mathematics with the aid of mathematical computer packages.

## Mentorships

Many different mentor arrangements are possible. It is important to find a quality mentor and to establish a good working relationship between the mentor and the gifted student. Mentors are often mathematics teachers, engineers, professors, or mathematics education students. Mentorship instruction should be individualized and fit the interests of both the student and the mentor. The mentor relationship should include the following elements: student commitment (to the relationship and to working); student self-assessment; a conceptual contract (what is to be

learned during the mentorship); a work plan (how it will be learned); and a final product.

Furthermore, in order to be more effective, the mentorship should have the following objectives: to make raising the student's level of expertise a goal; develop the student's ability to question; improve interpersonal and communication skills of the student; connect the mathematics to the larger world in which the student lives; set time-limit priorities on projects; and provide psychosocial support. Whereas the ideal tutor would be one who can take on both the role of academic teacher and the role of social developer, it is not always possible. Thus, it is advisable to use a double mentoring program that involves one mentor to work on academic material and another mentor to act as a socializing advisor.

## Examples of Specific Programs

Programs available for the gifted include the Julian C. Stanley Mentor Program (JCSMP), which is modeled after the Study of Mathematically Precocious Youth (SMPY) and is administered by the SMPY at the University of North Texas, and the Investigation of Mathematically Advanced Elementary Students (IMAES), also modeled after SMPY and run by the Connie Belin National Center for Gifted Education. The Texas Academy of Mathematics and Science at the University of North Texas is a program in which students take classes while finishing their final two years of high school and are given college standing as juniors upon entering many undergraduate institutions. Stanley, Lupkowski, and Assouline (1990) suggest mathematics clubs, competitions, and summer programs (such as Arnold Ross's program at the Ohio State University). National mathematics contests are also available, such as the American Mathematics Contests (AMC), the American Invitational Mathematics Examination (AIME), the USA Mathematics Olympiad (USAMO), the International Mathematics Olympiad (IMO), and the Mathematical Olympiads for Elementary Schools (MOES). There are also pullout programs, where students are allowed to leave their regular classes for periods of time to work on individual projects.

## PARENTAL AND SCHOOL COOPERATION

Parents also must take responsibility. They need to make contacts with the schools, develop a working relationship with appropriate personnel, and help to set up programming for their children. Parents must monitor their child's progress on a regular basis. Changes are often needed in programs; gifted students are often able to judge the appropriateness of their program very early on and parents must know to seek feedback. Educators should recognize parents as a source of expert information regarding their child. Because funding for intensive programming is difficult to get (either from the government or from private agencies), parents must be urged to learn about the legislation that governs their children's schooling.

Another important part of parental and school cooperation involves the reservations that parents have about their child's academic and social well-being when being accelerated. Parents need to be counseled about the intellectual and emotional advantages and disadvantages of the child's programming and be assured that they will be a part of the planning for their child. Parent meetings are a very important part of the programming.

## SOCIAL EFFECTS AND EMOTIONAL CHARACTERISTICS

There has been concern over possible negative social effects of various types of gifted programs. Extracurricular opportunities should be provided for gifted children to interact with agemates. Weiss, Haier, and Keating (1974) reported that mathematically gifted students are not socially maladjusted, but are instead solid and competent. They are more mature, responsible, dependable, and morally upright than average students and are usually characterized as independent, quick, sharp-witted, foresighted, versatile, and intelligent.

The possibility does exist, however, of some detrimental social effects for which to be prepared. Friendships involving highly gifted students are an important social factor to consider. A substantial number of gifted students have a positive outlook on their friendships, but many report that they have fewer friends than they would like and that they feel that being smart made it more difficult to form friendships. This is not necessarily a serious problem, but it does call for special sensitivity and guidance involving parents, educators, and sometimes psychologists (Janos, Marwood, and Robinson 1985).

Also, some highly gifted students feel heightened feelings of alienation and are hypersensitive; a student may experience problems in separating his or her identity from a parent's. These students have exaggerated views of adults' expectations, experience problems differentiating between their roles and those of their agemates, and use perfectionism to batter their own self-concepts. This tension may be caused by the uneven development between the intelligence (adultlike) and the emotions (childlike) in a highly gifted student. This gap often causes unrealistic goals and then emotional breakdowns when these goals are not achieved. Gifted students suffer from "inappropriate environments" (Roedell 1984, 129). These students often become bored when in their age-group classrooms, and this boredom is usually only partially alleviated by many gifted programs. Many of the pullout gifted programs (programs where students are "pulled out" of their regular classrooms) are only part-time and may not even be directed at the gifted student's talent (e.g., a mathematically gifted student may be asked to work on a project in the arts).

Because of superior information processing and understanding of what supreme achievements are, some gifted students consequently develop ambitions and self-concepts that are unattainable. This can lead either to accomplishing nothing or to attempts to achieve unattainable goals. In either case, low self-concept is the result. All of these pressures can cause an immense amount of stress on such a student. Retaining a family psychologist or counselor to help alleviate or even prevent some of the common problems is advisable where and when appropriate and economically feasible.

Much of the research about the social and emotional well-being of highly gifted students is directed at the vulnerabilities of these students. Scholwinski and Reynolds (1985) found that high-IQ children scored significantly lower on all of the anxiety scales of the Revised Children's Manifest Anxiety Scale (RCMAS) than average students. Overall, the highest anxiety category for high-IQ children was related to excess worry and sensitivity. Results of this study should help dispel the stereotype that high-IQ children are particularly anxious and unstable.

*See also* Competitions; Gifted, Special Programs; Techniques and Materials for Enrichment Learning and Teaching; Technology

## SELECTED REFERENCES

Benbow, Camilla P., Susan Perkins, and Julian C. Stanley. "Mathematics Taught at a Fast Pace: A Longitudinal Evaluation of SMPY's First Class." In *Academic Precocity: Aspects of Its Development* (pp. 51–78). Camilla P. Benbow and Julian C. Stanley, eds. Baltimore, MD: Johns Hopkins University Press, 1983.

Brown, Martha M. "Highly Gifted Students: The Needs and Potential of the Highly Gifted: Toward a Model of Responsiveness." *Roeper Review* 6(3)(1984):123–127.

Davidson, Janet E., and Robert J. Sternberg. "The Role of Insight in Intellectual Giftedness." *Gifted Child Quarterly* 28(2)(1984):58–64.

Grandgenett, Neal. "Roles of Computer Technology in the Mathematics Education of the Gifted." *The Gifted Child Today* 14(1)(1991):18–23.

Hirsch, Christian R., and Robert A. Laing, eds. *Activities for Active Learning and Teaching: Selections from the Mathematics Teacher.* Reston, VA: National Council of Teachers of Mathematics, 1987.

House, Peggy A., ed. *Providing Opportunities for the Mathematically Gifted, K–12.* Reston, VA: National Council of Teachers of Mathematics, 1987.

Janos, Paul M., Kristi A. Marwood, and Nancy M. Robinson. "Friendship Patterns in Highly Intelligent Children." *Roeper Review* 8(1)(1985):46–49.

Kirschenbaum, Robert J. "An Interview with Julian C. Stanley." *The Gifted Child Today* 15(4)(1992):42–45.

Krutetskii, Vadim A. *The Psychology of Mathematical Abilities in School Children.* Joan Teller, trans. Chicago, IL: University of Chicago Press, 1976. (Original work published 1968).

Nelson, Doyal, and Joan Worth. *How to Choose and Create Good Problems for Primary Children.* Reston, VA: National Council of Teachers of Mathematics, 1983.

Pohl, Victoria. *How to Enrich Geometry Using String Designs.* Reston, VA: National Council of Teachers of Mathematics, 1986.

Reys, Barbara. *Elementary School Mathematics: What Parents Should Know About Problem Solving.* Reston, VA: National Council of Teachers of Mathematics, 1982.

Roedell, Wendy C. "Vulnerabilities of Highly Gifted Children." *Roeper Review* 6(3)(1984):127–130.

Sachs, Leroy, ed. *Projects to Enrich School Mathematics: Level 2 and Level 3.* Reston, VA: National Council of Teachers of Mathematics, 1988.

Schaaf, William L. *The High School Mathematics Library.* Reston, VA: National Council of Teachers of Mathematics, 1987.

Scholwinski, Ed, and Cecil R. Reynolds. "Dimensions of Anxiety among High IQ Children." *Gifted Child Quarterly* 29(3)(1985):125–130.

Sobel, Max A., ed. *Readings for Enrichment in Secondary School Mathematics.* Reston, VA: National Council of Teachers of Mathematics, 1988.

Stanley, Julian C., Ann E. Lupkowski, and Susan G. Assouline. "Eight Considerations for Mathematically Talented Youth." *The Gifted Child Today* 13(2)(1990):2–4.

Stevenson, Frederick. *Exploratory Problems in Mathematics.* Reston, VA: National Council of Teachers of Mathematics, 1992.

Thornton, Carol A., and Nancy S. Bley, eds. *Windows of Opportunity: Mathematics for Students With Special Needs.* Reston, VA: National Council of Teachers of Mathematics, 1994.

Trowell, Judith M., ed. *Projects to Enrich School Mathematics: Level 1.* Reston, VA: National Council of Teachers of Mathematics, 1990.

Weiss, Daniel S., Richard J. Haier, and Daniel P. Keating. "Personality Characteristics of Mathematically Precocious Boys." In *Mathematical Talent: Discovery, Description, and Development* (pp. 126–139). Julian C. Stanley, Daniel P. Keating, and Lynn H. Fox, eds. Baltimore, MD: Johns Hopkins University Press, 1974.

Wolfle, Jane A. "Enriching the Mathematics Program for Middle School Students." *Roeper Review* 9(2)(1986): 81–85.

PATRICIA A. BROSNAN
JAMES A. FITZSIMMONS

# GIFTED, SPECIAL PROGRAMS

The investigation of the needs of students of high ability, in mathematics as well as in other areas, has its origins in cognitive psychology and so is a phenomenon largely of the twentieth century. The early studies of Francis Galton (1869) and Lewis M. Terman (1925, 1937) concerned high achievement in general and its possible hereditary component in particular ("hereditary genius"). The general trend of investigation since then has been toward more practical items: specific questions of identification and interventions on behalf of students who achieve a high level of understanding or creativity early and need more stimulation or guidance. There has been a more or less continuous tradition since Terman's time of interest in young people gifted in mathematics, but the most intense activity in the field began with the launching of the Soviet satellite *Sputnik*, dubbed a "scientific Pearl Harbor" by the first director of the National Science Foundation (NSF) (England 1982, 237). Since then, a rich array of programs has developed to serve the needs of high-ability students in a variety of ways.

## CENTRAL RESOURCES

Julian Stanley is among the pioneers in this movement. In 1971, to apply twenty years of work with high-ability students, he founded the Study of Mathematically Precocious Youth (SMPY) at Johns Hopkins University. Starting with students who were extremely precocious, Stanley and his co-workers explored (and continue to explore) problems of identification, the structure of intelligence, developmental questions, and types of pedagogical intervention. Work focused on the Scholastic Aptitude Test as a tool for identification (on the pre-high school level) and on acceleration, rather than enrichment, as a tool for intervention. Investigation also stressed individual counseling and tutoring when appropriate, as well as articulation with the educational system wherever possible. In 1991, the name of the program was changed to the Study of Exceptional Talent (SET) and was expanded to include work with students with high verbal as well as mathematical ability. SET also publishes a bimonthly journal called *Imagine,* which reports on activities of and opportunities for students served by the program.

Aside from the intellectual influence of Stanley's work on other researchers, his investigations have spawned numerous other programs of research and intervention. The Center for Talented Youth (CTY) at Johns Hopkins University took over the teaching and academic coaching functions of SMPY. Today, CTY runs a number of programs for enrichment and acceleration for students in junior high school and the early high school years. The acceleration programs are fast-paced (a year of subject matter in three weeks) and give students access to tools of thought and communication. Enrichment programs provide environments for intellectual exploration often missing in the usual school curriculum. There are programs for commuter students to the Johns Hopkins campus as well as summer programs on campuses throughout the country.

Another direct outgrowth of the SMPY program is the work of the Office of Pre-collegiate Programs for Talented and Gifted (OPPTAG) at Iowa State University. Here Camilla Benbow's work is a direct continuation of Stanley's. Benbow's unique contributions include a twenty-five-year longitudinal study of high-ability students. The depth of this database, together with the remarkable interest on the part of the subjects themselves in the study, have allowed Benbow and her colleagues to conduct investigations of factors that influence the careers of individuals identified early as having high mathematical ability. OPPTAG also offers a variety of Saturday and summer programs in mathematics, including activities for parents and teachers.

The Connie Belin National Center for Gifted Education at the University of Iowa is an important national resource. Programs include teacher training institutes and precollege programs for gifted students. In addition, the center conducts talent searches, hosts national academic contests, provides consultations for schools, and conducts research on the cognitive and affective needs of talented students.

Numerous regional centers provide resources similar to those already described. The Talent Identification Program (TIP) at Duke University also started as an offshoot of Stanley's work at Johns Hopkins. The Center for Talent Development at Northwestern University serves students in the midwestern area, and the Center for Gifted Studies at the University of Southern Mississippi serves that state and its neighbors. The National Research Center on the Gifted and Talented at the University of Connecticut concentrates on theoretical research and the development of teachers and teaching materials for high-ability students.

The Education Program for Gifted Youth at Stanford University implements a different model of service. Through interactive computer programs, this center offers advanced coursework to both elementary and high school students. Offerings run from early algebra and prealgebra through advanced placement work. Contact with instructors is maintained through e-mail question-and-answer sessions and telephone conversations.

## SUMMER PROGRAMS

Probably the earliest summer work with high-ability students was that of Arnold Ross. His summer program for teachers, started during the *Sputnik* era, grew into a student program, first at Notre Dame University, then the University of Chicago, and currently at Ohio State University. The Ross program is built around a core curriculum in number theory. This branch of mathematics includes results easily accessible to high school students with just a year or two of algebra, as well as theorems and conjectures that are the subject of current mathematical research. Number theory is particularly useful for the training of younger students because certain of its elementary results lead naturally to more general or more advanced techniques and concepts.

Central to Ross's concept in starting his program was the idea of an intellectual community. Early in the history of the program it was discovered that its alumni, working as "counselors" or "mentors" the following summer, could set a tone for achievement and purpose with students new to the program. This process has progressed to the point where a second program, PROMYS, has been started at Boston University by Glenn Stevens, an alumnus of Ross's program. Stevens's motivation for starting this program was to continue and expand Ross's work. The devoted following of its alumni and the strong tradition of successful summers have proven the effectiveness of this effort.

Several summer programs involving classwork have since become available for high-ability students of mathematics. The Hampshire College Summer Studies in Mathematics, established in 1971 by David Kelly, exposes students to a variety of mathematical topics in a less formal, but equally rigorous, atmosphere. Topics covered vary from year to year, and include a wide range of materials from outside the usual secondary or early college curriculum. Examples of topics covered include algebraic structures, graph theory, inequalities, and knot theory. The emphasis of the program is on having students see themselves as active creators of mathematics rather than as learners of mathematical facts.

The USA Mathematical Talent Search combines a summer program with a year-round competition by correspondence. Coordinated by Professor George Berzsenyi at the Rose-Hulman Institute of Technology, the program offers four rounds of five long-answer problems each year. Berzsenyi took as his model the problem section of *KOMAL,* the Hungarian problem-solving journal for high school students of mathematics and physics. Problems are chosen from across the mathematical spectrum. They require a minimum of background but a lot of thought. Students work on the problems by themselves and send in their written responses. The responses are graded and returned to the student. A Young Scholar's Summer Program is based on the competition and offers training in problem solving as well as such mathematical content as axiomatic set theory, graph theory, and operations research. The Talent Search is run with the support of the National Security Agency, the NSF, and the Consortium for Mathematics and its Applications (COMAP). The problem sets, together with answers and hints, are published in the COMAP journal.

An alternative to the classroom model for summer programs is the internship model. Although widespread in other fields, this model is not often used in mathematics, largely because of its labor-intensive nature. One successful example is the Research Science Institute at the Massachusetts Institute of Technology, a program founded by Admiral H. G. Rickover in 1984 and administered by the Center for Excellence in Education. This program places students in research laboratories in many fields, including apprenticeships in mathematics. A key element in the success of this sort of program is the choice of and articulation with mentors. These are often university faculty or graduate students, who agree with the interns on a set of problems to be pursued. Progress is monitored daily and intervention provided when required. The intern writes a paper summarizing his or her results at the end of the program and delivers an oral report to the institute as a whole. Information about summer programs, both on the classroom model and the internship model, may be found in the booklet *Directory of Student Science Training Programs for Pre-college Students,* published each year by Science Service.

The NSF also supports summer programs for high-ability students. NSF involvement in all aspects of education came slowly and with great reluctance (England 1982, 227–254) until the *Sputnik* era. At that time it became clear that federal intervention in precollege education was critical to the development of the country's scientific infrastructure and that the NSF was the appropriate agency to oversee this effort. The main thrust of this intervention consisted of summer institutes for teachers. However, a set of enrichment programs for high school students in all areas of science, called Summer Science Talent Programs, was set up. This program came to an end in the 1980s, but a similar set of programs was funded under the title Young Scholars Programs. While these programs serve high-ability students, and some of the above-mentioned programs receive NSF support, most summer programs supported by NSF target a specific population, such as the handicapped, females, or minority groups.

## MATHEMATICAL CONTENT

Among the difficulties of working with high-ability students is the extreme stratification of the mathematical profession. Mathematicians working in industry and in academia and teachers of mathematics in precollege environments have little chance to interact professionally. This has not been the case in eastern Europe; there for generations researchers with international reputations have considered it part of their professional duty to work with precollege students. Among other results of this culture has been a flowering of books and journals full of mathematical results accessible to precollege students.

The American situation, so different from that in eastern Europe, changed with the launching of *Sputnik*. The trauma of this event occasioned a deep interest in the Soviet system of education, including a flurry of translations from Russian and other East European sources. Since then, interest in writing curriculum and extracurricular materials has been uneven. The New Mathematical Library, published first by Random House and later by the Mathematical Association of America, has continued to produce high-quality materials on the high school level. Such publishers as Creative Publications, Dale Seymour Publications, and MathPro Press have taken the risk of making such materials available for sale despite the lack of a wide market.

Another aspect of the mathematical scene that has been lacking in America is a mathematical journal for students of high ability. Again, the lead has been taken by eastern Europe, where such journals exist in almost every country, sometimes on a very high level. The most important of these is *Kvant,* a student journal of mathematics and physics published by the Russian (formerly Soviet) Academy of Sciences. In 1988, the National Science Teachers' Association, with support from the NSF, negotiated an agreement to prepare an American version of *Kvant* in English. The resulting magazine, *Quantum,* contains mostly Russian contributions but includes articles by American teachers and mathematicians as well. Articles are accessible to high school students but go far beyond the usual curriculum. Their style is informal and interactive, with problems and exercises interspersed throughout the exposition. A second journal, *Mathematics and Informatics,* is edited by the Bulgarian Academy of Sciences and published by the Science and Technology Publishing Company of Singapore. This journal contains contributions from all over the world, in both mathematics and informatics (computer science).

## COMPETITIONS

Competitions in mathematics have had a long and distinguished history. An early version of Stokes' Theorem, the generalization of the fundamental theorem of calculus to higher dimensions, appeared on the tripos examinations in Victorian England. The algebraic solutions to equations of degrees 3 and 4 were stimulated in part by contests in the solution of equations held in the universities of Renaissance Italy. And in ancient times, Archimedes' Cattle Problem was originally posed as a challenge to another mathematician. Competition, both formal and informal, seems to be one way in which mathematical creativity asserts itself.

A network of contests on the local level extends throughout the nation. Some of these are run by colleges and universities, others by teachers and schools, and still others by profit-making organizations. On the national level, the Committee on American Mathematics Competitions, a joint effort of seven professional organizations, runs several contests. A junior high school contest consists of short-

answer questions. Three high school contests, culminating in the United States of America Mathematical Olympiad, form a pyramid leading to participation in the International Mathematical Olympiad.

The American Regions Mathematics League (ARML) also sponsors several events. An annual on-site competition, held simultaneously at two sites, draws 1,500 students and teachers from around the country. A long-answer, Olympiad-type contest is held in the high schools three times each year. An exchange program with Russia, although not itself a contest, includes competitive events: a team from Moscow competes every other year at the ARML contest.

The International Tournament of the Towns originated as a Russian competition. It offers two sets of long-answer questions each year. Students receive individual recognition, and the cities with the best sets of papers get prizes. The World Federation of National Mathematics Competitions publishes a journal with news about this and other international competitions.

The National Society of Professional Engineers runs two national contests and edits training and enrichment materials. The contests are organized by state, with a final event in Washington. The Junior Engineering Talent Search, or JETS, pits teams of students against each other in grappling with long-answer problem situations. Among competition opportunities for younger students, two of the most important are MathCounts, offering contests on the local, state, and national levels to middle school students, and Mathematical Olympiads for Elementary Schools, with 80,000 participants from around the world. The Mandelbrot competition is an offshoot of some of those described here, in the sense that it was started by mathematicians who had been stimulated by participation in mathematics contests as students. It offers both individual and team events to participants in the high schools.

Aside from timed-problem-solving competitions, there are many mathematics and science "fairs" throughout the country. These provide students with opportunities to share their work in mathematics on a deeper level than a timed competition might allow. The most highly organized network of science fairs is the International Science and Engineering Fairs, run by Science Service, an organization that also runs the highly prestigious Intel (formerly Westinghouse) Science Talent Search.

Although not a mathematics competition, SuperQuest, a contest in computer science, borrows from other disciplines. Its structure allows teams of students to use computers to grapple with difficult problem situations. The final stage of this contest involves a summer trip to one of the national supercomputer installations, where students take classes, are given access to the large machines, and work on their final competition projects.

The World Federation of National Mathematics Competitions, based in Australia, provides information on mathematics competitions in over seventy countries throughout the world. The Federation journal, *Mathematics Competitions,* includes articles and questions from member organizations.

## OTHER RESOURCES

Mu Alpha Theta is a national mathematics honors fraternity. Most of its activities are on a local level, with chapters in many schools. A summer "convention" brings thousands of students together for classes, contests, talks, and mathematical recreations. Mu Alpha Theta publishes a newsletter and a series of monographs on mathematical topics accessible to high school and middle school students.

COMAP develops applications-oriented curriculum materials at the college and precollege levels. These include instructional modules, texts, and teacher-training materials. COMAP also produces three journals, one for undergraduate mathematics, another for secondary mathematics, and a third for elementary school teachers. The USA Mathematical Talent Search questions appear in the COMAP secondary school journal *Consortium.*

Israel Moiseyevich Gelfand, a major figure in twentieth-century mathematics, has organized a correspondence school for students seeking more mathematics training than is available in local high schools. His work is based on three texts, published by Birkhäuser. Students are sent a series of assignments based on these texts. Their responses are graded centrally, and feedback and additional materials support the student's individual pace of progress. The work of Joseph Renzulli (1977) is important for teachers working with younger children. Renzulli's structures apply to all subject areas, including mathematics.

The National Association for Gifted Children is an advocacy group for children with gifts in a variety of areas. Based in Washington, DC, it publishes *The Gifted Child Quarterly,* a newsletter, and several references on the education of gifted children.

*See also* Cognitive Learning; Competitions; Gifted; Mu Alpha Theta

## SELECTED REFERENCES

Clendening, Corinne P., and Ruth Ann Davies. *Creating Programs for the Gifted.* New York: Bowker, 1980.

*Consortium.* Lexington, MA: COMAP, 1984– .

*The Elementary Mathematician.* Lexington, MA: COMAP, 1988– .

England, J. Merton. *A Patron for Pure Science: The National Science Foundation's Formative Years, 1945–57.* Washington, DC: National Science Foundation, 1982.

Galton, Francis. *Hereditary Genius.* London, England: Macmillan, 1869.

Gelfand, Israel M., Elena G. Glagoleva, and Aleksandr A. Kirillov. *The Method of Coordinates.* Boston, MA: Birkhäuser, 1990.

Gelfand, Israel M., Elena G. Glagoleva, and Emanuel E. Shnol. *Functions and Graphs.* Boston, MA: Birkhäuser, 1990.

Gelfand, Israel M., and Mark Saul. *Trigonometry.* Boston, MA: Birkhäuser, in press.

Gelfand, Israel M., and A. Shen. *Algebra.* Boston, MA: Birkhäuser, 1993.

Greenlaw, M. Jean, and Margaret E. McIntosh. *Educating the Gifted: A Sourcebook.* Chicago, IL: American Library Association, 1988.

*Imagine.* Baltimore, MD: Johns Hopkins University Press, 1994– .

Laubenfels, Jean. *The Gifted Student: An Annotated Bibliography.* Westport, CT: Greenwood, 1977.

*Mathematics and Informatics.* Singapore: Science and Technology Publishing Company, 1990– .

*Quantum.* Arlington, VA.: National Science Teachers' Association and Springer-Verlag New York, 1988– .

Renzulli, Joseph S. *The Enrichment Triad Model: A Guide for Developing Defensible Programs for the Gifted and Talented.* Mansfield Center, CT: Creative Learning, 1977.

Stanley, Julian C., William C. George, and Cecilia H. Solano, eds. *The Gifted and the Creative: A Fifty-Year Perspective.* Baltimore: Johns Hopkins Press, 1977.

Terman, Lewis M. *Measuring Intelligence.* Boston, MA: Houghton Mifflin, 1937.

———. *Mental and Physical Traits of a Thousand Gifted Children.* Genetic Studies of Genius, Vol. I. Stanford CA: Stanford University Press, 1925.

*UMAP Journal.* Lexington, MA: COMAP, 1980– .

**Sources of Further Information**

American Regions Mathematics League
711 Amsterdam Avenue
New York, NY 10025
212 666 5188

Center for Gifted Studies
University of Southern Mississippi
Southern Station Box 8207
Hattiesburg, MS 39406-8207
601 266 5236

Center for Talent Development at Northwestern
University
617 Dartmouth
Evanston IL 60208
708 491 3782

Connie Belin National Center for Gifted Education
College of Education
The University of Iowa
210 Lindquist Center
Iowa City, IA 52242

Consortium for Mathematics and Its Applications
57 Bedford Street, Suite 210
Lexington, MA 02173
800 772 6627
email: order@comap.com

Creative Publications
5040 West 111 Street
Oaklawn, IL 60453
800 624 0822

Dale Seymour Publications
PO Box 10888
Palo Alto, CA 94303
800 872 1100

Duke University Talent Identification Program
1121 West Main Street
Suite 100
Durham, NC 27701-2028
919 683 1400

Education Program for Gifted Youth
Ventura Hall
Stanford University
Stanford, CA 94305-4115
415 723 0512
email: raviglia@epgy.stanford.edu

Gelfand Outreach Program in Mathematics
Center for Mathematics, Science and Computer
Education
SERC Building, Room 239
Busch Campus, Rutgers University
Piscataway, NJ 08855-1179
908 932 0669
email: harriet@gandalf.rutgers.edu

Hampshire College Summer Studies in Mathematics
c/o David Kelly
Box NS, Hampshire College
Amherst, MA 01002-5001
413 582 5375
email: dkelly@hamp.hampshire.edu

*Imagine: Opportunities and Resources for Academically
Talented Youth*
Published by Johns Hopkins University
2715 North Charles Street
Baltimore, MD 21218-4319
800 548 1784

Junior Engineering Technical Society
1420 King Street, Suite 405
Alexandria, VA 22314-2715
703 548 JETS

Mandelbrot Competition
c/o Greater Testing Concepts
PO Box A-D
Stanford, CA 94309

MathCounts Foundation
1420 King Street
Alexandria, VA 22314
703 684 2831

Mathematical Association of America
1529 18th Street NW
Washington, DC 20036

Mathematical Olympiads for Elementary Schools
125 Merle Avenue
Oceanside, NY 11572

*Mathematics and Informatics*
c/o George Berzsenyi
Department of Mathematics, Box 121
Rose-Hulman Institute of Technology
Terre Haute, IN 47803-3999

MathPro Press
PO Box 713
Westford, MA 01886
508 649 3003

Mu Alpha Theta
University of Oklahoma
601 Elm Avenue, Room 423
Norman, OK 73019-0315
405 325 4489

National Association for Gifted Children
1155 15th Street NW, Suite 1002
Washington, DC 20005
202 785 4268

National Council of Teachers of Mathematics
1906 Association Drive
Reston, VA 22091

OPPTAG at Iowa State University
W172 Lago Marcino Hall
Ames, IA 50011
515 294 1772

*Quantum*
Springer-Verlag New York, Inc.
Journal Fulfillment Services, Inc.
PO Box 2485
Secaucus, NJ 07096-9813
800 777 4643

Research Science Institute
Center for Excellence in Education
7710 Old Springhouse Road
McLean, VA 22102
703 448 9062

Ross Summer Mathematics Program
Department of Mathematics
Ohio State University
100 Mathematics Building
231 West 18th Avenue
Columbus, OH 43210-1174
614 292 4975

Study of Exceptional Talent at Johns Hopkins Center
for Talented Youth
Johns Hopkins University
Baltimore, MD 21218
410 516 0309

SuperQuest
National Center for Supercomputing Applications
605 East Springfield
Champaign, IL 61820
217 244 4908

Tournament of the Towns
c/o ARML
711 Amsterdam Avenue
New York, NY 10025
212 666 5188

USA Mathematics Talent Search
Box 121
Rose Hulman Institute of Technology
Terre Haute, IN 47803
812 877 8391

World Federation of National Mathematics
Competitions
University of Canberra, PO Box 1
Belconnen ACT 2616 Australia

Young Scholars Program
National Science Foundation, Room 885
4201 Woodrow Wilson Boulevard
Arlington, VA 22230

MARK E. SAUL

# GOALS OF MATHEMATICS INSTRUCTION

Goals for education are shaped and dictated by society. Changes in societal trends and the economy impact the education system. In recent years, mathematics education, being subsumed under the education system, has been placed under tremendous pressure to change from all sectors of society—political, economic, and educational. Politicians at the federal, state, and local levels have emphasized the current and future importance of mathematics and science to society. Business leaders have called on educators to train workers who reason mathematically and who can work cooperatively. Mathematics educators, dissatisfied with the current status of mathematics education, have written standards to motivate changes in mathematics instruction and content. We will examine here the new goals for mathematics education that have emerged from this climate of social change and will describe the resulting new goals for students.

## NEW GOALS FOR SOCIETY

Historically, societies have established schools to transmit their culture to the young and provide citizens with the opportunity for self-fulfillment. For the agrarian/industrial society of the first half of this century, the goals of learning mathematics were to obtain basic mathematical skills for a lifetime of work in an industrial and agricultural economy and to educate a small elite who would go to college and on to professional careers. During the second half of the twentieth century, industrialized countries throughout the world have experienced a shift from industrial to information economies. Today's schools, however, still operate under the legacy of the industrial society of the past, which is no longer relevant to educating youth for the information age (National Research Council (NRC) 1989, 11). The availability of low-cost calculators, computers, and other technology has changed the nature of the physical, life, and social sciences and business, industry, and government. "Information is the new capital and the new material, and communication is the new means of production" (National Council of Teachers of Mathematics (NCTM) 1989, 3). New goals have been proposed by the education community to address these changes.

The new goals have been developed to address the changing needs of workers in the information age. Because work will be "less manual but more mental; less mechanical but more electronic; less routine but more verbal; less static but more varied" (NRC 1989, 11), schooling must be, too. In addition, work will require more facility in reasoning and communicating mathematically and less in calculating and applying routine procedures. Therefore, the "new social goals for education include (1) mathematically literate workers, (2) lifelong learning, (3) opportunity for all, (4) an informed electorate" (NCTM 1989, 3).

## 1. Mathematically Literate Workers

The mathematics to be learned today is significantly more diverse than it was when many current leaders and educators studied it. Mathematics involves far more than calculations—clarifying the problem, deducing the consequences, formulating alternatives, and developing appropriate tools are all a part of the mathematicians' craft. Statistics has blossomed from its roots in agriculture and genetics into a rich mathematical science that provides tools for analyses of uncertainty and development of forecasts in such diverse applications as consumer and stock market surveys, enhancing digital photographs, and policy analysis in every area of human affairs. Problems in computer science and social science have invigorated the discipline of discrete mathematics. New tools such as game theory and decision theory are being applied to the human sciences to assist in making choices and decisions (NRC 1989, 5).

## 2. Lifelong Learning

As colleges, universities, and continuing education attract a larger proportion of the population, schools are being pressured to prepare all students for some type of postsecondary study. The level of numeracy and literacy formerly associated with the few who entered higher education is now to be a goal for all. The facility with mathematics formerly required only of those entering scientific careers is now a necessary foundation for lifelong work in the information age (NRC 1989, 10).

From another perspective, students entering school now are looking at a working environment

that is changing with increasing frequency. Preparing for only one career is no longer efficient. Those entering the work force today expect to have at least three different careers in a working lifetime, and that number may increase. The instruction that students receive needs to give them the mathematical understanding and flexibility to adjust to their changing circumstances and to the new applications of mathematics that will arise.

## 3. Opportunity for All

One of the main goals of mathematics education is to increase the participation of *all students* and to improve and increase their mathematical education at every level. The board of directors of the National Council of Teachers of Mathematics (NCTM) has defined "all students" to mean "(1) students who have been denied access in any way to educational opportunities as well as those who have not; (2) students who are African American, Hispanic, American Indian, and other minorities as well as those who are considered to be a part of the majority; (3) students who are female as well as those who are male; (4) students who have not been successful in school and in mathematics as well as those who have been successful" (NCTM 1991, 4).

Too often in the past, students socially and economically well positioned have had more educational opportunities than those in minority subpopulations. Also, there has been a pervasive societal assumption that mathematics and the hard sciences are a male domain. One of the most quoted goals for mathematics education, attributed to Robert M. White, is that mathematics must become a "pump" instead of a "filter in the pipeline" (NRC 1989, 6). Filtering out minorities and women from the mathematically privileged has become a liability. Greater demands for mathematical understanding are being placed on all workers, and schools can no longer afford to teach only a few the mathematics that all will need. In addition, societal pressures that all be given equal economic opportunity place a responsibility on mathematics educators to provide all students with equal opportunity to learn mathematics. It is increasingly important to society that everyone be given the tools to excel.

Unfortunately, the American public tends to assume that differences in accomplishment in school mathematics are due primarily to differences in innate ability. Expectations are low, and parents and other members of the public are heard to say, "I could never do math." This is not true in Asian societies, for example, where the expectation is that learning mathematics is achieved by hard work (Stevenson and Stigler 1992). Therefore, a subgoal of creating opportunity for all is changing the public perception of who can do mathematics.

### 4. An Informed Electorate

In modern democratic societies, political and social decisions involve complex technical issues. Environmental issues, nuclear energy, defense spending, space exploration, social issues, and taxation are increasingly intermingled. In order to be politically informed on the issues of a complex technological society, citizens need to interpret complex and sometimes conflicting information (NCTM 1989, 4–5). These goals imply an educational system organized to serve as a resource for all citizens throughout their lives (NCTM 1989, 3).

## NEW GOALS FOR STUDENTS

In order to accomplish the aforementioned societal goals, goals for schools and students must also change. These overarching goals for students have been summarized in the *Curriculum and Evaluation Standards* (NCTM 1989) as providing students with *mathematical power.* "Mathematical power" is defined by NCTM as "the ability to explore, conjecture, and reason logically; to solve nonroutine problems; to communicate about and through mathematics; to connect ideas within mathematics and between mathematics and other intellectual activity" (NCTM 1991, 1). For NCTM, students who exhibit mathematical power have developed "personal self-confidence and a disposition to seek, evaluate, and use quantitative and spatial information in solving problems and in making decisions" (NCTM 1991, 1). Such students manifest "flexibility, perseverance, interest, curiosity, and inventiveness" (NCTM 1991, 1).

Five general goals for all students have been articulated in the K–12 curriculum and evaluation standards: "(1) that they learn to value mathematics, (2) that they become confident in their ability to do mathematics, (3) that they become mathematical problem solvers, (4) that they learn to communicate mathematically, (5) that they learn to reason mathematically" (NCTM 1989, 5) According to NCTM, these goals will require a change in the nature of the classroom, because "*what* a student learns depends to a great degree on *how* he or she has learned it"

(emphasis in original) (NCTM 1989, 5). Thus, classrooms must become "places where interesting problems are regularly explored using important mathematical ideas" (NCTM 1989, 5).

### 1. Learning to Value Mathematics

The accomplishment of this goal relates to the historical development of mathematics and the contribution of mathematics to society and to the culture. Today, theoretical mathematics has become more diverse and deeper in complexity and abstraction, and still more vital to our technological society. Historically, mathematical development has proceeded in the same way: practical problems and theoretical pursuits have stimulated and been intertwined each with the other. "It is the intent of this goal . . . to focus attention on the need for student awareness of the interaction between mathematics and the historical situations . . . and the impact . . . on our culture and our lives" (NCTM 1989, 7).

### 2. Becoming Confident in One's Own Ability

As they study mathematics, students need to feel themselves capable of applying their growing mathematical power to situations in their own lives. Whether they imagine themselves in such "real-world" situations as papering a wall or carpeting a floor or purchasing items at a store, they should be aware of and apply the appropriate mathematics. "To some extent, everybody is a mathematician and does mathematics consciously" (NCTM 1989, 6).

### 3. Becoming a Mathematical Problem Solver

Problem solving has been a basic goal of mathematics education for many years. Perhaps in different terms, problem solving was an emphasis of mathematics teaching in the "new math" era of the 1960s. Problem solving was central to the 1977 definition of the basics in mathematics education in the National Council of Supervisors of Mathematics statement (NCSM 1977), which foreshadowed the first recommendation of the *Agenda for Action*, the goals statement by NCTM: "Problem solving must be the focus of school mathematics in the 1980s" (NCTM 1980, 2)

The goal of problem solving has evolved to mean more than the ability to translate word problems into mathematical symbols. Problem-solving activities should not be limited to exercises that students can solve easily but should also include "problems that may take hours, days, and even weeks to solve . . . . Some problems should be open-ended with no right answer, and others need to be formulated" (NCTM 1989, 6). In addition, problem-solving activities should not only engage students independently but should also include activities that are best solved by small groups or by a class working cooperatively.

## 4. Learning to Communicate Mathematically

The goal of communication in mathematics reveals that doing mathematics is more than working with numbers, formulas, and equations. It also requires interpretation, argument, justification, and explanation. Group settings for problem solving are a rich medium for developing the ability to communicate mathematically. As students engage in solving a problem in concert with their peers, they exchange their interpretations of the problem, share their ideas for solving the problem, and explain their rationale for why a solution path is or is not fruitful. These interchanges may occur between student and student, students and teacher, and students and the whole class. Other rich mediums for developing mathematical communication are journal writing, written reports of solutions, and problem posing. In communicating ideas, the language of mathematics becomes more natural, and students learn to clarify, refine, and consolidate their thinking (NCTM 1989, 6).

## 5. Learning to Reason Mathematically

Gathering information, making and testing conjectures, building an argument to support (interim and final) conclusions, and evaluating the validity or efficacy of arguments are fundamental to learning mathematics. A demonstration of good reasoning should be rewarded more than finding correct answers (NCTM 1989, 6). Mathematical reasoning is not limited to formal proof, but rather includes all forms of argumentation by which students explain to themselves and their peers as well as their teachers why they feel justified in making certain decisions and choosing certain paths during their problem-solving process. It includes inductive as well as deductive forms of reasoning. Inductive reasoning includes thinking that proceeds from examining multiple, well-chosen examples in the problem-

solving process to making a conjecture that is then tested in other, more general cases. Deductive reasoning includes the application of already-tested arguments to specific situations.

## 6. Making Connections with Mathematics

Although not listed in the introduction to the NCTM *Curriculum and Evaluation Standards*, this is clearly a goal of mathematics education for all students. Connections include linking of conceptual knowledge with symbols and procedures, connecting topics within mathematics, and applying mathematics to other subjects and to the world outside the classroom. The goal of making connections in mathematics includes encouraging students to recognize and relate the various forms of a problem, whether it is presented as a diagram, a group of mathematical symbols, written or spoken text, or a concrete model. For example, when students are solving a problem from a real-world setting, they can begin the solution process by creating a mathematical interpretation or model. The mathematical model can often be expressed in more than one form, such as an equation and a graph. Using the mathematical model, students can then work toward a solution that often they must reinterpret into a form in keeping with the original problem situation.

## SUMMARY

As is evident from this discussion of the goals of mathematics education, there are two general and overarching goals: to increase participation in mathematics learning and to make mathematics learning more accessible. Both goals require a change in the public perception of mathematics and of those who use it. It appears that adults and youth alike find mathematics mystifying, difficult, and irrelevant. In defining the goals listed here, mathematics educators have tried to demystify mathematics and to make it relevant and accessible by tying it to the solution of real-world applications, as in mathematical modeling. They believe that helping people to see mathematics as relevant and useful will increase their disposition to learn the subject. This increased disposition to learn mathematics will be evident in increasing participation in mathematics learning and in changing attitudes toward mathematics.

*See also* Applications for the Classroom, Overview; Career Information; Goals 2000; Mathematics, Nature; Modeling; Standards, Curriculum

## SELECTED REFERENCES

National Council of Supervisors of Mathematics. "Position Paper On Basic Skills." *Arithmetic Teacher* 25(Oct. 1977):19–22.

National Council of Teachers of Mathematics. *An Agenda for Action: Recommendations for School Mathematics of the 1980s.* Reston, VA: The Council, 1980.

———. *Curriculum and Evaluation Standards for School Mathematics.* Reston, VA: The Council, 1989.

———. *Principles and Standards for School Mathematics.* Reston, VA: The Council, 2000.

———. *Professional Standards for Teaching Mathematics.* Reston, VA: The Council, 1991.

National Research Council. *Everybody Counts: A Report to the Nation on the Future of Mathematics Education.* Washington, DC: National Academy Press, 1989.

Stevenson, Harold W., and James W. Stigler. *The Learning Gap: Why Our Schools Are Failing and What We Can Learn From Japanese and Chinese Education.* New York: Summit, 1992.

DOUGLAS T. OWENS
GAYLE M. MILLSAPS

# GOALS 2000

Refers to the "Goals 2000: Educate America Act," passed by the U.S. Congress in 1994, which set a national agenda for education. In 1983, the report *A Nation at Risk* described the United States as a nation with a "rising tide of mediocrity" in its elementary and secondary schools, with students whose deficiencies threatened the nation's economy. It called for clear and rigorous academic standards to help remedy the situation. It was not until 1989, however, that President George Bush convened the nation's governors at an historic education summit, where they agreed to establish national education goals. A number of organizations were created to help move this agenda, including the National Education Goals Panel (NEGP) and the National Council on Education Standards and Testing (NCEST). The reports of these groups set the stage for the legislation that was enacted in 1994 during the administration of President Bill Clinton.

The Goals 2000 act set eight goals (NEGP Report 1994), as follows. By the year 2000: (1) all students will start school ready to learn; (2) the high school graduation rate will increase to at least 90%; (3) all students will leave grades 4, 8, and 12 having demonstrated competence in challenging subject matter including English, mathematics, science, foreign languages, civics and government, economics, arts, history, and geography; (4) the nation's teaching force will have access to programs for the continued improvement of their professional skills and the opportunity to acquire the knowledge and skills needed to instruct and prepare all American students for the next century; (5) U.S. students will be first in the world in mathematics and science achievement; (6) every adult American will be literate and will possess the knowledge and skills necessary to compete in a global economy and exercise the rights and responsibilities of citizenship; (7) every school in the United States will be free of drugs, violence, and the unauthorized presence of firearms and alcohol and will offer a disciplined environment conducive to learning; and (8) every school will promote partnerships that will increase parental involvement and participation in promoting social, emotional, and academic growth of children.

The mathematics and science emphasis of goal (5) was prompted by the results of a number of projects (LaPointe, Mead, and Phillips 1989; Mathematical Sciences Education Board 1989) that compared the mathematics and science achievement of students in industrialized nations and found U.S. students to rank near the bottom. In order to meet this goal, the nation would need clear standards comparable to those in high-achieving countries; science and mathematics educators would need to provide a different kind of preservice training to help teacher candidates facilitate broader and deeper mathematics understanding among an increasingly diverse group of students, and current teachers would need substantial professional development.

Goals 2000 required states to develop high standards for what students should know and be able to do and challenging assessments to measure their progress. It prompted intense activity in this area. By mid-1995, forty-nine states were engaged in setting standards (American Federation of Teachers 1995), though the content and rigor varied from state to state. Some states have developed standards that are clear and specific enough to provide guidance for teachers, curriculum developers, and others who will be using them. For example: "The student will differentiate between area and perimeter and identify whether the application of the concept of perimeter or area is appropriate for a given situation. (Grade 5)." Others produced standards that offered little guidance for becoming first in the world. For example: "Students should become mathematical problem solvers. To develop these abilities, students need the experience of working with diverse problem-solving situations." Some states plan to align state assessment with these standards, but many have no plan to hold students accountable for reaching them.

The initial plan contained substantial funding for professional development activities to enable teachers to teach to these standards, with 90% of the money targeted to local school districts and much of that to individual schools.

See also Federal Government, Role; State Government, Role

## SELECTED REFERENCES

American Federation of Teachers. *Making Standards Matter: A Fifty-State Progress Report on Efforts to Raise Academic Standards.* Washington, DC: The Federation, 1995.

LaPointe, Archie E., Nancy A. Mead, and Gary W. Phillips. *A World of Differences: An International Assessment of Mathematics and Science.* Princeton, NJ: Educational Testing Service, 1989.

Mathematical Sciences Education Board, National Research Council. *U.S. School Mathematics from an International Perspective: A Guide for Speakers.* Washington, DC: The Council, 1989.

National Commission on Excellence in Education. *A Nation at Risk: The Imperative for Educational Reform.* Washington, DC: Government Printing Office, 1983.

National Education Goals Panel. "The National Education Goals." In *The National Education Goals Report: Building a Nation of Learners 1994* (pp. 8–11). Washington, DC: Government Printing Office, 1994.

———. Web site: //www.negp.gov/

ALICE J. GILL

# GÖDEL, KURT FRIEDRICH
## (1906–1978)

Greatest logician of the twentieth century whose fundamental work in the 1930s completed the transformation of logic into a mathematical subject and established set theory as a sophisticated field of mathematics. Gödel was born in Brünn in the Austro-Hungarian Empire, now Brno in the Czech Republic. In 1930, he established the Completeness Theorem, that for the natural formalization of mathematical reasoning in terms of variables, "for all," and "there is," if a statement is *true* under every possible interpretation of the language, then it is *provable* according to specified rules. Its corollary, the Compactness Theorem, that a collection of statements is satisfiable exactly when every finite subcollection is, became pivotal for *model theory,* the mathematical theory of formal semantics.

In 1931, Gödel established his best-known result, the Incompleteness Theorem: any consistent rule-governed proof system that includes some basic arithmetic is *incomplete;* there is a true arithmetical statement that is not provable in the system. Gödel coded statements and proofs with numbers, a procedure now known as *Gödel numbering,* and developed a statement that asserts, in the coding, its own unprovability. The result established a crucial barrier to David Hilbert's program, an ambitious scheme for establishing the overall consistency of mathematics by finitary means and has since been associated with the limits of reasoning. The algorithmic techniques that Gödel developed led, in the hands of Alan Turing, Alonzo Church, and Stephen Kleene, to *recursion theory,* the mathematical theory of computability and the basis of contemporary theoretical computer science.

During 1935–1937, Gödel established the *relative consistency* of the Axiom of Choice and the Continuum Hypothesis in set theory, proving that these two assertions, though they go beyond the basic axioms of set theory, do not introduce any new contradictions. The Continuum Hypothesis asserts that every infinite set of real numbers is in one-to-one correspondence with either the set of natural numbers or else the set of all the reals. It was to articulate and establish this hypothesis that led Georg Cantor to create the subject of set theory. The Axiom of Choice asserts that for every set $S$ there is a function that chooses an element from every nonempty member of $S$. This axiom was a crucial part of the first axiomatization of set theory, by Ernest Zermelo. To establish his result, Gödel carried out a sort of Gödel numbering along a hierarachy to formulate an "inner model," and his work introduced formal definability techniques that have become a basic part of contemporary set theory. In 1940 Gödel was permanently installed at the Institute for Advanced Study in Princeton. From 1947 to 1951, he provided solutions to Einstein's field equations concerning time travel into one's past.

See also Axioms; Hilbert; Logic; Mathematics, Foundations; Sets

## SELECTED REFERENCES

Enderton, Herbert B. *A Mathematical Introduction to Logic.* New York: Academic, 1972.

Gödel, Kurt F. *Collected Works.* 3 vols. Solomon Feferman et al., eds. New York: Oxford University Press, 1986–1995.

Hofstadter, Douglas. *Gödel, Escher, Bach: An Eternal Golden Braid.* New York: Vintage, 1979.

Nagel, Ernest, and James R. Newman. "Göedel's Proof." In *The World of Mathematics* (pp. 1668–1695). James R. Newman, ed. New York: Simon and Schuster, 1956.

AKIHIRO KANAMORI

# GOLDBACH'S CONJECTURE

Famous unproven hypothesis first made in a letter of 1742 from Christian Goldbach (1690–1764) to Leonhard Euler. Commonly stated, every even number greater than 2 is the sum of two primes; that is, $4 = 2 + 2$; $6 = 3 + 3$; $8 = 3 + 5$; $10 = 3 + 7$; $12 = 5 + 7$; and so on. From that time, mathematicians, trained and amateur alike, have tried unsuccessfully to find either an even number not the sum of two primes, thus disproving the conjecture, or a proof that holds for all even numbers however large. In 1937, Ivan Vinogradov proved analytically a related conjecture that every "sufficiently large" odd number is the sum of at most three odd primes. Similarly, modern computers have checked increasingly larger even numbers without finding an exception to the rule. Therefore, despite over 250 years of determined effort, the validity of Goldbach's conjecture remains one of the most intriguing of a number of unresolved hypotheses in the theory of prime numbers.

*See also* Number Theory, Overview; Prime Numbers

### SELECTED REFERENCES

Barnett, Isaac A. *Some Ideas about Number Theory.* Washington, DC: National Council of Teachers of Mathematics, 1961.

Fey, James. "Prime and Composite Numbers." In *Historical Topics for the Mathematics Classroom* (pp. 65–66). 2nd ed. John K. Baumgart et al., eds. Reston, VA: National Council of Teachers of Mathematics, 1989.

WILLIAM D. JAMSKI

# GOLDEN SECTION

A division of a segment into two segments such that the whole is to the longer as the longer is to the shorter. If $a$ and $b$ denote the lengths of the longer and shorter segments and $g$ denotes the ratio, we then have $(a + b)/a = a/b = g$, and therefore $g = (1 + \sqrt{5})/2$. Called the *golden ratio,* $g$ is approximately 1.61803. Some authors call $1/g$ the golden ratio; note that $g^{-1} = (-1 + \sqrt{5})/2 = g - 1$.

The ruler and compass construction for the golden section is Proposition 10, Book IV, in Euclid's *Elements* (Euclid 1956). For a given segment $\overline{AB}$, let $\overline{BC} = \overline{AB}/2$ with segment $\overline{BC}$ perpendicular to segment $\overline{AB}$; let $D$ be on segment $\overline{AC}$ such that $\overline{CD} = \overline{CB}$; and let $E$ be on segment $\overline{AB}$ such that $\overline{AE} = \overline{AD}$. Then $E$ cuts $\overline{AB}$ in a golden section (Figure 1). Further, if $F$ is such that $\overline{BF} = \overline{EF} = \overline{EA}$, then $\triangle ABF$ is isosceles, with each base angle double the third angle. So $\angle BAF$ must be a 36° angle, and we can construct regular decagons and regular pentagons with a ruler and compass.

A diagonal and a side of a regular pentagon are in the golden ratio. The five diagonals of a regular pentagon form a figure called a *pentagram,* which was used as a sign of recognition in the secret society formed by the followers of Pythagoras. Any two segments from a pentagram that look as if they might be in the golden ratio actually are, as illustrated in the Disney classic *Donald in Mathmagic Land* (1959). A rectangle is "golden" if its adjacent sides are in the golden ratio. As an easy classroom illustration, a $5 \times 8$ index card provides a better approximation than a $3 \times 5$ card. Examples of golden rectangles are found in architecture from the Acropolis in Athens to the U.N. Secretariat in New York City. The golden ratio is also ubiquitous in art and sculpture based on the human form. For a construction of a golden rectangle starting with square *ABCD,* let $E$ be the midpoint of segment $\overline{BC}$ and let $F$ be such that $\overline{EF} = \overline{ED}$, with $C$ between $E$ and $F$. Then the rectangle *ABFG* is a golden rectangle (Figure 2). Furthermore, $\overline{DF}$ and $\overline{CF}$ are the lengths of sides of regular

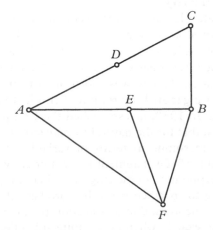

**Figure 1** *Ruler and compass construction for the golden section.*

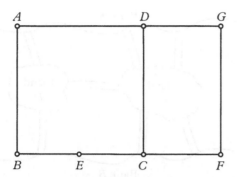

**Figure 2** *Construction of a golden rectangle.*

pentagons and regular decagons inscribed in a circle with radius *AB*.

*See also* Art; Enrichment, Overview; Recreations, Overview

## SELECTED REFERENCES

Baumgart, John K., et al. *Historical Topics for the Mathematics Classroom.* 2nd ed. Reston, VA: National Council of Teachers of Mathematics, 1989.

*Donald in Mathmagic Land.* Burbank, CA: Walt Disney, 1959. Animated film/video cassette.

Euclid. *The Thirteen Books of Euclid's Elements.* 2nd ed. 3 vols. Thomas L. Heath, trans. New York: Dover, 1956.

Eves, Howard W. *Introduction to the History of Mathematics.* 6th ed. Philadelphia, PA: Saunders College Publishing, 1990.

Huntley, Herbert E. *The Divine Proportion: A Study in Mathematical Beauty.* New York: Dover, 1970.

Newman, Rochelle, and Martha Boles. *The Golden Relationship: Art, Math and Nature.* Book 1: *Universal Patterns.* 2nd ed. Bradford, MA: Pythagorean Press, 1987.

GEORGE E. MARTIN

# GRAPH THEORY, OVERVIEW

A graph is a set of points in a defined space and a set of segments that connect some points. It is common to think of a graph as a curve on a plane such as a parabola. But graphs are more fundamental than simply curves on a plane.

A point of a graph is called a *node;* an interconnecting segment between two nodes is called an *edge* or *arc.* If we let $N$ denote the set of nodes and let $E$ denote the set of edges, we can say a graph $G$ is the ordered set $G = (N,E)$. We denote the cardinal number of set $N$ and set $E$ by $\#N$ and $\#E$, respectively. So, for $N = \{n_1, n_2, \ldots, n_j\}$ and $E = \{e_1,$

$e_2, \ldots, e_k\}$, $\#N = j$ and $\#E = k$. All graphs must contain at least one node; that is, $\#N \geq 1$. A null graph is one that has no edge, so $\#E = 0$.

Figure 1 shows the five-node, three-edge graph $G_1 = (N_1, E_1)$ where $N_1 = \{a, b, c, d, e\}$ and $E_1 = \{x, y, z\}$. If we describe each edge by the nodes at its two endpoints, we can rewrite the set $E_1$ as a set of *un*ordered pairs $E_1 = \{(a,d), (c,d), (b,b)\}$ such that $x = (a,d)$, $y = (c,d)$, and $z = (b,b)$. The edge $z$ is a *self-loop* or simply a *loop,* because it has the same endpoints. When $\#N$ and $\#E$ are both finite, the graph is a *finite graph;* when the graph has an infinite number of points, the graph is an *infinite graph.* The graph of Figure 1 is clearly finite.

It is possible to have more than one edge connecting the same two nodes. Such edges are *parallel edges.* Consider the five-node, five-edge graph of Figure 2 (see below). Here, $G_2 = (N_2, E_2)$, where $N_2 = \{a, b, c, d, e\}$ and $E_2 = \{s, t, u, v, w\} = \{(a,b), (b,c), (b,c), (c,d), (a,e)\}$. Edges $t$ and $u$ are parallel edges. A graph having at least one parallel edge and no loop is a *multigraph.* A graph displaying no parallel edge and no loop is a *simple graph.*

An interesting feature of a graph is a *path* going from one node of the graph to another node along its

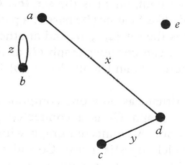

**Figure 1** *A five-node, three-edge graph.*

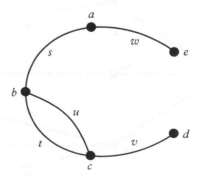

**Figure 2** *A five-node, five-edge graph.*

edges. We denote a *path* by a sequence of nodes. For example, we represent a path from node $n_1$ to node $n_2$ in graph $G = (N,E)$ by the sequence $(n_1, \ldots, n_i, n_{i+1}, \ldots, n_2)$, where each unordered pair $(n_i, n_{i+1})$ formed from the sequence is an element of $E$. From graph $G_1$ of Figure 1, the sequence $(a, d, c)$, with unordered pairs $(a,d)$ and $(d,c)$, members of $E_1$, represents a path from $a$ to $c$. Notice that no path exists from node $b$ to any other node other than itself because no such unordered pair is a member of $E_1$. Also, neither $(a,c)$ nor $(c,e)$ is a member of set $E_1$. A path is a *simple path* if no node appears more than once in the sequence describing it. The path $(a, d, c)$ from $G_1$ is a simple path.

A path that starts and ends at the same node is a *cycle* or *circuit*. From graph $G_2$ of Figure 2, the path $(b, c, b)$ is a cycle. It is a *simple cycle* because no node in the sequence, except the first and last nodes, appears more than once. For a given graph, the set of all nodes that have a path connecting them to each other is a *component* of the graph. Graph $G_2$ of Figure 2 contains only one component because for any two nodes in the graph we can describe a path between them. For example, for nodes $c$ and $e$, the sequence $(c, b, a, e)$ along edges $t$, $s$, and $w$ is one possible path. However, graph $G_1$ of Figure 1 contains three components. One component is the set $\{b\}$. Clearly, an isolated node like $b$ cannot be part of a path. Another component is the set $\{a, c, d\}$, and another is the set $\{e\}$. Thus, we can consider graph $G_1$ as the disjoint union of these sets and write $N_1 = \{b\} \cup \{a, c, d\} \cup \{e\}$.

A graph that has only one component is a *connected graph*; graph $G_2$ is a connected graph and graph $G_1$ is not. A connected graph with no cycles and no parallel edges is a *tree*. Consider the graph $G_3 = (N_3, E_3)$ of Figure 3, with $N_3 = \{n_1, n_2, \ldots,$

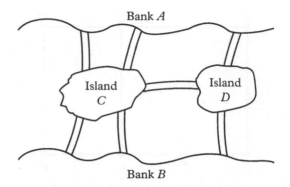

**Figure 4**  *The Königsberg bridge problem.*

$n_{10}\}$ and $E_3 = \{e_1, e_2, \ldots, e_9\}$. This tree contains ten nodes and nine edges. For any tree, $T = (N_T, E_T)$, $\#E_T = \#N_T - 1$; that is, the number of edges is always one less than the number of nodes.

It is useful at this point to reflect on the beginnings of graph theory. In the middle of the eighteenth century, Leonhard Euler addressed a problem created from a pastime of the people of Königsberg, Prussia, now Kaliningrad, Russia. The Pregel River passed through the city of Königsberg, forming two river banks, $A$ and $B$. Two small islands in the river, $C$ and $D$, connect to each other and to the river banks by seven bridges, as shown in Figure 4. People of the city would take a stroll and try to start and end at the same place while crossing each bridge exactly once. They soon began to realize that this goal was not possible.

Euler expressed the problem in mathematical terms. A point (a node) represented each river bank and each island; an arc (an edge) represented each bridge, as shown by the graph of Figure 5. The problem reduced to the following: find a closed path (a cycle or circuit) such that each edge is traversed exactly once. We call a path containing each edge of a

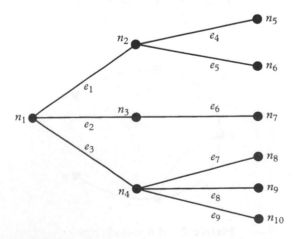

**Figure 3**  *A tree.*

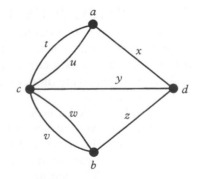

**Figure 5**  *A graph for the Königsberg bridge problem.*

graph an *Euler path* and call a closed path containing each edge of a graph an *Euler circuit*. The solution to the Königsberg problem hinges on the number of edges attributed to each node of the graph that models the problem. The number of edges having node $n_i$ as an endpoint is the *degree* of node $n_i$ and is denoted $\deg(n_i)$. We formalize the graph of Figure 5 as $K = (N_K, E_K)$, where $N_K = \{a, b, c, d\}$ and $E_K = \{t, u, v, w, x, y, z\}$. The degrees of each node are: $\deg(a) = 3$, $\deg(b) = 3$, $\deg(c) = 5$, and $\deg(d) = 3$. Euler proved that a connected graph $G$ contains an Euler circuit if and only if the degree of every node of $G$ is even. For the graph $K$ of Figure 5, no node is of even degree, so an Euler circuit cannot exist.

So far, the discussion of edges in a graph has implied that an edge can start or end from either of the two nodes that define it. This is not always the case. Sometimes an edge must be directed; that is, it can start only from one node and end at another. A graph that contains at least one directed edge is a *directed graph* or *digraph*. By convention, we define an edge in a digraph by the ordered pair $(n_i, n_j)$, meaning that the edge starts at $n_i$ and ends at $n_j$. An arrow with obvious direction represents a directed edge.

Figure 6 shows a digraph, $D = (N, E)$, of four nodes and three edges, where $N = \{a, b, c, d\}$ and $E = \{(a,b), (c,b), (c,d)\}$. With digraphs, it is important to distinguish arrows (directed edges) entering a node from those leaving a node. The number of edges directed toward the node $n_i$ is the *indegree* of the node, denoted $\deg^{in}(n_i)$. The number of edges directed away from the node $n_i$ is the *outdegree* of the node, denoted $\deg^{out}(n_i)$. For the digraph of Figure 6, $\deg^{in}(a) = 0$ and $\deg^{out}(a) = 1$, $\deg^{in}(b) = 2$ and $\deg^{out}(b) = 0$, $\deg^{in}(c) = 0$ and $\deg^{out}(c) = 2$, and $\deg^{in}(d) = 1$ and $\deg^{out}(d) = 0$. For a digraph, the degree of the node $n_i$ is the sum of the indegree and outdegree of $n_i$; that is, $\deg(n_i) = d^{in}(n_i) + d^{out}(n_i)$. For the digraph of Figure 6, $\deg(a) = 1$, $\deg(b) = 2$, $\deg(c) = 2$, and $\deg(d) = 1$. Note that the sum of the degrees of all nodes in a digraph (or a graph) must be even and equal to twice the number of edges. To traverse a path in a digraph, we must traverse the nodes along the edges of the digraph in the direction of the arrows of the edges. In Figure 6 there are three arrows between the nodes of the digraph. However, the digraph is not connected, in the sense that a path does not exist between any two nodes of the digraph. For example, node $a$ is not connected to node $c$ or node $d$ because the arrow from node $c$ to node $b$ is in the opposite direction. When a path does exist between any two nodes of a digraph, the di-

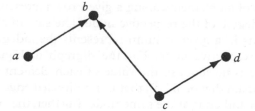

**Figure 6**   *A digraph of four nodes and three edges.*

graph is considered *strongly connected*. We can represent any graph as a digraph by replacing each edge with two opposite directed edges. The digraph of Figure 7 is the representation of the graph of Figure 2.

It is often convenient to represent a graph (or a digraph) by a matrix called the *adjacency matrix* of the graph. To do this, we represent each node as a row position with a corresponding column position. An entry or element of the matrix corresponds to the number of edges going from one node (the row position) to another node (the column position).

Consider the digraph of Figure 6. The adjacency matrix corresponding to the digraph is

$$
\begin{array}{c}
 \\ a \\ b \\ c \\ d
\end{array}
\begin{array}{cccc}
a & b & c & d \\
\end{array}
\left[
\begin{array}{cccc}
0 & 1 & 0 & 0 \\
0 & 0 & 0 & 0 \\
0 & 1 & 0 & 1 \\
0 & 0 & 0 & 0
\end{array}
\right]
$$

From the first row, the matrix entries show that only one directed edge starts from node $a$ and that it ends at node $b$. The second and fourth rows show that no directed edge starts from nodes $b$ and $d$, respectively. The third row indicates that two directed edges start from node $c$; one edge ends at node $b$ and the other edge ends at node $d$. Likewise, an entry in a column indicates the number of edges ending at a node. The

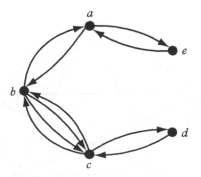

**Figure 7**   *A digraph representation of a graph.*

sum of all elements along a given row represents the outdegree of the respective node. The sum of all elements in a given column represents the indegree of the respective node. For the digraph of Figure 6, there is no loop, so the value of each element along the main diagonal is 0; that is, no directed edge starts from and ends at the same node. Further, the sum of all elements in the matrix represents the number of directed edges of the digraph.

For the graph of Figure 5 (the Königsberg graph), the corresponding adjacency matrix is

$$\begin{array}{c} \\ a \\ b \\ c \\ d \end{array} \begin{array}{cccc} a & b & c & d \\ \left[\begin{array}{cccc} 0 & 0 & 2 & 1 \\ 0 & 0 & 2 & 1 \\ 2 & 2 & 0 & 1 \\ 1 & 1 & 1 & 0 \end{array}\right] \end{array}$$

The value of each element along the main diagonal of the matrix is 0, so the graph contains no loop. Also, there is a symmetry about the main diagonal, suggesting the existence of a pair of opposite directed edges for each undirected edge of the graph. The sum of the elements in the first row $a$ (or first column) is 3, the degree of node $a$. Similarly, the sum of the elements in row $b$ is 3. In row $c$ the sum is 5, and in row $d$ the sum is 3. The sum of all elements in the matrix is 14. This means that there are 14 directed edges. Since the original graph has no directed edge, we conclude that the graph has seven undirected edges.

Graphs and digraphs are very important in the study of mathematics and computing. One use of graphs is to design computer circuits so that they are as simple as possible. To do this, we use the idea of a *planar graph*—a graph in which no edge overlaps with any other edge. The graph of Figure 8 is not planar because it is not possible to draw the graph without overlapping at least one edge. The graph of Figure 9(a) shows two overlapping edges. However,

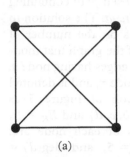
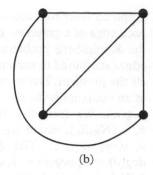

(a)          (b)

**Figure 9** *A planar graph (two views).*

the graph is planar because we can redraw the graph with no overlapping edge, as shown in Figure 9(b). A challenge in designing computer circuits is to simplify the wiring (edges) of the circuit; that is, make their graphs as planar as possible.

There are many more interesting applications of graphs, such as in molecular diagrams, communications, and transportation. Other uses are modeling the organization of data (Aho and Ullman 1992), describing the shortest-path algorithm (Crisler and Meyer 1993), and displaying algorithms that construct minimal spanning trees (Epp 1990). Other uses of graphs include solving coloring problems (Maurer and Ralston 1991), describing finite state machines (Gould 1988), and illustrating breadth-first and depth-first algorithms (Gersting 1987). Graphs have become a practical and theoretical tool in the study of mathematics and the sciences.

Teachers of mathematics at all levels can use graphs in their classrooms. In the elementary school, students can construct digraphs to show how they walked from one place to another. In the middle school, students can find out if certain given graphs can be made into planar graphs. At the high school level, we can emphasize the natural association between graphs and matrices. Graphs are fun and students can learn more mathematics by studying them.

*See also* Discrete Mathematics; Enrichment, Overview; Recreations, Overview

### SELECTED REFERENCES

Aho, Alfred V., and Jeffrey D. Ullman. *Foundations of Computer Science.* New York: Computer Science, 1992.

Crisler, Nancy, and Walter Meyer. *Shortest Paths. Geometry and Its Applications.* Lexington, MA: COMAP, 1993.

Epp, Susanna S. *Discrete Mathematics with Applications.* Belmont, CA: Wadsworth, 1990.

Gersting, Judith L. *Mathematical Structures for Computer Science.* New York: Freeman, 1987.

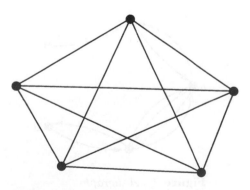

**Figure 8** *A nonplanar graph.*

Gould, Ronald. *Graph Theory*. Menlo Park, CA: Benjamin-Cummings, 1988.

Maurer, Stephen B., and Anthony Ralston. *Discrete Algorithmic Mathematics*. Reading MA: Addison-Wesley, 1991.

JOHN IMPAGLIAZZO

# GRAPHS

Geometric representations of numerical data or relationships. Two major types are *function graphs* and *statistical graphs*. The latter, sometimes called *charts,* are for the presentation of statistical data. Often used in business and journalism, charts include bar graphs, line graphs, pie charts, and a host of other specific types. Most computer spreadsheet programs include provisions for creating many kinds of charts from numerical data. In mathematics, a reference to graphs usually means function graphs, of which there are several kinds. Here, the graphs of explicit functions, implicit functions, and functions of two variables are discussed.

## GRAPHS OF EXPLICIT FUNCTIONS

A *function* is a rule, usually expressed as a formula, that assigns to each of a collection of numbers another number. If $f$ is the symbol representing the function, $x$ represents a number, and $y$ is the number that $f$ assigns to $x$, we call $y$ *the image of $x$ under $f$,* written $y = f(x)$, and read "$y$ equals $f$ of $x$." The collection of all numbers to which $f$ assigns images is called the *domain* of $f$, and the set of all images is called the *range* of $f$. For example, the function $f(x) = x^2 - 1$ assigns to each real number $x$ the number one less than the square of $x$. For instance, $f(3) = 8$, $f(1) = 0$, $f(-2) = 3$, and $f(-3) = 8$. The domain of $f$ is the set of all real numbers, and the range of $f$ is the set of real numbers not less than $-1$.

A means for representing functions geometrically was developed in the early seventeenth century by René Descartes and Pierre de Fermat. Because Descartes published the idea (*Discours de la Méthode* 1637) and Fermat's papers were published only posthumously, the resulting "coordinate geometry" is often called *Cartesian* geometry. Two lines in the plane are chosen, a horizontal line, called the *x-axis,* and a vertical line, called the *y-axis.* Their point of intersection is called the *origin* of the coordinate system. On each axis, the points are associated with the numbers of the real number system, zero usually cor-

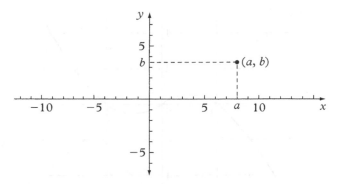

**Figure 1**  *Rectangular coordinate system.*

responding to the origin in each case. It is also customary to place the positive numbers to the right of the origin on the *x*-axis and above the origin on the *y*-axis.

There is a pair of numbers, called *coordinates,* associated with each point in the plane. Coordinates are represented as ordered pairs; the point with coordinates $(a,b)$ lies on the vertical line meeting the *x*-axis at the number $a$ and on the horizontal line meeting the *y*-axis at the number $b$. The first coordinate $a$ is called the *x-coordinate* or *abscissa,* and the second coordinate $b$ is called the *y-coordinate* or *ordinate* (see Figure 1). The *graph* of a function $f$ is the set of points $(x,y)$ in the Cartesian plane such that $x$ is in the domain of $f$ and $y = f(x)$. For example, the graph of $f(x) = x^2 - 1$ passes through the points $(-2,3)$, $(0,-1)$, $(1,0)$ and $(3,8)$ because the second coordinate in each case is the image of the first coordinate under $f$.

The usefulness of a graph is that it is a visual representation of a numerical relationship and the features of the relationship can be read from the graph. Where the graph of a function is above the *x*-axis, the function is positive; where the graph touches the *x*-axis, the function is zero. Where the graph rises as followed from left to right, the function is increasing; where it falls, the function is decreasing. Where the graph is increasing and curving upward, or concave upward, the function is increasing at an increasing rate; where the graph is increasing and concave downward, the function is growing at a decreasing rate. For example, the graph of the function $f(x) = x^3 - x$ (Figure 2) shows pictorially that the function has value zero at $x = -1$, $x = 0$, and $x = 1$; is positive for $-1 < x < 0$, and $x > 1$; and negative for $x < -1$ and $0 < x < 1$. The function is increasing for $x < -\sqrt{3}/3$ and for $x > \sqrt{3}/3$ and decreasing for $-\sqrt{3}/3 < x < \sqrt{3}/3$; concave upward for

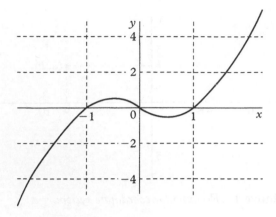

**Figure 2**  $y = x^3 - x$.

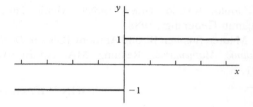

**Figure 4**  $f(x) = \dfrac{|x|}{x}$ *discontinuous at*
$x = 0$.

$x > 0$ and concave downward for $x < 0$; and grows large without bound as $x$ gets large without bound. Similarly, the graph of the function $f(x) = 1/(x^2 + 1)$ (Figure 3) shows that the function is always positive and never greater than 1, has its maximum value of 1 at $x = 0$, and approaches 0 as $x$ gets large.

For most functions defined by simple formulas, the graph will be a single curve in the plane; such a graph is called *continuous*. For some functions, the graph will be *discontinuous*, coming in several pieces or having points missing. For example, the function $f(x) = |x|/x$ is not defined at 0, has value 1 if $x > 0$, and has value $-1$ if $x < 0$ (Figure 4).

Creating the graph of a function is usually a matter of finding points on the graph and sketching the graph through the points, by hand or by machine. Graphing calculators and computer graphing software make the generation of accurate graphs much easier than drawing them by hand, but bring with

them their own terminology and limitations. Typically, before a graph can be drawn by machine, a formula defining the graph must be entered in a form readable by the computer. Then a viewing window must be chosen; only that portion of the graph that appears in that window will be drawn. If axes will appear in the window, the user may decide whether they will be drawn, whether there will be scale marks on the axes, whether they will be labeled, and whether the axes will be drawn through the origin or through some other point. Some machines allow the user to decide whether there will be gridlines drawn, what colors the graph, axes, and background will be, and so on. Most systems allow for drawing another graph without erasing the first, so that multiple graphs can be shown. On the other hand, a graph may be shown incorrectly on a machine, perhaps because of limited screen resolution. In general, the more flexible the graphing environment, the longer it takes to learn to use it.

## GRAPHS OF IMPLICIT FUNCTIONS

If $E(x,y)$ is an expression in $x$ and $y$, the equation $E(x,y) = 0$ is said to define $y$ *implicitly* as a function $f$ of $x$ if and only if the equation is satisfied when $y$ is replaced with $f(x)$. Typically, a single equation in $x$ and $y$ defines more than one function implicitly; for example, the equation $x^2 + y^2 = 1$ defines both of the functions $y = \sqrt{1 - x^2}$ and $y = -\sqrt{1 - x^2}$ implicitly. If an equation $E(x,y) = 0$ defines a function $f$ implicitly, the graph of $f$ is also called the *graph of the equation* $E(x,y) = 0$; it consists of the points $(x,y)$ such that $E(x,y) = 0$. It is usually drawn by finding the explicit functions defined by the equation and graphing them, or by using a computer, which samples many points and interpolates between points for which $E(x,y)$ is positive and points for which $E(x,y)$ is negative.

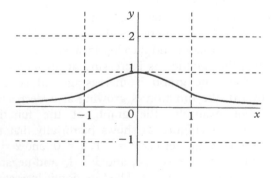

**Figure 3**  $y = \dfrac{1}{x^2 + 1}$.

## FUNCTIONS OF TWO VARIABLES

A *function of two variables* is a rule that assigns a number to a *pair* of numbers. If $x$ and $y$ are numbers and $z$ is the number assigned to the pair $(x,y)$ by a function $f$, we say $z$ is the *image* of $(x,y)$ and write $z = f(x,y)$. The set of all pairs $(x,y)$ for which $f$ assigns images is called the *domain* of $f$, and may be thought of conveniently as a set of points in the Cartesian plane. For example, the function $f(x,y) = x^2 + y^2$ is the function assigning to any pair of numbers the sum of their squares; its domain is the entire plane. On the other hand, the function $g(x,y) = \sqrt{1 - x^2 - y^2}$ assigns to a pair of numbers the square root of the difference between the sum of their squares and 1; its domain consists only of points for which $x^2 + y^2 \leq 1$; namely, the set of points on and inside the unit circle.

Because three numbers are involved with a function of two variables, a three-dimensional coordinate system is required to represent its graph. We take three lines in space that are mutually perpendicular to one another as axes; it is customary to take the *x-axis* and *y-axis* to be horizontal, so that the *z-axis* will be vertical. The plane containing the $x$- and $y$-axes is called the *xy-plane,* and the point of intersection of the axes is called the *origin.*

Two distinct orientations of the axes are possible. A *right-hand* system is one in which, if one were to point the first finger of the right hand in the direction of the positive $x$-axis and the second finger in the direction of the positive $y$-axis, then the thumb would be pointing in the direction of the positive $z$-axis. A right-hand system is customary, for the usual Cartesian plane is seen if one looks down on the $xy$-plane from a point on the positive $z$-axis (in a left-hand system, either the $x$- or $y$-axis would be oppositely oriented).

The *graph* of a function $f$ of two variables is the set of points $(x,y,z)$ in space for which $(x,y)$ is in the domain of $f$ and $z = f(x,y)$. It is convenient to think of $f(x,y)$ as a directed distance from the $xy$-plane at the point $(x,y)$. Thus, a function of two variables typically has as its graph a surface in space.

Representing a surface in space is most easily done by drawing a *wireframe* model of the surface. For example, the curve in which the $yz$-plane intersects a surface is called the *trace* of the surface in that plane. If several planes parallel to the $yz$-plane are chosen and the traces of the surface in them are drawn, we get a family of curves lying on the surface. If several planes parallel to the $xz$-plane are chosen, the traces of the surface in them form another family of curves on the surface. If both families are drawn,

the wireframe model of the surface appears and the shape of the surface is evident.

Computer graphics programs (see Newman and Sproull 1979, 325–351) for drawing graphs of functions of two variables typically start with the wireframe model. First, a point in space from which the graph will be viewed is chosen, and then several planes parallel to the $xz$- and $yz$-planes are chosen and the traces of the graph in those planes are drawn. Programs may offer coloring or shading to visually enhance the graph drawn; a part of the surface not visible from the chosen point of view can be hidden. The graph of $f(x,y) = x^2 - y^2$ is shown as a wireframe model, plain (Figure 5) and shaded (Figure 6), so that portions of the surface that could not be seen if the surface were opaque are not shown (*hidden surface* mode).

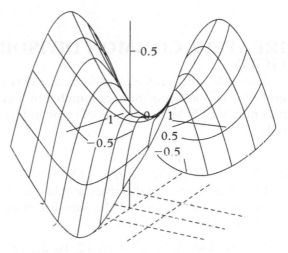

**Figure 5** $z = x^2 - y^2$ *wireframe, hidden surface drawing.*

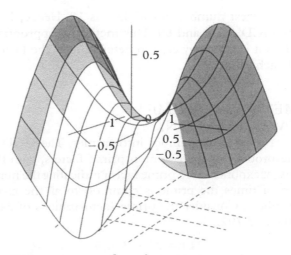

**Figure 6** $z = x^2 - y^2$ *shaded, hidden surface drawing.*

*See also* Coordinate Geometry, Overview; Data Representation; Functions of a Real Variable

## SELECTED REFERENCES

Boyer, Carl B. *A History of Mathematics.* 2nd ed. Revised by Uta C. Merzbach. New York: Wiley, 1989.

Garner, Lynn E. *Calculus and Analytic Geometry.* San Francisco, CA: Dellen, 1988.

Newman, William H., and Robert F. Sproull. *Principles of Interactive Computer Graphics.* New York: McGraw-Hill, 1979.

Spiegel, Murray R. *Schaum's Outline Series: Mathematical Handbook of Formulas and Tables.* New York: McGraw-Hill, 1994.

Struik, Dirk J. *A Source Book in Mathematics, 1200–1800.* Cambridge, MA: Harvard University Press, 1969.

LYNN E. GARNER

# GREATEST COMMON DIVISOR (GCD)

Also called the *greatest common factor,* the largest number that is a divisor of each number in a given set of numbers. Several methods exist for finding the GCD.

## METHOD I (LISTING FACTORS)

List all the factors of the given numbers and find the largest factor common to the lists. For instance, consider the factors of 48 and 60:

Factors of 48: 1, 2, 3, 4, 6, 8, 12, 16, 24, 48

Factors of 60: 1, 2, 3, 4, 5, 6, 10, 12, 15, 20, 30, 60

The largest number in *both* lists is 12. Hence, 12 is the GCD of 48 and 60. This method is appropriate when it is relatively easy to determine all the factors of each number.

## METHOD II (PRIME FACTORIZATION)

Factor each number into primes. The GCD is the product of the common prime factors, with the lowest exponent on a prime factor indicating the number of times the prime is common to all the given numbers. Consider the prime factorizations of 240, 504, and 792:

$$240 = 2^4 \times 3 \times 5$$
$$504 = 2^3 \times 3^2 \times 7$$
$$792 = 2^3 \times 3^2 \times 11$$

Notice that 2 is a common prime factor, occurring at least three times in each number. Likewise, 3 is a common prime factor, occurring at least once in each number. So, the GCD is $2^3 \times 3^1$, or 24. This method is particularly useful when it is difficult or cumbersome to list all the individual factors of the given numbers.

## METHOD III (EUCLIDEAN ALGORITHM)

Euclid's algorithm, described in Book 7 of the *Elements* (Euclid 1956), ca. 300 B.C., is used with pairs of numbers. First the larger number is divided by the smaller. If there is no remainder, then the smaller number is the GCD. If there is a remainder, it is divided into the previous divisor. This process is continued until there is no remainder. The last divisor is the GCD of the two numbers. The GCD of 24 and 90 is found by first dividing 90 by 24. The remainder is 18. Now 24 is divided by 18. The remainder is 6. Next 18 is divided by 6, leaving no remainder. The last divisor is 6, so 6 is the GCD of 24 and 90. Most number theory texts contain a proof of the Euclidean algorithm (Long 1987; Ore 1948). The proof requires a demonstration that the last divisor divides the original two numbers and that any common divisor of the original two numbers also divides the last divisor.

## METHOD IV (GEOMETRIC APPROACH)

The Euclidean algorithm can be modeled geometrically by creating a rectangle whose dimensions are the given pair of numbers. From one of the rectangle's edges, the largest square possible is removed. From the remaining rectangle, again the largest square possible is removed. This process is continued until the entire rectangle has been removed. The length of the side of the last square removed is the GCD of the numbers representing the dimensions of the original rectangle. (See Olson 1991 for student sheets to use with this approach.)

## APPLICATIONS

The GCD is used in a number of situations. If both the numerator and denominator of a fraction, either arithmetic or algebraic, are divided by their GCD, then the fraction is simplified to lowest terms in a single step. If the GCD of the numerator and denominator is 1, then the fraction is already in lowest terms. The GCD of two numbers can be used to find the least common multiple of the two numbers. If $d$ is the GCD of $a$ and $b$, then their least common multiple is $ab/d$. For instance, the GCD of 25 and 30 is 5 and their least common multiple is 150.

A linear Diophantine equation of the form $ax + by = c$ has integer solutions if and only if the GCD of $a$ and $b$ also divides $c$. Consider the following problem: "An adult ticket to an amusement park costs \$20 and a child ticket costs \$12. How many adult and child tickets can be purchased for \$240? for \$230?" If $a$ is the number of adult tickets and $c$ is the number of child tickets, then $20a + 12c = 240$ describes the problem for the first question. This equation has at least one integer solution because 4, the GCD of 20 and 12, divides 240. One solution is to purchase 9 adult tickets and 5 child tickets. However, $20a + 12c = 230$ has no integer solutions because the GCD of 20 and 12 does not divide 230.

*See also* Factors; Fractions

SELECTED REFERENCES

Euclid. *The Thirteen Books of Euclid's Elements*. 2nd ed. 3 vols. Thomas L. Heath, trans. New York: Dover, 1956.

Long, Calvin T. *Elementary Introduction to Number Theory*. 3rd ed. Englewood Cliffs, NJ: Prentice-Hall, 1987.

Olson, Melfried. "A Geometric Look at Greatest Common Divisor." *Mathematics Teacher* 84(Mar. 1991): 202–208.

Ore, Oystein. *Number Theory and Its History*. New York: McGraw-Hill, 1948.

Reimer, Wilbert, and Luetta Reimer. *Historical Connections in Mathematics*. Vol. 2. *Resources for Using History of Mathematics in the Classroom*. Fresno, CA: AIMS, 1993.

DENISSE R. THOMPSON

# GREEK MATHEMATICS

The development of mathematics as an organized scientific body of knowledge is generally credited to the ancient Greeks, beginning with the work of Thales of Miletus in the sixth century B.C. A substantial portion of the early history of Greek mathematics, in particular geometry, is documented in the *Eudemian Summary* of Proclus, which was written in the fifth century A.D. Proclus based his work on a comprehensive history of geometry (from the time of Thales to 335 B.C.) that was written by Eudemus, a pupil of Aristotle. Thales was probably the first Greek mathematician to utilize the method of deductive reasoning to draw geometric conclusions. He is credited with discovering and perhaps even prov-

ing that the base angles of an isosceles triangle are congruent and that an angle inscribed in a semicircle is a right angle.

The next major figure in the development of Greek mathematics is Pythagoras, a student of Thales, who was born on the island of Samos around 580 B.C. After extensive travel to both Egypt and the Orient, Pythagoras migrated to the Greek city of Crotona in southern Italy where he founded a society devoted to the study of mathematics, philosophy, and science. A basic tenet of this society was that numbers (the positive integers) rule the universe, that is, everything in the universe can be described in terms of the positive integers and their ratios. The Pythagoreans' mystical view of numbers accounted for their association of different numbers with both physical and abstract attributes. For instance, they associated the number 1 with males, the number 2 with females, and the number 3 with marriage. The number 4 was associated with justice, which accounts for the use today of the expression "square deal" to represent a fair transaction. Their mystical interest in numbers resulted in an obsessive preoccupation with astrological matters, but it was also responsible for their extensive development of number theory and the eventual discovery of irrational numbers. This discovery devastated the Pythagoreans, who now recognized that their philosophical views of numbers were both inadequate and inconsistent.

Unfortunately, this lack of acceptance of incommensurable (irrational) magnitudes by the Greeks was largely responsible for the early demise of Eudoxus's theory of equal ratios, an early anticipation by him of the real number system. In this theory, Eudoxus introduced the concept of magnitudes, which were continuous geometric entities such as segments, angles, areas, and so on, and used them to define the equality of two ratios. He succeeded in linking the notions of ratio and proportion to geometry. Eudoxus is also commonly credited with being the first Greek to utilize the method of exhaustion to find the perimeter and area of a given circle, thus anticipating the infinite processes of the calculus. This discovery came at about the same time that Greek mathematicians and philosophers were struggling with the concept of infinity and its paradoxes as promulgated by Zeno of Elea.

All the significant mathematical advances in geometry, number theory, and infinite processes from Thales's time to about 300 B.C. were assimilated into one monumental work, called the *Elements*, by Euclid. The thirteen books (chapters) that comprise the *Elements* demonstrate the ingenuity and skill of Euclid in organizing and systematizing the body of existing mathematical knowledge into a coherent and logical axiomatic system. Although the *Elements* does not meet the modern criteria of mathematical rigor for an axiomatic system, it served as the model for the development of future axiomatic systems.

Shortly after Euclid produced his *Elements,* there emerged the most significant personality in the history of Greek mathematics, Archimedes. His contributions to mathematics, physics, mechanics, hydrostatics, and many other fields demonstrated his versatility and acumen as a mathematician, scientist, and inventor. Among the numerous mathematical achievements attributed to Archimedes is his anticipation of the calculus, which resulted from his ingenious application of the method of exhaustion to finding areas and volumes. Using this method, Archimedes determined the area and volume of a given figure by enclosing it between two other figures, one inscribed and one circumscribed. By essentially using a limit process, Archimedes was able to "close in" on the desired area or volume. Examinations of treatises written by Archimedes indicate quite clearly that his analyses of area and volume problems utilize ideas very similar to modern-day integration methods in calculus. The death of Archimedes in 212 B.C. ended the remarkable 400-year classical period, beginning in the sixth century B.C., characterized as the "glory that was Greece."

The classical period of Greek mathematics provides a fertile source for integrating the historical development of mathematics into the school curriculum. Euclid's magnificent organization of the known mathematics of his time into a logical sequence of propositions offers students a precursor of modern mathematical systems, Appolonius's treatment of conic sections demonstrates a synthetic approach to the topic that is in contrast to the analytic approach exhibited in contemporary curricula, and Archimedes's work with circles shows students early formulations of integral calculus. Although the death of Archimedes marked the end of the classical period of Greek mathematics, it did not bring an end to the contributions of the ancient Greeks. The later Alexandrian period, spanning the first four centuries A.D., represents another flourishing of mathematical activity through the works of Claudius Ptolemy, Diophantus, Heron, Pappus, and several other notable Greeks.

Ptolemy (A.D. 100–170), through his two masterpieces the *Almagest* and *Geographika Syntaxis,* achieved for astronomy and geography what Euclid had accomplished for geometry, that is, a comprehensive and brilliantly organized rendering of those disciplines that remained the definitive works on those subjects for many centuries thereafter. Diophantus of Alexandria (ca. A.D. 250) produced a monumental work, *Arithmetica,* which carved a niche for the Greeks in the historical development of algebra. The *Arithmetica,* a collection of 184 problems and their solutions, transformed algebra from its rhetorical stage of verbal representation, dating back to the Egyptians and Babylonians, to the syncopated stage that incorporates the use of symbolic abbreviations to represent unknown quantities. This was indeed a major advance in the evolutionary development of algebra. The last great mathematician of Hellenistic times is the geometer Pappus (ca. A.D. 320). In his great work, *Mathematical Collections,* Pappus succeeded in consolidating all the major geometric knowledge of his time while simultaneously offering the mathematical world alternative proofs and approaches to well-known results dating back to the "golden age of Greece." Pappus's work also provided a critical record of the mathematical activity of ancient geometers that would have been lost forever. The remarkable achievements in Greek mathematics came to an end after the death of the prominent female mathematician Hypatia in A.D. 415.

*See also* Archimedes; Calculus, Overview; History of Mathematics, Overview; Number Theory, Overview; Women and Mathematics, History

## SELECTED REFERENCES

Baumgart, John K., et al., eds. *Historical Topics for the Mathematics Classroom.* 2nd ed. Reston, VA: National Council of Teachers of Mathematics, 1989.

Calinger, Ronald, ed. *Classics in Mathematics.* Englewood Cliffs, NJ: Prentice-Hall, 1995.

Dunham, William. *Journey Through Genius: The Great Theorems of Mathematics.* New York: Wiley, 1990.

Fauvel, John, and Jeremy Gray. *The History of Mathematics: A Reader.* London, England: Macmillan, 1987.

Heath, Thomas L. *A Manual of Greek Mathematics.* New York: Dover, 1963.

Stillwell, John. *Mathematics and Its History.* New York: Springer-Verlag, 1989.

Swetz, Frank, ed. *From Five Fingers to Infinity: A Journey Through the History of Mathematics.* Chicago, IL: Open Court, 1994.

———, et al., eds. *Learn from the Masters.* Washington, DC: Mathematical Association of America, 1994.

ANTHONY V. PICCOLINO

# GROUP THEORY

A *group* is a collection $G$ of elements $a, b, c, \ldots$, together with an operation that assigns a product $ab$ in $G$ to each pair of elements $a$, $b$ such that: (1) the operation is associative, meaning $a(bc) = (ab)c$ for all elements $a$, $b$, $c$ in $G$; (2) $G$ has an identity element, that is, there is a special element $e$ in $G$ such that $ea = ae = a$ for all elements $a$ in $G$; (3) every element $a$ in $G$ has an inverse, that is, an element $a^{-1}$ in the collection $G$ such that $aa^{-1} = a^{-1}a = e$.

Sometimes additive notation is used for the operation on a group, in which case $ab$ is written $a + b$ and $a^{-1}$ is written $-a$. A group is *commutative*, or *abelian* (in honor of Neils Henrik Abel (1802–1829)), if $ab = ba$ for all elements in the group. A *subgroup* of a group is any collection $H$ of elements in the group such that the product $ab$ and inverse $a^{-1}$ lie in $H$ whenever $a$ and $b$ lie in $H$.

The collection of integers, $\{\ldots, -2, -1, 0, 1, 2, \ldots\}$, is a group under the operation of addition. Addition assigns the sum $n + m$ to each pair of integers $n$, $m$ and is an associative operation since $n + (m + s) = (n + m) + s$ for all integers $n$, $m$, and $s$. The integer $0$ is the identity element since $0 + n = n + 0 = n$ for all integers $n$. The inverse of an integer $n$ is its negative, $-n$, since $n + (-n) = (-n) + n = 0$. The group of integers under addition is commutative since $n + m = m + n$ for all integers $n$, $m$. Similarly, the set of rational numbers (fractions), the set of real numbers, and the set of complex numbers are commutative groups under addition. The set of nonzero rational numbers is a group under multiplication. But the set of nonzero integers is not a group under multiplication since not every integer has an inverse that is an integer.

Groups arise frequently in geometry as symmetry groups, that is, as groups of rigid motions of the plane that transform a given geometric figure into itself. An equilateral triangle, when rotated 0°, 120°, or 240° counterclockwise about its center, returns to a position coincident with itself. The triangle may also be reflected about a line passing through any one of its three vertices and the midpoint of the opposite side and still be coincident with itself. The symmetry group of an equilateral triangle $ABC$ contains six elements, three rotations ($R_0$, $R_{120}$, $R_{240}$), and three reflections ($S_A$, $S_B$, $S_C$) and is a finite group of order 6. The six elements are multiplied by following one transformation by another. The multiplication table, or Cayley table, of the group is shown in Figure 1. The product $S_A S_B$, for example, is computed by beginning with $S_B$ and following it by $S_A$; the result is the rotation $R_{120}$. Thus, $S_A S_B = R_{120}$.

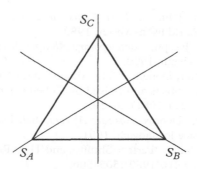

| $\circ$ | $R_0$ | $R_{120}$ | $R_{240}$ | $S_A$ | $S_B$ | $S_C$ |
|---|---|---|---|---|---|---|
| $R_0$ | $R_0$ | $R_{120}$ | $R_{240}$ | $S_A$ | $S_B$ | $S_C$ |
| $R_{120}$ | $R_{120}$ | $R_{240}$ | $R_0$ | $S_C$ | $S_A$ | $S_B$ |
| $R_{240}$ | $R_{240}$ | $R_0$ | $R_{120}$ | $S_B$ | $S_C$ | $S_A$ |
| $S_A$ | $S_A$ | $S_B$ | $S_C$ | $R_0$ | $R_{120}$ | $R_{240}$ |
| $S_B$ | $S_B$ | $S_C$ | $S_A$ | $R_{240}$ | $R_0$ | $R_{120}$ |
| $S_C$ | $S_C$ | $S_A$ | $S_B$ | $R_{120}$ | $R_{240}$ | $R_0$ |

**Figure 1** *Symmetry group of a triangle.*

Group theory explores the properties of groups, such as the relationship beween the number of elements in a finite group and the possible number of elements in each of its subgroups. Joseph Louis Lagrange (1736–1813) showed that the possible number of elements in a subgroup of a finite group must be a divisor of the number of elements in the group. The symmetry group of the triangle, for example, has six elements; according to Lagrange's theorem, it may have a subgroup containing three elements, but it cannot have a subgroup containing four elements. In fact, the three rotations $R_0$, $R_{120}$, and $R_{240}$ form a subgroup of three elements. Lugvig Sylow (1832–1918) proved several theorems about the existence and nature of subgroups of prime power order in a finite group, including the fact that if $p^n$ is the largest power of a prime $p$ that divides the number of elements in a group, then the group must have a subgroup containing $p^n$ elements. The symmetry group of the triangle, for example, must have a subgroup containing two elements since 2 is the largest power of 2 that divides 6. The two elements $R_0$ and $S_A$ form such a subgroup. Group theory has important applications in many areas of mathematics and the physical sciences (Dyson 1964; Kolata 1982; Mackey 1973; White 1967).

*See also* Abstract Algebra; Transformations

## SELECTED REFERENCES

Dyson, Freeman. "Mathematics in the Physical Sciences." *Scientific American* 211(1964):129–146.

Fraleigh, John. *A First Course in Abstract Algebra.* Reading, MA: Addison-Wesley, 1989.

Gallian, Joseph. *Contemporary Abstract Algebra.* Lexington, MA: Heath, 1990.

Hofstadter, Douglas. "The Magic Cube's Cubies Are Twiddled by Cubists and Solved by Cubemeisters." *Scientific American* 244(1981):20–39.

Kletzing, Dennis. *Abstract Algebra.* San Diego, CA: Harcourt Brace Jovanovich, 1991.

Kolata, Gina. "Perfect Shuffles and Their Relation to Math." *Science* 216(1982):505–506.

Mackey, George W. "Group Theory and Its Significance for Mathematics and Physics." *Proceedings of the American Philosophical Society* 117(1973):374–380.

White, J. Edmund. "Introduction to Group Theory for Chemists." *Journal of Chemical Education* 44(1967): 128–135.

DENNIS KLETZING

# GUIDELINES FOR SCIENCE AND MATHEMATICS IN THE PREPARATION PROGRAM OF TEACHERS OF SECONDARY SCHOOL SCIENCE AND MATHEMATICS, GUIDELINES FOR SCIENCE AND MATHEMATICS IN THE PREPARATION PROGRAM OF ELEMENTARY SCHOOL TEACHERS

Two sets of guidelines issued by the American Association for the Advancement of Science (AAAS) and National Association of State Directors of Teacher Education and Certification in 1961. The secondary school guidelines were revised in 1971 due to "extensive changes in the science and mathematics curricula [since 1961]." The original elementary school guidelines stressed the understanding of processes, concept development, interdisciplinary treatment, laboratory work, and in-service education. The original secondary guidelines emphasized the sequential nature of mathematics, the importance of a major in mathematics sufficient for graduate work, and curriculum recommendations. The guidelines of 1971 discussed humaneness, societal issues, curriculum recommendations (a major sufficient for graduate work, computing, mathematical modeling, written and oral communication of mathematics), the nature of learning, and adaptation for continued learning. Programs for junior high school teachers were to include elementary calculus, linear and abstract algebra, geometry, probability and statistics, number theory, and history of mathematics.

*See also* History of Mathematics Education in the United States, Overview

## SELECTED REFERENCES

American Association for the Advancement of Science. *Guidelines and Standards for the Education of Secondary School Teachers of Science and Mathematics.* Washington, DC: The Association, 1971.

Jones, Phillip S., ed. *A History of Mathematics Education in the United States and Canada.* 32nd Yearbook. Washington, DC: National Council of Teachers of Mathematics, 1970.

JAMES K. BIDWELL
ROBERT G. CLASON

# H

## HADAMARD, JACQUES SALOMON (1865–1963)

French mathematician (and mathematical physicist) whose first important contributions, which were in the area of functions of complex variables, dealt with the analytic continuation of a Taylor's series. Hadamard studied at the École Normale Supérieure, which awarded his doctoral degree in 1892. His teaching career began at the Lycée Buffon in Paris (1890–1893) and continued at a number of French institutions, often simultaneously, including the Sorbonne (1897–1909), Collège de France (1909–1937), and the École Polytechnique (1912–1937). He was elected to the Académie des Sciences in 1912 and was also an associate member of foreign academies such as the National Academy of Sciences of the United States and the Royal Society of London.

In 1896, Hadamard and Charles de la Vallée-Poussin (1866–1962), working independently, solved the problem concerning the distribution of prime numbers (posed by Carl Gauss in 1792 and now known as the Prime Number Theorem), demonstrating that the function $\pi(x)$, denoting the number of prime numbers less than $x$, is asymptotically equal to $x/(\ln x)$. Hadamard also influenced hydrodynamics, mechanics, and probability theory, as well as logic. He was interested in pedagogy and wrote on concepts in elementary mathematics. Like Henri Poincaré, he also wrote about the process of mathematical creation (1945).

*See also* Creativity; Number Theory, Overview; Prime Numbers; Series

### SELECTED REFERENCES

Hadamard, Jacques. *The Psychology of Invention in the Mathematical Field.* New York: Dover, 1945.

Mandelbrojt, S. "Jacques Hadamard." In *Dictionary of Scientific Biography* (pp. 3–5). Vol. 6. Charles C. Gillispie, ed. New York: Scribner's, 1972.

Maz'ya, Vladimir, and Tatyana Shaposhnikova. *Jacques Hadamard, a Universal Mathematician.* Providence, RI: American Mathematical Society, 1998.

Poincaré, Henri. "Mathematical Creation." In *The World of Mathematics* (pp. 2041–2050). Vol. 4. James R. Newman, ed. New York: Simon and Schuster, 1956.

LOUISE S. GRINSTEIN

SALLY I. LIPSEY

## HEALTH

Can be studied by many mathematical models. Our bodies, and those of all living organisms, are made up of cells. These are electrically polarized, meaning that electrical charges are separated, being electrically positive on one side of the cell membrane and negative on the other. The polarity of cells may change at various times, for example, before mechanical events such as muscle contraction. Indeed, the polarization and depolarization of certain cells in the heart give rise to its rhythmic contraction and, thus, the familiar heart beat. The latter is responsible for the pulse and the resulting flow of blood and other substances throughout the body. Muscles such as the heart contract and this contractile process is a mechanical event that may be viewed in mathematical models somewhat similar to those that apply to inanimate objects like springs. The blood vessels, which carry blood throughout the organism, are histologically distinguishable from the musculature of the heart and from the striated muscles responsible for willed movement and are classified as smooth muscle. Their elasticity is an important factor in the magnitude of the blood pressure, and their role as conduits for blood has invited a great deal of mathematical model building that is useful in understanding normal and abnormal flow patterns. The lungs deliver inhaled oxygen to the blood and, thus, to all the individual cells.

Control of all these processes is in part due to nerves, which conduct electrical impulses to muscles and other structures and to a myriad of chemicals and hormones. The organism is a complex organization of physicochemical processes; it is not surprising that many mathematical models are used to study this organization.

Fundamental to an understanding of cellular functions is the concept of the cellular receptor. This is a structure on or in the cell that binds specific chemical substances that reach it through the blood or the nerves. The receptor is thus a recognition site for a chemical. Heart muscle, for example, contains receptors for adrenalin and noradrenalin. When these molecules react with the receptor (termed a *beta* receptor) a series of intimate changes occur that cause the heart to beat rapidly and forcefully. The heart contains many other kinds of receptors as do virtually all other structures in the body. Many drugs act by mimicking the action of natural chemicals or by blocking the action of these chemicals. It is instructive, therefore, to illustrate the mathematical modeling of the interactions between chemical substances and cellular receptors. Such models have proved to be most useful in understanding the body's physiology and biochemistry and have provided a rational basis for the use of drugs to treat specific abnormal conditions.

The mathematical model for the interaction of a chemical, denoted $L$, and a receptor, denoted $R$, is one in which these bind in a reversible bimolecular reaction to form a complex: $L + R \longleftrightarrow LR$. Receptors ($R$) are generally fixed in number and the quantity of chemical ($L$) greatly exceeds this number so that the reaction is represented mathematically as $dx/dt = k_1L(R - x) - k_2x$, where $x$ is the quantity of the complex (bound receptor) and $k_1$ and $k_2$ are constants that uniquely characterize the chemical (or drug) and the receptor. It is this characterization that is so important in understanding the way cells or organ systems respond to chemical substances and drugs. This model equation and its mathematical solution have greatly aided the drug development process as well. An interesting empirical finding is that the increase in effect that generally is associated with larger doses of a drug or larger concentrations of a natural chemical has a special relationship to logarithms. Simply stated, provided the drug dose is not too low or too high, the magnitude of the drug-induced effect is found to be linearly related to the logarithm of the dose; that is, if a plot of effect against the logarithm of the dose is made (in Cartesian coordinates) the resulting graph is linear.

Mathematics is not only useful in these receptor reactions but also has wide usage in the drug delivery process. Most drugs are taken from absorptive sites, for example, from the oral cavity to the gastrointestinal tract and from there into the blood. Once in the blood, which delivers the drug molecule to all parts of the body, the quantity of drug will decrease due to its elimination through excretion by the kidney, chemical degradation in the liver, and possible losses from other biological pathways. But regardless of the pathway for ridding the body of the drug, the most common relation for quantity versus time is the decreasing exponential, which is expressed in terms of the drug's concentration $C$, its original concentration $C_0$, and the time $t$: $C = C_0e^{-at}$, where $a$ is a positive constant related to the rate of elimination and connects with the drug's half-life ($t_{1/2}$) according to the equation $t_{1/2} = \ln(2)/a$. This mathematical fact is of paramount importance in understanding how fast drugs leave the body and thus how much to give and when to give subsequent doses. In multiple dose regimens, the patient frequently takes the same dose $D$ every $T$ hr. It is desirable to know how concentrations build in patients with chronic illness who take a drug periodically. This question leads to the interesting relation for the peak amounts:

$$D(1 + f + f^2 + \cdots)$$

where $f = e^{-aT}$. Students of mathematics will recognize the geometric progression, a concept introduced in high school algebra. In this case, the ratio of any term to the preceding term is symbolized by $f$ and is related to the drug's exponential decay constant $a$ and the time interval between doses, $T$. Because $f < 1$, the infinite progression above converges and has the limit $D/(1 - f)$. Thus, drugs, which are eliminated exponentially, do not accumulate boundlessly when given repeatedly, but stay less than this upper limit.

Physiological processes are controlled. For example, the volume of blood per unit time pumped by the heart, the blood pressure, the body temperature, the concentration of many body chemicals, and many other phenomena depend on various feedback processes. Many principles of control system engineering have been used to study these physiological phenomena, and the mathematical methods used by control engineers have been enormously useful in these studies and in the development of the myriad of prosthetic devices used in modern medicine. Clinical practitioners also need an understanding of mathematical statistics in order to assess the efficacy of these and other treatments. Accordingly, almost all medical schools require that entering students be adequately prepared in mathematics.

*See also* Nursing

## SELECTED REFERENCES

Hoppensteadt, Frank C., and Charles S. Peskin. *Mathematics in Medicine and the Life Sciences.* New York: Springer-Verlag, 1992.

Rashevsky, Nicholas. *Mathematical Principles in Biology and Their Applications.* Springfield, MA: Thomas, 1961.

Tallarida, Ronald J., and Rodney B. Murray. *Manual of Pharmacologic Calculations with Computer Programs.* 2nd ed. New York: Springer-Verlag, 1987.

RONALD J. TALLARIDA

# HEXADECIMAL SYSTEM

A method of writing numbers in base 16, instead of the usual decimal system using base 10. The digits used in the hexadecimal system are 0, 1, 2, 3, 4, 5, 6, 7, 8, 9, A, B, C, D, E, and F, where A stands for 10, B for 11, C for 12, D for 13, E for 14, and F for 15. For example, the hexadecimal number 5D4 denotes the number $5 \times 16^2 + 13 \times 16 + 4$, which is the decimal number 1,492. We often add the base as a subscript, if a number is not written in decimal, so this hexadecimal would be written as $5D4_{16}$. Hexadeci-

mal numbers are also called *radix sixteen numbers*. The general hexadecimal number $r_m \ldots r_2 r_1 r_0$ represents $r_m \times 16^m + \cdots + r_2 \times 16^2 + r_1 \times 16^1 + r_0 \times 16^0$. (See Knuth 1973, Ch. 4.1 and Resnikoff and Wells 1984, Ch. 1, for the history of positional number systems.)

The hexadecimal system only became popular when digital computers started representing numbers internally in binary format. The advantage of the hexadecimal system is that it can represent these numbers more compactly than binary. It is easy to convert numbers between their hexadecimal and binary forms. Each hexadecimal digit corresponds to four binary digits; $0_{16} = 0000_2$, $1_{16} = 0001_2$, $2_{16} = 0010_2$, $3_{16} = 0011_2$, $4_{16} = 0100_2$, $5_{16} = 0101_2$, $6_{16} = 0110_2$, $7_{16} = 0111_2$, $8_{16} = 1000_2$, $9_{16} = 1001_2$, $A_{16} = 1010_2$, $B_{16} = 1011_2$, $C_{16} = 1100_2$, $D_{16} = 1101_2$, $E_{16} = 1110_2$, and $F_{16} = 1111_2$. Binary numbers can be converted to hexadecimal by grouping the binary digits in fours, starting at the binary point, and then replacing each group of four binary digits by the corresponding hexadecimal digit. For example, $110\ 1111\ 0010\ 1100_2 = 6F2C_{16}$.

It is a little more complicated to convert a whole number from the decimal form to the hexadecimal form. This is achieved by repeatedly dividing by the base 16 until there is no remainder. Consider the decimal number 3,236. Dividing by 16 gives $202 \times 16 + 4$. Then dividing the quotient 202 again by 16 gives $12 \times 16 + 10$. The process is continued in this way until there is no remainder. The remainders in these divisions are the hexadecimal digits.

$$
\begin{aligned}
3,236 &= 202 \times 16 + 4 \\
&= (12 \times 16 + 10)16 + 4 \\
&= 12 \times 16^2 + 10 \times 16 + 4 \\
&= CA4_{16}
\end{aligned}
$$

Arithmetic can be performed directly using hexadecimal numbers, but you have to know the addition and multiplication tables for all pairs of numbers up to 15. For example, $7_{16} + D_{16} = 7 + 13 = 20 = 14_{16}$ and $7_{16} \times D_{16} = 7 \times 13 = 91 = 5B_{16}$. Some calculators have options for performing hexadecimal arithmetic.

Just as fractional numbers can be represented in the decimal system by using expansions with a decimal (or radix) point, fractional numbers can be represented in the hexadecimal system using expansions with a radix point. The decimal equivalent of $2E.3F_{16}$ is thus $2 \times 16 + 14 + 3 \times 16^{-1} + 15 \times 16^{-2} = 46 + (63/256) = 46.24609375$.

Students should construct addition and multiplication tables for hexadecimal digits. They can also explore number systems in other bases and discuss the advantages and disadvantages of different bases.

See also Bases, Complex; Bases, Negative; Binary (Dyadic) System

SELECTED REFERENCES

Knuth, Donald E. *The Art of Computer Programming*. Vol. 2. *Seminumerical Algorithms*. Reading, MA: Addison-Wesley, 1973.

Resnikoff, Howard L., and Raymond O. Wells. *Mathematics in Civilization*. New York: Dover, 1984.

WILLIAM J. GILBERT

# HILBERT, DAVID (1862–1943)

The most prominent spokesperson for the "formalist" school of mathematicians who believe that the consistency and completeness of the set of rules by which mathematics is done are the most important considerations in deciding the ultimate basis for mathematical thought. Although Gödel's work established that such a set of rules could never be found, most practicing mathematicians today are indeed formalists who follow the Hilbertian tradition.

Hilbert was born and died in Germany. He initiated his university studies in mathematics at the University of Königsberg, the institution where Emmanuel Kant studied and taught and where the mathematician Carl G. Jacobi established a strong center for mathematics in the nineteenth century. After proving himself to be a mathematician of the first rank in various university positions, Hilbert was appointed in 1895 to a chair at Göttingen University, the most prestigious center for mathematics in the world from the time of Carl F. Gauss until the rise of Nazi Germany in the early 1930s. He remained there for the rest of his career.

At the turn of the century, Henri Poincaré and Hilbert were considered the two greatest living mathematical scientists; by the slimmest of margins, Hilbert was chosen to give the keynote address to the Second International Congress of Mathematicians at Paris in 1900. In that talk, he proposed twenty-three mathematical problems that he deemed would be the lifeblood of mathematical research in the first decades of the new century. His choices proved to be quite prophetic and fruitful to the mathematical community. For example, the second problem was to establish that the axioms of arithmetic are consistent. In 1931, Kurt Gödel surprisingly showed that this problem is unsolvable when he proved that in *no* logical system can the logical consistency of its axiom set *ever* be established by operating within the system itself.

As a mathematician, Hilbert is renowned for repeatedly mastering one field of mathematics for a substantial period of time, then turning his efforts to a quite different area of mathematical endeavor. Thus he began his studies in algebraic fields associated with algebraic language and number theory, moved on to foundational questions in geometry, then proceeded to make invaluable contributions to continuous mathematics (analysis), before doing research in theoretical physics.

See also History of Mathematics, Overview; Mathematics, Foundations; Philosophical Perspectives on Mathematics

SELECTED REFERENCES

Burton, David. *History of Mathematics, An Introduction*. 3rd ed. Dubuque, IA: Brown, 1995.

Eves, Howard W. *An Introduction to the History of Mathematics*. 6th ed. Philadelphia, PA: Saunders College Publishing, 1990.

Gillispie, Charles, ed. *Dictionary of Scientific Biography*. Vol. VI. New York: Scribner's, 1981–1990.

RICHARD M. DAVITT

# HISTORY OF MATHEMATICS, OVERVIEW

"The history of mathematics is exhilarating, because it unfolds before us the vision of an endless series of victories of the human mind." So wrote the noted twentieth-century historian of science George Sarton (1884–1956), who clearly recognized the importance of knowing the history of science and understanding its relationship to all human endeavors (Sarton 1958). In his writings, Sarton frequently singled out mathematics as a special discipline a true understanding of which required a historical background. In this respect, he echoed the sentiments of many early-twentieth-century educators who urged a stronger association of mathematics teaching with a history of the subject. In general, they felt that history enriched mathematics learning for students and led to a better understanding of just how mathematical concepts were conceived, formalized, and eventually applied. Perhaps the most active individual advocate of this movement in the United States was David Eugene Smith (1860–1944) of Teachers College, Columbia University. Smith's involvement with the learning and teaching of mathematics convinced him that in order to develop a good pedagogy for the subject, one must know and understand its history. He researched that history and lectured and wrote widely on its uses in teaching (Smith 1958). Perhaps influenced by Smith and his disciples from Teachers College, many classroom teachers during the first half of the twentieth century adhered to this principle. In its first yearbook (1926), the National Council of Teachers of Mathematics (NCTM), the principal professional support organization for teaching of mathematics in North America, featured an article (Austin et al. 1926, 198) describing the experiences of a teacher who challenged her mathematics students with such questions as "Who was Thales?" and "Where did the signs '+', '−', '×', '÷' and '=' come from?" Since that early period, the NCTM, through its series of publications and conference presentations, has supported the premise that the history of mathematics be reflected in the teaching of mathematics.

Recent suggested reforms intended to improve mathematics education include the recommendation that "students should have numerous and varied experiences related to the cultural, historical, and scientific evolution of mathematics so that they can appreciate the role of mathematics in the development of our contemporary society and explore relationships among mathematics and the disciplines it serves: the physical and life sciences, the social sciences, and the humanities. . . . It is the intent of this goal—learning to value mathematics—to focus attention on the need for student awareness of the interaction between mathematics and the historical situations from which it has developed and the impact that interaction has on our culture and our lives" (NCTM 1989, 5–6). Once again, the need for associating history with the teaching of mathematics is strongly advocated.

## SOME RATIONALES

Many reasons justify the use of history in teaching mathematics. One theory is that mathematics learning recapitulates history; that is, during a student's prolonged mathematics learning experience, concepts and techniques are identified and refined in a manner paralleling their historical evolution. For example, a child associates number with quantity before undertaking mathematical operations such as addition, learns addition before undertaking subtraction, and learns multiplication before division; a mastery of all these operations must be obtained be-

fore learning to compute square roots. There is a definite conceptual ordering, a hierarchy of concepts in the learning of mathematics that, in most cases, parallels historical orderings. Even within specific concepts, such historical and psychological orderings can be identified.

Historically, it appears that humans first conceived counting numbers, then went on to devise fractions, from which a rational number concept arose, followed by the concepts of irrational numbers and then negative numbers. Children today learn their number concepts in a similar order; and while the sequencing may be justified as progressing from the simple to the complex, historically it can be described as going from the "concrete to the abstract." Pedagogically, we also now realize that mathematics teaching and learning should proceed from the concrete to the abstract; thus historical experiences foreshadowed and predicted eventual pedagogy.

Mathematical concepts are interconnected and related. This web of relationships possesses a structure; we teach and learn along this structure. Students should be made aware of these connections in mathematics. They should recognize and appreciate relations between concepts, such as the operations of addition and subtraction as inverses; multiplication as repeated addition; $\sin \theta$, $\cos \theta$ as cofunctions. Studying the historical evolution of such concepts helps to achieve this.

While a historical approach to teaching can help structure and make more meaningful the learning of mathematics, that is, supply a logic and order to mathematical ideas, it can also demystify mathematics by supplying humanistic reasons for various procedures and techniques: a base 10 numeration system is popular because, quite simply, people always counted on their 10 fingers. However, far too often, teachers teach mathematics without ever talking about it, its origins and evolution; thus many students, and people in general, feel that mathematics is almost magical and that its form and content were conceived by isolated geniuses. Jacques Barzun, a respected commentator on the American educational scene, has noted this phenomenon: "I have more than an impression—it amounts to a certainty—that algebra is made repellent by the unwillingness or inability of teachers to explain why. There is no sense of history behind the teaching, so the feeling is given that the whole system dropped down ready made from the skies, to be used only by born jugglers" (Barzun 1945, 82).

This mystique lends an air of incomprehensibility to mathematics, encourages memorization, and fosters anxieties. An understanding of the human roots of mathematics—that its ideas and concepts were labored over by men and women, evolved over thousands of years, and were gradually refined into their present form—may alleviate this situation. The story of just how mathematical problems were recognized and solved, perhaps imperfectly at first but eventually corrected, is reassuring. Approximation has always been part of the mathematical scene, for example, the ancient Egyptians estimated the area of a circle by computing $(8/9)^2$ of the area of its circumscribing square, and the Babylonians and Chinese of the pre-Christian era approximated the area of a circular segment by using a trapezoid. For problem solvers, it is psychologically satisfying to realize the omnipresence of approximation techniques. Students should understand that people had to grapple with mathematical concepts and procedures and that many times the answers did not easily appear. History testifies that mathematics is not magical but the result of prolonged and persistent human involvement.

The history of mathematics contains features of the human drama: it encompasses adventure, intrigue, and conflicts, which if related and examined, can capture student interest and further motivate learning. Mystery abounds in Pythagorean number theory: the identification of perfect numbers, amicable numbers, and so on; the construction of golden rectangles and magic squares; ancient Indian attempts to square the circle; the cabalistic numerology of the Middle Ages and the properties of the transfinite. Adventure can be found in the deciphering of Mayan head glyph numerals, solving problems from the Rhind papyrus, examining a Babylonian tablet such as "Plimpton 322" with its listing of Pythagorean triplets and the resulting realization that the Babylonians of 2000 B.C. knew the Pythagorean Theorem, or following Johann Kepler's quest for the geometric orbits of planets. Intrigue abounds in such episodes as the Pythagorean discovery and concealment of irrational numbers, François Viète's (1540–1603) involvement with secret codes; Isaac Newton's (1642–1727) alchemy experiments; and the clouded identity of M. Le Blanc (1776–1831), benefactor to Carl Friedrich Gauss (1777–1855), the "Prince of Mathematicians." Conflicts are found in the Cardano-Tartaglia dispute (1545) over the solution of the cubic equations, the dual discovery of the calculus by Isaac Newton and Gottfried Wilhelm Leibniz (1646–1716), and the Janos Bolyai (1802–1860) and Gauss correspondence on non-Euclidean geometry, to name a few instances.

In discussing historical issues with students, ideas are sharpened and many significant questions arise. For example, what does it mean to invent a mathematical concept? Simon Stevin (1548–1620) is often credited with inventing decimal fractions; does this mean he was the first mathematician to use decimal fractions? Did Euclid (ca. 300 B.C.) originate the geometry described in his famous text? The issue of priority in mathematical accomplishment is worthy of examination. The ancient Egyptians used rectangular grids in their design of wall paintings for tombs and yet Cartesian coordinates are associated with the French mathematician René Descartes (1596–1650). Why is this? Similarly, the triangular array of numbers from which binomial coefficients can be obtained was employed for this purpose by Arabs in the eleventh century and, independently, by the Chinese in the thirteenth century and yet it bears the name "Pascal Triangle" in honor of Blaise Pascal (1623–1662). He certainly did not discover it, but, still, he seems to have received the credit for it. Why does history regard the accomplishments of one person or group of people and ignore those of others? Is history always fair?

The people of mathematics themselves are interesting. By learning about the individuals who contributed to the body of mathematical knowledge, their lives, their accomplishments, and their failures, students further realize that mathematics is made by real people much like themselves. The fraternity atmosphere, secret signs, and mystical rites of the early Pythagoreans, Zeno's paradoxes, and Hippocrates of Chios (ca. 440 B.C.) experiments with the quadrature of lunes, all appeal to students. Galileo Galilei's (1564–1642) challenge to established authority strikes a note of accord with contemporary young people. Gottfried Wilhelm Leibniz's (1646–1716) concern with unifying the Christians of Europe and converting the Chinese with the aid of binary arithmetic demonstrates a connection between mathematics and religion. Charles Babbage's (1792–1871) failure to construct his "analytic engine" illustrates some of the problems experienced by computer pioneers.

Often, even just a quotation attributed to a famous mathematician can serve as the basis for a meaningful classroom discussion: "The deep study of nature is the most fruitful source of mathematical discovery" (Joseph Fourier, 1768–1830); "If we wish to foresee the future of mathematics, our proper course is to study the history and present condition of the science" (Jules Henri Poincaré, 1854–1912); "God made the integers; all else is the work of man"

(Leopold Kronecker, 1823–1891). Do students agree with such statements? What was the context of the speaker's statement? Who was he or she? Quotations and the questions they elicit are yet another learning facet of the history of mathematics.

Further, the history of mathematics broadens students' perspectives by providing links between mathematics and the social, economic, or political conditions that influenced it. For example, what social factors slowed the indigenous development of Egyptian mathematics while, at about the same period of history, mathematics in the Greek city-states was blossoming? Why did Alexandria eventually become a center of mathematical accomplishment? How did societal sponsorship of regional universities and national academies provide stimulus to mathematical development? How did the rise of mercantile capitalism in medieval Europe affect mathematical thinking? Did the Confucian spirit diminish the development of mathematics in China? Why, in the field of mathematics, have women been traditionally repressed? Seeking answers for such queries in an open discussion exposes the interdependence of mathematics to many social, political, and cultural factors. Awareness of these connections also provides a lesson for teachers, that is, mathematics should not be taught in isolation from the world around it but must continually be referenced to people and their lives.

Historical considerations can also call attention to the multicultural history of mathematics—the diversity of peoples who have contributed to the growth of the science. For too many years the history of mathematics has been viewed in a limited Eurocentric light; that is, the "legacy of the Greeks" as transmitted through western European society was believed to be the basis of mathematics. Recent work by such researchers as George Gheverghese Joseph (1991) of Great Britain, Paulus Gerdes (1990) of Mozambique, and Claudia Zaslavsky (1973) of the United States is beginning to dispel this myth and reveal the mathematical accomplishments of many non-Western peoples. Still, several components of our traditional history of mathematics attest to culturally diverse origins. For example, from the Babylonians we inherited a system of circular and chronological measure based on 60: 360 degrees in a circle, 60 minutes in an hour, and so on. The early Egyptian use of geometry in constructing massive structures still fascinates. Chinese facility with computing algorithms resulted in the most accurate estimate for a value of $\pi$ in the ancient world; Zu Chongzhi (ca. A.D. 500) obtained a value

3.14159267. Hindu mathematicians did distinguished work with indeterminate equations. Islamic scholars contributed to the development of our system of trigonometric functions and broadened early knowledge of combinatorics. Our word *algebra* evolved from the Arabic *al-jabr, al* meaning "the" and *jabr* meaning "restoring." Our numeral system is called the "Hindu-Arabic" system, although some scholars believe it should be called the "Sino-Hindu-Arabic" system as they feel our numerals really originated in China. Mayan mathematics provides an example of a non-decimal base system. Historical surveys should consider the mathematics of traditional peoples, of the Americas and of sub-Saharan Africa. The history of mathematics is composed of a multicolored and textured fabric, a tapestry students can readily appreciate.

## STRATEGIES FOR INCORPORATING HISTORY

Historical material may be introduced into the instructional process as a natural adjunct to the classroom discussion, not separated from the theme of learning and understanding mathematics. Over the years, teachers have incorporated the history of mathematics into classroom lessons and assignments in various ways. Some successful strategies include: (1) providing historical origins of the concepts, procedures, and mathematical terms under consideration; (2) enriching lessons with brief anecdotes about the lives and work of relevant mathematicians; (3) building lessons around actual historical activities; (4) constructing and displaying timelines; (5) assigning actual historical problems as class and homework exercises; (6) employing appropriate visual aids: posters, films, and videos; (7) assigning miniresearch projects on particular mathematical accomplishments or mathematicians. Each of the strategies warrants a closer examination.

### Revelation of Historical Origins

Every concept and term that is used in mathematics has a history. Often, that history is itself revealing of the functional or intuitive foundations of the concept. For example, the word *mathematics* comes from the Greek *mathematikos*, which means "disposed to learning," indicating the high intellectual esteem the Greeks assigned to mathematical activity. In an algebraic context, the phrase "completing the square" is associated with symbolic manipulation whereby the roots of a quadratic equation can be isolated. For example, $x^2 + 14x - 32 = 0$ can be restated as $x^2 + 14x + 49 = 81$, from which is obtained $(x + 7)^2 = 81$ and $x = 2$, $x = -16$. However, the origins of this phrase lie in the geometric-algebraic solution procedures of early societies in which the quadratic expression was interpreted as a geometric square of which the partial length of its side was known. The problem then became one of finding the total length, or "completing the square." Note the term *quadratic* itself evolved from the Latin word *Quadrare*—"to make square." Similarly, the simple expression, "she is good at figures," a compliment on computational ability, can be traced back to the Pythagorean concern with figurate numbers, numbers that could imaginatively be assigned geometric shapes, figures. The common plus sign, +, began as a symbol used by German merchants to indicate the content of barrels and casks: a + indicated the barrel was full, − indicated a lacking; that is, it was partially full, some content was missing (Austin et al. 1926, 198). Such facts, while seemingly minor, animate the concepts under discussion and provide an association that may assist retention and increase appreciation of the concept or technique being learned.

## Anecdotes

A brief story can often enliven and enrich an otherwise bland lesson. An example is the story of the young Carl Friedrich Gauss. The mathematically precocious youth frustrated his mathematics master by solving his classroom exercises too quickly. One time the master gave the class a busywork problem, one he felt would keep them occupied for some time, to find the sum of the natural numbers from 1 to 100. Gauss almost instantly gave the correct answer, 5,050. The ten-year-old boy had noted a pattern:

$$
\begin{array}{l}
1 + \phantom{0}2 + \phantom{0}3 + \cdots + \phantom{0}99 + 100 \\
100 + 99 + 98 + \cdots + \phantom{0}2 + \phantom{00}1 \\
\hline
101 + 101 + 101 + \cdots + 101 + 101 = 100 \times 101
\end{array}
$$

As this was twice the required sum, he divided by 2 to obtain the correct answer, 5,050. This little scenario illustrates the importance of recognizing and, if possible, utilizing patterns in mathematics. The "Eureka episode," in which Archimedes (287–212 B.C.) discovers the principle of specific gravity while taking a bath, leaps up naked, and runs down the street shouting "Eureka!" ("I have discovered it"), interjects humor into a class while pointing out the problem-solving necessities of persistence and experimenta-

tion. Human drama is incorporated into a lesson by illustrating how a mathematical concept is refined or a greater precision achieved. Consider, for example, the numerical refinement for the value of $\pi$. In the ancient world of 1000 B.C., 3 was used as a value for $\pi$; by 240 B.C., Archimedes improved the value to 223/71; about A.D. 150, Ptolemy obtained 377/120; in A.D. 480, the Chinese mathematician Zu Chongzhi improved the value further to 355/113; in A.D. 1150, Bhaskara II found $\pi$ to be 3927/1250; and finally in about A.D. 1202, Fibonacci introduced a value for $\pi$ of 865/275. Stories can reflect the victory of the human spirit, as when Sofia Kovalevskaia (1850–1891) achieved recognition as a mathematician and was acclaimed the "Princess of Science," triumphing over the bias of her time against women. Knowledge of Srinivasa Ramanujan's (1887–1920) quest to have his mathematical achievements recognized despite social and economic disadvantages can inspire young mathematicians to overcome hardships.

## History-based Activities

Historically, many acts of mathematical discovery or accomplishment were achieved as a result of specific activities. Early mathematicians, in great part, learned by doing. Action-based strategies for learning and understanding are very much a part of the modern education scene. Students, with some guidance, can replicate successful historical activities and obtain the satisfaction of discovery for themselves. Unknown volumes of given solids can be obtained by dissection techniques, that is, the dissection of the solid, either physically or theoretically, into a collection of smaller, known volume solids, the sum of whose volumes will equal the whole. Volumes can also be obtained by Archimedean water displacement techniques. The ancient Chinese and Egyptians were known to utilize volume dissection experiments. Shadow measurements and computations formed the basis of early trigonometry. Eratosthenes (ca. 230 B.C.), a Greek astronomer-mathematician, measured the circumference of the earth by using two shadows; his experiment can readily be undertaken by a high school class and accomplished with great satisfaction and good accuracy. A variety of classical ruler-compass constructions can be undertaken by a geometry class. Methods of exhaustion or Buffon's (1707–1788) needle-dropping experiment can be used to obtain a numerical value for $\pi$. Students can perform computations in various ways: by employing early algorithms such as chessboard or gelosia multiplication and galley division or by making and using computational devices such as a set of Napier's bones or Genaille-Lucas rods. A set of Platonic solids can be built out of construction paper and manipulated to discover Euler's formula relating vertices, $V$, edges, $E$, and faces, $F$: $V + F - 2 = E$.

Such activities can be teacher directed or incorporated into individual learning tasks or modules, the use of which can encourage small-group, cooperative learning (Mitchell 1978). While commercial sets of historical activities exist (Swetz, *Learning*, 1994), frequently teachers prepare their own for classroom use.

## Timelines

The construction and use of historical timelines give a visual appreciation of the growth and development of mathematics. Timelines illustrate the evolutionary nature of mathematics. They place people and events in a perspective, one built upon dependence. Mathematical accomplishment is contingent upon the efforts of many people working over a long period of time. Timelines can be constructed by students as either a class or individual project. Their scope can vary from the general to the specific: the growth of mathematics from 600 B.C. to A.D. 1600, events in the development of algebra, famous women in mathematics, the history of "Fermat's Last Theorem," and so on. They can also be constructed around a nonmathematical social or historically significant event, thus providing an interdisciplinary as well as historical approach to mathematics learning. For example, students can list chronologically the mathematical accomplishments and events within 100 years of Columbus's voyage to the Americas (i.e., 1392–1592) or list the noted mathematicians alive during the American Revolution. A personal computer can assist in storing and processing data and actually generating timelines.

## Historical Problems and Exercises

Over 2000 years of accumulated mathematical activity have resulted in the production of a wealth of problems. Some of these problems, in altered forms, have become standard exercises in modern textbooks; however, a multitude remain in their original form. In undertaking the solution to such problems, students make many discoveries about mathematics

and the peoples and cultures that used that mathematics.

A problem from the Rhind papyrus (1650 B.C.) attests to the Egyptian society's dependence on grain: "Suppose a scribe says to you that 4 overseers have drawn 100 great quadruple *hekat* of grain, and their gangs consist of 12, 8, 6 and 4 men. How much does each overseer receive?" (Chace 1979). Babylonian problems from the Old Kingdom period (2000 B.C.) denote their users' concern with irrigation and channel construction and maintenance: "It is known that the digging of a canal becomes more difficult the deeper one goes. In order to compensate for this fact, differential work allotments were computed: A laborer working at the top level was expected to remove $\frac{1}{3}$ *sar* of earth in one day, while a laborer at the middle level removed $\frac{1}{6}$ *sar* and one at the bottom level, $\frac{1}{9}$ *sar*. If a fixed amount of earth is to be removed from a canal in one day, how much digging time should be spent at each level?" (Neugebauer and Sachs 1945, 57).

A Chinese grain problem from 100 B.C. indicates that mathematicians of that time could solve systems of linear equations and compute with negative numbers: "Three sheafs of good crop, 2 sheafs of mediocre crop, and 1 sheaf of a poor crop produce 39 *dou* of grain. Two sheafs of good, 3 of mediocre, and 1 of poor produce 34 *dou*. One sheaf of good, 2 of mediocre, and 3 of poor produce 26 *dou*. What is the yield for one sheaf of good crop? One of mediocre crop? One of poor crop?" (Van der Waerden 1983, 47). A Greek problem of A.D. 500 tells the reader about the life of the mathematician Diophantus and shows that students even then enjoyed a good riddle: "God granted him to be a boy for the sixth part of his life, and adding a twelfth part of this, He clothed his cheeks with down. He let him the light of the wedlock after a seventh part, and five years after his marriage, He granted him a son. Alas, late-born wretched child; after attaining the measure of half his father's life, chill Fate took him. After consoling his grief by the science of numbers for four years, he ended his life" (Eves 1990, 197). The reader is then expected to determine the duration of each part of Diophantus's life.

Students can also try their skill at solving some of the more classical or formal problems from the history of mathematics, such as performing the *"Pons asinorum"* or bridge of fools proof from the Middle Ages to show that the base angles of an isosceles triangle are congruent. Other interesting problems of this nature are finding the area under one arch of a cycloid and determining the quadrature of an ellipse with given semiaxes of length $a$ and $b$, ($\pi ab$). In a sense, solving such problems allows students to touch the past.

## Use of Visual and Tactile Teaching Aids

There are many learning/teaching aids commercially available (Swetz, *Learning,* 1994) or easily constructed that help in the instruction of a mathematical concept and also relate to a historical situation. A Chinese abacus can be employed to teach children place value. Computation algorithms can physically be carried out on a medieval counting table with plastic chips as counters. Some students are fascinated by demonstrations of early written algorithms such as multiplication by the cross and galley division. Archimedes's discovery of the volume relationships among a right circular cylinder, a hemisphere, and a cone, all having the same radii and height ($1:2:3$) can be replicated with appropriate volume containers. Plasticine clay can be readily shaped into a Babylonian tablet on which a pencil point imprint can make cuneiform markings by which numerals can be inscribed and later deciphered by students.

Poster sets containing portraits and descriptive information on the life and work of selected mathematicians are commercially available (Swetz, *Learning,* 1994). Facsimiles and copies of old mathematical illustrations stir student interest and videos and films recreate many notable events in the history of mathematics. There is a growing collection of such videos, especially those produced by the Open University in England.

## Reports

Student reports are a convenient and effective means of using history to enrich mathematics education. These reports can be assigned as an exercise combining mathematics with writing skills. Reports can be delivered orally before the class or serve as informational displays. Researching the lives and work of particular mathematicians seems to be an appealing task for students. Identifying a name such as Isaac Newton with a real person who had a specific date and place of birth (1643 in Lincolnshire, England), a record of schooling (Grantham public school and Trinity College, Cambridge), participation in family life, and an attraction to mathematics helps to demystify the subject. The dogged persistence of historical problem solvers such as Kepler in

determining the planetary orbits or William Rowan Hamilton (1805–1865) in solving the puzzle of his quaternions can be examined. Such revelations of perseverance can inspire modern problem solvers. Further, the combined findings of such reports help to promote the concept that mathematical achievement is a collective and diverse endeavor drawing on the talents of men and women from various social levels, nationalities, and cultures.

Besides focusing on individuals, reports can also center around relevant themes and issues in the history of mathematics; for example, multiplication in the sixteenth century, circle squares, solving the cubic equation, and irrational numbers. Also of interest are relationships between individuals: teacher and pupil (e.g., Pythagoras and Thales), or father and son (the Bolyais), or, more pointedly, specific associations such as Charles Babbage and Ada Lovelace, Augustus De Morgan and George Peacock, and Ramanujan and Hardy. Research questions can be open-ended and test student initiative, for example, "Name several mathematicians who were also recognized as poets." Omar Khayyam and Sofia Kovalevskaia are two examples.

Teachers can direct research effort by preparing lists of appropriate subjects and/or questions, for example, "Who was the Fibonacci who lent his name to the number sequence 1, 1, 2, 3, 5, . . . ?" Of course, school library holdings should provide basic references for such research projects.

## CONCLUSION

The history of mathematics is a bountiful resource that can enrich and strengthen mathematics learning. History provides human interest to mathematics; it associates concepts with their situational origins and the people who helped shape and formalize those concepts. Further, it reveals the universal and culturally diverse nature of mathematical involvement. Such revelations have helped to reshape perceptions of mathematics education and opened questions for scholarly debate and research such as, Do different people view mathematics in different ways and thus use it and learn it differently? This question, arising largely from a historical perspective, has spawned a new discipline, ethnomathematics, which rests on the premise that all people possess and articulate a natural mathematics.

Although numerous authorities, organizations, and committee reports advocate the use of the history of mathematics in teaching and learning, the benefits of this approach remain anecdotal and conjectural. There are few experimental research findings on the rewards of a historical approach to mathematics teaching. This field remains a fertile area for further investigation.

*See also* Babylonian Mathematics; Chinese Mathematics; Greek Mathematics; Indian Mathematics; Islamic Mathematics

## SELECTED REFERENCES

Austin, Charles M., et al. *A General Survey of Progress in the Last Twenty-five Years.* 1st Yearbook. National Council of Teachers of Mathematics. New York: Bureau of Publications, Teachers College, Columbia University, 1926. Reprinted. New York: AMS Reprint, 1966.

Barzun, Jacques. *Teacher in America.* Boston, MA: Little, Brown, 1945.

Baumgart, John K., et al. *Historical Topics for the Mathematics Classroom.* 2nd ed. Reston, VA: National Council of Teachers of Mathematics, 1989.

Boyer, Carl B. *A History of Mathematics.* 2nd ed., rev. Revised by Uta C. Merzbach. New York: Wiley, 1991.

Chace, Arnold Buffum. *The Rhind Mathematical Papyrus.* Reprint of 1927–29 ed. Reston, VA: National Council of Teachers of Mathematics, 1979.

Eves, Howard W. *An Introduction to the History of Mathematics.* 6th ed. Philadelphia, PA: Saunders College Publishing, 1990.

Gerdes, Paulus. "On Mathematical Elements in the Tchokive 'Sona' Tradition." *For the Learning of Mathematics* 10(1990):31–34.

Gillispie, Charles C., ed. *Dictionary of Scientific Biography.* New York: Scribner's, 1970–80.

Joseph, George G. *The Crest of the Peacock: Non-European Roots of Mathematics.* London, England: Tauris, 1991.

Katz, Victor. *A History of Mathematics: An Introduction.* New York: HarperCollins, 1993.

Mitchell, Merle. *Mathematical History: Activities, Puzzles, Stories and Games.* Reston, VA: National Council of Teachers of Mathematics, 1978.

National Council of Teachers of Mathematics. *Curriculum and Evaluation Standards for School Mathematics.* Reston, VA: The Council, 1989.

———. *Principles and Standards for School Mathematics.* Reston, VA: The Council, 2000.

Neugebauer, Otto, and Abraham Sachs. *Mathematical Cuneiform Texts.* New Haven, CT: American Oriental Society, 1945.

Sarton, George. *Ancient and Medieval Science during the Renaissance.* New York: Barnes, 1958.

Smith, David E. *History of Mathematics.* 2 vols. New York: Dover, 1958.

Swetz, Frank J. *From Five Fingers to Infinity: A Journey through the History of Mathematics.* Chicago, IL: Open Court, 1994.

———. *Learning Activities from the History of Mathematics.* Portland, ME: Walch, 1994.

Van der Waerden, Bartel L. *Geometry and Algebra in Ancient Civilizations.* New York: Copyright © Springer-Verlag, 1983.

Zaslavsky, Claudia. *Africa Counts: Number and Pattern in African Culture.* Boston, MA: Prindle Weber and Schmidt, 1973. New York: Hill, 1979; Revised 1999.

FRANK J. SWETZ

# HISTORY OF MATHEMATICS EDUCATION IN THE UNITED STATES, OVERVIEW

Prior to the end of the nineteenth century, mathematics as a school subject in the United States was cast in a supporting role, one whose purpose was to advance the more valued goals of gaining practical skills and developing good habits of mind. With the approach of the twentieth century came a growing recognition that mathematics itself was an important discipline. Eventually educators saw that mathematics provided for an array of practical applications and presented a domain to explore reasoning and thinking skills.

The history of mathematics education in the United States is outlined here beginning with a broad sketch of the colonial period and the first 100 years of the Republic. The last quarter of the nineteenth century, a period of significant change for mathematics and for education in the United States, is treated separately. Events in the twentieth century are examined in twenty-five-year intervals. The first interval, 1900 to 1925, was a period of vigorous committee activity that resulted in many reports urging various reforms in teaching mathematics. During the second quarter of the century, major world events—depression and war—affected the school population, and the social sciences exerted different influences on mathematics education. After World War II through the early 1970s, a new wave of reforms propelled mathematics into the forefront of national discourse on education. A section on recent trends in mathematics education since 1975 concludes the outline.

## THE COLONIAL PERIOD

During the colonial period in America, "getting an education" meant studying the classical curriculum, which included the mathematical topics of basic arithmetic, some geometry, and logic. Formal schooling was available at a few academic institutions such as the Boston Latin School and Harvard College, both established in the 1630s. A minute proportion of American youth attended formal schools at any level. Toward the end of the seventeenth century, Thomas Brattle, a professor at Harvard, was the first to exhibit a serious interest in mathematics instruction by publishing an account of crossing the fabled *pons asinorum* ("bridge of fools"), that is, proving that the base angles of an isosceles triangle are congruent. In 1729, Isaac Greenwood, Hollis Professor of Mathematics and Natural Philosophy at Harvard, wrote a treatise entitled *Arithmetic, Vulgar and Decimal*. Thus, the mathematics taught even in the colleges was at a basic level.

During the 1600s, elite grammar schools like Boston Latin prepared a small number of young men for the ministry or for further study at college. In the 1700s, academies were founded to provide further training for those in the expanding mercantile class and for those who did not intend to enter a college. One of the best known was Benjamin Franklin's Philadelphia Academy. In the academies, mathematics was viewed as a practical skill. A few individuals like Benjamin Banneker acquired mathematical skills on their own. General elementary schools also evolved, but their primary mission was to teach language skills, and consequently mathematics did not occupy a prominent position in the curriculum. As a result, formal mathematics instruction in colonial America provided only rudimentary arithmetic and computational skills to a very select group of the population.

## THE REPUBLIC DEVELOPS:
### 1800–1875

During the nineteenth century, the United States struggled to establish itself as a nation, expanded its boundaries across the continent, fought a civil war, absorbed an increasing number of immigrants, and turned its attention to the need for improved education. From 1800 to 1875, mathematical activity in the United States was limited in both volume and level of sophistication when compared with that of Europe. Thomas Jefferson, who encouraged the establishment of the Military Academy at West Point and who later founded the University of Virginia, was responsible in large part for whatever emphasis mathematics was given during this period.

The United States Military Academy opened in 1802, organized along the lines of the École Polytechnique, a famous technological institute established in Paris by Napoleon a few years earlier. Claude Crozet, a graduate of the École Polytechnique, joined the academy faculty to teach mathe-

matics. The curriculum at West Point focused on applications of mathematics and science useful for engineering and military needs. Lieutenant Colonel Sylvanus Thayer became head of the academy in 1817. During his sixteen years as superintendent, Thayer encouraged use of French instruction methods and textbooks, in part as a reaction against the colonial British traditions. The British college of the time aimed to educate "gentlemen" for the elite classes. This class orientation, combined with the fact that mathematics in Britain had languished in the post-Newton era, made the British model seem inappropriate for an emerging democracy.

The new French polytechnic model, on the other hand, prepared men to apply scientific knowledge to practical problems. Not surprisingly, French mathematics texts proved more useful for the sciences than did British texts. In addition, French texts benefited from that country's prodigious mathematical activity during the eighteenth century. For these reasons, translating French texts into English became an important function for college instructors who were ambitious for their students. Charles Davies, chairman of the mathematics department at West Point, published a translation of Adrien-Marie Legendre's *Geometry* that became the standard text for Euclidean geometry in the United States for the remainder of the nineteenth century. Davies also wrote texts used in many schools and colleges. Davies' *The Logic and Utility of Mathematics with the Best Methods of Instruction Explained and Illustrated* (1850) was the first mathematics book in the United States that included methods for teaching.

The University of Virginia had closer ties to the British mathematical community than did West Point. From 1827 to 1840, the British mathematician Charles Bonnycastle taught at Virginia using the best texts from England and France. James Joseph Sylvester taught briefly at Virginia in 1841 and went on to become an important contributor to mathematics in his native England.

Following the success of West Point, other independent technical institutes were established, notably Rensselaer Polytechnic in 1824 and the Massachusetts Institute of Technology in 1861. Beginning in 1847, industrialists such as Abbott Lawrence (textiles) and Joseph Sheffield (railroads) financed scientific schools at leading colleges like Harvard and Yale. Although these institutions contributed to a growing community of mathematically trained personnel, their purpose was to prepare individuals for careers in industry and engineering rather than to develop scientists and mathematicians. Despite these bright spots, most mathematics at American colleges during the first half of the nineteenth century did not go much beyond elementary algebra and geometry. Prior to the Civil War, few colleges included the calculus in their curricula, even for the most able students. The culmination of work in mathematics was often its application to astronomy or physics.

If strong American college programs in mathematics were rare before 1875, so were noteworthy American contributors to the field. Nathaniel Bowditch, a seaman interested in improving navigation, translated Pierre Laplace's *Celestial Mechanics* over a ten-year period in the 1830s with the assistance of Benjamin Peirce. Peirce later became a professor at Harvard and was among the first in the United States to recognize research as an important activity of a collegiate mathematics department. In 1870, Peirce completed *Linear Associative Algebra,* a significant advance in the emerging field of abstract algebra. Since no appropriate American journal for theoretical research existed at the time, Peirce collaborated with friends in the Coast and Geodetic Survey to publish the manuscript privately. His son, Charles Sanders Peirce, developed the philosophy of American pragmatism and made important contributions to the field of logic.

Schooling during the period 1800 to 1875 also was affected by institutions, by individuals, and, to an extent, by foreign influences. The numbers and types of schools continued to increase, sponsored by communities or by religious groups. The "common school," intended to develop character, self-discipline, and good citizenship, became a prevalent model. The curriculum of these schools was teacher-centered, since what was taught depended on what the teacher knew.

In colonial times, the "rule method" had been accepted widely as the proper pedagogy for teaching arithmetic calculations. Using this approach, the teacher set forth the rule, illustrated its use by example, and provided practice for the students. In 1825, Warren Colburn published an arithmetic handbook based on the methods of the Swiss educator Johann Pestalozzi. Colburn advocated an inductive approach using concrete objects and a discovery orientation for teaching basic ideas of quantity to children. Colburn's book gained wide popularity and remained in use for many years.

By the middle of the nineteenth century, attention to the pedagogy of arithmetic was superseded by what would become the dominant model for learning during the next fifty years: mental discipline. This view held that the mind was like a muscle, needing strong exercise to develop its potential. Proponents of this theory saw mathematics, with its precision and formal structure, as the ideal "calisthenics" for the mind. Although this idea placed mathematics in an important position in the curriculum, it also promoted the attitude that the usefulness of mathematics was in mental exercise rather than in its real-world applications.

## RISE OF THE UNIVERSITY AND GROWTH OF THE SCHOOLS: 1875–1900

The closing decades of the nineteenth century in the United States brought dramatic developments in education at all levels. The inauguration of research institutions such as Johns Hopkins University in 1876 and the University of Chicago in 1892 stimulated mathematical investigations. The founding of journals for research, the establishment of the American Mathematical Society, and mathematical interchanges with other countries were essential factors in the development of a vital community of mathematics scholars in the United States. As interest in research mathematics grew, more and better-trained professors encouraged the next generation of academics to pursue mathematics at the highest levels. With the rise of universities came increased attention to the preparation of students for college entrance. During the 1890s, a rapid growth in high school enrollments intensified this attention. Eventually, the field of mathematics education emerged as an area of special study when it was recognized that college entrance ultimately depended on the training of the teachers who prepared the students.

### University Events

A remarkable quarter century of mathematical growth in the United States began with the 1876 opening of Johns Hopkins University. Modeled on the German university, Johns Hopkins provided a center for the pursuit of scientific "truth" through scholarly research at the graduate level. J. J. Sylvester came from England to organize the graduate program in mathematics. By 1880, Johns Hopkins had conferred doctoral degrees for studies in several areas of mathematics, including non-Euclidean geometry. Still, American students interested in mathematics continued to travel to Europe, especially Germany, for advanced study.

In conjunction with the World Columbian Exposition of 1893, the American mathematical community convened an International Congress at Chicago. Felix Klein, the eminent German mathematician, delivered the inaugural address at the congress. Afterward, at Northwestern University, Klein gave a series of lectures that included information on German approaches to teaching mathematics. Klein's lecture series was arranged by Eliakim Hastings Moore, who had just moved to the newly opened University of Chicago. As part of his plan to develop a research-oriented department at Chicago, Moore employed two German mathematicians, Oskar Bolza and Heinrich Maschke. By 1900, the mathematics department at Chicago was the recognized leader in the United States, offering over thirty graduate-level courses.

## School Events

During this same quarter century, from 1875 to 1900, American public schools evolved into community institutions available to an expanding population. The mental discipline theory, with its emphasis on mental arithmetic, affected the curriculum, the texts, and the teaching in most schools. Viewed as an ideal exercise for the mind, arithmetic had shifted from the college to the secondary school curriculum during the middle of the nineteenth century. By the last quarter of the century, arithmetic was positioned firmly in the elementary school. Edward Brooks, who trained teachers in what was then known as a "normal school," published *Philosophy of Arithmetic*, one of the first texts intended to help elementary school teachers in teaching arithmetic. It reflected Brooks's orientation toward a synthetic, deductive approach to the subject.

At the high school level, the most widely used texts during the 1880s were those by George A. Wentworth, a teacher at Exeter Academy. Wentworth's books were used at both secondary and college levels well into the twentieth century. The mathematics taught in the expanding high schools had to combine requirements for college entrance with vocational and community needs. In the decade from 1890 to 1900, enrollment in public high schools more than doubled, with the percentage of students enrolled in some type of mathematics course (usually algebra, geometry, or trigonometry) rising substantially. The growth of high schools and the rise of the university required that college faculty and school administrators examine the secondary school curriculum. In 1893, the National Educational Association sponsored the Committee of Ten chaired by Charles Eliot, the president of Harvard

University. Special subcommittees assessed the entire school curriculum and recommended sweeping changes. The mathematics subcommittee, led by Simon Newcomb of Johns Hopkins University, suggested a restructuring of arithmetic to eliminate arcane material included solely for mental discipline purposes. The introduction of more concrete problems and inductive development of rules were encouraged. All students in high school, college bound or not, were to complete the first year of algebra. The courses in geometry and advanced algebra were to be taught side-by-side for two years. This report was the first of many studies on school mathematics conducted by various committees during the three decades from 1893 to 1923.

Two years after the Committee of Ten issued its report, the elementary education Committee of Fifteen presented its recommendations, including an increased emphasis on the importance of mathematics in the elementary school curriculum and the completion of arithmetic study by the end of sixth grade. The seventh and eighth grades, then still a part of the elementary school, were urged to begin the preliminary study of algebra.

## REPORTS AND REFORMS:
### 1900–1925

A new generation of philosophers and psychologists, including John Dewey and G. Stanley Hall, questioned the theory of "faculty psychology," with its pedagogic corollary of mental discipline. Dewey proposed that the study of number and arithmetic be motivated by the idea and the act of measurement. Dewey's work was cited in separate addresses given by the British mathematics professor John Perry in 1901 and by the American mathematician E. H. Moore in 1902. Both Perry and Moore emphasized the practical aspects of mathematics rather than mental discipline. They called for a realignment of the mathematics curriculum so that algebra, geometry, and physics would be taught together in a manner that emphasized applications. Each advocated a laboratory approach using concrete objects for teaching mathematics and science. Moore collaborated with Jacob William Albert Young, an assistant professor of the pedagogy of mathematics, in establishing a program at the University of Chicago to train mathematics teachers. The program accommodated undergraduate and graduate students and was based on Moore's ideas regarding the laboratory method and curricular "fusion" of mathematics and science.

In New York City, a different program for preparing mathematics teachers was initiated by David Eugene Smith at Teachers College, Columbia University. In 1900, one year before arriving at Teachers College, Smith wrote a seminal work in mathematics education, *The Teaching of Elementary Mathematics.* He intended it as a handbook for teachers of arithmetic, algebra, and geometry at the primary or secondary school levels. Smith emphasized the importance of an active role for the classroom teacher in deciding on content and methodology. He incorporated a historical perspective on both mathematics and mathematics teaching, and he advocated consideration of an international viewpoint on education. Smith developed a doctoral program in the history and teaching of mathematics, and in 1906, the first Ph.D. degrees in the new field of mathematics education were awarded at Teachers College. Along with the University of Chicago, Teachers College advanced the idea that the combined study of both mathematics and its pedagogy required different emphases than either mathematics or education alone.

In 1911, two prominent committees, one national and the other international, issued reports on the state of mathematics education in America. The National Committee of Fifteen on the Geometry Syllabus, chaired by Herbert Slaught of the University of Chicago, recommended inclusion of more concrete examples in geometry, elimination of topics like limits, and acceptance of informal proofs, while still maintaining emphasis on the logical structure of the subject. These suggestions, influenced by the Perry movement and by Moore's laboratory approach, were intended to remedy the difficulties encountered by many students in geometry.

The American Committee of the International Commission on the Teaching of Mathematics, chaired by David Eugene Smith, prepared extensive descriptions of mathematical instruction in the United States at virtually all levels and in all types of schools, elementary through university. Committees from many countries submitted reports at the fifth International Congress of Mathematicians held in 1912 at Cambridge, England. In comparing American secondary schools with those of Europe, the American commissioners—Smith of Columbia, William Osgood of Harvard, and Young of Chicago—identified the main issue as the need for teachers who were better prepared in mathematics. Regarding the curriculum, they recommended the shift of topics from college to secondary school so that analytic geometry and even the calculus might be introduced in the high school.

The concern for preparing teachers who could carry out the committee's recommendations for more sophisticated mathematics in the schools met opposition from the National Education Association's Commission on the Reorganization of Secondary Education. In its report of 1918, the committee asserted that each discipline must justify its place in the curriculum by explaining how it met "seven cardinal principles" they set forth for education. The seven cardinal principles viewed the purpose of schooling as the promotion of social harmony and civic virtue. To proponents of the cardinal principles, abstract mathematical subjects, especially algebra, did not appear to meet the goals of promoting citizenship and ethical character. A vigorous defense was mounted by regional professional mathematics teacher organizations formed in the previous decade, including the New England Association, the Middle States and Maryland Association, and the Central Association of Science and Mathematics Teachers. In 1920, the newly formed National Council of Teachers of Mathematics joined the debate.

The major document in mathematics education of this period was the 1923 Report of the National Committee on Mathematical Requirements. The committee's report identified three aims for mathematical instruction that reflected an accommodation of three viewpoints: practical or utilitarian aims, more general "disciplinary" aims, and cultural aims. The report also suggested a restructuring of the seventh through ninth grades into a junior high school. All junior high school students were to take a sequence of "general mathematics" courses that stressed fundamental notions of arithmetic, algebra, intuitive geometry, numerical trigonometry, and demonstrative geometry. Although not all of its reforms were enacted, the 1923 Report remained an influential document into the 1940s.

## SOCIAL CONCERNS, PSYCHOLOGICAL THEORIES, NATIONAL DEFENSE ISSUES: 1925–1950

The Great Depression (1929–1941) and the outbreak of World War II (1939) affected education in direct ways. Since job opportunities were scarce, the number of students continuing their education beyond the elementary school level increased sharply. Limited financial resources and growing school enrollments hindered adoption of the recommendations made in the 1923 Report. The junior high school concept and, to some extent, the "general mathematics" sequence were implemented. The 6-3-3 model—six years of elementary school, three years of junior high school, and three years of senior high school—prevailed in the public schools over the more traditional 8-4 model. Although many schools adopted a "general mathematics" curriculum for the junior high school, the format it took, especially at the ninth-grade level, was not entirely what the 1923 Report had envisioned. Faced with increasing numbers of students who had varying abilities and interests, schools chose different ways to include a course called "general mathematics" in their curricula. Rather than the survey of topics from algebra and geometry called for in the report, some schools simply retitled or reconfigured existing courses, such as a lower-level treatment of algebra or commercial arithmetic. More than half the schools at this time required mathematics in the ninth grade. After the ninth grade it was an elective subject and enrollments in courses beyond elementary algebra declined.

During this period, new psychological theories influenced school mathematics, especially at the elementary level. In the 1920s and 1930s, the connectionism of Edward L. Thorndike of Teachers College, Columbia University, dominated educational psychology. Mathematics was an ideal field for studying stimulus-response bonds that Thorndike and others sought to investigate. Attention was given also to questions of transfer of training from one field to another.

Other theories challenged the basis of connectionism. Gestalt psychology emphasized insight and discovery, not conditioned response, as the key to learning. William Brownell proposed a "meaning" theory, asserting that the real purpose of instruction in arithmetic was to develop students' abilities to think about, analyze, and describe quantitative situations. A "readiness" theory promoted patience in instruction, urging teachers to delay teaching a new skill until the child was mentally ready to learn it. This view led to a related theory of incidental learning: arithmetic skills were to be acquired not through direct instruction but as a result of natural experiences that provided opportunities to develop and use those skills.

In addition to these theories on the psychology of learning, views on the purpose of education and the role of the school in society influenced the way mathematics was taught. The progressive education movement influenced American education during the 1930s. Grounded in Dewey's conception of education for democracy and in the view that practical aspects of learning were of greatest importance, the movement challenged the mathematics curriculum and the effectiveness of teaching methods. Progressive education also supported the child study movement that put the child's needs at the center of decisions on curriculum and pedagogy. These social-utilitarian views themselves were to be challenged as a result of the nation's experience in World War II.

Assessment of mathematical skills took on new urgency with the 1941 entry of the United States into the war. Many recruits had not mastered even the rudiments of the secondary school mathematics needed to perform their military jobs. Concerns of the national defense establishment, combined with general concerns for basic mathematical literacy, led to a reconsideration of the high school mathematics curriculum in the 1950s.

## POSTWAR REFORMS: 1950–1975

After World War II ended in 1945, government reports anticipated requirements for the American workforce in the postwar years. The reports pointed out that the mathematical preparation available in schools was inadequate and outdated for careers in government, business, industry, or in the scientific and engineering professions. As a result, during the 1950s and 1960s the mathematics community initiated a series of curricular reforms that led to a number of curriculum projects labeled "new mathematics." A combination of government, foundation, and university resources supported many of these projects. The earliest project began in 1952 at the University of Illinois. Led by Max Beberman, the University of Illinois Committee on School Mathematics (UICSM) prepared materials for the high school, tested them in selected schools, and trained teachers in the appropriate use of the materials. The UICSM materials emphasized the logical structure of algebra as well as geometry. The reform of the school curriculum reflected the view held throughout the twentieth century by many mathematicians: mathematics, while certainly useful in applications for the sciences, was more than a tool; it had intrinsic and aesthetic value as a structural system itself.

At the end of the 1950s, the National Science Foundation supported the formation of the School Mathematics Study Group (SMSG) to produce prototype texts for school use. With Edward Begle of Yale University as the head of SMSG, mathematicians joined with secondary school teachers and supervisors to prepare texts that emphasized the structure of mathematics, with little attention to applications. In the 1960s, the Madison Project, headed by Robert Davis at Syracuse University, developed enrichment materials for the elementary school. Howard Fehr of Teachers College, Columbia University, led another group that published a textbook series intended to "unify" the study of algebra and geometry at the secondary level. These and other projects addressed the needs of college-bound students rather than the general student population. The availability of government and foundation funding, along with university support for these curriculum projects, contributed to a pronounced increase in research and graduate study in mathematics education during the 1960s and early 1970s.

Another characteristic of this period was the growing influence of psychologists, who saw in mathematics a discipline well suited to studies of cognition and learning. Jean Piaget's theory of the stages of intellectual development in children greatly influenced elementary school curriculum and pedagogy. Piaget, a Swiss psychologist, held that every child develops through a sequence of four stages: sensory-motor (birth to 2 years), preoperational (2 to 6 years), concrete operational (6 to 14 years), and formal operational (14 to adult). Piaget emphasized the role of concrete experiences in developing concepts of number and shape. He also emphasized matching teaching methods to the appropriate age and intellectual stage of the child. These two ideas were similar to those of earlier theorists who had emphasized inductive reasoning and "readiness" for learning.

Robert Gagné's theory of "learning hierarchies" advocated analysis of larger concepts into subordinate concepts, principles, and skills. In his view, students first must master these lower-level "tasks" in order to learn the higher-order concepts. At the top of his hierarchy Gagné placed two types of learning: the learning of principles and the solution of problems.

David Ausubel was another psychologist whose theories gained prominence during this period. Ausubel praised the use of discovery techniques for teaching preadolescents. For secondary and college-level students, however, he advocated the more "efficient" approach of careful exposition and demonstration by the teacher, followed by opportunities for students to solve appropriate problems. Ausubel's principle of the "advance organizer" urged the teacher to make efforts to place the strategy of the lesson before the students prior to instruction.

With the many changes proposed for the nation's schools, disagreement was to be expected. The wave of reforms met an undercurrent of resistance from various quarters, most vocally and vociferously from Morris Kline, a professor at New York University. Kline criticized the "new math" for neglecting traditional topics and applications. The new favored topics, Kline asserted, were of unproven value. Moreover, he contended, the treatment overemphasized rigor, abstraction, symbolism, and the self-sufficiency of mathematics. Kline's 1973 book *Why Johnny Can't Add* was cited by those who were skeptical of the modernized curriculum. The "back-to-basics" movement in the early 1970s grew out of dissatisfaction with the results of the new curriculum and its implementation, especially at the elementary school level. Test results did not demonstrate the efficacy of the "new math" as promised by reformers. Defenders of the new curriculum pointed out that traditional algorithmic and symbol manipulation skills were far easier to measure than were the higher-order attributes, such as understanding or problem-solving ability emphasized in new texts. Still, back-to-basics advocates swayed parents and administrators throughout the nation and the world.

## CURRENT DEVELOPMENTS: POST-1975

One result of the demise of the "new math" was increased concern for including in the curriculum applications of mathematics to real-world problems. By the end of the 1970s, the National Council of Teachers of Mathematics (NCTM) in its *Agenda for Action* identified problem solving as the main focus for the 1980s. Based largely on the earlier work of George Polya, the problem-solving movement generated new initiatives in curricular materials and influenced the direction of professional development for teachers. Topics in probability and statistics were incorporated into the curricula of secondary and elementary schools. Efforts of organizations like the Consortium for Mathematics and Its Applications (COMAP) further emphasized applications.

Technology has continued to influence both the curriculum and the pedagogy of mathematics. The increasing sophistication and decreasing cost of calculators and computers have made them important tools for introducing real-world problems into the classroom. Computer-related topics from discrete mathematics are finding their way into the school curriculum at all levels. In colleges and universities, a major effort begun in 1986 to reform the teaching of calculus has resulted in new curriculum projects that incorporate technology.

Reflecting broader national trends, issues such as standards, assessment, equity, multiculturalism, and interdisciplinary approaches are receiving increased attention in education. The NCTM sponsored three documents produced by committees of mathematics teachers and educators. In 1989, the first document, *Curriculum and Evaluation Standards for School Mathematics,* appeared as an outgrowth of the problem-solving movement and in response to the 1983 alarm *A Nation at Risk,* issued by the National Commission on Excellence in Education. The mathematics *Standards* spearheaded a nationwide drive toward updated, more-demanding standards for the core academic subjects. In 1991, *Professional Standards for Teaching Mathematics* described ways to prepare, support, and develop teachers who could implement the new curriculum standards envisioned in the initial document. The final publication in the trilogy, *Assessment Standards for School Mathematics* (1995), offered strategies for assessing student performance in meeting the expectations promulgated in the NCTM curriculum standards. Together, these three publications present the NCTM vision for mathematics education as the nation moves into the twenty-first century.

*See also* Begle; Cambridge Conference on School Mathematics (1963); College Entrance Examination Board (CEEB) Commission Report (1959); Commission on Postwar Plans; Committee of Fifteen on Elementary Education; Committee of Ten; Dewey; Mathematical Association of America (MAA), Historical Perspective; Moore; National Committee on Mathematical Requirements; National Council of Teachers of Mathematics (NCTM), Historical Perspective; Pestalozzi; School Mathematics Study Group (SMSG)

SELECTED REFERENCES

Begle, Edward G. "Some Lessons Learned by SMSG." *Mathematics Teacher* 66(1973):207–214.

Bidwell, James K., and Robert G. Clason, eds. *Readings in the History of Mathematics Education.* Washington, DC: National Council of Teachers of Mathematics, 1970.

Commission on the Reorganization of Secondary Education. *Cardinal Principles of Secondary Education.* U.S. Bureau of Education. Bulletin 1918, No. 35. Washington, DC: Government Printing Office, 1918.

International Commission on the Teaching of Mathematics. The American Report. *Report of the American Commissioners of the International Commission on the Teaching of Mathematics.* U.S. Bureau of Education. Bulletin 1912, No. 14. Washington, DC: Government Printing Office, 1912.

Jones, Phillip S., ed. *A History of Mathematics Education in the United States and Canada.* 32nd Yearbook. Washington, DC: National Council of Teachers of Mathematics, 1970.

Kilpatrick, Jeremy. "A History of Research in Mathematics Education." In *Handbook of Research on Mathematics Teaching and Learning* (pp. 3–38). Douglas A. Grouws, ed. New York: Macmillan, 1992.

National Committee on Mathematical Requirements of the MAA. *The Reorganization of Mathematics in Secondary Education.* Oberlin, OH: Mathematical Association of America, 1923.

National Education(al) Association. *Report of the Committee of Fifteen on Elementary Education.* New York: American Book, 1895.

———. *Report of the Committee of Ten on Secondary School Studies with the Reports of the Conferences Arranged by the Committees.* U.S. Bureau of Education. Bulletin 1893, No. 205. Washington, DC: Government Printing Office, 1893.

Tarwater, Dalton, ed. *The Bicentennial Tribute to American Mathematics, 1776–1976.* Buffalo, NY: Mathematical Association of America, 1977.

EILEEN F. DONOGHUE

# HOMEWORK

Assignments to be completed outside the normal mathematics class period. Historically, most of this assigned work was completed by students at home; hence the term *home study* and later the term *homework* became common (Langdon and Stout 1969). In contemporary schooling, completion of assigned work in school during study periods that are not a part of the regular mathematics class period is also called homework.

The purpose for assigning homework and the nature of homework assignments vary considerably. Homework has frequently been used to help students develop speed and accuracy in performing mathematical procedures such as adding and multiplying numbers and factoring algebraic expressions. Assignments designed for such skill building have often been lengthy, comprised of a single type of exercise, such as adding fractions with unlike denominators or factoring quadratic polynomials, and have only required students to mimic step-by-step procedures presented earlier by the classroom teacher (Stake and Easley 1978). This type of homework still prevails in many classrooms in spite of studies that show distributing practice over time is much more effective than giving massive blocks of practice exercises (see Darley, Glucksberg, and Kinchla 1991).

Research has shown that if students understand a procedure or algorithm, then it takes them less time to develop proficiency with the skill and that developing understanding also increases the likelihood that students will be able to decide when to use the skill appropriately in a problem-solving situation (Brownell 1947). Thus research supports teaching practice in which a mathematical skill is developed initially in a meaningful way and then homework is used to provide distributed practice to assist students in gaining proficiency with the skill.

In mathematics, studies of homework effects on student achievement have been mixed. Austin (1979) reported, for example, that a review of research on mathematics homework showed that homework was preferable to no homework and that the effects of homework were cumulative, although the value of drill-oriented homework was questionable. Parish (1976) found that homework improved mathematics grades without adverse effects on student attitude. Nevertheless, surveys of the vast homework research literature show mixed results, although most conclude that students who do homework perform better than students who do no or less homework (e.g., Cooper, "Synthesis," 1989). No doubt one reason for the differential effectiveness of homework across studies is that different definitions of homework were used and that it was implemented in very different instructional settings. It is worthwhile to note, however, that substantial gains in student mathematical performance were achieved in a large experimental research study in which regular homework assignments were a required part of an instructional program that also emphasized increased time for meaningful development of mathematical ideas (Good, Grouws, and Ebmeier 1983).

As part of the mathematics education reform movement of the 1990s, homework is being used for a variety of important purposes other than skill development. These uses include but are not limited to increasing understanding, demonstrating applications, and developing connections. Many teachers, for example, are using homework to provide opportunities for students to apply their mathematical skills to real-world problems. Using homework in this way alleviates the pressure of having to solve a problem in class in a fixed amount of time. This homework policy also allows time for students to think and reflect, two processes that are critical to the development of problem-solving ability.

Many educators have provided detailed suggestions for how to prepare homework assignments (e.g., Posamentier and Stepelman 1995). What seems most important from these recommendations is that homework assignments have a clear purpose, be of reasonable length, and require student thinking and that students be held accountable for their work.

*See also* Lesson Plans; Task Analysis

## SELECTED REFERENCES

Austin, Joe Dan. "Homework Research in Mathematics." *School Science and Mathematics* 79(Feb. 1979): 115–121.

Brownell, William A. "The Place of Meaning in the Teaching of Arithmetic." *Elementary School Journal* 47(Jan. 1947):256–265.

Cooper, Harris. *Homework.* White Plains, NY: Longman, 1989.

———. "Synthesis of Research on Homework." *Educational Leadership* 47(3)(1989):85–91.

Darley, John M., Sam Glucksberg, and Ronald A. Kinchla. *Psychology.* Englewood Cliffs, NJ: Prentice-Hall, 1991.

Doyle, Mary Anne E., and Betsy S. Barber. *Homework as a Learning Experience.* Washington, DC: National Education Association, 1990.

Good, Thomas L., Douglas A. Grouws, and Howard Ebmeier. *Active Mathematics Teaching.* New York: Longman, 1983.

Langdon, Grace, and Irving W. Stout. *Homework.* New York: Day, 1969.

Parish, Doris C. "A Comparative Study of the Effects of Required Drill Homework Versus No Homework on Attitudes Toward Achievement in Mathematics." Ed.D. diss. University of Houston, 1976.

Paschal, R. A., T. Weinstein, and Herbert J. Walberg. "Effects of Homework: A Quantitative Synthesis." *Journal of Educational Research* 78(2)(1984):97–104.

Posamentier, Alfred S., and Jay Stepelman. *Teaching Secondary School Mathematics.* Columbus, OH: Prentice-Hall, 1995.

Stake, Robert E., and Jack A. Easley. *Case Studies in Science Education.* Urbana, IL: University of Illinois Press, 1978.

DOUGLAS A. GROUWS

# HUMANISTIC MATHEMATICS

The proposition that mathematics issues from the same spirit as art, music, and poetry. "A mathematician, like a painter is a maker of patterns. The mathematician's patterns, like the painter's or the poet's, must be beautiful; the ideas, like the colours or the words, must fit together in a harmonious way. Beauty is the first test: there is no permanent place in the world for ugly mathematics" (Hardy 1967, 85).

According to Dorothy Buerk and Jackie Szablewski (1993), "mathematics has a public image of an elegant, polished, finished product which obscures its human roots. It has a private life of human joy, challenge, reflection, puzzlement, intuition, struggle and excitement. Mathematics IS a humanistic discipline, but the humanistic dimension is often limited to its private world. Mathematics students see the elegant mask, but rarely see this private world. . . . Their exposure to mathematics . . . often leaves them silent and insecure."

Thirty mathematicians, concerned with the distance between the public image and private life of mathematics, met in 1986 to examine mathematics as a humanistic discipline. Humanistic dimensions of mathematics discussed included: (1) an appreciation for the role of intuition, not only in understanding but in creating concepts that appear in their final version to be "merely technical"; (2) an appreciation of the human dimensions that motivate discovery—cooperation, the urge for holistic pictures; (3) an understanding of the value judgments implied in the growth of a discipline; logic alone never completely accounts for *what* is investigated, *how* it is investigated, and *why* it is investigated; and (4) a recognition of the need for new teaching and learning formats that will wean students from a view of knowledge as certain, to-be-received.

The *Humanistic Mathematics Network Journal* began publication in 1986 in order to share ideas. With support from the Exxon Education Foundation, the journal is now sent free to almost 2,000 readers all over the world. Among the goals of the humanistic mathematics movement are to remind teachers and professors that mathematics is more than facts and formulas to be memorized and that mathematics shares an inner life with music and poetry. The mathematician Hermann Weyl wrote, "My work always tried to unite the true with the beautiful; but when I had to choose one or the other, I usually chose the beautiful." (Rothstein 1995, 139).

## TEACHING AND LEARNING

It is common on the first day to give students detailed instructions about examinations, punctual and tardy homework, the exact percentage that each activity contributes to the final grade, and other information that is reminiscent of a truth-in-lending notice. Learning is presented as entries in a bookkeeping ledger (White 1985). However, there are other goals. Siu (1957) maintains that "in addition to thinking, the student should be provided with the education of feeling. He should not be led into the abject slavery of formal logic and rationality. . . . What is essential in education is a receptiveness to extrapolations into the totality of nature and a communion with her" (1957, 96, 99). Siu (1957, 87) also contrasts technical knowledge with "other" knowledge: "Although real, it is as imprecise as an exhilarating spring day. . . . It is not dispensed in measured doses. It is absorbed slowly and subconsciously into the moral fiber and intimate intuition of the person over a long period of time." The "other" knowledge can be taught only indirectly.

## MEETING THE CHALLENGE

It is a challenge for the teacher and students to become aware and to attend to that "other" knowledge. It requires the teacher to find convincing ways of carrying out such a program, and it requires the students to participate in a spirit that transcends the traditional goals of classroom learning. Such a program emphasizes questions over answers. "The cutting edge of knowledge is not in the known but in the unknown, not in knowing but in questioning. Facts, concepts, generalizations and theories are dull instruments unless they are honed to a sharp edge by persistent inquiry about the unknown" (Thompson 1977, 109) My questions are not how to solve a problem or prove a theorem but why the problem is solved as it is, or what is the meaning of the theorem, how it connects with other parts of the text. Answers are not as important as questions, especially questions that probe the limits of our knowledge. The textbook is not our master but an object of our critical thinking. If the class is successful, then the students will write or rewrite their own textbook. Albert Einstein felt that his imagination and curiosity were more important than his knowledge of certain facts (Moore 1969).

## LEARNING WITHOUT ANXIETY

Efforts to eliminate stress and anxiety from a class sometimes produce another kind of anxiety. Students may be skeptical that a class can be anxiety-free and may fear that they cannot learn in such an environment. Examples of independent attempts to structure anxiety-free and student-centered courses were a student-centered course in the calculus of variations (White 1974) and a course in thermodynamics at Purdue University (Mullen 1975). Both courses were inspired by Carl Rogers (1969, 1961). Although both groups of students had wide latitude to explore those pathways that were most appealing, the calculus of variations course was guided by several textbooks that the students used for reference, while the thermodynamics course was built around a series of projects. Both classes were essentially anxiety-free. In both classes the student consensus was that, although the course had been enjoyable and the students learned a lot, more might have been learned if the course had used conventional lectures and a text. Mullen wrote, "I doubt that they would have learned more . . . although they might have suffered more. . . ."

Two reasons why some students thought that they did not learn as much as they would have in a conventional format is that their learning seemed not to be work but rather play, or that learning occurred to satisfy a personal need rather than a teacher's demand. Both Rogerian experiments were guided by the teacher's vision of an educational situation that made the student the central figure. And in both the teacher was transformed by the experience because the teacher shared the students' learning and growth. The absence of anxiety also applies to the teacher, who can risk exploring new ideas and participate in the learning and growing process.

In the calculus of variations class, the students duplicated their notes, which were distributed to everyone. The final set of common notes totaled more than 150 pages. It paralleled a conventional text in having a beginning, middle, and end. Each student also wrote an individual in-depth paper on a particular aspect of the subject. The course started with no set syllabus and no prescribed text and ended with a coherent course that ranged deeper and wider than a comparable conventional course, a mathematics course with the same spirit of give and take, of intellectual excitement that one may imagine an idealized seminar in Shakespeare to have.

Thompson (1977) wrote, "The nature of intellectual ferment in a classroom designed to promote student inquiry is such that the instructor will always feel ambiguous about who is in charge. As a matter of fact, no one is in charge. If the end sought is the search rather than the found, it soon becomes apparent that no person can be in charge in the usual sense. What is really in charge is a way of behaving toward learning—an approach to a subject. Both the teacher and the taught are caught up in a mode of inquiry. The instructor becomes a guide to learning rather than the authority who dispenses questions and answers. The student becomes in large measure her own teacher because with the materials at hand she must search for meanings and in so doing raise questions appropriate to the relationships she wishes to investigate."

## TOWARD A DEFINITION

Perhaps as a legacy of the experience of most people in elementary school, there is the felt need for an unambiguous definition of humanistic mathematics. As with music and art, analogies and comparisons may be the best that we can hope for. Sherman Stein (1992) considers humanistic mathematics a style of teaching that eschews an emphasis on memorization and encourages students to be independent thinkers and communicators. Philip Davis (1993) examines those qualities that mathematics shares with literature. In part, he says that: (1) "Mathematics, like literature has metaphor;" (2) "Mathematics, like poetry, has ambiguity;" (3) "Mathematics possesses an aesthetic component which is strong and which is immediately apparent to the practitioner at the higher levels of the subject;" (4) Like "poetry (which according to T. S. Eliot, cannot be totally written down) mathematics cannot be totally formalized;" (5) "Mathematics has mystery and can convey awe;" (6) Mathematics has a history . . . Not "because it is something unique to mathematics . . . but because it is often asserted that the truths of mathematics are atemporal, and hence, stand outside of history."

Stravinsky said that the musician should find in mathematics a study "as useful to him as the learning of another language is to a poet." In discussing "the art of combination which is composition," the composer quoted the mathematician Marston Morse: "Mathematics are the result of mysterious powers which no one understands, and in which the unconscious recognition of beauty must play an important part. Out of an infinity of designs a mathematician chooses one pattern for beauty's sake and pulls it down to earth." Morse, Stravinsky says, could as well have been talking about music. It is not only in the clarity of things but in their beauty and mystery that the two arts join (Rothstein 1995).

Pure mathematics is "one of the humanities, because it is an intellectual discipline with a human perspective and a history that matters" (Tymoczko 1993, 11). Also, humanistic mathematics is not "just 'friendly math' or 'touchy feely' math. It is mathematics with a human face. . . ." Without humanistic mathematics, "educators may teach students to compute and to solve, just as they can teach students to read and to write. But without it educators can't teach students to love or even like, to appreciate or even understand mathematics."

*See also* Art; Cooperative Learning; Enrichment, Overview; Mathematics, Nature; Music; Philosophical Perspectives on Mathematics; Poetry; Recreations, Overview

SELECTED REFERENCES

Buerk, Dorothy, and Jackie Szablewski. "Getting Beneath the Mask, Moving Out of Silence." In *Essays in Humanistic Mathematics* (pp. 151–164). Alvin M. White, ed. Washington, DC: Mathematical Association of America, 1993.

Davis, Philip. "The Humanistic Aspects of Mathematics and Their Importance." In *Essays in Humanistic Mathematics* (pp. 9–10). Alvin M. White, ed. Washington, DC: Mathematical Association of America, 1993.

Hardy, Godfrey H. *A Mathematician's Apology.* Cambridge, England: Cambridge University Press, 1967.

Moore, A. D. *Invention, Discovery and Creativity.* Garden City, NY: Anchor, 1969.

Mullen, James G. "An Attempt at a Personalized Course in Thermodynamics." *American Journal of Physics* 43(4)(1975):354–360.

Rogers, Carl R. *On Becoming a Person.* Boston, MA: Houghton Mifflin, 1961.

———. *Freedom to Learn.* Columbus, OH: Merrill, 1969.

Rothstein, Edward. *Emblems of Mind, The Inner Life of Music and Mathematics.* New York: Times Books, 1995.

Siu, R. G. H. *The Tao of Science.* Cambridge, MA: MIT Press, 1957.

Stein, Sherman. "Toward a Definition of Humanistic Mathematics." *Humanistic Mathematics Network Journal* (7)(Apr. 1992).

Thompson, R. "Learning to Question." In *Teaching in Higher Education: Readings for Faculty.* Steven Scholl and S. E. Inglis, eds. Columbus, OH: Ohio Board of Regents, 1977.

Tymoczko, Thomas. "Humanistic and Utilitarian Aspects of Mathematics." In *Essays in Humanistic Mathematics* (pp. 11–14). Alvin M. White, ed. Washington, DC: Mathematical Association of America, 1993.

White, Alvin M. "Humanistic Mathematics: An Experiment." *Education* 95(2)(1974):128–133.

———. "Teaching Mathematics as Though Students Mattered." In *Teaching As Though Students Mattered.* Joseph Katz, ed. San Francisco, CA: Jossey-Bass, 1985.

ALVIN M. WHITE

# HUMOR

Related to mathematics in many ways. In *Mathematics and Humor* and *I Think, Therefore I Laugh,* John Paulus (1980, 1985) gives scholarly, and often humorous, treatments of the logic and mathematics of humor. He considers both mathematics and humor

forms of intellectual play. Humor also has a place in the mathematics classroom. Jokes, riddles, limericks, puns, cartoons, and so on as well as spontaneous humor can help establish rapport, ease tension, and facilitate learning.

Although research on the specific effectiveness of humor in educational settings is generally inconclusive (Ziv 1988; Zillman and Bryant 1983), a few inferences are possible (Coleman 1992). As one would expect, sarcastic humor has only detrimental effects—cognitive and affective. For older students, positive effects on achievement have been associated with use of humor that was relevant to understanding the concepts studied. Irrelevant humor seems helpful only with preadolescent students. Humor embedded in tests seems to have mixed results: it is likely to benefit the low-anxious students but be detrimental for the high-anxious. Humor in a textbook increases students' enjoyment of reading it. Effects of humor do not seem to be related to gender.

In this context, the following are some examples of using humor in teaching mathematics. Suppose you are beginning a unit on probability and statistics. Tell your students about the cautious homeowner who, after reading that 55% of all traffic accidents occur within five miles of one's home, decided to move. When the laughter subsides, ask what's wrong with the homeowner's thinking. Need something to liven up a discussion of numbers and infinity? Tell about the maiden who was being courted by two avid suitors and told them she would give her hand to whomever could name the greatest number. After a minute of deep concentration, the first young man yelled, "10 billion." Upon hearing this, the other fellow, looking disappointed and defeated, murmured simply, "Darn!" Often quotes from public figures can be the source of humor and associated mathematical discussion. Two of many are: "If crime went down 100%, it would still be 50 times higher than it should be"; "Baseball is 90% mental. The other half is physical" (Petras and Petras 1993). Have students explain why these are funny. Their explanations should reveal something about their understanding of fractions and percentages.

When using humor in teaching, be aware that there is a relationship among sense of humor, age, and level of cognitive development (McGhee 1979). For example, in Piagetian terms a youngster who does not yet conserve substance does not realize that a change in an object's shape (a hunk of modeling clay, for example) does not affect the amount of substance it contains. Such a youngster is not likely to see any humor in the story about Mr. Jones who had ordered a large pepperoni pizza and when asked by the waiter if he would like that cut into six slices or eight slices, replied, "Better just have six slices, I could never eat eight" (McGhee 1979, 158).

Comic strips and cartoons often contain humor with a mathematical twist. Sydney Harris (1990) has produced many cartoons that spoof scientists and mathematicians; and what mathematics student can restrain laughter when viewing the burly, black-hatted gunslinger in Gary Larson's (1982) Old West saloon demanding, "I asked you a question, buddy . . . What's the square root of 5,248?" *Mathematics and Humor* (Azzolino, Silvey, and Hughes 1978) is a plethora of cartoons, funny drawings, jokes, and puns specifically designed for the use of mathematics teachers. Here you will find a sketch of two rabbits musing "Fibonacci numbers surely are a hare-raising experience" (p. 18) and Tom Swifties such as, "'The concavity changes here,' said Tom with inflection" (p. 28).

If students see humor and mathematics as not mutually exclusive, they will also know that mathematics teachers can laugh. Conversely, avoiding humor would surely confirm the impression of Euclid embodied in the following limerick, attributed to Leo Moser:

In the Greek mathematical forum,
Young Euclid was present to bore 'em.
He spent most of his time
Drawing circles sublime
And crossing his pons asinorum.
(Azzolino, Silvey, and Hughes 1978, 49)

*See also* Humanistic Mathematics; Poetry

## SELECTED REFERENCES

Azzolino, Aggie, Linda Silvey, and Barnabas Hughes, eds. *Mathematics and Humor.* Reston, VA: National Council of Teachers of Mathematics, 1978.

Coleman, J. Gordon. "All Seriousness Aside: The Laughing-Learning Connection." *International Journal of Instructional Media* 19(3)(1992):269–276.

Harris, Sydney. *You Want Proof? I'll Give You Proof!* New York: Freeman, 1990.

Larson, Gary. *The Far Side Gallery.* Kansas City, MO: Andrews, McMeel, 1982.

Loomans, Diane, and Karen Kolberg. *The Laughing Classroom.* Tiburon, CA: Kramer, 1993.

McGhee, Paul. *Humor: Its Origin and Development.* San Francisco, CA: Freeman, 1979.

Paulus, John. *I Think, Therefore I Laugh.* New York: Columbia University Press, 1985.

———. *Mathematics and Humor.* Chicago, IL: University of Chicago Press, 1980.

Petras, Ross, and Kathryn Petras. *The 776 Stupidest Things Ever Said.* New York: Main Street Books (Doubleday), 1993.

Thaves, Bob. *Frank and Ernest: Batteries Not Included.* New York: Holt, Rinehart and Winston, 1983.

Zillman, Dolf, and Jennings Bryant. "Uses and Effects of Humor in Educational Ventures." In *Handbook of Humor Research* (pp. 173–193). Paul McGhee and Jeffrey Goldstein, eds. New York: Springer-Verlag, 1983.

Ziv, Avner. "Teaching and Learning with Humor: Experiment and Replication." *Journal of Experimental Education* 57(1)(1988):5–15.

WILLIAM JURASCHEK

# I

## INDEX NUMBERS

Statistical measures used by businesspersons, consumers, economists, and government officials that measure changes which have taken place in the prices, quantities, or values of various commodities. Although important studies of family expenditures and wholesale prices had been made by the United States government during the 1890s, the first Consumer Price Index, then called a *Cost of Living Index,* grew out of a decision by the Shipbuilding Labor Adjustment Board during World War I. The board's objective was to attain a fair wage scale in shipbuilding yards by adjusting wages when the cost of living increased generally. The Bureau of Labor Statistics began regular publication of this index in 1921 and, although often revised, has continued publication to the present.

In its simplest form, an index number is a *ratio expressed as a percentage.* Suppose that in the years 1990 and 1998 the average retail price of a certain type of cheese in the United States was, respectively, $5.00 and $5.75 per pound. Converting these prices into a ratio and then into a percentage, we find that in 1998 the price is 115.0% of what it was in the year 1990. This kind of index refers to a *single* commodity, but there are also index numbers that show the changes in such complicated areas as consumer prices in general, wholesale prices in general, and total industrial production. For instance, the previously mentioned Consumer Price Index, which is often cited in newspapers and on television, is intended to measure price changes in food, clothing, shelter, fuel, drugs, transportation fares, doctors' and dentists' services, and other goods and services that people buy for day-to-day living. This information is important to businesspersons and to others in determining the amount and direction of change in consumer prices.

The Wholesale Price Index is also prepared by the Bureau of Labor Statistics and serves the same purpose for wholesale prices. The Index of Industrial Production is published by the Federal Reserve Board, and it provides information concerning the *quantity* of industrial production in the United States. Many other indexes are published by private organizations.

During the past few decades, the use of index numbers has extended to many new areas of human activity. For example, psychologists measure intelligence quotients, which are index numbers comparing a person's intelligence with that of the average for a person of the same age. Health authorities prepare indexes that show changes in the adequacy of health care in hospitals and other health facilities. Boards of education construct indexes to measure the effectiveness of school systems. Sociologists construct indexes measuring population changes. The National Weather Service has devised a "discomfort index" to measure the combined effects of heat and humidity. The movements of price indexes are vital to millions of workers whose employment contracts include escalator clauses, which provide automatic wage increases when the Consumer Price Index rises by a certain amount. The incomes received by Social Security beneficiaries, retired military and government personnel, postal workers, and food stamp recipients are, by law, also affected by the Consumer Price Index.

*See also* Statistics, Overview

## SELECTED REFERENCES

Anderson, David R., Dennis J. Sweeney, and Thomas A. Williams. *Statistics for Business and Economics*. 5th ed. St. Paul, MN: West, 1993.

Chou, Ya-Lun. *Statistical Analysis for Business and Economics*. New York: Elsevier, 1989.

Freund, John E., Frank J. Williams, and Benjamin M. Perles. *Elementary Business Statistics: The Modern Approach*. 6th ed. Englewood Cliffs, NJ: Prentice-Hall, 1993.

Mendenhall, William, James E. Reinmuth, and Robert J. Beaver. *Statistics for Management and Economics*. Boston, MA: PWS-Kent, 1993.

BENJAMIN M. PERLES

# INDIAN MATHEMATICS

The earliest sources for the history of mathematics in India are Vedic works, the *Sulvasutras,* which probably date back to 1000 B.C. Among the mathematical ideas present in these texts (all related to the theoretical requirements for building altars) are a formula for the area of a circle, a method for calculating square roots, and a discussion of the Pythagorean Theorem, including a diagram in a special case that makes the result obvious.

It was in medieval times, however, that the Indians made their greatest contributions to mathematics, particularly in the fields of trigonometry, algebra, and combinatorics. In trigonometry, they took over the Greek chord function from the work of Hipparchus (ca. 150 B.C.) and replaced it with the sine, although for the Indians, the sine of an arc was not a ratio but half the length of the chord subtending double the arc in a circle of given radius. In fact, our word *sine* comes from a series of mistranslations of the Sanskrit *jya-ardha*, meaning "half-chord." Not only did the Indians produce extensive tables of sines through the use of geometrical theorems but they

also were able by the fifteenth century to calculate sines and cosines to any degree of accuracy through the use of power series. As part of their development of these series, they even worked out the differentials of the sine and cosine.

Much of medieval Indian algebra is detailed in the *Brahmasphutasiddhanta* of Brahmagupta (first half of the seventh century). For example, he presented the rules for operations on positive and negative integers: "The sum of two positive quantities is positive; of two negative is negative; of a positive and a negative is their difference. . . . In subtraction, the less is to be taken from the greater, positive from positive; negative from negative. When the greater, however, is subtracted from the less, the difference is reversed. . . . The product of a negative quantity and a positive is negative; of two negative, is positive; of two positive, is positive . . ." (Colebrooke 1817, 339). He also presented the quadratic formula—in words—in essentially the same form as we know it today, although he only managed to find a single solution. Bhaskara (1114–1185), on the other hand, discussed the conditions under which there were two (positive) solutions.

Both Brahmagupta and Bhaskara were particularly interested in solving indeterminate equations in integers. The former gave detailed procedures for solving the simultaneous congruences $N \equiv a \pmod{r}$ and $N \equiv b \pmod{s}$ and made a beginning toward solving the quadratic indeterminate equation $Dx^2 \pm b = y^2$, a special case of which is usually called the *Pell equation*. A complete solution to the latter problem was given by Bhaskara in his *Siddhantasiromani*. In particular, Bhaskara showed that the smallest solution to $61x^2 + 1 = y^2$ was $x = 226,153,980$, $y = 1,766,319,049$. This problem was, in fact, proposed by Pierre de Fermat in 1657 as a challenge problem to other European mathematicians.

Indian mathematicians from earliest times appear also to have been interested in combinatorial questions. Although we have no knowledge of how they derived the results, the basic formula for calculating the number of ways to choose $k$ objects from a set of $n$ was written out in words by Mahavira (ninth century) as well as Bhaskara and applied to numerous examples. Bhaskara also dealt with permutations with and without repetition.

Finally, our own number system, usually called the *Hindu-Arabic place value system*, probably has its origin in India, although the precise steps in its development are currently unknown. What is known is that in 662, Severus Sebokht, a Syrian priest, remarked that the Hindus have a valuable method of calculation "done by means of nine signs." Sebokht was clearly referring to a place value system, but did not mention the zero. Nevertheless, an Indian manuscript probably dating from the seventh century does use a dot as a symbol for zero as part of a place value system, and the same symbol also appears in the *Chiu-chih li,* a Chinese astronomical work compiled by Indian scholars for the Chinese emperor in 718.

*See also* Diophantine Equations; History of Mathematics, Overview

## SELECTED REFERENCES

Colebrooke, Henry T. *Algebra with Arithmetic and Mensuration from the Sanscrit of Brahmegupta and Bhascara.* London, England: Murray, 1817.

Datta, Bibhutibhusan, and Avadhesh N. Singh. *History of Hindu Mathematics: A Source Book.* Bombay, India: Asia, 1961.

Joseph, George Gheverghese. *The Crest of the Peacock: Non-European Roots of Mathematics.* London, England: Tauris, 1991.

Van der Waerden, Bartel L. *Geometry and Algebra in Ancient Civilizations.* Berlin, Germany: Springer-Verlag, 1983.

VICTOR J. KATZ

# INDIVIDUALIZED INSTRUCTION

The norm throughout history until about the middle of the nineteenth century. The individualization that occurred prior to the mid-1800s happened because relatively few students received a formal school-based education; the rest worked on the farms or in skill-oriented jobs that required little formal education. As the need increased for a better-educated workforce, schools adopted group instruction as the best model because it enabled large numbers of people to be educated efficiently. There was a resulting loss of one-to-one instruction. Those who did not succeed in school simply dropped out. The group approach proved to be less than satisfactory because of the huge waste in human capital.

Harold Shane (1962) noted that several generations of educators in the United States had been examining ways to modify the organizational structure of the schools in order to cope with individual student differences. In the decades immediately preceding 1960, the self-contained classroom in the elementary school was the norm. In those classrooms, students were assigned primarily on the basis of age and tended to receive the same instructional treatment, regardless of their individual differences. Some teachers tried to cope with and accommodate individual differences within the framework of their own classrooms by individualizing seatwork and homework, informal grouping, and special before- and after-school sessions. At the secondary level, individual differences were usually met by curriculum tracking and homogeneous grouping strategies within subject areas.

In the 1960s, the quality of schooling came under attack from a variety of sources, and educators and psychologists began to examine what schools were doing, especially the ways that student differences were being handled. A number of strategies were developed and implemented to try to cope with the vast array of individual differences in achievement, aptitude, and learning style. Some of the strategies focused on small-scale changes built on the efforts of individual teachers, while others focused on large-scale (schoolwide and systemic) initiatives.

Many of the efforts to accommodate individual differences focused on curricular and instructional models that were based on the work of John Carroll and Benjamin Bloom. In 1963, Carroll presented a model of school learning that challenged the time-honored assumption that schooling should provide a fixed amount of time (and no more) for all students to learn whatever was to be taught. His model asserted that learners would learn whatever was to be taught to the extent to which the learner spent the amount of time that he/she needed to learn the task! Bloom's (1968) research and inquiries built on Carroll's work and led to the development of mastery learning models, many of which were applied to school mathematics, especially at the elementary school level.

Maurice Gibbons (1970) developed a matrix that could be used to classify and describe the nature of different programs designed to individualize instruction. His system enables one to make fine distinctions among programs according to variables such as attendance, materials and methods, pace of study, curriculum goals, instructional environment, and the nature of assessment.

John Edling (1970) proposed a simpler scheme in which he identified four basic types of individualized instruction programs: (1) individually diagnosed and prescribed, (2) personalized, (3) self-directed, and (4) independent study. His categories were determined by the extent to which learning objectives and instructional materials and methods were fixed by the school or the extent to

**Table 1** *Types of Individualized Instructional Programs Based on Degree of Student Choice\* of Objectives and Methods and Materials* [Source: Division of Continuing Education, Oregon State University]

| Learning objectives | Methods and materials | |
|---|---|---|
| | Fixed | Optional |
| Fixed | Individually diagnosed and prescribed | Self-directed |
| Optional | Personalized | Independent study |

*Determined by the primary emphasis of each program.

which learners had choices. Table 1 illustrates Edling's category system.

In the individually diagnosed and prescribed programs, primary attention is given to differences in rate of learning, with minor attention given to student interests or learning style. Typically, the curriculum is highly structured with a clearly defined set of objectives and specified instructional materials for students to use as they work toward mastery of the objectives. Diagnostic tests are used to determine each student's knowledge in reference to the curriculum, and then the student is placed in the curriculum and given materials designed to expedite progress through the curriculum at his/her own rate.

Self-directed programs not only focus on differences in rate of learning but also attempt to accommodate differences in learning style. In classrooms or schools using this model, students often have access to a variety of instructional media and environments, such as textbooks, manipulative materials, games, computers, other students in one-to-one and small-group settings, and learning centers. Personalized programs also focus on rate of learning and attempt to accommodate individual differences in interests, while not paying much attention to learning styles. Programs following the independent study model consciously focus on individual differences in rate of learning, interests, and learning styles. The latter two models are not often found in programs that are used to individualize instruction in school mathematics, especially at the elementary school level where the curricular objectives are specified by some organizational authority.

The Individually Prescribed Instruction (IPI) project is perhaps the best known major systemic attempt to individualize instruction using Edling's "fixed objectives/fixed methods and materials" model. In the early versions of IPI, the objectives

were written in terms of desired pupil behaviors and ordered in a sequence to minimize gaps. Pupils' skills were diagnosed. Then the students were placed at an appropriate point on the instructional track. Instructional materials were developed to facilitate pupil progress with a minimum of teacher-pupil interaction, and immediate feedback was available to the pupils. Data were collected and analyzed to aid in the continuing modification of materials and procedures (Lindvall and Bolvin 1967, 236–250). The IPI program evolved into the Adaptive Learning Environments Model (ALEM) over a period of ten years, until it included increased attention to student selection of materials, grouping based on common instructional and social needs, and family involvement to integrate school and home learning (Glaser and Rosner 1975, 84–135). ALEM would be classified as a self-directed model.

Another of the early major systems for individualizing instruction illustrates both Edling's self-directed and personalized models. PLAN (Program for Learning in Accordance with Needs) used teaching-learning units (TLUs) that consisted of objectives, learning materials, and methods (Flanagan, Shanner, Brudner, and Marker 1975, 136–167). Each student's curriculum consisted of a series of TLUs that were based on an assessment of the student's interests, needs, and abilities. In the lower levels of the mathematics program, students had the option of using deductive or inductive approaches to learning the same objectives, and at the higher levels, they were able to select from, and study, a wide range of objectives and commercially available learning materials. A computer facility was used to collect, store, and process data regarding the performance and progress of each student, as well as the student's school history. It served as a basis for decision making in a joint teacher-student meeting about the next objectives that the student would study.

Teachers, principals, curriculum coordinators, and professors have collaborated in a number of ways on the classroom or school level in grass-roots attempts to increase the amount of instructional attention that can be focused on the needs of individual learners. Journals such as *School Science and Mathematics, Arithmetic Teacher, Mathematics Teacher, Educational Technology, Science Education,* and *Educational Leadership* contain literally hundreds of accounts of these small-scale efforts. Strategies used by these teachers included self-paced worksheets and programmed instruction, manipulative materials for use by individuals or small groups, filmstrips and videocassettes to help students who had not mastered the objectives, learning packets for use in independent study settings, within-class grouping, cross-age tutoring, performance contracts, learning centers, mathematics laboratories, computer-aided instruction, team teaching, and nongraded patterns (cutting across several classrooms or even including a whole school), among others. Many of the teachers developed with their colleagues simple-to-elaborate classroom management systems (both paper-based and computer-assisted) for tracking students and materials.

Although these efforts are laudable, many of them suffer from having too narrow a focus. Teaching is a complex act, and good teachers need to focus on a wide array of curricular and instructional variables in their attempts to improve learning and instruction. Nicely (1986) described an instructional model for teaching elementary school mathematics that was developed and modified over a period of seventeen years by more than 100 teachers. The flexible-grouping model was a practical but comprehensive instructional plan that attended to many of the critical concerns, including congruence among the written, taught, and tested curricula, basic skills, problem solving, and higher-order thinking in elementary school mathematics, enrichment activities, mastery learning, diagnostic-prescriptive teaching, direct instruction, productive time-on-task, and the use of effective teaching and learning strategies, in an attempt to accommodate many individual learner differences. Nicely notes that this model is not the answer to all the existing problems in improving elementary mathematics instruction, but rather is an illustration of one way to deal with the complex and interacting variables that face classroom teachers. In Edling's classification system, this flexible grouping plan would be an example of the individually diagnosed and prescribed model with considerable attention being given to self-directed and personalized models.

At the postsecondary level, perhaps the best-known attempt to individualize instruction was the personalized system of instruction (PSI), which was devised by Keller (1968). His plan organizes the mathematics courses into units that are presented in a linear sequence. The students must master each unit before going on to the next unit. Each unit has objectives, suggested instructional resources, study questions, and proctors to tutor course enrollees. The proctors are usually students who completed the course successfully in the not-too-distant past. Edling would classify PSI as a self-directed approach to individualizing instruction.

Until recently, the computer has been used largely for managing data regarding student performance, retrieving appropriate instructional resources, or providing instruction on or reinforcement of basic skills and content. But most of these systems have been narrowly designed. Kyle Peck (1994) has called for the development of computer-based technologies to help teachers in their modern roles in restructured performance-based school systems. The recent insights of cognitive science, when used in conjunction with the ever-increasing capabilities of the computer, provide powerful means for individualizing pace of instruction; selection of content and resources; networking possibilities with learners and teachers in other classrooms, schools, states, and across the world; and levels of inquiry and understanding that are presently available to relatively few fortunate individuals.

As communities and their schools become increasingly diverse and as student differences multiply, whole-class instruction will be increasingly less desirable. As the knowledge explosion continues and demands for an increasingly competent and continually learning workforce increase, there will be an increasing demand for instructional effectiveness and efficiency. J. D. Fletcher (1992) noted that instructional systems that respond to individual learners' needs by attending to content, pace, sequence, and learning style without raising costs will contribute significantly to effectiveness and efficiency. He contends that research needs to be conducted to determine how individualized systems of instruction can best be integrated with current instructional practices and institutions. Although the desirability, effectiveness, and promise of such systems are clear, ways to ensure their effective implementation in schools and classrooms on a wide scale are more elusive.

*See also* Cognitive Learning; Computer-assisted Instruction (CAI); Technology; Tutoring

## SELECTED REFERENCES

Bloom, Benjamin S. "Learning for Mastery." *UCLA Evaluation Comment* 1(2)(1968):1–12.

Carroll, John. "A Model of School Learning." *Teachers College Record* 64(8)(1963):723–733.

Edling, John. *Individualized Instruction: A Manual for Administrators*. Corvallis, OR: Continuing Education Publications, Oregon State University, 1970.

Flanagan, John C., William M. Shanner, Harvey J. Brudner, and Robert W. Marker. "An Individualized Instructional System: PLAN*." In *Systems of Individualized Education* (pp. 136–167). Harriet Talmage, ed. Berkeley, CA: McCutchan, 1975.

Fletcher, J. D. "Individualized Systems of Instruction." In *Encyclopedia of Educational Research* (pp. 613–620). Vol. 2. 6th ed. Marvin C. Alkin, ed. New York: Macmillan, 1992.

Gibbons, Maurice. "What Is Individualized Instruction?" *Interchange* 1(2)(1970):28–52.

Glaser, Robert, and Jerome Rosner. "Adaptive Environments for Learning: Curriculum Aspects." In *Systems of Individualized Education* (pp. 84–135). Harriet Talmage, ed. Berkeley, CA: McCutchan, 1975.

Keller, Fred S. "Goodbye, Teacher . . ." *Journal of Applied Behavior Analysis* 1(1968):79–89.

Lindvall, C. Mauritz, and John O. Bolvin. "Programed Instruction in the Schools: An Application of Programing Principles in Individually Prescribed Instruction." In *Programed Instruction* (pp. 217–254). 66th Yearbook, Part II. National Society for the Study of Education. Phil C. Lange, ed. Chicago, IL: University of Chicago Press, 1967.

Nicely, Robert F., Jr. "Synthesizing Research-Based Instructional Strategies for Effective Teaching." In *Mathematics Teaching and Learning* (pp. 29–36). 1986 Yearbook. Robert F. Nicely, Jr. and Thomas F. Sigmund, eds. University Park, PA: Pennsylvania Council of Teachers of Mathematics, 1986.

Peck, Kyle. "Performance-Oriented Educator's Toolkit (POET)." Unpublished proposal to move Pennsylvania toward Goals 2000 (pp. 1–7). University Park, PA: Institute for the Reinvention of Education, 1994.

Shane, Harold G. "The School and Individual Differences." In *Individualizing Instruction* (pp. 44–61). 61st Yearbook, Part I. National Society for the Study of Education. Nelson B. Henry, ed. Chicago, IL: University of Chicago Press, 1962.

ROBERT F. NICELY, JR.

# INEQUALITIES

Represent order relationships between two real numbers. One useful classical inequality is the triangle inequality, which states that the sum of the lengths of any two sides of a triangle is always greater than or equal to the length of the third side (West et al. 1982, 268). The symbols for order relationships include < (less than), ≤ (less than or equal to), > (greater than), and ≥ (greater than or equal to). Thomas Harriot, an English mathematician, first used the signs > and < in his algebra book *Artis analyticae praxis*, published after his death in 1631 (Eves 1983, 230). However, other mathematicians did not use these symbols immediately.

In denoting an order relationship, the arrow always points to the smaller number. For example, $2 <$ 3 and $3 > 2$ are equivalent relationships since 2 is smaller than 3, and 3 is larger than 2. On a number line, $2 < 3$ since 2 lies to the left of 3, and $3 > 2$ since 3 lies to the right of 2. Although $2 < 3$, the number line shows that $-2 > -3$. This example may help students understand that multiplying by a negative number reverses the inequality symbol.

An unconditional inequality is always true. Such inequalities include $5 > 2$ or $(x + 1)^2 + 3 > 2$. A conditional inequality is true for some values of $x$. For example, $x + 2 > 5$ is true for $x > 3$.

The National Council of Teachers of Mathematics (NCTM) recommends that students in grades 5–8 investigate inequalities informally (NCTM 1989, 102), while students in grades 9–12 be able to represent physical situations as inequalities and solve them (NCTM 1989, 150). Informal approaches include using materials such as a double-pan scale in which weights on one side are heavier than the weights on the other side (College Entrance Examination Board and Educational Testing Service 1990, 233) or Algebra Lab Gear manipulatives (Wah and Picciotto 1994, T21). The Project for Computer-Intensive Curricula in Elementary Algebra (1991) is a good source for physical representations of inequalities and how to use graphs to test and define rules for equivalent inequalities (9-55–9-79).

## SOLVING AN INEQUALITY IN ONE VARIABLE

This is similar to solving a linear equality in that the goal is to construct equivalent inequalities until the variable remains by itself on one side of the inequality. One can add or subtract equal quantities on both sides of the inequality. For example, if $x + 3 > 5$, then $x + 3 - 3 > 5 - 3$ or $x > 2$. It is also possible to multiply or divide equal positive quantities on both sides of the inequality. However, when multiplying (or dividing) equal negative quantities on both sides of the inequality, the direction of the inequality symbol reverses, as in the following:

$$2(5 - x) > 14$$
$$10 - 2x > 14$$
$$-2x > 4$$
$$x < -2$$

Set notation, interval notation, or intervals on a number line can be used to represent solutions to algebraic inequalities. In set notation, $x < -2$ is expressed as $\{x | x < -2\}$; in interval notation, $(-\infty, -2)$, and as a graph:

The open parenthesis, ), or open circle at $-2$ indicates that $-2$ is not included in the graph. When the solution set includes the endpoint, interval notation uses brackets. For example, $[-2, 5]$ represents all numbers between $-2$ and $5$, inclusive. As shown in the graph, either brackets or closed circles at $-2$ and $5$ indicate that they are included in the solution set:

The following method (McLaurin 1985) is useful when solving algebraic inequalities that have polynomials of degree greater than 1, absolute values, or rational expressions: (1) Replace the inequality sign with an equality sign, and solve the equations. (2) Use the solution(s) to divide the number line into distinct intervals. (3) Select a value from each interval and test it in the original inequality. (What is true for the tested value will be true for the entire interval.) (4) Write down the interval(s) that make the inequality true. If the original inequality is $\leq$ or $\geq$, include endpoints.

*Example of a Quadratic Inequality:* Solve $x^2 - x \leq 6$.

1. Solve $x^2 - x = 6$.

$$x^2 - x - 6 = 0$$
$$(x - 3)(x + 2) = 0$$
$$x = 3 \quad \text{or} \quad x = -2$$

2. The intervals that arise from these solutions are $(-\infty, -2)$, $(-2, 3)$, and $(3, \infty)$.

3. Many choices are possible, but $-3$, $0$, and $4$ are easy to substitute into $x^2 - x \leq 6$. For $x = -3$, $12 > 6$, so the inequality is false; for $x = 0$, $0 < 6$, so the inequality is true; and for $x = 4$, $12 > 6$, so the inequality is false.

4. The solution of the given inequality is $[-2, 3]$.

*Example of an Absolute Value Inequality* $|x + 1| > 3$.

1. Solve $|x + 1| = 3$, which is equivalent to solving $x + 1 = 3$ and $x + 1 = -3$. The solutions are $x = 2$ and $x = -4$.

2. The intervals are $(-\infty, -4)$, $(-4, 2)$, and $(2, \infty)$.

3. For $x = -5$, the inequality is true, $|-4| > 3$; for $x = 0$, the inequality is false, $|1| < 3$; and for $x = 3$, the inequality is true, $|4| > 3$.

4. The solution is $(-\infty, -4) \cup (2, \infty)$.

When solving inequalities having rational expressions, one may determine the intervals by factoring both numerator and denominator. The solution must exclude any values that make the denominator zero. This method can also be used for inequalities that have radical, logarithmic, exponential, or trigonometric functions (Dobbs and Peterson 1991).

Although algebraic solution techniques are limited to those problems that can be solved by factoring, the use of graphing calculators or computers can greatly expand the options. For example, if an inequality has the form $y - f(x) < k$, students can solve this inequality by graphing $y = f(x) - k$ and finding the values of $x$ where the graph lies under the $x$-axis. In the following illustration, $y = f(x) - k < 0$ for $(-\infty, a)$, $(b, c)$ $(d, \infty)$. (See Figure 1.)

## SOLVING INEQUALITIES IN TWO VARIABLES

The solution for an inequality in two variables is sometimes shown as a region on a rectangular coordinate graph. For a linear inequality in the form $ax + by < c$, or $ax + by > c$, one can graph the line and use a test point to determine on which side of the line the solution lies. Another method is to solve the inequality for $y$. The direction of the inequality determines if the solution set lies above or below the line. The graphing convention is to use a solid line if

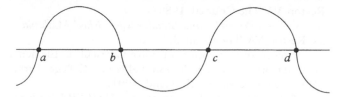

**Figure 1** $y = f(x) - k$.

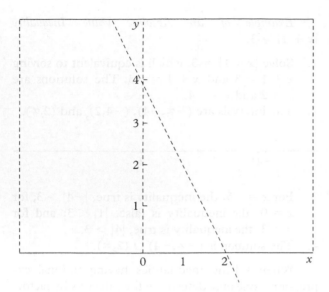

**Figure 2** $y < -2x + 4$, *including* (0,0).

the line is to be included; otherwise, a dotted line is used. For example, the solution set for the linear inequality $2x + y < 4$ is the set of all points in the region under the line $y = -2x + 4$ since $y < -2x + 4$.

*See also* Equalities; Linear Programming

SELECTED REFERENCES

Beckenbach, Edwin, and Richard Bellman. *An Introduction to Inequalities.* New York: Random House, 1961.

College Entrance Examination Board and Educational Testing Service. *ALGEBRIDGE.* Providence, RI: Janson, 1990.

Dobbs, David E., and John C. Peterson. "The Sign-Chart Method for Solving Inequalities." *Mathematics Teacher* 84(Nov. 1991):657–664.

Eves, Howard W. *An Introduction to the History of Mathematics.* 5th ed. Philadelphia, PA: Saunders College Publishing, 1983.

Lial, Margaret, et al. *Intermediate Algebra.* New York: HarperCollins, 1996.

McLaurin, Sandra C. "A Unified Way to Teach the Solution of Inequalities." *Mathematics Teacher* 78(Feb. 1985):91–95.

National Council of Teachers of Mathematics. *Curriculum and Evaluation Standards for School Mathematics.* Reston, VA: The Council, 1989.

———. *Principles and Standards for School Mathematics.* Reston, VA: The Council, 2000.

Project for Computer-Intensive Curricula in Elementary Algebra. *Computer-Intensive Algebra.* College Park, MD: The University of Maryland, 1991.

Wah, Anita, and Henri Picciotto. *ALGEBRA Teacher's Edition.* Mountain View, CA: Creative, 1994.

Waits, Bert K., and Franklin Demana. "A Computer-graphing Based Approach to Solving Inequalities." *Mathematics Teacher* 82(May 1989):327–331.

West, Beverly Henderson, et al. *The Prentice-Hall Encyclopedia of Mathematics.* Englewood Cliffs, NJ: Prentice-Hall, 1982.

ANN E. MOSKOL

# INSTRUCTIONAL MATERIALS AND METHODS, ELEMENTARY

Strategies or activities to help students learn. Materials and methods should be appropriate for the age and maturity of the students. For example, manipulatives such as Cuisenaire rods, unifix cubes, and base 10 blocks can assist in the teaching of basic concepts such as number sense and operations. A first-grade classroom could be using the Cuisenaire rods in the teaching of length and color, a second-grade classroom could use them to make addition sentences such as R + E = O or red plus blue = orange. A third-grade classroom could use them to teach subtraction such as O − R = E (orange − red = blue). A fourth-grade classroom could use them to teach multiplication and division such as light green times purple = 12 (three groups of four = 12 or four groups of three = 12). In the fifth grade, students could use them to learn fractions, for example, red = $\frac{1}{5}$ of orange. In the teaching of geometry, manipulatives such as tangrams, pentominoes, supertangrams, multilink cubes, and polydrons help teachers introduce new concepts. Software such as Tessel-Mania by MECC is a wonderful tool to help students understand tessellations. In the 1990s, calculators and computers were of great help in classroom instruction.

Problem solving is one of the reasons for studying mathematics. Solutions to many problems should be found by teamwork. Practice with the tools of problem solving and critical thinking is achieved by means of cooperative learning. The ability to communicate is further encouraged by journal writing, expository writing, and oral discussions.

Learning proceeds through the individual student's construction of understanding, not just by studying facts and rules from a textbook. Thus teaching is the facilitation of student construction of knowledge, not just the delivery of information. Using a variety of types of assessment (often called "authentic assessment"), including judgment of work collected in individual student portfolios, helps teachers understand what it is their students have achieved.

*See also* Calculators; Computers; Field Activities; Instructional Methods, Current Research; Instructional Methods, New Directions; Techniques and Materials for Enrichment Learning and Teaching

## SELECTED REFERENCES

Aichele, Douglas B., ed. *Professional Development for Teachers of Mathematics.* 1994 Yearbook. Reston, VA: National Council of Teachers of Mathematics, 1994.

Harmin, Merrill. *Inspiring Active Learning: A Handbook for Teachers.* Alexandria, VA: Association for Supervision and Curriculum Development, 1994.

Lewis, Anne C. "An Overview of the Standards Movement." *Phi Delta Kappan* 76(10)(1995):744–750.

McCarthy, B. *The 4 MAT System: Teaching to Learning Styles with Right/Left Mode Techniques.* Barrington, IL: EXCEL, 1981.

National Council of Teachers of Mathematics. *Curriculum and Evaluation Standards for School Mathematics.* Reston: VA: The Council, 1989.

———. *Principles and Standards for School Mathematics.* Reston, VA: The Council, 2000.

Reesink, Carole J., ed. *Teacher-Made Aids for Elementary School Mathematics: Readings from the "Arithmetic Teacher" and "Teaching Children Mathematics."* Vol. 3. Reston, VA: National Council of Teachers of Mathematics, 1998.

Steen, Lynn A. *On the Shoulders of Giants: New Approaches to Numeracy.* Washington, DC: National Academy Press, 1990.

BETTY M. ERICKSON

# INSTRUCTIONAL METHODS, CURRENT RESEARCH

The major underlying assumptions of the reform initiatives of the 1990s have focused on the way knowledge is acquired by individuals. The reform movement is based on a constructivist theory of learning, with origins in the work of Jean Piaget. This theory suggests that learning occurs when learners actively engage in making meaning and seeking understanding, rather than passively receiving knowledge from others. Teaching for construction of knowledge is characterized by several principles: (a) the student must be actively engaged in order to build understanding, (b) learning entails making sense of new situations based on previously constructed ideas, and (c) reflective thought is a critical component of the learning process.

Within the constructivist framework, research support has been established for several instructional strategies: placing an emphasis on problem solving, using multiple representations to explore a concept, encouraging the use of oral and written language by the learners to exchange mathematical ideas, using cooperative learning groups, and making assessment an ongoing and integral part of instruction.

## PLACING AN EMPHASIS ON PROBLEM SOLVING

Research on problem solving was abundant throughout the 1980s as was research on instruction in problem solving. Views about using problem solving in instruction have evolved from problem solving should be added to the curriculum to mathematics should be taught via problem solving (Schroeder and Lester 1989). Research that examines this second view, problem solving as an integral part of instruction, suggests that children are in fact learning mathematics in a more meaningful way with greater understanding when instruction is problem-centered (see, for example, Wood et al. 1993; Wood and Sellers 1996). Based on numerous studies with similar findings, the National Council of Teachers of Mathematics in its *Curriculum and Evaluation Standards for School Mathematics* (1989) has recommended problem-solving-based instruction for the improvement of mathematical learning.

## USING MULTIPLE REPRESENTATIONS TO EXPLORE A CONCEPT

Probably the most significant findings from research that inform practice are those involving the use of multiple representations of mathematical concepts. Jerome Bruner (1966) and Zoltan Dienes (1960) were early proponents of this approach, and researchers since then have continued to explore the effects of multiple representations on students' learning of mathematics (Janvier, Girardon, and Morand 1993). Representations that have been studied include, among others, concrete models, pictures, symbolic notation, and graphs and tables to display data. In particular, the advent of technology has facilitated the dynamic interchange of representations (Kaput 1992).

Manipulative materials, for example, have proven successful in promoting understanding of mathematical concepts when the instruction is designed in a way that helps students build connections between various forms of representation (see, for example, Fuson et al. 1997). Computer microworlds have been used to support students' learning of diverse mathematical topics by linking representations to their understanding, for example, in the teaching of rational numbers (Steffe and Wiegel 1994; Hunting et al. 1996), multiplicative structures (Confrey 1994), and algebra (Thompson 1989; Yerushalmy 1997), among many other topics. The research about students' interactions within the microworld environment supports the use of microworlds to develop students' mathematical understanding and reasoning skills and has greatly contributed to the knowledge base of how students build meaning and reason within the realm of the different mathematical topics.

The use of electronic tools for geometric construction (e.g., Geometric Supposer, Geometer's Sketchpad, Cabri Geometry) has also been the subject of recent research involving representations of mathematics concepts. In algebra instruction, researchers' attention has been on the positive impact on learning when students use graphing calculators or computer programs with graphing capabilities. Finally, some attention has also been given to the positive impact of spreadsheets, which can represent and manipulate data in various forms, on engaging students in learning a variety of mathematical concepts and developing problem-solving skills.

## ENCOURAGING THE USE OF ORAL AND WRITTEN LANGUAGE TO EXCHANGE MATHEMATICAL IDEAS

Written and oral language is a significant component of reflective thought and a critical component of the construction of knowledge. The articulation and interpretation of ideas in written or oral form require learners to organize their thoughts in a meaningful way, linking new ideas to previously learned ones.

Support for the use of oral communication as an integral part of the construction of ideas can be found in the work of Terry Wood, Paul Cobb, and Erna Yackel (1993). In teaching experiments conducted in a second-grade classroom, the researchers documented the power of oral communication among students and teachers in building understanding of mathematical ideas. Raffaella Borasi's (1992) work points to the importance of oral communication in assessing high school students' conceptual understanding of geometrical concepts. Her report of a teaching experiment carefully describes

how students learn mathematics as they negotiate meanings while pursuing a curriculum driven by their personal inquiries. "Writing to learn" strategies are the focus of the work of Joan Countryman (1992), who describes the use of student writing to assess understanding and to help students make connections between their prior knowledge and their understandings of new ideas. The use of literature to provide meaningful contexts for learning is suggested in the work of David Whitin (1995).

## USING COOPERATIVE LEARNING GROUPS

Although cooperative learning has received increasing support in much of the research literature, many questions remain unanswered with regard to its effectiveness (Good, Mulryan, and McCaslin 1992). Neil Davidson and Diana Lambdin Kroll (1991), however, have described many research results in which cooperative learning environments have led to the improvement of student achievement. In addition, many noncognitive positive effects of cooperative work have been reported in the research literature (Leikin and Zaslavsky 1997). These include findings that "cooperation promotes self-esteem, increased efforts to achieve, enhanced psychological health and caring relationships, and ability to take the perspective of another person" (Davidson and Kroll 1991, 363).

## MAKING ASSESSMENT AN ONGOING AND INTEGRAL PART OF INSTRUCTION

To support a constructivist framework for instruction, it is essential that teachers learn as much as possible about their students' prior knowledge as they plan experiences for them (Clarke 1997; Davis 1997). Within this perspective, alternative forms of assessment become a major component of a teacher's daily practices (Webb 1993). Different strategies are suggested for assessing students' understanding, including observations, open-ended tasks, assessment of group work, journals, interviews, portfolios, and project presentations. Meaningful assessment must be aligned with the curriculum, and so manipulative materials, group work, writing experiences, and the use of technology may each be involved in the assessment process.

An issue of significance to the discussion of aligning assessment practices to instruction is the relationship between instruction and mandated testing practices. Lynn Hancock and Jeremy Kilpatrick (1993) summarized research findings describing the impact of mandated tests on instruction. They reported on survey studies of teachers, who revealed how instructional practices can suffer as they strove to prepare students for success on standardized tests. Teachers reported often being driven to pay increased attention to the basic skills and numerical manipulations expected on standardized tests. Some studies report a negative impact of standardized tests on students (e.g., Smith and Rottenberg 1991) and on the work lives of teachers, as defined, for example, by an increase in demands on their time, in paperwork, in pressures for student performance, and in other areas that resulted in a general increase in levels of stress (e.g., Corbett and Wilson 1991).

*See also* Assessment of Student Achievement, Overview; Constructivism; Cooperative Learning; Test Construction; Writing Activities

## SELECTED REFERENCES

Borasi, Raffaella. *Learning Mathematics Through Inquiry*. Portsmouth, NH: Heinemann, 1992.

Bruner, Jerome. *Toward a Theory of Instruction*. Cambridge, MA: Belknap Press of Harvard University Press, 1966.

Clarke, Doug M. "The Changing Role of the Mathematics Teacher." *Journal for Research in Mathematics Education* 28(May 1997):278–308.

Confrey, Jere. "Splitting, Similarity and Rate of Change: A New Approach to Multiplication and Exponential Functions." In *The Development of Multiplicative Reasoning in the Learning of Mathematics*. Guershon Harel and Jere Confrey, eds. Albany, NY: State University of New York Press, 1994.

Corbett, H. Dickson, and Bruce L. Wilson. *Testing, Reform, and Rebellion*. Norwood, NJ: Ablex, 1991.

Countryman, Joan. *Writing to Learn Mathematics: Strategies that Work*, Portsmouth, NH: Heinemann, 1992.

Davidson, Neil, and Diana Lambdin Kroll. "An Overview of Research on Cooperative Learning Related to Mathematics." *Journal for Research in Mathematics Education* 22(Nov. 1991):362–365.

Davis, Brent. "Listening for Differences: An Evolving Conception of Mathematics Teaching." *Journal for Research in Mathematics Education* 28(May 1997):355–376.

Dienes, Zoltan. *Building Up Mathematics*. London, England: Hutchinson, 1960.

Fuson, Karen C., et al. "Children's Conceptual Structures for Multidigit Numbers and Methods of Multidigit Addition and Subtraction." *Journal for Research in Mathematics Education* 28(Mar. 1997):130–162.

Good, Thomas, Catherine Mulryan, and Mary McCaslin. "Grouping for Instruction in Mathematics: A Call for Programmatic Research on Small-Group Processes."

In *Handbook of Research on Mathematics Teaching and Learning* (pp. 165–196). Douglas A. Grouws, ed. New York: Macmillan, 1992.

Hancock, Lynn, and Jeremy Kilpatrick. "Effects of Mandated Testing on Instruction." In *Measuring What Counts: A Conceptual Guide for Mathematics Assessment* (pp. 149–174). National Research Council. Washington, DC: National Academy Press, 1993.

Hunting, Robert P., Gary Davis, and Catherine A. Pearn. "Engaging Whole-Number Knowledge for Rational Number Learning Using a Computer-Based Tool." *Journal for Research in Mathematics Education* 27(May 1996): 354–379.

Janvier, Claude, Catherine Girardon, and Jean-Charles Morand. "Mathematical Symbols and Representations." In *Research Ideas for the Classroom: High School Mathematics* (pp. 79–102). Patricia S. Wilson, ed. New York: Macmillan, 1993.

Kaput, James. "Technology and Mathematics Education." In *Handbook of Research on Mathematics Teaching and Learning* (pp. 515–556). Douglas A. Grouws, ed. New York: Macmillan, 1992.

Leikin, Roza, and Orit Zaslavsky. "Facilitating Student Interactions in Mathematics in a Cooperative Learning Setting." *Journal for Research in Mathematics Education* 28(May 1997):331–354.

National Council of Teachers of Mathematics. *Curriculum and Evaluation Standards for School Mathematics*. Reston, VA: The Council, 1989.

———. *Principles and Standards for School Mathematics*. Reston, VA: The Council, 2000.

Schroeder, Thomas L., and Frank K. Lester, Jr. "Developing Understanding in Mathematics via Problem Solving." In *New Directions for Elementary School Mathematics* (pp. 31–56). 1989 Yearbook. Paul R. Trafton, ed. Reston, VA: National Council of Teachers of Mathematics, 1989.

Smith, Mary L., and Claire Rottenberg. "Unintended Consequences of External Testing in Elementary Schools." *Educational Measurement: Issues and Practice* 10(4)(1991): 7–11.

Steffe, Leslie P., and Heide G. Wiegel. "Cognitive Play and Mathematical Learning in Computer Microworlds." In *Learning Mathematics: Constructivist and Interactionist Theories of Mathematics Development* (pp. 7–30). Paul Cobb, ed. Dordrecht, Netherlands: Kluwer, 1994.

Thompson, Patrick W. "Artificial Intelligence, Advanced Technology, and Learning and Teaching of Algebra." In *Research Issues in the Learning and Teaching of Algebra* (pp. 135–161). Carolyn Kieran and Sigrid Wagner, eds. Reston, VA: National Council of Teachers of Mathematics, 1989.

Webb, Norman, ed. *Assessment in the Mathematics Classroom*. 1993 Yearbook. Reston, VA: National Council of Teachers of Mathematics, 1993.

Whitin, David J. "Connecting Literature and Mathematics." In *Connecting Mathematics Across the Curriculum* (pp. 134–141). 1995 Yearbook. Peggy A. House, ed. Reston, VA: National Council of Teachers of Mathematics, 1995.

Wood, Terry, Paul Cobb, and Erna Yackel. "The Nature of Whole Class Discussion." In *Rethinking Elementary School Mathematics: Insights and Issues* (pp. 55–68). Terry Wood, Paul Cobb, Erna Yackel, and Deborah Dillon, eds. Reston, VA: National Council of Teachers of Mathematics, 1993.

———, and Deborah Dillon, eds. *Rethinking Elementary School Mathematics: Insights and Issues.* Reston, VA: National Council of Teachers of Mathematics, 1993.

Wood, Terry, and Patricia Sellers. "Assessment of a Problem-Centered Mathematics Program: Third Grade." *Journal for Research in Mathematics Education* 27(May 1996):337–353.

Yerushalmy, Michal. "Designing Representations: Reasoning about Functions of Two Variables." *Journal for Research in Mathematics Education* 28(July 1997): 431–466.

BEATRIZ S. D'AMBROSIO
EILEEN L. POIANI

# INSTRUCTIONAL METHODS, NEW DIRECTIONS

Uses of new technologies and possible transformations of the K–14 instructional environment. The impetus for change stems from several sources: technological change; the poor performance of American students on local, national, and international assessment tests; the decline in the number of mathematics, science, and engineering undergraduate majors and doctoral candidates in the United States; and the inability to meet the nation's workforce needs.

Research in mathematics education and in learning psychology has shown that instruction based on rote exercises limited students' development and ability to formulate creative strategies to solve new problems. As stated in *Everybody Counts* (National Research Council (NRC) 1989, 58–59): "Educational research offers compelling evidence that students learn mathematics well only when they *construct* their own mathematical understanding. To understand what they learn, they must enact for themselves verbs that permeate the mathematics curriculum: 'examine,' 'represent,' 'transform,' 'solve,' 'apply,' 'prove,' 'communicate.' This happens most readily when students work in groups, engage in discussion, make presentations, and in other ways take charge of their own learning." Proposed standards for curriculum, evaluation, and teaching recommend focusing instruction on "problem solving, communication, reasoning, and mathematical connections" in an active learning environment (National Council of Teachers of Mathematics (NCTM) 1989, 1991).

Recommendations for reform of American mathematics education have included the following: "shifting from authoritarian models based on 'transmission of knowledge' and 'drill and practice' to student-centered methods featuring 'stimulation of learning' and 'active exploration' . . ." (NRC 1991, 7). The Mathematical Sciences Education Board (MSEB) as well as statewide coalitions of mathematical sciences groups have tried to improve instruction by influencing school administrators, teachers, and policy makers to adopt new methodologies. At the level of the first two years of undergraduate mathematics courses (years 13 and 14), the calculus reform movement has encouraged the use of graphing calculators (NRC 1991). For two-year college and lower-division undergraduate offerings below the calculus level, *Crossroads in Mathematics* addressed the framework for course standards and instructional strategies that are consistent with other reform initiatives of the 1990s (American Mathematical Association of Two-Year Colleges 1995, 69).

## USES OF NEW TECHNOLOGIES

The philosophical basis for the move to technology is clear: "It is now possible to execute almost all of the mathematical techniques taught from kindergarten through the first two years of college on hand-held calculators. . . . the changes in mathematics brought about by computers and calculators are so profound as to require readjustment in the balance and approach to virtually every topic in school mathematics" (NRC 1990, 2). Calculators, computers, electronic communication, interactive television (ITV), CD-ROMs (compact disc read-only memory), and other new technologies pose challenges for instructors. The question arises: what types of innovations are appropriate and how does the use of technology change the nature of mathematics learning and teaching?

### Beliefs about Mathematics

The use of technology may change common perceptions. Studies have revealed that students believe mathematics consists of a collection of algorithms and that learning mathematics entails memorizing those algorithms (NRC 1989, 57). Unfortunately, many traditional school experiences with mathematics have reinforced this belief. The typical school experience consists of learning a set of rules, being shown an example or two that illustrate how to follow the rules, and then mimicking the solution by solving many problems just like the examples shown. The computer or calculator can help teachers shift students' focus from the algorithms to thinking about problem-solving strategies. The student of mathematics—whether college- or job-bound—needs to be proficient in identifying which algorithm to use, and when it is most efficient to solve by calculator, mental arithmetic, or pencil and paper.

## Calculators and Computers—Dynamic Exploration

The teacher can use a calculator with overhead projection capability while students use a companion calculator at a desk or table. Examples described in the NCTM *Standards* illustrate the use of calculators in problem solving. Actual, "real-life numbers" (as opposed to contrived "nice numbers" that can be handled by paper and pencil) can enliven a problem—from the size of a laser beam to the earth's distance from the Milky Way. Sharpening the student's ability to estimate answers is necessary when relying on technology. If the wrong calculator keys are pressed, a student will benefit from using estimation to detect a wrong answer. Developing mental arithmetic strategies for estimation can begin with simple examples in the elementary grades: estimating the number of seeds in a pumpkin, guessing the height of classmates, or finding how many boxes of a given size are needed to fill the classroom (NCTM 1989, 36–37).

In high school algebra classes, much time is spent on finding the roots of quadratics. Although proficient in calculating the roots, students quite often lack an understanding of how the graphs of quadratics relate to the technique of finding the roots. They are unable to identify real-world situations that can be represented by quadratic relationships, such as the golden ratio found in architecture and music $[(1 + \sqrt{5})/2]$ and the links among velocity, distance, and time. The use of computers or graphing calculators allows students to explore real data and the relationships among the variables. Students can easily graph the relationships to obtain visual evidence; engage in mathematical modeling and explore whether the equations they create produce the same graphs as those from the real data; then study the differences and refine the mathematical model accordingly. For example, modeling the spread of the HIV virus can be done in different grades using various levels of statistical sophistication. By being able to explain the meaning of the quadratic roots in a real-world context, students will

realize that few real-world problems have solutions that coincide with the simple mathematical models in many textbooks. This may lead students to think more deeply about the real situations and the mathematical models and patterns available to them. Students and teachers can then pursue together the mathematics appropriate to solving the problem, rather than searching for problems that lend themselves to the use of specific mathematical techniques (NRC 1989, 32–33; 1990, 12).

Existing commercial and home-grown software allows students to generate many examples of a concept quickly, so exploration and investigation of ideas become dynamic. For example, given an equation of a line, $y = mx + b$, what happens to the graphical representation if one constant is changed? This exploration can begin with hand-drawn graphs and move to software that illustrates the graphs for several different constants on one coordinate system. The process of dynamic explorations allows students to work within a context that encourages raising and testing conjectures that are intriguing and generate curiosity and thereby tends to make the mathematics more meaningful.

Sophisticated graphics are revolutionizing the ways students visualize all geometries. Software such as *Algebra Xpresser* (Hoffer 1991) allows students to link different representations for a mathematical idea. A change in a relationship among variables represented in a table, an equation, or a graph results in changes in the other representations. In a technology-rich environment, the student can explore how the graph of the equation $y = x + 5$ changes as the constant changes. Similarly, the student could "pick up" the line and move it around on a coordinate plane and watch how the equation changes to match new positions of the line. This encourages students to look for patterns in the various representations of relationships among variables and in how these representations are connected. It is particularly useful to have students predict how a change in one representation might appear in another (Kaput 1992).

## Electronic Communication and Interactive Television (ITV)

Computer availability permitting, mathematics classes in the United States and around the world can become "electronic pals" to enrich each other from a distance. The electronic bulletin board completely changes the nature and use of bulletin boards in the K–14 classroom. At each grade level, challenging problems drawn from journals such as *Teaching*

*Children Mathematics, Mathematics Teaching in the Middle School, Mathematics Teacher* (NCTM); *Math Horizons* (Mathematical Association of America); and *Pi Mu Epsilon Journals* can be proposed. Many newspapers and journals available electronically provide additional course material. Prizes for team and individual solutions can be mathematics books, computer software, or a math T-shirt. Communication can be in English or in another language if students are available to translate. Literally any topic in class can be explored over the World Wide Web and other electronic connections. Studying the mathematics of the spread of disease, the change in the national debt, or the impact of new drugs to fight cancer can be coupled with information on the history of the problem, the current research, and biographies with pictures of the people conducting the research.

Linked by satellite or fiber optics, interactive television can enable one classroom to interact with another a mile away or a world away. Tackling problems such as improving the quality of water or examining global air traffic from a mathematical point of view can engage several classes in a shared project. Showcasing successful instructional methods in mathematics classrooms can promote reform and provide new teachers with electronic mentors.

## CD-ROMs

This technology enables the student and teacher to page through references to see the historical context in which particular parts of mathematics were discovered and to see the mathematicians—men and women—who discovered them. The connectedness of mathematics to people, time, and society becomes a dimension of instruction.

## TRANSFORMATIONS OF THE INSTRUCTIONAL ENVIRONMENT

Engaging students actively in their exploration of mathematics can also involve the following strategies: cooperative and team learning and writing to learn.

## Cooperative and Team Learning

Many preparatory schools traditionally have centered their instruction around an oval "Harkness table," where a dozen or so students, with an instructor, converse and learn. Each student is stretched to challenge, speak, and excel. From kindergarten through college, in many places the mathematics learning environment is changing from familiar rows of static desks to clustered seating that fosters learning through conversation, writing, and other interactive situations. Dr. Ernest Boyer, former president of the Carnegie Foundation for the Advancement of Teaching, has pointed out that very few jobs of the twenty-first century will be solitary, so team learning is ideal preparation for team working (Saint Peter's College 1990).

For the student, the success of cooperative learning may be fostered by the teacher who encourages risk taking, allows for errors, and does not intimidate the math anxious. For the teacher/facilitator, adapting to new noise levels, sharing decision making and authority for knowledge, planning flexibly to permit student-initiated inquiry, and collaborating with students in the quest for knowledge present many challenges to the long-held image of the role of the teacher.

## Writing to Learn

Understanding mathematics requires good communications skills. Writing about how and what to learn, writing reflections on the learning process and on the content, and keeping a journal to promote logical organization—all are braided into the "writing-to-learn" approach that gained popularity in the 1980s and 1990s (Connolly and Vilardi 1989). This method changes the class tempo and, hence, the learning environment. From asking students to impersonate famous mathematicians to reflecting on the meaning of an algebra topic, from writing a mathematical autobiography to interviewing a mathematician, from describing how one will prepare for a test to writing a legacy of advice for the next class—informal writing can reinforce the process of understanding the mathematics. The capstone experience, which involves writing a paper that is planned and critiqued with the instructor, revised at least once, and presented to the class, further supports the writing-to-learn approach.

## Inclusion of Traditional Methods

The new opportunities in mathematics instruction offer students new paths to discovering mathematics and additional ways to implement techniques in problem solving, which George Pólya encouraged (Pólya 1971). Discovering mathematics can grow from constructing it, from using manipulatives as basic as blocks or as advanced as a graphing calculator. Group settings can magnify the effect. Flash-cards, even electronic ones, can still have a role in a course of instruction, as can lectures, question/answer sessions, self-paced instruction, and practice methods. Each of these traditional methods can have a place in the menu of instructional strategies in mathematics. It is the role of the teacher/facilitator to find the proper blend of the new and traditional methodologies that will serve each type of student most effectively.

The difficulties teachers and students have in finding a common place to start in a class of mixed achievement, preparedness, and cultures can be ameliorated by building the instruction around familiar patterns and connections. Examples of this strategy at different grade levels include measuring the heights of famous buildings and comparing them to how high students can jump, folding familiar materials such as cupcake papers into geometrical shapes, comparing data in a table to graphical representations, relating symmetry and spatial relations to art work, and using statistical methods in a local election (NCTM 1989, 32–35, 84–86, 146–149).

New perceptions about assessment may also directly impact the mathematics classroom. "Assessment in support of standards must not only measure results, but must also contribute to the educational process itself" (NRC 1993, vii). Assessment should be an integral part of the learning process, rather than its culmination. Assessment standards and methods can vary depending on their purpose. Possible purposes include monitoring students' progress, making instructional decisions, evaluating students' achievement, and evaluating programs. Several forms of alternative assessments are being used by innovative teachers and schools. For students, these include the use of portfolios, journals, performance-based tasks such as project-based presentations, direct communication between teacher and pupil, and student self-evaluation. These options afford opportunities for students to engage in meaningful learning experiences at the same time that they inform teachers and students of the nature of student learning. For planning and evaluating instruction, teachers are encouraged to use evidence of classroom learning in three ways: (1) to examine the effects of the tasks, discourse, and learning environment on students' mathematical knowledge, skills, and dispositions; (2) to make instruction more responsive to students' needs; and (3) to ensure that every student is gaining mathematical power (NCTM 1995, 45). To have an effect on instruction, assessment "should look first at the accuracy of the mathematics that is taught. In addition, an evaluation should determine whether that mathematical content is treated in a manner that is sensitive to the developmental level and mathematical maturity of the students" (NCTM 1989, 244).

*See also* Calculators; Computers; Constructivism; Cooperative Learning; Goals of Mathematics Instruction; Instructional Methods, Current Research; Research in Mathematics Education, Current; Software; Technology; Writing Activities

## SELECTED REFERENCES

American Mathematical Association of Two-Year Colleges (AMATYC). *Crossroads in Mathematics: Standards for Introductory College Mathematics before Calculus.* Memphis, TN: The Association, 1995.

Connolly, Paul, and Teresa Vilardi, eds. *Writing to Learn Mathematics and Science.* New York: Teachers College Press, 1989.

D'Ambrosio, Ubiratan. *Socio-cultural Bases for Mathematics Education.* Campinas, Brazil: University of Campinas, 1985.

Eisenhower National Clearinghouse for Mathematics and Science Education. *Guidebook to Excellence: A Directory of Federal Resources for Math and Science Education.* Stock no. 065-000-00641-3. Pittsburgh, PA: U.S. Department of Education, Office of Educational Research and Improvement, 1994.

Hoffer, Alan. *Algebra Xpresser.* Acton, MA: Bradford, 1991.

Kaput, James. "Technology and Mathematics Education." In *Handbook of Research on Mathematics Teaching and Learning* (pp. 515–556). Douglas A. Grouws, ed. New York: Macmillan, 1992.

National Council of Teachers of Mathematics. *Assessment Standards for School Mathematics.* Reston, VA: The Council, 1995.

———. *Curriculum and Evaluation Standards for School Mathematics.* Reston, VA: The Council, 1989.

———. *Professional Standards for Teaching Mathematics.* Reston, VA: The Council, 1991.

———. *Principles and Standards for School Mathematics.* Reston, VA: The Council, 2000.

National Education Goals Panel. *The National Education Goals Report: Building A Nation of Learners. 1997.* Washington, DC: Government Printing Office, 1997.

National Research Council. *Counting on You—Actions Supporting Mathematics Teaching Standards.* Washington, DC: National Academy Press, 1991.

———. *Everybody Counts: A Report to the Nation on the Future of Mathematics Education.* Washington, DC: National Academy Press, 1989.

———. *Measuring What Counts: A Conceptual Guide for Mathematics Assessment.* Washington, DC: National Academy Press, 1993.

———. *Reshaping School Mathematics: A Philosophy and Framework for Curriculum.* Washington, DC: National Academy Press, 1990.

Oaxaca, Jaime, and Ann W. Reynolds. *Changing America: The New Face of Science and Engineering—Actions.* Washington, DC: Task Force on Women, Minorities, and the Handicapped in Science and Technology, Dec. 1989.

Pea, Roy D. "Cognitive Technologies for Mathematics Education." In *Cognitive Science and Mathematics Education* (pp. 89–122). Alan H. Schoenfeld, ed. Hillsdale, NJ: Erlbaum, 1987.

Poiani, Eileen L. "Cultural Diversity and Bonding in Mathematics." In *American Perspectives on the Seventh International Congress on Mathematical Education* (pp. 56–57). John A. Dossey, ed. Reston, VA: National Council of Teachers of Mathematics, 1993.

Pólya, George. *How to Solve It: A New Aspect of Mathematical Method.* 2nd ed. Princeton, NJ: Princeton University Press, 1971.

Saint Peter's College. Satellite Television Symposium: *Teaching Strategies in the Urban Classroom into the 21st Century: K–College.* Jersey City, NJ: 1990.

Webb, Norman L., and Thomas A. Romberg, eds. *Reforming Mathematics Education in America's Cities.* The Urban Mathematics Collaborative Project. New York: Teachers College Press, 1994.

EILEEN L. POIANI
BEATRIZ S. D'AMBROSIO

# INTEGRATION

A term usually applied to two separate but related processes: indefinite integration and definite integration. The symbol $\int f(x)dx$ denotes the indefinite integral or antiderivative of a function, $f(x)$, and indicates the process of reversing differentiation. If $F(x) = \int f(x)\,dx$, then $dF/dx = f(x)$. That is, $F(x)$ is the function whose derivative with respect to $x$ is $f(x)$. The integral sign, $\int$, is simply a Roman S, alluding to the Greek S, $\Sigma$, which represents summation. The definite integral process involves summation, and the integral sign relates to the connection between the two processes.

There is no unique indefinite integral. Since the derivative of a sum is equal to the sum of the derivatives, and since the derivative of any constant $C$ is 0, if

$$\frac{dF(x)}{dx} = f(x)$$

then

$$\frac{d(F(x) + C)}{dx} = \frac{dF(x)}{dx} + \frac{dC}{dx} = \frac{dF(x)}{dx} + 0 = f(x)$$

This is emphasized by writing $\int f(x)\,dx = F(x) + C$.

Definite integration is a more involved process. Let $f(x)$ be a function defined on the interval $[a,b]$. A partition of the interval is a set $\{x_i\}$ so that $a = x_0 < x_1 < \ldots < x_n = b$. Let $\Delta x_i = x_i - x_{i-1}$ for $i = 1 \ldots n$. For any partition select a set $\{c_i\}$ so that $c_i$ is in $[x_i - x_{i-1}]$. Then a Riemann sum for this partition is $f(c_1)\,\Delta x_1 + f(c_2)\,\Delta x_2 + \cdots + f(c_n)\,\Delta x_n$ or $\sum_{i=1}^{n} f(c_i)\,\Delta x_i$. The definite integral of $f(x)$ over the interval $[a,b]$ is the limit, as max $\Delta x_i \to 0$, of all possible Riemann sums over $[a,b]$ and is denoted by $\int_a^b f(x)\,dx$. This limit must be the same regardless of the choice of the partition or set $\{c_i\}$. If $f(x)$ is continuous on $[a,b]$, then this limit exists.

For $f(x) > 0$, $f(c_i)$ represents the height of a rectangle over the base $\Delta x_i$, and $f(c_i)\,\Delta x_i$ represents the area of this rectangle and approximates the area under the graph of $y = f(x)$ over the interval $[x_i - x_{i-1}]$. The Riemann sum is an approximation to the area under the curve $y = f(x)$ over the entire interval $[a,b]$, and the definite integral represents the actual area under this curve for $x$ in $[a,b]$.

If $f(x)$ and $g(x)$ are functions on $[a,b]$ and $k$ is a constant, then $\int_a^b [f(x) \pm g(x)]\,dx = \int_a^b f(x)\,dx \pm \int_a^b g(x)\,dx$ (that is, the operation is linear in $f$ and $g$); $\int_a^b k\,f(x)\,dx = k\int_a^b f(x)\,dx$ (that is, constants may be factored out of the process); and $\int_a^b f(x)\,dx = \int_a^c f(x)\,dx + \int_c^b f(x)\,dx$. The last expression simply says that the area under the curve from $a$ to $b$ is the same as the sum of the areas under the curve from $a$ to $c$ and from $c$ to $b$. If $c > b$ this means that the area from $c$ to $b$ must be interpreted as the negative of the area from $b$ to $c$. Thus $\int_c^b f(x)\,dx = -\int_b^c f(x)\,dx$. Also, $\int_a^a f(x)\,dx = 0$. The integral $\int_a^x f(t)\,dt$ is the area under $y = f(x)$ from a fixed point, $a$, up to a variable $x$. As such, it is a function of $x$ and has a derivative, which is $(d/dx)\int_a^x f(t)\,dt = f(x)$ for continuous functions $f(x)$.

There is a connection, known as the Fundamental Theorem of Calculus, between the indefinite integral and the definite integral. If $f(x)$ is continuous on $[a,b]$ and $F(x)$ is an indefinite integral of $f(x)$, then $\int_a^b f(x)\,dx = F(b) - F(a)$. This connection makes the definite integral a practical tool of science and engineering. If a problem involves a continuous varying function $f(x)$, then the problem is broken up into many small ones on small subintervals and $f(x)$ is treated as constant there. The resulting solution is a Riemann sum. If the small subintervals are made smaller, then the solution tends to a definite integral. But the indefinite integral makes calculating this limit relatively simple. For example, if $f(x)$ represents velocity and $x$ represents time, then $f(c_i)\,\Delta x_i$ is approximately the displacement traveled over the interval $[x_{i-1}, x_i]$. The Riemann sum is the approximate displacement over the interval $[a,b]$ and $\int_a^b f(x)\, dx$ is the exact displacement.

The definite integral process is applied in many other situations. If $f(x,y)$ is a function of two variables $x$ and $y$ where $(x,y)$ varies over some region $R$ in the $xy$-plane, then the definite integral of $f(x,y)$ denotes a process similar to the one-dimensional definite integral. First the region $R$ is broken up into a set of small areas, $\Delta A_{i,j}$, and a set of points is selected so that $(x_i, y_j)$ is in $\Delta A_{i,j}$. The Riemann sum is formed: $\sum_{i=1}^n \sum_{j=1}^m f(x_i, y_j)\, \Delta A_{i,j}$. If this sum tends to one common value regardless of the mode of partitioning $R$ and regardless of the choice of points $(x_i, y_j)$, this common value is the integral of $f(x,y)$ over $R$ and it is denoted by $\iint_R f(x,y)\, dA$. This is also called a double integral of $f(x,y)$ over $R$. If $f(x,y)$ is continuous and $R$ is sufficiently simple, in the sense of being bounded by a finite number of continuous curves, then this integral always will exist. The integral of $f(x_1, x_2, \ldots, x_n)$ over a set $R$ in $n$-space is defined in a similar manner. If $f(x,y) > 0$, then the double integral of $f(x,y)$ over $R$ is the volume under the surface $z = f(x,y)$ above the region $R$. If the double integral of $f(x,y)$ over $R$ exists and $R$ is expressible as either $a \leq x \leq b, g(x) \leq y \leq h(x)$ or $c \leq y \leq d, r(y) \leq x \leq 1(y)$, then the double integral is computed as an iterated integral of either $\int_a^b [\int_{g(x)}^{h(x)} f(x,y)\, dy]\, dx$ or $\int_c^d [\int_{r(y)}^{1(y)} f(x,y)\, dx]\, dy$. When evaluating this type of integral, the innermost integral is computed and all variables other than that of the inner differential are treated as constants. Integrals in higher dimensional spaces are also equivalent to similar iterated integrals.

The one-dimensional integral also is present in 3-space as a "line integral." A curve $C$ in space is a one-parameter set $(x(t), y(t), z(t))$, where $t$ varies from $a$ to $b$. A partition of $[a,b]$ produces a partition of curve $C$. A point $c_i$ in $[t_{i-1}, t_i]$ corresponds to a point $p_i = (x(c_i), y(c_i), z(c_i))$ on a small segment $\Delta s_i$ of the curve. The line integral $\int_C f(x,y,z)\, ds$ is the limit of the sum $\sum_1^n f(p_i) \Delta s_i$ as $\Delta s_i$ shrinks to 0. If $f(x,y,z)$ is continuous and $C$ is sufficiently smooth so as to have continuously varying tangent lines, this integral equals

$$\int_a^b f(x(t), y(t), z(t)) \sqrt{\left(\frac{dx}{dt}\right)^2 + \left(\frac{dy}{dt}\right)^2 + \left(\frac{dz}{dt}\right)^2}\, dt$$

When $f(x,y,z)$ is 1, this is the length of $C$.

The double integral also extends to 3-space as a surface integral. A surface $S$ is a two-parameter set of points $(x(u,v), y(u,v), z(u,v))$, where $(u,v)$ range over a set $R$ in the $(u,v)$ space. A partition of $R$ causes a partition of $S$ into small $\Delta S_{i,j}$ surface area elements with $\Delta S_{i,j}$ corresponding to $\Delta R_{i,j}$. An ordered pair $(u_i, v_j)$ in $\Delta R_{i,j}$ corresponds to a point $p_{i,j} = (x(u_i, v_j), y(u_i, v_j), z(u_i, v_j))$ in $\Delta S_{ij}$. The surface integral $\iint_S f(x,y,z)\, dS$ is the limit of the sum $\sum_{i=1}^n \sum_{j=1}^m f(p_{i,j})\, \Delta S_{i,j}$ as $\Delta S_{i,j}$ approaches 0. If $f(x,y,z)$ is continuous and $S$ is sufficiently smooth so as to have continuously varying normal vectors to the surface, then this surface integral exists and equals

$$\iint_R f(x(u,v), y(u,v), z(u,v))$$

$$\times \sqrt{\left(\frac{\partial(y,z)}{\partial(u,v)}\right)^2 + \left(\frac{\partial(z,x)}{\partial(u,v)}\right)^2 + \left(\frac{\partial(x,y)}{\partial(u,v)}\right)^2}\, dA$$

where $\partial(x,y)/\partial(u,v) = x_u y_v - x_v y_u$. If $f(x,y,z)$ is 1, then this is the surface area of $S$.

The basic essence of integration is to replace a continuous function $f(p)$ over some domain by a function made up of a discontinuous set of constant approximations to this function. On each of the small subsets of a partition of the domain the approximation is constant. Sum these small approximations and then shrink the subsets of the domain. If the function and the domain are sufficiently smooth in the geometry of the domain space (basically has a boundary of continuous pieces that have continuously varying direction), then these sums tend to an integral. This integral can usually be expressed as an iterated integral of simple, one-dimensional, definite integrals that are then evaluated through the Fundamental Theorem of Calculus if antiderivatives can be found. Modern computer techniques can evaluate these integrals without recourse to the Fundamental Theorem of Calculus. There are other forms of integrals such as Stieltjes and Lebesgue integrals. The Stieltjes integral plays an important role in probability theory and can deal with a function that appears to have infinite jumps at isolated points but still has a finite area under the curve. Lebesgue integrals can handle situations where the function being integrated has a noncountable infinite number of discontinuities. Although these situations may seem difficult to visualize, they can arise in important theoretical discussions.

*See also* Calculus, Overview; Differentiation

## SELECTED REFERENCES

Anton, Howard. *Calculus with Analytic Geometry*. 5th ed. New York: Wiley, 1995.

Hughes-Hallett, Deborah, Andrew M. Gleason., et al. *Calculus*. New York: Wiley, 1992.

Marsden, Jerrold E., and Anthony J. Tromba. *Vector Calculus*. 3rd ed. New York: Freeman, 1988.

Repka, Joe. *Calculus with Analytic Geometry*. Dubuque, IA: Brown, 1994.

Stewart, James. *Calculus*. 3rd ed. Monterey, CA: Brooks/Cole, 1995.

Thomas, George B., and Ross L. Finney. *Calculus and Analytic Geometry*. 8th ed. Reading, MA: Addison-Wesley, 1992.

BRUCE H. STEPHAN

# INTEGRATION OF ELEMENTARY SCHOOL MATHEMATICS INSTRUCTION WITH OTHER SUBJECTS

Methods of teaching elementary mathematics in conjunction with other disciplines. The elementary curriculum is a rich source of opportunities for integrating mathematics with science, language arts, physical education, and so on. Connections can be made using a variety of models and/or techniques. At the core of integrated instruction lies a fundamental belief that knowledge is multidimensional and requires information from a variety of sources and that (1) the real world is integrated, not fragmented, (2) problems, whether childhood or adult, are not compartmentalized, and (3) problem solving requires the skills and knowledge of several subjects. This is not to say that organizing information into workable parts while focusing upon specific content knowledge is wrong. In fact, a variety of forms of organizational focus should be implemented to facilitate knowledge acquisition as efficiently and effectively as possible. The following models represent a context for understanding the possibilities for curriculum design. These models begin with the notion that integration is a goal driven by a central theme.

## MULTIDISCIPLINARY APPROACH

This curriculum design model focuses on separate disciplines tackling the same theme. The approach views the curriculum through the lens of a discipline that includes content from other disciplines to increase relevance. This model breaks down a few of the boundaries among subject areas but leaves the disciplines intact enough to allow teachers to continue to organize knowledge through the definition of the discipline.

For example, a social studies teacher could have the students read novels about the period of history being studied. This scenario maintains the integrity of the individual discipline (history) while providing an opportunity for students to immerse themselves in the period through historical fiction (literature). A teacher who decides to study bears as a part of the science curriculum may, in an effort to involve "bears" in every aspect of the curriculum, have the students count gummy bears during mathematics instruction. Using gummy bears is an engaging counting manipulative for early childhood students. Also, teddy bears can be used for categorizing, sorting, graphing, estimation, prediction, and weighing and measurement activities, thereby facilitating more mathematical thinking and actions. To further expand upper primary and intermediate students' understanding of bears, activities with meaningful mathematical involvement could involve students in collecting, analyzing, categorizing, graphing, and reporting data on a variety of potential research topics associated with bears (e.g., size, weight, body measurements, types, extinction rates, hibernation patterns, habitat destruction rates, etc.).

Mathematics teachers who subscribe to the integration approach want to recognize relationships among different mathematical topics and be able to use mathematics in other curriculum areas as well as daily life (National Council of Teachers of Mathematics (NCTM) 1989). The connections that they want children to explore should reflect authentic learning experiences rather than be the result of force-fitting mathematics into a thematic approach. In other words, they want to create mathematical connections within the theme. When teams design and organize an integrated curriculum, they answer the question: what is important to learn within different disciplines? When working with other content specialists, a "math" specialist can develop lessons that naturally connect mathematics. Teachers of one discipline can then connect content from another. Estimation, for example, can be explored in each of the content areas to show that all the content areas use estimation skills—money, distances, size, weight, time, dates, and so on.

## INTERDISCIPLINARY APPROACH

Instead of applying themes within subject areas, this approach emphasizes commonalties across disciplines; for example, decision making and problem solving involve the same principles regardless of discipline. The interdisciplinary approach emphasizes metacognition and learning how to learn and focuses on the question: how can curriculum encourage higher-order thinking? A greater reliance is placed on performance assessments that reach beyond the boundaries of the disciplines. The emphasis is on process rather than product. The following examples illustrate interdisciplinary approaches for integrating mathematics with other content areas.

## Mathematics and Science

Science often uses mathematics as a tool, including graphs, computation, measurement, patterning, estimation, problem solving, reasoning, geometry, statistics, and probability. Whole numbers, fractions, and decimals are all used in science. It is virtually impossible to gain a thorough understanding of research methodology or problem solving without accessing both domains of content knowledge.

## Mathematics and Children's Literature

This connection has become increasingly popular and complements the whole-language philosophy of language learning. Using children's literature provides enhanced mathematical understanding by focusing on the development of content-appropriate mathematics that emerges from the literature. For example, Marilyn Burns's delightful tale of *The Greedy Triangle* (1994) enables teachers to include a picture storybook that provides students with an opportunity to understand the nature of shapes, geometry, angles, sides, nomenclature, and so on.

## Mathematics and Language Arts

As the expressive skills (writing and speaking) and the receptive skills (reading and listening) develop, the breadth of language arts increases to include discipline-specific vocabulary development that is essential for holistic literacy development. Activities such as creating a **KWL** chart, wherein the teacher encourages the students to work collaboratively to brainstorm what they **K**now (prior knowledge), what they **W**ant to learn (questions), and what they have **L**earned about a specific theme or topic, including specific vocabulary, result in the "integration of language arts as children write and discuss their experiences in mathematics" (NCTM 1989, 35).

## Mathematics and Social Studies

Many disciplines make up the social studies curriculum. Mathematics connects with physical geography, especially the map skills of measurement and directionality. Cultural geography naturally extends to learning about mathematics in other countries and cultures. Economics education provides opportunities for computational and graphing skills as well as conceptual understandings.

## Mathematics and the Arts

The expressive arts include multiple opportunities for mathematical connections; for example, musical notation requires an understanding of fractions; geometry concepts are evident in art; physical education includes scoring and sequencing and the use of weights and measurements, including time and distance, as well as geometry: spheres and circles (balls, shots, rings, etc.), ovals (tracks, stadiums, etc.), quadrilaterals (playing fields), and so on.

## TRANSDISCIPLINARY APPROACH

The word *transdisciplinary* is used to emphasize a transcendence beyond the boundaries of traditional disciplines (Lauritzen and Jaeger 1994). The approach considers the broader question: how can we teach students to be productive citizens in the future? The skills emphasized are no longer specific to content areas, but rather involve skills such as change management, dealing with ambiguity, perseverance, and confidence. The emphasis is upon meaning and relevance through a life-centered approach. The teacher considers important concepts to be explored and together with the students determines the essential questions that provoke exploration and the best activities to facilitate the learning experience. The content and procedures of individual disciplines are transcended. The interconnections are so numerous and far-reaching they seem limitless: theme, strategies, and skills seem to merge when the theme is set in its real-life context.

For example, focusing on the theme "exploring our world" and using Kathy Lasky's *The Librarian Who Measured the Earth* (1994), the teacher can provide a springboard for children into the world of transdisciplinary curriculum. As the teacher reads the selection aloud, children record questions they have about Eratosthenes, Greece, Egypt, measuring the earth, biographies, circumference, and so on. These questions form the nucleus of exploration as students engage in research to seek answers—and yet more questions. It is at this point that they recognize the dissolution of disciplines in greater intellectual pursuits. The student is guided by the nature of discovery, experiencing intellectual disequillibrium, trying to form a hypothesis, testing the hypothesis, drawing upon past learning, reorganizing learning, discarding irrelevancies, forming new vistas of understanding. The teacher facilitates the discovery, teaching research methodology, providing resources or information about where they can be found, providing affirmation, encouraging the struggle that ambiguity creates, and asserting that the acquisition of knowledge demands that we spend time in intellectual unrest.

## CONCLUSION

The pursuit of intellectual experience has historically remained within individual disciplines. While the integrated approaches honor the importance of individual disciplines, they encourage authentic and meaningful learning that can help students apply and synthesize knowledge from more than one discipline (Gardner and Boix-Mancilla 1994). Integrated learning environments provide an arena for students to explore and construct genuine connections between higher-order thinking and the world around them.

*See also* Art; Books, Stories for Children; Consumer Mathematics; Enrichment, Overview; Music

### SELECTED REFERENCES

Burns, Marilyn. *The Greedy Triangle*. New York: Scholastic, 1994.

Gardner, Howard, and Veronica Boix-Mancilla. "Teaching for Understanding in the Disciplines—and Beyond." *Teachers College Record* (Winter 1994):198–218.

Lasky, Kathy. *The Librarian Who Measured the Earth*. Toronto, ONT, Canada: Little, Brown, 1994.

Lauritzen, Carol, and Michael Jaeger. "Language Arts Teacher Education within a Transdisciplinary Curriculum." *Language Arts* 71(12)(1994):581–587.

National Council of Teachers of Mathematics. *Curriculum and Evaluation Standards for School Mathematics*. Reston, VA: The Council, 1989.

———. *Principles and Standards for School Mathematics*. Reston, VA: The Council, 2000.

LYN TAYLOR
LIBBY QUATTROMANI

# INTEGRATION OF SECONDARY SCHOOL MATHEMATICS INSTRUCTION WITH OTHER SUBJECTS

Methods for coordinating the teaching and learning of mathematics with that of subjects outside mathematics. Until the late 1950s, educators considered mathematics an applied subject that should be taught for the main purpose of solving everyday problems in a variety of fields. But after the Soviet launching of the *Sputnik* satellite in 1957, serious discussion began about the nature of school mathematics. An important issue was whether mathematics should be "pure" or "applied." The proponents of pure mathematics believed that students would learn best if they were taught the theory and structure of mathematics, an approach that would make easier the transition to university-level mathematics, where more scientists, engineers, and mathematicians were to be educated for the space race (Farrell and Farmer 1988).

The heavy emphasis on theory in the "modern math" curriculum of the 1960s gradually eroded, however, leaving much of the structure but little of the content. This was the status of secondary school mathematics until the 1989 publication of the *Curriculum and Evaluation Standards for School Mathematics* by the National Council of Teachers of Mathematics (NCTM). An important goal set out by the *Standards* is a core curriculum, making applications of a variety of mathematical topics accessible to all students, not only those preparing for college. Students should be able to "apply integrated problem-solving strategies to solve problems from within and outside mathematics" and to "apply the process of mathematical modeling to real-world problem situations" (NCTM 1989, 137). Examples of mathematical modeling include the use of networks to represent airline routes, matrices to analyze production and sales, and statistical techniques in analyzing and predicting election results.

These new goals shifted the argument away from a choice between pure and applied mathematics to how to integrate mathematics with other subjects. The growing use of mathematics in many disciplines makes it essential that the types of mathematical content taught be broadened beyond the traditional areas of algebra and geometry. Statistics and probability are especially important in fields as diverse as economics and ecology. Business depends on data such as costs, profits, and market research. Naturalists use sampling techniques to determine wildlife populations in order to manage habitat.

Instructional strategies that have the most promise for integrating mathematics with other subjects are those in which students are able to work in small groups, independently, or across grade levels in order to explore problems that are drawn from real life. Planning and teaching in interdisciplinary teams is an important component of this approach. Each team member has expertise in a subject area and can serve as a resource and consultant. The team may be responsible for the instruction of a group of 100–125 students, who rotate among the team members for their course work, often working on an interdisciplinary unit that encompasses all of the subjects they are studying.

Special curriculum projects are the best source for materials. Examples of the better-known projects are the Consortium for Math and Its Applications (see Garfunkel 1988) and Project TIMS (Edwards 1993). These and other projects have developed a large collection of units and modules that apply mathematics to many subjects. Other sources include publications by the NCTM (see Austin 1991; Froelich, Bartkovich, and Foerster 1991; Swetz and Hartzler 1991).

Despite the impetus toward integration at the secondary level, the integration of mathematics with other subjects has not been and still is not a central concern in American education. The reforms of the 1990s, while emphasizing broader goals for mathematics, did not completely embrace the idea. Separate courses in algebra, geometry, and analysis are still the main pattern in secondary schools. Applications are mainly relegated to the end of the lesson or chapter rather than being the central focus or motivation for learning concepts and skills. In order to achieve true integration of mathematics with other subjects, however, curricula must be drawn from and built around specific, applied problems. Teachers from several disciplines need to work together to plan and coordinate the instruction. Currently, this interdisciplinary team planning and teaching is the exception and exists mainly in elementary and middle schools. For teachers and others who wish to use an integrated approach, the teaching strategies and materials are available but not in the mainstream, requiring teachers to spend additional preparation and planning time in order to teach integrated mathematics. The wide implementation of integrated secondary school mathematics would require a significant amount of curriculum development, teacher training, and staff restructuring.

*See also* Applications for the Classroom, Overview; Applications for the Secondary School Classroom; Biomathematics; Chemistry; Environmental Mathematics; Health; Physics; Teacher Development, Secondary School

### SELECTED REFERENCES

Austin, Joe D., ed. *Applications of Secondary School Mathematics, Readings from the Mathematics Teacher.* Reston, VA: National Council of Teachers of Mathematics, 1991.

Edwards, Leo. *Project TIMS (Teaching Integrated Math/Science).* Washington, DC: National Aeronautics and Space Administration, 1993.

Farrell, Margaret A., and Walter A. Farmer. *Secondary Mathematics Instruction: An Integrated Approach.* Providence, RI: Janson, 1988.

Froelich, Gary, Kevin G. Bartkovich, and Paul A. Foerster. *Connecting Mathematics: Addenda Series, Grades 9–12.* Reston, VA: National Council of Teachers of Mathematics, 1991.

Garfunkel, Solomon A., proj. dir. *For All Practical Purposes.* New York: Freeman, 1988.

National Council of Teachers of Mathematics. *Curriculum and Evaluation Standards for School Mathematics.* Reston, VA: The Council, 1989.

———. *Principles and Standards for School Mathematics.* Reston, VA: The Council, 2000.

Swetz, Frank, and Jefferson S. Hartzler, eds. *Mathematical Modeling in the Secondary School Curriculum: A Resource Guide of Classroom Exercises.* Reston, VA: National Council of Teachers of Mathematics, 1991.

GERALD KULM

# INTERACTIVE MATHEMATICS PROGRAM (IMP)

Four-year, problem-based mathematics curriculum for high school students. Written by four mathematics educators, the program is designed to meet the needs of both college-bound and noncollege-bound students.

Curriculum development of IMP began in 1989 in California as part of an effort to fulfill the vision of the *Curriculum and Evaluation Standards* of the National Council of Teachers of Mathematics (NCTM) (1989). Other influences include the work of Robert B. Davis (The Madison Project) at Rutgers University and the late William Johntz of Project SEED. Basic assumptions and features of the project include: (a) a shift from a skill-centered to a problem-centered curriculum; (b) broadening the scope of the secondary curriculum to include such areas as statistics, probability, and discrete mathematics; (c) changes in pedagogical strategies, including emphasis on communication and writing skills as well as access to appropriate technology; and (d) expansion of the pool of students who receive a "core" mathematics education. IMP differs from other problem-based instructional programs in that students are encouraged to develop skills within the context of large, complex problems built into a series of units.

## CURRICULUM DESIGN

IMP replaces the four-course sequence typical of most high school mathematics programs with an integrated course. Whereas much traditional mathematics teaching is structured around skills and concepts in isolation, IMP calls on students to work in context, experimenting with examples, looking for and articulating patterns, and making, testing, and proving conjectures.

Beginning with a list of concepts and skills, the developers worked on creating problem settings in which these factors could arise in a meaningful way. The curriculum consists of four- to eight-week units that are each organized around a central problem or theme. Over the course of four years, the curriculum revisits themes and concepts in a spiraling sequence, providing students with ongoing opportunities to develop mathematical understanding with increasing sophistication. Students who are unable to complete the full program and transfer to more traditional classes do not seem to have problems with the transition.

From 1989 to 1992, the first three years of the IMP curriculum were pilot-tested at three high schools in the San Francisco Bay area, and the fourth year of the curriculum was piloted from 1993 to 1994 in four schools. Materials have been revised, tested, and reviewed. Along with text booklets, students use graphing calculators and manipulatives to work on problems. The following is a summary of the curriculum:

### Year 1

The first-year curriculum contains an introduction to problem-solving strategies, the use of variables, and the meaning and use of functions and graphs, as well as concepts from statistics, geometry, and trigonometry. These mathematical ideas are set in varied contexts, such as the settlement of the American West, games of chance, Edgar Allen Poe's "The Pit and the Pendulum," and the measurement of shadows.

### Year 2

Students work with powerful mathematical ideas, including the chi-square ($\chi^2$) statistic, the Pythagorean Theorem, and linear programming, and learn a variety of approaches to solving equations. Problem contexts include statistical comparison of populations, the geometry of the honeycomb, and maximization of profits from a cookie store. There is also a miniunit on developing mathematical writing skills.

### Year 3

Students extend their understanding of material studied in preceding years of the curriculum while learning about and applying new topics, such as combinatorics, derivatives, and the algebra of matrices. A baseball pennant race, population growth, and decision making on land use provide some of the contexts for the mathematical concepts.

### Year 4

Fourth-year IMP has a more varied subject matter than a calculus-focused course and includes topics such as circular functions, computer graphics, and statistical sampling. Units build on the strong knowledge base of students who have completed three years in the program. Problem settings include a Ferris wheel circus act and election polling.

## ASSESSMENT

Opportunities for assessment of learning go beyond the traditional paper-and-pencil test to include self-assessment, student portfolios, oral presentations, written explanations, teacher observations, and group work.

To give students experience in working independently on substantive problems, IMP has incorporated Problems of the Week (POWs) into each unit. Some POWs are directly related to the specific mathematics of the unit under study; others are classic mathematics problems, independent of the unit.

## COLLABORATIVE LEARNING

The "interactive" aspect of IMP refers, in part, to the program's emphasis on students interacting with each other by working in groups. Together, students tackle problems that usually are too complex to be solved by any one individual. Students make written and oral presentations that help clarify their thinking and refine their ability to communicate mathematically. Even though students are allowed to collaborate on problems, evidence of individual interpretation can be seen in their written portfolio work.

## TEACHING CHALLENGES

Professional support for teachers is essential for the success of this program. In addition to the new content described above, teachers must also master new instructional strategies. They learn to incorporate the Socratic method of questioning in class, which allows them to probe for solutions rather than telling students the answer to the problem. Participating teachers attend summer and school-year workshops led by veteran IMP teachers and receive ongoing support from a community of other IMP teachers and resource staff. In these workshops, teachers learn mathematics in a style that mirrors the curriculum, working cooperatively and constructing their own ideas, along with working on classroom management issues.

## IMPLEMENTATION

Dissemination of the IMP curriculum is currently funded by a grant from the National Science Foundation. IMP is taught in many schools in such states as California, Colorado, Illinois, and Pennsylvania.

A five-year evaluation of the program is being conducted by Norman Webb at the Wisconsin Center for Education Research. This study will examine the long- and short-term effects of the curriculum on mathematics achievement of high school students in the program as compared to students enrolled in a traditional mathematics course sequence. The evaluation will also investigate post–high school careers and college experiences of IMP students, as well as outline professional development requirements for successful implementation of IMP within a school.

*See also* Curriculum Trends, Secondary Level; Research in Mathematics Education, Current

## SELECTED REFERENCES

Davis, Robert B. *Learning Mathematics, the Cognitive Science Approach to Mathematics Education.* Norwood, NJ: Ablex, 1984.

Fendel, Dan, et al. *Interactive Mathematics Program.* 4 vols. Berkeley, CA: Key Curriculum, 1996–1999.

National Council of Teachers of Mathematics. *Curriculum and Evaluation Standards for School Mathematics.* Reston, VA: The Council, 1989.

———. *Principles and Standards for School Mathematics.* Reston, VA: The Council, 2000.

National Research Council. *Everybody Counts: A Report to the Nation on the Future of Mathematics Education.* Washington, DC: National Academy Press, 1989.

DANIEL M. FENDEL
DIANE RESEK
LYNNE ALPER
SHERRY FRASER

# INTEREST

The charge you pay someone for allowing you to have the use of her/his money for a period of time. Suppose that you want the use of $1,000 for a year.

You could borrow the $1,000 from another person, the lender, and at the end of the year you would pay the lender back the $1,000. In addition you would pay the interest the lender charged for making it possible for you to use the money for a year. If the interest charge were $100, you would pay the lender a total of $1,100 at the end of the year when your loan came due. The interest charge, expressed as a percentage of the loan amount, is known as the *interest rate*. In this example the interest rate is 10% because the interest charge is 10% of the $1,000 you borrowed. Interest rates are usually quoted on a one-year basis and are referred to as *annual interest rates*.

Suppose you were to borrow the $1,000 for two years at an annual interest rate of 10%. You would pay back the loan at the end of the second year after having had the use of the $1,000 for two years. If your lender charged you $100 interest for each of the two years you would have been charged what is known as *simple interest*. At the end of the second year you would pay the lender back the $1,000 plus two years of interest at $100 per year, or a total repayment of $1,200.

The lender might charge you *compound interest* on the two-year loan. At the end of the first year you would owe the lender $1,100, that is, the $1,000 loan plus $100 in first-year interest. In the second year the lender would charge you 10% interest on the $1,100 you owed at the end of the first year, which would be $110. When you repaid the loan at the end of the second year, you would pay the lender back the $1,000 plus $100 interest for the first year plus $110 interest for the second year, for a total repayment of $1,210.

If it were a three-year loan, the interest charge for the third year would be 10% of the $1,210 you owed at the end of the second year, or $121. When you repaid the loan at the end of the third year you would have had the use of the $1,000 for three years. You would pay the lender back the $1,000 loan plus $100 in first-year interest plus $110 in second-year interest plus $121 in third-year interest, for a total repayment of $1,331. In borrowing the $1,000 at compound interest the total interest charges on the three-year loan would be $331. If you had been able to borrow the $1,000 at simple interest, the total interest charges would have been only $300. Compound interest is sometimes described as "being charged 'interest on the interest.'" Most loans made by banks and other kinds of lenders are made on a compound interest basis.

The compounding of interest can be expressed in mathematical form. Let $P_t$ be the future value of the loan principal after $t$ years; let $P_0$ be the present value of the loan principal; let $r$ be the annual interest rate; let $t$ be the number of years to maturity; and let $m$ be the number of times per year that interest is compounded. Thus $P_t = P_0(1 + r/m)^{mt}$. The value of an initial loan of \$100 after three years when the interest rate is 10% will be:

Annual compounding ($m = 1$):
$$P_3 = 100 \, (1 + .10/1)^3 = 133.10$$
Semiannual compounding ($m = 2$):
$$P_3 = 100(1 + .10/2)^6 = 134.00$$
Daily compounding ($m = 365$):
$$P_3 = 100 \, (1 + .10/365)^{1095} = 134.98$$
Continuous compounding ($m = \infty$) where
$$P_t = P_0 e^{rt} \colon P_3 = 100(2.71828\ldots)^{.30} = 134.99$$

*See also* Consumer Mathematics; Finance

## SELECTED REFERENCES

Brealey, Richard A., Stewart C. Myers, and Alan J. Marcus. *Fundamentals of Corporate Finance.* New York: McGraw-Hill, 1995, Chapter 3.

Ross, Stephen A., Randolph W. Westerfield, and Bradford D. Jordan. *Essentials of Corporate Finance.* Chicago, IL: Irwin, 1996, Chapter 5.

RALPH L. NELSON

# INTERNATIONAL COMPARISONS, OVERVIEW

Since education reflects the culture in which it takes place, one expects to find variation in schooling that mirrors the diversity of cultures around the world. Cross-national studies have documented and analyzed the content and organization of the school curriculum, how the curriculum is organized, and how teachers handle the subject matter of mathematics in their daily instruction. More recently, research has addressed more basic issues, such as how culture impinges upon conceptual development and affects the very nature of the mathematics that is being taught and learned (Smith, Stanley, and Shores 1971; Stake and Easley 1978; Bishop et al. 1993; D'Ambrosio 1994).

## CURRICULAR VARIATION

Variation in curricular content and organization occurs in what topics are covered, when and how they are covered, and with what intensity. In the early days of cross-national research in education, some subject matter (notably, mathematics) was assumed to be relatively impervious to cultural influences. Indeed, mathematics was the subject matter of choice for the First International Mathematics Survey (FIMS) undertaken by the International Association for the Evaluation of Educational Achievement (IEA) since, the researchers posited, there was "certain agreement internationally upon its aims, contents and methods" (Husén 1967, vol. 1, pp. 33–34).

A major, though somewhat unanticipated, finding of FIMS was that of considerable cross-national variation in the content of mathematics and in the degree of emphasis that was given various topics within school curricula. Consequently, the FIMS researchers concluded, in future international studies of educational achievement, "The educational system . . . would have to be seen much more closely in its social context than was the case when the First IEA Mathematics Study was designed" (Husén 1967, vol. 2, p. 307).

## The Royaumont Conference

In 1961, just three years prior to FIMS, an international curriculum survey was carried out by the Organization for European Economic Cooperation (OEEC). The findings were reported at a major international conference on school mathematics in Royaumont, France (OEEC 1961). The Royaumont Conference was a major event in the history of mathematics education since it brought together early in the "new mathematics era" leading international figures to share the many significant developments in school mathematics that were under way, especially in North America and Europe.

The Royaumont Conference established, independently of FIMS, that there was significant cross-national variation not only in the content of school mathematics but in the time in school (age, year, or grade) when various topics (such as algebra, geometry, and statistics) were first introduced. A secondary analysis (Hirstein 1980) revealed the "layer cake" approach to school mathematics that typifies the curriculum in the United States and in some Canadian provinces. The "layer cake curriculum" called for algebra and geometry in the first three years of high school as separate and sequential topics (typically, one year of algebra, one year of geometry, then another year of algebra). In virtually every other educational system around the world, school mathematics was integrated, with the various topics (algebra, geometry, and so on) presented together and developed in depth in successive years in school.

## The Second International Mathematics Study (SIMS)

In the 1980s, SIMS, a study of school mathematics in twenty countries, again under the aegis of IEA, undertook an analysis of the curriculum using secondary sources of information (typically, questionnaires about a country's school mathematics program that were completed by mathematics educational specialists in each country). SIMS identified substantial between-country differences in the school curriculum, especially in geometry and probability. Much less variation was found in algebra (Travers and Westbury 1989).

The countries also differed substantially in the emphasis they placed on various topics. It was noted, for example:

> The Japanese curriculum provides their seventh graders with an intensive introduction to algebra. The Belgian and French programs focus on geometry and extensive work on common fractions (decimal fractions are dealt with in earlier grades). Correspondingly dramatic growth in achievement during the school year takes place for these topics in Japan, France and Belgium. By contrast, the U.S. curriculum has little intensity, with little sustained attention paid to any aspect of mathematics. (McKnight et al. 1987, 94)

## PEDAGOGICAL VARIATION

A unique feature of SIMS was its focus on the classroom and how teachers handle the subject matter of mathematics in their day-to-day classroom instruction. This information was captured largely by use of a series of self-report "classroom process questionnaires" that were administered in eight countries, including France, Belgium, Canada, and the United States. There were five such questionnaires, one for each of the major content strands at the eighth year of schooling: algebra; common and decimal fractions; geometry; measurement; and ratio, proportion, and percent.

SIMS found differences in teaching methodology according to the concept taught, the representation of the content, the approach (formal or informal, more verbal or less), and the type of problems being addressed (Burstein et al. 1992; Robitaille and Travers 1992; McKnight and Cooney 1992). An especially novel analysis of the voluminous "classroom process" data was carried out by Robin, who found teachers in France and Belgium whose ". . . outlook favors mastery of deductive approaches, including 'a formal Euclidean approach based on an axiomatic system used to prove theorems.' There is also a rejection of trial-and-error as well as informal approaches" (Robin 1992, 227). On the other hand, Robin found, teachers from North America drew examples from everyday life and utilized student experiences to validate the mathematical propositions dealt with in class (Robin 1992).

In a separate, small-scale investigation, Sugiyama (1987) analyzed major middle school mathematics textbooks in Japan and the United States and found:

- More emphasis . . . in the books of Japan than in the United States on the meaning of mathematical operations and when it is appropriate to use them.
- In the United States, problem-solving strategies are provided in the textbook for the student to read. In Japan, problem solving is taught by the teacher.
- The level of difficulty of the content is greater in Japanese than in U.S. textbooks.
- In Japan, the teacher teaches all the content to all students. In the United States, students are expected to learn the material by reading the book and solving problems themselves (Sugiyama 1987, 238).

Stevenson, Lee, and Stigler (1986) studied the teaching and learning of mathematics in first- and fifth-grade classrooms in Japan, China, and the United States. The researchers found that Japanese mathematics classrooms were characterized by far more verbal explanation (by both teacher and students) than were the Chinese and U.S. classrooms:

> . . . the most common means of publicly evaluating student work in the Japanese classrooms was to ask a student who obtained the *incorrect* answer to a problem to put his work on the board, then to discuss, with the entire class, the process that led to the error. The most common form of evaluation in the American classrooms, by contrast, was simply to praise a student who answered the problem correctly (Stigler and Baranes 1988–89, 296).

The use of open-ended problems has been common in Japanese elementary schools for some time (see, for example, Shimada 1987). Recently, the use of "open-ended problems" in U.S. schools has been the subject of cross-national research and experimentation. In contrast to traditional classroom problems that are formulated to have one and only one correct answer ("closed problems"), the open-ended approach uses "incomplete" problems, which have a multiplicity of correct answers or approaches to solutions. Becker and his colleagues concluded, on the basis of their research, that the use of open-ended problems has great potential in the United States for engaging students in mathematical thinking (Becker et al. 1990, 17).

## MATHEMATICS EDUCATION AND CULTURE

The classroom is affected by many factors, including how mathematics is perceived as a subject. "Whether or not teachers can identify the particular nature of the subject, they must hold beliefs and values with respect to mathematics that influence how they teach. These will affect what content they select, whether they consider it accessible to all pupils, and how they choose to make it accessible to them" (Nickson 1992, 103).

Effects of cultural background may be seen in terms of language and traditions. Stigler and his associates investigated how children learn to count in Chinese, Japanese, Korean, and English (Stigler and Baranes 1988–89). Of special interest was the fact that counting in English is not completely regular (the existence of the teens tends to obscure the base-10 principle, with different number names used up to 20). In Chinese, rules for number names are completely regular and predictable. Twelve, in Chinese, is the word "ten-two." The researchers compared the performances of Chinese- and English-speaking four- and five-year-olds on a variety of counting tasks and found the Chinese scoring consistently higher than the American children on all tasks.

Joseph (1994) has commented on differences in mathematical traditions and in the cognitive structures of mathematics across cultures. The work of Geoffrey E. R. Lloyd (1990), Joseph notes, documents that the Chinese were only concerned with whether a certain formula or algorithm produced a correct solution and show little interest in the Greek notion of proof. Joseph also draws distinctions between the contrasting styles of mathematical argument in India and Western countries: "Proofs are social and cultural artifacts. They evolve in a particular social and cultural context. . . . this is important since we might tend to forget that part of finding out how a proof works includes finding out how well its intended audience (the author included) is prepared to follow it" (Joseph 1994, 189–190). Awareness of the central role of culture in mathematics learning

(e.g., Bishop 1988) has paved the way for concerted efforts to devise varieties of curricula and approaches to instruction that are designed to make mathematics accessible to the vast majority of students.

D'Ambrosio (1994) has established the study of "ethnomathematics," in which it is assumed that forms of mathematics vary as a consequence of being embedded in cultural activities whose purpose is other than "doing mathematics." D'Ambrosio and his colleagues (D'Ambrosio 1994, 10), an international group of specialists in mathematics and science education, developed a framework (below) that illustrates how differences in the Traditional (Western) and Confucian (Eastern) approaches to learning reflect each culture and lead to major differences in emphasis on how education is organized.

| Traditional (Western) Individual work | Confucian (Eastern) Group work |
|---|---|
| Children grouped by their abilities within their age groups | Children of all abilities are merged by age group, but slower students are given more time and extra help |
| Understanding leads to doing | Doing leads to understanding |
| Emphasis on concepts and insight leading to skills | Emphasis on graded skills exercises facilitating understanding of concepts and insight |

In the same report, D'Ambrosio reiterates the central role that must be played by mathematics in the preparation of effective, functional citizens for the twenty-first century and outlines fundamental reforms that are needed in order to achieve such ambitious goals. Cross-national studies such as those reported here provide a critically important part of the knowledge base on which more effective mathematics education programs can be built.

*See also* Congresses; International Organizations; International Studies of Mathematics Education

## SELECTED REFERENCES

Becker, Jerry P., and Tatsuro Miwa, eds. *Proceedings of the US-Japan Seminar on Mathematical Problem-Solving.* Carbondale, IL: Southern Illinois University Press, 1987; ERIC No. ED304315. Columbus, OH: ERIC 1987.

Becker, Jerry P., Edward A. Silver, Mary G. Kantowski, Kenneth J. Travers, and James W. Wilson. "Some Observations of Mathematics Teaching in Japanese Elementary and Junior High Schools." *Arithmetic Teacher* 38(Oct. 1990):12–21.

Bishop, Alan J. *Mathematical Enculturation: A Cultural Perspective on Mathematics Education.* Dordrecht, Netherlands: Kluwer, 1988.

——, et al. *Significant Influences on Children's Learning of Mathematics.* UNESCO Document Series, #47. Paris, France: UNESCO, Education Sector, 1993.

Burstein, Leigh, et al., eds. *The IEA Study of Mathematics III: Student Growth and Classroom Processes.* Oxford, England: Pergamon, 1992.

D'Ambrosio, Ubiratan, ed. *Science, Mathematics, Engineering and Technology Education for the 21st Century: Report of the 1992 NSF Summer Symposium on Science Education.* Arlington, VA: National Science Foundation, 1994.

Grouws, Douglas A., ed. *Handbook of Research on Mathematics Education.* New York: Macmillan, 1992.

Grouws, Douglas A., Thomas J. Cooney, and Douglas Jones, eds. *Effective Mathematics Teaching.* Reston, VA: National Council of Teachers of Mathematics, 1988.

Hashimoto, Yoshihiko. "Classroom Practice of Problem-Solving in Japanese Elementary Schools." In *Proceedings of the US–Japan Seminar on Mathematical Problem-Solving* (pp. 94–119). Jerry P. Becker and Tatsuro Miwa (eds.). Carbondale, IL: Southern Illinois University Press, 1987. ERIC No. ED304315, Columbus, OH: ERIC, 1987.

Hirstein, James J. "From Royaumont to Bielefeld: A Twenty Year Cross-national Survey of the Content of School Mathematics." In *Comparative Studies of Mathematics Curricula: Change and Stability 1960–1980* (pp. 55–89). Hans Steiner (ed.). Bielefeld, Germany: Institute for Mathematical Didactics, 1980.

Husén, Torsten, ed. *International Study of Achievement in Mathematics: A Comparison of Twelve Countries.* New York: Wiley, 1967.

Joseph, George G. "Different Ways of Knowing: Contrasting Styles of Argument in India and the West." In *Selected Lectures from the Seventh International Congress on Mathematical Education* (pp. 183–196). David F. Robitaille, David H. Wheeler, and Carolyn Kieran (eds.). Sainte-Foy, Quebec, Canada: Laval University Press, 1994.

McKnight, Curtis C., and Thomas J. Cooney. "Content Representation in Mathematics Instruction." In *The IEA Study of Mathematics III: Student Growth and Classroom Processes* (pp. 179–224). Leigh Burstein, et al. (eds.). Oxford, England: Pergamon, 1992.

McKnight, Curtis C., et al. *The Underachieving Curriculum: Assessing US School Mathematics from an International Perspective.* Champaign, IL: Stipes, 1987.

Nickson, Marilyn. "The Culture of the Mathematics Classroom, an Unknown Quantity." In *Handbook of Research on Mathematics Education* (pp. 101–114). Douglas A. Grouws (ed.). New York: Macmillan, 1992.

Organization for European Economic Cooperation (OEEC). *New Thinking in School Mathematics.* Paris, France: The Organization, 1961.

Robin, Daniel. "Teachers' Strategies and Students' Achievement." In *The IEA Study of Mathematics III: Student Growth and Classroom Processes* (pp. 225–258). Leigh Burstein, et al. (eds.). Oxford, England: Pergamon, 1992.

Robitaille, David F. "Contrasts in the Teaching of Selected Concepts and Procedures." In *The IEA Study of Mathematics III: Student Growth and Classroom Processes* (pp. 147–177). Leigh Burstein, et al. (eds.). Oxford, England: Pergamon, 1992.

Robitaille, David F., and Robert A. Garden, eds. *The IEA Study of Mathematics II: Contexts and Outcomes of School Mathematics.* Oxford, England: Pergamon, 1989.

Robitaille, David F., and Kenneth J. Travers. "International Studies of Achievement in Mathematics. In *Handbook of Research on Mathematics Teaching and Learning* (pp. 687–709). Douglas A. Grouws (ed.). New York: Macmillan, 1992.

Robitaille, David F., et al. *Curriculum Frameworks for Mathematics and Science.* Vancouver, British Columbia, Canada: Pacific Educational, 1993.

Saxe, Geoffrey B. "Effects of Schooling on Arithmetical Understandings: Studies with Oksapmin Children in Papua, New Guinea." *Journal of Educational Psychology* 77(1985):503–513.

Shimada, Shigeru. "Problem-Solving—The Present State and Historical Background." In *Proceedings of the US–Japan Seminar on Mathematical Problem-Solving* (pp. 5–32). Jerry P. Becker and Tatsuro Miwa (eds.). Carbondale, IL: Southern Illinois University Press, 1987; ERIC No. ED304315, Columbus, OH: ERIC, 1987.

Smith, Othanel B., William Stanley, and Harlan Shores. "Cultural Roots of the Curriculum." In *The Curriculum: Context, Design and Development* (pp. 16–19). Richard Hooper (ed.). Edinburgh, Scotland: Oliver and Boyd, 1971.

Stake, Robert E., and Jack A. Easley. *Case Studies in Science Education.* Washington, DC: Government Printing Office, 1978.

Stevenson, Harold W., Shin-Ying Lee, and James W. Stigler. "Mathematics Achievement of Chinese, Japanese, and American Children." *Science* 231(1986):693–699.

Stigler, James W., and Ruth Baranes. "Culture and Mathematics Learning." In *Review of Research in Education* (pp. 253–306). Vol. 15. Ernst Z. Rothkopf (ed.). Washington, DC: American Educational Research Association, 1988–89.

Stigler, James, Shin-Ying Lee, and Harold Stevenson. "Curriculum and Achievement in Mathematics: A Study of Elementary School Children in Japan, Taiwan and the United States." *Journal of Educational Psychology* 74(1982): 315–322.

Sugiyama, Yoshishige. "A Comparison of Word Problems in American and Japanese Textbooks." In *Proceedings of the US–Japan Seminar on Mathematical Problem-Solving* (pp. 228–253). Jerry P. Becker and Tatsuro Miwa (eds.). Carbondale, IL: Southern Illinois University Press, 1987; ERIC No. ED304315, Columbus, OH: ERIC, 1987.

Travers, Kenneth, and Ian Westbury, eds. *The IEA Study of Mathematics I: Analysis of Mathematics Curricula.* Oxford, England: Pergamon, 1989.

KENNETH J. TRAVERS
assisted by J. McKeown

# INTERNATIONAL ORGANIZATIONS

This entry provides a brief overview of a few of the international organizations that promote the study of mathematics education. The International Commission on Mathematical Instruction (ICMI), which has been in existence since the early 1900s, plays a major role in identifying general problems and trends in mathematics education, as well as seeking to bring international resources to address these issues. Following World War II, ICMI became a subcommission of the International Mathematical Union (IMU), and the officers of ICMI are now appointed by the general assembly of the IMU. The work of ICMI is carried out mainly through its International Congress on Mathematics Education (ICME), which emphasizes the importance of sharing among the member countries information about mathematics curricula and teaching practices. ICME I was held in Lyons, France, in 1969. ICME II was convened in Exeter, the United Kingdom, in 1972. Thereafter, the congresses have met quadrennially, with recent sites being Quebec (1992), Seville (1996), and Tokyo (2000). In 2004, ICME 10 will be in Denmark. These meetings are significant events, bringing together major figures in mathematics education from around the world and addressing such key issues as curricular trends, applications of technology to mathematics education, and developing theories of mathematics education. One important activity of ICMI between the congresses is the annual International Mathematical Olympiad, a competition that attracts young people from increasing numbers of countries around the world.

ICMI also carries out regional activities in mathematics education through two associations, the South East Asia Mathematical Society (SEAMS) and the Inter American Council on Mathematical Education (IACME). Both of these groups convene periodic meetings to address issues of particular relevance to their world regions.

The International Group for the Psychology of Mathematics Education (PME), founded in the mid-1970s, meets annually in various locations around the world to report on research on the psychology of mathematics education, broadly conceived. PME conducts annual conferences, with recent ones being: Tsukuba (1993), Lisbon (1994), Recife (1995), and Valencia (1996). A useful overview of the work of the group is provided by Nesher and Kilpatrick (1990).

The International Organization of Women in Mathematics Education (IOWME) was established in 1980 to provide a forum for those interested in the relationship between gender and the learning and teaching of mathematics. Business meetings are held at each ICME.

*See also* Competitions; Congresses

## SELECTED REFERENCES

Bishop, Alan J., et al. *Significant Influences on Children's Learning of Mathematics.* UNESCO Document Series, #47. Paris, France: UNESCO, Education Sector, 1993.

Kilpatrick, Jeremy. "International Commission on Mathematical Instruction." In *International Encyclopedia of Education, Supplement* (p. 432). Neville Postlethwaite and Torsten Husén (eds.). London, England: Pergamon, 1988.

Nesher, Pearla, and Jeremy Kilpatrick. *Mathematics and Cognition.* Cambridge, England: Cambridge University Press, 1990.

Tall, David. *Advanced Mathematical Thinking.* Boston, MA: Kluwer, 1991.

Union of International Associations, eds. *Yearbook of International Organizations.* Vol. I. Munich, Germany: Saur, 1995–96.

KENNETH J. TRAVERS

# INTERNATIONAL STUDIES OF MATHEMATICS EDUCATION

Range from studies of a few key educational variables in two or three countries to cooperative multinational ventures designed to characterize many aspects of mathematics education in each participating country. Generally speaking, the intent of these studies is to provide educators with new insights into their own systems by giving a view of educational approaches that are both similar to and different from their own. Such insights can suggest factors that influence student learning and thereby inform discussions regarding curriculum design, teaching, and policy making.

The most prominent studies are the large-scale undertakings of the International Association for the Evaluation of Educational Achievement (IEA) and the International Assessments of Educational Progress (IAEP). IEA was founded in 1959 with the purpose of comparing educational systems and student achievement worldwide. IAEP is a project of the Educational Testing Service (ETS) designed to compare student achievement in the United States to those of other countries. Many other researchers around the world have also contributed substantially to this field. While the studies all focus on mathematics education in several countries, important differences can be recognized in both the purpose and design of these studies, yielding a broad literature that contributes to education in many diverse ways.

## CHANGING CHARACTER OF COMPARATIVE STUDIES

Generally, at the turn of the nineteenth century, comparative education depended largely on the reports of travelers to foreign educational systems (Halls 1990). These early descriptions provided information about foreign education systems that readers could use to make comparisons with their own. The intent of these studies was both to support nationalistic competition among nations and to promote "cultural borrowing," the adoption of features found in the educational systems of other countries (Holmes 1981). More prevalent in work undertaken since 1950 are studies that make explicit comparisons of specific variables between systems. For example, Flanders studied the interactions of teacher influence, student attitudes, and student achievement in mathematics and social studies in the United States and New Zealand (Flanders 1965). Beyond those particular variables, however, there was no explicit comparison made regarding other features of the educational systems.

A more comprehensive approach was taken in IEA's First International Mathematics Study (FIMS), which made comparisons among many variables of the educational systems in twelve participating countries: Australia, Belgium, England, Federal Republic of Germany, Finland, France, Israel, Japan, the Netherlands, Scotland, Sweden, and the United States (see, for example, Husén 1967, vols. 1 and 2). This was followed in the early 1980s by the Second International Mathematics Study (SIMS) involving a total of twenty educational systems around the world (Travers and Westbury 1989; Robitaille and Garden 1989; Burstein 1992). In this study, the curriculum was considered to be an important variable, as were the personal, school, and broader social contexts of education.

SIMS made a significant advance beyond FIMS in that it focused on the study of mathematics curriculum at three levels: the *Intended Curriculum* (as intended by national and system-level authorities); the *Implemented Curriculum* (as interpreted and put into practice by teachers and presented to students); and the *Attained Curriculum*, the mathematics actually learned by students as evidenced by their achievement and attitudes. The Third International Mathematics and Science Study (TIMSS) also used this same "tripartite" model of the curriculum. International comparisons in terms of the intended, implemented, and attained curriculum are not only used in IEA's projects. They are also implicit in other studies such as those conducted by Harold Stevenson and his colleagues at the University of Michigan (e.g., Stigler, Lee, and Stevenson 1990).

## METHODOLOGIES

The two IEA studies (FIMS and SIMS) relied primarily on a series of questionnaires that were used to characterize student achievement and other variables. This also is the case with the IAEP studies that involved six countries in its 1988 survey (Robitaille and Travers 1992). Other studies use direct observations of classroom interactions and interviews, as well as questionnaires and written tests. For example, in the Flanders (1965) studies, observational data were analyzed quantitatively. Stevenson and his colleagues used interpretive analyses of the observational data as well as student interviews to characterize both students' mathematical knowledge and the cultural attributes that may influence the development of that knowledge (Stevenson, "Bracey's Broadsides . . . ," 1993). Still other contributions to comparative education in mathematics utilize research conducted independently in different countries to develop evidence for specific or general relationships. For example, Taylor (1991) assembled data from many countries to argue that out-of-school mathematics and school curricula tend to be unrelated in those cultures with a colonial history.

Different study designs emerge from attempts to balance the strengths of different approaches with the information desired. Projects on the scale of the IEA and IAEP studies make use of samples that are large enough to characterize dominant patterns among variables, but they cannot provide the in-depth social and cultural contexts of learning that may affect achievement differences (Stigler and Baranes 1988–89). On the other hand, a series of small-scale cross-cultural studies conducted by researchers at the University of Michigan provide insights into the social and cultural influences on achievement differences (Stevenson, "Why Asian . . . ," 1993; Stigler, Lee, and Stevenson 1990). These smaller studies, however, have the disadvantage of not utilizing nationally representative samples (Bracey 1993).

## INTENT AND INTERPRETATION

Some comparative studies are intended to give detailed descriptions of students' abilities in mathematics in different systems without attempting to suggest causes or solutions. Others go beyond this, developing broader causal and relational models of how educational factors influence each other. In the IEA studies, for example, the goal is to use natural variability among educational systems worldwide to gain insight into the effects of different ways of undertaking educational endeavors (Husén 1967). The apparent influences of different approaches to education can be suggestive of alternatives that might be tried in other systems. This intent is distinct from the cultural borrowing approach of early comparative international studies in its recognition that no strategy or process can be transplanted from one system to another without first being adjusted, sometimes profoundly, with respect to the context of the new setting.

Variability in culture and contexts of education makes interpreting the results from comparative studies a complex task. Since the countries have different national emphases, the common assessment questionnaire or test cannot be thoroughly representative of any particular country. Husén (1967), speaking about the first IEA study, has addressed this issue, stating "with some justifications, one might paradoxically say that the tests . . . are equally appropriate or inappropriate to all countries participating in the study" (vol. 1, p. 21). Flanders (1994), for example, has claimed that the SIMS assessment questionnaire was not representative of the United States "curriculum defined by students' texts" (p. 260). Even if the problem of common assessment is taken into account, interpretation remains complex due to several other factors.

All IEA studies have attempted to respond to some of these complexities by introducing a variable called "opportunity to learn" (OTL). OTL is an IEA innovation that measures whether or not students were taught the content needed to correctly respond to items in the assessment questionnaire. Since OTL is basically the measurement of the implemented curriculum, it is referred to as "implemented coverage" (Travers, Garden, and Rosier 1989, 9). IEA studies are predicated upon the notion that the achievement results from different countries should not be viewed as a "horse race" (see, for example, Robitaille et al. 1993). Rather, interpreting achievement as a function of OTL provides meaningful insights into mathematics education (see, for example, Travers and McKnight 1985). In SIMS, the nine-year-old Japanese children scored significantly higher than their counterparts in the United States in algebra. However, Japanese children also have a significantly higher OTL than did the U.S. children on these items. A more informative question thus becomes, Why do the Japanese children have greater OTL than the U.S. children? Although OTL is undoubtedly one of the most important variables in the interpretation of results on international assessments, accurate determinations of OTL, whether they are provided by teachers, students, or curriculum analysis, remain problematical (Freudenthal 1975; UNESCO 1993).

Complicating further the interpretation of the results of these international studies is that, especially with respect to the larger projects, popular media may neglect comparisons of OTL or cultural differences among systems and report the achievement results at face value (Robitaille and Travers 1992). Such simplified presentations greatly increase the risk of misinterpretation. Experience in this area has shown that achievement and other results have to be interpreted along with many other variables, and with specific purposes in mind.

## IMPACT

Comparative studies in mathematics education have impact in several areas of education including debates about educational policy, instructional methods, and the effects of sociocultural factors in education. To illustrate the range of impacts that even a single study can have, the SIMS results are discussed below in relation to how they have been used in specific applications. In many cases the primary goals of comparative studies are at the policy level. There are many examples of the direct effects of comparative studies in general, and SIMS in particular, on educational policy. In New Zealand "curriculum writers were specifically instructed to add emphasis to those subject areas found deficient in the New Zealand SIMS results compared to other OECD nations" (Plomp and Loxley 1993, 4). Plomp and Loxley go on to say that "perhaps the best example of how national policy makers and educators have benefited from international comparisons of achievement performance comes from the U.S.A. where analysis of SIMS data provided the Americans with many conclusions that are still having an impact on the way mathematics is organized in American secondary schools" (p. 5). Marklund (1989) describes, too, how the low achievement of Swedish students in SIMS fueled an ongoing policy debate, giving support to reforms that had been suggested but that had received little attention up to that time. By contrast, the high achievement of students in

Hungary was taken as an indicator of the success of recent curriculum changes and as a suggestion that attention should be paid to comparisons with "international standards, that is, to the level of developed countries and not to provincial, politically set criteria" (Báthory 1989, 49).

The effects of various instructional methods have also been examined using the SIMS findings. Tracking "appeared to be ineffective in identifying talent, biased against certain social groups, and restrictive of both OTL and growth in achievement" (Kilpatrick 1990, 420). Broader instructional emphases, too, can be addressed through comparisons of systems with tendencies toward, for instance, empirical verification or axiomatics as were found to be favored by English-speaking countries and by Belgium respectively (Kilpatrick 1990).

With respect to the sociocultural factors, SIMS results have been used to critically examine hypotheses regarding gender differences in mathematics achievement. Specifically, the SIMS data showed gender effects in the results from thirteen-year-olds, with boys outperforming girls most dramatically on items involving measurement and transformational geometry, and girls outperforming boys most markedly on items in whole-number computation, common fractions, and rational expressions in geometry (Robitaille 1989, Chap. 6). These results were interpreted as tending to support a notion that girls might be inherently superior in some areas of mathematics and boys in others. To consider these issues more closely, Hanna (1990) provided an analysis that, by looking at the levels of gender differences across countries, supports an argument that "contradicts those theories that attempt to explain male superiority in mathematics on the basis of biological differences," concluding that "such gender differences as are found among thirteen-year-olds are due to out-of-class experiences and to psychosocial processes rather than biological differences" (pp. 31–32). That this analysis would not have been possible using only within-country data highlights the opportunities created by such studies for addressing specific questions, especially in secondary analyses.

In addition to such secondary analyses, comparative studies provide methodological insights and innovations and often raise issues that are undertaken with more detailed studies. In some cases, the data pool that results from such large-scale studies offers opportunities to test and compare a variety of statistical procedures (see, for example, Muthén 1991; Muthén, Kao, and Burstein 1991) and theoretical models related to student learning (see, for example, Ethington 1991).

The models developed for international endeavors also often provide a basis for further research. In the Dominican Republic, for instance, SIMS provided a model for a national achievement survey that raised concerns regarding the mathematics learning of students in that country (Luna, Ganzalez, and Wolfe 1990; Crespo 1990). Based on these initial findings, a comprehensive program of research and educational innovation to improve mathematics education was undertaken in that country (Luna 1992). Important, too, in providing information regarding educational reform, is the fact that IEA studies have provided opportunity for people to make longitudinal comparisons by maintaining a common core of achievement items and addressing similar contextual variables (see, for example, Robitaille 1990).

*See also* International Comparisons, Overview

## SELECTED REFERENCES
Báthory, Zoltan. "How Two Educational Systems Learned from Comparative Studies: The Hungarian Experience." In *International Comparisons and Educational Reform* (pp. 45–50). Alan C. Purves (ed.). Alexandria, VA: Association for Supervision and Curriculum Development, 1989.

Bracey, Gerald W. "American Students Hold Their Own." *Educational Leadership* 50(5)(1993):66–67.

Burstein, Leigh, et al., eds. *The IEA Study of Mathematics III: Student Growth and Classroom Processes.* Oxford, England: Pergamon, 1992.

Crespo, Sandra. "Mathematics Achievement in the Dominican Republic: Grade 12." Master's thesis, University of British Columbia, 1990.

Ethington, Corinna A. "A Test of a Model of Achievement Behaviors." *American Educational Research Journal* 28(1)(1991):155–172.

Flanders, James R. "Textbooks, Teachers, and the SIMS Test." *Journal for Research in Mathematics Education* 25(3)(1994):260–278.

Flanders, Ned A. *Teacher Influence, Pupil Attitudes, and Achievement.* Washington, DC: U.S. Department of Health, Education, and Welfare, 1965.

Freudenthal, Hans. "Pupils' Achievement Internationally Compared." *Educational Studies in Mathematics* 6(1975):127–186.

Halls, William D., ed. *Comparative Education: Contemporary Issues and Trends.* London, England: Kingsley/UNESCO, 1990.

Hanna, Gila. "Mathematics Achievement of Boys and Girls: An International Perspective." *Ontario Mathematics Gazette* 28(3)(1990):28–32.

Holmes, Brian. *Comparative Education: Some Considerations of Method.* London, England: Allen and Unwin, 1981.

Husén, Torsten, ed. *International Study of Achievement in Mathematics: A Comparison of Twelve Countries.* 2 vols. Stockholm, Sweden: Wiley, 1967.

Kilpatrick, Jeremy. "Apples and Oranges Again." *Journal for Research in Mathematics Education* 21(5)(1990): 416–424.

Luna, Eduardo. "Dominican Republic: The Study on Teaching and Learning of Mathematics." *Prospects* 22(4)(1992):448–454.

Luna, Eduardo, Sara Ganzalez, and Richard Wolfe. "The Underdevelopment of Educational Achievement: Mathematics Achievement in the Dominican Republic Eighth Grade." *Journal of Curriculum Studies* 22(4)(1990): 361–376.

Marklund, Inger. "How Two Educational Systems Learned from Comparative Studies: The Swedish Experience." In *International Comparisons and Educational Reform* (pp. 35–44). Alan C. Purves (ed.). Alexandria, VA: Association for Supervision and Curriculum Development, 1989.

Muthén, Bengt O. "Multilevel Factor Analysis of Class and Student Achievement Components." *Journal of Educational Measurement* 28(4)(1991):338–354.

Muthén, Bengt O., C.-F Kao, and Leigh Burstein. "Instructionally Sensitive Psychometrics: Application of a New IRT-based Detection Technique to Mathematics Achievement Test Items." *Journal of Educational Measurement* 28(1)(1991):1–22.

Plomp, Tjeerd, and William Loxley. "International Comparative Assessment and Curriculum Reform." Paper presented to UNESCO International Forum for Project 2000+ on Science and Technology Education, Paris, July, 1993.

Robitaille, David F. "Achievement Comparisons Between the First and Second IEA Studies of Mathematics." *Educational Studies in Mathematics* 21(1990):395–414.

———. "Students' Achievements: Population A." In *The IEA Study of Mathematics II: Contexts and Outcomes of School Mathematics* (pp. 102–125). David F. Robitaille and Robert A. Garden (eds.). Oxford, England: Pergamon, 1989.

Robitaille, David F., and Robert A. Garden, eds. *The IEA Study of Mathematics II: Contexts and Outcomes of School Mathematics.* Oxford, England: Pergamon, 1989.

Robitaille, David F., Curtis McKnight, William H. Schmidt, Edward Britton, Senta Raizen, and Cynthia Nicol. *Curriculum Frameworks for Mathematics and Science.* Vancouver, British Columbia, Canada: Pacific Educational, 1993.

Robitaille, David F., and Kenneth J. Travers. "International Studies of Achievement in Mathematics." In *Handbook of Research on Mathematics Teaching and Learning* (pp. 687–709). Douglas A. Grouws (ed.). New York: Macmillan, 1992.

Stevenson, Harold W. "Bracey's Broadsides Are Unfounded." *Educational Leadership* 50(5)(1993):68.

———. "Why Asian Students Still Outdistance Americans?" *Educational Leadership* 50(5)(1993):63–65.

Stigler, James W., and Ruth Baranes. "Culture and Mathematics Learning." In *Review of Research in Education* (pp. 253–306). Vol. 15. Ernst Z. Rothkopf (ed.). Washington, DC: American Educational Research Association, 1988–89.

Stigler, James W., Shin-Ying Lee, and Harold W. Stevenson. *Mathematical Knowledge of Japanese, Chinese, and American Elementary School Children.* Reston, VA: National Council of Teachers of Mathematics, 1990.

Taylor, Nick. "Independence and Interdependence: Analytical Vectors for Defining the Mathematics Curriculum of Schools in a Democratic Society." *Educational Studies in Mathematics* 22(1991):107–123.

Travers, Kenneth J., Robert A. Garden, and Malcolm Rosier. "Introduction to the Study." In *The IEA Study of Mathematics II: Contexts and Outcomes of School Mathematics* (pp. 1–16). David F. Robitaille and Robert A. Garden (eds.). Oxford, England: Pergamon, 1989.

Travers, Kenneth J., and Curtis C. McKnight. "Mathematics Achievement in US Schools: Preliminary Findings from the Second IEA Mathematics Study." *Phi Delta Kappan* 66(Feb. 1985):407–413.

Travers, Kenneth J., and Ian Westbury, eds. *The IEA Study of Mathematics I: Analysis of Mathematics Curricula.* Oxford, England: Pergamon, 1989.

UNESCO. "Searching for Standards." In *World Education Report 1993* (pp. 75–89). Lucon, France: UNESCO, 1993.

DAVID F. ROBITAILLE
assisted by Ed Robeck and Hari Koirala

# INTERPOLATION

Estimating unknown results or making predictions within the range of data or information already available. Graphical methods of interpolation are most frequently used; in particular, estimation of results from a least-squares regression line (line of best fit). Here are the results of a simple physics experiment in which various masses are suspended from a spring (in random order) and the length of the spring measured:

| mass (g) | 100 | 200 | 300 | 400 | 500 | 600 | 700 |
|---|---|---|---|---|---|---|---|
| length (cm) | 22.9 | 27.3 | 32.2 | 36.5 | 42.0 | 47.0 | 53.2 |

The regression line estimated from these results has the equation $y = 17.29 + 0.05x$ (see Figure 1). Thus a prediction can be made for the length of the spring when a mass of 550 g is suspended from it, by substituting $x = 550$ into the equation to give an estimate of $y = 44.79$ cm. Graphical interpolation is also used when the median of a set of continuous data is estimated from a cumulative frequency curve or polygon.

*Linear interpolation* is a term used to refer to the calculation of intermediate values, such as the median in a set of data. This method assumes a uniform

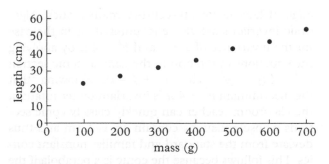

**Figure 1**

or linear distribution of results between points that are known and is therefore the equivalent of the graphical methods outlined above. This method may be used to estimate intermediate values from tables or to estimate the median directly from the data, without drawing a graph. The following data, for the heights of a group of pupils, are used as an example:

| Heights | Frequency | Upper limit for group | Cumulative frequency |
|---------|-----------|----------------------|----------------------|
| 150–154 | 3 | 154.5 | 3 |
| 155–159 | 4 | 159.5 | 7 |
| 160–164 | 6 | 164.5 | 13 |
| 165–169 | 3 | 169.5 | 16 |
| 170–174 | 2 | 174.5 | 18 |

There are 18 pupils here, so the median of their heights will be midway between the 9th and 10th values. The median must therefore lie in the class interval from 159.5 cm to 164.5 cm; and there are 6 pupils with heights in that interval. Assuming that we do not have access to the original raw data, we split the 5-cm interval into 6 equal spaces (see Figure 2):

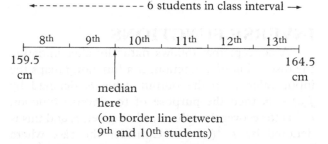

**Figure 2** *Calculation of the median in a set of data.*

The median must occur between 9 and 10, that is, 2/6 of the way through that interval giving:

$$\text{Median height} = 159.5 \text{ cm} + 2/6 \times 5 \text{ cm}$$
$$= 159.5 + 1.67$$
$$= 161.17 \text{ cm}$$

*See also* Least Squares: Median

SELECTED REFERENCES

Gravetter, Frederick J., and Larry B. Wallnau. *Statistics for the Behavioral Sciences.* St Paul, MN: West, 1992.

Hudson, Brian, and Mary Rouncefield. *Handling Data Core Y.9 Century Maths.* Cheltenham, England: Thornes, 1992.

Marsh, Catherine. *Exploring Data.* Cambridge, England: Polity, 1988.

Rouncefield, Mary, and Peter Holmes. *Practical Statistics.* London, England: Macmillan, 1989.

MARY ROUNCEFIELD

## INVARIANCE

Properties or values that remain unchanged as transformations occur and thus identified as invariants. Illustrations of invariants may pinpoint the obvious or they may single out unchanging relationships that are surprising. In the case of projecting one plane onto another plane, lines will be transformed into lines, but circles need not be transformed into circles (Figure 1, below). Hexagons, for example, will be transformed into hexagons, but their perimeters,

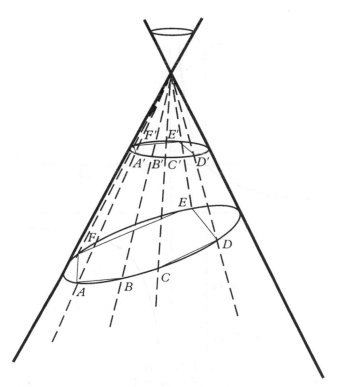

**Figure 1** *A projection that transforms a circle into an ellipse.*

angle measures, and areas will generally be altered. Rigid motions, which include translations and rotations, however, will preserve these properties of perimeters, angle measures, areas, and more. Fundamentally, Euclidean geometry is the study of rigid motions, whereas projective geometry is the study of those properties held invariant under projection.

Projective geometry describes how we see the world about us and arose as a consequence of problems in the practical realm of everyday living (as did Euclidean geometry and mathematics in general). For example, distortions occurred in the artistic representation of scenes. Still, the recognition of these painted scenes disturbingly raised the question of what visible features in the original were preserved on canvas. Whereas angles, areas, and shapes were distorted, other characteristics were not. Such a consideration of preserved features led ultimately to the formalizing of transformation theories and the notion of invariance.

As noted in Figure 1, conic sections are transformed into conic sections by projection and give rise to such far-reaching theorems as those of Blaise Pascal (1623–1662) and Charles Julien Brianchon (1783–1864). Other remarkable advances based on tranformations and invariants describe the cutting

edge of late-nineteenth-century mathematics. Algebraic invariants are also encountered, as in the case for the invariance of $a + c$ and $b^2 - 4ac$ by applying the equations of rotation to the general conic section $ax^2 + bxy + cy^2 + dx + ey + f = 0$. Knowing that the discriminant $b^2 - 4ac$ is invariant under rotation, the classroom teacher can quickly classify conic sections whose equations contain an $xy$ term and thus deviate from the standard and familiar nonslant conics. This follows because the conic is a parabola if the discriminant is zero, an ellipse if the discriminant is negative, and a hyperbola if the discriminant is positive. For example, the curve $xy + 4x - 3y = 0$ is a hyperbola as the discriminant $b^2 - 4ac$ is 1, a positive number (Figure 2).

The notions of invariance lend themselves to some of the most difficult of questions in modern mathematics, topology for example, and identify areas of deep challenge for mathematicians.

*See also* Conic Sections (Conics); Projective Geometry; Transformations

### SELECTED REFERENCES

Eves, Howard W. *College Geometry.* Boston, MA: Jones and Bartlett, 1995.

Greenberg, Marvin J. *Euclidean and Non-Euclidean Geometries: Development and History.* New York: Freeman, 1993.

Kline, Morris. *Mathematical Thought from Ancient to Modern Times.* New York: Oxford University Press, 1972.

Reid, Constance. *A Long Way from Euclid.* New York: Crowell, 1963.

Wallace, Edward C., and Stephen F. West. *Roads to Geometry.* Englewood Cliffs, NJ: Prentice-Hall, 1992.

RICHARD L. FRANCIS

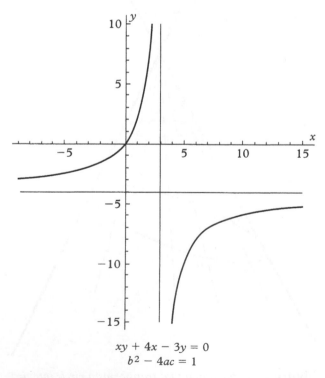

$$xy + 4x - 3y = 0$$
$$b^2 - 4ac = 1$$

**Figure 2** *Hyperbola.*

# INVERSE FUNCTIONS

A concept that follows naturally after initiating the discussion of a function. If a function $f$ maps the input value $a$ to the output value $b$, denoted by $f(a) = b$, then the purpose of the inverse function $f^{-1}$ is to recover the value $a$, when given $b$, and this is denoted by $f^{-1}(b) = a$. Consider the case where function $f$ is defined by the following: "Given any input value, first add 2 to it; multiply this result by 3; then finally subtract 4." Now, if the function value so obtained is 17 (i.e., $b = 17$), what was the input value? This value turns out to be 5 and is written $f^{-1}(17) = 5$. This kind of problem is common in introductory algebra courses. In this situation $f$ could be expressed algebraically by $f(x) = 3(x + 2) - 4$.

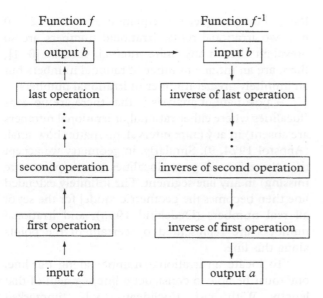

**Figure 1**  *Order of operations.*

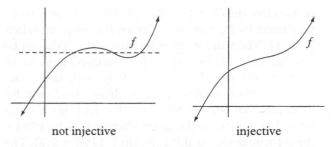

**Figure 2**  *Determining whether f is injective.*

Knowing that the function value is 17 gives $3(x + 2) - 4 = 17$. The algebra student adds 4 to each side of the equation, giving $3(x + 2) = 21$, then divides both sides by 3, $x + 2 = 7$, and finally subtracts 2 to yield $x = 5$. The following describes the behavior of this particular inverse function, $f^{-1}$: "Given any input value, first add 4 to it; divide this result by 3; then finally subtract 2." This gives $f^{-1}(x) = [(x + 4)/3] - 2$. What is important here is to realize that the operations performed by $f^{-1}$ are the inverses of the operations performed by $f$, and are applied in reverse order (see Figure 1).

Thus, if $f$ begins by adding 2 to the variable and ends by multiplying by 3, then $f^{-1}$ begins by dividing by 3, and ends by subtracting 2.

No problems occur if the operations defining $f$ are elementary enough (e.g., addition, multiplication). When the operations include powers and roots, the situation becomes more difficult. If an operation involves raising a quantity to the power $p$ (i.e., $x^p$), the inverse operation should be extracting the $p$th root ($x^{1/p}$). Difficulty can arise, as an even root must necessarily be positive. Thus, if $f(x) = x^2$ and $a = -2$, then $b = f(a) = 4$, and $f^{-1}(x) = \sqrt{x}$ gives $f^{-1}(4) = 2$, which is not equal to $-2$.

The difficulty here is that functions defined by odd powers are injective (this means two different input values produce different function values), while functions defined by even powers are not injective. A function can be determined to be injective by examining its graph. If the graph of $f$ intersects some horizontal line in at least two places, then $f$ is not in-

jective; otherwise, $f$ is injective (see Figure 2) (Nicodemi 1987). A continuous injective function must either be an increasing or a decreasing function. An important observation is that if function $f$, defined on domain $D$ with range $R$, is injective, then its inverse $f^{-1}$ exists.

Furthermore, $f^{-1}$ has domain $R$ and range $D$ and is defined by $f^{-1}(b) = a$, providing $f(a) = b$.

If $a \in D$, then $f^{-1}(f(a)) = a$, and if $b \in R$, then $f(f^{-1}(b)) = b$. Thus, the compositions $f \circ f^{-1}$ and $f^{-1} \circ f$ yield the identity functions on $R$ and $D$, respectively, and the graphs of $f$ and $f^{-1}$ are symmetric with respect to the diagonal line $y = x$ (see Figure 3) (Leithold 1990, 428).

Many functions are not injective over their natural domain. The quadratic $y = x^2$, for one, is not injective over the entire real line as is the case with many polynomials. Often, in this case, a restricted domain can be found for which $f$ is injective. For example, $f(x) = \sin(x)$ is not injective on $(\infty, -\infty)$, but

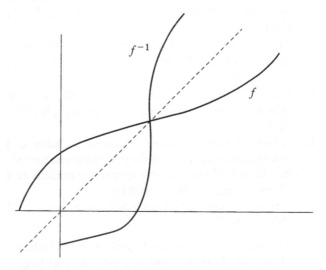

**Figure 3**  *Graphical symmetry of f and $f^{-1}$.*

is injective on $D = [-\pi/2, \pi/2]$. Furthermore, when restricted to $D$, $f$ assumes its entire range of values $[-1, 1]$. Then the inverse function, $\sin^{-1}(x)$, can be defined on $[-1, 1]$, with values (angles) lying in $D$. The inverse sine key, $\sin^{-1}$, along with $\cos^{-1}$ and $\tan^{-1}$, is common on today's calculators. It is found, for example, that $\sin^{-1}(0.9) = 1.1198$, and this is interpreted as 1.1198 is the angle (hence, it is 1.1198 radians) whose sine is 0.9 (i.e., $\sin 1.1198 = 0.9$). The inverse trigonometric functions play an important role in calculus and analysis for they serve to define many integrals (Varberg and Purcell 1992, 368–378).

*See also* Functions of a Real Variable

SELECTED REFERENCES

Leithold, Louis. *The Calculus with Analytic Geometry.* 6th ed. New York: Harper and Row, 1990.

Nicodemi, Olympia. *Discrete Mathematics.* St. Paul, MN: West, 1987.

Varberg, Dale, and Edwin Purcell. *Calculus with Analytic Geometry.* 6th ed. Englewood Cliffs, NJ: Prentice-Hall, 1992.

THOMAS P. DENCE

# IRRATIONAL NUMBERS

Real numbers that cannot be represented as quotients $a/b$ of two integers ($b \neq 0$) (Niven 1961). Their existence was first discovered by the Pythagoreans in Greece sometime between the sixth and third centuries B.C. (Boyer 1991, 71). Today, several classes of irrational numbers that were not imagined by the ancient Greeks are recognized:

1. integral $p$th roots of rational numbers $r$ that are not perfect $p$th powers (e.g., $r^{1/p} = \sqrt{2/3}$, $\sqrt[4]{4}$, or $\sqrt[10]{50}$);
2. values of trigonometric functions and of their inverses at most of their arguments (e.g., $\cos 7$, $\sin 10$, $\tan^{-1}(1/4)$);
3. values of logarithmic and exponential functions at most of their arguments (e.g., $\log_{10} 5$, $\ln(5/2)$, $(2/3)^{\sqrt{2}}$);
4. values of various other functions in advanced mathematics at most of their arguments, including Bessel functions, the gamma function, and elliptic integrals (Beyer 1984);
5. certain special numbers such as $\pi$, $e$, $\pi^{-1}$, $e^{-1}$ (Niven 1947).

The many classes of irrational numbers indicate that these numbers are widespread. Even in beginning mathematics they make their appearance effort-

lessly. For example, the equation $x^2 - 2x - 2 = 0$ has two irrational roots. Irrational numbers are so prevalent that in any finite interval, such as $[0, 1]$, there are an infinite number of rational numbers but an infinitely greater number of irrational numbers.

A postulate of analysis is that there are no gaps (localities where either rational or irrational numbers are absent) in any finite interval, no matter how small (Apostol 1974, 9). Similarly, in geometry we accept that there are no gaps (localities where points are missing) in any line segment. The infinitely extended line then becomes the geometric model for the set of all real numbers (Dedekind 1956), and irrational numbers are represented by certain of the points along this line.

To locate an irrational number along the line, one must be able to construct a line segment of that length. With just Euclidean tools (unmarked straightedge and compass), however, a line segment of irrational length can be constructed only if the length is a number in Class 1 above and $p$ is a power of 2, or if the length is some finite arithmetic combination of such numbers. Segments of length $\sqrt{1 + 2\sqrt{2}}$ or $\sqrt[4]{5}$ can be constructed, but a segment of length $\sqrt[3]{2}$ cannot. Consequently, trisection of an angle is not generally possible, because this is equivalent to the construction of a line segment whose length is the cube root of a number that is not a perfect cube (Dickson 1921). The squaring of the circle using Euclidean tools also is impossible, because this is equivalent to the construction of a line segment of length $\pi$, a number not even in Class 1.

Some properties of irrational numbers include the following:

1. The sum, difference, product, or quotient of two irrational numbers may be either rational or irrational (e.g., $\sqrt{2} \cdot \sqrt{2} = 2$, but $\frac{1}{2}\sqrt{2} + \frac{1}{2}\sqrt{2} = \sqrt{2}$).
2. If $x$ is irrational and $b \neq 0$ is rational, then $b + x$, $b - x$, $b \cdot x$, $b/x$, and $x/b$ are all irrational (e.g., $2 - \pi$ is irrational).
3. An irrational number raised to an irrational power may be either rational or irrational (see Jones and Toporowski 1973).
4. The decimal representation of an irrational number contains infinitely many decimal places, and the pattern is nonrepeating.

An irrational number can be approximated to any finite number of decimal places by the use of the general technique of infinite series (Dence and Dence 1994, 303). Alternatively, roots of any order can be estimated using Newton's method (Dence and Dence 1994, 86). Square roots, in particular, can also be estimated using continued fractions (Olds 1963, 51; Johnson 1989). A continued fraction convergent $a_n/b_n$ to any irrational number $\alpha$ is actually that rational number $a/b$, with $0 < b \leq b_n$, which minimizes the error $|\alpha - (a/b)|$ (Rose 1988, 120). Thus, 38/17 is a continued fraction convergent to $\sqrt{5}$, and it is the best rational approximation to $\sqrt{5}$ among fractions with denominators 17 or less.

*See also* Algebraic Numbers; Continued Fractions; Rational Numbers; Transcendental Numbers

## SELECTED REFERENCES

Apostol, Tom M. *Mathematical Analysis.* 2nd ed. Reading, MA: Addison-Wesley, 1974, pp. 9–11.

Beyer, William H. *CRC Standard Mathematical Tables.* 27th ed. Boca Raton, FL: CRC, 1984.

Boyer, Carl B. *A History of Mathematics.* 2nd ed., Revised by Uta C. Merzbach. New York: Wiley, 1991, pp. 71–74.

Castellanos, Dario. "The Ubiquitous $\pi$." *Mathematics Magazine* 61(1988):67–98, 148–163.

Dedekind, Richard. "Irrational Numbers." In *The World of Mathematics* (pp. 528–536). Vol. 1. James R. Newman (ed.). New York: Simon and Schuster, 1956.

Dence, Joseph B., and Thomas P. Dence. *A First Course of Collegiate Mathematics.* Malabar, FL: Krieger, 1994, pp. 86, 303–308.

Dickson, Leonard E. "Why It Is Impossible to Trisect an Angle or to Construct a Regular Polygon of 7 or 9 Sides by Ruler and Compasses." *Mathematics Teacher* 14(1921): 217–223.

Johnson, Kenneth R. "An Iterative Method for Approximating Square Roots." *Mathematics Magazine* 62(1989):253–259.

Jones, J. P., and S. Toporowski. "Irrational Numbers." *American Mathematical Monthly* 80(1973):423–424.

Niven, Ivan. *Numbers: Rational and Irrational.* New York: Random House, 1961, pp. 21–37, 38–51.

———. "A Simple Proof That $\pi$ Is Irrational." *Bulletin of the American Mathematical Society* 53(1947):509.

Olds, Carl D. *Continued Fractions.* New York: Random House, 1963, pp. 51–87.

Rose, Harvey E. *A Course in Number Theory.* Oxford, England: Oxford University Press, 1988, pp. 120–125.

JOSEPH B. DENCE

# ISLAMIC MATHEMATICS

By the middle of the eighth century A.D., the new religion of Islam had spread from its origins in Arabia westward through all of North Africa and into Spain and eastward nearly to India. Its outward spread slowed, whereupon the Islamic empire then split into several parts. In the eastern segment, the caliph al-Mansūr founded his new capital in Baghdad in 766; his successors al-Rashid and al-Ma'mūn developed the city into a commercial and intellectual center, establishing a major library, which collected mathematical and scientific manuscripts from east and west, and a research institute, the House of Wisdom, to which scholars from all parts of the caliphate were invited to translate the Greek and Indian works and then to create new knowledge. Over the next five hundred years, Islamic mathematicians (and others also writing in the Arabic language), in Baghdad and elsewhere in the Islamic domains, built on the Greek and Indian foundations and developed mathematical knowledge in decimal arithmetic, algebra, combinatorics, geometry, and trigonometry.

Muhammad ibn-Mūsā al-Khowarizmi (ca. 780–850), an early member of the House of Wisdom, wrote one of the first arithmetic texts in Arabic discussing the number system that had been developed in India, now known as the Hindu-Arabic place value system. This work was important not only in Islam but also in Europe, where Latin versions of it in the twelfth century helped spread the advantages of this system. Al-Khowarizmi's work, however, dealt only with place values for integers. Decimal fractions were first discussed, at least in a formal sense, in a work written in Damascus in 952 by Abu l-Hasan al-Uqlidisi. But it took another two hundred years for mathematicians to realize the importance of decimal fractions as approximations to real numbers. It was al-Samaw'al ibn Yahyā al-Maghribi (ca. 1125–1180) who noted that when one uses the standard algorithm to calculate square roots, in general one "obtains answers infinite in number, each of which is more precise and closer to the truth than that which precedes it."

Al-Khowarizmi also wrote the earliest Islamic algebra text, the *Hisāb al-jabr wa-l-muqābala,* in which he discusses in detail the solution of quadratic equations. For example, he shows that to solve the equation "squares and roots equal to numbers," or, in modern notation, $x^2 + bx = c$, one first halves the number of roots, squares that result, adds it to the constant term, takes the square root of that sum, and finally subtracts from the square root half the number of roots. One can translate al-Khowarizmi's verbal method, which he applies to numerous examples, into a version of the modern quadratic formula

$$x = \sqrt{\left(\frac{b}{2}\right)^2 + c} - \frac{b}{2}$$

To justify this procedure, al-Khowarizmi presented a geometric derivation of the result, beginning with the interpretation of the left side of the original equation as a square added to a rectangle.

Al-Khowarizmi's text was translated into Latin in the twelfth century and provided Europe with one of its introductions to algebra. But one of the earliest European writers on algebra, Leonardo of Pisa (ca. 1170–1240), based the algebraic portion of his *Liber abaci* on the algebra text of the Egyptian mathematician Abū Kāmil ibn Aslam (ca. 850–930).

Other Islamic authors developed the techniques of algebra further over the succeeding centuries. For example, Abū Bakr al-Karaji (d. 1019) worked out the algebra of exponents and its use in adding, subtracting, and multiplying polynomials. Al-Samaw'al in essence used negative exponents in dividing polynomials and showed how to continue the process indefinitely in the case where the division did not come out even. These men also developed the idea of inductive proof for dealing with certain arithmetic formulas, including those for the sum of the integral cubes and for the Pascal triangle of binomial coefficients. Ibn al-Haytham (965–1039) showed how to develop a formula for the sum of any integral powers and used the result for fourth powers to find the volume of the solid formed by revolving a parabola around its base. Omar Khayyam (1048–1131) used intersections of conic sections to solve cubic equations but was unable to find a strictly algebraic method. Omar's methods were improved upon by Sharaf al-Din al-Tūsi (d. 1213), who analyzed the solutions to cubics and developed functional techniques for determining the number of solutions to various types of cubics.

Islamic authors were very interested in combinatorics from as early as the eighth century, particularly in the question of how many words can be formed out of certain letters of the Arabic alphabet. Over the years the questions gradually became more abstract until in the thirteenth century, ibn al-Banna (1256–1321), from Marrakech in what is now Morocco, was able to give detailed derivations of formulas for both the number of combinations and the number of permutations of $k$ elements taken from a set of $n$. Similar ideas were being discussed by Hebrew scholars in Spain and southern France around the same time, but it is not known if there was any transmission of combinatorial ideas between the Jewish and Islamic mathematicians.

There was, however, transmission of some geometric ideas from Islamic scholars to European ones. In particular, the work of Nasir al-Din al-Tūsi (1201–1274), on "proving" the parallel postulate of Euclid, was adapted by his son in 1298 and published in Rome three hundred years later. It was then commented on by John Wallis in connection with his own work on the parallel postulate and eventually became the starting point for Saccheri and others in the development of non-Euclidean geometry. Many Islamic scholars also wrote commentaries on the tenth book of Euclid's *Elements* and attempted to erase the Greek distinction between number and magnitude by turning Euclid's "irrational" magnitudes into "numbers."

Islamic scholars absorbed Greek trigonometry, both from Greek sources and from the Indian adaptations of it, and then reworked it to fit their own needs. In particular, they developed all six modern trigonometric functions and proved new results useful for solving plane and spherical triangles. For example, Abū Nasr Mansūr ibn 'Iraq (d. 1030), derived several important theorems in spherical trigonometry, including the law of sines, which his student Muhammad ibn Ahmad al-Birūni (973–1055), used to determine the *qibla,* the direction of Mecca that one needed to face during prayer. Two centuries later, Nasir al-Din al-Tūsi wrote the earliest comprehensive work on plane and spherical trigonometry, independent of astronomy, the *Treatise on the Complete Quadrilateral,* in which is found for the first time the law of sines for plane triangles as well as systematic methods for solving plane and spherical triangles.

Although Islamic scholars had carried forward the mathematical enterprise in many areas, only a small portion of their work was translated into a European language in the late Middle Ages in time to influence European work on the same subject. But because subsequent European developments often resemble Islamic work, there are questions as to whether the Islamic material reached Europe through still unknown channels. Much further research needs to be done to learn the complete story of Islamic mathematics and its place in the general development of the subject.

*See also* History of Mathematics, Overview

## SELECTED REFERENCES

Berggren, J. Lennart. *Episodes in the Mathematics of Medieval Islam.* New York: Springer-Verlag, 1986.

Katz, Victor J. *A History of Mathematics: An Introduction.* New York: HarperCollins, 1993.

Rashed, Roshdi. *The Development of Arabic Mathematics: Between Arithmetic and Algebra.* A. F. W. Armstrong, trans. Dordrecht, Netherlands: Kluwer, 1994.

Youschkevitch, Adolf P. *Les Mathématiques Arabes (VIIIᵉ–XVᵉ siècles).* Paris, France: Vrin, 1976.

VICTOR J. KATZ

# ISOMORPHISM

Term used to identify structural sameness. Two mathematical structures are considered to be the same, or *isomorphic,* if there is a one-to-one correspondence between them. If the structures have one or more operations on them, the operations are required to match under the correspondence. An isomorphism of an algebraic structure with itself is called an *automorphism* of the structure. For example, two sets $S$ and $S'$ are isomorphic if there is a one-to-one correspondence $s \longleftrightarrow s'$ between their elements; each element $s$ in $S$ corresponds to one and only one element $s'$ in $S'$, and vice versa. The sets $S = \{1, 2\}$ and $S' = \{3, 4\}$ are isomorphic; $1 \longleftrightarrow 3, 2 \longleftrightarrow 4$. It does not matter that the particular elements in these sets are different; from the point of view of isomorphism, all that matters is that their elements can be paired in a one-to-one manner.

For algebraic structures such as groups, rings, and fields, which have addition and multiplication operations defined on them, the operations also must be taken into account. Two groups, $G$ and $G'$, are isomorphic if there is a one-to-one correspondence $g \longleftrightarrow g'$ between the elements $g$ in $G$ and $g'$ in $G'$ with the property that if $g_1 \longleftrightarrow g_1'$ and $g_2 \longleftrightarrow g_2'$, then $g_1 g_2 \longleftrightarrow g_1' g_2'$, where $g_1, g_2$ are in $G$ and $g_1', g_2'$ are in $G'$. If the operations on $G$ and $G'$ are written additively, the requirement for isomorphism is written $g_1 + g_2 \longleftrightarrow g_1' + g_2'$.

For example, let $G$ be the group of integers under the operation of addition, and let $G'$ be the group of even integers under addition. Pair each integer $n$ in $G$ with its double $2n$ in $G'$, $n \longleftrightarrow n' = 2n$. This pairing is a one-to-one correspondence and is an isomorphism, since, if $n \longleftrightarrow n' = 2n$ and $m \longleftrightarrow m' = 2m$, then $n + m \longleftrightarrow 2n + 2m = n' + m'$. From the point of view of group theory, the group of integers and the group of even integers are the same, although they have different elements.

For rings and fields, which have two operations, both operations must be taken into account. Two fields, $K$ and $K'$, are isomorphic if there is a one-to-one correspondence $x \longleftrightarrow x'$ between elements $x$ in $K$ and $x'$ in $K'$ with the property that if $x \longleftrightarrow x'$ and $y \longleftrightarrow y'$, then both $x + y \longleftrightarrow x' + y'$ and $xy \longleftrightarrow x'y'$, where $x$ and $y$ are elements in $K$, and $x', y'$ are corresponding elements in $K'$. The process of taking the conjugate of a complex number is an example of an automorphism of the field of complex numbers. If $z = a + bi$ is a complex number, its conjugate is the number $\bar{z} = a - bi$. The pairing $z \longleftrightarrow \bar{z}$ of a complex number with its conjugate is a one-to-one correspondence of the field $K$ of complex numbers with itself and is an isomorphism, since it has the property

that $z_1 + z_2 \longleftrightarrow \overline{z_1 + z_2} = \bar{z_1} + \bar{z_2}$ and $z_1 z_2 \longleftrightarrow \overline{z_1 z_2} = \bar{z_1}\,\bar{z_2}$ for all complex numbers $z_1, z_2$. Geometrically, complex conjugation reflects the complex number plane about the real axis.

*See also* Abstract Algebra; Fields; Group Theory

SELECTED REFERENCES

Fraleigh, John. *A First Course in Abstract Algebra.* Reading, MA: Addison-Wesley, 1989.

Gallian, Joseph. *Contemporary Abstract Algebra.* Lexington, MA: Heath, 1990.

Kletzing, Dennis. *Abstract Algebra.* San Diego, CA: Harcourt Brace Jovanovich, 1991.

DENNIS KLETZING

# ISRAEL

A country in which all aspects of educational policy are determined by the Ministry of Education of the national government. Education is compulsory until age 15. This includes six grades of elementary school and three grades of junior high school. The expectation is that adolescents will continue their education until age 18 either in academic tracks or in vocational tracks. Some of the vocational schools also have academic tracks. Others have mathematical programs that are quite minimal. At the end of the academic tracks, the student has to pass matriculation examinations that are created and evaluated by the Ministry of Education. Mathematics is one of the subjects that all students in the academic tracks must study. There are three mathematics programs, however, that are offered to the students and that are different from each other in scope. The five-unit program (five weekly hours of mathematics during the three years of high school) includes calculus, algebra, and plane Euclidean geometry and also the geometry and algebra of vectors. The four-unit program (four weekly hours) is quite similar to the five-unit program. The main difference between the two programs is in the level of exercises that are given as class or home assignment and also as matriculation examination items. The four-unit program does not include some harder topics, such as complex numbers, which the five-unit program does include. The three-unit program (three weekly hours of mathematics) has a minimal exposure to calculus, does not include such algebraic topics as the Gaussian method of solving linear equations and other topics that the four- and five-unit programs do include, and its matriculation examination items are simple and stereotypical.

## THE CURRICULUM

Historically, the secondary mathematics curriculum was a heritage of the English mandate that terminated in 1948 with the establishment of the state of Israel. During the next twenty-eight years, the curriculum underwent some insignificant changes. In 1976 the Ministry of Education approved a new secondary mathematics curriculum that had some revolutionary elements. The initiative to promote this curriculum came from some distinguished mathematicians at various universities who were not satisfied with the old curriculum. The revolutionary elements of the curriculum were: (a) canceling trigonometry as a separate topic and including it in analysis, the framework for studying all functions; (b) starting differentiation at the 10th grade (also in the three-unit program); (c) including complex numbers, the Gaussian method for solving linear equations, and geometry of vectors in the five-unit program; and (d) including recursion and linear programming in the three-unit program. The Ministry of Education could not force the schools to adopt the new curriculum, and as of 1995 only about 25% of the schools switched to the new curriculum. The main objection to the new curriculum came from the teachers. They claimed that the mathematics was too complicated to teach and the new books that were written for the curriculum were unsatisfactory. Thus there are actually two forms of the mathematics matriculation examination, one for the old curriculum as well as one for the new curriculum.

## THE MATRICULATION EXAMINATIONS

The matriculation examinations, as all types of mass examinations, have to be stereotypical, because the educational system must have models for the final examinations. Thus, learning very often becomes learning for the final examinations. It is quite common that teachers and students are busy solving problems from previous matriculation examinations and pay little attention to concepts and understanding. The emphasis, in many cases, is primarily on mathematical procedures and not on concepts and mathematical thinking.

## TEACHER PREPARATION

The preparation for high school mathematics teaching is done at the universities. The future teachers need a bachelor's degree in mathematics. Their preparation for teaching includes some general courses in education, a didactic course, and some practical work in schools. The teacher preparation for prospective junior high school mathematics teachers is done in teachers' colleges, where the students also study some mathematics, the level of which is much less than that required for a bachelor's degree. The main emphasis is on pedagogy. The mathematical aspects of the preparation of prospective elementary school teachers is included in the overall preparation they get in teachers' colleges.

## THE USE OF TECHNOLOGIES

As in many technological societies, there also are attempts to use modern technologies in the teaching of mathematics. At this stage, most of the these attempts are made within experimental frameworks. A variety of computer software for graphing and some computer software for learning Euclidean geometry in a dynamic way has been tested in several schools. There also are projects in which large parts of the curriculum are developed by means of the graphing calculator.

## MATHEMATICAL EDUCATION RESEARCH

The mathematical education research community in Israel is one of the biggest (per capita) in the world. Most of the work deals with cognitive aspects of learning mathematics, such as mathematical intuitions, misconceptions in mathematical thinking at all levels, and problem solving. Some of the studies are associated with development projects. The development projects are evaluated, including attempts to understand how students cope with the new learning materials and with the new technologies. The range of ages that are studied spreads from kindergarten to college. The research results are published in international research journals, mainly in English. Israel also participates in international comparative studies; results fall around the mean of the developed countries.

*See also* International Studies of Mathematics Education; Third International Mathematics and Science Study (TIMSS)

### SELECTED REFERENCE

Robitaille, David F., and Kenneth J. Travers. "International Studies of Achievement in Mathematics." In *Handbook of Research on Mathematics Technology and Learning* (pp. 687–709). Douglas A. Grouws (ed.). New York: Macmillan, 1992.

SHLOMO VINNER

# J

## JAMES, WILLIAM (1842–1910)

Philosopher and psychologist, and one of the originators of pragmatism. He was among the earliest to challenge beliefs held by faculty psychology theorists that components of mental functioning can be strengthened by exercise. James was suspicious of absolutes, orthodoxy, and dogmatism. His 1890 *Principles of Psychology* included a section on mathematics that challenged the then-current view that number and geometrical ideas are obtained by abstracting properties from physical reality. He saw physical reality, in which material continually divides and fuses, as too chaotic to form a basis for the unchanging numbers of abstract mathematics. As later paraphrased, James felt that "*number* is not (psychologically) got *from* things, it is put *into* them" (McLellan and Dewey 1895, 71). James found the basis of knowledge in "*constructiveness*" by which a child "accumulates a store of physical conceptions." He stressed the motivational importance of the child's interest in "sensible properties of material things" during the first seven or eight years (James 1904, 146).

*See also* History of Mathematics Education in the United States, Overview; Psychology of Learning and Instruction, Overview

### SELECTED REFERENCES

Bidwell, James K., and Robert G. Clason, eds. *Readings in the History of Mathematics Education*. Washington, DC: National Council of Teachers of Mathematics, 1970.

James, William. *The Principles of Psychology*. 2 vols. New York: Holt, 1890.

———. *Talks to Teachers on Psychology: And to Students on Some of Life's Ideals*. New York: Holt, 1904.

Jones, Philip S., ed. *A History of Mathematics Education in the United States and Canada*. 32nd Yearbook. Washington, DC: National Council of Teachers of Mathematics, 1970.

McLellan, James A., and John Dewey. *The Psychology of Number and Its Application to Methods of Teaching Arithmetic*. New York: Appleton, 1895.

JAMES K. BIDWELL
ROBERT G. CLASON

## JAPAN

Prior to the sixteenth century, formal mathematics was not a significant part of Japanese culture. Japan was an agricultural country, and cities and commerce were not yet developed. *Sangi* (calculating chips or pieces) were used for limited computation. The *Soroban* (Japanese abacus), adapted from the *swan-pan* (Chinese abacus), replaced *Sangi* in the sixteenth and seventeenth centuries. Its use was taught during the *Edo* period (1603–1868) in the *Terakoya* children's schools established in Buddhist temples for children of commoners. These schools provided basic training in reading, writing, and arithmetic and emphasized the use of the *Soroban* and calligraphy. The *Soroban* was very important for pupils in the commercial class, and the content of textbooks relevant to the *Soroban* dealt with basic arithmetic and financial problems.

*Wasan* was the Japanese mathematical system used to solve problems with algebraical symbols instead of calculating with *Sangi* and the *Soroban* (Kawajiri 1982). *Wasan* was fully developed during the *Edo* period following the end of the civil wars in 1603, when the economy expanded and cities were established. *Tokugawa Bakufu* (the central government) or *Han* (local government) hired *Wasan* scholars for the purpose of financial accounting, castle construction, and water irrigation. *Wasan* usually was taught in private, exclusive academies called *Shijuku*, organized by a distinguished scholar for instructing able disciples, both *Samurai* and commoners. One of the greatest *Wasan* scholars was Seki Kowa. He is said to have originated systems of linear equations ten years before Leibniz (Eves 1990). *Wasan* gradually was changed to focus upon aesthetic recreation rather than real-life applications. Mathematical problems were hung in front of temples and provided *Wasan* scholars with a form of intellectual recreation. *Wasan* scholars of this period preferred the study of intriguing geometric relations to practical scientific problems.

European mathematics gradually was introduced to Japan by the Chinese and Dutch, in spite of the "closed-door" policy of the *Edo* period. European arithmetic, trigonometry, and logarithms, however, were all regarded as being at a lower level than the mathematical amusements of the *Wasan* scholars. By the end of the *Edo* period, various types of private academies had developed, specializing in subjects such as medicine, *Kokugaku* (Japanese studies), and *Rangaku* (Dutch studies). During this period, Japan had increasing contact with other nations and felt foreign pressure to open its doors to the outside world.

## THE PERIOD FROM 1868 TO 1945

Changes in the Japanese political structure and in social organization occurred after the Meiji Restoration of 1868. *Monbusho* (the Ministry of Education, Science, and Culture) was established in 1871, and the fundamental *Code of Education* was decreed in 1872—the first comprehensive plan for the national education system. By the early 1900s, the Meiji government had established the system that remained the basis of Japanese education until 1945. School organization closely resembled the American system, beginning with a single track to provide for elementary, secondary, and higher education. In 1881, the system had a 6–4–2–4 sequence (elementary, lower middle, upper middle, and college) but changed to a 4–2–5–3–3 (ordinary elementary, higher elementary, middle, higher school, and college) in 1900. In order for Japan to meet the urgent needs for modernizing military, scientific, and technological activities, the Meiji government abandoned the *Wasan* tradition and adapted Western mathematics.

During this period, American and European textbooks on arithmetic, algebra, geometry, and trigonometry were translated into Japanese and used in classes. American textbooks written by Charles Davies and Horatio N. Robinson, British textbooks by I. Todhunter, and French textbooks by P. M. Bourdon and M. F. Delille were used in the elementary schools and middle schools. Later, textbooks by Japanese authors such as Rikitaro Fujizawa and Dairoku Kikuchi replaced these editions. The *Monbusho* controlled the writing and the publishing of textbooks. Methodology encouraged in the textbooks began with "an example of a calculation, followed by an exercise using the same type of calculation and then an application of the calculation" (Miwa 1992, 180).

*Sanjutsu* (arithmetic) was one of the core subjects from the first year in the elementary school. The instructional objectives were to train children in simple calculations, with stories from daily life, in order to help them to develop careful thinking skills. *Sugaku* (mathematics), consisting of arithmetic, algebra, geometry, and trigonometry, was a required subject in middle school. The instructional purposes were to give the students knowledge of numbers and quantities, to develop their arithmetic computation skills, to help them apply mathematics to the problems of everyday life, and to train their clear and logical thinking skills (Miwa 1992).

## THE PERIOD FROM 1945 TO 1960

After World War II, Japanese education was in disarray. In order to democratize Japan, a complete reform of the education system was undertaken. Some of the resulting changes included the institution of a 6–3–3–4 sequence (elementary, junior high, high school, and college); the introduction of university-based teacher education; and support for equal access to higher education. The first *Course of*

*Study,* composed with the help of the U.S. Occupation Army and reflecting the occupation policy, was introduced in 1947 under the urgent need for educational reform after the war. It was revised extensively in 1951.

Progressive education, the predominant influence in the United States in the 1930s and 1940s, was introduced directly to Japan. The new curricula derived from progressive education were characterized as experience-based and student-centered. They encouraged applying problem-solving methods to student experience and community activities. Pedagogical criticisms of progressive education were raised by scholars, educators, and teachers. For example, students studied addition or subtraction in a particular practical setting. As a result, they often did not have time to study the concepts of addition or subtraction intensively. Studies showed that student performance in arithmetic was poorer than in prewar days (Tooyama 1972). These negative reactions toward progressive education were part of a larger social and political transition that affected the course of Japan's history.

In 1958, *Monbusho* abandoned progressive curricula and completely revised the *Course of Study.* More systematic and thorough instruction in mathematics was required. Researchers developed various methods of teaching mathematics during that time; in particular, "Suido Houshiki," developed by Hiraku Tooyama, influenced mathematics education. "Suido Houshiki" is similar to a "top-down approach" to solving problems. A standard prototype is at the apex, and some related problems on the next levels. Thus, if a student has difficulty, the teacher can determine the nature of the problem by backtracking to the prototype example (Tooyama 1972).

## THE EARLY 1960s TO PRESENT

By the 1960s, postwar recovery and accelerating economic growth placed increased demands upon the Japanese education system. The rapid growth in elementary school enrollment increased the competition for entrance to both high school and universities. "New Mathematics" was introduced in Japan via revisions in the *Course of Study* in 1968, 1969, and 1970. New concepts, such as sets, structure, transformations, mappings, topology, and matrices became part of the mathematics curricula, along with novel terminology and a renewed focus on abstract concepts (Miwa 1992).

The 1964 First International Study of Educational Achievement indicated that Japanese students in the thirteen-year-old age group ranked highest in mathematics among those of twelve countries, including the United States, Australia, and several European countries. In addition, Japanese students were most positive in their attitudes toward mathematics. Teaching standards in the elementary grades are considered principally responsible for this success (Anderson 1975).

Today, elementary and junior high school education remains compulsory and free, with enrollment at nearly 100%. Teaching methods are left to the classroom teacher; however, most teachers utilize similar teaching strategies. "Teachers require children to purchase a 'math set'—a box of colorful, well-designed materials used for illustrating and teaching basic mathematical concepts" (Stevenson and Stigler 1992, 161).

Arithmetic in Japanese elementary schools consists of numbers and calculation, measurement, geometrical figures, and mathematical relations. Children in second grade learn the *kuku* mnemonic, which is a rhythmic song about the multiplication tables. After *kuku,* they study the *Soroban* as part of the third- and fourth-grade mathematics curriculum. They are taught the basics of addition, subtraction, multiplication, and division using this form of the abacus. To encourage "number sense," children are taught that experience with the *Soroban* operators represents intermediate results on their "mental abacus," that is, the students visualize the *Soroban* in their minds.

*Origami* or paper folding is used in teaching geometric concepts and vocabulary, serving as a motivational device. Using paper folding, several important geometric processes can be completed more easily than with compasses and rulers. Students learn concepts and skills involving polygons, angles, measurement, symmetry, and congruence.

Creative problem solving and mathematical processes are emphasized in both elementary and junior high schools (White 1987). Textbooks now are developed on the assumption that knowledge should be cumulative from semester to semester, in contrast to the "spiral" organization of Western texts. Japanese educators believe that if a concept or skill is taught well the first time, it is unnecessary to repeat it at a later grade. "Textbooks that contain short lessons, a limited number of practice problems, and practically no ancillary material make it possible for the class to cover every detail contained in every textbook" (Stevenson and Stigler 1992, 141).

After junior high school, students take entrance examinations for admission to high school. The influence of high school and the college entrance examination on mathematics is overwhelming. Mathematics is a central subject in the examination, and Japanese parents believe that higher education provides the way to a prosperous future. As a result, an entire alternative-school industry has been established to promote test achievement.

*Juku* (e.g., Kumon) are private after-school classes or "cram" schools that have an important role in mathematics education. The term *juku* referred originally to a study room at home. Eventually, however, the term denoted "a small private, practice-oriented, individualized school" distinguished by the teacher's personal style (Stevenson, Azuma, and Hakuta 1986, 83). *Juku* provide opportunities for students to obtain tutoring, remediation, and enrichment, as well as preparation for entrance examinations (Becker et al. 1990). Many Japanese students attend two or three evenings a week and on Saturday afternoons. Whenever school is not in session, full-time attendance at the *juku* is the norm (Lynn 1988).

## THE MATHEMATICS CURRICULUM

At present, the official mathematics curriculum in Japan is divided into three levels—the elementary school level, the junior high school level, and the high school level.

### Elementary School (Grades 1 through 6; Ages 7 to 12)

A new *Course of Study* began in 1992 (Miwa 1992). Mathematics is taught for four forty-five-minute periods per week in grade 1 and for five forty-five minute periods per week in grades 2 through 6. The objectives are to convey basic skills and knowledge of numbers, quantities, and geometric figures; to develop abilities and motivation to apply mathematics in daily life; and to train students in logical thinking and its applications. The content of arithmetic in elementary school is numbers and calculations, quantities and measurement, geometrical figures, and mathematical relations.

### Junior High School (Grades 7 through 9; Ages 13 to 15)

A new *Course of Study* (Monbusho, *Chigakuko* . . ., 1989) was put into effect in 1993. Mathematics is taught for three fifty-minute periods per week in grade 7 and for four fifty-minute periods in grades

8 and 9. The aims of teaching mathematics are to deepen students' understanding of basic concepts, principles and laws of numbers, quantities, and geometrical figures; to learn mathematical expressions and how to use them; to develop the ability to solve problems; and to develop mathematical, logical, and precise thinking. The content of mathematics in junior high school is limited to numbers and algebraic expressions, functions, geometrical figures, and probability and statistics.

### High School (Grades 10 through 12; Ages 16 to 18)

A new *Course of Study* (Monbusho, *Kotogako* . . ., 1989) was introduced in 1994. Mathematics courses account for sixteen credits in the curriculum. One credit is earned for attending thirty-five periods (fifty minutes each) during an entire school year. The areas of study are divided as follows: *Mathematics I*, four credits; *Mathematics II*, three credits; *Mathematics III*, three credits; *Mathematics A*, two credits; *Mathematics B*, two credits; *Mathematics C*, two credits. *Mathematics I* is required of all students in the first year; all later courses are optional. *Mathematics A, B,* and *C* include computer activities. *Mathematics A* introduces numerical computations using the computer. *Mathematics B* emphasizes basic knowledge about the function of the computer, programming, and algorithms. The main focus of *Mathematics C* is applied mathematics using computers. The purpose of teaching mathematics at the high school level is to deepen students' understanding of basic concepts, principles, and laws of mathematics and to develop their logical thinking skills. Specific topics include:

*Mathematics I:* quadratic functions, geometric figures and measure, treatment of numbers of cases, and probability and basic theorems;

*Mathematics II:* polynomial functions, plane figures, and equations in a plane;

*Mathematics III:* functions and limits, derivatives of functions, and integration;

*Mathematics A:* numbers and algebraic expressions, plane geometry, sequences, and calculation and computing;

*Mathematics B:* vectors, complex numbers, probability, binomial and normal distributions, and computing and calculation;

*Mathematics C:* matrix and linear algebra, curves, numerical computation, and statistics.

## THE FUTURE

Although Japan is a highly developed technological society, the use of handheld calculators traditionally has not been allowed in classes. The former national *Course of Study* did not recommend calculator use despite its reliance upon the *Soroban* as a teaching aid in elementary school. In addition, the attitudes of parents and elementary school teachers did not favor the use of calculators. The situation was similar with microcomputers. Although the use of computers was widespread in experimental schools, a large number of teachers and educators traditionally made no use of them (Becker et al. 1990).

In 1985, in response to the widespread adoption of microcomputer technology and its potential impact on schooling, *Monbusho* coordinated a national effort to develop computer skills. The new *Course of Study* (Monbusho, *Chugakuko* . . . , 1989; *Kotogako* . . . , 1989) proposes two changes in policy: teachers should make use of educational media, such as computers, to improve the effectiveness of instruction, and should make use of handheld calculators and computers, appropriate to the occasion, to improve the effectiveness of learning. Thus, the teacher is encouraged to use the computer as an educational "delivery system," while the student is urged to use calculators and computers as mathematical tools.

*See also* Abacus; International Comparisons, Overview; International Studies of Mathematics Education; Third International Mathematics and Science Study (TIMSS)

### SELECTED REFERENCES

Anderson, Ronald S. *Education in Japan: A Century of Modern Development.* Washington, DC: Government Printing Office, 1975.

Becker, Jerry P., Edward A. Silver, Mary G. Kantowski, Kenneth J. Travers, and James W. Wilson. "Some Observation of Mathematics Teaching in Japanese Elementary and Junior High School." *Arithmetic Teacher* 38(Oct. 1990):12–21.

Eves, Howard W. *An Introduction to the History of Mathematics.* 6th ed. Philadelphia, PA: Saunders College Publishing, 1990.

Japan Society of Mathematical Education (JSME). *Mathematics Education in Japan.* English summary parts of JSME Yearbooks 1–4. Tokyo, Japan: The Society, 2000.

Kawajiri, Nobuo. *Bakumatsu ni okeru yoroppa gaukujutsu juyo no ichidanmen: Uchida Itsumi to Takono Choei, Sakuma Shozan.* Tokyo, Japan: Tokai daigaku shunpan kai, 1982.

Lynn, Richard. *Educational Achievement in Japan: Lessons for the West.* Armonk, NY: Sharpe, 1988.

Miwa, Tatsuro. "Mathematics education in Japan." In *Studies in Mathematics Education* (pp. 179–190) Vol 8.

Robert Morris and Manmohan S. Arora (eds.). Paris, France: UNESCO, 1992.

Monbusho. *Chugakuko shidosho sugakuhen.* Osaka, Japan: Osaka Shoseki, 1989.

———. *Kotogako gakushu shido yoryo kaisetsu sugakuhen, reshuhen.* Tokyo, Japan: Gyosei, 1989.

Stevenson, Harold W., Hiroshi Azuma, and Kenji Hakuta. *Child Development and Education in Japan.* New York: Freeman, 1986.

Stevenson, Harold W., and James W. Stigler. *The Learning Gap: Why Our Schools Are Failing and What We Can Learn from Japanese and Chinese Education.* New York: Summit, 1992.

Tooyama, Hiraku. *Sugaku no manabikata, oshiekata.* Tokyo, Japan: Iwanami Bunko, 1972.

White, Merry. *The Japanese Educational Challenge: A Commitment to Children.* New York: Free Press, 1987.

MINORU ITO
BRUCE R. VOGELI

## JOINT COMMISSION REPORT (1940)

*The Place of Mathematics in Secondary Education: The Final Report of the Joint Commission of the Mathematical Association of America and The National Council of Teachers of Mathematics* (NCTM) (Reeve 1940). The report recognized the diversity of students in high schools that resulted from the increased percentage of high school-aged students attending school. It proposed tracked sequences of mathematics classes designed to accommodate students with varied needs for and interest in mathematical training. While the report gave consideration to the trends of the day—incidental learning, basing learning on students' felt needs, skills needed in everyday life, and knowledge necessary for living in a democracy—it questioned whether a mathematics curriculum should be built entirely around these trends. For a wider perspective, the report can be compared to the 1940 Progressive Education Association's *Mathematics in General Education,* which took a more radical view in accepting the then-current trends.

*See also* History of Mathematics Education in the United States, Overview; Progressive Education Association (PEA) Report

### SELECTED REFERENCES

Bidwell, James K., and Robert G. Clason, eds. *Readings in the History of Mathematics Education.* Washington, DC: National Council of Teachers of Mathematics, 1970.

Jones, Phillip S., ed. *A History of Mathematics Education in the United States and Canada.* 32nd Yearbook. Washington, DC: National Council of Teachers of Mathematics, 1970.

Reeve, William D., ed. *The Place of Mathematics in Secondary Education: The Final Report of the Joint Commission of the Mathematical Association of America and The National Council of Teachers of Mathematics.* 15th Yearbook. New York: Bureau of Publications, Teachers College, Columbia University, 1940.

JAMES K. BIDWELL
ROBERT G. CLASON

# JOINT POLICY BOARD FOR MATHEMATICS (JPBM)

A consortium of the American Mathematical Society, the Mathematical Association of America, and the Society for Industrial and Applied Mathematics. Formed in 1983 by those three societies, JPBM consists of a ten-member board appointed by the societies, providing a forum for the leadership of the societies to discuss issues in the mathematical sciences and take joint action when warranted. JPBM articulates policy directions and advocates courses of action concerning the mathematical sciences; represents mathematics to the White House, Congress, and the federal agencies; and increases public understanding of the role and importance of the mathematical sciences through national events and media relations. JPBM maintains a Washington office, funded by the three societies, to implement comprehensive programs in the areas of public information and government affairs.

JPBM'S principal outreach activity is Mathematics Awareness Week, held in late April each year to increase public understanding of and appreciation for mathematics. Each year a national theme is selected and theme materials are developed and distributed, using in recent years electronic vehicles. Activities for Mathematics Awareness Week are generally organized by college and university departments, institutional public information offices, student groups, and related associations and interest groups. They have included a wide variety of workshops, competitions, exhibits, festivals, lectures, and symposia.

*See also* American Mathematical Society (AMS); Mathematical Association of America (MAA)

SELECTED REFERENCES
JPBM, 1529 18th Street NW, Washington, DC, 20036. 202-234-9570, 202-462-7877 fax, jpbm@math.umd.edu. Information on Mathematics Awareness Week can be obtained from mharris@deans.umd.edu

and from the World Wide Web at http://forum.swarthmore.edu/maw/.

RICHARD HERMAN

# JULIA, GASTON (1893–1978)

A significant contributor to number theory, analysis, and the theory of Hilbert spaces. It is, however, his work in complex dynamics, which addresses the iteration of complex functions, for which he is most famous.

The iterates of a point $z$ in the complex plane under a function $f$ are the points $z$, $f(z)$, $(f(z))$, $(f(f(z)))$, etc. A simple example of iteration occurs with Newton's method for solving an equation of the form $f(z) = 0$. Let $N$ be the Newton's method function for $f$. Pick a point $z$ near a suspected solution of $f(z) = 0$, evaluate $N(z)$, plug the result back into $N$, and continue the process ad infinitum in hopes that the iterates of $z$ under $N$ converge to the solution.

Along with Pierre Fatou (1878–1929), who independently achieved many of the same results, Julia is considered the cofounder of the contemporary theory of complex dynamics. Both men made fundamental discoveries about what has come to be known as the Fatou and Julia sets and the theory of chaos. The Julia set of a complex function $f$ is the set of points whose iterates behave chaotically, while the Fatou set consists of those points whose iterates do not behave chaotically. For his achievements in complex dynamics, Julia won France's most prestigious mathematics prize in 1918.

Julia also was awarded the French Legion of Honor for bravery during World War I. He suffered a terrible wound to his face in the midst of a furious German attack, which left him permanently disfigured and temporarily mute. Yet he refused to be taken from the battlefield until after the Germans were repulsed.

*See also* Chaos; Fatou

SELECTED REFERENCES
Alexander, Daniel. "Civilized Mathematics." *Focus* 14(3)(June 1994):10–11.
———. *A History of Complex Analysis from Schröder to Fatou and Julia.* Wiesbaden, Germany: Vieweg, 1994.
Devaney, Robert. *Chaos, Fractals, and Dynamics: Computer Experiments in Mathematics.* Menlo Park, CA: Addison-Wesley, 1990.
Hervé, Michel. "Gaston Julia." In *Dictionary of Scientific Biography, Supplement.* New York: Scribner's, preprint.

DANIEL S. ALEXANDER

# K

## KEPLER, JOHANNES (1571–1630)

German astronomer and mathematician. He studied at the University of Tübingen, became a mathematics teacher in Graz (1594–1600), an imperial mathematician in Prague (1601–1612) and in Linz (1612–1626), and, finally, the collaborator of Albrecht Wenzel Eusebius von Wallenstein in Sagan (now Poland). His fifteen main works belong to the scientific world literature, especially his *Cosmographical Mystery* (1594), his *New Astronomy* (1609), which contains the first two Keplerian laws, his *Epitome of the Copernican Astronomy* (1617–1620), his most mature and most comprehensive publication, and his *Harmony of the World* (1619), which contains the third Keplerian law. These three laws read as follows: (1) the planetary orbits are ellipses—one of their focuses is occupied by the sun; (2) the radius vector from the sun to the planet covers in equal periods equal areas; and (3) the squares of the periods of revolution of two planets are to each other as the third powers of the great semiaxes of their orbits.

Kepler was authorized to use Tycho de Brahe's observational data in order to calculate the orbit of Mars. He succeeded in discovering its elliptic shape and the so-called law of area by applying infinitesimal methods. In such a way he became a forerunner of infinitesimal geometry. For example, a sphere is supposed to consist of infinitely many quadratic pyramids whose apices reach to the center of the sphere, while the infinitely small quadratic basal surfaces cover the surface of the sphere. In his *Harmony of the World,* he elaborated a geometrical foundation of consonance theory, being convinced that God's creation is full of harmonies. Consonant intervals are defined by means of sections of a circle. Such a section leads to a ratio of a consonant interval if and only if the parts of the circle form with the whole circle (for example, 2:5, 3:5), as well as with each other (2:3) ratios corresponding to regular polygons, which are constructable by ruler and compasses. Thus, there are exactly seven ratios (1/2, 2/3, 3/4, 4/5, 5/6, 3/5, 5/8) that define a consonant interval. The five regular polyhedra played a crucial role in his cosmology. As to physics, he introduced the concept of force into celestial physics and contributed considerably to the scientific development of optics. Kepler is of interest both to teachers and to students of mathematics, because he paved the way to the infinitesimal calculus, studied intensively the five regular polyhedra and tesselations as well, and demonstrated how mathematics is connected with music theory.

*See also* History of Mathematics, Overview; Music; Polygons

383

SELECTED REFERENCES

Caspar, Max, and Martha List, eds. *Bibliographia Kepleriana.* Munich, Germany: Beck, 1968.

Field, Judith Veronica. *Kepler's Geometrical Cosmology.* London, England: Athlone, 1988.

Hofmann, Joseph Ehrenfried. "Johannes Kepler als Mathematiker." *Praxis der Mathematik* 13(1971):287–293, 318–324.

———. "Uber einige fachliche Beiträge Keplers zur Mathematik." In *Internationales Kepler-Symposium, Weil der Stadt 1971, Referate und Diskussionen* (pp. 261–284) Fritz Krafft, Karl Meyer, and Bernhard Sticker (eds.). Hildesheim, Germany: Gerstenberg, 1972.

Kepler, Johannes. *Gesammelte Werke.* 20 vols. Max Caspar, Franz Hammer, and Volker Bialas (eds.). Munich, Germany: Beck, 1938–1988.

Stephenson, Bruce. *The Music of the Heavens. Kepler's Harmonic Astronomy.* Princeton, NJ: Princeton University Press, 1994.

EBERHARD KNOBLOCH

# KINDERGARTEN

Good mathematics programs for children at the kindergarten level can be developed using the recommendations of two professional organizations—the National Council of Teachers of Mathematics (NCTM) and the National Association for the Education of Young Children (NAEYC).

According to NAEYC, a curriculum for children ages 3–8 should have a theoretical basis consistent with current knowledge about how children learn. Its long-range goal, to prepare children to function productively in a democratic society, should be attained through meaningful learning in a variety of engaging situations. NAEYC guidelines describe a curriculum that provides a conceptual framework based on prior knowledge that respects the individuality of children. Such a curriculum would engage children in learning tasks and emphasize children's reasoning, decision-making abilities, and problem-solving abilities (Bredekamp and Rosegrant 1992). Among the characteristics listed by NAEYC as signifying high-quality early childhood programs are those in which developmentally appropriate programs are delivered by adults who have received specialized preparation. Acceptable curricula would include experiences that stimulate learning in all areas, supply appropriate activities for a wide range of developmental levels, and focus on learning as an interactive process between adults and children.

According to the NCTM, the kindergarten mathematics curriculum should encourage exploration and communication about mathematical discoveries. In such a curriculum, instruction and content are interwoven; adults actively but unobtrusively mediate children's discussions about relevant and important mathematics (Ramey et al. 1988). At all grade levels, four NCTM standards form an essential organizing core—(1) mathematics as problem solving, (2) mathematics as communication, (3) mathematics as reasoning, and (4) mathematical connections (NCTM 1989, 1991). The recommendations of both organizations flow from a constructivist framework—that is, a belief that children must build up meanings for themselves, as opposed to passively receiving them from others. The importance of encouraging children to think, reason, and experiment (Hayes, Palmer, and Zaslow 1990; NCTM 1989, 1991) is emphasized. Both professional groups recommend providing an environment in which children discuss and record mathematical ideas and in which caring and knowledgeable teachers respond to the individual strengths and needs of each child. In such programs, child–child interaction also is valued.

Both organizations assert that teaching practices are inextricably woven into the curriculum development process. The teacher's role is to choose significant learning tasks and to guide, support, encourage, pose questions to, and monitor the children at work. Children are expected to listen and respond to their peers and the teacher, and to make and to articulate judgments based on their experiences. Space, time, and materials are to be provided that facilitate children's learning and the development of social as well as intellectual skills. Both sets of guidelines also describe evaluation as an ongoing process that assures that each child progresses toward attainable goals. Errors are seen as a natural and inevitable part of the learning process and a valuable indicator of which learning experiences are the most appropriate next steps for this individual. Consistent with a constructivist philosophy, each set of guidelines deemphasizes standardized tests and underscores the validity of teacher judgment. The main questions of evaluation become "What does this child understand, and how does he or she feel about that?" rather than merely "What can this child (or class) do?"

The flavor of such a curriculum can be illustrated with specific examples from each of the K–4 Curriculum Standards (NCTM 1989). Many of the activities described below are further developed in *Kindergarten Book* (Burton et al. 1991): (1) children decide together how many pattern blocks each table should have if each child at the table needs five pattern blocks for a mathematics activity (Mathematics as Problem Solving); (2) children compare the number of connecting cubes (such as unifix cubes) each is able to grab with the right hand (Mathematics as Communication); (3) children make a train with attribute blocks such that each "car" is different from the preceding "car" in only one way (Mathematics as Reasoning); (4) children who are gathering leaves for a science display classify them into groups and then arrange the set of leaves by size (Mathematical Connections); (5) children guess how many paper clips are in a small box. The teacher records their guesses on a large chart, then has a volunteer count the collection. The children discuss which estimates came "close" (Estimation); (6) children sort a double-six set of dominoes according to whether the number of dots is equal to, less than, or greater than 7 (Number Sense and Numeration); (7) children take part in a "counting down" skit such as Five Little Pumpkins (Concepts of Whole Number Operation); (8) children model number stories involving addition and subtraction with pasta teddy bears (Whole Number Computation); (9) children examine the imprint of the faces of three-dimensional blocks that the teacher has stamped into a sand tray, then try to determine which geoblock made the "footprints" (Geometry and Spatial Sense); (10) children measure in steps the perimeter of large geometric figures traced on the floor with masking tape and record the measurements in their math journals (Measurement); (11) the class makes a bar graph using actual library books classified by size, topic, or some other criteria of interest to the children (Statistics and Probability); (12) children divide a set of counters so that each child gets a fair share (Fractions and Decimals); (13) children create a pattern using stuffed animals brought from home. They describe the pattern using directional and ordinal words and record the pattern on paper for a door display (Patterns and Relationships).

In the contemporary kindergarten curriculum, mathematics is relevant, engaging, and meaningful to children and involves active exploration of mathematical questions. It may appear as play, but it is play with a purpose.

*See also* Preschool; Standards, Curriculum

SELECTED REFERENCES

Bredekamp, Sue, and Teresa Rosegrant. *Reaching Potential: Appropriate Curriculum and Assessment for Young Children.* Washington, DC: National Association for the Education of Young Children, 1992.

Burton, Grace, Terrence Coburn, John Del Grande, Mary M. Lindquist, and Lorna Morrow. *Kindergarten Book.* Reston, VA: National Council of Teachers of Mathematics, 1991.

Cataldo, Christine Z. *Infant and Toddler Programs: A Guide to Very Early Childhood Education.* Reading, MA: Addison-Wesley, 1983.

Hayes, Cheryl D., John L. Palmer, and Martha J. Zaslow, eds. *Who Cares for America's Children? Child Care Policy for the 1990's.* Washington, DC: National Academy Press, 1990.

National Council of Teachers of Mathematics. *Curriculum and Evaluation Standards for School Mathematics.* Reston, VA: The Council, 1989.

———. *Principles and Standards for School Mathematics.* Reston, VA: The Council, 2000.

———. *Professional Standards for Teaching Mathematics.* Reston, VA: The Council, 1991.

Ramey, Craig T., Donna M. Bryant, Frances A. Campbell, Joseph J. Sparling, and Barbara H. Wasik. "Early Intervention for High Risk Children: The Carolina Early Intervention Program." In *Fourteen Ounces of Prevention: A Casebook for Practitioners* (pp. 32–43). Richard H. Price, Emory L. Cowen, Raymond P. Lorion, and Julia Ramos-McKay (eds.). Washington, DC: American Psychological Association, 1988.

GRACE M. BURTON

# KLEIN, FELIX (1849–1925)

Outstanding German mathematician who helped found the Mathematical Institute at the University of Göttingen, which became the world center for research until the rise of Hitler in 1933. In 1872 he presented his *Erlanger Programm,* which viewed the existing geometries in terms of their invariant properties under group transformations (e.g., rotations, translations, reflections). This reorganization of the field placed projective geometry at the base and Euclidean geometry as one branch. Klein's summer lectures to German mathematics teachers in the early 1900s were published by 1908. They were translated into English and printed in the United States in 1932 and 1939 as *Elementary Mathematics from an Advanced Standpoint.* These lectures were an early attempt to present school mathematics from a modern mathematical viewpoint. The geometry portion of the lectures was based on transformations.

*See also* Projective Geometry; Transformational Geometry

SELECTED REFERENCES

Jones, Phillip, ed. *A History of Mathematics Education in the United States and Canada.* 32nd Yearbook. Washington, DC: National Council of Teachers of Mathematics, 1969.

Klein, Felix. *Elementary Mathematics from an Advanced Standpoint*. New York: Dover, 1945.

JAMES K. BIDWELL
ROBERT G. CLASON

# KOVALEVSKAIA, SOFIA (1850–1891)

Recipient of the first doctorate in mathematics awarded to a female in modern times. Motivated to learn mathematics in childhood by her attempts to understand calculus lecture notes used as nursery wallpaper, Kovalevskaia was denied admission to universities in Russia because of her sex. She arranged a marriage of convenience in order to be able to travel abroad. She studied for a time in Heidelberg and then worked with Karl Weierstrass in Berlin. Her doctorate in mathematics was awarded in absentia by the university in Göttingen in 1874.

Unable to secure a position in mathematics at a university in Russia, Kovalevskaia spent a turbulent period involved in political activism and disastrous financial speculations. During this time, her marriage was consummated and a daughter was born. After a hiatus of seven years, Kovalevskaia returned to mathematical research. At the University of Stockholm, Gustav Mittag-Leffler was instrumental in securing for her the first university professorship for a woman in Europe in modern times. Kovalevskaia died tragically from pneumonia at the height of her creativity.

One of the three papers submitted for her doctoral dissertation contains a result basic to the theory of partial differential equations, now known as the Cauchy-Kovalevskaia Theorem. In 1888 Kovalevskaia was awarded the Prix Bordin of the French Academy for her paper on the rotation of a solid body about a fixed point. Other areas of her research included work on the refraction of light in a crystalline medium, the rings of Saturn, and the reduction of Abelian integrals to elliptic integrals. Elected a corresponding member of the Russian Imperial Academy of Sciences, she served as a liaison between the analytic school of Weierstrass and the applied work of the Russian mathematicians of the time.

Kovalevskaia also wrote a number of short stories and plays as well as a memoir of her childhood. She was a strong proponent of the full participation of women in all aspects of society, particularly education.

*See also* Women and Mathematics, History

## SELECTED REFERENCES

Cooke, Roger. *The Mathematics of Sonya Kovalevskaya*. New York: Springer-Verlag, 1984.

Koblitz, Ann Hibner. *A Convergence of Lives, Sofia Kovalevskaia: Scientist, Writer, Revolutionary*. Boston: Birkhäuser, 1983.

Kovalevskaia, Sofya. *A Russian Childhood*. Translated by Beatrice Stillman. New York: Springer-Verlag, 1978.

Rappaport, Karen D. "S. Kovalevsky: A Mathematical Lesson." *American Mathematical Monthly* 88(8)(1981): 564–574.

MARY W. GRAY

# L

## LADIES' DIARY (WOMAN'S ALMANACK) (1704–1841)

Eighteenth-century English magazine devoted largely to mathematical problems and puzzles, all contributed in verse. Males were the main contributors, although women were represented in significant numbers. The beginning of the *Ladies' Diary* coincides with the popularization of mathematics, part of the eighteenth-century fascination with the new science and technology of Newton's universe. The mathematical recreation magazines, of which the *Ladies' Diary* was one, were addressed to the nonmathematician, although they required a significant degree of mathematical sophistication of their readers, more so as time went on. As mathematical literacy spread in response to developing technology's requirements for more mathematically sophisticated workers, women, as homemakers, were left behind. This effect was reflected in the decline in the number of women contributors over the life of the publication.

*See also* Women and Mathematics, History

### SELECTED REFERENCES

Leybourn, Thomas. *The Mathematical Questions Proposed in the Ladies' Diary, 1704–1816*. London, England: Mawman, 1817.

Meyer, Gerald D. *The Scientific Lady in England 1650–1760*. Berkeley, CA: University of California Press, 1955.

Perl, Teri. "The Ladies' Diary . . . Circa 1700." *Mathematics Teacher* 70(1977):354–358.

———. "The Ladies' Diary or Woman's Almanack, 1704–1841." *Historia Mathematica* 6(1979):36–53.

TERI PERL

## LAGRANGE, JOSEPH LOUIS (1736–1813)

Born in Turin, Italy, with an early interest in literature rather than mathematics. However, a tract by Edmund Halley turned him toward mathematics. In 1754, he was appointed professor of geometry at the Royal School of Artillery in Turin. In 1766, he moved to Berlin where he was made director of the Mathematics Department of the Berlin Academy. In 1787, he took a professorship in Paris, and in 1795 became the first president of the École Polytechnique.

At the age of nineteen, he communicated to Euler the general idea of the calculus of variations. His initial triumph with the invention of the calculus of variations was followed by major achievements in analytical dynamics and solving the differential equations now known as Lagrange's equations of motion.

Students first meet Lagrange's work in calculus when they apply the Lagrange multiplier method to the problem of maximization or minimization subject to constraints. Besides these significant contributions to analysis, Lagrange made major discoveries in other branches of mathematics. He was the first to prove that every positive integer is a sum of four squares. He discovered the method of approximating algebraic numbers by continued fractions. For example,

$$\frac{1 + \sqrt{5}}{2} = 1 + \cfrac{1}{1 + \cfrac{1}{1 + \cfrac{1}{1 + \cdots}}}$$

The successive convergents of this continued fraction are

$$1/1, \; 2/1, \; 3/2, \; 5/3, \; 8/5, \; 13/5, \; 21/13, \ldots$$

These numbers arise by truncation of the infinite continued fraction, and they are (relative to the size of their respective denominators) the best fractional approximations of $(1 + \sqrt{5})/2$. Note that the sequence of numerators (also denominators) is the Fibonacci sequence.

In the theory of finite differences, Lagrange is remembered for Lagrange interpolation. Namely if $f(x)$ is a polynomial of degree $n$,

$$f(x) = \sum_{i=0}^{n} f(a_i) \prod_{\substack{j=0 \\ j \neq i}}^{n} \frac{(x - a_j)}{(a_i - a_j)}$$

This formula allows the construction of a polynomial $f(x)$ given only the $n + 1$ required values $f(a_0)$, $f(a_1), \ldots, f(a_n)$.

*See also* Continued Fractions; Fibonacci Sequence

### SELECTED REFERENCE
Bell, Eric Temple. *Men of Mathematics.* New York: Simon and Schuster, 1937.

GEORGE E. ANDREWS

# LANGUAGE AND MATHEMATICS IN THE CLASSROOM

The relationship between language and mathematics which has been analyzed by mathematics educators and researchers from different perspectives: the nature of the language–mathematics connection as expressed in terms of vocabulary, semantics, syntax, and symbols; the nature of communication in mathematics with respect to classroom discourse and the integration of the language arts; and the nature of sociocultural and linguistic influences based on analyses of mathematics achievement of nonnative speakers of English.

## THE LANGUAGE–MATHEMATICS CONNECTION

Vocabulary, semantics, syntax, and symbolism are elements of mathematical language. Although the vocabulary and symbolism of mathematics may be considered as aspects of semantics, these are discussed separately to highlight specific characteristics.

*Vocabulary* The language of mathematics is characterized by three types of vocabulary—general, technical, and special (Earle 1976, 17). *General vocabulary* words and symbols are from ordinary, everyday conversation—the words and symbols of natural language (e.g., "peace," "house," "$," "towel"). *Technical vocabulary* words and symbols are precise, concise, high in concept density, and specific to the study of mathematics (e.g., "subtrahend," "numerator," "vinculum," "$<$," "polygon"). The most difficult type of vocabulary is *special vocabulary,* because it includes words and symbols that have different meanings in natural language and in mathematics (e.g., "table," "base," "$\wedge$," "plane"). To add to this difficulty, some homonyms such as "sum–some," and "plane–plain" may cause confusion.

Encouraging students to invent their own terms is a frequently recommended technique that can be used to help students acquire mathematical vocabulary. By this technique, students build on their natural language to represent mathematical concepts in terms meaningful to them (Ellerton and Clements 1991; Fouch and Nichols 1959; Rubenstein 1996). Then the teacher helps the students to connect their inventions with the appropriate mathematical language. Students also may benefit from the study of prefixes, suffixes, and word roots (Milligan and Milligan 1983).

*Semantics* Mathematical vocabulary, terms, symbols, and notation do more than simply name or label—they house underlying concepts that are usually abstractions, not directly or easily represented as concrete objects. The meaning of the underlying, embedded concepts housed in words, terms, symbols, and notation make up "what" is to be communicated. This is the semantics of mathematical language, referred to as "understandings" by Hiebert (1984) and referred to as "deep structure" by Skemp

(1982). Opportunities to discover, experience, and discuss mathematical ideas informally build a foundation of understanding upon which formal and conventional expressions can be built. Instructional recommendations suggest that such experiences should precede the introduction of formal mathematical language (Curcio 1990; Skemp 1982).

*Syntax* The rules, procedures, and conventions of operating within the mathematical language-symbol system are analogous to the grammar of a language. They provide a framework for "how" to communicate within the system. This is the syntax of the mathematical language, referred to as "form" by Hiebert (1984) and referred to as "surface structure" by Skemp (1982). The surface structure transmits the meaning of the deep structure. Since the connection between the surface and deep structure is implicit, the learner's background knowledge is essential in understanding the meaning of the message. Difficulties students encounter in understanding mathematics are often related to their inability to interpret form with understanding, their lack of relevant background knowledge of the mathematics embedded in the vocabulary and the symbols, and their reliance on surface structure cues that may lead to misinterpretations.

Students who solve word problems using a "clue word" approach rely on surface structure cues that may impede or replace comprehension. For example, given "The sum of two numbers is 15. One of the numbers is 8. What is the other number?" Relying on the meaning of "sum" without analyzing its use in context, may lead students to add 15 and 8 and respond with "23." Another example from Clement's research (1982) is related to translating a mathematical relationship from words into symbols. Many college engineering students were found to use a word-order matching approach to translate "There are six times as many students ($S$) as there are professors ($P$)" into symbols. Students who attended only to surface structure and not to the meaning of the sentence translated the sentence as $6S = P$, instead of $6P = S$. Wollman (1983) found that providing college students with explicit instruction in "sentence meaning" increased the number of correct translations for similar examples.

*Symbols* According to Richard Skemp, "symbols are an interface between the inner world of our thoughts, and the outer, physical world" (1982, 28). As early as about 3000 B.C., the use of symbols for unknowns was introduced by the Babylonians (Kline 1972, 3, 10). Icons, ideographs, and pictorial images became part of the symbolic language of many cultures and were adapted for recording quantity, shape, and measures. It was not until the sixteenth century that scientific needs and improved methods of calculation contributed to a more formal development of symbolism. The scholars of the time did not, however, appreciate the power of the shorthand notation. The benefits of using symbols were not fully appreciated until the seventeenth century, when the use of symbolism was deliberate and the power and generality of its use became evident (Kline 1972, 262).

The advent of the "New Math" in the 1960s brought an increase in the classroom use of symbolism. For example, set and function notation, symbolic logic, vectors, matrices, and symbols for inverses found their way into school mathematics (Woodrow 1982, 298). As advances in modern technology facilitate mechanical procedures that involve manipulating symbols, the role of such devices as symbolic manipulators may need to be examined and the amount of time prescribed for mastering mechanical pencil-and-paper skills may change.

## COMMUNICATION IN MATHEMATICS

Educators often compare two particular formats of classroom discourse characterized by the terms *traditional*, or *inquiry-based*. In traditional mathematics classes, teachers rely predominantly on lecturing; students listen and take notes, copy notes from the board, or do examples. Among the many impressions that students receive from this format is that the only proper language is the language of the teacher or the textbook.

The "inquiry-based" classroom differs from the traditional classroom in that students have more opportunities to discuss mathematics with each other and the instructor and to express themselves orally, and in writing. Those who advocate the inquiry-based approach believe that it provides assurance that the students are actively involved constructing meanings and solving problems. The inquiry-based classroom seeks to incorporate methods of the language arts: clarifying, explaining, describing, conjecturing, and defending one's position. Advocates believe that encouraging students to defend their ideas, explain their reasoning, and describe their strategies by writing or speaking contributes to the development of their mathematical language (Countryman 1992; Pimm 1987). By listening to students' informal discussions as well as to a formal defense of their conjectures, teachers gain insight into students' level of understanding, according to Pirie (1996). Students who read each other's work experience the need to agree on the meaning of words, symbols, and the structure of visual displays. Development of critical reading skills also is essential for learning to read a mathematics textbook (Earle 1976; Shuard and Rothery 1984).

A language-experience approach in mathematics is characterized by building on or creating a common or shared experience related to the development of a particular mathematical concept. As children explain their interpretations and understandings orally and in writing, they listen to each other, share strategies, and record the big mathematical ideas developed during an activity. The teacher is able to assess the students' level of understanding from their oral or written explanations and the visual representations they create (Curcio 1990). Instructional decisions related to language and concept development are aided by such an assessment.

## SOCIOCULTURAL AND LINGUISTIC INFLUENCES

Research on the influence of sociocultural and linguistic factors on mathematics learning and achievement has provided some insight in understanding the successes and difficulties of students in different cultures. For example, the structure of the base-10 number system is more explicit in the Japanese language than it is in English, and, as a result, Japanese first graders understand place value concepts better than their American counterparts (Miura and Okamoto 1989). In her work with Black students in Washington, DC, Orr (1987) found that the differences between standard English and Black English vernacular interferes with Black students' performance in mathematics.

Learning mathematics in a nonnative language presents challenges to both the students and the teacher. Recommended instructional strategies build on developing and reviewing vocabulary and terms in context. Opportunities for reading, writing, and talking about mathematics contribute to developing communication skills and positive attitudes toward mathematics and improve understanding (Cuevas 1990).

## FINAL COMMENTS

Students develop mathematical language by using it in discussing, recording, reading, and writing about their observations, conjectures, strategies, and solutions. Teachers can help make the language–mathematics connection by comparing and contrasting natural and technical language, by creating a classroom climate that encourages student discourse, and by dealing with sociocultural and linguistic differences that may affect mathematics achievement.

*See also* Errors, Arithmetic; Language of Mathematics Textbooks; Reading Mathematics Textbooks

## SELECTED REFERENCES

Aiken, Lewis R., Jr. "Language Factors in Learning Mathematics." *Review of Educational Research* 42(1972): 359–385.

Clement, John. "Algebra Word Problem Solutions: Thought Processes Underlying a Common Misconception." *Journal for Research in Mathematics Education* 13(Jan. 1982):16–30.

Cocking, Rodney R., and Jose P. Mestre, eds. *Linguistic and Cultural Influences on Learning Mathematics*. Hillsdale, NJ: Erlbaum, 1988.

Countryman, Joan. *Writing to Learn Mathematics: Strategies That Work*. Portsmouth, NH: Heinemann, 1992.

Cuevas, Gilbert. "Increasing the Achievement and Participation of Language Minority Students in Mathematics Education." In *Teaching and Learning Mathematics in the 1990s* (pp. 159–165). 1990 Yearbook. Thomas J. Cooney (ed.). Reston, VA: National Council of Teachers of Mathematics, 1990.

Curcio, Frances R. "Mathematics as Communication: Using a Language Experience Approach in the Elementary Grades." In *Teaching and Learning Mathematics in the 1990s* (pp. 69–75). 1990 Yearbook. Thomas J. Cooney (ed.). Reston, VA: National Council of Teachers of Mathematics, 1990.

Earle, Richard A. *Teaching Reading and Mathematics*. Newark, DE: International Reading Association, 1976.

Ellerton, Nerida F., and M. A. (Ken) Clements. *Mathematics in Language: A Review of Language Factors in Mathematics Learning*. Geelong, Victoria, Canada: Deakin University Press, 1991.

Fouch, Robert S., and Eugene D. Nichols. "Language and Symbolism in Mathematics." In *The Growth of Mathematical Ideas, Grades K–12* (pp. 327–369). 25th Yearbook. Phillip S. Jones (ed.). Washington, DC: National Council of Teachers of Mathematics, 1959.

Hiebert, James. "Children's Mathematics Learning: The Struggle to Link Form and Understanding." *The Elementary School Journal* 84(May 1984):497–513.

———. "The Struggle to Link Written Symbols with Understanding: An Update." *Arithmetic Teacher* 36(Mar. 1989):38–44.

Kline, Morris. *Mathematical Thought from Ancient to Modern Times*. New York: Oxford University Press, 1972.

Milligan, Constance P., and Jerry L. Milligan. "A Linguistic Approach to Learning Mathematics Vocabulary." *Mathematics Teacher* 76(Oct. 1983):488–490.

Miura, Irene T., and Yukari Okamoto. "Comparisons of U.S. and Japanese First Graders' Cognitive Representation of Number and Understanding of Place Value." *Journal of Educational Psychology* 81(Mar. 1989):109–113.

Orr, Eleanor Wilson. *Twice as Less*. New York: Norton, 1987.

Pimm, David. *Speaking Mathematically: Communication in Mathematics Classrooms*. London, England: Routledge and Kegan Paul, 1987.

Pirie, Susan B. "Is Anybody Listening?" In *Communication in Mathematics, K–12 and Beyond* (pp. 105–115). 1996 Yearbook. Portia Elliot (ed.). Reston, VA: National Council of Teachers of Mathematics, 1996.

Rubenstein, Rheta. "Strategies to Support Mathematics Language Learning." In *Communication in Mathematics, K–12 and Beyond* (pp. 214–218). 1996 Yearbook. Portia Elliot (ed.). Reston, VA: National Council of Teachers of Mathematics, 1996.

Shuard, Hilary, and Andrew Rothery, eds. *Children Reading Mathematics*. London, England: Murray, 1984.

Skemp, Richard R. "Communicating Mathematics: Surface Structures and Deep Structures." *Visible Language* 16(Summer 1982):281–288.

Steinbring, Heinz, Maria G. Bartolini Bussi, and Anna Sierpinska, (eds.). *Language and Communication in the Mathematics Classroom*. Reston, VA: National Council of Teachers of Mathematics, 1998.

Wollman, Warren. "Determining the Sources of Error in a Translation from Sentence to Equation." *Journal for Research in Mathematics Education* 14(May 1983): 169–181.

Woodrow, Derek. "Mathematical Symbolism." *Visible Language* 16(Summer 1982):289–302.

FRANCES R. CURCIO

# LANGUAGE OF MATHEMATICS TEXTBOOKS

A very specialized language used by mathematicians in communicating with one another. This written language enables mathematicians to formulate definitions, to verify theorems, and, in general, to pose ideas precisely (Olson 1976). Among its special features, mathematical language has modes of abbreviation that are universally understood by mathematicians. As a result, mathematical text is densely packed with information.

Instructors and authors of textbooks use very much the same language in communicating to students at the university level. By doing so, they implicitly assume that students share the conventions of the language. Students, however, are quite inexperienced in the complex modes of analysis required for extracting the intended information from mathematical text. Not only are the concepts themselves foreign to the students but they are presented, to a significant extent, in a foreign language. Typically, instructors and authors of textbooks do not make an explicit, systematic effort to help students cope with this language.

In confronting a definitional passage such as the following one from a university textbook, the relatively untrained reader meets with difficulty. (Readers unfamiliar with vector spaces may wish to focus primarily on the italicized commentary below.)

## Subspaces

If $V$ is a vector space over the reals (or complexes), there are certain subsets of $V$, called *subspaces, that are again vector spaces under the same algebraic operations*. The purpose of this section is to study these objects. *(The terms "again" and "under" have a specialized usage here, departing from their typical English usage.)*

*Definition* If $V$ is a vector space and $H$ is a nonempty subset of $V$ having the properties:

(i) Whenever $x$ and $y$ belong to $H$, then $x + y$ belongs to $H$. *(The term "whenever" is used here in a specialized way as a "universal quantifier," signifying a generalization about ANY members $x, y$ of the set $H$.)*

(ii) If $x$ belongs to $H$ and $\alpha$ is a scalar, then $\alpha x$ belongs to $H$. *(There are implicit universal quantifiers here as well, associated with $x$ and $\alpha$. The intended meaning is, "if $x$ is any vector whatsoever belonging to the subset $H$ and $\alpha$ is any scalar whatsoever, then the vector $\alpha x$ also belongs to $H$.")*

$H$ is said to be a *subspace* of the vector space $V$ . . .

As an example in $R^3$, let $L$ be the set of vectors lying on some line passing through the origin . . . all vectors in $L$ are scalar multiples of a single nonzero vector in $L$, for definiteness, say $u$. *(The text provides a "generic" example of an entire family of lines rather than a single, specific, and, hence, more accessible example.)*

If $x$ and $y$ belong to $L$, we have $x = \alpha u$ and $y = \beta u$ for suitable scalars $\alpha$ and $\beta$. Then $x + y = (\alpha + \beta)u$. Thus, $x + y$ being a scalar multiple of $u$ must necessarily belong to $L$. Moreover, since $\gamma x = \gamma(\alpha u) = (\gamma \alpha)u$, we also see that if $x$ belongs to $L$ and $\gamma$ is a scalar, $\gamma x$ belongs to $L$. Thus having verified conditions (i) and (ii) in the definition of a subspace above, we see that $L$ is a subspace of $R^3$. . . . *(Only after the fact does the text mention that it has just provided a proof that the set $L$ is indeed a subspace of $R^3$. Only an experienced reader of mathematical text might overcome this poor writing practice.)*

**Exercises**

1. Which of the following subsets of $R^2$ are subspaces? (a) $\{(x,y) \mid x = 3y\}$
*(Use of "x" and "y" in this very first exercise differs entirely from use in the definition. The correspondences between ingredients in the exercise and their counterparts in the definitional passage do not leap out at the untrained reader. Routine exercises assume unexplicated skills that no previous coursework addresses.)*

When it comes to basing a proof on such a definitional passage, the nature of student difficulties emerges, revealing that they do not share the mathematicians' tacit understandings concerning mathematical objects, symbols, and definitions. Students are "outsiders" (Vygotsky 1962). Their abundantly flawed and ambiguous communication to their instructors makes this very clear. Students make idiosyncratic use of symbols that the textbook has defined for very specific use. For example, the textbook will use the symbol $M_{32}$ consistently to denote the vector space of all $3 \times 2$ real matrices, while a student will make use of the symbol to denote some specific matrix. Students' use of a symbol such as $R^5$ suggests that they are not aware that the object denoted by this symbol is a specific set, a fixed entity, and not simply one of some kind of entities. Even when they do realize that a set is at hand, they have great difficulty in describing what its members are, particularly in handling the quantifiers required for doing so.

The research leading to these findings probed students' understanding of definitions by analyzing text passages and analyzing students' writing in response to written quizzes. Quiz items included providing a definition of $R^5$, when the book had defined $R^n$, or $M_{59}$, when the book had defined $M_{mn}$. Quizzes also included questions about procedure: "Suppose it is claimed that a matrix S is symmetric. Write a procedure for verifying whether this claim is true. Explain what has to be done, including some English words in your explanation." These written quizzes, administered to students in several parallel sections of a linear algebra course, were followed by tape-recorded interviews with individuals about their responses (Rin 1982). The following is an interview excerpt concerning a standard course symbol, $M_{59}$.

*Interviewer:* So what's $M_{59}$?
*Student:* That's going to be a 5,9 matrix of things.
*Interviewer:* It's going to be a matrix?
*Student:* Well, $M_{59}$ M is . . . You can have $R_{59}$ or $Q_{59}$ or anything else.
*Interviewer:* Are you considering M?
*Student:* No, that's not the same. I'm using M as just a variable, like X or B or anything else. I'm not considering M to be a special space.
*Interviewer:* Okay, how about the book, though (opening book to appropriate page)?
*Student:* The book on the other hand appears to feel that M is a space, or is a collection. It's all of them. I was thinking of it more as a single one.
*Interviewer:* Okay, so what's $M_{59}$?
*Student:* $M_{59}$ would be the collection of all . . . 5 by 9 matrices.

At the same time, instructional practice generally is not designed to help students explicitly acquire the needed linguistic skills. Instructional practice may, in fact, be "weeding out" students unnecessarily. Exercises rarely probe directly student acquisition of course vocabulary. Nor do instructors initially supply definitions in wordy, verbose, redundant form, gradually leading students to cope with the abbreviated language of mathematics. Such wordy formulations, along with exercises specifically designed to probe and enhance student understanding of features of the language, can be effective in addressing difficulties.

## SPECIFIC SUGGESTIONS FOR CLASSROOM PRACTICE

At the secondary school level, for example, teachers can be mindful of being very explicit about quantifiers. When talking about the constant function $f$ defined by the formula "$f(x) = 5$ for any real number $x$," be sure to include the phrase "for any real number $x$" and *not* to leave it out, emphasizing that it is an integral part of the definition of the function $f$. Explain that the function $f$ takes the value 5 *no matter which number is chosen.* So for any number $x$ whatsoever, the value assigned to $x$ by the function $f$ is 5, whether the number $x$ is 2, $-4.778$, pi, 23 million, or whatever. This kind of redundant, wordy explanation may help prevent the misconception reflected in the statement "the derivative of any number is zero," uttered by a calculus student in connection with the (abbreviated) formula: $f'(5) = 0$.

Mathematics teachers at all levels can help students by creating opportunities to communicate and then correcting students' written and oral communication. We can assign exercises addressing specific linguistic skills that comprise subskills of proof writing or solution write-up. Students can use help, for example, with the linguistic conventions followed when introducing a symbol to denote a "typical" member $(a,b,c,d,e)$ of the set $R^5$. We also can provide practice in inferring a procedure based on a definition, as in the case of the symmetric matrix mentioned above. Opportunities for attending to linguistic practice abound. As with many human difficulties, amelioration begins with attention.

*See also* Language and Mathematics in the Classroom; Reading Mathematics Textbooks

### SELECTED REFERENCES

Olson, David R. "Towards a Theory of Instructional Means." *Educational Psychologist* 12(1)(1976):14–35.

Pimm, David. *Speaking Mathematically—Communication in Mathematics Classrooms.* London, England: Routledge and Kegan Paul, 1987.

Rin, Hadas. *Linguistic Barriers to Students' Understanding of Definitions in a College Mathematics Course.* Berkeley, CA: Ph.D. diss. University of California at Berkeley, 1982.

Vygotsky, Lev. *Thought and Language.* Cambridge, MA: MIT Press, 1962.

HADAS RIN

# LAPLACE, PIERRE SIMON (1749–1827)

An influential member of the Academy of Sciences, also *de facto* leader of the Bureau of Longitudes upon its founding in 1795, an important policy maker (though not a professor) at the *École Polytechnique* founded the year before, and influential member of numerous committees. Thus Laplace spent much of his career at the summit of the profession of mathematics and astronomy. His only equal was Joseph L. Lagrange (1736–1813).

Laplace's scientific career exhibits a continuous preoccupation with three connected areas: (1) he took up from Euler and others the mathematical astronomy of his time, with particular interest in details of the real and apparent motions of the planetary bodies and the issue of the stability of the solar system; (2) to this and other ends, he drew on

and enriched solutions of the partial differential equation now named after him. In rectangular coordinates,

$$\partial^2 V/\partial x^2 + \partial^2 V/\partial y^2 + \partial^2 V/\partial z^2 = 0$$

He also developed related theories such as the Legendre functions (known in the nineteenth century as "Laplace's functions"). He introduced special methods of summing series, especially finite but very long ones. He was an important pioneer in mathematical statistics, contributing to asymptotic theory and such major techniques as the Central Limit Theorem.

In addition to an unceasing flow of papers, he published influential books: *Traité de mécanique céleste* (five volumes, 1799–1825) and the more popular *Exposition du système du monde* (editions from 1796): and the *Traité analytique des probabilités* (editions from 1812) with the more popular *Essai philosophique sur les probabilités* (1814 and later editions).

### SELECTED REFERENCES

Gillispie, Charles C., ed. *Dictionary of Scientific Biography.* Vol. 15. New York: Scribner's, 1978, pp. 273–403.

Gillispie, Charles C. *Pierre-Simone LaPlace, 1749–1827: A Life in Exact Science.* Princeton, NJ: Princeton University Press, 1997.

IVOR GRATTAN-GUINNESS

# LEARNING DISABILITIES, OVERVIEW

Significant learning disorders affecting as many as 15–20% of school-aged children that interfere with the acquisition of certain academic skills, while sometimes also thwarting overall productivity in school. In most cases, underlying neurodevelopmental dysfunctions, subtle handicaps of brain function, perpetuate these disorders. The dysfunctions, which may be inborn or acquired, consist of weaknesses in such areas as attention, language, or memory that impede learning in specific skill domains, such as reading, spelling, or written output (Feagans and McKinney 1981). In this entry, we will examine specifically the neurodevelopmental dysfunctions that exert a negative effect on the mastery of mathematics. Such disorders of mathematical learning are thought to affect approximately 6% of school-aged children (Norman and Zigmond 1980). We will explore the manifestations and their frequent complications as well as the assessment and management of these conditions.

## COMMON FORMS OF DYSFUNCTION AFFECTING MATHEMATICAL LEARNING

Children with mathematics learning disorders comprise a very heterogeneous group. The reasons for their difficulty are likely to vary markedly, and therefore it is important to avoid glib generalizations about them. A multitude of different patterns of neurodevelopmental strength and weakness exist among them, and many such students contend with more than one dysfunction (Levine 1994; Levine, Lindsay, and Reed 1992). It is possible to classify their dysfunctions into three general categories. First, there are dysfunctions that are manifest in incomplete comprehension. Students who reveal these patterns find it hard and often unrewarding to interpret content to which they are exposed in a mathematics class or textbook. The second category includes a range of problems with memory. When children exhibit weaknesses within this category of dysfunction, they may understand mathematics but have serious problems retaining its facts or procedures (a classic form of so-called dyscalculia) or they may show other manifestations of inadequate memory capacity. Finally, there is a group of children who exhibit weaknesses of attention and/or problem solving. They may understand and remember relatively well, yet perform poorly when it comes to making use of efficient, systematic approaches to mathematical challenges. We will examine these three categories and the different neurodevelopmental dysfunctions that are found within them.

### Dysfunctions of Comprehension

Children in a mathematics class confront a constant and rapid flow of ideas and facts that must be well understood in order to be applied effectively. They must process novel concepts and procedures in depth and with speed. They need to understand the language of mathematics and also be able to visualize a wide range of entities, dimensions, and processes. Often students cannot succeed in meeting the comprehension challenges. Common dysfunctions are described next.

1. Incomplete conceptualization—Some children display a chronically tenuous grasp on the conceptual content of mathematics. They have a weak sense of the critical features that come together to form a concept in arithmetic. It may be hard for them to understand what a fraction is or how fractions relate to decimals and percentages. They may never fully master the concept of equation. Such students often make use of an extreme algorithmic approach to mathematics. That is, they imitate procedures without actually knowing what they are doing and exactly why they are doing it! They commonly overrely on rote memory and derive little satisfaction from mathematics. Many of them experience frustration in other courses that demand strong concept formation.

2. Language dysfunction—There are learners who have to struggle incessantly with the language demands of the curriculum. Mathematics language can be especially elusive (Durkin and Shire 1991). Students with language dysfunction may have trouble mastering the subject's technical vocabulary, words such as *denominator, subtrahend, coefficient,* and *hypotenuse.* On a syntactic level, these students are apt to have their greatest difficulty with the solution of word problems, as it can be hard for them to interpret the grammatical construction and arrive at the realization that information provided in the problem is not necessarily used in the order in which it is mentioned. Furthermore, these students can become hopelessly confused when a teacher provides a verbal explanation. Some of them prefer to learn exclusively through visual demonstration models rather than striving to process complex explanatory linguistics. Additionally, they may show serious and humiliating dysfluency when they try to explain to classmates or a teacher a particular concept or how they solved a problem. Many students with language dysfunctions have problems acquiring skills in reading and in written expression as well.

3. Ineffective visualization—Much of mathematics learning is enhanced through visual imagery. In fact, many competent mathematicians are highly effective at going back and forth from verbalization to visualization, using each to fortify the other. Some students, however, are virtually unable to visualize critical processes and concepts (Miles 1992). As a result, they have trouble appreciating and retaining the distinctive features of geometric forms or dimensions of measurement, and they endure confusion in trying to understand fully such entities as proportion, ratio, and place value. They are likewise prone to confusion over directionality, as a result of which they commonly proceed from left to right when they should be operating from right to left (as in long multiplication) or they work downward

when they should be working from the bottom up (as in subtraction). These students may attempt to learn mathematics exclusively as a language. Many of them have problems in certain scientific subjects that also demand visualization. Thus, it may be hard for them to picture planets rotating and revolving in the solar system or to capture the image of the spiraled configurations of DNA molecules.

4. Incomplete appreciation of symbolic representation—The curriculum in mathematics demands that students feel comfortable comprehending and applying a battery of abstract symbols, ranging from operational signs, to exponents, to algebraic unknowns. There are children who have a seeming inability to progress beyond concrete processing to the representation of entities or phenomena through symbols. Their slowness to internalize the symbol systems constitutes a substantial handicap in mathematics. They also may have problems dealing with abstraction, interpreting metaphors, and processing figurative symbolic language, such as might be encountered in an English class. They are often described as highly "concrete" in their thinking.

5. Diminished comprehension monitoring—There are some students who fail to monitor their understanding of the content of mathematics. Such individuals not only fail to understand but often they do not understand that they are not understanding! On occasions when they do realize that they do not understand, they may have substantial problems understanding exactly what it is they do not understand about the subject matter. As a result, they can fall behind in mathematics learning without realizing soon enough that they are accumulating significant gaps in their comprehension. These students often over-rely on rote memory and make use of extreme algorithmic approaches to problem solving.

## Dysfunctions of Memory

Mathematics learning imposes some of the most stringent demands on memory that a child experiences during the school years. There is a relentless requirement for the registration, consolidation, and rapid recall of facts and procedures. Some students lack facility with memory and may exhibit one or more specific forms of the following dysfunctions:

1. Imprecise or slow declarative (factual) recall—These children have great difficulty mastering and remembering mathematics facts. The multiplication tables are likely to be especially troublesome for them. They are apt to review their facts repeatedly only to be unable to recall them while taking a test. Often such children ultimately learn their facts but then exhibit delayed automatization, which is to say that they are unable to recall the facts rapidly and effortlessly by their middle school years. Consequently, work in mathematics becomes excessively labored and time-consuming. In addition, it can be hard to superimpose new knowledge over poorly automatized prior learning.

2. Deficient procedural recall—A group of children has particular problems with the recall of methods. It is hard for them to remember multistep processes in mathematics. They may confuse the order of operations or simply not remember how to apply specific procedures (such as reducing a fraction). Ultimately, these students, too, may show delayed automatization. Some display pervasive difficulties with the appreciation and recall of sequences and the preservation of serial order. Consequently, it may be hard for them to retrieve the order of the months of the year and to comply with multistep instructions.

3. Weak pattern recognition—Some students show an inability to recognize previously encountered wording or visual imagery in a mathematical problem. This shortcoming makes it hard for them to know what to do when. Their weak recognition memory may impede the ability to discern a pattern of words in a word problem that suggests the need for a particular type of solution or algorithm. They are likely to be misled by the superficial differences that overlie the patterns. For example, a student may not recognize that two word problems call for the same strategies because one of them is about a car and a truck, while the other describes a horse and a dog. Their underlying similarity (i.e., both deal with relative rates) eludes her or him. Reduced pattern recognition also may pervade the visual-spatial realm, as a student is unable to recognize a geometric form embedded within other forms. Some children with weak pattern recognition perform poorly with rule application, as they fail to detect the "if" in a situation that calls for the use of a particular "if . . . then" rule.

4. Limited active working memory—While working with computations and/or mental arithmetic, students must be able to suspend in memory the various components of what they

are working on while they are working on it. This requires active working memory capacity that is inadequate in certain children. They have problems holding one part of a task in mind while they are completing some other aspect of that task with the need then to return to that first part without having lost it from memory. Consequently, these students keep on forgetting what they are doing while they are doing it. While carrying a number, they are likely to forget what it was they had intended to do once they had finished carrying that number. Moreover, they frequently develop test-taking anxiety that interferes further with active working memory function. Some of these students also experience difficulty while writing reports, as they lose the ideas they were going to put in writing while pausing to decide whether or not to insert a comma in a sentence.

## Dysfunctions of Attention and Problem Solving

It is not unusual for a student's acquisition of mathematics skills to be compromised because she or he is unfocused, unsystematic, and/or disorganized in approaching the challenges of the subject. Varying combinations of difficulty with attention and overall problem-solving abilities commonly are found in such children.

1. Attention deficits—Impairments of attention are thought to be the most common of the neurodevelopmental dysfunctions. They have the potential to perpetuate underachievement across the curriculum. In some cases, attention deficits are accompanied by hyperactivity and behavior problems. In mathematics, the traits associated with attentional dysfunction can exact a substantial toll. These traits and their effects on mathematics learning and performance are summarized in Table 1.

2. Ineffective problem solving—Poor performance on problem-solving tasks is a common occurrence in students with difficulties in mathematics. There is some overlap in the signs of attentional dysfunction and those of ineffective problem solving. This is because their problems with concentration, their impulsivity, and their inattention to detail prevent many of these children from applying good problem-solving techniques. It is the case, however, that there exist many students who have no evidence of atten-

**Table 1** *Traits of attentional dysfunction and their potential impacts on mathematics performance*

| Trait | Impacts on mathematical performance |
|---|---|
| Mental fatigue | Trouble persisting, a tendency to "burn out" and lose focus while studying, listening in class, or completing assignments |
| Weak saliency determination | Difficulty determining what is most important or relevant; trouble isolating needed information in word problems; tendency to get distracted while studying or working |
| Superficial processing | Lack of attention to fine detail (such as operational signs or the difference between a 3 and a 5), preference for the "big picture" |
| Improper pacing | Tendency to make careless errors because work pace is too rapid or frenetic; in some cases, excessively slow pacing |
| Cognitive impulsivity | Failure to plan before undertaking a mathematics problem; a tendency to apply the first method that comes to mind instead of considering alternatives |
| Poor previewing | A reluctance to look ahead and predict or estimate answers or likely outcomes |
| Weak self-monitoring | Problems with quality control; a tendency to commit careless mistakes without noticing them both while and after solving problems |
| Performance inconsistency | A pattern of unpredictable, erratic accuracy and productivity from moment to moment and from day to day; highly inconsistent test scores, random error patterns due to variable concentration |
| Low reinforceability | Trouble learning from previous experience, so that what was highly successful one day may not be applied the next day |

tional dysfunction but manifest difficulties with their problem-solving skills in mathematics and sometimes in other content areas as well. The components of deficient problem solving are depicted in Table 2.

The various neurodevelopmental functions that are commonly affected in students who have trouble learning in mathematics are depicted in Figure 1.

**Table 2** *Components of deficient problem solving and their impacts on mathematical performance*

| Component | Impacts on mathematical performance |
| --- | --- |
| Diminished planning | A tendency to proceed with mathematical challenges without any preconceived approach or plans |
| Weak sense of "stepwisdom" | A lack of awareness that complex problems must be solved in a series of manageable incremental steps rather than randomly or all at once; a tendency to feel overwhelmed in mathematics |
| Failure to use strategies | An apparent inability to consider tactics to simplify mathematics tasks |
| Lack of alternative strategies | Difficulty making use of a second strategy when a first one is not working; a tendency to be rigid rather than flexible in problem solving |
| Trouble applying rules | Failure to make use of previously acquired results relevant to particular problem-solving challenges |
| Deficient reasoning | Weaknesses in certain forms of reasoning essential in mathematics: primarily, difficulty with proportional and analogical thinking skills |
| Poor on-line monitoring and regulation | Trouble detecting one's errors or noting that one is "off track" and then self-righting |

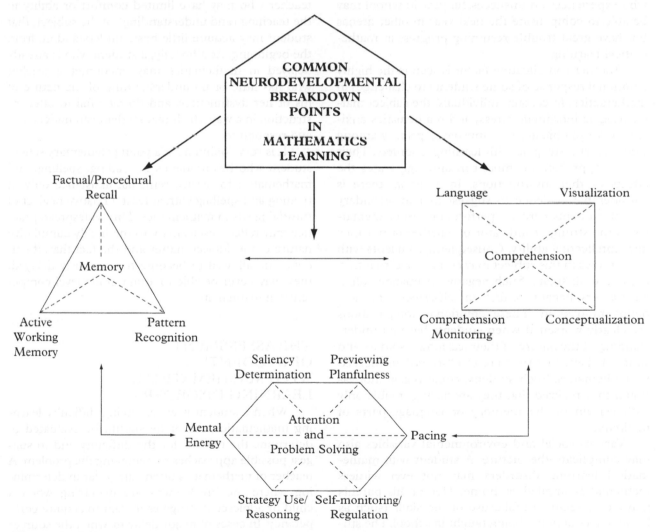

**Figure 1** *This diagram depicts the areas of neurodevelopmental function that are most likely to be impaired in students with mathematics learning disorders. Individual students may experience difficulty in one or more of these areas.*

## THE FREQUENT COMPLICATIONS OF MATHEMATICS LEARNING DISORDERS

Problems with the acquisition of mathematics skills and concepts carry with them a series of complicating factors that may perpetuate continuing failure and frustration. First, there are the complications introduced by the highly cumulative nature of the subject matter, as a result of which students whose skills become moderately delayed may fail to grasp and assimilate new inputs, thereby falling progressively behind over the years of study. They are likely to feel justifiably overwhelmed and disoriented in a mathematics class. Furthermore, they may come to feel that they can never catch up, that they are simply too hopelessly delayed. A student who experiences an unsuccessful year in school may be able to compensate the next year in other arenas but have great trouble resuming progress in mathematical learning.

Another complicating factor is seen in the highly emotional response of some students to challenges in mathematics. In certain individuals, the subject matter triggers inordinate stress, and mathematics anxiety and/or phobia are not unusual, especially among students who struggle with learning disorders. Their intense apprehension most certainly aggravates the effects of their dysfunctions. In addition, there is considerable evidence to suggest that in secondary school, students perceive performance in mathematics as the strongest indicator of whether or not they have intellectual ability. Consequently, students with poor grades in this subject may come to feel that they are not at all smart. Such negative evaluations often have a very negative effect on self-esteem and academic motivation. These emotional complications are clearly worsened when students have no understanding of the nature of their learning disorders and simply believe that they are too "stupid" to succeed in mathematics. Such students could regain motivation if they realized that they are having trouble only with certain of the memory or language parts of mathematics.

Various social and environmental variables also may complicate the picture. A student with mathematical learning disorders may not ever witness mathematics applied at home. Her or his parents may never seem to make use of the skills (such as balancing equations) being taught in school. The student may then rationalize that mathematics is somehow irrelevant ("Why do I have to learn this? . . . I'm never gonna use it"). There are some students who actually fear displaying competency in mathematics out of concern that they will be labeled by their peers as "nerds" or "dorks." This threat of social ostracism can actually prompt some students with mathematical learning disorders to rationalize their lack of motivation to improve. In some cases, recurring failure and embarrassment in a mathematics class can cause a student to develop disruptive or oppositional behaviors. Sometimes these misguided actions and facades of indifference or defiance represent defensive tactics on the part of a student who would much prefer to be perceived as a behavior problem than as a "dummy."

Finally, there are teaching factors that complicate the plight of a child with mathematical learning disorders. If a child with certain dysfunctions is taught early mathematical skills and concepts by a teacher who may have limited comfort or ability in the teaching (and understanding) of the subject, that student may acquire little basic understanding from the beginning. Additionally, a student who is already delayed in mathematics may encounter a teacher who has little or no understanding of the nature of his or her dysfunctions and thereby fail to offer instruction in a way which that student can understand and respond to.

It is very common for a young elementary school student who has problems in reading, spelling, and mathematics to receive remedial assistance only in reading and spelling with at least a relative neglect of tutorial needs in mathematics. This widespread practice may reflect insensitivity to the highly cumulative nature of the subject matter and the fact that, if students are allowed to become too seriously delayed, they may never be able to attain grade level competency in mathematics.

## THE ASSESSMENT OF STUDENTS WITH MATHEMATICAL LEARNING DISORDERS

When a student is experiencing difficulty learning mathematics, she or he should be evaluated to determine the reasons for the difficulty and to suggest possible approaches to managing the problem. A teacher of mathematics often can go far in determining where the breakdowns are occurring when a child or adolescent struggles in vain to acquire competency. In cases of major delay or when the sources of the difficulty remain elusive, a more formal evaluation is called for. The latter may be carried out by a psychologist or educational diagnostician with a strong background in mathematics learning disorders or, alternatively, by a multidisciplinary team that can examine closely the multiple educational, neurodevelopmental, emotional, and environmental issues that may be converging to deter learning in mathematics. The following sources of information can be used by teachers and clinicians in their efforts to understand these students:

1. Analyses of work samples—Teachers, in particular, should become adept at examining test papers and homework to determine and describe the underlying nature of a child's errors. For example, when the student attempts a word problem, does it look as if he or she cannot decipher the language of the problem? Or do the mistakes suggest the use of wrong algorithms? Is the student missing the wording pattern that suggests the need for a particular algorithm? Or can the student simply not remember how to undertake a particular operation (such as division)? Do errors reflect a lack of recall of basic arithmetic combinations? Does the student take too long to complete problems and fail to finish a quiz because of delayed automatization of facts and/or algorithms? Are there indications of too much reliance on rote recall with little or no understanding of concepts? Such error analyses can be highly revealing, as they suggest the presence of specific neurodevelopmental dysfunctions that are compromising performance.

2. Achievement testing—Standardized achievement tests can be helpful in identifying which particular mathematics skills are deficient and in comparing a student to others in the same grade level. Often so-called process-oriented individual achievement tests or dynamic assessment techniques yield more specific findings than traditional group tests, as they allow an examiner to observe a student in the process of problem solving and reasoning that facilitates observations of relevant parameters, such as attention, strategy use, anxiety level, and rate of processing. A keen observer can detect where the breakdowns are occurring, while the student is trying to compute or problem solve.

3. Questionnaires and interviews—Standardized questionnaires may shed light on the nature of a learning difficulty. These inventories may be completed by parents, teachers, or the student. They are generally designed to elicit information on the child's overall development and behavior as well as a range of other issues. One such set of instruments is called the ANSER System (see Levine 1994, and available from Educator's Publishing Service, Cambridge, Massachusetts), which enables a school or clinic to gather data on a student's health, behavior, patterns of learning, and areas of strength and strong interest. These questionnaires also are designed to screen for attention deficits, emotional difficulties, and specific developmental lags (such as problems with memory or language). In some cases, direct interviews with the student can yield helpful information about the reasons for mathematical difficulties. The Mathematics Interview is an example of this. A portion of this instrument is illustrated in Figure 2 (Levine 1994).

4. Cognitive testing—There can be direct testing of specific neurodevelopmental functions. Some of these may be assessed on traditional intelligence tests. In other cases, more detailed neuropsychological, neurodevelopmental, or language assessments may be indicated (Fleischner 1994).

5. Physical and neurological examination—It is important to rule out an underlying medical or neurological condition that may be contributing to a mathematics learning difficulty. A hearing problem, a visual impairment, a seizure disorder, or any number of chronic congenital and acquired conditions may need to be identified and treated in some cases.

The ultimate product of an evaluation should be a valid and clear description of a student's deficits in mathematics learning. Which skills and subskills are intact and which are delayed or incompletely automatized? There should then be a description of the student's observed neurodevelopmental strengths and weaknesses (such as in language ability and memory) as well as a rendering of any complicating factors (such as anxiety or a loss of motivation). The description should attempt to establish the connections between the neurodevelopmental dysfunctions and the difficulty in mathematics. It can then be used to develop an appropriate individualized plan to manage that student.

# THE MANAGEMENT OF STUDENTS WITH MATHEMATICS LEARNING DISORDERS

Managing the educational care of a student with mathematics learning disorders can be rewarding and successful. The management needs to be

| Memory | 0 | 1 | 2 | 3 |
|---|---|---|---|---|
| 11. It can take too long to remember math facts (like multiplication) on a test or on homework. | | | | |
| 12. Some kids have trouble remembering how to do things in mathematics. | | | | |
| 13. It can be much faster and easier to do hard math problems using a calculator. | | | | |
| 14. There are students who forget what they're doing in the middle of a math problem. | | | | |
| 15. Some kids can do much less math in their heads than other kids. | | | | |
| 16. A lot of students find that when they look at a math problem, it just doesn't look familiar to them. Every problem seems very different from all the others. | | | | |
| 17. Doing things in the right order is one of the hardest parts of mathematics. | | | | |
| 18. Some people have to stop and think about everything in math; none of it seems to be fast and automatic for them. | | | | |
| 19. In math it can be very hard to remember exact shapes when you hear their names. | | | | |
| 20. Some kids complain they understand how to do things in class and then they forget how on an assignment or a test. | | | | |
| **Attention** | 0 | 1 | 2 | 3 |
| 21. It is possible to make too many careless mistakes because they rush through stuff too much in math. | | | | |
| 22. It can be very hard to keep concentrating during math class. | | | | |
| 23. Some kids say they keep getting very tired when they do math. | | | | |
| 24. A student might complain that there are too many little details to think about in math. | | | | |
| 25. There are students who have trouble in math sometimes because they don't really stop and think and plan enough before they do a problem. | | | | |
| 26. It can be very hard to get started on math problems. | | | | |
| 27. There are students who have problems estimating; they have no idea about what the answer will be before they start a problem. | | | | |
| 28. Some students have minds that don't like subjects like math because there's only one correct answer all the time. | | | | |
| 29. For some people it is extremely boring to check over math work. | | | | |
| 30. Some kids are very inconsistent in math; they do really well sometimes and very poorly other times. | | | | |

**Figure 2** *Sample items from a mathematics interview.*
*These two parts of a mathematics interview (taken from Levine 1994) cover various aspects of memory and attention. The interview presents the child with these statements, and the child responds by stating the extent to which each pertains to him or her. The interview also contains sections on comprehension, problem solving, and attributions.*

planned and executed in a systematic manner. One can delineate three steps that represent optimal management. These include (1) demystification, (2) the use of bypass strategies, and (3) intervention at breakdown points.

Demystification

A student cannot strive to improve a function he does not have the word for. It is imperative that students with learning difficulties acquire an understanding of their problems as well as the language with which to discuss them rationally. The process of educating children about their learning disorders is called *demystification*. An adult (often a teacher) meets with the student and talks about his or her strengths and weak areas. Plain language is used interspersed with concrete examples and analogies. The discussion is nonaccusatory, highly supportive, and optimistic. The student is told of ways in which the learning disorder can be dealt with and is reassured that this problem is very different from mental retardation, that he or she is basically smart.

Demystification also can be offered to small groups of students. In this case, an adult leader conducts a discussion that covers various strengths and abilities that are needed to learn mathematics. Participants are then asked to share with each other their own strengths and weaknesses. Case studies of children with mathematics learning disorders and a self-assessment questionnaire are sometimes employed to facilitate and direct these kinds of discussion groups.

## The Use of Bypass Strategies

*Bypass strategies* are classroom techniques used to circumvent a student's neurodevelopmental dysfunctions, thus allowing for the continuing acquisition of knowledge despite the presence of a learning disorder. Such provisions are often essential if a student is not to show further deterioration and anxiety in mathematics classes. The common forms of bypass strategies are summarized in Table 3.

## Interventions at Breakdown Points

It is possible to work specifically to repair those aspects of mathematical learning that are deficient. This endeavor might entail directing management to one or more delayed skills and/or working on a neurodevelopmental dysfunction that is impeding attainment. Often these kinds of intervention need to be implemented outside of a regular classroom (such as in a tutorial setting or study skills program). Some forms of intervention at breakdown points, however, can be integrated effectively and unobtrusively within a mathematics class. Moreover, certain techniques that would help a child with a mathematics learning disorder might well be beneficial to all of the students in a room. The forms of intervention at breakdown points and examples are presented in Table 4.

The kinds of management techniques outlined here may need to be supplemented with other forms

**Table 3** *Forms and examples of bypass strategies*

| Bypass strategy | Example(s) |
|---|---|
| Alteration of rate | Allowing more time on tests; warning student in advance he will have to put a problem on the board tomorrow |
| Reduction of volume | Giving fewer problems on test (e.g., completing remainder at home); reducing length of a homework assignment |
| Change of format | Helping a student with language problems to learn through visual demonstration models or concrete manipulatives |
| Use of devices | Permitting a student to use a calculator; having a child keep a reminder card to check for errors after calculating |
| Modified testing format | Providing several examples and having students find errors rather than having to retrieve methods and facts during the test |
| Collaboration | Encouraging a child to work as a partner with one who is adept at mathematics |
| Adjustment of curriculum | Offering special mathematics courses for children with difficulties |

**Table 4** *Forms and examples of interventions at the breakdown points*

| Intervention | Example(s) |
|---|---|
| Scaffolding | Providing a part of a task or the plan for a task and having a student complete the remainder (which includes his area of weakness or frequent breakdown) |
| Ending tasks at the breakdown point | Giving students word problems and asking only that they circle the data needed to solve them (especially helpful to students with attention problems) |
| Achieving automatization of weak subskills | Devising games, using educational software to render math facts or procedures completely automatic |
| Staging | Helping a student come up with a written list of steps for solving a particular kind of problem in a stepwise fashion |
| Mapping | Having a student maintain an atlas or scrapbook of diagrams (maps) of key concepts and their critical features |
| Articulating and elaborating | Encouraging students to describe and teach concepts and procedures to others rather than just imitating or doing them |

of intervention that take place outside of a mathematics classroom. Such treatments include the following.

1. Tutorial support—A student may require remedial tutoring in mathematics as well as other subjects. To be effective, such assistance should be coordinated closely with the work of the regular mathematics teacher. The content should either be focused upon current class material or directed toward the mastery and automatization of earlier presented material. Additionally, students of all ages with mathematics learning disorders often can benefit from help in their overall organization and study skills.

2. Parental involvement—Optimally, it is always helpful for a mathematics teacher to have the parents as allies and collaborators with younger students. They can assist with homework, offer mock tests, and design games and incentives for the automatization of subskills. Parents helping a child need to be keenly sensitive to her or his embarrassment over the learning problem. They should not be overly critical of the child while offering help.

3. Special therapies—Some children may benefit from treatment programs aimed at improving specific neurodevelopmental functions. For example, a child enduring serious trouble with the verbal demands of the curriculum may gain from speech and language therapy. Others may require help with motor skills for writing or memory skills for studying.

4. Counseling—When anxiety, depression, an extreme loss of motivation, or substantial behavior problems complicate the picture, a student may require counseling from a mental health specialist. It is important that the classroom teacher be aware that this kind of intervention is taking place.

5. Medication—Some students with attentional dysfunction receive stimulant drug therapy (most commonly Ritalin or Dexedrine) that is aimed at increasing the strength of concentration and enabling them to be more reflective. This form of treatment needs to be very closely monitored by a knowledgeable physician. In particular, such intervention may help certain individuals become more able to focus upon the fine detail of mathematics and also better equipped to pace themselves and persist at problem-solving tasks. Students with extreme anxiety, depression, or serious behavior problems may be given a range of other psychopharmacological agents to enable them to function more adaptively. Classroom teachers should know when a student is taking one of these drugs. They should be knowledgeable about observable side effects in the classroom, and they should be active participants in the process of determining whether or not the treatment is effective.

Mathematics learning disorders above all need to be handled in such a way that they will not seriously damage the self-esteem and overall feelings of effectiveness of a growing child. In particular, public humiliation or excessive open criticism in front of peers is likely to have a long-term malignant effect. A supportive teacher who is willing to seek an understanding of the child's struggles, a caring mentor who can offer a flexible and compassionate approach while preserving the pride of a young child, is likely to achieve the most constructive and humane outcome.

*See also* Mathematics Anxiety; Psychology of Learning and Instruction, Overview; Remedial Instruction; Tutoring; Underachievers, Special Programs

## SELECTED REFERENCES

Durkin, Kevin, and Beatrice Shire, (eds.). *Language in Mathematical Education: Research and Practice.* Philadelphia, PA: Open University Press, 1991.

Feagans, Lynn, and Donald McKinney. "The Pattern of Exceptionality Across Domains in Learning-Disabled Children." *Journal of Applied Developmental Psychology* 1(1981):313–328.

Fleischner, Jeannette E. "Diagnosis and Assessment of Mathematics Learning Disability." In *Frames of Reference for the Assessment of Learning Disability* (pp. 441–458). G. Reid Lyon (ed.). Baltimore, MD: Brookes, 1994.

Levine, Melvin D. *Educational Care.* Cambridge, MA: Educator's Publishing, 1994.

———, Ronald S. Lindsay, and Martha S. Reed. "The Wrath of Math." *Pediatric Clinics of North America* 39(1992):525–536.

Miles, Tim. "Some Theoretical Considerations." In *Dyslexia and Mathematics* (pp. 1–22). Tim Miles and Elaine Miles (eds.). New York: Routledge, 1992.

Norman, Charles A., and Naomi Zigmond. "Characteristics of Children Labeled and Served as Learning Disabled in School Systems Affiliated with Child Service Demonstration Centers." *Journal of Learning Disabilities* 13(1980):542–547.

MELVIN D. LEVINE

# LEAST SQUARES

Mathematical method that finds the relation between a dependent variable and one or more independent variables, for which the sum of the squares of the residual (error) is minimum. The least-squares method was developed independently by Carl Friedrich Gauss in Germany, Adrien Marie Legendre in France, and Robert Adrain in America. Legendre (1805) was the first to publish it. Several excellent books on the theory of least squares and its applications were published in the late nineteenth and twentieth centuries. Particularly noteworthy is that of Merriman (1884) on the theory.

For example, suppose that the height ($ht$) of Monterey pine trees is a linear function of rainfall ($rf$) as given by $ht = a_0 + a_1 (rf)$, where $a_0$ and $a_1$ are constants. To determine $a_0$ and $a_1$, an experimenter grew one-foot Monterey pine trees in similarly controlled environments where variation in rainfall was controlled by providing different amounts of irrigation. The following table presents the amount of rainfall (in inches) and the measured height (in inches) of Monterey pine trees at the end of one year.

**Table 1**

| $rf$ (inches) | 13 | 19 | 11 | 17 | 27 | 23 |
|---|---|---|---|---|---|---|
| $ht$ (inches) | 21 | 26 | 20 | 24 | 34 | 32 |

This data contains experimental errors. From Table 1 and the model equation, we have:

$$a_0 + 13a_1 = 21 \qquad a_0 + 17a_1 = 24$$
$$a_0 + 19a_1 = 26 \qquad a_0 + 27a_1 = 34 \qquad (1)$$
$$a_0 + 11a_1 = 20 \qquad a_0 + 23a_1 = 32$$

We have more equations than unknowns. Such systems are known as overdetermined systems. One possibility is to determine $a_0$ and $a_1$ from a part of (1) by ignoring the rest. Since the data comes from the same source, it is difficult to know which equations contain large errors. Thus, we cannot justify determining $a_0$ and $a_1$ from a part of (1). It seems reasonable to select $a_0$ and $a_1$ such that the average error in these six equations is minimum. Since $a_0$ and $a_1$ do not satisfy (1) exactly, the error given in each equation of (1) is $21 - a_0 - 13a_1$, $26 - a_0 - 19a_1$, $20 - a_0 - 11a_1$, $24 - a_0 - 17a_1$, $34 - a_0 - 27a_1$, and $32 - a_0 - 23a_1$ for a given $a_0$ and $a_1$. There are many ways to define the average error, but the most convenient and often used is the sum of squares:

$$E^2 = (21 - a_0 - 13a_1)^2 + (26 - a_0 - 19a_1)^2$$
$$+ (20 - a_0 - 11a_1)^2 + (24 - a_0 - 17a_1)^2$$
$$+ (34 - a_0 - 27a_1)^2 + (32 - a_0 - 23a_1)^2$$

We wish to find $a_0$ and $a_1$ for which $E^2$ is a minimum. The values of $a_0$ and $a_1$ that make $E^2$ a minimum are called a *least-squares* solution of (1).

By means of partial differentiation, we can provide $a_0$ and $a_1$ for which $E^2$ is a minimum. As a result of differentiating $E^2$ first with respect to $a_0$ and then with respect to $a_1$, we have two equations in $a_0$ and $a_1$:

$$\left. \begin{array}{r} 6a_0 + 110a_1 = 157 \\ 110a_0 + 2198a_1 = 3049 \end{array} \right\} \qquad (2)$$

Solving (2) for $a_0$ and $a_1$, we get $ht = 8.91 + 0.94(rf)$. The graphs of the data points and $ht = 8.91 + 0.94(rf)$ are shown in Figure 1. This problem may be generalized to a least-squares solution of $m$ equations in $n$ unknowns where $m > n$.

The least-squares method can be used to fit a function to given data. The least-squares fit of a linear function is $P_1 (x) = a_0 + a_1 x = y$, where $a_0$ and $a_1$ are determined from the given data $(x_1, y_1)$, $(x_2, y_2), \ldots, (x_N, y_N)$ by the following formulas. (Note: One can also compute $a_0$ and $a_1$ of $ht = a_0 + a_1(rf)$ for Table 1 from these formulas.)

$$a_0 = \frac{(\sum_{i=1}^{N} y_i)(\sum_{i=1}^{N} x_i^2) - (\sum_{i=1}^{N} x_i y_i)(\sum_{i=1}^{N} x_i)}{N(\sum_{i=1}^{N} x_i^2) - (\sum_{i=1}^{N} x_i)^2}$$

$$a_1 = \frac{N(\sum_{i=1}^{N} x_i y_i) - (\sum_{i=1}^{N} y_i)(\sum_{i=1}^{N} x_i)}{N(\sum_{i=1}^{N} x_i^2) - (\sum_{i=1}^{N} x_i)^2}$$

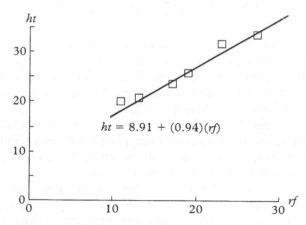

$$ht = 8.91 + (0.94)(rf)$$

**Figure 1**

Next, we consider an exponential form given by $g(x) = a_0 e^{a_1 x} = y$ to fit a given N data points $(x_1, y_1)$, $(x_2, y_2), \ldots, (x_N, y_N)$ with positive weights $w_1, w_2, \ldots, w_N$. Choosing different values of $w_i > 0$ allows us to attach varying degrees of importance to different data points. If all points are given equal importance, then we take $w_i = 1$ for $i = 1, 2, \ldots, N$. The least-squares procedure requires that we minimize $E^2(a_0, a_1) = \sum_{i=1}^{N} w_i (y_i - a_0 e^{a_1 x_i})^2$. Setting the partial derivatives of $E^2$ to zero and simplifying, we get

$$a_0 \left( \sum_{i=1}^{N} w_i e^{2 a_1 x_i} \right) - \sum_{i=1}^{N} w_i y_i e^{a_1 x_i} = 0 \quad \text{and}$$

$$a_0 \left( \sum_{i=1}^{N} w_i x_i e^{2 a_1 x_i} \right) - \sum_{i=1}^{N} w_i y_i x_i e^{a_1 x_i} = 0$$

This system is nonlinear in $a_0$ and $a_1$ and is solved by the methods of numerical solutions of systems of nonlinear equations (Patel 1994). Another common approach, however, is to transform the given exponential form, $g(x) = a_0 e^{a_1 x} = y$, into the logarithmic form $\ln g(x) = \ln a_0 + a_1 x$. Letting $\ln g(x) = Y$, $\ln a_0 = A$, $a_1 = B$, and $x = X$ leads to

$$Y = A + BX \tag{3}$$

This is a linear equation. The values of $A$ and $B$ are obtained by minimizing $E^2(A, B) = \sum_{i=1}^{N} w_i [Y_i - (A + B X_i)]^2$, where $X_i = x_i$ and $Y_i = \ln y_i$. It can be verified that the normal equations for (3) are given by

$$A \sum_{i=1}^{N} w_i + B \sum_{i=1}^{N} w_i X_i = \sum_{i=1}^{N} w_i Y_i \tag{4}$$

and

$$A \sum_{i=1}^{N} w_i X_i + B \sum_{i=1}^{N} w_i X_i^2 = \sum_{i=1}^{N} w_i X_i Y_i$$

Solving (4) for $A$ and $B$ gives the least-squares approximation of the transformed problem but not of the original problem. Many times, a nonlinear fit can be transferred into a linear fit by using a proper transformation (Maron 1987).

Consider a mixture of four radioactive materials. We know their rates of decay, but we do not know how much of each is in the mixture. Then our experimental data would behave like $y = a_0 e^{-\alpha_1 t} + a_1 e^{-\alpha_2 t} + a_2 e^{-\alpha_3 t} + a_3 e^{-\alpha_4 t}$. We extend the least-squares procedure to determine $a_0, a_1, a_2, a_3$. So far we considered polynomials and exponentials. We can generalize the least-squares procedure to fit $M$ lin-

early independent functions to a given set of N data points. Consider $\phi_1(x), \phi_2(x), \ldots, \phi_M(x)$ to be linearly independent functions and $w_1, w_2, \ldots, w_N$ to be weights. Let $(x_1, y_1), (x_2, y_2), \ldots, (x_N, y_N)$ be the given N data points. We fit $g(x) = \sum_{j=1}^{M} a_j \phi_j(x)$ by selecting $a_1, a_2, \ldots, a_M$ in such a way that $E^2(a_1, a_2, \ldots, a_M) = \sum_{i=1}^{N} w_i [y_i - \sum_{j=1}^{M} a_j \phi_j(x)]^2$ is minimum.

*See also* Curve Fitting; Linear Algebra, Overview

## SELECTED REFERENCES

Legendre, Adrien Marie. *Nouvelles Methodes pour la Determination des Orbites des Cométes.* Paris, France: Courcier, 1805.

Maron, Melvin J. *Numerical Analysis: A Practical Approach,* 2nd ed. New York: Macmillan, 1987.

Merriman, Mansfield. *A Text-Book on the Method of Least Squares.* New York: Wiley, 1884.

Patel, Vithal A. *Numerical Analysis.* Philadelphia, PA: Saunders College Publishing, 1994.

VITHAL A. PATEL

# LEBESGUE, HENRI (1875–1941)

Arguably the most influential of three young French mathematicians (Émile Borel and René Baire being the others) who around 1900 transformed mathematics through their incorporation of set theory into analysis. Lebesgue developed a theory of measuring sets that he used to create what is often called the Lebesgue integral, a fundamental reworking of the nineteenth-century conception of the integral. Lebesgue's work has been called the "first genuine theory of integration" (Hawkins 1975, ix).

Lebesgue's innovations were initially criticized by some of the more traditional members of the French mathematical community. It was perhaps for this reason that Lebesgue himself occasionally questioned the value of his own contributions. In time, however, Lebesgue's work came to be regarded as one of the most revolutionary mathematical developments of the last hundred years.

In addition to his research interests, Lebesgue strove to make higher mathematics accessible to his students. For example, he discussed the historical background of a particular topic as a means of explaining why it interests mathematicians. He also suggested that one way for a teacher to maintain a fresh point of view was to avoid reuse of previous lecture notes.

## SELECTED REFERENCES

Grattan-Guinness, Ivor. "Integral, Content and Measure." In *Companion Encyclopedia of the History and Philosophy of the Mathematical Sciences* (pp. 21–29). Ivor Grattan-Guinness (ed.). London, England: Routledge, 1994.

Hawkins, Thomas. "Henri Lebesgue." In *Dictionary of Scientific Biography* (pp. 110–112). Vol. 8. Charles C. Gillispie (ed.). New York: Scribner's, 1973.

———. *Lebesgue's Theory of Integration.* New York: Chelsea, 1975.

Kline, Morris. *Mathematical Thought from Ancient to Modern Times.* New York: Oxford University Press, 1972, pp. 1040–1051.

Lebesgue, Henri. *Measure and the Integral* (pp. 1–7). Kenneth O. May (ed.). San Francisco, CA: Holden-Day, 1966.

DANIEL S. ALEXANDER

# LEGENDRE, ADRIEN MARIE (1752–1833)

One of the pioneers of the theory of elliptic functions. Shortly after completing his studies at the Collège Mozarin in Paris, Legendre was appointed professor of mathematics at the École Militaire and, subsequently, at the École Normale. In 1782 his work on ballistics received the prize from the Berlin Academy, and he was appointed to the French Academy in 1783. During the French Revolution, he was a member of the council appointed to introduce the decimal system and of the commission to determine the length of the meter.

His mathematical work was broad and deep. In number theory, he formulated the law of quadratic reciprocity. A number $n$ is called a quadratic residue modulo the prime $p$ if there is a perfect square $m^2$ for which $p$ divides $n - m^2$. If $p$ and $q$ are two odd primes, the law of quadratic reciprocity states that $p$ and $q$ are each quadratic residues of the other or each not quadratic residues of the other precisely when at least one of $(p - 1)/4$ and $(q - 1)/4$ is an integer. Thus, 11 and 37 are each quadratic residues of the other, because 37 divides $14^2 - 11 = 37 \cdot 5$ and 11 divides $9^2 - 37 = 4 \cdot 11$; indeed while $(11 - 1)/4$ is not an integer, $(37 - 1)/4 = 9$ is. His work on spherical harmonics includes the development of the celebrated Legendre polynomials. He also was among the first to consider the statistical method of least squares.

*See also* History of Mathematics, Overview; Least Squares; Number Theory, Overview

## SELECTED REFERENCE

Gillispie, Charles C., ed. *Dictionary of Scientific Biography.* Vol. 8. New York: Scribner's, 1973, pp. 134–143.

GEORGE E. ANDREWS

# LEIBNIZ, GOTTFRIED WILHELM (1646–1716)

German mathematician who cofounded the calculus, along with Isaac Newton (1642–1727). Central to the theory each man developed were powerful symbolic tools for the computation of derivatives and integrals that reflected their discovery that determining the area under a curve was the inverse process of finding its tangent. Newton and Leibniz, however, developed their ideas independently and approached the calculus from very different perspectives. Indicative of these differences was Leibniz's invention of a flexible notational system still in use today. For example, the differential symbols $dy$ and $dx$ and the integral symbol $\int$ are due to Leibniz. Leibniz also was more concerned than Newton with the articulation of formulas for techniques such as integration by parts and the product rule, techniques Newton almost certainly knew but did not bother to state explicitly.

Although no one argues that Newton was the first to develop many fundamental notions of the calculus, Leibniz published many of these results before Newton, since the latter was apparently overly fearful of criticism from those who did not understand his work. This helped fuel a needless dispute over who deserved priority for the discovery of the calculus. In 1712, the British Royal Society, of which Newton was then president, actually charged Leibniz with plagiarizing Newton's work. Few, if any, historians now believe this charge to be true, but Leibniz's reputation was nonetheless damaged.

Leibniz is most famous for his numerous mathematical discoveries, including one of the earliest calculating machines and the first use of the concept of a function (Boyer 1991, 406), but he also made fundamental contributions to philosophy. He created a symbolic language called the *characteristica universalis* that he hoped would have the flexibility to represent complex arguments in all fields of thought. Although he fell short of this grandiose goal, his investigations were an important first step in the development of symbolic logic. It was perhaps this same awareness of the power of good symbolism that led him to develop such convenient notation for both the integral and the derivative.

Leibniz was in many respects a mathematical amateur. He studied philosophy in college, and like many seventeenth-century mathematicians, was a lawyer. He worked his entire adult life as a courtier for various German principalities. Although his diplomatic travels brought him into contact with mathematicians in Paris, London, and the Netherlands, he learned mathematics virtually on his own.

*See also* Calculus, Overview; Logic; Newton

## SELECTED REFERENCES

Aiton, Eric. *Leibniz, A Biography.* Bristol, England: Hilgar, 1985.

Boyer, Carl B. *A History of Mathematics.* 2nd ed., rev. Revised by Uta C. Merzbach. New York: Wiley, 1991, pp. 391–414.

Hall, Alfred Rupert. *Philosophers at War: The Quarrel Between Newton and Leibniz.* Cambridge, England: Cambridge University Press, 1980.

Hofmann, Joseph. *Leibniz in Paris, 1672–1676.* Cambridge, England: Cambridge University Press, 1974.

Hofmann, Joseph, et al. "Gottfried Wilhelm Leibniz." In *Dictionary of Scientific Biography* (pp. 149–168). Vol 8. Charles C. Gillispie (ed.). New York: Scribner's, 1973.

Katz, Victor. *A History of Mathematics: An Introduction.* New York: HarperCollins, 1993, pp. 428–493.

Kline, Morris. *Mathematical Thought from Ancient to Modern Times.* New York: Oxford University Press, 1972, pp. 342–383.

DANIEL S. ALEXANDER

# LENGTH, PERIMETER, AREA, AND VOLUME

Measurement is one of the principal real-world applications of mathematics. It bridges two critical realms of mathematics: geometry or spatial relations and real numbers. Done well, education in measurement can connect these two realms, each providing conceptual support to the other. Indications are, however, that this potential is usually not realized.

## STUDENT PERFORMANCE ON MEASUREMENT TASKS

Many students use measurement instruments or count units in a rote fashion and apply formulas to attain answers without meaning (Clements and Battista 1992). For example, less than 50% of seventh graders can determine the length of a line segment when the beginning of the ruler used to measure it is not aligned at the beginning of the line segment.

Only about 10% of seventh graders and 52% of entering secondary students can find the area of a square given the length of one of its sides. Misconceptions are common. Many students believe that the angle sum of a quadrilateral is the same as its area and that the area of a quadrilateral can be obtained by transforming it into a rectangle with the same perimeter. Research has not just identified problems. Studies also have contributed to our understanding of the critical mathematical aspects of measurement. More important, they have helped us understand how students think about measurement concepts and thus can guide educational efforts to ameliorate weaknesses in students' performance. These are discussed in the following sections. The final section discusses implications for teaching.

## BACKGROUND: MATHEMATICAL CONCEPTS OF MEASUREMENT

There is a fundamental difference between scientific and mathematical measurement. Scientific measurement is observational and always includes some degree of error. Mathematical measurement is based on certain foundational concepts. We will present these concepts briefly in the context of length measurement (for a discussion, see Wilson and Osborne 1988).

1. Number assignment: Given a pair of points $A$ and $B$, there is exactly one nonnegative number, $d(A,B) > 0$, that is the length of segment $AB$.

2. Comparison: If segment $AB$ is contained in segment $AC$, then $d(A,B) < d(A,C)$.

3. Congruence: Segment $AB$ congruent to segment $CD$ means that $d(A,B) = d(A,C)$.

4. Unit: There is a line segment that can be assigned the length 1.

5. Additivity: A line segment made by joining two distinct line segments has a length equal to the sum of the lengths of the joined segments. This is important, as now we know that the structure of addition can guide children's thinking about measurement, and vice versa.

6. Archimedean iteration: If a point $B$ is between points $A$ and $C$ on a line, then a counting number $n$ can be found so that $n \times d(A,B) > d(A,C)$. Thus, some whole number of copies or iterations of $AB$ laid end-to-end is all that is needed to get beyond point $C$. This notion is the basis for developing number lines and ruler use.

These ideas should be encountered by children in an informal, significant setting, not introduced formally.

The history of measurement is the story of a continuing effort to achieve standardization of measures and measurement processes; such understandings must be constructed by children over time.

## LENGTH

An early emphasis in research on measurement focused on conservation—the idea that a physical quantity does not change during certain transformations. Jean Piaget and his collaborators (Piaget and Inhelder 1967; Piaget et al. 1960) found that children younger than five years would judge length in terms of endpoints only; therefore, a line segment and a bent path with the same endpoints would be judged to have the same length. These researchers claimed that children achieve an understanding of linear and area measurement only at about age nine. Such understandings depend on mental operations such as subdivision and ordering.

If two equal-length strips are cut as shown in Figure 1, younger children will view one segment as longer. They may judge the left strip to be longer, because it has longer pieces, or they may judge the right strip to be longer, because it has more pieces (Carpenter and Lewis 1976). When they can coordinate both the subdivision and reconnection of the parts and the ordering of the positions of the parts, they can understand linear measurement, according to the Piagetian position. When first to third graders were given number line estimation problems to be solved with specially designed rulers, they progressively altered their estimation strategies from strictly sequential ones (that neglect the matter of distance between points on the number line) to ones that incorporate elements of proportional reasoning (Petitto 1990). Their acquisition of a concept of equal intervals required learning both the convention of equal intervals and conservation of length.

Children develop, for example, three types of strategies for solving different length problems: (1) some students, rather than segmenting lengths and connecting the number for the measure with the length of the line segment, applied general strategies such as visual guessing of measures; (2) most students drew hash marks, dots, or line segments to partition lengths; they needed to have perceptible units such as these to quantify the length; and (3) a few other students did not use physical partitioning; however, they did use quantitative concepts in discussing the problems, drew proportional figures, integrating the number for the measures with the lengths of the segments, and sighted along line segments to assign them a length measure. These students have created an abstract unit of length, a "conceptual ruler" that they can project onto unsegmented objects. Thus, children must create an abstract unit of length (Clements et al. 1997; Steffe 1991). This is not a static image, but rather an interiorization of the process of moving (visually or physically) along an object, segmenting it, and counting the segments.

This study also has implications for teaching. Working with length activities using the computer program Logo helped these students develop more sophisticated strategies (Clements et al. 1997; Clements and Meredith 1994). Other research indicates that Logo can help young children learn about measurement, because Logo's turtle graphics provides an arena in which young children may use units of varying size, define and create their own units, maintain or predict unit size, and create length rather than endpoint representations through either iterative or numeric distance commands (Clements and Battista 1992).

On and off computer, teachers cannot assume that children understand measure fully, even if they can complete textbook exercises involving reading a pictured ruler aligned with a pictured object. More worthwhile experiences include measuring with different size units and different materials, from paper clips to inch cubes to rulers. Students should also compare the measures of different lines, measure the same length with different units, and use computer programs that present measurement tasks with varying unit sizes (Clements and Meredith 1994).

## AREA, PERIMETER, AND VOLUME

Understanding of area measure, according to the Piagetian position, involves coordinating many ideas.

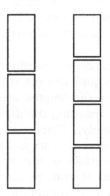

**Figure 1**

Children must understand that subtracting equal parts from equal wholes results in equal remainders and that decomposing and rearranging shapes does not affect their area. Important is the ability to coordinate two linear dimensions to build the idea of a two-dimensional space. This explains why understanding of area is often not fully developed until twelve to thirteen years of age.

Recent research has revealed that just structuring a two-dimensional region is not a simple task. For example, while most elementary students have no difficulty covering a region with tiles and finding area by counting individual tiles, many cannot *represent* the results of such actions in a drawing. These children do not interpret arrays in terms of rows and columns, which obstructs their learning about area measurement and probably makes formula use next to meaningless. Some elementary students can make an array drawing; however, many of these students still cannot apply their multiplication and linear measurement skills to determine the area of an array (Outhred and Mitchelmore 1992).

Research has identified "rules" that children use to make area judgments. For example, four- and five-year-olds match one side of figures to match their areas. They also use height and width rules to make area judgments. Children from six to eight years use a linear extent rule, such as the diagonal of a rectangle. Only after this age do most children move to multiplicative rules (Forman 1993). Elementary school children often confuse perimeter and area. For example, they believe that counting the units around a figure gives its area. Many adolescents can conserve area, but believe that the perimeter of the figure is also conserved.

Differentiating and coordinating these two measures is a difficult task. Teachers help when they offer many experiences comparing areas, encouraging children to use their own strategies (even one-by-one counting) rather than teaching rote rules. Children should also build different shapes (e.g., rectangles) with the same area, checking the perimeter for each. They should guess and check how many unit squares fit into various rectangular frames. When three dimensions are involved, it is no surprise that people have similar and sometimes greater difficulties with volume concepts. Again, informal educational opportunities, starting with counting, are important.

## PEDAGOGICAL IMPLICATIONS

While researchers and educators have taken a variety of positions, most agree that understanding measurement is an important conceptual area, as well as a skill. Teaching of measurement should build on children's intuitive spatial understanding and help children establish connections between this understanding and number. Counting preassigned units or multiplying dimensions in formulas alone will not result in quality learning. As we have seen, many students apply formulas without understanding how they work or what they mean. Such rote practice is especially harmful to students' development of understanding of the nature of units of measurement in one, two, and three dimensions. Experiences with qualitative comparison are indicated. Young children should have a variety of experiences comparing the size of objects in various dimensions; for example, finding all the objects in the room that are as long as their forearm. Connections should be made to number through counting, and then to arithmetic, including simple proportional relations such as doubling or halving, using real-world materials and problems.

Close observation of children's strategies for solving measurement problems is pedagogically useful. For example, teachers presenting children with length tasks such as sketching a rectangle with particular dimensions should observe if students are partitioning the lengths (Clements et al. 1997). Many students draw hash marks, dots, or line segments to partition or segment lengths but do not maintain equal-length parts. They need to have perceptible units to quantify the length. Continued presentation of such tasks, such as drawing a $10 \times 5$ cm rectangle, with an emphasis on equal internal partitioning and the creation of different units of length, will help these students. Eventually, students should use a variety of nonstandard and standard units to discover that the results of counting to yield a measure depend on the unit. (Note that some research indicates that standard instruments such as rulers may support reasoning more than nonstandard units such as paper clips. The traditional "nonstandard then standard" sequence may be overly simplistic.)

Students who do not partition lengths may be using one of two quite distinct strategies. Some can mentally partition the lengths using a "conceptual ruler" (i.e., they are using the third type of strategy identified previously). These students should be challenged with more difficult tasks (see Figure 2). If they do not show signs of this ability, they may be using the first kind of strategy; such students need to engage in partitioning and iterating lengths, continually tying the results of that activity to their counting.

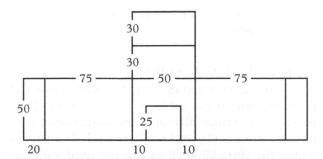

**Figure 2** *Students fill in all the missing measures (Clements et al. 1995).*

For example, they might draw a favorite toy, measuring the toy and drawing it using the same (and, later, a smaller) measure.

The research on area and volume suggest that even young children can base their judgments of quantity on several dimensions, but in many contexts, such as judging the amount of water in a glass, they use only one. When they do attend to multiple dimensions, they often use an additive rather than a multiplicative rule. In middle school, students start to use multiplicative reasoning, though even older students are not consistent in this regard; many still calculate the perimeter when they need to find the area. Instruction on formulas for area and perimeter may actually hinder the development of both concepts and skills in measurement. Unfortunately, such bare-bones instruction on formulas is all most mathematics textbooks include (Fuys et al. 1988).

Thus, for example, simple counting of units to find area (achievable by preschoolers) leading directly to teaching formulas is a recipe for instructional disaster. Instead, educators should build upon young children's initial spatial intuitions and appreciate the need for students to: (a) construct the idea of measurement units (including development of a measurement sense for standard units; e.g., finding common objects in the environment that have a unit measure); (b) have many experiences covering quantities with appropriate measurement units and counting those units; (c) spatially structure the object they are to measure (e.g., linking counting by groups to the structure of rectangular arrays; building two- and three-dimensional multiplicative concepts); (d) construct the inverse relationship between the size of a unit and the number of units used in a particular measurement; and (e) construct two- (and later three-) dimensional space and corresponding multiplicative relations. For example, ask students to

make boxes out of nets (2D patterns that fold into 3D shapes) and determine how many cubes fit in the boxes. Students also might determine how many tiles cover a given rectangle (sometimes the tiles could be square, other times they could be nonsquare rectangles). Students could make buildings (not prisms, but columns with different heights) and be asked to draw the buildings from different perspectives.

For middle school students, these basic ideas can be expanded to include explicit generalizations of measurement concepts and processes across different types of quantities (e.g., the foundational concepts of measurement); differences between scientific and mathematical measurement (scientific measurement is based on observation and always contains error); and the relationships between length, perimeter, area, and volume (e.g., what happens to area when the length of a side is doubled).

For most students, connections between geometric forms and numerical ideas are tenuous at best, even in situations designed to emphasize and develop these connections. Such lack of linkages would appear to limit the growth of number sense, geometric knowledge, and problem-solving ability. Studying more measurement and geometry may ameliorate this situation. Using situations that help students forge such links is also indicated. Students' integration of number and geometry was especially potent and synergistic in Logo environments (especially mathematically oriented versions of Logo, e.g., Clements and Meredith 1994). In most tasks, students should use measurement as a means for achieving a goal, not only as an end in itself.

*See also* Geometry, Instruction; Logo; Measurement; Systems of Measurement

## SELECTED REFERENCES

Carpenter, Thomas P., and Ruth Lewis. "The Development of the Concept of a Standard Unit of Measure in Young Children." *Journal for Research in Mathematics Education* 7(1976):53–58.

Clements, Douglas H., and Michael T. Battista. "Geometry and Spatial Reasoning." In *Handbook of Research on Mathematics Teaching and Learning* (pp. 420–464). Douglas A. Grouws (ed.). New York: Macmillan, 1992.

———, Joan Akers, Virginia Woolley, Julie Sarama Meredith, and Sue McMillen. *Turtle Paths.* Palo Alto, CA: Seymour, 1995.

Clements, Douglas H., Michael T. Battista, Julie Sarama, Sudha Swaminathan, and Sue McMillen. "Students' Development of Length Measurement Concepts in a Logo-based Unit on Geometric Paths." *Journal for Research in Mathematics Education* 28(1997):70–95.

Clements, Douglas H., and Julie Sarama Meredith. *Turtle Math*. Montreal, Quebec, Canada: Logo Computer Systems (LCSI), 1994.

Forman, Ellice. "Middle School Students' Understanding of Area, Perimeter, Surface Area, and Volume." Unpublished manuscript, University of Pittsburgh, 1993.

Fuys, David, Dorothy Geddes, and Rosamond Tischler. *The van Hiele Model of Thinking in Geometry among Adolescents*. Reston, VA: National Council of Teachers of Mathematics, 1988.

Outhred, Lynne, and Michael Mitchelmore. "Representation of Area: A Pictorial Perspective." In *Proceedings of the Sixteenth PME Conference* (pp. 194–201). Vol. II. William Geeslin and Karen Graham (eds.). Durham, NH: Program Committee of the Sixteenth PME Conference, 1992.

Petitto, Andrea L. "Development of Numberline and Measurement Concepts." *Cognition and Instruction* 7(1990):55–78.

Piaget, Jean, and Bärbel Inhelder. *The Child's Conception of Space*. New York: Norton, 1967.

———, and Alina Szeminska. *The Child's Conception of Geometry*. London, England: Routledge and Kegan Paul, 1960.

Steffe, Leslie P. "Operations that Generate Quantity." *Learning and Individual Differences* 3(1991):61–82.

Wilson, Patricia S., and Alan Osborne. "Foundational Ideas in Teaching about Measure." In *Teaching Mathematics in Grades K–8: Research Based Methods* (pp. 78–110). Thomas R. Post (ed.). Boston, MA: Allyn and Bacon, 1988.

DOUGLAS H. CLEMENTS
MICHAEL T. BATTISTA

# LESSON PLANS

A form of instructional planning, which is a primary responsibility of every classroom teacher. Often, the success or failure of the instructional experience for both teacher and student is directly related to the thoroughness of such planning from broad-based goals to seemingly trivial, procedural issues. A lesson plan is written to organize the flow of information among students and teachers. It is not designed to disseminate information exclusively from the teacher to the students; rather, a lesson plan can, and should, be designed with many potential forms of information flow in mind. The development of goals is an important first step in the lesson-planning process. When writing lesson plans with a mathematics focus, teachers may find it helpful to consider the five goals of the National Council of Teachers of Mathematics (NCTM 1989). These goals provide a framework for the development of long-term instructional planning as well as individual lesson plans.

## CURRICULUM MAPPING

Curriculum mapping includes both horizontal planning, within the grade level, among the disciplines; and vertical planning, the sequence of curriculum from grade level to grade level. As a part of long-term curriculum mapping, the need exists to develop individual lesson plans. Individual lesson plans provide a vehicle for the teacher to plan daily, short-term lessons either exclusive to a particular discipline or across two or more disciplines.

## A LESSON PLAN FORMAT
### Critical Elements of the Daily Lesson Plan

With the daily lesson plan in mind, the critical elements to include are: (a) *objectives/outcomes* for the lesson; (b) *key concepts, skills,* and/or *strategies* to be taught; (c) *essential questions* to be asked; (d) *methodology* to be used; (e) *resources* needed; (f) *procedure* for implementing instruction; (g) *closure;* (h) *assessment;* and (i) *lesson adaptations.* Furthermore, it is important that teachers have a clear rationale for choosing all the elements of the lesson plan and relating the learning to real-life needs and experiences.

The following sections include a more in-depth explanation of the critical elements of the daily, lesson-plan format and an example that can guide teachers in creating teacher-directed plans, student-centered plans, multiple grouping scenarios, mini-lessons, and so forth. It also can be modified for discovery and experiential learning. A continuous cycle of the critical lesson elements can be instantaneous or can occur over long periods of time. "For all of us, the ultimate goal is more coherent, organic, and integrated schooling for American young people" (Zemelman, Daniels, and Hyde 1993, 17).

### Objectives/Outcomes

Many state boards of education and school districts have mandated adherence to the achievement of outcomes rather than merely "covering the curriculum." In this entry, we use both the terms *objective* and *outcome* cautiously, and refer to objectives with intended learning outcomes (Gronlund 1995). We define "outcomes-based education" (OBE) as "focusing curriculum, instruction, and assessments on student success in achieving the knowledge skills and attitudes through standards-based outcomes"

(NCTM 1993, 232). In our discussion, any reference to outcomes is based upon a direct connection to these standards.

Setting student outcomes includes recognizing what the learner is to know and be able to do upon completion of the learning experience. This includes a broad-based understanding of the nature of learning, curriculum implementation, assessment, and classroom climate. It is important to examine each of these areas when developing meaningful outcomes. They should be very specific and valuable for teaching. The two broad categories that we have found to be particularly helpful are cognitive and social outcomes.

*Cognitive Objectives/Outcomes* Bloom's taxonomy of cognitive objectives contains six major classes, that is, knowledge, comprehension, application, analysis, synthesis, and evaluation (Bloom et al. 1956), and is widely used by educators planning lessons (Farrell and Farmer 1988). Figure 1 presents detailed descriptors that connect Bloom's taxonomy, Gardner's (1983) levels of thinking and seven comprehensive categories or "intelligences," and student performance. Operant verbs are the primary connectors to classify student performance and provide specific examples for day-to-day practice.

In the Second International Mathematics Study (SIMS), Weinzweig and Wilson (1977) adapted Bloom's six cognitive levels to four levels of cognitive mathematical capability expected of students: computation, comprehension, application, and analysis (NCTM 1993). Their classification replaced knowledge with computation and eliminated the synthesis and evaluation levels. In Figure 1 (Quattromani and Nathanson-Mejia 1996), Gardner's levels of thinking relate to Bloom's taxonomy and student performance followed by several mathematical examples. This matrix provides a visual framework for moving from concrete to abstract exemplars of intended student behavior, as well as for arranging educational behaviors from simple to complex. This is based on the idea that a simple behavior may be integrated with other equally simple behaviors to form a more complex behavior. Therefore, the following relationships include: (a) *retention of information* (linguistic and logical-mathematical) to *knowledge;* (b) *micro skills* (linguistic, logical-mathematical, spatial, bodily-kinesthetic, and musical) to *simple analysis, application,* and *comprehension* (Note: Analysis is further classified to be simple or complex); (c) *critical thinking* (linguistic, logical-mathematical, spatial, bodily-kinesthetic, and musical) to *evaluation* and *complex analysis;* (d) *creative thinking* (linguistic, logical-mathematical, interpersonal and intrapersonal) to *synthesis;* and (e) *problem solving/decision making* (linguistic, logical-mathematical, interpersonal and intrapersonal) to *evaluation, synthesis,* and *analysis.* We have determined that the inclusion of higher-level thinking skills and cognitive outcomes is crucial to a more comprehensive notion of instructional planning. The expert development of cognitive outcomes allows for a clearer understanding of the assessment of student performance. If teachers know what they are trying to assess, they are better able to determine the most appropriate assessment.

*Social Objectives/Outcomes* Creating a positive classroom environment is a major responsibility of teachers. This category encompasses those expectations for student affective, communicative, and physical needs with particular emphasis on classroom management issues. Creating a positive classroom environment requires that both teachers and students have an active role in discourse, particularly questioning, listening, clarifying, and problem posing (NCTM 1993). This includes determining grouping strategies, such as whole class, small group, collaborative, and/or cooperative. Emphasis must be placed upon student expectations for appropriate behavior in relationship to the educational task at hand. Additionally, the teacher, as instructional planner, has the responsibility to recognize student needs while facilitating the best possible pedagogy. "What students learn is fundamentally connected with HOW they learn it. What students learn—about particular concepts and procedures as well as about thinking mathematically—depends on the ways in which they engage in mathematical activity in their classrooms" (NCTM 1991, 21).

*Example* If you are introducing students to a new manipulative material it is important to allow them time for free play and exploration (Fuys and Tischler 1979). It is truly a natural process that satisfies human curiosity. Often, when exploration time is not planned, classroom management deteriorates. Furthermore, it is necessary to bring closure to this exploration. This may be accomplished by involving the class in a brief processing discussion of what they learned or a brief mathematics learning-log entry.

The major shifts described and woven into the *Professional Standards* (NCTM 1991) indicate the importance of not only cognitive outcomes but also social outcomes. They call for a shift:

- toward classrooms as mathematical communities (and) away from classrooms as simply a

**Figure 1** *Classification of student performances according to level of thinking*

| Levels of Thinking and Gardner's Intelligences | Relationship to Bloom's Taxonomy | Student Performances with Operant Verbs | Mathematical Examples |
|---|---|---|---|
| **Retention of Information**<br><br>**Intelligences:**<br>Linguistic<br>Logical-Mathematical | **Knowledge (Recall)** | Define<br>List<br>Name<br>Repeat<br>Label<br>Compute | **Define** equilateral<br>**List** five even numbers<br>**Name** six geometric shapes<br>**Repeat** oral directions<br>**Label** a given angle<br>**Add** 117 + 45 |
| **Micro Skills**<br><br>**Intelligences:**<br>Linguistic<br>Logical-Mathematical<br>Spatial<br>Bodily-Kinesthetic<br>Musical | **Analysis (Simple)** | Sort/categorize<br>Compare/contrast<br><br>Determine cause/effect<br><br>Determine main idea | **Classify** an unfamiliar organism according to phylum; and **compare** the genus and species.<br>**Determine** the causes of air pollution, and the effects.<br>**Determine** the theme of Greenpeace. |
| | **Application** | Apply rules or principles<br>Calculate | **Find** the area of a trapezoid.<br>**Calculate** the average of your quizzes. |
| | **Comprehension** | Paraphrase/restate<br>Describe<br><br>Give examples/sketch<br><br>Identify<br>Read symbols/Translate<br>Explain in own words/tell | **Restate** the definition of congruent.<br>**Describe** several assumptions of Euclidean geometry.<br>**Give an example** of a non-Euclidean problem/draw a **sketch.**<br>**Read** the weather map, **identify** the **symbols, explain** the predicted conditions. |
| **Critical Thinking**<br><br>**Intelligences:**<br>Linguistic<br>Logical-Mathematical<br>Spatial<br>Bodily-Kinesthetic<br>Musical | **Evaluation**<br><br><br><br>**Analysis (Complex)** | Evaluate<br>Judge<br>Assess<br>Criticize<br>Classify<br>Analyze<br>Distinguish<br>Compare/contrast | **Evaluate** the **accuracy, credibility, and fairness** of a statistical newspaper article, and offer a **critical review.**<br>**Compare and contrast** mathematics education in the United States, Russia, and Japan.<br>**Analyze** the outcomes. |
| **Creative Thinking**<br><br><br><br>**Intelligences:**<br>Linguistic<br>Logical-Mathematical<br>Interpersonal<br>Intrapersonal | **Synthesis** | Propose<br>Design<br>Create<br>Plan<br><br>Formulate<br><br>Construct/prepare | **Propose** a recycling plan.<br>**Design** a recycling container.<br>**Create** word problem.<br>**Plan** a campaign to increase public awareness about recycling.<br>Correctly **formulate** a chemistry experiment.<br>**Construct** a tetrahedron model. |
| **Problem Solving/ Decision Making**<br><br>**Intelligences:**<br>Linguistic<br>Logical-Mathematical<br>Interpersonal<br>Intrapersonal | **Evaluation**<br><br><br>**Synthesis**<br><br><br>**Analysis** | Evaluate solution/decision options<br><br>Propose a solution<br>Create decision options<br><br>Define the problem<br>Establish goals/criteria<br>Analyze resources | **Recommend** a solution based upon mathematical rationale.<br>**Create solution options** to local pollution problems.<br>**Describe** the duration, scope, and consequences of local air pollution. |

collection of individuals; for example, paired, problem-solving, or cooperative learning;

- toward logic and mathematical evidence as (student) verification (and) away from the teacher as the sole authority for right answers;
- toward mathematical reasoning (and) away from merely memorizing procedures; for example, cooperative problem solving, development of student-generated problems and solutions;
- toward conjecturing, invention, and problem solving (and) away from an emphasis on mechanistic answer finding; for example, storytelling and problem posing;
- toward connecting mathematics, its ideas, and its applications (and) away from treating mathematics as a body of isolated concepts and procedures (p. 3); for example, immersion in simulation to provide real-life mathematical scenarios.

## Key Concepts, Skills, or Strategies

Key concepts, skills, and/or strategies are the ideas that remain with us for a lifetime. They help students make connections to other learning experiences and provide a foundation for cognitive development. A teacher should choose appropriate concepts both carefully and wisely; for it is this decision that ultimately guides the teacher through the many other instructional decisions that lay ahead. If the key concepts, skills, and/or strategies are clearly identified, then both the teacher and students will remain on course by reflecting back to them during instruction and closure.

## Essential Questions

Essential questions are those questions that focus upon the key concepts, skills and/or strategies to drive instruction and encourage exploration. These questions are generated by both teachers and students. Bloom's taxonomy can help guide the development of essential questions to ensure that multiple levels are addressed and investigated.

## Methodology

The teacher determines which model of teaching (Joyce and Weil 1996) will be used and whether the lesson is open or closed-ended. Models of teaching include a variety of possibilities; for example, concept attainment, direct instruction, inquiry, inductive, simulations, role play, cooperative, and so on. A closed-ended lesson would not allow choice for

students and would usually be primarily teacher-centered, whereas an open-ended lesson allows considerable choice for students and is developed with students as the central consideration. This is not to say that closed-ended lessons are inappropriate, but, simply, that teachers should know the type of lesson they are developing and why. If there is little student choice involved in the lesson, then what will the teacher do to ensure that all students might participate to the fullest of their ability?

## Resources

Determine all the resources and materials (books, paper, multimedia) needed for both teacher and students to complete the entire assignment; the method for obtaining them; and the timeline for securing all materials. Schedule any opportunities for experiential learning, whether on-site or field experience, well in advance.

## Procedure

The procedure of the lesson appears to be linear sequential; however, the concept of a continuous cycle lesson is more congruent with the reflection and decision making that underlie best practice (Zemelman et al. 1993). Not all the preceding elements of a lesson plan need to be used all the time. Securing student engagement in the learning experience is paramount. Anything else that occurs is to further pique student interest and enhance the learning experience. The following list of lesson elements provides a series of attributes to consider during comprehensive instructional planning.

*Anticipatory Set* When developing an anticipatory set, the teacher needs to consider the following:

- what information students need to achieve the outcomes;
- focusing students' attention, practicing or reviewing previous learning, or creating interest in new learning;
- developing a mental set through an interesting activity. How it shall be given to students (teacher, book, film, records, demonstration or combination);
- whether the lesson should include teacher or student modeling that is visual, verbal, or hands-on.

*Check for Understanding* Asking effective questions during the instructional cycle. (See Johnson 1982, for a detailed discussion of the art of questioning.)

- posing questions to extend students' capacity for reasoning (NCTM 1991);
- determining whether the student has acquired knowledge;
- sampling-group response . . . signaling—agree, disagree, not sure.

*Provide Opportunities for Guided and Independent Practice* Important opportunity for students to develop fluency without a teacher present or with minimal teacher guidance.

- Teacher provides written or verbal assignment(s).
- Teacher determines: How much? How often? How well?
- Teacher provides clarification and extensions and/or creates a plan for remediation.

## Closure

Johnson (1982) believes that the last five minutes of a class are at least as important as the first five minutes. The teacher must plan for effective closure; to avoid ending with the traditional "working on homework" or "running out of time" as the bell rings. This is the moment to refer to the outcomes and key concepts; to restate or review them; to clarify the instruction; to facilitate discussion and provide comments; and/or reteach to small groups, if necessary. Consider whether the students can tell you what they learned and whether it is congruent with the objectives.

## Assessment

This is a very difficult and complex part of instructional planning. Assessment is the area that should provide information to students, teachers, and parents but often causes anxiety for all constituencies. The teacher should consider both process and product assessment, as well as contextualized versus decontextualized measures.

## Lesson Adaptations

Well-developed plans provide a guide for educators in their daily practice. "The teacher of mathematics should engage in ongoing analysis of teaching and learning . . . in order to . . . adapt or change activities while teaching" (NCTM 1991, 63). The teacher must be prepared to reflect on his/her lesson plan and adapt it as needed. Reflective practitioners need to be able to learn and improve their planning and teaching (NCTM 1991; Schon 1983; Zemel-

man, Daniels, and Hyde 1993), whether it be "on-the-spot," before your afternoon section, or before the next time you teach it. An effective way of discussing the main standards addressed in a lesson can include putting an overhead of the NCTM Standards (grades 2–4, 5–8, or 9–12) on your projector and asking the students to identify which standards were addressed in the lesson.

## EXAMPLE I: AN ELEMENTARY MATHEMATICS LESSON PLAN

### Objectives/Outcomes

*Cognitive Objectives/Outcomes*

a. The student will be able to identify a variety of geometric shapes in the environment.

b. The student will be able to use a camera, draw pictures, or cut and paste magazine pictures.

*Social Objectives/Outcomes*

a. The student will be able to interact with print during read aloud to make connections with the environment.

b. The student will be able to engage in appropriate cooperative learning behaviors using self-regulation and group processing skills.

### Key Concepts/Skills/Strategies

The *concept* for this lesson is geometric shapes. The *skills* focus on the identification and differentiation of shapes and the use of media. The primary *strategies* include prediction of story events and how they relate to the concepts, as well as working in whole-class and cooperative groups.

### Essential Questions

a. Where are geometric shapes in the environment?

b. What geometric shapes are found in the environment?

### Methodology

a. Teacher-directed, whole-class, read aloud, and discussion.

b. Student-directed, teacher-facilitated, cooperative groups.

### Materials

You will need the book *The Greedy Triangle* (Burns 1994), chart paper, markers, crayons, glue, scissors, construction paper, magazines, disposable cameras, an overhead transparency for the directions, and each student's learning log.

## Procedure

*Anticipatory Set* Introduce *The Greedy Triangle* (Burns 1994). Ask for predictions about the book. Record predictions on chart paper. Read the book aloud.

### Check for Understanding

- What does the title mean to you?
- Compare predictions with story line.
- Encourage predictions as you read.
- Record responses on chart paper.

### Guided and Independent Practice

- After read-aloud of *The Greedy Triangle*, the students form cooperative groups.
- Remind students of learning-behavior expectations during cooperative-group activities.
- Show an overhead transparency with directions for the activity.
- Each group reviews the directions and decides if there are any questions or concerns.
- The teacher asks for input or questions.
- The class leaves the room together and the students take along their learning logs to record shapes they find and where they are located.
- When the class returns to the room the teacher encourages the groups to reach consensus about the shapes they will photograph.
- The teacher demonstrates the use of the camera.
- After the groups have reached consensus, each group gets a camera.
- The class goes outside again to collect their photographs.
- After the photographs are developed, each group organizes their *Geometric Shapes Photojournals*, labels each photograph, and prepares to present their findings.

To audit the contributions of individual students, the teacher may color-code markers with students' names as they write descriptions, enter information in their learning logs, and initial the contributions of other group members.

## Closure

*Small Groups* Discuss their findings and write responses in learning logs.

*Whole Class* Each group describes their *Geometric Shapes Photojournal*. The students must be able to share why they chose their photographs. Each group member must describe his or her own photograph using relevant data and appropriate responses.

## Assessment
### Products

*Geometric Shapes Photojournal*

Develop a rubric to establish criteria and levels of acceptable performance.

*Mathematics Learning Logs*

Develop focus questions to aid students' exploration of the topic during the activity.

Review learning logs for comments, descriptions, and answers to focus questions.

Write responses to learning logs.

*Mathematics Portfolios*

Is a showcase of individual, student work related to mathematics.

Contains both student- and teacher-selected artifacts.

Provides an opportunity to show student growth over time.

### Process

*Learning-Behaviors Checklist*

Teacher observes group dynamics.

Teacher uses checklist of desired learning behaviors to record and rate observations of Social Objectives/Outcomes.

*Learning Logs*

Students record group and individual learning behaviors.

## EXAMPLE II: A SECONDARY MATHEMATICS LESSON PLAN
### Objectives/Outcomes

*Cognitive Objectives/Outcomes*

a. The student will be able to discover what the important variables are in a pendulum experiment and hypothesize the relationships between them.
b. The student will be able to determine the average period of a pendulum swing for six trials.

*Social Objectives/Outcomes*

a. The student will be able to work effectively with peers to gather data and discover relationships.
b. The student will be able to recognize and discuss the elements of connection among literature, science, and mathematics using *The Pit and the Pendulum* by Edgar Allan Poe.

## Key Concepts/Skills/Strategies

The *concepts* for this lesson are pendulums and variables. The *skills* focus on data collection, recognizing variables, and determining the relationships among variables. The *strategies* include using scientific method, group decision making, and presentation of findings.

## Essential Questions

a.  What are the important variables in a pendulum?
b.  What is/are the relationship(s) between/among them?

## Methodology

a.  Teacher-directed, whole-class, read-aloud, and discussion.
b.  Student-directed, teacher-facilitated, small, cooperative-group discovery.

## Materials

The teacher will need the book *The Pit and the Pendulum* by Edgar Allan Poe. Each group will need individual experiment logs, markers, chart paper, masking tape, one ball of string, three metal washers of assorted sizes, one tape measure, and one stopwatch.

## Procedure

*Anticipatory Set* Introduce *The Pit and the Pendulum* by Edgar Allan Poe, read an excerpt (see Interactive Mathematics Program 1996).

*Check for Understanding*

- Discuss the reading.
- What is the focus of the story?
- What is a pendulum?
- What is its significance?

## Guided and Independent Practice

- After read-aloud of *The Pit and the Pendulum,* the students form groups of four.
- Remind students of learning-behavior expectations during cooperative-group activities.
- Provide written directions for experiment for each group.
- Each small group will create a pendulum, practice swinging it (noticing anything of interest), and record observations on the chart paper.
- Students will create up to four (4) pendulums as well as vary the length of the string.

- Students pose mathematical relationships and see if they can verify their predictions.
- Students use experiment logs to record hypotheses, observations, revisions, and conclusions.
- Students will record interactions and impressions of group process as well as connections among mathematics, science, and literature.

## Closure

*Small Groups* Discuss findings and write responses on chart paper and experiment logs.

*Whole Class*

a.  Each small group will share one of their predictions, observations, revisions, and conclusions with the class.
b.  Debrief the group process, including the connections made among the disciplines.
c.  Provide opportunities for clarification of any areas of confusion and/or contributions by students.

## Assessment

*Products* Collect the experiment logs, record contributions, and write comments.

*Process* Using a behavior checklist, determine level of interactions by each student and the cooperative strategies of the small groups.

## Lesson Adaptations

The focus of teacher reflection, self-assessment, and adaptation centers upon his/her ability to engage the students in an integrated mathematics lesson using a variety of instructional modalities. All students, regardless of limitations, must be included to the fullest extent practicable. If the teacher is having difficulty finding solutions to particularly difficult student learning issues, he or she should consult with other teachers and resource personnel whose responsibility it is to collaborate with the classroom teacher for optimum student learning.

## SUMMARY

The lesson plan format we have suggested and described is adaptable for discovery, manipulative-based, activity-centered, teacher- or student-directed, and/or collaborative lessons. It also can be modified for students with special needs. We have found it to be an effective format for planning lessons to reach all students.

*See also* Assessment of Student Achievement, Overview; Questioning; Task Analysis

## SELECTED REFERENCES

Bloom, Benjamin S., Max D. Engelhart, Edward J. Furst, Walker H. Hill, and David R. Krath-Wohl, eds. *Taxonomy of Educational Objectives*. New York: McKay, 1956.

Burns, Marilyn. *The Greedy Triangle*. New York: Scholastic, 1994.

Farrell, Margaret A., and Walter A. Farmer. *Secondary Mathematics Instruction: An Integrated Approach*. Providence, RI: Janson, 1988.

Fuys, David J., and Rosamond W. Tischler. *Teaching Mathematics in the Elementary School*. Boston, MA: Little Brown, 1979.

Gardner, Howard. *Frames of Mind: The Theory of Multiple Intelligences*. New York: Basic Books, 1983.

Gronlund, Norman E. *How to Write and Use Instructional Objectives*. 5th ed. Engelwood Cliffs, NJ: Merrill, 1995.

Interactive Mathematics Program. *The Pit and the Pendulum*. Berkeley, CA: Key Curriculum, 1996.

Johnson, David R. *Every Minute Counts*. Palo Alto, CA: Seymour, 1982.

Joyce, Bruce, and Marsha Weil. *Models of Teaching*. 5th ed. Boston, MA: Allyn and Bacon, 1996.

National Council of Teachers of Mathematics. *Assessment Standards for School Mathematics, Working Draft*. Reston, VA: The Council, 1993.

———. *Curriculum and Evaluation Standards for School Mathematics*. Reston, VA: The Council, 1989.

———. *Principles and Standards for School Mathematics*. Reston, VA: The Council, 2000.

———. *Professional Standards for Teaching Mathematics*. Reston, VA: The Council, 1991.

Quattromani, Mary E., and Sally Nathanson-Mejia. *Developing Content Literacy for ESL and Native English Speaking Students*. Working manuscript, 1996.

Schon, Donald E. *The Reflective Practitioner: How Professionals Think in Action*. New York: Basic Books, 1983.

Vygotsky, Lev S. *Mind in Society: The Development of Higher Psychological Processes*. Michael Cole, Vera John-Steiner, Sylvia Scribner, and Ellen Souberman (eds.). Cambridge, MA: Harvard University Press, 1978.

Weinzweig, A. I., and J. W. Wilson. *Second IEA Mathematics Study: Suggested Tables of Specifications for the IEA Mathematics Tests*. Working Paper I. Wellington, New Zealand: IEA International Mathematics Committee, 1977.

Zemelman, Steven, Harvey Daniels, and Arthur Hyde. *Best Practices: New Standards for Teaching and Learning in America's Schools*. Portsmouth, NH: Heinemann, 1993.

LIBBY QUATTROMANI
LYN TAYLOR

# L'HOSPITAL, GUILLAUME FRANÇOIS ANTOINE, MARQUIS DE (1661–1704)

Author of the first textbook on calculus, *Analyse des Infiniment Petits,* published in Paris in 1696. The book consisted largely of the lecture notes of his teacher, Johann Bernoulli (Jean I), who, in return for a regular salary, sent his mathematical discoveries to L'Hospital to be used in any manner L'Hospital thought fit.

The *Analyse* contained the basic Leibnizian formulae for differentiation and applied them to problems such as the construction of tangents to curves and the determination of maximum and minimum points. The text also contained "L'Hospital's rule," a result actually discovered by Johann Bernoulli, which states that the limiting value of the ratio of two quantities, both of which approach zero, is the same as the limiting value of the ratio of their derivatives. L'Hospital also wrote a major work on conics, *Traité Analytique des Sections Coniques,* which was published posthumously in 1707. This, together with the *Analyse,* formed the standard texts on calculus and conics used throughout the eighteenth century.

## SELECTED REFERENCES

Boyer, Carl B. *A History of Mathematics*. 2nd ed. rev. Revised by Uta C. Merzbach. New York: Wiley, 1989.

Eves, Howard W. *An Introduction to the History of Mathematics*. 5th ed. Philadelphia, PA: Saunders College Publishing, 1983.

Newman, James R. *The World of Mathematics*. 4 vols. London, England: Allen and Unwin, 1960.

Struik, Dirk J. *A Concise History of Mathematics*. New York: Dover, 1948.

ROD HAGGARTY

# LIBERAL ARTS MATHEMATICS

College courses in mathematics appreciation for students in nonscience fields. Typically, these students are in the humanities or the less quantitative areas of the social sciences who have no specific mathematics requirements in their major but who are required to fulfill a collegewide requirement of intellectual breadth by completing one or two courses in mathematics or related fields such as computer science or statistics. A smaller but very significant pool of students consists of those with a liking for mathematics (often a result of happy experiences in school or solid successes with other college courses) who enroll out of pure interest. Courses that concentrate on techniques such as algebraic manipulation and programming rather than on concepts often are ruled out by such a requirement. At the same time, many courses that are rich in conceptual content, such as calculus and linear algebra, are usually beyond these students because of their inadequate technical background. Thus a new set of courses has been created.

There are two families of elementary mathematics courses that have existed for some time and that are often used as liberal arts mathematics courses: finite mathematics and number systems. Finite mathematics courses usually center around probability experiments in which the set of possible outcomes is finite, such as rolling dice, playing a lottery, or conducting a political preference poll. The analysis of such experiments requires careful logical reasoning, and the use of set notation and related devices such as tree diagrams is often helpful. Indeed, such courses are often entitled "Sets, Logic, and Probability." This course can serve as an excellent introduction to a course in elementary statistics; for this reason it is sometimes recommended for business students. But this dual use can alter its intellectual content in ways that run counter to the intent of the liberal arts requirement. The number systems course examines the foundation of the systems of integers, rational numbers, and, to some extent, real and complex numbers. The course is often a requirement for elementary education students. A major goal is to demonstrate that the "laws" of arithmetic such as $(-a)(-b) = ab$ are not arbitrary but arise as logical consequences of basic axioms. Enrichment material on numerical notation and computational tricks is often included. Without the motivating desire to be well prepared to teach mathematics in elementary school, however, many liberal arts students find it hard to sustain interest in such material for an entire term.

Topics from finite mathematics and number systems courses are often incorporated in modern liberal arts mathematics courses, but usually only as units or single chapters. The older courses may be considered as elementary explorations of mathematics *in depth;* the newer course is an exploration *in breadth.* Topics considered include, in addition to those mentioned above, elementary number theory, graph theory, fractal geometry, plane tessellation, systems of voting, methods of fair division, rigid motion and symmetry, gnomonic growth in nature, linear programming, direction fields, cryptanalysis, logic circuits, mathematical machines, transfinite numbers and their arithmetic, Markov chains, game theory, non-Euclidean geometry, four-dimensional geometry, and elementary group theory. Experienced teachers will often modify their syllabus as the course proceeds in response to questions and evidence of class enthusiasm. It is not uncommon to see five or more separate topics in a fourteen-week semester. Additionally, students are often required to find and explore a mathematical topic on their own—often one that is at least tangentially related to their major field. (The teacher must, of course, be well prepared with suggestions and advice.) This exploration may result in a five- to eight-page illustrated paper, which forms a part of the student's portfolio. In some versions of the course, the last two weeks are conducted like a miniature scientific meeting, at which the students give short illustrated lectures on their chosen topics.

At least part of the inspiration for liberal arts mathematics courses stems from the expository literature of popular mathematics aimed at the general public.

*See also* Cryptography; Discrete Mathematics; Game Theory; Group Theory; Logic

## SELECTED REFERENCES

Abbott, Edwin A. *Flatland, a Romance of Many Dimensions.* New York: Dover, 1952.

Gamow, George. *One, Two, Three . . . Infinity.* New York: Dover, 1988.

Garfunkel, Solomon A., proj. dir. *For All Practical Purposes.* 2nd ed. New York: Freeman, 1991.

Gudder, Stanley. *A Mathematical Journey.* New York: McGraw Hill, 1994.

Hogben, Lancelot. *Mathematics for the Million.* 4th ed. New York: Norton, 1968.

Newman, James R., ed. *The World of Mathematics.* 4 vols. New York: Simon and Schuster, 1956.

Stein, Sherman K. *Mathematics, the Man-Made Universe.* New York: Freeman, 1976.

Tannenbaum, Peter, and Robert Arnold. *Excursions in Modern Mathematics.* 2nd ed. Englewood Cliffs, NJ: Prentice Hall, 1995.

Wolfe, Harold E. *Introduction to Non-Euclidean Geometry.* New York: Holt, 1945.

ROBERT J. BUMCROT

# LIE, SOPHUS (1842–1899)

Norwegian mathematician who, in the 1870s, applied the abstract group concept to the study of differential equations. This marked the first important application of group theory, a subject created by Evariste Galois around 1830 in his investigations into the solution of equations, to a topic not directly related to the theory of equations. In turn, this led

Lie to develop the theory of continuous transformation groups, often called Lie groups. A simple example of a continuous group is the set of rotations of a circle through $t$ degrees as $t$ varies continuously between 0 and 360.

Like his countryman, Niels Abel, almost fifty years before him, Lie felt that living in Norway isolated him from the international mathematical community, so he decided to travel to Berlin and Paris. Unfortunately, the Franco-Prussian War broke out soon after he arrived in Paris in the summer of 1870. Since Lie was an avid hiker he decided to walk back to Norway via the French Alps. Soon after he began his journey, he was briefly jailed by the French under suspicion of being a German spy. He used his time in jail wisely, making several important mathematical discoveries. Upon his release he resumed his hike.

*See also* Group Theory

## SELECTED REFERENCES

Gray, Jeremy. "Differential Equations and Groups." In *Companion Encyclopedia of the History and Philosophy of the Mathematical Sciences* (pp. 470–474). Ivor Grattan-Guinness, ed. London, England: Routledge, 1994.

Hawkins, Thomas. "The Birth of Lie's Theory of Groups." *Mathematical Intelligencer* 16(2)(1994):6–17.

Stewart, Ian. "Lie Groups." In *Companion Encyclopedia of the History and Philosophy of the Mathematical Sciences* (pp. 761–765). Ivor Grattan-Guinness, ed. London, England: Routledge, 1994.

Yaglom, Isaak Moiseevich. *Felix Klein and Sophus Lie.* Boston, MA: Birkhäuser, 1988.

DANIEL S. ALEXANDER

# LIMITS

Let us consider the following algebraic relation given by $y = 3x - 2$. A table such as the following may be prepared:

| $x$ | $y$ |
| --- | --- |
| 0.5 | −0.5 |
| 0.75 | 0.25 |
| 0.9 | 0.7 |
| 0.99 | 0.97 |
| 0.999 | 0.997 |
| 0.9999 | 0.9997 |
| 1 | 1 |
| 1.001 | 1.003 |
| 1.010 | 1.030 |

As $x$ increases, the value of $y$ also increases. As $x$ approaches 1 more and more closely, the value of $y$ also approaches the value 1. In mathematical terminology, the limit, as $x$ approaches 1, of $3x - 2$ is 1. Symbolically, $\lim_{x \to 1} 3x - 2 = 1$. The limit is the value that the function approaches as the variable approaches a specific number. If a function becomes arbitrarily close to a unique number, $L$, as $x$ approaches $c$ from either side, then we say the limit of $f(x)$ as $x$ approaches $c$ is $L$ (Larson and Hostetler 1986). Augustus Louis Cauchy (1789–1857) was the first mathematician who was responsible for the expression that $f(x)$ becomes arbitrarily close to $L$ as $x$ approaches $c$.

As another example, consider the function $y = 1/x$. As $x$ increases in the positive sense, the value of $1/x$ becomes smaller and smaller. This can be symbolized as $\lim_{x \to +\infty} 1/x = 0$. From a graphical standpoint, the value of $y$ gets closer and closer to the $x$-axis, but never touches it; we say that the graph is asymptotic to the $x$-axis. As $x$ gets smaller and smaller in the negative sense, the value of $1/x$ also tends to zero, and $\lim_{x \to -\infty} 1/x = 0$. Therefore, $\lim_{x \to \infty} 1/x = 0$. There is also a formal, rigorous approach to limits that requires more sophisticated mathematics.

*See also* Asymptotes; Calculus, Overview; Continuity

## SELECTED REFERENCE

Larson, Roland, and Robert Hostetler. *Calculus with Analytic Geometry.* Alternate 3rd ed. Lexington, MA: Heath, 1986.

STEVEN J. BALASIANO

# LINEAR ALGEBRA, OVERVIEW

The study and solution of problems involving linear combinations. A linear combination is an expression built up from a set of mathematical objects using the operations of addition and scalar multiplication. For example, $3x + 2y + 4z$ is a linear combination of the variables $x$, $y$, and $z$. Basic linear algebra provides a simple and powerful tool for solving a wide variety of applied problems. Linear algebra applications are common in engineering and in the physical, biological, and social sciences. Often the same linear models are used for solving problems in such diverse areas as biology, business, ecology, economics, and demography.

Linear algebra is one of the fastest-growing areas of mathematics. It is a powerful and essential tool for pure as well as applied mathematics. In recent years there has been a major emphasis on linear algebra in the mathematics community. The Society of Industrial and Applied Mathematics (SIAM) has a special interest group in linear algebra that frequently sponsors linear algebra conferences. The International Linear Algebra Society (ILAS) was formed in 1989 to promote worldwide linear algebra activities. The Mathematical Association of America (MAA) supports contributed paper sessions on linear algebra education at their annual joint meetings with the American Mathematical Society.

The importance of linear algebra has increased dramatically with the advent of the computer age. Modern electronic computers are particularly well suited for linear algebraic computations. Almost all of the problems encountered in scientific computing are solved using linear techniques. Even nonlinear problems are solved using linear approximations. For example, one can approximate the solution to a nonlinear equation $f(x) = 0$ by successively finding the roots of a sequence of linear equations. Two well-known methods for doing this are Newton's Method and the Secant Method. Sophisticated applied problems such as the solution of boundary value problems or partial differential equations are solved numerically by translating the problem into a large system of linear equations. The solution to such a system can then be calculated using routines from any of the linear algebra software libraries.

Linear algebra did not appear in the undergraduate mathematics curriculum until the 1960s. It quickly became a required course for all mathematics majors. Originally the audience for this course consisted almost exclusively of majors; however, the audience has expanded with the growing importance of the subject. Now many departments other than mathematics either require linear algebra or list it as a recommended elective. Indeed, the vast majority of linear algebra students are from applied fields. The standard first course in linear algebra is taught at the sophomore level and includes the following topics: matrices and systems of linear equations, determinants, vector spaces, linear transformations and eigenvalues, orthogonality and least squares.

## SYSTEMS OF LINEAR EQUATIONS, MATRICES, AND DETERMINANTS

A linear equation is an equation of the form

$$a_1x_1 + a_2x_2 + \cdots + a_nx_n = b$$

where $x_1, x_2, \ldots, x_n$ are unknowns and $a_1, a_2, \ldots, a_n, b$ are real numbers. An $m \times n$ linear system is a system of $m$ linear equations in $n$ unknowns of the form

$$a_{11}x_1 + a_{12}x_2 + \cdots + a_{1n}x_n = b_1$$
$$a_{21}x_1 + a_{22}x_2 + \cdots + a_{2n}x_n = b_2$$
$$\vdots$$
$$a_{m1}x_1 + a_{m2}x_2 + \cdots + a_{mn}x_n = b_m$$

An $n$-tuple of real numbers $(x_1, x_2, \ldots, x_n)$ is a solution of the system if and only if these numbers satisfy all of the equations. The standard method for solving linear systems is called *Gaussian elimination*. It involves using elementary operations to reduce the system to an equivalent form that is simpler and easy to solve. The following three operations are used in the Gaussian elimination process:

*Operation 1.* Interchange the order of two equations.

*Operations 2.* Multiply an equation by a nonzero real number.

*Operation 3.* Add a multiple of one equation to another equation.

The numerical coefficients that define the left-hand side of the linear system form a rectangular matrix. With the proper definitions of matrix equality and matrix multiplication an $m \times n$ linear system can be interpreted as a matrix equation of the form $A\mathbf{x} = \mathbf{b}$, where

$$A = \begin{pmatrix} a_{11} & a_{12} & \ldots & a_{1n} \\ a_{21} & a_{22} & \ldots & a_{2n} \\ \vdots & & & \\ a_{m1} & a_{m2} & \ldots & a_{mn} \end{pmatrix};$$

$$\mathbf{x} = \begin{pmatrix} x_1 \\ x_2 \\ \vdots \\ x_n \end{pmatrix}; \qquad \mathbf{b} = \begin{pmatrix} b_1 \\ b_2 \\ \vdots \\ b_m \end{pmatrix}$$

Matrices with only one column, such as $\mathbf{x}$ and $\mathbf{b}$, are called *vectors*.

If $A$ and $B$ are two matrices with the same dimensions, they can be added by adding their corresponding entries. One also can multiply each of the entries of a matrix by a real or complex number. This operation is called *scalar multiplication*.

With each $n \times n$ matrix $A$, one can associate a real number, det $(A)$, called the determinant of $A$. The determinant can be used to characterize the conditions under which a matrix will have a multiplicative inverse. Simply stated, $A$ has an inverse if and only if $\det A \neq 0$. Although determinants were once a major area of mathematics, their importance has declined considerably. Now they serve mainly as a pedagogical tool in matrix theory.

## VECTOR SPACES, LINEAR TRANSFORMATIONS, AND EIGENVALUES

The set of all column vectors with $n$ entries is referred to as Euclidean $n$-space and is denoted by $R^n$. There are two basic operations that can be performed on the vectors in $R^n$. These operations are vector addition and scalar multiplication.

The operations of addition and scalar multiplication are commonly used not only in $R^n$ but in many different mathematical settings. Usually the same algebraic rules apply regardless of the setting. Thus, a general theory of mathematical systems involving these two operations has widespread application.

A *vector space* or *linear space* is a mathematical system consisting of a set of elements called *vectors* and two operations, vector addition and scalar multiplication. The operations must satisfy a set of eight axioms. In addition to the Euclidean spaces, one can define vector spaces of matrices as well as vector spaces of continuous functions. Abstract vector spaces whose elements are functions are often referred to as *function spaces*.

If every vector in the vector space $V$ can be written as a linear combination of some collection of vectors $\mathbf{x}_1, \mathbf{x}_2, \ldots, \mathbf{x}_n$, then those vectors are said to *span V*. If in addition the $n$ vectors are *linearly independent*, then the vectors are said to form *a basis* for $V$. The vectors are linearly independent if none of the vectors can be expressed as a linear combination of the other $n - 1$ vectors. There can be many different bases for a vector space, but any two bases must have the same number of vectors. If $V$ has a basis consisting of $n$ vectors, then $V$ is said to have dimension $n$. The basis vectors essentially define a coordinate system for the vector space. For example the standard basis for the two-dimensional vector space $R^2$ consists of the two vectors

$$\mathbf{e}_1 = \begin{pmatrix} 1 \\ 0 \end{pmatrix} \qquad \mathbf{e}_2 = \begin{pmatrix} 0 \\ 1 \end{pmatrix}$$

Any vector $\mathbf{x}$ in $R^2$ can be written as a linear combination of $\mathbf{e}_1$ and $\mathbf{e}_2$:

$$\mathbf{x} = \begin{pmatrix} x_1 \\ x_2 \end{pmatrix} = \begin{pmatrix} x_1 \\ 0 \end{pmatrix} + \begin{pmatrix} 0 \\ x_2 \end{pmatrix} = x_1\mathbf{e}_1 + x_2\mathbf{e}_2$$

Thus $x_1$ and $x_2$ are the coordinates of $\mathbf{x}$ with respect to the standard basis for $R^2$. If another basis such as

$$y_1 = \begin{pmatrix} 1 \\ 1 \end{pmatrix} \qquad y_2 = \begin{pmatrix} 1 \\ -1 \end{pmatrix}$$

is used, then we can write

$$\mathbf{x} = \begin{pmatrix} x_1 \\ x_2 \end{pmatrix} = \frac{x_1 + x_2}{2}\begin{pmatrix} 1 \\ 1 \end{pmatrix} + \frac{x_1 - x_2}{2}\begin{pmatrix} 1 \\ -1 \end{pmatrix}$$

Thus,

$$\mathbf{x} = c_1\mathbf{y}_1 + c_2\mathbf{y}_2$$

where

$$c_1 = \frac{x_1 + x_2}{2} \quad \text{and} \quad c_2 = \frac{x_1 - x_2}{2}$$

are the coordinates of $\mathbf{x}$ with respect to the basis vectors $\mathbf{y}_1$ and $\mathbf{y}_2$. The key to solving many applied problems is the choice of basis vectors.

A linear transformation $L$ from a vector space $V$ to a vector space $W$ is a mapping from $V$ to $W$ that preserves linear combinations, that is,

$$L(c_1\mathbf{x}_1 + c_2\mathbf{x}_2) = c_1L(\mathbf{x}_1) + c_2L(\mathbf{x}_2)$$

If $L$ is a linear transformation from $R^n$ to $R^m$, then there exists an $m \times n$ matrix $A$ such that $L(\mathbf{x}) = A\mathbf{x}$ for every $\mathbf{x}$ in $R^n$. Thus we can think of the matrix $A$ as a representation of the linear transformation $L$.

Of particular importance in applications are linear transformations mapping $R^n$ into itself. To best understand the effects of such a transformation one must choose the proper basis for $R^n$. In particular it is desirable to have a basis $\{\mathbf{x}_1, \mathbf{x}_2, \ldots, \mathbf{x}_n\}$ with the property that the image of each basis vector is some scalar multiple of that vector. Thus if the linear transformation is represented by the matrix $A$, then

$$A\mathbf{x}_i = \lambda_i\mathbf{x}_i \qquad i = 1, \ldots, n$$

The scalars $\lambda_i$ are called *eigenvalues* and the vectors $\mathbf{x}_i$ are called *eigenvectors*.

For example, if

$$A = \begin{pmatrix} 1 & -1 \\ 2 & 4 \end{pmatrix} \qquad \mathbf{x}_1 = \begin{pmatrix} 1 \\ -2 \end{pmatrix} \qquad \mathbf{x}_2 = \begin{pmatrix} -1 \\ 1 \end{pmatrix}$$

then

$$A\mathbf{x}_1 = 3\mathbf{x}_1 \quad \text{and} \quad A\mathbf{x}_2 = 2\mathbf{x}_2$$

Thus $A$ has eigenvalues $\lambda_1 = 3$ and $\lambda_2 = 2$. The eigenvectors $\mathbf{x}_1$ and $\mathbf{x}_2$ are natural basis vectors for the operator $A$, since the effect of $A$ on any linear combination of these vectors is easily determined:

$$A(c_1x_1 + c_2x_2) = c_1Ax_1 + c_2Ax_2 = 3c_1x_1 + 2c_2x_2$$

Eigenvalues arise naturally in a wide variety of applied fields including chemistry, economics, engineering, genetics, and physics. Eigenvalues can be interpreted physically as frequencies associated with vibrations. While a person may be able to tune a guitar without having studied the theory of eigenvalues, an engineer building earthquake-resistant structures must ensure that the natural frequencies or eigenvalues of the structure lie outside of earthquake bandwidths.

## INNER PRODUCTS, ORTHOGONALITY, AND LEAST SQUARES

If $A$ is an $m \times n$ matrix, then the $n \times m$ matrix formed by interchanging the rows and columns of $A$ is called the *transpose* of $A$ and is denoted $A^T$. For example, if

$$A = \begin{pmatrix} 1 & 2 & 3 \\ 2 & 4 & 6 \end{pmatrix}$$

then

$$A^T = \begin{pmatrix} 1 & 2 \\ 2 & 4 \\ 3 & 6 \end{pmatrix}$$

In particular, if $\mathbf{x}$ and $\mathbf{y}$ are vectors in $R^n$, then the product $\mathbf{x}^T\mathbf{y}$ is a $1 \times 1$ matrix or essentially a scalar.

$$\mathbf{x}^T\mathbf{y} = (x_1 x_2 \cdots x_n) \begin{pmatrix} y_1 \\ y_2 \\ \vdots \\ y_n \end{pmatrix}$$

$$= x_1 y_1 + x_2 y_2 + \cdots + x_n y_n$$

The real number $\mathbf{x}^T\mathbf{y}$ is called the *scalar product* of $\mathbf{x}$ and $\mathbf{y}$. The length of a vector $\mathbf{x}$, denoted by $\|\mathbf{x}\|$ is defined by $\|\mathbf{x}\| = \sqrt{\mathbf{x}^T\mathbf{x}}$. The distance between two vectors $\mathbf{x}$ and $\mathbf{y}$ is defined to be $\|\mathbf{x} - \mathbf{y}\|$, the length of the difference of the vectors.

Two vectors in $R^n$ are said to be *orthogonal* if their scalar product is equal to 0. Nonzero vectors in $R^2$ or $R^3$ can be represented geometrically by directed line segments through the origin. In this case, vectors will be orthogonal if and only if the line segments are perpendicular. Thus, orthogonality is just a generalization of the concept of perpendicularity. Scalar products can be generalized to arbitrary vector spaces by associating a real number with every pair of vectors $\mathbf{x}$ and $\mathbf{y}$. The real number is called the inner product of $\mathbf{x}$ and $\mathbf{y}$ and is denoted $\langle \mathbf{x}, \mathbf{y} \rangle$. The rule defining the inner product must have similar algebraic properties to the rule used for scalar products. Just as with the scalar product, the inner product can be used to define orthogonality, the length of a vector, and the distance between two vectors.

Least-squares problems are the most important application of orthogonality theory. Fitting curves to data gives rise to inconsistent $m \times n$ linear systems of the form $A\mathbf{x} = \mathbf{b}$ where $m$ is generally far greater than $n$. To obtain a least-squares fit, find the vector $\mathbf{x}$ that minimizes the distance between the vectors $A\mathbf{x}$ and $\mathbf{b}$. The least-squares solution $\mathbf{x}$ is characterized by the condition that the vector $\mathbf{r} = \mathbf{b} - A\mathbf{x}$ must be orthogonal to each of the column vectors of $A^T$. This condition is equivalent to the linear system

$$A^T A\mathbf{x} = A^T\mathbf{b}$$

## THE HISTORY OF LINEAR ALGEBRA

Gaussian elimination was developed by Carl Friedrich Gauss (1777–1855) around 1795 as part of his efforts to solve least-squares problems for celestial and geodesic computations. Early Chinese manuscripts have been uncovered that explain Gaussian elimination for $3 \times 3$ systems. These manuscripts predate Gauss by several centuries. The algebraic method of least squares was first published by Adrien-Marie Legendre in 1805; however, credit for this method is usually attributed to Gauss, who used it ten years earlier. The method was justified as a statistical procedure by Gauss in an 1809 paper. The first publication of Gauss-Jordan elimination occurred in a handbook on geodesy by Wilhelm Jordan. It is a common mistake to assign credit for the process to the famous mathematician Camille Jordan instead of the geodetist Wilhelm Jordan. Athloen and McLaughlin (1987) and Tucker (1993) have helped to set the record straight.

Matrix theory itself is an outgrowth of the theory of determinants that dates back to some work of Gottfried Leibniz (1646–1716) in 1693. Determinants were once a major area of mathematical research. Matrix notation and conventions were derived much later as an attempt to present the proper language for the study of determinants. In 1848, James J. Sylvester used arrays as devices to generate determinants. The term *matrix* was coined by Sylvester from the Latin word for "womb." The rules for matrix algebra stem from the work of Arthur Cayley in 1855. Surprisingly, matrix multiplication was not derived in connection with linear systems of equations. Instead, Cayley was studying transformations of the form

$$T_1: \quad x' = ax + by$$
$$y' = cx + dy$$

The transformation $T_1$ corresponds to a change from $xy$-coordinates to $x'y'$-coordinates. Let $T_1$ be represented by a $2 \times 2$ array of coefficients:

$$\begin{pmatrix} a & b \\ c & d \end{pmatrix}$$

Given a transformation $T_2$ corresponding to a second change of coordinates,

$$T_2: \quad x'' = ex' + fy'$$
$$y'' = gx' + hy'$$

substitute for $x'$ and $y'$ to find the composite transformation $T_2(T_1(x, y))$ representing a change from $xy$- to $x''y''$-coordinates

$$x'' = e(ax + by) + f(cx + dy)$$
$$= (ea + fc)x + (eb + fd)y$$
$$y'' = g(ax + by) + h(cx + dy)$$
$$= (ga + hc)x + (gb + hd)y$$

Cayley defined the product of the matrix representing $T_2$ times the matrix representing $T_1$ to be the matrix that represents the composite transformation $T_2(T_1(x, y))$. Thus,

$$\begin{pmatrix} e & f \\ g & h \end{pmatrix} \begin{pmatrix} a & b \\ c & d \end{pmatrix} = \begin{pmatrix} ea + fc & eb + fd \\ ga + hc & gb + hd \end{pmatrix}$$

Since the composite functions $T_2(T_1(x, y))$ and $T_1(T_2(x, y))$ are in general not equal, matrix multiplication is not commutative. In general, if $A$ and $B$ are $n \times n$ matrices, then $AB \neq BA$. Matrices have become the basic tool for scientific computing. Determinants, however, have proved to be useless for most computational purposes.

Early in this century, matrices were viewed as either a finite dimensional example of linear transformations or simply as a notational convenience. The significance of matrix algebra became more apparent when Werner Heisenberg used it in 1925 as a basic tool for his work in quantum mechanics.

In the last fifty years, matrix factorizations have become increasingly important. Numerical linear algebra is based on matrix factorizations, and many of the topics in the standard linear algebra course can be interpreted in terms of factorizations. The $LU$ factorization, introduced by Alan Turing in 1948, is now included in many beginning linear algebra courses.

Computing the $LU$ factorization of a matrix is equivalent to using Gaussian elimination to reduce the matrix to upper triangular form.

Other important factorizations are the Schur decomposition, $A = QTQ^*$, which is used for computing eigenvalues, and the $QR$ factorization and singular value decomposition $A = U \Sigma V^T$, which are used to solve least-squares problems. The singular value decomposition is one of the primary tools in numerical linear algebra. This factorization first appeared in an 1873 paper by the Italian differential geometer Eugenio Beltrami. It also appeared in an independent work published by Camille Jordan in 1874. J. J. Sylvester rediscovered the factorization in 1889/90. All three discoverers gave different proofs.

The most influential person in the development of numerical linear algebra has been James Hardy Wilkinson (1919–1986). Wilkinson worked with Alan Turing on the logical design of the Pilot Ace computer at the National Public Laboratory in Great Britain after World War II (1946–1948). In the following years, Wilkinson did pioneering work in numerical linear algebra and was instrumental in developing the LINPACK and EISPACK software libraries for matrix computations. Wilkinson is perhaps best known for a technique he invented called *backward error analysis*. In the early days of computing there was some doubt about the feasibility of using Gaussian elimination for computing solutions of linear systems. It was suspected that the process would be unstable in the sense that a build-up of roundoff error during the elimination process would yield inaccurate computed solutions. Based on his experiences computing solutions to ballistics problems during the war, Wilkinson was confident that the elimination method actually was stable. He was able to prove numerical stability using backward error analysis.

## CURRICULUM AND REFORM

The trend in linear algebra education is toward a more matrix-oriented course with a wide variety of applications. Modern courses also are taught from a more geometric point of view and incorporate the use of software. Still, the problem remains that mathematics students need to learn more linear algebra than the basics that are covered in a first course. It remains a challenge to the mathematics community to decide how to best fit a second linear algebra course into the curriculum.

Linear algebra courses in the 1960s were originally taught from an abstract algebra point of view. A second type of course emerged during the 1970s and

1980s with increasing emphasis on matrix theory and applications. There also has been a growing emphasis on geometry. Vector space theory is now generally taught from a geometric (rather than algebraic) point of view.

Because of the growing importance of linear algebra in the curriculum, a number of major reform efforts have been initiated in recent years. In 1990, a National Science Foundation (NSF) sponsored Linear Algebra Curriculum Study Group (LACSG) was formed. The group met at the College of William and Mary and drew up the following set of recommendations:

1. The syllabus and presentation of the first course must respond to the needs of client disciplines.
2. Mathematics departments should seriously consider making their first course in linear algebra a matrix-oriented course.
3. Faculty should consider the needs and interests of students as learners.
4. Faculty should be encouraged to utilize technology in the first linear algebra course.
5. At least one "second course" in matrix/linear algebra should be a high priority for every mathematics curriculum.

The majority of students taking linear algebra are not mathematics majors. Additionally, the majority of mathematics majors end up working in industry rather than going on to graduate school. In light of these observations, the LACSG felt that a more practical matrix-oriented course would best meet the needs of the students. The committee report (Carlson et al. 1993) went on to propose a core syllabus for a matrix-oriented course.

A difficulty with any elementary linear algebra course is that the course ends before one is able to cover many of the most important topics. There is simply too much essential material to fit into a single course. The obvious solution is to require a second course in linear algebra. This course could be applied, numerical, or theoretical in nature. Many of the larger universities offer second courses in linear algebra as electives and a growing number of these universities are making a second course required.

## USING TECHNOLOGY TO TEACH LINEAR ALGEBRA

Sponsored by NSF, the ATLAST (Augmenting the Teaching of Linear Algebra through the use of Software Tools) Project is the first major initiative to support the implementation of the recommendations of the LACSG. The goal of the project is to encour-

age and facilitate the use of software in the teaching of linear algebra. ATLAST has offered faculty workshops on the use of software in teaching linear algebra. Instructors attending the ATLAST workshops were required to design their own computing projects and test them in class. These were then submitted for inclusion in a database of computer exercises.

The following are some of the types of exercises recommended by the ATLAST Project:

1. *Discovery exercises.* By working through computer examples, students are able to observe interesting properties and discover theorems for themselves. Discovery exercises help students gain numerical intuition into matrix theory. After discovering a result, students are then asked either to prove or explain why the result will work in general.
2. *Application exercises.* Realistic applications often involve a significant amount of computation. The computer makes these types of applications accessible. If presented properly, applications can serve to motivate much of the course material.
3. *Geometrical exercises.* Modern software packages have sophisticated graphics capabilities that can be used to illustrate geometrically many of the concepts taught in linear algebra courses. For example, students can gain insight into how linear transformations work by graphing the various images of a figure under a wide variety of transformations. Most of the standard topics in a linear algebra course have geometric interpretations that can be illustrated on the computer.

*See also* Least Squares; Vector Spaces

## SELECTED REFERENCES

Athloen, Steven C., and Renate McLaughlin. "Gauss-Jordan Reduction: A Brief History." *American Mathematical Monthly* 94(1987):130–142.

Bell, Eric Temple. *The Development of Mathematics.* New York: McGraw-Hill, 1945.

———. *Men of Mathematics.* New York: Simon and Schuster, 1937.

Björck, Ake. "Least Squares Methods." In *Handbook of Numerical Analysis.* Vol. 1. Philippe G. Ciarlet and Jacques L. Lions (eds.). Amsterdam: Elsevier/North Holland, 1990.

Boyer, Carl B. *A History of Mathematics.* 2nd ed. Revised by Uta C. Merzbach. New York: Wiley, 1989.

Carlson, David, Charles R. Johnson, David Lay, and A. Duane Porter. "The Linear Algebra Curriculum Study Group Recommendations for the First Course in Linear Algebra." *College Mathematics Journal* 24(Jan. 1993): 41–46.

Gauss, Carl Friedrich. *Theory of the Motion of the Heavenly Bodies Moving about the Sun in Conic Sections* [1809]. Trans. Charles H. Davis. New York: Dover, 1963.

Legendre, Adrien-Marie. *Nouvelle Méthodes pour la Détermination des Orbites des Comètes*. Paris, France: Courcier, 1805.

Leon, Steven J. *Linear Algebra with Applications*. 4th ed. New York: Macmillan, 1994.

Tucker, Alan. "The Growing Importance of Linear Algebra in Undergraduate Mathematics." *College Mathematics Journal* 24(Jan. 1993):3–9.

STEVEN J. LEON

# LINEAR PROGRAMMING

Mathematical theory of minimization or maximization of a linear function (the *objective function*) subject to linear constraints *(limiting conditions)*. The most common form of linear program has the constraints described as inequalities, and the variables nonnegative, as follows.

*Example 1*   Maximize $\sum_{i=1}^{m} c_i x_i$ subject to

$$\sum_{i=1}^{n} a_{ij} x_i \leq b_j, \qquad j = 1, \ldots, n; x_i \geq 0$$

Although some real life situations may seem to require a different form of problem description, it is generally possible to reformulate the description so that it "fits" the form shown in Example 1. One may therefore consider the solution of linear program (1) without loss of generality. Thus, for example, one may have to minimize rather than maximize; minimizing a function, however, is equivalent to maximizing its negative. Some constraints may be equations; however, an equation may be replaced by two inequalities. Some negative variable values may be permitted; however, variables of unrestricted value may be described as the difference of two nonnegative variables.

A *solution* of a linear program is a set of $x_i$ that satisfy the $n$ constraints. When the $x_i$ are nonnegative, the solution is *feasible*. A feasible solution that maximizes (minimizes) the objective function is an *optimal solution*. The number providing the maximum or minimum value is called the *value* of the linear program. An elementary (graphical) approach for a two-variable problem is the "corner point" method. One graphs the system of inequalities (graphing each inequality separately on the same set of axes) and considers the overlapping regions. Each point in the region of intersections—called the *feasible region*—will satisfy the set of inequalities. Then one considers all corner points (the intersection of lines comprising

the boundaries of the feasible region). Those points whose coordinates maximize (or minimize, as appropriate) the value of the objective function are optimal solutions.

When large numbers of variables and constraints are involved, the *simplex method* is most commonly used. Computation required for this method can be tiresome; therefore several computer packages are available for this purpose (LINDO, MATLAB, and others). Using matrix notation, a linear programming problem may be described as follows.

*Example 2*   Maximize $Cx$ subject to $Ax \leq b$, $x \geq 0$. The *dual* is then the following problem.

*Example 3*   Minimize $b^T V$ subject to $A^T V \geq C^T$, $V \geq 0$, where superscript $T$ represents the *transpose* of the matrix.

Note that the dual of linear program (3) is (2). Basic to the theory of linear programming is the *Fundamental Duality Theorem*:

> If a linear program and its dual both have feasible solutions, then they both have optimal solutions, and both have the same value. If either program does not have a feasible solution, then neither has an optimal solution.

The concepts underlying the theory of linear programming date back to the early nineteenth century (Fourier 1826). The first use of the term *linear programming* in a published work was by George B. Dantzig (1949). The Duality Theorem was first noted by John von Neumann in private notes around 1947. A proof of duality based on von Neumann's notes was first published by Gale, Kuhn, and Tucker (1951). The application of the theory to economics was first noted by John von Neumann and Oskar Morgenstern (1944). The principal methods used—particularly the *simplex method*—were developed by Dantzig and others in the United States (1940s and 1950s). A fairly comprehensive treatment of the theory may be found in Goldman and Tucker (1956).

To illustrate the simplex method, consider the following linear program.

*Example 4*   Maximize $x + 2y$ subject to

$$x \geq 0, \qquad y \geq 0$$
$$3x + y \leq 22$$
$$2x + 3y \leq 24$$

First convert inequalities to equations by adding an appropriate nonnegative value to the smaller side of each inequality, yielding:

$$3x + y + s_1 = 22$$
$$2x + 3y + s_2 = 24$$

The quantities $s_1$ and $s_2$ are called *slack variables*. The resulting equations and the objective functions are placed in tableau form in the following way:

|    | $x$ | $y$ | $s_1$ | $s_2$ | constant |
|----|-----|-----|-------|-------|----------|
| 1. | 3   | 1   | 1     | 0     | 22       |
| 2. | 2   | 3   | 0     | 1     | 24       |
| 3. | −1  | −2  | 0     | 0     | 0        |

Note that each row contains the coefficients of the variables $(x, y, s_1, s_2)$ and the constant terms of each equation in the appropriate column, and the last row (the extra row) contains the negatives of the coefficients of the objective function. The rules for application of the method follow:

Rule 1. Find the most negative entry in the extra row of the tableau, and note the column in which the number occurs; this is the *pivot column*.

Rule 2. Form ratios of the value in the constant column to the positive elements of the pivot column. The row with the smallest such ratio is the *pivot row*.

Rule 3. Convert the entry in the pivot column and pivot row to 1 by multiplying each element of the pivot row by the reciprocal of the element (called the *pivot element*). In the example, the $y$ column is the pivot column and the pivot row is row 2 (the ratio 24/3 is smaller than 22/1), hence our pivot element is 3. Multiply row 2 by 1/3, yielding

| 2/3 | 1 | 0 | 1/3 | 8 |
|-----|---|---|-----|---|

This is now the pivot row.

Rule 4. Convert the remaining numbers in the pivot column to 0s by multiplying the pivot row by a suitable constant, such that adding the resulting values to the other row will produce a zero in the pivot column.

Multiplying the pivot row by −1 and adding that result to row 1 yields a zero in row 1, column $y$. Multiplying the pivot row by 2 and adding that result to row 3 produces a zero in row 3, column $y$. The new tableau is

|    | $x$  | $y$ | $s_1$ | $s_2$  | constant |
|----|------|-----|-------|--------|----------|
| 1. | 7/3  | 0   | 1     | −1/3   | 14       |
| 2. | 2/3  | 1   | 0     | 1/3    | 8        |
| 3. | 1/3  | 0   | 0     | 2/3    | 16       |

Continue to apply rules 1, 2, 3, and 4 until either every entry in the extra row is nonnegative (in which case no pivot column can be chosen) or every column that has a negative entry in the

extra row has no positive element. In the latter case, no solution exists; in the former, a solution exists and is obtained as follows: if a column is in 0–1 form (one 1 and the remaining entries zeros), assign to the variable of that column the constant in the row with the 1; if a column is not in 0–1 form, assign zero to the variable of that column. The value of the objective function is found in the entry of the extra row in the constant column. Thus in the example, since no negative values exist in the extra row, the solution has been found: $x = 0$, $y = 8$, and the value of the objective function is 16.

Note that the graphical method, although not very practical, is most beneficial for an intuitive understanding of linear programming. It is therefore suggested that, for initial classroom presentation, a simple problem be solved using both the corner point and simplex methods.

With regard to prerequisites, students should be familiar with matrix algebra. If the course is to include proofs of theorems, a knowledge of linear algebra and a "reasonable" degree of mathematical maturity are required. A course in linear programming would generally be taught by mathematics faculty but made available to students in other fields—particularly economics—with sufficient mathematical experience and maturity.

*See also* Linear Algebra, Overview

## SELECTED REFERENCES

Dantzig, George B. "Maximization of Linear Functions of Variables Subject to Linear Inequalities." In *Activity Analysis of Production and Allocation* (pp. 339–347). Tjalling C. Koopmans (ed.). New York: Wiley, 1951.

———. "Programming in a Linear Structure." *Econometrica* 17 (1949).

———, L. R. Ford, and D. R. Fulkerson. "A Primal-Dual Algorithm for Linear Programs." In *Linear Inequalities and Related Systems* (pp. 171–181). Harold W. Kuhn and Albert W. Tucker (eds.). Princeton, NJ: Princeton University Press, 1956.

———, and D. R. Fulkerson. "On the Max-Flow Min-Cut Theorem of Networks." In *Linear Inequalities and Related Systems* (pp. 215–221). Harold W. Kuhn and Albert W. Tucker (eds.). Princeton, NJ: Princeton University Press, 1956.

———, Alex Orden, and Philip Wolfe. "Generalized Simplex Method for Minimizing a Linear Form Under Linear Inequality Restraints." *Pacific Journal of Mathematics* 5(1955):183–195.

Fourier, Joseph. "Solution d'une Question Particulière du Calcul des Inégalites." *Bulletin des Sciences par la Société Philomathique de Paris* (1826):99–100.

Gale, David, Harold W. Kuhn, and Albert W. Tucker. "Linear Programming and the Theory of Games." In *Activity Analysis of Production and Allocation* (pp. 317–329). Tjalling Charles Koopmans (ed.). New York: Wiley, 1951.

Gass, Saul I. *Linear Programming.* New York: McGraw-Hill, 1958.

Goldman, A. J., and Albert W. Tucker. "Theory of Linear Programming." In *Linear Inequalities and Related Systems* (pp. 53–98). Harold W. Kuhn and Albert W. Tucker (eds.). Princeton, NJ: Princeton University Press, 1956.

Vajda, Steven. *The Theory of Games and Linear Programming.* New York: Wiley, 1956.

Von Neumann, John. "Über ein Ökonomisches Gleichungssystem und eine Verallgemeinerung des Brouwerschen Fixpunktsatzes." *Ergebnisse eines Mathematischen Kolloquiums* 8(1937):73–83. [Translated as "A Model of General Economic Equilibrium." *Review of Economic Studies* 13(1)(1945–1946):1–9.]

———, and Oskar Morgenstern. *Theory of Games and Economic Behavior.* Princeton, NJ: Princeton University Press, 1944.

RINA YARMISH

# LIOUVILLE, JOSEPH (1809–1882)

Associated with a number of famous theorems. Liouville was born on March 24, 1809 at Saint-Omer, Pas de Calais. In 1825 he entered the École Polytechnique in Paris. He began teaching at the Collège de France in 1837 and was appointed a Professor of Mathematics there in 1851. In 1836, he founded the *Journal de Mathématique,* which became known as *Liouville's Journal.* In 1839, he was elected to the Académie des Sciences. Liouville had a profound impact on a wide variety of mathematical topics. His name is associated with a number of famous theorems. For example, he was the first mathematician to discover transcendental numbers; the ones he constructed bear his name. An example of a Liouville number is

$$\sum_{n=1}^{\infty} \frac{(-1)^{n-1}}{2^{n!}}$$

He was a pioneer of the theory of integration in finite terms, a subject of intense recent interest in the development of computer algebra. His work in differential equations led to what today is called Sturm-Liouville theory. His theorem that a bounded complex function analytic everywhere is constant is ubiquitous in the theory of complex variables. His further contributions in number theory, Galois theory, mechanics, potential theory, and other areas are described in Lutzen (1990).

SELECTED REFERENCE
Lutzen, Jesper. *Joseph Liouville, 1809–1882, Master of Pure and Applied Mathematics.* New York: Springer-Verlag, 1990.

GEORGE E. ANDREWS

# LOBACHEVSKY, NIKOLAI IVANOVICH (1792–1856)

Along with Carl F. Gauss and Janos Bolyai, independently and simultaneously discovered a new version of plane geometry in which there exists more than one parallel to a given line through any point not on the line. This revolutionary theory of parallels was systematized by them in essentially the same form more than two millennia after Euclid had *assumed* the familiar parallel postulate, wherein there exists a unique parallel to a line through a point not on the line, and only after numerous others had attempted to *prove* it as an "intuitively obvious" *theorem* of plane geometry. Of the three, Lobachevsky deserves the most credit for the eventual acceptance of such a non-Euclidean geometry by the mathematical community and subsequent, fruitful applications of its concepts by the scientific community in areas of modern physics, including the Einstein theory of relativity. In 1829–1830, he was the first to publish these theories in an obscure Russian journal of the University of Kazan. Afterward, he worked diligently to see that his revolutionary ideas circulated within the mathematical community by translating them into French and German and publishing them in established journals. Gauss never published his ideas on the subject and Bolyai never had sufficient prestige in the scientific world to change the longstanding human belief that Euclidean geometry was *a priori* the only possible valid geometry.

*See also* Axioms; History of Mathematics, Overview; Non-Euclidean Geometry

SELECTED REFERENCES
Burton, David. *History of Mathematics: An Introduction.* 3rd ed. Dubuque, IA: Brown, 1995, pp. 536–540.

Eves, Howard W. *An Introduction to the History of Mathematics.* 6th ed. Philadelphia, PA: Saunders College Publishing, 1990, pp. 498–500.

Gillispie, Charles, ed. *Dictionary of Scientific Biography.* Vol 8. New York: Scribner's, 1981–1990, pp. 428–435.

RICHARD M. DAVITT

# LOGARITHMS

Credited to John Napier (1550–1617), a Scottish landowner who, in 1614, published *Mirifici logarithmorum canonis descriptio (A Description of the Wonderful Law of Logarithms)*. Although Jobst Bürgi, a Swiss watchmaker, independently discovered logarithms, he did not publish his results until 1620 and thereby lost his claim to credit (Boyer 1991, Yozwiak 1969). Earlier mathematicians, notably Archimedes (287–212 B.C.) and Michael Stifel (1486–1567), had observed connections between the product or quotient of terms in the geometric progression $1, r, r^2, r^3, \ldots$ and the associated sum or difference of terms in the arithmetic progression $0, 1, 2, 3, \ldots$ composed of their exponents. Napier, however, was the first to use this relationship to simplify the calculations needed to complete astronomical problems involving spherical trigonometry (Kline 1972). At that time, sines of angles were defined as lengths of lines. Hence, Napier originally defined logarithms geometrically as lengths of segments. In Figure 1, sine values from 10,000,000 to 0 are represented on segment $\overline{AB}$, a segment of fixed length taken to be 10,000,000; ray $\overrightarrow{DE}$ is of infinite length. Points $C$ and $F$ begin moving simultaneously with the velocity of point $C$ being numerically equal to the distance $\overline{CB}$ and the velocity of point $F$ being constant and equal to the initial velocity of point $C$. This results in successive lengths $x$ being in geometric progression, while successive lengths $y$ are in arithmetic progression. Napier defined the logarithm of $\overline{CB}$ to be $\overline{DF}$, or $\log x = y$, and used this definition to develop a table of the logs of sines of angles for successive minutes of arc with logarithm 10,000,000 = 0 (Boyer 1991; Kline 1972; Yozwiak 1969).

One of the first changes to Napier's logarithmic system was suggested by Henry Briggs (1561–1631), who associated logarithms with appropriate powers of 10. In 1624, he published logarithm tables for the numbers from 1 to 20,000 and from 90,000 to 100,000 (Boyer 1991). Logarithms with base-10 are called *common logarithms* and denoted with the abbreviation *log*. The modern connection of logarithms to exponents appeared in William Gardiner's *Tables of Logarithms*, published in 1742, and is credited to William Jones (Kline 1972). Euler used this definition in his 1748 treatise *Introductio in analysis infinitorum*.

Today, logarithms are generally introduced to students through one or more of the following approaches.

## Logarithms as Exponents

$y$ is the logarithm of $x$ with base $b (b > 0, b \neq 1)$, written $y = \log_b x$, if and only if $b^y = x$. For instance, $\log_3 81 = 4$ because $3^4 = 81$. This definition leads to the following properties:

(1) $\log_b 1 = 0$

(2) $\log_b b = 1$

(3) $\log_b b^n = n$

(4) $\log_b(xy) = \log_b x + \log_b y$

(5) $\log_b\left(\dfrac{x}{y}\right) = \log_b x - \log_b y$

(6) $\log_b x^n = n \log_b x$

Until the advent of calculator technology, these properties were applied, together with values in logarithmic tables, to transform complicated multiplication and division problems into addition and subtraction problems and root extraction problems into problems involving multiplication by a rational number. For instance, consider the problem:

$$N = \frac{435^2 \cdot 786}{\sqrt{239}}$$

Using the properties of logarithms and recognizing that the square root is equivalent to the 1/2 power, this problem can be rewritten as $\log N = 2 \log 435 + \log 786 - 0.5 \log 239$.

A table of logarithms for numbers between 1 and 10 could now be used. Each number may be written in scientific notation. The logarithm may be expressed as the sum of the *characteristic* and *mantissa*, where the characteristic is the exponent of 10 and the mantissa is found in the table as the logarithm of the number between 1 and 10.

Any positive number $b, b \neq 1$, can be used as the base of the logarithm although tables and calculator functions do not exist for all bases. To facilitate evaluation of logarithms, a change of base property, $\log_b x =$

$$\frac{\log_a x}{\log_a b}$$

can be used to write an equivalent expression using a base $a$ for which a table of values exists or for which

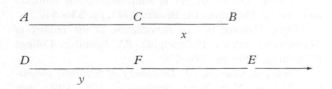

**Figure 1**

a logarithmic function is built into the calculator. Thus,

$$\log_5 36 = \frac{\log 36}{\log 5} \approx 2.23$$

## Logarithmic Functions as Inverses of Exponential Functions

The function $f$ defined as $f(x) = a^x$, where $a > 0$, $a \neq 1$, is the exponential function with base $a$. Then the inverse of the exponential function $f$ is the logarithmic function with base $a$, denoted $\log_a x$. Therefore, $f^{-1}(x) = \log_a x$.

## Relating Logarithms to Area

In calculus, the definite integral which gives the area bounded by the $x$-axis and the curve

$$y = \frac{1}{x}$$

from 1 to $x$ is often defined to be the natural logarithmic function, denoted by the abbreviation ln. That is,

$$\ln x = \int_1^x \frac{dt}{t}$$

Natural logarithms are logarithms with base $e = 2.7182818 \ldots$; the exponential function $y = e^x$ has the property that it is its own derivative and $e$ is the limit of

$$\left(1 + \frac{1}{n}\right)^n$$

as $n \to \infty$. The properties of this natural logarithmic function are determined based on integral properties and areas under the hyperbola; this leads to an integral definition for ln $x$ that is consistent with the algebraic definition for $\log_e x$.

## Applications

While linear scales are constructed so the difference of successive units is constant, logarithmic scales are constructed so the ratio of successive units is constant. Examples of logarithmic scales are the Richter scale for measuring the amplitude of earthquakes, the decibel scale for measuring the intensity of sound, the pH scale for measuring the acidity of solutions, the magnitude scale for measuring the intensity of light from celestial objects, and the scales on radio dials (Senk et al. 1996; Lindstrom 1988).

For example, an earthquake measuring 6.2 on the Richter scale has a measured amplitude that is 10 times greater than one measuring 5.2 and 100 times greater than one measuring 4.2. Continuous compounding, radioactive decay, and population growth are just some of the situations described by exponential equations. Solutions to such equations often involve the use of logarithms. For instance, $2000(1.06)^t$ gives the value of a \$2,000 investment after $t$ years when compounded annually at 6%. The investment triples when $6000 = 2000(1.06)^t$. The value of $t$ can be found by evaluating

$$\frac{\log 3}{\log 1.06}$$

so that the investment triples in roughly 18.9 years.

*See also* e; Equations, Exponential; Equations, Logarithmic

### SELECTED REFERENCES

Bolduc, Elroy J., Jr. "Student Construction of a Table of Common Logarithms." *School Science and Mathematics* 77(Feb. 1977):154–156.

Boyer, Carl B. *A History of Mathematics.* 2nd ed., rev. Revised by Uta C. Merzbach. New York: Wiley, 1991.

Deakin, Michael A. B. "A Numerical Approach to Natural Logarithms." *Mathematics Teacher* 66(Mar. 1973): 239–242.

Eves, Howard. *An Introduction to the History of Mathematics.* 4th ed. New York: Holt, Rinehart and Winston, 1976.

Hartman, Janet. "Approximating Logarithms Intuitively." *Mathematics Teacher* 74(Apr. 1981):276–277.

Hurwitz, Marsha. "Conjecturing with Logarithms." *Mathematics Teacher* 84(Jan. 1991):35–36.

Kalman, Dan, and Charles E. Mitchell. "Logarithms in the Year 10 A.C." *School Science and Mathematics* 81(Feb. 1981):124–130.

Kline, Morris. *Mathematical Thought from Ancient to Modern Times.* New York: Oxford University Press, 1972.

Kluepfel, Charles. "When Are Logarithms Used?" *Mathematics Teacher* 74(Apr. 1981):250–253.

Lindstrom, Peter A. "Earthquakes and Logarithms." In *Newsletter of the Consortium for Mathematics and Its Applications* (Winter 1988): HiMap Pull Out Section.

Rahn, James R., and Barry A. Berndes. "Using Logarithms to Explore Power and Exponential Functions." *Mathematics Teacher* 87(Mar. 1994):161–170.

Senk, Sharon L., et al. *Advanced Algebra.* 2nd ed. Glenview, IL: Scott Foresman, 1996.

Yozwiak, Bernard J. "Logarithms." In *Historical Topics for the Mathematics Classroom* (pp. 142–145). 31st Yearbook. John K. Baumgart et al. (eds.). Washington, DC: National Council of Teachers of Mathematics, 1969.

DENISSE R. THOMPSON

# LOGIC

Study of proof and rational argument begun by the ancient Greeks, notably among them Aristotle. The fundamental idea is that a proof, or "logical" argument, consists of a series of assertions, each one being either an initial assumption or else following "logically" from previous ones in the series, according to some "logical rules." The Aristotelian school examined correct arguments that may be formulated as a series of assertions known as subject-predicate propositions.

A *proposition* is a sentence that is either true or false. The subject-predicate propositions are those consisting of two entities, a subject and a property, or predicate, ascribed to that subject. The logical rules that Aristotle identified as the patterns that must be followed to construct a valid proof (using subject-predicate propositions) are known as *syllogisms*. These are rules for deducing one assertion from exactly two others. An example of a syllogism is

> All men are mortal.
> Socrates is a man.
> Socrates is mortal.

The idea is that the third assertion, the one below the line, follows "logically" from the previous two. Despite its simplicity, what makes Aristotle's contribution so significant is that he found general patterns in such examples and succeeded in classifying all seventeen of the logically valid syllogisms among the 256 possible patterns. Modern logic, which derives from the work of the Greeks, consists of two closely related branches: propositional logic and predicate logic. Both are highly algebraic. The use of algebra to study logic and rational argument was introduced by the English mathematician George Boole in the nineteenth century.

Propositional logic starts off with some "basic" propositions. The only thing known about these propositions is that they *are* propositions; that is, they are statements that are either true or false. A number of precisely stipulated rules permit combination of these basic propositions to produce more complex propositions. Arguments that consist of a series of such compound propositions can then be analyzed. This approach to logic is highly abstract, since the logical patterns uncovered are devoid of any context. The theory is independent of what the various propositions say. The patterns of reasoning uncovered in this manner are those that depend solely on the *logical form* of the various propositions, the way complex propositions are built up from simpler ones.

One way to combine propositions is by the operation of *conjunction*: given propositions $p$ and $q$, form the new proposition ($p$ and $q$). In general, all that can be known about this compound proposition is its truth status. Given the truth status of $p$ and $q$: if both $p$ and $q$ are true, then the conjunction ($p$ and $q$) will be true; if one or both of $p$ and $q$ are false, then ($p$ and $q$) will be false. The pattern is most clearly presented as a "truth table." The following "truth tables" give the truth pattern for conjunction and for three other operations on propositions, namely, disjunction, ($p$ or $q$), the *conditional*, ($p \rightarrow q$), and *negation, not p*. Reading along a row, each entry indicates the truth value of the compound proposition that arises from the truth values of the component propositions.

| $p$ | $q$ | $p$ and $q$ | $p$ or $q$ | $p \rightarrow q$ | not $p$ |
|-----|-----|-------------|-----------|-------------------|---------|
| T | T | T | T | T | F |
| T | F | F | T | F | F |
| F | T | F | T | T | T |
| F | F | F | F | T | T |

The last of these columns, *not p*, is self-explanatory. Two of the others require some comment. In everyday language, the word *or* has two meanings. It can be used in an exclusive way, as in the sentence, "The door is locked *or* it is not locked." In this case, only one of these possibilities can be true. Alternatively, *or* may be used inclusively, as in "It will rain *or* snow." In this case, there is the possibility that both will occur. In everyday communication, people generally rely on context to make the intended meaning clear. But in the logic of propositions, there is no context, only the bare knowledge of truth or falsity. To eliminate ambiguity, mathematicians made a choice when they were formulating the rules of propositional logic. They chose the inclusive version. (It is an easy matter to express the exclusive-or in terms of the inclusive-or and the other logical operations, so there is no loss in making this particular choice.)

There is no common English word that directly corresponds to the conditional. It is related to logical implication, so *implies* would be the closest common word. But the conditional does not fully capture the notion of implication. Implication involves some kind of *causality*; the assertion ($p$ implies $q$) suggests a connection between $p$ and $q$. Since the operations of propositional logic are defined solely in terms of truth and falsity, this method of definition cannot capture completely the notion of implication. The conditional [$p \rightarrow q$] reflects the two patterns of truth that arise from an implication, namely:

- if it is the case that *p implies q*, then the truth of *q* follows from that of *p*, and
- if it is the case that *p* is true and *q* is false, then it cannot be the case that *p implies q*.

These considerations give the first two rows of the truth table for the conditional. The remainder of the truth table, which concerns the two cases when *p* is false, is completed in a fashion that leads to the most useful theory.

The following table lists a number of basic properties of these definitions. The symbol ↔ denotes logical equivalence, the logician's version of equality. It means that the expressions on each side of the symbol ↔ have exactly the same truth table.

$$[p \text{ and } q] \leftrightarrow [q \text{ and } p]$$

$$[p \text{ or } q] \leftrightarrow [q \text{ or } p]$$

$$p \text{ and } [q \text{ and } r] \leftrightarrow [p \text{ and } q] \text{ and } r$$

$$p \text{ or } [q \text{ or } r] \leftrightarrow [p \text{ or } q] \text{ or } r$$

$$p \text{ and } [q \text{ or } r] \leftrightarrow [p \text{ and } q] \text{ or } [p \text{ and } r]$$

$$p \text{ or } [q \text{ and } r] \leftrightarrow [p \text{ or } q] \text{ and } [p \text{ or } r]$$

$$\text{not } [p \text{ and } q] \leftrightarrow [\text{not } p] \text{ or } [\text{not } q]$$

$$\text{not } [p \text{ or } q] \leftrightarrow [\text{not } p] \text{ and } [\text{not } q]$$

$$\text{not not } p \leftrightarrow p$$

$$[p \rightarrow q] \leftrightarrow [\text{not } p] \text{ or } q$$

Deductions in propositional logic are carried out using the following, single deduction rule, called *modus ponens*:

From $[p \rightarrow q]$ and *p*, infer *q*

This rule accords with the intuition that the conditional corresponds to the notion of implication. It should be stressed that the *p* and *q* here do not have to be simple, noncompound propositions. As far as *modus ponens* is concerned, these symbols may denote any proposition whatsoever. Indeed, throughout propositional logic, the algebraic symbols used almost invariably denote arbitrary propositions, simple or compound.

In propositional logic, a *proof* consists of a series of propositions such that each proposition in the series is either deduced from previous ones by means of *modus ponens* or else is one of the assumptions that underlie the proof. Though it does not represent all kinds of reasoning, not even all kinds of mathematical proof, propositional logic has proved to be extremely useful. In particular, to all intents and purposes, today's electronic computer is simply a de-

vice that can perform deductions in propositional logic.

The final step in analyzing the patterns involved in mathematical proofs was supplied by Charles Sanders Pierce, Guiseppe Peano, and Gottlob Frege at the end of the nineteenth century. Their idea was to take propositional logic and add further deductive mechanisms that depend on the nature of the propositions, not just their truth values. The resulting theory is called *predicate logic*. Predicate logic has no unanalyzed, "atomic" propositions. All propositions are regarded as built up from more basic elements. In other words, in predicate logic, a study of the patterns of deduction is preceded by, and depends upon, a study of certain linguistic patterns—the patterns of language used to form a proposition.

This system of logic takes as its basic elements not propositions but properties, or *predicates*. The more simple of these are the very same constituents of Aristotle's logic, predicates such as:

. . . is a man

. . . is mortal.

Predicate logic, however, allows for more complex predicates, involving two or more objects, such as:

. . . is married to . . .

that relates two objects (people), or

. . . is the sum of . . . and . . .

that relates three objects (numbers).

Predicate logic extends propositional logic, but the focus shifts from propositions to "sentences" (the technical term is *formula*). The actual rules for sentence construction are somewhat complicated to write down precisely and completely, but the following simple examples should give the general idea. In predicate logic, Aristotle's proposition *All men are mortal* is constructed like this:

For all *x*, if *x* is a man, then *x* is mortal.

This version has the mathematical advantage over the original in that the internal, logical structure is explicit. This structure becomes more apparent when the logician's symbols are used instead of English words and phrases.

First, the logician writes the predicate "*x* is a man" in the abbreviated form Man(*x*) and the predicate "*x* is mortal" in the form Mortal(*x*). The phrase *for all* may be abbreviated by the letter *A*, and the phrase *there exists* may be abbreviated by the letter *E*. Using this notation, *All men are mortal* looks like this:

$$Ax[\text{Man}(x) \rightarrow \text{Mortal}(x)]$$

Written in this way, all the logical constituents of the proposition and the underlying logical pattern are obvious.

A major, characteristic feature of logic as a mathematical discipline is that it separates the notion of a *statement about* a mathematical structure (such as the integers or the real numbers or the two-dimensional Euclidean plane) from the notion of a *property of* that structure. Study of the former is referred to as *syntax,* the latter as *semantics.* (In everyday terms, a sentence is a syntactic entity, the meaning of a sentence is a semantic notion.) For example, let $N$ be the structure whose domain is the set of natural numbers, and whose sole predicate is the two-place relation less than, $<$. The structure $N$ provides an *interpretation* for the formula $Ax\,Ey\,P(x, y)$.

The variables are taken to range over the set of all natural numbers and the predicate symbol $P$ is taken to denote the relation $<$. Under this interpretation, the formula above has a meaning. In fact, this formula makes an assertion about the structure $N$: It says that for every natural number $x$, there is another natural number $y$ that is greater than $x$. This assertion is indeed true of the structure $N$. On the other hand, if $M$ is the structure consisting of the finite set $\{1,2,3,4,5\}$ with the single two-place relation $<$ on that set, then $M$ also provides an interpretation for the formula above, but under this interpretation the formula makes a false claim about the structure. These ideas were formalized by the logician Alfred Tarski in the early part of the twentieth century. Tarski's definitions provided formally defined notions of "meaning" and "truth" for formulas.

The two key features of classical logic are (1) it captures the abstract patterns used in formal, mathematical reasoning and provides a precise, formal model of a mathematical proof; and (2) it distinguishes syntax (language) from semantics (meaning). More recent developments in logic, in large part spurred on by problems in computer science, have replaced the truth-based logic described above by an information-based logic. In these newer logical frameworks, the basic step of deducing one true statement from known true statements is replaced by the step of producing new information from available information.

*See also* Diagrams, Euler and Venn; Logic Machines

## SELECTED REFERENCES

Devlin, Keith. *Sets, Functions and Logic.* London, England: Chapman and Hall, 1992.

Mendelson, Elliott. *Introduction to Mathematical Logic.* New York: Van Nostrand, 1987.

KEITH DEVLIN

# LOGIC MACHINES

Devices for solving problems in logic, first used to solve the classical syllogism, a form of reasoning where a conclusion is drawn from two premises. An example is

Encyclopedias are useful.
<u>Useful books are valuable.</u>
Encyclopedias are valuable.

Charles Stanhope (1753–1816), an eighteenth-century British scientist, invented the first of these machines, the Stanhope Demonstrator. This mechanical device could solve numerical syllogisms (anticipating the work of Augustus De Morgan [1806–1871], who is credited with introducing them), as well as elementary problems in probability. An example of a numerical syllogism is

There are 10 people of whom 4 are women.
<u>8 people are young.</u>
At least 2 women are young.

William Stanley Jevons (1835–1882), the British logician and economist, constructed a logic machine in 1869 capable of rapidly identifying the valid lines of a truth table consistent with the premises of a problem. Sometimes referred to as a logical "piano," which it resembled, this device could not handle syllogisms well, but it became the prototype of nearly all the logic machines that followed as syllogistic reasoning gave way to propositional logic.

The front of the machine contains openings for the letters representing the sixteen combinations of four terms and their negatives in columns. The keyboard's twenty-one keys has a copula for equality as the center key. To its left and right, symmetrically arranged, are the four positive terms (uppercase letters) and the four negative terms (lowercase letters). The first key on the left is labeled *finis;* the last key on the right is labeled *full stop.* The second and twentieth keys are .|. for the inclusive "or."

Consider the proposition "All $A$ is $C$," which Jevons viewed as the equation "$A = AC$." Using the machine to operate on this equation involves pressing the A key on the left, followed by the copula, the A and C keys on the right, ending with the full stop. All combinations of terms that are inconsistent with this proposition will be eliminated automatically from the face of the machine, leaving only the conclusions that can be drawn.

The next advance was made by Allan Marquand (1853–1924), professor of art and archeology at Princeton University, who in 1881 with Charles G. Rockwood, Jr., a mathematical colleague, constructed a machine like Jevons's capable of handling the sixteen true-false arrangements of four terms. Although simpler to use than Jevons's machine, statements like "some *A* is *B*" could not be included. The breakthrough, incorporating "some" statements, came in 1910 when Charles P. R. Macaulay, an Englishman working in Chicago, built an easily operated compact four-term machine explicitly for propositional logic.

The first electric machine, a device for solving syllogisms, was constructed in 1936 by Benjamin Burack, a Chicago psychologist. But it was Allan Marquand who had drawn the first circuit design in 1885—for an electric version of his machine, a device never built. The first electric machine for propositional logic constructed in 1947 by two students at Harvard University, William Burkhart and Theodore A. Kalin, was a Jevons-type machine with a capacity of twelve rather than four terms. All of these special-purpose electric machines were supplanted ten years later by the general-purpose digital computer.

*See also* Logic

## SELECTED REFERENCES

Gardner, Martin. *Logic Machines and Diagrams.* 2nd ed. Chicago, IL: University of Chicago Press, 1982.

Jevons, William Stanley. *Pure Logic and Other Minor Works.* Robert Adamson and Harriet A. Jevons (eds.). Bristol, England: Thoemmes, 1991.

Marquand, Allan. "A New Logical Machine." *Proceedings of the American Academy of Arts and Sciences* 21(1885):303–307.

FRANCINE F. ABELES

# LOGO

Computer language designed to develop students' creative ability and problem-solving skills as they use the computer. The emphasis in this language usually involves "turtle graphics." It uses a "turtle"-shaped cursor for most of its operations, which draw a range of figures from simple line segments to complex diagrams involving variables. After learning no more than a few simple commands, the students are ready to explore many instructional ideas. All of this provides for a learning environment (microworld) that is motivational, interesting, and exciting. The students can create their own new commands or procedures, with new names and tasks, as they explore mathematical concepts. They are very eager to teach the "turtle" how to draw new designs and see the results on the screen and may spend long hours playing and learning.

Logo is a high-level language that has been specifically designed so that it may be used by small children as well as adults. This is possible because Logo allows students complete control of the learning environment. This control is provided as the students learn to make shapes and create other tasks and develop and name new procedures that can be used as commands (referred to as *primitives* in some cases) or as subprocedures for other procedures. The students have many options to manipulate and a lot of freedom when using them. Also, as they learn how to use Logo, the students get more and more sophisticated in the creation and use of these commands and procedures. This learning process is like teaching the "turtle" new words, and the students are actually building their own computer language. Logo has been specially useful in the area of mathematics. Because of this powerful feature, Logo is not a toy, or a language only for children. It has been used successfully to teach numbers, measurement, and geometry. For example, to explore geometry, students move the turtle around on the screen, make shapes, and draw pictures. Thus, students learn about sides and angles in a unique manner. They have to input the length of the sides and the size of the angles in order to control the "turtle." An example of the FORWARD (FD) command is the following (the small triangle in Figure 1 is called the turtle, and it has moved ten "turtle steps" forward leaving a trail):

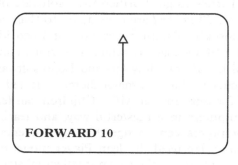

**FORWARD 10**

**Figure 1**

Also, a child can develop a procedure to draw a square (see Figure 2). In this procedure RT stands

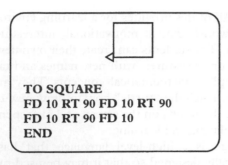

```
TO SQUARE
FD 10 RT 90 FD 10 RT 90
FD 10 RT 90 FD 10
END
```

**Figure 2**

for RIGHT, which commands the turtle to turn to the right a number of degrees, in this case 90 degrees. Every time the student enters the word SQUARE, the turtle will draw a square with sides equal to ten turtle steps—it becomes a new word in the Logo vocabulary.

Logo, whose name is derived from the Greek word for "reason," was introduced in 1968 and developed as more than a computer language under the leadership of Seymour Papert at the Massachusetts Institute of Technology (MIT). It stems from an earlier language called LISP (LISt Processing), which has been used by many researchers in the field of artificial intelligence. His original explorations during the development of Logo were with a robot turtle that could move around on the floor and draw pictures when certain commands were given. Seymour Papert (1980) described Logo as the name for a philosophy of education in a growing family of computer languages associated with it. This includes computer languages and software that provide the learning environment proposed by Papert: Terrapin Logo, Logo Writer, Logo Plus, Micro Worlds, WIN Logo, Object Logo, Harvard Associates Logo, and Mach Turtles Logo. Furthermore, software like the *Geometer's Sketchpad* are considered to share a similar philosophy. Another variation of Logo ideas is named LEGO/Logo. It refers to various combinations of LEGO building kits and Logo software. He also indicated that two major themes affected his research for ten years at MIT: Children can learn to use computers in a masterful way, and learning to use computers can change the way students learn everything. He used the Jean Piaget model of children as builders of their own intellectual structures to develop his ideas about learning with Logo. His emphasis was not on the computer but on learning. For him, the "turtle" was a computational "object-to-think-with." His vision of the learning environ-

ment demanded free contact between children and computers.

Logo is associated in many cases with exploratory learning or discovery learning with computers. This approach provides opportunities for experimentation with the computer, includes opportunities for peer interaction, and includes teacher support activities in the form of probing questions or other type of follow-up activities. For example, the teacher, after some introductory activities dealing with polygons, degrees, and Logo, can ask the students to find a way to "teach" the turtle how to draw a regular hexagon. They can explore different alternatives or solutions in small groups, and eventually share their findings. One of many possible solutions to this problem is the following:

```
TO HEXAGON
REPEAT 6 [FD 50 RT 60]
END
```

When this procedure is entered into the Logo program, it will draw a regular hexagon by repeating the command inside the brackets, move forward fifty turtle steps and turn right (RT) 60 degrees, six times.

Some educators and researchers have proposed a more guided discovery approach in which the teachers have a more significant and active role within the Logo learning environment (Clements 1986). This approach must include insightful teacher questions and guidance that will facilitate student reflection on the procedures developed, as well as proper interaction and communication between peers (this could include cooperative learning activities). In this type of setting, the teacher will ask probing questions, make sure the students are on the right track, and have specific objectives and outcomes in mind as the students explore.

One of the versions of Logo, created by Logo Computer Systems International, has paired a word-processing program with Logo. The program, called Logo Writer, has been used by teachers who encourage their students to write about their Logo projects. Also, programs in Logo, called "instant" or "single-stroke" Logo, can be written by teachers that permit students to press one key, possibly followed by RETURN, to command the turtle. The teacher could assign a key for each of the following actions: FORWARD 10, BACK 10, RIGHT 30, and LEFT 30. For example, the students can command the "turtle" to move ten "turtle steps" forward by just pressing the f key on the computer keyboard.

Logo computer programming has been described as an environment in which children can develop problem solving, inductive reasoning, logical thinking, and other mathematical ideas. Researchers (Abelson and diSessa 1981; Foster 1972; Ortiz and MacGregor 1991; Papert, Watt, diSessa, and Wier 1979) have found a connection between Logo programming and problem solving. In the last few years, educators have evinced new interest in Logo. Papert's books (1993; 1996) have been a factor in this renewed attention. Educators in general, however, still have not seen a widespread use of Logo in classrooms. Some of the reasons might be that an effective use of Logo in classrooms will require more changes in the way that mathematics and other subjects are taught, and also availability of computers to students. Standards for school mathematics recommended by The National Council of Teachers of Mathematics (NCTM 1989) and the efforts related to their implementation are steps in the right direction.

*See also* Geometry, Instruction; Software

## SELECTED REFERENCES

Abelson, Harold, and Andrea diSessa. *Turtle Geometry: The Computer as a Medium for Exploring Mathematics.* Cambridge, MA: MIT Press, 1981.

Clements, Douglas H. *Delayed Effects of Computer Programming in Logo on Mathematics and Cognitive Skills.* Paper presented at the Annual Meeting of the American Educational Research Association, San Francisco, CA., 1986.

Foster, Thomas E. *The Effect of Computer Programming Experiences on Student Problem Solving Behaviors in Eighth Grade Mathematics.* ED 038034. Cambridge, MA: ERIC Document Reproduction Services, 1972.

National Council of Teachers of Mathematics. *Curriculum and Evaluation Standards for School Mathematics.* Reston, VA: The Council, 1989.

———. *Principles and Standards for School Mathematics.* Reston, VA: The Council, 2000.

Ortiz, Enrique, and S. Kim MacGregor. "Effects of Logo Programming on Understanding Variables." *Journal for Educational Computing Research* 7(1)(1991):37–50.

Papert, Seymour. *The Children's Machine: Rethinking School in the Age of the Computer.* New York: Basic Books, 1993.

———. *The Connected Family: Bridging the Digital Generation Gap.* Marietta, GA: Longstreet, 1996.

———. *Mindstorms: Children, Computers and Powerful Ideas.* New York: Basic Books, 1980.

Papert, Seymour, D. Watt, Andrea diSessa, and S. Wier. *Final Report of the Brookline Logo Project.* Vol II. Memo No. 545. Cambridge, MA: Massachusetts Institute of Technology Artificial Intelligence Laboratory, 1979.

### Resources

CLIME Connections: (Council for Logo and Technology in Mathematics Education). Special interest group of the National Council of Teachers of Mathematics. Quarterly Newsletter can be ordered from 10 Bogert Ave., White Plains, NY 10606. Phone: (914) 946-5143.

Epistemology and Learning Group: In the forefront of Logo Research. Media-Lab E15-309, Massachusetts Institute of Technology, 20 Ames Street, Cambridge, MA 02139. Phone: (617) 253-7851.

Logo Foundation: Nonprofit educational organization incorporated in New York State. *Logo Update* is their newsletter, and is published three times yearly. Their Professional Development Services provide teachers with Logo Learning opportunities in various formats: publications, institutes, workshops, courses, users groups, on-site support, and telecommunications. Most of the Logo versions described in this article and other Logo products can be found through this foundation. Subscription is free from 250 West 57th Street, Suite 2603, New York, NY 10107-2603. Phone: (212) 765-4918.

Logo Special Interest Group: Subdivision of the International Society for Technology in Education (ISTE). Newsletter *(Logo Exchange)* can be ordered from 1787 Agate Street, Eugene, OR 97403. Phone: (503) 346-4414.

ENRIQUE ORTIZ

## LOW ACHIEVERS
*See* DEVELOPMENTAL (REMEDIAL) MATHEMATICS; LEARNING DISABILITIES, OVERVIEW; REMEDIAL INSTRUCTION; UNDERACHIEVERS; UNDERACHIEVERS, SPECIAL PROGRAMS

# M

## MACLAURIN, COLIN (1698–1746)

Most famous for his *Treatise of Fluxions* (1742), a version of the calculus. Maclaurin entered the University of Glasgow at age 11 and was appointed to a chair of mathematics at the University of Aberdeen at age 19, later moving to the University of Edinburgh. His *Treatise* was written in response to the criticisms by philosopher and cleric George Berkeley (1685–1753) of the foundations of Newton's theory of fluxions. Not only did Maclaurin's work demonstrate the rules of fluxions from a classical geometric point of view but it also contained the earliest analytic proofs of the rule for calculating power series of functions (Maclaurin series), the derivative tests for maxima and minima, and the Fundamental Theorem of Calculus. Maclaurin's other major text, *A Treatise of Algebra in Three Parts* (1748), considered algebra as a generalized arithmetic. It included rules and justifications for operating with negative quantities, used word problems in showing how to solve linear and quadratic equations, and contained the earliest discussion of Cramer's Rule for solving systems of linear equations.

*See also* Maclaurin Expansion

### SELECTED REFERENCES

Katz, Victor J. *A History of Mathematics: An Introduction.* New York: HarperCollins, 1993.

Turnbull, Herbert W. *Bi-centenary of the Death of Colin Maclaurin.* Aberdeen, Scotland: University Press, 1951.

———. "Colin Maclaurin." *American Mathematical Monthly* 54(1947):318–322.

VICTOR J. KATZ

## MACLAURIN EXPANSION

In the eighteenth century, mathematicians, including the great Leonhard Euler, developed the modern notion of "function." Their idea was to arithmetize the study of functions, including the calculus, which had previously been undertaken largely geometrically. Thus, they limited the operations they were willing to use in the construction and manipulation of functions to standard algebraic operations, taking limits, and summing infinite series. They studied transcendental functions, including exponential and logarithmic functions and the trigonometric functions, by representing these functions as infinite series. They then used the series representations to discover many properties of and relationships among the transcendental functions.

In 1715, Brook Taylor found a particular algorithm for constructing an infinite series representation of a general function. Taylor's series were central to a treatise on the calculus written in 1742 by Colin Maclaurin. Both Taylor and Maclaurin were students of Isaac Newton, who, together with Gottfried W. Leibniz, is credited with the development of modern calculus.

The Maclaurin expansion of a function is the Taylor series expansion of the function about the point $a = 0$; that is, a power series representation of the function that converges to the function for values of $x$ close to 0.

The Maclaurin series expansion of the function $f$ is the power series

$$f(0) + f'(0)x + \frac{f''(0)}{2!}x^2 + \frac{f'''(0)}{3!}x^3 + \cdots$$

Power series, including Maclaurin series, are commonly used to represent transcendental functions, such as the trigonometric functions, exponential functions, and logarithmic functions.

Geometrically, if you consider the polynomial formed by the first few terms of any Maclaurin series, that is, a partial sum of the series, you have a polynomial that is similar to the function represented in that (1) the function and the polynomial have the same value at zero, and (2) the graph of the function and the graph of the polynomial have similar shape near zero, since their first few derivatives at zero are the same. In this sense, the first term of the Maclaurin series is the best constant approximation of the function near zero; the first two terms give the best linear approximation of the function near zero, the first three terms give the best quadratic approximation of the function near zero, and so on.

An effective classroom demonstration is to show the graphs of the original function together with those of several partial sums for the Maclaurin expansion. Near zero, as you use more and more terms of the power series expansion, the graphs become nearly indistinguishable. However, farther away from zero, they are not very similar. This also indicates the importance of knowing the interval on which the power series converges to the function value, since the first few terms of the power series may be used to approximate the value of the function for appropriate values of the independent variable.

Maclaurin series expansions for some common functions are:

$$\sin(x) = x - \frac{x^3}{3!} + \frac{x^5}{5!} - \frac{x^7}{7!} + \cdots$$

$$\cos(x) = 1 - \frac{x^2}{2!} + \frac{x^4}{4!} - \frac{x^6}{6!} + \cdots$$

$$\exp(x) = 1 + x + \frac{x^2}{2!} + \frac{x^3}{3!} + \cdots$$

$$\ln(1 + x) = x - \frac{x^2}{2} + \frac{x^3}{3} - \frac{x^4}{4} + \cdots$$

The conditions under which the Maclaurin series converges to the function are given by Taylor's theorem, using $a = 0$: Let $f$ be a function with derivatives of all orders in some interval $(a - r, a + r)$. The Taylor series

$$f(a) + f'(a)(x - a) + \frac{f''(a)}{2!}(x - a)^2 +$$

$$\frac{f'''(a)}{3!}(x - a)^3 + \cdots$$

represents the function $f$ on that interval if and only if

$$\lim_{n \to \infty} R_n(x) = 0$$

where $R_n(x)$ is the so-called remainder in Taylor's formula. That is,

$$R_n(x) = \frac{f^{(n+1)}(c)}{(n + 1)!}(x - a)^{(n+1)}$$

and $c$ is some point in $(a - r, a + r)$.

The size of the remainder is the error of approximation when the value of a function is approximated using the first $n$ terms of the power series expansion.

The partial sums of the Maclaurin series for the function $f$ are called *Maclaurin polynomials*. The $n^{\text{th}}$ Maclaurin polynomial for $f$ consists of the first few terms of the Maclaurin series expansion and is a polynomial of degree $n$. These polynomials approximate the function $f$ near $x = 0$ in that the value of the function and that of each polynomial match at $x = 0$, and the first $n$ derivatives of the $n^{\text{th}}$ Maclaurin polynomial and the corresponding derivatives of the function $f$ have the same value at $x = 0$.

For example, the fourth Maclaurin polynomial for $f(x) = \ln(1 + x)$ is

$$P_4(x) = x - \frac{x^2}{2} + \frac{x^3}{3} - \frac{x^4}{4}$$

and $P_4(0) = f(0) = 0$, $P_4'(0) = f'(0) = 1$, $P_4''(0) = f''(0) = -1$, and so forth.

One suggested approach to introducing the topic of Maclaurin and Taylor series into the classroom is to begin with the Maclaurin polynomials. Ask the students to guess the best constant approximation to the function near zero. Then ask for the best linear approximation. Most, from their study of the calculus, will see that the tangent line is the best choice. Observe that the linear function they have suggested has the same derivative as the function they are trying to approximate and that both pass through the point $(0, f(0))$. Next, consider the best quadratic approximation. For this, we have the chance to ensure that the approximating curve bends in the right direction. Since the direction and the amount of "bend" are described by the second derivative, it is reasonable to write a quadratic function that has the same function value and the same values for the first and second derivatives at zero as the original function. Continuing in this manner, as a little computation will show, gives you precisely the Maclaurin polynomials. The Maclaurin series results when this process is carried out indefinitely.

*See also* Power Series; Series

## SELECTED REFERENCES

Edwards, Charles H., Jr. *The Historical Development of the Calculus.* New York: Springer-Verlag, 1979.

Hughes-Hallett, Deborah, Andrew M. Gleason, et al. *Calculus.* New York: Wiley, 1992.

Ithaca College Calculus Group. *Calculus, An Active Approach with Projects.* New York: Wiley, 1993.

Strang, Gilbert. *Calculus.* Wellesley, MA: Wellesley Cambridge Press, 1991.

DIANE DRISCOLL SCHWARTZ

# MAGIC SQUARES

Mathematical curiosities that have been known for over 4,000 years. A magic square of order $n$ consists of a sequence of distinct integers arranged in an $n$ by $n$ square array so that the sum of the numbers in every row, in every column, and in each main diagonal is the same. This sum is called the *magic constant* of the square. Usually, the integers used are the numbers from 1 to $n^2$, and the square is then called a *traditional magic square*. Magic squares exist for all orders larger than 2. For example, the nine integers in the first square in Figure 1 form a traditional magic square of order 3, with magic constant 15. The second square is a traditional magic square of order 4, with magic constant 34. The magic constant in a

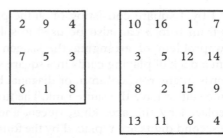

**Figure 1**

traditional magic square of order $n$ must be $n(n^2 + 1)/2$. This follows because the sum of all the numbers from 1 to $n^2$ is $n^2(n^2 + 1)/2$, and each of the $n$ rows must have the same sum. According to an ancient Chinese legend, the Emperor Yu, who lived around 2200 B.C., discovered a mystic turtle on the banks of the Yellow River. The markings on the turtle's back looked like dots that formed a third-order magic square. A treatise in Arabic on magic squares was written in the ninth century A.D. In 1514, Albrecht Dürer produced an engraving entitled "Melancholia" that contained a magic square of order 4, with the numbers 15 and 14 appearing as the two central numbers on the bottom row.

One method of constructing magic squares of order $n$ is to use numbers written in the base $n$. For example, the magic square of order 4 in Figure 1 was constructed from the first square in Figure 2, consisting of the numbers from 00 to 33 in base 4. The right (units) digits of each row, column, and diagonal form an arrangement of the four possible digits 0, 1, 2, 3. The left digits also form an arrangement of the four digits, and so the sum of each row, column, and diagonal must be the same. Furthermore, each number is a different combination of the four digits, and so the numbers in the square are 0 to 15, in some order. The magic square is now obtained by converting each number from base 4 to the decimal system and then adding one to each number.

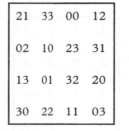

**Figure 2**

This 4 by 4 square consisting of all the pairs of elements from 0 to 3 can also be used to solve the well-know problem of arranging the sixteen court cards from a deck of playing cards in a square so that no two cards in any row, column, or diagonal belong to the same suit or have the same rank. The first digits are replaced by the ace, king, queen, and jack, while the second digits are replaced by the four suits. There are many other methods for constructing magic squares, and the methods for constructing odd- and even-ordered squares are usually quite different. Various methods of construction are given in Ball and Coxeter (1987, Ch. 7), Benson and Jacoby (1976), Eiss (1988), and Kraitchik (1953).

The left digits, written separately, of the first 4 by 4 square in Figure 2 and the right digits, written separately, are examples of an interesting mathematical object called a *Latin square* (see Figure 3). A Latin square of order $n$ is an $n$ by $n$ array of $n$ numbers arranged so that each number appears exactly once in each row and once in each column. The two Latin squares obtained in Figure 3 are called *orthogonal* because, when they are superimposed, each number of the first square occurs exactly once with each number of the second square. Latin squares are used in different mathematical areas such as designing statistical experiments and in coding theory (Stein 1976, 243–263). Any two orthogonal Latin squares of order $n$, using the numbers 0 to $n - 1$, can be combined in the manner above to produce a magic square, so long as each diagonal of both Latin squares contains every number exactly once.

In 1779, Leonhard Euler, in a paper called "On a new type of magic square," posed the famous problem of the thirty-six officers from six ranks and six regiments. He claimed that it was impossible to arrange these officers on parade in a 6 by 6 square so that each row and each column contained one officer from each rank and one from each regiment. For this reason, orthogonal Latin squares are sometimes called Euler squares. In 1899, it was proved, by exhaustive enumeration, that the problem of the thirty-six officers was insoluble. However, in 1959 it was shown that orthogonal Latin squares exist for all orders larger than 6 (Stein 1976, 243–263).

As an exercise in elementary addition, the class should construct all possible magic squares of order 3 using a given set of numbers. A comparison of the answers will lead to a discussion of the symmetries of a square. As an exercise in conversion from base 4, students should first try to solve the problem of arranging the sixteen court cards, or four different-colored cards containing four pictures. Then the class should construct Latin squares of order 4 and determine which are orthogonal. Finally, these orthogonal Latin squares can be used to solve the court card problem. Students could also be asked to show why there is no magic square of order 2.

*See also* Enrichment, Overview; Recreations, Overview

## SELECTED REFERENCES

Ball, Walter W. R., and Harold S. M. Coxeter. *Mathematical Recreations & Essays.* New York: Dover, 1987.

Benson, William H., and Oswald Jacoby. *New Recreations with Magic Squares.* New York: Dover, 1976.

Eiss, Harry E. *Dictionary of Mathematical Games, Puzzles, and Amusements.* New York: Greenwood, 1988.

Kraitchik, Maurice. *Mathematical Recreations.* New York: Dover, 1953.

Stein, Sherman K. *Mathematics: The Man-made Universe.* San Francisco, CA: Freeman, 1976.

WILLIAM J. GILBERT

# MANIPULATIVES

*See* ABACUS; ATTRIBUTE BLOCKS; CUISENAIRE RODS; DIENES; KINDERGARTEN; MIDDLE SCHOOL; MONEY; MONTESSORI; PATTERNING; PRESCHOOL; SPACE GEOMETRY, INSTRUCTION; STERN; SUBTRACTION

# MANIPULATIVES, COMPUTER GENERATED

Manipulatives and computer software, each important on their own, can make contributions to mathematics learning when used together. Manipulatives provide concrete experiences, computers add a next level of abstraction by connecting the concrete three-dimensional model or manipulative and the

| 2 | 3 | 0 | 1 |
|---|---|---|---|
| 0 | 1 | 2 | 3 |
| 1 | 0 | 3 | 2 |
| 3 | 2 | 1 | 0 |

| 1 | 3 | 0 | 2 |
|---|---|---|---|
| 2 | 0 | 3 | 1 |
| 3 | 1 | 2 | 0 |
| 0 | 2 | 1 | 3 |

**Figure 3**

symbolic, paper-and-pencil representation of the mathematical idea. The relationship between manipulatives and computer software can be organized into four categories: mirroring, modeling, managing, and manipulating.

## MIRRORING

*Mirroring* refers to using software that displays objects that look and are used in exactly the same way as the corresponding manipulatives. Such software has several advantages over manipulatives used alone. It provides an intermediate level of abstraction, and it can monitor student responses, an important feature with problem-solving activities such as attribute games, where the *range* of "correct" answers makes teacher monitoring difficult. Computer software can also add depth to the manipulative. In geoboard software, for example, students have access to many different types of on-screen geoboards. They can select from square, triangular, or circular boards and assign different numbers of pinpoints to each. Finally, software provides an "endless" supply of manipulatives, and the manipulatives are less messy and easier to clean up than the real thing since they vanish when the program is turned off.

## MODELING

*Modeling* refers to using the manipulative to visualize a concept that is being presented in the computer program. An example is a software program that asks users to identify the relative heights of an array of buildings given a top-down view and clues showing front, right, left, or back side views. A manipulative such as Cuisenaire rods set up to represent buildings of different heights to model the computer clues can be remarkably helpful in visualizing the computer information.

## MANAGING

Manipulatives can be controlled in much the same way as machines can be remotely controlled. In *LEGO TC Logo,* for example, the manipulative, in the form of a machine or robot-like construction built by the user, is managed, maneuvered, or manipulated under computer control. Science software that records data generated by a physical object such as a thermometer or a light meter that has been connected to the computer recording device is another example of the managing relationship.

## MANIPULATING

A new class of software/manipulative relationship includes the software itself as the manipulative. Here users "touch," stretch, and manipulate the shapes of figures on the screen using an input device such as a mouse to observe relationships and guide discovery. In a calculus program, for example, users observe the slope lines at each point on the screen-displayed curve as the curve is traced by the mouse. The curve itself can also be manipulated in various ways using the mouse (e.g., translated, stretched, etc.). These changes are reflected in the equation and tables of values that are simultaneously displayed in separate windows on the screen. A geometry program allows users to explore relationships and generalize about geometric constructs. For example, having constructed the angle bisectors of a triangle and observed that these lines meet at a single point, the user may manipulate and distort the shape of the triangle using the mouse, thus observing that this relationship remains constant for different triangles.

Understanding how manipulatives and computer software, two classes of educational tools, can work together allows maximum educational benefit to be derived from the use of both.

*See also* Software; Technology

### SELECTED REFERENCES

*Geometers' Sketchpad* (Macintosh and IBM-compatible computers). Berkeley, CA: Key Curriculum, 1991.

*LEGO TC Logo* (Apple and IBM-compatible computers). Enfield, CT: LEGO Dacta, 1992.

Perl, Teri. "Manipulatives and the Computer: A Powerful Partnership for Learners of all Ages." *Classroom Computer Learning* 10(Mar. 1990):20–22, 26–29.

TERI PERL

# MARKOV PROCESSES

Processes that owe their name to Andrei Andreevich Markov (1856–1922), the Russian mathematician who first investigated Markov chains in his analysis of the sequence of consonants and vowels in Pushkin's poem *Evgeny Onegin* (1831). Roughly, the idea is to express some level of memory in an evolving system. For a Markov process, it is assumed that the present state of the system is dependent only on its most recent previous state. We illustrate some of the concepts of Markov processes through the simplest process of this kind, namely, the Markov chain.

A homogeneous Markov chain $\{X_t\}$ is a sequence of random variables with discrete parameter (usually time) $t = 0,1,2, \ldots$, defined on the finite or infinite state space of the nonnegative integers by a transition probability matrix $P = \{p_{ij}\}$, where

$$p_{ij} = P\{X_{t+1} = j \mid X_t = i, X_{t-1} = k, X_{t-2}$$

$$= 1, \ldots, X_0 = m\} = P\{X_{t+1} = j \mid X_t = i\}$$

$$\sum_j p_{ij} = 1; \ i, j, k, l, \ldots, m \geq 0 \text{ integers}$$

We see that the probability that $X_{t+1} = j$ is conditional only on $X_t = i$. This expresses the Markov property that the value of $X_{t+1}$ depends only on the value of $X_t$, and not on earlier values. It should be noted that for a homogeneous Markov chain, the transition matrix $P$ is the same for all $t$.

Let us consider a simple example, that of the homogeneous Markov chain with two states, 0, 1, having the transition probability matrix

$$P = \begin{bmatrix} p_{00} & p_{01} \\ p_{10} & p_{11} \end{bmatrix}, \quad p_{00} + p_{01} = p_{10} + p_{11} = 1$$

To make this example more concrete, let us consider a game involving two coins labeled 0 and 1. Coin 0 is fair, so that it comes up with heads and tails with equal probabilities of 0.5. Coin 1 is biased, so that its probability of heads is 0.6 and of tails 0.4. We start with coin 0; if its throw results in heads, we switch to coin 1, while if we get tails we continue with coin 0. With coin 1, if we obtain heads we continue with it, but if we get tails we switch back to coin 0. We can continue the game indefinitely for $t = 1, 2, 3, \ldots$.

We now have the specific transition probability matrix

$$P = \begin{bmatrix} p_{00} = 0.5 & p_{01} = 0.5 \\ p_{10} = 0.4 & p_{11} = 0.6 \end{bmatrix}$$

for this system. We may ask, starting with coin 0, what is the probability of ending with coin 1 at the second throw? This could happen in two ways, (a) or (b):

|     | Start $t = 0$ | Throw $t = 1$ | Throw $t = 2$ |
|-----|---------------|---------------|---------------|
| (a) | Coin 0        | Coin 0        | Coin 1        |
| (b) | Coin 0        | Coin 1        | Coin 1        |

We can readily see that the probability of $X_2 = 1$, starting from $X_0 = 0$, is

$$p_{01}^{(2)} = P\{X_2 = 1 \mid X_0 = 0\} = [p_{00} \ p_{01}] \begin{pmatrix} p_{01} \\ p_{11} \end{pmatrix}$$

$$= (p_{00})(p_{01}) + (p_{01})(p_{11})$$

$$= (0.5)(0.5) + (0.5)(0.6)$$

$$= 0.55$$

In general, for $t \geq 1$, starting from $X_0 = 0$ so that $P\{X_0 = 0, X_0 = 1\} = [1 \ \ 0]$, we have that

$$P\{X_t = 0, X_t = 1 \mid X_0 = 0\}$$

$$= [1 \ 0] \begin{bmatrix} 0.5 & 0.5 \\ 0.4 & 0.6 \end{bmatrix}^t = [1 \ 0] P^t$$

Thus, if we take $t = 3$,

$$P\{X_3 = 0, X_3 = 1 \mid X_0 = 0\}$$

$$= [1 \ 0] \begin{bmatrix} 0.5 & 0.5 \\ 0.4 & 0.6 \end{bmatrix}^3 = [1 \ 0] \begin{bmatrix} 0.445 & 0.555 \\ 0.444 & 0.556 \end{bmatrix}$$

$$= [0.445 \ \ 0.555]$$

and $P\{X_3 = 0 \mid X_0 = 0\} = 0.445$, while $P\{X_3 = 1 \mid X_0 = 0\} = 0.555$. A similar result holds for the case of $n + 1$ states, when the initial probabilities are

$$P\{X_0 = 0, \ldots, X_0 = n\} = [a_0, a_1, \ldots, a_n] = a$$

with

$$\sum_{i=0}^{n} a_i = 1$$

and

$$P\{X_t = 0, \ldots, X_t = n\} = aP^t, P = \{p_{ij}\},$$

$$i, j = 0, \ldots, n$$

Markov chains can be nonhomogeneous over their parameter $t$, so that the transition matrix $P(t)$ is dependent on $t$; they may also have a continuous state space. An example of a Markov process with continuous state space and discrete parameter $t = 0, 1, 2, \ldots$, which occurs in time series, is the autoregressive sequence

$$X_{t+1} = \rho X_t + e_t, E(e_t) = 0, X_0 = x_0$$

with the serial correlation $\rho$ ($e_t$ mutually independent and $|\rho| < 1$).

More generally, one can define Markov chains with continuous parameter $t \geq 0$ and discrete or continuous state space. A simple example of a Markov chain in continuous time $t \geq 0$, with two states 0,1, that parallels the discrete time chain just described when $p_{00} = e^{-\lambda}$ and $p_{11} = e^{-\mu}$, $\lambda, \mu > 0$ can be defined by the continuous probabilities

$$\begin{bmatrix} p_{00}(t) = \dfrac{1}{\lambda + \mu}(\mu + \lambda e^{-(\lambda+\mu)t}) & p_{01}(t) = 1 - p_{00}(t) \\ p_{10}(t) = 1 - p_{11}(t) & p_{11}(t) = \dfrac{1}{\lambda + \mu}(\lambda + \mu e^{-(\lambda+\mu)t}) \end{bmatrix}$$

indicating the transitions from $i = 0, 1$ to $j = 0, 1$ in any time $t \geq 0$.

Markov processes are extremely useful as models of natural phenomena; for example, they have been used in birth and death models for a variety of biological population processes. They have also been used in Markov decision processes in operations research.

*See also* Probability, Overview

## SELECTED REFERENCES

Bartlett, Maurice S. *An Introduction to Stochastic Processes*. Cambridge, England: Cambridge University Press, 1960.

Doob, Joseph L. *Stochastic Processes*. New York: Wiley, 1953.

Feller, William. *An Introduction to Probability Theory and Its Applications*. Vol. 1. 3rd ed. New York: Wiley, 1968.

————. *An Introduction to Probability Theory and Its Applications*. Vol. 2. New York: Wiley, 1966.

Heyman, Daniel P., and Matthew J. Sobel. *Stochastic Models in Operations Research*. Vol. 1. New York: McGraw-Hill, 1982.

Ross, Sheldon M. *Stochastic Processes*. New York: Wiley, 1983.

JOSEPH GANI

# MATHEMATICAL ASSOCIATION OF AMERICA (MAA)

An association whose purpose is to "assist in promoting the interests of mathematics in America, especially in the collegiate field." Currently based in Washington, D.C., it was founded in 1915 in Columbus, Ohio. Although most MAA members are mathematics educators at the college and university level, anyone interested in mathematics is welcome to join. The association is governed by a board of governors, but much of its activity is centered in local sections to which each member automatically belongs depending on geographical region. There are also MAA student chapters on many college campuses.

## MEETINGS

The MAA holds its annual meeting in January and a summer meeting in August. These meetings, held jointly with the American Mathematical Society, serve as a forum for the latest developments in mathematical theory and applications in collegiate mathematics and the college curriculum. In addition, the local sections hold regular meetings and activities that involve both high school and college students.

## EDUCATION

The MAA deals with many aspects of mathematics education, especially those related to the teaching of mathematics at the collegiate level. It has launched many projects, of which a number are funded by government and private grants. Some examples are Teaching Mathematics with Calculators: A National Workshop with the National Council of Teachers of Mathematics (NCTM); Curriculum Action Project, which is reported in the MAA publication *Heeding the Call for Change: Suggestions for Curricular Action* (Steen 1992); Development of Software for Computer-Based Placement Tests; Development of Standards for Teachers of Mathematics; Interactive Mathematics Text Project; and Strengthening Underrepresented Minority Mathematics Achievement (SUMMA) Program, the first undertaking of the Office of Minority Participation in Mathematics.

In addition, the association conducts the MAA Placement Test Program (PTP); the Putnam Competition for college students; and the American Mathematics Competitions for junior and senior high school students, including the American Mathematics Contests AMC $\rightarrow$ 8, AMC $\rightarrow$ 10, and AMC $\rightarrow$ 12, the American Invitational Mathematics Examination (AIME), and the U.S.A. Mathematical Olympiad (USAMO). These competitions make up the selection process that talented high school students go through in order to participate in the International Mathematical Olympiad (IMO).

*See also* Mathematical Association of America (MAA), Historical Perspective

## SELECTED REFERENCES

*American Mathematical Monthly*. Washington, DC: MAA.

*Carus Mathematical Monographs (CARUS)*. Washington, DC: MAA. A monograph series.

*Classroom Resource Materials*. Washington, DC: MAA. A series.

*College Mathematics Journal*. Washington, DC: MAA.

*Dolciani Mathematical Exposition (DOL)*. Washington, DC: MAA. A book series.

*MAA Notes*. Washington, DC: MAA. A paperback series.

*Math HORIZONS*. Washington, DC: MAA.

*Mathematics Magazine*. Washington, DC: MAA.

National Research Council. *Everybody Counts: A Report to the Nation on the Future of Mathematics Education*. Washington, DC: National Academy Press, 1989.

————. *Moving Beyond Myths: Revitalizing Undergraduate Mathematics*. Washington, DC: National Academy Press, 1991.

*New Mathematical Library (NML)*. Washington, DC: MAA. A paperback series.

Steen, Lynn A., ed. *Heeding the Call for Change: Suggestions for Curricular Action*. Washington, DC: MAA, 1992.

LILY E. CHRIST

# MATHEMATICAL ASSOCIATION OF AMERICA (MAA), HISTORICAL PERSPECTIVE

Founded in 1915 and incorporated in 1920, the MAA was created primarily to support the continued publication of the *American Mathematical Monthly,* which had been founded in 1894. Originally, the MAA had four major functions all dealing with mathematics education at the undergraduate collegiate level: (1) to provide organized activity; (2) to form a medium of communication and a forum for exchange of ideas; (3) to furnish a place for publications of scientific articles and papers; and (4) to publish historical articles, books, reviews, notes and news (May 1972, 20). Although the association retains the undergraduate level as its primary focus, it also serves as a bridge between high school and collegiate mathematics. The MAA now also publishes an "intermediate" journal, *The College Mathematics Journal,* which includes articles for high school and early college. The *Monthly* maintains a section on collegiate mathematics education. The MAA sponsored the American High School Mathematics Examination, first given in 1950, as well as a junior high school contest. A program for secondary school lectures was begun in 1958.

*See also* Mathematical Association of America (MAA)

## SELECTED REFERENCES
Jones, Phillip S., ed. *A History of Mathematics Education in the United States and Canada.* 32nd Yearbook. Washington, DC: National Council of Teachers of Mathematics, 1970.

May, Kenneth O. *The Mathematical Association of America: Its First Fifty Years.* Washington, DC: MAA, 1972.

JAMES K. BIDWELL
ROBERT G. CLASON

# MATHEMATICAL EXPECTATION

Most simply expressed as *average* value. If we read that a teenager can expect to visit his or her dentist 1.9 times a year, a married woman can expect to have 2.4 children, or a person can expect to eat 10.6 pounds of cheese per year, it must be clear that we are not using the word *expect* in its colloquial sense. A person cannot very well go to the dentist 1.9 times or have 2.4 children, and it would be surprising, indeed, if we found someone who has eaten 10.6 pounds of cheese in any given year. Figures like these must be interpreted as *averages,* namely, as *mathematical expectations.*

Originally, the concept of a mathematical expectation arose in connection with games of chance, and in its simplest form it can be stated as follows:

If a person stands to win the amount $a$ with probability $p$ and the amount 0 with probability $1 - p$, his or her mathematical expectation is given by the product $a \cdot p$.

For instance, if we buy one of 2,000 raffle tickets for a color television set worth $640, the probability for each ticket is $p = 1/2,000$ and our mathematical expectation is $640 \cdot 1/2,000 = \$0.32$ (or at least the equivalent of this amount in merchandise). Note that 1,999 tickets will not pay anything at all, one of the tickets will pay $640, and on the average the 2,000 tickets will pay $640/2000 = \$0.32$ per ticket. This average is the mathematical expectation.

To illustrate how the concept of a mathematical expectation can be generalized, suppose that in the raffle just described there are also ten consolation prizes of a small calculator worth $12. Now we can argue that 1,989 tickets will not pay anything at all, one ticket will pay $640, ten tickets will each pay $12, and on the average the 2,000 tickets will pay $(640 + 10 \cdot 12)/2,000 = \$0.38$ per ticket. Note that this result is also given by

$$0 \cdot (1,989/2,000) + 640 \cdot (1/2,000) + 12 \cdot (10/2,000) = \$0.38$$

which is the sum of the products obtained by multiplying each "payoff" by the corresponding probability. Generalizing from this example, we can give the following, more general, definition of a mathematical expectation:

If there are $k$ payoffs, $a_1, a_2, \ldots, a_k$, of which one must occur, and their respective probabilities are $p_1, p_2, \ldots, p_k$, then the mathematical expectation is $a_1 \cdot p_1 + a_2 \cdot p_2 + \cdots + a_k \cdot p_k$.

Note that if we let $a_1 = a$, $a_2 = 0$, $p_1 = p$, and $p_2 = 1 - p$, we get the original definition.

The payoffs mentioned in the raffle ticket example are all values of random variables, and need not be cash amounts. In general, the *expected value* of a discrete random variable is the sum of the products obtained by multiplying each value of the random variable by the corresponding probability; symbolically, the expected value of the discrete random variable $X$, denoted by $E(X)$, is given by

$\Sigma x \cdot (P(X = x))$, where the summation extends over the finite number or countable infinity of values of the random variable. If $X$ is a continuous random variable, the summation must be replaced with an integral.

*See also* Mean, Arithmetic; Random Variable

SELECTED REFERENCES
Freund, John E. *Introduction to Probability.* New York: Dover, 1993.
Hoel, Paul G. *Introduction to Mathematical Statistics.* 5th ed. New York: Wiley, 1984.

JOHN E. FREUND

# MATHEMATICAL LITERACY

The ability to understand and communicate mathematical ideas in the language of mathematics. In common usage, we mean by illiteracy the inability to read or write. Knowledge of how spoken language is represented by written symbols, however, does not guarantee that one is able to understand what is read or to write in an understandable way. Comprehension and expression, whether written or oral, will be severely limited by a small vocabulary or by ignorance of the rules of the language's grammar and syntax. To comprehend serious literature, it is also necessary to understand the use of such literary devices as symbolism, metaphor, and allegory. Finally, there are the problems of allusion and context. It is often impossible to know what an author is talking about without having first read other works of literature.

These notions about the nature and extent of literacy are central to the way we teach English, but it is not generally appreciated that they should be applied to the teaching of mathematics as well. We need to understand that mathematics is a language. Not only does mathematics have a special symbolic notation and vocabulary but an entire grammar, syntax, and literature of its own. Mathematical objects, things like lines, shapes, and numbers, are entirely abstract and can be observed only with the human imagination. Unlike the case of physical objects, just giving names to imaginary objects does not advance the capabilities of existing language appreciably. In order to converse about the abstractions of space and numbers, the definitions and attributes of mathematical objects and of the operations that can be performed upon them must be codified with a consistent formalism; an entire new language must be created. Removed from the context of this parent language, the symbols and words of mathematics have no meaning. It is thus meaningless to speak of translating mathematics into nonmathematical language.

Creating new kinds of mathematical objects and inventing ways to talk about them is the realm of pure mathematics. To understand the nature of this activity, the essential role played by whimsy must be appreciated. Undaunted by centuries of naysayers, the creative genius Bernhard Riemann was able to develop a whole new system of geometry based on the whimsical supposition that there are *no* lines through a given point that are parallel to a given line. This example illustrates another characteristic of pure mathematics: it is often quite arbitrary. One can also construct a perfectly good geometry by imagining that there may be more than one parallel through a given point to a given line.

Using the abstract imaginings of pure mathematics to understand our physical universe is known as applied mathematics. The method of application can be described more precisely: mathematical language is used to construct metaphors that represent specific aspects of the observable world. Describing the earth as a sphere using Euclidean geometry or the universe as a spherical space using Riemannian geometry are examples of such metaphorical constructions. Many of the greatest insights in human history have been expressed in this form, the ideas of Isaac Newton, James Maxwell, and Albert Einstein to name but a few of the most notable. The fact that scientists prefer to call these ideas *laws, theories,* or *models* should not be a source of confusion; by nature they are all mathematical metaphors. The power of mathematical language is that the implications of these metaphors for a wide variety of specific circumstances can be worked out in great detail. The subtleties of the metaphor, however, will be apparent only to those who appreciate the subtleties of the language in which they are written.

These, then, are the goals of the quest for mathematical literacy: to appreciate the aesthetic and intellectual appeal of the language of pure mathematics and to grasp the meaning of metaphors constructed with it. Pursuit of proficiency in performing symbolic or numeric manipulations will produce little progress toward either of these goals. When students encounter mathematics outside the mathematics classroom, however, it is just this sort of proficiency that is demanded of them. For example, students are usually taught that an ideal gas is one that obeys a certain formula. While this formula can be manipulated to calculate the properties of gases in some situations, the more powerful insights about the behavior of gases are contained in the metaphor of ideality: the molecules of an ideal gas collide with the walls of the container but never hit each other. The details of how the gas formula can be obtained directly from this metaphorical picture require a bit more knowledge of mathematics than most students have acquired. Science educators condescend to give students the formula but keep the metaphor hidden behind a veil. This veil has generally kept the relevance of mathematics hidden from scholars in the humanities and social sciences, and their failure to include mathematics in the discussion of their disciplines is most unfortunate. Mathematics, for instance, has a great deal to say about the notion of "self-evident truths" invoked by Thomas Jefferson.

To most students, the enormous impact of mathematics on human intellectual history is not apparent. By this measure, only a small population of students manages to assimilate enough mathematical language to achieve a reasonable level of literacy. Correcting this widespread illiteracy will not require major increases in the concentrated study of mathematics; but from the very beginning, the subject must be taught as language invented for thinking about abstract concepts and integrated across the entire spectrum of curricula.

*See also* Language and Mathematics in the Classroom; Language of Mathematics Textbooks

SELECTED REFERENCES
Bullock, James O. "Literacy in the Language of Mathematics." *American Mathematical Monthly* 101(Oct. 1994):735–743.

King, Jerry P. *The Art of Mathematics.* New York: Plenum, 1992.

Snow, Charles P. *The Two Cultures and a Second Look.* 2nd ed. New York: Cambridge University Press, 1964.

Steen, Lynn Arthur, ed. *Why Numbers Count: Quantitative Literacy for Tomorrow's America.* New York: College Entrance Examination Board, 1997.

JAMES O. BULLOCK

## MATHEMATICAL SCIENCES EDUCATION BOARD (MSEB)

A subdivision (created in 1985) of the Center for Science, Mathematics, and Engineering Education of the National Research Council (NRC). The National Academy of Sciences, for which NRC is the main operating agency, appoints the twenty-one members of the MSEB for a term of office of three

years, which may be renewed. The board and its subcommittees study the current condition of mathematics education and promulgate policies affecting those issues deemed most significant. Activities of the MSEB include publishing books, booklets, and reports and holding convocations. The MSEB conducts many activities in conjunction with other organizations; for example, the symposium *The Nature and Role of Algebra in the K–14 Curriculum: A National Symposium,* held in 1997, was cosponsored by the National Council of Teachers of Mathematics.

*See also* Federal Government, Role; Mathematical Association of America (MAA); National Council of Teachers of Mathematics (NCTM); National Science Foundation (NSF); United States Department of Education; United States National Commission on Mathematics Instruction (USNCMI)

SELECTED REFERENCES
Mathematical Sciences Education Board. *Getting Started with Teachers.* Washington, DC: National Academy Press, 1997.

———. *Measuring Up: Prototypes for Mathematics Assessment.* Washington, DC: National Academy Press, 1993.

———. *MSEB 1996–97 Annual Report: Decisions that Count.* Washington, DC: National Academy Press, 1997.

———. *On the Shoulders of Giants: New Approaches to Numeracy.* Washington, DC: National Academy Press, 1990.

LOUISE S. GRINSTEIN
SALLY I. LIPSEY

## MATHEMATICAL SPECTRUM

A periodical printed in Britain, first published in the academic year 1968–69, by the Applied Probability Trust, a foundation that encourages mathematical research and teaching and provides a variety of student prizes for excellence in mathematics. The initial aim of *Mathematical Spectrum,* based on the ideas of the mathematicians Harry Burkill and Leon Mirsky and the statistician Joseph Gani, was to provide mathematically talented students in schools, colleges, and universities as well as their teachers with articles containing some mathematical challenge. It was intended to supplement the more standard mathematics texts and introduce students and teachers to research areas in mathematics through expository articles readily understandable by them. Initial editors Joseph Gani and Harry Burkill were followed in 1979–80 by the current editor, David Sharpe.

Typically, issues of *Mathematical Spectrum* have consisted of about six articles on pure and applied mathematics, statistics, and biomathematics, followed by a computer column, letters to the editor, problems and solutions, and book reviews. Most articles are written by school and university teachers, but students are also encouraged to contribute papers, which are often polished by the editor before publication. Prizes are awarded annually for student contributions, letters, and solutions to problems.

When the magazine was first started in the late 1960s, it had only a few competitors in its particular area of publication. With the increasing understanding of the role of mathematics in education and economic development, the number of magazines with similar (if not identical) aims has greatly increased. Among these are the Mathematical Association of America's *College Mathematics Journal* and *Math Horizons,* the National Council of Teachers of Mathematics' *Mathematics Teacher,* and the Teaching Statistics Trust's *Teaching Statistics.* However, *Mathematical Spectrum* continues to provide challenging mathematical fare for students and teachers.

*See also* Mathematical Association of America (MAA); National Council of Teachers of Mathematics (NCTM)

### SELECTED REFERENCE
*Mathematical Spectrum.* ISSN 0025-5653. Sheffield S3 7RH, England: Applied Probability Trust, Hicks Building, The University.

JOSEPH GANI

## MATHEMATICS, DEFINITIONS

"A *point* is that which has no part." "A *line* (curve) is breadthless length." "A *straight line* is a line which lies evenly with the points on itself." These definitions appeared (ca. 300 B.C.) in Euclid's *Elements* (Euclid 1956). They would make little sense if one did not already know what a point, curve, or line is. Euclid was, of course, presenting an axiomatic system and (presumably) believed that *all* its concepts should be formally defined. We now realize, with 2,000 years' hindsight, that we must accept some concepts as undefined. Such concepts are said to be *implicitly* defined by the axioms (Eves 1990).

Geometry did not begin with Euclid. His axiomatic system came after centuries of heuristic geometric thinking, in which no formal definitions of point and line were needed. In general, the historical order of development in mathematics is the reverse of the logical. Another example: "the derivative was first *used;* it was then *discovered;* it was then *explored* and *developed;* and it was finally *defined*" (Grabiner 1983). Pedagogy should take a page from history rather than from logic: definitions should come only when there is a clear need for then. For instance, mathematics got along without a formal definition of real numbers for two millennia.

For 3,000 years mathematicians developed algebra, geometry, trigonometry, analytic geometry, and calculus—all *before* the term *function* was formally defined. Of course mathematicians had encountered *instances*—tables, curves, equations, physical laws—of what later came to be known as functions. Students, too, should develop an "instinct for functionality" before they are given the definitions. Teachers can deal with formulas, plot graphs, and introduce tables without formally introducing functions.

But what *is* a function? Definitions have varied throughout history (Kleiner 1993). This (but not only this) argues for *tentative* definitions in the classroom, which may require revision with changing circumstances. One may begin with function as a formula and *perhaps* later introduce function as a set of ordered pairs. (The notion of formula, too, has evolved (Kleiner 1993).)

We conclude with some comments on the "logic" of definitions. Is one free to give definitions as one pleases? Up to a point. For example, define a "recahedron" as a ten-faced regular polyhedron. There is no such object, although this is far from obvious. Aristotle, Gottfried Wilhelm Leibniz, and Luitzen Egbertus Jan Brouwer all warned against defining a concept without demonstrating its existence (Euclid 1956; Eves 1990). Moreover, some mathematicians have insisted that to show the existence of the object being defined it is necessary to *construct* it. Proposition 1 of Book I of Euclid's *Elements* shows how to construct equilateral triangles; this has been interpreted (by some) as demonstrating the existence, hence legitimizing the definition, of an equilateral triangle as a triangle in which all three sides are equal (Euclid 1956). Another example: define a function $f$ by (1) $f(x) = 1$ if $x$ is a positive integer and there are $x$ successive zeros in the decimal expansion of $\pi$, (2) $f(x) = 0$ for all other real numbers $x$. Brouwer would not countenance such a definition since it is not constructive. For instance, what is $f(999)$? We do not know, and *may never know,* if there are or are not 999 consecutive zeros in the decimal expsansion of $\pi$ (Kleiner 1993).

Finally, "impredicative" definitions must be treated with great caution. For example, define a set

$N$ as $N = \{x : x \notin x\}$, where $x$ is a set. (The set $u$ of all infinite sets is an infinite set; hence, $u \in u$ and $u \notin N$.) It can be readily shown that $N \in N$ if and only if $N \notin N$—this is the famous Russell paradox (Eves 1990). The set $N$ is thus not well defined. Its definition is *impredicative* in the sense that the definition of membership in $N$ depends on $N$. Russell expressed this phenomenon in his *vicious circle principle:* no set $S$ is allowed to contain members $m$ definable only in terms of $S$, or members $m$ involving or presupposing $S$ (Eves 1990). Obey the vicious circle principle or you run the risk of a vicious circle.

*See also* Axioms; Functions; Paradox

## SELECTED REFERENCES

Euclid. *The Thirteen Books of Euclid's Elements.* 2nd ed. 3 vols. Thomas L. Heath, trans. New York: Dover, 1956.

Eves, Howard W. *Foundations and Fundamental Concepts of Mathematics.* 3rd ed. Boston, MA: PWS-Kent, 1990.

Grabiner, Judith V. "The Changing Concept of Change: The Derivative from Fermat to Weierstrass." *Mathematics Magazine* 56(1983):195–206.

Kleiner, Israel. "Functions: Historical and Pedagogical Aspects." *Science and Education* 2(1993):183–209.

ISRAEL KLEINER

# MATHEMATICS, FOUNDATIONS

A mathematical specialty that attempts to put every area of mathematics on a logical footing. Although the use of mathematics predates history, logical deductive reasoning in mathematics began with the Greeks in the fifth century B.C.

Aristotle (394–321 B.C.) recognized the necessity of the acceptance of "indemonstrable" principles to avoid endless arguments regarding the truth of statements (Wilder 1952). These accepted principles are assumed to be true and are called *axioms* (or *postulates*). Similarly, some terms ("primitive" words) must be accepted as undefined in the mathematical sense. A mathematical definition must be a complete characterization of the term being defined, unlike the dictionary definition of a term, which leads to other terms and eventually back to the term originally sought. The circularity of dictionary definitions is not acceptable in a mathematics definition. The stated undefined terms, definitions, and axioms together form the basis for what is called an *axiomatic* (or *deductive*) system. Rules of logic are then applied

to the basic statements to deduce other statements, called *theorems*. These theorems are used along with the basis to establish other theorems. This process (or deductive procedure) was likely quite common with Aristotle and his contemporaries. The publication of Euclid's *Elements* (about 300 B.C.) exerted tremendous influence in establishing this process as the accepted way to validate statements, and as a result had a great impact. It served as a model for rational thought, formulating arguments, and judging the validity of arguments not only in mathematics and scientific disciplines but also in other domains.

The revolutionary characteristic of this approach is that it deals with ideas, not physical entities. In proving theorems, one is divorced from anything tangible and deals with pure thought. There are distinct advantages in this. Axiomatic systems often apply in physical situations completely unrelated to the motivating reality; modification of an axiomatic system can lead to new mathematics and to refinement of mathematical concepts. Formulating the axioms very often gives direction to further study of any concept.

It is the sense of mathematics as dealing with ideas rather than physical entities that differentiates pure mathematics from the sciences. Mathematics is not an empirical science, and, therefore, the conclusions arrived at cannot be validated nor impugned by physical observations. The conclusions derived in an axiomatic system are accepted as true, and they are true for any application that fits the model, and so the standard set by the Greeks continued. Among many others, both Isaac Newton (1643–1727) in his *Principia* and Adrien Legendre (1752–1833) in his treatment of analytic numbers used this approach as the structure of their work (Wilder 1952).

During the late nineteenth century, mathematicians began to question seriously the "natural" logic of the Greeks. It was at this time that Guiseppe Peano (1858–1932), David Hilbert (1862–1943), and their contemporaries developed the modern form of the foundations of mathematics based on formal mathematical logic and couched in the language of mathematical logic. Peano employed this language in his work. As one of the originators, Peano emphasized the desirability of a minimum number of undefined terms and the independence of the axioms; that is, one should not assume something that can be established deductively from the other assumptions. Hilbert was concerned about the consistency of axiomatic systems (i.e., that the assumptions did not lead logically to contradictory statements), and he attempted to prove consistency of systems. It was believed that consistency was necessary to make the conclusions irrefutable. As a result of the leadership of Peano, Hilbert, and others, almost all areas of mathematics were then treated on an axiomatic, deductive basis involving mathematical logic and set theory.

The attempts by Hilbert and many others to establish the consistency of axiomatic systems were, however, doomed to failure. The brilliant mathematical logician Kurt Gödel (1906–1978) proved in a 1931 paper that it is impossible to prove (within the context of the system) that the system is consistent. The paper also established that in any consistent system, there are undecidable (as to true or false) statements (Nagel and Newman 1956). Gödel's theorem showed that the axiomatic method which had been accepted as the cardinal method of formal mathematics had limitations. The formal deductive process demonstrated, as its crowning achievement, proof of its inability to validate itself (Newman 1956). But despite this limitation, the foundations of mathematics contribute significantly to the understanding, communication, exploration, and extension of mathematical concepts.

There are distinct advantages to the formal deductive procedures of the foundations of mathematics. A particular axiomatic system can serve as a model in diverse branches of mathematics, for example. A theorem proved is, then, true in each of the diverse applications of the model. Also, altering an existing system leads to new concepts and new branches of mathematics. Axiomatic characterization of mathematical concepts that have evolved over time leads to further knowledge of the concepts (theorems) and gives direction for further study. There are also disadvantages to an overdependence on the process. If it should turn out that the logic upon which it is based has flaws, what can be said of the reliability of the theory developed? Also, strict adherence to the process tends to narrow the focus and impedes creative thought (Wilder 1952).

Because the mode of thought of the foundations of mathematics is essential to the understanding of mathematics and important to communication in mathematics and science, it is an integral part of the teaching of mathematics. An argument has been formulated that the reason teaching and learning mathematics and science are difficult is that the rational deductive thought process is not a natural human attribute (Cromer 1993). The premise is that since this mode of thought has evolved only in the Greek culture and in no other, it is not natural. Training in rational deductive thought, then, should accompany the teaching of mathematics and science.

*See also* Axioms; Logic; Mathematics, Definitions; Proof

## SELECTED REFERENCES
Cromer, Alan. *Uncommon Sense.* New York: Oxford University Press, 1993.

Kline, Morris. *Mathematics: The Loss of Certainty.* New York: Oxford University Press, 1980.

Nagel, Ernest, and James R. Newman. "Goedel's Proof". In *The World of Mathematics* (pp. 1668–1695). James R. Newman, ed. New York: Simon and Schuster, 1956.

Newman, James R. "Commentary on the Foundations of Mathematics." In *The World of Mathematics* (pp. 1614–1618). James R. Newman, ed. New York: Simon and Schuster, 1956.

Wilder, Raymond. *Introduction to the Foundations of Mathematics.* New York: Wiley, 1952.

HARALD M. NESS

# MATHEMATICS, NATURE

Encompasses both a systemized body of knowledge and a process that acts as a basis for scientific thinking. Mathematics studies numbers, shapes, arrangements, abstract objects, and the relations among these objects. We use mathematics to prove theorems, which in turn enable us to understand how and why things work. The goal of mathematics is to find the simple underlying ideas that explain seemingly complex problems. The symbolic language of mathematics allows precise, brief, and comprehensible communication of number, size, and relations. By organizing information systematically it gives us a powerful tool for examining and solving problems. This is done by focusing on the simple invariants that explain the key ideas and generalizing these ideas on an abstract level. Deductive reasoning, the method of applying the rules of logic to prove theorems in mathematics, makes the conclusion a logical consequence of the assumptions and previously developed theorems. Mathematics applies the deductive process to extend concepts that build on this foundation for continued research and investigation, which may be theoretical or applied in nature.

The Greek mathematics that developed between 750 B.C. and A.D. 200 gave us a structure that organized the discipline. Euclid's *Elements,* dating back to 300 B.C., sets forth the principles of geometry and a foundation to examine this mathematical structure. It begins with undefined terms (such as *point, line,* and *plane*) and definitions that attempt to describe in everyday language the mathematical objects being studied. Certain axioms about these objects are then postulated, for example, that there is a unique line joining any two distinct points. The purpose of these definitions and axioms is so that everyone would agree on precisely what objects are under study and what is being assumed about them. Next, from these axioms and definitions one derives theorems that describe relationships among the objects being studied. These theorems are proven by deductive reasoning or logic. This process is called the *axiomatic method of inquiry*, since it begins with axioms and then derives consequences from them.

The legacy of Greek mathematics in geometry, trigonometry, and astronomy provided the foundation for subsequent advances during the Renaissance. However, Greek mathematicians were limited by their inability to handle infinite processes. In the mid-1600s, Gottfried Leibniz and Isaac Newton independently developed the area of mathematics known as the calculus, which is a systematic way of handling infinite processes. Calculus addresses problems that have to do with motion and change, using time as a variable. Newton used calculus and the physical laws for gravitational attraction to develop models that explained planetary motion. Calculus has applications in almost all areas of science—in physics, for the study of mechanics, electricity, and magnetism; in chemistry, to describe the rates of chemical reactions; in biology, to model population growth; in economics, to forecast economic growth.

Advances in mathematics often occurred as a result of the need to describe and solve physical problems. Mathematics can be said to have two branches: pure and applied. Applied mathematics deals with solving problems that arise in areas such as business, astronomy, physics, chemistry, and computer science. Pure mathematics studies abstract concepts that may not seem to have any practical applications. Topology, abstract algebra, number theory and analysis are the main branches in that category. But the distinction between pure and applied mathematics is hard to make. Much of the pure mathematics, while developed on a theoretical and conceptual level, has proved to have important applications. Similarly, problems in applied mathematics often give rise to developments in pure mathematics.

Along with its rational and analytical aspects, mathematics is also a dynamic and ever-changing human activity. One reason mathematics has fascinated people over the centuries is the simplicity with which one can state the problems. The Four Color Problem is a famous problem which states that a map on a sphere can be colored with four colors so

that no adjacent countries have the same color. This problem was only recently proved with the aid of a computer. Another famous problem was Fermat's Last Theorem. Before he died, Pierre de Fermat left a conjecture in the margin of his manuscript that an equation of the form $x^n + y^n = z^n$ has positive integer solutions for $x, y,$ and $z$ only if $n = 2$. After 300 years and many futile attempts by mathematicians, the problem was finally proved.

Yet many old problems remain unsolved even after hundreds of years. For example, we call a pair of primes *twin primes* if they differ by 2. Thus, 5 and 7, 11 and 13, and 17 and 19 are twin primes. The "twin prime" conjecture states that there are infinitely many such twin primes, but this has not been proven. Like many problems in mathematics, it is easy to state but seemingly very difficult to prove. We still do not know if there are infinitely many twin primes. A theorem of Kurt Gödel goes one step further. Gödel proved that there are problems in number theory which can neither be proved nor disproved.

Mathematics contains simple but profound questions that can be described at all levels, not just the advanced. It is embedded in our culture, in, for example, the principles of perspective used in paintings since the Renaissance, in the geometry of architecture, and in the foundations of music theory. Its methods give structure to social and economic thought, and it gives science a language and tools with which to operate. Mathematics is a creative process, an art form, and an expression of the human mind motivated by insight, intuition, and a desire to understand the world in which we live.

*See also* Mathematics, Foundations; Number Theory, Overview; Philosophical Perspectives on Mathematics

## SELECTED REFERENCES

Adler, Alfred. "Mathematics and Creativity." In *Mathematics: People, Problems, Results* (pp. 3–10). Vol. II. Douglas M. Campbell and John C. Higgins, eds. Belmont, CA: Wadsworth International, 1984.

Courant, Richard, and Herbert Robbins. *What Is Mathematics?*. New York: Oxford University Press, 1941.

Halmos, Paul. "Mathematics as a Creative Art." In *Mathematics: People, Problems, Results* (pp. 19–29). Vol. II. Douglas M. Campbell and John C. Higgins, eds. Belmont, CA: Wadsworth International, 1984.

Hollingdale, Stuart. *Makers of Mathematics*. New York: Penguin, 1989.

Kasner, Edward, and James R. Newman. *Mathematics and the Imagination*. Redmond, WA: Microsoft, 1989.

Wilder, Raymond, "The Nature of Modern Mathematics." In *Learning and the Nature of Mathematics* (pp. 35–48). William E. Lamon, ed. Chicago, IL: Science Research Associates, 1972.

HIROKO K. WARSHAUER

MAX L. WARSHAUER

# MATHEMATICS ANXIETY

Fear and avoidance of mathematics. Many adults who are competent in other areas fear mathematics and do their best to avoid the subject. For some, the phobia is so strong as to induce physical reactions, such as nausea, headaches, and mental paralysis. "Math anxiety" became a topic of discussion in the mid-1970s in connection with women's efforts to enter fields that had previously been the domain of men. Failure to take the prerequisite mathematics courses in high school excluded them from the college majors that would lead to high-paying careers. People of working-class background as well as most African Americans, Hispanics, and Native Americans were also underrepresented in fields requiring some mathematical background (National Science Foundation 1990, vii–ix).

Many societal factors and beliefs are responsible for fear and avoidance of mathematics among large sections of the population (Zaslavsky 1994):

- The belief that one needs a "mathematical mind" to do mathematics is perpetuated by the media and accepted by parents and teachers. It is coupled with the myth that many women and members of certain ethnic/racial groups are biologically incapable of doing higher mathematics.

- Childhood socialization patterns encourage boys to be assertive and take risks, while girls are brought up to be cautious, a trait that serves them poorly when they are introduced to unfamiliar concepts or take timed multiple choice tests. Boys are more likely than girls to engage in activities that foster interest and skills in mathematics. Boys attribute their success to ability, while girls attribute success to luck.

- School mathematics stresses competition among students, emphasizes rote memorization of procedures and early introduction of abstract symbols and terminology, and has little relevance to the lives of the students. Low-income and minority students are most likely to be exposed to and suffer from this type of mathematics education.

- Because of the cumulative nature of the subject, the failure to grasp a key concept may cause an individual to fall behind to the point where it is difficult to catch up.

To accommodate students who suffer from fear and avoidance of mathematics, colleges and other organizations have set up programs and clinics. These might be one-session workshops, a full course with a title such as "Math Without Fear," or a clinic staffed by a mathematics instructor and a psychologist. Some programs emphasize math tutoring, some concentrate on the psychological aspects, while others, like "Mind Over Math," combine both (Kogelman and Warren 1978). In the mid-1970s, psychologist Sheila Tobias set up one of the first university clinics in order to serve female students after Wesleyan University in Connecticut had become coeducational. Working with mathematics instructors, she found that these women lacked confidence rather than competence (Tobias 1993).

Once people are aware of the problem, what are the remedies? Many facilitators begin by encouraging students to write their "math autobiographies," to review the incidents that led to their fear of mathematics. As they discuss their experiences with others in the group, they discover that they are not alone in their misery, that they are not necessarily stupid or to blame for their phobia. Further discussion centers on exposing the myths about the nature of mathematics—that it is mainly work with numbers, it involves a great deal of memorization, there is just one correct procedure and one correct answer for each problem, speed is of the essence, working with other people is cheating. Students learn that they do use mathematics in many ways in their daily lives and that they actually know more than they realize. They should feel proud of devising their own methods to solve problems, instead of being ashamed of using unorthodox procedures. People learn best when they construct their own knowledge rather than being told what to do.

Students may have felt lost because their learning styles were in conflict with those of their instructors. Some people do best with visual presentations, while others prefer aural or kinesthetic modes of learning. Many profit from the use of manipulative materials such as Cuisenaire rods or algebra blocks once they have overcome their inhibitions about using these aids. Students accustomed to their instructors' glib presentations may not realize that some problems may require a great deal of time and thought and that initial approaches may have to be discarded. Speed is not of the essence! Hard work, persistence, and good study habits will bring their reward.

Instructors can help students to learn various problem-solving techniques: listing the given facts and the relationships among them, making tables and drawing diagrams, trying to solve an easier problem of the same type, looking for errors instead of tearing up the paper. Once students have arrived at a solution and checked it, they should reflect on their methodology. Perhaps a different procedure would have been more efficient, something to remember for future use. If all else fails, they should not hesitate to ask for help from the instructor or tutor.

Educators investigating the affective aspects of learning find that working in collaborative groups improves the learning of most students. Cooperative learning is one key to the success of the Mathematics Workshop Program, initiated by Uri Treisman at the University of California at Berkeley and later adopted on many campuses. Students work individually on challenging problems, then join with others for further exploration. A faculty person is available as a last resort. Context is also important— experiencing and analyzing each problem to relate it to one's own world. An increasingly popular technique is for students to keep journals in which they write about their difficulties and successes, a running commentary on their progress (Zaslavsky 1994).

Many of the practices described here are equally useful in the elementary and secondary classroom. Encouraging students to talk and write about mathematics, to work in cooperative learning groups, to use manipulative materials, and to devise their own procedures are all ways of helping children to construct their own mathematical knowledge. Most important is that mathematics be interesting and relevant to their lives.

*See also* Minorities, Career Problems; Minorities, Educational Problems; Remedial Instruction; Techniques and Materials for Enrichment Learning and Teaching; Women and Mathematics, Problems

SELECTED REFERENCES
Kogelman, Stanley, and Joseph Warren. *Mind Over Math.* New York: Dial, 1978.
National Science Foundation. *Women and Minorities in Science and Engineering.* Washington, DC: The Foundation, 1990.
Tobias, Sheila. *Overcoming Math Anxiety.* Rev. ed. New York: Norton, 1993.
Zaslavsky, Claudia. *Fear of Math: How to Get over It and Get on with Your Life.* New Brunswick, NJ: Rutgers University Press, 1994.

CLAUDIA ZASLAVSKY

# MATHEMATICS EDUCATION, ISSUES

*See* ASSESSMENT OF STUDENT ACHIEVEMENT, ISSUES; BASIC SKILLS; DEVELOPMENTAL (REMEDIAL) MATHEMATICS; ETHNOMATHEMATICS; EVALUATION OF INSTRUCTION, ISSUES; GIFTED, SPECIAL PROGRAMS; LEARNING DISABILITIES, OVERVIEW; MATHEMATICS ANXIETY; MINORITIES, CAREER PROBLEMS; MINORITIES, EDUCATIONAL PROBLEMS; MULTICULTURAL MATHEMATICS, ISSUES; PHILOSOPHICAL PERSPECTIVES ON MATHEMATICS; UNDERACHIEVERS; WOMEN AND MATHEMATICS, PROBLEMS

# MATHEMATICS EDUCATION, STATISTICAL INDICATORS

A system of state indicators of the quality of science and mathematics education in public schools developed in cooperation with the state departments of education, federal agencies, and professional organizations by the Council of Chief State School Officers (CCSSO). The 1995 report on mathematics and science indicators (Blank and Gruebel 1995) is the third in a series of biennial reports on state indicators. The reports are primarily intended for use by policy makers and educators. The design, management, and reporting of indicators has been supported by the National Science Foundation (NSF) since the project was initiated in 1986.

The CCSSO strongly supports the development and use of state-level student assessments from the National Assessment of Educational Progress (NAEP) as an indicator of student learning in mathematics. The NAEP mathematics assessment is the best source for student achievement indicators that are comparable state-to-state and that adequately assess the range and depth of knowledge and student skills recommended by states and school districts. The NAEP assessment results and supporting questionnaires from students and teachers are based on a sample of 2,000 students per state at each assessed grade. Although the data do not provide a way for states to analyze student achievement for each school and district, the results are still extremely valuable as overall indicators. They provide a way to monitor state progress in student achievement, to assess education received by specific groups of students, and, most significant, to determine by state the relationship of student achievement to characteristics of schools, classroom practices, and teachers. The NAEP 1992 Mathematics Assessment results provide at least three indicators of student proficiency by state: (1) proficiency results by mathematics achievement level, (2) state trends in mathematics improvement, and (3) differences in proficiency by student race/ethnicity (Blank and Gruebel 1995).

## MATHEMATICS PROFICIENCY BY LEVEL AND STATE TRENDS

Summary statistics are reported for state-by-state and national results from the 1992 NAEP mathematics assessment in grade 8. The Proficient level on the NAEP assessment is defined, in brief, as: "Eighth grade students should apply mathematics concepts and procedures consistently to complex problems in the NAEP content areas—Numbers and Operations, Measurement, Geometry, Algebra and Functions, Statistics and Probability, and Estimation" (levels are defined by the National Assessment Governing Board) (Mullis et al. 1993). The percentage of students at or above the "Proficient" level is defined as a key indicator by the National Education Goals Panel (NEGP) in its annual reporting on Goal 3 (NEGP 1995).

Seven states had significant improvement from 1990 to 1992 in the percentage of students at or above the Proficient level, while eighteen states had significant improvement from 1990 to 1992 in the average mathematics proficiency scale score (statistical significance generally means at least 4 points on the NAEP scale of 0 to 500). In 1992, state percentages of students at or above Proficient varied from six states above 30% (Minnesota, Iowa, North Dakota, Wisconsin, Nebraska, Maine) to six states below 15% (Mississippi, Louisiana, Alabama, Arkansas, West Virginia, New Mexico). Nationally, 25% percent of grade 8 students scored at or above the Proficient level in 1992, and 4 percent were at or above the Advanced level, while 63% of grade 8 students scored at or above the Basic achievement level (thus, 37% of students scored below the Basic level). The percentage of grade 8 students above the Proficient level increased from 19% in 1990 to 25% in 1992, a significant improvement. Nationally, 18% of fourth graders were at or above the Proficient level. Nine states had one-fourth of grade 4 students at or above the Proficient level (Maine, Minnesota, Iowa, New Hampshire, Wisconsin, New Jersey, Connecticut, Massachusetts, Nebraska). Five states had 10% or less of grade 4 students at or above the Proficient level. These aggregate figures on mathematics proficiency can be disaggregated by the content areas of mathematics that were assessed at grades 4 and 8 in the 1992 NAEP: numbers/operations, measurement, statistics/probability, algebra/functions, geometry, and estimation.

National Center for Education Statistics reports show there was overall improvement in NAEP mathematics and science scores over the decade from 1982 to 1992 (Mullis et al. 1994). For example, the national average proficiency in mathematics had statistically significant, but modest, increases at grades 4, 8, and 12 over the period (e.g., grade 4 average mathematics proficiency scale score at 219 in 1982 and 230 in 1992). In science, approximately the same improvements were made at these grade levels (e.g., grade 4 average science proficiency at 221 in 1982 and 231 in 1992).

## NATIONAL TRENDS BY RACE/ETHNICITY AND GENDER

NAEP results by race/ethnicity show larger rates of improvement in proficiency scores for minority students than for whites from 1982 to 1992. In mathematics, African American students' proficiency at age 9 (grade 4) improved over 10 points (195 to 208), and science also improved significantly (187 to 200). A 13-point improvement on NAEP is substantial. In mathematics, thirteen-year-olds scored 43 points higher on average than nine-year-old students. Thus, a 13-point improvement for age 9 African American students' scores on NAEP could be interpreted as the equivalent of an improvement of more than one year of school. Scores for Hispanic students also increased in the decade—20 points at age-9 science and 10 points in age-9 mathematics. Even though minority scores on NAEP improved more than the scores of whites in the 1980s, however, the gap in mathematics and science proficiency between majority and minority students was still substantial.

No gender differences were found in average mathematics proficiency at age 9 and age 13 as of 1992 (Mullis et al. 1994). The 1992 national NAEP results in mathematics show slightly more males than females scoring at the Advanced level at grade 12, and the average proficiency for males is slightly higher than females at the 12th grade. The trend from 1982 to 1992 showed females had improved NAEP scores in mathematics relative to males at all levels.

*See also* Assessment of Student Achievement, Overview; Goals 2000; National Assessment of Educational Progress (NAEP)

## SELECTED REFERENCES

Blank, Rolf K., and Doreen Gruebel. *State Indicators of Science and Mathematics Education, 1995.* Washington, DC: Council of Chief State School Officers, 1995.

Mullis, Ina V. S., John A. Dossey, Eugene H. Owen, and Gary W. Phillips. *NAEP 1992: Mathematics Report Card for the Nation and the States.* Washington, DC: National Center for Education Statistics, 1993.

———, John A. Dossey, Jay R. Campbell, Claudia A. Gentile, Christine O'Sullivan, and Andrew S. Latham. *NAEP 1992 Trends in Academic Progress.* Washington, DC: National Center for Education Statistics, 1994.

National Education Goals Panel. *The National Education Goals Report.* Washington, DC: Government Printing Office, 1995.

ROLF K. BLANK

# MATHEMATICS IN CONTEXT (MiC): A CONNECTED CURRICULUM FOR GRADES 5 THROUGH 8

A comprehensive mathematics curriculum for the middle grades funded in 1990 by the National Science Foundation for development and field testing. MiC reflects the content and pedagogy suggested by the *Curriculum and Evaluation Standards for School Mathematics* (National Council of Teachers of Mathematics (NCTM) 1989), *Professional Standards for Teaching Mathematics* (NCTM 1991), and *Assessment Standards for School Mathematics* (NCTM 1995). The project made use of knowledge about the teaching and learning of mathematics to create a research-based but teacher-oriented set of materials (Romberg, *A Blueprint . . .* , 1992). The materials were field tested in several schools in Wisconsin, Iowa, Missouri, Tennessee, California, Florida, and Puerto Rico and were published by Encyclopedia Britannica Education Corporation in 1996 and 1997.

The development of the curriculum units reflects a collaboration between research and development teams at the Freudenthal Institute at the University of Utrecht, The Netherlands, and the National Center for Research in Mathematical Sciences Education at the University of Wisconsin. A total of forty units have been developed, ten for each grade 5 through 8. A "blueprint" for the contents of these units was prepared by an international advisory committee (Romberg, *A Blueprint . . . .* , 1992). Initial drafts of the units were prepared by Dutch researchers. The units were then modified by staff members at the University of Wisconsin to make them appropriate for U.S. students and teachers.

The units are unique in that they make extensive use of real-world contexts. From the context of tiling a floor, for example, flows a wealth of mathematical applications, such as similarity, ratio and proportion, and scaling. Units emphasize the interrelationships among mathematical domains, such as number, algebra, geometry, and statistics. The purpose of the units is to connect mathematical content both across mathematical domains and to the real world.

Because the philosophy underscoring the units is that of teaching mathematics for understanding, the curriculum has tangible benefits for both students and teachers. For students, mathematics ceases to be seen as a set of disjointed facts and rules. Rather, students come to view mathematics as an interesting endeavor that enables them to better understand their world (Gravemeijer 1994). Also, since all students should be able to reason mathematically, the activities within a unit are at multiple levels so that the able student can go into more depth yet the student having trouble can still make sense out of the activity (Van den Heuvel-Panhuizen 1995). For teachers, the rewards of seeing students excited by mathematical inquiry, their redefined role as guide and facilitator of inquiry, and their collaboration with other teachers result in innovative approaches to instruction, increased enthusiasm for teaching, and a more positive image with students and society (DeLange et al. 1993).

Hans Freudenthal called this approach *Realistic Mathematics Education,* and contrasted it with typical classroom instruction in the following way:

> [Traditional instruction presents] a view of the individual as a programmable computer, who, however, will never approach the performances typical of the computer. The education we are developing is determined by a different view of fellow man and by a different view of mathematics—not as subject matter, but as a human activity; allied to reality, close to the child, [and] socially relevant. Bring these characteristics together as [an activity] worthy of a human being: worthy of the human as a learner, teacher, guide to and creator of education (Freudenthal 1987).

*See also* Research in Mathematics Education

## SELECTED REFERENCES

DeLange, Jan, Gail Burrill, Thomas Romberg, and Martin van Reeuwijk. *Learning and Testing Mathematics in Context: The Case: Data Visualization.* Pleasantville, NY: Wings for Learning, 1993.

Freudenthal, Hans. "Mathematics Starting and Staying in Reality." In *Developments in School Mathematics Edu-*

cation Around the World (pp. 279–295). Vol. 1. Izaak Wirszup and Robert Streit, eds. Reston, VA: National Council of Teachers of Mathematics, 1987.

Gravemeijer, Koeno. *Developing Realistic Mathematics Education.* Culemborg, Netherlands: Technipress, 1994.

National Council of Teachers of Mathematics. *Assessment Standards for School Mathematics.* Reston, VA: The Council, 1995.

———. *Curriculum and Evaluation Standards for School Mathematics.* Reston, VA: The Council, 1989.

———. *Principles and Standards for School Mathematics.* Reston, VA: The Council, 2000.

———. *Professional Standards for Teaching Mathematics.* Reston, VA: The Council, 1991.

Romberg, Thomas A., ed. *A Blueprint for Math in Context: A Connected Curriculum for Grades 5–8.* Madison, WI: Wisconsin Center for Education Research, 1992.

———. "Problematic Features of the School Mathematics Curriculum." In *Handbook of Research on Curriculum* (pp. 749–788). Philip W. Jackson, ed. New York: Macmillan, 1992.

Van den Heuvel-Panhuizen, Marja. "Assessment and Realistic Mathematics Education." Ph.D. diss., University of Utrecht, Netherlands, 1995.

THOMAS A. ROMBERG

# MAYAN NUMERATION

Sophisticated numeration and hieroglyphic systems that were used to make astronomical predictions and to develop intricate calendars by the Mayan people. The Maya created a base-20 (vigesimal) numeration system for commercial dealings. They performed addition and subtraction, but there is no evidence of any real multiplication or division operations nor any evidence of the use of fractions or decimals. The Mayan people and their culture flourished in Central America and the Yucatan peninsula of southern Mexico during the classic period from A.D. 300 to A.D. 900. In addition to their accomplishments in astronomy, the Maya constructed impressive pyramids and temples. Remnants of this culture are found in the lowland ruins of extensive ceremonial centers that functioned as sites for religious activities, judicial courts, civic administration, and commercial markets. The pyramids that dominated these centers, with temples erected on top, were built from stone without the use of wheels or metal tools. The orientation of these temples together with the few surviving books suggest that the Maya were able to predict accurately the positions of planets and the dates of solar eclipses. Scribes who were proficient in arithmetic and calendric computations were accorded special status.

Mayan religious traditions emphasized the passage of time and the cyclical nature of life. In their calendar, numerals were identified with specific gods. The Maya divided the *haab,* their calendar for the seasonal year, into 18 *uinals* (months), each consisting of 20 *kins* (days). The 20 days in a *uinal* had distinct names: the first was always *imix* and the last (twentieth) was always *ahau.* A sequence of eighteen 20-day *uinals* (360 days) was called one *tun* (year), the usual unit for counting time. A *tun* was followed by *uayeb,* a period of 5 "unlucky" days that harbored likely misfortune—the Mayan equivalent of Halloween. Our own Gregorian calendar is not so regular. We divide the 365-day seasonal year into 12 months that vary in length from 28 to 31 days. We have only seven distinct names for days (our week: Sunday, Monday, . . . , Saturday), and these names are repeated in a continuous cycle throughout the months and years. Hence, Sunday (the first day of the week) is not always the name of the first day of a new month. For the Maya, twenty *tuns* form a *katun* and twenty *katuns* constituted a *baktun,* the largest unit of time commonly used. This aggregation of years into successive groups of 20 corresponded to their base-20 numeration system. In a similar way, we use successive groups of 10, the base of our numeration system, to aggregate years into decades, centuries, and millennia.

The Mayan numeration system used only three symbols for counting: a dot (●) representing 1, a dash (—) representing 5, and a shell-like figure (◯) representing 0. It was an additive system, so a dot and a dash together represented 1 + 5 = 6. Figure 1 shows how numbers from 1 to 19 were written. For larger numbers, the Maya incorporated positional notation into their numeration system. Their numerals were recorded vertically, with place values ascending from bottom to top. For calendar notations, the actual value of each place corresponded to the *haab* calculation. Since there were only 18 *uinals* in a *tun,* the third place had a value of 18 (20) in-

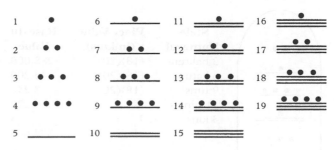

**Figure 1**

stead of $20^2$, as in a true base-20 system. The numeral on the stele (stone monument) drawn in Figure 2 represents 334,663 *kins* (days). This Mayan numeral can be translated into the modern base-10 numeral using the place value table shown in Figure 2.

An important innovation within the Mayan numeration system was the employment of the shell-like symbol to indicate that a particular place was not used, similar to the way zero functions as a place holder in modern numeration systems. However, for the Maya the shell symbol connoted "completion" of the place value rather than our notion of deficiency. Thus, the Mayan numeral for 20 *kins* (1 *uinal*) was

and for 360 *kins* (1 *tun*) it was

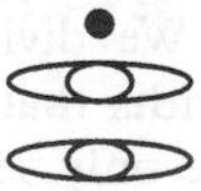

Inscriptions on stelae indicate that the Maya reckoned dates from a starting point equivalent on our calendar to August 10, 3113 B.C., more than thirty centuries before the classic period of Mayan culture. Apparently, the Maya associated this date with an important event in their creation myth. As a result, Mayan dates required rather large numbers. Our system uses the birth of Christ as a reference point for designating dates. For example, the classic period for Mayan civilization began around A.D. *(anno domini)*, 300, that is, 300 years after the birth of Christ, while the classic period for Greek civilization ended around 300 B.C., that is, 300 years before the birth of Christ.

The Maya devised a second calendar that was used as a sacred almanac to forecast the influence of various gods on a particular date. Mayan priests interpreted the relative potency of each god's influence and offered guidance on how to conduct daily life, somewhat like astrologers today. This religious calendar consisted of 260 days (a *tzolkin*) and also reflected the importance placed on a cyclical view of events. Although there were distinct names for each of the 20 days of a *uinal* (month), the days in the religious calendar also were numbered consecutively,

from 1 to 13. The first day of the calendar, *imix*, was numbered 1; the thirteenth day, *ben*, was numbered 13. On the fourteenth day, *ix*, the number cycle began again so that this day was numbered 1. Thus, the twentieth day of the calendar, *ahau*, was numbered 7. On the twenty-first day, designated 8 *imix*, the name cycle began again but the number cycle continued in order. The number 1 would not coincide again with a day named *imix* until the 20-day name cycle had been repeated 13 times, that is, until $20(13) = 260$ days (a *tzolkin*) had elapsed. At that time a new period began and the calendar was repeated. Due to leap years and variations in month length, our calendar, as noted earlier, is not as regular as the Mayan calendars. In 1995, the first day of the year, January 1, falls on the first day of the week, Sunday. January 1, 1996 falls on Monday. January 1 does not fall on a Sunday again until 2006; however, it falls on Monday again in 2001.

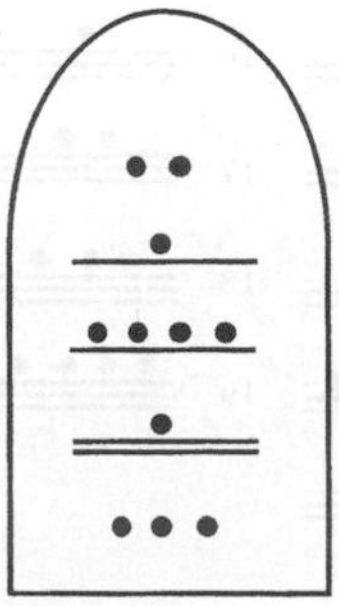

The Maya coordinated their seasonal year calendar *(haab)* and the religious calendar-almanac *(tzolkin)* into a longer cycle called the *Calendar Round*. The length of this cycle was determined by using the least common multiple of the length of the *haab* plus the *uayeb* (365 days) and the length of the *tzolkin* (260 days), that is, 18,980 days or fifty-two years. Hence, each day in the fifty-two-year cycle could be identified by the unique combination of its *tzolkin* designation (day name and number) and its *haab* designation (month name and day position). Completion of a calendar year or an entire Calendar Round cycle were occasions for festivals and ritual celebrations. In a similar way, we celebrate New Year's Day and mark the turn of the century and a new millennium.

In the classroom, students might compare our own calendar system with the system devised by the Maya. In our system, the week length is constant, cycling continuously every 7 days, but the month length varies. In the Mayan system, the day names and month length cycled regularly in groups of 20. Determining on what day of the week some particular date (e.g., July 4 or a student's birthday) will fall in some future year of our calendar presents a challenge for students that would not arise in the *haab* calendar. We count years in groups of 10 (decades and centuries), while the Maya counted years in groups of 20 *(katuns and baktuns)*. Such comparisons highlight the arbitrary nature of most divisions of time; however, the length of the day and the length of the seasonal year are virtually equivalent in the two systems. Teachers may wish to incorporate the numeration system and the calendars devised by the Maya into the curriculum as one aspect of a broader

| Stele Numeral | Place Value (in kins) | Base-10 Value |
|---|---|---|
| 2 baktuns | $(18)(20)^3$ | 288,000 |
| 6 katuns | $(18)(20)^2$ | 43,200 |
| 9 tuns | $(18)(20)$ | 3,240 |
| 11 uinals | 20 | 220 |
| 3 kins | 1 | 3 |
| | | 334,663 |

**Figure 2**

456

study of pre-Columbian and Hispanic colonial America, as part of a unit on multicultural aspects of mathematics, or as an example in the study of base numeration systems.

*See also* Calendars; Ethnomathematics

## SELECTED REFERENCES

Bidwell, James K. "Mayan Arithmetic." *Mathematics Teacher* 60(Nov. 1967):762–768.

Closs, Michael P. "Mathematicians and Mathematical Education in Ancient Maya Society." In *Selected Lectures from the Seventh International Congress on Mathematical Education* (pp. 77–88). David F. Robitaille, David H. Wheeler, and Carolyn Kieran, eds. Sainte-Foy, Quebec, Canada: Les Presses de l'Université Laval, 1994.

Gates, W. E., trans. and ed. *Yucatan Before and After the Conquest, by Friar Diego de Landa, etc.* Maya Society Publ. No. 20. Baltimore, MD: The Society, 1937.

Kinsella, John, and A. Day Bradley. "The Mayan Calendar." *Mathematics Teacher* 27(Nov. 1934):340–343.

Morley, Sylvanus G. *An Introduction to the Study of the Maya Hieroglyphs.* Washington, DC: Government Printing Office, 1915.

Smith, David Eugene. *History of Mathematics.* Vol. 2. Boston, MA: Ginn, 1925. Reprinted. New York: Dover, 1958.

Thompson, J. Eric S. *Maya Hieroglyphic Writing: An Introduction.* 2nd ed. Norman, OK: University of Oklahoma Press, 1960.

EILEEN F. DONOGHUE

# MEAN, ARITHMETIC

The most commonly used average, or typical, value. It is calculated by adding all the results in the data set and dividing the total by the number of results used. It was used by both the ancient Egyptians and the Babylonians as a summary measure. Thus, the mean of a simple list of numbers is found by $\bar{x} = (\Sigma x)/n$, where $\bar{x}$ stands for the sample mean and $\Sigma$ means total or sum. So the mean of the weights 2 kg, 3 kg, 2.5 kg, 6.1 kg, and 4 kg is the total of the results, 17.6 kg, divided by 5, giving a mean of 3.52 kg.

The mean can be found only for numerical data, either discrete or continuous. If data have been summarized into a frequency table, the frequency of each value or result must be taken into consideration: $\bar{x} = (\Sigma(f \cdot x))/n$, where $f$ stands for the frequency of each value, $f \cdot x$ is each result multiplied by its own frequency.

The mean is sometimes referred to as the "expected value," or $E(x)$. When a continuous function,

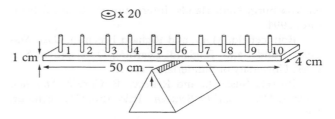

**Figure 1** *A simple balance.*

$f(x)$, is used to model the probability density function of a continuous variable, $E(x) = \int xf(x)$, where this integral is evaluated over the range of values that $x$ can take.

The mean is a balancing point in the data, so that $\Sigma(x - \bar{x}) = 0$ (and $\Sigma(x - \bar{x})^2$ is a minimum). Thus it may be compared to a center of gravity (Cassell 1989). This analogy may not be helpful to pupils unless they are given concrete experiences of finding centers of gravity. Cassell describes some simple apparatus that can be used with quite young pupils to demonstrate that the mean is a balancing point. He describes a simple balance on which weights can be used to represent data items (see Figure 1). The pivot can be moved along until the weights balance, and this balancing point is found to be the mean.

Pupils learn also that (a) the mean need not be one of the values in the data, and (b) the mean need not be a whole number, even if all the data are whole numbers.

The algorithm for finding the mean is easy to apply but does not make intuitive sense. Even younger children have intuitive ideas of representativeness, but these may be abandoned if they are taught the algorithm too early (Leon and Zawojewski 1991). Thus, Russell and Corwin (1989) recommend that the median and then the mode should be introduced to children as "typical" values well before they are taught to calculate the mean.

*See also* Mathematical Expectation; Measures of Central Tendency

## SELECTED REFERENCES

Cassell, David. "What Do We Mean by the Mean?" *Teaching Statistics* 11(2)(1989):38–39.

Hawkins, Anne, Flavia Jolliffe, and Leslie Glickman. *Teaching Statistical Concepts.* London, England: Longman, 1992.

Leon, Marjorie Roth, and Judith S. Zawojewski. "Use of the Arithmetic Mean: An Investigation of Four Properties." In *Proceedings of the Third International Conference on Teaching Statistics* (pp. 302–306). Vol. 1. David Vere-Jones, ed. Voorburg, Netherlands: International Statistical Institute, 1991.

Rouncefield, Mary, and Richard Freeman. *Basic Statistics.* Sunderland, England: Learning Development Services, University of Sunderland, 1993.

Russell, Susan J., and Rebecca B. Corwin. *Statistics: Middles, Means, and In-Betweens.* Palo Alto, CA: Seymour, 1989.

MARY ROUNCEFIELD

## MEAN, GEOMETRIC

A measure of central tendency that is used only rarely but does have a few specialized applications. The geometric mean is given as

$$G = \sqrt[n]{(x_1 x_2 x_3 \ldots x_n)}$$

or, for a frequency distribution:

$$G = \sqrt[n]{(x_1^{f1} x_2^{f2} x_3^{f3} \ldots x_n^{fn})}$$

If logarithms are used, the formulation is simpler:

$$\log G = (\sum f \log x)/n$$

where $x$ stands for values taken by the variable being measured, $f$ for the frequency of each result, and $n$ for the number of results.

This last expression provides an indication that the geometric mean is most useful in situations where (a) a logarithmic transformation has been used; (b) data may be modeled by an exponential or power function (e.g., population increase); or (c) index numbers or rates of change are to be averaged. For example, the population of a country may grow as an exponential function, particularly under idealized conditions, with stable birth and death rates and zero net migration. The geometric mean is more appropriate than the arithmetic mean to interpolate an estimate of the population at a date between two other dates for which the population is known.

*See also* Interpolation; Measures of Central Tendency

### SELECTED REFERENCE
Moroney, M. J. *Facts from Figures.* London, England: Pelican, 1990.

MARY ROUNCEFIELD

## MEAN VALUE THEOREMS

Theorems of both integral and differential calculus that are linked to the concept of average value.

The French mathematician Joseph-Louis Lagrange (1736–1813) discovered the first of the mean value theorems of differential calculus. It states that, for a real-valued function $f$ of a real variable that is continuous on the closed-interval $[a,b]$ and differentiable on the open interval $(a,b)$, there exists a number $c$ between $a$ and $b$ such that $f(b) - f(a) = f'(c)(b - a)$. The theorem does not give an algorithm to find $c$, nor does it assert that $c$ is unique (it need not be). This is a feature common to all the mean value theorems. Geometrically, Lagrange's Mean Value Theorem can be interpreted to mean that there is a line tangent to the graph of $f$ at some point $(c, f(c))$ that is parallel to the line passing through the points $(a, f(a))$ and $(b, f(b))$. The method of proof involves the use of a theorem by the French mathematician Michel Rolle (1652–1719), who established that if a differentiable function $f$ of a real variable assumes the same value at two points $a$ and $b$, then there is a point $c$ between $a$ and $b$ such that $f'(c) = 0$. (The Lagrange theorem applies this result to a function that measures the vertical distance between the graph of $f$ and a secant line through two points on the graph of $f$.)

A generalization of Lagrange's theorem was stated and proved by the French mathematician Augustin Cauchy (1789–1857). Cauchy's theorem states that, for two real-valued functions $f$ and $g$ of a real variable, continuous on the closed interval $[a,b]$ and differentiable on the open interval $(a,b)$, if $g(b) - g(a) \neq 0$ and $f'$ and $g'$ are not both zero at any point of the open interval $(a,b)$, there exists a number $c$ between $a$ and $b$ such that

$$[f(b) - f(a)]g'(c) = [g(b) - g(a)]f'(c)$$

Less well known is the work done by Bhaskara (1114?–1185), a twelfth-century mathematician of Ujjain, a center of learning in the north of India. In his book on astronomy, *Siddhanta Siromani,* written around 1150, Bhaskara, popularly known in India as Bhaskaracharya ("Bhaskara the Teacher"), provides an application of what would be called Rolle's Theorem in the Western tradition: he concluded that the differential of a certain expression is equal to zero when a planet is either at its furthest distance or its closest distance from the earth (Joseph 1991).

The Mean Value Theorem for integral calculus was derived by both Lagrange and Sylvestre-François Lacroix (1765–1843) in a form that makes it a restatement of Lagrange's theorem for differential calculus. Its statement is: let $f$ be a real-valued function of a real variable that is continuous over $[a,b]$. Then there exists a number $c$ between $a$ and $b$ such that

$$\int_a^b f(x)\,dx = f(c)(b-a)$$

For a continuous, positive-valued function $f$, this equation can be interpreted as follows: The left-hand side represents the area of the region $R$ below the graph of $f$ and above the $x$-axis between the vertical lines $x = a$ and $x = b$. The right-hand side represents the area of a rectangle of height $f(c)$ and width $b - a$. Thus, $c$ is a point where the area of a rectangle is equal to the area of $R$.

An extension of this theorem concerns two functions $f$ and $g$. Under suitable assumptions—$f$ continuous over $(a,b)$, $g$ nonnegative over $(a,b)$, and the existence of the integrals of $fg$ and $g$ over the interval $[a,b]$ is one common set of assumptions—there is a number $c$ in $[a,b]$ such that

$$\int_a^b f(x)g(x)\,dx = f(c)\int_a^b g(x)\,dx$$

The first of the mean value theorems for integral calculus appears as a corollary of this extension when the function $g$ is the identity function, $g(x) = x$. (Cauchy derived his own proof of the theorem as part of his more rigorous development of a theory of integrals.)

## MOTIVATION AND GOALS

The Mean Value Theorem of differential calculus plays a vital role in the proofs of many of the theorems in calculus, including the Fundamental Theorem, which states that, for a differentiable function $f$, $\int_a^b f'(x)\,dx = f(b) - f(a)$. Its use, therefore, tends to be theoretical rather than practical. The theorem for integral calculus is best illustrated geometrically, as suggested earlier. Both mean value theorems are intimately linked to the notion of an average value, as explained in the next section.

## PEDAGOGICAL APPROACHES

Students should understand how crucial the mean value theorems are in proving other theorems. Their more practical application involves some discussion. A common task that students are given in their study of the differential calculus is to determine the number of real zeroes a given function possesses. This involves an understanding of the Mean Value Theorem for differential calculus, and it is a good question to ask. Less useful is the typical "warm-up" drill that asks them to verify that a given function behaves the way the Mean Value Theorem says that it should behave over a given interval.

Some of the best word problems invoking the mean value theorems are those that involve the notion of an average value. Comparing instantaneous speed with the average speed of a vehicle over a certain time interval is one such example. Similarly for integral calculus, to calculate the average temperature for a certain period of time over which temperatures are known offers students the opportunity to explore what is meant by the expression "average temperature." It also offers teachers the chance to get their students to compare their answers with what is frequently done to calculate average temperature, that is, to take the arithmetic average of the high and low temperatures over a given period of time. What is said here of temperature, of course, can be said equally well of any measurable quantity over time or length or whatever independent variable is appropriate.

The Cauchy Mean Value Theorem is illustrated best in terms of the parametric representation of a plane curve by two functions, say $f$ for the $x$ coordinate and $g$ for the $y$ coordinate. The slope of the line joining one point $(f(a),g(a))$ on such a curve to a second point $(f(b),g(b))$ is equal to the slope of at least one tangent line to the curve at a point $(f(c),g(c))$ where $c$ lies between $a$ and $b$. For this illustration, $g'$ must not vanish between $a$ and $b$.

## RECOMMENDED MATERIALS FOR INSTRUCTION

Chalk and a board are useful to illustrate the mean value theorems although recent technological developments, such as the graphing calculator, provide ways to reinforce student understanding.

*See also* Calculus, Overview; Differentiation; Integration

## SELECTED REFERENCES

Boas, Ralph P. "Who Needs Those Mean-Value Theorems, Anyway?" In *A Century of Calculus* (pp. 182–186). Pt. 2. Tom M. Apostol et al., eds. Washington, DC: Mathematical Association of America, 1992.

Boyer, Carl B. *A History of Mathematics*. 2nd ed. Revised by Uta C. Merzbach. New York: Wiley, 1989.

Buck, R. Creighton. *Advanced Calculus*. 3rd ed. New York: McGraw-Hill, 1978.

Grabiner, Judith V. *The Origins of Cauchy's Rigorous Calculus*. Cambridge, MA: MIT Press, 1981.

Joseph, George Gheverghese. *The Crest of the Peacock*. London, England: Tauris, 1991.

Stein, Sherman K. *Calculus and Analytic Geometry*. New York: McGraw-Hill, 1987.

ROBERT J. FRAGA

# MEASUREMENT

Procedure for making comparisons to standard units. When a five-year-old raises her hand to register the height of a pile of blocks and then walks across the room to compare the position of her hand to the top of another pile of blocks, she is measuring. It is of little importance that she does not hold her hand steady as she crosses the room, nor that she has ever used the terms *measure, measurement, quantity, meter, metric, length*. She can measure, even though she cannot tell us how tall the piles are in inches, centimeters, or feet. She can measure, even though she has not yet learned how to use a measuring tape or ruler. The distance between the floor and her hand serves as an external standard for comparing the heights of the two piles of blocks. By calibrating the height of her hand to the height of the first pile and then comparing that height to that of the second pile, she can determine which of the two piles is higher, or whether they are equal in height. In doing so, she works with an implicit notion of transitivity: if the height of her hand is equal to the height of pile A but lower than the height of pile B, then pile B must be taller than pile A.

Although the child can measure, it would be an oversimplification to affirm that she has "acquired the concept of measurement," as if it were an all-or-none phenomenon. Measurement, like many other mathematical and scientific concepts, develops over a long period of time and becomes enriched as the child masters related concepts, symbolic representations, and techniques. To understand this, it can be helpful to consider the relations of measurement to counting, to division, and to the use of mathematical notation.

## COUNTING, ADDITIVE COMPOSITION, AND MEASUREMENT

Counting provides a means of determining the cardinality (e.g., number of fish in an aquarium) or ordinality (e.g., order of arrival of marathon runners) of a set of objects. Although counting would seem to be very distinct from measuring, the order of numbers plays an important role in comparing magnitudes and ordering measures, that is, measurements of magnitudes. One counts as one "paces off" the number of steps from one side of a room to the other. The total number of paces can serve to describe the width of the room. We say, for instance, that "the living room" is 13 paces across; "13 paces" is the measure of the room width. We can then mea-sure the width of another room, say, the dining room, and express its width in paces. Let us imagine the dining room is 10 paces across. By virtue of the order of the numbers 13 and 10, we know which room is wider. If we ignore, for the moment, *measurement error*, we can even express the magnitude of the difference in paces: the living room is 3 paces wider than the dining room. We could just as easily note that the dining room would be the same width as the living room were it widened by 3 paces.

Several characteristics distinguish this example from the one regarding the piles of blocks. In the room measurement example, there is a repeatable *unit of measure,* namely, the (magnitude of the) child's pace. This unit of measure, along with rules for its use, allows us to transform *unmeasured quantities* (the widths of the rooms, with no numbers implied) into *measured quantities,* or *measures*. Furthermore, the arithmetical operations of addition and subtraction allow us to express certain relations that would otherwise be left implicit. In this sense, addition and subtraction play an important role in the enrichment of the student's concept of measurement.

As long as the target quantity can be regarded as an integral multiple of the chosen unit of measure, measurement can be thought of as a "counting off." However, when one begins to pay attention to the quantity that remains after the maximum number of units is removed, a question arises: how should we measure the remaining quantity? One possibility is to introduce a smaller unit of measure. For example, a person might describe a room width as "13 paces plus 2 shoe lengths across."

When working with a conventional *system of measures,* such as the English measurement system, one can obtain more accurate values by measuring first in feet and then in inches (rather than working only with whole numbers of feet). The final measure is expressed as a composition based on both units. Both of these measures can be determined by using a measuring tape. A person's height, say 5 feet, 9 inches, can be thought of as entailing the following relations: height = $(5 \times \text{foot}) + (9 \times \text{inch})$. This example expresses height as integral multiples of two units of measure, foot and inch. There is nothing inherently wrong with this, but since the ratios of each unit to its nearest subunit vary throughout the system (the ratio of miles : yards = 1 : 1,760; the ratio of yards : feet = 1 : 3; the ratio of feet : inches = 1 : 12), computations can be unwieldy.

There is a major advantage to expressing measures in terms of only one unit (and decimal multiples and factors of this unit): there will be a natural correspondence between how numbers are written and how measurement units are interrelated. For example, 5.31 meters will signify 5 meters + 3 decimeters + 1 centimeter. Consider now how to interpret a decimal value in a nondecimal (non–base-10) measurement system. How much is "5.31 feet" or "5.31 years"? Students will sometimes mistakenly read these values as "5 feet, 31 inches" and "5 years and 31 days." But 0.31 feet is not 31 inches since 100 inches do not make 1 foot. Likewise, 0.31 years is not equivalent to 31 days since 100 days do not comprise 1 year. Conversions between decimal and multiunit representations require taking into account mismatches between ratios among units in the number system and units in the measurement system.

## SOME RELATIONS BETWEEN DECIMAL MEASUREMENT AND DIVISION

Measurement and division have much in common. The target quantity—the one being measured—can be thought of as a dividend, the unit of measure can be thought of as a divisor, and the result can be thought of as a quotient. The remainder quantity, if any exists, corresponds to the remainder of division. Accordingly, if we use a specified unit to measure some previously unmeasured length, the result will consist of a quotient and a remainder. Note that the length and the unit are physical quantities, whereas the quotient and the remainder are numbers with no units of measure. If the length is an integral multiple of the unit, then the remainder resulting from the measurement will equal zero and the quotient will state how many times the unit fits into the length.

Let us suppose that the ratio, length:unit, cannot be expressed as an integer. Suppose, for example that length:unit is equivalent to 17:5 (even though we are not aware of this relation before we have taken our measurements.) If we can presume that there is no error in measurement, then the following should take place (see Figure 1). We will first discover that the unit fits completely 3 times into the length. We remove or, so to speak, "subtract" 3 units from the length, thereby obtaining a remainder. We can now proceed to use $\frac{1}{10}$ of the unit as our secondary unit of measure. We find that $\frac{1}{10}$ unit fits exactly 4 times into the remainder. Summarizing in notation, we can state that the length = 3 × unit + 4 ($\frac{1}{10}$ × unit) = 3.4 unit. Had we known the precise ratio, length:unit, beforehand, we could have dispensed altogether with measurement, for 17 divided by 5 will yield 3.4. (Another way to express this is "length:unit = 17:5 = 3.4:1".)

In the preceding example, a target length is measured against a specified unit. The unit fits wholly into the target quantity 3 times. One-tenth of the unit fits 4 times into the remainder. We conclude that the target quantity measures 3.4 units. The ratio of the target quantity to the unit is 17:5. To understand this, imagine that the target quantity is 17 inches long and the unit is 5 inches long. (What is 17 divided by 5?) The *maximum common measure* of the length and the unit equals $\frac{1}{5}$ of the unit or $\frac{1}{17}$ of the length to be measured.

If we can ignore measurement error, there will be circumstances in which there will always be a remainder, no matter how many times we move to a smaller unit, $\frac{1}{10}$ of the previous unit. Imagine that the ratio of the target length to the unit of measure were 7:3. Measurement would in principle always leave a remainder. This can be appreciated by noting that 7 ÷ 3 yields the nonterminating decimal number 2.333333 . . . . The initial unit would fit 2 times into the total length being measured. Then $\frac{1}{10}$ of the unit would fit into the first remainder length 3 times, $\frac{1}{100}$ of the unit would fit 3 times into the second remainder, $\frac{1}{1,000}$ of the unit would fit 3 times into the third remainder, and so on *ad infinitum*. This is equivalent to the expression length = 2.333 . . . unit = 2 × unit + 3 × 0.1 unit + 3 × 0.01 unit + 3 × 0.001 unit . . . .

## OTHER KINDS OF MEASUREMENT

Just as one could measure by tenths one could measure by other bases. Nondecimal measurement will express the ratio of the target quantity to units in diverse ways. If the ratio length:unit happens to be 7:3, for example, and we measure, with no measurement error, "by thirds," we should discover that length = 2 × unit + 1($\frac{1}{3}$ × unit). Thus, barring measurement error, measurement terminates or not depending on the measurement base selected. This

Length to be measured

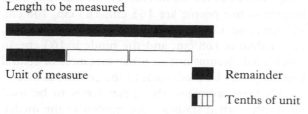

Unit of measure            ■ Remainder

                  ▥ Tenths of unit

**Figure 1**  *Measurement by base unit and tenths of unit.*

corresponds directly to the fact that $2\frac{1}{3}$ can be expressed as the terminating value $2.1_3$ in base 3 or as the nonterminating $2.3333\ldots$ in base 10.

The Euclidean algorithm can be thought of as a measurement procedure in which successive remainders serve as units of measure. If the ratio of the length:unit can be expressed as a rational number, then the Euclidean algorithm will terminate. All of the examples given so far will thus have finite solutions according to the Euclidean algorithm. For example, if target length:unit = 17:5, the unit will fit 3 times into length. The remainder, $R1$ will then fit twice into the unit and leave a second remainder, $R2$. $R2$ will fit into $R1$ exactly 2 times, leaving no remainder. The algorithm has thereupon terminated. (See Figure 2.)

The preceding example works with the same length and unit as in Figure 1. Again we note that the unit fits wholly into the target quantity 3 times, leaving a remainder, which we will call $R1$. $R1$ fits twice into the unit, leaving a second remainder, $R2$. $R2$ fits exactly twice into $R1$. The ratio of the target quantity to the unit must be 17:5. To understand this, consider $R2$ to have the value 1. How many times will $R2$ fit into the length? How many times will $R2$ fit into the unit? The results can be summarized as follows:

$$\text{length} = 3 \times \text{unit} + R1$$

$$\text{unit} = 2 \times R1 + R2$$

$$R1 = 2R2$$

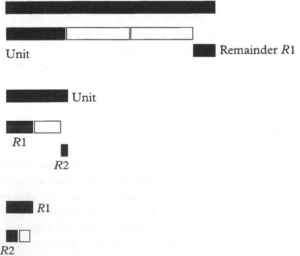

Length to be measured

Unit

Remainder $R1$

Unit

$R1$

$R2$

$R1$

$R2$

**Figure 2** *Measurement according to the Euclidean algorithm.*

$R2$ is the *maximum common measure* of the length and the unit. It is analogous to the idea of a *maximum common divisor.* $R2$ fits into length exactly 17 times; it fits into the unit exactly 5 times. That is to say, length:$R2$ = 17:1 and unit:$R2$ = 5:1. It follows that length = (17/5) units = 3.4 units.

This discussion has illustrated that measurement is tied to diverse concepts from arithmetic, number theory, and even algebra (the relations can be stated through first-order equations, power expansions, etc.). These links may not be immediately clear even to advanced students. One of the major challenges of mathematics education consists of creating situations and staging discussions in which students can begin to work out such relations.

*See also* Division; Systems of Measurement

SELECTED REFERENCES

Carraher, David W. "Learning about Fractions." In *Theories of Mathematical Learning* (pp. 241–266). Pearla Nesher, Leslie P. Steffe, Paul Cobb, Gerald A. Goldin, and Brian Greer, eds. Hillsdale, NJ: Erlbaum, 1996.

Fridman, L. M. "Features of Introducing the Concept of Concrete Numbers in the Primary Grades." In *Soviet Studies in Mathematics Education.* Vol. 6. *Psychological Abilities of Primary School Children in Learning Mathematics* (pp. 148–180). Leslie P. Steffe, ed., trans. Reston, VA: National Council of Teachers of Mathematics, 1991.

DAVID WILLIAM CARRAHER

# MEASURES OF CENTRAL TENDENCY

Used to summarize a set of data or results by giving "typical" values. A central or typical value is often required to answer such questions as: what should my pulse rate be? how much does a newborn baby weigh? my friend feels ill; is her temperature too high? how long does it take you to travel to school? The following are most often used as typical values: (1) the *mean,* or arithmetical average, (2) the result that occurs most often, called the *mode,* (3) a "middling" result, called the *median.* As an example, if the heights of five people are 163 cm, 163 cm, 168 cm, 170 cm, and 177 cm, the mean height is 168.2 cm, the median is 168 cm, and the mode is 163 cm. In specialized circumstances, other measures of central tendency may be used, such as the geometric mean.

Of these averages, the mean tends to be used most often when statistics are quoted in the media. However, if there are unusual values (outliers) in the data set, the median may be a better average to use, as it will be unaffected by those values. The mode is the only average that can be given for nonnumerical data, such as hair color or political affiliation. For most data sets, these three averages will give different answers. However, for a normal distribution, they will coincide.

*See also* Mean, Arithmetic; Mean, Geometric; Median; Mode

## SELECTED REFERENCES

Kennedy, Gavin. *Invitation to Statistics.* London, England: Robertson, 1983.

Marsh, Catherine. *Exploring Data.* Cambridge, England: Polity, 1988.

Moore, David S., and George P. McCabe. *Introduction to the Practice of Statistics.* New York: Freeman, 1993.

MARY ROUNCEFIELD

# MEDIA

The means of communication. Traditional media include storytelling, drama, sculpture, drawing, and painting, which involve small audiences. The invention of movable type and technologies such as electricity gave rise to today's prevalent media, including books, newspapers, magazines, film, radio, and, finally, television, arguably today's most influential medium. These media involve large audiences and exemplify mass media, that is, means of mass communication. More recently, the proliferation of computers has led to new media: multimedia, interactive media, and the Internet. The new media are essentially mass media but have some of the flavor of traditional media by including new modes for small-scale communication.

## NEW MEDIA

*Multimedia* refers to several media combined under the control of a computer. For example, a multimedia encyclopedia might have not only an article about sonata allegro form in music but much more. It could include excerpts of music from several time periods to illustrate the evolution of the form. The user would listen to the music and follow the scores (on the computer's screen) as they play (on the computer's speakers), pausing and repeating at will, even singling out one voice for analysis. Photographs of the composers and the performers might be included, but such an encyclopedia may also have video clips of the performers as they play. When the user finds an intriguing bit of information or name, the computer can instantly serve up all other references to it. Thus, a multimedia encyclopedia has elements of print, audio recording, television, and film but also provides complex links and ease of access.

Stimulated by the development of large-scale memory devices such as CD-ROMs, multimedia has blossomed into interactive media. The salient characteristic of interactive media is that a user affects not just the course of the experience but also its substance. A user not only reviews information but also contributes to it. An interactive encyclopedia encompasses a multimedia encyclopedia and more. For example, a user can comment on an item with text, picture, audio, video, or any combination of these. A comment can also take the form of a new link among existing items. Thus, a user modifies the body of the material and contributes to its complexity. Mathematics software is another example of interactive media. For example, graphing software, both on desktop computers and on handheld calculators, is becoming a common tool in algebra, geometry, and calculus classes. With it, a student can easily represent functions and data and so carry out experiments not otherwise feasible.

The Internet is an informal worldwide system of millions of communicating computers and networks of computers, ranging from single personal computers to large corporate and university computer systems. All one needs to join the Internet is an inexpensive computer accessory (a modem), software to manage the connection, and an account with one of the many companies providing access. As a communication system, the Internet resembles a telephone system, but the Internet is also a new form. Massive amounts of information (text, pictures, video, audio, etc.) pass rapidly among the computers under the idiosyncratic control of the individual users.

A major artifact of the Internet is the World Wide Web. The Web is a massive interactive encyclopedia, but with a high degree of collaboration. Tens of thousands of people contribute to the Web. Their contributions range from the most mundane and personal to tools for scholarly research and opportunities for diverting entertainment. The ease with which anyone can take on a role in the Web leads immediately to issues of standards and quality and, for some people, to issues of regulation. Users also have the problem of verifying content. By contrast, a newspaper typically has a reputation for accuracy established over time and is subject to criticism within its own medium. Reputation and criticism are useful to readers as they weigh the content of an article. Contributors to the Web may have no public reputation and may even be anonymous, and the Internet has no well-developed facilities for criticism. The whole mass is fluid, with people and organizations adding and deleting information, commenting on it, and constructing links throughout it.

Electronic mail (e-mail) is another significant feature of the Internet. An e-mail message goes in minutes to many people as easily as to a single person. The computers take care of routing, delivery, and storage. E-mail facilitates the formation of newsgroups, that is, groups of people linked by a common interest. Thousands of newsgroups exist and more appear every day. The members participate in a free-flowing discussion of the topic of choice with dozens, hundreds, or thousands of other members. Some groups are open to everyone; some are restricted in membership. For example, the participants in a summer institute in the teaching of geometry could form a newsgroup to maintain group cohesion and for group business, such as discussing mutual problems and successes. Thus, the Internet offers convenient one-to-many communication and makes it easy to gather people with like interests, even very narrowly conceived. Because the Internet is in a state of rapid development, its specific features and devices are certain to change.

## MEDIA AND INFORMAL EDUCATION

Media mirror and reinforce the collective consciousness. They are also powerful tools with which to affect people's attitudes and opinions and to educate, both informally and formally. Popular culture through the very media we are discussing reinforces negative public opinion about mathematics. Altering the general view of mathematics depends, of course, on the reform of mathematics teaching and learning in the schools. We can use the media to help ensure the success of that reform. However, harnessing the educational potential of the media is not a trivial problem. The creation, manufacture, and distribution of media products are complex and expensive activities. They take place in a commercial, free (although regulated) market. Under this system, the nature and size of the audience and the potential revenues are major factors. They influence in particular the content of the products offered and the amount of money invested in their production. Action and adventure sell movie tickets; education and mathematics do not. Open distribution venues such as mass-market magazines and newspapers and open-circuit television drive media production; they also set standards for the quality of production. Because audiences carry their expectations to other venues and because technical and creative innovations migrate to products for special markets, such as schools or museums, mass-market standards become, in effect, the norm.

Even though the standards and content of the open-media market dominate special-interest markets, there have been some successful educational applications under the rubric of informal education. Good examples can be found in the programming of the Public Broadcasting Service (PBS), which maintains a block of educational programming directed at children, including *Square One TV* (mathematics). There have been successes as well in the other mass media. For examples in radio, newspapers, and magazines as well as television, consult the proceedings of a conference on the popularization of mathematics held in 1989 at the University of Leeds (UK) (Howson and Kahane 1990).

## MEDIA AND SCHOOLS

Schools are a major venue for mass communication. Schools are relatively conservative in terms of the media they employ. Lectures (storytelling), demonstrations (drama), and books (text and pictures) continue to be the basis for most instruction. Newspapers and magazines play some role, radio and television have less of a role, and the new media far less still. Television was invented less than seventy years ago, and large-scale broadcast was introduced less than fifty years ago. Through PBS and its related consortia, instructional television (as contrasted with open-circuit broadcast television) matured in the last decade of the twentieth century. It provides large quantities of video and related print and services to teachers. Personal computers, which appeared less than twenty years ago, are already major cultural catalysts increasingly found in classrooms. Widespread, effective use of the new media is still to come, however, although perhaps in less time than was the case for television. Even so, these media are promising tools for educators.

New media impinge on schools in at least three areas: curriculum, teacher preparation and development, and enhancing the professional context for teachers. The new media can supplement curricular content in a significant way. The vast amounts of material available through them allow the infusion of fresh, current material into a course syllabus and provide the raw material for crosscurricular studies. The communication tools built into the new media facilitate wide-ranging collaboration, and they include the means of tailoring content to a much higher degree than is possible with other media. For example, films and videos help us to understand three-dimensional objects. Graphics software on a computer allows a student to set up a solid of revolution and to toy with the parameters. This is a powerful tool for developing skills in visualization.

Video enhances both preservice and continuing teacher education, especially by providing exemplars. On the other hand, teacher education must take notice of the pedagogical problems involved in the use of media in the classroom. Significant use requires familiarity with the media and media products. Most publishers in the school market also include videos among their products, both to enhance the implementation of their programs and as supplements to standard materials.

Mass media are also important for improving the professional context for teachers. But teachers need the opportunity and time to learn to use the media in their work and to plan how to use them effectively, which raises the issue of paying for staff time and materials. Easy access to vast quantities of this material points to a democratization of curriculum design and tailoring, but this raises issues of standards and of responsibility for curricular content. The communication facilities of the Internet may be the most promising aspect of the new media for teachers. Through them, teachers have the means for developing mutually supportive communities with similar interests. One important function of these communities is to supplement the word-of-mouth that helps a teacher sort through the masses of products available for the classroom.

Counterbalancing the potential are problems of curricular design, integration of supplementary materials, and developing pedagogy in the use of the media, among others. Fulfilling the media's promise for schools rests on at least three conditions: adequate investment in hardware (television sets, video players, computers, repair facilities, and so on); adequate investment in staff (continuing education, time to plan, technical support, and so on); and adequate investment in the experimentation necessary to develop applications. The National Science Foundation takes much of the last responsibility. Scarce resources force difficult choices, especially in the face of the pressing need to pursue educational equity.

## MEDIA AND MATHEMATICS

A comprehensive review of applications of various media to mathematics is not possible or even particularly valuable, given the rapid changes in product. Reviewing and critiquing the mass of products is one of the many functions of the professional organizations. The journals of the National Council of Teachers of Mathematics (NCTM) include review columns. The national and regional meetings of NCTM and the related local and state organizations provide a forum for examining and discussing products with colleagues. Listed here is a highly idiosyncratic collection of producers and products readers might find interesting or provocative. *Donald Duck in Mathmagic Land* is a classic film with a humorous view of mathematics. It is an early attempt to provide informal information about mathematics to a wide, general audience. Project Mathematics! produces a series of videos illustrating central ideas from the secondary curriculum. Titles include *The Pythagorean Theorem, Similarity,* and *Sines and Cosines.* The videos include excellent animations to illustrate the difficulties of their subjects. *For All Practical Purposes* is a video series accompanied by a coordinated textbook that includes more than twenty applications of mathematics in a lively presentation. *Channel One* broadcasts a daily news program into schools. This controversial project offers opportunities for considering data representation and analysis in timely contexts. The national newspaper *USA Today* employs innovative graphics that can be useful in school. *MATH TALK* videos derive from *Square One TV.* The former is specifically for use in schools, while the latter was primarily for home viewing. Each provides a humorous context for mathematics and problem solving.

*See also* Computer-assisted Instruction (CAI); Computers; Software; Technology

## SELECTED REFERENCES

*Channel One.* Nashville, TN: Whittle Communications.

Devlin, Keith. "Prime Time Television Discovers Mathematics." *Focus* 18(3)(1998):60.

*Donald Duck in Mathmagic Land.* Burbank, CA: Walt Disney, 1959.

Garfunkel, Solomon, proj. dir. *For All Practical Purposes.* New York: Freeman, 1988.

Howson, A. Geoffrey, and Jean-Pierre Kahane, eds. *The Popularization of Mathematics.* Cambridge, England: Cambridge University Press, 1990.

Project Mathematics! Pasadena, CA: California Institute of Technology, 1989–1995.

| Heights | Frequency | Upper limit of group | Cumulative frequency |
|---------|-----------|----------------------|----------------------|
| 150–154 | 3 | 154.5 | 3 |
| 155–159 | 4 | 159.5 | 7 |
| 160–164 | 6 | 164.5 | 13 |
| 165–169 | 3 | 169.5 | 16 |
| 170–174 | 2 | 174.5 | 18 |

*Square One TV MATH TALK*. New York: Children's Television Workshop, 1995.

*USA Today*. Arlington, VA: Gannett.

JOEL SCHNEIDER

# MEDIAN

The value which occurs exactly halfway through a set of data. It is a useful measure of central tendency as it is easily understood. The median can be calculated only for numerical data, or data that is at least ordinal, as the values must first be arranged in order. For example, the heights of five people are

165 cm, 180 cm, 178 cm, 142 cm, and 127 cm

These values must first be listed in order as

127 cm, 142 cm, *165 cm,* 178 cm, 180 cm

Thus, the median height is 165 cm. If another person, 151 cm tall, joins this group, the ordered list becomes

127 cm, 142 cm, *151 cm, 165 cm,* 178 cm, 180 cm

There is now no longer a middle value. The median is then found by taking the average of the two middle terms (giving 158 cm) as the central point of the data.

A stem and leaf diagram (Hudson and Rouncefield 1993; Marsh 1988) is a convenient way of arranging raw data in order. If the number of items ($n$) is odd, the middle value has its position at $(n + 1)/2$. So, for instance, the middle item out of 99 observations is the 50th. Where $n$ is even, there are two middle values, located at $n/2$ and $(n/2) + 1$. So for 100 values, the median is the average of the 50th and 51st.

The simplest method of finding the median of data arranged as a frequency table is to construct a cumulative frequency table and then draw a cumulative frequency graph. To locate the median for $n$ data items, a horizontal line is drawn across to the graph at $n/2$ on the cumulative frequency scale (Rouncefield 1988). As an example, the heights of 18 pupils, measured to the nearest cm, are organized as a frequency table. Two extra columns have been added to show the upper limit for each class interval (or group) and the cumulative frequency (running total).

The cumulative frequency graph is drawn by plotting the upper limit for each class interval on the horizontal axis, and the cumulative frequency on the vertical axis. The median is found by drawing a horizontal line across to the graph at $n/2 = 9$ and reading

off the corresponding height (see Figure 1). The value of the median estimated from the graph in this way is 161.25 cm. An alternative method is to use linear interpolation. If a distribution is modeled by a continuous probability density function $f(x)$, the median is that value of $x$ for which the area under the curve equals $\frac{1}{2}$.

It is recommended that pupils first use the median as a "typical value" for data they have collected themselves, in preference to other kinds of averages (Corwin and Friel 1989). Young pupils can start with small sets of data and move progressively from the use of an ordered list to a stem and leaf diagram to a cumulative frequency curve in the later secondary grades. The median is a "robust statistic" in that it is not affected by outliers or unusual results. Students can investigate this for themselves using computer software such as ELASTIC produced by Sunburst Communications (Rubin and Rosebery 1990). The median gives a more representative answer than the mean for data sets that are very skewed and that contain a relatively small number of very low or very high results (as in a typical distribution of incomes,

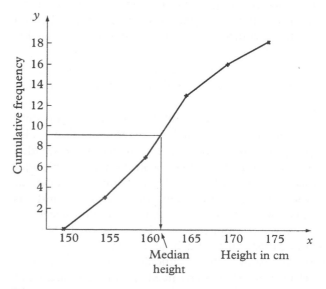

**Figure 1** *Cumulative frequency graph.*

where the mean may be affected by a small number of very high incomes).

*See also* Measures of Central Tendency; Probability Density

## SELECTED REFERENCES

Corwin, Rebecca B., and Susan N. Friel. *The Shape of the Data.* Palo Alto, CA: Seymour, 1989.

Hudson, Brian, and Mary Rouncefield. *Handling Data Core Y10/11. Century Maths Series.* Cheltenham, England: Stanley Thornes, 1993.

Marsh, Catherine. *Exploring Data.* Cambridge, England: Polity, 1988.

Rosebery, Ann, and Andee Rubin. "Teaching Statistical Reasoning with Computers." *Teaching Statistics* 12(2)(1990):38–42.

Rouncefield, Mary. "Now Can We Stop Using $(n + 1)/2$ for Finding the Median?" *Teaching Statistics* 10(1)(1988):20–22.

———, and Peter Holmes. *Practical Statistics.* London, England: Macmillan, 1989.

Rubin, Andee, and Ann Rosebery. "Teachers' Misunderstandings in Statistical Reasoning: Evidence from a Field Test of Innovative Materials." In *Training Teachers to Teach Statistics* (pp. 72–89). Anne Hawkins, ed. Voorburg, Netherlands: International Statistical Institute, 1990.

Vellman, Paul, and David Hoaglin. *Applications, Basics and Computing of Exploratory Data Analysis.* Boston, MA: Duxbury, 1981.

MARY ROUNCEFIELD

# MIDDLE SCHOOL

A school comprising the middle grades. Research has identified the middle grades (5–8) as crucial ones in the development of students' academic and social dispositions. Moreover, learning theorists have suggested that students' developmental stages should guide the design of the curriculum and the learning environment (Secada 1990). In response to these findings, school organizational patterns, variously called the *junior high school,* the *intermediate school,* or the *middle school,* have evolved. Over the last century, changing societal values and shifting curricular emphases have altered the intended functions of these schools. Consequently, it has not always been clear whether the school organizational pattern was designed to facilitate the students' transition to a high school subject-centered curriculum (the junior high school organization) or to acknowledge, in addition, the special cognitive, sociological, and physiological characteristics of the early adolescent (the middle school organization). Regardless of the school organizational pattern, it is universally accepted that students in early adolescence are in a transitional stage with unique psychosocial needs and cognitive styles that must be considered in the design of an appropriate mathematics curriculum (Silvey 1982).

## HISTORICAL BACKGROUND

The problem of articulation between elementary and secondary schools was first addressed at the turn of the century. The organizational structure of the twelve years of schooling at that time usually conformed to one of the following patterns: 8–4, 6–6, or 6–3–3. The end of World War I represented a new era in educational development. Enrollment in public schools began to rise rapidly and brought with it serious discussions about the design of the curriculum and the content appropriate for large numbers of students with diverse backgrounds and a broad range of needs. The transition from grade 8 in the elementary school to grade 9 in the high school was seen as too abrupt. A gradual introduction to secondary school subjects was needed, as well as greater flexibility of scheduling and provisions for student exploration (Johnson 1980). One solution was the establishment of junior high schools for grades 7 through 9; another was the creation of intermediate schools, or separate units within elementary schools, for grades 5 through 8. Beginning in California and Ohio in 1909, the junior high school was well established by the 1920s and had become the norm rather than the exception by the late 1930s.

During the period from 1920 to 1945, junior high school mathematics was shifting away from the practice (which had dominated earlier periods) of lengthy arithmetic computations, often presented in isolation from real applications. The 1926 National Committee on Mathematical Requirements (Koos 1927) recommended a more integrated mathematics curriculum. In particular, a *general mathematics* curriculum was recommended for the junior high level that included arithmetic applications, introductory algebra, informal geometry, statistics, and right-triangle trigonometry. Words such as *accuracy* and *drill* in the prefaces of textbooks were replaced with *meaning* and *understanding.* This trend, however, was short-lived. The textbooks of the 1930s and 1940s began to shift away from general mathematics to focus on mathematics for the consumer. Topics listed in the table of contents for an eighth-grade text, *Mathematics and Life* (Ruch, Knight, and Studebaker 1942), illustrate the trend: "Mathematics and the

Community; The Merchant and the Community; Taxes and Other Community Funds" (p. ix). The first half of the twentieth century had been a turbulent period in American history, encompassing an economic depression and world wars. At all levels education too, was in a state of flux. Consequently, efforts to enact sweeping reform in the mathematics curriculum failed (Jones 1970).

History has shown that school problems come into focus during periods of national crisis. The event that spawned major national reform in the mathematics curriculum was the U.S.S.R.'s launching of *Sputnik* in 1957. The curricular reform efforts that followed were aimed at the college-capable student. These innovations in mathematics education were collectively known as "new math." The junior high school mathematics textbooks of the "new math" era covered college preparatory concepts such as sets, algebraic properties of number systems, base arithmetic, and number theory. The "new math" trend lasted for approximately fifteen years.

In 1974, critics of the "new math" curriculum for elementary and middle grade levels complained that reform had produced programs that were highly theoretical, excessively formal, and too focused on deductive logic. Moreover, even though mathematics textbooks had been written to reflect the "new math" curriculum, questions arose as to the extent that actual reform had been implemented in the nation's schools. In order to answer these questions, the Conference Board of the Mathematical Sciences appointed a National Advisory Committee on Mathematical Education (NACOME 1975). The committee was charged with the task of preparing a complete overview and analysis of mathematics education (K–12) in the nation's schools. NACOME disclosed that junior high school mathematics curricular innovations were easy to implement, being unencumbered by college preparatory, Carnegie-unit requirements. However, for the same reason, they were more difficult to survey. NACOME pointed to the wide use of commercial texts designed for the "new math" curriculum as evidence of an altered junior high curriculum. However, there were no dependable data on the relative emphasis of new and traditional junior high school mathematical topics in the nation's schools.

By the 1980s, the National Council of Teachers of Mathematics (NCTM 1980) foresaw a reordering of priorities in the mathematics curriculum due to changes in societal needs and the availability of low-cost calculators and computers. In its position statement, *An Agenda for Action,* NCTM identified

problem solving as the primary goal of school mathematics at all levels; at the middle grade level, it should be "simple and embryonic" (NCTM 1980, 19) but should increase in sophistication throughout the high school years. The *Agenda* delineated societal goals to which the middle-grade mathematics curriculum should respond as follows:

- the preservation of mathematics as an important component of scientific culture,
- the development of future consumers of mathematics,
- the recognition and encouragement of mathematical talent,
- the development of users of elementary mathematical techniques, such as expressing relationships in a variety of ways, computing numerically, solving a broad range of problems, reasoning abstractly, and evaluating results (NCTM 1980).

These goals were considered paramount for teachers of middle-grade mathematics because this is the level at which students begin to make conscious decisions about whether to pursue mathematics further. If they choose not to pursue mathematics, this could be their culminating mathematical experience.

In the 1970s and 1980s, the majority of the nation's junior high schools were converted to middle schools of varying configurations. The physical design of the schools and the structure of the curriculum were intended to take into consideration the cognitive, social, and psychological characteristics of the early adolescent (see the next section, on learner characteristics). The middle school mathematics curriculum was the subject of NCTM's 1982 Yearbook (Silvey 1982). Articles in the yearbook responded to four prior calls for reform in school mathematics originating from NACOME (1975), the Mathematical Association of America (MAA 1978), the National Council of Supervisors of Mathematics (NCSM 1977), and NCTM (1980). The yearbook focused on models of effective mathematics instruction for middle school students (see the later section on curriculum). Clearly, mathematics for middle schools had become a prominent issue in mathematics education by the 1980s. The middle grades had been identified as the pivotal period of a student's education and recognized as a time of exploration, definitive academic decisions, and social choices.

At the same time, a national study of the state of America's educational system was being conducted by the National Commission of Excellence in Education. Results were brought to the forefront of public attention in its report, *A Nation at Risk* (1983), which disclosed disturbing trends of low student achievement in mathematics. The National Research Council (1989) confirmed these trends and cited significant reductions in the numbers of U.S. students entering mathematics-related careers. These alarming statistics culminated in a resounding call for national reform in the teaching and learning of mathematics at all levels. Thus, in 1987, NCTM enlisted the support of representatives from a wide range of professions and began to create a new set of curriculum guidelines.

NCTM published its new curriculum guidelines, *The Curriculum and Evaluation Standards for School Mathematics* (the *Standards*) in 1989. Unlike previous reform efforts, the *Standards* received endorsements from all of the major professional organizations pertaining to mathematics. This document proposed a major shift in the content and instructional practices in mathematics for the 1990s. The curriculum was revised to provide *all* students with broader-based mathematics content and to take advantage of the latest advances in technology. Instructional models were developed for three distinct levels: elementary, middle, and high school. By 1993, the majority of mathematics textbooks were revised to conform to the goals and content of the new curriculum. For the professional journals of the NCTM, the goals of the curriculum became thematic; and to serve the specific needs of the middle-grade mathematics teacher, the NCTM created a new journal, *Mathematics Teaching in the Middle School*.

## LEARNER CHARACTERISTICS

Relationships among instructional practice, curriculum, and students' developmental stages were not fully recognized until the 1970s. Prior to 1960, the dearth of research pertaining specifically to the educational needs of early adolescents was underscored by educational psychologist William Wattenberg (1965), who wrote: "The extent of our ignorance is symbolized by the fact that during the past four decades, . . . there have been exactly three books devoted to pre-adolescence" (p. 39). In 1978, Geneva Haertel's meta-analysis of research on early adolescence reported a lack of coordination between the formal reasoning abilities of the early adolescent and the sequencing and teaching of middle-grade mathematics. Haertel urged further study of relationships among cognitive ability, learning style, and instructional practice in mathematics. This was the

beginning of a series of studies that provided a conceptualization of human development that affected the design of the current curriculum for middle school mathematics.

The middle school years were the topics of the 79th Yearbook of the National Society for the Study of Education (Johnson 1980). In this volume, middle school students were described as energetic, restless, insecure, self-conscious, and responsive to peer pressure. Nearly all have attained what Jean Piaget (1974) termed the "concrete operational stage" and have begun to move into the "formal operational stage." Students in the fifth grade may still need concrete experiences to comprehend fully an array of mathematical topics. Eighth-grade students may have already progressed to the formal operational stage; that is, they are able to synthesize and consider hypothetical situations without corequisite concrete experiences. The yearbook pointed out that middle-grade students are busily forming attitudes and dispositions toward education, asking questions such as Who am I? In what directions will I go in my life? Thus, the educational experience of middle school students is of particular importance because the decisions they make at this level can dramatically affect their future. Failure to pursue mathematics may decrease career choices and the selection of college majors—particularly for females and minorities.

Elizabeth Fennema, an authority on gender issues in mathematics education, points out that even though females appear to receive the same educational opportunities as males to learn mathematics, evidence indicates lower achievement by females, lower enrollment in advanced mathematics classes by females, lower achievement in high-level mathematical tasks, and "more negative attitudes on the part of females toward the learning of mathematics" (Fennema 1982, 9).

Fennema's research indicates that the teaching and learning environment have a direct impact on students' achievement and attitudes toward mathematics. In the middle grades, the learning environments for males and females, while appearing to be the same, may differ a great deal. It has been shown that teachers generally interact more with males than with females and have differential expectations. Fennema pointed out several implications for the teaching behavior of middle school teachers that might help to alleviate these discrepancies. Specifically, she advises teachers to become blind to gender; maintain equal expectations; help students to develop feelings of confidence in doing mathematics; make students more aware of the usefulness of mathematics in choosing careers; become aware of students' attitudes toward mathematics; and combat the belief that mathematics is a male domain.

## THE LEARNING ENVIRONMENT

The characteristics of the learner and the nature of the content are chief considerations in providing conducive learning environments. Mathematics educators (Romberg and Carpenter 1986; Steen 1986) recommend that middle-grade students participate actively in the *process* of learning mathematics. Since these students are in need of social interaction and peer approval, an ideal learning environment will engage students both cognitively and socially. Thus, mathematics lessons that provide opportunities for students to work individually and in collaborative groups capitalize on the social characteristics of students while enabling them to take some responsibility for their own learning.

The *Professional Standards for Teaching Mathematics* (NCTM 1991) describes learning environments that enhance children's problem-solving skills as those that invite children to initiate questions and problems, make and investigate conjectures, listen to and question the teacher, use a variety of communication tools, and rely on mathematical evidence and deductive logic to determine the validity of arguments.

One way that students may be engaged both cognitively and socially is by assigning cooperative groups open-ended problems and extended projects. These projects should allow students to experience the connections of mathematics with other subjects. Their solutions to problems should engage students in investigations, requiring inductive and deductive reasoning, and oral and written communications. This way, students may become more independent, confident learners of mathematics and view it as a logical, problem-solving endeavor (NCTM 1980, 1989; Silvey 1982). To accommodate middle-grade students' transition to the formal operational stage, instruction should include multiple formats; that is, tactile, visual, and auditory instructional modes should be used to assist students in deriving the meaning of mathematical concepts (Furth and Wachs 1975). The role of the teacher should, therefore, shift from that of sole conveyer of knowledge to one who coordinates, guides, and helps students in their mathematical explorations.

## THE CURRICULUM

The middle-grade mathematics curriculum prior to the 1980s was viewed by many as "irrelevant and dull" (NCTM 1989, 65). Textbooks were repetitious, with the same topics repeated year after year; problems were contrived; and instruction focused on computation and drill. The prevailing belief of teachers was that children needed to learn basic arithmetic facts before they are able to apply their knowledge to solve real problems, so children were taught mechanics separate from each other and any real-world problems. The result was that children were often unable to use those skills to solve problems and acquired an attitude that "mathematics is meaningless drudgery" (Schwartz and Riedesel 1994, 69).

In contrast, problem solving is at the core of the new curriculum proposed by the NCTM. The four primary foci of mathematics instruction are identified as follows: (1) problem solving, (2) reasoning, (3) communications, and (4) connections. Rather than introducing problem solving after basic skills are mastered, the new curriculum recommends the opposite; that is, basic skills should be introduced in the context of solving problems. In this way, basic skills are properly situated, motivated, and relevant. Problem solving, therefore, is the driving force from which the content evolves in the new curriculum. The mathematics curriculum for grades 5–8, as defined by the *Standards* (1989), covers a broad range of topics spanning the following content strands: patterns and functions, measurement and geometry, probability and statistics, multiple modes of computation, estimation, algebra, number systems, and number theory. Attention also is given to the historical evolution of topics and the contributions of women and men to their development. The goal is for students to realize that mathematics is connected to every aspect of life, is a human invention, and has evolved from the efforts of many different people and cultures.

For students to see the relevance of mathematics, they need firsthand experience in applications taken from a variety of contexts. NCTM's 1982 Yearbook (Silvey 1982) presents a wealth of problem-solving ideas for integrating mathematics with sports, newspapers, out-of-doors measurement, and games. For instance, the information on sports trading cards may be used to create graphs, compare measures, and evaluate statistics. The newspaper is a ready resource of real-world problems dealing with ratios, averages, large numbers, and percentages. Out-of-door activities involving measurement of distances, diameters of tree trunks, and angles of elevation are excellent ways to engage students in the discovery of mathematical properties as they solve and analyze real problems.

## Projects

Numerous textbooks and enrichment resources describe open-ended projects appropriate for middle-grade mathematical investigations. The following is a sampling of projects adapted from multiple sources that are recommended for individual work or cooperative groups.

1. Collect and organize data from at least fifty people to discover if there is a relationship between the size of a person's (a) foot and height; (b) arm and leg length. Are there any implications for the design and marketing of clothing or furniture? Explain.

2. Select a baseball card. Use your calculator and the information on the back of the card to determine each of the following: In what year did the player get the most home runs? Least home runs? How old is the player? What was his best batting average in the years reported? Make a bar graph of the player's annual total hits for the years reported. Represent these data on a pie chart. What are the mean, median, and mode for the total number of hits for the years reported? Make a general statement regarding the performance of this player.

3. Take a square piece of 20 × 20 grid paper, scissors, and tape. Construct an open-ended box by following these instructions: (a) cut congruent square pieces from each of the four corners of the grid paper; (b) fold up and tape the sides to form an open-ended box. Determine the volume of the box. Next take a new 20 × 20 piece of grid paper and make a different box. Continue this process until you have found the box of maximum volume. Write a convincing argument supporting your conclusion.

4. Investigate which regular polygons tessellate the plane and which do not. Provide a mathematical explanation supporting your conclusion. Describe where tessellations occur in nature. Write a two-page paper describing the role of tessellations in Islamic culture and art.

5. Design a game of chance using coins, dice, or spinners. How do you win? Lose? Use probability to demonstrate whether this game is fair or unfair.

6. Write two different Logo (computer program) procedures for drawing a completed tangram puzzle: (a) write one procedure that employs the lengths of individual line segments; (b) write another procedure using only the coordinates of points.

7. Use the *Geometer's Sketchpad* (computer software) to construct the three medians of a triangle. What can you conclude about the areas of the interior regions of the triangle created by these medians?

8. Determine if it is possible for two shapes to have the same area and perimeter and be different in shape. Use your geoboard to investigate this problem.

9. Write a report describing the contributions of Ada Lovelace and/or Grace Hopper to the design and operation of computers. What is the role of binary arithmetic in the operation of computers?

Problems of this type are considered appropriate because they will engender responses on multiple levels and may form the context in which other mathematical explorations can occur. They are ideal for cooperative groups because, aside from the benefits students gain from working together academically, they may help develop students' skills in social interaction and sensitivity to cultures different from their own (Schwartz and Riedesel 1994).

## Manipulative Materials

A major focus of the new curriculum is on reasoning. Inductive and deductive reasoning may be nurtured with hands-on explorations, pictorial representations, and graphs. Students therefore need access to materials to use in these explorations. Many of the manipulative materials that are introduced in the lower grades to explore whole-number and basic geometric concepts can also be used in the middle grades to investigate properties of rational numbers, measurement, and geometric relationships. The following is a sampling of appropriate materials and associated topics for middle-grade instruction and investigation:

- circular sectors and number rods—to model fractions and decimals
- loop abacus—to demonstrate place value, regrouping, and arithmetic in other bases
- rulers, compasses, protractors and scissors—to use in measurement (length, perimeter, and area), geometry, and trigonometry
- tangrams, geoboards, dot and grid paper—to model fractions, decimals, percentages, probability, geometric concepts, area, and perimeter
- solid geometric models—to derive concepts of surface area and volume
- pattern blocks—to explore fractions, symmetry, patterns, and tessellations

- mirrors—to investigate symmetry, geometric constructions, and transformations
- balance beams and scales—to explore arithmetic relations, measurement, and algebraic equations
- algebra tiles—to model algebraic expressions, multiplication, and factoring
- spinners, dice, cards—to experiment with probability and statistics

There is no inherent value in manipulatives unless teachers use them effectively to motivate and clarify mathematical concepts. Specifically, manipulative materials may be used to construct physical models of mathematical concepts and to demonstrate the applications of mathematics in real-world situations.

The merit of hands-on explorations is supported by the historical evolution of mathematics itself. For example, Archimedes (287–212 B.C.), the greatest mathematician of ancient times, attributed the bulk of his discoveries to "diversions of geometry at play" (Edwards 1979, 29) and explained that "certain things first became clear to me by a mechanical method" (Edwards 1979, 68). Thus, hands-on explorations with manipulative materials may foster middle-grade students' own discoveries of mathematical relationships. Middle-grade teachers may use manipulative materials to bridge the gap between the concrete and symbolic modes of instruction that students experience at, respectively, the elementary and high school levels.

## Calculators

Another valuable tool for middle-grade instruction is the calculator (Bitter and Mikesell 1990). The new curriculum recommends that all middle-grade students "have access to an appropriate calculator" (NCTM 1989, 8). However, the appropriate use of calculators does not include using them as a substitute for learning basic arithmetic facts (Schwartz and Riedesel 1994). Nonscientific calculators have been developed specifically to serve the needs of the middle-grade mathematics students (e.g., The Explorer by Texas Instruments). Since rational numbers, probability, and statistics are content strands of the middle-grade curriculum, these calculators are designed with both decimal- and fraction-manipulation capabilities. In addition to relieving students of the burden of computation, calculators have been used effectively to help relate mathematical language, symbols, and representations (i.e., subtraction is the inverse of addition, multiplication is repeated addition, raising a number to a power is repeated multiplication, etc.) Calculators may also serve as

instructional tools for exploring patterns, verifying estimation, and enhancing problem-solving instruction (Coburn 1987).

Although classroom use of calculators remains a controversial issue, research has revealed that (1) calculators *do not* prevent students from learning and understanding mathematical operations, concepts, and skills; and (2) calculators *do not* impair students' learning and use of mental arithmetic skills (Shumway 1988). The key to using the calculator in classroom teaching is to identify when it provides the best approach for problem solution.

## Computers

The proposed NCTM curriculum recommends that every classroom be equipped with at least one computer for demonstration purposes. Additionally, it recommends that computer laboratories should be available for whole-class instruction. The *Standards* stipulate that "students should learn to use the computer as a tool for processing information and performing calculations to investigate and solve problems" (NCTM 1989, 8). At the middle-grade level, using interactive software that engages a student's curiosity, invites exploration and conjectures, and allows for validation is highly desirable. Although not exhaustive, examples of software that meet these criteria are as follows: electronic spreadsheets (e.g., ClarisWorks, EXCEL by Microsoft), the Logo language (e.g., Microworlds by LCSI, Terrapin Logo), graphics packages (e.g., Elastic Lines by Sunburst), and geometric exploratory programs (e.g., Geometric Supposers by Sunburst, The *Geometer's Sketchpad* by Key Curriculum Press). New ways of using technology for problem solving, computation, and visual representations of mathematical phenomena are evolving every day and should be examined for their potential for enhancing the teaching and learning of middle-grade mathematics.

## SUMMARY

The teaching and learning of middle-grade mathematics have undergone significant changes over the past century. The unique cognitive, psychological, and sociological needs of middle-grade students did not come into focus until the 1970s. It was then that learning theorists suggested that students' developmental stages should guide the design of the curriculum and the learning environment. These considerations became paramount in the design of the mathematics curriculum for the middle school.

The middle-grade years were identified as pivotal ones in which definitive decisions and social choices are made that often affect students' academic directions and career options. Changing societal needs and the advent of affordable technologies also affect the ways in which mathematics is taught and learned. These factors were considered in the development of NCTM's (1989) middle-grade mathematics curriculum guidelines. Previous curricular reform efforts have experienced only limited success. The future of middle-grade mathematics education and the implementation of the new curriculum guidelines are ultimately the responsibility of the classroom teacher.

*See also* Computers; Curriculum, Overview; Software; Teacher Development, Middle School; Teacher Preparation, Middle School; Technology

## SELECTED REFERENCES

Bitter, Gary, and Jerald Mikesell. *Using the Math Explorer Calculator.* Reading, MA: Addison-Wesley, 1990.

Coburn, Terrence. *How to Teach Mathematics Using a Calculator.* Reston, VA: National Council of Teachers of Mathematics, 1987.

Edwards, Charles H., Jr. *Historical Development of the Calculus.* New York: Springer-Verlag, 1979.

Fennema, Elizabeth. "Girls and Mathematics: The Crucial Middle Grades." In *Mathematics for the Middle Grades (5–9)* (pp. 9–19). 1982 Yearbook. Linda Silvey, ed. Reston, VA: National Council of Teachers of Mathematics, 1982.

Furth, Hans, and Harry Wachs. *Thinking Goes to School: Piaget's Theory in Practice.* London, England: Oxford University Press, 1975.

Haertel, Geneva D. "Literature Review of Early Adolescence and Implications for Science Education Programming." In *Early Adolescence: Perspectives and Recommendations* (pp. 93–203). Prepared by Paul DeHart Hurd. NSF Report SE-78-75. Washington, DC: Government Printing Office, 1978.

Johnson, Mauritz, ed. *Toward Adolescence: The Middle School Years.* 79th Yearbook of the National Society for the Study of Education. Chicago, IL: University of Chicago Press, 1980.

Johnston, William B., and Arnold E. Packer, eds. *Workforce 2000: Work and Workers for the Twenty-First Century.* Indianapolis, IN: Hudson Institute, 1987.

Jones, Phillip S., ed. *A History of Mathematics Education in the United States and Canada.* 32d Yearbook. Washington, DC: National Council of Teachers of Mathematics, 1970.

Koos, Leonard V. *The Junior High School.* Boston, MA: Ginn, 1927.

Mathematical Association of America (MAA). *PRIME—80: Proceedings on Prospects in Mathematics Education in the 1980s.* Washington, DC: The Association, 1978.

National Advisory Committee on Mathematical Education (NACOME). *Overview and Analysis of School Mathematics Grades K–12.* Washington, DC: Conference Board of the Mathematical Sciences, 1975.

National Commission on Excellence in Education. *A Nation at Risk: The Imperative for Educational Reform.* Washington, DC: Government Printing Office, 1983.

National Council of Supervisors of Mathematics. "Position Paper on Basic Skills." *Arithmetic Teacher* 25(Oct. 1977):19–22.

National Council of Teachers of Mathematics. *Addenda Series.* Reston, VA: The Council, 1991– .

———. *An Agenda for Action: Recommendations for School Mathematics of the 1980s.* Reston, VA: The Council, 1980.

———. *Curriculum and Evaluation Standards for School Mathematics.* Reston, VA: The Council, 1989.

———. *Mathematics Teaching in the Middle School.* Reston, VA: The Council. A journal.

———. *Principles and Standards for School Mathematics.* Reston, VA: The Council, 2000.

———. *Professional Standards for Teaching Mathematics.* Reston, VA: The Council, 1991.

National Research Council. *Everybody Counts: A Report to the Nation on the Future of Mathematics Education.* Washington, DC: National Academy Press, 1989.

National Science Board Commission on Precollege Education in Mathematics, Science and Technology. *Today's Problems, Tomorrow's Crises.* Washington, DC: National Science Foundation, 1982.

Piaget, Jean. *To Understand Is to Invent.* New York: Viking Press, 1948, 1974.

Romberg, Thomas A., and Thomas P. Carpenter. "Research on Teaching and Learning Mathematics: Two Disciplines of Scientific Inquiry." In *Handbook of Research on Teaching* (pp. 850–873). Merlin C. Wittrock, ed. New York: Macmillan, 1986.

Ruch, Giles M., Frederick B. Knight, and John W. Studebaker. *Mathematics and Life.* Bk. 2. Chicago, IL: Scott, Foresman, 1937, 1942.

Schwartz, James E., and C. Alan Riedesel. *Essentials of Classroom Teaching Elementary Mathematics.* Boston, MA: Allyn and Bacon, 1994.

Secada, Walter G. "The Challenges of a Changing World for Mathematics Education." In *Teaching and Learning Mathematics in the 1990s* (pp. 135–143). 1990 Yearbook. Thomas J. Cooney, ed. Reston, VA: National Council of Teachers of Mathematics, 1990.

Shumway, Robert. "Calculators and Computers." In *Teaching Mathematics in Grades K–8* (pp. 363–419). Thomas R. Post, ed. Boston, MA: Allyn and Bacon, 1988.

Silvey, Linda, ed. *Mathematics for the Middle Grades (5–9).* 1982 Yearbook. Reston, VA: National Council of Teachers of Mathematics, 1982.

Steen, Lynn. "A Time of Transition: Mathematics for the Middle Grades." In *A Change in Emphasis* (pp. 1–9). Richard Lodholz, ed. Parkway, MO: Parkway School District, 1986.

Wattenberg, William W. "The Junior High School—A Psychologist's View." *NASSP Bulletin* 49(1965):39–44.

SHARON WHITTON

# MINORITIES, CAREER PROBLEMS

Barriers to careers involving mathematics with respect to preparation as well as employment and career advancement to high levels. Adequate preparation in mathematics provides a solid foundation for a broad range of career choices, but minorities in the mathematical sciences can expect to encounter a host of barriers to their career options and to find few role models. Here, the term *minority* refers to people from racial and ethnic groups whose participation in mathematics is significantly less than their prevalence in the American population. According to the 1990 Census and National Science Foundation (NSF) data (NSF 1994), the underrepresented groups, which total 21.9% of the U.S. population, are African Americans, who form 12.1%; Hispanics, who are 9%; and American Indians/Alaskan Natives, who constitute 0.8%.

For the academic year 1990–91, there were 14,661 bachelor's and 3,615 master's degrees awarded in the mathematical sciences. Minorities however earned just 8.5% of these degrees at the baccalaureate and 5.1% at the master's level (Commission on Professionals in Science and Technology (CPST) 1994). Altogether, there were only 887 underrepresented minorities out of 19,992 mathematical sciences graduate students in 1991 (CPST 1994). Between July 1973 and June 1995, minorities earned 2.68% of 11,381 mathematical sciences doctorates awarded to U.S. citizens according to data compiled from the October 1974–December 1995 issues of the *Notices of the American Mathematical Society* (Fulton 1995). In 1990, minorities made up 14.4% of precollege teachers, while 28.6% of school-age children were members of a minority group. That same year, at postsecondary institutions African Americans were 5.0% of mathematical science faculty, Hispanics were 2.7%, and American Indians were 0.5%. In 1991, of academically employed African American Ph.D.s in the mathematical sciences, 11.6% were nontenure track versus 4.9% of whites. The 1992 pool of nonacademic employed mathematical and computer scientists was 6.8% African American and 3.1% Hispanic (CPST 1994).

In many ways, the barriers to careers in mathematics, as with barriers to education in general, are connected to economic and social conditions. Moreover, the elitist American attitude that "success in mathematics depends more on innate ability than hard work" (National Research Council 1991) has had very damaging effects on minorities in particular. In some locations, the practice of "tracking students" has resulted in disproportionate numbers of minority students being excluded from high-level mathematics classes, beginning in their early educational years. Predictions have been made that by the year 2000, well over one-third of public school children will be minority but only 5% of their mathematics teachers will come from minority populations (National Research Council 1989). Students who see few, if any, minority teachers in the mathematics classroom are accepting of the small numbers of minority students in advanced mathematics courses. Low expectations and the scarcity of positive role models create additional barriers to achievement.

As early as the fourth grade, minority students score poorly in mathematics, as measured by the National Assessment of Educational Progress (NAEP) (Mullis et al. 1993). NAEP scores show the same pattern for grades 8 and 12. The mathematics results for minorities planning to attend college are similar on the Scholastic Aptitude Test. Finally, on the quantitative and analytic reasoning portions of the Graduate Record Examination for prospective graduate students, means for minorities are less than those for the group as a whole.

The precise nature of the barriers in high school has been identified in a study (Oakes et al. 1990) for the RAND Corporation. Schools with a higher proportion of minority students tended to offer fewer sections of college preparatory or advanced courses. When schools were racially mixed, minority students were tracked into low-level classes more often than their white peers. Minorities generally had less access to "gatekeeping" courses such as algebra, geometry, and calculus.

This last finding is particularly significant because other research (Pelavin and Kane 1990) for the College Board showed there was no better predictor of college enrollment than high school geometry. For all ethnic groups, the rate of college attendance is about 80% when they have completed a year of geometry; the rates increase when they also plan to go to college. NAEP data similarly show that students who have taken more mathematics courses do better in grades 8 and 12 (Mullis et al. 1993). The Oakes study also found that low-income and minority students have less access to the best-qualified mathematics teachers. Indeed, low-ability students in more advantaged schools (high socioeconomic status, predominantly white, suburban) had a higher percentage of qualified teachers than high-ability classes in less-advantaged schools (low socioeconomic status, minority, inner city).

What are the barriers in college? Ethnic differences are minor when it comes to majoring in mathematics, science, or engineering. About 25% of college graduates from the high school class of 1980 majored in these fields whether they were white, African American, or Hispanic. Minority underrepresentation stems from the low college attendance and graduation rates experienced by minorities (National Center for Education Statistics 1990). For example, African Americans were 9.9%, Hispanics only 6.9% and American Indians just 0.9% of undergraduates in 1993 ("Affirmative Action" 1995).

In the fall of 1991, the historically Black colleges and universities (HBCUs) enrolled only 18% of African American undergraduates, an increase during the decade, yet they produced 48% of the African Americans earning bachelor's degrees in mathematics in the fall of 1989 (Hill, *Blacks in . . . Education,* 1992). Even more striking is the statistic that 31% of Ph.D.s earned by African Americans in mathematics during the years 1985–1990 went to HBCU graduates (Hill, *Undergraduate Origins of . . . Recipients,* 1992). This shows that minority students can achieve when opportunity is provided.

Efforts to overcome the barriers experienced by minorities have come from various segments of the community—academia, scientific societies, government, business, and industry. They range from informing children of the mathematical contributions of their ancestors through a multicultural curriculum to programs that financially support minorities who seek doctorates in mathematics. From those who believe that nothing less than a national comprehensive movement starting at early educational levels will significantly affect such entrenched problems, there have been calls to recruit more minority teachers at the precollege level. Both governmental and private organizations have explored recruitment strategies (Mathematical Sciences Education Board 1990; Quality Education for Minorities Project 1990). Institutions of higher education have organized enrichment programs for minority precollege students to supplement school mathematics curricula, and these programs seem particularly to impact their desire to attend college. The Mathematical Association of America (MAA) has compiled a handbook of such programs directed by mathematicians (MAA 1996).

Government agencies began to devote small percentages of their budgets to enhance the technical education of minorities. For example, the NSF broadened its agenda from primarily research to include a large education directorate. Many of its programs address increased access of minorities to mathematics and science. The National Aeronautics and Space Administration instituted a range of programs, many of which directly supported the postsecondary education of students designated as potential mathematicians and scientists. Similar programs from the Department of Defense, the National Institutes of Health, and others provide scholarships and support an enriched curriculum and preprofessional experiences for students in minority groups in an attempt to expand their career options.

Scientific societies slowly created programs to address the career development of students and the few minorities in their ranks. Early efforts came from the engineering schools and industries, leading to the National Action Council for Minorities in Engineering (1974), and from societies interested in the scientific development of specific populations, such as the American Indian Science and Engineering Society (1977). In the early 1970s, the National Association of Mathematicians emerged as an organization primarily concerned with the mathematics development of minority college students and young professionals. In the 1980s and 1990s, the older professional mathematical societies began to establish structures to address the lack of minority participation. A current initiative by the Conference Board of the Mathematical Sciences, which consists of fourteen national mathematics organizations, seeks to involve the entire mathematics community in this enterprise by establishing a Task Force on Minority Participation and Achievement and conducting, in May 1995, a related workshop for organization leaders.

Recently, a major ethnographic study (Seymour and Hewitt 1994) has examined reasons why science, mathematics, and engineering students change to nonquantitative majors. This study of students who have left or still remain in quantitative majors has underscored the need to change the academic culture in the areas of teaching and advising. Effectively staffed programs for minorities established within science and engineering departments help many to survive, but the best retention program for minorities seems to be improved learning experiences for all.

*See also* Minorities, Educational Problems

## SELECTED REFERENCES

"Affirmative Action: A Report." *Chronicle of Higher Education* 41(Apr. 28, 1995):A11–33.

Commission on Professionals in Science and Technology (CPST). *Professional Women and Minorities: A Total Human Resource Data Compendium.* Washington, DC: The Commission, 1994.

Fulton, John D. "AMS-IMS-MAA Annual Survey." *Notices of the American Mathematical Society* 42(12)(Dec. 1995):1504–1519.

Hill, Susan T. *Blacks in Undergraduate Science and Engineering Education.* Washington, DC: National Science Foundation, 1992.

———. *Undergraduate Origins of Recent Science and Engineering Doctorate Recipients.* Washington, DC: National Science Foundation, 1992.

Mathematical Association of America. "Strengthening Underrepresented Minority Mathematics Achievement (SUMMA)." *1996 Directory of Mathematics-based Intervention Projects.* Washington, DC: The Association, 1996.

Mathematical Sciences Education Board, National Academy of Sciences (MSEB). *Making Mathematics Work for Minorities: Framework for a National Action Plan, 1990–2000 and A Compendium of Program Proceedings, Professional Papers, and Action Plans.* Washington, DC: The Board, 1990.

Mullis, Ina V. S., John A. Dossey, Eugene H. Owen, and Gary W. Phillips. *NAEP 1992: Mathematics Report Card for the Nation and the States.* Washington, DC: National Center for Education Statistics, 1993.

National Center for Education Statistics, Office of Educational Research and Improvement, U.S. Department of Education (NCES). *Who Majors in Science? College Graduates in Science, Engineering, or Mathematics from the High School Class of 1980.* Washington, DC: Government Printing Office, 1990.

National Research Council. *Everybody Counts: A Report to the Nation on the Future of Mathematics Education.* Washington, DC: National Academy Press, 1989.

———. *Moving Beyond Myths: Revitalizing Undergraduate Mathematics.* Washington, DC: National Academy Press, 1991.

National Science Foundation. *Women, Minorities, and Persons with Disabilities in Science and Engineering: 1994.* Arlington, VA: The Foundation, 1994.

Oakes, Jeanne, et al. *Multiplying Inequalities: The Effects of Race, Social Class, and Tracking on Opportunities to Learn Mathematics and Science.* Santa Monica, CA: RAND, 1990.

Pelavin, Sol H., and Michael Kane. *Changing the Odds: Factors Increasing Access to College.* New York: College Entrance Examination Board, 1990.

Quality Education for Minorities Project. *Education That Works: An Action Plan for the Education of Minorities.* Cambridge, MA: Massachusetts Institute of Technology, 1990.

Seymour, Elaine, and Nancy M. Hewitt. *Talking about Leaving: Factors Contributing to High Attrition Rates among Science, Mathematics, and Engineering Undergraduate Majors.* Boulder, CO: Ethnography and Assessment Research, Bureau of Sociological Research, University of Colorado, 1994.

SYLVIA T. BOZEMAN
WILLIAM A. HAWKINS, JR.

# MINORITIES, EDUCATIONAL PROBLEMS

There is no shortage of studies and reports on the underrepresentation of minority students on the college-preparatory path, particularly in mathematics courses at the middle school and high school levels. Many reasons exist for that situation. It is clear, however, that lower academic expectations for minority students become a self-fulfilling prophecy. This entry focuses on barriers to minority student enrollment and success in college-preparatory mathematics courses at the precollege level, which have a direct bearing on the degree to which minority students are adequately represented in the science, engineering, and mathematics pipeline at the undergraduate and graduate levels. It includes a discussion of the obstacles and myths that impede minority student success in mathematics and concludes with a focus on EQUITY 2000, the College Board's national districtwide systemic education reform initiative. Because mathematics typically is the critical filter for the entire curriculum—driving enrollment patterns for students across all subject areas—EQUITY 2000 uses mathematics as a lever to reform the curriculum and change policies and programs *within entire school districts,* with the goal of closing the achievement gap between minority and nonminority, disadvantaged, and advantaged students.

## TRACKING

In spite of some progress made in the last thirty years, poor and minority and disadvantaged students in the United States are still frequently held to lower expectations than their majority and advantaged peers. There is a history of negative assumptions about their ability to learn, and regrettably there is still debate over their capacity to learn. The result is that far too many schools and school districts retain their outmoded and ineffective policies of tracking, whereby poor and minority students are herded into watered-down classes that limit their career and college options and often lead them down the path to a life in the permanent underclass.

A tracking policy makes assumptions and judgments about student abilities based on socioeconomic, racial, or ethnic status, resulting in many students being grouped into academic paths that lock them out of a chance to attend college, pursue a variety of worthwhile careers, or to become meaningful contributors to society. As a number of studies have shown, tracking almost always means that those students who need the most support to raise student performance get the least. The consequence is a two-tiered system characterized by the following:

- poor and minority students underrepresented in college-preparatory classes such as algebra and geometry and overrepresented in dead-end classes such as consumer math and general math;
- guidance counselors who presume that poor and minority students have neither the capabilities nor the inclinations to attend college, who guide students toward a course enrollment path based on low expectations, and who fail to provide adequate information to those students about college prerequisites and financial aid options; and
- teachers who fail to provide the necessary encouragement to poor and minority students because their expectations of those students' success are low or nonexistent.

The United States Office of Civil Rights has estimated that more than half of U.S. elementary schools have at least one "racially identifiable" classroom in their highest or lowest grade, meaning a classroom in which the proportion of students of a given race is substantially different from that of the school as a whole (Slavin and Braddock 1993). One consequence of those conditions is that minority and disadvantaged students enroll in college-preparatory courses such as algebra, geometry, and chemistry at rates far lower than their White and advantaged counterparts (Pelavin and Kane 1990).

## MYTHS ABOUT THE EDUCATION OF MINORITY STUDENTS

In the late 1980s, the Action Council on Minority Education of the Carnegie-supported Quality Education for Minorities (QEM) Project undertook a comprehensive examination of the educational status, needs, and possibilities of minority children, youth, and adults. The QEM study identified thousands of individual efforts on the parts of teachers, counselors, administrators, and parents to meet the needs of traditionally underserved students—efforts

that can serve as the starting point for the comprehensive, systemic reforms that are necessary to meet the students' needs. The 1990 QEM report set forth a plan for restructuring our educational system. The keys to this effort are twofold: (1) the elimination of tracking; and (2) the establishment of high expectations for all students.

The QEM report also identified ten myths about the education of minorities that serve as obstacles to their receiving a quality education. These myths are:

- learning is due to innate abilities and minorities are simply less capable of educational excellence than Whites;
- the situation is hopeless. The problems minority youths face, including poverty, teenage pregnancy, unemployment, drug abuse, and high dropout rates, are so overwhelming that society is incapable of providing effective responses;
- quality education for all is a luxury, since not all jobs presently require creativity and problem-solving skills;
- education is an expense and not an investment;
- equity and excellence in education are in conflict;
- all we need are marginal changes;
- minorities don't care about education;
- bilingual education delays the learning of English and hinders academic achievement;
- the problem will go away;
- educational success or failure is within the complete control of each individual, and in America anyone can make it.

Any effort to increase minority enrollment and achievement in college-preparatory courses must directly address these myths, beginning with an emphasis on raising standards and expectations for all students among teachers, counselors, principals, superintendents, and college faculty. Until these myths are successfuly challenged, classroom and school-based activities will not be restructured to provide equitable access to a quality education.

## TEACHERS AND COUNSELORS

A teamwork strategy between teachers and counselors that focuses on the academic development of every child can be a powerful tool for reaching students. Teachers and counselors are sometimes seen, however, as obstacles rather than facilitators in bringing about academic excellence for all students. This is reflected in the perceptions of teachers and counselors by other educators. In the spring of 1991,

EQUITY 2000 surveyed a cross-section of educators from K–12 and higher education to identify those perceptions. Among the highlights of that survey were the following.

## Perceptions about Mathematics
### Teachers

- math teachers fail half their students in algebra and then blame the students for the failing grade;
- math teachers build reasons to fail in projects that ask change of them;
- they act as if it is a "sign of weakness" to teach math to *all* students. They only want their "good" students to succeed and do not want to be bothered with the low achievers;
- math teachers don't want to teach students from diverse backgrounds as demographics change;
- of all the content areas, math teachers are least flexible;
- some math teachers discourage students from taking more math than the minimum math required for graduation.

## Perceptions about Counselors

- counselors don't encourage minority students to go to college;
- counselors don't encourage students—especially minority students—to take higher-level mathematics courses;
- counselors are "touchy-feely" people and suffer from math anxiety;
- in the interest of protecting kids, counselors shelter them from academic challenges;
- counselors tend to help students find the "easy way out" rather than putting them into the programs that are best for them;
- counselors do not see themselves empowered to make changes.

To be effective in multicultural classrooms, in addition to mastering the course content as a part of understanding how to deliver it effectively, teachers need to be sensitive to and knowledgeable about the diverse experiences and cultures of their students. They also need to create an atmosphere in their classrooms where the different cultures are valued, where instruction builds upon the strengths of students, and where expectations for *all* students are high. The Mid-Atlantic Equity Center of the American University in Washington, DC, has identified three variables that are crucial to a classroom climate where excellence is expected and equity is the norm; they include teacher expectations, the learning atmosphere, and social organization (Beane 1985).

## Teacher Expectations

A teacher's belief that every student can learn is communicated when that teacher

- expects the same academic goals for *all* students;
- holds all students accountable for the same standards of performance;
- uses questioning techniques that promote curiosity and inquiry and addresses different levels and types of questions to students at all achievement levels;
- gives all students equal time to contribute and reaches out to less-active students;
- provides all students with immediate and objective feedback and praises correct thinking and responses from all of them;
- responds to requests for extra help without embarrassing the students;
- encourages critical and analytical thinking by *all* students.

## Learning Atmosphere

To create an atmosphere where learning by all is valued, a teacher should

- encourage class discussion between females and males in a panel format;
- convey respect for different learning styles as the classroom norm;
- establish a relaxed but task-oriented environment;
- foster the ability of each student to feel control of his/her own future by listening to the ideas of students. Group work and cooperative learning are especially effective;
- identify and effectively use alternative strategies for reteaching the unit objectives to those students who have not mastered the skill;
- emphasize the practical application of the subject matter and invite role models from diverse backgrounds to attend and lead classes.

## Social Organization

Teachers, principals, and counselors can work jointly to establish an integrated learning environment by

- monitoring seating assignments to ensure that traditionally higher achievers are not always close to the teachers;
- using cooperative learning activities that promote achievement through mixed-ability grouping;
- encouraging study groups made up of students with different learning styles.

Expectations are powerful self-fulfilling prophecies. Teachers who care about their students, believe they can learn, build high expectations for them, are culturally literate about their backgrounds, and can relate to the worldviews and lives of their students, can meet the learning needs of those students, and raise their expectations of success. Classroom by classroom, until the entire school system is committed to reform aimed at *all* students meeting high standards, an emphasis should be placed on creating a constructive academic environment that places a high value on learning, nurtures high standards of respect and hard work, encourages creativity, and rewards academic excellence. Those who hold the key to creating excellence and equity are those who already know the extra effort required to bring about change in education—teachers, counselors, and administrators whose individual acts of courage and resolve can create a broad base of support for educational reform in schools.

## EQUITY 2000

Perhaps no program better reflects the type of districtwide reforms necessary to open the doors to college and entering the high-tech workplace for minority students than EQUITY 2000, the systemic educational reform initiative of the College Board. Established in 1990 and now operating at six sites in fourteen school districts serving nearly five hundred thousand students, EQUITY 2000 is a K–16 education reform program that uses mathematics as a lever to trigger districtwide changes in policy, curriculum, and instructional practices for the purpose of closing the gap in college-going and achievement rates between minority and nonminority, advantaged and disadvantaged students. EQUITY 2000 focuses on increasing academic achievement and building aspirations to go to college among *all* students, especially minority and disadvantaged students traditionally tracked out of the college-preparatory pipeline and placed in low-level courses, particularly in middle school and high school. One outcome of EQUITY 2000 is a substantial increase in the numbers of minority and disadvantaged students academically prepared for college. EQUITY 2000 is underway at six urban sites: Fort Worth, TX; Nashville, TN; Milwaukee, WI; Providence, RI; San Jose, CA (a consortium of nine districts); and Prince George's County, MD. Plans are underway now for an expansion of the program to districts across the country.

Recognizing that systemic education reform requires changes in policy, curriculum, instruction, guidance, and counseling, and community involvement, EQUITY 2000 touches upon all aspects of a district's operation through the six components of this model:

- policy changes to end tracking and raise standards, beginning with the requirement that *all* students complete algebra by the ninth grade and geometry by the tenth grade, and changes in the curriculum to reflect the standards promulgated by the National Council of Teachers of Mathematics (NCTM), as well as standards set in the other disciplines;
- ongoing professional development for teachers, counselors, and principals that focuses on improving instruction and counseling and on making the college option viable for *every* student;
- "safety net" programs, such as Saturday Academies and Summer Scholars Programs, that supplement and enrich regular coursework, leveling the playing field for students whose academic backgrounds and experiences may have shortchanged them in preparation for rigorous, college-preparatory work;
- empowering parents to become advocates for their children's education, so that parents of poor and minority students play a direct role in helping their children take the right courses and prepare for college;
- school-community partnerships that include links with colleges and universities, the business sector, and community-based organizations; and
- use of disaggregated student enrollment and achievement data to drive decisions and monitor the reform effort.

The impact of EQUITY 2000 at the sites has been significant, as suggested by the 1996 evaluation results:

- overall, enrollment trends in algebra indicate that each site is approaching its stated objective of 100% enrollment in algebra or higher by the ninth grade. Fort Worth, Milwaukee, and Providence have nearly reached that target. Four of the six sites show geometry enrollment rates in excess of 50% for all ethnic/racial groups, an impressive gain when placed in the context of the 30% enrollment rates that existed three to four years ago.

- passing rates in algebra average higher than 50%, with three sites above 60%. Many of the sites are now reporting enrollments in Algebra I or higher at rates exceeding 80% of all students in the district. Within that context, the passing rates across the sites—which on average are greater than 50%, and in some cases as high as 70%—are indeed impressive, and in many cases higher than the national average. Geometry passing rates are above 70% at all sites and exceed 83% at three sites. It is clear that thousands more students are taking and passing algebra and geometry at the sites than were doing so before EQUITY 2000.

The effects of EQUITY 2000 extend well beyond algebra and geometry, however, having transformed the mathematics curriculum across all grades to reflect higher standards for all students, and having triggered substantial reforms in science, English, and social studies curricula and instruction as well. At all of the sites, lower-level "watered-down" courses are being eliminated in other subject areas, and many more students are taking rigorous courses at all grade levels. The result is that thousands of students who might otherwise have been tracked out of the college pipeline now have the option of going to college.

EQUITY 2000 is guided by the belief that all students can achieve at high levels with high expectations, appropriate academic support, and a nurturing and motivating environment in the school. The power of EQUITY 2000 is its districtwide orientation—establishing as the norm in *every* school and *every* classroom that *all* students are expected to achieve at high levels. EQUITY 2000 increases the pool of disadvantaged and minority students who have the option to attend college and pursue studies in a wide range of fields, including those with a strong mathematics, science, and technology base.

*See also* Goals of Mathematics Instruction; Minorities, Career Problems; Multicultural Mathematics, Issues

### SELECTED REFERENCES

Beane, DeAnna Banks. *Mathematics and Science: Critical Filters for the Future of Minority Students.* Washington, DC: Mid-Atlantic Center for Race and Equity, American University, 1985.

Jones, Vinetta. "Preparing Teachers for Multicultural Classrooms: Can Research Help?" Conference Proceed-ings, Lida Lee Tall Third Annual Research Conference, Towson State University, Baltimore, MD, November 7, 1990.

——— "Views on the State of Public Schools." Paper presented at the Conference on the State of American Public Education, sponsored by Phi Delta Kappa, the Institute on Education Leadership and the Educational Excellence Network, February 4, 1993.

Pelavin, Sol, and Michael Kane. *Changing the Odds: Factors Increasing Access to the College.* New York: College Entrance Examination Board, 1990.

Quality Education for Minorities Project. *Education that Works: An Action Plan for the Education of Minorities.* Cambridge, MA: Massachusetts Institute of Technology, 1990.

Slavin, Robert E., and Jomills H. Braddock, III. "Ability Grouping: On the Wrong Track." A white paper presented at the EQUITY 2000 National Invitational Conference, Washington, DC, May 24, 1993.

VINETTA C. JONES

# MODE

The simplest kind of "average" or typical value. It is defined as the result or category that occurs with the highest frequency. The mode is especially suitable for categorical data, for which no other kind of average can be calculated. Even the youngest pupils can find the mode for data they have collected themselves. The following are all examples:

"The favorite ice-cream flavor in our class is strawberry."

"The mode for pets in our class is fish."

"The average wedding takes place on a Saturday."

"An alien visiting planet Earth would say that the average Earth inhabitant speaks Chinese."

The mode is not difficult to find. What is more difficult is to decide whether the mode in a particular instance provides any useful information about the data. The mode is sometimes called the *shopkeeper's average,* as the shopkeeper is most concerned about knowing which size garment is bought by most customers rather than knowing the mean (which need not be an actual size at all). For grouped numerical data, the modal class should be given, rather than any attempting to extract a single value, particularly if the raw data have been lost. Where data are modeled by a continuous probability density function, the mode appears as a maximum; that is, if $f(x)$ is the probability density function, the mode is that value of $x$ for which $f'(x) = 0$ and $f''(x)$ is negative.

*See also* Mean, Arithmetic; Measures of Central Tendency; Median

SELECTED REFERENCES

Barr, G. V. "Some Student Ideas on the Median and Mode." In *The Best of Teaching Statistics* (pp. 79–82). Peter Holmes, ed. Sheffield, England: University of Sheffield, Teaching Statistics Trust, 1988.

Corwin, Rebecca B., and Susan N. Friel. *The Shape of the Data.* Palo Alto, CA: Seymour, 1989.

Graham, Alan. *Investigating Statistics: A Beginners Guide.* London, England: Hodder and Stoughton, 1990.

MARY ROUNCEFIELD

# MODELING

Formulating a mathematical description of a process or situation. Mathematics is studied because it is a rich and interesting discipline, because it provides a set of ideas and tools that are effective in solving problems arising in other fields, and because it provides concepts useful in theoretical studies in other fields. When used in problem solving, mathematics may be applied to specific problems already posed in mathematical form, or it may be used to formulate such problems. When used in theory construction, mathematics provides abstract structures that may aid in understanding situations arising in other fields. Theory construction and problem formulation involve a process known as *mathematical model building*. Given a situation in a field other than mathematics or in everyday life, mathematical model building is the activity that begins with the situation and formulates a precise mathematical problem whose solution, or analysis in the case of theory construction, enables us to gain insight and understanding about the original situation.

Mathematical modeling usually begins with a situation in the real world, sometimes in the relatively controlled conditions of a laboratory and sometimes in the much less completely understood environment of meadows and forests, offices and factories, and everyday life. For example, a psychologist observes certain types of behavior in rats running in a maze, a wildlife ecologist studies the declining numbers of endangered sea turtles, or an economist seeks to understand the consequences of a specific tariff policy. The goal is to understand the observations and to predict future behavior. It is usual for the work to be based on detailed data, experience, and the recognition of similarities between the current situation and other situations that are better understood. This close study of the system is really the first step in model building, and much of the work must be done by someone who is familiar with the origin of the problem and the basic biology, economics, psychology, or whatever else is involved.

The next step, an important one that frequently determines the usefulness of the study, is to select those concepts to be considered as basic and to define them carefully. This step typically involves making idealizations and approximations whose purpose is to eliminate unnecessary information and to simplify that which is retained as much as possible. For instance, in a study of nesting sea turtles, an ecologist may decide that the color of a turtle is unimportant, but the age of the turtle and the extent of human use of the beach may be significant. This step of identification, idealization, and approximation is referred to as constructing a *real model*. This terminology is intended to convey that the context is still that of real things (animals, apparatus, etc.) but that the situation may no longer incorporate all features of the original setting. Returning to the sea turtles, the ecologist may construct a real model including turtles and beaches and the assumption that each year a turtle of a certain age always lays a specific number of eggs.

The third step (after forming a real model) is usually much less well defined and frequently involves a high degree of creativity. Keeping the real model in mind, the researcher attempts to identify the operative processes at work and to use a mathematical structure, mathematical notation and expressions, to describe these processes. The result is a *mathematical model* in which the real quantities and processes are replaced by mathematical symbols and relations (sets, functions, equations, etc.) and by mathematical operations (differentiation, matrix multiplication, etc.). Usually, much of the value of the study hinges on this step because an inappropriate identification between the real world and a mathematical structure is unlikely to lead to useful results. It should be emphasized that there may be several mathematical models for the same real situation. In such circumstances, it may be the case that one model accounts especially well for certain observations whereas another model accounts for others. There may not be a "best" model; the one to be used will depend on the precise questions to be studied.

After the problem has been transformed into symbolic terms, the resulting mathematical system is studied using mathematical ideas and techniques. The results of the mathematical study are theorems, from a mathematical point of view, and predictions, from the empirical point of view. The motivation for the mathematical study is not to produce new mathematics, that is, new abstract ideas or new theorems, although this may happen, but rather to produce new information about the situation being studied. In fact, it is likely that such information can be obtained by using well-known mathematical concepts and techniques. An important contribution of the study may well be recognizing a relationship between known mathematical results and the situation being studied.

The final step in the model-building process is the comparison of the results predicted using the mathematical model with the real world. The most desirable situation is that the phenomena actually observed are accounted for in the conclusions of the mathematical study and that other predictions based on the mathematics are subsequently verified by experiment. In fact, in many situations, an elaborate experiment is designed to determine whether the model gives predictions consistent with observations. In many cases, the agreement between predictions and observations is less than desirable, at least on the first attempt. Typically, among the conclusions of the mathematical work are some that seem to agree and some that appear to disagree with the outcomes of experiments. In such cases, every step of the process has to be reexamined. Has there been a significant omission in the step from the real world to the real model? Does the mathematical model reflect all the important aspects of the real model, and does it avoid introducing phenomena not observed in the real world? Is the mathematical work free from error?

It often happens that the model-building process proceeds through several iterations, each a refinement of the preceding, until finally an acceptable model is found. Pictorially, we can represent this process as in Figure 1. The solid lines in the figure indicate the process of building, developing, and testing a mathematical model as just outlined. The dashed line indicates an abbreviated version of this process that is often used in practice. The shortened version is particularly common in the social and life sciences, where mathematization of the concepts may be difficult. In either case, the steps in this process may be complex and there may be complicated interactions between them. However, for the purpose of studying the model-building process, this simplification is quite useful.

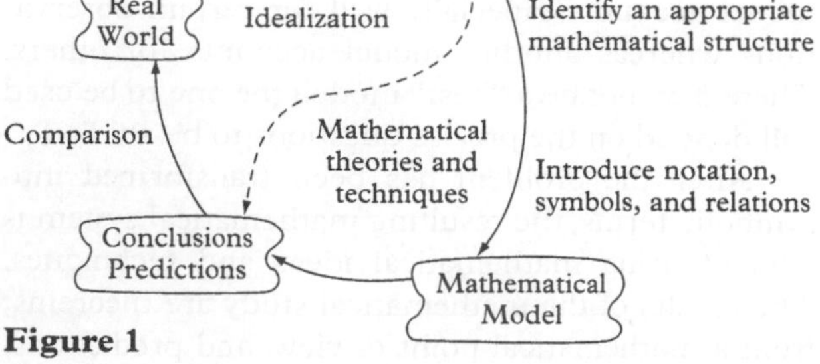

The use of mathematical models is now well established, and useful models have been developed for a wide variety of problems and situations. There are instances where models for important problems have been developed and refined for years or even centuries, and such instances may provide especially good illustrations of the model-building process. A study of planetary motion provides one of the best examples of the evolution of a mathematical model. Indeed, scientists from the Greeks to Einstein developed increasingly insightful models, and the resulting predictions have become incredibly accurate. The pictures obtained by the *Voyager* spacecraft are dramatic evidence of the validity of the model.

Examples of recent models that are important to society are those known as *linear programming models*. The problems leading to such models occur very frequently in the management sciences and fairly often in the social and life sciences. The model-building or problem-formulation phase is an essential first step in using linear programming methods, and we illustrate the ideas with a simple version of an example that has great historical importance, a version of one of the first problems solved using linear programming. The problem is one of selecting foods that satisfy certain dietary requirements in a way that minimizes the cost of the food.

The situation we consider is that of two foods and two dietary requirements; the techniques are essentially the same in more complicated situations. Suppose that we are to use broccoli and milk—hardly an appetizing diet, but a nutritious one—to meet dietary needs for calcium and iron and that we are to select amounts of these two foods that meet our needs at the least possible cost. The following data are available to us:

The diet must contain at least 0.8 gm calcium and 10 mg iron.

Each serving of broccoli contains 0.13 gm calcium and 1.3 mg iron, and each serving of milk contains 0.28 gm calcium and 0.2 mg of iron.

Broccoli costs $0.80 per serving, and milk costs $0.50 per serving.

**Figure 1**

We use $x$ to denote the amount of broccoli and $y$ to denote the amount of milk to be consumed, $x$ and $y$ each measured in units of servings. If there are 0.13 gm calcium in one serving of broccoli, then there are $0.13x$ gm in $x$ servings, and if there are 1.3 mg iron in one serving of broccoli, then there are $1.3x$ mg in $x$ servings. Likewise, there are $0.28y$ gm calcium in $y$ servings of milk, and $0.2y$ mg iron in $y$ servings of milk. Next, if we consume both $x$ servings of broccoli and $y$ servings of milk, then we have $0.13x + 0.28y$ gm calcium. The requirement that the diet provide at least 0.8 gm is, in our notation, $0.13x + 0.28y \geq 0.8$. A completely analogous argument leads to the inequality $1.3x + 0.2y \geq 10$, which represents the requirement that the diet contain at least 10 mg iron.

Looking at this discussion in the context of model building, notice that we have made several assumptions:

- The first assumption is that we actually know the calcium content of broccoli and milk. In the real world, there is likely to be variation in the vitamin content of foods, depending on how they are grown, processed, and stored. In some cases this variation can be quite large.
- Second, we are assuming that if we get 0.13 gm calcium from one serving of broccoli, then we get $0.13x$ gm from $x$ servings. This is a legitimate assumption for some values of $x$ but not for other values of $x$. As $x$ becomes large, it is impossible for a human to extract all nutrients available in the food.
- Next, we are assuming that if there is a certain amount of calcium available in the broccoli and another amount in the milk, then the sum of those two amounts is available in the combination of the two foods. This would be the case if the two foods were consumed independently; however, in some cases the presence of one food may affect the ability of the body to extract nutrients from other foods.

Next, turning to cost, we see that since each serving of broccoli costs \$0.80, $x$ servings should cost \$$0.80x$, and since each serving of milk costs \$0.50, $y$ servings of milk should cost \$$0.50y$. Therefore, the cost of a diet of $x$ servings of broccoli and $y$ servings of milk is \$$0.80x$ + \$$0.50y$. To find a lowest-cost diet, we should make $0.80x + 0.50y$ as small as possible. Again, we have made some assumptions:

- The statement that $x$ servings of broccoli cost \$$0.80x$ has the implicit assumption that there are no bulk discounts. Although the statement may

be true for the amounts of broccoli that an individual normally buys, there should be some savings if the purchaser is buying broccoli by the boxcar load!
- The statement that the cost of the diet is the sum of the costs of broccoli and of milk means that there are no discounts for purchasing both items. In fact, it is sometimes possible to negotiate a better price if you buy several items from one supplier.

This simple example illustrates the use of modeling in the formulation of a linear programming problem and the many assumptions that underlie such formulations. It is important to recognize that the reliability of solutions of linear programming problems as descriptions of the real world depends on the legitimacy of the assumptions, that is, on the validity of the model.

*See also* Linear Programming

SELECTED REFERENCES

Aris, Rutherford. *Mathematical Modelling Techniques.* London, England: Pitman, 1978.

Beltrami, Edward J. *Mathematics for Dynamic Modeling.* Boston, MA: Academic, 1987.

Sandefur, James T. *Discrete Dynamical Modeling.* New York: Oxford University Press, 1993.

Schiffer, Max M., and Leon Bowden. *The Role of Mathematics in Science.* Washington, DC: Mathematical Association of America, 1984.

DANIEL P. MAKI
MAYNARD THOMPSON

# MODERN MATHEMATICS

Best described by a comparison with classical mathematics, which is a mathematics of specified, individual objects that can be said to have an identity, or at least can be given an identity inside a theory. Thus, for example, in algebra over the real numbers, in traditional analytic geometry, or in classical analysis over the real or complex numbers, there is no abstractness as there is in modern mathematics. But many mathematicians of the classical period—for example, Christian Huygens (1629–1695), Leonhard Euler (1707–1785), Joseph Lagrange (1736–1813), Pierre Laplace (1749–1827), Bernhard Riemann (1826–1866), and Henri Poincaré (1854–1912)—already dealt with modern ideas, and classical mathematics has not disappeared. Classical subjects are still studied, but in a new form and with modern insights. Consider, for example, the famous statement of Pierre de Fermat (1601–1665) that says the equation $x^n + y^n = z^n$, where $x, y, z$ are positive integers, has no solutions for any integer $n$, $n > 2$. Only recently was progress made on proving this statement, using modern methods in a rather unexpected way.

Modern mathematics began to develop in the last decades of the nineteenth century on the basis of classical theories. This was not a process of mere causality—free creativity was important and led to a fundamental change in methods, contents, and forms. Out of this, essentially new directions developed. Two of the many famous mathematicians who contributed to this transformation should be mentioned: Georg Cantor (1845–1918), who was the initiator of set theory, which was of fundamental importance for coming developments; and David Hilbert (1862–1943), who introduced axiomatic methods, another fundamental step. These two concepts changed mathematics, and they can be traced in most modern developments in the field. In modern mathematics, attention is displaced from the properties of individual objects toward the study of collections, totalities, sets of various kinds, for example, point sets, families of curves, function spaces, and abstract sets defined by means of axiomatic methods. These sets may be finite or infinite.

The concept of structures defined on sets is a further development. In nearly any contemporary paper or book on mathematics, structural concepts find a place. A set $G$ is said to have the structure of a group (or be called a group) when there is defined on $G$ an internal operation (composition) such that with any two elements $a$, $b$ of $G$, there corresponds a uniquely determined element $c$ of $G$ satisfying some further conditions. It is called the *product* of $a$ and $b$ and may be symbolized as $c = a \star b$. When in a suitable way two compositions are defined—the additive and the multiplicative operation—the set is said to have the structure of a ring. A more specific composition leads to the concept of a field. Such operations may be defined in the form of a natural setting but also as symbols in axiomatic form.

Structures are a subject of study in algebra as well as in analysis and geometry. They have led to new concepts of "space," generalizing the classical Euclidean space in an essential way. This led to the development of the discipline of topology: the geometry of continuous maps. Extensive theories are developed on the properties of structures; for instance, the theory of groups and rings in an abstract sense, the concept of an algebra over a field. The concept of "algebra" is not limited to algebra in the classical sense over the real numbers $R$; algebra is no longer a unique concept. Vector spaces over a field, concrete or abstract, are introduced as additive structures with scalar multiplication. Maps between structures are frequent operations. They may concern maps of an internal kind as well as maps that are external. When internal or external structures are invariant under certain mappings, they are denoted as isomorphisms or homomorphisms. There is a broad field of applications for these concepts. Analysis over fields different from $R$ is an important aspect of modern mathematics. There is a very broad study of linear operators, which is the theory of maps $T$ satisfying the relation $T(x + y) = Tx + Ty$. A multiplicative example is the functional relation $\log xy = \log x + \log y$.

The study of structures in a general sense is to some extent characteristic of modern mathematics. These developments have had an impact on the form and content of educational programs. For example, vectors and elementary set theory are nowadays part of basic mathematical education.

*See also* Abstract Algebra; Cantor; Hilbert; Sets; Topology

## SELECTED REFERENCES

Dieudonné, Jean. *Panorama des Mathématiques Pures. Le choix bourbachique.* Paris, France: Gauthiers-Villars, 1977.

Monna, Antonie F. *Methods, Concepts and Ideas in Mathematics: Aspects of an Evolution.* CWI Tract. No. 23. Amsterdam, Netherlands: Mathematical Centre Amsterdam, 1986, 1989.

———. *The Way of Mathematics and Mathematicians; From Reality towards Fiction.* CWI Tract. No. 87. Amsterdam, Netherlands: Mathematical Centre Amsterdam, 1992.

Struik, Dirk J. *The Concise History of Mathematics.* 4th ed. New York: Dover, 1987.

Wilder, Raymond L. *Mathematics as a Cultural System.* Oxford England: Pergamon, 1981.

Young, Laurence. *Mathematicians and Their Times.* Amsterdam, Netherlands: North-Holland, 1981.

ANTONIE F. MONNA

# MONEY

An exchange unit for goods and services. Shells, beads, salt, animal hides, gold, and silver are forms of exchange units used in bartering and sales. In 600 B.C., Lydia (western Turkey) made coins as bean-shaped lumps of electrum, a natural mixture of silver and gold, and in A.D. 600, the Chinese developed paper money. In the United States, the Massachusetts Bay Colony was the first colony to make coins and to produce paper money. Two silver coins produced by the Massachusetts Bay Colony were the pine-tree shilling and the oak-tree shilling. Interesting facts about money are the numerical features of a Federal Reserve note, exchange rates of foreign currency, different features of foreign currency, the minting of coins, and the printing of paper money.

Play money coins and bills are useful in teaching a variety of mathematical skills. Students can count coins and bills to get a total value, or they can order a set of coins and bills by monetary value. They may explore foreign currency to determine differences in geometric shapes, variety of depicted drawings, coloration schemes, and use of watermarks. Some countries have bills of different sizes to help visually impaired people identify denominations; others depict historical events, important people, or nature on their currency. At an elementary level, organizing coins and bills visually and in chart form helps students learn the values. They may make tallies of each coin represented and do the required arithmetic to find the total value of these coins either with paper and pencil or by using a calculator. Playacting store by selling and buying items, by making change, and by trading reinforces number sense and estimation skills. Also, identification of various geometric shapes on currency improves visualization skills. At the middle school level, currency usage, as visual aids to learning, helps students with representation and comparison of decimal values and with an introduction to addition and subtraction of integers. Newspaper advertisements serve as an excellent resource for activities involving money. Students may devise a budget for monthly expenses using the classified section for prices of rents and supermarket advertisements for prices of food supplies.

Explorations in probability and statistics can evolve from finding all possible arrangements of a set of coins, such as a penny, a nickel, and a dime. Results can be displayed in graphs constructed by hand, on a graphing calculator, or with appropriate software. At high school and college levels, money is the basis of economics and calculus problems and instruction related to supply and demand, profit and loss, and maxima and minima. Completing income tax returns and balancing a checkbook are examples of lifelong applications of mathematics involving monetary computations.

*See also* Consumer Mathematics; Economics; Enrichment, Overview; Finance; Manipulatives, Computer Generated

## SELECTED REFERENCES

Axelson, Sharon L. "Supermarket Challenge." *Arithmetic Teacher* 40(Oct. 1992):84–88.

Bright, George W. "Teaching Mathematics with Technology: Simulations of Operating a Store." *Arithmetic Teacher* 36(Mar. 1989):52–53.

Chang, Lisa. "Multiple Methods of Teaching the Addition and Subtraction of Integers." *Arithmetic Teacher* 33(Dec. 1985):14–19.

Clason, Robert G. "How Our Decimal Money Began." *Arithmetic Teacher* 33(Jan. 1986):30–33.

Cook, Marcy. "Ideas: Spending Money." *Arithmetic Teacher* 36(May 1989):19–24.

Goodnow, Judy, and Shirley Hoogeboom. *Overhead Math: Manipulative Lessons on the Overhead Projector (Grades 3-6)*. Sunnyvale, CA: Creative, 1990.

Kitchen, Patricia, and Beth Piskora. "Cultivating Tomorrow's Crop of Clients—Kids." *American Banker* 156(Dec. 12, 1991):10.

Seuling, Barbara. *You Can't Count a Billion Bucks & Other Little Known Facts about Money*. New York: Doubleday, 1979.

Stevenson, Cathy L. "Teaching Money with Grids." *Arithmetic Teacher* 37(Apr. 1990):47–49.

REGINA BARON BRUNNER

# MONTESSORI, MARIA (1870–1952)

An Italian educator known for her work with primary-grade students. Montessori was trained as a physician (considered unbecoming for a woman at that time), graduating in 1896 at the top of her class, the first woman to obtain a medical degree from the University of Rome. In its approach to arithmetic, the Montessori method stresses cognitive development through the use of "didactic" (manipulative) materials. The materials include sticks for counting, cubes, rods, geometric solids, and sandpaper numerals pasted on cards. Montessori established schools in Italy and in various other countries. Her schools were popular in the United States during the early twentieth century, including the years 1913–1916, which she spent in this country. Following this period, the Montessori method lost popularity in the United States due, in part at least, to differences with educational philosophers of the time, such as William Kilpatrick and followers of John Dewey, who placed less emphasis on cognitive development and more on social interaction and fantasy play. Montessori schools became popular again in the United States beginning in the 1960s.

*See also* Cuisenaire Rods; Kindergarten; Manipulatives, Computer Generated; Preschool

## SELECTED REFERENCES

Hainstock, Elizabeth. *Essential Montessori, Updated Edition.* New York: New American Library, 1986.

Jones, Phillip S., ed. *A History of Mathematics Education in the United States and Canada.* 32nd Yearbook. Washington, DC: National Council of Teachers of Mathematics, 1970.

JAMES K. BIDWELL
ROBERT G. CLASON

## MOORE, ELIAKIM HASTINGS (1862–1932)

Eminent American mathematician at the University of Chicago, a distinguished analyst and algebraist, and a memorable teacher, many of whose students also became eminent. As retiring president of the American Mathematical Society in 1902, his presidential address focused on the teaching of school mathematics. This speech is considered the beginning of mathematics curriculum reform in the United States. After reviewing the state of pure and applied mathematics (particularly the work of John Perry of Britain), Moore offered "A Vision" for all elementary mathematics. He stressed a practical plan using applications and a laboratory approach—"a synthesis and development of the best pedagogic methods at present in use in mathematics and the physical sciences" (Moore 1967, 374). Moore later served on the National Committee on Mathematical Requirements of the Mathematical Association of America.

*See also* National Committee on Mathematical Requirements

## SELECTED REFERENCES

Bidwell, James K., and Robert G. Clason, eds. *Readings in the History of Mathematics Education.* Washington, DC: National Council of Teachers of Mathematics, 1970.

Jones, Phillip S., ed. *A History of Mathematics Education in the United States and Canada.* 32nd Yearbook. Washington, DC: National Council of Teachers of Mathematics, 1970.

Moore, Eliakim H. "On the Foundations of Mathematics." *Mathematics Teacher* 60(Apr. 1967):360–374.

JAMES K. BIDWELL
ROBERT G. CLASON

## MU ALPHA THETA

A national high school and junior college mathematics club cosponsored by the Mathematical Association of America and the National Council of Teachers of Mathematics. Mu Alpha Theta was founded in 1959 by Richard and Josephine Andree of the University of Oklahoma, Norman, Oklahoma, to recognize excellence in mathematics and encourage mathematics outside the classroom through the formation of mathematics clubs. Mu Alpha Theta promotes, encourages, and supports the formation of mathematics clubs at the local level. It provides a handbook for sponsors (teachers who serve as moderators), which includes a listing of "101 Ideas for Math Club Meetings." Each student member is provided four issues yearly of a publication called *The Mathematical Log,* which includes puzzle and contest problems; information about scholarships, mathematics contests and tournaments; a column about mathematics club activities; and other articles of mathematical interest. Student officers report on their delegate meetings (student sessions at the national convention), and all students are encouraged to submit articles and photographs.

Each August, Mu Alpha Theta hosts a national convention with three or four days of mathematics competition, including team tests, school tests, topic tests, ciphering competition, and chalk talk contests. The convention also has speakers, scholarship competition, and student delegate meetings. A high school is eligible to charter a Mu Alpha Theta chapter if it offers four years of college preparatory mathematics. The math club teacher or sponsor must have earned a college degree equivalent to a major in mathematics. Each school may induct students subject to its own standards for admission, but recommended standards include completion of at least four semesters of college preparatory mathematics (not to include such classes as general mathematics or consumer mathematics, etc.) and must have a grade point average of at least 3.0 on a 4-point scale.

*See also* Enrichment, Overview

## SELECTED REFERENCES

Andree, Richard, and Josephine Andree. *Cryptarithms.* 2 vols. Norman, OK: Mu Alpha Theta, 1978.

———. *Secret Ciphers.* 2 vols. Norman, OK: Mu Alpha Theta, 1979.

———. *Solving Ciphers.* 2 vols. Norman, OK: Mu Alpha Theta, 1979.

*Handbook for Sponsors.* Norman, OK: Mu Alpha Theta.

*The Mathematical Log.* Newsletter. Norman, OK: Mu Alpha Theta.

Reinthaler, Joan. *Mathematics and Music.* Norman, OK: Mu Alpha Theta, 1990.

Ruderman, Harry, ed. *Mathematical Buds.* Vols. 1–5. Norman, OK: Mu Alpha Theta, 1978– .

JOYCE M. BECKER

# MULTICULTURAL MATHEMATICS, ISSUES

Concern the nature and role of multicultural mathematics in and out of formal educational settings. Mathematics is a cultural product rooted in the ideological, sociological, emotional, and technological components of any culture (White 1959). Beliefs, customs, institutions, attitudes and behaviors, and technologies (or tools) impact the nature, goals, and construction of mathematical thought in any society or culture. Bishop (*Mathematical Enculturation,* 1988) has identified six fundamental activities in which all cultural groups engage: counting, locating, designing, measuring, playing, and explaining. Counting compares and orders discrete phenomena in the environment. It may consist of tallying, using concrete objects, or special words and names. Locating is the process of exploring, conceptualizing, and symbolizing one's environment. Designing is creating, modifying, conceptualizing, or symbolizing shapes or designs for use in that environment. Measuring uses objects and tools to quantify a variety of attributes of physical and nonphysical constructs. Playing is creating and participating in games or similar recreations governed by rules of engagement. Explaining is accounting for observed phenomena in the environment.

Different cultural groups may have very different ways of performing each of these activities. For example, many cultures rely on the use of written symbols for counting, but some cultures use body language or gestures to express numerals. Or some cultures employ complex systems of visual patterns that use simultaneous rather than successive methods of representing numerals and number sums (Gelman and Gallistel 1978; Davidson and Klich 1984; Saxe 1991). Each cultural group undoubtedly emphasizes or values one or more types of these fundamental activities over others. Activities may be performed exclusively of others or may be combined, such as measuring and designing. How any cultural group develops its mathematics depends, in great measure, on the conditions and constraints of its environment. That different cultures invent or create different mathematics is not in question; evidence that mathematics has a cultural history can be found in a large collection of research, including the works of Van Sertima (1986), Gerdes (1985), Closs (1986) and Zaslavsky (1973). The term *ethnomathematics* is frequently used to embody the perspective that mathematics is a cultural product and has a cultural history.

The term *ethnomathematics* is also used to describe systematic practices in mathematics and ways of knowing and thinking about mathematics that every cultural group possesses (D'Ambrosio 1985). Although preschool children possess some understanding of mathematics, their knowledge, ideas, experiences, and intuitions are largely ignored in their study of "school mathematics" (Nunes, Schliemann, and Carraher 1993; Stiff and Harvey 1988). For example, students' use of terms like *more* or *less* may not be consistent with that of school mathematics. In such cases, teachers and students may believe they are communicating when in fact they may each be using the terms very differently, leading to very different mathematical ideas about addition and subtraction. (See Cocking and Mestre 1988.) The failure to incorporate the mathematical understanding of young children into the teaching of mathematics raises important issues regarding the relevance of mathematics to their lives. That teachers do not take advantage of the mathematics that young children already know and bring to school is even more difficult to understand and explain when one compares the manner in which preschool students' existing language skills are exploited in the classroom to develop enhanced and more formal language skills. Similar contradictions in the way mathematics is taught in the schools compared to how many other subjects are taught raise important issues, many of which revolve around the ideas of *enculturation* and *acculturation.*

Enculturation is the process of helping young children adapt to the cultural patterns of their home and community. Acculturation is the process of helping young children (and adults) adapt to cultural patterns different from those of their home and community. Schools, among many other institutions, are usually given the responsibility of helping young people adapt to the predominant cultural patterns in the society. In a diverse society, many students are inducted into what is, for them, an alien culture by means of acculturation, often without their consent or the consent of their parents. It is not surprising, therefore, that families and communities become concerned about the effect that the acculturation process has on students' performance and success in schools (Bishop, "Mathematics Education," 1988). Issues related to the choices of learning environment, teacher, curriculum, and language are rooted in cultural perspectives and histories and may be difficult to resolve in schools serving diverse cultures.

There are many questions, for example, related to curriculum issues. What are the important, worthwhile concepts in mathematics that children should be taught? Which historical perspectives about the development of mathematics should be presented? What are the important concepts and ways of knowing that can be found in different cultures and should be combined into mainstream classroom instruction? How should a culturally diverse mathematics curriculum be structured? Questions related to language include the role of language in the development of mathematical skills and concepts. Should mathematics be a multilingual subject, in which meaning is created and expanded in the mother tongue of the child? How can the mathematical knowledge encoded in a student's language be made an integral part of the mathematics education of that child?

Ethnomathematical perspectives about formal mathematics education acknowledge that student values and experiences are an important component of any mathematics class. For example, "Western" perspectives value *control* over the environment, the use of *reason* and *logic* in decision making, and *openness* in the creation and examination of facts (Bishop, "Mathematics Education," 1988). Other cultures have different values. In many cultures, the individual and the group accept the forces of nature and strive to be one with it (rather than control it), reason and logic yield to authority and the wisdom of age and experience (often in opposition to "what seems reasonable"), and truths (rather than isolated facts) are closely guarded secrets that are shared only with the most worthy of students. Tate (1994) argues for *centricity* in mathematics classrooms as a way to remove barriers to the equitable mathematics education of diverse groups. The principles of centricity are at work when schools and teachers place diverse cultural groups at the center of the learning process and build upon their cultural and community experiences in making decisions about the learning environment, teacher selection, curricula development, and the role of language in instruction.

Students who choose to role play, make real-life assumptions, or work with a partner when attacking a mathematical problem may fail in schools because the way they view and approach many mathematics situations is often regarded by teachers and many fellow students as unimportant, misguided, or inappropriate (Stiff and Harvey 1988). Many more students fail in schools because school mathematics frequently lacks relevance to their lives and circumstances. Identifying genuine settings in which to pose mathematics that is relevant to the lives of diverse student populations establishes the teaching and learning of mathematics as an essential real-life activity. Teaching the "mathematics of phenomena" (Howson and Wilson 1986, 17) will help students better understand the complex social, political, economic, and technological environments that make up their world. It can also be used as a vehicle for engaging the cultural realities of students. For example, worthwhile mathematics for students besieged by an uncommon number of liquor stores in their community would focus upon "counting problems" in a real-life setting rather than the typical school setting in which counting is done simply for the sake of counting, and not used as a problem-solving tool! In this situation, students would be required to formulate mathematical incentives to get unwanted businesses to relocate (see Tate 1994). Students would have to develop mathematical arguments to strengthen their (political) position and counter the position of opponents.

Many issues related to creating mathematical power revolve around access to different mathematics curricula. Should access be differentiated based upon the characteristics of students or their future prospects? Should the standards of achievement be different for different groups of students? How are decisions about the tracking of students and the standards of attainment for students related to the cultural groups in which students belong? If the key to preparing students in mathematics is to begin school mathematics with students' ethnomathematical knowledge, issues related to teacher preparation and development arise. How should teachers be prepared to deal with the variety of student approaches to mathematical problems presented in school mathematics? Can school mathematics curricula achieve greater flexibility by reducing or modifying the scope and sequence of traditional mathematics instruction? Can teachers become effective curriculum builders for diverse student populations? And, if ethnomathematics is given a central role in the development of school mathematics, what are the best strategies for moving toward more formal mathematics teaching and learning? When should the transition begin?

*See also* Ethnomathematics; Minorities, Educational Problems; Women and Mathematics, Problems

## SELECTED REFERENCES

Bishop, Alan J. *Mathematical Enculturation: A Cultural Perspective on Mathematics Education.* Dordrecht, Netherlands: Kluwer, 1988.

―――. "Mathematics Education in Its Cultural Context." In *Mathematical Education and Culture* (pp. 179–191). Alan J. Bishop, ed. Dordrecht, Netherlands: Kluwer, 1988.

Closs, Michael P., ed. *Native American Mathematics.* Austin, TX: University of Texas Press, 1986.

Cocking, Rodney R., and Jose P. Mestre, eds. *Linguistic and Cultural Influences on Learning Mathematics.* Hillsdale, NJ: Erlbaum, 1988.

D'Ambrosio, Ubiratan. "Ethnomathematics and Its Place in the History and Pedagogy of Mathematics." *For the Learning of Mathematics* 5(Feb. 1985):44–48.

Davidson, Graham, and L. Z. Klich. "Ethnography, Cognitive Processes and Instructional Procedures." In *Cognitive Strategies and Educational Performance* (pp. 137–153). John R. Kirby, ed. Orlando, FL: Academic, 1984.

Gelman, Rochel, and C. R. Gallistel. *The Child's Understanding of Number.* Cambridge, MA: Harvard University Press, 1978.

Gerdes, Paulus. "Conditions and Strategies for Emancipatory Mathematics Education in Underdeveloped Countries." *For the Learning of Mathematics* 5(Feb. 1985):15–20.

Howson, A. Geoffrey, and Bryan Wilson, eds. *School Mathematics in the 1990s.* Cambridge, England: Cambridge University Press, 1986.

Nunes, Terezinha, Analucia D. Schliemann, and David W. Carraher. *Street Mathematics and School Mathematics.* Cambridge, England: Cambridge University Press, 1993.

Saxe, Geoffrey B. *Culture and Cognitive Development: Studies in Mathematical Understanding.* Hillsdale, NJ: Erlbaum, 1991.

Stiff, Lee V., and William B. Harvey. "On the Education of Black Children in Mathematics." *Journal of Black Studies* 19(Dec. 1988):190–203.

Tate, William F. "Race, Retrenchment, and the Reform of School Mathematics." *Phi Delta Kappan* 75(Feb. 1994):477–480, 482–484.

Van Sertima, Ivan. *Blacks in Science.* New Brunswick, NJ: Transaction, 1986.

White, Leslie A. *The Evolution of Culture.* New York: McGraw-Hill, 1959.

Zaslavsky, Claudia. *Africa Counts: Number and Pattern in African Culture.* Boston, MA: Prindle, Weber, and Schmidt, 1973. New York: Hill, 1979; revised 1999.

LEE V. STIFF

# MULTIPLICATION

Historically defined in terms of repeated additions of the same quantity: $3 \times 4 = 3 + 3 + 3 + 3$. In the seventh book of Euclid's *Elements*, multiplication is described as follows: "one number is said to multiply another when the number multiplied is so often added to itself as there are units in the number multiplying, and another number is produced." When the calculation of $3 \times 4$ is read as "3 multiplied by 4," the number 4 is the *multiplier* that determines how many times the *multiplicand*, 3, must be used in the addition process. Alternatively, $3 \times 4$ may be read as "3 times 4" or "3 lots of 4," in which case the number 3 is the *multiplier* and the number 4 is the *multiplicand*. In either case, the result, 12, is called the *product* of 3 and 4. Learning multiplication is aided by use of context and manipulative materials. For instance, children may determine the total number of crayons needed if three children are each to have four crayons. Manipulative materials include counters and cubes that children may use to model concrete situations.

To symbolize multiplication, a dot or cross ($\cdot$ or $\times$) is usually chosen. In his book *An Introduction to the History of Mathematics,* Howard Eves (1964) notes that "the cross symbol first appeared in an Appendix to the 1618 edition by Edward Wright of Napier's *Descriptio.*" The dot was adopted by the mathematician Gottfried Leibniz, who disliked the cross because it resembled too closely the letter x. However multiplication of whole numbers is defined and symbolized, extension of the definition to integers (zero, positive and negative whole numbers) and rational numbers (fractions) involves further considerations. What is meant, for example, by "$-4$ multiplied by $-3$" or by "3 multiplied by one quarter"?

## FRACTIONS

In practice, multiplication of fractions can present difficulties both in interpretation and in implementation. The meaning of $3 \times \frac{1}{4}$ and $\frac{1}{4} \times 3$ will be clear if the interpretation is that of "3 lots of one quarter" or "one quarter of 3" giving the product $\frac{3}{4}$. This use of the word *of* gives the key to interpreting calculations like "$\frac{1}{4} \times \frac{3}{4}$" as "one quarter of three quarters" (resulting in three sixteenths) but does not give a direct interpretation of decimal expressions like $0.2 \times 0.3$ since "0.2 of 0.3" has no application that will be familiar to children. By considering the fraction equivalents of 0.2 and 0.3, the product 0.06 may be calculated as follows: $\frac{2}{10} \times \frac{3}{10} = \frac{6}{100}$.

Clearly, the use of an electronic calculator to establish patterns of behavior when multiplying numbers gives sound experience of the patterns that are relevant. Operations like multiplying many different numbers by 5 or dividing different numbers by 10 will lead children to predict and generalize patterns that emerge in the solutions. The preceding result could be arrived at by considering $2 \times 3$, $2 \times 0.3$, $0.2 \times 3$, $0.2 \times 0.3$, $0.02 \times 0.3$, and so forth with a calculator and trying to explain the results. Through this sort of investigation, it may be seen that the position of the decimal point may be determined from the total number of decimal places in the original numbers.

## EXPRESSIONS FOR MULTIPLICATION

Calculating a multiplicative expression is often aided by recognition of the fact that multiplication is a *commutative* operation, so that $a \times b$ gives the same product as $b \times a$. The expression $4 \times 15$ may be calculated using 4 addends of 15 (i.e., $15 + 15 + 15 + 15$) or using 15 addends of 4 (i.e., $4 + 4 + 4 + \cdots + 4$). A variety of expressions may be used to interpret the multiplication symbol, including "15 multiplied by 4"; "4 times 15"; "15 times 4"; "4 lots of 15"; "15 lots of 4"; "4 fifteens"; "15 fours." It is important to be aware that each expression may be matched to a visual image involving repeated sets and that these images may differ but, because of the commutative law, the product will always be uniquely determined. As well as repeated sets for representing the product of two numbers, an arrangement of rows and columns to create a rectangular *array* will illustrate a Cartesian product and the fact that 3 rows of 4 elements will have the same total as 4 rows of 3 elements. (See Figure 1.)

## MULTIPLICATION TABLES

Multiplication facts are often learned as *multiplication tables,* in which the multiples of a given number are listed in order and may be learned by heart. The *7 times table,* for example, consists of the facts $1 \times 7 = 7, 2 \times 7 = 14, 3 \times 7 = 21, \ldots, 10 \times 7 = 70$. Quick access to any of these facts will help children in further calculations, but tables need to be taught in a meaningful way by encouraging use of relationships, for example, $6 \times 7$ will be twice $3 \times 7$. Understanding such relationships reduces the number of independent facts to be learned by heart.

Often neglected but very important are the 0 and 1 times tables:

$$1 \times 1 = 1 \qquad 0 \times 1 = 0$$
$$1 \times 2 = 2, \text{ etc.} \qquad 0 \times 2 = 0, \text{ etc.}$$

These give children particular pleasure to "learn" and contain fundamental results that will be used when multiplying multidigit numbers.

## MULTIDIGIT NUMBERS

Multiplication of multidigit numbers such as $132 \times 16$ will be easier if use is made of the *distributive* law, which enables this product to be "broken down." The distributive law states that $a \times (b + c) = (a \times b) + (a \times c)$. Using the distributive law, $132 \times 16$ becomes $132 \times 10$ added to $132 \times 6$, or

$$132 \times 16 = (132 \times 10) + (132 \times 6)$$

There are various algorithms that make use of this procedure in arranging a written calculation to record the subtotals and the final total (see Figure 2). The *gelosia* method is basically the same as the first written method, which is more popular today. It uses the cell arrangement as a convenient device for memorizing the different products that together make the final solution. This method has been known since the twelfth century, when it was used in India. It appears to have been carried from India to China and Arabia and subsequently to Italy, where it became known as the *gelosia* method because of its resemblance to the gratings used in front of win-

**Figure 1**  *Rectangular arrays to illustrate $3 \times 4$.*

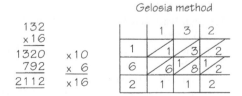

**Figure 2**  *Two pencil-and-paper methods for multiplication.*

dows. Children in school today are fascinated by such alternative written algorithms.

Both methods depend on the place value system used to represent numbers. Calculation of products involving larger numbers will depend on an understanding of multiplication by 10 and by powers of 10 (e.g., 100, 1000, 0.1 etc.). It is not always easy to provide practical materials like counters or cubes to illustrate the results of such multiplication, and a simple electronic calculator may be useful to help children investigate the patterns that emerge. From such investigations it may be tempting to conclude that "multiplying by 10 puts a zero on the end of any number." This is a very misleading conclusion, as can be seen when decimal fractions are involved; the result of multiplying 1.25 by 10 will not be to "put a zero on the end" but will "shift every number one place to the left" in the place value system of representation used for numbers. In all these calculations it is useful to calculate mentally an approximate solution that will act as a check that the calculated solution is about the right size. Note, for example, that $1.25 \times 10$ will be a bit bigger than $1 \times 10$, that is, 10. This will help avoid answers like 125 or 0.125.

Valuable shortcuts in multiplication are often possible using the *distributive* law. Take for example $132 \times 199$. This may be calculated as follows:

$$132 \times 199 = 132 \times (200 - 1)$$
$$= (132 \times 200) - (132 \times 1)$$

To calculate 199 times 132, first calculate 200 times 132 and then subtract 1 times 132. The distributive law becomes particularly important in algebra. Take, for example, the product $(x + 4)(x + 3)$. This becomes $x(x + 3) + 4(x + 3)$ when the distributive law is applied. It also enables multiplication to be defined for complex numbers:

$$[(a + ib)(c + id) = a(c + id) + ib(c + id)$$
$$= ac - bd + i(ad + bc)]$$

and for irrational numbers:

$(2 + \sqrt{3})(4 + \sqrt{5})$
$$= 2(4 + \sqrt{5}) + \sqrt{3}(4 + \sqrt{5})$$
$$= 8 + 2\sqrt{5} + 4\sqrt{3} + \sqrt{15}$$

Another useful rule that will help in calculations is the *associative* law, which enables products to be "broken down" in a different way. The associative law states

$$a \times (b \times c) = (a \times b) \times c$$

Use is made of the associative law when $132 \times 20$ is calculated as $132 \times 2 \times 10$:

$$132 \times 20 = 132 \times (2 \times 10) = (132 \times 2) \times 10$$

## JUSTIFYING THAT "THE PRODUCT OF 2 NEGATIVES IS POSITIVE"

When it comes to negative numbers, distinction must be made between the operation of subtraction and the idea of a negative number. At a more advanced level, one may use the ideas of the "additive inverses" and the "distributive law" to show $(-4) \times (-3) = 4 \times 3$, for example:

$$-4 + (+4) = 0 \quad \text{and} \quad (-4) \times (3) = -(4 \times 3)$$
$$\text{and so} \quad (-4) \times (-3) = -(-(4 \times 3))$$

Now $-(-(4 \times 3))$ is the additive inverse of $-(4 \times 3)$ and so is $(4 \times 3)$.

Children may find it helpful to construct charts showing the patterns of multiplication involving negative numbers like the following:

| | | |
|---|---|---|
| $3 \times 4 = 12$ | | $5 \times -3 = -15$ |
| $3 \times 3 = 9$ | | $4 \times -3 = -12$ |
| $3 \times 2 = 6$ | | $3 \times -3 = -9$ |
| $3 \times 1 = 3$ | | $2 \times -3 = -6$ |
| $3 \times 0 = 0$ | | $1 \times -3 = -3$ |
| $3 \times -1 = -3$ | | $0 \times -3 = 0$ |
| $3 \times -2 = -6$ | | $-1 \times -3 = 3$ |
| $3 \times -3 = -9$ | | $-2 \times -3 = 6$ |

Children who find it very difficult to remember facts may like to make and keep a personal copy of all the multiplication facts up to $10 \times 10$. Some pupils appreciate a table in four sections to show the results of multiplying both positive and negative numbers together.

## BEYOND NUMBERS

The language of multiplication is found in binary combinations of mathematical objects, such as matrices, vectors, and polynomials. Multiplication of matrices provides a particularly interesting operation because it does not satisfy the commutative law, that is, for two matrices $A$ and $B$, the products $A \times B$ and $B \times A$ are generally not equal—the order in which the elements are multiplied is crucial.

*See also* Arithmetic; Decimal System; Division; Estimation; Fractions

## SELECTED REFERENCES

Anghileri, Julia. *Children's Mathematical Thinking in the Primary Years.* London, England: Cassell, 1995.

Boyer, Carl B. *A History of Mathematics.* New York: Wiley, 1968.

Durkin, Kevin, and Beatrice Shire, eds. *Language in Mathematical Education: Research and Practice.* Philadelphia, PA: Open University Press, 1991.

Eves, Howard W. *An Introduction to the History of Mathematics.* Rev. ed. New York: Holt, Rinehart and Winston, 1964.

Post, Thomas R., ed. *Teaching Mathematics in Grades K–8: Research Based Methods.* 2nd ed. Boston, MA: Allyn and Bacon, 1992.

JULIA ANGHILERI

# MUSIC

The connections between mathematics and music are many and varied. Mathematics is used to count out musical rhythms, to study the sound waves that produce musical tones, to explain why we tune our instruments the way we do, and even to compose music. Quite often, people who are good at mathematics are musicians as well. The ancient Greeks had the belief, known as "the music of the spheres," that the universe was constructed based on the mathematical principles of harmony. The study of music in ancient times involved the study of the mathematical proportions that define musical intervals.

## COUNTING A RHYTHM

Music theory uses the *measure* and the *time signature* to set up the rules of rhythm for a particular piece of music. A piece is divided into measures that represent equal amounts of time. The time signature appears at the very beginning of the piece and is a symbol that resembles a fraction. The top number (like the "numerator") tells how many beats of equal length occur in a measure. The bottom number (like the "denominator") tells what kind of note gets one beat. The time signature does not have the properties of a fraction generally; that is, "reducing" a time signature changes the structure of the music.

The $\frac{4}{4}$ ("four four") time signature is the basis for naming the different kinds of notes. In $\frac{4}{4}$ time, a note that lasts for the whole measure is called a *whole note*. A note sustained for half of a measure is called a *half note,* and so on. Each successive power of one-half provides a new note to work with—quarter notes, eighth notes, sixteenth notes, thirty-second

notes, and so on. The pattern works the same way for spaces, or rests, in the rhythm, that is, whole rests, half rests, quarter rests, and so on. Adding a dot after any note increases its duration by one-half, creating, for example, a *dotted half note* or a *dotted quarter note.* This method also applies to rests.

In a time signature of $\frac{4}{4}$, the top number, 4, indicates that there are four beats of equal length per measure. Since the bottom number is 4, the *quarter note* gets one beat (if it were 2, the half note would get the beat, and so on). So a time signature of $\frac{4}{4}$ tells us that every measure must contain, altogether, the equivalent of four quarter notes. Using the fraction analogy, for this time signature the sum of the fractions represented by the notes must always be 1, because $\frac{4}{4} = 1$. Another common time signature is $\frac{3}{4}$. The quarter note still gets the beat (because there is a 4 on the bottom), but now there are only three beats per measure (because there is a 3 on top). That means each measure must contain the equivalent of three quarter notes; that is, the sum of the fractions represented by the notes must equal three-fourths. In $\frac{6}{8}$ time, each measure must contain the equivalent of six eighth notes, and the sum of the fractions must equal six-eighths.

The names for the notes are derived from $\frac{4}{4}$ time, so only in $\frac{4}{4}$ time does a quarter note actually take up one-fourth of the measure. A quarter note in $\frac{3}{4}$ time takes up one-third of a measure but is still called a quarter note for consistency in notation. Teachers should prepare for the possibility of confusion around this issue.

## SOUND WAVES

All sound travels in waves. On an oscilloscope, a sound wave is translated into a graph that is a type of sine curve. What makes one sound wave different from another and some sounds more musical than others? A musical *tone* lasts long enough and is steady enough to have pitch, loudness, and quality (timbre). These are related to the frequency, amplitude, and general shape of the tone's corresponding sine curve.

The *pitch* of a tone relates to the *frequency* (the number of vibrations that pass a given point per second) and represents how high or low a note sounds. Frequency is measured in *Hertz* (Hz), where 1 Hz = 1 vibration/second. It is the reciprocal of the period of the curve. The higher the frequency (the more vibrations per second), the higher the pitch; the lower the frequency (the fewer vibrations per second), the lower the pitch. The *loudness* or softness (dynamics) of a tone relates to the *amplitude* (the maximum displacement from the equilibrium). The larger the amplitude, the louder the tone; the smaller the amplitude, the softer the tone. Loudness is measured in decibels on a logarithmic scale. The *timbre* or quality of a tone distinguishes it from other tones with the same pitch and loudness. A piano, guitar, and French horn, for example, have distinctive sounds, and the corresponding sine curves have different general shapes; that is, they are more or less jagged and bumpy.

## MUSICAL PROPORTIONS

Pythagoras, a renowned Greek mathematician and philosopher, is reputed to have traveled to Mesopotamia during the sixth century B.C. and brought back many important mathematical ideas. Some of his findings may have inspired his experiments with the monochord. A monochord is a very simple instrument consisting of a string stretched taut over a moveable bridge. When the string is plucked on one side of the bridge, it vibrates and produces a tone. The pitch of the tone changes as the bridge is moved. Therefore, the pitch must be related to the length of the portion of the string that is vibrating. Shorten the string to exactly half its original length, and the new note will be an *octave* higher than the original. In a major scale, "do re mi fa sol la ti do," the first and last notes are an octave apart. The ratio of the new length (higher pitch) to the original length (lower pitch) is $\frac{1}{2}:1$, or $1:2$. In an octave interval, the ratio of frequencies of higher-pitched note to lower-pitched note is $2:1$. If the original tone (whole string) has frequency 440 Hz, the tone that sounds when half the string is plucked has frequency 880 Hz ($880:440 = 2:1$). Therefore, the frequency ratio and the string length ratio for an octave are reciprocals of one another. This is true for any interval.

In his experiments, Pythagoras found two other intervals he considered to be consonant (i.e., "sounding together"), the *fifth* and the *fourth*. The octave, fifth, and fourth have these names because they correspond to the eighth, fifth, and fourth notes of what is known as the Pythagorean diatonic scale. The fifth is the interval that results from shortening the monochord string to two-thirds of its original length. The string length ratio (new : old) is $\frac{2}{3}$ or $2:3$. The frequency ratio, then, should be $3:2$. If the original tone has frequency 440 Hz, the new tone has frequency 660 Hz, since $660:440 = 3:2$.

The final consonant interval, the fourth, is produced when the string is shortened to three-fourths of its original length. The string length ratio is $3:4$, and the frequency ratio is $4:3$; hence, the new tone must have frequency 586.67 Hz, since $586.67:440 = 4:3$.

Pythagoras' experiments with the monochord led to a method for tuning instruments with intervals in integer ratio, known as *Pythagorean tuning*. However, this tuning is rarely used today. Its intervals are unevenly spaced, making it sound different in every key; that is, the distances or intervals between the notes of the major scale are slightly different in every key. A method called *equal temperament*, which divides the octave into twelve evenly-spaced half-steps, is now employed in most parts of the Western world. Instead of using integer ratios, this method of tuning uses the irrational number $2^{1/12}$ as the frequency ratio between successive half-steps (take the twelfth root of the octave ratio $2:1$).

## COMPOSITION

Some of the techniques used in musical composition are related to plane geometry. Musical transformations can be compared to those *geometric transformations* called *isometries*, which relocate a rigid geometric figure in the plane while carefully preserving its size and shape. A geometric translation is like a musical transposition, a reflection across the $y$-axis is like a musical retrogression, a reflection across the $x$-axis is like a musical inversion, and so on. Some composers also use computers to create music. They derive rules from various compositional styles and apply them to randomly generated notes, rhythm, dynamics, and so on. Mathematicians have even used computers to create music based on the properties of fractals (mathematical objects that are self-similar on different scales), having as the goal music that is random enough to be interesting, yet not so random that the ear cannot discern a pattern.

*See also* Enrichment, Overview; Fractals; Transformations

## SELECTED REFERENCES

Ardley, Neil. *Mathematics, An Illustrated Encyclopedia.* New York: Facts on File, 1986.

Gardner, Martin. *Fractal Music, Hypercards, and More: Mathematical Recreations from Scientific American Magazine.* New York: Freeman, 1992.

Garland, Trudi H., and Charity V. Kahn. *Math and Music: Harmonious Connections.* Palo Alto, CA: Seymour, 1995.

Hutchins, Carleen Maley. *The Physics of Music: Readings from Scientific American.* San Francisco, CA: Freeman, 1978.

Levarie, Siegmund, and Ernst Levy. *Tone, A Study in Musical Acoustics.* Westport, CT: Greenwood, 1968.

McLeish, Kenneth, and Valerie McLeish. *The Oxford First Companion to Music.* London, England: Oxford University Press, 1982.

Reinthaler, Joan. *Mathematics and Music.* Norman, OK: Mu Alpha Theta, 1990.

Struik, Dirk, J. *A Concise History of Mathematics.* 4th ed. New York: Dover, 1987.

CHARITY VAUGHAN KAHN

# N

## NAPIER, JOHN (1550–1617)

A Scottish mathematician and inventor of logarithms born into a family of influential nobles and statesmen. Napier lived most of his life on the imposing family estate of Merchiston Castle, near Edinburgh, Scotland. He held no professional post, but kept busy with scientific and mathematical pursuits, developing a reputation as an intense, energetic amateur.

After twenty years of study and experimentation, Napier introduced the use of logarithms—an incredible breakthrough in simplifying computation. Logarithms made numerical calculations faster and more accurate by reducing multiplication and division to addition and subtraction. In one of his books on logarithms, Napier promoted the use of the decimal point as a way of simplifying the large numbers often required in mathematical tables. This soon became standard throughout Great Britain.

Napier developed another method for simplifying calculation known as *Napier's Rods* (see Figure 1). These moveable rectangular strips, often made of bones or ivory, were marked with multiplication tables. When arranged correctly, the rods made computations relatively easy.

In addition to his mathematical "hobbies," Napier invested time in agricultural experimentation, military science, and religious and political causes.

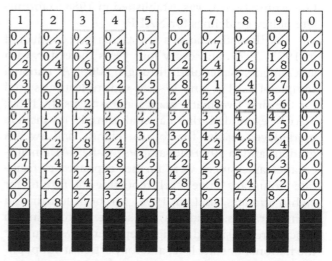

**Figure 1** *Napier's rods. [Louetta and Wilbert Reimer. "Connecting Mathematics with Its History" in National Council of Teachers of Mathematics, 1995 Yearbook, p. 106]*

*See also* Logarithms

### SELECTED REFERENCES

Abbott, David. *The Biographical Dictionary of Scientists: Mathematicians.* New York: Bedrick, 1986.

Baumgart, John K., et al. *Historical Topics for the Mathematics Classroom.* 31st Yearbook. Reston, VA: National Council of Teachers of Mathematics, 1989.

Boyer, Carl B. *A History of Mathematics.* 2nd ed. Revised by Uta C. Merzbach. New York: Wiley, 1989.

Burton, David M. *The History of Mathematics.* Boston, MA: Allyn and Bacon, 1985.

Eves, Howard W. *An Introduction to the History of Mathematics.* 6th ed. Philadelphia, PA: Saunders College Publishing, 1990.

Turnbull, Herbert W. *The Great Mathematicians.* New York: New York University Press, 1961.

LUETTA REIMER
WILBERT REIMER

# NATIONAL ASSESSMENT OF EDUCATIONAL PROGRESS (NAEP)

A Congressionally mandated survey of the educational achievement of U.S. students in a variety of subject areas, including mathematics, and changes in that achievement over time. To provide a context for the achievement results, NAEP also collects demographic, curricular, and instructional information from students, teachers, and school administrators. An independent board, the National Assessment Governing Board, was established in 1988 to formulate policy guidelines for NAEP.

NAEP mathematics assessments were conducted in 1973, 1978, and every four years thereafter, with assessments scheduled through 2008. A national sample of 9-, 13-, and 17-year-old students was used for the first two mathematics assessments. In 1982, NAEP expanded the national sample to include grade-level results for fourth, eighth, and twelfth graders, and in 1990 the samples for the NAEP mathematics assessment were further expanded to facilitate voluntary state-level reporting of results for eighth graders, and in 1992, for both fourth and eighth graders. Students are selected according to procedures designed to yield representative results at the national or state level and for particular subpopulations of students (e.g., gender, race/ethnicity, size/type of community). No additional preparation is necessary to take the NAEP assessments at any grade level.

The objectives that guide the development of the NAEP grade-level tests are determined through a legislatively mandated consensus process and take the form of frameworks delineating important content and process areas. For the mathematics assessment, the content areas typically include number, measurement, geometry, data analysis/statistics/probability, and algebra; the process categories are conceptual understanding, procedural knowledge, and problem solving. These frameworks are updated prior to each new assessment to reflect the most current thinking in school mathematics. The questions themselves are presented either in multiple choice or open-ended format.

Typically, each grade-level test consists of between 150 and 200 questions organized into non-overlapping sets of questions, called *blocks,* with each block having a fifteen-minute time limit. The number of questions per block varies depending on the format of the questions. Each student in the sample takes only three blocks of questions; and results are analyzed for the sample as a whole through use of Focused-Balanced Incomplete Block (BIB) Spiraling, a system that provides broad coverage of the subject being assessed while minimizing the assessment time required per student (see Mullis et al. 1993 for additional information on Focused-BIB spiraling).

After each NAEP administration, the National Center for Education Statistics (NCES) publishes the results in a series of widely disseminated reports that present summary statistics on student performance results and information from background questions (e.g., Mullis et al. 1991, 1993; Reese et al. 1997), on special topics such as problem solving (e.g., Dossey, Mullis, and Jones 1993), how mathematics is taught in school (Dossey et al. 1994), and on trends in student performance over time (e.g., Mullis et al. 1994; Campbell, Voelkl, and Donahue 1997). With respect to interpretive analyses of NAEP results, the National Council of Teachers of Mathematics (NCTM) has produced interpretive reports based on results from four of the first five NAEP mathematics assessments (Carpenter et al. 1978; Carpenter et al. 1981; Lindquist 1989; Kenney and Silver 1997).

*See also* Assessment of Student Achievement, Overview; Mathematics Education, Statistical Indicators

## SELECTED REFERENCES

Campbell, Jay R., Kristin E. Voelkl, and Patricia L. Donahue. *NAEP 1996 Trends in Educational Progress.* Washington, DC: National Center for Education Statistics, 1997.

Carpenter, Thomas P., Terrence G. Coburn, Robert E. Reys, and James W. Wilson. *Results from the First Mathematics Assessment of the National Assessment of Educational Progress.* Reston, VA: National Council of Teachers of Mathematics, 1978.

Carpenter, Thomas P., Mary Kay Corbitt, Henry S. Kepner, Jr., Mary Montgomery Lindquist, and Robert E. Reys. *Results from the Second Mathematics Assessment of the National Assessment of Educational Progress.* Reston, VA: National Council of Teachers of Mathematics, 1981.

Dossey, John A., Ina V. S. Mullis, Steven Gorman, and Andrew S. Latham. *How School Mathematics Functions.* Washington, DC: National Center for Education Statistics, 1994.

Dossey, John A., Ina V. S. Mullis, and Chancey O. Jones. *Can Students Do Mathematical Problem Solving?: Results from Constructed-Response Questions in NAEP's 1992 Mathematics Assessment.* Washington, DC: National Center for Education Statistics, 1993.

Kenney, Patricia Ann, and Edward A. Silver, eds. *Results from the Sixth Mathematics Assessment of the National Assessment of Educational Progress.* Reston, VA: National Council of Teachers of Mathematics, 1997.

———. *Results from the Seventh Mathematics Assessment of the National Assessment of Educational Progress.* Reston, VA: National Council of Teachers of Mathematics, 2000.

Lindquist, Mary Montgomery, ed. *Results from the Fourth Mathematics Assessment of the National Assessment of Educational Progress.* Reston, VA: National Council of Teachers of Mathematics, 1989.

Mullis, Ina V. S., John A. Dossey, Jay R. Campbell, Claudia A. Gentile, Christine O'Sullivan, and Andrew S. Latham. *NAEP 1992 Trends in Academic Progress: Achievement of U.S. Students in Science, 1969 to 1992; Mathematics, 1973 to 1992; Reading, 1971 to 1992; Writing, 1984 to 1992.* Washington, DC: National Center for Education Statistics, 1994.

Mullis, Ina V. S., John A. Dossey, Eugene H. Owen, and Gary W. Phillips. *NAEP 1992: Mathematics Report Card for the Nation and the States.* Washington, DC: National Center for Education Statistics, 1993.

———. *The STATE of Mathematics Achievement: NAEP's 1990 Assessment of the Nation and the Trial Assessment of the States.* Washington, DC: National Center for Education Statistics, 1991.

Reese, Clyde M., Karen E. Miller, John Mazzeo, and John A. Dossey. *NAEP 1996 Mathematics Report Card for the Nation and the States.* Washington, DC: National Center for Education Statistics, 1997.

PATRICIA ANN KENNEY

# NATIONAL CENTER FOR RESEARCH IN MATHEMATICAL SCIENCES EDUCATION (NCRMSE)

A research base for the reform of school mathematics funded from 1987 to 1996 by the U.S. Department of Education. The changes needed in the teaching and learning of mathematics in the United States are a consequence of several factors: development of new technologies; changes in mathematics itself; new knowledge about teachers, learning, teaching, and schools as institutions; and renewed calls for equity in learning mathematics regardless of race, class, gender, or ethnicity. To accomplish its mission of enabling developers and policy makers to make informed decisions for schools, the NCRMSE created national networks of scholars who collaborated on a long-range research agenda.

The center was organized into a director's office, a national advisory panel, and seven working groups. The national advisory panel advised the Center on the management of its research programs and reviewed its work. The working groups were:

*Learning/Teaching of Whole Numbers.* This working group investigated how children come to understand whole-number arithmetic, and how teachers influence the development of this understanding. Its research provided information on the critical variables in classroom instruction designed to develop children's learning of whole-number arithmetic with understanding. Its work is summarized in Hiebert et al. 1997 and Lampert and Blunk 1998.

*Learning/Teaching of Quantities.* This group examined what is known about the growth of student competence with number systems and about the instructional processes linking the procedural and conceptual aspects of these systems. In the middle grades, students require a complex conceptual basis for the quantitative notions they encounter and increased information on the semantics of symbol systems. Its work is summarized in Sowder and Schappelle 1995 and Sowder et al. 1998.

*Learning/Teaching of Algebra.* This group built a base of understanding for a reformed algebra and quantitative reasoning curriculum using a unified approach that begins with the idea of function. This approach incorporates electronic tools (calculators, graphing calculators, and computers) that support the manipulation of symbols and the examination of quantitative and spacial relationships. Its work is summarized in Kaput 1999.

*Learning/Teaching of Geometry.* This group investigated students' informal geometric notions about concepts such as shape, wayfinding, and measurement. The group's efforts yielded new knowledge about instruction in geometry. Its work is summarized in Lehrer and Chazan 1998.

*Learning/Teaching of Statistics.* This group examined ways that descriptive and inferential statistics and probability can effectively be integrated into the precollege mathematics curriculum. Students were able to use statistics to understand real-world phenomena, to interpret data summaries and displays of information, and to be critical of claims and arguments based on data. Its work is summarized in Lajoie 1997.

*Models of Authentic Assessment.* This working group identified a variety of models of assessment practices that are aligned, or in agreement with, the reform goals described in the *Curriculum and Evaluation Standards for School Mathematics* (National Council of Teachers of Mathematics 1989). The work of this group is summarized in Romberg 1995.

*Implementation of Reform.* This group examined the kinds of experiences, resources, and support systems teachers need if they are to carry out the reforms. The group identified ways schools can be organized to promote teaching for the development of mathematical power in all students, and how national, state, and local policies can influence teachers' beliefs and practices. Its work is summarized in Fennema and Nelson 1997; Secada et al. 1995.

Much of the work of NCRMSE was continued in 1996 by the newly funded (1996–2001) National Center for Improving Student Learning and Achievement in Mathematics and Science (NCISLA). This later research was based on the accomplishments of three previous projects: NCRMSE; Cognitively Guided Instruction, funded by the National Science Foundation (Carpenter, Fennema, and Franke 1996); and the Center on Organization and Restructuring of Schools (Newman and Wehlage 1995).

*See also* Research in Mathematics Education, Current; Van Hiele Levels

## SELECTED REFERENCES

Carpenter, Thomas, Elizabeth Fennema, and Megan Franke. "Cognitively Guided Instruction: A Knowledge Base for Reform in Primary Mathematics Instruction." *Elementary School Journal* (9)(1)(1996):3–20.

Carpenter, Thomas, and Richard Lehrer. "Teaching and Learning Mathematics with Understanding." In *Mathematics Classrooms that Promote Understanding.* Elizabeth Fennema and Thomas A. Romberg, eds. Mahwah, NJ: Erlbaum, 1997.

Fennema, Elizabeth, and Barbara Scott Nelson, eds. *Mathematics Teachers in Transition.* Mahwah, NJ: Erlbaum, 1997.

Hiebert, James, Thomas P. Carpenter, Elizabeth Fennema, Karen Fuson, Piet Human, Hanlie Murray, Alwyn Olivier, and Diane Wearne. *Classrooms for Understanding Mathematics.* Portsmouth, NH: Heinemann, 1997.

Kaput, James J. "Teaching and Learning a New Algebra." In *Mathematics Classrooms that Promote Understanding.* Elizabeth Fennema and Thomas Romberg, eds. (pp. 133–155). Mahwah, NJ: Erlbaum, 1999.

Lajoie, Susanne, ed. *Reflections on Statistics: Learning, Teaching, and Assessment in Grades K–12.* Mahwah, NJ: Erlbaum, 1997.

Lampert, Magdalene. "Why Study Mathematical Talk and School Learning?" In *Talking Mathematics: Studies of Teaching and Learning in School.* Magdalene Lampert and Merrie Blunk, eds. (pp. 1–14). New York: Cambridge University Press, 1998.

Lampert, Magdalene, and Merrie Blunk, eds. *Talking Mathematics: Studies of Teaching and Learning in School.* New York: Cambridge University Press, 1998.

Lehrer, Richard, and Donald Chazan, eds. *Designing Learning Environments for Developing Understanding of Space and Geometry.* Mahwah, NJ: Erlbaum, 1998.

National Council of Teachers of Mathematics. *Curriculum and Evaluation Standards for School Mathematics.* Reston, VA: The Council, 1989.

———. *Principles and Standards for School Mathematics.* Reston, VA: The Council, 2000.

Newman, Fred M., and Gary G. Wehlage. *Successful School Restructuring: A Report to the Public and Educators.* Madison, WI: Wisconsin Center for Education Research, 1995.

Romberg, Thomas A., ed. *Reform in School Mathematics and Authentic Assessment.* New York: State University of New York (SUNY) Press, 1995.

Secada, Walter, Elizabeth Fennema, and Lisa Byrd Adajian. *New Directions in Equity for Mathematics Education.* New York: Cambridge University Press, 1995.

Sowder, Judith T., Randolph Philipp, Bill Armstrong, and Bonnie Schappelle. *Middle Grades Teachers' Mathematical Knowledge and Its Relationship to Instruction.* Albany, NY: SUNY Press, 1998.

Sowder, Judith T., and Bonnie Schappelle, eds. *Providing a Foundation for Teaching Mathematics in the Middle Grades.* Albany, NY: State University of New York (SUNY) Press, 1995.

THOMAS A. ROMBERG

# NATIONAL COMMITTEE ON MATHEMATICAL REQUIREMENTS

A committee of university and secondary school representatives organized in 1916 by the Mathematical Association of America to study "the improvement of mathematical education and to cover the field of secondary and collegiate mathematics" (Bidwell and Clason 1970, 384). In 1923, the committee published its report, *The Reorganization of Mathemat-*

*ics in Secondary Education,* which affected mathematics education into the 1940s. The committee approved of creating the junior high school (grades 7–9) and making mathematics required through grade 9. The recommended curriculum consisted of arithmetic, intuitive geometry, algebra, numerical trigonometry, and other topics. For grades 10–12, the committee suggested plane geometry (with proofs), algebra, solid geometry, trigonometry, elementary statistics, and intuitive calculus. A chapter in the report was devoted to the informal function concept.

The curriculum advocated by the committee was conservative, and instruction was to have three aims: practical, disciplinary, and cultural. The most important aim was mental discipline, since it offered "insight into and control over our environment." This aim assumed the concept of "transfer of training" and included quantitative thinking (measurement, proportion, dependence, etc.), ability to think logically and with generalization, habits of mind associated with mathematics such as "a love for precision," and "functional thinking" about relationships.

*See also* Commission on Postwar Plans; Committee of Fifteen on Elementary Education; Committee of Ten; Committee on the Undergraduate Program in Mathematics (CUPM); Joint Commission Report (1940); Progressive Education Association (PEA) Report

## SELECTED REFERENCES

Bidwell, James K., and Robert G. Clason, eds. *Readings in the History of Mathematics Education.* Washington, DC: National Council of Teachers of Mathematics, 1970.

Jones, Phillip S., ed. *A History of Mathematics Education in the United States and Canada.* 32nd Yearbook. Washington, DC: National Council of Teachers of Mathematics, 1970.

JAMES K. BIDWELL
ROBERT G. CLASON

# NATIONAL COUNCIL OF SUPERVISORS OF MATHEMATICS (NCSM)

An organization for leaders in mathematics education, pre-K through adult, in the United States and Canada. Its purpose is to support mathematics education leadership at the school, district, college or university, state or province, and national levels through a cadre of well-trained, broadly informed, and perceptive leaders of mathematics education. NCSM believes that these leaders must be empowered and held accountable for overseeing and facilitating the implementation of substantive reform of school mathematics for all students. The approximately 2,500 members who have joined NCSM constitute a worldwide force of school, district, and state or province mathematics coordinators, supervisors, department chairs, and resource personnel, as well as mathematics education faculty at colleges and universities and mathematics editorial personnel from publishing companies.

Founded in 1968 and organizers of annual meetings held in conjunction with the annual meeting of the National Council of Teachers of Mathematics, NCSM has established a broad agenda of activity to (1) promote the importance of designated mathematics leadership at all levels, (2) provide ongoing experiences, information, and support to enable mathematics educators to assume leadership roles, (3) collaborate with leaders in business and industry, publishing, government, professional organizations, and boards of education to improve leadership in mathematics education, and (4) establish and maintain a communications network among mathematics leaders. Many credit the organization's 1976 "Position Paper on Basic Mathematical Skills" with beginning the end of the "back-to-basics" movement and launching the focus on problem solving as the core of school mathematics. More recently, NCSM has published and distributed source books for leaders in mathematics education to disseminate more broadly needed information on critical issues.

*See also* American Mathematical Association of Two-Year Colleges (AMATYC); Mathematical Association of America (MAA); National Council of Teachers of Mathematics (NCTM)

## SELECTED REFERENCES

National Council of Supervisors of Mathematics. "Position Paper on Basic Skills." Golden, CO: The Council, 1976. Reprinted in *Mathematics Teacher* 71 (Feb. 1978):147–152.

———. *Supporting Leaders in Mathematics Education: A Source Book of Essential Information.* Golden, CO: The Council, 1994.

STEVEN LEINWAND

# NATIONAL COUNCIL OF TEACHERS OF MATHEMATICS (NCTM)

A professional organization focusing on school mathematics. The membership of 130,000 comprises mainly K–12 teachers, mathematics supervisors, and college mathematics education instructors. Much of its work, supported by a capable headquarters staff, is carried out by volunteers. These efforts are complemented by the over 200 affiliated groups throughout the United States and Canada (Lindquist 1993). Since its founding in 1920, the NCTM has a rich history of providing guidance and leadership to mathematics education through guideline documents, position statements, journals, and other publications as well as through conferences and conventions. Working with the other mathematical organizations, the council has assumed the role of speaking to the public, to policy makers, and to the entire world of education on matters regarding mathematics learning and teaching.

The council presented its recommendations for mathematics for the 1980s in the document *An Agenda for Action* (Hill 1980). This document was based on an extensive survey of opinions, preferences, and priorities of diverse educational populations as well as on the knowledgeable advice from professionals in the field. It considered the lessons learned in the 1960s and 1970s: that change was not merely a matter of proposing new programs and that pressures from outside the educational establishment could not be ignored. This document clearly established problem solving as the focus of school mathematics, broadened the definition of basics, and expected consistent use of technology. It called for effective teaching, expanded ways of evaluating student learning, additional mathematical study by all students, professionalism of all mathematics teachers, and public support. It provided the underpinnings for the next major project of the council, the *Standards* (NCTM 1989, 1991, 1995).

Crosswhite, Dossey, and Frye (1989) described the challenges of developing the *Standards*, the first attempt of any teacher organization to specify national, professional standards for school curricula in its discipline. This was during a time when there was little encouragement from outside funders or outside support for standards. Crosswhite recalls the bold step of the NCTM Board in providing the funding, the most expensive single project that the council had ever undertaken. Looking back from the vantage point of the present, it was a courageous decision that has had positive repercussions. The council has grown both financially and in stature through its investment. More important, the lack of outside funding made the *Standards* the product of the profession, not of politics.

The *Curriculum and Evaluation Standards for School Mathematics* was released in 1989 after a two-year period of writing, gathering reactions, building consensus, and publishing. This document sets forth five goals for students: to value mathematics, to become confident in one's ability to do mathematics, to become a mathematical problem solver, to communicate mathematically, and to reason mathematically. The document acknowledges, more strongly than any previous position of the council, the necessity and importance of a comprehensive mathematics education for each and every student. No longer was mathematics to be considered the subject that would sort students and limit their future opportunities. The curriculum standards set broad guidelines for what mathematics students should know and be able to do. However, the document is not a prescription; its purpose was to encourage professional dialogue and reflection. The evaluation standards pointed the way to evaluating what students know and are able to do.

The second part of developing standards focused on teaching, evaluating teaching, and professional development of teachers. The *Professional Standards for Teaching Mathematics* was released in 1991. The underlying assumptions, that teachers are the key to change and that continued support of teachers is needed to ensure change, set a positive tone for the high expectations of excellence embodied in the document. To help all students learn the mathematics called for in the curriculum standards required monumental changes in teaching, such as reaching all students, helping students make sense of mathematics rather than learning only to mimic procedures, and encouraging discussion and reflection. Yet these recommendations were based on what successful teachers had been doing or realized needed to be done; thus, no matter how challenging, the council knew that these changes were possible.

No deep change in curricula or in instruction will happen unless assessment encourages that change. Additionally, the council recognized the responsibility of the profession in terms of accountability. Thus, the third phase of the council's work focused on assessment. The *Assessment Standards for School Mathematics* (1995) sets forth six standards that focus on assessing important mathematics, enhancing learning, ensuring equity, keeping the process open, making valid inferences, and establishing coherence. Each of these standards is illustrated in the different purposes of assessment—monitoring student progress, making instructional decisions, and evaluating students' achievement and programs.

The development of the *Standards* was a major stage in the council's continuing goal of improving the mathematics learning of every student and of supporting teachers in this effort. An updated version, *Principles and Standards for School Mathematics,* was published in 2000. The *Standards* have provided a focus for the council and have been influential in the educational reform movement of other disciplines. The council received the top honor of the American Society of Association Executives in 1995 for setting standards of excellence. The standards project has heightened the awareness of the council regarding all the work that is necessary in creating and documenting change. NCTM will continue its quest to provide support, to encourage research, to monitor, to make necessary adjustments, and to meet the needs of students and teachers. The diversity of the membership and of students along with the ever-changing needs for mathematics in society, however, make the development of the *Standards* only the first, albeit powerful, step in this process.

*See also* Mathematical Association of America (MAA); National Council of Supervisors of Mathematics (NCSM); National Council of Teachers of Mathematics (NCTM), Historical Perspective

### SELECTED REFERENCES

Crosswhite, F. Joe, John A. Dossey, and Shirley M. Frye. "NCTM Standards for School Mathematics: Visions for Implementation." *Arithmetic Teacher* 37(Nov. 1989):55–60.

Hill, Shirley. "President's Address: 58th Annual Meeting." *Arithmetic Teacher* 28(Sept. 1980):49–54.

Lindquist, Mary M. "Tides of Change: Teachers at the Helm." *Arithmetic Teacher* 41(Sept. 1993):64–68.

National Council of Teachers of Mathematics. *Assessment Standards for School Mathematics.* Reston, VA: The Council, 1995.

———. *Curriculum and Evaluation Standards for School Mathematics.* Reston, VA: The Council, 1989.

———. *Principles and Standards for School Mathematics.* Reston, VA: The Council, 2000.

———*Professional Standards for Teaching Mathematics.* Reston, VA: The Council, 1991.

MARY M. LINDQUIST

# NATIONAL COUNCIL OF TEACHERS OF MATHEMATICS (NCTM), HISTORICAL PERSPECTIVE

A national organization formed in 1920 for the improvement of secondary mathematics instruction, a task then controlled by college mathematicians. In 1920, the executive committee took over the publica-

tion of *The Mathematics Teacher,* formerly the journal of the Association of Teachers of Mathematics in the Middle States and Maryland. Publication of the NCTM yearbooks began in 1926. The NCTM has widened its role to grades K–12 as well as teacher preparation. Although other organizations led reform movements before 1940, the NCTM became a national leader in mathematics education reform after World War II. In 1950, the NCTM affiliated with the National Education Association and moved its headquarters to Washington, D.C. A second journal, *The Arithmetic Teacher,* was begun in 1954 and a third, *Journal for Research in Mathematics Education,* in 1970. In 1994, *The Arithmetic Teacher* was split into *Teaching Children Mathematics* and *Mathematics Teaching in the Middle Grades.*

*See also* National Council of Teachers of Mathematics (NCTM)

### SELECTED REFERENCES

Bidwell, James K., and Robert G. Clason, eds. *Readings in the History of Mathematics Education.* Washington, DC: National Council of Teachers of Mathematics, 1970.

Jones, Phillip S., ed. *A History of Mathematics Education in the United States and Canada.* 32nd Yearbook. Washington, DC: National Council of Teachers of Mathematics, 1970.

JAMES K. BIDWELL
ROBERT G. CLASON

# NATIONAL SCIENCE FOUNDATION (NSF)

Established in 1950 (Public Law 81-507) to support mathematics and science education. Support for education is the primary responsibility of the Directorate for Education and Human Resources (EHR), and support for education at the K–16 levels is concentrated in the Directorate's Division of Elementary, Secondary, and Informal Education, the Division of Undergraduate Education, the Division of Educational System Reform, and the Division of Human Resource Development. Other directorates also provide some support for mathematics education. Funds to support programs operated by EHR are appropriated annually by Congress.

Programs within EHR promote student, teacher, and faculty development, as well as improved public science literacy through the support of projects at the national, regional, and local levels. The directorate's five major goals are designed to provide quality science, mathematics, and technology education for *all* students. These goals are designed to ensure that (1) high-quality education in science, mathematics, and technology is available to every child in the United States, enabling all who have the interest and talent to pursue scientific and technical careers at all levels; (2) the educational pipelines carrying students to careers in science, mathematics, engineering, and technology yield sufficient numbers of individuals who can meet the needs of the technical workforce; (3) high-quality educational opportunities are available to those who select careers in science or engineering disciplines; (4) interested nonspecialists have opportunities to broaden their scientific and technical knowledge; and (5) the larger public has opportunities to develop an understanding of scientific and technological developments and processes.

The Division of Elementary, Secondary, and Informal Education operates grant programs that support activities to develop and disseminate effective instructional materials for prekindergarten to grade 12 (through the Instructional Materials Development program); to strengthen the effectiveness of teachers who are already in classrooms (through the Teacher Enhancement program); to prepare students for the demands of the advanced technological workplace; to provide research experiences for high-potential youth in grades 7–12 (through the Young Scholars program); and to increase public literacy (through the Informal Science Education program). The programs that provide support for these activities are described in *Elementary, Secondary and Informal Education, Program Announcement and Guidelines,* and examples of funded projects may be found in *Summary of Awards, Instructional Materials Development, Fiscal Years 1991–1994* and *Summary of Awards, Teacher Enhancement Active Awards, Fiscal Year 1994.*

The Division of Undergraduate Education focuses on improving science, mathematics, engineering, and technology education in universities and four- and two-year colleges by supporting the reform of curricula and laboratories (through the Course and Curriculum program and the Instructional Laboratory Improvement program); enhancing faculty (through the Faculty Enhancement program); improving the undergraduate preparation of K–12 science, mathematics, and technology teachers (through the Collaboratives for Excellence in Teacher Preparation program); and addressing advanced technician training (through the Advanced Technology Education program). Further information is available in the *Undergraduate Education Program Announcment and Guidelines.*

The Division of Human Resource Development supports activities from elementary through graduate education that are designed to increase the representation in science, mathematics, engineering, and technology of minorities, women, and persons with physical disabilities. Additionally, the division supports career and minority institutional development activities. For further information see the *Guide to Programs.*

The Division of Educational System Reform has responsibility for broad-based initiatives to stimulate major reform in K–12 science, mathematics, and technology education in states and in urban and rural areas. Examples of activities supported by these programs may be found in the *Statewide Systemic Initiatives in Science, Mathematics, and Engineering, 1994–1995.*

*See also* Federal Government, Role; National Science Foundation (NSF), Historical Perspective

## SELECTED REFERENCES

National Science Foundation. *Elementary, Secondary, and Informal Education, Program Announcement and Guidelines.* NSF 95-150. Arlington, VA: The Foundation, 1996.

———. *Guide to Programs.* NSF 95-138. Arlington, VA: The Foundation, 1995.

———. *Statewide Systemic Initiatives in Science, Mathematics, and Engineering 1994–1995.* NSF 94-175. Arlington, VA: The Foundation.

———. *Summary of Awards, Instructional Materials Development, Fiscal Years 1991–1994.* NSF 95-123. Arlington, VA: The Foundation.

———. *Summary of Awards, Teacher Enhancement Active Awards, Fiscal Year 1994.* NSF 96-18. Arlington, VA: The Foundation.

———. *Undergraduate Education Program Announcement and Guidelines.* NSF 96-10. Arlington, VA: The Foundation, 1995.

Public Law 81-507 (64 Stat. 149) (42 U. S. C. Sec. 1862–1887).

JOHN S. BRADLEY

# NATIONAL SCIENCE FOUNDATION (NSF), HISTORICAL PERSPECTIVE

First funded by Congress in 1950 to promote research and education in the sciences. The national curriculum movement "new mathematics" is dated by some from that time. Early grants were for establishing summer institutes and developing college-level text materials, but in 1957 NSF funded the Physical Science Study Committee (PSSC) to write secondary school teaching materials. In 1953, NSF began its support of summer and academic year in-service institutes for graduate study. Many secondary school teachers earned master's and doctoral degrees with these grants. This became a very influential program in the 1960s. NSF continues to fund programs for students and teachers (K–12) and research in mathematics curriculum development. During 1958–1972, NSF funded the School Mathematics Study Group to develop a national curriculum for the "new math" movement.

*See also* National Science Foundation (NSF)

## SELECTED REFERENCES

Jones, Phillip S., ed. *A History of Mathematics Education in the United States and Canada.* 32nd Yearbook. Washington, DC: National Council of Teachers of Mathematics, 1970.

Krieghbaum, Hillier, and Hugh Rawson. *An Investment in Knowledge.* New York: New York University Press, 1969.

JAMES K. BIDWELL
ROBERT G. CLASON

# NEUTRAL GEOMETRY

The geometry deducible from Euclid's first four postulates; also called *absolute* geometry. It thus consists of all theorems that can be deduced from the assumptions that (1) two points determine a straight line, (2) a straight line can be extended to any length, (3) there exists a circle with any point as center and any length as radius, and (4) all right angles are equal. No assumption is made about the uniqueness of parallel lines.

Neutral geometry can be thought of as the collection of all statements that are true in both Euclidean and hyperbolic geometry, since both of these start by assuming Euclid's first four postulates but then make opposite assumptions about the uniqueness of parallel lines. Theorems of neutral geometry include:

1. The congruence theorems for triangles, such as the side-angle-side theorem, which states that two triangles are congruent if they agree in two sides and the subtended angle.
2. An exterior angle of a triangle is greater than either opposite interior angle.

3. Given any point not lying on a given line, there exist one or more straight lines passing through the point parallel to the given line.
4. The sum of the angles of a triangle is equal to or less than a straight angle.

The first three of these are Euclid's Propositions 4, 16, and 31, respectively. Euclid's first twenty-eight propositions are all true in neutral geometry.

Many well-known statements, such as the Pythagorean Theorem and the formula for the area of a circle, are not theorems of neutral geometry. In neutral geometry it can be proved that the square of the hypotenuse of a right triangle is equal to or greater than the sum of the squares of the other two sides.

The purpose of neutral geometry is primarily pedagogical: to help one understand the possibility of non-Euclidean geometry, it helps to consider carefully exactly which of Euclid's propositions require the parallel postulate in their proofs. One of the discoverers of non-Euclidean geometry, Janos Bolyai, introduced the concept of neutral geometry to prepare his readers for his revolutionary discovery, and geometry teachers today would be well advised to do the same. The books listed in the "Selected References" should prove helpful.

*See also* Non-Euclidean Geometry

## SELECTED REFERENCES

Davis, Donald M. *The Nature and Power of Mathematics.* Princeton, NJ: Princeton University Press, 1993.

Greenberg, Marvin J. *Euclidean and Non-Euclidean Geometries: Development and History.* 2nd ed. New York: Freeman, 1980.

Trudeau, Richard J. *The Non-Euclidean Revolution.* Boston, MA: Birkhäuser, 1987.

DONALD M. DAVIS

# NEWTON, ISAAC (1642–1727)

The inventor of the calculus and largely self-educated in mathematics. About 1664, when Newton was an undergraduate at Cambridge University, where mathematics was not then greatly cultivated, he came upon the *Geometry* of René Descartes and then the writings of the other early masters of analytic geometry. Two problems seized his attention, drawing tangents to curves (related to what we now call differentiation) and finding the areas under curves (integration). Earlier mathematicians had developed algorithms for the areas under simple curves. Newton succeeded in extending this work by expanding equations of more complex curves into infinite series that he could integrate term by term. By 1668, he had learned how to find the area under any algebraic curve no matter how complex. Meanwhile, he had observed an inverse relation between his general algorithm for drawing tangents and the algorithm for finding areas. In 1666, in a context which considered that a curve is the path of a point in motion and an area is swept out by a line in motion, he set down the fundamental theorem of the calculus, the inverse relation of differentiation and integration. A tract that he dated October 1666 expounded what he came to call the Fluxional Method (from the past participle of the Latin verb *fluere,* "to flow") and what we know as the calculus.

Newton did not publish his work, however, although a small number of British mathematicians came to know about it. Ten years after his discovery, the German mathematician and philosopher Gottfried Wilhelm Leibniz, working independently from much the same sources, developed a similar method, which he named *the differential calculus.* Leibniz did publish his work, and the world of mathematics learned the calculus from him. Newton allowed himself to be convinced that Leibniz had stolen the method from him, and in the second decade of the eighteenth century the two men carried on a rancorous dispute over priority, which cast no credit on either.

Newton was also a monumental figure in the history of physics. His *Mathematical Principles of Natural Philosophy* (1687), which utilized the thought patterns of the calculus set in the idiom of geometry, defined the science of dynamics and used it to deduce the law of universal gravitation. His *Opticks* (1704) established the heterogeneity of light, with which he explained the phenomena of colors. He also devoted extensive time to theology, embracing heterodox beliefs similar to unitarianism. In 1696, Newton left Cambridge, the scene of all his scientific discoveries, to become first Warden and then Master of the Royal Mint in London, where he died after filling the post for nearly thirty years.

*See also* Fermat; Leibniz

## SELECTED REFERENCES

Boyer, Carl B. *The Concepts of the Calculus. A Critical and Historical Discussion of the Derivative and the Integral.* New York: Columbia University Press, 1939.

Hall, A. Rupert. *Philosophers at War: The Quarrel Between Newton and Leibniz.* Cambridge, England: Cambridge University Press, 1980.

Westfall, Richard S. *Never at Rest. A Biography of Isaac Newton.* New York: Cambridge University Press, 1980.

Whiteside, Derek T. *The Mathematical Papers of Isaac Newton.* 8 vols. Cambridge, England: Cambridge University Press, 1967–80.

RICHARD S. WESTFALL

# NOETHER, EMMY (1882–1935)

Played a central role in the development of abstract algebra through her own work and that of members of her circle. The change from a computational basis for algebra to an axiomatic approach is reflected in Noether's own research. Her 1908 doctoral dissertation at the University of Erlangen, Germany, on the theory of invariants was heavily computational. Subsequently, her papers established a conceptual framework for a unified theory of noncommutative algebras, and her work on ideals led to the concept of Noetherian rings. She also contributed to the study of relativity and particle physics by what is now known as *Noether's Theorem* and to the algebraization of topology.

The University of Erlangen, where Noether's father was a professor of mathematics, changed its policies to permit women to matriculate after she spent her first term there as an auditor. However, although she assisted her father in various ways, including giving lectures, she never had a position at the university. In 1915, she was invited by Felix Klein and David Hilbert to the University at Göttingen with the expectation that she would receive a position as *Privatdozent,* allowing her to lecture without being paid a salary. However, while Noether gave lectures announced under Hilbert's name, it took four years to overcome the opposition to having a woman on the faculty. It took another four years for Noether to achieve a paid position, although only as an untenured associate professor. In spite of her growing influence, which attracted algebraists from all over the world to work with her, she never became a professor at Göttingen.

In 1933, Noether was one of the first Jews to be dismissed from German universities. She emigrated to the United States and took a position as visiting professor at Bryn Mawr College and lectured at the Institute for Advanced Studies at Princeton, which gave refuge to many of her former colleagues. During her second year at Bryn Mawr she died after surgery.

*See also* Abstract Algebra; Women and Mathematics, History

SELECTED REFERENCES

Brewer, James W., and Martha K. Smith, eds. *Emmy Noether: A Tribute to Her Life and Work.* New York: Marcel Dekker, 1981.

Dick, Auguste. *Emmy Noether.* Boston, MA: Birkhäuser, 1981.

Srinivasan, Bhama, and Judith Sally, eds. *Emmy Noether in Bryn Mawr.* New York: Springer-Verlag, 1983.

MARY W. GRAY

# NON-EUCLIDEAN GEOMETRY

A term used primarily for the alternatives to the geometry developed by Euclid in his *Elements*. The two principal forms of non-Euclidean geometry are called *hyperbolic* and *spherical*. The former was developed independently by Nikolai Lobachevsky and Janos Bolyai in the late 1820s, while the latter was proposed by Bernhard Riemann in the 1850s.

One way of distinguishing these forms of geometry involves parallel lines, which are defined to be lines that never meet. In Euclidean geometry, there is exactly one line through a point parallel to a given line. In hyperbolic geometry, there are many lines through a point parallel to a given line. In spherical geometry, there are no parallel lines. Another way of distinguishing these forms of geometry is by the sum of the angles of triangles. In Euclidean geometry, the angle sum of a triangle equals a straight angle (180°). In hyperbolic geometry, this angle sum is less than a straight angle, while in spherical geometry it is greater than a straight angle.

Euclid's fifth postulate states that if two lines are cut by a transversal so that the sum of the interior angles on one side is less than a straight angle, then the lines must meet on that side. Many mathematicians sought either to replace this postulate with a simpler statement or to prove it by using just the other postulates. One simpler statement is that of John Playfair: there is at most one line through a point parallel to a given line. Another, devised by the great nineteenth-century German mathematician Carl Friedrich Gauss, says that there are triangles with arbitrarily large area. One noteworthy attempt to prove Euclid's fifth postulate was made in the 1700s by Gerolamo Saccheri. He considered quadrilaterals with two equal sides perpendicular to a third side. He proved that the other two angles could not be greater than 90° and hoped to prove similarly that they could not be less than 90°, for this would then imply that they were right angles, and from this he could deduce Euclid's fifth postu-late. He proved many consequences of the hypothesis that the vertex angles were less than 90° (acute) but eventually rejected this hypothesis because he felt that the behavior of lines at infinity implied by this hypothesis was unacceptable. Thus he claimed to have proved Euclid's fifth postulate, which he described in his book *Euclides ab omni naevo vindicatus (Euclid Freed of Every Flaw)*. If, instead, he had said that the consequences of the acute angle hypothesis seemed consistent, he might have been honored as the founder of non-Euclidean geometry.

That honor instead goes to the aforementioned Lobachevsky and Bolyai, whose treatises were published in 1829 and 1831, respectively. They both developed a large body of consequences of the assumption that there is more than one line through a point parallel to a given line. These consequences included trigonometric formulas, and both showed how units of length could be directly related to angles. The latter result gives an intrinsic length in hyperbolic geometry, a unit of length definable in purely geometric terms. Such a quantity does not exist in Euclidean geometry. Gauss had speculated about intrinsic length in letters written in the early years of the nineteenth century. From his letters, we find that he foresaw much of the hyperbolic geometry before Lobachevsky and Bolyai, but he did not publish any of these findings because he considered them too revolutionary. Lobachevsky, Bolyai, and Gauss all realized that one might try to determine which type of geometry applies to the physical universe by measuring the angle sums of large triangles, and Gauss tried to perform such a measurement from three mountaintops, with inconclusive results. They also realized that for most everyday measurements Euclidean geometry would be at least a very good approximation to reality.

Spherical geometry can be thought of as the geometry on a sphere, the surface of a ball, such as the earth. Here the "straight lines" are great circles, which are circles that separate the sphere exactly in half and are the shortest distances between points on them. This form of geometry fails to satisfy Euclid's second postulate, because straight lines can only be extended to a certain length before they intersect themselves. This form of geometry also has an intrinsic length, which can be thought of as the length of the longest straight line. In his famous 1854 lecture, *Über die Hypothesen, welche der Geometrie zu Grunde liegen (On the Hypotheses that Lie at the Foundations of Geometry)*, Riemann proposed spherical geometry as an alternative to Euclidean but also did much more,

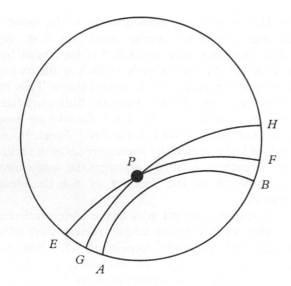

**Figure 1** *Poincaré's disk.*

## SELECTED REFERENCES

Davis, Donald M. *The Nature and Power of Mathematics.* Princeton, NJ: Princeton University Press, 1993.

Greenberg, Marvin J. *Euclidean and Non-Euclidean Geometries: Development and History.* 2nd ed. New York: Freeman, 1980.

Trudeau, Richard J. *The Non-Euclidean Revolution.* Boston, MA: Birkhäuser, 1987.

DONALD M. DAVIS

initiating the theory of higher-dimensional surfaces, called *manifolds.*

Lobachevsky and Bolyai did not prove that hyperbolic geometry is consistent, that is, free of contradictions. That required the models of hyperbolic geometry constructed within Euclidean geometry by Eugenio Beltrami in 1868 and Henri Poincaré in 1881. Beltrami's pseudosphere was a horn-shaped figure with constant negative curvature; portions of it modeled portions of hyperbolic geometry. Poincaré's disk (without boundary) modeled all of hyperbolic geometry. The "straight lines" in this model are circles that meet the boundary at right angles. In Figure 1, *EF* and *GH* are both straight lines passing through *P* and parallel to the straight line *AB*. Pleasing depictions of this model were created by the Dutch artist Mauritz C. Escher.

The significance of non-Euclidean geometry is that it freed people from the belief that space is necessarily Euclidean. It separated mathematics as the study of logical relationships from physics, which studied physical space. It also opened the way for other forms of geometry, such as differential geometry, which were essential to physicists like Albert Einstein. A good course in high school geometry should probably at least mention non-Euclidean geometry. The sphere as a model for spherical geometry and the Poincaré disk for hyperbolic geometry can be used to illustrate the subject without any technical details.

*See also* Bolyai; Gauss; Lobachevsky; Neutral Geometry

# NORMAL DISTRIBUTION CURVE

First observed by mathematical astronomers as the distribution of errors made in the measurement of the distances between heavenly bodies. This distribution was developed by the French mathematician Pierre S. Laplace (1749–1827) and first applied to astronomical errors by the German mathematician and astronomer Carl Friedrich Gauss (1777–1855). Subsequently, this function became known as the *Laplace-Gaussian curve* or, more recently, as the *normal curve.* The Belgian scientist Adolphe Quetelet (1796–1874) developed this function as a model for natural variations among biological populations. This line of inquiry was pursued with great enthusiasm by Francis Galton (1822–1911), who collected as many measurements of as many populations as he could in order to verify this principle of systematic variation. In 1889 he wrote, "Whenever a large sample of chaotic elements are taken in hand, and marshalled in the order of their magnitudes, an unsuspected and most beautiful form of regularity proves to have been latent all along."

The histograms reproduced in Figure 1 show the heights (in cm) of adult women and men in Great Britain and are drawn from information obtained from the United Kingdom Office of Population Censuses and Surveys. The graphs show quite clearly that males and females form two distinct populations with regard to height, but the two distributions are very similar in shape. They are both typical examples of the kinds of data that can be modeled by the normal distribution curve.

The normal distribution has the following density function:

$$f(x) = (1/\sigma\sqrt{2\pi})e^{-1/2[(x-\mu)/\sigma]^2}$$

where $\mu$ = population mean and $\sigma$ = population standard deviation. This function appears more simply if standardized measures are used. A standardized value is given as

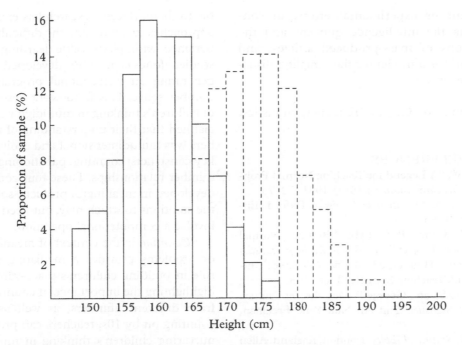

**Figure 1** *Heights of all adults aged 16–64 years in Great Britain: ———, females; — — —, males. [Source: Rouncefield and Holmes, Practical Statistics, 1989, Macmillan, reproduced with permission of Palgrave]*

$$Z = \left(\frac{x - \mu}{\sigma}\right)$$

This means that each value is expressed as a difference from the mean in units of the standard deviation. The function for the standardized normal distribution (see Figure 2) is

$$f(Z) = \left(\frac{1}{\sqrt{2\pi}}\right) e^{\frac{-z^2}{2}}$$

This function has mean = 0 and standard deviation = 1. The total area under the curve is equal to

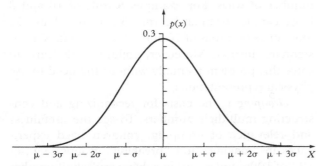

**Figure 2** *The shape of the normal distribution $X \sim N(\mu, \sigma^2)$. [Source: Rouncefield and Holmes, Practical Statistics, 1989, Macmillan, reproduced with permission of Palgrave]*

1, so areas under the curve may be used to estimate probabilities or proportions of the population. As integration of the function is laborious, tables are used to find areas under the standard normal curve.

The normal distribution curve is symmetrical, with the mean, median, and mode all coinciding at the center. It has several properties that can be usefully applied in the description of large populations:

a. 50% of the distribution occurs below the mean;
b. 68% occurs within one standard deviation either side of the mean;
c. Just over 95% occurs within two standard deviations of the mean.

Intelligence tests, for instance, are often set up so that the mean score of a large population will be 100 and the standard deviation 15. It is expected that 50% of the population will have intelligence test scores below the mean score of 100. Also, 68% of the population should achieve scores between 85 and 115, while just over 95% will have scores between 70 and 130. So a person with an intelligence test score of 130 or more is in the "top" $2\frac{1}{2}$% of such scores.

The normal distribution curve is widely used in biology as a model for the distribution of physical measurements in populations, as a model for errors in measurement or experimental errors, in constructs such as the intelligence quotient and the likely dimensions of mass-produced articles, and most importantly as a model for the sampling distribution of the mean.

*See also* Measures of Central Tendency; Statistics, Overview

## SELECTED REFERENCES

Gilchrist, J. W. "A Legend for Teaching Normal Probability Plotting." *Teaching Statistics* 15(3)(1993):72–75.

Kennedy, Gavin. *Invitation to Statistics.* Oxford, England: Robertson, 1983.

Maxwell, M. W., and B. C. Lyon. "Bivariate Normal Model: A Classroom Project." *The Best of Teaching Statistics* (pp. 135–139). Peter Holmes, ed. Sheffield England: University of Sheffield, Teaching Statistics Trust, 1988.

Moore, David S., and George P. McCabe. *Introduction to the Practice of Statistics.* 2nd ed. New York: Freeman, 1993.

Neave, H. R. *Statistical Tables.* London, England: Allen and Unwin, 1978.

Rouncefield, Mary, and Peter Holmes. *Practical Statistics.* London, England: Macmillan, 1989.

MARY ROUNCEFIELD

# NUMBER SENSE AND NUMERATION

According to authors of recent reforms for school mathematics, the major objective of elementary school mathematics (e.g., National Council of Teachers of Mathematics 1989, 1993). Although the expression "number sense" is a relatively new one, it is characterized as good intuition about numbers that enables one to judge whether numbers are used in "sensible" ways. More specifically, number sense encompasses an awareness and understanding of the meanings of numbers, their magnitudes, their relationships, and the relative effect of operating on numbers. Children with good number sense use numbers flexibly when mentally computing, estimating, and judging the reasonableness of results. This entry focuses on number sense in the context of whole-number numeration and place value, which involve those concepts, skills, and understandings necessary for naming and processing numbers greater than 10.

As far back as the 1930s, teachers have been concerned about nurturing reasonable understandings and uses of number among students. William Brownell (1935) addressed this in advocating that teachers should assist children to construct mathematical meanings and relationships that were sensi-

ble to them. Recent research has examined different approaches to addressing the difficulties children experience with place value learning. One of these studies (Jones et al. 1996) developed a framework for generating an instructional program in multidigit number sense. This framework described five levels of children's thinking in multidigit number sense and claimed that their understanding of multidigit numbers was multidimensional and evolved through four key constructs: counting, partitioning, grouping, and number relationships. These four constructs are best developed in meaningful problem-solving situations, the solutions to which may, but need not necessarily, involve a computational operation.

*Counting* in the context of meaningful situations or problems provides a mediating and continuing role in building children's place-value concepts. By highlighting the importance of counting on and back from different numbers, as well as counting and counting on by 10s, teachers can provide a basis for nurturing children's thinking in multidigit number sense. Repeated experiences with problems in which children count on by 10 are particularly useful. For example, "There are 34 candies in the jar. I bought 3 more rolls of 10 candies. If we put them in the jar, how many candies will there be then?"

*Partitioning* emphasizes part–part–whole relationships in place-value learning. Partitioning numbers in different ways helps children develop flexibility in representing and understanding multidigit numbers. The goal is for children to be able to "break numbers apart," that is, to partition numbers in both standard and nonstandard ways. Partitioning should occur in the context of real-world problems. For example, "You can buy loose candies or rolls of 10. If 68 are needed for a party, in how many different ways can they be bought?" Children with good number sense will readily solve this problem in a number of ways. For example, 6 rolls of 10 and 8 loose candies (standard form); 5 rolls of 10 and 18 loose ones or 3 rolls of 10 and 38 loose candies (nonstandard forms). Moreover, children will learn to solve this problem mentally without the need to use physical representations.

*Grouping* is the basis for recognizing and constructing multidigit numbers. To see the usefulness and relevance of grouping, children need experiences in "making" and "breaking" groups (e.g., of 10s, of 100s, and so on) in solving multidigit number problems. For example, "Grandma wants to buy some candies. A small bag costs 25¢, a big one 38¢. Grandma has only 70¢. Does she have enough to buy one of each?" Whether children solve this problem mentally or with manipulatives (e.g., unifix

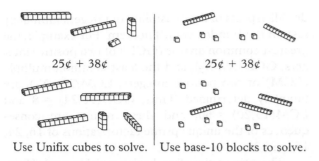

25¢ + 38¢          25¢ + 38¢

Use Unifix cubes to solve. | Use base-10 blocks to solve.

**Figure 1**

cubes, base-10 blocks, or dimes and pennies), there is a need to combine groups, including a group of 10 that is constructed from 1s. (See Figure 1.) Instead of determining an exact answer, children with good number sense may simply estimate by noting that five 10s are needed, but not enough more to make two more 10s. "So 70¢ is enough."

A real understanding of number includes the ability to think of a number *in relation to other numbers*. In particular, comparing the relative size of multidigit numbers, describing them in terms of "more" or "less," and ordering them are fundamental to number sense. Problems like the following are valuable for two reasons: (1) they involve two-step comparisons, which are both challenging and beneficial; (2) they provide a forum for listening and assessing children's thinking. For example, "The children estimated the number of candies in the jar. Cameron's estimate was 150; Stephanie's was 210. The actual count was 176. Who was closer? Explain your response." The four constructs: counting, partitioning, grouping, and number relationships can be used as the basis for developing a coherent, long-term number sense and numeration program. Ideally, this program should incorporate a problem-solving context and should be integrated with addition and subtraction computation.

Some of the implications that flow from recent research (e.g., Jones et al. 1996) are that multidigit number sense can be greatly enhanced if teachers:

Are sensitive to children's thinking in relation to the four key constructs—recognizing which are mastered and which invite further work.

Plan teaching and assessment experiences that focus on these constructs. For instance, counting might be incorporated with daily calendar work.

Engage children in collaborative problem-solving situations as the basis for exploring, developing, and applying multidigit number sense. Good problems are the key. For example: "One box has 68 wheels, another 54. We need 10 for

each toy. Is there enough for 100 toys? Explain."

Provide ongoing and systematic problem-solving experiences based on the four constructs, especially mental arithmetic and estimation, *prior to* the introduction of formal paper-and-pencil computational procedures. This will increase children's potential to be flexible as they work numerically.

Continue these problem-solving experiences, especially those that invite mental arithmetic and estimation, as a regular interface with paper-and-pencil computational work—so children retain their flexibility in dealing with number in many different ways.

These approaches are central to nurturing multidigit number sense in the early school program. They provide opportunities to build on children's prior knowledge, nurture their thinking through problem-focused experiences, and are consistent with a constructive orientation to learning.

*See also* Addition; Arithmetic; Decimal (Hindu-Arabic) System; Errors, Arithmetic; Estimation; Fractions; Subtraction

### SELECTED REFERENCES

Brownell, William A. "Psychological Considerations in the Learning and Teaching of Arithmetic." In *The Teaching of Arithmetic* (pp. 1–31). 10th Yearbook. National Council of Teachers of Mathematics. William D. Reeve, ed. New York: Teachers College Press, 1935.

Jones, Graham A., Carol A. Thornton, Ian J. Putt, Kevin M. Hill, A. Timothy Mogill, Beverly S. Rich, and Laura R. Van Zoest. "Multidigit Number Sense: A Framework for Instruction and Assessment." *Journal for Research in Mathematics Education* 27(3)(1996):310–336.

National Council of Teachers of Mathematics. *Curriculum and Evaluation Standards for School Mathematics.* Reston, VA: The Council, 1989.

———. *Number Sense and Operations.* Curriculum and Evaluation Standards for School Mathematics Addenda Series, Grades K–6. Reston, VA: The Council, 1993.

———. *Principles and Standards for School Mathematics.* Reston, VA: The Council, 2000.

CAROL A. THORNTON
GRAHAM A. JONES

# NUMBER THEORY, OVERVIEW

The study of the properties of numbers. A digit is missing from the number $N = 12345x$. When the $x$ is replaced by the correct digit, $N$ will be divisible by 9 (Note: In standard notation, 9 | $N$ is read "9 divides $N$.") What is the missing digit? This is recreational mathematics, suitable for use at a party. Yet behind every such problem there is an important concept or result in number theory (Beiler 1966). Just what is number theory? Arithmetic refers to the facts of addition, subtraction, multiplication, and division of integers and common rational numbers (e.g., $\frac{2}{3}$, $-\frac{8}{5}$). Deeper results concerning these numbers belong to number theory.

## HISTORY

Although there is an extensive history of the development and the writing of numbers in cultures worldwide (Flegg 1989), the systematic study of their properties began with the Greeks of antiquity. The first prominent student of number theory was Euclid (ca. 300 B.C.). Several sections of the *Elements* (Euclid 1956) deal with properties of numbers. A later Greek mathematician named Diophantus (ca. A.D. 250), who worked in Alexandria, Egypt (as did Euclid), did highly original work at the interface of number theory and algebra (Heath 1964). Arabic translations of Euclid's *Elements,* Diophantus's *Arithmetica,* and other Greek manuscripts reached Europe centuries later and stimulated Western interest in mathematics. The modern study of number theory begins with Pierre de Fermat (1601–1665). His famous Last Theorem (see later) has been a topic of interest in professional circles and in the popular press recently. The two most outstanding practitioners of number theory were Leonhard Euler (1707–1783) (Burckhardt 1983) and Carl Friedrich Gauss (1777–1855) (Bühler 1981). Between them they defined most of the field before 1900. Eminent number theorists in the twentieth century have included Godfrey H. Hardy (1877–1947), Waclaw Sierpinski (1882–1969), Viggo Brun (1885–1978), Ivan M. Vinogradov (1891–1983), Derrick H. Lehmer (1905–1991), and Paul Erdös (1913–1996). Today, number theory is a flourishing branch of mathematics with several divisions and little resemblance to the field in Fermat's time.

## PRIMES

Prime integers (or primes) are the building blocks of all integers. The Fundamental Theorem of Arithmetic, first proved by Gauss, says that a composite positive integer can be factored into primes in a unique way, except possibly for differences in the order of listing of the prime factors (Ore 1967). Thus, $42 = 2 \cdot 3 \cdot 7$, and no other set of primes will

do. All aspects of the divisibility of integers ultimately rest on the Fundamental Theorem. For example, the greatest common divisor (GCD) of two positive integers, GCD$(N_1, N_2)$, and the least common multiple (LCM) of two positive integers, LCM$(N_1, N_2)$, are uniquely determined. Thus, GCD$(16,24) = 8$ and LCM$(15,20) = 60$, and these results are consequences of the unique prime factorizations of 16, 24, 15, and 20.

The primes therefore have special appeal. However, there are many unanswered questions about them. They seem to occur irregularly. The Prime Number Theorem says that if $\pi(N)$ is the function that gives the number of primes in the interval $[2,N]$, then $\pi(N)$ is approximately $N/\ln N$ for large $N$, in the sense that

$$\lim_{N \to \infty} \frac{\pi(N)}{N/\ln N} = 1$$

To provide finer details on the occurrence of primes is extremely difficult. Also, our knowledge of very large primes is scanty. To show convincingly that a certain 100-digit integer is a prime is a formidable task, even on a large computer.

## CONGRUENCES

Gauss introduced the notation $a \equiv b$ (mod $n$), read "$a$ is *congruent* to $b$ modulo $n$," to mean that $n \mid (a - b)$; for example, $8 \equiv 2$ (mod 3). This deceptively simple symbolism has proved to be extremely fruitful, in part because congruences have many properties that parallel those of equalities. Some properties of the general congruences $a \equiv b$ (mod $n$) and $c \equiv d$ (mod $n$), and illustrated with the respective congruences $46 \equiv 4$ (mod 6) and $11 \equiv 5$ (mod 6), are tabulated here (Ore 1948).

| General property | Illustration |
|---|---|
| 1. $a + c \equiv b + d$ (mod $n$) | $46 + 11 \equiv 4 + 5$ (mod 6) |
| 2. $a^k \equiv b^k$ (mod $n$) ($k$ = positive integer) | $46^{10} \equiv 4^{10}$ (mod 6) |
| 3. $ka + k'c \equiv kb + k'd$ (mod $n$) ($k,k'$ = integers) | $85 \cdot 46 - 14 \cdot 11 \equiv$ $85 \cdot 4 - 14 \cdot 5$ (mod 6) |
| 4. If $k\mid a$ and $k\mid b$, then, $a/k \equiv b/k$ (mod $(n/\mathrm{GCD}(n,k))$) | $46/2 \equiv 4/2$ (mod $(6/\mathrm{GCD}(6,2))$), or $23 \equiv 2$ (mod 3) |

The illustrations for statements (2) and (3) certainly look nontrivial, yet can be written down immediately. Statement (4) is unique to congruences and has no parallel among equalities.

Congruence notation is just another way of symbolizing divisibility. For example, our opening question in this entry could be reworded as: for which digit $x$ is the congruence $12345x \equiv 0 \pmod 9$ valid? Congruence notation is suggestive of countless questions. Answers to some classic questions are stated in terms of congruences.

*Fermat's Theorem*

If $p$ is a positive prime and $a$ is any positive integer not divisible by $p$, then $a^{p-1} \equiv 1 \pmod p$.
Ex: $6^{12} \equiv 1 \pmod{13}$.

This amazing theorem defines a whole class of integers that are divisible by a given prime $p$.

*Euler's Theorem*

Let $GCD(a,n) = 1$. If $\phi(n)$ denotes the number of positive integers $k$, less than $n$, such that $GCD(k,n) = 1$, then $a^{\phi(n)} \equiv 1 \pmod n$.
Ex: $7^4 \equiv 1 \pmod{12}$, since $\phi(12) = 4$.

This theorem is even more amazing. It generalizes Fermat's Theorem to the case where the modulus is no longer a prime.

*Wilson's Theorem*

The congruence $(p - 1)! \equiv -1 \pmod p$ holds if and only if the positive integer $p$ is a prime.
Ex: $10! \equiv 1 \pmod{11}$, but $14! \not\equiv -1 \pmod{15}$.

This theorem is a foolproof criterion for primality but is useless practically because $(p - 1)!$ rapidly becomes astronomical.

A wider significance of Fermat's Theorem is that it has spawned much work on methods to identify large primes. Suppose $a = 2$ and $n$ is odd; then Fermat's Theorem, in its contrapositive form, would say that if $2^{n-1} \not\equiv 1 \pmod n$, then $n$ is not a prime. This statement is true. Its inverse, that $2^{n-1} \equiv 1 \pmod n$ implies $n$ is prime, is false, but the falsity seems to occur in only a small percentage of the cases. This is useful in trying to track down very large primes.

Another classic result involving congruences is due to Srinivasa Ramanujan (1887–1920) (Newman 1956; Kanigel 1991). Let $p(n)$ denote the number of ways that the positive integer $n$ can be partitioned into summands that are positive integers. For example, $p(5) = 7$ because $5 = 4 + 1 = 3 + 2 = 3 + 1 + 1 = 2 + 2 + 1 = 2 + 1 + 1 + 1 = 1 + 1 + 1 + 1 + 1$

*Ramanujan's Theorem*

For any integer $n \geq 0$ these congruences hold:

$$p(5n + 4) \equiv 0 \pmod 5$$
$$p(7n + 5) \equiv 0 \pmod 7$$
$$p(11n + 6) \equiv 0 \pmod{11}$$
$$\text{Ex: } p(9) = 30 \quad \text{and} \quad 30 \equiv 0 \pmod 5.$$

Partitions are a very active and fascinating subject in number theory. One can ask for the number of partitions of an integer under various interesting restrictions (e.g., even number of summands, all summands odd, no summand used more than twice, etc.). These kinds of problems are difficult (and therefore appealing) because additive questions tend to be intrinsically harder than multiplicative questions. Rigorous verification of Ramanujan's Theorem was a nontrivial affair.

## REPRESENTATION PROBLEMS

One class of additive number theory questions is the representation problems. These ask: which positive integers can be represented in a certain manner, and, for a given integer, in how many ways can this be done? Such problems date back to Euclid, who asked: which positive integers $n$ can be represented by

$$n = 1 + d_2 + d_3 + \cdots + d_r$$

where $1, d_2, d_3, \ldots, d_r$ are the $r$ positive divisors of $n$ (excluding $n$ itself)? Examples are $n = 6$ ($6 = 1 + 2 + 3$) and $n = 28$ ($28 = 1 + 2 + 4 + 7 + 14$). Such integers are called *perfect numbers*. Euclid proved that if $2^k - 1$ is a prime, then $n = 2^{k-1}(2^k - 1)$ is an even perfect number. Two millenia later, Euler proved that every even perfect number is of this form. There are at least thirty-five known perfect numbers; whether others exist is an open question. Nor is it known if there are any odd perfect numbers.

Which positive integers $n$ can be represented as the *sum of two squares* ($0^2$ is permitted)?

$$n = m_1^2 + m_2^2$$

Fermat answered this one. Any positive integer $n$ can be so represented unless $n$ has in its prime factorization an odd power of a $(4k + 3)$ prime. Thus, $245 = 5 \cdot 7^2 = 7^2 + 14^2$, but $702 = 2 \cdot 3^3 \cdot 13$ cannot be represented as the sum of two squares. A century after Fermat, it was shown by Joseph-Louis Lagrange (1736–1813) that every positive integer can be written as the *sum of four squares* ($0^2$ is permitted). Thus, we can represent 702 as $0^2 + 1^2 + 5^2 + 26^2$, and also as $1^2 + 5^2 + 10^2 + 24^2$. Far less is known about representations of numbers as sums of cubes, fourth powers, and so on. These higher powers might be expected to present difficulties. However, even the simpler-sounding representation problem of determining which positive even integer $n \geq 6$ can be represented as the sum of two odd primes (e.g., $8 = 3 + 5$) has proved to be quite difficult. Goldbach's Conjecture, named after Christian Goldbach (1690–1764), asserts that *all* even integers $n \geq 6$ can be so represented. This conjecture is still undecided after 250 years.

By far, the most notorious representation problem is the one referred to as *Fermat's Last Theorem*. For integer $n \geq 2$, the problem asks which perfect $n$th powers can be represented as sums of two other $n$th powers of positive integers?

$$z^n = x^n + y^n$$

The solutions to the problem were known completely when $n = 2$ (e.g., $5^2 = 3^2 + 4^2$, $13^2 = 5^2 + 12^2$) (Ore 1967). Fermat asserted (his Last Theorem) that for $n \geq 3$ there are *no* positive integers $z$ for which the above representation is possible. The Last Theorem had been investigated by computer up to quite large $n$, and no counterexamples had been found. This was suggestive but did not constitute a proof. Finally, in 1995, the Princeton mathematician Andrew Wiles proved that Fermat's conjecture is indeed correct.

## ALGEBRAIC AND TRANSCENDENTAL NUMBERS

Real numbers can be divided into two mutually exclusive groups. Those real numbers that can be found as the zeros of polynomials with integral coefficients

$$a_0 x^n + a_1 x^{n-1} + \cdots + a_n, \ a_i \ integers$$

are called *algebraic numbers*. Examples are $\frac{1}{2}$ (the zero of the polynomial $2x - 1$), $\sqrt{2}$ (a zero of the polynomial $x^2 - 2$), and $1 + \sqrt[3]{3}$ (a zero of the polynomial $x^3 - 3x^2 + 3x - 4$). Some algebraic numbers are rational and some are irrational. Real numbers that are not algebraic are called *transcendental numbers;* examples are $\pi$ and $e$. These numbers cannot be found as zeros of polynomials with integral coefficients.

The demonstration in the late nineteenth century that $\pi$ and $e$ were transcendental was regarded as a very significant accomplishment. In general, to show that any real number is a transcendental number is a difficult job, and even today no general methods are known for doing this. The search for criteria for transcendence is a major area of work in this part of number theory. We do know, however, that transcendental numbers are much more plentiful than are algebraic numbers. In fact, in a technical sense almost all real numbers are transcendental. But the trick is to distinguish the majority from the minority. It is easy to write down real numbers for which it is

not known whether they are algebraic or transcendental, for example, $\pi^{\sqrt{2}}$. A result obtained in 1934 and known as the Gelfond-Schneider Theorem states that $\alpha^\beta$ is transcendental if $\alpha$, $\beta$ are algebraic, $\alpha \neq 0$ or 1, and $\beta$ is irrational. Thus, the number $(\sqrt{2})^{\sqrt{2}}$ is transcendental; if either $\alpha$ or $\beta$ is not algebraic, then we can draw no conclusion. The theorem thus covers a minority of cases but has nevertheless stimulated much work.

## TECHNIQUES IN NUMBER THEORY

Elementary work in number theory, such as one might do in an introductory classroom setting or for recreation, involves only simple tools from arithmetic and algebra. Indeed, this is the way the discipline developed before 1800. Beginning in the middle of the nineteenth century, however, ideas and techniques from other areas of mathematics began to be applied to number theory, notably from analysis and abstract algebra. Today, a thorough knowledge of techniques in these fields is essential for advanced work in number theory.

For many (but not all) advanced problems in number theory a computer is also essential. For example, it has been indispensable in gathering data in the search for odd perfect numbers, in checking hypotheses on possible solutions to Fermat's Last Theorem, and in uncovering ever larger primes.

## NUMBER THEORY IN THE CLASSROOM

Unlike most branches of mathematics, number theory is approachable at a variety of levels. At the high school level, coverage of number theory could profitably focus on (a) elementary consequences of the Fundamental Theorem of Arithmetic, (b) exploring various aspects of congruences, and (c) using number theory as a vehicle to sharpen programming skills. Here are some sample student projects that could serve as a source of stimulation:

1. Develop a computer program for finding the prime factorization of an integer and then use it to work out the GCDs of several pairs of integers.
2. Read about the Euler $\phi$-function (LeVeque 1962, 42; McCoy 1965, 36). Then develop a computer program for evaluating $\phi(n)$ and use it to corroborate Euler's Theorem for several different values of $n$.

3. Find out what quadratic residues are (McCoy 1965, 85) and work them out for several moduli, prime and composite. Can you find any patterns of generalizations? Extend your investigations to cubic and quartic residues (Dence and Dence 1995).

4. A pseudoprime (to the base 2) is a *composite* positive integer $n$ such that $2^n \equiv 2 \pmod{n}$ (Rosen 1993). Find, by computer programming, the first three pseudoprimes. What happens if the base is changed to 3?

At the early undergraduate level, number theory coverage could include all of the above, plus some beginning exposure to abstraction, primarily through proof. The following major theorems in number theory are accessible at this level:

5. Read about continued fractions (LeVeque 1962, 73; Olds 1963). Learn the proof of Lagrange's Theorem, which says every quadratic irrational number $\alpha$ has a periodic continued fraction expansion. Then work out the continued fraction expansion of $\alpha = \sqrt{31}$, and prove that it is correct.

6. Work through the proof of Fermat's Theorem on the sum of two squares (Davenport 1983). Then write a program that will express a given positive integer, whenever possible, as the sum of two squares in as many ways as possible. Can you formulate a generalization?

*See also* Algebraic Numbers; Prime Numbers; Recreations, Overview; Transcendental Numbers

## SELECTED REFERENCES

Beiler, Albert H. *Recreations in the Theory of Numbers: The Queen of Mathematics Entertains.* New York: Dover, 1966.

Bühler, Walter K. *Gauss: A Biographical Study.* New York: Springer-Verlag, 1981.

Burckhardt, Johann J. "Leonhard Euler, 1701–1783." *Mathematics Magazine* 56(1983):262–273.

Davenport, Harold. *The Higher Arithmetic.* New York: Dover, 1983, pp. 115–120.

Dence, Joseph B., and Thomas P. Dence. "Cubic and Quartic Residues Modulo a Prime." *Missouri Journal of Mathematical Sciences* 7(1995):24–31.

Euclid. *The Thirteen Books of Euclid's Elements.* Thomas L. Heath, trans. 2nd ed. 3 vols. New York: Dover, 1956.

Flegg, Graham, ed. *Numbers Through the Ages.* London, England: Macmillan, 1989.

Heath, Thomas L. *Diophantus of Alexandria.* New York: Dover, 1964.

Kanigel, Robert. *The Man Who Knew Infinity: A Life of the Genius Ramanujan.* New York: Scribner's, 1991.

Kramer, Edna E. *The Nature and Growth of Modern Mathematics.* New York: Hawthorn, 1970.

LeVeque, William J. *Elementary Theory of Numbers.* Reading, MA: Addison-Wesley, 1962.

McCoy, Neal H. *The Theory of Numbers.* New York: Macmillan, 1965.

Newman, James R., ed. *The World of Mathematics.* Vol. 1. New York: Simon and Schuster, 1956.

Olds, Carl D. *Continued Fractions.* New York: Random House, 1963.

Ore, Oystein. *Invitation to Number Theory.* New York: Random House, 1967.

———. *Number Theory and Its History.* New York: McGraw-Hill, 1948.

Richards, Ian. "Number Theory." In *Mathematics Today: Twelve Informal Essays* (pp. 37–64). Lynn Arthur Steen, ed. New York: Springer-Verlag, 1978.

Rosen, Kenneth H. *Elementary Number Theory and Its Applications.* 3rd ed. Reading, MA: Addison-Wesley, 1993.

JOSEPH B. DENCE

# NUMERICAL ANALYSIS, METHODS

The various numerical approximations in use today for simulating physical systems on computing machines. A new scientific paradigm emerged in the twentieth century due to the availability (indeed proliferation) of modern computers. A computer works analogously to a person who has been given explicit instructions to follow and who has an infinite supply of paper and erasers. For example, suppose you wish to compute

$$(A + B) * 2$$

The following "program" will compute this using two pieces of paper, each of which has one of the numbers to use written on it ($A$ or $B$), erasers, and a person who knows how to add and multiply:

Read paper 1, name it $A$.

Read paper 2, name it $B$.

Add $A + B$.

Erase paper 2.

Write the sum $(A + B)$ on paper 2.

Read paper 2, name it $C$.

Compute $C * 2$.

Erase paper 2.

Write the product $(C * 2)$ on paper 2.

As in this example, a computer follows step-by-step instructions, keeping track of all intermediate results. Today, because we have the technology to construct high-speed general-purpose computers, the challenge is to define instructions for the computer in sufficient detail so as to allow the solution to be computed in as general a way as possible.

Scientists' metaphysical beliefs, coupled with the technology available, set the stage for the examination of certain problems. Computers have allowed us to develop and execute a significant number of models, for example. "A model is an abstract description of the real world; it is a simple representation of more complex forms, processes, and functions of physical phenomena or ideas" (Rubinstein 1975). A mathematical model reduces or simplifies complex, sometimes apparently disparate, physical phenomena, replacing them with a set of equations that we can examine and analyze. One of the many benefits of mathematical models is that the resulting equations highlight in which sense apparently disparate phenomena are similar. Unfortunately, these mathematical models are not usually easy to solve. In fact, most useful mathematical models do not have known solutions. This is why numerical approximations are used to estimate the solutions of mathematical systems.

Mathematical quantities are frequently approximated with algebraic expressions for calculations on a computer. In the 1600s, Bonaventura Cavalieri was among the first mathematicians to use a functional approximation equivalent to the modern practice of summing the leading terms of an infinite series (Hall 1981). This is one of the earliest uses of the discipline we now call *numerical analysis*. It is this approximation of the infinite with the finite that characterizes modern computer simulations of physical equations.

The idea is simple. When functions are equivalent to the infinite sum of a series, we can substitute the sum into an expression containing the function. Once all terms in an equation are written in terms of sums of infinite series, each sum can then be truncated to finite length and used to model the equation. This procedure can be used to cast complex integro-differential equations into an algebraic form that we know how to solve. When this is done, it is referred to as a *numerical method*.

Numerical techniques allow us to use computers to solve mathematical problems originally posed on a continuous domain. Having made such an approximation, the resulting numerical method has error. The analysis and quantification of this error is the backbone of numerical analysis: numerical analysis

provides bounds on the accuracy of computer simulations. Scientists cannot test every instance of every possibility; they must trust that the scientific community has examined representative samples in a mathematically consistent way. Computing machines are frequently used to increase the number of "experimental samples" we perform. Such "experiments" (i.e., simulations) typically apply a numerical method to compute the behavior of a physical system. Because computer models are, in general, less expensive and faster to execute than an experiment, they help us examine more possibilities.

The scientific computing paradigm was first used in the nineteenth century. Having attended lectures on Charles Babbage's mechanical computer in 1834 (Alic 1986) Ada Byron Lovelace, daughter of Lord Byron, prepared mathematical software for an automatic general-purpose digital computer. She used numerical approximations—in particular, the finite differences that we use later in this entry—to render continuous equations calculable on a digital computer. Babbage's published examples were algebraic; for example, he proposed that his calculating machine could calculate

$$a^6 + 5a^5b + 10a^4b^2 + 20a^3b^3 + 10a^2b^4 + 5ab^5 + b^6$$

Babbage developed the theory necessary to construct a general-purpose digital computer, transforming very basic automatic-processing technology into a computer of significant sophistication—comparable to the computers used in the 1950s. His machine performed "the calculation of the numerical quantities resulting from two or more successive operations" (Babbage 1837). However, the numbers are limited in their representation. Computers typically use a fixed number of places or digits for each number. Such truncation of numbers introduces error into the calculation. Computations with these approximate numbers introduce additional errors, and these errors sometimes accumulate and erode the accuracy of a result. In addition, errors are introduced when truncated sums of infinite series are used in place of functions, as just discussed. The nature of calculating on a digital computer is approximation.

Then how can we know whether a simulation, based as it is on approximations, is representative of the physical world? Simulations typically contain many simplifications and errors, only a few of which have been discussed here. Numerical analysis, the study of the error characteristics of numerical models, incorporates the details of a computer environment into a mathematical format. It quantifies the effects of the programming language, the machine architecture, the operating system, the mathematical model, the numerical approximation to the mathematical model, and even the quality of data. A model simulation is validated by analyzing its accuracy: the sources of error are tracked as the errors propagate. This requires knowledge of every aspect of the simulation—we formulate the mathematical errors due to each numerical approximation.

Scientific computing has been defined (Golub and Ortega 1993) as the "study of the techniques used to obtain approximate solutions to the problems that arise in science and engineering, especially the numerical solutions of mathematical models executed by computers." Hence it is the collection of tools, techniques, and theories used to numerically approximate the mathematical models that arise in science and engineering so as to render them suitable for solution on a computer. This approach is not new. Archimedes systematically constructed mathematical models and then utilized the scientific computing paradigm to make numerical approximations in order to predict or forecast future phenomena. Of course, without computing machines he had to make the calculations by hand.

The task of reducing a scientific question so that it can be approximated with a straightforward set of instructions that can be rendered in software is quite difficult and typically falls into the individual physical science disciplines. It requires first developing (or choosing) a mathematical model. Such a model may be constructed by analyzing experimental data until a set of governing equations has been derived that expresses the relationship between the data and the physical phenomena. For example, let's suppose that we want to model the Gulf Stream in the Atlantic Ocean in order to predict the location of the Gulf Stream at some future time. Such information is quite useful to ships, which can take advantage of (or avoid) this strong current. The equations for this phenomena are well studied, so choosing a set of equations is straightforward. Then, to convince ourselves of the usefulness of our computer model, we can perform checks against both the mathematical model and reality (data). These two steps occupy most of the numerical analysis done with numerical simulations. To check against reality, we might initialize our Gulf Stream model with data from a given date, advance the model in time to a prediction date, and then verify our prediction against data from the prediction date. There are other ways of verifying the model, including simple realism tests that include any known con-

straints; for example, we could check whether our ocean model predicts a departure from the climatological mean sea surface height. We could make sure that energy never becomes negative. Much more detailed data comparisons are, of course, typically performed. To check against a mathematical model, we might initialize our Gulf Stream simulation with a function that has a known solution (remember, these are hard to find). We advance the model in time until we have a prediction. We then compute the known solution at the forecast time and compare the actual location of the Gulf Stream against the model prediction.

Such detailed studies lead to a range of situations within which the model is applicable. We call this a *regime of applicability*. When computer simulations are executed within this regime of applicability, the predictions should be qualitatively correct. Outside the regime of applicability, the simulation predictions may not have a physical counterpart. It is very important to understand not only the physics predicted in the simulation but also when this physics can be directly related with confidence to real-world dynamics. Numerical analysis provides the information needed for us to have this confidence.

Many scientific disciplines have incorporated scientific computing into their research paradigms. These disciplines include, but are not limited to, genetic research; fluid dynamics, which includes oceanography; atmospherics; aerodynamics; astronomy; and the environmental sciences.

## NUMERICAL ANALYSIS MODELING PROJECT

Typically, simulations first express each term in the mathematical model by an equivalent series representation. The series can be in physical space, such as with a Taylor series, or in frequency (or an eigenfunction) space, as with a Fourier series. Depending on the choice of series approximations, different numerical methods are derived. In either case, such series contain, as the most general case, an infinite number of terms. Next, the mathematical model is approximated by truncating these infinite series to some finite number of terms. Most mathematical models (of equations) that describe physical dynamics contain derivatives (rates of change of a quantity) that must be approximated.

Let's consider the approximation of a time derivative using a Taylor series. The time derivative might represent the evolution of the Gulf Stream. The Taylor series allows us to calculate the value of an analytic function near a point, given all of the derivatives at that point:

$$f(t + \Delta t) = f(t) + \frac{f^{(1)}(t)}{1\,!} * \Delta t + \frac{f^{(2)}(t)}{2\,!} * \Delta t^2$$
$$+ \frac{f^{(3)}(t)}{3\,!} * \Delta t^3 + \cdots$$

Here $f$ is the function, and superscripts indicate the respective derivative.

Next, to approximate the derivative we can truncate the Taylor series after the term involving $f^{(1)}(t)/1!$, leaving the approximation

$$f(t + \Delta t) \approx f(t) + f^{(1)}(t)/1! * \Delta t$$

We have just introduced a source of error at this step. To use this approximation, we solve for the derivative $f^{(1)}(t)$ to see that

$$f^{(1)}(t) \approx (f(t + \Delta t) - f(t))/\Delta t$$

Wherever we use this approximation, we have truncated an infinite series in time into a finite number of terms from the same series. We stress again that it is this approximation of the infinite with the finite that characterizes modern computer simulations of physical equations.

An instructive yet simple computer class project simulates advection, a simple form of fluid flow that is an important part of the dynamics of the oceans. The equation is

$$\frac{\partial u}{\partial t} + a\,\frac{\partial u}{\partial x} = 0$$

Here $\partial/\partial t$ represents the changes in time while the $\partial/\partial x$ represents the changes in longitude along a given latitude. Disturbances in this system (such as the ripples in a pond when you toss in a pebble) are propagating at speed $a$. Using the Taylor series approximation technique, we can approximate each individual term:

$$\frac{\partial u}{\partial x} \approx \frac{u(x + \Delta x, t) - u(x, t)}{\Delta x}$$

at some given time, such as on May 15, 1992. Also let

$$\frac{\partial u}{\partial t} \approx \frac{u(x, t + \Delta t) - u(x, t)}{\Delta t}$$

at some given longitude $x = X$. Based on the Taylor series, such difference approximation methods are referred to as *finite differences*.

The nature of our approximations moves us from a continuous domain to a domain defined at only a finite number of locations. We have only simu-

lated the system at a finite number of space and time locations. Such a set of finite locations is referred to as a *grid*. Now assemble the entire equation:

$$\frac{u(x, t + \Delta t) - u(x, t)}{\Delta t}$$
$$+ a\,\frac{u(x + \Delta x, t) - u(x, t)}{\Delta x} = 0$$

The value of the function $u$ at some initial time $t = 0$ must be prescribed. These values are referred to as *initial conditions*. We must now decide what initial conditions to use, such as where to place the Gulf Stream. We might proceed by obtaining data on the location of the Gulf Stream at a specific date, for example. Now we can approximate the equation, at each $x = X$ location where we defined initial values, using the following model equation:

$$u(x, t + \Delta t) = u(x, t) - \Delta t * a\,\frac{u(x + \Delta x, t) - u(x, t)}{\Delta x}$$

This equation can be repeatedly solved for the next higher time step, advancing the equation forward in time.

Students can prepare a straightforward computer program to advance this system forward in time. Experiments with various values of $a$, $\Delta x$ and $\Delta t$ will allow students to define a regime of applicability. Such experiments are called *parameter-bracketing* experiments. To understand why the regime of applicability exists requires a college-level study of finite difference experiments; however, what is important at the level of this example is for students to understand that outside of this regime the model prediction results are invalid.

## SUMMARY

Scientific computing is an emerging paradigm, adding a new dimension—numerical methods—to scientific investigation. This type of investigation provides significant flexibility and cost benefits over the use of physical experiments only. Its usefulness requires a careful study of each numerical method, that is, numerical analysis. The paradigm has many applications, some of which are appropriate for high school students with elementary calculus skills, as illustrated in the example described here of finite difference approximations to simulate partial differential equations.

*See also* Modeling; Partial Differential Equations; Series

## SELECTED REFERENCES

Alic, Margaret. *Hypatia's Heritage.* Boston, MA: Beacon, 1986.

Babbage, Charles. "On the Mathematical Powers of the Calculating Engine." Unpublished ms., 1837. At Museum of History of Science, Oxford, England. In *The Origins of Digital Computers; Selected Papers* (pp. 19–54). 3rd ed. Brian Randell, ed. Berlin, Germany: Springer-Verlag, 1982.

Golub, Gene, and James M. Ortega. *Scientific Computing: An Introduction with Parallel Computing.* Boston, MA: Academic, 1993.

Hall, A. Rupert. *From Galileo to Newton.* New York: Dover, 1981.

Rubinstein, Moshe F. *Patterns of Problem Solving.* Englewood Cliffs, NJ: Prentice-Hall, 1975.

Shea, William R. *Galileo's Intellectual Revolution: Middle Period, 1610–1632.* New York: Watson, 1977.

A. LOUISE PERKINS

# NURSING

From the time of Florence Nightingale, mathematics has been an integral part of nursing practice. Today, the use of mathematics in the nursing profession can be divided into two general categories—the traditional clinical application and the recent trend to deliver more cost-effective health care. In most clinical health care settings, nurses are responsible for administering medications and infusion therapy, which often require dosage calculations. Nurses must know how to use the International System of Units (SI) and non-SI measurement systems. Calculations may include elementary arithmetic, ratio and proportion, and simple algebra. Clinical nurses also use measurement systems to obtain weight and height as well as fluid intake and output. These vital measurements help determine if clients (a more recent term for patients) are losing or retaining body water, an indication that they may be at risk for life-threatening complications.

Another example of the clinical application of mathematics is the measurement and documentation of vital signs—temperature, pulse rate, respiratory rate, and blood pressure. The nurse or other nursing staff member obtains and records these values on a graphic record or flow sheet using a manual or automated documentation system. The move toward the use of the Celsius scale for measuring temperature requires that nurses know the relationship between Fahrenheit and Celsius temperatures as well.

In addition to these examples of the traditional use of mathematics in the clinical setting, nurses in administrative and management roles use elementary arithmetic and principles of accounting to prepare budgets, analyze health care agency data, and compute staffing requirements based on client acuity levels. For example, if a unit in a hospital staffs at 6.5 nursing care hours per patient per day (PPD), the nurse manager can compute the number of staff needed for a given census on that unit using basic arithmetic. While this role for nursing administrators and managers will continue into the twenty-first century, nurses at *all* levels in any health care setting are expected to provide quality health care in a cost-effective manner. Heretofore, most nurses have not been required to focus on the cost of staffing, equipment, and supplies while providing care. However, as the United States continues to move toward managed care as its primary health care delivery system, nurses must know what it costs to care for each of their clients. Third-party payers, such as Medicare, medical assistance, and traditional insurance companies, pay a designated amount of money to providers for care of their clients. Therefore, to remain financially solvent, health care providers must reduce costs while continuing to deliver quality care.

To ensure that quality of care is not sacrificed as a result of the shift to cost reduction, nurses participate in the continuous quality improvement (CQI) process in their agencies. CQI requires that nurses collect client data; statistically analyze the data, using descriptive methods; identify potential or actual problems; develop and implement improvement plans; and follow up to ensure that the plans were successful in improving care. This process is another example of why and how nurses use mathematics as a part of their role in a managed care environment. In addition, advanced-practice nurses, especially at the doctoral level, must be prepared to design and implement clinical research in nursing, and nurses with master's degrees may assist with research projects as needed. These nurses must be familiar with research design, descriptive statistics, and inferential statistics.

*See also* Health

## SELECTED REFERENCES

Booker, Marilyn, and Donna D. Ignatavicius. *Infusion Therapy: Techniques and Medications.* Philadelphia, PA: Saunders, 1996.

Lipsey, Sally I., and Donna D. Ignatavicius. *Math for Nurses.* Philadelphia, PA: Saunders, 1994.

Tappen, Ruth M. *Nursing Leadership and Management: Concepts and Practice.* Philadelphia, PA: Davis, 1995.

DONNA D. IGNATAVICIUS

# O

## ONE-TO-ONE CORRESPONDENCE

The two-way mapping of one set onto another, element by element. Mathematical historians consider this type of mapping a precursor to the development of numeration systems and counting (Bunt, Jones, and Bedient 1976). Early herdsmen, it is believed, kept track of the number of their animals by assigning each one a pebble or a carved notch on a stick or bone (Bunt, Jones, and Bedient 1976, 2–3). This could be accomplished without actually counting in the sense of determining a number to represent "how many." In early childhood education, one-to-one correspondence is the basic principle underlying the concept of number and is a prerequisite to "rational counting," "conservation," and equivalence (Charlesworth and Lind 1995). Reys, Suydam, and Lindquist (1989) describe as a "rational counter" the child who can (1) say the names of the numbers in the correct order, (2) assign one and only one number-name to each object, (3) understand that the order in which the objects are counted does not matter, and (4) understand that the last number named also represents the total number of objects in the set (pp. 74–75). Jean Piaget (1965) coined the term *conservation* to explain the ability to recognize that the quantity (or number) of a certain set of objects remains the same when the objects are spread out as when they are moved closer together. The one-to-one matching of discrete sets may be what Piaget (1965) refers to as "provoked correspondence," where each cup has a saucer or each foot wears a shoe, and so on.

Accurate use of the one-to-one principle is more difficult for young children than it might appear to adults (Ginsburg 1977). Rote recitation of the number names in the correct order does not necessarily transfer to a deep understanding of number, quantity, and what it means "to count." In fact, these are but two of the four skills involved in rational counting (Reys, Suydam, and Lindquist 1989). Given a set of objects, very young children may not be able to deal with each object in the set, one at a time. The concept of one-to-one correspondence develops through children's everyday experiences (Charlesworth and Lind 1995). These include self-initiated activities, that is, a child places a toy animal on each block, and adult-initiated activities, that is, a teacher passes out one pair of scissors to each child. The concept also can be learned from fair-play situations such as resolving the conflict where one child has two toys and the other has none. (See Charlesworth and Lind 1995 for further examples.) The development of the one-to-one principle is a necessary prerequisite to rational counting and further understanding of number concepts (Reys, Suydam, and Lindquist 1989).

*See also* Arithmetic; Number Sense and Numeration

SELECTED REFERENCES

Bunt, Lucas, Phillip Jones, and Jack Bedient. *The Historical Roots of Elementary Mathematics.* Englewood Cliffs, NJ: Prentice-Hall, 1976.

Charlesworth, Rosalind, and Karen K. Lind. *Math and Science for Young Children.* 2nd ed. Albany, NY: Delmar, 1995.

Ginsburg, Herbert. *Children's Arithmetic: The Learning Process.* New York: Van Nostrand, 1977.

Piaget, Jean. *The Child's Conception of Number.* New York: Norton, 1965. Originally published in French in 1941 under the title *La Genèse du Nombre chez l'Enfant.*

Reys, Robert E., Marilyn N. Suydam, and Mary Montgomery Lindquist. *Helping Children Learn Mathematics.* 2nd ed. Englewood Cliffs, NJ: Prentice-Hall, 1989.

ROSALIND CHARLESWORTH
ELIZABETH SENGER

# OPERATIONS RESEARCH

Commonly abbreviated "O.R.," a scientific basis for executives' management decisions through mathematical or statistical models of their systems. Operations research began to be recognized as a distinct discipline early in World War II in England, where it is commonly called "Operational Research." Other names for O.R. include management science, decision analysis, and systems analysis. Most experts agree that certain parts of mathematics, such as linear and dynamic programming and queueing theory, and certain areas of applications, such as inventory theory and the theory of network flows, are part of O.R. More advanced topics are integer, nonlinear, and stochastic programming. Some authors include other topics that are important for management decision making, such as scheduling, game theory, reliability, and the theory of Markov Chains, in their O.R. books. The typical content may be seen, for example, in Anderson, Sweeney, and Williams (1985).

As implied above, the original applications of O.R. was to the conduct of World War II. Successful investigations were made of such problems as methods for finding and attacking U-boats, deployment of antiaircraft batteries, aircraft maintenance, and the optimum size of convoys (Morse and Kimball 1951). The ideas spread rapidly into many areas of business and government. The Operations Research Society of America was formed in 1952, and The Institute of Management Sciences in 1953. A key feature of O.R. is that both mathematical and computer techniques play a critical role. Analytic solutions give great

power and allow users to understand their dependence on parameters. On the other hand, the real situation is sometimes too complicated for analysis to be completely successful, and simulation is necessary to obtain specific results. It is then more difficult to be sure which assumptions are critical.

The twin motivations of correctness and practicality that motivate O.R. have meant that a number of famous O.R. problems have been important to theoretical computer science. The so-called traveling salesperson problem, in which one seeks the shortest closed loop through a series of $n$ locations, is a key example of a problem for which it is not known whether it can be solved in a number of steps that is polynomial in $n$. Similarly, the question whether linear programming requires a number of steps more than polynomial in the number of variables and constraints was unsolved for a long time. Practice has shown the most widely used solution method, the simplex algorithm of George Dantzig, to be quite efficient. An artificial example constructed by Victor Klee, however, required an unacceptably large number of steps for its solution. L. G. Khatchian in 1979 finally proved the problem to be polynomial, and Narendra Karmarkar in 1984 gave a practical algorithm for really large problems, with on the order of $10^6$ variables and constraints. Such problems are important, for example, in scheduling airline operations (Garfunkel 1994).

A first course in O.R. often requires algebra, geometry, discrete probability, and elementary statistics as prerequisites. Typically, very little if any calculus is needed initially, but there is frequent use of linear algebra and matrices. It often turns out to be practical to make the necessary probability and statistics part of the O.R. course. The best layout and use of networks for such purposes as communication or transport is sometimes taught as part of graph theory in a discrete mathematics course, or else the graph theory is developed as needed.

Applications of mathematics play an increasingly central role in all of school mathematics teaching, but many traditional applications require more knowledge of physical science than many students have acquired. Thus, examples from the world of decision making, which often depend less on technical knowledge and more on familiarity with everyday life, tend to be particularly popular. Here are some O.R. problems that are found in middle and high school curricula:

- The shortest tree connecting a given set of locations can be found by two different algorithms, attributed to Joseph Kruskal and Robert Prim, respectively. Both are easy to understand, to carry out, and to program. How do the two compare?

- The shortest path in a network between two given points $A$ and $B$ is easy to find, but no easier than the shortest paths from $A$ to all other points by essentially the same algorithm.

- Inventory control. In its simplest form, a store sells $k$ items per week at a price of $m$ dollars, it costs $h$ dollars a week to store an unsold item (this represents, for instance, the cost of capital), and it costs $p$ dollars for each order the store writes to its supplier. If it orders too infrequently, too much money is tied up in inventory; if it orders too frequently, ordering costs are overwhelming. What is the best strategy?

- What is the best sequence of operations for, say, cooking Thanksgiving dinner? Certain steps must precede certain others, and you know how long each step takes and how much help you have. When do you start and who does what when? Where are the bottlenecks?

Many O.R. problems, like the inventory problem above, have forces that pull in opposite directions. When the two desiderata are measured in the same units—dollars in that example—an optimization problem is easy to formulate. But what if the two desired features are measured in different units? For example, the logic used in scheduling a bank of elevators tries to minimize the energy consumption in running the elevators and also the passengers' waiting time! Now you can only minimize one quantity at once, and you do not know how to trade energy dollars for waiting time. One popular technique in such problems, which often occur in the real world, is to plot the boundary between possible and impossible pairs of values—in this case in a plane whose respective coordinate axes represent energy consumption and waiting time.

In addition to a variety of O.R. problems beginning to appear in school textbooks, a number of publications have been devoted specifically to such problems for use at the high school or early college level. (See Garfunkel 1994; North Carolina . . . 1987; and Perham and Perham 1992.)

Linear programming is perhaps the most widely used O.R. model, and O.R. courses tend to spend a lot of time on it. Unfortunately, the pictures that really show the student what is going on are easy to draw only in two dimensions, which means that the problem has only two or at most three variables. Easy

examples of linear programming involve transporting goods from several sources to several destinations (called transportation problems), the best way to allocate resources among several possible manufacturing schedules, or the cheapest way to meet a daily nutritional requirement. In none of these cases is it easy to give a realistic problem with only two variables! At least one fascinating real-world linear programming problem with only three variables has been published (Griffiths and Hassan 1978) and has been used with high school students.

*See also* Discrete Mathematics; Game Theory; Graph Theory, Overview; Linear Programming; Markov Processes; Modeling; Queueing Theory

## SELECTED REFERENCES

Anderson, David R., Dennis J. Sweeney, and Thomas A. Williams. *An Introduction to Management Science.* 4th ed. St. Paul, MN: West, 1985.

Garfunkel, Solomon, proj. dir. *For All Practical Purposes.* 3rd ed. New York: Freeman, 1994.

Griffiths, J. D., and E. M. Hassan. "Increasing the Shipping Capacity of the Suez Canal." *Journal of Navigation* 31 (May 1978):219–231.

Morse, Philip M., and George E. Kimball. *Methods of Operations Research.* New York: Wiley, 1951.

North Carolina School of Science and Mathematics, Department of Mathematics and Computer Science. *Contemporary Applied Mathematics Series.* Providence, RI: Janson, 1987.

Perham, Bernadette H., and Arnold E. Perham. *Topics in Discrete Mathematics.* 5 units. *Matrix Theory; Game Theory; Linear Programming Theory; Markov Chain Theory; Graph Theory.* Menlo Park, CA: Addison-Wesley, 1992.

HENRY O. POLLAK

# ORDINARY DIFFERENTIAL EQUATIONS

Physical laws involving rates of change (such as Newton's laws of motion) written in terms of differential equations. When there is one independent variable in a problem (such as a single spatial dimension or time), so-called ordinary differential equations are used. For example, the physical law that states that force is equal to the product of mass and acceleration can be written as $F = ma = mv' = mp''$ (here, $p$ represents position as a function of time $t$ and $v$ represents velocity).

An ordinary differential equation (ODE) is an equation in which the derivative of an unknown function is present. The order of the differential equation is the number of derivatives of the unknown function appearing in the equation. For only one unknown function, $u(x)$, an ordinary differential equation has the general form $F(x, u, u', u'', \ldots, u^{(n)}) = 0$. In a linear ordinary differential equation, the unknown function $u$ and its derivatives only appear to the zeroth and first powers. In this case, the equation has the general form $f_0(x)u + f_1(x)u' + \cdots + f_n(x)u^{(n)} = h(x)$. For example, the motion of an undamped pendulum near its equilibrium position (straight down) satisfies $u'' + mgu = 0$.

An ordinary differential equation can involve more than one unknown function, and there may be more than one ODE involving the unknown function(s). If so, one then has a system of ODEs. The most general system of ordinary differential equations has the form $\mathbf{v}' = \mathbf{f}(x,\mathbf{v})$, where $\mathbf{v}(x)$ is a vector of independent variables all depending on the single variable $x$ and $\mathbf{f}$ is a vector-valued function. For example, the equation of the pendulum described above could be written as $\{u' = v, v' = -mgu\}$, or as

$$\begin{pmatrix} u \\ v \end{pmatrix}' = \begin{pmatrix} v \\ -mgu \end{pmatrix}$$

Generally, an ODE has a single, unique solution. The form of the equation, the type of information given to the equation, and the actual information given, however, all influence the number of solutions. The number of equations needed to solve uniquely a system of ODEs may be more or less than the number of unknown functions; frequently it is the same.

An autonomous differential equation is one for which the independent variable does not explicitly appear. The most general autonomous ODE system has the form $\mathbf{v}' = \mathbf{f}(\mathbf{v})$; for a single independent variable the form is $F(u, u', u'', \ldots, u^{(n)}) = 0$. If $\mathbf{v}(x) = \mathbf{b}(x)$ and $u(x) = a(x)$ are solutions to these equations, then $\mathbf{v}(x) = \mathbf{b}(x + c)$ and $u(x) = a(x + c)$ are also solutions, for any constant $c$.

An initial value problem (IVP) for a differential equation is one for which the values of the dependent variable (and, perhaps, its derivatives) are specified at one value of the independent variable. A boundary value problem (BVP) is one for which the dependent variables (and, perhaps, its derivatives) are specified at more than one value of the independent variable. For example, consider the pendulum above. Specifying the pendulum position and velocity at one instant of time (say, $p(0) = 1$ and $v(0) = 0$, which corresponds to displacing the pendulum and letting it drop) results in an IVP. Specifying the pendulum position at two instants of time (say, $p(0) = 1$

and $p(1) = 1$, which corresponds to displacing the pendulum and giving it a sufficient push so that it returns to its starting position in a single unit of time), results in a BVP.

The solution to the above IVP is $u(t) = \cos(\sqrt{mg}t)$ while the solution to the above BVP is, usually,

$$u(t) = \cos(\sqrt{mg}t) + (((1 - \cos(\sqrt{mg}))/\sin(\sqrt{mg}))\sin(\sqrt{mg}t)$$

If $\sqrt{mg} = 2n\pi$, however, then the BVP has the solution $u(t) = \cos(\sqrt{mg}t) + A\sin(\sqrt{mg}t)$ for any value of $A$, and if $\sqrt{mg} = (2n + 1)\pi$, then the BVP has no solution.

In the study of differential equations, there is a common three-step process: (1) determine if the given problem has a solution, (2) determine if the problem has a unique solution, and (3) determine if the unique solution is stable. There are many definitions of stability, each applicable to different situations. The concept for each is the same: if small changes in the initial data (or the boundary data, or the coefficients in the equation, or some other parameter in the problem statement) result in corresponding small changes in the solution, then the equation is stable. A chaotic system is unstable to such a degree that small changes in the original problem statement results in huge changes in the solution.

Exact analytical solutions have only been found for limited classes of ordinary differential equations. Linear initial value problems are frequently solved by Laplace transform methods. Linear boundary value problems are frequently solved by Fourier transform methods. The most important technique for symbolically finding exact solutions is the method of Lie groups (which uses continuous transformation groups). Current computer algebra systems (such as Macsyma, Maple, or *Mathematica*) can solve symbolically a wide variety of first-order ordinary differential equations and linear second-order ordinary differential equations. Sometimes an approximate solution of an ordinary differential equation can be analytically obtained; frequently ordinary differential equations are solved numerically. Special schemes have been devised for nearly every type of ordinary differential equation. The major numerical methods include the finite element method and the method of finite differences.

Ordinary differential equations arise naturally in many physical models. Generally, in the construction of an equation whose solution represents the result of some physical process, each physical effect that is to be considered is reflected in one or more terms of the ordinary differential equation(s). For example, the ODE for the pendulum given above has a term corresponding to gravity and an inertial term. If the pendulum were damped, then the correct ODE would be $u'' + cu' + mgu = 0$ (which includes a term corresponding to damping effects). Alternately, ordinary differential equations are sometimes obtained during the analysis of partial differential equations (especially when using transform methods or separation of variables).

Because of the close connection with physical problems, the main subject of ODE research has traditionally been linear equations of the second order. Current interest centers on nonlinear equations and newly discovered phenomena such as chaos (chaos cannot appear in linear ODEs). Some of the more widely studied ordinary differential equations are the Bernoulli equation ($y' = a(x)y^n + b(x)y$), the Riccati equation ($y' = a(x)y^2 + b(x)y + c(x)$), the Airy equation ($y'' = xy$), the Bessel equation ($x^2y'' + xy' + (x^2 - n^2)y = 0$), and the Lorenz equations (which are chaotic).

*See also* Chaos; Modeling; Numerical Analysis, Methods

## SELECTED REFERENCES

Bender, Carl M., and Steven A. Orszag. *Advanced Mathematical Methods for Scientists and Engineers*. New York: McGraw-Hill, 1978.

Ince, Edward L. *Ordinary Differential Equations*. New York: Dover, 1964.

Zwillinger, Daniel. *Handbook of Differential Equations*. 2nd ed. Boston, MA: Academic, 1992.

DANIEL ZWILLINGER

# P

## PARADOX

A self-contradictory statement or argument (mathematical or logical), sometimes called a *fallacy*. Another type of paradox is a statement that is true but is unexpected or unlikely. An example of a fallacy is the intentionally incorrect "proof" of an obviously incorrect result. In such a paradox, the author gives reasons to support the validity of each step of the proof, yet the final result is certainly wrong. The intent is to challenge the reader to find out where the mistake has crept in. As an example, consider the following proof that $2 = 1$. Begin by taking any two numbers, $a$ and $b$, that are equal to each other. Then,

1. by assumption:      $a = b$
2. multiply by $a$:      $a^2 = ab$
3. subtract $b^2$:      $a^2 - b^2 = ab - b^2$
4. factor:      $(a - b)(a + b) = b(a - b)$
5. divide by $(a - b)$:      $a + b = b$
6. by step (1), replace $b$:      $a + a = a$
7. collect like terms:      $2a = a$
8. divide by $a$:      $2 = 1$

The mistake occurs in line 5, where the divisor, $(a - b)$, has the value 0, since $a = b$ from line 1. Dividing both sides of an equation by a common factor is usually a correct process, but not always; it can only be done when that factor is different from 0.

Many algebraic and arithmetical paradoxes involve dividing by 0, as shown in the example. Another common device used by paradox creators is taking square roots. If a reader naively accepts that $\sqrt{x^2} = x$ (instead of the correct statement $\sqrt{x^2} = |x|$), then the argument $2 = \sqrt{4} = \sqrt{(-2)^2} = -2$ might appear correct but is certainly a paradox, since the conclusion $2 = -2$ is ridiculous. The most common paradoxes in arithmetic involve either square roots or zero denominators. The best paradox authors are clever at concealing the false step. The educational value of studying paradoxes is that the student must not merely find the wrong step in the argument but should also confirm the majority of the steps as being correct.

There are also paradoxes in other areas of mathematics. In geometry, there are various proofs of results like "all triangles are equilateral" and "all angles are right angles." One has to track down the hidden defect. Perhaps it occurs when two lines are assumed to intersect, when there is a possibility that they could be parallel. Here is an interesting geometrical fallacy. Suppose $ABCD$ is a quadrilateral with $AB = CD$. We shall "prove" that $AD$ is parallel to $BC$. The method involves using a preliminary result; mathematicians call it a *lemma*. This lemma states that if in quadrilateral $ABCD$, $AB = CD$ and the base angles $A$ and $D$ are equal, as in Figure 1, then

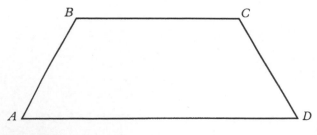

**Figure 1**

$AD \parallel BC$. The proof is adapted from Maxwell (1959). It uses the diagonals $AC$ and $BD$. Then there are two pairs of congruent triangles: $\triangle ABD \cong \triangle ACD$ and $\triangle ABC \cong \triangle DBC$. Choosing the correct pairs of matching angles from these triangles, we can quickly show that the interior angles $A$ and $B$ add up to half the sum of the total of the four interior angles of the quadrilateral. The latter sum is 360°. So for the two interior angles, $\angle A + \angle B = 180°$. Those are alternate angles for sides $AD$ and $BC$, and so $AD \parallel BC$ and the lemma is proved. No paradox so far. Next consider Figure 2, where we shall again assume $AB = CD$ but nothing about the base angles $A$ and $D$. In fact, we are going to "prove" that those angles are equal. The construction lines are shown in Figure 3.

We raise perpendicular bisectors to both $AD$ and $BC$ with feet $M$ and $N$. If they are parallel, we are

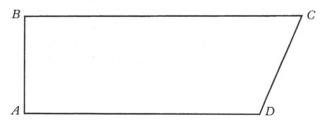

**Figure 2**

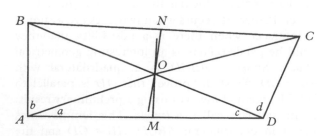

**Figure 3**

done. So assume not, and suppose that $O$ is the point of intersection. Draw the segments to the four vertices, $OA$, $OB$, $OC$, and $OD$. Again we have two pairs of congruent triangles: $\triangle AOM \cong \triangle DOM$ and $\triangle AOB \cong \triangle DOC$. For congruence proof, the first pair needs only the perpendicular bisector property of $OM$ as the locus of points equidistant from $A$ and $D$. The other pair of triangles uses the perpendicular bisector feature for both $OM$ and $ON$ and the given condition that $AB = CD$. Then, using the appropriate matching angles, $\angle a = \angle c$ and $\angle b = \angle d$. The final step is to add. Thus $\angle a + \angle b = \angle c + \angle d$, that is, the base angles $A$ and $D$ of the quadrilateral are equal. So the lemma applies, and the result $AD \parallel BC$ follows.

Paradoxes like the ones here appear to go back in mathematical experience for hundreds of years, and many paradoxes have no identifiable author.

## SOME RELATED CONCEPTS

Paradoxes should be distinguished from *errors*, which are cases where a mathematical proof is presented that has an unintended mistake. The latter is simply an embarrassment to the author. Another related but quite distinct idea is the case of a deliberately incorrect proof of a correct result presented in a way that is transparently mistaken. That kind of intentional mathematical amusement is called a *howler*. For example, consider the fraction 16/64. If we were to "cancel" the two "6" digits, one in the numerator and one in the denominator, we would get 1̸6/6̸4 = 1/4. This is a correct conclusion, but the steps leading to it are not. You should distinguish carefully between a paradox, which is a purportedly correct proof of a known false result, and a howler, which is an admittedly wrong proof of a true conclusion.

The field of logic has its own concept of paradox, different from the mathematical one. An example of a logical paradox is the sentence, "This statement is a lie." Is the sentence true? Suppose it is true. Then as a true statement, it must state a true fact. And that fact is that the statement is a lie. So it really is not true, contrary to what we supposed a few lines back. Well then, suppose instead the sentence in question is not true. Then it is a lie, and that is exactly what the statement says. So it really is true, contrary to the second trial assumption. There is no way that the sentence can be either true or not true. Logicians and mathematicians have been worried about that kind of paradox since the Englishman Bertrand Russell proposed it early in the twentieth century.

There is still no generally agreed upon way to resolve it. It is not a wrong result so much as it is an impossible one. That is a challenge more for philosophers and logicians than mathematicians.

A much more elementary logical paradox is one that is associated with the ancient Greek philosopher Zeno. He argued that since there are infinitely many points between the two ends $A$ and $B$ of a line segment, one cannot move from $A$ to $B$ because it must take an infinite amount of time to complete an infinite number of actions. Of course that argument is nonsense (and Zeno certainly knew it was) simply from the universal experience that people do go from here to there. Zeno's use of the paradox affected the development of Greek mathematics; the realm of number continued to have the property of discreteness, as before, but continuous magnitudes came to be treated differently, through geometry (Boyer 1968, 84). Centuries later, the formal mathematical explanation of Zeno's paradox utilized the theory of limits and convergence of infinite series (see Knopp 1928).

Finally, the term *paradox* can be used in its ordinary English-language meaning of something that is true but unexpected or unlikely. This does not have to be mathematical in nature, but there are some cases that are. For example, consider a group of 24 people whose birthdays are randomly distributed among the 365 days of the year. What is the probability that 2 of them have the same birthday (month and day)? Most people initially guess that the probability is quite low. In fact, it is better than 50%. This is one a teacher can make money on, if he or she can get students or colleagues to give good odds. Bet on finding a matching pair of birthday twins, and you will win more than half the time! It is also a stimulating educational example to show how the theory overwhelms the contrary human intuition.

Here is how the theory of the birthday-twin paradox works. Reverse the question, and ask for the probability that there are not any duplicate birthdates. With a "group" of one person, the probability of "no duplication" is 1. Consider now a second person in the group. The probability that the second birthdate does not match is 364/365. The reasoning is that there are 365 possible birthdates for person #2, and 364 of them are nonmatching. How about person #3? To have a nonmatch again, the birthday of that person must be among the 363 dates that do not match the birthdays of persons #1 and #2. Using the multiplication rule for independent probabilities, the event "no matching birthdays in a group of 3" is given by (364/365)(363/365). The same argument

continues to apply for a group of 4. The answer is (364/365)(363/365)(362/365). If the group size is $N$, the probability can be written as a fairly simply formula. It is $P = (365!)/[(365 - N)!(365^N)]$. Although the arithmetic becomes a bit burdensome, it can be done. If $N$ is 24 or greater, the probability of no match is less than 0.5; so the probability of having a match is better than 50%.

*See also* Logic; Probability Applications: Chances of Matches; Proof; Recreations, Overview

## SELECTED REFERENCES

Ball, Walter W. R. *Mathematical Recreations and Essays.* Harold S. M. Coxeter, ed. New York: Macmillan, 1947.

Boyer, Carl B. *A History of Mathematics.* New York: Wiley, 1968.

Bradis, Vladimir M., V. I. Minkovskii, and A. K. Kharcheva. *Lapses in Mathematical Reasoning.* J. J. Schorrkon, trans. New York: Macmillan, 1963.

Kempner, Aubrey J. *Paradoxes and Common Sense.* Princeton, NJ: Van Nostrand, 1959.

Kleiner, Israel, and Nitsa B. Movshovitz-Hadar. "The Role of Paradoxes in the Evolution of Mathematics." *American Mathematical Monthly* 101(1994):963–974.

Knopp, Konrad. *Theory and Applications of Infinite Series.* R. C. Young, trans. New York: Hafner, 1928.

Maxwell, Edwin A. *Fallacies in Mathematics.* Cambridge, England: Cambridge University Press, 1959.

BENJAMIN L. SCHWARTZ

# PARAMETRIC EQUATIONS

A plane curve in the Cartesian coordinate system may be given by a single equation such as $y = f(x)$, $x = g(y)$, or $f(x,y) = 0$. Examples: $y = x^2$ and $x = y^2 + y$ represent parabolas, while $x^2 + y^2 = 1$ represents a circle. In the more general method of giving the curve, two equations express $x$ and $y$ as a function in a third variable called a *parameter*. The two equations are called parametric equations, $x$ and $y$ are expressed in the form $x = f(t)$ and $y = g(t)$, where $f$ and $g$ are defined on the same interval $I$ and $t$ is the parameter of the curve. Each value of $t$ in $I$ determines a point $(x,y)$ in the $xy$-plane, and the set of all such points makes up the curve. For example, to sketch the curve given by $x = (t - 1)^2$ and $y = t + 1$, $-2 \leq t \leq 4$, one calculates first the values of $x$ and $y$ corresponding to several values of $t$. After plotting the points $(x,y)$ in a rectangular coordinate plane, one joins these points to get the curve shown in Figure 1. The arrows on the curve indicate the direction in which the curve is

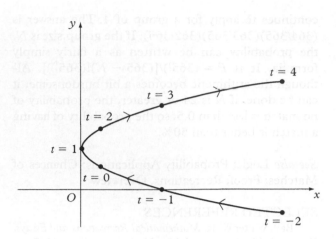

**Figure 1**

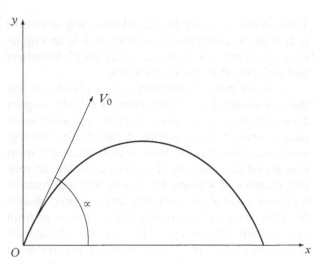

**Figure 2**

traced with increasing values of the parameter. To find the Cartesian equation of this curve, one eliminates the parameter $t$ from the parametric equations. Since $t = y - 1$, then $x = (y - 2)^2$, which represents a parabola.

In parametric equations, the parameter may represent time or may have a geometric interpretation, as illustrated by some of the following examples.

*Example* If a projectile is fired with an initial velocity of $v_0$ at an angle $\alpha$ above the horizontal and air resistance is disregarded, then it can be shown that the position of the projectile after $t$ seconds is given by the parametric equations: $x = (v_0 \cos \alpha)t$ and $y = (v_0 \sin \alpha)t - gt^2/2$, where $g$ is the acceleration due to gravity. Since $t = x/(v_0 \cos \alpha)$, then $y = x \tan \alpha - [g/(2v_0^2 \cos^2 \alpha)]x^2$. This shows that the trajectory of the projectile is part of a parabola that opens downward (Figure 2).

*Example* $x = a \cos \theta$, $y = b \sin \theta$; $0 < b < a$, $0 \le \theta < 2\pi$ are parametric equations of the ellipse $x^2/a^2 + y^2/b^2 = 1$. That is because $x/a = \cos \theta$ and $y/b = \sin \theta$ imply that $(x/a)^2 + (y/b)^2 = \cos^2\theta + \sin^2\theta = 1$. In Figure 3, the circle with the same center as the ellipse and radius equal to $a$ is called the *auxiliary circle* of the ellipse. By considering the intersections of the terminal side of $\theta$ with the ellipse and the auxiliary circle, each point $P(a \cos \theta, b \sin \theta)$ on the ellipse is associated with a point $P'$ $(a \cos \theta, a \sin \theta)$ on the auxiliary circle. This implies that the parameter $\theta$ represents the measure of the angle between $OP'$ and the positive direction of the x-axis.

Sometimes the easiest way to define a curve algebraically is to derive parametric equations for it.

*Example* Suppose a circle rolls without slipping along the x-axis. If $P$ is a point on the circle, then it traces a curve called the *cycloid*. Let us assume that $P$ was at the origin when the rolling started. In Figure 4, $P$ takes the position $(x,y)$. If $\angle ACP = \theta$ and the radius of the circle is $a$, then

$$x = OA - PB = a\theta - a \sin \theta$$

$$y = AC - BC = a - a \cos \theta$$

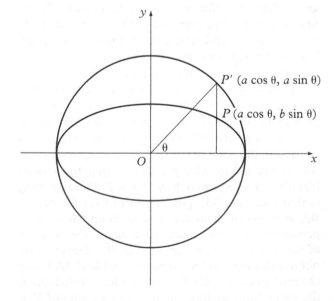

**Figure 3**

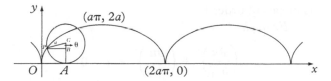

**Figure 4**

Thus, the cycloid is defined by the parametric equations

$$x = a(\theta - \sin \theta) \qquad y = a(1 - \cos \theta)$$

The maximum height of this cycloid is $2a$, which occurs at $\theta = (2n - 1)\pi$, where $n$ is an integer. At $\theta = 2n\pi$ the cycloid has cusps.

The history of the cycloid goes back to Galileo (1564–1642), who was the first to trace it. The scientist Christian Huygens proved that no matter where a particle is placed on an inverted cycloid, it takes the same time to slide to the bottom. In 1658, Huygens used this result in designing a pendulum clock whose oscillations are isochronous, that is, take exactly the same time (Boyer 1989).

The process of defining a curve by parametric equations is called *parametrization*. Parametrization of a curve is not unique, as illustrated by the case of the circle $x^2 + y^2 = a^2$. This circle can be defined parametrically by $x = a \cos \theta, y = a \sin \theta$; $0 \leq \theta < 2\pi$. Also, it can be defined by

$$x = a \sin 2\theta, y = a \cos 2\theta; \qquad 0 \leq \theta < \pi$$

or

$$x = t, y = \pm \sqrt{(a^2 - t^2)} \qquad -a \leq t \leq a$$

If a curve is given by the polar equation $r = f(\theta)$, then the easiest way to parametrize this curve is to make use of the relations $x = r \cos \theta$ and $y = r \sin \theta$. So we get, in this case, the parametric equations: $x = f(\theta) \cos \theta$ and $y = f(\theta) \sin \theta$, where now $\theta$ is playing the role of a parameter. For example, the cardioid $r = 1 + \cos \theta$ has the parametric equations $x = (1 + \cos \theta) \cos \theta$ and $y = (1 + \cos \theta) \sin \theta$, $0 \leq \theta < \pi$.

A curve in three-dimensional space may be defined by a set of three parametric equations in the form $x = f(t), y = g(t),$ and $z = h(t)$ for all $t$ in some interval $I$.

*Example* $x = \cos t, y = \sin t, z = t; 0 \leq t \leq 4\pi$ is a curve in space called a *circular helix*. This curve lies on the circular cylinder $x^2 + y^2 = 1$ (Figure 5a).

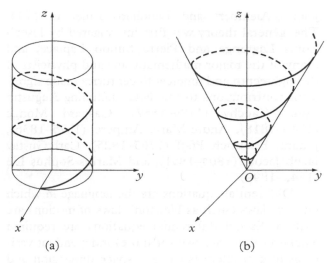

(a)            (b)

**Figure 5**

*Example* $x = t \cos t, y = t \sin t, z = t; t \geq 0$ is a curve in space called a *conical helix*. This curve lies on the right circular cone $z^2 = x^2 + y^2$ (Figure 5b).

For three-dimensional space curves, the formulas to calculate arc length and curvature are based on the parametric representation of these curves. Also, to study the path of an object moving in space, the parametric equations are the natural format to use. High school students learn very little about parametric equations. They should be encouraged to use their graphing calculators to explore new, interesting curves defined by parametric equations.

*See also* Conic Sections (Conics); Coordinate Geometry, Overview

### SELECTED REFERENCES

Boyer, Carl B. *A History of Mathematics*. Revised by Uta C. Merzbach. 2nd ed. New York: Wiley, 1989.

Mizrahi, Abe, and Michael Sullivan. *Calculus*. 3rd ed. Belmont, CA: Wadsworth, 1990.

Stewart, James. *Calculus*. 2nd ed. Pacific Grove, CA: Brooks/Cole, 1991.

Swokowski, Earl W. *Calculus*. 5th ed. Boston, MA: PWS-Kent, 1991.

AYOUB B. AYOUB

# PARTIAL DIFFERENTIAL EQUATIONS

Equations in which a partial derivative of an unknown function is present. These equations originated in the study of hydrodynamic problems by

Jean d'Alembert and Leonhard Euler in 1744. The general theory was first investigated by Joseph Louis Lagrange and Pierre Simon Laplace, and many of the major mathematicians and physicists of the eighteenth and nineteenth centuries made significant contributions to this field, including Augustin Louis Cauchy (1789–1857), Gaspard Monge (1746–1818), André-Marie Ampère (1775–1836), Johann Friedrich Pfaff (1765–1825), Carl Gustav Jacob Jacobi (1804–1851), and Marius Sophus Lie (1842–1899).

Differential equations are the language in which physical laws (such as Newton's laws of motion) are written. Partial differential equations are required whenever there are two (or more) independent variables in a problem (such as a space dimension and time). Generally, in the construction of an equation whose solution represents the result of some physical process, each different physical effect that is to be considered is reflected in the terms of the partial differential equation(s). For example, an equation describing a fluid mechanics problem may have one term that represents the effects of viscosity and one term that represents the effects of pressure.

The order of the differential equation is the number of derivatives of the unknown function appearing in the equation. For a single dependent variable $u(x) = u(x_1, \ldots, x_n)$, a partial differential equation of $k$th order has the general form

$$F\left(x_1, \ldots, x_n, u, \frac{\partial u}{\partial x_1}, \ldots, \frac{\partial u}{\partial x_n}, \ldots, \right.$$
$$\left. \frac{\partial^k u}{\partial x_1^k}, \ldots, \frac{\partial^k u}{\partial x_n^k}\right) = 0$$

A partial differential equation is called *linear* if $u$ and all derivatives of $u$ appear separately to the first degree. In this case, the equation has the general form

$$f_0(x)u + f_1(x)\frac{\partial u}{\partial x_1} + \cdots + f_n(x)\frac{\partial u}{\partial x_n} + f_{11}(x)\frac{\partial^2 u}{\partial x_1^2}$$
$$+ f_{12}(x)\frac{\partial^2 u}{\partial x_1 \partial x_2} + \cdots + f_{1\ldots1}(x)\frac{\partial^k u}{\partial x_1^k}$$
$$+ \cdots + f_{n\ldots n}(x)\frac{\partial^k u}{\partial x_n^k} = g(x)$$

There are no restrictions on the functions $\{g, f_0, \ldots, f_{n\ldots n}\}$. To cite some examples, consider $u(x) = u(x_1, x_2)$.

*Example*

$$x_1\frac{\partial u}{\partial x_1} + x_2\frac{\partial u}{\partial x_2} - 3u = 0$$

is linear, of order 1.

*Example*

$$\left(\frac{\partial u}{\partial x_1}\right)^2 + \left(\frac{\partial u}{\partial x_2}\right)^2 - u = 0$$

is nonlinear, of order 1.

When writing partial differential equations, we use the Laplacian operator (denoted $\nabla^2$) frequently; it is defined by

$$\nabla^2 = \sum_{i=1}^{n} u_{x_i x_i}$$

A partial differential equation (PDE) can involve more than one unknown function and there may be more than one involving the unknown function(s). One then has a system of partial differential equations. The number of solutions of a partial differential equation depends on the form of the equation, the type of information given to the equation (whether initial conditions, boundary conditions, or some combination), and the actual data. The number of equations needed to solve uniquely a system of partial differential equations may be more or less than the number of unknown functions; frequently, it is the same.

Partial differential equations are classified into three major classes: elliptic equations (which generally describe steady-state or potential phenomena), hyperbolic equations (which generally describe wave motion), and parabolic equations (which generally describe diffusive processes). Mixed-type equations switch from one type to another. For example, a PDE describing the air flow around an airplane wing may change type from hyperbolic in the supersonic region near the wing to elliptic in the subsonic region far from the wing.

As in ordinary differential equations, exact solutions have been found only for limited classes of partial differential equations. Frequently, partial differential equations are solved numerically. There are many numerical approximation schemes, and special schemes have been devised for nearly every type of partial differential equation. The two major numerical methods are the method of finite differences (in which each derivative of $u(x)$ is approximated by a linear combination of values of $u(x + \Delta x)$ and the resulting algebraic equations are solved) and the finite element method (in which a variation formulation of the equation is first obtained).

Because of their close connection with physical problems, the main subject of PDE research has traditionally been linear equations of the second order.

Current interest centers on nonlinear equations and newly discovered phenomena such as chaos. Some of the more widely studied partial differential equations are Schrödinger's equation (describing quantum mechanical systems), Einstein's equations (describing general relativity), the Navier-Stokes equations (describing fluid mechanical systems), the wave equation ($\nabla^2 u = u_{tt}$, which describes the propagation of sound), Helmholtz's equation ($\nabla^2 u = k^2 u$, which describes the resonate frequencies of a plate), and Laplace's equation ($\nabla^2 u = 0$, which describes the electromagnetic field surrounding an electrical cord).

*See also* Numerical Analysis, Methods; Ordinary Differential Equations

## SELECTED REFERENCES

Bender, Carl M., and Steven A. Orszag. *Advanced Mathematical Methods for Scientists and Engineers.* New York: McGraw-Hill, 1978.

Zauderer, Erich. *Partial Differential Equations of Applied Mathematics.* 2nd ed. New York: Wiley, 1989.

Zwillinger, Daniel. *Handbook of Differential Equations.* 2nd ed. Boston, MA: Academic, 1992.

DANIEL ZWILLINGER

## PASCAL, BLAISE (1623–1662)

When still a teenager, Pascal was introduced by his father, Etienne, to the mathematical circle in Paris around cleric and mathematical correspondent Marin Mersenne (1588–1648). He began his own mathematical researches before the age of 20 and soon built a small calculating machine. He is best known for his 1654 study of "Pascal's triangle," the array of integers representing either binomial coefficients or numbers of combinations that had been first studied in China, India, and the Middle East some 600 years earlier. Pascal developed and proved numerous properties of the triangle (sometimes explicitly using the principle of mathematical induction) and applied these properties to problems in probability and combinatorics.

Modern probability theory stems from Pascal's correspondence with Pierre de Fermat (1601–1665) in which they discussed the problem of the equitable division of the stakes in a game interrupted before its conclusion. Pascal further used his studies of the triangle to develop formulas for the sums of powers of the integers, which he then applied to derive the formula for integrating $x^k$. About the same time, he in-

vented the differential triangle, the infinitesimal triangle later applied by Gottfried Leibniz in his invention of the calculus, and used it in his evaluation of the definite integral of the sine and its powers.

*See also* Pascal's Triangle

## SELECTED REFERENCES

David, Florence N. *Games, Gods, and Gambling.* New York: Hafner, 1962.

Edwards, Anthony W. F. *Pascal's Arithmetical Triangle.* New York: Oxford University Press, 1987.

Taton, René. *L'oeuvre scientifique de Pascal.* Paris, France: Presses Universitaires, 1964.

VICTOR J. KATZ

## PASCAL'S TRIANGLE

A triangular array of integers usually written in the form of an equilateral triangle with 1s at the top vertex and running down the two slanting sides. It has an unlimited number of rows. Each integer in the interior of the triangle is the sum of the two integers immediately above it.

Motivation for the triangle is clear upon considering the problem of finding the coefficients in the binomial expansion $(x + y)^n$. For example, to find the coefficients of $(x + y)^3$ multiply $(x + y)^2$ by $(x + y)$. Abstract the multiplication process by ignoring $x$'s and $y$'s and treating them only as place holders, thus multiplying 1 2 1 by 1 1. Equivalently, take the 1 2 1 and write under it, shifted to the left, 1 2 1. Then add column by column, yielding 1 3 3 1. So if 1 2 1 are the coefficients of $(x + y)^2$, then 1 3 3 1 are the coefficients of $(x + y)^3$. This leads to an understanding of the process of generating Pascal's triangle.

We also can look at the problem of expanding $(x + y)^n$ in combinatorial terms. Since $(x + y)^3 = (x + y)(x + y)(x + y)$, to find the number of terms of the form $x^2 y$, say, find the number of ways to choose two $x$'s from the set of three (or, alternatively, one $y$ from the set of three) in the factorization. This shows that each coefficient in a binomial expansion is the number of combinations of choosing $r$ objects from a set of $n$, where $n$ is the exponent of $x + y$ and $r$ is either the exponent of $x$ or the exponent of $y$.

Thus the elements of Pascal's triangle, being the co-efficients in the binomial expansion, are also combinations. It was this relationship that was explained by Blaise Pascal in his 1654 paper, *Traité du Triangle Arithmétique*. In this paper Pascal also derived a number of properties of the triangle and related the elements to questions in probability, motivated by a gambling problem posed to him by the Chevalier de Méré.

It is not clear to what extent Pascal realized that this array of numbers was known earlier. The binomial coefficients were already known to Omar Khayyám in Persia and Bhaskra in India in the twelfth century. In the thirteenth century, Pascal's triangle was discovered by the Chinese, Yang Hui and Chu âShih-chieh, and rediscovered in Persia by Nasir-Ad Din At-Tusi, though this latter fact was not discovered until the 1960s (Boyer 1989, 231; Berge 1971, 6). In Europe, Pascal's triangle appeared in a work of Peter Apian in 1527 and again in 1544 in the *Arithmetica Integra* of Michael Stifel (Boyer 1989, 315–316).

Many interesting numerical facts are hidden in Pascal's triangle. It is therefore ideally suited to investigations by students. For example, it was observed only within the last few decades that the product of the six integers surrounding any interior element is always a perfect square (Hansell and Hoggatt 1971). There are many generalizations of Pascal's triangle. These include Leibniz's harmonic triangle and the Pascal pyramid (tetrahedron), which gives the coefficients in the expansion of $(x + y + z)^n$ (Pólya 1981, 88–89).

*See also* Combinatorics, Overview; Pascal

SELECTED REFERENCES

Berge, Claude. *Principles of Combinatorics*. New York: Academic, 1971.

Boyer, Carl B. *A History of Mathematics*. 2nd ed. Revised by Uta C. Merzbach. New York: Wiley, 1989.

Hansell, Walter, and Verner E. Hoggatt, Jr. "The Hidden Hexagon Squares." *Fibonacci Quarterly* 9(1971):120.

Pólya, George. *Mathematical Discovery: On Understanding Learning, and Teaching Problem Solving*. Combined ed. New York: Wiley, 1981.

GERALD L. ALEXANDERSON

# PATTERNING

Involves perceiving similarities and differences, differentiating between essential and nonessential features, and decoding a series rule. In patterning activities, students search for regularities, draw inferences, and create responses. Patterns can be classified by their repetends, that is, the part that repeats. The series *round button, square button, round button, square button* typifies an AB pattern. Other possible patterns include ABC, AABB, ABCD, and ABA, with the last one being the most difficult. While initial patterning activity often involves real objects or bodily movement (Baratta-Lorton 1976), patterns are also present in elementary mathematical activities. Some common examples of mathematical patterns include the even-odd pattern of the counting numbers, the "add-three" pattern of the "three table" in multiplication, the differences between square numbers, the graphs of functions, and designs constructed from repeated sequences of slides, flips, and turns.

Patterning activities are not equally difficult. The following examples of activities in an early childhood setting illustrate the levels of the established hierarchy of pattern processes:

*Reproduction:* Children copy the letters XXOXXO.

*Identification:* Children move in a pattern, then find others using the same form. (Jump, jump, hop is equivalent to step, step, slide.)

*Extension:* A child makes a train of connecting cubes using yellow, yellow, green as repetend. Another child adds to the train.

*Translation:* Children watch a clapping pattern, then transfer it to pattern blocks. Clap, clap, snap becomes triangle, triangle, hexagon.

*Interpolation:* In a chain of interlocking paper rings in two colors, one link is covered. The children guess its color.

Two-dimensional patterns such as those seen in fabric, tiling, and wallpaper can also be investigated. Searching for patterns in a hundreds board or an addition table is an especially rich mathematical activity. Patterns can also be colored on grid paper of various dimensions.

The inclusion of patterning at all grade levels is recommended for its power in fostering reasoning, encouraging a variety of responses, and developing spatial and number sense (National Council of Teachers of Mathematics 1989), and many mathematics educators, researchers, and teachers now recognize patterning as an engaging and powerful mathematical skill.

*See also* Tiling

## SELECTED REFERENCES

Baratta-Lorton, Mary. *Mathematics Their Way.* Menlo Park, CA: Addison-Wesley, 1976.

Burton, Grace M. *Towards a Good Beginning.* Menlo Park, CA: Addison-Wesley, 1985.

Coburn, Terrence, et al. *Patterns: Addenda Series, Grades K–6.* Reston, VA: National Council of Teachers of Mathematics, 1994.

National Council of Teachers of Mathematics. *Curriculum and Evaluation Standards for School Mathematics.* Reston, VA: The Council, 1989.

———. *Principles and Standards for School Mathematics.* Reston, VA: The Council, 2000.

Phillips, Elizabeth, et al. *Patterns and Functions: Addenda Series, Grades 5–8.* Reston, VA: National Council of Teachers of Mathematics, 1991.

GRACE M. BURTON

## PEANO, GIUSEPPE (1858–1932)

Best known for his pioneering work in symbolic logic and the axiomatic method (i.e., the Peano axioms for arithmetic). Peano graduated with "high honors" from the University of Turin in 1880 and went on to teach mathematics there for nearly fifty years, until his death. He always considered his work in analysis more important than his work in symbolic logic and was justifiably pleased with the positive reception of his textbooks on the differential and integral calculus, *Calcolo differenziale e principii di calcolo integrale* (1884) and his later *Lezioni di analisi infinitesimale* (1893). He also made notable contributions to actuarial mathematics and to rational mechanics.

Peano's publications in logic show, above all, the influence of George Boole (1815–1864), Ernst Schröder (1841–1902), Charles Sanders Peirce (1839–1914), and Hugh McColl (1837–1909). His *Arithmetices principia, nova methodo exposita* (1889) included his famous postulates for the natural numbers. The innovativeness of Peano's approach and symbolism (he introduced ∃, meaning "there exists," the notation for set membership, ε, and a new notation for universal quantification) had an immense influence, especially on Bertrand Russell (1872–1970).

Peano's admiration for Gottfried Wilhelm Leibniz's idea of a "universal characteristic" was reflected not only in his mathematical logic but in his support for an international language. Peano's proposal for an *"interlingua"* was *latino sine flexione* (Latin without grammar), which he published as early as 1903.

Later, he served as president of the Akademi Internasional de Lingua Universal (he was elected in 1908; two years later its name was changed to Academia pro Interlingua, with Peano remaining as president until he died in 1932).

Peano was also a great promoter of mathematics internationally. In 1891, he founded a new journal, *Rivista di matematica,* which stressed work in logic and foundations of mathematics. A year later he announced his *Formulario* project to state and prove virtually all of the basic theorems of mathematics using his own logical notation. By the time he gave up this project in 1908, the *Formulario* included some 4,200 theorems. Peano enjoyed a reputation as an inspiring teacher, and had a strong interest in pedagogy. He was a member of the Mathesis Society, and in 1914 supported a series of meetings for secondary teachers of mathematics in Turin (which continued through 1919).

*See also* Logic

## SELECTED REFERENCES

Cassina, Ugo. *Critica dei principii della matematica e questioni di logica.* Rome, Italy: Edizioni Cremonese, 1961.

———. *Dalla geometria egiziana alla matematica moderna.* Rome, Italy: Edizioni Cremonese, 1961.

Kennedy, Hubert C. *Giuseppe Peano.* Basel, Switzerland: Birkhäuser, 1974.

———. *Peano. Life and Works of Giuseppe Peano.* Dordrecht, Netherlands: Reidel, 1980.

———, ed. *Selected Works of Giuseppe Peano.* Toronto, ONT, Canada: University of Toronto Press, 1973.

Quine, Willard Van Orman. "Peano as Logician." *History and Philosophy of Logic* 8(1987):15–24.

Terracini, Alessandro, ed. *In memoria di Giuseppe Peano.* Cuneo, Italy: Liceo Scientifico Statale, 1955.

JOSEPH W. DAUBEN

## PEIRCE, CHARLES SANDERS (1839–1914)

Born in Cambridge, Massachusetts, Peirce was the son of the mathematician Benjamin Peirce (1809–1880). He attended Harvard University, obtaining an undergraduate degree in 1859, a master's degree in 1862, as well as an Sc.B. in chemistry in 1863. Peirce began working for the U.S. Coast (and Geodetic) Survey in 1861, under his father's direction. In 1876, he invented a new map projection, still found useful in 1946 for showing international air routes. This invention was the first application of elliptic function theory to conformal mapping for use in geographical studies (Eisele 1989, 45–58; 1972, 483). Some negligence in professional performance, and much transgression of the sexual mores of the time, forced his resignation in 1891; his remaining years were clouded by considerable poverty.

Peirce was a universal thinker of high caliber. He specialized in areas of mathematics, mechanics, physics, philosophy, history, and, above all, in logic. Peirce felt that mathematics study could develop the mind's ability to imagine, abstract, and generalize. In the area of logic, he united a modified version of George Boole's algebra of logic with Augustus De Morgan's logic of relations and produced a powerful algebraic means of developing propositional and predicate logic. While teaching at Johns Hopkins University from 1879 to 1884, he led a team of talented students, including Christine Ladd-Franklin (1847–1930), to pioneer quantification theory (a most important part of logical theory, concerned with the functors "there is . . ." and "for all . . ."). His work was taken up especially by Ernst Schröder (1841–1902), but this whole algebraic tradition was eclipsed by the mathematical logic of Giuseppe Peano (1858–1932) and his school and by Bertrand Russell (1872–1970). Much of Peirce's work is only now being published and evaluated.

*See also* Boole; De Morgan; Logic; Peano; Russell

SELECTED REFERENCES

Brent, Joseph. *Charles Sanders Peirce, A Life.* Bloomington, IN: Indiana University Press, 1993.

Eisele, Carolyn. "Charles Sanders Peirce." In *Dictionary of Scientific Biography* (pp. 482–488). Vol. 10. Charles C. Gillispie, ed. New York: Scribner's 1972.

———. "Thomas S. Fiske and Charles S. Peirce." In *A Century of Mathematics in America* (pp. 41–55). Pt. I. Peter Duren, ed. Providence, RI: American Mathematical Society, 1989.

Peirce, Charles S. *Writings of Charles S. Peirce. A Chronological Edition.* Max H. Fisch, ed. Bloomington, IN: Indiana University Press, 1982– . 30 volumes planned.

IVOR GRATTAN-GUINNESS

# PERCENT

A form used to write ratios. The term *percent* is derived from the Latin expression *per centum* and can be interpreted as "out of a hundred," "per hundred," "hundredths," and "by the hundred." Thus, "75 percent" means "75 out of a hundred," "75 per hun-

dred," or "75 hundredths." The expression "by the hundred" can help one understand the meaning of percent when the quantity is greater than 100. For example, a grade of 75 percent on a test with a maximum score of 200 points means that the candidate earned 75 out of each 100 points or 150 ($75 \times 2$ groups of 100) out of the 200. Percent can also be defined as the ratio of a number to 100. Thus, a score of 45 on a 50-question test can be interpeted as 45/50 which equals 90/100 or 90 percent. Fifteenth-century manuscripts on commercial arithmetic show abbreviated forms of percent, or *por cento,* as "per $\frac{o}{c}$" or "p $\frac{o}{c}$." By the middle of the seventeenth century, the symbol used was "per $\frac{o}{o}$," which was later abbreviated to "$\frac{o}{o}$" (Smith 1953). Today, percent is symbolized as $\%$ or %, as in 90%.

The term *percent* can, in some instances, be used interchangeably with the term *percentage,* which also means "parts per hundred." For example, if talking about highway conditions, one can say, "A percentage (or percent) of the highways is in good driving condition." As will be seen in the problem types described later, *percentage* also means the product of a percent and a quantity referred to as the *base.* The statistical term *percentile* denotes one of the partitioning marks that divide a set into 100 equal parts. A score, $S$, in the 80th percentile means that 80% of the scores in the sample were less than $S$ or that $S$ is among the top 20% of the scores. When the meaning of percent is understood, it is easy to write percents as fractions. Examples: 5% = 5/100; 85% = 85/100; 0.5% = 0.5/100. Percents can also be written as decimals. Examples: 68% = 68/100 = 0.68; 8% = 8/100 = 0.08; 1/2% = 1/2 ÷ 100 = 1/200 = 0.005. This method is equivalent to moving the decimal point two places to the left and "dropping" the percent sign.

Both fractions and decimals, in turn, can be expressed as percents. Procedures to achieve this are:

## Procedures to Change a Fraction to a Percent

*Method a* Change to an equivalent fraction with the denominator equal to 100:

$$\frac{2}{5} = \frac{40}{100} = 40\%$$

*Method b* Use a proportion or equivalent fraction:

$$\frac{4}{5} = \frac{a}{100} \quad \text{and } a = 80$$

*Method c* Use the division interpretation of fractions:

$$\frac{3}{12} = 3 \div 12 = 0.25 = 25/100 = 25\%$$

## Procedures to Change a Decimal to a Percent

*Method a* Change the decimal to a fraction and then to a percent:

$$0.8 = 8/10 = 80/100 = 80\%$$

*Method b* Multiply the decimal by 1 expressed as 100/100:

$$0.8 = 0.8 \left(\frac{100}{100}\right) = \frac{80}{100} = 80\%$$

This is equivalent to moving the decimal point two places to the right and "adding" the percent sign.

## USE OF PERCENTS

In commercial, banking, or statistical problems, percents are used to compare two quantities in the sense of a ratio. If a number $P$ *(percentage)* is compared to a number $B$ (the *base* or the *unit*), we write $P = R\%$ times $B$, where $R\%$ is the *rate* of comparison of $P$ to $B$. Thus, the equation $R\%$ times $B = P$ is the general equation for solving percent problems. Written as a proportion, it becomes $R/100 = P/B$ (Fujii 1965). Percent problems are characterized according to the unknown term in the comparison: the rate, the base, or the percentage. Examples of each type follow:

Type 1. Find the percentage (product).
Problem: What is 35% of 125?
Solution: $0.35(125) = P$ or
$35/100 = P/125$ and $P = 43.75$.

Type 2. Find the base (a factor).
Problem: 42 is 25% of what number?
Solution: $0.25B = 42$ or $25/100 = 42/B$
and $B = 168$.

Type 3. Find the rate (a factor).
Problem: What percent of 90 is 74?
Solution: $(R/100)(90) = 74$ or
$R/100 = 74/90$ and $R = 82.22$.

In instruction, care should be taken that students know the meaning of the prefix *per*, as in "4 aces per deck of cards" meaning "4 aces *out of* a deck of cards." In this situation, the word *deck* has a value of 52. It could be established that the expressions "4

out of 52," "4 parts of 52," and "4 per 52" are equivalent and can be written mathematically as 4/52. When students know the meaning of per (out of) and cent (hundred), they will more readily grasp the meaning of expressions such as "55 percent."

Students' inability to solve percent problems can be attributed to their lack of visual images or quantitative feel for numbers written as percents (Sobel and Maletsky 1988; Wiebe 1986). An attempt should be made to have students represent percents in concrete form, particularly in early work. The most common model recommended is geometric, in the form of circles, grids, and rods. Students can use the hundred block in the base-ten set to model different percents by covering it with unit cubes. A 10-by-10 grid paper can be colored by students to show particular percents (Cathcart, Pothier, and Vance 1994). Cuisenaire rods can be used to develop a concrete linear model for percentage (Erickson 1990). Money has been used successfully in developing percent sense (Rossini-Osiecki 1988). It is probable that the difficulty that students have in understanding percent problems stems from their lack of adequate knowledge of fractions and decimals. Also, developing proportional thinking skills is thought to be more beneficial to problem solving than the use of formal mechanical procedures (Post 1992).

*See also* Proportional Reasoning

### SELECTED REFERENCES

Angel, Allen R., and Stuart R. Porter. *A Survey of Mathematics*. Reading, MA: Addison-Wesley, 1989.

Cathcart, W. George, Yvonne M. Pothier, and James H. Vance. *Learning Mathematics in the Elementary and Middle Schools*. Toronto, ONT, Canada: Allyn and Bacon Canada, 1994.

Erickson, Dianne K. "Percentages and Cuisenaire Rods." *Mathematics Teacher* 83(Nov. 1990):648–655.

Fujii, John N. *Numbers and Arithmetic*. New York: Blaisdell, 1965.

Meserve, Bruce E., and Max A. Sobel. *Contemporary Mathematics*. 3rd ed. Englewood Cliffs, NJ: Prentice-Hall, 1981.

Post, Thomas R., ed. *Teaching Mathematics in Grades K–8: Research Based Methods*. 2nd ed. Boston, MA: Allyn and Bacon, 1992.

Rossini-Osiecki, Beverly A. "Using Percent Problems to Promote Critical Thinking." *Mathematics Teacher* 81(Jan. 1988):31–34.

Sobel, Max A., and Evan M. Maletsky. *Teaching Mathematics*. 2nd ed. Englewood Cliffs, NJ: Prentice-Hall, 1988.

Smith, David E. *History of Mathematics*. Vol. 2. New York: Dover, 1953.

Wiebe, James H. "Manipulating Percents." *Mathematics Teacher* 79(Jan. 1986):23–26.

YVONNE M. POTHIER

# PESTALOZZI, JOHANN HEINRICH (1746–1827)

A Swiss educator who considered observation and sense perception to be fundamental in learning. Pestalozzi's educational doctrine can be found in his *How Gertrude Teaches Her Children*. He felt that children learn from nature through their eyes, ears, and hands. Applied to arithmetic instruction, his method centered on object lessons and the use of concrete materials and avoided introducing children to abstraction and symbols prematurely. The large-scale introduction of Pestalozzian ideas into the United States is credited to Edward Sheldon, who introduced them in a school in Oswego, New York in 1859. In the specific area of arithmetic instruction, however, the credit belongs to Warren Colburn, whose *First Lessons in Arithmetic on the Plan of Pestalozzi with Some Improvements* (1822) was widely used throughout the rest of the nineteenth century.

*See also* Manipulatives, Computer Generated

## SELECTED REFERENCES

Colburn, Warren. *First Lessons in Arithmetic on the Plan of Pestalozzi, with Some Improvements.* Boston, MA: Cummings, Hilliard, 1822.

Jones, Phillip S., ed. *A History of Mathematics Education in the United States and Canada.* 32nd Yearbook. Washington, DC: National Council of Teachers of Mathematics, 1970.

Pestalozzi, Johann Heinrich. *How Gertrude Teaches Her Children: An Attempt to Help Mothers to Teach Their Own Children and an Account of the Method.* Lucy I. Holland and Frances C. Turner, trans. Syracuse, NY: Bardeen, 1898.

JAMES K. BIDWELL
ROBERT G. CLASON

# PHILOSOPHICAL PERSPECTIVES ON MATHEMATICS

Concerned with the main issues of the nature of mathematical objects and the justification of mathematical knowledge. These ancient and still controversial questions have educational implications that are seldom discussed in either the philosophical or the educational literature. One specific issue is whether infinite sets exist. Does the set of all natural numbers 1, 2, 3 . . . exist as a completed whole? Does the set of all points on a line segment exist as a completed whole? The problem of infinity goes back to Zeno (ca. 490–435 B.C.). It became central in the seventeenth century, when Gottfried Leibniz used infinitely small numbers to compute instantaneous velocities. (He thought of instantaneous velocity as an infinitesimal distance divided by an infinitesimal time interval.) Infinitesimal numbers are related to infinite numbers; a number in one class is the reciprocal of a number in the other class. Leibniz's competitor, Isaac Newton, computed velocities without using infinitesimals. His procedure depended essentially on the notion of limit, which was not clarified until the nineteenth century.

Aristotle introduced the distinction between the potential and the actual infinite. A family of objects is "potentially infinite" if it can be enlarged beyond any limit. The potential infinite is in Euclid's *Elements,* in the arbitrary extendibility of a line segment. This is a geometric counterpart to the indefinite continuability of the natural numbers. There is no last natural number. You can always add one more.

Few reject the potential infinite. Aristotle contrasted it with the "actual infinite"—an object obtained by *completing* infinitely many steps. That he thought self-contradictory. In a sense, he was right. Galileo Galilei discovered that in an infinite set Euclid's axiom "the whole is greater than the part" becomes false. The set of even numbers is a proper part of the set of all natural numbers, yet the two sets can be put into one-to-one correspondence. Every natural number is half of one even number; every even number is twice one natural number.

Despite the prohibition of Aristotle and virtually all mathematicians and philosophers prior to the 1880s, Georg Cantor brought the *completed* infinite into mathematics. His theory of infinite sets developed into one of the most fascinating and subtle mathematical subjects. Most remarkable is Cantor's "continuum hypothesis." He conjectured that every infinite subset of $R$, the set of all real numbers (also called *the continuum*), is one-to-one equivalent either to $N$ (the set of all natural numbers) or to $R$. Consequently, there would be no infinite cardinal number between aleph-null (the cardinality of $N$), and $c$, the cardinality of $R$. Cantor could not prove his conjecture. Extending Kurt Gödel's 1931 result, Paul J. Cohen in 1963 finally showed it to be undecidable. The axioms of set theory are insufficient to decide the question. You can take as an additional axiom, without introducing a contradiction, either the continuum hypothesis or its negation.

Platonism (also called *realism*) is a popular philosophy of mathematics. It says numbers, triangles, finite and infinite sets, and all objects of mathematical study have independent, timeless, immaterial, nonhuman, abstract existence. As a consequence, mathematical research is discovery, not creation. What is, always was and always will be, whether anyone discovers it or not. Platonism raises two obvious questions. Where and how do timeless, abstract objects exist? How do humans interact with immaterial abstractions? Plato said numbers and shapes exist in Heaven, where we meet them before we are born. No one repeats Plato's answer today. Answers are still being sought. The undecidability of Cantor's continuum hypothesis presents difficulties for some Platonists. They consider the continuum hypothesis a factual statement about real objects. It is either true or false. Some tried to find a plausible additional axiom to decide the question.

Why do so many mathematicians and mathematics teachers accept Platonism? The strongest reason is a universal experience of mathematical study and research: the answer to a problem is completely determined as soon as the problem is properly stated. In earlier times, a buttress of Platonism was its link to religion—in Plato, St. Augustine, René Descartes, and Gottfried Leibniz. The later logicism of Gottlob Frege and Bertrand Russell also was an influential brand of Platonism. But there has been little investigation of the pedagogic implications of Platonism. One indication was given by the famous English number theorist Godfrey H. Hardy (1877–1947). In expounding his Platonist view of mathematics, he compared the facts of mathematics to peaks in a distant mountain range. When he saw a peak, he pointed at it, and he hoped his pupil would then see it too.

A rival to Platonism is formalism. Formalism avoids the obscurity of Platonism by reducing mathematics to formal derivations from formal axiom systems. Mathematics is seen as just calculations (whether by hand or by machine does not matter). The leader of the formalists, David Hilbert, hoped that by thinking of mathematics as sequences of formal symbols, it would be possible to prove that mathematics is consistent. However, the famous incompleteness theorems of Kurt Gödel in the early 1930s seem to mean that Hilbert's consistency proof is unattainable. When formalism entered textbooks or classrooms, it was sometimes counterproductive. In some "new math" programs of the 1960s, the impression was sometimes given that mathematics is mainly formal axioms and derivations, thus obscuring the fact that mathematics is a purposeful activity connected to other fields of human activity.

Another well-known philosophy of mathematics is intuitionism, credited to the twentieth-century Dutch mathematician Luitzen E. J. Brouwer, and its American version, mathematical constructivism, due to Errett Bishop in the 1970s and 1980s. These viewpoints regard mathematics as thoughts in the head of individual mathematicians. Their principal dogma is rejection of "the law of the excluded middle" ("proof by contradiction") except for finite sets. They do not believe you prove A by disproving not-A. This doctrine rejects a significant part of standard classical mathematics. Many mathematicians are unwilling to make that sacrifice.

Recently, a new perspective of the philosophy of mathematics has emerged. Rooted in the works of Imre Lakatos and Ludwig Wittgenstein, its central idea is the obvious fact that mathematics is done by humans, for human purposes. It has been called "naturalism" by Philip Kitcher, "social constructivism" by Paul Ernest, "maverick" by William Aspray and Kitcher, "quasi-empiricism" by Lakatos and Thomas Tymoczko, and "humanism" by Reuben Hersh. This perspective often focuses explicitly on educational issues. Some current trends in U.S. mathematics education are: working in groups; problem solving; and writing as a regular part of mathematics assignments. For the humanist mathematics teacher, problem solving, group work, and use of language all are part of the nature of mathematics.

*See also* Cantor; Constructivism; Gödel; Humanistic Mathematics

SELECTED REFERENCES

Benacerraf, Paul, and Hilary Putnam, eds. *The Philosophy of Mathematics; Selected Readings.* Englewood Cliffs, NJ: Prentice-Hall, 1964.

Davis, Philip, and Reuben Hersh. *The Mathematical Experience.* Boston, MA: Birkhäuser, 1981; 2nd ed., with Elena Anne Marchisotto, 1995.

Ernest, Paul. *The Philosophy of Mathematics Education.* Washington, DC: Falmer, 1991.

Hardy, Godfrey H. "Mathematical Proof." *Mind* 38(1929):1–25.

Hersh, Reuben. "Fresh Breezes in the Philosophy of Mathematics." *American Mathematical Monthly* 102(1995): 589–594.

———. *What Is Mathematics, Really?* New York: Oxford University Press, 1997.

Lakatos, Imre. *Proofs and Refutations.* Cambridge, England: Cambridge University Press, 1976.

Rényi, Alfréd. *Dialogues on Mathematics.* San Francisco, CA: Holden Day, 1967.

Tymoczko, Thomas. *New Directions in the Philosophy of Mathematics.* Boston, MA: Birkhäuser, 1985.

White, Leslie. "The Locus of Mathematical Reality." *Philosophy of Science* 14(1947):289–303. Reprinted in *The World of Mathematics* (pp. 2348–2364). James R. Newman, ed. New York: Simon and Schuster, 1956.

REUBEN HERSH

# PHYSICS

Some physicists believe that physics began in the seventeenth century with Isaac Newton's (1642–1727) work on the laws of motion, the law of gravitation, and the nature of light (Gribbin 1984). It was common before the twentieth century for physicists also to be mathematicians, and Newton made his contributions to the development of the calculus while working on gravitational theory and the laws of motion. What is called a "law" in physics is not absolute, but is subject to change as new observations yield new theories. A physical law

> . . . is a statement, usually in the succinct and precise language of mathematics, of a relation that has been found by repeated experiment to hold among physical quantities and that reflects persistent regularities in the behavior of the physical world. A 'good' physical law has the greatest possible generality, simplicity and precision. The final criterion of a successful law of physics is how accurately it predicts experimental results. (Weidner and Sells 1980).

The process of using a physical theory to make predictions is comparable to what mathematicians call *modeling.* For example, qualitatively different models have been used to interpret the behavior of light, which is the visible part of the form of energy known as electromagnetic radiation. In the first century A.D., Heron of Alexandria proved the law of reflection of light: a ray of light that strikes a mirrored surface is reflected so that the angle of incidence is equal to the angle of reflection. These angles are shown in Figure 1. Figure 2 shows the elements of Heron's proof, which asserts that of all the light rays emanating from the source at $S,$ those that reach the observer at $O$ are the ones reflected at the point on the mirror surface such that the path from $S$ to $O$ via the mirror is shortest. This is the path shown as $SPO,$ where $S'$ is the image of $S$ in the mirror $MM'$ and $S'PO$ is a straight line. Any other path, such as $SRO,$

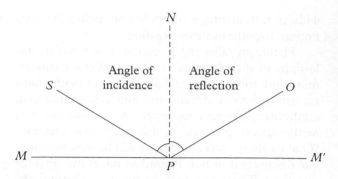

**Figure 1**

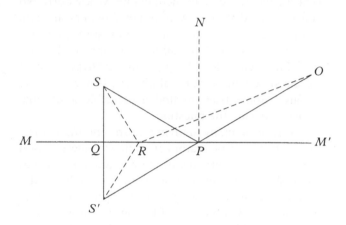

**Figure 2**

must be longer, by the triangle inequality. This proof treats light rays as streams of "corpuscles" or particles (Schiffer and Bowden 1984).

Newton's work in optics, also based on a particle model for light, included the study of refraction (the bending of light as it passes from one medium to another) and the development of the first reflecting telescope (Gribbin 1984). However, the Dutch scientist and mathematician Christian Huygens (1629–1695) proposed a wave model that could explain not only reflection and refraction of light but also properties such as interference, diffraction, and polarization, for which the corpuscular theory had no interpretation (Tipler 1976, 605). A familiar example of a wave is seen when a stone is dropped in a pool of water. The disturbance spreads out in concentric circles, though individual water molecules only move up and down, as illustrated in Figure 3. The mathematical analysis of waves makes natural use of sinusoidal functions, and the property of interference can be interpreted in terms of algebraic addition (or superposition) of

**Figure 3**

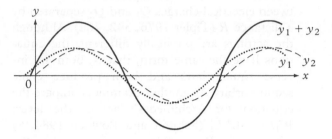

**Figure 4**

sinusoidal functions, as illustrated in Figure 4 (Young 1968, 17).

In the 1860s, it was shown in the laboratory that light is part of the electromagnetic spectrum that includes radio waves. After James Clerk Maxwell published his theory of electromagnetism in 1867, the wave model became the one more widely accepted. Maxwell's equations described light as a wave phenomenon and agreed well with experiment but left unanswered the question of what was "waving," so physicists began to search for such a medium that they called the "ether." Near the end of the nineteenth century, then,

> . . . it was generally believed that all natural phenomena could be described by Newton's laws, the laws of thermodynamics, and the laws of electromagnetism (as expressed in Maxwell's equations). There was nothing left for scientists to do but apply these laws to various phenomena and to measure the next decimal in the fundamental constants . . . (Tipler 1976).

There were basically two kinds of physical quantities in this "classical" physics, the continuous and the discrete. Quantities such as the speed or the momentum of a free particle, or the mechanical energy of a system of two particles, were continuous. Quantities such as the electric charge of a body, the rest mass of an atom, or the possible frequencies of vibration of a string fixed at both ends were discrete, yielding only

particular numbers, generally multiples of fundamental quantities or properties of the system (Weidner and Sells 1980, 87–88).

The world of physics was set into turmoil, however, around the turn of the century by a series of experiments that were completely at variance with the predictions of accepted theories in classical physics. In 1887, Albert Michelson and Edward Morley constructed an interferometer to measure the velocity of the earth relative to the ether. The ingeniously designed experiment, later verified by others, showed that no ether existed. This null result led to the theory of relativity, based on the following two postulates formulated by Albert Einstein (1879–1955) (Good 1968, 19):

1.  The principle of relativity: Physical laws are independent of the particular state of uniform motion of the observer.
2.  The constancy of the velocity of light: The speed of light in vacuum is $c \approx 3(10^8)$ m/sec regardless of the relative motion of the light source and the observer.

Also in the late nineteenth and early twentieth centuries, experiments were performed that were designed to measure electromagnetic radiation from a "black body" (a perfect absorber or emitter), the energy of electrons released from a clean metal surface receiving electromagnetic radiation (the photoelectric effect), and the light emitted by atoms (spectra). Such experiments examine the interaction of electromagnetic radiation with electrons. The results show that light energy can only take on values that are multiples of a basic "quantum" (a single quantum of light is called a *photon*). Further experiments with electrons—previously considered to be particles—revealed that they had wave properties as well. What has ensued from efforts to explain such unexpected results are probabilistic quantum theories involving wave and matrix mechanics and notions of wave-particle "duality" in which both photons and electrons are sometimes waves and sometimes particles but do not satisfy both models at the same time (Weidner and Sells 1980, sec. 5–4).

While relativity theory may be seen as an elaboration of Newton's classical laws of motion, quantum theories appear to violate all commonsense ideas of how the world behaves. Instead of the deterministic equations of classical physics, the quantum view involves stochastic equations and probability distributions. Erwin Schrödinger himself said about the quantum-mechanical equation that bears his name and works very well, "I don't like it and I'm sorry I ever had anything to do with it." (Gribbin 1984). To quote Richard Feynman (1985):

> . . . while I am describing to you *how* Nature works, you won't understand *why* Nature works that way. But you see, nobody understands that.
>
> . . .
>
> . . . It's a problem that physicists have learned to deal with. They've learned to realize that whether they like a theory or don't like a theory is *not* the essential question. Rather, it is whether or not the theory gives predictions that agree with experiment.

Feynman's work in quantum electrodynamics makes it possible to use a particle model for light (photons) and derive all the wave properties using quantum theory. Relativistic quantum mechanics and quantum field theory are needed to describe the experimentally observed interactions of fundamental particles. The list of fundamental particles no longer includes the proton and neutron of classical physics, since these are now known to be made up of smaller particles called *quarks,* but does include the photon and electron, as well as a variety of quarks, "gluons," and "antiparticles" (Okun 1987). Physicists and philosophers are still engaged in discussion and debate about the physical meaning of the mathematical theories of quantum physics (Gribbin 1984). In the list of references at the end of this entry, those by Feynman, Good, Gribbin, Okun, and Weidner and Sells provide some of the background of the debate.

Mathematics educators will find much useful material for the classroom in the literature of physics, only a small part of which is given in the list of references. In particular,

1. Throughout history, developments in mathematics and physics have reinforced each other. Abstract geometric theory comes to life when it is applied to optics, as we saw in Heron's proof of the law of reflection. As another example, the physical concepts of speed and velocity are historically authentic examples of ratios and vectors in mathematics. In fact, the idea of velocity as the limit of the ratio, change in distance/change in time as the change in time approaches zero, provides a natural connection to concepts of calculus such as a limit and a derivative (Tipler 1976, sec. 2–2).

2. The generality of mathematical structures can be seen from the variety of their physical settings.

For example, consider the following two equations:

$$F = GM_1M_2/R^2 \quad \text{and} \quad F = kQ_1Q_2/R^2$$

The first equation is Newton's law of universal gravitation, describing the gravitational force of attraction between bodies with masses $M_1$ and $M_2$ separated by a distance $R$; the second is Coulomb's law, which expresses the force between electrical charges $Q_1$ and $Q_2$ separated by a distance $R$ (Tipler 1976, 392, 702). Although the settings are physically different, the equations have the same form, that is, both are inverse square laws and both produce conic section orbits. As another example, compare the equation for Archimedes' law of the lever: $W_1L_1 = W_2L_2$ (Schiffer and Bowden 1984, 8) with that for Bernoulli's principle regarding the flow of an incompressible fluid: $A_1V_1 = A_2V_2$ (Tipler 1976, 244). Again, the form of the equations is the same, though the settings are different. The first equation states the condition for balance of weights $W_1$ and $W_2$ at distances $L_1$ and $L_2$, respectively, on opposite sides of a fulcrum (the heavier weight must be closer to the fulcrum). The second shows the relationship of the velocities $V_1$ and $V_2$ of a fluid flowing in a pipe when the pipe's cross-sectional area changes from $A_1$ to $A_2$ (water must flow faster in the narrower part of the pipe).

3. Since all physical theories are validated through experiment, the role of measurement in physics is absolutely fundamental. Mathematics curriculum materials have traditionally paid insufficient attention to the distinction between exact and measured numbers and to crucial information about units and precision used in producing measured numbers. It is often the case that measured numbers are obtained indirectly, either for convenience or by necessity. For example, it is easier to measure the lengths of the sides of a geometric figure and then use an appropriate formula to calculate its area than to actually measure the area. However, we must take into account the *precision* of the measured numbers; the area of a rectangle whose length and width respectively measure 8.3 cm and 6.4 cm is properly reported as 53 cm$^2$ rather than 53.12 cm$^2$, since the latter gives a false message about the accuracy to which the area is known. It is hard for a student to understand the effectiveness of a suitable approximation (Pólya 1977, 192–211), or the fact that not all digits produced by a hand-

held calculator are meaningful, when the student has consistently seen measured numbers treated as though they were exact numbers.

The intimate relationship between mathematics and physics has enriched both fields throughout history. Mathematics education can also be enhanced by utilizing such connections.

*See also* Applications for the Classroom, Overview; Applications for the Secondary School Classroom; Measurement; Modeling

## SELECTED REFERENCES

Asimov, Isaac. *The History of Physics.* New York: Walker, 1984.

Feynman, Richard P. *QED: The Strange Theory of Light and Matter.* Princeton, NJ: Princeton University Press, 1985.

Good, Roland H. *Basic Concepts of Relativity.* New York: Reinhold, 1968.

Gribbin, John. *In Search of Schrödinger's Cat: Quantum Physics and Reality.* New York: Bantam, 1984.

Okun, Lev B. $\alpha$, $\beta$ $\tau$, . . . Z: A Primer in Particle Physics. Chur, Switzerland: Harwood, 1987.

Pólya, George. *Mathematical Methods in Science.* Washington, DC: Mathematical Association of America, 1977.

Schiffer, Max M., and Leon Bowden. *The Role of Mathematics in Science.* Washington, DC: Mathematical Association of America, 1984.

Tipler, Paul A. *Physics.* New York: Worth, 1976.

Weidner, Richard T., and Robert L. Sells. *Elementary Modern Physics.* 3rd ed. Boston, MA: Allyn and Bacon, 1980.

Young, Hugh D. *Fundamentals of Optics and Modern Physics.* New York: McGraw-Hill, 1968.

BERNICE KASTNER

# PI

The ratio of the circumference (periphery) of a circle to its diameter. This ratio (symbolized by $\pi$) is the same for all circles, no matter what their size. Thus $\pi$ is a constant, approximately equal to 3.14159. Johann Lambert (1728–1777) proved that $\pi$ is irrational; thus its decimal equivalent is nonterminating and nonrepeating. It is also transcendental, as proved by Carl von Lindemann (1852–1939); it cannot be the root of any algebraic equation of finite degree with rational coefficients. The symbol $\pi$ first appeared in a work by William Jones (1675–1749) in 1706, but its adoption by Leonhard Euler (1707–1783) was responsible for its popularization (Cajori 1993, 9–10). The ratio $\pi$ played an essential role in the many attempts to square the circle with ruler and compass alone. These attempts were an irresistible challenge to many mathematicians and amateurs beginning with Anaxagoras (434 B.C.) and ending when Lindemann proved that it was impossible (1882). The number $\pi$ has come to play an important role not only in geometry but also in such areas as trigonometry, probability and statistics, and analysis.

As an introductory experiment, students may wrap a string tightly once around the top of a drinking glass, then measure the length of the string and the diameter of the glass. Dividing the first measurement by the second will yield a number slightly greater than 3. The same result will always happen regardless of the size of the glass. More advanced students may enjoy activities with Buffon's Needle and the Monte Carlo Method (Beckmann 1971, 158–165).

Because $\pi$ is irrational, the initial quest for approximate values was necessary for practical as well as scientific purposes. Before 1000 B.C., Babylonians used $3\frac{1}{8}$ for the value of the ratio, Egyptians used 256/81, and Chinese used 3. The Bible (1 Kings 7:23; 2 Chron. 4:2) suggests a value of 3. Over the centuries, improved approximations appeared. In the seventeenth century, for instance, Ludolph van Ceulen spent a major portion of his life calculating $\pi$ to 35 decimal places. By this time, the calculation was no longer for practical purposes, but simply a fascination. Using computers, in 1961 J. Wrench and Daniel Shanks found the value of $\pi$ correct to 100,625 decimal places, and in 1997, Y. Tamura and Y. Kanada extended the approximation to 51.5 billion digits (Blatner 1997).

Many infinite series for $\pi$ have been found since the first was published by Gottfried Leibniz (1646–1716). He used a series attributed to James Gregory (1638–1675):

$$\arctan x = x - \frac{x^3}{3} + \frac{x^5}{5} - \frac{x^7}{7} + \cdots$$

Letting $x = 1$, following Leibniz, gives arctan $1 = \pi/4$, and

$$\pi = 4(1 - \frac{1}{3} + \frac{1}{5} - \frac{1}{7} + \cdots)$$

(See Beckmann 1971, 132.) Euler published many expressions for $\pi$, including infinite series, infinite products, and continued fractions. Euler's work also led to the discovery of the relationship between $\pi$ and another important transcendental number, $e$; that is, the equation $e^{i\pi} = -1$.

*See also e;* Irrational Numbers; Transcendental Numbers

## SELECTED REFERENCES

Beckmann, Petr. *History of Pi.* New York: St. Martin's, 1971.

Berggren, Lennart, Jonathan Borwein, and Peter Borwein. *Pi: A Source Book.* New York: Springer-Verlag, 1997.

Blatner, David. *The Joy of π.* New York: Walker, 1997.

Boyer, Carl B. *A History of Mathematics.* Revised by Uta C. Merzbach. 2nd ed. New York: Wiley, 1989.

Cajori, Florian. *A History of Mathematical Notations.* Vol. 2. New York: Dover, 1993.

Eves, Howard W. *An Introduction to the History of Mathematics.* 6th ed. Philadelphia, PA: Saunders College Publishing, 1990.

Project Mathematics. "The Story of π." Pasadena, CA: California Institute of Technology, 1989. Video.

DEREK I. BLOOMFIELD

## PIKE, NICOLAS (1743–1819)

Author of *A New and Complete System of Arithmetic Composed for the Use of the Citizens of the United States* (1788), the first arithmetic text by an American author that was widely used in the United States. An earlier arithmetic (1729), attributed to Isaac Greenwood, was little used. Prior to Pike, texts more commonly available included English arithmetics of Edward Cocker, James Hodder, Thomas Dilworth, and a book by a Mrs. Slack under the pseudonym George Fisher; however, much arithmetic was taught without a text. Pike's text, which was not intended for young children, begins with multidigit computation and proceeds with a variety of applications, most of which pertain to business. The newly established federal money is given prominence, although both foreign and colonial monetary systems also appear, with methods for converting among them. Pike's *Arithmetic* contains sections on geometry, trigonometry, logarithms, algebra, permutations and combinations, circulating (repeating) decimals, the conic sections, physics problems, the Julian and Gregorian calendars, and finding the capacity of casks (gauging) and the tonnage of ships, including the tonnage of Noah's Ark.

*See also* History of Mathematics Education in the United States, Overview

## SELECTED REFERENCES

Bidwell, James K., and Robert G. Clason, eds. *Readings in the History of Mathematics Education.* Washington, DC: National Council of Teachers of Mathematics, 1970.

Jones, Phillip S., ed. *A History of Mathematics Education in the United States and Canada.* 32nd Yearbook. Washington, DC: National Council of Teachers of Mathematics, 1970.

Pike, Nicolas. *A New and Complete System of Arithmetic Composed for the Use of the Citizens of the United States.* NewburyPort, MA: Mycall, 1788.

JAMES K. BIDWELL
ROBERT G. CLASON

## PLANE GEOMETRY
*See* ANGLES; AXIOMS; CIRCLES; DISCOVERY APPROACH TO GEOMETRY; EUCLID; GEOMETRY, INSTRUCTION; GOLDEN SECTION; LENGTH, PERIMETER, AREA, AND VOLUME; LOGO; MATHEMATICS, DEFINITIONS; NEUTRAL GEOMETRY; PI; POLYGONS; PROOF; PROOF, GEOMETRIC; PYTHAGORAS; PYTHAGOREAN THEOREM; SIMILARITY; TILING; VAN HIELE LEVELS

*See also* Art; Birkhoff; Bolyai; Coordinate Geometry, Overview; Escher; Greek Mathematics; Islamic Mathematics; Lobachevsky; Mathematics, Foundations; Non-Euclidean Geometry; Software; Space Geometry, Instruction; Transformational Geometry; Transformations

## PLATO (427–347 B.C.)

One of the most influential philosophers not only in Greek antiquity but throughout the history of Western civilization. He held mathematics in high esteem, geometry in particular, as is evident from his statements that "God always geometrizes" and "Let no one ignorant of geometry enter here" (in referring to his academy). For Plato, studying geometry was a form of religious experience since he believed that geometric truths are in actuality manifestations of the cosmic spirit of the universe. He also recognized the important role that definitions played in mathematics and logical discourse. Consequently, historians attribute many of the definitions in Euclid's *Elements* to the efforts of the Platonic school.

It is commonly held that Plato was the first to specify that geometric constructions were to be performed using only what are now called Euclidean tools, namely the straightedge and collapsible compass. In his *Dialogues*, Plato related how the discovery of incommensurable magnitudes (i.e., irrational numbers) such as $\sqrt{2}$ greatly disturbed the Pythagoreans since their philosophy rested on the premise that everything in the universe can be expressed in terms of the natural numbers. In several of his other writings, Plato discussed Zeno's paradoxes, the theory of irrationals, treating numbers completely as abstractions, and the five regular polyhedra (Platonic solids). Historians credit Plato with discovering a mechanical solution for the problem of duplicating a cube, that is, constructing a cube whose volume is twice the volume of a given cube.

Ancient folklore relates that the gods sent a plague to the people of Athens, who then beseeched the oracle at Delos to see how they might appease the gods. They were told to construct an altar in honor of Apollo that was double the size of the existing altar. Plato purportedly interpreted this incident to mean that the gods were dissatisfied with the lack of scientific pursuit of geometry by the Greeks. He was responsible for making the quadrivium—arithmetic, geometry, music, and astronomy—an essential part of a liberal arts education in ancient Greece.

*See also* Irrational Numbers; Mathematics, Definitions; Paradox

SELECTED REFERENCES

Calinger, Ronald, ed. *Classics of Mathematics.* Englewood Cliffs, NJ: Prentice-Hall, 1995.

Kline, Morris. *Mathematics in Western Culture.* New York: Oxford University Press, 1953.

ANTHONY V. PICCOLINO

# POETRY

The mathematician Eric Temple Bell said that the human side of mathematics is mathematicians. The same might be said of poetry. Indeed, the poet reading a personal work conveys the pain and passion of experience that we take to represent humanness. We look for elements in the poem that will either remind us of our own lived experience or teach us something new about the life of another person. Mathematics, too, reminds us of the familiar and introduces us to the strange. We speak here not of simple calculations, but of real mathematics, the development of ideas that is the work of mathematicians. At the core, mathematics represents a struggle to make sense of information in the same way that poems represent solutions to problems. Both in mathematics and in poetry, symbols and patterns perform the work of constructing meaning and communicating knowledge and wisdom.

Of course, poetry shares with mathematics the dilemma posed by school learning. If only we could introduce our children to Shakespeare without making them read the sonnets and plays in class and take tests afterward. If only the obligatory unit on poetry offered every spring invited more students into the world of literature. If only the elegant patterns and shapes of mathematics prevailed in the minds of our students, rather than the baffling complexity of long division. Instead, the lessons of school lead too many people to dismiss poetry as irrelevant in the same way that they have rejected mathematics as useless, except for balancing checkbooks and measuring things. The instances are rare when a student hears the poet's voice or sees in a curve or a model the elegance that mathematics claims as central to the discipline. Many students give up on mathematics and on poetry long before they discover what both offer them in helping to make sense of the world and their lives. Here are missed opportunities.

What might students gain from exploring the intersection of two disciplines that use symbols to construct meaning? Two poems, a verse from William Blake's "Auguries of Innocence" (Blake 1996, 173) and the contemporary mathematician and poet JoAnne Growney's "A Mathematician's Nightmare," suggest what might be garnered from this arena.

> To see a world in a grain of sand,
> And a Heaven in a wild flower,
> Hold Infinity in the palm of your hand,
> And Eternity in an hour.

Writing in the eighteenth century, Blake reminds us that the infinite and the infinitesimal have long been sources of fascination to poets and to mathematicians. Zeno's paradoxes, infinite series, Hilbert's hotel, and questions about potential versus actual infinity continue to challenge mathematicians and students of mathematics.

In her poem, JoAnne Growney presents a bit of mathematics, the Collatz Conjecture. Start with any positive integer. If it is even, divide it by 2; if it is odd, increase it by half and round up to the nearest whole number. For example, starting with 18, divide by 2 to get 9; starting with 9, increase by 4.5 and round up to 14. Continue in this vein (18, 9, 14, 7, 11, 17, 26, 13, 20, 10, 5, 8, 4, 2, 1). The Collatz Conjecture, an unsolved problem, asserts that for any starting number the process will eventually lead to 1.

## A Mathematician's Nightmare

Suppose a general store—
items with unknown values
and arbitrary prices,
rounded for ease to
whole-dollar amounts.

Each day Madame X,
keeper of this emporium,
raises or lowers each price—
exceptional bargains
and anti-bargains.

Even-numbered prices
divide by two,
while odd ones climb
by half themselves—
then half a dollar more
to keep the numbers whole.

Today I pause before
a handsome beveled mirror
priced at twenty-seven dollars.
shall I buy or wait
for fifty-nine long days
until the price is lower?

Students might want to pursue the question raised about the $27 mirror or pose questions about prices of other items in the store.

Finding mathematics in poetry may seem easier than finding poetry in mathematics, but for some mathematicians, the mystery of $i$, the square root of $-1$, or Euler's remarkable equation $e^{i\pi} + 1 = 0$, suggest that poetry is there nonetheless. As students come to know and appreciate the wealth of ideas that reside in mathematics, perhaps we can help them see that a proof, the work of a mathematician, and a poem, the poet's medium, are more similar than not. The inspiration for both arises out of the desire to solve problems.

*See also* Humanistic Mathematics

### SELECTED REFERENCES

Bell, Eric T. *Men of Mathematics.* New York: Simon & Schuster, 1937.

Blake, William. *Selected Poetry.* Michael Mason, ed. New York: Oxford University Press, 1996.

Buchanan, Scott. *Poetry and Mathematics.* Chicago, IL: Midway Reprint, 1975.

Davis, Philip J., and Reuben Hersh. *The Mathematical Experience.* 2nd ed., with Elena Marchisotto. Boston, MA: Birkhäuser, 1995.

Growney, JoAnne. *Intersections.* Bloomsburg, PA: Kadet, 1993.

Stewart, Ian. *The Problems of Mathematics.* New York: Oxford University Press, 1987.

JOAN COUNTRYMAN

# POINCARÉ, JULES HENRI (1854–1912)

One of the greatest mathematicians, as well as a physicist, astronomer, and philosopher; often compared to Carl Friedrich Gauss (1777–1855). Poincaré's cousin, Raymond, was president of France during World War I. Poincaré graduated from the École Polytechnique in 1875. He obtained a degree in mining engineering and a doctorate degree in mathematics from the University of Paris in 1879. He taught at the University of Caen for two years and after that at the University of Paris, where he remained until his death. Poincaré was elected to the Académie des Sciences in 1887, becoming its president in 1906; and to the Académie Française in 1908.

Poincaré's view of science is known as "conventionalism," the freedom to set up many different theories for a set of experimental data (Miller 1996, 203). He contributed to such diverse areas as function theory (discovering the "automorphic functions" of one complex variable); Abelian functions and algebraic geometry; number theory; algebra; differential equations and celestial mechanics; mathematical physics; and algebraic topology as well as foundations of mathematics. In his work on the solar system, he anticipated chaos theory. In addition, he wrote literary, nontechnical works on mathematics, the philosophy of mathematics, and the psychology of mathematical creation.

*See also* Chaos; Creativity; Non-Euclidean Geometry

### SELECTED REFERENCES

Bell, Eric T. *Men of Mathematics.* New York: Simon and Schuster, 1986.

Boyer, Carl B. *A History of Mathematics.* 2nd ed. Revised by Uta C. Merzbach. New York: Wiley, 1991.

Dieudonné, Jean. "Jules Henri Poincaré." In *Dictionary of Scientific Biography* (pp. 51–61). Vol. 11. Charles C. Gillispie, ed. New York: Scribner's, 1975.

Miller, Arthur I. *Insights of Genius: Imagery and Creativity in Science and Art.* New York: Springer-Verlag, 1996.

Newman, James R. "Commentary on an Absent-Minded Genius and the Laws of Chance." In *The World of Mathematics* (pp. 1374–1379). Vol. 2. James R. Newman, ed. New York: Simon and Schuster, 1956.

Poincaré, Henri. "Chance." In *The World of Mathematics* (pp. 1380–1394). Vol. 2. James R. Newman, ed. New York: Simon and Schuster, 1956.

———. "Mathematical Creation." In *The World of Mathematics* (pp. 2041–2050). Vol. 4. James R. Newman, ed. New York: Simon and Schuster, 1956.

———. *Science and Hypothesis*. New York: Dover, 1905.

LOUISE S. GRINSTEIN<br>SALLY I. LIPSEY

# POISSON DISTRIBUTION

Credited to Siméon Denis Poisson (1781–1840), a French mathematician and mathematical physicist; given by the formula

$$f(x) = \lambda^x e^{-\lambda}/x! \qquad \text{for } x = 0, 1, 2, 3, \ldots$$

where $x$ is the number of independent occurrences of a rare event over a specified unit of time or space, $e = 2.7128\ldots$ (the base of the system of natural logarithms), and the parameter $\lambda$ is the expected, or average, number of occurrences. This formula can be used in many situations where a fixed number of successes per unit of time (or other kind of unit) is expected. For instance, if a town expects 2.4 automobile accidents per week, the number of accidents per week can be modeled as a Poisson variable with $\lambda = 2.4$; or where a baker may expect 10.7 raisins in a slice of raisin bread; or where a typist may be expected to make 1.5 errors per page.

*Example* Suppose that an airline terminal loses an average of seven pieces of luggage per day. Use the Poisson distribution with parameter $\lambda$ to determine the probability that on a certain day the airline terminal will lose

a. fewer than five pieces of luggage
b. at least five pieces of luggage

*Solution* Substituting $\lambda = 7$; $x = 4, 3, 2, 1,$ or $0$; $e^{-7} = 0.0009$ (from a table of values of $e^{-x}$) into the formula for the Poisson distribution with parameter $\lambda$, we get

a. $f(4) = (7^4 \cdot e^{-7}/4!)$
$= (2,401)(0.0009)/(4 \cdot 3 \cdot 2 \cdot 1) = 0.090$
$f(3) = (7^3 \cdot e^{-7}/3!)$
$= (343)(0.0009)/(3 \cdot 2 \cdot 1) = \quad 0.051$
$f(2) = (7^2 \cdot e^{-7}/2!)$
$= (49)(0.0009)/(2 \cdot 1) = \quad 0.022$
$f(1) = (7^1 \cdot e^{-7}/1!)$
$= (7)(0.0009)/1 = \quad 0.006$
$f(0) = (7^0 \cdot e^{-7}/0!)$
$= (1)(0.0009)/1 = \quad \underline{0.001}$
Total $\qquad 0.170$

Thus, the probability is 0.17 that the airline terminal will lose fewer than five pieces of luggage.

b. Subtracting the results of part (a) from 1.00, we get 0.83 for the probability that the airline terminal will lose at least five pieces of luggage.

As indicated in the example, the Poisson model is usually applied to counts of independent occurrences of a *rare* event, "rare" in the sense that, if a unit is divided into a very large number, $n$, of small components, the probability, $p$, that the event occurs in any one such component is exceedingly small. The distribution is therefore like that of a binomial variable, which counts occurrences in $n$ independent trials with probability $p$ of an occurrence per trial, but with a large $n$ and small $p$ and mean value $\lambda = np$. For this reason, the Poisson probability distribution function provides a good approximation to the binomial probability distribution function:

$$f(x) = n!/(x!(n-x)!)p^x(1-p)^{n-x}$$
$$\text{for } x = 0, 1, \ldots, n$$

where $n$ is large, $p$ small, and $np$ moderate in size. Although it is difficult to specify where $n$ is large and $p$ small, it is usually appropriate to use the Poisson formula shown next where $n$ is at least 100 and $np$ is less than 10. (Manual calculations of binomial probabilities can be laborious and time-consuming; using the Poisson approximation provides answers that are very nearly identical.) The Poisson approximation to the binomial distribution is given by the formula

$$f(x) = (np)^x \cdot e^{-np}/x! \qquad \text{for } x = 0, 1, 2, 3, \ldots$$

In this formula, $e = 2.7128\ldots$ and the values of $e^{-np}$ are read directly from a table. These can also be obtained from many handheld calculators or by using appropriate computer software.

*Example* It is known that 2% of the pearl buttons produced by a button manufacturer are discolored. Use the Poisson approximation to the binomial distribution to determine the probability that in a carton containing 200 randomly selected buttons, there will be exactly 3 discolored buttons.

*Solution* Substituting $np = (200)(0.02) = 4$, and $e^{-np} = e^{-4} = 0.018$ into the formula provided for the Poisson approximation to the binomial distribution, we get

$$f(3) = 4^3 \cdot e^{-4}/3! = 0.192$$

for this probability.

See also Binomial Distribution; e; Probability Distribution

SELECTED REFERENCES

Freund, John E., and Benjamin M. Perles. *Statistics: A First Course.* 7th ed. Upper Saddle River, NJ: Prentice-Hall, 1999.

Freund, John E., and Gary A. Simon. *Modern Elementary Statistics.* 9th ed. Upper Saddle River, NJ: Prentice-Hall, 1997.

McClave, James T., and P. George Benson. *A First Course in Business Statistics.* 5th ed. New York: Macmillan, 1992.

Sandy, Robert. *Statistics for Business and Economics.* New York: McGraw-Hill, 1990.

BENJAMIN M. PERLES

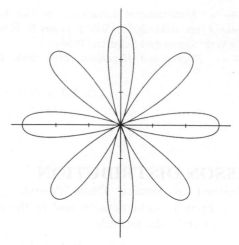

**Figure 1** *Rose curve, r = 3 cos 4θ.*

# POLAR COORDINATES

Specify how far a point is from the origin, and in which direction; first introduced by Isaac Newton (Boyer 1989, 456). In a rectangular coordinate system, a point is specified by its distances from the origin in directions parallel to the $x$-axis and parallel to the $y$-axis, similar to the street system in a city laid out in blocks. In some settings, however, such as at sea or in the air, a rectangular system is inefficient.

To understand a polar coordinate system and its relation to a rectangular system, consider a given point $P = (x,y)$ in the plane; the distance of $P$ from the origin is $r = \sqrt{x^2 + y^2}$. The direction in which the point $P$ lies is expressed as the measure of the angle formed by the positive $x$-axis and the *radial ray* from the origin through $P$. If $\theta$ is the angle, then the relationships $x = r \cos \theta$ and $y = r \sin \theta$ are satisfied. The point with rectangular coordinates $(x,y)$ then has *polar coordinates* $(r,\theta)$. (The angle $\theta$ is usually given in radians.) Because directions are not uniquely specified by angles, points do not have unique polar coordinates; the point $(r,\theta)$ is the same as the point $(-r, \theta + \pi)$ or the point $(r, \theta + 2\pi)$.

If $r$ is a function of $\theta$, say $r = f(\theta)$, then the *graph* of $f$ is the set of points $(r,\theta)$ satisfying the equation $r = f(\theta)$. Some common polar curves not easily described by rectangular equations are *rose curves, r = a cos nθ* (see Figure 1) and *limaçons, r = a + b cos θ* (see Figure 2). The *logarithmic spiral, r = ae^{kθ}* (see Figure 3) is found in nature, notably in the chambered nautilus.

See also Coordinate Geometry, Overview; Trigonometry

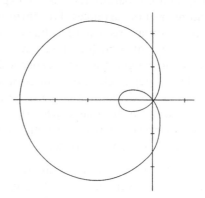

**Figure 2** *Limaçon, r = 3/2 − (5/2) cos θ.*

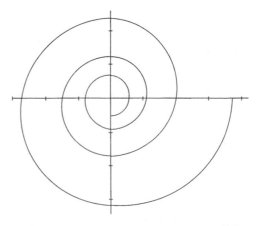

**Figure 3** *Logarithmic spiral, r = 2 e^{θ/10}.*

SELECTED REFERENCES

Boyer, Carl B. *A History of Mathematics.* 2nd ed. Revised by Uta C. Merzbach. New York: Wiley, 1989.

Garner, Lynn E. *Calculus and Analytic Geometry.* San Francisco, CA: Dellen, 1988.

Selby, Samuel M., ed. *CRC Standard Mathematical Tables.* Cleveland, OH: Chemical Rubber, 1991.

Spiegel, Murray R. *Schaum's Outline Series. Mathematical Handbook of Formulas and Tables.* New York: McGraw-Hill, 1994.

LYNN E. GARNER

## PÓLYA, GEORGE (1887–1985)

Author of writings on the use of problems in teaching mathematics and in the teaching of heuristics to solve problems. In fact, it has been said that Pólya is the father of the current emphasis on problem solving in mathematics instruction (Alexanderson 1987).

George Pólya was born in Budapest on December 13, 1887, one of five children. At the university, Pólya chose to study law, his father's profession, but he soon lost interest in it. After exploring various other areas (languages, literature, philosophy, and physics), he decided on mathematics. He received a Ph.D. degree from the University of Budapest in 1912, writing a dissertation on geometric probability. Over the course of his long and distinguished career as a mathematician, Pólya's interest in the act of problem solving led him to raise questions that yielded major mathematical research results and influenced the direction of several branches of mathematics in this century (real and complex analysis, number theory, probability, geometry, combinatorics, and mathematical physics).

As early as 1919 he demonstrated an interest in problem solving, heuristic reasoning, and mathematics teaching by publishing a paper on heuristics in the Swiss journal *Schweizerische Pädagogische Zeitschrift.* Although from that date on he continued to write papers on heuristics, it was not until 1945 that his book *How to Solve It* appeared. Its publication marks the beginning of his tremendous influence on mathematics education. *How to Solve It* was followed by other extremely influential works: *Mathematics and Plausible Reasoning* (1954) and two volumes of *Mathematical Discovery* (1962, 1965). Pólya published more than 250 articles and ten books and monographs, more than 70 of which were related to mathematics education.

*See also* Problem Solving, Overview

SELECTED REFERENCES

Alexanderson, Gerald, L., ed. *The Pólya Picture Album: Encounters of a Mathematician.* Boston, MA: Birkhäuser, 1987.

Curcio, Frances R., ed. *Teaching and Learning: A Problem-Solving Focus.* Reston, VA: National Council of Teachers of Mathematics, 1987.

Pólya, George. *How to Solve It: A New Aspect of Mathematical Method.* 2nd ed. Garden City, NY: Doubleday, 1957.

———. "On Learning, Teaching, and Learning Teaching." *American Mathematical Monthly* 70(1963):605–619.

———. *Mathematical Discovery: On Understanding, Learning, and Teaching Problem Solving.* Combined ed. New York: Wiley, 1981.

———. *Mathematics and Plausible Reasoning.* 2 vols. Princeton, NJ: Princeton University Press, 1954.

FRANK K. LESTER, JR.

## POLYGONS

Simple closed curves that consist of more than two line segments. The line segments are the sides of the polygon. The point where a pair of adjacent sides meet is called a *vertex* of the polygon. Each of these vertices is also a vertex of an angle of the polygon. A polygon has the same number of vertices and angles as it has sides. If the measure of each angle is less than 180°, the polygon is said to be *convex* (Figure 1a); otherwise it is called *concave* (Figure 1b). A polygon is classified according to the number of its sides and sometimes is given a special name, as indicated in the following table:

| Number of sides | Name of polygon | Number of sides | Name of polygon |
| --- | --- | --- | --- |
| 3 | Triangle | 7 | Heptagon |
| 4 | Quadrilateral | 8 | Octagon |
| 5 | Pentagon | 9 | Nonagon |
| 6 | Hexagon | 10 | Decagon |

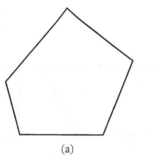
(a)

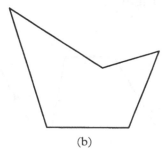
(b)

**Figure 1**

In general, a polygon with $n$ sides is called an $n$-gon, and the sum of the measures of its angles in degrees is equal to $180(n - 2)$.

A *triangle* may be classified according to its angle measures or the lengths of its sides as follows: (1) a right triangle, an obtuse triangle, or an acute triangle, if it has a right angle, an obtuse angle, or three acute angles, respectively; (2) equilateral, isosceles, or scalene, if it has three, two, or no equal sides, respectively. The elementary theorems on triangles are included in Euclid's *Elements*. Some important points related to a triangle $ABC$ are: (1) the *circumcenter O*, where the perpendicular bisectors of the sides meet; (2) the *centroid G*, where the medians of the triangle meet; (3) the *orthocenter H*, where the altitudes of the triangle meet. If $\triangle ABC$ is nonequilateral (Figure 2), the points $O$, $G$, and $H$ lie in a straight line called the *Euler line*, where $OG:GH = 1:2$ (Eves 1969). In the case of an equilateral triangle, the three points coincide.

A *quadrilateral* may have a special name if it has a certain property. For example:

1. A *trapezoid* is a quadrilateral with one pair of parallel lines (Figure 3(a));

2. A *parallelogram* is a quadrilateral with each pair of the opposite sides parallel (Figure 3(b));
3. A *rectangle* is a quadrilateral with four right angles (Figure 3(c));
4. A *rhombus* is a quadrilateral with four equal sides (Figure 3(d));
5. A *square* is a rectangle with four equal sides (Figure 3(e)).

Problems that dealt with triangles, rectangles, and trapezoids first appeared in ancient Egyptian papyri and Babylonian clay tablets that date back to ca. 2000 B.C. (Eves 1969). A theorem that applies to any quadrilateral $ABCD$ states that

$$AB \cdot CD + BC \cdot DA \geq AC \cdot BD$$

Equality holds if and only if the quadrilateral may be inscribed in a circle (Figure 4). The equality was discovered by Ptolemy of Alexandria (ca. 150), and the general case was found by Euler (1707–1783) (Hahn 1994). Ptolemy used his theorem to build up a table giving the lengths of the chords of the central angles, in a circle of radius 60, from $\frac{1}{2}°$ to 180° in steps of a half-degree. This table is equivalent to a table of sines (Eves 1969).

## REGULAR POLYGONS

In a regular polygon, all the sides are congruent and all the angles are congruent. A regular polygon can always be circumscribed by a circle. If the radius of the circumcircle is $r$ and the polygon has $n$ sides, then the perimeter of the polygon is $2nr (\sin \pi/n)$, and the area of the polygon is $nr^2 (\sin \pi/n \cos \pi/n)$. If $n$ is large enough, these two formulas should yield approximations for the circumference and the area of the circle of radius $r$. In 1795, Carl F. Gauss proved that a regular polygon with $n$ sides can be constructed by compass and straightedge if and only if $n = 2^r p_1 p_2 \ldots p_k$, where $r$ is a whole number and the $p_i$'s are distinct Fermat primes, that is, each of the form $2^{2^L} + 1$, $L$ being a whole number. This implies that a polygon with seven sides is not constructable

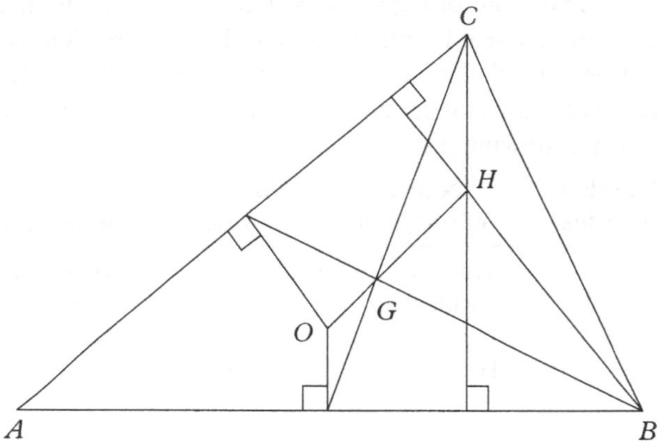

**Figure 2**

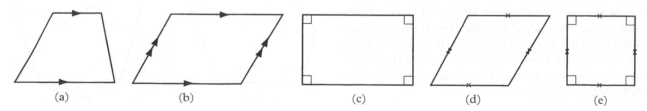

(a)    (b)    (c)    (d)    (e)

**Figure 3**

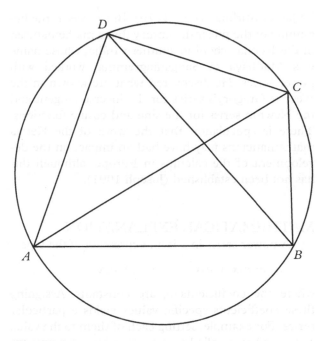

**Figure 4**

while a polygon with seventeen sides is constructable (Ore 1948).

In elementary school, students can learn about polygons and their properties through different activities such as matching geometric figures, drawing games, copying geoboard figures, and making solid figures. Students may use a protractor or *Logo* to explore the relation between angles of a polygon. Symmetries of regular polygons and using polygons to tile the plane are interesting topics for students. In high school, students should learn how to use deductive reasoning to prove some of the facts that they acquired informally.

*See also* Geometry, Instruction; Length, Perimeter, Area, and Volume

SELECTED REFERENCES

Billstein, Rick, Shlomo Libeskind, and Johnny W. Lott. *A Problem Solving Approach to Mathematics for Elementary School Teachers.* 6th ed. Reading, MA: Addison-Wesley, 1997.

Coxeter, Harold S. M., and Samuel L. Greitzer. *Geometry Revisited.* New York: Random House, 1967.

Del Grande, John, et al. *Geometry and Spatial Sense: Addenda Series, Grades K–6.* Reston, VA: National Council of Teachers of Mathematics, 1993.

Euclid. *The Thirteen Books of Euclid's Elements.* Vol. 2. Thomas L. Heath, trans. 2nd ed. New York: Dover, 1956.

Eves, Howard W. *An Introduction to the History of Mathematics.* 3rd ed. New York: Holt, Rinehart and Winston, 1969.

Hahn, Liang-shin. *Complex Numbers and Geometry.* Washington, DC: Mathematical Association of America, 1994.

Ore, Oystein. *Number Theory and Its History.* New York: McGraw-Hill, 1948.

AYOUB B. AYOUB

# POPULATION GROWTH

A phenomenon studied by the use of both discrete and continuous models. By a biological population we may mean the number of a particular species of animal in a particular region at a specific time. We may want to know how the size of this population changes with time, and what affects the size. So let us say that we look at the population at times $t(n)$. What will the population be the next time we look, at time $t(n + 1)$? The number of members in the population at time $t(n + 1)$ is the number at time $t(n)$ plus the number of births minus the number of deaths in the period from $t(n)$ to $t(n + 1)$. In the simplest population models, the birthrate $\beta$ and the death rate $\delta$ are constant. Then, if $p(t)$ is the population at time $t$, it would be modeled by the equation

$$p(t(n + 1)) = p(t(n)) + (\beta - \delta)(t(n + 1) - t(n))$$

We can think of the quantity $(\beta - \delta)$ as the population's growth rate. How would we make the model more realistic? We could have both the birth and the death rates depend on other factors, such as the age distribution of the population, competition for the food supply, harvesting, and possible genetic changes. We could also look simultaneously at two populations, in which one could be the food source of the other (this is called the "prey-predator model"). Another interesting situation is one in which the two populations compete for the same food supply.

The models we have discussed so far are called "discrete" because we examine the population at discrete times $t(n)$. The equations describing such discrete models are difference equations (Sandefur 1990). The mathematics of continuous models becomes relevant if we think of $t(n + 1) - t(n)$ going to zero and study the resulting differential equations (Hirsch and Smale 1974). For a model in which the times we consider are one unit apart, we call the population at time $n$ simply $p(n)$. Let the growth rate become smaller as the population increases—

a reasonable assumption if the territory or the food supply is limited. The equation now becomes

$$p(n + 1) = p(n) + rp(n)(A - p(n))$$

This is the so-called logistic model, which can show an amazing variety of behaviors, including a steady approach to a limiting population, periodicity, and even chaos. Natural populations give evidence of all of these. The differential equation governing the continuous model with the same assumptions does not lead to all of these behaviors.

The Italian mathematician Vito Volterra (1860–1940) was first led to study these problems on the basis of the populations of food fish and predators in the Adriatic during and after World War I. In recent years, computers have given us easy access to these models, and materials have become available for school and college study (Peitgen, Jürgens, and Saupe 1992).

*See also* Applications for the Classroom, Overview; Applications for the Secondary School Classroom; Biomathematics; Modeling

### SELECTED REFERENCES

Hirsch, Morris W., and Stephen Smale. *Differential Equations, Dynamical Systems, and Linear Algebra.* New York: Academic, 1974.

Peitgen, Heinz-Otto, Hartmut Jürgens, and Dietmar Saupe. *Fractals for the Classroom, Part One.* New York: Springer-Verlag, 1992.

Sandefur, James T. *Discrete Dynamical Systems: Theory and Applications.* Oxford, England: Clarendon, 1990.

HENRY O. POLLAK

# POWER SERIES

A series that first appeared prominently in the work of the French mathematician Joseph-Louis Lagrange (1736–1813). In fact, his approach to the calculus involved the expansion of a function in its Taylor series. Brook Taylor (1685–1731) was an English mathematician whose name has been given to the series expansion of a function despite the fact that this expansion was known to a Scottish mathematician, David Gregory (1627–1720), forty years before Taylor published his discovery in *Methodus Incrementorum* (1715). The special case of the Taylor series, which is known as the Maclaurin series, after the Scottish mathematician, Colin Maclaurin (1698–1746), was itself discovered (1730) more than ten years before Maclaurin's published result (1742)

by James Stirling (1692–1770). But it was at the beginning of the fifteenth century that a mathematician of the Kerala school in southern India, whose name was Madhava of Sangamagramma, worked with power series. He discovered what are known in the West as Gregory's series for the inverse tangent and the Newton series for the sine and cosine functions. There is speculation that the work of the Kerala mathematicians may have had an impact on the development of the calculus in Europe, although this has not been established (Joseph 1991).

## MATHEMATICAL EXPLANATION

A power series in $x$ is an expression of the type

$$a_0 + a_1x + a_2x^2 + \cdots + a_nx^n + \cdots$$

where the coefficients $a_i$ are constants. Assigning these coefficients specific values yields a particular series. For example, setting each of them to the value 1 gives what is called a *geometric* series. A power series in $x - a$ is obtained by replacing $x$ in the preceding expression by $(x - a)$. This involves no more than a change of coordinates, so the explanation given here will deal mostly with a power series in $x$.

The notion of convergence is crucial to the analysis of a power series. Let $S$ be a power series in $x$ and let $S_n$ be the sum of the first $(n + 1)$ terms of $S$:

$$S_n(x) = a_0 + a_1x + a_2x^2 + \cdots + a_nx^n$$

Thus $S_n$ is a polynomial of degree $n$. Suppose that, for a set $D$ of real numbers $x$, $S_n$ converges; that is, there exists a function $L$, defined on $D$, such that

$$\lim_{n \to \infty} S_n(x) = L(x)$$

for all $x$ in $D$. Then, by definition, $S$ converges to $L$ for $x$ in $D$. For example, the power series

$$S = 1 + x + x^2/2! + \cdots + x^n/n! + \cdots$$

converges to the exponential function $\exp(x) = e^x$ in $D = R$, the real line. The series

$$S = 1 + x + x^2 + \cdots + x^n + \cdots$$

converges to the function $L$ defined by the rule $L(x) = 1/(1 - x)$ in $D = (-1,1)$; $S$ does not converge outside $D$.

For any series $S$, there exists a radius $R$ of convergence of the series. $R$ may take the value 0, in which case $S$ converges only for the value $x = 0$; infinity, in which case $S$ converges for all values of $x$; or a positive value, in which case $S$ converges for all $x$ whose absolute value is less than $R$. For example, the radius of convergence of the geometric series

$(a + ax + ax^2 + \cdots + ax^n + \cdots)$ is 1. In practice, the radius of convergence of a series may be calculated by a ratio test.

The most important aspect of power series involves a reversal of the development just given. One starts with an infinitely differentiable function $f$ and asks for its power series representation. For a power series in $x$, this leads to the so-called Maclaurin series, in which the coefficients are determined by the formula

$$a_n = f^{(n)}(0)/n!$$

where $f^{(n)}(0)$ is the $n$th derivative of $f$, evaluated at $x = 0$. For a power series in $x - a$, the expression for $a_n$ is the same except that the derivative is evaluated at $a$ rather than at 0. This yields the so-called Taylor series for $f$. As for all power series, there is a radius of convergence $R$ associated with each Taylor (or Maclaurin) series. If $R$ is a positive real number, the series is said to represent the function $f$ over $(-R,R)$, which is called the *interval of convergence* for the series. The theory presented here can be extended to series of complex numbers, in which case the notion of the interval of convergence is replaced by that of the region inside the circle of convergence,

$$\{z|\ |z| < R\}$$

## MOTIVATION AND GOALS

The salient issues to address in a study of power series are those of representation and tolerance. The first of these provides the answer to the question: Where does a series represent a function? Students will need to determine the radius of convergence of the Taylor series. As stated, this corresponding "interval of convergence" is either a singleton (i.e., $\{0\}$ in the special case of the Maclaurin series or $\{a\}$ in the case of the Taylor series for a function expanded in terms of $x - a$); an interval of the type $(-R,R)$ for a Maclaurin series or $(a - R, a + R)$ for a Taylor series; or the entire real line.

The second issue arises when the function $f$ has a Taylor series representation $S$ about the point $a$ and a nonzero radius of convergence. What is required is that the absolute value of the difference

$$S(x - a) - S_n(x - a)$$

be bounded above by a given tolerance $T$ for a fixed (or for all) $x$ in the interval of convergence. The most common method of attack on this problem uses the Lagrange form of the difference. In effect, the lowest value of $n$ must be found such that the following inequality involving the Lagrange form of the difference is satisfied:

$$|f^{(n+1)}(c)\ [(x - a)^{n+1}/(n + 1)!]| \leq T$$

Here, $c$ is a number between $x$ and $a$ determined by an application of the Mean Value Theorem. Generally, an upper bound $u$ for $|f^{n+1}|$, restricted to the closed interval from $x$ to $a$ (or from $a$ to $x$, depending on which is smaller), is used in place of $f^{(n+1)}(c)$ in the preceding inequality to guarantee that the tolerance, $T$, is satisfied regardless of what $c$ is. For example, the Maclaurin polynomial of degree 5, which approximates the exponential function exp on the interval $[0,1]$, satisfies a tolerance of $T = 0.005$, since an upper bound $u$ for $\exp^{(6)}(x)$ on $I$ is $e$, and $e/6! < 0.00378$. Generally, of course, this value depends on the (fixed) value of $x$. If it does not, then the convergence of the Taylor polynomials $S_n(x - a)$ is said to be uniform.

Taylor series are no longer used extensively for approximation or evaluative purposes, but they continue to shed light on processes related to these. An example of this is provided by the problem of computing the zero of a function when the zero occurs on a "flat" part of the graph of the function. Consider specifically the case of the function $y = \cos x + \cosh x - 2$, which has a zero at $x = 0$. One popular computer algebra system (CAS), using Newton's method and a seed of $x = 4$, fails to converge to the prescribed accuracy and stops at $x = 0.057$. Insight into this failure is given by the Taylor series approximation of the function—it starts with $x^4/12$.

## PEDAGOGICAL APPROACHES

The issues of representation and tolerance presented here will absorb much of the students' time in their study of power series. The development of computer algebra systems and, more recently, of graphing calculators like the TI 92 or the HP 48G with symbolic manipulation capabilities has had an impact on the way students can cope with these issues. By plotting Taylor polynomials of increasingly higher degree, students will achieve a graphic sense of the way in which these polynomials approximate the given function over its interval of convergence. There is still pedagogic value, however, in students' determining the radius of convergence for a given series representation by use of the ratio test. Traditionally, attention has been paid, in the case where the radius $R$ is a nonzero real number, to the behavior of the series at the endpoints of the interval of convergence. This means that two particular series of constants have to be examined by the methods available to analyze such series. Teachers should keep in mind, however, that this is a detail that should not distract students' attention from the bigger picture of series representation.

Access to a CAS or a calculator with a symbolic manipulation capability assists students greatly to determine the degree of the Taylor polynomial required to meet a given tolerance. Technological assistance makes calculating derivatives less tedious, so the list of examples with which students can work is vastly expanded, and they can tackle "real-world" problems. An example of this is provided by approximations of the function $y = \exp(-x^2)$ and closely related functions such as the Gaussian. Such functions are used in the normal distributions encountered in statistics. A CAS helps significantly to determine bounds on the $n$th order derivatives of the Gaussian function and hence the degree of the approximating polynomial required to meet the given tolerance. Teachers who wish may also encourage their students to study graphically such functions as the Lagrange form of the difference cited here.

## RECOMMENDED MATERIALS FOR INSTRUCTION

Despite the obvious advantage that technological tools like a CAS provide, students are unlikely to abandon more traditional methods of paper and pencil, nor should they be discouraged from doodling with paper and pencil. The fact remains, however, that the impact of technology on the study of power series (and the whole of the calculus) is so great and the price of powerful equipment has become so low that teachers are urged to avail themselves of whatever lies within their budget and to experiment with its potential to enhance their students' understanding of the material studied.

*See also* Maclaurin Expansion; Mean Value Theorems; Series

## SELECTED REFERENCES

Apostol, Tom M. *Calculus.* Vol. 1. Waltham, MA: Blaisdell, 1967.

Bell, Eric Temple. *The Development of Mathematics.* New York: McGraw-Hill, 1940.

Boyer, Carl B. *A History of Mathematics.* 2nd ed rev. Revised by Uta C. Merzbach. New York: Wiley, 1991.

Buck, R. Creighton. *Advanced Calculus.* 3rd ed. New York: McGraw-Hill, 1978.

Joseph, George Gheverghese. *The Crest of the Peacock.* London, England: Tauris, 1991.

Stewart, James. *Calculus.* Monterey, CA: Brooks/Cole, 1987.

ROBERT J. FRAGA

# PRESCHOOL

An age group in which the teaching of mathematics can be based on young children's natural curiosity about the world. Preschool children explore freely, investigate, and learn early number concepts in spontaneous, natural ways—through the manipulation of objects and making discoveries. They are capable of concrete reasoning based on their past experiences. Mathematics plays a major role in the daily lives of preschool children: they are exposed to concepts of time through events such as breakfast, naptime, and bedtime; they live in a world where measures of weight, temperature, and capacity are daily evident; and they are exposed to coins and bills on a daily basis.

## EARLY INFLUENCES

Child study began late in the nineteenth century when Charles Darwin (1809–1882) studied his own son's behavior. Historically, it has been psychologists, rather than mathematicians, who have conducted research into how mathematical content is taught and learned, particularly at the elementary, early childhood, and preschool levels. A survey of research studies from 1880–1910 focusing on mathematics found thirteen psychological studies of the development of concepts and readiness.

Child-study advocates did their best work with preschool children, and many theorists contributed to the body of knowledge relating to preschool children and appropriate mathematics activities. Friedrich Froebel (1782–1852), for example, believed play was the foundation for a child's learning and that materials found in nature (cylinders, cubes, spheres, etc.) helped children explore the properties of matter. His beliefs are still considered today when plans are made for preschool mathematics programs. Johann Pestalozzi (1746–1827) believed that children from birth to age 5 learned best through direct experience involving physical activity. He was concerned about rote learning in schools and advocated that children's experiences be based on their interests (Brewer 1992). Maria Montessori (1870–1952), whose aim was to foster progression in the child's knowledge, believed children should handle real materials (such as puzzles, beads, and pattern blocks) as a means of learning if they were to complete real tasks (Brewer 1992). She believed that the transformation of mental structure occurs through active experience, and she used counting and recognition activities as well as puzzles in her work with young children in the early twentieth century. Montessori's work suggested that teachers be observers of children and that they help them build on previous experiences. Alfred Binet (1857–1911), while director of the first French psychological laboratory, suggested in 1899 that research on mathematics had to be done with children in schools and by teachers using observation, questionnaires, and experiments. G. Stanley Hall (1844–1924) launched a child-study movement because he believed children learned primarily in the same way as occurred during human evolution, that is, naturally, and that outside influences did not speed up the learning; thus, there was not much point in trying to influence the intellectual development of young children. Hall believed that formal teaching of arithmetic should be postponed until late in the school program and that the early years should be devoted to concrete experience requiring counting and measuring (Jones 1970).

## MORE RECENT INFLUENCES

Several theories of learning mathematics have had an influence on preschool mathematics, and there has been disagreement about what should be taught to children at various levels or ages and how the concepts should be taught. In 1935, William Brownell advocated a "meaning theory" for arithmetic, making connections between ideas, facts, and procedures. An "incidental learning theory" held that children learn arithmetic better if it is not systematically taught to them. The "readiness theory" stated that concepts or skills in mathematics should not be taught until children are mature enough to learn them. In 1929, Louis P. Benezet, superintendent of the schools in Manchester, New Hampshire, advocated that no arithmetic, other than estimating and using numbers in social situations, be taught until seventh grade. Two educators, John Dewey (1859–1952) and William Heard Kilpatrick (1871–1965), suggested that all mathematics that anyone needed could be learned through experience and that arithmetic should not be a separate subject in the elementary school curriculum.

Dewey believed the curriculum should actively involve children in real and meaningful activities because children learn by doing and that number concepts develop out of the child's activity. He also believed that the young child related parts to the whole and needed to make measurements as well as to count.

Francis Galton (1822–1911), Binet, Jean Piaget (1896–1980), Max Wertheimer (1880–1943), and Lev S. Vygotsky (1896–1934) have written about the mental abilities used in doing mathematics and have influenced our thinking about mathematics for the young child. Vygotsky and his followers believed that cognitive growth occurs through social interaction and conceived of the mind as a product of social life. They believed that the adult should provide for a sense of socially situated learning for preschool children. Children "learn to do things they observe others around them doing and for which they have some need in their own lives" (Brewer 1995, 28).

Jerome S. Bruner (1915– ) felt that children must manipulate objects in order to form images. He believed that there is a need to create connections between new and existing knowledge. Young children are in a stage of development Bruner called *enactive*, where they must physically manipulate objects. Zoltan Dienes (1916– ) suggested that free play would promote the wholeness of the mathematical structure. His ideas are based on the theory that learning can be integrated into one's personality and can become a personal fulfillment.

Caleb Gattegno (1911– ) developed Cuisenaire rods to be manipulated by children in order to help them learn mathematical concepts. He wrote arithmetic texts in which children used the rods to explore and learn numbers and measurement concepts. The method employed by Gattegno emphasized reasoning based on perceptual judgment.

## IMPLEMENTATION OF THEORIES AND RESEARCH

In the past, formal mathematics instruction employed such practices as memorizing number sequences and learning to do rote counting. Preschool children were taught to write numerals and to match numerals with drawings of sets. These concepts are considered to be abstract, and ones for which the preschool child was not developmentally ready. Because of what is known about preschool children and how they learn, these practices are now discouraged by some.

Before the 1960s, the term *arithmetic* was used for mathematics and children focused primarily on *acquiring* computational skills (numbers and operations on numbers) and *applying* computational skills. The acquisition and application of computational skills occurred in limited, though concrete, situations (Schultz, Colarusso, and Strawderman 1989). Beginning in the 1970s, *mathematics* became the focus of experiences for children. Mathematics included arithmetic, geometry, and measurement. Estimation and problem solving became an integral and essential component of early childhood mathematics. *Arithmetic* enables the preschool child to tell *how many* and *how much* and emphasizes an understanding of numbers and patterns. In *measurement* the child uses numbers to describe attributes of continuous things (length, area, volume, capacity, and time (Schultz, Colarusso, and Strawderman 1989). In the study of *geometry* the young child uses geometric models to explore shapes of varied dimensions and create changes in shape and size through *topological transformations* (Wolfinger 1994).

Constance Kamii, a follower of Jean Piaget, advocates the construction of two types of knowledge (in addition to social) that were proposed by Piaget —physical and logico-mathematical. She believes that physical knowledge is knowledge of objects that are "out there" and observable. Kamii defined logico-mathematical knowledge as consisting of the relationships that the subject creates and introduces into or among objects (Kamii, *Number,* 1982).

## MOTIVATION AND GOALS

The motivation and goals for preschool mathematics are influenced by the ideas of various theorists and how each believes that the learning of mathematics occurs. Constructivist theorists, whose ideas are based to a great extent on the works of Piaget, believe that motivation for cognitive activity (thinking and reasoning) comes primarily from within the individual, not from without. The learner constructs ideas by first manipulating objects and reflecting on the results, thereby reorganizing his or her mental constructs (Brewer 1995, 248). These theorists also believe that motivation comes from the environment and that children create their own representations of interaction with the world and build their own network of representation. Therefore, their goal is for children to interact with and become involved with the environment. For the constructivist, motivation occurs in small increments; that is, as the learner reorganizes, he or she has different motives. Motivation is inherent in the situation in which the child is involved. When the learner manipulates and reacts to the situation, new goals are created (Payne 1990).

It is now generally agreed that young children must have experiences in classifying, sorting, comparing, ordering, graphing, measuring and counting, and performing operations on numbers (Payne 1990). Preschool children aged 3 to 4 have no classification system. Before preschoolers can classify, they must have an understanding of the concepts of "alike" and "belonging," and of objects being put together that belong together. The adult (teacher or parent) must provide experiences for preschool children to learn to identify various objects, to describe their purpose, and to gain an understanding of location and position. Young children need concrete experiences that enable them to use the learned terms (*location, position,* etc.) and to apply the learned concepts, and they should have a variety of opportunities to handle, manipulate, and explore materials as they think about mathematical ideas. In addition, children should take part in activities designed to move them from dependency on the physical world to the abstract world of ideas.

## MATHEMATICAL EXPLANATIONS

To be successful, a learning experience must be accurately matched to the child's level of understanding. Appropriately designed, the content of preschool mathematics can enable children to investigate and expand their intuition (Charlesworth and Radeloff 1991). In addition, they can begin to establish a broader base for such informal mathematics knowledge. Young children are dependent on concrete observable events (physical knowledge) to help them "figure things out."

Mathematical experiences should also build on young children's interests. When tied to the experiences and interests of learners, mathematics serves a functional and authentic purpose. This consideration of the experiences and interests of preschool children aids teachers and curriculum planners in developing authentic mathematical experiences. Preschool mathematics topics and activities must be planned to stimulate children's mental action.

### Observation

Children learn to observe by observing. They must be encouraged to look for things in their environment and to note particular characteristics of the things observed. Children should be encouraged to use all their senses. They can observe fruits and vegetables, investigate toys, use "feeling" boxes, or experience sound and smell matching games. Experience charts, which record the child's own words and descriptions, can be made following observations. For example, after children have taken a walk around the school to observe the number of cars in the parking lot, windows on the school, and so on, the teacher can record the children's observations on a chart.

## Classification

Classification requires children to use their developing observational skills. Children look for similarities among objects—size, color, texture, shape or use—and try to group objects accordingly. Long blocks of time for free choice should be provided daily. During the exploratory stages, young children often begin with one trait and change to another in the middle of sorting a group of materials. They may also change the trait being used with every other object. It is appropriate, therefore, to provide a single, obvious characteristic—such as red cars, long blocks, large buttons, and so forth—to be classified by young children. The key to classification activities is to find a single characteristic that children can see easily, use the characteristic to classify, and then see if another characteristic can be found. Finding many ways in which to classify the same group of objects helps develop flexibility in the thinking of the preschool child.

## Comparing and Contrasting

In comparing and contrasting, children look for differences. When they build with blocks, they can make comparisons of the constructions. The ability to contrast objects can be developed in the same way as the ability to classify, but the focus is on how the objects are different. Concepts to be developed include more-less, big-small, up-down, and other spatial ideas that show differences in location. Children can compare volume through informal experiences of pouring water or sand.

## Patterning

Patterning is probably the most fun and certainly is the most important of mathematical skills for young children. Children use movement patterns by bending, turning, hopping, jumping, or running in a specific sequence. They can use sound patterns by using rhythm instruments. They can make patterns with blocks, investigate patterns in fabric, and color patterns onto fabric. Children can investigate the arrangements that can be made with a certain quantity of blocks or look for patterns that are formed in the veins of leaves, by snowflakes, when pebbles are dropped in a puddle, or listen for patterns in their favorite songs or poems. The greater the variety of patterns children experience, the more likely they are to see those patterns in more formal mathematics later (Payne 1990).

## Graphing

Graphs can be made by placing any objects (the children themselves or interesting leaves they find outdoors) into rows according to a common criteria and comparing those rows with each other on the basis of one-to-one correspondence. Lifesize graphs should be made for young children; for example, cash register tape can be placed on a chart to show height. Graphs made of real objects such as fruit or leaves allow children to develop the concepts of more, less, equal, same, and different. Graphs allow children to explore concepts about themselves and their classmates when they arrange themselves into lines on the basis of some characteristic—gender, favorite food, favorite story, number of brothers and/or sisters, number of teeth lost, number of children with birthdays in certain months, number of children wearing a certain color, having a certain hair color, having freckles.

## Informal Measurement

Understanding of measurement comes from experience. It occurs when children measure using any object that is more natural for them than rulers, yardsticks, or meter sticks. Children can classify and compare to find how many crayons will fill a box, how many lengths of a particular child it will take to reach from one side of the room to other, or how many children are needed on one end of the seesaw to balance the teacher on the other end. Pouring water into various-size containers, comparing the amount of rice needed to fill a jar to the amount of beans needed, finding out whether it takes syrup longer to flow down a ramp than water or oil, and finding how to get water to evaporate quicker, or an ice cube to melt faster are all appropriate experiences for preschool mathematics.

## Geometry

Preschool mathematics involves activities in which young children sort collections of objects.

They may sort solid geometric shapes or solids into two groups that are alike in some way and explain why they put blocks together as they did. These investigations of geometric shapes result in a variety of classifications and are a natural activity that young children engage in as they seek to make sense of their world. Topics appropriate for preschool mathematics include two- and-three dimensional geometry. The form or shape of an object is more significant than its color, and children are interested in shapes of things—objects in the environment, the shape of their bodies, shapes they can make with their bodies, and geometric shapes. Most things have a shape and the shapes of things help determine what they are or how they are different from other objects. Shapes should be found in numerous ways in the child's environment, and many opportunities for manipulation of the shapes should be provided. Preschool children should play with shapes through games, toys, art activities, songs, poems, and stories.

## Counting

Counting for young children begins with the memorization of the number names in their correct sequence. The next step is to use counting in natural and purposeful ways. It is natural for young children to count to find out how many more children like bananas than strawberries before a graph is made or to count how many more wheels a large truck has than a car when comparing and contrasting. Young children can count how many objects are in the blue group and how many are in the red group after classifying. They also count in games. When counting skills are developed through use, with objects and for a purpose, counting becomes meaningful and concepts of number develop.

## VARIETY OF PEDAGOGICAL APPROACHES

Successful investigation of mathematical concepts by preschool children depends on the use of manipulative materials as well as the atmosphere created in the classroom. When young children manipulate things in their world in ways that help them make sense of it, they are better able to organize the world into understandable parts. "Play with lots of thoughtful conversation is what math for young children should be" (Cartwright 1988, 44). Young children learn by constructing relationships based on their observations and experiences. Touching and

moving objects are essential parts of a child's construction of mathematical concepts. The teacher or parent must arrange a variety of free-play experiences in which children can interact.

Young children should be given opportunities to choose activities and to experiment with objects. It is widely held that these kind of experiences promote the preschool child's social and cognitive development. In addition, situations should be created that enhance thinking, analyzing, and reasoning (Eliason and Jenkins 1990). One way to accomplish this is by organizing the classroom as a mathematics laboratory. The mathematics laboratory is rich in materials for different types of activities and is child-centered, makes use of open-ended activities, and uses a multimedia-multisensory approach. Learning in the mathematics laboratory is based on the experiences that enhance concept formation, experiences in which preschool children investigate number concepts, make patterns, classify objects, make comparisons, and explore spatial and measurement concepts.

## RECOMMENDED ACTIVITIES AND MATERIALS FOR INSTRUCTION

The content and materials for preschool mathematics are determined by informal experiences that enable young children to investigate and expand their intuition. Activities and materials to be used for the development of mathematical concepts with young children should meet several criteria: (1) cause and effect should be immediate in time and space; (2) children should construct knowledge rather than be taught; (3) topics should allow young children to investigate both the forward and reverse results of an activity (children should put together and take apart, pour from one container to another, and then back into the original container); (4) activities should allow young children to discuss the "what" of an activity rather than the causal "why" of an activity; (5) activities should allow children to use forms of reasoning that are cognitively appropriate; and (6) activities should encourage appropriate oral development (Wolfinger 1994; Charlesworth and Radeloff 1991). Two types of instructional mathematical materials are those found in the environment (real life) and manipulative objects designed and specifically constructed as models to depict and model the mathematical concept being addressed. Manipulative objects and materials found in the environment are developmentally appropriate for use with preschool children. Preschool children can use manipulatives such as buttons to sort, classify, and count. Other materials in the environment that can be used for these activities include collections of beans, seeds, pasta, seashells, cotton balls, jar tops and lids, stickers, spools, straws, stirrers, tongue depressors, rocks, and craft sticks. Nuts, nails, screws, and bolts can be used for classification activities. Children can be encouraged to sort items according to shape, size, color or function. Bells can also be classified by size, color, shape, and according to sound. Scrap materials or textured papers can be sorted and classified according to how they feel or by color. Puzzles and puzzle pieces, both wooden and cardboard, are excellent for helping preschool children develop their eye-hand coordination as well as mathematical concepts. Empty cardboard boxes and cartons and other containers can be used to set up a play store so that children can gain an understanding of the concepts of money and economics. Objects such as buttons, screws, bolts, macaroni, and shells are also appropriate for one-to-one correspondence.

Simple games in which preschool children manipulate marbles, plastic chips, small blocks, and so forth are appropriate for allowing children to construct meaningful concepts in mathematics. Mathematically related experiences include those that allow children to roll objects at various angles. These experiences with objects from the environment allow preschool children to have direct multisensory experiences as they directly manipulate materials.

Measurement activities involving placing objects on the pans of a balance to see what happens are appropriate. Generally, however, the use of standard measuring instruments is not appropriate for preschool mathematics. Nor should written symbols be incorporated in preschool mathematics instruction, even though sometimes young children can and do read numbers. However, language use and vocabulary are important. The language of mathematics should be natural and mathematical "words" introduced in an informal manner. Young children can count aloud as they manipulate marbles, plastic chips, small blocks, or buttons.

Materials appropriate for preschool mathematics are used by children to explore a variety of possibilities rather than a single idea. The greater use and flexibility of the materials, the better.

*See also* Brownell; Bruner; Dewey; Dienes; Measurement; Montessori; Patterning; Pestalozzi; Psychology of Learning and Instruction, Overview

## SELECTED REFERENCES

Bredekamp, Sue, ed. *Developmentally Appropriate Practice* (Position Statement). Washington, DC: National Association for the Education of Young Children, 1986.

Brewer, Jo Ann. *Introduction to Early Childhood Education: Preschool Through Primary Grades.* Boston, MA: Allyn and Bacon, 1992, 1995.

Cartwright, Sally. "Play Can Be the Building Blocks of Learning." *Young Children* 43(5)(1988):44–46.

Charlesworth, Rosalind, and Deanna J. Radeloff. *Experiences in Math for Young Children.* Albany, NY: Delmar, 1991.

Eliason, Claudia, and Loa Jenkins. *A Practical Guide to Early Childhood Education.* Columbus, OH: Merrill, 1990.

Greenberg, Polly. "How and Why to Teach All Aspects of Preschool and Kindergarten Math Naturally, Democratically, and Effectively (For Teachers Who Don't Believe in Educational Excellence, and Who Find Math Boring to the Max)." Parts 1, 2. *Young Children* 48(May 1993):75–84; 49(Jan. 1994):12–18, 88.

Jones, Phillip S., ed. *A History of Mathematics Education in the United States and Canada.* 32nd Yearbook. Washington, DC: National Council of Teachers of Mathematics, 1970.

Kamii, Constance K. "Encouraging Thinking in Mathematics." *Phi Delta Kappan* 64(4)(1982):247–251.

———. *Number in Preschool and Kindergarten.* Washington, DC: National Association for the Education of Young Children, 1982.

———, with Georgia DeClark. *Young Children Reinvent Arithmetic: Implications of Piaget's Theory.* New York: Teachers College Press, 1985.

National Council of Teachers of Mathematics. *Curriculum and Evaluation Standards for School Mathematics.* Reston, VA: The Council, 1989.

———. *Principles and Standards for School Mathematics.* Reston, VA: The Council, 2000.

Payne, Joseph N., ed. *Mathematics for the Young Child.* Reston, VA: National Council of Teachers of Mathematics, 1990.

Schultz, Karen A., Ron P. Colarusso, and Virginia W. Strawderman. *Mathematics for Every Young Child.* Columbus, OH: Merrill, 1989.

Stone, Janet I. "Early Childhood Math: Make It Manipulative." *Young Children* 42(6)(1987):16–23.

Van de Walle, John A. "The Early Development of Number Relations." *Arithmetic Teacher* 35(6)(1988): 15–21, 32.

Williams, Connie, and Constance K. Kamii. "How Do Children Learn by Handling Objects?" *Young Children* 42(1)(1986):23–26.

Wolfinger, Donna M. *Science and Mathematics in Early Childhood Education.* New York: HarperCollins, 1994.

MARILYN S. NEIL

# PRIME NUMBERS

Integers $p > 1$ whose only positive divisors are 1 and $p$. Otherwise, $p$ is called a *composite* number. The notion of prime number is essential to every aspect of number theory. Primes have fascinated mathematicians since ancient times, and some fundamental results on primes were proved more than 2,000 years ago. The statement "There are infinitely many prime numbers," was first proved by Euclid. Euclid's original proof goes as follows. Suppose, on the contrary, that there are only finitely many prime numbers. Denote them by $p_1, p_2, \ldots, p_r$ and set $n = p_1 p_2 \ldots p_r + 1$. As $n$ differs from all $p_i$'s, $n$ is a composite number and as such has to be divisible by some $p_i$. However, since $p_i$ divides the term $p_1 \ldots p_r$, $p_i$ has to divide also the number 1, which is a contradiction.

It has also been known since ancient times that prime numbers form the building blocks of all integers, although the formal proof of this statement was derived much later. More precisely, every integer $n$ greater than 1 is either a prime or can be expressed as a product of primes. This representation of $n$ is unique except for the order of the factors. Because of the significance of the result, this statement is called the *Fundamental Theorem of Arithmetic*.

The oldest method for testing primality is the sieve of Eratosthenes. This method determines all primes below a given integer $n$. First all integers from 2 to $n$ are written down. Then composite numbers are systematically eliminated by crossing out multiples of primes, starting with 2, 3, 5, 7, 11, and so on. The procedure continues up to crossing out multiples of the largest prime smaller than $\sqrt{n}$, since any composite number $m < n$ has a factor smaller than $\sqrt{n}$. The integers that are left, those that did not fall through the sieve, are primes. As the sieve can be used only in the case when there is a list of all primes smaller than $\sqrt{n}$, different approaches are used for testing primality of large numbers.

## DISTRIBUTION OF PRIMES

One of the most interesting and, at the same time, most difficult questions concerning primes is their distribution. It can happen that two consecutive odd numbers, $p$ and $p + 2$, are primes; such primes are called *twin primes*. On the other hand, for any number $N$, there exist $N$ consecutive composite integers. In other words, the gap between two consecutive primes can be arbitrarily long. For example, the numbers $N! + 2, N! + 3, \ldots, N! + N$ provide $N - 1$ consecutive composite integers. Pafnuti

Chebyshev (1821–1894) proved a conjecture raised by Joseph Louis Bertrand (1822–1903), now known as Bertrand's postulate: for any natural number $n > 1$, there is a prime between the numbers $n$ and $2n$; that is, the gap between two consecutive primes $p < q$ is smaller than $p$.

The function $\pi(x)$, which is defined as the number of primes not exceeding $x$, is used to describe the distribution of prime numbers. It is a simple exercise to calculate $\pi(x)$ for small values of $x$, for example, $\pi(100) = 25$; that is, there are precisely 25 primes not bigger than 100. However, the values of $\pi(x)$ for very large $x$ are not known. Therefore, the focus is on obtaining a good estimate of the magnitude of $\pi(x)$, when $x$ grows over all bounds. A classical result, now called the *Prime Number Theorem*, claims that the limit of $\pi(x)/(x/\log x)$ equals 1 as $x$ approaches infinity, where $\log x$ is the logarithm to the base $e$. In other words, for very large $x$, $\pi(x)$ and $x/\log x$ hardly differ. Or, using probabilistic language, if a number about the size of $x$ is given, the probability is $1/\log x$ that the given number is prime. This statement was proved independently in 1896 by Jacques Hadamard (1865–1963) and Charles-Jean de la Vallée Poussin (1866–1962). It is an immediate consequence of the Prime Number Theorem that the series $\Sigma(1/p)$, where $p$ runs over all primes, is divergent. This means that primes are not that sparse as, for example, perfect squares.

## RECORDS

It is most likely that the part of number theory concerned with primes is the only area in all mathematics where mathematicians keep track of (some are even hunting for) records. An updated list of some of the records follows. A new record (largest) prime was found by David Slovinski and Paul Gage in 1992, who used a Cray-2 supercomputer. It took over 10,000 computer hours to locate the number $2^{756839} - 1$, but the primality of the number was verified in nineteen hours. The previous record prime was much smaller, "only" $391581 \times 2^{216196} - 1$. The record twin primes are $1,706,595 \times 2^{11235} \pm 1$, which were discovered in 1989 by a team headed by Bodo Parady. There are arithmetic progressions formed entirely by primes. The record is a progression of nineteen primes, where the first term is 8,297,644,387 and the common difference is 4,180,566,390, discovered by Paul A. Pritchard in 1985. The longest-known arithmetic progression of consecutive primes contains six terms and was found by Leon J. Lander and T. R. Parkin in 1967. The first term is 121,174,811, and the common difference equals 30. The largest value of $x$ for which $\pi(x)$ has been computed is $x = 4 \times 10^{16}$ and $\pi(4 \times 10^{16}) = 1,075,292,778,753,150$.

## APPLICATIONS

For a long time there were no significant applications of primes to real-life problems, but the situation has changed dramatically with the invention of the RSA cryptosystem. In essence, this approach to secret communication is a process by which a sender encodes the message so that an eavesdropper cannot understand it. The encoded message is then transmitted to a recipient, who, in order to be able to read it, needs to decode the message. Today, the need for secure communication is no longer purely the domain of the military. Transmission of various types of data (e.g., electronic banking and huge computer networks in general) requires a method of communication that guarantees that only authorized persons have access to some data files. Until recently, all known codes used the same secret key for both decoding and encoding. Both sender and recipient of the message therefore had to know the key. Any change of the key required a separate operation of distributing the key, which could be a technical problem (a spy and his "headquarters" are in mutually remote places) and a possible source of "leaked" information. In 1976, Whitfield Diffie and Martin Hellman came up with a revolutionary idea for changing the basic principle of secret communication by avoiding the distribution of the secret key. The effective use of this type of code, called RSA public cryptosystem, was developed by Ron Rivest, Adi Shamir, and Leonard M. Adleman. RSA requires choosing three numbers: two large primes, $p$ and $q$, and a number $r$ such that $r$, $p - 1$, and $q - 1$ are relatively prime; put $n = p \cdot q$. The numbers $r$ and $n$ are made public, and $p$ and $q$ are kept secret. A sender can encode a message using $r$ and $n$ by a simple method that is generally known. However, only the person possessing $p$ and $q$ is able to decode the message. Even the sender is not able to check whether the message was encoded correctly except by encoding it once more and comparing the results. It seems, at least at first glance, that the cryptosystem is of no practical use because knowing $n$ leads immediately to calculating $p$ and $q$, that is, to breaking the code. However, the trick is that although it is relatively simple to find large primes, it is practically impossible to factor large numbers. It is estimated that with the most advanced present technology, it would take 3.8 billion years to factor a 200-digit integer. On the other hand, it is a matter of minutes to test whether an arbitrary 100-digit number is a prime.

## OPEN QUESTIONS

There are many open problems concerning primes but only three, probably the most famous and popular ones, will be mentioned here. In 1742, Christian Goldbach conjectured that any integer $n > 5$ is the sum of three primes. This is equivalent to the statement that any even integer $n > 8$ is the sum of two primes. The best result known to date is that of Godfrey H. Hardy and John E. Littlewood, later proved also by Ivan M. Vinogradov using different techniques, stating that, starting with some number $n_0$, any odd integer that is bigger than the number $n_0$ is the sum of three primes. Although a lot of attention has been focused on twin primes, the fundamental question, whether there are infinitely many of them, is still open. Viggo Brun's famous result says that the series of reciprocals of all twin primes is convergent (compare with the corresponding result for primes), indicating that even if there were infinitely many twin primes then they would be relatively sparse. A result of Peter G. Lejeune Dirichlet (1805–1859) asserts if $a$ and $d$ are natural numbers that are relatively prime, then the arithmetic progression $a$, $a + d$, $a + 2d$, $a + 3d$, . . . , $a + nd$, . . . contains infinitely many primes. However, it is still not known if, for an arbitrary natural number $k$, there is at least one arithmetic progression consisting of $k$ prime numbers. According to a previously mentioned result of Pritchard, we are able to verify the statement for all $k < 19$.

*See also* Number Theory, Overview

## SELECTED REFERENCES

Burton, David M. *Elementary Number Theory.* Dubuque, IA: Brown, 1989.

Hewitt, Edwin. "The Riddle of Primes." *Mathematical Medley* 16(2)(1988):48–58.

Long, Calvin T. *Elementary Introduction to Number Theory.* 3rd ed. Englewood Cliffs, NJ: Prentice-Hall, 1987.

Ribenboim, Paulo. *The Book of Prime Number Records.* 2nd ed. New York: Springer-Verlag, 1989.

Rosen, Kenneth H. *Elementary Number Theory and Its Applications.* 3rd ed. Reading, MA: Addison-Wesley, 1993.

Underwood, Dudley. *Elementary Number Theory.* Boston, MA: Allyn and Bacon, 1980.

Vanden Eynden, Charles. *Elementary Number Theory.* New York: Random House, 1987.

PETER HORÁK

# PROBABILITY, OVERVIEW

Our world is full of uncertainty. Many of the words in our language imply degrees of certainty or uncertainty (e.g., *likely, certain, probable,* and *unlikely*). We frequently try to quantify the occurrence of uncertain events, asking questions such as: What are the chances that it will rain today? How likely am I to be involved in an automobile accident? What are the odds of winning the lottery? Answers to questions like these can be obtained using methods of probability. Probabilities are numerical values assigned to events so that a "0" means an event is not going to happen, and a "1" or "100%" means an event is certain to happen. Numbers in between indicate likelihood of occurrence based on their relative position between 0 and 1. A probability of 0.1 or 10% would indicate a low chance of an event occurring, and a probability of 0.99 or 99% would indicate a very strong chance of an event occurring.

Probability is a field of mathematics that attempts to describe randomness. Mathematicians do not think of "random" as meaning haphazard, but instead think of randomness as pertaining to phenomena that have uncertain individual outcomes but have a regular pattern of outcomes when examined over many repetitions (Moore 1991). For example, when tossing a fair coin, we do not know in advance if the coin will turn up heads or tails, but we do know that if the coin is tossed many times it will turn up "heads" about 50% of the time. Probability theory and methods are used in many diverse areas, for example, in computing the risks of disease in medical research, in computing life expectancy tables for actuarial purposes, and in establishing lotteries and games of chance.

## THE HISTORY OF PROBABILITY

Although games of chance and devices such as dice have existed for thousands of years, probability theory seems to have emerged in the 1600s. During that century, several mathematicians, including Blaise Pascal, Pierre Fermat, Christian Huygens, Gottfried Leibniz, Nicholas Bernoulli, and John Arbuthnot, examined the ways permutations and combinations could be used to solve gaming puzzles and to quantify uncertain outcomes of games of chance (Stigler 1986). Their methods involved estimating a probability of an event by taking a ratio between the number of mutually exclusive ways that an event could occur and the total of all equally likely, mutually exclusive outcomes (approaches now referred to as *theoretical* or *classical probability*). Leibniz is cred-

ited with suggesting that 1 and 0 be used to represent the extreme probabilities that an event will or will not happen. Other mathematicians extended this theory to look at repeated experiments involving random events and began to compute probabilities as the proportion of times an event occurred, now referred to as *relative frequency probabilities*. Bernoulli is credited with proposing that these relative frequencies, based on repetitions of an experiment (e.g., tossing a coin hundreds of times and calculating the percentage of times heads occur), would come close to theoretical probabilities of the same event (e.g., the probability of a head on a single toss being one-half or 50%) if the experiment was repeated a large number of times (Kennedy 1983).

Although the word *probability* was originally used to represent a physical property inherent to physical systems, it developed a second meaning in the 1950s. At that time, renewed interest in Bayes' Theorem led to a rethinking of the meaning of probability. Bayes' Theorem involves the estimation of the conditional probability that a particular causal event occurred, given information that a certain outcome has occurred, a type of "working backwards" approach to inference (Vogt 1993). "Bayesian" statisticians began the practice of examining beliefs prior to collecting empirical data and then revising these beliefs (probabilities) in light of the data (Folks 1981). A new definition of probability thus evolved, and the term came to be interpreted as a measure of belief in the truth of some statement.

Current computer technology has changed the way probabilities may be computed for complex events, allowing for simulations of data based on different probability models. Simulated data are used to calculate relative frequencies to estimate probabilities that are difficult to compute using traditional methods. For example, to study the probability of obtaining different lengths of runs of "heads" when tossing a coin many times, a model of a coin may be programmed and used to simulate data for hundreds of coin tosses. The computer can count the different lengths of runs and the proportion of times each run occurs in repeated simulations of the coin toss data.

## TEACHING PROBABILITY

Until the publication of the *Curriculum and Evaluation Standards for School Mathematics* by the National Council of Teachers of Mathematics (NCTM) in 1989, probability topics were not typically included in the K–12 curriculum in U.S. schools. Now, probability topics are being introduced as early as in the primary grades. Most instructional activities for elementary school students deal with simple explorations of chance, establishing ideas that some outcomes are more likely than others and that some things cannot be predicted in advance (NCTM 1989). The NCTM standards for middle school students include explorations of probability in real-world settings so that students model situations by carrying out experiments or simulations to determine relative frequency probabilities and construct sample spaces to calculate theoretical probabilities. High school students learn about random variables, discrete probability distributions, the normal curve, and additional ways to calculate theoretical and empirical probabilities (NCTM 1989). The Quantitative Literacy Materials developed for secondary level students offer a separate book on probability (Newman, Obremski, and Scheaffer 1986), which includes activities in which students generate data using random devices such as coins, spinners, dice, and cards and compare relative frequency probabilities to estimated theoretical probabilities. Diagrams, tables, and charts are also introduced and used as ways to display and calculate probabilities. A more traditional approach to learning probability may be found in secondary mathematics courses, which might include topics such as permutations, combinations, and traditional methods of computing theoretical probabilities.

Several materials have been produced to further explain, support, and give examples of how to implement the NCTM standards for probability. Activities in Zawojewski (1991) encourage a focus on reasoning by asking students to consider interpretation of medical test results, discuss strategies to use in card games, and identify certain and uncertain events using a story context. Hawkins, Jolliffe, and Glickman (1992) suggest a variety of random experiments to use with students to help them develop probabilistic concepts and to better distinguish between random and deterministic phenomena. Since the inclusion of probability and statistics in the precollege mathematics curriculum, materials have been written to help teachers think about the important concepts and ideas and to guide them in teaching these topics.

One concern in teaching probability is the abundant research documenting the difficulties students have understanding these concepts and using correct probabilistic reasoning (e.g., Garfield and Ahlgren 1988; Shaughnessy 1992). Some of these difficulties include a tendency for people to use a model of probability that leads them to make yes-or-no decisions about single events rather than looking at series of events (Konold 1989) or to identify a correct answer on a test because they know what the answer should be, but then fail to correctly apply that concept when solving a problem using a different context (Garfield 1995). Several psychologists have studied the relationship between students' natural intuitive beliefs about uncertain events and the formal, mathematical methods for calculating probabilities they learn in school (e.g., Kahneman, Slovic, and Tversky 1982; Fischbein, Nello, and Marino 1991). Although students may learn probability rules and procedures and may actually calculate correct answers on mathematics tests, research indicates that these same students rarely apply the methods they have learned when making their own judgments of the likelihood of uncertain events. This finding has led some educators to question the wisdom of trying to teach too much mathematical theory to students and has led them to provide students with more experiences gathering and interpreting data from random experiments to help them develop more correct, normative beliefs and intuitions about chance. This concern is apparent in *Benchmarks for Science Literacy* (American Association for the Advancement of Science 1993), which offers suggestions for the kinds of experiences students should have in order to develop correct ideas about probability and describes the kinds of understanding students should have at different grade levels (e.g., by the end of 8th grade, students should know that probabilities are ratios and can be expressed as fractions, percentages, or odds).

In addition to the curriculum materials developed for elementary- and secondary-level students, software programs are now available to help students solve probability problems by developing models to use in simulating and analyzing data. Programs available for secondary-level students include Probability Constructor (LOGAL 1995) and Prob Sim (Konold 1995). The Resampling Program (Simon and Bruce 1991) is available for students in college courses.

## RESOURCES FOR TEACHING

There are several articles that summarize research on teaching and learning probability, describe difficulties students have learning probability, and offer implications for teachers of statistics (e.g., Garfield and Ahlgren 1988; Garfield 1995; Shaughnessy 1992; Shaughnessy and Bergman 1993). Published proceedings from four different International Conferences on Teaching Statistics contain papers on probability teaching and learning at all educational levels and describe experiences of teachers and researchers around the world (e.g., Vere-Jones 1991). *Statistics for the Twenty-First Century* (Gordon and Gordon 1992) includes a few chapters on teaching probability and an important chapter on the psychology of learning probability that summarizes conceptual difficulties students have and suggests ways to help students learn to reason probabilistically (Falk and Konold 1992). At this date there is only one comprehensive collection of papers exclusively focused on probability education. This book covers the mathematical foundations of probability, historical perspectives, curriculum issues, the role of computers, and psychological perspectives on learning probability (Kapadia and Borovcnik 1991).

*See also* Normal Distribution Curve; Probability Applications: Chances of Matches; Probability Density; Probability Distribution; Probability in Elementary School Mathematics; Probability Problems, Computer Simulations

## SELECTED REFERENCES

American Association for the Advancement of Science. *Benchmarks for Science Literacy.* New York: Oxford University Press, 1993.

Falk, Ruma, and Clifford Konold. "The Psychology of Learning Probability." In *Statistics for the Twenty-First Century* (pp. 151–164). Florence Gordon and Sheldon Gordon, eds. Washington, DC: Mathematical Association of America, 1992.

Fischbein, Efraim, Maria Nello, and Maria Marino. "Factors Affecting Probabilistics Judgements in Children and Adolescents." *Educational Studies in Mathematics* 22(1991):523–549.

Folks, J. Leroy. *Ideas of Statistics.* New York: Wiley, 1981.

Garfield, Joan. "How Students Learn Statistics." *International Statistical Review* 63(1995):25–34.

———, and Andrew Ahlgren. "Difficulties in Learning Basic Concepts in Statistics: Implications for Research." *Journal for Research in Mathematics Education* 19(1988):44–63.

Gordon, Florence, and Sheldon Gordon, eds. *Statistics for the Twenty-First Century.* Washington, DC: Mathematical Association of America, 1992.

Hawkins, Anne, Flavia Jolliffe, and Leslie Glickman. *Teaching Statistical Concepts.* London, England: Longman, 1992.

Kahneman, Daniel, Paul Slovic, and Amos Tversky, eds. *Judgment Under Uncertainty: Heuristics and Biases.* Cambridge, England: Cambridge University Press, 1982.

Kapadia, Ramesh, and Manfred G. Borovcnik. *Chance Encounters: Probability in Education.* Dordrecht, Netherlands: Kluwer, 1991.

Kennedy, Gavin. *Invitation to Statistics.* Oxford, England: Robertson, 1983.

Konold, Clifford. "Informal Conceptions of Probability." *Cognition and Instruction* 6(1989):59–98.

———. Prob Sim Software. Santa Barbara, CA: Intellimation: Library for the Macintosh, 1995.

LOGAL. Probability Constructor Software. Pleasantville, NY: Sunburst, 1995.

Moore, David. *Statistics: Concepts and Controversies.* New York: Freeman, 1991.

National Council of Teachers of Mathematics. *Curriculum and Evaluation Standards for School Mathematics.* Reston, VA: The Council, 1989.

———. *Principles and Standards for School Mathematics.* Reston, VA: The Council, 2000.

Newman, Claire M., Thomas E. Obremski, and Richard L. Scheaffer. *Exploring Probability.* Palo Alto, CA: Seymour, 1986.

Shaughnessy, J. Michael. "Research in Probability and Statistics: Reflections and Directions." In *Handbook of Research on Mathematics Teaching and Learning* (pp. 465–494). Douglas Grouws, ed. New York: Macmillan, 1992.

———, and Barry Bergman. "Thinking About Uncertainty: Probability and Statistics." In *Research Ideas for the Classroom* (pp. 177–197). Vol. 3. *High School Mathematics.* Patricia Wilson, ed. New York: Macmillan, 1993.

Shulte, Albert, ed. *Teaching Statistics and Probability.* 1981 Yearbook. Reston, VA: National Council of Teachers of Mathematics, 1981.

Simon, Julian, and Peter Bruce. "Resampling: A Tool for Everyday Statistical Work." *Chance* 4(1)(1991):22–32.

Stigler, Stephen. *The History of Statistics: The Measurement of Uncertainty before 1900.* Cambridge, MA: Harvard University Press, 1986.

Vere-Jones, David. *Proceedings of the Third International Conference on Teaching Statistics.* Voorburg, Netherlands: International Statistical Institute, 1991.

Vogt, W. Paul. *Dictionary of Statistics and Methodology.* Newbury Park, CA: Sage, 1993.

Zawojewski, Judith S., et al. *Dealing with Data and Chance: Addenda Series, Grades 5–8.* Reston, VA: National Council of Teachers of Mathematics, 1991.

JOAN B. GARFIELD

## PROBABILITY APPLICATIONS: CHANCES OF MATCHES

This entry considers probabilities in several matching situations suitable for the mathematics level of the secondary school. The Fundamental Principle of Counting is the mathematical basis for these discussions.

A well-known probability experiment concerns the so-called Birthday Problem: in a randomly chosen group of $n$ persons, what is the probability that at least two persons share the same birthday anniversary? A direct method of solving this problem would be to calculate separately the probabilities of all possible distinct ways in which matches can occur and then sum these probabilities. This method, however, involves a substantial number of calculations. The following is an easier method: since $P$(at least one match) $+$ $P$(no matches) $= 1$, $P$(at least one match) $= 1 - P$(no matches). To calculate the probability of no matches, the Fundamental Principle of Counting must be used: suppose task A can be performed in $m$ ways and, after task A is completed, task B can be performed in $n$ ways. Tasks A and B can then be consecutively performed in $mn$ ways. This principle can be similarly extended to any number of consecutive tasks.

To solve the original problem, first calculate the probability that no two people (from a group of $n$ persons) share the same birthday. If $n > 365$, a match will necessarily occur since there are only 365 days in the year (ignoring February 29). Consequently, consider the case in which $n \leq 365$. If the $n$ persons were arranged in some order and each of them was to declare his/her birthday, then $365^n$ different $n$-tuples are possible. If all birthdates are considered to be equally likely, each of these $365^n$ $n$-tuples is equally likely as well. To find the number of $n$-tuples that contain no matches, again visualize the $n$ people declaring their birthdates in order. The first person may declare any of 365 days. To avoid a match, only 364 dates are available to the second person; 363 are available to the third person; 362 to the fourth person. The $n$th person has $(365 - n + 1)$ dates available. The number of ways in which the $n$ people can fail to have a match is $(365)(364)(363) \times (362) \ldots (365 - n + 1)$. The probability of no matches is thus

$$\frac{(365)(364)(363)(362) \ldots (365 - n + 1)}{365^n}$$

The probability of at least one match is then

$$1 - \frac{[(365)(364)(363)(362) \ldots (365 - n + 1)]}{365^n}$$

Surprisingly few people are needed in a group in order that the probability of at least one match exceeds 1/2. For 22 people, the probability of at least one match is

$$1 - \frac{[(365)(364) \ldots (344)]}{365^{22}}$$

$$= 1 - 0.5243 = 0.4757$$

However, the probability of at least one match for 23 people is

$$1 - \frac{[(365)(364)(363) \ldots (343)]}{365^{23}} = 0.5073$$

Let 23 be called the critical value in the Birthday Problem; that is, 23 is the smallest number of persons for which the probability of at least one match exceeds 1/2.

*A Challenge for Students* What is the smallest number of people needed in a group so that the probability of at least one match exceeds 90%? 99%?

Another birthday problem is sometimes confused with the problem just analyzed. Call this new problem "The Fixed Date Birthday Problem." Let a fixed date be selected and let a randomly chosen group of $n$ persons be visualized. What is the probability that at least one person has this fixed date as his/her birthday anniversary? There is no size restriction on $n$, since very large groups of people might still fail to match a fixed date. Again assume that each of the 365 days of the year (omit February 29) is equally likely as a birthday anniversary. If the experimental group consists of $n$ randomly chosen persons, the probability of any given person *not* having a birthday anniversary on the fixed date originally chosen is 364/365. Since these $n$ persons represent independent trials, the probability that none has a birthday anniversary on the fixed date originally chosen is

$$\frac{(364)^n}{(365)^n}$$

The probability that at least one person matches this date is then

$$1 - \frac{(364)^n}{(365)^n}$$

For example, if the experiment group consists of 23 persons, the probability that at least one person has his/her birthday anniversary on the fixed date is 0.0612, using the formula above with $n = 23$. For the original Birthday Problem the smallest value of $n$ for which the probability of at least one match exceeded 1/2 was 23. To calculate a similar result for the Fixed Date Birthday Problem, compute the smallest value of $n$ for which the probability of at least one person matching the fixed date exceeds 1/2. To solve this problem, solve the inequality:

$$1 - \frac{(364)^n}{(365)^n} > 0.5$$

Equivalently,

$$\frac{(364)^n}{(365)^n} < 0.5$$

Using logarithms,

$$n > \frac{\log_{10}0.5}{\log_{10}\frac{364}{365}}$$

Calculation reveals that $n > 252.65$. The smallest value of $n$ satisfying this condition is 253. By direct calculation, if $n = 252$, the probability of at least one person matching the fixed date is 0.4991, while if $n = 253$, the probability is 0.5005.

*A Challenge for Students* In the original Birthday Problem, what is the probability of at least one match if $n = 253$?

A similar type of experiment may be performed with ordinary (52-card) decks of playing cards: suppose that each of a group of $n$ students ($n \leq 52$) has a deck of well-shuffled playing cards. If $n > 52$, a match is inevitable. If at a given time each draws out a random card from his/her deck, what are the chances that at least two will draw exactly the same card (same suit, same numerical value)? This experiment may be considered a preferable illustration to the Birthday Problem since it may be repeated easily. Using the same reasoning as in the original Birthday Problem, the probability of at least one match resulting is

$$1 - \frac{(52)(51) \ldots (52 - n + 1)}{(52)^n}$$

*A Challenge for Students* For which value of $n$ does the probability of at least one match first exceed 1/2? (Ans. $n = 9$, $P = 0.5197$.)

This problem can be modified to the Fixed Card Problem: If a fixed card is designated in advance, what is the probability that at least one person will draw it? Using the same strategy as for the Fixed Date Birthday Problem, the probability that at least one person matches the fixed card is

$$1 - \frac{(51)^n}{(52)^n}$$

*A Challenge for Students* Construct a table comparing the probability ($P$) of at least one match in both card problems for several values of $n$. Continue this table until $P > 1/2$ for the Fixed Card Problem. When does this happen? Verify using logs.

Generalize the problem. Suppose that each of $n$ persons has a set of cards numbered consecutively from 1 to $r$. If each person selects a card at random from his/her deck, what is the probability of at least one match? The same reasoning as that used in the

**Table 1**

| r (number of cards held by each person) | Critical value for n (smallest number for which the probability of at least one match exceeds 1/2) |
|---|---|
| 2–5 | 3 |
| 6–9 | 4 ($r = 6$ is the same situation as if each person rolled one hexahedral die, and $r = 8$ if each person rolled one octahedral die) |
| 10–16 | 5 ($r = 12$ is the same situation as if each person rolled one dodecahedral die) |
| 17–23 | 6 ($r = 20$ is the same situation as if each person rolled one icosahedral die) |
| 24–32 | 7 |
| ⋮ | ⋮ |
| 100–116 | 13 |
| 117–134 | 14 |
| ⋮ | ⋮ |
| 310–340 | 22 |
| 341–372 | 23 ($r = 365$ yields the original Birthday Problem) |

two previous examples leads to the conclusion that the probability of at least one match is

$$1 - \frac{[r(r - 1)(r - 2) \ldots (r - n + 1)]}{r^n}$$

For a given value of $r$, what is the critical value for $n$; that is, what is the smallest value for $n$ so that the probability of at least one match exceeds 1/2? Table 1 depicts critical values for $n$ for specified values of $r$.

*A Challenge for Students* Construct a table similar to Table 1 for the Fixed Card Matching Problem. Compare the two tables.

The next group of problems involves the matching process but in a different setting. One method of packaging an allergy medication uses a 4 by 3 array. Each pill is formed by combining together white and colored half casings with the medications inside. In the packaging process, each identical pill is randomly placed with its colored side to the left or to the right. One such arrangement is shown in Figure 1.

A number of questions can be asked.

1. *How many different arrangements of orientations are possible on one card?* For each location a pill can be placed in two possible orientations. Since there are 12 pills per card, there are $2^{12}$ or 4,096 possible pill arrangements.

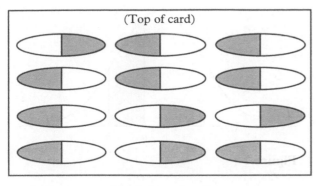

(Top of card)

(Shaded is colored)

**Figure 1**

**Table 2**

| Number of color-left pills on a card | Probability |
|---|---|
| 0 | 0.0002 |
| 1 | 0.0029 |
| 2 | 0.0161 |
| 3 | 0.0537 |
| 4 | 0.1208 |
| 5 | 0.1934 |
| 6 | 0.2256 |
| 7 | 0.1934 |
| ⋮ | ⋮ |
| 12 | 0.0002 |

2. *Suppose that a particular arrangement is specified in advance. For instance, the arrangement of Figure 1 is*

| R | L | L |
|---|---|---|
| L | L | L |
| L | R | R |
| L | R | L |

*What is the probability that the pills on a card that you purchase would be in that exact arrangement?* Since there are 4,096 possible arrangements on one card, the probability that a given arrangement will appear on a particular card is 1/4,096.

3. *Suppose that the number of pills in each of the two possible orientations (color-right or color-left) are counted, irrespective of their positions on the card. What are the probabilities of the various possible outcomes?* Consider the case shown in Figure 1; that is, consider 8 color-left and 4 color-right. The probability of any specific arrangement yielding these numbers is 1/4,096. The number of ways in which these numbers can occur is the number of ways of selecting 8 of the 12 positions in which to place the color-left pills. This can be done in $C(12,8)$, or 495 ways. The probability of this outcome is thus $495/4,096 \approx 12.08\%$. The probability of the 4 color-left and 8 color-right case is $C(12,4)/4,096 \approx 12.08\%$; this is the same probability as for the 8 color-left and 4 color-right case. Table 2 reports the probabilities for all 13 possible outcomes.

4. *Suppose that a box containing two cards of pills is purchased. What is the probability that both cards have the same arrangement of orientations?* Although this question appears different from question 2, it is, in fact, the same question. The first card can have any arrangement; it is then necessary for the second card to have that same arrangement. The probability that the second card will have the correct arrangement is then 1/4,096.

5. *Suppose that a box containing two cards of pills is purchased. What is the probability that the two cards have the same number of color-left pills regardless of their arrangement?* This could occur in each of 13 disjoint ways—both having 0 color-left, both having 1 color-left, both having 2 color-left, . . . both having 12 color-left. The probability is the sum of the squares of the separate probabilities, which sum to 0.1612, or 16.12%.

*A Final Challenge for Students* Find other situations in which the probability of matches can be analyzed. For instance, analyze lotteries and other gaming situations. The teacher may choose to present these concepts to the entire class. Another possible use is for enrichment in which individuals or small groups might research, extend, and explain them.

*See also* Probability, Overview

SELECTED REFERENCES

Duncan, David R., and Bonnie H. Litwiller. "Elevator Probabilities: Chances of Coincidences." *Mathematics Teacher* 84(Jan. 1991):64–65.

———. "A Question of Coincidences." *Mathematics Teacher* 78(May 1985):381–384.

Ginther, John L., and William A. Ewbank. "Using a Microcomputer to Simulate the Birthday Coincidence Problem." *Mathematics Teacher* 75(Dec. 1982):769–770.

Litwiller, Bonnie H., and David R. Duncan. "Maalox Lottery: A Novel Probability Problem." *Mathematics Teacher* 80(Sept. 1987):455–456.

———. "Matching Garage-Door Openers." *Mathematics Teacher* 85(Mar. 1992):217–219.

Miller, Charles D., Vern E. Heeren, and E. John Hornsby, Jr. *Mathematical Ideas.* New York: HarperCollins, 1994.

Spencer, Neville. "Celebrating the Birthday Problem." *Mathematics Teacher* 70(Apr. 1977):348–353.

BONNIE H. LITWILLER

DAVID R. DUNCAN

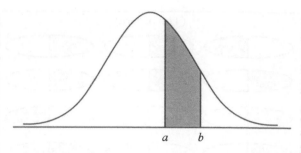

**Figure 2**

# PROBABILITY DENSITY

The term is borrowed from the language of physics, where the terms *weight* and *density* are used in very much the same way in which the terms *probability* and *probability density* are used in statistics. Actually, probability densities are functions that we use to define probability in connection with continuous random variables. When we deal with continuous random variables, the place of histograms is taken by continuous curves. As pictured in Figure 1, we can think of histograms with progressively narrower class intervals approaching the continuous curve and the areas of the rectangles representing the intervals approaching the corresponding areas under the curve.

Curves like these are the graphs of probability densities, and areas under curves like the one shaded in Figure 2 give the probabilities that continuous random variables will take on values in the corresponding intervals; in this case, the interval from *a* to *b*.

Many special kinds of probability densities are used in statistics, but most important, by far, is the one referred to as the *normal distribution*. The normal distribution was investigated first in the eighteenth century, when scientists observed a great degree of

regularity in errors of measurement. They found that the patterns they observed could be closely approximated by continuous curves, which they referred to as "normal curves of error" and attributed to the laws of chance. The graphs of normal distributions, like the two curves shown in the figures are all shaped like cross-sections of bells.

Other important probability densities (also called *continuous distributions*) are the *uniform*, *gamma*, and *beta* distributions and, of special relevance in statistical inference, the *exponential*, *chi-square*, *t*, and *F* distributions.

*See also* Normal Distribution Curve; Probability Distribution

## SELECTED REFERENCES

Freund, John E. *Mathematical Statistics.* 5th ed. Englewood Cliffs, NJ: Prentice-Hall, 1992.

Hastings, N. A. J., and J. B. Peacock. *Statistical Distributions: A Handbook for Students and Practitioners.* London, England: Butterworth, 1975.

Johnson, Norman L., and Samuel Kotz. *Continuous Univariate Distributions.* 2 vols. New York: Houghton Mifflin, 1970.

JOHN E. FREUND

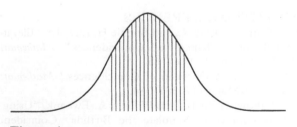

**Figure 1**

# PROBABILITY DISTRIBUTION

A function that associates a probability with each value within the finite or countably infinite range of a discrete random variable. Symbolically, the values of the function are the probabilities $P(X = x)$ that the random variable $X$ will take on the value $x$, with the domain of the function being the entire range of values of $X$. The probabilities $P(X = x)$ are usually denoted by symbols such as $f(x)$ or $g(x)$.

Probability distributions may be specified by actually displaying the values of the random variable together with the corresponding probabilities or, when possible, by means of formulas (equations) that enable us to calculate the probability $P(X = x)$ for any value of $x$ within the range of $X$. For instance, for three flips of a balanced coin there are the eight equally likely possibilities HHH, HHT, HTH, THH, HTT, THT, TTH, and TTT, where H stands for heads and T for tails. Counting the number of heads in each case, we arrive at the following probability distribution:

| Number of heads | Probability |
| --- | --- |
| 0 | 1/8 |
| 1 | 3/8 |
| 2 | 3/8 |
| 3 | 1/8 |

This probability distribution is also given by

$$f(x) = \binom{3}{x}/8 \qquad \text{for } x = 0, 1, 2, \text{ and } 3$$

where the *binomial coefficients* $\binom{3}{x}$ are, in fact, the coefficients 1, 3, 3, and 1 in $(a + b)^3 = a^3 + 3a^2b + 3ab^2 + b^3$.

The probability distribution for the number of heads in three flips of a balanced coin is a special case of the *binomial distribution*, which in general pertains to the number of "successes" in repeated trials. The binomial distribution applies also to the number of responses we may get to 500 questionnaires, the number of drivers wearing seat belts among 300 drivers stopped at a road block, or the number of cures among 80 patients injected with a new medication for a tropical disease. There are several other probability distributions that apply to repeated trials; among them are the *hypergeometric distribution*, the *geometric distribution*, and the *negative binomial distribution*. The *Poisson distribution* is a probability distribution that applies to the number of successes occurring during an interval of time rather than in repeated trials.

*See also* Binomial Distribution; Random Variable

## SELECTED REFERENCES

Hastings, N. A. J., and J. B. Peacock. *Statistical Distributions.* London, England: Butterworth, 1975.

Johnson, Norman L., and Samuel Kotz. *Discrete Distributions.* Boston, MA: Houghton Mifflin, 1969.

Olkin, Ingram, Leon J. Gleser, and Cyrus Derman. *Probability Models and Applications.* New York: Macmillan, 1980.

JOHN E. FREUND

# PROBABILITY IN ELEMENTARY SCHOOL MATHEMATICS

One of the recent reform recommendations for school mathematics (e.g., National Council of Teachers of Mathematics 1989; Lindquist et al. 1992; Zawojewski et al. 1991). This change can be ascribed to the technological needs of our age and to research that indicates (1) children enter school with some intuitive understandings of probability; and (2) these understandings increase with age and instruction (e.g., Kapadia and Borovcnik 1991; Shaughnessy 1992). Providing opportunities to link data and chance is the central part of the new focus.

Just as meteorologists, sporting coaches, insurance companies, engineers, and scientists, collect data to establish probabilities, so too can elementary school children collect data and carry out simulations to solve probability problems and explore games of chance. This approach capitalizes on the fact that many of the real-world problems that interest children involve probabilities that are determined by using or collecting data. For example, by finding what fraction of the evenings in the last two weeks parents allowed them to play their favorite video game, children can estimate their probability of being allowed to play it tonight.

## PEDAGOGICAL APPROACHES

A spirit of exploration and problem solving should permeate probability instruction in the elementary school. In fact, problem-solving explorations provide a natural setting for developing understanding of such key concepts as listing the outcomes of an experiment; determining the probability of an event; comparing the probability of two events; and finding conditional probabilities.

In attempting to solve problems involving chance, children should be encouraged to (1) predict probabilities based on their intuitive understanding of the problems, (2) design and carry out experiments or simulations to determine experimental probabilities, (3) analyze and interpret the results, and (4) compare their experimental probabilities with their original predictions. Our experiences suggest that analysis and reflection in collaborative groupings stimulate students to assess, modify, and extend their own probabilistic thinking. Moreover, we believe that it is through this process of review and reconstruction that children begin to challenge their own misconceptions and gradually build more formal thinking in both experimental and theoretical probability.

In carrying out simulations, students can determine experimental probabilities more precisely by comparing and pooling their data with others in the class. For example, in simulating the next two shots of a basketball player who has a 50% record in free-throw shooting, class results involving as many as 300 tosses of two coins are more likely to produce a better result than a small number of tosses obtained by a single group of students. Computer simulations are even more compelling because they enable the two coins to be tossed 1,000 times or more. When this process is repeated with similar problems, children are encouraged to develop a more pervasive and thoughtful perspective about randomness and probability.

The following suggestions may be helpful to teachers in generating worthwhile learning experiences in probability (Jones et al. 1997): (1) create problems that have game or real-world contexts familiar to the children, e.g., food, music, sport, travel, video games; (2) begin with problem situations that have a small number of outcomes and simple probabilities; (3) use problems that can be simulated with several probability devices (e.g., spinners, dice, coins, and random numbers); (4) generate extension problems that have a larger number of outcomes and require students to find more complex probabilities (e.g., three shots in a row from the 50% shooter); and (5) encourage collaboration and sharing in problem solving through small-group and follow-up class discussions. We have also found it valuable to have the students talk or write about their solution processes. Such communications open up opportunities for more insightful assessment of children's probabilistic thinking.

*See also* Probability, Overview; Probability Applications: Chances of Matches; Probability Problems, Computer Simulations

### SELECTED REFERENCES

Jones, Graham A., Cynthia W. Langrall, Carol A. Thornton, and A. Timothy Mogill. "A Framework for Assessing and Nurturing Young Children's Thinking in Probability." *Educational Studies in Mathematics* 32(2)(1997): 101–125.

Kapadia, Ramesh, and Manfred G. Borovcnik. *Chance Encounters: Probability in Education.* Dordrecht, Netherlands: Kluwer, 1991.

Lindquist, Mary M., with Jan Luquire, Angela Gardner, and Sandra Shekaramiz. *Making Sense of Data: Addenda Series, Grades K–6.* Reston, VA: National Council of Teachers of Mathematics, 1992.

National Council of Teachers of Mathematics. *Curriculum and Evaluation Standards for School Mathematics.* Reston, VA: The Council, 1989.

———. *Principles and Standards for School Mathematics.* Reston, VA: The Council, 2000.

Shaughnessy, J. Michael. "Research in Probability and Statistics: Reflections and Directions." In *Handbook of Research on Mathematics Teaching and Learning* (pp. 465–494). Douglas A. Grouws, ed. New York: Macmillan, 1992.

Zawojewski, Judith S., et al. *Dealing with Data and Chance: Addenda Series, Grades 5–8.* Reston, VA: National Council of Teachers of Mathematics, 1991.

CAROL A. THORNTON
GRAHAM A. JONES

# PROBABILITY PROBLEMS, COMPUTER SIMULATIONS

Revolutionized the way probability is taught. The theory of probability is famous for its many problems with seemingly unreasonable solutions, and so reasonable conjectures concerning these solutions are wrong (Székely 1986). Computer simulation allows theoretical results to be checked quickly, and the algorithms help us understand correct results better.

During World War II, Stanislas Ulam invented the Monte Carlo method, which enabled scientists to use probability theory to simulate certain phenomena. Using this method, Fano, Spencer, and Berger (1959) calculated how far atomic radiation can penetrate various barriers. The Monte Carlo method is useful in situations where deterministic equations are not known or cannot be solved easily. Probabilistic computer simulation is also useful when counting many objects. Consider a problem discussed by Mansheim and Baldridge (1987):

> One hundred people are lined up to enter a theater where the price of admission is $5.00. Sixty patrons have a $5 bill, the other 40 have a $10 bill. The cashier forgot the cash-box. What is the probability that all patrons can be admitted without delay?

Although phrased probabilistically, this is a counting problem. Not knowing a theoretical solution of this problem, one must examine $C(100,40) \approx 1.37*10^{28}$ different ways that sixty $5 bills and forty $10 bills can be arranged in a sequence and count those sequences that do not cause a delay. Or one could write a program that does the following: enter the numbers 1 through 100 into the computer and draw out 40 of these numbers at random, without replacement. If, for example, the number 29 is drawn, then we pretend that the customer in position 29 has a $10 bill; if the number 29 is not drawn, then the customer in position 29 has a $5 bill. We thus obtain what is called a *random sequence* of forty $10 and sixty $5 bills. We examine this sequence and determine whether, if real customers were lined up in the order given by the sequence, they could enter the theater without delay. If this happens, we call this sequence a "good" sequence. We now repeat this process, say, 1,000 times and count the number of "good" sequences we obtain. The ratio of the number of good sequences to the total number of sequences generated will be close to the correct answer: the probability that all customers can enter the theater without delay is 21/61. It is possible to adapt such techniques to the study of number theory and formulate conjectures that are almost surely true (Elliot 1979).

Difficulties do arise when random events are simulated by computers. Consider the following problem discussed in Flusser (1988):

A stick is randomly broken into three pieces. What is the probability that the pieces can be formed into a triangle?

It is easy to simulate this problem on a computer. This can be done as follows: all computers have a built-in pseudorandom number generator. Thus, if one gives the correct command, the computer produces a number between 0 and 1. The average user has no idea how this number was generated. If the command is repeated, a different number appears. It seems as though numbers are produced "at random" and that one number is as likely to appear as any other. Although these numbers are produced by deterministic means, they share most characteristics of uniformly distributed random numbers and are used as such. To simulate the "stick" problem, two random numbers, $a$ and $b$, both between 0 and 1, are generated. Assume that $a < b$ and that the stick is of unit length. The lengths of the three pieces are $a$, $b - a$, and $1 - b$, respectively. A triangle can be formed if and only if the length of the largest of these pieces is smaller than the sum of the lengths of the other two. In that case, one says one has "good" breaks. This experiment is repeated, say, 1,000 times and the ratio of the number of good breaks to the total number of breaks is calculated. This ratio will turn out to agree closely with the theoretical probability, 25%. But when 30 students were asked to break toothpicks at random, 23 (that is 77%) were able to form a triangle (Flusser 1988). It turns out that if someone breaks a stick "randomly," the breaks are more likely to occur near the middle; so the prob-

ability of getting "good" breaks is considerably higher that 25%. Computer simulation and theoretical computations yield compatible results because in both cases one makes the unrealistic assumption of uniformly distributed breaks. There now exist programs that enable users to generate random numbers from any desired distribution (Cooke, Craven, and Clarke 1982). But the major problem remains: "How does one choose the most appropriate distribution?" This, however, is not a mathematical problem.

A recent development is the use of computer graphics to illustrate the solution of probability problems. This technique does decrease the speed with which individual cases are generated and it does cut down the number of cases examined. It can, however, be used to aid the understanding of the mechanism by which random events are generated. Consider, for example, Bertrand's paradox. Bertrand (1907) asks for the probability that a randomly chosen chord of a circle will be smaller than the side of the inscribed equilateral triangle. In writing a program that draws a circle and picks a chord at random one soon discovers that first a mechanism that picks the chord must be specified. By varying the mechanism, programs can be written that yield results close to each of the values 1/2, 1/3, and 1/4. Such an exercise illustrates Bertrand's statement (1907, 5): "If the mechanism is not specified, the problem is not well posed." To understand any problem well enough to be able to solve it, whether theoretically or by simulation, the manner in which events are generated must be specified.

How can a teacher assigned to teach several units on probability to a junior or senior high school class use computer simulation to enhance students' experience? Having selected a problem whose solution students find counterintuitive, let them write a program that simulates the relevant events and let them study the results. It is vital that students write their own programs, or else the computer becomes a black box; students will not know what is going on and they will not believe the output. But if they write their own programs, they will have analyzed the problem and made it their own; they will have learned something.

*See also* Probability, Overview; Probability Applications: Chances of Matches

## SELECTED REFERENCES
Bertrand, Joseph Louis Francois. *Calcul des Probabilités.* 2nd ed. Paris, France: Gauthier-Villars, 1907. 3rd ed. New York: Chelsea, 1972.

Cooke, Dennis, A. H. Craven, and Geoffrey M. Clarke. *Basic Statistical Computing*. London, England: Arnold, 1982.

Elliot, Peter D. *Probabilistic Number Theory*. 2 vols. New York: Springer-Verlag, 1979.

Fano, Ugo, Lewis V. Spencer, and Martin J. Berger. "Penetration and Diffusion of X-rays." In *Handbuch der Physik* (pp. 660–817). Vol. 38, pt. 2. S. Flügge, ed. Berlin, Germany: Springer-Verlag, 1959.

Flusser, Peter. "Another Encounter with 'Rencontre'." *Mathematics and Computer Education* 23(3)(1989): 159–167.

———. "The Cookie Problem." *Iowa Council of Teachers of Mathematics Journal* 14(1985–1986).

———. "Theory, Simulation and Reality." *College Mathematics Journal* 19(May 1988):210–222.

———, and Dorothy Hanna. "Computer Simulation of the Testing of a Statistical Hypothesis." *Mathematics and Computer Education* 25(2)(1991):158–164.

Mansheim, Jan, and Phyllis Baldridge. "Three Methods of Attacking Problems in Discrete Mathematics, Part 2." *Mathematics Teacher* 80(Apr. 1987):282–288.

Székely, Gabor J. *Paradoxes in Probability and Mathematical Statistics*. Boston, MA: Reidel, 1986.

PETER FLUSSER

# PROBLEM SOLVING, OVERVIEW

The process that lies at the heart of all mathematical activity. In fact, there is widespread support for the notion that the ultimate aim of learning mathematics is to be able to solve problems. Further evidence of the importance of problem solving in mathematics is provided by the numerous books, monographs, and journal articles devoted to it. In addition, since the early 1970s problem solving has been the major focus of several conference reports and curriculum development efforts. Moreover, problem solving has been the subject of a substantial amount of research in recent years, perhaps more than any other topic in the mathematics curriculum. Despite the well-recognized importance of problem solving, its role in mathematics education is less clear. This lack of clarity can be attributed to two things: (1) over the years mathematics educators have given different meanings to "problem" and "problem solving"; and (2) problem solving is by its very nature an extremely complex type of human activity.

For many mathematics teachers the word *problem* describes any of the variety of mathematics tasks students perform in school: completing computational exercises, applying formulas, using algorithms, and solving story problems (sometimes called *word problems* or *verbal problems*). These problems are often classified as "routine problems" or exercises. However, for other teachers the word is reserved for a special type of mathematical activity. A problem of the latter type (often called *nonroutine*) is a task for which an individual or group wants or needs to find a solution but has no readily available procedure for getting one. An attempt is made to obtain a solution, using mathematical concepts, principles, and procedures.

There is general agreement that problem solving is the most complex kind of human learning, so complex as almost to defy description and analysis. Successful problem solving involves the process of coordinating one's knowledge, previous experiences, intuition, and various analytic and spatial abilities in an effort to determine a solution for a task where a procedure for determining the solution is not readily known. Of course, this does not describe what goes on when a person is actively engaged in solving a problem. One of the most intriguing aspects of problem solving is that two individuals can obtain the same solution to a problem using apparently different, but perfectly legitimate, methods. This characteristic of problem solving makes it difficult to decide on the best procedures to use in instruction.

Early in the twentieth century, psychologists began to develop models to explain problem-solving behavior in terms of cognitive processes. These efforts were the forerunners of what has come to be called *cognitive science*. Notable among the early attempts to develop these problem-solving models is the work of the European Gestalt psychologists from the 1920s through the 1950s (especially Karl Duncker, George Katona, and Max Wertheimer). The influence of the Gestaltists was prominent in the work of the early cognitive scientists Alan Newell and Herbert Simon (1972). They were concerned primarily with "puzzle" problems (e.g., the Tower of Hanoi problem), problems related to playing chess, and algorithmic problems but not with the sorts of mathematics problems of interest to teachers or education researchers. More recently, cognitive scientists have broadened their focus to include both mainstream mathematics problems and more open-ended types of problems. In addition, there has been recognition among cognitive scientists that learning in general, and problem solving in particular, takes place in a social context. Thus, today there is general acceptance within the cognitive science community that problem solving is considerably more complex than described by early cognitive models.

The ability to solve mathematics problems develops slowly over a very long period of time because success depends on much more than mathematical content knowledge, a fact that researchers interested in problem solving have only recently begun to realize. Today, researchers tend to focus their attention on questions in five broad, interdependent categories: (1) knowledge acquisition and utilization, (2) control, (3) beliefs, (4) affects, and (5) sociocultural contexts.

## KNOWLEDGE ACQUISITION AND UTILIZATION

Until rather recently the majority of research on mathematical problem solving has been devoted to the study of how knowledge involved in problem solving is acquired and utilized. "Knowledge" refers to both informal and intuitive knowledge as well as to formal knowledge. Included in this category are a wide range of resources that can assist the individual's mathematical performance. Especially important types of resources are the following: facts and definitions (e.g., 12 is a composite number, a rectangle is a parallelogram with four right angles), algorithms (e.g., the regrouping algorithm for subtraction), heuristics (e.g., drawing pictures, looking for patterns, working backwards), problem schemas (i.e., packages of information about particular problem types, such as distance-rate-time problems), and the host of other routine, but nonalgorithmic, procedures that an individual can bring to bear on a mathematical task. Since individuals understand, organize, represent, and ultimately utilize their knowledge in very different ways, for each individual attempting to solve a problem at a particular point in time, it is likely that at least some of the relevant mathematical concepts are at an intermediate stage of development. In such cases, problem solvers must adapt their concepts, as they understand them, to fit the problem situation. To the extent that they are able to make appropriate adaptations, they are successful in solving the problem.

## CONTROL

Even when individuals possess the knowledge and skills necessary to solve a particular problem, they are generally unsuccessful unless they are able to utilize these resources efficiently. Control refers to the marshaling and subsequent allocation of available resources to deal successfully with mathematical situations. More specifically, it includes decisions about planning, evaluating, monitoring, and regulating.

Two aspects of control processes have become increasingly popular as objects of research in recent years: knowledge about and regulation of cognition. The processes used to regulate one's behavior are often referred to as *metacognitive* processes, and these have become the focus of much attention within the mathematics education research community. In fact, recent research suggests that an important difference between successful and unsuccessful problem solvers is that successful problem solvers are much better at monitoring and regulating their activities than unsuccessful ones. A lack of control can have disastrous effects on problem-solving performance. And explicit attention to the metacognitive aspects of problem solving seems to bring the importance of monitoring behaviors to the forefront of students' awareness and can make a difference in their ability to make the most of the resources and skills they have at hand. For example, after a class has completed work on a problem the teacher might engage students in a discussion of what made the problem difficult for them, what they could have done to solve the problem more efficiently, and what they learned about their own strengths and weaknesses from having worked on the problem.

## BELIEFS

Beliefs constitute the individual's subjective knowledge about self, mathematics, the environment, and the topics dealt with in particular mathematical tasks. For example, many elementary school children believe not only that all mathematics story problems can be solved by direct application of one or more arithmetic operations but also that the operation to apply is determined by the "key words" in the problem. Beliefs shape attitudes and emotions and direct the decisions made during mathematical activity. Some research has focused on students' beliefs about the nature of problem solving as well as about their own capabilities and limitations.

## AFFECTS

Many individuals have very definite feelings related to the study of mathematics. It is not uncommon to hear people confide that they "always hated math" or "never felt confident about word problems." The affective domain, which includes individual feelings, attitudes, and emotions, is an important contributor to problem-solving behavior. However, until recently research in this area has been limited largely to examinations of the correlation between attitudes and mathematical problem-solving performance. Attitudes that have been shown to be related to problem-solving performance include motivation, interest, confidence, perseverance, willingness to take risks, and tolerance of ambiguity. Since about 1985, problem-solving researchers have become much more aware of the pervasive nature of affective variables, and consequently much more attention has been devoted to clarifying the nature of affective variables and to identifying and studying their impact on problem-solving teaching and learning.

## SOCIOCULTURAL CONTEXTS

Cognitive psychologists and others interested in mathematical problem solving now believe that human intellectual behavior must be studied in the context in which it takes place. That is to say, because human beings are immersed in a reality that both affects and is affected by human behavior, it is essential to consider the ways in which sociocultural factors influence cognition. Also, the development, understanding, and use of mathematical ideas and techniques grow out of social and cultural situations, often outside of school. The interactions that students have among themselves and with their teachers, as well as the values and expectations that are nurtured in school, shape not only what mathematics is learned but also how it is learned and how it is perceived. The wealth of sociocultural conditions that make up an individual's reality plays a prominent role in determining the individual's potential for success in solving mathematics problems both in and out of school.

The five research categories discussed here overlap much more than is suggested by the discussion (e.g., it is clearly not possible to completely separate affects, beliefs, and sociocultural contexts). And they not only overlap but they also interact in a variety of ways (e.g., beliefs influence affects, and they both influence knowledge utilization and control; sociocultural contexts have an impact on all the categories). The interdependence of these categories may explain why progress has been slow in developing a stable body of knowledge about how problem-solving ability develops, how individuals learn to be good problem solvers, and how problem solving should be taught.

Although progress has been slow, research does point to several conditions that influence problem-solving success. Among those that account for much

of the difficulty an individual can have with a particular problem, the following are particularly important:

1. the amount of information, number of variables, syntactical complexity, and the mathematical content of the problem;
2. the way the problem is posed for the problem solver and the context within which it is represented;
3. familiarity of the problem solver with acceptable solution procedures;
4. misleading incorrect solutions or solution procedures; e.g., a problem suggests a solution or procedure that is incorrect or ultimately of no help;
5. difficulty in locating reachable subgoals, generating complaints such as "I don't know how to get started," or "I don't know what to do first";
6. constraints arising from misconceptions, misunderstanding or overlooking information given in the problem;
7. affective factors associated with the problem solver's reaction to the problem, such as lack of motivation and perseverance, high degree of stress, and low tolerance for ambiguity—problems of this type are legion and are linked with the individual: one problem may be considered too "messy" to bother with, while another is not interesting because the individual does not enjoy solving that kind of problem.

Teachers who give conscious attention to helping students overcome obstacles such as these will be taking some very positive steps toward improving their students' problem-solving performance. Unfortunately, instruction that is restricted to helping students avoid error or difficulty is usually inadequate. What seems to be needed is instruction that is concerned not only with sources of difficulty but also with promoting the development of good problem-solving habits (e.g., perseverance, self-monitoring) and with the acquisition by students of a wide range of skills and strategies (e.g., making tables to organize data, estimating, looking for patterns, working backward, and solving a simpler problem). But how to help students become better problem solvers poses some perplexing questions because what causes changes in problem-solving behavior over time is not well understood.

Perhaps the most lucid thinking about mathematical problem solving has been done by the mathematician George Pólya (1957, 1981). For Pólya, there are four distinct, but interrelated, phases in the solution process: (1) understanding the problem, (2) devising a plan, (3) carrying out the plan, and (4) looking back. This model has been valuable to many teachers as a guide in organizing instruction, but it has been of less help in specifying the cognitive processes involved in successful problem solving.

A number of differing viewpoints regarding instruction in problem solving have been proposed. The most common of these are based on the thoughtful writings of Pólya, whose four-phase model, just described, has direct applicability to instruction. Attempts to develop instructional methods using Pólya's ideas typically have focused on teaching students various heuristic strategies, that is, planned actions or series of actions performed to assist in the discovery of a solution to a problem. There are other instructional approaches in addition to those patterned after Pólya's suggestions that have been used: (1) have students solve many problems without specific intervention by the teacher; (2) teach specific skills known to be essential to successful problem solving (for example, making tables, drawing diagrams, and translating from written form to equation form); (3) model good problem-solving behavior and have students imitate this behavior; and (4) some combination of the preceding.

A teacher using the first approach might select a large number of problems on the basis of certain criteria. For example, a fifth-grade teacher who wanted students to develop some facility in solving problems might use the following criteria: mathematical content must be no higher than fifth-grade level; problems must be interesting to the students; solution processes must be within the grasp of fifth graders; some problems should have more than one answer (others, no answer); there should be more than one way to solve each problem; and some problems should be related to others in the sense that they are similar to, build on, or are extensions of other problems. A collection of problems might be placed in a mathematics corner of the classroom, and students would be encouraged to work several problems during a given period of time. And from time to time the teacher would lead a class discussion about the students' attempts to solve certain problems.

Instruction aimed at developing specific skills useful in problem solving has been common in many mathematics textbooks. Skills that facilitate planning an attack, help in organizing relevant information, and otherwise assist the problem solver in using a strategy are usually emphasized. For example, instruction in making a table might begin with a demonstration by the teacher of how to make a table to solve a specific problem. The teacher might point out how the table helps to organize and keep track of information. Subsequently, students might practice reading and constructing tables. Finally, they might be asked to solve several problems for which solutions are made easier by making tables.

The third instructional approach, modeling good problem solving, is used from elementary school through graduate school. The teacher demonstrates how to solve a certain problem and directs the students' attention to salient procedures and strategies that enhance the solution of the problem. Students are then expected to solve problems using the processes modeled by the teacher.

Although research provides no clear direction, some combination of these approaches seems to be the most sensible one to use. Certainly, students will not improve their problem-solving skills unless they try to solve a wide range of types of problems. It also is the case that the likelihood of improved problem-solving performance is increased if students see good problem-solving behavior exhibited by their teacher. Finally, the acquisition of various specific skills and heuristics are likely to enhance problem-solving performance. In short, no single approach can be given an unqualified recommendation over the others. Instead, conscious attention to any or all of these approaches tends to have a positive effect on the problem-solving performance of students.

*See also* Competitions; Gifted; Pólya; Problem-Solving Ability and Computer Programming Instruction; Projects for Students; Psychology of Learning and Instruction, Overview

## SELECTED READINGS

Charles, Randall I., and Edward A. Silver, eds. *The Teaching and Assessing of Mathematical Problem Solving.* Reston, VA: National Council of Teachers of Mathematics, 1988.

Lester, Frank K. "Musings about Mathematical Problem Solving Research: 1970–1994." *Journal for Research in Mathematics Education* 25(6)(1994):660–675.

Newell, Alan, and Herbert A. Simon. *Human Problem Solving.* Englewood Cliffs, NJ: Prentice-Hall, 1972.

Pólya, George. *How to Solve It: A New Aspect of Mathematical Method.* 2nd ed. Garden City, NY: Doubleday, 1957.

———. *Mathematical Discovery: On Understanding, Learning, and Teaching Problem Solving.* Combined ed. New York: Wiley, 1981.

Schoenfeld, Alan H. "Learning to Think Mathematically: Problem Solving, Metacognition, and Sense Making in Mathematics." In *Handbook of Research on Mathematics Teaching and Learning* (pp. 334–370). Douglas A. Grouws, ed. New York: Macmillan, 1992.

———. *Mathematical Problem Solving.* Orlando, FL: Academic, 1985.

Silver, Edward A., ed. *Teaching and Learning Mathematical Problem Solving: Multiple Research Perspectives.* Hillsdale, NJ: Erlbaum, 1985.

FRANK K. LESTER, JR.

# PROBLEM-SOLVING ABILITY AND COMPUTER PROGRAMMING INSTRUCTION

A set of theories that problem-solving ability is enhanced by learning to program. In *Mindstorms: Children, Computers and Powerful Ideas,* Seymour Papert described a setting in which "children can learn to use computers in a masterful way, and that learning to use computers can change the way they learn everything else" (1980, 8). As research into programming's effect on cognitive skills proliferated, empirical findings did not always support its proposed potential. Discontinuities in the theory and practice of the transfer of problem-solving ability from computer programming to other domains have been attributed to lack of student achievement, the unfocused nature of targeted variables, and underdeveloped pedagogy (Johanson 1988).

Suggestions on how to focus examination of programming's effect on behavior have emerged from the problem-solving literature, where the cognitive and metacognitive skills associated with the solution of nonroutine mathematics problems have their counterparts in the programming task. Pólya's (1957) framework for analyzing mathematical problem-solving includes stages of understanding, planning, carrying out the plan, and looking back. Dalbey and Linn's model (1985) of the programming process consists of problem specification, design, coding, and debugging. Within these stages, students can employ heuristics, or methods frequently found to be useful in the solution of problems (Pólya 1957). Several research efforts have focused on particular methods thought to be capable of transfer from the programming domain. Learning computer programming was found to have positive effects on subgoal formation, systematic trial and error, analogical reasoning (Swan 1991), and enhancement and retention of the concept of variable (Ortiz and MacGregor 1991). Ehrlich, Abbott, Salter, and Soloway claim that learning to program can encourage a student to replace a descriptive approach to equations with an "active, procedural" one

(p. 5). Schoenfeld's framework (1985) for examining mathematical performance incorporates issues of control: decisions regarding planning, monitoring, and evaluating the solution process. Several studies have used observation and protocol analysis to investigate programming's effect on these metacognitive activities. Clements and Nastasi (1988) showed that elementary school students in a Logo environment exhibited a higher frequency of these activities than did a group involved in computer-assisted instruction. In contrast, Blume and Schoen (1988) showed eighth-grade programmers did not engage in planning more so than nonprogrammers.

Granting that a theoretical basis exists for problem-solving ability to be enhanced by learning to program, transfer may still be contingent upon the level of ability attained. Observation has shown a wide disparity in the pedagogy of programming, from rote learning to unguided discovery (Linn, Sloane, and Clancy 1987). The development of the Advanced Placement (AP) curriculum has helped standardize expectations of high school programming and made explicit many of the sought-after underlying processes (College Entrance Examination Board 1995). Incremental development and ongoing evaluation of programs became even more concrete with the recent incorporation into the AP curriculum of a "Case Study" (College Entrance Examination Board 1993), whereby students try to understand, assess, and go on to modify a comparatively large program.

*See also* Problem Solving, Overview

## SELECTED REFERENCES

Blume, Glendon W., and Harold L. Schoen. "Mathematical Problem-Solving Performance of Eighth-Grade Programmers and Nonprogrammers." *Journal for Research in Mathematics Education* 19(Mar. 1988):142–156.

Clements, Douglas H., and Bonnie K. Nastasi. "Social and Cognitive Interactions in Educational Computer Environments." *American Educational Research Journal* 25(Spring 1988):87–106.

College Entrance Examination Board. *Advanced Placement Course Description: Computer Science.* New York: The College Board, 1995.

———. *A Teacher's Manual for the "Directory Manager" Case Study.* New York: The College Board, 1993.

Dalbey, John, and Marcia Linn. "The Demands and Requirements of Computer Programming; A Literature Review." *Journal of Educational Computing Research* 1(1985):253–274.

Ehrlich, Kate, Valerie Abbott, William Salter, and Elliot Soloway. "Issues and Problems in Studying Transfer Effects of Programming." Unpublished paper.

Johanson, Roger P. "Computers, Cognition and Curriculum: Retrospect and Prospect." *Journal of Educational Computing Research* 4(1988):1–30.

Linn, Marcia, Kathryn Sloane, and Michael Clancy. "Ideal and Actual Outcomes from Precollege Pascal Instruction." *Journal of Research in Science Teaching* 24(5)(1987):467–490.

Ortiz, Enrique, and S. Kim MacGregor. "Effects of Logo Programming on Understanding Variables." *Journal of Educational Computing Research* 7(1)(1991):37–50.

Papert, Seymour. *Mindstorms: Children, Computers and Powerful Ideas.* New York: Basic Books, 1980.

Pólya, George. *How to Solve It: A New Aspect of Mathematical Method.* 2nd ed. Garden City, NY: Doubleday, 1957.

Schoenfeld, Alan H. *Mathematical Problem Solving.* Orlando, FL: Academic, 1985.

Swan, Karen. "Programming Objects to Think With: Logo and the Teaching and Learning of Problem Solving." *Journal of Educational Computing Research* 7(1991): 89–112.

RHONDA NEUBORN WEISSMAN

# PROGRAMMING LANGUAGES

Systems of symbolic expressions for algorithms that determine how a computer carries out tasks. The computer has been used as a tool for solving mathematics problems from the earliest days of its invention and for teaching and learning mathematics since the 1950s. It continues to be an integral part of mathematics education at all levels, as evidenced by the curriculum recommendations of the professional organizations in the field (Mathematical Association of America 1993; National Council of Teachers of Mathematics 1989).

Mathematics teachers and students use a variety of programming languages and software packages. Choices are dictated by the instructional level, the resources available, and the interests of the teacher and students. The programming language most often used in the elementary school classroom is Logo. On the secondary level, BASIC or Pascal is usually the first programming language introduced, sometimes followed by C, C++, LISP, or another high-level language. On the college level, Pascal, C, C++, Java, Ada, or LISP are the most widely used introductory languages. A distinction may be made between using programming as a tool for mathematics and programming as part of a computer science curriculum, but any of these languages can be used for either purpose. Spreadsheets such as Quattro Pro or Excel may be used as problem-solving and modeling tools for mathematics. Computer algebra systems such as Mathematica, Maple, and Derive, which are capable of doing graphics and manipulating formulas and symbols as well as numbers, are often used at the higher grade levels and the college level. Both spreadsheets and computer algebra systems can be programmed using their own languages, which include all the basic functions of any programming language, namely, data representation, operations, control structures, and input/output capabilities. The power of such systems has shifted emphasis in mathematics education away from traditional languages, but programming continues to be an important tool for expressing mathematical algorithms abstractly.

## LOGO

Logo was originally developed in the 1960s by a team at the consulting firm of Bolt, Beranek and Newman (Abelson and diSessa 1986). Its most famous advocate is Seymour Papert (Papert 1993), a professor who established the Logo Group at the Massachusetts Institute of Technology's Artificial Intelligence Laboratory in the 1970s. His objective was to create a learning environment in which mathematics is the natural language for communication, where children could become as fluent in mathematics as they are in their native languages. His philosophy of education was strongly influenced by the learning theories of Jean Piaget (1896–1980). Papert maintained that the computer could make the abstract concrete, enabling the child to progress from concrete to formal thinking at an earlier age. Papert saw Logo as the command language for "mathland," an environment in which children program a "turtle," a drawing instrument that may be a pen-carrying robot that can draw on a sheet of paper as it traverses the room under the child's command or a cursor that can draw a line on the screen as it moves to various screen locations. By developing his or her ability to control the turtle, the child becomes an active, self-directed learner. The student is initially taught a few very simple turtle graphics commands, such as *forward 20,* which moves the turtle forward 20 spaces, and *right 30,* which makes a 30-degree turn. By trial and error, the child will discover how to make the turtle draw simple geometric figures such as a square. These directions can be written as a procedure as follows:

```
to square
    repeat 4[forward 20 right 90]
end
```

The child may discover commands such as *left* and *back* independently and may learn to add conditions (*if* statements), variables, parameters, and even make the procedures recursive (i.e., able to call themselves). Explorations of the turtle's behavior make concrete some of the abstract concepts of geometry, making them available at a very early age. Inspired by Papert's work, elementary schools throughout the world incorporated Logo into the mathematics curriculum in the early 1980s. Although the early hope that Logo would revolutionize mathematics instruction has faded, many schools continue to teach Logo programming in mathematics classes and to use the extensions of the language.

## BASIC

BASIC, which stands for "Beginner's All-purpose Symbolic Instruction Code," was developed in 1964 by John Kemeny and Thomas Kurtz at Dartmouth College (Kemeny and Kurtz 1980). It was designed to be easy to learn and interactive, allowing the user to enter data at a terminal as opposed to providing data on punched cards or other batch media as required by earlier languages. The on-line editor identifies syntax errors in statements as they are entered, so users can correct errors immediately rather than having to wait for the entire program to be executed before any errors can be spotted. The original language consisted of only a few types of statements, allowing students to master the elements of the language quickly. These features made BASIC a popular language for instruction from its inception. Its popularity increased with the development of the microcomputer, since BASIC was provided with many operating systems.

Each BASIC statement is preceded by a line number, and the program is normally executed in line number order, unless the programmer transfers control by an *if* or a *GOTO* statement. The following simple program illustrates the use of the original BASIC statements:

```
10 REM This program allows you to enter a number and prints its double.
20 REM It repeats until the double is greater than 50.
30 PRINT "Enter a number"
40 INPUT n
50 PRINT "You entered"; n
60 LET m = 2 * n
70 IF m > 50 THEN 100
80 PRINT "Double your original number is"; m
90 GOTO 30
100 END
```

Despite its popularity, BASIC has well-recognized flaws, primarily due to its lack of structure. In structured design, programs are planned "top down," with a main program that has overall control and that causes subprograms, called *procedures*, to accomplish specific tasks. The procedures in turn may have subprocedures, so that the tasks of a large program are broken down into progressively smaller subtasks that are each easily accomplished and that fit together to solve the problem. Although early versions encouraged writing the entire program as a single unstructured list of instructions, current versions of BASIC include control structures that address these criticisms.

## PASCAL

The Pascal language, named in honor of the French mathematician and philosopher Blaise Pascal, was developed in 1968 by Niklaus Wirth in Zürich, Switzerland. It was designed as a tool for teaching students structured programming, using the "top-down" design approach. Because it encourages good programming habits and is easy to learn, it became the most widely taught first programming language in secondary schools and colleges. There is a published Pascal standard, but most popular Pascal compilers, such as Turbo Pascal, include extensions or enhancements of the language. A Pascal program consists of a heading and a block and ends with a period. The program heading consists of the reserved word *program*, followed by the name of the program and a list of the files the program uses. The block begins with a declaration section, listing the constants and variables that will be used and specifying the data type of each variable. Pascal has predefined types of *integer, real, boolean, char,* and *text,* as well as built-in data structures of *array, record, set,* and *file.* Programmers may create their own user-defined data types based on these types and structures by declaring them in a *type* declaration section before the variable declaration section. Procedures and functions, if any, appear next, followed by the main block of the program, which starts with the keyword *begin* and ends with the keyword *end,* followed by a period. The structure of procedures and functions mirrors that of the overall program. There is a standard syntax for each type of statement as well as standard keywords, operators, rules for identifiers, selection and looping mechanisms, parameter passing, recursion, and dynamic data structures. Statements are separated by a semicolon. Compound statements, which begin with the keyword *begin* and are terminated by *end,* may be used wherever a single statement may appear, allowing the programmer to group statements. The following simple program adds two integers:

```
program Simple;
var
    N1, N2, Sum: integer;
begin
    writeln('Enter two numbers');
    readln(N1,N2);
    Sum : = N1 + N2;
    writeln('The sum is', Sum)
end.
```

Most Pascal books describe widely used conventions for spacing, indentation, capitalization, documentation, and so on, which make programs easy to read and easy to modify (Savitch 1995). These elements can be used to create a carefully structured environment within which students can construct very powerful programs. However, the language lacks many features that are needed for commercial environments, so it has not been widely accepted outside the academic community.

## C++

C++ evolved from the programming language C, the most widely used language for writing professional software for microcomputers. C was developed by Dennis Richie at Bell Laboratories in 1972 for the purpose of writing and maintaining the UNIX operating system. C is an extremely powerful language, but C programs can be difficult to read and to modify. C++, an extension of C, was developed at Bell Laboratories in the early 1980s by C. Bjarne Stroustrup. Its major advantage over both C and previous languages is that it permits the use of a new programming paradigm called the *object-oriented approach* (Deitel and Deitel 1994). Here the programmer can define a data type and the set of operations that can be performed on it as an integrated object that is a model of items in the real world. Objects are reusable software components that can function as modules for building new software quickly and economically. Although it is more difficult for students to learn, C++ is quickly succeeding Pascal as the introductory programming language in colleges and universities because of the attraction of the object-oriented paradigm, and it has replaced Pascal as the underlying language on the Advanced Placement Test in Computer Science.

## JAVA

Java is both a programming language and a platform for running programs. The Java language is an object-oriented high-level language similar to C++. Java programs may be stand-alone applications that run on the Java platform, similar to other languages, or applets, which are Java programs that run within browsers on the World Wide Web. Java programs are compiled into an intermediate form called *byte codes,* which are then interpreted each time the program is run. This two-phase translation process makes it possible to compile a Java program on one machine and then run it on a variety of machines using different operating systems. With increasing use of the World Wide Web, the Java language is gaining in popularity.

## LISP AND SCHEME

LISP was developed in the late 1950s by John McCarthy as a language for symbolic computation. It has been used primarily in the field of artificial intelligence. The name is an acronym for LISt Processing, since the language uses the list as the basic data structure, and the program itself is a list. Traditional programming languages use an imperative paradigm, in which a program consists of a set of statements that the computer executes in sequence or by following control structures such as looping or branching. LISP uses a functional paradigm, based on the theory of recursive functions in mathematics. The original language consisted of functions for constructing and manipulating lists, defining and evaluating functions, and testing for equality (Steele 1990). Symbolic expressions, or s-expressions, are the basic elements used. An s-expression may be an atom or a list. An atom consists of letters, numbers, and some special characters. A list is a sequence of atoms or other lists, which are enclosed in parentheses and separated by blanks. Lists may be empty and may be nested. Prefix notation (usually called *Polish notation* in mathematics) is used, so that the first element represents a function to be performed on the remaining elements. The value of ( + 2 3) is 5, or 2 + 3. The arguments themselves may be functions, as in (+(* 4 5)( − 7 2)) which evaluates to 25, or (4*5) + (7−2). New functions are defined using *defun,* which stands for *define function,* as in

$$(defun\ square\ (x)\ (^{*}\ x\ x))$$

which defines the square of an argument, $x$. This function can then be used to create additional functions.

There are a number of dialects of LISP in existence, having a range of mathematical functions, program control structures, input/output functions, and functions for list manipulation, data structuring, function evaluation, and so on. Scheme is a variation that is widely used both for artificial intelligence programming and for teaching the concepts of computer science in introductory courses (Abelson, Sussman, and Sussman 1996).

## ADA

In the 1970s the Department of Defense initiated development of a new programming language, Ada, to support software development for its systems. The objective was to create software that was more reliable, easy to maintain, portable to other machines, and less costly (Booch, Bryan, and Petersen 1993). Software engineering methods, including top-down design, structured programming, and object-oriented programming, along with common support tools were used to help accomplish these objectives. Since the language was meant to be used to develop software for large, complex systems, it has features that help hide complexity. It uses functional decomposition, which means breaking a problem down into smaller, independent pieces, and data abstraction, which means that the details of how operations on data types are actually implemented are hidden from the programmer. Program modules are organized into packages, tasks, and subprograms, which are procedures and functions. The result is an "industrial-strength" general-purpose language that can be used for a wide variety of applications and systems. Ada is gaining acceptance among colleges and universities as a programming language for computer science instruction.

*See also* Computer Science; Logo; Software; Technology

### SELECTED REFERENCES

Abelson, Harold, and Andrea diSessa. *Turtle Geometry: The Computer as a Medium for Exploring Mathematics.* Cambridge, MA: MIT Press, 1986.

Abelson, Harold, Gerald J. Sussman, and Julie Sussman. *Structure and Interpretation of Computer Programs.* Cambridge, MA: MIT Press, 1996.

Arnold, Ken, and James Gosling. *The Java Programming Language.* 2nd ed. Reading, MA: Addison-Wesley, 1998.

Booch, Grady, Doug Bryan, and Charles G. Petersen. *Software Engineering with Ada.* 3rd ed. Redwood City, CA: Benjamin-Cummings, 1993.

Deitel, Harvey M., and Paul J. Deitel. *C++: How to Program.* Englewood Cliffs, NJ: Prentice-Hall, 1994.

Kemeny, John G., and Thomas E. Kurtz. *Basic Programming.* 3rd ed. New York: Wiley, 1980.

Mathematical Association of America. *Guidelines for Undergraduate Mathematical Sciences Programs.* Washington, DC: The Association, 1993.

National Council of Teachers of Mathematics. *Curriculum and Evaluation Standards for School Mathematics.* Reston, VA: The Council, 1989.

———. *Principles and Standards for School Mathematics.* Reston, VA: The Council, 2000.

Papert, Seymour. *Mindstorms: Children, Computers and Powerful Ideas.* 2nd ed. New York: Basic Books, 1993.

Sammet, Jean E. *Programming Languages: History and Fundamentals.* Englewood Cliffs, NJ: Prentice-Hall, 1969.

Savitch, Walter J. *Pascal: An Introduction to the Art and Science of Programming.* 4th ed. Redwood City, CA: Benjamin-Cummings, 1995.

Steele, Guy L., Jr. *Common Lisp: The Language.* 2nd ed. Bedford, MA: Digital, 1990.

CATHERINE M. RICARDO

# PROGRESSIVE EDUCATION ASSOCIATION (PEA) REPORT

*Mathematics in General Education* (1940), a report for the Commission on Secondary School Curriculum of the PEA that proposed building a mathematics curriculum around concrete problem situations which arise in ordinary life, an approach stressing incidental learning, motivation, and preparation for democratic living. The report documented a decrease in the percentage of students in traditional mathematics classes (classes that were primarily intended for college-bound students) at the same time as schools were experiencing an increase in total high school enrollments because the Depression kept many young people from finding jobs. Also cited was lack of student interest in mathematics. The report stressed mathematics as a tool for problem solving, and listed seven problem-solving concepts that are also unifying concepts in mathematics: (1) formulation and solution, (2) data, (3) approximation, (4) function, (5), operation, (6), proof, and (7) symbolism. Incidental learning, that aspect of the report which deemphasised learning based on mathematical structure, was challenged after World War II.

*See also* History of Mathematics Education in the United States, Overview

### SELECTED REFERENCES

Bidwell, James K., and Robert G. Clason, eds. *Readings in the History of Mathematics Education.* Washington, DC: National Council of Teachers of Mathematics, 1970.

Jones, Phillip S., ed. *A History of Mathematics Education in the United States and Canada.* 32nd Yearbook. Washington, DC: National Council of Teachers of Mathematics, 1970.

Progressive Education Association, Committee on the Function of Mathematics in General Education for the Commission on Secondary School Curriculum. *Mathematics in General Education.* New York: Appleton-Century, 1940.

JAMES K. BIDWELL
ROBERT G. CLASON

# PROJECTIONS, GEOMETRIC

Projections of an object into a plane in a parallel or perspective manner. Projecting a figure into a plane is analogous to making a shadow picture of the figure. Extend a line (an imaginary light ray) from a given point (an imaginary light source) to each point of the figure. The points where the lines of the "light rays" strike the plane form the projection of the figure into the given "image plane."

## PARALLEL PROJECTION

A projection is *parallel* if the imaginary light rays are parallel, that is, if the "light source" is infinitely far away. The shadow of an object under ordinary sunlight is a parallel projection. A parallel projection is *orthographic* if the image plane is perpendicular to the light rays (see Figure 1). A shadow on the ground is an orthographic projection if the sun is directly overhead. In coordinates, projection along vertical lines into the $x,y$-plane is given by the formula

$$f(x,y,z) = (x,y)$$

Three-dimensional figures become distorted when projected into a two-dimensional plane because the information contained in the third dimension is lost. For example, the projection $f(x,y,z) = (x,y)$ collapses the ellipsoid

$$x^2 + \left(\frac{y^2}{4}\right) + \left(\frac{z^2}{9}\right) = 1$$

onto the ellipse

$$x^2 + \left(\frac{y^2}{4}\right) = 1$$

Orthographic projections are commonly used in engineering drawings because they have the simplest formulas and therefore cause the simplest distortions. Parallel lines have parallel images under parallel projections, and a line segment parallel to the image plane projects to a line segment with the same

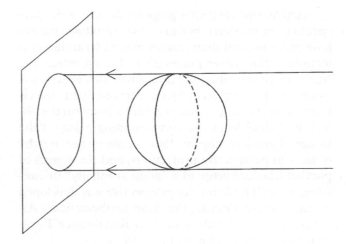

**Figure 1** *Orthographic projection.*

length. Draftsmen can easily calculate the exact length and position of a line segment in space if they are given its orthographic projections in two perpendicular image planes.

## PERSPECTIVE PROJECTIONS

If the "light source" is a finite distance from the image plane, then the projection is a *perspective projection.* The shadow of an object on the wall of a room that is illuminated by a single light bulb is a perspective projection (see Figure 2). A typical perspective projection, such as a projection into the $x,y$-plane from the point $(0,0,1)$ on the $z$-axis, is given by the formula

$$g(x,y,z) = (x/1 - z), y/(1 - z))$$

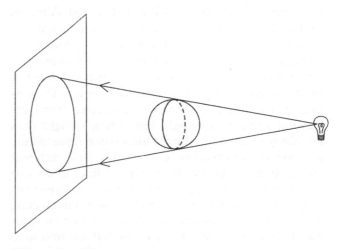

**Figure 2** *Perspective projection.*

Artists and computer graphics designers use perspective projections because the resulting images look more natural than images made by parallel projection. If the viewer places his or her eye where the light source was, then the image will look exactly the same as the original object, because light travels along the same line to the eye from a point on the object as it does from the corresponding point on the image. Far-away objects look smaller than nearby objects in perspective projections, and the images of parallel lines converge to a "point at infinity" or *vanishing point*. The theory of perspective was developed by the ancient Greeks, the Arab mathematician Alhazen (ca. A.D. 1000), and such Renaissance Europeans as Albrecht Dürer (1471–1528).

*See also* Projective Geometry

## SELECTED REFERENCES

Beskin, Nikolai Mikhailovich. *Images of Geometric Solids.* Moscow, Russia: Mir, 1985.

Bronowski, Jacob. *The Ascent of Man.* Boston, MA: Little, Brown, 1973.

*Encyclopaedia Britannica.* 15th ed. Chicago, IL: Encyclopaedia Britannica, 1984.

French, Thomas E., and Charles J. Vierck. *Engineering Drawing.* New York: McGraw-Hill, 1972.

Jennings, George A. *Modern Geometry with Applications.* New York: Springer-Verlag, 1994.

Pedoe, Daniel. *Geometry and the Liberal Arts.* New York: St. Martin's, 1976.

GEORGE A. JENNINGS

# PROJECTIVE GEOMETRY

Geometry that had its beginnings in the theory of perspective developed by Renaissance artists such as Leonardo da Vinci (1452–1519) and Albrecht Dürer (1471–1528). In order to make a drawing look realistic, shapes must be distorted appropriately. For example, the top of a circular barrel is not drawn as a circle, but as an ellipse whose eccentricity depends on the angle of view. A straight line, such as the corner of a building, however, is perceived from any viewing angle as a straight line. Thus, straight lines are always drawn as straight lines when representing a scene on canvas, but curves may have different shapes than in reality. Since parallel lines in a scene receding from the viewer appear to come together, they must be drawn in a picture as lines intersecting on the horizon. This gave rise to the notion that "parallel lines meet at infinity," an idea that was first used outside of art by Johannes Kepler in describing the

orbit of a comet and first developed systematically in the study of geometry by Girard Desargues (1639). Because points in the scene are "projected" onto the drawing, this type of geometry also came to be called *projective* geometry. Jean Victor Poncelet produced the first projective geometry book in 1822. Meanwhile, work by Garalamo Saccheri (1763), Janos Bolyai, Nikolai Lobachevsky, and Carl Friedrich Gauss (1820s) led to *hyperbolic* geometry, and work by Bernhard Riemann (1854) led to *elliptic* geometry. These geometries are called *non-Euclidean* because they contradict the parallel postulate; in hyperbolic geometry, there are many "parallels" to a line through a given point, while in elliptic geometry there are none. Advances in algebraic techniques during the nineteenth century such as the use of matrices and determinants and the development of vector spaces enabled Arthur Cayley (1882) to show that not only Euclidean geometry but also the two non-Euclidean geometries are special cases of projective geometry.

## AXIOMATIC FORMULATION

A *projective plane* is a collection of objects called *points* and objects called *lines* and a relation called *on* that does or does not hold between a given point and a given line such that:

Axiom 1. Given two distinct points, there is one and only one line on both of them.

Axiom 2. Given two distinct lines, there is a point on both of them.

Axiom 3. Given a line, there are at least three distinct points on it.

Axiom 4. Given a line, there is a point not on it.

Axiom 5. There is a line.

From these axioms, the following theorems can be proved:

Theorem 1: Given two distinct lines, there is one and only one point on both of them.

Theorem 2. Given a point, there are at least three distinct lines on it.

Theorem 3. Given a point, there is a line not on it.

Theorem 4. There is a point.

Notice that if the terms *point* and *line* are interchanged in Axiom 1, the result is Theorem 1. The same is true for Axiom 3 and Theorem 2, for Axiom 4 and Theorem 3, and for Axiom 5 and Theorem 4.

This symmetry between the axioms and these basic theorems leads to the *principle of duality,* which states that given any theorem about points and lines in a projective plane, its *dual,* obtained by interchanging the terms *point* and *line,* is also a theorem. Note that the dual of Axiom 2 is included in Axiom 1.

A *finite* projective plane (of order $n$) is a projective plane in which there are exactly $n + 1$ points on some line, for some positive integer $n$. It follows that there are exactly $n + 1$ points on each line, exactly $n + 1$ lines on each point, a total of $n^2 + n + 1$ points, and a total of $n^2 + n + 1$ lines. It is not fully known for which positive integers $n$ there exists a projective plane of order $n$. The study of finite planes leads to *combinatorial geometry.*

## ALGEBRAIC REPRESENTATION

The most common example of a projective plane is the *real* projective plane, usually described using *homogeneous coordinates.* In the real projective plane, a *point* is represented by an ordered triple $(x,y,z)$ of real numbers, not all zero, with triples $(x,y,z)$ and $(kx,ky,kz)$ representing the same point if $k$ is any nonzero real number. A *line* is represented by an ordered triple $[L,M,N]$ of real numbers, not all zero, with triples $[L,M,N]$ and $[kL,kM,kN]$ representing the same line if $k$ is any nonzero real number. The point $(x,y,z)$ is on the line $[L,M,N]$ if and only if $Lx + My + Nz = 0$. In the language of vector spaces, the real projective plane is obtained from the standard three-dimensional real vector space by calling each one-dimensional subspace a *point,* represented by any nonzero vector in it, and by calling each two-dimensional subspace a *line,* represented by any vector normal to it; a *point* is on a *line* in case the one-dimensional subspace is a subspace of the two-dimensional subspace. (Using such notions from vector space theory, projective spaces of arbitrary dimension can be constructed; a projective plane is a projective space of dimension 2.)

Two important results that may or may not hold in projective planes are the following:

*Theorem of Pappus:* If $L$ and $L'$ are two lines; $A$, $B$, and $C$ are three points on $L$; and $A'$, $B'$, and $C'$ are three points on $L'$, none of them being the point of intersection of $L$ and $L'$, then the three points $C'' = AB' \cap A'B$, $B'' = AC' \cap A'C$, and $A'' = BC' \cap B'C$ are collinear (see Figure 1).

*Theorem of Desargues:* If $ABC$ and $A'B'C'$ are two triangles such that lines $AA'$, $BB'$, and $CC'$ are concurrent, then the points $C'' = AB \cap A'B'$,

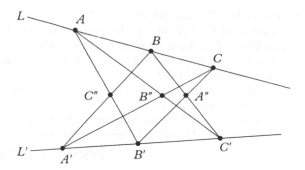

**Figure 1**

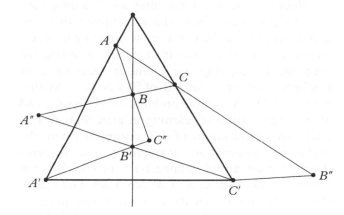

**Figure 2**

$B'' = AC \cap A'C'$, and $A'' = BC \cap B'C'$ are collinear (see Figure 2).

A projective plane in which the theorem of Desargues holds is called a *Desarguesian* plane; one in which the theorem of Pappus holds is called a *Pappian* plane. It turns out that the real projective plane is Pappian and every Pappian plane is Desarguesian. Moreover, every Pappian plane can be represented algebraically using homogeneous coordinates, perhaps over some field other than the real number field. On the other hand, a Desarguesian plane can be represented algebraically over a division ring (i.e., a ring whose nonzero elements form a group under multiplication) and thus is not necessarily Pappian. Non-Desarguesian projective planes exist that give rise to algebraic structures even more general than fields or division rings. A surprising theorem is that if a projective plane lies in a projective space of higher dimension, then the plane must be Desarguesian.

*See also* Bolyai; Gauss; Lobachevsky; Non-Euclidean Geometry

## SELECTED REFERENCES

Boyer, Carl B. *A History of Mathematics.* 2nd ed. Revised by Uta C. Merzbach. New York: Wiley, 1989.

Eves, Howard W. *Survey of Geometry.* Rev. ed. Boston, MA: Allyn and Bacon, 1972.

Garner, Lynn E. *Outline of Projective Geometry.* New York: North-Holland, 1981.

Wallace, Edward C., and Stephen F. West. *Roads to Geometry.* Englewood Cliffs, NJ: Prentice-Hall, 1992.

LYNN E. GARNER

# PROJECTS FOR STUDENTS

Recent models for teaching and learning have included, in some form, the assumption that students must create their own understanding of mathematical concepts, rather than merely adopting the language or copying the methods used by their teachers (National Council of Teachers of Mathematics 1989). A central problem of pedagogy, and of the organization of learning in general, is to reconcile the implications of this assumption with the traditional notion of curriculum: that students should master a well-defined body of knowledge at a predetermined pace and within a set time frame. One form of reconciliation of these two poles of thought is to categorize activities on a "scale" of student autonomy. Some activities are teacher dominated, others are directed solely by the student, and most can be described as partaking of both characteristics. The teacher must then choose the activities and levels of autonomy most suitable for the students in the class. On such a scale, the setting of independent projects for students belongs among the most "student-autonomous" activities possible. Indeed, one might say that any activity in which students themselves conduct investigations, set questions, and define goals is an independent project. In this sense, work on independent projects is among the most important mathematical experiences we can provide for students.

## WHAT CONSTITUTES AN INDEPENDENT PROJECT?

There is no clear-cut criterion for independence of student work. In a context of formal education, students cannot often be expected to generate their own with no input from teachers; this is not expected even on the graduate level. The problem is to minimize teacher intervention in order to maximize student productivity. In solving this problem, the teacher must make numerous subtle judgments. Should a well-defined problem be set, or should the student be freed to explore a whole area, then define the problem independently? Should the student be expected to absorb some mathematical background, or will he or she do better using what is already known? How much time will the student spend on the project? For how long will it hold the student's interest? What level of formality should be required in reporting the results of independent work? In making these judgments, a direct and personal relationship between student and teacher is critical. Similar judgments must be made in challenging groups of students with a mathematical investigation.

A first step is usually for the teacher to present concrete examples of mathematical content that will intrigue students. Several models for this are possible. Many schools have invested in "mathematical discovery" classes. These groups follow no set curriculum and may include students of different ages, abilities, or experiences. The usual school requirement of attendance in class may be restructured. The teacher presents a variety of problem situations or mathematical topics, on a variety of levels. Typically, teacher presentations fall far short of closure but rather open questions for exploration. At a certain point, the students choose topics for independent projects, either from among those presented or from other topics. When a student, or group of students, has found a topic for investigation, it is often expedient to adopt a second model for support, the tutorial model. In this scheme, teachers meet with individual students, or groups of students, to discuss progress, make suggestions, or correct errors. It is important to take the methodology, not just the content, of mathematical investigations as a subject of teaching. Students who have been exposed only to traditional curricula often do not understand the importance of observation or conjecture in mathematical discovery. These activities lie at the heart of mathematical investigation, on almost any level. It is a paradox of the discipline that most expositions of mathematical topics betray no trace of the creative process that produced the results. This paradox must be explained to the student, who will then experience it in his or her own work.

The following exercise is one of many which can provide practice in experimentation and conjecture. Many more can be found, for example, in the books by Pólya listed in the references.

*Exercise 1:* Evaluate $100 - (99 - (98 - (97 - (\cdots - (3 - (2 - 1)))) \ldots ))))$.

*Solution:* By direct calculation, we have:

$$1 = 1$$
$$2 - 1 = 1$$
$$3 - (2 - 1) = 2$$
$$4 - (3 - (2 - 1)) = 2$$
$$5 - (4 - (3 - (2 - 1))) = 3$$
$$6 - (5 - (4 - (3 - (2 - 1)))) = 3$$

So we can conjecture that the given expression is equal to 50, with an appropriate generalization for the corresponding differences for $N$ integers. The pattern can be formalized by defining a sequence $a_n$ as $a_1 = 1$ and $a_n = n - a_{n-1}$ for $n > 1$. An inductive proof of the pattern is immediate, and students can generalize by varying the sequence. An interesting follow-up can be furnished by asking students to evaluate $100^2 - (99^2 - (98^2 - (97^2 - (\cdots - (3^2 - (2^2 - 1^2)))) \ldots))))$.

## RESOURCES

However the guidance of independent research is managed, attention must be paid to providing students with resources of ideas and background. A brief list of bibliographies of mathematical sources is provided with this article. A teacher developing a program in independent research must become familiar with such sources. We are fortunate in mathematics in that we need few physical resources to do research. We do not need a well-equipped laboratory, animals, instruments, or reagents. Reasonable access to computers is helpful in many projects, and essential to some. But lack of enormous computer resources will not significantly hamper the learning process in a program of independent research. The most important sources of support, as in most educational endeavors, are human. For students doing independent research, this resource is critical and can take many forms. For example, the recruitment of a professor or researcher in mathematics who might be willing to guide students in independent research will allow them to go much further in their work than might otherwise be possible. A high school teacher who values the student's work, offering encouragement and guidance at regular intervals, can offer a different form of support. The creation of a peer support group sometimes makes an enormous contribution to the outcome of student research. Peers can be a sympathetic audience, offering intellectual and emotional support and giving productive criticism. The act of listening can be as useful for the audience as for the presenter. Finally, the role model

played by older or former students of a research program can be very important in attracting younger students to the program. An archive of research papers written by former students can generate ideas, and their bibliographies can save time in library work.

## TYPES OF PROJECTS

In the struggle to match a given student with a problem to work on, the teacher must consider a variety of factors. What is the student's background? Is any particular aspect of school mathematics of particular interest? Does the student have an interest in another field in which there may be applications of mathematical ideas? All these are important factors in guiding a student toward a particular topic. One criterion which often confuses both students and teachers is the notion of originality. Whether a topic or result has been explored before is essentially a matter of chance; with all the printed sources available through the years, and with student research becoming more popular and publicized, chances are good that any single result will have been achieved at some time or place earlier. On the graduate level, students spend a lot of time researching the literature to see what has been done already—and typically hold their breath while conducting their own research, lest someone publish the same result earlier. This factor should not be made important in precollege research programs. As a learning experience, independent research is valid whether or not the result achieved has been duplicated by someone else independently.

Topics for the most independent level of project might include generalizations of existing results, applications of mathematical methods to new areas or branches of art or science, or interesting examples and counterexamples. On another level, independent research might take the form of exploring an area of mathematics not usually discussed in high school curricula. The most meaningful cases of this experience occur when the student must synthesize his or her understanding of the new area from a variety of sources rather than simply digest a single chapter or article. An attempt simply to learn a well-worked-out part of mathematics may generate insights original to the student. On a third level, historical topics can offer another dimension to the mathematical understanding of students. Here the teacher must exercise caution. Many students are under the impression that historical research involves listing dates and names very precisely. It is often much more useful for students to research the original historical sources of mathematical ideas. For example, they can pick a mathematical exposition from Euclid or Gauss and explain it in modern terms. The result will contribute to the mathematical understanding of the student, as well as widening his or her view of how mathematics is created.

## DOING THE RESEARCH

Once a student has settled on a topic, he or she must be supported in doing the work. This usually means frequent discussion with a teacher or other mentor about how the investigation is proceeding. The mentor can play a variety of roles, sometimes encouraging the student on the path chosen, sometimes suggesting alternative approaches, but always shaping the project and tailoring it to the student's abilities and interests. Unfortunately, it frequently happens that the teacher or mentor will see a solution to a problem before the student. The mentor must then act with great restraint, guiding the student to see it without actually doing all the work. This is perhaps the most difficult part of the mentoring process.

## WRITING THE PAPER

Among the many facets of mathematics, perhaps the most elusive is the aspect of mathematics as a language. History shows that the struggle to express one's ideas mathematically can be as difficult an effort as developing these same ideas. For this reason, students engaged in independent research should be asked to write their results early and often. Having students read and criticize each other's written work is a technique useful both to teacher and student. Another way to teach writing is to take an historical source and rewrite the results in modern terms.

The following are exercises designed to help students grapple with the difficulties of mathematical exposition.

*Exercise 2:* Describe the following proposition using algebraic notation.

If a straight line be bisected, and a straight line be added to it in a straight line, the square of the whole with the added straight line and the square on the added straight line both together are double of the square on the half and of the square described on the straight line made up of the half and the added straight line as on one straight line (see Euclid 1956, Bk. II., Prop. 10).

*Solution:* If $X, Y$, and $Z$ are collinear, and $M$ is the midpoint of $XZ$, then $XZ^2 + ZY^2 = 2(XM^2 + MZ^2)$. This can be proved algebraically by letting $XZ = 2a$ and $YZ = b$. Then the statement is equivalent to the identity $(2a + b)^2 + b^2 = 2(a^2 + (a + b)^2)$.

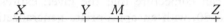

**Figure 1**

*Exercise 3:* Examine the following table, recording whatever patterns you may find:

| | |
|---|---|
| $3 \times 1 = 3$ | $2 \times 2 = 4$ |
| $4 \times 2 = 8$ | $3 \times 3 = 9$ |
| $5 \times 3 = 15$ | $4 \times 4 = 16$ |
| $6 \times 4 = 24$ | $5 \times 5 = 25$ |
| $7 \times 5 = 35$ | $6 \times 6 = 36$ |
| $8 \times 6 = 48$ | $7 \times 7 = 49$ |

Even very good students in ninth or tenth grade, who know the identity $(x + 1)(x - 1) = x^2 - 1$ often have trouble recognizing its application to positive integers or relating it to the pattern displayed here. Students frequently have to go about the process of "reinventing" algebra to make it serve the purpose of describing a pattern they see.

It is never too early for the student to write down results, even if they will have to be heavily revised later on. Techniques for mathematical exposition do not come naturally, and students must run into the obstacles of communicating their thoughts well in advance so that they have ample time to solve problems of communication. Larger issues of organization and of style will arise naturally and pose further difficulties for the student. The teacher or mentor must anticipate all these problems and work early with the student on their resolution.

## RECOGNITION AND MOTIVATION

Once the research paper is written, there are a variety of ways in which it can be used. Numerous mathematics and science fairs afford students an opportunity to exhibit their results and perhaps be recognized for them. Another way to provide closure to the experience of individual research is to organize a symposium of students in the same school or neighboring schools. Oral presentation of individual work provides an important learning opportunity, as students respond to questions by peers (or judges). In addition, the act of exhibiting one's own work can motivate the student to do more of his or her own research. Oral presentations require different skills from written work, skills that students typically find more difficult to acquire and that must be practiced for their own sake. Furthermore, the recognition by an audience of peers for original work done provides an excellent form of closure for one investigative experience and good motivation for the next.

*See also* Problem Solving, Overview; Techniques and Materials for Enrichment Learning and Teaching

## SELECTED REFERENCES

Brown, Stephen I., and Marion I. Walter. *The Art of Problem Posing.* Hillsdale, NJ: Erlbaum, 1983.

Euclid. *The Thirteen Books of Euclid's Elements.* Thomas L. Heath, trans. 2nd ed. 3 vols. New York: Dover, 1956.

Eves, Howard W. *An Introduction to the History of Mathematics.* 3rd ed. New York: Holt, Rinehart and Winston, 1969.

Gillman, Leonard. *Writing Mathematics Well.* Washington, DC: Mathematical Association of America, 1987.

Hess, Adrien, Glenn Allinger, and Lyle Andersen. *Mathematics Project Handbook.* Reston, VA: National Council of Teachers of Mathematics, 1989.

National Council of Teachers of Mathematics. *Curriculum and Evaluation Standards for School Mathematics.* Reston, VA: The Council, 1989.

———. *Principles and Standards for School Mathematics.* Reston, VA: The Council, 2000.

Pólya, George. *How to Solve It: A New Aspect of Mathematical Method.* 2nd ed. Garden City, NY: Doubleday, 1957.

———. *Mathematical Discovery: On Understanding, Learning, and Teaching Problem Solving.* Combined ed. New York: John Wiley, 1981.

———. *Mathematics and Plausible Reasoning.* 2 vols. Princeton, NJ: Princeton University Press, 1954.

———, and Jeremy Kilpatrick. *The Stanford Mathematics Problem Book.* New York: Teachers College Press, 1974.

Sachs, Leroy, ed. *Projects to Enrich School Mathematics: Level 2 and Level 3.* Reston, VA: National Council of Teachers of Mathematics, 1988.

Saul, Mark. "Teaching the Gifted in Mathematics." In *Mathematics Education in Secondary Schools and Two-Year Colleges* (pp. 301–319). Paul J. Campbell and Louise S. Grinstein, eds. New York: Garland, 1988.

Schaaf, William L. *A Bibliography of Recreational Mathematics.* Reston, VA: National Council of Teachers of Mathematics, 1970.

Solow, Dan. *How to Read and Do Proofs.* New York: Wiley, 1982.

———. *Reading, Writing and Doing Mathematical Proofs.* Palo Alto, CA: Seymour, 1984.

Stevenson, Frederick W. *Exploratory Problems in Mathematics.* Reston, VA: National Council of Teachers of Mathematics, 1992.

Trowell, Judith M., ed. *Projects to Enrich School Mathematics: Level 1.* Reston, VA: National Council of Teachers of Mathematics, 1990.

MARK E. SAUL

# PROOF

An argument one gives in order to convince others (and often one's self) of the correctness of one's assertion. In mathematics, unlike in many other areas, the standards of proof demand that *every* assertion be given a conclusive proof, that is, a proof beyond *any* doubt. It is the certainty provided by rigorous proof that sets mathematical knowledge apart from all other kinds of knowledge, including the sciences. In higher mathematics, proofs are absolutely essential because often theorems are nonobvious and even hard to believe. However, in lower levels of mathematics, there is often a tension between the demand to give a formal proof and the feeling that a proof is not necessary, particularly if the claim at hand seems intuitively clear and self-evident.

## MATHEMATICAL EXPLANATION OF THE NOTION OF PROOF

Mathematicians use proofs, mainly, to communicate to each other why they became certain of the truth of a newly found mathematical assertion. To achieve certainty, an ideal mathematical proof displays in a systematic way a finite sequential set of statements that leads from definitions, axioms (i.e., statements the truth of which is unquestioned in a given theory), and theorems (i.e., statements the truth of which has already been proved) to a conclusion, in such a way that as long as the axioms are accepted and the definitions are agreed upon, the conclusion is inevitable and its validity must be recognized. The word *axiom* comes from a Greek word meaning "to be thought worthy." An axiom is a statement the truth of which is to be accepted *without* argument or logical evidence because it is thought worthy as a starting point for further logical argument.

*Mathematical logic* is the area of mathematics that, among other things, treats the components of language used in a proof and the formal rules for deriving conclusions from premises. It determines exact standards for a proof, that is, the criteria that govern the acceptance or rejection of a proposed mathematical proof. The system of logic, which mirrors the majority of proofs found by practicing mathematicians, as well as those rediscovered by students at various stages of their mathematics-educational development, is called *second-order logic* or *predicate logic*. It is built on a simpler, and less expressive system called *first-order logic* or *sentential logic*.

Several deductive calculi can be designated for both of these systems. A commonly used one includes as its basic rule of inference the rule known as *The Law of Detachment* (or *Modus Ponens,* a Latin expression meaning "mode of thinking"). This rule refers to three independent propositions: $\alpha$, $\beta$, and *If $\alpha$, then $\beta$*. It permits us to infer the truth of $\beta$ from the truth of the proposition *If $\alpha$, then $\beta$*, given that the truth of the proposition $\alpha$ is also established. For example, let the statement: *If it is 7:00 A.M., then it is dark outside* be an instance of *If $\alpha$, then $\beta$*. Wherever this statement *is* indeed true, and whenever indeed the time is 7:00 A.M. ($\alpha$), one can be sure, without looking outside, that it is dark there ($\beta$), inferring it by the Law of Detachment. Other rules of inference can also be included in the deductive system.

Formally speaking, within this commonly used deductive calculus, a proof (or deduction) of a proposition $\alpha$ from a set of assumptions (or premises) $\Gamma$ is a finite sequence of propositions $\{\alpha_0, \alpha_1, \ldots, \alpha_n\}$ such that the last one, $\alpha_n$, is the proposition $\alpha$, to be proved. Any other proposition in the sequence is either an assumption (in the set $\Gamma$), or it is a logical axiom, or it is inferred by the Law of Detachment from two earlier propositions in the sequence. (See also Enderton 1972; Manin 1977; Suppes 1957.) Thus, a formal proof is a logical construction consisting of a chain of arguments that demonstrates explicitly how a set of assumptions necessarily implies a conclusion (see Types of Proofs later in this entry). Recently, this concept of proof as captured by mathematical logic has been challenged (see Historical Background later in this entry).

## MOTIVATION AND GOALS

It is not customary in mathematics to present proofs in a manner as detailed and rigorous as that developed by logicians. Usually, a proof omits trivial and routine steps and covers the unobvious transitions from the premises to the conclusions. The further away from formal proof one goes, the higher the risk that transitions, omitted based upon the assumption that they are obvious, are not that obvious to some of the audience. This may, in fact, lead one astray. However, sticking to formal proofs means putting down, explicitly, all the long and tiring details, which would not only obscure the general flow of the argument but would actually make it unreadable.

Although there are well-established techniques (if the prover has furnished enough details) for checking the correctness of a given proof, constructing a proof often demands ingenuity similar to that used for solving other kinds of mathematical problems. The insight it takes to construct a proof is closely related to the wisdom required afterward, of phrasing and composing it so as to achieve the right balance between formality and informality, taking into account the audience to whom the proof is addressed. Proofs in the usual mathematical sense can vary in style from being very intuitive, verbal, or even pictorial to highly symbolic and sophisticated. However, enough of the argument must be presented in order to permit anyone possessing the background to follow the line of reasoning with sufficient clarity.

## TYPES OF PROOFS

1. A proof can have many logical forms. In the simplest case, a proof is a sequence of direct deductions that starts with a deduction from accepted premises and leads step by step to the assertion to be proved, where it ends. Each one of the intermediate deductive steps follows necessarily from one or more of its predecessors or from a combination of them with a previously well-established statement. This form is called a *direct proof*. A short example of such a proof is

For all real numbers $a, b$,

$$(a + b)^2 = (a + b)(a + b) = a^2 + ab + ba + b^2$$
$$= a^2 + 2ab + b^2$$

Therefore, for all real numbers $a, b$,

$$(a + b)^2 = a^2 + 2ab + b^2$$

Here, the first equality is a definition, the next two equalities can be regarded as previously established facts, and the conclusion is derived by applying the transitivity property of equality.

2. Aristotle's logical principle of the "excluded middle" asserts that for any proposition $P$, either $P$ is true or *not-P* is true, and there is no third alternative. (The extendibility of this principle to the infinite was

challenged at the beginning of the twentieth century.) A proof either proceeds directly from premises to conclusion or makes use of Aristotle's principle of the "excluded middle" by showing that the negation of the conclusion leads to the negation of one or more of the premises, and therefore the negation of the conclusion must be false, that is, the conclusion is true. This is called an *indirect proof* or a *proof by reductio ad absurdum*, known also as *proof by contradiction* or *proof by negation*. To prove a proposition of the form *If α, then β*, an *indirect proof* starts with the assumption that the negation of *If α, then β*, that is, *α and not-β*, is true. The proof then proceeds to show how this assumption necessarily produces a contradiction, thus demonstrating the absurdity of the assumption. On the basis of the fundamental logical principle of the "excluded middle," the absurdity of *α and not-β*, establishes the truth of its contrary: *If α, then β*.

To illustrate this type of proof, consider the assertion: *If a prime is a whole number greater than 1 divisible only by 1 and by itself (α), then there are infinitely many different prime numbers (β)*. The proof of this assertion starts by assuming the contrary; namely, that given the definition of a prime number (α), there is only *a finite* number of distinct primes (not-β). Designate these primes by $p_1, p_2, p_3, \ldots, p_n$. This necessarily implies that one of them is the largest prime. Now, construct the sum of 1 and the (finite-by-assumption) product of all primes to get a new whole number, $W = 1 + p_1 \cdot p_2 \cdot p_3 \cdots \cdots p_n$. Is $W$ prime? Certainly not, as it is larger than any of the primes in the list. In particular, it is larger than the largest one. It therefore cannot itself be prime. However, $W$ cannot be anything else but prime since it is not divisible by any prime in the list (it leaves 1 as a remainder when divided by any prime), which implies that it is divisible by itself and by 1 only. We have reached an absurd situation by showing that $W$ *cannot* be prime and at the same time it *must* be prime. Because all the derivations in this line of argument are valid, this contradiction can only be a result of the initial assumption that the number of primes is finite. This falsifies that assumption, leaving as the only possible case the existence of infinitely many primes, which is what we set out to prove.

3. If the conclusion to be proved from the premises is the assertion that an object exists, a direct proof of this conclusion is usually referred to as *a constructive proof*, because the chain of deductions necessarily supplies an explicit procedure for exhibiting that object. For example, a constructive proof that a linear equation $ax + b = 0$ has a unique solu-

tion shows that $x = -b/a$ is the unique solution, where $a, b$, represent any two nonzero real numbers. If the existence proof is an indirect one, then it is referred to as a *nonconstructive proof*, because then we know only that asserting the object's nonexistence would lead to a contradiction, but we do not know how to directly exhibit that object. An example is the theorem usually attributed to Carl F. Gauss, stating that any polynomial of positive degree with complex coefficients must have a complex root. All the known proofs merely show that such a polynomial without any complex root must be of zero degree, concluding therefore that there must be a complex root to any polynomial of positive degree. Although there has been some progress recently toward writing down a procedure (algorithm) to exhibit a complex root, the general problem is far from being solved.

4. If the conclusion has the form of a general property that is asserted to hold for all natural numbers, then frequently there is a special technique, called *mathematical induction*, that can be employed to prove it. It is based upon the fifth axiom in Giuseppe Peano's (1858–1932) axiomatic foundations for the natural numbers. This axiom allows one to deduce the truth of a claim about all the natural numbers as a third step after the following two:

(i) establishing that the property holds for 1;
(ii) establishing the general conditional statement: for all $k \geq 1$, if the property holds for $k$, then it holds for $k + 1$.

For example, we shall prove by mathematical induction that the following is true for all natural numbers $n$:

If $m$ is a positive integer between two consecutive powers of 3 satisfying: $3^{n-1} < m \leq 3^n$, then $n$ is the number of weighings by a balance pan scale that suffice to guarantee the discovery of a false coin among $m$ identically looking coins of which one is known to be counterfeit (lighter).

*Step (i):* For $n = 1$ we have $3^0 < m \leq 3^1$ coins. For two coins it follows immediately that one weighing suffices. For three coins, put one coin on each pan and see if they balance. If they do, then the counterfeit is the third coin; if they do not, the counterfeit is the lighter coin.

*Step (ii):* Suppose $k$ weighings suffice for any number of coins $m$ that satisfies: $3^{k-1} < m \leq 3^k$. Let us prove that $k + 1$ weighings suffice for any number of coins $m$ that satisfies: $3^k < m \leq 3^{k+1}$.

Take any number of coins $m$ such that $3^k < m \leq 3^{k+1}$. Divide them into three subsets such that at least two have the same number of coins in them. (This is always possible.) The number of coins in each of the three subsets satisfies $3^{k-1} < m \leq 3^k$, and therefore, by the inductive hypothesis, $k$ weighings suffice to discover the false coin in each, if known to be there. To decide in which of the three the false coin indeed belongs, one more weighing suffices: put aside the subset with a different number of coins (or any of the three subsets if all have the same number). Put one of the two remaining (equal) subsets on each pan. If they balance, the false coin must be in the subset put aside. If not, it must be in the lighter subset. This completes the proof that $k + 1$ weighings suffice.

*Step (iii):* The establishment of these two facts completes the proof, as at this point, by Peano's fifth axiom, the statement is true for all $n$.

Note that in a proof by mathematical induction, step (i) can be proved for a number larger than 1. The proof then changes to regard a proposition about the subset of the natural numbers from that number on. Step (ii) can have another version in which the assumption is that the property holds for all natural numbers up to $k$. (See also Movshovitz-Hadar 1993.)

5. To prove that a certain general statement $\alpha$ is false, it is sufficient to exhibit one object that violates $\alpha$. In this case, it is customary to say that we have produced a refutation, or a *proof by counterexample*. For example, if $\alpha$ is "All prime numbers are odd," then the number 2 constitutes a counterexample to $\alpha$, and that makes $\alpha$ false. To disprove the proposition "If two triangles have two sides of one equal to two sides of the second and also have three angles of one equal to three angles of the second, then the triangles are congruent," it is sufficient to find one counterexample, such as the triangles with sides: 8, 12, 18, and 12, 18, 27. These triangles are similar; hence their angles are equal to each other. They also have two sides of one equal to two sides of the second. However, they are *not* congruent to one another, as the third sides are not equal. (Note that there is no correspondence between equal elements of these triangles, and therefore there is no inconsistency between this statement and the side-angle-side and angle-side-angle congruency theorems. The latter requires correspondence between equal elements to guarantee triangle congruency.)

Other types of proof have recently been introduced into mathematics. They are described in the next section. (See also Enderton 1972; Lakatos 1976, 1985; and Suppes 1957.)

## HISTORICAL BACKGROUND

The idea of proof was introduced in ancient Greece, in the sixth century B.C., as a part of the axiomatic method. From the time of its introduction, the concept of proof has been the bedrock of mathematics. By the turn of the fourth century B.C. to the third, Euclid had written the famous *Elements,* a thirteen-volume treatise applying the axiomatic method to geometry. It contained proofs—namely, deductions from explicitly stated axioms—for more than 400 propositions. Ever since, questions such as "What is a proof? What constitutes an acceptable proof?" have been occupying the minds of the mathematical community, which has never reached a consensus.

In the middle of the nineteenth century, George Boole published *The Mathematical Analysis of Logic,* followed by *The Laws of Thought,* laying the grounds for the formal study of mathematical reasoning. These publications instituted a school of thought that endeavored to establish a unification of logic and mathematics, reaching a climax in 1910 with Whitehead and Russell's *Principia Mathematica.* Paradoxically, this comprehensive study did not unite mathematicians over the basic questions concerning the nature of proof. In fact, at the beginning of the twentieth century, there was a split over the scope and limits of the notion of proof, dividing mathematicians into Formalists and Intuitionists, who hold irreconcilable views on the foundations of mathematics. The Formalists adhere to formal axiomatic systems as the foundations of mathematics. They believe in freeing any mathematical domain from contradictions through its establishment on a finite set of axioms from which all the theory can be logically derived. The Intuitionists deny the need for foundations in the form of axiomatic theories, since according to their view, mathematical entities are substantial objects in a realm of pure intuition and mathematical facts are objectively true statements describing existing realities. They admit only the mathematics developed through constructive proofs. Consequently, Intuitionists outlaw proofs by *reductio ad absurdum,* as these are based on the principle of the excluded middle.

Modern computers strengthened the experimental approach to mathematical proofs yet brought a new meaning to strict rigor. Benoit Mandelbrot in his address to the 7th International Congress on Mathematics Education said: "The slightest departure from absolute rigor makes it (the computer) scream 'error' at the programmer." In 1976, Kenneth Appel and Wolfgang Haken proved the Four Color

Theorem after more than 120 years of struggle by many mathematicians. In 1995, after about 350 years of false attempts, Andrew Wiles succeeded in showing that Pierre de Fermat was right in stating, in the margin of a book, that $x^n + y^n = z^n$ has no non-trivial solution in positive integers for $n \geq 3$. These proofs stretch over hundreds of pages and depend on other very long proofs. Moreover, they use mathematical concepts and techniques that can be followed by a very limited number of experts.

Doubts and uncertainty about such proofs have penetrated the mathematical community. Are such theorems indeed proved beyond the shadow of *any* doubt, as mathematics requires, or are they, rather like proof of guilt in a case of murder, proved beyond a *reasonable* doubt? To make things even more complicated, some assertions were shown to have such long proofs that, in practice, they cannot be displayed in writing at all. However, in some of such assertions, if we allow a very small probability of error, their ("probable") truth turns out to be provable with relative ease. As of the early 1970s, a proof beyond a reasonable doubt became acceptable in mathematics provided that the doubt can be guarded quantitatively. To distinguish this kind of proof from the proofs captured by mathematical logic, which in general are referred to as *deterministic proofs,* the notion of *probabilistic proof* was introduced (Rabin 1976). Already in 1947, the eminent Hungarian mathematician Paul Erdös used the probabilistic method to prove that a certain coloring of a graph exists (Alon and Spencer 1992). In 1985, another fascinating development occurred, with the introduction of *zero knowledge proof* by computer scientists. Its essence is to conclusively persuade another person that you deserve the credit for being able to prove something, without providing any details about the proof itself (except perhaps an upper bound for its length) (Goldreich 1988). In 1994, Intel's Pentium chip, the computer chip that is the heart of millions of personal computers, was found to have a mathematical flaw. In considering the doubts this discovery may shed on proofs such as that of the Four Color Theorem, Keith Devlin (1995) suggested a new definition of proof that reflects real, live mathematical proofs dealt with by mathematicians at the end of the twentieth century: "A proof is often just an argument that (i) has been accepted by a number of mathematicians whom the community at large feels it can trust on such matters, and (ii) has not yet been shown to be false" (see also Kleiner 1991).

## PEDAGOGICAL APPROACHES

Pedagogical practice varies between two extremes. On one end is the very formal approach, commonly exhibited in graduate-level university courses, where the underlying assumption is that students are mathematically self-motivated and the convention is that a presentation of a rigorous proof follows the presentation of theorems, one by one in a series. On the other extreme is the very informal and intuitive approach, dispensing with proofs altogether, at times. The latter is often associated with matter-of-fact acceptance of generalizations based upon a few particular examples, commonly occurring in elementary school. In between one finds attempts at provoking doubts and thereby making students feel the need to prove, stimulating students to look for and to construct proofs, leading classroom or group activities in which students negotiate the meaning of a plausible conjecture and phrase a reliable statement. These attempts are sometimes followed by providing students with mathematical experiences aimed at the acquisition of the language and skill for following another student's proof and for presenting their own proof, both orally and in writing.

There is a noticeable difficulty in striking the right balance between the development of intuitive understanding in mathematics and the enhancement of appreciation for the crucial role of proof and rigor in mathematics. The problem is to combine formal proving and informal explanation in a situation in which students first experience the need to convince others of the truth or falsity of a mathematical proposition and then reflect on what it is they ought to do to achieve it. This issue is associated with cognitive aspects of proof, such as abstraction, communication—oral and written—generalization, language, motivation, representation and visualization, and others. It has been the focus of many studies carried out by leading mathematics educators in recent years. (See, for example, Clements and Battista 1992; Hanna and Jahnke 1993; Harel 1996; Webb 1995.)

## RECOMMENDED MATERIALS FOR INSTRUCTION

It is through selecting the instructional materials, and through choosing the mode of operating in the classroom, that mathematics teachers at all levels convey the essence of mathematics to their students. Therefore, they should not only be familiar and comfortable with rigorous proofs but they also must be able to create a classroom atmosphere that tolerates lines of reasoning that may not stand up to strict rigor yet allow the growth of insight into mathematics in a gradual fashion, thereby strengthening students' mathematical intuition and understanding. The following bibliography contains recommended materials for instruction at various levels that can assist instructors in striking the right balance between formal and informal approaches to proof in mathematics.

In general, the habit of asking "why?" should be adopted by both learners and teachers of mathematics. It is at the heart of developing an understanding of the ways mathematical concepts and ideas are related to one another through reason and proof.

*See also* Axioms; Logic; Mathematics, Definitions

## SELECTED REFERENCES

Alon, Nogah, and Joel Spencer. *The Probabilistic Method.* New York: Wiley, 1992.

Billstein, Rick, Shlomo Libeskind, and Johnny W. Lott. *A Problem Solving Approach to Mathematics for Elementary School Teachers.* 5th ed. Reading, MA: Addison-Wesley, 1993.

Boole, George. *An Investigation of the Laws of Thought, on Which Are Founded the Mathematical Theories of Logic and Probabilities.* Cambridge, England: Macmillan, 1854. Reprint. Dover, 1951.

———. *The Mathematical Analysis of Logic, Being an Essay Towards a Calculus of Deductive Reasoning.* London, England: Walton and Maberley, 1848.

Brown, Stephen I., and Marion I. Walter. *The Art of Problem Posing.* 2nd ed. Hillsdale, NJ: Erlbaum, 1990.

Clements, Douglas H., and Michael Battista. "Geometry and Spatial Reasoning." In *Handbook of Research in Mathematics Teaching and Learning* (pp. 420–464). Douglas A. Grouws, ed. New York: Macmillan, 1992.

Devlin, Keith. "Editorial: Proof Beyond Reasonable Doubt." *Focus* 15(1)(1995):2–3.

Enderton, Herbert B. *A Mathematical Introduction to Logic.* New York: Academic, 1972.

Fendel, Dan, and Diane Resek. *Foundations of Higher Mathematics—Exploration and Proof.* Reading, MA: Addison-Wesley, 1990.

Fendel, Dan, et al. *Interactive Mathematics Program—4 vols.* Berkeley CA: Key Curriculum, 1996–1999.

Goldreich, Oded. "Randomness, Interactive Proofs, and Zero-Knowledge—A Survey." In *The Universal Turing Machine: A Half Century Survey* (pp. 377–405). Rolf Herken, ed. London, England: Oxford University Press, 1988.

Hadar, Nitsa. "Children's Conditional Reasoning. Part I. An Intuitive Approach to the Logic of Implication." *Educational Studies in Mathematics* 8(1977):413–438.

Hanna, Gila, and Niels Jahnke, eds. *Educational Studies in Mathematics, Special Issue on Proof* 24(4)(1993).

Harel, Guershon. "Proof Schemes." In *Research in Collegiate Mathematics Education.* Ed Dubinsky, Alan Schoenfeld, and Jim Kaput, eds. Providence, RI: American Mathematical Society, 1996.

Hughes-Hallett, Deborah, et al. *Calculus.* New York: Wiley, 1992.

Kleiner, Israel. "Rigour and Proof in Mathematics: A Historical Perspective." *Mathematics Magazine* 64(5)(1991): 291–314.

Lakatos, Imres. *Proofs and Refutations.* Cambridge, England: Cambridge University Press, 1976.

———. "What Does a Mathematical Proof Prove?" In *New Directions in the Philosophy of Mathematics* (pp. 153–162). Thomas Tymoczko, ed. Boston, MA: Birkhäuser, 1985.

Maher, Carolyn, and Amy M. Martino. "The Development of the Idea of Mathematical Proof: A Five-year Case Study." *Journal for Research in Mathematics Education* 27(2)(1996):194–214.

Mandelbrot, Benoit. "Fractals, the Computer and Mathematics Education." In *Proceedings of the Seventh International Congress on Mathematical Education* (pp. 77–98). Claude Gaulin, Bernard R. Hodgson, David H. Wheeler, and John C. Egsgard, eds. Sainte-Foy, Quebec, Canada: Les Presses de l'Université Laval, 1994.

Manin, Yu. *A Course in Mathematical Logic.* New York: Springer-Verlag, 1977.

Mason, John, with Leone Burton and Kaye Stacey. *Thinking Mathematically.* Menlo Park, CA: Addison-Wesley, 1985.

Movshovitz-Hadar, Nitsa. "Mathematical Induction—a Focus on the Conceptual Framework." *School Science and Mathematics* 93(8)(1993):408–417.

———. "School Mathematics Theorems—an Endless Source of Surprise." *For the Learning of Mathematics* 8(3)(1988):34–40.

National Council of Teachers of Mathematics. *Curriculum and Evaluation Standards for School Mathematics.* Reston VA: The Council, 1989.

———. *Principles and Standards for School Mathematics.* Reston, VA: The Council, 2000.

———. *Professional Standards for Teaching Mathematics.* Reston VA: The Council, 1991.

Nelsen, B. Roger. *Proofs Without Words—Exercises in Visual Thinking.* Washington DC: Mathematical Association of America, 1993.

Papert, Seymour. *Mindstorms: Children, Computers and Powerful Ideas.* New York: Basic Books, 1980.

Rabin, Michael O. "Probabilistic Algorithms." In *Algorithms and Complexity* (pp. 21–39). Joseph F. Traub, ed. New York: Academic Press, 1976.

Suppes, Patrick. *Introduction to Logic.* New York: Van Nostrand, 1957.

Webb, John H. *What's the Use of Mathematics?* New Series, no. 185. Cape Town, South Africa: University of Cape Town Printing Dept., 1995.

Whitehead, Alfred North, and Bertrand Russell. *Principia Mathematica*. 3 vols. New York: Cambridge University Press, 1910–1913.

NITSA MOVSHOVITZ-HADAR

# PROOF, GEOMETRIC

The process by which geometric ideas are tested and verified. Proof is the formal process by which the truth of mathematical ideas is demonstrated. It is based on an accepted set of rules. In this entry, common student geometric proof performance, research in the development of geometric proof, and suggestions for the teaching of geometric proof will be discussed.

## PROOF PERFORMANCE

Many recent studies on proof performance suggest that the ability to use formal deduction is lacking in many students who are taking or have taken secondary school geometry (Burger and Shaughnessy 1986; Usiskin 1982). For instance, asked to judge the mathematical correctness of inductive and deductive verifications of statements, 52% of students accepted an incorrect deductive argument as valid (Martin and Harel 1989). After finding or learning a correct proof for a statement, many students maintained that surprises were still possible and that further checks were desirable (Fischbein 1982). In a study of students aged 12 to 15, Peter Galbraith (1981) found that over a third of the students did not understand that counterexamples must satisfy the conditions of a conjecture but violate the conclusion; 18% felt that one counterexample was not sufficient to disprove a statement. Alan Schoenfeld (1986) described students as naive empiricists. In dealing with geometric constructions, for instance, they made a conjecture, then tested it by examining their construction. If the construction looked sufficiently accurate, the students were satisfied that the conjecture had been verified. In a formal study of students who studied geometric proof, Sharon Senk (1985) concluded that only about 30% of students in full-year geometry courses that teach proof reach a mastery level in proof writing.

Thus, the common misconceptions made by students are: they accept statements as true without sufficient mathematical evidence and at the same time doubt mathematically complete arguments; they are unable to create such proofs of statements that were not directly discussed in the class or text; and formal proofs do not seem to convince students of the truth of geometric statements. These errors suggest problems in students' learning of geometric proof and are the subject of much recent research.

## RESEARCH ON THE DEVELOPMENT OF PROOF UNDERSTANDING

Alan Bell (1976) claims that the need for proof occurs when a distinction is made between convincing oneself and convincing others of the truth of a mathematical idea. Jean Piaget believed students must be able to "see" things from an exocentric perspective and be capable of introspection before they establish the need for proof. He also believed that the development of reasoning occurs in three separate stages (Clements and Battista 1992). At the first stage, students fail to consider any but the most recent information. Their thinking is nonreflective and unsystematic. At the second stage, students consider prior action or previously acquired information to support or refute an idea. They begin trying to justify their predictions. There is an anticipatory character and purposefulness to searches for information. At the third stage, students can create general hypotheses that explain why things *must* occur. They progress beyond an empirically based belief that something is always true to making a logical conclusion that it must *necessarily* be true.

Studies on logical reasoning suggest that students' development of formal proof is related to their understanding of the language of logic, which develops with age. For example, until students reached a certain level of mathematical/logical maturity they misused both logical implication and logical connectors, and some of the misuse was regular (for example, a conditional statement interpreted as a biconditional statement). As students developed the language necessary for logical reasoning their proof performance increased.

According to Pierre van Hiele, the intuitive foundation of proof "begins with a pupil's statement that belief in the truth of some assertion is connected with belief in the truth of other assertions. The notion of this connection is intuitive: The laws of such a connection can only be learned by analysis" (1986, 124). Logic is created by analyzing and abstracting these laws, that is, by operating on the network of links between statements. Michael de Villiers (1987) concurs that deductive reasoning first occurs when the network of logical relations among properties of concepts is established. He further believes that proof is meaningless to students who do not doubt the validity of their own empirical observations. These students see proof as justifying the obvious.

Efraim Fischbein (1982) believes that intuition is an essential component of all levels of argument. Intuition is anticipatory: "While striving to solve a problem one suddenly has the feeling that one has grasped the solution even before one can offer any explicit, complete justification for that solution" (p. 10). For novices, intuition exists as the major component of an argument. For more advanced students, intuition plays the role of "advanced organizer" and is only the beginning of an argument. It not only convinces oneself of the truth of an idea but organizes the direction of more formal methods.

## IMPLICATIONS FOR INSTRUCTION

In much of school mathematics, intuition and empirical methods are discouraged, depriving students of an opportunity to develop a sound foundation for proof and its need. Students at a very early age should be encouraged to justify their mathematical ideas. They should be asked to show, demonstrate, or explain why they believe their mathematical thinking is correct in order to strengthen their conviction about these ideas. Students should also be encouraged to defend their thinking to groups of other students who hold opposing mathematical views. These types of discussions are beneficial in developing both an understanding of the need for and also the process of creating mathematical proof.

Teachers should use lessons appropriate for the developmental level of their students. Initially (which could be early in the elementary grades), teachers should provide activities that promote student exploration, have students make conjectures, and allow time for discussion between individual students. For example, students may be encouraged to use Logo to create as many different rectangles as they can. They would create a set of Logo commands for each rectangle and sketch pictures of the rectangles. Students would then share their sets of commands and be asked to conjecture about characteristics common to all sets of commands. Students would make conjectures and argue about their conceptions. At this point the students could go back and create more rectangles that fit their conjectures or create other rectangles that do not conform to other students' conjectures. The main purpose of such activities is to introduce the natural process of exploring, conjecturing, and verifying geometric ideas.

The next level of activities (which could occur in middle school) requires students to create more refined geometrical arguments. For example, students using *The Geometer's Sketchpad* may be asked to find evidence to support or refute given geometric statements (e.g., "The diagonals of a quadrilateral are congruent"). Students report their findings and discuss their reasons for their conclusions. At this level, teachers should encourage students to modify the statements (i.e., make them more or less specific) and indicate reasons for their modifications based on collected data. Here the main purpose of the activities is for students to realize or appreciate the type of evidence required to establish the truth of particular geometric ideas (e.g., that only one counterexample is necessary to disprove a statement).

If students have sufficient experiences with these activities, then dealing with formal geometric proof is achievable. To remain consistent with the prior activities, these new activities should keep the focus on students convincing other members of the class that their geometric ideas are correct. For example, students could write out in paragraph form an argument to support or refute the statement "Each median of an isosceles triangle cuts the original triangle into two congruent triangles." Have students exchange and edit each others' arguments. Teachers should emphasize the use of mathematical language and concepts and the need to clearly communicate and convince others with their own arguments.

The key to effective instructional activities that develop geometric proof in students is to focus on students' justification of ideas using increasingly sophisticated methods that make sense to them. Traditional methods that prematurely formalize the justification process, such as telling students to write down all the given data and then list any theorems that can be used, will only confuse students, causing them to memorize geometric proofs and to disconnect them from meaningful justification.

*See also* Proof

## SELECTED REFERENCES

Bell, Alan W. "A Study of Pupil's Proof-explanations in Mathematical Situations." *Educational Studies in Mathematics* 7(Feb. 1976):23–40.

Burger, W. F., and J. Michael Shaughnessy. "Characterizing the van Hiele Levels of Development in Geometry." *Journal for Research in Mathematics Education* 17(Jan. 1986):31–48.

Clements, Douglas H., and Michael T. Battista. "Geometry and Spatial Reasoning." In *Handbook of Re-

*search on Mathematics Teaching and Learning* (pp. 420–464). Douglas A. Grouws, ed. New York: Macmillan, 1992.

De Villiers, Michael D. "Research Evidence on Hierarchical Thinking, Teaching Strategies, and the van Hiele Theory: Some Critical Comments." Paper presented at the working conference for Learning and Teaching Geometry: *Issues for Research and Practice.* Syracuse, NY: 1987.

Fischbein, Efraim. "Intuition and Proof." *For the Learning of Mathematics* 13(2)(1982):9–18, 24.

Galbraith, Peter L. "Aspects of Proving: A Clinical Investigation of Process." *Educational Studies in Mathematics* 12(Feb. 1981):1–28.

Martin, W. Gary, and Guershon Harel. "Proof Frames of Preservice Elementary Teachers." *Journal for Research in Mathematics Education* 20(1)(1989):41–51.

Piaget, Jean. *Possibility and Necessity.* Vol. 2. *The Role of Necessity in Cognitive Development.* Helga Feider, trans. Minneapolis, MN: University of Minnesota Press, 1987.

Pólya, George. *Mathematics and Plausible Reasoning: Induction and Analogy in Mathematics.* Vol. 1. Princeton, NJ: Princeton University Press, 1954.

Schoenfeld, Alan H. "On Having and Using Geometric Knowledge." In *Conceptual and Procedural Knowledge: The Case of Mathematics* (pp. 225–264). James Hiebert, ed. Hillsdale, NJ: Erlbaum, 1986.

Senk, Sharon L. "How Well Do Students Write Geometry Proofs?" *Mathematics Teacher* 78(Sept. 1985): 448–456.

Usiskin, Zalman P. *Van Hiele Levels and Achievement in Secondary School Geometry.* Chicago, IL: University of Chicago, Department of Education, 1982.

van Hiele, Pierre M. *Structure and Insight.* Orlando, FL: Academic, 1986.

MICHAEL MIKUSA
MICHAEL T. BATTISTA

# PROPORTIONAL REASONING

A proportion is an equivalence of ratios. A proportion can thus be expressed as an equation, such as $2:3 = 4:6$. But since fractions express ratios, we can represent this equation as $\frac{2}{3} = \frac{4}{6}$. Since ratios also can stand for the operation of division, we can express the proportion by stating that $2 \div 3 = 4 \div 6$. These three equations are all mathematically equivalent. But research shows that students do not think of them as expressing the same ideas. Most students, for example, do not automatically think of division when working with fractions, and vice versa. Many will even deny that $2 \div 3$ yields $\frac{2}{3}$. They may insist, for example, that the result of the division must be expressed as a decimal fraction, such as $0.666. \ldots$

Ratios come in several varieties, and this variety has great implications for the nature of proportional reasoning. *Ratios of numbers,* or numerical ratios, concern the relative magnitude of two numbers (commonly, though not exclusively, integers). Each of the three preceding equations asserts the equivalence of two ratios of numbers. *Ratios of measured quantities,* or ratios of measures, are normally expressed in terms of units of measures. When the measures are given in the same unit, the ratio expresses a comparison of like things. For example, if John's height is 68 inches and Mary's height is 64 inches, then the ratio of their heights is 68 in. : 64 in. or 17 in. : 16 in. Since the units are the same, one sometimes suppresses or "cancels" the units, expressing the result as a ratio of numbers, $17:16$ in the present case. When the terms of a ratio entail units of unlike measures (e.g., 100 miles, 4 hours), division yields an *intensive quantity* such as 25 miles/hr. Intensive quantities emerged relatively recently in the history of mathematics. Although mathematicians of ancient Greece compared ratios of "time" with ratios of "distance," they did not use ratios composed of unlike measures. *Ratios of unmeasured quantities* are similar to ratios of measures, with the exception that the terms have not been expressed according to standard units of measure. The expression of the height to the base of a triangle as $H:B$ is an example of a ratio of unmeasured quantities.

Proportions are related to the linear functions, $y = ax + b$. The term $a$ is even known as a *constant of proportionality.* This constant bears important relations to many mathematical ideas. For example, if the graph of the function is drawn in a plane, the size of $a$ corresponds to the slope of the graph. If $a$ is represented as a fraction, such as $\frac{3}{5}$ the ratio of numerator to denominator describes the "rise over run." A slope of $\frac{3}{5}$ corresponds to the incline of the hypotenuse of a right triangle with height 3 units and base 5 units. The slope of a graph of a function at any point corresponds to the derivative of the function. In this sense, ratio and proportion are connected to fundamental ideas in calculus.

The psychological origins of ratio and proportion are rooted in perceptual judgment rather than arithmetical knowledge. By the age of 6, children are already quite sensitive to ratios of unmeasured quantities. They can note, for example, that a small clown has a relatively large nose. In terms of mathematical notation, this amounts to recognizing something like (clown-nose):(clown-face) > (normal-nose): (normal-face). Clearly, young children would no more understand such notation than they would understand expressions such as $2:3$ or $\frac{2}{3}$. Their knowledge of ratio and proportion is implicit, intuitive, and highly perceptual. So, although children develop a rudimentary, perceptually oriented understanding of ratio and proportion at an early age, their ability to express proportional relations through language, arithmetic statements, and algebraic equations takes a long time to develop and normally requires years of schooling. It is likely that the articulated, highly linguistic, and equation-bound problem-solving skills that schools require of the student do not fully substitute for the perceptual judgmental intuitions that develop in early childhood. Much of the working out of proportional relations entails reconciling what students intuitively understand about ratio and proportions with the formal representations they learn about in school. Strategies to help students strengthen their understanding and skills include encouragement of verbal activities with familiar ratios, comparison of ratios, and creation of pictures, tables, and graphs (Hoffer and Hoffer 1992).

*See also* Calculus, Overview; Coordinate Geometry, Overview; Division; Equations, Algebraic; Fractions; Rational Numbers

### SELECTED REFERENCES

Coxford, Arthur F., ed. *The Ideas of Algebra, K–12.* 1988 Yearbook. Reston, VA: National Council of Teachers of Mathematics, 1988.

Freudenthal, Hans. *Didactical Phenomenology of Mathematical Structures.* Dordrecht, Netherlands: Reidel, 1983.

Hart, Kathleen M. *Ratio: Children's Strategies and Errors.* Windsor, England: NFER-Nelson, 1984.

Hoffer, Alan R., and Shirley Ann K. Hoffer. "Ratios and Proportional Thinking." In *Teaching Mathematics in Grades K–8: Research-Based Methods* (pp. 303–330). Thomas R. Post, ed. 2nd ed. Boston, MA: Allyn and Bacon, 1992.

Schliemann, Analúcia, and David W. Carraher. "Proportional Reasoning in and out of School." In *Context and Cognition* (pp. 47–73). Paul Light and George Butterworth, eds. Hillsdale, NJ: Erlbaum, 1993.

Vergnaud, Gérard. "Multiplicative Structures." In *Acquisition of Mathematics Concepts and Processes* (pp. 137–174). Richard Lesh and Marsha Landau, eds. New York: Academic, 1993.

DAVID WILLIAM CARRAHER

# PSYCHOLOGY OF LEARNING AND INSTRUCTION, OVERVIEW

Mathematics teachers and educators often look to psychology for guidance, support, or even direct help but sometimes express disappointment in what they find. Mathematics is regarded as a difficult subject to learn, and teachers at all levels are only too aware of this. Thus they constantly strive to find ways of removing obstacles to learning and easing the process of acquiring knowledge. Although psychology is unlikely to offer detailed guidance concerning teaching particular subject matter to particular students at specific times, there is much that it does offer. Psychology provides valuable background information and knowledge that teachers can take into account and interpret in the context of their own courses and in relation to their own particular classes and pupils.

The particular issues of learning for which we might hope to find answers from psychology are many and various. All teachers are involved in preparing coherent sequences of lessons, so one of the most obvious issues is whether it is possible to enhance learning through the optimum sequencing of learning activities. Another concerns whether we might sometimes need to wait until students are ready. Yet another involves the extent to which it is possible or necessary for students to work things out for themselves rather than being informed by teachers. And another involves what kinds of ways of learning are most effective in terms of the quality, quantity, and retention of what is learned. These issues, and others, will all be followed up or referred to later in this entry. Of particular importance in all learning is the issue of transfer, and this is dealt with first.

## TRANSFER OF LEARNING

If our students could learn and apply only what they are taught, then the extent of their knowledge would remain within what they encounter in school and college, together with what they acquire in a more haphazard way from their environment. For many years now, it has been accepted that some transfer of ideas from one domain of knowledge to another does occur (Cormier and Hagman 1987). Such transfer might in theory be either "vertical"—say, from the mastery of addition, subtraction, multiplication, and division of numbers to the concept of binary operation—or "lateral"—from, say, mathematics to science or economics, or from algebra to geometry, or even perhaps from the multiplication of numbers to the multiplication of vectors and matrices (Gagné 1970). The important psychological issue here concerns the extent to which transfer can take place. Many years ago it was believed that any mental training would prove beneficial across broad areas of knowledge, though this view is not now generally accepted (Bigge 1982). It was believed, for example, that learning mathematics, or Latin, would make one a more logical thinker in any domain of knowledge; and even that exercising the mind on anything at all would provide benefits to intellectual development comparable to the benefits of physical exercise to athletic ability and sporting prowess. Currently, although it has been generally agreed for many years that some lateral transfer does occur, there is no unanimous agreement among psychologists about the extent to which it can happen and under what circumstances (Bigge 1982; Cormier and Hagman 1987).

Vertical transfer is, by comparison, relatively uncontroversial and is regularly assumed by teachers whenever learning is expected to proceed from the less abstract to the more abstract, but this does not mean to say that there is agreement about the optimum conditions for vertical transfer or the optimum sequence. Even within mathematics, however, there is scope for disagreement about whether, in relation to the earlier example, the step from numbers to vectors and matrices is lateral or vertical or whether it incorporates elements of both. This example clearly illustrates the difficulty of judging equivalence across different subject matter, even within the same school subject. However, it is clear to most mathematics teachers, for example, that the concept of proof in geometry does not appear to transfer automatically to proof in other domains of mathematics, never mind to proof in history or in a court of law, and one might justifiably inquire why. It seems that we still do not yet know enough about the conditions under which transfer of different kinds of learning is likely to occur. The best advice to mathematics teachers is not to assume that much will be transferred laterally and that even vertical transfer will not be automatic.

## ROTE LEARNING, MEANINGFUL LEARNING AND MEMORY

Students are expected to be able to retain and recall a considerable quantity of mathematical knowledge, ranging from simple facts like specialist words (e.g., *square*, *kilometer*, *height*) and symbols (e.g., 7, +, *dx*) to formulae (e.g., $A = 1/2h[a + b]$) and procedures such as those for the addition of fractions and the division of decimals. The three questions that immediately follow concern how to commit knowledge to memory in the first place, how to prevent forgetting, and how to retrieve from the vast store of data held in the mind.

Two contrasting methods available for attempting to achieve long-term storage in the mind may be referred to as rote learning and meaningful learning (Ausubel, Novak, and Hanesian 1978). To learn by rote is basically to learn by repetition or solely by frequent rehearsal. Meaningful learning ideally implies that any element of mathematical knowledge is not merely held and understood in isolation; it is comprehended in relation to other parts of the whole of mathematics, in that a kind of network of knowledge is built up in the mind. This is analogous to the tree-like networks of nerve cells and linkages that exist in the brain. Mathematics is a subject in which, if memory fails, it is often possible for those who have previously mastered the subject material to work out results and relationships again, seemingly by retreading paths through the network of knowledge held in the mind. Clearly, this is unlikely to happen unless the material was learned in a meaningful manner in the first place, so in this respect meaningful learning is preferable to rote learning. However, there is a clear efficiency element involved in being able to remember, because ready recall obviously allows more speedy progress to be made.

The currently accepted model of the mind incorporates the view that it includes facilities for both short-term and long-term storage (Slavin 1991). Short-term memory is all that is needed to remember a telephone number for long enough to press the keys, though even then constant repetition may be necessary to hold the number in mind (Klausmeier 1985). Only about seven items can be held in short-term memory, though "chunking" (grouping) of items to create composite, and preferably meaningful, items can sometimes extend the quantity. In learning mathematics, although procedures often only require numbers or other elements to be held in memory for a short time, it is likely that our ultimate objective involves long-term storage of the procedures, and indeed of ideas. A considerable amount of mathematical knowledge, ranging from number symbols and names of shapes to formulae like that for the area of a rectangle, is usually remembered without much effort because it is in almost daily use and is therefore rehearsed or reviewed constantly. Rarely used or more complicated knowledge, on the other hand, is more difficult to retain, as also is knowledge that has not been learned meaningfully. Additionally, the efficiency and capacity of our memory is believed to be a property of overall mental ability, so there are likely to be limits on what we can retain that are out of our control. For many people, however, their long-term memory seems to have virtually infinite capacity.

Learning by rote, or repetition, has seemingly always had a firm place in mathematics teaching. The "chanting" of multiplication tables was once virtually the only method used in order to fix the connections in the mind and also to achieve rapid recall. The method is no doubt still used, but perhaps not often nowadays as the only approach to learning tables. An element of rote learning seems unavoidable in the early stages of learning mathematics, when strings of letters and other symbols have to be associated with ideas like "twoness" ("two" and "2") and add ("+"), but before long it becomes clear that students seem to benefit when they can attach meaning to what they are being asked to learn. New material may often be taught meaningfully by relating it, or even mentally "attaching" it, to existing knowledge. Unlike rote practice, sometimes called drill, relating or attaching new ideas to existing knowledge is not an automatic procedure, and some reference to theories of learning is helpful in coming to understand how knowledge might best be extended or enhanced through meaningful learning. This issue will be followed up later.

Periodic review, which most students find essential as examinations draw near, is a standard means of temporarily preventing forgetting and frequently merely involves the straightforward repetition of facts, relationships, and procedures. Without periodic review, and particularly if the knowledge was learned by rote in the first place, it is normal to forget. Rote learning should not be equated with rehearsal, for some repetition and practice may be legitimate within an environment of meaningful learning. Retention can sometimes be assisted by acronyms and mnemonics, and these seem to be particularly common in trigonometry. "SOHCAHTOA" is an acronym that will be familiar to many readers, and "some officers have curly auburn hair to offer attraction" is an example of the many mnemonics available for helping to recall the three basic trigonometric ratios. In a sense, a mnemonic is an attempt to insert meaning when no other, more meaningful approach is available. It is never possible to guarantee or ensure recall, but the regular application and use of meaningfully learned knowledge help. Anxiety causes severe interference to the recall of knowledge, no matter how meaningfully it was learned, and this has serious implications when students take examinations.

Meaningful learning thus implies the comprehension of elements of mathematical knowledge in relation to other elements and to a larger whole, which might be thought of as a network of knowledge. The meaningful storage of knowledge within linked networks in the mind seems to permit easier recall, presumably because searching is able to proceed along the linkages, which is impossible with separately stored rote learning. Thus, for a young child, simple division without remainders might be completely meaningful, though perhaps only in the context of sharing. Eventually, however, the child ought to understand that division cannot always be carried out within the set of natural numbers and that what teachers know as "rational numbers" need to be introduced. Subsequently, this leads to other number sets like the integers, the irrational numbers, the real numbers, and for some students even the complex numbers, with the concept of division consequently becoming amended, broadened, or enhanced in light of a growing network of knowledge about number sets and operations. Even that is not the end of possible links with other knowledge, because division arises elsewhere, for example, in algebra. One teaching device for meaningful learning is to develop pictorial representations of networks of knowledge, sometimes known as concept maps or semantic networks (see Figure 1). These may be drawn and recorded, and periodically extended, by either teacher or students in a deliberate attempt to inject as much meaning as possible into how particular elements of knowledge relate to other elements, thereby supporting long-term retention.

However, although we might strive to achieve meaningful learning, we cannot guarantee that it will occur. Some elements of mathematics, like real numbers for example, seem to cause learning difficulties that cannot be overcome easily, and this phenomenon raises other learning issues like readiness. Many students do come to a meaningful understanding of real numbers over a period of time, almost as if they needed more time to consolidate, and perhaps make connections. Rote learning, however, would certainly not solve any difficulties in relation to learning ideas or concepts as difficult as real numbers, and it is necessary to persevere with trying to convey meaning.

## CONCEPTUAL LEARNING

The conceptual difficulties presented by real numbers are well known. Mathematics involves many concepts, some of which appear to be more difficult to learn than others. It is even difficult to capture the meaning of the term *concept* itself in a definition, and examples are often necessary, suggesting that mathematical concepts might also often be best learned from examples and counterexamples. Examples of concepts include everyday objects like

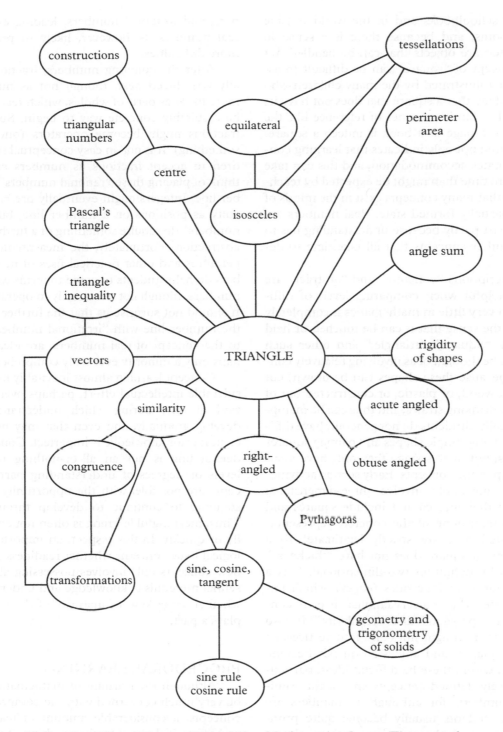

**Figure 1** *[Figure 3.1, page 27,* Learning Mathematics: Issues, Theory and Classroom Practice *(2nd ed., 1992) by Anthony Orton. Reprinted with permission.]*

"dog," "cup," and "chair," all of which exist in concrete form in many variations of shape, size, and appearance, and all of which seem to require time and experience before the concept becomes well formed. The concepts of "beauty," "justice," and "religion"

are even more difficult, presumably because they are more abstract. In mathematics, the concept of "triangle" appears to be relatively unproblematic for children, perhaps because examples and counterexamples seem to abound in the flat shapes students see in their schoolbooks and in the world outside their classrooms, and because there is a sense in which they too are objects that can be handled. Yet even the concept of "square" can be difficult to assimilate, a fact illustrated by the many children who vehemently deny that a square that does not have its sides parallel to familiar frames of reference like the boundaries of a page in a book is indeed a square. This illustration not only indicates that learning concepts necessitates accommodation, and this can take students more time than might be expected by teachers, but also that many concepts exist in the minds of us all in a partially formed state. Real numbers are difficult at least partly because understanding has to be built on other concepts that all take time to develop.

The descriptions "concrete" and "abstract" are sometimes helpful when comparing levels of difficulty, though very little in mathematics is completely concrete, in the sense that it can be touched or held in the hand. "Square," "triangle," and other such shapes might be thought of as involving relatively concrete ideas because these shapes can be drawn, cut out of paper, wood, or plastic, or constructed out of wire, rods, or drinking straws, and thus can be manipulated physically. Students do not generally have difficulty in classifying simple shapes and naming squares and triangles, but at this stage they only have a superficial grasp of the concepts. Early sorting activities with shapes inevitably involve three-dimensional solids, which therefore cannot include squares and triangles. In fact, none of the concrete representations suggested here are strictly legitimate! To a mathematician, a square does not have "thickness" because it is by definition two-dimensional. It is a regular polygon with four sides (edges), which immediately relates what is a very familiar shape to concepts such as "polygon" and "regularity." It also incorporates four straight lines and four vertices; so points, lines, planes, and angles all quickly become involved, and within these lie difficult ideas, particularly if precisely defined concepts are sought. Similarly with numbers, for although all numbers are abstractions, students usually become quite proficient at working with the counting numbers, almost as if they have become concrete entities. This suggests that what might be difficult and seem abstract at one stage of education might later become the relatively more concrete concepts on which subsequent abstractions can be built. Eventually, children may be able to think of the counting numbers as each having a position on a number line. The extension of this concept to incorporate integers, rational numbers, and irrational numbers, leading eventually to real numbers, is, however, likely to present many more difficulties.

After the counting numbers, fractions are usually introduced next, though not as numbers, but more likely as parts of wholes, which teachers find to be a suitably concrete way to begin. Subsequently, fractions might become operators (one-quarter *of* something). It is not an easy conceptual step for children to accept fractions as numbers and then to think of placing these "rational numbers" on a number line. Integers might eventually present less difficulty as positions on a number line, but "negative numbers" do require something of a further leap into abstraction. Fortunately, the measurement of temperatures and other practical uses of negative numbers do help students to come to terms with directed numbers, though not necessarily to operate on them. It should not surprise us that the further filling in of the number line with "irrational numbers," leading to the concept of real numbers, are ideas that perhaps only a minority ever truly comprehend.

Concept learning almost inevitably involves considerable intellectual effort, perhaps over a long period of time during which understanding might develop gradually, and even that may not be sufficient if prior knowledge is imperfect. Conceptual understanding is not an all-or-nothing matter, and levels or degrees of understanding particular concepts are possible, with the opportunity for understanding to continue to develop throughout life. Thus, meaningful learning is often not achieved easily or quickly. In this respect, an important psychological issue concerns whether readiness for learning new concepts only involves possessing all of the essential prerequisite knowledge and understanding or whether some kind of maturity of the intellect also plays a part.

## PROCEDURAL LEARNING

Although the learning of mathematics ought to be very much concerned with the comprehension of concepts, a considerable amount of teaching time and effort is devoted to the teaching of procedures. There are many such procedures in school mathematics, for example, long multiplication, long division, the addition, multiplication, and division of fractions, and the calculation of an arithmetic mean. Students are expected to remember how to carry out such procedures, even though they often have no means of learning them other than by rote, either because no attempt has been made to promote meaningful learning or because the ideas underlying the procedures are conceptually very difficult for the stage of development of the student. Thus, it should not be surprising that it is common for a procedure to be imperfectly remembered, or for two procedures, like addition and multiplication of fractions, to be confused.

In recent years, the term *instrumental understanding* has been accepted to describe the state of understanding limited to how to carry out a procedure correctly, while *relational understanding* describes the state of understanding both how and why the procedure works (Skemp 1976). Relational understanding is usually regarded as the ideal, but instrumental understanding is often the reality. A complicating factor is that, although some students can remember and accept instrumentally learned procedures both without complaint and without its handicapping progress, it seems that many others cannot. These latter students are eventually likely to reject mathematics as a meaningless collection of routines that they have to do but do not comprehend. It is very difficult to motivate students when what they are being required to learn conveys little meaning to them. However, some of the many who have to accept a procedure at an instrumental level when it is first introduced are eventually able to make sense of it and may even be able to link it to networks of knowledge in the mind. A major dilemma of learning mathematics thus concerns how much instrumentally learned knowledge is acceptable. It is difficult to suggest none, given the multiplicity of ways in which relational understanding and meaningful learning can develop, but once teachers accept some it is difficult to know where to draw the line.

## BEHAVIORISM

Conflicting theories about how students learn, and in consequence about how teaching should be conducted, have held sway at different times, certainly throughout the twentieth century. Educational pioneers of previous centuries have tried to promote methods that would enable children to discover knowledge and construct understanding for themselves, but the advent of mass education seems to have been associated more with behaviorist approaches than with any other. For many people, behaviorism is associated with conditioning and with experiments with rats, dogs, and pigeons. In the education of human beings, the essence of behaviorism is concerned with how to change human behavior, and in terms of intellectual behavior, is often associated

with the stimulus, response, reinforcement cycle. First, a stimulus is presented, often in the form of a simple question, then a response is demanded, and finally the response is reinforced, positively or negatively, depending on whether it was correct or not. The nature of the reinforcement could be intrinsic, perhaps simply the pleasure of achieving success or the disgrace of failing, or extrinsic, perhaps the award of a commendation, a "gold star," or a small gift, or a punishment if the response is incorrect. Early behaviorists such as Skinner and Thorndike (see Orton 1992) sought to encourage the formation of a strong "bond" or "connection" between a stimulus and the desired response, so the successful completion of a particular task would be rewarded in the expectation that the required behavior pattern would be repeated. The main means of strengthening the desired bonds was through practice, often called "drill." If the bond was between 2 + 3 (the stimulus) and 5 (the response), then instruction would need to be followed by the right amount of oral and written practice. The main responsibilities of the teacher would not only be to organize sufficient practice but also to arrange the material to be learned into the optimum sequence, so sequencing is an important issue within behaviorism. Behaviorist theory basically seems to be associated with instruction and depends on vertical transfer and equates learning with observed responses. It should be clear that mathematics teaching still frequently incorporates elements of behaviorism, and not only in striving for optimum sequencing of learning tasks. The most common method of teaching mathematics is exposition followed by practice of the skills and routines that have been demonstrated (Cockcroft 1982), and extrinsic rewards for good work are common. Teachers even sometimes still describe "2 + 3 = 5" as a "number bond."

## COGNITIVE APPROACHES TO LEARNING THEORY

In the 1960s and 1970s, what was known as "discovery learning" was enthusiastically promoted by educators such as Jerome Bruner (1973) as an alternative to behaviorism. It was suggested that this approach was more likely to promote meaningful learning than instruction based on demonstration and practice and indeed more likely to motivate. Bruner also encouraged teachers to accept that lateral transfer from one domain of knowledge to another could be a regular feature of learning. Put simply, the expectation of discovery learning was that students would come to understand more thoroughly if they discovered knowledge for themselves. One possible theoretical underpinning for learning by discovery was Gestalt theory (see Resnick and Ford 1984), which basically refuted the behaviorist view that knowledge was simply assimilated by practice and suggested that the mind develops principles of organization through which incoming data are interpreted, or in other words that structure was perceived in, or imposed on, perceptions. Gestalt theory also inspired the development of certain forms of manipulatives, considered by many to be all that was necessary to promote the discovery of particular mathematical ideas. We now know that such apparatus might often be necessary but that it may not be sufficient, because learning often does not automatically take place. A further possible theoretical underpinning for discovery learning was the work of Jean Piaget (see Orton 1992), perhaps most simply described as developmental psychology, which suggested that cognition is an aspect of human growth. Thus, maturation and interaction with, or within, a learning environment both influence the growth of understanding and knowledge.

Considerable debate has taken place over the years concerning the relative merits of discovery and reception (the intended outcome from instruction), partly perhaps because the interpretation of statements by proponents of discovery was often that everything could be, and therefore should be, discovered (see Shulman 1970 for an early discussion of this issue). Critics of discovery learning, such as Ausubel, were keen to assert that discovery is not always necessary or helpful, that accurate and meaningful learning can follow from high-quality exposition, that exposition does not imply merely instruction, and that it is ridiculous to think that there is sufficient time to wait while students discover the whole of mathematics at their own pace and in their own way. The impression that it was necessary for teachers to wait until discoveries occurred and that there was little that teachers could or should do to help led to accusations that such a policy was an abrogation of responsibility on the part of the teacher. Such interpretations are unfair to discovery learning, and in practice teachers have generally assumed that they did have a part to play. In fact, some early evidence (Shulman 1970) indicated that discovery was most rapid and effective when it was "guided" rather than "free," suggesting a vital and sensitive role for the teacher. As regards the impact of discovery learning, the view has been expressed (Cockcroft 1982) that most mathematics teachers, particularly at the secondary level, still depend heav-ily on exposition and instruction. This is perhaps because of its apparent efficiency, in that large numbers of students can be processed at the same time and in the same way, and partly one suspects because it seems easier than other methods. For many students, however, reception learning does not work well and often does not motivate sufficiently. Some teachers, particularly of young children, firmly believe that understanding needs to come from within the learner and often just simply cannot be imposed from without, an issue that will be taken up in the next section.

The work of Jean Piaget has been influential in many countries. His theory of intellectual development was based on results from innumerable experiments with children and students of all ages, described in many books, using methods of individual interviewing and often involving apparatus. This led him to hypothesize that children pass through several stages in their intellectual growth and that these stages are qualitatively different in nature. The implication assumed by some educators was that this placed limits on what could be expected of children, in that learning material was inappropriate if it required a level of thinking that was not yet available to the child. In terms of mathematics, this has led, for example, to suggestions that children are introduced to pencil-and-paper arithmetic too soon. The same theory could also be taken to explain why students have difficulty with topics like place value, ratio, real numbers, and algebra at the times when they are introduced. Clearly, this reminds us of the issue of readiness for learning. Piaget also believed that knowledge could not be delivered to children and that for relational understanding to take place it is often necessary for ideas to be discovered or worked out by the learner. It is important to realize that there have been many criticisms of Piaget's theory (see Orton 1992), not least of the suggestion that intellectual development occurs as a progression through discrete and identifiable stages. At the same time, educators could be accused of expecting too much direct guidance from a theory, considering the complexity of human learning, and perhaps even of trying to apply aspects of the theory inappropriately. Despite the criticisms, Piaget remains a giant in the history of educational psychology, and his work still exerts considerable influence, not least within constructivism (see next section).

## CONTEMPORARY THEORIES

Two contemporary schools of thought that seem to coexist without obvious conflict must be acknowledged, namely information processing and constructivism. The former, not surprisingly, concerns detailed study and analysis of how students solve problems and how they process data in the development of new knowledge. The latter concerns the claim that meaningful learning and relational understanding come from the efforts of the learner in making sense of external stimuli and in successfully integrating new information and ideas with existing knowledge. In the last resort, they would claim knowledge must be actively constructed by the learner, and this seems to suggest the possibilities of both vertical and lateral transfer. Some constructivists would go further and make the radical assertion that learning involves organizing and reorganizing one's own experiential world, not discovering a preexisting, objective, universally accepted world (see Lerman 1989).

Constructivism has implications for teaching, for example, that emphasis should be on creating the best learning environment for fostering the generation and construction of ideas, and this would frequently be an environment that encouraged activity. With younger students, this activity might often involve the use of manipulatives, but with older students the activity might be entirely mental. At any age, interaction through discussion among students would be encouraged, partly because the exchange of ideas frequently fosters learning better than does working alone and partly perhaps because the nonexistence of an objective world of knowledge suggests the social necessity of aiming for consensus. Here we see the influence of Lev Vygotsky (1978), as well as Piaget, on the development of constructivism. Exposition by the teacher would be regarded with suspicion by constructivists, on the grounds that it is often ineffective. However, the environment arranged by the teacher might involve discussion with the teacher and might involve the students carrying out a wide variety of tasks suggested by the teacher. It is important to emphasize that constructivist teaching is likely to make heavy demands on the teacher, who already has many burdens to cope with, particularly in inner-city schools, because of the constant need to provide the best learning environment for all students at all times. It is not controversial to maintain that, in the last resort, we all make sense of the world ourselves, for we are all doing it all the time, so in this regard some might say that constructivism tells us little that is new. But constructivism raises some important issues, including the role of a constructivist teacher in the classroom, the role and nature of discussions among pupils, and

how to provide help to busy teachers who might wish to develop alternative teaching methods. At a deeper level, it raises issues such as the nature of learning and of knowledge itself. A much fuller consideration of constructivism will be found in Davis, Maher, and Noddings (1990).

## PROBLEM SOLVING

There are three reasons why it is appropriate to conclude with a brief mention of problem solving. One reason is the current emphasis on problem-solving activities in mathematics education (National Council of Teachers of Mathematics 1989), another is that it allows us to connect information processing with constructivism, and the third is because it provides another means of comparing and judging the various learning theories. Problem solving is generally regarded as the pinnacle in any hierarchical classification of types of mathematics learning (Gagné 1970). By "problem solving" is meant the incorporation of novel problems or situations that require students to use previously acquired knowledge and expertise in an intelligent and insightful way in order to arrive at a solution or conclusion. It does not mean the typical practice problems and exercises that have always formed an element of mathematics textbooks. From the point of view of information processing, novel problems have always formed the raw material that students were expected to work on so that they would inevitably reveal their thought processes and provide information about the nature of their thinking. From the point of view of constructivism, in solving a novel problem students clearly must construct or develop meaning, and in a modest way they have to be creative. Problem solving would therefore seem to be implicit within a constructivist approach to teaching. In relation to theories of learning referred to earlier, problem solving formed an essential feature of Gestalt theory, and discovery implies the solution of whatever problems the learner faces, however simple they might be. Despite the work of Gagné (1970), however, behaviorism does not appear to encourage or perhaps even allow a problem-solving approach to learning mathematics.

Within current curriculum recommendations, it seems that teachers are being exhorted to use a problem-solving approach to mathematics teaching because it enables students to learn something of what it is to be a mathematician as much as because of contemporary views on learning. Whether a problem-solving approach to mathematics will be successful for all pupils, whether it will lead to better motivation or enhanced relational understanding or longer retention, remains to be seen. Given the difficulties in learning mathematics that weaker pupils have always experienced in the past, it would not be surprising if many experienced mathematics teachers are skeptical. However, advocates of the greater use of problem-solving approaches in learning mathematics do not necessarily claim that all pupils will benefit from the same problems at the same age. It is therefore essential that student performance and behavior continue to be monitored so that benefits and drawbacks can be measured and evaluated in relation to particular pupils and particular kinds of problems.

See also Constructivism; Learning Disabilities, Overview; Mathematics Anxiety; Problem Solving, Overview; Readiness; Spiral Learning

## SELECTED REFERENCES

Ausubel, David P., Joseph D. Novak, and Helen Hanesian. *Educational Psychology: A Cognitive View.* 2nd ed. New York: Holt, Rinehart and Winston, 1978.

Bigge, Morris L. *Learning Theories for Teachers.* 4th ed. New York: Harper and Row, 1982.

Bruner, Jerome S. *Beyond the Information Given.* London, England: Allen and Unwin, 1973.

Cockcroft, Wilfred H. *Mathematics Counts.* London, England: Her Majesty's Stationery Office, 1982.

Cormier, Stephen M., and Joseph D. Hagman, eds. *Transfer of Learning: Contemporary Research and Applications.* Orlando, FL: Academic, 1987.

Davis, Robert B., Carolyn A. Maher, and Nel Noddings. *Constructivist Views on the Teaching and Learning of Mathematics.* Reston, VA: National Council of Teachers of Mathematics, 1990.

Gagné, Robert M. *The Conditions of Learning.* 2nd ed. New York: Holt, Rinehart and Winston, 1970.

Klausmeier, Herbert J. *Educational Psychology.* 5th ed. New York: Harper and Row, 1985.

Lerman, Stephen. "Constructivism, Mathematics and Mathematics Education." *Educational Studies in Mathematics* 20(May 1989):211–223.

National Council of Teachers of Mathematics. *Curriculum and Evaluation Standards for School Mathematics.* Reston, VA: The Council, 1989.

———. *Principles and Standards for School Mathematics.* Reston, VA: The Council, 2000.

Orton, Anthony. *Learning Mathematics: Issues, Theory and Classroom Practice.* 2nd ed. London, England: Cassell, 1992.

Resnick, Lauren B., and Wendy W. Ford. *The Psychology of Mathematics for Instruction.* Hillsdale, NJ: Erlbaum, 1984.

Shulman, Lee S. "Psychology and Mathematics Education." In *Mathematics Education* (pp. 23–71). 69th Yearbook. National Society for the Study of Education. Pt. I.

Edward G. Begle, ed. Chicago, IL: University of Chicago Press, 1970.

Skemp, Richard R. "Relational Understanding and Instrumental Understanding." *Mathematics Teaching* 77(Dec. 1976):20–26.

Slavin, Robert E. *Educational Psychology: Theory and Practice.* 3rd ed. Boston, MA: Allyn and Bacon, 1991.

Vygotsky, Lev S. *Mind in Society: The Development of Higher Psychological Processes.* Michael Cole, et al., eds. Cambridge, MA: Harvard University Press, 1978.

ANTHONY ORTON

# PYTHAGORAS (ca. 580 B.C.–ca. 500 B.C.)

A Greek mathematician, philosopher, and astrologer. Few details are known of his life other than that he was born on the island of Samos and spent several years visiting Egypt and the Orient before returning home. It is believed that he subsequently migrated to Crotona, a Greek city in southern Italy, where he founded a society of individuals devoted to the study of mathematics, mysticism, philosophy, astrology, music, and the natural sciences. This brotherhood, known collectively as the Pythagoreans, contributed significantly to the body of mathematical knowledge associated with Greek antiquity. Central to the teachings of Pythagoras and his followers was the concept of number. The Pythagoreans used number to represent ideas, attributes, and even geometric entities. For example, they believed that 1 represents male, 2 represents female, and 3 represents marriage, the union of male and female. Four represented justice as evidenced by our modern-day reference to a fair and equitable situation as being a "square deal." The group is credited with developing a theory of proportion and discovering that the length of a diagonal in a unit square is an incommensurable quantity (irrational). It is commonly believed that the Pythagoreans were the primary movers in promoting the notion of geometry as a deductive science as opposed to the pre-Hellenic view of geometry as a collection of practical rules of thumb to solve mensurational problems. The Pythagoreans are also credited with discovering that musical harmonies depend on numerical ratios. This discovery subsequently inspired them to search for a similar "harmony of the spheres" to describe the motion of the planets. The Pythagoreans truly lived by their motto that "all is number."

See also Greek Mathematics; History of Mathematics, Overview; Music; Pythagorean Theorem

SELECTED REFERENCES

Baumgart, John K., et al. *Historical Topics for the Mathematics Classroom.* 2nd ed. Reston VA: National Council of Teachers of Mathematics, 1989.

Stillwell, John. *Mathematics and Its History.* New York: Springer-Verlag, 1989.

Swetz, Frank, ed. *From Five Fingers to Infinity: A Journey Through the History of Mathematics.* Chicago, IL: Open Court, 1994.

ANTHONY V. PICCOLINO

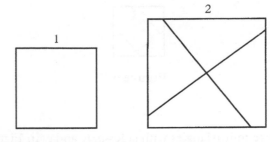

**Figure 2**

# PYTHAGOREAN THEOREM

Attributed to a group of Greek scholars known as the Pythagoreans and headed by Pythagoras (ca. 580 B.C.–ca. 500 B.C.). As a demonstration of the theorem, copy or trace and cut out the squares numbered 1 and 2 in Figure 2 and verify that they are the same size as squares 1 and 2 of Figure 1. Now cut square 2 on the lines into 4 pieces and see if you can arrange square 1 and the pieces of square 2 together in such a way as to cover square 3 exactly. (For the solution, see Figure 6.) This suggests that the sum of the areas of squares 1 and 2 is equal to the area of square 3. This is an instance of the Pythagorean Theorem.

The Pythagoreans saw that if they constructed squares on each of the sides of a right triangle, the sum of the areas of the squares on the two shorter sides (the legs) was equal to the area of the square on the longer side (the hypotenuse), see Figure 3. Formally, the theorem states:

In any right triangle, the sum of the squares of the two legs, $a^2 + b^2$, equals the square of the hypotenuse, $c^2$:

$$a^2 + b^2 = c^2$$

Here is a simple but elegant geometric proof of the Pythagorean Theorem: We are given two squares, each having length $a + b$ on a side and therefore of equal area. Each of the squares has been divided into squares and right triangles in a different fashion, as shown in Figures 4 and 5. From each of the squares

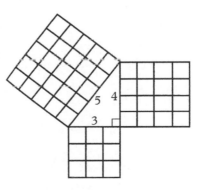

**Figure 3**   $3^2 + 4^2 = 25$
$9 + 16 = 25$

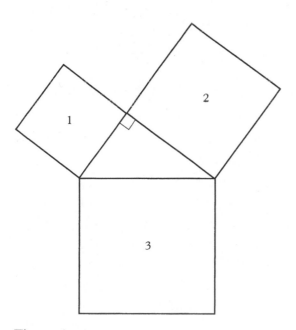

**Figure 1**

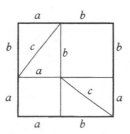

**Figure 4**

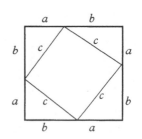

**Figure 5**

**Figure 6**

remove four triangles with sides *a, b,* and *c.* In Figure 4, the area that remains consists of two squares, one of area $a^2$ and the other $b^2$. In Figure 5, a square of area $c^2$ remains. We conclude that

$$a^2 + b^2 = c^2$$

where *a,b,* and *c* are the sides of a right triangle.

*Example 1:* Find the unknown side in a right triangle if a leg is 6 and the hypotenuse is 8.

*Solution:*

$$a^2 + b^2 = c^2$$

$$a^2 + 6^2 = 8^2$$

$$a^2 = 28$$

$$a = 2\sqrt{7} \approx 5.29$$

*Example 2:* If in triangle *ABC,* $a = 6$, $b = 8$, and $c = 10$, is the triangle a right triangle?

*Solution:* The converse of the Pythagorean Theorem is also true, that is, $c^2 = a^2 + b^2$ is true only for a right triangle.

$$10^2 = 6^2 + 8^2 \, ?$$

$$100 = 100$$

The triangle *is* a right triangle.

*See also* Pythagoras

## SELECTED REFERENCE

Fidkin, David. *An Introduction to the Pythagorean Theorem.* New York: Vantage, 1994.

DEREK BLOOMFIELD

# Q

## QUALITY CONTROL

Statistical quality control originated with the work of Walter Shewhart of Bell Laboratories in the 1920s. Although utilized in the United States and abroad, statistical quality control did not come into general use in the United States until World War II, when it became widely adopted in an effort to produce massive quantities of war materiel to exacting specifications. Following World War II, Japanese industry embarked upon a remarkable period of transformation. W. Edwards Deming, of the United States, is the best known of those who introduced statistical quality control to Japan, as well as bringing its industry a new theory of management that focuses upon quality. The *quality* of a product refers to some property such as the measurement of the outside diameter of an automobile piston ring, the weight of a bronze casting, the breaking strength of a piece of parachute cord, or the number of ounces of soda pop dispensed by a machine. Two apparently identical parts made from the same batch of raw materials, under controlled conditions, and produced only seconds apart, can nevertheless be different. These differences occur because it is seldom possible to duplicate the exact conditions existing at a given time, no matter how great an effort is made. Thus, any manufacturing process, however good, is charac-terized by a certain amount of variability that is of the same chance or random nature as the variations we might find between repeated rolls of a pair of dice.

Statistical control is usually achieved in a process by finding and eliminating another kind of variability called *assignable variation*. Under this heading we include variations in a process due to poorly trained operators, substandard raw materials, faulty machine settings, worn parts, and the like. Control charts are used to detect assignable variation. A control chart contains three horizontal lines: a central line to indicate the desired standard, an upper control limit, and a lower control limit. By plotting results from small samples that are taken periodically (say, each half-hour or hour), it is possible to check by such a chart whether the variation among samples may be attributable to chance or whether trouble in the form of assignable variation has entered the process. Specifically, if the sample data plotted on the chart falls above the upper control limit or below the lower control limit, one looks for trouble. Otherwise, the process is left alone.

Occasionally, however, specific configurations of the plotted points may also reveal trouble, even though these points fall between the upper and the lower control limits. Mean charts, median charts, and their corresponding range charts are commonly used control charts. Control charts are almost always constructed so that if a production process is in control (random variation only) the sequentially plotted samples fall between the control limits approximately 997 times out of 1,000. Thus, if a plotted point falls outside the control limits, we reject the premise that the variability is random and search for troublesome assignable variation. This procedure necessarily runs the risk of occasionally (about 3 times out of 1,000) looking for nonexistent trouble. The control chart tells us nothing about the risk of assuming that the process is under control when it is not. This type of analysis can be performed with operating characteristic curves, which we will not discuss here except to note that they are rarely used in practical production problems. In the control chart shown in Figure 1, sample 9 lies below the lower control limit and sample 10 lies above the upper control limit. These samples provide strong evidence that the process is out of control, and that production personnel should search for sources of assignable variation.

Acceptance sampling is a method of quality control that is used where parts or other inventory are purchased from a supplier, rather than manufactured. One method of checking whether a shipment conforms to specifications is 100% inspection, which is called *screening*. Screening is often impractical be-cause of its high cost or because it is impossible (for instance, if the testing is destructive). An assortment of sampling plans are available for utilization, and double and multiple sampling plans can be used to decrease the cost. In all sampling plans, there is a probability of rejecting a good lot (producer's risk) or accepting a bad lot (consumer's risk). If the producer and consumer specify these probabilities as well as the proportion of defectives above which a lot is considered to be bad, and the proportion of defectives below which a lot is considered to be good, it is possible to construct inspection plans that will meet these requirements. Multiple sampling plans produce the greatest savings because they begin with small samples and add further samples until a decision is reached. A multiple sampling plan might be presented as in Figure 2. Beginning with the first sample of size 25, we reject the lot if the number of defectives is 3 or more but do not accept the shipment even if there are no defects. If the lot has not been rejected, we proceed to the second sample of 25, combining the first and second samples. If the number of defectives is 1 or less, we accept the shipment, and if the number of defectives is 4 or more, we reject the shipment. If the number of defectives is 2 or 3, we continue with sample 3 and subsequent samples, if necessary, until the shipment is either accepted or rejected.

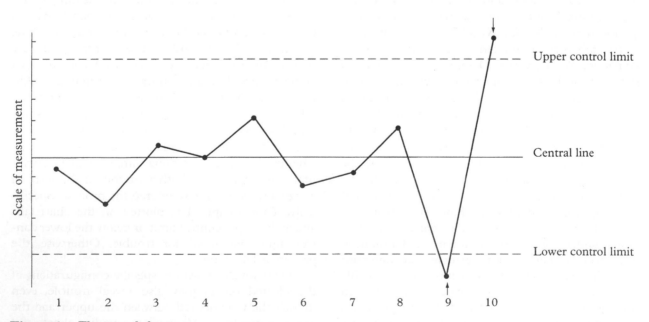

**Figure 1** *The control chart.*

**Figure 2** *Multiple sampling plan.*

| | | Combined samples | | |
|--------|----------------|------|----------------------|---------------------|
| Sample | Sample size | Size | Acceptance number | Rejection number |
| First | 25 | 25 | | 3 |
| Second | 25 | 50 | 1 | 4 |
| Third | 25 | 75 | 2 | 5 |
| Fourth | 25 | 100 | 3 | 6 |
| Fifth | 25 | 125 | 5 | 7 |
| Sixth | 25 | 150 | 6 | 7 |

*See also* Mean, Arithmetic; Median; Normal Distribution Curve; Standard Deviation; Statistics, Overview

### SELECTED REFERENCES

Deming, W. Edwards. *Out of the Crisis.* Cambridge, MA: Massachusetts Institute of Technology Center of Advanced Engineering Study, 1986.

Duncan, Acheson J. *Quality Control and Industrial Statistics.* 5th ed. Homewood, IL: Irwin, 1986.

Freund, John E., Frank J. Williams, and Benjamin M. Perles. *Elementary Business Statistics: The Modern Approach.* 6th ed. Englewood Cliffs, NJ: Prentice-Hall, 1993.

Miller, Irwin, and Marylees Miller. *Statistical Methods for Quality: With Applications to Engineering and Management.* Englewood Cliffs, NJ: Prentice-Hall, 1995.

BENJAMIN M. PERLES

# QUESTIONING

Critical to how students develop thinking skills. Knowing the correct questioning technique allows for the continuation of quality classroom discussion. The questions the classroom teacher asks his or her students help the students gather further information about the answer to the problem. Some teachers rarely give their students quality questions or the answers to quality problems, since once the answer is given, the students stop trying to solve the problem. Although good questioning techniques have been important to teachers at least since the time of Socrates, it was not until the late 1950s (when a team of researchers led by Benjamin Bloom developed a taxonomy, or classification system, of educational objectives) that the systematic organization of questioning began.

Myra and David Sadker (1990) investigated the state of questioning skills in our schools. From their review of a major study of classroom behavior, they reached the following conclusions:

1. 80 percent of classroom talk was devoted to asking, answering, or reacting to questions.
2. The state of questioning techniques in the classroom has remained basically unchanged for many years.
3. Teachers ask numerous questions. Primary school teachers ask $3\frac{1}{2}$ to $6\frac{1}{2}$ questions per minute. Elementary school teachers average 348 questions per day.
4. Although teachers ask a large number of questions, they generally show little tolerance in waiting for student replies. Typically, only one second passes between the end of a question and the next verbal interaction.
5. In classes where higher-order, thought-provoking questions are asked, students perform better on achievement tests.
6. The quality and quantity of student answers increase when teachers provide students with time to think.
7. The typical student asks approximately one question per month.

When working on cooperative learning with students, it helps to be a good listener and to circulate around the classroom. Teachers who are facilitators help students arrive at solutions to problems. Do students need answers to problems? Most definitely! Appropriate questioning will generate additional questions to help students arrive at answers to their questions. It may be helpful, for example, when a teacher asks: "Can you explain this or could you make a diagram to help with your explanation?" The teacher needs to know his or her students, to know when to interject and ask a question, and when to let the students ask the question.

What are the best types of questions? Should they have an immediate answer or should they spark additional interest in the question? One way to approach questioning is to think about what types of information are necessary to gather. If the class is receiving factual information, or opinions, then what types of questions should be asked? If the class has been able to come up with a hypothesis, based on their questions, then they have generated understanding. A teacher's role is not always to give the correct answers but to allow children the time and opportunity to build up their understanding of mathematics through contextualized problems that make sense to them and that provide opportunities for the teacher to gather information on student progress toward assimilating key mathematical ideas (Moon and Schulman 1995).

Students generate questions in the classroom, if the climate will allow this to happen. The classroom is an important place for questioning to occur. Questioning should take place almost constantly in every classroom. Teachers need to know that they can be challenged by the type of questions their students ask and that they are not expected to have an answer to every question. The competent teacher knows how to phrase questions and to use them to develop the learner's academic knowledge.

The use of skillful questioning techniques as a teaching tool has several special characteristics. The first is that the questioner is able seriously to consider students not as blank slates but as the repositories of knowledge and analytical ability, capable of contributing to a conceptual and substantive dialogue (Mackey and Appleman 1988). Second, clear questions are more productive than ambiguous ones. Third, if teachers wait after raising questions, giving students time to think rather than rushing to answer their own questions, the quality of students' answers is generally better (Mackey and Appleman 1988). Fourth, when students feel free to ask questions, more high-level learning takes place. Creating a classroom climate that is accepting of student responses and having students feel free to ask questions are critical to successful questioning techniques. Appropriate questioning techniques are important in the development of thinking skills. As John Dewey, the "father of progressive education," pointed out, questioning is an important form of thinking.

Developing good questioning techniques requires continuous self-evaluation. Improving questioning techniques can provide a more enjoyable teaching experience and a more enhanced learning experience to the students in the classroom today and in the workplace tomorrow. Teachers must develop an awareness of and be sensitive to how questions are framed. Teachers need to guide their students in developing good questioning techniques also since they will be the future problem solvers of the world. Questions should focus, guide, and facilitate learner thinking. A key to good teaching is the posing of the right question at the right time so that students will think. One of the most important aspects of the role of a teacher is to orchestrate the discourse in the classroom. The classroom teacher needs to pose interesting questions, ask students to clarify their ideas, monitor student participation, and decide when the teacher's input is needed (Moon and Schulman 1995).

*See also* Homework; Problem Solving, Overview; Projects for Students

## SELECTED REFERENCES
Dewey, John. *Democracy and Education.* New York: Macmillan, 1944.

Mackey, James, and Deborah Appleman. "Questioning Skill." In *Guide to Classroom Teaching* (pp. 146–165). Robert McNergney, et al. eds. Boston, MA: Allyn and Bacon, 1988.

Moon, Jean, and Linda Schulman. *Finding the Connections: Linking Assessment, Instruction, and Curriculum in Elementary Mathematics.* Portsmouth, NH: Heinemann, 1995.

Sadker, Myra, and David Sadker. "Questioning Skills." In *Classroom Teaching Skills* (pp. 112–148). James M. Cooper, ed. 4th ed. Lexington, MA: Heath, 1990.

BETTY M. ERICKSON

# QUEUEING THEORY

A queue is a waiting line. Queueing theory basically studies congestion in queues. Queueing problems include the following standard elements: the pattern in which customers arrive, the law governing the duration of service time, the number of available servers, and the rule by which the next customer to be served is selected—called the *queue discipline.* In the simplest models, both the time between customer arrivals and the time it takes to complete service are assumed to have a negative exponential distribution, there is one server, and the queue discipline is first-come, first-served. The characteristics of the system that are of principal interest include distributions of the number of customers in the queue, of the waiting time before service begins, and of the total time in the system.

The original impetus for queueing theory came from the telephone industry, with Agner K. Erlang in Denmark publishing the first major study in 1917. Nowadays, queueing is applied throughout business. There are waiting lines of customers in banks, supermarkets, and airports; cars must be kept moving past traffic lights and through tunnels; trucks wait to use loading docks; and both people and computers wait to use other computers. The problem is to provide an acceptable quality of service to customers at the lowest cost. The mathematical setting for the study of queues is probability theory. The derivation of most of the formulas requires calculus, and it adds greatly to the understanding of queues to be able to simulate them on a computer.

*See also* Modeling; Operations Research; Probability, Overview

## SELECTED REFERENCES

Gross, D., and C. M. Harris. *Fundamentals of Queueing Theory.* New York: Wiley, 1974.

Sloyer, Clifford, Wayne Copes, William Sacco, and Robert Stark. *Queues: Will This Wait Never End!* Providence, RI: Janson, 1987.

Takacs, Lajos. *Introduction to the Theory of Queues.* New York: Oxford University Press, 1962.

HENRY O. POLLAK

# R

## RAMANUJAN, SRINIVASA (1887–1920)

Important contributor to number theory. Born in southern India, Ramanujan became interested in mathematics at an early age and excelled. In 1904, he began studying at the Government College, where he was brilliant in mathematics but neglected English and physiology, and as a result, lost his scholarship. By 1913, without a college degree, Ramanujan seemed to be unable to rise from a low-paying job as a clerk in the Madras Port Trust. He had continued to study mathematics and, at the urging of friends, communicated some of his work to the famous English mathematician Godfrey H. Hardy.

Hardy was overwhelmed by the depth and quality of Ramanujan's achievements. Especially surprising to Hardy was the following evaluation of the continued fraction:

$$1/(1 + e^{-2\pi}/(1 + e^{-4\pi}/(1 + e^{-6\pi}/$$

$$(1 + \cdots )))) = \left( \sqrt{\frac{5 + \sqrt{5}}{2}} - \frac{\sqrt{5} + 1}{2} \right) e^{2\pi/5}$$

Hardy arranged for Ramanujan to come to England in 1914, where the two of them collaborated on path-breaking work in number theory and related topics. Perhaps their crowning achievement was the exact formula for the partition function. This formula asserts that the number of ways of writing an integer $n$ as unordered sums of positive integers is

$$\frac{1}{2\sqrt{2}} \frac{d}{dn} \left( \frac{\sinh \left( \pi \sqrt{\frac{2}{3} \left( n - \frac{1}{24} \right)} \right)}{\sqrt{n - \frac{1}{24}}} \right) + similar\ terms$$

In 1917, Ramanujan developed a consumptive illness. After convalescing in both England and India, he died back in Madras on April 26, 1920, at the age of 32. His life was chronicled in a NOVA program (Public Broadcasting Service) entitled *The Man Who Loved Numbers*.

*See also* Number Theory, Overview

### SELECTED REFERENCE

Kanigel, Robert. *The Man Who Knew Infinity: A Life of the Genius Ramanujan.* New York: Scribner's, 1991.

GEORGE E. ANDREWS

## RANDOM NUMBER

An integer that has the same probability of being selected as any other integer. In many areas of statistics, it is important to draw one or more integers at random from the sequence 0, 1, . . . , *n*. Typical values of *n* are 9, 99, . . . , 99999, but *n* may take on any value. Each of the numbers drawn must have the same probability $1/(n + 1)$ if it is to be labeled a random number. Historically, the generation of random variables dates back to the work of "Student" (W. S. Gossett), who used simulations in his work on the sampling distributions of the *t*-statistic and the correlation coefficient in 1908.

The generation of random numbers is equivalent to obtaining observations from the uniform probability distribution on the unit interval (0,1). This is now achieved rapidly by an arithmetic algorithm on a computer. The algorithm generates pseudorandom numbers by a deterministic mechanism approximating randomness. For example, Lehmer's linear congruential generator may be used (Knuth 1981). Several tables of random numbers generated in this way have been published, among them one by the RAND Corporation. Tests have been devised to check that these provide a reasonable approximation to exact randomness. Random numbers are often used to generate observations from a wide range of nonuniform probability distributions.

To illustrate a simple use of random numbers, consider the following example. Four children must be selected at random for a relay team from a group of 10, numbered 0, 1, . . . , 9. We use the excerpt from Abramowitz and Stegun's (1964) Table 26.11 of random numbers shown here to select them.

| | |
|---|---|
| 53479 | 81115 |
| 97344 | 70328 |
| 66023 | 38277 |
| 99776 | 75723 |
| 30176 | 48979 |

One may read the numbers across a row, or down a column; since the numbers are random, either direction is acceptable. Reading across the first row, the selection would be children 5, 3, 4, 7; reading down the first column it would be 5, 9, 6, 3, the duplicate 9 being neglected. Either selection is random.

*See also* Probability, Overview; Probability Distribution; Random Sample; Statistics, Overview

### SELECTED REFERENCES

Abramowitz, Milton, and Irene A. Stegun. *Handbook of Mathematical Functions.* Washington, DC: Government Printing Office, 1964.

Knuth, Donald E. *The Art of Computer Programming.* Vol. 2: *Seminumerical Algorithms.* 2nd ed. Reading, MA: Addison-Wesley, 1981.

Student. "The Probable Error of a Mean." *Biometrika* 6(1908):1–25.
———. "Probable Error of a Correlation Coefficient." *Biometrika* 6(1908):302–310.

JOSEPH GANI

## RANDOM SAMPLE

A sample that, in its selection, gives each element, measurement, or observation an equal chance of being selected. When feasible, random samples are preferred to other kinds of samples because they permit valid, or logical, generalizations.

For a more formal definition, we must distinguish between samples from finite and infinite *populations*. A finite population of size $N$ is a set of $N$ objects, and a sample of size $n$ from such a population is a subset of size $n$. Since there are $C(N,n) = N(N - 1) . . . (N - n + 1)/n!$ possible samples of size $n$ from a finite population of size $N$, such a sample is said to be random, if it is selected in a manner giving each possible sample the probability $1/C(N,n)$. For instance, if a finite population consists of the $N = 5$ elements $a$, $b$, $c$, $d$, and $e$ (which might represent the incomes of five persons), there are the $C(5,3) = 5 \cdot 4 \cdot 3/3! = 10$ different samples of size $n = 3$: *abc, abd, abe, acd, ace, ade, bcd, bce, bde,* and *cde*. If one of these samples is chosen in such a way that each one has the probability $1/C(5,3) = 1/10$ of being chosen, we can call it a random sample. In a situation like this, we could write each of the possible samples on a slip of paper, shuffle them thoroughly (say, in a hat), and then draw one without looking. In more complex situations, we can make use of a table of random numbers, a preprogrammed calculator, or special computer software.

Infinite populations arise when we deal with continuous random variables, countably infinite random variables (having as many possible values as there are whole numbers), or when we sample with replacement from a finite population. In that case, a random sample of size $n$ consists of the values of $n$ independent identically distributed random variables, and their common distribution is referred to as the *infinite population sampled*. It is customary to apply the term *random sample* also to the random variables themselves. For instance, if we get 2, 5, 1, 3, 6, 4, 4, 5, 2, 4, 1, and 2 in twelve rolls of a die, these numbers constitute a random sample if they are values of independent random variables having the same probability distribution:

$$f(x) = 1/6 \qquad \text{for } x = 1, 2, 3, 4, 5, \text{ or } 6$$

In practice, it can be difficult to judge whether data like these can actually be looked upon as a random sample.

*See also* Probability, Overview; Probability Distribution; Random Number; Statistics, Overview

### SELECTED REFERENCES

RAND Corporation. *A Million Random Digits with 100,000 Normal Deviates.* New York: Free Press, 1995.

Slonin, Morris J. *Sampling in a Nutshell.* New York: Simon and Schuster, 1973.

Williams, William H. *A Sampler on Sampling.* New York: Wiley, 1978.

JOHN E. FREUND

# RANDOM VARIABLE

In simplest terms, the modern concept, first found in Kolmogorov's monograph (1933), of a real-valued function defined on the points $\omega$ of a sample space $\Omega$, where these points represent the possible outcomes of some experiment. To each $\omega$ is attached a probability, so that for a discrete random variable (r.v.), namely one taking a set of discrete values $\{x_k; k = 0, 1, \ldots\}$, we have

$$P\{X(\omega) = x_k\} = p_k \geq 0, \quad \sum_k p_k = 1$$

and $\{p_k\}$ is known as the probability distribution of $X$. An example of this is the r.v. $X$ associated with the throw of a single six-sided die, for which the sample space is $\Omega = \{1,2,3,4,5,6\}$ and

$$P\{X = x_k\} = 1/6, \quad x_k = k = 1,2, \ldots, 6$$

Other commonly used discrete distributions are the binomial

$$p_k = \binom{n}{k} p^k (1 - p)^{n-k}, \quad 0 < p < 1,$$
$$k = 0,1, \ldots, n \text{ (a positive integer)}$$

and the Poisson

$$p_k = e^{-\lambda} \lambda^k/k!, \quad \lambda > 0, k = 0,1,2, \ldots$$

Tables of this last distribution for $0.02 \leq \lambda \leq 15$ can be found in Lindgren (1993).

A continuous r.v. $X$ takes values over the continuous range $a < x < b$, where $a$ may equal $-\infty$ and $b$ may equal $+\infty$. In this case, one can only make a probability statement about $X$ lying between two values $x_0 < x_1$, so that

$$P\{x_0 < X \leq x_1\} = F(x_1) - F(x_0)$$

Here, $F(x) = P\{X \leq x\}$, the distribution function (d.f.) of $X$, is nondecreasing, right continuous, and with at most a countable set of discontinuities. For details of the classical theory, the reader should consult a standard text on probability such as Loève (1977) or Grimmett and Stirzacker (1982).

Most continuous distributions used in practice are absolutely continuous functions, so that

$$F(x) = \int_{-\infty}^{x} f(u)\,du$$

where $f(x) \geq 0$ is defined as the probability density function (p.d.f.) of $X$. In this case

$$P(x_0 < X \leq x_1) = F(x_1) - F(x_0) = \int_{x_0}^{x_1} f(u)\,du$$

A common p.d.f. is

$$f(x) = \lambda e^{-\lambda x}, \quad x \geq 0, \lambda > 0$$

which defines a negative exponential r.v.; another is

$$f(x) = \frac{1}{\sqrt{2\pi}} e^{-(x^2/2)}, \quad -\infty < x < \infty$$

which defines the standard normal r.v. with mean 0 and standard deviation 1.

The graphs in Figure 1 illustrate both the p.d.f. $f(x)$ and the d.f. $F(x)$ of the standard normal r.v. $X$

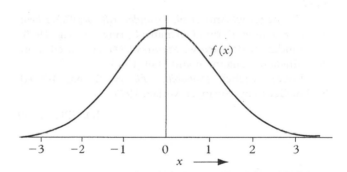

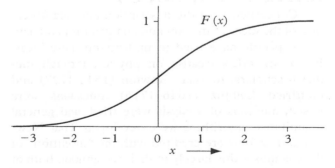

**Figure 1** *Graphs of $f(x) = \dfrac{1}{\sqrt{2\pi}} e^{-(1/2)x^2}$ and $F(x) = \displaystyle\int_{-\infty}^{x} f(u)\,du$.*

with mean 0 and standard deviation 1. The probabilities

$$F(x) = \frac{1}{\sqrt{2\pi}} \int_{-\infty}^{x} e^{-(x^2/2)} dx$$

are tabulated in Lindgren (1993) as well as in most other elementary texts. We provide an example to illustrate the use of these tables. The height of men $Y$ in a certain population is normally distributed with mean $\mu = 69$ in. and standard deviation $\sigma = 6$ in. What is the probability that a man selected at random has a height between 60 and 75 in.? First, we must reduce the normal r.v. $Y$ to a standard normal r.v. $X$ with mean 0 and standard deviation 1; this is done by the transformation

$$X = (Y - \mu)/\sigma = (Y - 69)/6$$

Thus, $Y = 60$ corresponds to $X = -1.5$, and $Y = 75$ to $X = 1$. Hence, $P\{60 < Y \le 75\} = P\{-1.5 < X \le 1\} = F(1) - F(-1.5) = 0.8413 - 0.0668 = 0.7745$.

*See also* Normal Distribution Curve; Probability, Overview; Probability Distribution; Statistics, Overview

## SELECTED REFERENCES

Grimmett, Geoffrey R., and David R. Stirzacker. *Probability and Random Processes.* Oxford, England: Clarendon, 1982.

Kolmogorov, Andrei N. *Grundbegriffe der Wahrscheinlichkeitsrechnung.* Berlin, Germany: Springer-Verlag, 1933.

Lindgren, Bernard W. *Statistical Theory.* 4th ed. London, England: Chapman and Hall, 1993.

Loève, Michel. *Probability Theory.* 2 vols. 4th ed. Berlin, Germany: Springer-Verlag, 1977.

JOSEPH GANI

# RATES IN CALCULUS

Concerned with the description and quantification of the ideas of motion and change. From ancient times, people have tried to understand these ideas. From Aristotle's treatise on physics, through medieval scholars, to Isaac Newton (1642–1727) and Gottfried Leibniz (1646–1716), questions were raised, methods of analysis were tried, and general quantitative algorithms were developed. That Newton and Leibniz are credited with the "invention" of the calculus is due largely to their recognition both of the significance of a particular quantitative way of studying change and of the inverse relationship between rate problems and area problems (the Fundamental Theorem of Calculus). Newton and Leibniz took what is essentially a geometric approach to studying change. It was left to the eighteenth-century mathematical community, led by Leonhard Euler (1707–1783), to introduce the modern concept of "function."

In calculus, the derivative of a function is a rate of change. Statements made about the rate of change of a quantity are often statements about derivatives. For example, the velocity of a car is the rate of change of the car's position with respect to time. Consider a car traveling at a nonconstant speed, whose distance from a starting point is given by the function $d = s(t)$. We can calculate the car's average velocity over various time intervals using the formula

$$\text{average velocity} = \frac{\text{(distance traveled)}}{\text{(time elapsed)}}$$

The question arises, can we find a meaningful way of quantifying the more elusive notion of velocity at an instant in time. To approximate the instantaneous velocity at some time $t_0$, we compute the average velocity for progressively shorter time intervals near $t_0$. If these approximations are approaching some limiting value, we define that limit to be the instantaneous velocity, $v(t_0)$. This velocity is the derivative, $s'(t_0)$, of the position function. In more formal terms, at any time $t$,

$$v(t) = s'(t) = \lim_{h \to 0} \frac{s(t + h) - s(t)}{h}$$

Geometrically, the derivative (velocity) at time $t$ is the slope of the tangent line to the position curve at time $t$.

The same process applies to finding the (instantaneous) rate of change of any quantity. That is, if $f$ is a function of $x$, and $h$ represents any small change in the value of $x$, then

$$f'(x) = \lim_{h \to 0} \frac{f(x + h) - f(x)}{h}$$

is defined to be the instantaneous rate of change of $f$ with respect to $x$, provided the limit exists. Thus, the derivative of a velocity function is the rate of change of velocity with respect to time, or acceleration. And, since the velocity, $v$, is the derivative of the position function then $(v(t) = s'(t))$, this means that the acceleration is the derivative of the derivative of the position function, or the second derivative, $s''(t)$. Note that the independent variable does not necessarily have to represent time. For example, if the function $a(x)$ represents the altitude in feet above sea level of a river as a function of the distance, $x$, in miles measured from the river's source, then $a'(x)$ is the rate at which the altitude is changing, measured in feet per mile. (Observe that, for the water in the river to flow, the value of $a'(x)$ could never be positive!) Similarly, if $p = g(s)$ represents the price of a commodity as a function of the number of units, $s$, supplied, then the derivative, $g'(s)$, represents the rate of change of price with respect to supply (i.e., the change in price that results from one unit increase in the supply).

So calculus, through derivatives, allows us to model rates of change of all sorts. Given enough information about a function, we can compute or approximate the rate of change of the function by computing or approximating its derivative. The opposite is also true. In many situations, we may have information about the rate of change of a quantity but not know how to describe the quantity explicitly. Calculus allows us to derive properties of the function from information about its derivative. If we can write an equation that is satisfied by the derivative of a function, we have what is known as a differential equation. Solving the differential equation (and possibly using other information about the function) will tell us what function describes the quantity. For example, knowing that the rate of growth of a population is proportional to the size of the population allows us to write the differential equation $f'(t) = kf(t)$. The solution of this differential equation is a family of exponential functions. Thus, we learn that the population is growing exponentially. If we also know the size of the population at two different times, we can derive the exact exponential function that describes the population.

*See also* Calculus, Overview

### SELECTED REFERENCES

Edwards, Charles H., Jr. *The Historical Development of the Calculus.* New York: Springer-Verlag, 1979.

Hughes-Hallett, Deborah, Andrew M. Gleason, et al. *Calculus.* New York: Wiley, 1992.

Ithaca College Calculus Group. *Calculus, An Active Approach with Projects.* New York: Wiley, 1993.

Strang, Gilbert. *Calculus.* Wellesley, MA: Wellesley Cambridge Press, 1991.

DIANE DRISCOLL SCHWARTZ

# RATIONAL NUMBERS

Numbers that can be expressed as $a/b$, where $a$ and $b$ are integers and $b$ does not equal zero. Some examples of rational numbers are: $\frac{4}{2}, \frac{-2}{1}, \frac{2}{3}$, and $\frac{19}{5}$. The rational numbers form a quotient field, which means that for both addition and multiplication, the set is closed, contains an identity, and inverses (except for zero in multiplication); each operation is associative and commutative; and the distributive property for multiplication over addition holds.

The rational numbers are dense, i.e., between any two rational numbers, no matter how close they are, there is an infinite number of rational numbers. For example, between $\frac{1}{4}$ and $\frac{1}{2}$ there is an infinite number of rational numbers. One way to explore this property is to select a rational number between $\frac{1}{4}$ and $\frac{1}{2}$, say $\frac{3}{8}$. Next, select a rational number between $\frac{3}{8}$ and $\frac{1}{2}$, say $\frac{7}{16}$. Continue in this way, narrowing the interval between the last selected rational number and $\frac{1}{2}$, until convinced that this process could continue indefinitely. Then repeat the process for two different rational numbers. Another property of rational numbers is that each number can be represented by an infinite number of fractions; these are said to constitute a class of equivalent fractions. The following fractions are equivalent and represent one rational number: $\frac{3}{4}, \frac{6}{8}, \frac{9}{12}, \ldots, \frac{3n}{4n}$, and $\frac{-3}{-4}, \frac{-6}{-8}, \frac{-9}{-12}, \ldots, \frac{-3n}{-4n}$, where $n$ is a counting number.

The rational numbers can be symbolized using either fractions or decimals. Whereas all fractions written in the form of a ratio of two integers, where the denominator does not equal zero, are rational numbers, not all decimals are rational numbers, since irrational numbers can also be written as nonrepeating decimals. Only those decimals that eventually terminate or repeat are rational numbers. Examples of rational numbers that terminate are $\frac{1}{2} = 0.5$, $\frac{3}{8} = 0.375$, and $\frac{32}{25} = 1.28$; some that repeat are $\frac{2}{3} = 0.6666\ldots = 0.\overline{6}$, $\frac{7}{6} = 1.1666\ldots = 1.1\overline{6}$, and $\frac{3}{7} = 0.428571428\ldots = 0.\overline{428571}$. To determine whether or not a rational number can be written as a terminating decimal, find the prime factors of its denominator when expressed as a fraction. When all of the prime factors are 2s and/or 5s the decimal terminates. Thus, in the preceding examples, since $8 = 2 \times 2 \times 2$ and $25 = 5 \times 5$, the numbers $\frac{1}{2}$, $\frac{3}{8}$, and $\frac{32}{25}$ terminate; whereas $6 = 2 \times 3$ and thus $\frac{2}{3}$, $\frac{7}{6}$, and $\frac{3}{7}$ do not terminate.

Fractions were used by many early cultures. As early as 2000 B.C. the Babylonians used sexagesimal fractions, i.e., fractions with denominators that were powers of 60 (Mainville 1969). The Egyptian mathematical system included a symbol for $\frac{2}{3}$ and a way of writing all unit fractions as a reciprocal of a whole number. All multiple fractions, except for $\frac{2}{3}$, were written as the sum of two or more unit fractions, e.g.,

$\frac{3}{4}$ was written as $\frac{1}{2}$ $\frac{1}{4}$, indicating the sum of the two unit fractions without the use of a plus sign. The Rhind Mathematical Papyrus, written in approximately 1650 B.C., includes addition, subtraction, multiplication, and division problems involving fractions (Chace 1986).

Most of the current educational research on rational numbers has been based on Thomas Kieren's theoretical work of analyzing the rational number construct into a number of different but related subconstructs. In his 1976 paper, Kieren lists and describes, in detail, seven interpretations of rational numbers: fractions, decimal fractions, equivalence classes of fractions, ratio numbers, operators or mappings, measures or points on a number line, and elements of a quotient field. These seven interpretations were revised a number of times, in later work, into his current model, which is comprised of the four subconstructs of quotients, measures, operators, and ratios (Kieren 1993). Behr, Post, and Lesh with other colleagues have systematically investigated students' rational number concepts. Based on this research they have revised Kieren's model to include a distinct part-whole subconstruct (Behr et al. 1983) and to further refine each of Kieren's subconstructs (Behr et al. 1993). Kieren (1993) and Behr et al. (1983) claim that for students to understand rational numbers, they must understand all of the interpretations of rational numbers and the interrelationships among them.

The study of rational numbers forms a substantial part of the curriculum in grades 4 through 8, but most of the focus is on computation with fractions and decimals rather than on the development of rational number concepts. This has been documented by clinical research and teaching experiments on the development of students' rational numbers concepts. One result of this research has been that researchers have increased awareness of the complexity of rational numbers concepts. Lacking an understanding of this set of numbers, most students rely on rules and procedures when dealing with rational numbers expressed as either fractions or decimals.

One recurring result in the many research studies that have been reported is that students inappropriately apply whole-number thinking to many rational-number tasks. For example, when comparing $\frac{1}{3}$ and $\frac{1}{4}$, many students will claim that $\frac{1}{4}$ is greater since 4 is greater than 3 (Behr et al. 1993). Elementary school students have a great deal of experience with whole numbers and this has influenced their concept of what a number is. They seem to try to assimilate rational-number experiences into their whole-number schema. This tendency is aided by the fact that three numbers are involved in the fraction symbol: two integers, the numerator and the denominator, and one rational number, identified by the relationship of the two integers. Most instruction on fractions focuses on counting to find the numerator and denominator of the fraction rather than on the underlying rational-number concepts and the fact that the fraction represents "a new number." Traditionally, the denominator and the numerator are taught as counts of "the number of parts in all" and "the number of parts that are shaded or different in some way." Most students do not perceive of these part-whole models as being representations of numbers, but rather that the associated fraction names the picture or part. Kerslake concluded that one difficulty that students have with fractions is that they "found it difficult to accept the fact that a fraction is a number" (Kerslake 1986, 91). Another difficulty with this definition of a fraction leads to students' inability to model fractions representing a number greater than 1, such as $\frac{5}{4}$ and $\frac{10}{3}$. One of Mack's students explained, "Fractions are a part of a whole. . . . They're always less than one whole" (Mack 1990, 22). This belief is common among elementary and middle school students.

Researchers recommend changing the way in which we characterize the denominator and numerator of a fraction. The focus should be on the denominator indicating the name of each part into which each physical unit has been partitioned and the numerator indicating the number of these parts being considered. If one or more regions are each partitioned into four equivalent parts, then each equivalent part is, initially, identified as being a fourth of one region and then the fourths are counted: one-fourth, two-fourths, three-fourths, four-fourths, five-fourths. . . .

Another factor that is responsible for students' focusing on a whole-number interpretation of fractions is that most students are introduced to fractional symbols before they have had the opportunity to learn the fractional terms. Pothier (1981) and Larson (1986) report that the fractional terms *half* and *quarter* are practically the only fractional terms used by primary grade students when engaged in fraction tasks. Prior to introducing students to fractional symbols, they need to have many opportunities for using fractional terms in real-world situations, such as cooking, crafts, and measuring activities. Only after students have developed a meaningful lexicon of fractional terms, such as *one-third, two-thirds, one-fourth, three-fourths, one-fifth, six-fifths,* etc., are they ready for the mathematical symbolism.

Pothier and Sawada (1983) describe the development of partitioning skills in students in grades K–3. They emphasize that partitioning is basic to the development of rational-number concepts. The students must be able to partition the whole to find fractional parts for themselves. Streefland (1993) describes how students' participation in fair-sharing activities helps them develop the concept of fractions as "mathematical objects." A sample activity would be for children to divide 3 pizzas fairly among 4 children. Students must also be able to do the reverse task of constructing a whole when given a fractional part. For example, when given $\frac{2}{5}$ of a rectangle, the student should be able to construct the whole rectangle.

The measure subconstruct as represented by associating rational numbers written as fractions with points on number lines is more difficult than the part-whole subconstruct for elementary and middle school students. One of the students' problems is identifying the unit (Larson, 1980; Bright et al. 1988). Students also have problems with the measure subconstruct when measuring to the eighth-inch. Larson (1991) reported that sixth graders could not consistently measure fourths and/or eighths of an inch. Students tried to remember the names of the parts rather than apply their rational-number knowledge to figure out the names of fractional parts. Students seldom counted the number of parts in an inch to determine the denominator; they either remembered the name of the part or they did not.

Rational-number concepts develop slowly over years. Premature symbolism and focus on abstract procedures interfere with students' understanding. Students need to be involved in instructional activities in which they clearly identify the unit, partition it into equivalent parts, and name and count the parts using fractional terms. Students need experience with using many different types of models (Cuisenaire rods, pattern blocks, circular pieces, number lines, rulers, sets, folded paper, etc.) to explore different interpretations of rational numbers (part-whole, division, ratio, operator, and measure) as well as different topics, such as improper fractions, ordering, equivalence, and estimation. Effective teaching activities involve real-world problems that students are interested in and are related to their current environment and culture.

*See also* Addition; Decimal (Hindu-Arabic) System; Fractions; Number Sense and Numeration

## SELECTED REFERENCES

Behr, Merlyn J., Guershon Harel, Thomas Post, and Richard Lesh. "Rational Numbers: Towards a Semantic Analysis—Emphasis on the Operator Construct." In *Rational Numbers: An Integration of Research* (pp. 13–47). Thomas P. Carpenter, Elizabeth Fennema, and Thomas A. Romberg, eds. Hillsdale, NJ: Erlbaum, 1993.

Behr, Merlyn J., Richard Lesh, Thomas R. Post, and Edward A. Silver. "Rational-Number Concepts." In *Acquisition of Mathematics Concepts and Processes* (pp. 92–127). Richard Lesh and Marsha Landau, eds. New York: Academic Press, 1983.

Bright, George W., Merlyn J. Behr, Thomas R. Post, and Ipke Wachsmuth. "Identifying Fractions on Number Lines." *Journal for Research in Mathematics Education* 19(1988):215–232.

Chace, Arnold B. *The Rhind Mathematical Papyrus.* Reston, VA: National Council of Teachers of Mathematics, 1986.

Kerslake, Daphne. *Fractions: Children's Strategies and Errors.* Windsor, England: NFER-Nelson, 1986.

Kieren, Thomas E. "On the Mathematical, Cognitive, and Instructional Foundations of Rational Numbers." In *Number and Measurement: Papers from a Research Workshop* (pp. 101–144). Richard Lesh, ed. Columbus, OH: ERIC/SMEAC, 1976.

———. "Rational and Fractional Numbers: From Quotient Fields to Recursive Understanding." In *Rational Numbers: An Integration of Research* (pp. 49–84). Thomas P. Carpenter, Elizabeth Fennema, and Thomas A. Romberg, eds. Hillsdale, NJ: Erlbaum, 1993.

Larson, Carol N. "Locating Proper Fractions on Number Lines: Effect of Length and Equivalence." *School Science and Mathematics* 80(1980):423–428.

———. "Primary Grade Students' Ability to Associate Fractional Terms and Symbols with Area and Set Models." In *Proceedings of the Eighth Annual Meeting of the North American Chapter of the International Group for the Psychology of Mathematics Education* (pp. 72–77). Glenda Lappan and Ruhama Even, eds. East Lansing, MI: Michigan State University, 1986. ERIC No. ED301443. Columbus, OH: ERIC, 1986.

——— "Sixth-graders' Knowledge of Fractional Parts of an Inch." In *Proceedings of the Thirteenth Annual Meeting of the North American Chapter of the International Group for the Psychology of Mathematics Education* (pp. 217–223). Vol. 2. Robert G. Underhill, ed. Blacksburg, VA: Virginia Tech, 1991.

Mack, Nancy K. "Learning Fractions with Understanding: Building on Informal Knowledge." *Journal for Research in Mathematics Education* 21(1990):16–32.

Mainville, Waldeck E., Jr. "Fractions." In *Historical Topics for the Mathematics Classroom* (pp. 135–137). 31st Yearbook. John K. Baumgart, et al., eds. Washington, DC: National Council of Teachers of Mathematics, 1969.

Pothier, Yvonne. "Listen, and You'll Know What I Mean." *Elements* 13(3)(1981):1–4.

———, and David Sawada. "Partitioning: The Emergence of Rational Number Ideas in Young Children." *Journal for Research in Mathematics Education* 14(1983):307–317.

Streefland, Leen. "Fractions: A Realistic Approach." In *Rational Numbers: An Integration of Research* (pp. 289–325). Thomas P. Carpenter, Elizabeth Fennema, and Thomas A. Romberg, eds. Hillsdale, NJ: Erlbaum, 1993.

CAROL NOVILLIS LARSON

# READINESS

The term describing the ability of young children to learn some fundamental concepts of mathematics. In 1960, with the publication of *The Process of Education* by Jerome S. Bruner, Jean Piaget's work became of widespread interest in mathematics education. *The Process of Education* was a report of a conference on mathematics education called "The Woods Hole Conference." A long quotation from a memorandum prepared by Bärbel Inhelder, one of Piaget's most important co-workers, defined the capacities for reason and logic of young children as follows:

> Basic notions in these fields are perfectly accessible to children of seven to ten years of age, provided they are divorced from their mathematical expressions and studied through material that the child can handle for himself (Bruner 1960, 43).

On the basis of the development of the mental operations of ordering and classification by seven years of age, the spirit was that school-aged children were ready to learn the fundamental structures of mathematics. This belief was the foundation for Bruner's famous concept of readiness to learn: "Any subject can be taught effectively in some intellectually honest form to any child at any stage of development" (Bruner 1960, 33). According to this concept, teachers could teach aspects of ordering and classification to children even before seven years of age. The key was to supply these children with material that they could handle and not require things like indirect measurement where children had to reason to compare, say, the length of two objects by using an intermediate measuring tool. After seven years of age, most children would be ready to reason in this way.

## FOUNDATION OF THE CONCEPT OF READINESS

A brief look at some essential aspects of Piaget's work helps to set a context for Bruner's idea of readiness to learn the fundamental structures of mathematics and for a discussion of a more contemporary view. In an autobiography, Piaget refers to an early insight he published in 1917 at the age of 21 that "never ceased to guide me in my variegated endeavors" (Piaget 1952, 241):

> My one idea, developed under various aspects in (alas!) twenty-two volumes, has been that intellectual operations proceed in terms of structures-of-the whole. These structures denote the kinds of equilibrium toward which evolution in its entirety is striving; at once organic, psychological and social (Piaget 1952, 256).

Piaget's structures-of-the-whole referred to how he believed, say, ordering operations were organized in thought. These were not mathematical structures like the structure of arithmetical operations. Rather, they were Piaget's understanding of how children organized their ordering activity that he abstracted from his observations of children. Starting from his fundamental idea, around 1920 Piaget

> noticed with amazement that the simplest reasoning task involving the inclusion of a part in the whole or the coordination of relations or the multiplication of classes (finding the common part of two wholes), presented for normal children up to the age of eleven or twelve difficulties unsuspected by the adult (Piaget 1952, 244).

This observation marked a fundamental shift in Piaget's career toward studies in genetic epistemology for which he became famous. Piaget subsequently formulated stages in the development of the structures-of-the-whole, and these stages were regarded as characterizing the development of logico-mathematical reasoning throughout childhood. He identified three broad stages: the preoperative stage from two to seven years, the concrete operational stage from seven to eleven years, and the formal operational stage occurring at the ages of eleven or twelve years (Piaget 1952, 247).

In 1960, readiness to learn mathematics was based on these three broad stages of intellectual development interpreted as if they explained children's objective mental reality. The attitude was that by understanding the stages, one could understand children's mental reality without engaging in intensive interaction with them. For example, according to Piaget, the child,

> reinvents for himself, around his seventh year, the concepts of reversibility, transitivity, recursion, reciprocity of relations, class inclusion, conservation of numerical sets, measurements, organization of spatial references (Piaget 1980, 26).

These concepts belonged to the stage of concrete operations. To understand children meant to understand that an operation like, say, assembling objects into classes could be reversed, a relation like "is a brother of" could be used to reason transitively, and so forth. Without these operations and their organizational structure, it was thought that the child would not be ready to learn what was being taught in school. Thus, the concept of readiness was made up of two complementary aspects. In any stage of development, the child was considered ready to learn those logical mathematical concepts that could be based on within-stage operations. But the child was considered not ready to learn logical mathematical concepts that were based on operations of the next stage. This idea was the basis for Bruner's (1960) concept of readiness to learn mathematics.

## A CONTEMPORARY CONCEPT OF READINESS

The shift in the concept of readiness that has occurred in mathematics education since 1975 (Steffe and Kieren 1994) follows from an emphasis on Piaget's genetic epistemology rather than on his structures-of-the-whole. The latter are now regarded as models Piaget made to account for his observations of children's ways and means of operating. For example, von Glasersfeld believes that Piaget's structures should be interpreted as a "conceptual tool for systematizing the investigator's experiences with subjects" (von Glasersfeld 1995, 71). We use the basic principles of Piaget's genetic epistemology in understanding children's mathematics rather than use Piaget's structures-of-the-whole in producing implications for education.

A rejection of the idea that the cognitive structures children gradually build are but reflections of the fundamental structures of mathematics is providing a new perspective on school mathematics. This rejection of the sharp separation between psychology and mathematics has led to accepting children's mathematical concepts and operations as legitimate parts of school mathematics. Including children's mathematics as a part of school mathematics is changing the earlier concept of readiness. Rather than question whether children are ready to learn some objective mathematical concept or operation, we now ask whether children are ready to construct some mathematical concept or operation that other children have been known to construct. The difference resides in the meaning of *construction* as well as what it is that the child constructs. To construct a

mathematical concept or operation means to more or less permanently modify a current concept or operation in the context of mathematical interaction. In this, the question is not whether the result of the modification matches aspects of fundamental mathematical structures. What is of interest is whether the result of the modification can be judged to be at a level of knowledge higher than the unmodified concept or operation.

A child is now considered ready to learn what other children have constructed who are like the particular child. What a child might learn could diverge in one of several directions, depending on the intentions and actions of the child's teacher. Acknowledging the importance of the teacher, however, does not mean that what a child learns is wholly a function of what the teacher does. A particular modification of a mathematical concept or operation cannot be caused by a teacher any more than nutriments cause plants to grow. Nutriments are used by plants for growth, but they do not cause plant growth.

What children make from their experiences, however, is open for observation, and a teacher can expect to experience regularities in these observations. These regularities, when abstracted and organized together for a group of similar children, can be regarded as a zone of potential construction for children like those for whom the regularities were experienced. Zones of potential construction are constellations of concepts a teacher constructs for the purpose of organizing and guiding future experiential encounters with children. Stages are still relevant, but the idea of a stage has changed from the developmental stages Piaget established to stages in the process of construction. For example, stages in the construction of children's counting schemes have been established (Steffe, Cobb, and von Glasersfeld 1988), but there is no assumption that these stages necessarily constitute more general developmental stages. Zones of potential construction can be formulated for a specific group of children across, for example, a six-month period or for the immediate future. Such possibilities only reflect the contemporary understanding of a zone of potential construction as a conceptual tool for systematizing the teacher's experiences when teaching children.

*See also* Bruner; Psychology of Learning and Instruction, Overview

### SELECTED REFERENCES

Bruner, Jerome S. *The Process of Education*. Cambridge, MA: Harvard University Press, 1960.

Piaget, Jean. "Jean Piaget." In *History of Psychology in Autobiography* (pp. 237–256). Edwin G. Boring, Herbert S. Langfeld, Heinz Werner, and Robert M. Yerkes, eds. Worcester, MA: Clarke University Press, 1952.

———. "The Psychogenesis of Knowledge and Its Epistemological Significance." In *Language and Learning: The Debate Between Jean Piaget and Noam Chomsky* (pp. 23–34). Massimo Piattelli-Palmarini, ed. Cambridge, MA: Harvard University Press, 1980.

Steffe, Leslie P., Paul Cobb, and Ernst von Glasersfeld. *Construction of Arithmetical Meanings and Strategies.* New York: Springer-Verlag, 1988.

———, and Thomas Kieren. "Radical Constructivism in Mathematics Education." *Journal for Research in Mathematics Education* 25(6)(1994):711–733.

Von Glasersfeld, Ernst. *Radical Constructivism: A Way of Knowing and Learning.* Washington, DC: Falmer, 1995.

LESLIE P. STEFFE

# READING MATHEMATICS TEXTBOOKS

More than just decoding but a process of constructing new mathematical understandings. The National Council of Teachers of Mathematics (1989) has issued comprehensive recommendations in the form of standards for changing the mathematics curriculum in grades K–12. "Mathematics as Communication" is listed as a standard at all grade levels and includes the recommendation that all students be able to read and write text representing mathematical knowledge and problem solving. Doing so, it is believed, will enable students to become independent learners of mathematics. Yet, there is general agreement that few students read their mathematical textbooks, for four reasons: the nature of text in mathematical textbooks, the norms of teaching mathematics, the understandings and approaches used to teach reading mathematics, and students' lack of appropriate reading skills (Borasi and Siegel 1990; Shuard and Rothery 1984; Cohen 1991; Nolan 1984). Feathers (1993) states that most reading in elementary school is with fictional narratives. The reading skills that students develop in reading fiction do not seem to help them in reading nonfiction since the style of the two types of text differs.

Shuard and Rothery (1984) analyzed British mathematics textbooks in order to identify the features of their written style that add to the difficulty of reading mathematical materials. The format of the British textbooks analyzed by Shuard and Rothery are similar enough to many mathematical textbooks published in the United States so that the features they describe can assist teachers and parents in eval-

uating the readability of current textbooks. Shuard and Rothery's analysis agrees with Shield's (1991) position that the technical writing style of most mathematical textbooks is usually dense—including diverse mathematical concepts and their relationships and using complex grammatical structures in a minimum number of statements. This results in a lack of redundancy that one usually finds in other text which can be helpful to the reader. Shuard and Rothery (1984) state that the text in mathematical textbooks includes explanations of concepts and methods, definitions of mathematical vocabulary and symbols, worked examples, and exercises. Explanations are not straightforward descriptions of the mathematical content; they usually involve examples where the student needs to provide the reasons for the sequence of steps in a mathematical procedure or relationship between included pictures, written statement, and mathematical equations. This necessitates that the reader stop to work out the examples with paper and pencil rather than passively read the text as one would do with a novel. Sometimes new information is presented in an indirect manner. For example, (1) new vocabulary that is very important to understanding the text may be incorporated into a sentence rather than introduced with a definition, and (2) rhetorical questions are often used to signal new ideas and, in contrast to other questions, are not expected to be answered by the student.

A second feature adding to the difficulty of mathematical text is that three types of vocabulary are used in mathematics: (1) terms that have the same meaning in "ordinary English" and in mathematics; (2) technical terms that are only used in mathematics, for example, *hypotenuse* and *parallelogram;* and (3) terms that have a different or more specialized meaning in mathematics than they do in "ordinary English," for example, *odd, product,* and *similar* (Shuard and Rothery 1984). Formal definitions of new vocabulary contain only the essential characteristics and are indicated by other technical terms (Shield 1991).

Due to the nature of mathematics, mathematical text contains many diverse mathematical symbols. Reading mathematical symbols differs from regular text in that there is no one-to-one correspondence between the mathematical symbols and the spoken equivalent. Many mathematical symbols can be read in many ways, and also the same idea can be represented by different symbols (Shuard and Rothery 1984). Some ways that the symbol $\frac{2}{3}$ can be read are as the fraction "two-thirds"; the ratio "two to three"; and the stated division "two divided by three." Also, the division, 2 divided by 3, can be written $2 \div 3$, $\frac{2}{3}$, and $3\overline{)2}$.

Mathematical text also contains various types of graphic language such as tables, graphs, diagrams, plans, maps, and pictorial illustrations. Different types of visual language require different eye movement from regular text, since some graphics are linear, some branching, some have two dimensions, and some are nonlinear. Other illustrations are decorative; some are related to the ideas presented in the text but are not essential; and others are essential. The reader needs to determine which illustrations are necessary in order to understand the passage being read. Other features of graphic materials are the conventions of shading, coloring, scale, and how motion is shown (Shuard and Rothery 1984).

Reading mathematical text is more than just decoding the words, symbols, and graphical materials; it is also following the ideas in a passage to construct new mathematical understandings. Shuard and Rothery (1984) maintain that "the text must have a clear 'storyline' or *flow of meaning* . . ." (p. 66). They suggest that teachers analyze text that causes difficulties by segmenting it into *meaning units;* each unit could be a sentence, illustration, example, definition, table, etc. The meaning units can be classified as being one of the following: (1) clear statements of information that are included in the text; (2) necessary mathematical information that is not explicitly stated in the text (rather, questions are asked and tasks provided from which the information is meant to be discovered); and (3) important pieces of background knowledge that is necessary for understanding the text but is omitted (the student must either infer the knowledge from the text or bring it from previous experience). Such an analysis helps identify missing crucial information, the number of new vocabulary words and how they are presented, and whether or not examples and graphic materials are adequately integrated into the written text. Bechervaise (1992) claims that linguistic features of mathematical language is one of the reasons why students do not read mathematics. He compares the graphophonics, syntactic, and semantic systems of mathematics to that of a foreign language and suggests that learning to read and to use the language of mathematics is analogous to learning a foreign language.

Considering the nature of mathematical text, it is not surprising that teachers report that students are not required to read their textbooks. In addition, few mathematics teachers are prepared to teach the reading of mathematics. Muth (1993) reports that middle school teachers, after completing a required content-area reading course in their teacher certification program, still felt unprepared to teach the reading of mathematics. The teachers claimed that reading mathematics was not specifically discussed in the course and hence were unsure as to the role of reading in learning mathematics and their role in teaching reading. These teachers used the textbook as a source of practice activities but not as a learning resource for students. Lehmann (1993) describes a special summer college mathematics program where one of the primary goals of the program was for students to read their mathematics textbook. In spite of this stated goal, the mathematics instructors were provided with few suggestions about what was involved in reading mathematics textbooks and techniques for designing teaching activities that would help students develop reading skills.

Reading teachers are as unlikely as mathematics teachers to teach reading of mathematics. Since the 1930s, the reading field has included a subfield identified as content-area reading. This specialty is designed to focus on reading in all the subject areas; however, most attention has been paid to the reading of science and social studies. Further, the reading field (much like the mathematics field) has, in the last twenty years, undergone a paradigm shift. Content-area reading experts attempted previously to delineate specific skills for each content area and recommended teaching activities, such as the use of teacher-directed reading activities; study and reading guides; various matching and completions activities involving terms and symbols, terms and definitions, terms and examples, etc.; various types of puzzles; and rewriting story problems (Nolan 1984).

In contrast, current theory suggests that the reading process is unitary (Goodman 1984) and that readers interact with the text to construct meaning. When applied to mathematics, this means that reading is a "mode of learning," a means to transform text into meanings mediated by experience, knowledge, and the context of the reading event (Siegel, Borasi, and Smith 1989). Within this perspective, students' prior experience, purpose for reading, and facility with the linguistic features unique to mathematics are critical for transacting with the text to construct meaning. In accord with this approach, Borasi and Siegel (1990) use the strategy "say something," in which students, in pairs, read the text and stop "at self-selected points to raise questions, make predictions, share an image or feeling, summarize, suggest alternatives, connect to other texts and contexts, and so on" (p. 11). Another technique, "cloning the author," involves students recording comments and questions on a set of index cards as they read. Students in pairs then make conceptual maps using the cards or classify the cards and then create labels for each category. Feathers (1993) uses brainstorming and predicting before reading to help students relate new information to previous knowledge and set up their own purposes for reading.

A current trend in mathematics education is to expose students to a broader range of reading materials concerning mathematics. These materials, which are easier to read and more user-friendly, can help students form a bridge from reading fiction to reading more technical materials. At the elementary and middle school levels, children's picture books, novels, mathematical problem-solving and puzzle books, and biographies of mathematicians are used in the teaching of mathematics. For example, the two novels, *The Phantom Tollbooth* (Juster 1961) and *The Toothpaste Millionaire* (Merrill 1972) both contain many mathematical references that are basic to the plot. The two volumes of *Mathematicians Are People, Too, Stories from the Lives of Great Mathematicians* (Reimer and Reimer 1990; 1995) were specifically written to interest students in mathematics as a human endeavor and to show that mathematics is more approachable than most people usually think. All major book stores and libraries have a section of adult and children's books on diverse mathematical topics written for the nonmathematician. Reading such books can lead to a greater appreciation of the role of mathematics in our lives and can develop a greater interest in mathematics.

*See also* Language and Mathematics in the Classroom; Language of Mathematics Textbooks; Textbooks

## SELECTED REFERENCES

Bechervaise, Neil. "Mathematics: A Foreign Language?" *Australian Mathematics Teacher* 48(June 1992):4–8.

Borasi, Raffaella, and Marjorie Siegel. "Reading to Learn Mathematics: New Connections, New Questions, New Challenges." *For the Learning of Mathematics* 10(Nov. 1990):9–16.

Cohen, Carl C. "Teaching and Testing Mathematics Reading." *American Mathematical Monthly* 98(Jan. 1991):50–53.

Feathers, Karen M. *Infotext Reading and Learning.* Markham, ONT, Canada: Pippin, 1993.

Goodman, Kenneth. "Unity in Reading." In *Becoming Readers in a Complex Society* (pp. 79–114). 83rd Yearbook. National Society for the Study of Education. Alan C. Purves and Olive S. Niles, eds. Chicago, IL: University of Chicago Press, 1984.

Juster, Norton. *The Phantom Tollbooth.* New York: Random House, 1961.

Lehmann, Jane N. *Reading Mathematics: Mathematics Teachers' Beliefs and Practices.* Doctoral diss., University of Arizona, 1993.

Merrill, Jean. *The Toothpaste Millionaire.* Boston, MA: Houghton Mifflin, 1972.

Muth, K. Denise. "Reading in Mathematics: Middle School Mathematics Teachers' Beliefs and Practices." *Reading Research and Instruction* 32(Winter 1993):76–83.

National Council of Teachers of Mathematics. *Curriculum and Evaluation Standards for School Mathematics.* Reston, VA: The Council, 1989.

———. *Principles and Standards for School Mathematics.* Reston, VA: The Council, 2000.

Nolan, James F. "Reading in the Content Area of Mathematics." In *Reading in the Content Areas: Research for Teachers* (pp. 28–41). Mary M. Dupuis, ed. Newark, DE: International Reading, 1984.

Reimer, Luetta, and Wilbert Reimer. *Mathematicians Are People, Too, Stories from the Lives of Great Mathematicians.* 2 vols. Palo Alto, CA: Seymour, 1990; 1995.

Shield, Malcolm. "Mathematics Textbooks: Much Maligned Much Neglected." *Australian Mathematics Teacher* 47(Oct. 1991):24–27.

Shuard, Hilary, and Andrew Rothery. *Children Reading Mathematics.* London, England: Murray, 1984.

Siegel, Marjorie, Raffaella Borasi, and Constance Smith. "A Critical Review of Reading in Mathematics Instruction: The Need for a New Synthesis." In *Cognitive and Social Perspectives for Literacy Research and Instruction* (pp. 269–277). 38th Yearbook. National Reading Conference. Sandra McCormick and Jerry Zutell, eds. Chicago, IL: The Conference, 1989.

CAROL NOVILLIS LARSON

# REAL NUMBERS

A system that has developed slowly over many centuries that includes the positive integers and the rational and irrational numbers. The positive integers or natural numbers (1, 2, 3, . . .) arose out of the necessity to count objects. The rational numbers ($p/q$, where $p$ and $q$ are integers, $q \neq 0$) were needed for commerce. Their existence can be traced as far back as the Babylonian cuneiform tablets (ca. 1850 B.C.) and the Rhind Papyrus (ca. 1650 B.C.). Pythagoras (ca. 560 B.C.–480 B.C.), or perhaps one of his students, discovered that $\sqrt{2}$ is irrational. The insight that there were numbers other than rational numbers occurred when the Pythagoreans tried to find the ratio of the length of the diagonal of a unit square to its side (Boyer 1991, 72). Theaetetus (ca. 417 B.C.–369 B.C.) demonstrated that there were an infinite number of irrationals (Asimov 1982, 17). It was not until ca. 300 B.C. that Euclid proved, using geometry, that $\sqrt{2}$ was irrational. It was many centuries later that Adrien Marie Legendre (1752–1833) proved that both $\pi$ and $\sqrt{\pi}$ were irrational. Julius W. R. Dedekind (1831–1916) showed that irrational numbers could be understood as "cuts" on a number line. A cut is a division of the rational numbers that separates them into two nonempty disjoint sets such that every element of the first set is less than every element of the second set. The point at which the cut is taken (i.e., the irrational number) is the upper bound of the first set and the lower bound of the second set.

Students need to develop a sense of numbers through concrete representations. Small and large numbers should be discussed, using examples from students' surroundings. To understand a very small number, students can be asked to try to measure, using a stopwatch, the amount of time a calculator takes to add two numbers once they are entered. To understand a large number, students can be asked to count the number of stars in the sky on a clear night. In the early grades, students may not be able to understand a nanosecond (one billionth of a second) but large numbers, such as a googol ($10^{100}$), are fascinating to some. They can write out the digits for a googol on the chalkboard and then discuss whether the number of raindrops that fall on their town or city is more or less than this number.

As rationals and then irrationals are introduced, a geometric representation of these numbers using a number line will help to give students a feeling for their size. Estimation and the concept of "between" are techniques for helping students order numbers. Have students locate the point on a number line that corresponds to a given number. Then have them discuss the set of numbers less than the number and the set of numbers greater than the number. Small numbers, such as those representing a nanosecond, can be discussed in the context of scientific applications. For example, it would take approximately 1 nanosecond for electricity to travel through a 1-foot-long piece of wire.

Most algebra textbooks present the real numbers as the union of the set of rationals (including zero, positive, and negative numbers) and irrationals. When studying polynomial equations and their solutions, students are interested to see that the reals can be divided into two different sets, namely, algebraic and transcendental numbers. A real number is an algebraic number if it is the solution of a polynomial equation with rational coefficients. Otherwise, it is a transcendental number: $\pi$, $e$ and $2^{\sqrt{3}}$ are examples of transcendental numbers.

A discussion of real numbers cannot be concluded without referring to the concept of "infinity." Although poetic license allows writers to substitute the word *infinity* for a large number, students in mathematics must be guided to understand the difference between finite and infinite. The number of objects in a set is said to be infinite if no matter how many are counted there is always one more. A more formal definition states that a set is infinite if there is a one-to-one correspondence between the elements of the set and a proper subset of itself. If asked to name an infinite set, students may mistakenly say the number of grains of sand on a beach. Although the number of grains of sand on a beach is very large, it is countable. For example, the beach at Coney Island, New York has been estimated to have $10^{20}$ grains of sand (Newman 1956, 2007). On the other hand, the set of natural numbers is infinite since no matter how large a number you choose, a larger number can be found by adding one to the chosen number. Newspaper and magazine articles can be used to illustrate how real numbers describe and quantify the world in which we live.

*See also* Algebraic Numbers; Fractions; Irrational Numbers; Rational Numbers; Transcendental Numbers

## SELECTED REFERENCES

Asimov, Isaac. *Asimov's Biographical Encyclopedia of Science and Technology.* Garden City, NY: Doubleday, 1982.

Boyer, Carl B. *A History of Mathematics.* 2nd ed. rev. Revised by Uta C. Merzbach. New York: Wiley, 1991.

Jacobs, Harold R. *Mathematics: A Human Endeavor.* 3rd ed. New York: Freeman, 1994.

Malcom, Scott P. "Understanding Rational Numbers." *Mathematics Teacher* 80(1987):518–521.

Newman, James R. *The World of Mathematics.* New York: Simon and Schuster, 1956.

Williams, Richard K. "On the Rationality of Exponentials and Logarithms." *Mathematics Teacher* 70(1977):750.

MONA FABRICANT

# RECREATIONS, OVERVIEW

Drawing from all branches of mathematics, problems range from modest challenges to cases still unsolved after centuries of trying. The educational attraction of recreational mathematics is that problems occur that students find understandable and intriguing; and from there the path leads to new areas of mathematical knowledge that the student is motivated to learn.

Those problems that are called *mathematical recreations* should have two features. First, the problem statement must be in a form easily understood without advanced mathematical training. Second, this simply stated problem should not have an immediately obvious solution. Adding a bookkeeper's column of figures, while elementary, is not recreational. We might say that a mathematical question should be considered recreational if it is easy to ask and hard to answer.

## LITERATURE SELECTIONS

To get the feel of the subject, a student should do additional reading. By far, the best books are those of Martin Gardner, based on his long-running column in *Scientific American*. Other names, equally famous and both from early in the twentieth century, are Henry E. Dudeney, an Englishman, and Sam Loyd, an American. Unlike Gardner, who is an omnivorous reader and eclectic collector, Loyd and Dudeney were themselves mathematicians who composed their own recreational problems. Their books are still available, fully as lively as Gardner's, but the mathematical level for solution of Loyd's or Dudeney's problems is generally considerably higher than Gardner's, often beyond school level.

## PERFECT NUMBERS

One way to classify integers is based on their divisors. In this section, the unit 1 is included as a divisor of $n$, but not $n$ itself. Find all the divisors of $n$; for example, the divisors of 10 are 1, 2, and 5. Add those divisors. If the sum of the divisors is

a. less than $n$, then $n$ is called *defective* (10 is defective, since $1 + 2 + 5 < 10$);
b. equal to $n$, then $n$ is *perfect;*
c. greater than $n$, then $n$ is *abundant.*

Perfect numbers have long been a topic of mathematical recreations. The first two are 6 (divisors 1, 2, and 3) and 28 (divisors 1, 2, 4, 7, and 14). There is a general method for finding perfect numbers. If a number $k$ of the form $k = 2^n - 1$ is prime (called a *Mersenne prime*), then $k(k + 1)/2$ is perfect. The Mersenne primes 3 and 7 give rise to the two examples 6 and 28. All such perfect numbers are even; and the proof that the only even perfect numbers are Mersenne-perfect is not too difficult (Ball 1947, 167). Less than twenty Mersenne primes are known, so that is also the number of known even perfect numbers. It is an open question how many there are;

conjecture is that there are infinitely many. How about *odd* perfect numbers? Almost nothing is known, including whether any exist. The little theory there is proves that an odd perfect number will have more than fifteen digits, so it will never be found by trial and error. Experimentation with small numbers suggests that odd numbers tend to be defective, but in fact for larger values, abundant numbers predominate among both even and odd integers. The smallest odd abundant number is 945. The literature of perfect numbers is mixed. Some of it involves number theory concepts well beyond school level; other writings are reasonably accessible. There are many extensions and variations. For example, a pseudoperfect number is one equal to the sum of some of its divisors.

## THE FALSE COIN PROBLEM

This problem first appeared in a mathematics journal in the mid-1940s and became an instant classic of recreational mathematics. A banker who has nine coins, identical in appearance, knows that one is counterfeit and weighs slightly less than the others. To detect weight differences, there is a balance scale with two pans that permits the user to determine which of the items in the pans is lighter or if the two sides are equal. For instance, the banker might weigh coins (A and B) against (C and D). In only two uses of the balance, can the banker identify the odd coin? Yes. Weigh (A, B, and C) against (D, E, and F). If one side is lighter, the odd coin is on that side. If the pans balance, the odd coin is among the other three. In any case, in one weighing, the uncertainty is reduced to three remaining coins. Weigh one of them against a second. (It can turn out that the odd coin is identified without ever being on the scale.)

Extensions of this problem abound. Suppose the banker knows that there is an odd coin but not the direction in which its weight differs from the good ones. Two weighings are not sufficient. If (A, B, and C) are lighter than (C, D, and E), it may be that the left side is light; but there is also the possibility that the odd coin is heavy and on the right side. Three weighings will do; but if we are going to be allowed that many weighings, we can accept a larger group of coins. With $n$ uses of the balance allowed, the number of coins can be up to $(3^n - 1)/2$. Solution theory leads unexpectedly into the ternary number system (suggested by the $3^n$ term), because with each weighing there are three possible outcomes: balance, left side heavy, right side heavy (Schwartz 1980, 131). Some other questions that may be considered are as follows. Suppose there is more than one odd coin. If there are two of identical weight (different from a good coin), how many coins can there be in the set when $n$ weighings are allowed? Suppose the two odd coins may be of different weights.

## INSTANT INSANITY

Many problems in mathematical recreations are based upon some kind of physical prop that may look like a toy but has mathematical features that a student can study. The best-known example, perhaps, is the Rubik Cube, another is the Meffert Pyraminx. Still a third, almost as famous, is Instant Insanity. Recall that Instant Insanity is a puzzle whose pieces are four cubes (see Figure 1). Each face is colored one of four colors, as shown: White, Red, Blue, or Green. The objective is to assemble the cubes into a square prism $1 \times 1 \times 4$ in such a way that each of the four $1 \times 4$ lateral faces displays all four colors. One key idea is that the colored faces may be considered in terms of pairs of opposite faces. For example, on Cube I, $B$ and $G$ are colors on a pair of opposite faces; and we need to consider only the paired set of colors $BG$. If the four cubes are assembled into a prism, the $G$ and $B$ faces on Cube I can be reversed in the arrangement by simply rotating the cube by 180°. Table 1 shows the pairs of colors from the four cubes.

The objective now is to find by inspection a set of pairs—four of them, one from each row. And in that set of four pairs, with eight face colors, there

### Table 1

| | Face pair | 1 | 2 | 3 |
|---|---|---|---|---|
| **Cube I** | | RR | BW | BG |
| **Cube II** | | RW | WG | BG |
| **Cube III** | | RW | RG | BW |
| **Cube IV** | | RG | BW | GG |

must be two each of each color, $W$, $R$, $G$, and $B$. Finding the pairs is trial and error, but it is far easier to inspect the table than to manipulate the cubes.

Here are some examples of trying to find a solution. Suppose a student starts in the table trying the pairs I-3 and II-3 in Cubes I and II, respectively. They both happen to be $BG$ pairs. So in the other two cubes, whatever choices are made must omit any $B$ or $G$ face. But in Cube IV that is impossible, so the initial trial choices must be a blind alley, and the student must start again. The first trial was an error! On the next trial, suppose the start is the pairs I-2 = $BW$, and II-3 = $BG$. Since neither of these trial choices includes an $R$ face, the remaining two choices must have two $R$'s. There is no $RR$ face available on either cube, so each of the last two face-pair choices must have one $R$. In Cube IV, that limits the choice to only IV-1. And then the final choice on Cube III has to be III-1. And that set of four pairs does have two each of every color, so it is a successful trial, not an error. (There is one other winning choice set.) That argument (Schwartz 1980, 3) may be a strain for students, but it should be ultimately within their reach. Mathematics is not supposed to be always easy; it is just supposed to be fun.

## OTHER PUZZLES WITH MOVING PIECES

A large class of puzzles use moving pieces, with the objective for the solver to arrange them into some pattern. Rubik's Cube, Meffert Pyraminx, and Instant Insanity have already been mentioned. Piet Hein's SOMA cube is another. There are also innumerable sliding block puzzles. The most famous, called the 15-Puzzle, is the Sam Loyd classic from the 1870s (Loyd 1959, 19). In virtually all of these cases, trial and error (about which more later) is the usual starting method. It may or may not lead to a more general mathematical theory. The 15-Puzzle, Rubik Cube, and Pyraminx have solutions involving group theory. Instant Insanity is effectively solved with some of the principles of combinatorics. But for many others, there has not been any overall

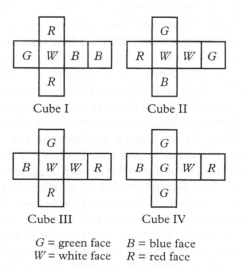

Cube I    Cube II

Cube III    Cube IV

$G$ = green face    $B$ = blue face
$W$ = white face    $R$ = red face

**Figure 1**    *Instant Insanity.*

theoretical idea that has been shown to apply. Mathematical recreations are simply unpredictable.

## FERMAT'S LAST THEOREM

The so-called Fermat's Last Theorem (FLT) may be the best example of a problem that is easy to state and hard to solve. Until it was finally resolved within the last few years, it was probably the most famous problem in all mathematics. Pierre de Fermat, a French mathematician stated it in 1637. The formulation is based on the familiar fact that three sides of a right triangle can all be integers, for example 3, 4, and 5. The Pythagorean property for such right triangles takes the form $3^2 + 4^2 = 5^2$. One simple generalization is to ask for other sets of integers ($a$, $b$, and $c$) with $a^2 + b^2 = c^2$. There are many solutions to that question. But a further generalization is to ask for sets of integers ($a$, $b$, $c$, and $n$) with $a^n + b^n = c^n$ and $n > 2$. That is what Fermat asked (in a note on a book margin) in the seventeenth century. (It was not really his last theorem.) It was not until 1995 that the American Andrew Wiles was finally able to prove that there is no solution.

## PATHS THROUGH GRAPHS

Nearly as spectacular, and more instructive as a recreational mathematics problem, is one that was solved by the Swiss mathematician Leonhard Euler in the early eighteenth century. It is conveniently presented graphically in Figure 2, which represents the

bridges of the town of Königsburg in Germany. The question asked is whether the graph is unicursal, that is, is it possible to traverse the graph, moving from vertex to vertex along the graph's edges, in such a way as to trace out every edge once and only once? The original formulation asked if a traveler could walk around Königsburg in a path that crossed each bridge once only. Euler found a simple and ingenious way to determine if a graph is unicursal. It relates to the degree of the vertices, that is the number of edges that begin (or end) at a vertex. A vertex that has an odd (or even) number of edges entering or leaving it is said to be of odd (or even, respectively) degree. If a graph has either zero vertices or two vertices of odd degree, then it is unicursal; otherwise not. The proof (Ball 1947, Chap. IX) is within the scope of high school students. A natural variation on the unicursal graph problem would seem to be the question of whether the spanning path can go just once through each vertex (rather than each edge). That version was asked in the 1880s by William Hamilton, an English mathematician and physicist; the question of characterizing a Hamiltonian circuit is still unsolved today. In recreational mathematics, there is no predicting in advance how profound will be the level of mathematics needed nor how educationally valuable any problem will be.

## THE FOUR-COLOR THEOREM

As a mathematical idealization, assume that a map of an area has countries that are each made up of only one region and border each other along some common length of boundary (more than a corner point). Can every such map be colored so that bordering countries have different colors and the total number of colors used altogether is four or fewer? This is the four-color problem, a long time competitor to FLT as the most famous problem in recreational mathematics. Its resolution in 1976 by Kenneth Appel and Wolfgang Haken was one of the high points of twentieth-century mathematics. The earliest presentation seems to come from about 1850 in London by an English graduate student to his teacher, Augustus de Morgan. For several decades, only desultory solution efforts were undertaken, but in 1879 A.B. Kempe published an alleged "proof" that turned out to be wrong and has become notorious in mathematical history as the Kempe catastrophe. For nearly 100 years more, some of the truly great names of modern mathematics struggled toward a valid derivation. A great deal of exciting mathematics was created, much of it (Saaty and

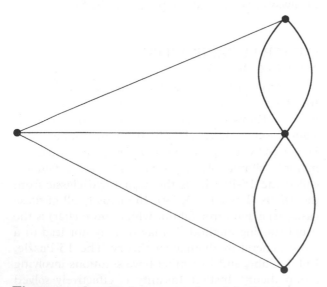

**Figure 2** *Graphical representation of the Bridges of Königsburg.*

Kainen 1986) within the scope of school readers today, in contrast to FLT literature. The final steps by Appel and Haken began with theoretical arguments that reduced the number of candidate map subregion arrangements to a large but finite number of cases that they called the "unavoidable set," with over a thousand members. Each of these they examined by using a computer program to prove four-colorability, a process that required hundreds of hours of computer time on the largest, fastest machines then available. (Today, those would be desktop toys!) So the four-color conjecture has now become the four-color theorem. Whether a less computationally top-heavy proof can be found remains an open issue.

## THE INTEGER CUBE

The final example is much less known than any of the preceding cases. After the FLT and four-color problem were solved recently, there was speculation in mathematical journals about the next "elementary" problem that would challenge the profession for another century or so. One of the suggestions is called *The Integer Cube,* which is not an accurate designation; but a more precise one like *The Integer Rectangular Parallelepiped* would not work. As with the FLT, the motivation for the integer cube starts with the observation that a rectangle can have integer values for the sides and diagonal. Can the same idea extend to three dimensions? Is there a rectangular box whose edges and diagonals are all integers? There would be three edge dimensions, three face diagonals, and one body diagonal through the center, a total of seven lengths in all. For a start, it is easy to make five of them integers. Take edges of length 4, 3, and 12. With that much progress so early, could you doubt that an integer cube will eventually be found? Most investigators are skeptical, but at present no one knows what kind of mathematical theory will be needed. Perhaps just Pythagorean triples. But do not count on it!

## TRIAL AND ERROR

In conclusion, we mention briefly several problems that are often considered to be recreational, but this writer thinks should not be. They are problems solved by simple extensive trial and error. SOMA cube construction is one example. Tangram figures make another. There is no underlying mathematical theory to trying one thing after another until you find something that works, and little educational

benefit either. One cannot be totally dogmatic. Trial and error as a starting method may lead to some underlying pattern, as with Rubik's Cube and Instant Insanity. But if it does not, then the problem is not even mathematics, even though it may be easy to state and hard to solve.

*See also* Enrichment, Overview; Graph Theory, Overview; Number Theory, Overview

SELECTED REFERENCES
    Ball, Walter W. Rouse. *Mathematical Recreations & Essays.* Harold S. M. Coxeter, ed. New York: MacMillan, 1947.
    Dudeney, Henry E. *The Canterbury Puzzles.* New York: Dover, 1958.
    Gardner, Martin. *AHA! Insight.* New York: Freeman, 1978.
    ———. *Knotted Doughnuts and Other Mathematical Entertainments.* New York: Freeman, 1986.
    ———. *Mathematical Carnival.* Washington, DC: Mathematical Association of America, 1989.
    ———. *Scientific American Book of Mathematical Puzzles and Diversions.* New York: Simon and Schuster, 1959.
    ———. *Second Scientific American Book of Mathematical Puzzles and Diversions.* New York: Simon and Schuster, 1961.
    ———. *The Unexpected Hanging.* New York: Simon and Schuster, 1969.
    *Journal of Recreational Mathematics.* Amityville, NY: Baywood.
    Loyd, Sam. *Mathematical Puzzles.* Selected and edited by Martin Gardner. New York: Dover, 1959.
    *Mathematical Gazette.* Leicester, England: Mathematical Association.
    *Mathematics & Informatics Quarterly.* Sofia, Bulgaria: Bulgarian Academy of Sciences, Institute of Mathematics; and Terre Haute, IN: Rose Hulman Institute of Technology.
    Phillips, Hubert ("Caliban"), S.T. Shovelton, and G. Struan Marshall. *Caliban's Problem Book.* New York: Dover, 1961.
    Saaty, Thomas L., and Paul C. Kainen. *The Four-Color Problem.* New York: Dover, 1986.
    Schwartz, Benjamin L. *Mathematical Solitaires & Games.* Farmingdale, NY: Baywood, 1980.

BENJAMIN L. SCHWARTZ

# RELATIVITY

The theory relating measurements that are made by moving observers. Imagine two observers traveling with constant velocity and moving away from one another with relative velocity $v$. Each holds up a clock, a 1 kg mass, and a meter stick that is parallel to a line formed by the observers. In 1904, Hendrik Lorentz published his famous "Lorentz Transformations" (Goldberg 1984, 99, 136), which predict that each observer will find that the other observer's clock runs slow, his meter stick has shrunk, and his mass has increased, as follows:

(i) Each observer will find that the other observer's hour is $1/\sqrt{1 - v^2/c^2}$ hours long,

(ii) the mass of the other observer's kilogram is $1/\sqrt{1 - v^2/c^2}$ kg, and

(iii) the other observer's meter is $\sqrt{1 - v^2/c^2}$ meters long. (In these formulas, $c \approx 3 \times 10^8$ meters per second is the speed of light.)

Formula (ii) led Einstein to the most famous equation in science, $E = mc^2$ (Goldberg 1984, 157, 136), which says that mass ($m$) can be converted into energy ($E$) and vice versa. This prediction was spectacularly confirmed when the first atomic bomb exploded at Alamagordo Air Force Base, New Mexico in July 1945.

In 1905, Albert Einstein published his special theory of relativity, which explained the Lorentz transformations and began a new era in physics. Special relativity is based on four ideas: the principle of relativity, which states the laws of physics are the same for all observers who are not accelerating; the Michelson-Morley experiments, which say that the speed of light is the same for all observers; the law of conservation of momentum, which says that the total momentum of the objects in a closed system does not change; and Einstein's careful analysis of what it means to say that two different events occur at the same time. Einstein (1961) explains his reasoning in simple terms using only common sense and high school algebra. Einstein worked for years to extend the theory of relativity to observers who were accelerating due to gravity or other forces. Hermann Minkowski recast special relativity as a theory about the geometry of "flat four-dimensional spacetime" in 1908. Building on these ideas, Einstein published his general theory of relativity in 1916, incorporating accelerated observers in a universe described as a "curved four-dimensional spacetime" and explaining gravity as an effect of the bending of space. These controversial ideas were confirmed when astronomers observed light being deflected by the sun's gravity during a 1919 solar eclipse, just as the general theory of relativity had predicted (Goldberg 1984, 174; Hoffmann 1983, 156).

*See also* Einstein

## SELECTED REFERENCES

Einstein, Albert. *Relativity, the Special and General Theory.* New York: Crown, 1961.

Goldberg, Stanley. *Understanding Relativity.* Boston, MA: Birkhäuser, 1984.

Hoffmann, Banesh. *Relativity and Its Roots.* New York: Freeman, 1983.

GEORGE A. JENNINGS

# REMEDIAL INSTRUCTION

Classes provided in many secondary schools and colleges to help low achievers in mathematics overcome deficiencies in their mathematics education. In these classes, teachers spend much time reviewing computational techniques, and students often work individually. Modeling computational skills takes a great deal of class time, so discussing applications and exploring the effects of changing the parameters in a problem often seem impractical. However, focusing on low-level computing skills does not help students learn to use mathematics to solve problems. The errors that remedial students make when trying to solve application problems indicate that they often base decisions on computing rules without understanding the meanings or relationships in applications.

In recent years, mathematics education reformers called for enriched curriculum for all students, including low achievers. The National Council of Teachers of Mathematics (1989) outlined appropriate mathematics curriculum goals in *Curriculum and Evaluation Standards for School Mathematics:* (a) stress conceptual understanding, (b) prepare students to communicate mathematically, (c) give students opportunities to develop mathematical reasoning abilities, and (d) teach students to solve both routine and nonroutine problems. Technology now makes using higher-level mathematics possible for those who have not mastered computational skills. Also, students can learn basic problem-solving techniques without actually computing solutions. If students in remedial classes do not have to master computational processes for solving problems, teachers can focus on concepts, relationships, effects of changing problem parameters, and experimenting.

This entry will discuss how teachers can change the focus of remedial classes from drill on computations and low-level applications to applications with interesting contexts, relationships, and decision making. Research will be discussed that supports the premise that low-ability mathematics students may benefit from spending less time learning computational procedures because their conceptual misunderstandings are often caused by focusing on attributes of symbols and computing rules. Suggestions will be given for using technology to develop lessons designed to help low-ability students understand concepts in ways that high-ability students understand them. Lesson suggestions will also be given to help students learn general problem-solving strategies while also learning mathematical concepts. Lessons that focus on understanding concepts, applications, and problem-solving strategies are more appropriate for group work and discussions as compared with computational-drill lessons.

## HISTORICAL CONTEXT

Before technology for computing was widely available, mathematics could only be used by those who could successfully perform the computations necessary to calculate solutions. Students in remedial mathematics classes often had not mastered the computational skills necessary to perform higher-order tasks; therefore, their curriculum has historically been dominated by low-level arithmetic skill learning. This was practical and consistent with educational learning theories. Robert Gagné's (1985) learning theory asserted that before higher-order tasks could be mastered, subordinate and prerequisite skills must be learned. The behaviorist theories of Burrhus F. Skinner (1968) and Edward Thorndike (1922) supported the practice of repetitive drills to reinforce rote learning.

## LOOKING FORWARD

Technology for doing simple arithmetic and complicated higher-level mathematical computations is now widely available. Calculators can manipulate algebraic expressions, do matrix operations, compute simple statistics, display graphs, and do calculus computations. People can use computer software to, for example, amortize loans, perform complicated statistical analyses, design electrical circuits, or model systems using operations research, even if they have no experience with the higher-level mathematics needed to do the computations that are involved in these applications. Students in remedial classes who have difficulty computing with numbers may be able to successfully learn to solve real-life applications using technological tools.

Many important problem-solving skills can be taught without the use of technology. Teachers can

design lessons with the goal of teaching problem-solving techniques, such as (a) breaking problems into component parts, (b) recognizing similar problems, (c) solving a simpler problem, (d) generating data and looking for a pattern, or (e) guessing a solution and revising the method based on the outcome. Students need to feel comfortable experimenting in order to become successful problem solvers. They should have the opportunity to work in groups so that they can discuss problem-solving strategies and learn to communicate about mathematics.

## LEARNING THEORY AND RESEARCH

Students in remedial classes often have understandings of numbers that do not help them become good problem solvers or that lead to incorrect interpretations of solutions. Brian Greer (1987) found that students who used the correct operations to solve application problems that involved whole numbers chose the wrong operations for similar application problems that involved fractions and decimals. For example, Greer reported that students divided to find the price of 0.923 kg of cheese costing $27.50 per kg and reported that they chose division because the answer would be smaller than the price per kg, and division gives smaller numbers for answers. Students chose the correct operations for solving problems if the model for whole numbers applied; they multiplied to find the price of 3 kg of cheese at $27.50 per kg. They did not recognize the structural similarities between these and other problems. Giving remedial students the opportunity to classify problems, without solving them, would help them learn to identify similar problems.

Warren Wollman (1983) found that students mistakenly believed variables were similar to units. For example, students may symbolize three feet as one yard, 3 ft = 1 yd, with ft and yd as units. They symbolize five times as many bats as cats, $5b = 1c$, incorrectly, as if $b$ and $c$ were units instead of variables that stand for numbers. Consequently, many students who can solve equations successfully are frequently unable to set up equations for solving problems. Lessons designed with the goal of teaching students the meaning of variables might help them distinguish variables from units.

Less able students tend to understand mathematics in ways that they cannot generalize. The counting models for whole-number addition and multiplication do not apply to other integers, fractions, or decimals, for example. As a consequence, students may learn to compute with integers, fractions, and decimals but not understand how to interpret the meaning of the computations. Practicing computing rules for different kinds of numbers is not likely to help students understand general interpretations of operations. Richard Skemp (1971) stressed the importance of building a general understanding to facilitate the assimilation of new concepts. For example, students should understand that models illustrate one way to interpret meaning but may not apply to all types of numbers and may not be the only possible interpretation. Students should be shown a general model for addition before learning to add other kinds of numbers. The number-line model for addition can be used to show the meaning of adding any kind of real numbers. Skemp argued that it is unlikely that most students will think of a model for understanding new concepts without help; students are more likely to try to incorrectly apply models they already understand.

Skemp warned that a conceptual model can be as powerful a hindrance as a help if it cannot be used to understand new concepts. A mathematical conceptual model must be adaptable so that it can change to accommodate new ideas and still be used to understand previous knowledge. Perhaps one reason remedial students experience math anxiety is because they try to generalize models incorrectly to interpret computations. Students would benefit from identifying an appropriate model for interpreting a problem before attempting to solve it.

## SUGGESTIONS FOR THE CLASSROOM

Strong evidence exists that using instructional computing tools can improve mathematical conceptual understanding at all levels of mathematics instruction (Heid 1988). Mathematics teachers can use technology to develop lessons for students in remedial classes to (a) illustrate mathematical relationships and concepts, (b) allow students to generate problems and experiment, (c) gain information mathematically and communicate the results. Students can learn problem-solving strategies without actually computing solutions. Students can learn from each other while working in groups on lessons that focus on generating and organizing data, discovering patterns, and identifying similarities among application problems.

### Illustrating Mathematical Relationships and Concepts

Remedial mathematics teachers can use spreadsheets to develop lessons for introducing algebraic concepts, such as the meaning of equals and of variables. Students can generate data using spreadsheets so that the focus of the lesson can be organizing data and discovering patterns and relationships, instead of computing. Students can enter the following formulas in consecutive cells of a spreadsheet:

| $= 3 * A1 + 2$ | $= 3 * (A1 + 2)$ | $= 3 * A1 + 6$ |
|---|---|---|

As students enter different numbers into the spreadsheet cell A1, the cells containing these formulas would automatically update with the correct calculation. The concept of variable is demonstrated when students change the number in cell A1. Students should make a table to keep track of the numbers they enter for A1 and the results:

| A | 3A + 2 | 3(A + 2) | 3A + 6 |
|---|---|---|---|
| | | | |
| | | | |
| | | | |

From the data they generate and organize, students may discover a pattern that would help them see that $3(A + 2) = 3A + 6$, regardless of the value of $A$ or whether $A$ is a whole number, integer, fraction, or decimal. Students who do not discover the pattern may see it when shown by other students. Lessons of this nature could be done using paper and pencil or symbolic calculators instead of spreadsheets. This lesson not only illustrates the meaning of variables and of equals but would teach students the problem-solving skill of generating and organizing data to help them discover patterns. This lesson could be modified and expanded to introduce the distributive property.

### Generating Problems and Experimenting

Student-generated problems are likely to result in numbers that lead to difficult calculations. If students are using calculators for doing computations, teachers need not worry that the resulting numbers involved in problems will be too cumbersome with which to work. Once solutions are found, exploring the effects of changing some of the parameters of problems is quick and simple. For example, students could use calculators, spreadsheets, or a variety of different software programs to compare the cost of borrowing or investing money for various lengths of time and at different interest rates. Many such comparisons are nonintuitive to inexperienced students. By using technology to learn how problem parameters affect solutions, students can learn to use mathematics as a tool for gaining information to make decisions.

## Gaining Information Mathematically and Communicating the Results

Teachers could design statistics lessons in which students gather data regarding topics that interest them. Students could use statistics software, spreadsheets, or calculators to compute simple statistics such as means, variances, and standard deviations. They could use software to graph frequencies for discrete data. For example, students might collect data regarding the number of hours male students study versus females. Many issues arise when working with actual data as opposed to textbook data. The importance of random sampling may be illustrated if they collect data only from their close friends and compare results with others. Students could collect observational data, such as tallying the number of male versus female drivers who are wearing seat belts.

Students could use graphing technology to compare results for different groups and appropriately illustrate results with graphs or tables. They could then write explanations to explain these results, which would teach them to express mathematical ideas. Students would learn to think critically when reading articles that generalize from statistical inferences drawn from data.

## Learning to Recognize Similarities Among Problems

Teachers can design lessons to help students acquire general problem-solving skills, such as recognizing similarities among problems. Teachers can create lessons that focus on classifying problems according to solution strategies. Teachers would need to first show students why a specific strategy would solve a given problem and discuss with them how to recognize when that strategy would apply to other problems. Students could then work in groups to discuss similarities in the structure of problems in order to classify them according to solution techniques.

Teachers could use the following problems as examples of different classifications:

a. A Christmas tree is decorated with red and green blinking lights. If the green lights blink on every 12 seconds and the red lights blink on every 18 seconds, how many seconds after they blinked together will they blink together again?
b. What is the total price of 5 pounds of jelly beans which cost $2.00 per pound?

The general method for solving problem (a) is finding the least common multiple of the numbers; the general method for solving problem (b) is multiplying price times the number of pounds purchased.

Student groups could then classify the following problems as to whether they would be solved like example problem (a) or (b):

1. Bill and Reah are running around a track together. If they begin at the same time and Reah completes the track every 8 minutes and Bill completes the track every 6 minutes, after how many minutes will they cross the starting point together?
2. Anne bought $\frac{1}{2}$ pound of hamburger, which cost $2.73 per pound. What was the total price?

After discussing how students classified the problems and why, student groups could write problems that would be solved like examples (a) and (b). Students could then use calculators to help them do the computations necessary to solve the problems. The focus of this type of lesson is on learning the problem-solving strategy of recognizing similar problems. Because computing is not the primary goal of this lesson, students are not likely to make the common errors that occur when deciding a strategy based on incorrectly applying computational models.

## Identifying Models

A common error among remedial students is incorrectly generalizing a model for interpreting a problem. Students would benefit from keeping a math journal in which they define different models, with examples, to which they can refer when interpreting an application. For example, several different fraction models exist including (a) the area model, in which a figure is divided into equal areas, some of which are shaded; (b) the set model, in which a set of identical objects make up one "whole" and some of the objects are shaded; (c) the set-grouping model, in which the set is divided into equal-sized groups and some groups are shaded; (d) the ratio model; and (e) the division model. Numerators and denominators have different interpretations for these models. Teachers could create lessons designed to help students recognize which model should be used to interpret a problem, such as asking students to classify problems by appropriate models. Then when students do solve problems, they could first indicate which model they are using to interpret it.

## CONCLUSION

Although calculators and computers can perform most of the calculations involved in the applications of mathematics, their influence on teaching in remedial mathematics classes has not been widespread. Teachers continue to teach the way they were taught and textbooks emphasize rote learning (Weiss 1989). Highly structured drill is the most popular way of teaching computational skills, and very little use of technology for developing higher-order thinking skills exists (Anderson 1993). If calculators and computers are used for computing, class time can be spent on concept-building activities and discussions. It becomes practical for remedial students to work together in a variety of ways, including activities in which they generate and solve their own problems. Investigating the effects of changing the parameters of problems, and comparing similar problems, becomes a more practical undertaking. When lessons focus on the structure of problems, discovering patterns or similarities, and basic problem-solving strategies, students will be less likely to make the common errors that result from focusing on computational rules. Class time made available by using technology can be spent effectively to further instructional goals.

*See also* Mathematics Anxiety; Software; Technology

## SELECTED REFERENCES

Anderson, Ronald E., ed. *Computers in American Schools 1992: An Overview.* Minneapolis, MN: University of Minnesota Press, 1993.

Gagné, Robert M. *The Conditions of Learning and Theory of Instruction.* 4th ed. New York: Holt, Rinehart and Winston, 1985.

Greer, Brian. "Nonconservation of Multiplication and Division Involving Decimals." *Journal for Research in Mathematics Education* 18(1987):37–45.

Heid, Kathleen M. "Resequencing Skills and Concepts in Applied Calculus Using the Computer as a Tool." *Journal for Research in Mathematics Education* 19(1988): 3–25.

National Council of Teachers of Mathematics. *Curriculum and Evaluation Standards for School Mathematics.* Reston, VA: The Council, 1989.

———. *Principles and Standards for School Mathematics.* Reston, VA: The Council, 2000.

Skemp, Richard R. *The Psychology of Learning Mathematics.* Middlesex, England: Penguin, 1971.

Skinner, Burrhus F. *The Technology of Teaching.* Englewood Cliffs, NJ: Prentice-Hall, 1968.

Thorndike, Edward L. *The Psychology of Arithmetic.* New York: Macmillan, 1922.

Weiss, Iris. *Science and Mathematics Education Briefing Book.* Chapel Hill, NC: Horizon Research, 1989.

Wollman, Warren. "Determining the Sources of Error in a Translation from Sentence to Equation." *Journal for Research in Mathematics Education* 14(3)(1983):169–181.

JANET L. JOHNSON

## RESEARCH IN MATHEMATICS EDUCATION

Although mathematics has been studied for centuries, mathematics education did not become a separate field of study until late in the nineteenth century. Disciplined inquiry, more formally known as research in mathematics education, is mostly a phenomenon of the twentieth century (Kilpatrick 1992). In fact, there is recently more research in mathematics education being done by a greater number of individuals than at any time in the past. Here we will outline some of the topics in mathematics education that are being investigated, describe some of the methodologies that are being used in mathematics education research, describe discussions about criteria for judging the quality of mathematics education research, and outline some of the sources where mathematics education research is reported.

## TOPICS BEING INVESTIGATED IN MATHEMATICS EDUCATION

The number of topics in mathematics education that have been researched is almost endless. The *Handbook of Research on Mathematics Teaching and Learning* (Grouws 1992), for example, contains twenty-nine chapters dealing with distinct topic areas for research in mathematics education. One topic that is just beginning to be investigated is the culture of the mathematics classroom. Questions include how and why some students are uncomfortable in a classroom where the teacher, rather than telling students if their answers are correct, expects them to decide for themselves if they are correct. Another research topic involves how to foster cooperation rather than competition when solving mathematics problems in the classroom. A third relatively new topic is that of teachers' and students' beliefs about mathematics. If teachers believe that mathematics is nothing more than rules to be memorized, how can they really teach problem solving? If students believe that "understanding" means giving a rule to solve a problem, will they ever really develop mathematical reasoning skills?

In addition to the fairly general research on mathematics classroom culture and beliefs, there is considerable research on teaching specific mathematical content. How does the wording of a subtraction problem influence the way a first-grade child interprets it? Why do students have so much trouble with fractions and ratios? Does teaching a graphical solution strategy for simultaneous linear equations help students to picture what a solution of simultaneous equations means?

Other major areas of research in mathematics education include the use of technology in instruction, ethnomathematics, assessment, individual differences in attitudes and achievement by gender, race, and social class, and international studies of achievement. Technology questions include effective use of graphing calculators to teach mathematics. Ethnomathematics includes how students see "out-of-school" mathematics differently than they see school mathematics.

These are examples of just a few of the hundreds of areas noted in Grouws (1992) where mathematics education research is taking place. Because the amount of research being conducted is so vast, it is difficult to keep abreast of developments in any one of the topic areas, let alone all of them. Understanding research, however, is one of the keys to both doing good research and to using the results of research to improve instruction (Research Advisory Committee 1995).

## RESEARCH METHODOLOGIES IN MATHEMATICS EDUCATION

One hundred years ago, "research" in mathematics education often meant describing methods of teaching specific topics. For example, reports were written about what sorts of applications of mathematics should be included in textbooks. By the early twentieth century, research in education in general tended to become more systematic and quantitative. In fact, the push toward quantitative research be-

came so strong that by the 1960s, most of the mathematics education research being done in the United States involved testing of students or teachers and then applying statistical tests to the data. A study by Zalman Usiskin (1972) of teaching high school geometry via transformations provides an example.

Usiskin studied thirty-six high school geometry classes. The eighteen classes comprising the experimental group were taught using a year-long curriculum based on transformational geometry. The control classes were taught using traditional Euclidean geometry textbooks. All students completed standard geometry achievement tests and standard mathematics attitude scales at the beginning and end of the study. Comparisons of test and attitude scale scores showed that students in the control classes did somewhat better than those in the experimental classes on "standard" geometry content. Attitude scale scores of both the experimental and control students declined over the period of the study, but there was no significant difference between the two groups.

Statistical research in mathematics still continues to produce useful information, but the predominance of statistical methodologies has faded. Diana Lambdin and Peter Kloosterman (1995) report that 63% of the manuscripts submitted to the *Journal for Research in Mathematics Education* in 1993 used some methodology other than statistics. One type of nonstatistical methodology that has become common is the clinical interview. Jan Mokros and Susan Jo Russell (1995), for example, interviewed seven students in each of grades 4, 6, and 8 to gain information on the students' conceptions of averages. The interviews, which included a series of open-ended problems about averages, were videotaped, audiotaped, and transcribed. By looking at the videotapes and transcripts, the researchers were able to see that most of the students exhibited inconsistencies in the way they thought about averages. The researchers found, however, that they were able to categorize the predominant way of thinking about averages for twenty of the twenty-one students. Although the complexity of the interview process restricted the number of students interviewed to twenty-one, a much greater understanding of the students' thinking processes was gained through the interviews than could have been possible through statistical research.

Another type of nonstatistical research that has been common outside of the United States for many years but is now becoming common in the United States is what is sometimes called a *teaching experiment*. Martin Simon and Glendon Blume (1994) used this technique to determine the extent to which twenty-six prospective elementary school teachers were able to understand the proportional relationship between area of a rectangular region and the lengths of the sides. As part of the experiment, Simon taught a mathematics content course while Blume helped in development of the course and observed each class session. Although a number of topics were covered in the course, Simon and Blume focused on data relating to the concept of multiplication. The two investigators reacted to videotapes of each class session independently and then shared their reactions with each other. Results of the study indicated that even though preservice elementary teachers have memorized the formula for area of a rectangle, they often do not have a sense of what area really involves. If true understanding of mathematics concepts such as area is going to be a goal of mathematics instruction, then more needs to be done to demonstrate the concept to all students, including preservice teachers. Otherwise, the teachers will continue to teach only what they know, and their students will be unlikely to gain true understanding.

A third common type of nonstatistical methodology is the case study of a small number of individuals. Melvin Wilson's (1994) study of what "Molly," a preservice secondary teacher, believed about teaching mathematics and understood about the concept of function exemplifies this type of research. In contrast to the teaching experiment, where the focus was on ways to teach a topic to help students master it, a case study may focus on how one or more individuals make sense of a topic. More specifically, a case study often includes substantial consideration of the characteristics of the individuals being studied as opposed to the characteristics of the instruction.

Similar to the interview and teaching experiment methodologies, the case study usually involves personal interviews, videotaping, and transcription. In Wilson's (1994) study, Molly was interviewed and observed throughout the ten-week period in which she was enrolled in a mathematics content course for secondary teachers. At the beginning of the study, Molly was asked to solve specific mathematics problems involving functions so that her notions about teaching could be analyzed in relation to her ability to work with functions. Findings of the study included a description of Molly's beliefs about mathematics and mathematics teaching. For example, Wilson found that Molly believed that teachers could have their classes do projects rather than worksheets, but only if the mathematics in the projects was related to what was being taught in the textbooks.

Molly also claimed to believe that understanding concepts was important, but that if students could get right answers, that was usually good enough.

Molly's instructor had hoped to instill the beliefs that mathematics projects were useful in a variety of situations and that understanding of mathematical concepts was important for long-term retention of skills. The study indicated that instruction had changed Molly's opinions on these issues, but she was still not thinking about them in the way the instructor had hoped. With respect to functions, the study showed that Molly evolved substantially in her understanding of the function concept and of the applicability of that concept in a variety of mathematical contexts. In his concluding remarks, Wilson pointed out that many of the recent reform documents in mathematics education urge that prospective mathematics teachers take more mathematics content courses. Molly, however, is a good example of someone who, even after considerable study of mathematics, had a limited conception of what it meant to understand mathematical ideas. In other words, Wilson argued, taking advanced mathematics courses does not necessarily help preservice teachers understand fundamental mathematics concepts better.

The three types of nonstatistical research mentioned are clearly only a sampling of the types of methodologies now being used in mathematics education research. In any field of research, the types of questions being asked should dictate the type of methodology used. Nonquantitative methods tend to be more useful than quantitative methods for determining the cognitive processes used by students and teachers, and it is certainly true that current trends in mathematics instruction focus on building instruction based on student thinking.

## CRITERIA FOR JUDGING THE QUALITY OF MATHEMATICS EDUCATION RESEARCH

One of the common criticisms of nonquantitative research is that there are no standards by which to judge it. From 1993 through 1995, members of the editorial board of the *Journal for Research in Mathematics Education (JRME)* discussed rigor in mathematics education research and the issue of whether there should be criteria for judging such research. Criteria for judging all educational research were suggested by Kenneth Howe and Margaret Eisenhart (1990) and helped editorial board members in the formulation of criteria specific to mathematics education. Although discussions are still taking place, and any criteria that do emerge will be subject to continual revision, Frank Lester and Diana Lambdin (in press) have summarized several criteria proposed by the board that appear to have considerable merit for judging the quality of both quantitative and qualitative research in mathematics education. Individually, these criteria guarantee little, but taken as a group, they appear to be good measures of research quality. Very briefly, the criteria are the following.

## Worthwhileness

Of all the criteria for judging the quality of mathematics education research, worthwhileness is clearly most important.

> Worthwhileness has to do with the potential of a research study for adding to and deepening our understanding of issues associated with mathematics teaching and learning. . . . Some key indicators of worthwhileness include: the study generates good research questions, the study contributes to the development of rich theories of mathematics teaching and learning, the study is clearly situated in the existing body of research on the question under investigation, and the study informs or improves mathematics education practice (Lester and Lambdin, in press).

In essence, worthwhileness confronts the issue of whether a specific research study provides mathematics educators with new knowledge. A study that does nothing to further understanding of mathematics learning or teaching is, in the eyes of the *JRME* editorial board, not particularly worthwhile.

## Goodness of Fit

Goodness of fit was defined by the *JRME* editorial board as the consistency between research questions and data collection and analysis techniques (see also Howe and Eisenhart 1990). Surveys, for example, are good for determining whether students like algebra better than geometry. Achievement tests are good for determining, in general, how many algebra and geometry facts students know. Surveys and tests, however, are usually not detailed enough to provide insight into why students like geometry or where their misunderstandings in geometry occur. To gain such insight, clinical interview methods are usually more effective. Questions such as how teachers' knowledge of geometry influence the way they

teach geometry can be answered best by observations of teachers. In short, goodness of fit between questions asked and research methods used is an important consideration in the quality of mathematics education research.

## Competence

Research reports should demonstrate that study planning, execution, and analysis were done in a competent manner. In a statistical experiment, it is important for instruments to measure what they purport to measure. For example, a study of mathematical problem solving that measures problem-solving ability based on a multiple choice test may be of lower quality than one where problem-solving ability is measured with open-ended, holistically scored problems. Investigators should be careful when generalizing beyond the populations that they have studied. Can findings about gifted students, for example, be generalized to average-ability students? When comparing two types of instruction, how sure can educators be that group differences resulted from the instruction rather than other factors?

In qualitative studies, competence issues, such as generalizability, are just as important, if not more important, than in quantitative studies. Moreover, because standards for competence are less clearly defined, investigators in qualitative studies need to be very careful to explain their methods. Data need to be thoroughly analyzed to make sure there is no evidence to refute conclusions that are made. Efforts should be made to collect data from a variety of sources and to explain the methods that were used to make sense of the data. Simply stated, research in mathematics education needs to be done in a way that makes sense to other investigators and to consumers of research.

## Openness

Investigators always bring their own perceptions, interpretations, and biases to their work. In qualitative and in quantitative studies, the researcher's expectations influence the type of data that are collected and the way that they are interpreted. For example, one investigator may decide that attitudes are not important in a study of problem solving and thus choose to ignore them. Another investigator may think that metacognition is too hard to measure and thus decide to ignore it. A third investigator may never have even considered attitudes or metacognition. All three investigators may have been acting competently based on their knowledge of important factors in the study of problem solving. What is important with respect to openness is that by carefully explaining what they did and why, the three investigators allow the reader of their research reports to decide if important variables had been omitted. Similar arguments can be made for being open about the population studied, the method of collecting data, and so forth.

## Ethics

Although a number of ethical questions could arise, two seem most important. First, because most research in mathematics education requires involvement of human subjects, it is essential that research be done in a way that minimizes risk and maximizes potential benefits for the individuals being studied. Investigators can reduce risk by doing all they can to protect the identity of subjects. They should also tell subjects, on a level that makes sense to them, about the purposes and findings of research they participate in. If subjects are asked to solve problems, going back and explaining the solutions to those who were unable to solve them is a good way for subjects to gain from participation in a study.

In addition to concern about human subjects, ethics concerns include adequate acknowledgment of individuals who contributed to the publication of a study. Specifically, graduate students, research assistants, cooperating teachers, and those who help in the conceptualization of the study and the writing of reports should be given appropriate recognition for their work.

## Credibility

Findings from research in mathematics education should be plausible and be based on data presented. Research sometimes yields surprising findings, but a good report contains possible explanations for those findings. Conclusions should follow from data presented in the study rather than from an author's preconceived notions about what the results might be. In addition, although clear exposition of ideas and reasoned argument for a point of view are essential elements of a good research report, consumers of research must be cautious of reports in which the eloquence of the writer masks flaws in the research design or conduct of the study.

In brief, the question of what constitutes high-quality research is a difficult one to answer. The preceding criteria (adapted from Lester and Lambdin,

in press) are the result of discussions at meetings of the *JRME* editorial board and of discussions at several national and international meetings. The criteria clearly overlap and are probably incomplete. However, manuscript reviewers and journal editors make implicit assumptions about quality when they accept or reject manuscripts. Making criteria for judging the quality of research explicit should help to provide an open process for accepting manuscripts for publication.

## SOURCES OF MATHEMATICS EDUCATION RESEARCH

Each year from 1971 to 1994, the *Journal for Research in Mathematics Education* has published an annotated bibliography, compiled by Marilyn Suydam and her colleagues, of research in mathematics education during the previous year. In 1994, Suydam and Brosnan found articles reporting research in mathematics education in sixty major English-language journals. Of these journals, five contained eight or more articles involving mathematics education research. The first three, *Journal for Research in Mathematics Education, Educational Studies in Mathematics*, and *Journal of Educational Psychology*, are intended primarily for researchers and thus many of the articles are fairly technical. The fourth and fifth journals, *School Science and Mathematics* and *Focus on Learning Problems in Mathematics,* are read by teachers as well as researchers and thus tend to include reports that are less complex than those in the first three journals. Less technical writing can also be found in a three-volume series about the implications of research for practice in mathematics education (Jensen 1993; Owens 1993; Wilson 1993).

*See also* Geometry, Instruction; Mathematics Education, Statistical Indicators; National Science Foundation (NSF)

## SELECTED REFERENCES

Grouws, Douglas A., ed. *Handbook of Research on Mathematics Teaching and Learning.* New York: Macmillan, 1992.

Howe, Kenneth, and Margaret Eisenhart. "Standards for Qualitative (and Quantitative) Research: A Prolegomenon. *Educational Researcher* 19(May 1990):2–9.

Jensen, Robert J., ed. *Research Ideas for the Classroom: Early Childhood Mathematics.* New York: Macmillan, 1993.

Kilpatrick, Jeremy. "A History of Research in Mathematics Education." In *Handbook of Research on Mathematics Teaching and Learning* (pp. 3–38). Douglas A. Grouws, ed. New York: Macmillan, 1992.

Lambdin, Diana V., and Peter Kloosterman. "Editorial." *Journal for Research in Mathematics Education* 26(May 1995):202–203.

Lester, Frank K., and Diana V. Lambdin. "The Ship of Theseus and Other Metaphors for Deciding What We Value in Mathematics Education Research." In *What Is Research in Mathematics Education?* Jeremy Kilpatrick and Anna Sierpinska, eds. Dordrecht, Netherlands: Kluwer (in press).

Mokros, Jan, and Susan Jo Russell. "Children's Concepts of Average and Representativeness." *Journal for Research in Mathematics Education* 26(Jan. 1995):20–39.

Owens, Douglas T., ed. *Research Ideas for the Classroom. Vol. 2. Middle Grades Mathematics.* New York: Macmillan, 1993.

Research Advisory Committee of the National Council of Teachers of Mathematics. "Research and Practice." *Journal for Research in Mathematics Education* 26(July 1995):300–303.

Simon, Martin A., and Glendon W. Blume. "Building and Understanding Multiplicative Relationships: A Study of Prospective Elementary Teachers." *Journal for Research in Mathematics Education* 25(Nov. 1994):472–494.

Suydam, Marilyn N., and Patricia A. Brosnan. "Research in Mathematics Education Reported in 1993." *Journal for Research in Mathematics Education* 25(July 1994):375–434.

Usiskin, Zalman P. "The Effects of Teaching Euclidean Geometry via Transformations on Student Achievement and Attitudes in Tenth-Grade Geometry." *Journal for Research in Mathematics Education* 3(Nov. 1972):249–259.

Wilson, Melvin R. "One Preservice Teacher's Understanding of Function: The Impact of a Course Integrating Mathematical Content and Pedagogy." *Journal for Research in Mathematics Education* 25(July 1994):346–370.

Wilson, Patricia S., ed. *Research Ideas for the Classroom: High School Mathematics.* New York: Macmillan, 1993.

PETER KLOOSTERMAN

## *REVOLUTION IN SCHOOL MATHEMATICS*

The title of a pamphlet prepared by the National Council of Teachers of Mathematics (NCTM) in 1961 to provide information for those interested in establishing new and improved mathematics programs reflecting the reforms of the "new math" of the late 1950s. The term *revolution* was applied because progress in mathematical research, automation, and computers was seen to be profound enough to warrant its use. The pamphlet described a number of the then current programs, including the School Mathematics Study Group, the University of Illinois Committee on School Mathematics, the University of Maryland Mathematics Project, the Boston College Mathematics Institute, the Ball State Teachers College Experimental Program, and the Developmental Project in Secondary Mathematics of Southern Illinois University. The pamphlet listed unifying themes of the programs, which went beyond such traditional topics as measurement and graphing to include "new math" topics such as set theory, mathematical structure (the pamphlet treated an example of a group of two elements in detail), nondecimal numeration systems, and probability and statistics. The pamphlet made it clear that NCTM did not specifically endorse any of the programs. A summary expressed the view that teachers were responsible for evaluating the new concepts for teachability and clarity of exposition but were not to be the sole determiners of what should be taught in schools.

*See also* Cambridge Conference on School Mathematics (1963); History of Mathematics Education in the United States, Overview; School Mathematics Study Group (SMSG)

### SELECTED REFERENCES

Jones, Phillip S., ed. *A History of Mathematics Education in the United States and Canada.* 32nd Yearbook. Washington, DC: National Council of Teachers of Mathematics, 1970.

National Council of Teachers of Mathematics. *The Revolution in School Mathematics.* Washington, DC: The Council, 1961.

JAMES K. BIDWELL
ROBERT G. CLASON

## RIEMANN, GEORG FRIEDRICH BERNHARD (1826–1866)

Made bold, imaginative, and far-reaching contributions to real and complex analysis, geometry, number theory, and topology. Riemann was "the man who more than any other influenced the course of modern mathematics" (Struik 1987). Born in a small village in Germany, Riemann studied mathematics with the great masters at Göttingen and Berlin and was later Peter Gustav Lejeune Dirichlet's successor at Göttingen. He was shy and modest and poor during much of his adult life and died of consumption while taking a "cure" in Italy. Riemann's publications were few, but all were ground-breaking.

Riemann defined the Riemann integral to investigate representability of functions by Fourier series. He introduced Riemann surfaces to give shape to his theory of complex functions, thus building bridges that linked analysis, geometry, and topology. It was in this context that he brought in fundamental topological notions. He introduced the Riemann-zeta function, defined by

$$\zeta(s) = \sum_{n=1}^{\infty} \frac{1}{n^s}$$

for complex $s$ with real part $> 1$ and extended to all complex $s$ by a process known as "analytic continuation," in order to study the distribution of primes. It is in this connection that he formulated the Riemann hypothesis, arguably still the most important unsolved mathematical problem, namely, that all the (nontrivial) roots of the Riemann-zeta function $\zeta(s)$ lie on the "critical line" $s = \frac{1}{2} + bi$.

He also broadened immensely the scope of geometry with his conception of Riemannian geometry, which proved vital sixty years later in Einstein's general theory of relativity. An elementary example of Riemannian geometry is the geometry of a sphere. Here the "straight lines" are the great circles of the sphere (e.g., the equator and the lines of longitude on the earth). The following are some properties of this geometry: there are no parallel lines, two points may determine more than one line, the sum of the angles of a triangle is greater than 180° and varies with the size of the triangle. This is, indeed, a "strange" geometry—quite different from Euclidean geometry but equally consistent (Burton 1991).

*See also* Non-Euclidean Geometry; Relativity

SELECTED REFERENCES

Bell, Eric Temple. *Men of Mathematics.* New York: Simon and Schuster, 1937.

Burton, David M. *The History of Mathematics: An Introduction.* 2nd ed. Dubuque, IA: Brown, 1991.

Freudenthal, Hans. "Riemann, Georg Friedrich Bernhard." In *Dictionary of Scientific Biography* (pp. 447–456). Vol. 11. Charles C. Gillispie, ed. New York: Scribner's, 1981.

Struik, Dirk J. *A Concise History of Mathematics.* 4th ed. New York: Dover, 1987.

ISRAEL KLEINER

# ROUNDING

Deals with the expression of numbers or measures through approximate values. People round whenever they use approximate, rather than exact, numerical information. Reasons for rounding include requirements of computation to achieve desired levels of accuracy or precision and such purposes as describing and comparing magnitudes, estimating, computing mentally, and judging the reasonableness of computed quantities. For example, a journalist may describe a $9,892,326.76 contract between two companies as a "10 million dollar deal."

Depending on the purpose of rounding, a given value can be rounded according to diverse criteria, as the examples in the table show.

| Original value | Rounded value | Criterion: rounded to | Explanation |
|---|---|---|---|
| 47.2392 inches | 47 in. | nearest inch | The original value, 47.2392 in., is between 47 and 48 inches but closer to the former value. So it is better to use 47 in. than 48 in. as an approximation. |
| | 4 ft | nearest foot | 47.2392 in. is closer to 4 ft (48 in.) than to 3 ft (36 in.) |
| | 47 1/4 in. | nearest quarter-inch | 47.2392 in. is closer to 47 1/4 in. (47.25 in.) than to 47 in. |
| | 47.2 in. | nearest tenth of an inch | 47.2392 in. is closer to 47.2 in. than to 47.3 in. |
| | 47.24 in. | nearest hundredth of an inch | 47.2392 is closer to 47.24 than to 47.23. |
| | 47.239 in. | nearest thousandth of an inch | 47.2392 is closer to 47.239 than to 47.240. |
| | 47.2392 in. | nearest ten-thousandth of an inch (or 4 places) | No rounding was done since the original value was presumed accurate to the nearest 0.0001 inch. |

Learning to round according to various criteria is helpful to elementary school children in their development of number sense.

How much rounding is acceptable depends on one's purposes. Consider π, the number of times the diameter of any circle fits into its circumference. π is sometimes expressed approximately by 3.14 or by the fraction, 22/7. For many purposes, such values will be sufficient. However, a manufacturer of precision ball bearings would find these approximations crude. A mathematician concerned with the properties of π itself—for example, how π can be expressed as a continued fraction—would view the value 3.14 as downright useless. Rather than adopt an arbitrary rule of thumb, such as "always round to two places!," we need to answer the question, "how accurate or precise does the answer have to be?"

*Rounding error* is the difference between the values before and after rounding. For example, if one rounds 2.53 to the nearest tenth, obtaining the rounded value, 2.5, there will be a rounding error of 0.03. The error, 0.03, represents a little more than 1% of 2.53. This error may be acceptable. But rounding error affects all subsequent calculations that rely on the rounded value. In computing compound interest on a large twenty-year loan, rounding error could lead to a miscalculation by thousands of dollars. In order to ensure accuracy, for example, to the nearest cent, it is necessary to keep information at least to tenths of cents.

*See also* Errors, Arithmetic; Estimation; Number Sense and Numeration; Pi

DAVID WILLIAM CARRAHER

# RUSSELL, BERTRAND ARTHUR WILLIAM (1872–1970)

Born in England, Russell entered Trinity College, Cambridge University, in 1890, where he was greatly influenced by George Edward Moore and Alfred North Whitehead. In 1893, he was seventh wrangler on part one of the mathematical tripos but a year later was awarded a first class with distinction in the natural sciences tripos. Based on his dissertation, *An Essay on the Foundations of Geometry*, Russell was offered a fellowship at Trinity College. Among his most important early publications was *A Critical Explanation of the Philosophy of Leibniz* (1899). In 1900, attending the International Congress of Philosophy in Paris, Russell met Giuseppe Peano and was immediately impressed by his emphasis upon symbolic logic. Russell had already discovered the paradox of classes (i.e., the class of classes that are not members of themselves). This, he showed,

doomed Gottlob Frege's *Grundgesetze der Arithmetik*, which tried to reduce mathematics to logic and prove thereby its self-consistency, just as it raised serious doubts about the foundations of set theory. Although Russell completed the first draft of *The Principles of Mathematics* at this time, it was not published until 1930.

Instead, in collaboration with his former teacher Whitehead, Russell set out to resolve the foundational questions he and other mathematicians were raising at the turn of the century. They sought to reduce all of mathematics to a symbolic logic to which a theory of types was applied in hopes of establishing a consistent framework from which the paradoxes of logic and set theory could be eliminated. This resulted in their monumental *Principia Mathematica*, published in three volumes between 1910 and 1913. Meanwhile, Russell had published a paper that was destined to become especially influential philosophically, "On Denoting," in *Mind* (1905). Increasingly, his concerns turned to issues in philosophy, and Ludwig Wittgenstein became a major influence on his thought. He began to travel widely and delivered the Lowell Lectures on "Our Knowledge of the External World" at Harvard University in 1914. Upon losing his position at Trinity College in 1916 over an antiwar pamphlet he had published, Russell was imprisoned two years later for sedition, having suggested the American army might be invited to break up strikes in England. During six months in prison, he wrote his *Introduction to Mathematical Philosophy* (1919), of which Anthony Quinton once said: "Comparing it with some of his later books, often spasmodic and casual, one might wish he had been imprisoned more often" (Quinton 1981, 904).

After World War I, Trinity College reinstated Russell and he continued to write on broad subjects and to lecture widely. In 1938, the University of Chicago invited him to return to the United States as a visiting professor, and thereafter he taught briefly at the University of California Los Angeles (UCLA) and at the City College of New York. After a contentious stint with the Barnes Foundation in Philadelphia, he spent the rest of World War II at Bryn Mawr, where he finished writing *A History of Western Philosophy* (1945), which was an international success and ensured Russell's financial security for the rest of his life. Invited back to Trinity College once more after the war (Russell was by now in his seventies), he became increasingly involved in antiwar and nuclear disarmament protests.

In 1955 Russell retired to Plas Penrhyn, Merionethshire, Wales. After four marriages and numerous affairs, he settled into a lionized old age. Although his contributions to mathematics and philosophy have been described as "less original" than those of either Wittgenstein or G. E. Moore, Russell was also a "listening-post for his age." Therein perhaps is the best explanation for his great success and popularity as a thinker and writer. Not only was he a member of the Royal Society of London (elected in 1906) but he won the Nobel Prize for literature in 1950. Russell died in Wales at age 98, in 1970.

*See also* Logic; Paradox; Philosophical Perspectives on Mathematics

## SELECTED REFERENCES

Clark, Ronald W. *The Life of Bertrand Russell.* New York: Knopf, 1976.

Crawshay-Williams, Rupert. *Russell Remembered.* London, England: Oxford University Press, 1970.

Gottschalk, Herbert. *Bertrand Russell; a Life.* New York: Roy, 1966.

Hendley, Brian Patrick. *Dewey, Russell, Whitehead: Philosophers as Educators.* Carbondale, IL: Southern Illinois University Press, 1986.

Park, Joe. *Bertrand Russell on Education.* Columbus, OH: Ohio State University Press, 1963.

Quinton, Anthony. "Bertrand Arthur William Russell." In *Dictionary of National Biography, 1961–1970* (pp. 901–908). E. T. Williams and C. S. Nicholls, eds. Oxford, England: Oxford University Press, 1981.

Russell, Bertrand. *Autobiography of Bertrand Russell.* 3 vols. Boston, MA: Little, Brown, 1969.

———. *Portraits from Memory and Other Essays.* New York: Simon and Schuster, 1956.

Sanisbury, Richard M. *Russell.* London, England: Routledge and Kegan Paul, 1979.

Tiles, Mary. *Mathematics and the Image of Reason.* London, England: Routledge, 1991.

Wood, Alan. *Bertrand Russell—The Passionate Sceptic. A Biography.* New York: Simon and Schuster, 1957.

JOSEPH W. DAUBEN

# RUSSIA

According to historical chronicles the first educational institutions in Russia were ancient Russian *uchilishcha* established after the adoption of Christianity in A.D. 988. Little is known about the curricula of these *uchilishcha*, which served children of landowners and merchants; however, historical records indicate counting and basic calculation skills were taught. The oldest known Russian mathematical work, by Kirik in 1136, utilized a unique Russian numerical system that permitted direct multiplica-

tion to obtain products rather than repeated addition. It is likely that this method was a form of Russian peasant multiplication frequently cited in Western textbooks (e.g., Musser and Burger 1991). Russian mathematicians of the time also knew how to calculate the beginning of Easter for each year. That problem and the problem of counting the days since Adam's birth were the most difficult considered. Based on the problems discussed and taught during the period from A.D. 900–1300, mathematical education in Russia seems comparable to that of other European countries of the time.

The next few centuries were marked by foreign invasions, especially by the Mongolians. As a result, education in general, and mathematics education in particular, experienced a severe decline. Fortunately, a new wave of "educational prosperity" began in the middle of the sixteenth century and reached its height at the beginning of the eighteenth century with the accession of Czar Peter the Great. As a result of Peter the Great's decree of January 14, 1701, a special School of Mathematical and Navigational Sciences was established in Moscow as the first public technical school in Europe. It accepted boys between the ages of twelve and seventeen of all "estates except bondservants." Many of the school's graduates became mathematics teachers for regular schools without further training. The school's enrollment of 200 pupils in 1703 increased to 500 by 1715. The curriculum, unique in Europe at that time, included arithmetic, geometry, trigonometry, astronomy, and navigation. The school was well equipped and even had its own observatory.

The first teachers at Peter the Great's School of Mathematical and Navigational Sciences were Leontyf Magnitsky and James Farvarson, a British "visiting professor." Magnitsky worked at the school for thirty-eight years and in 1703 published the famous textbook *Arithmetics,* which served as the main Russian mathematics textbook through the middle of the eighteenth century. Even today it is used by some teachers. The famous Russian scientist Michail V. Lomonosov studied from Magnitsky's *Arithmetics* and described it as "the gate to the sciences."

Russian public schools for girls were established during the eighteenth century to provide young women with a "liberal education." During this same period, additional schools for boys were opened. The curriculum of these schools included topics in arithmetic and geometry. The development of industry and trade, however, demanded more mathematical knowledge. In 1739, Euclid's *Elements* and in 1740,

Euler's *Universal Arithmetics* were translated into Russian for use in schools and universities.

The end of Peter the Great's reign in 1725 was followed by another century of decline. During the czarist regimes of the nineteenth century the curricula of public schools included arithmetic operations and some geometry, along with reading and writing. By 1850, some secondary schools required six to seven-and-a-half hours of mathematics instruction per week. At the end of the nineteenth century, mathematical education for the middle class in Russia had developed to a level comparable to that in Western European countries.

The significant progress in mathematical research at the end of the nineteenth century was responsible for initiatives in many nations to revise the mathematics curriculum. In Russia, several new publications promoted reform. In 1866, the Moscow Mathematical Society began publication of the *Proceedings of the Society*. In 1884, the famous Russian mathematician and educator I. S. Ermakov founded *The Magazine of Elementary Mathematics*. Both publications continue to this day.

In 1911–12 and 1913–14, the first and second Russian Congresses of Teachers of Mathematics were held and were attended by more than 1,000 participants. They dealt with the issues of mathematics as a science, methodology, psychology of teaching, manipulatives, and teacher preparation. They also discussed the results of Russia's participation in the International Commission on Mathematics teaching reform movement, influenced by Felix Klein of Germany and David Eugene Smith of the United States. The theme of the movement was that the school should pay more attention to "the art of using mathematical methods," not just to the expansion of the curriculum. In 1917, the Ministry of Education curtailed the changes recommended by the first and second congresses due to fears that the reforms promoted atheism and free thinking. The 1917 October Revolution and the subsequent Civil War prevented the third congress from being held.

After the October Revolution and the Civil War, the Bolshevik government attempted to provide education for all rather than for the select few. The demanding czarist curriculum, which included analytic geometry and introductory calculus in the eighth and ninth grades (ages 15–16), was too difficult for average students. In 1921, *The Mathematics Syllabus of the Unified Labor School* was published as a replacement for the older syllabus. The study of arithmetic was compressed into four years, and the function concept was emphasized as a major unifying theme. New teaching methods were introduced, including the Project Method, the Brigade Method, and the so-called System of General Development. These methods attempted to implement such dicta as "teach the whole child," "learn by doing," and "progressive" educational practice. Knowledge was classified not by subjects but by general themes, such as nature, labor, and society, complying in many respects with the educational philosophy of John Dewey. As a result of these reforms the level of educational achievement gradually decreased.

Governmental education decrees of 1931–32 returned Russian schools to a "class/lesson system" and to traditional subjects. Children entered school at age 6 or 7. Those completing the entire ten-year school graduated at age 17. The recommendations of the Teachers' Congresses of 1911–14 were resuscitated in part because industrialization in the 1930s and 1940s required the study of mathematics applications. Many new textbooks appeared during the next twenty years, including those of A. N. Kolmogorov, A. I. Markushevich, S. E. Lyapin, and G. M. Berman, as well as new editions of czarist-era texts by A. P. Kiselev and N. S. Rybkin. Many of these texts are still used as important references by Russian teachers.

The end of the 1950s and the beginning of the 1960s, when Chairman N. S. Khrushchev was the leader of the USSR, were the years of "polytechnic labor training." As part of the school curriculum, students were assigned to "polytechnic" or work-related activities. Mathematics courses were diluted as were those in many other subjects. The period of education was lengthened briefly from ten years to eleven years to permit time for polytechnic activities and work experience. The growth of computer technology, space research, and other scientific achievements during the Khrushchev era increased the need for highly qualified mathematicians. Khrushchev-era curricula, however, were not adequate to prepare such specialists; therefore, special schools for gifted students were established. Educators also emphasized out-of-class activities such as mathematics clubs and contests to promote recognition and development of mathematical talent.

After Khrushchev's departure, the new Brezhnev government appointed a commission led by A. N. Kolmogorov to revise the mathematics curriculum. The commission recommended a new curriculum that was implemented quickly, without careful research or field trial. The curriculum increased the time for mathematics in grades 1 through 8 to six hours per week, and in the high school to five hours per week. The theory of sets served as a unifying concept. Kolmogorov's curriculum was similar in many respects to "modern math" programs in the United States and was very innovative, but it was too abstract for many teachers and students accustomed to practical approaches. Teachers complained that it was not appropriate for all students. As a result, Kolmogorov's attempt to reform mathematics education in the USSR failed. Mathematics teaching returned to the ideas of the 1950s, based upon a unified curriculum for the entire country and only two approved textbooks.

A new wave of changes that began in the late 1980s was characterized by the inclusion of computer science as a separate subject and by attempts to provide different mathematics curricula for different students, such as mathematics for humanities, mathematics for mathematically gifted, mathematics for children with learning disabilities, etc. Many different textbooks appeared, including those by A. D. Aleksandrov, A. L. Verner, and V. I. Ryzhik for the junior high school and by M. I. Bashmakov for the high school. Teachers now can choose the text they prefer; however, schools cannot always provide the teacher's choice. The education system is now decentralized, and there is greater flexibility for experimentation by individual teachers. Current trends in Russian mathematics include the teaching of logic in the junior and senior high school, more attention to concepts than to facts and theorems, and emphasis on problem solving and the classroom use of calculators and computers. These trends encourage emphasis upon creative thinking skills rather than memorization of facts and algorithms.

*See also* China, People's Republic; England; France; Germany; Japan

## SELECTED REFERENCES

Cherkasov, R. S. "Otechestvennye Traditsii i Sovremennye Tendentsii v Razvitii Shkol'nogo Matematicheskogo Obrazovaniya." *Matematika v Shkole* 93(4)(1993): 73–77; 93(5)(1993):75–79; 93(6)(1993):75–76.

Curcio, Frances R. "The Development of Current Trends in Mathematics Education in Soviet Schools." In *Soviet Politics and Education* (pp. 361–372). Frank M. Sorrentino and Frances Curcio, eds., Lanham, MD: University Press of America, 1986.

Lavrovsky, Nikolai. *O Drevne-Russkikh Uchilishchakh.* Khar'kov, Russia: Universitet, 1854.

*Matematichesky Entsiklopedichesky Slovar'.* Moskva, Russia: Sovetskaya Entsiklopediya, 1988.

Musser, Gary L., and William F. Burger. *Mathematics for Elementary School Teachers: A Contemporary Approach.* 2nd ed. New York: Macmillan, 1991.

Vogeli, Bruce. *Soviet Secondary Schools for the Mathematically Talented.* Washington, DC: National Council of Teachers of Mathematics, 1968.

SERGEI MIKHELSON
BRUCE R. VOGELI

# S

## SCHOOL MATHEMATICS STUDY GROUP (SMSG)

Organized in 1958, through National Science Foundation funding, under the direction of Edward G. Begle of Yale University. SMSG moved with Begle to Stanford University in 1961. The initial charge was to provide a national reformed school mathematics curriculum based on the several experimental programs begun the previous decade. Begle was assisted by an advisory panel and a committee, dominated by mathematicians. Beginning with secondary school, SMSG eventually wrote sample texts for grades K–12, teacher's manuals, and support materials. The intent was to influence rather than dictate curriculum. The sample texts emphasized understanding through concepts of mathematical structure (the mathematician's view) rather than computational processing skills (the popular view). National acceptance of this approach created mass revisions in commercial text series. The advisory board last met in 1972 and issued a statement on objectives and minimum goals: objectives should be hortatory, feasible, and verifiable; they should represent minimum goals, should have a clear purpose and should be expressed in terms of student behaviors. No affective goals were offered. They advocated one pedagogical objective: "Teach understanding of a mathematical process before developing skill in the process" (SMSG 1972, 18).

*See also* Cambridge Conference on School Mathematics (1963); History of Mathematics Education in the United States, Overview; *Revolution in School Mathematics*

### SELECTED REFERENCES

Begle, Edward G. "SMSG: The First Decade." *Mathematics Teacher* 61(Mar. 1968):239–245.

Jones, Phillip S., ed. *A History of Mathematics Education in the United States and Canada.* 32nd Yearbook. Washington, DC: National Council of Teachers of Mathematics, 1970.

School Mathematics Study Group. *Newsletter* (38) (Aug. 1972).

JAMES K. BIDWELL
ROBERT G. CLASON

## SCHOOL SCIENCE AND MATHEMATICS ASSOCIATION (SSMA)

A not-for-profit organization of educators from the range of kindergarten through graduate school who have an interest in science and mathematics education. The Association, formerly the Central

Association of Science and Mathematics Teachers, has existed and published its journal since 1901.

SSMA's primary focus is the integration of science and mathematics content in K–12 schools. The association is interested in all kinds of dialogue relating to the improvement of instruction in the two subjects. Articles in its journal range from practical classroom activities to research reports. All research articles are written to be interesting and easily read and understood by the classroom teacher. The journal regularly includes feature articles, editorials, mathematical problems with solutions in subsequent issues, historical notes about the association, book and software reviews, association news, and an annual index (in the December issue). The association also publishes a newsletter for its individual members, *SSMArrt,* four times a year. Also, occasional monographs are produced in two series, Classroom Activities and Topics for Teachers.

Each year the association holds its annual meeting in conjunction with a local state science or mathematics teachers association or with a group of state associations. SSMA is heavily committed to helping the state organizations develop dialogue between the disciplines of science and mathematics.

*See also* School Science and Mathematics Association (SSMA), Historical Perspective

### SELECTED REFERENCES

Barufaldi, James P., ed. "Science and Mathematics for Early Childhood." *School Science and Mathematics* 92(Jan. 1982).

Berlin, Donna. *A Bibliography of Integrated Science and Mathematics Teaching and Learning Literature.* Bowling Green, OH: SSMA, 1991.

Duncan, David L., and Bonnie H. Litwiller. *Problem Solving with Number Patterns.* Bloomsburg, PA: SSMA, 1987.

Farmer, Walter A., and Margaret A. Farrell. *Activities for Teaching K–6 Math/Science Concepts.* Bloomsburg, PA: SSMA, 1989.

House, Peggy A. *Interactions of Science and Mathematics.* Bloomsburg, PA: SSMA, 1980.

Peterson, John C., ed. "Career Education." *School Science and Mathematics* 76(Apr. 1976).

*School Science and Mathematics.* Bloomsburg, PA: SSMA. A monthly.

Underhill, Robert G., ed. "Integrated Mathematics and Science." *School Science and Mathematics* 94(Jan. 1994).

Whitmer, John C. *Spreadsheets in Mathematics and Science Teaching.* Bloomsburg, PA: SSMA, 1992.

*Wingspread Conference Plenary Papers.* Bloomsburg, PA: SSMA, 1994.

DONALD L. PRATT

## SCHOOL SCIENCE AND MATHEMATICS ASSOCIATION (SSMA), HISTORICAL PERSPECTIVE

Organized as the Central Association of Science and Mathematics Teachers in April 1901, a time when mathematics teachers were often also science teachers. SSMA representatives have served on a variety of reform committees and commissions. In its activities and through its journal, *School Science and Mathematics,* the association continues to focus on teaching the interrelation of mathematics and science and the application of mathematics.

*See also* School Science and Mathematics Association (SSMA)

### SELECTED REFERENCE

Jones, Phillip S., ed. *A History of Mathematics Education in the United States and Canada.* 32nd Yearbook. Washington, DC: National Council of Teachers of Mathematics, 1970.

JAMES K. BIDWELL
ROBERT G. CLASON

## SCIENTIFIC NOTATION

Notation using negative and non-negative powers of 10 (see Figure 1) for convenient calculations with extremely large or small numbers, like the distances between stars or the dimensions of atoms.

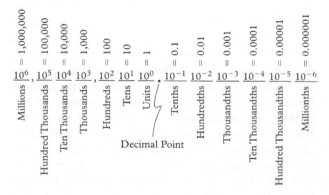

**Figure 1**

To write a number in scientific notation, we write it as the product of a number between 1 and 10 and a power of 10.

*Example 1* Write 6,237 in scientific notation.
Solution

$$6{,}237 = 6.237 \times 1000$$
$$= 6.237 \times 10^3$$

number between 1 and 10    power of 10

*Example 2* Write 0.006 in scientific notation.
Solution

$$0.006 = 6 \times 0.001 = 6 \times 10^{-3}$$

The following is a procedure for writing decimal-based numbers in scientific notation:

*To write a number in scientific notation:*

1. Start at the right of the first nonzero digit.
2. Count the number of digits from this starting point to the decimal point. (Remember, a decimal point is implied at the end of every whole number, even when it is not written.) This number will be the exponent, or power, of 10.
3. The sign of the exponent is positive (+) if you count to the right and negative (−) if you count to the left.
4. Write the given number with the decimal point after the first nonzero digit and multiply by the power of 10 found in steps 2 and 3.

*Example 3* Write 605,000 in scientific notation.
Solution

$$605{,}000 = 6\ 0\ 5\ 0\ 0\ 0$$

5 places to the right

$$= 6.05 \times 10^5$$

*Example 4* Change 0.000751 to scientific notation.
Solution  Start at the right of the 7 (the first nonzero digit) and count four places to the left to the decimal point, which yields an exponent of −4.

$$0.000751 = 0\ .\ 0\ 0\ 0\ 7\ 5\ 1$$

4 places to the left

$$= 7.51 \times 10^{-4}$$

To change a number from scientific notation to ordinary notation, just reverse the process.

*To change from scientific notation to ordinary notation:*

1. Move the decimal point as many places as the value of the exponent.
2. Move the decimal point to the right if the exponent is positive (+) and move it to the left if the exponent is negative (−). Add zeros if necessary.

*Example 5* Change $3.4 \times 10^5$ to ordinary notation.
Solution  Since the power of 10 is +5, we move the decimal point 5 places to the *right* (we must add four zeros).

$$3.4 \times 10^5 = 3.\ 4\ 0\ 0\ 0\ 0 = 340{,}000$$

*Example 6* $5.21 \times 10^{-4} = 0.0\ 0\ 0\ 5.\ 2\ 1 = 0.000521$
We moved the decimal point 4 places to the *left,* since the exponent is −4. We had to add three zeros.

*Example 7* Evaluate $120{,}000 \times 8{,}000{,}000$.
Solution

$$120{,}000 \times 8{,}000{,}000 = 1.2 \times 10^5 \times 8.0 \times 10^6$$
$$= (1.2)(8) \times 10^5 \times 10^6$$
$$= 9.6 \times 10^{11}$$

If you performed this calculation on most calculators, the answer would look like this: 9.6  11, which means $9.6 \times 10^{11}$.

*Example 8* Evaluate $(5{,}400{,}000)(700{,}000)$.
Solution

$$(5{,}400{,}000)(700{,}000) = (5.4 \times 10^6)(7.0 \times 10^5)$$
$$= (5.4)(7) \times 10^6 \times 10^5$$
$$= 37.8 \times 10^{11}$$

Notice that the answer is not in scientific notation since 37.8 is not a number between 1 and 10. To remedy this, rewrite 37.8 as follows:

$$37.8 \times 10^{11} = 3.78 \times 10^1 \times 10^{11} = 3.78 \times 10^{12}$$

Now the answer is in scientific notation.

*See also* Decimal (Hindu-Arabic) System

### SELECTED REFERENCE
Bloomfield, Derek I. *Introductory Algebra.* St. Paul, MN: West, 1994.

DEREK I. BLOOMFIELD

# SECONDARY SCHOOL MATHEMATICS

Current trends strengthen the vision of all high school students actively learning mathematics. The *Curriculum and Evaluation Standards for School Mathematics,* published by the National Council of Teachers of Mathematics (NCTM) (1989), recommended what mathematics students should learn in high school. Many curriculum reform projects, now in progress, are actually developing programs designed to bring the vision to reality. At the turn of the 1990s there was a greater variety of mathematics programs offered in senior high schools than at either the elementary or middle grade levels.

Most high school mathematics programs followed two divergent paths: the traditional college preparation program and a noncollege program. The latter path frequently subdivided into a general mathematics sequence and a remedial sequence. Any movement of students from one path to the other was almost always away from the college-bound program to the other program. Although many high schools often reported that they did not "track" students in mathematics, they did refer to past grades and standardized test scores to "advise" students about which mathematics classes to take. Schools usually had prerequisite requirements for enrollment in college preparation courses, beginning with Algebra I. This first class in algebra became the determining factor upon which course selection was based for the rest of the program. Students who were successful (high grades and/or high test scores) could continue in the college preparation program, the others transferred to the non-college-bound mathematics program for their remaining mathematics credits.

## RECENT HISTORY

Zalman Usiskin introduced the 1985 NCTM Yearbook by stating, "There have been, in the past thirty years, a revolution (the new math) and a rebellion (back-to-basics) in school mathematics. Recent reports from inside and outside mathematics education suggest that today we either are or should be embarking on, a second revolution" (Usiskin 1985, 1).

The greatest single event receiving credit for awakening the educational establishment to action in the late 1950s was, of course, the Soviet Union's successful launching of the space vehicle, *Sputnik,* in 1957. The United States reacted in several ways, but the reaction of particular interest here was the effort to reform mathematics and science education. The view then focused on the need for a new generation of engineers and scientists in order to "win" the race for space. The culmination of several federal grants to improve mathematics curriculum efforts became known as the "new math." Although many educators talk of the new math curriculum movement as though everyone understands what the term means, the term *new math* was itself not well defined in a mathematical sense. Of the content changes that accompanied the new math curriculum, some topics have continued on and are present in the "normal" mathematics curriculum today, and some have been left on the sidelines. The adjectives *rigorous, logical, abstract,* and *structured* are frequently identified with the new math. Formalism seemed to be a desired end in itself.

During the 1960s, many textbook publishers jumped on the new math bandwagon, and began to provide materials for teaching the new math programs. They proudly announced that their texts incorporated the recommendations of the School Mathematics Study Group, SMSG (a major force in the new math movement), and provided the structure of mathematics and the discovery of patterns. There was no emphasis on computational skills. In his book *Why Johnny Can't Add: The Failure of the New Math,* Morris Kline (1973) presented a view that although the "old" curriculum, pre-1960, was in need of reform, the new math movement made matters worse. Kline referred to the formalism, rigor, and precise language, also mentioned by others, as hindrances to student learning in mathematics. He made another important observation, that is, that mathematics had become a field of study in itself and had gotten away from its application to science and other areas. The new math seemed to work well for high-ability students, but the curriculum underwent some modification so that more students would be successful in it.

Because from 1964 through 1974 no report of national scope dealing with the secondary mathematics curriculum appeared, one could conclude that the profession seemed content with the revolution after the corrections. The general public, however, was not satisfied. Standardized test scores were dropping, and the decline was blamed on the curriculum (Usiskin 1985). During the later 1970s, textbook publishers began changing their emphasis. It was the failure of students to perform computational skills at a level that the public felt to be acceptable that received a great deal of attention in the press. A definite change in response to this perception was seen in the books brought forth in the late 1970s. The new math in particular was blamed for the decline of mathematics test scores. But the scores on the verbal sections of the SAT (SAT V) also dropped—even more than the mathematics scores (SAT M). In addition, the back-to-basics movement did not reverse the downward trend; rather, scores continued to drop. This may well have indicated that the causes of declining test scores involved more than the mathematics curriculum in the schools.

In 1977, NCTM published a short position statement on basic skills. The statement preceded the more frequently cited position paper adopted by the National Council of Supervisors of Mathematics (NCSM), and both were published in the October issue of *Arithmetic Teacher*. After agreeing that there was a genuine concern over falling mathematical achievement scores, the NCTM paper pointed out a danger in the back-to-basics curriculum if the basics were limited to an algorithmic, computational approach to mathematics. "We are deeply distressed, however, by the danger that the "back-to-basics" movement might eliminate teaching for understanding. It will do citizens no good to have the ability to compute if they do not know what computations to perform when they meet a problem. The use of the hand-held calculator emphasizes this need for understanding: one must know when to push what button" (NCTM 1977, 18). The NCTM statement provided a strong lead for the NCSM position paper, which identified the following basic skill areas:

1. problem solving,
2. applying mathematics to everyday situations,
3. alertness to the reasonableness of results,
4. estimation and approximation,
5. appropriate computational skills,
6. geometry,
7. measurement (both metric and customary systems),
8. reading, interpreting, and constructing tables, charts, and graphs,
9. using mathematics to predict,
10. computer literacy.

These topics became the focus for *An Agenda for Action* (NCTM 1980). The publication of these topical lists was designed to enhance the view of what comprised basic mathematics. During this time, the NCTM conducted a major survey of mathematics teachers concerning their perceptions of the mathematics curriculum. The survey was the focus of the *Priorities in School Mathematics* (PRISM) project. The results of this project supported the NCTM's earlier *An Agenda for Action* and was published in 1981 (NCTM 1981).

In 1985, Usiskin stated, "If our present curriculum were an experimental curriculum that we were

testing, we would be forced to pronounce it a failure for many of our students" (Usiskin 1985, 19). Such views within mathematics education set the stage for initiating the current reform efforts. The 1985 NCTM Yearbook, *The Secondary School Mathematics Curriculum,* helped pave the way for the curriculum and evaluation standards that the NCTM would be creating in the near future. The collected chapters formed both a vision of the difficulties associated with low achievement patterns in mathematics and some structure for what should be taught and ways to teach mathematics better than the current level of practice. The topics in this yearbook included views on the need for curriculum reform in mathematics, uses of technology in teaching mathematics, discrete mathematics, a mathematically integrated curriculum, and the need for strengthening the curriculum for non-college-bound students.

The Annenberg/CPB *Guide to Math and Science Reform* (1995) listed four major (funding in excess of $1 million) curriculum projects that were designed to impact the secondary mathematics curriculum. All of these projects are currently classified as ongoing, and most also have implications at the elementary level. Project *EQUALS: A Mathematics Equity Program* was begun in 1977 in California. A main goal of this project is to attract and retain women and minority students in mathematics. The program also focuses on the use of manipulative materials, cooperative learning, and authentic assessment, as well as promoting student awareness of potential career opportunities. Since its beginning, over seventy-thousand educators from forty states have participated in EQUALS training. There are currently training program sites in twenty-three states.

The *California Mathematics Project,* begun in 1982, focuses on assisting teachers with both the selection and development of curricular materials that can be used in the classrooms of those participating in the institutes. That same year, 1982, the *National Leadership Program* for teachers began. This effort was designed to improve teaching across the nation with a particular emphasis on the appropriate use of technology in teaching secondary mathematics.

The *University of Chicago School Mathematics Project* (UCSMP) began in 1983. The secondary component of this program has developed a curriculum for students from grade 7 *(Transition Mathematics Prealgebra)* through grade 12 *(Precalculus and Discrete Mathematics)*. The materials developed have been published by the Scott Foresman Company as a textbook series. One impact of this program has been a new emphasis on teaching the first course in algebra to average eighth-grade students.

## NAEP

The National Assessment of Educational Progress Test (NAEP) had shown some of the deficiencies that were and remain a great concern to mathematics educators. The conclusions from reviewing the early NAEP data seem to indicate that at all levels the students are doing much better when asked to simply demonstrate computational skills rather than to apply the same kinds of skills in problem situations, particularly two-step problems. The published NAEP test results from 1988 showed no improvement in this situation. The performance of students indicated a need to improve the mathematics curriculum for all students (Lindquist 1989). Thus the poor level of achievement in problem solving remained a major concern.

## NCTM Efforts

During the late 1980s, NCTM was drafting and editing standards for the mathematics curriculum. While many forces were behind the decision to create these standards, disappointment in the standardized test scores was certainly one factor. The most significant impact on reform in mathematics was the publication by NCTM of the *Curriculum and Evaluation Standards for School Mathematics* (1989).

"Historically, the purposes of secondary school mathematics have been to provide students with opportunities to acquire mathematical knowledge, skills, and modes of thought needed for daily life and effective citizenship, to prepare students for occupations that do not require formal study after graduation, and to prepare students for postsecondary education, particularly college" (NCTM 1989, 123). The fourteen high school standards form a structure for the topics to be covered in a mathematics core curriculum intended for all high school students. College-bound students are expected to learn additional topics in mathematics and with a deeper level of understanding.

1. *Mathematics as problem solving.* A focus on the process of problem solving remained at the center of the curriculum.
2. *Mathematics as communication.* This involves students being able to express their understandings both orally as well as in writing.
3. *Mathematics as reasoning.* Reasoning involves students making and testing conjectures, judging the validity of arguments, and making valid arguments themselves.
4. *Mathematical connections.* This involves making connections between various branches of mathematics as well as between mathematics and other disciplines.
5. *Algebra.* Algebra is that part of the mathematics curriculum where the use of variables to represent situations is a main focus.
6. *Functions.* The students learn to model real-world events and represent equations with graphs of functions.
7. *Geometry from a synthetic perspective.* The focus is on application of geometric properties of figures used to model problem situations.
8. *Geometry from an algebraic perspective.* This involves more of a coordinate geometry approach and encompasses transformations and working with vectors.
9. *Trigonometry.* The focus for all students is the application of trigonometry to solving problems involving triangles and periodic situations using sine, cosine, and tangent functions.
10. *Statistics.* All students should expand their abilities in using statistics to include drawing inferences from data.
11. *Probability.* Students should have an understanding of randomness and an ability to use simulation to generate or estimate probabilities.
12. *Discrete mathematics.* Because of computers, many problem situations can be modeled using sequences or recurrence relations. Other topics include enumeration problems and finite probabilities.
13. *Conceptual underpinnings of calculus.* This includes the traditional precalculus topics such as examination of relative maxima and minima of functions. Also included would be the process of infinite sequences and series as well as areas under a curve.
14. *Mathematical structure.* Here students are expected to understand the relationship of the real number system to other number systems and relate various systems with the problem situations that can be represented by each. This includes efforts to bring together seemingly unrelated areas of mathematics in their problem-solving applications.

## Core Curriculum

The concept of a core curriculum for all senior high students, proposed in the standards, is probably the most dramatic departure from the traditional curriculum. There are suggestions as to how the differing needs and backgrounds of students can be considered in the core program, and there are additional lists of topics that are specified for those students preparing for college study. Some school districts have already eliminated the general mathematics curriculum option from their course offerings and expect all students to take Algebra I.

## Other Trends

Shortly after the publication of the *Curriculum and Evaluation Standards for School Mathematics, Reshaping School Mathematics: A Philosophy and Framework for Curriculum* (1990) was published by the National Research Council. This book was designed to provide support for the reformation of the mathematics curriculum in general and for the NCTM standards in particular. While many teachers eagerly embraced the NCTM standards, many wondered just how they could alter what they were currently doing in order to do a "better" job of teaching mathematics. To help fill the need for materials that secondary mathematics teachers could use, the NCTM developed the high school Addenda Series. These books were created to provide activities that would be consistent with the standards and actively involve students. The high school Addenda Books include *Algebra in a Technological World, Connecting Mathematics, A Core Curriculum: Making Mathematics Count for Everyone, Data Analysis and Statistics across the Curriculum,* and *Geometry from Multiple Perspectives.*

## MORE RECENT PROJECTS

Of the eighty-three curriculum projects, cited in the Annenberg *Guide,* that contained some emphasis on secondary mathematics and that began in 1990 or later, fifty-one had an exclusive emphasis on the secondary grades (Annenberg/CPB 1995). It is noteworthy that since the publication of the NCTM standards the number of funded curriculum projects has risen dramatically. Many of these curriculum projects are tied to ongoing reforms in the curriculum and teaching of science.

Beginning in 1991, the National Science Foundation (NSF) began the funding of systemic reform projects that were coordinated at the state level and included both mathematics and science. There were twenty-four state-level projects (including Puerto Rico) in progress in 1994. The NSF featured the programs from Nebraska, Delaware, Michigan, Connecticut, South Dakota, Maine, and Montana in its publication *Foundation for the Future* (NSF 1994). In

addition to the state systemic initiatives, the NSF also featured several local programs in the booklet.

The Gateways III Conference, funded by the NSF, focused on several high school projects in 1994. Of the seven projects represented, four are also listed in the Annenberg *Guide.* The represented projects included *Applications/Reform in Secondary Mathematics* (ARISE), which was designed to be a 9–11 curriculum with a strong applications overview. The *Core-Plus Mathematics Project,* being developed in Michigan, planned a complete first three-year high school mathematics curriculum and then a fourth year transition to a college mathematics course. As the name implies, the *Connected Geometry Project,* being developed in Newton, Massachusetts, centered around geometric investigations and links with other topics and other subject disciplines. The *Interactive Mathematics Program* (IMP), from California, is a program that planned to feature the extensive use of nonroutine problems in a curriculum that would also feature group investigations, use of graphing calculators, extensive writing, and alternative methods of assessment. The *MATH Connections: A Secondary Mathematics Core Curriculum Initiative,* developed in Hartford, Connecticut, planned to use a unified approach to learning mathematics and blend the learning around a series of thematic threads. The *Systemic Initiatives for Montana Mathematics and Science* and the *University of Chicago School Mathematics Program* also were involved (Gateways III 1994).

The Center for Occupational Research and Development (CORD) has developed curricula in mathematics, sciences, technology, and communications that are specifically designed for those high school students who do not plan to pursue a university (four-year college) degree after high school. Of particular interest here is the *Applied Mathematics* program. The program features activities that integrate topics from algebra, geometry, and basic trigonometry in ways that involve hands-on learning and are related to the world of work. Much of the original support for this program came from the vocational and technical departments at state departments of education. This program might provide a solid link between high school academic mathematics teachers and the technical/vocational teachers. The skills covered in the two-year *Applied Mathematics* program have been correlated to the *Curriculum and Evaluation Standards,* and the *Applied Mathematics* program includes all of the topics that should be covered by all students. The State of Florida's Department of Education has determined that the completion of both Applied Mathematics I and Applied Mathematics II is equivalent to completion of Algebra I.

Continuing the view that mathematics reform is for all students, Webb and Romberg edited *Reforming Mathematics Education in America's Cities* (1994). This book reflects the study of collaboration as a process through the Urban Mathematics Collaborative Project. This continuing project began in 1984 with support from the Ford Foundation and has had an impact on eleven urban centers across the nation.

*See also* Curriculum Trends, Secondary Level; Teacher Development, Secondary School; Teacher Preparation, Secondary School

### SELECTED REFERENCES

Annenberg/CPB Math and Science Project. *The Guide to Math and Science Reform.* No. 3, Spring 1995. See web site http://www.learner.org.

Gateways III Conference. "Moving Forward." Washington, DC: National Science Foundation, October 6–9, 1994.

Hirsch, Christian R., ed. *The Secondary School Mathematics Curriculum.* 1985 Yearbook. Reston, VA: National Council of Teachers of Mathematics, 1985.

Kline, Morris. *Why Johnny Can't Add: The Failure of the New Math.* New York: St. Martin's, 1973.

Lindquist, Mary M., ed. *Results from the Fourth Mathematics Assessment of the National Assessment of Educational Progress.* Reston, VA: National Council of Teachers of Mathematics, 1989.

National Council of Supervisors of Mathematics. "Position Paper on Basic Skills." *Arithmetic Teacher* 25(Oct. 1977):19–22.

National Council of Teachers of Mathematics. *An Agenda for Action: Recommendations for School Mathematics of the 1980s.* Reston, VA: The Council, 1980.

———. *Assessment Standards for School Mathematics.* Reston, VA: The Council, 1995.

———. *Curriculum and Evaluation Standards for School Mathematics.* Reston, VA: The Council, 1989.

———. "Position Statement on Basic Skills." *Arithmetic Teacher* 25(Oct. 1977):18.

———. *Principles and Standards for School Mathematics.* Reston, VA: The Council, 2000.

———. *Priorities in School Mathematics: An Executive Summary of the PRISM Project.* Reston, VA: The Council, 1981.

———. *Professional Standards for Teaching Mathematics.* Reston, VA: The Council, 1991.

National Research Council. *Reshaping School Mathematics: A Philosophy and Framework for Curriculum.* Washington, DC: National Academy Press, 1990.

National Science Foundation. *Foundation for the Future: The Systemic Cornerstone.* Washington, DC: The Foundation, 1994.

Usiskin, Zalman. "We Need Another Revolution in Secondary School Mathematics." In *The Secondary School Mathematics Curriculum* (pp. 1–21). Christian R. Hirsch and Marilyn J. Zweng (eds.). 1985 Yearbook. Reston, VA: National Council of Teachers of Mathematics, 1985.

Webb, Norman L., and Thomas Romberg, eds. *Reforming Mathematics Education in America's Cities, The Urban Mathematics Collaborative Project.* New York: Teachers College Press, 1994.

RICHARD A. AUSTIN

# SERIES

An expression of the form

$$a_1 + a_2 + \cdots + a_n + \cdots = \sum_{j=1}^{\infty} a_j$$

An example of such a series is given by $1 + 1/2^2 + 1/3^2 + 1/4^2 + \cdots + 1/n^2 = \sum_{n=1}^{\infty} 1/n^2$, a series the sum of whose terms is known to equal $\pi^2/6$. Occasionally the word *series* refers to a finite sum of the type $a_1 + a_2 + \cdots + a_n$, although this is less common than its use in referring to infinite sums.

Classical Greek mathematics touched upon the subject of series, although much of its development was deferred until modern times. The Parisian scholar Nicole Oresme (1323?–1382) used a graphical procedure to sum several infinite series, and he proved that the harmonic series $1 + (1/2) + (1/3) + \cdots + (1/n) + \cdots$ diverged. A contemporary of Oresme, the Indian mathematician Madhava of Sangamagramma (1340?–1425), obtained a value of $\pi$, accurate to eleven decimal places, by using an infinite-series expansion for the inverse tangent function (Joseph 1991). Madhava is one of the mathematicians associated with the so-called school of Kerala, which flourished for two centuries in the south of India. Another of the mathematicians associated with the Kerala school is Nilakantha (1445–1555), whose works include infinite-series expansions. Nilakantha used geometric arguments to establish the validity of the formula

$$\sum_{n=1}^{N} \frac{n(n+1)}{2} = \frac{N(N+1)(N+2)}{6}$$

In his book, the *Tantrasangraha,* he goes further than Western mathematicians such as Isaac Newton (1642–1727) and James Gregory (1638–1675) by providing error approximations for truncations of the series expression for $\pi$ (Roy 1993). Two Chinese writers of the Sung Dynasty (900–1279), Shen Kao and Yang Hui, studied series in a geometric fashion paralleling the method of Nilakantha.

Newton is generally credited with dealing so successfully with series that mathematicians in Europe finally overcame a distaste for infinite processes that dated back to the days of classical Greek mathematics. Infinite series also played a major role in the work of Gottfried Leibniz (1646–1716), the European co-parent (with Newton) of the calculus. Work on series was done throughout the eighteenth and nineteenth centuries by Europe's great analysts, and space does not permit doing justice to the development of the subject in those two hundred years. Examples of some of the accomplishments of those years include a use of the ratio test by Carl Friedrich Gauss (1777–1855) to establish the convergence of his hypergeometric series (although the ratio test had been developed much earlier) and the work of the French mathematician Augustin-Louis Cauchy (1789–1857) on characterizing the convergence of a series in $n$ space in terms of the behavior of the "tail" of that series: $\sum a_n$ converges if and only if

$$\lim_{m,n \to \infty} (a_m + a_{m+1} + \cdots + a_n) = 0$$

This criterion is occasionally easier to check than the definition of convergence itself.

## MATHEMATICAL EXPLANATION

It is convenient to divide series into two categories: those that are finite and those that are infinite. Of these, the second is by far the more important, and even the word *series* is commonly restricted to refer only to the second category. Nonetheless, arithmetic series are examples of the first category, and their sums can be expressed in closed form:

$$\sum_{n=1}^{N} \{a + (n-1)\,d\}$$

$$= a + (a + d) + \cdots + \{a + (N-1)\,d\}$$

$$= \frac{N}{2} \{2a + (N-1)\,d\}$$

A formula for the sum of a finite geometric series also can be given:

$$\sum_{n=1}^{N} ar^{n-1} = a + ar + ar^2 + \cdots + ar^{N-1}$$

$$= \frac{a\,(1 - r^N)}{1 - r}$$

If $|r| < 1$, this formula yields a sum for an infinite geometric series in which the upper limit of summation $N$ is replaced by infinity, but a word about what is meant by an infinite series and the expression "the sum of an infinite series" is required first.

An infinite series is an expression of the sort

$$a_1 + a_2 + \cdots + a_n + \cdots$$

Associated with this expression is a sequence of partial sums $\{S_n\}$ where

$$S_n = a_1 + a_2 + \cdots + a_n$$

An infinite series is said to converge to $L$ if

$$\lim_{n \to \infty} S_n = L$$

If the sequence of partial sums for a series fails to have a limit, we say that the series diverges. Thus the series

$$1 + (1/2) + (1/4) + (1/8) + \cdots$$
$$+ (1/2^{n-1}) + \cdots$$

converges to 2 and the series

$$1 + (-1) + 1 + (-1) + \cdots + (-1)^{n+1} + \cdots,$$

whose partial sums oscillate between $+1$ and 0, diverges.

## MOTIVATION AND GOALS

The first of the principal questions to address in studying (infinite) series is whether a given series converges or diverges. Next, if a series does converge, then to what sum does it converge? To know the sum is to know how to write the series in closed form; working in the opposite direction, one can learn something about a number such as $\pi$ if that number is expressed as a series. Another question arises from the fact that new series can be created from a given series or set of series. Do these new series converge or diverge? For example, replacing each of the terms in a given series $S$ by its absolute value yields a new series of nonnegative terms, $A$. If $A$ converges, then $S$ is said to converge absolutely. If a series converges absolutely, then it converges according to the definition given in the preceding section.

A series that converges but that does not converge absolutely is said to converge conditionally. One of the remarkable facts about a conditionally convergent series, proved by Georg Bernhard Riemann (1826–1866), is that a rearrangement of its terms yields a new series with possibly a different convergence behavior. In fact, given any real number $r$, a conditionally convergent series $S$ can be rearranged so that it converges to $r$. For example, the alternating harmonic series,

$$1 - (1/2) + (1/3) - (1/4) + \cdots$$

converges to ln 2; however, given a number like $\pi$ a suitable rearrangement of the terms of this series produces a new series that converges to $\pi$. This is done by "stockpiling" the positive and the negative terms in two separate series: since these two series diverge, a new series, which converges to $\pi$, can be created by drawing first from one, then from the other of these two series. The switch from the one to the other occurs so that the partial sums of the new series oscillate around $\pi$.

Students in an elementary calculus course deal primarily with determining whether a series converges. They learn various tests for convergence, of which the ratio test and the integral test are likely to be the ones most heavily employed, at least for series of positive terms. Alternating series $\sum(-1)^{n+1}a_n, a_n > 0$, converge if two conditions are satisfied:

$$a_{n+1} \leq a_n, \text{ for all } n \quad \text{and} \quad \lim_{n \to \infty} a_n = 0$$

Having established that a given series converges, students frequently go on to the next series in the list and do not concern themselves with the value to which the given series converges. Also, students can easily develop negative attitudes toward divergent series in the belief that such series are "good for nothing." This is certainly not the case: For example, the harmonic series

$$1 + (1/2) + \cdots + (1/n) + \cdots$$

that diverges can be used to establish the fact that a stack of identical blocks can be built up in such a way that the stack extends as far to one side as one chooses without tumbling down (Fraga 1993).

## PEDAGOGICAL APPROACHES

Developments in technology have made it possible for students to produce, without much effort, estimates of sums for certain types of convergent series. What is required is an appropriate software package, such as *Mathematica* or Maple, or a calculator such as those produced by Texas Instruments and Hewlett-Packard. For example, the sum of a convergent series of positive terms, $\sum_{n=1}^{\infty}a_n$, can be approximated by the integral test if there is a continuous function $f$ such that $f(n) = a_n$. The "tail" of the series can be bounded below and above by integrals of $f$:

$$\int_N^{\infty} f(x)dx \leq \sum_{n=N}^{\infty} a_n \leq \int_{N-1}^{\infty} f(x)dx$$

Students may be able to determine the value of the index $N$ required to meet a given tolerance by calculating the integrals (although to do so may pose difficulties). They can then use a computer or a calculator to find the approximation:

$$\sum_{n=1}^{N} a_n$$

Students should have the opportunity to see how series arise in different contexts (Fraga 1993). They should *not* spend most of their time on what are essentially drill questions. For example, the concept of a proof without words arises naturally in asking students to establish that

$$1 + 2 + \cdots + n = n(n + 1)/2$$

by considering a sequence of $n \times (n + 1)$ dominoes sliced from one corner to the opposite corner:

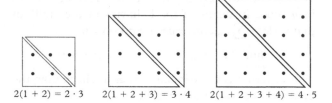

$$2(1 + 2) = 2 \cdot 3 \qquad 2(1 + 2 + 3) = 3 \cdot 4 \qquad 2(1 + 2 + 3 + 4) = 4 \cdot 5$$

**Figure 1**

By constructing figures $T_0, T_1, \ldots, T_n, \ldots$, where $T_0$ is an equilateral triangle and $T_{n+1}$ is obtained by replacing the middle third of each edge of $T_n$ by an outwardly directed equilateral triangle, students create a sequence of figures, the limit of which is called the Koch Snowflake (Kline 1972). Questions about the area and perimeter of the Snowflake involve an investigation of geometric series. Resolution of Zeno's Paradox, where a hare and a tortoise run a race in which the tortoise has a head start but the hare runs ten times as fast as the tortoise without apparently ever catching up, also involves an application of geometric series.

## RECOMMENDED MATERIALS FOR INSTRUCTION

It is important for students to approximate the sum of a convergent series, not simply to know that a given series converges. For this reason, some number-crunching capacity should be available to them. A number of software packages supportable on microcomputers and inexpensive calculators bring this capacity within students' grasp. Some kind of technological tool should be incorporated in the structure of an elementary calculus course.

*See also* Maclaurin Expansion; Paradox; Power Series

SELECTED REFERENCES

Boyer, Carl B. *A History of Mathematics.* New York: Wiley, 1968.

Buck, R. Creighton. *Advanced Calculus.* 3rd ed. New York: McGraw-Hill, 1978.

Fraga, Robert, ed. *Calculus Problems for a New Century.* Washington, DC: Mathematical Association of America, 1993.

Joseph, George Gheverghese. *The Crest of the Peacock.* London, England: Tauris, 1991.

Kline, Morris. *Mathematical Thought from Ancient to Modern Times* (pp. 1020–1021). New York: Oxford University Press, 1972.

Roy, Ranjan. "Anticipations of Calculus in Medieval Indian Mathematics." In *Readings for Calculus* (pp. 18–21). Underwood Dudley, ed. Washington, DC: Mathematical Association of America, 1993.

Stewart, James. *Calculus.* Monterey, CA: Brooks/Cole, 1987.

ROBERT J. FRAGA

# SETS

A mathematical discipline developed in the late nineteenth century to provide a theoretical foundation for mathematics. At that time, mathematicians, including Julius W. R. Dedekind (1831–1916), Karl Weierstrass (1815–1897), Heinrich E. Heine (1821–1881), and Georg Cantor (1845–1918), began to think about the properties of numbers in terms of classes or collections of numbers. For example, Dedekind showed how one could extend the rational numbers to the reals by regarding each irrational number as the point separating two classes, or in modern terms, sets, of rational numbers. Thus, the irrational number, $\sqrt{2}$, is the point that divides the set of all positive rational numbers whose squares are less than 2 from the set of all positive rational numbers whose squares are greater than 2. Later, Cantor developed the theory of sets as a full mathematical discipline.

Set theory continues to be a basic tool of much of modern theoretical and applied mathematics, and its language and notation are used extensively. Among other areas, elementary probability, often found in the secondary school curriculum, and areas of abstract algebra such as group theory rely heavily on the ideas and results of set theory. At the elementary school level, textbooks teach students to reduce fractions to lowest terms by having them form the set of all divisors of the numerator and the set of all the divisors of the denominator. The intersection of the two sets is the set of all common divisors of the numerator and denominator. The largest element of this last set is the greatest common divisor that is used to reduce the fraction.

A set is a collection of objects, called *elements*. Any object, call it $x$, may or may not belong to a given set $A$. If $x$ belongs to the set $A$, we also say that $x$ is an element of $A$, and we write $x \in A$. Otherwise, $x$ is not an element of $A$, and we write $x \notin A$. One way to describe a set is by listing its elements. $A = \{-3, 2, 7\}$ means that $A$ is the set whose elements are the three integers $-3$, 2, and 7. $\mathcal{J} = \{3, 4, 5, \ldots, 12\}$ means that $\mathcal{J}$ is the set of all integers from 3 to 12 inclusive. $N = \{1, 2, 3, \ldots\}$ means that $N$ is the set of all positive integers. The set $\{\}$, also denoted $\varnothing$, is the empty set or null set. $\varnothing$ contains no elements. You can also describe a set by specifying a property such that any object that has the property is an element of the set, and any object that does not have the property is not an element of the set. The notation $A = \{x \in N | x < 6\}$ (read "the set of all $x$ in the set of natural numbers such that $x$ is less than 6") means that $A$ is the set of all natural numbers that are less than 6. This way of defining a set is called the *rule method*, and the notation is called *set-builder notation*.

Besides the relationship of belonging that may or may not exist between an element and a set, there are some relationships that may exist between pairs of sets. If every element of one set, $A$, is also an element of another set, $B$, then we say that $A$ is a subset of $B$, and we write $A \subset B$. For example, $\{1, 3\}$ is a subset of $\{1, 2, 3, 4\}$. The empty set, $\varnothing$, is a subset of every set, and every set is a subset of itself. If two sets have exactly the same elements, then we say that the sets are equal, and we write $A = B$. If sets $A$ and $B$ are equal, then it must be that every element of $A$ is an element of $B$, and every element of $B$ is an element of $A$—that is, each set is a subset of the other. If $A$ is a subset of $B$, but $A$ is not equal to $B$, then $A$ is called a proper subset of $B$. In informal treatments of set theory such as one would encounter in mathematics courses through the first two years of college, one considers any well-defined objects as legitimate elements for sets. (This is not the case for more formal treatments.) In particular, the elements of a set can be other sets. The set whose elements are the distinct subsets of a given set, $A$, is called the power set of $A$, denoted $P(A)$. For example, if $A = \{1, 2\}$, then $P(A) = \{\varnothing, \{1\}, \{2\}, \{1, 2\}\}$.

In any discussion involving sets, we work within a universe of discourse, or universal set, which contains all the elements of all the sets in the discussion. Thus, every set being considered is a subset of this universal set, and every element is an element of the universal set. Different contexts call for different choices of the universal set. Working with sets only in the context of a universe of discourse avoids some logical problems that were first discovered in the early twentieth century. Up to that time, it was thought that any collection that one could conceivably describe could be a set. In 1902, Bertrand Russell pointed out a paradox (Russell's paradox) that showed that there are descriptions (involving self-reference) that cannot define sets. One version of Russell's paradox considers the following description: Let $C$ be the collection of all sets, $A$, such that $A$ is not an element of $A$. One could argue that such sets exist, since the empty set cannot be an element of itself. On the other hand, if $A$ is the set of all interesting ideas, we might argue that $A$ is itself an interesting idea. Thus, $A$ is an element of itself. The paradox arises when we ask the question, "Is $C$ an element of $C$?" If $C$ is an element of $C$, then, by definition of the collection $C$, $C$ is not an element of $C$. On the other hand, if $C$ is not an element of $C$, then $C$ satisfies the criterion for membership in the collection, $C$. This paradoxical result shows that the description we gave for the collection, $C$, cannot define a set. In defining a set, we must provide a condition such that, for any candidate for membership, either the object satisfies the condition and is therefore a member of the set, or it fails to satisfy the condition and thus is not a member of a set. This is what is meant by the requirement that sets must be "well defined."

There are many different ways to create new sets from sets that have already been defined. If $A$ and $B$ are two sets, then if you collect all the elements of $A$ together with all the elements of $B$ into a single set, you have constructed the union of $A$ and $B$, which is denoted $A \cup B$. For example, $\{1, 2, 3\} \cup \{2, 4, 6\} = \{1, 2, 3, 4, 6\}$. Thus, we can make the definition:

$$A \cup B = \{x \in U | x \in A \text{ or } x \in B\}$$

Similarly, the intersection of $A$ and $B$ is the set of all the elements that $A$ and $B$ have in common, and is denoted $A \cap B$. For example, $\{1, 2, 3\} \cap \{2, 4, 6\} = \{2\}$. We define:

$$A \cap B = \{x \in U | x \in A \text{ and } x \in B\}$$

If two sets have no elements in common, they are disjoint. That is, two sets are disjoint if and only if their intersection is the empty set. The complement of the set $A$ is the set of all elements of the universal set that are not elements of $A$. The complement of $A$ is often denoted $A'$. If $A = \{1, 2, 3, 4\}$ and the universal set, $U$, is $\{1, 2, \ldots, 10\}$, then $A' = \{5, 6, 7, 8, 9, 10\}$. Again, this gives a definition:

$$A' = \{x \in U | x \notin A\}$$

A very useful device for visualizing sets and set operations is the Venn diagram. In a Venn diagram, sets are represented by regions in the plane, and ways of combining sets can easily be represented as combinations of regions.

The study of set theory leads quite naturally to several related ideas. Counting the number of subsets of a finite set is a natural thing to try. The number of $r$-element subsets of an $n$-element set is simply $C(n,r)$—the number of combinations of $n$ objects taken $r$ at a time. When we teach the topic of combinations, we often point out that the values of $C(n,r)$ can be represented in a triangular array known as Pascal's triangle:

$$
\begin{array}{ccccccccccc}
 & & & & & 1 & & & & & \\
 & & & & 1 & & 1 & & & & \\
 & & & 1 & & 2 & & 1 & & & \\
 & & 1 & & 3 & & 3 & & 1 & & \\
 & 1 & & 4 & & 6 & & 4 & & 1 & \\
1 & & 5 & & 10 & & 10 & & 5 & & 1
\end{array}
$$

$$\text{etc.}$$

Each row of the triangle is obtained by adding pairs of entries from the preceding row. Furthermore, the values of the numbers $C(n,r)$ can be read from the triangle. If we begin counting the rows of the triangle with zero, then, for example, the fourth row, 1 4 6 4 1, contains the values of $C(4,0)$, $C(4,1)$, $C(4,2)$, $C(4,3)$, and $C(4,4)$. Now, we can use the triangle to make guesses about the number of subsets of a set. Adding the numbers in each row of Pascal's triangle leads to the (correct) guess that the total number of subsets of an $n$-element set is $2^n$.

Another interesting extension of the study of sets is the consideration of infinite sets. According to a definition first proposed by Dedekind, a set $A$ is infinite if there is a one-to-one correspondence between the elements of $A$ and the elements of some proper subset of $A$. Thus, the set $N$ of natural numbers is infinite, because there is a one-to-one correspondence between the elements of $N$ and the elements of the set $E$ of all even natural numbers. (Just match each natural number with its double.) Infinite sets come in different "sizes." Any set that is in one-to-one correspondence with the set of natural numbers is called *countably infinite,* or simply *countable.* It is not hard to show a good secondary school class that the set of all integers is countable and (more amazing) that the set of all rational numbers is countable. The latter demonstration is the result of Cantor's mathematical work and uses a "diagonal" counting method. The positive rational numbers are arranged in an infinite rectangular array, and the elements are counted by following the arrows:

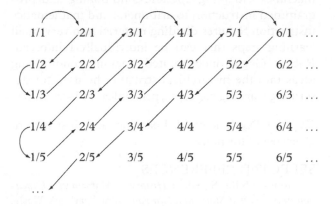

Duplicate values can be skipped in the count. A bit more subtle, but also interesting to students who have caught the spirit of working with infinite sets, is the fact that the set of real numbers is not countable (Smith, Eggen, and St. Andre 1990; Schwartz 1997).

Another interesting direction to take a class that is studying set theory is to compare the theory of sets and the results of set operations with the logical system known as the statement or predicate calculus. The definitions of the set operations are closely related to the basic logical connectives. For example, the intersection of sets $A$ and $B$ is the set of all elements that are in $A$ *and* also in $B$. Thus, the set operation *intersection* is associated with the logical connective *and*. The properties of sets that one can derive from the basic definitions are fully analogous to the properties of statements that govern the logical connectives. Each system, in turn, is an example of a more abstract structure called a Boolean algebra.

*See also* Cantor; Logic; Paradox; Real Numbers

SELECTED REFERENCES

Boyer, Carl B. *A History of Mathematics.* New York: Wiley, 1968.

Kolman, Bernard, and Robert C. Busby. *Discrete Mathematical Structures for Computer Science.* Englewood Cliffs, NJ: Prentice-Hall, 1984.

Mizrahi, Abe, and Michael Sullivan. *Finite Mathematics with Applications for Business and Social Sciences.* 6th ed. New York: Wiley, 1992.

Schwartz, Diane Driscoll. *Conjecture and Proof: An Introduction to Mathematical Thinking.* Philadelphia, PA: Saunders, 1997.

Smith, Douglas, Maurice Eggen, and Richard St. Andre. *A Transition to Advanced Mathematics.* 3rd ed. Belmont, CA: Brooks/Cole, 1990.

DIANE DRISCOLL SCHWARTZ

# SIMILARITY

Refers to geometric figures having the same shape but not necessarily the same size. Similar figures are scale models of each other. One can produce similar figures by enlarging or reducing a given figure with a photographic enlarger. (The technical term for enlargement is *dilation* or *homothety*.) If one enlarges a given triangle $\triangle ABC$ by a factor of $r$, one obtains a similar triangle $\triangle A'B'C'$ whose sides are $r$ times as long as the corresponding sides of $\triangle ABC$:

$$r = \frac{A'B'}{AB} = \frac{B'C'}{BC} = \frac{C'A'}{CA} \tag{1}$$

Therefore, similar triangles have proportional sides. Enlarging a figure does not change its angles, so corresponding angles of similar triangles are equal:

$$\angle A = \angle A', \quad \angle B = \angle B', \quad \angle C = \angle C' \tag{2}$$

One can also prove the converse: if either condition (1) or condition (2) holds, then the triangles are similar.

Proportionality underlies most practical uses of similar triangles.

*Example* A surveyor marks a point $A'$ at the base of the cliff, walks one hundred feet away to a point $B'$, takes a sighting to a point $C'$ at the top of the cliff directly above $A'$, and finds that $\angle A'B'C'$ measures 53°. How high is the cliff?

*Solution:* $\triangle A'B'C'$ is a right triangle with $A'B' = 100$ ft, $\angle A' = 90°$, $\angle B' = 53°$, and $\angle C' = (180° - 90° - 53°) = 37°$. Draw a similar triangle $\triangle ABC$ then measure the lengths of its sides. For instance, if $BC = 1$ ft, one finds that $AB \approx 0.60$ ft and $AC \approx 0.80$ ft. Since $(A'C'/AC) = (A'B'/AB)$, the result becomes

$$(A'C'/0.80 \text{ ft}) \approx (100 \text{ ft}/0.60 \text{ ft})$$

and

$$A'C' \approx (0.80)(100)/(0.60) \text{ ft} \approx 130 \text{ ft}$$

Trigonometry is based on the proportionality of similar triangles. A table of sines and cosines amounts to a table of the lengths of legs of right triangles △ABC with hypotenuse $BC = 1$ because $AB = \cos B$ and $AC = \sin B$. One can replace scale drawings with sines and cosines in problems involving similar right triangles because every right triangle is similar to one whose hypotenuse has length = 1. For example, in the problem above $AB = \cos 53° \approx 0.60182$ and $AC = \sin 53° \approx 0.79864$, so

$$A'C' \approx ((0.79864)\ (100\ \text{ft})/(0.60182)) \approx 132.70\ \text{ft}$$

Because trigonometric functions are known to arbitrarily high precision, one generally achieves more accurate answers with trigonometric functions than by measuring drawings.

Similar triangles find many applications in calculus, physics, and engineering, in related rates problems, in calculations of areas and volumes, and in resolving systems of forces. Similar triangles may also occur in non-Euclidean geometries, but the situation there is quite different. For example a "triangle" in spherical geometry has edges that are arcs of great circles. Two spherical triangles on a given sphere are similar if and only if they are congruent, for if one enlarges a given spherical triangle then the enlarged spherical triangle will fit only on an enlarged sphere.

A special kind of similarity is called "self-similarity." A figure is self-similar if parts of the figure have the same shape (after enlargement) as the whole figure. Such figures are called *fractals*.

*See also* Fractals; Plane Geometry; Trigonometry

SELECTED REFERENCES

Clemens, C. Herbert, and Michael A. Clemens. *Geometry for the Classroom.* New York: Springer-Verlag, 1992.

Hogben, Lancelot T. *Mathematics for the Million* (pp. 123–125). 4th ed. New York: Norton, 1983.

Jennings, George. *Modern Geometry with Applications.* New York: Springer-Verlag, 1994.

Thompson, James E. *Geometry for the Practical Worker.* 4th ed. New York: Van Nostrand Reinhold, 1982.

GEORGE A. JENNINGS

# SKINNER, BURRHUS FREDERIC (1904–1990)

A neobehaviorist or stimulus-response associationist psychologist whose learning theory followed Edward L. Thorndike's general thrust. According to Skinner's theory of learning, called *operant conditioning,* a response to a stimulus is made more frequent through reinforcement (repetition of the pattern in which a stimulus is followed by a response). Skinner was particularly successful in using operant conditioning to train animals. He trained pigeons to play a modified game of tennis, rats to obtain food using marbles in a vending machine, and dogs to retrieve bones by depressing a pedal. Skinner advocated teaching through programmed instruction using teaching machines. Adapting operant conditioning and programmed instruction to arithmetic and mathematics instruction requires dividing material into very small learning steps that can be individually reinforced. Robert Gagné incorporated operant conditioning ideas into the hierarchical structure he used for describing learning concepts, principles, and structures.

*See also* Psychology of Learning and Instruction, Overview; Thorndike

SELECTED REFERENCES

Jones, Phillip S., ed. *A History of Mathematics Education in the United States and Canada.* 32nd Yearbook. Washington, DC: National Council of Teachers of Mathematics, 1970.

Skinner, Burrhus F. *Science and Human Behavior.* New York: Macmillan, 1953.

JAMES K. BIDWELL
ROBERT G. CLASON

# SMITH, DAVID EUGENE (1860–1944)

A prolific writer whose works included very popular arithmetic text series and texts on teaching methods, as well as a two-volume *History of Mathematics* (1923 and 1925). Smith taught at Michigan State Normal College at Ypsilanti, was the principal at Brockport (New York) State Normal School, and was a professor at Teachers College, Columbia University from 1901 to 1926. He had influential roles in many professional organizations including the National Council of Teachers of Mathematics (NCTM), the Mathematical Association of America, and the American Mathematical Society. Smith grew up during the era of faculty psychology and lived through both the rise of William James's pragmatism and the rise of connectionist psychology. He did not become an advocate of any of these educational theories but remained eclectic, selecting what seemed valuable from a variety of sources.

*See also* History of Mathematics Education in the United States, Overview

SELECTED REFERENCES

Bidwell, James K., and Robert G. Clason, eds. *Readings in the History of Mathematics Education*. Washington, DC: NCTM, 1970.

Jones, Phillip S., ed. *A History of Mathematics Education in the United States and Canada*. 32nd Yearbook. Washington, DC: NCTM, 1970.

Smith, David E. *History of Mathematics*. New York: Dover, 1958.

JAMES K. BIDWELL
ROBERT G. CLASON

# SOFTWARE

Term for computer programs that determine the operation of hardware (computers and the associated physical equipment such as printers and scanners). Operating systems, programming languages, application packages, and content-specific programs are all software. Few people question the great potential that computer technology provides to enhance mathematics teaching and learning. Since the computer can do nothing without instructions from software, the extremely important role played by software is apparent.

## EARLY THINKING ABOUT COMPUTER SOFTWARE

Since the early days of computer use, there has been a great variety of computer software with evolving quality for mathematics education. An early categorization for computer uses in education was proposed in *The Computers in the School: Tutor, Tool, Tutee* (Taylor 1980). Actually, this book also helped frame the early thinking about computer software in education; that is, that software functions as a tutor, as a tool, and also as a tutee.

### Software Functions as a Tutor

As a tutor, software includes *drill and practice software, tutorials,* and *demonstration software.* The early uses of drill and practice software in mathematics education started in the late 1960s, involving simple arithmetic calculations and promising the benefits of immediate feedback, efficient record keeping, and motivation. Since then, drill and practice software has grown in capability and sophistication. *Math Blasters,* for example, is a drill and practice arithmetic facts program that uses a video game format, intended to be highly motivational to

students (see Simonson and Thompson 1994). This kind of software has been criticized by many mathematics educators because it wastes the power of the computer, encourages emphasis on lower-level skills, and does not promote active thinking. It is not consistent with recommendations for reform in mathematics education, in which skills are to be developed as tools for computing and problem solving but not as an end in themselves.

Computer tutorials are programs designed to teach skills and develop knowledge. Early computer tutorials in arithmetic and algebra simply used rote teaching and were very simple in design. Recent attempts to improve computer tutorials involve applying the principles of artificial intelligence to develop intelligent tutoring systems. The *Geometry Tutor* (Anderson 1988) is such a system. It makes instructional decisions based on the student's knowledge and difficulties so far and considers how to present information understandably, how to query the student, how the student answers, and what information should be reinforced. It has been piloted in classrooms in the Pittsburgh area, and the researchers have reported the participating students' achievement gains of about one standard deviation or one letter grade (Anderson 1992; Wertheimer 1990). The major complaints against this software are that it uses too many menus and is sometimes slow and that the proof structure is excessively rigid and without alternatives. Although quality tutorials such as the *Geometry Tutor* pay close attention to user–system interaction, numerous mathematics educators suggest that tutorials present information to students in very much the same way as teachers in traditional classrooms, where students may not be constructing knowledge by themselves to a sufficient degree.

Demonstration software is designed to illustrate concepts or present lessons. A good demonstration program is *Probability Theory* (1988). It includes topics such as *Venn Diagram, Simulate,* and *Markov Chain.* One sample activity is to simulate rolling a pair of regular dice. The user is prompted to enter the number of rolls. The computer quickly shows the data organized into four columns on the screen: the sum of the numbers on the dice; the frequency of getting the sum (in the sample space); the experimental probability of the sum; and the distribution bar graph resulting from the experiment (a theoretical bar graph is also on the screen for making comparisons). Some demonstrations like this are very clear and informative and may well serve a specific purpose.

## Software Functions as a Tool

Tool software is computer software used as a tool to facilitate the teaching and learning of most school subjects. Word processors, spreadsheets, database managers, graphing utilities, statistics analysis programs, and symbolic manipulators are all examples of tool software used in schools, especially in mathematics classrooms. Specific examples of early tool software include WordStar (a word processor), VisiCalc (a spreadsheet program), and dBASE I (a database management system). These packages are by no means mathematics-content-driven. Teachers and students, however, can use them to accomplish tasks related to mathematics teaching and learning. With VisiCalc, for example, a teacher might construct and save his or her gradebook, whereas students might solve some real-world problems such as creating a budget and calculating interest on a loan.

## Software Functions as a Tutee

Instead of tutoring students to learn the content required by the current curriculum, tutee software allows students to "teach" the computer to do the things they want to accomplish. In order to do so, students must learn a programming language. BASIC and Logo, as well as Pascal and FORTRAN, have been programming languages traditionally taught and used in mathematics classrooms. The advocates of student programming believe that analyses required in writing a program deepen student understanding of the underlying mathematics. However, "the search for evidence that programming experience influences mathematical behavior has not produced consistent or striking results" (Fey 1989, 260).

## CURRENT USES OF COMPUTER SOFTWARE

In recent years, there have been fundamental changes in software development. First of all, along with the rapid changes in computer hardware, software environments have become more flexible and user-modifiable. Many things that previously had to be programmed as needed can now be done easily with prepackaged instructional software. Many mathematics educators suggest that programming by students has become much less important and should not be emphasized any more. However, others still believe that programming can help students understand how to communicate with the computer and enable them to see more wide-ranging ways of creating and using software. These educators recommend, therefore, that both programming and using instructional software be stressed.

A new category of computer software, called innovative exploratory software, has emerged and many software packages in this category have been created. The impetus for the emergence of this new category of software came from the call for reform by the National Council of Teachers of Mathematics (NCTM) (1989) and other organizations as well as individual mathematics educators. It was fostered by funding from agencies that realized the importance of developing more powerful software, by the growth of computer hardware, and by the fact that research based on the uses of software categorized by Taylor (1980) had shown mixed results. Williams and Brown (1991) attribute these mixed results to researchers' preoccupation with a *comparative research paradigm,* which views technology as an experimental variable, independent of instructional context. They indicate that this research model provides very limited information for the reform of instruction and suggest that research should, instead, focus on the investigations of innovative ways of using technology to enhance and change teaching and learning. The creation and use of innovative exploratory software embodies such investigations. Olive (1990) reported a series of beginning efforts along this line.

The innovative exploratory software involves subject-specific exploratory tools, exploratory microworld environments, mentally engaging computational and analytical environments, multipurpose tools that can be innovatively used in mathematics instruction, thought-provoking games, multimedia software, and assessment applications.

## Subject-Specific Exploratory Tools

*The Geometric Supposers* (Schwartz and Yerushalmy 1985) is an early example of this kind of software. Developed for students to conjecture about Euclidean geometry, it provides students with a computerized set of geometric construction and measurement tools. With these tools, students can conveniently construct various kinds of geometric figures such as points, lines, triangles, quadrilaterals, and circles as needed and quickly collect a wealth of visual and numerical data, such as the measurements of related segments and angles as well as the calculation results about those measurements. By observing the relationships revealed by the data, it is natural for students to make conjectures and then collect more data to test the conjectures. Thus, *The Geometric Sup-*

*posers* has potential to facilitate students' inductive learning of geometry. It runs on both Apple and DOS-based computers.

In comparison to the *Supposers, The Geometer's Sketchpad* (GSP) (Jackiw 1991) is another tool for the student to explore a wider range of geometric phenomena. The features of GSP include the following: (a) Its constructions and measurements are dynamic. All objects are movable by dragging, and the related measurements and calculations change automatically and instantly (see Figure 1). (b) The construction done in GSP is not only a geometric object itself (e.g., a triangle) but represents and can easily be changed into any object of the same type (another triangle). This helps the user see the invariance and get a much more global view about the construction. (c) Its capability of tracing locus and animation enhances students' ability for visualization. (d) It emphasizes the spirit of transformation geometry. Reflections, translations, rotations, and dilations are easily available. (e) It can record and "play back" any geometric construction process by the "script" function. This feature not only saves students' time and effort from frequent repetitions of the "starting from scratch" process but also helps develop students' abstract thinking. A script of constructing the centroid of a triangle, for example, applies to an abstract tri-

angle, that is, a triangle with any shape and any size. (f) It can be used together with multimedia software like *Adobe Premier* to capture action sequences as QuickTime (an integration of audio and video) movies, enhancing GSP's demonstration power. (g) It is very intuitive and user-friendly. All students, including young children, can use it to explore geometric properties and relationships with little difficulty but with great interest based on their intuition and imagination. (h) It enables the user to create innovative learning environments (electronic systems with computer interface, dynamic graphics, and special tools for constructing mathematical ideas). By creating a coordinate system, for example, even algebraic relationships and their graphing can be explored within GSP (Lin and Hsieh 1993). Because of these features, GSP has been widely used in mathematics classrooms.

There is another geometry software package called *Cabri Geometry* (Laborde 1990), which is similar to GSP in most of the aspects just mentioned. While *Cabri Geometry* does not have the functions of making QuickTime movies and doing calculations on the measurements made, it seems better than GSP in creating and flexibly using vectors. Both packages have versions for Macintosh computers and for Microsoft Windows.

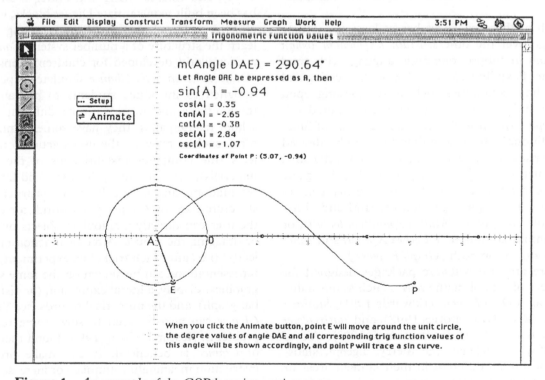

**Figure 1**   *An example of the GSP learning environment.*

In addition to geometry software, there are exploratory algebra software packages. *Math Exploration Toolkit* (Fey and Heid 1985) is the earliest software of this type. *Algebra Xpresser* (Hoffer 1991), *Graph Explorer* (Abrams 1992), and *Math Connections* (Rosenberg 1993) are more recent packages. The features possessed by more than one of these packages include the following: (a) All of them are function graphers that can generate graphs for a variety of mathematical functions quickly and accurately, providing students with powerful facilities to explore the behaviors, patterns, and characteristics of those functions. (b) All of them allow simultaneous, multiple graphs, but some limit the number of graphs at one time (three graphs for *Math Connections* and two graphs for *Graph Explorer*). (c) All of them have capabilities for zooming (adjusting the viewing window of the graph) and tracing (moving the cursor from one point to the next along the graph). These capabilities are excellent for *Algebra Xpresser* and *Graph Explorer* (easy, dynamic, and reversible), fair for *Math Connections,* but very limited and difficult to accomplish for *Math Exploration Toolkit.* (d) While *Math Exploration Toolkit* and *Math Connections* can only graph functions in rectangular form, *Algebra Xpresser* and *Graph Explorer* are able to graph functions in rectangular, polar, and pamametric forms. (e) *Math Exploration Toolkit* and *Algebra Xpresser* can also be used to do symbolic manipulations.

Each of these packages has at least one unique advantage over the others. *Math Exploration Toolkit* has a built-in subprogramming language with good power and flexibility; *Algebra Xpresser* has scaling and axis-labeling capabilities such as auto scaling upon zooming and easy rescaling upon user-entered values; *Graph Explorer* is able to graph a family of functions with one entry (more than two graphs allowed in this sense); and *Math Connections* allows dynamic transformations on graphs (every graph is draggable and the corresponding function will be changed accordingly). These packages run on IBM and IBM-compatible computers *(Math Exploration Toolkit),* or on Macintosh computers *(Algebra Xpresser and Math Connections),* or on both *(Graph Explorer).*

There are also software packages designed for other areas of school mathematics such as the statistics packages *Data Insights* (Edwards 1991), *Statistics Workshop* (BBN Laboratories 1992), and *StatExplorer* (Professional Systems International 1993). All of these software packages (geometry, algebra, statistics, etc.), though different in their content areas or degrees of sophistication, suggest an inductive approach based on experimentation, observation, data collecting, and conjecturing. Such an approach gives students the opportunity to engage in mathematics as active constructors rather than as passive recipients of others' mathematics knowledge (Olive 1990). In addition, with these software packages, students can learn some interesting and important mathematics ideas that they are not likely to learn in the traditional classroom. For example, students can use the features of GSP ("animation," "trace locus," etc.) to simulate the movement of planets and satellites in the solar system.

## Exploratory Microworld Environments

The term *microworld* was created by Papert in 1980 to name the exploratory learning environment of Logo programming. A mathematical microworld, according to Thompson (1987), is a system composed of objects, relationships among objects, and operations that transform objects and relationships. In practice, it incorporates a graphical display that depicts a visualization of the microworld's initial objects. This display, in conjunction with operations of the microworld's objects, constitutes a model of the concept(s) being proposed to the student. Examples of mathematical microworlds include *Blocks* (Thompson 1993), *CandyBar* (Olive and Steffe 1994), and *Chance* (Jiang and Potter 1994). *Blocks* has been built to instantiate Dienes blocks, which are effective manipulatives for helping young children learn the structure of a number system. *CandyBar* is a microworld developed for children's construction of rational numbers. *Chance* simulates experiments, designed to introduce students to probability. Although these microworlds involve different topics of school mathematics, they have some common features, such as allowing the user's actions on the objects (the computer instantiations of the physical materials); using multiple, dynamic, and linked representations; and having built-in construct-support structures. In the *Chance* microworld, for example, the user can directly manipulate objects by clicking or dragging the mouse. After each action on an object(s) (i.e., after each trial of an experiment), several representations can be shown on the same screen: a graphical display, a literal expression (or distribution bar graph), and the numerical records (see Figure 2). *Chance* can also be used to solve some real-world problems. Research (Jiang 1993) found that the use of *Chance* is beneficial. It can make probability instruction meaningful, stimulate or increase motivation, help students overcome their learning difficul-

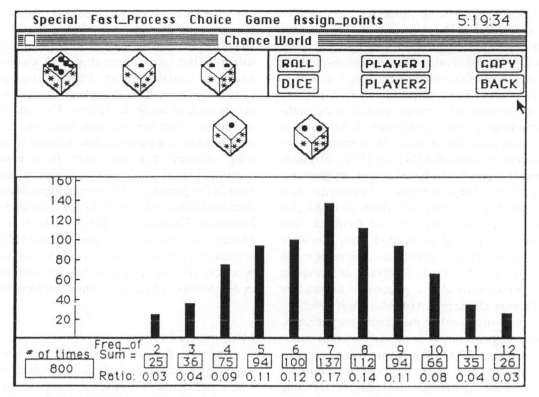

**Figure 2**  *An example of the* Chance *microworld environment. [Reproduced with permission from Zhonghong Jiang and Walter Potter]*

ties, help students correct their misconceptions about probability, and save instruction time.

## Mentally Engaging Computational and Analytical Environments

An example of this kind of software is *Mathematica* (Wolfram 1991), a sophisticated combination of tool and exploratory environments. It encompasses a larger portion of mathematics than any software already mentioned. The wide range of mathematical functions it performs are drawn from such areas as algebra, statistics, matrices, calculus, and analytic geometry. It solves equations and integrates and differentiates, to name a few. It has tools for explorations in statistical analyses and visualizing transformations on sets of data or generalizing patterns in function theory. Its graphing capabilities help the user explore various types of graphs such as parametric plots, three-dimensional plots, density plots, and contour plots. The user may also benefit from the visual effects of transformations and animation on three-dimensional graphs. *Mathematica* also has a built-in programming language, allowing the user to construct his or her own learning and problem-solving environments. It is structured for exploration

with menus and outlines of the options available on screen, yet the documentation, extensive as it is, requires substantial study and assimilation. The learning time for unguided learners is substantial, but the power for mathematical and scientific investigation is vast. *Mathematica* can only satisfactorily be used on mainframe, mini, and advanced microcomputers (microcomputers with large RAM, large storage, and fast processors). *Derive, Theorist,* and *Maple* are also excellent software in this subcategory. They may not be as comprehensive as *Mathematica*, but have potential for students to explore a variety of mathematical concepts.

## Multipurpose Tools That Can Be Innovatively Used

Through continuous upgrading, the current versions of spreadsheet, database, graphics, word processing, and desktop publishing software (*Microsoft Excel* 5.0, *Quattro Pro* 5.0, *WordPerfect* 6.0, *PageMaker* 5.0, etc.) have become much more powerful than the old versions. Although their origin was not in education, their multifunction features and capabilities can be applied to enhance mathematics teaching and learning in ways that emphasize active involvement, hands-on activities, critical thinking, collaborative learning, and visualization. Current word processors and desktop publishers, for example, have excellent features such as graphic and table capabilities, linking compatibility with other software packages, as well as spelling, thesaurus, and grammar checkers. With these features, the teacher can create mathematics worksheets containing graphics that either help motivate students' interest in their work or help make mathematics concepts meaningful (Wiebe 1992). Students can efficiently prepare drafts of major assignments (involving graphical representation of problems, data collection and analysis, etc.), save them on disks, give the disks to peers and teachers for feedback, and make as many revisions as needed until the work meets quality standards, without the trouble of much redrafting. The habit of draft, feedback, and revision is one of the recommended expectations for student work (California Department of Education 1992).

In comparison to other multipurpose packages, spreadsheet programs are more widely used in mathematics classrooms. The following are some examples of what a spreadsheet (*Microsoft Excel*, for example) can do to facilitate mathematics learning: (a) By using special features and commands, students can use an inductive and discovery approach to explore the properties of number sequences such as Fibonacci sequences and related topics; (b) by using features and commands similar to these, a table-approach in studying functions can be introduced so that students can numerically explore the solution of an equation, the behavior of a function, the meaning of square roots, maximum and minimum values, and other mathematical ideas; (c) by using the worksheet and the chart correspondingly, students can choose to plot points in a coordinate system, visualize the graph of a function (or functions), and compare the table representation with the graph of the function(s); (d) by using the built-in random number generator and other features, students can simulate probability experiments and model real-world probability problems (see McClintock and Jiang 1997).

### Thought-Provoking Games

In the 1980s, a series of computerized games such as *Green Globs* (Dugdale 1982), *Guess My Rules* (Barclay 1985), and *The King's Rule* (O'Brien 1985), no matter whether their mathematics emphasis was explicit or not, played a role in helping students develop inference strategies and mathematical problem-solving abilities. Based on increasingly advanced computer technology, a new group of computer games has been designed and made available in recent years. For instance, *Safari Search* (O'Brien 1991) is particularly thought-provoking for younger students but also interesting and challenging to adults. It contains twelve different strategy games, with varying levels of complexity and strategy. The games such as Intuit the Iguana, Find the Flamingo, and Search Out the Seal may have one or two animals hidden in a matrix. Each is based on an underlying strategy that may vary from counting to complex logical combinations of operations. To win each of the games, students must hypothesize, collect data, and make judgments. *Green Globs and Graphing Equations* (Dugdale and Kibbey 1996), running on Macintosh computers, has significantly enhanced the original version of *Green Globs*. It is easy to use and produces very nice graphics that are better than those in *Mathematica*, Maple, or other graphing software.

### Multimedia Software

Software for mathematics education is increasingly expanding its functions. It has been combined with other media, allowing the computer to manipulate videodisc players, CD-ROM drives (internal or external), VCRs, video and audio editors, and so on. Two examples of multimedia software used in mathematics education are *Measurement in Motion* (Learning in Motion 1994) and *Jasper Software* (Learning Technology Center 1992). With *Measurement in Motion,* students define measurements, make predictions, enter formulas, and run QuickTime movies, which are on either a hard disk or a CD-ROM, to take measurements. These measurements generate charts and graphs from which students can interpret results of the investigation. Students can also make digitized QuickTime movies of their own from VCR tapes or cameras using built-in or add-on audiovisual systems. *Jasper Software* is a computer program to control interactively the operation on a videodisc series called *The Adventures of Jasper Woodbury* by the staff of the Learning Technology Center at Vanderbilt University, which is designed to develop several cognitive skills in students, especially mathematical problem-solving skills. Multimedia software has brought in an integration of graphics, video, sound, animation, text, random access to information, and storage of large amounts of data.

### Assessment Applications

Many of the software packages already mentioned can be used to assess student learning. Those that can capture learning processes as they occur are especially appropriate for assessment. For instance, GSP's scripts capture the way a student does constructions, explores problem situations, or creates a geometric object. *Math Exploration Toolkit* has similar capabilities. These software features allow assessment that is ongoing, naturalistic, and performance-based.

Other approaches to assessment with computers include some software packages that have been developed specifically for assessment purposes. These packages either make current assessment practice more efficient or help change the assessment system. *Mathematics TestBuilder* (1993), for example, combines word processing, database usage, and layout capabilities with a complete set of mathematical symbols and diagrams, allowing teachers to design questions and create test sheets flexibly and easily. Two other examples are *Learner Profile* (Morgan Media 1994) and *Electronic Portfolio* (1995), which feature new assessment methods, emphasizing evidence to be gathered from multiple sources. With *Learner Profile*, teachers can use an electronic notepad or a bar code scanner to record student learning as it happens and transfer data between that device and the computer electronically. These data, where appropriate, can be transferred to spreadsheets for analysis, and reports can then be printed to show evidence of students' mathematical understanding. With *Electronic Portfolio*, teachers can maintain well-organized student portfolios by showing process and best pieces in the same portfolio, tracking progress in students' written or oral presentations, linking their own comments (text or sound) to students' work, and even adding QuickTime video to portfolios, to capture student performances and presentations.

## EQUITY AND USES OF COMPUTER SOFTWARE

Issues of computer equity have received considerable attention since the 1980s when computer use was rapidly increasing in schools. Research studies have revealed that poor, minority, and female students have less access to computers both at home and at school (see Sutton 1991). With regard to type of software use, researchers have found that poor and minority students were more likely to use drill and practice software than were middle-class and white students, and females outnumbered males in word processing but were underrepresented in programming (see Sutton 1991). These findings, however, are apparently more related to early uses of software. A number of authors have discussed the gender stereotyping of software. Many software packages, especially in the early days of development, were gamelike, with much aggression and violence. Many authors suggested that this content was inherently unappealing to girls, but there is only limited research data to support the belief (see Sutton 1991). One example of such research studies was provided by Malone (1981). He explored gender differences for an educational game called *Darts*, which was designed to teach students about fractions. He found that fifth-grade girls did not like the fantasy of a dart (actually an arrow) bursting balloons. The concern for the assumed gender bias in software content has led some manufacturers to produce special software such as *Berenstein Bears* introduced by Sega (see Kantrowitz 1994). The intent of this type of software is to stimulate girls' interest and to complement the way that girls like to learn (see Sutton 1991).

## SUMMARY AND EXPECTATIONS

Since the 1960s, and especially since the late 1970s when the microcomputer became available, the creation and use of mathematics software has developed from its early uses (tutor, tutee, and tool) to the innovative uses represented by the subject-specific exploratory software like *The Geometer's Sketchpad*. As the hardware technology gets more advanced, the innovative software packages are becoming more powerful. Although these packages differ from each other in mathematical content, the range of application, and/or the degree of sophistication, they all have the potential of encouraging students' active involvement and supporting student-directed, construction-oriented learning. In the classroom where software is effectively used, there is a fundamental change in the teacher's role. The teacher is no longer solely a lecturer who transmits the textbook knowledge to students, but functions as an organizer of learning activities and as a catalyst to stimulate critical thinking. The teacher also uses the computer technology for his or her own mathematical learning and becomes a learner along with the students (see Olive 1990).

Research studies on how the use of software packages can enhance mathematics teaching and learning as well as those on the related equity issues should be seriously designed and conducted. Up to now, the efforts put on these types of research are far short of what should be demanded. Only high-quality research findings can convincingly identify the strong features of the software that we should keep and expand and the limitations of the software that we should try to avoid. Since most of the existing software packages deal with middle school, high school, and college-level mathematics, more attention should be paid to developing software for elementary school mathematics. If more sophisticated multimedia software is created, it should be thoroughly tested and studied in research. While the student's active construction of knowledge is often a recommended goal, help for students who need guidance more than others in their mathematical explorations should also be taken into consideration. Although mathematics software may provide rich environments in which students study and do mathematics, the selection and use of the software determines its effectiveness for an individual. The responsibility of teachers and professors is to make use of mathematics software so that the focus stays on mathematics, not on the software itself. The contrast between teaching the functions and operations of software and teaching the mathematics with the aid of the software might well be worth our consideration and study. In learning all the options available in a program, it is important to keep the main goal in mind: learning about and appreciating the power of the mathematics that a particular program offers.

*See also* Enrichment, Overview; Techniques and Materials for Enrichment Learning and Teaching

## SELECTED REFERENCES

Anderson, John R. "Intelligent Tutoring and High School Mathematics." In *Intelligent Tutoring Systems*, Second International Conference, ITS '92 Montreal, Canada, June 1992 Proceedings (pp. 1–10). Claude Frasson, Gilles Gauthier, and Gordon I. McCalla, eds. Berlin, Germany: Springer-Verlag, 1992.

California Department of Education. *Mathematics Framework for California Public Schools*. Sacramento, CA: The Department, 1992.

Dugdale, Sharon. "Green Globs: A Microcomputer Application for Graphing of Equations." *Mathematics Teacher* 75(Mar. 1982):208–214.

Fey, James T. "Technology and Mathematics Education: A Survey of Recent Developments and Important Problems." *Educational Studies in Mathematics* 20(1989): 237–272.

Jiang, Zhonghong. *Students' Learning of Introductory Probability in a Mathematical Microworld*. Unpublished doctoral dissertation, University of Georgia, 1993.

———, and Walter D. Potter. "A Computer Microworld to Introduce Students to Probability." *Journal of Computers in Mathematics and Science Teaching* 13(2)(1994):197–222.

Kantrowitz, Barbara. "Men, Women, Computers." *Newsweek* (May 18, 1994):48–52.

Lin, Pao-ping, and Che-jen Hsieh. "Parameter Effects and Solving Linear Equations in Dynamic, Linked, Multiple Representation Environments." *Mathematics Educator* 4(Winter 1993):25–33.

Malone, Thomas W. "Toward a Theory of Intrinsically Motivating Instruction." *Cognitive Science* 4(1981): 333–369.

McClintock, Edwin, and Zhonghong Jiang. "Spreadsheets: Powerful Tools for Probability Simulations." *Mathematics Teacher* 90(Oct. 1997):572–579.

National Council of Teachers of Mathematics. *Curriculum and Evaluation Standards for School Mathematics*. Reston, VA: The Council, 1989.

———. *Principles and Standards for School Mathematics*. Reston, VA: The Council, 2000.

Olive, John. *Technology and School Mathematics*. Paper available at Athens, GA: University of Georgia, 1990.

Papert, Seymour. *Mindstorms: Children, Computers, and Powerful Ideas*. New York: Basic Books, 1980.

Simonson, Michael R., and Ann Thompson. *Educational Computing Foundations*. New York: Macmillan, 1994.

Sutton, Rosemary E. "Equity and Computers in the Schools: A Decade of Research." *Review of Educational Research* 61(4)(1991):475–503.

Taylor, Robert. *The Computers in the School: Tutor, Tool, Tutee*. New York: Teachers College Press, 1980.

Thompson, Patrick. "Mathematical Microworlds and Intelligent Computer-assisted Instruction." In *Artificial Intelligence and Instruction: Applications and Methods* (pp. 83–109). Gregg Kearsley, ed. Reading, MA: Addison-Wesley, 1987.

Wertheimer, Richard J. "The Geometry Proof Tutor: An Intelligent Computer-based Tutor in the Classroom." *Mathematics Teacher* 83(Apr. 1990):308–317.

Wiebe, James. *Computer Tools and Problem Solving in Mathematics*. Wilsonville, OR: Franklin, Beedle, 1992.

Williams, Carol J., and Scott W. Brown. "A Review of Research Issues in the Use of Computer-Related Technologies for Instruction: An Agenda for Research." In *Educational Media and Technology Yearbook 1991* (pp. 26–46). Brenda Brenyan-Broadbent and R. Kent Wood, eds. Englewood, CO: Libraries Unlimited, 1991.

### Selected Software References

Abrams, Joshua P. *Graph Explorer*. Glenview, IL: Scott, Foresman, 1992.

Anderson, John R. *Geometry Tutor*. Pittsburgh, PA: Carnegie Mellon University, Department of Psychology, 1988.

Barclay, Tim. *Guess My Rules*. Pleasantville, NY: HRM Software, 1985.

BBN Laboratories. *Statistics Workshop*. Pleasantville, NY: Sunburst, 1992.

Dugdale, Sharon, and David Kibbey. *Green Globs and Graphing Equations*. Pleasantville, NY: Sunburst, 1996.

Education Development Center. *MathFINDER Macintosh Version 1.1.* Armonk, NY: Learning Team, 1993.

Edwards, Lois A. *Data Insights.* Pleasantville, NY: Sunburst, 1991.

*Electronic Portfolio.* Jefferson City, MO: Scholastic, 1995.

Fey, James T., and M. Kathleen Heid. *Math Exploration Toolkit.* College Park, MD: University of Maryland, 1985.

Hoffer, Alan. *Algebra Xpresser.* Acton, MA: Bradford, 1991.

Jackiw, Nicholas. *The Geometer's Sketchpad.* Berkeley, CA: Key Curriculum, 1991.

Laborde, Jean-Marie. *Cabri Geometry.* France: Université de Grenoble 1, 1990.

Learning in Motion. *Measurement in Motion,* 1994.

Learning Technology Center. *Jasper Software.* Nashville, TN: Vanderbilt University, 1992.

*Mathematics TestBuilder.* Acton, MA: Bradford, 1993.

Morgan Media. *Learner Profile.* Pleasantville, NY: Sunburst, 1994.

O'Brien, Thomas C. *The King's Rule.* Pleasantville, NY: Sunburst, 1985.

———. *Safari Search.* Pleasantville, NY: Sunburst, 1991.

Olive, John, and Leslie P. Steffe. *CandyBar.* (Also called *TIMA: Bars*). Acton, MA: Bradford, 1994.

*Probability Theory.* Hanover, NH: True Basic, 1988.

Professional Systems International. *StatExplorer.* Glenview, IL: Scott, Foresman, 1993.

Rosenberg, Jon. *Math Connections.* Pleasantville, NY: Sunburst/Wings for Learning, 1993.

Schwartz, Judah, L., and Michal Yerushalmy. *The Geometric Supposers.* Pleasantville, NY: Sunburst, 1985.

Thompson, Patrick. *Blocks.* San Diego, CA: San Diego State University, 1993.

Wolfram, Stephen. *Mathematica.* Champaign, IL: Wolfram, 1991.

ZHONGHONG JIANG

# SPACE GEOMETRY, INSTRUCTION

Traditionally, the investigation of solid geometric shapes such as prisms, pyramids, cones, cylinders, and spheres, as well as the mathematical space in which these shapes are considered. The major curricular goals for three-dimensional (3D) geometry can be grouped into three clusters: (a) how objects, groups of objects, and space itself are organized or *structured;* (b) how objects in 3D space are *measured;* and (c) how objects are *described.* Moreover, consistent with the recommendations of the *Curriculum and Evaluation Standards for School Mathematics* (National Council of Teachers of Mathematics 1989),

another important, but often-neglected, general goal is the development of spatial visualization skills. As students progress through the grades, their structuring, measuring, describing, and visualizing must become increasingly sophisticated, especially for those who must deal with the demands of subjects such as 3D analytic geometry, calculus, and many topics in science and engineering.

The study of space geometry, like other areas of geometry, can beneficially utilize the van Hiele theory (Clements and Battista 1992) as a framework for choosing appropriate instructional tasks and gauging the development of students' thinking. This theory suggests that, in grades K–14, students' thinking about 3D geometry should progress from reasoning holistically about solids based on their appearance (level 1), to identifying the components of solids and informally describing solids in terms of their properties (level 2), to logically relating properties of solids and thereby hierarchically classifying them (level 3), and finally to reasoning formally and constructing proofs within an axiomatic system for 3D geometry (level 4).

## STRUCTURE

The structure of a 3D object refers to how its components are put together, how they are spatially organized and interrelated. The van Hiele theory suggests that structural knowledge can be developed by having students first become familiar with common solids as wholes, next by examining their components and investigating how these components are spatially related, and finally by describing these relationships as properties of solids.

Although reaching van Hiele level 2 in 3D geometry is a reasonable goal for the upper elementary grades, and perhaps level 3 by eighth grade, these levels are rarely reached in traditional curricula. For instance, Gutiérrez, Jaime, and Fortuny (1991) found that although 38 out of 41 preservice teachers and 1 out of 9 eighth graders had completely acquired van Hiele level 1, only 11 of the teachers and none of the eighth graders had completely acquired level 2, and 3 of the teachers and none of the eighth graders had completely acquired level 3.

A commonly employed instructional activity is to have students learn about the components of 3D shapes by identifying their faces, vertices, and edges, as well as tracing and comparing their faces. The goal of these activities should be for students to start mentally constructing the structure of solids as they decompose them into their constituent parts. When

Bourgeois (1986) utilized such an approach with third graders, the students' ability to select solids that were "pictured" by nets improved. However, although students who employed the analytic strategy of counting faces of the solids or nets were quite successful in associating solids to nets, very few students employed such an analytic strategy and thus could have been considered to be moving into van Hiele level 2. A similar, but mentally more demanding, activity is to have students make nets for solids. This requires students not only to examine the parts of solids but how they are related. Jean Piaget's research indicates that children have difficulty imagining nets for solids before the ages of 8 or 9 (Bourgeois 1986).

Another way to have students investigate the structure of 3D shapes is for them to examine the cross sections of the shapes, another level 2 activity. Davis (1973) implemented a 25-minute training session in which sixth-, eighth-, and tenth-grade students cut styrofoam solids and were asked questions about the cross sections. He then had the students select the correct drawing of cross sections of the solids. The number of students (out of 30) who scored 87.5% or higher was 2 for sixth grade, 15 for eighth grade, and 21 for tenth grade. The author suggested age 14 or higher as the mastery level for the concept of cross-sectioning geometric solids.

Battista and Clements' (1996) work with third graders gives us further insight into how students think about the components of solids. As these students enumerated edges of polyhedra that they had built out of sticks and connectors, they frequently miscounted the edges (which they noticed because their counts were inconsistent with previous counts or the counts of other students). It was not until they organized their counts (for example, they might count the edges on the top, then the bottom, then the lateral sides of a prism) that they were able to correctly enumerate the edges. These students imposed a different spatial structure on the parts of the solid as a result of attempting to count them. In fact, the counting task forced students to reflect on, and many times alter, their structuring of the objects. That reflection on counting acts should help students spatially structure a situation is consistent with Piaget's theory that the mental construction of shapes consists of the coordination and abstraction of the child's mental actions in dealing with the shapes (Piaget and Inhelder 1956). Thus, the key to helping students structure 3D shapes and move to higher van Hiele levels is the presentation of problems in which students concretely investigate parts of shapes and reflect on how these parts are related.

## MEASUREMENT

Measurement, particularly of volume and surface area, has traditionally been a major curricular emphasis in 3D geometry. The research indicates, however, that students are not learning this material effectively. As we shall see later, a major reason for this seems to be students' meager understanding of the structure of 3D objects.

A typical volume problem presents students with a diagram of a 3D cube array and asks them to determine its volume or the number of cubes in it. Ben-Chaim, Lappan, and Houang (1985) found that less than 50% of middle grade students could solve such problems. The results of the second National Assessment of Educational Progress (NAEP) showed that less than 40% of 17-year-olds could solve problems of this type (Hirstein 1981).

Typical errors that students in grades 5–8 made on these problems include counting the cube faces shown in the diagram, and perhaps doubling that number, and counting the number of cubes showing in the diagram, and perhaps doubling that number (Ben-Chaim, Lappan, and Houang 1985). Ben-Chaim, Lappan, and Houang suggested that (a) the counting-the-faces error was due to students dealing with the picture strictly as a two-dimensional object; (b) students who counted cubes showed an awareness of the three-dimensionality of the depicted object; (c) students who did not double their counts did not seem to visualize the hidden portions of the objects; and (d) many students had difficulty relating isometriclike drawings to the rectangular solids they represent. They concluded that these errors were related to students' visualization ability.

Hirstein (1981), on the other hand, attributed many of the errors students make on this type of problem to confusion between volume and surface area. In fact, Enochs and Gabel (1984) found that preservice elementary teachers' understanding of these two concepts was poor and that these students did not really understand either concept. Indeed, whereas 77% of these students were certain that the volume of a rectangular solid could be found by multiplying the length, width, and height, only 44% were certain that the volume was equal to the area of the base times the height, and only 58% were certain that the volume could be found by counting the number of unit cubes that filled the solid. The situation was worse for surface area, with only 34% knowing that it could be determined by counting the number of unit squares on the outside of the solid. Apparently, many of these students had memorized formulas for obtaining quantities that they did not understand. Because of students' fuzzy notions about both concepts, and because the concepts have traditionally been taught one after the other, it is easy to see why students' ideas about them are confused and intermingled. The authors suggest that these concepts be taught with a "hands-on" approach and that formulas be mentioned only at the end of exploration.

Battista and Clements (1996) provide a more elaborate description of students' solution strategies and errors in dealing with 3D cube arrays. Their data suggest that many students are unable to enumerate the cubes in a 3D array because they cannot coordinate the separate views of the array and integrate them to construct one coherent mental model of the array. Up to 64% of third graders and 33% of fifth graders showed evidence of this type of thinking. Ben-Chaim, Lappan, and Houang (1985) reported data suggesting that about 39% of fifth–eighth graders might be using this type of thinking. Eventually, as students become capable of coordinating views, they restructure the arrays. Those who complete a global restructuring utilize layering strategies. Those in transition utilize strategies that indicate a local, piece-by-piece, structuring.

## DESCRIPTION

One of the major goals of 3D instructional units should be to promote the development of the basic concepts and language needed to reflect on and communicate about spatial relationships in 3D environments. Such knowledge, along with visualization skills, is not only critical in mathematics but in science and engineering. Students must explore various ways to pictorially represent solids and learn to correctly visualize 3D configurations described in diagrams and different types of building instructions.

Mitchelmore (1980) found four stages of sophistication in children's drawing of 3D solids. In the first stage, the figure is drawn as a single, orthogonal face or as a general holistic outline. In the second stage, several visible faces are drawn but not properly depicted or connected, or hidden faces are included. In stage 3, drawings attempt to represent a single point of view and to depict depth. In stage 4, parallel edges on the object are represented by parallel lines in the drawing. The sophistication of students' drawings increased with grade level.

Cooper and Sweller (1989) examined students' ability to interpret various representations of simple three-dimensional objects. For all ages studied, objects, perspective drawings, and verbal descriptions were more easily interpreted than layer plans, orthogonal views, and coordinates. Grade 9 and grade 11 students were more successful than grade 7 students in interpreting layer plans, orthogonal views, and coordinate descriptions. Although other research suggests that middle school students often have difficulty with perspective drawings, 85% of this study's grade 7 students correctly interpreted such drawings.

Ben-Chaim, Lappan, and Houang (1989) gave students three weeks of instruction on representing solids and constructing solids from drawings. To assess their ability to communicate spatial information, students were given 3D models made up of 10 small cubes taped together and were to devise a description of what their building looked like. Before instruction, only 26% of students gave correct building descriptions; after instruction, 83% gave correct descriptions.

Again, thinking about the van Hiele levels can be helpful. Students' work with describing 3D shapes must be coordinated with their conceptual development concerning those objects. For example, to expect students who are at van Hiele level 1 to give verbal descriptions that describe the properties of shapes, rather than the vague, holistic descriptions characteristic of this level, simply encourages students to rotely memorize such descriptions.

## CURRICULAR RECOMMENDATIONS

At the primary level, students need experiences building with 3D objects. For instance, they can build a house or castle out of wooden geometric solids. Or one student might make a simple figure out of connecting cubes and another student could try to duplicate it. As students progress, they can also start thinking about the parts of shapes. For instance, students can use ink to stamp the faces of solids on cards; other students can try to identify which solids made the cards.

At the intermediate level, students can build polyhedra from sticks and connectors. They can identify the shapes of shadows made by geometric solids and explore notions of perspective by examining how various shapes look from different angles. They can make boxes that contain a given number of cubes. They can predict how many cubes fit in boxes, then check their answers with cubes. Given various kinds of diagrams of cube configurations, including pictures of several orthogonal views, they can build the configurations with connecting cubes. Toward the end of this period, they can be introduced to the formal concept of volume (initially, students should focus on enumerating cube arrays).

As students move into junior high, they can continue many of the activities started in the intermediate grades. But they also can deal with more abstract situations. They need to make sense of volume in wider contexts. For instance, they need to understand the relationship between linear and volume measurements and to make sense out of situations that involve rational, not just whole number, measurements.

At the high school level, students should study 3D coordinate systems, 3D transformations, cross sections of solids, and more complex volumes, such as those for nonpolygonal solids and shapes that must be decomposed. Problems that require interpretation of increasingly complex drawings should continue to be given to students.

As students move through the grade levels, they should be presented with tasks that encourage more and more visualization. However, because visualization ability differs widely among individuals, at all grade levels some students will require the presence of actual physical materials. It is essential to recognize that visual ability can be improved by giving students many opportunities to manipulate physical objects and to reflect on these manipulations.

*See also* van Hiele Levels

SELECTED REFERENCES

Battista, Michael T., and Douglas H. Clements. "Students' Understanding of Three-dimensional Rectangular Arrays of Cubes." *Journal for Research in Mathematics Education* 27(3)(1996):258–292.

Ben-Chaim, David, Glenda Lappan, and Richard T. Houang. "Adolescents' Ability to Communicate Spatial Information: Analyzing and Effecting Students' Performance." *Educational Studies in Mathematics* 20(1989): 121–146.

———. "Visualizing Rectangular Solids Made of Small Cubes: Analyzing and Effecting Students' Performance." *Educational Studies in Mathematics* 16(1985): 389–409.

Bourgeois, Roger D. "Third Graders' Ability to Associate Foldout Shapes with Polyhedra." *Journal for Research in Mathematics Education* 17(1986):222–230.

Clements, Douglas H., and Michael T. Battista. "Geometry and Spatial Reasoning." In *Handbook of Research on Mathematics Teaching.* Edited by Douglas A. Grouws. New York: Macmillan, 1992, pp. 420–464.

Cooper, Martin, and John Sweller. "Secondary School Students' Representations of Solids." *Journal for Research in Mathematics Education* 20(1989):202–212.

Davis, Edward J. "A Study of the Ability of School Pupils to Perceive and Identify the Plane Sections of Selected Solid Figures." *Journal for Research in Mathematics Education* 4(1973):132–140.

Enochs, Larry G., and Dorothy L. Gabel. "Preservice Elementary Teachers' Conceptions of Volume." *School Science and Mathematics* 84(1984):670–680.

Gutiérrez, Angel, Adela Jaime, and Jose M. Fortuny. "An Alternative Paradigm to Evaluate the Acquisition of the van Hiele Levels." *Journal for Research in Mathematics Education* 22(1991):237–251.

Hirstein, James J. "The Second National Assessment in Mathematics: Area and Volume." *Mathematics Teacher* 74(1981):704–708.

Mitchelmore, Michael C. "Prediction of Developmental Stages in the Representation of Regular Space Figures." *Journal for Research in Mathematics Education* 11(1980):83–93.

National Council of Teachers of Mathematics. *Curriculum and Evaluation Standards for School Mathematics.* Reston, VA: The Council, 1989.

———. *Principles and Standards for School Mathematics.* Reston, VA: The Council, 2000.

Parzysz, Bernard. "Knowing vs Seeing. Problems of the Plane Representation of Space Geometry Figures." *Educational Studies in Mathematics* 19(1988):79–92.

Piaget, Jean, and Bärbel Inhelder. *The Child's Conception of Space.* London, England: Routledge and Kegan Paul, 1956.

MICHAEL T. BATTISTA
MICHAEL MIKUSA
DOUGLAS H. CLEMENTS

# SPACE MATHEMATICS

The achievements of the space program owe much to mathematics. In the study of the solar system, large numbers are rounded and scientific notation is used. Calculations of center of mass, center of pressure, and trajectories are used in studying rocketry. Geometry is used in calculating the orbital paths of satellites and planets and in the measurement of the earth.

One of the first measurements of the Earth was made by a Greek mathematician, Eratosthenes, over two thousand years ago. He noted that at noon on the first day of summer the sun appeared to be directly overhead at the city of Syene in Egypt. At the same time, in the city of Alexandria a pole would cast a shadow such that the angle of the Sun's rays to the pole measured about 7.2°. By setting up a proportion using the given angle and the known distance between the two cities,

$$\frac{5{,}000 \text{ stadia}}{7.2°} = \frac{\text{circumference of the Earth}}{360°}$$

Eratosthenes obtained the circumference of the Earth, about 250,000 stadia or 25,000 miles (Soroka 1967, 66).

Describing the motion of a spacecraft requires describing the location and the orientation of the spacecraft and changes that occur during flight. This can be accomplished by setting up suitable coordinate systems. The orientation of the spacecraft in relationship to direction of flight is best described by a spacecraft-based system, while ground control is best suited by an Earth-based coordinate system. This system based upon two different coordinates requires the ability to make transformations between coordinate systems (Kastner 1985, 11).

The space program produces a vast amount of data. This data can then provide students with an opportunity to use their mathematical abilities to analyze real-world situations. Using the same data that the scientists used, students can decide whether or not to delay the launch of the *Challenger* on January 28, 1986. They can thus develop an understanding of the concept of probability modeling. This exercise allows for the discussion of alternative interpretations of the same data that can lead to two different decisions, launch or delay. The students are given a set of data points and asked to fit a smooth curve to the given points. They are then asked to estimate final values given various initial points. Afterward, the students are given the definitions for the variables in their models. They are asked to use their models to determine whether the temperature would allow the launch as scheduled. A discussion then will arise as to why the *Challenger* was launched (Tappin 1994).

Space mathematics lends itself greatly to interdisciplinary activities. One such activity is based on a journey to Mars. The social sciences are responsible for who goes to Mars, while the science, mathematics, and technology classes determine how to get to Mars. Physical requirements are left to the biosciences, nutrition is the responsibility of home economics, and health and physical education plan exercise routines for the crew (Litowitz and McLeod 1993). The entire school is involved, and space-related mathematics permeates all subjects.

Although space science is complex and usually beyond what is taught in high school, many aspects can be understood using high school mathematics. Many topics taught in the primary levels can be enhanced by adding a space theme to make them more realistic. There are not many books devoted entirely to space mathematics, but there are quite a few resources available for space-related subjects.

The National Aeronautics and Space Administration (NASA) teacher resource network offers NASA-produced teacher guides and other space education materials for teachers (Kastner 1985).

*See also* Applications for the Classroom, Overview; Applications for the Secondary School Classroom; Data Analysis; Modeling

## SELECTED REFERENCES
Kastner, Bernice. *Space mathematics: A Resource for Secondary School Teachers.* Washington, DC: NASA, 1985.

Litowitz, Len, and Lisa McLeod. "The Journey to Mars." *TIES* (Nov./Dec. 1993):32–36.

Soroka, John. "Measurement, a Window to the Universe." In *The Shapes of Tomorrow* (pp. 66–68). U.S. Committee on Space Science Oriented Mathematics (eds.). Washington, DC: Government Printing Office, 1967.

Tappin, Linda. "Analyzing Data Relating to the *Challenger* Disaster." *Mathematics Teacher* 87(Sept. 1994): 423–426.

JAMES E. PRATT

# SPIRAL LEARNING

A means of organizing learning through a structured mathematics curriculum. A sequence of learning experiences is constructed so that the learner works through the main concepts of mathematics at the elementary level, before tackling the same concepts again and again but at successively more sophisticated and advanced levels. Such a curriculum is said, by geometric analogy, to have a spiral shape. The approach has also been called the *concentric method.* Before the term *spiral curriculum* was used, educators argued for the progressive development of mathematical knowledge, with earlier learning revisited and extended. Such ideas go back to antiquity and were promoted more recently by progressive educators such as Johann Friedrich Herbart and John Dewey. According to Geoffrey Howson (1983), Herbart strongly influenced the form of the mathematics curriculum at the beginning of this century. He quotes Charles Godfrey, the author of well-known British geometry books, as follows, from a publication of 1911. ". . . [T]he Herbartian allows the mental content to grow by laying hold of ideas that will hook onto the old. . . . And the old will include not only old knowledge but old experience" (Howson 1983, 4). Similarly, Dewey describes the need for ". . . the progressive development of what is already experienced into a fuller and richer and also more organized form that gradually approximates that in which subject matter is presented to the skilled, mature person" (Dewey 1963, 73–74).

More recently, the spiral learning approach, and its name, is associated with the cognitive psychologist and curriculum theorist Jerome Bruner (1915– ). His idea of a spiral curriculum was based on the need for integration, continuity, and progression in learning. Traditionally, the mathematics curriculum was divided into separate segments, such as arithmetic, geometry, and algebra, which were studied in sequence. Bruner argues that the same concepts should recur throughout the curriculum in progressively more complex or sophisticated terms.

His claim is that "any subject can be taught effectively in some intellectually honest form to any child at any stage of development" (Bruner 1960, 33). Bruner's claim rests on his theory of mental representation, which is a development of the learning theory of Jean Piaget (1972). Bruner's theory is that there are three modes in which knowledge is mentally represented: enactive, iconic, and symbolic, which learners develop in this order. Enactive understanding depends on some action, such as using compasses to draw a circle, or a ruler for measuring. This leads to knowledge in the iconic mode, which is understanding through mental pictures or spatial imagery. For example, the concept of a circle is linked in many peoples' minds with a mental image of a circle. This leads to knowledge in the symbolic mode. Formal mathematical topics like rational numbers, ratio and proportion, and algebra can only be fully represented and understood in this mode. Bruner claims that we use all three modes, but that the earlier modes are more accessible and result in more vivid experiences and more memorable learning than the later modes. Hence spiral learning should first introduce children to mathematical concepts in enactive mode. Later, they should revisit them in both enactive and iconic forms. Subsequently they should encounter the concepts in all three modes.

Bruner's theory has useful consequences for the teaching of mathematics. It suggests a sequence of ways of representing knowledge to young learners. For example, the concept of rational number is developed through the physical cutting of objects into equal parts (enactive); the shading of "pizza" diagrams (iconic); finally, the use of rational number symbols and notation (symbolic). Bruner's theory of representation has been criticized, however, for understressing distinctions within the symbolic category. Pamela Liebeck (1984) argues that the communication of mathematical ideas in spoken language and in written mathematical symbolism must be sharply distinguished for instructional purposes, although they both fall within Bruner's symbolic category, and that representation in spoken language often precedes children's use of iconic representations.

The principle of spiral learning is independent of Bruner's theory of representation and can accommodate more refined theories of representation, such as those of Gerald Goldin (1987). Research has shown that a spiral learning approach leads to general but slow progress with exercises and problems of increasing complexity across the range of mathematical topics and with larger and more difficult number types in calculations (Hart 1981). Furthermore, in certain topics, such as the multiplication and division of rational numbers, when a spiral approach is not applied (the skills are not revisited), performance deteriorates. This confirms the value of spiral learning as a general principle of curriculum design, although it can be applied in many different ways. For the elementary school teacher the implication is that concepts should be introduced by means of activities, speech and discussion, and pictorial representations before moving on to written symbolism and that the concepts and the modes of representation should be revisited periodically. The implication for all teachers is to periodically revisit concepts and practice skills to maintain and extend previous learning.

*See also* Bruner; Dewey; Psychology of Learning and Instruction, Overview

## SELECTED REFERENCES

Bruner, Jerome. *The Process of Education.* Cambridge, MA: Harvard University Press, 1960.

Dewey, John. *Experience and Education.* New York: Macmillan, 1963. [Originally published 1938.]

Goldin, Gerald. "Cognitive Representational Systems for Mathematical Problem Solving." In *Problems of Representation in the Teaching and Learning of Mathematics* (pp. 125–145). Claude Janvier, (ed.). Hillsdale, NJ: Erlbaum, 1987.

Hart, Kathleen, ed. *Children's Understanding of Mathematics: 11–16.* London, England: Murray, 1981.

Howson, A. Geoffrey. *A Review of Research on Mathematical Education.* Pt. C. *Curriculum Development and Curriculum Research.* Windsor, England: NFER-Nelson, 1983.

Liebeck, Pamela. *How Children Learn Mathematics.* London, England: Penguin, 1984.

Piaget, Jean. *The Principles of Genetic Epistemology.* Wolfe Mays, trans. London, England: Routledge and Kegan Paul, 1972.

PAUL ERNEST

# SPORTS

Contains phenomena understandable by mathematical concepts taught at the secondary school level. Numbers are everywhere in sports. How many points were scored? How fast did she run? How high did he jump? Who has the higher batting average? In compiling and analyzing such sports-related numbers, the importance of arithmetic computations is well recognized. Less well known, however, is the role that mathematics, beyond mere computations, can play in gaining a better understanding of certain phenomena that occur in sports.

## CIRCUMFERENCE OF A CIRCLE

The first example is an illustration of how I was able to use mathematics to solve a sports-related problem that was unexpectedly posed to me a couple of years ago. I had just finished working out at the local high school track when a newly arriving runner approached me and asked the following question: "How many times around the track for one mile?" I answered "four times essentially," having said "essentially" since four times around is really 1600 meters, which is about 10 yards short of a mile. He replied, "How about if I run in the outside lane?" I thought about his question for a moment. The track is made up of two straightaways and two turns. By his staying in the outside lane, the only places on the track where his run would pick up extra distance are on the turns. Realizing that the turns on a track are semicircles, I proceeded to walk off, at about 1 yard per stride, the distance from the inside of lane 1 to the inside of lane 6 (this track had a total of only six lanes). This distance came out to be about 6 yards. I then mentally computed that if $r$ is the radius of each of these semicircles, then one lap run in lane 6 covers $2\pi(r + 6) - 2\pi r = 12\pi$ (or about 38) yards more than one lap run in lane 1 (see Figure 1). Thus, I told him that if he were to run four laps in lane 6, he would cover approximately 150 yards more than if he

ran the four laps in lane 1, giving him a total of just under 1.1 miles. I then thought, "I can use this same idea to figure out how far from the inside of lane 1 I would have to move out in order to run a total distance of 10 km on the track in only 24 laps, instead of the 25 it would take me to do it in lane 1."

By presenting situations like the one described above, teachers can lead their students to discover how mathematical formulas may be used to solve some interesting real-life problems. Teachers might want to pose the original problem here for students to work on in groups, including the "10 km in 24 laps" problem as an extension. Appropriate guidance could be given as needed. Additional extension problems that students should find interesting are the following:

1. We know that by running four laps in the inside of lane 1 on a standard 400-m track, the total distance covered is not quite one mile. How far from the inside of lane 1 would you have to move out in order to run exactly one mile in four laps of a 400-m track? (*Note:* The answer is a surprisingly small amount.)

2. A belt is placed snugly around the Earth's equator. How much longer would the belt have to be in order for it to fit around the equator at a distance of exactly 1 meter above the ground? (Assume that Earth is a sphere.)

3. If you walked all the way around the Earth's equator, starting and ending at the same point, how much farther would your head have traveled than your feet? (Assume again that Earth is a sphere.)

## FRACTIONS

In the 1963 major league baseball season, Frank Howard of the Los Angeles Dodgers had a batting average of 0.273. This average was computed by dividing 114, the total number of hits he obtained that year, by 417, his total number of official at-bats, and rounding off the decimal obtained to three places. That same year, his teammate Wally Moon had a batting average of 0.262. In the 1964 season, their respective averages were 0.226 and 0.220. Thus, Frank Howard had a higher batting average than Wally Moon over each of the two seasons. Can we therefore conclude that his batting average was higher than Wally Moon's when the two seasons are combined? While at first glance, this conclusion may appear to be correct, a closer look at the numbers reveals otherwise (see Figure 2 [Stevens]).

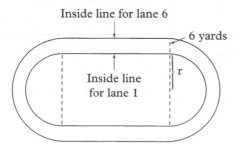

**Figure 1**  *How much farther do you run in lane 6?*

| | **Wally Moon** | | | | **Frank Howard** | | |
|---|---|---|---|---|---|---|---|
| | Hits | At-bats | Batting average | | Hits | At-bats | Batting average |
| 3 | 90 | 343 | 0.262 | | 114 | 417 | 0.273 |
| 4 | <u>26</u> | <u>118</u> | <u>0.220</u> | | <u>98</u> | <u>433</u> | <u>0.226</u> |
| | 116 | 461 | 0.252 | | 212 | 850 | 0.249 |

**Figure 2**  *Who was the better hitter?*

This phenomenon is commonly referred to as Simpson's paradox (Beckenbach 1979; Friedlander 1992; Simpson 1951). In order to analyze it, consider a batter who gets $a$ hits in $b$ at-bats in one season for a batting average of $a/b$, followed by $c$ hits in $d$ at-bats the next season for a batting average of $c/d$. The combined batting average over the two seasons is then $(a + c)/(b + d)$. It is not hard to show that, given any two positive fractions $a/b$ and $c/d$, the fraction $(a + c)/(b + d)$ must lie between them. That is, if $(a/b) \leq (c/d)$, then

$$(a/b) \leq (a + c)/(b + d) \leq (c/d)$$

Thus we see that, between any two given positive fractions, we can always find another positive fraction simply by adding the two numerators and then dividing the sum obtained by the sum of the two denominators. The fact that we can always find a positive fraction that lies between two given positive fractions is called the *density property* of the positive fractions.

To understand how Simpson's paradox can occur, note that the combined batting average $(a + c)/(b + d)$ is really a weighted average of the two batting averages $(a/b)$ and $(c/d)$. That is, depending on the relative sizes of $b$ and $d$, if $(a/b) \leq (c/d)$, then $(a + c)/(b + d)$ could end up being anywhere in

$[a/b, c/d]$, the interval between $a/b$ and $c/d$. If $b$ is very small in comparison to $d$, then $(a + c)/(b + d)$ will be closer to $c/d$, while if $b$ is very large in comparison to $d$, then $(a + c)/(b + d)$ will be closer to $a/b$. In the example of Figure 2, the interval $[a/b, c/d]$ for Wally Moon is $[26/118, 90/343] = [0.220, 0.262]$, with $(a + c)/(b + d) = 116/461 = 0.252$, while the interval $[a/b, c/d]$ for Frank Howard is $[98/433, 114/417] = [0.226, 0.273]$, with his $(a + c)/(b + d) = 212/850 = 0.249$. We see that, since 343 is so much larger then 118, Wally Moon's combined batting average of 0.252 over the two seasons is much closer to 0.262 than to 0.220, while since 433 and 417 are so close to each other, Frank Howard's combined batting average of 0.249 is about halfway between 0.226 and 0.273. Picturing these averages on the number line (see Figure 3), we see that Simpson's paradox was able to occur in this instance because Wally Moon's interval [0.220, 0.262] overlaps Frank Howard's interval [0.226, 0.273]. The reason the intervals overlap is that, even though Frank Howard had a higher batting average then Wally Moon over each of the two seasons, Wally Moon's batting average of 0.262 in his best season of the two is higher than Frank Howard's average of 0.226 in his worst season of the two. If the two intervals had not overlapped, then, regardless of how many at-bats they each would have

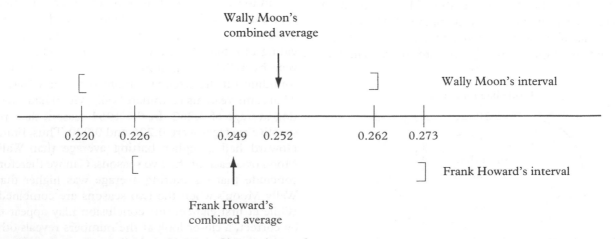

**Figure 3**  *Overlapping intervals in Simpson's paradox.*

had in each of the two seasons, Frank Howard's interval would have been situated entirely to the right of Wally Moon's interval, and therefore his combined average over the two seasons would have been higher than Wally Moon's.

For students who are not familiar with how batting averages are computed in baseball, teachers can present these ideas using test scores. For example, if a student scores 30% on a 20-point quiz and 90% on an 80-point exam, then the student has obtained $6 + 72 = 78$ points out of a possible $20 + 80 = 100$ points, for a combined average on the two tests of 78%. This represents the weighted average of the two scores $6/20 = 0.30$ and $72/80 = 0.90$. Since 80 is so many more points than 20, the combined average of 78% lies much closer to the 90% test score than it does to the 30% quiz score.

By analyzing either batting averages or test scores as here, students are not only able to gain a concrete understanding of the density property of positive fractions, but are also able to give meaning to the operation $\oplus$ defined by $a/b \oplus c/d = (a + c)/(b + d)$ for positive fractions $a/b$ and $c/d$. In addition, they can be led to discover that the density property can be generalized to any number of positive fractions as follows: if $a_1/b_1, a_2/b_2, \ldots, a_n/b_n$ are positive fractions such that $a_1/b_1 \leq a_2/b_2 \leq \ldots \leq a_n/b_n$, then

$$\frac{a_1}{b_1} \leq \frac{a_1 + a_2 + \cdots + a_n}{b_1 + b_2 + \cdots + b_n} \leq \frac{a_n}{b_n}$$

In fact, depending on the relative sizes of $b_1, b_2, \ldots, b_n$, the weighted average

$$(a_1 + a_2 + \cdots + a_n)/(b_1 + b_2 + \cdots b_n)$$

of the $n$ fractions could turn out to lie anywhere in the interval $[a_1/b_1, a_n/b_n]$. For this reason, Simpson's paradox can occur for any number of fractions. For example, Batter A could have a higher batting average than Batter B over each of several seasons, and yet not have a higher overall batting average than Batter B when the seasons are combined. As with the case for two seasons, a necessary condition that must be satisfied in order for Simpson's paradox to have a chance of occurring over several seasons is that the lowest of Batter A's seasonal batting averages be less than the highest of Batter B's seasonal averages.

Two examples of real-life occurrences of Simpson's paradox in this more general setting have appeared in the literature. The first one involves an unwarranted charge of sex discrimination that was leveled in 1973 at the University of California–Berkeley due to the overall rate of admission to graduate school being substantially less for female applicants than for male applicants. Yet, when examined on a department-by-department basis, female rates exceeded male rates in every instance (Bickel, Hammel, and O'Connell 1975). The second example details how the federal personal income tax rates decreased in each income category between 1974 and 1978, while at the same time the overall tax rate increased from 14.1% to 15.2% over those years (Wagner 1982).

## PROBABILITY AND COUNTING

The following example is illustrative of how teachers can find interesting mathematical problems in newspapers, magazines, or other media sources. In a 1989 letter to the editor in *Sports Illustrated* (Heuertz 1989) (see Figure 4), a reader asked what is the probability that, in any given season, the Big Eight Conference's final football standings will have a "perfect order finish" (that is, a finish where, after each of the eight teams has played each of the other seven exactly once in a round-robin tournament, as they do each year in Big Eight football, the first place team winds up with a 7–0 record, the second place team finishes 6–1, third place is 5–2, and so on down to the last place team, which finishes with a record of 0–7).

As the letter and the response to it pointed out, there were perfect order finishes in the Big Eight standings in 1986 and 1988. In addition, there was a perfect order finish in 1989. Is it surprising that, in three out of four seasons, the standings ended in a

A Letter:
## PERFECT ORDER
I was interested to see the perfect mathematical progression in the Big Eight's final conference standings in football this season.

| Won | Lost | Team |
|-----|------|------|
| 7 | 0 | Nebraska |
| 6 | 1 | Oklahoma |
| 5 | 2 | Oklahoma State |
| 4 | 3 | Colorado |
| 3 | 4 | Iowa State |
| 2 | 5 | Missouri |
| 1 | 6 | Kansas |
| 0 | 7 | Kansas State |

What is the probability of such a progression happening?
MATT HEUERTZ
*Wheaton, Ill.*

Sports Illustrated's Response:
If you could assume that all teams had a 50–50 chance of winning each game they played and that every game produced a winner, the probability of this progression occurring in an eight-team conference would be 1 in 6,667. However, because in real life skill comes into play and some teams are always stronger than others, such perfect progressions actually occur far more frequently than that. In fact, the Big Eight's final standings were also in perfect progression in 1986. –ED.

**Figure 4** *Perfect order finish in Big Eight football. [Reprinted courtesy of* Sports Illustrated: *Letters to the Editor column, January 9, 1989, copyright © 1989, Time Inc. All rights reserved.]*

perfect order finish? According to *Sports Illustrated*'s response to the reader, it is very surprising, since it claimed that the probability of a perfect order finish among eight teams in a round-robin tournament is 1 in 6,667 (assuming that all teams have a 50–50 chance of winning each game they play and that every game produces a winner). To see how the magazine might have come up with that figure, note that since each of the eight teams plays each of the other seven exactly once, there are $(8 \times 7)/2 = 28$ games played in the entire Big Eight season (we must divide by 2 in order to avoid counting each game twice). Since each game has two possible winners, there are $2^{28}$ total possible outcomes of all these games.

One of the $2^{28}$ possible outcomes is shown in the letter in Figure 4; namely, the actual perfect order finish of the 1988 Big Eight season, with Nebraska winning all seven of its games, Oklahoma winning all but the Nebraska game, and so on down to Kansas State losing all seven of its games. Another of the $2^{28}$ possible outcomes of all the Big Eight games would be a perfect order finish where Nebraska and Oklahoma switched places, so that Oklahoma would now have a perfect 7–0 record, Nebraska would be 6–1, winning all but the Oklahoma game, and all other teams would have the same record as shown in Figure 4. Note that if the season ends in a perfect order finish, then each team had to have lost to all of those who finished above it in the standings, and had to have beaten all of those who finished below it. Thus, given a particular perfect order finish, the outcomes of all the games are determined by the order in which the teams finished in the final standings. Hence, each possible perfect order finish in the Big Eight corresponds to exactly one of the $2^{28}$ possible outcomes of all the games. Therefore, the probability, $p_8$, that a perfect order finish will occur in a Big Eight football season is given by $p_8 = N/2^{28}$, where $N$ is the number of ways that a perfect order finish can occur. We can find $N$ by noting that there are eight choices for the team having the 7–0 record. Once this team is chosen, there are seven choices left for the team having the 6–1 record. After we choose this team, there are now six choices left for the team having the 5–2 record, and so on down to one choice remaining for the team having the 0–7 record. Thus we see, from the multiplicative counting principle, that

$$N = 8 \cdot 7 \cdot 6 \cdot 5 \cdot 4 \cdot 3 \cdot 2 \cdot 1 = 8!$$

so that $p_8 = 8!/2^{28} \doteq 0.00015$, which is about 1 in 6,657.6 (so *Sports Illustrated*'s figure of 1 in 6,667 must have been a slight misprint!).

For $n$ teams in a round-robin tournament, the same reasoning can be used to show that the probability $p$ that a perfect order finish will occur is given by $p_n = n!/2^{C(n,2)}$, where $n! = n \cdot (n - 1) \cdot (n - 2) \cdot \ldots \cdot 2 \cdot 1$ and $C(n,2) = n(n - 1)/2$ is the number of ways to choose two teams from $n$. (In general,

$$C(n,k) = n \cdot (n - 1) \cdot \ldots \cdot (n - (k - 1))/k!$$

is the number of ways to choose a set of $k$ teams from $n$, where the order in which we choose the $k$ teams does not matter.)

The question arises as to why, if it is so unlikely that a perfect order finish would occur among eight teams in a round-robin tournament, there were three out of four such finishes in the Big Eight from 1986 through 1989. The answer lies in the fact that, in real life, it is not often the case that all teams have a 50–50 chance of winning each game they play. In fact, in the late 1980s, this was most definitely not the case in Big Eight football. In 1986 and 1987, Nebraska and Oklahoma were dominant teams, each being very likely to beat all of the other six teams in the conference, while Kansas and Kansas State were inferior teams, each being very likely to lose to all of the other six conference opponents. By 1989, the mix had changed a bit, with Colorado and Nebraska being the two dominant teams and Missouri and Kansas State being the inferior ones. Thus, for all practical purposes, in the late 1980s, the probability of a perfect order finish in the Big Eight football standings equaled the probability that the middle four teams would achieve a perfect order finish among themselves. Assuming these four teams were more or less equally matched, this number is simply $p_4 = 4!/2^{C(4,2)} = 0.375$. If we let 0.375 represent the probability of a perfect order finish in Big Eight football during the late 1980s, then $1 - 0.375 = 0.625$ is the probability that such a finish would not occur. Therefore, the probability of a perfect order finish in at least three out of four seasons in those years is given by

$$C(4,3) \cdot (0.375)^3 \cdot (0.625)^1 + C(4,4) \cdot (0.375)^4 \doteq 0.15$$

Thus, while still rather unlikely, it is perhaps not quite so surprising that perfect order finishes occurred with such frequency in the Big Eight football standings in the late 1980s.

*See also* Applications for the Classroom, Overview; Fractions; Probability, Overview; Techniques and Materials for Enrichment Learning and Teaching

SELECTED REFERENCES

Beckenbach, Edwin F. "Baseball Statistics." *Mathematics Teacher* 72(1979):351–352.

Bickel, P. J., E. A. Hammel, and J. W. O'Connell. "Sex Bias in Graduate Admissions: Data from Berkeley." *Science* 187(1975):398–404.

Friedlander, Richard J., "Ol' Abner Has Done It Again." *American Mathematical Monthly* 99(1992):845.

Heuertz, Matt. Letter to the Editor. *Sports Illustrated* 70(1989):6.

Simpson, E. H. "The Interpretation of Interaction in Contingency Tables." *Journal of the Royal Statistical Society* ser. B, 13(2)(1951):238–241.

Stevens, Christine. Personal communication.

Wagner, Clifford H. "Simpson's Paradox in Real Life." *American Statistician* 36(1)(1982):46–48.

RICHARD J. FRIEDLANDER

# STANDARD DEVIATION

An important summary statistic used to describe the variability in a data sample taken from a population. The well-known sample mean is a measure of central tendency, but the standard deviation describes the amount of variability in a set of numbers. Both are important statistics that should be computed for any set of data. The sample standard deviation is computed as follows:

Step 1: Compute the arithmetic mean, $\overline{X}$. This is just the sum of all the data values in the sample, divided by the sample size $n$.

Step 2: Add up the squared deviations from the sample mean, that is, (data value − mean)$^2$, and then divide by $(n - 1)$. This is commonly called the *sample variance*.

Step 3: Take the square root of the result from step 2. This number is the standard deviation, denoted by $s$.

The three steps outlined above can be summarized by a single statistical formula for $s$,

$$s = \sqrt{\sum_{i=1}^{n} (X_i - \overline{X})^2/(n - 1)}$$

where $X_i$ represents a data value and $\overline{X}$ is the sample mean. As an example, consider the sample of size $n = 5$ given by (12, 10, 6, 14, 8). The mean is $\overline{X} = 50/5 = 10$, and the squared deviations are, respectively (4, 0, 16, 16, 4). The calculations of step 2 yield a sample variance of 10, and so $s$, the standard deviation, is $\sqrt{10} = 3.16$. Once the standard deviation has been computed, it is important to think about what this number means. The data sample is collected from a population, and the value of $s$ represents an estimate of the population standard deviation. In the example here, the statistic $s = 3.16$ can be loosely thought of as an "average value" of the squared deviations of the data points from the mean. It is actually uncommon for a researcher to compute $s$ by hand, since statistical packages for computers and calculators include the computation of $s$ as part of their basic routines for data analysis.

The field of statistics is sometimes referred to as "the study of variability," since it is often the goal of a statistician to reduce standard deviation (or variability) by discovering its many causes. In certain situations, such as computer chip production, the goal is to produce consistent parts satisfying certain dimension specifications. In this case, a small value of standard deviation is preferred.

As a final example, consider a company that manufactures balsam wood airplanes. Purchasers like it when their planes fly far, so maximizing flight distance is an obvious goal. Two designs for a new plane are being considered, and five flights are made for each design. The average flight distances, given in feet, are given below:

Design I: 51.0 feet     Design II: 54.0 feet

Looking only at means, one might conclude that Design II should be chosen for the new model. However, the two standard deviations for Design I and II are 4.95 and 20.0, respectively! This shows that the flight distances for Design II are highly variable relative to Design I and that no strategic decisions should be made based solely on the value of $\overline{X}$. The raw data for the ten flights,

Design I:    58, 45, 53, 48, 51

Design II:   40, 44, 42, 88, 56

confirm this, showing a possible outlier (88) in the Design II data. In fact, the company should collect more data and carry out a larger experiment before it makes any decisions based solely on $\overline{X}$.

*See also* Mean, Arithmetic; Statistics, Overview

SELECTED REFERENCES

Moore, David S. *Statistics: Concepts and Controversies.* San Francisco, CA: Freeman, 1985.

Spurrier, John D., Don G. Edwards, and Lori A. Thombs. *Elementary Statistics Laboratory Manual: Mac and MS/DOS Versions.* Belmont, CA: Duxbury, 1995.

LORI A. THOMBS

# STANDARDIZATION OF VARIABLES

The conversion of values into standard units. If values belong to different sets of data, it may be difficult or even impossible to make direct meaningful comparisons. For instance, if a student scores 68 and 77 in midterm examinations in history and French, it is true that 68 is less than 77, but the 68 may be an A and 77 may be a C. Comparisons like this become even more complicated when data are given in different scales of measurement; say, in inches and in pounds.

In situations like this, we often *standardize* the variables with which we are concerned; that is, we convert the observed values into *standard units* by means of the equation

$$z = (x - \bar{x})/s$$

where $\bar{x}$ and $s$ are the mean and the standard deviation of the set of data to which the values belong. If the set of data constitutes a finite population, we substitute $\mu$ and $\sigma$ for $\bar{x}$ and $s$. In standard units, $z$ tells us how many standard deviations a value lies above or below the mean of the set of data to which it belongs; thus, it does not depend on the scale of measurement. Had we known in the preceding example that the students in the history class averaged 54 with a standard deviation of 8, while those in the French class averaged 71 with a standard deviation of 6, we could have argued that the 68 is $z = (68 - 54)/8 = 1.75$ standard deviations above average for the history class, while the 77 is only $z = (77 - 71)/6 = 1.00$ standard deviation above average for the French class. Relatively speaking, the 68 is thus "better" than the 77.

Standardization is also used in connection with random variables. If a random variable $X$ has the mean $\mu$ and the standard deviation $\sigma$, then

$$Z = (X - \mu)/\sigma$$

is referred to as the corresponding *standardized random variable* (and its distribution as the corresponding *standardized distribution*). For example, normal distributions constitute a two-parameter family (depending on $\mu$ and $\sigma$), but standardized, there is only one *standard normal distribution*. Since normal curves of all shapes can be standardized as indicated here, the (widely available) table of the standard normal distribution applies to a great variety of problems involving normal distributions.

*See also* Normal Distribution Curve; Standard Deviation

## SELECTED REFERENCES

Freund, John E., and Frank J. Williams. *Dictionary/Outline of Basic Statistics.* New York: Dover, 1991.

Kendall, M. G., and W. R. Buckland. *A Dictionary of Statistical Terms.* 4th ed. New York: Longman, 1982.

JOHN E. FREUND

# STANDARDS, CURRICULUM

The mathematics curriculum in the United States has undergone numerous changes. Curriculum movements during the twentieth century were influenced by a variety of concerns, including students' mathematical preparation for the workplace, threats to national security, enrollment patterns in advanced mathematics courses, and, more recently, the proliferation of technological tools in classrooms. Each movement altered the mathematics curriculum in significant ways, generating controversy as to the content considered essential for all students, methods of presentation, and the academic environment in the nation's schools. To gain perspective and insight into current curricular trends, it may be worthwhile to consider the evolution of the mathematics curriculum. A brief review of history may clarify the methods, content, and nature of the discipline as well as provide a frame of reference for contemporary movements in the United States. It will be seen that many of the so-called innovative curriculum movements of today have attributes similar to curriculum movements of the past.

## HISTORY OF CURRICULUM DEVELOPMENTS
### Colonial America

The earliest settlements of colonial America had no formal educational facilities. When schools did develop around 1635, their primary focus was on reading and writing. This usually included the reading and writing of numerals and, therefore, counting. Mathematics per se was rarely mentioned. In the seventeenth and eighteenth centuries, the arithmetic that was taught was limited to rote computation painstakingly recorded in students' notebooks. Generally, arithmetic was not taught to students less than age 11. The educational philosophy of the time is best portrayed in the following excerpt from the preface of Thomas Dilworth's *The Schoolmaster's Assistant: Being a Compendium of Arithmetic Both Practical and Theoretical:*

I believe it [arithmetic] . . . a task too hard for children; and therefore the best way of instructing them in it is . . . to give them a general notion of it, in the easiest manner, and next to enlarge upon it afterward, if there be time; otherwise it must be done by themselves. (quoted in Jones and Coxford 1970, 14)

Thus, mathematics instruction as it exists today was nonexistent. Mathematics curriculum is often reflected in the textbooks in use at the time (Grouws 1992). Until the 1700s, the United States had been dependent upon Europe for all of its textbooks. This changed in 1729 when a Harvard professor, Isaac Greenwood, wrote the first mathematics text published in the United States. This book, titled *Arithmetick, Vulgar and Decimal,* covered the gamut of arithmetic from operations with whole numbers, common fractions, decimals, and percents to applications in construction and commerce. It was used primarily as a college text (Jones and Coxford 1970). Significant changes occurred around the mid-eighteenth century. Yale University made arithmetic an entrance requirement. Princeton and Harvard followed suit soon thereafter.

## Nineteenth Century

The period from 1821 to 1894 saw a rise in formal instruction at all levels. Practical, philosophical, and scientific forces influenced the mathematics curriculum. The demands of the Industrial Revolution, developing technologies, and westward expansion called for more advanced mathematics instruction at all levels. Most states began to enact laws requiring towns to establish schools. Arithmetic, algebra, and geometry began to emerge in these schools. By the middle of the century, Horace Mann, chairman of the Massachusetts Board of Education, and founder of the first public normal (teacher preparation) school, steered the United States on the path to free public education. During this period, arithmetic instruction moved down from the high school into the elementary school, and algebra had become a mainstay of the high school curriculum. At Harvard, Yale, and Princeton, algebra had become an admission requirement. Even non-college-bound students preparing for newly developed technologies, surveying, and navigation demanded more preparation in arithmetic, geometry, and trigonometry.

Besides the practical applications of mathematics, the subject itself was valued for mental discipline. During this period, the mind was believed to function like a muscle, that is, the more it was exercised, the stronger it became. This was known as the *mind-is-a-muscle theory.* Mathematics problems were considered mental exercises that strengthened the functioning of the mind (Brooks 1883). The mathematics curriculum reflected this philosophy by emphasizing memorization and computation. The pedagogy followed a rule-recitation method in which the teacher conveyed the rules and students were expected to apply them in a plethora of exercises. Similar teaching methods were employed throughout most of the nineteenth century. A major critic of the rule-recitation method was Warren Colburn. Colburn's famous 1830 lecture entitled "Teaching of Arithmetic" encouraged teachers to guide students through a series of questions and activities so students could "discover" the mathematical principles for themselves (Jones and Coxford 1970, 26). He beseeched teachers to emphasize reasoning rather than rules and to use realistic examples in their instructional methodologies. Although Colburn has been praised by historians for his inductive teaching principles, his approach was overshadowed by the more prevalent rule-recitation methodology.

## Early Twentieth Century

A major force in the mathematics curriculum in the early twentieth century was pressure to educate all students. Enrollment in elementary schools increased by 50% in the years spanning 1900 to 1930 (Farrell and Farmer 1988). Mathematics instruction was criticized for dull pedagogy, tedious drill, and little transfer of learning. Testing of draftees in World War I disclosed a large proportion of U.S. citizens deficient in mathematical skills (Farrell and Farmer 1988). This prompted one of the first systematic analyses of the mathematics curriculum in the United States.

In the late 1920s, a Committee of Seven, consisting of school principals and superintendents from midwestern cities, surveyed students to discover at what age specific mathematical topics were mastered. Analyses of these findings culminated in a set of recommendations concerning the optimal mental age at which topics should be taught. This report formed the basis for the scope and sequence of the school mathematics curriculum for more than twenty years (Washburne 1931). In a national effort to strengthen the mathematics profession, as well as to enhance the curriculum in schools, professional societies were organized in the early part of the century. The Mathematics Association of America (MAA), composed primarily of professional mathematicians, was organized in 1915 and the National Council of Teachers of Mathematics (NCTM) was established in 1920.

The period between World War I and World War II saw conflicts over ways to improve the mathematics curriculum. Conflict was generated with regard to emerging psychological theories of how students learn. Two opposite ends of the spectrum were the theories of Edward Thorndike (1874–1949) and John Dewey (1859–1952). From the 1920s to the 1940s, *stimulus-response* psychological theory, espoused by Thorndike, influenced the mathematics curriculum. Thorndike called his educational philosophy *Connectionism*. Education was conceived as building up connections between specific skills through repeated practice and reward of correct associations (Thorndike 1922). Concepts to be learned were broken down into a series of sequential, discrete steps and taught independently. Instruction called for memorization and proficiency in algorithmic processes. Consequently, Thorndike's textbooks in arithmetic and algebra consisted predominantly of drill and practice routines.

At the other extreme was John Dewey's and William Kilpatrick's philosophy that students should construct their own knowledge and learn mathematics through projects that students choose for themselves. They contended that formal mathematics instruction should be eliminated because all the mathematics that one needed to know could be learned through experience and *incidental learning* (Jones and Coxford 1970). This philosophy brought about the *Progressive Movement* (1923–1952) in education. As a result of this practice, however, in many schools very little mathematics was ever covered during the school day.

Seeing the weaknesses in both Thorndike's and Dewey's theories, William Brownell (1895–1977) advocated another learning theory which he termed *meaning theory*. This was similar to *Gestalt (or field) theory*. The main tenet of Gestalt theory is that understanding of the organizing characteristics of a structure is essential to comprehend the whole. Proponents of this theory called for the curriculum to stress reasoning and understanding (Brownell 1935). In mathematics, this meant that greater emphasis was placed on the development of relationships, patterns, and insight rather than on drill and practice. The value of drill was acknowledged but it ranked below the primary goal of understanding. Consequently, in the early twentieth century, algorithmic processes were no longer considered the only factors in designing and sequencing the curriculum.

## Mid-Twentieth Century

During the 1940s and 1950s, a *social-utility theory* was superimposed on the stimulus-response curriculum of the prior two decades (Wilson 1948). This was an effort to reorient the school curriculum around topics that were socially useful. Practical applications were the focus of mathematics instruction. The social-utility theory, however, was short-lived and frequently applied in a negative sense, that is, to determine which topics should be dropped from the curriculum. For instance, because students would have no occasion outside of the classroom to take $\frac{1}{3}$ of $\frac{1}{2}$ of 9 hours and 16 minutes, tasks such as these were dropped from the curriculum (Caldwell and Courtis 1925). Despite all of the so-called reforms in the curriculum and teaching of mathematics, reform was to occur at an extraordinarily leisurely pace. But suddenly, a momentous event occurred that was to jolt the confidence of the United States and spawn vigorous activity in the mathematics and science communities. It was the launching of a Russian satellite named *Sputnik*.

## Curriculum Revolution of the 1950s and 1960s

In the mid-1950s and 1960s, the launching of *Sputnik* by the Soviet Union was a monumental awakening for the United States. Its national security and world status were at risk; suddenly, there was concrete evidence that another country could obtain technological superiority. As a result, the National Science Foundation (NSF) committed millions of dollars toward strengthening the nation's mathematics and science education programs. The result was a major mathematics curriculum revision. Mathematicians were the leaders of this movement. They developed a curriculum based on what they perceived to be the weaknesses of the previous curriculum. One of their primary concerns was the widening gap between the school and college mathematics curricula. Consequently, their intent was to update the content for grades K–12 (College Entrance Examination Board 1959, i). Thus, the new mathematics curriculum was concerned with the needs of the subject rather than its social utility. The focus was on unifying mathematical themes such as set theory, logic, and structure. The discovery approach was resurrected and drill and computation deemphasized. The movement emphasized deductive logic, precision of language, and symbolism (Howson, Keitel, and Kilpatrick 1981). Concern for practical applications was reserved for the sciences and engineering.

Since much of the content and language was new to everyone, the curriculum was frequently termed "new math." Some of the products of the new math movement were textbooks that reflected a clearer picture of the true nature of mathematics; teachers' manuals accompanying textbooks that included a variety of supplementary materials; and the implementation of Piagetian theory in the creation of developmentally appropriate curriculum units for the elementary grades (Farrell and Farmer 1988). Also, a wealth of curriculum ideas and materials were published in professional journals and yearbooks of the National Council of Teachers of Mathematics. In practice, mathematics teachers did add the precise language and symbolism and reduced the focus on drill. Many of the teachers like most of the nation, however, did not fully understand the goals of the curriculum and, consequently, were unable to implement it effectively (Fey 1979; National Advisory Committee on Mathematical Education 1975).

Numerous curriculum projects are associated with the new math movement. By 1970, there were thirty-two American curriculum projects in mathematics. These are outlined in the *Seventh Report of the International Clearinghouse on Science and Mathematics Curricular Developments* (Lockard 1976). Six of the more prominent projects are as follows: the University of Illinois Committee on School Mathematics (this project began before *Sputnik* but continued into the 1960s), the University of Illinois Arithmetic Project, the School Mathematics Study Group, the Madison Project, the Secondary School Mathematics Curriculum Improvement Study, and the Comprehensive School Mathematics Program. The focus of the majority of these projects was on preparation of the college-bound student. Many recommended discovery learning, laboratory activities, and visual illustrations.

Since the impetus for the curriculum revolution was the space race that was initiated by the Russian *Sputnik,* most of the curriculum projects of the 1960s were geared toward educating future mathematicians, scientists and engineers. Yet, within the same decade, the United States began to experience race riots brought about by school desegregation issues. Consequently, near the end of the decade, national attention began to be diverted toward meeting the needs of disadvantaged students. Federal funds were no longer allocated as freely to curriculum projects for the college-bound student. The U.S. citizen began to demand that schools be held accountable for teaching *all* of the students.

# CONTEMPORARY MATHEMATICS CURRICULA
## Back to Basics in the 1970s

The motto of the 1970s was "back-to-basics." This was in reaction to the abundant symbolism and rigor of the new math movement. The emphasis shifted to accountability and assessment. Content reverted to concerns of social utility. Behavioral psychologists influenced the curriculum to the extent that it was sequenced and partitioned into hierarchical levels. In many places, instruction was defined by measurable behavioral objectives. Students were placed on individual continuous progress programs in which progress to higher content levels was contingent upon mastery of preceding levels.

In practice, the back-to-basics movement in mathematics resulted in a set of discrete, narrowly defined curriculum objectives, often lacking continuity and application. Problem solving was limited to contrived word problems often occurring at the end of mastery units. Ultimately, the National Council of Supervisors of Mathematics (NCSM 1977) issued a position paper that called for a broader definition of basic skills. In addition to the usual content strands of geometry, measurement, and computational skills, this position paper called for strands including problem solving, applications, estimation, reading tables and graphs, probability, and computer literacy. This broader interpretation of basic skills continued into the 1980s (NCSM 1989).

## Agenda of the 1980s

The National Council of Teachers of Mathematics' (NCTM) position paper, *An Agenda for Action*, identified problem solving as the major focus of the mathematics curriculum for the 1980s. And, for the first time, the role of calculators and computers in school mathematics was acknowledged. Specifically, NCTM recommended that "mathematics programs take full advantage of the power of calculators and computers at all grade levels" (NCTM 1980, 8). Furthermore, the needs of society were recognized in the recommendation that "more mathematics study be required for all students, and a flexible curriculum with a greater range of options be designed to accommodate the diverse needs of the student population" (NCTM 1980, 17). By the 1980s, the United States was being rapidly transformed from an industrial nation to an information society due in large measure to the proliferation of low-cost calculators, computers, and enhancements in electronic communication. Since mathematics as a discipline contributed significantly to the emergence of these technologies, mathematical literacy was becoming increasingly important for national productivity.

By the mid-1980s, it was apparent that the recommendations of *An Agenda for Action* had not come to fruition. Public concern had been heightened by a series of reports, the most influential of which was *A Nation at Risk* (National Commission on Excellence in Education 1983), which associated problems in school with the nation's economic difficulties and educational inequities. Disturbingly, national and international assessments disclosed marked declines in student achievement in mathematics. Students from the United States ranked near the bottom on international assessments in mathematical problem solving and achievement (National Commission on Excellence in Education 1983). The National Research Council (NRC) conducted a study of mathematics education from kindergarten through graduate school in the latter part of the 1980s in which sharp declines were reported in the percentage of U.S. citizens pursuing mathematics-related careers (NRC 1989). Other studies reported significant underrepresentation of women and minorities in scientific-related careers (NSF 1988).

Historically, mathematics instruction in the United States had come in two distinct tracks: basic mathematics for the majority, advanced mathematics for the college-bound. This dichotomy of the curriculum served to systemically eliminate substantial numbers of students from a growing number of technologically dependent professions (NRC 1989). It was clear that all students needed to be more literate in more advanced, broader-based mathematics (National Commission on Excellence in Education 1983; National Science Board Commission on Precollege Education in Mathematics, Science, and Technology 1982). Instruction in mathematics needed to be substantially revised to include new technologies and to increase its accessibility to a greater percentage of the population (National Science Board Commission on Precollege Education in Mathematics, Science, and Technology 1982).

## Curriculum and Evaluation Standards for School Mathematics

Recognition of these phenomena culminated in the release of NCTM's *Curriculum and Evaluation Standards for School Mathematics* (NCTM 1989). These represent the latest efforts to provide a mathematics program to address current and anticipated needs. Of course, the articulation of mathematics

curriculum standards does not signify a move toward a nationally dictated curriculum. In the U.S. tradition, individual states control their own educational systems and attendant curricula. Rather, the standards are intended as statements "that can be used to judge the quality of a mathematics curriculum . . . [They are] . . . facilitators of reform" (NCTM 1989, 2). *The Curriculum and Evaluation Standards for School Mathematics* (hereafter referred to as the *Standards*) articulates new societal and student goals for school mathematics. Societal goals are identified as (1) mathematically literate workers, (2) lifelong learners, (3) an informed electorate, and (4) opportunity for all. New curricular goals for students include (1) learning to value mathematics, (2) becoming confident in their mathematical ability, (3) becoming mathematical problem solvers, (4) learning to communicate mathematically, and (5) learning to reason mathematically.

The curriculum standards are intended to provide a common core of mathematics content for all students. Basic assumptions are that (1) instruction is to be developed via problem-solving situations that engage students actively in doing mathematics rather than just listening and responding to questions, (2) computers and calculators are to be used regularly as problem-solving and computational tools, and (3) a broad range of mathematics is to be covered. There are four overarching foci of the mathematics curriculum applicable at all levels: problem solving, communication, reasoning, and mathematical connections. The document also includes specific content strands for elementary, middle grade, and secondary levels (NCTM 1989). Included are illustrations of ways in which these standards may be met. Teaching strategies are suggested that are based on research pertaining to students' cognitive development and learning modalities. Finally, instructional examples are provided that reflect the processes and products of mathematics (NCTM 1989, 17).

The following examples are excerpts from the document that illustrate the types of problems that should be part of the curriculum. For the sake of comparison, all of the problems chosen for inclusion here were taken from the content area of geometry.

*Secondary-Level Problem* A container manufacturing company has been contracted to design and manufacture cylindrical cans for fruit juice. The volume of each can is to be 0.946 liters. In order to minimize production costs, the company wishes to design a can that requires the smallest amount of material possible. What should the dimensions of the can be? (NCTM 1989, 134)

*Middle-Grade Problem*   The class is divided into small groups. Each group is given square pieces of grid paper and asked to make boxes by cutting out pieces from the corners. Each group is given $20 \times 20$ grid paper. Students cut and fold the paper to make boxes sized $18 \times 18 \times 1$, $16 \times 16 \times 2, \ldots, 2 \times 2 \times 8$. They are challenged to find a box that holds the maximum volume and to convince someone else that they have found the maximum. (NCTM 1989, 80)

*Elementary Problem*   A dog is tied to a 5-meter rope at the middle of the side of a garage. The side of the garage is 10 meters long. Make a sketch and use centimeter grid paper to estimate the area and shape of the ground on which the dog can walk. (NCTM 1989, 50)

The *Standards* recommend that at all levels, the applications of mathematics be emphasized so that students will begin to realize that mathematics is applicable to everyday situations and to other curricular areas. Furthermore, students are expected to work cooperatively in constructing mathematical models, creating problem-solving strategies, and communicating results. Although somewhat controversial, the *Standards* were written with the expectation that paper-and-pencil calculations would be deemphasized and that calculators would be used regularly for tedious computations. Similarly, computers are viewed as important tools for mathematical explorations and problem solving.

*Evaluation Standards*   The designers of the *Standards* realized that instruction and evaluation are intimately related. Consequently, consideration of evaluation was deemed necessary in bringing the *Standards* to fruition. Thus, the document concludes with a set of evaluation standards that correlate with the overall vision for curriculum.

*Addenda Series*   To aid in the implementation of the curriculum, the NCTM has published an Addenda Series of softbound books for grades K–6, 5–8, and 9–12 that focus on specific content strands. Each book illustrates problems and methodologies that are developmentally appropriate for the indicated grade level.

In summary, the *Curriculum and Evaluation Standards for School Mathematics* represents the most recent effort in curriculum revision in school mathematics. It is intended as a vehicle for improving the teaching and learning of school mathematics. It promotes an environment that takes full advantage

of technological tools and in which the learning of mathematics becomes an active, constructive process based on solving real problems and communicating ideas. The *Standards* are expected to have a dramatic impact on the school mathematics curriculum. Much depends on the commitment of schools, teachers, and the general public to the basic tenets and methodologies advocated in the *Standards*.

## ONGOING CURRICULAR CONCERNS

Curriculum change is usually a slow process. Moreover, it is evident that curricular innovations occur in cycles; that is, curricular ideas of one generation recur in other generations. Mathematics as a discipline is continually changing: new mathematics is developed; new uses are discovered; new technological developments make certain mathematical procedures obsolete, while opening up new avenues for exploration and application. Clearly, the mathematics of the future is unpredictable. A curriculum is needed that will prepare all citizens to reason mathematically and to be able to use mathematics in a wide variety of situations.

*See also* Curriculum Trends, Elementary Level; Curriculum Trends, Secondary Level; History of Mathematics Education in the United States, Overview; *Revolution in School Mathematics;* Secondary School Mathematics

SELECTED REFERENCES
Brooks, Edward. *Mental Science and Methods of Mental Culture.* Lancaster, PA: Normal, 1883.

Brownell, William A. "Psychological Considerations in the Learning and the Teaching of Arithmetic." In *The Teaching of Arithmetic* (pp. 1–31). 10th Yearbook of the NCTM. William D. Reeve, (ed.). New York: Teachers College, Columbia University, 1935.

Caldwell, Otis W., and Stuart A. Courtis. *Then and Now in Education, 1845–1923: A Message of Encouragement from the Past to the Present.* Yonkers-on-Hudson, NY: World Book, 1925.

College Entrance Examination Board. *Appendices.* New York: The Board, 1959.

Connors, Edward A. "A Decline in Mathematics Threatens Science—and the U.S." *Scientist* 2(22)(1988): 9–12.

Farrell, Margaret, and Walter Farmer. *Secondary Mathematics Instruction: An Integrated Approach.* Providence, RI: Janson, 1988.

Fey, James T. "Mathematics Teaching Today: Perspectives from Three National Surveys." *Mathematics Teacher* 72(1979):490–504.

Grouws, Douglas A., ed. *Handbook of Research on Mathematics Teaching and Learning*. New York: Macmillan, 1992.

Howson, A. Geoffrey, Christine Keitel, and Jeremy Kilpatrick. *Curriculum Development in Mathematics*. Cambridge, England: Cambridge University Press, 1981.

Johnston, William B., and Arnold E. Packer, eds. *Workforce 2000: Work and Workers for the Twenty-First Century*. Indianapolis, IN: Hudson Institute, 1987.

Jones, Phillip S., and Arthur F. Coxford. "Mathematics in the Evolving Schools. In *A History of Mathematics Education in the United States and Canada* (pp. 11–89). 32nd Yearbook. Phillip S. Jones (ed.). Washington, DC: National Council of Teachers of Mathematics, 1970.

Lockard, J. David, ed. *Seventh Report of the International Clearinghouse on Science and Mathematics Curricular Developments, 1970*. College Park, MD: University of Maryland, Science Teaching Center, 1976.

National Advisory Committee on Mathematical Education (NACOME). *Overview and Analysis of School Mathematics Grades K–12*. Washington, DC: Conference Board of the Mathematical Sciences, 1975.

National Commission on Excellence in Education. *A Nation at Risk: The Imperative for Educational Reform*. Washington, DC: Government Printing Office, 1983.

National Council of Supervisors of Mathematics. "Essential Mathematics for the 21st Century: The Position of the National Council of Supervisors of Mathematics." *Arithmetic Teacher* 37(Sept. 1989):44–46.

————. "Position Paper on Basic Skills." *Arithmetic Teacher* 25(Oct. 1977):19–22.

National Council of Teachers of Mathematics. *An Agenda for Action: Recommendations for School Mathematics of the 1980s*. Reston, VA: The Council, 1980.

————. *Curriculum and Evaluation Standards for School Mathematics*. Reston, VA: The Council, 1989.

————. *Principles and Standards for School Mathematics*. Reston, VA: The Council, 2000.

National Research Council. *Everybody Counts: A Report to the Nation on the Future of Mathematics Education*. Washington, DC: National Academy Press, 1989.

National Science Board Commission on Precollege Education in Mathematics, Science, and Technology. *Today's Problems, Tomorrow's Crises*. Washington, DC: National Science Foundation, 1982.

National Science Foundation. *Women and Minorities in Science and Engineering*. Washington, DC: The Foundation, 1988.

Reys, Robert, Marilyn Suydam, and Mary Lindquist. *Helping Children Learn Mathematics*. 4th ed. Boston, MA: Allyn and Bacon, 1995.

Thorndike, Edward L. *The Psychology of Arithmetic*. New York: Macmillan, 1922.

Washburne, Carleton. "Mental Age and the Arithmetic Curriculum: A Summary of the Committee of Seven Grade Placement Investigations to Date." *Journal of Educational Research* 23(Mar. 1931):210–231.

Wilson, Guy M. "A Social Utility Theory as Applied to Arithmetic: Its Research Basis and Some of Its Implications." *Journal of Educational Research* 41(Jan. 1948): 321–337.

SHARON WHITTON

# STATE GOVERNMENT, ROLE

Public education in the United States has been primarily the responsibility of individual states. The federal Constitution makes no mention of education, and the Tenth Amendment reserves unspecified powers to the states. All state constitutions accordingly direct state legislatures to provide for a state system of public education (Valente 1994). Although at times the federal government may support state initiatives through grants and other means, all public education is the responsibility of the states. State policymakers at all levels determine the design, delivery, and assessment of mathematics instruction in the state. These policymakers include the governor, state legislators, state boards of education, and local district boards of education. The chief state school officer and the local school district superintendent have major responsibility for implementation of state educational policy. They are assisted by a host of administrators, instructional supervisors, directors or consultants, and, most importantly, the classroom teachers. In addition to setting education policy, the states are also obligated to fund public schools.

A national interest in the performance of America's students in mathematics has prevailed for over forty years. Sparked by the Soviet Union's launch of the Earth satellite *Sputnik* on October 4, 1957, the United States immediately became concerned that it had lost its competitive edge over the USSR in space exploration. Congress responded by passing the National Defense Education Act of 1958 to provide federal support for mathematics and science education and foreign language instruction. Federal grants administered through the states supported programs at the elementary, secondary, and higher education levels, including increased resources for teaching, improvements in curricula, and teacher training.

The current period of public concern about mathematics education dates back to 1983 with the publication of *A Nation at Risk: The Imperative for Educational Reform* (National Commission on Excellence in Education 1983), which was a scathing assessment of public school education in America.

Among the many criticisms reported was that the College Board's Scholastic Aptitude Tests (SAT) demonstrated a virtually unbroken decline of nearly 40 points in average mathematics scores from 1963 to 1980. This growing concern about declining academic achievement provided motivation for the President's and National Governors' Conference in Charlottesville, Virginia, in 1990 where agreement was reached on a set of six National Education Goals. The fourth goal stated that by the year 2000, U.S. students would be first in the world in science and mathematics achievement.

The states' role in the attainment of this fourth goal was summarized by the participants in a set of recommendations for state policymakers that was the outcome of a conference on improving mathematics and science education convened by the U.S. Secretary of Education in 1991. The conferees included governors, state legislators, representatives of state boards of education and state departments of education, and chief state school officers. They were joined by representatives of national science and mathematics organizations and the U.S. Department of Education. They conceded that to become first in the world in mathematics achievement by the year 2000 was an ambitious goal that could only be attained if state policymakers acted on the imperative to develop world-class standards in mathematics supported by curriculum frameworks, assessment systems, and staff capacity to meet the standards. The conferees issued a set of ten recommendations for state policymakers to (1) create public awareness of the crisis in mathematics and science education, (2) establish a vision for what students should know and be able to do, (3) set high standards for all students, (4) focus on long-range objectives, (5) plan for systemic reform, (6) develop an awareness of exemplary practices, (7) leverage all existing resources to create high standards, (8) develop policies to assure teacher competence in the delivery of new curriculum, (9) promote change in school organization as needed to meet the goals, and (10) invest in staff development (U.S. Department of Education 1992).

Educational standards and goals require measurement to assess their degree of attainment by students, schools, and school districts. During the 1991 annual convention of the National School Boards Association (NSBA), which includes state and local boards of education, the association president identified student assessment as one of the five major issues of school board concerns. The policy making delegate assembly passed a resolution to support efforts to improve the assessment of education. It urged its member boards to assure that assessments are effective and that they test students' mastery of skills (Penfield 1991). Inspired by *A Nation At Risk* and the National Education Goals, governors have led their states in a myriad of education reform initiatives with significant attention to mathematics education. Their goal is to raise students' academic achievement, with mathematics as one of the critical subject areas for improvement.

The National Center for Education Statistics (NCES) has attempted to monitor the impact of state reforms in education. Its research reveals that states have increased their high school graduation requirements and that over 80% now have higher coursework standards for high school graduation. In over 90% of the states, competency tests results are being used to evaluate individual student performance and the performance of entire schools. Although not a common practice, use of minimum grade point averages is the basis in some states for making decisions about promotion, graduation, and participation in extracurricular activities. At least 85% of the states have developed policies increasing instructional time. Some states have moved to standardize their curriculum, with 12% mandating course content, 13% setting learning objectives to be included in local districts' curricula, and 16% deciding the number and types of courses to be taught in the schools throughout the state.

As stated earlier, sometimes the federal government supports state initiatives through grants. Recent examples include the National Science Foundation (NSF) Statewide Systemic Initiative Program (SSI) and the Dwight D. Eisenhower Mathematics and Science Education Program. Acceptance of these grants reinforces the states' efforts in program development and implementation in mathematics education. The NSF recognizes that the states are the appropriate avenue for curriculum reform and it encourages the involvement of governors along with their staffs in the thirty states that are recipients of the NSF grants. The Eisenhower Program provides block grants to states for the improvement of teachers' skills in mathematics, science, and foreign language instruction and to increase the number of students enrolled in these subjects. Again, acceptance of the grants and implementation strategies rest with the individual states. It is important to note here that only a fraction of the states' education budget flows from the federal government.

In any discussion of the role of the states in mathematics education, attention has to be given to the local school districts. Many persons hold the view that the unit of change and reform in educational practices, including mathematics education, is the local school district and, more specifically, individual schools. Local district boards of education are accountable for operationalizing state education policy. However, in some states, they also have the authority to set policies that may exceed state standards and/or state requirements. One example is local school districts increasing course requirements in mathematics as a condition for a high school diploma or students being required to take specific courses. In this recent spate of school reform, states and local districts are placing greater emphasis on increasing standards and academic achievement by requiring all students to take more challenging courses in mathematics. The course most often cited is algebra. Where states do not practice statewide adoption of texts and other instructional materials, local school districts have choices in these areas, subject to approval by the state. In matters of personnel selection, state certification requirements prevail; however, local staff development programs are designed to meet specific changing needs created by new standards and curricula. Some local districts also develop curriculum frameworks subject to approval by the state.

In summary, the role of the states in mathematics education is to set policies for program implementation in the three broad areas of curriculum, staff capacity, and student achievement. The area of curriculum includes the establishment of state standards and goals for mathematics instruction, the approval of curriculum frameworks by grade level, approval of texts and other instructional materials, and the establishment of high school graduation requirements. Staff capacity applies to both preservice and inservice teachers. The state determines certification requirements in mathematics education teacher preparation programs, sets hiring practices such as the requirement to pass a state test, sets standards for renewal of teaching certificates, and sponsors staff development programs for experienced teachers. Along with setting policies for what will be taught and the qualifications of those who will teach, the state also sets standards for student achievement in mathematics. The awarding of a high school diploma is often predicated on meeting the state's standard of course requirements and satisfactory academic performance on state proficiency tests. In recent years, state policy makers have responded to encouragement to expand their collaboration with business and industry, parents, and other stakeholders in the community to gain their participation and support in the attainment of the state's educational goals. All of these efforts are il-

lustrative of the states' responsibility in the mathematics education of America's children.

*See also* Eisenhower Program; Federal Government, Role; National Science Foundation (NSF)

SELECTED REFERENCES

National Center for Education Statistics (NCES), U.S. Department of Education. *Overview and Inventory of State Requirements for School Coursework and Attendance.* Washington, DC: Government Printing Office, 1992.

National Commission on Excellence in Education, U.S. Department of Education. *A Nation At Risk: The Imperative for Educational Reform.* Washington, DC: Government Printing Office, 1983.

Penfield, Arlene R. "Five Major Issues of School Board Concern." *School Board News* 2(10)(1991):2, 7.

U.S. Department of Education. Office of Educational Research and Improvement (OERI). *Improving the Math and Science Curriculum: Choices for State Policy Makers.* Washington, DC: Government Printing Office, 1992.

Valente, William D. *Law in the Schools.* New York: Macmillan, 1994.

LEE ETTA POWELL

# STATISTICAL INFERENCE

As taught in secondary schools and at introductory college levels, essentially means that students are introduced to point and confidence interval estimation and to hypothesis testing. While descriptive statistics provide summaries of data that have been observed, inferential methods allow extrapolation beyond the observed sample *statistics* to estimated population *parameters*.

Hypothesis testing was developed by Jerzy Neyman and Karl Pearson in the early 1930s. Subsequently, a controversy emerged between "decision makers" and "significance testers." The main distinction between the two perspectives relates to whether or not a null hypothesis can be *accepted*. Ronald A. Fisher (1890–1962), who is thought of as being the father of statistical experimental design, believed that hypotheses could be either *rejected* or *not rejected*; they could not, however, be *accepted*. In Fisher's view, *scientific* inquiry should *not* be based on statistical decision making, in which the problem is seen as a matter of choosing which of two alternatives to accept. Moore (1991, Chapter 8) provides an interesting discussion of the distinctions between the significance-testing and decision-making approaches to statistical inference. He demonstrates how, in testing hypotheses, we in fact mix significance tests and decision rules. Lindgren (1968, Chapter 6) also compares and contrasts the two approaches in action.

Statistical *estimation* techniques, which underlie methods of hypothesis testing using *test statistics* (see later), are generally more informative because the *confidence-interval* approach provides an insight into the degree of precision that is achieved. A similar approach can be used when two samples are being compared to see if they are likely to have been drawn from the same population. In fact, sometimes we *know* that the samples were originally drawn from the same parent population, but we are interested in whether this still appears to be the case after we have intervened in some way.

For example, we may wish to establish whether a particular drug improves stroke victims' reaction times. Our research design might be based on randomly assigning stroke patients to one of two groups. The intention of this would be to try to establish broadly similar pretreatment groups, or at least to avoid selective bias in their constitution. (Now would be a good opportunity to invite the students to comment on the merits and demerits of this experimental design. Given the inherent variability of stroke patients' reaction times, and of the severity of impairment to their manual dexterity, it might be better to preassess the patients' reaction times and then use a matched group design.) Patients in Group A should be given a *placebo* drug (rather than no drug at all), and patients in Group B should be given the drug that is being tested. No one directly involved with the patients, or with their assessment, should know which patient is receiving which drug. This is called a "double-blind" situation. It is used in order to preclude experimenter effect. After the period of treatment, all the patients would be assessed on a reaction-time task. This would give us two samples of reaction-time measurements, one for each group of patients.

Classical statistical inference commences with a Null Hypothesis ($H_0$), which is tested against an Alternative Hypothesis ($H_1$). In this case, these might be:

$H_0$: Mean reaction time$_A$ = Mean reaction time$_B$
$H_1$: Mean reaction time$_A$ ≠ Mean reaction time$_B$

Although inferential statistics may *appear* to involve comparisons between a sample and a population, or between two samples, in reality the inferences are based on comparing hypothetical *populations* (Hawkins et al. 1992), and the statistical hypotheses always relate to *population parameters*. Students are prone to muddle statistics and parameters. Teachers need to be very careful to keep drawing the distinction and to ensure that they themselves do not refer to "statistic" when they mean "parameter." Colorcoding in presenting the concepts to students can help and can also be extended to highlight the distinctions between raw data distributions and derived sampling distributions of statistics.

In the reaction-time example, the parameters to be compared are $\mu_A$ and $\mu_B$, the means of the two populations of reaction times, for patients receiving placebo and drug treatments, respectively. These can only be estimated, as we have not measured the entire populations of reaction times. We could use our two sample means as *point* estimates of the two population means. In order to allow for sampling variability, however, we should construct two *interval* estimates. These enable us to identify, say with 95% confidence, the limits within which each population mean lies. If the resulting confidence intervals overlap, $H_0$ is not rejected (which is *not* the same as saying that it is accepted). On the evidence of the data that we have collected, we have no reason to believe that we are now dealing with more than one population of reaction times. Any observed difference between the estimated population means is small enough so that it may possibly have resulted from sampling variability, that is, from chance. If the confidence intervals do not overlap, however, $H_0$ is rejected. It is then assumed that, at the chosen level of significance (0.05 or 5% in this case), the samples no longer come from the same population.

Instead of using this kind of estimation approach, we can compute a *test statistic* that compares the sample data that we have observed with what we could have expected to observe if the Null Hypothesis pertained. The particular formula that is used for a test statistic depends on what parameter(s) we are investigating (e.g., difference between two population means, proportions, etc.) and also on the experimental design that is adopted (e.g., independent or matched groups). The following test statistic could be used in the example on reaction times:

$$\text{test statistic} = \{(\overline{X}_A - \overline{X}_B) - (\mu_A - \mu_B)\}/S_{\overline{X}_A - \overline{X}_B}$$

where

$\overline{X}_A - \overline{X}_B$ = Difference between sample means

$\mu_A - \mu_B$ = Difference between hypothesized population means

$S_{\overline{X}_A - \overline{X}_B}$ = Sample estimate of the standard error of the difference between two means

In *this* case, $(\mu_A - \mu_B) =$ zero under the Null Hypothesis that the two population means are equal. The numerator is therefore simplified to the difference between the two sample means. Provided that there has been an element of *random* selection or allocation in the sampling process, the measure of sampling variability in the denominator is assumed to reflect the variability that might be expected *purely as a result of chance*. The bigger this is, the bigger the difference between sample means has to be for us to think that the drug has had an effect, and *vice versa*. It is easier to spot a pitchfork, than a needle, in a haystack! Once again, however, it is important to show students how a different sample design can lead to a reduction in the size of the denominator (which is analogous to shrinking the haystack).

We can use a statistical table of the test statistic's probability distribution to tell us how likely it is that a value as large as our test statistic could occur as a result of chance alone. If that probability is very low, say less than 5%, we would reject the Null Hypothesis "at the 0.05 level of significance." We would decide that the samples are not now from the same population of reaction times and, hence, that the drug appears to have had an effect.

If there is an obvious direction in which the population means are likely to differ, a *one-tailed* rather than a *two-tailed* probability value can be used to indicate the critical value for the test statistic, which determines whether the Null Hypothesis will be rejected. A one-tailed test would be appropriate if, for example, the Alternative Hypothesis had been specified as

$H_1$: Mean reaction time$_A$ > Mean reaction time$_B$

Then, instead of our test statistic having to exceed a value of 1.96 (assuming that we are using Standard Normal Distribution or large-sample $t$-Distribution tables) before we can reject the Null Hypothesis at the 0.05 level of significance, it would only have to exceed 1.645. Furthermore, our conclusion would be that the drug appeared to have *reduced* the patients' reaction times (as opposed to *changed* them).

It is often the case that teachers are too quick to introduce their students to methods of hypothesis testing based on *test statistics*. Students can benefit from a more thorough grounding in estimation procedures. This will enable them to see the underlying mechanisms in operation and help them to realize that effective significance testing, and hence statistical inference in general, is a matter of skilled judgment and not merely an apparently automatic computational process, guaranteed to yield "correct" or useful inferences.

Classical inference involves the following table of choices and their associated possibilities for error. Balancing the error rates requires the statistician to use great skill and judgment at the time when the research design and the inference procedures are chosen. Apart from sequential and Bayesian approaches, which entail some degree of retrospection, in classical statistics the research process must be completely specified (including the hypotheses, and hence the conclusions that can be drawn) *before* any practical investigation commences. More recent Exploratory Data Analysis (EDA) procedures relax this requirement, allowing the data analysis itself to suggest new directions and hypotheses. They also need fewer distributional assumptions.

*Statistician's decision based on sample of observations*
(classical inference paradigm)

|  |  | Do not reject $H_0$ | Reject $H_0$ |
|---|---|---|---|
| **True state of population** | $H_0$ | Correct | Type 1 error |
|  | $H_1$ | Type 2 error | Correct |

Our students should be given the experience of designing their own studies and working with real data, but they will benefit more from such opportunities if they have to present their work to their peer group. This will make them justify their choice of methodology and discuss the reliability and validity of their findings with reference to precision and bias, and so on. They will have to address issues associated with the possibility of arriving at an incorrect inference, including the probabilities of those errors indicated in the decision table. When the students have reached this level, it would be appropriate to make them aware that classical hypothesis testing, although still widely practiced, has a major limitation. What the researcher *really* wants to know is

the "probability of error" $= Pr(H_0$ is true $\mid$ reject $H_0)$

The classical approach, however, does not provide this information, because it is based on

the "probability of making a mistake"

$$= \text{significance level}$$

$$= \alpha$$

$$= \text{Pr (type 1 error)}$$

$$= \text{Pr (reject } H_0 | H_0 \text{ is true)}$$

Other inferential procedures may therefore be preferred, but these lie outside the scope of most introductory-level courses. It is always useful for students to evaluate critically other people's published research, particularly papers in which tests of significance have been used without quoting confidence intervals. Students should be encouraged to work out the missing confidence limits where possible, to see whether their view of the reported research findings changes as a result.

Simulations allow students to experiment with statistical concepts. One cheap and easy practical experiment (which can be brought into the classroom to demonstrate point and confidence interval estimation) is that of pebble sampling. From a population of pebbles, individually identified by numbers painted on them, students are required to draw samples in order to estimate the mean weight of the population. If the class is big enough, each student can draw one simple random sample of a given size, and the results can be collected together to yield the sampling distribution of the means. Repeating the exercise using different sampling procedures, for example, stratified sampling, systematic sampling, and so on, teaches the students about these procedures. It also provides simple comparison graphs of the resulting sampling distributions and demonstrates to students the influence of sample design on the precision of their estimates of the population parameter. The effect of changing the size of the sample can also be studied, allowing students to see the interaction between precision and accuracy. This "concrete" example helps to illuminate other, more abstract, simulations that can then be tried on the computer!

*See also* Statistical Significance; Statistics, Overview

### SELECTED REFERENCES

Fisher, Ronald A. *Statistical Methods and Scientific Inference.* 3rd ed. New York: Hafner, 1973.

Hawkins, Anne, Flavia Jolliffe, and Leslie Glickman. *Teaching Statistical Concepts.* London, England: Longman, 1992.

Lindgren, Bernard W. *Statistical Theory.* 2nd ed. New York: Macmillan, 1968.

Moore, David S. *Statistics: Concepts and Controversies.* 3rd ed. San Francisco, CA: Freeman, 1991.

ANNE HAWKINS

# STATISTICAL SIGNIFICANCE

A sample statistic (e.g., the proportion of "Yes" votes in a straw poll), obtained from a *single* survey or randomized experiment, can provide a *point estimate* of a population parameter. If we were to repeat the sampling process over and over again, however, the observed value of the statistic would vary. Relying upon a single-sample point estimate is therefore not entirely satisfactory, and we would be better off to allow for its likely variability in making our estimate. This variability can be defined in terms of the *probability distribution* of the statistic. Thus, we can take account of likely sampling variability, even though we only take a *single* sample of observations.

By applying our knowledge of the sampling distribution of the statistic to the data that we have collected, we can establish upper and lower boundaries to our estimate of the population parameter. This *interval estimate* has a certain probability of including the true population parameter. The probability is known as the *confidence level* and is often set at either 95% or 99%. In the former case, we are willing to risk that in 5% of the times that we use this method, our confidence interval will miss the true population parameter. If this risk is too great, we may prefer to use the 99% level of confidence, whereby we expect to have only a 1% chance of this kind of error. However, other things being equal, the latter interval would be considerably wider and therefore less precise.

Alternatively, we may already have under consideration a particular value that we think may be the true value of the unknown parameter. This is usually referred to as the *null hypothesis, $H_0$*. We may then wish to find out, on the basis of a single sample of observations, whether $H_0$ could be true, that is, whether the statistic calculated from these observations supports our hypothetical value of the unknown parameter or contradicts it. In other words, we may wish to *test* the null hypothesis. We again refer to the probability distribution of the statistic, but we now focus on the variability of the statistic under the assumption that $H_0$ is true. If, in this distribution, the observed value of the statistic is found to be outside the range of what would have occurred from chance variation alone, we call the result *statistically significant* and reject the null hypothesis, although the null hypothesis could very well be true. The probability that the statistic falls outside this range, and therefore the risk that such an error is committed, is known as the *significance level* of the test; it is denoted by $\sigma$, which is usually taken as 5% or 1%.

As may be evident from this discussion, there is a close relationship between a confidence interval for an unknown parameter and testing a null hypothesis about that parameter. For example, to establish whether the removal of an expensive additive from paint increases its mean drying time from the original standard, we can take a sample of $n$ drying time increases, calculate the sample mean $\bar{x}$ and sample deviation $s$, and obtain the 95% confidence limits for the population mean increase in drying time as $\bar{x} \pm 1.96\, s/\sqrt{n}$, where 1.96 is the normal distribution deviate that encloses the central 95% of the probability. (Different formulae apply to the confidence limits for different parameters, based on the associated statistics and their distributions, but the principle remains the same.) These limits can be calculated from the sample, and the resulting interval may or may not contain the unknown population mean increase in drying time. In fact, there is a 5% chance that it does not. The interval can also be used to test the null hypothesis that the mean increase in drying time is some specified value, say zero, at the significance level of 5% (which is the complement of the confidence level) by simply checking to see if the calculated interval contains the value zero. If it does not, the null hypothesis is rejected. This is due to the fact that a 95% confidence level does not contain the value of the population mean specified by $H_0$ (in this case, zero) if and only if the sample mean falls outside the range of values that are likely to occur under $H_0$; in other words, if and only if it is *statistically significant* at the 5% level.

The fact that a result has statistical significance does not guarantee that it is an *important* result. Increasing the sample size and/or decreasing the variability of the population in question (e.g., by only working with one color of paint, so as to control for the possible effect of other additives on drying times) narrows the confidence interval, which makes it easier to find *significance*. Spurious significance can result, however, if, for example, an enormous sample size is used or if the investigation is focused on a very homogeneous, but not very representative, subsection of the population.

The width of the confidence interval, that is, the precision of the estimate, can also be influenced by the level of significance adopted. Although 0.1 and 0.01 levels are sometimes used, the 0.05 level is the most common. The reason for this probably dates back to a personal preference expressed by Ronald A. Fisher (1926), a statistician who was enormously influential in the development of statistical inferential techniques. This conventional acceptance of the "significance" (or relevance) of the 0.05 *level of significance,* has led to a tendency for academic journals to publish only research studies in which the results reach this particular level of significance. In fact, research studies that show results that are *not statisti-*

*cally significant* may be just as important for determining future research directions, because they provide information about the interventions that are *not* effective.

Random sampling, or random allocation, is necessary if we wish to avoid unsuspected sources of bias and therefore to obtain a fair measure of *chance* variability. It is important for students to realize that if a *biased* sampling method is used, a more precise confidence interval is less likely to be *accurate* in the sense of "correct." This is because there is less scope for including the true population parameter in a confidence interval based on a false starting point than if the estimate were to be less precise (Hawkins, Jolliffe, and Glickman 1992). The precision of the interval estimate should be manipulated so that a result that is found to have *statistical* significance will also have *meaningful* significance. In the example involving paint drying times, what would be meaningful should be established in advance by asking the paint manufacturer how big a change in drying time would be tolerated by his customers. Likely variance in the population can be estimated from existing data on drying time. A significance level can be selected that reflects the perceived "cost" to the manufacturer of a chance outcome being erroneously identified as "significant." The sample size needed to achieve a correspondence between statistical and meaningful significance can then be calculated, because sample size is the only unknown quantity in that part of the confidence interval formula that determines the limits, and hence the precision:

$$z_\alpha \sigma/\sqrt{n}$$

where

$z_\alpha$ = standardized normal deviate corresponding to the chosen level of significance.

$\sigma$ = square root of the population variance. The *sample* variance ($s^2$ computed with an $n - 1$ denominator) is an unbiased estimate of $\sigma^2$, so $\sigma$ can be replaced by $s$.

$n$ = sample size.

It is important for students to gain *practical* insights into the concept of the confidence interval, from which their understanding of significance levels can be developed. They need to experience concrete examples before moving on to more abstract computer simulations. While these can undoubtedly take the drudgery out of repeated sampling, and so on, students often have difficulty when they try to relate what they see on the computer to real-world situations. Of the many software simulations that are available, those that do not merely show sampling distributions being obtained from parent populations are probably more useful, particularly those that involve simulations of real sampling contexts like crops subjected to different fertilizers or populations of poultry reared under different conditions. The software should be highly interactive, so as to allow the students to experiment with different kinds of sampling methods (including those that are biased) and to manipulate things like sample size. They can then observe the resulting changes in the precision and accuracy of their interval estimates and monitor the effects on the outcome of their hypothesis tests.

See also Normal Distribution Curve; Probability, Overview; Random Sample; Standard Deviation; Statistical Inference; Statistics, Overview

## SELECTED REFERENCES
Fisher, Ronald A. "The Arrangement of Field Experiments." *Journal of the Ministry of Agriculture of Great Britain* 33(1926):504. [Quoted in "On Rereading R. A. Fisher." By Leonard J. Savage. *Annals of Statistics* 4(1976):471.]

Hawkins, Anne, Flavia Jolliffe, and Leslie Glickman. *Teaching Statistical Concepts*. London, England: Longman, 1992.

Moore, David S. *Statistics—Concepts and Controversies*. 3rd ed. San Francisco, CA: Freeman, 1991.

ANNE HAWKINS

# STATISTICS, OVERVIEW

A discipline separate from mathematics, with its own concepts and types of reasoning. The word *statistics* usually makes people think of numbers and mathematics, however. We are surrounded by statistics: results of opinion polls, batting averages, tables and graphs of financial data, and risks of contracting diseases. Statisticians work with *data*, which usually consist of sets of numbers, although qualitative information (such as type of car) may be transformed into a numerical code (e.g., 1 = Ford, 2 = Mazda, etc.). Statistical work typically involves collecting or producing data, organizing and analyzing data, representing data in graphs, and drawing conclusions from data (Moore 1991). It is because statistical work often involves numbers, symbols, and equations that it is often viewed as mathematics. Statistics however, did not originate within mathematics (Moore 1992). While mathematicians may study numbers as abstract concepts without a context, statisticians study numbers only in the context of what

these numbers represent in the world (e.g., weight measurements, test scores, or stock prices).

Statistics is sometimes thought of more broadly as including the study of probability (which deals with mathematical models of chance) and not just as the steps of collecting, analyzing, displaying, and drawing conclusions from data. Europeans typically use the word *stochastics* for this broader definition (Garfield and Ahlgren 1988). Most statistics curricula include the study of both data analysis and probability and connect these areas when samples of data are used to make estimates, inferences, or generalizations about unknown populations of data. For example, it is common to take a sample of voters and ask them questions about their intentions to vote for a particular candidate for governor and then to use the percentage of voters in this sample who support the candidate to make inferences about the preferences of all voters in the state.

## THE HISTORY OF STATISTICS

The word *statistics* seems to come from the Latin words *status* (meaning "state") and *statista* (meaning "statesman") (Folks 1981). How do the terms *state* and *statesman* relate to our modern understanding of statistics? Perhaps because the first statistical studies consisted of gathering descriptive information on 158 states for Alexander the Great. Other early uses of statistics date back to William the Conqueror, who ordered a statistical survey of England in 1085, early censuses of the Americas conducted by the Spaniards in the 1500s, and analyses of births and deaths in seventeenth-century England (Folks 1981). The first statistician is thought to be the Englishman John Graunt, who wrote the book, *Natural and Political Observations on the Bills of Mortality* in the mid-1600s. Graunt collected and analyzed mortality rates and birth rates in order to make estimates of the size of London's population. He also used samples to make estimates about future growth rates for the city's population (Kennedy 1983).

The word *statistics* did not appear in print until the late 1700s. Over the next two hundred years, many different disciplines began to use statistical methods. Probability models also were developed that led to a variety of methods for making statistical inferences (Stigler 1986). In the late 1800s, statistics were developed and used in the study of heredity by Charles Darwin and Francis Galton. In the late 1890s, Karl Pearson is credited with founding modern statistics and the field of biometrics. Ronald Fisher's work in the early 1900s, particularly in the area of experimental design, laid the foundation for much of the statistical work done today. Pearson and Fisher not only contributed substantially to modern statistical ideas and procedures but were famous for arguing over their different views of statistics and procedures for making statistical inferences. While Pearson and his colleagues viewed statistics as a decision-making science and established a system of conditions for formulating a null hypothesis and determining when it is safe to reject or not reject this hypothesis, Fisher thought of statistics as the science of studying and learning from data, thereby forming opinions and making inferences but not making decisions (Folks 1981).

The introduction of computers followed by rapid technological advances led to dramatic changes in analyzing and displaying data. One step in a data analysis that formerly could take a team of statisticians one week to compute by hand can now be accomplished in seconds on a computer. New forms of data analysis utilizing computer technologies evolved along with a different orientation toward analyzing and graphically representing data, referred to as Exploratory Data Analysis (EDA). The first book on this topic was published by John Tukey (1977), and now many of the methods he introduced are included in most introductory statistics courses. In contrast to previous methods of representing and analyzing data, EDA offers better tools to explore and understand data and to develop knowledge about what the data represent, often by using many different representations of the data to suggest new hypotheses and models (Hawkins, Jolliffe, and Glickman 1992).

## TEACHING STATISTICS

Until the National Council of Teachers of Mathematics (NCTM) published the *Curriculum and Evaluation Standards for School Mathematics* (1989), statistical topics were not typically included in the K–12 curriculum of U.S. schools. Now, most mathematics textbooks for the primary grades include some statistics topics. Separate curriculum materials are also available for teaching statistical ideas and methods at most grade levels. Most notable are the *Used Numbers* curriculum (e.g., Russell and Corwin 1989) for primary grades and the Quantitative Literacy Materials (e.g., Landwehr and Watkins 1986) for secondary grades.

Most instructional activities for elementary school students deal with collecting, tabulating, and graphing data. Students learn how to ask questions, collect and summarize samples of data, and use the data to answer their questions. They may learn numerical measures such as mean, median, mode, and range and simple graphing methods such as line plots, stem and leaf plots, bar graphs, histograms, and pie charts.

Secondary curriculum materials offer more techniques for analyzing, describing, and making inferences from data. Students may learn how to compute more measures of central tendency (also referred to as *measures of center*) and variability, construct and interpret boxplots and scatterplots, develop models to fit bivariate data sets, graph and interpret time series data, and make statistical inferences. In contrast to elementary mathematics curricula that are integrating statistics topics with mathematical topics, at the secondary level statistics may be taught as a separate course or unit where it must compete with algebra, geometry, and other mathematics courses.

Since the inclusion of statistics in the precollege curriculum, materials have been written to help teachers think about the important concepts and ideas and to guide them in teaching these topics. Whereas the NCTM standards provided the first guide for the types of experiences students should have (e.g., in grades K–4, the mathematics curriculum should include experiences with data analysis and probability so that students can collect, organize, and describe data), *Benchmarks for Science Literacy,* published four years later (American Association for the Advancement of Science 1993), offers descriptions of the basic ideas all students should develop in different grade levels (e.g., by the end of fifth grade students should know that a summary of data includes where the middle is and how much spread is around it). Several materials have been produced to further explain, support, and give examples of how to implement the NCTM standards (see Burrill 1994; Zawojewski 1991; Burrill et al. 1992). Other publications have appeared that offer advice for teachers and suggestions for learning activities (see Shulte 1981; Hawkins, Jolliffe, and Glickman 1992).

In addition to the statistics curriculum materials developed for elementary and secondary level students, software programs are now available to help students learn basic ideas of statistics and to easily enter, analyze, and graph data. These software programs are accompanied by instructional activities and suggestions for teachers and include *Statistics Workshop* (distributed by Wings for Learning/Sunburst) and *DataScope* (distributed by Intellimation).

At the college level, introductory courses may be offered in different departments. While the more traditional mathematical statistics courses are usually offered by departments of mathematics, courses in basic statistical ideas and methods may be offered in separate statistics departments as well as in departments of psychology, sociology, business, and health science. Introductory statistics courses are typically distinguished as being calculus-based (offering more mathematical theory) or service courses (offering more data analysis and applications). Topics in introductory courses usually include methods of organizing and graphing univariate and bivariate data, measures of center and variability, basic probability concepts and problems, sampling distributions, correlation and regression analysis, and one- and two-sample statistical tests.

## RESOURCES FOR TEACHING

In addition to the references and curriculum materials listed here, there are currently four regular publications for teachers and researchers interested in statistics education. The British journal *Teaching Statistics* publishes articles, data sets, and teaching activities for students aged 9–19. The electronic *Journal of Statistics Education* is available free of charge on the Internet and publishes articles primarily of interest to college statistics teachers. The *Statistics Teacher Network* is a newsletter for teachers primarily at the secondary level and is produced by the Joint Committee of the American Statistical Association and NCTM. The International Study Group for Research on Teaching and Learning Statistics produces a newsletter three times each year summarizing current research around the world (Garfield and Green 1988).

*See also* Data Analysis; Data Representation; Teacher Preparation in Statistics, Issues

## SELECTED REFERENCES

American Association for the Advancement of Science. *Benchmarks for Science Literacy.* New York: Oxford University Press, 1993.

Burrill, Gail, ed. *Teaching Statistics: Guidelines for Elementary through High School.* Palo Alto, CA: Seymour, 1994.

Burrill, Gail, et al. *Data Analysis and Statistics across the Curriculum.* Reston, VA: National Council of Teachers of Mathematics, 1992.

Folks, J. Leroy. *Ideas of Statistics.* New York: Wiley, 1981.

Garfield, Joan. "How Students Learn Statistics." *International Statistical Review* 63(1995):25–34.

———, and Andrew Ahlgren. "Difficulties in Learning Basic Concepts in Statistics: Implications for Research." *Journal for Research in Mathematics Education* 19(1988):44–63.

Garfield, Joan, and David Green. "Probability and Statistics Study Group: Looking back and Looking Forward." *Teaching Statistics* 10(1988):55–58.

Gordon, Florence, and Sheldon Gordon, eds. *Statistics for the Twenty-First Century.* Washington, DC: Mathematical Association of America, 1992.

Hawkins, Anne, ed. *Training Teachers to Teach Statistics.* Voorburg, Netherlands: International Statistical Institute, 1990.

Hawkins, Anne, Flavia Jolliffe, and Leslie Glickman. *Teaching Statistical Concepts.* London, England: Longman, 1992.

Kennedy, Gavin. *Invitation to Statistics.* Oxford, England: Robertson, 1983.

Landwehr, James, and Ann Watkins. *Exploring Data.* Palo Alto, CA: Seymour, 1986.

Moore, David S. *Statistics: Concepts and Controversies.* 3rd ed. New York: Freeman, 1991.

———. "Teaching Statistics as a Respectable Subject." In *Statistics for the Twenty-First Century* (pp. 14–25). Florence and Sheldon Gordon, eds. Washington, DC: Mathematical Association of America, 1992.

National Council of Teachers of Mathematics. *Curriculum and Evaluation Standards for School Mathematics.* Reston, VA: The Council, 1989.

———. *Principles and Standards for School Mathematics.* Reston, VA: The Council, 2000.

Pereira-Mendoza, Lionel, ed. *Introducing Data Analysis in the Schools: Who Should Teach It and How?* Voorburg, Netherlands: International Statistical Institute, 1993.

Russell, Susan Jo, and Rebecca Corwin. *The Shape of the Data.* Palo Alto, CA: Seymour, 1989.

Shaughnessy, J. Michael. "Research in Probability and Statistics: Reflections and Directions." In *Handbook of Research on Mathematics Teaching and Learning* (pp. 465–494). Douglas Grouws (ed.). New York: Macmillan, 1992.

———, and Barry Bergman. "Thinking About Uncertainty: Probability and Statistics." In *Research Ideas for the Classroom* (pp. 177–197). Vol. 3. *High School Mathematics.* Patricia Wilson (ed.). New York: Macmillan, 1993.

Shulte, Albert, ed. *Teaching Statistics and Probability.* 1981 Yearbook. Reston, VA: National Council of Teachers of Mathematics, 1981.

Stigler, Stephen. *The History of Statistics: The Measurement of Uncertainty before 1900.* Cambridge, MA: Harvard University Press, 1986.

Tukey, John. *Exploratory Data Analysis.* Reading, MA: Addison-Wesley, 1977.

Zawojewski, Judith S., et al. *Dealing with Data and Chance: Addenda Series, Grades 5–8.* Reston, VA: National Council of Teachers of Mathematics, 1991.

JOAN B. GARFIELD

# STERN, CATHERINE
## (1894–1973)

An elementary school educator, who was born in Germany and received her Ph.D. degree there in mathematics and physics in 1918. Six years later, motivated by the educational needs of her own children, Stern opened a Montessori school in Breslau. She came to the United States in 1938 and became a citizen in 1944. She believed in the Gestalt psychology of Max Wertheimer and was an early advocate of learning numbers through the manipulation of linear blocks made from colored cubes placed in frames. Three was represented by a white block containing 3 cubes; 36, by three dark-blue blocks of 10 cubes each and a red block of 6 cubes. Her approach, "structural arithmetic," was based on measurement more than counting. Children, aged 2 and up, manipulated these blocks and other materials to do whole number, fraction, and decimal calculations. Stern held that, "in experimenting with the materials, the child develops insight into the structure of the number system and the meaning and behavior of numbers" (Stern 1949, 287). Her ideas are found in the work of later educators such as Zoltan P. Dienes and Caleb Gattegno (Cuisenaire rods).

*See also* Cuisenaire Rods; Dienes

## SELECTED REFERENCE

Stern, Catherine. *Children Discover Arithmetic.* New York: Harper, 1949.

JAMES K. BIDWELL
ROBERT G. CLASON

# STOCHASTIC PROCESSES

Refers to observations taken from random phenomena. Most natural or social phenomena such as (1) daily rainfall precipitation, (2) the value of a stock on the first day of each month, (3) the number of telephone calls initiated during an hour in a town, (4) the continuous record of temperature at a weather station, and (5) the elevation above sea level along a mountain road are stochastic (from the Greek word for "random") or have a stochastic component. Observations on (1) or (2) lead to values at discrete time points. Observations on (3) taken continuously over the hour lead to a step function with a unit jump for each new call initiated. Observations on (4) provide a continuous graph of the temperature, while on (5) they record the elevation continuously with distance along the road. Each of these represents a stochastic process.

Technically, a stochastic process is a family of random variables $\{X(t); t \in T\}$ on a common probability space, where the parameter space $T$ is usually a subset of the real line $R$. Most often $t$ indicates time as in (1), (2), (3), and (4), but it can also be distance as in (5). One usually writes $\{X_t; t = 0, 1, 2, \ldots\}$ for a stochastic process with $t$ discrete, and $\{X(t); t \geq 0\}$ for one with $t$ continuous. Both $X_t$ and $X(t)$ can take either discrete or continuous values. Possible realizations of processes (2), (3) and (4) are illustrated in Figures 1–3.

The concept of a stochastic process dates back to the early 1900s when, for example, Agner Erlang (Brockmeyer, Halstrom, and Jensen 1948) had begun to use the properties of the Poisson process

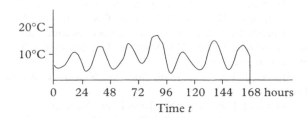

**Figure 1**   *Monthly value of a stock $X_t$ over a year.*

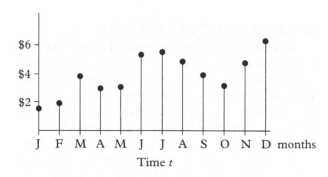

**Figure 2**   *Number of telephone calls $N(t)$ in an hour.*

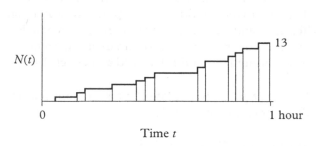

**Figure 3**   *Temperature record $X(t)$ over a week.*

$\{N(t); t \geq 0\}$ to characterize the number of telephone calls initiated during a period of time $(0, t)$. For such a process,

$$P\{N(t) = n\} = e^{-\lambda t}(\lambda t)^n/n! \qquad n = 0, 1, 2, \ldots$$

where the mean number of calls made per unit time is $\lambda > 0$. The Poisson process is often used to describe other arrival processes, such as the number of customers arriving at a bank counter in the interval of time $(0, t)$.

As a simple example, suppose that a bank has two counters each served by a teller. Each customer spends three minutes being served at a counter, and we may assume that the number $N(t)$ of customers arriving at the bank follows a Poisson process with $\lambda = 1$ customer arriving per minute. We can ask what is the probability that both tellers will be busy when a customer arrives just after 10:02 A.M., given that both tellers were free at 10:00 A.M. Counting the time $t$ in minutes, with $t = 0$ at 10:00 A.M. and $t = 2$ at 10:02 A.M., for both tellers to be busy just after $t = 2$, two or more customers must have arrived in the interval $(0, 2)$. None will have left since the service time is three minutes; thus

$$P\{N(2) \geq 2\} = \sum_{n=2}^{\infty} e^{-2}2^n/n!$$
$$= 1 - e^{-2}(1 + 2)$$
$$= 0.59$$

One might conclude, from the fact that the probability of being served is $0.41 < 0.5$, that the number of tellers is too small to service the incoming customers. If there had been three or four tellers, then

$$P\{N(2) \geq 3\} = 1 - e^{-2}(1 + 2 + 2^2/2!) = 0.32$$
$$P\{N(2) \geq 4\} = 1 - e^{-2}(1 + 2 + 2^2/2! + 2^3/3!)$$
$$= 0.14$$

and in these cases, the probability that the new customer could be served is 0.68 and 0.86, respectively, both greater than 0.5.

The Poisson process is the most basic stochastic process. Another that is often used to model random movement in the plane is the Wiener process (Brownian motion). Markov processes are used to describe phenomena that involve dependence on earlier states. Much of the work in stochastic processes is concerned with identifying the process that will approximate most closely to a particular set of observed data and then using the theoretical properties of the process for purposes of prediction.

*See also* Markov Processes; Poisson Distribution; Statistics, Overview

## SELECTED REFERENCES

Bartlett, Maurice S. *An Introduction to Stochastic Processes.* Cambridge, England: Cambridge University Press, 1960.

Brockmeyer, E., H. L. Halstrom, and Arne Jensen. "The Life and Works of A. K. Erlang." *Transactions of the Danish Academy of Technical Sciences* (2)(1948).

Brockwell, Peter, and Richard A. Davis. *Time Series: Theory and Methods.* New York: Springer-Verlag, 1987.

Doob, Joseph L. *Stochastic Processes.* New York: Wiley, 1953.

Feller, William. *An Introduction to Probability Theory and Its Applications,* vol. 1. 3rd ed. New York: Wiley, 1968.

———. *An Introduction to Probability Theory and Its Applications,* vol. 2. New York: Wiley, 1966.

Todorovic, Peter. *An Introduction to Stochastic Processes and Their Applications.* New York: Springer-Verlag, 1992.

JOSEPH GANI

# STUDENT TEACHER AUTONOMY, ELEMENTARY SCHOOL

The concept that teachers need to be autonomous themselves in order to teach for autonomy. Jean Piaget has commented that,

> The goal in intellectual education is not to know how to repeat or retain ready-made truths (a truth that is parroted is only a half-truth). It is in learning to master the truth by oneself at the risk of losing a lot of time and going through all the roundabout ways that are inherent in real activity (1974, 106).

This quote captures the essence of autonomy or self-regulation. Autonomy is the ability to decide for oneself without having to be told by others. Drawing from Piaget, Constance Kamii defined autonomy as

> . . . the ability to think for oneself and to decide between right and wrong in the moral realm, and between truth and untruth in the intellectual realm, by taking relevant factors into account, independently of reward and punishment (1992, 9).

Piaget's theory of constructivism explains how individuals actively construct knowledge by connecting new ideas with previous ones in an attempt to make sense of the world. This view of knowledge is different from the view of knowledge espoused by the cultural transmission or behavioristic paradigm in which knowledge is thought to be an internalized replica of external reality. Both Piaget and Kamii view autonomy as an aim of education. Kamii has researched constructivist theory in mathematics education and concluded that mathematics should not be taught as a set of rules and procedures to be memorized. Instead, teachers should develop an understanding of how children actively construct mathematical ideas and then facilitate that process. Constructivist teaching for autonomous development engages children in goal choosing, problem posing and solving, and assessing understanding.

Kamii's research (1985; 1989; 1994) confirms that children reinvent arithmetic autonomously when teachers serve as guides and do not force children to conform or memorize arithmetic "facts" and algorithms. The children in Kamii's constructivist primary mathematics program reinvented addition, subtraction, multiplication, and division by being encouraged to think about problems and devise their own approaches to solutions. For example, instead of memorizing that 9 plus 6 equals 15, children invented various approaches to solving the problem such as taking 1 from 6 which becomes 5 and making the 9 a 10, then adding the 10 to the 5 to get 15. Kamii found ". . . that the children who had three years of constructivist arithmetic generally did better than the traditionally instructed children both in logical and in numerical reasoning" (Kamii 1994, 207). The children in Kamii's constructivist mathematics programs were more autonomous compared to children who had been taught arithmetic in the traditional way. They had been encouraged to exchange points of view with others and to construct their own procedures for adding, subtracting, multiplying, and dividing. Not only did they invent their own procedures for solving problems, they were better able to explain their ideas than the traditionally taught children who were made to memorize. Constructivist teachers provided choices, reduced adult authority, promoted an exchange of points of view, and encouraged decision making.

The focus on autonomy as an aim for education is related to the recommendations in the *Standards of the National Council of Teachers of Mathematics* (NCTM) (NCTM 1989). While the *Standards* do not specifically identify autonomy as an aim, they do cite as goals for mathematics education problem solving, communication, and reasoning. All of these require a level of student self-regulation or autonomy. The *Standards* call for assessment evidence that students can formulate problems, apply a variety of strategies to solve problems, verify and interpret results, and generalize solutions. All are activities requiring student self-regulation. In addition, the NCTM *Professional Standards for Teaching Mathematics* recommends moving away from the teacher as sole authority for right answers and shifting toward student involvement in conjecturing, inventing, and problem solving. Autonomous students are more likely to be able to invent problems as well as solutions and less likely to rely on teachers for direction (Kamii 1994).

In order for autonomy to be an aim in the mathematics education of children, it must also be an aim of teacher education programs that prepare the teachers who will foster autonomy in children. Teacher education programs must also shift from a procedural approach to teaching mathematics to treating mathematics as sense-making. Teachers cannot teach for autonomy unless they themselves are autonomous. Autonomous teachers construct personally meaningful professional knowledge through a process of questioning what is educationally appropriate for children and making their best judgments after considering how children learn. Autonomous teachers are less likely to follow mandates from authorities without questioning whether what is being mandated is educationally appropriate. For example, an autonomous teacher can reason that using workbooks and photocopies that call for right answers would not be appropriate for children who learn best through posing their own problems and inventing their own solutions. Autonomous teachers are less likely to rely on others to mandate practices because they are able to make judgments for themselves.

Teacher education programs that foster autonomy in their students focus on two approaches: coursework in the program emphasizes increasing preservice teachers' understanding of children's understanding of mathematics and other areas (DeVries and Zan 1994), and the program experiences encourage preservice teachers to reconnect with themselves as learners in order to appreciate how children learn (Duckworth 1987). Fosnot (1989) and Schifter and Fosnot (1993) describe examples of programs that foster autonomy in preservice and inservice teacher development. The difficulties encountered in implementing such programs are due to the entrenchment of the behaviorist approach in both elementary school teaching and teacher education programs in higher education and to the difficult work involved in rethinking what it means to teach and learn. Such difficulties appear to be outweighed by the advantages of increased mathematical understanding in students who are able to reason for themselves.

Rheta DeVries and Betty Zan say that teachers who understand how children develop will be better able to foster that development. Teachers can listen carefully to children and engage them in activities that will give them insights into how children think. For example, they may play games with children in order to understand how children think numerically within the context of a game. Eleanor Duckworth describes teaching as research and puts student teachers in the role of researchers so they will better understand how they themselves learn. Student teachers pose problems, for example, such as inquiring into the meaning of place value and how they came to their understanding of it. Student teacher self-reflection promotes understanding what it is like to be a learner of elementary mathematics again. Teachers who have had a chance to act on a "wonderful idea" will be able to recognize and encourage the wonderful ideas that children have and want to explore.

Environments that foster autonomy place the responsibility for learning with the student teacher. David Boud (1988) calls for a teacher education environment in which students take the initiative in identifying their learning needs, setting goals for themselves, working collaboratively with others, creating problems to tackle, determining criteria to apply to their work, and engaging in self-assessment.

Student teachers can be encouraged to plan their own objectives and share the rationale for their practices with their supervising teachers and with other student teachers. This provides a chance for them to begin to question and challenge each other and to develop some confidence in their professional knowledge. Teacher education programs that help student teachers reflect on their goals and communicate them to supervising teachers will prevent the mindless replication of inappropriate practices such as having students memorize arithmetic "facts" without understanding (Castle and Rahhal 1992). Student teacher/supervising teacher negotiations over what student teachers do during the student teaching period occur very early in the experience (during the first two weeks) and regulate student teacher planning and implementation of learning activities (Grant and Castle 1990). Student teachers who are not able to negotiate activities into the daily schedule of their supervising teachers are likely to merely carry out the traditional practices of their supervising teachers (Castle and Meyer 1991). Course assignments aimed at promoting negotiations between a student teacher and a supervising teacher might include role playing situations between student teachers and supervising teachers, class discussions of what it means to negotiate and constructively resolve conflicts, an introductory letter from the student teacher to the supervising teacher describing the student teacher's expectations and goals for student teaching, a written plan of agreement or contract between a student teacher and supervising teacher detailing the student teacher's responsibilities during the student teaching period, and a final self-assessment statement that assesses the degree to which the student teacher perceived that activities were implemented during student teaching.

The degree to which a student teacher exerts autonomy has been found to be a function of student teacher initiative to discuss expectations with the supervising teacher, supervising teacher willingness to negotiate changes with student teacher, and student teacher initiative to self-assess and make accommodations versus need for external feedback from others (Castle and Meyer 1991). Evidence of student teacher autonomy can be observed in student teacher journal entries, written teaching plans, and in implementation of plans and projects. Examples that student teachers are fostering autonomy in children might include that they have planned and implemented child-initiated activities and provided opportunities for children to exert control over their learning experiences. Student teachers can learn to assess themselves by the extent to which they are autonomous in teaching and foster autonomy in children. Such reflection will lead to a deeper understanding of professional responsibility. Autonomy in children can be assessed by the extent to which children can pose their own problems and invent approaches to problems, are capable of making decisions without relying on adult authority or reward systems, ask questions when they do not understand, and work independently.

*See also* Constructivism

SELECTED REFERENCES

Boud, David. *Developing Student Autonomy in Learning.* 2nd ed. New York: Nichols, 1988.

Castle, Kathryn, and Douglas B. Aichele. "Professional Development and Teacher Autonomy." In *Professional Development for Teachers of Mathematics* (pp. 1–8). 1994 Yearbook. Douglas B. Aichele (ed.) Reston, VA: National Council of Teachers of Mathematics, 1994.

Castle, Kathryn, and Jane Meyer. "Student Teacher Autonomy: Negotiation of Student Teaching Experiences." *Journal of Early Childhood Teacher Education* 12(1)(Winter 1991):8.

Castle, Kathryn, and Kelly Rahhal. "Moving toward Developmentally Appropriate Practice in Primary Teaching." *Journal of Early Childhood Teacher Education* 13(1)(Winter 1992):3–6.

DeVries, Rheta, and Betty Zan. *Moral Classrooms, Moral Children.* New York: Teachers College Press, 1994.

Duckworth, Eleanor. *The Having of Wonderful Ideas and Other Essays on Teaching and Learning.* New York: Teachers College Press, 1987.

Fosnot, Catherine Twomey. *Enquiring Teachers Enquiring Learners, A Constructivist Approach for Teaching.* New York: Teachers College Press, 1989.

Grant, Kay, and Kathryn Castle. "Theory into Practice: Student Teachers' Construction of Knowledge." *Journal of Early Childhood Teacher Education* 11(1)(Winter 1990):13–14.

Kamii, Constance. "Autonomy as the Aim of Constructivist Education: How Can It Be Fostered?" In *Project Construct a Curriculum Guide—Understanding the Possibilities.* Deborah G. Murphy and Stacie G. Goffin (eds.). Jefferson City, MO: Missouri Department of Elementary and Secondary Education, 1992.

Kamii, Constance, with Georgia DeClark. *Young Children Reinvent Arithmetic.* New York: Teachers College Press, 1985.

———, with Linda Joseph. *Young Children Continue to Reinvent Arithmetic—2nd Grade.* New York: Teachers College Press, 1989.

———, with Sally Livingston. *Young Children Continue to Reinvent Arithmetic—3rd Grade.* New York: Teachers College Press, 1994.

National Council of Teachers of Mathematics. *Curriculum and Evaluation Standards for School Mathematics.* Reston, VA: The Council, 1989.

———. *Principles and Standards for School Mathematics.* Reston, VA: The Council, 2000.

———. *Professional Standards for Teaching Mathematics.* Reston, VA: The Council, 1991.

Piaget, Jean. *To Understand Is to Invent.* New York: Viking, 1974 (Paris, France: UNESCO, 1948).

Schifter, Deborah, and Catherine Twomey Fosnot. *Reconstructing Mathematics Education: Stories of Teachers Meeting the Challenge of Reform.* New York: Teachers College Press, 1993.

KATHRYN CASTLE

# SUBTRACTION

An operation inverse to addition and that may be taught in tandem with it. A simple subtraction statement consists of the minuend (the original quantity from which an amount will be subtracted), the subtrahend (the quantity to be subtracted), and the remainder (the difference between the quantities). For each subtraction statement, the corresponding addition statement may be given (e.g., $7 - 4 = 3$ and $3 + 4 = 7$).

## VARIETY OF PEDAGOGICAL APPROACHES

Some important considerations in the teaching of subtraction involve strategies for learning subtraction facts, techniques for simple and multidigit subtraction, and applications to word problems.

### Subtraction Facts

Different subtraction facts can be generated from single-digit subtrahends. Three different types of instructional activities are used to teach these facts: activities for understanding, relationship activities, and mastery activities. Activities to promote understanding include concrete or pictorial explorations of how the difference between two numbers is found. These explorations help children acquire number concepts by constructing them from within. The relationship activities are designed to reduce the memory requirements for identifying facts by capitalizing on the relationships inherent in the number system. A common activity used to demonstrate numerical relationships involves number families. Students learn that 2, 4, and 6 are a number family from which both subtraction and addition facts can be derived ($6 - 2 = 4$; $6 - 4 = 2$; $2 + 4 = 6$; and $4 + 2 = 6$). Therefore, for each number family students learn, they can generate solutions to four facts. Through the use of number families, children learn how addition and subtraction are inverse operations. Mastery activities are those activities that promote automaticity. When students are able to solve simple subtraction facts automatically, they, in turn, are better able to solve more complex subtraction problems. Activities such as fact games, timings, and flash card practice are included in this category. The relationship between teaching and learning is a function of the instructional environment. Classrooms that use different instructional approaches aimed at promoting conceptual understanding strengthen this relationship (Hiebert and Wearne 1993).

### Subtraction Problem-Solving Strategies

Subtraction problems may be categorized as those requiring regrouping or exchanging (formerly called "borrowing") and those that do not. Both types of problems may involve single-digit or multidigit numerals.

*Simple Subtraction* Subtraction may be taught to students using a variety of strategies. Often, subtraction is introduced to students through the use of counting strategies. Students may be encouraged to *count up* to determine the answer (for $8 - 5 = $ _____, students begin with 5 and count up to 8) or *count down* to determine the answer (for 8 − 5 = _____, students count backwards from 8). Frequently, subtraction counting strategies are combined with the use of manipulatives or counters whereby concrete objects represent numbers in the sequence. Another problem-solving strategy for subtraction involves the use of a number line. Students are taught to solve problems by making unit jumps either forward or backward on the number line. A third approach to introducing students to subtraction is through the use of line drawing. Students are taught to draw the number of lines represented by the minuend and then to cross out the number of lines represented by the subtrahend. The lines remaining show the difference.

*Multidigit Subtraction* This requires regrouping or exchanging. In order to solve the problem 73 − 56 = _____, a 10 must be traded from the seven 10s in the minuend and added to the three 1s to make thirteen 1s in the 1s column. At the same time, the 10 must be subtracted from the seven 10s to leave six 10s. In order to perform similar multidigit subtraction problems, students must understand several concepts. First, they must understand the concept of place value (Hiebert and Wearne 1992). That is, they must understand that 10s numbers can be traded for 1s, 100s for 10s, and so on. Another important skill in multidigit subtraction is understanding *when* regrouping is required. When complex subtraction is introduced, many students have difficulty determining whether regrouping is necessary. Finally, students must be able to rewrite problems after trading so that the problems can be solved easily. An important component skill required by multidigit subtraction is vertical alignment of columns. In order for students to complete multidigit subtraction problems successfully, numbers first must be lined up underneath each other in their respective columns. Vertical column alignment is especially important when students are working with numbers that have a different number of digits, for example, 976 − 49. Column grids or graph paper often are employed initially to promote accurate column alignment.

Another method of subtracting is known as the Austrian, or addition-carry, method. In the example 73 − 56 = _____, a group of 10s is taken from the 10s column and added to the 1s column. In the next step, since a 10 was taken away for use in the 1s column, a 10 is *added* to the five 10s in the subtrahend, making it six 10s. The primary difference between the Austrian method and the traditional method is that when the 10 is traded, the 10 is not subtracted from the 10s column in the minuend, it is added to the 10s column in the subtrahend.

Students often are taught that subtraction can be checked by adding the difference and the subtrahend (17 + 56 = 73). If the resulting sum is equal to the subtrahend, the subtraction operation was performed correctly. Not only is the checking activity important for self-monitoring but it also strengthens the concept of addition and subtraction as inverse operations. Estimating by rounding the answer before subtracting is another method for determining whether the solution is reasonable (e.g., to solve 73 − 56 students solve 70 − 60 = 10 then compare their computation to the estimated answer).

## Word Problems

Several different types of word problems may be solved using subtraction. These include action, classification, and comparison problems. Action problems refer to those word problems that contain a verb (breaks, loses, spends) that implies the subtraction operation (e.g., Taryn has $15.00. She *spends* $4.50 on pencils. How much money does she have left?). Classification problems do not contain an action verb but refer to two or more subordinate quantities and a superordinate quantity (e.g., There are 14 children in the class. Seven are girls. How many boys are in the class?). Comparison problems require students to determine the difference between two quantities (Sara is forty years old. David is thirty-six years old. How much older is Sara than David?). A variety of approaches is used to teach students to solve word problems. One approach is to have students act out the events in the problem. A second approach involves having students represent the quantities in the problem with concrete objects or a pictorial representation. Another approach is teaching students to use a number line. Estimation also is used in teaching students to solve word problems. Primary students solve word problems using a variety of strategies before formal instruction. These invented strategies often continue to be used after formal instruction begins. Instruction is most beneficial when it builds on children's natural problem-solving abilities (Carpenter and Moser 1984).

## RECOMMENDED MATERIALS FOR INSTRUCTION

A variety of materials may be used to teach students the concepts of subtraction. These materials include concrete objects such as blocks or unifix cubes, Cuisenaire rods, or other objects that may be used as counters. Students also are taught to use pictorial representations and the number line. For more complex subtraction, students may be introduced to computers or a calculator. Finally, teachers frequently use a variety of mathematical games, such as dominoes, to motivate students and to provide additional practice opportunities.

*See also* Addition; Algorithms; Basic Skills

## SELECTED REFERENCES

Carpenter, Thomas P., and James M. Moser. "The Acquisition of Addition and Subtraction Concepts in Grades One through Three." *Journal for Research in Mathematics Education* 15(May 1984):179–202.

Fuson, Karen C. "More Complexities in Subtraction." *Journal for Research in Mathematics Education* 15(May 1984):214–225.

———. "Research on Whole Number Addition and Subtraction." In *Handbook of Research on Mathematics Teaching and Learning* (pp. 243–275). Douglas A. Grouws (ed.). New York: Macmillan, 1992.

———. "Teaching Children to Subtract by Counting Up." *Journal for Research in Mathematics Education* 17(May 1986):172–189.

Geary, David C. *Children's Mathematical Development.* Washington, DC: American Psychological Association, 1994.

Hiebert, James, and Diana Wearne. "Instructional Tasks, Classroom Discourse and Students' Learning in Second Grade Arithmetic." *American Education Research Journal* 30(Summer 1993):393–425.

———. "Links between Teaching and Learning Place Value with Understanding in First Grade." *Journal for Research in Mathematics Education* 23(Mar. 1992):98–122.

Silbert, Jerry, Douglas W. Carnine, and Marcy L. Stein. *Direct Instruction Mathematics.* Columbus, OH: Merrill, 1990.

Steinberg, Ruth M. "Instruction on Derived Facts Strategies in Addition and Subtraction." *Journal for Research in Mathematics Education* 15(Nov. 1985):337–355.

MARCY L. STEIN
RONDA SIMMONS

## SUPPES, PATRICK (1922–)

Author of several books on logic and probability and a professor emeritus of philosophy, statistics, education, and psychology at Stanford University, where he joined the faculty in 1950. He received his doctorate in mathematics from Columbia University in 1950. For many years, he worked on elementary mathematics education. In 1958, with Newton S. Hawley, he developed geometrical ideas and constructions with grade 1 students. He was interested in theories of learning mathematics as well as mathematical theories of learning for grades K–12. In the 1960s, he advocated a behavioral theory of conditioned response as opposed to discovery learning. In 1963, he became the director of the Stanford Program in Computer-Assisted Instruction (CAI) for mathematics and other subjects (K–9). This was one of the earliest and most extensive explorations of CAI in the United States. The project utilized the computer in three ways: drill and practice, tutorial, and dialogue. The tutorial program used the computer in a complete instructional sequence. The dialogue portion was theoretical only, dealing with voice interaction, which was not possible at that time. Suppes was chief executive officer of Computer Curriculum Corporation from 1967–1990.

*See also* Computer-Assisted Instruction, (CAI)

## SELECTED REFERENCES

Hawley, Newton S., and Patrick Suppes. *Geometry for Primary Grades.* San Francisco, CA: Holden-Day, 1960.

Jones, Phillip S., ed. *A History of Mathematics Education in the United States and Canada.* 32nd Yearbook. Washington, DC: National Council of Teachers of Mathematics, 1970.

Seyfert, Warren C., ed. *The Continuing Revolution in Mathematics.* Washington, DC: National Council of Teachers of Mathematics, 1969.

Suppes, Patrick. *Language for Humans and Robots.* Cambridge, MA: Blackwell, 1992.

JAMES K. BIDWELL
ROBERT G. CLASON

## SYMMETRY

A feature that many geometric figures possess. The two basic types of symmetries of plane figures are exemplified in Figures 1(a) and 1(b). Figure 1(a) consists of two identical halves, each of which is a mirror image of the other. Figure 1(b) consists of several identical pieces equally spaced around a point.

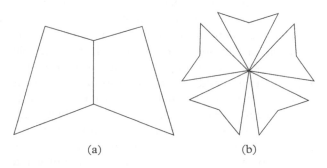

(a)                    (b)

**Figure 1**

In mathematical terms, a symmetry of a plane figure is an isometry, namely, a congruence transformation that maps the figure onto itself. The two isometries used to define symmetries are the reflection in a line and the rotation about a point.

*Line symmetry:* A plane figure F is said to have line symmetry if and only if there exists a line $l$ in the plane of F such that a reflection in $l$ will map F onto itself. The isosceles triangle $ABC$ in Figure 2(a) is symmetric with respect to the perpendicular bisector of $AB$.

*Rotational symmetry:* A plane figure F is said to have rotational symmetry of angle $\alpha \neq 0$ if and only if there exists a point $O$ in the plane of F such that a rotation of $\alpha < 360°$ with center $O$ will map F onto itself. Figure 2(b) represents an equilateral triangle with its sides $AB$, $BC$, and $CA$ extended by the same length. A rotation of $120°$ or $240°$ about the centroid $O$ of $ABC$ will map the figure onto itself. Hence, this figure has two rotational symmetries. Because of the extensions of the sides, the figure does not have any line symmetry.

*Point symmetry:* A plane figure F has point symmetry if and only if there exists a point $O$ in the plane of F such that a rotation of $180°$ (called a *half-turn*) with center $O$ will map F onto itself. So, the point symmetry is a special case of the rotational symmetry. The parallelogram $ABCD$ in Figure 2(c) has a point symmetry where the point of symmetry is the midpoint of the diagonal $AC$.

The square in Figure 2(d) has four reflections in lines $\ell_1$, $\ell_2$, $\ell_3$, $\ell_4$, two rotational symmetries of $90°$ and $270°$ about $O$, and is symmetric with respect to center $O$. In addition, it has the identity transformation I that maps every point of the square on itself. Thus the square has eight symmetries.

When a figure has two line symmetries whose lines of reflection are perpendicular, it will have a point symmetry, as is the case with the square. The reason is that the result of applying reflections in two lines intersecting in $O$ is a rotation about $O$ through twice the angle between the two lines.

The circle distinguishes itself from all plane curves by having an infinite number of symmetries. It is symmetric with respect to each of its diameters and with respect to its center. It also has rotational symmetry about its center through any angle.

*Translational symmetry:* A plane figure F has translational symmetry if and only if there exist a vector $\vec{v}$ in the plane of F such that a translation $T_{\vec{v}}$ will map F onto itself. It can be shown that F must extend to infinity (see Figures 3(a) and 3(b)).

The following tests for symmetry are used when figures are defined analytically:

1. If a curve is given by the Cartesian equation $f(x,y) = 0$, then the graph will be symmetric
   (i) with respect to the x-axis if $f(x,-y) = 0$ is equivalent to $f(x,y) = 0$, as is the parabola $y^2 = x$.

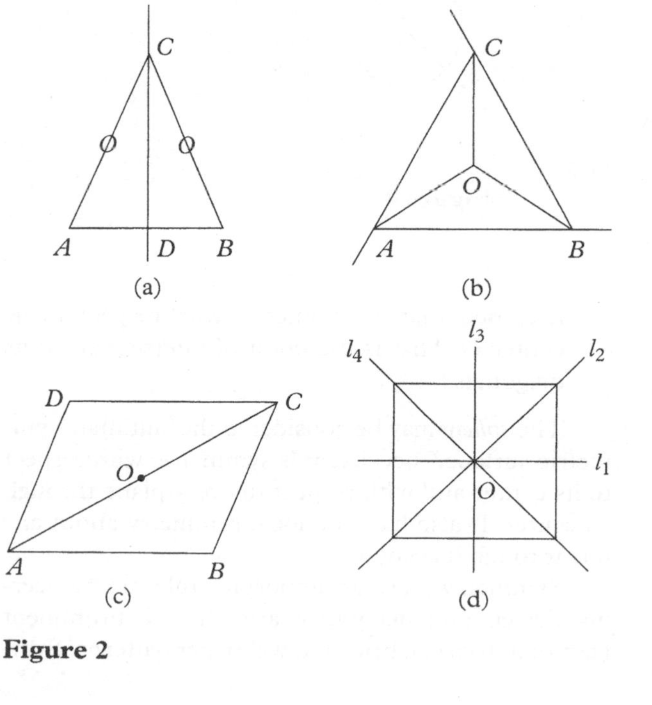

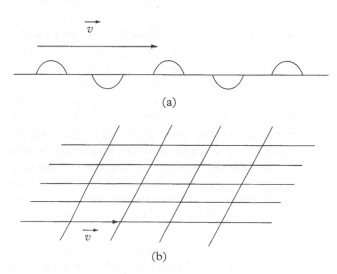

Figure 2

Figure 3

(ii) with respect to the y-axis if $f(-x,y) = 0$ is equivalent to $f(x,y) = 0$, as is the parabola $y = x^2$.

(iii) with respect to the origin if $f(-x,-y) = 0$ is equivalent to $f(x,y) = 0$, as is the cubical parabola $y = x^3$.

Because the ellipse $x^2 + 4y^2 = 4$ is symmetric with respect to each of the x-axis and y-axis, then it is symmetric with respect to its center (0,0).

(iv) with respect to the line $y = x$ if $f(y,x) = 0$ is equivalent to $f(x,y) = 0$, as is the hyperbola $xy = 1$. This curve is also symmetric with respect to the origin.

2. If a curve is given by the polar equation $f(r,\theta) = 0$, then the graph will be symmetric

   (i) with respect to the polar axis if $f(r,-\theta) = 0$ is equivalent to $f(r,\theta) = 0$, as is the cardioid $r = 2(1 + \cos\theta)$ (Figure 4(a)).

   (ii) with respect to the vertical line through the pole if $f(r,\pi - \theta) = 0$ is equivalent to $f(r,\theta) = 0$ as is the circle $r = \sin\theta$ (Figure 4(b)). In general, the graph will be symmetric with respect to the line $\theta = \alpha$ if $f(r,2\alpha - \theta) = f(r,\theta)$.

   (iii) with respect to the pole if $f(r,\pi + \theta) = 0$ is equivalent to $f(r,\theta) = 0$, as is the lemniscate $r^2 = 4\sin 2\theta$ (Figure 4(c)). This curve is also symmetric with respect to the line $\theta = \pi/4$.

In three-dimensional space, as in the plane, figures exhibit varying degrees of symmetry. The types of symmetry in this case are:

*Plane symmetry,* in which one-half of the figure is a mirror image of the other half with respect to the plane. Figure 5(a) depicts a right square pyramid that is symmetric with respect to each of four planes, two of which are *EAC* and *EBD*.

*Line symmetry* (or axial symmetry or rotational symmetry), in which a rotation through some angle about the line will map the figure onto itself. For the previous pyramid, a rotation about the line *EO* through 90° or 180° or 270° will map the pyramid onto itself. Also, the cone in Figure 5(b) is symmetric with respect to its axis *EO*. A rotation through any angle about the axis will map the cone onto itself. The cone is also symmetric with respect to any plane that contains its axis. Such a surface is called a *surface of revolution.*

*Point symmetry,* in which the point bisects each segment passing through it and having its ends on the figure. Figure 5(c) represents a paral-

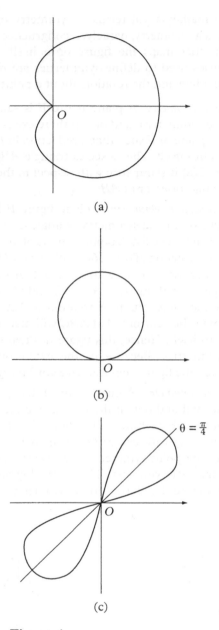

(a)

(b)

(c)

**Figure 4**

lelepiped that is symmetric with respect to its center *O*. That is the point of intersection of its diagonals.

The *sphere* may be considered the "ultimate symmetric surface," because it is symmetric with respect to its center and with respect to every plane through its center. It also has rotational symmetry about any line through its center.

Symmetry plays an important role in engineering design and decorative arts. It is a prominent feature in frieze, fabric, and wallpaper patterns. Sym-

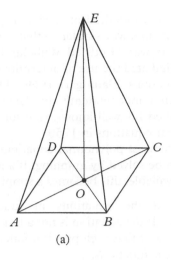

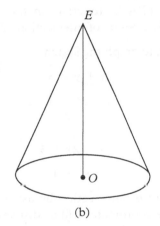

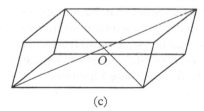

**Figure 5**

metry considerations are basic to the study of crystals and molecules in chemistry and the atoms in physics. Nature is inclined to produce symmetrical forms, as can be seen in a sunflower, a leaf, a starfish, and a snowflake.

Students in elementary schools start their experience with symmetry through paper-folding activities. Also, tracing paper can be used to explore the rotational symmetries of plane figures. Later, students may use *Logo* and *PreSuppose* software to investigate the different types of symmetry. Teachers may ask their students to classify polygons, including tri-

angles and quadrilaterals, on the basis of symmetry. The activity of constructing pyramids, prisms, and the Platonic polyhedra introduces the opportunity to examine the symmetries of these figures. Some students may be interested in creating artworks involving symmetry. At the college level, the topic of symmetry groups of geometric figures should be part of the teacher preparation curriculum.

*See also* Art; Coordinate Geometry, Overview

### SELECTED REFERENCES

Billstein, Rick, Shlomo Libeskind, and Johnny W. Lott. *A Problem Solving Approach to Mathematics for Elementary Teachers.* 6th ed. Reading, MA: Addison-Wesley, 1997.

Brown, Richard G. *Transformational Geometry.* Special ed. Palo Alto, CA: Seymour, 1973.

Geddes, Dorothy, et al. *Geometry in the Middle Grades: Addenda Series, Grades 5–8.* Reston, VA: National Council of Teachers of Mathematics, 1992.

Musser, Gary L., and William F. Burger. *Mathematics for Elementary Teachers: A Contemporary Approach.* 2nd ed. New York: Macmillan, 1991.

Stewart, James. *Calculus.* 3rd ed. Pacific Grove, CA: Brooks/Cole, 1995.

AYOUB B. AYOUB

## SYSTEMS OF EQUATIONS

Consists of two or more equations in two or more variables. A system of two linear equations is used in the business world to model the equilibrium price (the price where supply of the goods equals the demand for the goods). One equation represents the demand for the goods and the other equation represents the supply. The equilibrium price is found at the intersection of the two lines.

The unique solution, if it exists, of a pair of equations in two unknowns is represented by a single point on the Cartesian coordinate system that is the intersection of the two lines defined by the equations. The unique solution, if it exists, of three linear equations in three unknowns is a single point in a three-dimensional graphing system that is the intersection of the three planes defined by the three equations.

A system of equations in two unknowns can be solved by graphing the equations, solving for one variable in terms of the other in one of the equations and substituting into one of the other equations, or by eliminating one of the variables using addition or subtraction of equality.

It is easier to use the substitution method when one of the variables has a coefficient of 1, that is,

$$3x + 2y = 12$$

and

$$2x + y = 7$$

The method of eliminating one variable using addition and subtraction is helpful when all of the variables have coefficients greater than 1, that is,

$$4x - 2y = 14$$

and

$$2x + 3y = 3$$

Other methods of solving include Cramer's Rule using determinants, Gaussian elimination using expanded matrices, and matrix equations using properties of matrix operations (see the section on calculators).

To solve the system

$$3x + 2y = 8$$

$$4x - 3y = 5$$

using Cramer's Rule requires three determinants, $D_x$, $D_y$, and $D$:

$$D = \begin{vmatrix} 3 & 2 \\ 4 & -3 \end{vmatrix}$$

$$D_x = \begin{vmatrix} 8 & 2 \\ 5 & -3 \end{vmatrix}$$

$$D_y = \begin{vmatrix} 3 & 8 \\ 4 & 5 \end{vmatrix}$$

$D_x$ and $D_y$ are formed by replacing the $x$ or $y$ coefficients with the constant terms. The solutions are found by dividing the value of the appropriate determinant by the value of $D$:

$$x = D_x/D \quad \text{and} \quad y = D_y/D$$

To use Gaussian elimination on the same system, you would write an expanded matrix by adding the constants to each row, and then perform matrix operations on the rows or columns until the identity matrix for multiplication occurs in the first two columns.

$$\text{Transform} \begin{pmatrix} 3 & 2 & 8 \\ 4 & -3 & 5 \end{pmatrix} \text{ into } \begin{pmatrix} 1 & 0 & 2 \\ 0 & 1 & 1 \end{pmatrix}$$

The solution using either method is (2,1).

In the seventeenth century, the Japanese mathematician Seki Kowa systematized the old Chinese calculating sticks method of solving a system of linear equations in a way similar to that using determinants, but it was Gottfried Wilhelm Leibniz who formally originated the use of determinants with systems of equations (Baumgart 1969). Determinants were invented independently by Gabriel Cramer, who published his well-known rule for solving systems of linear equations in 1750.

All systems of polynomial equations in two unknowns can be solved by graphing. If a graphics calculator is available, this becomes a simple process:

1. Write all the equations in function form: $y = f(x)$. If the equation is not a function rewrite it in parts so that each part is a function.
2. Graph the functions.
3. Use the TRACE function on the calculator to identify the points of intersection, if they exist.

For example, to graph the system

$$2x + y = 3$$
$$y = x^2 - 4$$

use

$$y_1 = -2x + 3$$
$$y_2 = x^2 - 4$$

Find the point of intersection by using the TRACE function. The coordinates will be displayed along the bottom of the graph screen.

Most calculators that have graphics capabilities also have matrix operations built into the calculator. Therefore, linear systems with dimensions (the number of unknowns and equations) greater than two can be solved using these functions.

*Example*

$$3x + 2y + z = 5$$
$$2x - 3y + z = 8$$
$$x + 3y + 2z = 1$$

$$\begin{pmatrix} 3 & 2 & 1 \\ 2 & -3 & 1 \\ 1 & 3 & 2 \end{pmatrix} \begin{pmatrix} x \\ y \\ z \end{pmatrix} = \begin{pmatrix} 5 \\ 8 \\ 1 \end{pmatrix}$$

$$A \cdot X = B$$

$$X = A^{-1}B$$

where A is the coefficient matrix, X the variable matrix, and B the constant matrix.

The calculator is used to find the inverse of matrix A, multiply it by matrix B, and obtain the matrix of the unknowns, X. As calculators with these capabilities become more widely used, teachers can emphasize the various applications of systems of equations rather than the mechanics of their solutions. Many businesses use matrix form to display their data. Any time comparisons are made on groups of data, matrices can be used, that is, businesses can use the concept to compare sales in different stores or production on different assembly lines to analyze the materials needed and supplies available.

*See also* Determinants

## SELECTED REFERENCES

Baumgart, John K. "The History of Algebra." In *Historical Topics for the Mathematics Classroom* (pp. 233–260). 31st Yearbook. John K. Baumgart et al. (eds.). Washington, DC: National Council of Teachers of Mathematics, 1969.

Fey, James T., ed. *Calculators in Mathematics Education.* 1992 Yearbook. Reston, VA: National Council of Teachers of Mathematics, 1992.

Fleming, Walter, and Dale Varberg. *College Algebra: A Problem-Solving Approach.* Englewood Cliffs, NJ: Prentice-Hall, 1988, pp. 310–354.

Hall, Bettye C., and Mona Fabricant. *Algebra 2 with Trigonometry.* Englewood Cliffs, NJ: Prentice-Hall, 1993, pp. 144–247.

National Council of Teachers of Mathematics. *Curriculum and Evaluation Standards for School Mathematics.* Reston, VA: The Council, 1989.

———. *Principles and Standards for School Mathematics.* Reston, VA: The Council, 2000.

BETTYE C. HALL

# SYSTEMS OF INEQUALITIES

A set of two or more mathematical sentences indicating order relationships, symbolized by $>$, $\geq$, $<$, $\leq$, or $=$. The solution set of the system, known as the *feasible set* or *feasible region,* is the intersection of the solution sets of the individual inequalities comprising the system and is generally represented graphically.

To analyze systems of inequalities in two variables, begin by graphing each inequality on the same coordinate plane, using different markings to shade the solution set for each inequality. The feasible region is the intersection of the shadings of all the inequalities. The vertices of the feasible region correspond to the intersections of the boundaries of the inequalities. For instance, the system

$$\begin{cases} 2x + y > 4 \\ x - 3y < 6 \end{cases}$$

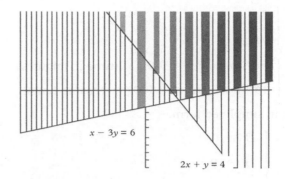

**Figure 1**

is graphed in Figure 1. The feasible set consists of the ordered pairs in the region marked by the overlapped shadings, indicating the intersection of the two inequalities. For the system

$$\begin{cases} x^2 + y^2 \leq 1 \\ x - y \geq 4 \end{cases}$$

there are no ordered pairs satisfying both inequalities simultaneously; hence, the solution is the empty set. Technological advances, such as the graphing calculator, have facilitated the study of systems of inequalities in the classroom. Using technology, several systems can be graphed in the time needed to graph one system by hand.

## LINEAR PROGRAMMING

Systems of linear inequalities are particularly useful in operations research, a field begun during World War II to analyze military operations. Today, there are economic, health, and environmental situations in which an optimal solution is required. Such problems can be modeled with systems of inequalities and solved using *linear programming* techniques.

For instance, suppose a student makes bracelets and hair bows. Each bracelet requires $1.50 in materials and 30 minutes to make; each bow requires $0.50 in materials and 20 minutes. In a week, the student can only spend $15 and 6 hours making these items. Assuming that all items made are sold and that a bracelet has a profit of $2.50 and a bow a profit of $1.00, how many of each item should be made to maximize the profit? If $B$ is the number of bracelets and $H$ is the number of bows, then a system describing the problem is

*703*

$$\begin{cases} B \geq 0 \\ H \geq 0 \\ 30B + 20H \leq 360 \quad \text{(time in minutes)} \\ 150B + 50H \leq 1500 \quad \text{(cost in cents)} \end{cases}$$

or, in simplified form,

$$\begin{cases} B \geq 0 \\ H \geq 0 \\ 3B + 2H \leq 36 \quad \text{(time in minutes)} \\ 3B + H \leq 30 \quad \text{(cost in cents)} \end{cases}$$

The task is to find the solution that maximizes the *linear objective function,* namely the profit function

$$P = 250B + 100H \quad \text{(in cents)}$$

For different values of $P$, the graphs of

$$P = 250B + 100H$$

are parallel lines, only some of which intersect the feasible region. The line that intersects the feasible region farthest from the origin contains the solution that maximizes the profit. According to a theorem proved in 1826 by Joseph Fourier (Senk et al. 1996), this optimal solution (either a maximum or a minimum depending on the situation) always occurs at one of the vertices of the feasible region, so the objective function need only be evaluated at these points. The following table gives the value of the profit function at each vertex of the feasible region.

| Number of bracelets | Number of bows | Profit (in cents) |
|---|---|---|
| 0 | 0 | 0 |
| 0 | 18 | 1800 |
| 8 | 6 | 2600 |
| 10 | 0 | 2500 |

Hence, making 8 bracelets and 6 bows generates the maximum profit.

*See also* Linear Programming; Operations Research

### SELECTED REFERENCES

Senk, Sharon L., et al. *Advanced Algebra.* 2nd ed. Glenview, IL: ScottForesman, 1996.

Ulep, Soledad A. "An Intuitive Approach in Teaching Linear Programming in High School." *Mathematics Teacher* 83(Jan. 1990):54–57.

Vest, Floyd. "Eco-Powered Programs: Using Linear Programming in Environmental Issues." *Newsletter of the Consortium for Mathematics and Its Applications* 40(Winter 1991):HiMAP Pull Out Section.

———. "Managing Your Money: Linear Programming II." *Newsletter of the Consortium for Mathematics and Its Applications* 42(Summer 1992):HiMAP Pull Out Section.

DENISSE R. THOMPSON

# SYSTEMS OF MEASUREMENT

Groups of commonly accepted, interrelated units that have been standardized by governments and scientific associations. They are used to answer questions such as "How tall are you?," "How many pounds did the baby weigh?," "How long will it take to get there?," and "How cold is it outdoors?" As they measure, people choose appropriate attributes, units, and tools for each measuring task, then use the unit to find out how many units comprise the object or event to be measured. For example, length is an attribute used to measure a person's height; centimeters might be used as the unit of measurement. A tape measure might be used as a tool to determine how many centimeters are needed to approximate the height.

The metric system, used throughout the world, is based on the meter, a measure of length, and includes related units such as cubic meters and liters to measure volume and capacity and grams to measure weight or mass. The units of the customary system, or inch-pound system, which is used widely in everyday life in the United States, include units such as inches to measure length, cubic inches and quarts to measure volume and capacity, and pounds to measure weight.

## HISTORICAL BACKGROUND

Historical records show that people have used measurement systems for thousands of years. Human activities such as keeping track of time and possessions, making things to desired sizes, and trading and selling products demanded measurement ideas. Communication was clarified by adoption of units and terms that could be used as a common basis for conversation and business transactions.

As measurements evolved, units were often based on body parts. The cubit, mentioned in the Bible and used in many early Mediterranean cultures, was the distance from a man's elbow to the end of his middle finger. The Romans used the term *unica,* the width of a thumb, for small measures of length; they transported the unit to lands they conquered. The English modified the term to *inch* and also defined its length with physical objects—the inch came to mean the length of three grains of barley placed end to end.

In local areas and kingdoms, measurements were often based on those of a particular person such as a king or lord. For example, about A.D. 1000 in England, the yard was defined as the distance from King Henry I's nose to the end of his outstretched thumb. Other stories of the often sporadic development of measurement units and their names are interesting. The term *week* developed as a measure of time based on the ancient Hebrew belief that the world was created in six days and the seventh day, or Sabbath, was a day to rest. The ancient Egyptians named the days for planets that were known at the time; the Romans modified the names using "Sun-day" and "Monday" to recognize the names of the sun and moon.

A variety of tools are necessary to use measurement systems, so along with measurement concepts and terms, people developed measurement devices. For example, knotted ropes helped ancient Egyptians as they measured lengths of land and building materials, and records from about 4000 B.C. show that Egyptians used balance scales to weigh grain, gold, and other products. As early as 1400 B.C., ancient Chinese and Babylonians used shadow clocks to keep track of time; by A.D. 1100, accurate mechanical clocks had developed in European and Asian societies.

As people traded and communicated with others in wider geographic locales and as science developed, the need for standardization of measurement became more apparent. In the 1600s, scientists and mathematicians began work on a simple, precise, measurement system with logically related units—one that could be used worldwide. In 1670, the Frenchman Gabriel Mouton proposed a decimal system, intended to be based on a measure of length—a fraction of the Earth's circumference; thus the metric system began to develop, spearheaded by members of the French Academy of Sciences. It was made legal in France in 1840, and its use spread to the scientific and engineering community throughout the world. The metric system is used for commercial and household purposes in most of the world's industrialized countries.

In 1975, the United States Congress passed the Metric Conversion Act, which called for voluntary changeover to the metric system. Although many efforts have been made to this end, both the metric and customary system are in use in the United States; both are taught in schools. The customary measurement system is used by the majority of people for everyday life purposes, while scientists, engineers, and medical personnel use the metric system in their work. Much industrial changeover to metrics has occurred, especially for products that are to be sold in international markets. Labels for most products—groceries, office products, and drugs—sold in the United States are marked with both customary and metric system units.

Commonly stated advantages of the metric system include its decimal base, which makes conversions among units easy. The system also uses uniform prefixes, whether applied to meters, liters, or grams. Commonly used metric prefixes are shown here:

| Prefix | Increase or decrease in unit |
| --- | --- |
| milli- | 0.001 (one thousandth) |
| centi- | 0.01 (one hundredth) |
| deci- | 0.1 (one tenth) |
| deka- | 10 (ten) |
| hecto- | 100 (one hundred) |
| kilo- | 1000 (one thousand) |

Another advantage of the metric system is that its units are logically related. A cubic centimeter and milliliter are defined as being the same amount. A cubic decimeter or liter of water at 4° Celsius is defined as having a mass of exactly 1 kilogram. So each cubic centimeter of cold water is a milliliter, and has a mass of 1 gram. Thus, measures of length, volume, capacity, temperature, and mass are linked in a direct way.

## GOALS FOR TEACHING MEASUREMENT SYSTEMS

Documents such as the National Council of Teachers of Mathematics' (NCTM) *Curriculum and Evaluation Standards for School Mathematics, K–12* (1989) help educators set goals for students as they learn measurement and measurement systems. For grades K–4 and 5–8, a measurement standard is one of thirteen presented for each range of grades. Although the *Standards* authors do not devote explicit attention to study of *systems* of measurement in grades K–4, they do advocate that students participate in experiences to gain background: students should "understand attributes of length, capacity, weight, area, volume, time, temperature, and angle." Young children typically work with estimation and measurement in both the metric and customary systems as they "develop the process of measurement and concepts related to units of measurement" (p. 51).

For grades 5–8, the *Standards* authors recommend more explicit attention to measurement systems: "In grades 5–8, the mathematics curriculum should include extensive concrete experiences using measurement" so that students can "understand the structure and uses of measurement systems" (p. 116). The *Standards* (1989) also emphasize use of measurement and measurement systems in realistic and problem situations involving rates and other derived and indirect measurements and formulas. As they work, students should select appropriate systems and units for each task.

The NCTM *Standards* do not include a separate section for measurement in recommendations for grades 9–12. The *Standards* authors, however, assume student understanding of measurement concepts, units, and systems as they advocate that students explore attributes of 2D and 3D figures; participate in problem solving using rates, speeds, and distances; analyze measurement data; and work with trigonometric functions.

## UNITS FOR SYSTEMS OF MEASUREMENT

Any physical object has a variety of attributes that might be measured, selecting appropriate units from a measurement system: length, area of surfaces, weight, and temperature, for example. Many objects have angles; volume or capacity are attributes that apply to three-dimensional objects. Time applies to duration or order of events.

*Length,* a linear or one-dimensional concept, concerns distance from one point to another. Length addresses the question "how long?" Perimeter is the distance around a figure. If the figure is a polygon, perimeter may be determined as the sum of the lengths of the sides. If the figure is a circle or other curve, the distance around it is called the circumference. Lengths are commonly measured in metric system units such as kilometers, meters, centimeters,

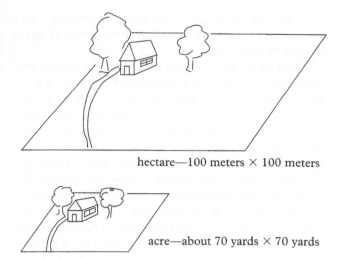

hectare—100 meters × 100 meters

acre—about 70 yards × 70 yards

**Figure 2**   *The hectare and acre are measures of area.*

and millimeters, and in customary system units such as miles, yards, feet, and inches (see Figure 1).

*Area,* a measure of surface, is expressed in square units such as square meters, square centimeters, or square miles. For measures of land, hectares and acres are often used (see Figure 2). Surface area is a measure applied to polyhedrons and cylinders—it is the sum of the areas of all the faces or curved surfaces. Students can picture surface area as the area that would be covered if the outside of a solid was painted.

*Capacity* and *volume* are three-dimensional measurement concepts; they describe how much space an object fills. Units for capacity include liters and milliliters in the metric system and gallons, quarts, pints, and cups in customary measurements (see Figure 3). Volume is measured in cubic units—cubes filling a space with no gaps and no overlapping. Commonly used units for volume include cubic meters and cubic centimeters and cubic inches and cubic feet.

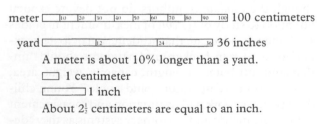

meter — 100 centimeters

yard — 36 inches

A meter is about 10% longer than a yard.

☐ 1 centimeter

☐ 1 inch

About 2½ centimeters are equal to an inch.

**Figure 1**   *Relative sizes of commonly used units of length.*

1 liter

1 quart

**Figure 3**   *The liter and the quart are measures of capacity that are very close in size.*

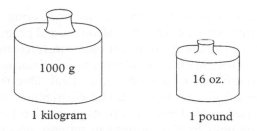

**Figure 4** *A kilogram is more than twice a pound.*

*Weight* and *mass* address the question "how heavy?" Weight is a measure of gravitational pull on objects; mass is a measure of quantity of matter. Though the terms are often used interchangeably in everyday life, some educators begin to draw a distinction between them in upper elementary grades or middle school. Some units for measurement of weight or mass include metric system units, such as grams and kilograms, and customary units, such as ounces and pounds (see Figure 4).

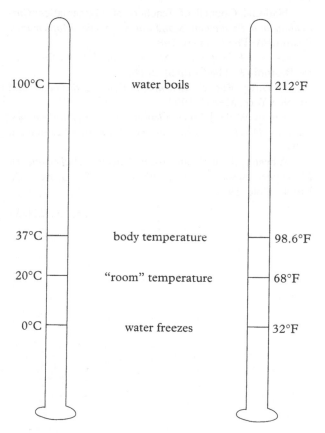

**Figure 5** *Both the Celsius and Fahrenheit scale use the term degree as a unit of temperature, but the size of the degree varies from scale to scale.*

Degrees are commonly used units for *temperature,* a measure of heat. The metric system uses degrees Celsius, a 100-degree scale between the freezing and boiling points of water. Customary units for temperature are degrees Fahrenheit, based on a 180-degree difference between the boiling and freezing points of water (see Figure 5).

*Time,* a measurement concept used to locate events in history or to compare durations of events, is measured with various devices such as clocks, timers, and calendars. Units such as seconds, minutes, hours, days, weeks, months, and years apply to time and are used in both the metric system and customary measurements.

*Angle,* a measure of rotation or turning, is the union of two rays with a common endpoint. Angles are typically measured with a protractor in both the metric and customary measurement systems. Angle measures are based on the idea that a complete rotation is 360 degrees. The use of 360 degrees seems to date from the Babylonian culture of about 4000 B.C.

## PEDAGOGICAL APPROACHES: BUILDING UNDERSTANDING OF MEASUREMENT SYSTEMS

The authors of the NCTM *Standards* (1989) stress the need for students to learn about measurement and measurement systems through actual experience, not merely by watching the teacher or examining pictures in a book. For young children, measurement experiences and development of vocabulary should start with direct comparison of attributes—having two children stand side by side and comparing their height, for example. Next children should order objects or events using a measurement attribute—perhaps placing five rocks in order from heaviest to lightest in weight.

A next stage is using nonstandard or arbitrary units as children learn about the measurement process. For example, children might use body parts (hands or feet) or readily available materials such as interlocking cubes to determine the length of a table or pencil, or they might use equal-sized paper wedges to discover how many of the wedges are needed to "fill in" an angle.

Finally, students should use standard units— widely agreed-on ones—from measurement systems. Since measurement is always approximate, in reporting results, educators should encourage children to use language that reflects this idea—phrases such as "about 7 centimeters," "between 20 and 21 pounds," and "almost 500 milliliters" are appropriate.

As they work with standard units, students must build concepts of measurement "benchmarks"—mental pictures of what a meter looks like, how long an hour seems in different situations, and what a kilogram feels like. To this end, educators might ask children to find many examples, and some nonexamples, of benchmarks. Children might collect various objects that are about a meter in length; they might then choose a few objects that are shorter and longer than a meter. Picturing benchmarks helps learners estimate and judge the reasonableness of results; it leads to understanding of related units within a system. Whether they use the metric system or customary measurements, students should be encouraged to think within the system, picturing units by using measurement benchmarks. Conversion between systems should be minimized. When it is necessary, conversion can be accomplished by using conversion charts or tables; any needed computations should be done on a calculator.

Finally, it is easy to connect measurement systems to engaging, realistic, problem-solving situations, to other mathematics topics, and to other school subjects. Educators can take advantage of the practicality and appeal of learning about measurement systems as they help students understand the need for measurement and clear communication about sizes, times, and other conditions such as heat.

## RECOMMENDED MATERIALS FOR INSTRUCTION

Since real experiences are essential as students learn about the measurement systems, materials are needed for instruction. Students should learn to use a variety of tools such as rulers, tape measures, scales, graduated cylinders and measuring cups, thermometers, clocks and calendars, and protractors as they explore the units for each measurement system. Tools made by students can supplement commercially manufactured ones. Real objects and events from the classroom and environment should be used as students learn to measure, select appropriate units from measurement systems, understand relationships among units within a measurement system, and refine their uses of measurement in problem solving.

*See also* Measurement; Temperature Scales

## SELECTED REFERENCES

*American Metric Journal.* American Metric Society. Bimonthly.

Billstein, Rick, Shlomo Libeskind, and Johnny W. Lott. *A Problem Solving Approach to Mathematics for Elementary School Teachers.* 4th ed. Redwood City, CA: Benjamin/Cummings, 1990.

Higgins, Jon L., ed. *A Metric Handbook for Teachers.* Reston, VA: National Council of Teachers of Mathematics, 1977.

National Council of Teachers of Mathematics. *Curriculum and Evaluation Standards for School Mathematics.* Reston, VA: The Council, 1989.

————. *Principles and Standards for School Mathematics.* Reston, VA: The Council, 2000.

Souviney, Randell. *Learning to Teach Mathematics.* 2nd ed. New York: Merrill, 1994.

Van de Walle, John A. *Elementary School Mathematics: Teaching Developmentally.* 2nd ed. New York: Longman, 1994.

Wheeler, Ruric E., and Ed R. Wheeler. *Mathematics for Elementary School Teachers.* 9th ed. Pacific Grove, CA: Brooks/Cole, 1995.

JEAN M. SHAW

# T

## TASK ANALYSIS

Describes the components of a task and how they are related to each other. In their comprehensive *Handbook of Task Analysis Procedures,* Jonassen, Hannum, and Tessmer (1989) describe various uses of task analyses in instructional design, such as constructing concept hierarchies (see Gagné 1977) and sequencing instructional events, both in traditional instruction and computer-assisted instruction (CAI). In mathematics education, the task to be analyzed may be as specific as an algorithm for calculating the sum of two two-digit numbers or as broad as teaching algebra or chaos theory.

A strategy often suggested for deciding how to teach a procedure—frequently an algorithm—is to analyze the procedure to determine all its components and then teach the components in some appropriate sequence, building up to the complete procedure. Task analysis will reveal which components are prerequisites for others, and this knowledge is used to sequence the instruction. Consider one of the algorithms for calculating the sum of two fractions as analyzed in Figure 1.

According to this analysis, students should first become proficient at listing multiples of numbers, then at choosing a least common multiple, then at creating equivalent fractions with a given denominator, and so on. Upon mastering each subtask, students are prepared to perform the algorithm in its entirety.

Performing a task analysis highlights prerequisites and possible stumbling blocks, but following it is no guarantee a complex procedure will be easily learned. Students must understand the conceptual underpinnings (Hiebert 1986). For the procedure in Figure 1, for example, students must understand what fractions are and how to write them, they should associate some physical meaning with adding fractions, and they must know what equivalent fractions are. For the mechanics of the procedure they must know what a multiple of a number is and what a numerator and a denominator are.

Prescriptions for performing mathematical procedures presented in textbooks are almost always based upon a task analysis of the procedure. The design of programmed learning units requires task analysis, as does the composition of a computer program. Also, formal lesson plans are a form of task analysis. The lesson plan formats suggested in Farrell and Farmer (1988) and Kennedy and Tipps (1994) are useful guides for analyzing a lesson.

*See also* Algorithms; Lesson Plans

*Task analysis for calculating $\frac{2}{3} + \frac{1}{4}$*

| | |
|---|---|
| 1. List multiples of first denominator (on paper or mentally). | 3, 6, 9, 12, . . . |
| 2. List multiples of second denominator (on paper or mentally). | 4, 8, 12, . . . |
| 3. Select least common multiple (least common denominator). | 12 |
| 4. Divide least common denominator by first denominator. | $12 \div 3 = 4$ |
| 5. Multiply this number times the first numerator. | $4 \cdot 2 = 8$ |
| 6. Write as numerator of fraction equivalent to $\dfrac{2}{3}$. | $\dfrac{8}{12}$ |
| 7. Repeat steps 4, 5, and 6 to get a fraction equivalent to $\dfrac{1}{4}$. | $\dfrac{3}{12}$ |
| 8. Write common denominator as denominator of the result. | $\dfrac{\phantom{8}}{12}$ |
| 9. Calculate sum of the numerators of the equivalent fractions. | $8 + 3 = 11$ |
| 10. Write this sum as the numerator of the result. | $\dfrac{11}{12}$ |

**Figure 1**

SELECTED REFERENCES

Farrell, Margaret A., and Walter A. Farmer. *Secondary Mathematics Instruction: An Integrated Approach*. Providence, RI: Janson, 1988.

Gagné, Robert M. *The Conditions of Learning*. 3rd ed. New York: Holt, Rinehart and Winston, 1977.

Hiebert, James, ed. *Conceptual and Procedural Knowledge: The Case of Mathematics*. Hillsdale, NJ: Erlbaum, 1986.

Jonassen, David, Wallace Hannum, and Martin Tessmer. *Handbook of Task Analysis Procedures*. New York: Praeger, 1989.

Kennedy, Leonard M., and Steve Tipps. *Guiding Children's Learning of Mathematics*. 7th ed. Belmont, CA: Wadsworth, 1994.

WILLIAM JURASCHEK

# TEACHER DEVELOPMENT, ELEMENTARY SCHOOL

At the elementary level, seeks to extend, remedy, or reinforce professional performance of teachers of mathematics. Development activities do not aim at the personal enrichment of teachers, but at helping them become more skilled, effective, or informed concerning professional practices, student learning, and the curriculum. Teacher development contrasts with teacher preparation in that the former serves practicing teachers, while the latter occurs, typically in a college setting, prior to employment as an educator. Development is not as time-delimited as preparation; it may occur periodically throughout one's career or it may be a continuous process, varying in intensity and emphasis as a teacher's needs evolve.

Whatever the level or subject area, professional development can be accomplished *only* by the individual teacher; it is not something that is done *to* a teacher but a process in which an educator chooses to engage, ultimately for the benefit of students. Professional development may be undertaken by an individual teacher or it may be a group activity, supported by agencies such as the National Council of Teachers of Mathematics (NCTM), school districts, state or federal offices, teachers' organizations, publishers, or foundations, to name but a few. Individual professional development is illustrated by the teacher who undertakes a program of reflection, reading, private study, and self-assessment, not involving colleagues or courses. An ideal of many teacher-preparation programs and professional associations is that teachers become "lifelong learners."

Professional development for teachers of mathematics at the elementary level may address what is often minimal or inadequate college-level study in mathematics. Even if a teacher's original preparation were exemplary, no preservice program can impart the knowledge and skills required for a lifetime of professional practice. Active and thoughtful participation in such professional organizations as NCTM, its state-level affiliates, or other associations can inspire, reinforce, and give direction to teachers' professional self-improvement. This participation is likely to include reading the appropriate journal(s), active participation at professional meetings, and communication with colleagues across the state and nation.

## EXTERNAL INFLUENCES

Inevitably, changes occur from year to year in the children we teach, the texts and curricular materials we employ, the expectations of the public and our profession, state mathematics curriculum requirements, and research in mathematics education. All teachers of mathematics must refine old and develop new skills, insights, and knowledge to remain effective for our students. As a prime example, the recommendations in NCTM's *Standards* entail many changes in teaching practices, subject matter, and even the organization of our schools and classrooms, for which virtually all inservice teachers need some form of professional development. *Reshaping School Mathematics* (National Research Council 1990) gave special attention to two "fundamentally important issues" rooted in the NCTM *Standards* and pivotal in teacher development:

- changing perspectives on the need for mathematics, the nature of mathematics, and the learning of mathematics;
- changing roles of calculators and computers in the practice of mathematics.

In addition, it was asserted that "[p]articular attention must be paid to in-service training of elementary school teachers . . . ." In support of the NCTM *Standards,* they emphasize that a "primary goal of elementary school mathematics is to develop number sense." This development entails students' "abilities to reason effectively with numerical information," which in turn comprises representation, operations, and interpretation (National Research Council 1990). For many or most practicing teachers, some type of development program is a practical necessity if they are to understand and implement such goals effectively, comfortably, and across the span of grades and student needs.

The National Research Council pointed out that teachers need more than sound preparation and freedom from unreasonable constraints, they need comprehensive programs of staff development that enable them to accept responsibility for their own professional development. Teachers should be

- Experimenting with alternative approaches and strategies in the classroom.
- Discussing issues in mathematics, mathematics teaching, and learning with colleagues.

- Learning with students, for students, and from students.
- Maintaining knowledge of contemporary mathematical practice.
- Proposing, designing, and evaluating mathematics programs for students and for professional development.
- Participating in workshops, courses, and other educational opportunities specific to mathematics.
- Participating in school and community efforts to effect positive change in mathematics education (National Research Council 1990).

*Professional Development for Teachers of Mathematics* was devoted to this topic (Aichele 1994). This volume included essays on teacher autonomy, a K–12 professional-development framework, lessons from research, teacher-led action research, teachers as assessors, teacher empowerment and leadership, and discussions of several specific programs.

With NCTM cooperation, the *PBS Middle School Math Project* offered an innovative opportunity for professional development during the 1994–95 school year. PBS (the Public Broadcasting Service) offered professional development opportunities linked closely to the NCTM *Standards* via distance learning courses, videoconferences, and electronic learning forums. The project also included an electronic resource center with e-mail, bulletin boards, discussion forums, and databases of resources for teaching mathematics. This project was the first to be launched by PBS's MATHLINE, a discipline-based educational service.

## MATHEMATICAL MASTERY AND TEACHER DEVELOPMENT

The *Guidelines for the Continuing Mathematical Education of Teachers* (Mathematical Association of America 1988) focused on teacher development through graduate courses or master's degree programs. It also addressed inservice and staff development programs that are not credit-bearing. The emphasis was on teachers' mastery of mathematics. But adult-level or academic knowledge of the mathematical content of the curriculum is insufficient for teachers at any level. Both professional preparation and development programs need to address teachers' *pedagogical knowledge of the subject matter* in the curriculum. This knowledge informs the judgments and decisions that teachers make every day about the selection, organization, and assessment of the mathematical ideas, skills, and applications we provide to our students. As mathematical topics are introduced into elementary-level classes, teachers need to understand the mathematics both as educated adults and as professional educators.

In *Everybody Counts*, after noting that "[t]oo often, elementary teachers take only one course in mathematics, approaching it with trepidation and leaving it with relief," it was stated that elementary-level teachers of mathematics "need to learn the history of mathematics and its impact on society, for it is only through history that teachers will come to know that mathematics changes and to see the difference between contemporary and ancient mathematics" (National Research Council 1989).

## CONCLUSION

Teacher development has been an important goal since mathematics teaching began. Long a neglected activity, it began to receive national attention in 1957 as the nation feared the scientific ascendancy of the former Soviet Union, symbolized by *Sputnik,* the first artificial Earth satellite. The National Defense Education Act of that era provided many teachers with advanced study in mathematics and spawned the "new math." The educational reform reports, which first appeared in the early 1980s and continued for many years, led to substantive innovations (as well as to recriminations and finger pointing) in mathematics education. Among the most significant changes were those presented in NCTM's *Standards.* The groups and forces pressing for improvement in American education are diverse and powerful; few if any fail to impinge on what mathematics teachers *do or know.* Teacher education, both pre- and inservice, should respond and serve the needs of America's mathematics teachers and their students.

*See also* Teacher Preparation, Elementary School, Issues

## SELECTED REFERENCES

Aichele, Douglas B., ed. *Professional Development for Teachers of Mathematics.* 1994 Yearbook. Reston, VA: National Council of Teachers of Mathematics, 1994.

Mathematical Association of America. *Guidelines for the Continuing Mathematical Education of Teachers.* Washington, DC: The Association, 1988.

National Board for Professional Teaching Standards. *What Teachers Should Know and Be Able to Do.* Detroit, MI: 1994.

National Council of Teachers of Mathematics. *Curriculum and Evaluation Standards for School Mathematics.* Reston, VA: The Council, 1989.

———. *Principles and Standards for School Mathematics.* Reston, VA: The Council, 2000.

National Research Council. *Counting on You: Actions Supporting Mathematics Teaching Standards.* Washington, DC: National Academy Press, 1991.

———. *Everybody Counts: A Report to the Nation on the Future of Mathematics Education.* Washington, DC: National Academy Press, 1989.

———. *Reshaping School Mathematics: A Philosophy and Framework for Curriculum.* Washington, DC: National Academy Press, 1990.

WILLIAM J. MCKEOUGH

# TEACHER DEVELOPMENT, K–12: CALIFORNIA MATHEMATICS PROJECT (CMP)

Inservice development program in California for school mathematics teachers. Reform in mathematics education creates a serious challenge to the mathematical teaching profession; the expressed changes require teachers to teach in ways that they were not taught and, to a large degree, have never experienced. Thus, the chance that widespread implementation of reform ideas is embraced is very small unless carefully planned professional development opportunities for teachers are designed.

One program that has been implemented in California and is adaptable to any part of the country is CMP. CMP subscribes to the idea that all teachers can become effective mathematics teachers. Presently, California has seventeen regional CMP sites that serve as headquarters for professional mathematical development activities for area teachers. Because California is quite large geographically and very diverse in population density, these sites sometimes serve areas as large as the entire state of Ohio or as narrow as one urban school district. The mathematics project is funded by the State of California and is part of a larger professional development model that includes projects in each of the disciplines required for high school graduation. The California Subject Matter Projects are externally evaluated by Inverness Research Associates. They continue to be shown as highly cost-effective models of professional development that positively influence the teaching practices of participants and are making a significant contribution to the statewide educational reform effort (St. John et al. 1995). Each CMP site has responded to local needs and conditions and hence the individual site programs have become as diverse as the populations they serve. The sites, however, share several key features and beliefs that bind them together and help to realize unified mathematics reform throughout California.

Most sites start with an intensive three- to four-week summer program. (Teachers apply to become part of the program and district support is required, often in the form of released days during the academic year.) Teachers receive a stipend for participation and units of credit are often available from the hosting university at the participant's expense. During the academic year, there are follow-up activities (both schoolday and weekend commitments) that bring the teachers together to continue their professional growth. Examples of these activities include grade-level specific curriculum workshops, classroom action research groups, seminars on specific topics such as assessment techniques or presentation skills, and "administrator days," where teachers and their administrators jointly attend sessions usually centered around site-based change and teacher support. After the first year of participation, participants continue to be involved in a host of ongoing professional development activities that the individual sites conduct, such as peer support groups, reading and research groups, and alumni mini-institutes. Many teachers subsequently become involved as staff for new cadres as all sites subscribe to a "teachers training teachers" model. Sites balance their efforts to bring in more new participants, support continuing members, and reach out to the larger population of teachers.

One common belief of the mathematics projects is that the desired student actions asked for by the reform (examining, representing, transforming, planning, designing, conjecturing, solving, evaluating, proving, and communicating) are, for many teachers, very different from what they have experienced as mathematics learners themselves. Hence, it would not be seen as an obtainable goal of teachers if they did not have the opportunity to experience mathematics in this way. Research continues to support the point that teachers most often teach the way they are taught. Therefore, *all sites of the CMP spend considerable time putting teachers in the role of active learners of mathematics.* Although the mathematical content varies from site to site, they all incorporate in-depth studies in mathematics. The mathematical content covered is secondary to the positioning of teachers in a constructivist's learning environment, that is, one where they will be involved in their own meaning-making of mathematical ideas and not just told procedural mathematical and/or pedagogical techniques. Changing teachers' beliefs about mathemat-

ics teaching and learning requires giving them new experiences in mathematical thinking and conceptual understanding (Hyde 1989).

Many of these experiences are accomplished with K–14 teachers working together. Indeed, even though sites often have grade-level breakout groups as part of their programs, substantial mathematical content is experienced with integrated teacher populations. In the beginning, participants are often somewhat skeptical of working with teachers of all grade levels (primary teachers fear the mathematical expertise of the secondary teachers, and secondary teachers feel that any mathematical content that is not at the level they teach is not appropriate for them). The strengths of having all teachers together, however, should not be underestimated. Together they can get multiple perspectives and views of mathematical ideas: primary teachers see how concepts spiral up through the curriculum and are needed in future years, and secondary teachers continually observe many varied solution techniques for problems for which they have only used (and taught) one prescribed formula. How mathematics curriculum reform is implemented in the classroom depends largely on teachers' images of the mathematics they are teaching (Bauersfeld 1980; Cooney 1985; Thompson 1984).

Perhaps the most significant shared belief of the mathematics projects (as well as other successful teacher development programs) is that they recognize that teachers are part of a school culture. Recognizing that schools have cultures is crucial to any professional development model; it implies that if any substantial change is going to take place through professional development programs, those programs must critically examine the attitudes, beliefs, values, and practices of the social community of the school. Lasting changes in mathematics classrooms result only when teachers confront their beliefs about what mathematics is, what it means to do mathematics, and how mathematics is taught and learned (House 1994).

Hence the projects never lose sight of the fact that teachers are human beings with beliefs and values that must be recognized. Teachers cannot be expected to actively participate in efforts in which they do not believe. Thus, the projects constantly engage themselves and the participants in self-reflections of their beliefs and endeavor to establish a climate for risk taking on the part of the teachers. Providing opportunities for reflecting and planning and for expressing and working through feelings about mathematics learning and teaching will increase the likelihood of teachers developing new understandings, challenging beliefs and assumptions, and changing rigid and unproductive practices (Weissglass 1994).

It cannot be overstated that effective professional development happens over time. The old incessant model of professional development, that is, providing one-shot workshops by outside experts, while entertaining, does very little in changing the climate or actions in individual teacher's classrooms. Teacher change takes time (Fullan 1990; Duffy and Roehler 1986; Romberg and Price 1983), and any vision of change must be seen by teachers as achievable, useful, and desirable. One-shot inservice models almost always fall short in these domains.

Finally, each site of the CMP wants all teachers in its service area to regard the CMP as their "professional home." CMP believes that the most productive way to facilitate change is to place special emphasis on nurturing the professional growth of teacher leaders. Instrumental in the success of establishing CMP as an effective professional development program is its broad definition of teacher leader. A teacher leader is not to be interpreted narrowly as one who leads an inservice workshop. Rather, a teacher leader in mathematics may be one who, for example,

- tries new assessment techniques in the mathematics classroom and shares the experience with the teacher next door,
- works to provide resources at the school level, thereby raising the level of support for the mathematics program in the school,
- becomes a mentor teacher in mathematics, helping other teachers to improve mathematics instruction,
- serves on school/district mathematics committees and works toward meaningful improvements in the mathematics program,
- connects to the larger network of teachers seeking improvement in mathematics learning and encourages other teachers to become actively involved,
- gets involved in a curriculum project helping to develop, pilot, and evaluate new mathematics materials,
- participates in state or national assessment programs helping to design, score, and evaluate new mathematics assessments,
- learns a new technology, uses it in the classroom, and helps other teachers to use it,
- works with diversity/equity programs to make mathematics accessible to all students,

- learns techniques of teaching ESL/LEP (English as a Second Language/Limited English Proficient) students and works to apply these techniques to mathematics teaching,
- develops presentation skills in order to share their knowledge and experiences with colleagues at mathematics meetings, workshops, institutes, etc.,
- becomes a spokesperson for improving mathematics learning and makes presentations to school administrators, school boards, parents, community groups, government, and the media (CMP 1994).

This list is not meant to be exhaustive, but rather only broad enough to give meaning to the idea of teacher leader. The list has become instrumental to each site in the design of its programs to meet the needs of its teachers. The success of CMP in developing teacher leaders has been measured by Inverness Research Associates. Besides the ability to contribute to other teachers in both formal and informal ways, there have been other important outcomes that have been measured. Among them are an increase in pedagogical skills, an increase in content knowledge, and a reconceptualizing of the discipline. But perhaps the best measure of success has been a look at the classrooms of participants. Not only is there evidence of students involved in reformed aligned activities with evidence of multiple modes of learning but also students of these teachers were judged as having more love and appreciation of the discipline than students of the traditional approach (St. John et al. 1995).

*See also* Teacher Development, Elementary School; Teacher Development, Middle School; Teacher Development, Secondary School

## SELECTED REFERENCES

Bauersfeld, Heinrich. "Hidden Dimensions in the So-Called Reality of a Mathematics Classroom." *Educational Studies in Mathematics* 11(Feb. 1980):23–42.

California Mathematics Project. *Vision Statement.* San Diego, CA: The Project, 1994.

Cooney, Thomas J. "A Beginning Teacher's View of Problem Solving." *Journal of Research in Mathematics Education* 16(Nov. 1985):324–336.

Duffy, Gerald, and Laura Roehler. "Constraints on Teacher Change." *Journal of Teacher Education* 37(Jan./Feb. 1986):55–58.

Fullan, Michael G. "Staff Development, Innovation and Institutional Development." In *Changing School Culture through Staff Development* (pp. 3–25). Yearbook. Bruce Joyce (ed.). Alexandria, VA: Association for Supervision and Curriculum Development, 1990.

House, Peggy A. "Empowering K–12 Teachers for Leadership: A Districtwide Strategy for Change." In *Professional Development for Teachers of Mathematics* (pp. 214–226). 1994 Yearbook. Douglas B. Aichele (ed.). Reston, VA: National Council of Teachers of Mathematics, 1994.

Hyde, Arthur A. "Staff Development: Directions and Realities." In *New Directions for Elementary School Mathematics* (pp. 223–233). 1989 Yearbook. Paul R. Trafton (ed.). Reston, VA: National Council of Teachers of Mathematics, 1989.

Romberg, Thomas A., and Gary G. Price. "Curriculum Implementation and Staff Development as Cultural Change." In *Staff Development* (pp. 154–184). Yearbook of the National Society for Study in Education, pt. 2. Gary A. Griffin (ed.). Chicago, IL: University of Chicago Press, 1983.

St. John, Mark, Kathleen Dickey, Barbara Heenan, Judy Hirabayashi, Kathy Medina, and Katherine Ramage. *Evaluating a Statewide Professional Development System.* Report no. 9. Inverness, CA: Inverness Research Associates, 1995.

Thompson, Alba G. "The Relationship of Teachers' Conceptions of Mathematics Teaching to Instructional Practice." *Educational Studies in Mathematics* 15(May 1984):105–127.

Weissglass, Julian. "Changing Mathematics Teaching Means Changing Ourselves: Implications for Professional Development." In *Professional Development for Teachers of Mathematics* (pp. 67–78). 1994 Yearbook. Douglas B. Aichele (ed.). Reston, VA: National Council of Teachers of Mathematics, 1994.

WILLIAM FISHER

# TEACHER DEVELOPMENT, MIDDLE SCHOOL

A crucial element of systemic change in the middle grade mathematics curriculum. The *Professional Standards for Teaching Mathematics* suggests that "teachers are key figures in changing the ways in which mathematics is taught and learned in schools" (National Council of Teachers of Mathematics [NCTM] 1991, 2). Indeed, teachers are being challenged to question traditional views about knowing, learning, and teaching mathematics. In many classrooms, major changes will need to take place in order to implement these recommendations. Many teachers will find that the curriculum and related instructional approaches currently being recommended are very different from what they experienced as mathematics students. "Teachers must examine their own assumptions about learning mathematics and face contradictions between the linear way in which mathematics is presently taught and what we know

about the dynamic and robust way in which it is actually learned" (Rasch 1994, 12).

The constructivist view of learning mathematics is at the heart of the current call for reform. A teacher's view of learning has a profound impact on his/her instructional practices. *Everybody Counts* suggests that

> Effective teachers are those who can stimulate students to *learn* mathematics well only when they *construct* their own mathematical understanding. To understand what they learn, they must enact for themselves verbs that permeate the mathematics curriculum: "examine," "represent," "transform," "solve," "apply," "prove," "communicate." This happens most readily when students work in groups, engage in discussion, make presentations, and in other ways take charge of their own learning (National Research Council 1989, 58–59).

Such teachers will create a classroom environment where teacher and students work together on significant mathematical tasks; the teacher will no longer simply present information to students. Emphasis will be placed on mathematical reasoning and problem solving instead of memorizing facts and procedures. "Problem situations that establish the need for new ideas and motivate students should serve as the context for mathematics in grades 5–8" (NCTM 1989, 66). The goal of teacher development programs for middle grade mathematics teachers, then, will be to help teachers create this kind of learning environment for their students.

First, teacher development programs must focus on crucial issues related to the middle grade mathematics curriculum and provide teachers with relevant information and experiences. Teachers should be given opportunities to

1. further their preparation in mathematics at a level appropriate for teaching middle grade mathematics;
2. analyze their views about the nature of mathematics, middle grade students as learners of mathematics, and the teaching of middle grade mathematics;
3. observe examples from real classrooms of a wide variety of instructional strategies, student activities, and curriculum materials;
4. interact with other teachers as they investigate and discuss questions concerning middle grade mathematics teaching and learning;
5. implement specific instructional techniques, student activities, and curriculum materials with their students.

Second, these programs should emphasize collaboration and innovation and be targeted specifically for middle grade mathematics teachers. Schools, colleges and universities, professional organizations, and state departments of education should work together to provide teachers with ongoing professional development activities.

Teachers, mathematics educators, and state education departments should develop specific standards for certification of middle grade mathematics teachers. Courses of study should be developed specifically for the middle grade mathematics teacher. Guidelines such as those suggested by the Mathematical Association of America (MAA) should be used in developing such courses (MAA 1991). At the state level, professional organizations should encourage state education departments to view middle grade certification as more than just a total number of mathematics hours.

Colleges and universities, together with school systems, should develop undergraduate and graduate programs that focus on the needs of middle grade mathematics teachers. For current middle grade teachers, this may mean retraining in mathematics in order to satisfy middle grade certification standards. Some of these teachers may need to take a number of mathematics and mathematics education courses in order to meet state guidelines, while other teachers may need only one or two courses. Such programs have already been developed or are currently in development at a number of colleges and universities, such as Portland State University, Oklahoma State University, Illinois State University, and Central Missouri State University. In these programs, courses in calculus, geometry, modern algebra, statistics and probability, and instructional strategies are designed specifically for both pre- and inservice mathematics teachers.

Needs of other teachers could be met by providing appropriate inservice workshops developed by local schools, university mathematics educators, and state departments of education. Summer workshops that focus on helping practicing teachers further develop their instructional skills and resources can serve as valuable experiences for these teachers. Universities and state departments of education can help such teachers provide inservice experiences for other teachers in their local districts. "Drive-in" conferences for teachers in a given region of a state can also be provided by universities and state departments of education.

Innovative ways of involving teachers in professional development experiences should be investigated. Distance learning technologies could be used to provide inservice workshops and university courses for teachers who are in locations that are geographically isolated or who otherwise are unable to attend such programs on college campuses. The use of distance learning technologies can make it possible for relatively large numbers of teachers to participate in such activities without having to travel great distances. Courses and workshops using a distance learning format have already been used in a variety of locations around the country. On a national level, the PBS Mathline Middle School Math Project has offered professional development opportunities via distance learning courses, videoconferences, and electronic learning forums. Mathline also provides an electronic math teacher resource center with e-mail, bulletin boards, discussion forums, and databases of resources for teaching mathematics. As an example of a state project, Central Missouri State University has developed two inservice courses designed specifically for practicing middle school mathematics teachers. These courses, Teaching Middle Grade Mathematics and the History of Mathematics were delivered by satellite to school sites throughout the state of Missouri. The Missouri Middle Mathematics and Show Me projects have given middle school mathematics teachers opportunities to investigate reform curriculum materials developed by National Science Foundation-sponsored projects.

Innovative instructional materials and other resources need to be developed and made available for middle grade mathematics teachers. Textbook publishers should begin to develop text materials that are compatible with the kinds of classrooms envisioned in the current recommendations for reform. Some beginning efforts have been made in this area and should be encouraged by members of the mathematics education community. Professional organizations can also provide materials that teachers can use to supplement and enrich existing middle grade programs. The NCTM *Addenda Series* is an example of such materials. In recent years, a number of projects have developed instructional materials that support innovative, alternative curricula for middle grade mathematics programs (*Mathematics in Context Project, Six through Eight Mathematics Project* (STEM), *Connected Mathematics Project, Middle School Mathematics through Applications Project,* and *Seeing and Thinking Mathematically Project*). The materials developed in these projects make extensive use of problems set in realistic contexts, adaptations from real applications, thematic modules, technology, and connections to other subject areas. To be effective, schools and universities should allocate time and financial resources to the process of teacher development. "Because teachers need time to learn and develop this kind of teaching practice, appropriate and ongoing professional development is crucial. Good instructional and assessment materials and the latitude to use them flexibly are also keys to the process of change" (NCTM 1991, 3).

Finally, individual teachers should take the primary responsibility for their professional development and should be willing to experiment with alternative instructional strategies in their classrooms. Teachers should take the initiative to participate in inservice workshops and courses, read professional publications, and actively participate in developing programs of study, curriculum materials, and policy. Teachers may be called on to assume different roles, including coach and coinvestigator, as well as making use of alternative forms of assessment. In order to implement the kinds of changes in curriculum and instructional methods recommended by the various calls for reform, middle grade mathematics teachers should understand the theoretical foundation underlying these recommendations. They should understand how middle grade students think about and learn mathematics. They need to investigate teaching strategies that will help students build on elementary school mathematics and yet lay the foundation for mathematics concepts and skills that will be encountered in later grades. Middle grade teachers should have opportunities to observe other teachers demonstrating a variety of teaching methods and, further, be able to practice these methods in an environment that provides support and relevant feedback. Ultimately, teachers' knowledge and beliefs will determine how much flexibility they have in implementing a variety of mathematics curricula and teaching methods. Teacher development should be a career-long, ongoing process.

*See also* Teacher Preparation, Middle School; Teacher Preparation, Secondary School

### SELECTED REFERENCES

*Connected Mathematics Project.* University of North Carolina—Chapel Hill. Chapel Hill, NC.

Epstein, J. L., and D. J. MacIver. *Education in the Middle Grades: Overview of National Survey of Practices and Trends.* Report No. 45. Baltimore, MD: Johns Hopkins University, Center for Research on Elementary and Middle Schools, 1990.

Fitzgerald, William M., and Mary Day Bouck. "Models of Instruction." In *Research Ideas for the Classroom.* Vol.

2. *Middle Grades Mathematics* (pp. 244–258). Douglas T. Owens (ed.). New York: Macmillan, 1993.

Mathematical Association of America. *A Call for Change: Recommendations for the Mathematical Preparation of Teachers of Mathematics.* Washington, DC: The Association, 1991.

*Mathematics in Context Project.* Wisconsin Center for Education Research. Madison, WI.

*Middle School Mathematics through Applications Project.* Institute for Research on Learning. Palo Alto, CA.

National Council of Teachers of Mathematics. *An Agenda for Action: Recommendations for School Mathematics of the 1980s.* Reston, VA: The Council, 1980.

———. *Curriculum and Evaluation Standards for School Mathematics.* Reston, VA: The Council, 1989.

———. *Principles and Standards for School Mathematics.* Reston, VA: The Council, 2000.

———. *Professional Standards for Teaching Mathematics.* Reston, VA: The Council, 1991.

National Research Council. *Everybody Counts: A Report to the Nation on the Future of Mathematics Education.* Washington, DC: National Academy Press, 1989.

Rasch, Kathe. "The Imperative for Quality Middle School Mathematics Curriculum and Instruction." *Midpoints* 4(Spring 1994):1–16.

*Seeing and Thinking Mathematically Project.* Educational Development Center. Newton, MA.

*Six through Eight Mathematics Project.* University of Montana. Missoula, MT.

Sowder, Judith, and Bonnie Schappelle, eds. *Providing a Foundation for Teaching Mathematics in the Middle Grades.* Albany, NY: State University of New York Press, 1995.

U.S. National Research Center. *Third International Mathematics and Science Study.* Report No. 7. East Lansing, MI: The Center, 1996.

TERRY GOODMAN

# TEACHER DEVELOPMENT, SECONDARY SCHOOL

Seeks to enhance, remedy, or reinforce professional performance by helping teachers become more skilled, effective, or informed concerning professional practices, student learning, and the curriculum. Teacher development is a central component of the reform of American education; they are complementary efforts. Teacher development contrasts with teacher preparation in that the former addresses the needs of practicing teachers, while the latter prepares preprofessionals to enter teaching with an initial credential or license. Professional development can be achieved *only* by an individual teacher: it is not something done *to* a teacher but a process in which an educator chooses to engage, ultimately for the benefit of students. Although professional development may be undertaken by an individual teacher, it is more commonly pursued with others. It may take the form of a graduate program, possibly leading to an advanced degree, or it may be long- or short-term training under the aegis of agencies such as the National Council of Teachers of Mathematics (NCTM), school districts, state or federal offices, teachers' organizations, or foundations, to name but a few. Individual professional development is illustrated by the teacher who undertakes a program of reflection, study, and self-assessment, not involving colleagues or courses.

## CONTEXT

Reform of secondary education has been a significant national concern for the last hundred years, beginning with the *Report of the Committee of Ten on Secondary School Studies* in 1893. When the former Soviet Union launched *Sputnik,* the first Earth-orbiting satellite, in 1957, there were loud demands for reform and improvement in American secondary education—particularly in mathematics, science, and foreign languages. In 1959, James Bryant Conant wrote *The American High School Today,* a study of the comprehensive high school, with a twenty-one-item "checklist for reform." Since then, many reports have been issued, including *A Nation at Risk: The Imperative for Educational Reform* (National Commission on Excellence in Education 1983); *Curriculum and Evaluation Standards for School Mathematics* (NCTM 1989); *Professional Standards for Teaching Mathematics* (NCTM 1991); and *Assessment Standards for School Mathematics* (NCTM 1995). Although the "reform literature" is not (indeed, need not be) internally consistent in its diagnoses and prescriptions, every report and recommendation has carried an implicit call for teacher development. No teacher preparation program can equip an educator for a lifetime of successful practice, nor can preservice education anticipate the needs of an experienced professional years after graduation. For many mathematics teachers, the demographic characteristics of the students they will teach throughout their careers will differ significantly from those of even recent years. The texts and curricular materials we employ, the emerging importance of technologies such as computers and (graphing) calculators, our modes of student assessment, state mathematics curriculum requirements, research findings in mathematics education, the expectations of parents, the public, and our profession all continue to change and to place new expectations and responsibilities on mathematics teachers. These "outside"

influences create a fundamental need for teacher development. "Burnout," a phenomenon common to many professions, affects mathematics teaching as well. The fact that a teacher may spend over forty years performing essentially the same role and tasks is a contributing factor. Teachers' traditional lack of effective control over much of their professional lives exacerbates matters. The fact that patent excellence in teaching rarely brings any greater reward than does marginal performance can be corrosive to some teachers. These "inside" influences only increase the need for teacher development.

*Everybody Counts: A Report to the Nation on the Future of Mathematics Education* [National Research Council (NRC) 1989] offered a powerful and far-ranging critique of then-current practices in mathematics education. Each point could form an agenda for a teacher development program. For example:

> Teachers . . . almost always present mathematics as an established doctrine to be learned just as it was taught. This "broadcast" metaphor for learning leads students to expect that mathematics is about right answers rather than about clear creative thinking (p. 57).

*Everybody Counts* also identified many areas of opportunity or concern in which many teachers would benefit from inservice education. Among them were a constructivist approach to student learning; constrictive state-certification requirements; teacher "insecurity" and "rigidity" with respect to personal mastery of the subject matter; teachers' flawed experiences in studying mathematics themselves and the negative effects on their own teaching; excessive reliance on textbooks; abuses of standardized testing, which can subvert meaningful learning; and a lack of serious investment in high-quality learning by non-college-bound and—especially—"disadvantaged groups" (pp. 60–73).

## SOME VIEWS AND VISIONS OF MATHEMATICS TEACHER DEVELOPMENT

The *Guidelines for the Continuing Mathematical Education of Teachers* (Mathematical Association of America 1988) focused on teacher development through graduate courses or master's degree programs. It also addressed inservice and staff development programs that are not credit-bearing. Its emphasis was on teachers' knowledge of mathematics.

*Reshaping School Mathematics* (NRC 1990) addressed teacher development and enhanced professionalism and offered useful guidelines for professional development: "Teachers need more than sound preparation and freedom from unreasonable constraints. They need comprehensive programs of staff development that enable them to accept responsibility for their own professional development." Among the points made were the necessity for

- Experimenting with alternative approaches and strategies in the classroom.
- Proposing, designing, and evaluating mathematics programs for students and for professional development.
- Participating in workshops, courses, and other educational opportunities specific to mathematics.
- Participating in school and community efforts to effect positive change in mathematics education.

*Professional Development for Teachers of Mathematics* (Aichele 1994) dealt with teacher development in mathematics. It included essays on teacher autonomy, a K–12 professional development framework, lessons from research, teacher-led action research, teachers as assessors, teacher empowerment and leadership, and discussions of several specific programs. Active long-term participation in such professional organizations as NCTM, its state-level affiliates, or other associations can inspire, reinforce, and give direction to teachers' professional self-improvement. This participation should include reading the appropriate journal(s), active participation at professional meetings, and communication with colleagues across the state and nation, for example by participation in forums on the Internet and World Wide Web. The Internet's resources for mathematics teachers are extensive and include access to educational software, research findings, and other teachers' plans. Formal graduate study in mathematics or education can offer teachers excellent opportunities for professional development.

## A SINGULAR OPPORTUNITY FOR RECOGNITION

The (nongovernmental) National Board for Professional Teaching Standards offers voluntary Board certification to accomplished teachers. Its origins are in *A Nation Prepared: Teachers for the 21st Century* (Carnegie Forum . . . 1986). Its mission was the establishment of "high and rigorous standards for what accomplished teachers should know and be able to do," along with creation of assessment and certification processes, all "to advance educational reforms

for the purpose of improving student learning in American schools" (National Board for Professional Teaching Standards 1995, 1). National Board certification offers both a powerful professional stimulus for self-development and the recognition of one's peers for successful professional achievement under demanding criteria.

## SOME CONSTRAINTS

Both mathematics education reform and teacher professional development—with their predictable false starts and lack of efficient structure—will be challenging. Numerous recommendations have been made over many years; billions of local, state, federal and private-sector dollars have been invested before now; and many thoughtful, dedicated educators have labored over many decades. The results of past efforts and recommendations are not great. The key external resources, always scarce, are time and money. The federal government's financial investment in teacher development in mathematics seems to have a negative slope, and state-level commitments are sure to be uneven. As a group, teacher unions have not been leaders in professional development, but their potential influence is great. Whether we as a profession and as a nation have the will to challenge and change teachers' traditionally passive roles and to modify teacher-centered, "stand-and-deliver" models of teaching is unclear. What is not unclear or uncertain is the imperative for mathematics teacher development as a driving force within a comprehensive reform of American education.

*See also* Teacher Preparation, Secondary School

### SELECTED REFERENCES

Aichele, Douglas B., ed. *Professional Development for Teachers of Mathematics.* 1994 Yearbook. Reston, VA: National Council of Teachers of Mathematics, 1994.

Carnegie Forum on Education and the Economy. Task Force on Teaching as a Profession. *A Nation Prepared: Teachers for the 21st Century.* Washington, DC: The Forum, 1986.

Conant, James Bryant. *The American High School Today: A First Report to Interested Citizens.* New York: McGraw-Hill, 1959.

Mathematical Association of America. *Guidelines for the Continuing Mathematical Education of Teachers.* Washington, DC: The Association, 1988.

National Board for Professional Teaching Standards. *An Invitation to National Board Certification.* Detroit, MI: 1995.

————. *What Teachers Should Know and Be Able to Do.* Detroit, MI: 1994.

National Commission on Excellence in Education. *A Nation at Risk: The Imperative for Educational Reform.* Washington, DC: Government Printing Office, 1983.

National Council of Teachers of Mathematics. *Assessment Standards for Teaching Mathematics.* Reston, VA: The Council, 1995.

————. *Curriculum and Evaluation Standards for School Mathematics.* Reston, VA: The Council, 1989.

————. *Principles and Standards for School Mathematics.* Reston, VA: The Council, 2000.

————. *Professional Standards for Teaching Mathematics.* Reston, VA: The Council, 1991.

National Education Association. *Report of the Committee of Ten on Secondary School Studies with Reports of the Conferences arranged by the Committees.* Washington, DC: Government Printing Office, 1893.

National Research Council. *Counting on You: Actions Supporting Mathematics Teaching Standards.* Washington, DC: National Academy Press, 1991.

————. *Everybody Counts: A Report to the Nation on the Future of Mathematics Education.* Washington, DC: National Academy Press, 1989.

————. *Reshaping School Mathematics: A Philosophy and Framework for Curriculum.* Washington, DC: National Academy Press, 1990.

WILLIAM J. MCKEOUGH

# TEACHER PREPARATION, COLLEGE

Mainly accomplished in mathematical sciences graduate departments and occasionally in colleges of education. More than two hundred departments offer the Ph.D. degree in pure and applied mathematics, statistics, operations research, or variously named computational and applied programs. Some six hundred departments offer a master's degree in mathematics; of those, there are a few specialized programs for two-year college teachers. (See Case 1994; Case and Huneke 1993; National Academy of Sciences 1992.) College mathematics teachers prepared in colleges of education generally find work only in two-year colleges except as mathematics education specialists working with K–12 teacher preparation in four-year college and university mathematics departments. Although some with doctorates in mathematics find two-year college employment, the minimal requirement of mathematical training required to teach at a two-year college is generally a master's degree in mathematics, and that is the most frequent qualification found (Shell 1990). Preparation is affected by hiring considerations, most re-

cently the lack of positions due to budget problems of public and private institutions (National Academy of Sciences 1995). Regional accreditation agency requirements also affect preparation because hiring is restricted; for example, institutions in the Southeast may assign only teachers who have successfully completed eighteen graduate hours in mathematics (Case and Huneke 1993, 95).

Although there are strong programs (specifically intended to produce college mathematics teachers) that lead to a degree designated Doctor of Arts, the degree has never been widely offered. Many such programs have in recent years been reworked and are now the sole offering, or options in, programs leading to the Ph.D. degree. Some departments of mathematics have recently added a dissertation option for specialization in collegiate mathematics education (National Academy of Sciences 1992, 30; Case and Blackwelder 1992, 418, and 1993, 807–809).

There are serious problems in the educational system of the United States that produce students often ill prepared to benefit from college teaching. The full participation of the faculty and graduate students of doctoral departments in the exciting explosion of mathematical research since *Sputnik,* a very good thing in itself, often diverts energy and attention from teaching and from preparation for teaching. With today's complex academic situation, it is appropriate—arguably, necessary—that a component of a graduate student's preparation be aimed at effective teaching. Only one or two universities included such a component twenty years ago. Now there are diverse models with similar goals. In addition, the major professional societies concerned with collegiate mathematics teaching have recognized this need and have produced reports and recommendations intended to be applied by preparatory departments and, in some cases, to be helpful also to individual beginning mathematicians and their mentors. Governmental agencies have given consideration to the various problems of preparation need and employment; the Committee on Science, Engineering, and Public Policy has taken a comprehensive view of changes affected by the present climate and made recommendations (National Academy of Sciences 1995).

## MATHEMATICAL CONTENT AND BREADTH

Wide diversity of mathematical content and of breadth of exposure to various areas is found in highly recognized doctoral mathematical sciences programs. Masters programs vary from a traditional pure mathematics program to specialized programs actually designed for industrial or government applications; a few have as their goal broad preparation for multiple demands of two-year college teaching. Recommendations for college teacher preparation that were produced over twenty years ago listed topics for specific analysis, algebra, and geometry courses. Although a recent report from the American Mathematical Association of Two Year Colleges (AMATYC) does list seventeen areas, it is not intended that each teacher study each topic at the graduate level. The report states that undergraduate and graduate courses should complement each other with "courses chosen broadly" from those areas (Foley 1992, A–4). Current recommendations for doctoral programs generally and teaching preparation in particular follow the practice that has emerged, with a diversity of areas and emphases. Over a five-year period (1989–1994), the practice was surveyed and recommendations published by the joint Committee on Preparation for College Teaching (CPCT), established in 1987 by the presidents of the American Mathematical Society, the Mathematical Association of America, and the Society for Industrial and Applied Mathematics. It was concluded that it was neither reasonable nor desirable to list essential inclusions in doctoral programs; any such list would rapidly be made obsolete by changes in the vitality and usefulness of the separate topics, as well as fail to reflect a consensus of the community of concerned scholars (Case 1994). In 1994, this joint study group affirmed its early thesis: "Breadth of knowledge forms the background for teaching a variety of courses, for advising students, and for recognizing different kinds of talent in mathematics. . . . It is essential if one is to be discerning in evaluating peers, hiring faculty, organizing curricula, and managing resources in support of student learning" (Mathematical Association of America 1989).

## PEDAGOGICAL INFORMATION AND TEACHING ISSUES

Although the excellence of U.S. and Canadian doctoral mathematical education is recognized by research mathematicians throughout the world, systematic preparation for teaching and other professional functions distinct from research is often lacking. For some doctoral departments there is immediacy of need for some pedagogical understanding because many mathematics graduate students themselves teach undergraduates, usually the most basic freshman courses. Attention is given in such

departments to the problem of teaching assistant competency (Case 1989; Foley 1992; Case and Huneke 1993). The longer preparatory view calls for a serious consideration in graduate school of potential future teaching responsibilities and pedagogical methods, with an understanding of the tripartite role of the professoriate and its requirements for distinguished scholarship, for service including administrative duties, and for classroom teaching. Recent reports address this concern carefully (e.g., Foley 1992, A5, 6) or in a general way (Conference Board 1992; National Academy of Sciences 1992; National Academy of Sciences 1995). A "generic" model of recommended activities is described in the joint CPCT report (Case 1994, 38–41).

A number of doctoral programs have recently made provisions for their students to study, think, and learn more about teaching methods and issues, usually revolving around information developed in a teaching seminar. In some cases, there is intensive effort over one or two semesters, and in others the activities, required or optional, are spread throughout the student tenure. Beginning mathematicians on postdoctoral appointments are sometimes included in these plans. The exact form, style, scope, and content of such a professional seminar depends on its institutional setting, participant needs and interests, and faculty leadership. Crucial success factors are that adequate time be allotted for this portion of the duties of the faculty director; appropriate enthusiasm for the project be maintained by the faculty participants; at least some of the faculty involved are highly respected senior mathematics faculty members who direct the research of graduate students; evaluations of strong and weak points are regularly conducted, with subsequent modifications of plans and activities; and careful meshing of activities and information with the knowledge, abilities, and needs of the particular individual students and postdoctoral participants. Eight very different successful plans in departments that cooperated in a project assisted by a grant from the Fund for the Improvement of Postsecondary Education have been presented (Case 1994, 45–88), and others have been more recently implemented.

## RESPONSIBILITY FOR TEACHING PREPARATION

Although the responsibility for graduate departments of assuring strong preparation of their students is apparent, the practical situation dictates that "hiring departments"—those making postdoctoral, visiting, or tenure-track first appointments—must share this responsibility. Postdoctoral fellows as well as advanced graduate students are an appropriate audience for formal and informal preparation programs. These hiring departments should accept responsibility that there be caring mentors as well as formal activities such as a seminar considering the broad range of faculty responsibilities. Hiring departments themselves will be better served when their departmental and college needs are understood. Unfortunately, unwritten rules sometimes exist that must be taken into account for career success and hence should not be secret (e.g., the grade distributions expected, examination customs).

At the present time there are many beginning mathematicians who must seek adequate teaching preparation outside formal programs. When there is no seminar or organized activities, senior faculty serving as mentors have a special responsibility. The novice faculty member must discover what is expected, both explicated and not. Even for those graduate students and new Ph.D.s who have an organized seminar, it is necessary to begin habits of self-evaluation and professional development in the teaching program as well as the research portion of the career; early acceptance of personal responsibility will serve mathematicians well through the changes of an academic career. (See "Continuing Education" in Foley 1992, A–4; "Whose Responsibility" and "How to Support Young Mathematicians" in Case 1994, 32–37.) Helpful activities include reading current literature; attendance at professional meetings; broadening mathematical knowledge; talking with other graduate students and faculty about teaching methods and controversies; observing classes of both professors and of other beginning mathematicians who are trying new methods, have won teaching awards, or are considered because of word-of-mouth to be especially effective; trying new ideas in the classroom; and developing interaction and communication skills with peers and students.

*See also* Developmental Mathematics Program and Apprenticeship Teaching

## SELECTED REFERENCES

Case, Bettye Anne, ed. *Responses to the Challenge: Keys to Improved Instruction by Teaching Assistants and Part-Time Instructors.* Washington, DC: Mathematical Association of America, 1989.

———. *You're the Professor, What Next?* Washington, DC: Mathematical Association of America, 1994.

Case, Bettye Anne, and M. Annette Blackwelder. "Doctoral Department Retention, Expectations, and Teaching Preparation." *Notices of the American Mathematical Society* 40(7)(1993):803–812.

———. "The Graduate Student Cohort." *Notices of the American Mathematical Society* 39(5)(1992):412–418.

Case, Bettye Anne, and J. Philip Huneke. "Programs of Note in Mathematics." In *Preparing Graduate Students to Teaching: A Guide to Programs that Improve Undergraduate Education and Develop Tomorrow's Faculty* (pp. 94–104). Leo M. Lambert and Stacey L. Tice, eds. Washington, DC: American Association of Higher Education, 1993.

Conference Board of the Mathematical Sciences. *Graduate Education in Transition.* Washington, DC: The Board, 1992. [Reprinted in *Notices of the American Mathematical Society* 39(5)(1992):398–402 and *Newsletter, Association for Women in Mathematics* (July/Aug. 1992):22–28.]

Foley, Gregory D. "Guidelines for the Academic Preparation of Mathematics Faculty at Two Year Colleges: Report of the Qualifications Subcommittee of the American Mathematical Association of Two-Year Colleges." Washington, DC: The Association, 1992. [Reprinted in Case 1994, A1–6.]

Joint Policy Board for Mathematics, Committee on Professional Recognition and Rewards. *Recognition and Rewards in the Mathematical Sciences.* Providence, RI: American Mathematical Society, 1994.

Madison, Bernard L., and Therese A. Hart. *A Challenge of Numbers: People in the Mathematical Sciences.* Washington, DC: National Academy Press, 1990.

Mathematical Association of America, Committee on Preparation for College Teaching. "How Should Mathematicians Prepare for College Teaching?" *Notices of the American Mathematical Society* 36(10)(Dec. 1989): 1344–1346.

National Academy of Sciences. *Educating Mathematical Scientists: Doctoral Study and the Postdoctoral Experience in the United States.* Washington, DC: The Academy, 1992.

———. *Reshaping the Graduate Education of Scientists and Engineers.* Washington, DC: The Academy, 1995.

Shell, Terry. "Do You (or Your Students) Want to Teach in a Two-Year College? Some Guidelines for Prospective Two-Year College Teachers." *Undergraduate Mathematics Education Trends* 2(4)(1990). [Reprinted in Case 1994, B–I 28–29.]

BETTYE ANNE CASE

# TEACHER PREPARATION, ELEMENTARY SCHOOL, ISSUES

Primarily center around the level of mathematical competency possessed by elementary teachers. Central to this issue are questions concerning the quantity and quality of the mathematical preparation prospective teachers receive during their college courses. Proposed changes in teaching and learning, current certification procedures, and attitudes of society and professionals toward mathematics are related issues. Traditionally, mathematics education, with its emphasis on rote memorization and prescribed algorithms, did little to promote critical thinking or problem-solving skills. Publications such as those of the National Council of Teachers of Mathematics (NCTM) (1989, 1991) and the Mathematical Association of America (MAA) (1991) address this deficiency and attempt to provide a forum for appropriate mathematical experiences for preservice elementary teachers.

Several factors are responsible for the inadequacy of the American mathematics education system. One of these is the quantity of mathematical experiences preservice elementary teachers receive in their college instruction. In many college and university programs only one or two courses in mathematics are required. These courses are often only a discussion of topics found in the elementary school curriculum or deal with mathematical content that is unrelated to elementary school mathematics. The publications of the NCTM (1991) and MAA (1991) make recommendations for improving course offerings.

If elementary teachers prepare students poorly in mathematics, it may be due in large part to deficiencies in their own preparation. It also may be due to the notion that poor comprehension and performance in mathematics are socially acceptable. Teachers may harbor negative views toward mathematics that are particularly damaging because they may be communicated, either consciously or otherwise, to impressionable young children. In order to teach mathematics effectively, elementary teachers need to possess an understanding of mathematical topics that extends beyond those they will actually be teaching. They need to develop both confidence and competence in order to teach mathematics in an appropriate manner. They need to be given opportunities to explore, in depth, more advanced mathematical topics in order to be able to assess the insights or difficulties of their students. They also must be aware of the multitude of activities, materials, and opportunities available, which will enrich their students' mathematical knowledge and vocabulary.

## CHANGES IN TEACHING AND LEARNING

The classroom envisioned in the NCTM reform documents (1989, 1991) is very different from the traditional view. In the new view, the role of the teacher will change from one of lecturer or dispenser of information to one of facilitator. Students will be viewed as learners who construct their own knowledge rather than as passive receivers of information. Thus, a strong mathematics background for elementary teachers remains essential, as is psychological and pedagogical knowledge.

It is believed that teachers tend to teach the way they were taught (Cipra 1991). The dominant mode of instruction in mathematics departments at institutions of higher learning has been the lecture method. Although this method may be effective for some, it is not appropriate as the sole teaching model for students who will be teaching in K–12 settings. Many reformers (Cipra 1991) believe that we cannot expect elementary teachers to teach in a manner that is not modeled for them at the college and university level. Aspects of teaching in a reformed classroom that may not be modeled include use of nonroutine problems as the focus of instruction without provision of procedures; adaptation to local concerns; different ways of organizing the classroom discussion; extended use of discussion, building on student ideas; and informal assessment methods to guide instructional decisions (Clarke 1997, 280).

## CERTIFICATION

Lack of consistency in the certification procedures mandated by individual states is another concern in regard to the mathematical preparation of elementary teachers. There is wide variation in state requirements for teacher certification, and this variation limits the validity of certification as a qualification measure (Blank 1988). Colleges and universities are aware of their state's certification requirements, and each offer a program that minimally satisfies the required number of courses. Elementary school teachers typically receive teaching credentials for grades K–6 or K–9 and are usually prepared as generalists without a specialization in any particular subject. Although this situation is being changed in some states that are requiring students to have a dual major (education and another academic area), it does not necessarily mean that preservice teachers will select mathematics as their second major.

## ATTITUDES

Another factor contributing to the poor mathematical preparation of elementary teachers is the attitude of university policymakers, many of whom are directly or indirectly involved in the process of planning and approving the course of study for preservice teachers. On the one hand, many university mathematics faculty are actively involved with the preparation of mathematics teachers and work in collaboration with mathematics education professors. On the other hand, a segment of the mathematics community does not believe that courses for elementary teachers warrant as much concern as courses designed to prepare students for upper-level or graduate school programs in mathematics. In many institutions, the mathematics courses for elementary teachers are taught by professors, adjunct faculty, and graduate students who have a minimal interest in the mathematical content of the courses or have little experience in or knowledge of the structure and workings of an elementary classroom. They unfortunately view the teaching of such a course as merely service to the mathematics department (Garfunkel and Young 1992). These courses often delve into an understanding of the nature of mathematics, however, demonstrate the interrelationships of different branches of mathematics, and relate how mathematics is connected to other academic disciplines. A more qualified and experienced instructor will, in general, have a much richer perspective and more personal insights to offer the prospective teacher. Relegating these courses to disinterested instructors deprives the prospective teacher of this expertise.

## CONCLUSION

The nature of teaching is complex and demanding, as is the task of preparing teachers. The direction of most teacher education programs is set by two national organizations—the National Association of State Directors of Teacher Education and Certification (NASDTEC) and the National Council for Accreditation of Teacher Education (NCATE)—as well as individual state governments. To resolve issues and effect changes in mathematics education, these organizations must work together with teacher educators, educational policy makers, mathematics supervisors, university mathematicians, mathematics education and education professors, and classroom teachers.

*See also* Curriculum Trends, Elementary Level; Evaluation of Instruction, Issues; Evaluation of Instruction, Overview; Goals of Mathematics Instruction; History of Mathematics Education in the United States, Overview; Psychology of Learning and Instruction, Overview; Teacher Development, Elementary School

## SELECTED REFERENCES

Austin, Rick. "Mathematics Teaching and Teachers in the Year 2000." *The Clearing House* 62(1)(1988):23–25.

Blank, Rolf. "Improving Research on Teacher Quality in Science and Mathematics: Report of a Symposium of Scientists, Educators, and Researchers." *Journal of Research in Science Teaching* 25(3)(1988):217–224.

Campbell, Patricia F., and Martin L. Johnson. "How Primary Students Think and Learn." In *Seventy-Five Years of Progress: Prospects for School Mathematics* (pp. 21–42). Iris M. Carl (ed.). Reston, VA: National Council of Teachers of Mathematics, 1995.

Cipra, Barry. *On the Mathematical Preparation of Elementary School Teachers: Report of a Conference Held at the University of Chicago.* Washington, DC: National Science Foundation, 1991.

Clarke, Doug M. "The Changing Role of the Mathematics Teacher." *Journal for Research in Mathematics Education* 28(3)(1997):278–308.

Garfunkel, Solomon A., and Gail S. Young. *In the Beginning: Mathematical Preparation for Elementary School Teachers.* Lexington, MA: COMAP, 1992, p. 3.

Mathematical Association of America. *A Call for Change: Recommendations for the Mathematical Preparation of Teachers of Mathematics.* Washington, DC: The Association, 1991.

National Council of Teachers of Mathematics. *Curriculum and Evaluation Standards for School Mathematics.* Reston, VA: The Council, 1989.

———. *Principles and Standards for School Mathematics.* Reston, VA: The Council, 2000.

———. *Professional Standards for Teaching Mathematics.* Reston, VA: The Council, 1991.

BEVERLY J. FERRUCCI

# TEACHER PREPARATION, MIDDLE SCHOOL

Comprises the knowledge and experience that provides the best preparation for teaching a modern curriculum. Teaching mathematics is a complex process and effective teachers must know mathematics, know how middle school students learn mathematics, and be able to use a variety of instructional strategies and related resources. Middle school mathematics teachers must understand the specific concepts and procedures found in the middle school curriculum. Middle school mathematics teachers need to see the curriculum as an integrated whole, yet recognize that many topics—geometry, measurement, probability, statistics, functions, and estimation—are important in themselves. Further, these specific topics provide contexts in which students can apply the computational skills they have developed earlier. Beyond this content knowledge, teachers need to investigate the "why" and "what if" questions in mathematics in order to be able to emphasize such investigations with students. For example, teachers need to know far more about division of fractions than the rule "invert and multiply." They need to investigate why this algorithm "works," what concrete and picture models can be used to develop meaning for the procedure, how to interpret the answer from such a calculation (especially the remainder), and alternative strategies/algorithms for dividing fractions.

Communicating mathematics involves understanding of patterns, definitions, conjectures, relationships, and generalizations. Recognition of how mathematics topics "fit together" and how mathematics relates to other subject areas will help teachers develop a more comprehensive understanding of middle school mathematics. Perhaps these teachers need to conceptualize the curriculum in terms of the "big ideas" in middle school mathematics. In *On the Shoulders of Giants*, five possible strands of mathematics are suggested: dimension, quantity, uncertainty, shape, and change (Steen 1990). Middle school mathematics teachers need to have opportunities to explore the breadth, depth, and richness of mathematics.

Prospective middle school mathematics teachers should be aware of current recommendations concerning curriculum and pedagogical reform (National Council of Teachers of Mathematics (NCTM) 1989, 1991) and research related to such recommendations (U.S. National Research Center 1996). Further, they should have opportunities to investigate a variety of curriculum materials, including reform materials such as those developed by National Science Foundation (NSF)-funded projects *(Mathematics in Context, Middle School Mathematics through Applications, Seeing and Thinking Mathematically, Six through Eight Mathematics, Connected Mathematics)*. Teachers' views of mathematics will influence the way they teach mathematics. "Their conceptions of mathematics shape their choice of worthwhile mathematical tasks, the kinds of learning environments they create, and the discourse in their classrooms" (NCTM 1991, 132).

Middle school mathematics teachers need to have an understanding of current learning theories and research and how these relate to the learning and teaching of mathematics. Understanding that learning requires the students to be actively engaged in the learning process and that their students construct meaning for new ideas by connecting these with what they already know (Confrey 1990) will in-

fluence how teachers teach. Many in the mathematics education community suggest that this constructivist view of learning is central to a teacher's understanding of learning (NCTM 1989, 1991; Owens 1993). Further, teachers should know what research has to say about issues such as the role of calculators and computers in learning mathematics, the role of language in learning mathematics, and the development of key mathematics ideas (i.e., proportional reasoning, spatial skills, etc.). Finally, teachers must be knowledgeable about a variety of instructional approaches and resources. Teachers need to be able to create relevant, worthwhile mathematical tasks for their students. "Students' opportunities to learn mathematics are a function of the setting and the kinds of tasks and discourse in which they participate. What students learn, about particular concepts and procedures as well as about thinking mathematically, depends on the ways in which they engage in mathematical activity in the classroom" (NCTM 1991, 21). We want teachers who are able to make use of a variety of models, cooperative learning, simulations, projects, guided discovery, questioning techniques, diagnosis and remediation, and technology in developing instructional activities.

When prospective teachers engage in these kinds of activities as learners, they are more likely to better understand how to use them with their students. Investigating areas of rectangles, all with a fixed perimeter, can help these teachers focus on relationships between area and perimeter, "maximizing" the area for a fixed perimeter, and appropriate models (geoboards, grid paper, functions, graphs, etc.) for representing this problem. Later, many of these same ideas can be used to investigate relationships between volumes and surface areas of three-dimensional figures. Likewise, carrying out a variety of hands-on probability experiments can help teachers investigate the meaning and applications of experimental versus theoretical probability, sample spaces, probability trees, compound probability, geometric probability, and expected value. The primary goal for middle school mathematics teachers, then, is to create learning environments that are conducive to learning. "Students' learning of mathematics is enhanced in a learning environment that is built as a community of people collaborating to make sense of mathematical ideas. It is a key function of the teacher to develop and nurture students' abilities to learn with and from others . . ." (NCTM 1991, 58).

Preparing mathematics teachers with the knowledge and experiences described here requires a collaborative effort involving schools and school systems, colleges and universities, professional organizations, and policy makers. Schools and teacher training institutions should strive to create programs that are designed specifically for preparing middle school mathematics teachers. In *A Call for Change: Recommendations for the Mathematical Preparation of Teachers of Mathematics* (Mathematical Association of America 1991), a program for middle school mathematics teachers that includes specific core courses is described. Each of these courses is to be designed for middle school teachers and should make use of manipulative materials, models, and technological tools and emphasize active learning. A number of similarly designed elective courses in such a program could be used to provide opportunities for students to be involved in practicum (classroom observations, tutoring, student teaching, apprentice teaching) experiences. Schools could contribute to such a program by making practicing teachers available to serve as mentors for preservice teachers and by working with colleges and universities to schedule practicum opportunities.

Colleges and universities already have or can begin now to develop programs of study that will adequately prepare middle school mathematics teachers. These programs should be relevant enough to reflect current needs and flexible enough to address tomorrow's challenges. At Central Missouri State University, for example, students are involved with courses that are designed specifically for the preparation of middle school mathematics teachers. The students in this program develop a broad mathematics background through courses in precalculus, calculus, geometry, modern algebra, problem solving, teaching strategies/practicum, and educational computing. A probability and statistics course is being developed for this program. Throughout the program, students make use of a wide variety of physical models, calculators, and computers and engage often in individual and small group investigative activities and projects. In each course students seek to relate the content and instructional models they are encountering to the curriculum and appropriate instructional strategies for the middle school. Examples from the middle school mathematics curriculum are used in each course, and students work extensively with reform curriculum materials such as those found in the NSF-funded curriculum projects. Students also are given opportunities to interact with practicing teachers and middle school students through various practicum experiences as well as participate in local, state, and national professional organizations. The experiences students have in such a program will greatly influence their own view of learning and teaching mathematics. Mathematics educators are also creating learning environments for these preservice teachers. "The experiences teachers have in these learning environments form expectations—implicitly or explicitly—of what constitutes good mathematics instruction. Such experiences provide the core from which teachers will eventually build learning environments for their own students" (NCTM 1991, 128).

*See also* Teacher Development, Secondary School; Teacher Preparation, Elementary School, Issues; Teacher Preparation, Secondary School

## SELECTED REFERENCES

Confrey, Jere. "What Constructivism Implies for Teaching." In *Constructivist Views on the Teaching and Learning of Mathematics* (pp. 107–122). Robert B. Davis, Carolyn A. Maher, and Nel Noddings (eds.). Reston, VA: National Council of Teachers of Mathematics, 1990.

*Connected Mathematics Project.* University of North Carolina-Chapel Hill. Chapel Hill, NC.

Mathematical Association of America. *A Call for Change: Recommendations for the Mathematical Preparation of Teachers of Mathematics.* Washington, DC: The Association, 1991.

*Mathematics in Context Project.* Wisconsin Center for Education Research. Madison, WI.

*Middle School Mathematics through Applications Project.* Institute for Research on Learning. Palo Alto, CA.

National Council of Teachers of Mathematics. *An Agenda for Action: Recommendations for School Mathematics of the 1980s.* Reston VA: The Council, 1980.

———. *Curriculum and Evaluation Standards for School Mathematics.* Reston, VA: The Council, 1989.

———. *Principles and Standards for School Mathematics.* Reston, VA: The Council, 2000.

———. *Professional Standards for Teaching Mathematics.* Reston, VA.: The Council, 1991.

National Research Council. *Everybody Counts: A Report to the Nation on the Future of Mathematics Education.* Washington, DC: National Academy Press, 1989.

———. *Reshaping School Mathematics: A Philosophy and Framework for Curriculum.* Washington, DC: National Academy Press, 1990.

Owens, Douglas T., ed. *Research Ideas for the Classroom.* Vol. 2. *Middle Grades Mathematics.* New York: Macmillan, 1993. Reston, VA: The Council, 1993.

Rasch, Kathe. "The Imperative for Quality Middle School Mathematics Curriculum and Instruction." *Midpoints* 4(Spring 1994):1–16.

*Seeing and Thinking Mathematically Project.* Educational Development Center. Newton, MA.

Silvey, Linda, ed. *Mathematics for the Middle Grades (5–9).* 1982 Yearbook. Reston, VA: National Council of Teachers of Mathematics, 1982.

*Six through Eight Mathematics Project.* University of Montana. Missoula, MT.

Steen, Lynn Arthur, ed. *On the Shoulders of Giants— New Approaches to Numeracy.* Washington, DC: National Academy Press, 1990.

U.S. National Research Center. *Third International Mathematics and Science Study.* Report No. 7. East Lansing, MI: The Center, 1996.

TERRY GOODMAN

# TEACHER PREPARATION, SECONDARY SCHOOL

A system that has evolved from a master–apprentice system early in the nineteenth century, through a period dominated by two-year normal schools, to the current, degree-oriented model. In the last century, as states sought to assure the quality of public educational systems, the education of prospective teachers emerged as a public policy issue. Institutions included "pedagogy" or "didactics" with other "practical arts." Today, teacher preparation programs are offered nationwide by hundreds of institutions at both the baccalaureate and master's degree levels. The undergraduate preparation of secondary mathematics teachers typically culminates in a liberal arts degree based on a concentration in mathematics and a program of study in professional education. If teacher preparation occurs at the master's degree level, an equivalent undergraduate mathematical preparation must have been achieved earlier. Many states have mandated the scope of study in mathematics required of prospective teachers, typically within an undergraduate major. It is common for states to require prospective teachers to pass a battery of norm-referenced tests covering "general education" and professional knowledge.

Guidelines for the mathematical content of the preparation of secondary-level mathematics teachers have been developed by several national organizations, including the National Council of Teachers of Mathematics (NCTM), the Committee on the Undergraduate Program in Mathematics (CUPM) of the Mathematical Association of America (MAA), and the Association of State Supervisors of Mathematics (ASSM). Broadly, they prescribe at least thirty semester hours in mathematics, an undergraduate major, and may specify a calculus sequence, probability and statistics, and such other courses as abstract algebra, discrete mathematics, geometry, the history of mathematics, linear algebra, and number theory. The *Guidelines for the Preparation of Teachers of*

*Mathematics* (NCTM 1993) have been incorporated into the requirements of the National Commission on Accreditation in Teacher Education (NCATE). To achieve this organization's accreditation in mathematics education, colleges must meet the NCTM *Guidelines.*

NCTM's *Guidelines* address the preparation of teachers of mathematics at three levels: K–4, 5–8, and 7–12.

. . . [T]he following themes, as suggested in the *Curriculum and Evaluation Standards for School Mathematics,* should be prominent in all mathematics-teacher preparation programs.

Problem solving in mathematics

Communication in mathematics

Reasoning in mathematics

Mathematical connections (both within the discipline and to its uses in the world around us).

In addition, mathematical experiences for all teachers should foster—

the disposition to do mathematics;

the confidence to learn mathematics independently;

the development and application of mathematical language and symbolism;

a view of mathematics as a study of patterns and relationships;

perspectives on the nature of mathematics through a historical and cultural approach.

NCTM's *Guidelines* make significant, complementary recommendations concerning the pedagogical component of teacher education, including field-based experiences.

The most prevalent model of professional preparation includes campus-based study in adolescent development, the social-historical foundations of education, and methods of teaching mathematics. These early courses may include a field-based experience; the prospective teachers become participant-observers in local schools, thus strengthening their understanding and mastery of professional practices and preparing them for their culminating professional courses, supervised student teaching, extending over one and sometimes two semesters. At this stage, prospective teachers work daily with practicing ("cooperating") teachers, their mentors, chosen for their professional proficiency and ability to work effectively with prospective teachers. College supervisors make periodic supervisory visits as the student (or "cadet") teachers assume increasing responsibility for classes. A weekly on-campus or on-site student-teaching seminar is usual, during which all mathematics student teachers meet with campus-based supervisors. In some institutions, a "professional semester" is required to complete student teaching, during which the participants spend a full college semester at a secondary school, take no other courses on campus, and spend full days in the schools.

Reform of secondary education has been a significant national concern for the last hundred years, beginning with the *Report of the Committee of Ten on Secondary School Studies* in 1893. When the Soviet Union launched the space satellite *Sputnik* in 1957, there were loud demands for improvement in academic programs, particularly in mathematics, science, and foreign languages. Many influential reports have appeared since then, including *A Nation at Risk: The Imperative for Educational Reform* (National Commission on Excellence in Education 1983), several international comparisons of student achievement that reflected adversely on American mathematics education, *Curriculum and Evaluation Standards for School Mathematics* (NCTM 1989), *Professional Standards for Teaching Mathematics* (NCTM 1991), *Assessment Standards for Teaching Mathematics* (NCTM 1995), and *Principles and Standards for School Mathematics* (NCTM 2000). Each of these has had material implications for teacher preparation. *Everybody Counts: A Report to the Nation on the Future of Mathematics Education* identified a number of concerns affecting teacher preparation programs including:

- lack of a constructivist approach to student learning,
- teachers' "insecurity" and "rigidity" with respect to their personal mastery of the subject matter,
- teachers' flawed experiences in studying mathematics themselves and the consequent negative effects on their own teaching, and
- teachers' excessive reliance on text books. (National Research Council 1989, 60–73)

Because there is relatively little research on teacher education (Cooney 1994, 613), teacher preparation programs differ from institution to institution. Debates about what is the best way to prepare teachers cannot be resolved on the basis of definitive research findings.

No teacher preparation program can equip an educator for a lifetime of successful practice, nor can preservice education anticipate the needs of an experienced professional years or decades after graduation. Ultimately, successful programs have graduates who are committed to being lifelong learners and are professionally responsive to such influences and issues as

- the changing demographics of students and our nation;
- emergent research in mathematics education,
- the changing role of federal and state governments in support of precollege mathematics education, and
- developing technologies such as (graphing) or symbol-processing calculators and computer programs based on neural networks or artificial intelligence that transcend earlier "drill-and-practice" approaches.

Among the important perennial and unresolved issues surrounding mathematics teacher preparation are

- definition and assessment of "excellence" in teaching mathematics, modulated for beginning teachers, career professionals, and teacher-leaders, whether or not in formal roles;
- methods of recognizing and rewarding excellent professional performance in mathematics education;
- formation and continuous updating of an accessible, reliable, and comprehensive knowledge base in mathematics education;
- identification and dissemination of methods and curricular materials that serve the needs of students with special needs, for example, limited English proficiency, handicapping conditions, cultural or social backgrounds different from those of "majority" populations; and
- recruitment and retention of mathematics educators from traditionally underrepresented groups.

Although teacher preparation programs can do much to enhance the quality of the education they offer future teachers of mathematics (and through them, the education offered to generations of mathematics students), they are nonetheless subject to significant influences over which they—of necessity—have little control. For example, the compensation offered to beginning and experienced teachers, the value society attaches to education and educators, the extent of governmental support for the spectrum of mathematics education activities, and the competition with other professions to recruit mathematically trained and dedicated individuals are all beyond the control of teacher education programs. Nonetheless, these and many other "external" influences exert strong leverage on the preparation of prospective teachers of mathematics.

*See also* Teacher Development, Secondary School

## SELECTED REFERENCES

Association of State Supervisors of Mathematics. *The Coordinated Implementation of National and State-by-State Reform in School Mathematics: The Report of the First Annual ASSM Planning Conference.* Chicago, IL: 1989.

Brown, Catherine, and Hilda Borko. "Becoming a Mathematics Teacher." In *Handbook of Research on Mathematics Teaching and Learning* (pp. 209–239). Douglas A. Grouws (ed.). New York: Macmillan, 1992.

Cooney, Thomas J. "Research and Teacher Education: In Search of Common Ground." *Journal for Research in Mathematics Education* 25(6)(1994):608–636.

Mathematical Association of America. *A Call for Change: Recommendations for the Mathematical Preparation of Teachers.* Washington, DC: The Association, 1991.

National Board for Professional Teaching Standards. *What Teachers Should Know and Be Able to Do.* Detroit, MI: 1994.

National Commission on Excellence in Education. *A Nation at Risk: The Imperative for Educational Reform.* Washington, DC: Government Printing Office, 1983.

National Council of Teachers of Mathematics. *Assessment Standards for Teaching Mathematics.* Reston, VA: The Council, 1995.

———. *Curriculum and Evaluation Standards for School Mathematics.* Reston, VA: The Council, 1989.

———. *Guidelines for the Preparation of Teachers of Mathematics.* Reston, VA: The Council, 1993.

———. *Principles and Standards for School Mathematics.* Reston, VA: The Council, 2000.

———. *Professional Standards for Teaching Mathematics.* Reston, VA: The Council, 1991.

National Education Association. *Report of the Committee of Ten on Secondary School Studies with Reports of the Conferences Arranged by the Committees.* Washington, DC: Government Printing Office, 1893.

National Research Council. *Counting on You: Actions Supporting Mathematics Teaching Standards.* Washington, DC: National Academy Press, 1991.

———. *Everybody Counts: A Report to the Nation on the Future of Mathematics Education.* Washington, DC: National Academy Press, 1989.

———. *Reshaping School Mathematics: A Philosophy and Framework for Curriculum.* Washington, DC: National Academy Press, 1990.

WILLIAM J. MCKEOUGH

# TEACHER PREPARATION, SECONDARY SCHOOL, REFORM PROJECT IN CALIFORNIA

Includes state standards for the acceptance of mathematics teacher credentialing programs. Much time and energy has been spent in the past two decades trying to change the mathematics education of our children. Our nation has moved from an industrial economy to an informational one, vastly changing the current and future needs of our society. More and more jobs call for us to work with our minds rather than with our hands, to solve problems rather than only to follow directions, and to communicate our mathematical thinking to others.

Few people disagree with the need for change in the mathematics curriculum, and many of our nation's K–12 schools have begun to implement new programs. The effort to revise mathematics education in our schools is immense, but it simply cannot be avoided. As a nation we must invest whatever it takes to prepare our children for the future. A blueprint for such curricular changes has been charted in many mathematics reform documents, such as *Everybody Counts* (National Research Council (NRC) 1989), *Reshaping School Mathematics: A Philosophy and Framework for Curriculum* (NRC 1990), and *Curriculum and Evaluation Standards for School Mathematics* (National Council of Teachers of Mathematics (NCTM) 1989).

All of the efforts to accomplish new goals for mathematics education, however, will be lost and remain relatively unsuccessful if we do not, at the same time, address the education of future mathematics teachers. Unfortunately, many of our prospective secondary mathematics teachers now receive a preparation at our institutions of higher education that is antiquated in both content and pedagogy. *Moving Beyond Myths: Revitalizing Undergraduate Mathematics* (NRC 1991, 17) reports,

The way mathematics is taught at most colleges—by lectures—has changed little over the past 300 years, despite mounting evidence that the lecture-recitation method works well only for a relatively small proportion of students. . . . Moreover, the syllabi of many undergraduate mathematics courses and the template-style textbooks are detached from the life experiences of students and are seen by many students as irrelevant.

The authors go on to report,

Since virtually all mathematics teachers received the majority of their mathematics instruction in traditional lecture courses, it is not surprising that lecturing has continued to be the most common way that mathematics is taught, both in high school and in college (p. 24).

Leading professional organizations like the NCTM and the Mathematical Association of America (MAA) are trying to play a leadership role in establishing a vision and setting a course of direction for the preparation of mathematics teachers. They have published resources that could (and should) influence programs at colleges and universities: *Professional Standards for Teaching Mathematics* (NCTM 1991) and *A Call for Change: Recommendations for the Mathematical Preparation of Teachers of Mathematics* (MAA 1991). Each is only a set of recommendations, however, that can either be accepted or rejected.

The State of California also recognizes the need for change in teacher preparation programs. So as to not rely on the happenstance of institutions embracing the recommendations just cited, in 1992 the Commission on Teacher Credentialing embarked upon the implementation of new program quality standards that must be met by any institution in the state that wishes to have an approved mathematics teacher credentialing program.

There are seventeen standards for the curriculum and content of subject matter programs in mathematics:

| | |
|---|---|
| Standard 1: | Program Philosophy and Purpose |
| Standard 2: | Mathematics as Problem Solving |
| Standard 3: | Mathematics as Communication |
| Standard 4: | Mathematics as Reasoning |
| Standard 5: | Mathematical Connections |
| Standard 6: | Mathematics with the Use of Technology |
| Standard 7: | Algebra |
| Standard 8: | Geometry |
| Standard 9: | Functions and Calculus |
| Standard 10: | Number Theory |
| Standard 11: | Mathematical Systems |
| Standard 12: | Statistics and Probability |
| Standard 13: | Discrete Mathematics |
| Standard 14: | History of Mathematics |
| Standard 15: | Equity and Diversity in the Program |
| Standard 16: | Delivery of Instruction in the Program |
| Standard 17: | Field Experiences in the Program (Commission on Teacher Credentialing 1992) |

The first six standards parallel very closely the beginning standards in the *Curriculum and Evaluation Standards for School Mathematics* (NCTM 1989) designed for K–12 programs. When reviewing institutional applications for mathematics credential programs, the commission requires that these standards (or rather, evidence of them) must be visible in all of the content standards (Standards 7–14) that follow. The last three standards relate to essential features of the overall curriculum; they require that mathematics be put in contexts in which students can understand the mathematics they are learning as well as ideas about how people learn mathematics.

Developing effective mathematics teacher preparation programs requires focusing on these beginning and ending standards. Sufficient mathematical content has always been in credential programs (although there is a need to modernize the content through the inclusion of topics like discrete mathematics). What is appropriately being carefully scrutinized now, however, is the delivery of instruction and connections between content areas. Prospective teachers must be given opportunities to learn mathematics in a style that is consistent with the ways in which they will be expected to teach.

One can get a better sense of the difference in expectations if we look more closely at individual standards. For example, in reviewing Standard 16, Delivery of Instruction in the Program, the commission judges whether a program meets this standard by considering the extent to which the program

- emphasizes learning to understand mathematics, not just following rules and procedures;
- provides opportunities for students to experience a variety of instructional formats such as small collaborative groups, individual explorations, peer instruction, and whole-class discussions facilitated by the students;
- provides opportunities for students to be actively involved in learning mathematics through tactile, visual, and auditory modalities;
- provides opportunities for students to develop and reinforce mathematical concepts and skills through open-ended situational lessons;
- provides opportunities for students to understand ways in which assessment can be con-

nected with instruction through use of portfolios, group and individual performance tasks, observations, and interviews. (Commission on Teacher Credentialing 1992)

The teaching of mathematics is receiving attention in the entire mathematical community. Professional organizations, such as the MAA, American Mathematical Society, Society of Industrial and Applied Mathematics, and the NCTM, advocate that mathematicians learn about learning; mathematicians need to think as deeply about how to teach as about what to teach (National Research Council 1991). This is a positive step; many university mathematicians find themselves not sufficiently knowledgeable in this area, but not without interest. "Since teachers teach much as they were taught, university courses for prospective teachers must exemplify the highest standards for instruction. . . . Prospective teachers should learn mathematics in a manner that encourages active engagement with mathematical ideas" (NRC 1989, 65–66).

In short, modern teacher preparation programs should focus on providing prospective teachers opportunities to examine and revise their assumptions about the nature of mathematics, how it should be taught, and how students learn mathematics. Further, these experiences should be embedded in the undergraduate mathematics courses, not delayed until fifth-year student-teaching experiences or specific mathematics pedagogy courses.

The following are factors to consider for each course in any preparation program:

1. University mathematicians need to model good mathematics teaching: to pose worthwhile mathematical tasks, create mathematical learning environments where risk taking is encouraged, and support the understanding that mathematics is an ongoing activity that uses a wide variety of tools and techniques, rarely insisting on only one way that things can be done.
2. Prospective teachers need to study mathematical concepts and the connections between them, reason mathematically, solve both traditional and nonroutine problems, communicate mathematical ideas both orally and in writing; understand the role of mathematics in culture and society, and be able to use the latest technology.
3. Finally, prospective teachers must have the opportunity to see, through examples, and have the chance to practice themselves how to orchestrate mathematical discourse. They must see how mathematical goals can be accomplished by

working both independently and collaboratively, how evaluation is a tool to determine what students know and how to modify teaching accordingly, how to foster a sense of a mathematical community, and how the needs of different sections of the student population can be reached so that mathematical power is an obtainable goal for all students.

*See also* Teacher Preparation, Secondary School

## SELECTED REFERENCES

Commission on Teacher Credentialing. *Mathematics Teacher Preparation in California: Standards of Quality and Effectiveness for Subject Matter Programs.* Sacramento, CA: The Commission, 1992.

Intersegmental Coordinating Council. *K–12 School Reform: Implications and Responsibilities for Higher Education.* Sacramento, CA: The Council, 1993.

Mathematical Association of America. *A Call for Change: Recommendations for the Mathematical Preparation of Teachers of Mathematics.* Washington, DC: The Association, 1991.

National Council of Teachers of Mathematics. *Curriculum and Evaluation Standards for School Mathematics.* Reston, VA: The Council, 1989.

———. *Principles and Standards for School Mathematics.* Reston, VA: The Council, 2000.

———. *Professional Standards for Teaching Mathematics.* Reston, VA: The Council, 1991.

National Research Council. *Everybody Counts: A Report to the Nation on the Future of Mathematics Education.* Washington, DC: National Academy Press, 1989.

———. *Moving Beyond Myths: Revitalizing Undergraduate Mathematics.* Washington, DC: National Academy Press, 1991.

———. *Reshaping School Mathematics: A Philosophy and Framework for Curriculum.* Washington, DC: National Academy Press, 1990.

WILLIAM FISHER

# TEACHER PREPARATION IN STATISTICS, ISSUES

Center around both the lack of statistical knowledge and specialized pedagogical training. In 1982, the Cockcroft Committee looked at United Kingdom (U.K.) mathematics education and reported that "some teachers, especially those who completed their degree courses some years ago, do not have sufficient knowledge of probability and statistics to teach in these areas effectively" (Cockcroft 1982). Furthermore, few teachers, including those whose degrees or other courses had contained statistics, had received any training in how to teach statistics. These observations were confirmed in a survey of 349 teachers of statistics, either within mathematics or within an applied discipline or as a subject in its own right (Hawkins, Jolliffe, and Glickman 1992; Pereira-Mendoza 1993). Approximately 20% of respondents had not studied statistics beyond secondary school level, and 6% claimed that they had had *no* formal statistical training. Those with lower levels of statistical training tended to be older, and more experienced, teachers. Time that is served teaching a subject cannot compensate, however, for the lack of relevant training in that subject.

With the exception of those who were teaching applied disciplines, *applications* and *scientific methodology* seemed to pose more problems for teachers than *statistical techniques*. If statistics were seen to be a mainstream subject, however, its teachers would need a stronger theoretical background. Many respondents stated that *scientific methods* were "not applicable" to them, despite their awareness of the need for more applied, project-based statistics teaching. Scientific methodology, however, is not only concerned with how data are collected. It also must be taken into account when data are evaluated and processed. If teachers think that this is "not applicable," how will their students acquire appropriate statistical understanding? Methods and materials are needed that will incorporate this aspect of statistics into all areas of statistical education and convince teachers and examination boards of its relevance.

Proceedings of successive International Conferences on Teaching Statistics (Grey et al. 1983; Davidson and Swift 1986; Vere-Jones 1991; ISI/INSEA 1994), scientific meetings of the International Association for Statistical Education (IASE), the International Statistical Institute (ISI) conferences and roundtables, and International Conferences on Mathematical Education (ICMEs) indicate that the problems are not unique to the United Kingdom. It is now common for developed countries to have "statistics for all" policies, but typically they lack the infrastructures for meaningfully implementing associated curriculum developments.

It is even harder to introduce statistics into the curriculum of developing countries. Apart from the lack of resources and the difficulties for teachers who wish to gain access to those which exist, in many such countries the mathematics curriculum is often "locked into place, by ill-informed bureaucracy, by stifling effects of competitive entrance examinations, and by the overriding consideration that changing

curricula must take second place to extending the availability of education to a larger proportion of the population" (Hawkins 1990, vi–vii). The International Statistical Institute, with sponsorship from organizations like UNESCO, is trying to alleviate teachers' difficulties in such countries, but there is still a long way to go, particularly where statistical literacy for all is seen by a government to be "politically inconvenient."

National professional statisticians' organizations also play an important role. For example, the U.K. Royal Statistical Society runs teachers' workshops and conferences, facilitates teacher networks, has associate school membership, provides teacher-training bursaries and, like the American Statistical Association, has established a National Centre for Statistical Education, based at the University of Nottingham.

At both pre- and inservice levels of teacher training, statistical education suffers, either by being part of mathematics education, where pressures of time (and inclination) marginalize it, or by being in competition with mathematics education. With more training resources and an increased demand for teachers of core curriculum subjects, students tend to opt for courses in mathematics or science education. The failure to recruit enough students, and hence to justify sufficient numbers of academic staff, impedes the development of a pedagogic discipline or research basis for statistical education in its own right. It is difficult, therefore, for students who want to specialize in statistical education to find enough relevant courses, resources, or training support. Research developments in teaching and learning statistics and probability are accruing, and gradually this circular situation is easing. Much needs to be done, however, to assure that the statistical trainers are themselves good exponents of statistical education. Few university-level statistics lecturers have had any preparation in teaching methods, and some statistical education trainers have had little training in statistics!

The ISI's Round-Table Conference (Hawkins 1990) recommended that the crisis in statistical education should be tackled first at the inservice level. Lack of finances or time allowances for inservice release, however, makes this difficult for teachers. National Statistics Competitions, like those run in the U.K., Hong Kong, and the United States, provide opportunities for giving teachers informal training in statistical project work. The Sheffield Hallam University's Postgraduate Diploma in Statistics and Statistical Education is a distance learning program, using correspondence tuition and summer schools to enhance teachers' statistical expertise as well as their awareness of research, curriculum, and pedagogic issues. At the fourth International Conference on Teaching Statistics, a number of other distance learning approaches to training were reported, including multimedia developments (ISI/INSEA 1994).

Teacher networks, such as *The Statistics Teacher Network* in the United States, also help. Funded by the National Council for Teachers of Mathematics, it produces a regular newsletter, providing a two-way communication channel between statistics teachers, researchers, and policy makers. Another useful network is the International Study Group for Research on Learning Probability and Statistics Concepts. This group has a tradition of informal exchanges concerning members' activities, research, and relevant news items through a newsletter, edited by Joan Garfield. The U.K.-produced journal *Teaching Statistics*, incorporating *IASE Matters*, also provides a communication channel for teachers. IASE publishes the *IASE Review* and the education section of the *International Statistical Newsletter*, both of which are distributed free to IASE members, along with papers in the *International Statistical Review*. With the increasing availability of computer resources in schools and colleges, the relative isolation of most statistics teachers could soon be remedied. Access to inservice training can also be improved by further exploitation of e-mail, bulletin boards like EDSTAT-L, and the electronic *Journal of Statistics Education*, with its archive of resources and software for statistics teachers.

A U.K. inservice initiative, recommended by the Cockcroft Committee (1982) and the Schools Council Project on Statistical Education, is that there should be a statistical coordinator in each school (or locality) who could promote inservice statistical education within that context. Although this has not been widely implemented, a coordinator's handbook is available (Holmes and Rouncefield 1990).

One style of teacher preparation involves putting teachers into the role of their students. At ICME-6, Shaughnessy (1988) described two Oregon projects, one for inservice and one for preservice teachers, to demonstrate the potential of computer simulations in the teaching of probability and statistics concepts. It was hoped that teachers would be encouraged to use the same approach with their own students.

In contrast, Hans Schupp (Hawkins 1990, 270–276) described an action research program, giv-

ing teachers responsibility for the development of curricula, teaching methods, and materials. This long-term project had the merit of stimulating the teachers' commitment. Subsequently, they became trainers, passing on their expertise to colleagues. A similar *cascade* model of inservice training, using teachers as trainers, was used in the U.S. Quantitative Literacy Project (Hawkins 1990, 219–227).

Changes in assessment methods and content can motivate and provide inservice training by changing people's attitudes toward statistics. What is taught and what is assessed, however, are interdependent. One is unlikely to change without the other. Some methods of assessment can themselves provide a context for inservice training. Ken Sharpe (Hawkins 1990, 283) outlined a scheme for teacher *consensus moderation* of students' statistical coursework, whereby teachers from several schools meet to assess their students' work. As a spinoff from this, through hearing about what is being achieved in other schools, the teachers themselves learned more about practical statistical work. Sharpe maintained that this was also "a powerful force for changing teachers' expectations."

### Useful Contacts

American Statistical Association, 1429 Duke Street, Alexandria, VA 22314-3402, USA (e-mail: asainfo@asa.mhs.compuserve.com)

EDSTAT-L, send one-line e-mail message to listserv@jse.stat.ncsu.edu—Subscribe edstat-1 Firstname Lastname.

IASE, International Association for Statistical Education, 428 Prinses Beatrixlaan, 2270 AZ Voorburg, Netherlands (e-mail: isi@cs.vu.nl)

*International Study Group on Learning Probability and Statistics Concepts.* Joan Garfield, Department of Educational Psychology, 210 Burton Hall, 178 Pillsbury Drive S.E., University of Minnesota, Minneapolis, MN 55455, USA. (e-mail: jbg@vx.cis.umn.edu)

*Journal of Statistics Education,* send two-line e-mail message to archive@jse.stat.ncsu.edu—Send index, send access.methods, or send post mail to Journal of Statistics Education, Department of Statistics, Box 8203, North Carolina State University, Raleigh, NC 27695-8203, USA

Royal Statistical Society, 12 Errol Street, London, England, EC1Y 8LX, U.K. (e-mail: rss@bristol/ac/uk)

*Statistics Teacher Network.* Jerry Moreno, Mathematics Department, John Carroll University, University Heights, OH 44118, USA (e-mail: moreno@jcvaxa.jcu.edu)

*Teaching Statistics.* Centre for Statistical Education, Department of Probability and Statistics, University of Sheffield, Sheffield, England, S3 7RH. (e-mail: p.holmes@sheffield.ac.uk)

## SELECTED REFERENCES

Brunelli, Lina, and Giuseppe Cicchitelli, eds. *Proceedings of the First Scientific Meeting of the International Association for Statistical Education.* Perugia, Italy: Università di Perugia, 1994.

Cockcroft, Wilfred H. *Mathematics Counts.* London, England: HMSO, 1982.

Davidson, R., and J. Swift, eds. *Proceedings of the Second International Conference on Teaching Statistics.* Victoria, BC, Canada: University of Victoria, 1986.

Grey, D. R., Peter Holmes, V. Barnett, and G. M. Constable, eds. *Proceedings of the First International Conference on Teaching Statistics.* Sheffield, England: University of Sheffield, Teaching Statistics Trust, 1983.

Hawkins, Anne, ed. *Training Teachers to Teach Statistics.* Voorburg, Netherlands: International Statistical Institute, 1990.

Hawkins, Anne, Flavia Jolliffe, and Leslie Glickman. *Teaching Statistical Concepts.* London, England: Longman, 1992.

Holmes, Peter, and Mary Rouncefield. *From Cooperation to Coordination (A file for coordinating school statistics).* Sheffield, England: Centre for Statistical Education, University of Sheffield, 1990.

ISI/INSEA. *Proceedings of the Fourth International Conference on Teaching Statistics.* Voorburg, Netherlands: International Statistical Institute, 1994; Rabat, Morocco: National Institute of Statistics and Applied Economics, 1994.

Pereira-Mendoza, Lionel, ed. *Introducing Data Analysis in the Schools: Who Should Teach It and How?* Voorburg, Netherlands: International Statistical Institute, 1993.

Robitaille, David F., David H. Wheeler, and Carolyn Kieran, eds. *Selected Lectures from the Seventh International Congress on Mathematical Education.* Sainte-Foy, Quebec, Canada: Les Presses de l'Université Laval, 1994.

Shaughnessy, J. Michael. *Computer Simulations of Probability Challenges: From Data to Theory.* Paper presented at the Sixth International Conference on Mathematical Education, Budapest, 1988.

Vere-Jones, David, ed. *Proceedings of the Third International Conference on Teaching Statistics.* Voorburg, Netherlands: International Statistical Institute, 1991.

ANNE HAWKINS

# TEACHING INTEGRATED MATHEMATICS AND SCIENCE PROGRAM (TIMS)

A teacher development and curriculum development project that works to enhance the teaching of mathematics and science in elementary schools. Schools in urban areas are a major focus of TIMS activities. Part of the Institute for Mathematics and Science Education (IMSE) at the University of Illinois at Chicago (UIC), TIMS was started by Howard Goldberg, a UIC particle physicist, and UIC professor Philip Wagreich. TIMS has received over $10 million in external funding from the National Science Foundation (NSF), the State of Illinois Scientific Literacy and Eisenhower funds, as well as direct support from school districts for professional development activities.

TIMS activities include curriculum development, staff development and outreach, and research.

*Curriculum Development* TIMS has developed two sets of nationally recognized curriculum materials, the TIMS Elementary Mathematics Curriculum and the TIMS Laboratory Experiments. The former is the result of six years and $5 million of testing and development. Funded by NSF with the goal of developing curriculum materials to support reform of teaching and learning of mathematics, this K–5 mathematics curriculum was published in 1997 as Math Trailblazers: A Mathematical Journey Using Science and Language Arts. The major premise underlying Math Trailblazers is that mathematics is best learned through active involvement in solving meaningful problems. Science provides an ideal setting for active learning by channeling a child's natural curiosity into systematic investigations of the real world.

*TIMS Laboratory Experiments* These are integrated mathematics/science experiments for students in grades 1–8 that are based on seventeen years of work by UIC mathematicians, scientists, and educators. More than 140 experiments help students learn scientific concepts and skills through engaging, hands-on activities. Underlying the experiments are the beliefs that children are natural scientists and that mathematics is best learned through active involvement in solving problems. *TIMS Laboratory Experiments* have four essential parts. First, children design their experiments and draw pictures of them. Second, they collect data and organize it into data tables. Third, they graph the data and look for patterns. Finally, they use the patterns they have discovered to answer questions.

*Staff Development and Outreach* Since the 1970s, TIMS has worked with elementary school teachers to foster professional development. In recent years, these activities have been funded by the Illinois Scientific Literacy Program and local school districts. In January 1996, TIMS began TIMS 95, a four-year, $3 million, NSF-funded effort with two components: (1) to establish Chicago-area, university–school partnerships to improve mathematics instruction through staff development; and (2) to create a set of resource materials for others to use in similar professional development efforts.

*Research* TIMS research efforts focus on the process of teacher change with the goal of developing more effective strategies for school change. A program of formative and summative evaluation investigates the effectiveness of the teacher enhancement program and of the resource materials (Goldberg and Wagreich 1990; Wagreich 1990).

*See also* Teacher Development, Elementary School

## SELECTED REFERENCES

Goldberg, Howard, and Philip Wagreich. "Focus on Integrating Science and Math." *Science and Children* 26(5)(1989):22–24.

———. "A Model Integrated Mathematics Science Program for the Elementary School." *International Journal of Educational Research* 14(2)(1990):193–214.

*Math Trailblazers: A Mathematical Journey Using Science and Language Arts.* Dubuque, IA: Kendall/Hunt, 1997.

Wagreich, Philip. "Teaching Integrated Math and Science: A Curriculum and Staff Development Project for the Elementary School." In *Mathematicians and Education Reform: Proceedings of July 6–8, 1988 Workshop* (pp. 63–130). Naomi Fisher, Harvey Keynes, and Philip Wagreich (eds.). Providence, RI: American Mathematical Society with Mathematical Association of America, 1990.

JAY BECKER

# TECHNIQUES AND MATERIALS FOR ENRICHMENT LEARNING AND TEACHING

Used traditionally to provide programming options for high-ability students. It is important to realize that providing challenging high-end learning opportunities in the mathematics classroom does not mean providing the *same* experiences for all students. Rather, the academic, social, and emotional needs of the individual learner must be recognized and met. An enrichment program focusing on authentic, high-end learning opportunities can be used to accomplish this goal.

## AUTHENTIC LEARNING

Authentic learning has as its main focus application—applying knowledge, thinking skills, and interpersonal skills to solving real-world problems. It consists of investigative activities that lead to the development of creative products in which students assume roles as first-hand writers, computer technicians, financial analysts, architects, or other professionals. The main purpose of these activities is to create situations where students are thinking, feeling, and acting in the same manner as these professionals in their preparation and delivery of products and services. These kinds of experiences serve as vehicles through which students apply their knowledge, thinking skills, creativity, and task commitment to real-world problems that are of interest to them (Renzulli 1994).

Enrichment learning and teaching is a form of authentic learning based on the premises that each learner is unique in interests, abilities, and learning styles; learning is more effective when students enjoy what they are doing; and learning is more meaningful and enjoyable when content and process are learned within the context of a real and present problem. In accordance with a constructivist approach, enrichment programs should emphasize that learning is most effective when the focus is on students' actions and understanding rather than the teacher's instructions. The teacher operates as a facilitator of learning, enabling students to assume the responsibility for their learning and thus experience the challenges, exhilaration, and honest satisfaction that come only from creation and consequent ownership of one's work.

When the focus of mathematics instruction is on the act of learning and the role of the student, learners become actual participants in the real world of mathematics. Students learn how to analyze, criticize, and select from alternative sources; how to think effectively about problems; and how to confront, clarify, and act creatively upon problems. Mathematics that is approached through a focus on real-world problems will not only help all students to become confident problem solvers but also encourage those students with special talents in mathematics to develop their interests and emerge as future leaders in the field.

## THE ENRICHMENT TRIAD MODEL

Enrichment learning and teaching is exemplified in the Enrichment Triad Model (Renzulli 1977), originally developed in the early 1970s as an alternative to didactic models for talent development. In the following years, numerous research studies (Renzulli and Reis 1994) and field tests in schools have provided opportunities for the development of large

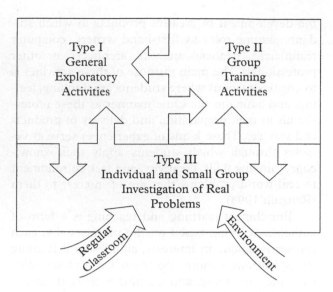

**Figure 1** *The Enrichment Triad Model. [Reprinted with permission]*

amounts of practical know-how that allow teachers to implement the model with relative ease. The three types of enrichment in the model are designed to replace dependence and passive learning with independence and engaged learning. They are depicted in Figure 1.

Type I enrichment consists of general exploratory experiences that are designed to expose students to topics and areas of study not ordinarily covered in the regular curriculum. These experiences might involve a field trip, a guest speaker, a film strip, an interest center, or a newspaper article. Type II enrichment consists of group training in thinking and feeling processes, learning-how-to-learn skills, research and reference skills, and written, oral, and visual communication skills. Some Type II activities might include learning observational and data-gathering skills, learning how to present data via graphs and computer displays, learning statistical analysis to interpret data gathered in a survey, and learning how to use a protractor and compass to construct geometric transformations. Type III enrichment consists of first-hand investigations of real problems, which often utilize the how-to skills learned in Type II training. The essential elements of Type III activities include a personal frame of reference for the student, a focus on advanced knowledge, a sense of audience for the product developed, and authentic evaluation. Some examples of Type III activities include original tessellations to be submitted as the cover of a school mathematics newspaper, a pictorial exhibit of the Fibonacci numbers in na-

ture, a land survey of school property conducted for the town planning and zoning department, and a statistical analysis of student attendance and tardiness trends using data from school records for the last twelve months.

Type I and Type II activities can be viewed as initial explorations that might entice a student to pursue a Type III investigation in a particular area of interest. They are usually pursued with a large group of students, while Type III activities are self-selected by individuals or members of an enrichment cluster (which will be described later). The interaction of the three types also includes "backward arrows" (i.e., the arrows shown in the figure that lead back from Type III to Type I, etc.). In many cases the advanced work of students in Type III activities can be used as Type I and Type II experiences for other students. With a strong emphasis on the student-as-learner and the investigation of real problems, the Enrichment Triad Model embodies the essence of enrichment learning and teaching.

## CURRICULAR MODIFICATIONS

In order to incorporate real-world problem solving into the mathematics curriculum, revision is necessary. One way to approach this revision is to implement the process of curriculum compacting (Reis and Renzulli 1992). This process eliminates repetitive drill and practice, thereby providing time for the incorporation of new material into the curriculum, such as probability and statistics and the pursuit of independent study projects. Compacting begins with pretesting in order to identify learning objectives already mastered. Students are allowed to "test out" of certain academic exercises and move on to new material. It gives students a window of time in which to pursue enrichment activities. A key feature of this process is that students will be given some freedom to make choices about the topic and methods through which the topic will be explored. This is especially appropriate for mathematically talented students who reach understanding sooner and often possess an intuitive grasp of concepts without the need for step-by-step explanation and repetitive examples.

### A Sample Mathematics Unit: The Concept of Area

To introduce the concept of area and provide a preassessment opportunity for the teacher to observe students in action as they deal with area concepts prior to instruction, the following problem situation is presented to the entire class. Students will discuss the problem in groups and then report their findings to the class.

> We have been given the opportunity to paint our classroom. We can select the color(s). Although the Parent-Teacher Organization (PTO) will pay for the supplies, we must decide what we need to buy and then purchase everything. Where do we begin? What do we need to know?

The teacher uses this activity to observe students, listen to their conversations, and get a feel for their understanding of the concepts associated with area. Following this activity, a pretest or posttest, taken from the textbook or created by the teacher, is administered to determine the mastery level of each student. The teacher then places students into groups according to their appropriate learning stage. The groups are differentiated by being involved in different types of investigations. A sample of activities pursued by four different groups might include the following:

1. students who will benefit from the entire unit of instruction might begin the unit by exploring the meaning of area and its relationship to perimeter (a problem situation to spark this exploration could be: How can we determine how much space the new swing set will take up on the playground and how much fencing will be needed to enclose it?);

2. students who already understand the definition of area might start their study by discovering the formulas for the area of common polygons (a problem situation they might consider in dealing with octagons is: How can the police department determine the amount of wood necessary to purchase for constructing 12 new stop signs?);

3. students who have mastered the concepts in no. 1 and no. 2 might begin this unit by investigating the area of circles, semicircles, and pie-shaped wedges (an interesting problem for these students to consider would be: How can we compare the cost effectiveness of purchasing pizzas based on the size of the pizza and the number of people being served?);

4. finally, the most advanced students might use their time to examine the area of irregular shapes (a real-world problem for study might be: How can we determine square footage of an irregularly shaped real estate plot for tax assessment?).

## Interest Development Centers

Utilizing the extra time found by eliminating repetitive and previously mastered curriculum, students may engage in enrichment activities in the form of interest development centers and individual or small group projects. Interest development centers make available a variety of Type I experiences. One "trigger" for motivating students, these centers provide a wide variety of higher-level thinking activities and also serve as springboards for independent projects and the formation of enrichment clusters, small groups of students who come together to explore a particular mathematics topic of mutual interest. Some examples of interest development centers for this unit on area include:

- Hands-On Center where students explore tangram activities with area; learn how to construct geometric transformations with a protractor and compass; create tessellations to use as book covers, fabric patterns, and wrapping paper; and combine area with coordinate graphing by playing "Where's the Rectangle?," a game in which students use coordinate points to try to guess the location of a hidden rectangle on a geoboard, given the area of the rectangle (see Downie, Slesnick, and Stenmark 1981).

- Architecture Center where students create scale drawings of their bedrooms and the classroom with the purpose of rearranging furniture, make scale models of solar houses (see Gould 1986), design graphs to show cost trends in building homes for the past fifty years, create a timeline to chart architectural periods and styles, and design a community of the future using LEGO blocks.

- Reading and Writing Corner where students can read about famous graphic artists, architects, and mapmakers, explore how-to books on creating three-dimensional polyhedra and tessellations, become authors by writing original poetry, riddles, puzzles, how-to books, and stories connected to area. (For example, students may write essays to answer the question: if I were a square in the land of circles, how could I find my "space"? In this essay, students would have the opportunity to compare and contrast the areas and perimeters of squares and circles while developing creative writing skills.)

- Mapping Center where students make treasure maps to locate hidden objects in the classroom or on school grounds, change an aerial picture into a street map, create games using latitude and longitude, and explore *Where Are You?*, a challenge about cartography (Science-by-Mail 1992).

- Computer Center where students use software programs such as MacDraw for designing and participate in National Geographic Kids Network, an online computer network that allows students to apply mapping skills to locate and communicate with other students around the world doing firsthand research on such topics as solar energy, acid rain, and weather forecasting.

## Independent Projects

Some of the activities in the interest development centers may spark motivation for a student to pursue an independent project or Type III investigation. These projects are investigative activities and artistic productions in which the student assumes the role of a firsthand inquirer—thinking, feeling, and acting like a practicing professional. The products that the student creates develop research skills and provide an opportunity to use authentic methodology. They are primarily directed toward bringing about a desired impact on an audience, whether it be fellow students, town officials, professionals in a specific field, or senior citizens. It is the student that comes up with the idea for the investigation and designs it. The teacher functions as a facilitator, pointing the student in the direction of resource persons and materials or providing direction in learning methodology to conduct the investigation. Some independent investigations that might arise from the centers listed earlier could include:

- Contracting local community officials for needed surveying or design projects. This could lead to designing an outside space such as a garden area or play space on school grounds or in a local park.
- Creating an architectural slide show of homes, museums, and buildings in the local area and presenting the show to the historical society.
- Creating a photography exhibit of geometry in the real world and displaying it at the local library.
- Researching the life of a famous cartographer, geometer, or architect and, dressed as the famous person, presenting the biography to the class as a first-person narrative.
- Redesigning a veterinarian's office space. This would include a visit to the veterinarian's office and consultation with the doctor and office staff on their needs.

## Enrichment Clusters

The Type I activities that students explore in the interest development centers also can be the impetus for the creation of an enrichment cluster. Enrichment clusters are small groups of students who come together with a teacher or facilitator at specially designated time blocks during the school day to pursue a common interest in a particular topic (Renzulli 1994), perhaps sparked by an activity in the interest development center. The main rationale for participation is that students and teachers *want* to be there. These clusters are ideal places to implement Type III enrichment with a small-group investigation that yields a product or service. Some clusters that might emerge from the previously mentioned interest centers include The Young Architects' Guild, The Cartography Society, The Computer Designing Club, The National Geographic Kids Network Team, and The Tessellation Fabric Company. In these clusters, the teacher functions as a methodological resource person and an agent for connecting student products with appropriate audiences. Acting as a guide, the teacher helps students focus on the four essential questions that must be addressed in a Type III investigation:

1. What are the different roles that are necessary to produce the product or service? In other words, what roles will students in the cluster have to perform (e.g., interviewer, writer, computer expert, graphic artist, calculation analyst) in order to get the product produced?
2. What are the resources used by professionals in this particular mathematics field to produce high-quality products (e.g., what resources do architects use to design innovative office spaces?)?
3. Who are the people in our community interested in the product or service we will produce/provide (e.g., What is the marketing strategy that we will use? Who should we contact that might be interested in a new design for their office space or building?)?
4. What steps need to be taken to ensure that our product or service will have an impact on our audience (e.g., How will we present our ideas for the new office space to our clients? How can we sell them on these ideas?)?

The goals of Type III enrichment, whether it be an independent project or the product or service undertaken by those involved in enrichment clusters, are larger than the methodology students learn or the products they produce. The hope is that students begin to think, feel, and act like creative and productive individuals and that they develop an attitude that reinforces the work of effective people: I can do . . . I can be . . . I can create (Renzulli 1994).

*See also* Enrichment, Overview; Games; Projects for Students; Recreations, Overview; Writing Activities

## SELECTED REFERENCES

*America 2000: An Education Strategy.* Washington, DC: U.S. Department of Education, 1991.

Carroll, Joseph M. *The Copernican Plan.* Andover, MA: Regional Laboratory for Educational Improvement, 1989.

*Claris MacDraw II 1.1.* Mountain View, CA: Claris, 1987–1989. [Computer software]

Downie, Diane, Twila Slesnick, and Jean Kerr Stenmark. *Math for Girls and Other Problem Solvers.* Berkeley, CA: University of California, Lawrence Hall of Science, 1981.

Gould, Alan. *Hot Water and Warm Homes from Sunlight.* Berkeley, CA: Great Explorations in Math and Science (GEMS), 1986.

House, Peggy A., ed. *Providing Opportunities for the Mathematically Gifted K–12.* Reston, VA: National Council of Teachers of Mathematics, 1987.

National Council of Teachers of Mathematics. *Curriculum and Evaluation Standards for School Mathematics.* Reston, VA: The Council, 1989.

———. *Principles and Standards for School Mathematics.* Reston, VA: The Council, 2000.

National Geographic Society. *National Geographic Kids Network.* Washington, DC: The Society, 1989.

Reis, Sally M., and Joseph S. Renzulli. "Using Curriculum Compacting to Challenge the Above-Average." *Educational Leadership* 50(Oct. 1992):51–57.

Renzulli, Joseph S. *The Enrichment Triad Model: A Guide for Developing Defensible Programs for the Gifted and Talented.* Mansfield Center, CT: Creative Learning, 1977.

———. *Schools for Talent Development: A Practical Plan for Total School Improvement.* Mansfield Center, CT: Creative Learning, 1994.

———, and Sally M. Reis. "Research Related to the Schoolwide Enrichment Model." *Gifted Child Quarterly* 38(1994):2–14.

Science-by-Mail. *Where Are You?* Boston, MA: Museum of Science, 1992.

Sheffield, Linda J. *The Development of Gifted and Talented Mathematics Students and the National Council of Teachers of Mathematics Standards.* Storrs, CT: National Research Center on the Gifted and Talented, 1994.

M. KATHERINE GAVIN
JOSEPH S. RENZULLI

(Note: the work reported herein was supported under the Javits Act Program (Grant No. R206R00001) as administered by the Office of Educational Research and Improvement, U.S. Department of Education. The opinions expressed in this entry do not reflect the position or policies of the Office of Educational Research and Improvement or the U.S. Department of Education.)

# TECHNOLOGY

No longer as controversial an issue as it was in the past in mathematics education. People have been exploring new uses of technology to enhance mathematics education. Among various kinds of technologies, computer technology receives the most attention. Other technologies, however, are also playing an increasingly important role. This entry concentrates on such technologies as calculators, computer networks, and educational multimedia.

## CALCULATOR TECHNOLOGY

Moving from the simplest four-operation calculator, to the calculator with memory, to the scientific calculator, to the programmable calculator, and finally to the graphing calculator, this technology is evolving very fast. Just a few years ago, people still debated the influence of calculators on objectives and on teaching arithmetic. Now, however, fewer people deplore the use of calculators, and graphing calculators, which have assumed many of the functions of microcomputers, are widely used in mathematics classrooms. Graphing calculators like the TI-82, TI-83, and TI-85 are actually a combination of algebra graphing software and a simplified version of a spreadsheet, allowing for the teaching of mathematics graphically, numerically, and analytically. Students and teachers can easily graph functions in rectangular, parametric, polar, or sequential form. For each graph, they can use zooming in, zooming out, or the tracing capability to explore the local details, the more global behavior, or the relationship between an independent variable and a dependent variable. A numerical approach is available by using the table feature of the TIs to investigate functions, equations, and number sequences visually as well as interactively. Programs can be constructed to perform specific tasks (e.g., sorting a list of numbers) and automatically saved for solving problems later on. Teachers and students can explore more content areas, including those that until recently were difficult for school mathematics—for example, curve fitting (see Wallace 1993; Moskowitz 1994; Goetz and Kahan 1995). Putting the power of a computer laboratory in the palm of one's hand, the TI-92 combines all the features of the TI-82 with interactive geometry, symbolic manipulation, and three-dimensional graphing. Designed for students to collect data from the real world and send the data to the calculator for statistical and functional analyses, the Calculator-Based Laboratory (CBL) system is compatible with the TI-82, TI-83, TI-85, and TI-92. The system usually involves a CBL unit and some supplemental devices that "plug in" to the unit. One of these devices may be a digital thermometer that reads temperature at intervals and stores the sequence of temperatures as a table in the calculator.

Casio, Sharp, Hewlett-Packard, and other kinds of graphing calculators are capable of operations similar to those of TIs but have their own nice features. For example, the Casio CFX-9850G PLUS uses onscreen, icon-driven menus, providing easy access to advanced functions combined with exclusive Casio color screens. Dual Graphing allows students to display two graphs on a single, large screen for side-by-side comparison. Dynamic Graphing will animate to reflect the range of results when one enters different variables. Multiple graphs can be overwritten and color-coded to their corresponding functions. It is significant that graphing calculators have become so inexpensive as to be affordable for most students (college students as well as high and middle school students). At the lower middle and elementary school level, simpler calculators like the four-function solar-powered calculator or scientific calculator are used to focus attention on the problem-solving process (Wilson 1991), help assess conceptual understanding (Carter, Ogle, and Royer 1993), or offer an early introduction to technology (Spiker and Kurtz 1992). Some special calculators have been designed for students with disabilities. One example is Talk Calculators designed for blind students.

## COMPUTER NETWORKS

Computer Networks provide powerful and affordable ways for communicating and accessing data. A computer laboratory can be a more efficient and effective learning environment when the computers are linked in a network. It would be ideal if the computers are both locally networked so that user-generated files, application programs, and hardware devices such as printers and scanners could be shared, and linked to the Internet, the world's largest network. The most significant uses of computer net-

works in mathematics education are those of telecommunications and the World Wide Web. Among different types of telecommunications, the most important one is electronic mail (e-mail), which is used to exchange text (letters, memos, etc.) between two or more individuals via computers. This application provides easy and convenient communication among teachers and students, who may be far away from each other. Since the attachment feature of e-mail allows for sending a whole document as it is, some mathematics educators even use e-mail to submit articles for publication and grant proposals for funding. All universities have free network access. Many states allow teachers to gain access to network services at low or no cost. Teachers in many school districts have been using e-mail to talk to each other and exchange ideas. Telecommunications also can be used in the classroom. For example, a Tennessee mathematics teacher coordinated a network project called "Ask A Teen" to teach high school statistics. Through e-mail, he and his students established an online interaction with a mathematics and science teacher in Iceland and his students. The Americans designed a survey covering topics such as morals, drug use, school, religion, and entertainment and e-mailed it to the teacher in Iceland. The two teachers administered the survey to their mathematics classes, answers were exchanged by e-mail, and an online meeting took place over the Internet. The instructors taught the statistics content (sample size, sampling methods, normal distribution, and other topics); students analyzed the results and tested for significant differences between the two groups. The students were able to understand the related mathematical ideas and developed a more positive attitude toward statistics (see Sanders 1996). An important benefit of telecommunications is that "when communicating via keyboard and computer screen, the stereotypes children have based on appearance, accent, and the like disappear" (Wiebe 1992, 218).

Since its first site was established in 1969, the Internet currently serves more than a billion users. Before the World Wide Web was instituted in 1991, the major computer tools accessing the Internet were Gopher, File Transfer Protocol (FTP), and Telnet. In recent years, the World Wide Web has become the most popular highway on the Internet. Students use a variety of web browsers to find relevant mathematics problems from a number of web sites such as http://forum.swarthmore.edu and http://www.mathpro.com/math.study. They solve the problems and share their solutions with other users of the Internet. Since the World Wide Web offers a huge amount of information that is frequently updated, students can collect various kinds of real-world data through the web, analyze the data, and learn related mathematics concepts and functional relationships. Mathematics teachers can find documents and articles ranging from the National Council of Teachers of Mathematics (NCTM) position papers to lesson plans written by experienced teachers and other information such as professional meeting announcements and introductions to mathematics software. Both teachers and students can download useful application programs and other files easily. For example, to do CBL experiments, a mathematics class can download from the web site http://www.ti.com TI-83 CBL programs and TI-GraphLinks that allow transferring the programs on the computer to the TI-83 calculator. Some mathematics educators have built their own home pages to share innovative mathematics reform ideas with colleagues and students all over the world. An excellent example of such home pages is http://jwilson.moe.uga.edu.

## COMPUTER-BASED MULTIMEDIA

A multimedia system adds to a basic computer system the computer coordination of graphics, video, animation, text, speech, and sound. The most popular uses of computer-based multimedia in mathematics education are related to CD-ROM and interactive video (videodisc technology).

CD-ROM (Compact Disc Read Only Memory) provides a low-cost method for storing an enormous amount of data (up to 663 megabytes) on a small compact optical disc (4.75 inches in diameter). The applications for CD-ROM in mathematics education, therefore, may include the storage of huge databases and multimedia programs involving a rich array of text, graphics, animations, audio, and video. *MathFINDER,* developed by the Education Development Center, for example, is a CD-ROM that stores a database of more than fifteen thousand pages, including the *Curriculum and Evaluation Standards for School Mathematics* (NCTM 1989), original teacher and student materials from thirty major curriculum projects, and articles that address the implementation of the standards. It helps teachers to understand new proposals for curriculum and instruction with a wealth of concrete examples.

Another example is ActivStats (Velleman 1998), which presents a complete introductory statistics course on CD-ROM. It integrates video, simulation, animation, narration, text, interactive experiments,

web access, and Data Disk, a full-function graphical interface statistics package, into a learning environment. It helps students discover and understand basic principles of exploring data, probability, and inference.

Videodiscs are read with a light beam and hence are not affected by repeated use. Unlike a videotape, any part or single image of the video on a videodisc can be accessed immediately without the need of fast forward or rewind. The videodisc also guarantees higher-resolution video images and uses dual stereo sound tracks. These features make videodisc technology very significant to mathematics education. An example of using this technology to develop students' mathematical power is *The Adventures of Jasper Woodbury,* a videodisc series by the staff at Vanderbilt University's Learning Technology Center. Based on the center's "Knowledge Construction" model, this series provides cognitive activities in the form of narrative stories that end with challenging problems for students. The random access of the data needed to solve the challenges can be controlled through a computer or a noncomputer device (a handheld remote controller, a barcode reader, or the front panel of the videodisc player). If a computer is involved, a higher-level interaction between students and the video (interactive video) can be achieved. *The World of Number* videodisc series from the National Curriculum Council of England and Wales is another example of videodisc technology in mathematics education. It includes six videodiscs that fit the standard curriculum, featuring students discussing and narrating the mathematics of its episodes. The "lessons" range in sophistication and content. For example, measurement problems show footage of traffic designed to allow measuring distances and time and to determine speeds of cars. *Perspectives,* a visual geometry activity provides different perspectives on objects and encourages students to solve visualization problems.

## VIDEOTAPE TECHNOLOGY AND TELEVISION

Videotape technology is relatively inexpensive. The mathematics phenomena embedded in real-world situations, as well as the students' data collection activities, may be recorded on videotapes for later classroom discussions on problem posing and problem solving. Examples of available videotapes are those produced by California State Polytechnic University, Pomona: *The Theorem of Pythagoras, Sines and Cosines I, Sines and Cosines II, Sines and Cosines*

*III, The Story of Pi,* and *Similarity.* Each has a blend of theory and applications, historical components, visualization segments, and animation. Although they naturally lack interactivity, they serve well in teaching traditional topics in somewhat visually and animation-enhanced ways. Television programs containing mathematical content are also used to present problem situations to students. Two examples are the television series *Square One TV* and *Newton's Apple.* The former has produced a substantial number of problem-solving episodes that require attention and critical thinking in a somewhat contrived "real-world" context. The latter is a science and mathematics series that features mathematical and scientific phenomena and explains, through expert guests, the principles of the phenomena.

## DISTANCE EDUCATION AND TELECONFERENCES

Distance education is the phrase used to describe education that takes place over audiovisual networks set up between remote sites. As an important part of distance education, teleconferences are becoming a prominent means of communicating and sharing ideas in the large community (statewide, interstate, or nationwide) of mathematics education. Satellite Educational Resources Consortium (SERC) has presented many teleconferences, some of which involve mathematics teaching, learning, assessment, and inservice teacher education. The teleconference series *Mathematics and Technology: The Transition Years,* produced by Stevens Institute of Technology and New Jersey Network, is an example. The four ninety-minute teleconferences held in February and March 1994 dealt with the integration of computers in prealgebra, algebra, and geometry classes. A national audience, including hundreds of teachers, participated in these teleconferences.

## OTHER TECHNOLOGIES

Still other technologies, including some newly introduced ones, are used in mathematics education. An example is electronic notepads, which have been used for assessment purposes. Newton messagepads (handheld information managers) are one kind of electronic notepad, introduced in the summer of 1993 by Apple Computer and aimed at revolutionizing the personal mobile computing environment. With a Newton messagepad, a teacher or a student can take notes in "electronic ink" to quickly write down his or her ideas. Its handwriting recognition capabilities for both printing and cursive writing

make it easy to translate the written notes into "typed" document files. If used with a computer having assessment software (e.g., *Learner Profile* developed by Morgan Media), it can receive descriptions of desired student learning from the computer, record students' actual learning process, and later transfer the recorded information back to the computer for analysis, to show evidence of students' mathematics understanding.

## A VISION OF THE FUTURE

The integration of new technologies with mathematics education may promote the goals of developing student autonomy in learning and student abilities of problem solving, mathematical reasoning, and communicating. Computer technology may continue to be the most important technology for mathematics education, but it may be combined with other technologies as well. Where multimedia workstations are equipped for teachers and students, the power of the computer is integrated with that of interactive video, CD-ROM, videotape, television, overhead projector, LCD panel, and/or other technologies. The graphing calculators and the CBL systems may also be used in the mathematics classroom more widely, and more emphasis may be put on telecommunications, the use of the Internet, and distance education.

*See also* Calculators; Computers; Media; Software

## SELECTED REFERENCES

Carter, Pamela L., Pamela K. Ogle, and Lynn B. Royer. "Learning Logs: What Are They and How Do We Use Them?" In *Assessment in the Mathematics Classroom* (pp. 87–96). 1993 Yearbook. Norman L. Webb (ed.). Reston, VA: National Council of Teachers of Mathematics, 1993.

Goetz, Albert, and Jeremy Kahan. "Surprising Results Using Calculators for Derivatives." *Mathematics Teacher* 88(Jan. 1995):30–33.

Kaput, James J., and Thomas A. Romberg. *Exploiting New Technologies for Reform in Mathematics Education.* Madison, WI: National Center for Research in Mathematical Science Education, 1989.

Moskowitz, Stuart. "Investigating Circles and Spirals with a Graphing Calculator." *Mathematics Teacher* 87(Apr. 1994):240–243.

National Council of Teachers of Mathematics. *Curriculum and Evaluation Standards for School Mathematics.* Reston, VA: The Council, 1989.

———. *Principles and Standards for School Mathematics.* Reston, VA: The Council, 2000.

Sanders, Mark. "Teaching Statistics with Computer Networks." *Mathematics Teacher* 89(Jan. 1996):70–72.

Spiker, Joan, and Ray Kurtz. "Teaching Primary Grade Mathematics Skills with Calculators." In *Computer Tools and Problem Solving in Mathematics* (pp. 226–229). By James Wiebe. Wilsonville, OR: Franklin, Beedle, 1992.

Sutton, Rosemary E. "Equity and Computers in the Schools: A Decade of Research." *Review of Educational Research* 61(4)(1991):475–503.

Velleman, Paul. *ActivStats*. Reading, MA: Addison-Wesley Interactive, 1998.

Wallace, Edward C. "Exploring Regression with a Graphing Calculator." *Mathematics Teacher* 86(Dec. 1993):741–743.

Wiebe, James. *Computer Tools and Problem Solving in Mathematics*. Wilsonville, OR: Franklin, Beedle, 1992.

Wilson, Katherine J. "Calculators and Word Problems in the Primary Grades." *Arithmetic Teacher* 38(Sept. 1991):12–14.

ZHONGHONG JIANG

## TEMPERATURE SCALES

Refers to the three most widely used scales (Celsius, Fahrenheit, and Kelvin), which are all centigrade scales. *Centigrade* is the name used to indicate that there are 100 grades, or "degrees," between two temperatures measured under distinct consistent conditions. Although the same word is used to indicate a difference in temperature on all three scales, the relative sizes of the degrees are not the same. Generally, temperature scale inventors used the temperatures of readily available substances, such as water, to devise their scales. Students should study temperature scales to recognize their arbitrary nature and to motivate the need for agreement in measuring hotness.

The Celsius scale was known as the centigrade scale until 1948, when an international conference on weights and measures renamed the scale for its inventor, the Swedish astronomer Anders Celsius (1701–1744). Celsius used fresh water as his indicator and named the boiling point of fresh water as 100°C and the freezing point of fresh water as 0°C (Asimov 1982). William Thomson, Lord Kelvin (1824–1907), a Scottish mathematician and physicist, used the same scale as Celsius but started his at 0 K, absolute zero where there is no heat at all in any substance. Hence, 273.15 K corresponds to 0° on the Celsius scale and 373.15 K maps to 100°C. These Kelvin measures were determined empirically through applications of Charles Law.

However, far earlier, the German physicist Gabriel Daniel Fahrenheit (1686–1736) created his temperature scale—also a centigrade scale (Asimov

1982)! Apparently, he chose the freezing point of sea water as 0°F but measured his own body temperature and named it 100°F! Oddly enough, his temperature must have been elevated from the "normal" body temperature of about 98°F. The United States has been depending on the metric system of weights and measures since 1866; however, the customary system of weights and measures in the United States still employs the Fahrenheit scale in daily weather reports and cooking recipes. The relative measure of freezing and boiling points of fresh water in this temperature scale are 32°F and 212°F. It is interesting to note that 1°C is about 1.8°F.

Following are some suggestions of exercises for students in mathematics classes:

*Elementary:* Read Ernest Hemingway's "A Day's Wait." This is a short story that plays on the confusion between normal body temperature on the Celsius and Fahrenheit scales. Discuss the need for standards in measurements and the confusion that might arise without standards.

*Middle School:* Have students devise their own temperature scales. Note the estimated temperatures for (1) needing a coat to go outside, (2) being warm enough to go swimming, (3) freezing and other temperatures.

*Algebra:* Graph the function $C = (5/9)(F - 32)$, for converting degrees Fahrenheit to degrees Celsius, and its inverse. Write the functions and their inverses for converting from degrees Celsius to degrees Kelvin and from degrees Fahrenheit to degrees Kelvin. Discuss the relationship of the two lines, noting their intersection. Offer this alternate conversion formula and have students determine why these two seemingly different equations generate the same result: $F + 40 = 1.8 (C + 40)$.

*See also* Systems of Measurement

SELECTED REFERENCE

Asimov, Isaac. *Asimov's Biographical Encyclopedia of Science and Technology*. Garden City, NY: Doubleday, 1982.

KATHLEEN M. HARMEYER

## TEST CONSTRUCTION

A process of designing and assembling questions or tasks to measure outcomes of selected educational objectives. Before 1850, whether it was an admission test to enter the university or a class test, most testing was done orally. The method was inefficient and often unfair. During the second half of the nineteenth century, many oral tests were replaced by written essay examinations. At the beginning of the twentieth century, research on psychometric assessment led to the production of standardized tests. A test is standardized if it has been used and revised until it produces uniform results under given conditions. It has the advantage of allowing teachers to compare the performance of their students with others who have taken the same test. The Stanford-Binet test for fundamental arithmetic and the Douglass Diagnostic Tests for elementary algebra are examples of tests used around 1920 (Thorndike and Hagen 1969, 1–7). Since 1947, the Educational Testing Service, created with a grant from the Carnegie Foundation, has been the leader in standardized testing in the United States. After 1960, the use of testing was extended to policy and program evaluation. In Europe, most university admission tests are not standardized and are made up primarily of open-ended questions.

## QUALITIES OF A GOOD TEST

A good test should be *reliable* and *valid;* in other words, it should test accurately what it is designed to test. *Timing* and level of *difficulty* should be planned carefully. Furthermore, if it is a class test, it should *help the student learn.*

### Reliability

The reliability of a test refers to whether it determines that student A's score represents what student A actually knows. If the errors in measurement are significant, the test is not reliable. The reliability of a test is established by giving it to the same groups of people on different occasions and comparing the results. Temporal errors, such as uncomfortable testing conditions or distracting noises, should be eliminated. *Scoring standards* for all questions must be clearly defined, and questions whose scoring might be subjective should be graded by two independent scorers to minimize the error.

### Validity

A test is valid if it satisfies the following two criteria:

- it covers the *content* of the unit or course;
- the *relative importance* of each content area is preserved.

Teachers involved in the course are the best judges of validity. In addition, the National Council of Teachers of Mathematics (NCTM) recommends ". . . assessment techniques focusing on *students' problem solving, reasoning and disposition toward mathematics* as well as on their understanding of content" (Meiring et al. 1992, vii).

### Correct Timing and Proper Difficulty

The objective of any test should define whether power or speed is more important. It should also define whether or not students will be using calculators and if so what types of calculators. The timing and difficulty levels are planned accordingly.

### Good Teaching Tool

Not only does a test tell the student what he or she knows or does not know, it should also *help the student learn*. The type of questions used should reflect our understanding of how children learn. The use of related questions makes tests a better teaching tool and gives students a chance to synthesize their work (Shafroth 1993, 12).

### Norm-Referenced Test

This feature is needed when it is desirable to compare the performance of each individual to that of others in the *same category*. The reference group might consist of all ninth graders in the nation or all people having completed certain training. The *standard score* is usually the positive or negative distance of the raw score from the mean, expressed in terms of the standard deviation. For instance, if the raw score is 50, the mean 40, and the standard deviation 5, the standard score will be equal to $+2$ since it is two standard deviations larger than the mean. The standard score can be used to determine the *percentile*, which indicates what percentage of the total group has done less well than the student who made a raw score of 50. Most norm-referenced tests, such as the *Comprehensive Test of Basic Skills* (1989), are commercial tests.

## CLASSIFICATION OF TESTS BY OBJECTIVE

The most important step in writing a test is a clear definition of what the *objectives* of the test are.

### Tests Used to Measure Achievement

The goal of an achievement test is to find out what students have learned or achieved so far. Many tests, such as the College Board Achievement Tests and most class tests, fall into this category. Some simply measure the basic arithmetic and algebraic skills of the student; others explore the understanding of more complex concepts such as functions, limits, etc. The results can be used for different purposes:

- to assign a grade to students;
- to give college credit, as is the case of the *College Level Examination Program* tests (CEEB, "College Level," 1990) and the Advanced Placement tests of the College Board (CEEB and ETS 1989);
- to diagnose students' weaknesses;
- to help teachers evaluate their teaching;
- to place students in the appropriate class.

## Tests Used as Diagnostic Tools

Achievement tests can be used as diagnostic tests, provided that they include several questions on each area to be tested and results of scoring permit easy recognition of the topics needing additional instruction. An example of such an instrument is the *Diagnostic Instrument in Pre-Algebra Concepts and Skills* (Beeson and Adams 1980). It is composed of nine independent parts, ranging from basic arithmetic to solution of linear equations. With each part is associated a set of prescriptive materials. The user must do the following: first identify the student's weakness, then teach the concept involved, and finally provide the appropriate drill material.

## Tests to Measure Efficacy of Instruction or Curriculum

These tests, often called *evaluation tools,* are used to judge the program. They are designed to portray the achievement of a *group of students* rather than the achievement of an individual student and so must be constructed differently. Along with multiple choice and written response items, they should include activities that measure the students' problem-solving skills. The results are used to modify the curriculum and improve the teaching. For instance, the National Assessment of Educational Progress (NAEP) has worked since 1989 "toward alignment with the NCTM Standards." It designed tests to assess students in five content areas and three categories of mathematical ability for three different grades. The NAEP tests were administered to students in grades 4, 8, or 12 from 10,000 schools. The results were used to compare achievements of students in the United States in 1992 and 1990, to assess the differ-

ences in achievement for various demographic groups, and to determine the level of performance for each grade level in each content area (Dossey, Mullis, and Jones 1993, 8–15).

## Tests Used for Placement

When achievement tests are used for placement purposes the placement strategy should be defined before testing takes place. It is often convenient to divide the test into several independent parts, which are scored separately. An example of tests used for this purpose are the *Descriptive Tests of Mathematics Skills* (CEEB 1990). Entering freshmen might take two of the following tests: *Elementary Algebra Skills, Intermediate Algebra Skills,* and *Functions and Graphs* to determine which college course they may register for.

## Tests Used as Predictors

Whether their purpose is for admission to college or for guidance in placing an employee, tests are not always totally reliable as predictors. For example, the *Scholastic Aptitude Test* is used to determine whether a student is likely to be successful in college even though some variables, such as willingness to study or determination to succeed, are not measured by the test. Research has shown that "even a valid measure of academic aptitude cannot be expected to provide perfect or even very good prediction in some circumstances" (Baird 1983). In designing a predictor-test, the test maker must decide whether the ability to solve difficult problems or speed is more important, decide what prior knowledge is needed, and design questions to test this knowledge. Finally, if reasoning ability is important, questions assessing the student's problem-solving skills and logical thinking should be included.

## Tests Used for Competition

The Committee on the American Mathematics Competitions is responsible for most of the national mathematics competitions in the United States. It appears that the stiffer the competition, the fewer the number of questions and the harder each question. Questions are based on the material that the student should have learned, but most of the time they require creativity and ingenuity. For instance, in the following question (Committee on the American Mathematics Competitions, *42d Annual,* 1991),

The sum of all real $x$ such that $(2^x - 4)^3 + (4^x - 2)^3 = (4^x + 2^x - 6)^3$ is
(a) 3/2 (b) 2 (c) 5/2 (d) 3 (e) 7/2

the student who thinks of replacing $(2^x - 4)$ by $a$ and $(4^x - 2)$ by $b$ will arrive at the answer much faster than the student who expands the binomials and trinomial in their original form.

## Tests Used as Part of the Curriculum

Tests designed to help students learn, such as take-home exams, open-book exams, or group-solution exams, are *learning tests*. All tests are *motivating tools*, but the results of a learning test should also guide future studying. To make it a *valuable learning experience,* the test should be interesting, should guide the student through logical steps to conclusions that are not obvious, and should be followed by a class discussion of difficult questions.

## TYPES OF QUESTIONS

Most tests will include different types of questions. The appropriate mix depends on the objective of the test.

## True-False, Completion, and Matching Items

These are common in teacher-made tests but should be used sparingly since they encourage rote memorization. The questions should be very clear to avoid any possibility of a misunderstanding. An example of a completion question is

Complete the following statement by citing the appropriate postulate:
Given $a, b, c$ real numbers, $a(b + c) = (b + c)a$ is an application of _____ .

## Multiple Choice Questions

This type of question is used most frequently on standardized mathematics tests. These tests have the advantage of being easy to grade, and are objective, norm-referenced, accurate in the scoring, and efficient for placement or diagnostic testing. Multiple choice questions are limited in scope, however, and often are not appropriate for testing problem-solving and critical-thinking skills. They should be written to measure accurately the knowledge of the subject matter; therefore, it should be difficult to answer them correctly on some other basis. A question test-

ing students' understanding of both the derivative of a function and the normal to a curve is the following (CEEB and ETS 1989):

If $x + 7y = 29$ is an equation of the line normal to the graph of $f$ at the point $(1,4)$, then $f'(1) =$
(a) 7 (b) 1/7 (c) $-1/7$ (d) $-7/29$ (e) $-7$

## Short Written-Response Questions (Constructed-Response Questions)

These can be written by teachers or taken from test banks provided by the textbook publisher. They can be graded as right or wrong or assigned partial credit when it seems appropriate, although that may encourage students' carelessness. If the answer is numerical and no partial credit is awarded, they can be computer-graded. For example, on the American Invitational Mathematics Examination, answers can be any integer from 000 to 999. For the question: "For how many real numbers $a$ does the quadratic equation $x^2 + ax + 6a = 0$ have only integer roots for $x$?," the answer is 010 (Committee on the American Mathematics Competitions, *Ninth Annual,* 1991, 5).

## Extended-Response Questions

These were included on the 1992 NAEP Test (Dossey, Mullis, and Jones 1993). They require students to explain their reasoning and are graded on a scale of 0 (Blank), 1 (Incorrect), 2 (Minimal), 3 (Partial), 4 (Satisfactory), and 5 (Extended). It is very important to indicate accurately ahead of time how the scoring will be done on questions of this type. Otherwise, it is easy to grade them unfairly. An example of an extended-response question for grade 12 is: "Why should the square of a positive integer ending in 5 end in 25?" It tests students' problem-solving skills. The extra credit awarded when the statement is "extended" encourages students' creativity.

## Proofs

Short proofs, such as "Show that the equation of the tangent to the parabola $y = x^2 - 4x + 7$ at the point $A(1,4)$ is $y = 6 - 2x$," or more formal proofs, such as those used in geometry, are a valuable tool for teaching students logical thinking and problem-solving skills. Since their main purpose is to teach reasoning, proofs should be straightforward and neither a trick nor knowledge of more advanced mathematics should facilitate the solution. The following proof was part of the University Entrance Examination in Portugal in 1990:

$u_1 = 0, v_1 = 1, u_{n+1} = (u_n + v_n)/2, v_{n+1} = (u_n + 3v_n)/4, n = 1, 2. \ldots$. Prove the following:

- $v_n - u_n = 1/4^{n-1}$, for all $n \in N$;
- $u_n$ is an increasing sequence, $v_n$ is a decreasing sequence;
- Both sequences converge (Shafroth 1993, 13).

Proofs can be graded on a scale similar to the scale used for extended-response questions or divided into several parts graded independently. In France and Germany, they represent 30% of the questions on the university entrance examination for mathematics or science majors.

## Graphs, Constructions and Visualization

These reinforce students' understanding of fundamental concepts such as "function," "maximum," "slope" and allow the investigation of interesting and realistic problems. If students are required to draw a graph, they should be provided with graph paper and required to use it. Some questions should require students to draw or recognize graphs on paper without the use of a calculator, while others can make use of the power of a graphing calculator. An example of the latter is the following:

Describe how you would use your graphing calculator to determine whether or not $\sin x = \sqrt{(1 - \cos^2 x)}$ is an identity" (Thompson and Senk 1993, 174).

## Open-ended Questions

These are becoming increasingly important in assessments to evaluate critical thinking and communication skills. They have been used since 1987 in the Netherlands on the Mathematics A examination, which is administered nationwide and emphasizes the application of mathematics to other subjects. Open-ended questions are also included in the new grade 12 California Assessment Program test where students have 12 to 15 minutes per question (Pandey 1992, 126). "For the teacher, open-ended questioning provides a better insight into students' understanding." An example is the following:

You want to teach your little cousin how to subtract two-digit numbers. Write out what you will say, and give examples of problems (Stenmark 1991, 22).

## STEPS IN CONSTRUCTION OF TEACHER-MADE TESTS

The following are basic steps in writing a test:

- Determine clearly what the *objectives* of the test are.
- Define the test *content* and the relative importance of each area. For instance, if the test is on solution of linear equations and inequalities and their applications, one could decide that 50% of the questions will involve solution of linear equations, 15% will be on linear inequalities, and the remainder, 35%, on applications.
- Define the amount of *time* allotted to the test, the level of *difficulty* of the test, and whether any tool will be allowed.
- Design the *test format*. Determine the number of questions and types of questions to be used. Decide what percentage of the test will be *routine* questions (multiple choice or short answers) and what percentage will test students' *understanding of concepts and problem-solving ability,* which require more extended responses. For instance, on the 1992 NAEP test, one-third of the questions and about one-half the time were devoted to constructed-response questions.
- *Write the questions.* Make use of computer banks for short-answer and multiple choice questions. Beware of the fact that problems in test banks provided by the publisher are usually almost identical to problems in the book and often do not test problem-solving skills. Teacher-written multiple choice questions should be checked by another teacher. If the test is to be administered to large numbers of students, it is worthwhile to *field test* the questions before incorporating them in a test.
- *Assemble the questions,* ordering them in such a way that the beginning of the test will be easy. This will boost students' confidence. Place the harder, time-consuming questions at the end.
- Determine the *grading scale* and include it on the test. Sometimes, it is also useful to indicate the *time* allotted to each question.
- *Review the test.* It should be reviewed by someone other than those who wrote it.
- After administering the test, *analyze* the results by drawing the frequency distribution. If the number of students is large and their ability to learn normally distributed, compare the results of the test to a normal curve and compute the *percentile* for each student. A *test-item analysis* is also very useful to determine the questions that

were poorly phrased or too difficult (Adkins 1974, 97–110).

Finally, to write tests efficiently, teachers should take advantage of the *technology* available. They can use a commercial test-generator to write their examination by picking questions from given test banks or use a test-builder or a word processor with mathematical capability to create their own questions. Most publishers provide a computer software package to teachers adopting one of their books. EXPTEST is an example of an excellent software program that can be used both as a test-builder or a test-generator (Smith 1995).

*See also* Assessment of Student Achievement, Overview

## SELECTED REFERENCES

Adkins, Dorothy C. *Test Construction.* 2nd ed. Columbus, OH: Merrill, 1974.

Baird, Leonard L. *Predicting Predictability: The Influence of Student and Institutional Characteristics on the Prediction of Grades.* College Board Report No. 83–30. New York: College Entrance Examination Board, 1983.

Beeson, B. F., and Sam Adams. *Adston Mathematics Skills Series: Diagnostic Instrument in Pre-Algebra Concepts and Skills (Junior High Level).* Educational Testing Service Collection, microfiche, 1980.

College Entrance Examination Board (CEEB). *College Level Examination Program (CLEP).* Princeton, NJ: 1990.

———. *Descriptive Tests of Mathematics Skills (DTMS).* Princeton, NJ: 1990.

——— and Educational Testing Service (CEEB and ETS). *The Entire 1988 AP Calculus BC Examination and Key.* Princeton, NJ, 1989.

Committee on the American Mathematics Competitions. *42d Annual American High School Mathematics Examination* (AHSME). Washington, DC: Mathematical Association of America, 1991.

———. *Ninth Annual American Invitational Mathematics Examination* (AIME). Washington, DC: Mathematical Association of America, 1991.

*Comprehensive Test of Basic Skills* (CTBS). 4th ed. New York: McGraw-Hill, 1989.

Dossey, John A., Ina V. S. Mullis, and Chancey O. Jones. *Can Students Do Mathematical Problem Solving? Results from Constructed-Response Questions in NAEP's 1992 Mathematics Assessment.* Washington, DC: National Center for Education Statistics, 1993.

Meiring, Steven, P., Rheta N. Rubenstein, James E. Schultz, Jan de Lange, and Donald L. Chambers. *A Core Curriculum—Making Mathematics Count for Everyone.* Grades 9–12. Reston, VA: National Council of Teachers of Mathematics, 1992.

Pandey, Tej. "Test Development Profile of a State-Mandated Large-Scale Assessment Instrument in Mathematics." In *Mathematics Assessment and Evaluation: Imperatives for Mathematics Educators* (pp. 100–127). Thomas Romberg, ed. Albany, NY: State University of New York Press, 1992.

Shafroth, Chantal. "A Comparison of University Entrance Examinations in the United States and in Europe." *Focus* 13(3)(Jun. 1993):1, 11–14.

Smith, Simon. *EXPTEST Version 6.0.* Monterey, CA: Brooks/Cole; Belmont, CA: Wadsworth; Boston, MA: PWS-Kent, 1995.

Stenmark, Jean Kerr, ed. *Mathematics Assessment: Myths, Models, Good Questions and Practical Suggestions.* Reston, VA: National Council of Teachers of Mathematics, 1991.

Thompson, Denisse, and Sharon L. Senk. "Assessing Reasoning and Proof in High School." In *Assessment in the Mathematics Classroom* (pp. 167–176). 1993 Yearbook. Norman L. Webb, ed. Reston, VA: National Council of Teachers of Mathematics, 1993.

Thorndike, Robert, and Elizabeth Hagen. *Measurement and Evaluation in Psychology and Education.* 3rd ed. New York: Wiley, 1969.

CHANTAL SHAFROTH

# TEXTBOOKS

The "earliest textbooks in colonial America dealt with the learning of English, the inculcation of moral and religious beliefs, *the teaching of arithmetic and numbers* (emphasis added), and the presentation of basic academic subjects and salable skills in less academic areas" (Squire 1992). James Squire goes on to note that during the eighteenth and nineteenth centuries, textbooks to help students learn about particular subjects emerged as a way of influencing what was learned in schools and colleges.

## ROLE OF TEXTBOOKS

Almost from their inception in American education, textbooks have been criticized (Squire 1988). The major basis for the concern is the heavy reliance on textbooks by most teachers of mathematics. J. B. Edmonson (1931) reported that during the first third of the twentieth century, textbooks were a major factor in determining what mathematics was taught. That this reliance on textbooks continued throughout the remainder of the century has been affirmed by mathematics educators (Johnson and Rising 1967) and the National Science Foundation (Brandt 1978). As recently as 1984, Steven

Willoughby, then president of the National Council of Teachers of Mathematics (NCTM), asserted that the "most important factor in determining what mathematics is taught is the textbook used" (Willoughby 1984).

This heavy reliance on textbooks led to continual examination of their status. In a thorough analysis of textbooks in school and society, Richard Venezky (1992) noted that there were three major attempts during the twentieth century to examine the nature and use of textbooks. The first analysis was reported in the 1931 yearbook of the National Society for the Study of Education (NSSE), *The Textbook in American Education* (Whipple 1931), in which chapter authors examined the preparation and selection of textbooks. The second study was reported by Lee J. Cronbach (1955) in *Text Materials in Modern Education*. He noted that at that time there was no comprehensive view of what textbooks should do, how they should contribute to schooling, or their strengths and weaknesses. The third major attempt identified by Venezky was another yearbook by the NSSE. The contributors to the 1990 yearbook of the NSSE—*Textbooks and Schooling in the United States*—examined the use and control of textbooks and their content, especially the strategies used by government and special interest groups to influence the content of textbooks (Elliott and Woodward 1990).

## TEXTBOOK EVALUATION IN THE EARLY TWENTIETH CENTURY

During the first third of the twentieth century, a number of attempts were made to evaluate textbooks. In 1918, Alfred Hall-Quest listed standards for textbook adoption that incorporated precise measures of the textbook's physical format. A flurry of publications in the 1920s attempted to develop justifiable standards for selection of textbooks. C. R. Maxwell listed items to be rated on an interval scale as excellent, good, fair, or poor (Maxwell 1921). The items were content-oriented and reflected the idea that the book should develop ideas for the student. R. H. Franzen and F. B. Knight developed a similar rating scheme to evaluate textbooks regarding interest, comprehension, permanent value of subject matter, value of method, and mechanical elements. Numerical values were assigned to the variables, and the value of each text could, theoretically, be expressed in one figure (Franzen and Knight 1922). John Fowlkes developed a form for evaluating text-

books that included a section containing the topics covered and the percentage of the total pages devoted to each topic. He included many items that required subjective evaluations without criteria for evaluation. He attempted to overcome the judgmental weaknesses by having evaluators give instances from the textbook to support the judgments (Fowlkes 1923).

## TEXTBOOK EVALUATION IN THE 1960S

During the 1960s, considerable attention was devoted to curriculum and materials evaluation in most content areas. In many cases, materials analysis systems were designed to help teachers, administrators, and curriculum personnel determine the extent to which the instructional resources dealt with the major issues in American education at that time—the development of higher-order thinking skills and the involvement of students in their learning. For example, in science, William Romey (1968) reported a scheme for quantitatively rating science textbooks or chapters from science textbooks on the basis of their content. The variables that he suggested for analysis included statements of fact, stated conclusions or generalizations, definitions, various types of questions, and directions.

William Stevens and Irving Morrissett developed a social science curriculum analysis system that asked questions about the expected student performance in the cognitive domain using the framework described by Benjamin Bloom (1956). Their system asks to what extent the student is called upon to perform processes that involve knowledge acquisition, comprehension, application, analysis, synthesis, and evaluation. It also asks if the author worded specific learning objectives in such a fashion that the verbs demonstrate student action behavior that is clearly observable and/or measurable. Although this scheme was developed for social science curricula, it asked questions about cognitive processes that have applicability in mathematics and other content areas.

An instrument that was designed by Maurice Eash (1969) to assess curriculum materials in a number of content areas (including mathematics) included items designed to assess student involvement and development of cognitive skills. Eash felt that materials emphasizing cognitive development skills usually focus on thinking processes such as understanding, discriminating, utilizing, chaining, and evaluating as opposed to specific subject products.

A number of schemes were developed specifically to determine the extent to which mathematics textbooks and instructional materials reflected the curricular values of the time. For example, in Pennsylvania, Project PRIMES (Pennsylvania Retrieval of Information for Mathematics Education System) was developed to serve as a resource to school districts to facilitate their selection of elementary mathematics textbooks. One of the objectives of the project was to analyze, store, and retrieve information related to elementary school mathematics curriculum materials (Creswell and Berger 1967). The analysis scheme included variables such as content and expected student behavior and was designed to provide an accurate description of the content of a textbook so that textbook adoption committees could determine the extent to which a text was congruent with their locally designed curriculum.

About the same time, the NCTM charged a committee with the responsibility of developing a set of criteria to aid teachers in the selection of a textbook to meet their needs. The committee developed an instrument to aid in making qualitative judgments, not in quantifying the characteristics of texts. The instrument was designed to help the user in the decision-making process, not to make the decision. The committee rejected the notion of using a single number to measure the quality of a textbook for use in a given institution but rather relied on the use of subjective judgments based on careful consideration of criteria that are relevant to the evaluation (Peak et al. 1965). The committee recognized the importance of students' engaging in higher-level cognitive tasks such as generalization, evaluation, synthesis, application, and thinking and included criteria to assess them. In 1982, the NCTM published *How to Evaluate Mathematics Textbooks* as a follow-up to the work done by the 1965 committee. That publication focused its questions on the extent to which there is conceptual development of content, opportunities for problem solving, and student involvement in higher-order intellectual activities.

## AN EXAMPLE OF IN-DEPTH TEXTBOOK ANALYSIS

Since mathematics textbooks exert such a powerful influence on what is taught in thousands of classrooms across the United States, some textbook analysis and evaluation systems moved away from generic assessments and toward specific assessments that would pinpoint exactly what it is that textbooks contribute to student learning. These analysis systems vary depending on the goals to be assessed. Robert F. Nicely, Jr. (1970) and several of his colleagues (G. Bradley Seager, Robert G. Dilts, and John S. Mowbray) at the University of Pittsburgh developed a way to analyze mathematics textbooks (at the individual problem level) in terms of content and the way it was developed and the level of cognitive involvement offered to the learners so that individual textbooks could be accurately described in great detail for comparison with each other and with local, state, and/or national standards. Applying this system to the complex numbers sections of secondary mathematics textbooks that had been published over a period of three decades, Nicely (1991) reported that the relative emphasis on lower-order cognitive behaviors (recall, iterate, compare, categorize, and illustrate) increased from about 84% in the 1960s to more than 93% in the 1970s, and then decreased to 90% in the 1980s. There was a concomitant decrease in the emphasis on higher-order cognitive behaviors (apply, explain, analyze, synthesize, prove, evaluate) —from 16% in the 1960s to 7% in the 1970s—and then an increase to 10% in the 1980s. The more recent texts showed a slight increase, although not a return to the 1960s level.

Nicely's analysis system was also applied to the decimal portion of four elementary school mathematics series that were published in the mid-1980s. A significant majority (72%–100%) of the problems posed for students in all series and at all grade levels was found to be at the lower end of the cognitive scale. And virtually all the lower-level situations simply required students to iterate (repeat a given procedure)! The second most frequent cognitive behavior was apply. All four series offered enrichment activities in grades 4, 5, and 6, but most of these enrichment problems required students only to iterate (Nicely, Fiber, and Bobango 1986).

In 1990, Janice Engelder and Nicely reported on the application of the analysis system to the exponential and logarithmic functions chapter in five recent algebra II with trigonometry textbooks. They found that each textbook had problems in at least six of the cognitive levels, but only one textbook used all levels. At least 80% of the problems in each textbook in this study fell into lower levels, such as recall, iterate, compare, categorize, and illustrate. Higher-order cognitive behaviors, such as apply, analyze, hypothesize, and prove, occurred at a low of 3% to a high of 20%, with the average of the combined categories being 14%. This study showed more higher-order cognitive activity required of students compared to the complex numbers analyses that were conducted on books published in the 1980s, but the textbooks were still not at the level of those published in the 1960s.

In order to meet the challenge of the *Curriculum and Evaluation Standards for School Mathematics* (NCTM 1989), mathematics teachers need instructional resources that provide opportunities for students to develop and practice behaviors that lead to the acquisition of higher-order thinking skills. Analyses indicate that most textbooks are inadequate in providing such opportunities. In the event that no textbooks can be found that will be likely to help students acquire these behaviors, teachers must develop or acquire other instructional materials that do foster such thinking. Teachers with expertise in framing questions and guiding discussions will be able to help their students operate at the desired intellectual levels. Problems that parallel real life and the use of calculators and computers as problem-solving tools can stimulate interest and new ideas for solving problems. If publishing companies and textbook and software authors focus on the design and development of high-quality curricular and instructional materials, and school districts and institutions of higher education develop and deliver high-quality staff development programs, teachers will have more resources and skills to help their students become effective problem solvers.

## TEXTBOOKS AND THE CHANGING COMMUNITIES

As communities and their schools become increasingly diverse, school curricula must address the needs of the changing population. In a study conducted in the United States, Susan Chipman and Veronica Thomas (1984) found that race and gender differences in students' interest in mathematics and science exist; in particular, they indicate a lower level of interest for African American and female students as compared to Caucasian male students.

During the 1991–92 school year, Iris Striedieck and Robert Nicely (1993) devised and implemented a plan to examine selected current (copyright 1991) eighth-grade mathematics textbooks to determine if and how females and minorities were depicted in illustrations. In examining these illustrations, they attempted to determine (1) the percentage of the relative portrayal of female and minority characters and compare these percentages with U.S. and Pennsylvania population statistics, (2) the nature (active or passive) of the roles that the female and minority characters portrayed, and (3) the overall image of fe-

males and minorities. While females were underrepresented in illustrations as compared to population statistics, representations of minorities generally exceeded their population statistics. Additionally, both groups were shown as active individuals in greater proportions than male Caucasians.

## NATIONAL CURRICULUM

In the United States, the issue of a national curriculum in mathematics (as well as in other subject areas) periodically surfaces. Donald Freeman et al. (1983) examined elementary school mathematics textbooks that were published in the late 1970s and found that claims for a national curriculum could only be made at a relatively high level of generality. The data from the aforementioned studies of selected secondary textbooks by Nicely and his colleagues can be interpreted to support Freeman's conclusions. Studies such as Freeman's need to be replicated and extended to new levels of the curriculum.

## RESOURCES AND THE FUTURE

There are numerous recent resources that are available to teachers, curriculum specialists, and researchers who are interested in delving into the many issues that surround the analysis, evaluation, selection, and influence of mathematics textbooks. This entry gives a hint of what is available. The NSSE yearbooks (Whipple 1931; Elliott and Woodward 1990), the *Handbook of Research on Curriculum* (Venezky 1992), and the *Encyclopedia of Educational Research* (Squire 1992) are good places to start a systematic inquiry. In addition, the *Handbook of Research on Mathematics Teaching and Learning* (Grouws 1992) sheds some light on international perspectives regarding mathematics textbooks. *A Guide for Reviewing School Mathematics Programs* (Blume and Nicely 1991) outlines ways that curriculum and materials selection committees in school districts can use the *Guide* to determine the extent to which textbooks and other instructional materials focus on the important elements of the *Curriculum and Evaluation Standards for School Mathematics* (NCTM 1989).

In our rapidly changing and increasingly complex society, learning throughout life is becoming more and more important. The instructional options for learners are increasing rapidly, and the textbook and other print media are at risk of being excluded as electronic forms of communication are enhanced and new developments are added to the mix. Electronic publishing, "interactive" textbooks, and the like hold potential for major alterations to the types of instructional materials with which students interact. The potential for teachers to personalize an electronic textbook to the specifications of an individual classroom or to modify its content to include topics that a class recently studied but that were not addressed in the original "electronic" textbook could lead to vast changes in the way teachers might use textbooks. Another new direction is the electronically modifiable textbook—one in which the student reads and then interacts with the text by providing additional examples, exploring additional cases, making conjectures, or drawing conclusions from the text and then recording the results in the student's own personalized text. As textbooks, whether print or electronic, evolve, great care needs to be taken to ensure that they are of high quality and meet the needs of a sophisticated society.

*See also* Language of Mathematics Textbooks; Media; Reading Mathematics Textbooks

## SELECTED REFERENCES

Bloom, Benjamin S., et al., eds. *Taxonomy of Educational Objectives. Handbook I: Cognitive Domain.* New York: McKay, 1956.

Blume, Glendon W., and Robert F. Nicely, Jr. *A Guide for Reviewing School Mathematics Programs.* Reston, VA: National Council of Teachers of Mathematics, 1991.

Brandt, Ronald. "NSF Study Finds Teaching 'By the Book' in U.S. Schools." *Association for Supervision and Curriculum Development (ASCD) News Exchange* 20(5)(1978):1–2.

Chipman, Susan F., and Veronica G. Thomas. *The Participation of Women and Minorities in Mathematical, Scientific and Technical Fields.* Washington, DC: Howard University Institute for Urban Affairs and Research, 1984.

Creswell, Doris E., and Emanuel Berger. *The Development of an Information System for Elementary School Mathematics Curriculum Materials: A Feasibility Study.* Harrisburg, PA: Department of Public Instruction, 1967.

Cronbach, Lee J., ed. *Text Materials in Modern Education.* Urbana, IL: University of Illinois Press, 1955.

Eash, Maurice J. "Assessing Curriculum Materials: A Preliminary Instrument." *Educational Product Report* 2(Feb. 1969):18–24.

Edmonson, J. B. "Introduction." In *The Textbook in American Education.* 30th Yearbook, Pt. II. National Society for the Study of Education (pp. 1–6). Guy Montrose Whipple, ed. Bloomington, IL: Public School Publishing, 1931.

Elliott, David L., and Arthur Woodward, eds. *Textbooks and Schooling in the United States.* 89th Yearbook, Pt. I.

Chicago, IL: National Society for the Study of Education, 1990.

Engelder, Janice W., and Robert F. Nicely, Jr. "Mathematics Textbooks and the NCTM Standards: Discrepancies and New Directions." In *Implementing New Curriculum and Evaluation Standards* (pp. 9–17). Glendon W. Blume and M. Kathleen Heid, eds. 1990 Yearbook. University Park, PA: Pennsylvania Council of Teachers of Mathematics, 1990.

Fowlkes, John G. *Evaluating School Textbooks.* New York: Silver, Burdett, 1923.

Franzen, Raymond H., and Frederic B. Knight. *Textbook Selection.* Baltimore, MD: Warwick and York, 1922.

Freeman, Donald J., Therese M. Kuhs, Andrew C. Porter, Robert E. Floden, William H. Schmidt, and John R. Schwille. "Do Textbooks and Tests Define a National Curriculum in Elementary School Mathematics?" *Elementary School Journal* 83(5)(1983):501–513.

Grouws, Douglas A., ed. *Handbook of Research on Mathematics Teaching and Learning.* New York: Macmillan, 1992.

Hall-Quest, Alfred L. *The Textbook.* New York: Macmillan, 1918.

Johnson, Donovan A., and Gerald A. Rising. *Guidelines for Teaching Mathematics.* Belmont, CA: Wadsworth, 1967.

Maxwell, Charles R. *The Selection of Textbooks.* Boston, MA: Houghton Mifflin, 1921.

National Council of Teachers of Mathematics. *Curriculum and Evaluation Standards for School Mathematics.* Reston, VA: The Council, 1989.

———. *How to Evaluate Mathematics Textbooks.* Reston, VA: The Council, 1982.

———. *Principles and Standards for School Mathematics.* Reston, VA: The Council, 2000.

Nicely, Jr., Robert F. *Development of Procedures for Analyzing Materials for Instruction in Complex Numbers at the Secondary Level.* Ph.D. diss., University of Pittsburgh, 1970.

———. "Higher-Order Thinking Skills in Mathematics Textbooks: A Research Summary." *Education* 111(4)(1991):456–460.

———, Helen Fiber, and Janet C. Bobango. "The Cognitive Content of Elementary School Mathematics Textbooks." *Arithmetic Teacher* 34(2)(1986):60–61.

Peak, Phillip, et al. "Aids for Evaluators of Mathematics Textbooks." *Mathematics Teacher* 58(May 1965): 467–473.

Romey, William D. *Inquiring Techniques for Teaching Science.* Englewood Cliffs, NJ: Prentice-Hall, 1968.

Squire, James R. "Studies of Textbooks: Are We Asking the Right Questions?" In *Contributing to Educational Change* (pp. 127–169). Philip W. Jackson, ed. Berkeley, CA: McCutchen, 1988.

———. "Textbook Publishing." In *Encyclopedia of Educational Research* (pp. 1414–1420). Vol. 4. 6th ed. Marvin C. Alkin, ed. New York: Macmillan, 1992.

Stevens, W. William, Jr., and Irving Morrissett. "A System for Analyzing Social Science Curricula." New York: Educational Products Information Exchange (EPIE) Institute, Teachers College, Columbia University, n.d.

Striedieck, Iris M., and Robert F. Nicely, Jr. "Females and Minorities in Mathematics Textbooks: The Hidden Message." In *Alternative Assessment in Mathematics* (pp. 58–67). Glendon W. Blume and M. Kathleen Heid, eds. 1993 Yearbook. University Park, PA: Pennsylvania Council of Teachers of Mathematics, 1993.

Venezky, Richard L. "Textbooks in School and Society." In *Handbook of Research on Curriculum* (pp. 436–461). Philip W. Jackson, ed. New York: Macmillan, 1992.

Whipple, Guy Montrose, ed. *The Textbook in American Education.* 30th Yearbook, Pt. II. National Society for the Study of Education. Bloomington, IL: Public School Publishing, 1931.

Willoughby, Steven S. "Mathematics for 21st Century Citizens." *Educational Leadership* 41(Dec. 1983–Jan. 1984):45–50.

ROBERT F. NICELY, JR.

# TEXTBOOKS, EARLY AMERICAN SAMPLES

The *Sumario Compendioso* by Brother Juan Diez, a mathematical work prepared in 1556 in Mexico City, was the first textbook in any secular subject printed in the Western Hemisphere. Another historically significant early textbook is *Arithmetick, Vulgar and Decimal*

## Reglas ordinarias.

**Y** El que baftantemente tengo puefto por donde fin hazer quenta fe pueda faber el valor de qualquier varra o tejo' de plata o oro por diferente ley y pefo que tenga y el valor delos ynterefes que fe a coftumbran a dar por qualquier plata o oro hafta treynta por ciento. y affi mifmo el valor de qualefquier pefos de plata corriente comprados de enfayado razonando el ynteres de ocho a veynte por ciento juntamente con todo lo mas necefario dela nueva Efpaña có las reduciones de pefos ducados y coronas, de aqui adelante pondre algunas reglas delas necefarias en los reynos del Peru juntamente con algunas quiftiones para curiofos entre las quales van algunas del arte mayor referuadas al algebra: las quales conlo demas fino fuere tal como conuiene recebid la voluntad y fea caritatiuamente emendado de lafalta que tuuiere.

### Common Rules

Now that I have sufficiently explained, without doing the actual computing, how the value can be found of any ingot or bar of silver or gold of whatever standard or weight, and how to find the amount of commission up to thirty per cent which it is customary to give for any gold or silver; and, in the same way, how to ascertain the value of divers weights of silver currency bought s assayed, reckoning the commission from eight to twenty per cent, together with all else that is necessary in regard to the reduction of pesos, ducats, and crowns in New Spain,—from here on, I shall set forth some of the necessary rules which are used in the kingdom of Peru, together with certain problems for those who are interested, among which are certain parts of the *arte mayor* pertaining to algebra. If these with the rest do not entirely meet with the approval of the reader, may he accept my good intentions, and, in as kindly a spirit as possible, excuse the mistakes which I may have made.

**Figure 1** [*Source: Smith 1921*]

¶ Quiſtiones poꝛ losnumeros q̃drados

¶ Para auer de baȝer qualquier pꝛegunta que te fuere dmandada de numeros quadrados,es neceſſario que ſepas que es numero q̃drado y poꝛque ſe llama quadrado,y que es numero cubo y poꝛque ſe llama cubo.

¶ Numeros quadrados ſe llaman y ſon aquellos que nacen dla multiplicacion o ſon produȝidos de algun numero en otro ſemejante como.4,9,16,ꝛc.q̃ el,4,nace del,2,multiplicado poꝛ ſi meſino diȝiendo,2,veȝes,2,ſon,4,y el,9.nace del,3, poꝛel. meſino conſiguiente poꝛque,3,veȝes,3,ſon,9, delos quales numeros los lineales como el,2,o el,3,ſon las rayȝes;

### Problems Relating to Square Numbers

In order to know how to solve any problem that is given to you relating to square numbers it is necessary to know what a square number is and why it is called a square, and also what a cube number is and why it is called a cube. A square number is a number that is derived by the multiplication of a number by itself, as is the case with 4, 9, 16, &c. The 4 comes from multiplying 2 by itself, as when we say that 2 times 2 is 4; and the 9 is the product of 3 multiplied by itself, because 3 times 3 is 9. Of such numbers the lineals like the 2 or the 3 are called the roots.

**Figure 2**  *[Source: Smith 1921]*

¶ Si quiſieres ſaber vna cantidad de ducados quantos peſos ſon ſaca el ſeſmo de los ducados,y lo que reſtare ſera lo que buſcas.

¶ Si quiſieres ſaber tantos peſos quantas coꝛonas ſon ſin baȝerlo poꝛ marauedis,multiplica los peſos poꝛ nueue y parte poꝛ ſiete, y el aduenimiento ſeran coꝛonas.

¶ Exemplo.

¶ 56.ps.multiplicapoꝛ.9.ſon.504.parte poꝛ.7.vien̄.72.coꝛonas.

¶ Para baȝer dlas coꝛonas peſos,multiplica poꝛ .7. y parte poꝛ.9

If you wish to find out how many pesos there are in a number of ducats, subtract one sixth of the number of ducats from the number of ducats. If you wish to find how many crowns there are in a certain number of pesos, without reducing to maravedis, multiply the pesos by nine and divide by seven, and the quotient will be crowns.

### Example

56 pesos multiplied by 9 is 504. Divide by 7 and the result is 72 crowns. To reduce crowns to pesos, multiply by 7 and divide by 9.

**Figure 3**  *[Source: Smith 1921]*

by Isaac Greenwood. Printed in Boston in 1729, it was the first English-language arithmetic text written in the Western Hemisphere. Like arithmetic texts today, many of the problems in early texts dealt with then prevalent commercial topics—buying and selling gold and silver or purchasing food and clothing.

The author of the *Sumario Compendioso* was born in the province of Galicia, Spain, in the late 1400s. He traveled with Cortés and the Spanish conquistadors as a "chaplain" and as a missionary to the New World. Diez' journey with the Spanish fleet to Yucatan in 1518 is recorded in a work entitled *Itinerario*. In Juan Diez' introduction to the "Common Rules" of computation, he shows that he believed them important in conducting business in "New Spain" (see Figure 1). Diez' definitions of squares and square roots are very similar to the ones found in modern textbooks (see Figure 2). Figure 3 uses units of money that were common in the sixteenth century in the New World. It is interesting to interpret Diez' rules in terms of modern fraction and decimal symbolism. For example, the rule for converting ducats to pesos would be stated today as "the number of pesos = 5/6 of the number of ducats."

Isaac Greenwood (1702–1745), on the other hand, was born in New England. He attended Harvard College and, in 1724, received a Master of Arts degree. Greenwood traveled to England, where he

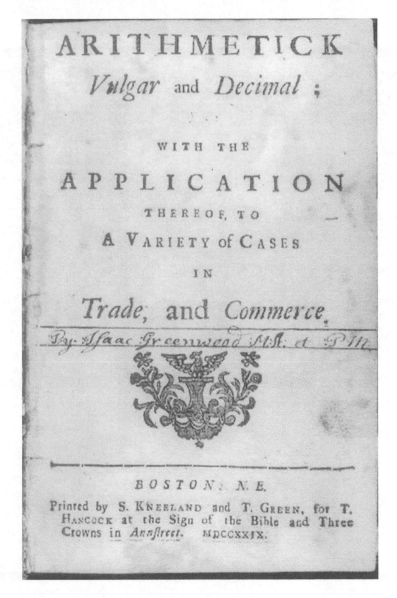

**Figure 4**  [*Source: George A. Plimpton Collection, Columbia University*]

studied Isaac Newton's work, including the method of "fluxions" (differential calculus). In 1728, Harvard appointed Greenwood the first Hollis Professor of Mathematics and Natural Philosophy. The Greenwood *Arithmetick* was the first to use a workbook format (see Figure 4). Since the copy from which these plates were made is that of the American patriot Samuel Adams, the handwritten entries in the workbook may be his own (see Figure 5).

Like Brother Juan Diez, Greenwood chose for his problems measures related to eighteenth-century life in colonial America. Apothecary measure—grains, scruples, drams, and ounces—were used to measure herbs, spices, and medicines. The unit "one scruple" was a very small amount, approximately the equivalent of a pinch of salt. Because of the care taken to measure in scruples, the word entered the English language to indicate careful consideration. People are scrupulous if they consider matters carefully. An activity that combines history and mathematics can be based upon a worksheet imitating a page from *Arithmetick, Vulgar and Decimal* as shown in Figure 6.

*See also* Enrichment, Overview; Ethnomathematics; Integration of Elementary School Mathematics Instruction with Other Subjects; Multicultural Mathematics, Issues; Systems of Measurement

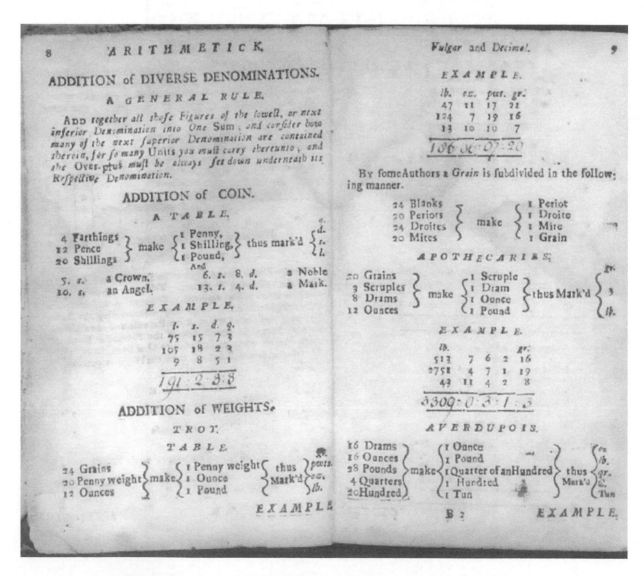

**Figure 5**   [*Source: George A. Plimpton Collection, Columbia University*]

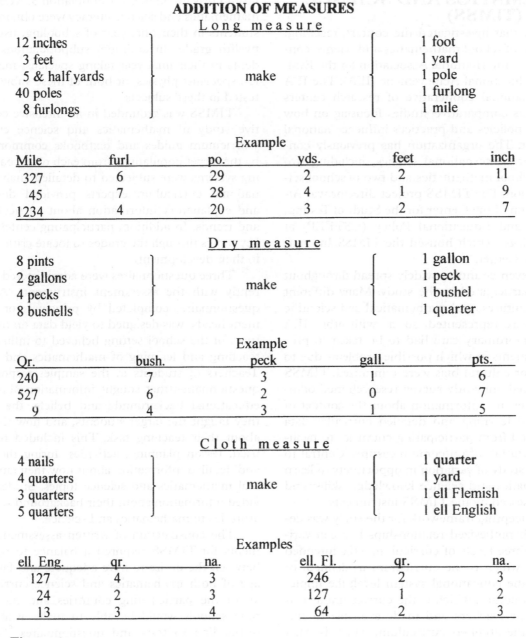

## ADDITION OF MEASURES

### Long measure

| | | |
|---|---|---|
| 12 inches | | 1 foot |
| 3 feet | | 1 yard |
| 5 & half yards | make | 1 pole |
| 40 poles | | 1 furlong |
| 8 furlongs | | 1 mile |

#### Example

| Mile | furl. | po. | yds. | feet | inch |
|---|---|---|---|---|---|
| 327 | 6 | 29 | 2 | 2 | 11 |
| 45 | 7 | 28 | 4 | 4 | 9 |
| 1234 | 4 | 20 | 3 | 2 | 7 |

### Dry measure

| | | |
|---|---|---|
| 8 pints | | 1 gallon |
| 2 gallons | make | 1 peck |
| 4 pecks | | 1 bushel |
| 8 bushells | | 1 quarter |

#### Example

| Qr. | bush. | peck | gall. | pts. |
|---|---|---|---|---|
| 240 | 7 | 3 | 1 | 6 |
| 527 | 6 | 2 | 0 | 7 |
| 9 | 4 | 3 | 1 | 5 |

### Cloth measure

| | | |
|---|---|---|
| 4 nails | | 1 quarter |
| 4 quarters | make | 1 yard |
| 3 quarters | | 1 ell Flemish |
| 5 quarters | | 1 ell English |

#### Examples

| ell. Eng. | qr. | na. | | ell. Fl. | qr. | na. |
|---|---|---|---|---|---|---|
| 127 | 4 | 3 | | 246 | 2 | 3 |
| 24 | 2 | 3 | | 127 | 1 | 2 |
| 13 | 3 | 4 | | 64 | 2 | 3 |

**Figure 6**

## SELECTED REFERENCES

Greenwood, Isaac. *Arithmetick, Vulgar and Decimal: With the Application Thereof to a Variety of Cases in Trade and Commerce.* Boston, MA: Kneeland and Green, 1729. Available in the George Arthur Plimpton Collection, Rare Book and Manuscript Library, Columbia University.

Karpinski, Louis C. *History of Arithmetic.* Chicago, IL: Rand, McNally, 1925.

Smith, David Eugene. *The Sumario Compendioso of Brother Juan Diez.* Boston, MA: Ginn, 1921. Only four copies of the original Diez work have survived, all of which are in museums and unavailable to teachers.

———, and Jekuthiel Ginsburg. *A History of Mathematics in America before 1900.* Chicago, IL: Open Court, 1934.

EILEEN F. DONOGHUE

# THIRD INTERNATIONAL MATHEMATICS AND SCIENCE STUDY (TIMSS)

A study that investigated the context, teaching, and learning of school mathematics and science conducted by the International Association for the Evaluation of Educational Achievement (IEA). The IEA is an international cooperative of research centers that conducts comparative studies focusing on how educational policies and practices influence national achievement. The organization has previously conducted several international studies, including two studies of school mathematics and two of school science education. The TIMSS project director was Albert E. Beaton of the Center for the Study of Testing, Evaluation, and Educational Policy (CSTEEP) at Boston College, which housed the TIMSS International Study Center.

Forty-seven countries widely spread throughout the world participated in the study. Many different languages, cultures, and mathematical and scientific traditions were represented, so, as with other IEA studies, extraordinary care had to be taken to produce instruments in which possible problems due to translation or cultural bias were minimized. TIMSS utilized a predominantly survey research methodology, but extensive information about the context of teaching and learning and detailed curricular data were obtained from participating countries to facilitate interpretation of outcome measures. Central to this was the study of variation in opportunity to learn the mathematics and science knowledge, skills, and attitudes assessed by the TIMSS instruments.

The conceptual framework for the study was derived from hypothesized relationships between variables from three levels of curriculum—the intended curriculum, which is the curriculum as defined, or implied, at the educational system level; the implemented curriculum, which is the curriculum as it is interpreted by teachers and made available to students; and the attained curriculum, which is what mathematics and science students have learned and what attitudes they have acquired toward these subjects during their schooling years. The primary goal was to identify those variables in the intended and implemented curricula that contribute to differences in the attained curriculum so that policy to effect improvements in mathematics and science education could be formulated. Target populations consisted of all students in the two adjacent grades containing a majority of nine-year-olds for Population 1, all students in the two adjacent grades containing a majority of thirteen-year-olds for Population 2, and

students in their final year of schooling in both public and private schools for Population 3. Measures of mathematics and science literacy were obtained from students in their final year of schooling, usually the twelfth grade. In addition, subpopulations of students in their final year taking specialist mathematics, specialist physics, or both of these subjects were tested in these subjects.

TIMSS was grounded in an intensive comparative study of mathematics and science curricula. Curriculum guides and textbooks commonly used by the target populations from each of the participating systems were subjected to detailed analyses, and national curriculum experts provided descriptive and explanatory information about curricular aims and trends. In addition, participating centers traced key topics through the grades to locate critical points in their development.

Three questionnaires were administered concurrently with the assessment instruments. A school questionnaire, completed by principals or department heads, was designed to yield data on those features of the school setting believed to influence the teaching and learning of mathematics and science. Teachers of students in the samples responded to questionnaires that sought information about their educational backgrounds and beliefs, the content they taught the target students, and how they went about their teaching task. This included resources used, lesson planning, activities during the lesson, and detailed information about how they taught several mathematics and science topics. Students provided information about their backgrounds and their attitudes to mathematics and science.

The construction of written assessment instruments for TIMSS required a balance to be struck between the desire to have adequate and fair coverage of both mathematics and science curricula for all of the participating countries and the limited time schools would be able to make available for responding to tests and questionnaires. This balance was achieved by adopting a test design for each population level that incorporated rotation of test forms, with clusters of items rotated among forms, to maximize coverage of curricula within acceptable testing time. A substantial proportion of the written tests consisted of free-response items, and these were coded to provide information about common misconceptions among learners and different approaches to solving problems as well as for correctness.

Other aspects of TIMSS of special interest to educators include assessment of levels of mathematics and science "literacy" among students as they reach the end of their schooling and the use of "hands-on" performance assessment tasks to complement the more traditional assessment methods. These tasks, designed to measure such things as ability to design an investigation, generate data, interpret this data, and generalize from the data, were administered to subsamples of nine-year-olds and thirteen-year-olds in the study.

*See also* Assessment of Student Achievement, Overview; Test Construction

## SELECTED REFERENCES

Beaton, Albert E., Michael O. Martin, Ina V. S. Mullis, Eugene Gonzalez, Theresa A. Smith, and Dana L. Kelly. *Science Achievement in the Middle School Years: IEA's Third International Mathematics and Science Study (TIMSS)*. Chestnut Hill, MA: CSTEEP, 1996.

Beaton, Albert E., Ina V. S. Mullis, Michael O. Martin, Eugene J. Gonzalez, Dana L. Kelly, and Theresa A. Smith. *Mathematics Achievement in the Middle School Years: IEA's Third International Mathematics and Science Study (TIMSS)*. Chestnut Hill, MA: CSTEEP, 1996.

Harmon, Maryellen, Theresa A. Smith, Michael O. Martin, Dana L. Kelly, Albert E. Beaton, Ina V. S. Mullis, Eugene J. Gonzalez, and Graham Orpwood. *Performance Assessment in IEA's Third International Mathematics and Science Study (TIMSS)*. Chestnut Hill, MA: CSTEEP, 1997.

Howson, A. Geoffrey. *Mathematics Textbooks: A Comparative Study of Grade 8 Texts*. Vancouver, BC, Canada: Pacific Educational, 1995.

Martin, Michael O., Ina V. S. Mullis, Albert E. Beaton, Eugene J. Gonzalez, Theresa A. Smith, and Dana L. Kelly. *Science Achievement in the Primary School Years: IEA's Third International Mathematics and Science Study (TIMSS)*. Chestnut Hill, MA: CSTEEP, 1997.

Mullis, Ina V. S., Michael O. Martin, Albert E. Beaton, Eugene J. Gonzalez, Dana L. Kelly, and Theresa A. Smith. *Mathematics Achievement in the Primary School Years: IEA's Third International Mathematics and Science Study (TIMSS)*. Chestnut Hill, MA: CSTEEP, 1997.

———. *Mathematics and Science Achievement in the Final Year of Secondary Schooling*. Chestnut Hill, MA: CSTEEP, 1998.

Robitaille, David F., ed. *National Contexts for Mathematics and Science Education: An Encyclopedia of the Education Systems Participating in TIMSS*. Vancouver, BC, Canada: Pacific Educational, 1997.

———, and Robert A. Garden, eds. *Research Questions and Study Design*. Vancouver, BC, Canada, Pacific Educational, 1996.

Robitaille, David F., Curtis C. McKnight, William H. Schmidt, Edward Britton, Senta Raizen, and Cynthia Nicol. *Curriculum Frameworks for Mathematics and Science*. Vancouver, BC, Canada: Pacific Educational, 1993.

Schmidt, William H., Curtis, C. McKnight, G. A. Valverde, R. T. Houang, and D. E. Wiley. *Many Visions, Many Aims*. 2 vols. Dordrecht, Netherlands: Kluwer, 1997.

ROBERT A. GARDEN

# THORNDIKE, EDWARD L. (1874–1949)

A behavioral psychologist and a connectionist whose experimental studies starting in the early twentieth century were instrumental in discrediting the faculty psychology theory of Edward Brooks and the theory that learning in one area can be automatically transferred to another area. Thorndike's psychology, which centers around stimulus-response learning bonds, was based on carefully designed statistical studies. If the objective were to perform multidigit addition quickly and accurately, Thorndike's analysis would determine what bonds were required, how many repetitions were required to establish each bond, and how drill should be spaced to maintain these bonds. Thorndike's three-volume *Educational Psychology* (1913) was a major contribution to the psychology of learning. The application of his theory to arithmetic teaching is set out in his *The Psychology of Arithmetic* (1922). Thorndike also wrote a series of arithmetic texts, *The Thorndike Arithmetics* (1917), which stressed common adult-life problems but which showed the effects of connectionism less than some texts that appeared in the 1920s.

*See also* Psychology of Learning and Instruction, Overview

## SELECTED REFERENCES

Bidwell, James K., and Robert G. Clason, eds. *Readings in the History of Mathematics Education*. Washington, DC: National Council of Teachers of Mathematics, 1970.

Jones, Phillip S., ed. *A History of Mathematics Education in the United States and Canada*. 32nd Yearbook. Washington, DC: National Council of Teachers of Mathematics, 1970.

Thorndike, Edward L. *The Psychology of Arithmetic*. New York: Macmillan, 1922.

JAMES K. BIDWELL
ROBERT G. CLASON

# TILING

A mathematical technique used by our ancestors to make mosaic patterns well before they invented writing. The words *tiling*, *tessellation*, and *mosaic* have the same mathematical meaning. To tile the plane, space, or any region means to cover that region with shapes such that no point is left uncovered and no point lies in the interior of two of the shapes. Intuitively, for the plane this means cutting the plane into pieces, allowing only edges to overlap. The Dutch artist Maurits C. Escher (1898–1972) is renowned for his imaginative use of tiling patterns.

Classic Islamic culture displayed implicit knowledge of some mathematical aspects of tiling through its extensive ornamentation, in particular at the Alhambra in Granada, Spain. However, little precedes the work of the last 100 years other than Johann Kepler's organized examination of tilings in 1619. The only possible edge-to-edge tilings of the plane by a single regular polygon are tilings by the equilateral triangle, the square, or the regular hexagon. (This was known by the Pythagoreans and is shown by considering how many vertex angles of some regular polygon can fit around one point.) These three tilings are called *regular* tilings. Kepler knew that there are precisely 11 edge-to-edge tilings by sets of regular polygons such that all vertices of the tiling "look alike." But Kepler's observation concerning these so-called Archimedean tilings was forgotten for 300 years.

Finding all possible tilings determined by sets of regular polygons is more involved than generally realized. Grünbaum and Shephard (1977) described the history and a comprehensive development. Every triangle tiles the plane. (Any triangle, together with its image obtained by rotation of 180° in the plane of the triangle about the midpoint of any one side, forms a parallelogram. Parallelograms stack to tile the plane.) A polygon is convex if the segment joining any two points in the interior of the polygon lies entirely in the interior. Surprisingly, every quadrilateral, whether convex or not, tiles the plane. (Any quadrilateral, together with its image obtained by rotation of 180° in the plane of the quadrilateral about the midpoint of any one side, forms a six-sided figure that has opposite sides congruent and parallel. Such hexagonal figures always stack to tile the plane.) The cataloging of those convex pentagons that tile the plane is not settled, although the list for convex hexagons was determined in 1918. No single convex polygon with more than six sides can tile the plane (Niven 1978).

Given a polygon that does tile the plane, how many essentially different tilings can be made with this polygon? For example, a square tiles the plane in infinitely many ways. (Start with the regular graph-paper tiling of the plane by a square but slide one row slightly; with admittedly dull results, there are infinitely many possible ways to slide rows to obtain new tilings.) On the other hand, a regular hexagon tiles the plane in exactly one way, the regular tiling frequently seen on bathroom floors. What are the possibilities between these two extremes? George E. Martin (1991) gives examples of polygons that tile the plane in exactly $n$ ways for $0 < n < 11$. It is unknown if there is a polygon that tiles the plane in exactly $n$ ways for any integer $n$ with $n > 10$.

Penrose tiles are a pair of tiles that together tile the plane but only in ways such that the resulting tilings are not repetitive under translation. Penrose tiles were created by Roger Penrose and announced by Richard K. Guy (1976). Martin Gardner (1977) made their existence known to a vast readership in his celebrated column "Mathematical Games." It is unknown whether there exists a single polygon that tiles the plane only in such ways that the resulting tilings are not repetitive under translation, although the analogous three-space problem has been solved by the production of such a shape.

Tiling activities fit into the mathematics curriculum at every level—from the first formulating of the attributes of shapes and space by elementary students, to the exploration of symmetry in nature by secondary students, to the facing of the unsolved research problems mentioned here by advanced students. Numerous books, posters, templates, tiling shapes, and other tiling materials for K–12 and higher can be found in the mathematics catalogs of such companies as Dale Seymour Publications and Creative Publications.

*See also* Art; Escher; Polygons; Symmetry

## SELECTED REFERENCES

Creative Publications. *Teachers' Catalog.* Mountain View, CA: Creative, 1995. Offers *Pentominoes* (plastic sets, p. 79), *Pattern Blocks* (plastic or wood, p. 74), *Tessellation Tracer* (plastic template, p. 75), *Tessellation Exploration Pack* (plastic shapes and template, p. 75), *Stained Glass Tessellation Posters* (11 posters, p. 118), and related teaching supplies.

Gardner, Martin. "Mathematical Games, Extraordinary Nonperiodic Tiling that Enriches the Theory of Tiling." *Scientific American* 236(Jan. 1977):110–121.

———. *Penrose Tiles to Trapdoor Ciphers.* New York: Freeman, 1989.

Grünbaum, Branko, and Geoffrey C. Shephard. *Tilings and Patterns.* New York: Freeman, 1987.

———. "Tilings by Regular Polygons." *Mathematics Magazine* 50(Nov. 1977):227–247.

Guy, Richard K. "The Penrose Pieces." *Bulletin of the London Mathematical Society* 8(Mar. 1976):9–10.

Klarner, David A., ed. *The Mathematical Gardner.* Belmont, CA: Wadsworth, 1981.

Martin, George E. *Polyominoes, A Guide to Puzzles and Problems in Tiling.* Washington, DC: Mathematical Association of America, 1991.

Niven, Ivan. "Convex Polygons that Cannot Tile the Plane." *American Mathematical Monthly* 85(Dec. 1978): 785–792.

Schattschneider, Doris. *Visions of Symmetry: Notebooks, Periodic Drawings, and Related Work of M. C. Escher.* New York: Freeman, 1990.

Seymour Publications. *Secondary Mathematics.* Palo Alto, CA: Seymour, 1994. Catalog offers *Pentominoes* (plastic sets, p. 27), *Mathmaster* (plastic template, p. 40), *Tessellations* (poster, p. 44), *Tessellation Teaching Masters* (overhead transparencies, p. 44), *Escher Neckties* (clothing, p. 47), *The Alhambra Past and Present* (video, p. 43), and related teaching supplies and books.

GEORGE E. MARTIN

# TIME

A concept best taught through three stages of instruction. Of what do you think when someone talks about "teaching students to tell time"? Most educators recall clocks made of paper plates and endless numbers of student worksheets with faces of analog and digital clocks. Perhaps the reason many students have difficulty learning to tell time is due to the methods utilized (i.e., paper clocks and worksheets) instead of developmentally appropriate activities. In order for students to understand the concept of time, three distinct stages of instruction are used. First, the students experience time concretely, through tactile and kinesthetic activities that involve the whole child. Second, students encounter representational forms through visuals. Finally, after the presentation of concrete and varied forms, students experience the abstract concept of identifying and writing times on clock faces. Integrating the concept of time into thematic units, children's literature, and the students' daily routines is also instrumental in understanding this complex concept.

Concrete explorations that involve holistic learning and integrate various mathematics concepts are necessary for students to understand how an analog clock works. A set of attribute blocks (six shapes in three colors, two sizes, and two levels of thickness for a total of sixty blocks) is used to arrange a circle where each adjacent block differs from its neighbor by only one attribute. Higher-order thinking, problem solving, reasoning, and communication are all essential to completion of the task. Students discuss the generally circular shape they have created, objects the circle resembles, and the relationship between this shape and a clock face. Next, the students place number cards (1–12) at the appropriate places on their clock faces so as to match the hours on a real clock. One student walks around the circle, stopping on each multiple of 5. A 1 is placed next to the block the first time the student stops; the second time, a 2; and so forth until the numerals 1 through 12 are placed around the circle (Andrade 1992). The students practice walking around the circle and counting by 5s as they step on a numeral card, thus building a foundation for multiplication and division as well as telling time. This activity can progress to students forming the hour hand and minute hand with their bodies, telling time to the hour, half-hour, and 5-minute intervals. The total movement expended during these activities will help students remember what they have done as telling time is integrated into other areas of the curriculum.

One way to incorporate representational forms of time and integrate the curriculum is through children's literature. A personal understanding of the various concepts is built on the kinesthetic foundation when literature is used to reinforce time-telling skills. There are numerous children's literature selections that are routinely associated with the concept of time. Stories that were not necessarily designed to develop time-telling skills are also excellent resources for developing personal understanding of various time concepts. *Clean Your Room, Harvey Moon!* (Cummings 1991) is a humorous story about a boy who is told to clean his room on Saturday morning. Naturally, he would rather be watching cartoons. After procrastinating, he hides everything under the rug. As various times are mentioned, the teacher demonstrates for young students how to display these times on a one-handed clock (Van de Walle 1994). The clock could be read using terms such as *approximately* and *almost* to enable the children to gain an understanding of time to the hour before modeling with a clock where the movement of the minute and hour hands is coordinated.

The abstract stage of telling time should be presented to students only after kinesthetic and representational activities that incorporate personal experiences. For example, students can identify the time of their favorite TV program, model it on a manipulative clock, draw the time on an analog clock, and then translate the time to a digital clock by writing the time (4:00 P.M.). The abstract concepts of A.M. and P.M. may be introduced with either the analog or digital model. Given the integrated use of both digital and analog models in modern society, both should be addressed through additional conversion activities such as finding times for movies in the newspaper or making time zone conversions. In addition, as A.M. and P.M. are discussed, the concept of the 24-hour clock should be introduced. The more abstract concepts should follow the concrete instruction provided through kinesthetic, tactile, visual, and auditory experiences. Teaching about telling time should allow all students multiple opportunities for learning to tell time.

*See also* Books, Stories for Children

### SELECTED REFERENCES

Andrade, Gloria S. "Teaching Students to Tell Time." *Arithmetic Teacher* 39(Apr. 1992):37–41.

Cummings, Pat. *Clean Your Room, Harvey Moon!*. New York: Bradbury, 1991.

Van de Walle, John A. *Elementary School Mathematics: Teaching Developmentally.* New York: Longman, 1994.

RUBY BOSTICK MIDKIFF
RONALD W. TOWERY

## TIME SERIES ANALYSIS

Refers to the analysis of statistical data that are collected, observed, or recorded at regular intervals of time. We search for observable regularities and patterns that are so persistent that they cannot be ignored. If we subsequently base our forecasts on such regularities and patterns, we are simply expressing our belief that what has happened in the past will, to a greater or lesser extent, happen again in the future. Although analysis of time series is generally employed in connection with business and economic data, it also applies to such periodic events as traffic counts on a certain highway, a dieter's daily caloric intake, or a person's heartbeat recorded on a computer printout. In business and economics, the analysis of time series is of greater practical importance since the patterns that are thus revealed form the basis of forecasts affecting the decisions made by business managers and government officials. The future is unpredictable, but to make the best forecasts, we must summarize and analyze what has happened in the past.

New and highly sophisticated computer models have been developed, but traditional methods continue to be employed because of their relative simplicity and because they produce quite adequate results. Forecasting is a valuable and widely used technique, where mathematical models are combined with business judgment and common sense.

These forecasts must be reexamined at intervals of time and adjusted if trends or seasonal patterns have changed. For example, long-term forecasts that predicted rapidly increasing sales for music recorded on audiocassettes became valueless when compact disks were introduced. Hundreds of millions of CDs were sold and, instead of increasing, sales of audiocassettes declined. Businesspeople who knew the music industry adjusted their forecasts to anticipate and reflect the new reality.

In traditional time series analysis, the movements or fluctuations of a time series are classified into four basic types of variation that, superimposed and acting in concert, account for the changes in the series over a period of time. These four components are secular trend, seasonal variation, cyclical variation, and irregular variation. It is assumed that there is a multiplicative relationship among them, namely, that any particular value in a series is a product of the factors which can be attributed to these components. If we designate an actual value (such as actual sales) as $Y$ and designate the four components of a time series as $T$, $S$, $C$, and $I$, we can say that $Y = T \times S \times C \times I$.

The secular (or long-term) trend of a time series is the smooth or regular movement of a series over a fairly long period of time. In other words, we might say that the trend of a time series reflects the overall sweep of its development, or that it characterizes the gradual and persistent pattern of its changes. For example, if we examine the population of the United States by age, we find that from the year 1820 to 1990, the median age of the population increased gradually and persistently from 16.7 years to 33.0 years. Based on this information, it has been predicted that by the year 2030 the median age will increase to about 42 years. Linear trends are trends that may be described by straight lines drawn on a graph, but trends of many time series can be described only by means of more complicated types of mathematical curves.

Seasonal variation consists of any kind of variation that is of a persistent nature, provided that the length of its repeating cycle is not more than one year. Although its name implies a connection with the seasons of the year, it includes any type of variation that is of a periodic nature. Thus, examples of seasonal variation include the number of passengers carried by buses in a certain community during various hours of the day and night, the number of patrons at a fast-food restaurant during various days of the week, and the sales of ski boots or bathing suits during the various months of the year. Almost any type of retail trade, manufacturing, agricultural, fishing, and other economic activity can be expected to have seasonal variation because of customs (holidays, for example), natural phenomena (such as weather), and other factors.

Cyclical variations are often called *business cycles.* These are up-and-down movements of a time series, which differ from seasonal variations in that they extend over longer periods of time and are often attributed to general economic conditions. Periods of prosperity, recession, depression, and recovery are sometimes considered to be the four phases of a complete business cycle. Many theoretical explanations of the business cycle exist, and there are differences of opinion concerning the causes of business cycles.

Irregular variations of time series are those that are random or those caused by special events such as wars, revolutions, elections, bank failures, fires, floods, and earthquakes. If the events are random occurrences, then, in the long run, such fluctuations will average out. If they are due to special events, these should be recognized and eliminated before the other three components are investigated and forecasts of future activity are made.

Calculated trend and seasonal factors can be multiplied for each period to obtain normal values. Normal values are forecasts of actual values that are not affected by cyclic or irregular values.

*Example 1:* The treasurer of a corporation has determined that a straight-line trend best describes the amount and direction of the firm's sales. Projecting the trend for quarterly (three-month) periods over the next two years, the treasurer obtains 100, 102, 104, 106, 108, 110, 112, and 114 thousands of dollars. The treasurer has also determined that the seasonal indexes for quarters 1, 2, 3, and 4 are 70, 115, 130, and 85. Use this information to forecast normal sales for each of the eight quarters.

*Solution:* Since seasonal indexes are expressed as percentages (with the percent sign omitted), we must change the indexes to proportions and then multiply by the corresponding trend values. The forecasted values are displayed in the last column of Table 1.

*Example 2:* If the actual sales for Example 1 are 77.0, 123.2, 135.2, 88.3, 71.8, 115.1, 136.0, and 95.9 thousands of dollars, use this data and the corresponding normal data to determine the cyclical-irregulars (the cyclicals and irregulars combined).

*Solution:* Divide the actual data by the normal data to obtain the cyclical-irregulars, expressed as proportions. Then multiply each proportion by 100 to obtain the cyclical-irregulars as percentages. Cyclical irregulars are usually expressed as percentages without a percent sign. The cyclical irregulars are displayed in the last column of Table 2. We may observe that the cyclical irregulars move downward

**Table 1**

| Year | Quarter | Trend $T$ (in thousands) | Seasonal $S$ | Normal $T \cdot S$ (in thousands) |
|------|---------|--------------------------|--------------|-----------------------------------|
| 1 | 1 | $100 | 0.70 | $ 70.0 |
| 1 | 2 | 102 | 1.15 | 117.3 |
| 1 | 3 | 104 | 1.30 | 135.2 |
| 1 | 4 | 106 | 0.85 | 90.1 |
| 2 | 1 | 108 | 0.70 | 75.6 |
| 2 | 2 | 110 | 1.15 | 126.5 |
| 2 | 3 | 112 | 1.30 | 145.6 |
| 2 | 4 | 114 | 0.85 | 96.9 |

**Table 2**

| Year | Quarter | Actual data $T \cdot S \cdot C \cdot I$ (in thousands) | Normal $T \cdot S$ (in thousands) | Cyclical-irregulars $\dfrac{T \cdot S \cdot C \cdot I}{T \cdot S} \cdot 100$ |
|------|---------|--------------------------------------------------------|-----------------------------------|------------------------------------------------------------------------------|
| 1 | 1 | $77.0 | $70.0 | 110.0 |
| 1 | 2 | 123.2 | 117.3 | 105.0 |
| 1 | 3 | 135.2 | 135.2 | 100.0 |
| 1 | 4 | 88.3 | 90.1 | 98.0 |
| 2 | 1 | 71.8 | 78.6 | 95.0 |
| 2 | 2 | 115.1 | 126.5 | 91.0 |
| 2 | 3 | 136.0 | 145.6 | 93.4 |
| 2 | 4 | 95.9 | 96.9 | 99.0 |

from quarter 1 of year 1 through quarter 2 of year 2, after which they rise for the final two quarters.

Readers who wish to pursue further study of traditional time series analysis can do so without prior statistical training if they are prepared in elementary algebra.

*See also* Modeling; Statistics, Overview

### SELECTED REFERENCES

Berenson, Mark L., and David M. Levine. *Statistics for Business and Economics.* 2nd ed. Englewood Cliffs, NJ: Prentice-Hall, 1993.

Daniel, Wayne W., and James C. Terrell. *Business Statistics: Basic Concepts and Methodology.* 6th ed. Boston, MA: Houghton Mifflin, 1991.

Freund, John E., Frank J. Williams, and Benjamin M. Perles. *Elementary Business Statistics: The Modern Approach.* 6th ed. Englewood Cliffs, NJ: Prentice-Hall, 1993.

Picconi, Mario J., Albert Romano, and Charles L. Olson. *Business Statistics: Elements and Applications.* New York: HarperCollins, 1993.

BENJAMIN M. PERLES

# TOPOLOGY

An area of pure mathematics that deals with properties of objects which are not affected by continuous deformation. Thus, geometric properties such as length and curvature are not of interest to a topologist, but the number of holes in an object is a topological property. A topologist is sometimes said to be a person who does not know the difference between a coffee cup and a doughnut, since if these are made out of elastic clay, one can be continuously deformed to the other without tearing or pinching together disjoint regions. Indeed, the hole in the handle of the cup corresponds to the hole in the doughnut, while the cupped part can be flattened and rounded. Another phrase commonly applied to topology is "rubber sheet geometry," since the properties that are of interest to a topologist are those that would be relevant to a stretchable rubber sheet.

An excellent example of topology is knot theory, which deals with how to tell whether one given knot (with the ends pinched together) can be deformed to another. Think of the knots as rubber bands, so that the length is not important. Actually, the knots themselves are all equivalent to a circle; the issue is whether a deformation of all of space can be achieved that sends one knot to the other. The subtlety of the subject can be seen by the example in Figure 1 of a knot that appears complicated but can

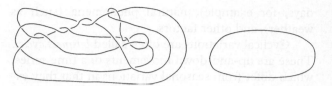

**Figure 1** *Knots.*

actually be manipulated into the trivial knot shown on the right. Knot theory became popular in the late nineteenth century when it was believed that a substance called *ether* pervaded all of space. Lord Kelvin (1824–1907) hypothesized that atoms were knots in the ether, with different knots corresponding to different atoms. Using ad hoc methods, mathematicians created an exhaustive list of all knots with 10 or fewer crossings (when pictured in the usual two-dimensional projection). But it was not until the 1920s that a systematic way of studying knots was devised by a Princeton mathematician, James Alexander (1888–1971). He gave a method of associating a polynomial with each knot diagram in such a way that equivalent knots would have the same polynomial. Thus, if two knots have distinct polynomials, we can conclude that one cannot be deformed to the other. The Alexander polynomial is not a perfect invariant, however, in that certain inequivalent knots have the same polynomial.

Since 1980, deep relationships of knot theory with quantum field theory in physics and with recombinant DNA experiments in molecular biology have been discovered. These have been related to major theoretical discoveries in knot theory during the same period, including the discovery of new polynomials to distinguish knots and the proof by Cameron Gordon and John Leucke that knots are determined by their complements. The complement of a knot refers to all points of space except the knot. The Gordon-Leucke result states that if the complements of two knots are homeomorphic, then the knots are equivalent. *Homeomorphic* is a central concept in topology. Two objects (often called *spaces*) are said to be homeomorphic if there is a one-to-one correspondence between points of one and points of the other in such a way that nearby points of one correspond to nearby points of the other. In general topology, there is not always a concept of distance, and an abstracted form of "nearby," the system of neighborhoods of each point, is used. The system of neighborhoods is called the *topology* of the space. If a set is equipped with a measure of the distance between its points, called a *metric*, then a typical neighborhood of a point is the set of all points within some distance $\epsilon$ of the point.

Among the goals of topology is the development of methods of showing that spaces are not homeomorphic. One such method uses the fundamental group of a space. This is a measure of the types of loops within the space, two loops being considered equivalent (or *homotopic*) if one can be continuously deformed to the other. For example, the fundamental group of the circle corresponds to the set of integers, since a loop can wrap around any number of times and in either direction. The fundamental group of a space is an algebraic object associated with each space in such a way that homeomorphic spaces have equivalent (isomorphic) fundamental groups. The Alexander polynomial of a knot diagram is derived from a study of the fundamental group of the complement of the knot.

Many of the spaces studied by topologists are difficult to envision concretely. For example, for any positive integer, $n$, one defines $n$-dimensional space, $R^n$, to be the set of ordered lists of real numbers $(x_1, \ldots, x_n)$, the $n$-dimensional ball, $B^n$, to be the points of $R^n$ such that $x_1^2 + \cdots + x_n^2 \leq 1$, and the $(n-1)$-dimensional sphere, $S^{n-1}$, to be the points satisfying $x_1^2 + \cdots + x_n^2 = 1$. Topologists work with $R^n$, $B^n$, and $S^{n-1}$ for an arbitrary large integer $n$, sometimes thinking by analogy from the cases $n = 2$ or 3 or often just working with some of the algebraic constructions associated with the space. Another large space considered by topologists is the set of all continuous real-valued functions defined on the closed interval from 0 to 1. Different topologies can be considered for this set. For example, the distance between functions $f$ and $g$ can be defined to be the largest value of $|f(t) - g(t)|$, or it might be defined to be $(\int_0^1 (f(t) - g(t))^2 dt)^{1/2}$. An example of a topological space whose topology does not come from a metric is given by the set of all real-valued functions defined on the real line, with a typical neighborhood of a point $f$ being determined by specifying for each number $t$ a positive number $\epsilon_t$ and taking all functions $g$ such that for all real numbers $t$, $|g(t) - f(t)| < \epsilon_t$. A question addressed by topologists and answered quite satisfactorily by Paul Urysohn in 1924 was to give conditions on a topological space which guarantee that its topology is induced by a metric. Such a result is called a *metrization theorem*.

Some theorems in topology that have had important applications are fixed point theorems. One of these is the Brouwer Fixed Point Theorem, proved in 1911; it is a focal point of most first courses in topol-

ogy. It involves continuous functions $f$ from $B^n$ to $B^n$; such an $f$ associates to each point $x$ of the ball a point $f(x)$, also in the ball, in such a way that if $x$ and $x'$ are close together, then so are $f(x)$ and $f(x')$. The theorem states that every such $f$ must have a fixed point, that is, a point $x$ for which $f(x) = x$. This says roughly that if you shake up a ball of liquid, then some particle will end up where it started. Applications of fixed point theorems occur in differential equations, game theory, and economics. In each case, a fixed point theorem guarantees the existence of a solution of an equation, which might be an optimal strategy in a game or a stable economic situation.

Although topology was not really an organized subject until the work of Henri Poincaré in the 1890s, some topological ideas were introduced by Leonhard Euler in the eighteenth century. One was the Königsberg Bridge problem, in which Euler showed that it was impossible to traverse the seven bridges of Königsberg so that each bridge is crossed just once. This is a problem in topology because it does not depend upon the lengths of the bridges, but only on their relative locations. Another topological result of Euler deals with convex polyhedra. A polyhedron is a solid with plane faces and straight edges arranged so that every edge joins two vertices and is also the common edge of two faces. It is convex if every two points in the solid can be connected by a line lying totally within the solid. Euler proved that for any convex polyhedron, the formula $V - E + F = 2$ holds, where $V$, $E$, and $F$ are, respectively, the number of vertices, edges, and faces of the polyhedron. This formula has been generalized in many ways by modern topologists, leading to various types of Euler characteristics, all of which involve a sum with alternating plus and minus signs of numbers related to the various dimensions.

An important class of topological spaces consists of manifolds. The idea of manifold was introduced by Bernhard Riemann in 1854. An $n$-dimensional manifold (or $n$-manifold, for short) is a space such that every point has a neighborhood which is homeomorphic to $R^n$. Some examples of two-dimensional manifolds are $S^2$, which can be thought of as the surface of a ball, and the torus, which can be thought of as an inner tube or the surface of a doughnut. Both of these manifolds are compact and orientable. *Compact* means roughly that it is bounded and contains all possible limit points, while *orientable* means that it is possible to choose, in a continuous way, a direction for little circles around all the points. It was proved in the late 1800s that every compact orientable 2-manifold is homeomorphic to a sphere or the surface of an *n*-holed doughnut for some positive integer *n*. Moreover, none of these 2-manifolds is homeomorphic to one another, as can be established by comparing their fundamental groups. This classification theorem for compact orientable 2-manifolds is the type of result for which topologists strive. One roadblock for a classification of compact orientable 3-manifolds is the Poincaré Conjecture. This conjecture, posed in 1904, states that a compact, orientable 3-manifold with trivial fundamental group must be homeomorphic to the sphere $S^3$. This conjecture remains unresolved and is the most famous outstanding problem in topology. The Poincaré Conjecture admits a natural extension to *n* dimensions for any positive integer *n*. For $n \geq 5$, this conjecture was proved in 1959 by Stephen Smale, and for $n = 4$ it was proved by Michael Freedman in 1981.

Building upon Freedman's work, Simon Donaldson proved in 1982 that $R^4$ admits more than one differentiable structure. This discovery was shocking because it was known that for every *n* except 4, $R^n$ admitted a unique differentiable structure. What this means is that there is a space *X* which is homeomorphic to $R^4$ but for which a homeomorphism cannot be chosen that is smooth, or differentiable, when using the identifications of neighborhoods in *X* with $R^4$. Donaldson's result is of interest to physicists, as it sheds light on the possible structures for four-dimensional space-time. A more basic concept in topology that has been widely applied by physicists, especially quantum physicists, is the notion of fiber bundle, developed by Norman Steenrod in the 1950s. A fiber bundle over a space *X* is another space *E* that we think of as sitting above *X* and projecting down to it in a manner which is called *locally trivial*. This means that above small portions *U* of *X*, *E* looks like a product of *U* with some other space.

A famous example of a fiber bundle is the Möbius band, pictured in Figure 2. This is obtained by making a half-twist in a strip of paper and then gluing the ends together. The Möbius band projects in a natural way to the circle formed from the line that ran along the middle of the strip. Over any incomplete portion *U* of that circle, this looks just like the product of *U* with an interval, but this is not the case globally. The Möbius band is an example of a manifold with boundary; whereas interior points have neighborhoods homeomorphic to $R^2$, boundary points have neighborhoods homeomorphic to half of $R^2$. It is the simplest example of a nonorientable manifold, since it is impossible to continuously assign directions to circles encircling its points.

Point-set, or general, topology is the study of properties, such as compactness or connectedness, that are directly related to the neighborhood systems of the spaces. Combinatorial topology is the study of spaces that are built up from simplices, which are *n*-dimensional generalizations of triangles. The study can be considered combinatorial since the space is totally determined by the way in which the simplices intersect one another. Algebraic topology is the study of the kind of information about spaces that can be obtained by associating to them algebraic gadgets such as the fundamental group.

Topology as a rigorous subject is generally not taught until the fourth year of the undergraduate mathematics major curriculum or the beginning of graduate school. An enrichment course for talented high school students could contain some topics in topology, such as the idea of a metric space and the construction of a Möbius band.

*See also* Graph Theory, Overview

## SELECTED REFERENCES

Adams, Colin C. *The Knot Book*. New York: Freeman, 1994.

Croom, Fred H. *Principles of Topology*. Philadelphia, PA: Saunders, 1989.

Hocking, John G., and Gail S. Young. *Topology*. Reading, MA: Addison-Wesley, 1961.

Munkres, James R. *Topology: A First Course*. Englewood Cliffs, NJ: Prentice-Hall, 1975.

Sieradski, Allan J. *An Introduction to Topology and Homotopy*. Boston, MA: PWS-Kent, 1992.

DONALD M. DAVIS

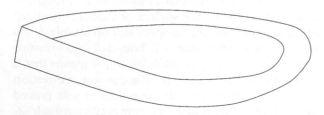

**Figure 2** *Möbius band.*

# TRANSCENDENTAL AND ALGEBRAIC FUNCTIONS

Defined by analogy with the concept of a transcendental number. Just as a transcendental number is one that is not algebraic, so a transcendental function is one that is not algebraic. A function *f* of a single variable *x* is *algebraic* if there is a way to combine the function with polynomials in $x$ to yield an expression that is identically zero. Consider for example the function $f$ given by

$$f(x) = (x + 2)/\sqrt{(x^2 + x + 1)}$$

We have

$$(f(x))^2 = (x^2 + 4x + 4)/(x^2 + x + 1)$$

$$(x^2 + x + 1)(f(x))^2 = x^2 + 4x + 4$$

$$(x^2 + x + 1)(f(x))^2 - (x^2 + 4x + 4) = 0$$

The last identity shows that $f$ is an algebraic function. Of course, for any function $g$, expressions such as $g - g$ and $0g$ are identically zero. We must not allow such "trivial" combinations. The precise definition may be stated as follows: a function of $x$ is algebraic if there exists a polynomial $P$ in another variable, say $y$, where the coefficients of $P$ are themselves polynomials in $x$ and at least one of these coefficients is not the zero polynomial, such that when $y$ is replaced by $f(x)$ the resulting expression $P(f(x))$ is identically zero. For the function

$$f(x) = (x + 2)/\sqrt{(x^2 + x + 1)}$$

we saw above that such a polynomial $P$ is

$$P(y) = (x^2 + x + 1)y^2 - (x^2 + 4x + 4)$$

A function is called *transcendental* if it is not algebraic. Such functions "transcend" the operations of arithmetic, in the sense just explained.

It is a good exercise in manipulative algebra to ask students to find polynomials $P(y)$ that prove various given functions $f(x)$ to be algebraic. For example, that

$$f(x) = \sqrt{(x - 2)} - \sqrt{(3 + x^2)}$$

is algebraic is shown by substitution of $f(x)$ for $y$ in the polynomial

$$P(y) = y^4 - 2(x^2 + x + 1)y^2 + x^4$$
$$- 2x^3 + 11x^2 - 10x + 25 = 0$$

Such exercises can easily be presented without reference to the formal definition of algebraic function or use of the second variable $y$.

The first transcendental functions encountered in the curriculum are the exponential and circular functions and their inverses. These functions occur in intermediate algebra and trigonometry and are consequently often called the *elementary transcendental functions*. If $b$ is a positive number different from 1, the *exponential function base* $b$ is $\exp_b(x) = b^x$. The inverse of this function is the *logarithm function base*

$b$, defined by $\log_b(x) = y$, where $\exp_b(y) = x$. The precise definition of $b^x$ for all numbers $x$ cannot be given without the use—directly or indirectly—of the limit concept (see later). The most satisfactory approach at the precalculus level may be to give the definition only for rational exponents $x = m/n$. This definition may be motivated as follows. (1) If $m$ and $n$ are positive integers, then

$$b^m b^n = (\text{product of } m \ b\text{'s}) \cdot (\text{product of } n \ b\text{'s})$$
$$= \text{product of } (m + n)b\text{'s} = b^{m+n}$$

and

$$(b^m)^n = \text{the product of } n \text{ factors, each a product of } m \ b\text{'s}$$
$$= \text{the product of } mn \ b\text{'s} = b^{mn}.$$

Suppose we want these laws to hold not just for positive integer exponents but for all rational number exponents. (2) Then $b^0 b = b^{0+1} = b$; so $b^0 = 1$. (3) Then $b^{-n}b^n = b^{-n+n} = b^0 = 1$; so $b^{-n} = 1/b^n$. (4) Also, $(b^{1/n})^n = b$; so $b^{1/n} = \sqrt[n]{b}$. (5) Then $b^{m/n} = (b^{1/n})^m$. Of course, this leaves some important facts unproved, namely, that with this definition we actually have $b^r b^s = b^{r+s}$ and $(b^r)^s = b^{rs}$ for all rational exponents $r$, $s$. These proofs are best postponed to second-term calculus or even to advanced calculus. The logarithm may then be "defined" as above, ignoring the lack of definition of $b^x$ for irrational $x$.

For the circular functions, consider a unit circle in a plane with Cartesian coordinates, having center $(0, 0)$ and radius 1. Given a number $x$, start at the point $(1, 0)$ and move along the circle a distance of $|x|$, in the counterclockwise direction if $x$ is positive, clockwise if $x$ is negative, to reach a point $P(x)$. Then the *sine* of $x$, denoted $\sin(x)$, is the second coordinate of $P(x)$ and the *cosine* of $x$, denoted $\cos(x)$, is the first coordinate of $P(x)$. Figure 1 shows an example of this definition where $0 < x < \pi/2$.

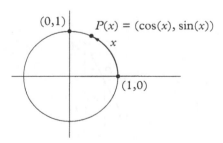

**Figure 1**

The other circular functions are defined in terms of sine and cosine as follows:

*tangent:*    $\tan(x) = \sin(x)/\cos(x)$

*cotangent:*    $\cot(x) = \cos(x)/\sin(x)$

*secant:*    $\sec(x) = 1/\cos(x)$

*cosecant:*    $\csc(x) = 1/\sin(x)$

Of the inverses of the circular functions, the most important are those of sine and tangent, defined by:

$$\sin^{-1}(x) = y, \quad \text{where } -\pi/2 \leq y \leq \pi/2$$
$$\text{and} \quad \sin(y) = x$$

and

$$\tan^{-1}(x) = y, \quad \text{where } -\pi/2 < y < \pi/2$$
$$\text{and} \quad \tan(y) = x$$

These functions occur frequently in integral calculus.

The circular functions were first employed in problems of surveying and measurement, both on or near the Earth's surface and in astronomy. In these problems, the functions are often called the *trigonometric functions*. The inverse tangent provides a convenient example of a one-to-one correspondence between a bounded and an unbounded set. For instance,

$$f(x) = 1/2 + (1/\pi)\tan^{-1}(x)$$

is a one-to-one function from the set of all real numbers onto the set of real numbers that are greater than 0 and less than 1.

A discussion of the origin of the names for the circular functions can be a worthwhile brief digression in a precalculus course to show the ancient roots of mathematical ideas that are still in use. Part of the origin may be seen in Figure 2. The tangent segment *TA*, which touches the unit circle at *T* (Latin *tangare*, to touch) has length $\tan(x)$, and the secant segment *OA*, which cuts the circle at *P* (*secare*, to cut) has length $\sec(x)$. But the $\sin(x)$ is equal to half the length of the chord *PQ*, so why isn't "sine" called "half-chord" in the original language of its creation? According to Howard Eves (1969, 196) that is exactly what it was first called (in Sanskrit) by Aryabhata the Elder (ca. 475–ca. 550), but a curious sequence of misreadings through Arabic into Latin led Gherardo of Cremona (ca. 1150) to translate it as *sinus*, the word for "bay."

Many authors omit parentheses with the elementary transcendental functions, writing $\log_b x$, $\cos x$, $\tan^{-1}x$ and so forth. But this notation can cause problems with calculators and computer software and so is avoided in some recent texts. An older notation for the inverse circular functions, *arcsin*, *arctan*, and so on, is still in use. As with transcendental numbers, the proof that the elementary transcendental functions are really transcendental is rarely included in the undergraduate curriculum.

Since the point $(\cos(x), \sin(x))$ is on the unit circle, we have the identity $(\sin(x))^2 + (\cos(x))^2 = 1$ for all $x$. (With omitted parentheses this becomes $\sin^2 x + \cos^2 x = 1$.) It is also apparent from the unit circle definition that, for all $x$,

$$\sin(x + 2\pi) = x$$
$$\sin(x + \pi) = \sin(-x) = -\sin(x)$$
$$\sin((\pi/2) - x) = \cos(x)$$

From this last result we see that

$$\cot(x) = \tan((\pi/2) - x) \quad \text{and}$$
$$\csc(x) = \sec((\pi/2) - x) \text{ (see Figure 3)}$$

Thus the prefix "co-" in the circular functions is seen to be derived from "complementary angle."

The set of algebraic functions is closed under function composition, that is, an algebraic function of an algebraic function is an algebraic function. This is not the case with transcendental functions. For ex-

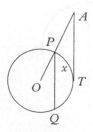

**Figure 2**

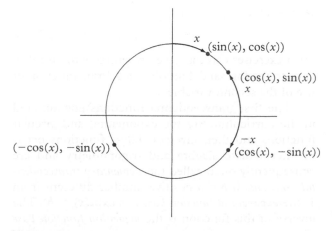

**Figure 3**

ample, the sine and inverse sine functions are transcendental, but their composition is the identity function, which is algebraic.

The attainment of skill with algebraic manipulation of the circular functions is still regarded by many teachers as valuable preparation for calculus, principally because of its use in techniques of integration. But the increased use of computer and calculator algebra systems, which carry out such manipulations as efficiently as their predecessors did arithmetic, together with the decrease in emphasis on algebraic manipulation in favor of graphical and numerical computation approaches to problems, are contributing to a decline in the importance of these techniques in the calculus curriculum. Still, some thoughtful educators remain wary of too much reliance on such devices.

Calculus offers two approaches to the definitions of the exponential and logarithm functions. The approach favored by most authors considers the function $L$ defined for $x > 0$ by

$$L(x) = \int_1^x dt/t$$

It is first shown that for all positive numbers $a$ and all rational numbers $r$, $L(a^r) = rL(a)$. Thus, $L(x)$ behaves like a logarithm $\log_e(x)$, where $e$ is the number for which $L(e) = 1$. It is natural then to define $\log_e(x) = L(x)$, for all positive numbers $x$. It follows from its definition that this function is strictly increasing and that its range is the set of all real numbers, so it has a unique inverse function $L^{-1}$, which is denoted $\exp_e$. Since $\exp_e(r) = e^r$ for all rational numbers $r$, it is natural to define $e^x = \exp_e(x)$ for all numbers $x$. For any positive $b$ different from 1, we then define $b^x = \exp_b(x) = \exp_e[x \log_e(x)]$ and $\log_b$ as the inverse of $\exp_b$. It can be shown that for $x > 0$, $\log_b(x) = \log_e(x)/\log_e(b)$; and this result could be used as an alternative definition of $\log_b$. The function $\exp_e$ is usually written simply as $\exp$. The *natural logarithm* of a positive number $x$, $\log_e(x)$ is written as $\ln(x)$ in the first calculus courses and on most calculators, but as $\log(x)$ in higher mathematics. Parentheses are often omitted: $\ln x$.

It is an instructive exercise in the definition of the definite integral together with the use of a small computer or a programmable calculator to show, using upper and lower sums, that $\int_1^{2.71} dx/x < 1$ and $\int_1^{2.72} dx/x > 1$, and hence that $2.71 < e < 2.72$. In fact, $e = 2.71828 \ldots$ is a transcendental number.

Another approach to these functions uses the precalculus definition of $b^r$ for rational numbers $r$, together with the fact that every real number $x$ is the limit of a sequence $(r_n)$ of rational numbers, to define

$$b^x = \lim_{n \to \infty} b^{r_n}$$

This approach is well suited to work with hand calculators, where for example $5^\pi$ may be approached through the sequence

$$(5^3, 5^{3.1}, 5^{3.14}, \ldots, 5^{3.14159}, \ldots)$$

A proof that, for all rational sequences $(r_n)$ converging to $x$, the sequences $b^{r_n}$ converge to the same number is usually omitted at this level. This approach is also useful motivation for the limit theorem

$$\lim_{h \to \infty} (1 + 1/h)^h = e$$

This result yields the formula for the value of an investment of $P_0$ dollars for $y$ years at $I$ percent per year compounded continuously:

$$P = P_0 e^{yI/100}$$

For example, \$10,000 at 8% per year compounded quarterly for 30 years grows to $10,000(1 + 0.08/4)^{4(30)} = \$107,651.63$, while compounded continuously it would become $10,000e^{0.08(30)} = \$110,231.76$.

In calculus, the transcendental functions are revealed as solutions to differential equations, as illustrated next. From the unit circle definition of sine, $\cos(x) < \sin(x)/x < \sec(x)$ for $-\pi/2 < x < \pi/2$, $x \neq 0$. This implies that

$$\lim_{x \to 0} \frac{\sin x}{x} = 1$$

from which are obtained the derivative formulas

$$(d/dx)\sin(x) = \cos(x)$$
$$(d/dx)\cos(x) = -\sin(x)$$

The derivative formulas for the other circular functions and the inverse circular functions are obtained from these. The results most often used are

$$(d/dx)\tan(x) = (\sec(x))^2$$
$$(d/dx)\sec(x) = \sec(x)\tan(x)$$
$$(d/dx)\tan^{-1}(x) = (1 + x^2)^{-1}$$
$$(d/dx)\sin^{-1}(x) = (1 - x^2)^{-1/2}$$

From the derivative formulas for sine and cosine it follows that the general solution to the differential equation for undamped vibration, $y'' = k^2 y$, where $k \neq 0$, is given by

$$y = A \sin(kx) + B \cos(kx)$$

where $A$ and $B$ are constants. It may also be worth class time to show how this formula may be written in the form of a *sine wave* $y = C \sin(kx + b)$, $C > 0$

and to derive the relations $C = \sqrt{(A^2 + B^2)}$, $b = \tan^{-1}(B/A)$ and to clarify the relation of the parameters $C$, $k$, $b$ to amplitude, frequency, wavelength, and phase shift for this extremely important curve.

From the integral definition of the logarithm and the Fundamental Theorem of Calculus comes the formula

$$(d/dx)\log_e(x) = 1/x$$

from which it can be shown that

$$(d/dx)\exp_e(x) = \exp_e(x)$$

It then follows that the general solution of the differential equation for growth or decay, $y' = ky$, where $k \neq 0$, is $y = Ce^{kx}$, where $C$ is a constant.

The differential equation $y'' - k^2y = 0$ may be used to introduce the *hyperbolic functions* $\sinh(x) = (e^x - e^{-x})/2$ and $\cosh(x) = (e^x + e^{-x})/2$. The general solution of this equation is

$$y = A\sinh(x) + B\cosh(x)$$

These functions are the fundamental trigonometric functions of hyperbolic non-Euclidean geometry, although this fact is rarely developed in the standard undergraduate curriculum. The graph of the hyperbolic cosine is the well-known *catenary*, the curve formed by a hanging cable suspended from two points. The four other hyperbolic functions are defined by analogy with the circular functions; for example, the hyperbolic tangent:

$$\tanh(x) = \frac{\sinh(x)}{\cosh(x)}$$

Identities with the hyperbolic functions may be obtained algebraically from properties of the exponential function. For example, omitting parentheses:

$$\sinh(x + y) = \sinh x \cosh y + \cosh x \sinh y$$

$$\cosh^2 x - \sinh^2 x = 1$$

The introduction of Taylor's Formula, which usually occurs about midway through the calculus course sequence, leads to useful power series expansions for several elementary transcendental functions. Among the most notable are these:

$$\exp(x) = \sum_{n=0}^{\infty} x^n/n!$$

$$\sin(x) = \sum_{n=0}^{\infty} (-1)^n x^{2n+1}/(2n + 1)!$$

$$\cos(x) = \sum_{n=0}^{\infty} (-1)^n x^{2n}/(2n)!$$

that converge for all $x$, and

$$\tan^{-1}(x) = \sum_{n=0}^{\infty} (-1)^n x^{2n+1}/(2n + 1)$$

$$\ln(1 - x) = -\sum_{n=1}^{\infty} x^n/n$$

that converge for $-1 \leq x < 1$. Using either a classroom computer with a monitor or a graphing calculator attached to an overhead projector, striking demonstrations of these and other results can be given by displaying partial sums of these series with, say, 1, 2, 4, 8, and 16 terms for various values of $x$. The different rates of convergence for these series become clear.

A number of trancendental functions are first encountered in more advanced courses in differential equations, engineering mathematics, and statistics, and consequently are often called the *higher transcendental functions*. Important examples include the *Gamma Function,* which is the extension of the integer factorial function

$$n! = n(n - 1)(n - 2) \cdots 2 \cdot 1$$

to all positive real numbers, the *Bessel Functions,* which arise in the study of vibrating membranes, and the *Error Function,* which occurs in the normal distribution of probability and statistics. Some of these functions are expressed as integrals and some are expressed as power series. A discussion of the Gamma Function,

$$\Gamma(x) = \int_0^{\infty} t^{x-1}e^{-t}dt, \qquad x > 0$$

concluding with a calculation of $(1/2)! = \sqrt{\pi}/2$ is an attractive enrichment topic for a strong calculus course.

*See also* Algebraic Numbers; Calculus, Overview; Exponents, Algebraic; Logarithms; Power Series; Transcendental Numbers; Trigonometric Functions, Elementary; Trigonometry

## SELECTED REFERENCES

Bateman Manuscript Project. *Higher Transcendental Functions.* New York: McGraw-Hill, 1953–1955.

Carico, Charles C. *Exponential and Logarithmic Functions.* Belmont, CA: Wadsworth, 1974.

Demana, Franklin. *Precalculus Mathematics: A Graphing Approach.* Reading, MA: Addison-Wesley, 1992.

Drooyan, Irving, and Walter Hadel. *Trigonometry: An Analytic Approach.* New York: Macmillan, 1967.

Eves, Howard W. *An Introduction to the History of Mathematics.* 3rd ed. New York: Holt, Rinehart and Winston, 1969.

Miller, Kenneth S., and John B. Walsh. *Elementary and Advanced Trigonometry.* New York: Harper, 1962.

Robison, John V. *Modern Algebra and Trigonometry.* New York: McGraw-Hill, 1966.

ROBERT J. BUMCROT

# TRANSCENDENTAL NUMBERS

A complex number $\alpha$ that is not algebraic, that is, if $\alpha$ is not the root of any polynomial equation with integer coefficients. Loosely speaking (and as the name implies), the algebraic numbers can be defined "algebraically" in terms of the usual arithmetic operations of addition, subtraction, multiplication and division (for example $\sqrt{2}$ is defined as the positive real number that when multiplied by itself gives the integer 2). The transcendental numbers, in contrast, require "analytic" descriptions that involve limiting operations. The numbers $e = 2.71828\ldots$ and $\pi = 3.14159\ldots$ are known to be transcendental.

The number $e$ was proved not to be algebraic by Charles Hermite in 1873, and $\pi$ by Ferdinand Lindemann in 1882. The transcendentality of $\pi$ proves the impossibility of the classical Greek geometry problem of "squaring the circle," namely, finding by straightedge and compass constructions of a square whose area is the same as the area of a given circle. Straightedge and compass constructions produce algebraic numbers involving at most a sequence of successive square roots, since the point(s) of intersection of two lines, a line and a circle, or two circles (the result of a single straightedge and compass construction) can be obtained with (at worst) the use of the quadratic formula. If the circle could be squared, the side of the square would involve $\sqrt{\pi}$, which would therefore be algebraic, and so its square, namely $\pi$, would also be algebraic.

Almost every complex number is transcendental, in the sense that the algebraic numbers are countable (it is theoretically possible to give a 1–1 correspondence between the algebraic numbers and the integers 1, 2, 3, . . .), whereas the real numbers and hence also the complex numbers are not countable. In a probabilistic sense, a complex number chosen "at random" has probability 1 of being transcendental.

Even though there are "many more" transcendental numbers than algebraic numbers, proving that a *specific* complex number is transcendental is generally quite difficult. There is no general criterion (other than the definition) to determine whether a given complex number is transcendental. One result (due to Joseph Liouville) that gives an infinite number of specific transcendental real numbers relates to how well a given real number can be approximated by fractions: a real number $\alpha$ is said to be *approximable to order n* if there is some positive constant $C$ and infinitely many rational numbers $p/q$ with $|\alpha - p/q| < C/q^n$. There should be infinitely many $p/q$ such that the difference between $\alpha$ and $p/q$ is small relative to the $n^{th}$ power of the denominator $q$. For example, suppose $\beta$ is the so-called golden ratio $(1 + \sqrt{5})/2$ and $F_n$ is the $n^{th}$ Fibonacci number defined by $F_0 = 0, F_1 = 1$, and the recursion $F_{n+2} = F_n + F_{n+1}$ (i.e., each successive Fibonacci number is the sum of the previous two—the first few are 0, 1, 1, 2, 3, 5, 8, 13, . . .). Then it can be proved that $|\beta - F_{n+1}/F_n| < 1/F_n^2$ for every $n > 0$, showing explicitly that $\beta$ is approximable to order 2 (with $C = 1$). On the other hand, it is known that $\beta$ is not approximable to any higher degree; this is a consequence of the theorem of Liouville, which states that if $\alpha$ is algebraic of degree $n$, then $\alpha$ is not approximable to any order greater than $n$ (note that $\beta$ satisfies $\beta^2 - \beta - 1 = 0$, so that $\beta$ is algebraic of degree 2). It also follows from Liouville's theorem that if there are an infinite number of extremely good rational approximations $p/q$ to $\alpha$, then $\alpha$ must be transcendental. For example, if for any positive integer $n$ there exists a rational number $p/q$ with $|\alpha - p/q| < 1/q^n$, then, in particular, $\alpha$ is approximable to order $n$ for any $n$, so $\alpha$ cannot be algebraic. As an explicit example, the number $\alpha = 0.110001000\ldots = 1/10^1 + 1/10^2 + 1/10^6 + 1/10^{24} + \ldots$ (the exponents being the values of the factorial function $n! = n(n-1)(n-2)\ldots(3)(2)(1)$) is transcendental because the first $N$ terms of the series for $\alpha$ give an extremely good rational approximation $p/q$ to $\alpha$ (satisfying $|\alpha - p/q| < 1/q^{N-2}$, for example).

Other numbers that are known to be transcendental include $e^\pi$, log 2, log 3/log 2, and $2^{\sqrt{2}}$. On the other hand, it is not known, for example, whether $\pi + e$ or $\pi^e$ are transcendental.

Also of interest is the question whether a given set of complex numbers depend algebraically on each other (for example, $\pi$ and $\sqrt{\pi}$ are both transcendental but one can be obtained from the other algebraically). This leads to the notion of *algebraic independence* and generalizes the notion of being transcendental (which is equivalent to the assertion that the powers of $\alpha$ are algebraically independent). The study of this area of mathematics is referred to as *transcendental number theory* and incorporates techniques from real and complex analysis and abstract algebra and, like algebraic number theory, is an active area of current mathematical research.

*See also* Algebraic Numbers; Logarithms; Pi.

## SELECTED REFERENCES

Courant, Richard, and Herbert Robbins. *What Is Mathematics?* London, England: Oxford University Press, 1969.

Hardy, Godfrey H., and Edward M. Wright. *An Introduction to the Theory of Numbers.* 4th ed. Oxford, England: Clarendon, 1965.

DAVID S. DUMMITT

# TRANSFORMATIONAL GEOMETRY

A new approach to the study of Euclidean geometry arising from Felix Klein's definition of a geometry in 1872 (Coxeter and Greitzer 1967). The underlying concept of this approach is a transformation (mapping) of a plane onto itself, by which is meant a one-to-one function whose domain and range both are the entire plane. The transformations that relate to the Euclidean geometry are called *isometries* and *similarities*.

An *isometry* is a transformation that maps each pair of points $A$ and $B$ to a pair $A'$ and $B'$ such that $AB = A'B'$. This implies that an isometry preserves the distances between points and the measures of angles. It also maps lines to lines and parallel lines to parallel lines. As a consequence of these properties, an isometry maps a geometric figure to a congruent one, and therefore isometries are sometimes called *congruence* transformations. The basic isometries are translations, rotations, and reflections.

A *translation* is an isometry that maps each point $A$ to the point $A'$ such that the vector $\overrightarrow{AA'}$ is equal to a given vector $\vec{v}$. Such a translation may be denoted by $T_{\vec{v}}$. In Figure 1(a), $T_{\vec{v}}(\triangle ABC) = \triangle A'B'C'$. Hence $\triangle ABC \cong \triangle A'B'C'$ and consequently the area of the parallelogram $ABB'A' =$ area of rectangle $ACC'A' = AA' \cdot AC$.

A *rotation* of $\alpha$ with center $O$ is an isometry in which $O$ is mapped to itself, and any other point $A$ is mapped to $A'$ with $OA = OA'$ and $\angle AOA' = \alpha$. Such rotation may be denoted by $C_{o,\alpha}$. In Figure 1(b), $C_{o,\alpha}(\triangle OAB) = \triangle OA'B'$. Hence $\triangle OAB \cong \triangle OA'B'$ and the angle between $AB$ and $A'B'$ equals $\alpha$.

A *half-turn* is a rotation of 180° and may be denoted by $H_o$, where $O$ is the center of rotation. In Figure 1(c), $ABCD$ is a trapezoid, $O$ is the midpoint of $AB$, and $H_o(ABCD) = BD'C'A$. Hence, $ABCD$ and $BD'C'A$ are congruent and have the same area. Consequently, the area of $ABCD =$ half the area of the parallelogram $D'CDC'$, which

$$= (1/2)(D'C)(\text{height of trapezoid})$$

$$= (\text{average of bases})(\text{height of trapezoid})$$

A reflection in a line $\ell$ is an isometry that maps any point $A$ not on $\ell$ to a point $A'$ such that $\ell$ is the perpendicular bisector of $AA'$. Any point on $\ell$ will be mapped to itself, which means that $\ell$ is invariant.

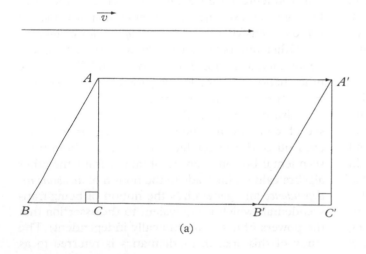

(a)

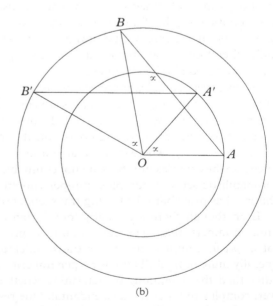

(b)

**Figure 1**

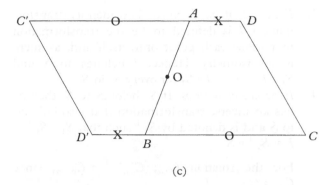

(c)

**Figure 1**

Such reflection may be denoted by $R_\ell$. In Figure 2(a), $R_\ell (\triangle ABC) = \triangle A'BC'$. While $\triangle ABC$ is oriented in counterclockwise sense, its image $\triangle A'BC'$ is oriented in clockwise sense. This shows that reflections reverse the orientation. For this reason, reflections are classified as opposite isometries. On the other hand, translations and rotations preserve the orientation and hence are called *direct* isometries.

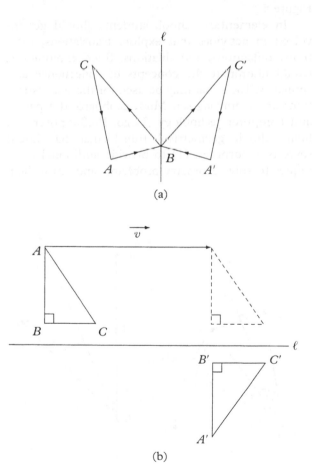

(a)

(b)

**Figure 2**

When two isometries are applied in succession, the equivalent transformation, referred to as the *composite* of the two isometries, is again an isometry. This concept will be used to define another opposite isometry. A *glide reflection* is the composite of a translation $T_{\vec{v}}$ followed by a reflection $R_\ell$, where $\ell$ is parallel to $\vec{v}$. This glide reflection may be denoted by $R_\ell \circ T_{\vec{v}}$. In Figure 2(b), $R_\ell \circ T_{\vec{v}}(\triangle ABC) = \triangle A'B'C'$. It can be shown that if $A$, $B$ and their images $A'$, $B'$ are given where $AB = A'B'$, then there are exactly two isometries, one direct (translation or rotation) and one opposite (reflection or glide reflection) that maps $AB$ onto $A'B'$.

Isometries form the basis for studying the topics of symmetry of figures, Frieze patterns, and tilings of the plane (Martin 1982).

The transformations called *similarities* include the isometries as special cases. A similarity is a transformation that maps each pair of points $A,B$ to a pair $A',B'$ such that $A'B' = kAB$, where $k$ is a positive number called the *ratio of similarity*. If $k = 1$, the similarity is an isometry. If $k \neq 1$, the similarity does not preserve distance, but still preserves the angle measure, maps lines to lines, and parallel lines to parallel lines. As a result, a similarity maps a geometric figure to a similar one. The basic similarities are dilation and dilative rotation (which are direct similarities) and dilative reflection (which is an opposite) similarity. Each one of these similarities has an invariant point called the *center of similarity*.

A dilation with center $O$ and ratio $k$ is a similarity that maps each pair of points $A,B$ to a pair $A',B'$ such that $OA' = kOA$ and $OB' = kOB$. This similarity may be denoted by $S_{o,k}$. In Figure 3(a), $S_{o,k}(\triangle ABC) = \triangle A'B'C'$. The triangles $ABC$ and $A'B'C'$ are similar, and the corresponding sides are parallel.

A dilative rotation of $\alpha$ with center $O$ and ratio $k$ is a similarity that is the composite of dilation of ratio $k$ and a rotation of $\alpha$ with the same center $O$. This similarity may be denoted by $S_{o,k,\alpha}$. In Figure 3(b), $S_{o,k,\alpha}(\triangle ABC) = \triangle A'B'C'$.

If $\alpha = 180°$, the rotation will be a half-turn and the similarity will be a dilation that may be denoted by $S_{o,-k}$. In Figure 3(c), $S_{o,-k}(\triangle ABC) = \triangle A'B'C'$, where the corresponding sides are parallel in opposite directions.

A dilative reflection with center $O$ on line of reflection $\ell$ and ratio $k$ is a similarity that is the composite of a dilation of ratio $k$ with center $O$ and a reflection in line $\ell$. This similarity may be denoted by $S_{o,k,\ell}$. In Figure 3(d), $S_{o,k,\ell}(\triangle ABC) = \triangle A'B'C'$. It can be shown that if $A,B$ and their images $A',B'$ are

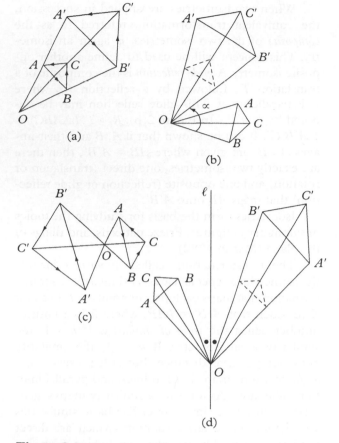

**Figure 3**

given where $A'B' = kAB$, $k \neq 1$, then there are exactly two similarities, one direct (dilation or dilative rotation) and one opposite (dilative reflection) that maps $AB$ onto $A'B'$.

The set $S$ of all similarities (including isometries) together with the operation of composition of two transformations form an algebraic structure called a *group* satisfying the following properties:

1. *Closure:* If $S_1$ and $S_2$ belong to $S$, that is, $S_1$ and $S_2$ are similarities, then their composite $S_2 \circ S_1$ is also a similarity. For example, the composite of the reflections $R_{\ell_1}$ and $R_{\ell_2}$, where the lines of reflections $\ell_1$ and $\ell_2$ intersect at $O$ and the angle between them has measure $\alpha$, is the rotation of $2\alpha$ about $O$, that is, $R_{\ell_2} \circ R_{\ell_1} = C_{o,2\alpha}$ (see Figure 4(a)). As another example, the composite of the dilative reflections $S_{o,k_1,\ell}$ and $S_{o,k_2,\ell}$ is the dilation with center $O$ and ratio $k_1 k_2$, that is, $S_{o,k_2,\ell} \circ S_{o,k_1,\ell} = S_{o,k_1 k_2}$, (see Figure 4(b)).
2. *Associativity:* If $S_1$, $S_2$, and $S_3$ belong to $S$, then $(S_1 \circ S_2) \circ S_3 = S_1 \circ (S_2 \circ S_3)$.

3. *Existence of an Identity:* The identity transformation $I$ is defined to be the transformation that maps each point onto itself and, as such, is an isometry. Hence, $I$ belongs to $S$ and $S_1 \circ I = S_1 = I \circ S_1$ for every $S_1$ in $S$.
4. *Existence of Inverses:* If $S_1$ belongs to $S$, then $S_1$ has an inverse transformation that also belongs to $S$ and is denoted by $S_1^{-1}$ such that $S_1 \circ S_1^{-1} = I = S_1^{-1} \circ S_1$.

For the rotation $C_{o,\alpha}$; $(C_{o,\alpha})^{-1} = C_{o,-\alpha}$, since $C_{o,\alpha} \circ C_{o,-\alpha} = I$.

For the reflection $R_\ell$; $(R_\ell)^{-1} = R_\ell$, since $R_\ell \circ R_\ell = I$.

For the dilation $S_{o,k}$; $(S_{o,k})^{-1} = S_{o,1/k}$, since $S_{o,k} \circ S_{o,1/k} = I$.

It can be shown that the set of all isometries form a subgroup of the group of similarities. Both the group and its subgroup are noncommutative, that is, $S_1 \circ S_2 \neq S_2 \circ S_1$, in general. In the case of the rotation $C_{o,90°}$ and the translation $T_{\vec{v}}$, the composites $T_{\vec{v}} \circ C_{o,90°}$ and $C_{o,90°} \circ T_{\vec{v}}$ are different, as illustrated in Figure 5.

In elementary school, students should get involved in activities that explore translations, rotations, reflections, and dilations. Such explorations would illuminate the concepts of congruence and similarity. Tools that may be used for these investigations are a drawing set, Mira, geoboard, dot paper, and computer technology (Logo, PreSupposer). In high school, geometrical transformations should serve two purposes: (1) to provide additional techniques to solve geometry problems, and (2) to help

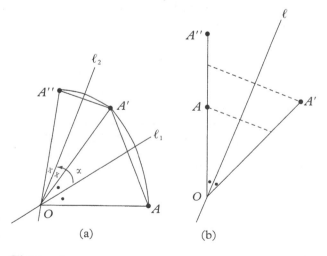

**Figure 4**

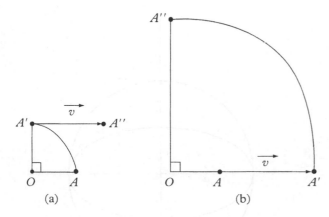

**Figure 5**

students gain more understanding of the concept of a function when they visualize the action of a transformation on a set of points. It would be helpful at this stage to extend the concept of a transformation to a coordinate plane setting. At the college level, transformation geometry should be part of a course designed for prospective mathematics teachers. Such a course should include the synthetic and analytic (using matrices or complex functions) treatments of the geometric transformations. In addition, this course should introduce the topic of transformation groups.

*See also* Abstract Algebra; Group Theory

## SELECTED REFERENCES

Billstein, Rick, Shlomo Libeskind, and Johnny W. Lott. *A Problem Solving Approach to Mathematics for Elementary Teachers.* 6th ed. Reading, MA: Addison-Wesley, 1997.

Brown, Richard G. *Transformational Geometry.* Special ed. Palo Alto, CA: Seymour, 1973.

Coxeter, Harold S. M., and Samuel L. Greitzer. *Geometry Revisited.* New York: Random House, 1967.

Eves, Howard W. *A Survey of Geometry.* Rev. ed. Boston, MA: Allyn and Bacon, 1972.

Martin, George E. *Transformation Geometry.* New York: Springer-Verlag, 1982.

Yaglom, Isaac M. *Geometric Transformations.* 2 vols. New York: Random House, 1962; 1968.

AYOUB B. AYOUB

# TRANSFORMATIONS

The term usually referring to the application of functions that change an equation, expression, or a geometric figure. The purpose of a transformation is to facilitate some process, such as finding a root, computing an integral, or solving a geometric problem. For example, to evaluate $\int_1^2 x(x-1)^3 dx$, one may use the transformation $x = u + 1$ to change the integral to $\int_0^1 (u+1)u^3 \, du$, which could be easily computed. The transformation used in this example is applied to one variable, so it is one-dimensional.

## GEOMETRIC TRANSFORMATIONS

A transformation $F$ defined on a plane is a one-to-one correspondence from the set of points in the plane onto itself.

1. Each point $P$ of the plane has a unique image $Q$ under $F$.
2. Each point $Q$ of the plane is the image of a unique point $P$ under $F$. As a result, the points of the plane are paired under $F$ where each pair consists of a point and its image. Reversing the roles of the point and its image yields another transformation, called the *inverse* of $F$, which is denoted by $F^{-1}$, where $F^{-1}(Q) = P$ if and only if $F(P) = Q$.

The set of transformations that relates to the Euclidean geometry is the set of linear transformations. These are given in terms of Cartesian coordinates by the equations:

$$x' = ax + by + c, \qquad y' = mx + ny + k$$

where the coefficients are real and $an - bm \neq 0$. The reason for this condition is to guarantee that each point of the plane be the image of a unique point. This can be seen when we solve the equations for $x$ and $y$ to get the inverse transformation:

$$x = [1/(an - bm)](nx' - by' + bk - cn)$$
$$y = [1/(an - bm)](-mx' + ay' + cm - ak)$$

A linear transformation maps lines on lines and parallel lines on parallel lines and preserves the ratio between segments on the same line or on parallel lines. Linear transformations may also be called *collineations* or *affine* transformations. In Figure 1, the square with vertices $O(0, 0)$, $A(2,0)$, $B(2,2)$, and $C(0,2)$ is transformed by $x' = x + y$ and $y' = 2x - y$ to the parallelogram with vertices $O(0,0)$, $A'(2,4)$, $B'(4,2)$, and $C'(2,-2)$. The center $(1,1)$ of the square is mapped on the center $(2,1)$ of the parallelogram. It is obvious that this transformation reversed the orientation from counterclockwise for the square $OABC$ to clockwise for the parallelogram $OA'B'C'$.

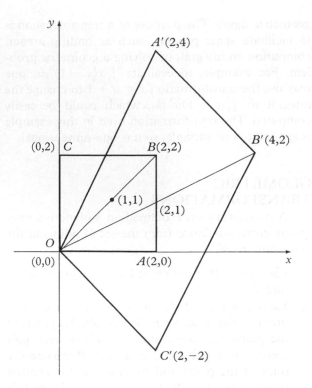

**Figure 1**

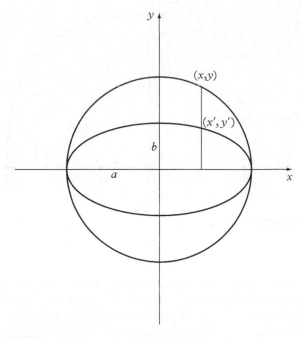

**Figure 2**

Figure 2 depicts the case of the circle $x^2 + y^2 = a^2$, which is mapped on the ellipse

$$\frac{(x')^2}{a^2} + \frac{(y')^2}{b^2} = 1$$

by the affine transformation $x' = x$ and $y' = (b/a)y$. Notice that for a geometric transformation, only one coordinate system is used, that is, the $x'$-axis and the $x$-axis are the same, and also the $y'$-axis and the $y$-axis are the same.

Special cases of linear transformations are isometries and similarities. Isometries are transformations that preserve distances. The basic isometries are translations, rotations, and reflections. A translation by a vector $\vec{v} = (g,h)$ is defined by the equations $x' = x + g$, $y' = y + h$. A rotation of angle $\alpha$ about the origin is given by the equations $x' = x \cos\alpha - y \sin\alpha$, $y' = x \sin\alpha + y \cos\alpha$. A reflection in the line $y = \tan\alpha$ is given by the equations $x' = x \cos 2\alpha + y \sin 2\alpha$, $y' = x \sin 2\alpha - y \sin 2\alpha$ (Ganz 1969).

The rotations and translations preserve orientation and are called *direct* isometries, while the reflections reverse the orientation and are called *opposite* isometries. When two isometries are applied in succession, then the equivalent transformation, referred to as the *composite* of the two isometries, is again an isometry. If the first isometry to be applied is $F$ and the second is $E$, their composite is denoted by $EF$. The composition of transformations is not commutative, that is, $EF \neq FE$, in general.

*Example:* Let us apply a rotation of 180° (called a *half-turn*) about the origin. This rotation is defined by the equations $x' = -x$ and $y' = -y$. And let us follow this by the application of a translation defined by $x'' = x' + 2$ and $y'' = y'$. Then the composite isometry will have the equations $x'' = -x + 2$ and $y'' = -y$. Consider the triangle with vertices $O(0,0)$, $A(1,0)$, and $B(0,2)$. Figure 3 illustrates $\Delta OAB$ and its image $\Delta O''AB''$ under this isometry. On the other hand, if we first apply the translation defined by $x' = x + 2$ and $y' = y$, followed by the rotation defined by $x'' = -x'$ and $y'' = -y'$, then the composite isometry will have the equations $x'' = -x - 2$ and $y'' = -y$. The image of $\Delta OAB$ under this isometry is $\Delta O''A''B''$, as depicted in Figure 4. Each one of the composite isometries is a new half turn, the first about $(1,0)$ and the second about $(-1,0)$.

*Example:* We will show that the composite of two reflections into two perpendicular lines is a half-turn about the point of intersection of the two lines. Let the two lines be the $x$-axis and the $y$-axis. The reflection in the $x$-axis is given by $x' = x$ and $y' = -y$. The second reflection, in the $y$-axis, is given by $x'' = -x'$ and $y'' = y'$. Hence the composite isome-

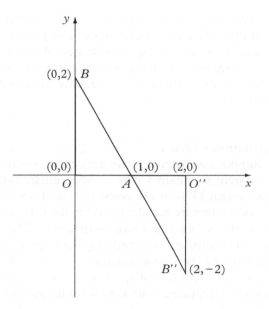

**Figure 3**

try is defined by $x'' = -x$ and $y'' = -y$, which represents a half-turn about the origin.

The representation of isometries is summarized by the following theorem: all isometries have the form

$$x' = ax + by + c \quad \text{and} \quad y' = \pm(-bx + ay) + d$$

where $a^2 + b^2 = 1$; the plus sign corresponds to a direct isometry and the minus sign corresponds to an opposite isometry. Conversely, all transformations

represented by such equations are isometries (Ganz 1969).

Similarities are transformations that carry each pair of points $A$ and $B$ to a pair $A'$ and $B'$ such that $A'B' = kAB$, where $k$ is a positive number called the *ratio of similarity*. Isometries are special cases of similarities when $k = 1$. A characteristic of similarities is that it preserves angle measures. The basic similarities are the dilations. A dilation with center $(0,0)$ and ratio $k$ is given by the equations $x' = kx, y' = ky$. Figure 5 depicts a triangle and its image under a dilation. The composite of a dilation and an isometry is a similarity that may be identified as a dilative rotation or a dilative reflection. The fundamental theorem of similarities states the following: every similarity transformation with ratio $k$ has the equations $x' = ax + by + c$ and $y' = \pm(-bx + ay) + d$, where $a^2 + b^2 = k^2$; the plus sign corresponds to a direct similarity and the minus sign corresponds to an opposite similarity. Conversely, all the transformations represented by such equations are similarities (Ganz 1969).

*Example:* Consider the two isosceles right angle triangles $OAB$ and $O'A'B'$ shown in Figure 6, with the points $O(0,0)$, $A(3,0)$, $B(0,3)$, $O'(2,8)$, $A'(5,4)$, and $B'(6,11)$. Since the two triangles are similar, we can find two similarities, one direct and one opposite, which will map $OAB$ onto $O'A'B'$. In each case, we will find the center of similarity, that is, the point which is mapped on itself. Let the equations of the direct similarity be

$$x' = ax + by + c, \quad y' = -bx + ay + d$$

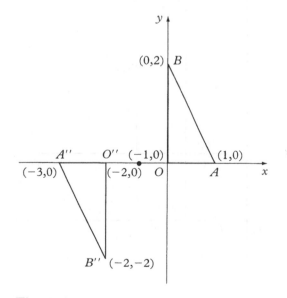

**Figure 4**

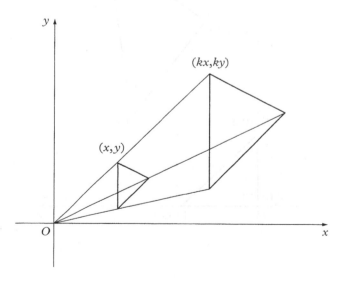

**Figure 5**

This similarity should map (0,0) on (2,8) and (3,0) on (5,4). Thus, $2 = c$, $8 = d$, and $5 = 3a + c$, $4 = -3b + d$. Hence, $a = 1$ and $b = 4/3$. Consequently, the equations of the similarity are $x' = x + (4/3)y + 2$ and $y' = -(4/3)x + y + 8$.

Now we set $x' = x$ and $y' = y$ and get $x = x + (4/3)y + 2$ and $y = -(4/3)x + y + 8$, which implies that the center of similarity is $(6,-(3/2))$. For the opposite similarity, we use the equations $x' = ax + by + c$ and $y' = -(-bx + ay) + d$. This similarity should map (0,0) on (2,8) and (3,0) on (6,11). Thus, $2 = c$, $8 = d$, and $6 = 3a + c$, $11 = 3b + d$. Hence, $a = (4/3)$ and $b = 1$. Consequently, the equations of the similarity are $x' = (4/3)x + y + 2$ and $y' = x - (4/3)y + 8$. If we set $x' = x$ and $y' = y$, we get $x = (4/3)x + y + 2$ and $y = x - (4/3)y + 8$, which implies that the center of similarity is $(-(57/8),(3/8))$.

## TRANSFORMATION OF COORDINATES

In analytic geometry, when two rectangular coordinate systems are imposed on the plane, a typical point $P$ has two pairs of coordinates, such as $(x,y)$ and $(x',y')$. The relation between these coordinates is expressed by the transformation equations in which the individual coordinates of one pair are expressed in terms of the coordinates of the other pair. A common reason for changing from one coordinate system to another is to simplify the equation of a curve.

### Translation of Axes

Suppose we translate the axes of an $xy$-coordinate system to obtain a new $x'y'$-coordinate system whose origin $O'$ is at the point $(h,k)$ and whose $x'$-axis and $y'$-axis are parallel to and in the same direction as the $x$-axis and $y$-axis, respectively. The two pairs of coordinates, $(x,y)$ and $(x',y')$, of a point $P$ are related by the translation equations: $x = x' + h$, $y = y' + k$, or, equivalently, $x' = x - h$, $y' = y - k$ (Figure 7). Translation of axes is useful in studying curves such as

$$y^2 - 4y - x + 5 = 0$$

Completing the square in $y$ gives $(y - 2)^2 = x - 1$, and so, with respect to a new coordinate system with origin at (1,2), the curve has the simple equation $(y')^2 = x'$, which represents a parabola (Figure 8).

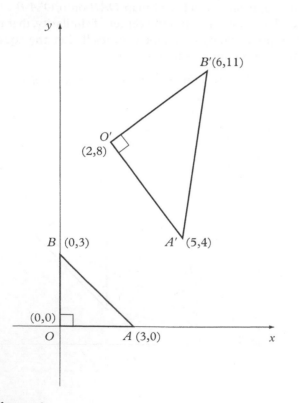

**Figure 6**

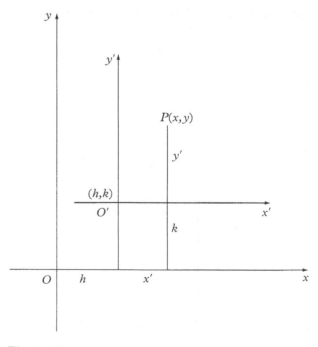

**Figure 7**

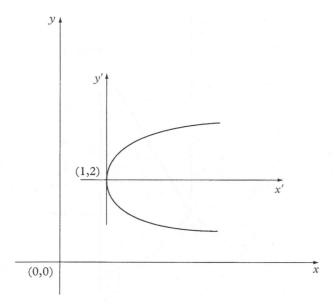

**Figure 8**

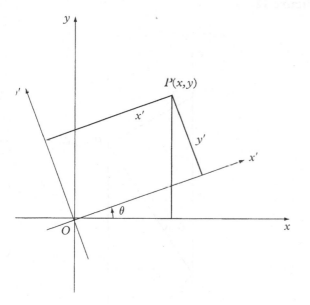

**Figure 9**

## Rotation of Axes

Suppose we rotate the axes of an $xy$-coordinate system about the origin through an angle $\theta$ to obtain a new $x'y'$-coordinate system (Figure 9). The two pairs of coordinates, $(x,y)$ and $(x',y')$, of a point $P$ are related by the rotation equations: $x = x' \cos \theta - y' \sin \theta$, $y = x' \sin \theta + y' \cos \theta$ (Anton 1995). Rotation of axes is helpful in identifying the conic represented by a quadratic equation, such as

$$17x^2 - 12xy + 8y^2 = 20$$

Let us rotate the axes by an angle $\theta$ such that the $xy$-term will be eliminated. Substituting for $x$ and $y$ from the rotation equations, we get

$$17(x'\cos \theta - y'\sin \theta)^2 - 12(x'\cos \theta - y'\sin \theta) \times$$

$$(x'\sin \theta + y'\cos \theta) + 8(x'\sin \theta + y'\cos \theta)^2 = 20$$

Setting the coefficient of $x'y'$ equal to 0 and simplifying the equation, we get $2 \tan^2\theta - 3 \tan \theta - 2 = 0$ or

$$(\tan \theta - 2)(2 \tan \theta + 1) = 0$$

which implies that $\tan \theta = 2$ or $\tan \theta = -(1/2)$. We can use either $\theta$ to rotate the axes. If we use $\theta = \tan^{-1} 2$, we get the simple equation $(x')^2 + 4(y')^2 = 4$, which represents an ellipse (Figure 10).

## Transformation to Polar Coordinates

Suppose a polar coordinate system is imposed on a plane with an $xy$-coordinate system such that the pole coincides with the origin and the polar axis aligns with the positive direction of the $x$-axis. The two pairs of coordinates $(x,y)$ and $(r,\theta)$ of a point are related by the transformation equations: $x = r \cos \theta$, $y = r \sin \theta$ or $r^2 = x^2 + y^2$, $\tan \theta = y/x$ (Figure 11). These equations may be used to investigate curves such as $r = 2a \cos \theta$. If we multiply by $r$, we get $r^2 = 2ar \cos \theta$. Now making use of the previous rela-

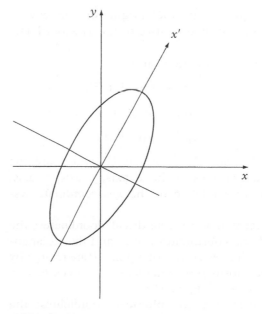

**Figure 10**

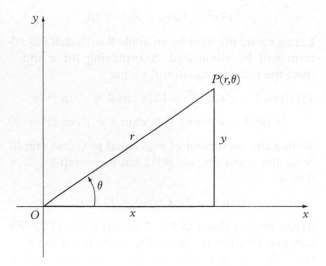

**Figure 11**

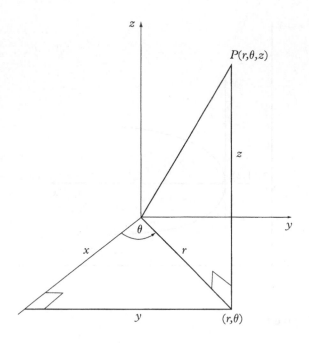

**Figure 12**

tions, the equation becomes $x^2 + y^2 = 2ax$ or $(x - a)^2 + y^2 = a^2$, which represents a circle of radius $a$ and center $(a,0)$. Also, the polar transformation is used to change a complex number such as $\sqrt{3} + i$, to the form $2 [\cos (\pi/6) + i \sin (\pi/6)]$.

## Transformation of Coordinates in Space

1. The translation equations are

$$x = x' + h, \qquad y = y' + k, \qquad z = z' + \ell$$

where $(h,k,\ell)$ is the origin of the $x'y'z'$-coordinate system relative to the $xyz$-coordinate system.

2. The rotation equations are

$$x = \ell_1 x' + \ell_2 y' + \ell_3 z',$$
$$y = m_1 x' + m_2 y' + m_3 z',$$
$$z = n_1 x' + n_2 y' + n_3 z'$$

where $(\ell_1, m_1, n_1)$, $(\ell_2, m_2, n_2)$, $(\ell_3, m_3, n_3)$ are the direction cosines of the $x'$-axis, $y'$-axis, $z'$-axis, respectively, relative to the $xyz$-coordinate system.

3. Transformation to cylindrical coordinates: the cylindrical coordinates $(r,\theta,z)$ and the rectangular coordinates $(x,y,z)$ of a point $P$ are related by the transformation equations $x = r \cos \theta$, $y = r \sin \theta$, $z = z$ (Figure 12).

4. Transformation to spherical coordinates: the spherical coordinates $(\rho,\theta,\phi)$ and the rectangular

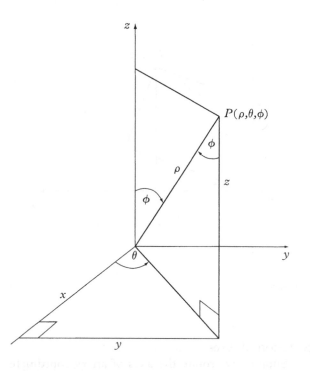

**Figure 13**

coordinates $(x,y,z)$ of a point $P$ are related by the transformation equations $x = \rho \sin \phi \cos \theta$, $y = \rho \sin \phi \sin \theta$, $z = \rho \cos \phi$ (Figure 13).

Students in high school do not learn enough about transformations. This topic provides a powerful tool to manipulate and solve geometric problems and students, especially the mathematically inclined, should be encouraged to make use of it to give them a breadth of view in geometry.

*See also* Coordinate Geometry, Overview; Polar Coordinates; Space Geometry, Instruction; Transformational Geometry

### SELECTED REFERENCES

Anton, Howard. *Calculus with Analytic Geometry.* 5th ed. New York: Wiley, 1995.

Ganz, David. *Transformations and Geometries.* New York: Appleton-Century-Crofts, Meredith, 1969.

Martin, George E. *Transformation Geometry.* New York: Springer-Verlag, 1982.

AYOUB B. AYOUB

# TRIGONOMETRIC FUNCTIONS, ELEMENTARY

Historically, functions developed to solve problems related to astronomy (Katz 1993). The word *trigonometry* comes from the Greek meaning "triangle measurement." There are two different, but consistent, approaches to defining the trigonometric functions, one using right triangles and one using arcs. The former method is convenient for many applications, while the latter allows for more general situations.

Given a right triangle $ABC$ with acute angles $\alpha$, $\beta$, and right angle $\gamma$, we define the sine of an angle $\alpha$ by

$$\sin \alpha = \text{opp/hyp}$$

where opp = length of side opposite angle $\alpha$ and hyp = length of hypotenuse. Since angle $\beta$ is the complement of angle $\alpha$, we define the *co*sine of $\alpha$ to be sin $\beta$, so that

$$\cos \alpha = \text{adj/hyp}$$

where adj = length of side adjacent to angle $\alpha$.

The four other trigonometric functions can be defined either in terms of the triangle $ABC$ or in terms of the sine and cosine.

$$\tan \alpha = \text{opp/adj} = (\sin \alpha)/\cos \alpha$$
$$\cot \alpha = \text{adj/opp} = (\cos \alpha)/\sin \alpha$$
$$\sec \alpha = \text{hyp/adj} = 1/\cos \alpha$$
$$\csc \alpha = \text{hyp/opp} = 1/\sin \alpha$$

By their nature, these definitions are only valid for angles $\alpha$ that satisfy $0 < \alpha < 90°$. For angles outside of this range, the right triangle interpretation is meaningless. Let $\theta$ represent any angle and visualize a line drawn in the Cartesian plane with initial side along the $x$-axis and terminal side moving counterclockwise from the initial side to an angle $\theta$. We let $\alpha_\theta$ denote the smallest nonnegative angle that this terminal side makes with the $x$-axis. It is clear that $\alpha_\theta$ satisfies $0 \leq \alpha_\theta \leq 90°$. We can then define sin $\theta$ and cos $\theta$ in terms of sin $\alpha_\theta$ and cos $\alpha_\theta$ as follows.

If the terminal side of $\theta$ lies in the second quadrant, $\sin \theta = \sin \alpha_\theta$ and $\cos \theta = -\cos \alpha_\theta$, while if it lies in the third quadrant, $\sin \theta = -\sin \alpha_\theta$ and $\cos \theta = -\cos \alpha_\theta$. Finally, if the terminal side of $\theta$ lies in the fourth quadrant, $\sin \theta = -\sin \alpha_\theta$ and $\cos \theta = \cos \alpha_\theta$. The minor difficulty with this is that we have not yet talked about the trigonometric functions when the angle is either 0° or 90°. We defined sin 0° = 0 and sin 90° = 1. With these definitions we conclude

$$\cos 0° = 1$$
$$\tan 0° = 0$$
$$\sec 0° = 1$$
$$\cot 0° \text{ undefined}$$
$$\csc 0° \text{ undefined}$$

Similar results follow for cos 90°, tan 90°, sec 90°, cot 90°, and csc 90°.

As mentioned earlier, an alternative approach to defining the trigonometric functions involves arcs. Consider a circle of radius 1 centered at the origin. Visualize a vertical real line whose origin is at the point $(1,0)$ with positive direction up. If we "wrap" this real line around the circle counterclockwise (assuming no thickness to the line), we find that to each real number $\theta$ on the line there corresponds a unique point $(x_\theta, y_\theta)$ on the circle. Notice that $\theta$ represents the length of the arc on the circle from $(1,0)$. (See Figure 1.) We define the sine and cosine of the real number $\theta$ by

$$\sin \theta = y_\theta \quad \text{and} \quad \cos \theta = x_\theta$$

As before, the other four trigonometric functions can then be defined in terms of sin $\theta$ and cos $\theta$.

This definition allows us to define the trigonometric functions as functions of real numbers instead of angles. If we allow the negative portion of the line to wrap around the circle in the clockwise direction, we can similarly define the trigonometric functions for the negative values of $\theta$. Notice that if $\alpha_\theta$ represents the angle with initial side along the positive $x$-axis that subtends the arc of length $\theta$, then our previous definitions are consistent so that $\sin \alpha_\theta = \sin \theta$, $\cos \alpha_\theta = \cos \theta$, and so on. (See Figure 2.)

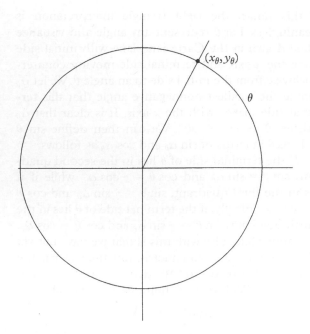

**Figure 1**

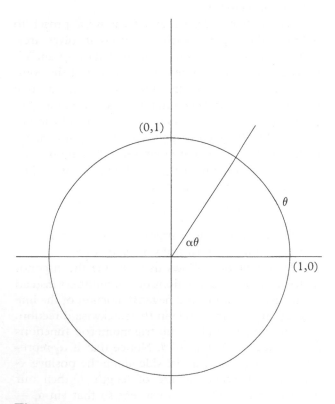

**Figure 2**

We observe that this definition of the trigonometric functions reveals a periodic characteristic since $x_\theta$ and $y_\theta$ repeat themselves every $2\pi$ (the circumference of the unit circle). In other words, for any value of $\theta$ for which each of the following is defined we have

$$\sin(\theta + 2\pi) = \sin\theta \qquad \cos(\theta + 2\pi) = \cos\theta$$
$$\tan(\theta + 2\pi) = \tan\theta \qquad \cot(\theta + 2\pi) = \cot\theta$$
$$\sec(\theta + 2\pi) = \sec\theta \qquad \csc(\theta + 2\pi) = \csc\theta$$

(Actually, $\tan\theta$ and $\cot\theta$ have a shorter period, namely $\pi$.)

These results allow us to evaluate the trigonometric functions for all real numbers as long as we know the values for $\theta$ in the interval $[0,2\pi)$.

*See also* Trigonometry

## SELECTED REFERENCES

Fleming, Walter, Dale Varberg, and Herbert Kasube. *Algebra and Trigonometry: A Problem Solving Approach.* 4th ed. Englewood Cliffs, NJ: Prentice-Hall, 1992.

Katz, Victor. *A History of Mathematics: An Introduction.* New York: HarperCollins, 1993.

HERBERT E. KASUBE

# TRIGONOMETRIC IDENTITIES

Equations which hold true for all values of the variables in their domains. Consider the sentence "The earth is the third planet from the sun." This is always true, whereas a statement such as "He is six feet tall" is true only when the "He" who is referred to actually satisfies the description. The same idea can be applied to mathematical sentences.

The *domain* of an expression in one variable is defined to be the set of all values of the variable for which the expression "makes sense." For example, the domain of the quadratic expression $x^2 + 2x + 1$ is the set of all real numbers, while the domain of $1/(x^2 - 1)$ consists of all real numbers other than $\pm 1$. An equation is called an *identity* if it becomes a true statement for all values of the variable in its domain. For example,

$$x^2 + 2x + 1 = (x + 1)^2$$

is an identity since it is true for all real values of the variable. In addition, the equation

$$1/(x^2 - 1) = 1/[(x - 1)(x + 1)]$$

is an identity even though it is not valid when $x = \pm 1$.

In trigonometry, there are some fundamental identities that play major roles in the subject. Our first group of identities are collectively called *Pythagorean Identities* because of their connection to the Pythagorean Theorem.

Consider the right triangle with legs $a$ and $b$ and hypotenuse $c$. If $\alpha$ is the angle opposite the side of length $a$, we conclude:

$$\sin \alpha = \frac{a}{c} \text{ and } \cos \alpha = \frac{b}{c}$$

From the Pythagorean Theorem we know that $a^2 + b^2 = c^2$. Using the expressions for $\sin \alpha$ and $\cos \alpha$ above we see that

$$\sin^2 \alpha + \cos^2 \alpha = \frac{a^2}{c^2} + \frac{b^2}{c^2} = \frac{a^2 + b^2}{c^2} = \frac{c^2}{c^2} = 1$$

While the argument above applies only to angles $0 < \alpha < \pi/2$, it can be shown that if $x$ is any real number, then

$$\sin^2 x + \cos^2 x = 1 \tag{1}$$

Dividing both sides of (1) by $\cos^2 x$, we obtain

$$\tan^2 x + 1 = \sec^2 x \tag{2}$$

Alternatively, dividing both sides (1) by $\sin^2 x$, we obtain

$$1 + \cot^2(x) = \csc^2 x \tag{3}$$

Notice that equations (2) and (3) have restricted domains where some of the trigonometric functions are undefined, but the equations are true whenever the functions exist. Hence, they are truly identities. Identities (2) and (3) can also be motivated via right triangle trigonometry and the Pythagorean Theorem using an argument similar to the one given here.

Another important class of trigonometric identities are called the *double-angle formulas*. The proofs that they are identities can be found in the literature (Fleming, Varberg, and Kasube 1992)

$$\sin 2x = 2 \sin x \cos x \tag{4}$$
$$\cos 2x = \cos^2 x - \sin^2 x \tag{5}$$
$$\cos 2x = 1 - 2 \sin^2 x \tag{5'}$$
$$\cos 2x = 2 \cos^2 x - 1 \tag{5''}$$

Notice that equations (5') and (5") follow directly from (5) by applying (1).

Similarly, we obtain *half-angle formulas* that yield

$$\sin\left(\frac{x}{2}\right) = \pm\sqrt{\frac{1 - \cos x}{2}} \tag{6}$$

and

$$\cos\left(\frac{x}{2}\right) = \pm\sqrt{\frac{1 + \cos x}{2}} \tag{7}$$

In each case, the appropriate sign is applied depending on the value of $\frac{x}{2}$ (Recall that $\sin \alpha > 0$ when $0 < \alpha < \pi$ and $\sin \alpha < 0$ when $\pi < \alpha < 2\pi$, while $\cos \alpha > 0$ when $(-\pi/2) < \alpha < (\pi/2)$ and $\cos \alpha < 0$ when $(\pi/2) < \alpha < (3\pi/2)$). Relying on the fact that the other four trigonometric functions can be expressed in terms of the sine and cosine, we need not state corresponding double- and half-angle formulas for them.

Trigonometric identities can be useful in simplifying expressions involving trigonometric functions or in determining the exact value of certain trigonometric functions. For example, using the half-angle formulas and the fact that $\cos 45° = \sqrt{2}/2$, we can determine the exact values for $\sin 22.5°$ and $\cos 22.5°$ (note: a $+$ sign is used since $0° < 22.5° < 90°$):

$$\sin 22.5° = \sin\left(\frac{45°}{2}\right) = \sqrt{\frac{1 - \cos 45°}{2}}$$

$$= \sqrt{\frac{1 - \frac{\sqrt{2}}{2}}{2}} = \sqrt{\frac{2 - \sqrt{2}}{2}}$$

$$\cos 22.5° = \cos\left(\frac{45°}{2}\right) = \sqrt{\frac{1 + \cos 45°}{2}}$$

$$= \sqrt{\frac{1 + \frac{\sqrt{2}}{2}}{2}} = \sqrt{\frac{2 + \sqrt{2}}{2}}$$

These identities also have applications in the calculus of the trigonometric functions, especially integration. There are many other trigonometric identities (Fleming, Varberg, and Kasube 1992), but those mentioned above are some that are most commonly encountered.

Suppose one wishes to verify an alleged trigonometric identity such as

$$(\sin^2 \theta)/\cos \theta = \sec \theta - \cos \theta$$

Start with the left-hand side (since it appears to be the more complicated) and obtain

$$(\sin^2 \theta)/\cos \theta = (1 - \cos^2 \theta)/\cos \theta = \sec \theta - \cos \theta$$

When introduced to trigonometric identities in class few students see their relevance at first. Using examples that illustrate the application of such identities to exact evaluation of some trigonometric functions can help alleviate this problem. If students are headed (eventually) to a calculus course, then trigonometric identities play a major role in certain techniques of integration (Stewart 1995).

*See also* Trigonometric Functions, Elementary; Trigonometry

### SELECTED REFERENCES

Fleming, Walter, Dale Varberg, and Herbert Kasube. *Algebra and Trigonometry: A Problem Solving Approach.* 4th ed. Englewood Cliffs, NJ: Prentice Hall, 1992.

Stewart, James. *Calculus.* 3rd ed. Monterey, CA: Brooks/Cole, 1995.

HERBERT E. KASUBE

# TRIGONOMETRY

In essence, unites many aspects of arithmetic, algebra, and geometry. It underlies such activities as determining the size of the earth, planning the placement of trees when landscaping, calculating the distance between cities on the earth's surface, and reducing noise in an airplane's passenger cabin.

## HISTORY

The word *trigonometry* means "the measure of the parts of a triangle" and is derived from the Greek words *trigon,* meaning "triangle," and *metron,* meaning "a measure." Like other branches of mathematics, trigonometry has evolved over a long period of time. Although somewhat sketchy records exist, it is known that spherical (three-dimensional) trigonometry was used by the Greek astronomer Hipparchus as early as the second century B.C. to help determine the earth's size as well as its relative distance from the sun and the moon. Around the same time, the Alexandrian astronomer Ptolemy completed a table of chords of circles that was roughly equivalent to our currently known table of sines (Boyer 1991, 158). Plane (two-dimensional) trigonometry was part of astronomy (spherical trigonometry) until the fifteenth-century European astronomer Johann Mueller, known as Regiomontanus, consolidated trigonometric material. Mueller's work helped to establish trigonometry as a discipline independent of astronomy. In the sixteenth century, the French mathematician François Viète computed extensive trigonometric tables and discovered many trigonometric identities algebraically (Boyer 1991, 307).

## RIGHT TRIANGLE TRIGONOMETRY

As new problems arise in commerce, surveying, and navigation, new mathematics is devised to solve

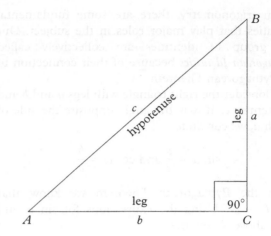

**Figure 1**

those problems. Plane trigonometry originally dealt with solving right triangles. However, it has been extended into a number of other areas. To solve a triangle means to use the known relationships between side lengths and angle measures to calculate the unknown side lengths or angle measures. In any right triangle $ABC$, the angles are lettered $A$, $B$, and $C$, where $\angle C = 90°$. The lengths of the sides opposite those angles are lettered $a$, $b$, and $c$, as shown in Figure 1. Using relationships and symbols from geometry, the following relationships always hold:

$$\angle A + \angle B + \angle C = 180°$$
$$\angle A + \angle B = 90°$$

Also, the Pythagorean Theorem states that $c^2 = a^2 + b^2$.

Six fundamental relationships between the length of sides and angle $A$ are defined as follows:

sine (sin) $A$ = (leg opposite angle $A$)/hypotenuse = $a/c$

cosine (cos) $A$ = (leg adjacent to angle $A$)/hypotenuse = $b/c$

tangent (tan) $A$ = (leg opposite angle $A$)/(leg adjacent to angle $A$) = $a/b$

cotangent (cot) $A$ = (leg adjacent to angle $A$)/(leg opposite angle $A$) = $b/a$

secant (sec) $A$ = hypotenuse/(leg adjacent to angle $A$) = $c/b$

cosecant (csc) $A$ = hypotenuse/(leg opposite angle $A$) = $c/a$

Three pairs of reciprocal relationships exist between the six trigonometric relationships. Note that no denominator can equal zero.

$$\sin A = 1/\csc A \qquad \cos A = 1/\sec A$$
$$\tan A = 1/\cot A$$

In addition to the preceding relationships, a third set of relationships hold based on the concept of complementary angles. (Two acute angles are complements if their sum is 90°.) The value of any defined trigonometric relationship of an acute angle equals the value of the cofunction of its complement:

$$\sin A = \cos(90° - A) = \cos B$$
$$\cos A = \sin(90° - A) = \sin B$$
$$\tan A = \cot(90° - A) = \cot B$$
$$\cot A = \tan(90° - A) = \tan B$$
$$\sec A = \csc(90° - A) = \csc B$$
$$\csc A = \sec(90° - A) = \sec B$$

The values of the six trigonometric relationships depend only on the size of angle $A$, not on the size of the right triangle. Figure 2 shows two similar right triangles with the same acute angle $A$. Thus the ratios of the corresponding sides are equal. From the smaller right triangle,

$$\sin A = (B'C')/(AB')$$

while in the larger triangle

$$\sin A = (BC)/(AB)$$

Since triangle $AB'C'$ is similar to triangle $ABC$,

$$(B'C')/(AB') = (BC)/(AB)$$

No matter which right triangle is used to compute $\sin A$, the value is the same. A similar argument can be used for the other five trigonometric relationships.

From the proportionality among similar triangles, tables of trigonometric ratios for various angles have been computed. The resulting tables, or calculators, that generate these values can be used to solve practical problems. For example, where possible, solar heating panels are placed on roofs facing due

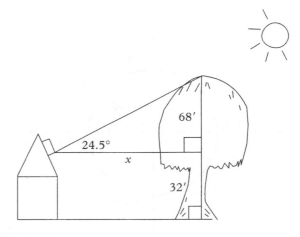

**Figure 3**

south in order to absorb maximum energy. Right triangle trigonometry can be used to determine where trees should be planted so as not to block sunlight striking the panels. For instance, suppose that on December 21, the winter solstice, the angle the sun makes with the horizontal at noon is the smallest noontime angle that occurs the entire year for locations in the northern hemisphere. Suppose you live in Chicago, Illinois, where the angle of the sun at noon on December 21 is 24.5°. Your house has solar panels on the roof with the bottom edge of the panel 32 feet above the ground. How far from your house should a maple tree be planted if you expect the tree to have a maximum height of 100 feet (see Figure 3)?

$$\tan 24.5° = 68/x$$
$$x = 68/0.4557$$
$$x = 149.2$$

Thus, the tree should be planted about 150 ft from the house.

## CIRCULAR FUNCTION TRIGONOMETRY

A *relation* is a set of ordered pairs of numbers. In a relation the set of replacements for the first variable is called the *domain*. The set of replacements for the second variable is called the *range*. When each member $x$ of the domain is paired with exactly one member $y$ of the range, the ordered pairs $(x, y)$ represent a special type of relation called a *function*.

The numerical value of a trigonometric function varies with the size of the angle. Consider the unit circle (radius 1) centered at the origin of the $x$-$y$ coordinate plane and an angle in standard position (vertex at the origin); see Figure 4. The initial side

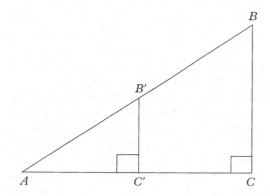

**Figure 2**

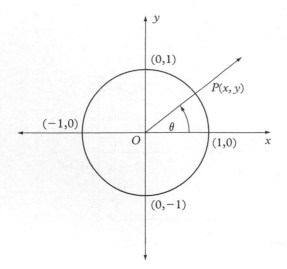

**Figure 4**

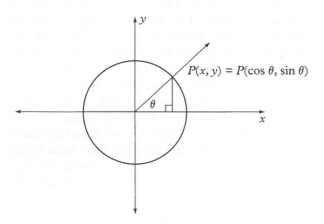

**Figure 5**

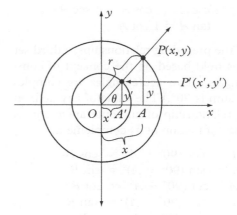

**Figure 6**

Figure 6). When a perpendicular is drawn from point $P$ to the $x$-axis, a right triangle is formed. This triangle is called a *reference* triangle. Because $P(x,y)$ and $P'(x',y')$ are corresponding vertices of the similar right triangles $POA$ and $P'OA'$, the side lengths are $x,y,r$ and $x',y'$, and 1, respectively. Since the lengths of the corresponding sides of similar triangles are proportional, $x'/1 = x/r$ and $y'/1 = y/r$. Thus, $x' = \cos \theta$ and $y' = \sin \theta$, $\cos \theta = x/r$, and $\sin \theta = y/r$.

The previous relationships hold for any angle in any quadrant and can be generalized as follows: for an angle $\theta$ in standard position, if $P(x,y)$ is any point (other than the origin) on the terminal side of $\theta$, then

$$\cos \theta = x/r \quad \text{and} \quad \sin \theta = y/r$$
$$\text{where } r = \sqrt{(x^2 + y^2)}$$

## Degree Versus Radian Measure

Two different measurement systems are commonly used to express the measure of an angle. By definition, the measure of an angle in circular function trigonometry is the amount and direction of rotation. Since one complete revolution of an initial ray on a circle is 360° (the initial and terminal rays coincide), 1 degree is 1/360 of a complete rotation. Note that a counterclockwise rotation is considered positive while a clockwise rotation is negative.

Angle measures can also be expressed using radians. If a central angle of a circle intercepts an arc that has the same length as the radius of the circle, the measure of the angle formed is 1 radian, abbreviated 1 rad (see Figure 7). In general, if $r$ is the radius of a circle and $s$ is the length of an arc intercepted by a central angle $\theta$, then the radian measure of $\theta$ can be found by dividing $s$ by $r$, that is, $\theta = s/r$ (see Figure

($x$-axis) intersects the circle at $(1,0)$, while the ray $OP$ (rotated counterclockwise from the $x$-axis) intersects the circle at $P(x,y)$. As $\theta$ changes, the coordinates of $P(x,y)$ will change. However, for each angle $\theta$, there is a set of unique ordered pairs $(\theta,x)$ and $(\theta,y)$. These two functions, respectively, are named the cosine function $\{(\theta,x)\}$ and the sine function $\{(\theta,y)\}$, that is, $x = \text{cosine } \theta$ and $y = \text{sine } \theta$ (see Figure 5). Because the cosine and sine functions are defined using a unit circle, they are often called *circular functions*. Since the unit circle has a radius of 1, values for $x$ and $y$ can range between $-1$ and 1. That is,

$$-1 \le \cos \theta \le 1 \quad \text{and} \quad -1 \le \sin \theta \le 1$$

The values of $\cos \theta$ and $\sin \theta$, for any angle $\theta$, can be found if a point $P(x,y)$ on the terminal side of $\theta$ is known, even if point $P$ is not on the unit circle (see

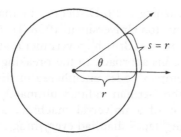

**Figure 7**  *Note: θ = 1 radian.*

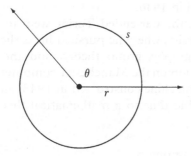

**Figure 8**  *Note: θ = $\frac{s}{r}$ radians.*

8). If no unit is specified, it is assumed that radian measure is being used. Since one-half revolution in radians equals π radians, π radians = 180°.

Radian measure can be used to calculate the distance between points on the earth's surface. Suppose, for example, that the distance along the earth's surface between Oslo, Norway and the equator was needed. An atlas indicates that the latitude of Oslo, Norway is 60° north of the equator (represented as 60°N). For practical purposes, assume that the earth is a sphere with a radius of 3,964 miles and $E$ is a point on the equator due south of Oslo (see Figure 9). Since 60° = π/3 radians, ∠OCE = π/3. By definition,

$$\angle OCE = OE/3,964 = \pi/3$$

Using π ≈ 3.14, $OE$ = (3,964)(3.14)/3 ≈ 4,149 miles. Therefore, along the earth's surface, Oslo, Norway is approximately 4,150 miles from the equator.

## Periodic Data and Graphing

Data or graphs that repeat in a pattern are said to be *periodic*. Knowing that a function such as the sine function or the cosine function is periodic simplifies its graphing. Angles that are less than 0 or greater than 2π are coterminal (angles in standard position whose terminal sides coincide) with angles

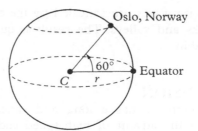

**Figure 9**  *Note: r = 3,964 mi.*

lying between 0 and 2π. The values of the circular functions repeat after one complete clockwise or counterclockwise rotation. Thus, to graph the sine function $y = \sin x$, first construct a table of values for $0 \le x \le 2\pi$ (see Figure 10). Note that when $y = \sin x$ is graphed on the x-y coordinate plane, each point on the graph has coordinates of the form $(x, \sin x)$, where $x$ is given in radians (see Figure 11). The cosine function is graphed in a similar manner.

Scientists have discovered that sound waves travel in patterns that mirror sine waves and cosine waves. Engineers, working to make airplane cabins quieter analyze sound using mathematical equations whose graphs illustrate the periodic nature of sound. These equations are constructed from the sine and cosine functions. Throughout the cabin, microphones pick up various sounds. A computer combines cabin sounds with engine speed information to produce sound waves that are the exact opposite of the undesirable noise. Technically, the computer-generated sound waves are 180° "out of phase" with the original undesirable noise. By projecting the suppression noise through the cabin speakers, the unde-

| x | 0 | $\frac{\pi}{6}$ | $\frac{\pi}{4}$ | $\frac{\pi}{3}$ | $\frac{\pi}{2}$ | $\frac{2\pi}{3}$ | $\frac{3\pi}{4}$ | $\frac{5\pi}{6}$ | π | $\frac{7\pi}{6}$ | $\frac{5\pi}{4}$ | $\frac{4\pi}{3}$ | $\frac{3\pi}{2}$ | $\frac{5\pi}{3}$ | $\frac{7\pi}{4}$ | $\frac{11\pi}{6}$ | 2π |
|---|---|---|---|---|---|---|---|---|---|---|---|---|---|---|---|---|---|
| sin x | 0 | 0.5 | 0.71 | 0.87 | 1 | 0.87 | 0.71 | 0.5 | 0 | −0.5 | −0.71 | −0.87 | −1 | −0.87 | −0.71 | −0.5 | 0 |

**Figure 10**

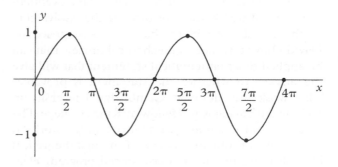

**Figure 11**

sirable noise waves are canceled by the corresponding peaks and valleys. The result is quiet in the plane's cabin.

## CONCLUSION

Trigonometry has a long and varied history. Even with the advent of high-speed and accurate computers and calculators, trigonometry remains an important discipline that models the real-world phenomena of periodic data, provides an important communication tool, and can be used to solve applied and theoretical problems.

*See also* Transcendental and Algebraic Functions; Trigonometric Functions, Elementary; Trigonometric Identities

### SELECTED REFERENCES

Boyer, Carl B. *A History of Mathematics.* 2nd ed., rev. Revised by Uta C. Merzbach. New York: Wiley, 1991.

DeMeis, Rick. "Noise Reduction—Quiet in the Sky." *Popular Science* 245(6)(Dec. 1994):35.

Hayden, Jerome D., and Bettye C. Hall. *Prentice Hall Trigonometry.* 2nd ed. Englewood Cliffs, NJ: Prentice-Hall, 1993.

Hirsch, Christian R., and Harold L. Schoen. *Trigonometry and Its Applications.* Mission Hills, CA: Glencoe/McGraw-Hill, 1990.

Hirsch, Christian R., Marcia Weinhold, and Cameron Nichols. "Trigonometry Today." *Mathematics Teacher* 84(Feb. 1991):98–106.

Ryan, Merilyn, Marvin E. Doubet, Mona Fabricant, and Theron D. Rockhill. *Prentice Hall Advanced Mathematics—A Precalculus Approach.* Englewood Cliffs, NJ: Prentice-Hall, 1993.

JEROME D. HAYDEN

# TURING, ALAN MATHISON (1912–1954)

One of the great mathematicians of the twentieth century and well known for proving the insolvability of the decision problem that had been posed by David Hilbert: is there a mechanical process that can be applied to a mathematical statement that will give the answer to whether the statement is provable? This work was done at Cambridge University in 1936 when Turing was a fellow at Kings College. The machine concept that he employed in his work on this decision problem was one of the first theoretical ideas leading to the modern stored-program electronic digital computer.

After earning a Ph.D. degree in mathematical logic at Princeton University in 1938, Turing became employed by the British government at Bletchley Park, where his approach to the breaking of the immensely complex codes enciphered on Enigma machines by the German military ultimately led to the construction of a universal machine, a computer with a single "tape" that had everything stored on it: instructions, data, and results of operations. The first such machine was the modified Electronic Numerical Integrator and Calculator, known as the ENIAC, completed in 1946.

When the war ended, Turing went to Manchester University, where he pursued his earlier interests in cryptography, group theory, and the Riemann zeta-function on the Manchester computer, a stored-program machine constructed in 1948, and devoted considerable time to a mathematical theory of embryology.

*See also* Computers

### SELECTED REFERENCES

Hodges, Andrew. *Alan Turing: The Enigma.* New York: Simon and Schuster, 1983.

Turing, Alan M. "Can a Machine Think?" In *The World of Mathematics* (pp. 2099–2133). Vol. 4. James R. Newman, ed. New York: Simon and Schuster, 1956.

Van Rootselaar, B. "Turing, Alan Mathison." In *Dictionary of Scientific Biography* (pp. 497–498). Vol. 13. Charles C. Gillispie, ed. New York: Scribner's, 1976.

FRANCINE F. ABELES

# TUTORING

Undertaken to improve the performance and increase the mathematical knowledge of one or a few "tutees," or learners. The advantage of a tutoring situation for the learner is that the tutor focuses continuous attention on the learner alone, addressing his or her unique needs without having to counterbalance those needs with the needs of other learners in a larger group.

Most often, a tutor is an older student or adult with extensive content knowledge and experience, but sometimes tutors are the learners' peers who have only just mastered the content themselves. Peer tutoring within the context of the school has been promoted by some programs and educators as a way of supporting the learning of students who are struggling by providing individualized help from classmates. Because peer tutors only recently struggled with the mathematical content themselves, they may be particularly sensitive to potential points of confusion and may be able to help classmates identify and address problematic areas. In the process of helping their classmates, peer tutors' own conceptual understanding is enhanced as they review and verbalize mathematical ideas. It is usually expected that the roles of tutor and tutee will at times be reversed, as different students may master different ideas more quickly during the course of instruction (Grossman 1985).

During the last two decades of the twentieth century, much computer software has been developed that provides personalized tutoring. The simplest of these programs provide extended drill and practice while noting and recording student responses. Patterns of student errors determine entry into branched paths for additional practice on tasks designed to address specific misunderstandings or gaps. Other programs provide hints for students as they work to solve complex problems in algebra and geometry (Kaput 1992). Research suggests that classroom use of computer-based tutoring programs can enhance learning as well as affect the social processes among students and teacher in positive ways (Schofield, Eurich-Fulcer, and Britt 1994).

The effectiveness of all tutoring relies on an accurate determination of the depth and breadth of the tutee's understanding of targeted mathematical concepts as well as the underlying ideas and skills necessary to support understanding. Since mathematical ideas are interrelated, misunderstandings and knowledge gaps often stand in the way of further learning (e.g., a student who does not "see" that a unit or whole may refer to a quantity other than 1 may well be unable to visualize "two-thirds of seven"; the task of subtracting quantities from 1,000 becomes an exercise in memorizing procedures that are devoid of meaning if a student is confused about the concept of place value). Tutees themselves are rarely aware of their own gaps or spotty knowledge; the tutor can discover deficits by asking the tutee to solve problems or do a series of exercises while simultaneously describing the procedures being used and explaining the reasoning behind them.

Once points of confusion are identified, the tutor should help the tutee acquire the necessary knowledge base as well as build self-confidence by choosing developmentally appropriate problems and working through them together. Problems should be meaningful to the tutee so that his or her experience and intuitions can be called upon to inform solutions and new ideas can be related to that which is already "owned" by the tutee. Asking questions such as "why did you do that?," "could you do it a different way?," and "can you make up a similar problem?" encourages the tutee to strategize solutions, fosters the expectation that procedures and their applications should be based on mathematical reasoning, and presents opportunities to discuss connections among related concepts.

The tutee should be highly engaged in the activities to derive the most benefit from the tutoring process; instructional practices should encourage active participation in discussions, hands-on modeling, and problem-solving strategizing rather than consist of lengthy verbal explanations or extended descriptions of step-by-step procedures by the tutor. Tutees must be participants in the process of tailoring the learning environment to meet their own needs. Indeed, research has shown that more questions are asked during a mathematics tutoring session than during a mathematics class and within each setting, a higher percentage of the questions are asked by tutees (20%) than by students (4%) (Graesser and Person 1994).

Tutoring experiences can enhance a tutee's understanding of mathematical content and, as well, provide a context for exploration of and reflection on personal learning strengths and weaknesses.

*See also* Manipulatives, Computer Generated; Questioning; Software

### SELECTED REFERENCES

Graesser, Arthur C., and Natalie K. Person. "Question Asking During Tutoring." *American Educational Research Journal* 31(Spring 1994):104–137.

Grossman, Anne S. "Mastery Learning and Peer Tutoring in a Special Program." *Mathematics Teacher* 78(Jan. 1985):24–27.

Kaput, James J. "Technology and Mathematics Education." In *Handbook of Research on Mathematics Teaching and Learning* (pp. 515–556). Douglas A. Grouws, ed. New York: Macmillan, 1992.

Schofield, Janet Ward, Rebecca Eurich-Fulcer, and Cheri L. Britt. "Teachers, Computer Tutors, and Teaching: The Artificially Intelligent Tutor as an Agent for Classroom Change." *American Educational Research Journal* 31(Fall 1994):579–607.

LYNDA B. GINSBURG

# U

## UNDERACHIEVERS

A categorization of students sometimes based on inequitable school policies and practices, peer pressure, negative attitudes, and misguided priorities. Children's lives are affected when they are tracked into elementary school classes in which teachers promote only memorization and computational skills. Furthermore, elementary school children, placed into low-level tracks based on the use of a standardized and possibly biased test of their ability and achievement in reading and mathematics, will likely remain in these tracks throughout their schooling (Goodlad 1984).

Beginning at this educational level, the unfair policies of grouping or tracking disproportionately affect economically disadvantaged groups of underachievers (Jones 1993; Oakes 1992). The less-challenging mathematics programs designed for these students focus solely on low-level and rote-oriented basic skills, omitting the challenges offered to students who come from more advantaged backgrounds (National Research Council 1989; Steen 1987). There is evidence to suggest that inferior programs are the norm for economically disadvantaged African American, Hispanic American, and American Indian students; they are not expected to excel in academically oriented courses that promote higher-order thinking skills (National Research Council 1989).

Far too often, many of these students are not provided a truly equal opportunity to learn substantive mathematics. Minority and/or female low achievers have historically been underserved by schools in this country (Darling-Hammond 1994; Secada 1992).

According to information published by the Office of Educational Research and Improvement (1992), other general educational policies, such as those aimed at easing the obstacles to learning and graduation, seem to reward students who make only a minimal effort at learning. For example, over the last twenty-five years, the rate of high school completion has risen dramatically, yet average achievement scores have gone down. It is questioned whether increased graduation rates have been attained only by reducing the standards of competence required of all students. Whether their academic records are deserving or not, high school students can almost always find colleges that will accept them. Students may thus find few educational incentives to achieve.

Because of peer pressure, students with strong mathematics aptitude may hide their interests and consciously restrain their achievement. High achievement in mathematics may not be socially acceptable, particularly at the high school level, where students make their first choices concerning curriculum. Students who develop and maintain negative dispositions about studying and learning mathematics may reduce or cease their efforts to achieve. They often give mathematics assignments low priority on their schedules of academic and extracurricular activities (e.g., other academic work, sports, community service, employment, and family responsibilities) that compete for their time and attention. In general, mathematics is often perceived by a large segment of the school population as too difficult, uninteresting, and of little use beyond basic computational skills (National Research Council 1989).

Too few students in the nation's schools are achieving at a high level in mathematics, as evidenced by increasing numbers of students who enroll in remedial or developmental coursework in college and who are ill-prepared to cope with the demands of today's workplace or to lead a productive life (National Research Council 1991). The National Research Council (NRC), in its document *Everybody Counts,* reports that "students enrolled in advanced high school mathematics courses come disproportionately from white upper- and middle-class families" (NRC 1989, 21).

Madison and Hart concluded in *A Challenge of Numbers* that if current academic and demographic trends continue, fewer students will study mathematics and select mathematics-based occupations. The segment of our society, white males, that has traditionally trained for the now-burgeoning mathematics-based careers, is shrinking in size and can no longer meet our nation's needs. While the need for technologically trained workers is increasing, the prospective supply is not. As the college-age population is shrinking, enrollments in remedial courses in secondary-level arithmetic, algebra, and geometry have shown a tremendous increase. Yet from entry into ninth grade through receipt of the doctoral degree, nearly one-half fewer students study mathematics each year. The problem is more acute in that white males now dominate the science and engineering work force, but their numbers are rapidly decreasing for the college-age group. By the year 2010, African American and Hispanic Americans together will constitute 30% of the traditional college-age population. But at all levels, few African Americans and Hispanic Americans study or practice mathematics (Madison and Hart 1990).

A goal of school mathematics programs is to prepare *all* students to function effectively in a skilled work force. Many of today's jobs demand ability to use quantitative reasoning and problem-solving skills. More than 85% of new entrants in the work force are expected to be female or minority. To promote excellence and to ensure access to high-quality educational opportunities for these students, new ways of motivating and retaining students in mathematics are being devised.

The National Council of Teachers of Mathematics (NCTM) underscores the importance of the need to rethink and restructure school mathematics when it reminds us:

> Today's society expects schools to insure that all students have an opportunity to become mathematically literate, are capable of extending their learning, have an equal opportunity to learn, and become informed citizens capable of understanding issues in a technological society (NCTM 1989, 5).

This organization has served as a national model for the education community by establishing standards for school mathematics. In its documents *Curriculum and Evaluation Standards, Assessment Standards for School Mathematics,* and *Professional Standards for Teaching Mathematics,* NCTM has set forth a new vision for school mathematics. The vision calls for dramatic changes in mathematics content, methods of assessment, pedagogical practices, and assumptions about who can learn mathematics. For all students, the *Curriculum and Evaluation Standards* advocates less emphasis on operations and computation in the curriculum and more emphasis on fostering the conceptual understanding of mathematical ideas and their application to rich problem-solving situations (NCTM 1989). An important consideration in promoting the achievement of high standards by all students is that they have an equal opportunity to learn high-level content (O'Day and Smith 1993). The underlying assumptions upon which such broad-based curricula should be established are (1) respect for cultural and linguistic differences and (2) high expectations that all students can learn.

The Chapter 1 (now Title 1) Program of the Elementary and Secondary Education Act (ESEA) has been reauthorized to eliminate all requirements to administer norm-referenced, multiple choice tests to students who are eligible to participate in the program ("Standardized Testing" 1994). The new legislation, now called Title 1, Part A, establishes new goals and policies for promoting the achievement of high standards, higher-order thinking skills, and the understanding of advanced mathematical concepts for youths who participate in the program. With the elimination of the requirement for standardized testing, the new law not only offers flexibility to states and districts in the design and implementation of new forms of assessment; it also supports the development of world-class standards for content and student performance ("Standardized Testing" 1994).

Similarly, under the Goals 2000: Educate America Act, P.L. 103-227, federal grants are being offered to states to create and implement world-class standards and to develop strategies to help students achieve them ("Goals 2000" 1994). Goals 2000 supports improved teacher training and increased parental involvement in the educational process. Goals 2000 asserts that "American students will be first in the world in math and science achievement." However, if this goal is to be attained, world-class instruction in mathematics must also be the norm for the nation's economically disadvantaged, underachieving students. Jones reminds us that ". . . if we eliminate tracking and create a level playing field—and then provide the support system for *all* students—then all students can succeed" (1993, 1).

Thus the score on a standardized test should not serve as the single source by which schools and teachers judge a student's intellectual potential. Researchers and educators suggest that traditional forms of standardized multiple choice tests do not provide a complete assessment of students' mathematical competencies and abilities to apply concepts and skills in rich, real-life problem settings (Worthen and Spandel 1991). To validly appraise students' strengths and needs, and also foster students' progress toward high expectations, the *Assessment Standards for School Mathematics* states that it is necessary to gather and utilize information from a variety of balanced and equitable sources. These sources include the students themselves, teachers, district committees, state or provincial mathematics supervisors, and major test publishers (NCTM 1995). The NRC likewise advises that "to assess development of a student's mathematical power, a teacher needs to use a mixture of means . . . (since) only broad based assessment can reflect fairly the important higher order objectives of mathematics curricula" (NRC 1989, 90). Schools must have comprehensive profiles on students in order to determine appropriate resources and to organize instruction in a way that fosters mathematical competence and intellectual growth (Darling-Hammond 1984; NCTM 1989). In addition, recent research has shown that the educational and social interests of underachieving students would be better served by offering them challenging academic experiences (Nieto 1994).

Substantive educational reform requires ongoing professional development for teachers to help them confront complex challenges related to teaching, program planning, and assessment. The ultimate goal should be to enable them to implement strategies to support *all* students in the achievement of mathematical literacy. *Professional Standards for Teaching Mathematics* attests to the revolutionary approach needed in mathematics education if all students are to gain mathematical power. Specifically, the *Professional Standards* calls for "the creation of a curriculum and an environment, in which teaching and learning are to occur, that are very different from much of current practice" (NCTM 1991, 1). The complexities of the issues to be addressed offer an even greater challenge when organizing instruction for students who have special needs. Teachers need support in implementing curricular changes and practices, including those which

- create a variety of critical-thinking and problem-solving experiences that reflect realistic applications of mathematics in a multicultural, global, and technologically advanced society;
- assign tasks that actively engage all students to deepen their understanding of mathematical concepts and applications;
- utilize pedagogical practices that promote equity so every student will have an equal opportunity to achieve high standards and to develop reasoning and thinking skills; and
- support increased involvement of families in the mathematics education of their children.

Key aspects of change must be addressed, however, if schools are to develop programs that help students become empowered and open to positive change. Recent studies have made teachers more aware of certain policies and practices that may be especially beneficial to underachievers, such as

- small classes that are conducive to effective teacher time management and classroom management (Bracey 1987),
- collaborative assignments given to heterogeneous (high/low) achieving student pairs, rather than homogeneous pairs, for the advantage of both categories of achievers (Bracey 1994).

Nieto concluded from her research that "students have a lot to teach us about how pedagogy, curriculum, ability grouping, and expectations of ability need to change so that greater numbers of young people can be reached" (1994, 423–424).

If the reform movement is to effect positive change, we must embrace these notions: (1) teachers are ultimately the catalysts of change, and (2) teachers must be prepared to support the development of students' mathematical power by seeking pedagogical practices that are most effective for actively engaging students in mathematical investigations that promote curiosity, reasoning, understanding of concepts, and problem-solving ability. Teachers must also employ classroom evaluation techniques that are different from those in current use. The NCTM argues against using only traditional paper-and-pencil tests and calls for expanding classroom instruction and assessment methods to include performance-based activities, collaborative learning, small-group discussion, observations, projects, portfolios, journal writings, role-playing, reporting, and open-ended questions:

> Paper-and-pencil tests, although one useful medium for judging some aspects of students' mathematical knowledge, cannot suffice to provide teachers with the insights they need about their students' understandings in order to make instruction as effectively responsive as possible (NCTM 1991, 63–64).

Parents serve as the child's first teachers and, if viewed as partners, they are invaluable resources in the educational process throughout schooling. Parent involvement can impact positively on achievement, enhance the child's self-esteem, and promote a more positive disposition toward school and learning. This involvement is critical for economically disadvantaged groups of low achievers since they often feel excluded from the educational process (Comer 1980). In an effort to mobilize support and increase public awareness of the need for a skilled work force, it is important to devise more efficient ways of informing parents and the community of the need to restructure school mathematics instruction and redesign assessment practices. When parents are involved as partners in the educational process, efforts by school administrators and teachers to improve performance, achievement, and attitudes toward learning are more likely to be achieved (Goldstein and Campbell 1991). The input of parents and community into the testing and assessment process might also help to eliminate ethnic, racial, and gender bias in testing (Worthen and Spandel 1991).

In an interview dealing with new perspectives on teaching and learning (Marriott 1990), Comer asserts that "most administrators [and teachers] are not trained to recognize that a student's academic and behavioral lapses may be linked to the cultural and social gap between home and school." Schools that have developed strategies for enhancing parent involvement based on Comer's school-based management team model have modified the negative experiences of at-risk students by improving their dispositions and attitudes toward school and learning. Research studies have shown that schools in which Comer's program has been implemented excelled in reading and mathematics for the lowest achievers (Marriott 1990).

A productive relationship between students and teachers and between home and school are key aspects of a successful learning and educational experience (Seeley 1981, 11). To facilitate parent involvement in the educational process, parents need to understand the major issues pertaining to testing, school structure, school operations, what, how, and to whom mathematics is taught, and what methods of assessment are being used in the instructional process. The level of parents' involvement in schools is intimately related to the success of their children in those schools (Goldberg 1990). Joyce Epstein, in an interview dealing with parents and school (Brandt 1989), lists five major types of parent involvement in schools, with their different outcomes and impacts upon parents, teachers, and students. Of these, two types of involvement very useful in devising effective partnerships between the home and school mathematics are

- Type 2—the "basic obligations of schools" to communicate with parents in the home regarding school programs and children's progress; and
- Type 4—"parent involvement in learning activities at home," or parent-initiated activities and child requests for help, or instructions from teachers to parents to help children at home "on learning activities that are coordinated with the children's classwork" (Brandt 1989, 25).

In type 2 involvement, more effective means of communication from the school to the home may be designed so that parents' involvement may be enhanced. For example, the use of mathematics portfolios may serve as one means by which students' work may be sent home for review and joint home assignments, then returned with comments and feedback to the teachers. Work with portfolios could be structured in such a way that both parents and students have input into the contents of the portfolios through projects designed for joint participation of parents and children in some aspects of its development (see Stenmark, Thompson, and Cossey 1986 for ideas). Through the use of such portfolios, parents can be helped to understand school programs better, and the school-home partnership can be enhanced.

Type 4, or parent involvement in learning activities in the home, is the kind of involvement for which parents often seek help. Type 4 involvement by parents can be enhanced when teachers make suggestions to parents on how to help their children at home to promote improved learning at school. Teachers can provide parents with information on the kinds of mathematical skills taught at each grade level. This could be structured so that students must discuss schoolwork at home with their parents at regular intervals. Goldstein and Campbell (1991) have devised a form to be used by parents to provide an assessment of mathematics activities performed by students at home. Epstein also notes some positive student outcomes as the result of the enhancement of type 4 involvements (Brandt 1989). These include completion of homework, increased student self-concept as a learner, and the enabling of students to get more practice, leading to greater achievement in the skills area.

Schools and teachers will benefit from devising strategies to inform parents and the community about the various changes taking place in school mathematics. Members of both the school and the neighboring communities have a critical role to play in promoting mathematical power for *all* students.

*See also* Underachievers, Special Programs

## SELECTED REFERENCES

Bracey, Gerald W. "Achievement in Collaborative Learning." *Phi Delta Kappan* 76(Nov. 1994):254–255.

————. "Small Is Beautiful—and Effective." *Phi Delta Kappan* 68(May 1987):703.

Brandt, Ronald. "On Parents and Schools: A Conversation with Joyce Epstein." *Educational Leadership* 47(2)(1989):24–27.

Comer, James. *School Power: Implications of an Intervention Project.* New York: Free Press, 1980.

Darling-Hammond, Linda. *Beyond the Commission Reports: The Coming Crisis in Teaching.* R-3177-RC. Santa Monica, CA: Rand, 1984.

————. "Performance-based Assessment and Educational Equity." *Harvard Educational Review* 64(1)(1994): 5–30.

"Goals 2000 Will Shape State, Local School Reform." *Education Daily* (Spec. Suppl.) 27(102)(May 27, 1994).

Goldberg, Mark F. "Portrait of James P. Comer." *Educational Leadership* 48(1)(1990):40–42.

Goldstein, Sue, and Francis A. Campbell. "Parents: A Ready Resource." *Arithmetic Teacher* 38(6)(Feb. 1991): 24–27.

Goodlad, John I. *A Place Called School: Prospects for the Future.* New York: McGraw-Hill, 1984.

Jones, Vinetta. "Views on the State of Public Schools." A paper presented to the Conference on the State of American Public Education. Sponsored by Phi Delta Kappa Institute on Educational Leadership and the Educational Excellence Network, February 4–5, 1993.

Madaus, George F. "A Technological and Historical Consideration of Equity Issues Associated with Proposals to Change the Nation's Testing Policy." *Harvard Educational Review* 64(1)(Spring 1994):76–95.

Madison, Bernard, and Therese A. Hart. *A Challenge of Numbers: People in the Mathematical Sciences.* Washington, DC: National Academy Press, 1990.

Marriott, Michel. "The Home's Link to School Success. A New Road to Learning: Teaching the Whole Child." *New York Times* (June 13, 1990):A1.

National Council of Teachers of Mathematics. *Assessment Standards for School Mathematics.* Reston, VA: The Council, 1995.

————. *Curriculum and Evaluation Standards for School Mathematics.* Reston, VA: The Council, 1998.

————. *Principles and Standards for School Mathematics.* Reston, VA: The Council, 2000.

————. *Professional Standards for Teaching Mathematics.* Reston, VA: The Council, 1991.

National Research Council. *Counting on You: Actions Supporting Mathematics Teaching Standards.* Washington, DC: National Academy Press, 1991.

————. *Everybody Counts: A Report to the Nation on the Future of Mathematics Education.* Washington, DC: National Academy Press, 1989.

Nieto, Sonya. "Lessons from Students on Creating a Chance to Dream." *Harvard Educational Review* 64(4)(Winter 1994):392–426.

Oakes, Jeannie. "Can Tracking Research Inform Practice?" *Educational Researcher* 21(4)(1992):12–21.

————. "Opportunity to Learn: Can Standards-Based Reform Be Equity-Based Reform?" In *Seventy-five Years of Progress. Prospects for School Mathematics* (pp. 78–98). Iris M. Carl, ed. Reston, VA: National Council of Teachers of Mathematics, 1995.

O'Day, Jennifer A., and Marshall S. Smith. "Systemic Reform and Educational Opportunity." In *Designing Coherent Education Policy: Improving the System* (pp. 250–312). Susan H. Fuhrmann, ed. San Francisco, CA: Jossey-Bass, 1993.

Office of Educational Research and Improvement. *Hard Work and High Expectations: Motivating Students to Learn.* Washington, DC: U.S. Department of Education, Office of Educational Research and Improvement, 1992.

Secada, Walter G. "Race, Ethnicity, Social Class, Language, and Achievement in Mathematics." In *Handbook of Research on Mathematics Teaching and Learning* (pp. 623–660). Douglas Grouws, ed. New York: Macmillan, 1992.

Seeley, David. *Education through Partnerships: Mediating Structures and Education.* Cambridge, MA: Ballinger, 1981.

"Standardized Testing Dropped from ESEA." *FairTest Examiner* 8(4)(Fall 1994):1, 9.

Steen, Lynn Arthur. "New Goals for College Mathematics." *Mathematics in College* (Spring-Summer 1987):5–10.

Stenmark, Jean, Virginia Thompson, and Ruth Cossey. *Family Math.* Berkeley, CA: Lawrence Hall of Science, University of California, 1986.

Wheelock, Anne. *Crossing the Tracks: How "Untracking" Can Save America's Schools.* New York: New Press, 1992.

Worthen, Blaine R., and Vicki Spandel. "Putting the Standardized Test Debate in Perspective." *Educational Leadership* 48(5)(1991):65–69.

LUCILLE CROOM
JUNE L. GASTÓN

# UNDERACHIEVERS: SPECIAL PROGRAMS

Efforts to enable more students to achieve mathematical competence. The mathematics "pipeline" can be either a pump or a filter. Unfortunately, it often serves as a filter to decrease the number of eligible and capable students completing rigorous academic curriculums that lead to lucrative occupations. Females and minorities are usually filtered out of this pipeline. Many of these students are among those labeled as underachievers in mathematics and discouraged from pursuing advanced study of the subject. In major international studies of student performance, American students have consistently scored well below their peers in other highly industrialized countries. Businesses, educational institutions, and the military have been burdened with extensive and exorbitant demands for remedial mathematics instruction. Due to academic and demographic trends, many well-paying positions in the burgeoning fields of science and technology cannot be filled (National Research Council 1989).

Clearly, the filtering process must be reversed. Mathematical competency must be a viable goal for all students, especially underachievers. These students must reevaluate their attitudes and priorities with respect to the subject and realize that mathematical empowerment can offer limitless access to rewarding careers. It is crucial that all students develop critical thinking and problem-solving skills and view themselves as competent, capable learners of the subject. In our highly competitive technological and global society, a concerted effort by educational institutions, community organizations, professional organizations, businesses, government, and the media is necessary to facilitate the attempts of all students to become mathematically proficient members of an effective citizenry and work force. There are many exemplary efforts to open the mathematics pipeline and improve access and equity in both educational and occupational areas. Many such public and privately funded efforts are currently underway.

Designed to enable more students to achieve appropriate mastery of either a traditional or an innovative mathematics curriculum, the reform efforts include programs with teacher support and development components, teacher and student involvement in the development of curricular content and methods, teacher mentoring and student mentoring, a technological dimension, collaborative learning techniques and restructured classrooms, summer academies and institutes, and parental involvement. The activities may be institutional, districtwide, statewide, or national in scope. All of these programs have addressed the issue of underachievement in mathematics.

## EFFORTS BY SCHOOLS
### Mathematics at Garfield High School, Los Angeles, CA

Jaime Escalante uses a sports team approach to capture his students' interests and motivate them to succeed. He gained national attention when his traditionally categorized low achievers from the East Los Angeles barrio made nontraditional academic gains in mathematics. With Escalante's coaching, they progressed from the foundations of mathematics through Advanced Placement (AP) calculus. Escalante's mathematics program has held national attention because it illustrated a way to motivate students who are not all mathematically gifted to endure the rigorous three-year secondary mathematics curriculum that culminates with AP calculus. He recruited students who have *ganas* (desire), not necessarily talent (Escalante and Dirmann 1990). The movie *Stand and Deliver* documents the success of eighteen of Escalante's first AP calculus students in 1982 and its impact on the mathematics curriculum at Garfield High School. It also documents Escalante's pedagogical expertise (Heyman 1990). Funds from a variety of community resources helped Escalante to continue to expand his program to serve a maximum number of minority students. Former graduates are invited back to encourage, counsel, and tutor new program participants (Escalante and Dirmann 1990).

## Math at East Harlem Tech (P.S. 72), New York, NY

Kay Toliver believes in teaching mathematics through listening, speaking, touching, and writing. She gives of her free time to help with classwork or homework questions and to be an attentive listener and advisor. She emphasizes that teacher expectations are vital to student success, especially for students who are regularly told they are disadvantaged and who may have the impression that mathematics is a subject beyond their abilities. She reminds her students that they must abandon negative attitudes toward the subject. Enforcing high expectations, she insists that with dedication and hard work anyone can learn mathematics. Some of her students have poor mathematics backgrounds and come from unstable families and unsafe neighborhoods; however, she does not let these drawbacks give them an excuse for failure. She makes sure that they have the supplies they need to do their work and that they have a safe, supportive learning environment.

Students have a greater opportunity to learn when they attend school regularly, change schools and are tardy less frequently, and cut classes less often. In order to overcome the chronic lateness and absenteeism common in many inner-city schools, Toliver focuses on capturing student attention and interest with special techniques. Her goal is to get students actively involved in creative activities that require their critical thinking and analytical skills. Sometimes she integrates mathematics with other subjects or uses costumes to seize their attention. Generally, the students are so preoccupied with mathematical activities that discipline is not a problem. Her outdoor Math Trail problems help students see the relevance of mathematical concepts in the urban environment. They also record their daily impressions in mathematics journals. Her assessment techniques include performance-based measures involving writing and drawing diagrams.

To facilitate teacher efforts, students need to have supportive parents and to attend schools that support academic achievement. Toliver thus enlists the support of the parents of her students. These parents, who may feel unsure of their own abilities and unneeded because their children no longer require parental assistance in mathematics, are encouraged to participate in Family Math sessions. The joint learning helps to open the lines of mathematical communication between the teacher, parents, and students. Kay Toliver has also initiated an annual Math Fair for students of all grades at East Harlem Tech. Due to the popularity and success of her endeavors, she has begun to work with other master mathematics teachers to disseminate information concerning successful classroom techniques.

## MOVE-IT! Mathematics, University of Houston-Victoria, Texas

Professor Paul Shoecraft believes that mathematics should be enjoyable. Classroom ideas that promote positive attitudes toward elementary-level mathematics are evident in the pedagogy and curriculum of his MOVE-IT! (Math Opportunities, Valuable Experiences—Innovative Teaching) program. Among the key pedagogical techniques are the skillful use of manipulatives to introduce mathematical concepts. Another technique facilitates the memorization of basic facts by focusing on counting skills, that is, counting up, counting down, and skip counting.

Algebra and geometry are also a part of the MOVE-IT! curriculum. Algebra is first introduced through the balance beam to model the basic concept of equality and to illustrate the equivalence of different equations. Students learn to analyze relevant data and to represent the results using vertical and horizontal bar graphs. They also learn to graph coordinates on the Cartesian plane. Employing manipulatives such as geoboards and elastic bands, students can determine the area of irregularly shaped geometric figures.

Shoecraft's program works for children from a variety of socioeconomic backgrounds. Using the MOVE-IT! mathematics program in Texas, some school districts have eliminated ability grouping and tracking. An important ingredient in the success of the program is the belief that, given a multiplicity of learning approaches, every child can learn mathematics (Nielsen 1990).

## INTERVENTION EFFORTS
### The Algebra Project

Robert P. Moses, a mathematician and civil rights leader, developed the Algebra Project in the 1980s. The project helps inner-city and minority middle school students gain access to higher-level mathematics. A Transition Curriculum helps sixth graders make the conceptual shift from arithmetic to algebraic thought processes. Seventh and eighth graders study algebra, supplementing the regular textbook with project modules.

The project curriculum engages students in co-operative learning activities that also develop abstract thinking skills. It calls for students to follow a five-step process in which they use real-life situations as tangible references for mathematical ideas. As the first step in the process, students experience an event, such as a field trip. They then create a model or pictures of the event and informally write about it. Their event descriptions are then formalized so that the event is appropriately depicted. Finally, students develop a symbolic representation of the event using mathematical concepts.

The Algebra Project is being used in Atlanta, Boston, Chicago, and other urban school districts across the country, where it has benefited from the support of community organizations. It is also implemented in rural areas such as the Mississippi Delta. Many parents have assisted with the implementation of the project and have provided classroom support to teachers (Silva et al. 1990; Kamii 1990; Jetter 1993).

## The Professional Development Program

Uri Treisman developed the Professional Development Program (PDP) at the University of California at Berkeley to reduce calculus failure rates of minority students with strong academic records. The program was developed in response to Treisman's research showing that certain socially reclusive tendencies of these students impeded their learning of calculus, the gateway to a variety of science and engineering majors. PDP "substituted for remedial efforts an approach to learning based on faculty involvement, academic challenge, collaborative learning, and growth of a student community. This approach to learning, which has always been a central part of the educational philosophy of the Historically Black Colleges and the private liberal arts colleges, has now been replicated in special programs in more than fifty universities" (National Research Council 1991, 34).

The workshop activities focus on the utilization of students' strengths, not the remediation of their weaknesses. It provides a supportive educational environment in which participants are challenged, expected to excel, and afforded ways to do so. Students are encouraged to work in study groups in an environment that favors mathematics accomplishment. Each two-hour PDP session is scheduled twice a week and supplements regular coursework. The sessions focus on special sets of rigorous and usually nonroutine problems designed to provide theoretical insight, reveal conceptual and procedural deficiencies, and help participants learn computational tricks and shortcuts that are usually known to only the best calculus students.

PDP workshop leaders, who may also be former mathematics graduate students, are chosen for their pedagogical expertise and content mastery. The leaders elicit mathematical communication and active participation in problem solving by giving hints and clues and encouraging experimentation. In this way, the workshop leaders serve as facilitators who help students hone and cultivate their analytical skills. Working in small, collaborative groups, students find the socioacademic environment supportive and challenging; they freely ask questions and share ideas with peers.

The effectiveness of Treisman's workshop is well documented. The grades, course completion, and graduation rates of minority students in mathematics courses and curriculums have shown dramatic increases. Many workshop students went on to study graduate-level mathematics (Jackson 1989; Asera 1990).

## The C³ Summer Institute

Uri Treisman, targeting low-achieving prealgebra students, also developed the C-cubed (Content, Competence, Confidence) Summer Institute. C³ has two objectives: to help students make the transitions from elementary to high school mathematics and to involve teachers in the revision of the prealgebra/algebra curriculum. The C³ Summer Institutes run in conjunction with regular summer school programs. Using certain curricular materials, teachers work with students in morning problem-solving workshops. Afternoon mathematics seminars are designed to familiarize teachers with the broader implications of the specially developed materials. In these seminars, groups of four teachers who cooperatively teach a morning course work with a mathematician to develop the curricular content of that morning course. The workshop structure and the use of collaborative learning by both students and teachers support the common resolution of participants to study realistic applications of mathematics. Funded by the National Science Foundation, the project has been successfully piloted in several California public schools (Nielsen 1990).

## Special Elementary Education for the Disadvantaged: Project SEED

Project SEED was initiated in 1963 by William F. Johntz, a mathematician and psychologist, to supplement the regular elementary school mathematics program. The goals of the nationwide program are to (a) improve mathematics achievement; (b) build academic confidence and self-esteem; (c) improve critical-thinking, problem-solving and communication skills; (d) increase the number of students from target schools who take advanced mathematics courses in secondary school; and (e) provide inservice training for classroom teachers.

Project SEED provides instruction in four periods a week for 14–18 weeks. The program pedagogy stresses mathematical communication. SEED students are engaged in rapid, interactive dialogue promoting active participation, understanding, and enthusiasm for mathematics. The Socratic, group-discovery approach requires that they answer a series of questions to discover certain mathematical principles or concepts. An important project component, inservice training for classroom teachers, involves instruction on mathematics content and methodology, observation of SEED lessons, and participation in one-on-one conferences, group workshops, and practice sessions.

Experienced Project SEED mathematics specialists may disseminate the project by recruiting local mathematicians and training them to become the specialists for newly organized SEED programs. Once assigned to classes, all mathematics specialists participate in three to four staff development workshops per week; they also do frequent observations and critiques of their peers. Although no special materials are needed for replication, there are internal curriculum guidelines and methodology for the specialists who teach the mathematics classes (National Council of Teachers of Mathematics 1993).

## QUASAR

The Learning Research and Development Center at the University of Pittsburgh has developed the Quantitative Understanding: Amplifying Student Achievement and Reasoning (QUASAR) Project to improve student participation and performance in mathematics. Using a coherent set of general principles as guides for reform, QUASAR adddresses the needs of schools serving economically disadvantaged children. Instructional practices at all sites feature increased emphasis on mathematical discourse, the application of mathematics to problems that are meaningful to students, and the use of physical and mental models to establish connections for abstract principles. Recognizing the importance of tying reform efforts to local conditions, QUASAR encompasses curriculum development and modification, staff development and ongoing teacher support, classroom and school-based assessment design, and outreach to parents and the community (Silver 1994).

## EFFORTS BY HIGHER EDUCATION

### MISS, California State University

The Mathematics Intensive Summer Session (MISS) is a one-month residential program to help high school girls master college preparatory mathematics at the second-year algebra level and above. The long-range goal is to encourage more female students to successfully complete the calculus courses that allow university majors such as mathematics, science, and engineering. The project targets high school juniors whose general academic backgrounds support college aspirations but whose mathematics performance does not.

The daily routine includes six hours of mathematics instruction, problem solving, and assessments. Since the program is residential, the students engage in activities to broaden their experiences after completing the daily routine. These activities include weekly computer-facility workshops, cultural and social activities, and conferences with guest speakers discussing a wide range of topics, including career options for women in science and mathematics and developing self-esteem. The MISS parent support group meets periodically, sponsors social events for the students, and works with the students to encourage their academic progress and continuing participation in the program (Pagni 1993).

### SKILL, South Dakota School of Mines and Technology (SDSM&T)

Since 1989, the SKILL (Scientific Knowledge for Indian Learning and Leadership) project, has exposed Native American students to extracurricular scientific and mathematical experiences. Two components of the program include a four-week residential program for new participants and a one-week reunion program. The four-week program is planned to increase students' mathematics and science skills and interests, confidence in their academic abilities, and awareness of the importance of mathematics and science in various careers. The one-week program is designed to motivate Native American students to consider careers in science, engineering, mathematics, and computer science.

The SKILL program, originally begun as an after-school program for fourth-grade Native Americans, expanded to include programs for other elementary, junior high, and high school students and training for teachers and parents. SDSM&T officials "have noted a better environment for Native American engineering and science students, an increase in the number of Native Americans declaring science and engineering majors, and positive changes in attitudes between students and faculty" (Gay 1994).

## Spelman College, Atlanta, Georgia

Spelman mathematics students are encouraged to think and communicate critically, logically, and creatively and to develop competence in decision making, problem solving, and quantitative reasoning. Such competencies provide students with those skills that prepare them for graduate and professional study as well as future employment.

The success of the mathematics and natural science program at Spelman is due to the special attention given to students that builds their confidence in their own ability to master mathematics. All natural science students participate in an eight week summer program prior to the beginning of their first year, during which study skills are developed and role models are established. This careful mentoring is continued throughout the undergraduate program and develops into opportunities for research experiences and special honors sections. The faculty devotes a great deal of energy to advising, since student motivation is the most powerful factor in learning (National Research Council 1991, 10).

The Center for Scientific Applications of Mathematics (CSAM) was established to advance Spelman's educational and research programs in science and mathematics and to develop new undergraduate models in those areas. CSAM has a unified program utilizing student and faculty research teams, interdisciplinary curricular efforts, a modern scientific infrastructure, and connections to other educational and industrial institutions. The center seeks to present mathematics as a unifying force, linking the natural sciences. Through this program, Spelman seeks to support a climate that produces outstanding African American women who enter scientific careers. CSAM's outreach program aims to enhance the teaching of science and mathematics at the secondary level in order to elicit early scientific interests in minority female students.

*See also* Family Math; Minorities, Career Problems; Minorities, Educational Problems; Remedial Instruction; Underachievers

## SELECTED REFERENCES

Asera, Rose. "Professional Development and a Study in Adaptation. *UME Trends* 2(4)(1990):1, 2, 4, 7, 8.

Escalante, Jaime, and Jack Dirmann. "The Jaime Escalante Mathematics Program." *Journal of Negro Education* 59(3)(1990):407–423.

Gay, Susan. "Projects: Scientific Knowledge for Indian Learning and Leadership (SKILL)." *Mathematics Teacher* 87(1)(1994):56.

Heller, Rachelle S., and C. Diane Martin. *Model Programs to Attract Young Minority Women to Engineering and Science: Report of a Working Conference.* Washington, DC: National Science Foundation, 1994.

Heyman, Ernest L. "Intensive Care: Observations of the Teaching of Jaime Escalante." *Kappa Delta Pi Record* 26(3)(1990):90–91.

Jackson, Allyn. "Minorities in Mathematics: A Focus on Excellence not Remediation." *American Educator* 13(1)(1989):22–27.

Jetter, Alexis. "Mississippi Learning." *New York Times Magazine* (Feb. 21, 1993):28–35; 50–51; 64; 72.

Kamii, Mieko. "Opening the Algebra Gate: Removing Obstacles to Success in College Preparatory Mathematics Courses." *Journal of Negro Education* 59(3)(1990): 392–406.

Keith, Sandra, and Philip Keith, eds. *Proceedings of the National Conference on Women in Mathematics and the Sciences.* St. Cloud, MN: St. Cloud State University, 1990.

Madison, Bernard, and Therese A. Hart. *A Challenge of Numbers: People in the Mathematical Sciences.* Washington, DC: National Academy Press, 1990.

*Mathematics, Science, and Technology Education Programs That Work: A Collection of Exemplary Educational Programs and Practices in the National Diffusion Network.* Washington, DC: U.S. Department of Education, Office of Educational Research and Improvement, 1994.

Matyas, Marsha L., and Linda S. Dix, eds. *Science and Engineering Programs: On Target for Women?* Washington, DC: National Academy Press, 1992.

National Council of Teachers of Mathematics. *Reaching All Students with Mathematics.* Reston, VA: The Council, 1993.

National Research Council. *Everybody Counts: A Report to the Nation on the Future of Mathematics Education.* Washington, DC: National Academy Press, 1989.

———. *Moving Beyond Myths, Revitalizing Undergraduate Mathematics Education*. Washington, DC: National Academy Press, 1991.

National Science Foundation Statewide Systemic Initiative Equity Focus Group. *Equity Framework in Mathematics, Science, and Technology Education* (draft). Newton, MA: Education Development Center, 1994.

Nielsen, Robert. "Anyone Can Learn Math: New Programs Show How." *American Educator* 14(1)(1990):29–34.

Pagni, David. "Projects: Targeting Girls: MISS." *Mathematics Teacher* 86(1)(1993):95–96.

Silva, Cynthia M., Robert P. Moses, Jacqueline Rivers, and Parker Johnson. "The Algebra Project: Making Middle School Mathematics Count." *Journal of Negro Education* 59(3)(1990):375–391.

Silver, Edward. *Illustrative Case: The QUASAR Project.* Pittsburgh, PA: University of Pittsburgh, Learning Research and Development Center, 1994.

Toliver, Kay. "The Kay Toliver Mathematics Program." *Journal of Negro Education* 62(1)(1993):35–45.

LUCILLE CROOM
JUNE L. GASTÓN

# UNITED STATES DEPARTMENT OF EDUCATION

The federal agency charged with improving teaching and learning in the U.S. school system. The role of the federal government in matters of education has had a bumpy history. Education's status has shifted from being a separate federal department to being an office or bureau housed within another federal department then back again to being a federal department—this time with cabinet status. Moneys to fund the federal education enterprise have gone through periodic fluctuations. The very existence of a federal education agency has been challenged repeatedly. The turbulent path of the federal role in education that extends even to today was well analyzed in 1992 by the National Research Council:

> Education itself has been a battleground of interest groups since at least the 1870s, and the conflicts naturally spill over onto research issues. There are several reasons for those conflicts. Americans put great faith in education as a means to upward mobility and a good life: they believe it can make a difference, they expect it to, and they are upset when it does not achieve that goal. Americans also hold deep-seated and differing values about both the goals and the means of education. When research findings or innovative programs contradict values and beliefs, the results are often dismissed, and the enterprise that produced them is criticized. For instance, an elementary school mathematics curriculum that encourages children to explore, conjecture, and challenge is considered by many scientists as an important investment for the future of science, but some parents see it as lacking discipline and encouraging disrespect (Atkinson and Jackson 1992, 108–109).

Although there was considerable disagreement over whether the U.S. Constitution permitted such a federal role, several developments led to the establishment of the U.S. Department of Education—without cabinet rank—in 1867. These developments included the concern for the imbalance in adequate public schooling in various regions of the country, the neglect of public schools during the Civil War, the growth of teacher professionals and their newly established National Teachers Association, the passage of the Morrill Act and the creation of the U.S. Department of Agriculture, and the establishment of the Freedman's Bureau, which pointed the way for other federal institutions to help disenfranchised and disadvantaged ex-slaves in the South (Vinovskis, in press).

Two years later, the department was reduced to the Bureau of Education (renamed the Office of Education in 1929), housed within the Department of Interior, whose major function was the gathering and dissemination of educational data. In 1929, President Hoover created a fifty-two-member National Advisory Committee on Education. Its 1931 report emphasized the haphazard way the federal government had become involved in educational matters and recommended that although the federal government should not control or regulate state or local education, it should aid schools through general grants to states, continue to gather statistical data, and engage in educational research. During the Depression, federal support for education increased through such programs as the Civilian Conservation Corps (CCC), the National Youth Administration (NYA), and the Works Progress Administration (WPA). In 1939, the Office of Education was transferred from the Department of Interior to the Federal Security Agency, which in 1953 became the Department of Health, Education, and Welfare.

However, it was not until the launching of *Sputnik* in 1957 that the Office of Education became particularly concerned with mathematics and science education. With passage of the Elementary and Secondary Education Act (ESEA) in 1965, the Office of

Education grew from a small, somewhat inconsequential federal agency into one of the largest federal government institutions. Its role changed from mainly that of collecting data to taking a proactive role in American education. And in 1979, Congress established the U.S. Department of Education, complete with cabinet status.

## THE FEDERAL ROLE IN MATHEMATICS EDUCATION

The U.S. Department of Education has made major contributions to the improvement of the learning and teaching of mathematics in this country. These contributions lie in four main categories: research and development; teacher preparation and development in mathematics (and science); educational stimulation for low-achieving poor children at risk of academic failure in mathematics (and reading); and the gathering, analysis, and reporting of a broad range of both national and international data on the status and progress of education. The department has also engaged in dissemination activities and scholarship programs in mathematics.

### Research and Development

Until 1954, when a Congressional act allowed the Office of Education to engage in cooperative research with colleges, universities, and state education departments, research done by the department was in-house. The primary goal of the federal research effort in the 1950s and early 1960s was to provide an academically enriched education for those students gifted in mathematics, science, and engineering.

In the early 1960s, President Lyndon Johnson created a Taskforce on Education whose 1964 report envisioned research as a key to the future of education. As a result of this report, the Bureau of Research was created to support research conducted outside the Office of Education. Funds were provided for research projects and programs designed to expand knowledge about the educational process, to develop new curricula and improve educational programs, to disseminate the results of these efforts to educators and to the public, and to train researchers in the field of education. The emphasis was shifting to reflect the vision of the Great Society: to provide all students with basic skills.

In 1972, the National Institute of Education (NIE) was created to expand the quality and quantity of educational inquiry. Since 1980, research and development efforts have been housed in the department's Office of Educational Research and Improvement (OERI).

Since NIE days, the department has funded significant research efforts in mathematics education, efforts that helped shape current reform in mathematics education. In the past, a considerable amount of research focused on students' misconceptions of computational algorithms. Usually before entering school, children develop counting strategies. Yet in early experiences with arithmetic, children often learn incorrect computational algorithms by rote, trying to make sense out of procedures that they did not understand. This line of research led to studies of how children best learn mathematical concepts.

Many of these research efforts have emanated from two of the department-funded National Educational Research Centers. The University of Pittsburgh-based Learning Research and Development Center (LRDC) was engaged in research on an understanding of the nature of the cognitive structures and mental processes that support learning in general and domain-specific learning in particular. Among LRDC's accomplishments was the development of a reasoning-based mathematics program designed to help children build on their mathematical intuitions. The Wisconsin Center, currently called the National Research and Development Center on Achievement in School Mathematics and Science, has also been studying how children learn mathematics. From this study arose the Cognitively Guided Instruction program, an offshoot of the taxonomy of word problems in addition and subtraction developed earlier. The center has spent considerable effort examining the school mathematics curriculum with respect to advances in mathematics and technology and new knowledge of how children learn mathematics. New curricular designs include exploring "big mathematical ideas" in depth rather than many ideas shallowly; integrating the learning of concepts from algebra and geometry into the elementary school curriculum (a "strands" versus a "layer cake" curriculum); deemphasizing rote learning; relegating more tedious tasks to the calculator or computer; engaging students in inquiry; experimenting, gathering and interpreting data; tapping the resources of technology to create, display, and share data, graphs, three-dimensional objects, and motion; using mathematics to solve real-world problems; and engaging students in meaningful out-of-school mathematics activities. The mission of the newly funded Wisconsin Center is to craft, implement in schools, and validate a set of principles for the design of classrooms that promotes understanding in mathematics and science.

Another result of President Johnson's Taskforce on Education was the establishment of regional educational laboratories. The laboratories were modeled after the national laboratories of the Atomic Energy Commission, with improvement and innovation in the education of the nation's children being considered at least as important an endeavor as the maintenance of our national defense. There are currently ten regional educational laboratories that provide technical assistance and dissemination to support school improvement activities throughout the nation and territories.

However, federal funding for educational research and development has always been on shaky grounds, dependent on the interests of a Congress that wants quick results. As Congressman John Brademas noted:

> I hope we will get away from the slot machine mentality that too many of us in Congress have who think that if you put a research nickel in on Monday you will get a quarter's worth of results out on Friday (U.S. Congress 1971, 98).

## Teacher Preparation and Development in Mathematics

The bulk of activities in teacher preparation and development in mathematics (and science) has occurred through the Dwight D. Eisenhower Education Program, enacted to respond to the nation's need for an increased understanding of mathematics and science by its students and a concern that there continue to be an adequate supply of mathematicians, scientists, and engineers to support our economic security and national defense. In order to accomplish this, it was decided that the thrust should be on improving the skills of teachers and the quality of mathematics and science instruction in elementary and secondary schools. The Eisenhower program is the largest federal program to address this priority. The Eisenhower State Program is a formula grant program that awards grants through the states to local school systems and universities in support of sustained and intensive high-quality professional development. The Eisenhower National Program provides discretionary grants for model teacher development and other programs related to improved student learning of mathematics and science. In addition to professional development activities, the Eisenhower National Program has funded ten

Regional Mathematics and Science Education Consortia to provide information and technical assistance to help states and school districts improve their mathematics and science programs. These consortia train and provide technical assistance to classroom teachers, administrators, and other educators to help them adapt and use exemplary instructional materials, teaching methods, curricula, and assessment tools. From its inception in 1984 until 1995, the Eisenhower programs supported teacher development in mathematics and science only. Since then, legislation has supported teacher development in all core subject areas.

## Educational Stimulation for Low-achieving Poor Children at Risk of Academic Failure in Mathematics

In 1965, Congress enacted a program to address economic inequity by improving educational opportunities for the children of poverty. Title 1 of Improving America's Schools Act (IASA), as it is now called, serves approximately 5.5 million students, basically in reading and mathematics. In 1992, Congress appropriated approximately $6.1 billion to support more than 5 million children under this program. In the past, the program relied heavily on "pull-out" instruction, that is, pulling children of poverty out of their regular classrooms for special, step-by-step instruction. However, instruction that addressed students' skill deficits in a narrow, sequential fashion could be found in both pull-out and in-class Title 1 settings. Still, according to a 1993 report, 97% of teachers reported an emphasis on reinforcing of concepts taught in regular class (drill and practice) and only 21% reported an emphasis on the development of higher-order thinking skills. Moreover, gains on standardized tests did not move Title 1 students ahead substantially toward the achievement levels of more advantaged students. (U.S. Department of Education 1993). Reports mandated by the "1992 National Assessment of Chapter 1 Act" recommended changes to the new Title 1 Program. Thus, Title 1 now requires "more advanced" skills be taught, and new provisions in the law encourage performance standards for Title 1 schools that are keyed to curricular frameworks and aligning Title 1 testing with regular school testing.

## Educational Statistics

Although the federal role in education has always placed emphasis on the gathering of statistical information, this role increased with the establishment of the National Center for Educational Statistics (NCES) in 1965. Housed within the U.S. Office of Education, NCES collects data on a broad range of indicators of the condition of education; its publications often impact public policy concerning the learning and teaching of mathematics in our schools and lead to further research and development in mathematics education.

Critical to mathematics education is its National Assessment of Educational Progress (NAEP), which collects statistical data related to the learning and teaching of several subject areas, among them mathematics. NAEP is both a product and a source of education research and development. Since 1973 NAEP has gathered information about the levels of student proficiency in mathematics and the related practices of teachers. These data have been collected, analyzed, and reported through various governmental publications. (See Mullis et al. 1993 and 1994; Dossey, Mullis, and Jones 1993.) For example, data show that average mathematics proficiency on the NAEP test has improved between 1973 and 1992 at the three ages tested: 9, 13, and 17. During the past decade, average proficiency in mathematics has improved for all three racial groups identified—white, black, and Hispanic—with differences between white and minority students narrowing. Nonetheless, the proportions of students demonstrating success with more complex mathematical tasks remains low, particularly for those subpopulations of students historically considered at risk.

In the past, NAEP mathematics assessments were based solely on multiple choice items. Based on the concern for teaching and learning advocated in such documents as the *Curriculum and Evaluation Standards for School Mathematics* (National Council of Teachers of Mathematics (NCTM) 1989), about one-third of the 1992 NAEP mathematics questions and approximately one-half of the students' response time were devoted to questions asking students to construct their own response. In general, there were lower levels of performance on these constructed-response questions than on the multiple choice items.

NCES also participates in international studies. It contributed to the 1982 Second International Mathematics Study (SIMS) and to the 1995 Third International Mathematics and Science Study (TIMSS). TIMSS focused on three distinct populations of students, defined in the United States as third and fourth graders, seventh and eighth graders, and twelfth graders. The main collection of data occurred in the spring of 1995. Of particular interest in

TIMSS are the videotaped classroom observational studies of eighth-grade mathematics and science teaching in Japan, Germany, and the United States and ethnographic case studies of key educational policy topics in those countries. Topics of study in the case histories include implementation of national standards, methods of dealing with ability differences, adolescents' attitudes toward school, and the daily lives and working environment of teachers. Such comparative data will certainly stimulate further research in the learning and teaching of mathematics and will undoubtably impact policy decisions.

## Other Activities

The department has a variety of dissemination activities that impact the learning and teaching of mathematics. Among such activities are the Educational Resources Information Center (ERIC), the Eisenhower National Clearinghouse for Mathematics and Science Education, and the National Diffusion Network (NDN). The department is currently creating an Expert Panel in Mathematics and Science to identify promising and exemplary programs and practices and disseminate this information to the mathematics and science education communities.

The department has also played a major role in interagency committees created under the President's Office of Science and Technology Policy (OSTP). The Committee on Education and Training (CET) and its predecessor, the Federal Coordinating Council for Science, Engineering, and Technology (FCCSET), were created to promote the federal government's leadership role by mobilizing national support for reform, by establishing national goals and federal long-range strategic plans, by initiating model reform efforts and a research and development agenda, and by bringing to bear the great scientific and technical resources managed by federal agencies.

*See also* Eisenhower Program; Federal Government, Role; National Assessment of Educational Progress (NAEP); Third International Mathematics and Science Study (TIMSS)

## SELECTED REFERENCES

Atkinson, Richard C., and Gregg B. Jackson, eds. *Research and Education Reform: Roles for the Office of Educational Research and Improvement*. Washington, DC: National Academy Press, 1992.

Dossey, John A., Ina V. S. Mullis, and Chancey O. Jones. *Can Students Do Mathematical Problem Solving?* Washington, DC: National Center for Education Statistics, 1993.

Florio, David H. "U.S. Department of Education." In *Encyclopedia of Educational Research* (pp. 1980–1989). Vol. 4. 5th ed. Harold E. Mitzel, ed. New York: Macmillan, 1982.

Goldberg, Milton, and Joseph C. Conaty. "U.S. Department of Education." In *Encyclopedia of Educational Research* (pp. 1470–1477). Vol. 4. 6th ed. Marvin C. Alkin, ed. New York: Macmillan, 1992.

Mullis, Ina V. S., John A. Dossey, Jay R. Campbell, Claudia A. Gentile, Christine O'Sullivan, and Andrew S. Latham. *NAEP 1992 Trends in Academic Progress.* Washington, DC: National Center for Education Statistics, 1994.

Mullis, Ina V. S., John A. Dossey, Eugene H. Owen, and Gary W. Phillips. *NAEP 1992 Mathematics Report Card for the Nation and the States.* Washington, DC: National Center for Education Statistics, 1993.

National Council of Teachers of Mathematics. *Curriculum and Evaluation Standards for School Mathematics.* Reston, VA: The Council, 1989.

————. *Principles and Standards for School Mathematics.* Reston, VA: The Council, 2000.

Sproull, Lee, Stephen Weiner, and David Wolf. *Organizing an Anarchy: Belief, Bureaucracy, and Politics in the National Institute of Education.* Chicago, IL: University of Chicago Press, 1978.

U.S. Congress, House, Select Subcommittee on Education. *To Establish a National Institute of Education.* Washington, DC: Government Printing Office, 1971.

U.S. Department of Education. *Reinventing Chapter 1: The Current Chapter 1 Program and New Directions.* Washington, DC: U.S. Department of Education, 1993.

Vinovskis, Maris A. *Changing Views of the Federal Role in Educational Statistics and Research.* (In press.)

CAROL E. B. LACAMPAGNE

# UNITED STATES NATIONAL COMMISSION ON MATHEMATICS INSTRUCTION (USNCMI)

Functions as liaison to the International Commission on Mathematical Instruction (ICMI) of the International Mathematical Union (IMU). USNCMI advises the National Academy of Sciences-National Research Council (NAS-NRC) in all matters pertaining to ICMI and encourages two-way communication with the U.S. mathematical sciences community on international activities in mathematics education. The USNCMI helps promote the advancement of mathematical sciences education in the United States and throughout the world and implements appropriate U.S. participation in ICMI activities. Members are appointed by the NAS-NRC.

USNCMI's recent activities include encouraging the National Council of Teachers of Mathematics (NCTM) to create a tax-exempt fund to support mathematics education activities in developing countries; advising, evaluating, and monitoring the Third International Mathematics and Science Survey (TIMSS); publicizing and encouraging participation in the First (October 1993, Moscow) and Second (October 1–7, 1995, Moscow) U.S./Russia Joint Conferences on Mathematics Education; coordinating and communicating with the Mathematical Sciences Education Board (MSEB), the Conference Board of the Mathematical Sciences (CBMS), the Mathematical Association of America (MAA), NCTM, and the American Mathematical Association of Two-Year Colleges (AMATYC) on future programs to discuss issues in international mathematics education; helping organize and publicize the August 16–20, 1994 ICMI-China Regional Conference on Mathematics Education in Shanghai; working with NCTM in the application for and administration of a travel grant program to the Eighth International Congress on Mathematical Education (ICME-8) in Seville, Spain in 1996; encouraging ICMI to undertake efforts in mathematical education that parallel those on which the IMU is embarking through IMU's World Math Year 2000 (WMY 2000) project and suggesting there be a WMY 2000 preplanning session at ICME-8.

Until November 1992, USNCMI was called the U.S. Commission on Mathematical Instruction (USCMI). USCMI was established in its current form by the NRC in 1953. When the NAS-NRC's Board on Mathematical Sciences (BMS) was established in November 1984, USCMI became a standing committee of BMS. In November 1992, USCMI adopted its current title. USNCMI reports annually to the BMS and to the Conference Board on Mathematical Sciences in addition to periodically organizing sessions at NCTM, MAA, and AMATYC annual meetings to keep the mathematical sciences community informed on international developments in mathematics education.

*See also* Federal Government, Role

SELECTED REFERENCES

Jackson, Allyn. "Reports from BMS." *Notices of the American Mathematical Society* 41(4)(1994):314.

Willoughby, Stephen S. "The United States Commission on Mathematical Instruction." *BMS Program Update* (Fall 1991):2.

JOHN R. TUCKER

# UNIVERSITY OF CHICAGO SCHOOL MATHEMATICS PROJECT (UCSMP)

A project that began in 1983 when the departments of mathematics and education at the University of Chicago received a six-year grant from the Amoco Foundation for the purpose of improving school mathematics in grades K–12. This effort has involved a broad range of activities, including the translation of mathematics education literature from other countries, the development of curriculum and teacher-training materials at both the elementary and secondary levels, and the critical evaluation of all project materials and models. UCSMP has created a mathematical sciences curriculum including substantial work in the mathematics of computer science and statistics. The curriculum brings the real world into the classroom, using calculators, computers and other available technology; a multitude of applications; and an emphasis on problem solving. It has less repetition and review at the elementary and junior high school levels than most existing texts in the United States, in an attempt to raise expectations for U.S. students based on the standards achieved by their counterparts in other educationally advanced countries.

In addition to arithmetic and its applications, the UCSMP *Everyday Mathematics* curriculum for grades K–6 investigates, informally but systematically, the basics of data gathering and analysis, probability, geometry, and algebra (Bell et al. 1995). For example, first-grade students create a Shapes Museum, collecting and organizing three-dimensional shapes such as cylinders and rectangular prisms. Beginning in second grade, students reinforce algebra skills by playing a game called Name That Number: given five numbers and a target number, students use any operations and as many of the five numbers as possible to arrive at the target number. Third-grade students participate in a year-long project gathering, graphing, and analyzing data on the daily sunrise and sunset. Fourth-grade students use a *World Tour* book to "visit" countries around the world, applying mathematics and geography concepts at each stop. Fifth-grade students record results of experiments on a probability meter, a poster-sized display of probabilities expressed as fractions, decimals, and percents from 0 to 1. Overall, the program attempts to take advantage of the young child's desire to explore and learn, instructs teachers in the creation of a mathematics-rich environment in the classroom, and makes a gradual transition from manipulatives to abstract concepts. Math Tools for Teachers, for example, is a UCSMP workshop series designed to enhance K–3 generalist teachers' knowledge of and strategies for teaching mathematics.

The six-year UCSMP secondary curriculum is designed for average students to begin in the seventh grade; better-prepared students are encouraged to begin earlier and more poorly prepared students later. The curriculum is distinguished by an abundance of applications and by its wider scope, with geometry, algebra, and some discrete mathematics in all courses and statistics and probability integrated into the study of algebra and functions. For example, lessons in the first algebra course are devoted to exponential growth and its use in compound interest; fitting a line and its equation to data; and using a quadratic equation to describe the path of a thrown ball. Lessons in the geometry course include scheduling round-robin tournaments, dart game probabilities, and optimal locations for taking pictures, among many others. In addition to the customary applications to physics and navigation, lessons in later courses examine such problems as deciding the authorship of historical documents, determining blood alcohol levels, and calculating test score distributions. Projects in every chapter provide opportunities for students to extend and explore ideas of their own choosing.

UCSMP materials and models have undergone extensive development, and their evaluations have resulted in numerous published reports. These studies indicate generally that UCSMP students significantly outperform comparison students on the broader range of content covered in the UCSMP curriculum while holding their own against their counterparts on traditional content. An estimated 3 million students across the nation were using UCSMP materials in 1995–96. Full-day inservice workshops at both the elementary and secondary level are held each summer on the University of Chicago campus and around the country for teachers and supervisors who will be using the materials in the upcoming school year. A two-day secondary-level conference open to both users and prospective users is held each year in November. UCSMP sponsored international conferences on mathematics education in 1985, 1988, and 1991, with proceedings published by the National Council of Teachers of Mathematics (NCTM). A number of its own translations of textbooks and related materials from such countries as Japan and Russia have also been published. Funding for UCSMP has come from the Amoco Foundation, National Science Foundation, Ford

Motor Company, Carnegie Corporation of New York, GE Foundation, GTE Corporation, Citicorp/Citibank, Exxon Education Foundation, and from the commercial publishers of UCSMP elementary and secondary materials, Everyday Learning Corporation and ScottForesman, respectively.

*See also* Curriculum, Overview; Curriculum Trends, Secondary Level

## SELECTED REFERENCES

Bell, Max, et al. *Everyday Mathematics K–5.* Evanston, IL: Everyday Learning, 1995.

Flanders, James R. "How Much of the Content in Mathematics Textbooks Is New?" *Arithmetic Teacher* 35(Sept. 1987):18–23.

Hirschhorn, Daniel B. "Implementation of the First Four Years of the UCSMP Secondary Curriculum." Ph.D. diss., University of Chicago, 1992.

———, Denisse R. Thompson, Zalman Usiskin, and Sharon L. Senk. "Rethinking the First Two Years of High School Mathematics with the UCSMP." *Mathematics Teacher* 88(8)(Nov. 1995):640–647.

Kodaira, Kunihiko. *Japanese Grade 7–9 Mathematics.* Hiromi Nagata, trans. Chicago, IL: UCSMP, 1992.

Krutetskii, Vadim A., ed. *Soviet Studies in Mathematics Education.* Vol. 8. *Issues in the Psychology of Abilities.* Joan Teller, trans. Chicago, IL: UCSMP, 1992.

McConnell, John W., et al. *Algebra.* 2nd ed. Glenview, IL: ScottForesman, 1996.

Moro, Mariia I., et al. *Russian Grade 1 Mathematics.* Robert H. Silverman, trans. Chicago, IL: UCSMP, 1992.

Moro, Mariia I., and Mariia A. Bantova. *Russian Grade 2 Mathematics.* Robert H. Silverman, trans. Chicago, IL: UCSMP, 1992.

Pcholko, A. S., et al. *Russian Grade 3 Mathematics.* Robert H. Silverman, trans. Chicago, IL: UCSMP, 1992.

Peressini, Anthony L., et al. *Precalculus and Discrete Mathematics.* Glenview, IL: ScottForesman, 1992.

Pyshkalo, Anatolii M., ed. *Soviet Studies in Mathematics Education.* Vol. 7. *Geometry in Grades 1–4: Problems in the Formation of Geometric Conceptions in Primary School Children.* Joan W. Teller, trans. Chicago, IL: UCSMP, 1992.

Rubenstein, Rheta N., et al. *Functions, Statistics and Trigonometry.* Glenview, IL: ScottForesman, 1992.

Sconiers, Sheila. *MathTools for Teachers K–3: Teacher Development Program.* Chicago, IL: University of Chicago Press, 1986.

———, Mary Fullmer, and Lydia Polansky. *Everyday Teaching for Everyday Mathematics.* Evanston, IL: Everyday Learning, 1996.

Senk, Sharon L., et al. *Advanced Algebra.* 2nd ed. Glenview, IL: ScottForesman, 1996.

Thompson, Denisse R. "An Evaluation of a New Course in Precalculus and Discrete Mathematics." Ph.D. diss., University of Chicago, 1992.

*UCSMP Newsletter* 1–18. Chicago, IL: UCSMP, 1987–1996.

Usiskin, Zalman. "The Beliefs Underlying UCSMP." *The Elementary Mathematician* (Winter 1988):7–12.

———. "If Everybody Counts, Why Do So Few Survive?" In *Reaching All Students with Mathematics* (pp. 7–22). Gilbert Cuevas and Mark Driscoll, eds. Reston, VA: National Council of Teachers of Mathematics, 1993.

———. "Lessons Learned from the Chicago Mathematics Project." *Educational Leadership* 50(8)(May 1993):14–18.

———. "The Sequencing of Applications and Modelling in the University of Chicago School Mathematics Project (UCSMP) 7–12 Curriculum." In *Applications and Modelling in Learning and Teaching Mathematics* (pp. 176–181). Werner Blum, et al., eds. London, England: Horwood, 1989.

———, et al. *Geometry.* 2nd ed. Chicago, IL: UCSMP, 1993.

———, et al. *Transition Mathematics.* 2nd ed. Glenview, IL: ScottForesman, 1995.

Wirszup, Izaak. "Education and National Survival: Confronting the Mathematics and Science Crisis in American Schools." *Educational Leadership* 41(4)(Dec. 1983/Jan. 1984):4–11.

———, and Robert Streit, eds. *Developments in School Mathematics Education Around the World.* 3 vols. Reston, VA: National Council of Teachers of Mathematics, 1987–1992.

ZALMAN USISKIN

# UNIVERSITY OF ILLINOIS COMMITTEE ON SCHOOL MATHEMATICS (UICSM)

The first project to create "new mathematics" for secondary schools. It began in 1952 as an effort to prepare high school graduates for college engineering programs. Max Beberman (1925–1971), who received his Ph.D. degree from Columbia University in 1953, was named the project director in 1955, and through the project became a national spokesman for curriculum reform. UICSM was funded by the Carnegie Foundation, and the guiding principles were that curriculum should be consistent, intrinsically exciting to students, and avoid excessive drill. The mathematics content stressed precise language and abstract structure but was to be learned by discovery methods. The term *pronumeral* for *variable* was created by the project. In 1956, the grade 9–12 courses included (then) innovative content such as the field properties, inequality relations, equations and inequalities and their graphs, sets and relations, functions, mathematical induction, exponential and logarithmic functions, complex numbers, and circular functions. In the 1960s, instructional units on fractions using the function notions of "stretchers" (numerator factors) and "shrinkers" (denominator factors) and an informal transformational geometry were developed.

*See also* History of Mathematics Education in the United States, Overview; School Mathematics Study Group

### SELECTED REFERENCES

Beberman, Max. *An Emerging Program of Secondary School Mathematics.* Cambridge, MA: Harvard University Press, 1958.

Bidwell, James K., and Robert G. Clason, eds. *Readings in the History of Mathematics Education.* Washington, DC: National Council of Teachers of Mathematics, 1970.

Jones, Phillip S., ed. *A History of Mathematics Education in the United States and Canada.* 32nd Yearbook. Washington, DC: National Council of Teachers of Mathematics, 1970.

JAMES K. BIDWELL
ROBERT G. CLASON

# UNIVERSITY OF MARYLAND MATHEMATICS PROJECT (UMMaP)

A project begun in 1957 and directed by the mathematician John R. Mayor. It was sponsored initially by the Carnegie Foundation and subsequently by the National Science Foundation. This project was the first to write "new mathematics" materials for the junior high school. It later wrote texts for prospective and inservice elementary teachers. Materials were written cooperatively by forty junior high school teachers and five university mathematicians. After classroom testing, two texts were published in the early 1960s. Language, mathematical structure, and number-numeral symbol distinctions were stressed, and statistics, probability, logic, and trigonometry were included. In 1961, the following chapter titles, considered unusual at the time, appeared in the seventh-grade text: "Systems of Numeration," "Symbols," "Factoring and Primes," "Mathematical Systems," "Logic and Number Sentences." The project stressed the psychology of learning and used Robert Gagné as a consultant. UMMaP was a consultant for the School Mathematics Study Group on early junior high school materials.

*See also* Gagné; History of Mathematics Education in the United States Overview; *Revolution in School Mathematics;* School Mathematics Study Group; University of Illinois Committee on School Mathematics

### SELECTED REFERENCES

Jones, Phillip S., ed. *A History of Mathematics Education in the United States and Canada.* 32nd Yearbook. Washington, DC: National Council of Teachers of Mathematics, 1970.

Seyfert, Warren C., ed. *The Continuing Revolution in Mathematics.* Washington, DC: National Council of Teachers of Mathematics, 1968.

JAMES K. BIDWELL
ROBERT G. CLASON

# V

## VAN HIELE LEVELS

A model describing five levels of thinking abilities through which students pass as they learn geometry. Pierre M. van Hiele and his wife, Dina van Hiele-Geldof, formulated the model in the 1950s as a way to explain the difficulties they observed their secondary school students encountering, especially in understanding the nature of proof (van Hiele 1957/1984, 1986; van Hiele and van Hiele-Geldof 1958). They argued that secondary geometry involves a relatively "high level of thinking" and that many students did not have sufficient experience in thinking at the prerequisite lower levels. The five ability levels are described next with examples drawn from the angle sum of triangles.

> *Level 0 (Visual)*. Identify, name, compare and operate on geometric figures according to their appearance as a whole; e.g., identify and measure angles of a given triangle.
>
> *Level 1 (Analysis)*. Analyze figures in terms of their components and relationships among components and empirically discover properties of a class of shapes (e.g., by folding, measuring) and think of a shape in terms of its properties (e.g., identify angles of equal measure in a grid of parallel lines; see Figure 1) and observe that the alternate-interior angles formed by a saw (i.e., $X_1$,

$X_2$, $X_3$, $X_4$) are equal in measure and that the corresponding angles formed by a ladder (i.e., $Y_1$, $Y_2$, $Y_3$, $Y_4$) are also equal in measure. Measure angles of triangles and discover that their sum is 180°, or tear off the angles of a triangle and arrange them to form a straight line to show that their sum is 180°.

> *Level 2 (Informal deduction)*. Interrelate previously discovered properties by giving or following informal arguments; e.g., in studying families of parallel lines in a triangular grid, use "saws" to explain why angle $A$ and angle $X$ have equal measures and angle $B$ and angle $Y$ also have equal measures; also to explain why the sum of the angles of the triangle must therefore be 180° (see Figure 2). Summarize the explanation by a "family tree" that shows the "ancestors" of the sum of the angles equal 180°, perhaps also later relate this to the angle sums for quadrilaterals and pentagons (see Figure 3).
>
> *Level 3 (Formal deduction)*. Prove theorems deductively and understand the role of postulates (axioms), definitions, theorems, and proof; e.g., give a deductive proof that the angle sum of a triangle is 180°, using the parallel postulate, definitions, and previously proven theorems.
>
> *Level 4 (Rigor)*. Establish theorems in different postulational systems and compare and analyze

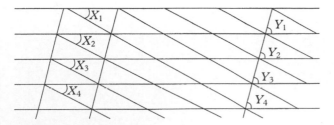

**Figure 1** *Grid, level 1.*

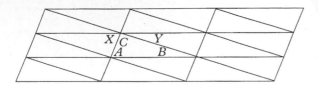

**Figure 2** *Grid, level 2.*

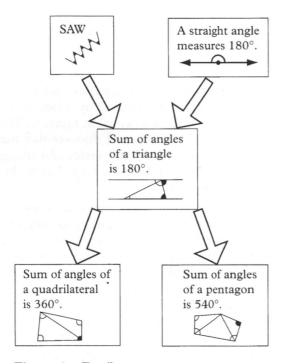

**Figure 3** *Family tree.*

process. There are "jumps" in the learning curve that reveal the presence of discrete levels of thinking. Second, the levels are sequential and hierarchal. Students cannot achieve one level without having passed through the previous levels. A third feature is that the levels are characterized by differences in objects of thought. At level 0, a student operates on individual shapes—for example, saying that a cutout shape is a square because it "looks like a box." At level 1, the student focuses on classes of shapes and discovers their properties—for example, sorting a collection of cutout shapes into two sets, squares and nonsquares, and then stating properties for squares. At level 2, properties and relations are objects of thought, and students use informal arguments to establish logical connections among them, as illustrated by the family tree in Figure 3.

A fourth feature is that "each level has its own linguistic symbols and its own system of relations connecting these symbols. A relation which is 'correct' at one level can reveal itself to be incorrect at another. Think, for example, of the relation between a square and a rectangle" (van Hiele 1957/1984, 246). At level 0, a child would say that a given square is not a rectangle because it does not look like one. However, at level 2, a student would use *square* and *rectangle* in a way that connects them to their properties and informal definitions. The student could now establish the correct relation between square and rectangle—that is, explain why all squares are rectangles but not all rectangles are squares. A student's metacognitive language can also have different meanings at different levels. For example, at level 0, a student uses *why* or *because* to mean looking at a specific shape; whereas a student at level 1 interprets *why* as experimenting with a class of shapes to make generalizations. At level 2, *why* means giving an informal deductive explanation using properties. As the van Hieles have noted, many failures in teaching geometry result from a language gap between teacher and student, with the teacher using language at level 2 or 3, for example, but the student hearing the language at level 0 or 1.

A fifth feature is that progress through the levels is more dependent upon instruction than on maturation and that certain types of instructional experiences can facilitate (or impede) progress within a level or to a higher level. For example, Dina van Hiele-Geldof (1957/1984) in a year-long teaching experiment guided twelve-year-old students through the levels for various topics such as angles and angle-sums, enabling them to reach level 2 thinking. Elementary and middle school students in the United

systems such as Euclidean and non-Euclidean geometries; e.g., examine angles and the angle sum of triangles in a non-Euclidean geometry, creating and proving theorems and comparing the results with those in Euclidean geometry.

The van Hiele model has five features. First, according to the van Hieles, learning is a discontinuous

States and abroad have also shown similar progress through carefully crafted instruction that embodied the van Hiele model (e.g., Fuys, Geddes, and Tischler 1988; Pyshkalo 1968). On the other hand, progress beyond level 0 can be constrained by curriculum that creates gaps in students' levels of thinking. For example, research on geometry material in U.S. textbooks for grades 3–8 revealed that most material required only level 0 thinking. Also, the limited material at levels 1 and 2 often included exercises and test questions that did not ask students to explain at those levels, thereby reducing the level of thinking needed to respond to level 0 (Fuys, Geddes, and Tischler 1988). As a result, students are unprepared for tenth-grade geometry, presented mostly at level 3.

The van Hieles proposed five sequential phases involving the content and method of instruction to facilitate progress from one level to the next. The phases are described next with examples for moving from level 0 to level 1 for the topic of angles of a triangle.

*Information:* The teacher acquaints students with the topic by means of materials that causes them to discern a certain structure (e.g., identifying angles and triangles in everyday life, constructing them out of paper or connecting sticks). The teacher uses the activities as a context to learn how students think and talk about the topic.

*Guided Orientation:* Students are presented with carefully crafted tasks designed to reveal various relations involving that topic (e.g., measuring and drawing angles, measuring sides and angles of triangles). The teacher guides students in appropriate explorations.

*Explicitation:* Students are guided to become explicitly aware of their geometric conceptualizations and to describe them, first in their own language and then in relevant mathematical terminology introduced by the teacher (e.g., saying "the triangle has two angles that have the same number of degrees" or "the angles have different numbers of degrees").

*Free Orientation:* Students learn through more complex tasks that can be done in different ways to explore relations involving that topic (e.g., using software to construct various triangles and measure their sides and angles).

*Integration:* Students summarize what they learned about the topic and formulate an overview of the new network of objects and rela-

tions they have learned (e.g., create a list of observations about triangles in general and special types of triangles such as right triangles, triangles with two equal sides).

Teachers can design activities sequenced in this way to provide students with what the van Hieles call "an apprenticeship" in geometry, which enables them to construct their understanding of the topic at the next level. Textbook materials that embody the van Hiele levels and phases have been produced for elementary, middle, and secondary school students. The model has also been incorporated into materials for teacher preparation, both in mathematics textbooks and in books on methods of teaching mathematics.

Since the 1980s, there has been considerable research on the van Hiele model itself and on its application to learning and instruction in geometry both in the United States and abroad (Clements and Battista 1992). Research generally supports the sequential and hierarchial nature of the levels but not the discontinuous jumps in learning between levels, since many students were found to be "in transition" between levels 0 and 1 or levels 1 and 2 (Burger and Shaughnessy 1986; Fuys, Geddes, and Tischler 1988). Moreover, students have been found to think at different van Hiele levels for different topics. The percentage of a student's successful performance on tasks at each level, or "degree of acquisition" for each level, has been used to create a profile that gives a better picture of a student's thinking in terms of levels than just classifying the student as thinking at a particular level (Gutierrez, Jaime, and Fortuny 1991). Van Hiele (1986) has suggested applying the model in other areas of mathematics (e.g., arithmetic, algebra) and other subjects (e.g., chemistry) as well as modifying the model from five levels to three: the visual, the descriptive, and the theoretical. Recent research has investigated the model in a variety of mathematical contexts such as plane figures and plane isometries in the middle school, LOGO-based instruction on plane geometry in grades 7–8, calculator-assisted solving of three-dimensional problems in college calculus, and non-Euclidean geometry in a course for prospective teachers. Research has also examined ways to assess a student's level (e.g., through journal writing, multiple choice instruments, interviews) and the relationship of the level of thinking to factors such as knowledge of mathematics vocabulary, performance in conjecturing and proof writing, geometry achievement, and Piagetian stages. The van Hiele model has also been investigated along with other theoretical perspectives in the design of learning environments that promote spatial and geometric thinking (Lehrer and Chazan 1998).

Thus, the van Hiele model has been useful for formulating and addressing theoretical and practical issues about learning and instruction in geometry. It offers mathematics educators and teachers at all grade levels a valuable framework for reflecting on how students think in geometry and how instruction can support the development of geometric thinking.

*See also* Geometry; Instruction

## SELECTED REFERENCES

Burger, William F., and J. Michael Shaughnessy. "Characterizing the van Hiele Levels of Development in Geometry." *Journal for Research in Mathematics Education* 17(Jan. 1986):31–48.

Clements, Douglas H., and Michael T. Battista. "Geometry and Spatial Reasoning." In *Handbook of Research on Mathematics Teaching and Learning* (pp. 420–464). Douglas A. Grouws, ed. New York: Macmillan, 1992.

Crowley, Mary L. "The van Hiele Model of Development in Geometric Thought." In *Learning and Teaching Geometry: K–12* (pp. 1–16). 1987 Yearbook. Mary M. Lindquist, ed. Reston, VA: National Council of Teachers of Mathematics, 1987.

Fuys, David, Dorothy Geddes, and Rosamond Tischler. *The van Hiele Model of Thinking in Geometry among Adolescents.* Reston, VA: National Council of Teachers of Mathematics, 1988.

Geddes, Dorothy, Julianna Bove, Irene Fortunato, David Fuys, Jessica Morgenstern, and Rosamond Welchman-Tischler. *Geometry in the Middle Grades. Addenda Series, Grades 5–8.* Reston, VA: National Council of Teachers of Mathematics, 1992.

Gutierrez, Angel, Adela Jaime, and Jose M. Fortuny. "An Alternative Paradigm to Evaluate the Acquisition of the van Hiele Levels." *Journal for Research in Mathematics Education* 22(May 1991):237–251.

Lehrer, Richard, and Dan Chazan, eds. *Designing Learning Environments for Developing Understanding of Space and Geometry.* Mahwah, NJ: Erlbaum, 1998.

Pyshkalo, Anatolii M. *Geometry in Grades 1–4. (Problems in the Formation of Geometric Conceptions in Pupils in the Primary Grades).* Alan Heffer, ed.; Izaak Wirszup, trans. Chicago, IL: University of Chicago Press, 1981. Original in Russian. Moscow, Russia: Prosveshchenie, 1968.

Van Hiele, Pierre M. *"A Child's Thought and Geometry."* In *English Translation of Selected Writings of Dina van Hiele-Geldof and Pierre M. van Hiele* (pp. 243–252). David Fuys, Dorothy Geddes, and Rosamond Tischler, eds. Brooklyn, NY: Brooklyn College, 1984. Original in French. *Bulletin de l'Association des Professeurs de Mathématiques de l'Enseignment Public* (1957):198–205. ERIC No. ED287697. Columbus, OH: ERIC, 1984.

——— *Structure and Insight.* Orlando, FL: Academic, 1986.

———, and Dina van Hiele-Geldof. "A Method of Initiation into Geometry at Secondary Schools." In *Report on Methods of Initiation into Geometry* (pp. 67–80). Hans Freudenthal, ed. Groningen, Netherlands: Wolters, 1958.

Van Hiele-Geldof, Dina. "The Didactics of Geometry in the Lowest Class of Secondary School." In *English Translation of Selected Writings of Dina van Hiele-Geldof and Pierre M. van Hiele* (pp. 1–220). David Fuys, Dorothy Geddes, and Rosamond Tischler, eds. Brooklyn, NY: Brooklyn College, 1984. Original in Dutch. Unpublished doctoral dissertation, University of Utrecht, 1957. ERIC No. ED287697. Columbus, OH: ERIC, 1984.

DOROTHY GEDDES
DAVID FUYS
C. JAMES LOVETT
ROSAMOND WELCHMAN

# VARIABLES

Symbols that represent unspecified elements of a set. Given the set (or domain) of counting numbers, for instance, the value of the expression $5x + 8$ depends on the counting number chosen to replace the variable $x$. For example, if $x$ is replaced by 2, then $5x + 8 = 18$. An important use of the variable is to represent the unknown in an equation. If $5x + 8 = 28$, then $x$ is unknown until the equation is solved. When symbolizing a fixed, but unspecified, quantity throughout a discussion, the variable (whose domain is now a single number) is called a *constant*. In elementary grades, variables may be denoted by shapes, such as squares or triangles. (Thus, in the case of $5 + \square = 6$, students are asked to fill in the square with a number.) In higher mathematics, variables are used to symbolize mathematical entities other than numbers, such as matrices or vectors.

Before the sixteenth century, the use of variables occurred unsystematically. The Rhind Papyrus (ca. 1650 B.C.) contained problems involving the solution of simple linear equations. The words *aha* or *heap* were used to designate the unknown (Boyer 1991, 15). The word *shai* was used by Mohammed ibn-Musa al-Khwarizmi (A.D. 825) to represent the unknown in his book *Al-Jabr wa'l muqabalah* (Boyer 1991, 228). Diophantus (A.D. 210–290) needed the concept of a variable to develop his theories on Diophantine equations. The symbolic notation for variables was developed in the sixteenth century. The idea of using letters to represent variables and constants is attributed to François Viète (1540–1603) (Asimov 1982, 90); he used vowels for the variables and consonants for the constants. René Descartes

(1596–1650) introduced the modern, customary notation of $x$, $y$, and $z$ for variables and $a$, $b$, and $c$ for constants (Asimov 1982, 117).

When asked to define "variable" in one word, some mathematicians gave the following responses: symbol, placeholder, pronoun, parameter, argument, pointer, name, identifier, empty space, void, reference, instance (Schoenfeld and Arcavi 1988, 420–421). Students should be led progressively from the idea of a variable as a placeholder to the understanding of a variable as representing elements of a set over a given domain. In particular, students need to be made aware that variables have different meanings in different contexts. For example, in a geometry problem $x$ may be used to represent the length of a side of a triangle, whereas in a linear programming problem $x$ may be used to represent the number of items to be manufactured. It is important to stress that within a particular equation the same letter in different positions must represent the same quantity but that changing the name of the variable wherever it appears in the equation does not change the quantity it represents. It is best to indicate exactly what a variable represents in the given problem.

The power of mathematical thinking lies in its capacity to generalize; variables are facilitators of this strength. There are many possible generalizations, including those that come from introducing different numbers of variables in a given discussion. Higher mathematics involves the study of single and multivariate functions. A function of one variable is a relationship between two sets of elements such that for each element in the first set there corresponds exactly one element in the second set. The variable representing the elements in the first set is called the *independent* variable. The variable representing the elements in the second set is called the *dependent* variable. In a multivariate function, one variable is determined by or dependent upon more than one other variable. For example, the area of a triangle is determined by the base length and the height.

*See also* Algebra, Introductory Motivation; Algebra Curriculum, K–12; Functions; Language and Mathematics in the Classroom

## SELECTED REFERENCES

Asimov, Isaac. *Asimov's Biographical Encyclopedia of Science and Technology.* Garden City, NY: Doubleday, 1982.

Boyer, Carl B. *A History of Mathematics.* 2nd ed. rev. Revised by Uta C. Merzbach. New York: Wiley, 1991.

Newman, James R. *The World of Mathematics.* New York: Simon and Schuster, 1956.

Schoenfeld, Alan H., and Abraham Arcavi. "On the Meaning of Variable." *Mathematics Teacher* 81(1988): 420–427.

Wagner, Sigrid. "What Are These Things Called Variables?" *Mathematics Teacher* 76(1983):474–479.

MONA FABRICANT

# VARIANCE AND COVARIANCE

In everyday English, any observed diversity or discrepancy that distinguishes things that would otherwise be considered equivalent. Many processes are subject to natural variation. For example, seeds from the same pod, grown in identical conditions, do not all produce plants of the same height. A machine at a bottling plant does not dispense exactly the same amount of liquid to each bottle. In statistics, *variance* has a related, but more specific, technical meaning. A set of data may usefully be summarized by its location (point of central tendency) and its spread (dispersion). Rather than relying on *range* (maximum value minus minimum value, which is peculiarly sensitive to the presence of extreme values), the dispersion is usually evaluated as some aggregate of the (positive and negative) deviations of the observations ($X_i$) from the point of central tendency (e.g., mean $= \bar{X}$). An intuitively appealing way of obtaining a representative value for these deviations would be to take a simple arithmetic average of the *modulus* (absolute value) of each deviation, that is, simply to ignore the direction of deviation. This measure of dispersion is called the *mean arithmetic deviation*.

The *variance*, however, is defined as the average of the *squared* deviations from the arithmetic mean of the data, which is another way of losing the positive or negative signs of the deviations. If the purpose is merely one of providing a descriptive statistic of the observed data, the divisor is the number, $n$, of deviations:

$$\text{variance} = \frac{\sum_{i=1}^{n} (X_i - \bar{X})^2}{n}$$

If, however, the purpose is to draw inferences about a population of data from which only a sample has been observed, it can be shown that the sample variance (as shown in the equation) gives a biased estimate of the population parameter. Therefore, in order to obtain an unbiased estimate of the population variance, the divisor in the formula is adjusted. The population variance, $\sigma^2$, is estimated by

$$s^2 = \frac{\sum_{i=1}^{n} (X_i - \bar{X})^2}{n - 1}$$

There is a common misconception that the square root of the sample variance, the *standard deviation,* yields an unbiased estimation of the population standard deviation. This is not the case. The unbiased estimation is based on the variances and not on the square roots of these values (Hawkins, Joliffe, and Glickman 1992).

Whether the variance can reliably be used as a summary statistic will depend on the reliability of the point of central tendency as a representative value. If data are to be summarized in this way, they should have an inherent homogeneity and symmetry about them. If the distribution is markedly skewed or has untypical outliers, the arithmetic mean, and hence the variance which is computed with reference to it, will be distorted from its assumed "centralness." This problem is further exacerbated in computing the variance by *squaring* the deviations, as this gives excessive weight to any extreme values.

Students should be given an opportunity to experiment with different-shaped distributions and to judge the effectiveness of the arithmetic mean and the variance as representations of central tendency and dispersion, respectively. In the Reasoning Under Uncertainty project (Rubin, Rosebery, and Bruce 1988) some attractive software, ELASTIC, was produced which allows students to manipulate data points and observe the changes that occur in the summary statistics shown on a graph. The main advantage of this software over many others is that it is particularly interactive and does not rely on the computer screen being redrawn each time an alteration is made.

Just as original data points show variability, so too will the summary statistics (e.g., arithmetic means) of successive samples. It is therefore necessary for us to be able to describe the *sampling distributions* of such statistics. Once again, we do this in terms of their central tendency and dispersion. In the absence of repeatedly drawing samples of size *n,* it is possible to estimate the *variance of the sample mean* based on a single sample. The sampling must be unbiased, that is, random, and the observations must be independent. Both empirically and mathematically, the *Central Limit Theorem* can be derived. This shows that the variance of the sample mean is $\sigma^2/n$, where the sample variance, $s^2$, with an $(n-1)$ denominator, can provide an unbiased estimate of the population variance, $\sigma^2$. It is therefore possible to obtain, on the basis of a single sample, a baseline measure of variation in the mean that could be expected by chance.

Because students tend to confuse population, sample, and sampling distributions (and hence have difficulty in distinguishing between their associated summary descriptions), a color-coding system can be helpful when introducing these concepts. In general, teachers should try to provide some sort of practical examples for the class before proceeding to computer simulations of the Central Limit Theorem. Students can collect their own population data (where the population of interest is chosen by the class) from which they can take repeated samples and look at the characteristics of the sampling distributions that they obtain. However, even if the workload is shared, this can be too time-consuming and teachers may prefer to use preexisting data. The *Journal of Statistics Education* archive has a growing number of data sets that may be of use.

There is a great deal of software available for simulating repeated sampling from a variety of differently shaped parent distributions. A number of useful pieces of simulation software are referenced in Hawkins, Jolliffe, and Glickman (1992), but the field of software development is extremely fluid and teachers should not be deterred from looking for newer products. It is important to ensure that the chosen software is really interactive and that students do not end up gazing mesmerized at the computer screen while the resampling process unfolds before their eyes. Students need opportunities to take on a more investigative, "what-if?," approach to computer simulations. If a piece of simulation software precludes this, then it is less than adequate as a teaching tool, and moreover it has been designed in a way that fails to capitalize on the real potential of the computer as a classroom resource.

The square root of the variance of the sample mean, $\sigma/\sqrt{n}$, known as the *standard error,* or *sampling error,* of the mean, plays a crucial role in *statistical inference* and estimation procedures. It is, however, just an example of a descriptive measure being applied to a particular sampling distribution. All summary statistics would be subject to sampling variability and can therefore have sampling errors associated with them. Students should experiment with sampling distributions of a variety of statistics (not just the arithmetic mean), for example, the standard error of the proportion of successes, or the standard error of the difference between two arithmetic means. This last measure of sampling variability would underpin hypothesis testing in a situation where two groups of data are to be compared (e.g., to discover if tomato plants treated with fertilizer A produce crops that are ready for harvesting earlier than those treated with fertilizer B, that is, have a shorter mean growing time). For any random variable, its mean may be expressed as its *expectation*. The use of expectation is of value because it represents a *weighted* sum, which is more appropriate, for example, when we are dealing with grouped data or data conforming to a probability distribution. Thus, the population variance may be regarded as the expectation of the squared deviations for a single variable:

$$\sigma^2 = \sum (x - \mu)^2 \cdot p(x) = E(X - \mu)^2$$

It is possible to extend the idea of variance to bivariate contexts, where we are interested in the degree to which two variables increase or decrease together. One measure of this is the *covariance*. If $X$ tends to be large when $Y$ is large, and small when $Y$ is small (e.g., the respective scores on IQ and reading comprehension tests), then the covariance is positive. However, if large values of $X$ tend to correspond with small values of $Y$ and vice versa (e.g., life expectancy at birth and infant mortality, by countries of the world), their covariance is negative. If the variables are independent (i.e., they do not tend to increase or decrease together), then their covariance is zero.

Just as the variance has been defined as an expectation in a univariate context, the covariance can be defined as the expectation of the product of deviations of two variables from their means:

$$\text{cov}(X,Y) = E[(X - \mu_X)(Y - \mu_Y)]$$

Since $\text{cov}(X,X) = \text{var}(X)$, it is not too surprising that the notation $\sigma_{XY}$ is often used for covariance. For a sample, the covariance also has a formulation that is analogous to that given for the sample variance:

$$S_{XY} = (\sum (X_i - \overline{X})(Y_i - \overline{Y}))/(n - 1)$$

Covariance is not a good measure of the concordance between two variables because it is sensitive to the scale of measurement that is adopted. For example, if $X$ is measured in seconds rather than minutes, then each $X$-deviation (and also the covariance, $\sigma_{XY}$) will be increased by a factor of 60. The *correlation coefficient* is a modification that overcomes this problem because it provides a measure of concordance that is independent of the units in which $X$ and $Y$ are measured. Dividing by the standard deviations ($\sigma_X$ and $\sigma_Y$) of $X$ and $Y$ has the effect of *standardizing* the units:

correlation coefficient, $\rho_{XY} = \text{cov}(X,Y)/(\sigma_X \sigma_Y)$

If $X$ and $Y$ are independent, then $\text{cov}(X,Y)$ equals zero, implying that independent variables are not correlated. However, the converse, namely that a zero correlation implies independence, does not hold (Lindgren 1968). It might merely mean, for example, that the relationship between the two variables is not *linear*.

It is important for students to approach the concept of covariance through graphics and exploratory data analysis. This will give them a better insight into the *statistical* meaning of the terms *relationship* and *dependence*. In everyday language we use these terms in a way that is much looser. Graphical methods also allow students to consider *multivariate* contexts that are more representative of the real world and to see how false correlations can arise from the interactions between variables or from nonsymmetrical distributions. This helps them to distinguish between *covariance* and *causality* (see Edwards 1976). Challenging the students' subjective perceptions of covariance by replicating some of the published studies can be an interesting classroom activity. Jennings, Amabile, and Ross (1982) describe some experiments on informal assessment of covariance to see whether people are more influenced by the appearance of the data or by their expectations of the variables themselves. Cleveland, Diaconis, and McGill (1982) also describe studies about what factors influence people's perceptions of covariance (correlation) as shown in scattergrams.

*See also* Correlation; Measures of Central Tendency; Statistical Inference; Statistics, Overview

## SELECTED REFERENCES

Cleveland, William S., P. Diaconis, and Robert McGill. "Variables on Scatterplots Look More Highly Correlated When the Scales Are Increased." *Science* 216(1982):1138–1141.

Edwards, Allen L. *An Introduction to Linear Regression and Correlation.* San Francisco, CA: Freeman, 1976.

Hawkins, Anne, Flavia Jolliffe, and Leslie Glickman. *Teaching Statistical Concepts.* London, England: Longman, 1992.

Jennings, Dennis L., Teresa M. Amabile, and Lee Ross. "Informal Covariation Assessment: Data-based versus Theory-based Judgments." In *Judgment Under Uncertainty; Heuristics and Biases.* Daniel Kahneman, Paul Slovic, and Amos Tversky, eds. New York: Cambridge University Press, 1982.

*Journal of Statistics Education.* Send two-line e-mail message to archive@jse.stat.ncsu.edu, send index or send access.methods, or send post mail to Journal of Statistics Education, Department of Statistics, Box 8203, North Carolina State University, Raleigh, NC 27695-8203, USA.

Lindgren, Bernard W. *Statistical Theory.* 2nd ed. New York: Macmillan, 1968.

Rubin, Andee, and Ann S. Rosebery. "Teachers' Misunderstandings in Statistical Reasoning; Evidence from a Field Test of Innovative Materials." In *Training to Teach Teachers Statistics* (pp. 72–89). Anne Hawkins, ed. Voorburg, Netherlands: International Statistical Institute, 1990.

———, and Bertram Bruce. *Reasoning Under Uncertainty*. Report No. 6851. Cambridge, MA: Bolt, Beranek, and Newman, 1988.

ANNE HAWKINS

# VECTOR SPACES

A theory providing a common framework for problems in many areas of mathematics. Vectors may be introduced as ordered pairs $(x_1, x_2)$ or ordered triples $(x_1, x_2, x_3)$ of real numbers that represent directed line segments in a Cartesian coordinate system. Two vectors are added by adding their corresponding coordinates. One can multiply a vector by a real number by multiplying each coordinate of the vector by that number. This multiplication is called *scalar multiplication*. In the Cartesian coordinate system the sum of two vectors is represented by a directed line segment corresponding to the diagonal of a parallelogram, and scalar multiplication is viewed as an operation that scales a vector along a line through the origin. Two- and three-dimensional vectors provide a geometric model for the various types of forces that are encountered in physics. In fact, the algebra of three-dimensional vectors was developed by the American physicist Josiah W. Gibbs in his treatise *Vector Analysis*, published in 1881. An earlier work by Hermann Grassman, published in German (1844), gave a detailed treatment of $n$-dimensional vector spaces. The significance of this work was not fully appreciated until many years later.

An $n$-tuple of real numbers can be thought of as a vector in an $n$-dimensional space. In linear algebra the standard notation is to represent the $n$-tuple as a column vector rather than a row vector. The set of all column vectors with $n$ entries is referred to as Euclidean $n$-space and is denoted by $R^n$. The operations of addition and scalar multiplication on $R^n$ are defined in the usual way. Thus, if $\mathbf{x}$ and $\mathbf{y}$ are vectors in $R^n$ and $c$ is a real number or scalar, then

$$c\mathbf{x} = c\begin{pmatrix} x_1 \\ x_2 \\ \vdots \\ x_n \end{pmatrix} = \begin{pmatrix} cx_1 \\ cx_2 \\ \vdots \\ cx_n \end{pmatrix}$$

and

$$\mathbf{x} + \mathbf{y} = \begin{pmatrix} x_1 \\ x_2 \\ \vdots \\ x_n \end{pmatrix} + \begin{pmatrix} y_1 \\ y_2 \\ \vdots \\ y_n \end{pmatrix} = \begin{pmatrix} x_1 + y_1 \\ x_2 + y_2 \\ \vdots \\ x_n + y_n \end{pmatrix}$$

It is possible to generalize further by defining the operations of addition and scalar multiplication on the set of all sequences of real numbers. In this case, the vector space is infinite dimensional. An infinite sequence is a function whose domain is the set of all natural numbers. In general, one can multiply any function by a scalar and add any two functions that have the same domain. Thus, if $f$ and $g$ are functions that are defined on the interval $[a, b]$, then the sum $f + g$ is defined to be the function whose value at any point $x$ in $[a, b]$ is given by $f(x) + g(x)$. If $c$ is a scalar, then the scalar product $cf$ is defined to be the function whose value at any point $x$ is $cf(x)$. Vector spaces whose elements are functions are called *function spaces*. They play an important role in the study of differential equations. In particular, the solution set of a linear homogeneous differential equation will be a vector space of functions.

The latter half of the nineteenth century marked the development of modern abstract algebra, a movement in mathematics to define and classify mathematical systems according to their algebraic structure. The two operations of addition and scalar multiplication appear in a wide variety of mathematical settings. It was natural, then, to define a general mathematical system based on these operations, as follows.

*Definition* A *vector space (linear space)* is a mathematical system consisting of a set of elements $V$, called *vectors*, and two operations, vector addition and scalar multiplication. The operations must satisfy the following axioms:

A1. $\mathbf{x} + \mathbf{y} = \mathbf{y} + \mathbf{x}$

A2. $(\mathbf{x} + \mathbf{y}) + \mathbf{z} = \mathbf{x} + (\mathbf{y} + \mathbf{z})$

A3. There exists a vector $\mathbf{0}$ in $V$ such that $\mathbf{x} + \mathbf{0} = \mathbf{x}$ for all $\mathbf{x}$ in $V$

A4. For each $\mathbf{x}$ in $V$ there exists an element $-\mathbf{x}$ such that $\mathbf{x} + (-\mathbf{x}) = \mathbf{0}$

A5. $c(\mathbf{x} + \mathbf{y}) = c\mathbf{x} + c\mathbf{y}$

A6. $(c + d)\mathbf{x} = c\mathbf{x} + d\mathbf{x}$

A7. $(cd)\mathbf{x} = c(d\mathbf{x})$

A8. $1 \cdot \mathbf{x} = \mathbf{x}$

The structure of a vector space can be enlarged by defining an *inner product* on the set of vectors. An inner product is a special way of assigning a scalar to each pair of vectors. The standard inner product on $R^n$ is formed by multiplying the corresponding entries of the vectors and then adding up all of the products. The inner product associated with the pair $(\mathbf{x},\mathbf{y})$ is denoted by $< \mathbf{x},\mathbf{y} >$. If $\mathbf{x}$ and $\mathbf{y}$ are vectors in $R^n$, then

$$< \mathbf{x},\mathbf{y} > = x_1 y_1 + x_2 y_2 + \cdots + x_n y_n$$

In the context of an inner product one can define the length of a vector as well as the distance between two vectors. The theory of inner product spaces is used to solve least-squares data fitting problems and least-squares approximation problems.

*See also* Abstract Algebra; Linear Algebra, Overview; Vectors

## SELECTED REFERENCES

Bell, Eric Temple. *The Development of Mathematics.* New York: McGraw-Hill, 1945.

Boyer, Carl B. *A History of Mathematics.* 2nd ed. Revised by Uta C. Merzbach. New York: Wiley, 1989.

Leon, Steven J. *Linear Algebra with Applications.* 4th ed. New York: Macmillan, 1994.

STEVEN J. LEON

# VECTORS

Represented by straight line segments in any direction, having both direction and magnitude. Some examples are a river flowing at 20 m.p.h. or an airplane flight at 150 m.p.h. in a southerly direction. Vectors can be denoted by either letters of the alphabet in bold type $(\mathbf{a})$ or by placing an arrow above the name of the line segment $(\overrightarrow{AB})$. If $P_1$ is the point of origin and $P_2$ the endpoint of the vector $\mathbf{R}$, the length of $P_1 P_2$ is the absolute value or magnitude of the vector and is written $R = |\mathbf{R}|$. If the point of the origin and the endpoint of a vector are the same, the vector is the null vector. A null vector is the only vector to have a length of zero and no direction. It is also called the *identity* vector. If the vector is in a plane, then it has two components, for example, $(x,y)$; but if the vector is in space, then it has three components, for example, $(x,y,z)$. To calculate the length of a vector in a plane, the Pythagorean Theorem is used with its components:

$$R = |\mathbf{R}| = \sqrt{[(x_2 - x_1)^2 + (y_2 - y_1)^2]}$$

William Rowan Hamilton (1805–1865) was the first person to use the word *vector* with regard to

quaternions, a complex system of ordered quadruples giving the scalar magnitude and three directions in space of a scalar field. He established that vectors could be added, so proving that a plane vector is a real number pair. Based on this, it was found that vectors are associative, distributive, and have an identity element. This meant that vectors could be manipulated and used to solve problems, such as finding, for example, the resultant of two or more forces acting in the same plane.

In 1873, a British physicist, James Clerk Maxwell, introduced vector analysis. Vector analysis deals with vector-valued functions of one or several variables and applies the concepts and methods of the differential and integral calculus. Thus, the use of vectors is extended to the fields of pure and applied mathematics and to theoretical physics, where equations of motion, forces, and friction are studied. As an example of these topics, consider a moving particle Q with components of velocity of 5 yd per second in the direction of the positive $x$-axis and 6 yd per second in the direction of the positive $y$-axis. The velocity vector of Q can be written as $(5\mathbf{i} + 6\mathbf{j})$ yd per second. By multiplying the vector by the number of seconds traveled, we obtain the position vector of the particle after $t$ seconds. If a wind increases its speed, the vector of the wind plus the vector of the particle will give its new magnitude and speed. A second particle, R, is introduced with a velocity vector $(\mathbf{i} - 2\mathbf{j})$; then the scalar or dot product, $\mathbf{q} \cdot \mathbf{r} = |\mathbf{q}| \cdot |\mathbf{r}| \cos t$, where $t$ is the angle between $\mathbf{q}$ and $\mathbf{r}$, will give a scalar quantity.

In an elementary school classroom, one of the simplest ways to introduce vectors is to start with a square grid on the floor, using carpet tiles to move from one point to another—initially moving to the right and up, then to the left and down, and finally a combination of the two, for example, right and down. Discuss with the group how they could describe the movements on paper. To build on this, use a game situation involving two players with dice, each starting at the bottom-left corner. The first player to reach the top-right corner is the winner. This can then progress to moving two-dimensional shapes around a grid, introducing the idea of translation.

*See also* Physics; Vector Spaces

## SELECTED REFERENCES

Ashurst, F. Garth. *Founders of Modern Mathematics.* London, England: Muller, 1982.

Bridgeman, Tony, P. C. Chatwin, and Charles Plumpton. *Vectors.* Basingstoke, England: Macmillan, 1983.

Coulson, Archibald E. *An Introduction to Vectors.* London, England: Longman, 1967.

Glenn, John, and Graham Littler, eds. *A Dictionary of Mathematics.* London, England: Harper and Row, 1984.

Graham, Ted. *Mechanics.* London, England: Collins Educational, 1995.

Hamilton, Alan G. *A First Course in Linear Algebra.* Cambridge, England: Cambridge University Press, 1987.

Nunn, Gordon. *Modern Mathematics.* Plymouth, England: Macdonald and Evans, 1978.

Pascoe, L. C. *Teach Yourself Modern Mathematics.* New York: Hodder and Stoughton, 1970.

Schiller, John J., and Marie A. Wurster. *College Algebra and Trigonometry: Basics Through Precalculus.* Glenview, IL: Scott, Foresman, 1987.

Selkirk, Keith. *Longman Mathematics Handbook.* Harlow, England: Longman, 1991.

LINDA JACKSON

# VON NEUMANN, JOHN
## (1903–1957)

Probably the most powerful *pure* and *applied* mathematician, theoretical physicist and computer scientist of the twentieth century. Born in Hungary on December 28, 1903, he died prematurely in Washington, D.C. on February 8, 1957. When he was a young boy, his teachers recognized that he had an astounding memory and prodigious talent for mathematics and science. By the age of nineteen he was already an established mathematician, publishing his results in professional journals. After a distinguished university career at Berlin and Hamburg between the world wars, von Neumann, like many of his scientific contemporaries, found it necessary to emigrate to the United States in 1930. He was soon invited to join the newly established Institute for Advanced Study at Princeton, New Jersey, where he became the youngest permanent member.

Von Neumann made important contributions to such diverse fields as logic and set theory, mathematical analysis, abstract algebra, quantum mechanics, and numerical analysis. During World War II, he participated in various scientific projects related to the war effort; in particular, he was a consultant on the construction of the atomic bomb at Los Alamos. After the war, he was a member of numerous government boards and committees, including the Atomic Energy Commission. During this period he became one of the leaders in the burgeoning field of computing machines, introducing the important concept of the stored computer program. He helped design reliable machines using unreliable components and self-reproducing machines or automata. His influence on modern-day computers is instantly recognizable. Additionally, he founded the mathematical field of game theory by introducing the concept of "strategy" and designing a mathematical model that made this concept amenable to mathematical analysis. He eventually worked with Oscar Morganstern in demonstrating how mathematical game theory has numerous, far-reaching applications in the field of economics. Today, game theory is a full-fledged mathematical discipline with important real-world applications.

*See also* Game Theory

## SELECTED REFERENCES

Burton, David M. *History of Mathematics: An Introduction.* 3rd ed. Dubuque, IA: Brown, 1995.

Eves, Howard W. *An Introduction to the History of Mathematics.* 6th ed. Philadelphia, PA: Saunders College Publishing, 1990.

Gillispie, Charles, ed. *Dictionary of Scientific Biography.* Vol. 14. New York: Scribner's, 1981–1990.

RICHARD M. DAVITT

# WEIERSTRASS, KARL (1815–1897)

The greatest analyst of his time—"the father of modern analysis." Born in Germany, Weierstrass studied law and commerce but did not complete his degree. He taught high school for fifteen years, doing mathematical research in his spare time. In 1856, he was appointed assistant professor at the University of Berlin.

Weierstrass contributed to all the branches of mathematical analysis: calculus, differential and integral equations, calculus of variations, and real and complex analysis. He published little; his ideas were spread through his excellent lectures and through his many students. His work is characterized by attention to foundations and by scrupulous logical reasoning. "Weierstrassian rigor" has come to denote strict standards of rigor. Weierstrass was largely responsible for the "arithmetization of analysis," basing it on (rigorous) arithmetic rather than (intuitive) geometric reasoning. This included a rigorous construction of the real numbers (different from Dedekind cuts and Cauchy sequences) and adoption of epsilon-delta arguments. Another of his important legacies was the rigorous development of complex analysis, based on power series.

*See also* History of Mathematics, Overview; Kovalevskaia; Real Numbers

## SELECTED REFERENCES

Bell, Eric Temple. *Men of Mathematics.* New York: Simon and Schuster, 1937.

Biermann, Kurt-R. "Weierstrass, Karl Theodor Wilhelm." In *Dictionary of Scientific Biography* (pp. 219–224). Vol. 14. Charles C. Gillispie, ed. New York: Scribner's, 1981.

Dugac, Pierre. "Éléments d'analyse de Karl Weierstrass." *Archive for History of Exact Sciences* 10(1973): 41–176.

ISRAEL KLEINER

# WEYL, HERMANN (1885–1955)

One of the most universal mathematicians of his generation, excelling at both pure and applied mathematics. Born near Hamburg, Germany on November 9, 1885, Weyl died in Zürich, Switzerland on December 8, 1955. He was undoubtedly the most gifted student of David Hilbert and helped extend the more than century-long reign of the University of Göttingen (initiated by Carl F. Gauss) as the premier center for mathematical research in the world. When Hilbert retired in 1930, Weyl was recruited from a professorship at the University of Zürich to head Göttingen's famed Mathematical Institute. However, political circumstances in Nazi Germany caused him to leave Göttingen in 1933 and accept a permanent position at the Institute for Advanced Study at Princeton, New Jersey, where he was a valued colleague of Albert Einstein and John von Neumann.

Weyl was equally at home in discrete and continuous mathematics, often ingeniously blending results from these quite disparate fields. Diverging from the beliefs of his mentor, Hilbert, he became a strong advocate for the intuitionist school of mathematicians, who insist that all mathematics be based on finite constructive methods and eschew indirect proofs such as the famous proof of Euclid that the number of primes is infinite.

*See also* Discrete Mathematics; Philosophical Perspectives on Mathematics; Proof

## SELECTED REFERENCES

Burton, David. *History of Mathematics: An Introduction.* 3rd ed. Dubuque, IA: Brown, 1995.

Eves, Howard W. *An Introduction to the History of Mathematics.* 6th ed. Philadelphia, PA: Saunders College Publishing, 1990.

Gillispie, Charles, ed. *Dictionary of Scientific Biography.* New York: Scribner's, 1981–1990.

RICHARD M. DAVITT

# WHITEHEAD, ALFRED NORTH (1861–1947)

Mathematician and philosopher who co-wrote with Bertrand Russell *Principia Mathematica.* Born in England, Whitehead finished his education at Trinity College, Cambridge (in 1883) as fourth wrangler in the mathematical tripos. The following year he was elected a fellow of Trinity, where he taught for the next twenty-five years. His first substantial book, *A Treatise on Universal Algebra* (1898), drew on William Rowan Hamilton's algebra of quaternions, Hermann Grassmann's calculus of extensions, and George Boole's symbolic logic. Over the next decade, Whitehead undertook a major project with his former student Bertrand Russell (1872–1970). They had both been impressed by Giuseppe Peano's axioms of arithmetic, and his idea that mathematics could be reduced to formal logic. Hoping to eliminate the paradoxical results from logic and set theory that had been discovered by Cesare Burali-Forti (1861–1931), Russell, and others at the turn of the century, Russell and Whitehead wrote *Principia Mathematica* (1910–1913). This was intended to provide a self-consistent, formally logical foundation for mathematics.

Later, Whitehead devoted himself increasingly to philosophy, integrating his mathematical ideas with the "process philosophy" of Henri Bergson, Samuel Alexander, and Conway Lloyd Morgan. In 1910, Whitehead left Cambridge for London and taught for a time at both University College (1911–1914) and Imperial College (1914–1924) before accepting a professorship in the Philosophy Department at Harvard University (1924–1927). In 1929, he published *Process and Reality, an Essay in Cosmology,* based on the Gifford Lectures he had given at Edinburgh in 1927–28. Whitehead died in Cambridge, Massachusetts, December 30, 1947.

*See also* Boole; Russell

## SELECTED REFERENCES

Dunkel, Harold Baker. *Whitehead on Education.* Columbus, OH: Ohio State University Press, 1965.

Hendley, Brian Patrick. *Dewey, Russell, Whitehead: Philosophers as Educators.* Carbondale, IL: Southern Illinois University Press, 1986.

Whittaker, Edmund T. "Alfred North Whitehead." *Dictionary of National Biography. 1941–1950* (pp. 952–954). L. G. Wickham Legg and E. T. Williams, eds. London, England: Oxford University Press, 1959.

———. "Alfred North Whitehead." *Obituary Notices of Fellows of the Royal Society* 17(1948):281–296.

JOSEPH W. DAUBEN

# WIENER, NORBERT (1894–1964)

Regarded as one of the premier mathematicians of the twentieth century. Wiener made substantial contributions to the fields of harmonic analysis, stochastic processes, and the mathematical theory of control systems. The son of a university professor, he was born in Columbia, Missouri. Wiener was able to read and write by the age of three, attended Harvard University as a teenager, and earned the Ph.D. degree in mathematics before the age of nineteen. Most of his professional career was spent at the Massachusetts Institute of Technology, where he remained until his retirement as a professor in 1960.

During World War II, Wiener's work on wave filters was applicable to conditions of antiaircraft fire. The work focused on the question: If the course of an airplane is difficult to predict because of "noisy" radar data, how can an electric circuit be used to filter out the noise? Clearly, this is a problem whose solution is central in communication theory, a branch of electrical engineering. But Wiener is probably best known for his work in cybernetics, the study of control and communication in the animal and the machine, that was the outgrowth of an extensive collaboration over many years with the physiologist Arturo Rosenblueth of the Harvard Medical School. Wiener was especially interested in neurophysiology because of its relation to pure mathematics, statistics, and electrical engineering, and saw its importance both as an interesting subject and also as an area that bridged a number of different disciplines. Wiener was a great admirer of the German mathematician Gottfried Wilhelm Leibniz (1646–1716), whom he described as the "patron saint of cybernetics."

*See also* Leibniz; Stochastic Processes

## SELECTED REFERENCES

Wiener, Norbert. *Cybernetics.* 2nd ed. Cambridge, MA: MIT Press, 1961.

——. *The Human Use of Human Beings; Cybernetics and Society.* Boston, MA: Houghton Mifflin, 1950.

RONALD J. TALLARIDA

# WOMEN AND MATHEMATICS, HISTORY

Characterized by slow, but increasing in modern times, progress in terms of recognition and total numbers of women in the field. Why, the question often arises, are there so few women in mathematics? Of course, the same question might be asked about the fields of art, music, drama, politics, or science. Essentially, the answer is the same for each: the position of women has not through the ages been conducive to creative work. As women participate more fully in all aspects of society, as they enjoy the same benefits as do men, the sheer numbers of women in all fields should rise. The few at the top, the few who are remembered for centuries, are a small percentage of those—men and women—in any field. As the pool of those working in a discipline comes to include more women, one can expect to find more women among the top handful who achieve a sort of immortality—in mathematics as elsewhere.

Although many women mathematicians have successfully combined research with major family responsibilities, others have suffered from their inability to secure uninterrupted periods for thought. Mary Somerville (1785–1872) complained that her household and visitors felt free to make demands on her time that would have been unthinkable if addressed to a man; Sofia Kovalevskaia wished for a "wife" to take care of her daughter and her household; in fact, in spite of financial hardship, she and her daughter frequently traveled with a nursemaid.

But in spite of hardships, in spite of their small numbers, there have been women whose achievements have received recognition. The contribution of the first of these is apocryphal in nature. It is said that when Queen Dido sought refuge for her exiled people by sailing into a harbor in what is now Tunisia, the indigenous people offered her the amount of land that could be surrounded by the hide of a steer. Cutting a hide into narrow strips and sewing them together into a long rope, she confronted the *problem of Dido:* what shape encompasses the maximum area for a fixed boundary length? The solution of a circle is said to have led to the original design of the city of Carthage.

Hypatia (ca. A.D. 370–415), known predominately as a teacher and expositor, is the next woman mathematician to achieve notoriety, primarily because she was killed in Alexandria in the fourth century at the hands of a Christian mob who objected to the paganism they saw in her and her school's preservation of Greek learning, including mathematics. Not until the eighteenth century do women mathematicians next gain recognition. Maria Agnesi's (1718–1799) best-known contribution was the study of a particular curve, whose name was mistranslated from the Italian *versiera* as "witch," leading to the appellation "witch of Agnesi," associating the mathematician somehow with the black arts. She also made an important contribution to mathematics education by publishing *Istituzioni Analitiche,* the first calculus text intended for young people.

The eighteenth and nineteenth centuries saw a number of women become known for their exposition of mathematics—Emilie du Chatelet (1706–1749), Caroline Herschel (1750–1848), Mary Somerville, and Ada Byron King, Countess of Lovelace (1815–1852). Given the circumstances in which they found themselves, these women frequently relied on a close relative or other mentor for support. The fathers of Hypatia and Agnesi were mathematicians. Herschel had a brother who was an astronomer. Although she did important work in the field, she was frequently denied access to observatories. Du Chatelet was a longtime companion of François de Voltaire. On the other hand, Somerville's father once forbade her to study mathematics, on the grounds that a woman's delicate health would not support such abstruse study. Beliefs died hard, for Somerville herself was said to have felt that her encouragement of her daughter's intellectual precocity contributed to the child's early death.

Lovelace's father, the poet Lord Byron, had no mathematical connections, but her mother had studied mathematics and her husband's position as a Fellow of the Royal Society gained her access to various resources. The Earl of Lovelace had interests in architecture, history, economics, education, literature, agronomy, and even statistics. Somerville, Herschel, and Lovelace moved in the same circles; in fact, it was Somerville who introduced Lovelace to Charles Babbage, the "father" of computing machines. Lovelace is often credited with publishing the first computer program, although in fact what she wrote was more a description, including the first published flowchart, of how Babbage's analytical machine would solve a problem such as the computation of the Bernoulli numbers. Somerville learned much of her mathematics by relying on informal tutelage from a number of noted mathematicians, such as William Wallace and Pierre-Simon Laplace; this was the only method of training for aspirants of either sex at that time. She in turn was helpful to many others, including women, particularly Ada Lovelace. Somerville achieved her greatest fame when she was denounced from the pulpit of Salisbury Cathedral for her translation of Laplace's work on celestial mechanics.

The existence of a supportive atmosphere or the opportunity for the exchange of ideas is a recurring theme in the lives of women mathematicians. Correspondence with Carl Friedrich Gauss and others was important to Sophie Germain (1776–1831); when she failed to get constructive criticism, her work suffered. Early on, Sofia Kovalevskaia's (1850–1891) father encouraged her study of mathematics, and it was when she communicated with other mathematicians, in Berlin and later in Paris and Stockholm, that she was able to produce excellent mathematics. Emmy Noether's (1882–1935) father was himself a mathematician, as was her brother, and she herself was the center of a circle of mathematicians at Göttingen.

Until very recently, women mathematicians have had little opportunity to secure professional positions, even when they produced important creative work. At Oxford, Somerville College is named for Mary Somerville, but Oxford has never had a female professor of mathematics, nor has Cambridge. Dame Mary Cartwright was "knighted," headed a college, obtained important results in the field of complex analysis, but never achieved a professorship. Kovalevskaia could not get a job in Russia, Germain had no position in France, Noether could not secure a substantial appointment in Germany.

Even when women made substantial contributions, recognition was difficult. Germain's name is missing from the list of French Academy of Sciences prize winners on the Eiffel tower, even though her prize-winning work on elasticity was crucial to the construction of the edifice. Many asserted that Kovalevskaia's best work was really due to Karl Weierstrass (1815–1887), even though he deemed her the "most gifted" of a distinguished array of his students. Noether was often referred to using the masculine article, "der Noether," or as the "father of modern algebra," as if a woman could not achieve what she did. Grace Chisholm Young (1868–1944), the first woman to receive a Ph.D. degree in mathematics in Göttingen other than *in absentia*, collaborated extensively with her husband William, under whose lone name the publications generally appeared. Writing more candidly than many, he said, "Mine the laurels now . . . yours the knowledge only." Florence Nightingale (1820–1910) made substantial contributions to the development of medical statistics, although it is not what she is remembered for.

Since Christine Ladd-Franklin was denied a Ph.D. degree from Johns Hopkins because of her sex in 1882, things have improved. Ladd-Franklin was best known for her theory of color vision, first developed while she was working in the laboratory of Hermann von Helmholtz (1821–1894) in Berlin. Hopkins finally awarded Ladd-Franklin her earned degree in 1926. The percentage of Ph.D. degrees in mathematics in the United States awarded to women held steady at 6% from 1930 until the late 1960s, when it started a steady rise, reaching over 20% in the early 1990s. However, part of the rise is attributable to the decline in the number of men receiving Ph.D. degrees in mathematics. In the countries of northern Europe the percentage of women receiving Ph.D. degrees in mathematics remains in the single digits, whereas in the Mediterranean region the percentage is between 20 and 30%. In every country the percentage of women declines as one moves up in academic rank, with very few women in tenured professorships. In 1994, there were fewer than five women professors of mathematics in the United Kingdom and about the same number in all of Scandinavia, reflecting the small number of Ph.D. degrees in mathematics awarded to women or to men in these countries. In the United States, the percentage of tenured full professors at doctorate-granting institutions who are women is small, with the actual number at each of the "top ten" institutions in single digits. This narrowing of the pipeline begins after the baccalaureate; in most countries, the percentage of the first degrees in mathematics awarded to women is close to 50%.

In spite of the narrowing of the pipeline, women are making notable achievements. In 1932, Noether, an algebraist, was the first woman to give a plenary address at an International Congress of Mathematicians; the next woman plenary speaker, geometer Karen Uhlenbeck, addressed the congress in 1990. The 1994 Congress, also in Zürich where Noether spoke, had two women plenary speakers, Marina Ratner and Ingrid Daubechies, a major contributor to the theory of wavelets. Two American women mathematicians, Uhlenbeck and biomathematician Nancy Koppel, have been awarded MacArthur "genius" fellowships, and there are two women mathematicians, Ratner and Uhlenbeck, in the U.S. National Academy of Sciences, an all-time high. When the first woman, Julia Robinson, was elected to the academy, she had never had a regular faculty position, but the honor accorded her prompted an immediate offer from the University of California at Berkeley.

That women appear with regularity among speakers at conferences and as officers of mathematical organizations is due in part to the efforts of the Association for Women in Mathematics (AWM), founded in 1971 to foster the study of mathematics by women and girls and to assist in the professional development of women mathematicians. Among the most active of all disciplinary groups devoted to the advancement of women, AWM sponsors conferences, makes awards to outstanding students and young faculty, and carries on a number of other advocacy activities.

In the last twenty-five years, much attention has been given to the differences in achievement of females and males on standardized mathematics tests once they reach adolescence. In fact, such a difference is not evident worldwide. Even in the United States the gap between the mean scores of males and females has lessened as women have begun to take more mathematics courses in high school. Formerly, many women were precluded not only from taking advanced mathematics courses but from entering careers in science and many other fields by their failure to take sufficient mathematics in high school. The major difference that still exists, and which has been the subject of much controversy, is on a specific standardized examination, the Scholastic Assessment Test (SAT); the most striking characteristic of its results is the heavy predominance of males at the top end of the scale. In spite of the fact that the SAT underpredicts college achievement for women while overpredicting for men, the SAT is used extensively for admissions and the award of financial aid, to the detriment of women. There are those who theorize that women are genetically inferior to males in mathematical ability, particularly that aspect of mathematical talent that relies heavily on spatial perception. Their belief is that women's thinking is more left-brain dominated so that they excel at verbal skills, whereas men are more right-brain influenced and thus more likely to do well in areas requiring visualization skills. One gender difference for which there does appear to be some ancedotal evidence is that creative women mathematicians tend to do their best work at a later age than do men.

It should be noted that not only is there an observed correlation between high mathematics ability and sex but there is also a positive correlation with asthma, left-handedness, and a number of other conditions. What is often neglected in this search to associate mathematics with gender stereotyping is that average test scores say very little about any individual's potential in mathematics, or in any other field.

In the 1970s the term *math anxiety* arose as a label for a supposedly psychological malady. Said to afflict primarily women, it inspired a cottage industry of those who suggested methods of curing the anxiety. Often, the recovered victim was still unable to do mathematics, although unworried about this deficiency.

The view that women cannot do mathematics as currently structured holds that the traditions of mathematics itself—or, more generally, science—are the cause of the dearth of women at the top of the profession. Its advocates assert that these traditions must change in order for women to participate fully. Exactly how they should change is not clear. On the other side of this issue, most women research mathematicians seek a change in the attitudes of some in the profession and certainly greater public acknowledgment of the potential and the achievements of women, but they are firm in their belief that women are quite capable of producing excellent mathematics without any transformation of the field itself.

It is often asked whether women are attracted to particular fields within mathematics. In fact, in the 1960s and 1970s there were a disproportionate number of women in a few fields, including algebra. Some conjectured that Noether had provided a powerful role model; however, because the number of women was so small, the influence of several mathematicians who had an unusually large number of women doctoral students skewed the statistics. It is also often thought that women are more likely to work in theoretical rather than applied areas. Currently, however, the percentage of women among those receiving Ph.D. degrees in applied areas is somewhat higher than the overall percentage of women among Ph.D. recipients. Moreover, statistics has always attracted a sizeable number of women, the achievements of Florence N. David (1909–1994), Gertrude Cox (1900–1978), and Elizabeth Scott (1917–1989) being particularly noteworthy; women have headed the U.S. Bureau of Labor Statistics and the Bureau of the Census. Thus, the enlightened view is that women mathematicians are much like men mathematicians and that the mathematics women do is much like the mathematics men do.

*See also* Germain; Kovalevskaia; Mathematics Anxiety; Noether; Women and Mathematics, Problems

## SELECTED REFERENCES

American Mathematical Society. "Women in Mathematics." *Notices of the American Mathematical Society* 38(7)(1991):701–754.

Brewer, James W., and Martha K. Smith, eds. *Emmy Noether: A Tribute to Her Life and Work*. New York: Marcel Dekker, 1981.

Damarin, Suzanne K. "Teaching Mathematics: A Feminist Perspective." In *Teaching and Learning Mathematics in the 1990s* (pp. 144–151). 1990 Yearbook. Thomas J. Cooney, ed. Reston VA: National Council of Teachers of Mathematics, 1990.

Fennema, Elizabeth, and Gilah C. Leder, eds. *Mathematics and Gender*. New York: Teachers College Press, 1990.

Gray, Mary W. "Association for Women in Mathematics —A Personal View." *Mathematical Intelligencer* 13(1991): 6–11.

Grinstein, Louise S., and Paul J. Campbell, eds. *Women of Mathematics*. Westport, CT: Greenwood Press, 1987.

Harding, Sandra. *Whose Science? Whose Knowledge?* Ithaca, NY: Cornell University Press, 1991.

Henrion, Claudia. *Women in Mathematics: The Addition of Difference*. Bloomington, IN: Indiana University Press, 1997.

Katz, Victor J. *A History of Mathematics*. New York: HarperCollins, 1993.

Kenschaft, Patricia C., and Sandra Keith, eds. *Winning Women into Mathematics*. Washington DC: Mathematical Association of America, 1991.

Linn, Marcia C., and Janet S. Hyde. "Gender, Mathematics, and Science." *AWM Newsletter* 23(3)(1993):17–23; (4)(1993):14–17.

Rosser, Phyllis. *The SAT Gender Gap: Identifying the Causes*. Washington DC: Center for Women Policy Studies, 1989.

Springer, Sally P., and Georg Deutsch. *Left Brain, Right Brain*. San Francisco, CA: Freeman, 1981.

Stein, Dorothy. *Ada, a Life and a Legacy*. Cambridge, MA: MIT Press, 1985.

Tobias, Sheila. *Overcoming Math Anxiety*. New York: Norton, 1978.

MARY W. GRAY

# WOMEN AND MATHEMATICS, INTERNATIONAL COMPARISONS OF CAREER AND EDUCATIONAL PROBLEMS

Characterized by similar cultural and gender biases in both developed and developing countries. Many factors influence educational and academic choices as well as career and life opportunities. In a particularly comprehensive model of academic choice, Eccles et al. (1985) cited important influences on paths selected, including the cultural climate within which choices are made; past events; the behaviors, attitudes, and expectations of significant socializers; individual perceptions of these attitudes and expectations; goals and self-image—including the desire to be a high achiever—task-specific beliefs and personal values placed on those tasks; and differential aptitudes. Thus, an individual's interpretation of reality, as well as more objective measures of the environment, are important determinants of future actions. Educational experiences and qualifications in turn facilitate or limit long-term occupational prospects.

## IMPORTANCE OF MATHEMATICS

The extent to which mathematics acts as a critical filter, particularly in many Western countries, for entry into college and university courses, further training, apprenticeships, and occupations—both technical and nontechnical—was publicized by Sells (1973). More recent data (U.S. Department of Labor 1991) confirm that those with a limited mathematics background continue to be barred from many occupations. In the United States, the underrepresentation of women and minorities in advanced mathematics courses and the fields of mathematics, science, and engineering has been described as a problem of national scope (Clewell, Anderson, and Thorpe 1992).

### Educational Experience

In countries such as the United States, the United Kingdom, and Australia, slightly more females than males remain in full-time education until the end of secondary schooling. In these countries, staying at school per se is not an area of disadvantage to females. Anomalies appear, however, when enrollments in intensive mathematics courses are considered. For example, in the United Kingdom, in the academic year 1991–92, three times more male than female students were enrolled full-time in the physical sciences, and males outnumbered females 2.5:1 in the mathematical sciences and 6.4:1 in engineering and technology. In contrast, females outnumbered males 2:1 in education courses (Central Statistical Office 1994). Large-scale testings such as the National Assessment of Educational Progress (NAEP) data continue to indicate that more American males than females take the more advanced high school mathematics courses.

Class, ethnicity, and gender affect educational opportunities in developing countries as well as in the more industrialized nations. Boys from poor families will, in general, be less likely to complete secondary schooling than girls from economically and socially advantaged homes. Yet in many developing, economically poor countries—particularly in Africa and South Asia—females are less likely than males to complete or even attend high school (Graham-Brown 1991). In rural areas, illiteracy rates among women continue to be high. In these countries stable military and economic environments, as well as educational experience, are important determinants of adult career opportunities. Although school enrollment data and high school completion rates are published in developing countries, separate statistics for boys and girls are typically unavailable or unreliable (Graham-Brown 1991). Mid-1980s estimates of seven male students for every female student at university in Africa suggest, however, that fewer females than males are now engaged in the most prestigious careers for which that level of education is a prerequisite. In science and related fields, women were underrepresented even more markedly. They comprised 10% of students in natural science programs and merely 1.4% of engineering students. As a consequence, there are few females in occupations that require these qualifications.

## Performance

Consistent between-gender differences in mathematics performance, when they are found, are dwarfed by much larger within-group differences. Data gathered in two International Studies of Mathematics Achievement completed to date illustrate this clearly. Differences in favor of boys were found in all populations in verbal and computational scores in each of the countries that participated in the first study (Husén 1967). However, girls in some countries performed better than boys in other countries. Similar findings emerged from the second study (Hanna 1994). Differences in educational methods adopted in different countries, socioeconomic variables, or other environmental influences, rather than genetic factors, are the most likely explanations for these between-country differences. Alternative forms of testing in mathematics, such as project work, extended investigations, or problems expected to take several weeks, are being used increasingly. These data illustrate that apparent gender differences in performance can be an artifact of the format of assessment used. Yet long-term perceptions that mathematics and related careers are more appropriate for males than females linger.

## CULTURAL MILIEU

Various indicators, in addition to educational experience, can be used to describe the broad cultural climate in which decisions about educational paths and careers are made.

## Health

In many countries, life expectancy at birth is higher for females then for males (e.g., 81 vs. 73 in Canada and 79 vs. 73 in the United States for those born in 1990). These projected life spans represent a considerable increase for both groups from those of earlier centuries. The more favorable estimates for females are not reflected in employment opportunities: contemporary statistics indicate that more males than females are engaged in full-time paid employment.

## Occupational Indicators

According to census data published in 1992, in Canada females comprise approximately 51% of the population but 41% of the work force. Some 58% of these are concentrated in two major occupational groups: clerks and salespersons. (Perhaps surprisingly, this figure was less in 1971: 55%). Women are far more likely than men to hold part-time positions: 25% of all female employees compared with 10% of the working males. Comparable data are found in other developed countries. Observation of these differences is likely to reinforce, and perpetuate, gender stereotyping of occupations and contribute to the perceptions of students, their parents, and teachers of gender-linked career paths.

Employment opportunities are generally bleak for women in less-developed countries. In Zimbabwe, for example, women represent less than 10% of those in the formal employment sector (Graham-Brown 1991).

## PERSONAL VARIABLES— ATTITUDES AND BELIEFS

Possible differences in the internal beliefs of females and males, and their effects on subject and career choice, have attracted considerable research activity (see, e.g., Leder 1992 for a review). The diversity of countries in which this work has taken place is illustrated by the following selection of findings from studies conducted during the past decade.

*Australia:* Males (in grade 7) held more stereotyped views about mathematics than females, indicated a greater enjoyment of mathematics, and appeared more confident about their achievement in mathematics even though no gender differences in mathematics were found.

*Bophuthatswana:* While no significant gender difference was found for mathematics anxiety, males were significantly more positive about mathematics than females (age group: 17–19). A positive correlation was found for both groups between attitudes to mathematics and achievement.

*Hong Kong:* Compared with females (in grade 7), males were more stereotyped about mathematics and considered themselves to have more natural ability. Students' perception of ability, usefulness of mathematics, and extent of creativity in mathematics were related to levels of mathematics achievement.

*Kenya:* Males were found to have more positive attitudes to mathematics than females (primary and secondary school students).

*The Netherlands:* Achievement, attitudes toward mathematics, and recognition that mathematics was required for the preferred vocations were significant predictors of the decision to study mathematics (in high school). The relative importance of the factors differed for males and females.

*Nigeria:* Girls from single-sex schools had more positive attitudes toward mathematics than those taught in coeducational schools.

*Papua New Guinea:* No gender differences in mathematics achievement were found. (The sample comprised high school students.) However, a higher proportion of males overestimated earlier mathematics achievement, believed their future job would involve mathematics, and intended to continue studying mathematics.

*South Africa:* Males (in grade 9) scored significantly higher than females on confidence, usefulness of mathematics, mathematics as a male domain, and perceived mother's, father's, and teacher's attitudes to mathematics. Intention to continue with mathematics was predicted by an attitudinal variable for females but not for males.

*Thailand:* Males (aged 13) held significantly more positive attitudes toward mathematics than females, even after controlling for achievement and parental support. Males also believed more strongly than females that mathematics was a male domain.

*Turkey:* As freshmen, females believed more strongly than males that mathematics was enjoyable, useful, and worth doing. Two years later, no gender differences were found: females' scores had decreased.

*USA:* Summaries (meta-analyses) of well over 100 separate samples (involving students aged 5–25) indicated that males were more stereotyped than females about mathematics as a male domain. Other gender differences in beliefs and attitudes about mathematics were typically small but persistent, with males' attitudes being more positive.

*(West) Germany:* Compared with females (aged 14–19), males were more interested in mathematics, in doing well in the subject, and were more traditional in gender-role expectations.

*Zambia:* Males (in grades 8–12) were more confident than females about mathematics, less anxious, enjoyed mathematics more and perceived mathematics to be more useful. Gender differences were greatest at the higher grade levels.

Data from these representative studies, conducted in developed as well as less-developed countries, indicate that there are small, subtle, but consistent and pervasive differences in the internal beliefs of males and females that mirror prevailing societal beliefs, expectations, and achievement behaviors.

## CONCLUSION

Feminist research, which has grown in volume and stature in recent years, has maintained strongly that males are privileged over females in a variety of cultural systems. Females grow up, live, and act in a society shaped by a structure based on the supremacy of males. Traditionally, males' beliefs, experiences, and behaviors have been accepted as the norm. For example, for many years, the content and illustrations of mathematics and science textbooks used in schools and universities reinforced that these were male rather than female pursuits. Gilligan (1982) and Belenky et al. (1986), among others, have attempted to provide a theoretical but research-based framework which argues that females (as a group) construct knowledge in subtly different ways from males (as a group). The implications of feminist epistemologies for conceptions of mathematics and the way it is taught continue to be explored. The impact of current patriarchal structures on the life and occupational chances of girls and women also continue to be scrutinized. The limitations of attempts to improve conditions for females in an environment that is otherwise left unchanged are increasingly being recognized. Instead, it is argued, there needs to be a redistribution of power, women's experience must be made central to mathematics, and the worth of women's pursuits and life goals must be acknowledged and celebrated. Meanwhile, mathematical qualifications remain an important mediator of career opportunities for those living in industrial societies. In developing countries, economic pressures, geographic isolation, health, attitudes to contraception and traditional duties are further crucial variables. In both settings, through academic choices and/or societal circumstances, the educational and career opportunities of females are, in general, more limited than those of males.

*See also* National Assessment of Educational Progress (NAEP); Women and Mathematics, Problems

## SELECTED REFERENCES

Belenky, Mary Fields, Blythe McVicker Clinchy, Nancy Rule Goldberger, and Jill Mattuck Tarule. *Women's Ways of Knowing: The Development of Self, Voice and Mind.* New York: Basic Books, 1986.

Central Statistical Office. *Annual Abstract of Statistics 1994.* London, England: Her Majesty's Stationery Office, 1994.

Clewell, Beatriz C., Bernice T. Anderson, and Margaret E. Thorpe. *Breaking the Barrier: Helping Female and Minority Students Succeed in Mathematics and Science.* San Francisco, CA: Jossey-Bass, 1992.

Eccles, J., Terry F. Adler, Robert Futterman, Susan B. Goff, Caroline M. Kaczala, Judith L. Meece, and Carol Midgley. "Self-perceptions, Task Perceptions, Socializing Influences, and the Decision to Enroll in Mathematics." In *Women and Mathematics: Balancing the Equation* (pp. 95–121). Susan F. Chipman, Lorelei R. Brush, and Donna M. Wilson, eds. Hillsdale, NJ: Erlbaum, 1985.

Gilligan, Carol. *In a Different Voice.* Cambridge, MA: Harvard University Press, 1982.

Graham-Brown, Sarah. *Education in the Developing World.* New York: Longman, 1991.

Hanna, Gila. "Cross-Cultural Gender Differences in Mathematics Education." *International Journal of Educational Research* 21(1994):58–68.

Husén, Torsten, ed. *International Study of Achievement in Mathematics: A Comparison in Twelve Countries.* 2 vols. Stockholm, Sweden: Almquist and Wiksell, 1967.

Leder, Gilah C. "Mathematics and Gender: Changing Perspectives." In *Handbook of Research in Mathematics Education* (pp. 597–622). Douglas A. Grouws, ed. New York: Macmillan, 1992.

Sells, Lucy. *High School Mathematics as the Critical Filter in the Job Market.* ERIC No. ED080351. Berkeley, CA: University of California, 1973.

U.S. Department of Labor. *Dictionary of Occupational Titles.* Washington, DC: Government Printing Office, 1991.

GILAH C. LEDER

# WOMEN AND MATHEMATICS, PROBLEMS

Investigated over time from differing theoretical frameworks. Females, from their high school years and beyond, do not participate or achieve in mathematics at the same rate as males. Recent studies have demonstrated that the schools have played a major role in perpetuating sex-role stereotyping. At the same time, there have been attacks on programs that promote the inclusion of females in fields that traditionally have been dominated by males (Sadker and Sadker 1994). These viewpoints, along with the wide publicity given the American Association of University Women's (AAUW) report *Shortchanging Girls, Shortchanging America,* have placed the issue of females' performance and participation in mathematics in the public forum (AAUW 1990). This entry will review the present state of knowledge regarding gender issues and mathematics education, as well as the new directions that current research is exploring.

Females often have been described as unable to perform as well as males in mathematical and scientific fields. Although the increase in the number of years of required study of mathematics in high school forces college-bound females to study more mathematics, as in the past, once the study of mathematics is optional, females often opt to discontinue their work in mathematics. They also score significantly lower on standardized tests given in high school and beyond. Based on data from 1992–93, females earned 47% of the mathematics degrees at the bachelor's level, 40% at the master's level, and 24% at the doctoral level. What is it about mathematics and mathematics education that even those females who do well and achieve at high levels choose to discontinue their studies?

## TRADITIONAL RESEARCH

Traditional, positivist research has identified many possible causes for females' lower achievement and participation in mathematics (Fennema and Leder 1990; Meyer and Fennema 1992; Leder 1992; Sadker and Sadker 1994). The results from this research relates mathematics achievement to numerous variables. These include biological factors such as genetic makeup or hormonal balance and "math anxiety," which is perceived as occurring more frequently in women, causing females' lower achievement levels. This research provides considerable evidence that perception of mathematics as a male domain discourages females from pursuing this area of study. Ironically, results of a National Assessment of Educational Progress (NAEP) study indicate that teenage boys see mathematics as a more appropriate area for males, while teenage girls perceive it as an appropriate area for females. Given the importance of teenage boys' views to teenage girls, it is understandable why many choose not to excel or go on in mathematics. Another important factor that is correlated with achievement in mathematics is whether one believes that mathematics will be useful to you in your future. At the eighth-grade level, where critical decisions are made about what mathematics is to be studied, males have a much more positive view of the relevance of mathematics to their future than do females (Meyer and Fennema 1992; Leder 1992). Research also indicates that teachers treat females and males differently in class, paying more attention to males, asking males questions that involve higher-level thinking than those asked of females, and praising males for their work. Teachers are more likely to praise female students for the neatness of their work or its appearance (Sadker and Sadker 1994).

Two major internal factors that are related to one's achievement in mathematics are one's perception of self as a learner of mathematics, with its re-

lated causal attribution for success and failure, and whether or not individuals possess autonomous learning behaviors. Causal attribution theory examines personal explanations for our success or failure. Two basic components, locus of control (whether the factor is internal or external) and stability (whether the condition we ascribe for our success or failure will be constant or can change), are examined (see Figure 1). One's ability is stable and internal, that is, for a given task at a given time, if no outside factor can influence one's ability. A task's difficulty is determined by someone else, and therefore its difficulty level is external and stable. One's effort is internal and unstable for deciding how hard one will work at any given time. A factor that is both external and unstable is how lucky one is or whether one can receive help.

Males tend to say that they do well in mathematics because they have the ability (internal and stable). If they do not do well in mathematics, they attribute their lack of success to something like the teacher was unfair (external and unstable). Females tend to say that they do well in mathematics because they were lucky or that the teacher was so good that they could not help but do well (external and unstable). They blame their lack of success in mathematics on lack of ability (stable and internal). Thus, males think that they are the source of their success and that their failure is someone else's fault, so they have hope for future success. On the other hand, females ascribe their success to factors over which they have no control and their failure to unchangeable conditions (Meyer and Fennema 1992).

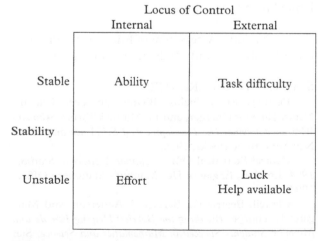

**Figure 1** *Causal attribution.*

The research on autonomous learning behaviors ties together many of the factors previously discussed (Fennema and Leder 1990). Individuals who possess autonomous learning behavior (ALB) choose to engage in high-level mathematical tasks, prefer to work independently, and persist with a task when it is perceived as difficult. Individuals who have ALB do well in mathematics. They perceive mathematics as useful, have confidence in their ability to do mathematics, and think it is appropriate for themselves, whether male or female, to do well in mathematics (sex role congruency). In addition, they attribute success to internal factors such as ability and effort and attribute failure to unstable factors, that is, they feel in control in academic situations.

## TRADITIONAL REMEDIES

The possible causes found by a traditional approach to the research on gender differences in mathematics proceeded from the assumption that if we could identify those individuals who succeeded in mathematics (primarily white males) and compare them with those who were not successful in mathematics (primarily females and nonwhite males), we could isolate factors that "cause" this difference. It was then assumed that if we could change the "unsuccessful" so that they would have the characteristics of the "successful," they too would succeed in mathematics. Given these basic assumptions, the solution to the gender problem in mathematics became making females behave as do these successful males. Such an assumption denies any substantive difference in the cognitive processes of females and males. It assumes that gender is not a salient variable in determining behavior and any effect related to gender is, and should be, easily modifiable. It also declares that males (the successful ones) are the norm; that how these successful males do mathematics is the only way of successfully doing it and that the road to success is to behave as these successful males behave (Secada 1995; Banks 1994; Eichler 1988).

## THEORETICAL FRAMEWORK: DIFFERENT VOICES

Recent work provides a theoretical, research-based foundation for a different starting point. Gilligan (1982) found that moral development in women followed a different path and was based on different values. She made the case that women were speaking "in a different voice." How women viewed the world and what they valued were not better or worse than how men viewed the world and what they valued; the two value systems were different. She called for honoring both women's and men's values and for listening to both their voices. This freed both women and men to look at how and what they value and decide which way is right for each individual. The acceptance of women's different voice with respect to the moral dimension led researchers to ask if the same is true for the cognitive.

In *Women's Ways of Knowing*, Belenky et al., (1986) document that the way women come to know is different from how men come to know. (Note: The expression "to know" is used in various works in epistemology and psychology to denote one's view of reality and what meaning one gives to the results of one's intuitive or learned knowledge. It is not synonymous with the process of learning or any particular pedagogical technique.) In particular, what is called the *procedural* way of knowing has the most profound impact for mathematics classrooms. In this procedural way, there are two approaches, a *separate* way that uses an arbitrary system for validating knowledge and a *connected* way that relies more on experience. *Women's Ways of Knowing* provides teachers with a new way of looking at how their women students learn. In mathematical research, proofs that are not deductive are viewed as less valuable than proofs of other kinds. There are still mathematicians waiting for the elegant, deductive proof of the Four Color Theorem even though mathematicians using computers have combined high-speed calculations with probabilistic considerations and deductive processes to justify their conclusions. More importantly, in most cases we would not know what to set out to prove (separate knowing) if we did not first know things through induction (connected knowing). Given that most women are connected knowers and most men are separate knowers, is it any wonder, with so much emphasis by research mathematicians on separate ways of knowing, that few women pursue mathematics and mathematics-related fields? Damarin (1990) describes a feminist perspective that sees the content of the school curriculum and modes of instruction as having been determined by male values and experiences.

The above-mentioned studies support a change in the starting point for discussions of gender and mathematics. They also prompt our looking for different modes of instruction and changes in the approach to the discipline itself. Although much of this work has focused on white females, there is evidence that the results apply to females of different cultures and races (Secada 1992).

## NEW PERSPECTIVES
## FOR REMEDIES
### The Discipline of Mathematics

There are fundamental and subtle issues that need to be addressed to make mathematical discussions more hospitable to females. Damarin (1990) discusses the language used in describing mathematics as being alienating to females. Problems are *tackled* and content is *mastered;* faculty *torpedo* students' proofs and students present *arguments*. Students are expected to *defend* their solutions rather than work together to improve them. Such confrontational language creates an environment inhospitable to many females, particularly adolescents for whom getting along with peers is so important. Yet a subtle change in the language of discourse can make being in a mathematics class more comfortable for females. Could we not expect students to integrate their new solution strategies into their set of strategies (rather than *master* new strategies) or encourage them to write more elegant proofs?

Another issue is how do mathematicians actually work? Mathematicians use both connected and separate knowing, but most importantly it is the connected knowing that leads to the need for separate knowing. Only after they have completed their experiential and intuitive knowing do they (sometimes) put on their separate knowing hat and prove what they have discovered using deductive proof. Yet mathematics is usually presented to students in the classroom as the result of separate knowing. Students rarely see how the theorems they are asked to prove evolved and why there was a need for these theorems.

Finally, as the discipline of mathematics is explored, the issue of what constitutes a proof, what is sufficient evidence so that something is known to be true or valid in mathematics, needs to be addressed. If females are more likely to be connected knowers, then there is a need to reexamine the emphasis on deduction over all other ways of knowing mathematics (Barrow 1992). The availability of high-speed computers enables mathematicians to prove things by examining many more cases than before. In addition to computer-aided proofs, there are other types of proofs that are sufficient and valid ways of knowing in mathematics that play to the connected style of knowing.

Mathematics instructors, on an elementary level, can use experiential and intuitive knowing to prove mathematical theorems. Consider the following theorem: The sum of two odd numbers is even. The traditional proof requires algebraic manipulation, using

the closure, commutative, and associative laws for addition and the distributive law for multiplication over addition. Generally, this is what is expected of individuals who are asked to prove the stated theorem. An alternative proof involves the students modeling even and odd numbers using manipulative materials such as egg cartons or Cuisenaire rods. Since an even number is a multiple of 2 or a pair of whole numbers, no matter what even number we want, we can picture or create a model for that number by stapling together as many egg cartons, or parts of egg cartons, as needed. The generalizability of the model is essential. Similarly, odd numbers are modeled by taking an even number and adding 1 or, in terms of egg cartons, leaving one extra egg compartment on the model of the sum. To add two odd numbers, put the models for two odd numbers together. What happens physically with the egg carton models is what happens abstractly in the deductive proof. This approach lets learners use models and their own experiences to prove the theorem. The generalizability of the model, the fact that any even or odd number can be represented with the egg cartons, is what makes this approach a proof. The representation of even and odd numbers is based on the abstract notion that even numbers are multiples of 2 (i.e., pairs of egg holders) and odd numbers are pairs plus 1.

The use of these models ties learning to one's own experience. It is connected knowing. Proofs using approaches such as these are as valid as deductive proofs and can be used, where possible, to make mathematics accessible to more individuals. Such proofs, however, are not possible or appropriate at some advanced levels of mathematics.

An examination of the discipline of mathematics should include the issue of the absence of women in discussions of who did mathematics. The lack of recognition of the presence and accomplishments of women mathematicians can be handled simply. Fortunately, there are several materials (Osen 1974; Perl 1978; Perl 1993) on women mathematicians' contributions to mathematics.

### The Methodology of the Mathematics Classroom

Mathematics classes traditionally have been taught as if the teacher and the textbook are the authorities, the sources of all knowledge, and it is the students' job to absorb that knowledge from those sources, not construct it on their own. To promote females' achievement and participation, instructors should balance their role as question or problem poser and source of answers, creating a more egalitarian environment. Students need to generate their own knowledge and connect it with the knowledge of other students. Students should learn from each other and share that learning in small groups, with the entire class, and with the instructor. A classroom that is a community of learners enables students to use their strengths and experiences to learn and become better students by learning how others learn. Such a classroom is based on a feminist model of pedagogy (Luke and Gore 1992).

Mathematics needs to be taught as a process, not as a universal truth handed down by the elders. Mathematical knowledge is not some predetermined entity. It is created anew for each learner, and all students can experience this act of creation. This creation may be figuring that if you forget what 7 + 8 is, you can figure out the answer by doing 7 + 7 + 1. The use of an imitation model of teaching, in which the impeccable reasoning of the teacher of "how a proof should be done" is presented to the students for them to mimic, is not an effective means of learning for women. This approach raises particular issues for women if the teacher is male.

In a community of learners, an instructor can design learning activities that enable students to use their experiences, either "real-world" or classroom-based, to enable them to learn. Writing can be used to help females learn mathematics (Buerk 1992). Learners can be actively involved in doing mathematics, causing them to engage in inquiry and reflect on their work. Alternative methods of solutions can be encouraged, where finding another way to solve a problem would be more valued than solving a similar problem in the same way. The emphasis would be on generating hypotheses rather than proving stated theorems. Given the work which indicates that for women connecting with people is important to their performance and decision making, these activities can be done cooperatively rather than competitively or in isolation from others (Gilligan 1982). Simply creating cooperative experiences, though, is not sufficient. Females are often at a disadvantage in mixed-sex groups, receiving less information and less likely to have their views prevail. Instructors should monitor cooperative learning experiences to prevent males from dominating the interaction and to use these opportunities to help females learn to lead, assert, and present their views while respecting the input and roles of the others in the group (Sadker and Sadker 1994).

*See also* Mathematics Anxiety; National Assessment of Educational Progress(NAEP); Women and Mathematics, History; Women and Mathematics, International Comparisons of Career and Educational Problems

SELECTED REFERENCES
American Association of University Women. *Shortchanging Girls, Shortchanging America.* Washington, DC: The Association, 1990.

Banks, James A. *An Introduction to Multicultural Education.* Boston, MA: Allyn and Bacon, 1994.

Barrow, John D. *Pi in the Sky: Counting, Thinking, and Being.* New York: Oxford University Press, 1992.

Belenky, Mary Fields, Blythe McVicker Clinchy, Nancy Rule Goldberger, and Jill Mattuck Tarule. *Women's Ways of Knowing: The Development of Self, Voice, and Mind.* New York: Basic Books, 1986.

Buerk, Dorothy. "Women's Metaphors for Math." In *Math and Science for Girls: A Symposium Sponsored by the National Coalition of Girls' Schools (June 16–21, 1991)* (pp. 90–100). Anna Pollina and Louise Gould, eds. Concord, MA: National Coalition of Girls' Schools, 1992.

Damarin, Suzanne K. "Teaching Mathematics: A Feminist Perspective." In *Teaching and Learning Mathematics in the 1990s* (pp. 144–151). 1990 Yearbook. Thomas J. Cooney, ed. Reston, VA: National Council of Teachers of Mathematics, 1990.

Eichler, Margit. *Nonsexist Research Methods: A Practical Guide.* Boston, MA: Unwin Hyman, 1988.

Fennema, Elizabeth, and Gilah C. Leder, eds. *Mathematics and Gender.* New York: Teachers College Press, 1990.

Gilligan, Carol. *In a Different Voice.* Cambridge, MA: Harvard University Press, 1982.

Leder, Gilah C. "Mathematics and Gender: Changing Perspectives." In *Handbook of Research on Mathematics Teaching and Learning* (pp. 597–622). Douglas A. Grouws, ed. New York: Macmillan, 1992.

Luke, Carmen, and Jennifer Gore. *Feminisms and Critical Pedagogy.* New York: Routledge, 1992.

Meyer, Margaret R., and Elizabeth Fennema. "Girls, Boys, and Mathematics." In *Teaching Mathematics in Grades K–8: Research Based Methods* (pp. 443–464). 2nd ed. Thomas R. Post, ed. Boston, MA: Allyn and Bacon, 1992.

Osen, Lynn M. *Women in Mathematics.* Cambridge, MA: MIT Press, 1974.

Perl, Teri. *Math EQUALS: Biographies of Women Mathematicians and Related Activities.* Menlo Park, CA: Addison-Wesley, 1978.

———. *Women and Numbers: Lives of Women Mathematicians plus Discovery Activities.* San Carlos, CA: Wide World Publishing/Tetra, 1993.

Sadker, Myra, and David Sadker. *Failing at Fairness: How America's Schools Cheat Girls.* New York: Scribner's, 1994.

Secada, Walter G. "Race, Ethnicity, Social Class, Language, and Achievement in Mathematics." In *Handbook of*

*Research on Mathematics Teaching and Learning* (pp. 623–660). Douglas A. Grouws, ed. New York: Macmillan, 1992.

————. "Social and Critical Dimensions for Equity in Mathematics Education." In *New Directions for Equity in Mathematics Education* (pp. 146–164). Walter G. Secada, Elizabeth Fennema, and Lisa Byrd Adajian, eds. New York: Cambridge University Press, 1995.

JUDITH E. JACOBS

# WOMEN AND MATHEMATICS EDUCATION (WME)

An association affiliated with the National Council of Teachers of Mathematics (NCTM) for the promotion of the mathematics education of girls and women. One aim of WME is to work cooperatively with other organizations to further common goals, such as effecting change in the mathematics education community in general with regard to issues of women and mathematics and monitoring publications and programs for sexist assumptions, behavior, and language.

WME promotes leadership among women in the broad mathematics education community. Its members strive to effect change by encouraging research in the area of women and mathematics, by emphasizing the need for student and teacher programs that help promote the study of mathematics by females at all levels, and by facilitating communication among special-interest groups, such as those interested in technology and gender. At annual and regional NCTM meetings, WME presents sessions focusing on women and mathematics. For instance, at the NCTM annual meeting in 1995, Judith E. Jacobs and Joanne Rossi Becker, founding members of WME, conducted a minicourse entitled "Creating a Gender-Equitable Multicultural Classroom Using Feminist Pedagogy." At the 1994 NCTM annual meeting, WME activities included a panel discussion by Ruth Cossey, Marieta Harris, and Diana Lambdin on "Assessment and Gender Equity: Issues and Challenges."

The *WME Newsletter* published three times per year serves as a general communications link among educators concerned with issues of women in mathematics. This publication includes relevant articles, announcements, and resources such as video and book reviews and research by WME members on gender issues affecting kindergarten to adult learners. Additionally, WME members develop programs, activities, and/or materials promoting awareness of the need for females to continue the study of mathematics.

*See also* Association for Women in Mathematics (AWM); Women and Mathematics, History; Women and Mathematics, International Comparisons of Career and Educational Problems; Women and Mathematics, Problems

## SELECTED REFERENCES

Olson, Judith. "Connecting the Past with the Future: Women in Mathematics and Science." Videotape series of 4 tapes. Macomb, IL: Western Illinois University, 1993–1994.

————, and Robin Thorman. *Selected Bibliography: Resources for Gender Equity in Mathematics and Technology.* South Hadley, MA: Women and Mathematics Education, 1991.

REGINA BARON BRUNNER

# WRITING ACTIVITIES

Not a traditional tool for mathematics teachers but strongly recommended by the National Council of Teachers of Mathematics (1991) to fulfill the stated goal of mathematics as communication. Reform movements in education created programs such as "Writing Across the Curriculum" and "Writing to Learn." Many state mathematics education documents also emphasize writing as an effective teaching tool.

Writing helps students clarify thinking and solidify understanding (Fernanadez 1994; Clarke et al. 1993). Writing also helps the teacher understand the student's thinking (Miller 1992; Watson 1991). There are indications that students who can write about concepts can use them correctly (Grossman et al. 1993); that writing improves student attitudes about mathematics (Davision and Pearce 1990); and that writing improves student performance (Lesnak 1993; Davision and Pearce 1990).

The writing experience can be divided into three phases: prewriting, writing, and postwriting. Prewriting is the phase in which teacher and students set the stage and develop ideas about the topic, through discussion or a discovery activity. Prewriting is extremely important as it helps students develop something worth writing about. If one portion of the writing experience is to be deleted, it is better to delete postwriting than prewriting.

In the writing phase, students clarify thinking by putting thoughts on paper. Writing is a cyclical process of thinking, writing, thinking, rewriting, thinking, rewriting, and so on. During the beginning of the writing stage, students need to write their thoughts, rough as they may be, and develop their thinking. This initial stage should be free from the stress of grammar and spelling constraints, which are important later but not during this creative portion of the process.

Postwriting is rewriting or editing. In this phase, students edit and correct grammar and spelling. Students can read each other's papers to help with editing. Depending on the importance of the paper, editing can be thorough or cursory. Editing is facilitated by peer review and teacher comments.

The ubiquitous suggestion, "write about a famous mathematician," is not an example of using writing as a learning tool. Successful activities usually ask students to explore their thinking and clarify understanding through writing on topics they already understand, as opposed to writing and teaching themselves at the same time. Within the writing assignment, it is wise to indicate the intended audience, such as another student, a child, the school newspaper, an English teacher, a diary, the teacher, or even a pet. Many teachers write some assignments along with their students in order to model the importance and challenge of writing and to understand the students' experiences. If students are allowed as a group to edit the teacher's writing, the teacher's model shows students that writing is a process for everyone. It also can demonstrate that the finished product varies considerably from the original and may reduce the feeling of intimidation that students have.

## SUGGESTED ASSIGNMENTS

What is the difference between an algebraic expression and an algebraic sentence?

Explain how to determine if a fraction can be reduced.

Discuss the difference between division problems and fractions.

Write numbered steps for adding fractions.

Explaining your reasoning, give an example of a situation where it would be appropriate to round $9\frac{1}{3}$ to 10.

Explain how you would estimate the number of M&M's in a fishbowl.

The sum of the angles in a triangle is 180°; write a convincing argument that a four-sided figure has an angle sum of 360°.

Explain how to factor a trinomial, using numbered steps and an example.

Discuss what we covered today; end with a question.

Compare and contrast the meanings of the word *similar* in English and in mathematics.

Write a letter to your English teacher telling her what you learned in Chapter 5.

Find three examples of misleading numbers in a newspaper and explain why they are misleading.

*See also* Assessment of Student Achievement, Issues; Assessment of Student Achievement, Overview; Enrichment, Overview; Instructional Methods, Current Research; Instructional Methods, New Directions

## SELECTED REFERENCES

Clarke, David J., et al. "Probing the Structure of Mathematical Writing." *Educational Studies in Mathematics* 25(3)(1993):235–250.

Davision, David, and Daniel Pearce. "Perspectives on Writing Activities in the Mathematics Classroom." *Mathematics Education Research Journal* 2(1)(1990):15–22.

Fernandez, Maria L., Nelda Hadaway, and James W. Wilson. "Problem Solving: Managing It All." *Mathematics Teacher* 87(3)(1994):195–199.

Gray, Virginia. *The Write Tool to Teach Algebra.* Berkeley, CA: Key Curriculum, 1993.

Grossman, Frances Jo, et al. "Did You Say 'Write' in Mathematics Class?" *Journal of Developmental Education* 17(1)(1993):2–4, 6, 35.

Lesnak, Richard J. "Using Linguistics in the Teaching of Developmental and Remedial Algebra." ERIC No. ED366383. Columbus, OH: ERIC, 1993.

Miller, L. Diane. "Teacher Benefits from Using Impromptu Writing Prompts in Algebra Classes." *Journal for Research in Mathematics Education* 23(4)(1992):329–340.

National Council of Teachers of Mathematics. *Principles and Standards for School Mathematics.* Reston, VA: The Council, 2000.

———. *Professional Standards for Teaching Mathematics.* Reston, VA: The Council, 1991.

Watson, Jane. "Research for Teaching." *Australian Mathematics Teacher* 47(4)(1991):18–19.

Zollman, Alan, and Deneese L. Jones. "Accommodating Assessment and Learning: Utilizing Portfolios in Teacher Education with Preservice Teachers." ERIC No. ED368551. Columbus, OH: ERIC, 1994.

VIRGINIA GRAY

# Y

## YOUNG, JACOB WILLIAM ALBERT (1865–1948)

A mathematics educator of the David E. Smith era. After receiving his Ph.D. degree in group theory from Clark University in 1892, he joined the mathematics faculty at the University of Chicago. Named assistant professor and then associate professor of the pedagogy of mathematics, he remained at Chicago through 1926. Young served on the American Mathematical Society's Committee on the Definition of College Entrance Requirements in Mathematics (1903) and the International Commission on the Teaching of Mathematics (1911–12). He wrote a methods book, *The Teaching of Mathematics in the Elementary and Secondary Schools* (1907).

*See also* History of Mathematics Education in the United States, Overview

### SELECTED REFERENCES

Bidwell, James K., and Robert G. Clason, eds. *Readings in the History of Mathematics Education*. Washington, DC: National Council of Teachers of Mathematics, 1970.

Jones, Phillip S., ed. *A History of Mathematics Education in the United States and Canada*. 32nd Yearbook. Washington, DC: National Council of Teachers of Mathematics, 1970.

JAMES K. BIDWELL
ROBERT G. CLASON

# Y

# Z

## ZENO (ca. 490–ca. 435 B.C.)

A Greek philosopher known for the paradoxes he posed dealing with infinity as well as the relation between discrete and continuous concepts. Born in the southern Italian city of Elea, a Greek colony, Zeno was said to have been executed there because of political activity. A member of the Eleatic school of philosophy, he was a disciple of the noted Eleatic philosopher, Parmenides, with whom, it is thought, he visited Athens about 450 B.C. Among Zeno's most famous paradoxes are *Dichotomy, Arrow,* and *Achilles and the Tortoise.* The statements of the paradoxes are known only as quoted by Aristotle.

*Dichotomy* and *Arrow* force a rethinking of the concept of motion. *Dichotomy* is based on the assumption that a line segment is infinitely divisible. To move from the initial to the terminal point of a line segment requires that its midpoint must first be reached. But first, the one-quarter point must be reached, which in turn implies that the one-eighth point must be reached, then one-sixteenth, one-thirty-second, . . . , and so forth, generating an infinite number of points. Thus, motion is impossible because there is no first fractional part that can be covered and motion cannot begin. *Arrow* is based on the assumption that time consists of indivisible instants. At any particular instant, the moving arrow is in a specific position. Since this is true at every instant, the arrow is at a standstill at all times.

*Achilles and the Tortoise* postulates that if Achilles runs faster than the tortoise but the tortoise has a head start, say equal to the distance from $a$ to $b$, Achilles can never overtake the tortoise. For while Achilles proceeds from $a$ to $b$, the tortoise will go from $b$ to $c$, and while Achilles proceeds from $b$ to $c$, the tortoise will go from $c$ to $d$, and so on. The explanation of this paradox lies in the fact that motion is measured by space intervals per unit of time, not by numbers of points. If Achilles takes times $t_1, t_2, t_3, \ldots, t_i, \ldots$, in seconds, to go from $a$ to $b$, $b$ to $c$, $c$ to $d$, . . . , then he will overtake the tortoise in the time equal to the sum of the $t_i$ provided the result is finite. If $t_1 = 1/2$, $t_2 = 1/4$, $t_3 = 1/8$, $t_4 = 1/16$, . . . , the sum of the $t_i$ is $1/2 + 1/4 + 1/8 + \cdots = 1$ second.

Zeno's paradoxes have been discussed extensively through the years. The questions he raised made Greek mathematicians avoid the use of infinitesimals. His work influenced the development of calculus and set theory. Such later mathematicians as Bernhard Bolzano (1781–1848), Richard Dedekind (1831–1916), and Georg Cantor (1845–1918) attempted to resolve the difficulties that Zeno had raised.

*See also* Cantor; Dedekind; Paradox

## SELECTED REFERENCES

Ball, W. W. Rouse. *A Short Account of the History of Mathematics.* New York: Dover, 1960.

Boyer, Carl B. *A History of Mathematics.* 2nd ed. rev. Revised by Uta C Merzbach. New York: Wiley, 1991.

Eves, Howard W. *An Introduction to the History of Mathematics.* 6th ed. Philadelphia, PA: Saunders College Publishing, 1990.

Kline, Morris. *Mathematical Thought from Ancient to Modern Times.* New York: Oxford University Press, 1972.

Kramer, Edna E. *The Nature and Growth of Modern Mathematics.* Princeton, NJ: Princeton University Press, 1981.

Newman, James R. *The World of Mathematics.* 4 vols. New York: Simon and Schuster, 1956.

Von Fritz, Kurt. "Zeno of Elea." In *Dictionary of Scientific Biography* (pp. 607–612). Vol. 14. Charles C. Gillispie, ed. New York: Scribner's, 1981.

LOUISE S. GRINSTEIN

SALLY I. LIPSEY

# ZERO

An important ingredient of positional notation and once simply a placeholder, that is, a symbol for an empty space or absence of quantity. Considered as a real number, 0 is neither positive nor negative. It is greater than every negative number, less than every positive number, and is represented on the real number line as the point of division between the positive and negative numbers. In set language, 0 is the cardinal number of the empty (null) set. The set of complex numbers also contains the number 0, denoted by $0 + 0i$.

There is evidence that, over many centuries, forerunners of 0 were used as placeholders in a variety of cultures. Babylonians used a slanted wedge (ca. 700 B.C.); Greeks, an O (possibly from omicron, ca. 300 B.C.); Mayans, zero glyphs in many styles (ca. 300 B.C.–ca. A.D. 900), and Indians a small superscript ° (ca. A.D. 900). It was convenient to have a method of clarifying the difference between, for example, 37 and 307. But because of fear of fraud, or just superstition, its adoption was not widespread until the Middle Ages. The invention of the printing press (A.D. 1438) helped to establish the uniqueness of the symbol. General use of 0 was motivated by and contributed to the growth of commerce as the greater efficiency of recorded place value notation over the counting board or abacus came to be appreciated. Gradually the perception of 0 changed. What was just a placeholder or symbol for *nothing* began to receive recognition as a bona fide number, and, like

other numbers, could be used for purposes of measurement and counting. As a measurement, 0 is a well-known choice for the temperature (on the Celsius scale) of water at its freezing point, and for representing the latitude of the equator and the longitude of Greenwich. On the Kelvin temperature scale, the theoretical point at which all motion ceases is called *absolute zero*.

## BASIC PROPERTIES OF THE NUMBER 0

If $a$ is a real number, the following properties hold. (When $a$ has an imaginary component, analogous considerations exist for the complex plane.)

*Addition*

$$a + 0 = 0 + a = a$$

Thus, in a mathematical structure such as a group, in which 0 is an element and addition is an operation, 0 is the *additive identity*.

*Subtraction*

$$a - 0 = a \text{ but } 0 - a = -a$$

*Multiplication*

$$a \times 0 = 0 \times a = 0$$

*Division* If $a \neq 0$, $0/a = 0$ but $a/0$ is undefined. No value $c$ can be chosen such that $a/0 = c$ because $0 \times c = 0 \neq a$. Sometimes the expression $1/0 = \infty$ is used but this is simply a shorthand way of writing (for $x > 0$) $\lim_{x \to 0^+} 1/x = +\infty$, that is, $1/x$ increases without bound as $x \to 0^+$.

If $a = 0$, $a/0 = 0/0$ is indeterminate because its value depends on the source of the expression. For example, $0/0$ sometimes arises in attempting to find by substitution the value of $\lim_{x \to 0} kx/x$, where $k \neq 0$. In this case, $x \to 0$ but $x \neq 0$ and $kx/x$ may be simplified first so that $\lim_{x \to 0} kx/x = \lim_{x \to 0} k = k$. Some cases in which $0/0$ appears may be resolved by use of *L'Hospital's Rule*. This rule states that, subject to certain continuity and differentiability restrictions, the limit of the ratio of two quantities both of which approach 0 is equal to the limit of the ratio of their derivatives. Thus,

$$\lim_{x \to 0} (\cos x - e^x)/5x = \lim_{x \to 0} (-\sin x - e^x)/5 = -1/5$$

*Exponentiation* Let $a > 0$. By definition, $a^0 = 1$. Also, $0^a = 0$ but $0^0$ is indeterminate. It is reasonable to define $a^0$ to have the value 1 because $\lim_{x \to 0} a^x = 1$. This may be demonstrated in elementary algebra classes by construction of an illustrative table of decreasing values (based on division by 2) including: $2^4 = 16$, $2^3 = 8$, $2^2 = 4$, $2^1 = 2$, $2^0 = 1$. An alternative method is to evaluate $a^n/a^n$ in two ways and compare the results: $a^n/a^n = 1$ and $a^n/a^n = a^{n-n} = a^0$.

We may show that $0^0$ is indeterminate by evaluating such limits as $\lim_{x \to 0^+} x^0 = \lim_{x \to 0^+} 1 = 1$ and $\lim_{x \to 0^+} 0^x = \lim_{x \to 0^+} 0 = 0$.

## SOME CONCEPTS INVOLVING THE NUMBER 0
### *Algebra* and the *Zeros of a Function*

If a function $f(x)$ has the value 0 when $x = k$, then $k$ is called a *zero* of $f(x)$. In other words, a zero of a function $f(x)$ is a solution (root) of the equation $f(x) = 0$. The expression is sometimes confusing because in most cases a zero of a function does not have the value 0. For instance, let $f(x) = (x - 1)(x - 2)$. Then $x = 1$ and $x = 2$ are solutions (roots) of the equation $(x - 1)(x - 2) = 0$. Therefore 1 and 2 are the zeros of $f(x)$. Zeros of a function may be found by direct algebraic methods or approximated by iterative procedures. They may also be found or approximated by graphing the function and determining the points of intersection of the graph with the $x$ axis.

The zeros of the derivative function $f'(x)$ are *critical values*, the values of $x$ at which $f(x)$ may have relative or global maxima or minima, or changes in concavity. The zeros of a second derivative function $f''(x)$ are values of $x$ where a point of inflection may exist.

Learning how to find the zeros of a function led to advances in techniques for solving equations, and thus facilitated the solution of problems. From the time of the Babylonians to the Middle Ages, equations were solved by trial and error, or, as in the case of quadratic equations for example, by classification into such types as $x^2 + px = q$ and $x^2 + q = px$. With the new perception of the number 0 came the realization that equations in different classifications could be solved by one method: appropriately arranging all terms on the left hand side of the equation and setting the resulting expression equal to 0.

### *Linear Algebra* and the *Null Vector*

A vector whose initial and terminal points coincide is called the null vector and may be represented by $\mathbf{0}$. In the plane, $\mathbf{0} = (0,0)$. In 3-dimensional space, $\mathbf{0} = (0,0,0)$, and in $\mathbb{R}^n$, $\mathbf{0}$ is an n-tuple of zeros, that is, each of the n components is 0. The null vector has no direction and its length is 0. Given any real number $c$ and any vectors $\mathbf{x}$ and $\mathbf{y}$, the usual operations of scalar multiplication ($c \cdot \mathbf{x}$) and addition ($\mathbf{x} + \mathbf{y}$) yield unsurprising results when $\mathbf{0}$ is involved. Thus, $c \cdot \mathbf{0} = \mathbf{0}$ and $\mathbf{x} + \mathbf{0} = \mathbf{0} + \mathbf{x} = \mathbf{x}$. In other words, $\mathbf{0}$ is the additive identity element of a vector space.

Analogous properties hold for any null matrix, that is, a matrix each of whose elements is 0. A surprising property of matrix algebra involves nonzero divisors of a null matrix. In ordinary arithmetic or algebra, the product of two factors is 0 if and only if at least one of the factors is 0. In matrix algebra, however, the product of two nonzero matrices may be a null matrix. For example, let $A = \begin{pmatrix} 0 & 0 \\ 3 & 3 \end{pmatrix}$ and $B = \begin{pmatrix} 0 & 2 \\ 0 & -2 \end{pmatrix}$. Then $AB = \begin{pmatrix} 0 & 0 \\ 0 & 0 \end{pmatrix}$.

Matrices are often used in game theory for *payoff* tables. If, in a two-person game, the payoff for one of the players is the negative of the payoff for the other player, the game is called a zero-sum game. In a zero-sum game, one player's gain is necessarily the other player's loss.

### *Further Concepts Involving* 0

The ramifications of the development of 0 from concrete placeholder to abstract concept are impossible to give fully. Students will find numerous occasions to appreciate its role. A significant influence in the development of calculus, 0 is necessary for the concept of instantaneous change, and occurs often in the evaluation of limits. Ubiquitous in all of mathematics, 0 is one of the significant numbers whose relationship was memorialized by Euler when he established that $e^{i\pi} + 1 = 0$.

*See also* Arithmetic; Calculus, Overview; Decimal (Hindu-Arabic) System; Exponents, Algebraic; Game Theory; Linear Algebra, Overview; Mayan Numeration

### SELECTED REFERENCES
Barnett, Janet Heine. "A Brief History of Algorithms in Mathematics." In *The Teaching and Learning of Algorithms in School Mathematics* (pp. 69-77). Lorna J. Morrow and Margaret J. Kenney, eds. 1998 Yearbook. Reston, VA: National Council of Teachers of Mathematics, 1998.
Boyer, Carl B. *A History of Mathematics.* 2nd ed. rev. Revised by Uta C. Merzbach. New York: Wiley, 1991.

Cajori, Florian. *A History of Mathematical Notations.* Vol. I: *Notations in Elementary Mathematics.* New York: Dover, 1993.

Eves, Howard. *An Introduction to the History of Mathematics.* 6th ed. Philadelphia, PA: Saunders College Publishing, 1990.

Kaplan, Robert. *The Nothing That Is: A Natural History of Zero.* New York: Oxford University Press, 1999. Website for notes and bibliography: www.oup-usa.org/sc/0195128427/.

Larson, Roland, and Robert Hostetler. *Calculus with Analytic Geometry.* 3 vols. 3rd ed. Lexington, MA: Heath, 1986.

Reid, Constance. *From Zero to Infinity.* 4th ed. Washington, DC: Mathematical Association of America, 1992.

LOUISE S. GRINSTEIN
SALLY I. LIPSEY

# Index

Page numbers in **boldface** indicate entry titles.

## A

AAAS. *See* American Association for the Advancement of Science
abacus, **1–2**, 114
  Chinese, 1, 377
  as earliest calculator, 87
  Japanese, 1, 377–378, 379
  loop, 471
  spike, 188
Abel, Niels Henrik, **2**, 4, 108, 279, 311, 419
Abelian groups, 2, 311
ability grouping, **2–4**
  cooperative learning and, 152–153
  tracking, 2, 3, 474, 476–477, 791
absolute geometry. *See* neutral geometry
absolute zero, 838
abstract algebra, **4–7**, 19, 82–83
  Dedekind's contribution, 189
  field theory in, 4, 7, 22
  Galois's contribution, 4, 5, 272
  group theory, 4–5, 311–312
  ideals, **6–7**
  isomorphism, 375
  *Linear Associative Algebra* (Peirce), 325
  Noether's contribution, 504–505
  rings, 4, **5–7**
  at secondary level, 177
  set theory and, 653
  vector spaces, 7, 816–817
abstractions, Platonic, 537, 543

Abu Bakr ibn Aslam, 374
Academia pro Interlingua, 533
acceleration
  as in motion, 93–94, 614–615
  as programs for students, 214–215
  arithmetic, 40
  enrichment vs., 285
achievement tests, 744–745, 791
  *See also* assessment of student achievement *headings*
*Acta Eruditorum* (Bernoulli), 73
ActivStats (CD-ROM), 741
actuarial mathematics, **7–9**
  as career, 104
  De Moivre's work, 189
  probability and, 8, 560
Ada (programming language), 133, 578
Adaptive Learning Environment Models (ALEM), 340
Addenda Series (NCTM), 649, 681, 716
adding machines, mechanical, 87–88
addition, **9–12**
  binary, 74
  commutative, 11, 311
  in Egyptian mathematics, 211, 212
  field properties, 10–11, 254
  properties of, 254
  zero's properties in, 838
  *See also* subtraction
addition-carry (Austrian) method of subtraction, 697
additivity, as mathematical concept of measurement, 406
*Adobe Premier* (software), 659

Adrain, Robert, least squares, 403
adult education, **12–13**
  distance learning and, 711, 716, 732–733, 742
advanced placement program (AP), **13–14**, 120
  examinations and tests, 13–14, 745
  high school programming, 574
Advanced Placement Tests of the College Board, 13, 745
*Adventures of Jasper Woodbury, The* (videodisc series), 662, 741
advising, **14–16**
African Americans
  career problems, 474–476
  math anxiety, 451
  NAEP mathematics and science scores, 1982–1992, 453
  Spelman College women's program, 800
  textbook depiction of, 751
  underachievers, 791, 792
African customs as applied to the mathematics classroom, **16–17**, 211
*Agenda for Action, An* (NCTM; 1980), 175, 221, 329, 468, 500, 647, 679, 680
Agnesi, Maria, 821
Ahmose, 212
AIME. *See* American Invitational Mathematics Examination
Airy equation, 523
Akademi Internasional de Lingua Universal, 533

Alabama, mathematics proficiency scores, 1992, 453
Alaskan Natives, career problems, 474–476
Aleksandrov, A. D., 642
Alexander, James, 764
Alexander, Samuel, 820
Alexander the Great, first statistical studies, 689
algebra
    Boolean, 19, 82–83, 190, 655
    coordinate geometry, 154
    equalities, 223–224
    factoring in, 20
    Fundamental Theorem of, 127, 268, 277
    geometric approach, 19
    Greek mathematics and, 310
    history of, 19, 224, 310
    ideals theory and, 7
    Indian mathematics and, 338
    instruction, 20, 497, 647
    introductory motivation, 17–19
    of logic, 82
    matrix, 6, 19, 214, 422–423
    in projective geometry, 581
    of propositions, 82
    of sets, 82
    software study, 349, 660
    of switching circuits, 82, 83
    symbolic, 190
    tensor, 19
    transcendental functions, 766–771
    of vectors, 19
    zeros of a function in, 839
    *See also* abstract algebra; algebraic numbers; linear algebra
algebra, introductory motivation, **17–19**
algebra curriculum, K–12, **19–21**
    NCTM standards, 648
    software study packages, 349, 660
    symposium (1997) on, 446
    teaching methods, 20, 497, 647
algebraic independence, 771
algebraic numbers, **21–23**, 128
    number theory, 189, 278, 512
    real numbers and, 623
    theory of, 23
    transcendental, 21, 766–771
*Algebra in a Technological World* (Addenda Book), 649
*Algebra Lab, The* (manipulative program), 223
Algebra Project, The (Illinois), 797–798
*Algebra Xpresser* (software), 349, 660
algorithms, **23–24**
    algebra curriculum and, 20
    Chinese mathematics and, 319
    discrete mathematics and, 202, 203
    Euclidean, 23, 308, 462
    Euler's use of, 235
    Sieve of Eratosthenes, 228
    task analysis and, 709
Alhazen, 580

*Alice in Wonderland* (Carroll), 107, 108
*Alice T. Schafer Mathematics Prize*, 56, 224, 812
*Al-jabr w'al muqabalah* (al-Khowarizmi), 224, 235, 373–374, 812
*Almagest* (Ptolemy), 310
alternative curricula, middle school, 716
Alternative Hypothesis ($H_1$), 685, 686
alumni associations, 15
Amabile, Teresa M., 815
AMATYC. *See* American Mathematical Association of Two-Year Colleges
*AMATYC News, The* (journal), 24
*AMATYC Review, The* (journal), 24
American Arithmometer Company, 88
American Association for the Advancement of Science (AAAS), 178
    *Benchmarks for Science Literacy* (1993), 690
    elementary and secondary school teacher preparation guidelines, 312
American Association of University Women (AAUW), *Shortchanging Girls, Shortchanging America*, 827
American High School Mathematics Examination, 444
*American High School Today, The* (Conant), 718
American Indians. *See* Native Americans
American Indian Science and Engineering Society, 475
American Invitational Mathematics Examination (AIME), 125, 286, 443, 746
    contest, 61–62
American Mathematical Association of Two-Year Colleges (AMATYC), 15, **24–25**
    college teacher preparation, recommendations for, 721
    and the United States National Commission on Mathematics Instruction, 805
*American Mathematical Monthly*, 444
American Mathematical Society (AMS), 15, **25–26**, 56
    college teacher preparation recommendations, 721
    Joint Policy Board for Mathematics, 382
    linear algebra education, 420
American Mathematical Society (AMS), historical perspective, **26**, 325
American Mathematics Contests, 25, 61, 124, 286, 443
American Regions Mathematics League (ARML), 125, 291
American Statistical Association (ASA), **26–27**
    *Statistics Teacher Network* (newsletter), 691
American University, Mid-Atlantic Equity Center, 478

Ames, A. F., 197
Amoco Foundation, 806
Ampère, André-Marie, 530
AMS. *See* American Mathematical Society
analytic geometry, 154, 190, 265–266
    conic sections, 142–144
    Descartes' contribution, 142, 190
    Fermat's contribution, 251
    *See also* coordinate geometry, overview
Analytical Engine (Babbage calculator), 65, 87
Analytical Society, 65
*Analytical Theory of Heat, The* (Fourier), 257
analytical trigonometry, 189
Anaxagoras, 541
Anderson, David R., 520
Andree, Richard and Josephine, 486
angles, **27–30**
    half-angle formulas, 783
    measurement of, 27, 784–786
animation, 89, 131
*Annuities upon Lives* (De Moivre), 189
Anscombe data sets, 162
antiparticles, 540
anxiety, math. *See* mathematics anxiety
AP. *See* advanced placement program
Apian, Peter, 532
Apollonius, 59, 142, 251, 310
Appel, Kenneth, 588–589, 626, 627
applications. *See* actuarial mathematics; applications for the classroom, overview; applications for the secondary school classroom; biomathematics; business mathematics; chemistry; coding theory; consumer mathematics; cryptography; economics; engineering; environmental mathematics; finance; health; interest; linear programming; modeling; nursing; operations research; physics; population growth; quality control; queueing theory; relativity; space mathematics; sports; temperature scales; time; time series analysis
applications for the classroom, overview, **30–34**
    goals of mathematics instruction, 296–297
applications for the secondary school classroom, **34–36**
applied mathematics, 30–32, 450
    Center for Occupational Research and Development program, 649
    consumer mathematics as, 148–149
Applied Probability Trust (United Kingdom), 446
apprentice teaching, 195–197
approximation theory, continued fractions and, 150–151

approximations, 540 *See also* estimation; extrapolation; interpolation; rounding

Arabic mathematics. *See* Islamic mathematics

Arbuthnot, John, 560

arcs
  calculus and, 98
  graphs and, 301

Archimedean iteration, 406

Archimedes, **36–37**, 91, 277
  anticipation of calculus by, 310
  Eureka episode, 320
  mathematical models and numerical approximations, 515
  numerical value of pi, 320
  as seventeenth-century mathematics influence, 251
  and value of hands-on exploration, 472

architecture, 16–17

area
  Archimedes's analyses of, 310
  in calculus, 95–98
  of circle, 115
  Egyptian formulas, 212
  geometric computations, 66
  measurement, 407–408, 706
  *See also* length, perimeter, area, and volume

Aristotle, **37**, 63, 272, 447, 448, 614
  on actual vs. potential infinite, 536
  principle of excluded middle, 587
  quotations of Zeno's paradoxes, 837
  syllogistic logic, 197

arithmetic, **37–43**
  basic skills, 69–72
  in binary system, 74, 254
  calculations in, 65
  drill, 39, 40, 123, 170
  equalities, 223
  errors, 228–230
  exponents, 242–244
  Fundamental Theorem of, 268, 510, 558
  history of teaching, 39–41, 676–677
  meaning theory of learning, 39, 83–84
  measurement and, 460
  Smith textbook, 656
  software, 657
  Thorndike textbook, 759
  University of Illinois Arithmetic Project, 679
  zero in, 838
  *See also* addition; division; multiplication; subtraction

*Arithmetica* (Diophantus), 243, 251, 310, 510

*Arithmetica Integra* (Stifel), 532

*Arithmetices principia, nova methodo exposita* (Peano), 533

*Arithmetick, Vulgar and Decimal* (Greenwood), 324, 677, 753, 755, 756

arithmetic mean. *See* mean, arithmetic

arithmetic series, 651

*Arithmetic Teacher*, 501, 647

Arkansas, mathematics proficiency scores, 1992, 453

*Arrow* (Zeno paradox), 837

*Ars conjectandi* (Bernoulli), 73

art, **43–47**
  elementary school mathematics integration into, 355
  Escher's woodcuts and lithographs, 231
  perspective, 580
  symmetry in, 700
  tiling, 759–761

Artin, Emil, 7

*Artis analyticae praxis* (Harriot), 342

ASA. *See* American Statistical Association

Ask a Teen (network project), 740

Aspray, William, 537

assessment. *See* assessment of student achievement, issues; assessment of student achievement, overview; College Entrance Examination Board (CEEB) Commission Report (1959); competitions; evaluation of instruction, issues; evaluation of instruction, overview; international studies of mathematics education; mathematics education, statistical indicators; National Assessment of Educational Progress (NAEP); test construction

assessment of instruction. *See* evaluation of instruction, issues; evaluation of instruction, overview

assessment of student achievement, issues, **47–50**
  evaluation of instruction, issues and, 236–237
  learning disabilities and, 398–399
  marked declines in mathematics (1980s), 680
  as school board major concern, 683

assessment of student achievement, overview, **50–56**
  in Canada, 102
  classroom impact of new perceptions of, 351
  current research in instructional methods and, 346–347
  Elementary and Secondary Education Act, 792–793
  *EQUALS: A Mathematics Equity Program*, 647
  Interactive Mathematics Program, 358
  international mathematics education, studies of, 366
  lesson plans and, 414
  National Center for Research in Mathematical Sciences Education, 498
  NCTM recommended methods, 794

software packages for, 662–663
statistical indicators, 452–454
student portfolios, 102
teacher-made tests, 102
technological advances in, 742
test construction, 743–748
Third International Mathematics and Science Study, 758–759
*See also* National Assessment of Educational Progress

*Assessment Standards for School Mathematics* (1995; NCTM), 48, 50, 330, 500, 718, 728, 793

Association for Computing Machinery (ACM), 131–132

Association for Teaching Aids in Mathematics (ATAM), 57, 217

Association for the Improvement of Geometrical Teaching (England), 217

Association for Women in Mathematics (AWM), 15, **56**, 823

Association of State Supervisors of Mathematics (ASSM), 727

Association of Teachers of Mathematics (ATM), 57, 217

associations. *See* resources

associative law
  addition, 10–11
  multiplication, 491

Assouline, Susan G., 285, 286

astronomics. *See* space mathematics; space race

astronomy
  chaos theory and, 109
  elementary trigonometic functions, 781–782
  history of trigonometric use in, 784
  Mayan numeration, 455–457
  measurement in, 219
  normal distribution curve, 506
  Pythagorean "harmony of the spheres" and, 602
  space mathematics, 668–669

asymptotes, **57–59**

athletics. *See* sports

Athloen, Steven C., 422

ATLAST (Augmenting the Teaching of Linear Algebra through the use of Software Tools), 424

ATM. *See* Association of Teachers of Mathematics

attentional dysfunction, 396, 400, 402

attribute blocks, **59–60**, 188

Attribute Games and Problems, 60

"Auguries of Innocence" (Blake), 543

Augustine, St., 537

Austin, Joe Dan, 331

Australia
  mathematics education studies, 365
  women and mathematics, 824–825, 826

Austria, advanced placement program, 13

Austrian (addition-carry) method of subtraction, 697
Ausubel, David, 147, 329, 600
authentic learning, 735
automorphism, 375
autonomous learning behavior, 829
autonomy, of elementary school student teacher, 693–696
averages
  mathematical expectation and, 44–45
  mean, arithmetic, 457–458
  measures of central tendency, 462–463
  median, 466–467
  mode, 480–481
awards for students and teachers, **60–63**
  congresses and, 141
AWM. *See* Association for Women in Mathematics
axes, translations, 778–779
axiomatic method, 78, 448, 450
  abstract algebra and, 5–6
  Dedekind's contribution, 5, 7, 189
  Euclid's contribution, 234–235, 588
  Peano's contribution, 5, 533, 587, 588
  projective geometry, 580–581
  proof and, 585–590
  *See also* axioms
Axiom of Choice, 299
axioms, **63–64**
  definition of, 585
  foundations of mathematics and, 63, 448
  of Gödel, 299, 316, 449
  nature of mathematics and, 450
  postulates vs., 235
  for projective geometry, 580–581
  proof and, 585–586, 588
  *See also* axiomatic method

**B**
Babbage, Charles, **65**, 87, 318, 514, 822
Babylonian mathematics, **65–67**
  base 60-system fractions, 260, 615
  circles, 115
  circular and chronological system of measurement, 319
  fractions, 615
  introduction of symbols for unknowns, 389
  mean, 457
  pi value, 66, 541
  placeholder, 838
  Plimpton 322 tablet, 318
  problems using polygons, 548
  real numbers, 622
back-to-basics movement
  arithmetic teaching and, 40
  as backlash to curriculum modernization, 171, 174, 177, 178, 329
  basic skills teaching and, 69–72
  calculator use as argument against, 220
  as ineffective against declining test scores, 647
  limitations of, 679

backward error analysis, 423
Bagel problem-solving program, 130
Banna, ibn al-, 374
Banneker, Benjamin, 16, 324
Barnsley, Michael, 259
Barzun, Jacques, 317
base 10, 315, 509
  ancient Egyptian and Indian systems, 187
  blocks, 243, 344
  origins of, 38
  *See also* decimal system
base 16, 315–316
base 60, 260, 615
bases, complex, **67–68**
bases, negative, 67, **68**
  conversion of decimals to, 68
Base-10 Blocks, 243, 344
Bashmakov, M. I., 642
BASIC (programming language), 202, 576, 658
*Basic Geometry* (Birkhoff and Beatley), 78
basic skills, **69–72**, 228–230
  addition as, 9–11
  curriculum trends, 174
  drill and practice, 69–70, 123, 170
  instructional materials, 70–71
  manipulatives in, 70–71
  NCTM position statement on (1977), 174, 647, 679
  subtraction as, 696–698
Battista, Michael T., 666, 667
Bayes' Theorem, 560
Beasley, John D., 275
Beatley, Ralph, 78
Beaton, Robert E., 758
Beberman, Max, 201, 328, 807
Becker, Jerry P., 362
Becker, Joanne Rossi, 832
Begle, Edward Griffith, **72–73**, 328, 643
behaviorism
  back-to basics movement and, 647
  Brownell's views on, 84
  "discovery learning" challenge to, 177, 599–600
  operant conditioning, 271, 656
  overview, 599
  Skinner's theories, 271, 599, 656
  stimulus-response theory, 69, 599, 678, 759
  Thorndike's theories, 759
Belenky, Mary Fields, 827, 829
Belgium
  international comparisons of mathematics curriculum, 361
  mathematics education studies, 365
Bell, Alan, 591
Bell, Eric Temple, 543
bell curve, 58
  *See also* normal distribution curve
Bell Laboratories, 605
Bell numbers, 121
Beltrami, Eugenio, 80, 423, 506
Benbow, Camilla P., 285, 289

Ben-Chaim, David, 666, 667
*Benchmarks for Science Literacy* (AAAS; 1993), 690
Benezet, Louis, 553
*Berenstein Bears* (software), 663
Bergson, Henri, 820
Berkeley, George, 437
Berman, G. M., 641
Bernoulli, Daniel, 73, 210
Bernoulli, Jakob, 73, 76
Bernoulli, Johann (Jean I), 73, 235
Bernoulli, Johann II, 73
Bernoulli, Nicolaus III, 73
Bernoulli equation, 523
Bernoulli family, **73–74**, 75, 77, 235, 560
Bernoulli numbers, 73
Bernoulli random variables, 75
Bernoulli's law of large numbers, 73
Bertrand, Joseph Louis, 558
Bertrand's paradox, 558, 568
Berzsenyi, George, 124, 290
Bessel equation, 523
Bessel Functions, 214, 770
Bhaskara II, 338, 458, 532
  value of pi, 320
biased sample, statistical significance and, 688
Bible, pi value in, 541
binary (dyadic) system, **74–75**, 254
  conversion to hexadecimal, 315
  Leibniz calculator and, 87
binary quadratic forms, 278
Binet, Alfred, 553
Binet, Jacques P. M., 192
binomial coefficients, 76, 120–121, 532
binomial distribution, **75**
binomial expansions, 114
binomial theorem, **75–77**
  Bernoulli (Jakob) proof, 73
  extension to complex exponents, 2
  as fundamental mathematical theorem, 268, 269
biomathematics, **77–78**
  environmental mathematics and, 221–223
  population and, 77, 549–550
biomedical engineering, 214
biometrics, 689
biquadratic reciprocity, 278
Birkhoff, George David, **78–79**
Birthday Problem (probability matches), 562, 563, 564
Biruni, Muhammad ibn Ahmad al-, 374
Bishop, Alan J., 487
Bishop, Errett, 537
bits, 74
black body radiation, 539
Black English vernacular, 390
Blake, William, 543
blind students, talk Calculators for, 740
block grants, 683
*Blocks* (mathematical microworld), 660
blocks, base 10, 243, 344

Bloom, Benjamin, 339, 749
and taxonomy of cognitive objectives, 411, 412, 413, 607
Blunk, Merrie, 497
Board on Mathematical Sciences (BMS), **79**
Bolyai, Janos, **79–80**, 235
non-Euclidean geometry development and, 318, 503, 505, 506, 580
Bolzano, Bernhard, 108, 267, 837
Bolza, Oskar, 326
Bombelli, Raphaello, 127
bonds, 255
Bonnycastle, Charles, 324
books, stories for children, **80–81**
integration with elementary school mathematics, 355
about time, 761
Boole, George, 6, 19, **81–82**, 190, 534
as influence on Peano, 533
mathematical reasoning groundwork, 588
as Whitehead influence, 820
Boolean algebra, 19, **82–83**, 190, 655
Peirce's adaptation, 534
Bophuthatswana, 826
Borasi, Raffaella, 346
Boston College, Third International Mathematics and Science Study, 758
Boston Latin School, 323, 324
Boston University, mathematically gifted program, 289
Boud, David, 695
Bourdin, Louis Pierre Marie, 186
Bourdon, P. M., 378
Bourgeois, Roger D., 666
Bowditch, Nathaniel, 325
box plot, 181, 182, 185, 186
Boyer, Ernest, 350
Boyle's Laws, 57, 58
brachystochrone, 73
Brademas, John, 803
Brahmagupta, 338
*Brahmasphutasiddhanta* (Brahmagupta), 338
Brattle, Thomas, 324
Brianchon, Charles Julien, 370
Bridge-It (game), 274, 275–276
"bridges of Königsberg" problem, 302–303, 616
Brigade Method, 641
Brooks, Edward, **83**, 123, 326, 759
Brouwer, Egbertus Jan, 447
Brouwer, Luitzen E. J., 537
Brouwer Fixed Point Theorem, and topology, 765
Brown, Martha M., 285
Brown, Scott W., 658
Brownell, William Arthur, 39, **83–84**, 328, 508, 553, 678
Brueckner, Leo, 39
Brun, Viggo, 510, 559
Bruner, Jerome, **84–85**, 345, 553

discovery learning and, 599
readiness concept, 618, 619
spiral learning and, 670
Buerk, Dorothy, 332
Buffon, Georges-Louis Leclerc de, 320
Buffon's Needle, 541
Bulgarian Academy of Sciences, 291
bulletin boards, electronic, 733
*Bunches and Bunches of Bunnies* (Mathews), 243
Burali-Forti, Cesare, 820
Bureau of Labor Statistics, Consumer Price Index, 337, 338
Burgi, Jobst, 226
Burkhardt, Hugh, 32
Burkill, Harry, 446
burnout, teacher, 718
Burns, Marilyn, 355, 414, 415
Burroughs, William, 87–88
Bush, George, 298
business mathematics, **85–86**
spreadsheets, 131
spreadsheet software, 472, 661, 662
systems of equations, 701, 703
time series analysis, 762–764
butterfly effect, 109
bypass strategies, for learning disabilities, 401
Byron, Augusta Ada, Countess of Lovelace, 65, 514, 821, 822
bytes, 74

**C**

cabalistic numerology, 318
*Cabri Geometry* (software), 659
calculators, **87–90**
availability of, 32, 48
in calculus, 97
computational knowledge for use of, 647
in consumer education, 149
as counterargument to back-to-basics movement, 220–221
debates over classroom role of, 220
for estimating sums of convergent series, 652
for exponents, 243–244
graphing. *See* graphing calculator
instructional materials and methods, 344, 349
in Japanese mathematics curriculum, 381
for middle-school instruction, 472
national workshop on teaching with, 443
for primary grades instruction, 71
programmable, 88, 97
recommendations for use, 89, 224, 679, 681
scientific, 20, 88, 97
symbolic manipulation capabilities, 552
systems of equations and, 702–703
technology of, 88, 97, 739–740
calculus, overview, **90–100**
Archimedes anticipation of, 310
Bernoulli family publications on, 73

binomial theorem and, 76
in business mathematics, 85
Cauchy's contribution, 108, 151
continuity and, 151–152
development of, 37
differential, 73, 210, 214, 268, 504, 533
economics and, 210
Euler's contribution, 235
Fermat's contribution, 251
first textbook for young people, 821
first textbook on, 417
Fundamental Theorem of, 268, 352–353, 437, 459
higher education teaching reform movement, 329
integral, 73, 98, 214, 268, 533
integration and, 97, 352–353
Lagrange's contribution, 387–388, 550
Leibniz-Newton dispute, 189
Leibniz's contribution, 90, 268, 405–406, 438, 450, 504, 531, 614
length of arcs in, 98
limit of a function and, 151
limits in, 91–92
mean value theorems, 268, 458–459
NCTM standards, 648
Newton's contribution, 90, 251, 268, 318, 405, 438, 450, 503–504, 614
Peano textbooks, 533
rates, 614–615
secondary level teaching, 177
tensor, 213
transcendental and algebraic functions, 766–771
of variations, 73, 387
Zeno's paradoxes as influence on, 837
zero's significance in, 838
calendars, **100**
Mayan numeration, 455–457
California
*EQUALS* project, 175
EQUITY 2000 program, 479
field testing of Mathematics in Context curriculum, 454
junior high school establishment, 467
Mathematics Project, 647, 712–715
secondary school teacher preparation reform, 729–731
special programs for underachievers, 796, 798, 799–800
teacher development, 712–715
teacher preparation reform project, 729–731
*See also* University of California, Berkeley
California Assessment Program test, 747
California Mathematics Project (CMP), 647, 712–715
California State Polytechnic University, Pomona, 741–742
California State University, Mathematics Intensive Summer Session, 799
California State University, Northridge, 195

*Call for Change: Recommendations for the Mathematical Preparation of Teachers of Mathematics, A* (1991), 726, 730
Cambridge Conference on School Mathematics (1963), **100–101**
Cambridge University, 216
Campbell, Francis A., 795
Canada, **101–102**
  factors in women's educational and career paths decisions, 825
  "layer cake curriculum," 360
  National Council of Supervisors of Mathematics, 499
  National Council of Teachers of Mathematics, 500
  teaching practice, 101–102
*CandyBar* (mathematical microworld), 660
Cantor, Georg, **102–103**, 259
  continuum hypothesis, 103, 152, 299, 536, 537
  role in modern mathematics, 484
  and Zeno's paradoxes, 837
Cantor sets, 103, 152, 259, 653, 655
capacity, as measurement system, 706
capital gains or losses, 256
Cardan, Jerome, 19
Cardano's formula, 22
Cardano-Tartaglia dispute, 318
career information, **103–107**
  actuarial science, 8–9, 104
  advice sources, 14–16
  choice of classes and, 106–107
  computer science, 105
  consulting/industrial mathematics, 104–105
  cryptology, 105
  employment decline during 1980s, 680
  employment examples, 104–106
  employment problems faced by minorities, 474–476
  engineering, 105
  in *EQUALS: A Mathematics Equity Program*, 647
  finance, 105
  goals of mathematics instruction and, 295–296
  operations research, 105
  predictions on future opportunities, 792
  research mathematics, 105
  retailing/marketing, 105–106
  statistics, 106
  teaching, 106
  *See also* minorities, career problems
Carnegie Corporation of New York, 807
Carnegie Foundation, 807, 808
Carpenter, Thomas, 176
Carroll, John, 339
Carroll, Lewis/Charles Lutwidge Dodgson, **107–108**, 198
Cartan, Eli, 6
Cartesian coordinates, 154, 318
Cartesian geometry, 305
Cartesian plane, 91, 154
Cartwright, Mary, 822

cascade model, of inservice training, 733
Casualty Actuarial Society, 8
Cauchy, Augustin-Louis, **108**, 192, 268, 419, 530
  axiomatic systems, 5
  calculus refinements by, 108, 151
  convergence of series, 651
  Fourier competition with, 257
  Mean Value Theorem for differential calculus, 458, 459
Cauchy-Kovalevskaia Theorem, 386
causal attribution theory, 828
causality, covariance distinguished from, 815
Cavalieri, Bonaventura, 514
Cayley, Arthur, 5, 6, 19, 422, 423, 580
Cayley table, 5
CD-ROMs, 350, 463, 741
cellular functions, mathematical models of, 314
Celsius, Anders, 743
Celsius temperature scale, 707, 743, 838
Center for Gifted Studies, Northwestern University, 289
Center for Gifted Studies, University of Southern Mississippi, 289
Center for Occupational Research and Development, 35
  curricula for non-college-bound students, 649
Center for Statistical Education, 26
Center for Talented Youth, 289
Center on Organization and Restructuring of Schools, 498
Central Association of Science and Mathematics Teachers. *See* School Science and Mathematics Association
Central Limit Theorem, 393
Central Missouri State University, 716, 726
central tendency. *See* measures of central tendency
certification
  of actuaries, 9
  of elementary school teachers, 723
  by National Board for Professional Teaching Standards, 719
Ceulen, Ludolph van, 541
*Challenge of Numbers, A* (Madison and Hart), 792
chance. *See* probability, overview; probability applications: chances of matches
*Chance* (mathematical microworld), 660
*Channel One*, 465
chaos, **108–109**, 249, 544
  Fatou's and Gaston's work, 382
  functions, 264
  ordinary differential equations and, 523
*characteristica universalis*, 405
Charles Law, Kelvin measures and, 743
charts, 305
  *See also* data analysis; data representation

Chatelet, Emilie du, 821
Chazan, Donald, 497
Chebyshev, Pafnuti, 558
chemistry, **110–111**
  chaotic processes, 109
Cheng Chuan Zhang, 114
Cheng Jian Gong, 114
Cheng Xing Shen, 114
China, People's Republic, **111–113**
  mathematics teaching and learning, international comparison of, 361, 362
Chinese mathematics, **113–114**, 318
  abacus use, 1, 377
  actuarial science, 8
  algorithms and, 319
  calendars and, 100
  combinatorics, 120–121
  fundamental theorems, 269
  Pascal's triangle discovery, 532
  pi value, 541
Chipman, Susan, 751
*Chiu-chih li* (Chinese astronomical work), 339
chords, 114, 115
Christina, queen of Sweden, 190
Christofferson, Halbert, 201
Chu Shih-chieh, 532
Chuck Miller Memorial Scholarship Award, 25
Church, Alonzo, 299
circles, **114–115**
  circumference, 115, 671
  radius, 27, 114
  ratio of circumference to diameter. *See* pi
circular function trigonometry, 189, 785–788
circumference, 115, 671
  ratio to diameter. *See* pi
Citicorp/Citibank, 807
Civilian Conservation Corps (CCC), 801
Civil Rights, U.S. Office of, 477
ClarisWorks (software), 472
classical inference, 686
classification, 555, 607
*Clean Your Room, Harvey Moon!* (Cummings), 761
Clement, John, 389
Clements, Douglas H., 574, 666, 667
Cleveland, William S., 815
Clinton, Bill, 298
clocks. *See* time
close-ended lessons, 413
Closs, Michael P., 487
closure, 10
clubs. *See* Mu Alpha Theta
CMP. *See* California Mathematics Project; Connected Mathematics Project; Cooperative Mathematics Project
coalitions, **115–116**
Cobb, Paul, 346
Cockcroft Committee, 731, 733
Cockcroft Report (1982), 217
Cocker, Edward, 542

coding theory, **117–118**, 254
  discrete mathematics and, 203
cognitive learning, **118–119**, 599–600
  computer programming's effect on, 574
  function of, 146
  gifted programs, 288–294
  Montessori method of, 485
  Piaget's developmental stages in, 284–285, 329, 600
  problem solving and, 570–573
Cognitively Guided Instruction program, 498, 802
Cohen, Paul J., 536
Colburn, Warren, **119**, 325, 536, 677
collaborative learning. *See* cooperative learning
Collatz Conjecture, 543
College Entrance Examination Board (CEEB)
  achievement tests, 744–745
  advanced placement test, 13, 745
  EQUITY 2000 education reform initiative, 476, 478, 479–480
  test construction, 745
College Entrance Examination Board (CEEB) Commission Report (1959), **120**
College Level Examination Program, 745
*College Mathematics Journal, The*, 444, 447
colleges and universities
  advanced placement credits, 13
  community college teacher awards, 63
  computer time-sharing activities, 128
  courses for middle grade mathematics teachers, 716
  curriculum trends, 171–173, 691
  evaluation of instruction, 238
  female vs. male degree recipients in mathematics, 822, 828
  history of U.S. mathematics education, 323–326
  math anxiety programs and clinics, 451
  minority students, 474–475, 476–480
  placement counseling, 14
  SAT scores as student success predictors, 745
  teacher preparation, 720–722, 726
  underachiever special programs, 799–800
  women Ph.D. recipients in mathematics, 822
  *See also* College Entrance Examination Board; curriculum trends, college-level; *specific institutions*
Columbia University, Teachers College, 250, 327
COMAP. *See* Consortium for Mathematics and its Applications
combinations, probability and, 560, 561
combinatorial geometry, 581
combinatorics, **120–122**
  as discrete mathematics, 121, 122, 202
  Indian mathematics and, 33
  Islamic mathematics and, 374

recreational puzzles, 625
Comer, James, 794
Commission on Postwar Plans, **122**
Commission on Teacher Credentialing, standards set by, 730–731
Committee of Fifteen on Elementary Education, 83, **123**
  curriculum recommendations (1895), 326
Committee of Seven (1920s survey), 677
Committee of Ten, **123**, 168
  National Education Association sponsorship of, 326
  secondary school studies report, 718, 728
Committee on Education and Training (CET), 804
Committee on Preparation for College Teaching (CPCT), recommendations, 721
Committee on Science, Engineering, and Public Policy, 720
Committee on the American Mathematics Competitions, 291, 745
Committee on the Mathematical Education of Teachers (COMET), 124
Committee on the Undergraduate Program in Mathematics (CUPM), **123–124**, 727
commodities, index numbers, 337–338
common denominator, 261
communication
  in mathematics, 297, 389–390
  NCTM standards, 648, 680
commutative algebra, 7
commutative groups, 2, 11, 311
commutative operations
  addition, 11, 311
  fractions, 260
  multiplication, 490
compacting, curriculum, 736
Compactness Theorem, 299
comparative studies, cross-cultural. *See* international studies of mathematics education
comparison, as mathematical concept of measurement, 406
competency testing, 683
competition, in mathematical games, 274
competitions, **124–127**
  international, 364
  Mathematical Association of America sponsorship of, 443
  for mathematically gifted, 286, 291–292
  Mu Alpha Theta sponsorship, 486
  in statistics, 732
  tests used for, 745–746
completed infinite, 536
Completeness Theorem, 299
completing the square, 225–226
completion questions, in test construction, 746

complex bases. *See* bases, complex
complex dynamics, 246, 382
complex exponents, 2, 241
complex function theory, 108, 109, 246
complex numbers, 108, **127–128**, 189, 215
  in abstract algebra, 6
  algebraic, 21
  field of, 255, 268
  noninteger, 67
compound interest, 7
  *See also* finance; interest
comprehension dysfunction, 394–395
Comprehensive School Mathematics Program, 679
comprehensive school movement in England, 217
*Comprehensive Test of Basic Skills* (norm-referenced test), 744
computer-assisted instruction (CAI), **128–131**, 194, 220
  for basic skills, 69, 71
  drill and practice with, 128, 129, 131, 170
  individualization, 341
  manipulatives generation, 440–441, 789
  Suppes's contribution to, 128, 698
  task analysis uses, 709
computer-based placement tests, 443
computer-managed instruction (CMI), 129
computer science, **131–134**
  careers in, 105
  finite fields and, 254
  programming languages, 131, 575–578, 657, 658
Computer Science Accreditation Commission (CSAC), 132
Computer Society of the Institute of Electrical and Electronics Engineers (IEEE-CS), 132
computers, 20, 31, **134–141**, 463–464
  as algorithm machines, 23–24
  availability of, 32, 48
  binary system and, 74, 87
  chaos theory and, 109
  consumer education and, 149
  creativity and, 165
  curriculum applications, 33
  data analysis and, 690
  design, 19
  duo-docking systems, 138, 139
  early designs, 65
  in elementary school, 139, 344
  in enrichment programs, 220, 738
  factoring large numbers with, 36
  functions and, 264
  as geometric ideas study tool, 280
  graphing capabilities, 48, 307
  increased use in mathematics education, 134–135
  in Japanese mathematics curriculum, 381
  learning laboratories and, 134
  linear algebra and, 420

mathematically gifted use of, 286
mathematical proofs and, 588–589
measurement activities, 407, 409
in middle school, 472
multimedia software, 138, 741
NCTM recommendations, 472, 679, 681
and new directions in instructional methods, 349
and numerical analysis methods, 513–517
operating systems, 132
pi value extension calculation with, 541
simulations by, 181
simulations of probability problems, 568–569
teacher use of, 134, 136–138, 139
technology of, 740–741
von Neumann's contribution, 818
*See also* CD-ROMs; Internet and World Wide Web; software
*Computers in the School: Tutor, Tool, Tutee, The* (Taylor), 657
Computing Sciences Accreditation Board (CSAB), 132
Conant, James Bryant, 718
concentric method, 669
conditional equalities, 223
*Conditions of Learning, The* (Gagné), 271
Conference Board of the Mathematical Sciences, 32
    Task Force on Minority Participation and Achievement, 475
    United States National Commission on Mathematics Instruction and, 805
confidence-interval approach, 685
confidence level, 687, 688
congresses, **141–142**, 364
    *See also headings beginning with* International Congress
congruence
    definition of, 277
    as mathematical concept of measurement, 406
    and number theory, 510–511
conic sections (conics), **142–144**
    invariance and, 369–370
    L'Hospital's contribution, 417
    *See also* coordinate geometry, overview
Connected Geometry Project, 649
connected graphs, 302
connected knowing, 829, 830
Connected Mathematics Project (CMP), **144–145**, 179, 716, 725
Connecticut
    mathematics proficiency scores, 1992, 453
    systemic reform projects, 649
    Wesleyan math anxiety clinic, 451
    *See also* University of Connecticut; Yale University
*Connecting Mathematics* (Addenda Book), 649
connectionism, 84, 328, 678, 759

connections, mathematical, NCTM standards, 680
Connie Belin National Center for Gifted Education, 286, 289
conservation of momentum, law of, 628
conservation, Piaget's use of term, 519
*Consortium* (journal), 292
Consortium for Mathematics and its Applications (COMAP), 35–36, 290, 292, 329, 357
constant of proportionality, 593
constructions
    classical Greek problems, 255, 542–543
    spatial reasoning, 280, 667–668
constructivism, **145–148**
    Colburn's theories as, 119
    current research on instructional methods, 345–348
    data representation and, 186
    Dewey's view of mathematics and, 197
    discovery method and, 146, 201
    learning environment and, 601
    middle school teacher preparation and, 725
    motivation and goal-setting and, 554
    Piaget's theory of, 146, 345, 554, 601, 693–694
    problem-solving teaching and, 146, 601
    social, 537
consumer mathematics, **148–149**
    economic analysis and, 210
    interest computation and, 259–260
    money and, 484–485
Consumer Price Index, 337, 338
contests, for mathematically gifted, 286
continued fractions, **149–151**
    irrational numbers and, 373
    Lagrange's work, 388
    for pi, 541
continuity, **151–152**
    *See also* continuum hypothesis
continuous distribution, 566
continuous functions, 151, 266, 267
continuous graphs, 306
continuous transformation groups (Lie groups), theory of, 419
continuum hypothesis, 103, 152, 299, 536, 537
control charts, 605–606
control systems, 820
conventionalism, 544
convergence, of a series, 651–652
Cooper, Martin, 667
cooperative games, 273
cooperative learning, **152–153**, 201, 202, 221
    and art, 44
    collaborative, 152, 194, 195–196, 793
    creativity and, 164–165
    *EQUALS: A Mathematics Equity Program*, 647
    group problem solving as, 195–196
    instructional methods, 346, 350

Interactive Mathematics Program, 358
math anxiety and, 452
middle school open-ended mathematics projects, 471
NCTM *Curriculum and Evaluation Standards for School Mathematics*, 681
Cooperative Mathematics Project (CMP), 176
coordinate geometry, overview, **154–159**
    cylindrical coordinate system, 158
    Descartes' contribution, 190
coordinates, homogeneous, 581
coordinates, polar. *See* polar coordinates
coordinates, transformations of, 778–780
Copernicus, 272
core curriculum, 356, 648–649
*Core Curriculum: Making Mathematics Count for Everyone, A* (Addenda Book), 649
Core-Plus Mathematics Project (CPMP), **159–161**, 649
    assessment in, 160
correlation, **161–164**, 244
    variance and covariance, 813–816
Corwin, Rebecca B., 457
cosine
    transcendental and algebraic functions, 767–768, 769–770
    in trigonometry, 784–785, 786
    videotapes about, 465, 741–742
*Cosmographical Mystery* (Kepler), 383
Cossey, Ruth, 832
Cost of Living Index, 337
Council of Chief State School Officers (CCSSO), 452–454
counseling. *See* advising; career information
countably infinite set, 654
counting
    in arithmetic, 38
    combinatorics and, 120–122
    international comparisons, 362
    measurement and, 460
    of musical rhythms, 492
    in number sense and numeration, 508
    as preschool activity, 556
    probability applications, 562, 563
    sports application, 673–674
    zero and, 838
Cournot, Antoine A., 210
*Cours d'Algèbre Supérieure* (Serret), 5
*Course of Study* (Japan), 378–379, 380, 381
covariance. *See* variance and covariance
Cox, Gertrude, 824
C++ (programming language), 577
CPMP. *See* Core-Plus Mathematics Project
Cramer, Gabriel, 192, 702
Cramer's Rule for solving linear equations, 437, 702
Creative Publications, tiling materials by, 760

creativity, **164–166**
  Hadamard on, 313
  Poincaré on, 544
criterion-referenced testing, 129
*Critical Explanation of the Philosophy of Leibniz, A* (Russell), 639
Cronbach, Lee J., 749
*Crossroads in Mathematics*, 348
Crosswhite, F. Joe, 500
Crozet, Claude, 324
cryptography, 31, 65, **166–167**
  careers in, 105
  RSA system, 167, 559
C3 Summer Institute, 798–799
cube duplication, 543
cube root, 113
cubic equations, 224
cubit, historical background of term, 704
Cuisenaire, Emile-Georges, 59, 167
Cuisenaire Company, 167
Cuisenaire rods, **167–168**, 198, 230, 245, 344, 441, 553, 692, 697
  for arithmetic study, 40, 42
  for fractions study, 261
Cuisenaire's Metric Blocks, 243
cumulative frequency graph, 466
curriculum. *See* addition; adult education; advanced placement programs; algebra curriculum; basic skills; calculus, overview; computer science; curriculum, overview; curriculum trends, college level; curriculum trends, elementary level; curriculum trends, secondary level; discrete mathematics; division; geometry instruction; integration of elementary school mathematics instruction with other subjects; integration of secondary school mathematics instruction with other subjects; kindergarten; liberal arts mathematics; measurement; middle school; multiplication; preschool; probability in elementary school mathematics; programming languages; secondary school mathematics; standards, curriculum; subtraction
curriculum, overview, **168–170**
  business mathematics, 85–86
  calculator implementation, 89–90
  in Canada, 101, 360
  classification skills, 169
  classroom applications, 30, 31, 33
  computer-assisted instruction and, 131
  enhancement, 218–221
  enrichment, 218–221, 736–739
  evaluation tools, 745
  humor used in, 334–335
  incorporation of history into, 319–322
  innovative alternatives for middle school programs, 716
  international studies of, 365, 366

issues in, 170–171
  in Japan, 380
  lesson plans and, 410
  National Committee on Mathematical Requirements recommendations, 499
  problem-solving as basis, 578
  reform and reform needs, 171, 792–794
  reform in France, 262
  reform suggestions, 317
  in Russia, 640–641
  spiral learning, 669–670
  standards, 676–682
  Teaching Integrated Mathematics and Science Program materials, 734
  technology and, 170
  test construction, 746
  Third International Mathematics and Science Study, 758–759
  University of Chicago School Mathematics Project, 806–807
  *See also* curriculum; *Curriculum and Evaluation Standards for School Mathematics*
Curriculum Action Project, 443
*Curriculum and Evaluation Standards for School Mathematics* (NCTM; 1989), 38, 178, 500, 628, 680–681, 718, 728, 729
  available on MathFINDER CD-ROM, 741
  on computer use by students, 472
  Connected Mathematics Project and, 144
  core curriculum recommendation, 356
  evaluation of instruction issues and, 236
  high school mathematics recommendations, 646
  impact on mathematics reform, 648
  Interactive Mathematics Program and, 357
  on mathematical power, 296
  on measurement and measurement systems, 705–706, 707
  as model of teaching practices reforms, 169
  as outgrowth of problem-solving movement, 330
  on preparation program for teaching mathematics, 727
  on problem-solving-based instruction, 345
  proposed major shift in content and instructional practices, 469, 792
  on spatial visualization skills development, 665
  standards for K–12 programs compared with Commission on Teacher Credentialing standards, 730
  on statistics inclusion in curriculum, 690
  student self-regulation or autonomy capabilities, 694
  on teacher development, 711, 712

  on ten basic skill areas, 174, 647, 679
  text of, 175
curriculum issues, **170–171**
  ethnomathematics modifications, 233
curriculum trends, college-level, **171–173**
  liberal arts mathematics, 417–418
  mathematics entrance requirements in United States, 677
  statistics courses, 691
curriculum trends, elementary-level, **174–177**
  enrichment programs, 218–219
  evaluation of instruction, 236
  exponents, presentation, 243
  instructional materials and methods, 344–345
  integration with other subjects, 354–356
  kindergarten, 384–385
  lesson plan example, 414–415
  measurement and measurement systems, 705
  number sense and numeration as major objective, 508–509
  probability study, 560–561, 567–568
  standards, 676–682
  statistics study, 690
  student teacher autonomy, 693–696
  symmetry study, 701
  teacher development, 710–712
  teacher preparation, 722–724
  TIMS Elementary Mathematics Curriculum, 734
curriculum trends, secondary-level, 159–160, **177–179**, 646–650
  enrichment programs, 219
  evaluation of instruction, 238
  exponents presentation, 243
  Interactive Mathematics Program, 357–359
  probability study, 177, 561
  standards, 676–682
  teacher development, 717–720
  teacher preparation, 727–729
  University of Chicago School Mathematics Project, 806–807
curvature, 278
  of surfaces, 280
curve-fitting, **180**
curves
  approximation of, 99, 180
  closed, 114
  definition of, 529
  normal distribution and, 566
  parametric equations and, 529
  plane, 73
  probability density and, 566
  rose, 546
  self-similar, 259
  space-filling dragon, 67
  three-dimensional, 529
customary system (inch-pound system) of measurement, 704, 705
cybernetics, 820–821

cycloid, 73
cyclotomy, 278

# D

D'Alembert, Jean Le Rond, 73, **181**, 530, 268
Dale Seymour Publications, tiling materials by, 760
Dalton Plan, 2
Damarin, Suzanne K., 829, 830
D'Ambrosio, Ubiratan, 362
Dantzig, George B., 425, 520
*Darts* (software), 663
Darwin, Charles, 689
data analysis, **181–184**
  statistical, 689–691
  *See also* data representation
*Data Analysis and Statistics across the Curriculum* (Addenda Book), 649
databases, 131
Data Disk, 741
*Data Insights* (software), 660
Data Processing Management Association (DPMA), 132
data representation, **184–186**
  spreadsheets, 131
  spreadsheet software, 472, 661, 662
*DataScope* (software), 690
Daubechies, Ingrid, 823
David, Florence N., 824
Davidson, Janet E., 284
Davidson, Neil, 152, 346
Davies, Charles, **186–187**, 324, 378
Davis, Edward J., 666
Davis, Philip, 333
Davis, Robert, 201, 329, 357
dBASE I (database management system), 658
*De Algebra Tractatus: Historicus and Practicus* (Wallis), 223
Deborah and Franklin Tepper Haimo Awards, 63
decay formula, exponential, 227
decimal system, 27, 38, **187–188**
  addition and, 9
  Babbage calculator, 87
  decimal measurement, 461
  division, 207
  Islamic mathematics and, 373
  origin in India, 338–339
  percent conversion from, 534, 535
  rational numbers, 615
  scientific notation, 644–645
  zero and, 838
  *See also* base 10
decision analysis. *See* operations research
Dedekind, Julius W. R., 623
  *See* Dedekind, Richard
Dedekind, Richard, 5, **189**
  Dedekind cuts, 189, 623
  sets, 653, 654
  theory of ideals, 6–7
  and Zeno's paradoxes, 837

deductive reasoning
  axiomatic, 78, 235, 448, 450, 533
  in discovery approach to geometry, 201
  history of, 448
  nature of mathematics and, 449
Defense Department, U.S., 475
definite integral, 352
definitions. *See* mathematics, definitions
Delaware, 649
Delille, M. F., 378
Deming, W. Edwards, 605
Democritus, 211
De Moivre, Abraham, 9, **189**
De Moivre's Theorem, 189, 241
De Morgan, Augustus, 6, **189–190**, 216, 534, 626
demystification, 400–401
Denning, Peter, 132
denominator, 260
  continued fractions, 149–150
  ways to find common, 261
derivatives
  applications of, 93–95
  calculating, 92
  in calculus, 92–98
  falling bodies and, 93–94
  growth and decay, 95
  maxima and minima in, 95
  meaning of, 93
  zeros of, 838
*Derive* (software), 661
Desargues, Girard, 580, 581
  Theorem of, 581
Desarguesian plane, 581
Descartes, René, 154, **190–191**, 268, 537
  analytic geometry, 142, 154, 190, 251, 266
  Cartesian coordinates, 154, 318
  exponent notation, 226, 240, 243
  geometrical plane, 91
  geometrical representation of functions, 305
  notations for variables, 266, 813
Descartes' Rule of Signs, 220
*Descriptive Tests of Mathematics Skills* (1990), 745
desktop publishing software, 661, 662
detachment, law of, 586
determinants, **191–193**
  Leibniz's work, 422
  systems of equations and, 702
deterministic proof, 589
Deutsch, Morton, 152
Deutsche Mathematiker Vereinigung, 103
Developing Mathematical Processes (DMP), **193–194**
developmental (remedial) mathematics, **194–195**
Developmental Mathematics Program and Apprenticeship Teaching, **195–197**
*Development of Ability in Arithmetic, The* (Brueckner), 39
Development of Software for Computer-Based Placement Tests, 443

Development of Standards for Teachers of Mathematics project, 443
development of teachers. *See* teacher development *headings*
deviation. *See* standard deviation
Devlin, Keith, 589
DeVries, Rheta, 695
Dewey, John, 112, 168, **197**, 608
  arithmetic teaching influence, 39, 197, 326
  incidental learning philosophy, 553, 678
  reflective thinking method, 201
  spiral learning and, 669–670
Diaconis, S. P., 815
*Diagnostic Instrument in Pre-Algebra Concepts and Skills* (Beeson and Adams), 745
diagnostic tests, 745
diagrams, Euler and Venn, **197–198**
*Dialogues* (Plato), 543
diameter, 114
  Egyptian formula, 212
  ratio to circumference. *See* pi
diatonic scale, 493
*Dichotomy* (Zeno paradox), 837
didactics of mathematics, 282
Diderot, Denis, 181
Dido, queen of Carthage, 821
Dienes, Zoltan, 59, 188, **198**, 345, 553, 692
Dienes blocks, 40, 42, 198, 660, 692
Diez, Brother Juan, 753, 755
difference engine (Babbage calculator), 65
differential calculus, 214, 268, 504
  Bernoulli contributions, 73
  mathematical economics theory and, 210
  Peano textbooks, 533
  *See also* calculus, overview
differential equations. *See* ordinary differential equations; partial differential equations
differential triangle, 531
differentiation, **198–200**
  in calculus, 97, 214, 268, 504
  mean value theorems, 458–459
  *See also* calculus, overview
Diffie, Whitfield, 166, 559
"Digital Display Technology: A Comprehensive Tool for Education" (NASA project), 252
digraph (directed graph), 303
Dilworth, Thomas, 542, 676–677
Diophantine equations, 150, **200–201**, 309
Diophantus, 19, 124, 200–201, 321, 812
  algebraic symbols inception, 224
  *Arithmetica*, 243, 251, 310, 510
directed graph (digraph), 303
*Directory of Student Science Training Programs for Pre-college Students*, 290
Dirichlet, Peter G. Lejeune, 559
Dirichlet pigeonhole principle, 121, 637
disabilities
  calculator design for, 740
  *See also* learning disabilities

disadvantaged students
   EQUITY 2000 for, 476, 478, 479–480
   underachievers, 791–796
   *See also* minorities, educational
      problems
discontinuous graphs, 306
*Discourse on the Method of Rightly
      Conducted Reason* (Descartes), 154,
      190, 240, 243
discovery approach to geometry, **201–202**,
      434
discovery learning, 177, 201–202,
      599–600
discrete mathematics, **202–204**
   combinatorics as, 121, 122, 202
   NCTM standards, 648
*Disquisitiones Arithmeticae* (Gauss),
      277–278
*Dissertatio de Arte Combinatoria* (Leibniz),
      120
distance formula. *See* coordinate
      geometry, overview
distance learning
   statistics and statistical education
      program, 732–733
   teacher development opportunities,
      711, 716
   technology of, 742
distributive fields, 254
distributive law
   in fractions, 260
   multiplying multidigit numbers, 490,
      491
   probability, 263, 539, 545, 566–567
divergent series, 78, 651–652
dividends, 255, 256
divisibility tests, **204–205**
   greatest common divisor, 308–309
division, **205–208**
   fractions, 207, 261–262
   measurement and, 461
   polynomial long division, 190
   by zero, 207–208, 838
   zero's properties in, 838
*Doctrine of Chances* (De Moivre), 189
Dodgson, Charles Lutwidge. *See* Carroll,
      Lewis
domain, 266, 785
Dominican Republic, 367
*Donald Duck in Mathmagic Land* (Disney
      movie), 300, 465
Donaldson, Simon, 766
Dossey, John A., 500
double-angle formulas, 783
Douglas, Ronald G., 172
Douglass Diagnostic Tests (elementary
      algebra), 744
drill and practice
   arithmetic, 39, 40, 123, 170
   basic skills, 69–70, 123, 170
   computer-assisted, 128, 129, 131, 170
   curriculum standards, 677, 678, 679
   elimination in curriculum compacting,
      736

software for, 657
   stimulus-response theory, 599, 749
duality, principle of, 581
Duality Theorem of Linear Programming,
      273
Dudeney, Henry E., 624
Duke University, gifted program, 289
Duncker, Karl, 570
Dürer, Albrecht, 439, 580
Dwight David Eisenhower Professional
      Development Act of 1988, 250
Dwight D. Eisenhower Mathematics and
      Science Education Program, 683
   *See also* Eisenhower program
dyadic system. *See* binary (dyadic) system
dynamical (time dependent) systems,
      theory of, 109
dynamical systems theory, 109
dyscalculia, 394
dysfunctions. *See* learning disabilities

**E**
*e*, **209–210**
   pi relationship, 541
   transcendental nature of, 209, 512
earthquake intensity
   exponents and, 244
   logarithmic equations, 228
Eash, Maurice, 749
East Harlem Tech (P.S. 72; N.Y.C.), 797
Eccles, J., 824
*École Normale* (Paris), 257
*École Polytechnic* (Paris), 257
economics, **210–211**
   money and, 484–485
   time series analysis, 762–764
   *See also* finance; interest
edges, graphs and, 301
Edkins, Joseph, 114
Edling, John, 339–340, 341
Edmonson, J. B., 748
EDSTAT-L (bulletin board), 733
Educate America Act of 1994 (Goals
      2000), 298–299, 793
Education Acts of 1870, 1902, 1944, and
      1988 (England), 217
educational problems. *See* learning
   disabilities, overview; mathematics
   anxiety; minorities, educational
   problems; underachievers;
   underachievers, special programs;
   women and mathematics,
   problems; women and
   mathematics, international
   comparisons of career and
   educational problems
*Educational Psychology* (Thorndike), 759
Educational Resources Information
      Center (ERIC), 804
Educational Services, Inc., 60
educational statistics, 803–804
*Educational Studies in Mathematics* (ESM;
      journal), **211**
Educational Testing Service (ETS)

advanced placement tests, 13
international studies of mathematics
      education, 364
and standardized testing, 744
Education and Human Resources (EHR),
      501–502
Education Department, U.S. *See* United
      States Department of Education
Education Program for Gifted Youth,
      Stanford University, 289
Edyth May Sliffe Awards, 62–63
*École Normale* (Paris), 257
Egyptian mathematics, **211–212**, 318
   algebraic equations, 224
   base-10 system, 187
   circles, 115
   exponents, 242–243
   fractions, 211–212, 259–260, 615–616
   geometry, 212, 319
   mean, 457
   pi value, 541
   problems using polygons, 548
eigenvalues, 192, 193, 421–422, 423
eigenvectors, 421
8-4 educational model, 328, 467
Eighth International Congress on
      Mathematical Education, 805
Einstein, Albert, **212–213**
   $e = mc^2$ equation, 628
   equations for general relativity, 531
   and Gödel's solutions to field equations,
      299
   on imagination and curiosity, 332
   relativity theories, 79, 212, 278, 531,
      628, 638
   Weyl association, 820
Einstein's equations, 531
Eisenhower National Clearinghouse for
      Mathematics and Science
      Education, 214, 804
Eisenhower Program, **213–214**
   Dwight D. Eisenhower Mathematics
      and Science Education Program,
      683
   funding for Teaching Integrated
      Mathematics and Science Program,
      734
   teacher preparation and development,
      803
   *See also* Dwight David Eisenhower
      Professional Development Act of
      1988
EISPACK software library, 423
ELASTIC (software), 466, 814
elasticity studies, 280
Elastic Lines (software), 472
electrical engineering, 215
electromagnetism, law of, 539
electronic bulletin boards, 733
electronic calculators, 88–90
electronic communication, 134, 349–350
electronic journals, 691, 733
electronic mail (e-mail), 464, 740
electronic notepads, 742

Electronic Numerical Integrator and Calculator (ENIAC), 788
*Electronic Portfolio* (software), 663
Elementary and Secondary Education Act of 1965 (ESEA), 792, 801
*Elementary Mathematics from an Advanced Standpoint* (Klein), 385
elementary school mathematics. *See* curriculum trends, elementary-level; teacher development, elementary school; teacher preparation, elementary school, issues
*Elementary Treatise on Determinants, An* (Dodgson), 108
*Elements* (Euclid), 63, 255, 300, 309–310, 542, 548
  alternative geometries, 505–506
  axiomatic method and proofs, 588
  deductive procedure use, 448
  elementary theorems on triangles, 538
  Fundamental Theorem of Arithmetic proof, 268
  importance of, 234–235
  Islamic scholars on tenth book of, 374
  mathematics definitions, 447
  on multiplication, 488
  on potential infinites, 536
  on principles of geometry, 449
  on properties of numbers, 510
  Russian translation of, 640
  theorems, 115
elements, in sets, 653
*Elements of Geometry* (Playfair), 27
Eliot, Charles, 326
ellipses, 142,
  reflective properties of, 143
  *See also* conic sections
elliptic differential equations, 530
elliptic function theory
  Abel's contribution, 2
  Legendre's contribution, 405
  Peirce's application to mapping, 534
elliptic geometry, 580
e-mail, 464, 740
Emerging Scholars Program, 172
*Emmy Noether Lectures*, 56
Employee Retirement Income Security Act of 1974, 8
ENC. *See* Eisenhower National Clearinghouse for Mathematics and Science Education
enculturation, multicultural mathematics and, 487–488
Encyclopedia Britannica Education Corporation, 454
*Encyclopedia of Educational Research* (Aikin ed.), 751
*Encyclopédie* (Diderot), 181
Engelder, Janice, 750
engineering, **214–216**
  as career, 105
  National Action Council for Minorities in Engineering, 475

National Society of Professional Engineers, 291
orthographic projections, 579
England, **216–218**
  advanced placement program, 13
  Applied Probability Trust, 446
  international comparison of women's mathematics education and career problems, 824–825
  in international studies comparisons, 365
  *Mathematical Spectrum* (journal), 446–447
  national curriculum, 216, 741
  percentage of women receiving Ph.D. degrees in mathematics, 822
  Royal Statistical Society, 732
  teacher preparation in statistics, 731–732, 733
  *See also* Oxford University; University of Leeds; University of Nottingham
English language, 362
enhanced curriculum, 218–221
ENIAC (Electronic Numerical Integrator and Calculator), 788
Enochs, Larry G., 666
enrichment. *See* art; books, stories for children; calendars; career information; competitions; enrichment, overview; Escher, Maurits C.; Fibonacci sequence; games; Goldbach's conjecture; golden section; humanistic mathematics; humor; Logo; magic squares; media; Mu Alpha Theta; music; paradox; Pascal's triangle; poetry; probability applications: chances of matches; projects for students; recreations, overview; techniques and materials for enrichment learning and teaching; tiling; writing activities
enrichment, overview, **218–221**
  acceleration vs., 285
  arithmetic programs, 40
  computers and, 220, 738
  FAMILY MATH and *MATEMÁTICA PARA LA FAMILIA* program, 246–248, 286–287, 794–795
  middle school open-ended projects, 471
  techniques and materials, 218–219, 735–739
enrichment clusters, 738–739
Enrichment Triad Model, 735–736
enumeration theorem, 121
environmental mathematics, **221–223**
  *See also* population growth
Eötvös, Lóránd, 124
epidemiology, 77
*Epitome of the Copernican Astronomy* (Kepler), 383
Epstein, Joyce, 794–795
equalities, **223–224**

  *See also* equations, algebraic; inequalities
equal ratios, Eudoxus's theory of, 309
*EQUALS: A Mathematics Equity Program*, 175, 246, 247, 248, 647
equations
  cubic, 224
  differential, 81, 95, 202, 223, 530, 615
  differing degrees, 19
  Diophantine, 149–151, 200–201, 309
  in Egyptian mathematics, 212
  equalities, 223–224
  equivalent, 224
  exponential, 226–227
  first-degree, 212, 225
  higher-order, 19, 114
  linear, 19, 113, 114, 154, 701, 702
  logarithmic, 227–228
  of motion, 215, 387
  nonlinear, 20
  ordinary differential, 521–523
  parametric, 88, 527–529
  partial differential, 73, 181, 529–531
  polar, 88
  quadratic, 19, 66, 144, 200, 228
  quartic, 111, 124, 224
  second-degree, 212, 225
  second-order linear differential, 215
  simultaneous, 137
  stochastic, 539
  systems of, 701–703
  theory of, 272
  wave, 531
  *See also* equations, algebraic
equations, algebraic, 18, **224–226**, 240–242, 272
  Abel's studies, 2
  equalities, 223, 224
  field theory, 7
  groups and, 4–5
  polynomial, 2, 127, 226, 623
  real numbers and, 623
  solving with radicals, 5
  zeros of a function and, 839
equations, exponential, **226–227**
equations, logarithmic, **227–228**
equation-solving capabilities, graphic calculators, 88–89
equilibria, 110
EQUITY 2000 (educational reform initiative), 476, 478, 479–480
equivalent fractions, 261
Eratosthenes, **228**, 320, 558, 668
Erdös, Paul, 510, 589
Erlang, Agner K., 608, 692
*Erlanger Programm* (Klein), 385
Ermakov, I. S., 641
Ernest, Paul, 537
error, paradox distinguished from, 526
error, sampling. *See* standard error
Error Function, 770

errors, arithmetic, **228–231**
 in kindergarten, 384
 rounding, 639
Escalante, Jaime, 796
Escher, Maurits Cornelius, **231**, 506, 760
ESM. *See Educational Studies in Mathematics*
*Essai philosophique sur les probabilités* (Laplace), 393
*Essay on Probabilities, and on Their Application to Life Contingencies and Insurance Offices* (De Morgan), 189
estimation, **231–232**
 in consumer mathematics, 148–149
 in preschool, 554
 in subtraction, 697
ethnicity and race
 international comparisons of women's mathematics education and career problems, 825
 multicultural mathematics issues, 487–489
 NAEP mathematics and science scores, 1982–1992, 453
 testing bias and, 794
 *See also* ethnomathematics; minorities, career problems; minorities, educational problems; *specific groups*
ethnomathematics, **232–234**
 described, 362
 and issues in multicultural mathematics, 487–489
 Mayan numeration, 455–457
Euclid, **234–235**, 318, 623
 algorithms, 23, 308, 462
 axiomatic method, 588
 coordinate geometry, 156
 definitions attributed to Platonic school, 542
 dominance in English mathematics education, 217
 elementary theorems on triangles, 548
 Fifth Postulate (Parallel Postulate), 63, 80, 235, 374, 505
 first four postulates, 503
 Fundamental Theorem of Arithmetic proof, 268
 geometric approach to algebra, 19
 irrational numbers, 623
 limerick about, 335
 number theory, 510, 511
 parallel postulate of, 80, 374
 potential infinite and, 536
 prime number proof, 558
 proofs, 588
 *See also Elements*; Euclidean geometry
*Euclid and His Modern Rivals* (Dodgson), 108
Euclidean geometry, 6, 79, 255
 alternatives, 505–506
 angles, 27
 axioms, 63, 64
 Cartesian coordinates and, 154

Fifth Postulate, 63, 235, 505
 and geometry instruction, 279
 Legendre text, 324
 neutral geometry and, 503
 postulates, 503
 as projective geometry, 580
 proof, 591–592
 software use, 658–659
 standard nineteenth-century U.S. text, 324
 theorems, 115, 156
 *See also* non-Euclidean geometry
Euclidean tools, 543
*Euclides ab omni naevo vindicatus (Euclid Freed of Every Flaw)* (Saccheri), 505
*Eudemian Summary* (Proclus), 309
Eudemus, 309
Eudoxus, 37, 211, 235, 309
Euler, Leonhard, **235–236**, 387, 483
 combinatorics, 120, 121
 diagrams to represent syllogisms, 197
 Diophantine equations and, 200
 on *e* as irrational, 209
 equation using zero, 838
 function concept, 235, 263, 614
 Goldbach's conjecture in letter to, 300
 Königsberg bridge problem solution, 302–303, 626
 magic square problem, 440
 on natural logarithm, 241
 number theory, 510, 511
 partial differential equations and, 530
 perfect numbers, 511
 pi symbol popularization, 235, 241, 541
 quadratic reciprocity law, 277
 relation equation, 269, 544
 Russian translation of text, 641
 topological ideas introduced by, 765
Euler circuit, 303
Euler path, 303
Euler relation, 269
Euler-Cotes formula, 235
evaluation of instruction, issues, **236–238**
evaluation of instruction, overview, **238–240**
 evaluation tools, 745
*Everybody Counts: A Report to the Nation on the Future of Mathematics Education* (1989), 348, 712, 715, 718, 728, 729, 792
Everyday Learning Corporation, 807
Eves, Howard, 489
EXCEL (software), 472
excluded middle, law of the, 537, 586–587
expanded notation, 187
expanding functions, 265–266
experimental programs, *See* Connected Mathematics Project (CMP); Core-Plus Mathematics Project (CPMP); Developing Mathematical Processes (DMP); Developmental Mathematics Program (DMP) and

Apprenticeship Teaching; Interactive Mathematics Program (IMP); Mathematics in Context (MiC); National Center for Research in Mathematical Sciences Education (NCRMSE); research in mathematics education; teacher development, K–12, California Mathematics Project (CMP); teacher preparation, secondary school, Reform Project in California; Teaching Integrated Mathematics and Science Program (TIMS); University of Chicago School Mathematics Project (UCSMP)
experiments, 734, 741
Exploratory Data Analysis (EDA), 686, 690
exploratory learning, 222
exponential decay formula, 227
exponential notation, 145
exponents, algebraic, **240–242**
 as continuous function, 151
 function, 227
 logarithms, 235, 241
 notation, 226, 240, 243
 transcendental and algebraic functions, 766–771
 zero, 838
 *See also* equations, exponential
exponents, arithmetic, **242–244**
*Exposition du système du monde* (Laplace), 393
EXPTEST (software), 748
extended-response test questions, 746
extension theory, 5
extrapolation, **244**
Exxon Education Foundation, 807
Exxon K-3 Mathematics Specialist Program, 175

**F**
factors, **245–246**
 in algebraic expressions, 20, 246
 greatest common divisor, 308–309
 prime, 245, 308
Fahrenheit, Gabriel Daniel, 743
Fahrenheit temperature scale, 707, 743
falling bodies, computation of speed of, 93–94
False Coin Problem, 624–625
FAMILY MATH and *MATEMÁTICA PARA LA FAMILIA*, **246–249**, 286–287, 794–795
Farmer, Walter A., 709
Farr, William, 35
Farrell, Margaret A., 709
Farvarson, James, 640
Fatou, Pierre, 109, **249**, 382
Fawcett, Harold, 64, 201
feasible set (feasible region), 703
Federal Communications Commission, 210

Federal Coordinating Council for Science, Engineering, and Technology (FCCSET), 804
federal government, role, 79, **249–250**
  economic analysis, 210
  Eisenhower program, 213–214
  field activities sponsorship, 252
  Goals 2000: Educate America Act, 298–299, 793
  minorities education, 475
  teacher preparation, 720
  United States Department of Education, 193, 213, 801–805
  United States National Commission on Mathematics Instruction, 805
  *See also* National Science Foundation; *specific legislation and organizations*
Fehr, Howard Franklin, **250–251**, 329
Fennema, Elizabeth, 176, 469–470, 498
Fermat, Pierre de, 154, **251**, 560, 589, 626
  challenge problem of, 338
  geometric representation of functions, 305
  number theory, 510, 511
  Pascal correspondence, 531
  positive integers, 511
  probability theory, 251, 531
Fermat primes, 548
Fermat's Last Theorem, 6, 22–23, 251, 281, 450, 510
  description of, 626
  Wiles solution, 512, 626
Feynman, Richard, 540
fiber bundle, 766
Fibonacci, Leonardo, 251–252
Fibonacci sequence, 78, **251–252**
  in binomial theorem, 76
  Lagrange and, 388
  software for study of, 662
  transcendental numbers and, 771
field activities, **252–254**
  enrichment, 219
field theory, 7, 189
fields, 189, **254–255**
  abstract algebra, 7
  algebraic numbers, 21–23, 128
  of complex numbers, 255, 268
  isomorphism, 375
  properties of addition, 10–11, 254
Fields Medals, 141
15-Puzzle, 625
Fifth Postulate (Parallel Postulate), 63, 80, 374
  non-Euclidean geometric contradictions, 235, 505, 580
finance, **255–256**
  as career, 105
  consumer mathematics and, 148–149
finite fields, 254
finite graphs, 301
finite mathematics courses, as liberal arts staple, 418
finite planes, 581
finite projective plane, 581

finite series, 651
Finland, international studies of mathematics education, 365
First International Mathematics Survey (FIMS), 360, 365
First International Study of Educational Achievement, 379
*First Lessons in Arithmetic on the Plan of Pestalozzi with Some Improvements* (Colburn), 119, 536
Fisher, George (pseud. for Mrs. Slack), 542
Fisher, Ronald A., 163, 684, 688, 689–690
Five Colleges' Calculus in Context curriculum, 172
Fixed Date Birthday Problem (probability matches), 563, 564
fixed point theorems, 765
Flanders, Ned A., 365, 366
Fletcher, J. D., 341
Florida, 454, 649–650
Florida State University, 271
flowcharts, 220
  algorithms and, 23
Focus-Balanced Incomplete Block (BIB) Spiraling, and NAEP mathematics assessment, 496
Folium of Descartes, 58
*For All Practical Purposes* (video series), 465
Ford Motor Company, 806–807
forecasting, and time series analysis, 762–764
formal proof, 586, 591
formalism, 588, 646
  Hilbert and, 316, 537
*Formulario* project, 533
formulas, **256–257**
FORTRAN (programming language), 131, 658
Fortuny, Jose M., 665
Fosnot, Catherine Twomey, 694
*Foundations for the Future* (1994), 649
foundations of mathematics. *See* mathematics, foundations
Four Color Theorem, 134, 220, 450, 589, 626–627
Fourier, Jean Baptiste Joseph de, **257–258**, 522
Fourier, Joseph, 318, 704
  *See also* Fourier, Jean Baptiste Joseph de
Fourier equation, 257
Fourier series, 214, 257
fractals, 67, 203, **258–259**
  chaotic processes, 109
  definition of, 258, 260
  musical composition using properties of, 493
  similarity and, 656
Fraction Bars, 261
fractions, **259–262**
  addition of, 9
  for batting averages analysis, 671–673

continued, 149–151
division of, 207, 261–262
in Egyptian mathematics, 211–212, 259–260, 615–616
exponents, 243
greatest common divisor, 308–309
multiplication of, 261–262, 489–490
percent conversion from, 534–535
rational numbers as, 615–616
sexagesimal, 615
sports applications, 671–673
whole-number interpretations, 616
Fraenkel, Abraham, 6
France, **262–263**
  international comparisons of mathematics curriculum, 361
  international studies of mathematics education, 365
Franklin, Benjamin, 39, 168
Freedman, Michael, 766
Freeman, Donald, 751
Frege, Gottlob, 103, 537, 639
French Revolution, 247
Frenet, Celestin, 2
Freudenthal, Hans, 211, 454
Freudenthal Institute, 454
Frieze patterns, 773
Froebel, Friedrich, 552
Fry, Thornton C., 31
Frye, Shirley M., 500
Fujizawa, Rikitaro, 378
Fuller, Thomas, 16
function graphs, 305
functions, **263–265**, 785
  algebraic, 209, 266
  alternating symmetric, 192
  asymptotic, 266, 267
  Bessel, 214, 770
  circular, 189, 785–788
  complex, 108, 109, 246
  of complex variables, 215
  continuous, 151, 266, 267
  counting, 120
  data sets and, 267
  definition of, 447
  derivative of, 614
  domain of, 266
  elliptic, 534
  as Euler's focus in calculus, 235, 263, 614
  expanding, 265–266
  explicit, graphs of, 305–306
  exponential, 151, 227
  Fourier, 257
  generation of, 120
  graphs of implicit, 306
  hyperbolic, 73
  inverse, 370–372
  logarithmic, 151
  modern idea of, 437
  "nice," 151
  partition, 611
  probability density and, 566
  probability distribution, 545

properties of, 266
range of, 266
rational, 151
"salt and pepper," 152
sinusoidal, 539
smoothness of, 151
trigonometric, 73, 235, 264, 266
zeros of, 838
functions of a real variable, **265–268**, 813
of two variables, 307
function spaces, 421
Fund for the Improvement of
Postsecondary Education, 247, 721
Fundamental Duality Theorem, 425
fundamental operations
addition as, 9
arithmetic, 38
*See also:* basic skills; division;
multiplication; subtraction
Fundamental Principle of Counting
(probability matches), 562, 563
Fundamental Theorem of Algebra, 127,
268, 277
Fundamental Theorem of Arithmetic,
268, 510, 558
Fundamental Theorem of Calculus, 268,
352–353, 437, 459
fundamental theorems of mathematics,
**268–269**
of algebra, 127, 268, 277
of arithmetic, 268, 510, 558
of calculus, 268, 352–353, 437, 459, 614
*See also* binomial theorem; Pythagorean
Theorem

**G**
Gabel, Dorothy L., 666
Gage, Paul, 558
Gagné, Robert Mills, **271**
theory of learning hierarchies, 271, 329
as University of Maryland Mathematics
Project consultant, 808
use of operant conditioning ideas, 271,
656
Gale, David, 275, 425
Galilei, Galileo, 77, **271–272**, 318, 536
Galois, Evariste, 255, **272**
abstract algebra and, 4, 5, 272
field theory, 7
group theory and, 418
polynomial equations and, 278
Galton, Francis, 163, 288, 553
normal distribution curve and, 506
statistics used by, 689
Game of Life, 35
game theory, 202, **272–274**
use by economists, 210
von Neumann and, 210, 272, 274, 276,
818
games, **274–277**
for algebra teaching, 17
application of African customs, 17
applications for secondary school
classroom, 34, 35

computerized, 662
for enrichment learning and teaching,
737
game theory, 202, 210, 272–274, 818
recreational mathematics, 623–627
subtraction and, 698
games of chance, 273, 444–445, 471
Gamma Function, 770
Gani, Joseph, 446
Gardner, Howard, 411, 412
Gardner, Martin, 624, 760
Garfield, Joan, 733
Garfield High School (Los Angeles), 796
Gateways III Conference, 649
Gattegno, Caleb, 59, 167, 553, 692
Gauss, Carl Friedrich, **277–279**
Binomial Theorem proof, 269
Bolyai correspondence, 318
Dedekind as student of, 189
distribution of prime numbers and, 313
Euclid's Fifth Postulate and, 235, 505
on Euler's mathematical contribution,
235
field of complex numbers, 254–255,
268
Fundamental Theorem of Arithmetic
and, 268
Germain correspondence, 281, 822
importance in mathematics, 36, 108,
235, 819
least squares and, 278, 403
non-Euclidean geometry development
and, 80, 235, 318, 505, 580
normal distribution curve applied to
astronomical errors by, 506
number theory and, 277, 510
patterns in mathematics and, 320
Poincaré compared with, 544
polynomial of positive degree theorem,
587
ratio test for convergence of
hypergeometric series, 651
regular polygon construction, 548
*See also* Gaussian elimination
Gaussian (normal) distribution, 278
Gaussian elimination, 278, 420, 422, 423,
702
Gaussian integers, 6, 67, 278
GCD. *See* greatest common denominator
GE Foundation, 807
Gelfand, Israel Moiseyevich, 292
Gelfond-Schneider Theorem, 512
gelosia method of multiplication, 490–491
gender. *See* women and mathematics
*headings*
General Educational Development test, 12
*Geographika Syntaxis* (Ptolemy), 310
*Geometer's Sketchpad, The* (software), 280,
471, 472, 659, 660, 663
geometric mean. *See* mean, geometric
*Geometric Supposers* (software), 472,
658–659
Geometric Supposer Series (software),
135

geometric transformations, 493, 775–778
*Geometry* (Legendre), 324
geometry
absolute, 503
algebra and, 19
alternative plane, 80
analytic, 142–144, 190, 251, 265–266
Archimedes and, 37
area computation, 66
axioms, 63, 235
Birkhoff postulates, 78
Birkhoff textbook, 78–79
Cartesian, 305
combinatorial, 581
concrete, 123
conic sections, 142–144
Connected Geometry Project, 649
coordinate, 154–159
discovery approach to, 201–202
Egyptian mathematics and, 212, 319
elliptic, 580
formal, 123
Greek mathematics and, 309, 310
hyperbolic, 580
lattice, 78
Lobachevsky's work, 427
neutral, 503
non-Euclidean. *See* non-Euclidean
geometry
paradox in, 527
pi's importance in, 541
Plato and, 542–543
postulates of, 63, 78, 235, 448, 505
practicality of, 32
preschool activities, 555–556
projections, 579–580
projective, 370, 385–386, 580–581
proof, 591–592
Pythagorean theorem, 602
solid, 37, 144
space, 665–668
symmetry groups, 311
transformational, 17, 385–386,
722–775
*See also* angles; circles; Euclidean
geometry; fractals; geometry,
instruction; graphs and graphing;
length, perimeter, area, and
volume; polygons
*Geometry* (Legendre), Davies translation,
324
geometry, instruction, **279–281**
measurement and, 409
National Center for Research in
Mathematical Sciences Education
Study, 497
NCTM standards, 648
*Perspectives* (visual geometry activity),
741
preschool activities, 555–556
software, 472, 657, 658–659
in solar system studies, 668
space geometry, 665–668
van Hiele levels, 809–812

*Geometry, The* (Descartes), 190
*Geometry from Multiple Perspectives*
  (Addenda Book, NCTM), 649
geometry of absolute space. *See* non-
  Euclidean geometry
*Geometry Tutor* (software), 657
Georgia, 797–798
Gerdes, Paulus, 319, 487
Germain, Sophie, **281**
Germany, **281–283**
  advanced placement program, 13
  international comparisons of women's
    mathematics education and career
    problems, 826
  in international studies comparisons,
    365
  Third International Mathematics and
    Science Study videotaped
    classroom observational studies in,
    804
*Gesellschaft fuer Didaktik der Mathematik
  (GDM)*, 282
Gestalt theory, 570, 600, 678
Gibbons, Maurice, 339
Gibbs, Josiah W., 816
gifted, **283–288**
gifted, special programs, 285–286,
  **288–294**
*Gifted Child Quarterly, The* (journal), 292
Gilligan, Carol, 827, 829
Ginsburg, Herbert, 146
Girard, Albert, 268
Glasersfeld, Ernst von, 146
Glickman, Leslie, 814
gluons, 540
*Goals for School Mathematics* (1963
  report), 100
goals of mathematics instruction, 41, 165,
  **294–298**
Goals 2000, **298–299**, 793
Gödel, Kurt Friedrich, 269, **299–300**,
  536
  on axiomatic systems, 299, 316, 449
  Incompleteness Theorem, 103, 299,
    537
  number theory proof, 450
Gödel numbering, 299
Godfrey, Charles, 669
Goldbach, Christian, 209, 300, 512, 559
Goldbach's conjecture, **300**, 512
Goldberg, Howard, 734–735
Golden Mean, 220
golden ratio, 300
golden section, 252, **300–301**
Goldin, Gerald, 670
Goldman, A. J., 425
Goldstein, Sue, 795
Gordon, Cameron, 764
Gorenstein, Daniel, 5
Gossett, W. S., 612
Gou Gu Theorem, 113
government. *See* federal government, role;
  mathematics education, statistical
  indicators; state government, role

Graduate Record Examination (GRE)
  advising and, 15
  minority students' scores, 474
*Graduate Studies in Mathematics* (AMS
  report), 25
Grandgenett, Neal, 286
*Graph Explorer* (software), 660
graph of the equation, 306
graph theory, overview, **301–305**
  combinatorics relationship, 121
  discrete mathematics and, 202, 203
  Euler's solution of "bridges of
    Königsberg" problem and, 626
graphical interpolation, 368
graphics software, 472
graphing calculator, 88–89, 739–740
  for coordinate geometry problems, 154
  for data analysis, 181
  for environmental mathematics, 222
  for exponential equation, 227
  for functions of real variable, 267
  integration into curriculum, 48, 221
  for logarithmic equations, 20, 21, 228
  for polynomial equations, 224, 226
  for remedial mathematics, 194
  special features, 88–89, 739–740
  for systems of inequalities, 703
graphs, **305–308**
  area calculation, 95–98
  data representation, 184, 185, 186
  elementary school use of, 690
  paths through, 626
  polynomial equations, 226, 702
  preschool activities with, 555
  software, 463
  in test construction, 747
  of velocity, 266
Grassman, Hermann, 816, 820
Graunt, John, 689
gravity, 538, 540, 628
Gray, Mary, 56
GRE. *See* Graduate Record Examination
Great Britain. *See* England
Great Depression (1930s), 327–328
greatest common divisor (GCD), **308–309**
*Greedy Triangle, The* (Burns), 355, 414, 415
Greek mathematics, **309–310**
  Archimedes, 36–37
  Egyptian mathematics and, 211
  Eratosthenes, 228, 320, 558, 668
  Euclid, 234–235
  exponents and, 243
  harmony of the spheres and, 492, 602
  irrational numbers discovery, 372
  logical deductive reasoning, 448
  nature of mathematics and, 449–450
  number theory, 510
  perspective theory, 580
  placeholder, 838
  proof concept, 588
  Pythagoras, 602
  ratios, 593
  Zeno, 837
  Zeno's paradoxes and, 527

*Green Globs* (computerized game), 135,
  662
*Green Globs and Graphing Equations*
  (computerized game), 662
Greenwood, Isaac, 324, 542, 677,
  755–756
Gregory, David, 550
Gregory, James, 269, 541
grouping (instructional)
  ability, 2, 3
  in cooperative learning, 152, 153
  *See also* tracking
grouping (mathematical)
  in arithmetic, 38
  Egyptian numeral system, 211
  in number sense and numeration, 508
group theory, **311–312**
  Abel's work, 2
  abstract algebra and, 4–5
  coding theory and, 117
  Galois's work, 418
  isomorphism and, 375
  Lie's work, 418–419
  recreational puzzles, 625
  research in, 5
  set theory and, 653
Growney, JoAnne, 543–544
Grünbaum, Branko, 760
*Grundgesetze der Arithmetik* (Frege), 639
GSP software. *See Geometer's Sketchpad,
  The*
GTE Corporation, 807
*Guess My Rules* (computerized game), 662
guided learning, 271
*Guide for Reviewing School Mathematics
  Programs, A* (Blume and Nicely),
  751
Guidelines for Science and Mathematics
  in the Preparation Program of
  Elementary School Teachers, **312**
Guidelines for Science and Mathematics
  in the Preparation Program of
  Teachers of Secondary School
  Science and Mathematics, **312**
*Guidelines for the Continuing Mathematical
  Education of Teachers* (MAA), 711,
  716, 718
*Guidelines for the Preparation of Teachers of
  Mathematics* (NCTM; 1993), 727
*Guide to Math and Science Reform* (1995),
  175
*Guide to Math and Science Reform*
  (Annenberg/CPB 1995), 647, 649
Guilford, Joy P., 118
Gutiérrez, Angel, 665
Guy, Richard K., 760

**H**
Hadamard, Jacques Salomon, **313**, 558
Haertel, Geneva, 469
Haier, Richard J., 287
Haken, Wolfgang, 588–589, 626, 627
half-angle formulas, 783
half-turn, 772, 776

Hall, G. Stanley, 326, 553
Halley, Edmund, 9, 387
Hamilton, William Rowan, 6, 127, 626, 820
Hammurabi, 224
Hampshire College Summer Studies in Mathematics, 290
Hancock, Lynn, 347
*Handbook of Research on Curriculum* (Jackson ed.), 751
*Handbook of Research on Mathematics Teaching and Learning* (Grouws ed.), 632, 751
*Handbook of Task Analysis Procedures* (Jonassen, Hannum, and Tessmer), 709
*Hands-On Math* (software), 71
hands-on mathematics and science programs. *See* field activities
Hanna, Gila, 367
Hardy, Godfrey H., 510, 537, 559, 611
Harkness table, 350
harmonic triangle, 532
harmonies, musical, 592, 602
harmony of the spheres, 492, 602
*Harmony of the World* (Kepler), 383
Harper, William Rainey, 141
Harriot, Thomas, 342
Harris, Lauren J., 280
Harris, Marieta, 832
Harris, Sydney, 335
Harris, William T., 123
Hart, Therese A., 792
Harvard University
    Center for Cognitive Studies, 84
    in history of mathematics education, 323–324
    mathematics entrance requirements, 677
    Russell's delivery of Lowell Lectures, 639
    Whitehead's professorship, 820
Hausdorff, Felix, 259
Hawkins, Anne, 814
Hawkins, David, 59
Hawley, Newton S., 698
health, **313–315**
    index numbers and, 337
    Nightingale's medical statistics, 822
    nursing's use of mathematics and, 517
    women's educational paths and careers, 825
*Heeding the Call for Change: Suggestions for Curricular Action* (Steen ed.), 443
Hein, Piet, 625
Heine, Heinrich E., 653
Heisenberg, Werner, 423
Hellman, Martin, 166, 559
Helmholtz's equations, 531
Herbart, Johann Friedrich, 669
Hermite, Charles, 209, 771
Heron of Alexandria, 538
Herschel, Caroline, 821
Hersh, Reuben, 537

hexadecimal system, **315–316**
hidden surface mode, 307
Hiebert, James, 388, 389, 497
Hiele-Geldof, Dina van, 64
    *See also* van Hiele-Geldof, Dina
Hiele, Pierre van, 64
    *See also* van Hiele, Pierre M.
hieroglyphics, in Mayan numeration, 455–457
high school mathematics. *See* secondary school mathematics
higher education. *See* colleges and universities
higher transcendental functions, 770
Hilbert, David, **316**
    consistency goal, 103, 299
    as formalist, 316, 537
    Gödel Incompleteness Theorem and, 103, 299
    modern mathematics and, 316, 448–449, 484
    Noether association, 504
    Paris Congress presentation, 141, 316
    ring theory and, 6
    Turing's work and, 788
    Weyl as student, 819, 820
Hilbert's hotel, 543
Hilbert spaces, 382
Hill, Shirley, 32
Hill codes, 167
Hindu-Arabic place value system. *See* decimal system
Hindu mathematics. *See* Indian mathematics
Hipparchus, 338, 784
Hippocrates of Chios, 318
Hirstein, James J., 666
Hispanics
    career problems, 474–476
    math anxiety, 451
    NAEP mathematics and science scores, 1982 to 1992, 453
    underachievers, 791, 792
histograms, 185
history of mathematics, overview, **316–323**
    Abel, Niels, 2, 4, 108, 279, 311, 419
    Archimedes, 36–37
    Aristotle, 37, 63, 272, 447, 448, 614
    Babbage, Charles, 65, 87, 318, 514, 822
    Babylonian mathematics, 65–66
    Bernoulli family, 73
    Bolyai, Janos, 79–80
    Boole, George, 81–82
    Brownell, William A., 83–84
    Cantor, Georg, 102–103
    Carroll, Lewis, 107–108, 198
    Cauchy, Augustin-Louis, 108
    Chinese mathematics, 113–114, 269, 318
    Colburn, Warren, 119
    D'Alembert, Jean, 181
    Davies, Charles, 186–187
    Dedekind, Richard, 5–7, 189, 623, 653, 654, 837

De Moivre, Abraham, 189
De Morgan, Augustus, 189–190
Descartes, René, 190
Dewey, John, 197
Diophantus, 200–201
Dodgson, Charles (Lewis Carroll), 107–108
Egyptian mathematics, 211–212
    in England, 216–218
Eratosthenes, 228, 320, 558, 668
Euclid, 234–235
Euler, Leonhard, 235
Fatou, Pierre, 109, 249, 382
Fermat, Pierre de, 251
Fourier, Jean, 257–258, 522
Galilei, Galileo, 77, 271–272, 318, 536
Galois, Evariste, 4, 5, 7, 255, 272, 278, 418
Gauss, Carl Friedrich, 277–279
Germain, Sophie, 281, 822
Greek mathematics, 309–310, 372, 689
Hadamard, Jacques S., 313, 558
Hypatia, 310, 821
Indian mathematics, 261, 319, 320, 338–339
Islamic mathematics, 260, 268, 319, 373–374, 760
Julia, Gaston, 109, 249, 382
Kepler, Johannes, 383–384, 580, 760
Kovalevskaia, Sofia, 386, 822
*Ladies' Diary*, 387
Lebesque, Henri, 404–405
Legendre, Adrien Marie, 405
Leibniz, Gottfried Wilhelm, 405–406
L'Hospital, Guillaume, 192, 417
Lie, Sophus, 418–419
Liouville, Joseph, 427, 771
Lobachevsky, Nikolai Ivanovich, 427
Maclaurin, Colin, 192, 437, 550
Mayan numeration, 319, 455–457
modern mathematics, 483–484
Napier, John, 495–496
Newton, Isaac, 503–504
Noether, Emmy, 7, 504–505, 822, 823
Pascal, Blaise, 531–532
Peano, Giuseppe, 533
Peirce, Charles S., 6, 325, 533–534
Plato, 542–543
Poincaré, Jules H., 316, 318, 544–545
Pythagoras, 602
Ramanujan, Srinivasa, 235, 320, 511, 611
Riemann, Georg Friedrich Bernhard, 637–638
    in Russia, 640–642
    statistical theory, 689
Turing, Alan M., 299, 423, 788
Weierstrass, Karl, 819
Weyl, Hermann, 819–820
Whitehead, Alfred N., 103, 588, 639, 820
Wiener, Norbert, 820–821
Zeno, 837
zero perceptions, 66, 187, 211, 838

*See also* Einstein, Albert; Gödel, Kurt; Hilbert, David; Lagrange, Joseph L.; Laplace, Pierre S.; Russell, Bertrand; von Neumann, John; women and mathematics, history

*History of Mathematics* (Smith), 656

history of mathematics education in the United States, overview, **323–330**

Begle, Edward G., 72–73, 328, 643

Birkhoff, George D. 78–79

Brooks, Edward, 83

Brownell, William Arthur, 83–84

Bruner, Jerome, 84

Cambridge Conference on School Mathematics, 100

Colburn, Warren, 119

College Entrance Examination Board (CEEB) Commission Report, 120

Commission on Postwar Plans, 122

Committee of Fifteen on Elementary Education, 83, 123

Committee of Ten, 123, 168

Committee on the Undergraduate Program in Mathematics, 123–124, 727

curriculum standards, 676–682

Davies, Charles, 186–187

Dewey, John, 197, 326

Dienes, Zoltan P., 198

early American textbook samples, 753–757

Fehr, Howard F., 250

first new math program for secondary schools, 807–808

Gagné, Robert M., 271

Guidelines for Science and Mathematics, 312

James, William, 197, 377

Joint Commission Report (1940), 381–382

Montessori, Maria, 485–486, 552, 553

Moore, Eliakim H., 326, 486

National Committee on Mathematical Requirements, 498–499

National Council of Teachers of Mathematics, historical perspective, 39, 501

National Science Foundation, historical perspective, 502–503

Pestalozzi, Johann H., 119, 536, 552

Pike's arithmetic textbook, 542

Progressive Education Association (PEA) Report, 578–579

*Revolution in School Mathematics*, 637

School Mathematics Study Group, 328, 643

School Science and Mathematics Association, historical perspective, 644

secondary education reform, 718

Smith, David Eugene, 656–657

Stern, Catherine, 692

Suppes, Patrick, 128, 698

University of Illinois Committee on School Mathematics 328, 679, 807–808

University of Maryland Mathematics Project, 808

Young, Jacob William Albert, 835

*See also* Klein, Felix; psychology of learning and instruction, overview; textbooks; Thorndike, Edward L.

*History of Western Philosophy, A* (Russell), 639

Hodder, James, 542

Hodgkin, Alan, 77

homeomorphism, 764, 765

homework, **330–331**

homogeneous coordinates, 581

Hong Kong, 826

Hoover, Herbert, 801

Horner's Method, 220

Houang, Richard T., 666, 667

*How Gertrude Teaches her Children* (Pestalozzi), 536

howler, paradox vs., 526

*How to Evaluate Mathematics Textbooks* (NCTM; 1982), 750

Hua Luo Geng, 114

Hull, William P., 59

humanistic mathematics, **331–334**, 537

poetry and, 543–544

*Humanistic Mathematics Network Journal*, 332

humor, **334–335**

Hungary, 367

Hurkle problem-solving program, 130

Husén, Torsten, 366

Hutton, Charles, 216, 217

Huxley, Andrew, 77

Huygens, Christian, 483, 538, 560

hyberbolic geometry, 580

hydraulic pressure, Bernoulli principle of, 73

hydrodynamics, 73

*Hydrodynamica* (Bernoulli), 73

hydrostatics, 37

Hypatia, 310, 821

hyperbolas, 142, 143

equilateral, 57, 58

reflective properties, 144

hyperbolic differential equations, 530

hyperbolic functions, 770

hyperbolic geometry, 503, 505, 506, 580

hypothesis testing, 684, 698

## I

*IASE Review*, 733

IBM computer, 128

Ibn al-Haytham, 374

ICMI-China Regional Conference on Mathematics Education, 805

ideal, definition of, 6–7

identities, 252

Illinois

Algebra Project, 797–798

Arithmetic Project, 679

curricular reform projects, 328, 679

Scientific Literacy, Teaching Integrated Mathematics and Science Program funding, 734

*See also* University of Chicago; University of Illinois

Illinois State University, 716

imaginary numbers, 189

*See also* complex numbers

*Imagine* (journal), 289

IMP. *See* Interactive Mathematics Program

Improving America's Schools Act of 1965, 803

IMU. *See* International Mathematical Union

inch, historical background of term, 704–705

inch-pound system (customary system) of measurement, 704, 705

incidental learning theory, 39, 84, 553, 678

incommensurable magnitudes. *See* irrational numbers

Incompleteness Theorem, 103, 299, 537

indefinite integration, 352

independent projects. *See* projects for students

index numbers, **337–338**

Indian mathematics, 319, **338–339**

combinatorics, 120–121

equality sign, 224

Kerala school, 550, 650

origins of fraction symbolization, 261

placeholder, 838

Ramanujan, Srinivasa, 235, 320, 511, 611

value of pi, 320

zero origination, 38

*See also* decimal system

indirect proof, 235, 587

intuitionist rejection of, 820

individualized instruction, 193–194, **339–342**

Individually Prescribed Instruction (IPI) project, 340

inductive proof, 73

industrial mathematicians, 104–105

inequalities, **342–344**

systems of, 703–704

*See also* equalities

inference

basic rule of, 586

*See also* statistical inference

infinite continued fractions, 150

infinite graphs, 301

infinite numbers

algebraic, 21

infinitesimals relationship, 536

infinite processes, 90, 309

infinite products, for pi, 541

infinite series, 543, 651

completed vs. potential, 536

convergence of, 2

for pi, 541

*See also* Maclaurin expansion; series

infinite sets, 654
  continuity and, 152
  as philosophical question, 536, 537
infinitesimals, 220, 536, 837
infinitesimal triangle, 531
infinity
  Aristotle and, 37, 536
  and asymptotes 57–59
  concept of, 37
  Continuum Hypothesis, 299
  in division by 0, 207–208, 838
  and $e$, 209
  Eudoxus and, 309
  parallel-line convergence, 580
  as philosophical question, 536
  potential, 37
  real numbers and, 623
  vanishing point, 580
  See also Cantor, Georg; Maclaurin
    expansion; paradox; series; Zeno
informal measurement, 555
information age, 136, 294–298
inference, basic rule of, 586
Inhelder, Bärbel, 618
innovative exploratory software, 658
inquiry-based classroom discourse,
  389–390
inservice teacher training, 718
  in statistics, 733
  See also teacher development headings
instantaneous change concept, 838
Instant Insanity (puzzle), 625, 627
Institut fuer Didaktik der Mathematik der
  Universität Bielefeld (IDM), 282
Institute for Advanced Study, Princeton
  University, 504, 820
Instructional materials. See abacus; ability
  grouping; adult education; algebra,
  introductory motivation; algebra
  curriculum, K–12; attribute blocks;
  calculators; computer assisted
  instruction (CAI); computers;
  cooperative learning; creativity;
  Cuisenaire rods; discovery approach
  to geometry; errors, arithmetic;
  estimation; field activities;
  geometry, instruction; homework;
  individualized instruction;
  instructional materials and
  methods; instructional methods,
  current research; instructional
  methods, new directions; language
  of mathematics in the classroom;
  language of mathematics
  textbooks; length, perimeter, area,
  and volume; lesson plans; Logo;
  manipulatives, computer
  generated; money; number sense
  and numeration; patterning; Pólya,
  George; probability problems,
  computer simulations; problem-
  solving, overview; problem-solving
  ability and computer programming
  instruction; questioning; reading

mathematics textbooks; remedial
  instruction; software; space
  geometry, instruction; Stern,
  Catherine; task analysis;
  technology; textbooks; textbooks,
  early American samples; tutoring;
  Van Hiele levels; writing activities;
  see also entries under headings of
  mathematical topics
instructional materials and methods,
  elementary, 344–345
  addition, 11
  algebraic equations, 224–225
  arithmetic, 41–42
  basic skills, 70–71
  decimal system, 188
  for measurement systems, 708
  programmed materials, 220
instructional methods, current research,
  345–348
  cooperative learning groups, 346, 350
  goals of, 294–298
  See also evaluation of instruction, issues;
    evaluation of instruction, overview
instructional methods, new directions,
  348–352
  writing activities, 832–833
insurance. See actuarial mathematics
Integer Cube, 627
integers
  exponents, 242
  Gaussian, 6, 67, 278
  in group theory, 311
  positive, 242, 533, 536, 622
integral calculus, 73, 98, 214, 268, 533
integrals, 73, 257
  probability, 189
integrated circuit, 88
integrated learning, 170
integration, 97, 352–354
  indefinite and definite, 352
  Mean Value Theorems, 458–459
integration of elementary school
  mathematics instruction with other
  subjects, 354–356
integration of secondary school
  mathematics instruction with other
  subjects, 356–357
Intel (formerly Westinghouse) Science
  Talent Search, 291
intelligence quotient. See IQ
intelligence tests, normal distribution
  curve, 507
interactive learning, 169
Interactive Mathematics Program (IMP),
  33, 179, 357–359, 649
Interactive Mathematics Text Project, 443
interactive multimedia, 130–131, 463
interactive television (ITV), 349–350
interactive video, 741
Inter American Council on Mathematical
  Education (IACME), 364
interdisciplinary programs. See
  biomathematics; environmental

mathematics; integration of
  elementary school mathematics
  instruction with other subjects;
  integration of secondary school
  mathematics instruction with other
  subjects
interdisciplinary teams. See team teaching
interest, 359–360
  actuarial mathematics and, 7, 8
  consumer mathematics and, 148–149
  economics and, 210
  finance and, 255
interest development centers, enrichment
  learning, 737–738
International Actuarial Association, 8
International Assessments of Educational
  Progress (IAEP), 364
International Association for Statistical
  Education (IASE), 732, 733
International Association for the
  Evaluation of Educatonal
  Achievement (IEA)
  mathematics studies, 360, 361, 364,
    365, 366
  Third International Mathematics and
    Science Study, 365, 758–759
International Commission on
  Mathematical Instruction (ICMI),
  364
  United States National Commission on
    Mathematics Instruction as liaison
    to, 805
International Commission on the
  Teaching of Mathematics,
  American Committee, 327
international comparisons, overview,
  101–102, 360–363
  See also Canada; China, People's
    Republic; England; France;
    Germany; international
    organizations; international studies
    of mathematics education; Israel;
    Japan; Russia; Third International
    Mathematics and Science Study
    (TIMSS); women and
    mathematics, international
    comparisons of career and
    educational problems
International Conferences on
  Mathematical Education,
  732
International Conferences on Teaching
  Statistics, 561–562, 732,
  733
International Congress of Mathematicians
  (1932), 823
International Congress of Mathematics
  (1900; Paris), 141, 316, 639
International Congress of Mathematics
  (1912), 327
International Congress on Mathematics
  Education (ICME), 141, 364
International Congress on the Teaching of
  Mathematical Modelling, 32

International Editorial Board, 211
International Group for the Psychology of Mathematics Education (PME), 364
International Linear Algebra Society (ILAS), 420
International Mathematical Olympiad, 286, 291, 364, 443
International Mathematical Union (IMU), 141, 364, 805
International Organization of Women and Mathematics Education (IOWME), 364
international organizations, **364**
    congresses, 141–142
International Science and Engineering Fairs, 291
International Statistical Institute, 732
*International Statistical Newsletter*, 733
*International Statistical Review*, 733
International Studies of Mathematics Achievement, 825
international studies of mathematics education, **364–368**
    Canada and, 101–102
    Germany and, 281–283
    International Commission on the Teaching of Mathematics and, 327
    Israel and, 375–376
    Japan and, 377–381
    methodologies of, 365, 367
International Study Group for Research on Learning Probability and Statistics Concepts, 733
International Study Group for Research on Teaching and Learning Statistics, 691
International Tournament of Towns, 125, 291
internet and world wide web, 463–464, 465
    as electronic communication, 134, 349–350, 464, 740
    electronic journals, 691, 733
    mathematics education applications, 350, 740–741
    PBS Mathline Middle School Math Project, 716
    as tool for teacher development, 719
internships, for mathematically gifted, 290
interpolation, **368–369**
Interstate Commerce Commission, 210
interval estimate, 687
interventions
    for learning disabilities, 401–402
    for underachievers, 797–799
*Introductio in Analysin Infinitorum* (Euler), 235, 241
*Introduction to Mathematical Philosophy* (Russell), 639
*Introduction to the History of Mathematics, An* (Eves), 489
intuitionism, 537, 588, 820
invariance, **369–370**

Inverness Research Associates, 712, 714
inverse functions, **370–372**
Investigation of Mathematically Advanced Elementary Students (IMAES), 286
Iowa
    mathematics proficiency scores, 1992, 453
    university program for the gifted, 289
Iowa State University, 289
IQ (intelligence quotient)
    to determine mathematically gifted, 285, 287
    index numbers and, 337
    normal distribution curve, 507
irrational exponents, 241, 242
irrational numbers, 309, **372–373**
    Dedekind cuts, 189, 623
    discoveries of, 372, 543, 622–623
    *e*, 209
    pi, 541, 623
    Plato on, 543
    real numbers and, 622–623
Islamic mathematics, 319, **373–374**
    Binomial Theorem, 268
    fraction notation, 260
    perspective, 580
    tiling, 760
    zero use, 187
    *See also* decimal system
isometry, 699, 772, 776–777
    musical composition and, 493
isomorphism, **375**
Israel, 365, **375–376**
*Issues in Mathematics Education* (AMS), 25
*Istituzioni Analitiche* (Agnesi), 821
*I Think, Therefore I Laugh* (Paulus), 334

**J**

Jacobi, Carl Gustav Jacob, 192, 316, 530
Jacobs, Judith E., 830
Jacquard, Joseph, 87
Jaime, Adela, 665
James, William, 197, **377**
Japan, **377–381**
    abacus use, 1, 377–378, 379
    in international comparisons of mathematics curriculum, 361, 362
    in international studies of mathematics education, 365, 366
    quality control, 605
    sociocultural and linguistic influences on mathematics learning and achievement, 390
    Third International Mathematics and Science Study videotaped classroom observational studies in, 804
    translations of University of Chicago School Mathematics Project, 806
*Jasper Software* (multimedia software), 662
Java (programming language), 133, 577
Jefferson, Thomas, 324
Jena Plan, 2

Jennings, Dennis L., 815
Jiang Ze Han, 114
Johns Hopkins University, 325
    denial of Ph.D. to woman student, 822
    programs for the gifted, 288, 289
Johnson, C. E., 172
Johnson, David R., 414
Johnson, Lyndon, Taskforce on Education creation by, 802
Johntz, William, 357, 799
Joint Commission Report (1940), **381–382**
Joint Policy Board for Mathematics (JPBM), **382**
Jolliffe, Flavia, 814
Jones, William, 541
Jordan, Camille, 5, 422, 423
Jordan, Wilhelm, 422
Joseph, George Gheverghese, 319, 362
*Journal de Mathématique (Liouville's Journal)*, 427
*Journal for Research in Mathematics Education*, 501
*Journal fuer Mathematik-Didaktik* (German journal), 282
*Journal of Statistics Education* (electronic journal), 691, 733
JPBM. *See* Joint Policy Board for Mathematics
*juku* (Japanese private after-school classes), 380
Julia, Gaston, 109, 249, **382**
Julian C. Stanley Mentor Program (JCSMP), 286
Junior Engineering Talent Search (JETS), 291
junior high school
    enrichment program, 220
    establishment of, 467, 499
    first new math project, 808
    *See also* middle school

**K**

Kamii, Constance, 554, 693, 694
Kanada, Y., 541
Kaput, James J., 497
Karmarkar, Narendra, 520
Katona, George, 570
Keating, Daniel P., 287
Keller, Fred S., 341
Kelly, David, 290
Kelly, George, 147
Kelvin, Lord (William Thomson), 743, 764
Kelvin temperature scale, 743, 838
Kemeny, John, 202, 576
Kempe, A. B., 626
Kendall, Maurice G., 163
Kennedy, Leonard M., 709
Kenya, 826
Kepler, Johannes, **383–384**, 580, 760
Kerala school, 550, 650
Key Curriculum Press, 280, 471, 472, 659, 660, 663

Khatchian, L. G., 520
Khayyám, Omar, 269, 374, 532
Khruschev, Nikita S., 641
Khowarizmi, Muhammad ibn Musa abu Djafar al-, 24, 170, 224, 235, 373–374, 812
Kikuchi, Dairoku, 378
Kilpatrick, Jeremy, 347
Kilpatrick, William Heard, 553, 678
kindergarten, **384–385**
*Kindergarten Book* (Burton et al.), 384–385
kinetics
  D'Alembert's principle, 181
  theory of gases, 73
King, Ada Byron. *See* Byron, Augusta Ada, Countess of Lovelace
*King's Chessboard, The* (Birch), 243
*King's Rule, The* (computerized game), 662
Kiselev, A. P., 641
Kitcher, Philip, 537
Klee, Victor, 520
Kleene, Stephen, 299
Klein, Felix, 80, 326, **385–386**, 641, 772
  Noether association, 504
Kline, Morris, 40, 329, 646
knot theory, topology and, 764
knowledge, definition of, 571
Koch Snowflake, 652
Kolmogorov, A. N., 641, 642
*KOMAL* (Hungarian journal), 290
Königsberg bridge problem, 302–303, 626
Koppel, Nancy, 823
Korea, 362
Kovalevskaia, Sofia, **386**, 822
Kowa, Seki, 378, 702
Krelth, Kurt, 222
Kroll, Diana Lambdin, 346
Kronecker, Leopold, 6, 103, 319
Krull, Wolfgang, 7
Kruskal, Joseph, 521
Krutetskii, Vadim A., 283
Kuhn, Harold W., 425
Kummer, Ernst Eduard, 6, 103
Kürschák, József, 124
Kürschák Mathematical Competition, 124
Kurtz, Thomas, 202, 576
*Kvant* (student journal), 291
KWL charts, 355

**L**

Lacroix, Sylvestre-François, 458–459
Ladd-Franklin, Christine, 534, 822
*Ladies' Diary* (*Woman's Almanack*) (1704–1841), **387**
Lagrange, Joseph Louis, 278, **387–388**, 483
  determinants, 192
  Fourier as student, 257
  function of a real variable, 266
  Germain's work and, 281
  group theory, 311
  Mean Value Theorem, 458–459
  partial differential equations, 387–388, 530

positive integers, 511
power series, 550
Lajoie, Susanne, 498
Lakatos, Imre, 537
Lambdin, Diana, 832
Lambert, Johann, 541
Lampert, Magdalene, 497
Lander, Leon, 558
language and mathematics in the classroom, **388–391**
  current research on instructional methods, 346
  effects of language dysfunction on mathematical learning, 394
  integration in elementary school, 355
  mathematical literacy and, 445–446
  multicultural mathematics and, 488
  variables and, 813
language of mathematics textbooks, **391–393**
Laplace, Pierre Simon, **393**, 483, 530
  determinants, 192
  Fourier as student, 257
  normal distribution curve, 506
  translation of Somerville work, 822
Laplace-Gaussian curve. *See* normal distribution curve
Laplace transforms, 214, 522
Lappan, Glenda, 666, 667
Larson, Gary, 335
Lasky, Kathy, 355
Latin square, 440
lattice geometry, 78
Law of Detachment, 586
Lawrence, Abbott, 324
laws. *See specific topics*
laws of chance. *See* probability, overview; probability application: chances of matches
*Laws of Thought, The* (Boole), 588
*Learner Profile* (software), 663
learning
  arithmetic, 83–84
  authentic, 735
  autonomous behavior, 829
  Bruner's four "theorems" of, 84
  cognitive. *See* cognitive learning
  collaborative, 194, 195–196, 358, 793
  comparing and contrasting on, 555
  conceptual, 596–598
  connectivity theorem, 84
  construction theorem, 84
  contrast and variation theorem, 84
  cooperative, 102, 201, 202, 221
  counting in, 556
  discovery, 177, 599–600
  to estimate, 231
  exploratory, 222
  guided, 271
  integrated, 170
  intellectual development and, 84
  interactive, 169
  meaningful, 595–596
  meaning theory of, 39, 83–84, 553, 678

notation theorem, 84
observation in, 554–555
patterning in, 242, 243, 532, 555
problem-solving ability and, 574
procedural, 598–599
psychology of, 594–602
remedial instruction, 629–630
retention, 596
rote, 595–596
sense perception and, 536
social aspects of, 147
spiral, 669–670
stages in, 198
team, 350
theories of, 145, 553, 594–602, 599–600
transfer of, 146, 594–595
transmission model, 146
*See also* psychology of learning and instruction, overview
learning disabilities, overview, **393–402**
*See also* minorities, educational problems
Learning Research and Development Center (LRDC), 802
learning tests, 746
least squares, 244, **403–404**
  Gauss's work, 278, 403
  interpolation and, 368
  Legendre's work, 278, 403, 405
  linear algebra and, 422
Lebesgue, Henri, **404–405**
Lebesgue integrals, 353, 404
LeBlanc (pseudonym of Sophie Germain), 281. *See also* Le Blanc, M.
Le Blanc, M., 318. *See also* LeBlanc
Legendre, Adrien Marie, 251, **405**
  axiomatic system, 448
  Davies translation of *Geometry*, 186, 324
  Fourier association, 257
  least squares, 278, 403, 422
  proof of pi and pi squared as irrational numbers, 623
  quadratic reciprocity law, 277
Legendre functions, 393
Legendre polynomials, 405
LEGO TC Logo, 441
Lehmer, Derrick H., 510
Lehmer's generator, 612
Lehrer, Richard, 497
Leibniz, Gottfried Wilhelm, **405–406**
  Bernoulli (Jakob and Johann) as student, 73
  calculator invention, 87
  calculus development, 90, 251, 268, 405–406, 438, 450, 504, 531, 614
  on combinatorics, 120
  concept definition, 447
  on connection between mathematics and religion, 318
  determinants, 192, 422, 702
  first systematic use of binary system, 74
  geometric diagrams for syllogistic logic, 197

harmonic triangle, 532
infinite series, 651
infinite series for pi, 541
infinitesimal numbers, 536
introduction of *function* as term, 266
matrix theory origins and, 422
multiplication symbol adoption, 489
Newton dispute on calculus invention, 189
Platonism and, 537
probabilities notation, 560
"universal characteristics" concept, 533
as Wiener influence, 821
Lejeune-Dirichlet, Peter G., 257
length, perimeter, area, and volume, **406–410**, 706
*See also* area; volume
Leningrad Mathematical Olympiad, 125
Leonardo da Vinci, 580
Leonardo of Pisa, 374
lesson plans, **410–417**
task analysis as a form of, 709
Leucke, John, 764
Lewin, Kurt, 152
L'Hospital, Guillaume François Antoine, Marquis de, 192, **417**
*Liber abaci* (Leonardo of Pisa), 374
liberal arts mathematics, **417–418**
*Librarian Who Measured the Earth, The* (Lasky), 355
Lie, Marius Sophus, 530. *See also* Lie, Sophus.
Lie, Sophus, **418–419**. *See also* Lie, Marius Sophus.
Liebeck, Pamela, 670
light
law of reflection of, 538
perspective projections, 579–580
quantum theory and, 540
speed of, 628
limaçons, 546
limits, **419**, 527
zero and, 838
Lindemann, Carl von, 541
Lindemann, Ferdinand, 771
Lindquist, Mary Montgomery, 519
linear algebra, overview, 31, **419–425**
business mathematics and, 85
coding theory and, 117
in college curriculum, 173
computer science and, 132
coordinate geometry and, 154
determinants, 192–193
engineering and, 214
null vector and, 838
revisions in, 173
vector spaces, 420, 421, 816–817, 838
Linear Algebra Curriculum Study Group (LACSG), 424
*Linear Associative Algebra* (Peirce), 325
linear equations, 19, 113, 114, 154, 701, 702
Diophantine, 200–201
linear interpolation, 368–369

linear programming, 157, 202, **425–427**
matrix games and, 273
modeling and, 482–483
operations research and, 520, 521
systems of inequalities and, 703
linear space, 421
linear transformations, 421, 775–776
line symmetry, 699, 700
*LINK* (ASA), 26
LINPACK (software library), 423
Liouville, Joseph, **427**, 771
*Liouville's Journal (Journal de Mathématique)*, 427
Li Shan Lan, 114
LISP (programming language), 577–578
literacy in mathematics. *See* mathematical literacy
literature, children's. *See* books, stories for children
Littlewood, John, 559
Liu Hui, 113
Lloyd, Geoffrey E. R., 362
Lloyd Morgan, Conway, 820
Lobachevsky, Nikolai Ivanovich, 64, 80, 235, **427**, 505–506, 580
local school districts. *See* school districts, local
logarithmic spirals, 546
logarithms, **428–429**
chemistry and, 110
continuity and, 151
*e* and, 209
equations, 227–228
as exponents, 235, 241
functions, 151
graphing calculator use for, 20, 21, 228
Napier's contribution, 209, 241, 495–496
transcendental and algebraic functions, 766–771
logic, **430–432**
attribute blocks and, 59, 60
Boolean algebra and, 82–83
compound statements in, 202
creativity and, 165
diagrams for syllogistic, 197–198
discrete mathematics and, 203
Dodgson's (Lewis Carroll) work in, 107–108
first-order, 586
foundations of mathematics and, 448–449
Gödel's contribution, 299
inference, basic rules of, 586
language of, 591
Leibniz's contribution, 197, 405–406
mathematical, 585–586
paradox in, 526–527
Peano's contribution, 533, 820
Peirce's contribution, 534
predicate, 586
proof and, 585–586, 588–589
Russell and Whitehead's *Principia Mathematica* on, 588, 639, 820

second-order, 586
sentential, 586
symbolic, 533, 639
in young children, 618
*Logic and Utility of Mathematics with the Best Methods of Instruction Explained and Illustrated, The* (Davies), 186–187, 324
logic machines, **432–433**
Logo (programming language), **433–435**, 575–576, 658, 660
geometry instruction and, 280
measurement activities, 407, 409
for middle-school instruction, 471, 472
problem-solving ability and, 574
symmetry studies, 701
Lomonosov, Michail V., 640
London Mathematical Society, 189
London University, 216
long division, 190, 207
loop abacus, 471
Lorentz, Hendrik, 628
Lorentz Transformations, 628
Lorenz, Edward, 109
Lorenz equations, 523
Lotka, Alfred, 77
Louise Hay Award, 56
Louisiana mathematics proficiency scores, 1992, 453
Lovelace, Ada Byron. *See* Byron, Augusta Ada, Countess of Lovelace
low achievers. *See* developmental (remedial) mathematics; learning disabilities, overview; remedial instruction; underachievers; underachievers, special programs
Lowell Lectures, 639
Loxley, William, 366
Loyd, Sam, 624, 625
Lubby, William, 227
*LU* factorization, 423
Lupkowski, Anne E., 285, 286
Lyapin, S. E., 641

**M**
MAA. *See* Mathematical Association of America
MacDraw (software), 738
Maclaurin, Colin, 192, **437**, 550
Maclaurin expansion, **437–439**
Maclaurin series, 550, 551
Macsyma (software), 522
Madhava of Sangamagramma, 550, 650
Madison, Bernard, 792
Madison Project, 329, 357, 679
magic squares, **439–440**
Magnitsky, Leontyf, 640
Maine
mathematics proficiency scores, 1992, 453
systemic curriculum reform projects, 649
Malone, Thomas W., 663
Malthus, Thomas, 226

management science. *See* operations research

Mandelbrot, Benoit, 109, 258, 588

Mandelbrot competition, 125, 291

Mandelbrot sets, 258, 259

manifolds, topological spaces and, 765

manipulatives
abacus, 1–2, 114
for algebraic equations, 225
for art projects, 43
attribute blocks, 59–60, 188
for basic skills, 70–71
to correct arithmetic errors, 230
Cuisenaire rods, 167–168
current research on instructional methods, 345–346
Davies's recognition of, 187
Dienes, Z. P., 198
for discovery learning, 600
Egyptian numerals as, 211
elementary school use of, 344
enrichment, 220
in *EQUALS: A Mathematics Equity Program*, 647
for grouping, 508–509
for introducing exponents, 243
Japanese origami, 379
kindergarten use of, 384–385
as math anxiety aids, 451
middle school use of, 471–472
money as, 485
Montessori method use of, 485
Napier's rods, 495
patterning, 242, 243, 532, 555
Pestalozzi's advocacy of, 536
preschool use of, 557
in space geometry, instruction, 667–668
Stern's work on, 692
for subtraction, 697
for symmetry study, 701
for teaching about time, 761
*See also* Cuisenaire rods; instructional materials and methods; technology

manipulatives, computer generated, **440–441**
for tutoring, 789

Mann, Horace, 677

Mansūr, Abū Nasr ibn 'Iraq, 374

Maple (software), 522, 652, 661

map projection, 533–534

marketing careers, 105–106

Marklund, Inger, 366

Markov, Andrei Andreevich, 441

Markov chain
operations research and, 520
software used in study of, 657

Markov processes, **441–443**, 693

Markushevich, A. I., 641

Martin, George E., 760

Maryland
EQUITY 2000 program, 479
University of Maryland Mathematics Project, 808

Maschke, Heinrich, 326

mass, as system of measurement, 707

Massachusetts
Algebra Project, 797–798
history of mathematics education in, 323, 324
mathematics proficiency scores, 1992, 453
Wiener's association, 820
*See also* Boston College; Boston University; Harvard University

Massachusetts Institute of Technology
first educational use of computers, 128
founding of, 324
Logo Group, 575
Research Science Institute, 290

mass action law, 199

matching items, in test construction, 746

materials for instruction. *See* instructional materials and methods

*Math Blasters* (software), 657

*Math Connections* (software), 660

*MATH Connections: A Secondary Mathematics Core Curriculum Initiative*, 649

MathCounts, 291

MATHCOUNTS competition, 60–61, 125

*Mathematica* (software), 522, 652, 661

*Mathematical Analysis of Logic, The* (Boole), 588

Mathematical Association (MA; England), 217

Mathematical Association of America (MAA), **443**
American Mathematical Society and, 26
Association for Women in Mathematics and, 56
awards and, 63
*Call for Change: Recommendations for the Mathematical Preparation of Teachers of Mathematics*, 726, 730
as careers in mathematics advice source, 15
*College Mathematics Journal*, 444, 447
college teacher preparation recommendations, 721
Committee on Discrete Mathematics, 203
Committee on the Undergraduate Program in Mathematics, 123
on elementary school teacher preparation, 723
founding of, 677–678
*Guidelines for the Continuing Mathematical Education of Teachers*, 711, 716, 718
handbook on enrichment programs for minority precollege students, 475
Joint Policy Board for Mathematics, 382
on linear algebra education, 420
*Math Horizons*, 447
Mu Alpha Theta cosponsorship, 486
National Committee on Mathematical Requirements, 498

reform in mathematics education advocacy, 468
on secondary school teacher preparation, 727, 730
United States National Commission on Mathematics Instruction and, 805

Mathematical Association of America (MAA), historical perspective, **444**

mathematical biology. *See* biomathematics

*Mathematical Collections* (Pappus), 310

mathematical connections, NCTM standards, 648

mathematical expectation, **444–445**

mathematical job market. *See* career information

mathematical literacy, **445–446**

*Mathematical Log, The* (publication), 486

mathematical modeling. *See* modeling

Mathematical Olympiads for Elementary Schools (MOES), 286, 291

*Mathematical Principles of Natural Philosophy* (Newton). *See Principia*

mathematical reasoning. *See* reasoning

Mathematical Sciences Education Board (MSEB), 116, **446**
emphasis on new instructional methodologies, 348
United States National Commission on Mathematics Instruction and, 805

*Mathematical Spectrum* (periodical), **446–447**

mathematical structure, NCTM standards, 648

*Mathematical World* (AMS publication), 25

"Mathematician's Nightmare, A" (Growney), 543–544

mathematics, definitions, **447–448**
foundations of mathematics and, 448–449
proof, 585–590

mathematics, foundations, **448–449**

mathematics, nature, **449–451**

Mathematics Action Group, 56

*Mathematics and Humor* (Azzolino, Silvey, and Hughes), 335

*Mathematics and Humor* (Paulus), 334

*Mathematics and Informatics* (journal), 291

*Mathematics and Life* (Ruch, Knight, and Studebaker), 467–468

*Mathematics and Technology: The Transition Years* (teleconference series), 742

mathematics anxiety, **451–452**
learning disabilities and, 396, 398, 401
women and, 823, 826, 828

*Mathematics Competitions* (journal), 292

mathematics education, statistical indicators, **452–454**

mathematics education goals. *See* goals of mathematics instruction

Mathematics in Context (MiC): A Connected Curriculum for Grades 5 through 8, **454–455**, 716, 725

*Mathematics in General Education* (1940 PEA report), 381, 578

Mathematics Intensive Summer Session (MISS), California State University, 799
Mathematics Learning Centers, 42
mathematics proficiency scores, 1992. *See under specific states.*
*Mathematics Teacher* (journal), 447, 501
*Mathematics Teaching in the Middle Grades* (journal), 501
*Mathematics Teaching in the Middle School* (journal), 469
*Mathematics TestBuilder* (software), 663
*Mathematik in der Schule* (German journal), 282
*Mathematikunterricht, Der* (German journal), 282
*Math Exploration Toolkit* (software), 660, 663
MathFINDER (CD-ROM), 741
*Math Horizons* (journal), 447
Math Labs, 42
MATHLINE (educational service), 711, 716
MATHTALK (videos), 465
*Math Trailblazers: A Mathematical Journey Using Science and Language Arts* (TIMS curriculum), 734
matrices, 202
  in game theory, 838
  null, 838
  square, 191
matrix algebra, 6, 19, 214
  linear algebra and, 422–423
  null vector and, 838
matrix equations, 702
matrix games, 272–273
matrix notation, and linear programming, 425
matrix theory, 191, 192–193
Maxwell, James Clerk, 215, 539, 817
May, Robert, 109
Mayan numeration, 319, **455–457**
  placeholder, 838
McCarthy, John, 577
McColl, Hugh, 533
McGill, Robert, 815
McLaughlin, Renate, 422
McLellan, James A., 197
McLellan, John, 39
mean, arithmetic, **457–458**
  deviation, 813
  as measure of central tendency, 462–463
  standard deviation, 675
mean, geometric, **458**
meaningful learning, 595–596
meaning theory of learning, 553, 678
  in arithmetic, 39, 83–84
mean value theorems, 268, **458–459**
measurement, **460–462**
  history of, 219
  informal, 555
  length, perimeter, area, and volume, 406–410, 706
  in nursing, 517

in physics, 540–541
in preschool, 555
solar, 219
in space geometry, 666–667
systems of, 704–708
zero's use in, 838
*Measurement in Motion* (multimedia software), 662
measures of central tendency, **462–463**, 690
  mean, arithmetic, 457–458
  mean, geometric, 458
  median, 466–467
  mode, 480–481
media, **463–466**
  interactive, 130–131
  technological advances in, 741–742
  *See also* multimedia
median, **466–467**
  interpolation and, 368–369
  as measure of central tendency, 462–463
medicine. *See* health; nursing
Meffert Pyraminx, 625
memorization, 677, 678, 694
  of arithmetic functions, 38
  of basic skills, 71–72
  psychology of learning and, 595–596
memory dysfunction, as learning disability, 395–396
Menaechmus, 142,
mental discipline theory, 326, 677
mental representation theory, 670
*Mental Science and Methods of Mental Culture* (Brooks), 83
mentorships, for mathematically gifted, 286
Méré, Chevalier de, 532
Merriman, Mansfield, 403
Mersenne, Marin, 531
Mersenne primes, 625
Merseth, Katherine K., 33
Mesopotamian mathematics, 65–67
meteorology, chaos theory and, 109
method of exhaustion, 309, 310
methods for instruction. *See* instructional materials and methods
*Methodus Incrementorum* (Taylor), 550
Metric Conversion Act of 1975, 705
metric system, 704, 705
metrization theorem, 765
Michelson, Albert, 539
Michelson-Morley experiments, 628
Michigan
  reform projects, 649
  *See also* University of Michigan
*Microsoft Excel* (spreadsheet software), 661, 662
microworld, definition of term, 660
Microworlds (software), 472
Mid-Atlantic Equity Center, American University, 478
middle school, **467–474**
  Connected Mathematics Project, 144
  curriculum, 470–472
  enrichment program, 220

Mathematics in Context (MiC): A Connected Curriculum for Grades 5 through 8, 454–455, 716, 725
  teacher development, 715–717
  teacher preparation, 724–727
Middle School Mathematics through Applications Project, 716, 725
mind-is-a-muscle theory, 677
"Mind Over Math," 451
*Mindstorms: Children, Computers and Powerful Ideas* (Papert), 574
Minimum Performance Test, 69
Minkowski, Hermann, 628
Minnesota, mathematics proficiency scores, 1992, 453
minorities, career problems, **474–476**
  math anxiety and, 451–452
  special programs for underachievers, 799–800
minorities, educational problems, **476–480**
  math anxiety and, 451–452
  NAEP mathematics and science scores, 1982–1992, 453
  special programs for underachievers, 796–801
  Strengthening Underrepresented Minority Mathematics Achievement (SUMMA) Program, 443
  textbook depictions of, 751
Min Si He, 114
mirroring, computer-generated manipulatives and, 441
Mirsky, Leon, 446
*Miscellanea Analytica* (De Moivre), 189
misere Nim (game), 277
Mississippi
  mathematics proficiency scores, 1992, 453
  program for the gifted, 289
Missouri, 454
Mitchelmore, Michael C., 29, 667
Möbius band, 766
mode, **480–481**
  as measure of central tendency, 462–463
modeling, 204, **481–483**
  classroom applications, 31, 32, 33
  computer-generated manipulatives and, 441
  definition of, 538
  as numerical analysis method, 513–517
model theory, 299
modern mathematics, **483–484**
  foundations of mathematics and, 448–449
module, 189
*Modus Ponens* (Law of Detachment), 586
MOES. *See* Mathematical Olympiads for Elementary Schools
momentum, law of conservation of, 628
money, **484–485**
  consumer mathematics and, 148–149
  finance and, 255–256
Monge, Gaspard, 257, 530

Montana, 649
Monte Carlo Method, 9, 541, 568
Montessori, Maria, **485–486**, 552, 553
Moore, David S., 684–685
Moore, Eliakim Hastings, 326, **486**
Moore, George Edward, 639, 640
Moore, Robert, 137
moral development theory, 829
Morgenstern, Oskar, 210, 425
Morley, Edward, 539
Morra (game), 274, 276
Morrissett, Irving, 749
Morse, Marston, 333
mortality tables, 8, 9
mosaic tile patterns, 759–761
Moscow Mathematical Society, 641
Moser, Leo, 335
Moses, Robert P., 797–798
motion
    equations of, 215, 387
    laws of, 215, 538
    straight-line, 214
    vertical, 214
    Zeno's paradoxes on, 837
motivation. *See* Advanced Placement
    Program (AP); advising; awards for
    students and teachers; career
    information; competitions; goals of
    mathematics instruction; Goals
    2000; humanistic mathematics;
    mathematical literacy
Mouton, Gabriel, 705
MOVE-IT Mathematics (underachievers
    program), 797
*Moving Beyond Myths: Revitalizing
    Undergraduate Mathematics* (NRC;
    1991), 729–730
MSEB. *See* Mathematical Sciences
    Education Board
Mu Alpha Theta, 292, **486–487**
Mueller, Johann, 784
Muir, Thomas, 192
Mullen, James G., 333
multicultural mathematics, issues,
    **487–489**
    educational problems of minorities and,
    476–480
    *See also* ethnomathematics
multidigit numbers
    addition of, 10
    number sense and numeration,
    508–509
multimedia, 463–464
    computer-based, 741
    interactive, 130–131, 463
    software, 138, 662
multiple choice questions, in test
    construction, 746
multiple representations, of mathematical
    concepts, 345–346
multiplication, **489–492**
    binary, 74
    complex bases and, 67
    field properties of, 254

of fractions, 261–262, 489–490
group theory and, 311
inverse division, 207
scalar, 420, 816
zero's properties in, 838
Murray, James, 77
music, 383–384, **492–494**
    Pythagorean harmonics, 493, 602

**N**
NAEP. *See* National Assessment of
    Educational Progress
Napier, John, 209, 227, 241, **495–496**
Napierian logarithm, 209, 241
Napier's rods, 495
Napier-Stokes equations, 531
Napoleon Bonaparte, 257
NASA. *See* National Aeronautics and
    Space Administration
Nash, John, 276
Nastasi, Bonnie K., 574
National Academy of Sciences, 446
    women mathematicians, 823
National Academy of Sciences-National
    Research Council (NAS-NRC), 805
National Action Council for Minorities in
    Engineering, 475
National Adult Literacy Survey, 12
National Advisory Committee on
    Education, 801
National Advisory Committee on
    Mathematical Education
    (NACOME), 468
National Aeronautics and Space
    Administration (NASA), 252, 475,
    669
National Alliance of State Science and
    Mathematics Coalitions
    (NASSMC), 116
National Assessment of Educational
    Progress (NAEP), **496–497**
    assessment approach, 804
    on gender and achievement in
    mathematics, 828
    on males vs. females enrolled in
    advanced high school mathematics
    courses, 825
    on minority students' mathematics
    scores, 474
    statistical indicators of mathematics
    education and, 452–454
    test construction and, 745, 746, 747
    test results and, 648, 666
    test results by state, 453
National Association for Gifted Children,
    292
National Association for the Education of
    Young Children (NAEYC), 384
National Association of Mathematicians,
    475
National Association of State Directors of
    Teacher Education and
    Certification (NASDTEC), 312
    teacher education programs, 724

National Board for Professional Teaching
    Standards, 719
National Center for Education Statistics
    (NCES)
    establishment of, 804
    monitoring of state educational reforms
    by, 683
    on NAEP mathematics and science
    scores, 1982–1992, 453
    publication of NAEP mathematics
    assessments by, 496
National Center for Improving Student
    Learning and Achievement in
    Mathematics and Science
    (NCISLA), 498
National Center for Research in
    Mathematical Sciences Education
    (NCRMSE), **497–498**
National Commission on Accreditation in
    Teacher Education (NCATE), 727
National Commission on Excellence in
    Education, *A Nation at Risk* report,
    330, 468–469, 680, 682–683, 718,
    728
National Committee of Fifteen on the
    Geometry Syllabus, 327
National Committee on Mathematical
    Requirements, **498–499**
    1926 recommendations, 467
National Council for Accreditation of
    Teacher Education (NCATE), 724
National Council of Supervisors of
    Mathematics (NCSM), 69, 174, **499**
    basic skills position paper, 647
    definition of basics in mathematics
    education, 296
    reform in mathematics education, call
    for, 468
National Council of Teachers of
    Mathematics (NCTM), 35,
    **499–501**
    Addenda Series, 176, 649, 681, 716
    *Agenda for Action* (1980), 175, 221, 329,
    468, 500, 647, 679, 680
    arithmetic goals, 39, 41
    *Assessment Standards for School
    Mathematics* (1995), 48, 50, 330,
    500, 718, 728, 793
    basic skills position statement (1977),
    174, 647, 679
    calculator use recommendations, 89, 224
    Commission on Postwar Plans, 122
    computer use recommendations, 472,
    679, 681
    Core-Plus Mathematics Project,
    159–160
    curriculum reform, 172
    *See also Curriculum and Evaluation
    Standards for School Mathematics*
    (1989)
    enrichment yearbooks, 220
    equalities study recommendations,
    223–224
    establishment of, 678

Fehr presidency, 250
goals of mathematics instruction, 295, 296
*Guidelines for the Preparation of Teachers of Mathematics* (1993), 727
inequalities study recommendations, 342
international conferences on mathematics education proceedings publication, 806
interpretive reports on NAEP mathematics assessments, 496
Joint Commission Report (1940), 381–382
journals, 447, 469, 501
on kindergarten mathematics curriculum, 384
lesson-plan writing recommendations, 410
mathematically gifted programs, 285
mathematics education reform advocacy, 468, 792
Mu Alpha Theta cosponsorship, 486
*New Directions for Elementary School Mathematics*, 175
preparation of mathematics teachers and, 327, 730
*Principles and Standards for School Mathematics* (2000), 501, 728
problem-solving-based instruction recommendation, 345, 500
*Professional Standards for Teaching Mathematics* (1991), 239, 330, 411, 413, 470, 500, 694, 715, 718, 728, 730, 793
on reflecting history of mathematics in mathematics curriculum, 317
*Revolution in School Mathematics* (1961), 637
standards publication, 69, 116, 175
statistics newsletter copublication, 691
teacher preparation, elementary school, 723
teacher preparation, secondary school, 727
on test validity, 744
textbooks selection criteria, 750
United States National Commission on Mathematics Instruction joint effort, 805
Women and Mathematics Education as affiliate of, 832
writing activities recommended by, 832
yearbooks, 175, 468, 470, 501, 647
National Council of Teachers of Mathematics (NCTM), historical perspective, 39, **501**
National Council on Education Standards and Testing (NCEST), 298
National Curriculum Council of England and Wales, videodisc series available from, 741
National Defense Education Act of 1958, 682

National Diffusion Network (NDN), 804
National Education Association (NEA)
Commission on the Reorganization of Secondary Education, 168, 327
Committee of Fifteen, 83, 123, 326
Committee of Ten, 123, 168, 326
National Council of Teachers of Mathematics affiliation with, 501
National Education Goals Panel (NEGP), 298, 453, 683
National Education Research Centers, 802
National Geographic Kids Network (online network), 738
National Institute of Education (NIE), creation of, 801
National Institutes of Health, 475
*National Leadership Program*, 647
National Longitudinal Study of Mathematical Abilities, 72
National Research and Development Center on Achievement in School Mathematics and Science, 802–803
National Research Center on the Gifted and Talented, University of Connecticut, 289
National Research Council, 41, 79
on assessment of students' mathematical power, 793
curriculum reform and, 176
*Everybody Counts: A Report to the Nation on the Future of Mathematics Education* (1989), 348, 680, 712, 715, 718, 728, 729, 792
on federal government's role in education, 801
on low student achievement and reduced numbers entering mathematics-related careers, 469
Mathematical Sciences Education Board, 116, 446
*Moving Beyond Myths: Revitalizing Undergraduate Mathematics* (1991), 729–730
*Reshaping School Mathematics; A Philosophy and Framework for Curriculum* (1990), 649, 711, 718–719, 729
on teacher development, 711, 712
National School Boards Association (NSBA), 683
National Science Foundation (NSF), 26, **501–502**
calculus awards, 172
Connected Mathematics Project, 144
Core-Plus Mathematics Project, 159–160
curriculum projects, 176, 177, 178
Developing Mathematical Processes, 193–194
funding for Cambridge Conference on School Mathematics (1963), 100
funding for FAMILY MATH and *MATEMÁTICA PARA LA FAMILIA*, 247

funding for Interactive Mathematics Program dissemination, 359
funding for mathematics and science education programs, 1950s and 1960s, 678
funding for Mathematics in Context, 454
funding for reform materials development, 725
funding for systemic reform projects, 649
funding for Teaching Integrated Mathematics and Science Program, 734
funding for University of Chicago School Mathematics Project, 806
funding wind-down, 250
grant coordination, 250
grant support of state initiatives, 683
Linear Algebra Curriculum Study Group sponsorship, 424
new media applications investments, 465
Statewide Systemic Initiatives by, 176
summer programs for high-ability students, 290
support for School Mathematics Study Group formation, 328
support for USA Mathematical Talent Search, 290
teacher training and, 177
University of Maryland Mathematics Project sponsorship, 808
National Science Foundation (NSF), historical perspective, **502–503**
National Security Agency, 290
National Society for the Study of Education (NSSE), 469, 749, 751
National Society of Professional Engineers, 291
National Teachers Association, 801
National Youth Administration (NYA), 801
*Nation at Risk: The Imperative for Educational Reform, A* (1983), 171, 178, 221, 330, 468–469, 680, 682–683, 718, 728
Goals 2000 and, 298
*Nation Prepared: Teachers for the 21st Century, A* (1986), 719
Native Americans
career problems, 474–476
math anxiety, 451
SKILL project for, 799–800
underachievers, 791
*Natural and Political Observations on the Bills of Mortality* (Graunt), 689
natural numbers (positive integers), 242, 536
Peano's postulates, 533, 587, 588
real numbers and, 622–623
zero and, 838
naturalism, 537
Nature and Role of Algebra in the K-14 Curriculum, The: A National Symposium (1997), 446

nature of mathematics. *See* mathematics, nature

NCES. *See* National Center for Education Statistics

NCSM. *See* National Council of Supervisors of Mathematics

NCTM. *See* National Council of Teachers of Mathematics

*n*-dimensional manifolds, theory of, 278

NEA. *See* National Education Association

Nebraska
  mathematics proficiency scores, 1992, 453
  systemic reform projects, 649

negative bases. *See* bases, negative

negative exponents, 240–241, 243

negative numbers
  multiplication of, 491
  zero as greater than, 838

Nelson, Barbara Scott, 498

Netherlands
  international comparison of women's mathematics education and career problems, 826
  in international mathematical studies comparisons, 365
  Mathematics in Context curriculum units, 454
  open-ended test questions, 747

Neumann, John von. *See* Von Neumann, John

neutral geometry, **503**

*New and Complete System of Arithmetic Composed for the Use of the Citizens of the United States, The* (Pike), 542

*New Astronomy* (Kepler), 383

Newcomb, Simon, 326

*New Directions for Elementary School Mathematics* (NCTM; 1989), 175

Newell, Alan, 570

New Hampshire, mathematics proficiency scores, 1992, 453

New Jersey, mathematics proficiency scores, 1992, 453

New Jersey Network, *Mathematics and Technology: The Transition Years* (teleconference series), 742

new math, 32, 72, 100, 123, 169
  back-to-basics backlash against, 171, 174, 177, 329
  enrichment materials and, 220
  first project for junior high school, 808
  first project for secondary schools, 807–808
  formalism and, 537
  history of, 468, 502, 503, 646
  impetus for initiating, 40
  Kline's criticism of, 329
  products of movement, 679
  symbolism and, 389

New Mathematical Library, 290–291

New Mexico, mathematics proficiency scores, 1992, 453

newsgroups, Internet, 464

Newton, Isaac, **503–504**
  axiomatic system, 448
  binomials and, 76
  calculus development, 90, 251, 268, 318, 405, 438, 450, 503–504, 614
  as chaotic set influence, 109
  dispute with Leibniz on calculus invention, 189
  estimating roots of any order, 372–373
  exponent notation, 240–241
  fractional and negative exponents, 243
  law of universal gravitation, 540
  laws of motion, 538, 539
  laws of motion and ordinary differential equations, 521
  mathematical greatness of, 36
  optics, 538
  physics inception and, 538
  polar coordinates, 546
  series, 650, 651
  theory of fluxions, 437
  velocities computation, 536

Newton messagepads (handheld information managers), 742

Newton-Raphson method (chemistry), 111

*Newton's Apple* (television series), 742

Newton's Method, 420

New York Mathematical Society, 26

New Zealand
  comparative study with United States, 365
  international SIMS study, 366

Neyman, Jerzy, 684

Nicely, Robert F., Jr., 341, 750, 751

Nieto, Sonya, 793

Nigeria, 826

Nightingale, Florence, 822

Nilakantha, 650

Nim (game), 274, 275, 277

node, graphs and, 301, 302, 303, 304

Noether, Emmy, 7, **504–505**, 822, 823

Noetherian rings, 504

Noether's Theorem, 504

nonagon, 547

noncooperative games, 273

nondecimal measurement, 461

non-Euclidean geometry, **505–506**, 544
  axioms and, 63, 64
  Bolyai-Gauss development of, 79–80, 318, 503, 505–506
  elliptic, 580
  Euclid's postulates and, 235, 505
  hyperbolic, 503, 505, 506, 580
  Islamic mathematics and development of, 374
  Lobachevsky development of, 80, 427, 505, 580
  as projective geometry, 580
  Riemann's contribution, 78, 235, 445, 505–506, 580, 638
  spherical, 505–506

nonzero-sum games, 273

normal curve. *See* normal distribution curve

normal distribution curve, 75, **506–508**, 566
  standardization of variables and, 676

norm-referenced tests, 744

North Dakota, mathematics proficiency scores, 1992, 453

Northwestern University, program for the gifted, 289

Norway, advanced placement program, 13

notation
  for calculus, 91
  Euler's standardization of, 235
  for fractions, 260
  Mayan positional, 454
  scientific, 145, 644–645, 668
  systematized, 235
  Viète's first symbolic, 190, 240, 812–813
  zero's positional importance, 838

notebook computers, 138

NSF. *See* National Science Foundation

nuclear bomb, Einstein $e = mc^2$ equation and, 628

nuclear wastes, exponential decay formula for, 227

Null Hypothesis ($H_0$), 685, 686, 687, 690

null vector, 838

number assignment, as mathematical concept of measurement, 406

number lines
  addition and, 11
  point named as fractions, 261
  rationals and irrationals representations, 623
  subtraction counting strategies and, 697

numbers
  algebraic, 21–23, 623
  Bernoulli, 73
  binary, 74–75, 87
  cardinal, 103
  complex, 21, 67, 108, 127–128, 189, 215, 255, 268
  composite, 558
  cubic, 243
  Dewey's view of, 197
  Egyptian system, 211
  Fibonacci, 76, 78
  imaginary, 189
  index, 337–338
  infinite, 21, 536
  multidigit, 10, 508–509
  natural, 242, 533, 536, 622
  negative, 19, 66
  perfect, 511, 624
  powers of. *See* exponents, arithmetic
  prime. *See* prime numbers
  Pythagorean view of, 37
  random, 611–612
  rational, 22, 145, 200, 242, 254, 260, 615–617
  real, 103, 128, 189, 622–623
  representations, 145
  Stirling, 121
  transcendental, 21, 209, 623
  transfinite ordinal, 103

number sense and numeration, **508–509**
  manipulatives used to teach basic
    concepts of, 344
  one-to-one correspondence, 519–520
number systems, 6
  African, 16–17
  bases, 2, 10, 16, 38, 60, 66, 74–75, 113,
    187, 215
  binary, 74–75, 87, 215, 254, 315
  calculators and, 88
  characteristics of, 260
  commutative, 6
  complex, 19
  fractions and, 260
  general structure theory, 6
  hexadecimal, 67, 68
  Hindu-Arabic, 38
  hypercomplex, 6
  liberal arts courses on, 418
  Mayan, 319, 455–457
  operations of, 144
  positional, 67
  properties of, 144
  rational, 112
  real, 100, 123
  sexagesimal, 27, 66
number theory, overview, 251, **509–513**
  algebraic, 7, 189, 278, 512
  combinatorics and, 120
  Diophantine equations, 200–201
  Euclidean algorithm proof, 308
  Euler's contribution, 235
  Fermat's contribution, 251
  Gauss's contribution, 277, 510
  Hadamard's contribution, 313
  Julia, Gaston, 382
  Legendre's contribution, 405
  for mathematically gifted, 289
  origin in Greek mathematics, 309
  Ramanujan's contribution, 611
  transcendental numbers and, 512
Number War (game), 274–275, 277
numerator, 260
numerical analysis, methods, **513–517**
  correlation, 159–163
nursing, 517

**O**

octagon, 547
Office of Educational Research and
  Improvement, 791
Office of Minority Participation in
  Mathematics, 443
Office of Pre-collegiate Programs for
  Talented and Gifted (OPPTAG),
  Iowa State University, 289
Ohio, 467
Ohio State University, mathematically
  gifted program, 286, 289
Ohm's law, 215
Oklahoma State University, 716
Olive, John, 658
*One Million* (Hertzberg), 243
one-to-one correspondence, 38, **519–520**

*On the Hypotheses that Lie at the
  Foundations of Geometry* (Riemann),
  505–506
*On the Shoulders of Giants* (Steen ed.), 725
open admissions, 194
open-ended lessons, 413
open-ended problems, 361–362
open-ended projects, middle school, 471
open-ended questions, in test
  construction, 747
operant conditioning, 271, 656
operations research, **520–521**
  career information, 105
  systems of inequalities, 703–704
opportunity to learn (OTL), 366
*Opticks* (Newton), 504
optics, 538
order of operations, inverse functions and,
  371
order relationships, inequalities, 342–344
ordinary differential equations, **521–523**
Oresme, Nicole, 243, 265–266, 650
Organization for European Economic
  Cooperation (OEEC),
  international curriculum survey,
  360
origami, 379
orthogonality, 422
orthographic projections, 579
Osgood, William, 327
outcomes-based education, definition of,
  410–411
Oxford University, 216, 822

**P**

*PageMaker 5.0* (software), 661
paper-folding activities, for symmetry
  studies, 701
Papert, Seymour, 574, 575, 660
Pappian plane, 581
Pappus, 251, 310
  Theorem of, 581
Papua New Guinea, 826
parabolas, 142
  differential equations, 530
paradox, **525–527**
  Bertrand's, 558, 568
  Cantor's theory engendering, 103
  of classes, 639
  Petersburg, 73
  Russell's, 448, 654
  Simpson's, 672, 673
  Zeno's, 318, 527, 543, 652, 837
Parady, Bodo, 558
parallel lines, 580
parallelogram, 548
Parallel Postulate. *See* Fifth Postulate
parallel projection, 579, 580
parametric equations, 88, **527–529**
parents
    FAMILY MATH and *MATEMÁTICA
      PARA LA FAMILIA*, 246–249,
      286–287, 794–795
  and gifted students, 286–287

as partners in educational process,
  794–795
Paris conference (1900). *See* Second
  International Congress of
  Mathematics
Parish, Doris C., 331
Parkhurst, Helen, 2
Parkin, T. R., 558
Parmenides, 837
partial convergents, continued fractions,
  150
partial differential equations, 73, 181,
  **529–531**
  Lagrange's contribution, 387–388, 530
partitioning, 202, 260, 261, 617
  functions, 611
  in number sense and numeration, 508
PASCAL (programming language),
  576–577, 658
Pascal, Blaise, 87, 318, 370, **531**, 532,
  560, 576
  probability theory, 251, 531
Pascal pyramid, 532
Pascal's triangle, 76, 78, 268, 318, 374,
  **531–532**
  in curriculum enrichment, 220
  sets and, 654
  Yang Hui Triangle as, 114
path, graphing, 301–302, 626
pattern blocks, 471
patterning, 242, 243, **532–533**
  preschool, 555
  *See also* tiling
pattern recognition, learning disabilities
  and, 395
Paulus, John, 334
payoff tables, game theory, 838
PBS Middle School Math Project, 711,
  716
Peacock, George, 6, 190
Peano, Giuseppe, **533**, 534
  axiomatic foundation for natural
    numbers, 5, 587, 588
  modern mathematics and, 448–449
  as Russell and Whitehead influence,
    533, 639, 820
PEA Report. *See* Progressive Education
  Association (PEA) Report
Pearson, Karl, 684, 689, 690
  sample correlation coefficient, 161–163
Peck, Kyle, 341
pedagogical approaches. *See* instructional
  materials and methods
peer evaluation of instruction, 237
Peirce, Benjamin, 6, 190, 325, 533, 534
Peirce, Charles Sanders, 6, 325, **533–534**
Pell, John, 200
Pell's equation, 200, 251, 338
Pennsylvania
  history of mathematics education, 324
  Project PRIMES, 750
  *See also* University of Pittsburgh
Penrose, Roger, 760
Penrose tiles, 760

pentagon, 547
pentagram, 300
percent, 69, 145, **534–536**
perfect numbers, 511, 624
perimeter, measurement of, 407–408
Perkins, Susan, 285
Perl (programming language), 133
permutations, 5, 560, 561
Perry, John, 326
Perry movement (England), 217
Persian mathematics, 532
personalized system of instruction (PSI), 341
perspective, theory of, 580
perspective projections, 579–580
*Perspectives* (visual geometry activity), 741
Pestalozzi, Johann Heinrich, 119, **536**, 552
Petersburg paradox, 73
Petersen, Peter, 2
Peterson, Penelope, 176
Peter the Great, 640
Pfaff, Johann Friedrich, 530
phenomenon of whispering galleries, 143
Philadelphia Academy, 324
philosophical perspectives on mathematics, **536–538**
    Dewey's contribution, 197
    humanistic mathematics, 331–334, 537
    Russell's contribution, 639–650
    Whitehead's contribution, 820
    Zeno's paradoxes, 318, 527, 543, 652, 837
    *See also* constructivism
*Philosophy of Arithmetic* (Brooks), 83, 326
photon, 539, 540
physics, **538–541**
    Bernoulli (Daniel) as father of mathematical, 73
    chaotic processes, 109
    Einstein special theory of relativity and, 628
    exponential equations, 226
    Noether's work, 504
    quantum mechanics, 264
    quantum theory, 212, 540
    vectors, 817–818
pi, **541–542**
    in Babylonian mathematics, 66, 641
    concept of, 66
    first use of symbol $\pi$, 235, 541
    as irrational number, 541, 623
    as transcendental number, 512, 541, 771
    value of, 320, 541, 650
    videotape about, 742
Piaget, Jean, 575
    on autonomy, 693–694
    Bruner's theories and, 618, 670
    coining of term *conservation*, 519
    constructivism theory, 146, 345, 554, 601, 693–694
    definition of angle, 28
    geometry instruction theory, 280
    influence on education by, 40

on judging length, 407
and Logo, 434
on one-to-one correspondence, 519
preschool education theory and, 553, 554
psychology of learning and teaching, 600, 601
readiness concept, 618
stages of cognitive development theory, 40, 284–285, 329, 679
on stages of reasoning development, 591, 618
and Van Hiele levels, 280
pictorial representations, for subtraction, 697
Pike, Nicolas, **542**
Pirie, Susan B., 390
*Pit and the Pendulum, The* (Poe), 415, 416
placeholder, zero and zero forerunners, 838
Placement Test Program (PTP), 443
placement tests, 745
*Place of Mathematics in Secondary Education: The Final Report of the Joint Commission of the Mathematical Association of America and The National Council of Teachers of Mathematics, The*, 381–382
place value
    arithmetic, 38
    decimal system, 187, 188
    difficulties in learning, 508
    subtraction and, 697
    *See also* abacus; addition; Cuisenaire rods; Dienes; number sense and numeration; rounding
PLAN (Program for Learning in Accordance with Needs), 340
planar graphs, 304
plane, projective, 580, 581
plane curves, 73
plane geometry. *See* angles; axioms; circles; discovery approach to geometry; Euclid; geometry, instruction; golden section; length, perimeter, area, and volume; Logo; mathematics, definitions; neutral geometry; pi; polygons; proof; proof, geometric; Pythagoras; Pythagorean theorem; similarity; tiling; Van Hiele levels
    *See also* art; Birkhoff; Bolyai; coordinate geometry, overview; Escher; Greek mathematics; Islamic mathematics; Lobachevsky; mathematics, foundations; non-Euclidean geometry; software; space geometry, instruction; transformational geometry; transformations
plane symmetry, 700
Plato, 211, **542–543**

Platonic solids (polyhedra), 543
Platonism, 537
Playfair, John, 27, 505
Plomp, Tjeerd, 366
Poe, Edgar Allan, 415, 416
poetry, **543–544**
Poincaré, Jules Henri, 483, **544–545**
    chaos processes and, 109
    hyperbolic geometry models constructed by, 506
    Poincaré Conjecture, 766
    significance in history of mathematics, 316, 318
    topology, 765, 766
Poincaré, Raymond, 544
point and confidence interval estimation, 684
point estimate, 687
point symmetry, 699, 700
Poisson, Simeon-Denis, 257, 545
Poisson distribution, 75, **545–546**, 567
Poisson process, 692–693
polar coordinates, 73, 156, 157, **546–547**
    transformations, 779–780
polar equations, 88
Pólya, George, 121, **547**, 572–573, 574
Pólya enumeration theorem, 121
polygons, **547–549**
    tiling, 760
polyhedra, 543
polynomial equations. *See* equations, algebraic
polynomial long division, 190
polynomials
    in abstract algebra, 5, 6
    characteristic, 192–193
    in chemistry, 110–111
    coding theory and, 117
    cubic, 22
    degree of, 22
    irreducible, 22
    Legendre, 405
    quartic, 22
    rational coefficients of, 22
    roots of, 192
    Taylor, 551, 552
Poncelet, Jean Victor, 580
*pons asinorum* (bridge of fools), 324
population growth, **549–550**
    biomathematics and, 77
    Malthusian exponential theory, 226
portion cups and chips, 225
Portland State University, 716
positional notation
    in Mayan numeration, 454
    zero's importance in, 838
positive integers. *See* natural numbers
postulates, 63, 448
    axioms vs., 235
    Euclid's, 63, 80, 235, 374, 503, 505
potential infinite, 536
*Power of Ten* (Morrison and Morrison), 243
power series, **550–552**

practice. *See* drill and practice
pragmatism, 197
*Praxis der Mathematik* (German journal), 282
precision, 540, 638–639
predicate logic, 586
predictions, 481–482
predictor tests, 745
preparation of teachers. *See* teacher preparation *headings*
preschool, **552–557**
  and kindergarten curriculum, 384
  Montessori method, 485–486, 552, 553
Presidential Awards for Excellence in Mathematics Teaching, 62
President's and National Governor's Conference on National Education Goals (1990), 683
PreSuppose (software), 701
Prim, Robert, 521
prime factorization, 245, 308
prime numbers, **558–559**
  abstract algebra and, 6
  definition of, 268
  Fundamental Theorem of Arithmetic and, 268
  Gauss's contribution, 279, 313
  Goldbach's conjecture, 300
  Hadamard's contribution, 313
  number theory and, 510, 511
  Sieve of Eratosthenes, 228
  twin primes, 450
Prime Number Theorem, 313, 558
Princeton University
  Institute for Advanced Study, 504, 820
  mathematics entrance requirements, 677
*Principia* (Newton), 448, 504
*Principia Mathematica* (Russell and Whitehead), 103, 588, 639, 820
*Principles and Standards for School Mathematics* (NCTM; 2000), 179, 501, 728
*Principles of Mathematics, The* (Russell), 639
*Principles of Psychology* (James), 377
*Priorities in School Mathematics* (PRISM) project, 647
Pritchard, Paul, 558
probabilistic proof, 589
probability, overview, **560–562**
  actuarial mathematics and, 8, 560
  applications for classroom, 31
  bell curve of, 58
  binomial coefficients and, 76
  classical, 560
  De Moivre's contribution, 189
  De Morgan's contribution, 189
  discrete mathematics and, 202, 203
  distributions, 263, 539, 545
  at elementary level, 560–561, 567–568
  Fermat's contribution, 251, 531
  Gauss's contribution, 278
  Markov processes and, 442

  microworlds for, 660
  NCTM standards, 648
  Pascal-Fermat correspondence on, 531
  probability integral, 189
  queueing theory and, 608
  at secondary-level, 177, 561
  set theory and, 653
  in sports, 673–674
  statistics and, 689
  theories of, 82, 251, 531
*Probability and its Engineering Uses* (Fry), 31
probability applications: chances of matches, **562–566**
  birthday-twin paradox, 527
probability density, **566**
  median and, 466
probability distribution, **566–567**, 687
probability in elementary school mathematics, 560–561, **567–568**
probability problems, computer simulations, **568–570**
*Probability Theory* (software), 657
problem of Dido, 821
Problems of the Week (POWs), 358
problem solving, overview, 69, 547, **570–573**
  with addition, 10
  with algorithms, 23
  applications for the classroom, 32
  in arithmetic, 40
  in back-to-basics movement, 679
  calculators and computers used in, 349
  in Canadian education, 102
  college-level curriculum, 172
  concepts as unifying mathematics concepts, 578
  constructivism and, 146, 601
  creativity and, 164–165
  Dewey's focus on, 197
  divisibility testing and, 204
  dysfunctions in, 396–397
  emphasis as instructional method, 344–345
  with fractions, 261–262
  in groups, 195–196
  information processing and, 601
  as main focus of NCTM mathematics program proposal, 296–297, 329, 468, 470, 500, 679, 680
  math anxiety and, 452
  NCTM standards, 470, 648
  and number sense and numeration, 508
  PEA report recommendations, 578
  Pólya's writings on, 547, 572–573, 574
  poor achievement levels, 648
  recreational, 623–627
  skills development, 247
  software, 130
  *Square One TV* episodes about, 742
  steps for, 197
  with subtraction, 696–697
problem-solving ability and computer programming instruction, **574–575**
procedural learning, 598–599

*Process and Reality, An Essay in Cosmology* (Whitehead), 820
*Process of Education, The* (Bruner), 84, 618
Proclus, 234, 309
*Professional Development for Teachers of Mathematics* (Aichele ed.), 711, 719
Professional Development Program, University of California at Berkeley, 798
professional organizations. *See* resources
*Professional Standards for Teaching Mathematics* (NCTM; 1991), 239, 330, 411, 413, 470, 500, 694, 715, 718, 728, 730, 793
Professors Rethinking Options in Mathematics for Prospective Teachers, 172
proficiency scores. *See* under *state headings*
programmed materials, 220
programming languages, 131, **575–578**
  Ada, 133, 578
  BASIC, 202, 576, 658
  C++, 577
  FORTRAN, 131, 658
  Java, 133, 577
  LISP, 577–578
  Logo, 433–435, 471, 472, 575–576, 658, 660
  Pascal, 576–577, 658
  Perl, 133
  problem-solving ability and instruction in, 574–575
  Scheme, 577–578
  software and, 657, 658
programming, linear, 85, 202
progressions
  arithmetic, 85
  geometric, 85
Progressive Education Association (PEA) Report, **578–579**
  Joint Commission Report (1940) compared with, 381
progressive education movement, 168, 328, 678
  in Japan, 379
Project for Computer-Intensive Curricula in Elementary Algebra, 224, 342
projections, geometric, **579–580**
projective geometry, **580–582**
  invariance and, 370
  Klein's work, 385–386
  Theorem of Desargues, 581
  Theorem of Pappus, 581
projective plane, 580
Project Mathematics!, 465
Project PRIMES (textbooks selection), 750
Project SEED (program for the disadvantaged), 357, 799
projects for students, **582–585**
  in enrichment learning, 736–738
  middle school open-ended, 471
Project TIMS. *See* Teaching Integrated Mathematics and Science Program
PROMYS, Boston University, 289

proof, **585–591**
  in cognitive learning, 118
  by contradiction, 537, 587
  foundations of mathematics and, 449
  indirect, 235, 587
  inductive, 73
  paradox vs. error or howler, 526
proof, geometric, **591–593**
proofs, in test construction, 746–747
proportion, 69, 145, 235
  theory of, 37
proportional reasoning, **593–594**
  percent, 534–535
propositions, 82, 235
"Psychological Considerations in the
    Learning and Teaching of
    Mathematics" (Brownell), 83
*Psychology of Arithmetic, The* (Thorndike),
    759
psychology of learning and instruction.
    *See* ability grouping; advising;
    cognitive learning; constructivism;
    cooperative learning; goals of
    mathematics instruction;
    humanistic mathematics; learning
    disabilities, overview; psychology of
    learning and instruction, overview
psychology of learning and instruction,
    overview, **594–602**
  behaviorism, overview, 599
  *See also* behaviorism *as separate heading*
  Brook's contribution, 83
  Brownell's contribution, 83–84
  Bruner's contribution, 84
  cognitive approaches, 118–119,
    599–600
  constructivism and, 145–148, 601
  in curriculum standards, 678
  Dewey's theories, 197, 678
  gifted development and, 284–285
  International Group for the Psychology
    of Mathematics Education, 364
  James's contribution, 377
  readiness concept, 618–619
  Skinner's theories, 656
  spiral learning, 669–670
  stimulus-response theory, 69, 599, 678
  Thorndike's theories, 678, 759
  University of Maryland Mathematics
    Project and, 808
  women's problems with mathematics
    and, 829
  *See also* Piaget, Jean
*Psychology of Number and Its Application to
    Methods of Teaching Arithmetic, The*
    (Dewey), 197
Ptolemy, Claudius
  Greek mathematics and, 310
  table of chords of circles, 784
  value of pi, 320
Public Broadcasting Service (PBS)
  educational programming, 464
  professional development opportunities
    for teachers, 711, 716

Puerto Rico, field testing of Mathematics
    in Context curriculum in, 454
punched card concept, 87
Purdue University, 332–333
pure mathematics, 31, 32, 114, 450
Putnam Competition, 443
puzzles. *See* recreations, overview
Pythagoras, 211, **602–603**
  irrational numbers discovery, 372, 543,
    622
  monochord experiments, 493
Pythagorean diatonic scale, 493
Pythagorean identities, 783
Pythagorean number theory, 318
  Indian mathematics and, 338
  shape classification, 243
Pythagoreans (brotherhood), 318, 602
  irrational numbers discovery, 372, 543,
    622
  pentagram use, 300
  *See also* Pythagorean Theorem
Pythagorean Theorem, 66, 113, 268, 503,
    **603–604**, 783
  for calculating vector length in a plane,
    817
  videotape on, 465, 741
Pythagorean tuning, 493

**Q**

qibla, 374
Qin Jui Shao, 113, 114
*QR* factorization, 423
quadratic functions, 263–264
quadratic reciprocity laws, 277–278, 405
quadric equations, 19, 66, 144, 200, 228
quadrilateral, 547, 548
quality control, **605–607**
Quality Education for Minorities (QEM)
    Project, 477
quantification theory, 533, 534
Quantitative Literacy Materials, 690
Quantitative Literacy Program, 26
*Quantum* (magazine), 291
quantum mechanics, 264
quantum theory, 212, 539–540
quarks, 540
QUASAR (Quantitative Understanding:
    Amplifying Student Achievement
    and Reasoning) project, 799
quaternions, 6, 817
*Quattro Pro 5.0* (software), 661
questioning, **607–608**
  tutoring and, 789
Quetelet, Adolphe, 596
queueing theory, **608–609**
  and operations research, 520
QuickTime video, 662, 663
Quinton, Anthony, 639

**R**

race. *See* ethnicity and race
radians, 786
radicals, 4
  cyclotomic equation and, 278

radius, 27, 115
radix sixteen numbers, 315
Rahn, Johann, 206
Ramanujan, Srinivasa, 235, 320, 511, **611**
Ramsey's Theorem, 121–122
RAND Corporation, 474
random number, **611–612**
random sample, **612–613**
  statistical significance, 612–613, 688
random variable, 75, **613–614**
  standardization and, 676
range, 266, 785
rates in calculus, **614–615**
ratio, 145
  equivalence of, 593
  law of, 35
  of measured quantities, 593
  musical harmonies and, 602
  numerical, 593
  percent as form of, 534–535
  pi, 541
  proportional reasoning and, 593–594
  roots of, 243
  of unmeasured quantities, 593
  varieties, 593
rational exponents, 241
rational functions
  as continuous, 151
  real numbers and, 622–623
rational numbers, **615–618**
  in abstract algebra, 5, 6, 272
  definition of, 260, 615
  in Diophantine equations solutions,
    200
  exponents, 242–244
  irrational numbers and, 372
  microworld for, 660
  real numbers and, 622–623
  *See also* decimal system; fractions
Ratner, Marina, 823
reaction-diffusion theory, 77
readiness, **618–619**
  theory, 553
reading mathematics textbooks, **620–622**
  language of, 391–393
real exponents, 241
realism. *See* Platonism
Realistic Mathematics Education, 454
real model, 481
real numbers, 103, 128, 512, **622–623**
  abstract algebra, 6
  Birkhoff postulates and, 78
  Dedekind cuts, 189
  inequalities, 342–344
  zero as positive/negative division point,
    838
  *See also* irrational numbers; rational
    numbers
*Real World and Mathematics, The*
    (Burkhardt), 32
reasoning
  as basic skill, 69
  curriculum standards and, 680
  developmental stages in, 591

goals of mathematics instruction and, 297

as major focus of new mathematics curriculum, 471–472

proportional, 593–594

in secondary school mathematics, 648

by young children, 618

*See also* deductive reasoning; logic; proof; proof, geometric

Reasoning Under Uncertainty project, 814

Reckoning Machine (Leibniz calculator), 87

Recorde, Robert, 216, 217, 223, 224

recreations, overview, **623–627**

paradox, 526, 527

problem solving and, 570

*See also* games

rectangle, 548

golden, 300–301

rectangular coordinates, 212

recursion theory, 159, 299

Redfield, J. Howard, 121

*reductio ad absurdum* method, 235, 587, 588

Reeve, William, 201

reflection, 772–773, 774, 776

reflective thinking, 201

*Reforming Mathematics Education in America's Cities* (Webb and Romberg eds.), 650

reform literature, 468–469

regulatory agencies, U.S., 210

Reimann, Georg, 78, 235

relation, 599, 785

relativity, **627–628**

Einstein general and special theories of, 79, 212–213, 278, 628, 638

Noether's work, 504

non-Euclidean geometry and, 79, 638

quantum mechanics, 539, 540

remedial instruction, 69, **628–632**

developmental mathematics, 194–195

programs for underachievers, 796–801

*See also* Developmental Mathematics Program and Apprenticeship Teaching

Rensselaer Polytechnic, 324

Renzulli, Joseph, 292

*Reorganization of Mathematics in Secondary Education, The* (1923 report), 498–499

*Report of the Committee of Ten on Secondary School Studies* (1893), 718, 728

*Report of the National Committee on Mathematical Requirements* (1923), 327

representation problems, and number theory, 511–512

research in mathematics education, **632–637**

National Center for Research in Mathematical Sciences Education, 497–498

research mathematics, careers in, 105

*Reshaping School Mathematics: A Philosophy and Framework for Curriculum* (1990), 649, 711, 718–719, 729

resources. *See* American Mathematical Association of Two-Year Colleges (AMATYC); American Mathematical Society (AMS); American Statistical Association (ASA); Association for Women in Mathematics (AWM); Association of Teachers of Mathematics (ATM); Board on Mathematical Sciences (BMS); coalitions; congresses; *Educational Studies in Mathematics (ESM);* Eisenhower Program; Joint Policy Board for Mathematics (JPBM); Mathematical Association of America (MAA); Mathematical Sciences Education Board (MSEB); *Mathematical Spectrum;* National Council of Supervisors of Mathematics (NCSM); National Council of Teachers of Mathematics (NCTM); National Science Foundation (NSF); School Science and Mathematics Association (SSMA); United States Department of Education; United States National Commission on Mathematics Instruction

retailing careers, 105–106

retention of learning, 596

*Revised Children's Manifest Anxiety Scale* (RCMAS), 287

*Revolution in School Mathematics* (NCTM; 1961), **637**

Reynolds, Cecil R., 287

Reys, Robert E., 519

Rhind papyrus (1650 B.C.), 242–243, 321, 622, 812

Rhode Island, EQUITY 2000 program, 479

rhombus, 548

Riccati equation, 523

Richie, Dennis, 577

Richter scale, 228, 244

Rickover, H. G., 290

Riemann, Georg Friedrich Bernhard, 64, 267, 483, **637–638**

conditionally convergent series proof, 651

$n$-dimensional manifolds theory, 278, 765

non-Euclidean geometry development, 78, 235, 445, 505–506, 580, 638

trigonometric series thesis, 258

Riemann sum, 352, 353

Riemann surfaces, 637

right triangle, 27

trigonometry, 784–785

*See also* Pythagorean Theorem

ring theory, 4, 5–7, 189, 504

*Rivista di matematica* (journal), 533

Robin, Daniel, 361

Robinson, Horatio N., 378

Robinson, Julia, 823

Rogers, Carl, 332

Rolle, Michel, 458

Rolle's Theorem, 458

Roman abacus, 1

Romberg, Thomas A., 498, 650

Romey, William, 749

Rose-Hulman Institute of Technology, 290

Rosenblueth, Arturo, 821

Ross, Arnold, 286, 289

Ross, Lee, 815

rotation

of angle, 27, 28, 29

transformational geometry, 772, 774, 776

rotational symmetry, 699

rote learning, 595–596. *See also* memorization

rounding, **638–639**

estimation, 232

Royal Military Academy (Woolwich, England), 216

Royal Statistical Society (England), 732

Royaumont Conference, 360

Rubik Cube, 625, 627

rule method

of definition of sets, 653

of teaching mathematics, 325

rule-recitation method, of memorization and computation, 677

Russell, Bertrand Arthur William, 103, 537, **639–640**

logic, 103, 534, 588

logical paradox, 526

paradox, 448, 654

Peano as influence on, 533, 639

Platonism and, 537

on sets, 654

Whitehead association, 103, 639, 820

Russell, Susan J., 457

Russell's paradox, 448, 526, 654

Russia, **640–642**

abacus use, 1

University of Chicago textbooks translations, 806

U.S. Joint Conferences on Mathematics Education, 805

Russian (formerly Soviet) Academy of Sciences, 291

Rybkin, N. S., 641

Ryzhik, V. I., 642

## S

Saccheri, Gerolamo, 505, 580

Sadker, Myra and David, 607

*Safari Search* (computerized game), 662

sample variance, 675

*See also* random sample; standard error

sampling, acceptance, 606, 607

*Sand Reckoner, The* (Archimedes), 37

Sarton, George, 316

SAT. *See* Scholastic Aptitude Test

Satellite Educational Resources Consortium (SERC), 742
scalar multiplication, 420, 816
scalar product, 422
Scandinavia, percentage of women mathematics Ph.D. recipients, 822
scatterplots, 161
Schappelle, Bonnie, 497
Scheme (programming language), 577–578
Schifter, Deborah, 694
Schoenfeld, Alan, 33
Scholastic Aptitude Test (SAT)
    as college success predictor, 745
    decline in scores, 646–647, 683
    female vs. male scores, 823
    as mathematical giftedness determinant, 285, 288–289
    minority students' scores, 474
Scholwinski, Ed, 287
school districts, educational reform roles, 683–684
Schoolmaster's Assistant: Being a Compendium of Arithmetic Both Practical and Theoretical, The (Dilworth), 676–677
School Mathematics Project (England), 217
School Mathematics Study Group (SMSG), 643
    Begle directorship, 72
    Birkhoff postulates adoption, 78–79
    curriculum trends and, 174
    National Science Foundation support for, 328
    new math movement and, 679
    recommendations to textbook publishers, 646
    University of Maryland Mathematics Project as consultant to, 808
School of Mathematical and Navigational Sciences (Russia), 640
School Science and Mathematics (journal), 644
School Science and Mathematics Association (SSMA), 643–644
School Science and Mathematics Association (SSMA), historical perspective, 644
Schröder, Ernst, 109, 533, 534
Schrödinger, Erwin, 539–540
Schrödinger's equation, 531, 540
Schupp, Hans, 733
Schur decomposition, 423
Schwartz, Benjamin L., 276
Schwartz, Richard, 222
Science and Technology Publishing Company of Singapore, 291
Science Service, 291
Scientific American (magazine), 624
scientific method, 37, 190, 271
scientific notation, 145, 644–645
    in solar system studies, 668

Scotland, 365
Scott, Elizabeth, 824
Scott Foresman Company, 647, 807
Sebokht, Severus, 338–339
Secada, Walter, 498
secant, 115
Secant Method, 420
Second International Congress of Mathematics (1900; Paris), 141, 316, 639
Second International Mathematics Assessment, 178
Second International Mathematics Study (SIMS), 361, 365, 366–367, 411, 804
secondary school mathematics, 646–650
    barriers for minorities, 474
    classroom applications, 34–36
    integration with other subjects, 356–357
    interactive program, 357–359
    Joint Commission Report (1940), 381–382
    lesson plan example, 415–416
    measurement/measurement systems studies, 706
    National Committee on Mathematical Requirements recommendations, 499
    reforms, 718, 728, 729
    statistics studies, 690
    teacher development, 717–720
    teacher preparation, 727–729
    See also curriculum trends, secondary-level
Secondary School Mathematics Curriculum Improvement Study, 250, 679
Secondary School Mathematics Curriculum, The (NCTM Yearbook), 647
Seeing and Thinking Mathematically Project, 716, 725
self-contained classrooms, 339
self-directed educational programs, 340
Sells, Lucy, 824
semantics, 388–389
sentential logic, 586
separate knowing, 829, 830
sequencing of learning materials, 271, 329, 599
series, 650–653
    divergent, 78, 651–652
    Fourier, 257–258
    Hadamard's work, 313
    Maclaurin expansion, 437–439
    numerical analysis methods, 514
    Taylor, 266, 313, 438, 515–516, 550, 551
    See also time series analysis
Serret, Joseph-Alfred, 5
set-builder notation, 653
sets, 653–655
    algebra of, 82
    Boolean, 82–83

Cantor, 103, 259, 653, 655
    chaotic, 109
    combinatorics, 120–122
    Dedekind, 189
    discrete mathematics, 202
    finite, 537
    Fourier, 257
    Gödel, 299
    infinite, 536, 537
    Lebesgue, 404–405
    in modern mathematics, 484
    of real numbers, 103
    theories of, 102–103, 203, 210
    transfinite, 102–103
    Zeno's paradoxes as influence on, 837
"seven bridges of Königsberg" problem, 121, 626
Seventh Report of the International Clearinghouse on Science and Mathematics Curricular Developments (Lockard; 1976), 679
Shane, Harold, 339
Shanks, Daniel, 541
Sharpe, David, 446
Sharpe, Ken, 733
Shaughnessy, J. Michael, 733
Sheffield, Joseph, 324
Sheffield Hallam University, 732–733
Sheldon, Edward, 536
Shen Kao, 650
Shephard, Geoffrey C., 760
Shewhart, Walter, 605
Shin-Ying Lee, 361
Shoecraft, Paul, 797
shopkeeper's average. See mode
Shortchanging Girls, Shortchanging America (AAUW report), 827
short written-response questions (constructed-response questions), 746
Siddhanta Siromani (Bhaskara), 338, 458
Sierpinski, Waclaw, 510
Sieve of Eratosthenes, 228
significance level, 687, 688
Silverman, David L., 277
SIM (game), 274, 275
similarity, 655–656
    transformations, 772, 773–774, 776, 777
    videotape about, 465, 742
Simon, Herbert, 570
simple graphs, 301
Simpson's paradox, 672, 673
SIMS. See Second International Mathematics Study
simulation
    computer, 130, 657
    in consumer mathematics, 149
    software, 657
simultaneous equations, 137
sine
    Leibniz evaluation of, 531
    transcendental and algebraic functions and, 767–768, 769–770

in trigonometry, 784–785, 786
videotapes about, 465, 741–742
Siu, R. G. H., 332
6-6 educational model, 467
6-3-3 educational model, 328, 467
Six through Eight Mathematics Project (STEM), 716, 725
*Si Yuan Yu Jian (Precious Mirror)*, 114
Skemp, Richard R., 388, 389
SKILL (Scientific Knowledge for Indian Learning and Leadership) project, South Dakota School of Mines and Technology, 799–800
skills. *See* basic skills
Skinner, Burrhus Frederic (B. F.), 271, 599, **656**
slack variables, 426
Slaught, Herbert, 327
Slocomb, William, 170
Slovinski, David, 558
Smale, Stephen, 109, 766
Smith, David Eugene, 316–317, 327, 641, **656–657**
SMSG. *See* School Mathematics Study Group
social constructivism, 537
social science, mathematics and, 31
social studies, integration of elementary school mathematics, 355
Society for Industrial and Applied Mathematics, 15
  college teacher preparation recommendations, 721
  Joint Policy Board for Mathematics, 382
  linear algebra conferences, 420
Society of Actuaries, 8
Socrates, 201, 607
Socratic method
  in Interactive Mathematics Program, 358
  purpose of, 201
software, **657–665**
  *Algebra Xpresser*, 349, 660
  for basic skills teaching, 71
  computer-assisted instruction, 128, 129–131, 349
  computer-based multimedia, 130, 741
  computer-based placement tests, 443
  computer-generated manipulatives, 440–441
  for convergent series instruction, 652
  for elementary instruction, 344
  enrichment programs, 220
  for geometry instruction, 280
  graphing, 463
  as interactive media, 463
  for linear algebra study, 424
  for matrix computations, 423
  measurement activities, 407, 409
  for median studies, 466
  for middle-school instruction, 472
  for ordinary differential equations study, 522

for problem-solving, 130
spreadsheets, 131, 472, 658, 661, 662
for statistics basics study, 690
for symmetry studies, 701
for test-building or test-generating, 748
tutorial, 128, 129, 131, 657, 658, 789
*See also* computers
solar system
  chaos theory and, 109
  measurement, 219
  space mathematics, 668–669
solid geometry, 37, 144
Solon, 211
SOMA cube, 625, 627
Somerville, Mary, 821, 822
*Soroban* (Japanese abacus), 377–378, 379
sound waves, 492–493
South Africa, 826
South Asia
  women's mathematics studies and career opportunities, 825, 826
  *See also* Indian mathematics
South Dakota, systemic reform projects, 649
South Dakota School of Mines and Technology (SDSM&T), SKILL project, 799–800
South East Asia Mathematical Society (SEAMS), 364
Soviet Union. *See* Russia
Sowder, Judith T., 497
space geometry, instruction, **665–668**
  conic sections and, 142–144
space mathematics, **668–669**
  *See also Sputnik* launching
space race, 174, 177, 646, 679, 682
spatial reasoning, 280
spatial skills
  measurement and, 408, 409
  visualization development as NCTM goal, 665
Spearman, Charles Edward, 163
special programs
  FAMILY MATH and *MATEMÁTICA PARA LA FAMILIA*, 246–249, 286–287, 794–795
  for gifted, 285–286, 288–294
  for underachievers, 796–801
  *See also* enrichment, overview; learning disabilities, overview; minorities, educational problems
Spelman College, 800
Spencer, Herbert, 39
spherical geometry, 505–506
spike abacus, 188
spiral learning, **669–670**
sports, **671–675**
Sprague-Grundy theory, 275, 276
spreadsheets, 131
  software, 472, 658, 661, 662
*Sputnik* launching
  effect on U.S. mathematics and science education, 40, 468, 646, 678, 682, 712, 718, 728, 801

new market for physicists and mathematicians following, 103
*Square One TV* (PBS television series), 464, 465, 742
square root, 66, 113
  continued fractions and, 373
  paradox, 527
SSMA. *See* School Science and Mathematics Association
*Stand and Deliver* (film), 796
standard deviation, **675**
standard error (sampling error), 814
standardization of variables, **676**
standardized tests
  achievement assessment, 48
  achievement by learning disabled assessment, 399
  decline in scores, 1960s and 1970s, 646
  elimination under Elementary and Secondary Education Act, 792–793
  gender and scores on, 828
  history of, 744
  instructional methods research and, 346–347
  mathematics scores, 174
*Standards* (NCTM documents). *See Assessment Standards for School Mathematics* (1995); *Curriculum and Evaluation Standards for School Mathematics* (1989); *Principles and Standards for School Mathematics* (2000); *Professional Standards for Teaching Mathematics* (1991)
standards, curriculum, **676–682**
  goals of mathematics instruction, 296–297
  Goals 2000: Educate America Act, 298–299
  kindergarten, 384–385
Stanford-Binet tests
  to determine mathematical giftedness, 285
  for fundamental arithmetic, 744
Stanford University
  program for the gifted, 289
  School Mathematics Study Group, 643
Stanley, Julian C., 285, 286, 288
state government, role, **682–684**
  Goals 2000: Educate America Act, 298–299
  *See also specific states*
State Mathematics Coalition Project, 116
State of Illinois Scientific Literacy, Teaching Integrated Mathematics and Science Program funding, 734
Statewide Systemic Initiative Program (SSI), 683
*StatExplorer* (software), 660
statistical graphs, 305
statistical inference, **684–687**
  variance and covariance in, 814
statistical significance, **687–689**
  random sample, 612–613, 688

statistics, overview, **689–691**
 ActivStats (CD-ROM), 741
 American Statistical Association, 26
 Ask a Teen network project, 740
 careers in, 106
 exploratory data analysis, 31
 Gauss's contribution, 278
 median-based, 181
 National Center for Research in
  Mathematical Sciences Education
  study, 497–498
 NCTM standards, 648
 quality control, 605–606
 software study packages, 660
 teacher preparation issues, 690–691,
  731–734
 *See also* binomial distribution;
  correlation; curve-fitting; data
  analysis; data representation;
  extrapolation; index numbers;
  interpolation; least squares;
  Markov Processes; mathematical
  expectation; mean, arithmetic;
  mean, geometric; measures of
  central tendency; median; mode;
  normal distribution curve; Poisson
  Distribution; probability density;
  probability distribution; random
  number; random sample; random
  variable; standard deviation;
  standardization of variables;
  statistical inference; statistical
  significance; stochastic processes;
  time series analysis; variance and
  covariance
Statistics Teacher Network, 733
*Statistics Teacher Network* (newsletter), 691
*Statistics Workshop* (software), 660, 690
Steen, Lynn, 222
Steenrod, Norman, 766
Stein, Sherman, 333
Steinitz, Ernst, 7
STEM. *See* Six through Eight
  Mathematics Project
stem-and-leaf plot, 181, 185
Stern, Catherine, **692**
Sternberg, Robert J., 284
Stevens, Glenn, 289
Stevens, William, 749
Stevens Institute of Technology, 742
Stevenson, Harold W., 361, 365
Stevin, Simon, 318
Stieltjes integrals, 353
Stifel, Michael, 532
Stigler, James W., 361, 362
stimulus-response theory, 69, 599, 678
 Brownell's opposition to, 84
 Thorndike's advocacy, 759
Stirling, James, 550
Stirling numbers, 121
stochastic processes, 539, 689, **692–693**
stocks and bonds, 255
stoichiometry, 110
Stone, Marshall, 100

story problems. *See* word problems
strategy, in game theory, 272
strategy stealing, 276
Stravinsky, Igor, 333
streaming, 3
Strengthening Underrepresented Minority
  Mathematics Achievement
  (SUMMA) Program, 443
Striedieck, Iris, 751
Stroustrup, Bjarne, 577
structuralism, 146
Student Mathematics League, 24
student teacher autonomy, elementary
  school, **693–696**
Study of Exceptional Talent (SET), 289
Study of Mathematically Precocious Youth
  (SMPY), 286, 288
Sturm, Charles, 257
Sturm-Liouville theory, 427
subtraction, **696–698**
 in Egyptian mathematics, 212
 zero's properties in, 838
 *See also* addition
Su Bu Qing, 114
"Suido Houshiki" (mathematics teaching
  method), 379
*Sulvasutras*, 338
*Sumario Compendioso* (Diez), 753, 755
summer programs, for mathematically
  gifted, 289–290
Summer Science Talent Programs, 290
Sunburst Communications
 ELASTIC software, 466, 814
 software for middle-school instruction,
  472
SuperQuest, 291–292
superscript notations. *See* exponents,
  arithmetic
Suppes, Patrick, 128, **698**
Suydam, Marilyn N., 519
*swan-pan* (Chinese abacus), 377
Sweden, international comparisons of
  mathematics education, 365,
  366
Sweeney, Dennis J., 520
Sweller, John, 667
*Syllabus of Plane Algebraic Geometry, A*
  (Dodgson), 108
syllogisms, diagrams for, 197–198
Sylow, Ludvig, group theory, 311
Sylvester, James Joseph, 192, 324, 325,
  422, 423
symbolic algebra, 190, 240
symbolic logic, 103, 533, 639, 820
symbols
 algebraic, 190, 224, 240
 in Egyptian mathematics, 211, 212
 for equality, 223
 Euler's contribution, 235
 in integration, 352
 language in teaching mathematics and,
  389
 learning disabilities and, 395
 in Mayan numeration, 455, 456

 for multiplication, 488
 for order relationships, 342
 origins of, 38, 320
 placeholders, 838
 poetical, 543
 systems of inequalities and, 703
 variables and, 812–813
 for zero, 38, 187, 339
symmetry, 159, **698–701**, 773
symmetry groups, 311
syntax, 389
*Systemic Initiatives for Montana
  Mathematics and Science*, 649
systems analysis. *See* operations research
systems of equations, **701–703**
systems of inequalities, **703–704**
systems of measurement, **704–708**
 length, perimeter, area, and volume,
  406–410
 measurement, 460–462
 temperature scales, 743
Szablewski, Jackie, 332

**T**
*Tagungen fuer Didaktik der Mathematik*,
  282
Talent Identification Program (TIP),
  Duke University, 289
Talk Calculators, 740
Tamura, Y., 541
Tangram figures, 627
*Tantrasangraha* (Nilakantha), 650
task analysis, **709–710**
 hierarchical organization, 271
Tate, William F., 488
Taylor, Brook. *See* Taylor series
Taylor, Nick, 365
Taylor, Robert, 658
Taylor polynomials, 551, 552
Taylor's Formula, 770
Taylor series, 266, 313, 438, 515–516,
  550, 551
teacher development, elementary school,
  **710–712**
teacher development, K–12: California
  Mathematics Project (CMP), 647,
  **712–715**
teacher development, middle school,
  **715–717**
teacher development, secondary school,
  **717–720**
 *EQUALS: A Mathematics Equity
  Program*, 647
 *National Leadership Program*, 647
Teacher Education and Certification
  (1961), 312
teacher evaluation. *See* evaluation of
  instruction, issues; evaluation of
  instruction, overview
teacher networks, 733
teacher preparation, college, **720–722**
 curriculum trends, 172–173
 field-based, 172
 symmetry studies, 701

*See also* Developmental Mathematics Program and Apprenticeship Teaching

teacher preparation, elementary school, issues, **722–724**

student teacher autonomy, elementary school, 693–696

teacher preparation, middle school, **724–727**

teacher preparation, secondary school, **727–729**

Addenda Series (NCTM), 649, 681, 716

California Mathematics Project, 647

teacher preparation, secondary school, reform project in California, **729–731**

teacher preparation in statistics, issues, 690–691, **731–734**

Teachers College, Columbia University, 250, 327

teaching apprenticeships, 195–196

teaching as career, 106

*See also* National Council of Teachers of Mathematics

teaching awards, 62–63

teaching capability, 179

*Teaching Children Mathematics* (journal), 501

teaching in teams, 356–357

Teaching Integrated Mathematics and Science Program (TIMS), 179, 357, **734–735**

*Elementary Mathematics Curriculum Project*, 176, 734

teaching-learning units (TLUs), and individualized instruction, 340

Teaching Mathematics with Calculators: A National Workshop, 443

*Teaching of Elementary Mathematics, The* (Smith), 327

*Teaching of Mathematics in the Elementary and Secondary Schools, The* (Young), 835

*Teaching Statistics* (British journal), 447, 691, 733

Teaching Statistics Trust, 447

team learning, 350

team teaching, 356–357

techniques and materials for enrichment learning and teaching, 218–219, **735–739**

abacus as, 1–2

field activities, 252–254

gifted programs, 285–286

*See also* manipulatives

technology, **739–743**

basic skills and, 71

calculator, 87–90, 739–740

*See also* calculators *as separate heading*

communication, 136

computer-based multimedia, 741

computer networks, 740–741

digital, 138

distance education and teleconferences, 742

field study, 253–254

growth of, 171

impact on curriculum, 170, 329

impact on instructional methods, 329, 348–352

interactive, 130–131, 363

job market and, 103–104

mathematically gifted programs and, 286

middle-school instruction and, 472

multimedia, 363–364, 741

research using, 139–140

space race and, 171

for student achievement assessment, 48

for teaching linear algebra, 424

videotape and television, 464, 741–742

*See also* computer-assisted instruction; computers; graphing calculator; internet and world wide web; software

telecommunications, 740

teleconferences, 742

telescope, development of, 272

television

instructional, 464

interactive (ITV), 349–350

professional development opportunities for teachers, 711, 716

videotape technology and, 741–742

temperature scales, 707, **743**

nurses' use of, 517

zero in, 838

Tennessee

EQUITY 2000 program, 479

field testing of Mathematics in Context curriculum, 454

Terman, Lewis M., 288

Terrapin Logo (software), 472

TesselMania (software), 344

test construction, **743–748**

advanced placement, 13–14

bias issues, 794

computer-based placement tests, 443

current research, 346–347

MAA computer-based placement tests project, 443

Third International Mathematics and Science Study, 758

test statistics, 685–686

Texas

EQUITY 2000 program, 479

mathematically gifted program, 286

underachievers programs, 797

*See also* University of Houston-Victoria; University of North Texas

Texas Academy of Mathematics and Science, 286

*Textbook in American Education, The* (Whipple, ed.), 749, 751

textbooks, **748–753**

Birkhoff's *Basic Geometry*, 78

Brownell's arithmetic texts, 83

calculus, 417, 821

Colonial Era. *See* textbooks, early American samples

as curriculum direction determinant, 220

in England, 216, 217

first calculus textbook, 417

first calculus textbook for young people, 821

first in English, 216

first mathematics text published in the United States, 677

first widely used arithmetic text by American author, 542

Interactive Mathematics Text Project, 443

in Japan, 378, 379

Japanese compared with American, 361

language of, 391–393

for middle school mathematics teachers, 716

1930s and 1940s, 467

1960s and 1970s, 646

national curriculum, 751

new math era for junior high students, 468

Pike's arithmetic texts, 542

reading, 620–622

resources and the future, 751–752

secondary school mathematics, 647

Smith's arithmetic series, 656

Thorndike's on teaching, 759

textbooks, early American samples, **753–757**

*Textbooks and Schooling in the United States* (Elliott and Woodward, eds.), 749

*Text Materials in Modern Education* (Cronbach, ed.), 749

Thailand, women's mathematics education and career problems, 826

Thales, 114, 211, 309

Thayer, Sylvanus, 324

Theaetetus, 622

Theorem of Desargues, 581

Theorem of Pappus, 581

theorems, 448, 450, 481

mean value, 268, 458–459

neutral geometry, 503

for projective geometry, 580–581

*See also* fundamental theorems of mathematics; *specific theorems*

*Théorie des Fonctions Analytiques* (Lagrange), 266

*Theorist* (software), 661

theory of relativity. *See* relativity

theory of special relativity. *See* relativity

thermodynamics, 110, 539

Third International Mathematics and Science Study (TIMSS), 365, **758–759**, 804, 805

Thomas, Veronic, 751

Thompson, D'Arcy, 77

Thompson, Patrick, 660

Thompson, R., 333

Thomson, William, Lord Kelvin, 743, 764
Thorndike, Edward L., 39, 69, 656, **759**
  behaviorism, 599, 759
  connectionism, 328, 678, 759
*Thorndike Arithmetics, The* (Thorndike), 759
three-dimensional (3D) geometry. *See* space geometry, instruction
tiles, algebra, 224, 225
tiling, 231, **759–761**, 773
time, **761–762**
  art and, 46
  as system of measurement, 707
timelines, in teaching history of mathematics, 321
time series analysis, **762–764**
TIMS. *See* Teaching Integrated Mathematics and Science Program
TIMS Elementary Mathematics Curriculum Project, 176, 734
TIMS Laboratory Experiments, 734
TIMSS. *See* Third International Mathematics and Science Study
Tipps, Steve, 709
Tobias, Sheila, 451
Todhunter, I., 378
Toliver, Kay, 797
tool software, 658
Tooyama, Hiraku, 379
topology, 78, **764–766**
  modern mathematics and development of, 484
*Tortoise* (Zeno paradox), 837
*Toward a Theory of Instruction* (Bruner), 84
Tower of Hanoi problem, 570
tracking
  as ability grouping, 2, 3
  minority students and, 474, 476–477
  underachievers and, 791
traditional classroom discourse, 389
traditional methods, inclusion in new directional instruction methods, 350–351
*Traité analytique des probabilités* (Laplace), 393
*Traité analytique des sections coniques* (L'Hospital), 417
*Traité de dynamique* (D'Alembert), 181
*Traité de mécanique céleste* (Laplace), 393
*Traité des substitutions et des équations algébriques* (Jordan), 5
*Traité du triangle arithmétique* (Pascal), 532
transcendental and algebraic functions, **766–771**
  asymptotes, 58
transcendental numbers, **771–772**
  algebraic, 21, 766–771
  *e* as, 209, 512
  Liouville's discovery of, 427
  and number theory, 512
  and real numbers, 623
  theory of, 771
  transcendental and algebraic functions, 766–771

transfer of learning, 146, 594–595
transformational geometry, 17, **772–775**
  Klein's work, 385–386, 772
transformations, **775–781**
  invariance and, 369–370
  musical, 493
transistors, 88
translation, 772, 774, 776
translational symmetry, 699
trapezoid, 548
*Treatise of Algebra in Three Parts, A* (Maclaurin), 437
*Treatise on Algebra, A* (Peacock), 6
*Treatise on the Complete Quadrilateral* (al-Tusi), 374
*Treatise on Universal Algebra, A* (Whitehead), 820
Treisman, Uri, 452, 798
triangle
  area of, 192
  harmonic, 532
  inequality, 342
  in polygons, 547, 548
  right triangle trigonometry, 784–785
  symmetry group of, 311
  *See also* Pascal's triangle; Pythagorean Theorem
trigonometric functions, elementary, **781–782**
  transcendental and algebraic functions, 768
trigonometric identities, **782–784**
trigonometric series, 257–258
trigonometry, **784–788**
  analytical, 189
  classroom applications, 32
  De Moivre's contribution, 189
  Fourier's contribution, 257–258
  Indian mathematics, 338
  Islamic mathematics, 374
  NCTM standards, 648
  pi's importance in, 541
  similarity, 655–656
  *See also* trigonometric functions; trigonometric identities; vectors
true-false questions, in test construction, 746
Tucker, Alan, 422
Tucker, Albert W., 120, 425
Tukey, John, 690
Turing, Alan Mathison, 299, **788**
  *LU* factorization, 423
Turing machine, 24
Turkey, women's mathematics education and career problems, 826
turtle math, 27
Tusi, Nasir al-Din al-, 374, 532
Tusi, Sharaf al-Din al-, 374
tutoring, **788–789**
  for learning disabled students, 402
  software for, 128, 129, 131, 657, 658
twin prime conjecture, 450
Tymoczko, Thomas, 537

**U**
*Über die Hypothesen, welche der Geometrie zu Grunde liegen* (Riemann), 505–506
UCSMP. *See* University of Chicago School Mathematics Project
Uglidisi, Abu l-Hasan al-, 373
Uhlenbeck, Karen, 823
UICSM. *See* University of Illinois Committee on School Mathematics
Ulam, Stanislas, 568
UMMaP. *See* University of Maryland Mathematics Project
underachievers, **791–796**
underachievers, special programs, **796–801**
UNESCO, 732
Unified Science and Mathematics in the Elementary School, 33
unifix cubes, 344, 509, 697
unit, as mathematical concept of measurement, 406
United Kingdom. *See* England
United States Department of Education, **801–805**
  Developing Mathematical Processes, 193–194
  Eisenhower program, 213–214
  Fund for the Improvement of Postsecondary Education, 247
  *See also* National Assessment of Educational Progress
United States Military Academy, 324
United States National Commission on Mathematics Instruction (USNCMI), **805**
United States Quantitative Literacy Project, 733
*Universal Arithmetics* (Euler), 641
universal characteristics (Leibniz concept), 533
universities. *See* colleges and universities; *specific institutions*
University of California, Berkeley, 452
  Lawrence Hall of Science, 246, 248
  Professional Development Program, 798
University of Chicago, 101, 325, 326, 639, 835
University of Chicago School Mathematics Project (UCSMP), 175, 647, 649, **806–807**
University of Connecticut, 289
University of Erlangen, 504
University of Göttingen, 277, 316, 504
  Mathematical Institute, 385, 819
University of Houston-Victoria, 797
University of Illinois
  Arithmetic Project, 679
  curricular reform projects, 328, 679
University of Illinois, Chicago
  Teaching Integrated Mathematics and Science Program, 734–735
  TIMS Elementary Mathematics Curriculum Project, 176

University of Illinois Committee on School Mathematics (UICSM), 328, 679, **807–808**
University of Iowa, 289
University of Königsberg, 316
University of Leeds, 464
University of Maryland Mathematics Project (UMMaP), **808**
University of Michigan, 365
University of North Texas, 286
University of Nottingham, National Centre for Statistical Education, 732
University of Pittsburgh
  Learning Research and Development Center, 802
  QUASAR project, 799
University of Southern Mississippi, 289
University of Virginia, 324
University of Wisconsin
  Developing Mathematical Processes, 193–194
  Mathematics in Context curriculum units, 454
unsolved problems, 627
Urysohn, Paul, 765
USA Mathematical Talent Search, 290, 292. See also U.S. Mathematical Talent Search (USAMTS)
USA Mathematics Olympiad (USAMO), 62, 286, 291, 443
Used Numbers curriculum, 690
Usiskin, Zalman, 633, 646, 647
U.S. Mathematical Talent Search (USAMTS), 124, 125. See also USA Mathematical Talent Search
USNCMI. See United States National Commission on Mathematics Instruction
U.S./Russia Joint Conferences on Mathematics Education, 805

**V**
validity test, 744
Vallée-Poussin, Charles de la, 313
Van Hiele, Pierre M., 279–280, 591, 809–812. See also Hiele, Pierre van
Van Hiele-Geldof, Dina, 809–812. See also Hiele-Geldof, Dina van
Van Hiele levels, **809–812**
  Piaget and, 280
  space geometry, 665, 666, 667
Van Sertima, Ivan, 487
Vanderbilt University, 662, 741
Vandermonde, Alexandre T., 192
variables, **812–813**
  function of two, 307
  functions of a real, 265–268, 813
  quality control and, 605
  random, 75, 613–614, 676
  solving inequalities with one or two, 342–344
  standard deviation and, 675
  standardization of, 676

variance and covariance, **813–816**
variations, calculus of, 73
vector analysis, 215, 817
Vector Analysis (Gibbs), 816
vector spaces, 424, 580, **816–817**
  field theory and, 7
  in linear algebra, 420, 421, 816–817, 838
  zero as additive identity, 838
vectors, 202, **817–818**
  algebra of, 19
  calculus, 214
  in coordinate geometry, 155–156
  definition of, 420
velocity, 214, 215
  computations, 536
  escape, 215
  graph of, 266
  of light, 539
Venezky, Richard, 749
Venn, John, 197
Venn diagrams
  sets and, 654
  software for study of, 657
  syllogistic logic, 197
Verner, A. L., 642
video, 465, 741–742
Viète, François
  codes, 318
  equation solutions, 224
  symbolic notation, 190, 240, 812–813
  trigonometric tables, 784
Vinogradov, Ivan M., 300, 510, 559
VisiCalc (spreadsheet software), 658
visual aids, 220
visualization, 667–668
  in algebra, 224–225
  in arithmetic, 38
  in test construction, 747
vocabulary, 388
Voltaire, 181
Volterra, Vito, 77, 550
volume
  Archimedes's analyses of, 310
  in calculus, 98
  measurement of, 407–408, 706
Von Neumann, John, 425, **818**, 820
  game theory, 210, 272, 274, 276, 818
Vygotsky, Lev, 147, 553, 601

**W**
Wagreich, Philip, 734–735
Wallenstein, Albrecht Wenzel Eusebius von, 383
Wallis, John, 223
Wasan (Japanese mathematical system), 378
Washburne, Carleton, 2
wave equations, 531
waves, sound, 492–493
Webb, Norman, 359, 650
Wedderburn, Joseph Henry Maclagan, 6
week, historical background of term, 705
Weierstrass, Karl, 103, 259, 267, 653, **819**
  axiomatic systems, 5

continuity definition, 151
  theory of functions, 266
weight, as system of measurement, 707
Weinzweig, A. I., 411
Weiss, Daniel S., 287
Wentworth, George A., 326
Wertheimer, Max, 553, 570
Wesleyan University, math anxiety clinic, 451
Westinghouse Science Talent Search (now Intel), 291
West Virginia, mathematics proficiency scores, 1992, 453
Weyl, Hermann, **819–820**
What's Happening in the Mathematical Sciences? (AMS; 1993), 25
Where Are You? (cartography challenge), 737–738
Whetstone of Witte, The (Recorde), 223, 224
White, Robert, 295
Whitehead, Alfred North, 103, 588, 639, **820**
whole numbers
  fraction study in relation with, 261
  NCRM study on learning and teaching of, 497
Wholesale Price Index, 337
Why Johnny Can't Add: The Failure of the New Math (Kline), 329, 646
Wiener, Norbert, **820–821**
Wiener process (Brownian motion), 693
Wiles, Andrew, 512, 589, 626
Wilkinson, James Hardy, 423
Williams, Carol J., 658
Williams, Thomas A., 520
William the Conqueror, 689
Willoughby, Steven, 749
Wilson, J. W., 411
Wilson, Melvin, 634
Wilson's Theorem, 511
Winnetka Plan, 2
wireframe model, 307
Wirth, Niklaus, 576
Wisconsin
  EQUITY 2000 program, 479
  field testing of Mathematics in Context curriculum, 454
  Mathematics in Context curriculum units, 454
  mathematics proficiency scores, 1992, 453
  National Research and Development Center on Achievement in School Mathematics and Science, 802–803
  See also University of Wisconsin
Wisconsin Center for Education Research, 359
Wittgenstein, Ludwig, 537, 639, 640
WME. See Women and Mathematics Education
WME Newsletter, 832
Wolfle, Jane A., 285
Wollman, Warren, 389

women and mathematics, history, **821–824**
  Association for Women in Mathematics, 15, 56, 823
  Byron, Augusta Ada, Countess of Lovelace, 65, 514, 821, 822
  Germain, Sophie, 281
  Hypatia, 310, 821
  Kovalevskaia, Sofia, 386, 822
  *Ladies' Diary (Woman's Almanack)* (1704–1841), 387
  Montessori, Maria, 485–486
  Noether, Emmy, 504–505
women and mathematics, international comparisons of career and educational problems, 367, **824–827**
  International Organization of Women in Mathematics Education, 364
women and mathematics, problems, 821, **827–832**
  bias in software content, 663
  bias in testing, 794
  math anxiety, 451–452
  middle school teaching and learning environment, 469–470
  NAEP mathematics and science scores, 1982–1992, 453
  Scholastic Aptitude Test scores, 823
  textbook depictions, 751
Women and Mathematics Education (WME), **832**
*Women's Ways of Knowing* (Belenky et al.), 829
Wood, Terry, 346
Woods Hole Conference, 618
*WordPerfect 6.0* (software), 661

word problems, 11, 34, 389, 570, 697
word processing software, 131, 658, 661–662
WordStar (software), 658
Works Progress Administration (WPA), 801
World Federation of National Mathematics Competitions, 291, 292
World Math Year 2000 project, 805
*World of Number, The* (videodisc series), 741
World War II, 327–328, 605, 820
World Wide Web. *See* Internet and World Wide Web
Wrench, J., 541
writing activities, **832–833**
  in enrichment learning and teaching, 737
  new directions in instructional methods, 350
Wylie, Alexander, 114

**X**
Xong Qing Lai, 114
Xu Bao Lu, 114

**Y**
Yackel, Erna, 346
Yahya al-Maghribi, Samaw'al ibn al-, 373, 374
Yale University, 324
  mathematics entrance requirements, 677
Yang Hui, 113, 114, 532, 640
Yang Hui Triangle, 114
Yang Wu Zhi, 114

Young, Grace Chisholm, 822
Young, Jacob William Albert, 326, 327, **835**
Young Scholars Program, 290
Yu Da Wei, 114

**Z**
Zambia, international comparison of women's mathematics education and career problems, 826
Zan, Betty, 695
Zariski, Oscar, 7
Zaslavsky, Claudia, 319, 487
Zeno, 309, 536, **837**
Zeno's paradoxes, 318, 527, 543, 652, 837
*Zentralblatt fuer Didaktik der Mathematik* (German journal), 282
Zermelo, Ernst, 103, 299
zero, **838**
  in Arabic mathematics, 187
  division by, 207–208
  early concepts, 66, 838
  in Egyptian mathematics, 211
  evolution of, 38
  exponents, 243
  in paradox, 527
  symbols for, 339
0.05 level of significance, 688
zero-sum games, 272, 838
Zhou Wei Liang, 114
Zhu Cong Zhi, 113
Zhu Shi Jie, 114
Zimbabwe, 826
Zu Chongzhi, 319
  value of pi, 320